COMPREHENSIVE ORGANIC CHEMISTRY

Volume 1

Stereochemistry, Hydrocarbons, Halo Compounds, Oxygen Compounds

Edited by J. F. Stoddart, University of Sheffield

Volume 2

Nitrogen Compounds, Carboxylic Acids, Phosphorus Compounds

Edited by I. O. Sutherland, University of Liverpool

Volume 3

Sulphur, Selenium, Silicon, Boron, Organometallic Compounds

Edited by D. Neville Jones, University of Sheffield

Volume 4

Heterocyclic Compounds

Edited by P. G. Sammes, The City University, London

Volume 5

Biological Compounds

Edited by E. Haslam, University of Sheffield

Volume 6

Author, Formula, Subject, Reagent, Reaction Indexes

Edited by C. J. Drayton, Pergamon Press, Oxford

COMPREHENSIVE ORGANIC CHEMISTRY

The Synthesis and Reactions of Organic Compounds

Volume 5 Biological Compounds

Edited by E. HASLAM
UNIVERSITY OF SHEFFIELD

PERGAMON PRESS
OXFORD · NEW YORK · TORONTO · SYDNEY · PARIS · FRANKFURT

U.K.	Pergamon Press Ltd., Headington Hill Hall, Oxford OX3 0BW, England
U.S.A.	Pergamon Press Inc., Maxwell House, Fairview Park, Elmsford, New York 10523, U.S.A.
CANADA	Pergamon of Canada, Suite 104, 150 Consumers Road, Willowdale, Ontario, M2 J1P9, Canada
AUSTRALIA	Pergamon Press (Aust.) Pty. Ltd, P.O. Box 544, Potts Point, N.S.W. 2011, Australia
FRANCE	Pergamon Press SARL, 24 rue des Ecoles, 75240 Paris, Cedex 05, France
FEDERAL REPUBLIC OF GERMANY	Pergamon Press GmbH, 6242 Kronberg-Taunus, Pferdstrasse 1, Federal Republic of Germany

First edition 1979

British Library Cataloguing in Publication Data
Comprehensive organic chemistry
1. Chemistry, Organic
I. Barton, *Sir*, Derek
547 QD251.2 78-40502

ISBN 0-08-021317-0
ISBN 0-08-021319-7 (set)

Printed in Great Britain by A. Wheaton & Co. Ltd., Exeter

Foreword

During more than a century, the development of organic chemistry has been associated with extensive documentation. Vast numbers of textbooks, monographs, and reviews have been published with the objective of summarizing and correlating the results obtained by many thousands of organic chemists working in academic and industrial research laboratories. However, out of this colossal literature there is but a relatively small number of textbooks and multi-volumed works which have become generally accepted as representing real steps forward in the presentation of our subject.

During the classical era of organic chemistry (1820–1940), textbooks which had a profound influence on the teaching of the subject included, for example, works by Armstrong (1874), van't Hoff (1875), Roscoe–Schorlemmer (1878), Richter (1888), Gattermann (1895), van't Hoff–Werner–Eiloart (1898), Meyer–Jacobson (1902), Schmidt–Rule (1926), Karrer (1928), Freudenberg (1933), Richter–Anschütz (1935), and Gilman (1938). These texts provide an opportunity to comment on the relationship between the history of organic chemistry and its associated publications. The *Treatise on Chemistry* by Roscoe and Schorlemmer consisted of three volumes (5343 pages) published in nine parts over the period 1878–1892: the major component was Volume III (6 parts, 3516 pages) which was devoted to organic chemistry. Another instructive example is the important work *Lehrbuch der Organischen Chemie*, produced by Victor Meyer and Paul Jacobson. The increase in size from the edition (1735 pages) published during 1902–1903 to the edition (5115 pages) published over the period 1913–1924 is striking.

Many have expressed concern about the problems of maintaining effective contact with the expanding literature of organic chemistry, but few have allowed themselves to become involved with attempted solutions. The decision to publish Comprehensive Organic Chemistry was not taken lightly. The absence of a work reflecting the current rapid development of modern organic chemistry has been lamented by many eminent chemists, including the late Sir Robert Robinson (1886–1975) who played an important role in the initiation of this project shortly before his death. Comprehensive Organic Chemistry was conceived, designed, and produced in order to meet this deficiency. We realised that the current rate of growth of organic chemistry demanded speedy publication and, furthermore, that its interaction with other subjects including biochemistry, inorganic chemistry, molecular biology, medicinal chemistry, and pharmacology required the collaboration of many authors. The selection of topics to be included in order to justify the work as being comprehensive has not been easy. We recognize that some areas of organic chemistry have not been given the detailed treatment which can be justified, but we have done our best to meet the expectations of the majority of readers. In particular, we have not made a special section for Theoretical Organic Chemistry. This is not because of any lack of appreciation on our part of the importance of Theory. It is because a correct treatment of Theory cannot be made comprehensible in an abbreviated form. It is also because Theory changes with time more rapidly than the facts of the subject. Theory is better treated in our view in specialist monographs. The same arguments apply equally to the fundamental subject of Stereochemistry. Any comments regarding errors and omissions will be appreciated so that they can be dealt with in future editions.

The contents of each volume have been brought together so as to reflect what are judged to be the truly important facets of modern organic chemistry. The information is presented in a concise and logical manner with mechanistic organic chemistry being adopted to provide a constant and correlative theme. The dominating intention of the Editorial Board has been to ensure the publication of a contribution to the literature of organic chemistry which will be genuinely useful and stimulating. Emphasis has therefore been given throughout to the properties and reactions of all the important classes of organic compounds, including the remarkable array of different compounds prepared by synthesis as well as natural products created by biosynthesis. Of course, the study of natural products provided the original foundation stones on which modern synthetic organic chemistry now firmly stands.

As a major presentation of modern organic chemistry, Comprehensive Organic Chemistry will be doubly useful because we have provided, in a separate volume, an extensive index. Not only have the contents of the work been indexed in the ordinary way, but we have also added a substantial number of additional references from the original literature. These do not appear in the text itself. Thus, the reader who wishes to obtain additional information about reactions and reagents mentioned in the text will quickly be able to consult the original literature. The Index volume has been prepared by a team from Pergamon Press.

Our debt to the Authors and to the Volume Editors is considerable. We are very grateful to all our colleagues for the efficient way in which they have tried to meet the challenges (and the deadlines!) which have been presented to them. We hope that the Authors have enjoyed their association with this venture. In a lighter vein, we also trust that their feelings are different from the statement 'this task put system into my soul but not much money into my purse' attributed to Henry Edward Armstrong (1848–1937) after he had written his *Introduction to Organic Chemistry* in 1874.

We are delighted to acknowledge the masterly way in which Robert Maxwell, the Publisher, and the staff at Pergamon Press have supported the Volume Editors and the Authors in our endeavour to produce a work which correctly portrays the relevance and achievements of organic chemists and their contributions to knowledge by research.

D. H. R. Barton
Chairman

W. D. Ollis
Deputy Chairman

Contents

Preface to Volume 5

"History is not just a catalogue of events put in the right order like a railway timetable. History is a version of events. Between the events and the historian there is a constant interplay. The historian tries to impose on events some kind of a rational pattern: how they happened and even why they happened".

A. J. P. Taylor, *Essays in English History.*

So it is that citric acid (von Scheele, 1784), quinic acid (Hofmann, 1790), morphine (Seturner, 1805), glycine (von Braconnet, 1820), geraniol (Jacobsen, 1871), peltogynol (Robinson, 1935), cephalosporin C (Abraham, 1955), and mevalolactone (Folkers, 1956) is not merely a record of the discovery and isolation of particular organic compounds from natural sources arranged in the right chronological order, but in a different context it portrays a constant theme and influence on the way, over the past two centuries, organic chemistry has developed as a scientific discipline. The investigations of the chemistry of natural products have not only been an essential element in man's endeavours to unravel the mysteries of the living world, but at the same time these studies have constantly refreshed and enriched the very fabric of organic chemistry itself, imparting new ideas and stimulating new directions in which the subject may grow. The strength of organic chemistry lies in its rich diversity. For those for whom it is above all an enabling science, to be used rather than admired for its own intrinsic elegance, the greatest excitements derive from looking outside its conventional boundaries. For many these interests take them within the compass of biology, where a multitude of properties and problems await an explanation and description at the molecular level. It is entirely appropriate therefore that a volume devoted to the organic chemistry of biological compounds should be included in the series Comprehensive Organic Chemistry.

Despite the series' collective title, Volume 5 is not primarily intended to be a comprehensive text embodying data on the structure and chemistry of all natural products as they are known. Rather its intention is to present, with sufficient background information and perspective, topics of major interest and importance in contemporary biological organic chemistry and so to give the reader some of the style and flavour of current developments in these areas of research. It is clear that within the past two decades new horizons have appeared in the study of natural products which strengthen its kinship and its ties to biochemistry. Increasingly, attention is drawn to the fundamental relationships which are seen to exist between structure, the keystone of organic chemistry, and biological function.

The present text sets out to reflect these ideas and relationships. In so far that we may have succeeded in this task the volume editor's thanks are due almost entirely to the many contributors not only for the chapters which they have composed but also for their helpful ideas and suggestions to improve the format and content of the volume. No less significantly, deadlines were met and friendships preserved.

The editors's thanks are also due to the other members of the editorial board for their comments and criticisms and to Dr. Colin Drayton of Pergamon Press for his considerable help and advice during the production of this text.

Sheffield E. HASLAM

Preface to Volume 5

[illegible]

[illegible]

[illegible]

[illegible]

[illegible]

Contributors to Volume 5

Professor M. Akhtar
Department of Biochemistry, University of Southampton

Dr. G. C. Barrett
Department of Science, Oxford Polytechnic

Dr. G. M. Blackburn
Department of Chemistry, University of Sheffield

Dr. G. Britton
Department of Biochemistry, University of Liverpool

Dr. J. D. Bu'Lock
Department of Chemistry, University of Manchester

Dr. B. W. Bycroft
Department of Chemistry, University of Nottingham

Professor D. T. Elmore
Department of Biochemistry, Queen's University of Belfast

Dr. B. T. Golding
Department of Molecular Sciences, University of Warwick

Professor F. D. Gunstone
Department of Chemistry, University of St. Andrews

Dr. J. R. Hanson
Department of Chemistry, University of Sussex

Dr. P. M. Hardy
Department of Chemistry, University of Exeter

Dr. E. Haslam
Department of Chemistry, University of Sheffield

Dr. R. B. Herbert
Department of Chemistry, University of Leeds

Dr. P. Hodge
Department of Chemistry, University of Lancaster

Professor L. Hough
Department of Chemistry, Queen Elizabeth College, University of London

Dr. D. W. Hutchinson
Department of Molecular Sciences, University of Warwick

Dr. P. M. Jordan
Department of Biochemistry, University of Southampton

Dr. J. F. Kennedy
Department of Chemistry, University of Birmingham

Dr. A. J. Kirby
University Chemical Laboratory, University of Cambridge

Dr. P. F. Knowles
Astbury Department of Biophysics, University of Leeds

Dr. G. Lowe
Dyson Perrins Laboratory, University of Oxford

Professor D. A. Rees
Unilever Research, Sharnbrook, Bedford

Dr. A. C. Richardson
Department of Chemistry, Queen Elizabeth College, University of London

Dr. R. C. Sheppard
M.R.C. Laboratory of Molecular Biology, Cambridge

Professor R. Thomas
Department of Chemistry, University of Surrey

Dr. R. T. Walker
Department of Chemistry, University of Birmingham

Dr. C. A. White
Biochemistry Department, Lea Castle Hospital, Kidderminster

Professor H. C. S. Wood
Department of Chemistry, University of Strathclyde

Contents of Other Volumes

PART 21

BIOLOGICAL CHEMISTRY: INTRODUCTION

21.1
Biological Chemistry: Introduction

E. HASLAM
University of Sheffield

21.1 PROLOGUE

Changes in the meaning of words is a process that goes on ceaselessly, inevitably, and imperceptibly. When first used by Berzelius at the beginning of the nineteenth century to signify a particular branch of chemistry, the description *organic* clearly implied the study of compounds which existed naturally as constituents of living matter. By the end of the century the word organic had, in the chemical context, long ceased to indicate a substance that is formed only in living systems, and when asked to define organic chemistry Roscoe (1871) described it as 'the chemistry of the carbon compounds' and Schorlemmer (1894) as 'the chemistry of the hydrocarbons and their derivatives'— definitions which are as acceptable today as when they were first made. This transformation was a reflection of the tremendous advances made during the nineteenth century in our understanding of the chemistry of the element carbon. It was in addition a recognition of the apparently limitless breadth of the subject. Since those days the changing emotional content with which the words 'natural product' have been endowed illustrates the changing views which chemists have expressed on the role which the study of such substances have played in the development of organic chemistry. Such studies have nevertheless continued with unabated vigour for the past century and a half. They are for many, if not all, a vital component of the whole subject and underline the continuing and intimate relationship of organic chemistry to biology.

Over the past 150 years the increasing volume of research in organic chemistry has helped to amass a vast body of information, and a continual problem for those who seek to keep abreast of all its developments is to place particular discoveries in their right context and perspective and to correctly assess their real significance to the whole. The contents of this volume have therefore been assembled not so much to be fully comprehensive as to be comprehensible, to reflect what are judged to be the truly important facets of the present state of biological organic chemistry. For a chemist, structure is all important and experience has shown that knowledge of structure is the only secure basis on which advances in biological chemistry can be made. Consistent with this attitude, considerable emphasis has been laid on the structure, synthesis, and reactivity of the organic compounds characteristic of all living systems — *primary metabolites* such as carbohydrates, lipids, amino-acids, peptides and proteins, nucleosides, nucleotides, and nucleic acids. Structure, however, should not be an end in itself, and the future of investigations in these areas must increasingly involve the chemist and use this chemical information as a starting point for the study of problems that lie deep in the realms of biology. The remarkable flowering of molecular biology in the 1960s means, for example, that many of the phenomenological problems of biology are now more clearly defined as problems for exponents of both the behavioural and molecular sciences. A typical case in point is that of enzymes. It is possible to study these substances simply as biological entities with a view solely to understand their influence upon the processes which they mediate. However, a much more strongly chemical approach is developed in this text to illustrate the extent of present knowledge concerning the ways in which the catalytic action of enzymes is achieved. Similar topics are likewise developed where appropriate elsewhere.

Traditionally, one of the most productive areas of enquiry for the organic chemist has been that of the chemistry of the so-called *secondary metabolites* such as phenols, quinones, terpenes, alkaloids, and the various pigments which organisms — particularly plants and micro-organisms — produce but for which no clear biological function has been identified. Studies of the chemistry of terpenes, for example, was instrumental in the very early development of synthetic methods (Scheme 1, W. H. Perkin's synthesis of (±)-α-terpineol, 1904), and in the recognition of one of the most commonly encountered molecular rearrangements in organic chemistry (Scheme 2, Wagner–Meerwein rearrangement).

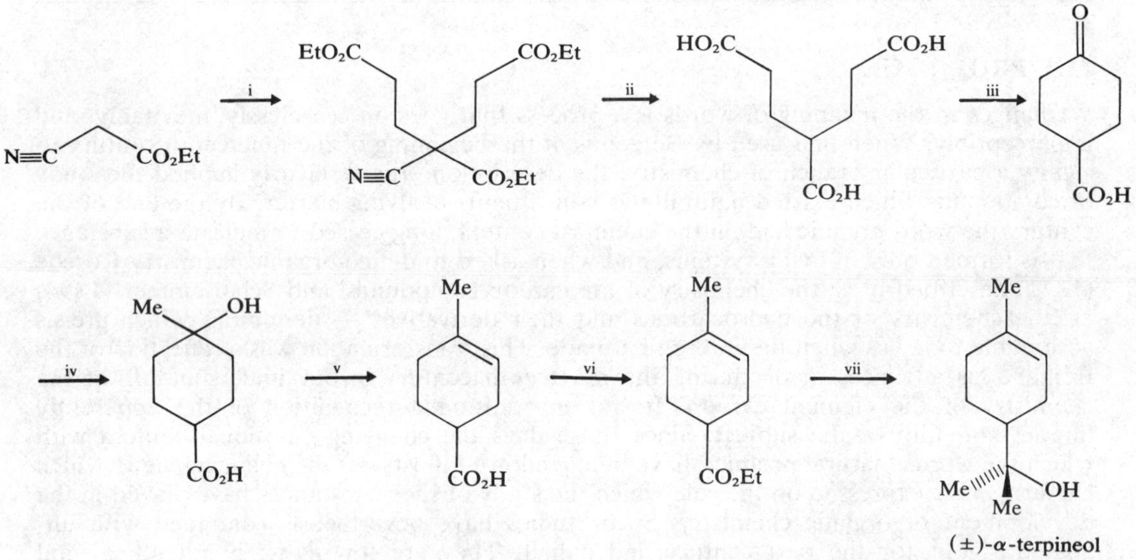

i, $ICH_2CH_2CO_2Et$, OEt^-; ii, H^+; iii, Ac_2O; iv, MeMgI; v, HBr; OH^-; vi, H^+, EtOH; vii, MeMgI.

SCHEME 1 Synthesis of (±)-α-terpineol — W. H. Perkin (1904)

The principles of conformational analysis enunciated in 1950 by D. H. R. Barton,[1] which changed the whole basis of the approach to the stereochemical analysis of carbon compounds, were based on ideas from chemical physics. There is little need to remind the serious student of the subject that the author chose examples to illustrate these new concepts in their most striking way from the chemistry of terpenes and steroids, nor that the article was entitled 'The conformation of the steroid nucleus'!

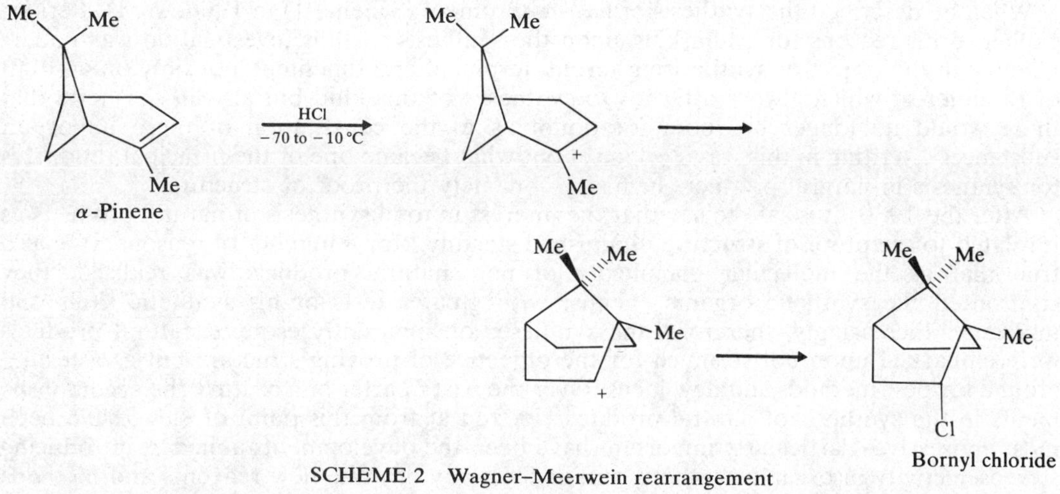

SCHEME 2 Wagner–Meerwein rearrangement

There does remain, however, a fundamental and as yet unresolved problem concerning these lavishly accumulated secondary metabolites. Simply stated it is — what is their biological function? In this connection some consider these substances as waste products of metabolism, although many are toxic to the organisms which produce them unless they are dissipated into the environment (*e.g.* the volatile monoterpenes produced by many plants) or are sequestered harmlessly in the organism itself. Others regard secondary metabolites, and in particular those found in plants, to be important factors in the co-evolution of plants, animals, and insects. The view is expressed that the selective pressures of animal and insect depredation has determined the qualities of chemical production by plants, and has led to the metabolism of particular products which repel animals and other herbivores and thus relieve the animal depredation. Whichever view is correct, and biologists clearly differ on this point at present, a first step towards an understanding of the function of secondary metabolites in an organism is to understand their biological origins. The organic chemist has made significant progress in this area during the past two decades and the routes by which many secondary metabolites are biosynthesized are now clearly delineated. These studies are reviewed in the final sections of this volume and they will provide a firm factual basis for further investigations of this problem.

Historically, organic chemistry has its origins in the study of substances obtained from living matter, but as far as its relationship to biology is now concerned the wheel has almost come full circle. The ample evidence of the ways in which research into fundamental problems in biology are now proceeding, and to which some reference has been made, leaves little doubt that the influence of the discipline of organic chemistry upon biology is likely to grow rather than diminish in the future. Organic chemistry will thus return to nourish and enrich the soil from which it has itself developed.

21.2 BIOGENETICALLY PATTERNED SYNTHESES

A very large part of current activity in the whole sphere of organic chemistry is concerned with synthesis and it is pertinent to note briefly the relatively recent influences which the chemistry of natural products have had upon synthesis. In particular the concept of biogenetic-type or biogenetically patterned synthesis is worthy of further amplification. For the organic chemist, if one takes Robinson's early landmark in this field (the synthesis of tropinone[2]) as the immaculate conception, then the period of gestation can be said to have been relatively long and uneventful, for it is only relatively recently that the ideas have begun to bear forth the promised fruit of earlier years.

When he described the synthesis of (±)-α-terpineol (Scheme 1) in 1904, W. H. Perkin[3] explained his reasons for embarking upon the synthesis . . . 'this investigation was undertaken with the object of synthesising terpin, terpineol and dipentene not only on account of the interest which always attaches to syntheses of this kind but also in the hope that there would no longer be room for doubt as to the constitution of these important substances'. Writing in this way Perkin set out what became one of the principal rationales for synthesis in natural product chemistry — namely the proof of structure.

After the 1930s it is safe to say that the interest in total synthesis of natural products as it related to the proof of structure diminished steadily,[4] for a number of reasons. It is also true that as the molecular complexity of new natural products was realised, they confronted the synthetic organic chemist with greater tests of his synthetic skills and ingenuity. Increasingly, therefore, the syntheses of apparently esoteric natural products were embarked upon not so much for the objective of proving structure but as a testing ground for new methods and new ideas. Over the past quarter of a century the accomplishments in the synthesis of natural products, looked at from this point of view, have been most impressive. Particularly important have been the developments aimed at introducing stereoselectivity in organic synthesis and at the array of quite new reagents and methods which have been introduced into the armamentarium of the chemist. Examples of many of these features are found in syntheses, such as those of strychnine, the β-lactam antibiotics, chlorophyll, vitamin B-12, and the prostaglandins.

J. N. Collie, one of the pioneers of biosynthetic studies and of the biogenetic-type synthesis of natural products, wrote in 1893 . . . 'the attempt to artificially produce naturally occurring substances, and to imitate in the laboratory some of the many processes which are perpetually being carried on around us in nature, has always been one of the chief aims of the organic chemist'. Nevertheless, some 70 years later van Tamelen was able to remark on the almost complete lack of similarity between the synthetic pathways adopted by the organic chemist to synthesize intricate natural molecules and the methods and routes which are presumed to be employed in nature for the production of these compounds. The now famous synthesis of tropinone by Robinson, contrasting as it did with the lengthy conventional synthesis of the same material announced by Willstätter, was the first example which demonstrated the intrinsic elegance of synthetic methods based on the philosophy of building natural molecules under mild conditions from components which simulate or suggest those used in nature. The full potential of these ideas has only been slowly exploited, and the relatively modest achievements made in this direction were reviewed by van Tamelen[6] in 1961. However, the acquisition since 1950 of much greater information concerning biosynthetic pathways has, predictably, led to much greater activity in this field. Some of the recent examples of the use of these concepts in the planning and the execution of organic syntheses are discussed below.

Biogenetic-type or biogenetically patterned syntheses of natural products are generally understood to mean syntheses which utilize, at least in their principal aspects, pathways and reactions which are modelled upon those used in nature for the biosynthesis of the self-same molecules. Generally this comparison does not extend to reagents and to reaction conditions. In this sense those syntheses such as Robinson's of tropinone, which may be achieved under reaction conditions which also closely parallel the physiological ones, are perhaps best categorized as both biogenetically and physiologically patterned.

Collie's early investigations which led ultimately to the recognition of the polyketide hypothesis were based,[5,7] in part, on the reactions of dehydracetic acid (**1**); its transformation with alkali, presumably *via* the intermediate (**2**) to the known natural product (**3**) (orcinol), and with the action of acid and alkali to the naphthol (**4**) (Scheme 3), were the most significant. Subsequent studies of biogenetic-type syntheses of the polyketide class of metabolite have been hampered more by the relative inaccessibility and intrinsic lability of the appropriate poly-β-carbonyl precursors than by difficulties in their subsequent transformation to aromatic products typical of the polyketide class of metabolite.

Numerous syntheses of monocyclic aromatic compounds, which involve free or partially protected β-tetracarbonyl compounds as precursors, have been reported. Scott and

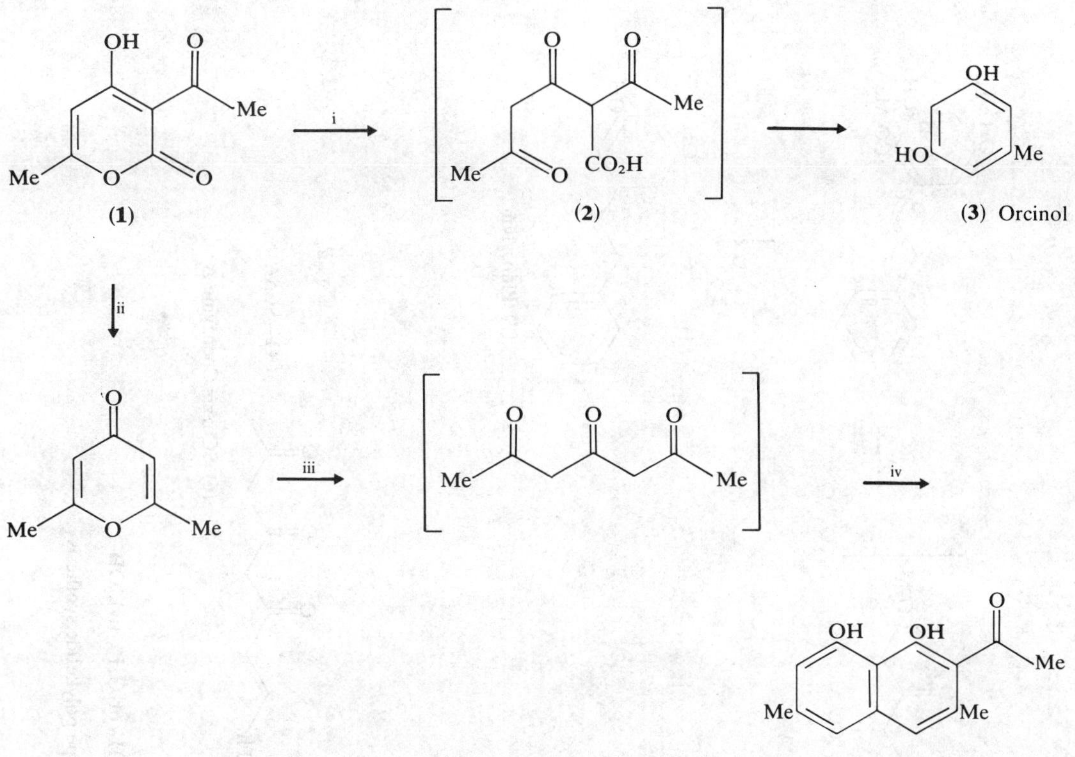

(1) (2) (3) Orcinol (4)

i, OH^-; ii, H^+; iii, OH^-; iv, OH^-.

SCHEME 3 Collie's biogenetic polyketide syntheses

Money and their collaborators, for example, have adopted[8] the original idea of Collie and later Birch to protect the labile β-keto-compound in the form of a pyrone ring. Using the dipyrone (**5**) — potentially equivalent to the biological precursor (**6**) — they were able to convert this with different basic treatments to the natural stilbene pinosylvin (**7**) and the flavanone (±)-pinocembrin (**8**) (Scheme 4).

A notable feature of much of this work has been the different directive effects observed in cyclizations exhibited by magnesium ions, and Crombie[9] has provided an explanation of many of these phenomena in terms of the selective chelating effects of the magnesium ions with the poly-β-keto-precursors.

Until comparatively recently the corresponding use of the higher poly-β-carbonyl compounds and their derivatives in biogenetically patterned synthesis had not led to any successful syntheses of naphthalene- or anthracene-based polyketide natural products. However, very recently biogenetic-type syntheses of polycyclic polyketide metabolites using partially protected β-hexa- and β-hepta-ketones were announced by Harris and his collaborators.[10,11] Successful syntheses of 6-hydroxymusizin (**9**) and barakol (**10**) (Scheme 5) and eleutherinol (**11**) (Scheme 6) were achieved by this group. Eleutherinol (**11**), it will be noted, provided initially a very effective demonstration of the validity of the polyketide theory of biogenesis, when application of the theory led to a successful revision of its structure.[12]

The control of the cyclization process to give the desired product becomes a formidable problem with the higher poly-β-carbonyl compounds. However, Harris and his collaborators have successfully overcome these problems in biogenetically patterned syntheses of the natural anthraquinones emodin (**12**) and chrysophanol (**13**) (Scheme 7). The two syntheses follow closely similar routes after the first condensation step.

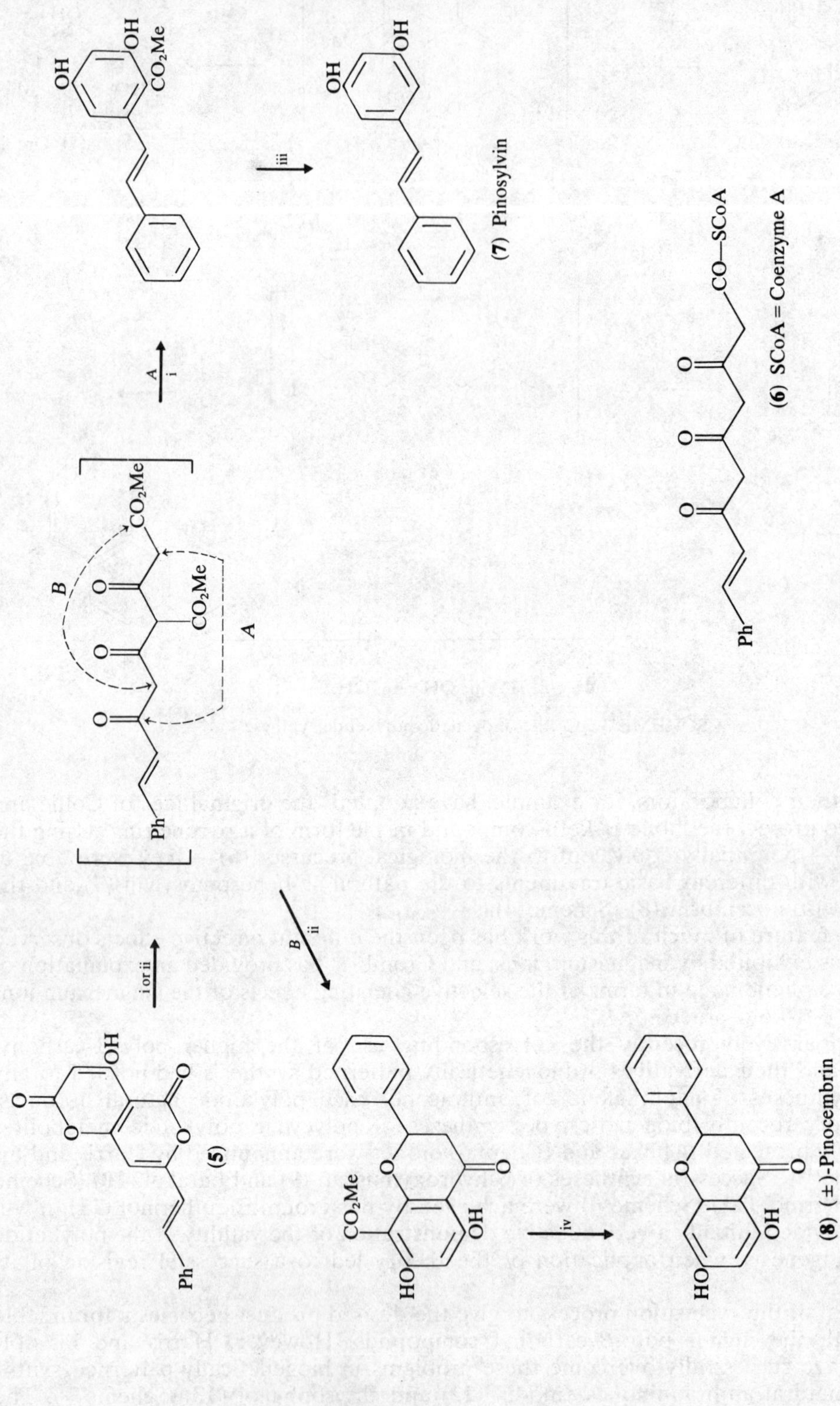

i, KOH; ii, $Mg(OMe)_2$; iii, H^+ or OH^-; (iv), OH^-.

SCHEME 4 Biogenetic-type polyketide synthesis

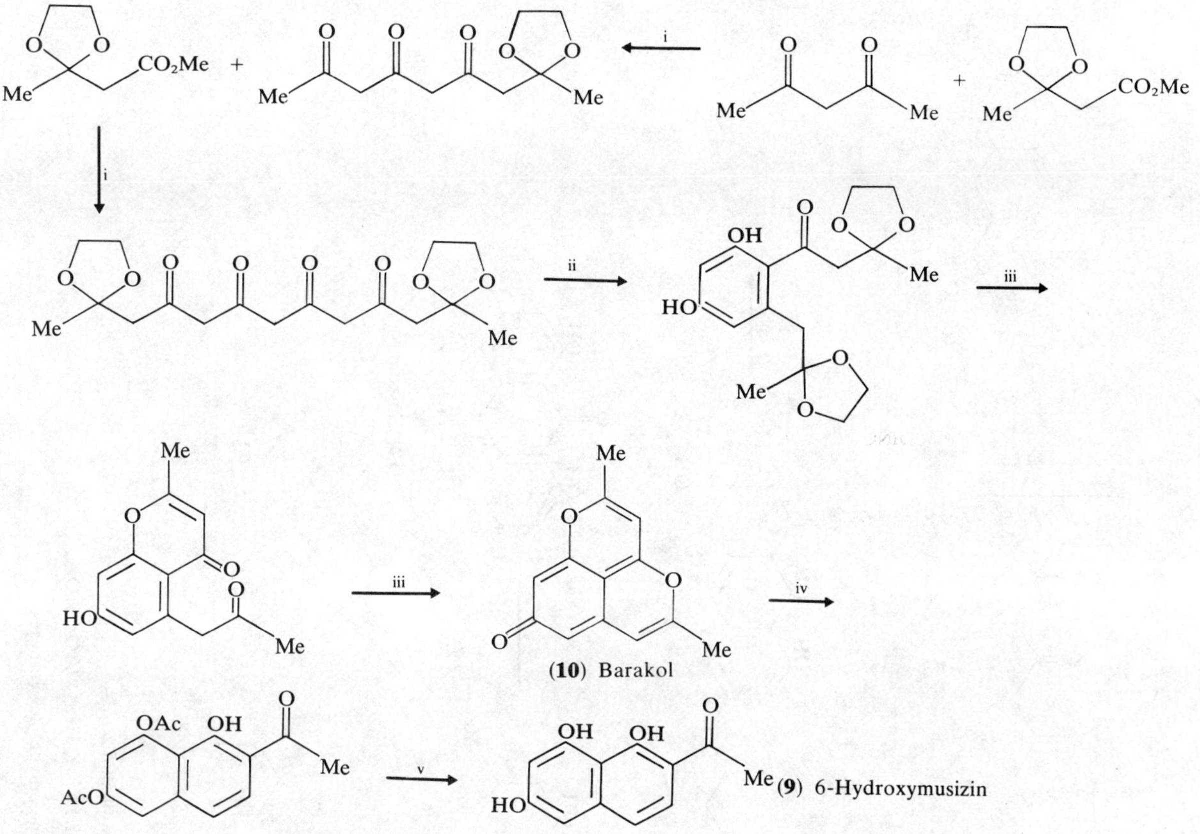

i, $LiNPr^{i}_2$; ii, Pr^{i}_2NH; iii, H^+; iv, Ac_2O, pyridine; H^+; v, OH^-.

SCHEME 5 Biogenetic-type polyketide syntheses

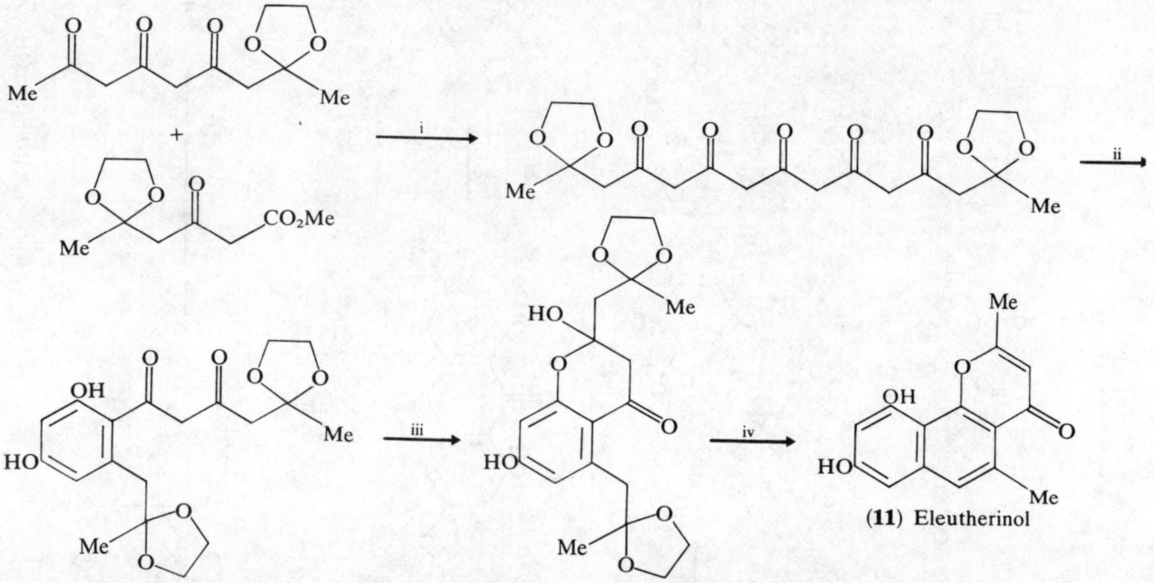

i, $LiNPr^{i}_2$; ii, Et_3N; iii, spontaneous; iv, H_3O^+, Pr^{i}_2NH, CF_3CO_2H.

SCHEME 6 Biogenetic-type polyketide synthesis

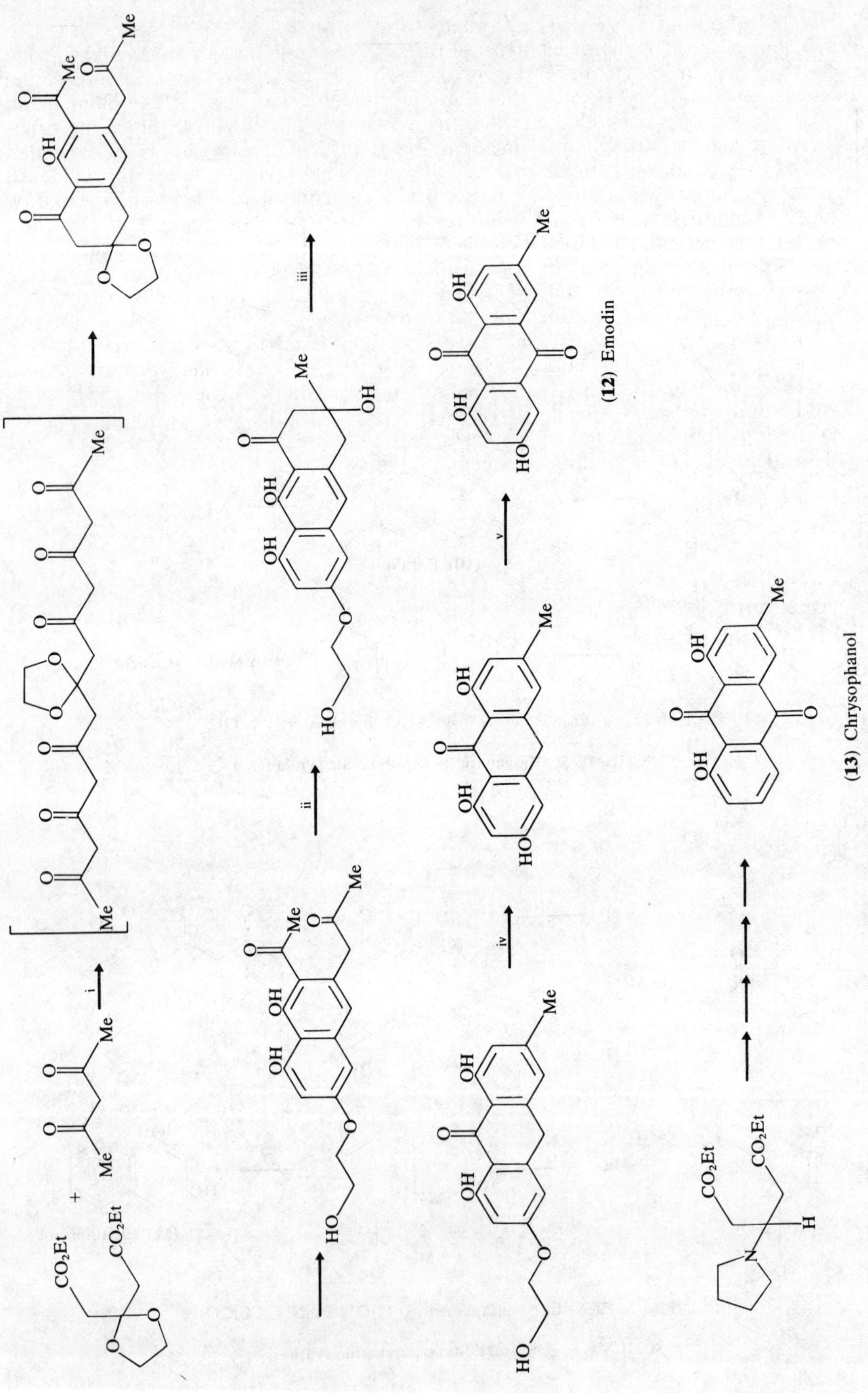

i, $LiNPr^i_2$; ii, OH^-; iii, pyridine or H^+; iv, HI; v, CrO_3, HOAc.

SCHEME 7 Biogenetic-type polyketide syntheses

Although Robinson once remarked[13] that 'we must not allow the biosynthetic tail to wag the chemical dog', many of his ideas in the field of biosynthesis were seminal. None more so than his concept that the steroid skeleton may be derived by a folding of the hydrocarbon squalene, and his idea that the morphine alkaloids arise in the plant by an oxidative coupling of a phenolic benzyltetrahydroisoquinoline precursor. Since the original conception and its subsequent refinement,[14] a great deal of effort has been expended to elucidate the actual biosynthetic pathway of morphine and finally to use this evidence to devise a biogenetically patterned synthesis. The biosynthetic pathway was delineated by the brilliant experimental work of several investigators, notably Battersby and Barton. The crucial step in this metabolic pathway is the *para–ortho* oxidative coupling of reticuline (**14**) to give the dienone salutaridine (**15**). Many previous attempts to bring about this coupling have, with the single exception of that of Barton and his associates, usually given the *para–para* product ($\leqslant 4\%$) or the *ortho–para* product ($\leqslant 53\%$). Barton and his group reported a yield of 0.03% of (±)-salutaridine by ferricyanide oxidation of (±)-reticuline. Schwartz and Mami[15] have now effected the vital oxidative coupling in a less than miniscule yield (11–23%) using thallium trifluoracetate on the *N*-trifluoracetyl or *N*-ethoxycarbonyl derivatives of (±)-*N*-nor-reticuline. With this step accomplished, and with the completion of earlier relays, a successful biogenetic-type synthesis of the morphine alkaloids is thus complete (Scheme 8).

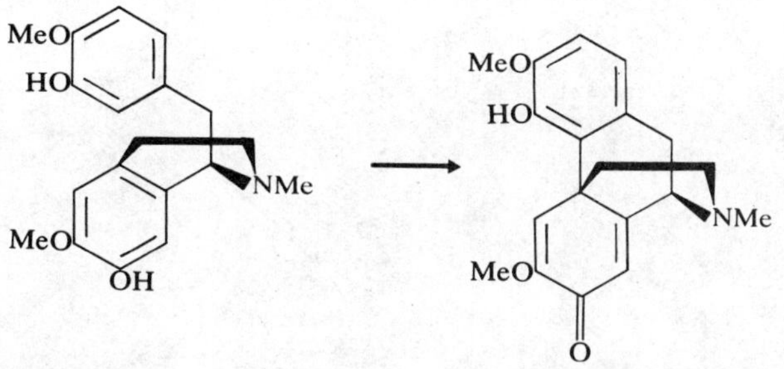

(**14**) (**15**)

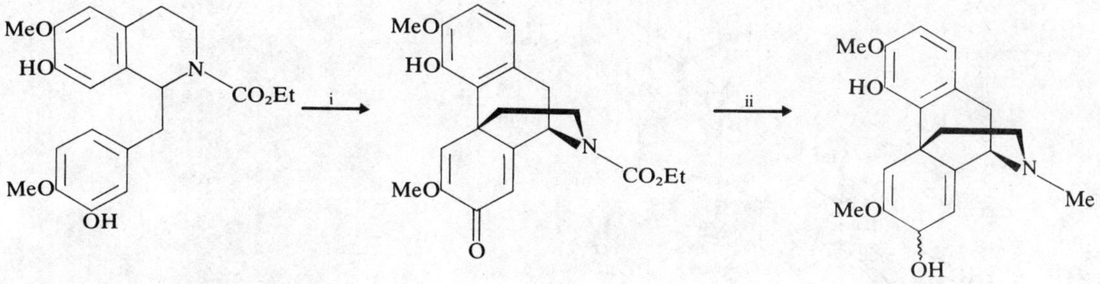

N-Ethoxycarbonylnor-reticuline

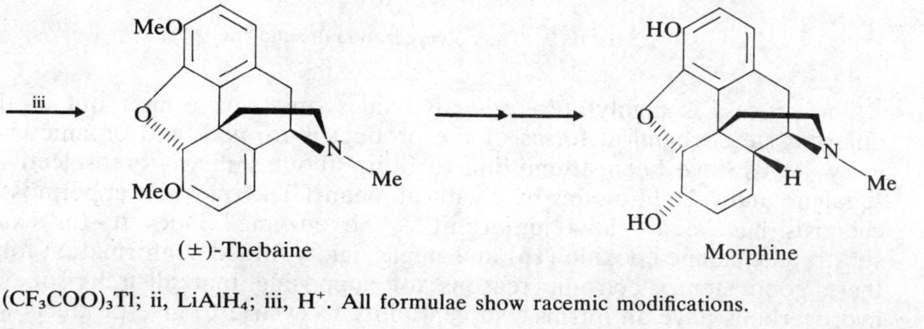

(±)-Thebaine Morphine

i, $(CF_3COO)_3Tl$; ii, $LiAlH_4$; iii, H^+. All formulae show racemic modifications.

SCHEME 8 Biogenetic-type synthesis of morphine alkaloids

One of the intellectually most satisfying achievements in the area of biogenesis has been the determination of the pathway by which cholesterol is metabolized from acetyl coenzyme A. During the controversies which surrounded the proof of the constitution of cholesterol, Robinson first suggested[16] that the cholesterol carbon skeleton could be identified with the hydrocarbon squalene less three carbon atoms. Subsequently, of all the stages involved in the biogenesis of cholesterol, that particular step in which squalene, *via* the epoxide (**16**), is cyclized to give lanosterol (**17**) has intrigued organic chemists most of all (Scheme 9).

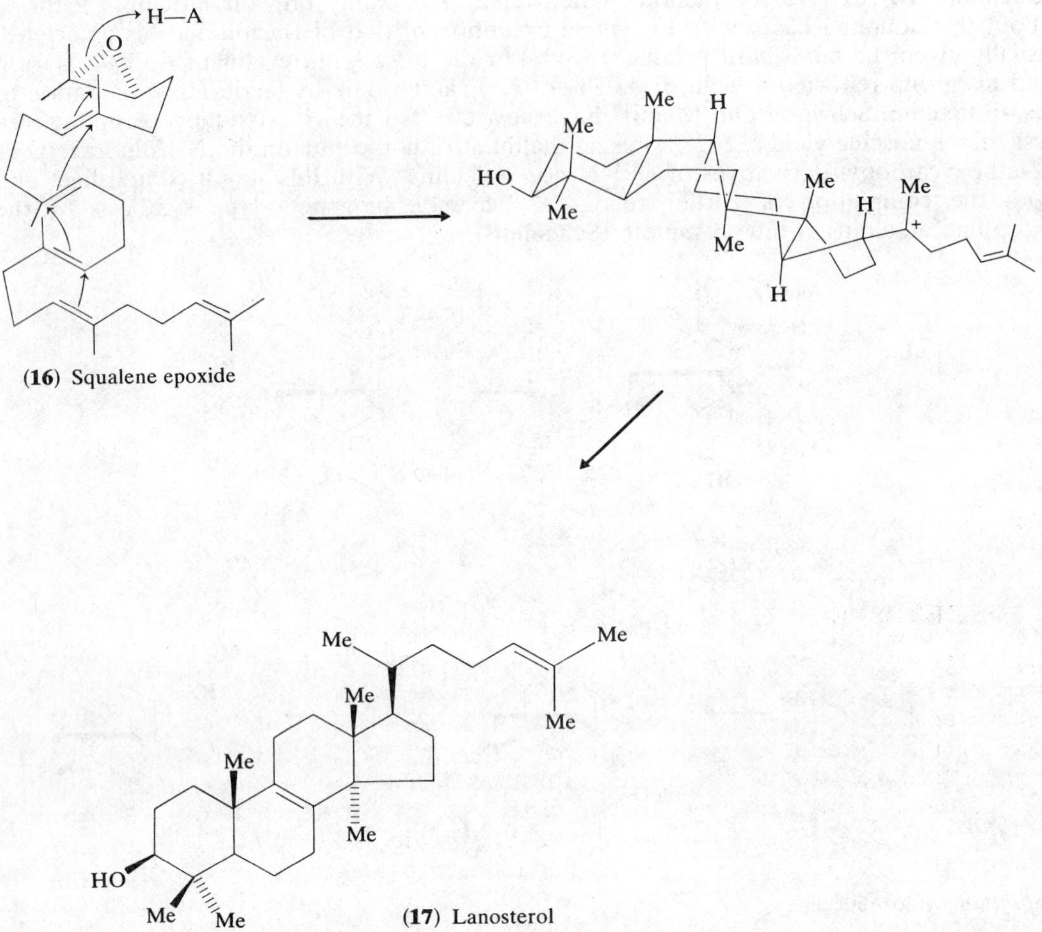

(**16**) Squalene epoxide

(**17**) Lanosterol

SCHEME 9 The polycyclization of squalene epoxide

The process is highly stereoselective since only one isomer out of the 128 possible different stereochemical forms of the product is formed, and organic chemists since the early 1950s have been attempting to bring about similar stereoselective cyclizations of squalene and related olefins by chemical means. The question uppermost in the minds of chemists has been — how important is the enzyme? Does it, for example, hold the substrate squalene epoxide (**16**) in a single rigidly folded conformation for reaction, or are there good stereoelectronic reasons for supposing that all-*trans*-squalene and similar hydrocarbons have an intrinsic susceptibility to cyclize to give products with the 'natural' relative configuration — a concept set forth independently by Stork and Eschenmoser.[17,18]

A minor flood of publications[19] has followed the discussion of this hypothesis in the literature, and with them a number of biogenetic-type syntheses of terpenes and steroids have been successfully completed. These have used as their basis the inherent stereoselectivity of many polyolefin cyclizations, which has been established by model experiments. Although squalene 2,3-epoxide (**16**) itself on acid treatment ($SnCl_4$ in benzene) yielded[18] two non-steroidal tricyclic systems (**18**) and (**19**), acid-catalysed cyclization (picric acid in nitromethane) of the epoxydiol (**20**) derived from squalene gave the natural triterpene (±)-malaricanediol (**21**; yield 7%).[20]

Johnson has similarly used the principle ingeniously[21,22] in the successful generation of the D-homosteroid (**22**) and of (±)-16,17-dehydroprogesterone (**23**) (Scheme 10).

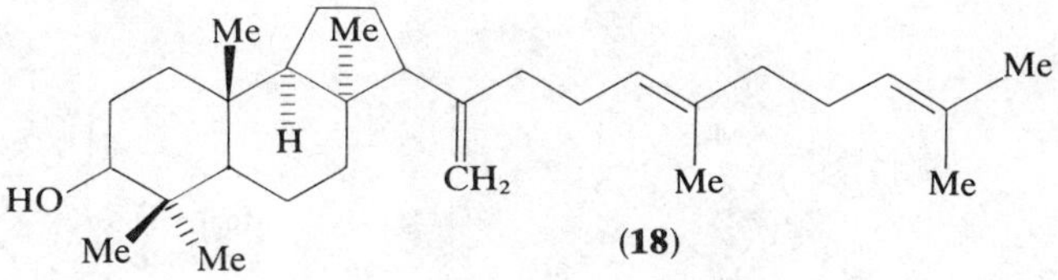

(**18**)

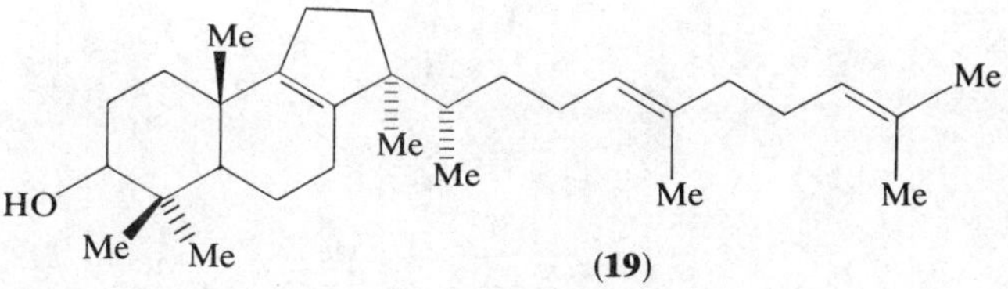

(**19**)

21.3 STEREOSPECIFICITY OF ENZYME REACTIONS

Enzymes are the catalysts for biological reactions. Their catalytic power is often quite prodigious and, coupled with their specificity for the substrate, this enables an organism to organize and select, for a given molecule, just one discrete metabolic pathway out of the many possible chemical reactions which that molecule and its progeny may undergo. The specificity of the enzyme for a particular substrate may be structural or stereochemical in origin. Structural specificity may be quite distinctive or, alternatively, it may be relatively broadly based, as for example is shown by the hydrolytic enzymes of the digestive system. Stereospecificity is a feature of the enzymic catalysis of reactions involving enantiomers and was first noted well over a century ago, but over the past 30 years there has developed a very extensive field of research into the often much more subtle stereochemical characteristics of enzyme reactions. This area of investigation emerged during the 1940s and developed from a number of quite unrelated observations in biochemistry, in particular the interesting controversy which arose concerning the status of citric acid (**24**) as an intermediate in the Krebs or tricarboxylic acid cycle.

Early studies of the Krebs cycle using labelled carbon dioxide[23] showed that the labelled carbon atom introduced into the metabolite α-ketoglutarate (**25**) was present in the carboxyl group • adjacent to the carbonyl group. Primarily because of the localization of the label, the reaction sequence was assumed to by-pass citrate and to proceed by way of isocitrate (**26**) to α-ketoglutarate. It was argued that the symmetrical molecule citrate would of necessity behave as though the isotopic label were distributed equally in the two equivalent carboxyl groups and would thus give α-ketoglutarate (**27**) labelled in both of its carboxyl groups (Scheme 11).

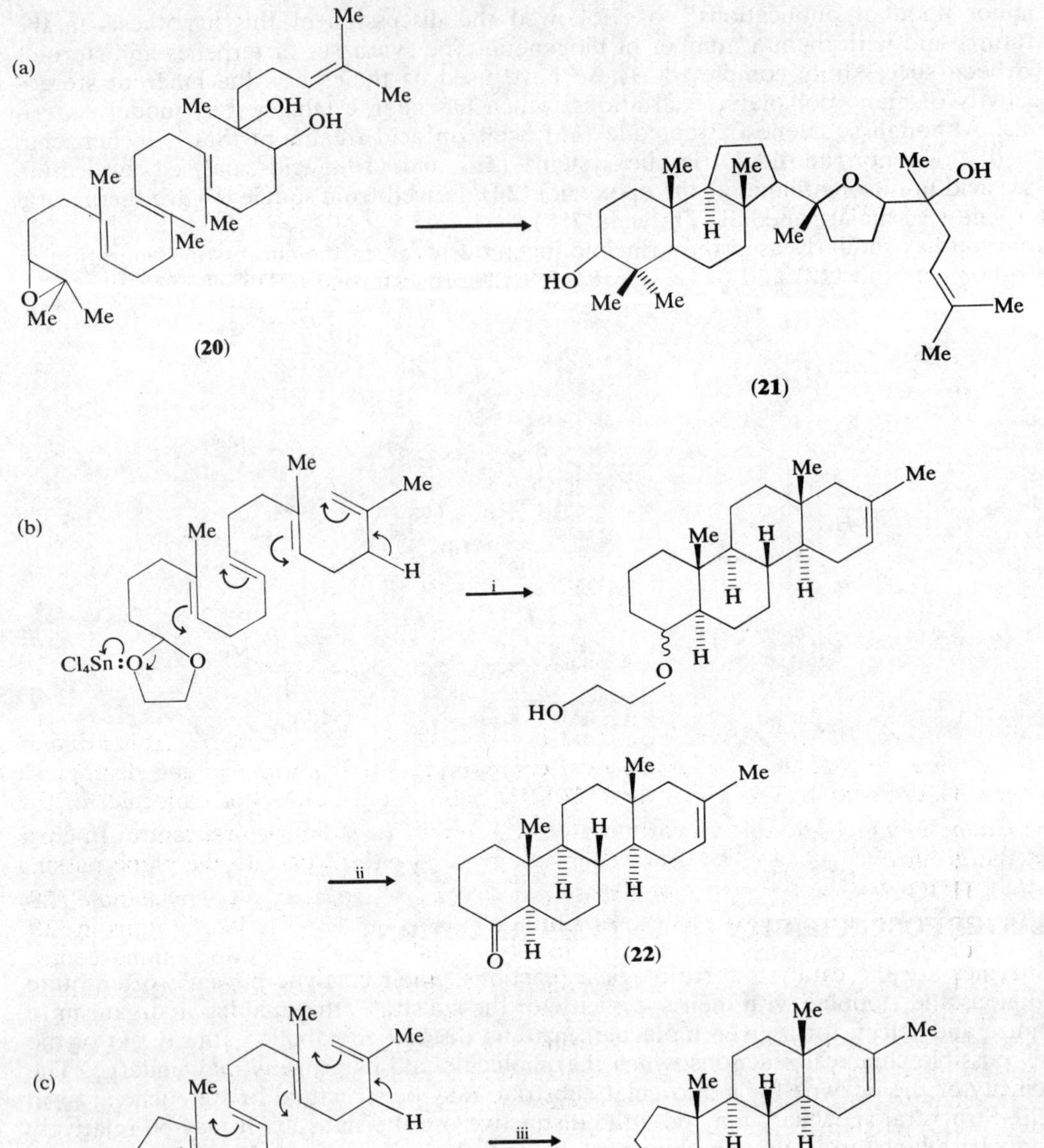

i, $SnCl_4$; ii, tosyl chloride, pyridine; NaI; Zn; CrO_3; iii, CF_3CO_2H; iv, OsO_4, H_2S; $Pb(OAc)_4$; v, KOH. All formulae represent racemic modifications.

SCHEME 10 Biogenetic-type triterpene and steroid synthesis

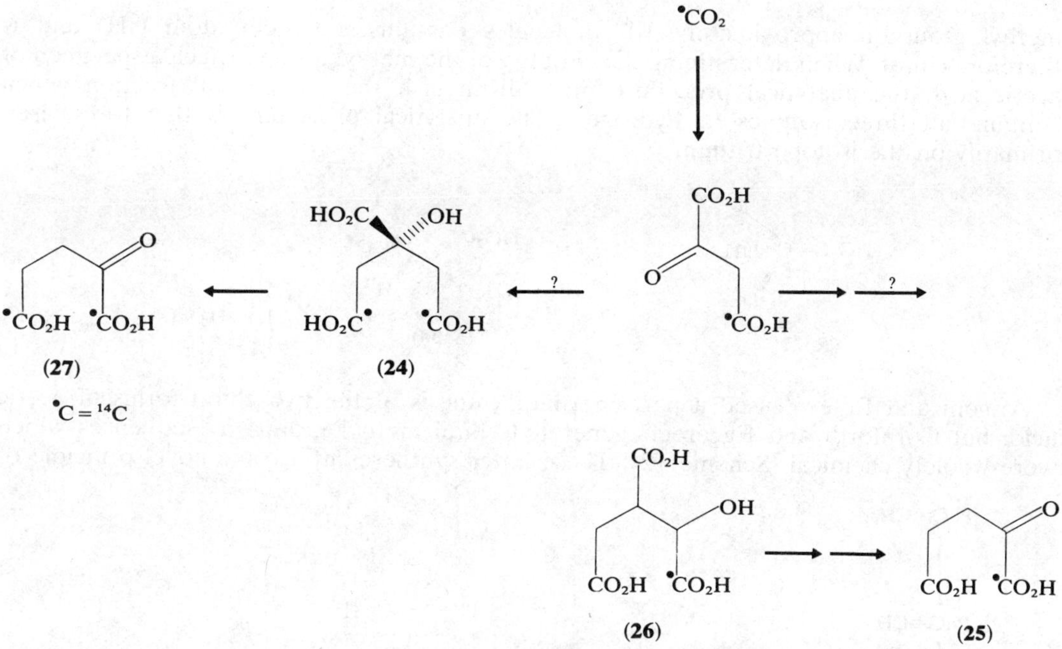

SCHEME 11 Early studies of the tricarboxylic acid cycle

Ogston,[24] however, first pointed out that it is possible for an asymmetric reagent, such as an enzyme, to react with a symmetrical compound such as citric acid and distinguish between its identical groups. Citric acid (**24**) is a member of the class of molecules of the type $\overset{*}{C}aabc$ in which the starred carbon atom is referred to as a prochiral centre. In citric acid the groups *a* (—CH_2CO_2H) are enantiotopic and with respect to the plane passing through HOOC—$\overset{*}{C}$—OH are related as object to mirror image. In L-phenylalanine (**28**) the two groups *a* are protons and the prochiral centre is at C-3. The two protons in (**28**) have a diastereotopic relationship due to the chirality at the C-2 α-amino-centre. Diastereotopic groups are, in principle, distinguishable by physical methods and both enantiotopic and diastereotopic groups are differentiated in a chiral environment such as is provided by an enzyme. Many experimental observations in the field of enzyme chemistry have amply confirmed these conclusions. It is an area of research in which the organic chemist has been at the forefront and has made impressive and imaginative contributions to almost every aspect of this work.

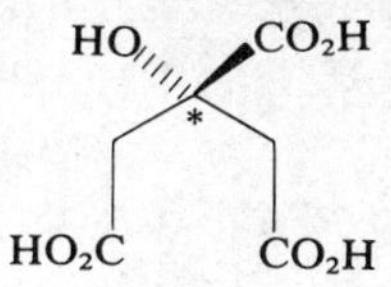

(**24**)

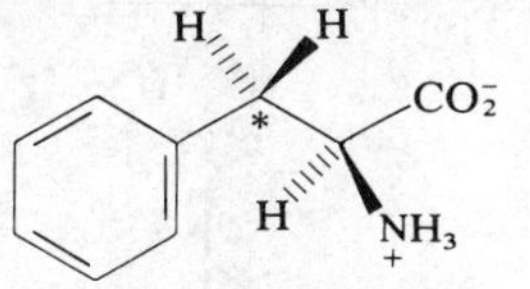

(**28**)

Undoubtedly one of the outstanding investigations in this field and, indeed, in the whole of recent research in organic chemistry, has been the study of the stereochemical course of enzyme-catalysed reactions which involve the interconversion of methyl and methylene groups.[25] Two groups independently chose very similar approaches to this most subtle of stereochemical problems.[26,27] Samples of acetic acid were generated in which the methyl group has, by isotopic substitution (^1H, ^2H, ^3H), either the *R* (**29**) or the *S* (**30**) configuration. Although almost all the methyl groups of such samples contain ^1H and ^2H, only one methyl group in approximately 10^6 molecules contains a tritium atom (^3H) and is therefore chiral. When determining the chirality of the methyl group in such a specimen of acetic acid, the analytical procedure must distinguish that very small fraction which contains all three isotopes of hydrogen. The analytical procedure is therefore based primarily on the isotope tritium.

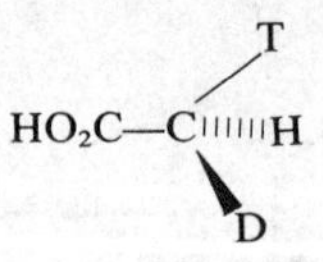

(*R*, **29**)

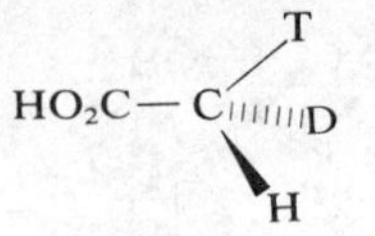

(*S*, **30**)

{D = ^2H, T = ^3H}

Arigoni and Retey[27] used a part-enzymic synthesis of the two chiral forms of acetic acid, but Cornforth and Eggerer[26] generated chiral methyl groups by sequences which were wholely chemical (Scheme 12). These latter syntheses utilize the novel principle of

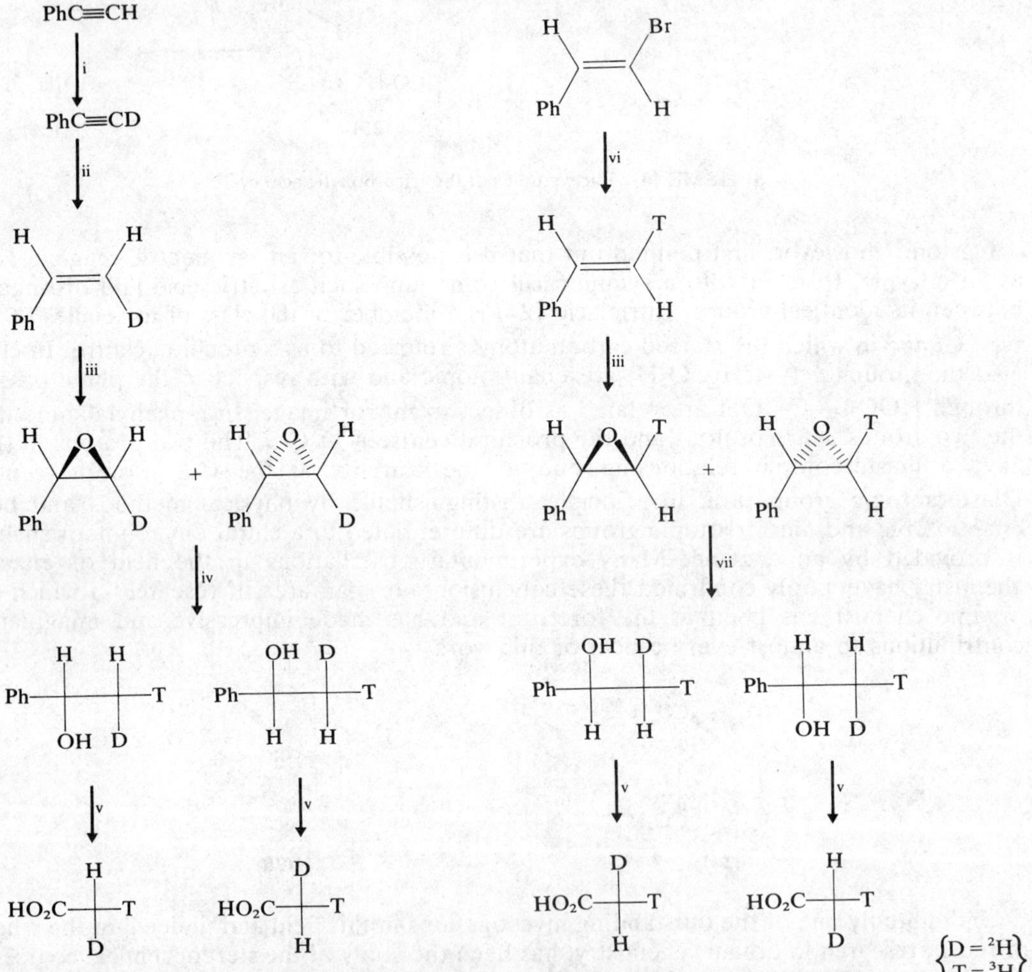

(*S*, **30**) (*R*, **29**) (*R*, **29**) (*S*, **30**)

i, EtMgBr, D_2O; ii, N_2H_2; iii, *m*-chloroperbenzoic acid; iv, lithium borotritiide; resolution of acid phthalates with brucine; v, CrO_3, H^+; CF_3CO_3H; OH^-; vi, BuLi, ^3HOH; vii, lithium aluminium deuteride; resolution of acid phthalates with brucine.

SCHEME 12 Chemical synthesis of chiral acetic acids

resolving the two chiral methyl groups by building them into a racemic diastereoisomer. This racemate is resolved at the alcohol group and subsequently the carbinol carbon atom is transformed to the carbonyl group of the acetic acid. In later work (Scheme 13), Arigoni[28] developed an interesting chemical synthesis of both forms of chiral acetic acid in which the sense of chirality of the methyl group is dictated by the absolute stereochemistry of the intermediate (**31**) and by the strict geometrical requirements imposed upon the two transition states that lead to its formation.

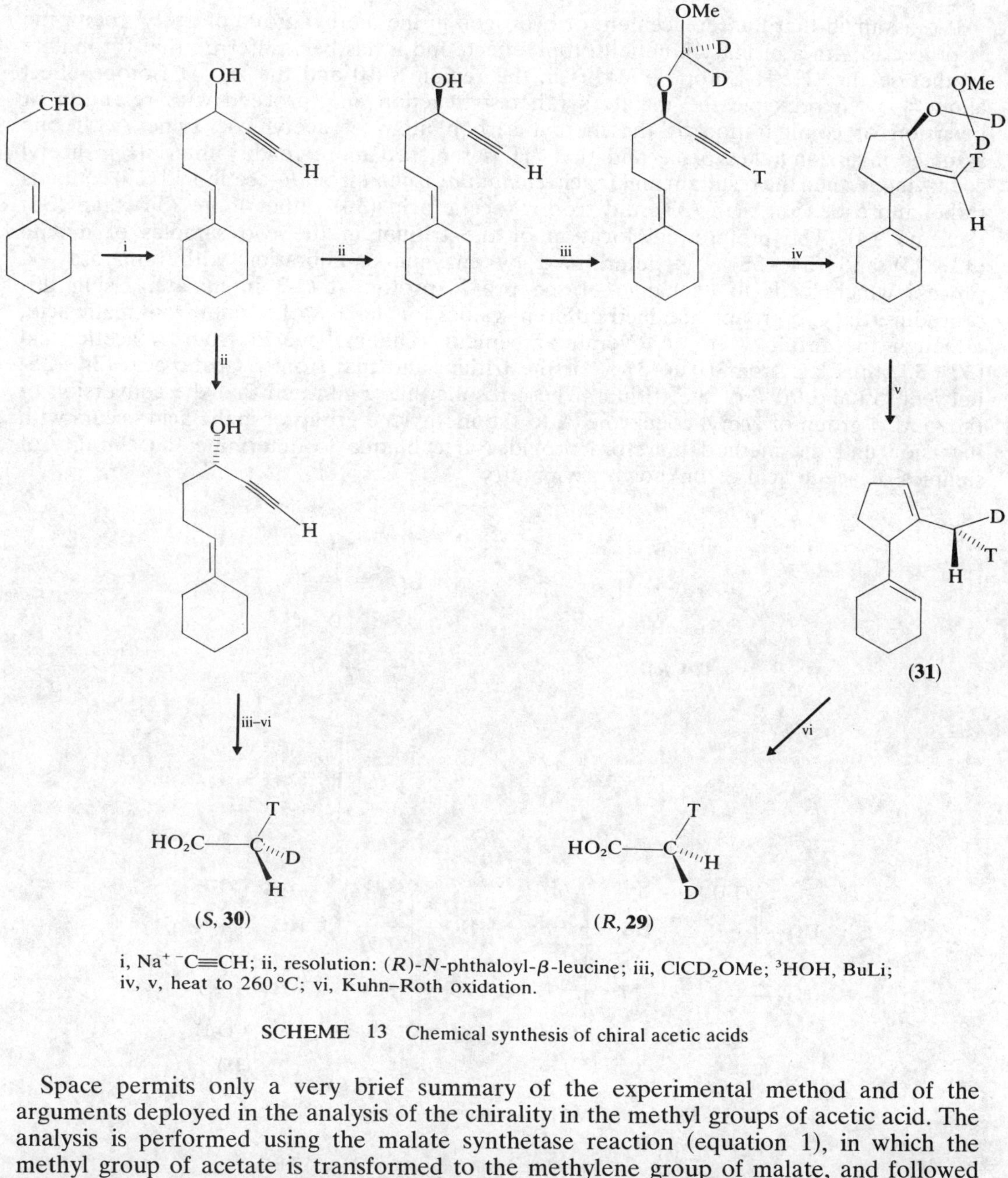

i, Na^+ $^-C{\equiv}CH$; ii, resolution: (*R*)-*N*-phthaloyl-*β*-leucine; iii, $ClCD_2OMe$; ^3HOH, BuLi; iv, v, heat to 260 °C; vi, Kuhn–Roth oxidation.

SCHEME 13 Chemical synthesis of chiral acetic acids

Space permits only a very brief summary of the experimental method and of the arguments deployed in the analysis of the chirality in the methyl groups of acetic acid. The analysis is performed using the malate synthetase reaction (equation 1), in which the methyl group of acetate is transformed to the methylene group of malate, and followed the fate of the isotope tritium. Both the *R* and *S* forms of acetic acid (**29** and **30**) were condensed as their coenzyme A derivatives with glyoxylic acid to give samples of malate.

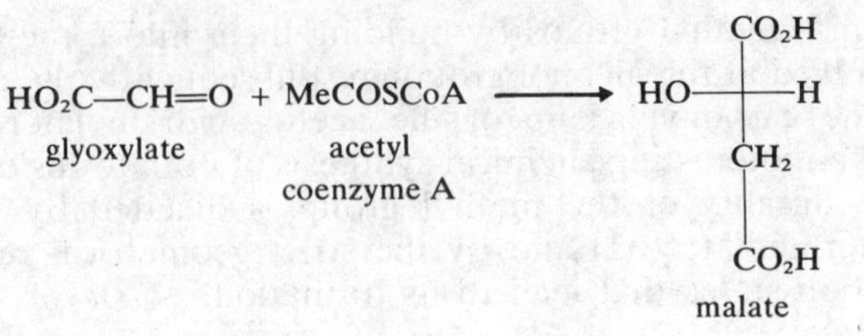

(1)

It is assumed that the replacement of hydrogen in the methyl group of acetyl coenzyme A proceeds with a primary kinetic isotope effect, and it has been inferred that for malate synthetase the $^1H/^3H$ isotope effect is in the region 8–10 and the $^2H/^3H$ isotope effect about 4–5. In principle the malate synthetase reaction may proceed with retention or inversion of configuration at the methyl carbon atom of acetyl coenzyme A. If one assumes inversion takes place and that ^1H is replaced more readily than ^2H in acetyl coenzyme A, then the resultant makic acid *containing tritium* from *R*-acetic acid (**29**) contains rather more of (**32**) than (**33**) and from *S*-acetic acid (**30**) rather more (**34**) than (**35**) (Scheme 14). The preferential location of the tritium in the two samples of malate (**32**+**33**) and (**34**+**35**) was determined by enzymic equilibration with fumarase—a process which leads to exchange of the pro-*R* proton at C-3 in malate. Using this procedure the two groups obtained different values for the loss of tritium from malic acid, although the results were in overall agreement. The malic acid from *R*-acetic acid (**32**+**33**) thus lost from 10 to 33% of the tritium and that from *S*-acetic acid (**34**+**35**) between 69 and 90% of the tritium. These results thus confirmed that the conversion of the methyl group of acetyl coenzyme A to the methylene group of malic acid occurs with inversion and the method therefore provides a technique to determine the chirality of samples of acetic acid of unknown asymmetry.

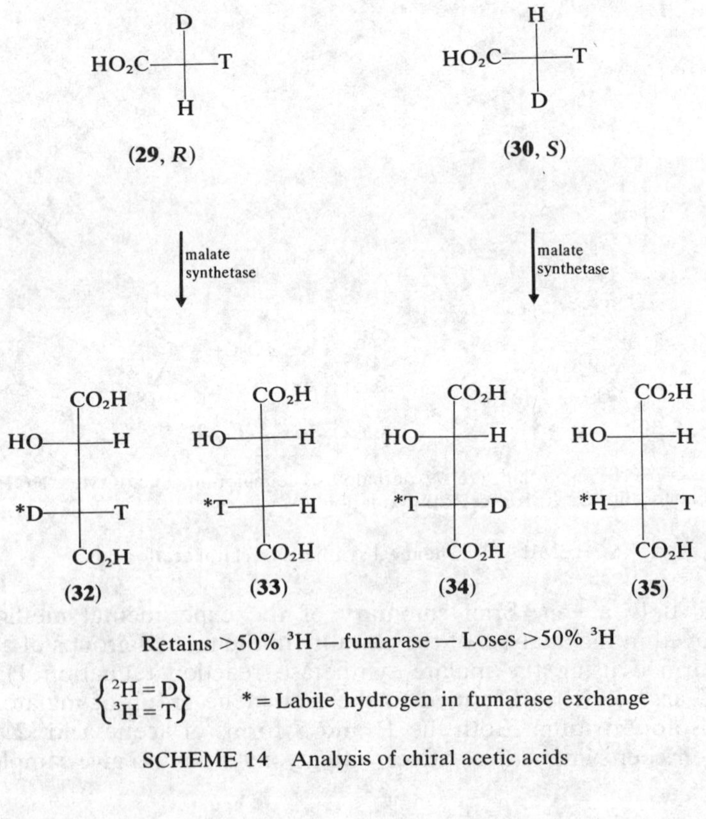

SCHEME 14 Analysis of chiral acetic acids

The principles of these investigations have been applied to a variety of analogous enzyme reactions. Typical of these is the investigation of the stereochemistry of the 3-hydroxy-3-methylglutaryl CoA lyase[29] (equation 2).

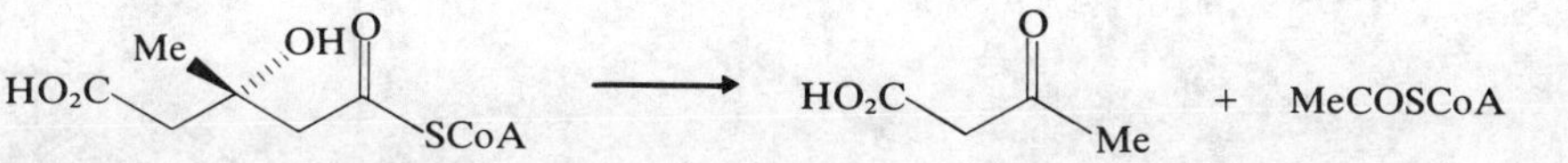

(2)

(SCoA = Coenzyme A)

Samples of 2*S*,3*S* (with 2*R*,3*R*) and 2*R*,3*S* (with 2*S*,3*R*) 3-hydroxy-3-methyl[2-$^3H_1$]glutaryl CoA were prepared from appropriately tritiated mevalonic acids (Scheme 15). Only the 3*S*-isomer is active in the enzymic degradation, and the chirality of the methyl group of the acetyl group of the acetyl coenzyme A derived from the lyase reaction was determined using the malate synthetase method of analysis. These results

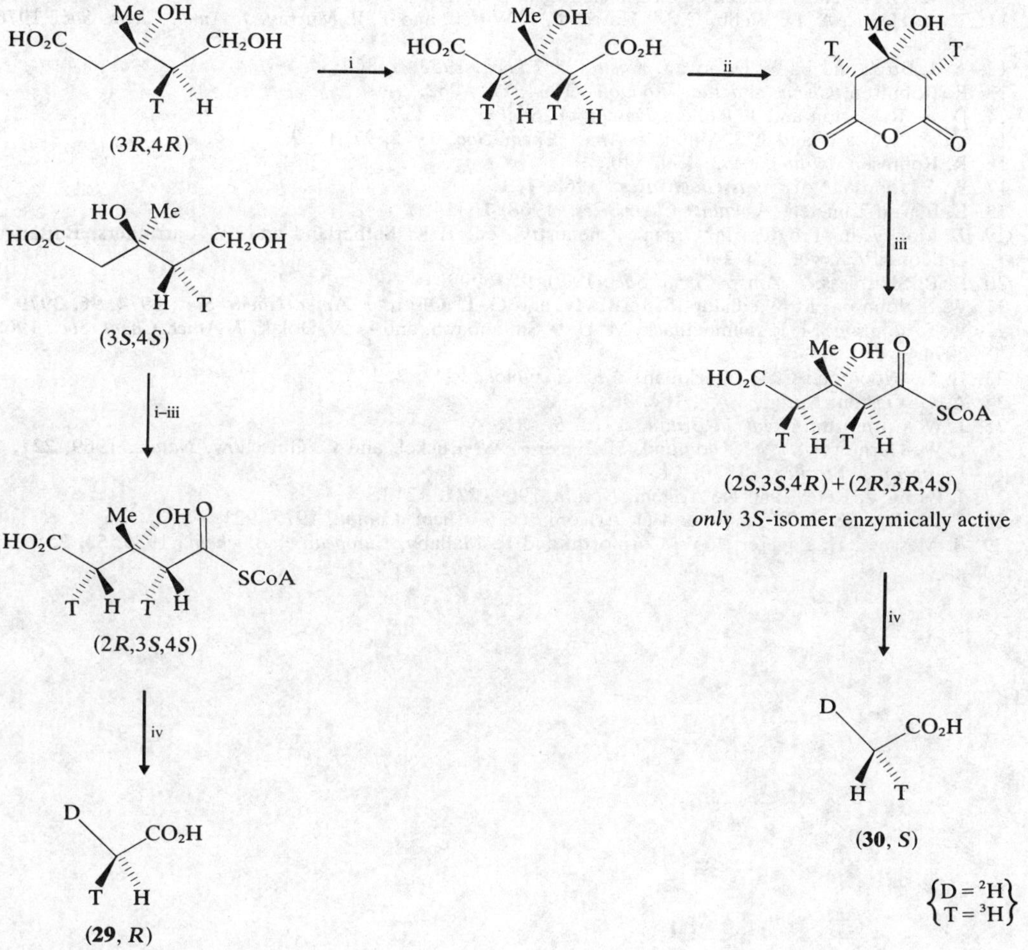

i, $Zn(MnO_4)_2$; ii, Ac_2O; iii, CoASH; iv, 3-hydroxy-3-methylglutaryl CoA lyase, D_2O.

SCHEME 15 Analysis of the stereospecificity of 3-hydroxy-3-methylglutaryl CoA lyase showed that the acetic acid derived from the substrate with the 2*S*,3*S* configuration had the *S* configuration and that *R*-acetic acid was derived from the alternative 2*R*,3*S* configuration of the substrate. It was concluded that the lyase reaction is therefore stereospecific and proceeds with inversion.

References

1. D. H. R. Barton, *Experentia*, 1950, **6,** 316.
2. R. Robinson, *J. Chem. Soc.*, 1917, 762.
3. W. H. Perkin, *J. Chem. Soc.*, 1904, 654.
4. R. Robinson, *J. Chem. Soc.*, 1936, 1079.
5. J. N. Collie and W. S. Meyers, *J. Chem. Soc.*, 1893, 63, 122, 329.
6. E. E. van Tamelen, *Fortschr. Chem. org. Naturstoffe*, 1961, **19,** 242.
7. J. N. Collie, *J. Chem. Soc.*, 1907, **91,** 1806.
8. T. Money, *Chem. Rev.*, 1970, **70,** 553.
9. L. Crombie, D. E. Games, and M. H. Knight, *Chem. Comm.*, 1966, 355.
10. T. M. Harris and P. J. Wittek, *J. Amer. Chem. Soc.*, 1975, **97,** 3270.
11. T. M. Harris, A. D. Webb, C. M. Harris, P. J. Wittek, and T. P. Murray, *J. Amer. Chem. Soc.*, 1976, **98,** 6065.
12. A. J. Birch and F. W. Donovan, *Austral. J. Chem.*, 1953, **6,** 360.
13. R. Robinson, *Chem. and Ind. (London)*, 1934, **12,** 1062.
14. D. H. R. Barton and T. Cohen, *Fetschr. A. Stoll*, 1957, 117.
15. M. A. Schwartz and I. S. Mami, *J. Amer. Chem. Soc.*, 1975, **97,** 1239.
16. R. Robinson, *Nature*, 1932, **130,** 540.
17. W. S. Johnson, *Accounts Chem. Res.*, 1968, **1,** 1.
18. E. E. van Tamelen, *Accounts Chem. Res.*, 1968, **1,** 111.
19. T. Money, in 'Progress in Organic Chemistry', ed. J. K. Sutherland and W. Carruthers, Butterworths, London, 1973, vol. 8, p.29.
20. K. B. Sharpless, *J. Amer. Chem. Soc.*, 1970, **92,** 6999.
21. W. S. Johnson, K. Wiedhaup, S. F. Brady, and G. L. Olson, *J. Amer. Chem. Soc.*, 1974, **96,** 3979.
22. W. S. Johnson, M. F. Semmelhack, M. U. S. Sultanbawa, and L. A. Dolak, *J. Amer. Chem. Soc.*, 1968, **90,** 2994.
23. H. G. Wood and C. H. Werkman, *Adv. Enzymol.*, 1942, **2,** 135.
24. A. G. Ogston, *Nature*, 1948, **162,** 963.
25. J. W. Cornforth, *Chem. in Britain*, 1970, **6,** 431.
26. J. W. Cornforth, J. W. Redmond, H. Eggerer, W. Buckel, and C. Gutschow, *Nature*, 1969, **221,** 1212; *European J. Biochem.*, 1970, **14,** 1.
27. J. Lüthy, J. Rétey, and D. Arigoni, *Nature*, 1969, **221,** 1213.
28. C. A. Townsend, T. Scholl, and D. Arigoni, *J.C.S. Chem. Comm.*, 1975, 921.
29. B. Messner, H. Eggerer, J. W. Cornforth, and R. Mallaby, *European J. Biochem.*, 1975, **53,** 255.

PART 22

NUCLEIC ACIDS

22.1
Introduction

G. M. BLACKBURN
University of Sheffield

"A very distinguished organic chemist, long since dead, said to me in the late eighties: 'The chemistry of the living? That is the chemistry of protoplasm; that is superchemistry; seek, my young friend, for other ambitions'." (F. G. Hopkins, 1933)[1]

'Dear Colleague, drop the idea of large molecules; organic molecules with a molecular weight higher than 5000 do not exist. Purify your product . . . then it will crystallise and prove to be a low molecular weight compound'. (H. Wieland to H. Staudinger, *ca.* 1929)[2]

The promptings of a Professor of Anatomy and Physiology at Basle to his nephew, Friedrich Miescher, stimulated the young man to begin an investigation of the chemical constitution of pus cells. In 1868, these were obtained by washing used bandages from a local surgical clinic. The study of the chemistry of nucleic acids had begun!

Miescher isolated a phosphorus-containing substance which was not digested by the proteolytic enzyme pepsin and he named it 'nuclein'.[3] He next looked at the chemistry of sperm from Rhine salmon. There he found a salt-like combination of a nitrogen-rich organic base, 'protamine', and a phosphorus-rich material which 'assumed the role of an acid'. He assigned this second material the empirical formula $C_{29}H_{49}N_9O_{22}P_3$ and also named it nuclein.[4]

In some respects, Miescher's work must be deemed premature. The limited criteria then available for chemical purity and identity, compounded in the nominal congruity of these two substances, led to controversies and confusions which took many years to resolve.

Independently, progress was being made in two other related fields of endeavour. In 1847, Liebig isolated a substance from beef muscle extracts[5] which he called 'inosinic acid'. It was shown later to contain phosphorus as phosphoric acid and, 60 years after its discovery, a five-carbon pentose sugar. Levene and Jacobs initially named this 'carnose' and subsequently identified it as the previously unknown substance, D-ribose,[6] whence they identified the full structure for inosinic acid as hypoxanthine-riboside 5′-phosphate (**1**).

Contemporaneously, the development of aniline dyes, following William Perkin's invention of mauveine in 1856, stimulated a systematic study of the staining of biological specimens. In general, it was observed that cell nuclei were deeply stained by dyes of basic character. This property led Flemming[7] to coin the description 'chromatin' for that substance in the cell nucleus from which nuclein was derived. Such work led on to the recognition of rod-like segments of chromatin, visible only at critical stages in cell division, which were believed to be the bearers of the hereditary material and for which the name 'chromosomes' was adopted.[8] The vital relationship between this cytological work and Miescher's studies was perceived by E. B. Wilson:[9]

> 'Now chromatin is known to be closely similar to, if not identical with, a substance known as nuclein ($C_{29}H_{49}N_9P_3O_{22}$ according to Miescher) which analysis shows to be a tolerably definite chemical compound of nucleic acid and albumin. And thus we reach the remarkable conclusion that inheritance may, perhaps, be effected by the physical transmission of a particular compound from parent to offspring'.

From this point on, alas, the chemical investigation of nucleic acids played second fiddle to that of proteins for nearly 50 years.

22.1.1 EARLY STRUCTURAL STUDIES ON NUCLEIC ACIDS

Following their work on inosinic acid, Levene and Jacobs showed that guanylic acid, which was isolated by alkaline treatment of pancreatic nucleoprotein, had the related structure guanine-ribose phosphate (**2**) and was dephosphorylated on mild hydrolysis to guanosine. This product and the related one, inosine, derived by similar hydrolysis of (**1**), were called 'nucleosides' while their phosphate esters, (**1**) and (**2**), were named 'nucleotides'.

At that time, chemical studies dealt mainly with nucleic acids derived either from yeast or from calf thymus. Partial alkaline hydrolysis of yeast nucleic acid in ammonia solutions led to the four nucleosides adenosine (**3**), guanosine (**4**), cytidine (**5**), and uridine (**6**).

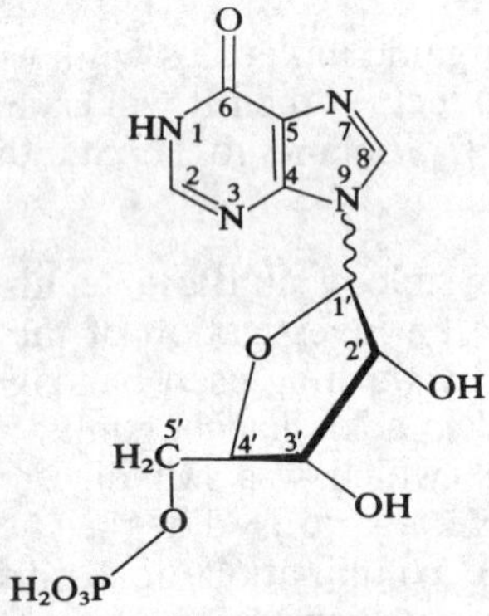

(**1**)

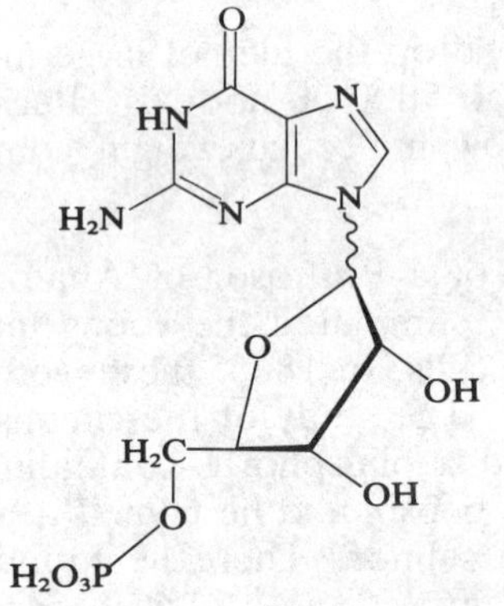

(**2**)

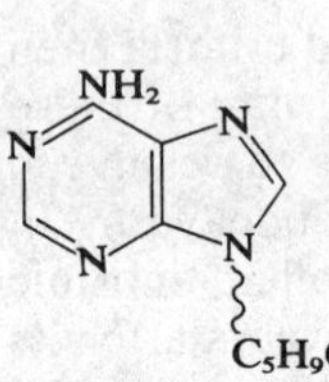

(**3**)

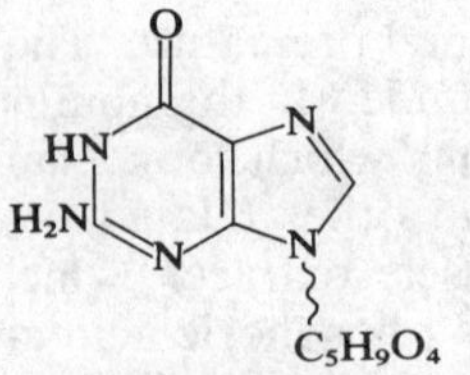

(**4**)

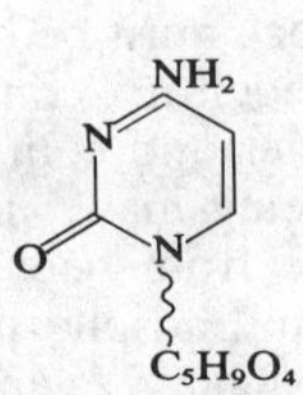

(**5**)

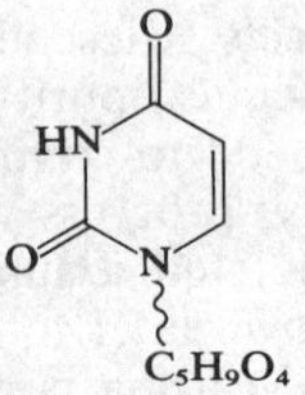

(**6**)

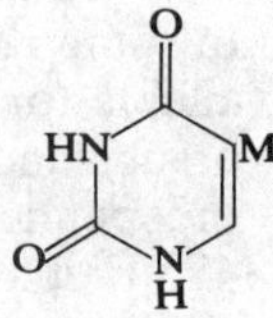

(**7**)

Thymus nucleic acid proved to be resistant to alkaline hydrolysis and the structures of the nucleosides derived from it thus defied analysis for a further 20 years. Then Levene[10] recognized the sugar as 2-deoxy-D-ribose and thereby accounted for its unusual property of restoring the colour to Schiff's reagent. Thymus nucleic acid also yielded four heterocyclic bases, adenine, guanine, cytosine, and — instead of uracil — thymine (**7**). These two features established the distinction between the deoxyribonucleic acids, which at that time were thought (like thymus nucleic acid) to be of animal origin, and the ribonucleic acids, then considered to be a characteristic component of plant tissues.

The belief in their similarity was stronger than the awareness of their difference. The fortuitous presence of approximately equal quantities of adenosine and guanosine in nucleic acid obtained from yeast or from wheat germ, coupled with an erroneous determination of molecular weight, misled Levene to suggest the unifying concept of the tetranucleotide structure for both deoxyribo- and ribo-nucleic acids. This was formulated[11] as follows:

```
phosphate–sugar–base
               |
       phosphate–sugar–base
                      |
              phosphate–sugar–base
                             |
                     phosphate–sugar–base
```

If this conclusion seems strange today, when nucleic acid species are known to contain from eighty to over a million nucleotide residues, it is salutary to recall that in the late 1920's Hermann Staudinger was at work refuting bald criticisms from the chemical establishment of his radically new ideas on polymer structure.[2] Organic chemistry was not then ready for the macromolecular structure analysis which the nucleic acids demanded.

22.1.2 STRUCTURE OF DNA

22.1.2.1 Isolation and characterization of DNA

Nucleic acids are materials very difficult to handle. They are sensitive to cleavage by enzymes (ribonuclease can be detected on fingertips), by extremes of pH and temperature, and even by mechanical shearing forces. The length of a DNA* molecule for the common bacterium *Escherichia coli* is about 1 mm while its diameter is some 2 nm. Thus simple stirring or even incautious pipetting of DNA solutions usually results in a large reduction in molecular weight. Consequently, most early preparations of DNA yielded fragmentary material of low molecular weight.

One widely used method for the preparation of protein-free DNA calls for the rupture of the cell wall or membrane and then uses an anionic detergent, such as sodium dodecyl sulphate, to separate the protein as a precipitate, leaving the nucleic acid in solution. The DNA is generally obtained by the cautious addition of ethanol to a salt solution of the nucleic acid. The gelatinous precipitate can then be either spooled on a glass rod or drawn into fibres.[12]

The DNA obtained from calf thymus glands by such a procedure was shown to have a molecular weight of 5–10 million as determined by light scattering, diffusion, sedimentation, and viscosity measurements.[13] On the one hand, such material is vastly superior to that with which the primary structure of DNA was solved. On the other, it is material wholly inadequate for the pursuit of the structure of the chromosome.

* The abbreviations DNA for deoxyribonucleic acid and RNA for ribonucleic acid were generally adopted in the 1950s.

22.1.2.2 The primary structure of DNA

Chemical studies on the primary structure of DNA were interwoven with, but often lagged behind, those on RNA. They are given priority here in order to preserve a unity of discussion of primary, secondary, and tertiary structure for DNA in which the importance of the secondary structure of DNA is dominant.

The early impediment to structural studies on DNA lay in its resistance to selective hydrolysis. Acidic cleavage affects the bonds between purines and deoxyribose faster than those of the phosphate diesters. Yet such diesters are resistant to alkaline hydrolysis. [Neighbouring group participation from the 2′-hydroxyl group is the essential feature of the alkaline hydrolysis of RNA (Chapter 22.3)]. Thus the use of an arsenate-inhibited phosphatase enzyme to fragment DNA by Klein and Thannhauser[14] in 1935 was an important achievement. It made available the four deoxyribonucleotides in crystalline form. Deoxyadenylic acid (**8**), deoxyguanylic acid (**9**), deoxycytidylic acid (**10**), and deoxythymidylic* acid (**11**) could each be hydrolysed to the corresponding deoxyribonucleoside by the action of an enzyme specific for 5′-phosphate esters. This and much other evidence suggested that the structures assigned (**8**)–(**11**) were those appropriate. Nonetheless, the rigorous proof of their correctness was later to be provided by the synthetic studies of the Cambridge school under the leadership of Alexander Todd.

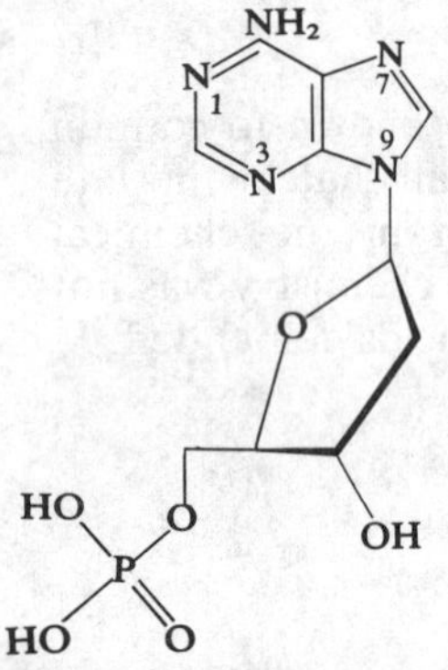

(**8**)

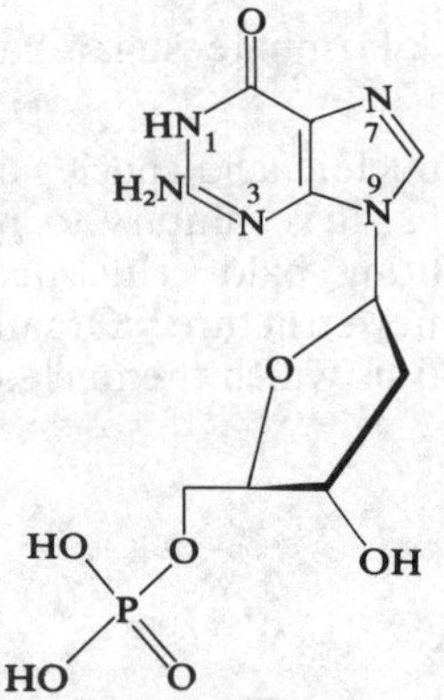

(**9**)

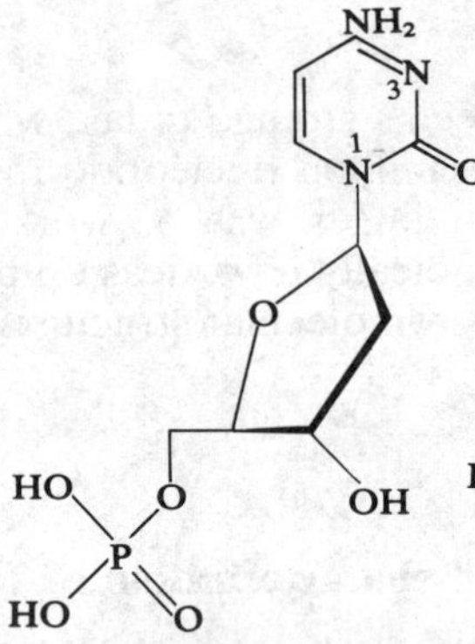

(**10**)

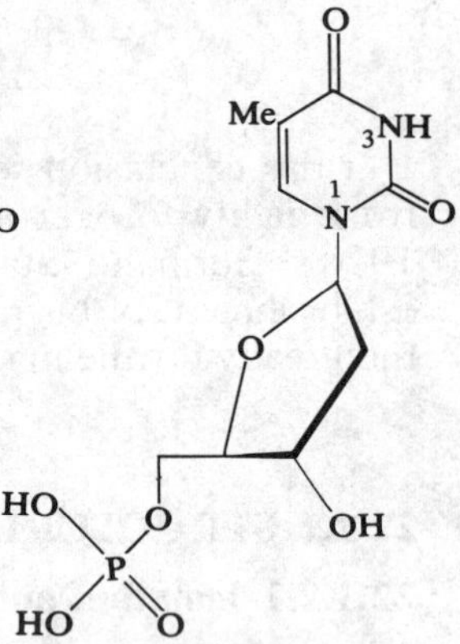

(**11**)

In the event, the syntheses of the 2′-deoxyribonucleosides were far more difficult than those of their ribose counterparts, both on account of the scarcity of 2-deoxyribose and because of the relative instability of its derivatives required for glycosidic bond formation. Thus the first synthetic achievement in this area was the direct conversion of ribonucleosides into deoxyribonucleosides, illustrated for the transformation of ribothymidine into deoxyribothymidine (**12**) (Scheme 1)[15] (see also Section 22.2.3.4).

The position of attachment of the sugar to the heterocyclic rings was determined by unambiguous synthesis in the ribo-series (Section 22.1.3.2). In the deoxyribonucleosides it was identified as N-1 for the pyrimidines†, as a result of direct chemical interconversions, and as N-9 for the purines by ultraviolet spectroscopic comparison with 9-methyladenine and 9-methylguanine.[16,17]

The configuration at the point of attachment of the heterocyclic base to the anomeric carbon of deoxyribose was established elegantly by the formation of a cyclonucleoside (**14**) on heating 3′-*O*-acetyl-5′-*O*-toluene-*p*-sulphonyldeoxyadenosine (**13**) in acetone.[18]

*Until the discovery of ribothymidine in transfer RNA (Chapter 22.2), *deoxythymidine* was commonly referred to as *thymidine*. Thus, (**11**) was originally named thymidylic acid.

† A now-discarded numbering system for pyrimidines used the reverse assignments for N-1 and N-3 to those in current use. Thus the older literature associates the glycosidic bond with N-3 for the pyrimidines.

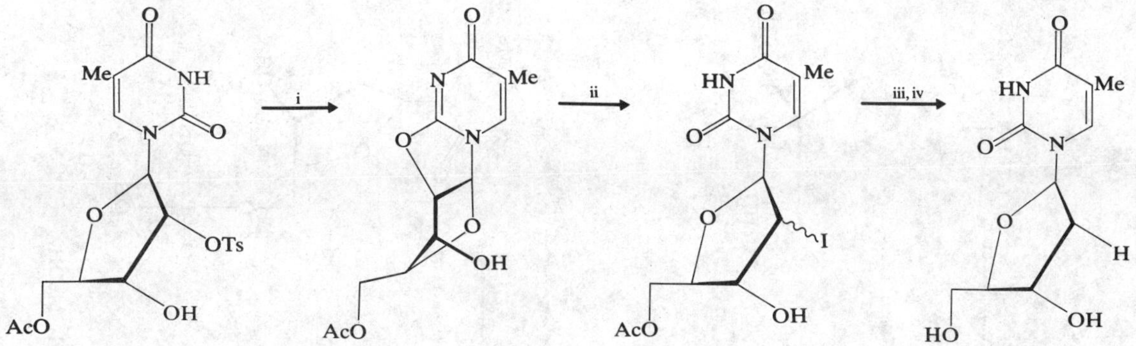

(12)

i, NaN_3: ii, NaI, H^+; iii, NH_3, EtOH; iv, H_2, Pd.

SCHEME 1

Such an intramolecular process cannot occur in the case of an α-glycoside. In the pyrimidine series, a similar proof for the β-glycosidic linkage was achieved by the transformation of 3′-iodo-2′,3′-dideoxyribothymidine (**15**) into cyclodeoxythymidine (**16**). On acidic hydrolysis, this substance gave thymine (**7**) and 2-deoxy-D-xylose (**17**).[19]

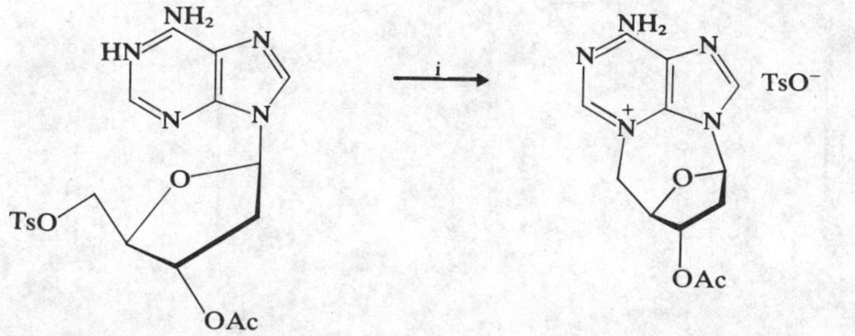

(13) (14)

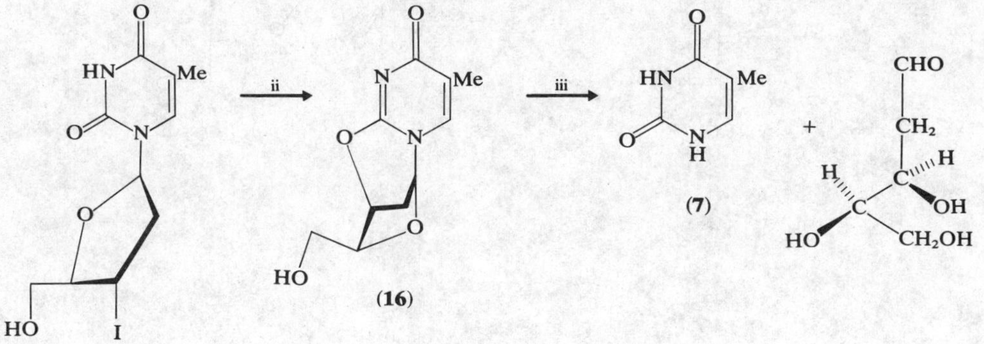

(15) (16) (7) (17)

i, heat; ii, AgOAc, MeCN; iii, H_3O^+.

SCHEME 2

The key feature in the identification of the site of phosphorylation in the deoxyribonucleotides lay in the regioselectivity of triphenylmethyl (trityl) chloride for reaction with primary rather than with secondary hydroxyl groups. It underpinned the specific syntheses of deoxythymidine 3′-phosphate (**18**) and 5′-phosphate (**19**) from deoxythymidine.[20] The 5′-phosphate proved to be identical with the mononucleotide which had been obtained by Thannhauser.[14]

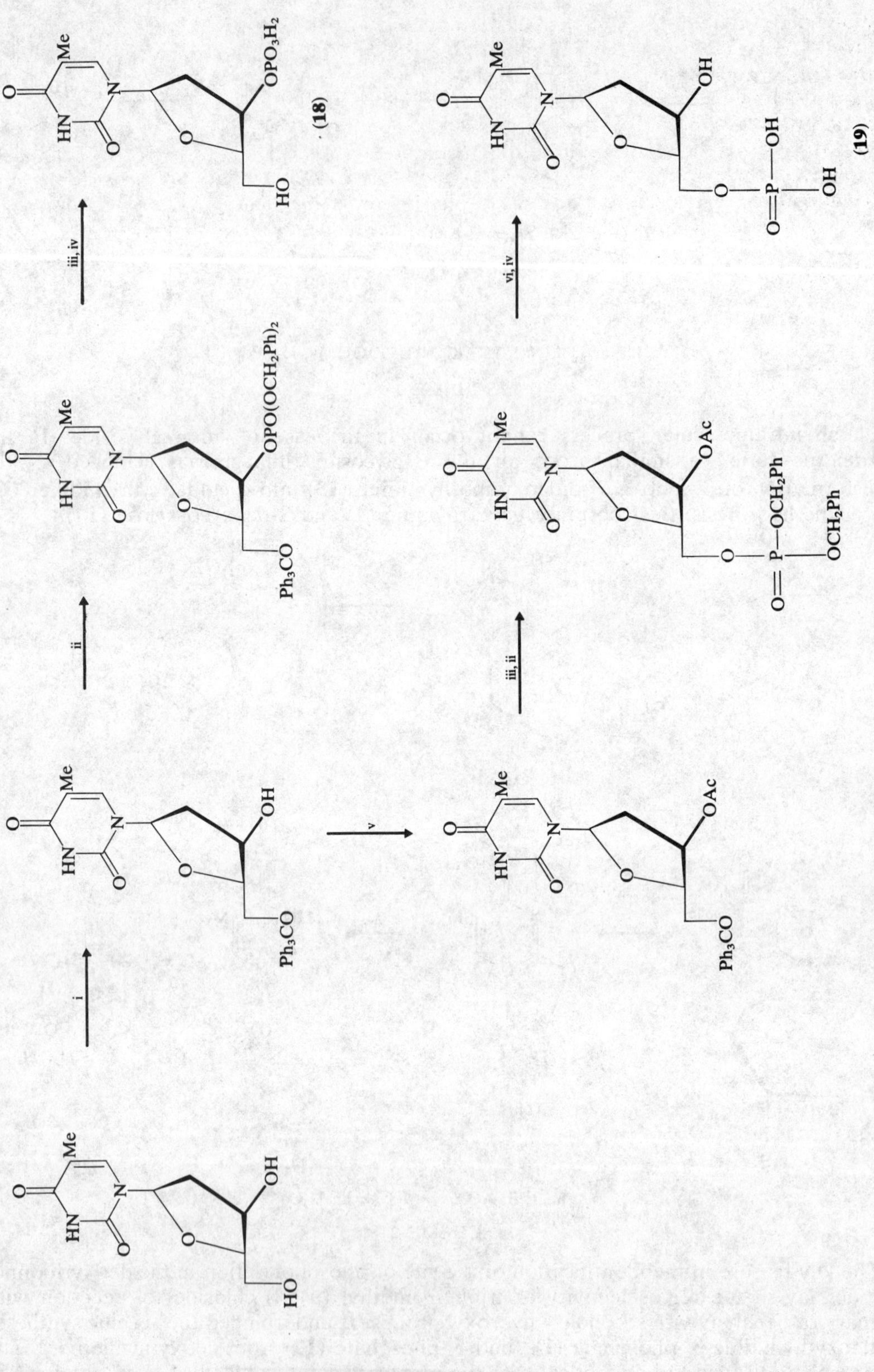

i, Ph_3CCl, pyridine; ii, $(PhCH_2O)_2POCl$; iii, H^+; iv, H_2, Pd; v, Ac_2O, pyridine; vi, NH_3, MeOH.

SCHEME 3

While deoxycytidine 3′- and 5′-phosphates could be prepared in an analogous fashion,[21] the well-known lability of the purine glycosidic bond to acidic hydrolysis posed problems in the purine nucleotide syntheses. These were solved by a painstaking separation of the 3′- and 5′-monoacetates of deoxyadenosine from the mixture obtained as a result of partial deacetylation of 3′,5′-di-*O*-acetyldeoxyadenosine (**20**).[20] They were identified by mild hydrolysis of the former to give the known 3-*O*-acetyldeoxyribose (**21**). The isomeric monoacetates were separately phosphorylated and protecting groups removed to give deoxyadenosine 3′-phosphate (**22**) and 5′-phosphate (**8**). Similar procedures were used to achieve syntheses of deoxyguanosine 3′-phosphate and 5′-phosphate (**9**).[22]

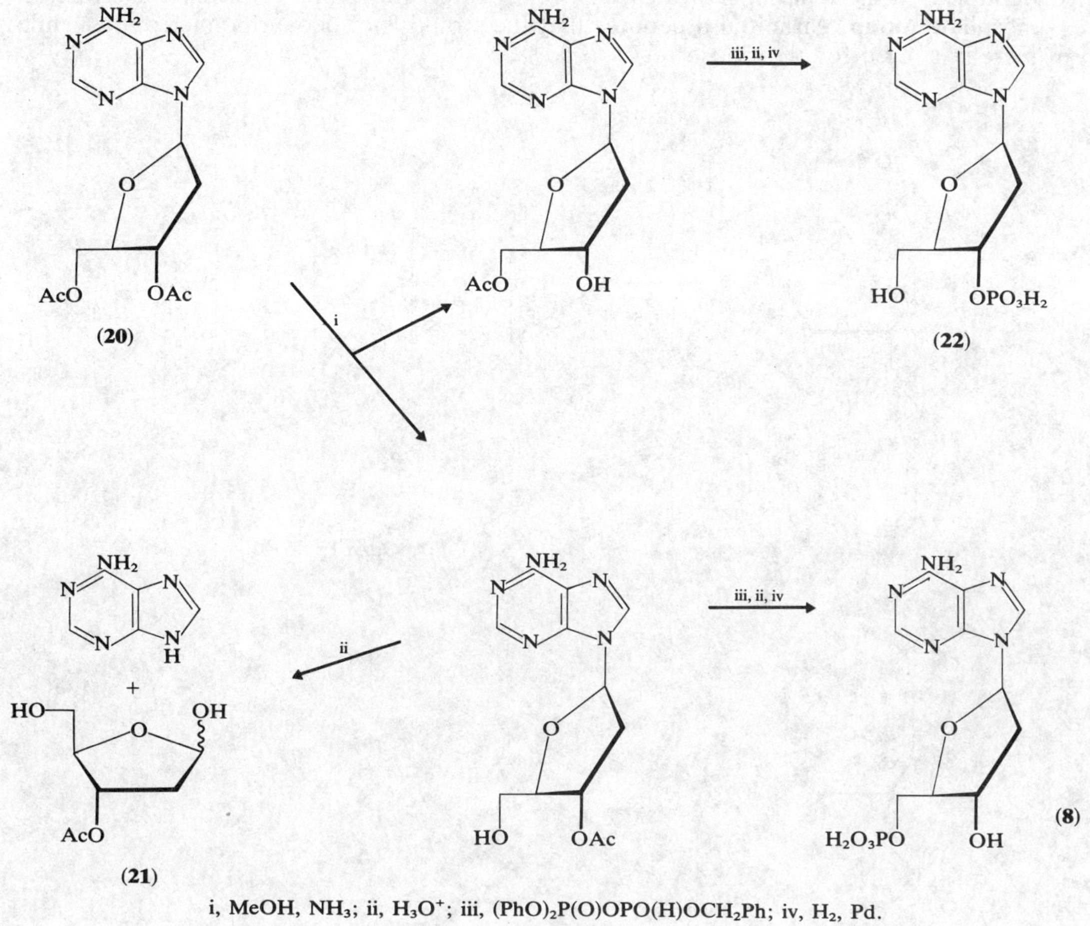

i, MeOH, NH_3; ii, H_3O^+; iii, $(PhO)_2P(O)OPO(H)OCH_2Ph$; iv, H_2, Pd.

SCHEME 4

These synthetic achievements provided the direct identification of the 5′-deoxynucleotide products of enzymic hydrolysis of DNA which had been made available by the work of Klein and Thannhauser.[14] They also defined the structures of the 3′-deoxynucleotides which were later to be obtained from DNA by enzymatic digestion using a micrococcal nuclease.[23] The 3′,5′-diphosphates of deoxycytidine and deoxythymidine were also synthesized[24] and used to prove the structures of the diphosphates which had been obtained as minor acidic hydrolysis products of DNA.

The general primary structure of DNA was thus to hand. The monomers each possessed two hydroxyl groups per nucleoside; the stability of DNA to hydrolysis was exactly that expected for a simple phosphate diester (half-life 116 days at 20 °C, pH 7); and DNA molecules were known to be polymers of high molecular weight. Since Levene's time, Staudinger had won the battle for macromolecular structure and estimates of the molecular weight of DNA had risen to well over 1 million with successive improvements in the methods available for its preparation in pure form.

Thus the overwhelming evidence available in the early 1950s showed that DNA consisted of a linear polynucleotide in which one nucleoside is linked *via* its 5′-hydroxyl group through a phosphate diester unit to the 3′-hydroxyl group of one neighbour and by a 3′ to 5′ phosphate linkage to the other neighbour, as in (**23**). The cumbersome, full diagrammatic representation was quickly abbreviated to a simpler form (**24**), reminiscent of Fischer projections for sugars, which in turn gave way to the condensed form (**25**).* By convention,[25] p as a prefix indicates a 5′-phosphate group and as a suffix indicates a 3′-phosphate group. An initial d denotes that the strand is of deoxyribonucleotides while an r prefix is used for the ribose series.

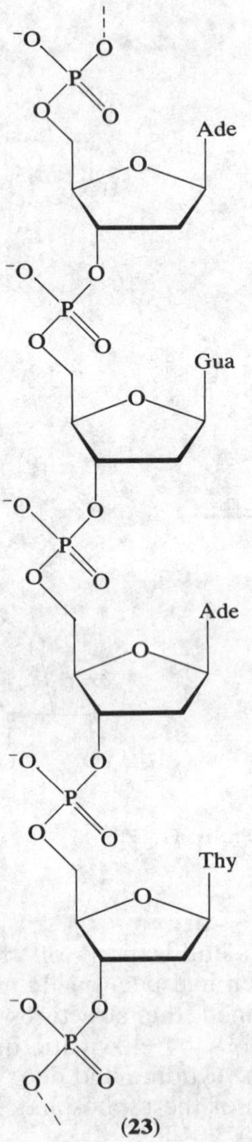

(**23**)

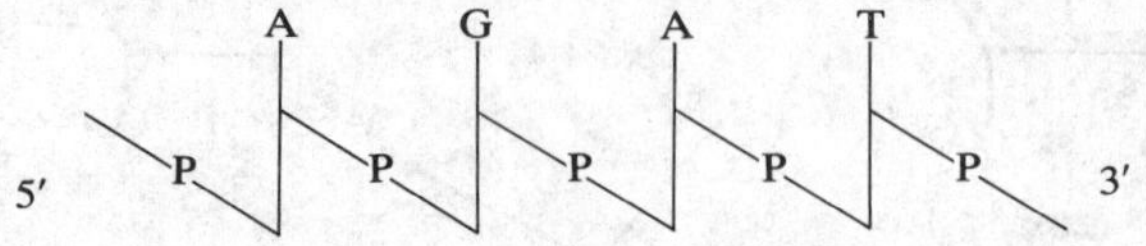

(**24**)

5′ . . . dpApGpApTp . . . 3′

(**25**)

* By convention, DNA strands are drawn with the 3′→5′ phosphate diester linkage from left to right. This leaves a free 5′-OH group at the left-hand end and a 3′-OH group at the right-hand end of the strand.

Structural chemists cautiously reserved judgement for a long time on the uniqueness of the 3′→5′ phosphate linkage in DNA. It was conceivable that chemical methods might have failed to detect a rare 5′→5′ or 3′→3′ linkage or even the presence of an occasional pyrophosphate group. Such caution has proved to be unnecessary. Today there is no evidence for any internucleotide linkage in DNA other than the 3′→5′ phosphate diester bond.

In principle, the investigation of the primary structure of DNA should have led on logically, as in the case of polypeptide chemistry, to analysis of the sequence of the four monomers in DNA strands. In practice this task had to be shelved for over 20 years until pure, discrete DNA species became available. In the event the whole future of biology was transformed by the explosive impact of the discovery of its secondary structure.

22.1.2.3 The secondary structure of DNA

The tetranucleotide hypothesis for the structure of DNA waxed in the 1920s, waned in the 1930s, and died in the 1940s. Three lines of inquiry provided the evidence which demanded a new concept for DNA secondary structure.

First, the polymeric nature of DNA was perceived in greater detail than could be explained by gel-like aggregates of tetranucleotides. Its physical properties were those expected of a thin, rod-like molecule whose length exceeded its width by a factor which grew from 300 to 700 as preparations of DNA improved.[26] Gulland recognized that the diversity of biological specificity shown by DNA demanded a greater variety of nucleic acid structure than that which might possibly be provided by a repeating tetranucleotide unit that could, after all, vary only in length.

Secondly, Astbury turned aside from his work on protein structure to make X-ray diffraction measurements on stretched DNA fibres. He concluded 'A spacing of 3.3_4 Å along the fibre axis corresponds to that of a close succession of flat or flattish nucleotides standing out perpendicularly to the long axis of the molecule to form a relatively rigid structure'.[27] He also thought that the true repeat period along the DNA axis must be at least 17 times the thickness of a nucleotide. His evidence also conflicted with the tetranucleotide hypothesis which predicted a repeat of 13 Å ($=3.3_4 \times 4$).

Thirdly, and decisively, Chargaff applied paper chromatography to the quantitative analysis of the base compositions of nucleic acids (see Table 1), using one of the first ultraviolet spectrophotometers commercially available for his analyses. The accuracy he achieved was comparable with that available for the amino-acid content of proteins and more than good enough to demolish the tetranucleotide hypothesis.

TABLE 1
Base Ratios obtained by Chargaff[a] in 1949

DNA source	*Adenine*	*Thymine*	*Guanine*	*Cytosine*
Calf thymus	1.7	1.6	1.2	1.0
Beef spleen	1.6	1.5	1.3	1.0
Yeast	1.8	1.9	1.0	1.0
Tubercle bacillus	1.1	1.0	2.6	2.4

[a] E. Chargaff, E. M. Vischer, R. Doniger, C. Green, and F. Misani, *J. Biol. Chem.*, 1949, **177**, 405.

From these and more extensive analyses of base ratios, three conclusions were drawn. First, the amount of purines was always equal to that of pyrimidines. Secondly, the ratio adenine : thymine and the ratio guanine : cytosine was always close to unity. Thirdly, the ratio (A+T)/(G+C) varied widely between DNAs from different species, but was constant for DNA from different organs within the same species. The tetranucleotide hypothesis predicted equal proportions for all four bases and was now totally discredited.

The events which brought together Watson and Crick to solve the problem of DNA secondary structure make compelling reading.[28,29] The strength of their partnership lay both in its variety of scientific experience and in the candid, almost rude, mutual criticism which enabled the two partners to discard one after another model for the structure of DNA — *until* they created one in conformity with all the available data.

That process drew on the results of many different research teams. The X-ray diffraction patterns obtained by Wilkins and by Franklin showed that DNA fibres were highly crystalline and could be characterized by an *A-form* structure at 70% relative humidity and a 'wet' *B-form* structure at above 90% humidity.[30] The data on the *B*-form strikingly showed that DNA was a helix with a 3.4 Å separation between nucleotides and a 34 Å repeat of helical conformation. Watson deduced that the number of nucleotides per unit crystallographic cell was better in accord with a double-stranded helix than with a triple helix. Crick, seeing that the *A*-form was monoclinic C^2, recognized that the helical chains must run antiparallel to produce diad axes perpendicular to the helix axis. While Linus Pauling proposed a helix with the bases outside the phosphate–sugar backbone, Watson and Crick settled for a structure with the sugar–phosphate backbone on the *outside* of the helix. Crick, seeing that the *A*-form was monoclinic $C2$, recognized that the helical chains the bases fit in?

Acid–base titration studies strongly indicated that inter-base hydrogen bonding existed. Moreover, X-ray structures of the hydrochlorides of adenine and guanine showed a multiplicity of such hydrogen bonds. Watson initially made models pairing like-with-like bases in which thymine (**26**) and guanine (**27**) were employed in the *enolic* tautomeric forms. That was, after all, how they were conventionally depicted in standard textbooks of the time.[31] (Such homo-pairing (**28**) could not be generated for the bases in their correct, *keto*-tautomeric forms).

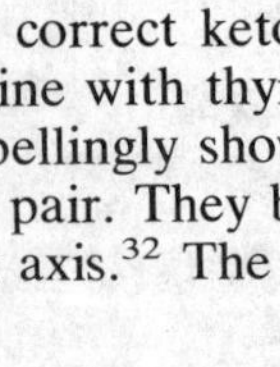

(**26**)

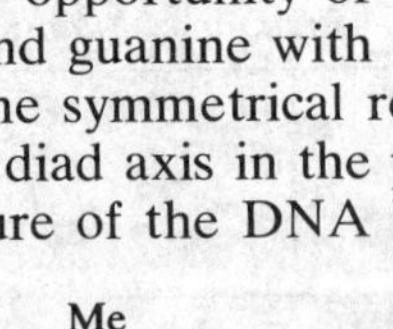

(**27**)

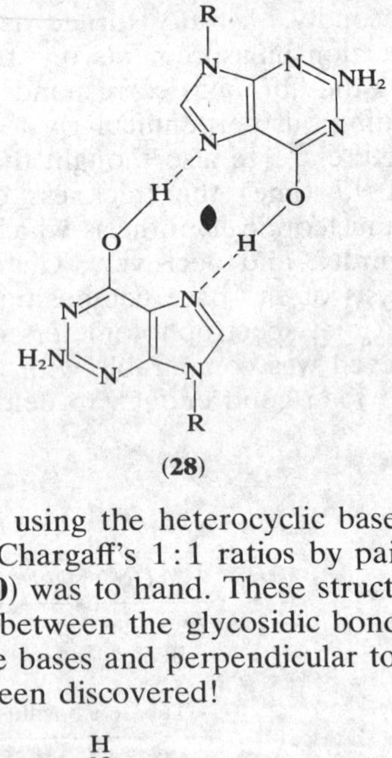

(**28**)

As soon as Watson was appraised of the necessity of using the heterocyclic bases in their correct keto-forms, the opportunity of explaining Chargaff's 1 : 1 ratios by pairing adenine with thymine (**29**) and guanine with cytosine (**30**) was to hand. These structures compellingly showed the same symmetrical relationship between the glycosidic bonds in each pair. They both have a diad axis in the plane of the bases and perpendicular to the helix axis.[32] The core structure of the DNA helix had been discovered!

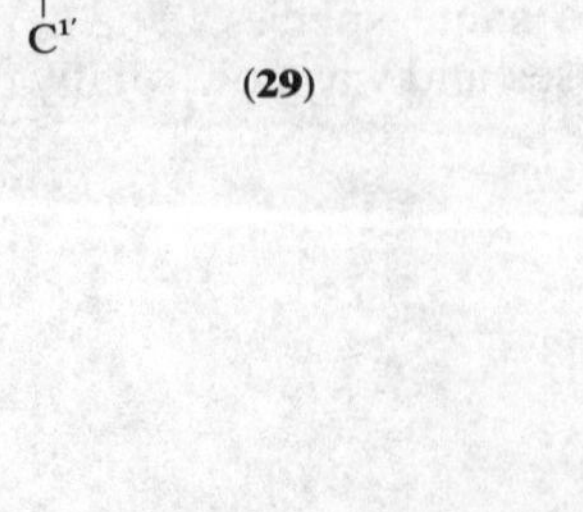

(**29**)

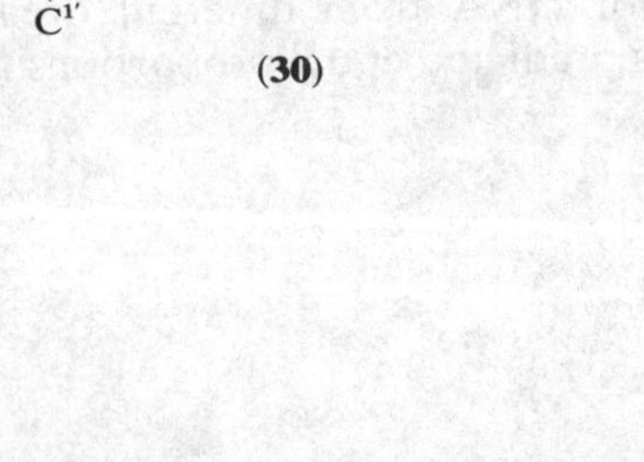

(**30**)

The original Watson–Crick model for the paracrystalline *B*-form of DNA has been refined by Wilkins (Figure 1).[30] Its principal features are an intertwined, right-handed, double helix which has ten residues per 34 Å period. The base pairs are virtually perpendicular to the helical axis and the sugar rings nearly planar (generally considered to be in the *C*2-*endo* conformation) and at right angles to the bases. The phosphate residues lie on a cylindrical surface some 9 Å radius from the helix axis which runs between the hydrogen bonds and link pairs of bases.

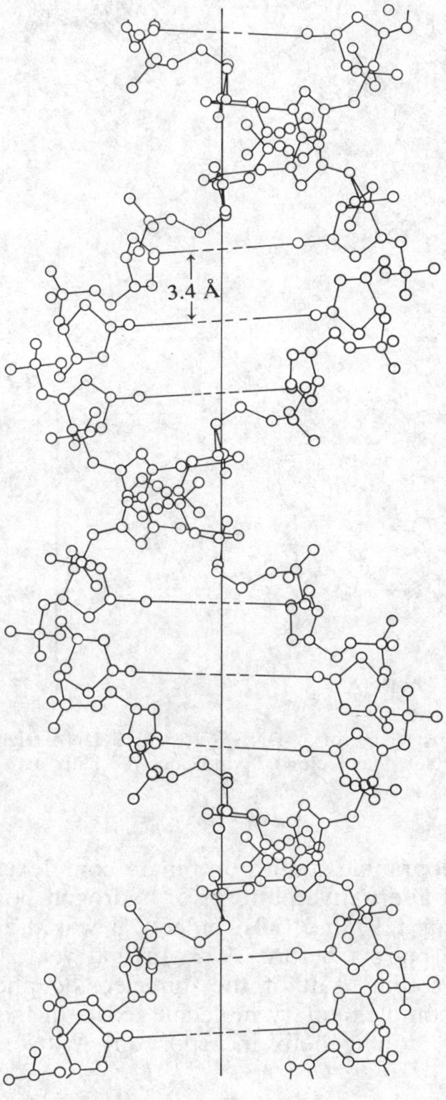

Figure 1 Projection of the *B*-conformation of the DNA molecule (Reproduced from Wilkins,[30] by courtesy of Elsevier)

For the more highly crystalline *A*-conformation of DNA, the base-pairs are tilted at 20° from the plane normal to the helical axis, there are 11 nucleotides per turn of 28 Å, and the double, right-handed helix is 22 Å wide (Figure 2).

Much of the ensuing debate over the correctness of this structure has hinged on the pattern of hydrogen bonding between complementary base pairs, adenine : thymine and

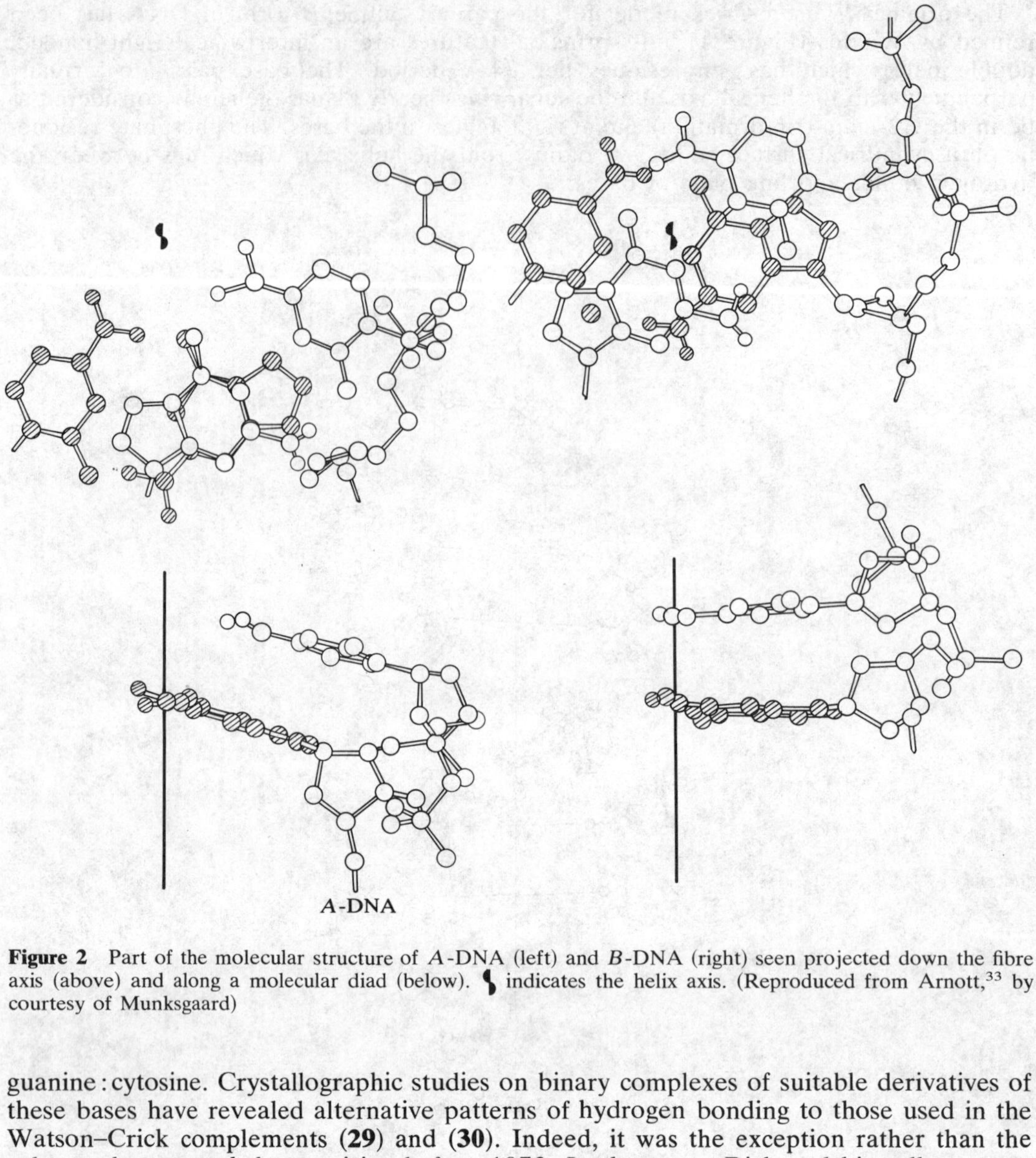

Figure 2 Part of the molecular structure of *A*-DNA (left) and *B*-DNA (right) seen projected down the fibre axis (above) and along a molecular diad (below). ⟆ indicates the helix axis. (Reproduced from Arnott,[33] by courtesy of Munksgaard)

guanine : cytosine. Crystallographic studies on binary complexes of suitable derivatives of these bases have revealed alternative patterns of hydrogen bonding to those used in the Watson–Crick complements (**29**) and (**30**). Indeed, it was the exception rather than the rule to observe such base-pairing before 1973. In that year, Rich and his colleagues at M.I.T. crystallized the disodium salt of the dinucleoside phosphate of adenosine and uridine, rApU. This self-complementary molecule exists in the crystal as a segment of a right handed, antiparallel, double helix incorporating Watson–Crick hydrogen bonding (Section 22.1.3.4).

Of the fundamental structural and biological accuracy of the pairing of A with T and G with C there has been no doubt. This complementarity lies at the heart of the correlation between structure and function of nucleic acids (Chapter 22.5) It is also an essential feature of a recently-suggested, alternative secondary structure for DNA which attempts to resolve one problem not answered by the Watson–Crick model. That is no less than the fearsome topological implications of strand-separation during the course of biological replication of a fully-interwound double helix (Section 22.5.1.1).

This *side-by-side* structural model[34] (Figure 3) envisages an oscillating strip which turns in a right-handed double helix for 17 Å and then reverses the helical sense to a

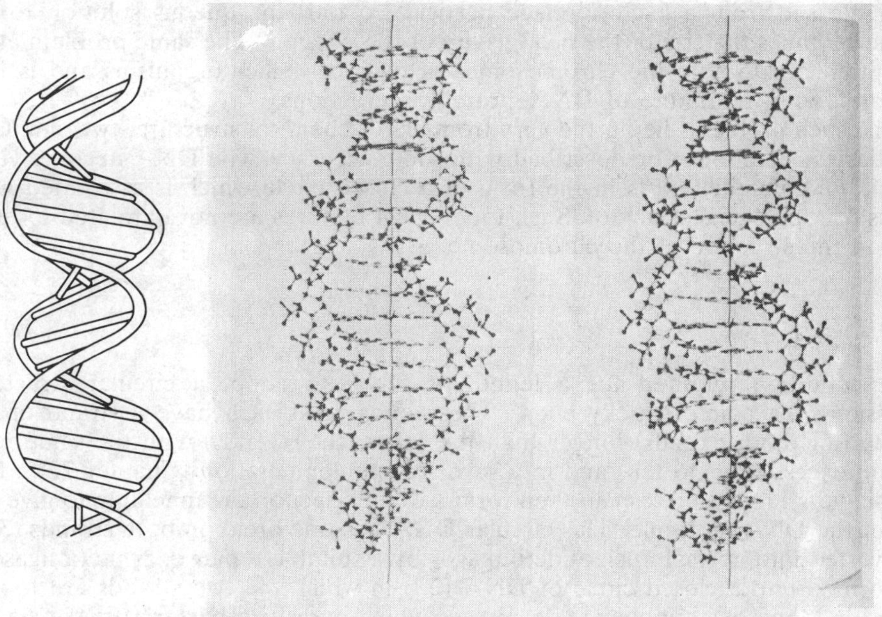

Figure 3 A simplified representation of the side-by-side model of DNA secondary structure (left) and a stereoview of the detailed model (right).* (Reproduced from Rodley *et al.*,[34] by courtesy of the US National Academy of Science)

left-handed progression for a further 17 Å. It thereby maintains the general features of the DNA *B*-form (!), a 34 Å period of 10 residues, and Watson–Crick paired bases in planes normal to the helix axis. It imposes detailed changes on the conformations of the sugar ring, of the C-4′—C-5′ bond, and of the phosphate ester linkages in order to achieve the left-handed helical sections and also the tight turns which occur at the transitions from right-to-left and left-to-right sense of helicity. Its inventors claim that this model has reasonable steric contacts and would provide a fibre X-ray diffraction pattern not unlike that of the *B*-form of DNA.

This model appears to offer two structural advantages. It should be capable of separation into two single strands by a simple 'unzipping' action without any of the topological constraints which attend a fully interwound double helix. It is also capable of a greater degree of flexibility to bending than the stiff, rod-like behaviour of the Watson–Crick double helix. In this respect it may figure in the formation of a 'kinky helix' which has been described[35] as a possible feature of chromatin structure.

In following the historical development of the subject so far, undue emphasis has been placed on the hydrogen bonding between bases. While this feature undoubtedly is responsible for the complementation of bases in many different situations, it alone does not hold the two strands of the helix together. The current consensus is that the dominant helix-forming factor is the stacking of the planar, quasi-aromatic bases on top of one another and that the consequent solvent-exclusion, or hydrophobic, forces are the main support of the double helix.

22.1.2.4 The tertiary structure of DNA

The image of DNA presented in the previous section requires considerable development to bring it closer to the reality of DNA in the living cell. For instance, the *E. coli* bacterium is 2 μm long but envelopes a DNA chromosome that is 1.1 mm long. Also, the

* This stereo-pair diagram is intended for viewing left-eye to left diagram and right-to-right (*parallel viewing*).

DNA released from a bacteriophage particle expands in aqueous solution to occupy a volume 15 times the size of the head cavity of the phage.[36] The same problem attends the description of DNA in the chromosomes of cells of higher organisms and is intricately associated with the nature of DNA–protein interactions.

While such problems lie on the very frontiers of chemical structure, two simpler aspects of tertiary structure can be described with some accuracy. The DNA in some viruses and bacteria undoubtedly exists in the form of a closed circle which is associated with novel aspects of topological structure. Such topological features are not unrelated to developing ideas on the structure of the chromosome.

(*i*) *Circular DNA*

The condition required for a length of DNA to adopt a circular structure is its possession of a pair of 'sticky ends'. These ends must each have a strand of the same polarity (*e.g.* both 5′-ends) longer than the other (the 3′-ends) with the same number of bases in excess (four to ten) and in a reverse complementation (Chapter 22.5, Figure 7). Such single-stranded pieces can then form a duplex section on annealing to give a circular form of the DNA molecule. This circular DNA has one break in both strands (**31**) which, if converted into normal nucleotide linkages by a suitable repair enzyme (a ligase), can be transformed into a closed circle of DNA (**32**) in which the two strands are topologically bonded.[37] Many bacteriophage (*i.e.* viruses which parasitize bacteria) possess such sticky ends and anneal to give 'Hershey circles'. Closed circular DNA is found in the native state in many viruses, bacteria, and mitochondria, in which form it is biologically active,[38] and can be produced by controlled chemical synthesis (Chapter 22.4).

Two principal features of tertiary structure result from such topological bonding: the DNA strands cannot be separated without breaking open either or both of them; the number of times one circular strand winds around the other when the closed circle is constrained to lie in a plane is invariant. This number is the *topological winding number*, α. A second parameter, β, known as the *duplex winding number*, is the ratio of the number of base pairs in the closed circle, *n*, to the number of pairs per turn of the unconstrained helix, *m*. If *m* changes as a result of change in pH, ionic strength, or temperature, then β also changes. Such a change creates conformational strain which can be relieved by the creation of twists or *supercoils* as a topological feature of the tertiary structure.

The number of such supercoils is called the *superhelix winding number*, τ, and is simply related to the topological winding number by the relation $\tau = \alpha - \beta$. Behaviour of this sort is illustrated as follows. In the closed circle (**32**) α, β, and τ are all zero.[39] On winding three right-handed and three left-handed turns into the duplex (**33**) there is no net change in β $(+3-3=0)$. When the three left-handed turns are converted into three superhelical left-handed turns (**34**), a strainless circle remains in which now $\beta = +3$ and $\tau = -3$. As an alternative transformation, the three left-handed turns in (**33**) can be exchanged for three *right-handed, interwound* superhelical turns which *do not rotate around each other* (**35**) and again α remains zero. Now if one strand in (**35**) is nicked to enable the three duplex turns to unwind and then is resealed, β changes from $+3$ to 0 and thus α changes from 0 to -3. When this new configuration is constrained to lie in a plane, there will be a topological transformation of its three interwound superhelical turns into three left-handed duplex turns.

Such superhelical DNA is readily observed by electron microscopy. It appears that closed circular DNA from natural sources has one superhelical turn for every 300 or so

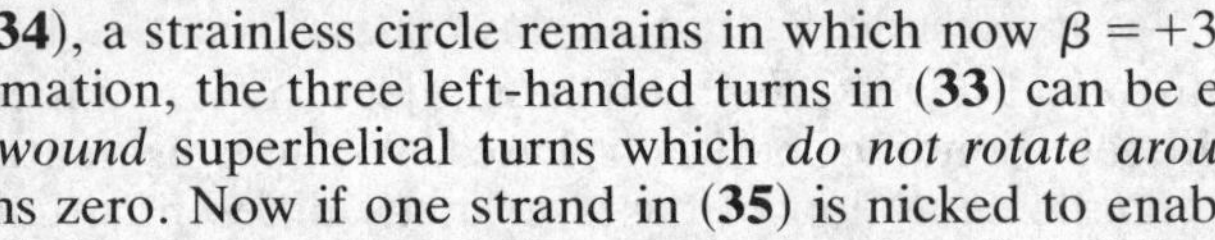

(**31**) (**32**) (**33**) (**34**) (**35**)

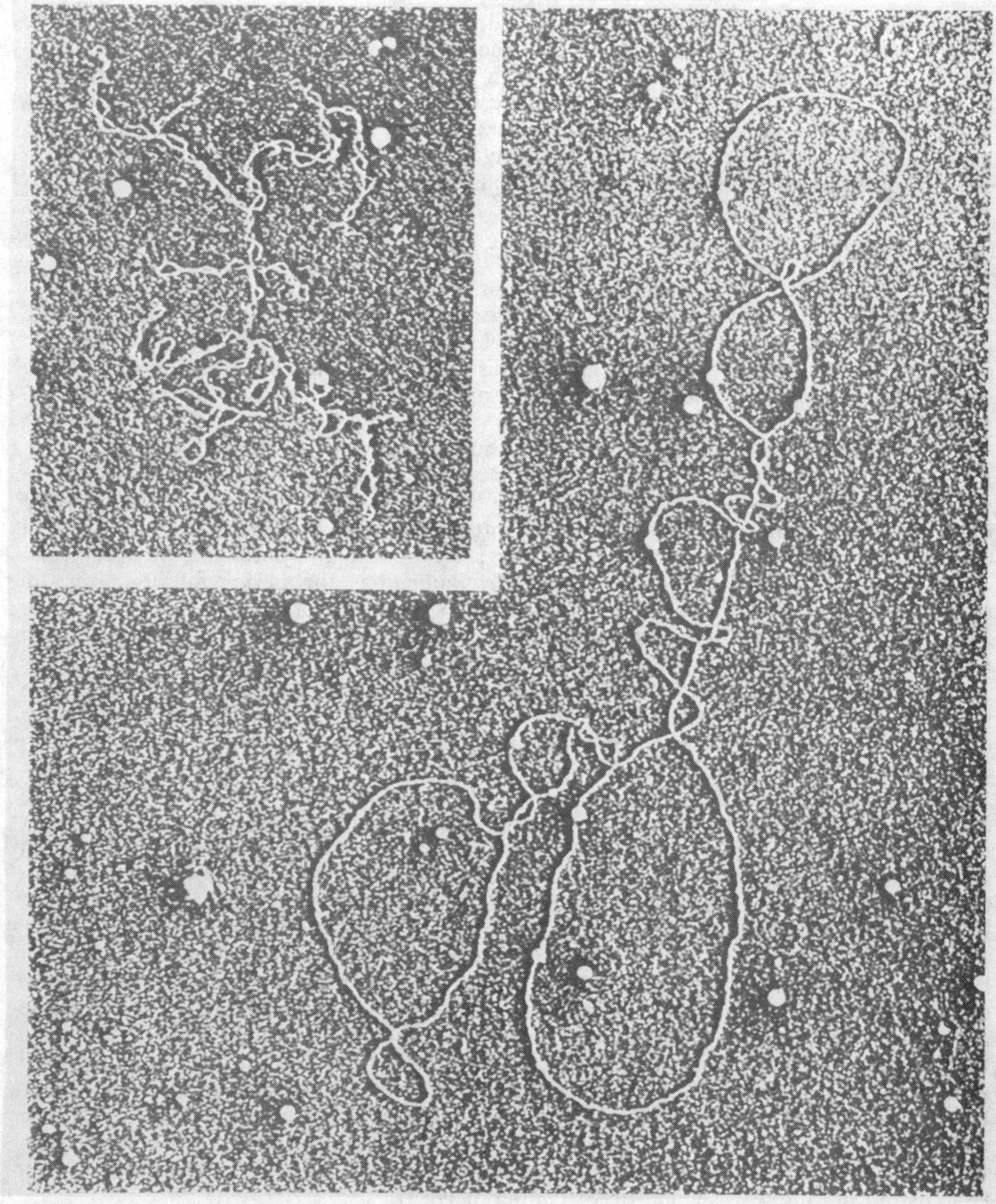

Figure 4 Electron micrographs of closed circular DNA molecules. Examples of twisted (128 crossings, high-salt preparation) and relatively untwisted (22 crossings, low-salt preparation) circular DNA from bacteriophage λ. (Reproduced from Bode and MacHattie,[42] by courtesy of Academic Press)

base pairs. The number of supercoils changes on alkaline titration of the DNA or on intercalation of it by water-soluble aromatic dyes. Both of these processes unwind the DNA double helix[39–41] and thus diminish β. That change can be monitored easily by the change in sedimentation velocity[37] of the circular DNA or directly by electron microscopy. A fine example of the latter (Figure 4) shows the change from a low-salt preparation of circular DNA with 128 crossings to a high-salt preparation of the same DNA with only

22 crossings. These sorts of change have led to the suggestion that the DNA has no superhelical turns *in vivo* at the time when the nicked Hershey circle (**31**) is sealed to a closed circle (**32**), but it is effectively in a high-salt environment. Thus when the DNA is isolated *in vitro* at low ionic strength, the helical pitch of the duplex decreases and results in an increase in α, which is manifest in an increase in the number of supercoils. It has been argued that this hypothesis explains why circular DNAs from a variety of sources show a relatively constant number of supercoils per unit length.[37,39]

(*ii*) *Chromosomal DNA*

The highest state of organization of DNA in the cell is the chromosome. At the onset of cell division the chromosomes assume a condensed form, which was seen under the microscope over a century ago. The 46 chromosomes of the human cell have a total length of 200 μm yet contain about 2 m of DNA. This gives a packing ratio of 10 000 fold to be explained by the tertiary structure of DNA and its interaction with the chromosomal proteins.

Chromatin, the material which makes up the chromosomes, contains DNA and some RNA, histone proteins of five varieties, and non-histone proteins (up to 400 varieties in mammalian cells) as well as some lipid. The fashion in which these are organized has begun to fall into place since 1974 and recent ideas can be discussed in four stages of developing complexity: the nucleofilament, the solenoid, the unit fibre, and the chromatid.

The 'string of beads' model for the *nucleofilament* as the basic unit of chromatin structure was devised by Kornberg.[43] It assimilates the ratios of the five histone proteins found associated with DNA, the nature of their aggregation, and the fact that certain nuclease enzymes cleave most of the chromatin into pieces about 200 base-pairs long. This model has a basic structural unit called a nucleosome. It is composed of two molecules each of the histone proteins H2A, H2B, H3, and H4 associated with a coil form of 200 DNA base pairs. It seems that only about one in four of these nucleosomes is also associated with non-histone proteins.

First suggestions were for the DNA to be coiled around the outside of the protein core of the nucleosome to form a bead (**36**) which was linked to adjacent beads by a short length of free DNA. More recently it was argued that the DNA helix lies in an ordered state between two identical tetramers of the H2A, H2B, H3, and H4 proteins. These can open 'like a book' to permit copying of the DNA without dislodging the histone proteins. Electron micrographs of chromatin beads cannot discriminate between these models.

Because the beads are some 10 nm in diameter and the DNA associated with each must be some 70 nm long, a packing factor of 7 is achieved. This must involve the DNA helix in either smooth curvature or tight bending. Crick has suggested that the second of these alternatives may be energetically more favourable and has described a way in which a kink in the DNA helix might contribute to its tertiary structure.[35] It simply involves the unstacking of bases at one point in the helix and rotating the appropriate C-4′—C-5′ bonds in the two strands from the *gauche* to the *trans* conformation. The result is to form a bend in the helix of about 98° towards the side of the minor groove. Clearly, eight such kinks at intervals of 20 base pairs could wind 200 base pairs of DNA around one protein bead as in (**36**) and give second-order coiling of DNA.

The form of aggregation of these beads involves third-order coiling of DNA. Both electron micrographs[44] and neutron diffraction studies[45] show a repeating 11 nm unit in the chromatin filament. The model suggested for these supercoils is a *solenoid* with a diameter about 31 nm, having some six nucleosomes per turn (**37**). It is held together by the fifth histome protein, H1, of which there is about one molecule per bead. In turn, this gives a further packing factor of 6.

It is conceivable that there is only one more level of coiling intermediate between the solenoid and that half of the dividing chromosome known as a *chromatid.* The most recent intermediate proposed is the *unit fibre.*[46] By very careful manipulation, dividing human chromatids can be teased apart to give unit fibres 2–10 μm long and 400 nm thick. They appear to have a thick-walled tube, such as might be formed by a solenoid coiling tightly upon itself in a spring-like fashion. It seems possible that these coils are susceptible to differential staining by basic dyes and thereby give rise to the banded pattern seen in stained chromosomes (**38**). Whatever the biological implication, the unit fibre achieves a further packing factor of 40 and leaves only a final factor of 6 to attain the overall condensation of DNA found in the chromosome (**39**), *i.e.* a factor of 3 for the chromatid. It says nothing for the contribution of the non-histone, regulatory proteins nor for the lipid and RNA involved, but it represents a remarkable achievement in the structural analysis of DNA.

In summary, the emerging picture of the chromatid has a single molecule of DNA running along its entire length. This interacts with four histone proteins to form nucleosomes (**36**) which, by interaction with the fifth protein, wind into a solenoid (**37**) that may also contain non-histone proteins. The further organization of the solenoid depends on the phase of the cell cycle. It is diffuse at prophase but highly condensed during cell division, possibly into unit fibres (**38**) which are primary components of the condensed chromosome (**39**).

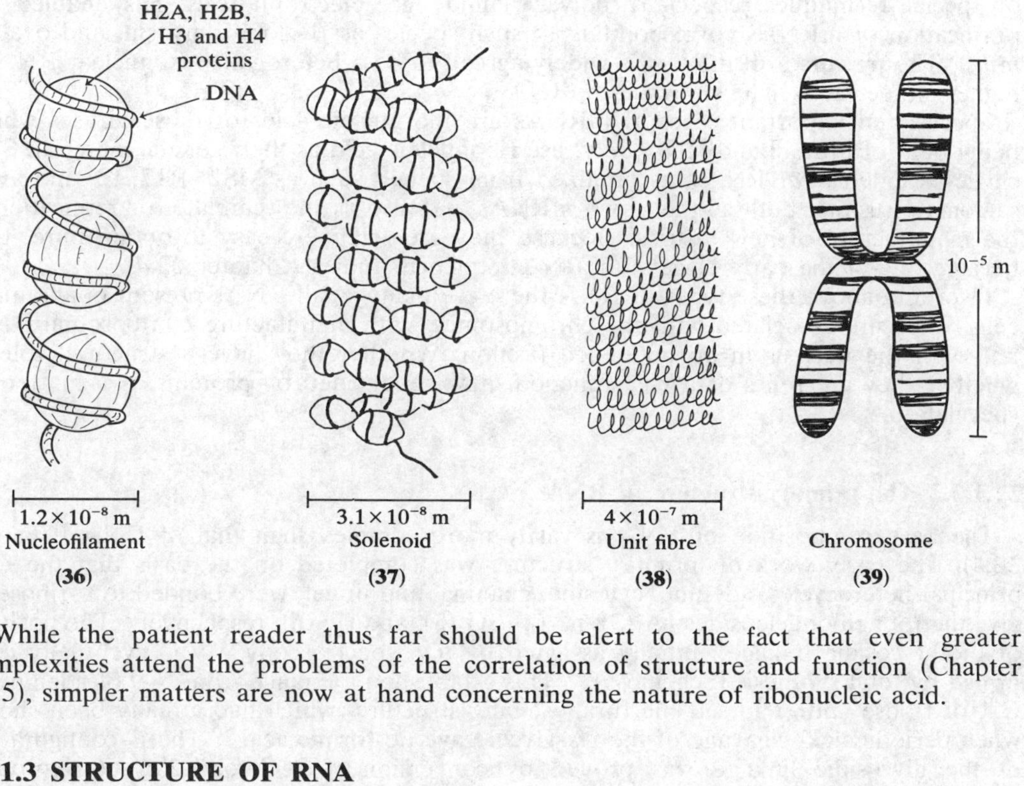

1.2×10^{-8} m Nucleofilament **(36)** | 3.1×10^{-8} m Solenoid **(37)** | 4×10^{-7} m Unit fibre **(38)** | Chromosome **(39)**

While the patient reader thus far should be alert to the fact that even greater complexities attend the problems of the correlation of structure and function (Chapter 22.5), simpler matters are now at hand concerning the nature of ribonucleic acid.

22.1.3 STRUCTURE OF RNA

It is justifiable to discuss DNA as though only that which originates in the cell nucleus in eukaryotes is of prime significance. However, the varieties of important RNA species are many and have individual character.

22.1.3.1 Isolation and characterization of RNA

Most cell species have three principal varieties of RNA, all of which are copied from chromosomal DNA according to the principle of base pairing. They have a common primary structure but differ in sequence of residues, in length, and in higher-order structure.[47]

The bulk of RNA is found as a major structural component of ribosomes — the particles associated with the manufacture of proteins (Section 22.5.3.2). This is *ribosomal RNA*, rRNA. Ribosomes can be obtained readily by centrifugation and then dissociated into protein (40%) and rRNA (60%). Bacterial cells provide three major RNA species which are designated according to their sedimentation behaviour, 5S, 16S, and 23S rRNA (the corresponding mammalian RNA is 5S, 18S, and 28S). In terms of length, they have 120, 1600, and 3000 residues, respectively. They are single-stranded species and covalently continuous (with the possible exception of 28S rRNA).

The first species of RNA to be characterized was *transfer RNA*, tRNA, which accounts for 10–15% of the total cellular RNA and is found in solution in the cell cytoplasm. There can be up to 100 million tRNA molecules in a mammalian cell of some 50 different varieties, each 73–85 nucleotides long. They are readily isolated, purifiable by chromatographic techniques, and in some cases can be crystallized.

The most transient form is that RNA which acts as the *messenger* to carry information from the chromosome to the site of protein biosynthesis. Such mRNA accounts for only 3–5% of the cellular RNA, is heterogeneous in size (around 10^3 nucleotides), and has a short lifetime (varying from $t_{\frac{1}{2}}$ of 1 min in bacteria to $\frac{1}{2}$ day in human tumour cells). The use of special techniques, especially polyacrylamide gel electrophoresis, has enabled the purification of mRNAs corresponding to such species as β-globin, myosin, and ovalbumin.[48] The precursor of mRNA is widely agreed to be a heterogeneous, nuclear RNA of higher molecular weight known as hnRNA.

Special and important cases of mRNAs are the nucleic acids of those viruses which, being built of protein and RNA only, use ribonucleic acid as their genetic material. Such single-stranded *viral* RNAs, as obtained from species such as MS2, R17, f2, and avian sarcoma virus, act both as their own mRNAs and also as the template for replication in the manufacture of new viruses. Because they are relatively easy to obtain pure, they became one of the early targets for RNA sequence studies (Chapter 22.4).

Over and above these types of RNA there are many small RNAs present in all animal cells. They are associated with the chromosomes, with manufacture of ribosomal RNA, and with the nucleus in an undefined fashion. Whether they have a structural role or whether they are relics of a past function, now superseded by proteins, is a matter for speculation.

22.1.3.2 The primary structure of RNA

The base composition of RNA is vastly more complex than that of DNA (Chapter 22.4). The early work on primary structure was completed on the basis that the four principal heterocycles, adenine, cytosine, guanine, and uracil, were bonded to D-ribose to give the four ribonucleosides rA (**3**), rC (**4**), rG (**5**), and rU (**6**), respectively. The position of the glycosidic linkage, initially assigned by u.v. spectroscopy,[16] was eventually confirmed by total synthesis (Scheme 5).[49] This established a linkage from N-9 of adenine to C-1 of ribose and endorsed the furanose ring structure, which had already been shown when periodic acid cleavage of the *cis*-glycol gave no formic acid.[50] The β-configuration of the glycosidic linkage was proved by conversion of 5′-*O*-tosyl-2′,3′-*O*-isopropylideneadenosine (**40**) into the cyclonucleoside (**41**).[51]

The structures of the pyrimidine nucleosides were approached differently. On methylation, uridine (**6**) gave a mono-*N*-methyluridine (although uracil gives 1,3-dimethyluracil), which hydrolysed to give 3-methyluracil (**42**). Thus, uridine must have the ribose attached at N-1. Cytidine (**5**) is directly transformed into uridine on deamination with nitrous acid (Chapter 22.5), which rounds off the determination of the structures of the ribonucleosides.

The actual shape of the nucleosides does not seem to have been considered important until Furberg determined the crystal structure of cytidine, which showed that the plane of the sugar ring was at right angles to the plane of the cytosine base.[52] The inclusion of this feature in the work of the DNA model builders was an important advance.

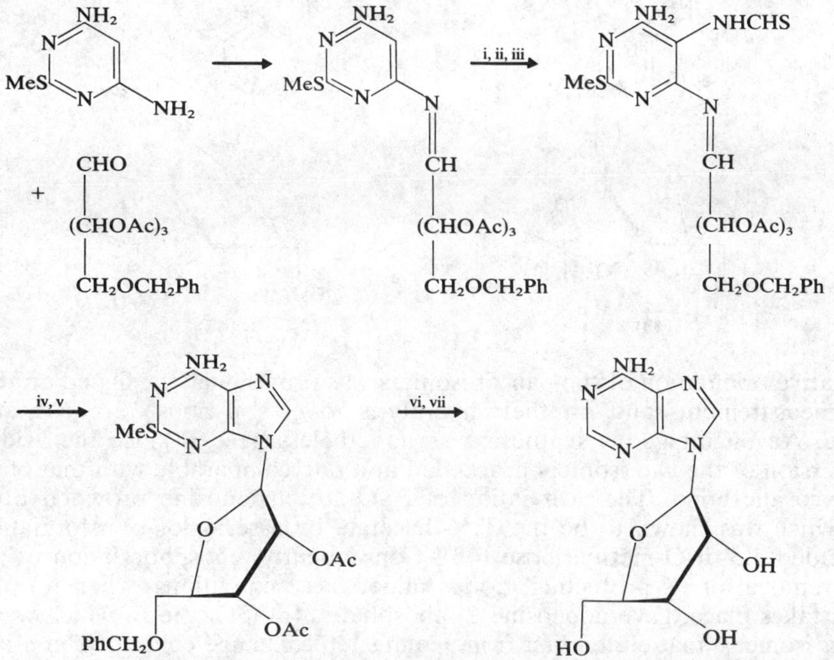

i, ArN$_2^+$; ii, Zn, AcOH; iii, HCSSH; iv, EtONa, EtOH; v, Ac$_2$O; vi, Raney nickel; vii, MeONa, MeOH.

SCHEME 5

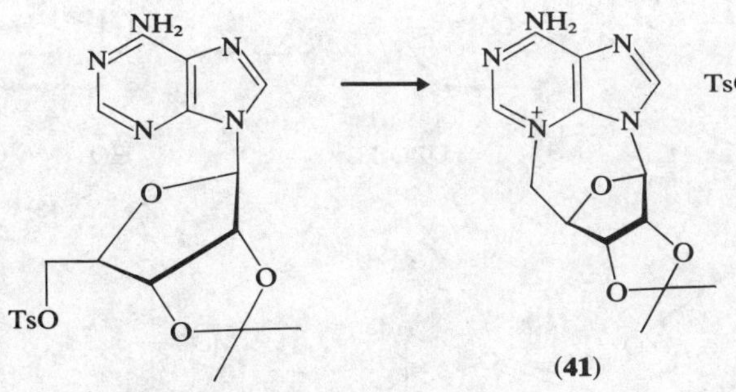

(40) **(41)**

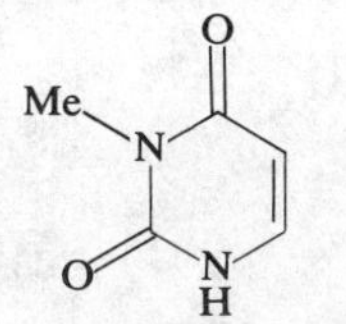

(42)

The structures of the ribonucleotides followed quickly from the availability of the isopropylidene protecting group for the *cis*-diol at positions-2′ and -3′ of the ribofuranose residue. Typically, phosphorylation of isopropylideneadenosine (**43**) by means of dibenzyl phosphorochloridate, with subsequent removal of protecting groups, provided adenosine 5′-phosphate,[53] which proved to be identical with the adenine nucleotide isolated in 1927 by Embden and Zimmermann.[54]

The four nucleotides, prA, prC, prG, and prU, prepared by this route were, however, different from the 'four mononucleotides' obtained by alkaline hydrolyis of RNA. Early work had, indeed, assigned to them ribonucleoside 3′-phosphate structures but Cohn, by means of ion exchange chromatography, was able to separate each of the 'four mononucleotides' into pairs of isomers.[55] In the case of the two adenylic acids, several methods were used to establish that they were the 2′- and 3′-phosphates of adenosine, including their mutual conversion into the 2′,3′-cyclic phosphate (**44**) by treatment with trifluoroacetic anhydride.[56]

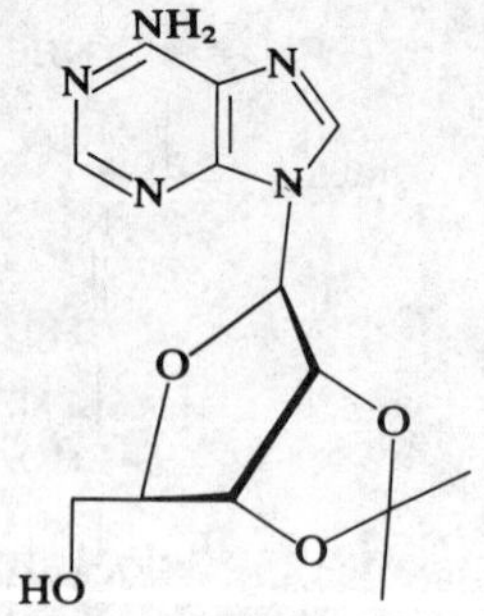

(**43**)

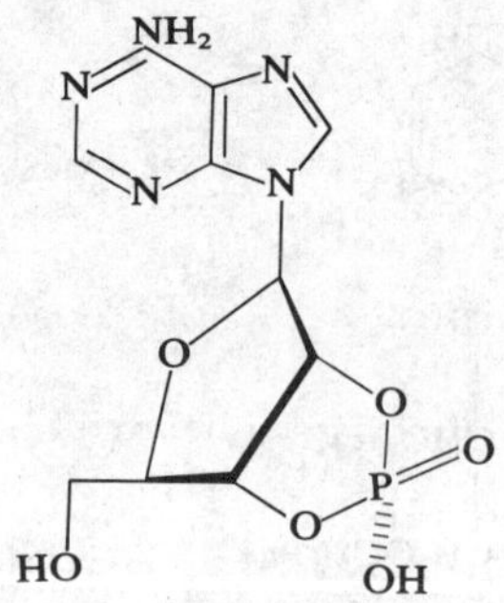

(**44**)

The relative orientation of the pair of isomers was provisionally assigned on the basis of physical measurements and on their hydrolysis to give D-ribose 2-phosphate and 3-phosphate. An unambiguous synthesis was nonetheless needed since the acid-catalysed interconversion of the two isomers proceeded at a rate comparable with that of hydrolysis of the glycosidic bond. The acetylation of 5′-*O*-acetyladenosine provided a crystalline product which was shown to be the 3′,5′-diacetate by a series of transformations which finally yielded 3,5-di-*O*-methylribose (**45**). Consequently, phosphorylation of this diacetate and removal of its protecting groups under basic conditions, when no phosphorus migration takes place, gave adenosine 2′-phosphate (**46**) (Scheme 6) which was identical with the *a* isomer[57] that eluted first from Cohn's ion exchange column. The *b* isomer was shown to be adenosine 3′-phosphate, rAp, by X-ray crystallography.[58] A related synthesis showed that uridylic acid *a* was the 2′-phosphate and Furberg[59] later completed the crystallographic analysis of cytidylic acid *b*, which proved it to be cytidine 3′-phosphate, rCp.

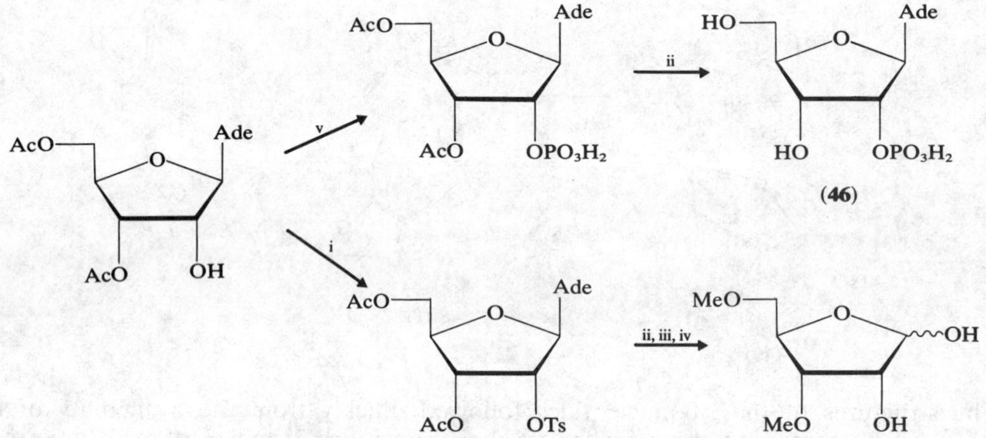

i, TsCl, pyridine; ii, NH_3, MeOH; iii, Ag_2O, MeI; iv, acid; v, phosphorylation.

SCHEME 6

At the corresponding stage of analysis, the polymeric structure of DNA was virtually self evident. However, for RNA the chemical evidence so far presented is compatible with three polymeric structures. The internucleotide linkage can be 5′→3′, or 5′→2′, or a mixture of both in variable proportions. This problem had to be resolved by the use of specific enzymatic hydrolysis.

Ribonuclease had been isolated from pancreas glands and crystallized in 1940. It attacks RNA by cleaving the ester bond at phosphates attached to the 3′-position of pyrimidine nucleosides (see Chapter 22.3). The nature of its action was appreciated by Markham and Smith, who showed that the 2′,3′-cyclic phosphates of uridine and cytidine (**47**) were the initial products of RNase action on RNA and, in turn, were slowly hydrolysed to the 3′-phosphates in a subsequent enzymatic step.[60] No comparable enzyme was then available with a like specificity for purine residues (the takadiastase enzyme was discovered later), but a phosphodiesterase from spleen was to hand and it acted as an exonuclease* to give all four ribonucleoside 3′-phosphates. Such information was enhanced by the demonstration that these enzymes were equally well able to hydrolyse the benzyl phosphate esters of the 3′-nucleotides (**48**), but would not operate on the corresponding benzyl esters of the 2′-nucleotides (Scheme 7). It was therefore clear that the internucleotide linkage involved the 3′- and not the 2′-position.

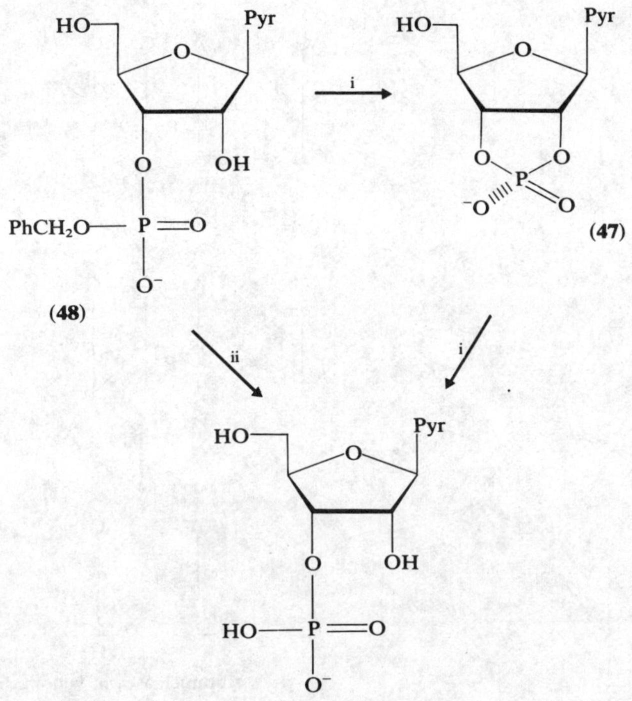

i, ribonuclease; ii, H_2, Pd.

SCHEME 7

The evidence supporting the use of the 5′-position stemmed from the work of Gulland and Jackson[61] and was completed by Cohn and Volkin.[62] Using arsenate to inhibit an unwanted phosphomonoesterase activity, they exposed RNA to an intestinal phosphodiesterase and obtained all four ribonucleoside 5′-phosphates. The same result was achieved later using a snake venom diesterase. This work confirmed the general primary structure of RNA as having only 5′→3′ linkages (**49**). The pattern of enzymatic hydrolysis is illustrated using a condensed structural representation for RNA (Scheme 8).

The availability of the 2′-hydroxyl group makes chain branching a formal possibility in RNA. However, no evidence has been presented to support that possibility. RNA, like DNA, appears to be a linear polynucleotide with invariant 5′→3′ internucleotide linkages.

One major difference between RNA and DNA is the large number of modified bases which are found in the former. This situation is particularly pronounced in the case of

* *endo*-Attack takes place in the middle of a chain; *exo*-attack cleaves residues from terminal positions on a chain.

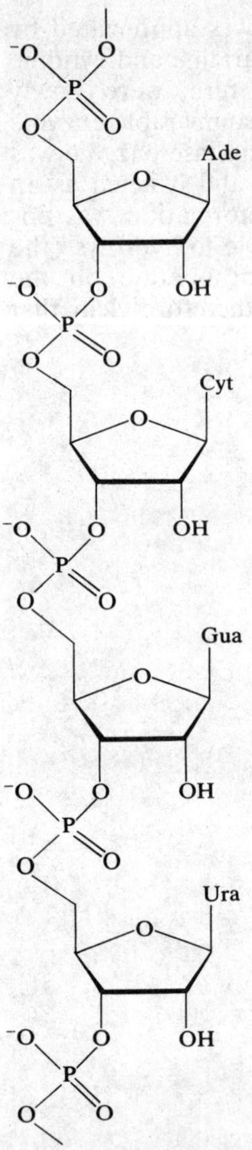

(49)

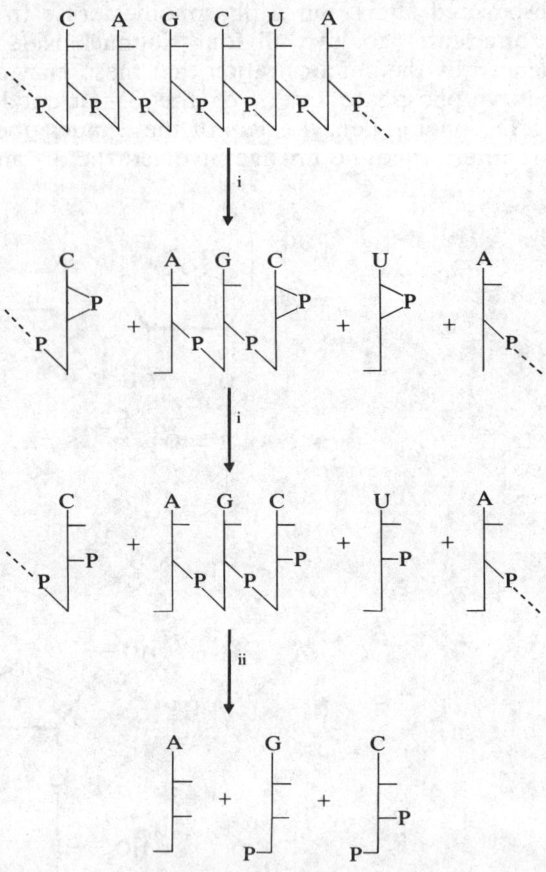

i, ribonuclease; ii, venom diesterase.

SCHEME 8

transfer RNA. The discussion of the primary sequence of determination of RNAs is therefore deferred until after the presentation of the chemistry of the minor nucleosides (Chapter 22.2). However, some of the results of such sequence studies (Chapter 22.4) must be assumed in the following discussion of the secondary and tertiary structure of RNA.

22.1.3.3 The secondary structure of RNA

While most naturally occurring RNA species are single-stranded nucleic acids, a small proportion of viruses, such as Reovirus, possess RNA in a double-helical form. These RNAs have a base composition in which A = U and G = C and they show a marked resistance to digestion by ribonuclease unless they are first denatured by heating. Fibres can be drawn from solutions of such RNA or from preparations of synthetic double-stranded polymers such as poly(rA) · (rU) and then used for X-ray diffraction studies.[63] The X-ray data reveal that double-stranded RNA adopts a helical structure which resembles A-DNA quite closely (about 10° of tilt of the plane of the base-pairs and 11–12 bases per turn of the helix). This conformation for *A*-RNA, like that of *A*-DNA, appears to be dictated by the puckering of the sugar ring in the C-3-*endo* conformation. It is thus apparent that uracil can complement adenine as effectively as does thymine in hydrogen-bond formation.

A second type of double helix containing RNA was discovered in 1961. It is created by the 'hybridization' of a strand of RNA with a strand of DNA of complementary base sequence.[64] The importance of such RNA:DNA hybrids has expanded enormously and today they are a key feature of sequence determination and of gene manipulation (Sections 22.4.4 and 22.5.4). Such hybrids are readily formed by incubation of a solution of the two single-stranded species at about 25 °C below the temperature at which the duplex is half-dissociated. They are more stable than DNA:DNA duplexes of corresponding base structure.[65] Such hybrids also appear to have the *A*-DNA conformation.

The stability of such helices is generally thought to arise from the stacking of bases, as in DNA. Indeed, single-stranded ribopolynucleotides have a more ordered structure than their deoxyribo-counterparts. Poly(rA) exists as a highly-extended rod in solution at room temperature. A study of the free energy values for the dimeric association of species such as rA_nU_n, rA_nCGU_n, and rA_nUGU_n has enabled estimates to be made for the energy of stacking of purine:pyrimidine pairs. The pairs C · G, A · U, and G · U give ΔG_0 values of -10, -5, and $0\ \mathrm{kJ\ mol^{-1}}$, respectively.

By the addition of a positive energy penalty for unpaired bases in loops and elsewhere, it is possible to devise a computer program to calculate the relative stabilities of feasible secondary structures for single-stranded RNA molecules of known base sequence.[66] It now appears that such molecules with a randomly-generated base sequence can be arranged into secondary structures in which about half of all the bases are paired in helical regions and that such structures are expected to be thermodynamically stable.[67]

Nature is no less efficient! Since Holley provided the first[68] base sequence for a tRNA molecule in 1965, now known to contain a spurious G · C pair,[69] followed by the analysis of tRNA sequences at up to a dozen *per annum*, a common 'cloverleaf' structure has been applied to them all. This puts 40 bases into paired, helical regions out of the number (73–79) available for this purpose. Such conformations are typically represented by the cloverleaf for $tRNA^{Phe}$ from yeast (**50**). This secondary structure at first appeared to be

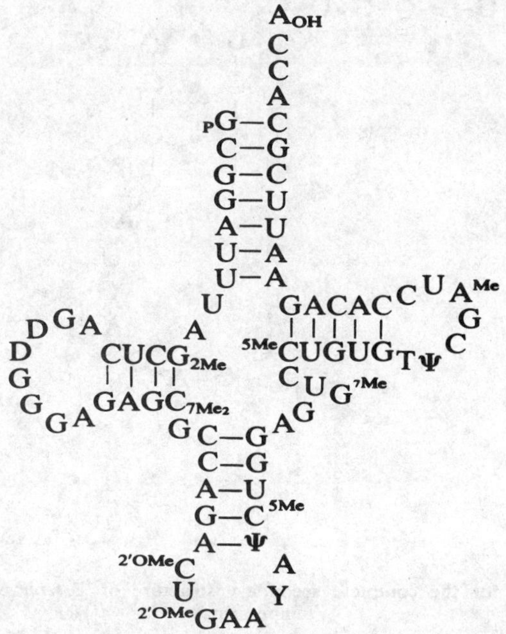

(50) $tRNA^{Phe}$ (Yeast)

speculative in nature, but each new tRNA sequence discovered fitted the same pattern. Material evidence has now been provided from the X-ray crystallographic determination of the structure of $tRNA^{Phe}$ (Section 22.1.3.4) and therefore makes vastly more credible the sweeping proposals now advanced for the secondary structure of larger species such as rRNA and viral RNA.

The 120-nucleotide chain of 5S ribosomal RNA from *E. coli* was the first achievement of a more rapid method of sequence determination[70] than that employed by Holley. Its solution properties indicate a higher-order structure of some rigidity with many base pairs, but its sequence permits of too many possible secondary structures for a clear-cut choice. In this case, variable-temperature, high-resolution n.m.r. spectroscopy has been applied as a discriminator. The data thus obtained[71] appear to favour a structure of 28 base-pairs, of which 17 are of the most stable C · G type, located in five helical regions (**51**).

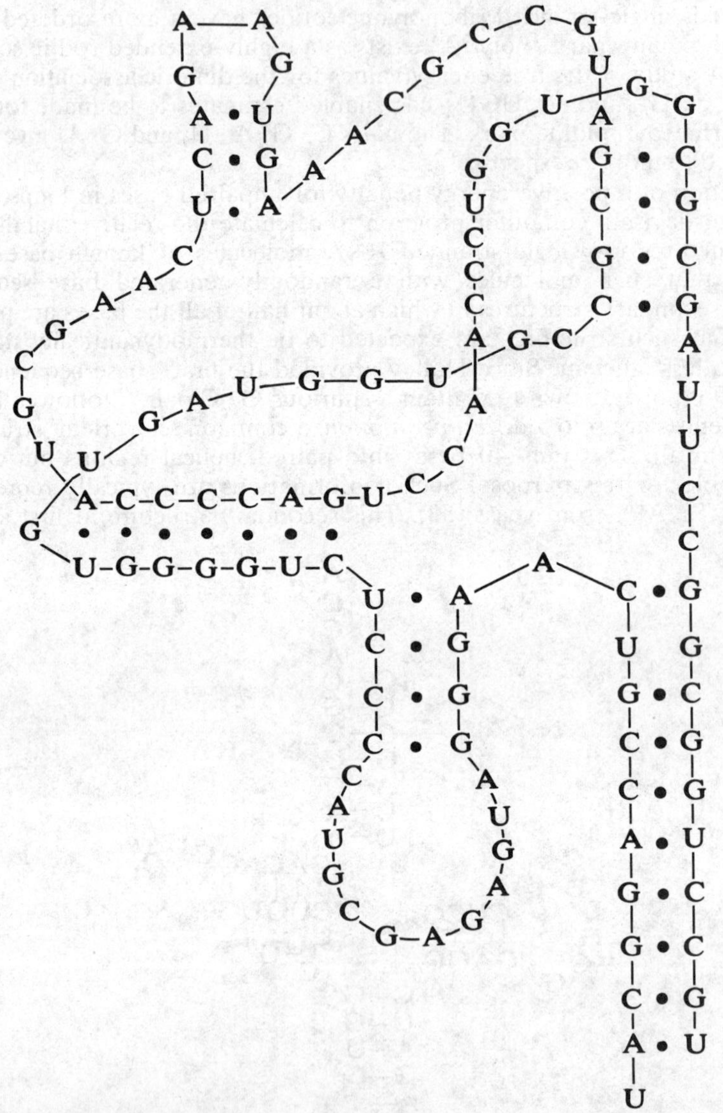

Proposed model for the complete secondary structure of *E. coli* 5S RNA at room temperature.

(51)

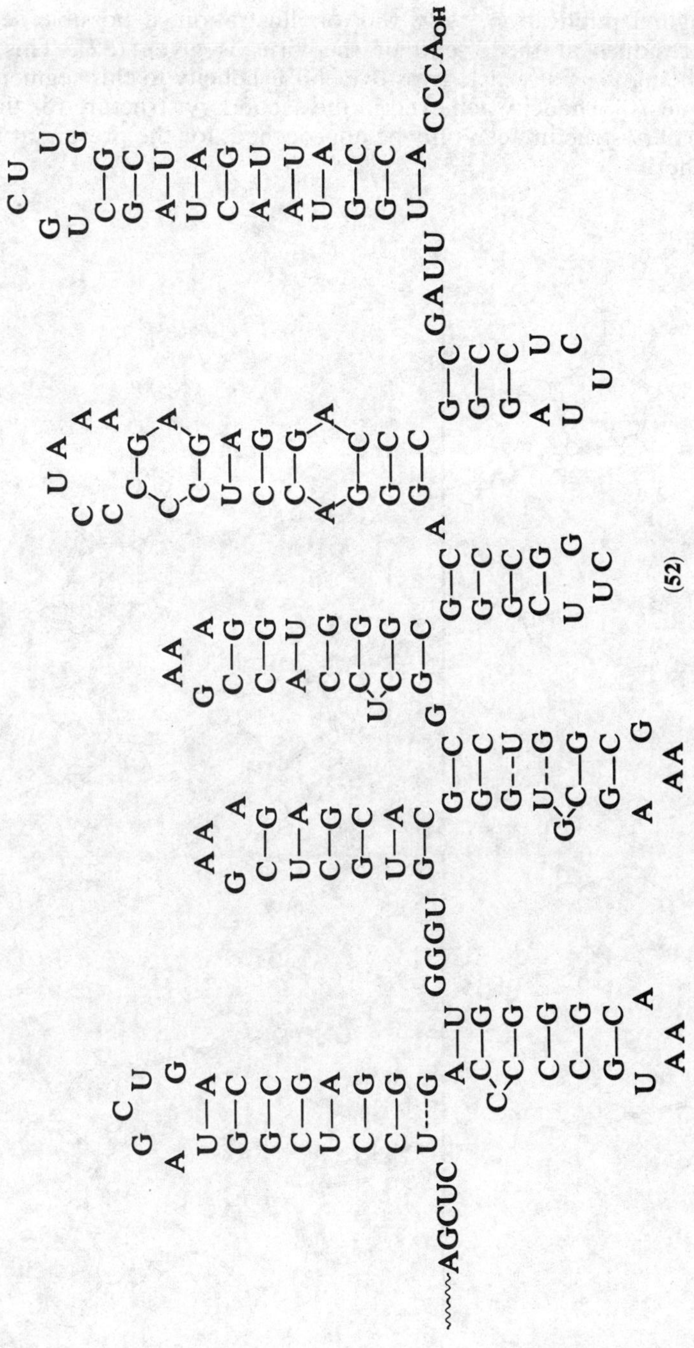

(52)

The proliferation of primary sequences for mRNAs from eukaryotic cells (Section 22.4.4) creates enormous possibilities for speculation about RNA secondary structure. These are, however, dwarfed in comparison with the complexities possible for viruses. The first complete nucleotide sequence for a bacteriophage, MS2, was announced in 1976. It has a 3569 nucleotide chain and the secondary structure for only half of it occupies a double page of the original publication.[72] By way of illustration, a possible secondary structure for the 179 residues at the 3′-end of the virus is given (**52**). This has 55 base-pairs, 43 of them being G · Cs, which gives over 60% helicity to this segment of the structure. Even with some 70% helicity achieved in the secondary structure for the whole virus, the problem of tertiary structure can only be approached, for the present at least, by the X-ray crystallographer.

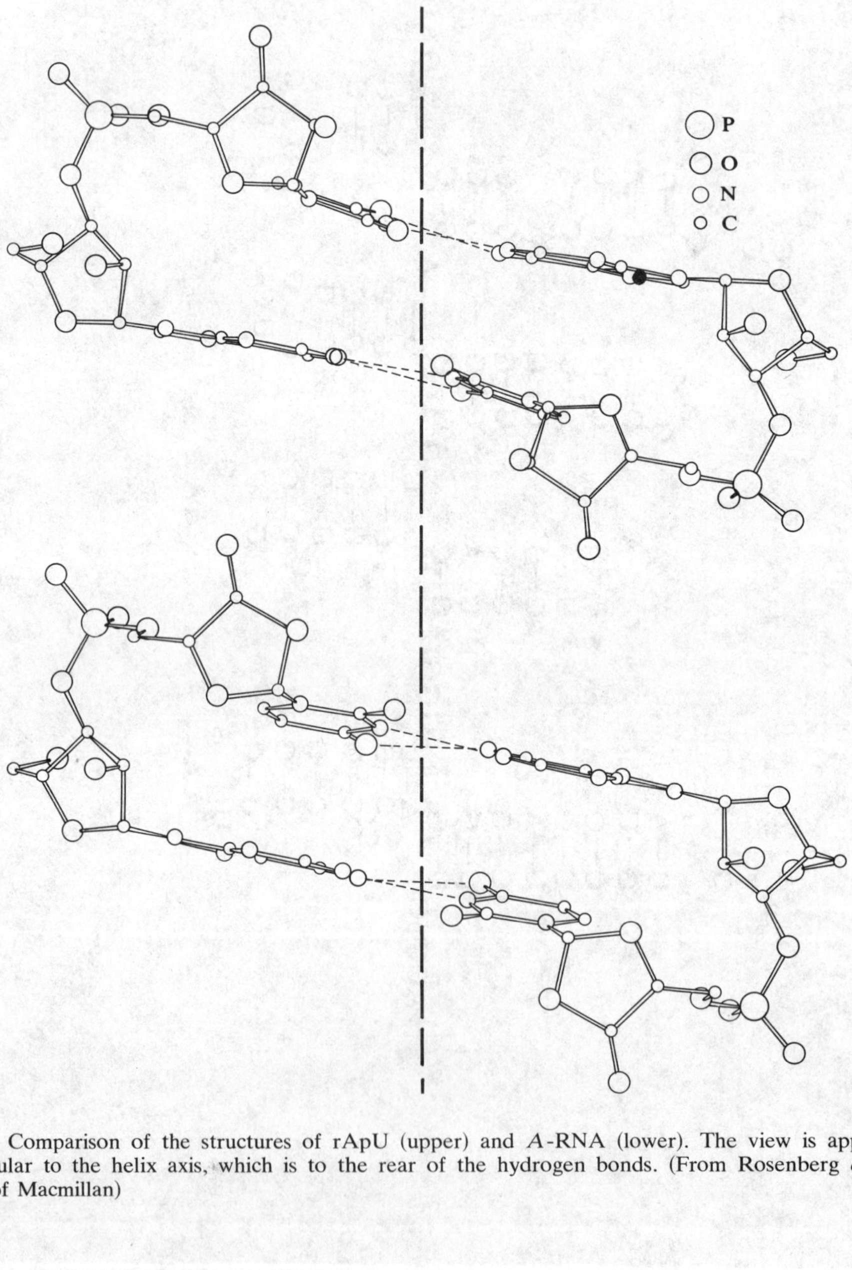

Figure 5 Comparison of the structures of rApU (upper) and *A*-RNA (lower). The view is approximately perpendicular to the helix axis, which is to the rear of the hydrogen bonds. (From Rosenberg *et al.*,[73], by courtesy of Macmillan)

22.1.3.4 The tertiary structure of RNA

Physicochemical measurements on single-stranded RNA species in solution leave no doubt that such molecules have tertiary structures, often compact ones. Multiple hydrogen bonding between bases is readily detected by high-resolution n.m.r. spectroscopy. Nonetheless, as in the domain of protein structures, the possibilities are too varied to be conquered by any other than crystal-structure analysis. Two major contributions have been made for the tertiary structure of RNA.

Both of the dinucleoside phosphates rApU[73] and rGpC[74] are self-complementary in base-pairing potential. They both form well-defined crystals which give X-ray diffraction data to 1 Å resolution. These data reveal that both species exist as short sections of right-handed, antiparallel double helixes in which the ribose–phosphate–ribose backbones are linked by Watson–Crick hydrogen bonds between the paired bases, $\begin{matrix}\text{ApU}\\ \vdots\ \vdots\\ \text{UpA}\end{matrix}$ and $\begin{matrix}\text{GpC}\\ \vdots\ \vdots\\ \text{CpG}\end{matrix}$. The structure for ApU bears a striking resemblance to that proposed for the *A*-form of RNA (Figure 5).

This achievement was soon overshadowed by the completion of crystallographic structures for yeast $tRNA^{Phe}$. The initial announcement of 3 Å data on an orthorhombic crystalline form[75] was followed by one based on monoclinic crystals of the same tRNA species which, extended to 2.5 Å, produced the more accurate picture of the tertiary structure of the molecule.[76] At this resolution the backbone can easily be traced and shows the cloverleaf pattern arranged into an 'L'-shaped tertiary conformation. It is achieved by stacking the 'dihydro U' stem on the anticodon stem and the 'TψC' stem on the acceptor stem. This places the anticodon loop at one extremity of the 'L'-shape and the 3′-end at the other. The tertiary organization is maintained by additional base-pairs and base-triplets which are usually *not* of Watson–Crick type. Several of these are associated with features of the primary base sequence which have been recognized as near-invariant in large numbers of tRNAs. Most significantly, they stack so as to exclude water molecules (*e.g.* G15 · C48 and U8 · A14). The extensive use of the 2′-hydroxyl group as either a donor or acceptor of a hydrogen bond is an additional notable feature (Figure 6).

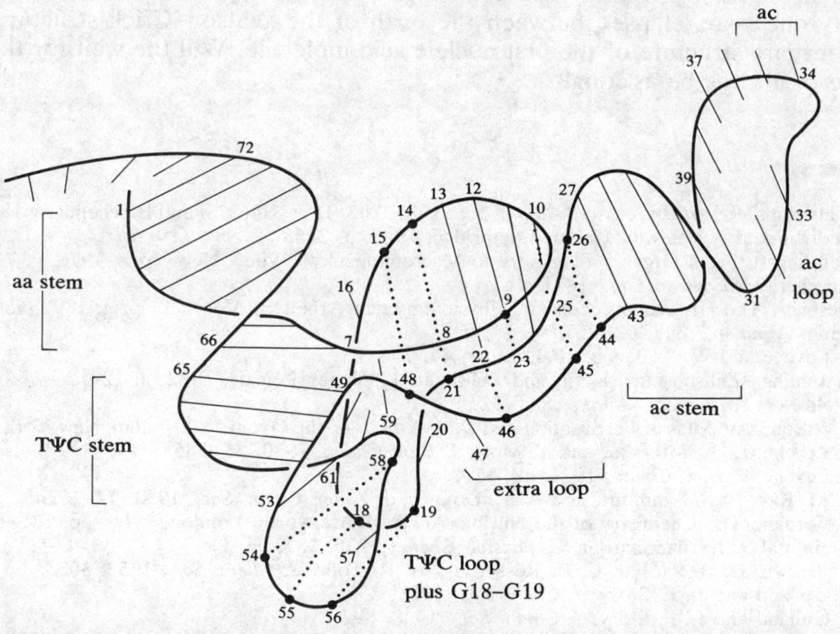

Figure 6 A schematic diagram of chain folding and tertiary interactions between bases in yeast $tRNA^{Phe}$. The ribose–phosphate backbone is shown as a continuous line. Base-pairs in the helical sections are represented by long light lines and non-paired bases by shorter lines. Base-pairs additional to the 21 in the cloverleaf structure are represented by dotted lines. (From Ladner *et al.*,[76] by courtesy of the US National Academy of Science)

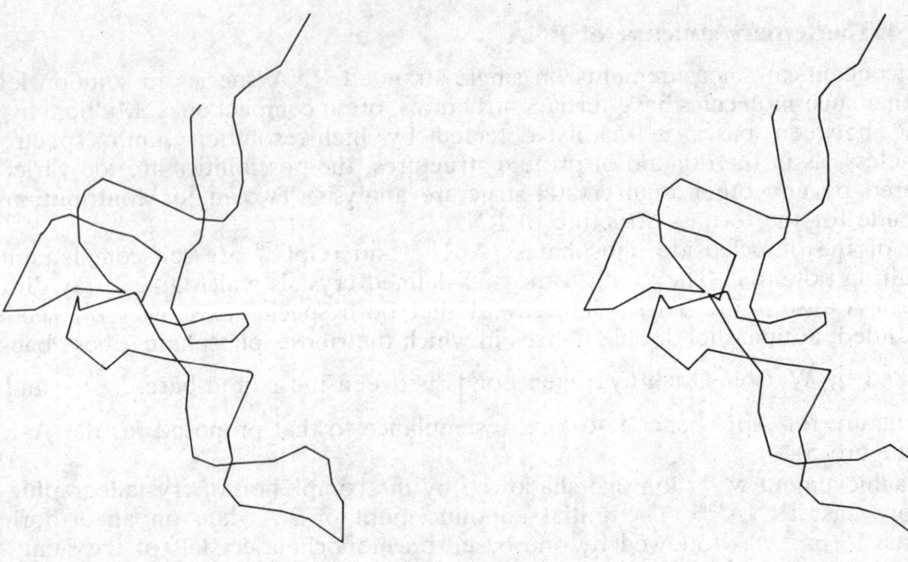

Figure 7 Stereo-diagram of the tertiary structure of $tRNA^{Phe}$ giving the C-1′ backbone positions (*parallel viewing*). (From Kim,[77] by courtesy of Academic Press)

For the benefit of those readers capable of resolving stereopair diagrams, a 3-D projection of the structure based on the 3 Å data is also provided (Figure 7).

The constant anxiety of molecular biologists that crystallographic tertiary structures may not survive dissolution in water has been somewhat allayed for $tRNA^{Phe}$ by the spectroscopic analysis at 360 MHz of the hydrogen-bonded protons of the molecule in solution. Such protons characteristically resonate at 10–15 δ and provide temperature-dependent data under both CW and FT conditions. Analysis of all the data thereby obtained in solution provides *post-hoc* support for the crystal structure conformation.[78]

Twenty-one years elapsed between the birth of the Watson–Crick structure for DNA and the tertiary structure of the first nucleic acid molecule. Will the wait for the first total structure of a virus be as long?

References

1. F. G. Hopkins, *Report Brit. Assoc. Advmt. Sci.*, 1933, **103,** 1; in 'Hopkins and Biochemistry 1861–1947,' ed. J. Needham and J. Baldwin, Heffer, Cambridge, 1949, p. 245.
2. H. Staudinger, 'From Organic Chemistry to Macromolecules,' Wiley, New York, 1970, p. 79.
3. F. Miescher, *Med. chem. Unt.*, 1871, 441.
4. F. Miescher, 'Die Histochemischen und Physiologischen Arbeiten,' Vogel, Leipzig, 1897, vol. II, p. 66.
5. J. Liebig, *Annalen*, 1847, **62,** 257.
6. P. A. Levene and W. A. Jacobs, *Ber.*, 1911, **44,** 746.
7. W. Flemming, 'Zellsubstanz, Kern, und Zelltheilung,' Vogel, Leipzig, 1882, p. 129.
8. W. Waldeyer, *Arch. mikr. Anat.*, 1888, **32,** 1.
9. E. B. Wilson, 'An Atlas of Fertilisation and Karyokinesis of the Ovum,' Macmillan, New York, 1895, p. 4.
10. P. A. Levene, L. A. Mikeska, and T. Mori, *J. Biol. Chem.*, 1930, **85,** 785.
11. P. A. Levene, *J. Biol. Chem.*, 1921, **48,** 119.
12. E. R. M. Kay, N. S. Simmons, and A. L. Dounce, *J. Amer. Chem. Soc.*, 1952, **74,** 1724.
13. D. O. Jordan, 'The Chemistry of the Nucleic Acids,' Butterworth, London, 1960, pp. 181–193.
14. W. Klein and S. J. Thannhauser, *Z. physiol. Chem.*, 1935, **231,** 96.
15. D. M. Brown, D. B. Parihar, C. B. Reese, and A. R. Todd, *J. Chem. Soc.*, 1958, 3035.
16. J. M. Gulland and L. F. Story, *J. Chem Soc.*, 1938, 259.
17. J. M. Gulland and L. F. Story, *J. Chem. Soc.*, 1938, 692.
18. W. Anderson, D. H. Hayes, A. M. Michelson, and A. R. Todd, *J. Chem. Soc.*, 1954, 1882.
19. A. M. Michelson and A. R. Todd, *J. Chem. Soc.*, 1955, 816.
20. A. M. Michelson and A. R. Todd, *J. Chem. Soc.*, 1953, 951.
21. D. H. Hayes, A. M. Michelson, and A. R. Todd, *J. Chem. Soc.*, 1955, 808.

22. A. M. Michelson and A. R. Todd, *J. Chem. Soc.*, 1954, 34.
23. L. Cunningham, *J. Amer. Chem. Soc.*, 1958, **80,** 2546.
24. C. A. Dekker, A. M. Michelson, and A. R. Todd, *J. Chem. Soc.*, 1953, 947.
25. Abbreviations used are in accord with the recommendations of the IUPAC–IUB Commission, *J. Biol. Chem.*, 1970, **245,** 5171.
26. J. M. Gulland, *Cold Spring Harbor Symp. Quant. Biol.*, 1947, **12,** 95.
27. W. T. Astbury and F. O. Bell, *Nature*, 1938, **141,** 747.
28. J. D. Watson, 'The Double Helix,' Weidenfelt and Nicholson, London, 1968.
29. R. Olby, 'The Path to the Double Helix,' Macmillan, London, 1974.
30. M. H. F. Wilkins, 'Comprehensive Biochemistry,' ed. M. Florkin and E. H. Stotz, Elsevier, Amsterdam, 1963, vol. 8, chapter IV.
31. *Cf.* J. N. Davidson, 'The Biochemistry of Nucleic Acids,' Butterworth, London, 2nd edn., 1953.
32. J. D. Watson and F. H. C. Crick, *Nature*, 1953, **171,** 737.
33. S. Arnott, S. D. Dover, and A. J. Wonacott, *Acta Cryst.*, 1969, **B25,** 2192.
34. G. A. Rodley, R. S. Scobie, R. H. T. Bates, and R. M. Lewitt, *Proc. Nat. Acad. Sci. U.S.A.*, 1976, **73,** 2959.
35. F. H. C. Crick and A. Klug, *Nature*, 1975, **255,** 530.
36. E. Kellenberger, *Adv. Virus Res.*, 1961, **8,** 1.
37. M. J. Waring, *Ann. Reports Chem. Soc.*, 1967, **64B,** 493; 1968, **65B,** 551.
38. D. R. Helinski and D. B. Clewell, *Ann. Rev. Biochem.*, 1971, **40,** 899.
39. J. Vinograd, J. Lebowitz, and R. Watson, *J. Mol. Biol.*, 1968, **33,** 173.
40. W. Bauer and J. Vinograd, *J. Mol. Biol.*, 1968, **33,** 141.
41. W. Keller, *Proc. Nat. Acad. Sci. U.S.A.*, 1975, **72,** 4876.
42. V. C. Bode and L. A. MacHattie, *J. Mol. Biol.*, 1968, **32,** 673.
43. R. D. Kornberg, *Science*, 1974, **184,** 868.
44. J. T. Finch and A. Klug, *Proc. Nat. Acad. Sci. U.S.A.*, 1976, **73,** 1897.
45. K. Ibel, B. Carpenter, J. Baldwin, and E. M. Bradbury, *Nucleic Acids Res.*, 1976, **3,** 1739.
46. F. H. C. Crick, A. L. Bax, and J. Zeuthen, *Proc. Nat. Acad. Sci. U.S.A.*, 1977, **74,** 1595.
47. *Cf.* R. H. Burdon, 'RNA Biosynthesis,' Chapman and Hall, London, 1976.
48. G. Brawerman, *Ann. Rev. Biochem.*, 1974, **43,** 621.
49. G. W. Kenner, C. W. Taylor, and A. R. Todd, *J. Chem. Soc.*, 1949, 1620.
50. B. Lythgoe and A. R. Todd, *J. Chem. Soc.*, 1944, 592.
51. V. M. Clark, A. R. Todd, and J. Zussman, *J. Chem. Soc.*, 1951, 2952.
52. S. Furberg, *Acta Chem. Scand.*, 1950, **4,** 751; *Acta Cryst.*, 1950, **3,** 325.
53. J. Baddiley and A. R. Todd, *J. Chem. Soc.*, 1947, 648.
54. G. Embden and M. Zimmermann, *Z. physiol. Chem.*, 1927, **167,** 137.
55. W. E. Cohn, 'Currents in Biochemical Research,' ed. D. Green, Interscience, New York, 1956, p. 460 and refs. therein.
56. D. M. Brown, D. I. Magrath, and A. R. Todd, *J. Chem. Soc.*, 1952, 2708.
57. D. M. Brown, G. D. Fasman, D. I. Magrath, and A. R. Todd, *J. Chem. Soc.*, 1954, 1448.
58. D. M. Brown, G. D. Fasman, D. I. Magrath, A. R. Todd, W. Cochran, and M. M. Woolfson, *Nature*, 1953, **172,** 1184.
59. E. Alver and S. Furberg, *Acta Chem. Scand.*, 1959, **13,** 910.
60. R. Markham and J. D. Smith, *Biochem. J.*, 1952, **52,** 552, 558, 565.
61. J. M. Gulland and E. M. Jackson, *J. Chem. Soc.*, 1938, 1492.
62. W. E. Cohn and E. Volkin, *Nature*, 1951, **167,** 483; *J. Biol. Chem.*, 1953, **203,** 319.
63. S. Arnott, *Progr. Biophys. Mol. Biol.*, 1970, **21**, 265.
64. B. D. Hall and S. Spiegelman, *Proc. Nat. Acad. Sci. U.S.A.*, 1961, **47,** 137; C. L. Schildkraut, J. Marmur, J. Fresco, and P. Doty, *J. Biol. Chem.*, 1961, **236,** PC3.
65. D. E. Kennell, *Progr. Nucleic Acid Res.*, 1971, **11,** 259.
66. C. Delisi and D. M. Crothers, *Proc. Nat. Acad. Sci. U.S.A.*, 1971, **68,** 2682.
67. J. Gralla and C. DeLisi, *Nature*, 1974, **248,** 330.
68. R. W. Holley, J. Apgar, G. A. Everett, J. T. Madison, N. Marquisee, S. H. Merrill, J. R. Penswick, and A. Zamir, *Science*, 1965, **147,** 1462.
69. J. R. Penswick, R. Martin, and G. Dirheimer, *F.E.B.S. Letters*, 1975, **50,** 28.
70. G. G. Brownlee, F. Sanger, and B. G. Barrell, *Nature*, 1967, **215,** 735; *J. Mol. Biol.*, 1968, **34,** 379.
71. D. R. Kearns and Y. P. Wong, *J. Mol. Biol.*, 1974, **87,** 755.
72. W. Fiers, R. Contreras, F. Duerinck, G. Haegeman, D. Iserentant, J. Merregaert, W. Min Jou, F. Molemans, A. Raeymaekers, A. Van den Berghe, G. Volckaert, and M. Ysebaert, *Nature*, 1976, **260,** 500.
73. J. M. Rosenberg, N. C. Seeman, J. J. P. Kim, F. L. Suddath, H. B. Nicholas, and A. Rich, *Nature*, 1973, **243,** 150.
74. R. O. Day, N. C. Seeman, J. M. Rosenberg, and A. Rich, *Proc. Nat. Acad. Sci. U.S.A.*, 1973, **70,** 849.
75. S. H. Kim, F. L. Suddath, G. J. Quigley, A. McPherson, J. L. Sussmann, A. H. J. Wang, N. C. Seeman, and A. Rich, *Science*, 1974, **185,** 435.
76. J. E. Ladner, A. Jack, J. D. Robertus, R. S. Brown, D. Rhodes, B. F. C. Clark, and A. Klug, *Proc. Nat. Acad. Sci. U.S.A.*, 1975, **72,** 4414.
77. S. H. Kim, *Progr. Nucleic Acid Res.*, 1976, **17,** 181.
78. R. Römer and V. Varadi, *Proc. Nat. Acad. Sci. U.S.A.*, 1977, **74,** 1561.

22.2
Nucleosides

R. T. WALKER
University of Birmingham

Although this chapter is in a series devoted to organic chemistry, the fact that it appears in a section entitled 'Nucleic Acids' means that, in general, this subject will be dealt with from that standpoint, such that the peculiar syntheses and preparations of the many nucleoside analogues which have been reported will not be mentioned unless they contain something of direct interest to a person working in the nucleic acid field. As the subject of nucleic acids covers so many separate disciplines, it is more important in this text to emphasize those aspects of the chemistry of the many known antibiotics and other antimetabolites which are nucleosides, rather than to catalogue many compounds with heterocycles containing all possible arrangements of nitrogen atoms and containing different sugar configurations, which have often only been made as an intellectual challenge and which very often have no particularly interesting biological properties.

Perhaps a definition of a nucleoside would be advantageous, as we have moved a long way from the original definition, which was concerned with the carbohydrate derivatives of purines and pyrimidines obtained by hydrolysis of the nucleic acids. Nowadays, all compounds of synthetic or natural origin which contain a heterocyclic base, linked through nitrogen or carbon to the C-1 position of a sugar, can be regarded as nucleosides, but even that definition excludes 1,3-dideazauridine,[1] which looks suspiciously like a derivative of resorcinol. The common naturally-occurring nucleosides are adenosine (**1**), deoxyadenosine (**2**), guanosine (**3**), deoxyguanosine (**4**), cytidine (**5**), deoxycytidine (**6**), uridine (**7**), and thymidine (**8**). All are β-*N*-glycosides of either a purine or pyrimidine and D-ribose or 2-deoxy-D-ribose. The currently-accepted numbering system is given, although older publications use a different system for the pyrimidine heterocycle.

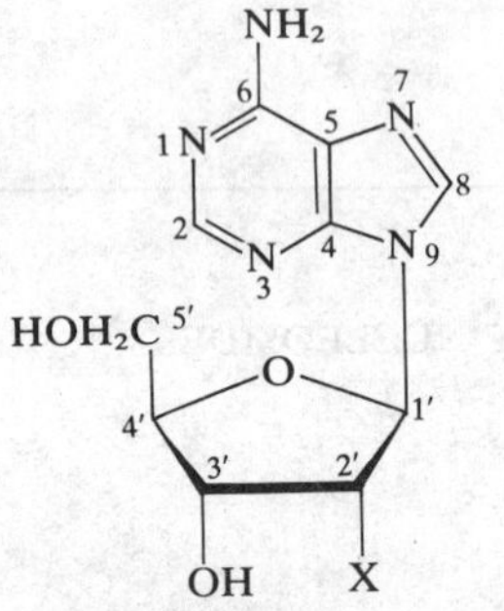

(**1**) X = OH
(**2**) X = H

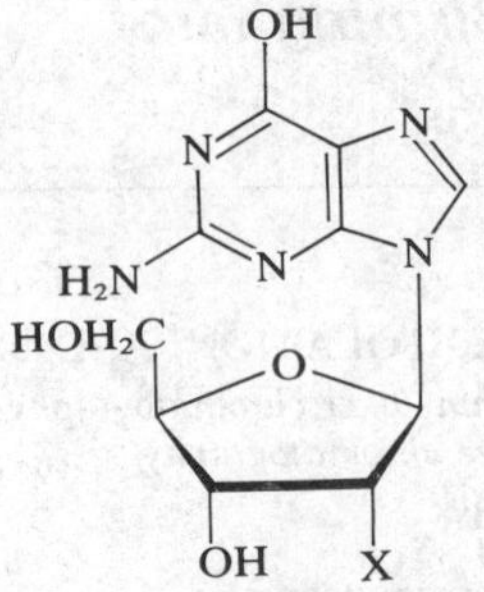

(**3**) X = OH
(**4**) X = H

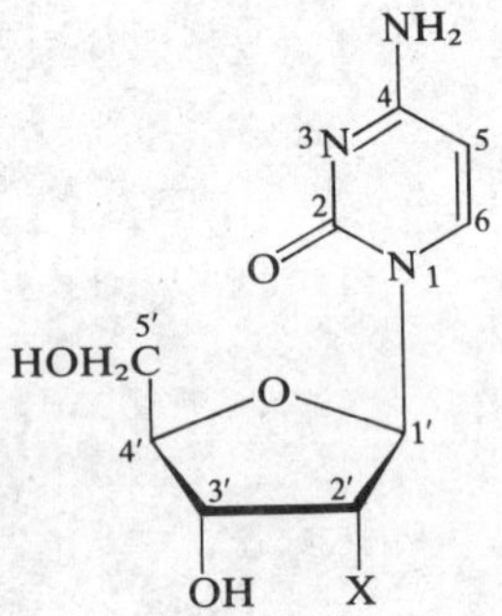

(**5**) X = OH
(**6**) X = H

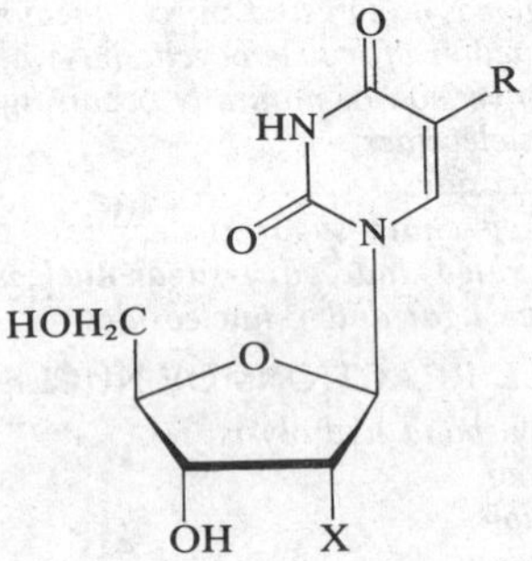

(**7**) X = OH, R = H
(**8**) X = H, R = Me

There is no lack of general textbooks in this field,[2–14] there are two continuing series,[15,16] which have occasionally been known to include some chemical articles, there is a journal[17] devoted to the field, and a compilation of abstracts[18] which has catalogued some 50 000 references in the past six years. The highlights in each year, with particular reference to chemistry, are included in the *Annual Reports of the Chemical Society*, and the *Annual Reviews of Biochemistry* often cover indivdual subjects in greater depth.

The emphasis in nucleoside research has changed drastically over the years. Originally, the main objective was to prove the structure of naturally-occurring compounds obtained from nucleic acid hydrolyses, and whereas this is still sometimes the case with the many modified nucleosides found in tRNA, more often than not the aim is the synthesis of some potential antibiotic or nucleoside analogue which it is hoped will have medicinal use. Underlining this change in emphasis it was recently reported at a Symposium on the Chemistry, Biology and Clinical Uses of Nucleoside Analogs[19] that there would be 'hardly a major field of medicine that will not feel the impact of a significant contribution from the field of nucleosides and nucleotides within the next 25 years'.

This chapter will thus start with a brief description of the methods available for the isolation and separation of nucleoside mixtures (usually from natural sources). General methods of nucleoside synthesis will be dealt with in some detail and this will be followed by a description of the syntheses of certain specific nucleoside types, *e.g. C*-nucleosides, cyclonucleosides, *etc.* Nothing more than a catalogue of the physical methods used to determine nucleoside structures can be given, although leading references will be listed. The chemical reactions considered will be restricted to those of particular interest to the nucleic acid chemist; the general reactions of the sugar residues and the heterocycles will be found elsewhere in their relevant chapters.

22.2.1 OCCURRENCE, ISOLATION, AND SEPARATION OF NUCLEOSIDES

One has only to make an absolute statement on the occurrence of nucleosides for it rapidly to be proved incorrect. Thus, other then the eight major nucleosides (**1**)–(**8**), most classes of nucleic acids so far distinguished have been shown to contain modified nucleosides to a greater or lesser degree. 5-Methyldeoxycytidine has long been known to be produced in hydrolysates from plant and mammalian DNA;[7] each bacterial restriction enzyme system is thought to contain a methylase which methylates specific base residues in the host DNA in the sequence recognized by the restriction endonuclease,[7] thus giving rise to the potential occurrence of modified nucleosides from bacterial DNA. Phage DNAs often contain large amounts of modified nucleosides and it is very likely that more surprises are in store as the composition of the nucleic acids of other phages is determined.[7]

RNA is still the best source of modified nucleosides, particularly tRNA, where the number of modified nucleosides seems to increase at a rate equal to or greater than the number of tRNAs whose base sequence is known.[8,20] The most unusual nucleosides so far identified which are derived from a polynucleotide structure must be Y (**9a**)[20] and Q* (**9b**),[21] although the former has only been identified as its base and there is some question about the identity of the nucleoside.[22] The role of many of these modified nucleosides is not understood except that the probable partial roles of a large enough number have been identified to make it clear that few, if any, of the modifications are random events and it is likely that their presence confers some advantage upon the organism.

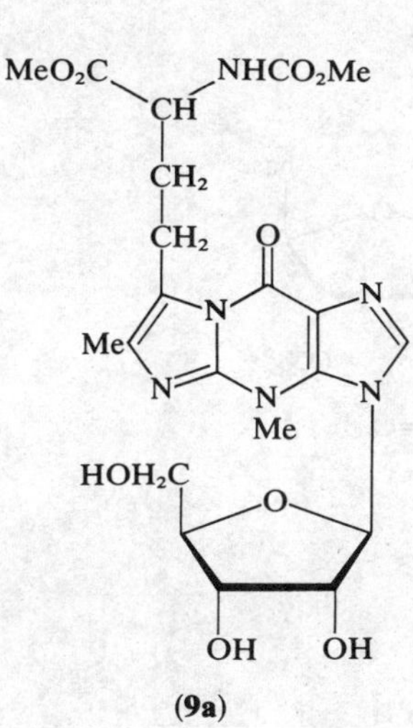

(**9a**)

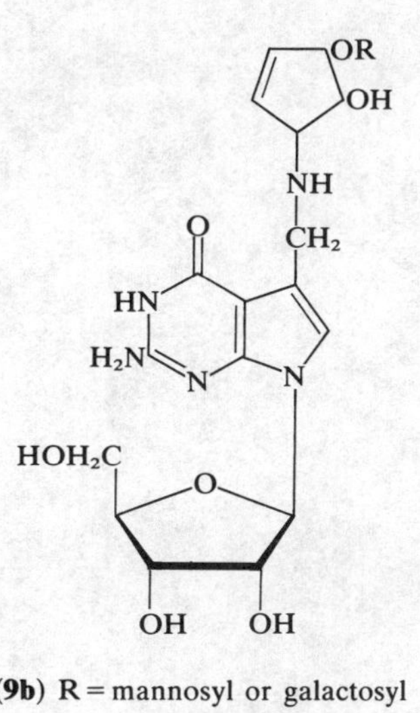

(**9b**) R = mannosyl or galactosyl

Ribosomal RNA also contains modified nucleosides; again their role is not known, although the replacement of two N^6-dimethyladenosines by adenosine in *Escherichia coli* rRNA causes the bacterium to be resistant to the drug Kasugamycin. Whether this is the primary result of the modification or just some side effect is not known, but this is not really very surprising as the role of rRNA in the functioning of the ribosome is only just beginning to be understood. Until recently, it could be stated that mRNA does not appear to contain any modified nucleosides. However, it is now well documented that many eukaryotic and viral mRNAs are 'capped', *i.e.* they contain the sequence $m^7G^{5'}ppp^{5'}(N_{mp})_{1-2}Np$ . . . (**10**) at their 5′-termini. Again there is at present some doubt about the purpose of these modified nucleosides, but it has provided the organic chemist with the incentive to synthesize nucleosides and nucleotides modified in this way.[23]

Obviously, many nucleosides are present as intermediates in the biosynthesis and breakdown of nucleotides and polynucleotides. In addition to the so-called spongonucleosides, strictly applied to the modified purine nucleosides obtained from the Caribbean

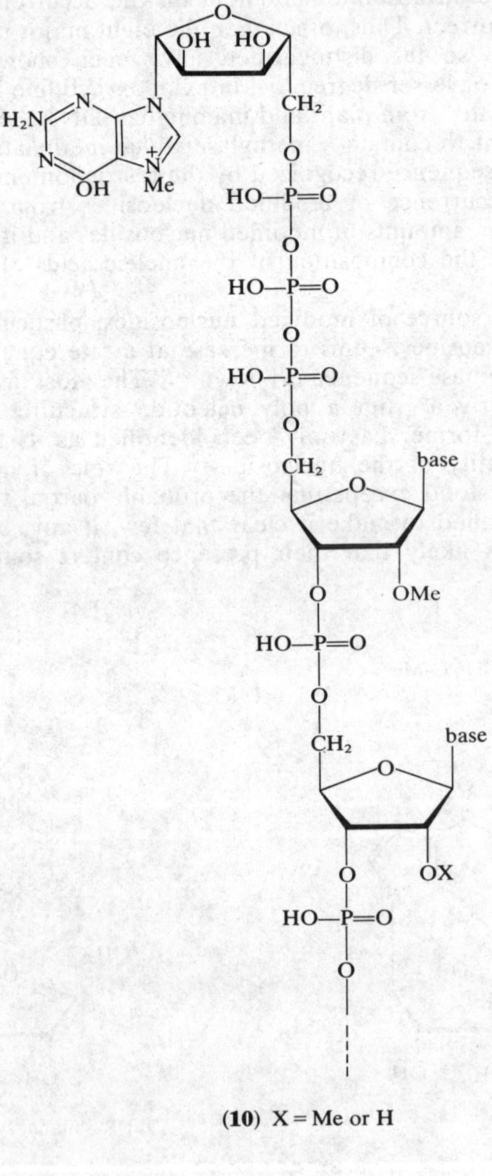

(**10**) X = Me or H

sponge, *Cryptotethya crypta*, which are arabinose derivatives, many antibiotics are nucleoside derivatives, often having modified sugar residues, and these will be discussed in greater detail later. Nucleosides are relatively easy to isolate from the chemical or enzymatic hydrolysates of naturally-occurring polynucleotides; the conditions and practical details for this can be found in the general textbooks on nucleic acids,[2,7,24] and all commercially-available samples of the principal nucleosides are obtained in this way. When attempting to isolate large quantities of such nucleosides, rather crude fractionation techniques have to be used which inevitably depend upon differential solubility and ion-exchange methods. For the isolation of small amounts of modified nucleosides, either from a natural source or a chemical synthesis, many more-efficient methods are available and these are discussed individually below. Finally, it must be remembered that nucleoside isolation is often achieved *via* dephosphorylation of nucleotides,[25] the isolation and separation of which are outside the scope of this section.

22.2.1.1 Paper and thin-layer chromatography

These methods were among the first to be used in this field and much of the literature dealing with this subject is rather old or difficult to locate; most people seem to have their own particular preferences for solvents, *etc.*, often for no strict scientific reasons. Three useful articles contain the R_F values of many nucleosides in several solvent systems[26] and a recent review[27] covers the practical techniques. Cellulose column chromatography is rarely used as the separations achieved never seem to approach those obtained on paper or thin-layers. However, the only real advantage of cellulose as the supporting material, apart from its neutrality, is that ultraviolet-absorbing components can be detected on cellulose plates at very low concentrations without having to add a fluorescent compound to the cellulose (which is destroyed by acidic solvents), whereas much more stable thin-layers can be obtained from silica, on which separations similar to those obtainable on paper are possible if the same solvents are used. Silica t.l.c. also has the advantage that the Dische cysteine–sulphuric acid spray can be used for the detection of deoxynucleosides and in fact the plates can be heated for a sufficient time that ribonucleosides may be detected, thus turning this reagent into a general one for all nucleosides. Preparative-layer chromatography (plates for which can be prepared by pouring a slurry of the support on to the plate and tapping it gently, thus obviating the need for expensive spreading apparatus) and column chromatography (particularly with silica) are often used for separating larger quantities of nucleosides. Short fat columns, packed with thin-layer grade silica gel and run under slight pressure, give very fast and efficient separations under conditions similar to those to used to achieve the same separation on a thin-layer plate.[28] Preparative-scale partition chromatography (counter current distribution) has been used for nucleoside separation, but is rather time-consuming and wasteful of solvents.

22.2.1.2 Ion-exchange chromatography

This technique has the advantage over partition and absorption chromatography when it is used for the fractionation of large quantities of a material that is soluble in water. Obviously the main use of ion-exchange chromatography is in the isolation and fractionation of polynucleotides and nucleotides, but the pK_a values of the heterocyclic bases are such that, between pH 4–10, some of the nucleosides can at least bear a partial charge and can hence be separated by such techniques. There are two main types of support used for the resins, either polystyrene or cellulose, and both have been used widely for nucleoside separations. Perhaps the most widely used method is that due to Dekker.[29] Since the sugar hydroxyl groups of nucleosides are known to have pK_a values in the range 13–17, if a strongly basic anion exchange resin is used with an eluant of aqueous methanol, various nucleoside derivatives and even anomeric mixtures can be separated.[29]

22.2.1.3 Electrophoresis

Few separations of a preparative nature can be achieved by electrophoresis of nucleosides, but the technique is often used to monitor the course of a reaction. It is particularly useful in the detection of ring *N*-substituted pyrimidine nucleosides, but the pH at which any significant migration is possible is often rather extreme and so the method is of limited use. One application of particular use in the nucleoside field is the electrophoresis of ribonucleosides in borate buffer at pH 10, under which conditions a stable complex between the borate and the *cis*-diol grouping is formed and results in the migration of ribonucleosides towards the anode.[30]

22.2.1.4 Gel filtration

The commercially-available gel exclusion media Sephadex G10 and G15, and Biogel P2, can be used for nucleoside fractionation. The advantage of these materials is that the substances eluted do not 'tail', recovery is quantitative, and salt gradients do not have to be used. It is unlikely that any chemical degradation of labile nucleosides will occur under the conditions of gel chromatography, something which is always liable to occur with ion-exchange and adsorption chromatography.[27]

22.2.1.5 Gas-phase chromatography

25 Years after the invention of this technique, the applications in the nucleoside field are still very limited, because of the lack of volatility of most nucleoside derivatives. The trimethylsilyl ethers of nucleosides have the required volatility and can be prepared by the action of *N,O*-bis(trimethylsilyl)acetamide on a nucleoside.[31] Each nucleoside gives only one peak when injected at 260 °C. Hexamethyldisilazane and/or trimethylsilyl chloride can also be used to prepare these ethers. A review of gas-phase chromatographic analysis of nucleic acid derivatives covers the practical details of silylation and the methods used to separate such derivatives.[31] A more recent article lists most of the relevant references in this field and presents a detailed analysis of structure–retention relationships for many nucleosides.[32] *N*-Trimethylsilylimidazole has also been used to prepare the trimethylsilyl ethers of nucleosides without having to maintain strictly anhydrous conditions with pyrimidine nucleosides; the sugar hydroxyl groups are silylated quantitatively at room temperature, but the exocyclic amino-group and O-4 of 4-keto-nucleosides are resistant. A general procedure has been described for the separation of anomeric *O*′-trimethylsilyl nucleosides on a highly polar column which can be adapted for analytical or preparative purposes.[33] This paper also reviews the other methods available for the separation of isomeric mixtures of nucleosides. Pyrimidine deoxyribonucleosides could not be separated in this way. When attempts were made to silylate the heterocyclic moiety to increase the volatility of the derivative by raising the temperature of silylation, degradation occurred. However, the pentose derivative produced retains the nucleoside configuration at the anomeric centre and this method can therefore be used to identify and distinguish between α- and β-nomers. With purine nucleosides a somewhat higher temperature for the initial silylation reaction is required.

This appears to be a technique whose use in the nucleoside field can only increase in the future and its somewhat limited use at the moment is mainly due to the ease in separating nucleosides by other techniques.

22.2.1.6 High-pressure liquid chromatography

This is, of course, not a technique which can strictly be isolated by itself as, depending on the column packing, this is only an extension of methods previously described. However, the method has such overwhelming advantages in the increase and speed of resolution that it is advantageous to consider separately the uses to which this technique has been put. Ion exchange, adsorption, and gel exclusion packings have all been used and the use of these techniques has been described by Leonard when applied to the critical separation of some cytokinins (adenosine derivatives), including the separation of *cis*- and *trans*-isomers.[34] This technique is still in its infancy in the nucleic acid field, but it is surely destined to play an ever increasing role in the separation of nucleoside mixtures, and as a gel filtration column to desalt nucleoside solutions recovered from ion-exchange eluates.

22.2.2 PHYSICOCHEMICAL PROPERTIES

While a detailed description of techniques is beyond the scope of these volumes, organic chemists, and those working in the nucleoside field are no exception, have become increasingly reliant on physical methods. The work can be divided conveniently into three categories: (a) theoretical studies; (b) spectroscopic methods used for the identification and assay of nucleosides; and (c) molecular conformation studies. In the space available, only a list of recent reviews and some leading references to the most important of these methods can be given.

Theoretical studies, led by the Pullmans[35] and others,[9,13,14] have been mainly concerned with the calculation of the distribution of partial charges on nucleosides and the calculation of the dipole moments of such molecules. Considerable success has been achieved in obtaining values close to the experimentally-determined ones.

Review articles dealing with spectroscopic techniques include those on i.r.,[10,12–14] Raman,[10,13,14] u.v.,[8,10,12–14,37] o.r.d.,[10,12,14] c.d.,[10,13,36,37] n.m.r.,[12,14,36,37] e.s.r.,[13] n.q.r.,[14] and e.p.r.[13] Other electronic properties of nucleosides such as charge distribution and ionization constants[12] have also been reviewed. Apart from u.v. and n.m.r. techniques, the most widely used method for the identification of nucleosides is mass spectrometry. This technique has been reviewed,[10,12,38,39] and some modern refinements which are particularly suitable for obtaining spectra from small amounts of non-volatile, labile substances have been described.[34,40] The thermodynamic properties of nucleosides which can be used to understand the interaction of nucleic acid components with themselves have also been examined.[14]

The conformational studies of nucleosides inevitably involve the use of X-ray diffraction and much of the work has been done and reviewed by Arnott and co-workers[41] and Sundaralingam and co-workers.[12,19,42] N.m.r.,[43] c.d.,[44] and o.r.d.[12] techniques have also been used to determine the preferred conformations of nucleosides in solution.

22.2.3 CHEMICAL SYNTHESIS OF NUCLEOSIDES

Only 20 years ago,[45] the synthesis of nucleosides in a review article could be dismissed in the sentences ‘Although synthesis of all the natural ribonucleosides has been effected, the methods involve relatively complex operations and for most biochemical studies, the preparation of the ribonucleosides and deoxyribonucleosides from natural sources is recommended. Should synthetic material be necessary, the original literature should be consulted’. This was followed by a list of *two* references! Many people would still agree that the methods are relatively complex and the subject is more of an art than a science, but progress is such that many synthetic nucleosides are now commercially available.

There are several general texts dealing with nucleoside synthesis and, in general, little reference will be made to the original literature. This field has tended to be dominated by a rather select group of scientists and it is hoped that these general texts and a few references to these key people’s work will help readers not only to be aware of the situation at the time of writing but also to follow future developments. The general texts include one written by Michelson[3] and published in 1962, which is a comprehensive account of the knowledge available at the time. The most recent review is by Goodman,[46] which covers the literature to the end of 1971. There is also a laboratory manual entitled 'Synthetic Procedures in Nucleic Acid Chemistry', but this contains a rather disappointing and apparently random selection of nucleoside syntheses solicited from workers in the field with little or no attempt to exert any editorial control. Thus it contains the preparation of a collection of syntheses of strange analogues interspersed with some very useful and more generally-applicable preparations.[12]

Until recently there has been little attempt to try to bring any order into the confused area of the mechanism of nucleoside synthesis by condensation reactions. However, this is now being rectified[47,48] and although two of the articles are not readily available, both can be thoroughly recommended to anyone interested in the science rather than the art of nucleoside synthesis.

22.2.3.1 Condensation reactions

(i) Koenigs–Knorr (heavy-metal) procedure

The original method of this type was the reaction of acetobromoglucose with methanol in the presence of silver carbonate to give the methyl glycoside. This was followed by the use of silver salts of some halogenated purines with the same halogenose to give the blocked nucleoside. The first major advance in this reaction in nucleoside synthesis was due to Davoll and Lowy,[49] who showed that the chloromercuri-derivatives of certain purines gave much better yields of nucleosides and this was extended further to the pyrimidine field after the introduction of reproducible procedures for the preparation of the necessary mercuri-pyrimidine derivatives.[50] The mechanism of the reaction has been reviewed by Fox.[47] With pyrimidines, the initial product is an *O*-glycoside which then reacts with a further mole of sugar halide (in the presence of HgX_2) to form nucleosides (Scheme 1).

The rate of conversion of *O*-glycosides into nucleosides catalysed by HgX_2 is dependent upon the ease of cleavage of the C-2—C-1′ bond. It is possible that the HgX_2 assists in this cleavage by forming a π-complex with the heterocyclic moiety. In a modification of this method,[51] the free base can be condensed with the halogenose in nitromethane in the presence of mercuric cyanide; the mechanism of the reaction is very similar.

In the case of purine nucleoside formation,[52] it is now thought that the initial product is an N-3 glycosyl derivative which rearranges by an intermolecular reaction to give (usually) the N-9 product[10] (Scheme 2). The ultimate site(s) to which the glycosyl residue migrates depends upon the thermodynamic stability of the product(s). The mercuric cyanide–nitromethane procedure may also be applied to purine nucleoside synthesis.

From the mechanisms given, it is clear that in general only 1′,2′-*trans*-nucleosides should result from such reactions when halogenoses bearing a 2-acyloxy-substituent are used. This (as will be seen later) is the important advantage of this method when compared with the other methods available, as in practice the stereochemical control ensures the production of β-ribonucleosides. For deoxynucleoside syntheses, anomeric mixtures are produced and little can be done to influence the relative proportions of each isomer in the final product.

(ii) The Hilbert–Johnson (and silyl) procedures

(a) With alkoxypyrimidines. The synthesis of 1-methyluracil by Hilbert and Johnson in 1930, by the reaction of 2,4-dimethoxypyrimidine with methyl iodide and subsequent

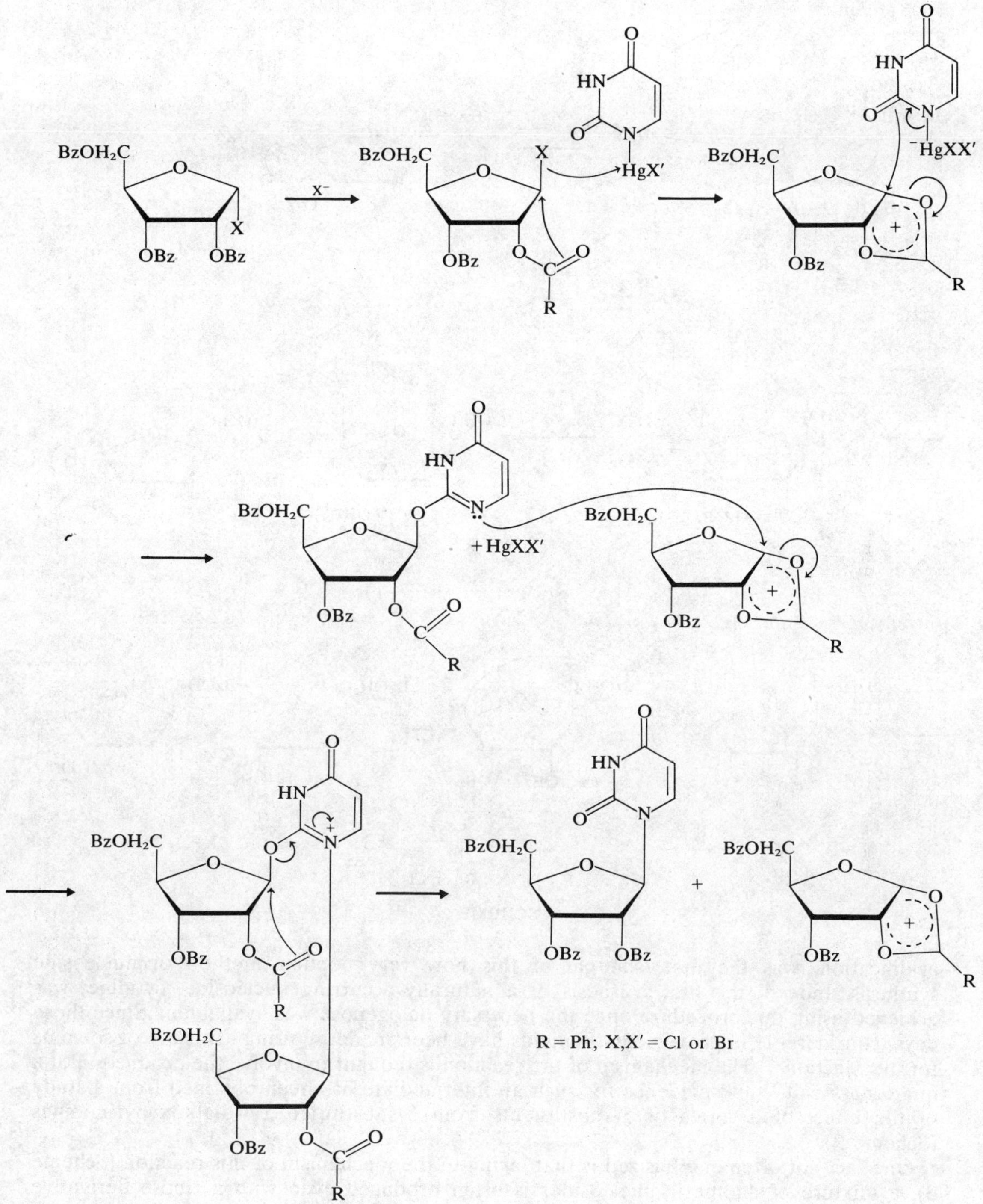

SCHEME 1

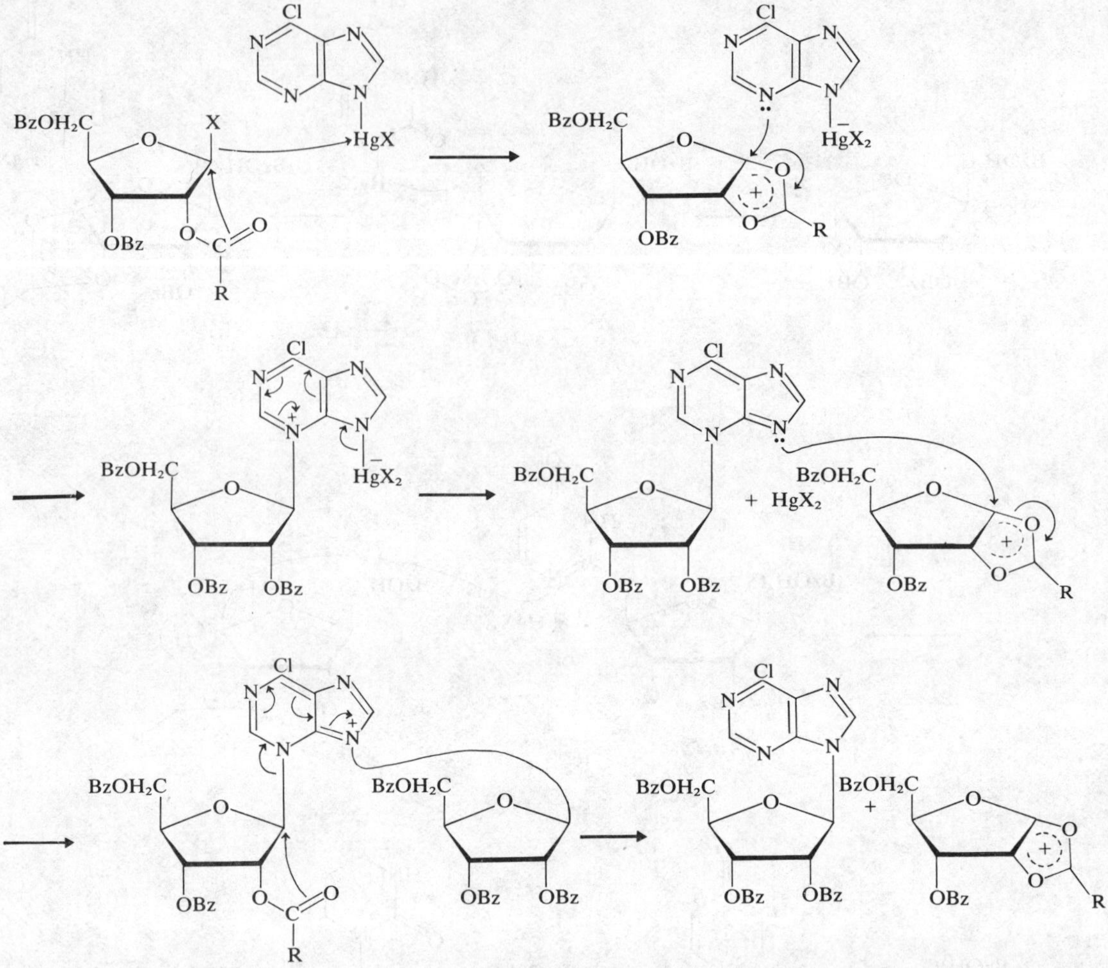

R = Ph; X = Br or Cl

SCHEME 2

acidification, was the first example of this now very popular method of nucleoside synthesis. Indeed, the first synthesis of a naturally-occurring nucleoside, cytidine, was achieved using this procedure once the necessary halogenose was available.[53] Since those days, considerable improvements in yields have been made by using a variety of solvents for the reaction.[47] The mechanism of the reaction is thought to involve the production of a quaternary salt,[54] and evidence for such an intermediate has been obtained from a study of the effect of a series of 5-substituents from 5-substituted 2,4-dialkoxypyrimidines (Scheme 3).[55]

One fact not often emphasized is that owing to the mechanism of this reaction (Scheme 3), a mixture of anomeric nucleosides is often produced even with a ribose derivative bearing a 2-acyloxy-substituent. The relative contribution of carboxonium ion to the overall reaction determines the relative amount of 1′,2′-*cis*-nucleoside and, although in many cases only the 1′,2′-*trans*-nucleoside has been isolated from such reactions, it cannot be presumed that such a product will be produced without the necessary structural assignment being made. There is at least one documented case where the identity of the product from such a reaction was assumed to be the β-anomer and which led to confusion in the literature.[56]

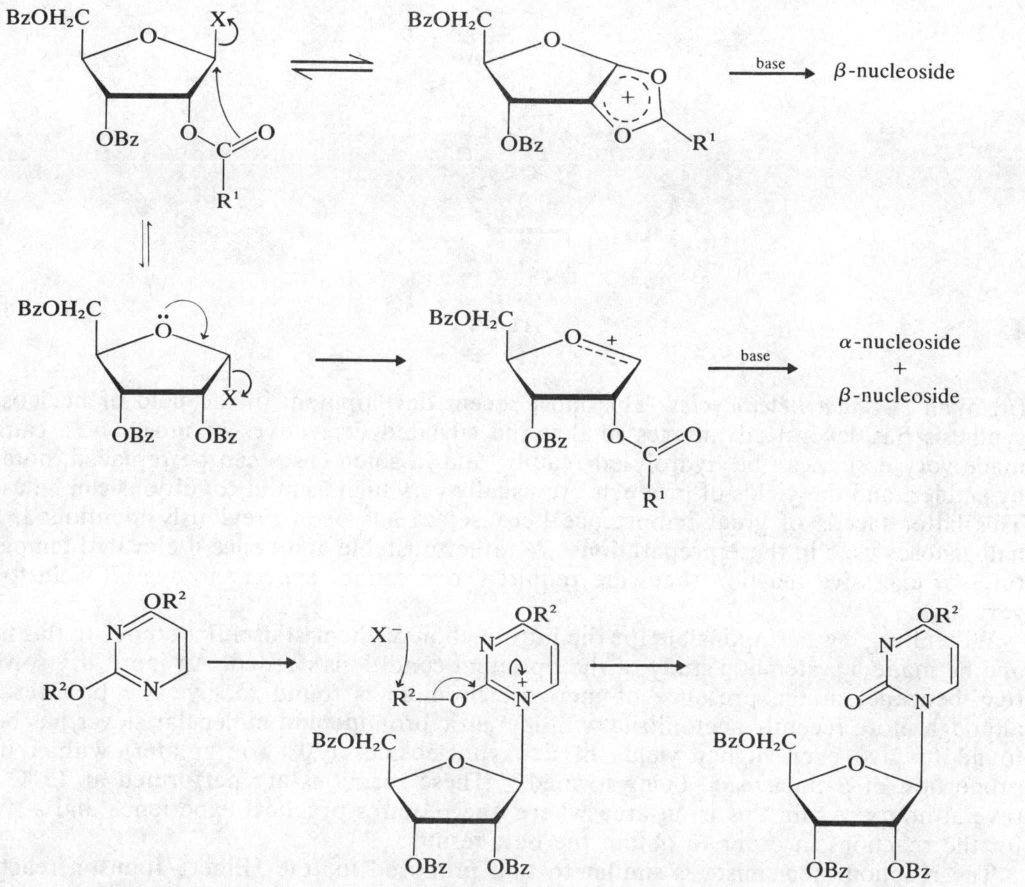

R¹ = phenyl; R² = alkyl; X = Cl or Br

SCHEME 3

Although both α- and β-nucleosides can be expected as products from Hilbert–Johnson reactions, it is also possible, by altering the reaction conditions, to alter the proportion of each which is formed. This in some measure accounts for the enormous diversity of reaction conditions reported in the literature and which to the newcomer suggests that the subject is more of an art than a science. The most useful additive seems to be mercuric bromide, which not only increases the rate of reaction but also affects the stereochemistry of the products so that more β-nucleoside is produced. This is particularly important in the synthesis of deoxynucleosides because the halogenose used is rather unstable and any increase in reaction rate will increase the yield of both anomers, although in practice proportionately more β-anomer generally appears to be formed. 3,5-Di-*O*-(*p*-nitrobenzoyl)-2-deoxy-α-D-*erythro*-pentofuranosyl chloride can be obtained crystalline, and reaction of this with 2,4-dialkoxypyrimidines has sometimes been found to give predominantly the α-nucleoside with retention of configuration. This has been explained as being caused by the formation of the intermediate (**11**), which would then undergo attack at C-1 by the nucleophile.

With halogenoses bearing a 2-acyloxy-function, it is postulated that the mercury assists in the formation of the 1,2-acyloxonium ion, and in the reaction with 2-deoxyhalogenoses it assists in the dissociation of the halide ion to form the carboxonium ion. Thus in the presence of mercuric bromide, the Hilbert–Johnson and the mercuri-procedures are very much alike.

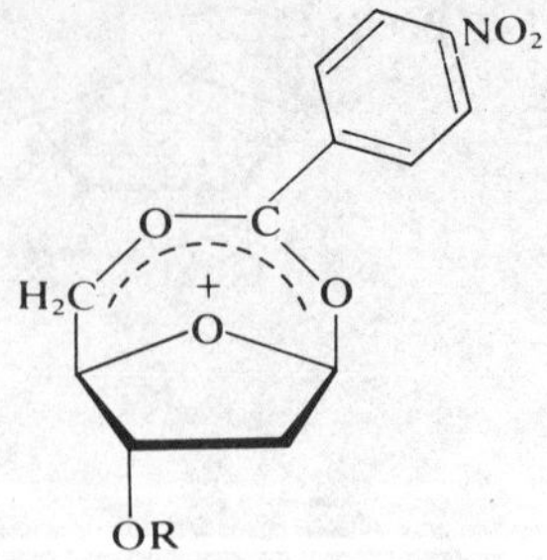

(11)

(*b*) *With silylated heterocycles.* This most recent development in the field of nucleoside synthesis has several advantages in that the silylated derivatives of most bases can be made very easily, can be hydrolysed readily, and in some cases can be replaced, notably by amines, and the yields of products are usually very high as mild conditions can be used. This latter fact is of great importance because, as has been previously mentioned, the halogenoses used in these preparations are rather unstable and hence if elevated temperatures or excessive reaction times are required, one cannot expect the overall yield to be very high.

Wittenburg[57] was responsible for the introduction of the first useful methods in this field and he made a systematic study of the optimum conditions needed. An inert, dry solvent (*e.g.* benzene) in the presence of mercuric bromide is found to give the best results, although more recently acetonitrile with mercuric bromide and molecular sieves has been found to give even higher yields of deoxynucleosides (70% or greater) with a high proportion of β-nucleoside being formed.[58] These reactions are performed at 25 °C for several hours. Again this is an area where one requires practical experience and a 'feel' for the reactions in order to obtain the best results.

The reaction mechanism is similar to that proposed for the Hilbert–Johnson reaction (Scheme 2), with the leaving group being trimethylsilyl instead of ethyl. The ratio of anomeric products produced has also been found to depend upon whether or not trimethylsilyl chloride is formed during the condensation.[59] Under conditions which favour its rapid removal, then only β-nucleoside is formed, while conditions which favour its retention in the reaction mixture favour production of the α-anomer. It appears that the α-anomer is formed by direct S_N2 displacement of the β-halogenose, which is formed by trimethylsilyl chloride-catalysed anomerization of the α-halogenose (Scheme 4).

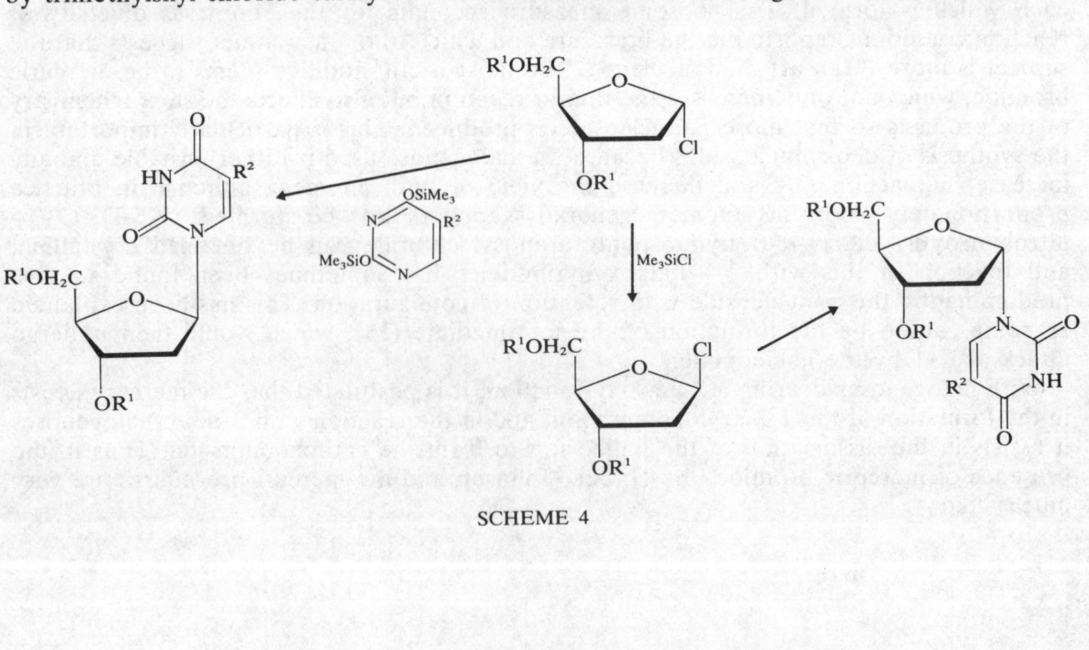

SCHEME 4

Another advantage of the silyl procedure is that it can be applied to the synthesis of purine nucleosides. Initial products are the N-3 glycosyl derivatives, which subsequently rearrange to give the N-9 derivative. However, unlike the mercury procedure for purine nucleoside synthesis, dissociation of halide from the halogenose is not assisted by an electrophile, which means that sometimes mixtures of anomeric nucleosides are obtained.[60]

One of the most interesting and useful developments in this field is due to the work of Vorbrüggen and co-workers, who obtained nucleosides from the Friedel–Crafts-catalysed reaction of a peracylated sugar and a silylated heterocycle in 1,2-dichloroethane or acetonitrile as solvent.[48,61] The authors[48] regard the reaction as a modification of the Hilbert–Johnson reaction, whereas Fox[47] considers the reaction to be related to the fusion reaction of Helferich which is discussed below. The advantage of the method is that the sugar derivative is readily available and is stable, the reactions proceed rapidly (from a few minutes to a few hours) at room temperature and in very high yield (>90%). There is so far only one reported exception to the product formed being exclusively the β-nucleoside. The catalyst usually used is stannic chloride, although recently trimethylsilyl perchlorate or trimethylsilyl trifluoromethanesulphonate, $Me_3SiSO_3CF_3$, have been found to be superior catalysts. In addition, stannic chloride often leads to the production of colloids which are difficult to work with. Little has been done in the deoxynucleoside field, although it depends whether the Vorbrüggen modification is taken to be the use of a peracylated sugar or the Friedel–Crafts catalyst, or both. Some deoxynucleosides have been prepared in high yield from the normal 2-deoxyhalogenose using a Friedel–Crafts catalyst, but a mixture of anomers in the ratio of 1:1 seems to be the normal product. The mechanism of ribonucleoside formation has been studied in some detail and is given in Scheme 5.

It can be argued that under these thermodynamically-controlled and at least partially-reversible conditions, the silylated base can only attack to give exclusively β-nucleoside. Both catalysts are regenerated in this procedure and so there should be no need for their presence in molar amounts. Vorbrüggen,[48] however, pointed out that the interaction of catalysts with silylated bases had not been studied. He was encouraged to do so by the discovery that silylated 5-nitrouracil reacted with peracylated sugar very rapidly (5 min) in the presence of only 0.2 equivalents of $SnCl_4$ (97% yield), whereas 5-methoxyuracil required 1.4 equivalents and in 2.5 h only gave a 53% yield of the β-nucleoside, a 27% yield of α-nucleoside, and 13% of the N-1,N-3 bis-riboside. It was found that with increasing electron-releasing capacity of the 5-substituent, more stable and hence less reactive donor–acceptor complexes between the heterocycle and catalyst result eventually in one equivalent of catalyst being neutralized; thus more than one equivalent of catalyst is required for nucleoside formation. Solvents like acetonitrile which diminish the stability of such complexes and weak Lewis acids such as $Me_3SiSO_3CF_3$ are preferred.

In acetonitrile, silylated 5-methoxyuracil gave a 90% yield of β-nucleoside with 1.4 equivalents of stannic chloride in 12 h; using the catalyst trimethylsilyl trifluorosulphonate, the same yield could be obtained in 1,2-dichloroethane in 4 h using 1.1 equivalents of catalyst. N.m.r. evidence showed that the catalyst gave an N-1 σ-complex and thus only the free silylated base reacts with the 1,2-acyloxonium ion to give the desired N-1 nucleoside. The N-1 σ-complex reacts only very slowly to give the N-3 nucleoside. The reversibility of these reactions has been demonstrated by heating a mixture of the N-1, N-3, and N-1,N-3 bis-glycosides of 6-methyluracil in acetonitrile in the presence of $Me_3SiSO_3CF_3$, whereupon only the thermodynamically most stable product, benzoylated 6-methyluridine, was present. This reaction will be referred to again when transglycosylation reactions are considered.

(c) *Miscellaneous reactions.* There are several nucleoside synthesis reactions which may be regarded as related to the Hilbert–Johnson procedure. Among these is the reported

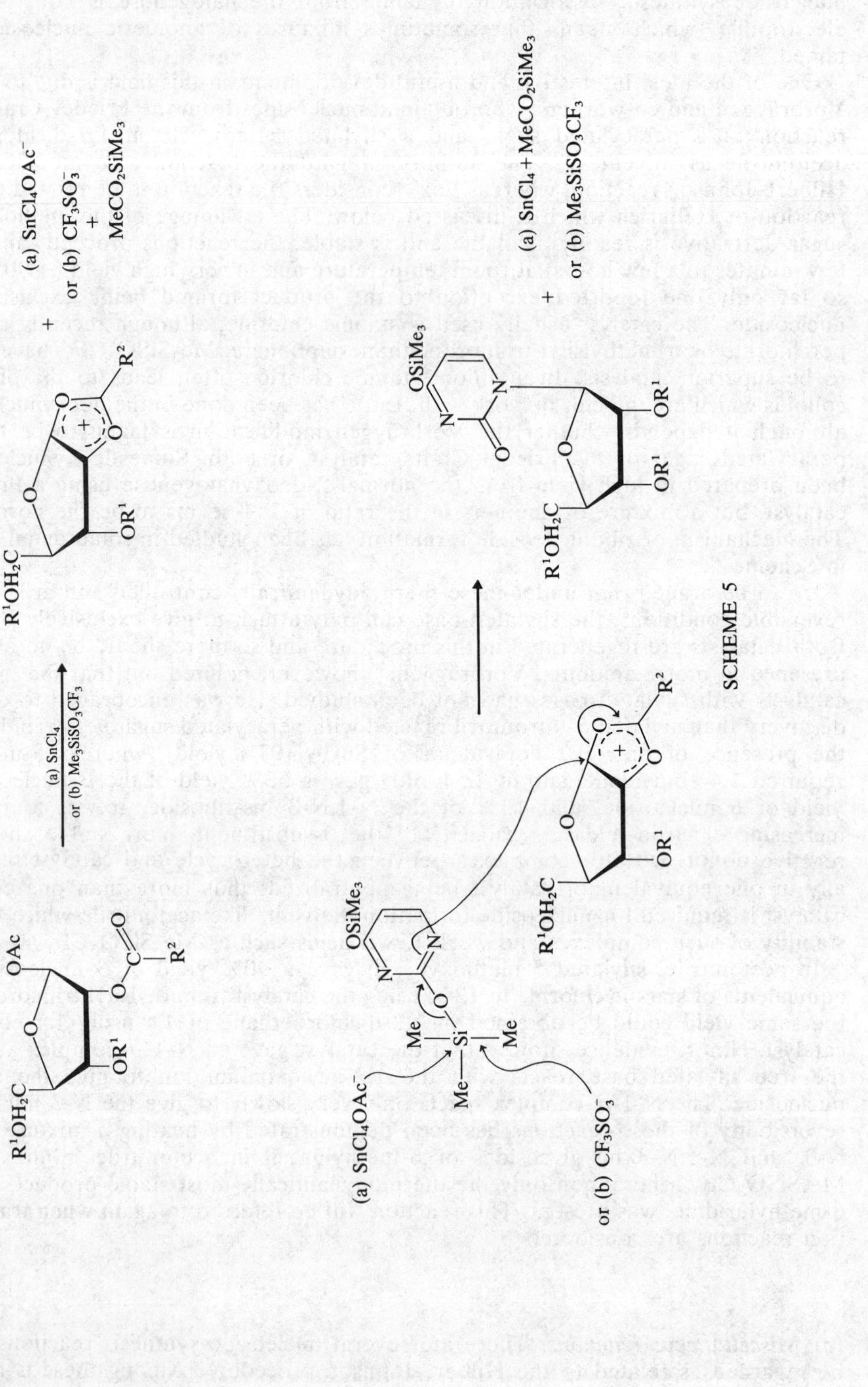
(a) SnCl4
or (b) Me3SiSO3CF3
(a) SnCl4OAc−
or (b) CF3SO3−
+
MeCO2SiMe3
(a) SnCl4 + MeCO2SiMe3
or (b) Me3SiSO3CF3
OSiMe3

SCHEME 5

synthesis of adenosine (and the N-3 glycoside) by the direct reaction of adenine with tri-*O*-benzoyl-D-ribofuranosyl bromide in acetonitrile or dimethylformamide.[62] *N*-2-Acetylguanine has also been ribosylated in dimethylacetamide to give a mixture of *N*-7- and *N*-9-ribosylguanine derivatives.[63] A comprehensive account of these reactions has been published.[46]

(*iii*) *The Helferich method* (*fusion procedure*)

The original method of Helferich for arylglycosyl synthesis was first used by Sato *et al.*[64] for the preparation of purine nucleosides. The method involves the fusion of a peracylated sugar (*e.g.* tetra-*O*-acetyl-β-D-ribofuranose) with various fusible purines under vacuum in the presence of a catalytic amount of toluene-*p*-sulphonic acid for 10–20 min. The yields vary widely (from 3 to 74%) and anomeric mixtures are produced. Many different catalysts have been investigated and purines which fuse well give the highest yields. The main advantage of the method is that free purine bases and commercially-available, stable, peracylated sugars can be used. Nevertheless, the Vorbrüggen modification of the Hilbert–Johnson reaction has many advantages and nowadays is almost always used for pyrimidine nucleoside synthesis, although these can be prepared *via* a fusion reaction.[65] A related non-catalysed reaction has been reported[66] and, unlike the mercuri-procedure, there is no evidence for the intermediate formation of *N*-3-glycosylpurines.

Some pertinent data[66] which can be explained mechanistically are that in a non-catalysed fusion using a β-sugar, only β-nucleoside was formed; with the α-sugar an anomeric mixture was produced. Under sulphamic acid catalysed conditions, the β-sugar gave β-nucleoside when fused for a short time, but an anomeric mixture if the fusion was left for longer than 10 min. Pure β-nucleoside could be converted into an anomeric mixture on fusing with catalyst for 10 min. These data have been rationalized by Fox[47] in the mechanism given in Scheme 6.

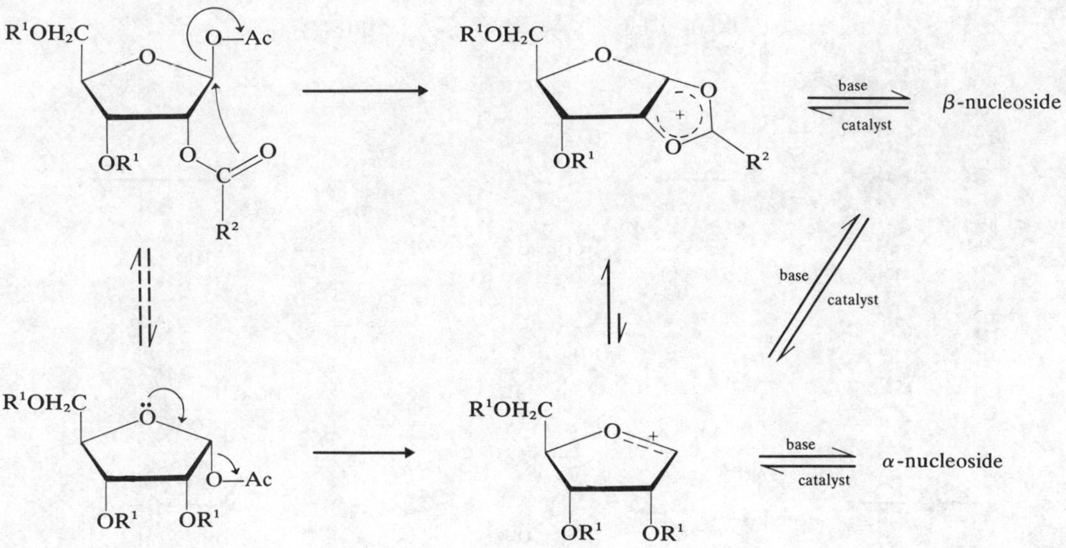

SCHEME 6

22.2.3.2 Transglycosylation and miscellaneous reactions

This section includes methods of nucleoside synthesis which either do not belong clearly under one of the other headings used or they have an equal right to belong to several. Also, although the mechanism of transglycosylation reactions are very closely linked to the condensation reactions previously discussed, from a practical point of view it is better to deal with them separately.

Several transglycosylation reactions have been used for nucleoside synthesis and, in this context, transglycosylation will refer only to the transfer of a sugar residue (usually from a nucleoside) to another heterocyclic base and not to the transfer of the sugar from one position of the base to another, as O→N and N→N glycosyl migrations have already been covered as an integral part of nucleoside condensation reactions.

In order to facilitate the cleavage of the glycosyl linkage, pyrimidine nucleosides have been acylated at the basic and glycosyl moieties and then heated with purine bases in the presence of acid catalysts.[67] Thus tetra-acetylcytidine, when reacted with 6-benzamidopurine in xylene in the presence of mercuric bromide, gave an anomeric mixture (mainly β) of the blocked α- and β-adenosines. Many other purine and imidazole bases can be substituted for the 6-benzamidopurine, and tetra-acetyluridine can be used as the glycosyl donor.

In a recent further investigation of this method,[68] it has been found that the glycosyl group of a naturally-occurring polyoxin (**12**) could be transferred to a purine, thus enabling these analogues to be prepared for the first time. Previous attempts to isolate the glycosyl residue had failed owing to the lability of the highly-strained furanose ring of this nucleoside. Thus conversion of the octosyl acid A (**12**) to a cytosine derivative (**13**) (cytosine derivatives give better yields) and transglycosylation of this in the presence of the trimethylsilyl derivative of 6-benzamidopurine with trimethylsilyl perchlorate or trimethylsilyl trifluoromethanesulphonate as catalyst in acetonitrile–dichloroethane under reflux for several hours gave, after deblocking, the crystalline nucleoside (**14**) in 50% overall yield. The adenine analogue of polyoxin C (**15**) has also been prepared using this method and the transglycosylation reaction should prove to be an excellent method for the preparation of nucleoside analogues of modified sugars such as amino-deoxy- and uro-sugars, whose derivatives are not easy to prepare by other standard procedures.

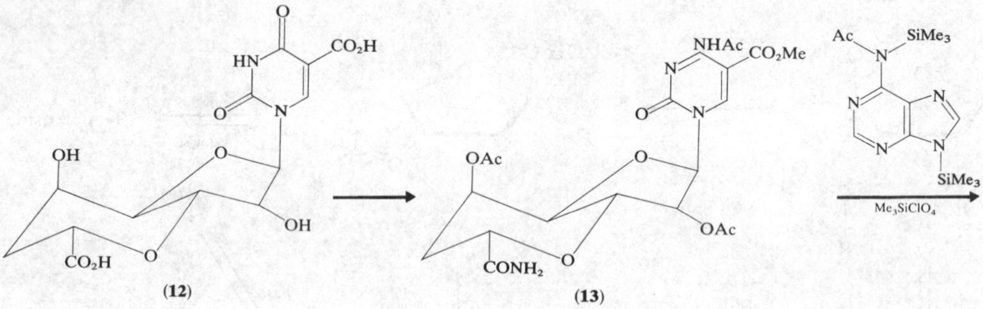

(**12**) (**13**)

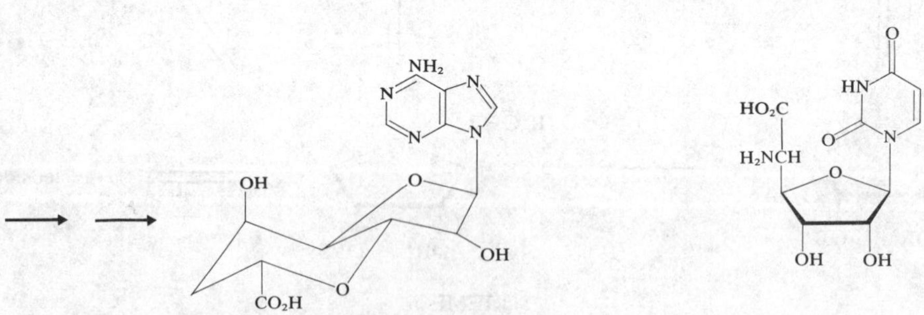

(**14**) (**15**)

The first transglycosylation reaction involving the exchange of one sugar residue (ribose) for another (glucose) has been reported.[69] The mercuric cyanide catalysed reaction of acetobromoglucose (**16**) with 2′,3′-*O*-isopropylideneinosine in nitromethane gives reasonable yields (70 °C, 1 h, 39% N-7, 20% N-9) of the glucosylhypoxanthines. It is assumed that the reaction proceeds by attack of the 1,2-acetoxonium ion, formed by the action of the catalyst on the glycosyl halide, which then reacts to give a bis-glycosylated species (**17**) followed by elimination of the ribose moiety and an N-7 → N-9 glucosyl shift. The transglycosylation reactions which are now available are certain to play an ever-increasing role in the production of purine nucleoside analogues from naturally-occurring pyrimidine nucleosides or suitable sugar derivatives.

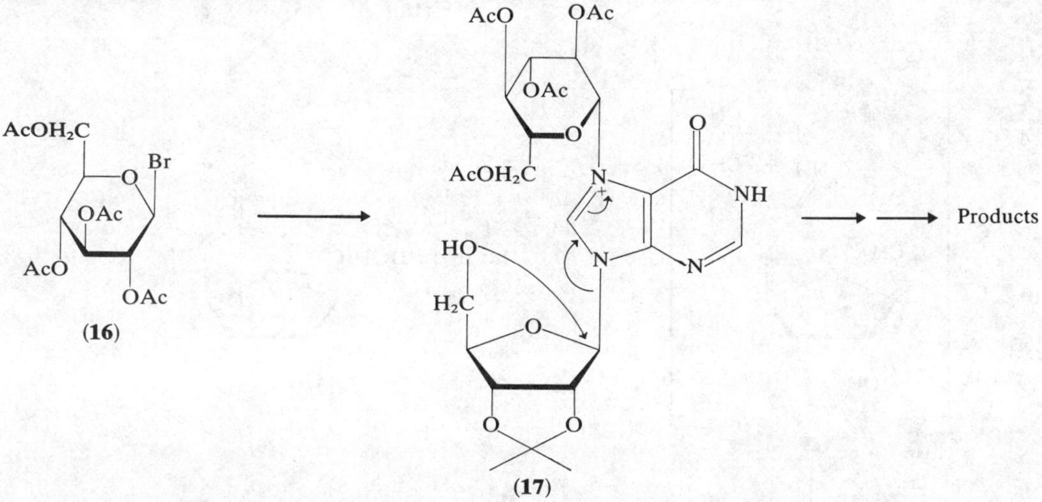

(**16**) (**17**)

Some years ago it appeared that the synthesis of adenosine and deoxyadenosine could be achieved merely by condensing the base, adenine, with the unprotected sugars D-ribose and 2-deoxy-D-*erythro*-pentose in 'polyphosphoric ester'.[70] However, the reaction with the deoxy-sugar was repeated independently and of the six major components in the product, which even so only accounted for less than 20% of the starting material, both deoxyadenosine and its α-anomer were present and the latter predominated.[71] Other similar reactions have confirmed that this synthesis is neither useful nor stereospecific.

Sanchez and Orgel[72] published a novel synthesis of cytosine arabinoside in which a base is formed from a glycosylated intermediate. As will be seen, others have used this method for the synthesis of cyclonucleosides and L-nucleosides. The key reaction is that between arabinose, cyanamide, and ammonia to give an amino-oxazoline (**18**); needless to say, this reaction was initially proposed as a potential prebiotic route to nucleosides. The amino-oxazoline is then treated with cyanoacetylene which, presumably *via* (**19**) and the O^2,2′-cyclonucleoside, is hydrolysed to cytosine arabinoside. Although photochemical-induced isomerization to give cytidine was achieved, the yield was certainly not of commercial interest! The amino-oxazoline has also been condensed with methyl acrylate to give O^2,2′-cyclo-5,6-dihydrouridine (**20**), a compound that is very difficult to prepare in any other way.[73]

Using L-arabinose as starting material, Holý[74] reacted the amino-oxazoline (**21**) with methyl propiolate to prepare O^2,2′-cyclo-L-uridine (**22**). A four-step synthesis yielded 2′-deoxy-L-uridine (**23**) in good yield and enabled the 2′-deoxy-L-nucleosides of thymine and cytosine to be prepared.

22.2.3.3 Construction of a heterocyclic base after glycosylation

Methods other than the direct condensation reactions for the preparation of nucleosides will inevitably often be found to be used either as a method of proof of structure or for the preparation of analogues (often aza- or deaza-nucleosides). It is not intended to

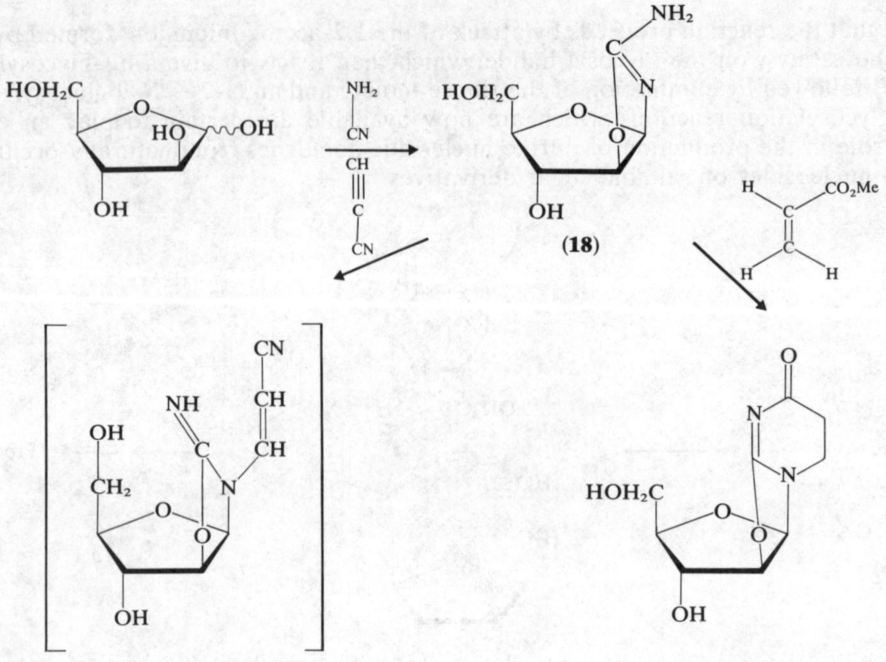

(18)

(19)

(20)

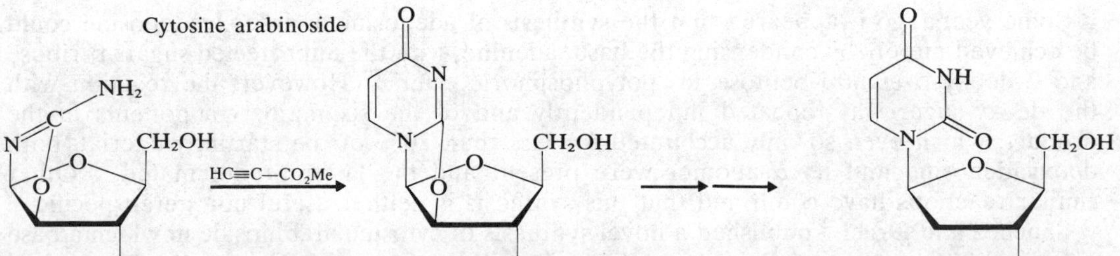

(21) (22) (23)

describe in detail any significant number of nucleoside analogue syntheses and the following discussion may often appear to be of more historical than practical interest, although the main fundamental types of preparation are covered and some emphasis is placed upon more modern methods which are useful for large-scale syntheses. It must be remembered, however, that nowadays most of the naturally-occurring nucleosides and many analogues are synthesized by direct condensation of a heterocyclic base with a sugar.

(i) *Pyrimidines*

Glycosyl azide intermediates have been used as precursors in pyrimidine nucleoside synthesis. 2,3,5-Tri-*O*-benzoyl-β-D-ribofuranosyl chloride reacts with sodium azide to give the glycosyl β-azide, which can be reduced to an anomeric mixture of ribosylamines. Unfortunately, this preparation is rather tedious, the ribosylamines readily mutarotate, and the compound usually has to be made *in situ* and used immediately to avoid $O \rightarrow N$ migration of the 2-*O*-benzoyl group.[75] However, recently the preparation in high yield

(80%) of a stable crystalline ribofuranosylamine derivative, 2′,3′-*O*-isopropylidene-D-ribofuranosylamine toluene-*p*-sulphonate, has been described.[76] This is easily prepared from D-ribopyranosylamine, acetone, 2,2-dimethoxypropane and toluene-*p*-sulphonic acid. The composition of the anomeric mixture varies with the solvent used; in chloroform, little α-anomer can be detected whereas in dimethyl sulphoxide, 25% of the α-anomer is present. The stable ribofuranosylamine derivative has been used to prepare the α- and β-anomers of 5-cyanouridine and 5-acetyluridine and the β-anomers of 5-methyl-2-thiouridine and uridine, although in the latter preparations the production of a small amount of the α-anomer could not be excluded.

Tri-*O*-benzoyl-D-ribofuranosylamine (**24**) has been used to prepare uridine by reaction with β-ethoxy-*N*-ethoxycarbonylacrylamide (**25**), 2-thiouridine (and hence uridine) by reaction with β-ethoxyacryloyl isothiocyanate (**26**), and 5-methyl-2-thiouridine by reaction with β-methoxy-α-methylacryloyl isothiocyanate (**27**).[77] Only β-nucleoside products were reported, even though the ribofuranosylamine used is an anomeric mixture.

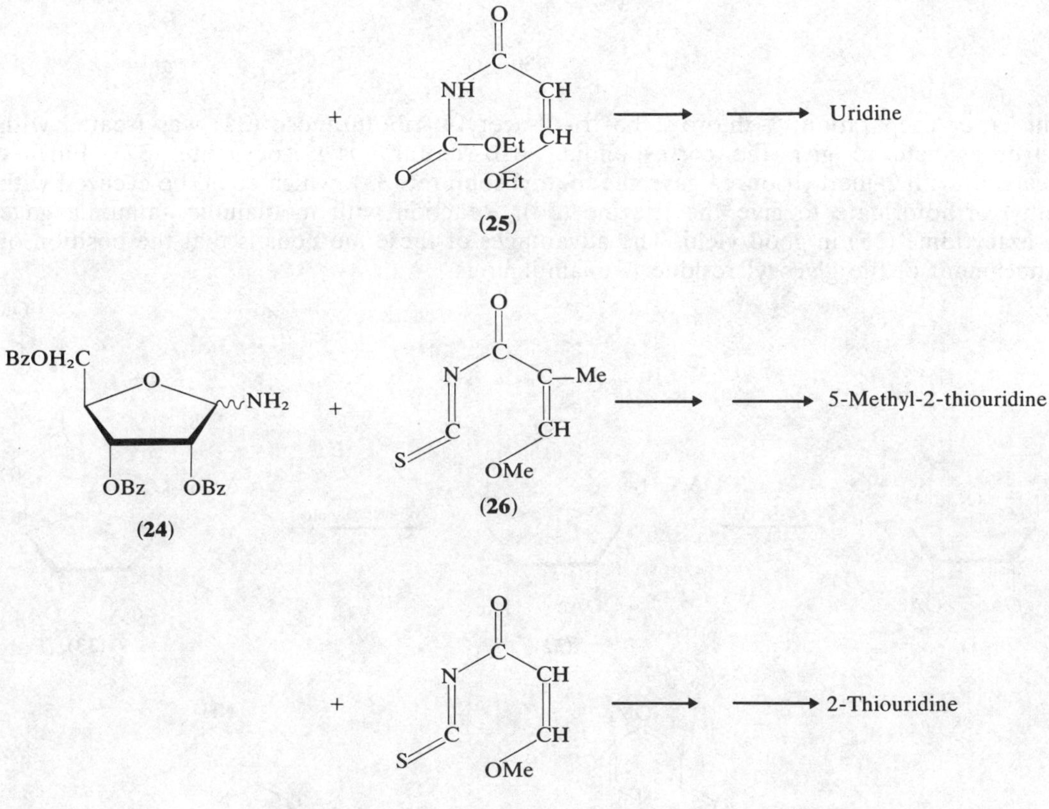

Glycosylureas have also been used as intermediates in pyrimidine nucleoside synthesis. An anomeric mixture of 3,5-di-*O*-benzyl-2-deoxy-D-ribofuranosylureas can be obtained from the reaction of 1-*O*-acetyl-3,5-di-*O*-benzyl-2-deoxy-D-ribofuranose with urea in acetic acid in the presence of HCl.[78] The pure β-anomer (**28**) can be obtained crystalline and subsequent reaction with β-ethoxyacryloyl chloride (**29**) or β-ethoxy-α-methacryloyl chloride (**30**), followed by ring closure with ammonia and hydrogenolysis to remove the benzyl groups, results in the synthesis of 2′-deoxyuridine and thymidine, respectively.

In related syntheses, Šorm and co-workers, who were responsible for much of the pioneering work in this field in the 1960s, prepared 1-β-D-ribofuranosyl-5,5-diethylbarbituric acid[79] and 5-azacytidine[80] from a 1-ureido-sugar and from (**32**), respectively. In

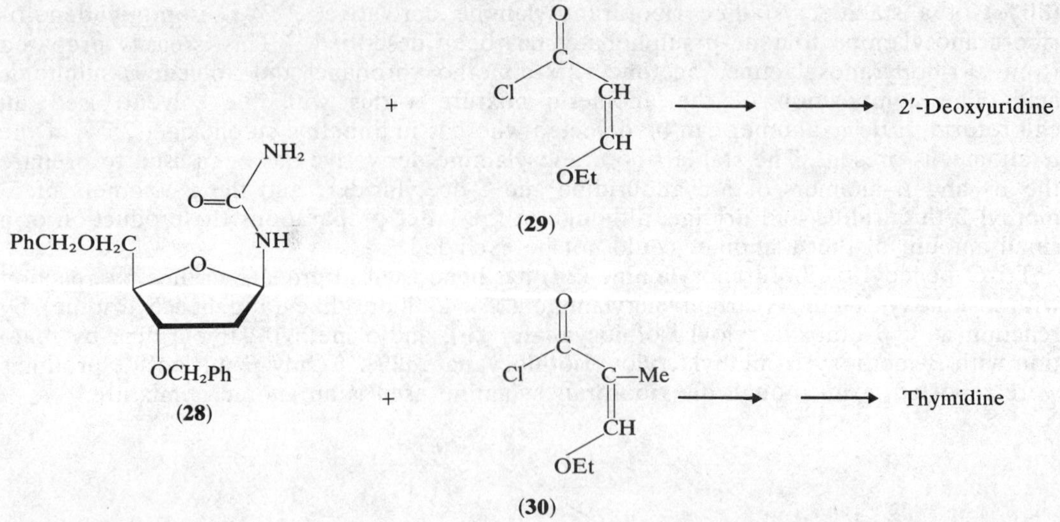

(28) (29) (30)

the latter preparation, 1-chloro-2,3,5-tri-*O*-acetyl-D-ribofuranose (**31**) was treated with silver cyanate to give the corresponding β-D-ribofuranosyl isocyanate (**32**). Further reaction with 2-methylisourea gave the methylisobiuret (**33**), which could be cyclized with ethyl orthoformate to give the triazine (**34**). Reaction with methanolic ammonia gave 5-azacytidine (**35**) in good yield. The advantages of these methods is that the position of attachment of the glycosyl residue is unambiguous.

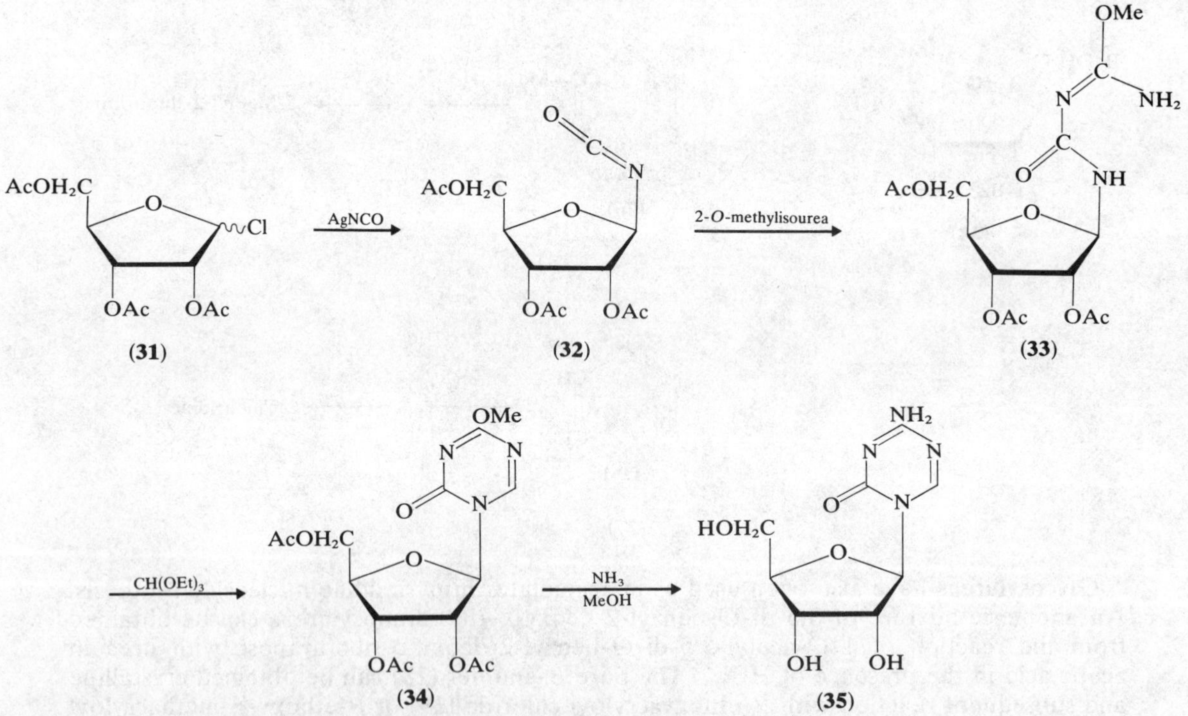

(31) (32) (33) (34) (35)

(*ii*) *Purines*

Purine nucleoside synthesis *via* pyrimidines was first used by Todd and his co-workers to assign unambiguously the position of the glycosyl residue in the naturally-occurring purine nucleosides. Starting from 4,6-diamino-2-methylthiopyrimidine and condensing it with 2,3,4-tri-*O*-acetyl-5-*O*-benzyl-D-ribose, adenosine was produced after a seven-stage synthesis.[81] The method has also been applied to the synthesis of 9-substituted 6-chloropurines[82] and could presumably be adapted to the synthesis of nucleosides from the ribosylamine previously mentioned, but there is little incentive to investigate this type of synthesis as so many better methods are now available.

Purine nucleoside synthesis, *via* imidazole glycosides particularly for the synthesis of analogues, is very widely used and the literature up to 1966 has been carefully reviewed.[83] Xanthosine was first synthesized in this manner, once again by Todd and co-workers.[81] Silver 4,5-dimethoxycarbonylimidazole (**36**) was condensed with 2,3,5-tri-*O*-acetyl-D-ribofuranosyl chloride to give the substituted imidazole nucleoside (**37**), which on treatment with ammonia yielded 1-D-ribofuranosylimidazole-4,5-dicarboxamide (**38**). The pyrimidine ring was closed using a modified Hofmann reaction with alkaline hypobromite to give xanthosine. This method was later modified by Baddiley and co-workers,[84] who prepared the imidazole nucleoside (**37**) directly by reaction of 2,3,5-tri-*O*-acetyl-D-ribofuranosyl azide with the dimethyl ester of acetylenedicarboxylate. It has been found[34] that traces of the *N*-7-glycosyl as well as the desired *N*-9-glycosyl derivatives are produced in these reactions.

A variation of this method has been used by Robins and co-workers to make specifically the *N*-7-glycosylpurine nucleoside.[85] The imidazole nucleoside (**39**) was formed in an acid-catalysed fusion reaction and then treated with potassium cyanide to give (**40**), which was reduced to the amino-nitrile (**41**) and cyclized with dimethoxymethyl acetate followed by deacetylation to give 7-(β-D-ribofuranosyl)adenine (**42**). The corresponding guanine derivative has also been made by a similar procedure.

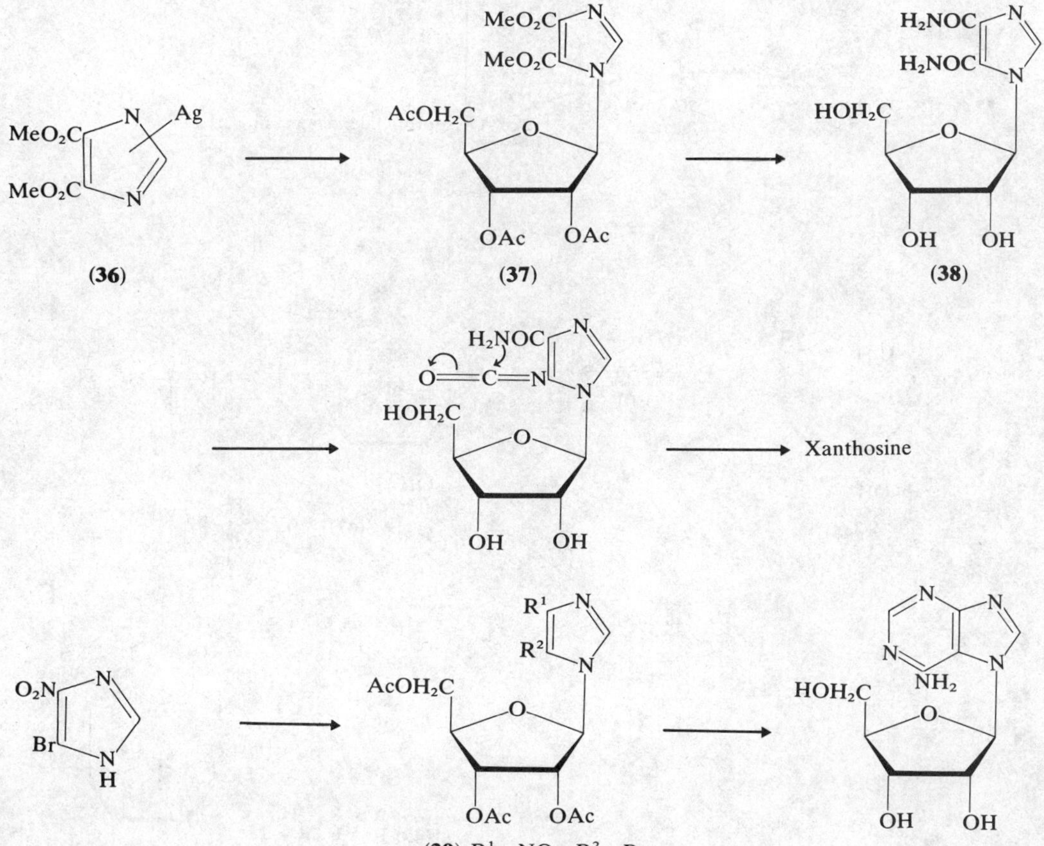

(**36**) (**37**) (**38**)

(**39**) $R^1 = NO_2$, $R^2 = Br$
(**40**) $R^1 = NO_2$, $R^2 = CN$
(**41**) $R^1 = NH_2$, $R^2 = CN$

(**42**)

5-Amino-1-β-D-ribofuranosyl-4-imidazolecarboxamide (AICA-riboside) produced by fermentation[86] has been the focus of intensive research as the starting material for purine nucleoside syntheses. Some years ago, Yamazaki and co-workers reported two methods of synthesis of guanosine from AICA-riboside (**46**), and recently two further modifications to one of these routes to guanosine and a preparation of inosine have been detailed. One of

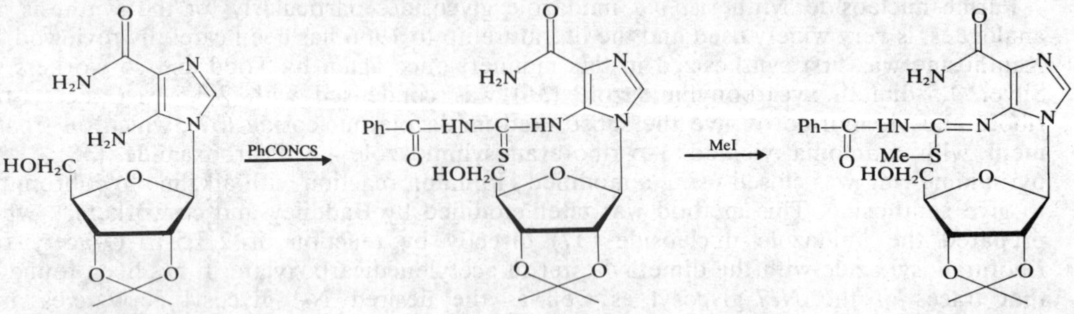

(**43**) (**44**) (**45**)

2′,3′-O-Isopropylidene-AICA-riboside

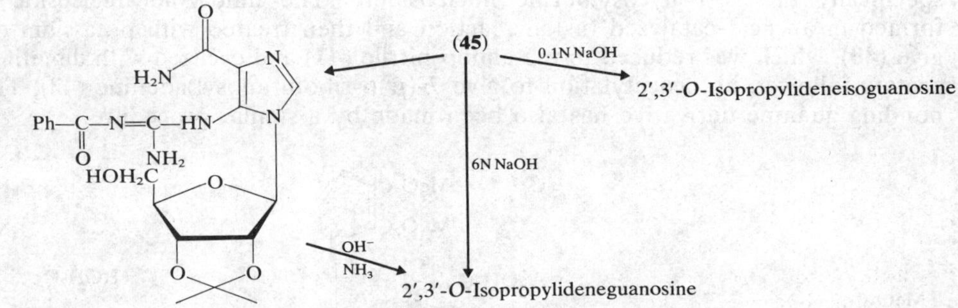

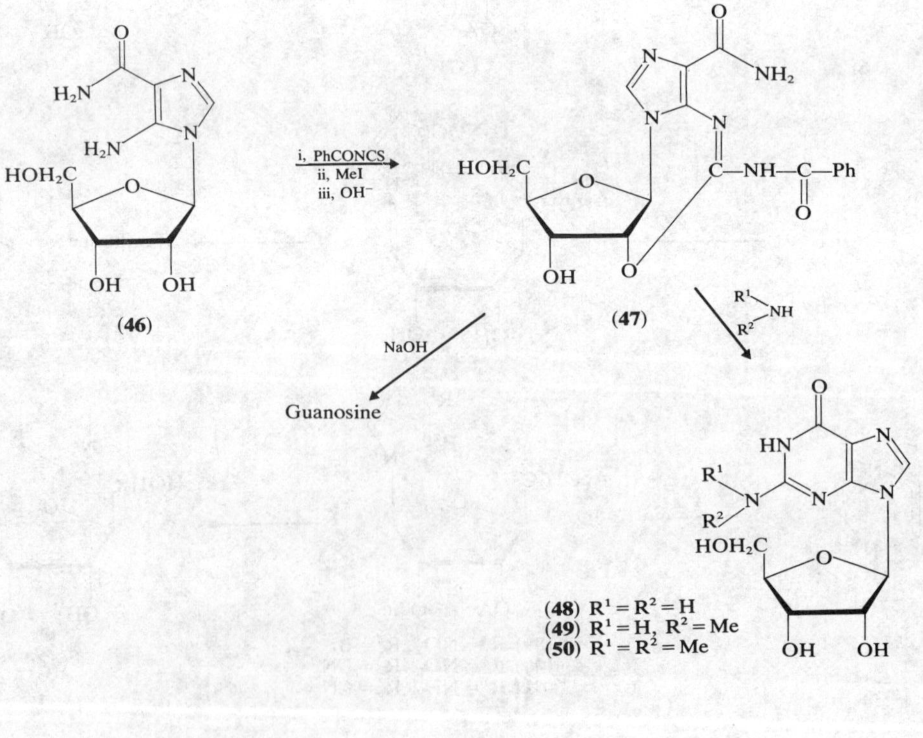

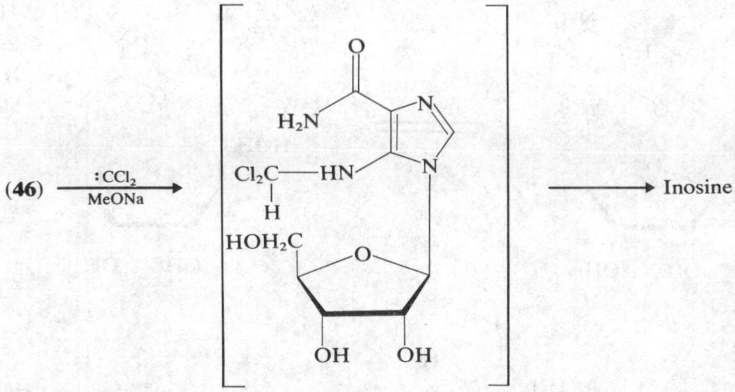

the early methods described the treatment of 5-(N'-benzoyl-5-methylisthiocarbamoyl)-amino-1-(2′,3′-*O*-isopropylidene-β-D-ribofuranosyl)-4-imidazolecarboxamide (**45**) with ammonia, followed by ring closure with alkali. The second method concerned the reaction of AICA-riboside with sodium methyl xanthate followed by oxidation and subsequent amination. The latter preparation proved the more convenient of the two when done on a large scale. One of the new routes comes from an examination of the ring closure of (**45**) in alkaline solution. It has been found that treatment of (**45**) with 6 N NaOH results in a high yield (50% overall without intermediate isolation) of 2′,3′-*O*-isopropylideneguanosine, and treatment of (**45**) with 0.1 N NaOH gives a reasonable yield of 2′,3′-*O*-isopropylideneisoguanosine. It has also been found that the intermediate corresponding to (**45**), produced from the free nucleoside rather than the isopropylidene derivative, when allowed to stand at an alkaline pH forms a cyclonucleoside of novel structure (**47**). This product can be converted into xanthosine, guanosine (**48**), N^2-methylguanosine (**49**), or N^2,N^2-dimethylguanosine (**50**) by treatment with NaOH, ammonia, methylamine, and dimethylamine, respectively.

Inosine has been produced from AICA-riboside (**46**) by reaction with the carbene formed from chloroform, carbon tetrachloride, or hexachloroethane in the presence of sodium methoxide.[36]

22.2.3.4 Interconversion of naturally-occurring nucleosides

This is necessarily a heterogeneous collection of reactions and the contents could just as well be considered as chemical reactions of nucleosides. However, there is some merit in collecting these methods together, particularly now that the present emphasis in this subject is on analogue synthesis.

(*i*) *Deamination*

Methods for the conversion of adenine and cytosine nucleosides to hypoxanthine and uracil nucleosides, respectively, have been available for a long time (indeed, much commercial DNA contains deoxyuridine residues at the expense of deoxycytidine residues). Nitrous acid is the usual reagent used and is particularly useful for the preparation of hypoxanthine-containing nucleosides. In the pyrimidine series, cytosine nucleosides are usually more difficult to prepare than uracil nucleosides and so their deamination is of less synthetic use.[87]

Another mild method for the deamination of cytosine to uracil derivatives involves the use of bisulphite (5 h, pH 5, 37 °C). This adds reversibly across the 5,6-double bond to give the 5,6-dihydro-6-sulphonate, which rapidly deaminates and subsequently releases bisulphite on treatment with alkali (Scheme 7). The synthetic uses of this reaction have been discussed.[88]

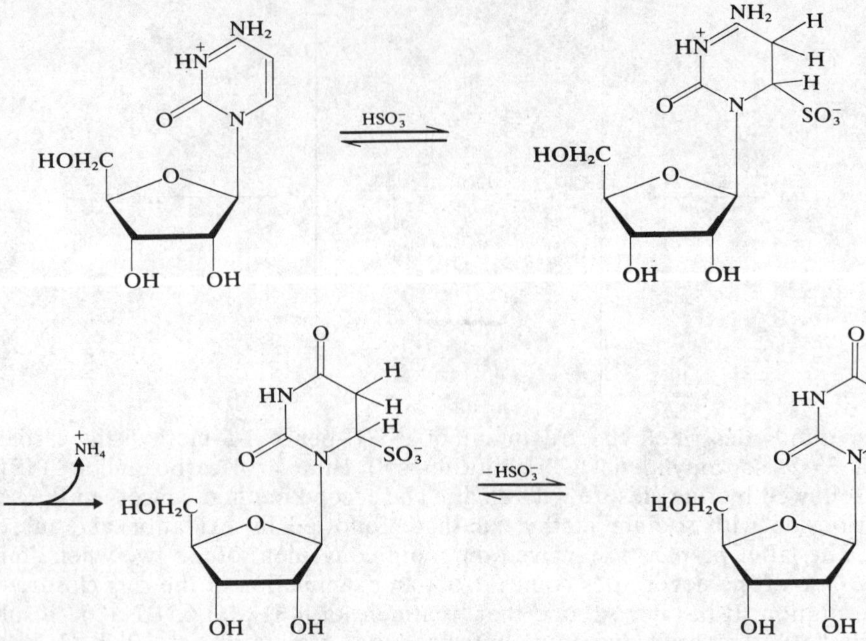

SCHEME 7

(*ii*) *Amination*

There are two well-established methods in the literature[89] for the preparation of cytosine nucleosides starting from the corresponding uracil derivatives. Fox has used the reaction of phosphorus pentasulphide, which results in the conversion of the 4-oxo-group to a thiol group. This can subsequently be replaced by treatment with ammonia. Alternatively, the 4-chloro-derivative can be prepared either by using the Vilsmeier–Haak reagent[47] (chloromethylenedimethylammonium chloride) or phosphorus oxychloride, and the resulting 4-chloro-group can subsequently be replaced by treatment with ammonia.

Phenyl phosphorodiamidate (**51**) has been reported to convert tautomeric oxo-hydroxy-groups directly to amino-groups. Although it has been used in the conversion of 6-methyluracil to 2,4-diamino-6-methylpyrimidine,[90] the yield of cytidine from uridine is not good.

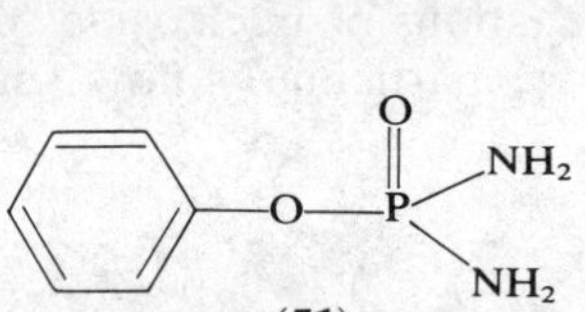

(**51**)

Vorbrüggen and co-workers have reported a one-step conversion of uridine to cytidine and the method has also been used to convert thymidine to 5-methyl-2′-deoxycytidine and inosine to adenosine (even at the nucleotide level).[91] Persilylation, of free or acylated uridines, activates the 4-position and subsequent treatment with ammonia gives the corresponding cytosines (80% yield). The activated intermediates formed during the Hilbert–Johnson reaction of silylated uracils can be converted analogously with ammonia into the corresponding cytidines. Thymidine requires the addition of catalytic amounts of ammonium sulphate (to give 5-methyl-2′-deoxycytidine, in 61% yield) and inosine gives adenosine (72% yield). The mechanism of the reactions has been discussed and the purine amination is formulated as a typical acid catalysed addition–elimination reaction (Scheme 8).

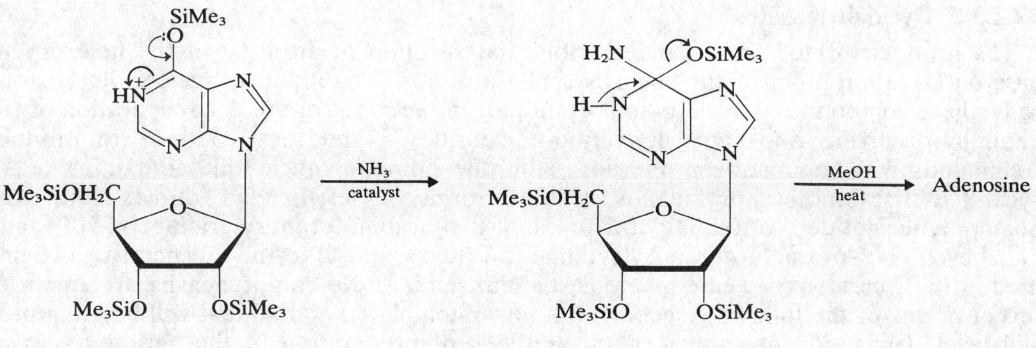

SCHEME 8

(*iii*) *Ribonucleosides* → 2′-*deoxyribonucleosides*

There are several advantages in devising a good synthetic method for this conversion. The preparation of deoxynucleosides is always more complicated than the equivalent ribonucleoside as the sugar is less stable (and more expensive) and it is often impossible to avoid the production of a mixture of α- and β-anomers. Many ribonucleoside preparations, by contrast, are rapid, quantitative, and give only the β-nucleoside. Three methods are available for the ribo → deoxy conversion. The first requires the preparation of the O^2,2′-cyclonucleoside (**52**) by the reaction of uridine with diphenyl carbonate.[92] The crude product is benzoylated directly with benzoyl cyanide in an aprotic solvent and further converted to the 2′-chloro-2′-deoxynucleoside (**53**) by the action of anhydrous hydrogen chloride in DMF. This gives exclusively the ribo-product, which can be reduced with tri-n-butyltin hydride to the blocked deoxynucleoside, from which the benzoyl protecting groups can be removed to give the free deoxynucleoside. Another method is somewhat similar in that 2′,3′-*O*-benzylideneuridine gives 3′-*O*-benzoyl-2′,5-dibromo-2′-deoxyuridine on reaction with *N*-bromosuccinimide, presumably *via* the O^2,2′-cyclonucleoside.[93] This can be converted to 2′-deoxyuridine by catalytic reduction and debenzoylation. The yields, however, are not as high as in the previous method.

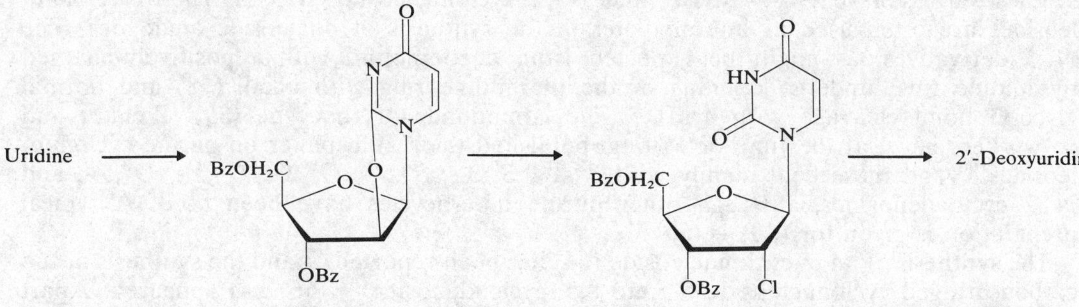

(**52**) (**53**)

Recently, another method of 2′-deoxynucleoside synthesis has been proposed which depends upon the reduction of the acetyl derivatives of adenosine-2′,3′-thiocarbonate with tri-n-butyltin hydride in the presence of azobis-isobutyronitrile in dimethylacetamide.[94] Removal of the blocking groups resulted in the production of a separable mixture of the 2′-deoxyadenosine (60%) and 3′- deoxyadenosine (31%).

22.2.3.5 Cyclonucleosides

The product formed from the intramolecular reaction of an atom of the heterocyclic base on a carbon atom of the sugar is called a cyclo- (or anhydro-) nucleoside. Historically, these compounds played an important part in the proof of the β-configuration of the naturally-occurring ribo- and deoxyribo-nucleosides,[95] and they continue to produce compounds with interesting properties. Thus the homopolynucleotide containing S^8,2'-cyclo-9-β-D-arabinofuranosyladenosine (**54**) forms a left-handed duplex with the homopolynucleotide containing 6,2'-cyclo-1-β-D-arabinofuranosyluridine (**55**),[96] and O^2,2'-cyclo-1-β-D-arabinofuranosylcytidine (**56**) (cyclo-ara-C to the biochemist!) is being used in the treatment of acute myeloblastic leukaemia.[97] No comprehensive treatment of this or of any of the following sections can be contemplated and all that will be attempted will be a description of some general methods of preparation of the various types of compounds that have been synthesized, together with some key references. The early work on cyclonucleosides has been reviewed in detail[3] and a review of more recent work is available.[46]

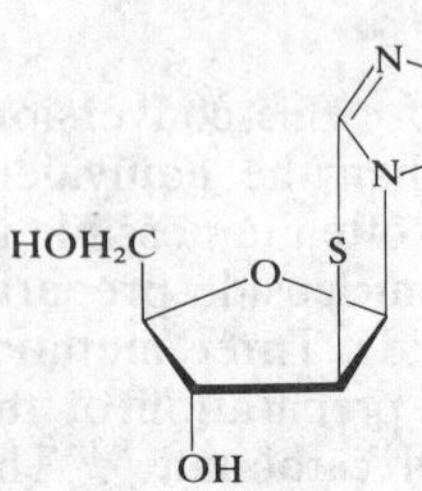

(**54**)

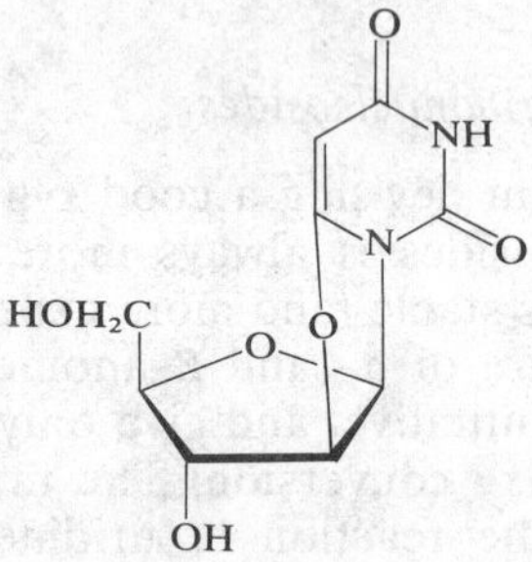

(**55**)

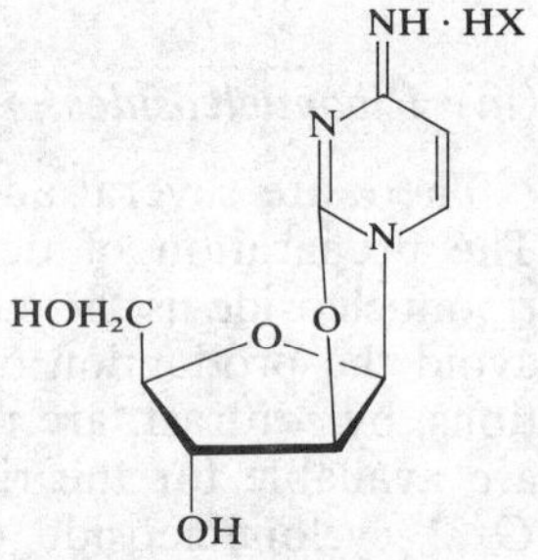

(**56**)

(i) *Purine cyclonucleosides*

The N^3,5'-cyclonucleoside of adenosine (**57**) was the first cyclonucleoside to be fully characterized (by X-ray)[95] and can be prepared easily by heating a solution of 5'-*O*-toluene-*p*-sulphonyl-2',3'-*O*-isopropylideneadenosine. Although the unprotected cyclonucleoside can be prepared from 5'-*O*-toluene-*p*-sulphonyladenosine, this reaction does not proceed as easily as with the isopropylidene derivative because cyclic acetal formation causes a change in the conformation of the sugar ring and brings the 5'-carbon atom close to the heterocyclic ring. Similar cyclonucleosides of other purine nucleosides have been synthesized, as have N^3,4'- (**58**)[98] and N^3,3'-cyclonucleosides (**59**).[99] These cyclonucleosides are often used as intermediates in the synthesis of nucleoside analogues. The N^3,5'-derivatives, as might be expected from a compound with a positively-charged pyrimidine ring, undergo opening of the pyrimidine ring with alkali (**58**) and normal glycosyl bond cleavage with acid to give compounds of the type (**59**). Ikehara and co-workers have, at the time of writing, published their 29th paper on purine cyclonucleosides[100] and these deal mainly with S^8,2'-, S^8,3'-, S^8,5'-, O^8,2'-, O^8,3'-, O^8,5'-, and N^8,2'-cycloadenosines, although other purine heterocycles have been used. A typical preparation is given for (**60**) → (**63**).

The synthesis of an α-cyclonucleoside (**64**) has been reported[101] and the synthesis of the carbon-bridged cyclonucleoside 5'-keto-8,5'-cycloadenosine[102] (**65**) has appeared. Apart from the work from Ikehara's group, little has been reported on the chemical reactions of purine-8-cyclonucleosides.

(ii) *Pyrimidine cyclonucleosides*

2',3'-*O*-Isopropylidene-O^2,5'-cyclocytidine tosylate was the first pyrimidine cyclonucleoside to be prepared,[95] and this was achieved merely by heating the isomeric covalent

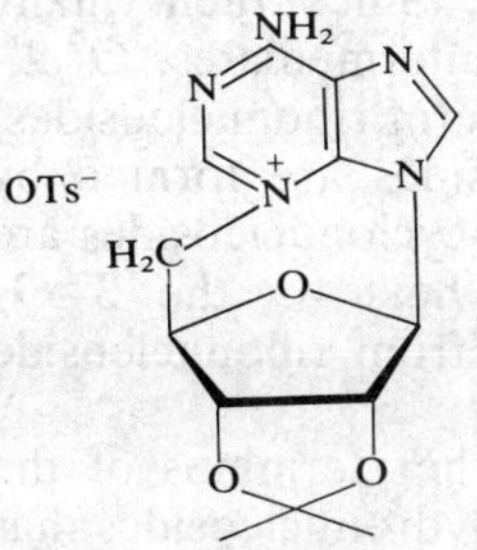

(57)

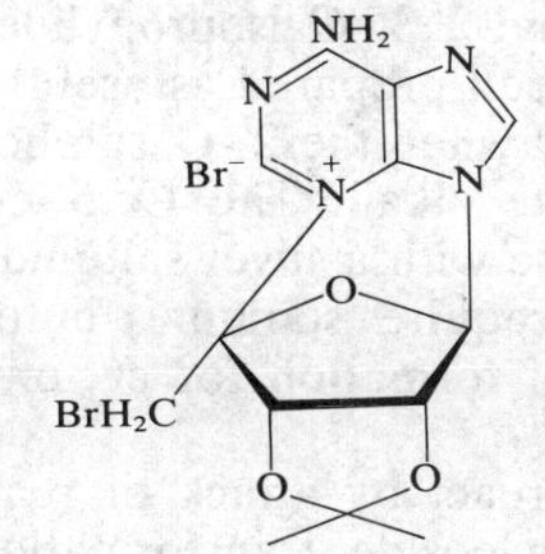

(58)

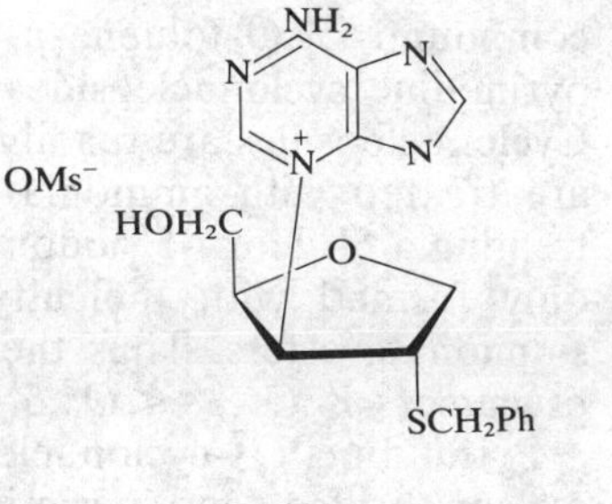

(59)

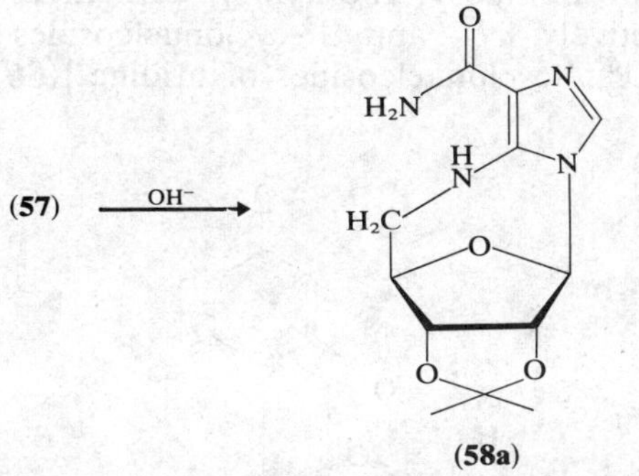

(58a)

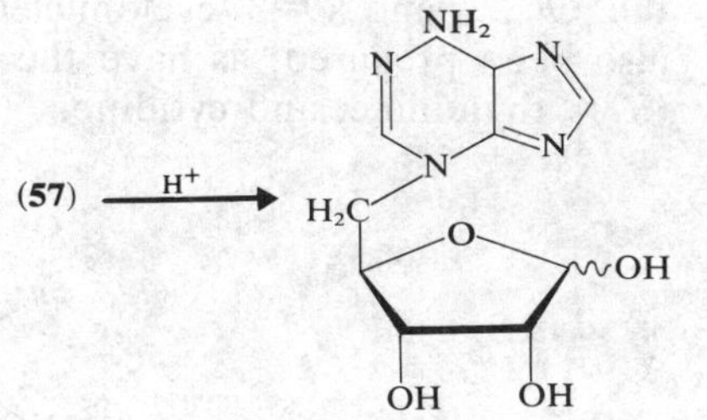

(59a)

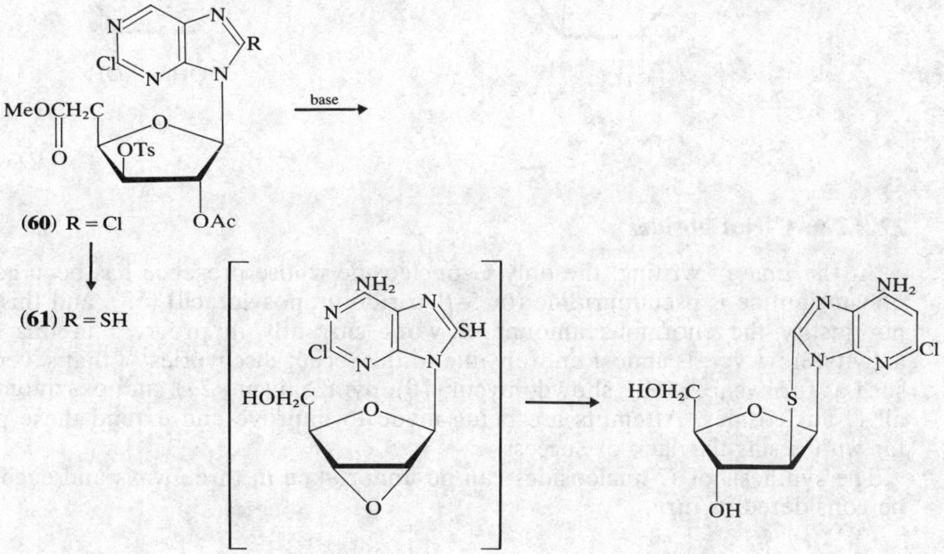

(60) R = Cl

(61) R = SH

(62)

(63)

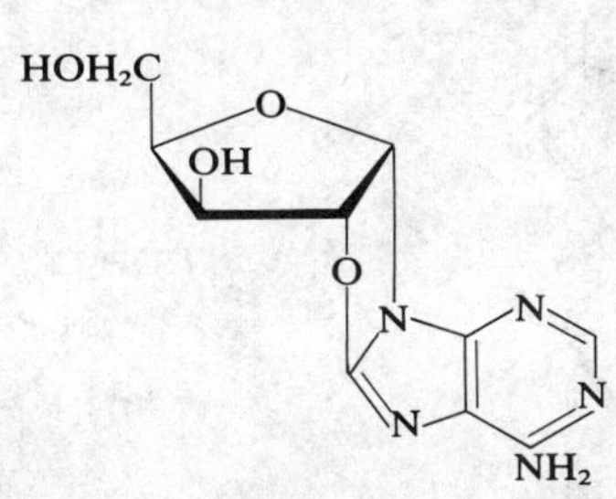

(64)

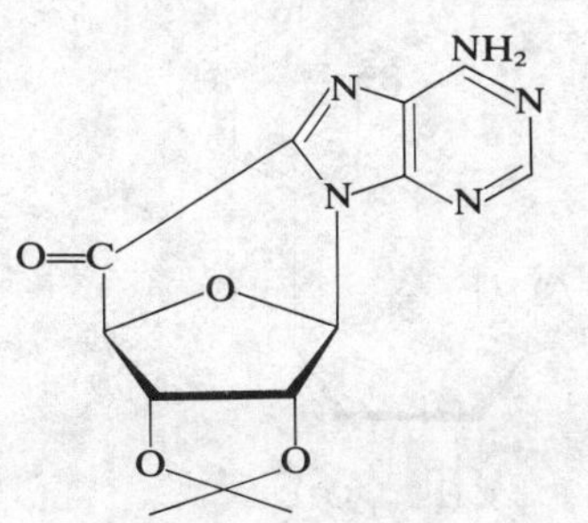

(65)

compound 5′-*O*-toluene-*p*-sulphonyl-2′,3′-*O*-isopropylidenecytidine. Since then, many pyrimidine cyclonucleosides have been prepared as useful synthetic intermediates. O^2,2′-Cyclonucleosides are readily formed when the 2′-*O*-sulphonate esters (of ribonucleosides) are treated with ammonia or dilute alkali. The O^2,5′-cyclonucleosides are formed by treating a 5′-deoxy-5′-iodonucleoside with a silver salt and the O^2,3′-cyclonucleosides are only formed with difficulty and require sodium t-butoxide and heat on the 3′-*O*-sulphonate esters. Thus the ease of formation for cyclonucleosides from ribonucleoside precursors is O^2,3′-< O^2,5′-< O^2,2′-.

Pyrimidine O^2-cyclonucleosides react by attack of reagent at either terminus of the oxygen bridge, and pyrimidine nucleoside transformation *via* anhydronucleosides has been reviewed.[103] Hydrolysis of the cyclo-linkage can be achieved with either acid or alkali and is accompanied by the production of arabinosyl and xylosyl derivatives from the O^2,2′- and O^2,3′-cyclonucleosides, respectively. N^2- and S^2-cyclonucleosides have also been prepared, as have the O^6,2′- and O^6,5′-cyclonucleosides of uridine [(**66**) and (**67**)], thymidine, and cytidine.

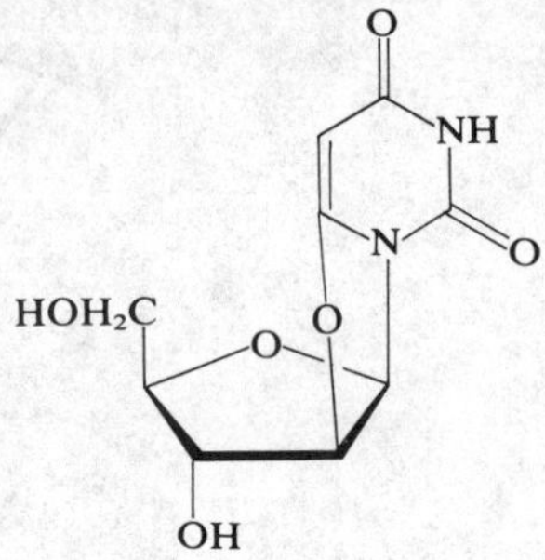

(**66**)

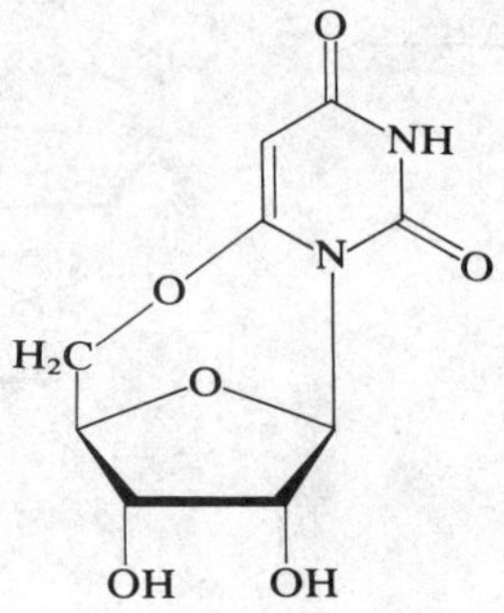

(**67**)

22.2.3.6 *C*-Nucleosides

At the time of writing, the only *C*-nucleoside whose presence has been confirmed in a polynucleotide is pseudouridine (or 5-β-D-ribofuranosyluracil) (**68**), and this alone would not justify the enormous amount of work currently in progress in this field.[104] This activity, however, is almost entirely due to the recent discoveries[105] that several antibiotics such as formycin A (**69**), showdomycin (**70**), pyrazomycin (**71**), and oxazinomycin (**72**) are all *C*-nucleosides. Attempts are being made to improve and extend these properties, so far with a singular lack of success.

The synthesis of *C*-nucleosides can be undertaken in three ways and each of these will be considered in turn.

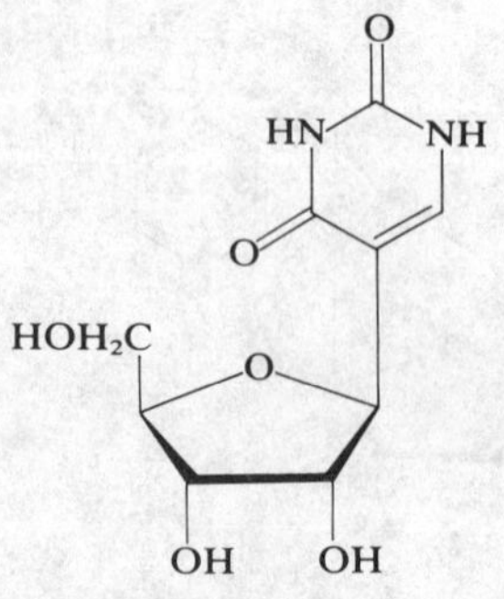

(**68**)

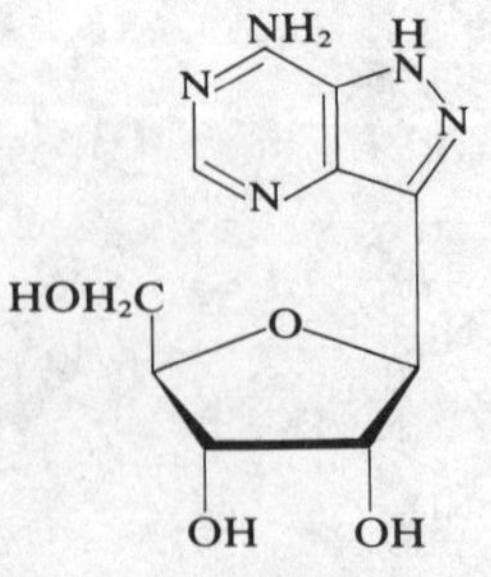

(**69**)

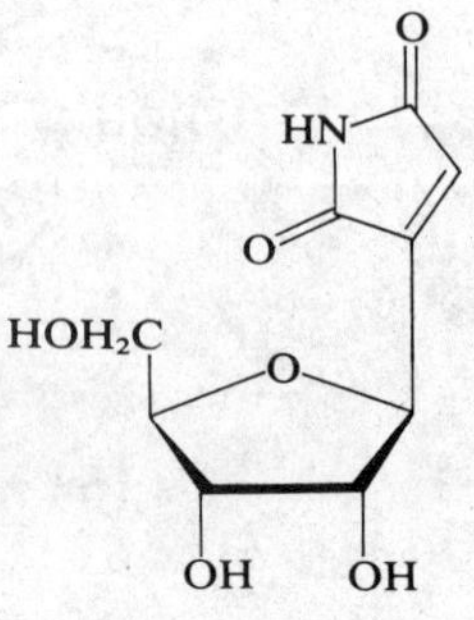

(**70**)

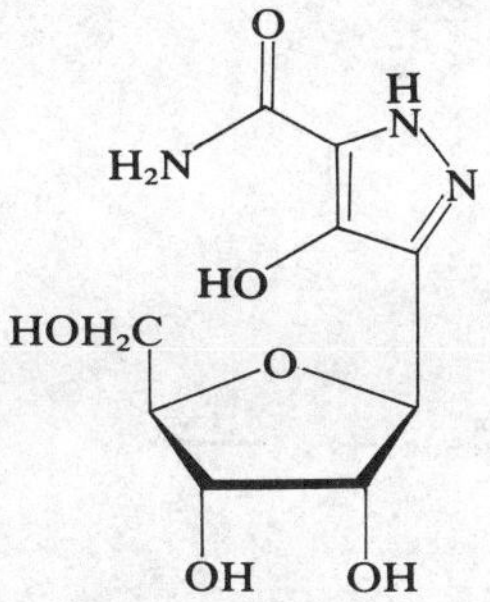

(71)

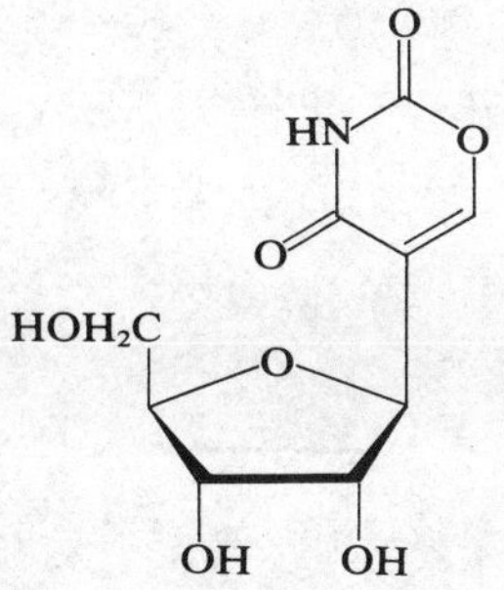

(72)

(i) Condensation reactions

The most well-known *C*-nucleoside, pseudouridine, has been synthesized *via* a direct condensation reaction. The best yields (10%) are obtained from the condensation of 2,4-di-t-butoxy-5-lithiopyrimidine with 2,4:3,5-di-*O*-benzylidene-D-ribose followed by removal of the blocking groups.[106] In general, however, condensation methods are rarely used for *C*-nucleoside synthesis as more predictable and versatile syntheses are required.

(ii) Conversion of existing C-nucleosides

Once again, this method is not widely used, although pseudouridine has been prepared from oxazinomycin (**72**) and the conversion of pseudouridine to 5-(β-D-ribofuranosyl)-6-azauracil is given in Scheme 9.

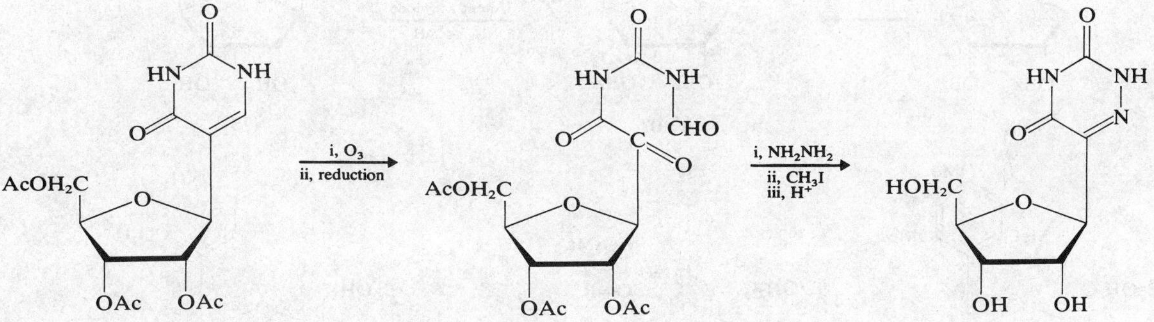

SCHEME 9

(iii) Elaboration of heterocycles from suitably functionalized anhydroalditols

This is by far the most useful approach and only a few typical examples of the many methods available can be given here.

The 1-diazo-sugar has been used as a precursor of formycin A analogues, following 1,3-dipolar cycloaddition of methylenedicarboxylate[107] (Scheme 10).

The 1-cyano-sugar (**73**) has also been used, either by Tronchet *et al.*[108] with 1,3-dipolar cycloaddition of acetylene to give iso-oxazoles and pyrazoles, or by Moffatt and co-workers[109] to give the amidoxime (**74**) by reaction with hydroxylamine, followed by cyclization to give either oxadiazoles (**75**) or oxadiazolines.

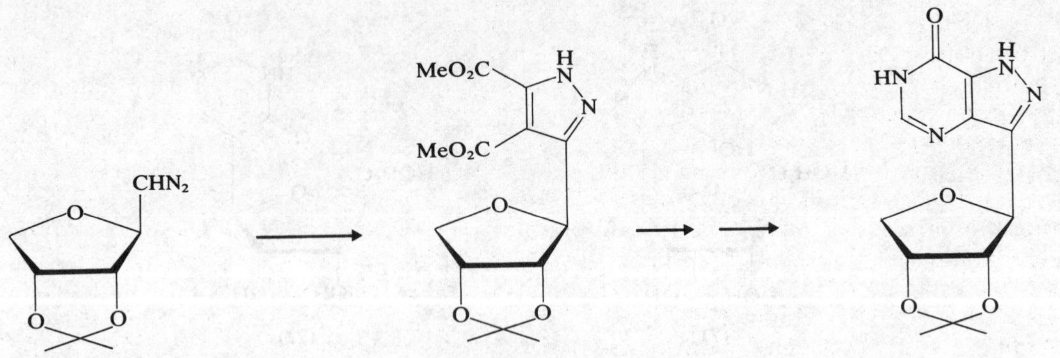

SCHEME 10

Ozonolysis of the 1-arylriboside (**76**) was used as the first step in the synthesis of showdomycin by Šorm and co-workers.[110] The reaction proceeds *via* the pyruvate ester (**77**), the maleate ester (**78**), and the maleamide (**79**).

A method which has received much attention recently involves the direct condensation of aldofuranosyl halides with stabilized carbanions. Hanessian and co-workers[111] were the first to show that such sugar derivatives could be reacted with dialkyl sodiomalonates to give moderate yields of the *β*-*C*-glycosides (**80**), but a non-participating group at C-2 is required to prevent the formation of compounds of the type (**81**) when 2,3,5-tri-*O*-acyl-*β*-D-ribofuranosyl halides are used.

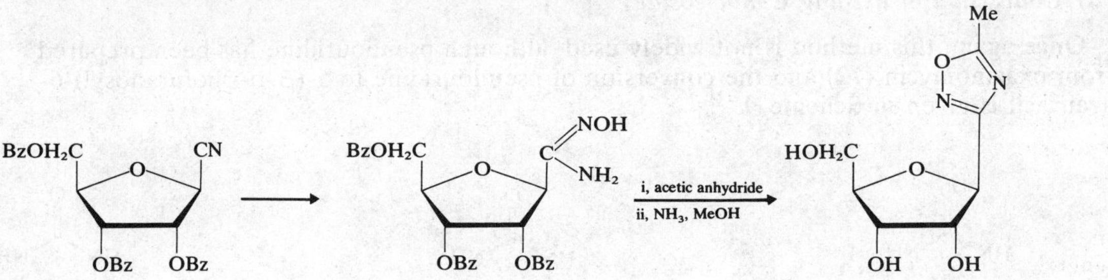

(**73**) (**74**) (**75**)

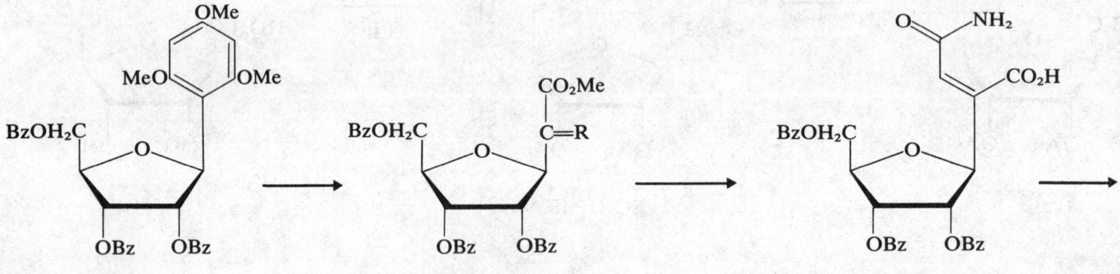

(**70**)

(**76**) (**77**) R = O

(**78**) R = $CHCO_2Et$ (**79**)

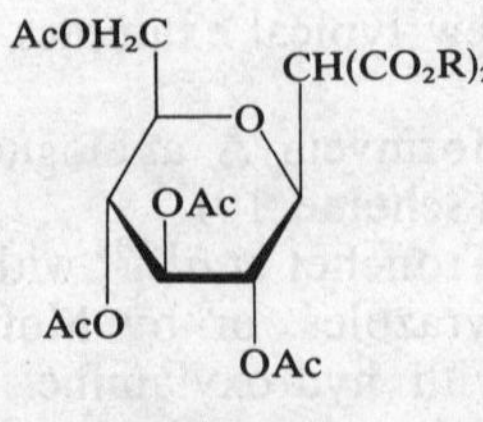

(**80**)

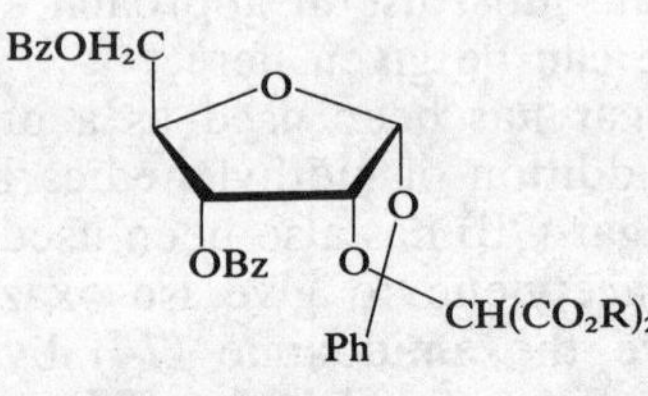

(**81**)

Fox and co-workers[112] have detailed the synthesis of 2,3-*O*-isopropylidene-5-*O*-trityl-β-D-ribofuranosyl chloride in good yield, and this is an ideal starting material for the synthesis of β-D-ribofuranosyl-*C*-glycosides. Moffatt and co-workers[113] have examined in detail the reaction of 2,3-*O*-isopropylidene sugars with stabilized ylides and have shown that, contrary to expectation (and the presumption of Ohrui and Fox[112]), the sterically more-hindered α-isomers, in which the anomeric substituent and isopropylidene moieties are *cis* disposed to one another, are the thermodynamically most-stable products, whereas the desired *trans*-isomer is the kinetic product. The addition of both phosphoranes and sodiomalonates have been studied. Finally, cycloaddition to glycosylethynes has been used by Buchanan *et al.*[114] (to give pyrazoles and a pyrazalone), by Fox and co-workers[115] (to give triazoles), by Moffatt and co-workers (to give pyrazoles and triazines), and by Arakawa *et al.*[116] (to give pyrazoles).

22.2.3.7 Branched-chain-sugar nucleosides

Impetus has been provided for the synthesis of branched-chain-sugar nucleosides by the reports that 2′- and 3′-*C*-methyladenosine and 3′-*C*-methylcytidine are effective antiviral agents in mice and also that two 2′-*C*-nitromethylhexopyranosylpurines are active against KB tumour cells. There are two main methods for preparing such nucleosides: firstly by synthesis of a suitable carbohydrate precursor followed by a conventional nucleoside condensation reaction, and secondly by the attack of nucleophiles on 2′- or 3′-keto-nucleosides. Thus most of the chemistry involved is standard carbohydrate chemistry and, as such, is covered in Part 26 of this publication. It will suffice, therefore, to give just a few examples here.

Walton and co-workers[117] reported the preparation of the sugars required for the synthesis (by the heavy-metal procedure) of 2′-*C*- and 3′-*C*-methyladenosine (**82**) and, by a similar condensation reaction, Albrecht and Moffatt synthesized 6-dimethylamino-9-[3-deoxy-3-nitro (and -amino)]methyl-β-D-ribofuranosylpurine,[118] (**83**) and (**84**). 2′-*C*-Nitromethyluridine has been prepared *via* a Friedel–Crafts-catalysed condensation of silylated uracil with a suitable sugar derivative (**85**) (Scheme 11). The chemistry and biochemistry of branched-chain sugars has recently been reviewed.[119]

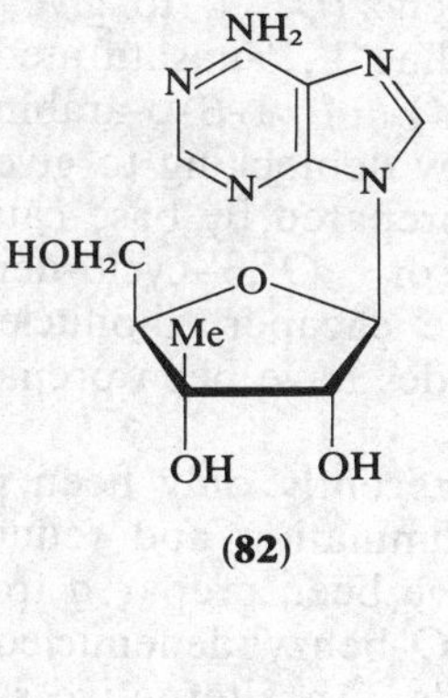

(**82**)

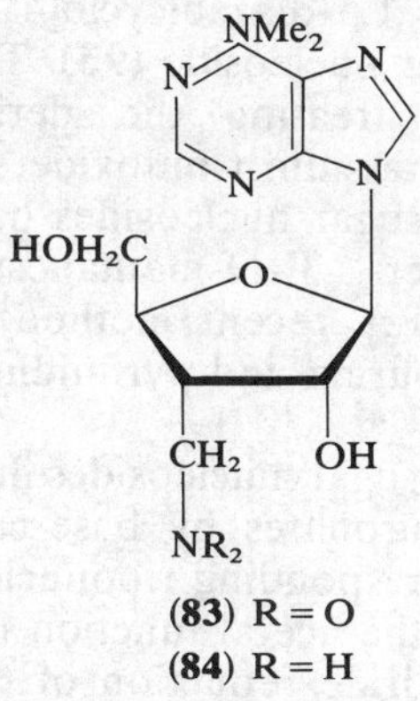

(**83**) R = O
(**84**) R = H

Rosenthal and co-workers[120] have exploited the alternative approach and prepared 9-(2′-*C*-nitromethyl)-β-D-lyxofuranosyladenine (**88**) by the ruthenium tetroxide oxidation of the nucleoside derivative (**86**), followed by reaction of the keto-derivative (**87**) with nitromethane in the presence of sodium methoxide.

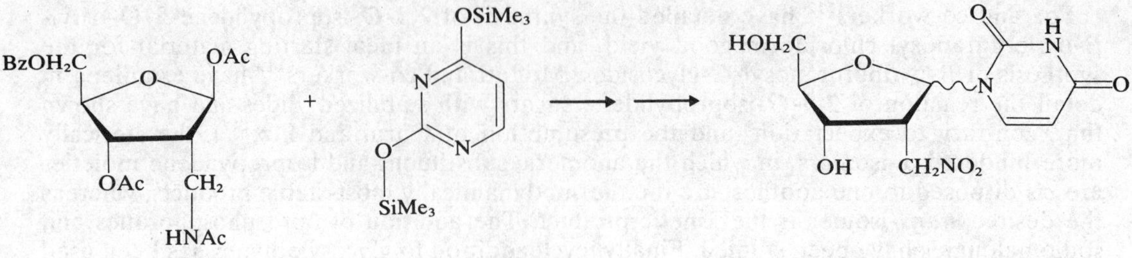

(85)

SCHEME 11

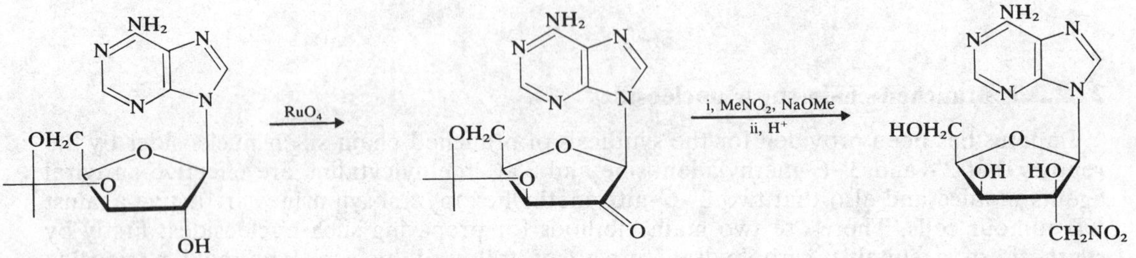

(86) **(87)** **(88)**

22.2.3.8 Unsaturated- and epoxy-sugar nucleosides

Nucleosides containing unsaturated sugars have attracted considerable interest, particularly since antibiotics such as augustmycin A (decoyinine) (**89**) have been shown to contain such structures. The compounds are also obviously of use as intermediates in the synthesis of other analogues and a compilation of recent references concerning this aspect is of value.[121]

1′,2′-Unsaturated-sugar purine and pyrimidine nucleosides are the most recently synthesized members of this group. The former were prepared from 2′,3′-*O*-methoxyethylideneadenosine (**91**), which with sodium iodide and pivaloyl chloride in pyridine gave, among other products, the 2′-iodo-derivative (**92**).[122] This was treated with permanganate to remove the 3′-enol ester group, which was then silylated and the elimination was performed with 1,5-diazabicyclo[4,3,0]non-5-ene (DBN) to give, after deblocking, the unsaturated-sugar nucleoside (**93**). The pyrimidine 1′,2′-unsaturated nucleosides have been prepared by treating the derivative (**94**) of 1-β-D-arabinofuranosyl-O^2,2′-cyclouridine with potassium t-butoxide, followed by deblocking to give (**95**).[123]

2′,3′-Unsaturated-sugar nucleosides have been prepared by base-catalysed elimination reactions of either 3′-*O*-methanesulphonyl or O^2,3′-cyclo-derivatives of 2′-deoxynucleosides.[124] A recent method[124] uses the cheaper ribonucleosides as starting material, and both purine and pyrimidine nucleosides have been prepared as outlined in Scheme 12.

3′,4′-Unsaturated-sugar nucleosides have until recently only been prepared from 2′-deoxynucleoside 5′-uronates by base-catalysed elimination and reduction of the ester function.[124] The corresponding ribonucleosides have been prepared from the base catalysed elimination of the acetal function from 2′,3′-*O*-benzylidenenucleoside-5′-aldehydes, followed by borohydride reduction of the aldehyde. An alternative synthesis of (**97**) in good yield has been achieved by the treatment of (**96**) with DBN in acetonitrile at 80 °C.[124]

Pyrimidine 4′,5′-ribo- and 2′-deoxyribo-nucleosides have been prepared from the corresponding nucleosides *via* dehydrohalogenation of the appropriately acylated 5′-iodo-5′-deoxynucleosides using either silver fluoride in pyridine or DBN.[125] The adenine

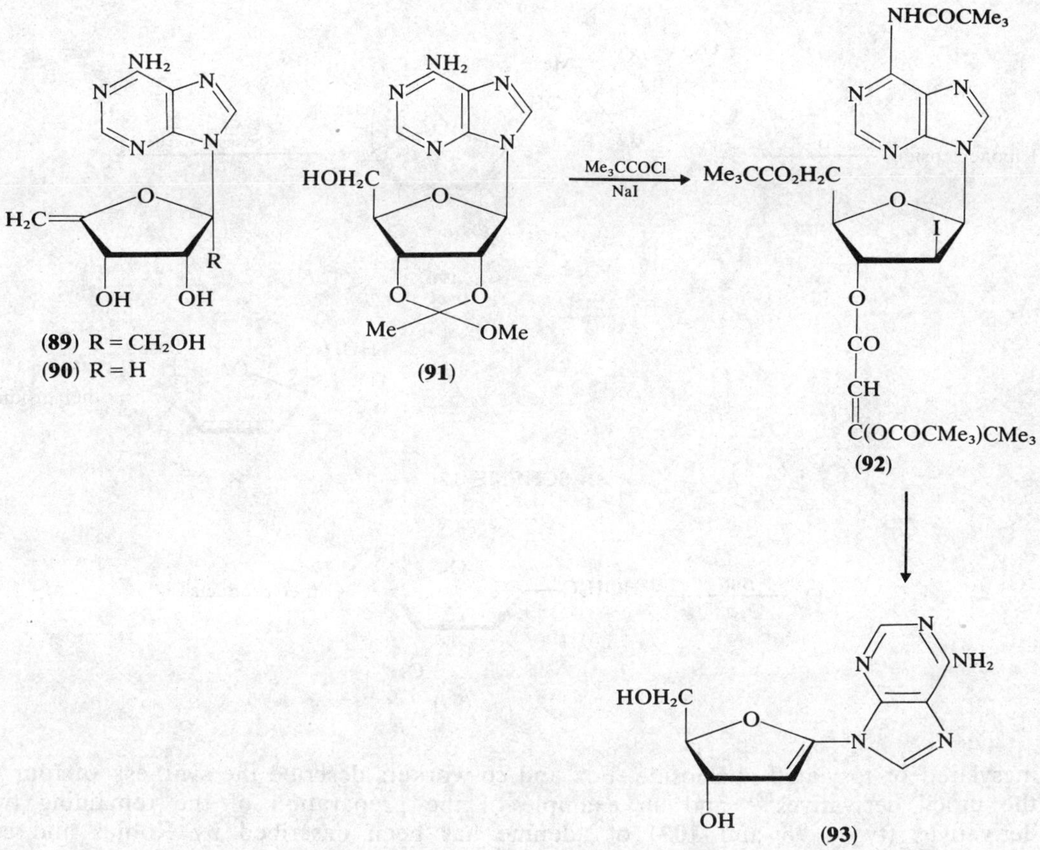

(**89**) R = CH_2OH
(**90**) R = H

(**91**)

(**92**)

(**93**)

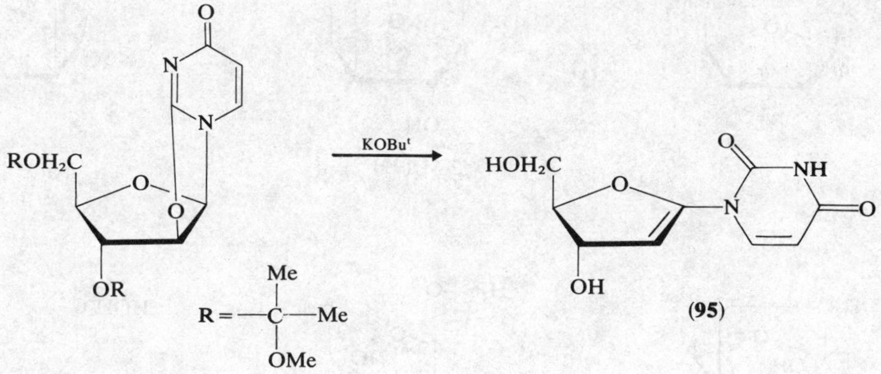

(**94**)

(**95**)

analogue (**90**) can be prepared by the reaction of 2′,3′-*O*-ethoxymethylidene-5′-*O*-(*p*-toluylsulphonyl)adenosine with potassium t-butoxide followed by a mild deblocking procedure.

As a result of investigations into the chemistry of simple ribofuranoside derivatives, methods for the synthesis of nucleoside epoxides have been available for a long time.[126] There are six possible epoxypentofuranosyl derivatives for each heterocyclic base, (**98**)–(**103**), and most of these can be obtained from the action of base on a suitably protected

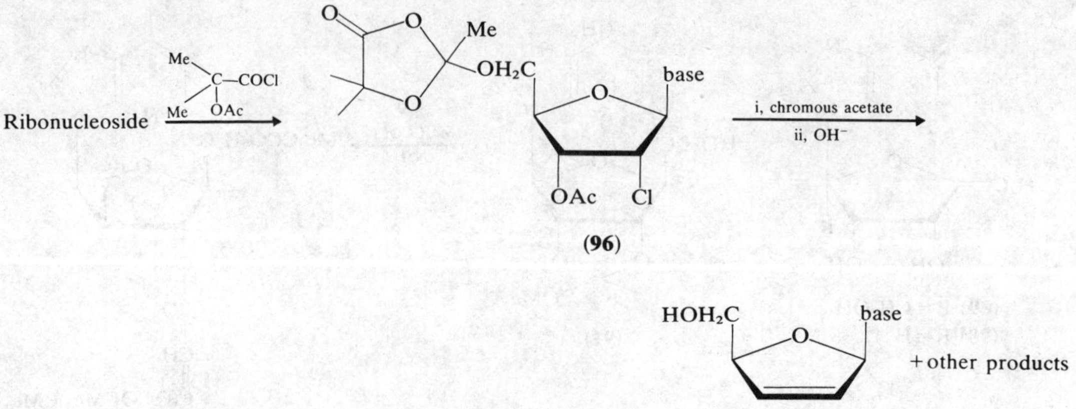

SCHEME 12

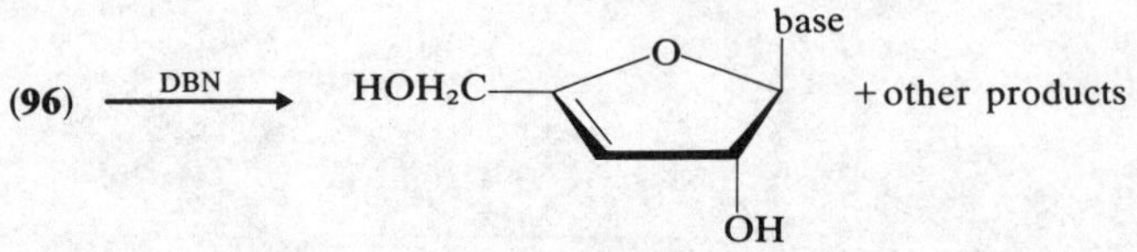

(97)

mesylated or tosylated nucleoside. Fox and co-workers describe the synthesis of four of the uracil derivatives[127] and an example of the preparation of the remaining two derivatives (types **98** and **103**) of adenine has been described by Robins and co-workers.[128]

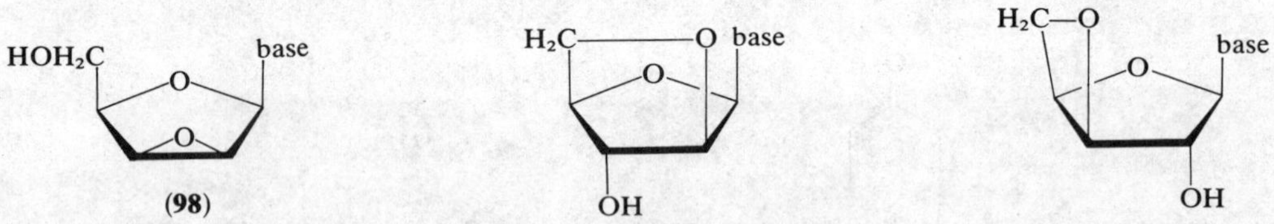

(98) **(99)** **(100)**

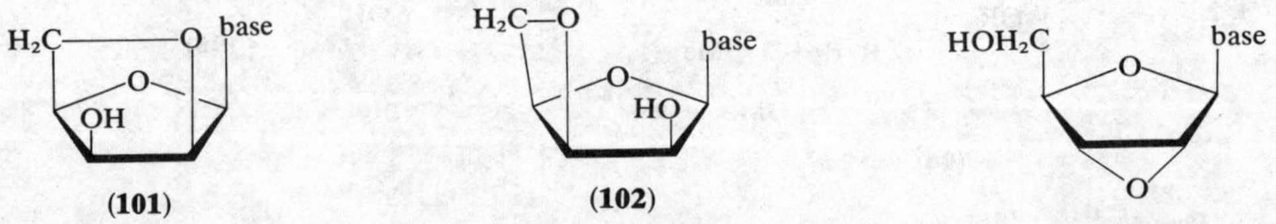

(101) **(102)** **(103)**

22.2.3.9 Acyclic-sugar and L-nucleosides

Almost all of the research into each of these two types of nucleoside has been done by one group of scientists. Acyclic-sugar nucleosides have been prepared by Wolfrom and Horton,[19,129] the L-nucleosides by Šorm and Holý.[130]

Acyclic-sugar nucleosides have all been prepared by condensing a suitable acyclic sugar (usually the dithioacetal of a glucose derivative) with a silylated heterocycle to give derivatives (**104**) as shown. Adenine, cytosine, uracil, and thymine derivatives have been made. An epimeric mixture at C-1′ is produced and it is possible to separate and to identify these by physical methods. The compounds so formed are potential antimetabolites and the conformations adopted by the acyclic sugar residue have been studied extensively.

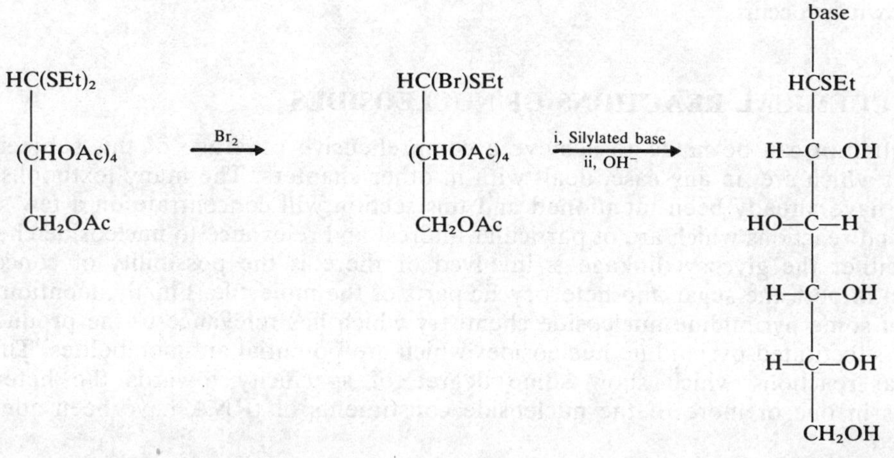

(**104**)

Synthetic procedures have been developed for the chemical synthesis of both purine and pyrimidine L-ribonucleosides and for pyrimidine 2′-deoxy-L-ribonucleosides, but there is no general method available for the synthesis of purine 2′-deoxy-L-ribonucleosides in significant quantities from readily-available starting materials.

Both purine and pyrimidine L-ribonucleosides have been prepared by the reaction of the sodium salt of the heterocycle in DMF with 2-*O*-tosyl-5-*O*-trityl-L-arabinose (**105**), which is obtainable from L-arabinose (a readily available L-sugar) in several steps (Scheme 13). Alternatively, pyrimidine L-ribonucleosides and the pyrimidine 2′-deoxy-L-ribonucleosides can be obtained from the O^2,2′-anhydro-L-uridine, which is prepared from L-arabinose *via* the amino-oxazoline (**18**) as previously described.

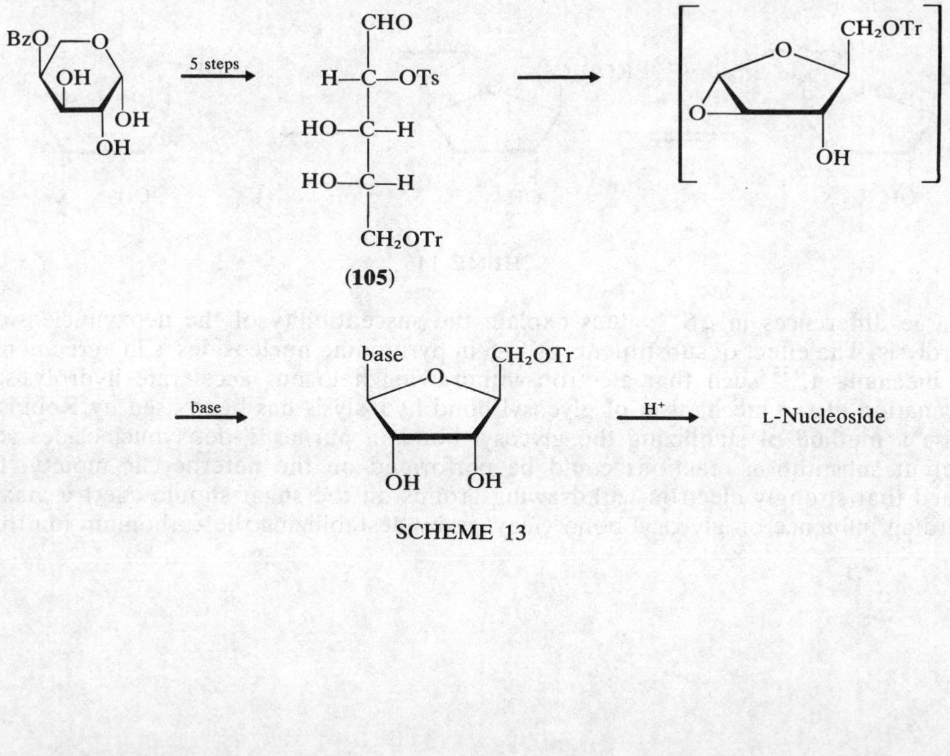

SCHEME 13

In bacterial cell walls there is a strong barrier to L-nucleoside penetration, which is governed by a permease protein system so that L-nucleosides neither penetrate nor interfere with the penetration of other nucleosides. Surprisingly, evidence for the synthesis of di- and tri-nucleotides has been obtained from the action of polynucleotide phosphorylase from *Micrococcus luteus* on the 5′-diphosphate of L-adenosine.[131] L-Nucleosides have also been given to rats, and although there is no significant incorporation into DNA or RNA, some nucleoside diphosphate is produced. Similarly, the application of L-uridine results in the production of L-cytidine *in vivo*, a process which in the natural D-series was not known to occur.

22.2.4 CHEMICAL REACTIONS OF NUCLEOSIDES

No attempt can be made to achieve a comprehensive coverage of this subject, many areas of which are, in any case, dealt with in other chapters. The many textbooks on the subject have already been mentioned and this section will concentrate on a few selected topics and reactions which are of particular interest and relevance to nucleoside chemistry. Thus, either the glycosyl linkage is involved or there is the possibility of concomitant reaction in both the sugar and heterocyclic parts of the molecule. Finally, mention will be made of some pyrimidine nucleoside chemistry which has relevance to the production of some 5-substituted pyrimidine nucleosides which are potential antimetabolites. The many chemical reactions which show some degree of specificity towards the heterocyclic residues in one or more of the nucleoside constituents of tRNA have been adequately reviewed.[8]

22.2.4.1 Glycosyl bond hydrolysis

In recent years there has been some progress towards an explanation for the well-known difference in rate of glycosyl bond hydrolysis between the ribo- and deoxyribonucleosides. The former are more stable than the latter and, in each class, the purine bases are hydrolysed much more rapidly than are the pyrimidines. Some kinetic data are now available for the common nucleosides over a wide pH range and the data can now be rationalized in a mechanism involving a pre-equilibrium protonation of the heterocyclic ring, followed by a rate-limiting ionization of the glycosyl bond and not, as is quoted in most textbooks, *via* a Schiff's base intermediate following initial protonation of the sugar ring-oxygen (Scheme 14).[132,133]

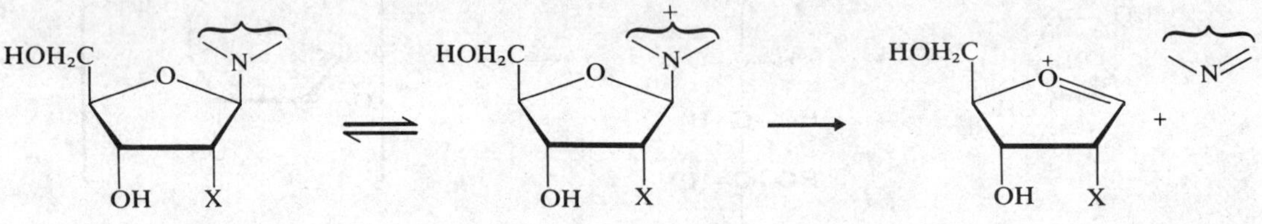

SCHEME 14

Large differences in $\Delta S^{\ddagger}$ values explain the susceptibility of the deoxynucleosides to hydrolysis. The effect of substituents at C-5 in pyrimidine nucleosides is in agreement with this mechanism,[133] such that electron-withdrawing halogens accelerate hydrolysis. This explanation of the mechanism of glycosyl bond hydrolysis has been used by Robins[134] to devise a method of stabilizing the glycosyl bond in purine 2′-deoxynucleosides so that different substitution reactions could be performed on the heterocyclic moiety. It was argued that strongly electron-withdrawing groups on the sugar should exert a maximum retarding influence on glycosyl bond cleavage by destabilizing the carbonium ion formed.

Thus 2′-deoxyinosine (**106**) was converted into the 3′,5′-bis-*O*-trifluoroacetate (**107**). This could then be converted into the 6-chloro-derivative (**108**) in 81% yield, whereas, under identical conditions, 3′,5′-di-*O*-acetyl-2′-deoxyinosine (**109**) gave over 90% of the base, hypoxanthine.

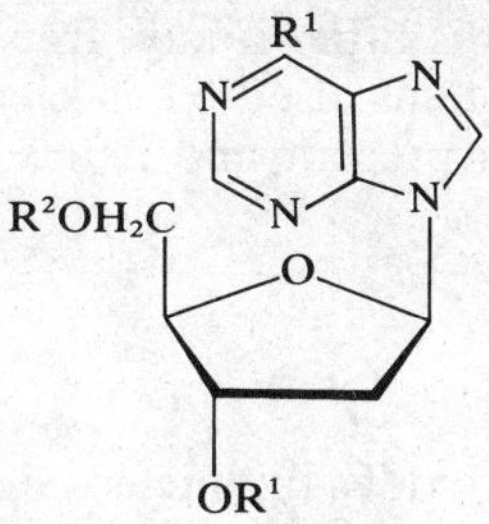

(**106**) $R^1 = OH$, $R^2 = H$
(**107**) $R^1 = OH$, $R^2 = CF_3CO$
(**108**) $R^1 = Cl$, $R^2 = CF_3CO$
(**109**) $R^1 = OH$, $R^2 = MeCO$

22.2.4.2 Oxidation

(*i*) *Sugar group*

5′-Carboxylic acids of nucleosides (**110**) have been prepared by using Pt and oxygen[135] or the chromium trioxide–pyridine complex[136] on the free nucleosides. The yields with the latter method are not good, however, owing to the concomitant oxidation of the C-3′ hydroxyl group followed by a base-catalysed β-elimination of the heterocyclic base from the 3′-ketonucleoside. Oxidation of 2′,3′-*O*-isopropylideneadenosine with potassium permanganate under basic conditions gives good yields of the 5′-carboxylic acid,[137] but this method is restricted to adenine nucleosides because, as will be seen later, all other heterocyclic groups of common nucleosides are readily oxidized by permanganate.

Nucleoside-5′-aldehydes (**111**), which are of greater synthetic use than the 5′-carboxylic acids, have been prepared by the action of the Pfitzner–Moffatt reagent[138] (*N,N*′-dicyclohexylcarbodi-imide–dimethyl sulphoxide) on the 2′,3′-*O*-isopropylidene nucleoside derivatives; 3′-*O*-acetylthymidine undergoes a similar reaction. The aldehydes so produced are rather unstable. The 5′-aldehydes of adenosine and uridine have also been produced following hydrolysis of the photolysis product of the corresponding nucleoside-5′-azides.[139] Although it has not apparently been used as yet in the nucleoside field, it would appear that pyridinium chlorochromate would also be an effective method of producing 5′-aldehydes and the 2′- and 3′-ketonucleosides.[140] As already mentioned, 3′-ketonucleosides are unstable in basic conditions and hence the preparation of '3′-ketouridine' (**112**) necessitates the action of the Pfitzner–Moffatt reagent (or, better, using dimethyl sulphoxide with either acetic anhydride or phosphorus pentoxide) on 2′,5′-di-*O*-trityluridine. The application of similar methods has resulted in the preparation of 2′-ketouridine and other pyrimidine and purine ketonucleosides.

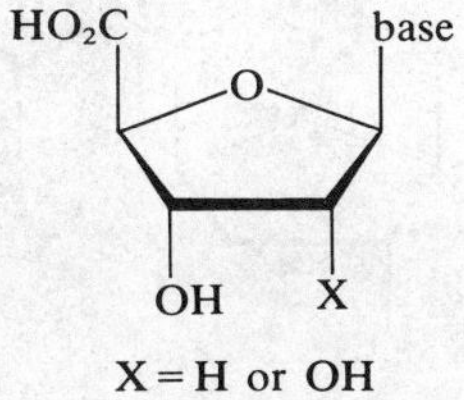

X = H or OH

(**110**)

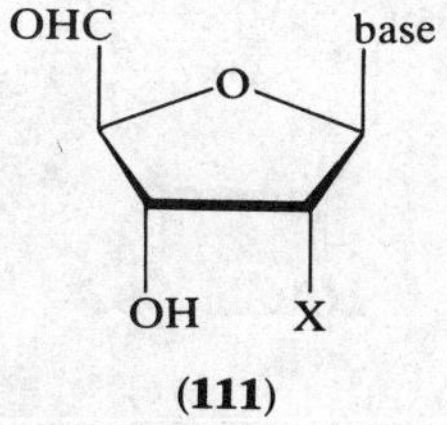

(**111**)

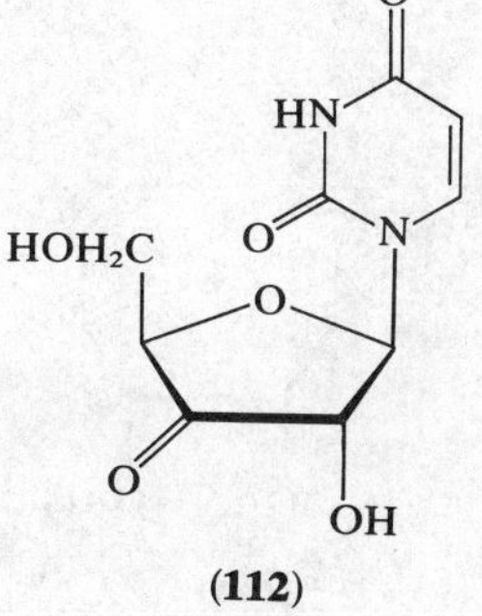

(**112**)

The reaction of periodate with a nucleoside to give the so-called 'nucleoside dialdehyde' is a well-known reaction and has been used, among other things, for the production of amino-sugar nucleosides,[141] and as the basis of methods for the determination of base sequence[142] and base composition[143] of polyribonucleotides. It was originally thought that the dialdehydes were not stable, but recently the dialdehydes from the four common ribonucleosides have been isolated in a solid and stable form.[144,145] However, the structure of these compounds is not clear. There is little, if any, free aldehyde group and although they are homogeneous in all t.l.c. systems used, it is clear from their n.m.r. spectra that a mixture of sugar configurations is present and it has been suggested that polymers are produced.[145]

(*ii*) *Heterocyclic group*

The action of hydrogen peroxide or organic peroxides on nucleosides leads to *N*-oxides, some of which have been found to be carcinogenic.[146] Thus, adenosine with 30% hydrogen peroxide in acetic acid gives the 1-*N*-oxide (**113**). Owing to their acid lability, monoperphthalic acid is required for the production of the corresponding derivatives of the purine deoxynucleosides. Cytidine N^3-oxide (**114**) is produced by the action of *M*-chloroperbenzoic acid in acetic acid on cytidine.

Adenine nucleosides are relatively inert to the action of alkaline potassium permanganate, whereas the other nucleosides are oxidized at varying rates to a complex mixture of products. 2′,3′,5′-Tri-*O*-acetylguanosine gives 2,3,5-tri-*O*-acetylribofuranosylurea (**115**) in low yield and guanidine.[147] 2′,3′,5′-Tri-*O*-benzoylcytidine gives *N*-(2,3,5-tri-*O*-benzoylribofuranosyl)-*N*′-oxalylurea[148] (**116**), and thymine gives 5,6-dihydroxy-5,6-dihydrothymidine[149] (**117**).

Similar products are undoubtedly formed from the oxidation of other pyrimidine nucleosides, but they have proved to be too unstable to be isolated. Unless steps are taken to protect the sugar hydroxyl groups, degradation of the sugar moiety occurs with elimination of the base. The final nitrogen-containing products of oxidation are the glycosylureas, glycosylbiurets (from cytidine), and guanidine (from guanine).

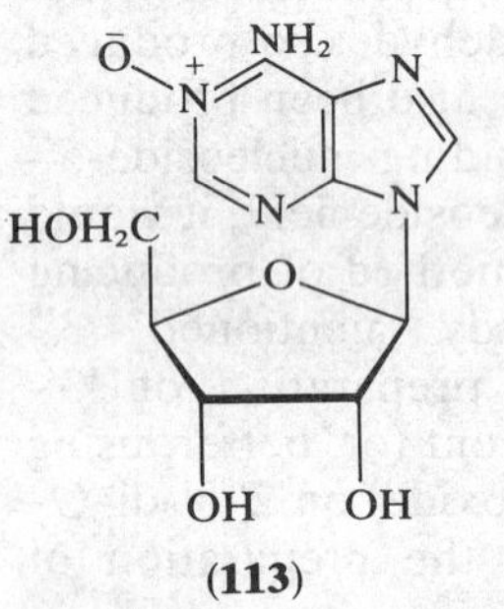

(**113**)

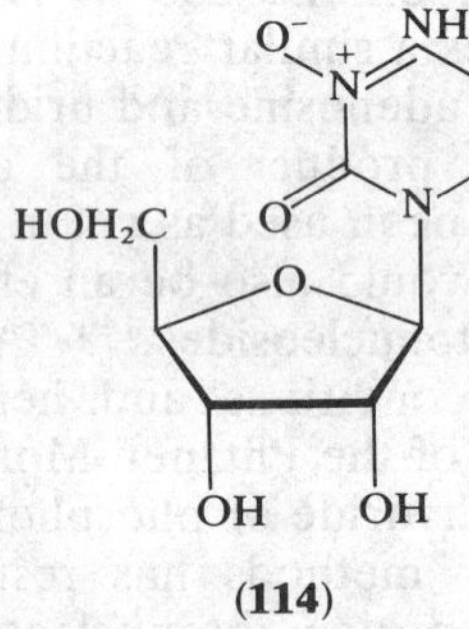

(**114**)

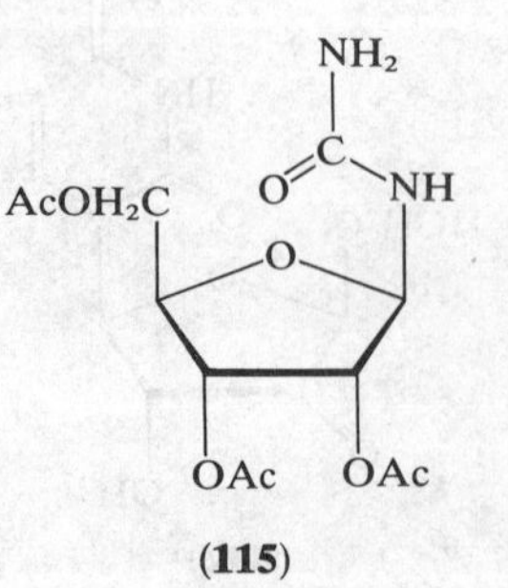

(**115**)

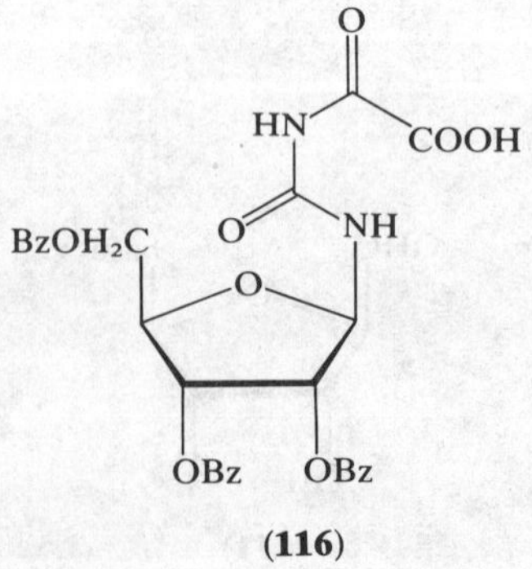

(**116**)

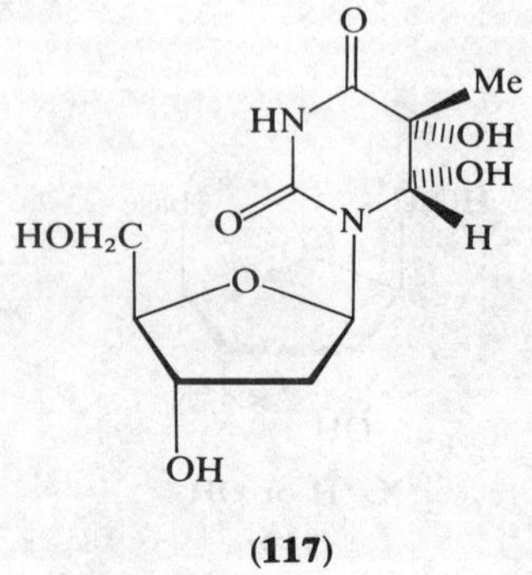

(**117**)

Osmium tetroxide oxidizes thymine residues much faster than the other bases, and 5,6-dihydroxy-5,6-dihydrothymidine has also been isolated from this reaction.[149] Compound (**118**) and the cyclic osmate ester have been isolated from the reaction of osmium tetroxide with 3′,5′-di-*O*-acetylthymidine in benzene.[150] Potassium peroxidisulphate, $K_2S_2O_8$, has a different specificity in that guanosine and deoxyguanosine are oxidized 500 times faster than any other nucleoside to give products similar to those obtained from a permanganate oxidation.[151]

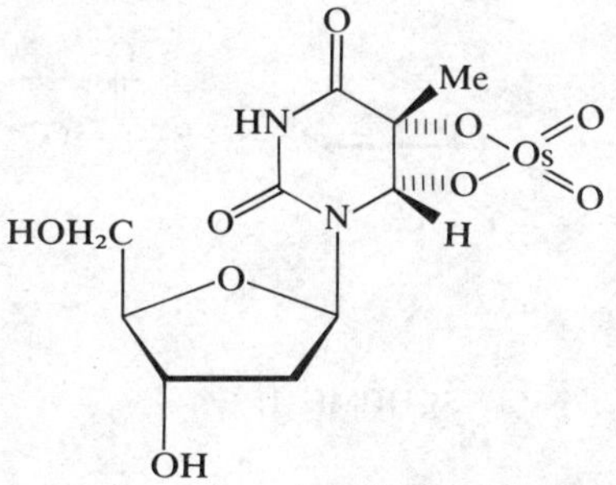

(**118**)

22.2.4.3 Acylation

Acylated derivatives of nucleosides and nucleotides are widely used in polynucleotide synthesis and the use of these derivatives will be found in that section. Khorana has pioneered the way in the preparation of suitable base-acylated derivatives and details of his syntheses have been published.[152,153] In general, exocyclic amino-groups are acylated under conditions which acylate the sugar hydroxyls as well, and the latter can be deacylated under mild conditions to leave the *N*-acylated nucleosides. The action of benzoyl chloride in pyridine on uridine gives 2′,3′,5′-tri-*O*-benzoyluridine, but with an excess of the reagent, N^3-benzoyl-2′,3′,5′-tri-*O*-benzoyluridine is produced. The *N*-benzoyl group is resistant to acidic hydrolysis. Cytidine and adenosine on vigorous acylation give pentabenzoyl derivatives and there is evidence to suggest that two acyl groups are attached to the exocyclic nitrogen atom.[154] N^4-Acetylcytidine can be prepared directly from cytidine by the selective method described by Fox and co-workers.[155] This has been modified to give more consistent results and the preparation of other N^4-acylated cytidines has been described.[156] N^4-Benzoyl-2′,3′,5′-tri-*O*-benzoylcytidine (or, better, N^4-acetyl-2′,3′,5′-tri-*O*-benzoylcytidine) can be selectively N^4-deacylated under mild acidic conditions and, contrary to previous reports, no deamination or glycosyl bond cleavage could be detected.[157] The former compound can also be selectively N^4-debenzoylated by treatment with phenol.[158] In a similar way, N^6,N^6-dibenzoyl-2′,3′,5′-tri-*O*-benzoyladenosine could be selectively N^6-debenzoylated[158] or selectively 2′-*O*-debenzoylated, the latter reaction being accomplished by hydrazinolysis.[159] Hydrazine also selectively 2′-*O*-debenzoylates N^2-benzoyl-2′,3′,5′-tri-*O*-benzoylguanosine and 2′,3′,5′-tri-*O*-benzoylinosine. Partial selectivity for the 2′-position can also be obtained with fully benzoylated uridine and cytidine.

The protection of alcoholic hydroxyl groups as esters is a widely used procedure, particularly in the carbohydrate and steroid fields,[160] and mention will only be made of one or two derivatives which have found particular use in nucleoside chemistry. Acetyl and benzoyl protecting groups have been used most frequently and they can be removed by alkaline hydrolysis or ammonolysis. Base-sensitive methoxy- and aryloxy-acetyl groups,[161] 2,2,2-tribromoethoxycarbonyl[162] (removed by reduction), 3-benzoylpropionyl[163] (removed by hydrazinolysis), laevulinyl[164] (removed by sodium borohydride or hydrazine in pyridine–acetic acid[165]), and dihydrocinnamyl esters[166] (removed by chymotrypsin) have all been used. The use of tosyl and mesyl esters in the preparation of cyclonucleosides and epoxynucleosides has already been mentioned. Some selectivity for the 5′-hydroxyl group can be achieved with tosyl chloride and the same selectivity has been claimed for the treatment of thymidine with diethyl azodicarboxylate, triphenylphosphine, and benzoic acid in hexamethylphosphoric triamide (Scheme 15).[167]

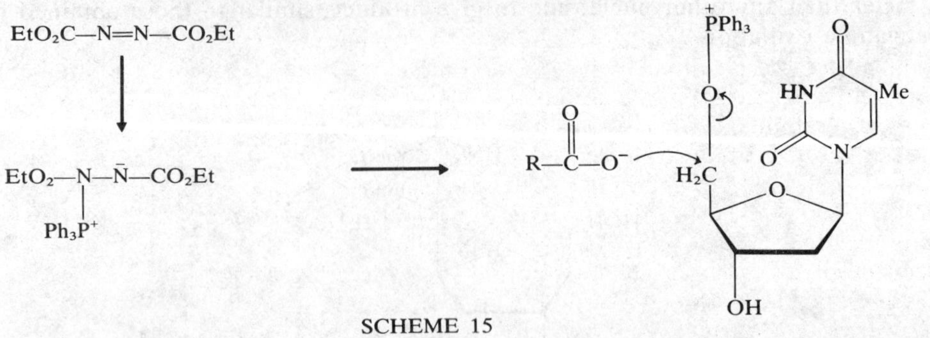

SCHEME 15

The pivaloyl (trimethylacetyl) group is also selective for the 5′-hydroxyl group. The reaction of ribonucleosides with dibutyltin oxide gives crystalline 2′,3′-*O*-(dibutylstannylene)nucleosides (**119**),[168] which on reaction with acyl chlorides and anhydrides give 3′-*O*-acyl derivatives, while reaction with tosyl chloride selectively yields the 2′-tosylnucleosides. Monoacetylation of 5′-*O*-substituted derivatives of uridine and adenosine gives a mixture of the 2′-*O*- and 3′-*O*-acetates. An equilibrium mixture favouring the 3′-*O*-acetate can be prepared by boiling the mixture in pyridine and by crystallization; an almost quantitative yield of 3′,5′-di-*O*-acetyluridine can be obtained by the partial acetylation of 5′-*O*-acetyluridine.[169]

The specific preparation of a 2′(3′)-*O*-monoacylnucleoside is best performed by the acidic hydrolysis of the corresponding 2′,3′-cyclic orthoesters (Scheme 16).[170] Tetra-acetoxysilane, $Si(OAc)_4$, reacts with 5′-*O*-acetyluridine to give 3′,5′-di-*O*-acetyluridine as the major product.[171]

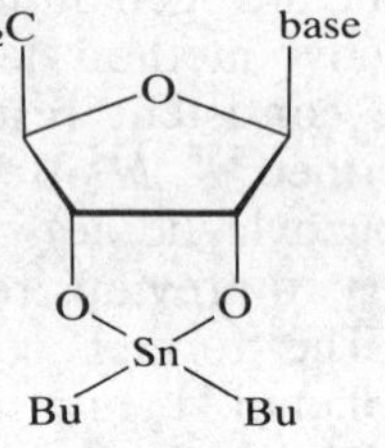

(**119**)

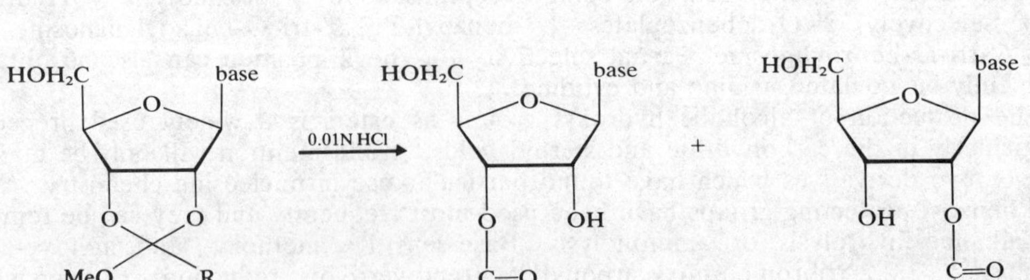

SCHEME 16

22.2.4.4 Alkylation

Because of the relevance to the mutagenicity and carcinogenicity of alkylating agents, there is an enormous amount of data available on the alkylation of nucleosides. Space, however, permits only a brief survey of the methods available for the production of alkyl derivatives of synthetic utility, and for other aspects of the subject the relevant review articles should be consulted.[133,172]

There are potentially so many sites in a nucleoside molecule which can be alkylated that, unless precautions are taken, a very complex mixture of products is obtained. This means that the isolation of any particular compound in reasonable yield, despite the use of Dowex OH^- columns,[29] is more a matter of luck than planning. Even the reasons for the differences found in the extent of alkylation of the various positions in a nucleoside are not fully understood. If one considers the ring nitrogen atoms which are available for alkylation in the purine and pyrimidine nucleosides, the order of basicity of these positions, as measured by pK_a, is not directly related to their relative ease of alkylation. It has been pointed out[173] that protonation is an equilibrium process, whereas alkylation is irreversible and thus subject to kinetic control and the influence of steric factors.

Thus, alkylation of nucleosides with diazomethane, alkyl halides, or dimethyl sulphate gives a complex mixture of *O*′-, *O*-, ring *N*-, and exocyclic *N*-alkylated products, although in general the most reactive site towards esters of strong acids in guanosine is N-7, in adenosine is N-1, and in cytidine is N-3, whereas uridine reacts only very slowly. However, under strongly alkaline conditions with dialkyl sulphates, *O*′-alkylated products only are produced,[174] and although complex mixtures of such compounds can still result, this is one of the easiest ways to produce such *O*′-alkylated nucleosides. This reaction has been used for adenine- and cytosine- (and hence uridine- by deamination) containing nucleosides, and also for the reaction of an alkyl halide on 2′,3′-*O*-isopropylidene-uridine,[175] to give almost entirely *O*′-alkylated products. This has been rationalized in terms of the hardness and softness of the reactive centres and correlation of the reaction at a hard centre with an S_N1 type mechanism and at a soft centre with an S_N2 reaction. Strong aqueous alkaline reactions with alkylating agents are of the S_N1 type, which thus favours reaction at the hard *O*′-positions.

Diazomethane in the presence of catalytic amounts of $SnCl_2$ also gives exclusively 2′-*O*- and 3′-*O*-alkylated nucleosides from a ribonucleoside, but the method, although very convenient, is limited to monoalkylation of the *cis*-hydroxyls of nucleosides.[176] Selective *O*′-alkylation of ribonucleosides can also be achieved by the action of alkyl halides on the intermediate tin derivative previously described.[168] 2′-*O*-(*o*-Nitrobenzyl)uridine has also been prepared *via* the tin derivative; this group can be removed photochemically and has been used in oligoribonucleotide synthesis.[177]

Other ethers of synthetic use are the benzyl ethers, which can be removed by hydrogenation, and trityl ethers (and their *p*-methoxy-derivatives), which are selective for the 5′-hydroxyl group and are susceptible to a range of mild acidic hydrolyses.

Silyl ethers have been used to give nucleoside *O*′-alkylsilyl ethers.[178] For ribonucleosides, the most useful reagents are t-butyldimethylsilyl chloride and tri-isopropylsilyl chloride, and from uridine the 2′- and 2′,5′-protected nucleosides have been prepared. Removal of the silyl groups is achieved by the action of Bu^n_4NF in tetrahydrofuran, which in general does not affect other acid- or base-labile protecting groups.

22.2.4.5 Acetal formation

These derivatives are often used as blocking groups in oligoribonucleotide synthesis. Like ordinary alcohols, nucleosides react with vinyl ethers under acid-catalysed conditions. The first such reaction to be described was the reaction of dihydropyran with uridine-3′,5′-cyclic phosphate (**120**) to give 2′-*O*-tetrahydropyranyl-3′,5′-cyclic phosphate

(**121**).[179] This was then converted into 5′-*O*-trityl-2′-*O*-tetrahydropyranyluridine-3′-phosphate (**122**), which was used in the first oligoribonucleotide synthesis. The tetrahydropyranyl group can be removed with dilute acid. One drawback is that a mixture of diastereoisomers is formed, which often means that crystallization of the product is difficult. This has been overcome by the use of 4-methoxy-5,6-dihydropyran (**123**).[180] This, being a symmetrical group, does not give diastereoisomers and it can be removed very easily under acidic conditions. However, the starting material is not readily available. Reactions with several other vinyl ethers have been described.[9]

Probably the most widely used reaction is the preparation of 2′,3′-*O*-isopropylidene derivatives of ribonucleosides (**124**) by the reaction of a free nucleoside with acetone under acidic catalysis. Many different such cyclic acetals have been prepared, covering a wide range of stability to acidic hydrolysis.[9]

The reaction of benzaldehyde in the presence of trifluoroacetic acid with a ribonucleoside leads to the production of a diasteriomeric mixture of 2′,3′-*O*-benzylidene-nucleosides (**125**). These are readily hydrolysed under mild acidic conditions and again, by introducing substituents into the aromatic ring, the rate of hydrolysis can be altered by a factor of 1000.[9]

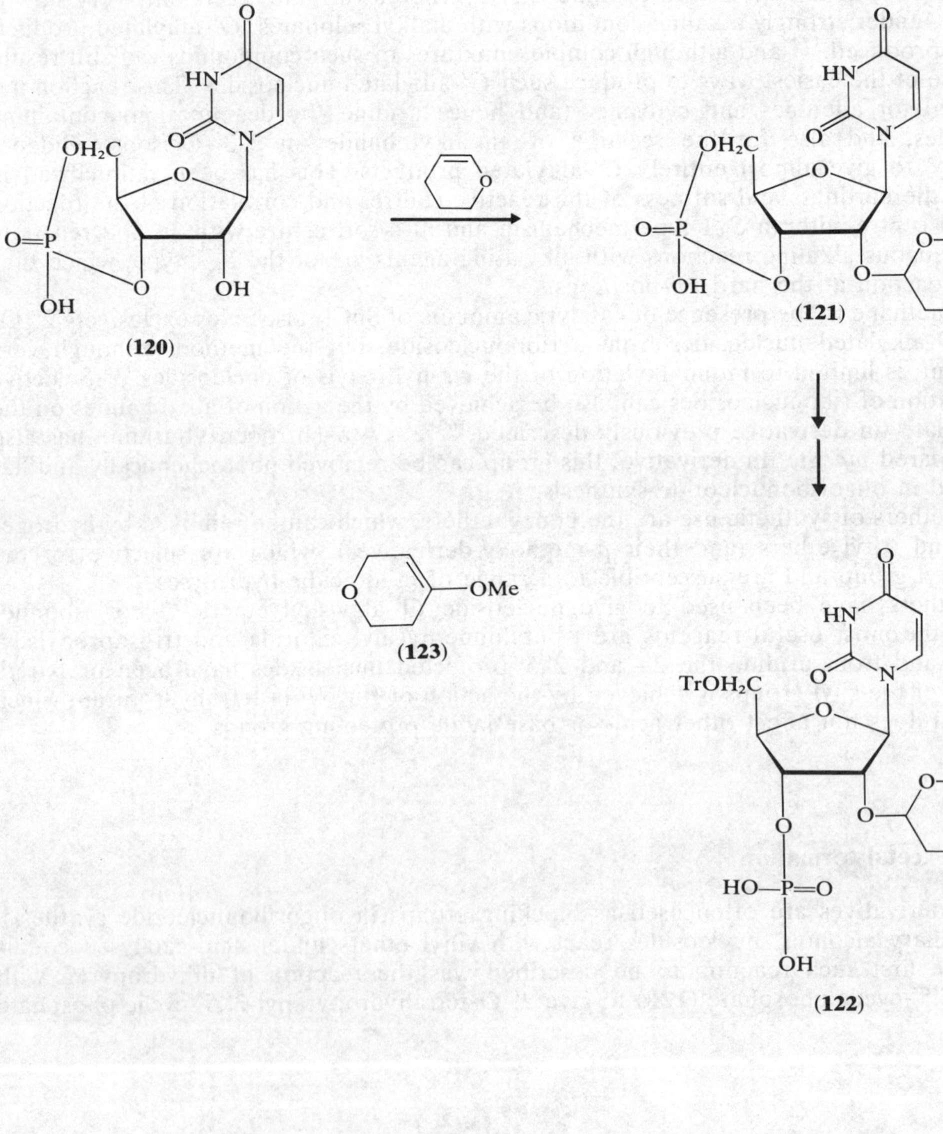

(**120**) (**121**) (**123**) (**122**)

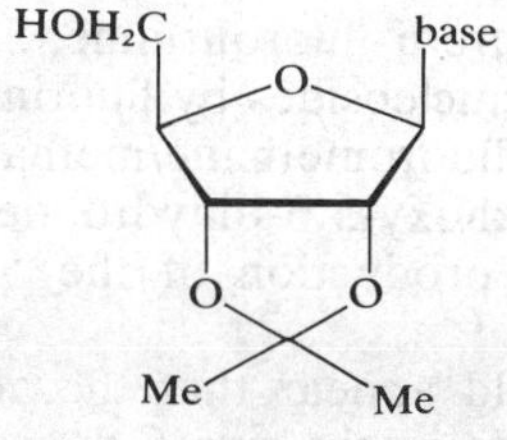

(124)

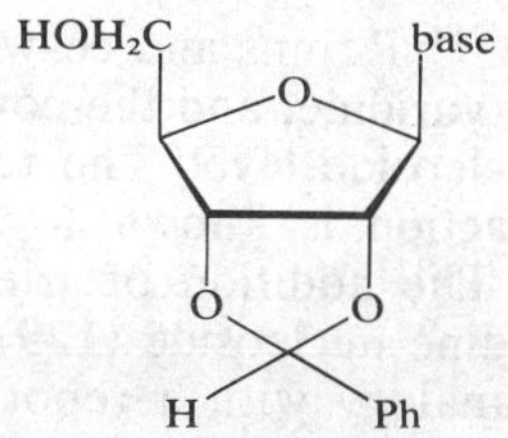

(125)

22.2.4.6 Synthesis and properties of some pyrimidine (particularly 5-substituted) nucleosides

A variety of 5-substituted pyrimidine nucleosides exhibit a distinct antiviral activity in mammalian cell cultures and experimental animals. This section will deal with the synthesis and properties of such compounds and, in doing so, much of the chemistry of pyrimidine nucleosides will be covered. The potential uses in medicine for nucleoside analogues will be examined.

5-Mercuri-[181] and 5-palladium-uridines[182] have been used as the precursors for the introduction of substituents into the 5-position of the heterocyclic ring. Thus, 5-mercurated uridines could be converted to the 5-tritium labelled uridine by reduction with sodium borotritiide. The action of *N*-bromosuccinimide resulted in the production of the 5-bromo-derivative and iodine gave the 5-iodonucleosides. The organo-palladium intermediates have to be generated *in situ* from the mercurinucleosides and have been used for the introduction of carbon chains at the C-5 position. Thus, 5-ethyl-2′-deoxyuridine and 5-allyluridine could be synthesized in 68 and 78% yields, respectively. The reaction of uridine, but not 5-bromouridine, with a perfluoroalkylcopper complex gives the 5-(perfluoroalkyl)uridine, but no mechanism has been suggested for this reaction.[183]

The C-5 position of pyrimidine nucleosides is of course the most aromatic position in the molecule, and electrophilic aromatic substitution reactions might be expected to yield C-5 derivatives. However, as will be seen later, because of the ease with which addition to and subsequent elimination from the 5,6-double bond occurs in pyrimidine nucleosides, it is often difficult to distinguish between these mechanisms, and, in any case, the exact mechanism of many of the reactions has not been studied in any detail.

Thus, 5-nitrouridine can be obtained from a suitably protected uridine under classical aromatic nitration conditions[184] and is presumably an example of this type of reaction. Hydroxymethylation of uridine with formaldehyde gives a 20% yield of 5-hydroxymethyluridine,[9] but a better method is the base-catalysed hydroxymethylation of 2′,3′-*O*-isopropylideneuridine.[185] Apart from a comment that the reaction is reversible, no-one seems to be prepared to commit themselves to a mechanism for this reaction.[9]

The mechanism of halogenation is even more confused. Under non-aqueous conditions, chlorination and bromination take place very readily to give the 5-halogenopyrimidine nucleosides, but iodination requires heating with nitric acid for several hours. Because of this and the relative lability of the glycosyl bond in cytosine-containing nucleosides, iodination of these has to be achieved with iodine in chloroform in the presence of HIO_3, but reaction to give 5,6-dihydro-products is also found to occur.[9] *N*-Haloimides and ICl also give 5-iodo-derivatives. All these reactions, when performed under anhydrous conditions, are presumed to proceed by electrophilic aromatic substitution mechanisms, but in the presence of water, addition across the 5,6-double bond occurs and the 5-halogeno-6-hydroxy-5,6-dihydropyrimidine nucleosides can be isolated. Thus, bromination of uridine with bromine water gives 5-bromo-6-hydroxy-5,6-dihydrouridine (**126**), but on heating the elements of water can be eliminated and the 5-bromo-derivative (**127**) isolated.

A direct synthesis of 5-fluoropyrimidine nucleosides obviously has tremendous commercial importance, and using the reagent trifluoromethyl hypofluorite developed by

Barton,[186] Robins and co-workers[134] have managed to prepare 5-fluorouridine, 5-fluoro-2′-deoxyuridine, and the corresponding cytosine-containing nucleosides by fluorination at the nucleoside level. The reaction is performed in trichlorofluoromethane/methanol and the reaction is known to proceed *via* the 5-fluoro-6-methoxy-5,6-dihydro derivative (**128**). The addition of triethylamine is necessary for the production of the 5-fluoropyrimidine nucleoside (**129**).

By analogy with a reported reaction with uracil, it would appear that the action of fluorine diluted with nitrogen in glacial acetic acid on uridine might give 5-fluorouridine with the intermediate production of the 5-fluoro-6-hydroxy-5,6-dihydro-derivative.[187]

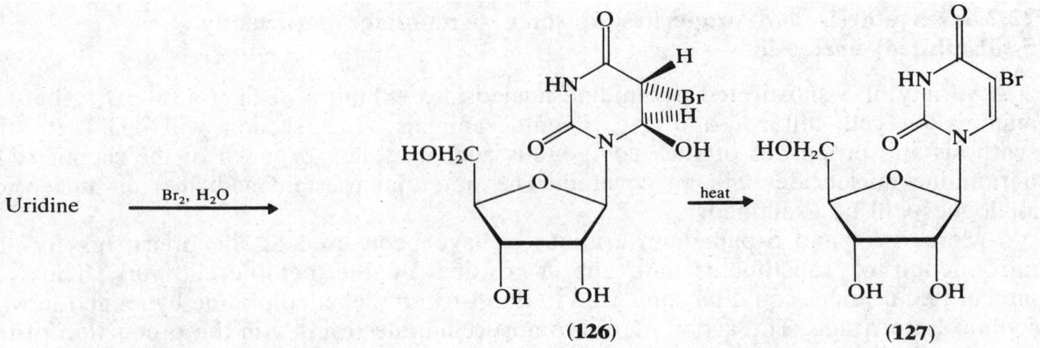

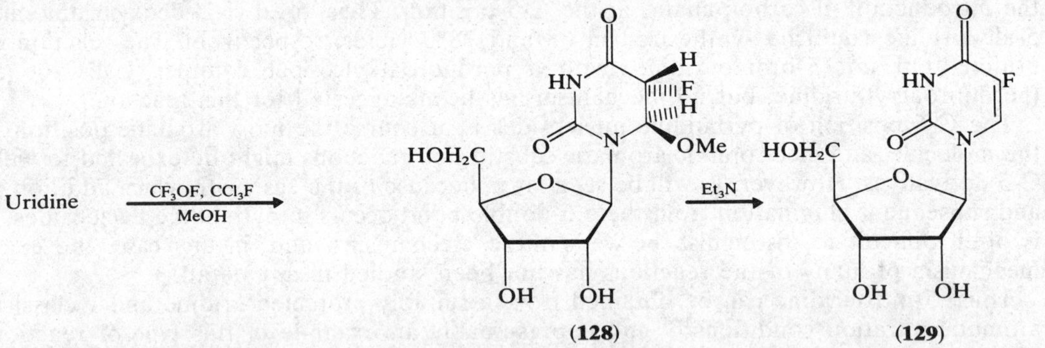

Direct thiocyanylation of uridine and 2′-deoxyuridine has been achieved with the pseudo-halogen chlorothiocyanogen, ClSCN.[188] 5-Mercaptopyrimidines can be easily prepared by reduction of the thiocyano-derivatives with dithiothreitol. 5-Mercaptopyrimidine nucleosides have also been produced directly by the reaction of methyl hypobromite in methanol on the pyrimidine nucleoside. This is subsequently reacted with sodium hydrosulphide in dimethylacetamide. Once again the reaction goes *via* the 5,6-dihydro-derivative, although the actual course of the reaction with uracil- and cytosine-containing nucleosides is different (Scheme 17).

Two well-documented reactions of a pyrimidine nucleoside, both of which involve the production of 5,6-dihydro-intermediates, are the reactions of cytosine derivatives with hydroxylamine, and uridine and cytidine derivatives with bisulphite. Both subjects have been authoritatively reviewed[88,133] and the action of bisulphite has already been detailed (p.75). The reaction of hydroxylamine with cytosine moieties can be rationalized in terms of Scheme 18.

The reaction with bisulphite has also been used for isotope exchange at C-5 in uridine and cytidine and the dehalogenation of 5-bromouridine. Cysteine can also cause isotope exchange at C-5,[189] and dehalogenation of 5-bromouridine at pH 8.9,[190] and these reactions are also thought to proceed *via* a 5,6-dihydro-intermediate.

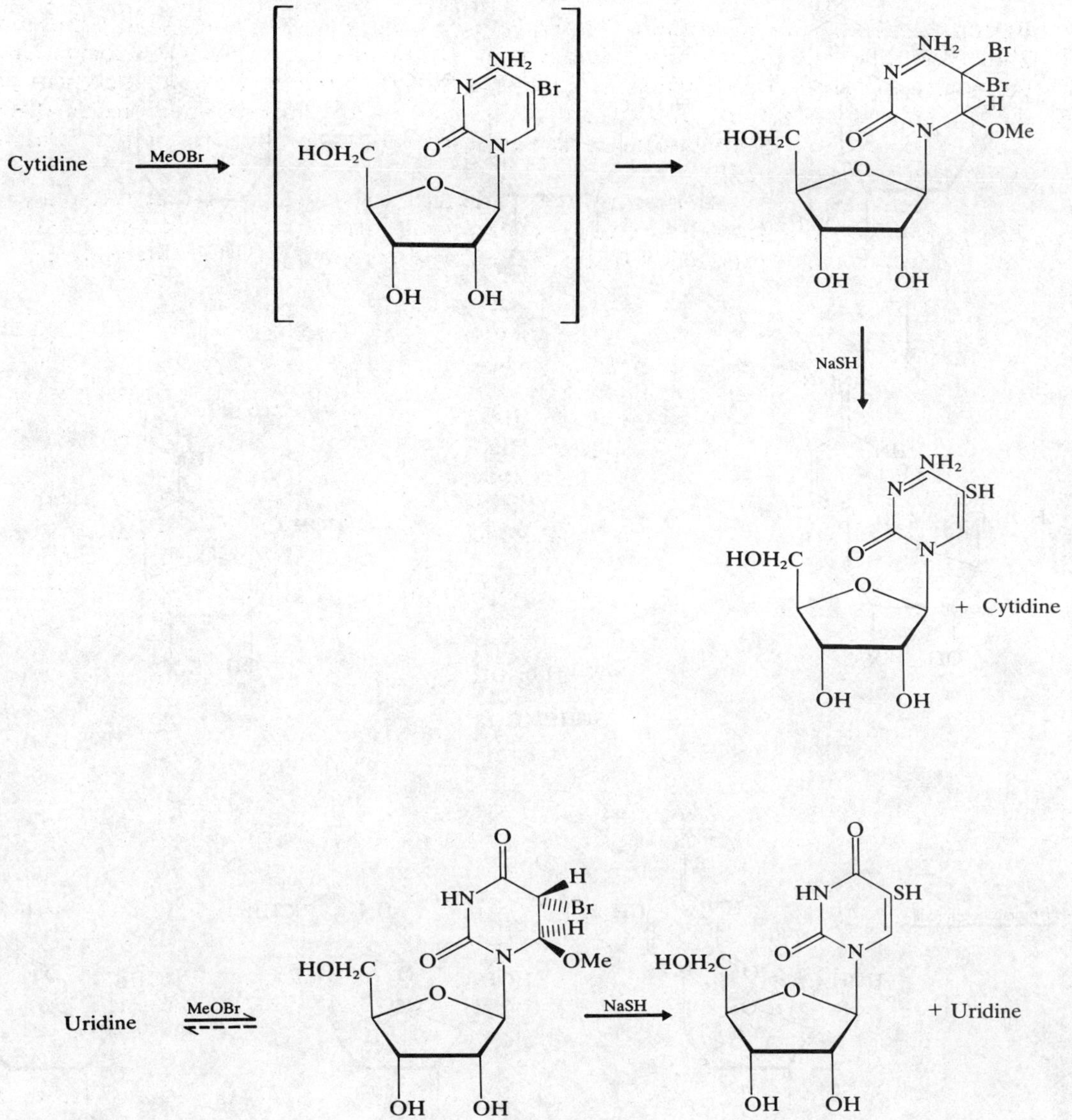

SCHEME 17

5-Substituted pyrimidine nucleosides have also been prepared by the chemical modification of other nucleosides. For example the trimethylsilyl ether of 5-iodo-2′-deoxyuridine has been converted to 5-cyano-2′-deoxyuridine by the action of cuprous cyanide in dry pyridine.[191]

5-Bromouracil-containing nucleosides have been converted to 5-aminonucleosides by treatment with ammonia[192] and have been oxidized to 5-hydroxynucleosides with lead oxide.[193] 5-Nitrouridine can be reduced to 5-aminouridine.[46] Thymidine has been used as the starting material for the production of 5-formyl-, 5-hydroxymethyl-, and 5-carboxy-2′-deoxyuridine (Scheme 19).[194]

In a similar reaction, treatment of 5-ethyl-2′-deoxyuridine with *N*-bromosuccinimide resulted in the production of 5-(1-bromoethyl)-2′-deoxyuridine, from which HBr could be eliminated with base to give 5-vinyl-2′-deoxyuridine. 5-Formyluridine can be prepared by the oxidation of 5-hydroxymethyluridine,[195] 5-acetyluridine can be reduced to 5-(1-hydroxyethyl)uridine, and 5-(2,2-dibromovinyl)-2′-deoxyuridine has been reacted with

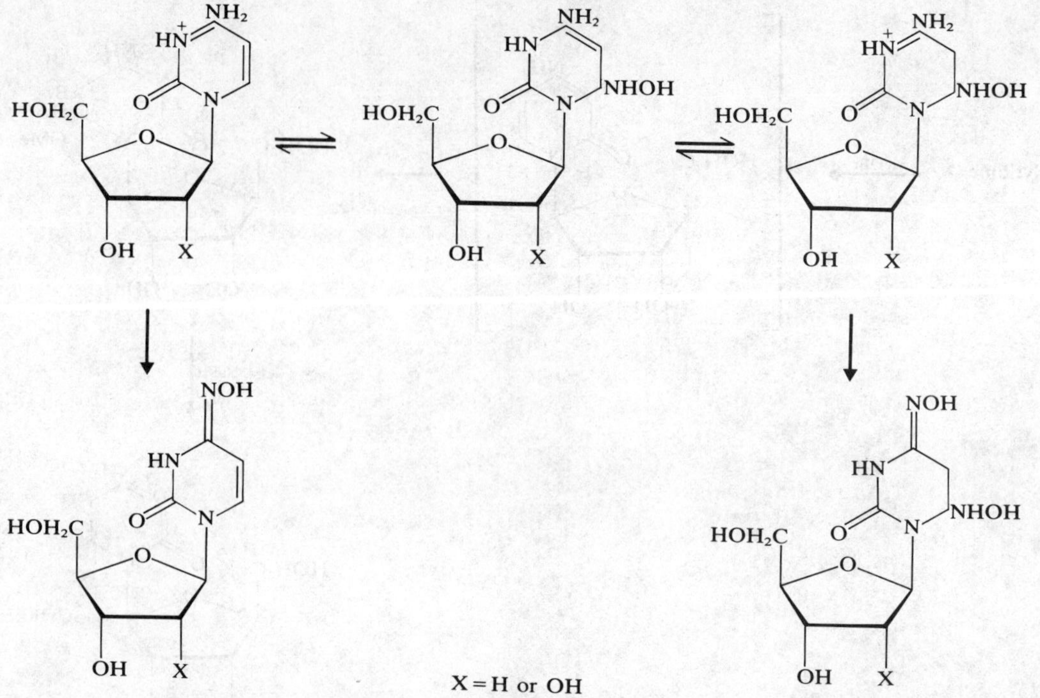

X = H or OH

SCHEME 18

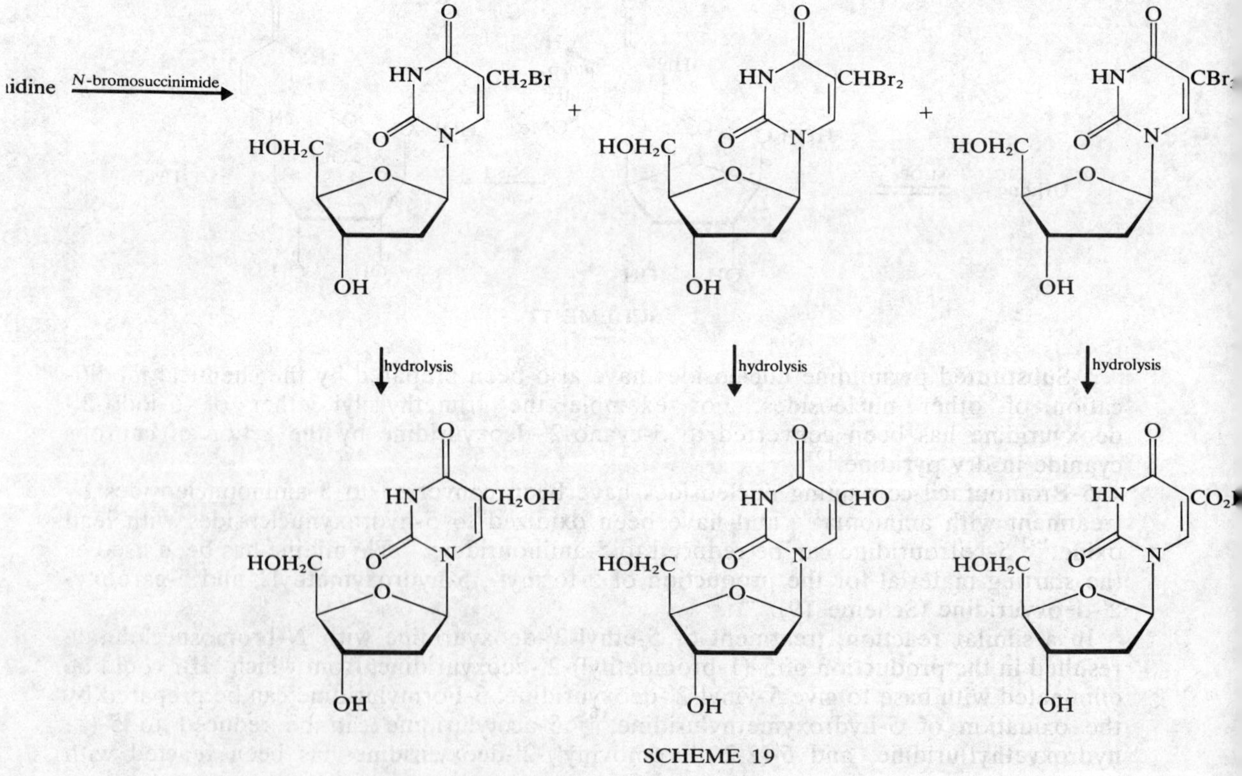

SCHEME 19

phenyl-lithium to give 5-ethynyl-2′-deoxyuridine.[196] Many 5-substituted pyrimidine nucleosides have been prepared by direct condensation of a sugar with a pre-formed base. These include 5-cyclopropyluridine,[197] several 5-alkylated (for example, isopropyl, t-butyl, and n-hexyl) and 5-halogenated 2′-deoxyuridines,[198] 5-hydroxymethyl-2′-deoxyuridine,[199] 5-benzyloxymethyl-2′-deoxyuridine,[200] 5-azauridine, 5-acetyl-2′-deoxyuridine, 5-vinyl-2′-deoxyuridine,[201] 5-ethynyl-2′-deoxyuridine,[196] 5-dibromovinyl-2′-deoxyuridine,[196] and 5-trifluoromethyl-2′-deoxyuridine.[202]

Many of these 5-substituted nucleosides, particularly the 5-substituted 2′-deoxyuridines, are thymidine analogues and much work has been reported on investigating the antiviral properties of these compounds.[19,203–208] However, in 1974[203] 5-iodo-2′-deoxyuridine was one of only three compounds officially approved for limited clinical application in antiviral chemotherapy and is used in the treatment of herpes simplex corneal keratitis. There are several potential ways to inhibit viral growth in cells and these include inhibition of virus adsorption, inhibition of virus penetration and uncoating, and finally inhibition of intracellular events. It is this last property with which we are most interested when discussing thymidine analogues, but even here the mode of action of many of the present analogues is not known and many apparently similar compounds have very different effects. The problem is best illustrated by reference to the 5-halogeno-2′-deoxynucleosides. 5-Fluoro-2′-deoxyuridine is not incorporated into DNA and acts, upon its conversion to its 5′-monophosphate, as a powerful inhibitor of thymidylate synthetase. This nucleoside is of course at present not used as an antiviral agent, but is an antineoplastic drug. There are several reports[209,210] which suggest that it has no antiviral properties, but recently[208] it has been shown to be as effective as (and less toxic than) 5-iodo-2′-deoxyuridine in protecting primary rabbit kidney cells from the cytopathic effect of vaccinia and herpes simplex virus. 5-Iodo- (and 5-bromo-) 2′-deoxyuridine are readily incorporated into host and viral DNA and have been shown to inhibit a number of enzymes involved in DNA synthesis. These compounds, however, are assumed to exert their primary action after their incorporation into DNA by preventing the normal flow of information from this DNA. 5-Trifluoromethyl-2′-deoxyuridine, after phosphorylation, is also a potent inhibitor of thymidylate synthetase, but is also incorporated into DNA of both virus and host[19] and is thought to affect the transcription of viral late mRNA.

5-Iodo- and 5-bromo-2′-deoxyuridine are also mutagenic. The lower pK_a of 5-iodo-2′-deoxyuridine is thought to permit a greater probability of base-pair errors after incorporation into DNA because it is in the anionic form to a significantly greater extent.[204] It is also known that the glycosyl bond of the 5-halogenonucleosides is more labile and the presence of such nucleosides in DNA is known to increase the lability of the DNA to irradiation. Apart from increasing the mutation rate in bacteria, 5-iodo-2′-deoxyuridine is known to induce the formation of virus particles in cell culture, produce chromosomal damage, and affect embryonic development and differentiation. Thus there is obviously room for an improved antiviral agent and whereas the ID_{50} value, which is a measure of the concentration of the drug required to inhibit virus-induced cytopathogenicity, is a useful index, the therapeutic index, which is the ratio of host DNA synthesis ID_{50} to virus ID_{50} (*i.e.* a measure of the usefulness of the drug within its toxic limit), is probably of greater value.

5-Ethyl-2′-deoxyuridine has a high therapeutic index because of its low toxicity and as it is not mutagenic, it could prove to be a useful addition to the list of antiviral agents. This nucleoside, however, is known to be incorporated into host and viral DNA, but 5-cyano-2′-deoxyuridine, which also has a very high therapeutic index because it is completely non-toxic, has been shown not to be incorporated into bacterial or viral DNA.[191] Presumably because it is not incorporated into DNA, 5-fluoro-2′-deoxyuridine also has a very high therapeutic index.

Recently,[204] a list of ten desirable features for the properties of an antiviral agent were listed to be: ease of preparation, good solubility at physiological pH, chemical and metabolic stability, sufficient non-polarity to avoid problems of cell transport, no incorporation into the DNA of the uninfected cell, no immunosuppression, no activation of virus, no teratogenic, mutagenic, or carcinogenic effects. 5-Iodo-2′-deoxyuridine clearly does not have several of these properties and 5-fluoro-, 5-ethyl-, and 5-cyano-2′-deoxyuridine obviously have potential as antiviral agents; promising results have also been reported for a range of other 5-alkyl-2′-deoxyuridines, including 5-allyl-, 5-propyl-, and 5-vinyl-2′-deoxyuridine,[206] and these are thought to exert their effect by inhibiting virus-specific thymidine kinases.

It is only a matter of time before some useful antiviral agents which are nucleoside derivatives become available and, with increasing knowledge of the mode of action of those agents already discovered, it is becoming increasingly possible to predict, scientifically, new lines of fruitful investigation. Thus the combination of the known properties of 5-iodo-2′-deoxyuridine and 5′-amino-5′-deoxythymidine in the compound 5′-amino-2′,5′-dideoxy-5-iodouridine has apparently resulted in the production of a compound which has the potent viral inhibiting characteristics of 5-iodo-2′-deoxyuridine but at the same time is presumably not capable of being incorporated into DNA and is thus much less toxic.[204] Whether the ideal antiviral agent will ever be prepared is a matter for conjecture, but it is likely that in the future many of the candidates for this enviable position will be nucleosides.

References

1. R. A. Sharma, M. Bobek, and A. Bloch, *J. Medicin. Chem.*, 1975, **18,** 473.
2. 'The Nucleic Acids', ed. E. Chargaff and J. N. Davidson, Academic Press, New York, 1955, vol. I.
3. A. M. Michelson, 'The Chemistry of Nucleosides and Nucleotides', Academic Press, New York, 1963.
4. D. O. Jordan, 'The Chemistry of the Nucleic Acids', Butterworths, Washington, 1960.
5. J. H. Parish, 'Principles and Practice of Experiments with Nucleic Acids', Longman, London, 1972.
6. W. Guschlbauer, 'Nucleic Acid Structure', Springer Verlag, Berlin, 1976.
7. J. N. Davidson, 'The Biochemistry of the Nucleic Acids', Chapman and Hall, London, 8th edn., 1976.
8. R. H. Hall, 'The Modified Nucleosides in Nucleic Acids', Columbia, New York, 1971.
9. N. K. Kochetkov and E. I. Budovskii, 'Organic Chemistry of Nucleic Acids', Parts A and B, Plenum, London, 1971.
10. 'Basic Principles in Nucleic Acid Chemistry', ed. P. O. P. Ts'O, Academic Press, New York, 1974, vols. I and II.
11. P. Langen, 'Antimetabolites of Nucleic Acid Metabolism', Gordon and Breach, New York, 1975.
12. 'Synthetic Procedures in Nucleic Acid Chemistry', ed. W. W. Zorbach and R. S. Tipson, Wiley, New York, 1973, vols. 1 and 2.
13. 'Physico-chemical Properties of Nucleic Acids', ed. J. Duchesne, Academic Press, London, 1973, vols. 1 and 2.
14. V. A. Bloomfield, D. M. Crothers, and I. Tinoco, Jr., 'Physical Chemistry of Nucleic Acids', Harper and Row, New York, 1974.
15. 'Progress in Nucleic Acid Research and Molecular Biology', ed. W. E. Cohn, Academic Press, New York.
16. 'Progress in Biophysics and Molecular Biology', ed. J. A. V. Butler and D. Noble, Pergamon, Oxford.
17. 'Nucleic Acids Research', Information Retrieval Ltd., London, vol. 1, 1974.
18. 'Nucleic Acids Abstracts', ed. A. Williamson, Information Retrieval Ltd., London, vol. 1, 1971.
19. 'Chemistry, Biology and Clinical Uses of Nucleoside Analogues', ed. A. Bloch, *Ann. New York Acad. Sci.*, 1975, 255.
20. B. G. Barrell and B. F. C. Clark, 'Handbook of Nucleic Acid Sequences', Joynson–Bruvvers Ltd., Oxford, 1974.
21. H. Kasai, K. Nakanishi, R. D. Macfarlane, D. F. Torgerson, Z. Ohashi, J. A. McCloskey, H. J. Gross, and S. Nishimura, *J. Amer. Chem. Soc.*, 1976, **98,** 5044; N. Okada, N. Shindo-Okada, and S. Nishimura, *Nucleic Acids Res.*, 1977, **4,** 415.
22. C. B. Reese and N. Whittall, *Nucleic Acids Res.*, 1976, **3,** 3439.
23. M. J. Robins, M. MacCoss, and A. S. K. Lee, *Biochem. Biophys. Res. Comm.*, 1976, **70,** 356.
24. 'Methods in Enzymology', ed. L. Grossman and K. Moldave, Academic Press, New York, 1967, vol. 12, pt. A; 'Microbial Production of Nucleic Acid Related Substances', ed. K. Ogata, S. Kimoshita, T. Tsuneda, and K. Aida, Wiley, Chichester, 1976.
25. S. Nishimura, in 'Progress in Nucleic Acid Research and Molecular Biology', ed. J. N. Davidson and W. E. Cohn, Academic Press, New York, 1972, vol. 12, p. 49.
26. K. Fink, R. E. Cline, and R. M. Fink, *Anal. Chem.*, 1963, **35,** 389; K. Fink and W. S. Adam, *J. Chromatog.*, 1966, **22,** 118; H. Rogg, R. Brambilla, G. Keith, and M. Staehelin, *Nucleic Acids Res.*, 1976, **3,** 285.
27. S. Zadražil, Ref. 12, vol. 2, chapter 9.
28. B. J. Hunt and W. Rigby, *Chem. and Ind.* (*London*), 1967, 1868.
29. C. A. Dekker, *J. Amer. Chem. Soc.*, 1965, **87,** 4027.

30. J. D. Smith, Ref. 2, Chapter 8.
31. A. E. Pierce, Ref. 12, vol. 2, chapter 3.
32. S. E. Hattox and J. A. McCloskey, *Anal. Chem.*, 1974, **46,** 1378.
33. T. D. Kulikowski and D. Shugar, *Acta Biochim. Polon.*, 1974, **21,** 169.
34. D. L. Cole, N. J. Leonard, and J. C. Cook in "Proceedings of International Conference on Recent Developments in Oligonucleotide Synthesis and Chemistry of the Minor Bases of tRNA", Poznań, Poland, 1974, pp. 153–174.
35. B. Pullman and A. Pullman, in 'Progress in Nucleic Acid Research and Molecular Biology', ed. J. N. Davidson and W. E. Cohn, Academic Press, New York, 1969, vol. 9, p. 327.
36. A. M. Jeffrey, K. W. Jennette, S. H. Blobstein, I. B. Weinstein, F. A. Beland, R. G. Harvey, H. Kasai, I. Miura, and K. Nakanishi, *J. Amer. Chem. Soc.*, 1976, **98,** 5714.
37. A. M. Jeffrey, S. H. Blobstein, I. B. Weinstein, F. A. Beland, R. G. Harvey, H. Kasai, and K. Nakanishi, *Proc. Nat. Acad. Sci. U.S.A.*, 1976, **73,** 2311.
38. H. Kasai, Z. Ohashi, F. Harada, S. Nishimura, N. J. Oppenheimer, P. F. Crain, J. G. Liehr, D. L. von Minden, and J. A. McCloskey, *Biochemistry*, 1975, **14,** 4198.
39. J. G. Liehr, D. L. von Minden, S. E. Hattox, and J. A. McCloskey, *Biomedical Mass Spectrometry*, 1974, **1,** 281.
40. R. D. Macfarlane and D. F. Torgerson, *Science*, 1976, **191,** 920.
41. S. Arnott, in 'Progress in Biophysics and Molecular Biology', ed. J. A. V. Butler, Pergamon, New York, 1970, vol. 21, p. 265.
42. 'Structure and Conformation of Nucleic Acids and Protein–Nucleic Acid Interactions', ed. M. Sundaralingam and S. T. Rao, University Park Press, Baltimore, 1975.
43. W. Guschlbauer and T. D. Son, *Nucleic Acids Res.*, *Special Publication No. 1*, 1975, S85.
44. A. Andre and W. Guschlbauer, *Nucleic Acids Res.*, 1974, **1,** 803.
45. A. R. Todd, in 'Methods in Enzymology', ed. S. P. Colowick and N. O. Kaplan, Pergamon, Oxford, 1957, vol. 3, p. 811.
46. L. Goodman, Ref. 10, vol. 1, chapter 2.
47. K. A. Watanabe, D. H. Hollenberg, and J. J. Fox, *J. Carbohydrates, Nucleosides Nucleotides*, 1974, **1,** 1.
48. H. Vorbrüggen, Ref. 34, p. 428; U. Niedballa and H. Vorbrüggen, *J. Org. Chem.*, 1976, **41,** 2084.
49. J. Davoll and B. A. Lowy, *J. Amer. Chem. Soc.*, 1951, **73,** 1650.
50. J. J. Fox, N. Yung, J. Davoll, and G. B. Brown, *J. Amer. Chem. Soc.*, 1956, **78,** 2117.
51. N. Yamaoka, K. Aso, and K. Matsuda, *J. Org. Chem.*, 1965, **30,** 149; K. A. Watanabe and J. J. Fox, *J. Heterocyclic Chem.*, 1969, **6,** 109.
52. B. Shimizu and M. Miyaki, *Chem. and Ind.* (*London*), 1966, 664; B. Shimizu and M. Miyaki, *Chem. Pharm. Bull.* (*Japan*), 1970, **18,** 1446.
53. G. A. Howard, B. Lythgoe, and A. R. Todd, *J. Chem. Soc.*, 1947, 1052.
54. T. L. V. Ulbricht, *J. Chem. Soc.*, 1961, 3345.
55. M. Prystaš and F. Šorm, *Coll. Czech. Chem. Comm.*, 1966, **31,** 3990 and references cited therein.
56. M. Roberts and D. W. Visser, *J. Amer. Chem. Soc.*, 1952, **74,** 668; F. Farkaš, L. Kaplan, and J. J. Fox, *J. Org. Chem.*, 1964, **29,** 1469.
57. E. Wittenburg, *Z. Chem.*, 1964, **4,** 303; E. Wittenburg, *Chem. Ber.*, 1968, **101,** 1095.
58. L. Otvös, A. Szabolcs, J. Sági, and A. Szemzó, *Nucleic Acids Res.*, *Special Publication No. 1*, 1975, S49.
59. T. J. Bardos, M. P. Kotick, and C. Szantay, *Tetrahedron Letters*, 1966, 1759.
60. T. Nishimura and B. Shimizu, *Chem. Pharm. Bull.* (*Japan*), 1965, **13,** 803.
61. U. Niedballa and H. Vorbrüggen, *J. Org. Chem.*, 1974, **39,** 3672 and references cited therein.
62. M. J. Covill, H. G. Garg, and T. L. V. Ulbricht, *Tetrahedron Letters*, 1968, 1033.
63. B. Shimizu and M. Miyaki, *Chem. Pharm. Bull.* (*Japan*), 1967, **15,** 1066.
64. T. Simadate, Y. Ishido, and T. Sato, *Chem. Abs.*, 1962, **56,** 11 692.
65. W. Pfleiderer and R. K. Robins, *Chem. Ber.*, 1965, **98,** 1511.
66. A. Hosono, K. Fujii, T. Tada, H. Tanaka, Y. Ohgo, Y. Ishido, and T. Sato, *Bull. Chem. Soc. Japan*, 1973, **46,** 2818.
67. B. Shimizu and M. Miyaki, *Tetrahedron Letters*, 1968, 855.
68. T. Azume, K. Isono, P. F. Crain, and J. A. McCloskey, *Tetrahedron Letters*, 1976, 1687.
69. F. W. Lichtenthaler and K. Kitahara, *Angew. Chem. Internat. Edn.*, 1975, **14,** 815.
70. G. Schramm, H. Groetsch, and W. Pollman, *Angew. Chem.*, 1961, **73,** 619.
71. J. A. Carbon, *Chem. and Ind.* (*London*), 1963, 529.
72. R. A. Sanchez and L. E. Orgel, *J. Mol. Biol.*, 1970, **47,** 531.
73. C. M. Hall, G. Slomp, S. A. Mizsak, and A. J. Taylor, *J. Org. Chem.*, 1972, **37,** 3290.
74. A. Holý, *Tetrahedron Letters*, 1971, 189.
75. J. Baddiley, J. G. Buchanan, R. Hodges, and J. F. Prescott, *J. Chem. Soc.*, 1957, 4769.
76. N. J. Cusack, B. J. Hildick, D. H. Robinson, P. W. Rugg, and G. Shaw, *J. C. S. Perkin I*, 1973, 1720.
77. G. Shaw, R. N. Warrener, M. H. Maguire, and R. K. Ralph, *J. Chem. Soc.*, 1958, 2294.
78. J. Šmejkal, J. Farkaš, and F. Šorm, *Coll. Czech. Chem. Comm.*, 1966, **31,** 291.
79. M. Sprinzl, J. Farkaš, and F. Šorm, *Coll. Czech. Chem. Comm.*, 1967, **32,** 4280.
80. A. Piskala and F. Šorm, *Coll. Czech. Chem. Comm.*, 1964, **29,** 2060.
81. G. W. Kenner, C. W. Taylor, and A. R. Todd, *J. Chem. Soc.*, 1949, 1620.
82. M. Ikehara, E. Ohtsuka, S. Kitagawa, K. Yagi, and Y. Tonomura, *J. Amer. Chem. Soc.*, 1961, **83,** 2679.

83. L. B. Townsend, *Chem. Rev.*, 1967, **67,** 533.
84. J. Baddiley, J. G. Buchanan, and G. O. Osborne, *J. Chem. Soc.*, 1958, 3606.
85. R. J. Rousseau, R. K. Robins, and L. B. Townsend, *J. Amer. Chem. Soc.*, 1968, **90,** 2661.
86. A. Yamazaki, M. Okutsu, and Y. Yamada, *Nucleic Acids Res.*, 1976, **3,** 251; see also the two preceding papers and references cited therein.
87. See J. R. Tittensor and R. T. Walker, *European Polymer J.*, 1968, **4,** 39.
88. See H. Hayatsu, in 'Progress in Nucleic Acid Research and Molecular Biology', 1976, vol. 16, p. 75.
89. See J. Brokěs and J. Beránek, *Coll. Czech. Chem. Comm.*, 1974, **39,** 3100.
90. N. A. Arutunyan, V. I. Gunar, and S. I. Zavyalov, *Izvest. Akad. Nauk S.S.S.R., Ser. khim.*, 1970, 904.
91. H. Vorbrüggen, K. Krolikiewicz, and U. Niedballa, *Annalen*, 1975, 988; H. Vorbrüggen, K. Krolikiewicz, and U. Niedballa, *Ann. New York Acad. Sci.*, 1975, **255,** 82.
92. See A. Holý and D. Cech, *Coll. Czech. Chem. Comm.*, 1974, **39,** 3157.
93. M. M. Ponpipom and S. Hanessian, *Canad. J. Chem.*, 1972, **50,** 2530, and references cited therein.
94. D. H. R. Barton and R. Subramanian, *J. C. S. Chem. Comm.*, 1976, 867.
95. V. M. Clark, A. R. Todd, and J. Zussman, *J. Chem. Soc.*, 1951, 2952; W. Andersen, D. H. Hayes, A. M. Michelson, and A. R. Todd, *ibid.*, 1954, 1882.
96. M. Ikehara and T. Tezuka, *J. Amer. Chem. Soc.*, 1973, **95,** 4054.
97. J. H. Burchenal, V. E. Currie, M. D. Dowling, J. J. Fox, and I. H. Krakoff, *Ann. New York Acad. Sci.*, 1975, **255,** 202.
98. G. Lünzmann and G. Schramm, *Biochim. Biophys. Acta*, 1968, **169,** 263.
99. A. P. Martinez, W. W. Lee, and L. Goodman, *J. Org. Chem.*, 1966, **31,** 3263.
100. M. Ikehara and M. Muraoka, *Chem. Pharm. Bull.* (*Japan*), 1976, **24,** 672.
101. M. Ikehara, Y. Nakahara, and S. Yamada, *Chem. Pharm. Bull.* (*Japan*), 1970, **18,** 2441.
102. P. J. Harper and A. Hampton, *J. Org. Chem.*, 1972, **37,** 795.
103. J. J. Fox, *Pure Appl. Chem.*, 1969, **18,** 223.
104. S. Hanessian and A. G. Pernet, *Adv. Carbohydrate Chem. Biochem.*, 1976, **33,** 111.
105. R. J. Suhadolnik, 'Nucleoside Antibiotics', Wiley Interscience, New York, 1970; K. Gerzon, D. C. de Long, and J. C. Cline, *Pure Appl. Chem.*, 1971, **28,** 489.
106. D. M. Brown, M. G. Burdon, and R. P. Slatcher, *J. Chem. Soc. C*, 1968, 1051.
107. See E. M. Acton, K. J. Ryan, D. W. Henry, and L. Goodman, *Chem. Comm.*, 1971, 986.
108. J. M. J. Tronchet and A. Jotterand, *Helv. Chim. Acta*, 1971, **54,** 1131.
109. D. B. Repke, H. P. Albrecht, and J. G. Moffatt, *J. Org. Chem.*, 1975, **40,** 2481.
110. L. Kalvoda, J. Farkaš, and F. Šorm, *Tetrahedron Letters*, 1970, 2297.
111. S. Hanessian and A. G. Pernet, *Canad. J. Chem.*, 1974, **52,** 1280.
112. H. Ohrui and J. J. Fox, *Tetrahedron Letters*, 1973, 1951; R. S. Klein, H. Ohrui, and J. J. Fox, *J. Carbohydrates, Nucleosides Nucleotides*, 1974, **1,** 265.
113. H. Ohrui, G. H. Jones, J. G. Moffatt, M. L. Maddox, A. T. Christensen, and S. K. Byram, *J. Amer. Chem. Soc.*, 1975, **97,** 4602.
114. J. G. Buchanan, A. R. Edgar, M. J. Power, and G. C. Williams, *J. C. S. Chem. Comm.*, 1975, 501.
115. F. G. De Las Heras, S. Y.-K. Tam, R. S. Klein, and J. J. Fox, *J. Org. Chem.*, 1976, **41,** 84.
116. K. Arakawa, T. Miyasaka, and N. Hamamichi, *Nucleic Acids Res., Special Publication No. 2*, 1976, S1.
117. S. R. Jenkins, B. A. Arison, and E. Walton, *J. Org. Chem.*, 1968, **33,** 2490, and references cited therein.
118. H. P. Albrecht and J. G. Moffatt, *Tetrahedron Letters*, 1970, 1063.
119. H. Grisebach and R. Schmid, *Angew. Chem. Internat. Edn.*, 1972, **11,** 159.
120. A. Rosenthal, M. Sprinzl, and D. A. Baker, *Tetrahedron Letters*, 1970, 4233.
121. J. Zemlicka, J. V. Freisler, R. Gasser, and J. P. Horwitz, *J. Org. Chem.*, 1973, **38,** 990.
122. M. J. Robins and R. A. Jones, *J. Org. Chem.*, 1974, **39,** 113.
123. M. J. Robins and E. M. Trip, *Tetrahedron Letters*, 1974, 3369.
124. See T. C. Jain, I. D. Jenkins, A. F. Russell, J. P. H. Verheyden, and J. G. Moffatt, *J. Org. Chem.*, 1974, **39,** 30.
125. See J. P. Verheyden and J. G. Moffatt, *J. Org. Chem.*, 1974, **39,** 3573.
126. C. A. Dekker and L. Goodman, in 'The Carbohydrates, Chemistry and Biochemistry', ed. W. Pigman and D. Horton, Academic Press, New York, 1970, vol. IIA, chapter 29.
127. I. L. F. Doerr, J. F. Codington, and J. J. Fox, *J. Org. Chem.*, 1965, **30,** 467, and following papers.
128. M. J. Robins, Y. Fouron, and R. Mengel, *J. Org. Chem.*, 1974, **39,** 1564.
129. See D. Horton, K. Blieszner, and R. A. Markovs, 'Proceedings of International Conference on the Synthesis, Structure and Chemistry of tRNA and their Components', Poznań, Poland, 1976, p. 68.
130. See A. Holý, Ref. 129, p. 134.
131. J. Simúth and A. Holý, *Nucleic Acids Res., Special Publication No. 1*, 1975, S165.
132. See R. T. Walker, *Ann. Reports Chem. Soc.*, 1972, **69B,** 531.
133. See D. M. Brown, in 'Basic Principles in Nucleic Acid Chemistry', ed. P. O. P. Ts'O, Academic Press, New York, 1974, vol. II, chapter 1.
134. M. J. Robins, *Ann. New York Acad. Sci.*, 1975, **255,** 104.
135. G. Moss, C. B. Reese, K. Schofield, R. Shapiro, and A. R. Todd, *J. Chem. Soc.*, 1963, 1149.
136. A. S. Jones, A. R. Williamson, and M. Winkley, *Carbohydrate Res.*, 1965, **1,** 187.
137. R. P. Schmidt, U. Schloz, and D. Schwille, *Chem. Ber.*, 1968, **101,** 590.
138. K. E. Pfitzner and J. G. Moffatt, *J. Amer. Chem. Soc.*, 1963, **85,** 3027.

139. D. C. Baker and D. Horton, *Carbohydrate Res.*, 1972, **21,** 393.
140. E. J. Corey and J. W. Suggs, *Tetrahedron Letters*, 1975, 2647.
141. H. A. Friedman, K. A. Watanabe, and J. J. Fox, *J. Org. Chem.*, 1967, **32,** 3775.
142. K. Randerath, E. Randerath, L. S. Y. Chia, R. C. Gupta, and M. Sivarajan, *Nucleic Acids Res.*, 1974, **1,** 112.
143. E. Randerath, C. T. Yu, and K. Randerath, *Anal. Biochem.*, 1972, **48,** 172.
144. A. S. Jones, A. F. Markham, and R. T. Walker, *J. C. S. Perkin I*, 1976, 1567.
145. F. Hansske, M. Sprinzl, and F. Cramer, *Bioorganic Chem.*, 1974, **3,** 367.
146. G. B. Brown, in 'Progress in Nucleic Acid Research and Molecular Biology', ed. J. N. Davidson and W. E. Cohn, Academic Press, New York, 1968, vol. 8, p. 209.
147. A. S. Jones and R. T. Walker, *J. Chem. Soc.*, 1963, 3554.
148. R. S. Goody, A. S. Jones, and R. T. Walker, *Tetrahedron*, 1971, **27,** 65.
149. P. Howgate, A. S. Jones, and J. R. Tittensor, *J. Chem. Soc. C*, 1968, 275.
150. P. J. Highton, B. L. Murr, F. Shafa, and M. Beer, *Biochemistry*, 1968, **7,** 825.
151. R. C. Moschel and E. J. Behrman, *J. Org. Chem.*, 1974, **39,** 1983, 2699.
152. H. G. Khorana, K. L. Agarwal, H. Buchi, M. H. Caruthers, N. K. Gupta, K. Kleppe, A. Kumar, E. Ohtsuka, U. L. RajBhandary, J. H. van de Sande, V. Sgaramella, T. Terao, H. Weber, and T. Yamada, *J. Mol. Biol.*, 1972, **72,** 209 and following papers.
153. H. G. Khorana, K. L. Agarwal, P. Besmer, H. Buchi, M. H. Caruthers, P. J. Cashion, M. Fridkin, E. Jay, K. Kleppe, R. Kleppe, A. Kumar, P. C. Loewen, R. C. Miller, K. Minamoto, A. Panet, U. L. RajBhandary, B. Ramamoorthy, T. Sekiya, T. Takeya, and J. H. van de Sande, *J. Biol. Chem.*, 1976, **251,** 565 and following papers.
154. P. A. Lyon and C. B. Reese, *J. C. S. Perkin I*, 1974, 2645.
155. B. A. Otter and J. J. Fox, in 'Synthetic Procedures in Nucleic Acid Chemistry', ed. W. W. Zorbach and R. S. Tipson, Wiley, New York, 1973, vol. 1, p. 285.
156. R. C. Bleaney, A. S. Jones, and R. T. Walker, *Tetrahedron*, 1975, **31,** 2423.
157. R. S. Goody and R. T. Walker, *J. Org. Chem.*, 1971, **36,** 727.
158. N. Nakazaki, N. Sakairi, and Y. Ishido, *Nucleic Acids Res., Special Publication No. 2*, 1975, S9.
159. Y. Ishido, N. Nakazaki, and N. Sakairi, *J. C. S. Chem. Comm.*, 1976, 832.
160. C. B. Reese, in 'Protective Groups in Organic Chemistry', ed. J. F. W. McOmie, Plenum, London, 1973, p. 95.
161. C. B. Reese and J. C. M. Stewart, *Tetrahedron Letters*, 1968, 4273.
162. A. F. Cook, *J. Org. Chem.*, 1968, **33,** 3589.
163. R. L. Letsinger and P. S. Miller, *J. Amer. Chem. Soc.*, 1969, **91,** 3356.
164. A. Hassner, G. Strand, M. Rubenstein, and A. Patchornik, *J. Amer. Chem. Soc.*, 1975, **97,** 7327.
165. J. H. van Boom and P. M. J. Burgers, *Tetrahedron Letters*, 1976, 4875.
166. H. S. Sachdev and N. A. Starkovsky, *Tetrahedron Letters*, 1969, 733.
167. O. Mitsunobu, J. Kimura, and Y. Fujisawa, *Bull. Chem. Soc. Japan*, 1972, **45,** 245.
168. D. Wagner, J. P. H. Verheyden, and J. G. Moffatt, *J. Org. Chem.*, 1974, **39,** 24.
169. D. M. Brown, G. D. Fasman, D. I. Magrath, and A. R. Todd, *J. Chem. Soc.*, 1954, 1448.
170. H. P. M. Fromageot, C. B. Reese, and J. E. Sulston, *Tetrahedron*, 1968, **24,** 3533.
171. K. Kondo, T. Adachi, and I. Inoue, *Nucleic Acids Res., Special Publication No. 2*, 1975, S5.
172. P. Brookes, *Life Sci.*, 1974, **16,** 331.
173. R. Shapiro, *Ann. New York Acad. Sci.*, 1969, **163,** 624.
174. Z. Kazimierczuk, E. Darzynkiewicz, and D. Shugar, *Biochemistry*, 1976, **15,** 2735 and references cited therein.
175. A. S. Jones, R. K. Patient, and R. T. Walker, *J. Carbohydrates, Nucleosides Nucleotides*, 1977, **4,** 301.
176. M. J. Robins, S. R. Naik, and A. S. K. Lee, *J. Org. Chem.*, 1974, **39,** 1891.
177. E. Ohtsuka, S. Tanaka, and M. Ikehara, *Nucleic Acids Res.*, 1974, **1,** 1351.
178. K. K. Ogilvie, K. L. Sadana, E. A. Thompson, M. A. Quilliam, and J. B. Westmore, *Tetrahedron Letters*, 1974, 2861 and following paper.
179. M. Smith, D. H. Rammler, I. H. Goldberg, and H. G. Khorana, *J. Amer. Chem. Soc.*, 1962, **84,** 430.
180. B. E. Griffin and C. B. Reese, *Tetrahedron*, 1969, **25,** 4057.
181. R. M. K. Dale, E. Martin, D. C. Livingstone, and D. C. Ward, *Biochemistry*, 1975, **14,** 2447.
182. D. E. Bergstrom and J. L. Ruth, *J. Amer. Chem. Soc.*, 1976, **98,** 1587.
183. D. Cech, R. Wohlfeil, and G. Etzold, *Nucleic Acids Res.*, 1975, **2,** 2183.
184. I. Wempen, I. L. Doerr, L. Kaplan, and J. J. Fox, *J. Amer. Chem. Soc.*, 1960, **82,** 1624.
185. K. H. Scheit, *Chem. Ber.*, 1966, **99,** 3884.
186. D. H. R. Barton, *Pure Appl. Chem.*, 1970, **21,** 285.
187. D. Cech, L. Hein, R. Wuttke, M. v. Janta-Lipinski, A. Otto, and P. Langen, *Nucleic Acids Res.*, 1975, **2,** 2177.
188. T. Nagamachi, J.-L. Fourray, P. F. Torrence, J. A. Waters, and B. Witkop, *J. Medicin. Chem.*, 1974, **17,** 403.
189. Y. Wataya, H. Hayatsu, and Y. Kawazoe, *J. Biochem.*, 1973, **73,** 871.
190. F. A. Sedor and E. G. Sander, *Biochem. Biophys. Res. Comm.*, 1973, **50,** 328.
191. R. C. Bleackley, A. S. Jones, and R. T. Walker, *Nucleic Acids Res.*, 1975, **2,** 683.
192. M. Friedland and D. W. Visser, *Biochim. Biophys. Acta*, 1961, **51,** 148.

193. M. Roberts and D. W. Visser, *J. Amer. Chem. Soc.*, 1952, **74,** 668.
194. D. Bärwolff and P. Langen, *Nucleic Acids Res., Special Publication No. 1*, 1975, S29.
195. R. E. Cline, R. M. Fink, and K. Fink, *J. Amer. Chem. Soc.*, 1959, **81,** 2521.
196. J. Perman, R. A. Sharma, and M. Bobek, *Tetrahedron Letters*, 1976, 2427.
197. I. Bašnak and J. Farkaš, *Nucleic Acids Res., Special Publication No. 1*, 1975, S81.
198. L. Otvös, A. Szabolcs, J. Sági, and A. Szemzó, *Nucleic Acids Res., Special Publication No. 1*, 1975, S49.
199. V. S. Gupta and G. L. Bubbar, *Canad. J. Chem.*, 1971, **49,** 719.
200. M. P. Mertes and M. T. Shipchandler, *J. Heterocyclic Chem.*, 1971, **8,** 133.
201. R. A. Sharma and M. Bobek, *J. Org. Chem.*, 1975, **40,** 2377.
202. C. Heidelberger, D. G. Parsons, and D. C. Remy, *J. Amer. Chem. Soc.*, 1962, **84,** 3597.
203. D. Shugar, *F.E.B.S. Letters*, 1974, **40,** S48.
204. W. H. Prusoff and D. C. Ward, *Biochem. Pharmacol.*, 1976, **25,** 1233.
205. E. De Clercq. P. F. Torrence, J. A. Waters, and B. Witkop, *Biochem. Pharmacol.*, 1975, **24,** 2171.
206. Y.-C. Cheng, B. A. Domin, R. A. Sharma, and M. Bobek, *Antimicrobial Agents and Chemotherapy*, 1976, **10,** 119.
207. E. De Clercq and D. Shugar, *Biochem. Pharmacol.*, 1975, **24,** 1073.
208. E. De Clercq, personal communication.
209. E. C. Herrmann, *Appl. Microbiol.*, 1968, **16,** 1151.
210. T. Y. Shen, J. F. McPherson, and B. O. Linn, *J. Medicin. Chem.*, 1966, **9,** 366.

22.3

Nucleotides and Related Organic Phosphates

D. W. HUTCHINSON
University of Warwick

22.3.1 GENERAL PROPERTIES OF NATURALLY OCCURRING PHOSPHATES

Nucleotides are phosphate esters of nucleosides, the phosphoric acid being attached to one of the hydroxyl groups of the sugar residue of the nucleoside. The phosphoric acid can be attached to the 2′,-, 3′-, or 5′-hydroxyl groups in ribonucleosides, while 2′-deoxynucleosides can be esterified on the 3′- or 5′-hydroxyl groups of the sugar. Mononucleotides (*e.g.* adenosine 5′-phosphate, AMP*) are esters of orthophosphoric acid and hence contain one phosphorus atom per nucleoside molecule. The majority of common mononucleotides are monoesters of phosphoric acid, although nucleoside diesters (*e.g.* adenosine 3′,5′-cyclic phosphate, cAMP) are known. Mono- and di-esters of pyrophosphoric acid (*e.g.* adenosine diphosphate, ADP or nicotinamide-adenine dinucleotide, NAD^+) also occur naturally, as do monoesters of tripolyphosphoric acid (*e.g.* adenosine triphosphate, ATP). Nucleotides are optically active due to their sugar residues, and absorb strongly in the ultraviolet on account of their constituent heterocyclic bases. The last property is frequently used for the detection and estimation of nucleotides, for example during chromatography. Structures of some common nucleotides are given in (**1**)-(**9**), and some of their physical properties are given in Table 1.

Mono- and di-esters of phosphoric acids are acidic, water-soluble compounds and their acidic nature governs the choice of chromatographic method used for their analysis. Paper chromatography was one of the earliest chromatographic techniques which were applied to the separation of nucleotides, and this technique has been in use for many years.[1] Thin

* Detailed abbreviations of common nucleotides are given in the Instructions to Authors of the *Biochemical Journal.*

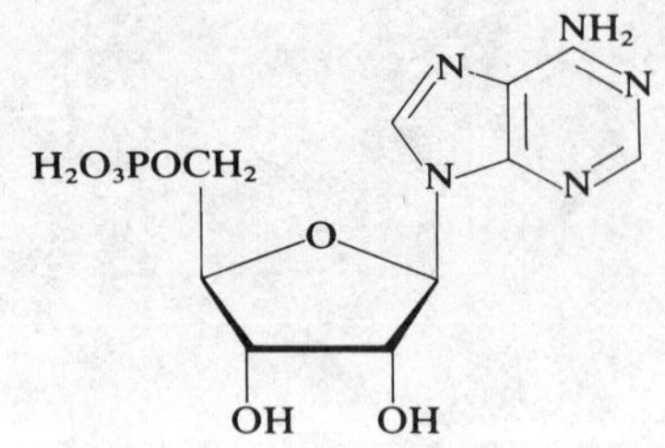

(1) 5′-AMP

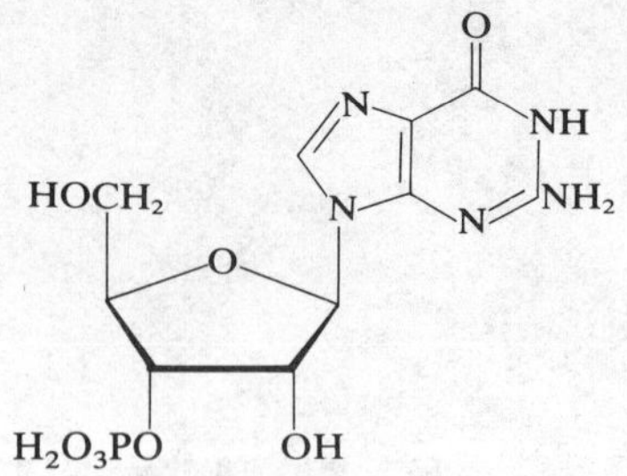

(2) 3′-GMP

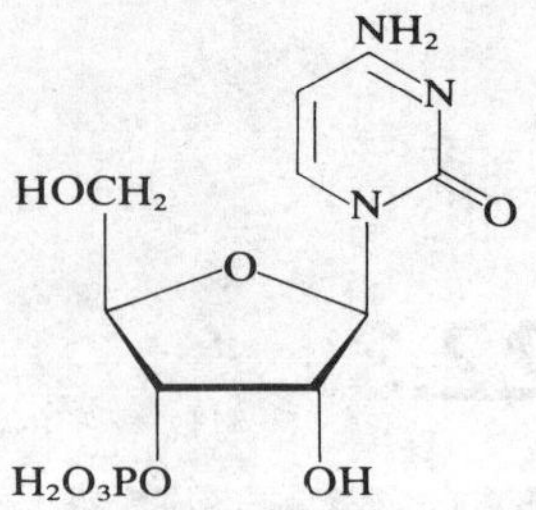

(3) 3′-CMP

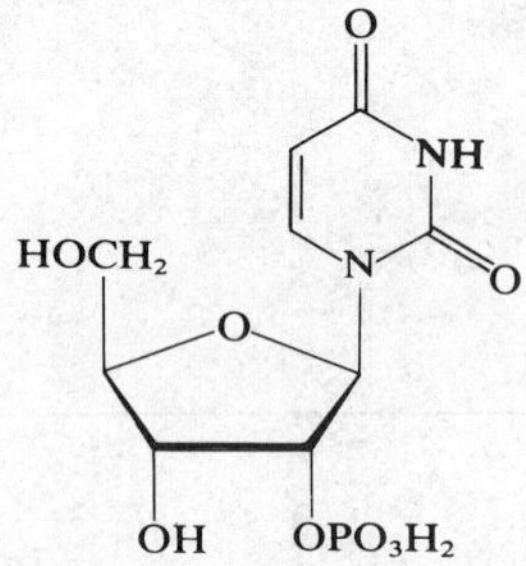

(4) 2′-UMP

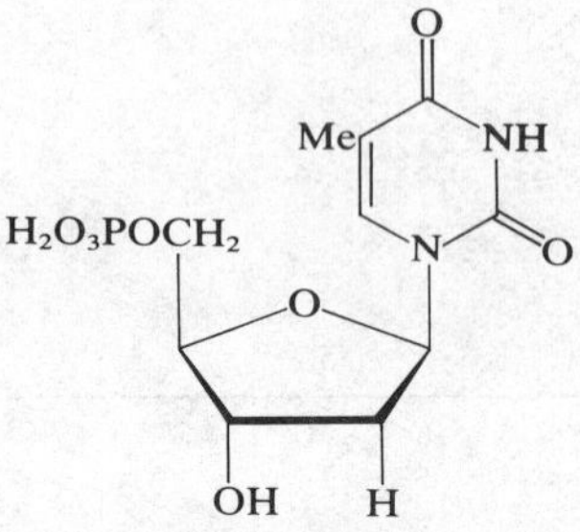

(5) 5′-dTMP

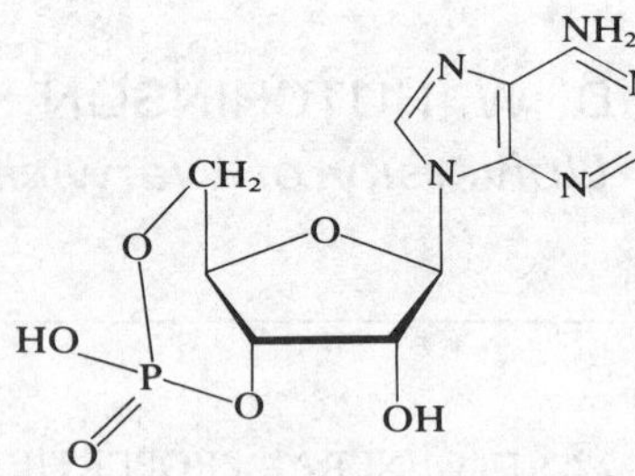

(6) cAMP

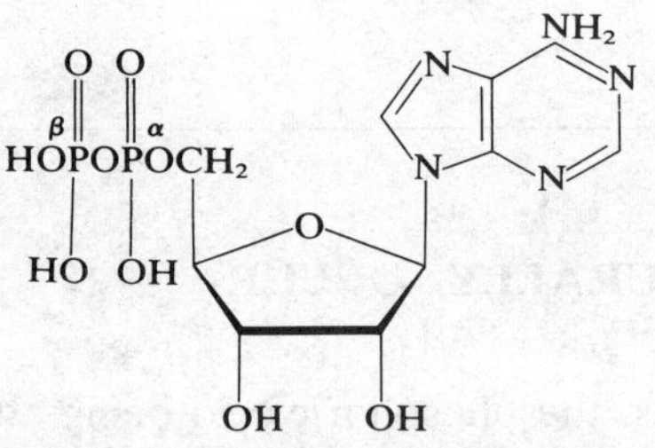

(7) ADP

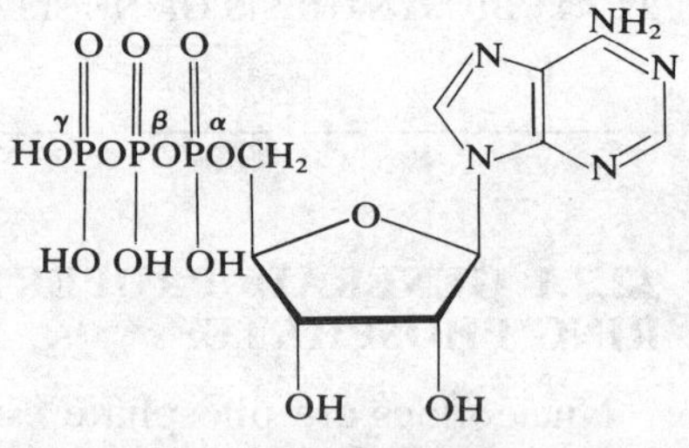

(8) ATP

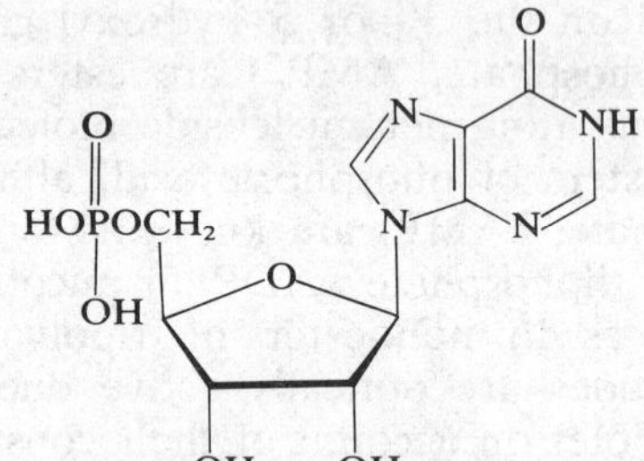

(9) IMP

layer chromatography using cellulose or silica gel provides another method for the rapid analysis of nucleotide mixtures.[2]

The application of ion-exchange chromatography to the analysis of nucleotide mixtures followed directly from the application of this technique to the separation of inorganic ions. Chronologically, cation exchange chromatography was examined before anion exchange, but the latter technique has proved to give the most satisfactory separation of phosphate esters.[3] Adsorption of a mixture of phosphates on to an anion exchange resin, followed by

TABLE 1

Physical Properties of Some Nucleotides[a]

Name	Empirical formula	M.p. (°C)	$[\alpha]_D$	Ultraviolet spectroscopic data					
				pH	λ_{max}/nm	$\varepsilon \times 10^{-3}$	pH	λ_{max}/nm	$\varepsilon \times 10^{-3}$
Adenosine 5′-phosphate (**1**)	$C_{10}H_{14}N_5O_7P$	192	−26.0	2	257	15.0	11	259	15.4
Guanosine 3′-phosphate (**2**)	$C_{10}H_{14}N_5O_8P$	175–180	−57.0	1	257	12.2	10.8	257	11.25
Cytidine 3′-phosphate (**3**)	$C_9H_{14}N_3O_8P$	233	+27.1	2	279	13.0	12	272	8.9
Uridine 2′-phosphate (**4**)	$C_9H_{13}N_2O_9P$	190–191	+22.3	2	262	9.9	13	261	7.3
Thymidine 5′-phosphate (**5**)	$C_{10}H_{15}N_2O_9P$	—	+7.3	2	267	10.2	7	267	10.2
Adenosine 3′,5′-cyclic phosphate (**6**)	$C_{10}H_{12}N_5O_6P$	219–220	−51.3	2	256	14.5	6	260	15.0
Adenosine 5′-diphosphate (**7**)	$C_{10}H_{15}N_5O_{10}P_2$	215 (dec)	—	2	257	15.0	11	259	15.4
Adenosine 5′-triphosphate (**8**)	$C_{10}H_{16}N_5O_{13}P_3$	218 (dec)	—	2	257	14.7	11	259	15.4

[a]Taken from 'Handbook of Biochemistry selected data for Molecular Biology', 2nd edn., ed. H. A. Sober, The Chemical Rubber Co., Cleveland, Ohio, 1970.

elution with an increasing concentration of acid or salt, is a standard method for the isolation and analysis of nucleotides. More recently, ion-exchange celluloses and dextrans have been developed as these have the advantage of increased selectivity over polystyrene resins.[4] Gas–liquid chromatography has been little used for the detection and analysis of nucleotides,[5] mainly on account of their involatile nature and the need for derivatization before analysis.

The first mononucleotide to be isolated was inosinic acid (IMP, **9**), which was obtained from meat hydrolysates by Liebig in 1847, some 20 years before the isolation of nucleic acids from pus cells by Miescher. The relationship between mononucleotides and nucleic acids became apparent in the first half of the twentieth century, largely as a result of work by Levene and others.[6] Inosinic acid is not of common occurrence and was formed during Liebig's isolation by the deamination of AMP, which itself was not isolated until 1927 when it was obtained from muscle. Since then the common nucleotides of adenine, cytosine, guanine, thymine, and uracil have all been isolated, as well as many minor nucleotides, *e.g.* those derived from pseudouridine, dihydrouridine, and methylated adenosines or guanosines.

The first nucleotide coenzyme to be discovered was nicotinamide-adenine dinucleotide (NAD^+, **10**), which was observed early in the twentieth century by Harden and Young as

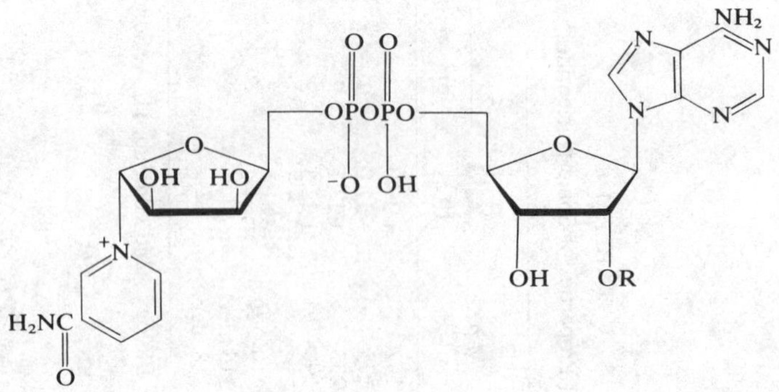

(10)

NAD^+ R = H
$NADP^+$ R = PO_3H_2

the heat-stable cofactor of alcoholic fermentation. Following the development of radioactive tracers and mild isolation techniques, *e.g.* ion-exchange chromatography, very many other coenzymes have been found.[7] These participate in biological oxidation/reduction, group transfer, and polymer synthesizing reactions. These coenzymes will be discussed in more detail later in this chapter. Other important, naturally occurring phosphate esters, such as those which are constituents of cell membranes (phospholipids and teichoic acids) or which are involved in the biosynthesis of natural products such as terpenes or steroids, will not be discussed here but will be dealt with in the appropriate chapters of this volume.

Nucleotides are monomer units of nucleic acids, the latter being formed by the enzymic polymerization of nucleoside triphosphates. This reaction is discussed in detail in Chapter 22.4. One of the most important naturally occurring mononucleotides is AMP, which, together with ADP and ATP, plays an important role in intermediary metabolism. ADP and ATP are acid anhydrides and can play an important part in biological processes by accepting or donating phosphoryl residues. For example, in the presence of the appropriate enzyme, ATP and flavin mononculeotide (FMN) interact to give the oxidation/reduction coenzyme flavin-adenine dinucleotide (FAD, **11**).

$$\text{FMN} + \text{ATP} \rightleftharpoons \text{FAD} + \text{inorganic pyrophosphate}$$

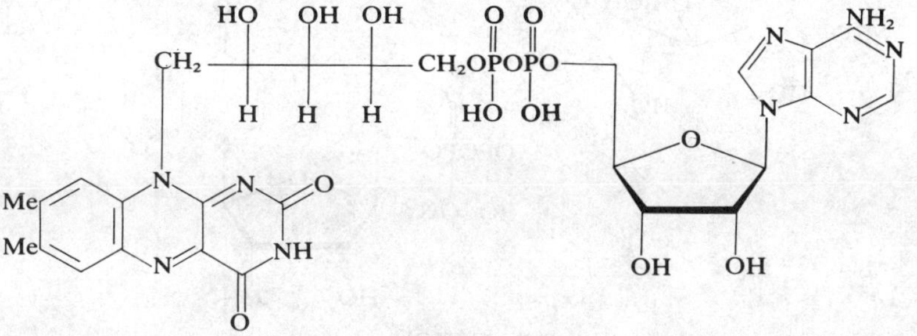

(11) FAD

The nicotinamide coenzymes NAD^+ and its 2′-phosphate $NADP^+$ are also oxidation/reduction coenzymes and are involved in the dehydrogenation of alcohols to carbonyl compounds (equation 1). Hydride transfer takes place from the alcohol, *e.g.* ethanol or lactate to the 4-position of the nicotinamide ring, generating reduced nicotinamide-adenine dinucleotide (NADH). Two classes of nicotinamide-dependent dehydrogenases are known which differ in the stereospecificity of the hydride transfer from C-4 in the NADH. L-Lactate dehydrogenase and liver alcohol dehydrogenase are two common enzymes which transfer hydride to the nicotinamide to give the *R*-configuration at C-4.[8] The function of the adenosine and pyrophosphoryl residues in these coenzymes is to bind the coenzyme to the apoenzyme protein in the most advantageous manner for the hydride transfer to take place.

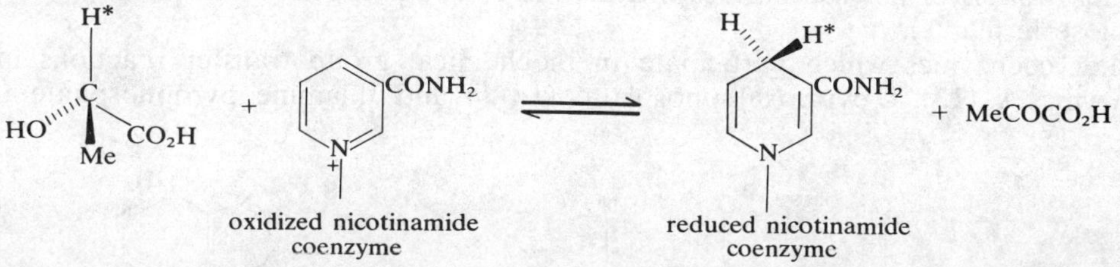

(1)

Nicotinamide coenzymes can generally be disassociated from their apoenzymes without undue difficulty; however, flavin coenzymes (**11**) are much more tightly bound to their apoenzymes and frequently can only be removed following extensive denaturation. Flavoproteins catalyse a wide variety of dehydrogenation reactions, *e.g.* the reduction of NAD^+, the oxidation of purines such as xanthine, the oxidation of amino-acids, and the dehydrogenation of succinate. In the last reaction, two hydrogen atoms *trans* to one another are removed to give fumarate:

$$\text{Succinate} + \text{FAD} \rightleftharpoons \text{fumarate} + \text{FADH}_2$$

Until 1970 it was believed that the reduction of the flavin moiety occurred in two one-electron steps and that semiquinones were intermediates. However, it has recently been suggested[9] that the oxidation/reduction observed in reductions catalysed by flavoproteins (equation 2) may not be due to direct hydrogen transfer but may be the consequence of a multi-step reaction sequence not involving free radicals; studies on substrate analogues support this view.

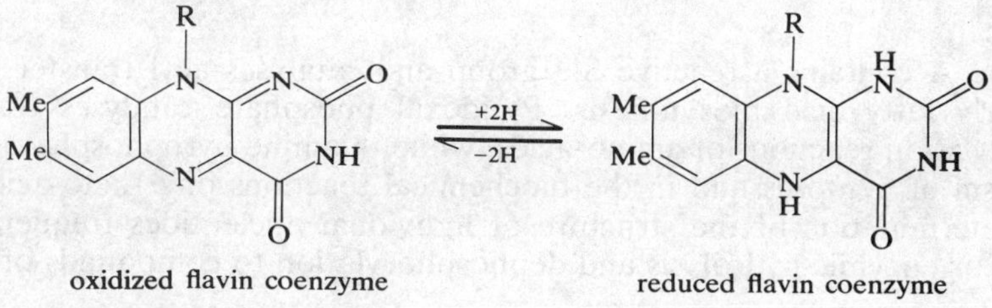

(2)

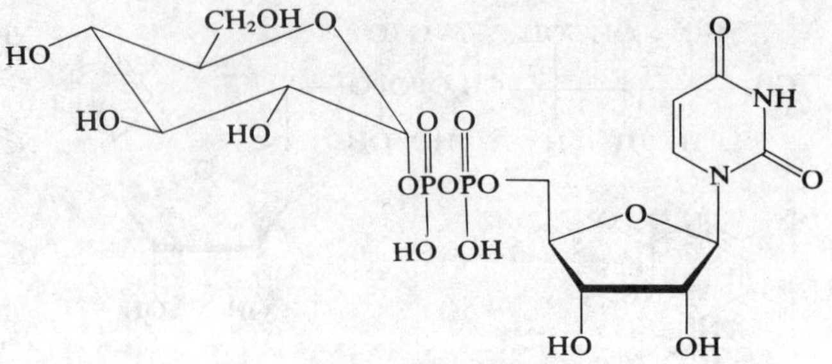

(12) UDPGlc

Another function of nucleotide coenzymes in biological reactions is to catalyse group transfer reactions. Thus, nucleoside diphosphate sugars (*e.g.* uridine diphosphate glucose, UDPGlc, **12**) play a major role in the synthesis of oligo- and poly-saccharides. This can be illustrated by considering the biosynthesis of sucrose in plants.[10] One of the major routes involves the conversion of D-fructose 6-phosphate into sucrose phosphate, followed by the almost irreversible hydrolysis of the latter to sucrose:

$$\text{UDPGlc}+\text{fructose 6-phosphate} \rightleftharpoons \text{UDP}+\text{sucrose phosphate}$$

In this reaction, the glucose–pyrophosphate bond is cleaved, with the transfer of the glucosyl residue to the fructose phosphate. UDPGlc is also the precursor of starch in plants and glycogen in yeast. Other nucleoside diphosphate sugars which are involved in the biosynthesis of polysaccharides include ADPGlc (glycogen in bacteria) and GPDGlc (cellulose in plants).

Other coenzymes which participate in biochemical group transfer reactions include Coenzyme A (**13**),[11] pyridoxal phosphate (**14**),[12] and thiamine pyrophosphate (**15**).[13]

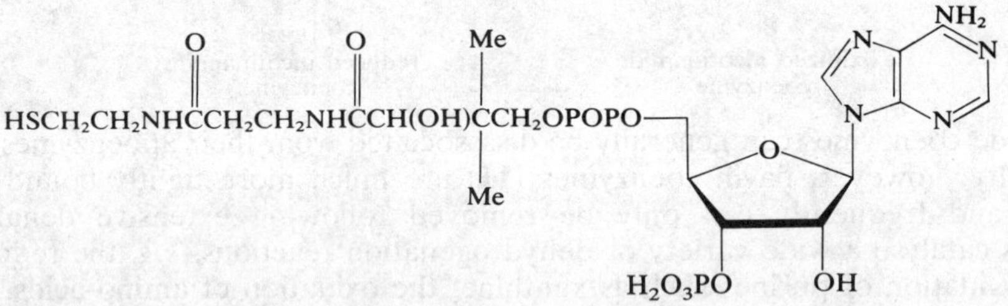

(13) CoA

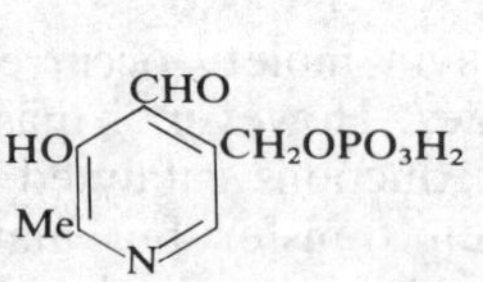

(14)

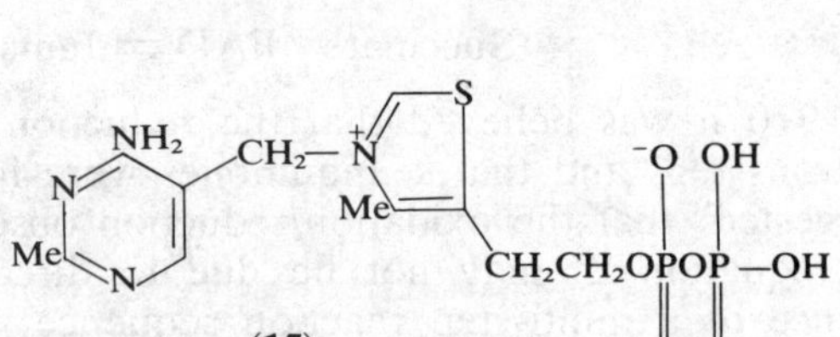

(15)

Coenzyme A contains a reactive SH group and catalyses acyl transfer reactions *in vivo*, particularly fatty acid biosynthesis. Pyridoxal phosphate catalyses transamination and decarboxylation reactions of amino-acids, while thiamine pyrophosphate is involved in the metabolism of pentoses and in the biochemical reactions of α-keto-acids.

The determination of the structure of individual nucleotides frequently makes use of chemical or enzymic hydrolysis and dephosphorylation to compounds of known structure.

Historically, the heterocyclic bases and constituent sugars of the common nucleotides were soon recognized, and the remaining problem was the location of the phosphoryl residue on the sugar.[14] This was determined comparatively easily for the 5′-nucleotides and for deoxynucleoside 3′-phosphates. However, the ready interconversion of ribonucleoside 2′- and 3′-phosphates caused problems for the early workers. The unequivocal location of the phosphoryl residues in these nucleotides followed the application of ion-exchange chromatography to the separation of the components of the hydrolysate of yeast RNA. Two isomeric pairs *a* and *b* were obtained for the four common nucleotides. Careful hydrolysis of adenylic acids *a* and *b* gave ribose 2- and 3-phosphates, respectively, which were identified by their reduction to ribitol phosphates. Ribitol 3-phosphate is optically inactive, while an optically active compound was obtained by the reduction of ribose 2-phosphate. The structures of the other nucleotide isomers was obtained in a similar manner, and the structures of nucleotide coenzymes were also determined by hydrolysis.

The application of modern techniques, particularly radioactive labelling, enzymic hydrolysis, and spectroscopy to the structure determination of nucleotides available in very limited amounts is well illustrated by the following example.[15] A nucleotide involved in the control of RNA synthesis in *Escherichia coli* was isolated by ion-exchange chromatography and shown to contain guanine by ultraviolet spectroscopy. The presence of ribose was also inferred by a specific colour reaction. When the *E. coli* was grown in a medium containing ^{32}P-labelled orthophosphate, the nucleotide was obtained radioactively labelled. Hydrolysis of the labelled nucleotide by a phosphomonoesterase revealed that it contained four phosphate residues per guanosine molecule, and it was considered unlikely that any of these phosphate residues were attached to the heterocyclic base as this would cause a significant change in the u.v. spectrum of the guanine. The nucleotide was stable to periodate and hence some of the phosphate residues must be attached to the 2′- or 3′-hydroxyls of the sugar. The mobility of the nucleotide during ion-exchange chromatography at different pH values indicated that there were four primary and two secondary phosphoryl dissociations. From the above evidence three isomeric structures could be envisaged: pppGp, ppGpp, and pGppp (or their 2′-isomers). Snake venom phosphodiesterase which cleaves between the α- and the β-phosphate residues of nuclcoside 5′-polyphosphates degraded the nucleotide to a triphosphate, and hence the structure must be guanosine 3′,5′-dipyrophosphate (**16**) or the 2′,5′-dipyrophosphate. Final proof of the structure of (**16**) came from its enzymic synthesis.[16] Using an impure enzyme system from *E. coli*, the terminal pyrophosphoryl residue of ATP was transferred to GDP to give (**16**). In two separate experiments using ATP labelled with ^{32}P in either the β- or the α-positions, ppG^{32}pp and ppGp^{32}p were obtained. Hydrolysis of ppG^{32}pp with a pyrophosphatase gave the known guanosine 3′,5′-diphosphate labelled in the 3′-position, and hence (**16**) was the 3′,5′-dipyrophosphate.

The conformations of mono- and oligo-nucleotides in solution and in the solid state have been studied by a number of techniques.[17] For example, optical rotatory dispersion and circular dichroic spectroscopy show that oligonucleotides exist in solution in the *anti*- (**17**, **18**) rather than in the *syn*-conformation (**19**, **20**), and X-ray diffraction measurements show that this tends to be the preferred conformation for mononucleotides in the crystalline state.[18] From the change in the chemical shift of the H-8 proton with pH in the ^{1}H nuclear magnetic resonance spectra of purine nucleoside 5′-phosphates, it has been deduced[17] that these nucleotides are in the *anti*-conformation as there is little change in the chemical shift with pH for the 3′-phosphates. The lanthanide probe technique indicates that the nucleoside moiety in AMP is in the *anti*-conformation both in water and in DMSO, although the torsion angle differs in the two solvents.[19] An X-ray diffraction determination of the structure of disodium ATP shows that the nucleoside moiety is in the *anti*-conformation and that the phosphate chain is folded back towards the adenine residue.[20] The proximity of the phosphate chain to the adenine ring can also be shown by ^{1}H n.m.r., as paramagnetic ions form complexes with ATP in which the signal due to H-8 is considerably broadened.[21]

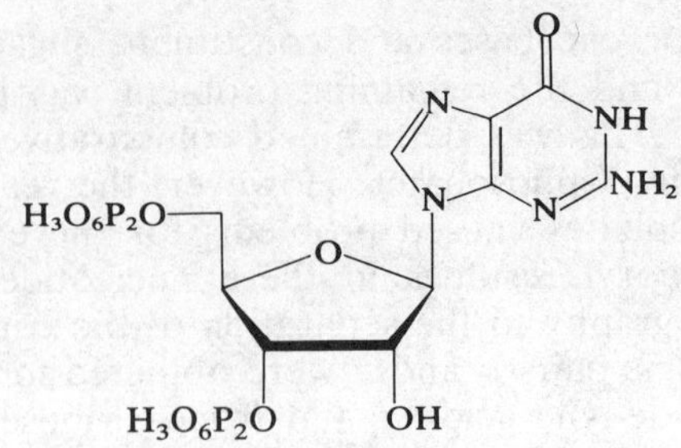

(16)

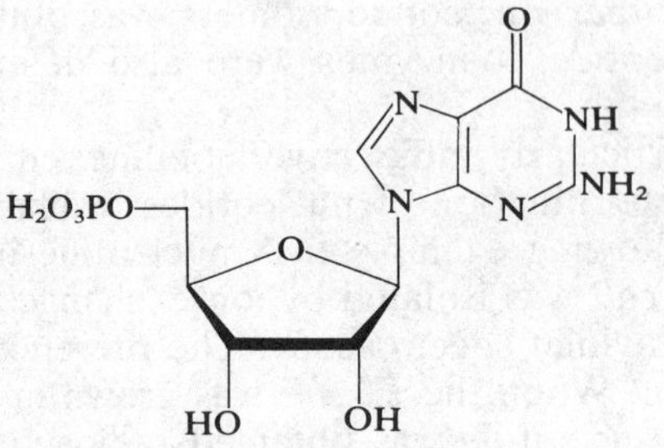

(17) *anti*-pG

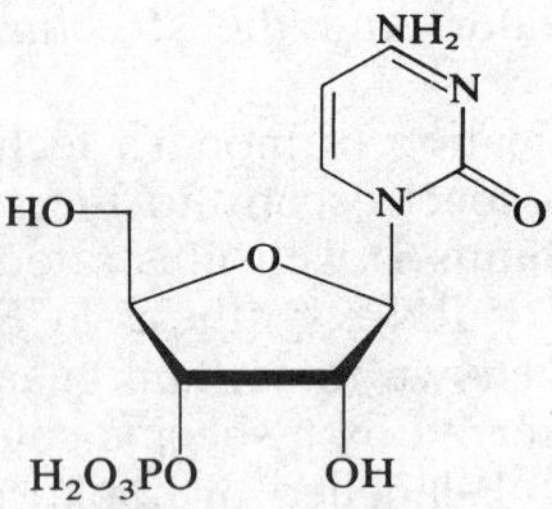

(18) *anti*-Cp

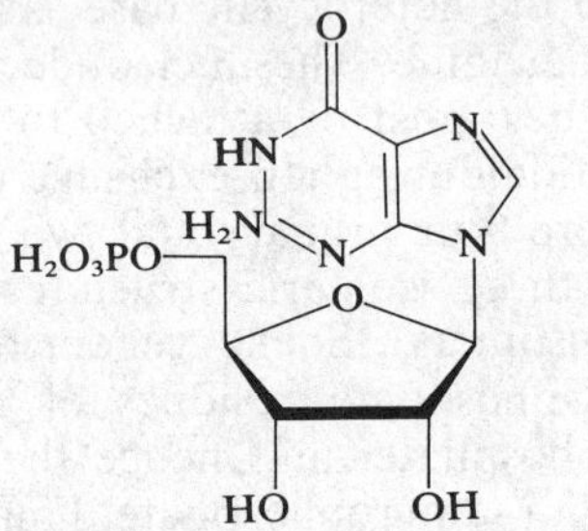

(19) *syn*-pG

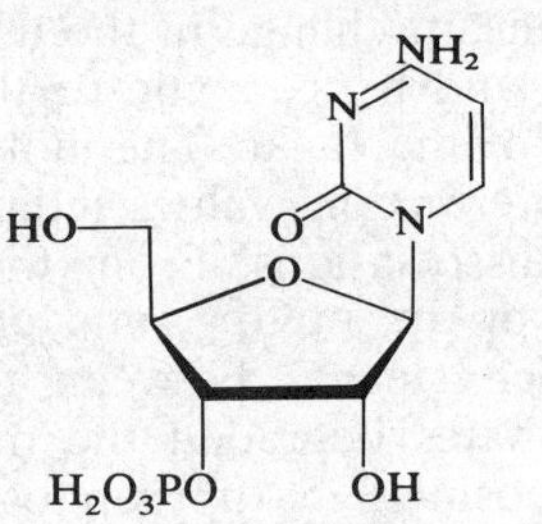

(20) *anti*-Cp

22.3.2 NUCLEOPHILIC REACTIONS OF PHOSPHATE ESTERS AND ANHYDRIDES

Since the reactions of the sugars and the heterocyclic bases in nucleosides have been discussed at length earlier (Chapter 22.2), only reactions involving the phosphorus atom in nucleotides will be described in this section. Thus, the chemical and enzymic hydrolysis of mono- and poly-nucleotides will be covered in some detail, while other hydrolysis reactions of nucleic acids, *e.g.* depurination in acid, will only be mentioned because of the increased lability of the phosphodiester chain resulting from this reaction.

Most simple mono- and di-alkyl phosphates are hydrolysed only slowly by acid or base[22] and this is the case with DNA. RNA, on the other hand, is much more susceptible to hydrolysis, breaking down at an appreciable rate in acid and hydrolysing rapidly in base. This increased reactivity of RNA is due to neighbouring group participation by the 2′-hydroxyl groups of the ribose residues. This is also observed with other phosphates. Thus, esters of glycerol phosphoric acids hydrolyse much more rapidly than esters of simple alicyclic phosphoric acids such as ethyl phosphoric acid. The initial products of hydrolysis of RNA are ribonucleoside 2′,3′-cyclic phosphates (**21**), and these are cleaved rapidly to the 2′- and 3′-phosphates. The ready cleavage of the ribonucleoside 2′,3′-cyclic phosphates is a general property of phosphodiesters in a five-membered ring, and these are hydrolysed 10^7–10^8 times faster than open-chain phosphodiesters under comparable conditions.[23] The enhanced reactivity of these cyclic phosphates is due in the main to ring strain, which can be released when a trigonal bipyramidal intermediate is formed during the hydrolysis reaction. In the case of ethylene phosphate (**22**) for example, addition of water to the phosphoryl group produces a quinquecovalent intermediate (**23**) in which the ethylene group spans one apical and one equatorial position. Since groups may enter and leave the quinquecovalent intermediate (**23**) only at apical positions, if pseudorotation[24] of this intermediate occurs to give (**24**) either ring fission or incorporation of oxygen from solvent can take place. Moreover, ring fission occurs with exclusive P—O cleavage. Ribonucleoside 2′,3′-cyclic phosphates can form similar quinquecovalent intermediates during acid hydrolysis and pseudorotation could place either C-2′—O or C-3′—O in an apical position, explaining the observed mixture of isomers of nucleotides formed during the hydrolysis.

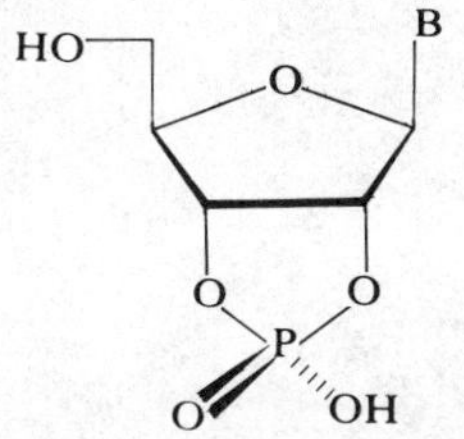

(**21**)

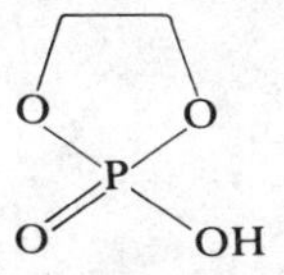

(**22**)

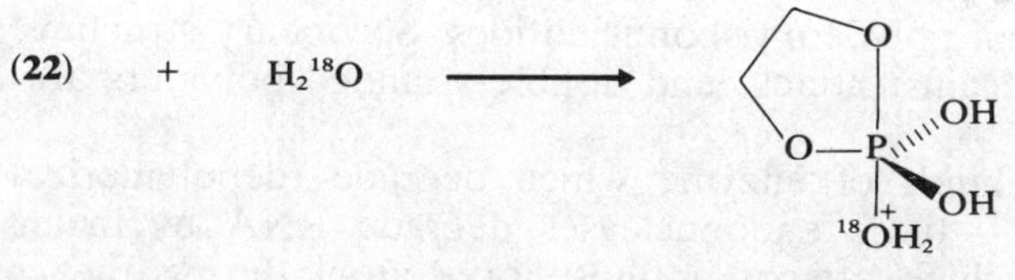

(**23**)

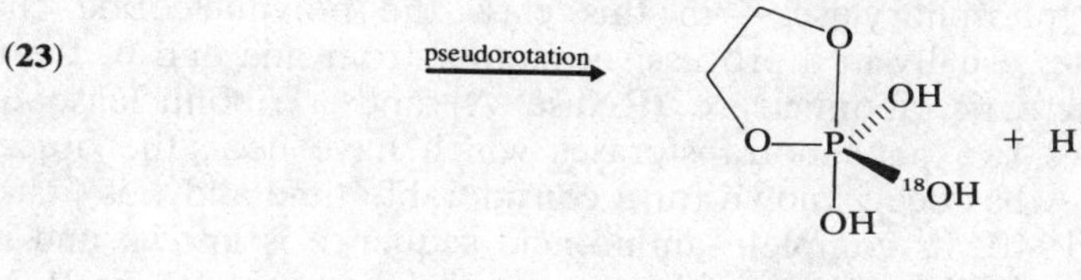

(**24**)

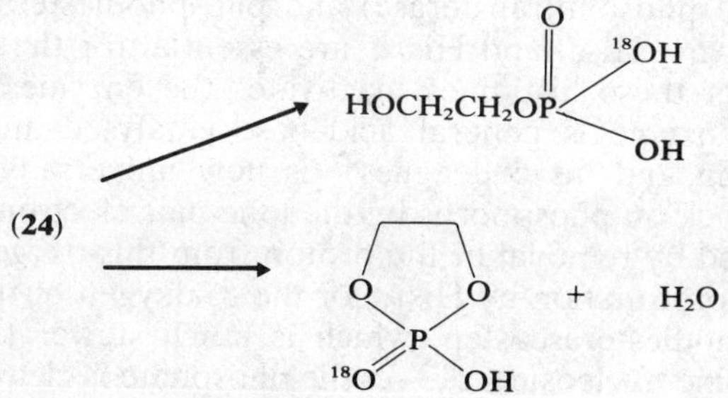

The base-catalysed hydrolysis of polyribonucleotides or ribonucleoside 2′,(3′)-phosphodiesters may proceed by intermediates similar to (**23**). No isomerization of nucleoside 3′-phosphodiesters to 2′-phosphodiesters occurs during partial base-catalysed hydrolysis. This may be because pseudorotation of the quinquecovalent intermediate is inhibited owing to the presence of the negatively charged P—O$^-$ rather than the P—OH ligands. It is energetically unfavourable to place negatively charged ligands in apical positions during pseudorotation; hence the 3′-phosphates predominate as the products of the base-catalysed hydrolysis of RNA. Alternatively, this reaction can be regarded as a simple S_N2(P) displacement proceeding with an 'in-line' mechanism (equation 3).

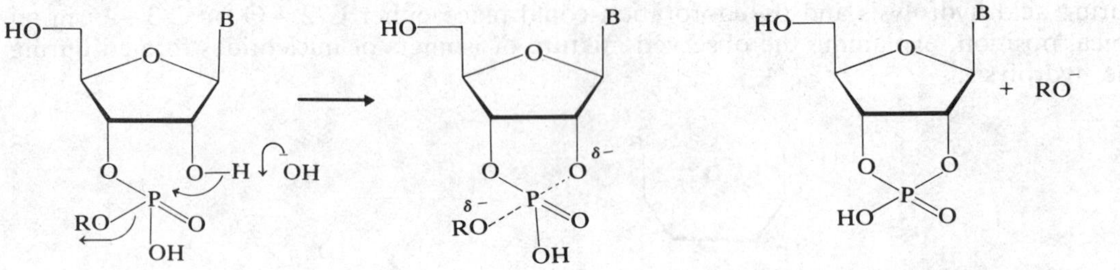
(3)

In-line displacement

Metal ions can act as electrophilic catalysts during the hydrolysis of phosphate esters and here presumably the reaction proceeds through the formation of a metal complex. Once again the presence of a neighbouring hydroxyl function has a pronounced rate-enhancing effect, and polyribonucleotides are much more susceptible to metal-ion catalysed degradation than polydeoxyribonucleotides. Secondary structure appears to protect the polynucleotide against attack, and double-stranded polymers are attacked relatively slowly.[25]

There are three kinds of enzyme which degrade (depolymerize) nucleic acids: (i) phosphotransferases — these endonucleases degrade RNA by means of a 2′-hydroxyl group of a ribose which attacks a phosphorus atom during the chain-breaking step, thereby transferring the phosphorus atom to this 2′-group from the 5′-hydroxyl of the adjacent nucleotide; (ii) phosphodiesterases — there are a large number of these enzymes (which can be endo- or exo-nucleases), differing in their specificity for bases and sugar residues; (iii) phosphorylases — in this class the polynucleotide chain is cleaved by orthophosphate, usually in a processive manner from one end of the chain.

Bovine pancreatic ribonuclease (RNase A) and a ribonuclease of microbial origin (RNase T_1) are two phosphotransferases which have been the objects of considerable study. RNase A has been known for a considerable time and was first isolated crystalline by Kunitz in 1940. Its complete amino-acid sequence is known and its crystal structure has been determined by X-ray diffraction.[26] The enzyme cleaves RNA specifically after pyrimidine residues in the polynucleotide chain, producing oligonucleotides which terminate in pyrimidine nucleoside 3′-phosphates. The hydrolysis of RNA is a two-step process consisting of chain cleavage (phosphotransferase) and phosphodiesterase reactions. Two histidine residues in the enzyme His_{12} and His_{119} are essential for the catalytic reaction, and methoxycarbonylation of these histidines inactivates the enzyme (equation 4). The cleavage of RNA by the enzyme is general acid–base catalysed and the mechanism originally proposed by Rabin and his colleagues[27] is now universally accepted. In the phosphotransferase step, attack on phosphorus by the lone pair electrons of the 2′-oxygen of the ribose ring is facilitated by removal of the proton from this oxygen by His_{12}. Chain cleavage is accompanied by protonation by His_{119} of the 5′-oxygen of the next nucleotide in the chain. In the phosphodiesterase step, which is much slower than the phosphotransferase step, the pyrimidine nucleoside 2′,3′-cyclic phosphate is cleaved by the enzyme

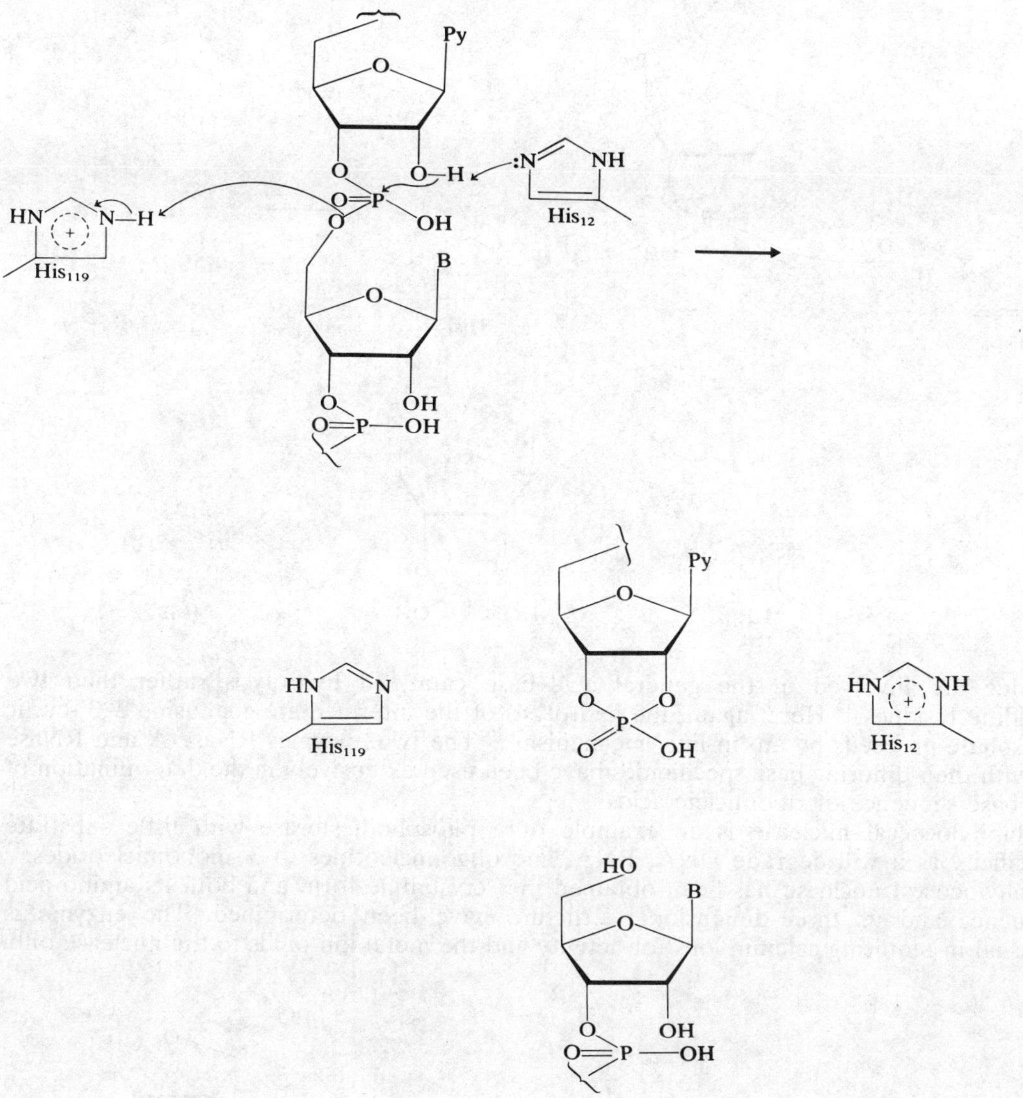

(4)

specifically to the 3′-phosphate in the presence of water (equation 5). This reaction is also subject to general acid–base catalysis by the two histidine residues in the ribonuclease.

The phosphodiesterase step proceeds by an 'in-line' mechanism and this has been shown unequivocally for uridine 2′,3′-cyclic phosphorothioate.[28] Of the two diastereoisomers of this cyclic phosphorothioate, one (**25**) can be obtained crystalline as its triethylammonium salt. This isomer is hydrolysed by RNase A and the kinetic parameters for this hydrolysis are similar to those for uridine 2′,3′-cyclic phosphate. When (**25**) was hydrolysed in ^{18}O-enriched water to the 3′-phosphorothioate and the product reconverted into the cyclic phosphorothioate by a method with known 'in-line' stereochemistry, the amount of incorporation of isotope into the crystalline salt was in excellent agreement with the value required for 'in-line' opening of the cyclic phosphorothioate by the ribonuclease (Scheme 1).

Ribonuclease T_1 was discovered much more recently than RNase A; its amino-acid sequence is known, but a complete crystal structure has not yet been published. RNase T_1 is an endonuclease which cleaves RNA after guanosine residues and its mechanism of action is very similar to that of RNase A except that a histidine and a glutamic acid

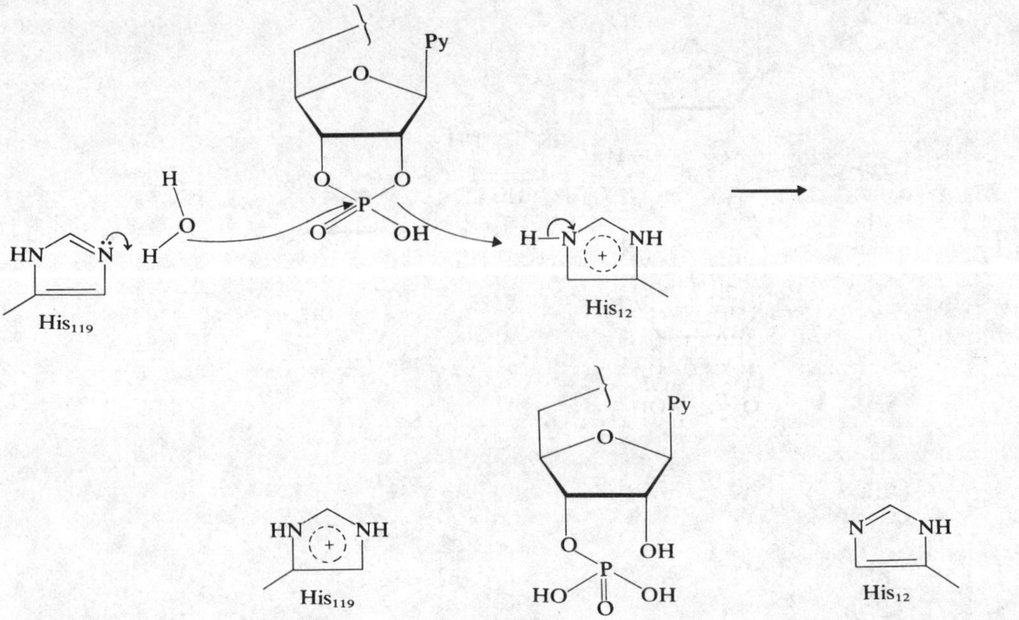

(5)

residue are involved in the general acid–base catalysed hydrolysis rather than two histidine residues.[29] Here, again, the hydrolysis of the intermediate guanosine 2′,3′-cyclic phosphate proceeds by an 'in-line' mechanism.[30] The two enzymes RNase A and RNase T_1 with their differing base specificities have been used extensively in the determination of the base sequence of ribonucleic acids.

Staphylococcal nuclease is an example of a phosphodiesterase with little substrate specificity as it will degrade DNA, RNA, and oligonucleotides to 3′-mononucleotides.[31] Staphylococcal nuclease has been obtained in a crystalline form and both its amino-acid sequence and its three-dimensional structure have been determined. The enzyme is unusual in requiring calcium ions for activity and the metal ion binds to the nuclease only

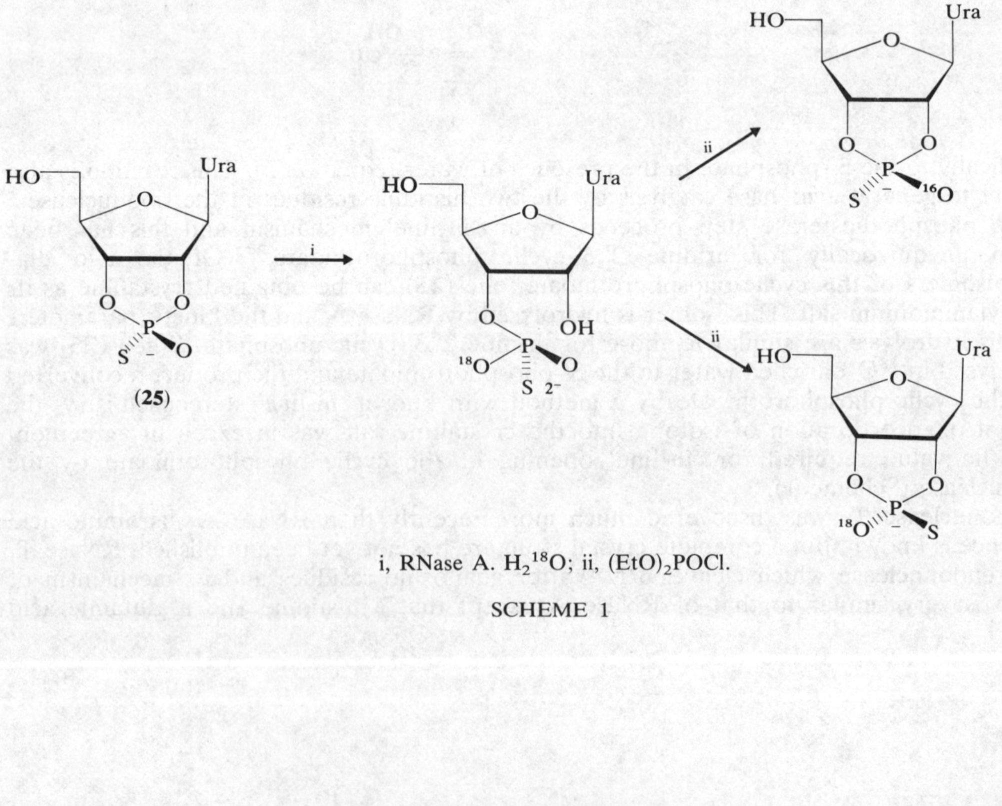

i, RNase A. $H_2{}^{18}O$; ii, $(EtO)_2POCl$.

SCHEME 1

in the presence of substrate or an inhibitor, *e.g.* thymidine 3′,5′-diphosphate. Staphylococcal nuclease acts as an endonuclease, but once chain-breaks have been made, it can act as an exonuclease. An unusual feature of the hydrolysis of thymidine 3′-phosphate 5′-(4-nitrophenyl) phosphate by the nuclease is that thymidine 3′-phosphate and 4-nitrophenyl phosphate are the sole products.[32] No 4-nitrophenol is formed, as might be expected in the chemical hydrolysis of this substrate. 4-Nitrophenoxide would be expected to be a far better leaving group in this hydrolytic reaction than the 5′-oxyanion of thymidine, and there is no kinetic evidence for a group acting as a general acid which would enable the neutral nucleoside to be eliminated. It has been suggested that the hydrolytic path can be explained by assuming the formation of a quinquecovalent intermediate which cannot pseudorotate freely owing to steric constraints imposed by the enzyme, particularly by co-ordination to the calcium ion. Thus, the better leaving group (4-nitrophenoxide) cannot assume an apical position in the intermediate. When expulsion of the thymidine 5′-oxyanion from an apical position occurs, it is followed by protonation in a non-rate-limiting step.

Polynucleotide phosphorylase, which is an example of the third class of polynucleotide-degrading enzyme, will degrade polyribonucleotides in the presence of inorganic orthophosphate to give nucleoside 5′-diphosphates in a reversible process.[33] Although this enzyme has been known for over 20 years, its biological function is still unknown. Polynucleotide phosphorylase was first recognized as an enzyme which could polymerize nucleoside 5′-diphosphates:

$$\text{ADP} \underset{\text{Mg}^{2+}}{\overset{\text{polynucleotide phosphorylase}}{\rightleftharpoons}} \text{poly(A)} + \text{P}_i$$

and with the pure enzyme from certain bacteria, *e.g. E. coli*, a primer is necessary for polymerization to occur without an initial time lag. Polynucleotide phosphorylase has not yet been obtained pure and there is no information on amino-acid residues which might be involved in its active site. However, despite the lack of knowledge as to its detailed mechanism of action, the enzyme is used extensively for the preparation of polyribonucleotides containing natural or synthetic base residues.

Chemical depurination of a polynucleotide can be followed by chain scission. For example, diphenylamine in aqueous formic acid will degrade DNA to oligonucleotides with the release of some inorganic orthophosphate (Scheme 2),[34] a reaction which has been used extensively in the sequence determination of nucleic acids.

One product of the reaction is levulinic acid, and presumably the aldehyde group of the deoxyribose residues which are exposed by the depurination reacts with the diphenylamine to form a Schiff base which then eliminates the phosphoryl residue from the 5′-position. Degradation of depurinated DNA with alkali does lead to chain scission, but this method is less satisfactory than the acid-catalysed reaction described above as there are many side reactions. The terminal base can be eliminated in an amine-catalysed reaction from RNA following periodate oxidation (Scheme 3). The amine catalyses the removal of the modified nucleoside at the 3′-end of the RNA and, once eliminated, further oxidation of the fragment liberates the free base.[35]

Adenosine triphosphate occupies a central role in intermediary metabolism and Lipmann introduced the concept of 'energy-rich phosphate bonds' in 1941 to explain why the standard free energy of hydrolysis of ATP and certain other phosphates, e.g. creatine phosphate, appeared to be so much higher than the standard free energy of hydrolysis of other phosphates such as AMP.[36] This concept has frequently been used in discussions of the reactions of ATP,[37] and it has been claimed on a number of occasions that ATP can 'store' energy released by degradative metabolic processes and can use the stored energy as required to drive synthetic reactions. Recently the concept of energy-rich phosphate bonds has been critically reassessed,[38] and it has been pointed out that the Lipmann concept is appropriate only for closed systems which contain energy-linked reactions. Since real organisms are open systems, this concept of energy-rich bonds cannot strictly be applied, and while phosphate esters can be arranged in order of decreasing standard free

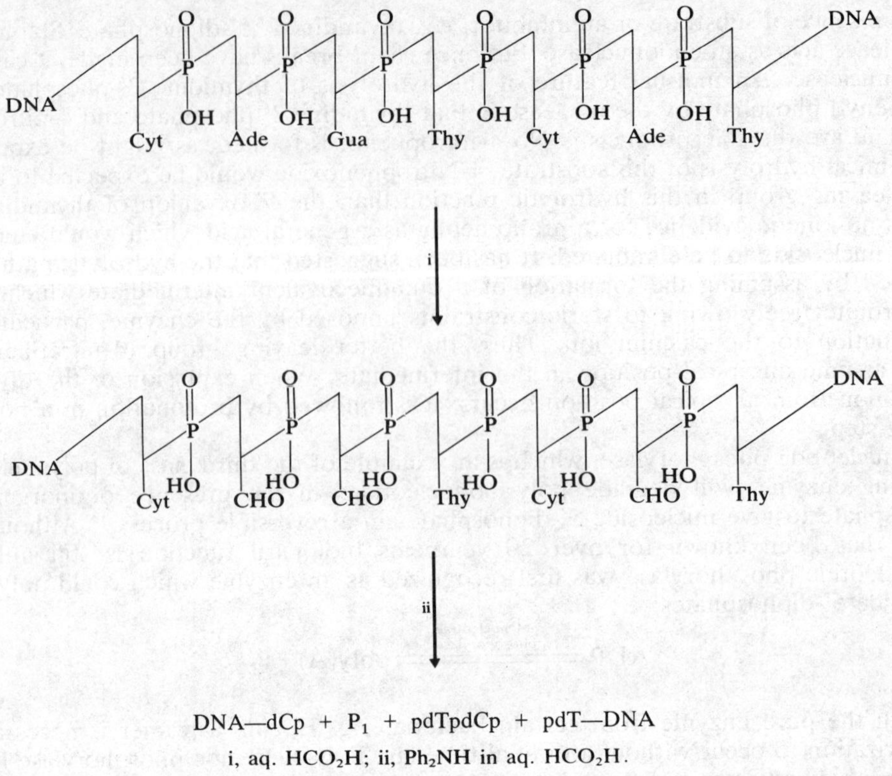

SCHEME 2 Degradation of DNA by formic acid and diphenylamine.

energy of hydrolysis, this is only an indication of the direction of transphosphorylation in a closed system.

The vital role of ATP in metabolic processes is due to its capability to phosphorylate substrates to form esters or mixed anhydrides which can then be hydrolysed to the required products. The vast majority of phosphoryl transfer reactions involving ATP take place at either the α- or the γ-phosphorus atoms when either an orthophosphoryl or an adenylyl residue is transferred [Scheme 4, pathways (i) and (iii)]. Very few reactions of

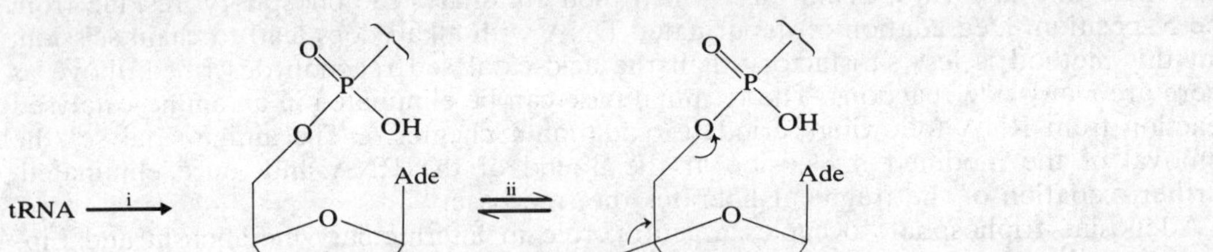

SCHEME 3

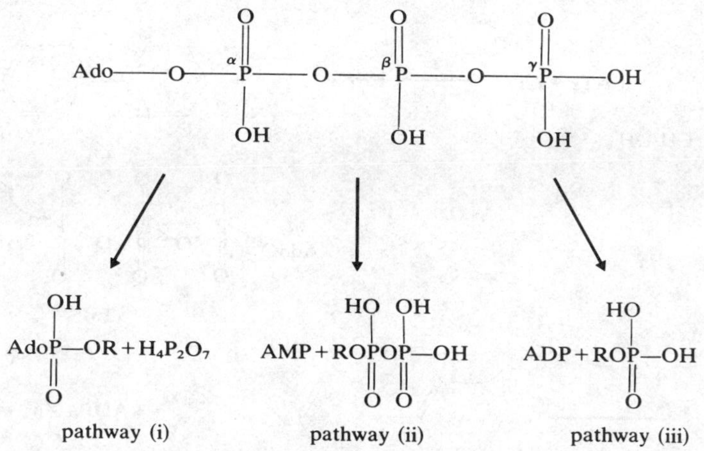

Possible pathways for ATP cleavage

SCHEME 4

ATP take place at the β-phosphorus atom, leading to pyrophosphoryl transfer [pathway (ii)]. One reason for this difference in reactivity may be due to differences in electron density at the phosphorus atom, and ^{31}P n.m.r. shows that the α- and γ-phosphorus atoms are deshielded with respect to the β-phosphorus atom.[39] Phosphoryl and nucleotidyl transfer reactions can take place by mechanisms varying between two extremes: (i) a dissociative (S_N1) mechanism in which the departure of a nucleophile leaves triply co-ordinate metaphosphate as the reactive species; or (ii) an S_N2 mechanism involving a quinquecovalent intermediate, as has been mentioned earlier for nucleases. Protonation, metal co-ordination or esterification inhibit reactions taking place by the S_N1 mechanism, presumably because of decreased availability of lone-pair electrons on oxygen for $d_\pi - p_\pi$ bonding with phosphorus in the metaphosphate intermediate. On the other hand, an S_N2 displacement should be accelerated by charge neutralization such as metal co-ordination, presumably because electron withdrawal from phosphorus is increased.[40] Not only can metal chelation affect the electrophilicity of phosphorus, but it can also induce strain in the ground state and can favourably influence pseudorotation. Metal ions can, in addition, co-ordinate simultaneously with the entering ligand and the phosphoryl group under attack and hence increase the nucleophilicity of the entering ligand. A example of the effect of a metal ion on the mechanism of an enzymic reaction has already been mentioned for Staphylococcal nuclease, when the metal ion appears to influence the pseudorotation of a quinquecovalent intermediate.

The chemical hydrolysis of ATP to ADP and inorganic phosphate is accelerated by divalent metal ions, and Cu(II) ions are the most effective of those studied so far.[41] The adenine base also plays an important part in the hydrolysis by co-ordinating through N-7, and Cu(II) ions have little effect on the hydrolysis of CTP to CDP when this extra co-ordination cannot take place. Another example of a metal ion-promoted dephosphorylation of ATP is the zinc ion-catalysed phosphorylation of 2-hydroxymethyl-1,10-phenanthroline.[42] The phosphoryl transfer is absolutely dependent on the presence of zinc ions and presumably proceeds through a ternary complex (**26**) involving the phenanthroline, ATP, and a zinc ion. Once (**26**) is formed the stereochemistry of the complex may bring the primary alcohol function of the phenanthroline close to the α-phosphoryl group of the ATP and hence phosphorylation is assisted (equation 6). It has been suggested that this reaction is a model for many biological phosphoryl transfer reactions from ATP.

Recent ^{31}P n.m.r. studies show that copper ions bind to the β- and γ-phosphoryl groups of ATP, but magnesium ions bind only to the β-phosphoryl group,[39] in contrast to

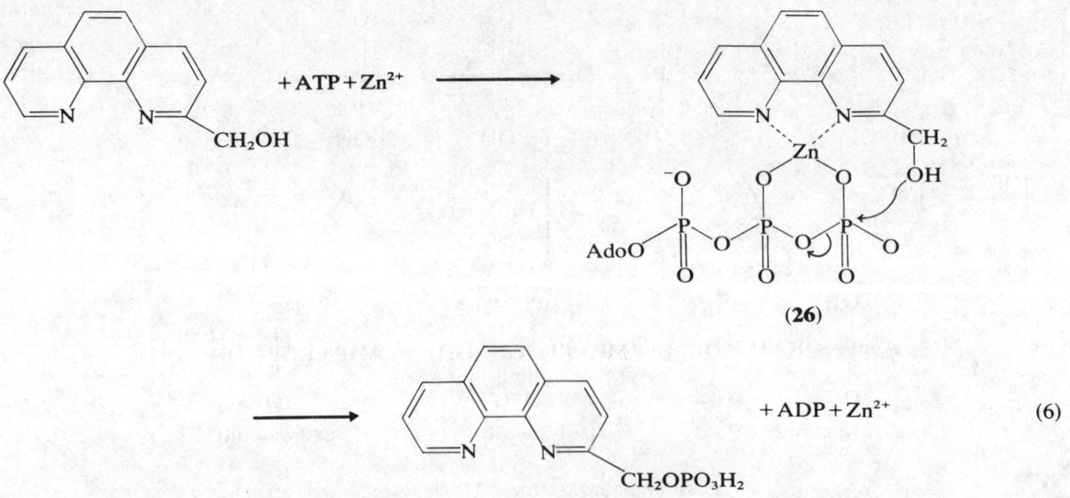

(**26**)

(6)

conclusions reached in an earlier study.[21] Manganese(II), nickel(II), and cobalt(II) ions bind to all three phosphoryl groups of ATP, as well as N-7 of the adenine base. This difference in co-ordination behaviour between Mn(II) and Mg(II) as observed by ^{31}P n.m.r. must be taken into account in model studies when Mn(II) ions are used as paramagnetic probes for enzyme–substrate interactions, and mechanistic conclusions reached from studies with Mn(II) ions may not be valid *in vivo*. For example, in the reaction complex of many enzymic phosphoryl and nucleotidyl transfer reactions which require magnesium ions for full activity, co-ordination of the metal ion to the β-phosphoryl group of ATP may occur together with neutralization of charge on either the α- or the γ-phosphoryl groups by a positively charged amino-acid side-chain in the enzyme, or by another metal ion as in the case of membrane-bound ATPase,[40] which are dependent on monovalent ions as well as Mg(II) ions for full activity. Examples of enzymic reactions of nucleoside polyphosphate esters which involve phosphoryl, pyrophosphoryl, or nucleotidyl transfer are given in Table 2.

TABLE 2

Some Enzymic Phosphoryl, and Nucleotidyl Transfer Reactions involving Nucleoside Polyphosphate Esters[a]

Phosphoryl transfer	*Pyrophosphoryl transfer*	*Nucleotidyl transfer*
ATP: D-fructose 1-phosphotransferase	ATP: thiamine pyrophosphotransferase	ATP: NMN$^+$ adenylyltransferase
ATP: riboflavin 5′-phosphotransferase	ATP: D-ribose 5-phosphate pyrophosphotransferase	ATP: FMN adenylyltransferase
ATP: AMP phosphotransferase (adenylate kinase)		UTP: α-D-glucose 1-phosphate uridylyltransferase
ATP: creatine phosphotransferase		GTP: α-D-mannose 1-phosphate guanylyltransferase
ATP: acetate phosphotransferase		CTP: choline phosphate cytidylyltransferase
ATP: pyruvate phosphotransferase (pyruvate kinase)		Nucleoside diphosphate: polynucleotide nucleotidyl-transferase (polynucleotide phosphorylase)
etc.		Nucleoside triphosphate: RNA nucleotidyltransferase (RNA polymerase)
		Deoxynucleoside triphosphate: DNA nucleotidyltransferase (DNA polymerase)
		etc.

[a] Taken from M. Dixon and E. C. Webb, 'The Enzymes', Longmans, London, 1964, 2nd edn.

There are few well-authenticated reactions of ATP in which a pyrophosphoryl residue is transferred to a substrate. Two well-studied examples are the synthesis of D-ribose 5-phosphate 1-pyrophosphate from D-ribose 5-phosphate by an enzyme which has been isolate from many sources, and the synthesis of thiamine pyrophosphate from thiamine by an enzyme which has been isolated from both animal tissue and yeast. Both enzymes require magnesium ions as cofactor and if the β-phosphoryl group of the ATP co-ordinates with a magnesium ion, as demonstrated by ^{31}P n.m.r., then nucleophilic attack should lead to pyrophosphoryl transfer. However, further work is needed to elucidate the exact mechanisms of action of these enzymes.

Nucleotidyl transfer can occur with nucleoside polyphosphate esters other than ATP. For example, RNA and DNA polymerases catalyse nucleotidyl transfer from the common ribo- and deoxyribo-nucleoside triphosphates, while the polymerization reaction catalysed by polynucleotide phosphorylase is an example of an enzymic reaction in which nucleotidyl residues are transferred from ribonucleoside 5′-pyrophosphates. The biosynthesis of nucleoside diphosphate sugars is carried out by nucleotidyl transferring enzymes, *e.g.* UDP-glucose is formed by UTP-α-D-glucose 1-phosphate uridylyltransferase:

$$\text{UTP} + \alpha\text{-D-glucose 1-phosphate} \rightleftharpoons \text{UDPGlc} + \text{pyrophosphate}$$

DNA ligase is an enzyme that can join DNA chains to each other and rather unusual nucleotidyl transfer reactions occur during this repair process. When the ligase from bacteriophage T_4 is incubated with ATP, an adenylate-ligase is formed in which the adenylate residue is joined to the ε-amino-group of the protein by means of a phosphoramidate bond.[43] The phosphoramidate then reacts with a 5′-phosphoryl group of a 'nicked' DNA to make a pyrophosphate which is then cleaved by the 3′-hydroxyl group of the deoxyribonucleoside on the other side of the 'nick' to complete the internucleotide line and liberate AMP. The DNA ligase from *E. coli* is an even more unusual enzyme as the adenylate residue is transferred from NAD^+ to the ε-amino-group of the active lysine (Scheme 5). This is the only occasion observed so far when NAD^+ functions as a cofactor for adenylate transfer rather than in its more usual capacity as an oxidation/reduction coenzyme.

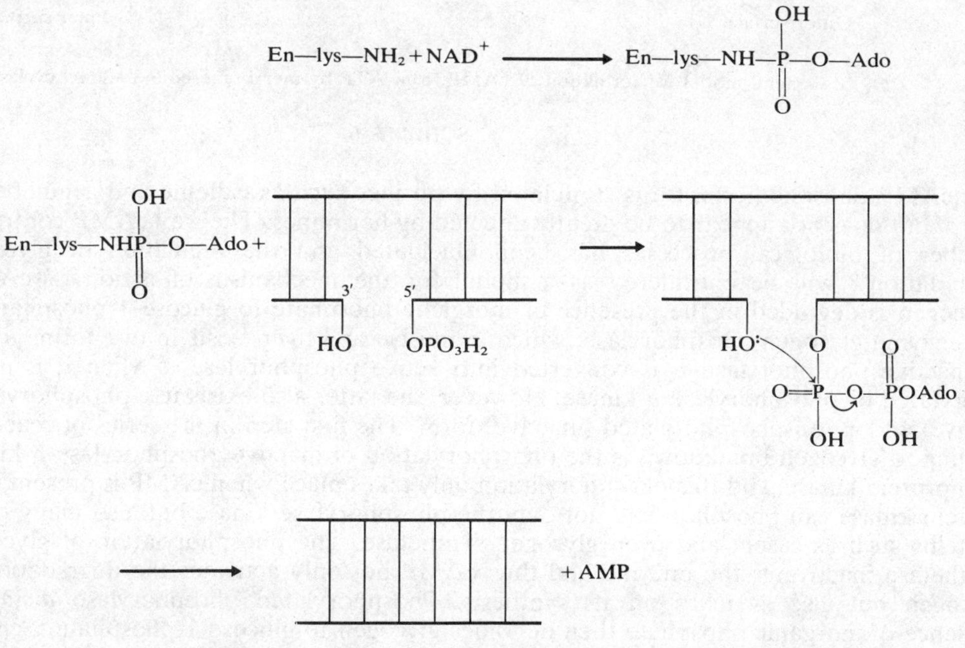

Schematic representation of repair of double stranded DNA by DNA-ligase from *E. coli*

SCHEME 5

Adenosine 3′,5′-cyclic phosphate (cAMP, **6**) is well established as an intracellular second messenger controlling the actions of many hormones.[44] Guanosine 3′,5′-cyclic phosphate is another second messenger controlling a variety of biological processes and other cyclic phosphates, *e.g.* cCMP, may also be involved in biological control. In tissues from a wide variety of sources, catecholamines and peptide hormones (*e.g.* insulin or glucagon) act *via* an enzyme, adenylate cyclase,[45] which converts ATP into cAMP. The level of cAMP in a cell at any given moment is controlled by two enzymes, adenylate cyclase and a specific phosphodiesterase which inactivates cAMP by cleaving it to 5′-AMP:

$$\text{ATP} \xrightarrow{\text{adenylate cyclase}} \text{PP}_i + \text{cAMP} \xrightarrow{\text{phosphodiesterase}} 5'\text{-AMP}$$

In general, adenylate cyclase is membrane-bound and is stimulated by hormones which bind to a receptor site on the outside of the membrane and activate the catalytic (cyclizing) site on the inside of the membrane which converts ATP to cAMP (Scheme 6).

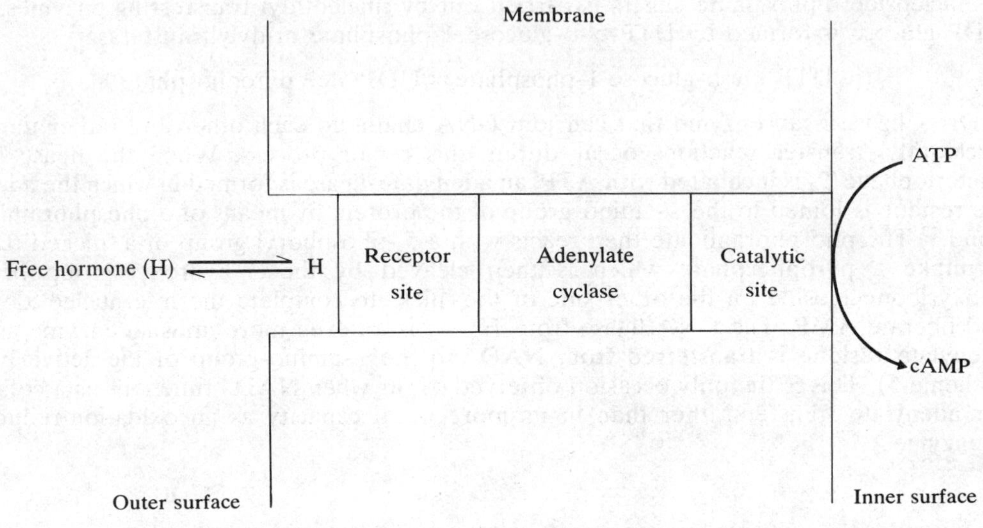

General model for formation of cAMP from ATP by membrane-bound adenyl cyclase

SCHEME 6

The cAMP phosphodiesterase is stimulated by purines such as caffeine and a number of drugs, but does not appear to be greatly affected by hormones. The way cAMP controls a number of biological processes has been elucidated and the regulation of glycogen degradation[46] will be considered as a model for the mechanism of action of cAMP. Glycogen is degraded in the presence of inorganic phosphate to glucose 1-phosphate by the enzyme glycogen phosphorylase, which in muscle and liver exists in two forms *a* and *b*. Inactive phosphorylase *b* is converted into active phosphorylase *a* when it is phosphorylated by phosphorylase *b* kinase. However, the latter also exists in a phosphorylated (active) and a non-phosphorylated (inactive) form. The first step in the series of reactions leading to glycogen breakdown is the phosphorylation of inactive phosphorylase *b* kinase by a protein kinase, and this phosphorylation only takes place when cAMP is present. The protein kinase can phosphorylate not only the phosphorylase kinase but also many other proteins such as casein and even glycogen synthetase. The phosphorylation of glycogen synthetase inactivates the enzyme and thus cAMP not only activates the degradation of glycogen but also switches off its synthesis. Phosphorylated phosphorylase *a* in the presence of inorganic phosphate then degrades glycogen to glucose 1-phosphate (Scheme 7). By the 'cascade' process outlined above, the effect of a small amount of cAMP is amplified and hence cAMP can exert a very fine control on glycogen degradation.

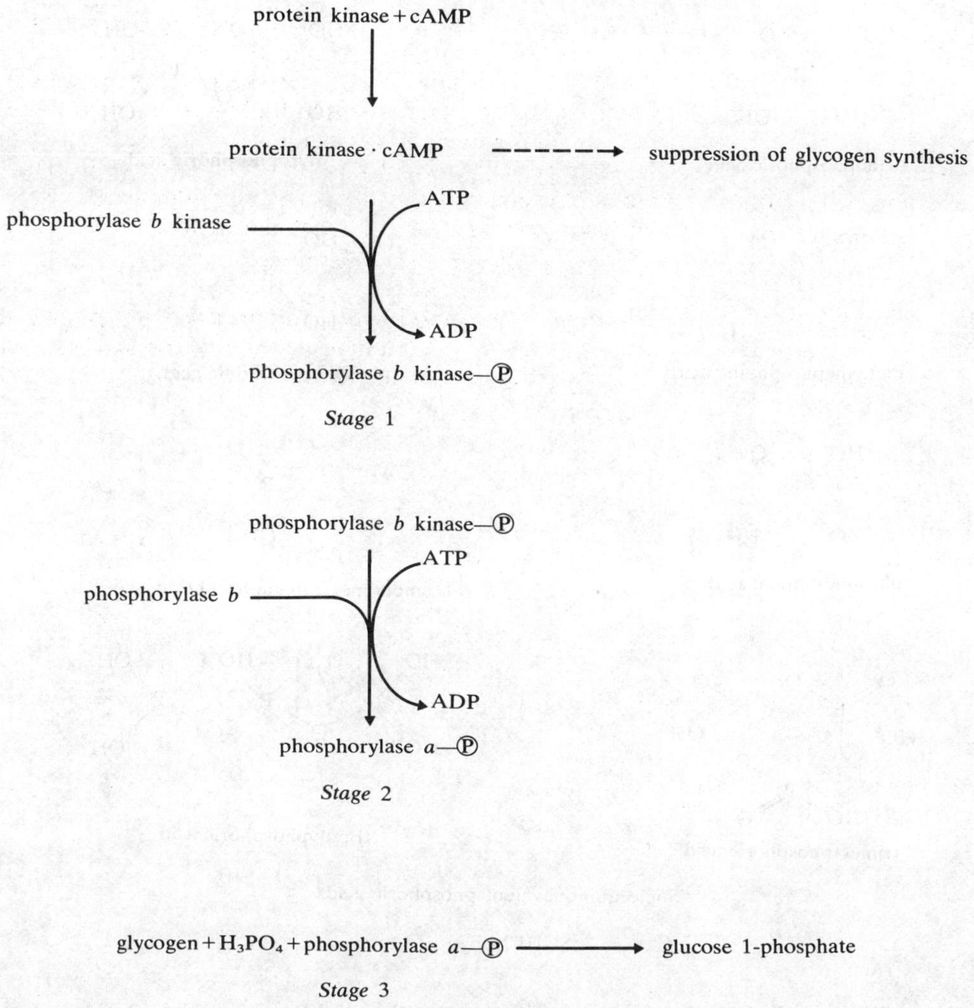

SCHEME 7 Regulation of degradation and synthesis of glycogen by cAMP

22.3.3 PREPARATION OF PHOSPHORIC ACIDS, AMIDES, ANHYDRIDES, AND ESTERS

All naturally occurring organophosphorus compounds are derivatives of quinquevalent phosphorus and are usually esters of orthophosphoric acid or condensed polyphosphoric acids (Scheme 8), although some phosphonates do occur naturally. Orthophosphoric acid (H_3PO_4) contains four oxygen atoms associated with the phosphorus atom and phosphorylation of an alcohol (nucleoside) takes place with the transfer of the phosphorus atom and only three oxygen atoms to the substrate alcohol. Phosphorylation is but one example of acylation and acyl transfer reactions from carboxylic acid derivatives have been well studied.[47] Two main reaction pathways have been observed, either a unimolecular decomposition to an acylium ion [Scheme 9, pathway (i)] or a synchronous displacement reaction can occur which may or may not involve a tetrahedral intermediate [pathways (ii) and (iii)]. Similarly, phosphorylation reactions can occur by initial unimolecular decomposition to metaphosphate [pathway (iv)], although few unambiguous cases of

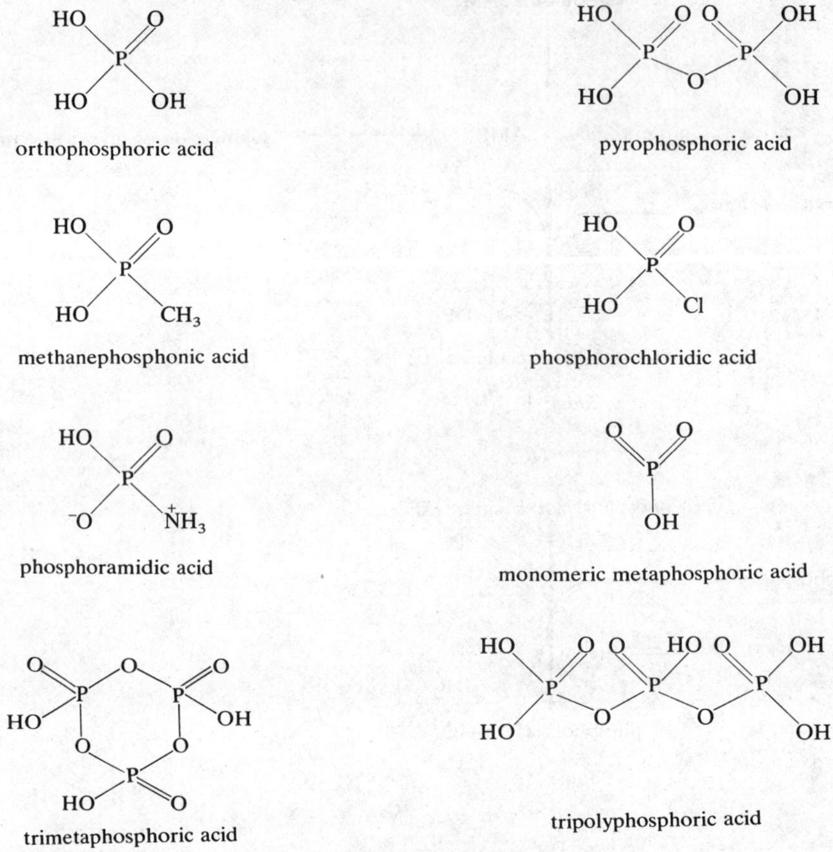

Some quinquevalent phosphoric acids

SCHEME 8

metaphosphate-mediated reactions have been recorded.[48] Most phosphorylation reactions proceed through displacement reactions which may or may not involve quinquecovalent intermediates [pathways (v) and (vi)]. These displacement reactions have already been discussed in the context of polynucleotide hydrolysis. There is little unequivocal evidence in the literature on whether quinquecovalent adducts are intermediates in many phosphorylation reactions, although phosphorylating agents which must act by the formation of such adducts have recently been developed.[49]

The simplest phosphoryl transfer reactions make use of displacement of halide from carbon. When the silver salt of a phosphoric acid or ester is treated with an acyl or alkyl halide, silver chloride is precipitated and the phosphorylated derivative is obtained. This method has been used to make dinucleoside phosphates, phospholipids, and acyl phosphates, but has largely been superseded as yields are generally low and the reaction products impure. Alternative methods of phosphorylation involve the activation of the orthophosphoric acid, and a major problem is how to carry this out selectively and under mild conditions.[50–52]

The activation of phosphoric acid was initially achieved by the use of phosphorochloridates, which are acyl halides. Phosphoryl chloride in the presence of a tertiary base was introduced many years ago for the synthesis of nucleotides and other sugar phosphates. This is still the reagent of choice for the synthesis of symmetrical aryl phosphotriesters (Scheme 10). Care must be taken to avoid the formation of by-products such as mono- and di-esters, and substantial yields of alkyl chloride can be formed during the

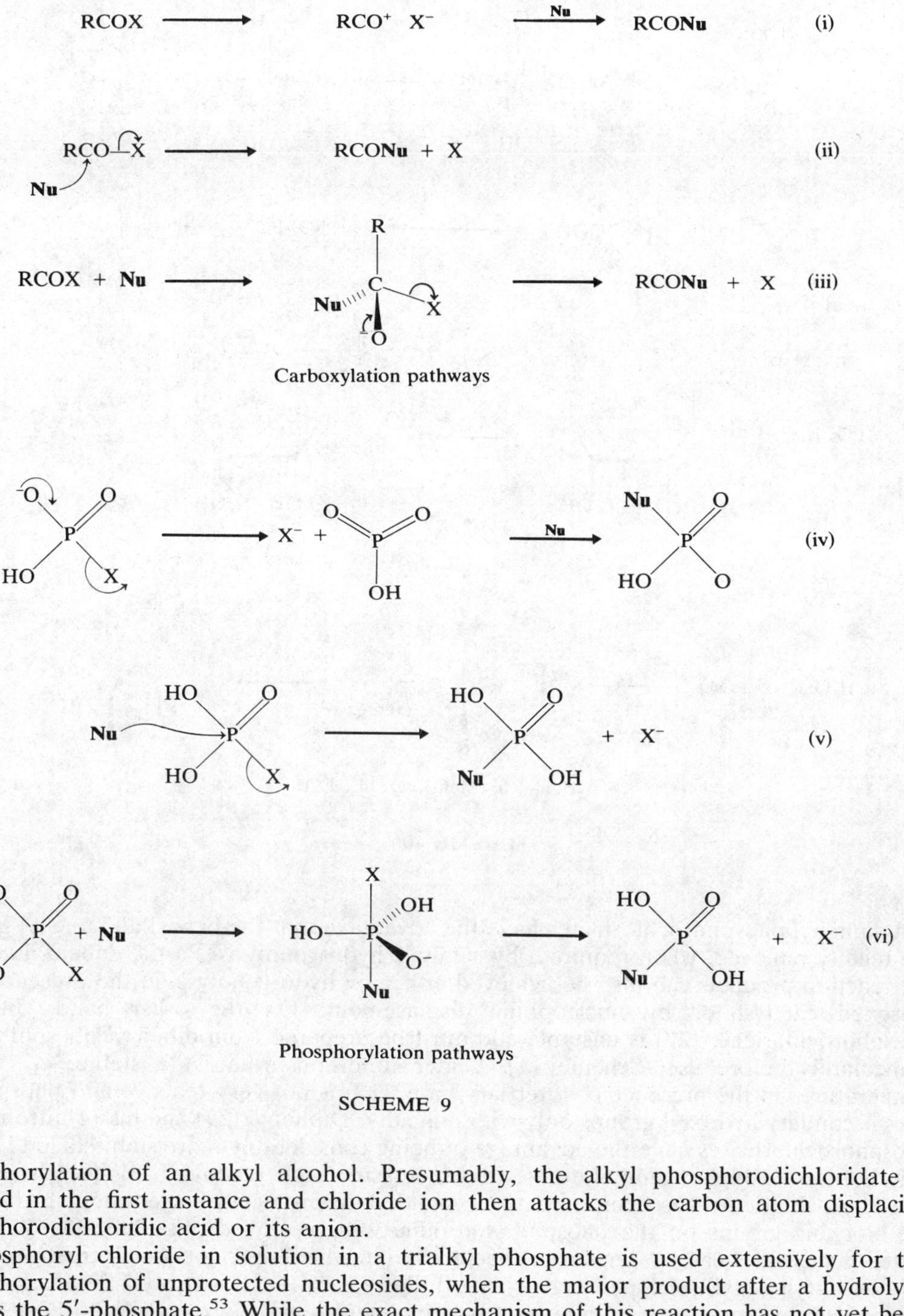

SCHEME 9

phosphorylation of an alkyl alcohol. Presumably, the alkyl phosphorodichloridate is formed in the first instance and chloride ion then attacks the carbon atom displacing phosphorodichloridic acid or its anion.

Phosphoryl chloride in solution in a trialkyl phosphate is used extensively for the phosphorylation of unprotected nucleosides, when the major product after a hydrolytic step is the 5′-phosphate.[53] While the exact mechanism of this reaction has not yet been elucidated, the formation of a bulky complex between the phosphoryl chloride and the solvent may be the cause of steric selectivity. The unwanted formation of di- and tri-esters during nucleotide synthesis can be overcome by using phosphoromonochloridic acid [$(HO)_2POCl$] as the phosphorylating agent. Whilst the parent acid is highly unstable, diesters of phosphoromonochloridic acid are stable and have been used extensively as phosphorylating agents. A number of diesters have been developed to take advantage of different methods of removing the protecting groups at the end of the synthesis. Dibenzyl

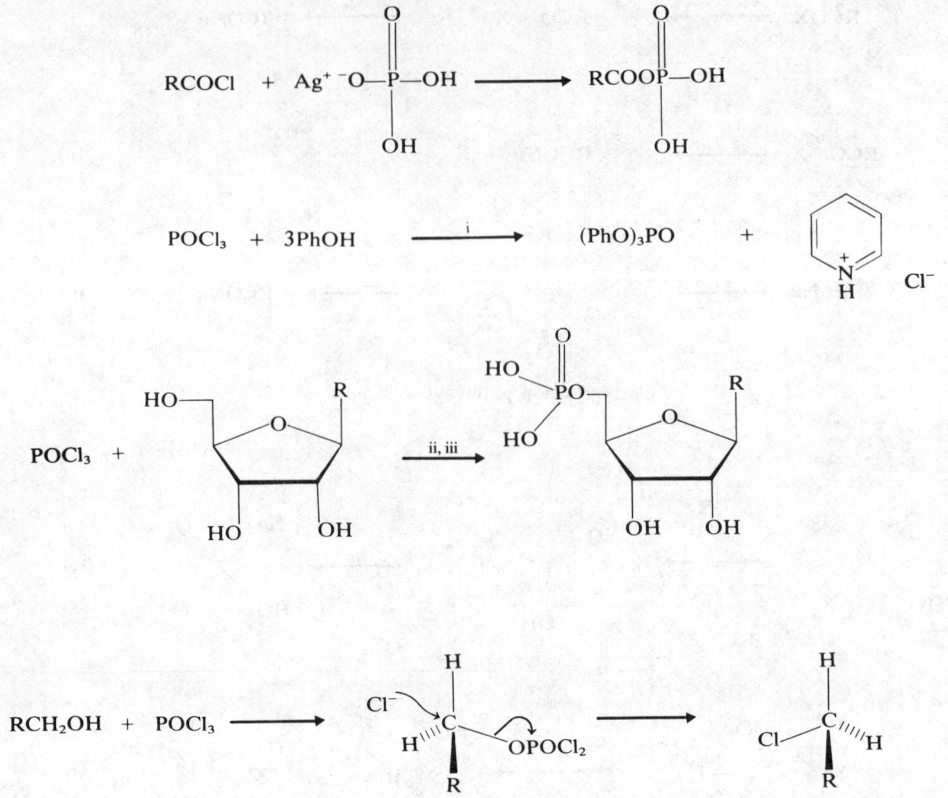

i, pyridine; ii, $(MeO)_3PO$; iii, H_2O.

SCHEME 10

and diphenyl phosphorochloridates have the advantage that the benzyl and phenyl groups are readily removed, when required, by catalytic hydrogenolysis. Partial debenzylation of the reaction products can also be achieved either by hydrogenolysis in the presence of a poisoned catalyst or by nucleophilic displacement. On the other hand, dibenzyl phosphorochloridate (**27**) is unstable and must be prepared from dibenzyl phosphite (**28**) immediately before use (Scheme 11). Under conditions when it is stable, *i.e.* at low temperatures in the presence of a tertiary base, (**27**) is not very reactive and phosphorylates secondary hydroxyl groups only with difficulty. Diphenyl (**29**) and bis-(4-nitrobenzyl) phosphorochloridates have the advantage of being considerably more stable than (**27**) and may be used at higher temperatures, thus increasing their reactivity. Bis-(2-nitrobenzyl) phosphorochloridate has been used to phosphorylate protected nucleosides; in this case the blocking groups on the phosphoryl residue were removed photolytically.[54] Unprotected nucleosides can be converted specifically into the 5′-nucleotides by treatment with bis-(2-t-butylphenyl) phosphorochloridate followed by acid hydrolysis.[55] Here again, steric hindrance must play an important part in determining the selectivity of the reaction. Other phosphorohalidates with acid-labile blocking groups are bismorpholino phosphorobromidate (**30**) and bisanilino phosphorochloridate (**31**), although the blocking groups are best removed from substrates phosphorylated with (**31**) by treatment with an alkyl nitrite.[56] Bis-(2,2,2-trichloroethyl) phosphorochloridate is a phosphorylating agent which has found considerable use for the preparation of nucleotides and phospholipids, the 2,2,2-trichloroethyl groups being removed from the products by zinc dust and acetic acid (Scheme 11).[57]

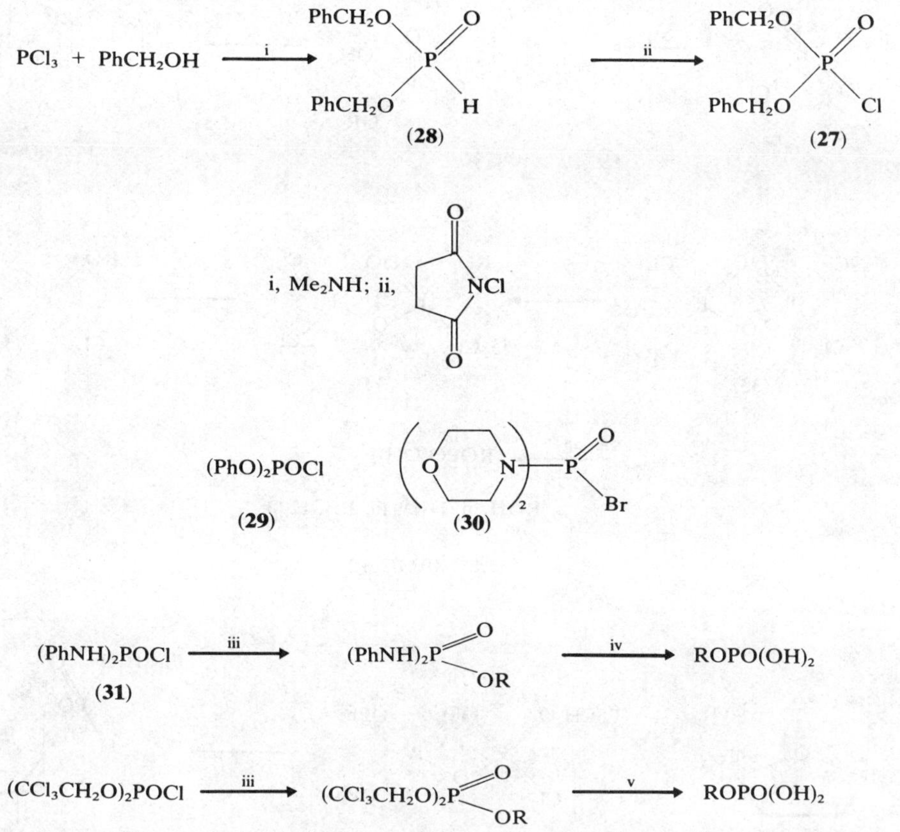

iii, ROH; iv, EtONO, H_2O; v, Zn, HOAc

SCHEME 11

1,2-Phenylene phosphorochloridate (**32**) has been proposed as a reagent for the preparation of monoalkyl phosphates and nucleotides. Cyclic triesters formed from (**32**) are easily hydrolysed with ring-opening, and the 2-hydroxyphenyl residue is readily removed by oxidation of the ring-opened product (Scheme 12).[58] Unprotected nucleosides can be phosphorylated specifically in the 5′-position by pyrophosphoryl chloride (**33**) in solution in a phenolic solvent.[59] Presumably the P—O—P linkage in the initial reaction product (**34**) is cleaved hydrolytically before chloride ion is displaced, and hence the 5′-phosphate is the major product rather than the pyrophosphate.

Complex phosphorochloridates can be used to prepare dinucleoside phosphates and polyphosphate esters, but the synthetic routes are generally laborious and this method has been superseded by the carbodi-imide or phosphoroamidate routes which are discussed more fully below. An example of the phosphorochloridate route is the synthesis of UTP.[60] A mixed anhydride, *O*-benzylphosphorus *O,O*-diphenyl phosphoric anhydride (**35**) prepared by treating a salt of monobenzyl phosphite with (**29**), is a phosphonylating agent and will react with 2′,3′-*O*-isopropylidene uridine to give a secondary phosphite. Chlorination of the latter gives uridine 5′-benzyl phosphorochloridate, which will react with a salt of tribenzyl pyrophosphoric acid to give the fully esterified nucleoside triphosphate (**36**) (Scheme 13). Hydrogenation of (**36**) gives rise to UTP. A major drawback of such syntheses is that fully esterified intermediates, *e.g.* (**36**), are very easily hydrolysed. Furthermore, unsymmetrically substituted polyphosphates rearrange readily to symmetrical products, and both these factors contribute to cause low yields. Triesters of pyrophosphoric acid, however, are more stable than tetraesters and this difference in reactivity

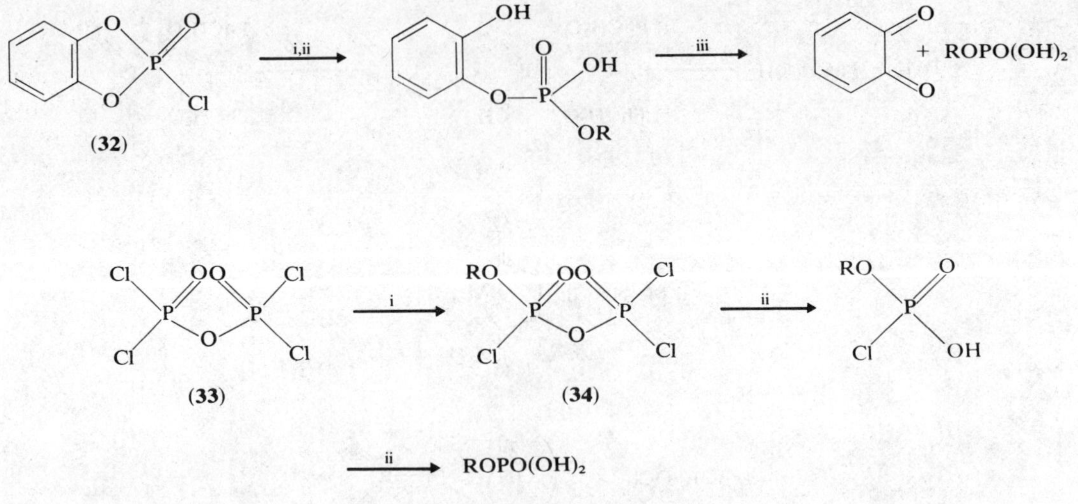

i, ROH; ii, H_2O; iii, Br_2, H_2O.

SCHEME 12

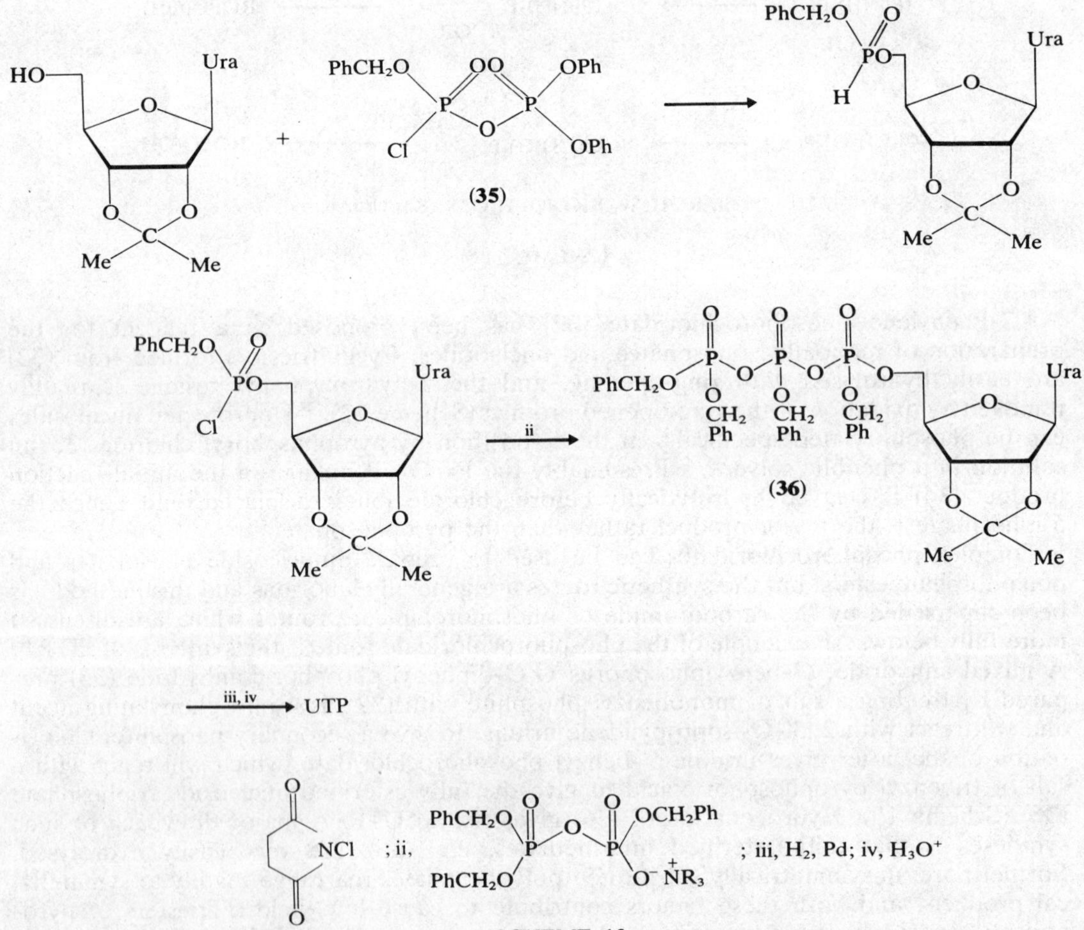

SCHEME 13

has enabled a simple synthesis of nucleoside polyphosphates to be developed. Diphenyl phosphorochloridate will react with salts of nucleotides and other phosphomonoesters to give triesters of pyrophosphoric acid. These triesters will undergo an exchange reaction with a second phosphomonoester to give a diester of pyrophosphoric acid and diphenyl phosphate. Using this technique, UDPGlc has been synthesized from UMP and α-D-glucose 1-phosphate (Scheme 14).[61] Tetraphenyl and tetrakis-(4-nitrophenyl) pyrophosphates, which can easily be prepared from the corresponding phosphodiesters, are

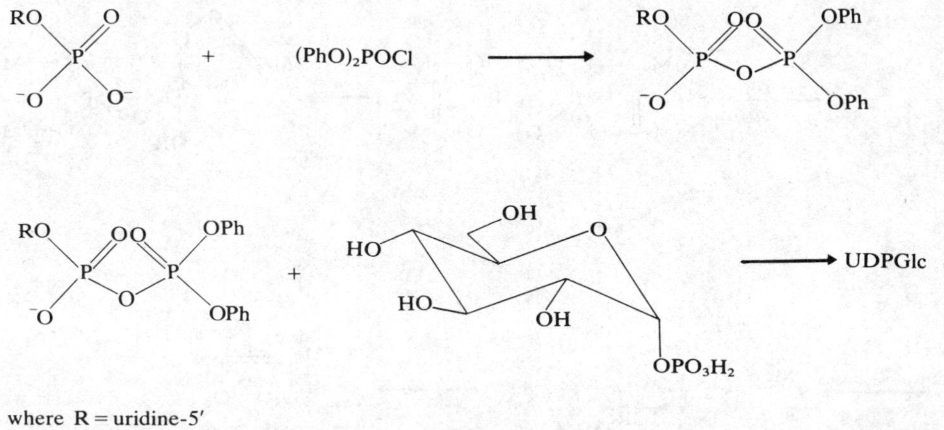

SCHEME 14

powerful phosphorylating agents and can be used to phosphorylate alcohols. Acetoin enediol cyclopyrophosphate (**37**) will selectively phosphorylate primary alcohols in the presence of secondary alcohols.[49] This reagent can be prepared from the oxyphosphorane (**38**) in the following manner (Scheme 15). Treatment of biacetyl with trimethyl phosphite gives (**38**), which can be hydrolysed under controlled conditions to methyl acetoinenediol cyclophosphate (**39**). Demethylation of (**39**) with pyridine followed by treatment of the demethylated salt with phosgene leads to (**37**). Successive reaction of (**37**) with two different alcohols gives the unsymmetrical phosphodiester and quinquecovalent derivatives are plausible intermediates. Hydroxide ion-catalysed hydrolysis of the dimethylacetoin residue from the triester occurs rapidly, being at least 10^6 times faster than the hydrolysis of trimethyl phosphate. While (**37**) has only been applied to the synthesis of model phosphodiesters, it is suggested that it could be used for the preparation of oligonucleotides.[49] However, (**37**) like all tetraesters of pyrophosphoric acid, is very sensitive to water and can only be used under strictly anhydrous conditions, which is a disadvantage in the nucleotide field when substrates are frequently soluble only with difficulty in anhydrous solvents.

Arene sulphonyl chlorides with phosphomonoesters form mixed anhydrides which are phosphorylating agents, and these mixed anhydrides have been used extensively by Khorana and his co-workers for the preparation of oligo- and poly-nucleotides, culminating in the synthesis of the structural gene for alanine tRNA from yeast.[62] A disadvantage of this method is that the sulphonyl chloride will sulphonate alcoholic functions and hence reduce the yield. This tendency for sulphonation can be overcome if arene sulphonyl halides bearing bulky groups, *e.g.* tri-isopropylbenzenesulphonyl chloride (TPS) or sulphonyl chlorides immobilized on an inert support, are used. Dinucleoside phosphates can be prepared without difficulty from suitably protected monomeric units using arenesulphonyl chlorides, and these reagents have been used to synthesize phospholipids. Furthermore, protected deoxyribonucleosides can be block-polymerized in the presence of phenyl phosphoric acid and TPS to give oligonucleotides with phosphotriester bonds. The phenyl groups can be removed by alkaline hydrolysis and, once other protecting groups are

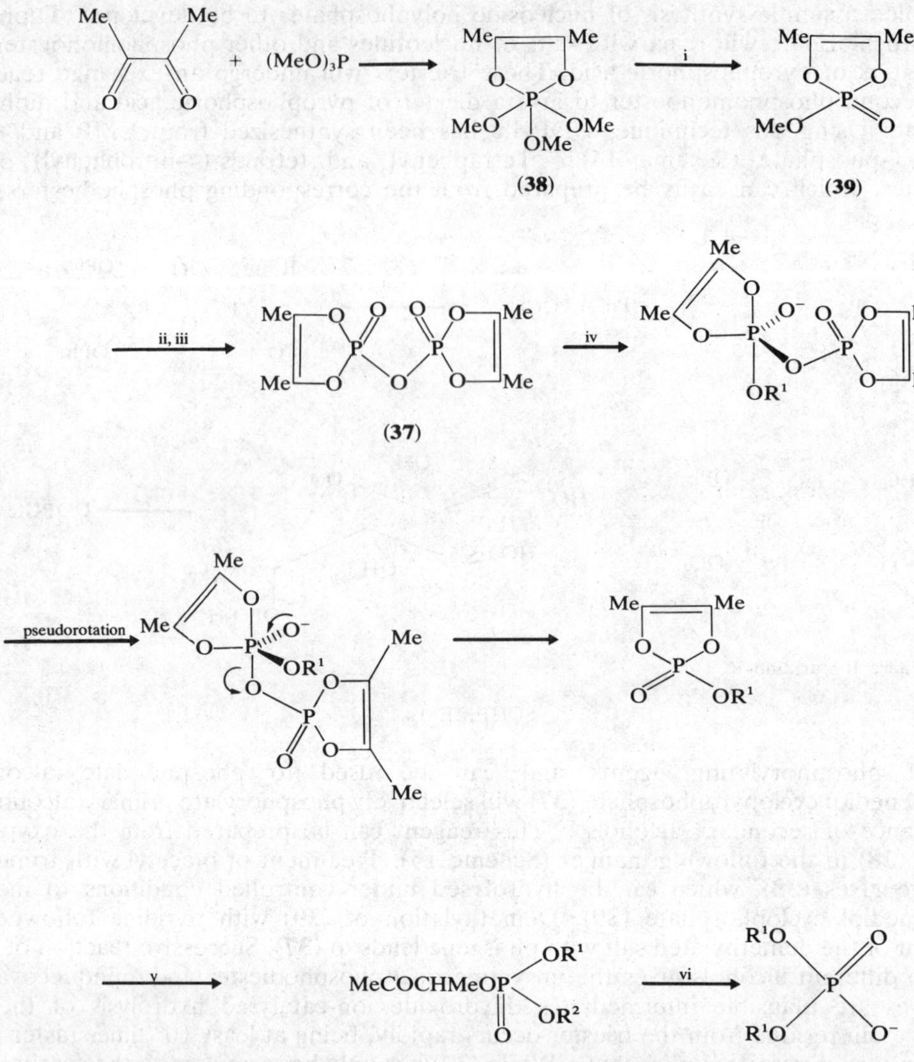

i, H_2O in benzene; ii, pyridine; iii, $COCl_2$; iv, R^1OH; v, R^2OH; vi, OH^-.

SCHEME 15

removed, the oligonucleotide is obtained.[63] This method cannot be used for the preparation of oligoribonucleotides as chain cleavage would occur during the reaction with aqueous base. However, fluoride ion will remove phenyl groups from phosphotriesters under conditions when the internucleotide link should be stable, making this a promising synthetic approach (Scheme 16).[64] Arenesulphonyl triazoles and tetrazoles also appear to be excellent condensing agents for oligonucleotide synthesis.[65]

As mentioned above, the phosphorohalidate route to oligonucleotides and nucleoside polyphosphates frequently involves complex synthetic procedures and yields are frequently low owing to exchange reactions and hydrolysis of fully esterified intermediates. The direct activation of phosphoric acids should be more attractive as this should reduce the number of steps in the reaction sequence. The activation of phosphoric acids by arenesulphonyl chlorides is one such reaction but here, again, moisture must be rigorously excluded from the reaction for optimum yields. An alternative method for the direct activation of phosphoric acids is by the use of carbodi-mides, a method which can tolerate

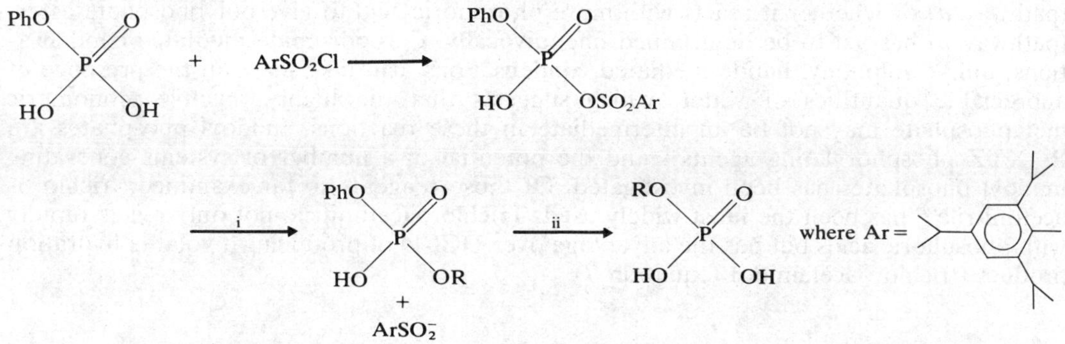

i, ROH; ii, F⁻.

SCHEME 16

the presence of appreciable quantities of water. The protection of extra acidic functions on the phosphomonoester is unnecessary and condensation reactions take place rapidly at room temperature. If dicyclohexylcarbodi-imide (DCCD) is used, its hydration product (dicyclohexylurea) is highly insoluble in most solvents and can easily be removed at the end of the synthesis.[50] The mechanism of formation of phosphate esters by DCCD is still a matter of conjecture. When phosphoric acids are treated with excess DCCD in anhydrous solvents, trimetaphosphates are produced,[66] presumably by the trimerization of monomeric metaphosphates. On the other hand, although the ^{31}P n.m.r. evidence for the formation of nucleoside monometaphosphates during the reaction between nucleotides and arenesulphonyl chlorides has been claimed,[67] neither mono- nor tri-metaphosphates were observed when nucleotides were treated with DCCD in pyridine. Signals due to a complex mixture of products were observed, including those due to a trisubstituted pyrophosphate. When DCCD reacts with a phosphoric acid, an imidoyl phosphate (**40**) is formed initially (Scheme 17), but whether this breaks down to monomeric metaphosphate

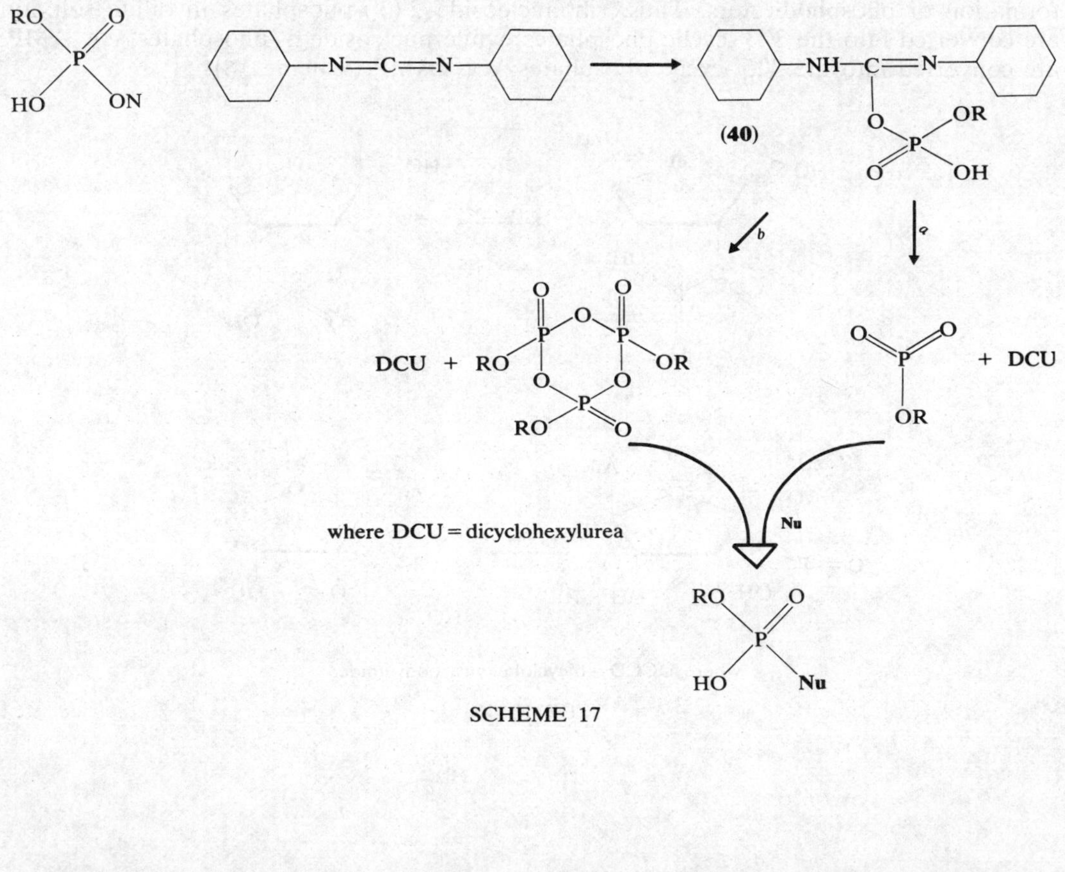

SCHEME 17

(pathway *a*) or whether it reacts with more phosphoric acid to give polyphosphate esters (pathway *b*) has yet to be determined unequivocally. Carbodi-imide-mediated condensations, unlike sulphonyl halide-mediated condensations, can take place in the presence of appreciable quantities of water, which suggests that the highly reactive monomeric metaphosphate may not be an intermediate in these reactions. Imidoyl phosphates are P−XYZ phosphorylating agents[52] and the potential of a number of systems generating imidoyl phosphates has been investigated. Of those reagents so far examined, trichloroacetonitrile[68] has been the most widely used. Trichloroacetonitrile not only reacts rapidly with phosphoric acids but has the advantage over DCCD of producing a volatile hydration product (trichloroacetamide) (equation 7).

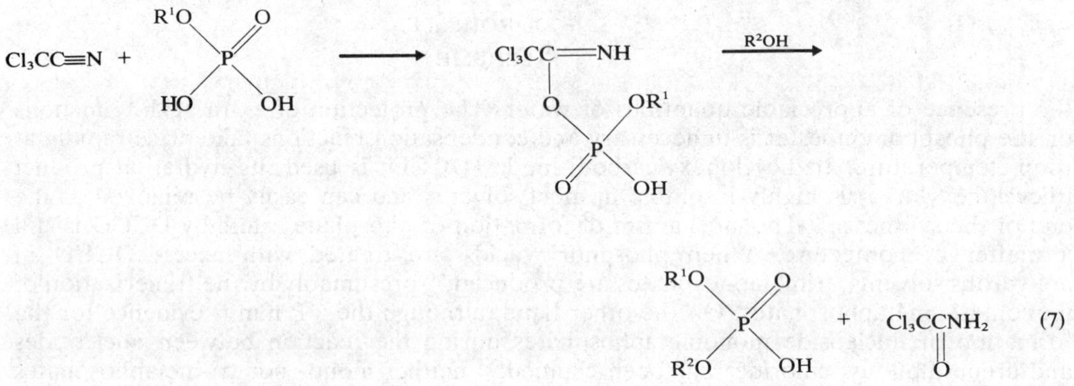

(7)

Despite the uncertainty as to the exact mechanism of the phosphorylation reaction, its ease of use has resulted in DCCD being applied to the synthesis of many phosphodiesters and oligonucleotides.[50] Carbodi-imides are also valuable reagents for the intramolecular formation of phosphodiesters. Thus, ribonucleoside 2′(3′)-phosphates in dilute solution are converted into the 2′,3′-cyclic phosphates, while nucleoside 5′-phosphates, *e.g.* AMP, are converted into the 3′,5′-cyclic phosphates, *e.g.* cAMP (Scheme 18).

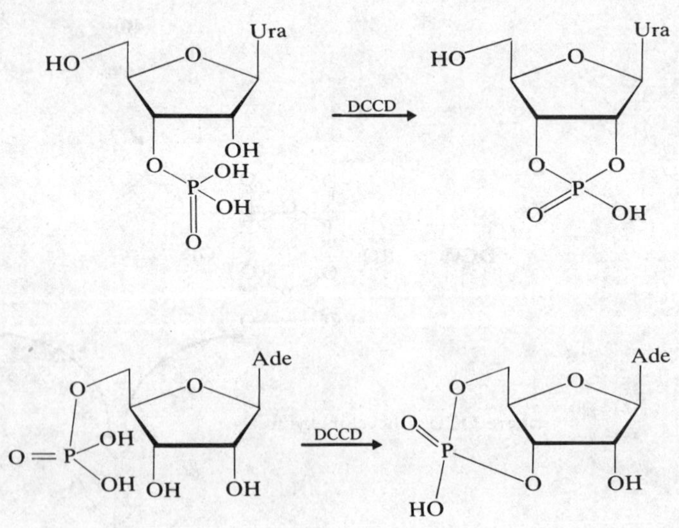

where DCCD = dicyclohexylcarbodi-imide,

SCHEME 18

Nucleoside polyphosphates can be prepared using DCCD, but when two different phosphoric acids are brought into reaction with a carbodi-imide, random formation of the two symmetrical and one asymmetrical products occurs, leading to low yields. Moreover, when orthophosphoric acid is one of the reactants, polymeric products, *e.g.* nucleoside tri- and tetra-phosphates, can be formed, again reducing yields. Thus, carbodi-imides have largely been replaced by other more specific methods for pyrophosphate synthesis, *e.g.* phosphoramidates. Satisfactory yields of asymmetric P^1,P^2-diesters of pyrophosphoric acid can be obtained in some special instances. In contrast to the synthesis of UDPGlc by the DCCD route, when very low yields were obtained, cytidine diphosphate choline can be prepared in reasonable yield by this method and no P^1,P^2-dicholine pyrophosphate is formed.[69] Choline phosphate is a zwitterion under the conditions of the condensation and hence is a stronger acid than CMP. Choline phosphate is consequently the poorer nucleophile for carbon and CMP adds on to the DCCD with little competition from the choline phosphate. As phosphoryl groups are good nucleophiles for phosphorus there is little difference in reactivity of choline phosphate or CMP towards the imidoyl phosphate (or its reaction products) and both CDP-choline and P^1,P^2-dicytidine pyrophosphate are formed in the second step of the synthesis (Scheme 19).

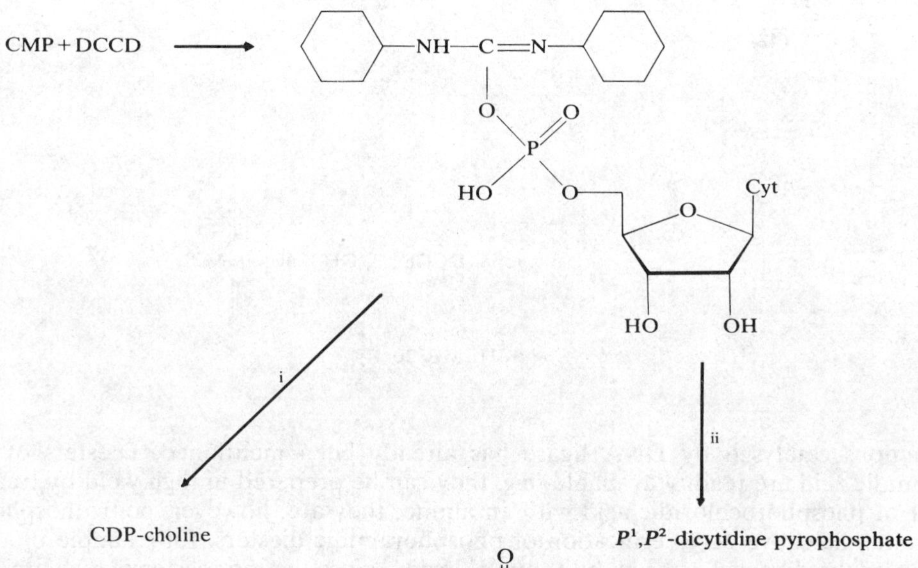

i, $Me_3\overset{+}{N}CH_2CH_2OP(O)(OH)—O^-$; ii, CMP.

SCHEME 19

2-Cyanoethyl phosphate (**41**) is a more convenient reagent than orthophosphoric acid for the DCCD-mediated phosphorylation of alcohols.[70] When (**41**) is used, a phosphodiester is formed which does not react further to give polyphosphates. The 2-cyanoethyl group can be removed when required by alkaline hydrolysis. Another class of alkaline-labile blocking group for phosphates is derived from *S*-substituted 2-mercaptoethanols (**42**). The latter can be oxidized to sulphones under mild conditions and the sulphones are removed rapidly by alkaline hydrolysis (Scheme 20).[71]

Some organophosphorus compounds containing a P—N bond can function as phosphorylating agents and will react selectively with phosphoryl groups to afford polyphosphates. Phosphoramidic acid and its derivatives are the most synthetically useful reagents of this class, and the intervention of an *N*-phosphorolysine residue in the series

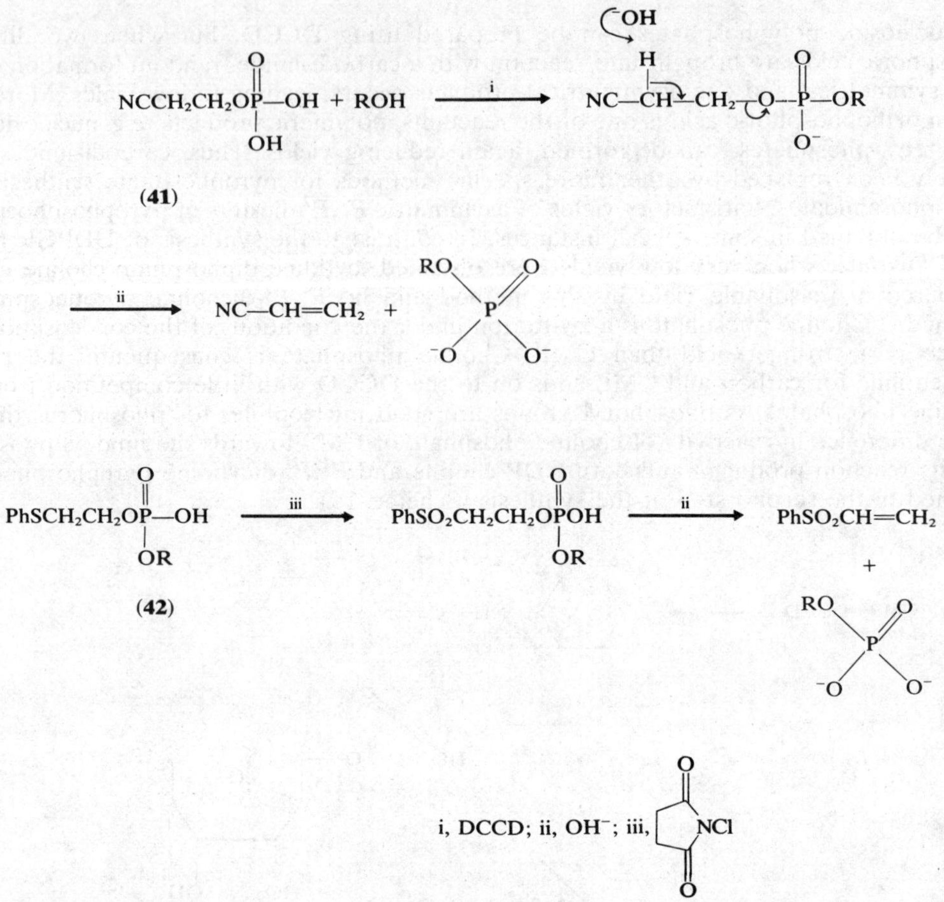

SCHEME 20

of reactions catalysed by DNA ligase has already been mentioned. Diesters of phosphoramidic acid are readily available, *e.g.* they can be prepared in high yield by treating a diester of phosphorochloridic acid with an amine; they are, however, poor phosphorylating agents. Selective de-esterification of phosphoramidic diesters, for example by debenzylation of the dibenzyl ester with lithium chloride, gives the monoesters which are potent phosphorylating agents. A much more convenient route to phosphoramidic monoesters, which is particularly applicable to the preparation of nucleoside phosphoramidates, is the reaction between DCCD, a phosphomonoester (nucleotide), and an amine (Scheme 21).[72]

When a monoester of phosphoramidic acid is heated, a pyrophosphate is formed in a reaction which follows second-order kinetics (Scheme 22).[73] This observation, together with the known selectivity of phosphoramidic monoesters for phosphoryl groups rather than alcohol functions, suggests that phosphoryl transfer from phosphoramidates occurs by a direct displacement reaction rather than by a route which involves monomeric metaphosphates. The nature of the departing amine plays an important part in determining the rate of phosphoryl transfer. For example, monophenyl phosphate will react almost quantitatively with adenosine 5′-phosphoropiperidate in pyridine after one hour at room temperature, while less than 10% reaction occurs under the same conditions with a phosphoramidate derived from a weaker base, adenosine 5′-phosphoranisidate.[72] Nucleoside phosphoromorpholidates offer advantages of reactivity and solubility in organic solvents and are the reagents of choice for the synthesis of unsymmetrical pyrophosphates such as NAD^+, FAD, Coenzyme A, and nucleoside diphosphate sugars.[51] They have also

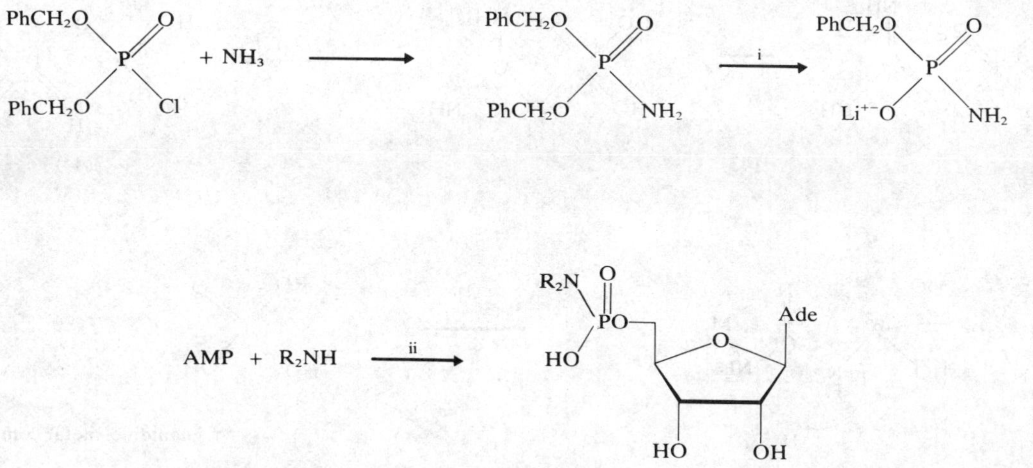

i, LiCl; ii, DCCD;

SCHEME 21

been used extensively to prepare nucleoside di- and tri-phosphates containing natural or atypical bases, *e.g.* 5-chlorocytidine diphosphate.[74]

In contrast to phosphoramidic acids derived from simple amines, *N*-acylphosphoramidates can phosphorylate alcohols under neutral conditions in high yield.[75] Acylation of the nitrogen atom presumably reduces the p_π–d_π contribution to the P—N bond and hence increases the reactivity of the phosphoramidate. Phosphoroguanidates (**43**) are much more stable than phosphoramidates. They are resistant to hydrolysis and do not function as phosphorylating agents under conditions when phosphoramidates readily participate in phosphoryl transfer. This marked contrast in reactivity can be attributed to differences in the nature of the P—N bond in the two classes of compound. For phosphoroguanidates, p_π–d_π bonding is not significantly reduced on protonation as the formal positive charge can be distributed over the whole guanidinium residue (**44**). If electrons are withdrawn from the guanidinium residue either by acylation or by the formation of a metal complex, the p_π–d_π contribution to the P—N bond is greatly reduced and phosphoryl transfer can now occur (Scheme 23).[52]

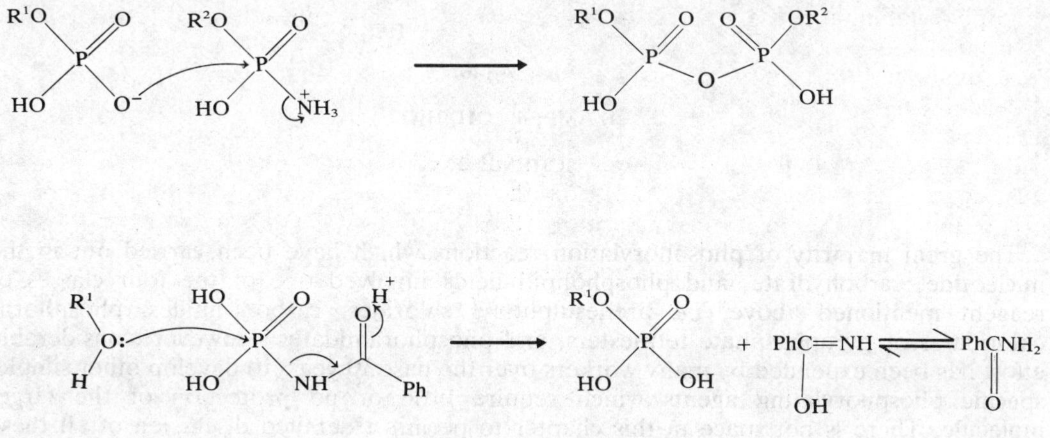

SCHEME 22

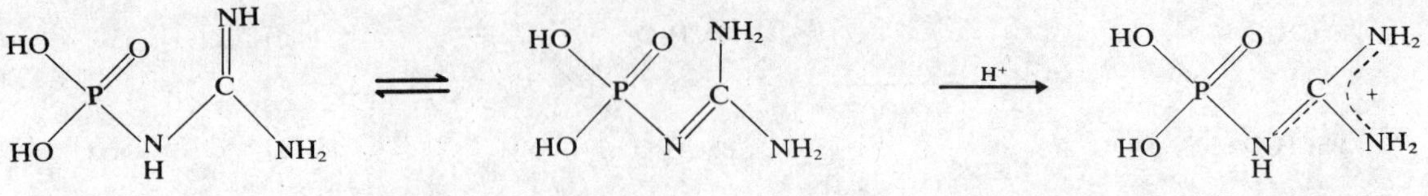

(43) (44)

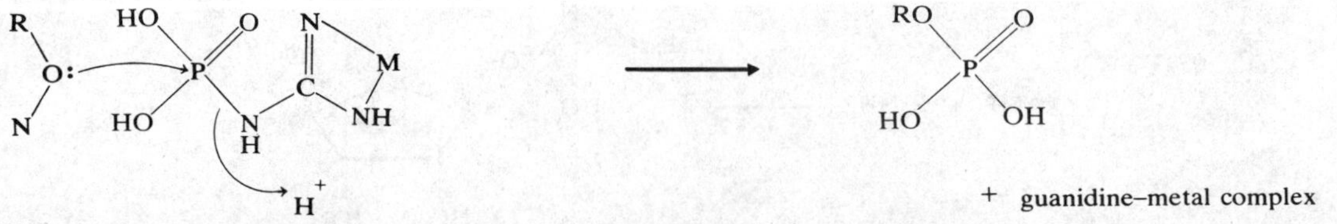

SCHEME 23

Phosphoroimidazolidates are more reactive than simple phosphoramidates and can phosphorylate alcohols, amines, and phosphoric acids.[76] *N,N'*-Carbonylbisimidazole (**45**) reacts vigorously with phosphomonoesters such as AMP to give phosphoroimidazolidates, and a simple preparation of nucleoside polyphosphates based on this reaction has been devised.[77] 2-Cyanoethyl phosphoroimidazolidate (**46**), which can be prepared from (**41**) and (**45**), has been suggested[78] for the synthesis of α-^{32}P-labelled nucleoside diphosphates of high specific radioactivity (Scheme 24).

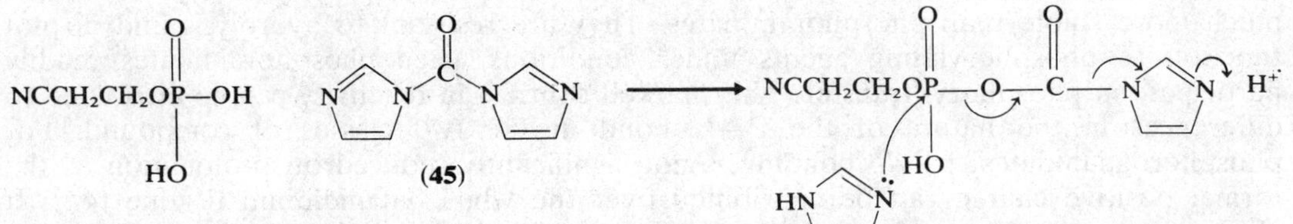

(45)

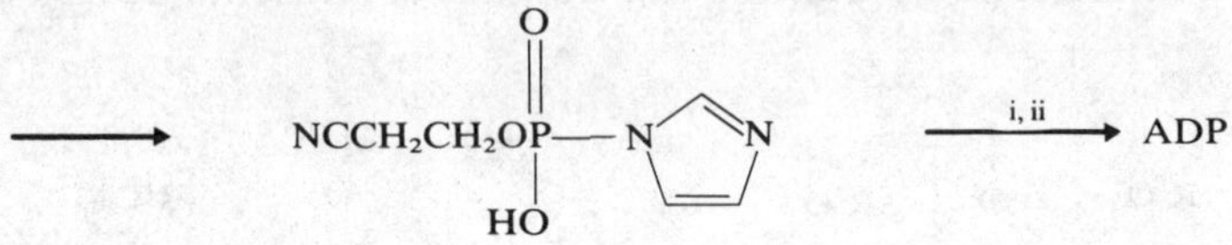

(46)

i, AMP; ii, ^{-}OH, H_2O.

SCHEME 24

The great majority of phosphorylation reactions which have been carried out in the nucleotide, carbohydrate, and phospholipid fields involved one of the four classes of reagent mentioned above, *i.e.* arenesulphonyl chlorides, carbodi-imides, phosphorochloridates or pyrophosphate tetraesters, and phosphoramidates. However, considerable effort has been expended by many workers over the past 30 years to develop other simple, specific phosphorylating agents which require little or no protection of the target molecule. There is not space in this chapter to permit a detailed discussion of all these reagents, but some which show promise will be mentioned below.

The selective phosphorylation by a bulky phosphorylating agent of the 5′-hydroxyl group of a nucleoside has already been mentioned. Another phosphorylating agent making use of steric hindrance for selectivity is formed by mixing triphenylphosphine, diethyl azodicarboxylate, and a phosphodiester.[79] This mixture will phosphorylate unprotected nucleosides selectively to 5′-nucleotides and will phosphorylate octan-2-ol with inversion at C-2 (Scheme 25)[80]. This suggests that the S_N2 displacement of triphenylphosphine from a phosphoroxyphosphonium ion is the final step in this reaction.

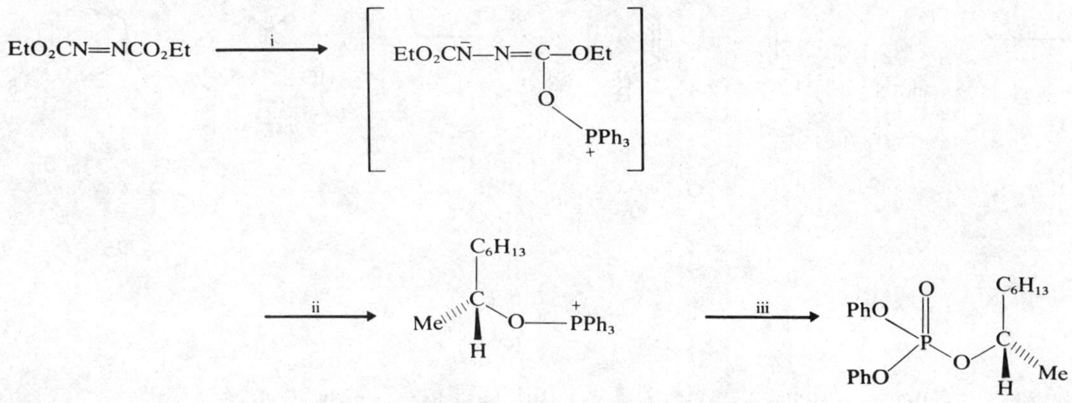

i, Ph_3P; ii, octan-2-ol; iii, $(PhO)_2P(=O)—O^-$.

SCHEME 25

As illustrated above, nucleotide and oligonucleotide synthesis can be carried out in the absence of acids or bases by oxidation/reduction condensations.[81] A mixture of triphenylphosphine and 2,2′-dipyridyl disulphide can act as a condensing agent in an oligonucleotide synthesis which is accompanied by the formation of triphenylphosphine oxide and two equivalents of pyridine-2-thione. These reagents also convert nucleoside 2′(3′)- and 5′-phosphates to the corresponding cyclic phosphodiesters. It is postulated that interaction between triphenylphosphine and the disulphide affords the phosphonium salt (**47**), which then reacts with the phosphomonoester to give the phosphoroxyphosphonium salt (**48**). Phosphorylation reactions promoted by phosphonium and carbonium ions are well known,[82] and hence (**48**) should be a potent phosphorylating agent (Scheme 26). Complex molecules such as Coenzyme A have been prepared by this route, which appears to have considerable synthetic potential.

Nucleosides can undergo transesterification with trialkyl phosphites to give the 2′,3′-*O*-cyclic phosphite esters (**49**), which can be de-alkylated to the cyclic phosphonites (**50**). Oxidation of the latter with hexachloroacetone yields the nucleoside 2′,3′-cyclic phosphates (Scheme 27).[83] Some ring opening of the cyclic phosphite can occur during de-alkylation and hence nucleoside 2′(3′)-phosphates are formed during the oxidation step. This is one of the more successful applications of phosphite oxidation to the synthesis of nucleotides.

Nucleoside 5′-phosphoric di-n-butylphosphinothioic anhydrides (**51**), which are formed quantitatively in the reaction between nucleotides and di-n-butylphosphinothionyl bromide, can readily be isolated as stable solids.[84] While these anhydrides are not hydrolysed at all rapidly by water, they are activated by oxidizing agents such as silver acetate and in the presence of salts of ortho- or pyro-phosphoric acids give ADP, ATP, *etc.* in excellent yields (Scheme 28). One great advantage of this synthetic route to nucleoside polyphosphates is that alcohol functions need not be protected and unwanted polyphosphates are not formed.

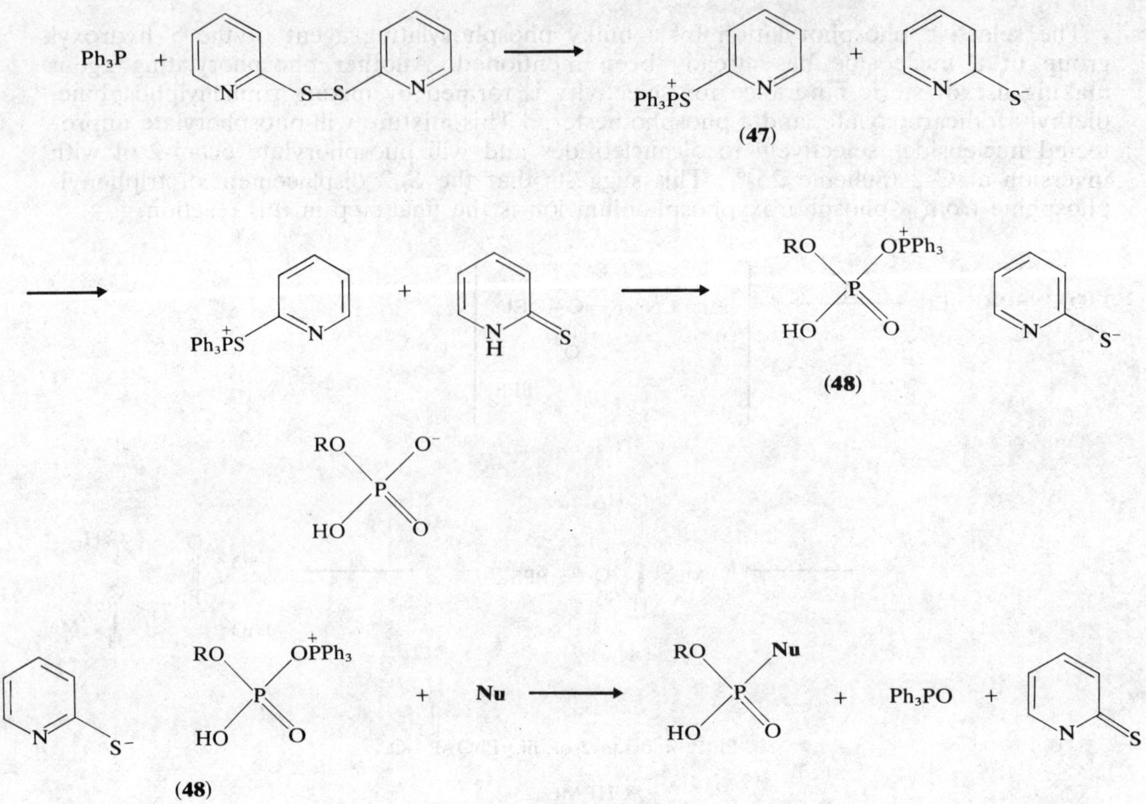

SCHEME 26

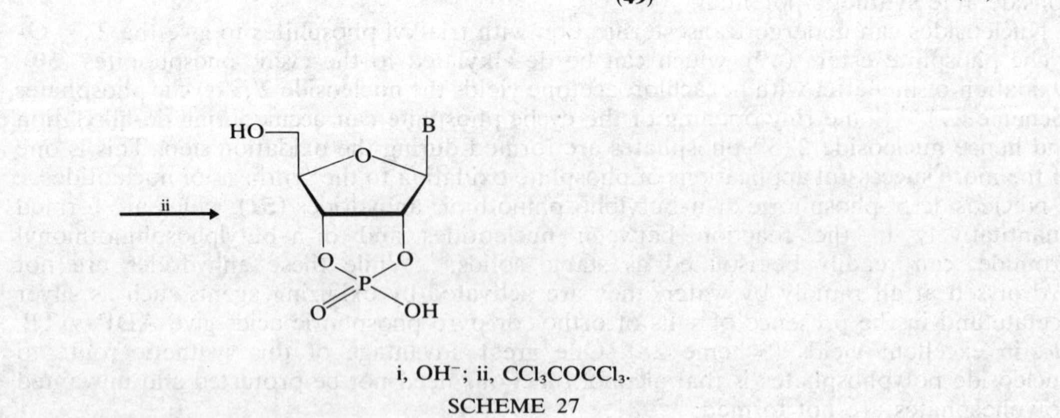

i, OH^-; ii, CCl_3COCCl_3.

SCHEME 27

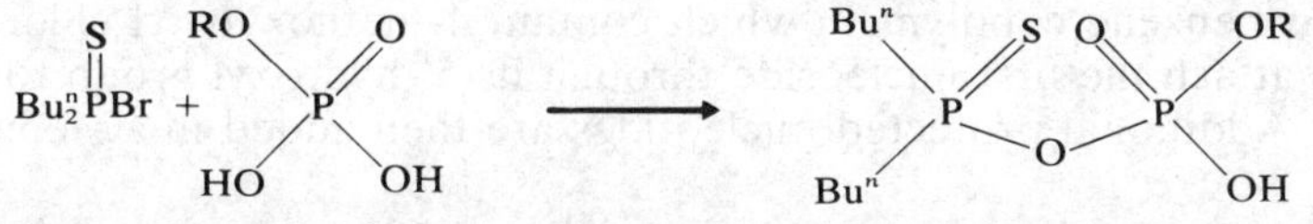

(51)

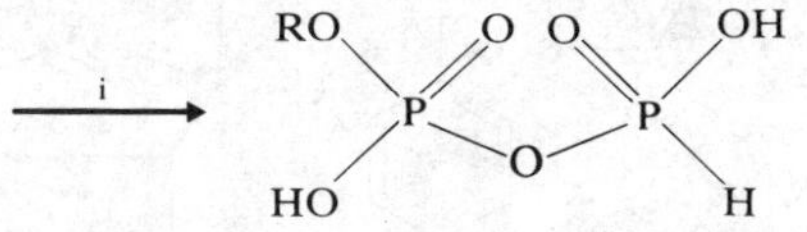

i, AgOCOMe, $R_3\overset{+}{N}\overset{-}{O}PO_3H_2$.

SCHEME 28

Modified nucleotides in which a phosphoryl oxygen atom has been replaced by a sulphur, nitrogen, or carbon atom have been used extensively to study enzyme mechanisms. Nucleoside phosphorothioates have been particularly widely used,[85] and the investigation of the mechanism of ribonuclease hydrolysis with uridine 2′,3′-cyclic phosphorothioate has been mentioned earlier in this chapter. ATP analogues with methylene groups in place of oxygen atoms between the α,β or β,γ phosphorus atoms can give information on phosphoryl and nucleotidyl transfer in biological systems. For example, ribose 5-phosphate 1-methylenediphosphonate (**52**) has been isolated from the reaction between ribose 5-phosphate and 5′-adenylyl methylenediphosphonate, which is catalysed by 5-phosphoribosyl pyrophosphate synthetase from Ehrlich ascites tumour cells (Scheme 29).[86] This is confirmation of the observation made earlier in this chapter that a pyrophosphoryl residue is transferred in a single step in this enzymic reaction. Polynucleotides and nucleoside triphosphates containing phosphoramidate bonds have been prepared,[87] but these have not been used to any extent to study enzyme mechanisms owing to the lability of the P—N bond under mildly acidic conditions.

The chemical synthesis of polymers with defined sequences of monomer subunits can be simplified enormously by attaching one end of the growing polymer to an insoluble support, as purification of the polymer after each chemical step can easily be achieved by filtration. This method has been very popular in the peptide field and the repetitive steps can be automated.[88] The solid-phase synthesis of polynucleotides has not been as successful as the solid-phase synthesis of polypeptides, mainly because of difficulties in achieving quantitative yields in the successive steps. Arenesulphonyl chlorides are the most useful reagents for forming the internucleotide link, although care must be taken to achieve anhydrous conditions to ensure maximum yields. Polystyrene and cross-linked

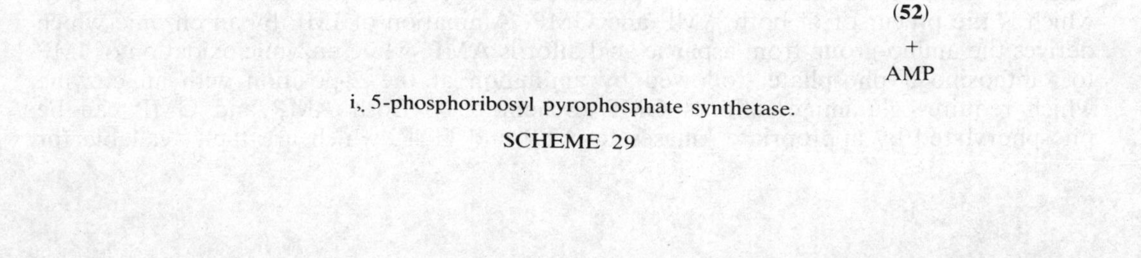

(52)

+

AMP

i, 5-phosphoribosyl pyrophosphate synthetase.

SCHEME 29

styrene–divinylbenzene copolymers which contain 4-methoxytrityl chloride residues have been used to attach the first nucleoside through its 5′-hydroxyl group to the solid support (equation 8). Additional protected nucleotides are then added in a stepwise fashion using

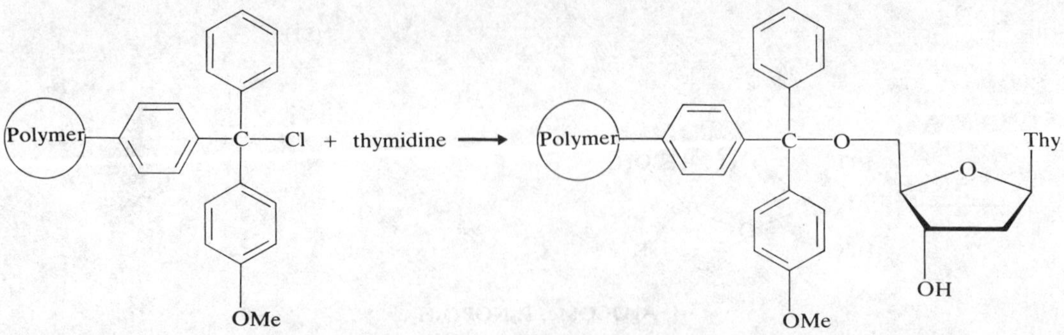

(8)

arenesulphonyl chlorides or DCCD and the polymer is cleaved from the support with acid at the end of the reaction.[89] Succinylated polystyrene is another support which has been attached to the 5′-hydroxyl groups of nucleosides for solid-phase synthesis,[90] and in this case the oligonucleotide is released from the support by alkaline hydrolysis. Silica, Sephadex LH 20, and poly(ethylene glycol) have all been tried as insoluble supports in polynucleotide synthesis with reasonable success.[91] Evidence has been presented from reactions carried out on solid supports that the co-operative interaction of at least three phosphoryl residues is needed before carbodi-imide-mediated phosphorylations can occur.[92] A similar interaction of three or more phosphoryl residues also appears necessary in solid-phase reactions promoted by arenesulphonyl chlorides. This is evidence for the participation of polyphosphates rather than monomeric metaphosphate in these phosphorylations.

As a footnote to the section on the synthesis of nucleotides, a simple enzymic method should be mentioned. A crude phosphotransferase obtained by homogenizing wheat shoots will use 4-nitrophenyl phosphate as phosphoryl donor to phosphorylate specifically the 5′-position of nucleosides in virtually quantitative yield after a few hours.[93] The reaction can be carried out on a large scale and no protection of other reactive groups is necessary.

22.3.4 BIOSYNTHESIS OF NUCLEOTIDES

As has been discussed earlier in this chapter, the chemical synthesis of nucleotides is usually carried out by phosphorylating a suitably protected nucleoside.[101] This has led to considerable effort being expended in a search for protecting groups for hydroxyl[94] and exocyclic amino-functions.[95] In living organisms, major pathways to both purine and pyrimidine nucleotides start with the phosphorylated sugar ribose 5-phosphate (**53**). In pigeon-liver extracts the purine is built up in the following manner.[96] ATP pyrophosphorylates (**53**) to produce ribose 5-phosphate 1-pyrophosphate (**54**), which is then aminated enzymically by glutamine to β-D-1-aminoribose 5-phosphate (**55**). The imidazole ring of the purine is then synthesized in a number of steps, culminating in the formation of 1-β-D-ribofuranosyl-5-aminoimidazole 5′-phosphate (**56**). The pyrimidine ring is then built on to the imidazole to produce inosine 5′-phosphate (**9,** IMP) (Scheme 30), which is the precursor of both AMP and GMP. Amination of IMP by an enzyme which derives the amino-group from aspartic acid affords AMP, while enzymic oxidation of IMP to xanthosine 5′-phosphate, followed by amination at the 2-position with an enzyme which requires glutamine, affords GMP (Scheme 31). Both AMP and GMP can be phosphorylated by appropriate kinases to ATP and GTP, which are then available for

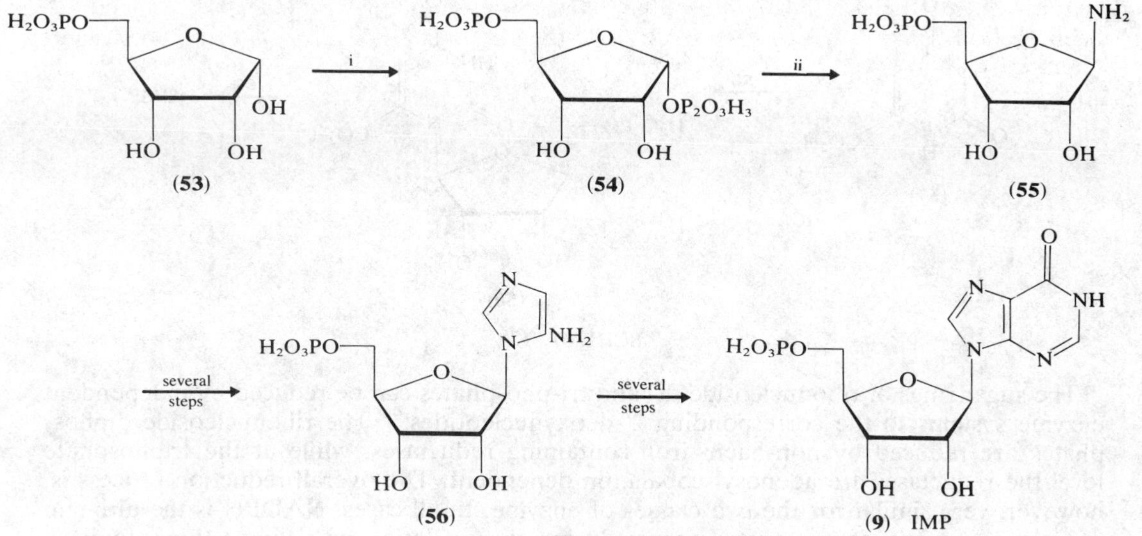

i, ATP; ii, glutamine.

SCHEME 30

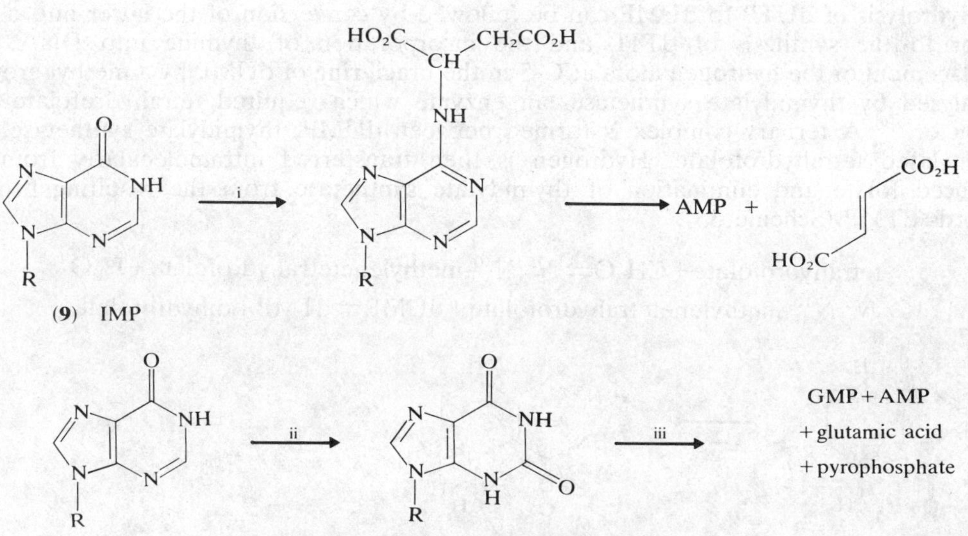

where R = β-D-ribofuranosyl 5-phosphate.

i, glutamine, GTP; ii, oxidase; iii, glutamine, ATP.

SCHEME 31

nucleic acid and coenzyme synthesis. Interference with the above pathway, for example by drug treatment, can override natural control systems and purine nucleotide biosynthesis from preformed purines then becomes important.[97]

Orotidine (**57**) and its 5′-phosphate (**58**) are key intermediates in pyrimidine biosynthesis.[98] The pyrimidine ring in (**57**) is synthesized enzymically from carbamyl phosphate and aspartic acid; the glycosidic bond in (**58**) is formed when (**57**) reacts with (**54**) in the presence of orotidine 5′-phosphate pyrophosphorylase (Scheme 32). Decarboxylation of (**58**) affords UMP. After UMP has been phosphorylated to UTP by an appropriate kinase, amination of the pyrimidine ring can occur, resulting in CTP. Two different enzymes for this process have been detected. In bacteria, ammonia is the source of nitrogen, while in animals the nitrogen is obtained from glutamine.

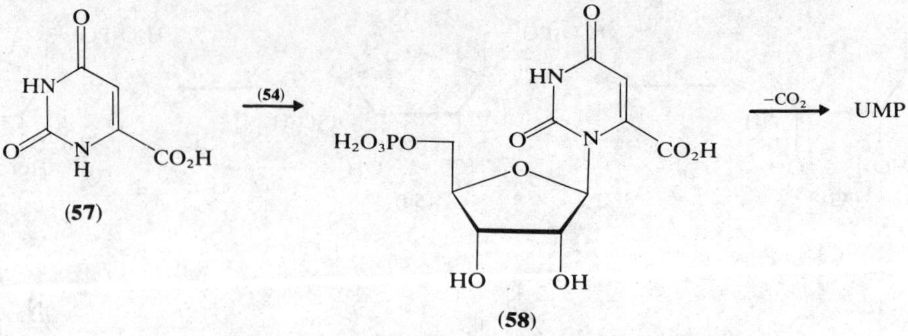

SCHEME 32

The sugar rings of ribonucleoside di- and tri-phosphates can be reduced by independent enzyme systems to the corresponding 2′-deoxynucleotides.[99] The ribonucleoside diphosphates are reduced by non-haem iron-containing reductases, while at the triphosphate level the reductases are adenosyl-cobalamin dependent. The overall reduction process is, however, very similar for the two classes of enzyme. In all cases, NADPH is the ultimate reductant and hydrogen transfer occurs during the oxidation of reduced thioredoxin, a sulphur-containing protein. Furthermore, the 2′-hydroxyl group is replaced in all cases with retention of configuration.

Hydrolysis of dUTP to dUMP can be followed by conversion of the latter into dTMP prior to the synthesis of dTTP and the incorporation of thymine into DNA. The replacement of the hydrogen atom at C-5 in the uracil ring of dUMP by a methyl group is catalysed by thymidylate synthetase, an enzyme which required tetrahydrofolate as a cofactor.[100] A ternary complex is formed between dUMP, thymidylate synthetase, and formylated tetrahydrofolate. Hydrogen is then transferred intramolecularly from the reduced folate and elimination of thymidylate synthetase from the resulting product affords dTMP (Scheme 33).

$$\text{tetrahydrofolate} + CH_2O \rightleftharpoons N^5,N^{10}\text{-methylenetetrahydrofolate} + H_2O$$
$$N^5,N^{10}\text{-methylenetetrahydrofolate} + \text{dUMP} \rightleftharpoons \text{dTMP} + \text{dihydrofolate}$$

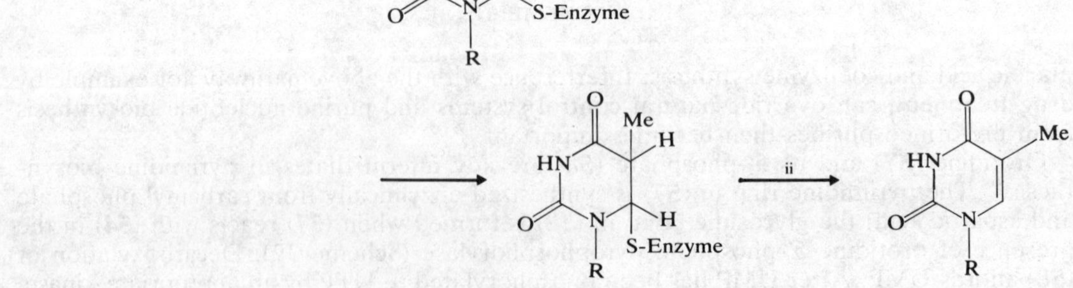

where R = β-D-2-deoxyribofuranosyl 5-phosphate.
i, dUMP, thymidylate synthetase; ii, −thymidylate synthetase.

SCHEME 33

A number of nucleotides which contain modified bases have been detected in tRNA and these appear to arise from the modification of bases in a precursor tRNA. Modification occurs at the macromolecular level and the conformation of the precursor tRNA plays a part in determining which bases are modified.

References

1. G. R. Wyatt, in 'The Nucleic Acids', ed. E. Chargaff and J. N. Davidson, Academic Press, New York, 1955, vol. I, p. 243.
2. H. K. Mangold, in 'Thin Layer Chromatography', ed. E. Stahl, Springer Verlag, Berlin, 1965, p. 440.
3. E. Heftmann, 'Chromatography', Reinhold, New York, 1961, p. 554.
4. S. R. Ayad, 'Techniques of Nucleic Acid Fractionation', Wiley-Interscience, London, 1972, p. 64.
5. J. L. Wiebers and J. A. Shapiro, *Biochemistry*, 1977, **16,** 1044; D. R. Burgard, S. P. Perone and J. L. Wiebers, *ibid.*, 1977, **16,** 1051.
6. P. A. Levene and L. W. Bass, 'Nucleic Acids', A.C.S. Monograph No. 56, Reinhold, New York, 1931.
7. D. W. Hutchinson, 'Nucleotides and Coenzymes', Methuen, London, 1964.
8. J. W. Cornforth, *Quart. Rev. Chem. Soc.*, 1969, **23,** 125.
9. M. Akhtar and D. C. Wilton, *Ann. Rep. Chem. Soc.(B)*, 1973, **70,** 98.
10. H. Nikaido and W. Z. Hassid, *Adv. Carbohydrate Chem. Biochem.*, 1971, **26,** 351.
11. P. Goldman and P. R. Vagelos, in 'Comprehensive Biochemistry', ed. M. Florkin and E. H. Stolz, Elsevier, Amsterdam, 1964, vol. 15, p. 71.
12. H. C. Dunathan, *Adv. Enzymol.*, 1971, **35,** 79.
13. L. O. Krampitz, 'Thiamin Diphosphate and Its Catalytic Functions', Dekker, New York, 1970.
14. A. M. Michelson, 'The Chemistry of Nucleosides and Nucleotides', Academic Press, London, 1963, p. 98.
15. M. Cashel and B. Kalbacher, *J. Biol. Chem.*, 1970, **245,** 2309.
16. J. Sy and F. Lipmann, *Proc. Nat. Acad. Sci. U.S.A.*, 1973, **70,** 306.
17. P. O. P. Ts'o, in 'Fine Structure of Proteins and Nucleic Acids', ed. G. D. Fasman and S. N. Timasheff, Dekker, New York, 1970, p. 49.
18. G. A. Jeffrey and M. Sundaralingam, *Adv. Carbohydrate Chem. Biochem.*, 1976, **32,** 353.
19. C. D. Barry, J. A. Glasel, A. C. T. North, R. J. P. Williams, and A. V. Xavier, *Biochem. Biophys. Res. Comm.*, 1972, **47,** 166.
20. O. Kennard, N. W. Isaacs, J. C. Coppola, A. J. Kirby, S. G. Warren, W. D. S. Motherwell, D. G. Watson, D. L. Wampler, D. H. Chenery, A. C. Larson, K. A. Kerr, and L. R. di Sanseverino, *Nature*, 1970, **225,** 333.
21. M. Cohn and T. R. Hughes, Jr., *J. Biol. Chem.*, 1962, **237,** 176.
22. J. R. Cox, Jr. and O. B. Ramsay, *Chem. Rev.*, 1964, **64,** 317.
23. F. H. Westheimer, *Accounts Chem. Res.*, 1968, **1,** 70.
24. I. Ugi, D. Marquarding, H. Klusacek, P. Gillespie, and F. Ramirez, *Accounts Chem. Res.*, 1971, **4,** 288.
25. G. L. Eichorn, in 'Inorganic Biochemistry', ed. G. L. Eichorn, Elsevier, Amsterdam, 1973, vol. 2, p. 1210.
26. F. M. Richards and H. W. Wyckoff, in 'The Enzymes', ed. P. D. Boyer, Academic Press, New York, 1971, 3rd edn., vol. 4, p. 647.
27. A. Deavin, A. P. Mathias, and B. R. Rabin, *Nature*, 1966, **211,** 252.
28. D. A. Usher, D. I. Richardson, Jr., and F. Eckstein, *Nature*, 1970, **228,** 663.
29. T. Uchida and F. Egami, in 'The Enzymes', ed. P. D. Boyer, Academic Press, New York, 1971, 3rd edn., vol. 4, p. 205.
30. F. Eckstein, H. H. Schulz, H. Rüterjans, W. Haar, and W. Maurer, *Biochemistry*, 1972, **11,** 3507.
31. C. B. Anfinsen, P. Cuatrecasas, and H. Taniuchi, in 'The Enzymes', ed. P. D. Boyer, Academic Press, New York, 1971, 3rd edn., vol. 4, p. 177.
32. B. M. Dunn, C. DiBello, and C. B. Anfinsen, *J. Biol. Chem.*, 1973, **248,** 4769.
33. T. Godefroy-Colburn and M. Grunberg-Manago, in 'The Enzymes', ed. P. D. Boyer, Academic Press, New York, 1972, 3rd edn., vol. 7, p. 533.
34. K. Burton, in 'Methods in Enzymology', ed. L. Grossman and K. Moldave, Academic Press, New York, 1967, vol. 12A, p. 222.
35. M. Uziel, *Biochemistry*, 1973, **12,** 938.
36. F. Lipmann, *Adv. Enzymol.*, 1941, **1,** 99.
37. E. R. Stadtman, in 'The Enzymes', ed. P. D. Boyer, Academic Press, New York, 1973, 3rd edn., vol. 8, p. 1.
38. B. E. C. Banks and C. A. Vernon, *J. Theoret. Biol.*, 1970, **29,** 301.
39. T.-D. Son, M. Roux, and M. E. Ellenberger, *Nucleic Acids Res.*, 1975, **2,** 1101.
40. A. S. Mildvan and C. M. Grisham, *Structure and Bonding*, 1974, **20,** 1.
41. P. E. Amsler and H. Sigel, *European J. Biochem.*, 1976, **63,** 569.
42. D. S. Sigman, G. M. Wahl, and D. J. Creighton, *Biochemistry*, 1972, **11,** 2236.
43. I. R. Lehman, *Science*, 1974, **186,** 790.

44. G. A. Robison, R. W. Butcher, and E. W. Sutherland, 'Cyclic AMP', Academic Press, New York, 1971.
45. T. Braun and L. Birnbaumer, in 'Comprehensive Biochemistry', ed. M. Florkin and E. H. Stolz, Elsevier, Amsterdam, 1975, vol. 25, p. 65.
46. I. Pastan and R. L. Perlman, *Nature New Biology*, 1971, **229,** 5.
47. D. P. N. Satchell, *Quart. Rev. Chem. Soc.*, 1963, **17,** 160; D. P. N. Satchell, *Chem. and Ind. (London)*, 1974, 683.
48. C. H. Clapp, A. Satterthwait, and F. H. Westheimer, *J. Amer. Chem. Soc.*, 1975, **97,** 6873.
49. F. Ramirez, *Synthesis*, 1974, 90; F. Ramirez, J. F. Marecek, and I. Ugi, *J. Amer. Chem. Soc.*, 1975, **97,** 3809.
50. H. G. Khorana, 'Some Recent Developments in the Chemistry of Phosphate Esters of Biological Interest', Wiley, New York, 1961.
51. D. M. Brown, in 'Advances in Organic Chemistry: Methods and Results', ed. R. A. Raphael, E. C. Taylor, and H. Wynberg, Interscience, New York, 1963, vol. 3, p. 75.
52. V. M. Clark and D. W. Hutchinson, in 'Progress in Organic Chemistry', ed. Sir James Cook and W. Carruthers, Butterworths, London, 1968, vol. 7, p. 75.
53. M. Yoshikawa, T. Kato, and T. Takenishi, *Tetrahedron Letters*, 1967, 5065; F. Eckstein, M. Goumet, and R. Wetzel, *Nucleic Acids Res.*, 1975, **2,** 1771.
54. M. Rubinstein, B. Amit, and A. Patchornik, *Tetrahedron Letters*, 1975, 1445.
55. J. Hes and M. P. Mertes, *J. Org. Chem.*, 1974, **39,** 3767.
56. W. S. Zielinski and J. Smrt, *Coll. Czech. Chem. Comm.*, 1974, **39,** 2483.
57. F. Eckstein and K. H. Scheit, *Angew. Chem. Internat. Edn.*, 1967, **6,** 362.
58. T. A. Khwaja, C. B. Reese, and J. C. M. Stewart, *J. Chem. Soc. (C)*, 1970, 2092; T. A. Khwaja and C. B. Reese, *Tetrahedron*, 1971, **27,** 6189.
59. K.-I. Imai, S. Fujii, K. Takanohashi, Y. Furukawa, T. Masuda, and M. Honjo, *J. Org. Chem.*,1969, **34,** 1547.
60. G. W. Kenner, A. R. Todd, R. F. Webb, and F. J. Weymouth, *J. Chem. Soc.*, 1954, 2288.
61. A. M. Michelson, *Chem. and Ind.* (*London*), 1960, 1267.
62. H. G. Khorana, K. L. Agarwal, H. Büchi, M. H. Caruthers, N. K. Gupta, K. Kleppe, A. Kumar, E. Ohtsuka, U. L. RajBhandary, J. H. van de Sande, V. Sgaramella, T. Terao, H. Weber, and T. Yamada, *J. Mol. Biol.*, 1972, **72,** 209.
63. R. Arentzen and C. B. Reese, *J. C. S. Perkin I*, 1977, 445.
64. K. K. Ogilvie, S. L. Beaucage, and D. W. Entwistle, *Tetrahedron Letters*, 1976, 1255.
65. N. Katagiri, K. Itakura, and S. A. Narang, *J. Amer. Chem. Soc.*, 1975, **97,** 7332; J. Stawinski, T. Hozumi, and S. A. Narang, *Canad. J. Chem.*, 1976, **54,** 670.
66. G. Weimann and H. G. Khorana, *J. Amer. Chem. Soc.*, 1962, **84,** 4329.
67. D. G. Knorre and V. F. Zarytova, *Nucleic Acids Res.*, 1976, **3,** 2709.
68. F. Cramer and G. Weimann, *Chem. Ber.*, 1961, **94,** 996.
69. E. P. Kennedy, *J. Biol. Chem.*, 1956, **222,** 185.
70. G. M. Tener, *J. Amer. Chem. Soc.*, 1961, **83,** 159.
71. K. L. Agarwal, M. Fridkin, E. Jay, and H. G. Khorana, *J. Amer. Chem. Soc.*, 1973, **95,** 2020.
72. J. G. Moffatt and H. G. Khorana, *J. Amer. Chem. Soc.*, 1961, **83,** 649.
73. V. M. Clark and S. G. Warren, *J. Chem. Soc.*, 1965, 5509.
74. M. A. W. Eaton and D. W. Hutchinson, *Biochemistry*, 1972, **11,** 3162.
75. C. Zioudrou, *Tetrahedron*, 1962, **18,** 197.
76. H. Schaller, H. A. Staab, and F. Cramer, *Chem. Ber.*, 1961, **94,** 1621.
77. D. E. Hoard and D. G. Ott, *J. Amer. Chem. Soc.*, 1965, **87,** 1785.
78. R. H. Symons, *Biochim. Biophys. Acta*, 1970, **209,** 296.
79. O. Mitsunobu, K. Kato, and J. Kimura, *J. Amer. Chem. Soc.*, 1969, **91,** 6510.
80. O. Mitsunobu and M. Eguchi, *Bull. Chem. Soc. Japan*, 1971, **44,** 3427.
81. T. Mukaiyama, *Angew. Chem. Internat. Edn.*, 1976, **15,** 94.
82. H. kaye and Lord Todd, *J. Chem. Soc.*(*C*), 1967, 1420; V. M. Clark, D. W. Hutchinson and P. F. Varey, *ibid.*, 1969, 74.
83. A. Holý and F. Šorm, *Coll. Czech. Chem. Comm.*, 1969, **34,** 1929.
84. T. Hata, K. Furusawa, and M. Sekine, *J. C. S. Chem. Comm.*, 1975, 196.
85. F. Eckstein, *Angew. Chem. Internat. Edn.*, 1975, **14,** 160.
86. A. W. Murray, P. C. L. Wong, and B. Fredrichs, *Biochem. J.*, 1969, **112,** 741.
87. R. L. Letsinger, J. S. Wilkes, and L. B. Dumas, *J. Amer. Chem. Soc.*, 1972, **94,** 292.
88. J. M. Stewart and J. D. Young, 'Solid Phase Peptide Synthesis', Freeman, San Francisco, 1969.
89. H. Hayatsu and H. G. Khorana, *J. Amer. Chem. Soc.*, 1967, **89,** 3880.
90. K. F. Yip and K. C. Tsou, *J. Amer. Chem. Soc.*, 1971, **93,** 3272.
91. H. Köster, and K. Heyns, *Tetrahedron Letters*, 1972, 1527, 1531, 1535.
92. G. M. Blackburn, M. J. Brown, M. R. Harris, and D. Shire, *J. Chem. Soc. (C)*, 1969, 676.
93. J. Giziewicz and D. Shugar, *Acta Biochim. Polon.*, 1975, **22,** 87.
94. C. B. Reese, in 'Protective Groups in Organic Chemistry', ed. J. F. W. McOmie, Plenum, London, 1973, p. 95.
95. R. I. Zhdanov and S. M. Zhenodarova, *Synthesis*, 1975, 222.
96. J. M. Buchanan, in 'The Nucleic Acids' ed. E. Chargaff and J. N. Davidson, Academic Press, New York, 1960, vol. III, p. 303.

97. A. W. Murray, D. C. Elliott, and M. R. Atkinson, *Progr. Nucleic Acid Res. Mol. Biol.*, 1970, **10,** 87.
98. G. W. Crosbie, in 'The Nucleic Acids' ed. E. Chargaff and J. N. Davidson, Academic Press, New York, 1960, vol. III, p. 323.
99. H. P. C. Hogenkamp and G. N. Sando, *Structure and Bonding*, 1974, **20,** 24.
100. T. K. Bradshaw and D. W. Hutchinson, *Chem. Soc. Rev.*, 1977, **6,** 43.
101. V. Amarnath and A. D. Broom, *Chem. Rev.*, 1977, **77,** 183.

22.4

Nucleic Acids: Structure Determination and Synthesis

G. M. BLACKBURN
University of Sheffield

Four subjects relating to the determination of structure and synthesis of nucleic acids are to be described. These deal with the determination of sequence and the chemical synthesis of both DNA and RNA. The order of presentation, synthesis before sequence, and DNA before RNA, is somewhat arbitrary but can be justified in part by historical priorities.

22.4.1 CHEMICAL SYNTHESIS OF DEOXYRIBO-OLIGONUCLEOTIDES

The task of synthesis of small DNA segments of DNA has been associated with the name of Har Gobind Khorana for over 20 years. During that time, the targets of synthesis have changed while the methods used to attain them have stayed sensibly of the same lineage. Initially, the preparation of ribotrinucleoside diphosphates provided the means of unravelling the genetic code (Chapter 22.5) and was followed by the block copolymerization of deoxyribo-oligonucleotides, which confirmed the working validity of the code. Next, with the determination of the sequence of the first tRNA species, the synthesis of its DNA complement, gene-synthesis, became a definable objective. Now, recombinant DNA technology makes it possible in principle to manufacture multiple copies of a synthetic piece of DNA and insert them into bacteria adjacent to any chosen piece of DNA. Goals beyond these are, at present, as much the province of the politician as of the scientist.

22.4.1.1 Synthesis *via* phosphodiesters

The general procedure developed[1] for the preparation of oligomers from 4 to 18 nucleotides long involves the stepwise or block condensation of a 5′-phosphate with a unit having a free 3′-hydroxyl group. The condensing agent is usually an arenesulphonyl chloride. Dicyclohexylcarbodi-imide does, however, present fewer problems for isolation and purification of the product in the special case of dinucleoside phosphate formation. In order to avoid unwanted condensation reactions, the residue bearing the 3′-OH group usually carries an acid-labile protecting group at the 5′-end. Conversely, the 5′-phosphate is commonly acetylated on the 3′-OH group. The amino-functions of the three bases would react directly with the arenesulphonyl chloride and are therefore protected by acylation, as previously discussed. Such a synthesis is illustrated (Scheme 1) by the preparation of the protected trimer (**1**), which can be further built up into the octamer (**2**) by block addition of dpCpC and dpGpGpA.

If, alternatively, the trimer (**1**) is treated with acid to remove the monomethoxytrityl residue and then phosphorylated at the 5′-position, a block copolymer can be manufactured by deacetylation and treatment with a condensing agent. It is usual to add a little of the 5′-protected 3′-OH trimer (**3**) as a chain initiator. Such a synthesis leads to the formation of the polymer dGpApA$(pGpApA)_n$. The reader should be able to work out, in due course, the structure of the mRNA which would be made from transcription of this polymer and, hence, the amino-acids which should be incorporated in protein biosynthesis directed by such a synthetic message (Chapter 22.5).

Syntheses of this type have been used to prepare fragments of up to some 20 units long, which Khorana[2] believes to be near to the maximum chain length that can be purified effectively. From that point on, chain extension calls for the augmentation of purely chemical methods by the use of enzymes.

The key catalyst here is the enzyme DNA ligase. It has the function of repairing a 'nick' in one strand of a DNA duplex which lies between the 5′-phosphate of one residue and the 3′-hydroxyl group of its neighbour — the *acceptor.* Some ligases require ATP as cofactor; others use different coenzymes for the task. In order to bring the 5′-phosphate of the *donor* segment into juxtaposition with the acceptor segment, both of them are annealed with a third segment, the *splint,* which has consecutive base sequences complementary to the donor and acceptor fragments in the opposite polarity (Scheme 2). In practice, a minimum of four hydrogen-bonded pairings has to be made between each partner. Providing that ambiguities in the complementation patterns are avoided, two or three such ligations can be made simultaneously, as illustrated for the formation of the duplex section (**4**). This block actually corresponds to residues 30 to 60 of $tRNA^{Ala}$ as sequenced by Holley (see Section 22.1.3.3).

While the synthesis of the DNA corresponding to the sequence of that tRNA was still incomplete, Khorana embarked on a second task — the synthesis of the gene for the 76-nucleotide $tRNA^{Tyr}_{Su^{3+}}$ from *E. coli* (which was selected mainly for reasons associated with its special biological properties). After work on the project had started, newer developments showed that this tRNA is derived from a precursor RNA molecule with 126 residues, to which objective the synthesis was reprogrammed. By methods similar to those described, Khorana's team prepared 26 fragments, varying from four to thirteen deoxynucleotides in length, and grouped them into four duplexes, 1–26, 27–56, 57–89, and 90–126, each with the appropriate sticky ends. These were then welded together by appropriate use of kinase phosphorylation, annealing, and ligation.[3]

That achievement was but the penultimate step in the grand design. This gene can only express its biological activity if it is flanked by a 21 base-pair terminator sequence and by a 59 base-pair promoter segment. The determination of the sequences of these fragments was technically more difficult than that of the precursor tRNA but, once accomplished, the final segments were duly manufactured and tailor-fitted on to the earlier molecule to give the 206 base-pairs of the final objective, finished in 1976 (Figure 1).

Not only is it unquestionably the greatest attainment yet accomplished in synthetic

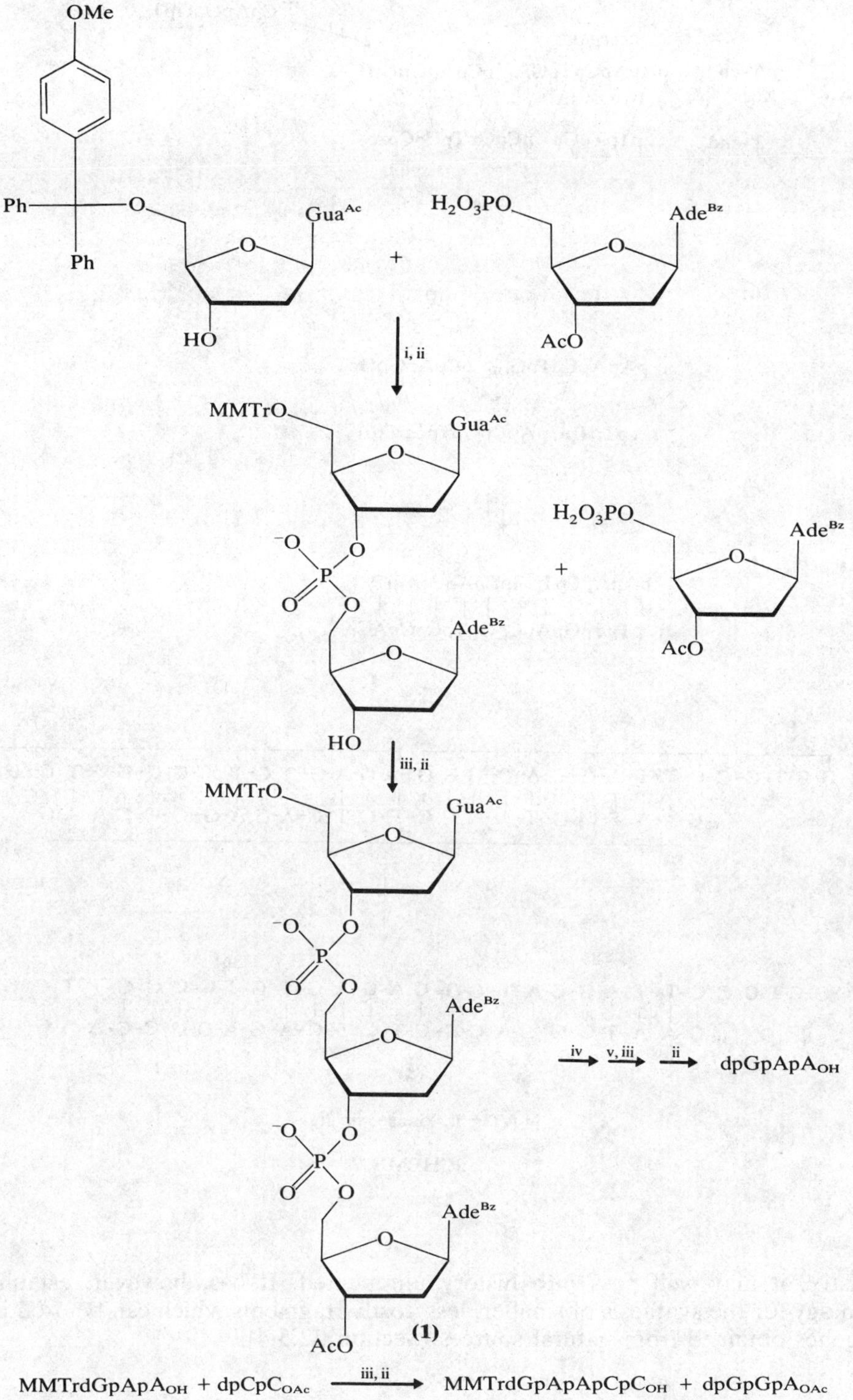

(1)

$MMTrdGpApA_{OH} + dpCpC_{OAc}$ —iii, ii→ $MMTrdGpApApCpC_{OH} + dpGpGpA_{OAc}$

(3)

—iii, ii→

$MMTrdGpApApCpCpGpGpA_{OH}$

(2)

i, DCCD; ii, OH^-; iii, MS or TPS; iv, CF_3CO_2H; v, $NCCH_2CH_2OPO_3H_2$

SCHEME 1

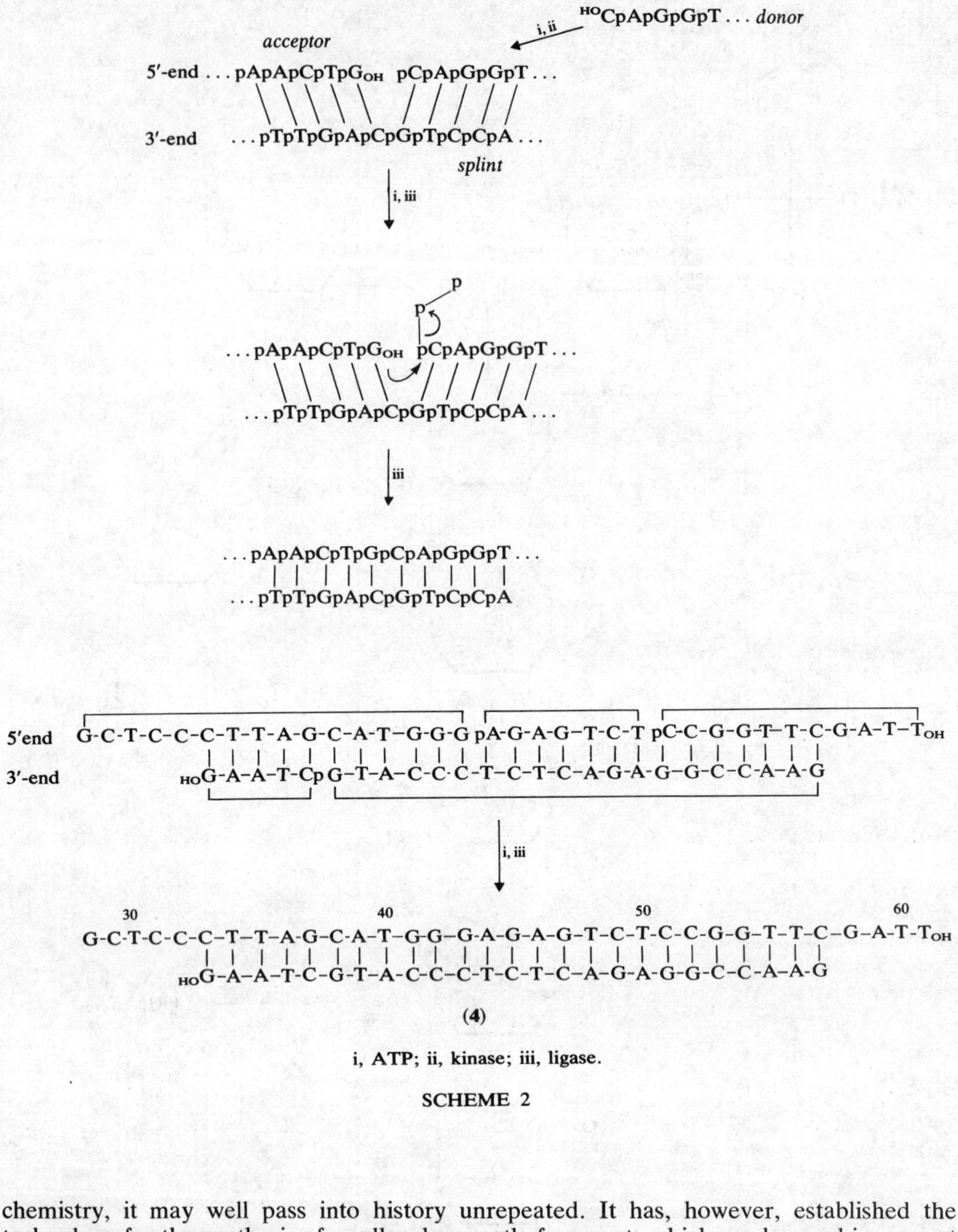

(4)

i, ATP; ii, kinase; iii, ligase.

SCHEME 2

chemistry, it may well pass into history unrepeated. It has, however, established the technology for the synthesis of smaller, less-costly fragments which can be used in concert with genes obtained from natural sources (Section 22.5.4).

22.4.1.2 Synthesis *via* phosphotriesters

Phosphodiester synthesis gives products which are polyanions. This fact imposes two major problems on such work: solubility in non-polar solvents is negligible, and purification is limited to separation-by-charge. Both of these difficulties can be circumvented by

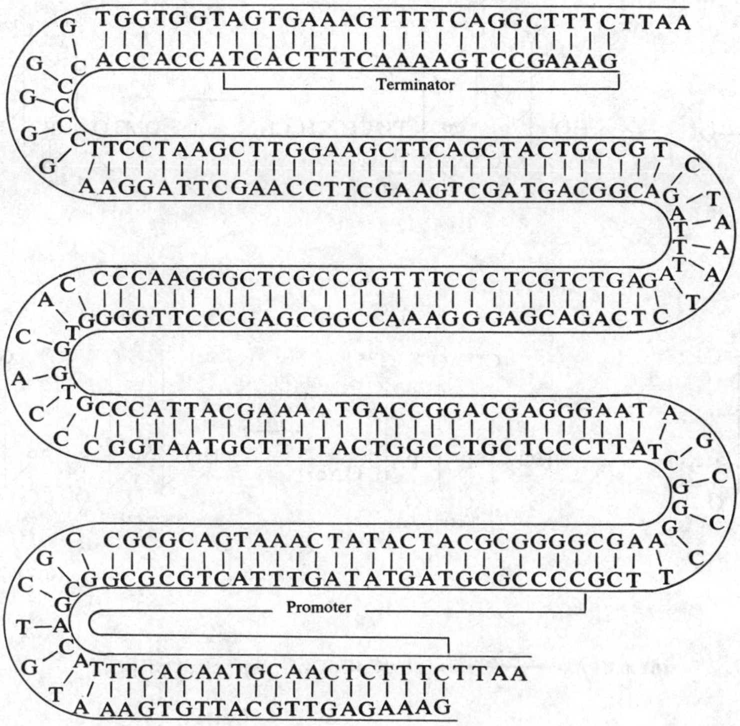

Figure 1 The completed gene for tRNA^{Tyr} showing the control sequences and single-stranded ends used for its insertion into a virus for biological testing.

resort to phosphate triesters. They can be prepared by the condensation of a suitably protected nucleoside 3′-phosphate diester with a 5′-hydroxyl group of an acceptor nucleotide. The arenesulphonic acid chlorides and certain active sulphonyl-triazoles or -tetrazoles are suitable reagents for this task. The products formed are neutral molecules, soluble in organic solvents, and can be purified by chromatographic techniques, especially h.p.l.c. on silica gel. After the completion of synthesis of the nucleotide fragment, the phosphate triesters can be cleaved to diesters without difficulty and without further hydrolysis of the internucleotide phosphate diester linkage. Various syntheses have used 2-cyanoethanol,[4] 2,2,2-trichloroethanol,[5] or one of several phenols[6] to esterify the phosphate group.

The method is illustrated here by reference to one of three syntheses of self-complementary, decameric DNA fragments which, in duplex form, are substrates for cleavage by restriction enzymes[7] (Sections 22.4.3 and 22.5.4). The dinucleotide dCpCp is protected by 4-chlorophenol on the phosphates, by benzoyl residues on the bases, and by dimethoxytritylation of the 5′-hydroxyl group. The acceptor is the 5′-OH group of the protected trinucleotide dGpGpA. Condensation is effected by means of a sulphonyl tetrazolide and then the base-labile 2-cyanoethyl residue on the 3′-terminal phosphate is selectively removed under mild alkaline conditions to give the protected pentamer (**5**). A similar condensation of the protected dTpCp with dCpGpG gives the pentamer (**6**). These pentamers are then coupled to give the decameric compound which, on removal of all protecting groups with ammoniacal pyridine, followed by mild acid treatment, was purified by cellulose t.l.c. treatment to afford 25 mg of the product (**7**) as a homogeneous duplex (Scheme 3). It is a substrate for the restriction enzyme *Bam* I.

Synthetic decameric duplexes of this sort are intended to be butt-jointed to chemically

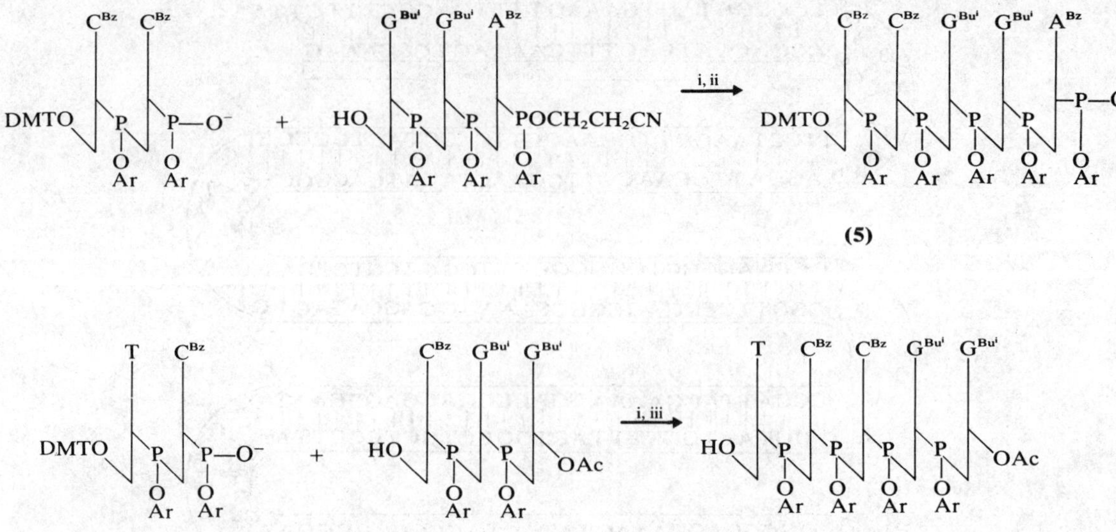

(5)

(6)

(6) + (5) $\xrightarrow{\text{i, iv, iii}}$ dCpCpGpGpApTpCpCpGpG$_{OH}$

(7)

i, TPST, pyridine; ii, NaOH, EtOH, pyridine; iii, 80% AcOH; iv, NH_4OH, pyridine; Ar = 4-chlorophenyl; TPST = 2,4,6-tri-isopropylbenzenesulphonyl tetrazolide.

SCHEME 3

or enzymatically prepared DNA genes by ligation. When they are then subjected to the sequence-specific endonuclease action of a restriction enzyme (Section 22.4.3) they are transformed into sticky ends. Since identical sticky ends can be made by the action of the same restriction enzyme on a bacterial chromosome, the opportunity can be created to introduce an artificial gene into a chromosome (Section 22.5.4).

22.4.1.3 Synthesis using solid-phase supports

The idea of a solid-phase support for the sequential synthesis of a polymer of defined sequence was developed by Merrifield for peptide work (Chapter 23.6). Early attempts to apply it and cognate technology to the synthesis of oligonucleotides were relatively unsuccessful.[1] Put simply, the growing oligonucleotide seemed to get knotted within the polymer support to make its end inaccessible for further extension: yields fell drastically before the number of nucleotides reached double figures. That situation appears to have been overcome by resort to a new type of polyamide support system which was developed for peptide work in the first instance.[8]

The first nucleotide is bonded to the polymer matrix by esterification of its 5′-phosphate with a 2-thioethanol group. At the end of the sequence of reactions extending the polynucleotide chain, this linkage is oxidized to give a sulphone, which can be detached from the completed nucleotide by a β-elimination process under basic conditions (Scheme 4).[9]

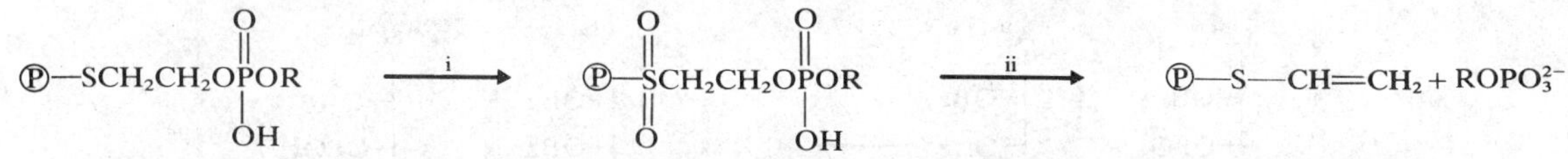

i, *N*-Chlorosuccinimide; ii, mild alkali.

SCHEME 4

The techniques used for nulceotide elongation are essentially the same as those described for the phosphate diester approach, with a few technical embellishments. 3′-Acetyldeoxynucleoside 5′-phosphates are condensed on to the growing polymer using tri-isopropylbenzenesulphonyl chloride followed by deacetylation in the usual way. This routine is repeated to add the second, third, *etc.* residues. At this stage, the only purification procedure involved is thorough washing of the insoluble polymer to remove excess reagents and by-products. On completion of the chain, base treatment releases a mixture of oligonucleotides which ideally should be homogeneous. In practice, the consequence of incomplete condensation at each step leads to an accumulation of errors and oligonucleotides missing one, two, or three residues are also formed. As long as the complete chain is not too big, this mixture can be resolved by careful chromatography. By this means the heptanucleotide dpCpApGpTpGpApT was prepared in 15–20% yield, free of hexa-and penta-nucleotides.

22.4.2 CHEMICAL SYNTHESIS OF RIBO-OLIGONUCLEOTIDES

The presence of the 2′-hydroxyl groups in the ribose series presents complications which have greatly retarded synthetic progress in this field relative to that for deoxyribo-oligonucleotide synthesis. In spite of the development of special protecting groups (Chapter 22.2) and other finesses, most objectives can be attained better by the combination of chemical deoxyoligonucleotide synthesis with enzymatic amplification and transcription into RNA.[10]

22.4.2.1 Synthesis *via* phosphodiesters

The preferred route for ribonucleotide synthesis has been influenced by the ready availability of 3′-phosphomonoesters from ribonuclease digestion of RNA. In the chemical synthesis of the 64 possible ribotrinucleoside diphosphates,[11] a 3′-phosphate bearing the acid-labile monomethoxytrityl group was condensed with a free 5′-hydroxyl group of an acceptor nucleotide. Selective deprotection of the 5′-hydroxyl group and a repeat of the sequence of operations gave the product, illustrated for the preparation of the codon rGpCpA (Scheme 5).

This incremental addition of residues to the 5′-end of the growing chain is the same as the mode adopted for triester synthesis in the deoxyribonucleotide series. It has been incorporated into a block condensation synthesis of the nonanucleotide rCpGpUpCpCpApCpCpA$_{OH}$ by the addition of the protected trimer rCpGpUp on to the product of joining rCpCpAp with rCpCpA$_{OH}$, all in suitably protected form.[12]

22.4.2.2 Synthesis *via* phosphotriesters

The advantages for purification by resource to triester synthesis in the ribonucleotide field barely outweigh the extra problems involved. These are created firstly by the risk of

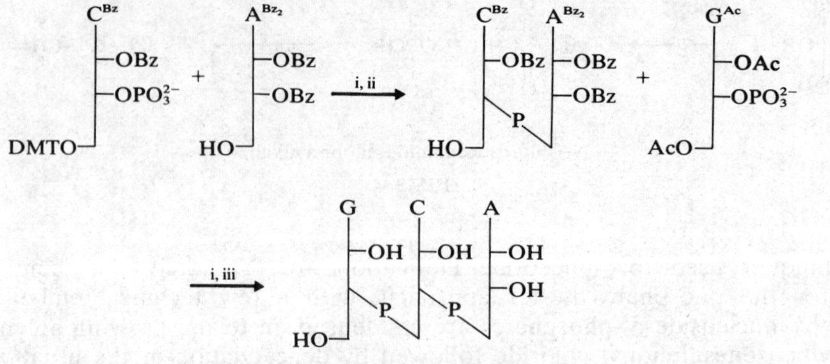

i, DCCD or TPS; ii, H^+; iii, NH_3, MeOH. DMT = 'dimethoxytrityl' = bis-*p*-anisylphenylcarbinyl.

SCHEME 5

phosphoryl migration from 3′- to 2′-positions during the final conversion of phosphate triesters into diesters and, secondly, by the need to manipulate separately *four* species of protecting groups — on the base, the phosphate, and the 2′- and 5′-hydroxyl groups. Nonetheless, the fact that a nonanucleotide has been made in this way speaks volumes for the detailed, careful development of protecting groups in this field of synthesis.

Routes have been developed for chain extension at both 5′- and 3′-ends. Neither as yet appears to have established a superiority over the other. The synthesis of a sequence from the anticodon loop for $tRNA_f^{Met}$ from *E. coli* provides a fine example of methods using stepwise and block additions to the 3′-hydroxyl group.[13] The procedures employed are similar to those used for the preparation of the simple trimer, rUpUpU (Scheme 6).

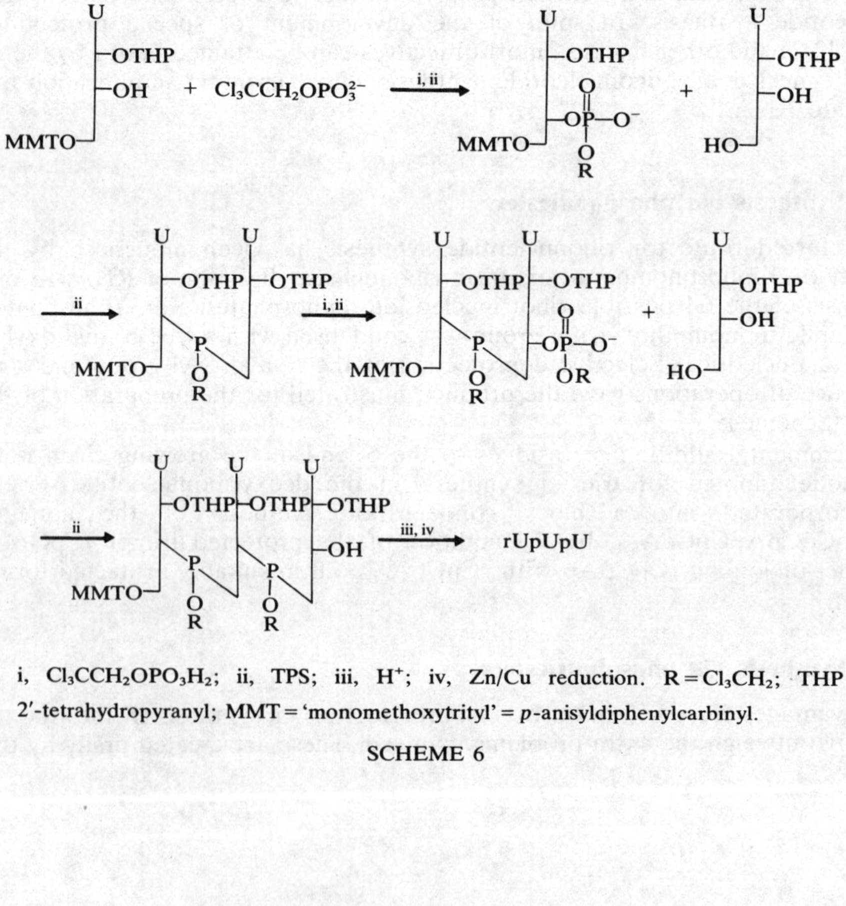

i, $Cl_3CCH_2OPO_3H_2$; ii, TPS; iii, H^+; iv, Zn/Cu reduction. R = Cl_3CH_2; THP = 2′-tetrahydropyranyl; MMT = 'monomethoxytrityl' = *p*-anisyldiphenylcarbinyl.

SCHEME 6

The following features are notable: (i) the trichloroethyl protecting group for phosphate is easily removed by reduction using a zinc/copper couple in neutral medium; (ii) the tetrahydropyranyl protecting group is adopted for the 2′-hydroxyl group; (iii) tri-isopropylbenzenesulphonyl chloride promotes the condensation of trichloroethyl phosphate on to the secondary hydroxyl group of a 2′,5′-protected nucleoside to form a phosphodiester; but (iv) the same condensing agent shows good regioselectivity for the primary 5′-hydroxyl group in the formation of the phosphotriester. Apparently, steric hindrance between the 2′-*O*-tetrahydropyranyl group and the sulphonylphosphoric anhydride intermediate excludes condensation at the secondary alcohol and thus obviates the need for a protecting group for the 3′-hydroxyl position.

This regioselectivity has resulted in a degree of similarity between the above synthesis and schemes which add monomers on to the 5′-end.[1,6] Both approaches share the same problem of slower rates of reaction for the formation of phosphotriesters compared with phosphodiesters. A hybrid synthesis has thus been developed to combine the better features of both strategies. In it, the internucleotide linkage is formed by a conventional phosphodiester procedure and is then transformed into a triester using 2-cyanoethanol and tri-isopropylbenzenesulphonyl chloride before the addition of the next monomer.[14]

22.4.2.3 Synthesis using enzyme-catalysed phosphorylation

Two types of condensation can be carried out effectively by enzymatic methods: the addition of a single residue to the 5′-terminus of an oligonucleotide, and the head-to-tail joining of two chains by a ligase enzyme. The action of DNA-dependent RNA polymerase is described later (Section 22.5.2).

The enzyme polynucleotide phosphorylase has been much used to condense a ribonucleotide 5′-pyrophosphate on to the 3′-hydroxyl group of a dinucleoside phosphate or larger acceptor. One disadvantage of the method used is the production of inorganic phosphate during the course of the reaction, which promotes the reverse process, *i.e.* the partial phosphorolysis of the 3′-terminal residue. This has been circumvented by the removal of phosphate from the reaction mixture as soon as it is formed.[15]

The unwanted, repetitive addition of monomers on to the 3′-end can be forestalled by the use of 2′(3′)-*O*-isovalerylribonucleoside 5′-pyrophosphates as donors in the condensation reaction. The bulky protecting group blocks further addition of a monomer on to the 3′-terminus.[16] In this way, two or even three residues can be added sequentially to give products of heptanucleotide size, but the method is most useful for the addition of a single residue, often having a modified base function, on to a preformed oligomer.

Although the DNA ligase used in Khorana's gene synthesis calls for juxtaposition of the acceptor and donor elements by the use of a splint, a less demanding RNA ligase has been isolated from *E. coli* infected with bacteriophage T4. It catalyses the phosphorylation of the 3′-hydroxyl group of a ribo-oligonucleotide by the 5′-phosphate of a donor. The enzyme appears to exhibit no base specificity and units as small as trimers can be linked together without a splint. The dodecamer $r(Ap)_5Cp(Up)_5U$ has been made in excellent yield by ligating the two hexamers $r(Ap)_5C$ and $rp(Up)_5U$.[17]

While synthetic achievements in the RNA field still lag a long way behind those for DNA, a major problem for investigators in both fields of endeavour has been the identification of targets of sufficient value to justify the expenditure of prodigious effort in achieving them. The early lead in purification and sequence determination of RNA species over DNA molecules presented the DNA genes for tRNAs as the first major objectives for synthesis. By contrast, the impossibility of reading the genetic code backwards to define a unique RNA sequence from that of a particular protein has restricted the goals for RNA synthesis to smaller fragments which often contain specialized minor bases.

This seemingly limited future has been transformed at a stroke. The remarkable breakthrough in DNA sequence determination must ensure an ample supply of worthwhile synthetic objectives for DNA construction for the next period of development of the field. Moreover, the progression of genetic engineering may well depend on the availability of synthetic fragments of DNA duplexes for specific tasks.

22.4.3 SEQUENCE DETERMINATION FOR DNA

Even in 1973, an effective approach to DNA sequence determination was the transformation of the task into an RNA sequence problem! Four years later, a complete *volte face* has inverted the dependency. Two methods are responsible for this about-turn.

Sanger and Coulson have devised a method for DNA sequence analysis which is based on enzymatic copying of a single-stranded DNA species.[18] Maxam and Gilbert's method relies on chemical modification of the four DNA bases and operates on single- or double-stranded DNA species with equal effect.[19] Both methods utilize the autoradiographic detection of ^{32}P-labelled oligonucleotides which have been resolved as a function of length alone by electrophoresis of denatured fragments on polyacrylamide gels. In application, their success owes much to recent developments in the enzymology of nucleic acids, especially to the use of restriction enzymes to cleave DNAs and of reverse transcriptase to make cDNA, which is a complement of an RNA template. The following descriptions of the techniques will, however, presuppose the availability of a homogeneous DNA sample of appropriate length.

Sanger's 'plus and minus' technique provides sequence information in two confirmatory modes. The *minus* technique involves copying from a single-stranded DNA using DNA-dependent polymerase or from an RNA species using reverse transcriptase. The newly-synthesized strands are made radioactive by ^{32}P incorporation. Fragments of different lengths, each of which stops before a given base, are prepared by *limiting* the supply of that base. As a result of running four parallel syntheses, deficient in A, C, G, and T respectively, stoppages occur in front of the different bases (−A, −C, −G, and −T). Side-by-side electrophoresis of these four samples provides a progression of bands on the autoradiograph from which the sequence of some 100–150 residues can be read directly. The *plus* procedure involves exonucleolytic degradation of the synthetic DNA fragments by removing bases one-at-a-time from the 3′-terminus. This calls for a modified DNA polymerase I which can be halted at specific positions along the degradative sequence by the *addition* of one of the four nucleoside triphosphates. Once again, by running four reactions in parallel, each augmented by one of the nucleotides (+A, +C, +G, or +T), degradation can be stopped in front of one of the bases (A, C, G, or T) and, again, gel electrophoresis gives a pattern of bands from which the sequence can be read directly. This methodology is illustrated in Figure 2.

Major successes have been achieved by the application of the plus and minus technique to determination of the sequence of cDNA[20] derived from human β-globin mRNA and to the first complete sequence[21] of a DNA bacteriophage, ϕX 174.

The Maxam-Gilbert technique for DNA sequence determination achieved the particular distinction of being widely adopted in advance of publication.[19] It is essentially tied to the use of four controlled chemical degradations, each one of which shows a greater or lesser degree of selectivity for one of the four bases. It works equally well with single- and double-stranded DNA provided that only one strand is radioactively labelled with ^{32}P. That is achieved by terminal phosphorylation using γ-^{32}P-labelled ATP and either a 3′-terminal kinase or a 5′-terminal one. This procedure puts a ^{32}P-label at both ends of a duplex, but on alternate strands. Therefore, cleavage by a restriction enzyme provides separable fragments, each having a single label and with a potential overlap sequence relating the two (depending on the choice of restriction enzyme). In four separate reactions this material is subjected to one of the four chemical degradations under mild conditions, so that only about one base in 50–100 is modified. Cleavage of the strand at the modified position provides a mixture of fragments, each ending at a particular base. Just as in the 'plus and minus' technique, gel electrophoresis and autoradiography provides a pattern of bands from which the base sequence can be read, up to 150 residues at a time.

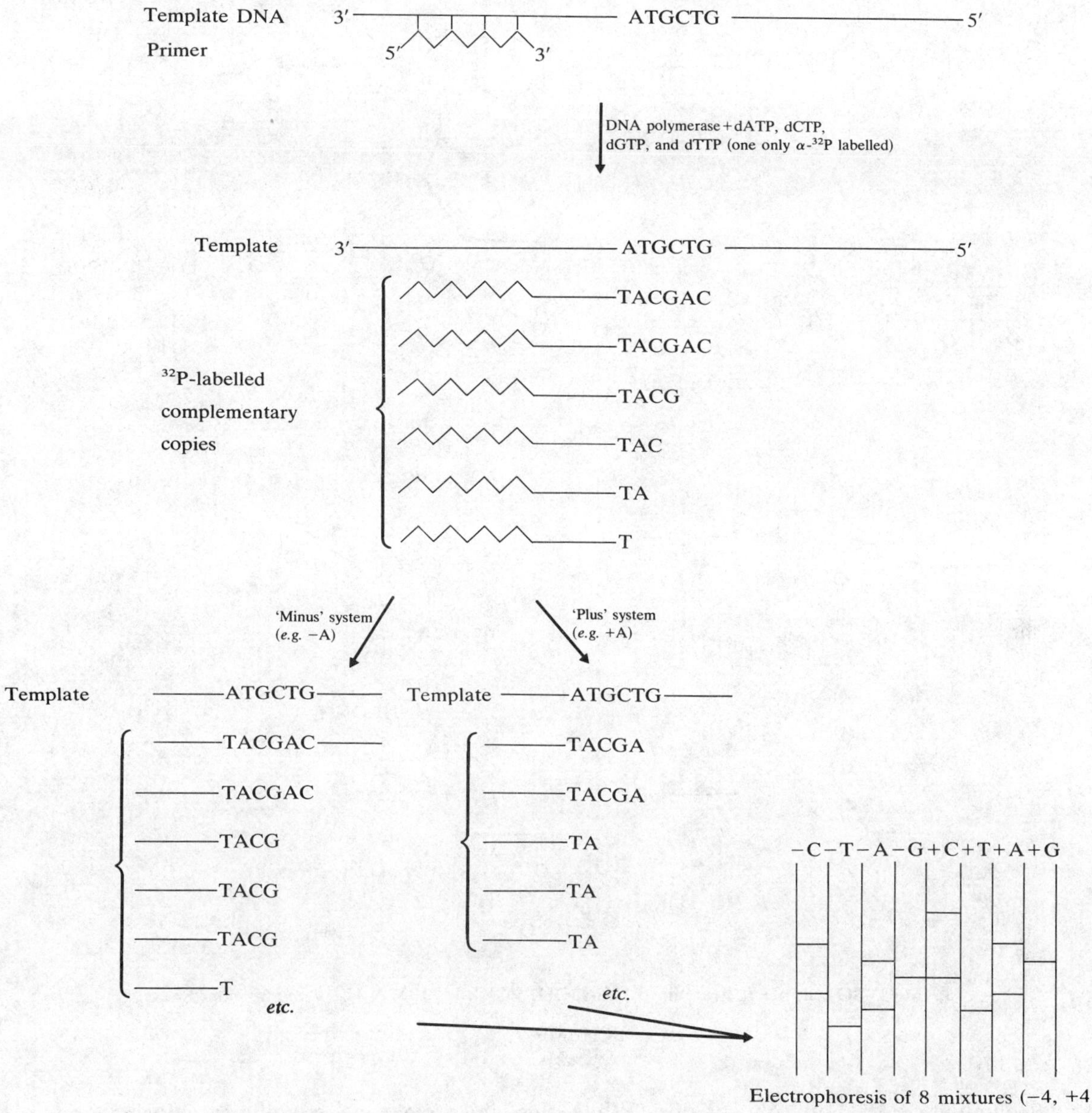

Figure 2 The principle of the 'plus and minus' method for DNA sequence determination.[18]

The specific cleavages are achieved as follows. Treatment of DNA with dimethyl sulphate methylates guanines at N-7 and adenines[22] at N-3. The glycosidic bonds of such derivatives are easily hydrolysed and subsequent alkaline treatment cleaves the phosphates from the free deoxyribose (Scheme 7). Guanine is methylated five times faster than adenine, so this routine gives a stronger cleavage at guanine (G > A). That situation is readily reversed to give a strong adenine/weak guanine cleavage by using dilute acidic conditions for hydrolysis of the glycosidic bond, which suppresses hydrolysis of 7-methylguanosine. Subsequent alkaline fission of the phosphates gives principal cleavage at adenine residues (A > G).

Both pyrimidines are attacked by hydrazine[23] (Chapter 22.2), which cleaves open the base to give a ribosylurea. The DNA backbone is then split using piperidine to cause

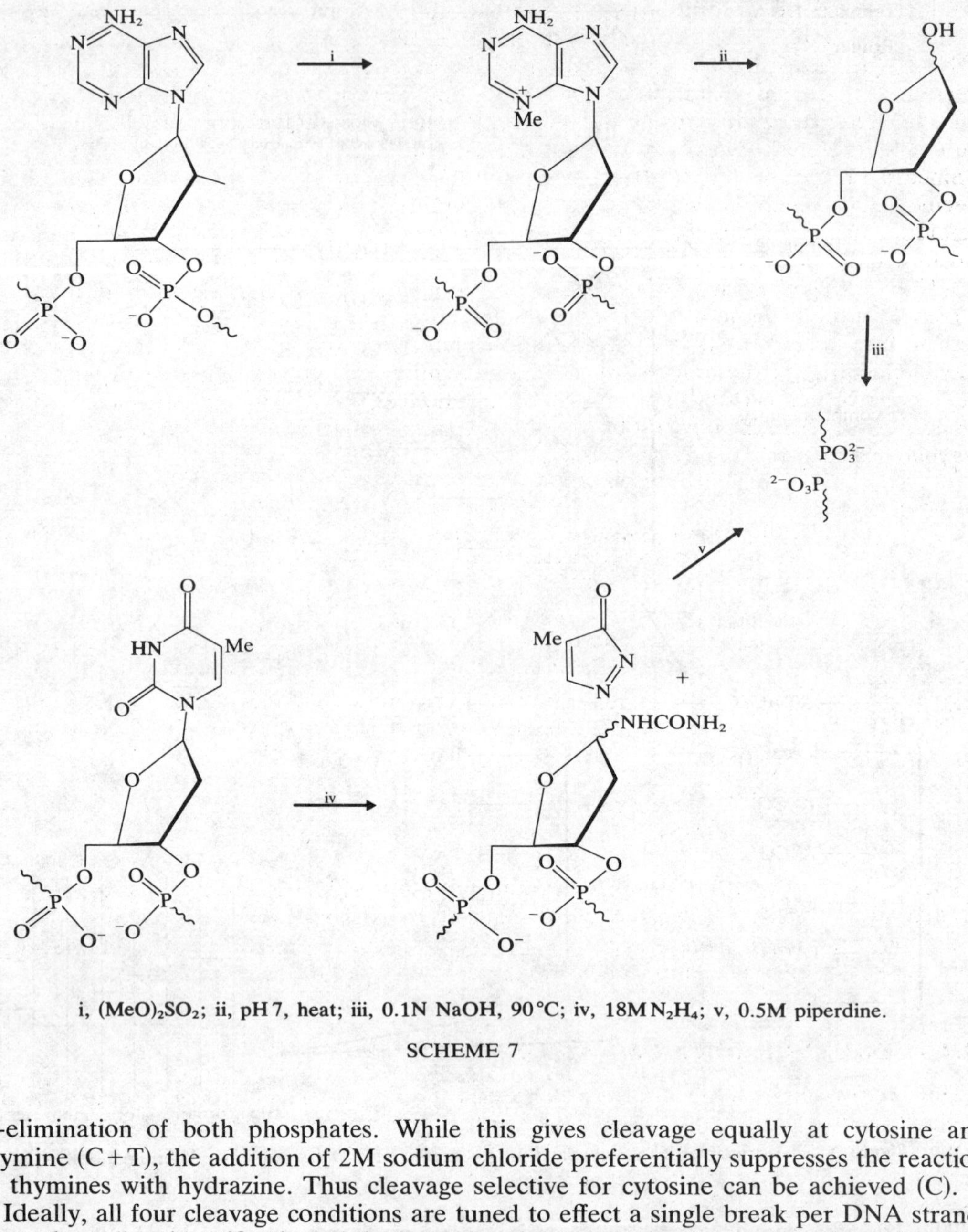

i, $(MeO)_2SO_2$; ii, pH 7, heat; iii, 0.1N NaOH, 90°C; iv, 18M N_2H_4; v, 0.5M piperdine.

SCHEME 7

β-elimination of both phosphates. While this gives cleavage equally at cytosine and thymine (C+T), the addition of 2M sodium chloride preferentially suppresses the reaction of thymines with hydrazine. Thus cleavage selective for cytosine can be achieved (C).

Ideally, all four cleavage conditions are tuned to effect a single break per DNA strand. They then allow identification of the bases by direct analysis of an autoradiograph of the four samples run side-by-side on gel electrophoresis. Adenine shows as a darker band in the second reaction (A > G) than in the first, guanine as a darker band in the first (G > A) than the second. Thymine appears as a band in only the third reaction (C+T), while cytosine also appears as the only band in the fourth reaction (C).

The minor limitations of this technique are offset by the duplicate information which can be gained from independent analyses of the complementary strands. The use of enzymes is limited to terminal phosphate labelling and to restriction cleavage of strands into suitable lengths, around 100–150 residues, with overlaps. The method has been particularly applied to the sequence of bases in control sections of genes, *e.g.* the 223 base-pair duplex for the 5S RNA gene in baker's yeast shows both the promoter and terminator sites for transcription.[24] Another fine example of its use is the complete primary structure for rabbit β-globin mRNA, tackled by reverse transcription into cDNA.[25] In the course of amplification of the cDNA by cloning of bacterial plasmids (Section 22.3.4), the 13 residues from the 5′-end were lost. Fortunately, their sequence was available from a study using the 'plus and minus' technique![26] Together they provided the complete gene sequence of 589 nucleotide pairs.

Since a crucial part of the technology of DNA sequence determination lies in the specific use of nuclease enzymes, a brief comment on their properties is called for. Reference has already been made to *endonucleases,* which cut chains into smaller chains, and to *exonucleases,* which clip off terminal residues without exhibiting any base specificity. The *restriction nucleases* are bacterial endonucleases which recognize specific base sequences, usually some 4 to 6 residues long and possessing a centre of symmetry.[27] The enzyme thus acts to break *both* strands in a symmetrical relation to this centre (Table 1). The complexities of the different sequences recognized regulates the frequency with which they appear in large DNA molecules. For instance, *Eco R1*, from *E. coli*, makes a single scission of λ viral DNA (5000 nucleotide pairs) while *Hae* III, from *Haemophilus aegyptius*, breaks the same DNA 17 times.

TABLE 1
Recognition Sequence for Four Restriction Enzymes and the Number of Targets for each in Two Viral DNAs[a]

	↓	↓	↓	↓
Recognition sequences	GTPyPuAC	GAATTC	GGCC	AAGCTT
	········	········	······	········
	CAPuPyTG	CTTAAG	CCGG	TTCGAA
	↑	↑	↑	↑
	Hind II	*Eco R1*	*Hae* III	*Hind* III
DNA targets				
λ(MW $\approx 3.2\times10^7$)	34	5	750	6
SV40 (3×10^6)	5	1	17	6

[a] Arrows indicate points of cleavage.

The selection of appropriate enzymes thus provides pieces of convenient size to use for DNA sequence analysis and also gives adequate opportunities for overlapping sequences. In addition, the types of cleavage made by species such as *Hind* III can be employed to generate sticky ends for joining DNA fragments by ligation (Section 22.5.4).

22.4.4 SEQUENCE DETERMINATION FOR RNA

The relative ease of purification of species of tRNA and of rRNA, along with the availability of endonucleases for RNA with different base specificities, enabled such sequences to be first determined in 1965. The early methods of Holley[28] and Zachau[29] were soon displaced by techniques developed by Sanger for the sequence determination of 5S RNA.

There are three stages to this classic sequence analysis. Firstly the RNA is cleaved into fragments of moderate size (*ca.* 50 base pairs) by limited RNase digestion. Secondly these discrete fragments are sequenced by total base analysis and further fragmentation into small pieces which can be sequenced directly. Finally the larger fragments are put into order by means of overlap sequences. Such overlaps are generated as a result of the different patterns of cleavage given by pancreatic RNase (pyrimidine specific) and Takadiastase T_1 (guanosine specific). The difficult separation of the many fragments produced by such limited nuclease action is best achieved by a two-dimensional separation ('fingerprinting') combining ionophoresis on cellulose acetate at pH 3.5 in one direction with ionophoresis on DEAE cellulose paper at acidic pH.[30] In all cases, fragments are detected non-destructively by autoradiography of ^{32}P-labelled nucleotides.

That feature has been turned to very good advantage in *homochromatography*, a technique in which the radioactive oligonucleotide fragments are developed on a chromatogram using a solution containing a non-radioactive mixture of fragments of all sizes, produced by RNase digestion of crude RNA. On DEAE or polyethylenimine cellulose t.l.c. plates, the smaller polyanions are bumped along by the larger ones when the cationic support phase is saturated with nucleotide fragments. The radioactive sample thus separates into a series of frontal zones which are cleanly detected by autoradiography.[31]

Valuable as this technique has been, the discovery of the enzyme *reverse transcriptase* in 1972 has now been applied to revolutionize RNA sequence determination. Most eukaryotic mRNAs possess a 3′-terminal poly(A) sequence (Section 22.5.2.1) which on annealing to an oligodeoxythymidylic acid, poly(dT), creates a primer sequence for the initiation of DNA synthesis by this enzyme. The mRNA is thus transcribed into cDNA, which can either be used directly for sequence analysis or converted into a double-stranded DNA by further action of the reverse transcriptase. At the present time, this approach has become the dominant technique for RNA sequence determination.[32]

In its turn, this technique may be displaced by a direct determination of RNA sequences along the lines of the Maxam–Gilbert DNA method. It is possible to map separately adenines, guanines, and pyrimidines in RNA by electrophoretic size-fractionation of alkaline and endonuclease digests on acrylamide gels.[33] At such time as uridine can be differentiated from cytidine, the method should be as powerful and rapid as that currently used for DNA analysis.

Wide and varied benefits have accrued from RNA sequence analysis. Over 100 tRNA sequences have been determined.[34] The sequence of yeast $tRNA^{Phe}$ has been as essential in complementing the X-ray data analysis to derive the 3-D structure (Section 22.1.3.4, Figures 6 and 7), as has proved the case for determination of protein structures. Even more important are the insights created into the relationship between structure of nucleic acids and biological function. The detail known of the workings of transcription and translation depend in much depth on sequence information of RNA species. A simple example is the derivation of tRNA molecules from precursors by excision of parts in a maturation process — very much akin to the formation of insulin from preproinsulin (Figure 3).

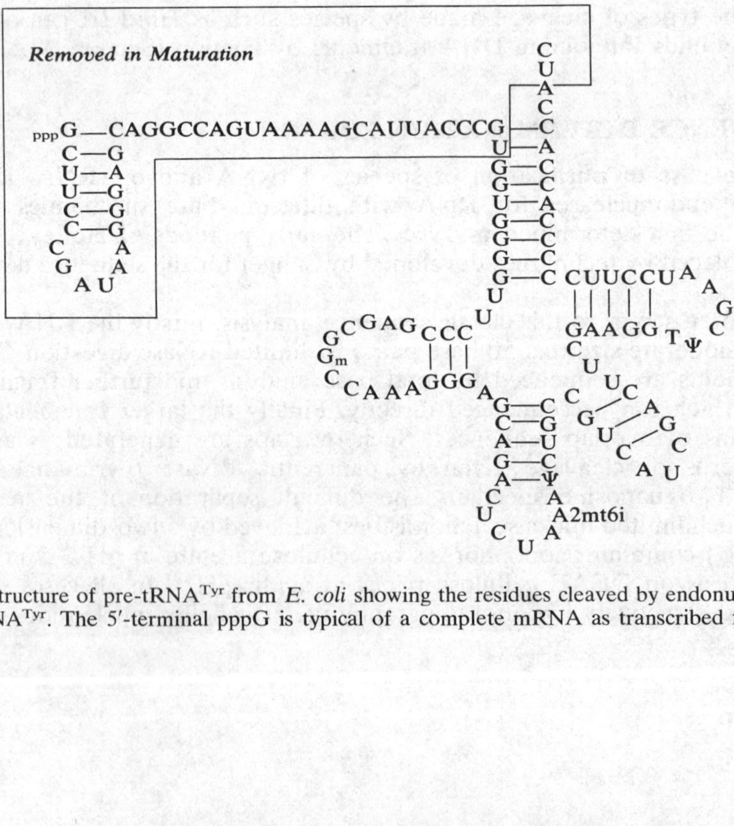

Figure 3 The structure of pre-$tRNA^{Tyr}$ from *E. coli* showing the residues cleaved by endonucleolytic action to give mature $tRNA^{Tyr}$. The 5′-terminal pppG is typical of a complete mRNA as transcribed from DNA.

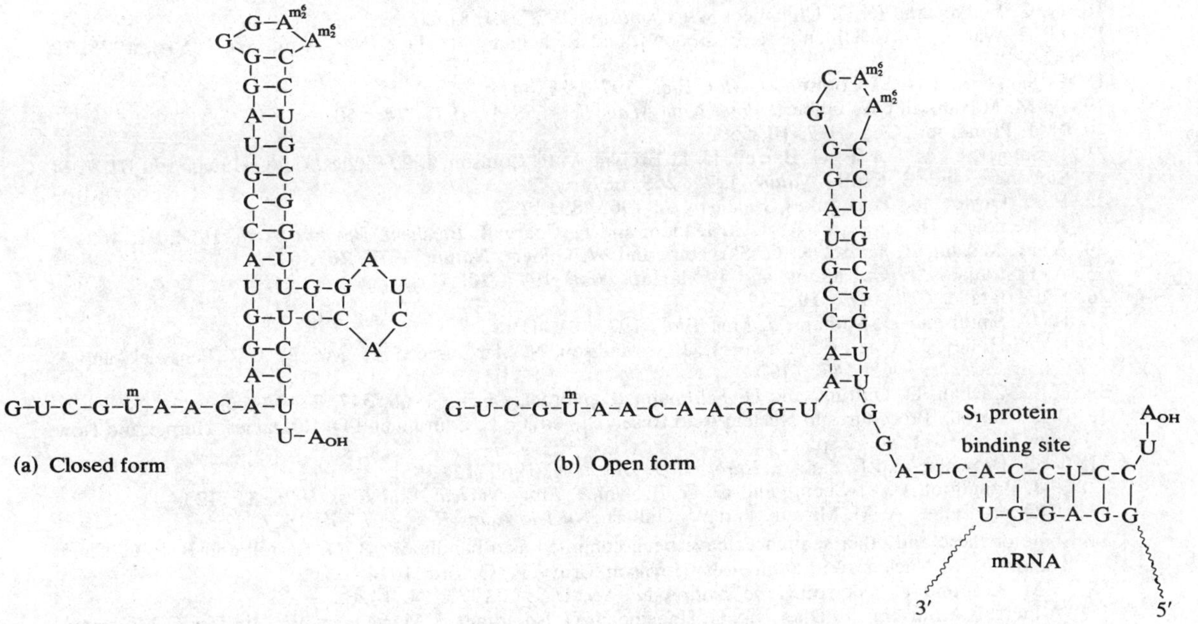

Figure 4 Closed (a) and open (b) conformations for the 3′-terminus of *E. coli* 16S rRNA showing possible mRNA attachment.

In the field of ribosomal RNA, sequences determined near the 3′-end of 16S RNA may possibly adopt a closed conformation which switches to an open form in the presence of a messenger RNA to allow pairing of complementary RNA sequences and identification of the point on the messenger for initiation of protein synthesis (Figure 4).

The sequence[25] of the first mammalian messenger RNA has provided a direct comparison between the structure of the genetic code and its practical utilization that has revealed an unexpected discrimination in choice between degenerate codons for the same amino-acid. Sequence analysis of heterogeneous RNA from the cell nucleus is providing the first steps in relating such hnRNA to mRNA.[35] Finally, the heroic determination of the whole sequence of residues in the RNA of virus MS2 provides a bridge between structural chemistry and life itself.[36]

Such attainments herald a new era in the understanding of structure–function relationships in nucleic acids, which forms the subject of Chapter 5 of this discussion.

References

1. H. Kössel and H. Seliger, *Fortschr. Chem. org. Naturstoffe*, 1975, **32,** 297.
2. H. G. Khorana, *Biochem. J.*, 1968, **109,** 709.
3. H. G. Khorana *et. al.*, *J. Biol. Chem.*, 1976, **251,** 565 *et seq.*
4. R. L. Letsinger and K. K. Ogilvie, *J. Amer. Chem. Soc.*, 1970, **91,** 3350.
5. F. Eckstein and I. Rizk, *Chem. Ber.*, 1969, **102,** 2362.
6. C. B. Reese, *Chim Org. Phosphore* (*Colloq. Internat. Centre Nat. Recherche Sci.*), 1970, **182,** 319.
7. R. H. Scheller, R. E. Dickerson, H. W. Boyer, A. D. Riggs, and K. Itakura, *Science*, 1977, **196,** 177.
8. M. J. Gait and R. C. Sheppard, *Nucleic Acids Res.*, 1977, **4,** 1135.
9. F. Brandstetter, H. Schott, and E. Bayer, *Tetrahedron Letters*, 1974, 2705.
10. D. S. Jones, S. Nishimura, and H. G. Khorana, *J. Mol. Biol.*, 1966, **16,** 454.
11. R. Lohrmann, D. Söll, H. Hayatsu, E. Ohtsuka, and H. G. Khorana, *J. Amer. Chem. Soc.*, 1966, **88,** 819.
12. E. M. Ohtsuka, S. Ubasawa, S. Morioka, and M. Ikehara, *J. Amer. Chem. Soc.*, 1973, **95,** 4725.
13. T. Neilson and E. S. Werstiuk, *J. Amer. Chem. Soc.*, 1974, **96,** 2295.
14. J. Smrt, *Tetrahedron Letters*, 1972, 3437.
15. J. J. Sninsky, G. N. Bennett, and P. T. Gilham, *Nucleic Acids Res.*, 1974, **1,** 1665.

16. G. C. Walker and O. C. Uhlenbeck, *Biochemistry*, 1975, **14,** 817.
17. G. C. Walker, O. C. Uhlenbeck, E. Bedows, and R. I. Gumport, *Proc. Nat. Acad. Sci. U.S.A.*, 1975, **72,** 122.
18. F. Sanger and A. R. Coulson, *J. Mol. Biol.*, 1975, **94,** 441.
19. A. M. Maxam and W. Gilbert, *Proc. Nat. Acad. Sci. U.S.A.*, 1977, **74,** 560.
20. N. J. Proudfoot, *Cell*, 1977, **10,** 559.
21. F. Sanger, G. M. Air, B. G. Barrell, N. L. Brown, A. R. Coulson, J. C. Fiddes, C. A. Hutchinson III, P. M. Slocombe, and M. Smith, *Nature*, 1977, **265,** 687.
22. P. D. Lawley and P. Brookes, *Biochem. J.*, 1963, **89,** 127.
23. A. Temperli, H. Türler, P. Rust, A. Danon, and E. Chargaff, *Biochem. Biophys. Acta*, 1964, **91,** 462.
24 A. M. Maxam, R. Tizard, K. G. Skryabin, and W. Gilbert, *Nature*, 1977, **267,** 643.
25. A. Efstratiadis, F. C. Kafatos, and T. Maniatis, *Cell*, 1977, **10,** 571.
26. F. E. Baralle, *Cell*, 1977, **10,** 549.
27. H. O. Smith and D. Nathans, *J. Mol. Biol.*, 1973, **81,** 419.
28. R. W. Holley, J. Apgar, G. A. Everett, J. T. Madison, N. Marquisee, S. H. Merrill, J. R. Penswick, and A. Zamir, *Science*, 1965, **147,** 1462.
29. H. G. Zachau, D. Dütting, and H. Feldmann, *Z. physiol. Chem.*, 1966, **347,** 212 *et seq.*
30. B. G. Barrell, 'Procedures in Nucleic Acid Research,' ed. G. L. Cantoni and D. R. Davies, Harper and Row, New York, 1971, vol. 2, p. 751.
31. G. G. Brownlee and F. Sanger, *European J. Biochem.*, 1969, **11,** 395.
32. N. J. Proudfoot, C. C. Cheng, and G. G. Brownlee, *Prog. Nucleic Acid Res.*, 1976, **19,** 123.
33. H. Donis-Keller, A. M. Maxam, and W. Gilbert, *Nucleic Acids Res.*, 1977, **4,** 2527.
34. Some of these and other sequences have been compiled as a handbook; B. G. Barrell and B. F. C. Clark, 'Handbook of Nucleic Acid Sequences,' Joynson–Bruvvers, Oxford, 1974.
35. V. M. Kish and T. Pederson, *Proc. Nat. Acad. Sci. U.S.A.*, 1977, **74,** 1426.
36. W. Fiers, R. Contreras, F. Duerinck, G. Haegeman, D. Iserentant, J. Merregaert, W. Min Jou, F. Molemans, A. Raeymaekers, A. Van den Berghe, G. Volckaert, and M. Ysebaert, *Nature*, 1976, **260,** 500.

22.5

Nucleic Acids: Structure and Function

G. M. BLACKBURN
University of Sheffield

The foregoing discussion of the structure and synthesis of nucleosides, nucleotides, and nucleic acids is capable of standing on its own as a prime sector of natural product chemistry. The deeper significance of and inherent links between its several parts can only be appreciated by full consideration of the relationship between structure and function of nucleic acids. Whatever are the pretensions of the rest of this part of Volume 5 to be comprehensive, this chapter must be unashamedly partial and selective, not least in its emphasis on the structural aspects of the molecules involved.

The primary statement relating structure and function is

'DNA makes DNA, makes RNA, makes Protein'.

It asserts that the hereditary substance is DNA. This molecular material is ultimately responsible for the precise transmission of information from parent to progeny cells, and for the control of the totality of chemical activity within the normal cell, as mediated by its catalytic proteins. From the point of view of a geneticist, chromosomes contain a discrete, linear nucleic acid, regions of which, called genes, are each responsible for the formation of a specific cellular product. These gene products are either polypeptides or structural RNA molecules, ribosomal and transfer ribonucleic acids.

This simple picture holds good for most sorts of cell, but it has to be modified in the special cases of two sorts of virus. Firstly, some viruses, like polio and bacteriophage MS2, have a chromosome which is a strand of RNA. Their progeny are formed without the intervention of DNA.[1] Secondly, certain tumour viruses also have RNA as their genetic material but, in the infected host cell, it is transcribed into DNA and from this *provirus* progeny RNA strands are transcribed later. For obvious reasons, the enzyme associated with this special activity has been called reverse transcriptase.

These relationships can be accommodated in the following scheme, which can now be examined in some detail:

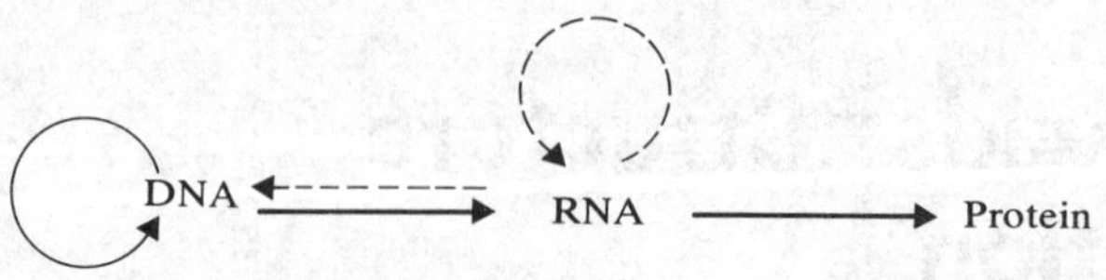

22.5.1 REPLICATION OF NUCLEIC ACIDS

22.5.1.1 Replication of double-stranded DNA

When a cell divides, the two daughter cells both contain chromosomes in which the DNA is virtually an exact copy of the DNA molecule(s) of the parental DNA. This is the simplest observation of replication. In practice, one each of the strands of the parental duplex appears in each daughter cell along with one newly-made complementary strand. The enzyme which brings about this new synthesis is *DNA polymerase III*, which matches incoming deoxynucleotides, as their 5′-triphosphates, by Watson–Crick pairing to the bases on both of the unwound parental strands. This pattern is known as semiconservative replication (Figure 1).

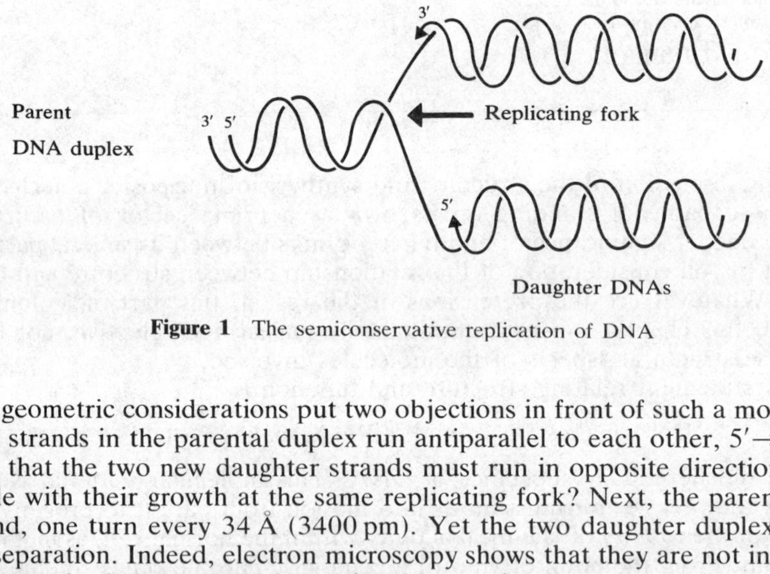

Figure 1 The semiconservative replication of DNA

Simple geometric considerations put two objections in front of such a model. In the first place, the strands in the parental duplex run antiparallel to each other, 5′→3′ and 3′→5′. It follows that the two new daughter strands must run in opposite directions. How is this compatible with their growth at the same replicating fork? Next, the parental strands are interwound, one turn every 34 Å (3400 pm). Yet the two daughter duplexes are capable of ready separation. Indeed, electron microscopy shows that they are not interwound even when they are in the process of formation from a circular parental duplex! What sort of swivel process allows the unwinding of the parental duplex? *It has to accommodate an apparent rotational velocity in* E. coli *of nearly* 100 *rev* s^{-1}.

The evidence suggests that a symmetrical but complex pattern of discontinuous replication takes place on both strands. The unwinding of the double-stranded DNA is promoted by binding of numerous protein particles which become attached to the parental DNA, at a chosen initiating site, and contrive to expose the separated, complementary strands of the DNA ready for new synthesis. Next, a species of RNA polymerase makes a short length of RNA, some 20 to 25 nucleotides long, which is complementary to and associated with the parental DNA strand. This acts as a primer for DNA polymerase III, a large, bulky enzyme, which now makes a new DNA strand of some 1000 residues in continuity with the RNA. Such synthesis takes place in a $5' \rightarrow 3'$ direction by the condensation of deoxynucleotide 5′-triphosphates on to the 3′-terminal hydroxyl group and on *both* strands of the parental DNA. Since the enzyme works close to reversibility, it gives the maximum thermodynamic control to the selection of the correct, incoming deoxyribonucleoside by base-pairing with one in the existing strand. In this way, a number of pieces, called Okazaki fragments, are aligned on each parental strand (Figure 2).

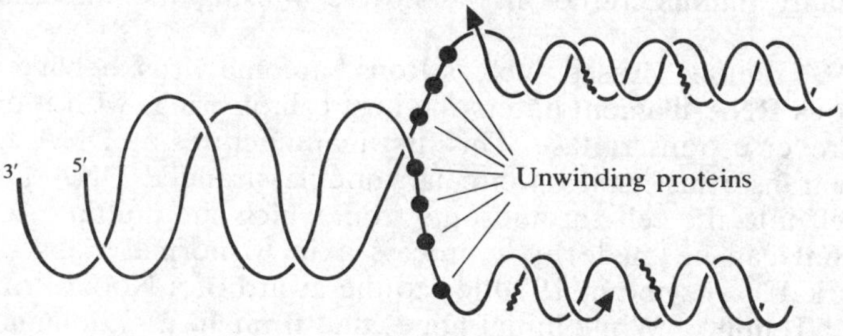

Figure 2 Details of the replication of DNA, showing RNA primers, ∿∿, at 5′-ends of new DNA fragments with DNA polymerase III effecting synthesis in a $5' \rightarrow 3'$ direction, →, and unwinding proteins, ●—●

In order to close the gaps between the newly made pieces of DNA, DNA polymerase I accomplishes two tasks. First, it extends the synthesis of incomplete Okazaki fragments up to the adjacent RNA segments and, secondly, it digests away these RNA segments by acting as an exonuclease on their 5′-ends. As a result of these processes, both daughter strands are married to new DNA with complete base complementation. However, there exist single-strand breaks in the new DNA. This is precisely the feature which is repaired by ligase enzymes which make the new chains whole.

The intriguing feature of this complicated performance is the complete removal of those RNA fragments which appear to be essential for DNA replication to start. Why? It has been suggested that it is a deliberate device to eliminate an unacceptable frequency of mistakes in pairing which might well be greater at the initiation of base-complementation than in its steady-state continuation.

Less attention has been directed to the problem of deconvoluting the parental DNA ahead of the replicating fork. One suggestion is that a single parental strand is 'nicked' to allow conformational freedom for the DNA to unwind and is then sealed before the separation of the parental strands. Until a better answer is forthcoming, contemplation of this problem is somewhat eased by the 'oscillating helix' model presented as an alternative to the Watson–Crick structure for DNA.[15]

The foregoing description of replication in eukaryotic systems applies equally well to the circular DNA of bacteria with one modification. Various results show that replication of bacterial and viral circular DNA duplexes begins at some point on the chromosome and then proceeds simultaneously in both directions away from that point. Thus an original 'eye-shaped' intermediate (-⊂⊃-) converts to a -⊂⊃—— shape on reaching the end of a linear molecule, but converts to two circles in the case of a bacterial chromosome.

22.5.1.2 Replication of viral nucleic acid

Viral nucleic acids are of four types: double- and single-stranded and DNA or RNA. The double-stranded DNA viruses replicate in much the same way as bacteria. Single-stranded DNA viruses, like ϕX 174, initiate their life cycle by injecting their circular DNA into a host *E. coli* bacterium in which the infectious 'plus' strand acts as template for the manufacture of a complementary 'minus' strand to give a circular duplex. From this, new copies of the (+) strand are rolled off continuously and cut into identical lengths before joining of the 3′- to 5′-ends to form a closed, single-stranded circle.

Most RNA viruses have a single-stranded nucleic acid, *e.g.* tobacco mosaic virus. This is generally replicated by a pattern of events similar to that described for ϕX 174 *via* the intervention of a (−) RNA strand. A lesser number of RNA viruses, such as Reovirus, have double-stranded RNA in which the base ratios obey Chargaff's rules. Replication proceeds directly to RNA daughter molecules, probably *via* a semiconservative scheme without intervention of DNA. Such replication requires the presence of a special RNA replicase which is usually manufactured in the host cell using the infecting RNA as messenger.

A small group of RNA tumour viruses, such as Rous Sarcoma virus, behave in a unique way. When the infectious RNA filament enters the host cell, it brings with it one or more copies of the enzyme reverse transcriptase. This first manufactures a cDNA complement to the RNA and then transforms it into a circular, double-stranded DNA *provirus.* The provirus integrates itself into the cell chromosome from which site multiple copies of the infectious RNA filament can be made by a process akin to normal transcription. The discovery of this remarkable enzyme in 1970 led to the award of a Nobel Prize to David Baltimore and Howard Temin.[2,3] While initial hopes that it might assist diagnosis of viral cancer have waned, its value to general aspects of molecular biology is substantial.

22.5.2 TRANSCRIPTION OF DNA INTO RNA

DNA base sequences predetermine the sequence of amino-acids in protein synthesis. Two separate stages participate in that control. First, the *transcription* of information from DNA to mRNA takes place in the cell nucleus. Secondly, the *translation* of mRNA information into protein occurs in the cytoplasm.

22.5.2.1 Biosynthesis of mRNA: transcription

mRNA is manufactured in the cell by the condensation of ribonucleoside 5′-triphosphates on to the free 3′-hydroxyl group of the nascent RNA chain. Thus, as in DNA, biosynthesis takes place in a 5′→3′ direction (Figure 3). A single enzyme, *RNA polymerase*, catalyses this process and demands a DNA template to work from. The template may be single- or double-stranded and it alone dictates the sequence of bases incorporated, according to normal patterns of complementation. When double-stranded DNA is used, only one particular strand is adopted by RNA polymerase to act as a template. If, however, the DNA is damaged or if its strands are separated, then both can act as templates. The possible origin of this selectivity will be described later (Section 22.5.2.2).

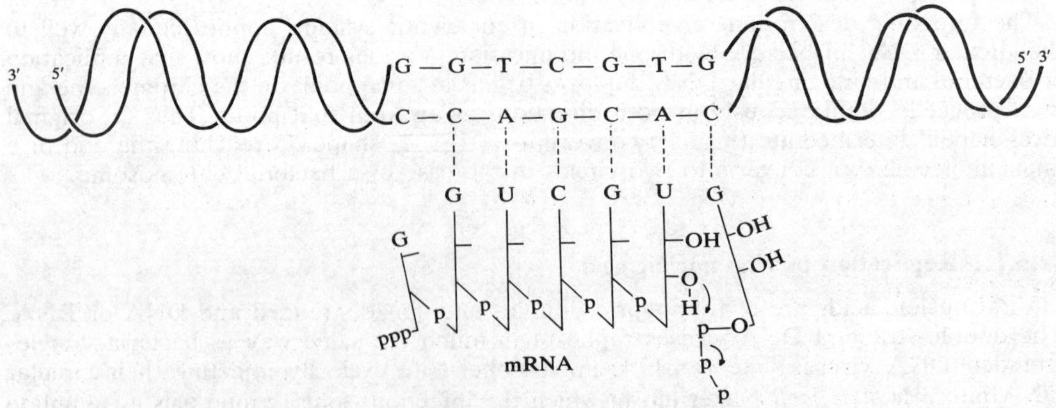

Figure 3 5′→3′ Pattern of biosynthesis of mRNA

Three things happen to eukaryotic mRNA after it has been transcribed from DNA. Before it leaves the cell nucleus, the 5′-end of the mRNA is 'capped' by condensation on to it of guanosine 5′-triphosphate (Scheme 1). This cap is now methylated by means of *S*-adenosylmethionine to place methyl groups at N-7 of the terminal guanine base and on the 2′-hydroxyl group of the 5′-residue of the mRNA to give the characteristic terminal unit, illustrated for capped Reovirus, m^7GpppG_mC . . . (**1**).

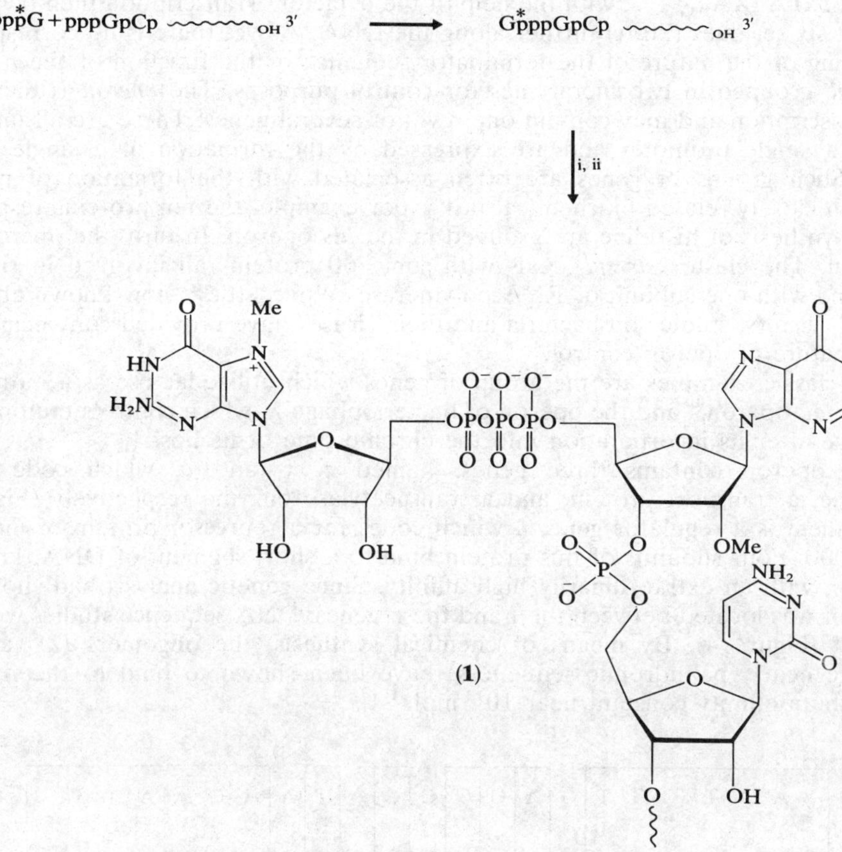

(**1**)

i, *S*-Adenosylmethionine; ii, methylase enzymes.

SCHEME 1

Tracts of poly(A) of up to 200 residues are now tagged on to the 3′-end of the capped mRNA and the product, m^7GpppG_m~~~~A_{200} is then exported from the cell nucleus into the cytoplasm, where further methylation may occur at N-6 of internal adenine residues. Completed mRNAs are about 1500 residues long, excluding the variable poly(A) end. There is some evidence that the much larger hnRNA found in the cell nucleus, which also can be capped and has poly(A) tracts, is a formal precursor of mRNA.

22.5.2.2 Control of transcription

There is vastly more DNA in the normal cell chromosome than is ever used for transcription. Two levels of control seem to be applied to it. The chromosomal proteins must be deemed to permit transcription of some DNA and not of the rest. That situation would appear to be invariant for all cells of the species. On the other hand, different genes are used in cells having different functions within a species and within the same cell at different stages of its life cycle or in response to changes in its habitat. This is described as gene regulation and possibly works by control of the access of RNA polymerase to chromosomal DNA.

RNA polymerase is a very large enzyme (MW $\sim 5 \times 10^6$) which can coopt additional subunits for specific tasks: the σ subunit for starting mRNA synthesis and the ρ subunit for releasing the mRNA at the end of synthesis. The results of DNA sequence studies suggest that RNA polymerase recognizes a *promoter* site, which has a sequence approximating to . . . TAT$\overset{A}{G}$AT$\overset{A}{G}$ . . . , with the help of the σ factor. Transcription then begins at a point about six residues (20 Å) further along the DNA. As yet there is no corresponding understanding of the nature of the terminator sequence or the function of the ρ factor.

Genes are grouped in two hierarchies for control purposes. The *operon* is the unit of mRNA transcription and may contain one, two, or several genes. These are all under the control of a single promoter and are expressed by the formation of a single mRNA molecule. Such groups of genes are often associated with the formation of products employed in closely related biochemical tasks. For example, the ten proteins responsible for the biosynthesis of histidine are grouped in the *his* operon. In turn, the operons may be clustered. The cluster *str-spc* deals with some 60 proteins all involved in ribosome structure and with one subunit of RNA polymerase. While little is now known about the function of clusters, studies on bacteria and their viruses have provided convincing detail about the nature of operon control.

The two classic examples are the group of genes which utilize lactose as a nutrient for *E. coli*, the *lac* operon,[4] and the operon of bacteriophage λ, which controls multiplication of the phage after its incorporation into the chromosome of its host.[5]

The *lac* operon contains three genes, named *z*, *y*, and *a*, which code for β-galactosidase, a transport protein, and a transacetylase enzyme respectively (Figure 4). Preceding them is a regulator gene, *i*, which codes for a repressor protein of molecular weight 38 000. Four subunits of this protein bind to a short segment of DNA, known as the *operator*, with an extraordinarily high affinity. Since genetic analysis had shown that this operator was located between the *i* and the *z* genes, DNA sequence studies were able to locate it (Figure 4). By means of chemical synthesis, the oligomers (**2**)[6] and (**3**)[7] (which have nearly palindromic sequences) have been shown to bind to the repressor protein with an affinity constant near 10^{13} mol^{-1} l.

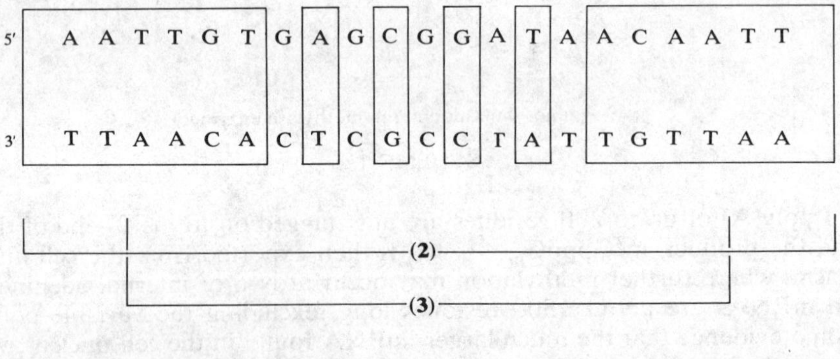

The *promoter* region lies between the *z* gene and the operator and has two functions. Residues 53 to 84 provide a binding site for a complex formed between a catabolic gene activator protein, CAP, and cyclic AMP. The level of cyclic AMP rises when the bacterium experiences a shortage of metabolic products derived from glucose. Thus the level of binding of the cAMP:CAP complex to the promoter region rises and, in an unknown manner, facilitates the binding of RNA polymerase at an adjacent site (Figure 4). Thus *positive control* is exerted by increasing the frequency with which RNA polymerase latches on to the promoter region when the bacterium gets hungry!

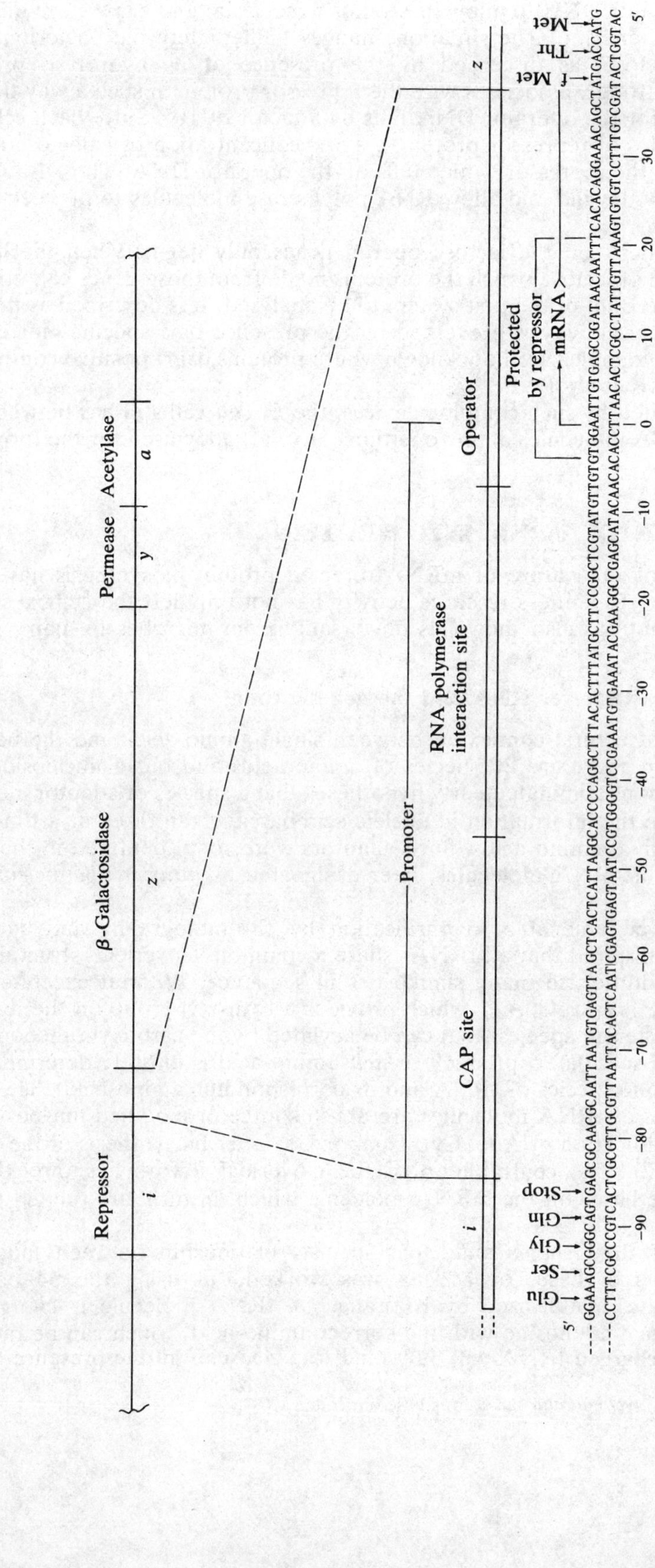

Figure 4 The *E. coli lac* operator showing the relation of the *i*, *z*, *y*, and *a* genes to the promoter and operator sites. The DNA sequence for the control region is expanded in the lower part of the diagram

There is a block between the RNA polymerase and the locus for initiation of mRNA biosynthesis and it is, literally, in the shape of the repressor protein. This cannot be displaced from the operator DNA to allow its strands to separate and expose the initiatior site to the transcription enzyme. The situation changes immediately the bacterium finds itself supplied with lactose, as discerned by the presence of its hydrolysis product, galactose. This hexose strongly associates with the repressor protein in such a way that the affinity of the complex for the operator DNA falls by a factor of 10^4. Since each cell only possesses a few molecules of repressor protein, a small concentration of galactose acts as an inducer and binds to the repressor, which falls off the operator DNA. The DNA of the operator region can now unwind and allow RNA polymerase molecules to proceed to the site of mRNA initiation.

In summary, biosynthesis on the lactose operon genes only begins when the hungry bacterium finds a source of lactose which the proteins made from those genes can utilize.

The foregoing is an account of the first system to be analysed. It is described as *negative control* because biological activity is prevented by the presence of a specific molecule — the repressor protein. Examples are now known where proteins exert positive control and even positive *or* negative control.

Even though it is limited by such complex devices, the *E. coli* cell still has between one and two thousand mRNA molecules at any one time. How are they used for the formation of proteins?

22.5.3 TRANSLATION OF mRNA INTO PROTEIN

The broad outlines of the nature of mRNA-directed protein biosynthesis have been recognized for a decade. Relentless research activity has not only left the central scheme relatively unchanged, but has also shown its universality from microbes to man.

22.5.3.1 The function of transfer RNA and the genetic code

There is no evident structural connexion between single amino-acids and the bases of nucleic acids. Moreover, there are 20 species of amino-acids and but 4 nucleoside base types. These considerations stimulated early hypotheses that a species of adaptor molecule should exist to correlate the information in nucleic acid bases, taken three-at-a-time, with the structures of individual amino-acids. Such adaptors were soon identified in the shape of small relatively-soluble RNA molecules, later designated as *transfer ribonucleic acids*, tRNAs.

Various species proved amenable to purification by chromatography and, later, to sequence analysis. It transpired that all tRNAs share a common 'cloverleaf' structure (see Chapter 22.1, **50**) in addition to many similarities in sequence. Without exception, the 3′-end has the sequence ... $pCpCpA_{OH}$ which provides a hydroxyl group on the terminal adenosine (2′ or 3′ in different species) that can be acylated by the carboxyl function of an amino-acid. (As will be seen later, precisely *which* amino-acid is entirely determined by an enzyme specific for one species of tRNA and its corresponding amino-acid). The loops and non-helical residues of tRNA molecules are a rich source of modified nucleosides — invariably made by modification of Ade, Cyt, Gua, or Ura *after* biosynthesis of the tRNA molecule (Chapter 22.2). The central loop of the cloverleaf always has three bases* complementary to three bases on the mRNA molecule which, in turn, are related to one particular amino-acid.

The identification of the *genetic code*, that special relationship between individual amino-acids and triplets of bases or *codons*, was worked out using the 64 possible trinucleoside diphosphates synthesized by Khorana.[8] A tRNA molecule is *charged* by acylation of its 3′-terminal adenosine with the correct amino-acid, which can be radioactively labelled. Such a charged tRNA will only bind to a *ribosome* in the presence of the

* The frameshift mutant of *S. typhimurium* has a four-base anticodon CCCC.

appropriate codon. It turned out that 61 of the codons promoted physical binding of an aminoacyl tRNA to ribosomal particles and three did not. These were later identified as signals for termination which have no corresponding tRNA.

The 64 assignments (Table 1) have been verified by *in vitro* protein synthesis using synthetic ribonucleoside messengers derived from DNAs of repeating sequence, such as $dp(GpGpAp)_nGpGpA_{OH}$ (Chapter 22.4, Scheme 1).† They now permit the full correlation between a given protein and its corresponding mRNA, as for instance the 3′-end of the *i* gene with the amino-acid sequence . . . Glu-Ser-Gly-Gln-COOH (Figure 4).

TABLE 1
The Genetic Code

First position	*Second position*				*Third position*
(5′-*end*)	U	C	A	G	(3′-*end*)
U	Phe	Ser	Tyr	Cys	U
	Phe	Ser	Tyr	Cys	C
	Leu	Ser	Stop	Stop	A
	Leu	Ser	Stop	Trp	G
C	Leu	Pro	His	Arg	U
	Leu	Pro	His	Arg	C
	Leu	Pro	Gln	Arg	A
	Leu	Pro	Gln	Arg	G
A	Ile	Thr	Asn	Ser	U
	Ile	Thr	Asn	Ser	C
	Ile	Thr	Lys	Arg	A
	Met	Thr	Lys	Arg	G
G	Val	Ala	Asp	Gly	U
	Val	Ala	Asp	Gly	C
	Val	Ala	Glu	Gly	A
	Val	Ala	Glu	Gly	G

22.5.3.2 The role of the ribosome

The ribosome is the essential machine in protein biosynthesis. It is an aggregate of over 50 proteins and three major rRNA species which combine to give a smaller particle (MW $\approx 10^6$), called the S30 particle for bacteria, and a larger one (MW $\approx 2\times 10^6$), known as the S50 particle for bacteria. It is the smaller particle which binds mRNA species and contains the 16S rRNA (MW $\approx 5\times 10^5$) for which the 3′-tail sequence, . . . $CAUUA_{OH}$, appears to be the complement for a sequence in mRNAs near to the point at which protein synthesis starts.

Protein biosynthesis begins when the 30S particle forms a ternary complex with the messenger RNA, using the initiator codon AUG, and a special tRNA molecule. This has been charged with methionine and then formylated on the amino-group of that amino-acid (*N*-formyl-met-$tRNA_f^{Met}$). The larger ribosomal subunit now adds on (and other protein factors) to give the initiation complex.

The 50S ribosomal particle has two sites for accepting the L-shaped molecules of tRNA. One of them, called the *A*-site, only accepts tRNAs charged with free amino-acids. The other, *P*-site, adjacent to it, accepts tRNAs from the *A*-site providing that the charging group has an amino-function which is part of an amide linkage — either as a result of formylation in the initial case or as a peptide bond in the general case (Figure 5).

† This DNA should produce the messenger $rppp(UpCpCp)_nUpCpC_{OH}$ which leads to the incorporation of serine, leucine, and proline into three species of protein.

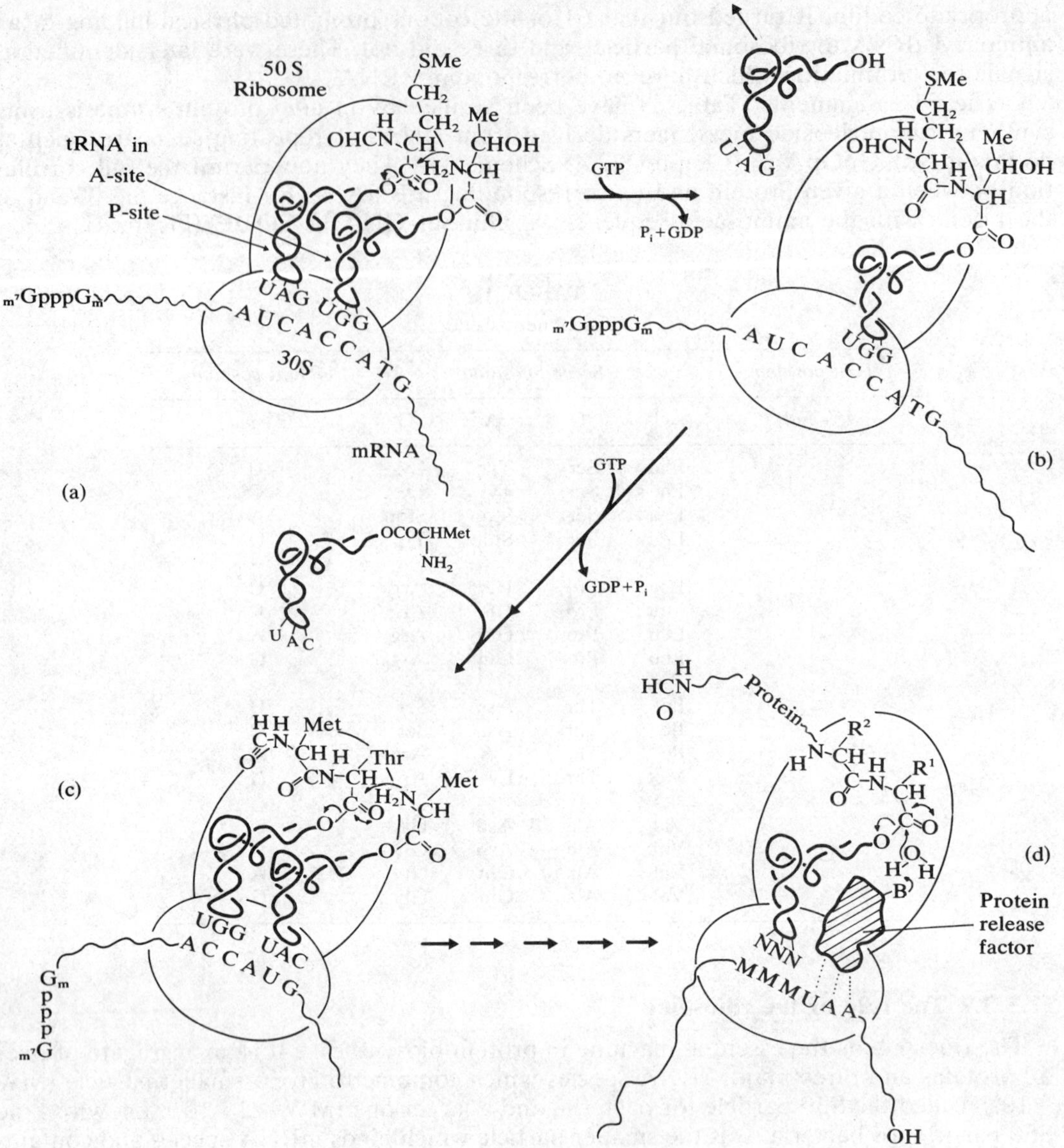

Figure 5 Scheme of initiation (a), elongation (b–c), and termination (d) steps in protein biosynthesis. The horizontal line represents the messenger RNA: tRNAs are L-shaped blocks.

The fmet-tRNA$_f^{Met}$ moves into the *P*-site, shifting with it the mRNA to expose the second triplet of bases in the *A*-site. The charged tRNA with an anticodon of complementary bases (thr-tRNA^{Thr} in the case of the β-galactosidase protein in Figure 4) binds in the *A*-site. Its amino-function can now act as a nucleophile and attack the active ester bond (Figure 5a) of the charged tRNA in the *P*-site to create the first peptide bond of the new protein. The released tRNA$_f^{Met}$ drops off the ribosomes (Figure 5b) so that the peptidyl-tRNA^{Thr} can shunt into the *P*-site, again shifting with it the mRNA. Thus the mRNA advances by precisely three residues for each peptide bond made and keeps the triplet code *in phase*. The third charged amino tRNA can now enter the vacant *A*-site in response to the third codon (AUG and met-tRNA$_f^{Met}$ for β-galactosidase) and a second ester aminolysis (Figure 5c) creates the second peptide bond.

This process is driven by the free energy of hydrolysis of two molecules of GTP per cycle (-60 kJ mol^{-1}). It continues until the *A*-site encounters one of the three terminator codons, UAA, UAG, or UGU (Figure 5d). No tRNA species exists with an anticodon complementary to these codons, so that protein synthesis cannot continue. Instead, protein factors recognize these sequences, enter the *A*-site, and catalyse the hydrolysis of the active ester linkage of the peptidyl-tRNA in the *P*-site. Thus the newly made protein is released from the tRNA of its last residue, the ribosome drops off the mRNA, and synthesis of that protein is complete.

In the economy of nature, revealed by the electron microscope, translation of DNA into mRNA and transcription of mRNA into protein are going on in concert. Fir-tree-like structures can be seen which are thought to show a DNA trunk at whose apex the first of a series of RNA polymerase molecules is starting to make mRNA. As these progress down the trunk, branches of increasing length of mRNA sprout sideways, along which are bound numerous ribosomal particles, all busily making protein! In molecular terms, mRNA synthesis proceeds in the $5' \rightarrow 3'$ direction with ribosomes at intervals of 25 nm making protein in the $N \rightarrow C$ direction.

22.5.3.3 The genetic code in use

Three facets of the use of the genetic code are revealed from a study of structure and function: its degeneracy, its duplicity, and its mutability.

The *degeneracy* of the code is a statistical necessity. There are more triplets ($4^3 = 64$) than amino-acids. This degeneracy is not distributed uniformly (Table 1). Arginine, leucine, and serine all enjoy six codons while tryptophan makes do with one. As we have seen, the codon AUG doubles up for initiation of protein synthesis using $tRNA_f^{Met}$ and for insertion of methionine into internal positions using $tRNA_m^{Met}$.

In the mRNA of bacteriophage MS2, some 90% of the 3569 residues are translated to make three proteins. The use of the codon GUU significantly exceeds that of $GG\left\{\begin{matrix}A\\C\\G\end{matrix}\right.$ for incorporation of glycine and UAC is definitely preferred over UAU (32:9) for tyrosine. Such preferences[9] may indicate a similar preference for these codons in the host bacterium, *E. coli.* However, the situation is more extreme for human β-globin mRNA, where 75% of the nucleotides are translated. The preferences for CUG (leucine), GUG (valine), and AGG (arginine) are significant at the 0.001% level (χ^2 test). Possibly in this case the correlation is with the tRNA complement of the specialized reticulocytes where globin is made.

In other respects the use of the degeneracy of the code is more uniform, and it may well be that the freedom of choice so frequently available in the third position of a codon (Table 1) is manipulated in viral mRNAs to maximize the proportion of complementary base-pairing in hairpin loops and thus increase the stability of secondary and/or tertiary structure.[10,11]

The *duplicity* of the genetic code has been suspected for some time, largely because certain viruses like ϕX 174 do not have enough nucleotide residues to provide a linear correspondence for all of the amino-acids which are encoded in their chromosome. The nine proteins of ϕX 174 contain some 1926 amino-acids, which is about 10% in excess of the coding capacity of its 5375 nucleotides.

The amazing answer is that the same sequence of bases is used in one phase to code for one protein and in a second phase to code for a second protein! Thus, the gene for protein B lies within the gene for protein A and gene E lies entirely within gene D. This is illustrated for the terminal sequences of the latter two proteins[12] (Figure 6). The end of these genes almost certainly overlaps the ribosome binding site for the J protein which, remarkably, is out of phase with both proteins D and E.

This hypercompression of information may be typical of viruses and may occur in bacteria, but its existence is not yet defined for higher organisms. Many eukaryotic species use only a minor part of the DNA in their chromosomes and, even then, many genes are

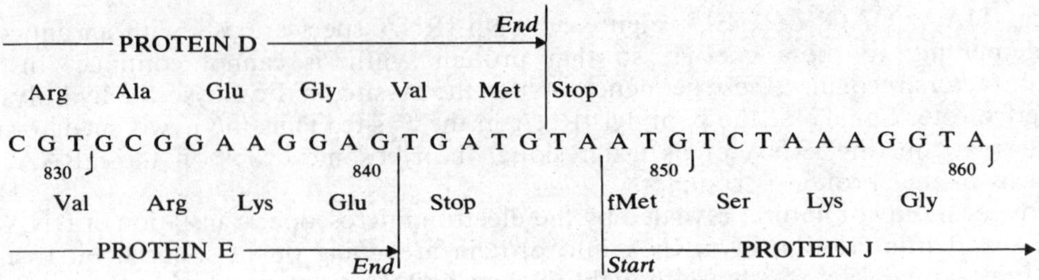

Figure 6 DNA sequence[12] for bacteriophage ϕX 174 showing residues 827 to 860 along with their functional assignments.

repeated to an amazing extent. One species of South African frog has no less than 20 000 identical copies per cell of the gene for one of its rRNAs.

The balance between the creative opportunities of change in specific bases in DNA (creating mutations) and the accuracy of protein synthesis (sustaining the life of the organism) is at the heart of evolution. The enzymes which charge a tRNA with its specific amino-acid have an error rate as low as $1:10^4$ for homologous amino-acids. In replication, the accuracy is even greater and estimates of errors rarely exceed 1 in 10^7.

This situation can be altered drastically by the influence of chemicals, some simple, some complex, on DNA prior to replication or on RNA prior to translation. One example will suffice.

Both the environmental pollutants, nitrous acid and sulphur dioxide, can promote the deamination of cytosine to uracil (Scheme 2). Such a modification can have three types of consequences on protein synthesis, as seen by inspection of the genetic code (Table 1). Firstly, changes of C to U within the third position of the codon would have no effect on amino-acid incorporation in any of the 16 cases. Secondly, a change from C to U in the first position of the code can change the CAA codon (glutamine) to UAA (stop) and thus bring synthesis of a particular protein to an untimely end. Equally, a change from CAU (histidine) to UAU (tyrosine) could displace a catalytically active residue by an inactive one. For a protein with a vital role in the cell, both such changes would be lethal and no progeny could survive replication of a thus modified DNA strand. Thirdly, some such changes could introduce an amino-acid with a functional side-chain, such as cysteine (UG^{U}_{C}) or serine (UCX) in place of a very different function, arginine (CG^{U}_{C}) or proline (CCX), respectively, and thereby create new protein species of modified structure and potentially creative catalytic power.

Whether higher organisms evolved by accumulation of many such point mutations over

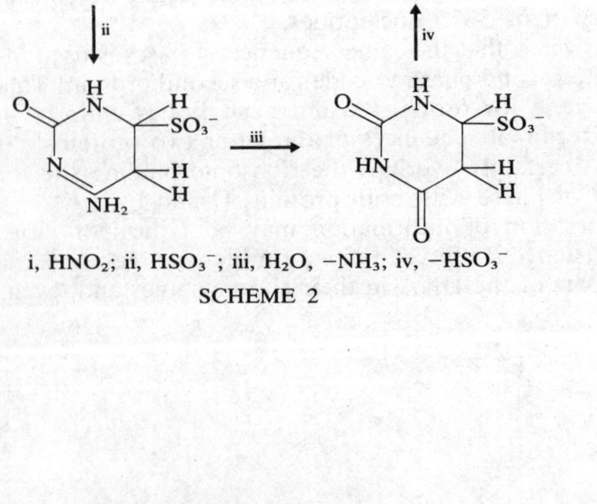

i, HNO_2; ii, HSO_3^-; iii, H_2O, $-NH_3$; iv, $-HSO_3^-$

SCHEME 2

the centuries or rather by wholesale swapping of larger DNA segments remains for future determination. What is clear now is that the powerful combination of knowledge of structure and function with chemical synthesis makes it possible to contemplate genetic engineering.

22.3.4 GENETIC ENGINEERING

Two of the fundamental objectives of genetic engineering are the correction of genetic defects, such as sickle-cell anaemia (a point mutation in globin), and the augmentation of

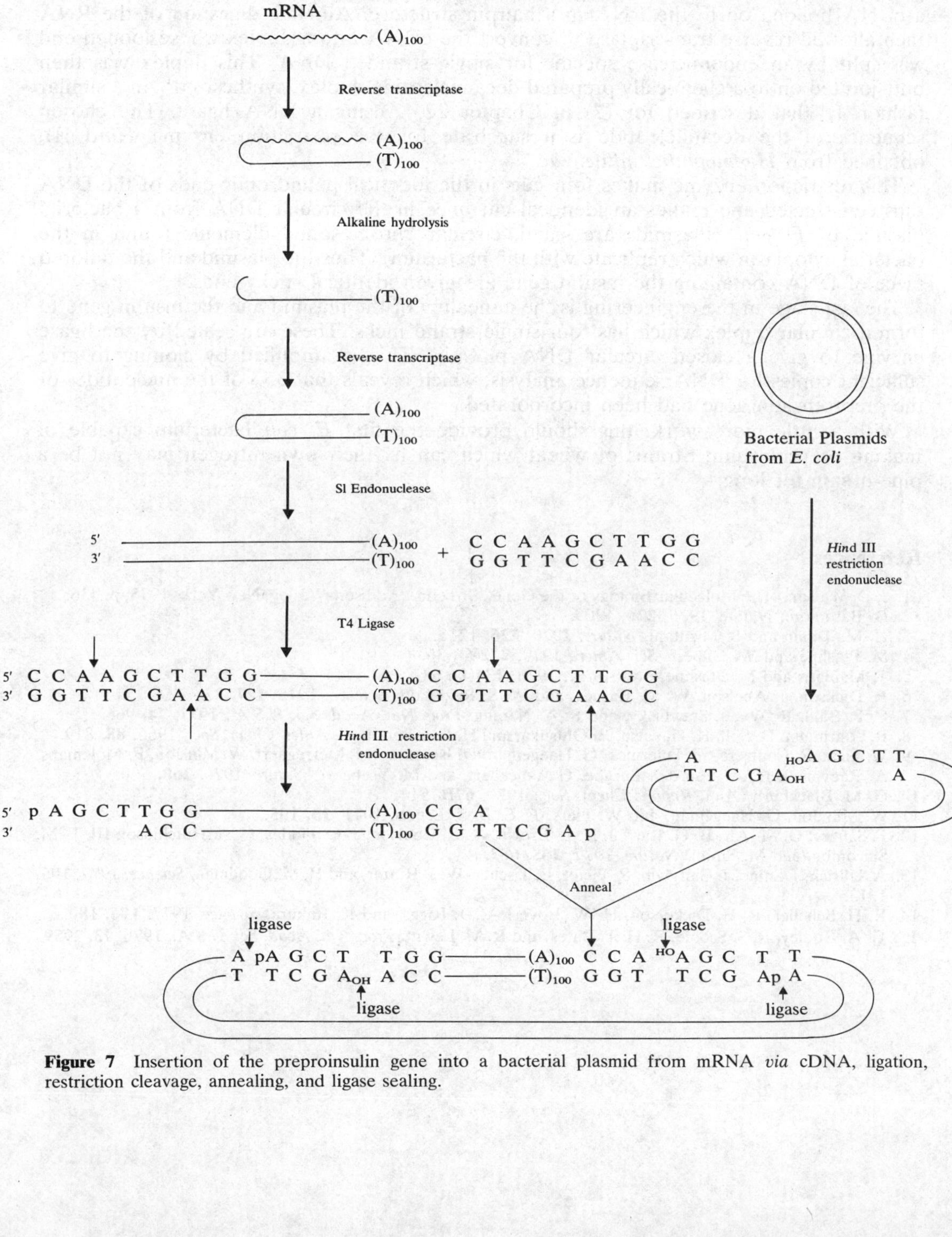

Figure 7 Insertion of the preproinsulin gene into a bacterial plasmid from mRNA *via* cDNA, ligation, restriction cleavage, annealing, and ligase sealing.

sound genes by others, such as the insertion of a nitrogenase gene into wheat chromosomes. It now appears that a hybrid of such goals is to hand: the insertion of an insulin gene into *E. coli* bacteria.[13]

The insulin hormone has two peptide chains, A (20 residues) and B (30 residues). These are produced from a single protein precursor, preproinsulin, in which 23 of its 108 amino-acids precede the B peptide and 35 join the B peptide to the A peptide. The mRNA for this protein thus has at least 327 nucleotides.

Rutter[13] and his colleagues isolated a mixture of mRNAs which contained a predominant species of 450 nucleotides corresponding to preproinsulin and having a 3′-tract of poly(A). This mRNA was used (Figure 7) as a template for reverse transcriptase to make a cDNA bonded on to the RNA in a hairpin structure. Alkaline digestion of the RNA then allowed reverse transcriptase to convert the cDNA into a duplex whose looped end was split by an endonuclease specific for single-stranded DNA. This duplex was then butt-jointed on to a chemically prepared decanucleotide duplex, synthesized[14] in a similar fashion to that described for (**7**) in Chapter 22.4, using a DNA ligase. The chosen sequence of the decanucleotide is a substrate for the restriction enzyme *Hin*d III, obtained from *Haemophilus influenzae.*

This restriction enzyme makes four cuts in the identical palindromic ends of the DNA thus constructed and makes an identical cut *once* in the circular DNA from a bacterial plasmid of *E. coli.* (Plasmids are small, circular chromosomal elements found in the bacterial cytoplasm which replicate with the bacterium). Thus the plasmid and the tailored piece of DNA containing the insulin gene are given identical sticky ends.

The last stage in the engineering is the annealing of the plasmid and the insulin gene to form a circular duplex which has four single-strand nicks. These are sealed by the ligase enzyme to give a closed circular DNA plasmid. This is amplified by cloning to give sufficient copies for DNA sequence analysis, which reveals that 353 of the nucleotides of the preproinsulin gene had been incorporated.

With a little more work, this should provide the first *E. coli* bacterium capable of making preproinsulin! Strains of wheat which can fix their own nitrogen may not be a pipe-dream for long.

References

1. J. D. Watson, 'The Molecular Biology of the Gene', 3rd edn., Addison-Wesley, New York, 1975, p. 436.
2. D. Baltimore, *Nature,* 1970, **226,** 1209.
3. H. M. Temin and S. Mizutani, *Nature,* 1970, **226,** 1211.
4. M. Ptashne and W. Gilbert, *Sci. Amer.,* 1970, **222(6),** 36.
5. T. Maniatis and M. Ptashne, *Sci. Amer.,* 1976, **234(1),** 64.
6. R. Dickson, J. Abelson, W. M. Barnes, and W. S. Reznikoff, *Science,* 1975, **187,** 27.
7. C. P. Bahl, R. Wu, J. Stawinsky, and S. A. Narang, *Proc. Nat. Acad. Sci. U.S.A.,* 1977, **74,** 966.
8. R. Lohrmann, D. Söll, H. Hayatsu, E. Ohtsuka, and H. G. Khorana, *J. Amer. Chem. Soc.,* 1966, **88,** 819.
9. W. Fiers, R. Contreras, F. Duerinck, G. Haegeman, D. Iserentant, J. Merregaert, W. Min Jou, F. Molemans, A. Raeymaekers, A. Van den Berghe, G. Volckaert, and M. Ysebaert, *Nature,* 1976, **260,** 500.
10. G. M. Blackburn, *Ann. Reports Chem. Soc.,* 1970, **67B,** 514.
11. W. Min Jou, G. Haegeman, and W. Fiers, *F. E. B. S. Letters,* 1971, **13,** 105.
12. F. Sanger, G. M. Air, B. G. Barrell, N. L. Brown, A. R. Coulson, J. C. Fiddes, C. A. Hutchinson III, P. M. Slocombe, and M. Smith, *Nature,* 1977, **265,** 687.
13. A. Ullrich, J. Shine, J. Chirgwin, R. Pictet, E. Tischer, W. J. Rutter, and H. M. Goodman, *Science,* 1977, **196,** 1313.
14. R. H. Scheller, R. E. Dickerson, H. W. Boyer, A. D. Riggs, and K. Itakura, *Science,* 1977, **196,** 180.
15. G. A. Rodley, R. S. Scobie, R. H. T. Bates, and R. M. Lewitt, *Proc. Nat. Acad. Sci. U.S.A.,* 1976, **73,** 2959.

PART 23

PROTEINS: AMINO-ACIDS AND PEPTIDES

23.1
Introduction

E. HASLAM
University of Sheffield

23.1.1 THE NATURE OF PROTEINS

'Meine Herren!
Da die Proteïnstoffe bei allen chemischen Processen im lebenden Organismus auf die eine oder andere Weise betheiligt sind, so darf man von der Aufklärung ihrer structur und ihrer Metamorphosen die wichtigsten Aufschlüsse für die biologische Chemie ewarten . . . Nur über den Zeitpunkt, wo ein Zusammenwirken von Biologie und Chemie erfolgreich sein werde gingen und gehen noch heute die Ansichten auseinander'.

Emil Fischer, 'Untersuchungen über Aminosäuren, Polypeptide und Proteïne', *Ber.*, 1906, **39,** 530.

'Giant molecules therefore were supposed not to exist — only clusters of ordinary small molecules forming particles of undefined mass.
Today we have, I think, definitive proof that this view is wrong. Investigations along different lines have given the result that the proteins are built up of particles possessing the hall-mark of individuality and therefore are in reality giant molecules'.

Th. Svedberg, Opening address to a Royal Society Discussion on 'The Protein Molecule', *Proc. Roy. Soc.*, 1939, **127B,** 17.

The authorship of the word protein and its introduction into the vocabulary of science is generally attributed to the Dutch chemist Gerhardus Johannes Mulder in 1838:

'There is present in plants as well as animals a substance which . . . performs an important function in both. It is without doubt the most important of the known components of living matter, and it would appear that, without it, life would not be possible. This substance has been named protein (Gr. *proteios*, first)'.

There nevertheless seems little doubt that in using this word Mulder had adopted a suggestion made to him by the great Swedish chemist Berzelius without making any specific acknowledgement to him.[1] Mulder himself had become interested in the study of albuminous substances and he suggested that the proteins were compounds of phosphorus and sulphur with an organic radical $C_{40}H_{62}N_{10}O_{12}$. Although Liebig and others soon provided evidence to demolish Mulder's theory of protein structure, no further attempts to arrive at a notion of protein constitution were made for very many years.

Proteins are amongst the most complicated and the largest of molecules which nature produces; in addition they remain some of the most labile and difficult to purify. These properties presented almost insurmountable technical difficulties to early workers in this field, and added to these was the unfortunate association of protein chemistry with the concept of colloids. Proteins for a long time were thought of as types of ill-defined colloids, the molecules of which defied isolation and definition. However, as the criteria of purity for proteins improved there emerged the molecular concept in protein chemistry and the realisation that proteins were after all merely large complicated molecules whose solutions were no-less molecular than those of their component amino-acids. Recognition of this view was due in no small way to the work of Svedberg, Sorensen, and others.

Against this background of enormous experimental difficulties and conceptual uncertainties, Fischer[2] embarked in 1899 on his classical work which decisively influenced all subsequent work in the field of protein chemistry. He set himself a bold objective, namely the synthesis of protein-like organic substances. His approach was characteristically direct and uncomplicated:

> 'Whereas cautious colleagues fear that a rational study of this class of substance will encounter insuperable difficulties, because of their awkward physical properties, other optimistically-inclined observers, among whom I number myself, believe that one could at least attempt to besiege the virgin fortress with all the present day resources; since only through daring can the limits of the potential of our methods be determined'.

At the commencement of his work about a dozen amino-acids were recognized as cleavage products of proteins, and others were recognized shortly afterwards. Fischer embarked on several projects: the synthesis of amino-acids and the resolution of their racemic forms; the study of amino-acid derivatives in order to facilitate the separation of amino-acid mixtures; and, most significantly, the recombination of two or more amino-acids to give compounds which he called 'polypeptides'. The magnitude of Fischer's approach to the problems of polypeptide synthesis made an immediate impact and tended to overshadow the pioneering work of others, such as Curtius, in this field. In 1901 he announced the synthesis of glycyl-glycine, the simplest dipeptide, and in 1907 an 18 amino-acid polypeptide leucyl-triglycyl-leucyl-triglycyl-leucyl-octaglycyl-glycine. Even by the standards of today this was a remarkable synthetic achievement. The synthesis of glycyl-glycyl-glycine illustrates the general process employed by Fischer (Scheme 1); it is, in its strategy and principles, a remarkable forerunner of the methods used today for peptide synthesis (see Chapter 23.6).

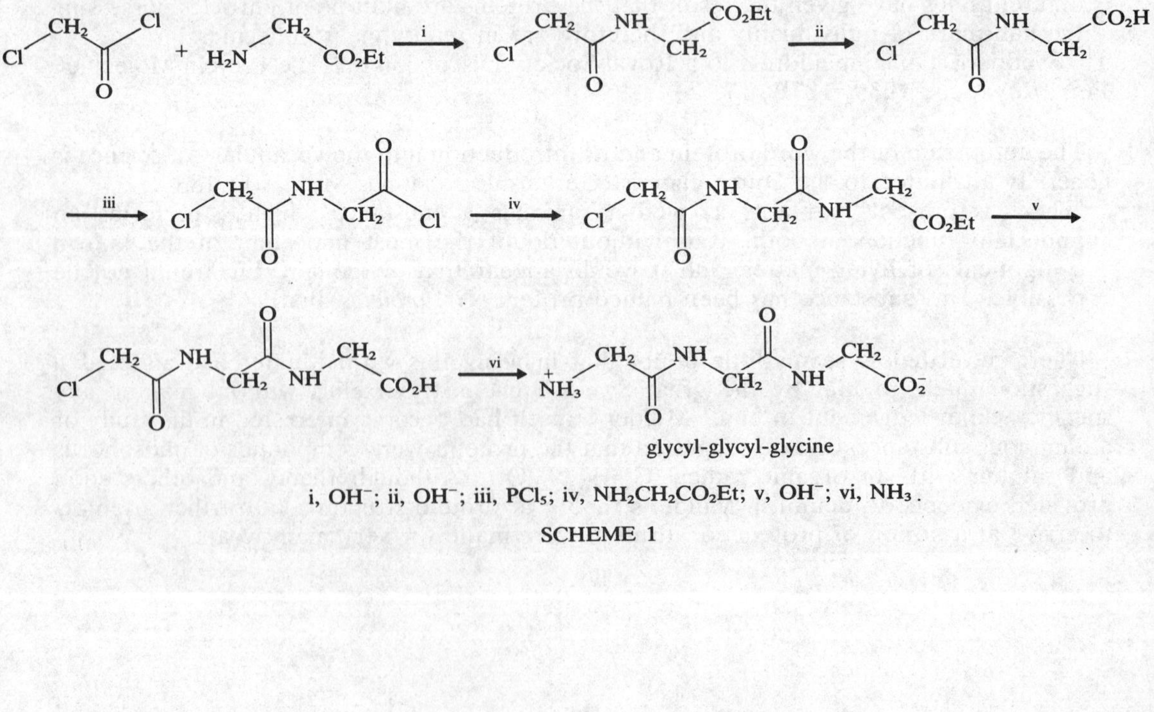

i, OH^-; ii, OH^-; iii, PCl_5; iv, $NH_2CH_2CO_2Et$; v, OH^-; vi, NH_3.

SCHEME 1

The great similarities in chemical and physical properties between the synthetic polypeptides of Fischer and some proteins lent further support to the suggestion put forward earlier by Fischer and independently by Hofmeister in 1902 of the peptide theory of protein structure. This theory proposed that protein molecules were built up only of chains of α-amino-acids (and later, of course, α-imino-acids were included) bound together by peptide (amide) bonds between their α-amino and α-carboxyl groups (see **1**). Fischer himself allowed that other possible modes of linkage between amino-acids might be possible in the protein molecule, and combined with continuing doubts about the size and complexity of natural proteins it gave rise to a variety of speculations[3] in the period 1920–1940 concerning alternative modes of linkage between amino-acids. Sanger[4] wrote in 1952 that probably the best evidence in support of the peptide theory of protein structure was indeed that, since its announcement in 1902, no facts had been found to contradict it, and Sanger himself provided some of the first decisive proof of this theory by his elucidation of the complete structure of the protein hormone insulin.

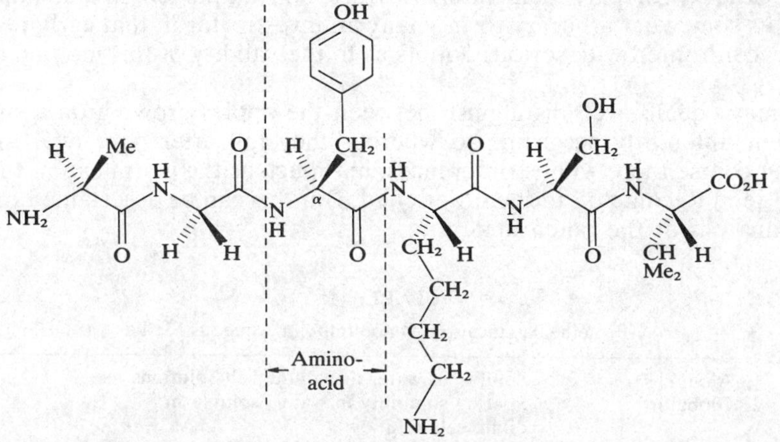

(1) Polypeptide structure

Neither Fischer nor many of the protein chemists who followed for the next 20 years were able to solve satsifactorily what had been one of the paramount problems of research into proteins. This was how could one prepare from protein extracts homogeneous chemical substances, and the answers to these problems did not arrive until the advent of chromatographic methods in the 1940s.[5] Since their introduction there has been a dramatic improvement in the quality of protein preparations and of the analytical data appertaining to proteins and this has formed the basis for the relative explosion which has occurred since 1950 in our knowledge of their structure and function. These methods and their applications may be found in a range of texts.[6–8] Pre-eminent amongst these techniques are probably the development of the Spackman–Stein–Moore apparatus for the automated complete amino-acid analysis of proteins, and the use of the armamentarium of powerful physical methods such as X-ray crystallography and nuclear magnetic resonance to examine three-dimensional structure.

Early workers were clearly disconcerted to discover the apparent size to which the polypeptide chain might grow in some proteins in order to accommodate the estimates of molecular weight which had been made. Several concluded[3] that 'some configuration must be present which acts in such a way as to render the molecule far more compact than would be supposed on the simplest and most obvious premises'. The great excitements associated with research on biopolymers such as proteins and nucleic acids and the growth of molecular biology have come from the recognition that such constraints on shape and size do indeed exist. The determination of the precise three-dimensional structures of proteins by crystallographic techniques, and, in some cases, the investigations which have shown the discrete changes in conformation which proteins undergo as they are involved in biological reactions, have pointed the way to solutions of that most important relationship in biological chemistry, namely that between structure and function to which Fischer alluded in 1906.

23.1.2 CLASSIFICATIONS, FUNCTION, AND STRUCTURE

Proteins occur in a considerable range of forms and exhibit a correspondingly wide range of physical properties. Early classifications were based on one of these physical criteria — their solubility in aqueous media. One such classification (British Physiological Society, 1907) illustrates (Table 1) how protein solubility varies from the very soluble globulins and albumins to the virtually insoluble scleroproteins such as α-keratin and sclerotin. This form of classification is purely empirical in nature, although several of the names associated with particular categories are still employed in the literature today. Because of its somewhat arbitrary form, many an investigator in that early period was, not surprisingly, confronted with serious doubts as to the validity of this method of characterization[9]:

> '... one may equally well distinguish between the apples growing on a single tree and name them differently according to whether they fall after one, two, three or more shakes. Of course there will be differences, inasmuch as the worm eaten, the big and the small will tend to come off in the order cited. But this can be ascertained only through a detailed analysis of the fallen material'.

TABLE 1

Protein Classification by Solubility in Aqueous Media

1. Albumins	Soluble in water and dilute salt solutions
2. Globulins	Sparing solubility in water; soluble in dilute salt solutions
3. Prolamins	Insoluble in water; soluble in 50–90% ethanol
4. Glutelins	Insoluble in water, dilute salt solutions, and aqueous ethanol; soluble in dilute acids and alkalis
5. Scleroproteins	Insoluble in aqueous media

Another classification which has been employed is based upon the products of hydrolysis. Simple proteins are defined as those which yield only amino-acids (or their degradation products) on hydrolysis. On the other hand, conjugated proteins produce not only amino-acids but also other organic or inorganic molecules (prosthetic groups). The classification is then based on the chemical nature of the various prosthetic groups and typical categories which result are nucleoproteins, glycoproteins, lipoproteins, phosphoproteins, flavoproteins, and metalloproteins. Thus the nucleoproteins in ribosomes are combined with nucleic acids by multiple ionic linkages, glycoproteins are covalently linked with carbohydrates, lipoproteins are associated with lipids and cholesterol by various non-covalent forces, and metal ions are bound to metalloproteins by ionic or dative covalent bonds. This classification nevertheless serves to illustrate the undoubted complexity of many proteins as they occur in living matter.

However, as more and more has been learnt concerning the nature of proteins and their individual biological roles, classifications have generally changed to become based upon properties which relate to their great functional diversity and versatility. Whilst the protein content of whole tissues varies widely, the protein within cells normally constitutes over half of the cell's dry weight. Several distinctive groups may be discerned: enzymes, which serve as catalysts for the biochemical reactions in the cell; the food storage proteins; the structural proteins; transport proteins; contractile proteins; antibodies; toxins; hormones; and regulatory proteins. It is possible on a somewhat broader interpretation of their biological functions to classify these into three major categories (Table 2) — storage proteins, structural or mechanical proteins, and proteins which express their various biological properties by combination or association with ions or other molecules. As with most classifications, overlap between different categories exists. Myosin, for example, is in most of its characteristics a typical fibrous structural protein, but it also possesses enzymic activity which is associated with its function in muscle action.

TABLE 2
Proteins — Biological Function

1. **Storage Proteins**
   Albumin (egg white), Casein (milk)
   Zein (corn), Gliadin (wheat)
   Ferritin (spleen)
2. **Structural or Mechanical Proteins**
   α-Keratin and its derivatives (hair, hoof, horn)
   Collagen (bone and tendon), Fibroin (silk)
   Elastin (ligaments), Sclerotin (insects)
   Glycoproteins (cell wall)
   Myosin (muscle), Fibrinogen–Fibrin (blood)
3. **Proteins which associate with other molecules or ions**
   (i) *Enzymes* (*e.g.* Chymotrypsin, Ribonuclease, Lysozyme, Snake Venoms)
   (ii) *Hormones and Regulatory Proteins* (*e.g.* Insulin, Glucagon, Gastrin, Adrenocorticotrophic Hormone, Histones and Nuclear Proteins)
   (iii) *Immunoproteins* (*e.g.* IgG Globulins)
   (iv) *Transport proteins* (*e.g.* Myoglobin, Haemoglobin, Cytochromes, Membrane Carrier Proteins)
   (v) *Photo- and Chemo-receptors* (*e.g.* Opsin)

Proteins besides displaying a wide range of physical and biological characteristics also exhibit a very wide range of molecular weights. Insulin, the protein hormone, has a molecular weight of about 6000 and the enzymes ribonuclease and lysozyme have molecular weights in the range 12 000–14 000. These protein molecules stand at the lower limits of a range which extends through haemoglobin (64 500), hexokinase (96 000), tryptophan synthetase (117 000), and horse γ-globulin (150 000) to the protein component of tobacco mosaic virus with a molecular weight in the region of 4 000 000. Most proteins with molecular weights in excess of 20 000–30 000, however, possess a *quaternary structure* and usually consist of a number of individual sub-units which associate by means of non-covalent forces. Haemoglobin thus has four units, each of which is composed of a separate polypeptide chain, and tobacco mosaic virus has well over 2000 separate polypeptide chains. In passing it is worth noting that although there is apparently considerable variation in the size of these sub-units in different proteins, many, for example human haemoglobin, correspond approximately in size to the sub-unit of 17 600 (a polypeptide chain of ~150 amino-acids) originally proposed by Svedberg[10] and which gave rise to the Svedberg law of multiples. The manner in which the sub-units of proteins associate to form the fully active biological complex is generally referred to as the *quaternary* structure of the protein. The study of sub-unit interactions is a very important aspect of contemporary molecular biology since these interactions undoubtedly play significant roles in many biological reactions. The, by now, classical examples of the influence of such properties on biological phenomena are the allosteric behaviour of enzymes and the cooperativity of oxygen binding in haemoglobin. The tetrameric structure of haemoglobin contains four oxygen binding sites and the structure is so organized that when one of these sites is oxygenated the subtle changes in conformation of the polypeptide chain around that site are transmitted to a second site so as to increase its affinity for oxygen. As oxygenation occurs at each of the first three sites, then so this effect occurs at the next site. This behaviour has important consequences for the characteristics of haemoglobin as an oxygen carrier.

Discussion of the structure and conformational properties of individual polypeptide sub-units conventionally falls under three headings — *primary*, *secondary*, and *tertiary* structure — following the suggestion of Linderstrøm-Lang.[11] Organisms use a basic set of 20 amino-acids for protein synthesis (see Chapter 23.2). Unusual amino-acids which occasionally occur in protein structures frequently arise by chemical modification of haem prosthetic group (residues 70–80) have remained constant throughout the evolutionary process, whereas other parts of the molecule which appear to have little significance particular residues after assembly. Thus, for example, the hydroxyproline and hydroxylysine residues in collagen are derived from proline and lysine groups in the polypeptide chains originally constructed. The specific sequence of amino-acids which is incorporated into a particular polypeptide chain is referred to as its *primary structure* and is genetically determined. Using only the 20 different amino-acids the possible permutations are astronomical. Thus for a polypeptide chain of 300 amino-acid residues, approximately 10^{500} structures are possible and for haemoglobin there exists approximately 1.35×10^{167} possible arrangements of the amino-acids in its four polypeptide chains. The polypeptide chain in its biologically active state may exist in a complex highly convoluted (but highly ordered) shape in which particular amino-acid side-chains, which are widely separate in the overall primary structure, are brought into close spatial juxtaposition. One view is that it is nevertheless the primary structure which finally determines — by the interplay of interactions within the molecule and with the surrounding aqueous medium — this overall shape (*tertiary structure*) of the protein. This view suggests that the primary structure and hence its determination, by chemical and physical means, is all important (see Chapter 23.3).

Dramatic changes in properties may result from the substitution of one amino-acid for another in a protein molecule, and the alteration of a glutamic acid residue to valine in one of the four polypeptide chains of haemoglobin changes its properties drastically and results in the disease sickle cell anaemia. Other amino-acid changes, however, may have little or no effect on biological activity, and an interesting example of these effects is seen amongst the various cytochrome *c* molecules isolated from organisms which have a quite different evolutionary status.[12] The cytochromes act in biological oxidations as a means of electron transfer and one of these molecules cytochrome *c* can be readily solubilized and isolated. The complete amino-acid sequence has been determined for the protein from over 40 species and comparisons have been made between the various sequences and with the three-dimensional (X-ray) model of horse heart cytochrome *c*. Cytochrome *c* does not appear to have undergone radical evolutionary change, but some sites (notably positions 70–80 in the 104 amino-acid sequence) are quite invariant whilst in others wide changes appear to be tolerated. It is significant that the amino-acid residues which surround the in the electron-transfer reaction are occupied by six or more different amino-acids in different species.

Some of the most exciting developments in science as a whole over the last two decades have stemmed from the elucidation by X-ray crystallographic techniques of the three-dimensional structures of many biopolymers and in particular the globular proteins. These structures have in many cases formed the basis for the construction of working models and hypotheses which explain how these proteins carry out their specific function in the biological system. Particular mention in this respect may be made of the haemoglobin and myoglobin molecules[13,14] and of the enzymes lysozyme[15] and chymotrypsin.[16] Although in principle it should be possible one day to predict the overall shape or conformation (*tertiary structure*) which a particular polypeptide chain would adopt in a given environment, such times seem some distance away. Nevertheless, some of the important factors which help to determine polypeptide conformations may be delineated. Pre-eminent amongst these is the interaction of the polypeptide with the aqueous medium of the cell — the ability of the water to solvate the several hydrophilic amino-acid side-chains and the corresponding need to sequester in the interior of the protein structure the hydrophobic amino-acid side-chains. In addition there are the constraints of the chemical structure of the polypeptide itself — the *trans* planar peptide bond and the fact that around any particular peptide linkage the chain can only assume conformations which result from rotation about the ($N—C_\alpha$) or ($C_\alpha—C{=}O$) bonds. To these factors may then be added the proviso that the particular favoured conformation(s) once formed will be stabilized by the formation of weak non-covalent bonds between suitably placed groups (ionic, hydrogen-bonds, charge-transfer complexes involving aromatic rings, hydrophobic interactions) and also by the creation of disulphide links between appropriately placed cysteine residues. Where the amino-acid residues within the gross structure adopt a regular repeating relationship to one another, these features are termed *secondary structure*; the α-helix (as it occurs in keratin) and the rippled β-pleated sheet (as it occurs in fibroin) are two particular examples of secondary structure normally characteristic of fibrous proteins but which have been found as stabilizing features of the conformations of globular proteins.[17]

References

1. H. Hartley, *Nature*, 1951, **168,** 244.
2. E. Fischer, *Ber.*, 1906, **39,** 530; 1907, **40,** 1754.
3. H. B. Vickery and T. B. Osborne, *Physiol. Rev.*, 1928, **8,** 393.
4. F. Sanger, *Adv. Protein Chem.*, 1952, **7,** 1.
5. A. J. P. Martin and R. L. M. Synge, *Adv. Protein Chem.*, 1945, **2,** 1.
6. 'Analytical Methods in Protein Chemistry', ed. P. Alexander and R. J. Block, Pergamon, Oxford, 1960–1961, 3 vols.
7. 'Methods in Enzymology', ed. C. H. W. Hirs, Academic Press, New York, 1967, vol. XI.
8. 'The Proteins', 2nd edn., ed. H. Neurath, Academic Press, New York, 1963.
9. H. E. Schultze and J. F. Heremans, 'Molecular Biology of Human Proteins', Elsevier, Amsterdam, 1966.
10. Th. Svedberg and B. Sjögren, *J. Amer. Chem. Soc.*, 1928, **50,** 3318.
11. K. Linderstrøm-Lang, Symposium on Peptide Chemistry, Chemical Society Special Publication No. 2, London, 1955, p. 1.
12. R. E. Dickerson, *Sci. Amer.*, 1972, **226**(4), 58.
13. M. F. Perutz, *Nature*, 1970, **228,** 726.
14. J. C. Kendrew, *Sci. Amer.*, 1961, **205**(6), 96.
15. D. C. Phillips, *Sci. Amer.*, 1966, **215**(5), 78.
16. D. M. Blow, *Accounts Chem. Res.*, 1976, **9,** 145.
17. R. E. Dickerson and I. Geis, 'The Structure and Action of Proteins', Harper-Row, New York, 1969.

23.2

Amino-acids Found in Proteins

P. M. HARDY
University of Exeter

The scope of this chapter is limited to amino-acids which have been firmly established as protein constituents. Peptides synthesized by non-ribosomal means, which contain a wider variety of amino-acids, are discussed in Chapter 23.4. Non-protein amino-acids in their free form occur in such profusion and structural diversity that it is difficult to generalize on their constitution and synthesis. Thompson *et al.*[1] have reviewed this field, and subsequent work is conveniently abstracted in an annual report on amino-acid chemistry;[2] for details of such compounds the reader is referred to these sources. The chemical reactions of α-amino-acids, other than those of analytical interest, are dealt with in Chapter 9.6, as are β-, γ-, and ω-amino-acids.

The earlier work on protein amino-acids has been discussed comprehensively in the three-volume compendium of Greenstein and Winitz,[3] and material covered in this treatise will only be briefly referred to and specific references will not be given. In the main, therefore, this chapter deals with advances made in the chemistry of the amino-acid constituent of proteins since 1961.

23.2.1 STRUCTURAL SURVEY

23.2.1.1 The genetically coded amino-acids

The amino-acid residues incorporated into proteins during synthesis on the ribosome are limited to those which are coded for by the sequence of bases in messenger RNA and

TABLE 1

The Genetically Coded Protein Amino-acids listed in Chronological Order of Isolation*

1. Glycine	$CH_2(NH_2)CO_2H$	Gly
2. Leucine	$Me_2CHCH_2CH(NH_2)CO_2H$	Leu
3. Tyrosine		Tyr
4. Serine	$HOCH_2CH(NH_2)CO_2H$	Ser
5. Glutamic acid	$HO_2CCH_2CH_2CH(NH_2)CO_2H$	Glu
6. Aspartic acid	$HO_2CCH_2CH(NH_2)CO_2H$	Asp
7. Phenylalanine	$PhCH_2CH(NH_2)CO_2H$	Phe
8. Alanine	$MeCH(NH_2)CO_2H$	Ala
9. Lysine	$H_2N(CH_2)_4CH(NH_2)CO_2H$	Lys
10. Arginine	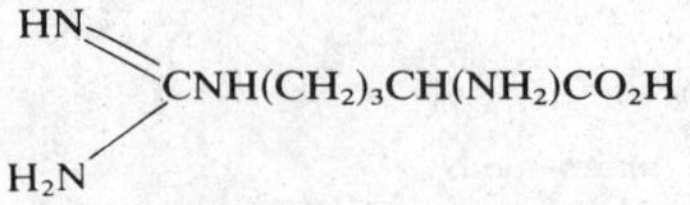	Arg
11. Histidine	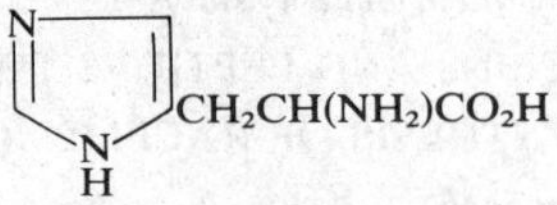	His
12. Cysteine	$HSCH_2CH(NH_2)CO_2H$	Cys
13. Valine	$Me_2CHCH(NH_2)CO_2H$	Val
14. Proline	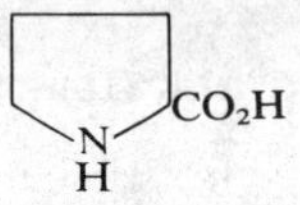	Pro
15. Tryptophan	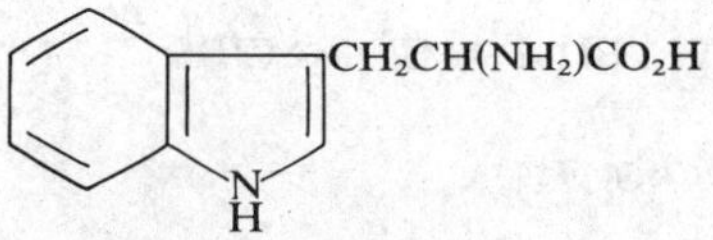	Trp
16. Isoleucine	$EtMeCHCH(NH_2)CO_2H$	Ile
17. Methionine	$MeSCH_2CH_2CH(NH_2)CO_2H$	Met
18. Asparagine	$H_2NCOCH_2CH(NH_2)CO_2H$	Asn or Asx
19. Glutamine	$H_2NCOCH_2CH_2CH(NH_2)CO_2H$	Gln or Glx
20. Threonine	$MeCH(OH)CH(NH_2)CO_2H$	Thr

* Symbols according to the instructions to authors in *The Biochemical Journal*, 1978.

have specific transfer RNA molecules. The process of translation only involves 19 α-amino-acid residues and the α-imino-acid proline (Table 1). It should be noted, however, that it is possible to induce the protein-synthesis mechanisms of bacteria to accept at least a small proportion of an amino-acid analogue in place of the normal amino-acid. For example, a lower homologue of proline (azetidine-2-carboxylic acid) has been incorporated into *E. coli* protein to replace one quarter of the proline.[4] However, such substitution only occurs in the absence of an adequate supply of the normal amino-acid.

The structures of the 20 normal amino-acid protein components (isolated from hydrolysates) had all been determined by 1935; the first to be characterized was glycine in 1820 by Braconnot, and the last was threonine. Although cysteine is incorporated as such, the functional forms of many peptides and proteins contain the oxidized product cystine, in which the disulphide cross-links may be either intra- or inter-molecular. Apart from glycine, all the coded protein amino-acids are optically active, and of identical chirality at the asymmetric α-carbon atom. By analogy to the usual nomenclature of the carbohydrates, they are commonly referred to as being of the L-configuration, L-serine being the reference compound of the series. With the exception of cysteine, this corresponds in all cases to the *S*-configuration in the Cahn–Ingold–Prelog convention; the position of the sulphur in cysteine is such as to render L-cysteine of the *R*-configuration. Isoleucine and threonine have further centres of asymmetry at the β-carbon atoms; the stereoisomers found in proteins are (2*S*,3*S*)-2-amino-3-methylvaleric and (2*S*,3*R*)-2-amino-3-hydroxybutyric acids, respectively.

23.2.1.2 Other amino-acids found in protein acid hydrolysates

The constituents of protein hydrolysates are not always confined to the genetically-coded amino-acids thyroxine and 3,3′,5-tri-iodothyronine are the two thyroid hormones, now well established,[5] and chemical or heat treatment of proteins prior to hydrolysis can give rise to artefacts. The unusual amino-acids arising under physiological conditions can be divided into two groups. The first group comprises compounds produced by substitution of relatively small groups into the normal protein constitutents (Table 2). These all appear to involve enzyme-induced reactions and the substituents introduced are principally *C*-hydroxyl, side-chain *N*-methyl, or halogen into the aromatic ring of tyrosine.

The amino-acids thyroxine and 3,3′,5-tri-iodothyronine are the two thyroid hormones, and seem to arise from oxidative coupling and iodination of two molecules of tyrosine

TABLE 2
Substituted Amino-acids found in Protein Acid Hydrolysates[a]

Date of isolation	*Amino-acid*		*Source*
1962	3-Hydroxyproline		Collagen
1940	δ-Hydroxylysine		Gelatin
1969	3,4-Dihydroxyproline		Diatom cell wall[b]
1902	4-Hydroxyproline		Gelatin
1959	ε-*N*-Methyl-lysine		{*Salmonella* flagellin; Calf thymus histone}
1967	ε-*N*-Dimethyl-lysine		Calf thymus histone
1968	ε-*N*-Trimethyl-lysine		Several histones
1967	3-Methylhistidine		Rabbit muscle actin
1968	N^G-Methylarginine		Calf thymus histone
1971	{$N^GN'^G$-Dimethylarginine; N^G,N^G-Dimethylarginine}		Bovine encephalitogenic protein
1970	$N^\varepsilon N^\varepsilon N^\varepsilon$-Trimethyl-δ-hydroxylysine		Diatom cell wall[c]
1948	3-Iodotyrosine		Thyroglobulin
1931	3,5-Di-iodotyrosine		Thyroglobulin
1951	3-Bromotyrosine		Gorgonian scleroprotein
1972	3-Chlorotyrosine		{Whelk scleroprotein[d]; Locust cuticular protein}
1972	3,5-Dichlorotyrosine		*Limulus* cuticle[e]
1971	3′-Bromo-5-chlorotyrosine		Whelk scleroprotein[f]
1972	Thyroxine		Thyroglobulin
1953	3,3′,5-Tri-iodothyronine		Thyroglobulin

[a] References are only given where the compounds are not discussed in Ref. 5. [b] T. Nakajima and B. E. Volcani, *Science*, 1969, **164,** 1400. [c] T. Nakajima and B. E. Volcani, *Biochem. Biophys. Res. Comm.*, 1970, **39,** 28. [d] S. Hunt, *FEBS Letters*, 1972, 24, 109; S. D. Anderson, *Acta Chem. Scand.*, 1972, **26,** 3097. [e] S. Welinder, *Biochim. Biophys. Acta*, 1972, **297,** 491. [f] S. Hunt, *Biochim. Biophys. Acta*, 1971, **252,** 301.

(Chaper 30.3). Thyroglobulin is a glycoprotein of molecular weight 660 000, which comprises the bulk of the colloid in thyroid follicles, and serves as a storage vehicle for the two hormones as well as providing a matrix for their synthesis. Rabbit thyroglobulin contains the iodinated amino-acids in the following amounts (residues per molecule): 3-iodotyrosine 9.6, 3,5-di-iodotyrosine 9.5, thyroxine 2.3, and tri-iodothyronine 0.6.[6] Although chlorotyrosines can be formed under certain conditions on hydrolysing proteins in hydrochloric acid, 3,5-dichlorotyrosine was isolated after basic hydrolysis and 3-chlorotyrosine after enzymic hydrolysis, clearly establishing their presence in the intact protein. The presence of hydroxyl substituents is particularly associated with the structural proteins collagen and elastin, while *N*-methylation is characteristic of the histones.

Deliberate modification of amino-acid residues in proteins is, of course, widely used in studying protein structure, but will not be considered here. Artefacts due to heating proteins or treating with chemicals for other reasons are also not uncommon. For example, the scorching of milk proteins leads to the formation of the acid-stable compounds pyridosine (**1**) and furanosine (**2**)[7] through the action of glucose on the ε-amino-group of lysine. The use of glutaraldehyde to cross-link protein chains also involves lysine, and the concomitant formation of the pyridinium derivative (**3**) has been demonstrated.[8]

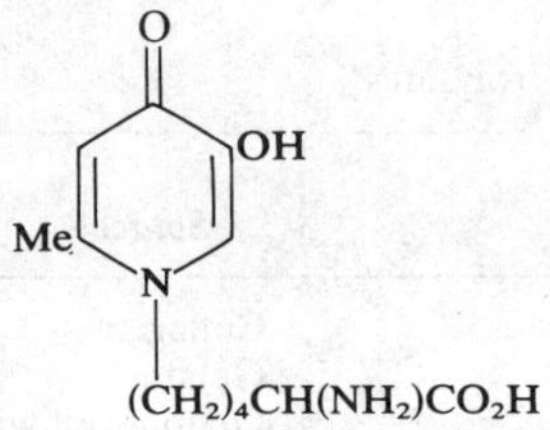

(**1**)

(**2**)

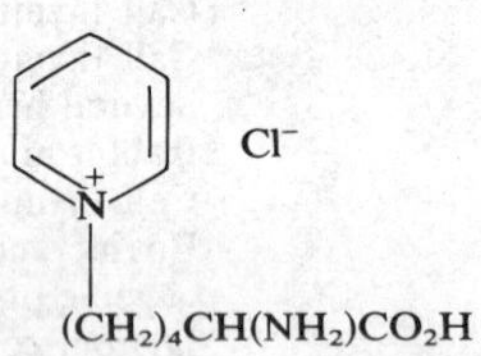

(**3**)

The second, and so far rather select, group of acid-stable naturally modified amino-acids are those in which two normal residues have become cross-linked (Table 3). The proteins in which they occur are rendered less soluble by their formation. The first step in the process leading to desmosine and isodesmosine is the enzymic oxidation of lysine to allysine (α-amino-adipic semialdehyde), three residues of which condense with an unchanged lysine residue to form a reduced pyridine ring which then undergoes oxidation to the final product. Dityrosine, originally isolated from the rubber-like protein from the wing-hinge ligament of locusts, has also been isolated from elastin;[9] it can be synthesized from tyrosine by the reaction of horseradish peroxidase and hydrogen peroxide. Treatment of proteins with alkali can lead to the formation of dehydroalanine from cysteine; addition of the thiol group of another cysteine residue to the double bond generates the acid-stable difunctional amino-acid lanthionine (Scheme 1). This was originally isolated from alkali-treated wool, but has also been found in ovokeratin from chick eggshell membranes after exposure to base.

Nucleophilic attack of lysine or ornithine (**5**) (itself a degradation product of arginine with alkali) on dehydroalanine similarly gives rise to lysinoalanine (**4a**) and ornithoalanine (**4b**).[10]

TABLE 3
Cross-linked Amino-acids found in Protein Hydrolysates

Amino-acid	*Formula*	*Source*
Dityrosine	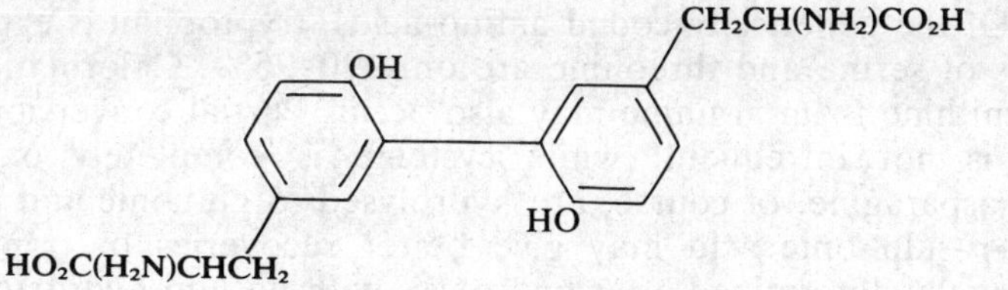	Resilin[9]
Desmosine	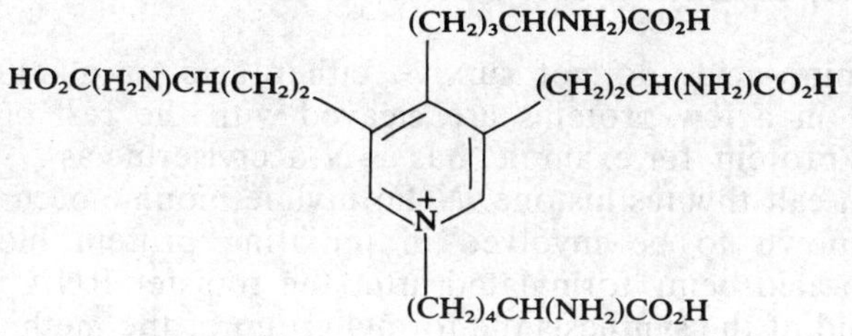	Elastin[a]
Isodesmosine	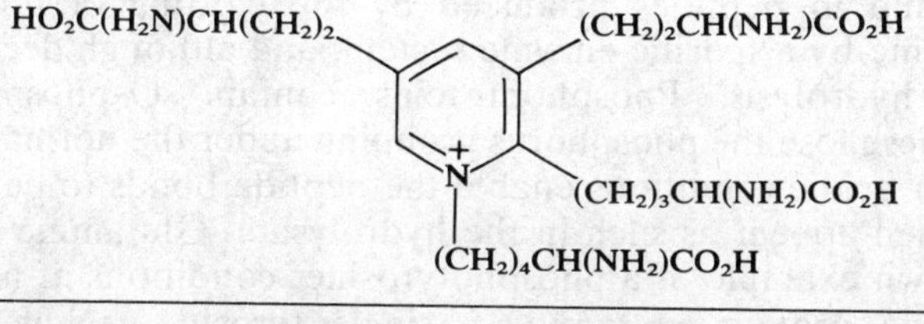	Elastin[a]

[a] J. Thomas, D. F. Elsden, and S. M. Partridge, *Nature*, 1963, **200**, 651.

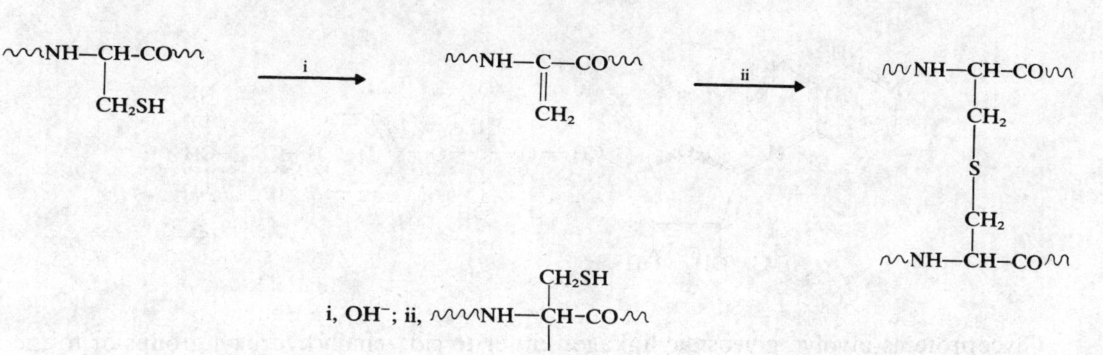

SCHEME 1

$HO_2CCH(NH_2)(CH_2)_nNHCH_2CH(NH_2)CO_2H$

(**4**) a; $n = 4$
b; $n = 3$

$H_2N(CH_2)_3CH(NH_2)CO_2H$

(**5**)

23.2.2 THE HYDROLYSIS OF PROTEINS AND PEPTIDES TO AMINO-ACIDS

Although hydrochloric acid (6 M, 24 h, 120 °C evacuated sealed tube) is normally used to break proteins down into their constituent amino-acids, it is not a method free of side-reactions. Of the genetically-coded amino-acids, tryptophan is extensively degraded, while recoveries of serine and threonine are only 90–95%. Chlorination of tyrosine and formation of ornithine from arginine may also occur. Partial conversion of methionine to its sulphoxide is not infrequent, while cysteine is completely oxidized to cystine. Glutamine and asparagine, of course, are hydrolysed to glutamic and aspartic acids. The use of toluene-*p*-sulphonic acid may give better recoveries of tryptophan,[11] but this amino-acid is usually determined after hydrolysis with barium hydroxide. Besides causing racemization, however, alkaline hydrolysis results in high losses of serine, threonine, cysteine, and arginine.

Many modified amino-acids do not survive either acid or alkaline hydrolysis. The *N*-acyl groups found in a few proteins are cleaved with the rest of the amide bonds. Tobacco mosaic virus protein, for example, has α-*N*-acetylserine as *N*-terminal, and ε-*N*-acetyl-lysine occurs in calf thymus histone. *N*-Formylmethionine occurs in several bacterial proteins; it is known to be involved in initiating protein biosynthesis in these organisms, the amino-acid being formylated after the transfer-RNA is charged with the amino-acid. At the end of the synthesis the formyl group of the methionine residue is not always released.[12] A number of peptide hormones have a pyroglutamyl residue (pyrrolidonecarboxylic acid) as *N*-terminal (spontaneous cyclization of *N*-terminal glutamine giving rise to this), which opens on hydrolysis to give glutamic acid. Tyrosine-*O*-sulphate, first found in peptides produced by the clotting of bovine fibrinogen, is generated from tyrosine by a specific enzyme system, and although decomposed by acid is stable to alkaline hydrolysis. Phosphoproteins contain *O*-phosphoserine and *O*-phosphothreonine; these lose the phosphorus grouping under the normal conditions of acid hydrolysis, but milder acidic conditions enable the peptide bonds to be cleaved selectively and these residues then appear as such in the hydrolysate. Glutamine synthetase from *E. coli* is the single known example of a phosphotyrosine; conditions of nitrogen surfeit lead to biosynthesis of the protein containing a single tyrosine residue *O*-substituted by adenylic acid (**6**).[13]

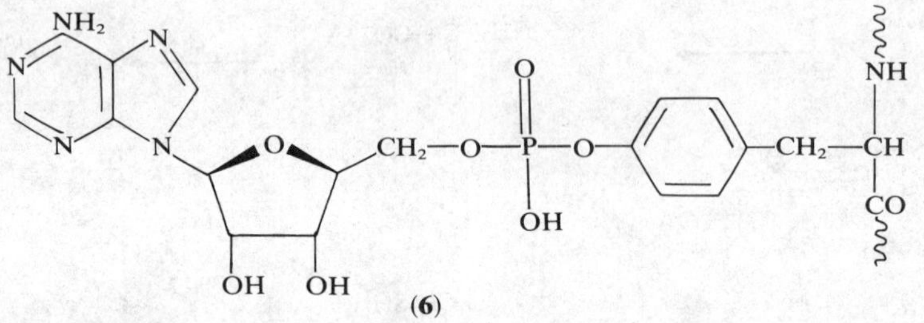

(**6**)

Glycoproteins involve glycosidic linkages either to side-chain hydroxyl groups or to the nitrogen of asparagine, although one case of *S*-substitution is known in a urine protein, digalactosylcysteine having been isolated from this compound;[14] in all cases, however, the carbohydrate is lost on mild acid hydrolysis. Enzymic removal of such sugar residues prior to acid hydrolysis is often beneficial; the presence of carbohydrates during the latter process can lead to amino-acid destruction.

The calcium binding sites of several proteins have been found to be relatively rich in residues of γ-carboxyglutamic acid. It occurs, for example, at 10 positions in the first 33 residues of the 582 residue prothrombin sequence, in each case as closely positioned doublets. Carboxylation of the glutamic acid residues in this protein appears to be mediated by vitamin K.[15] As a malonic acid derivative, decarboxylation occurs on acid hydrolysis, but it can be isolated after alkaline hydrolysis,[16] although it gives a rather weak ninhydrin reaction. Citrulline (**7**), long known to be an intermediate in the alkaline degradation of arginine to ornithine, has been established as a constituent of the medulla cell protein of mammalian hair and porcupine quill.[17] It is presumably derived from protein-bound arginine, and shows up in acid hydrolysates as ornithine. The insolubility of these same proteins is ascribed to the presence of ε-(γ-glutamyl)lysine cross-links which likewise do not survive the conditions of peptide bond cleavage.[18]

$$NH_2CONH(CH_2)_3CH(NH_2)CO_2H$$

(**7**)

The cross-links of collagen, which have an essential role in determining its physical properties, are due to the interaction of α-aminoadipic semialdehyde, produced from lysine by lysyl oxidase, with other amino-acid residues. None of the products is stable to acid as they are Schiff bases, but after reduction with sodium borohydride several bifunctional products have been isolated, *e.g.* lysinonorleucine (**8**), the trifunctional aldol-histidine (**9a**), and the tetrafunctional histidino-hydroxymerodesmosine (**9b**).[19] A similar series of products is formed from δ-hydroxylysine. Isolation of amino-acid residues unstable to extremes of pH has in most cases been accomplished after enzymic hydrolysis. Bacterial proteases with a wide specificity of action can alone often effect almost complete cleavage of peptides to the component amino-acids. Use of a mixture of four very specific enzymes (chymotrypsin, trypsin, prolidase, and aminopeptidase M) covalently bound to Sepharose has been found a convenient and mild alternative to acid hydrolysis.[20]

$$HO_2C(NH_2)CH(CH_2)_3NH(CH_2)_4CH(NH_2)CO_2H$$

(**8**)

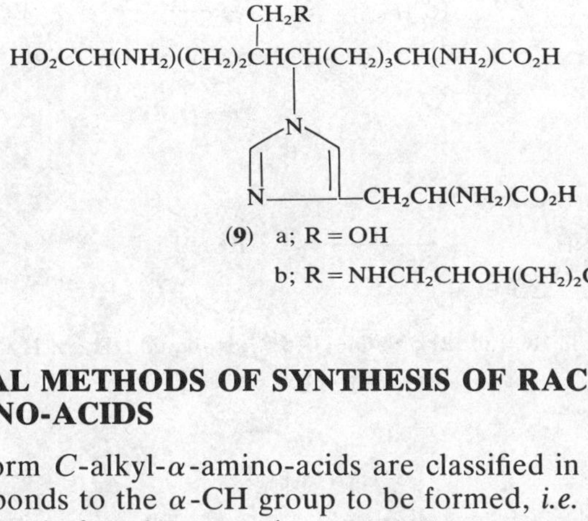

(**9**) a; R = OH

b; R = $NHCH_2CHOH(CH_2)_2CH(NH_2)CO_2H$

23.2.3 GENERAL METHODS OF SYNTHESIS OF RACEMIC *C*-ALKYL-α-AMINO-ACIDS

Reactions to form *C*-alkyl-α-amino-acids are classified in this chapter according to the last of the three bonds to the α-CH group to be formed, *i.e.* as aminations, *C*-alkylations, or carboxylations. A fourth group is necessary to encompass methods in which the amino-acid derivative is generated either by a rearrangement or where it is difficult to distinguish the order of bond formation. Space considerations preclude a completely comprehensive coverage of the topic; the reactions discussed were chosen on the basis of their proven utility or because they illustrate the diversity of synthetic approaches to α-amino-acids.

23.2.3.1 Amination of carboxyl components

Reductive amination of α-keto-acids (asymmetric hydrogenolytic transamination) is an important general method falling into this category, but as it is a synthesis involving prochiral intermediates it is dealt with in Section 23.2.4.1. The other classical method of this type is the straightforward nucleophilic attack of an amino-component on an α-halocarboxylic acid substrate. In its simplest form an α-chloro- or α-bromo-carboxylic acid is treated with an excess of ammonia (Scheme 2). To prevent dialkylation, ammonia

$$BrCHRCO_2H + NH_3 \longrightarrow H_2NCHRCO_2H$$

SCHEME 2

may be replaced by hexamethylenetetramine; hydrolysis of the initial product is then necessary. A variation using an α-halo-ester and potassium phthalimide (Scheme 3) has

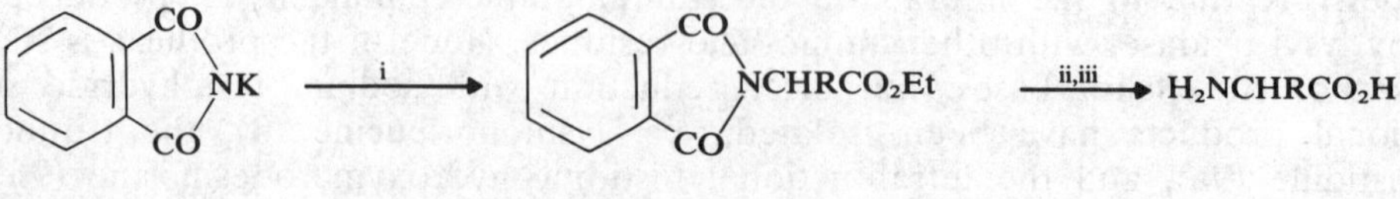

i, $BrCHRCO_2Et$; ii, N_2H_4; iii, OH^-.

SCHEME 3

been employed for the preparation of α,ω-diamino-acids. As an alternative to direct α-halogenation of a carboxylic acid to prepare the starting material, an alkyl halide can be used in conjuction with ethyl acetoacetate (Scheme 4). More recently, amination of sodiomalonate derivatives with chloramine (Scheme 5)[21] and of a carbene ester with

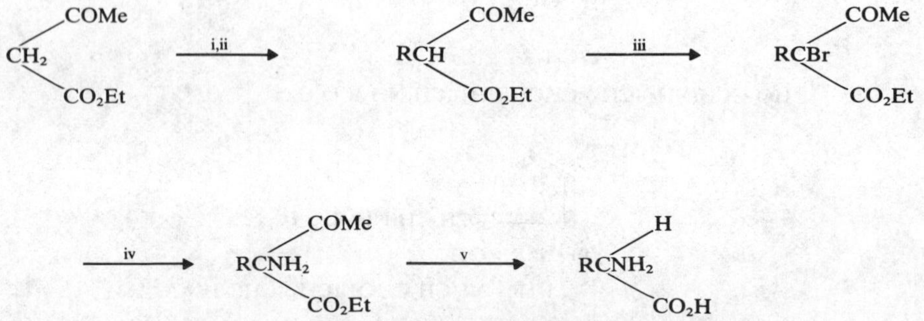

i, NaOEt; ii, RCl; iii, Br_2, NaOH, 0–5 °C; iv, liquid NH_3; v, H_3O^+.

SCHEME 4

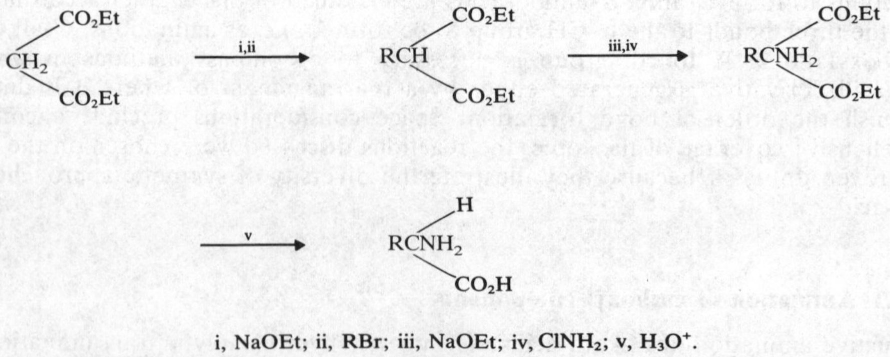

i, NaOEt; ii, RBr; iii, NaOEt; iv, $ClNH_2$; v, H_3O^+.

SCHEME 5

amines (Scheme 6)[22] have been developed as amino-acid syntheses. In the latter method the side-chain originates as an α-keto-ester, and if optically active groups are present (R^1 and R^2), there is a modest degree of asymmetric induction.

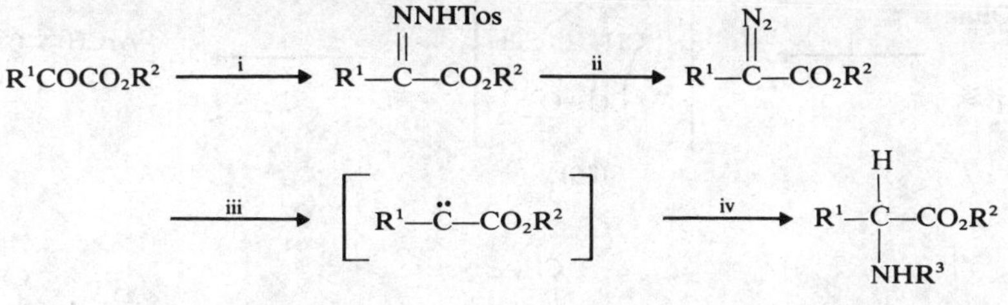

i, $TosNHNH_2$; ii, NEt_3, CH_2Cl_2; iii, CuCN at 50 °C; iv, R^3NH_2.

SCHEME 6

23.2.3.2 *C*-Alkylation of α-amino-acid derivatives

Two long-established methods are based on the introduction of a side-chain into derivatives of glycine. The first of these, originated by Sorensen in 1903, utilized in its original form the alkylation of ethyl *N*-phthalimidomalonate (Scheme 7); this reagent,

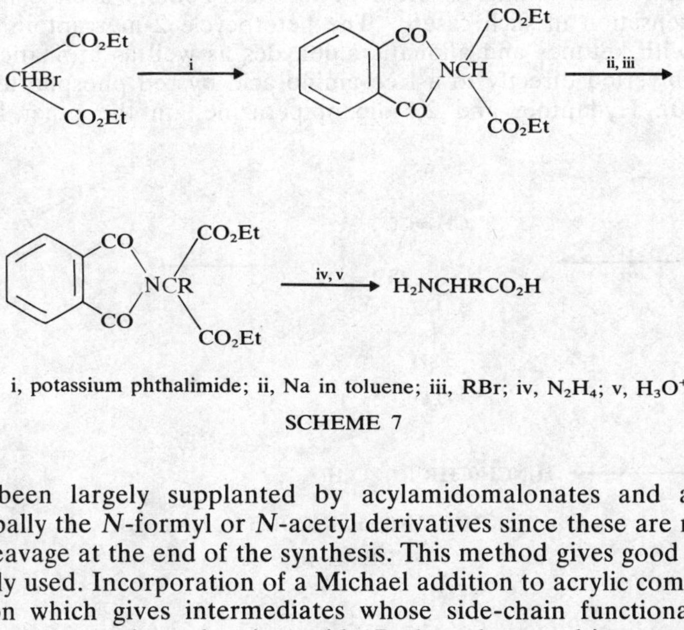

i, potassium phthalimide; ii, Na in toluene; iii, RBr; iv, N_2H_4; v, H_3O^+.

SCHEME 7

however, has been largely supplanted by acylamidomalonates and acylamidocyano-acetates, principally the *N*-formyl or *N*-acetyl derivatives since these are more susceptible to hydrolytic cleavage at the end of the synthesis. This method gives good yields in general and is still widely used. Incorporation of a Michael addition to acrylic compounds (Scheme 8) is a variation which gives intermediates whose side-chain functional group can be further elaborated to a variety of amino-acids. Perhaps less used in recent years has been

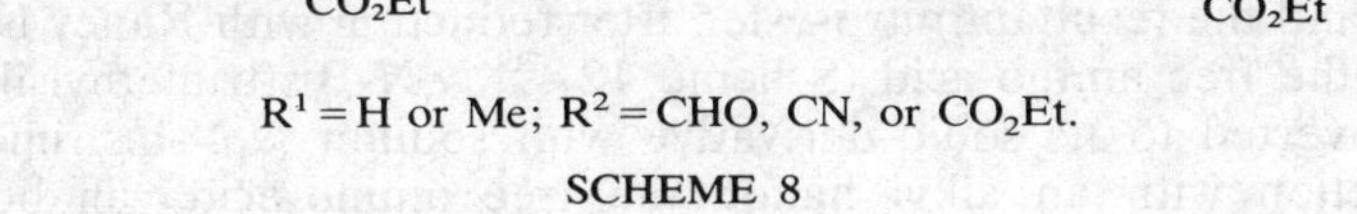

$R^1 = H$ or Me; $R^2 = CHO$, CN, or CO_2Et.

SCHEME 8

the classical azlactone synthesis (Scheme 9). This works well with aromatic aldehydes, but until recently two hydrolysis steps were necessary as the unsaturated azlactone is not directly reducible by hydrogenation over palladium or platinum, or by red phosphorus and

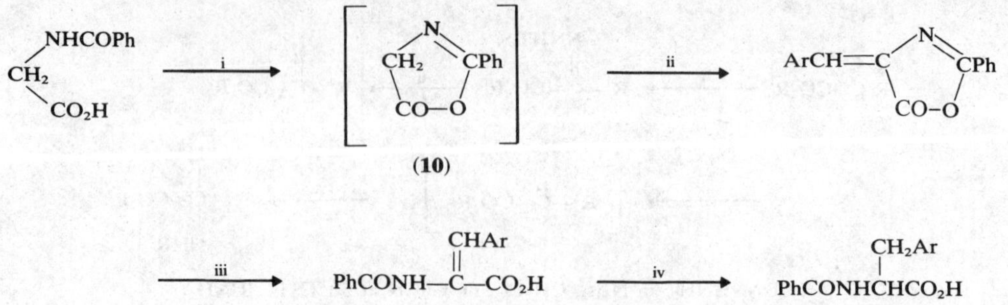

i, Ac_2O, NaAc; ii, ArCHO; iii, OH^-, H_2O; iv, Pd, H_2.

SCHEME 9

hydriodic acid. However, in 1972 it was shown that treatment with Raney nickel and hydrogen (2–4 atom) in alcoholic ammonia caused reductive solvolysis of the azlactone to the acylamino-acid amide in one step.[23] The saturated azlactone (**10**) is not normally isolated, but this is necessary if aliphatic aldehydes or ketones are to be used, otherwise yields are very low; lead acetate is often better than sodium acetate in catalysing the subsequent condensation in such cases.[24] The heterocycle 2-mercaptothiazol-5-one (**11**) condenses well with ketones and aliphatic aldehydes as well as aromatic aldehydes, the product being converted directly to a free amino-acid by red phosphorus and hydriodic acid (Scheme 10). Hydantoin and 2,5-dioxopiperazine, on the other hand, will only

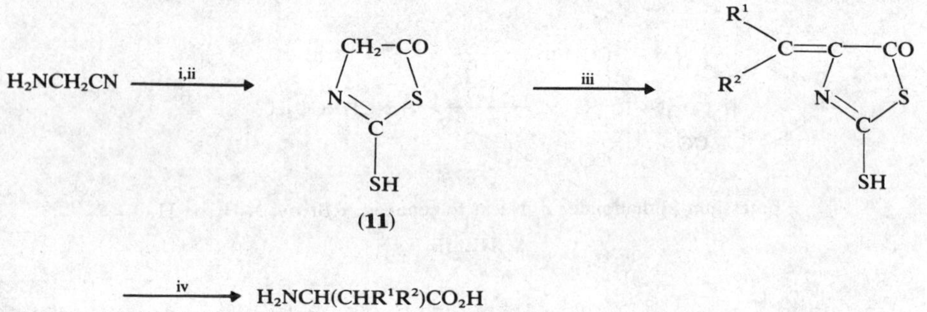

i, CS_2, Me_2CO; ii, HCl; iii, R^1COR^2; iv, red P, HI.

SCHEME 10

condense with aromatic aldehydes; the condensation products are converted to free amino-acids by reduction followed by hydrolysis. 3-Phenylhydantoins, however, have been used in a general amino-acid synthesis in which the side-chain is introduced by alkylation of an enolate (**12**) stabilized as its magnesium chelate (Scheme 11).[25]

Two recent general *C*-alkylation methods do not involve heterocyclic intermediates. Methyl nitroacetate in the presence of sodium methylate can undergo electrophilic attack by alkyl halides, and the resultant nitro-ester after reduction with Raney Ni and hydrogen is hydrolysed to the free amino-acid (Scheme 12).[26] *N,N*-Bis(trimethylsilyl)glycine ethyl ester can be converted to its sodio-derivative with sodium *N,N*-bistrimethylsilylamide; again, after reaction with an alkyl halide the free amino-acid can be generated by

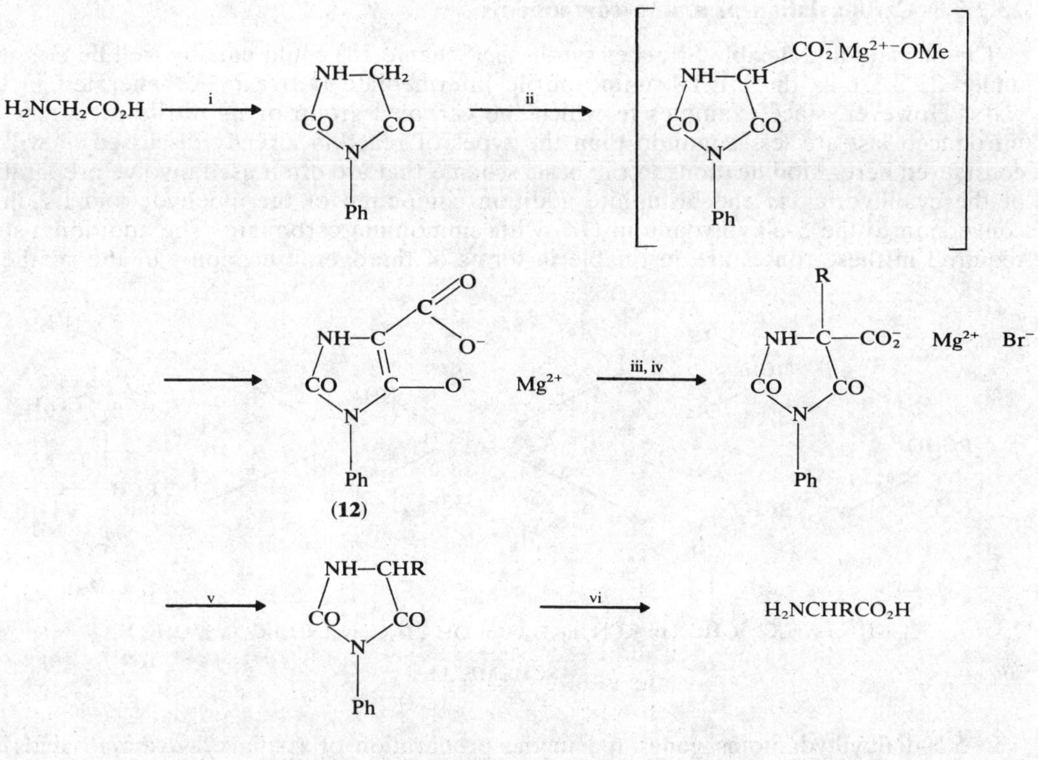

i, PhNCO; ii, $(MeOCO_2^-)_2Mg^{2+}$; iii, NaOMe; iv, RBr; v, H_3O^+; vi, $Ba(OH)_2$.

SCHEME 11

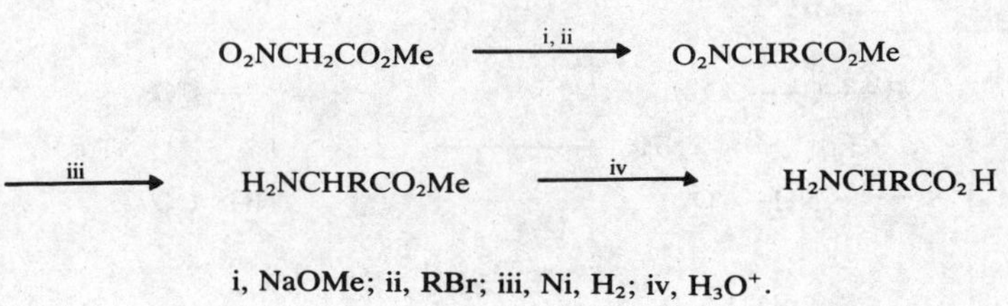

i, NaOMe; ii, RBr; iii, Ni, H_2; iv, H_3O^+.

SCHEME 12

hydrolysis, the trimethylsilyl groups being rapidly cleaved by dilute aqueous or ethereal hydrogen chloride.[27] A final reaction worth mentioning is the addition of Grignard reagents to the imines of glyoxylates (Scheme 13). If optically active protecting groups are used [*e.g.* $R^1 = PhCH(Me)$— and $R^2 = (-)$-menthyl] there is some asymmetric induction; L-alanine was thus made in 63.5% optical purity and 45% yield using this procedure.[28]

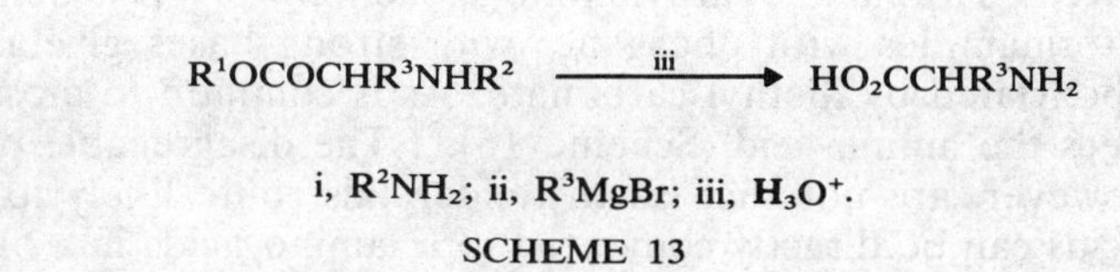

i, R^2NH_2; ii, R^3MgBr; iii, H_3O^+.

SCHEME 13

23.2.3.3 Carboxylation of amino-components

The efficient and flexible Strecker synthesis (Scheme 14) could equally well be classified under 23.2.3.1 as the crucial amino-nitrile intermediate (**13**) can be generated in two ways. However, since examples in which the carboxyl group or its nitrile precursor are introduced last are less common than the types of reaction already discussed, it will be considered here. Modifications to the basic scheme that are often used involve preparation of the cyanhydrin *via* the bisulphite addition compound of the aldehyde and its direct conversion to the 5-alkylhydantoin (**14**) with ammonium carbonate. The additional steps required in these routes are justifiable in terms of the overall economy of the synthesis.

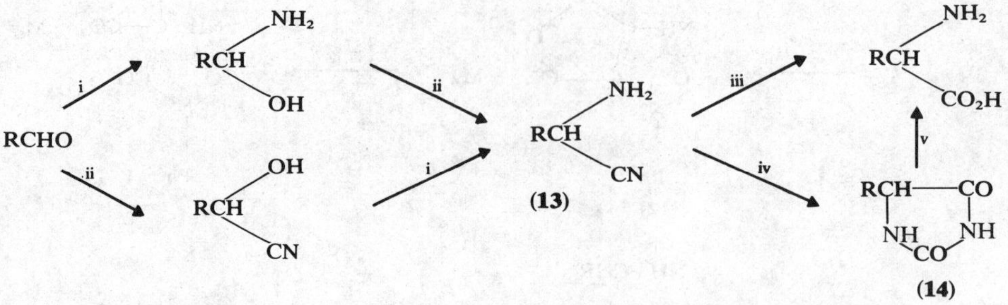

i, NH_3 or NH_4Cl; ii, HCN or KCN; iii, H_3O^+or OH^-, H_2O; iv, $(NH_4)_2CO_3$; v, OH^-, H_2O.

SCHEME 14

The 5,5-dialkylhydantoins generated in the preparation of *C*-dialkyl-α-amino-acids are often resistant to hydrolysis. Tosylation of these intermediates, however, gives products from which the desired amino-acids can more easily be released (Scheme 15).[30]

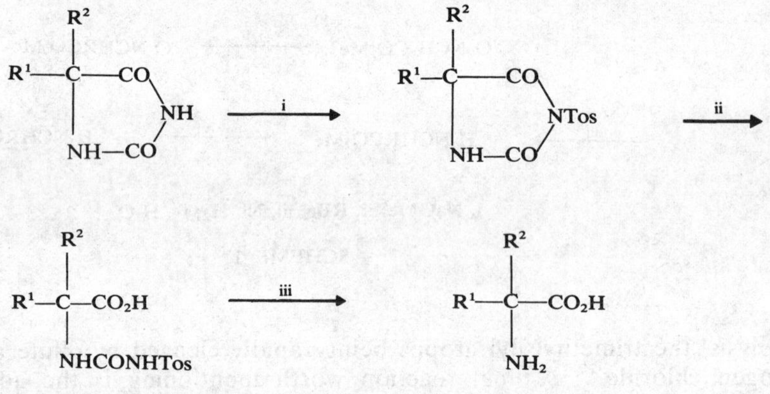

i, TosCl, KOH; ii, OH^-, H_2O; iii, 10% aq. HCl, reflux 1 h.

SCHEME 15

Two novel carboxylation reactions have been developed recently. Isocyanides, available from the improved Hofmann carbylamine reaction or by dehydration of *N*-monosubstituted formamides with phosgene, with strong bases give rise to carbanions which can be carboxylated by diethyl carbonate. As is common to most methods, a final hydrolytic step frees the amino-acid (Scheme 16).[29] The disagreeable physical properties of isocyanides, however, are not such as to render this route likely to become popular. Free carboxylic acids can be directly converted to α-amino-acids in a one-step procedure

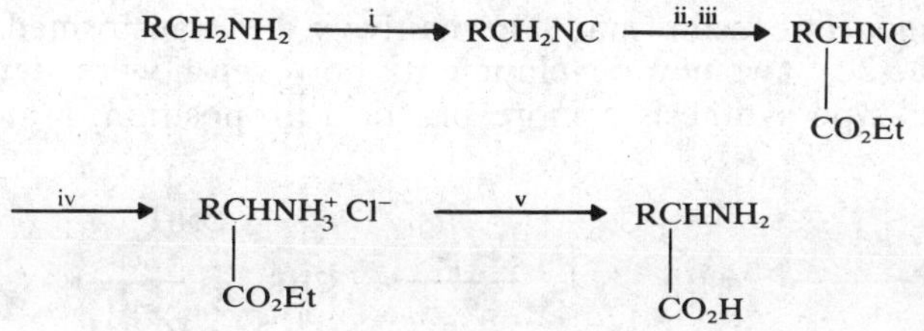

i, $CHCl_3$, OH^-; ii, NaH; iii, $(EtO)_2CO$; iv, HCl, H_2O; v, OH^-, H_2O.

SCHEME 16

by the amination of α-lithiated salts (Scheme 17). The best yields were obtained using *O*-methylhydroxylamine, but even these were rather modest; isovaleric acid, for example, gave valine in 33.9% yield.[31]

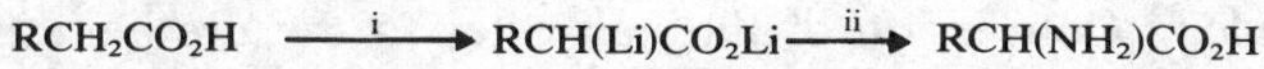

i, $LiNPr^i_2$; ii, NH_2OMe.

SCHEME 17

23.2.3.4 Rearrangement methods

The Curtius rearrangement has been used in the preparation of amino-acids since 1921, but the original form of its use (Scheme 18), in which the partial hydrolysis of the diester (**15**) is often difficult, was largely superseded by the Darapsky procedure (Scheme 19).

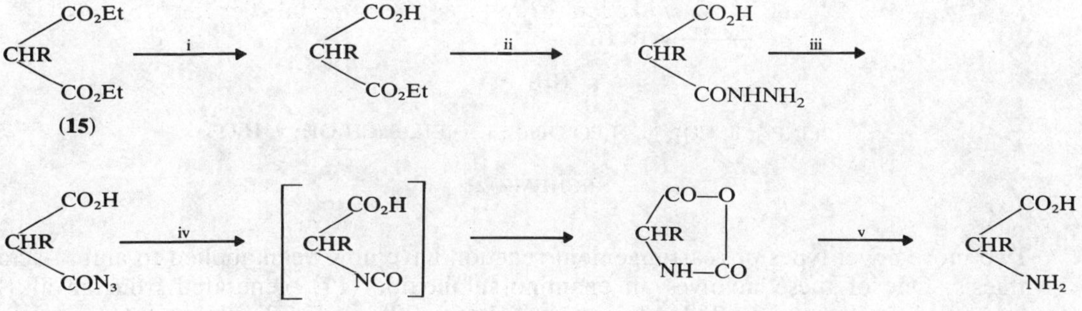

i, KOH; ii, N_2H_4; iii, HNO_2; iv, heat; v, H_3O^+.

SCHEME 18

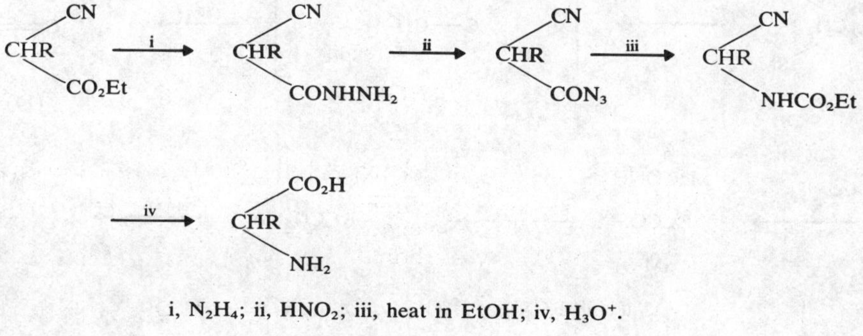

i, N_2H_4; ii, HNO_2; iii, heat in EtOH; iv, H_3O^+.

SCHEME 19

The alkylated cyanoacetic ester may alternaltively be transformed using a Hofmann degradation (Scheme 20). Two new developments, however, now render a modified form of the original Curtius-type synthesis a more practical proposition. Malonic half-esters can

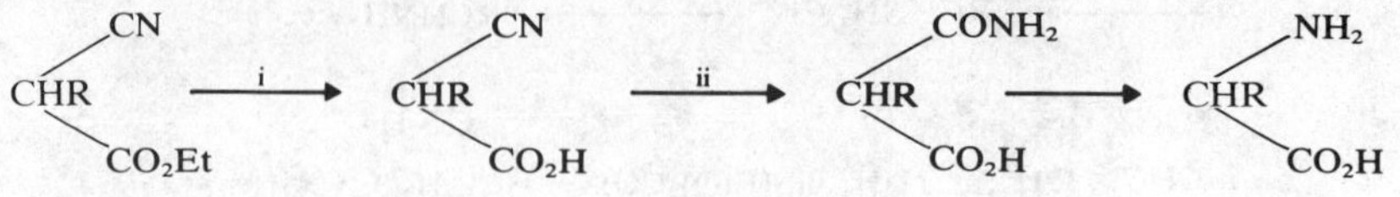

i, OH^-, H_2O; ii, conc. H_2SO_4; iii, Br_2, NaOH.

SCHEME 20

conveniently be prepared by α-carboxylation of simple esters; the ester is added to n-butyl-lithium and di-isopropylamine in tetrahydrofuran and the resulting mixture treated with carbon dioxide.[32] Conversion of the half-ester to the urethane (**16**) can be accomplished using diphenyl phosphorazidate, but the alcohol must be added an hour after this reagent otherwise the diester is the only product (Scheme 21).[33] However, since carboxylic acids can be directly aminated (Scheme 17), it remains to be seen if this longer route is a viable alternative.

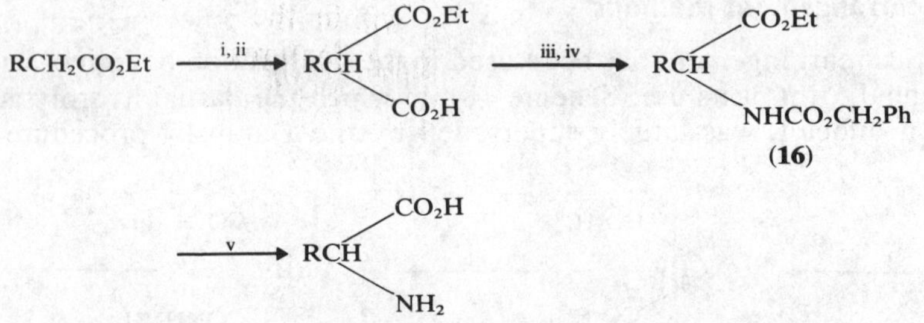

(**16**)

i, $LiNPr^i_2$; ii, CO_2; iii, $N_3PO(OPh)_2$; iv, NEt_3, $PhCH_2OH$; v, H_3O^+.

SCHEME 21

Two more novel types of rearrangement reaction have now been applied to amino-acid synthesis. One of these involves an enamino-sulphoxide (**17**), generated from an alkyl cyanide and methyl thiomethyl sulphoxide (Scheme 22), and a desulphurization step is required before a final hydrolytic stage,[34] The other concerns the rearrangement of an

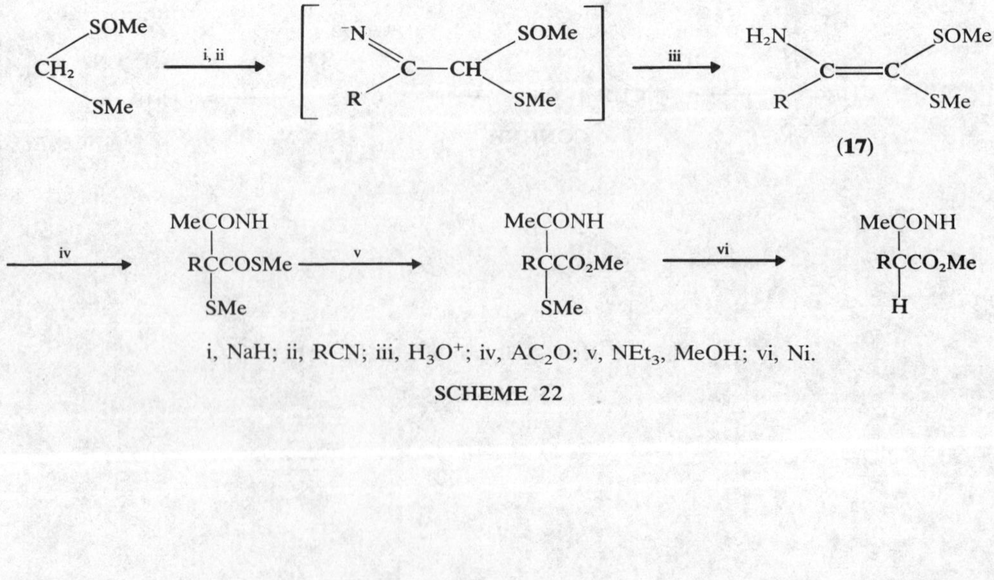

(**17**)

i, NaH; ii, RCN; iii, H_3O^+; iv, AC_2O; v, NEt_3, MeOH; vi, Ni.

SCHEME 22

N-chloroimidate to an amino-orthoester; alkoxyazirine intermediates are thought to be involved in this step, the mechanism resembling the Neber reaction (Scheme 23).[35] In a few syntheses of amino-acids it is difficult to distinguish the order in which bonds to the

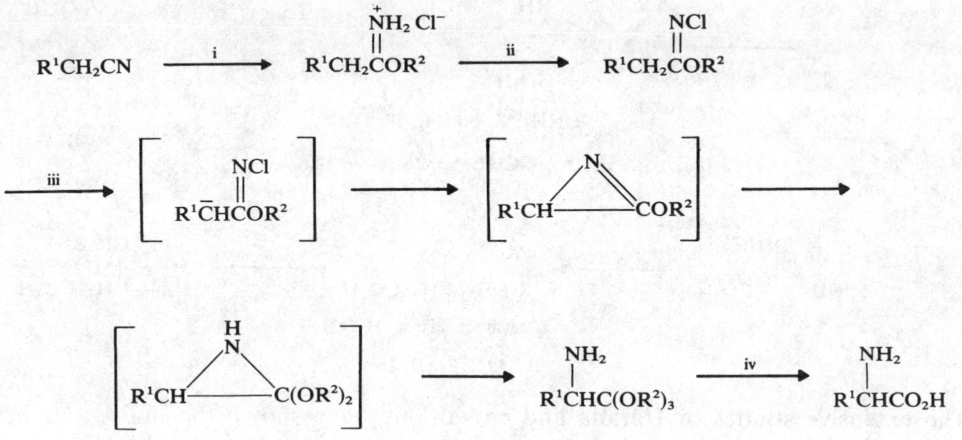

i, R^2OH, HCl; ii, NaOCl; iii, $NaOR^2$, R^2OH; iv, H_3O^+.

SCHEME 23

α-carbon atom are formed. Examples of this are the cobalt carbonyl catalysed reactions shown in Scheme 24, where the carboxyl and amino-functions are introduced in the same reaction.[36]

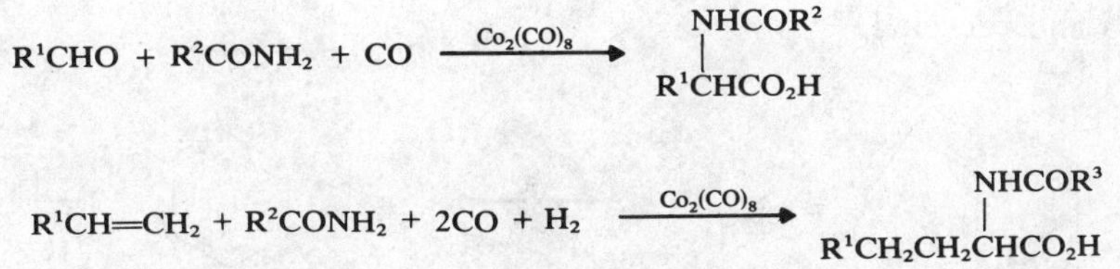

SCHEME 24

23.2.4 PREPARATION OF ENANTIOMERS OF *C*-ALKYL-α-AMINO-ACIDS

23.2.4.1 Asymmetric synthesis

For most synthetic purposes, amino-acid enantiomers are required. Although, in general, resolution of racemic products is still the predominant route to such stereoisomers, a good deal of progress has been made towards developing asymmetric syntheses which can compete with these established methods. The most widely studied reactions of this type are those in which the α-*C* asymmetric centre is generated during reduction of a double bond. This may be either the C=N of the Schiff base of an α-keto-acid (Scheme 25; asymmetric hydrogenolytic transamination) or the C=C bond of an enamine (Scheme 26), the latter usually being generated *via* an azlactone. In the case of the imine reduction, the asymmetric inducer is most often located in the starting amine, whilst asymmetric hydrogenation catalysts have been the principal inducers as far as acylamido-acrylic acids are concerned.

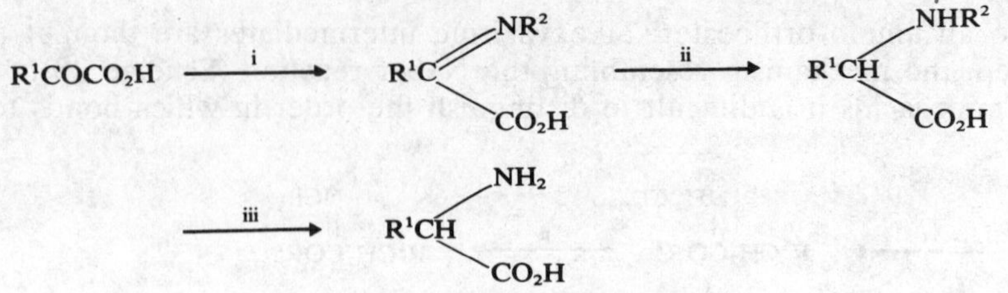

i, R^2NH_2; ii, Pd, H_2; iii, H_3O^+.

SCHEME 25

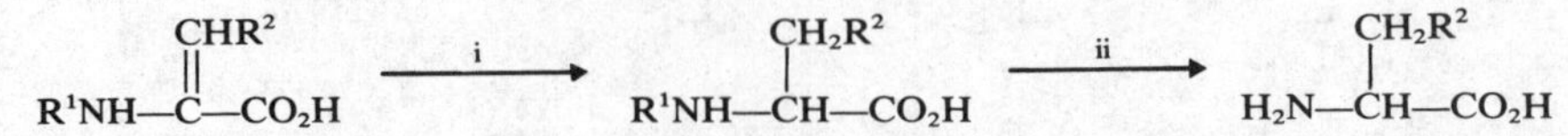

i, catalyst, H_2; ii, H_3O^+.

SCHEME 26

The extensive studies of Harada and co-workers have shown the limitations in using simple amines as asymmetric inducers in imine reduction. Chiral α-alkylbenzylamines enable L-alanine to be prepared in 60% optical purity, but catalytic reduction must be carried out at low temperature. At 17 °C the product is totally racemic, and inversion occurs at higher temperature, levelling out above 50 °C to 43% optically pure D-alanine. The degree of asymmetry induced also decreases with increasing side-chain size.[37] The problem of conformational mobility in the substrate was brilliantly solved by Corey and co-workers through the use of a cyclic imine (Scheme 27). In this way D-alanine of 96% optical purity has been prepared, the method constituting the first really practical asymmetric synthesis. The *N*-aminoindolenine inducer (**18**) can be regenerated from the product (**19**) and recycled.[38]

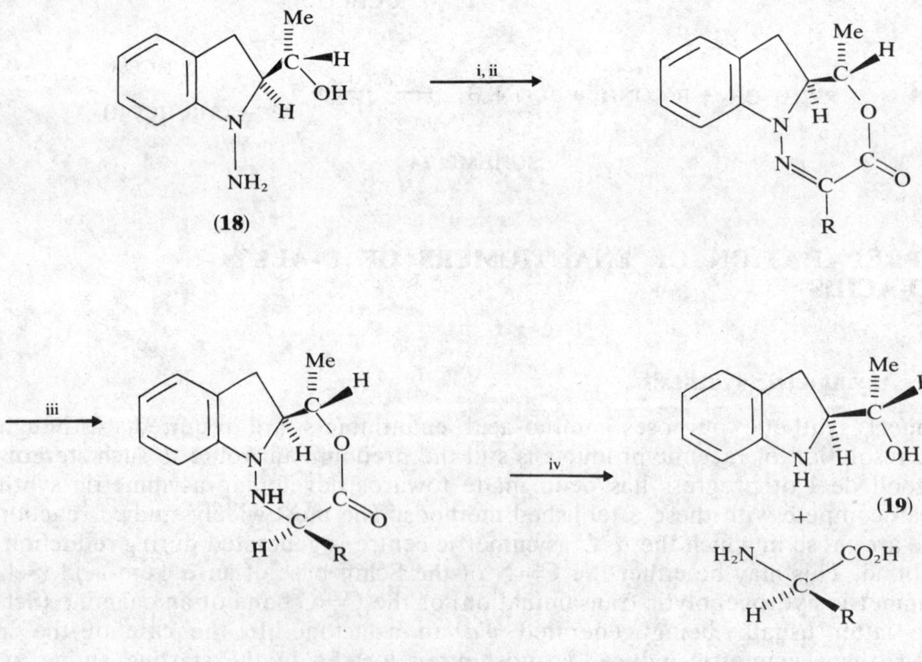

i, $RCOCO_2Me$; ii, NaOMe, PhH; iii, Al/Hg; iv, Pd, H_2.

SCHEME 27

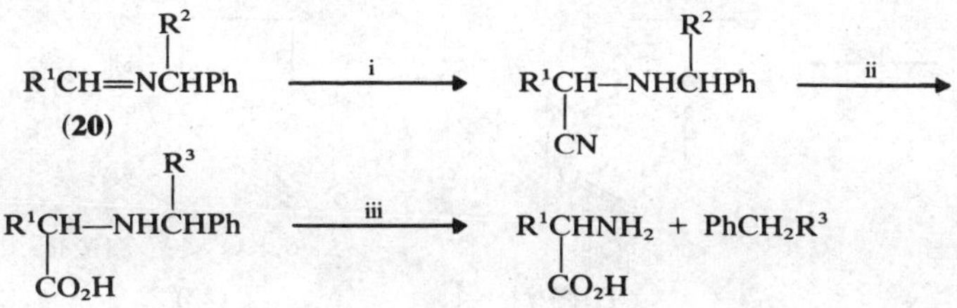

i, HCN; ii, H_3O^+; iii, Pd, H_2.

SCHEME 28

The use of optically active α-alkylbenzylamines in the Strecker synthesis to make the imine (**20**) (Scheme 28) only affords amino-acids of optical purity up to 58%,[39] but alkylation of similar imines of ethyl gloxylate (Scheme 29) has resulted in optical yields of up to 95% if the imine is complexed to enneacarbonyldi-iron. Complexation affords a mixture of diastereoisomers; the predominant one (**21**) is isolated and treated with alkyl bromide.[40] The optical inducer, as in all transamination reactions apart from the ingenious route of Corey, is not recoverable; the asymmetry of the product amino-acid is only gained at the expense of its destruction.

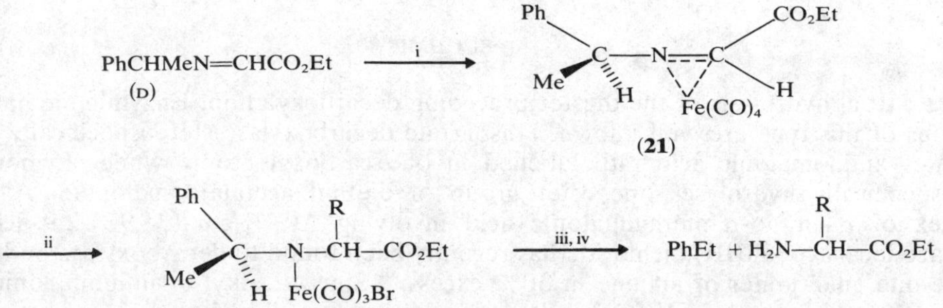

i, $Fe_2(CO)_9$; ii, RBr; iii, Pd, H_2; iv, OH^-, H_2O.

SCHEME 29

As far as α-acylamino-acrylic acids are concerned, the most efficient chiral catalysts yet discovered are of the homogeneous variety and based on rhodium. Products of 95% optical purity have been obtained using a catalyst prepared from $[Rh(cyclo\text{-}octa\text{-}1,5\text{-}diene)Cl]_2$ and 1,2-bis-(*O*-anisylphenylphosphino)ethane, the phosphorus atom itself being the chiral centre. The optical yield is not sensitive to temperature or pressure.[41] The azlactones normally used to make the α-acylaminoacrylic acids are not themselves reducible by catalysts of this type. If the aminoacrylic acid is part of a dioxopiperazine ring containing an optically active amino-acid as the other component, an optically inactive heterogeneous catalyst can allow highly stereospecific asymmetric hydrogenation. Use of (*S*)-proline (**22**) in this way results in nearly 90% optical efficiency (Scheme 30), and the chiral inducer is recoverable. The required side-chain is introduced as an α-keto-acid, the unsaturated dioxopiperazine (**24**) being generated *via* a 5-hydroxydioxopiperazine (**23**); reduction is effected in the presence of Adams catalyst to give the (*S*,*S*)-cyclodipeptide in essentially quantitative yield. By this means L-alanine was obtained in *ca.* 60% yield from pyruvic acid.[42] *N*-Substitution markedly reduces the efficiency of the hydrogenation step, especially for non-aromatic amino-acids;[43] azlactones are therefore ruled out as precursors in the preparation of the unsaturated dioxopiperazines.

Asymmetric synthesis by the popular alkylation of acylamidomalonic ester route (Scheme 7) necessarily involves chiral induction to occur during either the decarboxylation

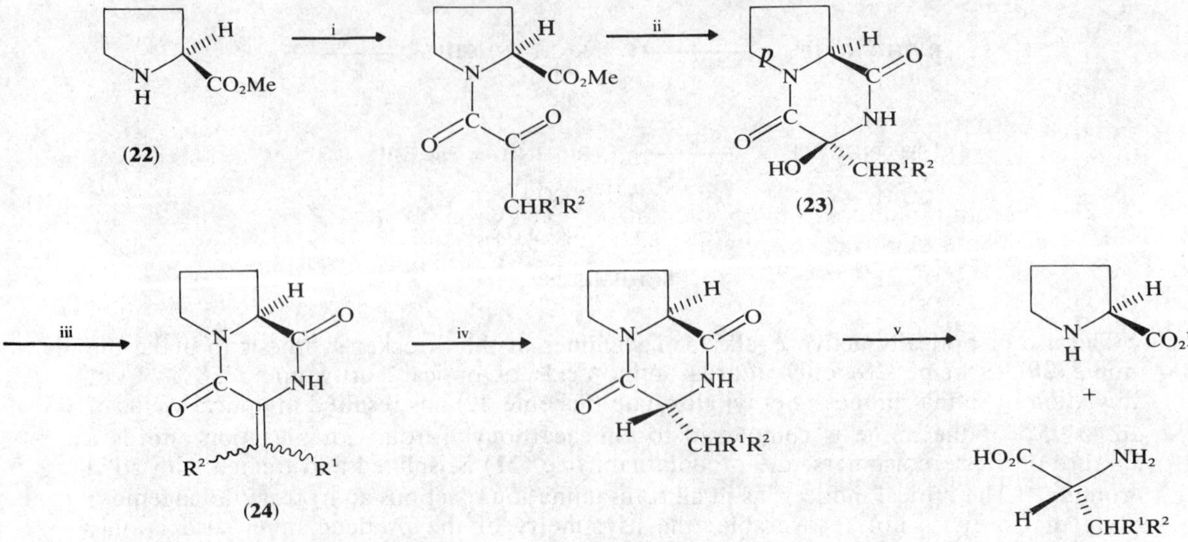

i, $RCOCO_2H$, dicyclohexylcarbodi-imide; ii, NH_3, $MeOCH_2CH_2OMe$; iii, CF_3CO_2H; iv, PtO, H_2; v, H_3O^+.

SCHEME 30

or on a partial hydrolysis of the diester preceding decarboxylation. Enzymic mediation of reactions of this type are well known; L-aspartate decarboxylase stereospecifically decarboxylates aminomalonic acid radiolabelled in one carboxyl group, while chymotrypsin stereospecifically hydrolyses one ester group of diethyl acetamidomalonate. A cobalt complex of α-amino-α-methylmalonic acid involving $\Lambda(-)_{436}$-α-(2*S*,9*S*)-2,9-diamino-4,7-diazodecanecobalt(III) dichloride has recently been found to decarboxylate on heating to give one enantiomer of alanine in 30% excess.[44] Since α-alkyl-α-aminomalonic acids can be obtained from their *N*-formyl diethyl esters by careful hydrolysis,[45] a first step has been taken in transforming this well-established synthetic route for preparation of racemic amino-acids into a stereospecific one.

23.2.4.2 Resolution of racemates

The early methods of resolving amino-acid racemates were based on selective crystallization, but the availability of relatively pure enzymes, notably the acylases, has led to enzymic methods substantially supplanting these. In recent years there has been a further shift towards the development of column chromatographic methods because of the practical convenience flow methods offer. In this discussion the accent is on methods that are of preparative use rather than those only useful on an analytical scale.

Most crystallization methods involve the preparation of diastereoisomeric salts, usually from the *N*-acyl-DL-amino-acid and an optically active base. The synthetic mixture of enantiomers is allowed to react with an optically active base such as brucine, strychnine, or 1-phenylethylamine in a solvent and at concentrations so chosen that one of the two diastereoisomeric salts formed crystallizes out. If necessary, the product is recrystallized until optically pure. The more soluble diastereoisomer can likewise be concentrated in the solution. The salts obtained are then decomposed back to the amino-acid. If the acyl group is appropriately chosen, the product may be directly usable in synthesis. *N*-Benzyloxycarbonyl-DL-amino-acids, for example, can be resolved with natural (−)-ephedrine in many cases. When no β-methyl side-chain is present, the salt of the D-isomer is deposited; when such a group is present, the L-isomer salt comes out of solution preferentially (phenylalanine is the one exception).[46] However, in common with many methods of resolution it is not generally applicable, failing with tyrosine, tryptophan, or glutamic acid. Crystallization methods, of course, often require conditions which vary with the particular amino-acid involved.

Amino-acids with basic side-chains, *e.g.* histidine and lysine, may be resolved without acylation by salt formation with optically active camphoric- or camphor-sulphonic acids. DL-Lysine has also been resolved using an optically active metal complex. Slow addition of dodecatungstophosphoric acid to a concentrated solution of $K(-)_{5461}[Co(EDTA)]$ (1 mol) and the racemic amino-acid hydrochloride (2 mols) gives an initial precipitate of the tungstophosphate of D-lysine. If enough acid is added, the salt of its antipode eventually precipitates.[47]

Derivatives of racemic amino-acids can also be resolved by preferential crystallization from supersaturated solutions on addition of a seed crystal of one antipode. One of the most general ways of doing this is using aromatic sulphonate salts. Both stereoisomers can be obtained in this way.[48]

The most widely applied of the enzymic methods of resolution utilizes a mammalian kidney acylase to hydrolyse the L-enantiomer of *N*-acetyl- or *N*-chloracetyl-DL-α-amino-acids. The resulting free L-amino-acid can be simply separated from the *N*-acyl-D-antipode by solvent extraction or ion-exchange chromatography. Acylase I is specific for the neutral non-aromatic amino-acids and glutamic acid, acylase II for aspartic acid, and acylase III for the aromatic amino-acids. Proline is not susceptible to the acylases, but has been resolved with an amidase. Recently a continuous flow resolution using a column packed with kidney acylase bound to DEAE-cellulose has been reported.[49] *N*-Acetyl-DL-methionine (0.02 M solution at pH 7.0) is completely resolved on a 1 × 4 cm column of this polymer-bound enzyme at a flow rate of 3 ml per hour at 37 °C. This method also avoids the inhibition of the enzyme which occurs on build-up of the free L-amino-acid when the reaction is carried out in solution.

Use of amino-acid oxidases which catalyse the decomposition of one enantiomer to an α-keto-acid (Scheme 31) has been extensive, but usually confined to analytical work. It

$$\text{DL-}NH_2CHRCO_2H \xrightarrow{\text{i}} RCOCO_2H + NH_3 + H_2O_2 + \text{unsusceptible enantiomer}$$

i, D- or L-amino-acid oxidase, O_2.

SCHEME 31

has the advantage of needing no formation of derivatives, but the susceptible enantiomer is destroyed and some amino-acids are resistant to attack. L-Amino-acid oxidase is found in some snake venoms, and its D-counterpart is found in mammalian kidney. Enzymes, of course, may be used to form derivatives as well as to fragment substrates. In 1937 Bergmann and Fraenkel-Conrat found that papain catalyses peptide-bond formation, the first example being the formation of an anilide from *N*-benzyloxycarbonylglycine. This enzyme is specific for L-amino-acids, and formation of arylhydrazides from racemic *N*-acylamino-acids has been particularly studied. One problem has been the fact that some attack on the D-enantiomer does occur; however, study of a range of substituted arylhydrazides has now established that *o*-fluorophenylhydrazine gives hydrazides containing 99.9% of L-amino-acid, alanine being used as the test amino-acid.[50] (Phenylhydrazine by comparison gives only 88.2% L-phenylhydrazide.) This method is therefore now of practical use in amino-acid resolution; it has the additional merit that the derivative crystallizes out from the aqueous solution as the reaction proceeds.

Of the column chromatographic methods of resolution, g.l.c. methods have received the most attention. Amino-acid derivatives can be resolved either by using an optically active stationary phase or by forming diastereoisomeric derivatives. As far as the latter are concerned, *N*-trifluoroacetyl-L-prolyl-DL-amino-acid alkyl esters have been fully explored, *N*-trifluoroacetyl-L-proline having the the advantage of being resistant to racemization during formation of the dipeptide.[51] The most successful optically active stationary phases have been dipeptide derivatives such as cyclohexyl *N*-trifluoroacetyl-L-α-aminobutyryl-L-α-aminobutyrate,[52] although *N*-lauroyl-L-valine-t-butylamide[53] has produced comparable results. It appears that best results are obtained when three hydrogen bonds can be formed between the stationary phase and the amino-acid derivative (**25**), imposing a

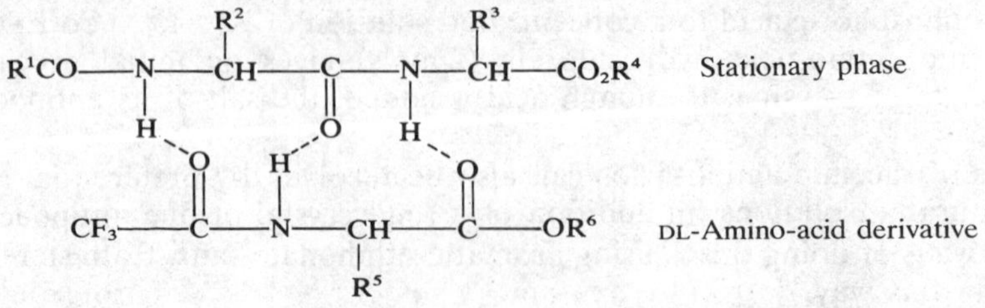

(25)

conformation on the 'solute', the handedness of which is determined by the configuration of the amino-acids in the stationary phase. The solutes cease to be mirror-images and become 'conformational' diasteroisomers.[54] Stationary phases containing *N*-acylproline will be in this case less efficient as it has no NH group with which to form hydrogen bonds.

G.l.c. methods, however, as might be expected, have only found use for analytical purposes. Low-pressure liquid chromatography on columns is much more suitable for preparative work, but as yet no system has been generally applied to amino-acid resolution. Optically active amino-acids in the form of their metal complexes have proved useful resolving agents, although only a few substrates have been fully evaluated. The coupling of L-proline to a copolymer of chloromethylated styrene and divinylbenzene followed by treatment with copper sulphate gives a column packing which will simply resolve DL-proline. If the racemate is applied in water, L-proline is eluted with this solvent, but D-proline only emerges after elution with 1 M ammonium hydroxide.[55] It may not, of course, be convenient to use an enantiomer of the amino-acid to be resolved, but this is not always a necessary condition of separation. When D-methionine DL-sulphoxide was used as the polymer-bound asymmetric agent, DL-isoleucine could be resolved in a similar way, the D-isomer eluting with water and the L-isomer with molar ammonia.[56]

The covalent attachment of a chiral cyclic polyether ionophore to a macroreticular divinylbenzene-crosslinked polystyrene (**26**) has also produced a useful asymmetric column packing for amino-acid resolution. This forms more stable complexes with the

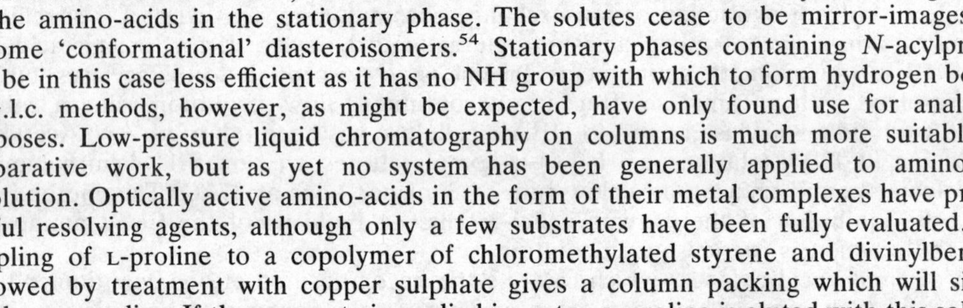

(26)

R-enantiomers of all amino-acids studied, with the exception of phenylalanine. A number of racemic amino-acids have successfully been resolved on this material as their perchlorate or hexafluorophosphate salts using chloroform–acetonitrile mixtures as solvent. The column, however, must be run at 0 °C as chiral recognition decreases with increase in temperature.[57] One example only has been reported of the use of affinity chromatography for amino-acid resolution. Tryptophan has been resolved using a column packed with agarose coupled to bovine serum albumin through a succinoylaminoethyl bridge. Using borate buffer (pH 2.0) containing 1% dimethyl sulphoxide, D-tryptophan is eluted; the L-antipode only appears after changing the solvent to 0.1 M acetic acid.[58]

23.2.5 ISOTOPICALLY LABELLED α-AMINO-ACIDS

23.2.5.1 Hydrogen isotopes

Tritium-labelled amino-acids have been very widely studied. Compared with other radioisotopes, tritium has the advantage of being relatively cheap and high specific activity can be incorporated. However, there may be less certainty about the position of the label than with isotopes of carbon or nitrogen, and in some molecular environments it can be relatively labile. Only soft β-radiation is emitted. Labelling with tritium has recently been comprehensively reviewed in a monograph,[59] and readers are referred to this excellent source for more extensive background information.

The labelling of amino acids by direct exchange reactions is finding less favour nowadays because of its random and often non-uniform nature and the tendency for concomitant racemization to occur. The original general procedure of Wilzbach, first published in 1956, involves simple contact between an organic compound and tritium gas for a period varying from days to weeks. The radiation of the tritium induces exchange between ^1H and ^3H, but relatively low specific activities are achieved and side-reactions frequently require the product to be extensively purified. Better results are often obtained by adding as catalysts those metals which function as hydrogenation catalysts. The nature of the metal and the particular conditions under which it is used can markedly alter the labelling pattern. For example, when phenylalanine is treated with Pt–^3H_2O at 135 °C, 73% of the label is found in the aromatic nucleus and 27% in the side-chain,[60] but use of PdO/$BaSO_4$–$^3H_2$ in glacial acetic acid or basic aqueous solution gives material with 92% of the label incorporated residing at the benzylic position.[61]

The aromatic rings of amino-acids such as phenylalanine, tyrosine, and tryptophan can be specifically labelled by catalytic halogen–tritium replacement in the presence of base. Tyrosine residues can be incorporated into peptides before iodination (3,5-substitution), and some natural peptides have been labelled by iodination of their tyrosine residues and subsequent exchange of iodine for tritium. An example of the latter is the synthesis of 23-[3,5-$^3H_2$–Tyr]-β-corticotropin-(1–24)-tetracosapeptide by iodination of a protected 11–24 fragment containing a free α-amino-group, coupling to a derivative of the 1–10 fragment, and finally incorporating tritium using a mixture of Pd/C–Rh/$CaCO_3$ as the catalyst, the calcium carbonate here acting as the required base.[62] It should be noted, however, that [3,5-$^3H_2$]tyrosine loses tritium by exchange in acid solution. Ring-tritiated phenylalanine is stable to 5% hydrochloric acid under reflux, but does undergo tritium loss on heating in >80% sulphuric acid.

One of the most useful ways of introducing tritium into free amino-acids is by the reduction in the presence of tritium of unsaturated precursors. If the dehydroamino-acid is α,β-unsaturated, as in the azlactone synthesis of amino-acids (Scheme 9), [2,3-$^3H_2$]amino-acids are generated. In general, however, more stable radiolabelled amino-acids are produced by the reduction of the alkene linkages in the side-chain remote from the α-asymmetric centre. For example, [4,5-$^3H_2$]leucine may be prepared in this way by the acetamidomalonate route from 1-chloro-2-methylprop-2-ene or [3,4-$^3H_2$]proline from 3,4-dehydroproline. Because of the auto-decomposition of tritiated compounds on storage, it is often useful to prepare a suitable unlabelled intermediate which can be hydrogenated with tritium immediately before use, especially where peptides are concerned. An example of this approach is the use of t-butoxycarbonyl-L-tryptophyl-L-4-dehydronorleucyl-L-aspartyl-L-phenylalanine amide as a precursor for the gastrin analogue BOC-Trp-NorLeu-Asp-Phe-NH_2.[63]

Specific labelling at the α-position of amino-acids may be accomplished by the decarboxylation of α-acetaminomalonic acid derivatives (Scheme 7) in an acid solution of a tritiated solvent. Alternatively, amino-acids may be labelled at this position directly under conditions which cause racemization at the α-C atom, *e.g.* strong alkali or on refluxing with acetic anhydride in acetic acid. However, the use of [α-or β-^{3}H]-labelled amino-acids is better avoided when many biological investigations are being carried out. Tritium exchange at these positions can occur through transamination reactions (Scheme 32); the loss of tritium from the β-position of amino-acids is in fact used as a method for

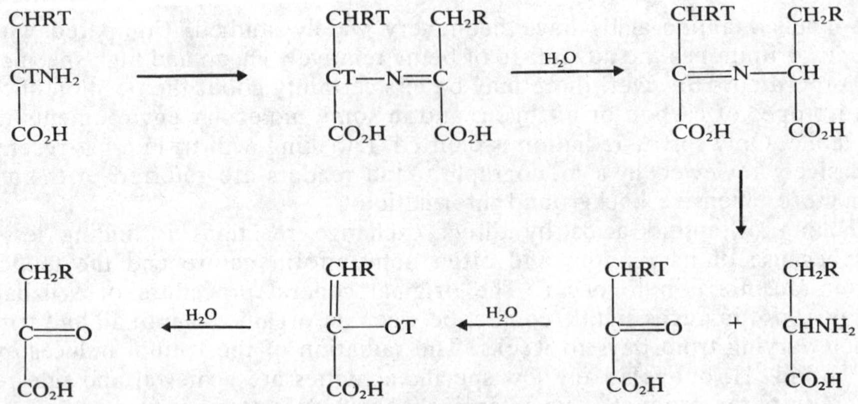

T = tritium (^{3}H)
SCHEME 32

the assay of transaminases. Treatment of α,β-tritiated α-amino-acids with amino-acid oxidases or renal acylase may result in considerable loss of activity, so caution must be exercised when using enzymes for resolving racemic radiolabelled amino-acids.

Stereoselective labelling of the β-methylene groups of aromatic amino-acids has been developed in connection with studies of the mechanism of biochemical reactions affecting the β-centre. The azlactone synthesis (Scheme 9) affords (*Z*)-oxazolinones from [α-^{3}H]aromatic aldehydes, and, after hydrolysis to the corresponding (*Z*)-α-benzoylamino-cinnamic acids, these compounds undergo *cis* addition on catalytic hydrogenation to yield a mixture of (βR)-L- and (βS)-D-[β-^{3}H]amino-acids (Scheme 33). These can be racemized at the α-centre by 10 M hydrochloric acid at 180 °C without loss of tritium, and resolution of the *N*-chloroacetyl derivates with carboxypeptidase [only the α-(*S*)-isomers are attacked] enables all four stereoisomers to be obtained.[64] The generation of side-chain

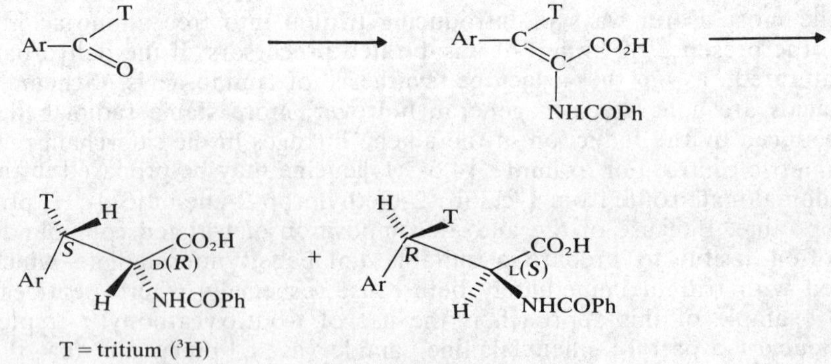

SCHEME 33

functional groups from precursors can be used to introduce tritium at these sites, often at a conveniently late stage in the synthesis. For example, [6-^{3}H]lysine and [5-^{3}H]ornithine can be prepared by sodium borotritide reduction of δ-cyanonorvaline and γ-cyano-α-aminobutyric acid, respectively.[65]

Deuterium-labelling methods are very similar to those used for tritium except that as it is a stable isotope, radiation-induced exchange cannot be used for its introduction. Its limitations parallel those of tritium, but its lack of radioactivity simplifies experimental work and short synthetic procedures are rather less crucial.

23.2.5.2 Carbon and nitrogen isotopes

Like tritium, as a long-lived radioisotope carbon-14 has proved a most versatile radiolabel. The primary source of this radioisotope is $^{14}CO_2$, and the bulk of the shorter chemical syntheses of ^{14}C-labelled amino-acids generate products with isotopic substitution in either the functional groups or in positions close to them. Routes to the [1-^{14}C]- and [2-^{14}C]-labelled alkyl halides, carboxylic acids, α-keto-acids, and aldehydes required for the most widely-used amino-acid syntheses are outlined in Scheme 34. Routes to [2-^{14}C]glycine and ethyl [2-^{14}C]acetamidocyanoacetate are depicted in Scheme 35. The

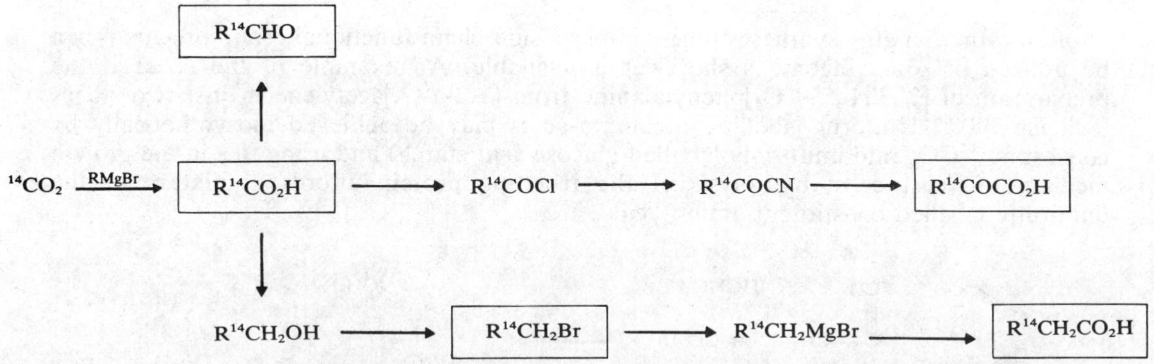

Intermediates for amino-acid syntheses are boxed.

SCHEME 34

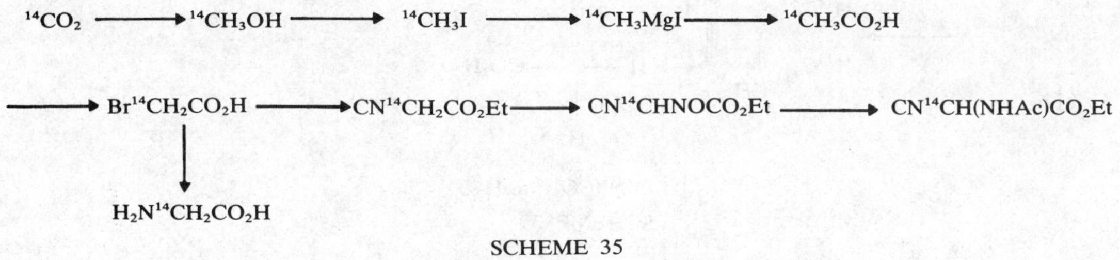

SCHEME 35

shortest popular route to [1-^{14}C]-labelled amino-acids is the Strecker synthesis (Scheme 14) using K^{14}CN. In some cases α-keto-acids may be preferable to aldehydes for this purpose (Scheme 36).[66] The Strecker synthesis may also be used to make [2-^{14}C]amino-acids from $^{14}CO_2$ in only five steps in one reaction vessel without isolation of intermediates (Scheme 37).[67] Isotopic substitution at the 3-position is often best achieved by the use of [1-^{14}C]-labelled aldehydes in the azlactone synthesis (Scheme 9). Labelling at γ or higher positions in the side-chain requires introduction of the isotope at an early stage

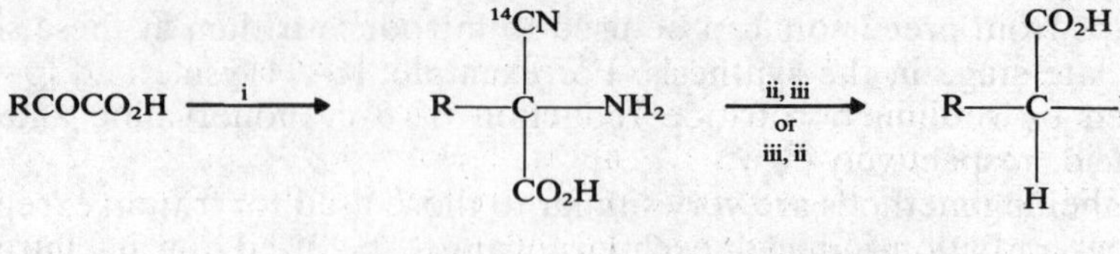

i, $K^{14}CN$, NH_4Cl; ii, decarboxylate; iii, H_3O^+.

SCHEME 36

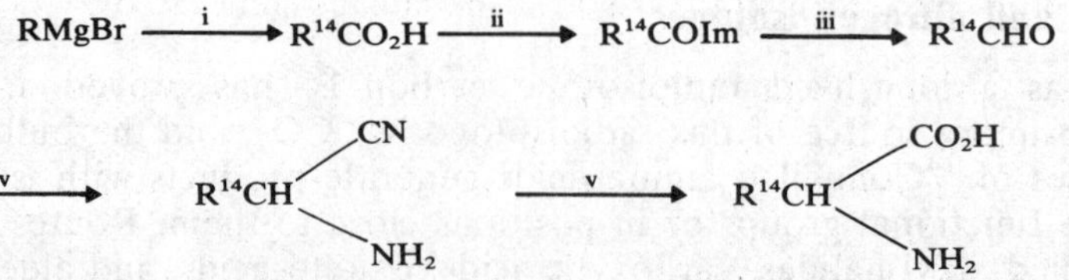

i, $^{14}CO_2$; ii, carbonylbis-imidazole; iii, $LiAlH_4$; iv, NH_4Cl, KCN; v, H_3O^+.

SCHEME 37

of often rather lengthy syntheses unless either a side-chain functional group precursor can be utilized or some ingenious short-cut is available. An example of the latter is the preparation of [2′,3′,4′,5′-$^{14}C_4$]phenylalanine from [1,2-$^{14}C_2$]acetylene in just two stages (Scheme 38).[68] Uniform labelling of amino-acids may be achieved biosynthetically by converting $^{14}CO_2$ into uniformly labelled glucose (*via* starch) and using this in the growth medium of yeast cells; hydrolysis of the resulting protein affords a mixture of the uniformly-labelled constituent amino-acids.

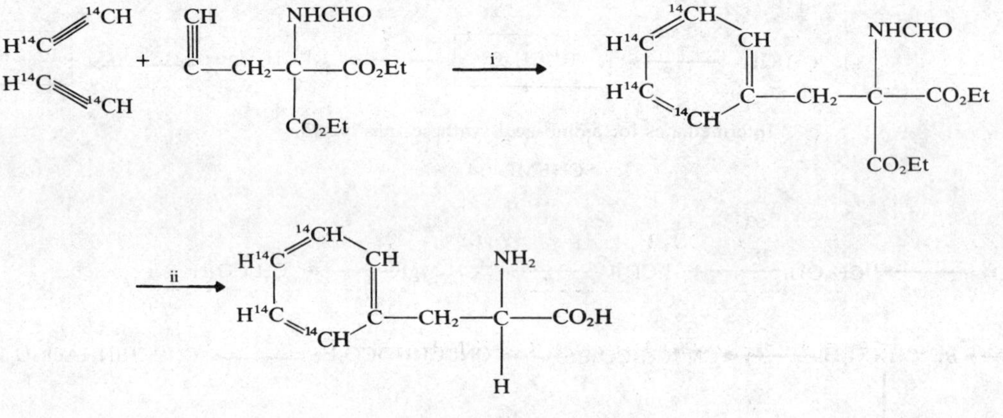

i, $(Ph_3P)_2Ni(CO)_4$; ii, H_3O^+.

SCHEME 38

Carbon-13 is not radioactive, but unlike ^{12}C and ^{14}C its nuclear properties are such that it may be detected by n.m.r. spectroscopy. Modern Fourier transform techniques enable ^{13}C chemical shifts to be obtained from samples without isotopic enrichment; the natural abundance (1.1%) suffices with the sensitive instruments now available.[69] However, boosting the labelling reduces the sample size required (and may therefore particularly aid biological work) and if carbon–carbon coupling constants are required, highly enriched compounds are necessary. Uniformly ^{13}C-labelled amino-acids have been prepared biosynthetically using the alga *Spirilina maxima* grown in the presence of $NaH^{13}CO_3$. The algal protein was hydrolysed, initially by the enzyme pronase and then by 2*M* sulphuric acid, and the resultant 85% ^{13}C-enriched amino-acid mixture fractionated on an ion-exchange column.[70] Specifically ^{13}C-labelled amino-acids have also been synthesized as tools to solve mechanistic problems. A notable example is the preparation of valine containing a $^{13}CH_3$ group in place of one of the methyl groups of its isopropyl side-chain to enable an n.m.r. study to be made of the incorporation of valine into penicillin. This problem has elicited such interest that three groups have independently developed syntheses of (2*S*)-[4-^{13}C]valine. These syntheses are outlined in Schemes 39–41,[71–73] and illustrate the number of steps required to specifically substitute a carbon atom in an alkyl side-chain remote from a functional group. Two of the products are racemic at the *α*-*C* atom, but in the other case this centre is stereospecifically generated by using the enzyme *β*-methylaspartase.

Carbon-11 has a short half-life (20.3 min) but can be detected *in vivo* whereas ^{14}C cannot. One example of its use is in the preparation of [1-^{11}C]-DL-*α*-phenylalanine by the carboxylation of an isocyanide (Scheme 42). The product was obtained in good yield within 40 minutes of introducing $^{11}CO_2$ into the reaction sequence.[74] The longest lived radioisotope of nitrogen is nitrogen-13 (half-life 10 min). Amino-acids containing this label have been synthesized from $^{13}NH_3$ using enzymes immobilized on porous derivatized

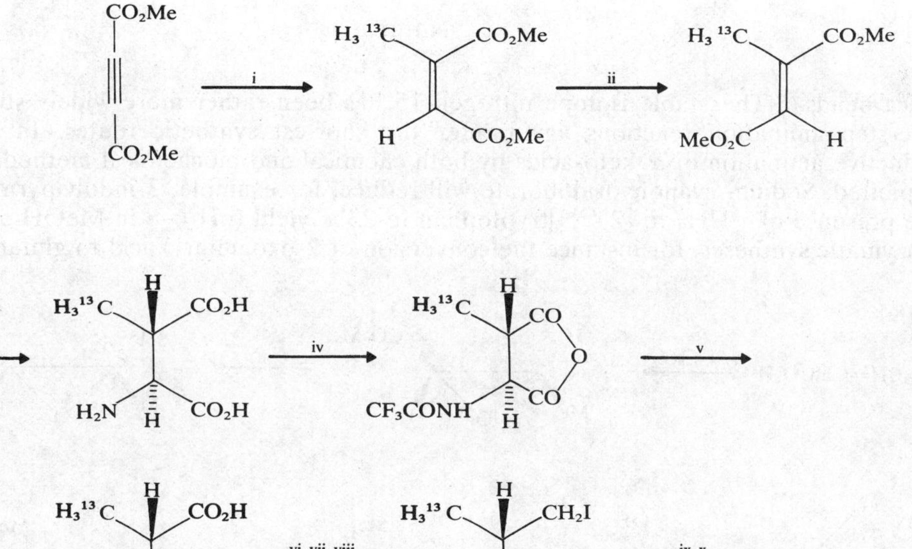

i, ($^{13}CH_3$)$_2$Cu; ii, *hν*, Br_2 (trace); iii, *β*-methylaspartase; iv, $(CF_3CO)_2O$; v, MeOH; vi, B_2H_6 at 0 °C; vii, $MeSO_2Cl$; viii, NaI; ix, Pd/C, H_2; x, H_3O^+.

SCHEME 39

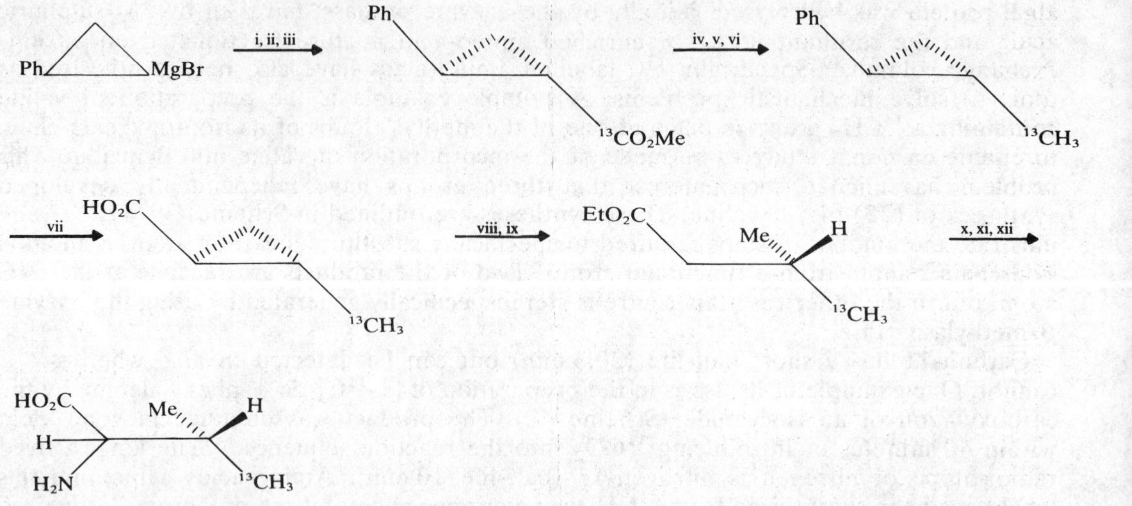

i, $^{13}CO_2$; ii, CH_2N_2; iii, $KOBu^t$; iv, $LiAlH_4$; v, $MeSO_2Cl$; vi, $LiAlH_4$; vii, O_3; viii, $MeCHN_2$; ix, Na, liquid NH_3; x, MeOH, KOH; xi, Br_2, PCl_3; xii, NH_3.

SCHEME 40

silica beads.[75] The stable isotope nitrogen-15 has been rather more widely studied, and one-step amination reactions again offer the shortest synthetic routes. In particular, reductive amination of α-keto-acids by both chemical and biochemical methods has been exploited. Sodium cyanohydridoborate will reduce, for example, 3-indolylpyruvic acid in the presence of $^{15}NH_3$ to [2-^{15}N]tryptophan in 23% yield (pH 6–8 in MeOH at 25 °C).[76] Enzymatic syntheses, for instance the conversion of 2-oxoglutaric acid to glutamic acid in

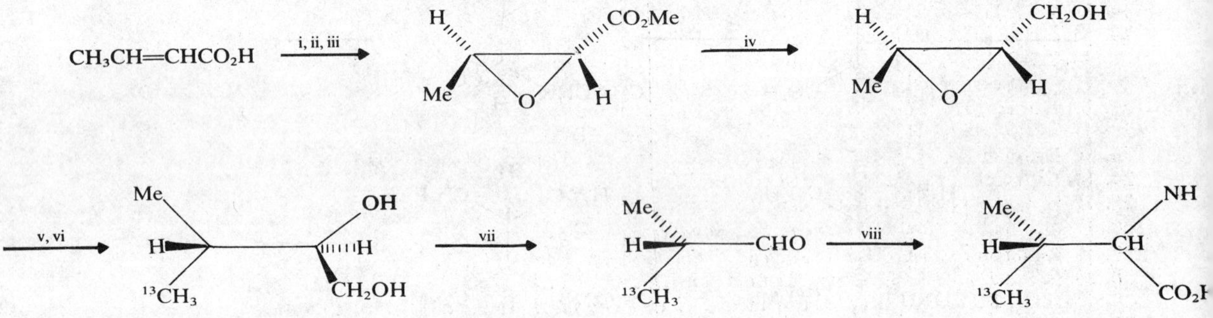

i, H_2O_2, Na tungstate; ii, resolve with brucine; iii, CH_2N_2; iv, $NaBH_4$; v, MeLi; vi, $^{13}CH_3Li$; vii, HIO_4; viii, Strecker synthesis.

SCHEME 41

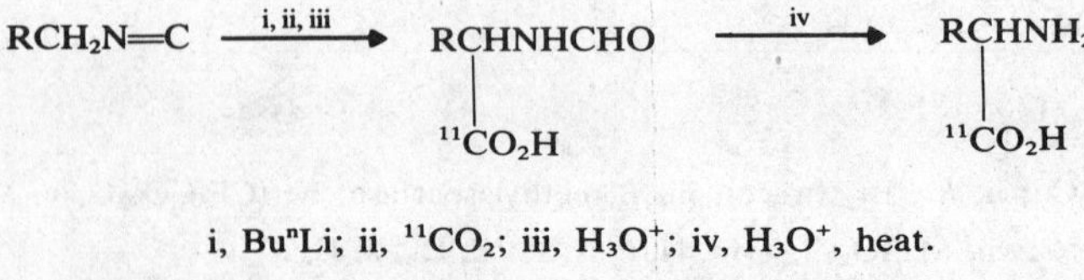

i, Bu^nLi; ii, $^{11}CO_2$; iii, H_3O^+; iv, H_3O^+, heat.

SCHEME 42

the presence of $^{15}NH_4Cl$ and reduced nicotinamide adenine dinucleotide phosphate, have the advantage of yielding the L-amino-acid rather than the racemate. In this latter reaction NADPH can be continuously regenerated by coupling the first reaction to the oxidation of glucose 6-phosphate, which serves to drive the synthesis of L-glutamic acid to completion.[77] Direct amination of α-halocarboxylic acids is less attractive as it requires a large excess of ammonia, which is wasteful of the nitrogen isotope. [^{15}N]Phthalimide has found some use in amino-acid synthesis, *e.g.* in the preparation of [2-^{15}N]lysine.[78]

References

1. J. F. Thompson, C. J. Morris and I. K. Smith, 'New Naturally Occurring Amino-acids', in *Ann. Rev. Biochem.*, 1969, **38,** 137.
2. 'Amino-acids, Peptides, and Proteins', Specialist Periodical Reports, The Chemical Society London, 1969 and succeeding years.
3. J. P. Greenstein and M. Winitz, 'Chemistry of The Amino-Acids', Wiley, New York, 1961, vols. 1–3.
4. L. Fowden and M. H. Richmond, *Biochim. Biophys. Acta*, 1963, **71,** 459. Analogue incorporation has been reviewed by M. H. Richmond, *Bacteriol. Rev.*, 1962, **26,** 398.
5. See, for example, H. B. Vickery, 'A Review of Amino-acids described since 1931 as Components of Native Proteins', in *Adv. Protein Chem.*, 1972, **26,** 81.
6. J. T. Dunn, *J. Biol. Chem.*, 1970, **245,** 5954.
7. P. A. Finot, R. Viani, J. Bricout, and J. Mauron, *Experientia*, 1969, **25,** 134; K. Heyns, J. Heukeshoven, and K.-H. Brose, *Angew. Chem. Internat. Edn.*, 1968, **7,** 628.
8. P. M. Hardy, G. J. Hughes, and H. N. Rydon, *J.C.S. Chem. Comm.*, 1976, 157.
9. F. W. Keeley and F. S. Labella, *Biochim. Biophys. Acta*, 1972, **263,** 52.
10. K. Ziegler, I. Melchert, and C. Lürken, *Nature*, 1967, **214,** 404.
11. T.-Y. Liu and Y. H. Chang, *J. Biol. Chem.*, 1971, **246,** 2842; tryptophan has been reviewed by A. Fontana and C. Toniolo, in 'Progress in the Organic Chemistry of Natural Products', Springer-Verlag, New York, 1976, vol. 33, p. 309.
12. R. Haselkorn and L. B. Rothman-Denes, *Ann. Rev. Biochem.*, 1971, **40,** 411.
13. R. L. Heinrikson and H. S. Kingdon, *J. Biol. Chem.*, 1971, **246,** 1099.
14. C. J. Lote and J. B. Weiss, *FEBS Letters*, 1971, **16,** 81.
15. H. R. Morris, A. Dell, T. E. Petersen, L. Sottrup-Jensen, and S. Magnusson, *Biochem. J.*, 1976, **153,** 663.
16. P. A. Price, A. S. Otsuka, J. W. Poser, J. Kristaponis, and N. Raman, *Proc. Nat. Acad. Sci. U.S.A.*, 1976, **73,** 1447.
17. P. M. Steinert, H. W. J. Harding, and G. E. Rogers, *Biochim. Biophys. Acta*, 1969, **175,** 1.
18. H. W. J. Harding and G. E. Rogers, *Biochim. Biophys. Acta*, 1972, **257,** 37.
19. M. L. Tanzer, T. Housley, L. Berube, R. Fairweather, C. Franzblau, and P. M. Gallop, *J. Biol. Chem.*, 1973, **248,** 393. See also the review by W. Traub and K. Piez, *Adv. Protein Chem.*, 1971, **25,** 243.
20. H. P. J. Bennett, D. F. Elliott, B. E. Evans, P. J. Lowry, and C. McMartin, *Biochem. J.*, 1972, **129,** 695.
21. M. Horiike, J. Oda, Y. Inouye, and M. Ohno, *Agric. Biol. Chem. (Japan)*, 1969, **33,** 292.
22. J.-F. Nicoud and H. B. Kagan, *Tetrahedron Letters*, 1971, 2065.
23. A. Badshah, N. H. Khan, and A. R. Kidwai, *J. Org. Chem.*, 1972, **37,** 2916.
24. M. Crawford and W. T. Little, *J. Chem. Soc.*, 1959, 729.
25. H. L. Finkbeiner, *J. Amer. Chem. Soc.*, 1964, **86,** 961.
26. E. Kaji and S. Zen, *Bull. Chem. Soc. Japan*, 1973, **46,** 337.
27. K. Rühlmann and G. Kuhrt, *Angew. Chem. Internat. Edn.*, 1968, **7,** 809.
28. J.-C. Fiaud and H. B. Kagan, *Tetrahedron Letters*, 1970, 1813.
29. K. Hiroi, K. Achiwa, and S.-I. Yamada, *Chem. Pharm. Bull. (Japan)*, 1968, **16,** 444.
30. K. Matsumoto, M. Suzuki, and M. Miyoshi, *J. Org. Chem.*, 1973, **38,** 2094.
31. S.-I. Yamada, T. Oguri, and T. Shioiri, *J.C.S. Chem. Comm.*, 1972, 623.
32. M. W. Rathke and A. Lindert, *J. Amer. Chem. Soc.*, 1971, **93,** 2318.
33. K. Ninomiya, T. Shioiri, and S.-I. Yamada, *Chem. Pharm. Bull. (Japan)*, 1974, **22,** 1398.
34. K. Ogura and G.-I. Tsuchihashi, *J. Amer. Chem. Soc.*, 1974, **96,** 1960.
35. W. H. Graham, *Tetrahedron Letters*, 1969, 2223.
36. H. Wakamatsu, J. Uda, and N. Yamakami, *Chem. Comm.*, 1971, 1540.
37. K. Harada and K. Matsumoto, *J. Org. Chem.*, 1969, **33,** 4467; 1968, **32,** 1794; K. Harada and T. Yoshida, *Bull. Chem. Soc. Japan*, 1970, **43,** 921; *Chem. Comm.*, 1970, 1071.
38. E. J. Corey, H. S. Sachdev, J. Z. Gougoutas, and W. Saenger, *J. Amer. Chem. Soc.*, 1970, **92,** 2488.
39. K. Harada and T. Okawara, *J. Org. Chem.*, 1973, **38,** 707; K. Harada, T. Okawara, and K. Matsumoto, *Bull. Chem. Soc. Japan*, 1973, **46,** 1865.
40. J. Y. Chenard, D. Commereuc, and Y. Chauvin, *J.C.S. Chem. Comm.*, 1972, 750.
41. W. S. Knowles, M. J. Sabacky, B. D. Vineyard, and D. J. Weinkauff, *J. Amer. Chem. Soc.*, 1975, **97,** 2567.
42. B. W. Bycroft and G. R. Lee, *J.C.S. Chem. Comm.*, 1975, 988.
43. H. Poisel and U. Schmidt, *Chem. Ber.*, 1973, **106,** 3408.

44. R. C. Job and T. C. Bruice, *J. Amer. Chem. Soc.*, 1974, **96,** 809.
45. J. Thanassi and J. S. Fruton, *Biochemistry*, 1962, **1,** 975.
46. K. Oki, K. Suzuki, S. Tuchida, T. Saito, and H. Kotake, *Bull. Chem. Soc. Japan*, 1970, **43,** 2554.
47. R. D. Gillard, P. R. Mitchell, and H. L. Roberts, *Nature*, 1968, **217,** 949.
48. S. Yamada, M. Yamamoto, and I. Chibata, *J. Org. Chem.*, 1973, **38,** 4408.
49. T. Barth and H. Maskova, *Coll. Czech. Chem. Comm.*, 1971, **36,** 2398.
50. J. L. Abernethy, E. Albano, and J. Comyns, *J. Org. Chem.*, 1971, **36,** 1580.
51. B. Halpern and J. W. Westley, *Tetrahedron Letters*, 1966, 2283; H. Iwase, *Chem. Pharm. Bull.* (*Japan*), 1974, **22,** 2075.
52. W. Parr and P. Y. Howard, *Angew. Chem. Internat. Edn.*, 1972, **11,** 529.
53. B. Feibush, *Chem. Comm.*, 1971, 544.
54. B. Feibush, *Tetrahedron*, 1970, **26,** 1361.
55. S. V. Rogozhin and V. A. Davankov, *Chem. Comm.*, 1971, 490.
56. S. V. Rogozhin, V. A. Davankov, I. A. Yamskov, and V. P. Kabanov, *J. Gen. Chem., U.S.S.R.*, 1972, **42,** 1605.
57. G. D. Y. Sogah and D. J. Cram, *J. Amer. Chem. Soc.*, 1976, **98,** 3038.
58. K. K. Stewart and R. F. Doherty, *Proc. Nat. Acad. Sci. U.S.A.*, 1973, **70,** 2850.
59. E. A. Evans, 'Tritium and its compounds', Wiley, New York, 2nd edn., 1974.
60. M. C. Clifford, E. A. Evans, A. E. Kilner, and D. C. Warrell, *J. Labelled Compounds*, 1975, **11,** 435.
61. E. A. Evans, H. C. Sheppard, J. C. Turner, and D. C. Warrell, *J. Labelled Compounds*, 1974, **10,** 569.
62. D. E. Brundish and R. Wade, *J.C.S. Perkin I*, 1973, 2875.
63. C. S. Pande, J. Rudick, L. Ornstein, I. L. Schwartz, and R. Walter, *Mol. Pharmacol.*, 1969, **5,** 227.
64. G. W. Kirby and M. J. Varley, *J.C.S. Chem. Comm.*, 1974, 833; G. W. Kirby and J. Michael, *J.C.S. Perkin I*, 1973, 115 and refs. therein.
65. I. Mezo, M. Havranek, I. Teplan, J. Benes, and B. Tanacs, *Acta Chim. Acad. Sci. Hung.*, 1975, **85,** 201.
66. I. Teplan, I. Mezo, L. Bursics, and J. Marton, *Acta Chim. Acad. Sci. Hung.*, 1969, **60,** 301.
67. L. Pichat, P. N. Liem, and J.-P. Guermont, *Bull. Soc. chim. France*, 1971, 837.
68. L. Pichat, P. N. Liem, and J.-P. Guermont, *Bull. Soc. chim. France*, 1972, 4224.
69. J. B. Stothers, 'Carbon-13 n.m.r. spectroscopy', Academic Press, New York, 1972.
70. S. Fermandjian, S. Tran-Dinh, J. Savrda, E. Sala, R. Mermet-Bouvier, E. Bricas, and P. Fromageot, *Biochim. Biophys. Acta*, 1975, **399,** 313.
71. H. Kelunder, C. H. Bradley, C. J. Sih, P. Fawcett, and E. P. Abraham, *J. Amer. Chem. Soc.*, 1973, **95,** 6149.
72. D. J. Aberhart and L. J. Lin, *J.C.S. Perkin I*, 1974, 2320.
73. N. Neuss, C. H. Nash, J. E. Baldwin, P. A. Lemke, and J. B. Grutzner, *J. Amer. Chem. Soc.*, 1973, **95,** 3797.
74. W. Vaalburg, H. D. Beerling-van der Molen, and M. G. Woldring, *Nuclear Medicine*, 1975, **14,** 60.
75. M. B. Cohen, L. Spolter, C. C. Chang, N. S. MacDonald, J. Takahashi, and D. D. Bobinet, *J. Nuclear Medicine*, 1974, **15,** 1192.
76. R. F. Borch, M. D. Bernstein, and H. D. Durst, *J. Amer. Chem. Soc.*, 1971, **93,** 2877.
77. W. Greenaway and F. R. Whatley, *J. Labelled Compounds*, 1975, **11,** 395.
78. J. Mizon and Ch. Mizon, *J. Labelled Compounds*, 1974, **10,** 229.

23.3
Peptides and the Primary Structure of Proteins

D. T. ELMORE
The Queen's University of Belfast

23.3.1 GENERAL STRATEGY FOR DETERMINING THE PRIMARY STRUCTURE OF PROTEINS

In general, proteins are composed of the 20 amino-acids in Table 1 of Chapter 23.2. These are the only amino-acids which can be coded for by triads of bases in messenger ribonucleic acid (mRNA). When other amino-acids are present in proteins (*e.g.* cystine, hydroxyproline), they are formed by post-translational modifications (*e.g.* oxidation of two cysteine residues, hydroxylation of proline).

The general strategy for determining the primary structure of a protein comprises several steps. It is necessary (a) to carry out a quantitative analysis of a total hydrolysate in order to determine the relative molar proportions of amino-acids present (Section 23.3.2); (b) to determine the molecular weight by suitable physical methods in order to calculate the number of residues of each amino-acid present;[1–3] (c) to determine the number of polypeptide chains present in the molecule by chromatographic or electrophoretic separation or by quantitative analyses of amino-terminal (*N*-terminal) and carboxyl-terminal (*C*-terminal) residues (Section 23.3.4); (d) to separate the chains if more than one is present in the molecule, bearing in mind that chains may be linked covalently through a disulphide bond (Section 23.3.3) or non-covalently to form a quaternary structure (Section 23.3.7); (e) to cleave each chain, preferably by specific methods (Sections 23.3.5 and 23.3.6), into fragments of convenient size for sequencing by, for example, the Edman method of stepwise degradation (Section 23.3.4). It should be noted that under step (c) the *N*-terminal residue(s) may be acylated and the *C*-terminal residue(s) may be present as amide. End-group analysis (Section 23.3.4) will yield negative results in these cases. An additional problem arises with those proteins which contain covalently bound groups (prosthetic groups) which are non-peptide in structure. If present, the position and mode of attachment of a prosthetic group must be determined.

Determination of the structure of insulin, the first protein to be sequenced, was commenced by Sanger and his colleagues before the Edman method of stepwise degradation had been published. Sanger relied heavily on the use of partial acid hydrolysis to cleave the two chains of insulin into fragments that were small enough to be characterized by techniques then available. Although the two chains contain only 21 and 30 amino-acids respectively, many fragments were formed and the amount of experimental work involved in their separation, identification, and sequential location occupied Sanger for several years and earned him a Nobel prize. With modern techniques the same project would today occupy at most a few weeks. Current strategy aims to cleave a protein into a small number of relatively large fragments. Cleavage may be achieved chemically (Section 23.3.5) or enzymatically (Section 23.3.6). It is customary to use more than one method of cleavage so that different fragments may be obtained whose sequences overlap. After separation of the small number of fragments isolated, the latter may be further cleaved by an appropriate method and the smaller fragments separated before sequencing by the Edman method. If the initial fragments are of a suitable size, sequencing can be carried out without the need to effect additional cleavage reactions. This strategy succeeds because (a) several powerful methods for separating peptides are available and (b) stepwise degradation of a polypeptide can be carried on for about 40 stages by improvements to the Edman method.

The importance of the foregoing approach to the determination of the primary structure of proteins can best be illustrated by some hypothetical examples. Suppose that a protein can be subjected to limited cleavage to give three fragments (A, B, C):

Gly—Lys	Ala—Arg	Ser—Asp
A	B	C

As soon as the *N*- and *C*-terminal residues of the original protein and of the three fragments have been identified, the order of the fragments in the original molecule is known provided that the terminal residues differ. If the original protein giving rise to the fragments A, B, and C has *N*-terminal Gly and *C*-terminal Asp, it must have the partial structure:

Gly—Lys-Ala—Arg-Ser—Asp
←A→ ←B→ ←C→

The problem would be only slightly more complicated if two of the fragments had the same *N*-terminal or *C*-terminal residue as the original protein. Determination of a partial sequence of perhaps two or three residues at the *N*- or *C*-terminus of the original protein and comparison with the corresponding sequences of the fragments would serve to identify the fragments originating from the end(s) of the protein. If a protein gives rise to more than three fragments, however, the two outside fragments could be identified, but the order of the inner fragments would be unknown. In this more usual case, it is necessary to produce another set of fragments by an alternative method of cleavage in order to identify overlaps. Let us consider a hypothetical case of a peptide which has *N*-terminal Val and *C*-terminal Leu and which produces by one method of cleavage four fragments E, F, G, and H whose amino-acid sequences are determined to be as follows:

Val-Gly-Ser-Lys	E
Ala-Ser-Phe-Gly-Lys	F
Asp-Gln-Tyr-Ala	G
Tyr-Gly-Leu	H

Since only E has *N*-terminal Val and only H has *C*-terminal Leu, it is clear that these peptides originate from the ends of the original peptide. The order in which F and G occur in the original peptide is indeterminate at this stage and so we could write a partial structure for the peptide as follows:

E(F, G)H

where the parentheses and comma indicate that the order of the enclosed fragments is unknown. Suppose that the original peptide can be selectively cleaved by an alternative procedure to give four different fragments W, X, Y, Z whose sequences are determined to be as follows:

Val-Gly-Ser-Lys-Ala-Ser-Phe	W
Gly-Lys-Asp-Gln-Tyr	X
Ala-Arg-Tyr	Y
Gly-Leu	Z

The sequence -Lys-Ala- in fragment W indicates that F follows E. Similarly, the sequence -Lys-Asp- in fragment Y indicates that G follows F. Hence, the original peptide has the sequence:

Val-Gly-Ser-Lys-Ala-Ser-Phe-Gly-Lys-Asp-Glu-Tyr-Ala-Arg-Tyr-Gly-Leu

This simple example demonstrates the value of obtaining overlapping sequences using alternative methods of cleavage.

As indicated above, post-translational modifications may occur to some amino-acid residues. In some cases this causes no difficulties; indeed, if only one or two residues are modified, their position(s) in the neighbouring sequences may be more readily identified. In other cases, post-translational modifications may cause complications in the determination of the primary structure of proteins. The most common case is the oxidation of two cysteine residues to give a cystine residue containing an intramolecular or intermolecular disulphide bond. Methods for dealing with this problem are given in Section 23.3.8. Another interesting case concerns the vitamin K-dependent carboxylation of glutamic acid residues in several of the blood-clotting factors. The additional carboxyl group is situated on the γ-carbon atom. Consequently, acidic hydrolysis effects decarboxylation and it is not surprising that γ-carboxyglutamic acid was not discovered in prothrombin until 1973. In glycoproteins, carbohydrate residues are usually attached to the side chains of asparagine or serine and their presence may interfere with the cleavage by proteinases during the determination of primary structure. In addition, the number of hexose units attached at a particular site may vary from molecule to molecule, giving rise to multiple fragments containing the same sequence of amino-acids.

23.3.2 AMINO-ACID ANALYSIS OF PROTEINS

Analytical data for some amino-acids can be obtained on proteins without need for hydrolysis. For example, the tryptophan content can be determined by magnetic circular dichroism measurements,[4] by spectrofluorimetric measurements on the reduced protein in the presence of sodium dodecyl sulphate,[5] or spectrophotometrically in 6M guanidine hydrochloride.[6] Since tryptophan and some other amino-acids are destroyed during hydrolysis of proteins with HCl, such techniques appear attractive. Improved techniques for hydrolysing proteins with minimal destruction of amino-acids have recently been described, however, and it seems likely that the rapid, mechanized, and sensitive methods available for determination of all the amino-acids present in proteins will ensure that these have the widest use.

Treatment of proteins with 6M HCl *in vacuo* at 110 °C effects hydrolysis of peptide bonds, but simultaneously destroys tryptophan, hydrolyses asparagine and glutamine to aspartic acid and glutamic acid respectively, and partially destroys serine, threonine, and cyst(e)ine. Peptide bonds between amino-acids with bulky side chains such as Ile and Val are rather resistant to hydrolysis. It is common, therefore, to hydrolyse samples of protein for 1, 2, and 3 days and to extrapolate the values for amino-acids such as Ser and Thr to zero time and for Ile and Val to infinite time. In the case of cyst(e)ine, it is preferable either to oxidize it to cysteic acid or convert it into either *S*-carboxymethylcysteine or *S*-4-pyridylethylcysteine before hydrolysis (Section 23.3.3), since all these derivatives are stable. Commonly, black humin may be formed, especially if the protein contains carbohydrate. After hydrolysis, hydrochloric acid must be removed to avoid interference with the subsequent separation of amino-acids.

Hydrolysis of proteins with 3M toluene-*p*-sulphonic acid or 4M methanesulphonic acid[7,8] containing 0.2% tryptamine *in vacuo* at 110 °C for up to 3 days gives good recovery of amino-acids including tryptophan, although carbohydrate may interfere. Tryptophan can also be determined after alkaline hydrolysis of proteins, but arginine, cyst(e)ine, serine, and threonine are all destroyed. The total amide content due to asparagine and glutamine can be determined by hydrolysis with 10M HCl at 37 °C for 10 days followed by assay for ammonia by the microdiffusion technique. Separate determination of asparagine and glutamine can be accomplished by prior esterification of free carboxyl groups by treatment with methanol–acetic anhydride followed by reduction of the ester groups with lithium borohydride. After acid hydrolysis, aspartic and glutamic acids are determined in the form of γ-hydroxy-α-aminobutyric acid and δ-hydroxy-α-aminovaleric acid, respectively; the contents of asparagine and glutamine are obtained by subtraction of these values from the contents of aspartic and glutamic acids obtained by total hydrolysis of the unmodified protein. Complete enzymic hydrolysis of proteins without destruction of amino-acids can be achieved using a mixture of conjugates of 'Sepharose' with trypsin, chymotrypsin, prolidase, and aminopeptidase M.[9]

When hydrolysis of a protein has been carried out, the mixture of amino-acids must be separated and quantitatively determined. Gas–liquid chromatography has the attraction of speed and high sensitivity, especially when coupled to mass spectrometry.[10] It is necessary, of course, to convert free amino-acids into more volatile derivatives for gas–liquid chromatography, and therein lies the rub. Most published methods involve two reactions, esterification of the carboxyl group and acylation of the amino-group. It is necessary that both reactions shall proceed as nearly quantitatively as possible and that the derivatives shall be separable. The publication of several hundred papers over the space of 20 years is indicative of the difficulties encountered. The carboxyl group has usually been converted into simple alkyl esters ranging from methyl to pentyl, while *N*-trifluoroacetyl or *N*-heptafluorobutyryl groups have been popular for blocking the amino- or imino-group because they can be sensitively detected by the electron capture detector in g.l.c. Difficulty in acylating the guanidino-group in arginine and the thermal instability of cysteine derivatives owing to β-elimination reactions are examples of the problems involved. The technique with appropriate references has been briefly reviewed.[11]

The most widely used method for separating the amino-acids in a protein hydrolysate is ion-exchange chromatography. Sulphonated polystyrene resin crosslinked with divinylbenzene is the preferred ion-exchange medium. At low pH values, amino-acids exist as cations in solution:

$$\overset{+}{N}H_3CHRCOO^- \underset{}{\overset{H^+}{\rightleftharpoons}} \overset{+}{N}H_3CHRCOOH$$

In this form they bind to the anionic groups of the sulphonated resin. Elution can be achieved either by raising the pH and displacing the above equilibrium from right to left or by increasing the ionic strength so that the added cations compete with amino acids for binding sites on the resin. Aspartic acid, glutamic acid, and cysteic acid, which is formed if cyst(e)ine residues have been oxidized (Section 23.3.3), are most easily eluted since they are dibasic acids. Conversely, lysine and arginine are eluted with greatest difficulty, since they each have an extra protonated group in the side-chain. Between these extremes, retention of amino-acids depends heavily on hydrophobic interactions between the side chains of the amino-acids and the aromatic structure of the ion-exchange resin. Not surprisingly, the aromatic amino-acids are rather strongly bound by hydrophobic forces and emerge just before lysine and arginine. On the other hand, the presence of a neutral polar group such as hydroxyl or amide weakens hydrophobic interactions so that amino-acids such as serine, threonine, asparagine, and glutamine emerge before leucine, isoleucine, and valine.

The resolution of the amino-acids, *R*, depends on the diameter of the particles of the ion-exchange resin, the length of the column and the linear eluent flow velocity[12] according to the relationship:

$$R \propto \frac{1}{d}\sqrt{\frac{Z}{2U}}$$

where *d* is the mean particle diameter, *Z* is the length of the column, and *U* is the flow rate. High resolution requires a long column of fine resin, and this means that the column offers a high resistance to flow. This factor, together with the need to elute the amino-acids at reproducible times after commencement of chromatography, necessitates the use of a constant displacement reciprocating pump in order to control the flow of eluent at a constant and suitable rate.

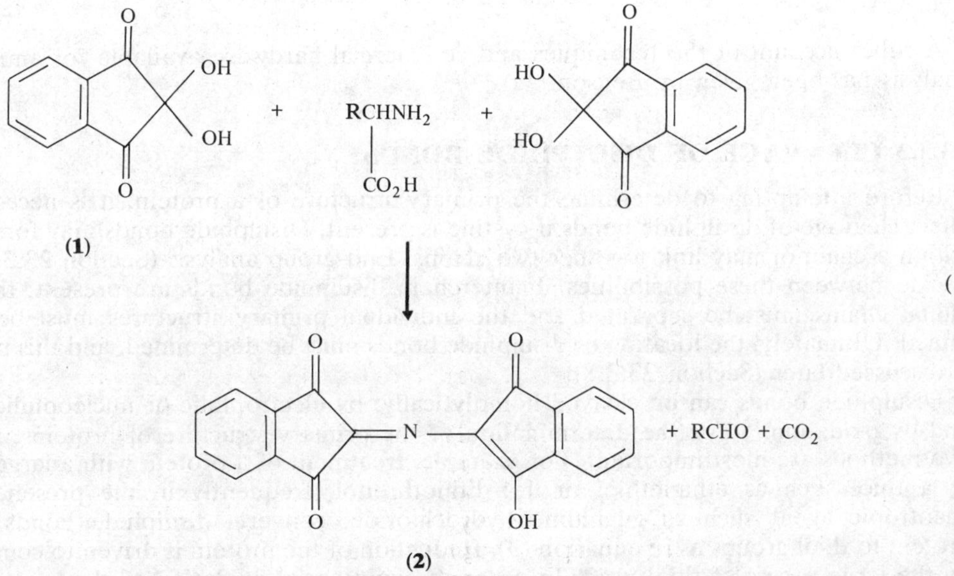

(1)

(1)

(2)

For detection and quantitative determination of the amino-acids, the eluate is mixed with a solution of ninhydrin (**1**) and the mixture is passed through a heating coil to bring about the reaction in equation (1). The product (Ruhemann's purple) (**2**) is spectrophotometrically assayed at 570 nm and the absorbance is output to a two-pen recorder so that each amino-acid gives a near Gaussian peak. Imino-acids (proline, hydroxyproline) give products which absorb maximally at 440 nm. Consequently, an additional spectrophotometer is required with the output connected to the second pen of the recorder. The peak areas may be determined by manual computation or, alternatively, the data output can be fed to an integrator or a minicomputer for processing. The ninhydrin solution is prepared in a mixture of an organic solvent such as dimethyl sulphoxide or 2-methoxyethanol with an acetate buffer at pH 5.2 and contains in addition a reducing agent such as stannous chloride or hydrindantin. The latter prevents oxidative side reactions which affect the precision of the assay. The various amino-acids give slightly different spectrophotometric yields and a standard mixture of amino-acids is run routinely to check the overall precision of the assay. The sensitivity of the analysis is about 1 nmol. Attempts have been made to improve the sensitivity by developing spectrofluorimetric assays. One potential reagent is 4-phenylspiro[furan-2(3*H*),1′-phthalan]-3,3′-dione (**3**) ('fluorescamine'), which is non-fluorescent and which reacts with amines and amino-acids at pH 9–11 to give *N*-substituted derivatives of 3,5-diphenyl-5-hydroxy-2-pyrrolin-4-ones (**4**) (equation 2). The latter are excited at 390 nm and emit maximally at 475 nm. While the increased sensitivity relative to the spectrophotometric method is attractive at first sight, the reagent is expensive, imino-acids do not give fluorescent products, and the pH of the acidic eluate must be substantially raised for reaction to occur. An alternative fluorogenic reagent is *o*-phthalaldehyde, although this also suffers from the last two disadvantages.

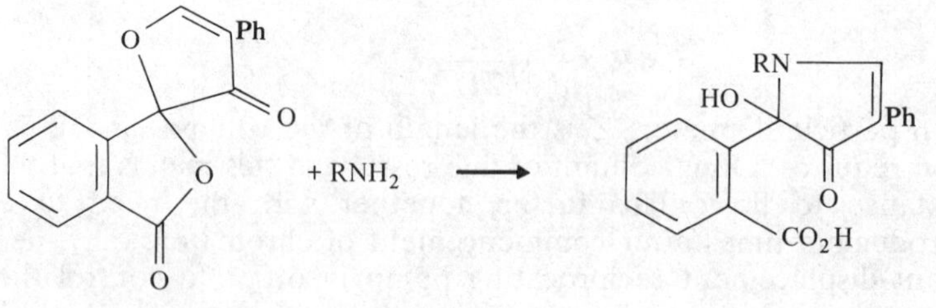

(2)

(**3**) (**4**)

A fuller account of the techniques and commercial hardware available for amino-acid analysis has been given by Benson.[13]

23.3.3 CLEAVAGE OF DISULPHIDE BONDS

Before attempting to determine the primary structure of a protein, it is necessary to effect cleavage of disulphide bonds if cystine is present. Disulphide bonds may form loops within a chain or may link together two chains. End-group analysis (Section 23.3.4) may decide between these possibilities. If interchain disulphide bonds are present, the individual chains must be separated and the individual primary structures must be determined. Ultimately, the location of disulphide bonds must be determined, and this problem is discussed later (Section 23.3.8).

Disulphide bonds can be cleaved homolytically, by electrophilic or nucleophilic attack and by oxidation.[14] For the determination of the primary structure of proteins, the last two methods are most important. For example, treatment of a protein with a large excess of a thiol such as ethanethiol or 1,4-dithiothreitol, frequently in the presence of a chaotropic agent such as guanidine hydrochloride, converts disulphide bonds in the protein to thiol groups as in equation (3). Reduction of the protein is driven to completion by the large excess of thiol used. In order to avoid aerial oxidation of the protein thiol groups, it is customary to block them by treatment with iodoacetic acid, 4-vinylpyridine, or aziridine (Scheme 1). The modified protein gives rise to *S*-carboxymethylcysteine (**5**), *S*-2-(4′-pyridyl)ethylcysteine (**6**), or *S*-2-aminoethylcysteine (**7**), respectively, on hydrolysis to amino-acids for quantitative analysis.

$$\begin{array}{ccccccccccc}
R^1 & & R^3 & & & & R^1 & & R^3 & & \\
| & & | & & & & | & & | & & \\
NH & & NH & & & & NH & & NH & & SCH_2CH_2OH \\
| & & | & & & & | & & | & & | \\
CHCH_2SSCH_2 & \!\!\!\!\!\! & CH & + & 2\,HSCH_2CH_2OH & \rightleftharpoons & CHCH_2SH & + & CHCH_2SH & + & SCH_2CH_2OH \\
| & & | & & & & | & & | & & \\
CO & & CO & & & & CO & & CO & & \\
| & & | & & & & | & & | & & \\
R^2 & & R^4 & & & & R^2 & & R^4 & &
\end{array} \qquad (3)$$

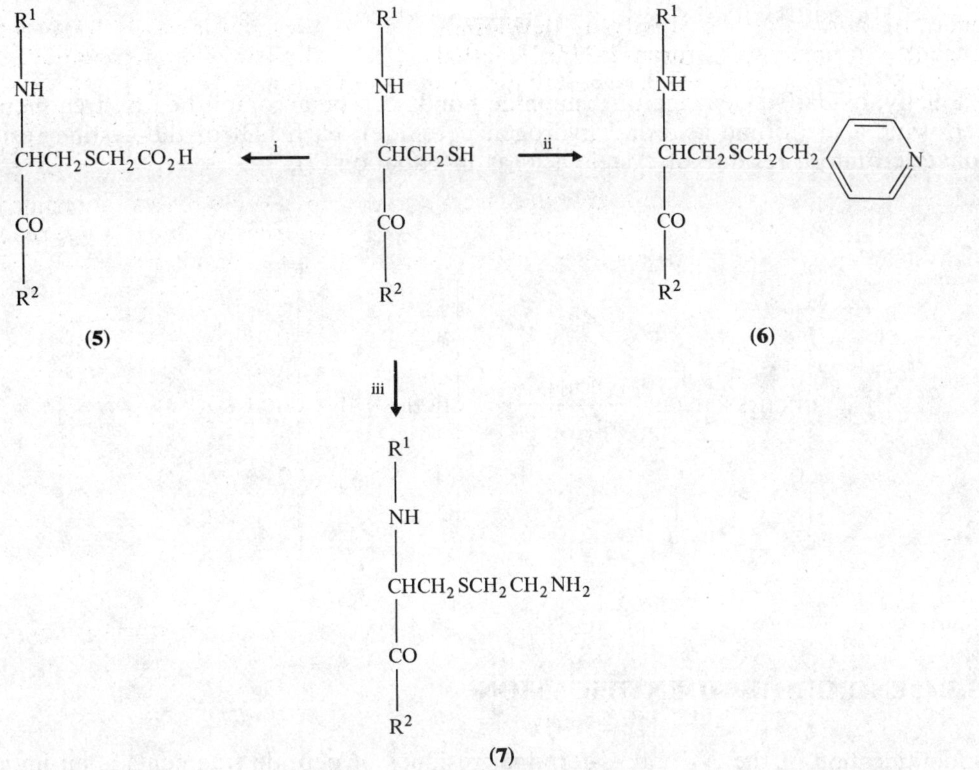

i, ICH_2CO_2H; ii, 4-vinylpyridine; iii, aziridine

SCHEME 1

Nucleophilic cleavage of the disulphide bond can be effected by sulphite ion as in equation (4). The cleavage is favoured by addition of Hg^{2+} or Ag^{+} to remove the thiolate ion formed. Alternatively, and more usually, conversion of cystine residues into *S*-sulphocysteine residues is completed by addition of Cu^{2+} ions, which reoxidize the liberated thiol groups to disulphide for further cleavage by sulphite ion as in equation (5). The overall reaction is summarized by equation (6).

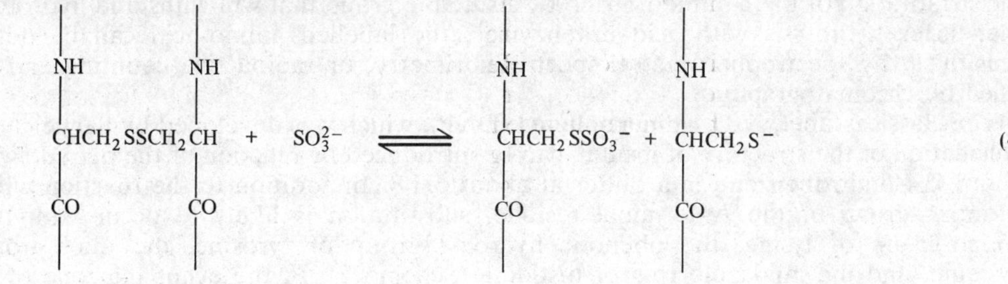

(4)

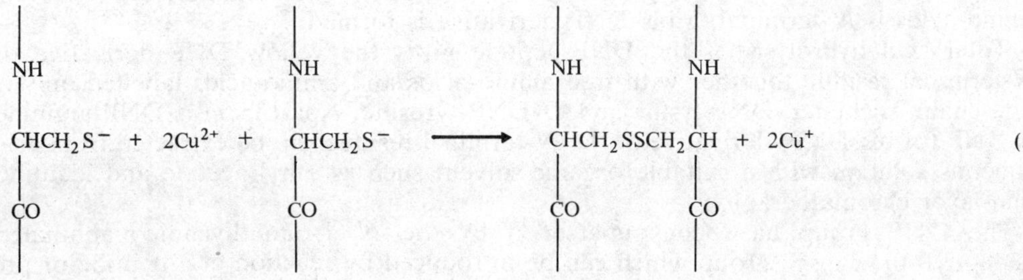

(5)

$$R^1SSR^2 + 2Cu^{2+} + 2SO_3^{2-} \longrightarrow R^1SSO_3^- + R^1SSO_3^- + 2Cu^+ \qquad (6)$$

Finally, oxidative cleavage of disulphide bonds can be accomplished by treatment with performic acid (formic acid and hydrogen peroxide); each half of the cystine residue is converted into a residue of cysteic acid as in equation (7).

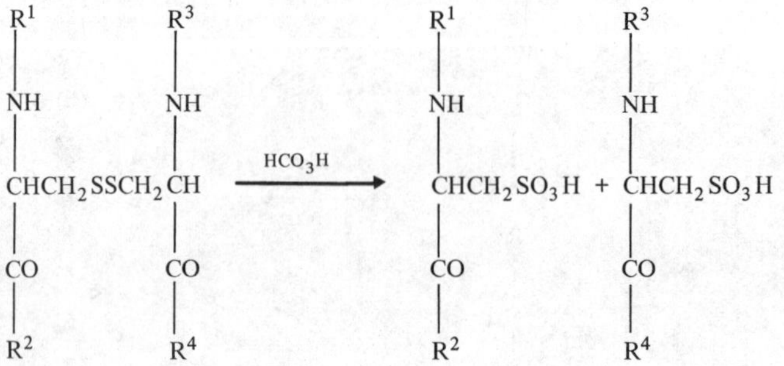

(7)

23.3.4 END-GROUP IDENTIFICATION

Identification of the *N*- and *C*-terminal residues of peptide fragments is an important part of the process of elucidating the primary structure of a protein. When this procedure can be combined with a method of stepwise degradation from one end of the molecule as described below, it is easy to understand the exponential growth in our knowledge of protein sequences in the last 20 years.

23.3.4.1 Identification of *N*-terminal groups

One method for identifying the *N*-terminal amino-acid in a peptide or protein involves the substitution of the α-amino-group by a suitable group that will withstand hydrolysis so that, after hydrolysis with acid or enzymes, the labelled amino-acid can be detected sensitively by spectrophotometry, spectrofluorimetry, or radioactive counting and identified by chromatography.

The classical label, *N*-2,4-dinitrophenyl (DNP), which was developed by Sanger[15] in the elucidation of the structure of insulin, can be introduced by reaction of the peptide with 1-fluoro-2,4-dinitrobenzene in a buffer at about pH 8. In addition to the reaction with the α-amino-group of the *N*-terminal residue, substitution is likely to occur with the ε-amino-group of lysine, the phenolic hydroxyl group of tyrosine, the thiol group of cysteine, and the imidazole ring of histidine (Scheme 2). In the event that one of these amino-acids is *N*-terminal, a bis-DNP derivative is formed.

Total acid hydrolysis of the DNP-peptide gives the yellow DNP derivative of the *N*-terminal residue together with free amino-acids and amino-acids labelled only in the side-chain, such as ε-DNP-lysine and *O*-DNP-tyrosine. Apart from α-DNP-arginine, the α-DNP (or bis-DNP) derivative of the *N*-terminal residue can be extracted from acidified aqueous solution with a suitable organic solvent such as ethyl acetate and identified by thin-layer chromatography.

The DNP group has been superseded by the *N*-(1-dimethylaminonaphthalene-5-sulphonyl) or 'dansyl' group which can be introduced by reaction of a peptide or protein

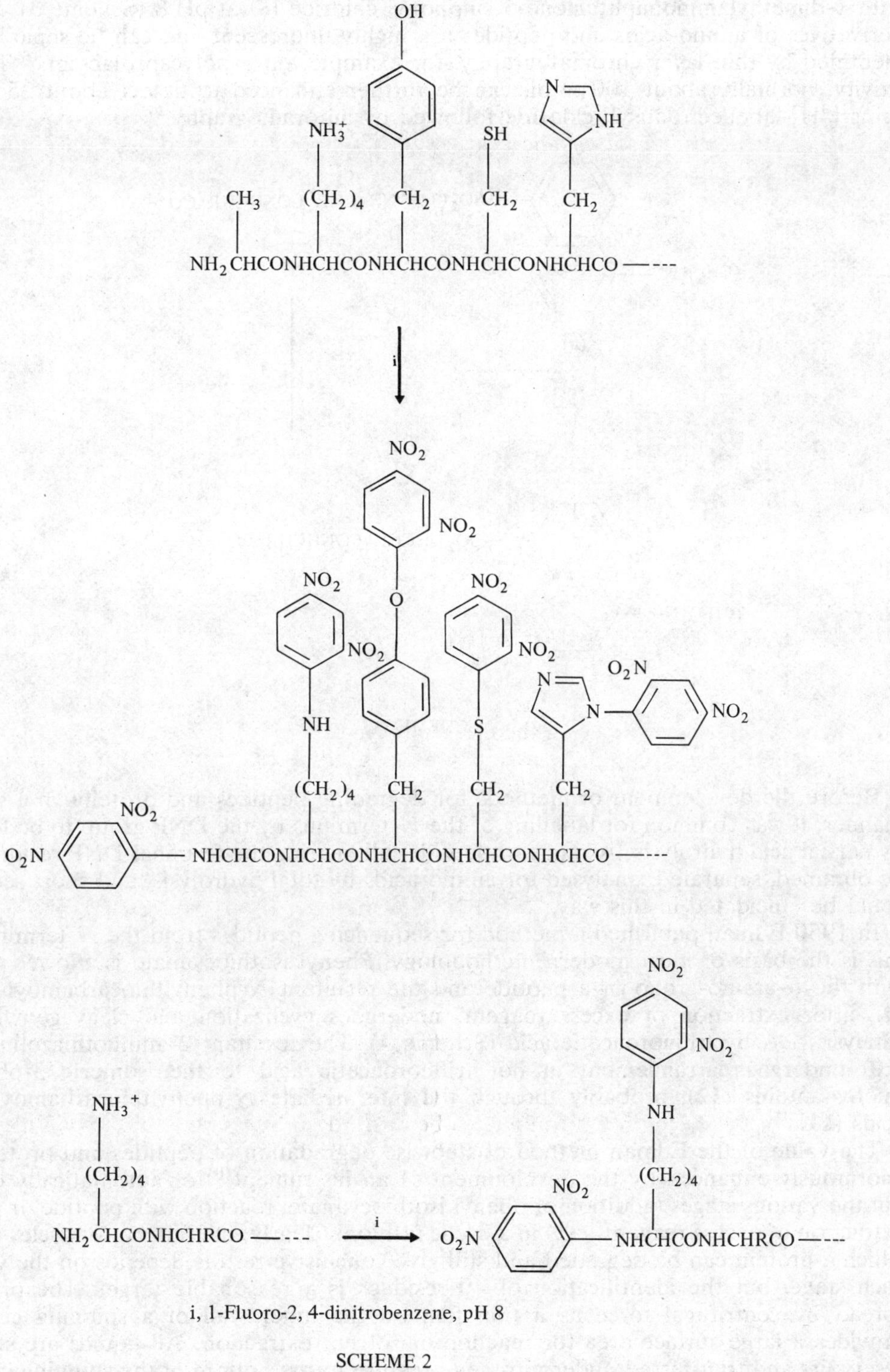

i, 1-Fluoro-2, 4-dinitrobenzene, pH 8

SCHEME 2

with 1-dimethylaminonaphthalene-5-sulphonyl chloride (**8**) at pH 8 (Scheme 3). 'Dansyl' derivatives of amino-acids and peptides are highly fluorescent and can be separated and identified by thin-layer chromatography, for example, on ε-polycaprolactam.[16] The sensitivity, normally about 100 pmol, can be further enhanced to detect about 25 fmol by using [^{3}H]-labelled 'dansyl' chloride followed by autoradiography.[17]

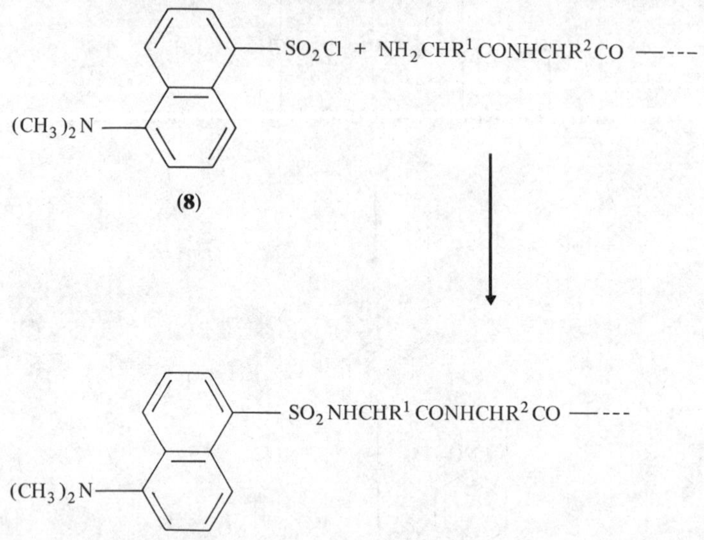

SCHEME 3

Before the development of methods for degrading peptides and proteins in a stepwise manner, it was common for labelling of the *N*-terminus by the DNP group to be followed by partial acid hydrolysis. A range of peptides all bearing the terminal DNP residue could be obtained, separated, analysed for amino-acids by total hydrolysis, and short sequences could be elucidated in this way.

In 1950 Edman published a method for sequencing peptides from the *N*-terminus, and this is the basis of most modern methodology. Phenyl isothiocyanate is allowed to react with the α-amino-group of a peptide and the resultant *N*-phenylthiocarbamoyl-peptide (**9**), after extraction of excess reagent, undergoes cyclization and cleavage with acid catalysis, *e.g.* by trifluoroacetic acid (Scheme 4). The resultant 2-anilinothiazolin-5-ones (**10**) undergo rearrangement in hot trifluoroacetic acid to the isomeric 3-phenyl-2-thiohydantoins (**12**), probably through the intermediate *N*-phenylthiocarbamoylamino-acids (**11**).

The value of the Edman method of stepwise degradation of peptides and proteins was enormously enhanced by the development of an instrument[18] for automatically carrying out the various stages (addition of phenyl isothiocyanate, reaction with peptide or protein, extraction of excess reagent, *etc.*) in a cyclic fashion. Clearly, the number of cycles through which a protein can be degraded and still give conclusive results depends on the yields at each stage, but the identification of 40 residues is a reasonable target. The protein is spread by centrifugal force as a thin film on the inner wall of a spinning cup. This provides a large surface area for reaction or solvent extraction. All liquids are stored in reservoirs and transferred under nitrogen pressure to the bottom of the spinning cup. The resultant 2-anilinothiazolin-5-ones from each degradation step are collected separately by fraction collector for subsequent identification. The amount of peptide or protein undergoing degradation must be enough (1–2 mg) to form a stable film in the spinning cup. The method has been reviewed in detail.[19]

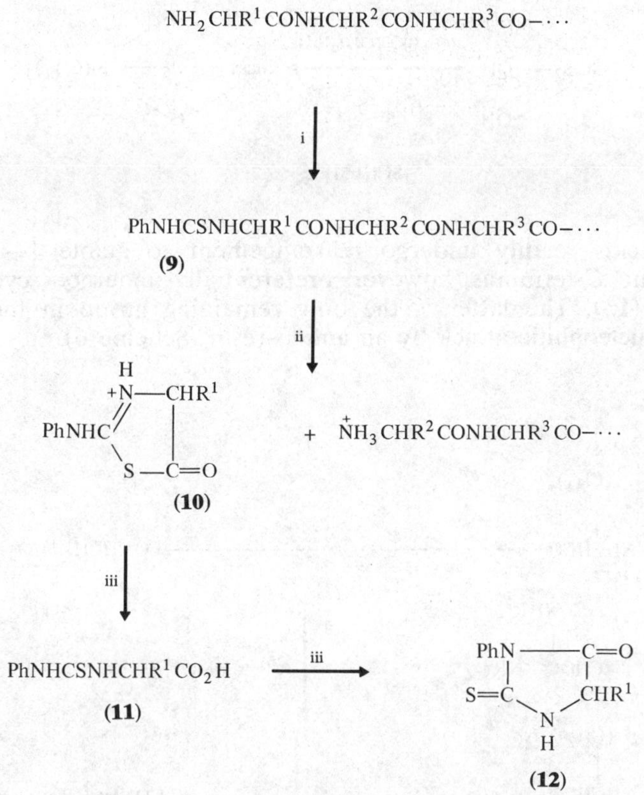

i, PhN=C=S, pH 8; ii, CF_3CO_2H; iii, hot CF_3CO_2H

SCHEME 4

The development of the solid-phase method of peptide synthesis (Chapter 23.6) led to a search for a parallel procedure in stepwise degradation. Separation of the 2-anilinothiazolin-5-ones from residual peptide or protein is thereby simplified. Since the first automatic sequencer was devised for solid-phase work,[20] several commerical machines have become available[21] and much effort has been expended in testing various solid supports and in developing methods for chemically attaching peptides to them for sequence analysis. In the earliest methods aminopolystyrene was used, but it does not swell in aqueous media so that the peptide to be degraded may have to be coupled in the presence of organic solvents in which it is poorly soluble. Secondly, coupling of large peptides or proteins to the support often proceeds in only poor yield. An alternative inert support is porous glass, which has been treated with 3-aminopropyltriethoxysilane (Scheme 5). The Si—O bonds are somewhat labile to acid, so that losses are considerable at each degradation step with peptides which are attached at only one point. A polydimethylacrylamide-based resin with β-alanyl substituents, which was developed for solid-phase peptide synthesis, seems to offer some advantages over both the foregoing supports.[22] Coupling of the peptide to the resin and its subsequent degradation proceed in high yield.

All of the above solid supports have free amino-groups and these are the sites for the attachment of peptides or proteins for sequential analysis. Three main methods are used. In the first of these, the peptide is treated with *N*-ethyl-*N'*-(3-dimethylamino)-propylcarbodi-imide in the absence of a nucleophile. The carboxyl groups in the peptide initially give *O*-acylisoureas and those derivatives situated on the side chains of aspartic

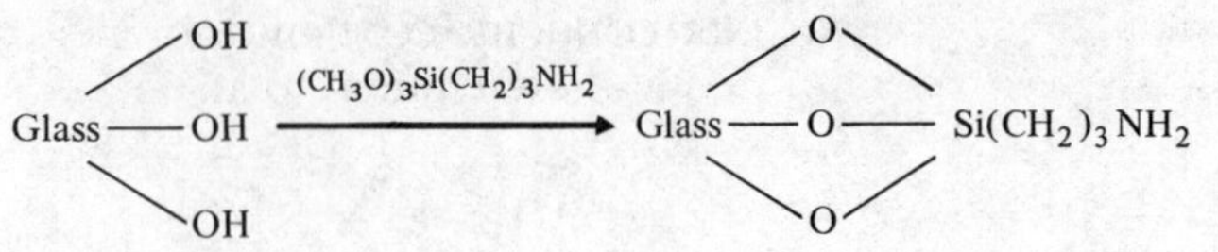

SCHEME 5

and glutamic acids readily undergo rearrangement to stable *N*-acylureas. The *O*-acylisourea at the *C*-terminus, however, preferentially undergoes cyclization to give an oxazolin-5-one (**13**). The latter is the only remaining group in the peptide which is susceptible to nucleophilic attack by an amino-resin (Scheme 6).

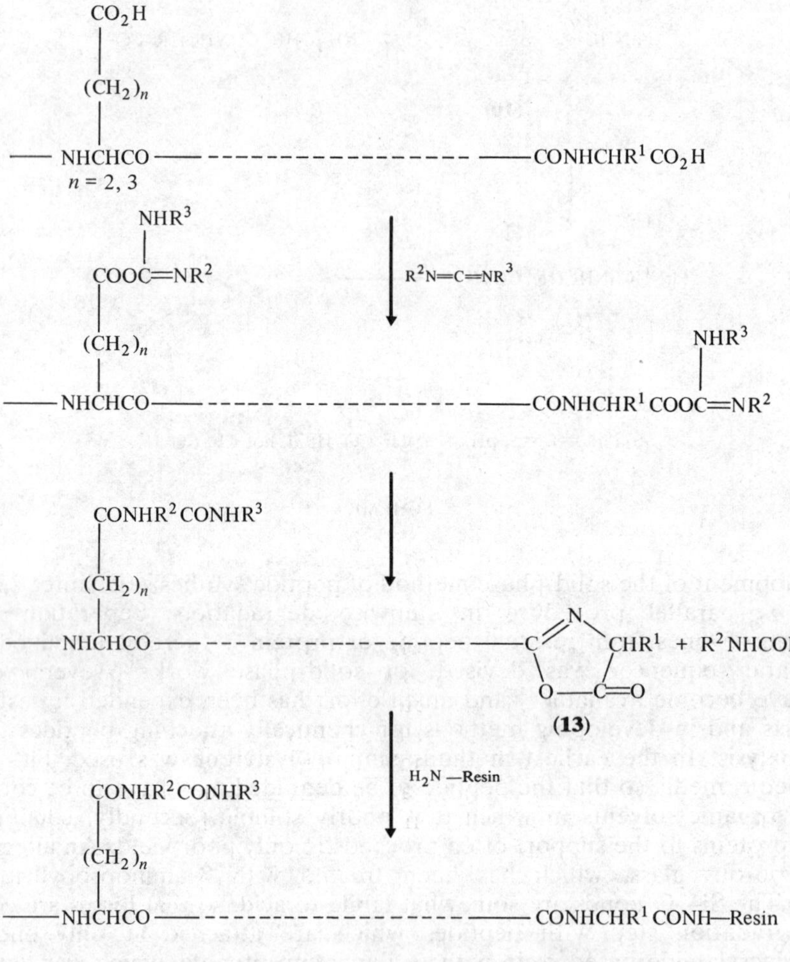

SCHEME 6

Apart from the variable yields achieved, Asp and Glu residues appear as derivatives in the subsequent Edman degradation.

Secondly, after cleavage of a protein with cyanogen bromide (Section 23.3.5), all peptides except that from the *C*-terminal region terminate with a residue of homoserine or its lactone. Complete lactonization can be achieved by treatment with trifluoroacetic acid and then the petptides undergo covalent attachment to an amino-resin (Scheme 7).

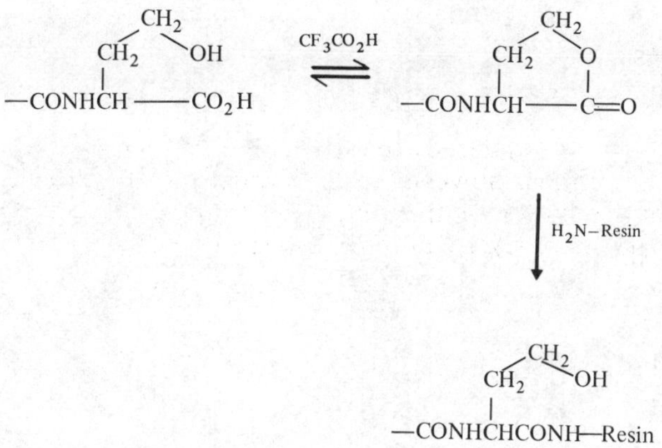

SCHEME 7

Finally, perhaps the most commonly used method, especially for peptides derived from trypsin-catalysed hydrolysis of a protein, uses *p*-phenylene di-isothiocyanate as coupling agent. ε-Amino-groups of lysine or of *S*-(2-aminoethyl)cysteine and the α-amino-group at the *N*-terminus react with one isothiocyanate group of *p*-phenylene di-isothiocyanate when a 50- to 100-fold molar excess of reagent is added to a peptide. Coupling is completed by adding the amino-resin as in Scheme 8. Some tryptic peptides will terminate with Arg rather than Lys and conversion into ornithine is accomplished by removing the guanidino-group with 50% hydrazine at 75 °C for 15 min. Thereafter, coupling to the resin with phenylene di-isothiocyanate proceeds as for peptides of lysine. Since the α-amino-group has already undergone thiocarbamoylation, the first treatment with phenyl isothiocyanate is omitted. Treatment with acid effects the first cleavage, but the 2-anilinothiazolin-5-one corresponding to the first residue remains attached to the resin. The *N*-terminal residue must therefore be identified by some other method such as the manual Edman technique. Likewise, no thiohydantoin corresponding to Lys is detected by this method. If this method of coupling is used with tryptic peptides, lysine or ornithine is *C*-terminal and no gaps in the sequence determination should occur after the *N*-terminal residue. A peptide derived from a protein by some other type of cleavage may contain several lysine residues in the chain but not at the *C*-terminus. Coupling to the resin involves all or most of the lysine residues and so several gaps may occur in the Edman degradation. When the stepwise degradation reaches the last lysine residue, any remaining peptide falls off the resin and is lost.

Irrespective of the particular variation of the Edman procedure employed, the 2-anilinothiazol-5-one corresponding to the *N*-terminal amino-acid is collected at each cycle and then converted into the 3-phenyl-2-thiohydantoin (**12**) by heating with trifluoroacetic acid. Several methods[23] are available for the identification of the 2-thiohydantoin. Before convenient methods for separating 2-thiohydantoins had been developed, it was common practice to hydrolyse them to amino-acids so that the amino-acid analyser could be used for their identification and quantitative determination. The hydrolysis, however, inevitably destroys some of the thiohydantoins (*e.g.* those corresponding to Ser and Thr) and more direct methods are now favoured.

Nearly all the 3-phenyl-2-thiohydantoins can be separated by two-dimensional thin-layer chromatography on silica gel plates. A useful pair of solvent systems are chloroform–ethanol (98 : 2 v/v) and chloroform–ethanol–methanol (88.2 : 1.8 : 10 v/v). If

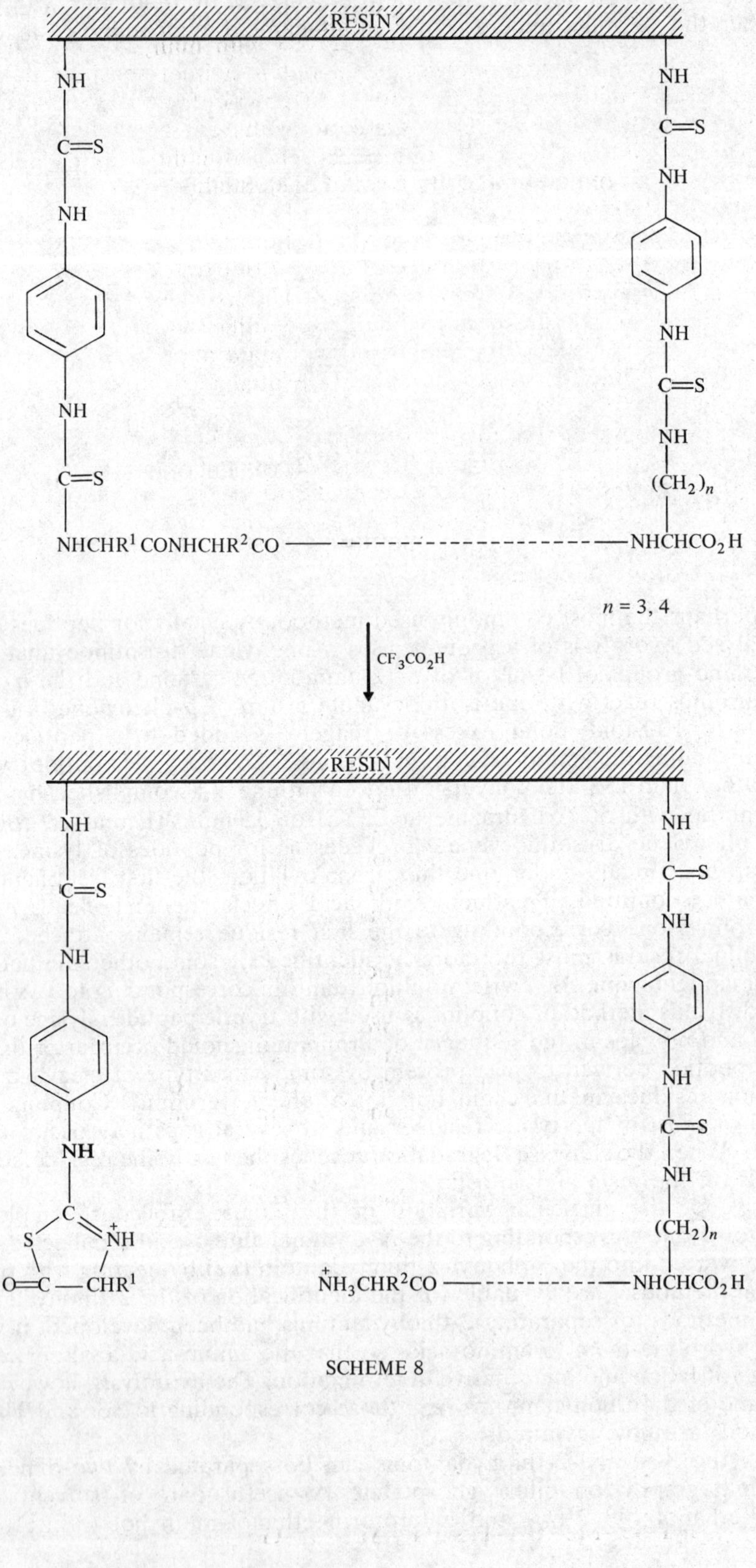
RESIN
NH
C=S
NH
NH
C=S
NHCHR1 CONHCHR2CO
NH
C=S
NH
NH
C=S
NH
(CH2)n
NHCHCO2H
n = 3, 4
CF3CO2H
RESIN
NH
C=S
NH
NH
C
S
NH
O=C
CHR1
NH3CHR2CO
NH
C=S
NH
NH
C=S
NH
(CH2)n
NHCHCO2H

SCHEME 8

the silica gel contains a fluorophor, absorbing spots can be seen in ultraviolet light (254 nm). Alternatively, the plates can be sprayed with ninhydin or with iodine–azide solution. Thin-layer chromatography is cheap, but is neither sensitive nor quantitative unless phenyl [^{35}S]isothiocyanate is used for the stepwise degradation. In the latter case, detection can be achieved by autoradiography or sufficient unlabelled 2-thiohydantoins are applied to the plate to detect by one of the above methods and then the spots are removed and assayed with a scintillation counter. The method is suitable with 0.5 nmol of 3-phenyl-2-thiohydantoin.

Gas–liquid chromatography using a flame-ionization detector and integrator offers the opportunity of sensitive and quantitative analysis. Both free 3-phenyl-2-thiohydantoins and their silylated derivatives may be separated. Unfortunately, the derivatives corresponding to arginine and cysteic acid cannot be identified by this technique. If methyl isothiocyanate is used in place of phenyl isothiocyanate in the Edman degradation, the resulting 3-methyl-2-thiohydantoins and their silylated derivatives can be more easily resolved by g.l.c.

The rapid development of high-performance liquid chromatography (HPLC) in biochemical analysis has resulted in considerable efforts to apply it to the separation of 3-phenyl-2-thiohydantoins. The substantial capital cost of the equipment is offset by the convenience of continuous and quantitative monitoring by u.v. absorption. Complete separation on one column, however, is difficult; while most of the thiohydantoins can be resolved by HPLC on a column of silica gel using, for example, a gradient of increasing concentration of methanol in heptane–chloroform (1 : 1 v/v), the derivatives of acidic amino-acids are difficult to elute. One solution involves using a second column containing a hydrophobic adsorbent in reverse phase HPLC.

Short sequences at the *N*-terminus of proteins can often be identified by enzymatic degradation. Several aminopeptidases are commercially available which differ in their specificities. Leucine aminopeptidase from porcine kidney, as its name suggests, preferentially hydrolyses leucine and similar amino-acids with hydrophobic side-chains. Although velocities of cleavage of *N*-terminal amino-acids cover a wide range, hydrolysis is likely to continue up to but not including an amino-acid which precedes proline. Because of the wide variation in rates of cleavage, caution is required in interpreting the results of degradations catalysed by this enzyme. It is particularly important to remove samples for analysis at regular intervals during degradation. For example, a protein with the *N*-terminal sequence Ala-Leu-Leu . . . would probably yield more alanine than leucine in the early stages of degradation, but the situation would be reversed as time progressed.[24] Another enzyme isolated from porcine kidney, aminopeptidase M, is much less specific and probably more useful for sequencing proteins.

Dipeptidyl aminopeptidase or cathepsin C has an unusual specificity and removes dipeptide units from the *N*-terminus of a protein. Hydrolysis continues until either Lys or Arg becomes *N*-terminal or Pro is in positions 2 or 3 in the chain. Thus five dipeptides are removed from β-corticotropin:

Ser-Tyr↓Ser-Met↓Glu-His↓Phe-Arg↓Trp-Gly↓Lys-Pro-Val . . .

and two dipeptides are removed from angiotensin II:

Arg-Arg↓Val-Tyr↓Ile-His-Pro-Phe

The liberated dipeptides can be separated by ion-exchange chromatography similar to the techniques used for amino-acid analysis, but with a stream-splitter fitted so that fractions can be collected for further analysis. An alternative set of dipeptides can be obtained by removing the *N*-terminal residue by one cycle of the Edman method before exposure to the dipeptidyl aminopeptidase. Thus if the *N*-terminal Ser residue is removed from β-corticotropin, dipeptides will be released as follows:

Tyr-Ser↓Met-Glu↓His-Phe↓Arg-Trp-Gly-Lys-Pro-Val . . .

The isolation of Ser-Tyr, Ser-Met, Glu-His, and Phe-Arg in the first degradation and Tyr-Ser, Met-Glu, and His-Phe in the second degradation is sufficient to give the complete sequence of the first eight residues. When a resistant bond is encountered, removal of one or two residues by the Edman procedure should produce a product that is susceptible to further degradation by the enzyme.

23.3.4.2 Identification of *C*-terminal groups

Determinations of the *C*-terminal residue of peptides and proteins are less satisfactory than for the *N*-terminal residue. Numerous chemical methods have been proposed, but none has stood the test of time. Fortunately, two pancreatic enzymes, carboxypeptidase A and carboxypeptidase B, can provide a limited amount of information about the *C*-terminal sequence. Carboxypeptidase A preferentially cleaves amino-acids with aromatic side-chains, but most other amino-acids are cleaved more slowly. *C*-Terminal Pro is resistant. This enzyme is useful for identifying the *C*-terminal sequences of peptides produced by cleavage with α-chymotrypsin (Section 23.3.6) since these are likely to have an aromatic amino-acid at the *C*-terminus. The difficulty in interpreting the results of a prolonged digestion of a peptide or protein with carboxypeptidase A parallels that encountered using aminopeptidases at the *N*-terminus (Section 23.3.4.1). Carboxypeptidase B is much more specific than carboxypeptidase A and cleaves only Lys or Arg. These residues will be *C*-terminal in peptides produced by hydrolysis of a protein by trypsin. These enzymes can be used alternately or in conjunction in order to achieve more extensive degradation. Several other carboxypeptidases are known which are less specific and one of these, carboxypeptidase Y, from yeast has recently been applied to the determination of *C*-terminal sequences.

23.3.5 SELECTIVE CHEMICAL METHODS FOR CLEAVING PEPTIDE BONDS

Although several methods have been described for effecting selective cleavage of certain peptide bonds in the interior of a protein chain, for the purpose of producing fragments suitable for sequential analysis by the Edman procedure, few have found wide acceptance. The most popular method uses cyanogen bromide to bring about cleavage of methionyl peptide bonds (Scheme 9). The methionyl residue is converted into homoserine and its lactone, which becomes the *C*-terminal residue of all fragments except that at the *C*-terminus of the original protein. Because Met is a relatively rare amino-acid, the fragments tend to be quite large but suitable for extensive sequential analysis by the modern variants of the Edman procedure. Moreover, the presence of homoserine lactone at the *C*-terminus provides a suitable locus for covalent attachment to an insoluble support for solid-phase sequence analysis (Scheme 7).

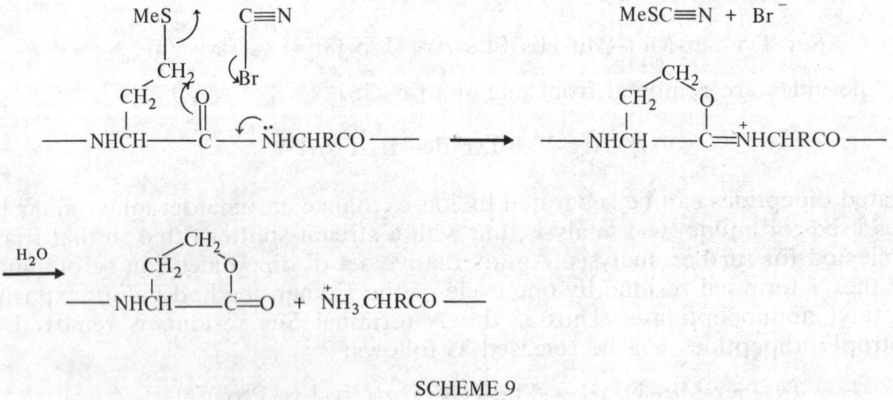

SCHEME 9

Tryptophanyl peptide bonds and, less readily, those involving the carbonyl group of tyrosyl residues are cleaved by *N*-bromosuccimide. In both cases, bromination is followed by nucleophilic attack by carbonyl-oxygen and displacement of bromide ion to give a spirolactone (Schemes 10 and 11).

Specific chemical methods for the cleavage of peptide bonds involving other amino-acids have been developed, but have not been widely applied. Now that the Edman procedure can be applied to large protein fragments, the most useful methods of cleavage are those which involve the rarest amino-acids, and Trp and Met are in this category.

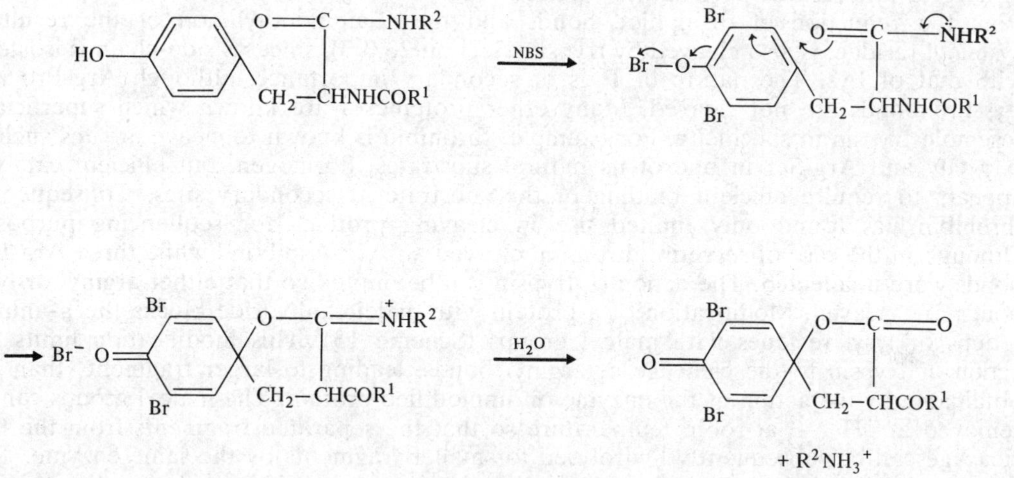

SCHEME 10

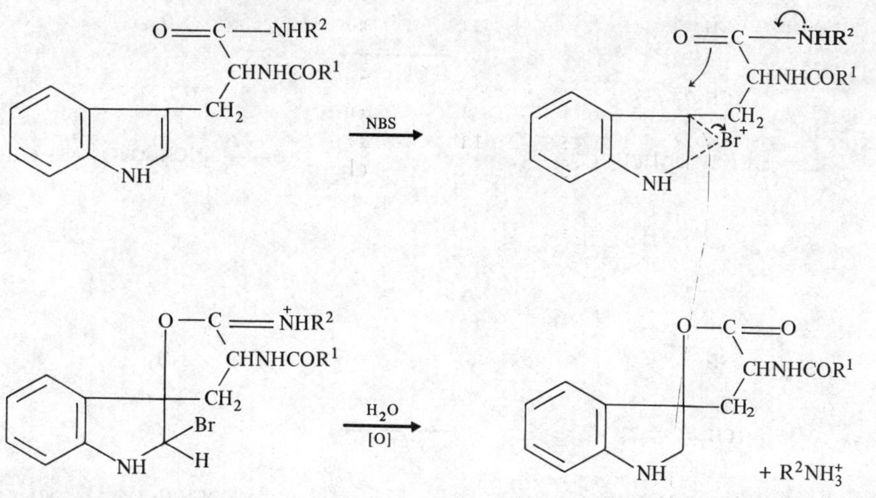

SCHEME 11

23.3.6 ENZYMATIC METHODS FOR CLEAVING PEPTIDE BONDS

The use of exopeptidases such as aminopeptidases and carboxypeptidases for determining the sequences of amino-acids near the *N*- and *C*-termini of proteins has been described in Section 23.3.4. Endopeptidases are proteolytic enzymes which cleave selectively some peptide bonds at points removed from the ends of the protein molecule. Endopeptidases differ enormously in specificity. Usually, the amino-acid residues on either side of a scissile peptide bond are the most important determinants of specificity of proteolytic enzymes. Thus, trypsin cleaves peptide bonds in which the carbonyl group of an Arg or Lys residue participates (Scheme 12). A thialysyl peptide bond, which results from the reduction of disulphide bonds and the *S*-aminoethylation of the resultant cysteinyl residue, is also cleaved by trypsin (Section 23.3.3), since its side-chain is isosteric with that of Lys. The nature of R is of secondary importance, although Arg-Pro and Lys-Pro bonds are not cleaved. Many other proteinases are known which superficially resemble trypsin in specificity. For example, thrombin is known to cleave at sites such as Arg-Gly and Arg-Ser in one of its natural substrates, fibrinogen, but efficient catalysis appears to require efficient binding of the substrate at secondary sites. Consequently, thrombin has found only limited use in cleaving proteins for sequencing purposes, although in the case of secretin, thrombin cleaved an Arg-Asp bond while three Arg-Leu bonds were unaffected. The action of trypsin can be limited so that either arginyl or lysyl bonds are cleaved. Modification of a protein with maleic anhydride blocks the ε-amino-groups of lysyl residues with maleyl groups (Scheme 13). This modification limits the action of trypsin to the cleavage of arginyl bonds, leading to larger fragments than are obtainable by the action of the enzyme on unmodified protein. The maleyl groups can be removed at pH 2–3 at room temperature so that the separated fragments from the first cleavage can be subsequently hydrolysed to smaller fragments by the same enzyme. This procedure helps in the determination of the order of tryptic peptides in the original protein. Alternatively, the ε-amino-groups of lysyl residues can be trifluoroacetylated by reaction with *S*-ethyl trifluorothioacetate (Scheme 14). After cleavage of arginyl bonds, *N*-trifluoroacetyl groups can be removed by treatment with aqueous piperidine at 0 °C.

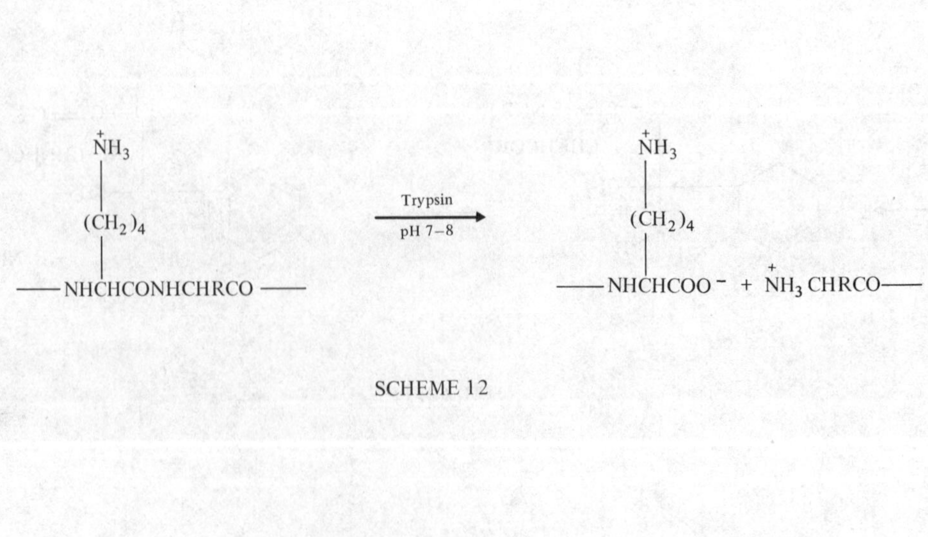

SCHEME 12

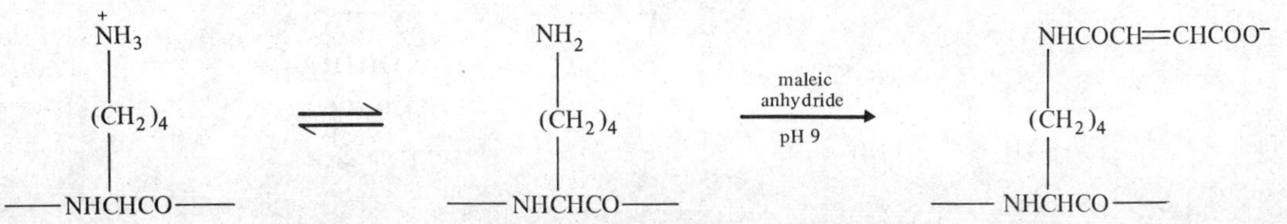

SCHEME 13

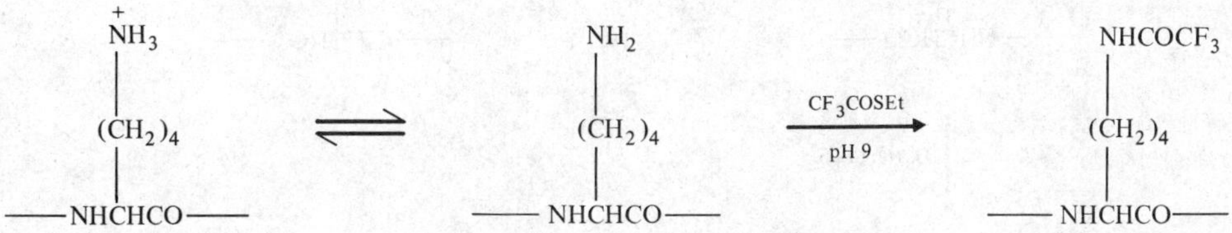

SCHEME 14

Conversely, arginyl side-chains can be blocked by a variety of 1,2- or 1,3-biscarbonyl compounds such as cyclohexane-1,2-dione or the carbanion of nitromalondialdehyde (Scheme 15). Neither of the derivatives (**14, 15**) are hydrolysed by trypsin. The modifying group is removed from (**14**) by treatment with hydroxylamine at pH 7.0 for 7–8 h at 37 °C. Reduction of (**15**) with sodium borohydride yields (**16**), which is sensitive to trypsin in spite of the presence of the substituent on the side-chain.

When the primary structure of a protein has been elucidated, it is frequently unnecessary to carry out a complete study of the amino-acid sequence of the homologous proteins from closely related species. Rapid results may be obtained by using the trypsin 'fingerprinting' technique. The protein of known structure and the homologous protein are separately hydrolysed with trypsin and the peptides are separated, usually by two-dimensional analysis using electrophoresis or chromatography. If there are only a few differences in sequence, most of the peptides will appear at identical positions on the two-dimensional 'fingerprint'. Those few peptides which appear to be different can be eluted and subjected to amino-acid analysis and Edman degradation. This technique has been especially useful in the study of abnormal haemoglobins, which usually differ from normal haemoglobin at only one locus.

α-Chymotrypsin has also found wide application in sequence analysis of proteins. Its substrate specificity is different from that of trypsin and is less restricted. Peptide bonds involving the carbonyl groups of Trp, Tyr, and Phe are rapidly cleaved while peptide bonds involving the carbonyl groups of smaller hydrophobic amino acids (Leu, Ile, Met, Val) are cleaved more slowly. Occasionally, unexpected cleavages occur such as, for example, the hydrolysis of Asn^{148}-Ala^{149} in the autocatalytic activation of chymotrypsinogen by α-chymotrypsin. α-Chymotrypsin is particularly useful for obtaining a different population of peptides from that obtained by cleavage with trypsin or cyanogen bromide so that overlapping sequences can be elucidated.

Microbial proteinases differ enormously in specificity; some are used for producing peptides suitable for sequential analysis by the Edman method whereas others have broad specificities, so that they are useful for producing small peptides to determine the position of disulphide bonds (Section 23.3.8) or the point of attachment of covalently bound coenzyme or carbohydrate. Proteinases from *Myxobacter* AL-1 (proteinase II) or *Armillaria mellea* cleave peptide bonds containing the —NH— group of lysyl residues.[25,26] This is a potentially useful way of determining overlapping sequences when trypsin is the main

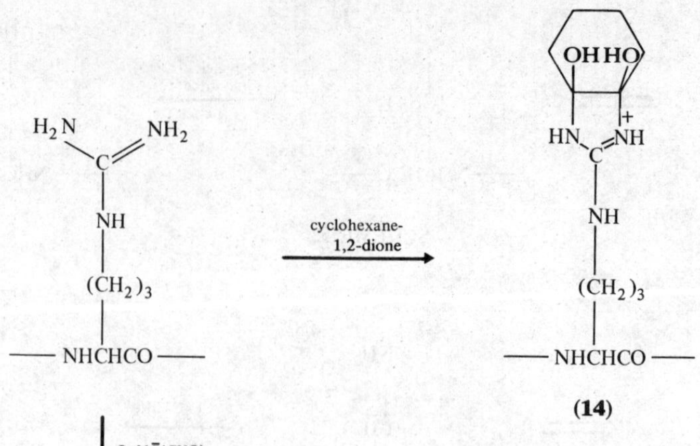

SCHEME 15

tool for obtaining fragments for sequential analysis. A further proteinase from *Staphylococcus aureus* cleaves peptide bonds containing the carbonyl group of aspartic or glutamic acid.[27] None of these enzymes has been extensively used in sequencing proteins, presumably because they are not commercially available. Thermolysin, which is a thermostable proteinase from *B. thermoproteolyticum,* is commercially available and in common use. It is an endopeptidase with a specificity similar to that of carboxypeptidase A; it cleaves peptide bonds containing the —NH— group derived from amino-acids with aromatic or hydrophobic aliphatic side-chains. Because it is an endopeptidase it does not hydrolyse residues at or near the ends of the polypeptide chain. It is useful for identifying overlaps when chymotrypsin has been used to provide the main fragments for sequential analysis.

Other proteinases such as pepsin, papain, and some microbial enzymes have broad specificities and generally produce too many fragments for general use in sequential analysis. As indicated above, however, they are useful for obtaining small peptides containing some particular feature such as a disulphide bond.

23.3.7 SEQUENCING BY MASS SPECTROMETRY

Peptides are not sufficiently volatile to be studied directly by mass spectrometry using electron impact to achieve ionization. The first attempts to apply mass spectrometry to sequence determination involved preliminary acylation of amino-group(s) and esterification of carboxyl group(s). The mass spectra of these derivatives contain peaks whose *m*/*e* ratios indicate that cleavage occurs on either side of carbonyl groups. Cleavage at C—N bonds given acylium ions, —NHCHRC≡O⁺, whereas cleavage at C—C bonds yields aldiminium ions, —$\overset{+}{N}$H═CHR. There is a tendency for additional fragmentation to occur, leading to loss of side chains of some amino-acids including valine, leucine, asparagine, serine, threonine, and cysteine.

The situation was improved considerably by permethylation of the *N*-acetylated peptide. This pretreatment not only increases volatility, but the mass spectra are considerably simplified because acylium ions usually predominate. Several methods for permethylation have been tried, but the method of choice employs a mixture of methyl iodide, sodium hydride, and dimethyl sulphoxide.[28,29] By restricting reaction time to 1–3 min and avoiding a large excess of methyl iodide, methylation of amide nitrogen, both in the peptide chain and in the side-chains of asparaginyl and glutaminyl residues, is achieved with minimal quaternization of nitrogen and sulphur atoms in the side chains of histidine, arginine, methionine, and cysteine. If it occurs, quaternization decreases volatility.

It is an obvious extension of the above technique to use a combination of gas–liquid chromatography and mass spectrometry to separate and identify small peptides produced by hydrolysis of a protein with proteinases. For example, the dipeptides liberated from a protein by the action of dipeptidyl aminopeptidase (Section 23.3.4.1) can be derivatized, separated by g.l.c. and identified by mass spectrometry. The alternative set of dipeptides obtained after an initial cycle of the Edman degradation can be treated similarly.

It has been found possible to identify mixtures of small peptides obtained from enzyme digests of proteins after a partial separation by gel filtration and ion-exchange chromatography. Comparison of mass spectra obtained at different source temperatures and by alternative use of permethylation and perdeuteriomethylation usually allows correct allocation of peaks to their parent peptides.

There is promise that technical developments will lead to a much wider use of mass spectrometry for sequencing proteins. The use of field ionization and field desorption ionization gives rise to rather simple spectra with a high abundance of the molecular ion. Since the sample does not have to be vaporized for field desorption mass spectrometry, the molecular weight of peptides can be determined without the need for acylation and permethylation. A different advantage accrues from the use of chemical ionization. As well as acylium ions, an alternative fragmentation pattern gives rise to ions from the *C*-terminal region (Scheme 16). This provides information about the *C*-terminal region of a peptide that is complementary to that obtained by electron impact. Finally, the mass spectrometer can be interfaced with a digital computer through an analog to digital converter. Programs are available which search the array of fragment masses for an *m*/*e* ratio which accurately corresponds to the *N*-terminal amino-acid and its *N*-acyl substituent. The programme then searches for a fragment ion whose *m*/*e* ratio corresponds accurately to the sum of two amino-residues in the form of a *N*-acylated dipeptide and so on as far as the analysis can be continued. Most of the computer-assisted analysis has used a high-resolution mass spectrometer and consequently the hardware is extremely expensive.

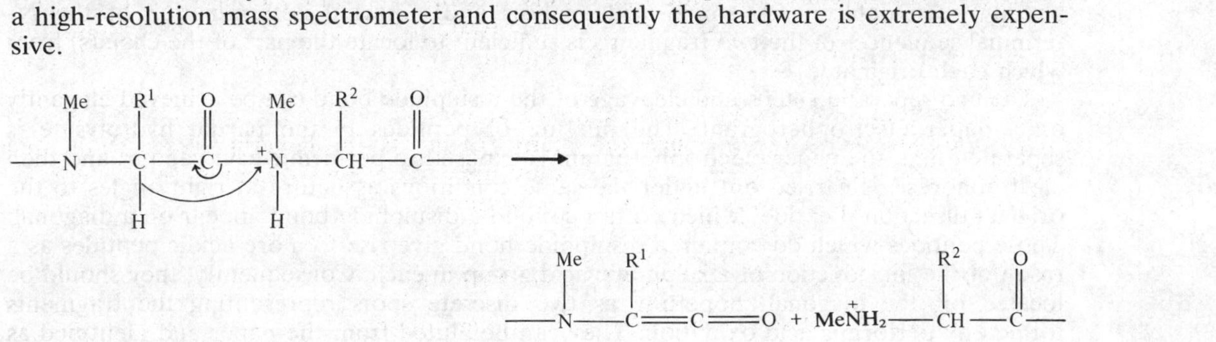

SCHEME 16

To summarize, mass spectrometry is provocatively promising,[30] but the volume of results obtained by the Edman degradation has led to some conservatism. Mass spectrometry, however, would be highly competitive if the tedium of isolating pure peptides could be avoided by further development of the method for sequencing mixtures of peptides with data-capture and -processing by on-line computer.

23.3.8 LOCATION OF DISULPHIDE BONDS

Although disulphide bonds are responsible in part for the secondary and tertiary structures of a protein, their location forms part of the process of determining the primary structure.

In their determination of the primary structure of insulin, Sanger and his colleagues encountered considerable problems because of the proclivity of disulphides to undergo exchange reactions under both acidic and mildly alkaline conditions. In strongly acidic solution, as used for partial acid hydrolysis, the following reactions (equations 8–9) may occur. Addition of a thiol tends to drive the first reaction from right to left and thus inhibits exchange.

$$\mathrm{RSSR + H^+ \rightleftharpoons RS^+ + RSH} \tag{8}$$

$$\mathrm{R^1S^+ + R^2SSR^2 \rightleftharpoons R^1SSR^2 + R^2S^+} \tag{9}$$

$$\mathrm{R^2SSR^2 + R^1S^- \rightleftharpoons R^1SSR^2 + R^2S^-} \tag{10}$$

$$\mathrm{R^1SSR^2 + R^1S^- \rightleftharpoons R^1SSR^1 + R^2S^-} \tag{11}$$

Under mild alkaline conditions, such as might be used during hydrolysis of a protein with trypsin, traces of thiolate ions catalyse the reactions (10) and (11). This type of exchange reaction could occur if the protein undergoing hydrolysis contains both cysteine and cystine. Provided the proteinase catalysing the hydrolysis does not contain an essential thiol group, the thiol groups in the substrate can be blocked during hydrolysis by addition of an appropriate thiol reagent such as *N*-ethylmaleimide.

In order to locate disulphide bonds, the protein is partially hydrolysed to relatively small fragments with a proteinase of broad specificity such as pepsin and under conditions designed to minimize exchange reactions. The peptides are separated and their amino-acid contents determined to ensure that each is homogeneous and contains only one disulphide bond. The disulphide bond in each peptide is cleaved (Section 23.3.3) and the two fragments are separated. Assuming that the primary structure of the protein has already been determined, amino-acid analysis and a partial determination of the *N*-terminal sequences of the two fragments is sufficient to locate the part of the chain(s) from which each originated.

The two separation steps and cleavage of the disulphide bond can be achieved elegantly on a paper electropherogram. The mixture of peptides in the partial hydrolysate is separated first, the paper electropherogram is exposed to performic acid vapour, and then electrophoresis is carried out under the same conditions as before at right angles to the original direction. Peptides which do not contain a disulphide bond appear on a diagonal. Those peptides which do contain a disulphide bond give rise to more acidic peptides as a result of the introduction of a sulphonic acid group in each. Consequently, they should be located off the diagonal, hopefully as two discrete spots representing the fragments formed by performic acid oxidation. They can be eluted from the paper and identified as above.

23.3.9 PRIMARY STRUCTURE AND EVOLUTION OF PROTEINS

Messenger ribonucleic acids (mRNA) code for 20 different amino-acids in protein biosynthesis. Some amino-acids such as Trp, Met, and His, however, occur much less frequently than others. This is explicable from the genetic code since, for example, there is only one codon for Trp and six for Ser. Coding for Ser is said to be degenerate. Just as languages make differential use of the letters or characters of alphabets, so proteins might be regarded as biological messages written in a 20-character alphabet of amino-acids. The analogy can be pressed a little further; written languages, apart from low-level computer languages, contain considerable redundancy. This is an advantage because every communication channel is subject to noise and, if redundant material is present in a message, there is a good chance that the main import of the message will survive. If every amino-acid in a protein were essential for manifestation of biological activity, a single genetic mutation, which may be regarded as biological noise, would be lethal and further evolution would be impossible. We now know that many proteins contain redundant amino-acids. For example, although β-corticotropin contains 39 amino-acid residues, a synthetic eicosapeptide comprising the *N*-terminal region of the hormone has almost full biological activity. Significantly, variations in primary structure in β-corticotropin from different species occur in the *C*-terminal region of the molecule. We also know that the secondary and tertiary structure of a protein is largely determined by the sequence of amino-acids. For example, if the four disulphide bonds of bovine pancreatic ribonuclease are reduced, although the product is inactive, the conformation must be fairly close to that of the native protein, since aerial reoxidation leads to recovery of most of the activity. Obviously, no appreciable scrambling of disulphide bonds occurs, as would be expected if the reduced enzyme assumed a widely different conformation. It is reasonable to conclude that proteins contain some amino-acid residues which are essential for biological activity, some may be redundant, and perhaps the majority determine the overall conformation and hence the spatial relationship between the essential residues. Consequently, a mutation which does not lead to a gross change in the chemical nature and size in the replacement amino-acid may well not impair and may indeed improve the biological efficiency of the protein molecule concerned. Such conservative replacements are favoured by the genetic code; thus UCU, UCC, UCA, and UCG all code for Ser whereas ACU, ACC, ACA, and ACG all code for Thr. Thus a base change in the first member of the triad only replaces one β-hydroxamino-acid by another. The third base in these triads does not affect the amino-acid introduced into the protein. Not all mutations are conservative and harmless. Replacement of Glu by Val at the sixth residue in the β-chain of haemoglobin is the cause at molecular level of sickle cell anaemia. The abnormal haemoglobin in its deoxygenated state tends to form insoluble polymers with consequent haemolysis. Mutations in proteins occur at regular rates which depend on the particular protein. Fibrinogen has evolved fairly rapidly, whereas cytochrome *c* has evolved much more slowly. Nevertheless, the number of amino-acid differences between homologous proteins from different species provides a method of dating the divergence of the species and thereby constructing a phylogenetic tree. Cytochrome *c* proteins from man and Rhesus monkey differ at only one amino-acid residue, whereas the proteins from man and a moth (*Samia cynthia*) differ at 31 locations.

Biological evolution has not only involved the replacement of amino-acid residues in the corresponding molecules from different species. It is obvious that gene doubling has occurred at various times, leading to the divergent evolution of homologous proteins from the same ancestral gene. Thus, about 40% of the amino-acid sequences of trypsin and chymotrypsin are identical and approximately another 10% represent conservative replacements. Not surprisingly, both molecules have similar conformations and biological functions.

In contrast, protein molecules are known whose primary, secondary and tertiary structures are almost totally different yet possess similar biological activity. A classical example is the case of chymotrypsin and the bacterial proteinase subtilisin. Despite the gross differences in structure, the essential amino-acids at the active sites of the enzymes are identical and perform their catalytic function by a similar mechanism. Such proteins are believed to have arisen by convergent evolution.

23.3.10 CORRELATION OF PROTEIN STRUCTURE AND BIOLOGICAL ACTIVITY

Several methods are available for attempting to identify the amino-acid residues which are responsible for the biological activity of proteins. Firstly, the protein can be subjected to limited degradation, especially at the *N*- and *C*-termini using aminopeptidases and carboxypeptidases, respectively. For example, up to three residues can be removed from the *C*-terminus of ribonuclease using carboxypeptidase and activity is retained. More extensive degradation in this region, however, leads to inactivation.

Secondly, the protein can be subjected to chemical modification in the side chains of the constituent amino-acids. Obviously, experimental results will be simpler to interpret if the modifying reagents are relatively specific so that only one or a few sites are affected. For example, the site at which the coenzyme, pyridoxal phosphate, binds to aminotransferase was easily identified. An aldimine, formed by condensation of the coenzyme with the ε-amino-group of a lysine residue, was reduced with sodium borohydride so that it survived hydrolytic degradation and could be identified. Likewise, enzymes with an essential thiol group such as alcohol dehydrogenase, 3-phosphoglyceraldehyde dehydrogenase, and papain are usually inhibited by reaction with *p*-chloromercuribenzoic acid or iodoacetic acid. Specificity of modification of proteins can be enhanced if the structure of the reagent simulates that of a molecule (substrate, coenzyme, allosteric effector) which normally binds to the protein. This technique is often referred to as affinity labelling.[31] Two examples will suffice to illustrate the technique. Glycyl-L-arginine (**17**) is a competitive inhibitor and poor substrate for carboxypeptidase B and therefore binds at the active site of the enzyme. α-*N*-Bromoacetylagmatine (**18**) irreversibly inhibits the enzyme and the phenolic hydroxyl group of Tyr^{248} is alkylated.[32] The second example concerns the enzyme oestradiol 17β-dehydrogenase. 3-Bromoacetoxyoestrone (**19**) reacts covalently with the human placental enzyme and can be reversibly reduced and oxidized as a 'fixed' substrate.[33] In contrast, 16α-bromoacetoxyoestradiol 3-methyl ether (**20**) irreversibly inhibits the enzyme by alkylating the imidazole nucleus of a histidine residue.[34]

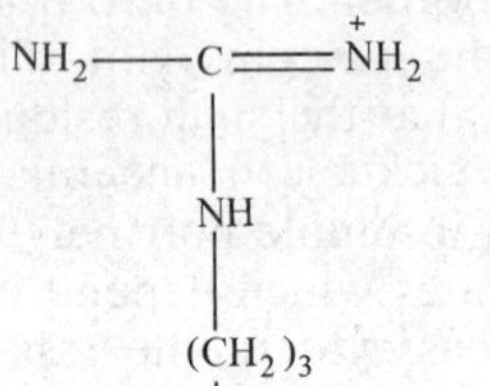

(17)

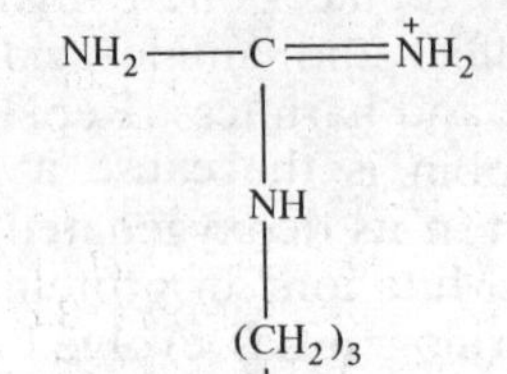

(18)

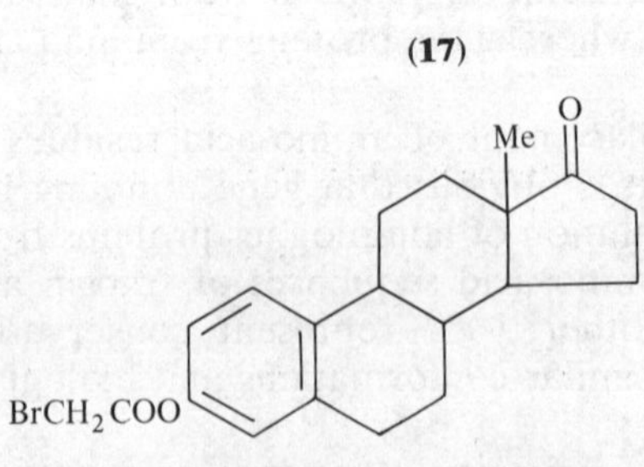

(19)

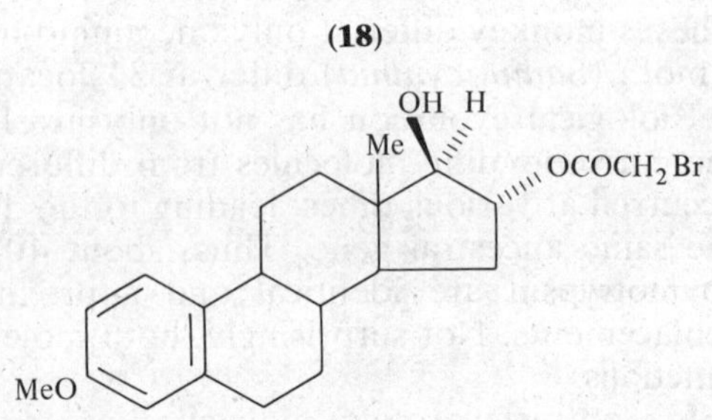

(20)

In the third method, chemical synthesis (Chapter 23.6) of analogues, especially of peptide hormones, has provided much useful information concerning the relationship between structure and biological activity. Gastrin, a heptadecapeptide amide from gastric antral mucosa, has the *C*-terminal sequence -Gly-Trp-Met-Asp-Phe-NH_2. Synthesis of this peptide amide has revealed that it possesses the full activity of the intact hormone. It is not necessary to confine the synthesis to include only the 20 amino-acids normally found in proteins. A fragment of β-corticotropin has been synthesized with α-aminobutyric acid in place of Met^4 and this was active, yet oxidation of Met^4 to the sulphoxide in the native hormone destroyed activity. It may be concluded that the polarity of the side chain of this residue is more important than its precise chemical structure.

Finally, useful information can be derived from comparison of the sequences of homologous proteins. The case of β-corticotropin has been cited above (Section 23.3.9), but many others are known. For example, the sequences around the essental serine and histidine residues in trypsin, chymotrypsin, elastase, thrombin, and plasmin are highly conserved. In addition, the sequence Asn-Pro-Lys-Lys-Tyr-Ile-Pro-Gly-Thr-Lys-Met (residues 70–80) is present in cytochrome *c* from numerous sources. Conservation of such sequences is a strong indication that they represent essential parts of the protein structure.

References

1. G. K. Ackers, in 'The Proteins', 3rd edn., ed. H. Neurath, R. L. Hill, and C.-L. Boeder, Academic Press, New York, 1975, vol. 1, chapter 1.
2. K. Weber and M. Osborn, in 'The Proteins', 3rd edn., ed. H. Neurath, R. L. Hill, and C.-L. Boeder, Academic Press, New York, 1975, vol. 1, chapter 3.
3. K. E. van Holde, in 'The Proteins', 3rd edn., ed. H. Neurath, R. L. Hill, and C.-L. Boeder, Academic Press, New York, 1975, vol. 1, chapter 4.
4. G. Barth, E. Bunnenberg, and C. Djerassi, *Analyt. Biochem.*, 1972, **48,** 471.
5. D. M. Kirschenbaum, *Analyt. Biochem.*, 1971, **44,** 159.
6. H. Edelhoch, *Biochemistry*, 1967, **6,** 1948.
7. T. Y. Liu and Y. H. Chang, *J. Biol. Chem.*, 1971, **246,** 2842.
8. R. J. Simpson, M. R. Neuberger, and T. Y. Liu, *J. Biol. Chem.*, 1976, **251,** 1936.
9. H. P. J. Bennett, D. F. Elliott, B. E. Evans, P. J. Lowry, and C. McMartin, *Biochem. J.*, 1972, **129,** 695.
10. F. P. Abramson, M. McCaman, and R. McCaman, *Analyt. Biochem*, 1974, **57,** 482.
11. A. Darbre, *Biochem. Soc. Trans.*, 1974, **2,** 70.
12. P. B. Hamilton, D. C. Bogue, and R. A. Anderson, *Analyt. Chem.*, 1960, **32,** 1782.
13. J. R. Benson, in 'Instrumentation in Amino Acid Sequence Analysis', ed. R. Perham, Academic Press, New York, 1975, chapter 1.
14. P. C. Jocelyn, 'Biochemistry of the SH group', Academic Press, New York, 1972, chapter 5.
15. F. Sanger, *Bull. Soc. Chim. biol.*, 1955, **37,** 23.
16. K. R. Woods and K. T. Wang, *Biochim. Biophys. Acta*, 1967, **133,** 369.
17. S. R. Burzynski, *Analyt. Biochem.*, 1975, **65,** 93.
18. P. Edman and G. Begg, *European J. Biochem.*, 1967, **1,** 80.
19. M. D. Waterfield and J. Bridgen, in 'Instrumentation in Amino Acid Sequence Analysis', ed. R. Perham, Academic Press, New York, 1975, chapter 2.
20. R. A. Laursen, *European J. Biochem.*, 1971, **20,** 89.
21. R. A. Laursen, A. G. Bonner, and M. J. Horn, in 'Instrumentation in Amino Acid Sequence Analysis', ed. R. Perham, Academic Press, New York, 1975, chapter 3.
22. E. Atherton, J. Bridgen, and R. C. Sheppard, *FEBS Letters*, 1976, **64,** 173.
23. J. Bridgen, A. P. Graffeo, B. L. Karger and M. D. Waterfield, in 'Instrumentation in Amino Acid Sequence Analysis', ed. R. Perham, Academic Press, New York, 1975, chapter 4.
24. R. J. DeLange and E. L. Smith, in 'The Enzymes' 3rd edn., ed. P. D. Boyer, Academic Press, New York, vol. 3, chapter 3.
25. M. Wingard, G. Matsueda, and R. S. Wolfe, *J. Bacteriol.*, 1972, **112,** 940.
26. S. Doonan, H. J. Doonan, R. Hanford, C. A. Vernon, J. M. Walker, F. Bossa, D. Barra, M. Carloni, P. Fasella, F. Riva, and P. L. Walton, *FEBS Letters*, 1974, **38,** 229.
27. J. Houmard and G. R. Drapeau, *Proc. Nat. Acad. Sci. U.S.A.*, 1972, **69,** 3506.
28. S.-I. Hakomori, *J. Biochem. (Japan)*, 1964, **55,** 205.
29. E. Vilkas, E. Lederer, and J. C. Massor, *Tetrahedron Letters*, 1968, 3089.

30. H. R. Morris and A. Dell, in 'Instrumentation in Amino Acid Sequence Analysis', ed. R. Perham, Academic Press, New York, 1975, chapter 5.
31. E. Shaw, *Physiol. Rev.*, 1970, **50,** 244.
32. T. H. Plummer, Jr. and M. T. Kimmel, *J. Biol. Chem.*, 1969, **244,** 5246.
33. E. V. Groman, R. M. Schultz, and L. L. Engel, *J. Biol. Chem.*, 1975, **250,** 5450.
34. J. A. Katzenellenbogen, T. S. Ruh, K. E. Carlson, H. S. Iwamoto, and J. Gorski, *Biochemistry*, 1975, **14,** 2310.

23.4

Naturally Occurring Low Molecular Weight Peptides

B. W. BYCROFT
University of Nottingham

Naturally occurring low molecular weight peptides can be conveniently divided, on structural and biosynthetic grounds, into two distinct groups: those with structures related to protein and those with structural features atypical of protein.

The distinction in the former group between peptides and proteins is problematic, and historically has been based on molecular size. The ability to pass through a natural dialysis membrane normally defines the upper limit for a peptide and this corresponds to a molecular weight of *ca.* 10 000 or approximately 100 amino-acid residues. For the purpose of this review, low molecular weight peptides are considered to range from dipeptides to peptides containing approximately 50 amino-acid residues. The major interest in this area over the last quarter of a century has centred largely on peptides with hormonal activity and there have been notable advances in this field, which represents an important interface between biology and chemistry.

The latter group of peptides is structurally much more diverse and encompasses all the compounds containing two or more amino-acids linked by an amide bond but which possess some structural features not characteristic of protein. These include unusual amino-acids, *i.e.* those not found in protein, amino-acids with the D-configuration, or, at a higher oxidative level, unusual amide linkages, *e.g.* γ-glutamyl peptides, ester linkages (depsipeptides) and various cyclic structures. These peptides have been mainly isolated from micro-organisms and many possess substantial biological activity. Some have been noted because of their toxicity to plants and animals, while others have found application as antibacterial, antitumour, and antiviral agents. Ionophoric peptides have been used as valuable tools to study ion transport across natural and artificial membranes.

It is perhaps significant that the individual members of both these groups of peptides have been isolated and characterized because of some specific biological activity. In the future, it is probable that more sophisticated biological assay techniques or detailed chemical analysis of the biological material will reveal an even greater variety of structural types than is already apparent.

As well as the structural differences, there is increasing evidence that the biosynthesis of both sets of peptides differ substantially. The majority of peptide hormones are derived by enzymatic cleavage of larger protein molecules (prohormones), the biosynthesis of which takes place at the ribosomes and is under rigid genetic control. For the biosynthesis of the peptide antibiotics, the evidence clearly points to a non-ribosomal enzymatic process which is less discriminating and which often leads to the production by the same organism of mixtures of closely related antibiotics differing only at one of the amino-acid sites.

The conjugates of both groups of peptides with sugars, lipids, and nucleosides have been excluded from this review.

23.4.1 LOW MOLECULAR WEIGHT PEPTIDES RELATED TO PROTEIN

In view of the breadth of this topic, discussion is limited to structural and biosynthetic aspects; the principles of chemical synthesis employed in this area are reviewed elsewhere in this volume. The considerable number of primary structures of small peptides so far characterized are conveniently tabulated in a valuable publication.[1] The rapidly accumulating sequence data have provided interesting clues to evolutionary and functional interrelationships among peptides. In many cases, peptides which perform functionally identical tasks in related organisms exhibit subtle differences in their primary structure. The species specificity of a peptide is due to the rigid genetic control of the biosynthesis of the protein from which it is derived.

The past two decades have seen great advances in the understanding of protein biosynthesis.[2,3] A general outline of the process is shown in Scheme 1 (see also Chapter 22.5). The primary structure of a protein, *i.e.* the amino-acid sequence, is encoded as a sequence of bases in the deoxyribonucleic acid (DNA) of a gene within the nucleus of the cell. The expression of this genetic information is mediated by messenger ribonucleic acid molecules (mRNA) on to which the information is transcribed by complementary pairing of the bases between the mRNA and DNA. The actual construction of the protein takes place on the ribosomes, which are cellular structures distributed throughout the cytoplasm of the cell. Complementary interaction between the mRNA and the aminoacyl transfer RNAs (there is a specific transfer RNA for each of the 20 protein amino-acids) during the course of the synthesis ensures that the genetic information is translated into a precise amino-acid sequence. The extraordinarily high fidelity[4] of this decoding process makes certain that functional peptides and protein are reproduced accurately from generation to generation.

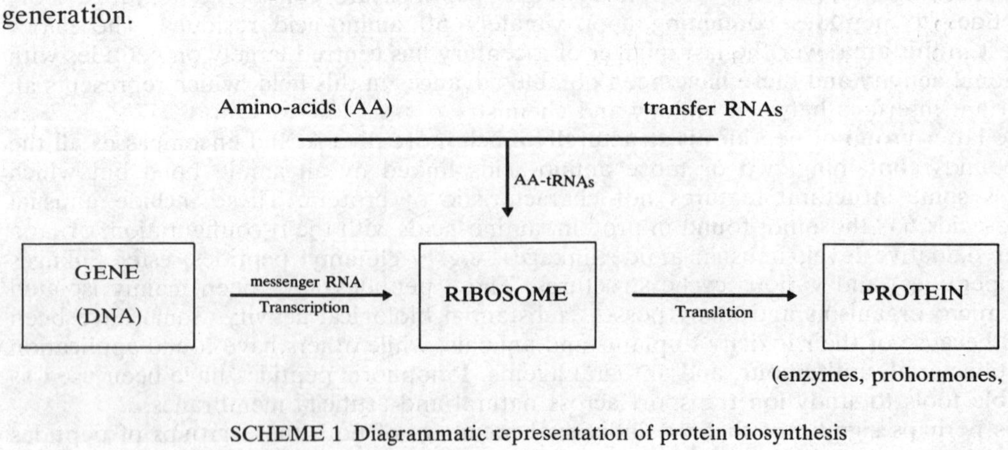

SCHEME 1 Diagrammatic representation of protein biosynthesis

The variations of the primary structure of peptides serving homologous functions among different organisms is illustrated with the vertebrate neurohypophysial hormones (Table 1). Oxytocin and vasopressin have occupied a unique position in the development of peptide chemistry since the classic pioneering investigations of du Vigneaud in the early 1950s.[5] Recent studies on the distribution of these hormones have shown that the common structural pattern characterized by nine amino-acid residues with the disulphide bridge connecting the 1 and 6 positions is general, and that variations of the amino acids are limited to the 3, 4, and 8 positions. The differences have been ascribed to genetic changes from a common neurohypophysial hormone in the course of the evolution of vertebrates.[6]

TABLE 1
Structure of Vertebrate Neurohypophysical Hormones*

(a)	Cys—Tyr—Ile—Ser—Asn—Cys—Pro—Gln—Gly—NH_2 1 2 3 4 5 6 7 8 9 glumitocin	Cys—Tyr—Ile—Gln—Asn—Cys—Pro—Arg—Gly—NH_2 1 2 3 4 5 6 7 8 9 vasotocin
(b)	Cys—Tyr—Ile—Ser—Asn—Cys—Pro—Ile—Gly—NH_2 isotocin	vasotocin
(c)	Cys—Tyr—Ile—Gln—Asn—Cys—Pro—Ile—Gly—NH_2 mesotocin	vasotocin
(d)	Cys—Tyr—Ile—Gln—Asn—Cys—Pro—Leu—Gly—NH_2 oxytocin	Cys—Tyr—Phe—Gln—Asn—Cys—Pro—Arg—Gly—NH_2 arginine-vasopressin

* (a) Cartilagineus fishes (rays); (b) bony fishes; (c) amphibians; (d) mammals (except the porcine hormones, which are oxytocin and lysine-vasopressin).

It is of considerable interest to note that variation of the basic structures by chemical synthesis has led to some 200 analogues.[7] Substitution in positions other than 3, 4, or 8 leads to inactive or weakly active compounds and only the 4 and 8 positions can be replaced by an alternative amino-acid without dramatic loss of activity. Indeed, [4-Thr]-oxytocin (**1**) is a highly active synthetic analogue which has so far not been encountered in nature. The considerable effort expended in modifying the biological activity of these natural hormones has been rewarded recently with the introduction of deamino-[8-D-Arg]-vasopressin (**2**) as a specific drug for the treatment of diabetes insipidus.[8] The problems of the intrinsic metabolic instability of peptides have been overcome, in this particular case, by the introduction of a D-amino-acid residue.

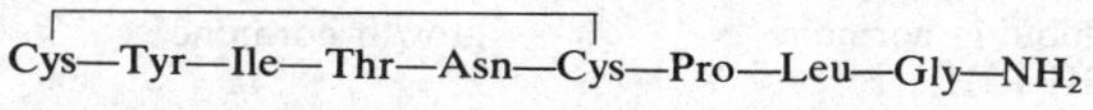

(**1**)

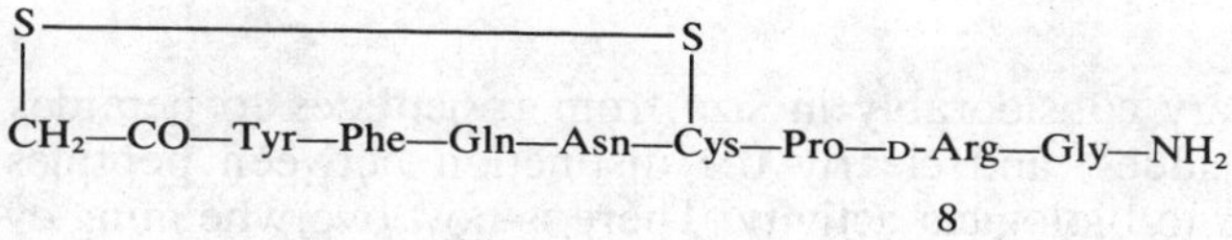

(**2**)

23.4.1.1 Peptide hormones

The hormones secreted by the endocrine glands are for the most part excitatory in their actions; they stimulate the growth and development, or the functional activity of certain tissues. Some of the important peptide hormones which have been fully characterized are listed in Table 2, together with the gland that secretes them and a brief outline of the physiological function. The list is not intended to be comprehensive, but to give an overall view, as well as clarifying the rather confusing nomenclature which has arisen in this area. For a more detailed treatment of the endocrinology, structure–activity relationships, and biochemistry the reader is referred to recent reviews.[9–11]

TABLE 2

Peptide Hormones[a]

Gland	*Hormone*	*Principal function*
Pancreas	Insulin (51)	Regulates metabolism of carbohydrate, fat, and protein
Gastrointestinal tract	Secretin (27)	
	Cholecystokinin-pancreozymin (CCK) (33)	Influences secretions of the gastrointestinal tract
	Gastrin (17)	
	Vasoactive intestinal peptide (VIP) (28)	
Placenta	Placental lactogen (PL) (191)	Prolactin and growth activities
Thyroid and para-thyroid	Parathyroid hormone (PH) (84)	Regulates metabolism of calcium and phosphorus
	Calcitonin (32)	
Pituitary:		
anterior lobe	Adrenocorticotrophin (ACTH) (34)	Controls activity of adrenal cortex
	Growth hormone (GH) (191)	Controls growth of skeleton and tissue
	Prolactin (198)	Influences growth of mammary glands and induces secretion of milk
intermediate lobe	α and β melanophore stimulating hormone (MSH) (13 & 9)	Disperses melanin granules in the skin
posterior lobe	Vasopressin (9)	Acts as antidivetic and vasoconstrictor
	Oxytocin (9)	Stimulates contractions of smooth muscle
Hypothalamus	Thyrotropin releasing hormone (TRH) (thyroliberin) (3)	Control the release and and synthesis of hormones of the anterior pituitary
	Luteinizing hormone releasing hormone (LRH) (gonadoliberin) (10)	
	Growth hormone release inhibiting hormone (GH-RIH) (somatostatin) (14)	Inhibits the release of growth hormone

[a] Abbreviations and number of amino-acid residues in parentheses.

The molecules vary considerably in size from tripeptides to peptides with more than 190 amino-acid residues, and clearly the distinction between peptides and proteins is arbitrary in relation to biological activity. There is now overwhelming evidence that most of the peptide hormones are derived from larger prohormones.[10] This process is well documented for insulin (**3**), and the proinsulins together with the cleaving enzymes have been isolated (Scheme 2). The insulins are unusual in that they consist of two peptide chains[12] linked to each other by two disulphide bridges, and it had long been a puzzle how living systems constructed these molecules. Similarly, the effectiveness of partial and total syntheses has been limited by the lack of specificity in the oxidative coupling of the two chains.[13]

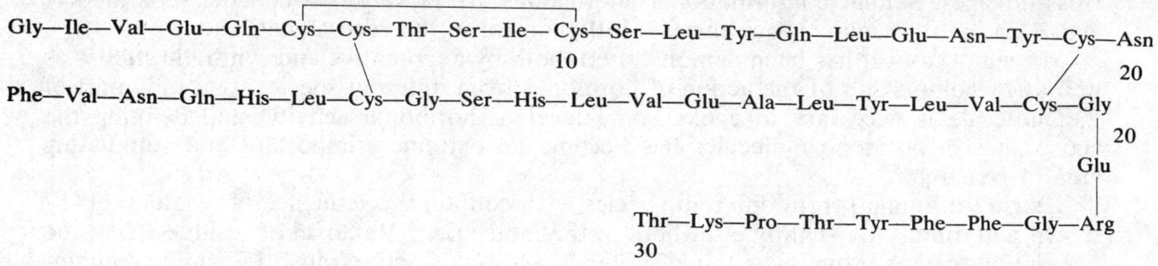

(**3**)

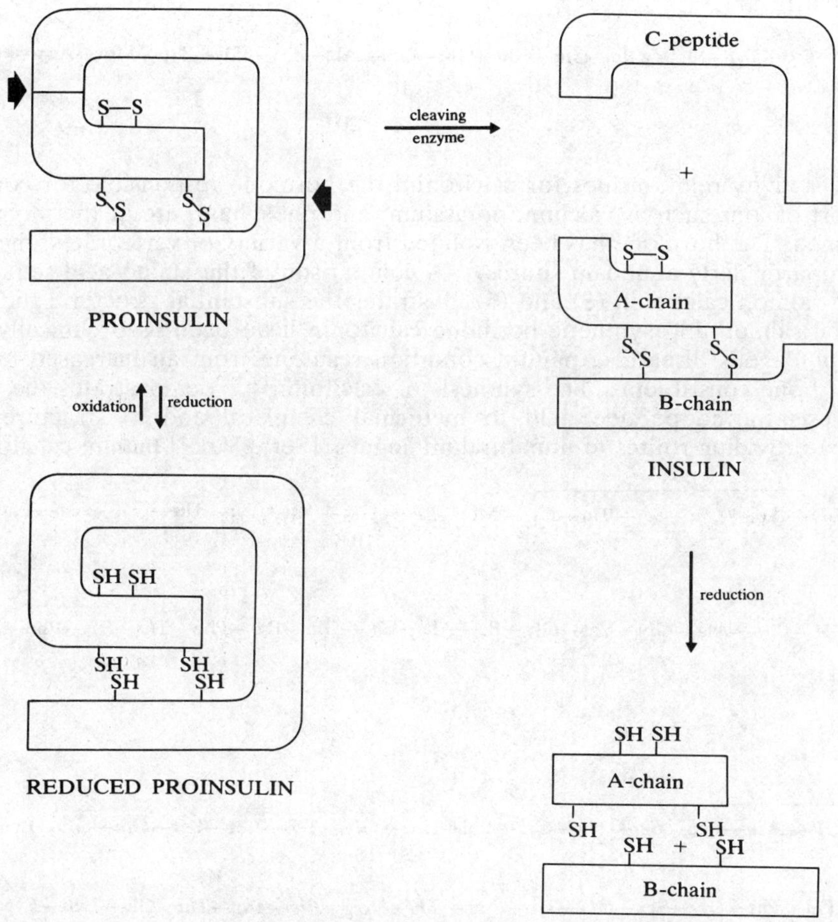

SCHEME 2 The derivation of insulin from proinsulin

These observations find a convenient explanation in the structure of the proinsulins. The molecules consist of a single peptide chain, and insulin together with an inactive fragment (which has been termed the C-peptide) are formed by proteolytic action at the two sites indicated (Scheme 2). The reduced form of proinsulin can be reoxidized in high yield to the prohormone. The geometry of the reduced molecule which is inherent in the primary structure of the peptide chain controls the specific formation of the disulphide cross-linkages. Similar conformational interactions are lacking between the separated A and B chains of insulin; hence there is little control in the recombination.

In many cases it has been demonstrated, both by degradative and synthetic studies as well as by comparison of the action of hormones from different species, that only part of the molecule is necessary to evoke some level of hormonal activity, and defining the active sites in hormone molecules has become an extremely important and stimulating area of research.[14]

Gastrin (**4**; human) from different species each contain the sequence of residues 11–17 shown and differ only slightly elsewhere in the molecule.[15] Removal of residues from the *N*-terminus causes progressive loss of potency. However, the tetrapeptide amide containing the sequence 14–17 still retains the full range of physiological activity but is only about one-tenth as potent as the natural hormone.[16] So far no special function has been demonstrated for the residues 1–13, which contain the unusual sequence of five adjacent glutamic acid residues.

Glu—Gly—Pro—Tyr—Met—Glu—Glu—Glu—Glu—Glu—Ala—Tyr—Gly—Trp—Met—Asp—PheNH_2

10 17

(4)

Structure–activity relationships for calcitonin, the hormone responsible for controlling the transport of ions such as calcium, potassium, and phosphate, are at the moment less clearly defined. The hormone has been isolated from a variety of vertebrates, the salmon providing a particularly abundant source.[17] A comparison of the amino-acid sequence for human and salmon calcitonin (**5**) and (**6**) illustrates the substantial structural differences. Both natural salmon and synthetic hormone calcitonin have been used clinically for the treatment of Paget's disease, a painful condition resulting from an increased metabolic turnover of bone constituents. The synthesis of calcitonin[17,18] demonstrates the value of total synthesis in the peptide field to medicinal chemistry and to structure–activity studies,[14] by providing routes to unnatural analogues. [Ser^{29},Val^{31}]-human calcitonin has,

Cys—*Gly*—*Asn*—*Leu*—*Ser*—*Thr*—*Cys*—Met—*Leu*—*Gly*—Thr—Tyr—Thr—*Gln*—Asp—Phe—Asn

10

NH_2—*Pro*—Ala—*Gly*—Val—*Gly*—Ile—Ala—*Thr*—Gln—*Pro*—Phe—*Thr*—His—Phe—*Lys*

30 20

(5)

Cys—*Gly*—*Asn*—*Leu*—*Ser*—*Thr*—*Cys*—Val—*Leu*—*Gly*—Lys—Leu—Ser—*Gln*—Glu—Leu—His

10

NH_2—*Pro*—Thr—*Gly*—Ser—*Gly*—Thr—Asn—*Thr*—Arg—*Pro*—Tyr—*Thr*—Gln—Leu—*Lys*

30 20

(6)

(Residues common to both species in italics)

for example, some five times the activity of the human hormone and longer duration of action.

The isolation and characterization of peptide hormones is normally painstaking and laborious work; none more so than the hypothalamic hormones.[19] The hypothalamus is a region of the brain tissue which operates a fine control on the endocrine system, influencing the release or inhibition of hormones from the anterior lobe of the pituitary gland. Only nanogram quantities of hormones are present in the tissue of a single animal. The initial investigations concerning the thyrotropin releasing hormone (TRH) involved the prodigious task of processing the extracts of hundreds of thousands of porcine hypothalami and even this gave only partially purified material. The amino-acids found in the hydrolysate were initially regarded as imputities, and it was only when the three amino-acids histidine, proline, and glutamic acid were found in equimolar quantities in sufficiently pure preparations that the peptide nature of the molecule was suspected. The six possible isomeric tripeptides were synthesized, but all were found to be biologically inactive. Subsequently, several lines of enquiry led to a consideration of a pyroglutamic acid residue and a terminal amide, and these observations finally resulted in the synthesis of (**7**), which proved to be identical with the natural TRH.[20,21] Thus the synthesis of the hormone and the determination of its constitution were achieved simultaneously.

A further factor, the luteinizing hormone releasing hormone (LRH), was later shown to be a decapeptide (**8**) which also possesses an *N*-terminal pyroglutamic acid residue.[22,23] More recently a number of other releasing and inhibitory factors have also been isolated and are in various stages of characterization. Growth hormone release inhibiting hormone (GH-RIH) or somatostatin (**9**) has been fully characterized[24] and the synthesis[25] of the

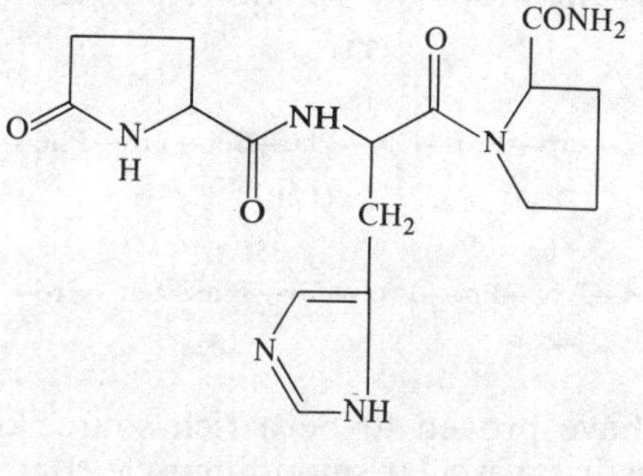

(**7**)

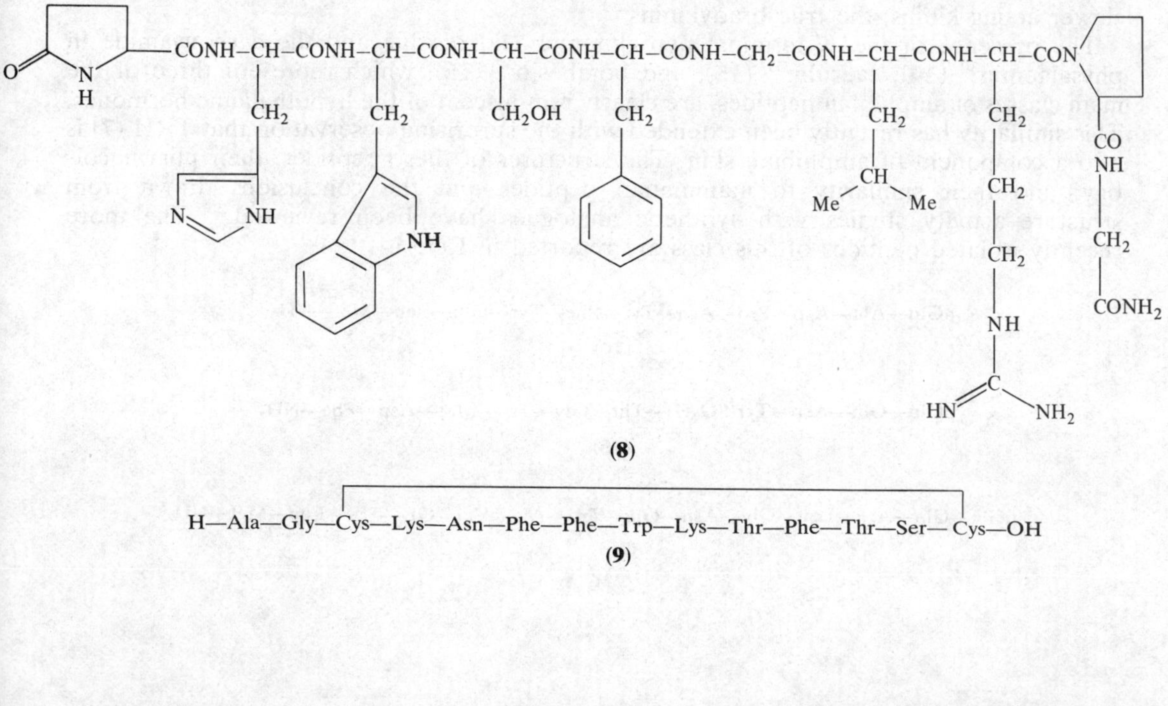

(**8**)

(**9**)

hormone along with several analogues have been described.[14] No species specificity has yet been reported for the hypothalamic hormones.

There is evidence that the TRH may be anomalous in regard to its biosynthesis. The tripeptide can be formed *in vitro* from the precursor amino-acids and hypothalamic extracts. The synthesis is not inhibited by protein-blocking antibiotics but it is inhibited by the addition of iodoacetamide or mercuric chloride, which inactivates most enzyme systems.[26] This suggests that the synthesis may be carried out enzymatically by a TRH synthetase, perhaps similar to the glutathione synthetase (see Section 23.4.2.1).

23.4.1.2 Vasoactive peptides

It has long been known that several vasoactive peptides are produced by the action of proteolytic enzymes on inactive proteins present in the blood plasma. These include the closely related peptides angiotensin I (**10**) and II (**11**), bradykinin (**12**), and khallidin (**13**). Much of the biochemistry associated with the release of these peptides has been extensively investigated[27,28] and is now well understood. The compounds are synthetically readily accessible and therefore have been the subject of intensive studies in a search for potential antihypertensive agents.[27]

Asp—Arg—Val—Tyr—Ile—His—Pro—Phe—His—Leu

(10)

Asp—Arg—Val—Tyr—Ile—His—Pro—Phe

(11)

Arg—Pro—Pro—Gly—Phe—Ser—Pro—Phe—Arg

(12)

Lys—Arg—Pro—Pro—Gly—Phe—Ser—Pro—Phe—Arg

(13)

The skin of amphibia have proved to be a rich source of peptides which also exert powerful effects on mammalian vascular smooth muscle. Because of the prompt nature of their contractile action they have been termed tachykinins, to distinguish them from the slower acting kinins, the true bradykinins.[29]

The presence of the *N*-terminal pyroglutamic acid residue and the carboxyamide in physalaemin[30] (**14**), caerulin[31] (**15**), and bombesin[32] (**16**), which represent three of the main classes of amphibian peptides, are clearly reminiscent of the hypothalamic hormones. This similarity has recently been extended with the surprising observation that TRH (**7**) is also a component of amphibian skin. The structures of these peptides, their pharmacology, and their similarity to mammalian peptides and the conclusions drawn from structure–activity studies with synthetic analogues have been reviewed.[33] The more recently isolated peptides of this class are reported in Ref. 14.

pGlu—Ala—Asp—Pro—Asn—Tyr—Phe—Tyr—Glu—Leu—Met—NH_2

(14)

pGlu—Gln—Asp—Tyr(SO_3H)—Thr—Gly—Trp—Met—Asp—Phe—NH_2

(15)

pGlu—Gln—Arg—Leu—Gly—Asn—Glu—Trp—Ala—Val—Gly—His—Leu—Met—NH_2

(16)

23.4.1.3 Peptides acting on the nervous system

Recent results in a number of fields point to the widespread presence and action of small peptides in nervous tissue. In particular the long postulated and much sort after natural opiates have been isolated from pig brain and shown to be a pair of simple pentapeptides[34] (**17**) and (**18**). The existence of such physiological transmitter-like molecules has long been suspected in order to explain the function of the morphine receptors which are significant in transmission of pain, but for which morphine and related compounds are not the usual agonists. It is of interest to note that in both cases the *N*-terminal amino-acid is tyrosine, and this has invoked considerable discussion concerning the topology of the peptides and their relationship to that of morphine.[35]

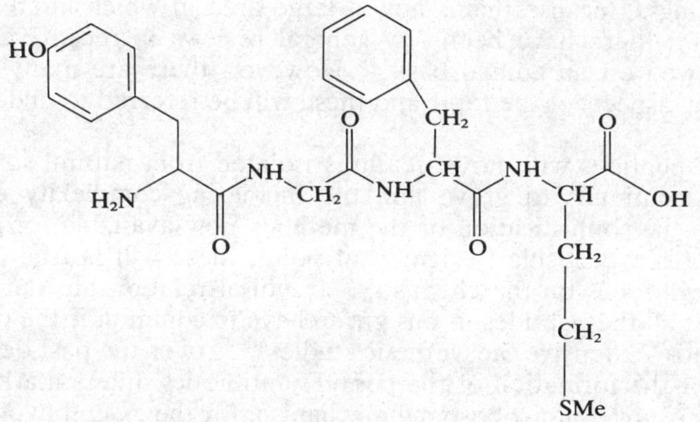

(**17**)

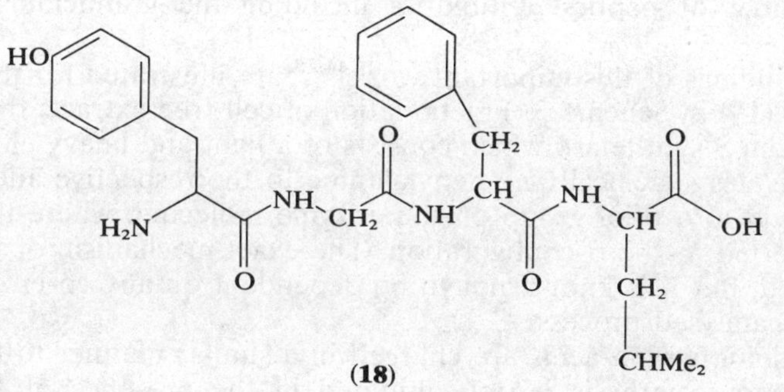

(**18**)

The hypothalamic hormones have been shown to act on the central nervous system. For example, injections of low doses of TRH (**7**) into the cerebral ventricle of cats causes a substantial fall in body temperature.[36] This has led to the proposal that a neurosecretory process operates which carries chemical signals to the higher centres of the brain, and has opened a new area of medicinal chemistry. Analogues of hypothalamic hormones are now being made and tested both for hormonal and CNS activity.

Perhaps the studies in this area which have attracted most general attention, however, concern the peptides isolated from the brains of animals conditioned to respond to certain stimuli. There are several claims that these peptides are capable of transferring the conditioned response when injected into control animals.[37] Scotophobin is the best known of these substances. The biological results have been the subject of some controversy and recent chemical data have not helped resolve the confusion. A synthetic peptide with the sequence originally proposed has been found to be inactive, but the activity of the natural peptide has been verified.[38] The field is undoubtedly in the early stages of development, but obviously has considerable interest with respect to the chemistry and biology, and perhaps significant implications in much wider spheres.

23.4.2 LOW MOLECULAR WEIGHT PEPTIDES WITH STRUCTURAL FEATURES ATYPICAL OF PROTEIN

As outlined in the introduction, this is a structurally diverse group of compounds, and there are well-defined classes of compounds within the group. One major class which is undoubtedly modified peptides, namely the penicillins and cephalosporins, has been omitted, because this has now become such a large field that it warrants a completely separate discussion (see Chapter 23.5).

In an attempt to maintain a coherent pattern, the group has been subdivided into small linear peptides, cyclic dipeptides (2,5-dioxopiperazine), cyclic homodetic and heterodetic, large, and miscellaneous peptides; the various classes of related compounds are described under these headings. Because this is now a large area in which interest is not confined solely to chemistry, there have been few general reviews in recent years, although the literature is reviewed on an annual basis.[39] However, there are many excellent reviews covering individual aspects of the topic and these will be referred to under the appropriate heading.

The number of peptides with novel features isolated from natural sources, particularly micro-organisms, continues to grow, and the increasing complexity of some of these molecules reflects the sophistication of the methods now available for structure elucidation. Unless there is a particularly significant point, these will not be described and the major attention will focus on the chemistry, structural relationships and biosynthesis.

The feature that all the peptides in this group have in common is the presence of one or more peptide bonds. Extensive biosynthetic studies[40–43] over the past decade indicate that the mechanism for the formation of the peptide antibiotics differs markedly from that of protein synthesis. A novel non-ribosomal mechanism for the assembly of amino-acids with the sequence determined by protein templates is emerging, and has been investigated in detail for a number of peptide antibiotics including the gramicidins, tyrocidine, and bacitracin.

The principal findings of this important work[44,45] are illustrated for the cyclic decapeptide gramicidin S (**19**) in Scheme 3. Fractionation of cell-free extracts from *Bacillus brevis* affords a gramicidin S synthetase which consists of a light and heavy enzyme system. The light enzyme activates specifically L-phenylalanine to the respective adenylate and transfers the amino-acid to a thiol group on the enzyme molecule, where the chirality at the α-centre is converted to the D-configuration. The exact mechanism of this process is not clearly understood, but it is known not to be dependent on any co-enzyme factor and is probably a base-catalysed process.

The four remaining amino-acids are charged in a similar manner to the heavy enzyme system as thioesters. Synthesis is only initiated in the presence of the phenylalanine charged light enzyme, whereupon the D-phenylalanine is transferred from the light to the heavy enzyme with concomitant formation of a peptide bond between the phenylalanine and the proline. The further elongation of the enzyme bound peptide chain up to the pentapeptide occurs by the serial addition of the respective amino-acids as shown. Following the addition of the final amino-acid leucine, gramicidin is released from the enzyme complex. The exact mechanism of the head-to-tail cyclization of the two pentapeptides is not yet clear.

The process of the chain elongation is considered to involve the participation of a phosphopantetheine residue which has been detected in the heavy enzyme system. It is proposed that the pantetheine acts as a swinging arm thiol donor, accepting and donating the growing peptide chain to and from the amino-acid binding sites by a transthiolation process, until the terminal amino-acid is reached. The similarity with the synthesis of fatty acids has been pointed out and a detailed comparison of the two processes made.[46]

Although the system requires the presence of all five amino-acids to operate, it is possible to replace a number of them by analogues, *e.g.* the light enzyme can be charged with tyrosine or tryptophan and will subsequently afford the corresponding substituted gramicidin. Acceptable quantities of antibiotics can be obtained from the cell-free

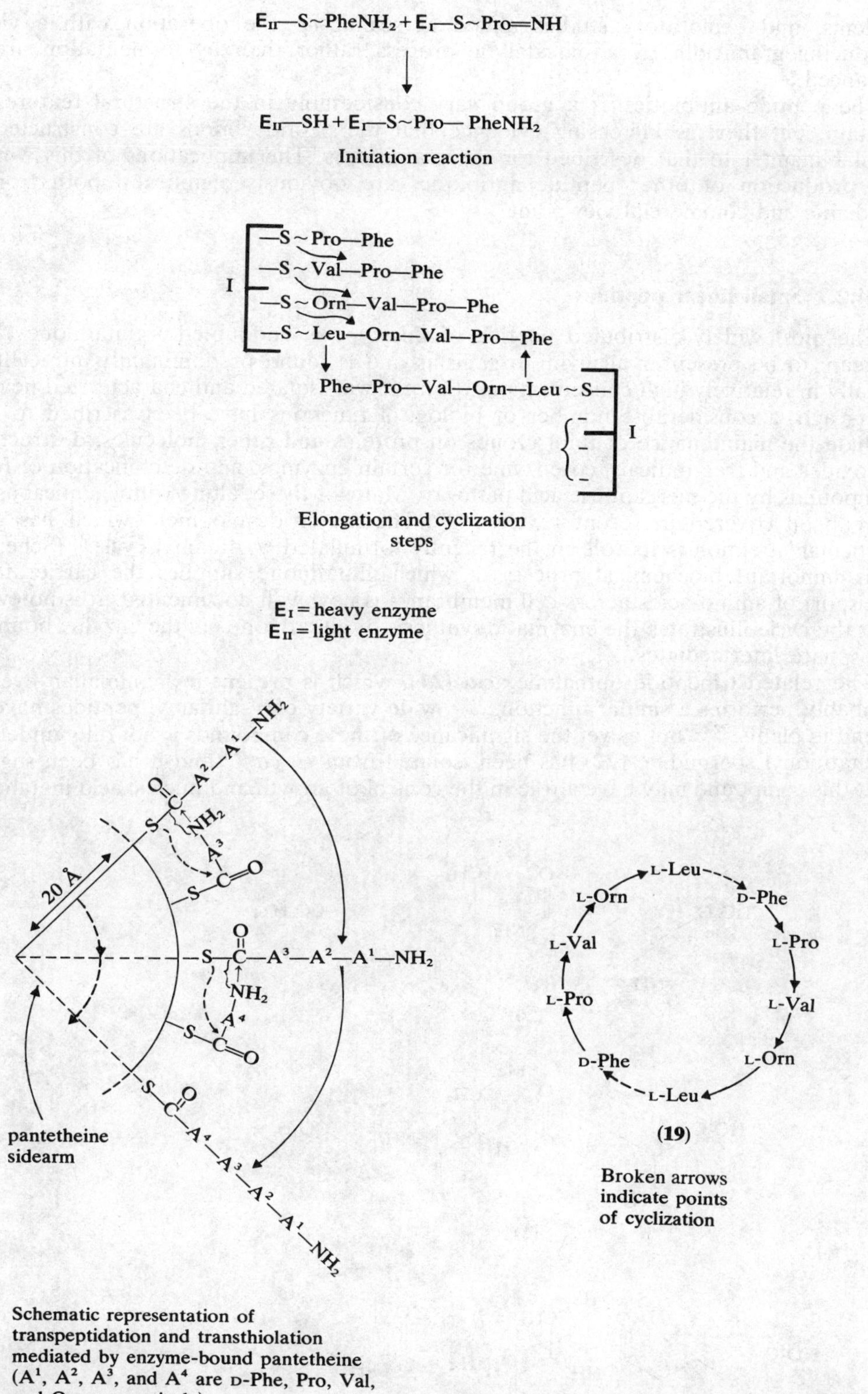

SCHEME 3 Non-ribosomal biosynthesis of gramicidin S (Reproduced with permission from Lipmann[45])

systems, and exploratory studies aimed at scaling-up the operation with a view to producing gramicidin by a biocatalytic process rather than by fermentation are well advanced.[47]

The peptide antibiotics as a group vary considerably in the structural features they contain, but there is increasing evidence that the peptide bonds are constructed in a similar manner to that described for the gramicidins. The implications of this work for the production of other peptide antibiotics are obviously significant, both from the academic and commercial viewpoint.

23.4.2.1 Small linear peptides

The most widely distributed peptide of this type is undoubtedly glutathione (**20**). It appears to be present in all living organisms and is found predominantly intracellularly, usually in relatively high concentrations. Since it was isolated and characterized nearly 50 years ago, a considerable number of biological functions have been ascribed to it and include the maintenance of thiol groups on proteins and other molecules, destruction of peroxides and free radicals, co-enzyme for certain enzymes, and detoxification of foreign compounds by the mercapturic acid pathway. Many of these, along with chemical aspects, have been covered in recent reviews.[48,49] The major development which has gained particular attention is its role in the recently formulated γ-glutamyl cycle[50] (Scheme 4). This important biochemical process in which glutathione supplies the carrier for the transport of amino-acids across cell membranes is now well documented. It is noteworthy that the cycle illustrates the enzymatic synthesis of glutathione *via* the enzyme bound acyl phosphate intermediates.

The related tripeptide opthalmic acid (**21**), which is present in mammalian eyes lens, probably performs a similar function.[51] A wide variety of γ-glutamyl peptides have been found in plants,[52,53] but as yet the significance of these compounds is not fully understood. Glutathionyl spermidine (**22**) has been isolated from *E. coli*[54] and it has been suggested that this compound may have a role in the control of growth and nucleic acid metabolism.

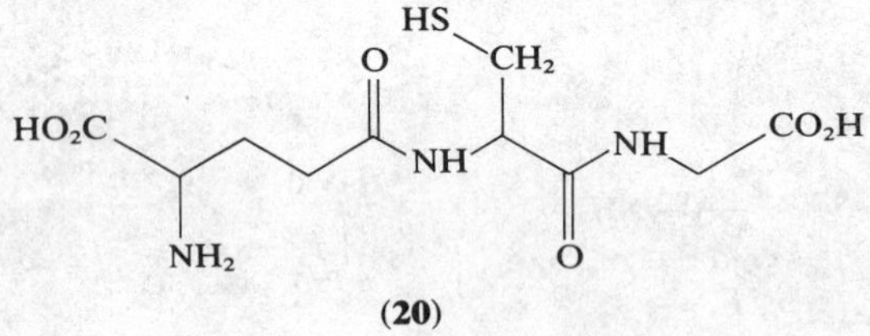

(20)

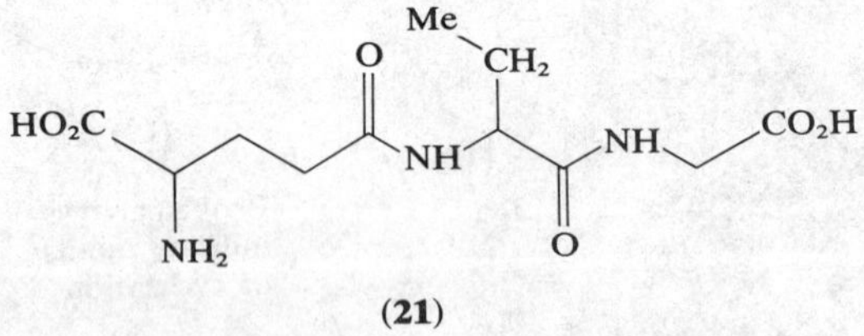

(21)

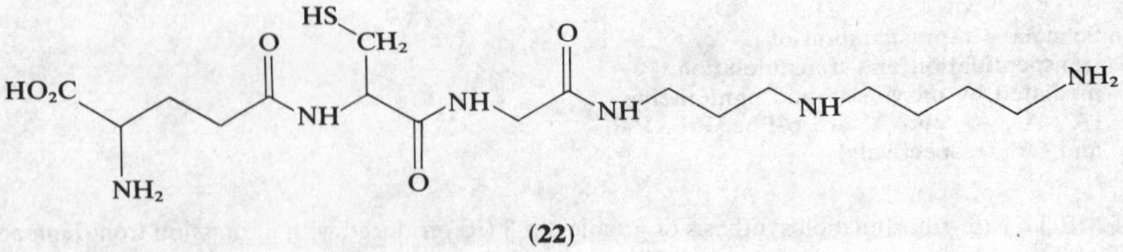

(22)

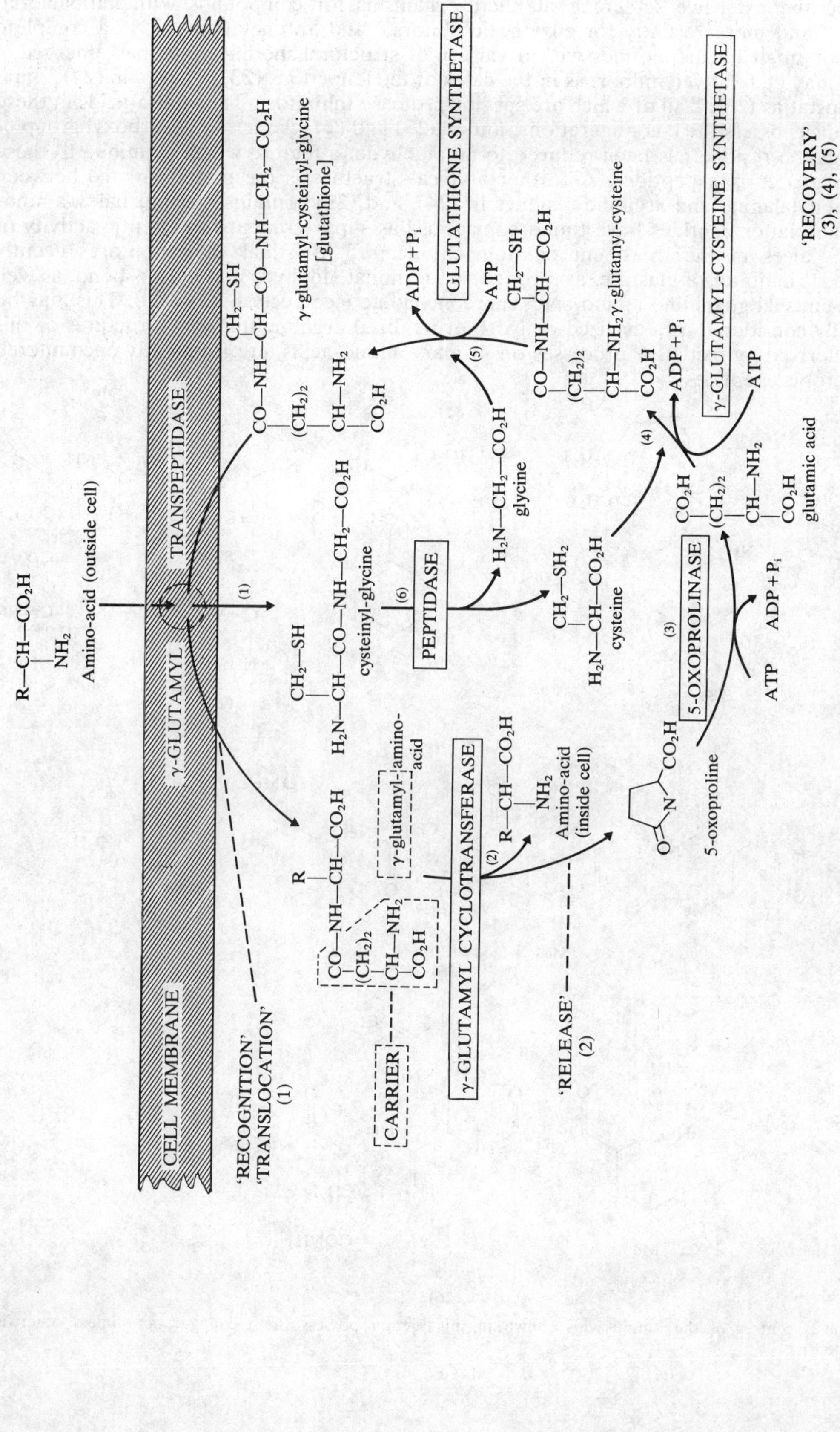

SCHEME 4 The glutathione cycle
(Reproduced with permission from A. Meister, *Science*, 1973, **33**, 180)

From the extensive screening of micro-organisms for compounds with antibacterial activity, and more recently for enzyme inhibitors[55] and antimetabolites,[56,57] a complete range of small linear peptides with a variety of structural modifications has emerged.[57] These may be relatively minor, as in the cases of the leupeptins (**23**),[58] antipain (**24**),[59] and the pepstatins (**25**),[60] all of which are specific protease inhibitors. The amino-acids in these molecules possess the L-configuration,* and in (**23**) and (**24**) the terminal carboxyl group of the arginine residue has been reduced to an aldehyde, a feature which is unique to these naturally occurring peptides. A further atypical structure is the ureido-linkage between the phenylalanine and arginine residues in (**24**), and (**25**) contains the unusual γ-amino-acid. The latter peptides have found clinical use as suppressants of the peptic activity of gastric juices in the treatment of stomach ulcers.[55] Elastinal (**26**), a more recently isolated[61] inhibitor of elastase, also contains a terminal aldehyde and ureido-bond, as well as the unusual guanidino amino-acid, capreomycidine (see Section 23.4.2.3). This may be formally considered as a cyclized dehydro or oxidized arginine unit and structures of this type, derived by oxidative processes on primary amino-acids, are frequently encountered in microbial peptides.[62]

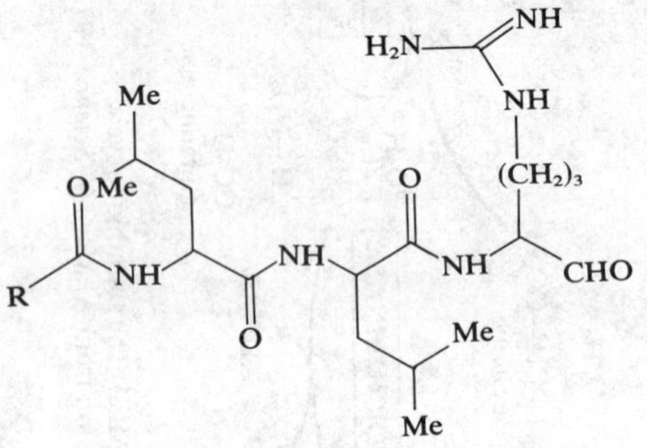

(23)

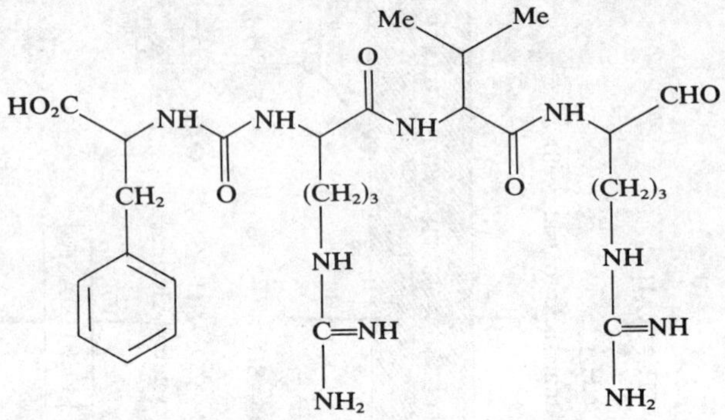

(24)

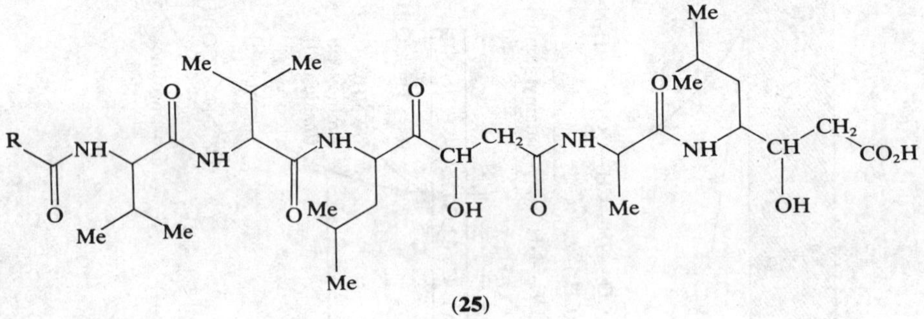

(25)

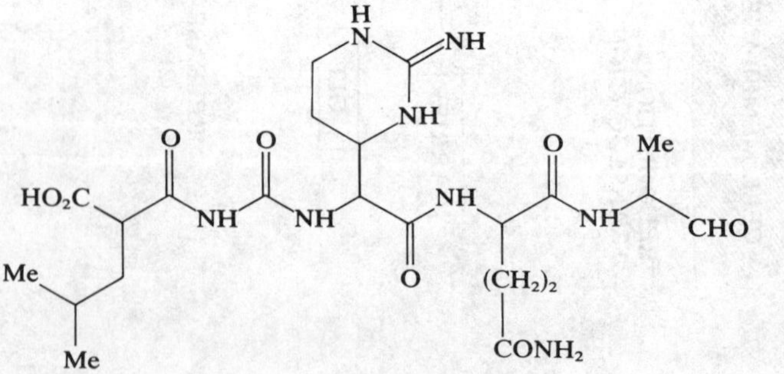

(26)

* The α-centres of the amino-acids shown in this section possess the L-configuration unless otherwise indicated.

It has been pointed out that amino-acids, with both primary and unusual structures, often occur with the D-configuration. For primary amino-acids with two chiral centres, the inversion of configuration normally occurs only at the α-centre,[63] as demonstrated in the simple tripeptide (**27**) which incorporates an *allo*-D-threonine residue.[64] Stravidin[65] (**28**) exemplifies two further anomalous features, namely the frequent occurrence of *N*-methylated amino-acids, although rarely with D-configuration, and non-protein amino-acids.

Non-protein or unusual amino-acids abound in microbial peptides. In some the derivation from primary amino-acids is apparent, while for others it is more obscure, and in some cases may represent independent pathways. Whether the modifications occur on the amino-acids before their incorporation into peptides, as apparently happens with most D-amino-acids, or take place on the peptide itself, is not well defined, but it is likely that both processes operate.

The α-hydrazino-acid 1-amino-D-proline, present in linatine (**29**), is presumably formed before it is coupled with glutamic acid. Studies on the mode of action of this anti-metabolite have shown that the γ-glutamyl residue acts as a carrier of the D-amino acid which blocks pyridoxal-mediated enzyme systems.[66] The basic peptide[67] (**30**) can be formally derived from the corresponding tripeptide (**31**) by oxidation of the cysteine, cyclization of the thiol with the neighbouring peptide bond to form the thiazole, and decarboxylation. The former of these events, if not all, most likely occurs at the peptide stage.

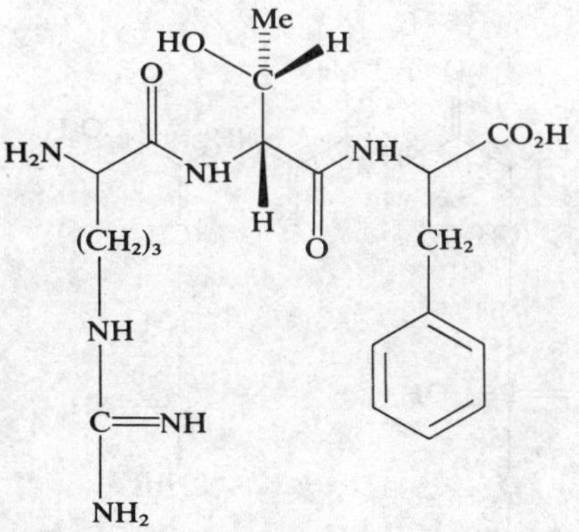

(**27**)

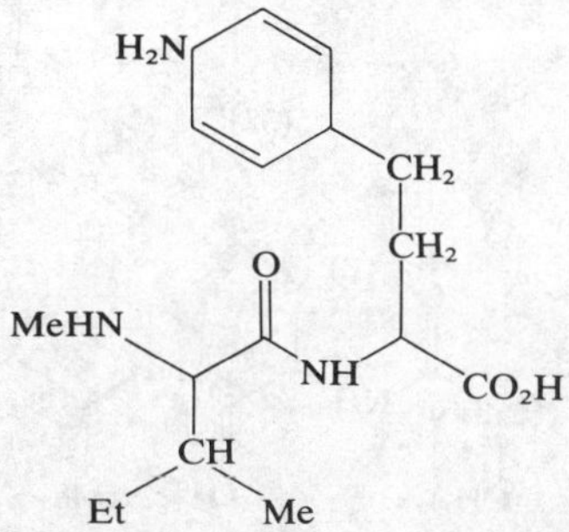

(**28**)

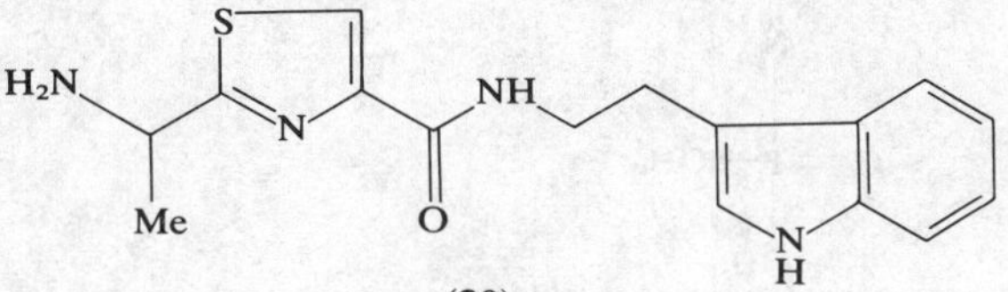

(**30**)

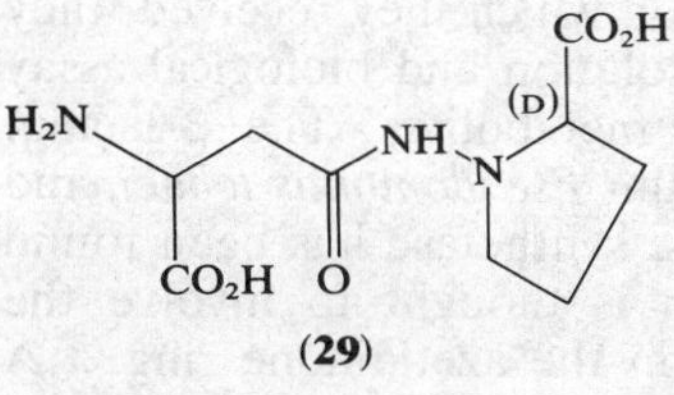

(**29**)

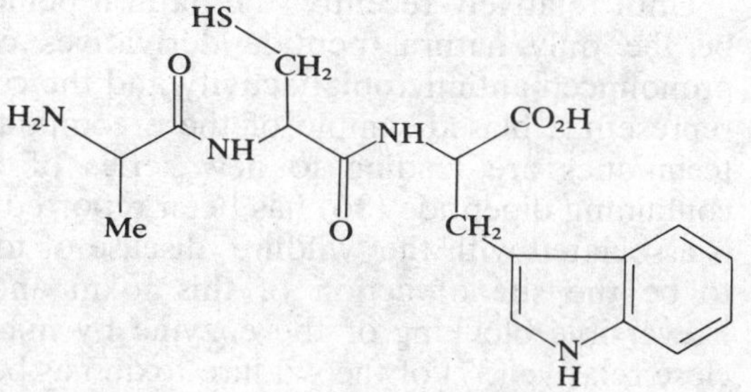

(**31**)

The presence of diazoketone groups in naturally occurring compounds was treated with some scepticism when it was first reported. However, the structures of a number of amino-acids[68] and the peptides duazomycin[69] A and B (**32**) have since been authenticated. The phosphine[70] and sulphoximine[71] peptide derivatives (**33**) and (**34**) are both specific inhibitors of glutamine synthetase, and the structural similarities of the *N*-terminal amino-acids, both of which occur in the free state, with glutamine are apparent. Of particular interest is the finding that both tripeptides are more potent microbial growth inhibitors than the constituent amino-acids, signifying the importance of peptides as carriers of amino-acid analogues. The novel related compound (**35**) is supposedly the toxin responsible for the symptoms of bean halo-blight.[72]

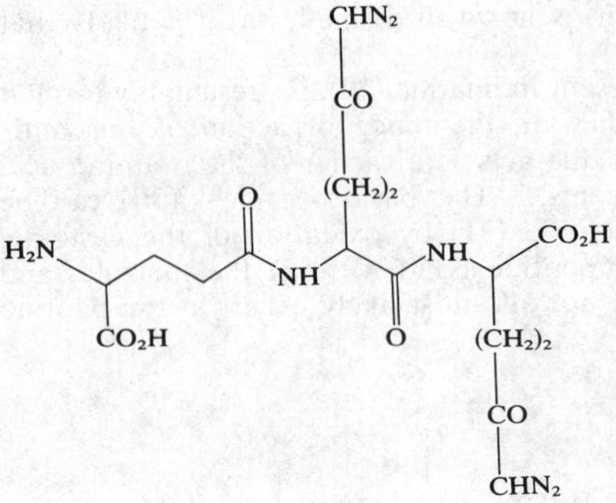

(**32**)

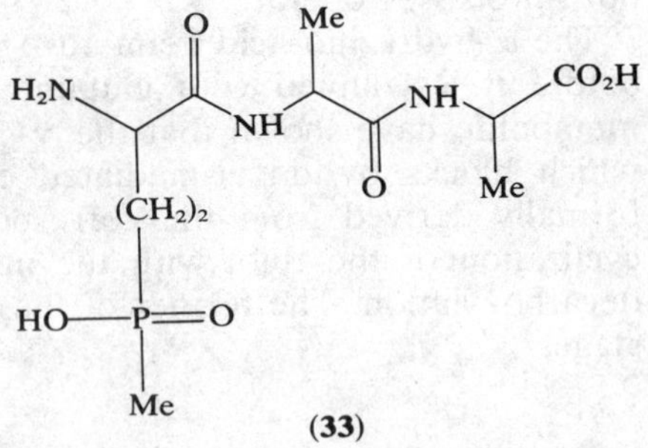

(**33**)

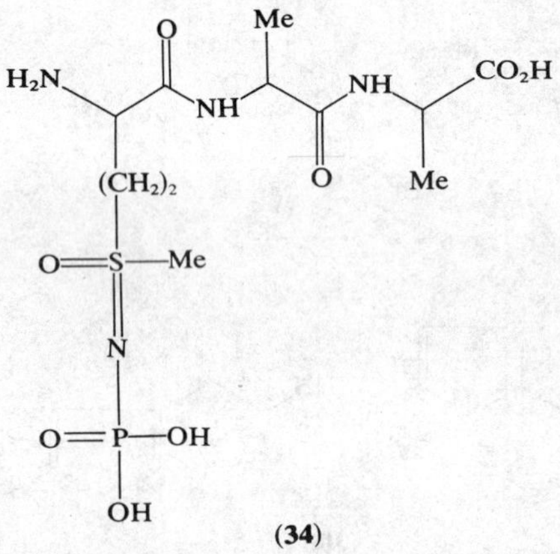

(**34**)

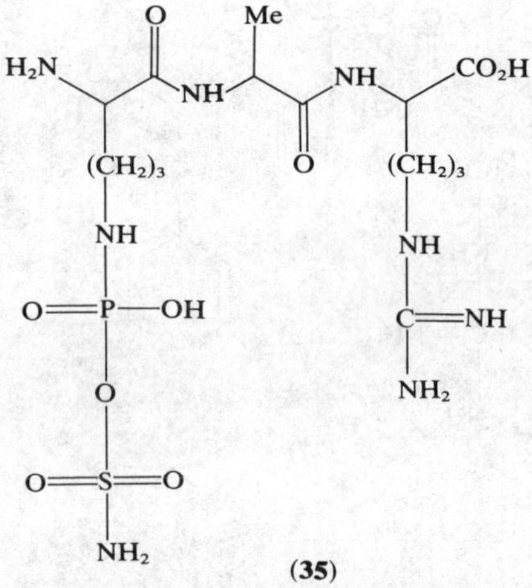

(**35**)

Until relatively recently, the natural penicillins and cephalosporins were considered to be the only natural peptide derivatives containing β-lactam rings. Because of their pronounced antimicrobial activity and the consequent attention which they received, they represent a biased sample of these compounds. Improved isolation and biological assay techniques are leading to new series of these interesting metabolites. The β-lactam containing dipeptide (**36**) has been reported from the bacterium *Pseudomonas tobaci*, and is associated with the 'wildfire' disease of tobacco. Glutamine synthetase has been found to be the site of action of this toxin and the mechanism is thought to involve the irreversible blocking of the enzyme by nucleophilic attack on the azetidinone ring.[73] A close relative (**37**) of the wildfire toxin has been identified from a *Streptomyces* species and also blocks glutamine synthetase.[74]

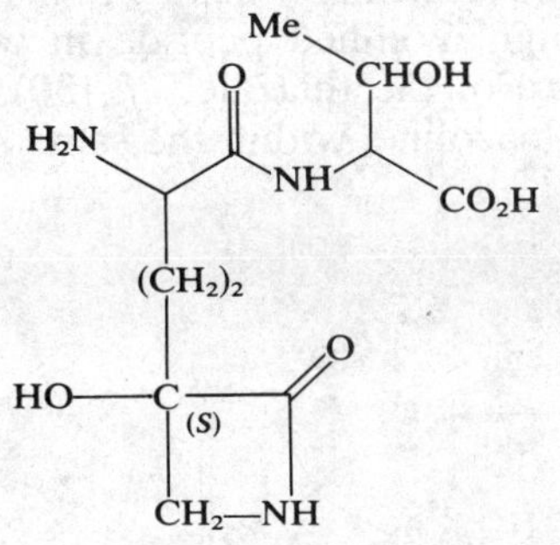

(36)

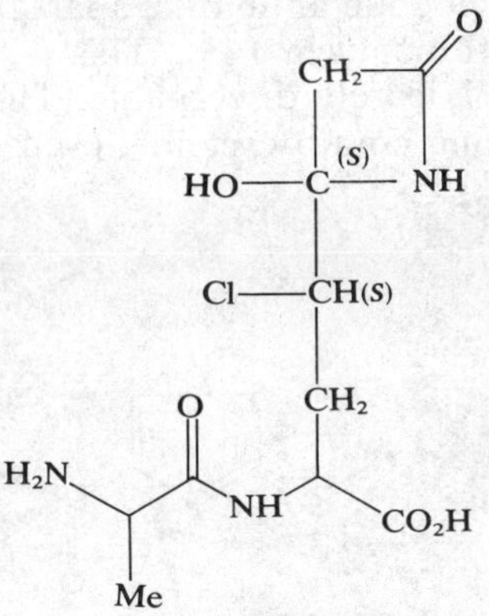

(37)

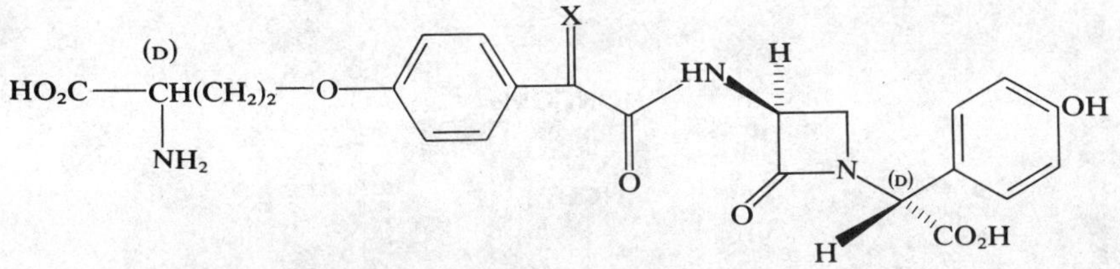

(38) X = NOH (*syn* with respect to the acylamino-group)
(39) X = NOH (*anti* with respect to the acylamino-group)
(40) X = H, NH_2
(41) X = O

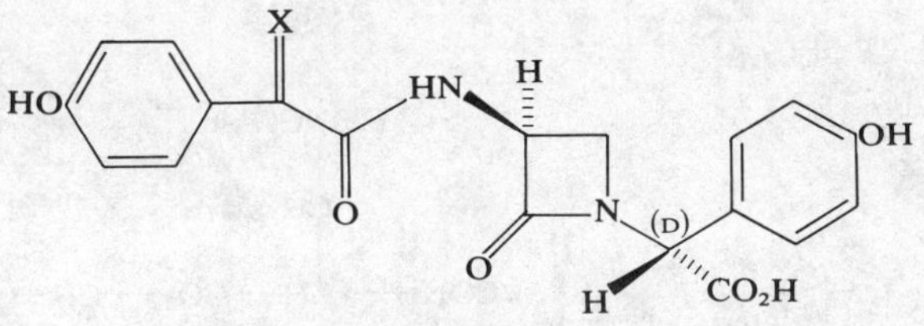

(42) X = NOH (*syn*)
(43) X = NOH (*anti*)
(44) X = H, NH_2

The nocardicins, a further group of monocyclic β-lactam antibiotics from the bacterium *Nocardia uniforms*, represent one of the most exciting new developments in this area. The major component nocardicin A has broad spectrum activity against Gram-negative bacteria and low toxicity. The structures of nocardicin A (**38**), as well as the minor components (**39**)–(**44**) from the fermentation, have been established and confirmed by synthesis. All aspects of the chemistry of these novel compounds have been covered in an excellent article in the proceedings of a recent symposium on β-lactam antibiotics.[75] The presence of D-*p*-hydroxyphenylglycine in the nocardicins is particularly noteworthy, since it is rarely found in microbial peptides, but more so because it is one of the most effective side-chains in the semi-synthetic penicillins. Once again Nature has precedented human efforts.

The only other peptide antibiotic, apart from the nocardicins, which contains a terminal oxime group is althiomycin (**45**). This is also a high modified peptide in which basic peptide structure is barely discernable. The presence of the thiazole, *cf.* (**30**), a masked α,β-dehydrocysteine, and D-cysteine, cyclized to a thiazoline, within the same molecule is interesting.[76,77]

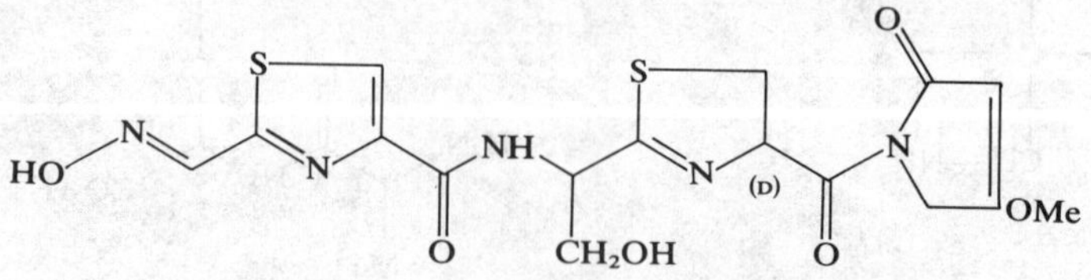

(**45**)

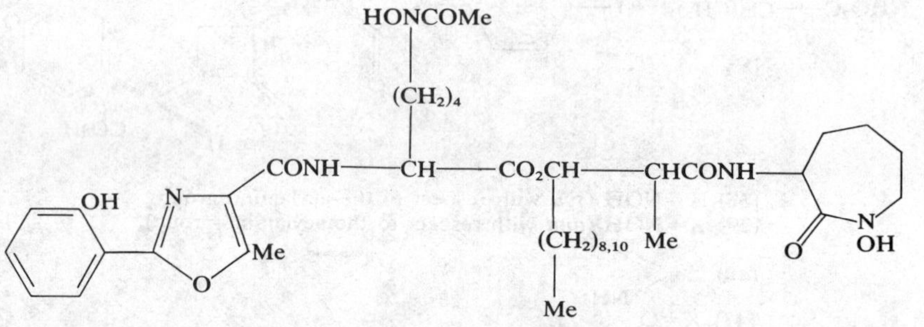

(**46**)

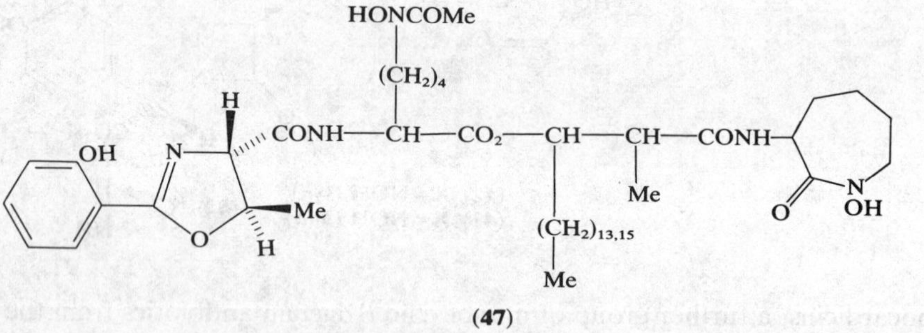

(**47**)

The corresponding oxazoles and oxazolines derived from serine or threonine are less common, but have been observed in members of the mycobactin family.[78] These are a group of bacterial iron-complexing hydroxamic acid derivatives which are reported to be growth factors for mycobacteria. Nocobactin[79] (**46**) resembles mycobactin M[80] (**47**) in structure with the exception that it has an oxazle in place of an oxazoline ring. The hydroxamic acids structures, which are *N*-hydroxylated derivatives of ornithine, are important in the complexing of iron and are also present in the cyclic peptides of the siderochrome family (see Section 23.4.2.3).

23.4.2.2 2,5-Dioxopiperazines (cyclic dipeptides)

2,5-Dioxopiperazines are amongst the most numerous and ubiquitous of peptide derivatives found in nature. They also differ in some aspects of their chemistry from other cyclic peptides, and therefore it has been considered expedient to treat them as a separate group. An excellent review which comprehensively deals with all aspects relating to natural and synthetic 2,5-dioxopiperazines, and covers the literature up to the end of 1973, offers an important source of information on these compounds.[81] No attempt to give an exhaustive coverage has been made, but the salient points considered of importance in relation to naturally occurring peptides are outlined.

Considerable care has to be taken in the isolation of these compounds and their derivatives, since they are readily formed from peptides and proteins on enzymatic or chemical hydrolysis, as well as thermolysis. Indeed, it appears that the characteristic bitter taste of a number of aged alcoholic beverages and roasted cocoa beans are due, in part, to *cyclo* dipeptides formed by one or other of these processes. However, their careful isolation, particularly from extracts and culture filtrates of micro-organisms, testifies to their genuine metabolic character.

A selection of microbial cyclic dipeptides derived from primary L-amino-acids for which there is sufficient evidence to show that they are true natural products are shown in Table 3. Many have also been isolated from a variety of other organisms, *e.g.* cyclo-Gly-L-Pro has recently been identified in the extracts of the starfish *Luidia clatharata.*

TABLE 3
Some Naturally Occurring 2,5-Dioxopiperazines derived from Primary L-Amino-acids

cyclo-L-Pro-L-Leu	*Aspergillus fumigatus*
cyclo-L-Pro-L-Val	*Aspergillus ochraceus*
cyclo-L-Pro-L-Phe	*Rosellinia necatrix*
cyclo-L-Pro-Gly	Yeast extract
cyclo-L-Pro-L-Tyr	Yeast extract
cyclo-L-Ala-L-Leu	*Aspergillus niger*
cyclo-L-Phe-L-Phe	*Streptomyces noursei*

It is interesting to note the relatively large number of proline derivatives in this group of natural products. This is perhaps not too surprising, when it is considered that, in order to form dioxopiperazines, the dipeptide precursors have to adopt a *cis* geometry about the peptide linkage, and it is well established that the activation energy for the *cis–trans* isomerism of *N*-acylproline derivatives is considerably lower than for other peptide bonds. Unlike most other cyclic peptides, the two peptide bonds in dioxopiperazines possess the *cis* configuration and this does influence the topology and chemistry of these molecules.

The presence of many of the peptide modifications observed in the preceding section are apparent in cyclic dipeptides (Table 4). α,β-Didehydroamino-acid residues are common structural features in (**48**), (**50**), and (**51**), as are *N*-methyl groups in (**49**) and (**50**), and *N*-hydroxy-groups in (**51**) and (**52**).[82–89] The mechanism of oxidation (dehydrogenation) has been investigated in a number of cases, including mycelianamide (**51**), and shown to involve the specific loss of the *pro-S*-methylene hydrogen from labelled amino-acids.[90]

Careful biosynthetic studies have also demonstrated that rhodotorulic acid (**52**) is metabolized from ornithine, which undergoes *N*-hydroxylation before acetylation and dimerization.[91] The cycloserine dimer is interesting in that both the amino-acid residues have the D-configuration. The chlorine-containing metabolite (**54**) is stated to have substantial antitumour activity, but, as yet, little is known about its biosynthesis.

Pulcherrimic acid (**55**) is representative of a number of naturally occurring hydroxypyrazine derivatives which can be considered structurally as highly oxidized forms of the

TABLE 4

Some Simple Modified 2,5-Dioxopiperazines

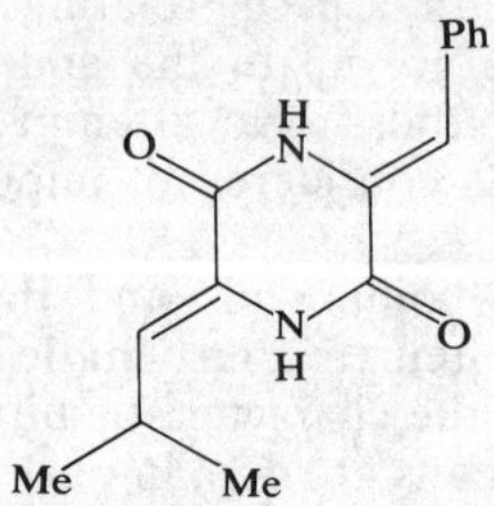

(**48**)
Albonoursin[82]
Streptomyces noursei

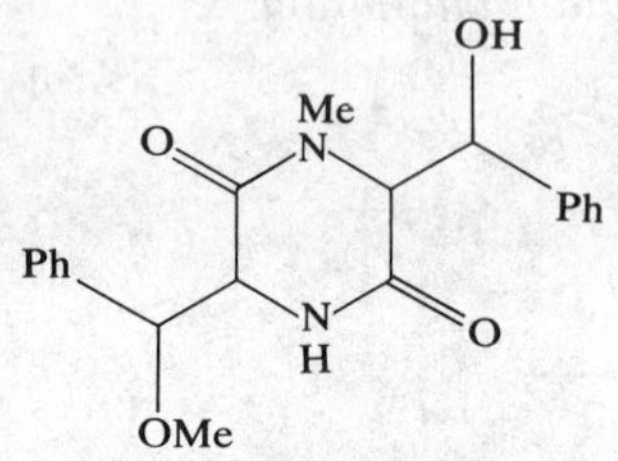

(**49**)
Picroroccellin[83]
Roccella fuciformis

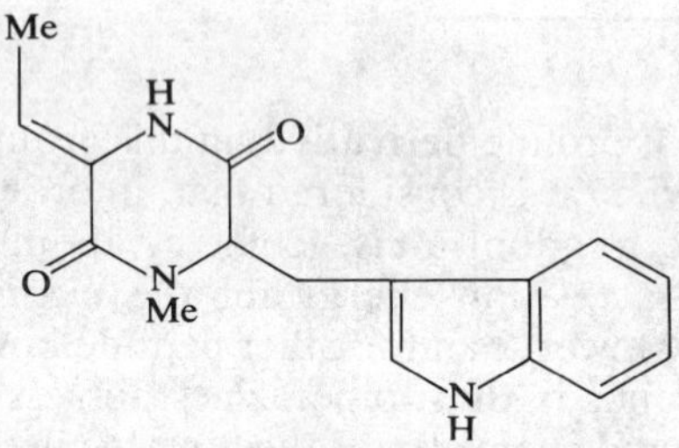

(**50**)[84]
Streptomyces spectabilis

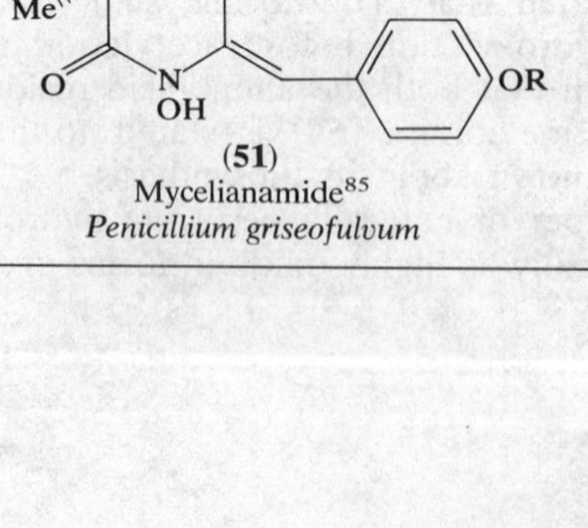

(**51**)
Mycelianamide[85]
Penicillium griseofulvum

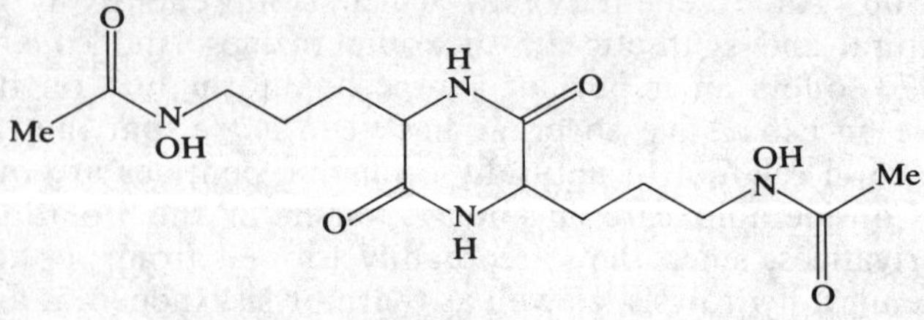

(**52**)
Rhodotorulic acid[86]
Rhodotorula pilimanae

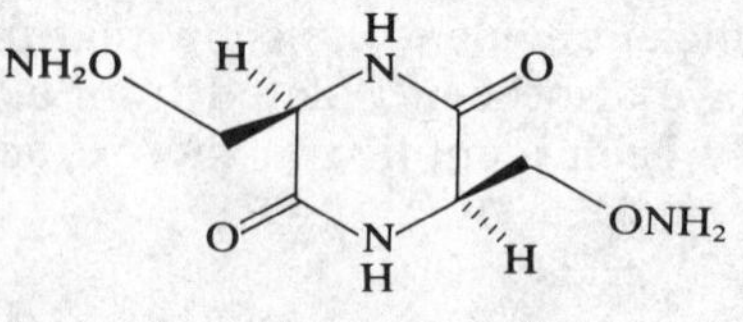

(**53**)
Cycloserine dimer[87]
Streptomyces orchidaceus

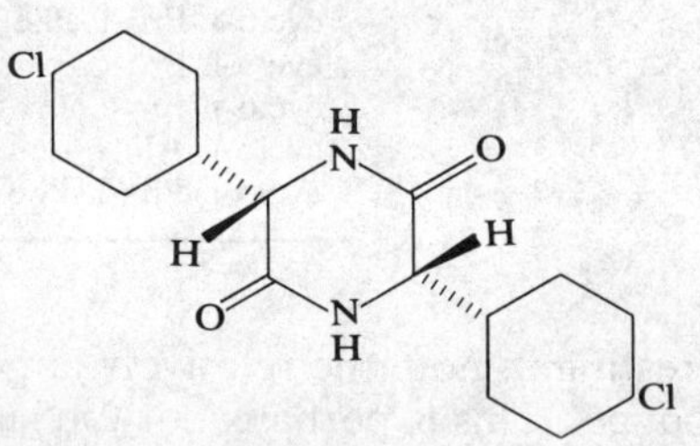

(**54**)[88]
Streptomyces griseoluteus

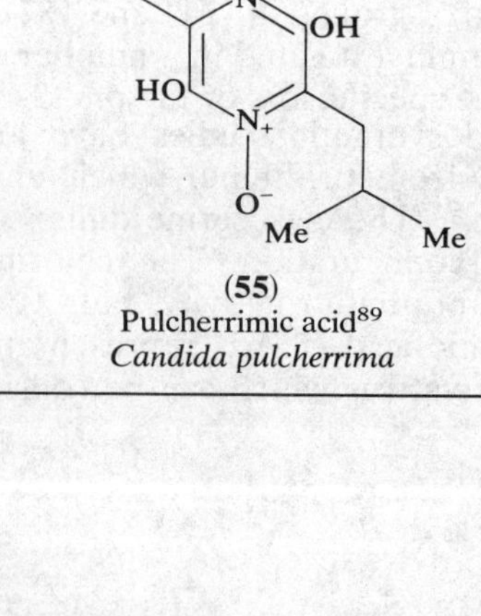

(**55**)
Pulcherrimic acid[89]
Candida pulcherrima

corresponding dioxopiperazine. Labelling experiments have shown that cyclo-L-Leu-L-Leu is converted into (**55**) by *C. pulcherrima*, but in general there is evidence that pyrazines are not derived from these precursors.[92]

A variety of dioxopiperazines containing an isoprenylated tryptophan unit and some of which contain interesting new ring systems have been characterized over the past 15 years. A cross-section of the various types of these metabolites (**56**)–(**66**) is shown in Table 5.[93–100] Echinulin (**64**) was historically the first of this new group to be isolated, but since then a number of simpler derivatives have been identified, as exemplified by (**56**)–(**60**).[93–96] The presence of α,β-didehydroamino-acid systems is once again noticeable, and there is evidence that the dehydrogenation occurs at the cyclo-dipeptide level with the loss of the *pro-S*-methylene hydrogen.

The isoprenylation of these compounds also occurs in the cyclized state, and recent studies have revealed that cyclo-L-Tyr-L-Ala is converted into (**58**) by a cell-free extract of *A. amstelodami* in the presence of 3,3-dimethylallyl pyrophosphate. The rapid formation of (**58**) was shown to be followed by its further incorporation into echinulin (**64**),[94] thus establishing the order in which the dimethylallyl groups are introduced into the indole nucleus. The alkylation pattern of the benzenoid ring is quite normal, although the inversion of the C-5 unit at position-2 is a novel feature. The mechanism of this inversion is thought to involve initial alkylation on the indole nitrogen atom followed by an acid-catalysed rearrangement. The isolation of (**66**)[100] lends support to this proposal. The inverted C-5 unit at position-3 in roquefortine[96] (**60**) also probably results from an *N*-alkylation followed by a Claisen-type rearrangement.

The remaining metabolites in Table 5 are derived by subsequent transformations of the indole residue and interaction of the dioxopiperazine ring with either the C-5 unit or the indole itself. These proposals are summarized in Scheme 5. Although they are to some extent speculative, they do offer a coherent interpretation which is chemically plausible. Particularly noteworthy is the proposed novel cycloaddition reaction for the formation of brevianamide A (**63**). Chemical precedent for such reactions now exists.[81]

The ergot peptides represent a further group of dioxopiperazine metabolites containing a modified isoprenylated tryptophan residue, which in this case is lysergic acid. Ergot, the sclerotia of the fungus *Claviceps purpurea* found on rye crops and grasses, has a long history both as a cause of social disaster and as a drug. Bread made from contaminated rye was the cause of the poisoning which is now referred to as ergotism, but was previously known as Holy Fire or St. Anthony's Fire. Clinical use of these peptides is restricted to the treatment of migraine and the control of haemorrhage after child birth.

The chemistry and biosynthesis of lysergic acid is a specialist subject and beyond the scope of this article, and the reader is referred to recent reviews.[101,102] The ergot peptides (**67**) are amides of lysergic acid which occur as interconvertible pairs epimeric at C-8. The amines are cyclic tripeptides, the biosynthesis of which is still somewhat obscure. They are known to be derived from the respective amino-acids and an *N*-acyl cyclic dipeptide intermediate is implicated since the recent isolation of (**68**) from an ergot producing organism.[103] This is claimed to be a shunt product formed by the epimerization of the key intermediate. The subsequent cyclization could occur either by the addition of an α-hydroxyl group to the peptide bond of the dioxopiperazine ring (**69**) or, alternatively, by the addition of the peptide amide to an acylimine (**70**). The former process was adopted, in principle, for the chemical synthesis of these compounds.[101]

The final group of metabolites covered in this section are the sulphur-bridged dioxopiperazines (Table 6), and these can be further divided into those related to gliotoxin and those of the sporidesmin type.[104]

Gliotoxin (**71**), a highly antifungal and antibacterial compound, was first isolated in 1936 and the chemistry extensively investigated over the next two decades. The correct structure was finally proposed[105] in 1958 and subsequently confirmed by X-ray analysis. The sporidesmins were detected as a result of a correlation of facial eczema of cattle and sheep with fungal infection of pastures. The structure of sporidesmin (**72**) based on chemical and X-ray methods followed.[106] The related compound with a tetrasulphide

TABLE 5
Some 2,5-Dioxopiperazines containing Isoprenylated Tryptophan Units

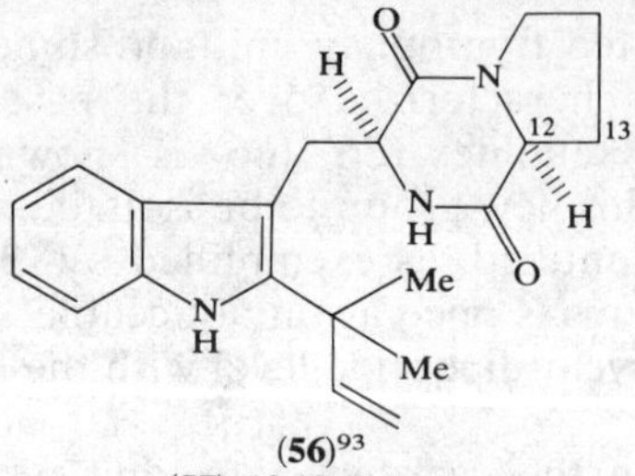

(**56**)[93]
(**57**) 12,13-Dehydro
Aspergillus ustus

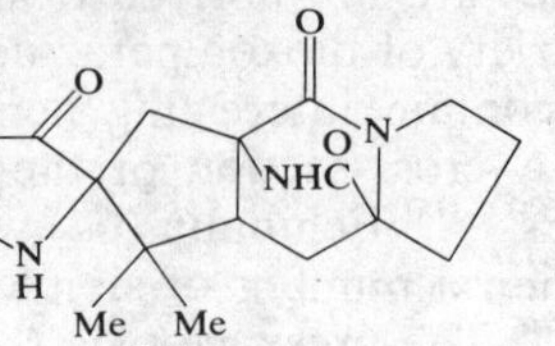

(**63**)
Brevianamide A[97]
P. brevicompactum

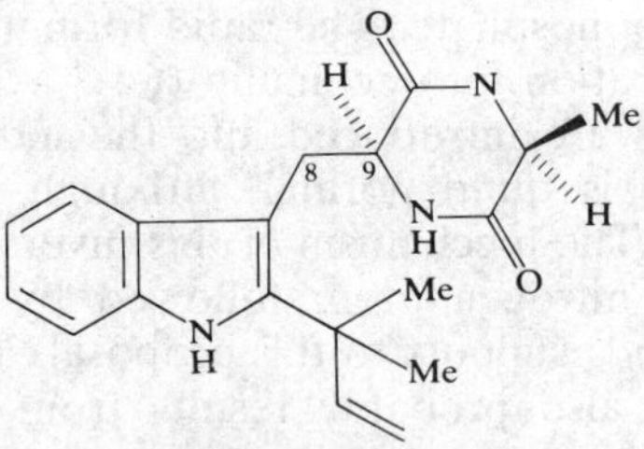

(**58**)[94]
(**59**) 8,9-Dehydro[95]
Aspergillus amstelodami

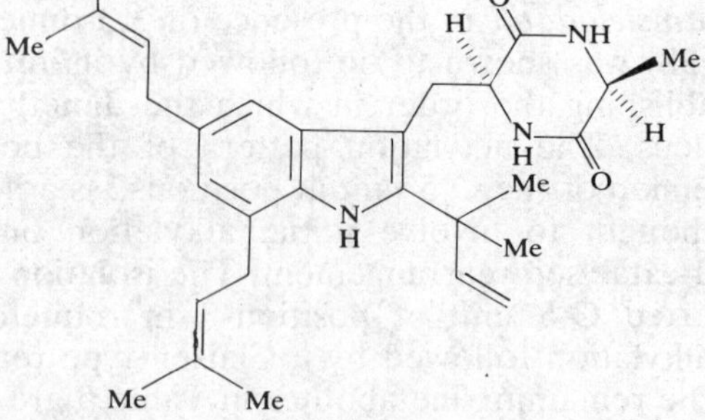

(**64**)
Echinulin[98]
A. echinulatus
A. amstelodami

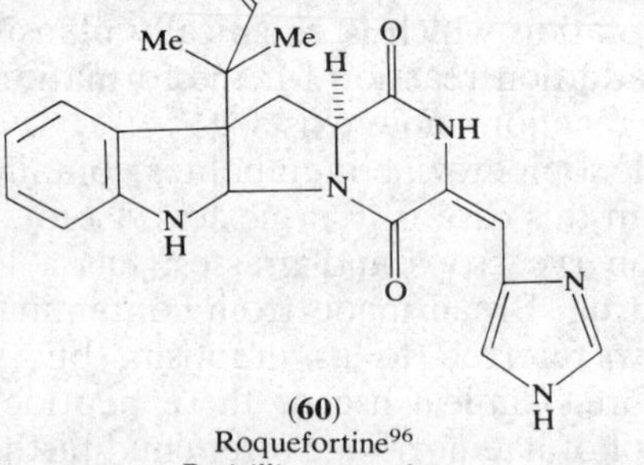

(**60**)
Roquefortine[96]
Penicillium roqueforti

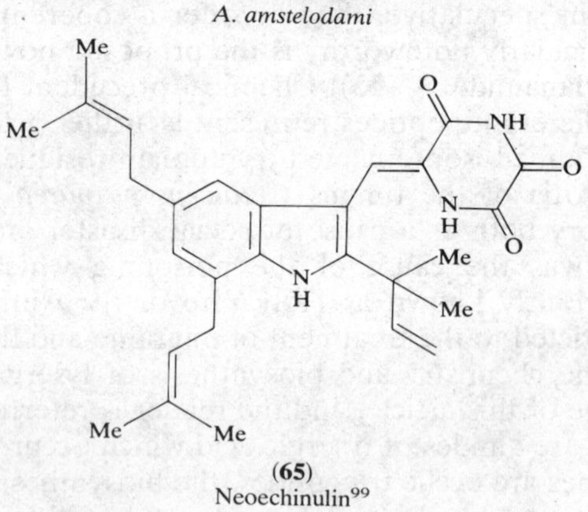

(**65**)
Neoechinulin[99]
A. amstelodami

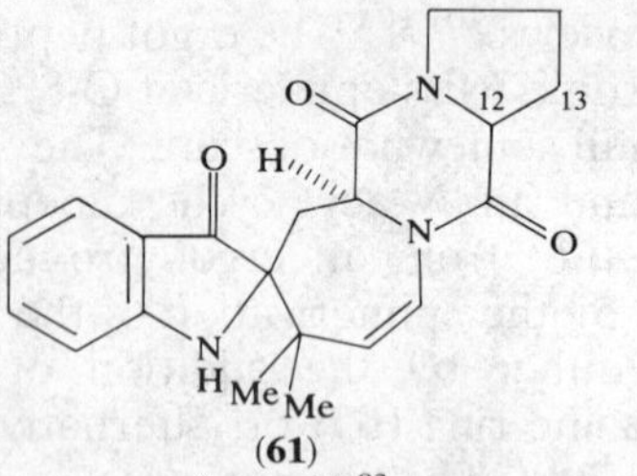

(**61**)
Austamide[93]
and 12,13-dehydro
Aspergillus ustus

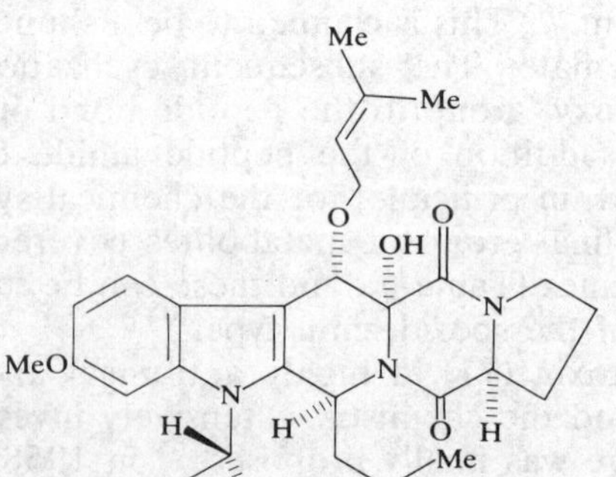

(**66**)
Fumitremorgin[100]
P. verruculosum

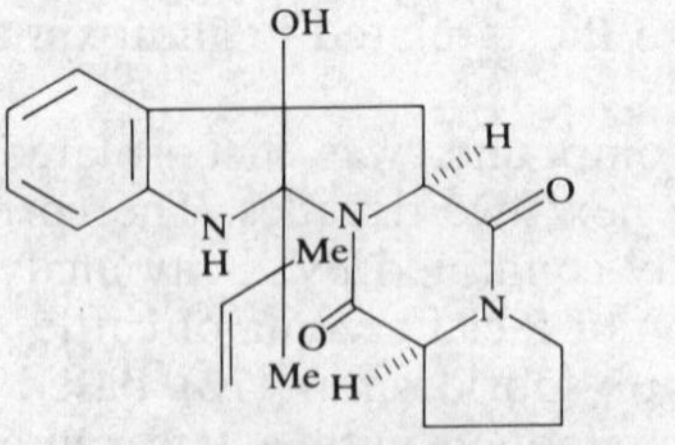

(**62**)
Brevianamide E[97]
P. brevicompactum

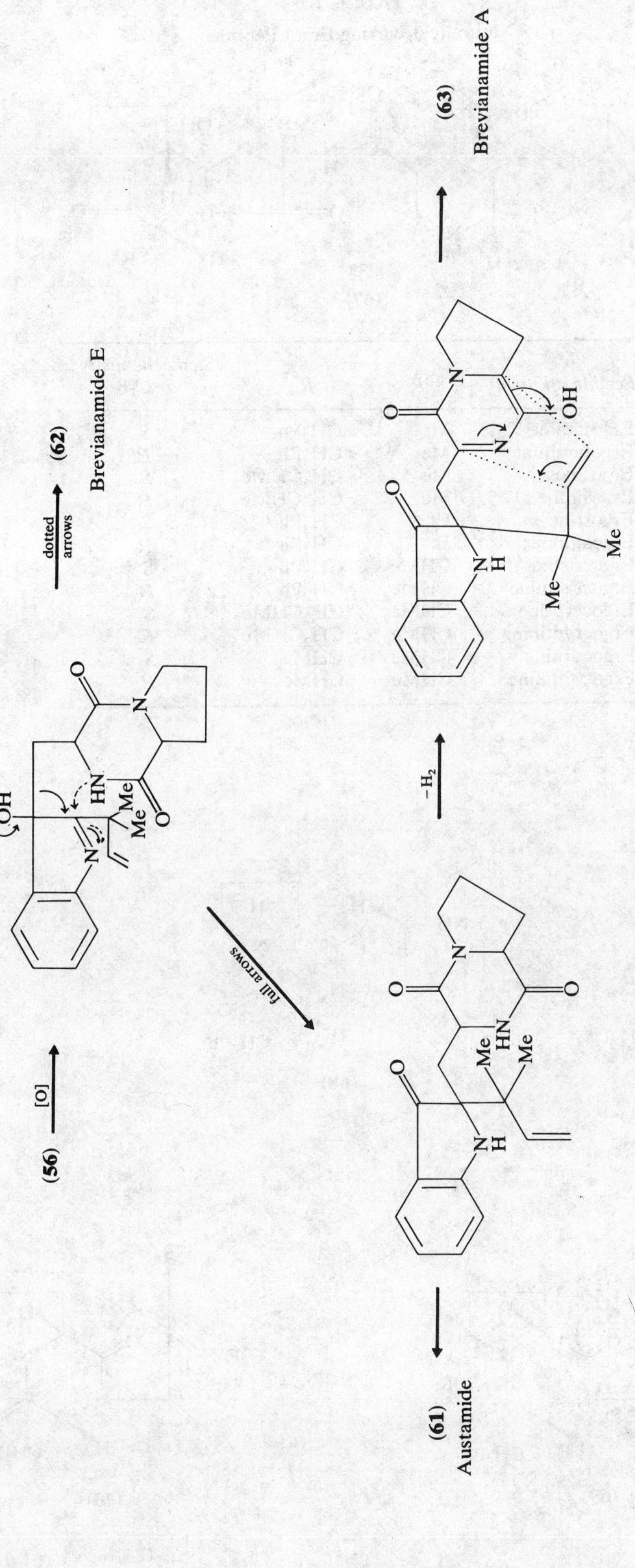

SCHEME 5 Proposed biogenesis of the brevianamides

TABLE 6
Naturally occurring Ergot Peptides

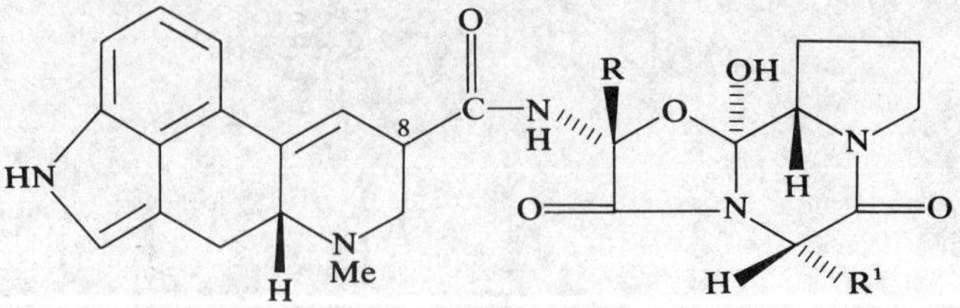

(67)

Peptide alkaloid	*R*	*R*[1]	*Stereochemistry at C*-8
Ergotamine	Me	CH_2Ph	*S*
Ergotaminine	Me	CH_2Ph	*R*
Ergosine	Me	CH_2CHMe_2	*S*
Ergosinine	Me	CH_2CHMe_2	*R*
Ergostine	Et	CH_2Ph	*S*
Ergostinine	Et	CH_2Ph	*R*
Ergocristine	$CHMe_2$	CH_2Ph	*S*
Ergocristinine	$CHMe_2$	CH_2Ph	*R*
Ergocryptine	$CHMe_2$	CH_2CHMe_2	*S*
Ergocryptinine	$CHMe_2$	CH_2CHMe_2	*R*
Ergocornine	$CHMe_2$	$CHMe_2$	*S*
Ergocorninine	$CHMe_2$	$CHMe_2$	*R*

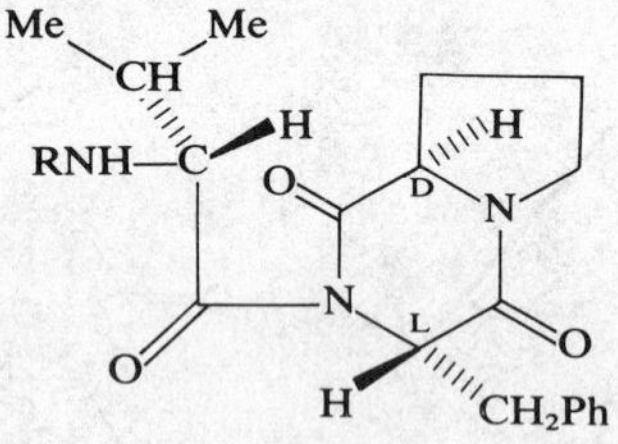

(68)

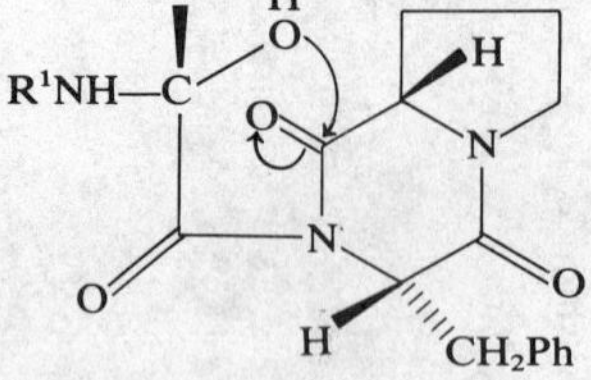

(69)

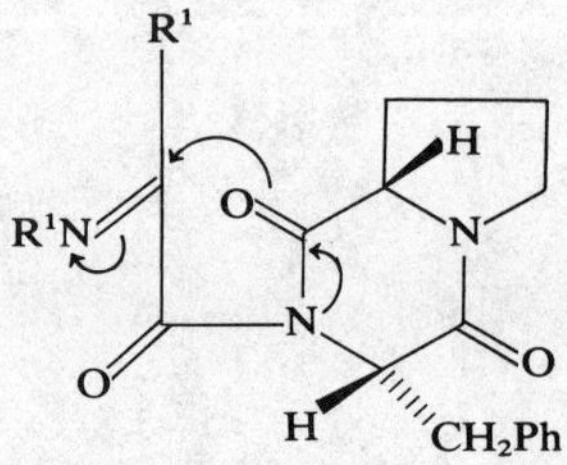

(70)

TABLE 7

Some Naturally Occurring Sulphur-bridged Dioxopiperazines

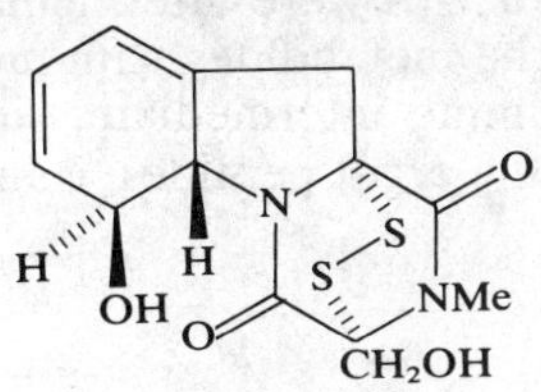

(**71**)
Gliotoxin[105]
Trichoderma viride

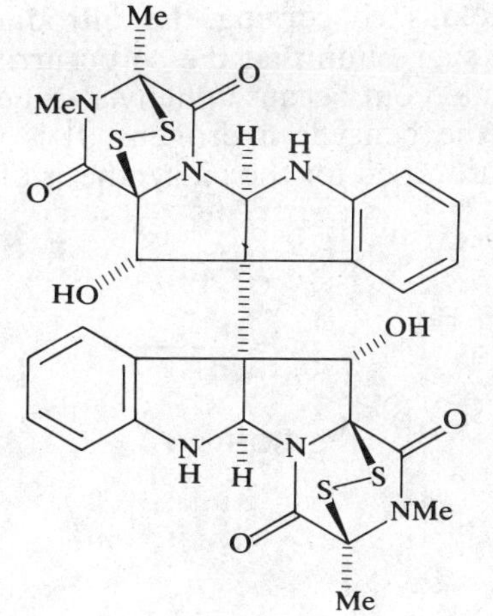

(**75**)
Verticillin A[109]
Verticillium sp.

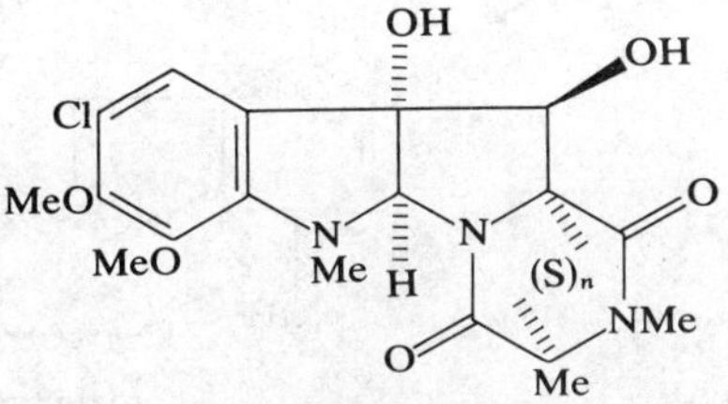

(**72**)
Sporidesmin B[106], $n = 2$
(**73**)
Sporidesmin G[107], $n = 4$
Pithomyces chartarum

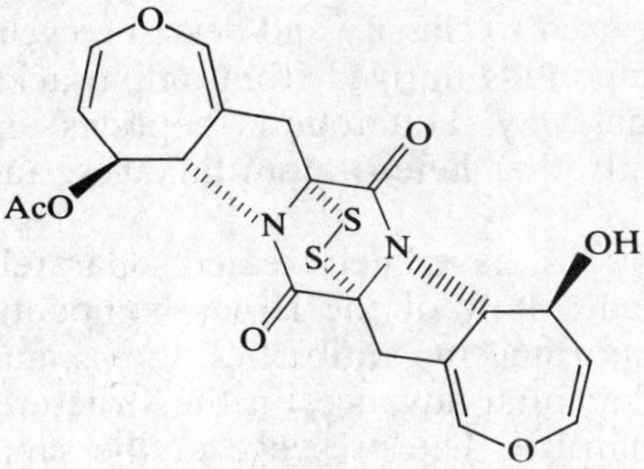

(**74**)
Aranotin[108]
Arachniotus aureus

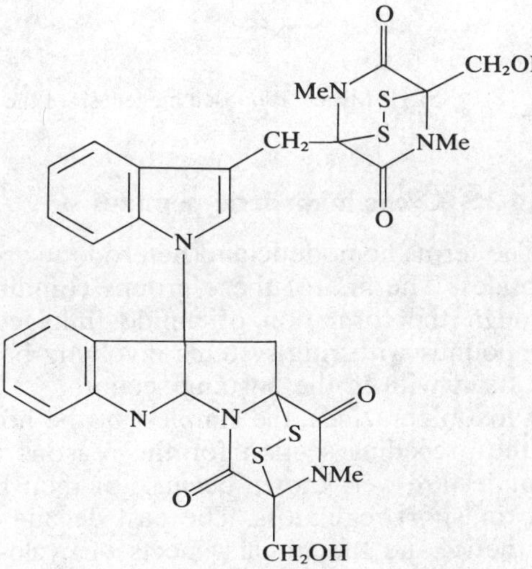

(**76**)
Chetomin[110]
Chaetomin cochliodes

bridge (**73**) was subsequently isolated and characterized.[107] Aranotin (**74**) and related compounds, which bear a considerable resemblance to gliotoxin, were identified as a result of their antiviral activity.[108]

Several more complex dimeric metabolites have been isolated recently and are illustrated by (**75**) and (**76**). Verticillin A (**75**) has, in addition to antimicrobial activity, considerable anti-tumour properties. It is interesting to note that (**75**) has the opposite chirality to sporidesmin with respect to the dithiopiperazine rings. The structure of the related dimer chetomin has recently been revised to (**76**), but the configuration remains to be elucidated.[110]

Cyclo-L-Phe-L-Ser has been shown to be an efficient precursor of gliotoxin,[111] and further labelling studies are consistent with the proposed mechanism for the formation of the dihydro-aromatic systems of the gliotoxin type *via* a benzene oxide intermediate, and *via* the benzene oxide–oxepin equilibrium for the aranotin-type ring system (Scheme 6).[112] The details concerning the introduction of the sulphur bridges remain obscure. An earlier suggestion that the co-occurring dethiodehydropiperazines were intermediates has been ruled out because phenylalanine is incorporated into the metabolites without loss of either the benzylic hydrogens. It is probable that an acylimine intermediate, similar to those proposed for the biosynthesis of brevianamide A and the ergot peptides, is involved.

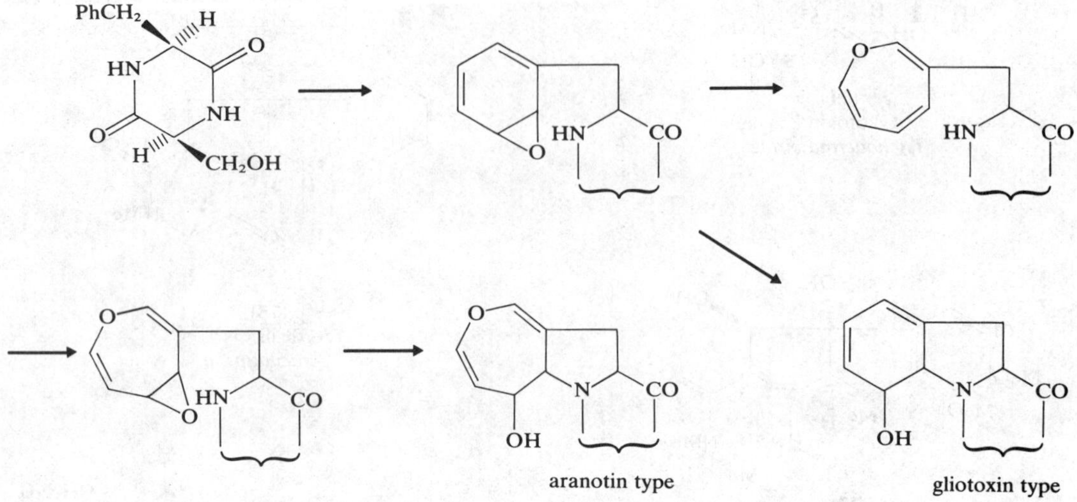

SCHEME 6 Proposed biogenesis of the dihydroaromatic rings in aranotin and gliotoxin

23.4.2.3 Cyclic homodetic peptides

The terms homodetic and heterodetic are commonly used to classify and describe cyclic peptides. The first of these groups contains cyclic compounds derived from amino-acids through the formation of amide linkages in the usual way. Heterodetic peptides are compounds with ring systems involving both amide and other hetero-atom linkages, and are dealt with in the ensuing section.

Dioxopiperazines, the simplest of the homodetic peptides, have been treated separately in the preceding section for the reasons already stated. Many of the larger homodetic peptides are well known because of their biological importance as antibiotics, toxins, and ion transport regulators. The past decade has seen substantial advances in the structure, synthetic, and functional aspects of cyclo-peptide chemistry. The presence of the same sort of atypical features within the amino-acid residues, as observed in the linear and cyclic dipeptides, is manifest, and these, as well as the aspects outlined above, have been recently reviewed.[113]

Currently, owing to the increasing application of physical methods, principally X-ray analysis, n.m.r., and optical (o.r.d. and c.d.) spectroscopy, the emphasis has shifted to a consideration of the topology of these important molecules and its relationship to their biological function.[114–116] Yet another and equally significant reason for this change of emphasis is the greater rigidity of the cyclopeptides with respect to their linear counterparts, which reduces the number of interconverting forms, and to some extent facilitates analysis. Nevertheless, they still possess some degree of flexibility and frequently the crystal conformation differs from that adopted in solution. Clearly detailed discussion of conformation is beyond the bounds of this review, but the salient points which relate to the chemical or biological properties of the molecules are noted.

As far as the author is aware, no cyclic homodetic tripeptide has been isolated to date, and only a few naturally occurring tetrapeptides have been fully characterized. The amino-acid sequence of tentoxin (**77**), a fungal phytotoxin, had been the subject of some controversy until it was resolved by an X-ray analysis of the dihydro-derivative.[117] The reduction of the didehydrophenylalanine unit results in the stereospecific formation of the D-isomer of *N*-methylphenylalanine, and both the *N*-methyl amino-acid units in the crystal lattice conformation of dihydrotentoxin adopt *cis* peptide bonds. There is increasing crystallographic and solution evidence that proline and *N*-methyl amino-acids in cyclic peptides often possess *cis* amide bonds, and there is speculation that this may be one of their functions in microbial peptides. The advantage of *cis* amides, particularly in small peptides, is the considerable release of ring strain which they allow. It was therefore surprising that the X-ray analysis of the dihydro-derivative of chlamydocin (**78**), a highly cytostatic cyclic tetrapeptide, revealed an all-*trans* conformation for the peptide bonds.[118] This structure represents the first naturally occurring or synthetic cyclotetrapeptide for which this has been observed, although each of the peptide units is significantly distorted from planarity. Other points of interest concerning this molecule are the previously unknown L-α-aminodecanoic acid and the α-methylalanine, which is rarely encountered in small peptides.

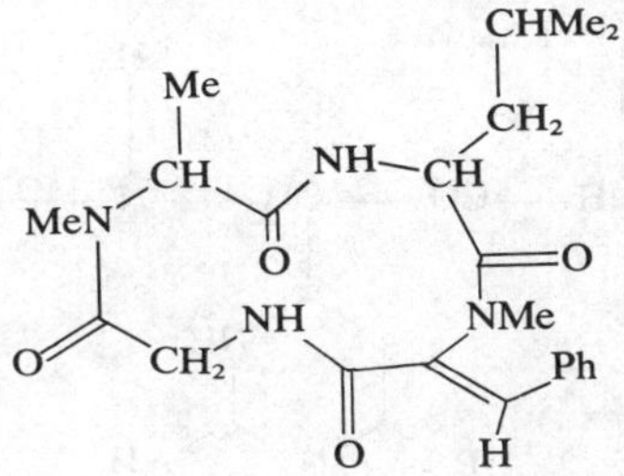

(**77**)

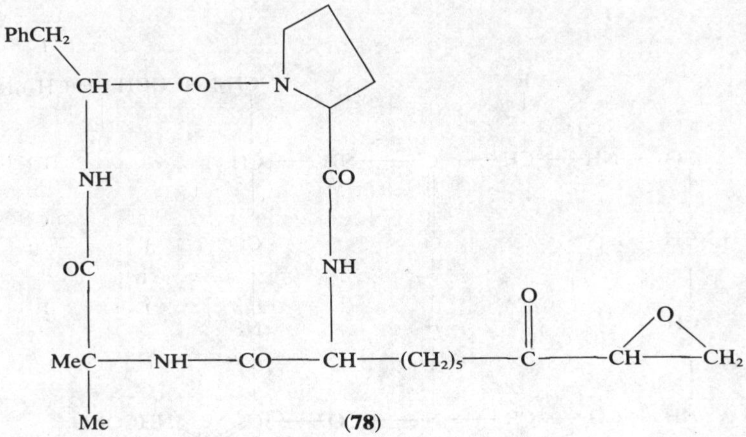

(**78**)

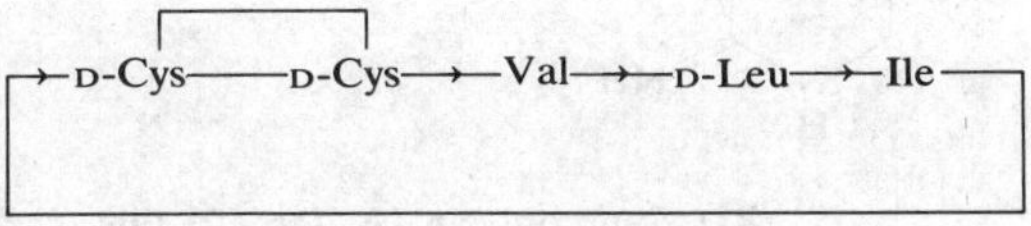

(**79**)

The malformins, a family of cyclic pentapeptide mycotoxins, derive their name from the malformations that they produce in germinating bean plants. The toxins were isolated some 15 years ago and extensively investigated over the intervening period. The structure originally proposed for the main component, malformin A, has recently been re-examined and the alternative structure (**79**) formulated and subsequently confirmed by synthesis.[119] This and other examples offer the salutary lesson for all natural product chemists that small peptides do represent challenging structural problems and cannot always be solved merely by the application of spectroscopic methods and the minimum of chemical investigation.

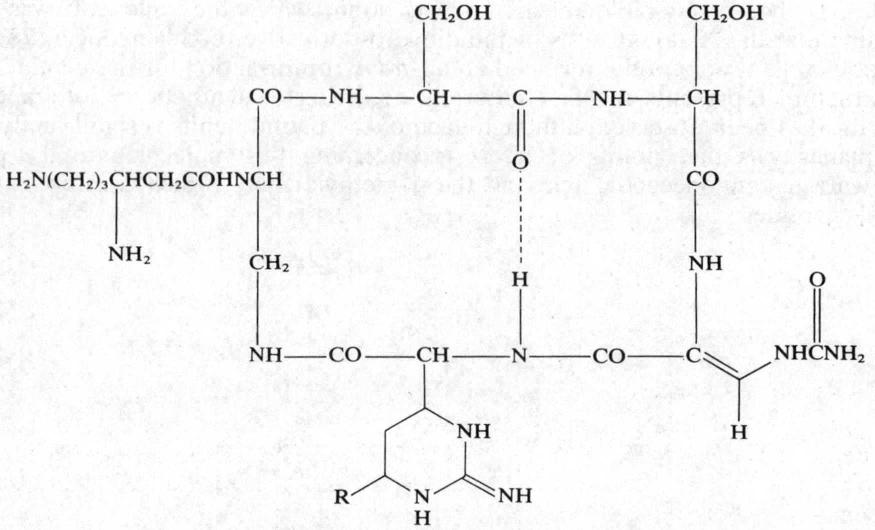

(**80**) Tuberactinomycin O, R = H
Viomycin, R = OH

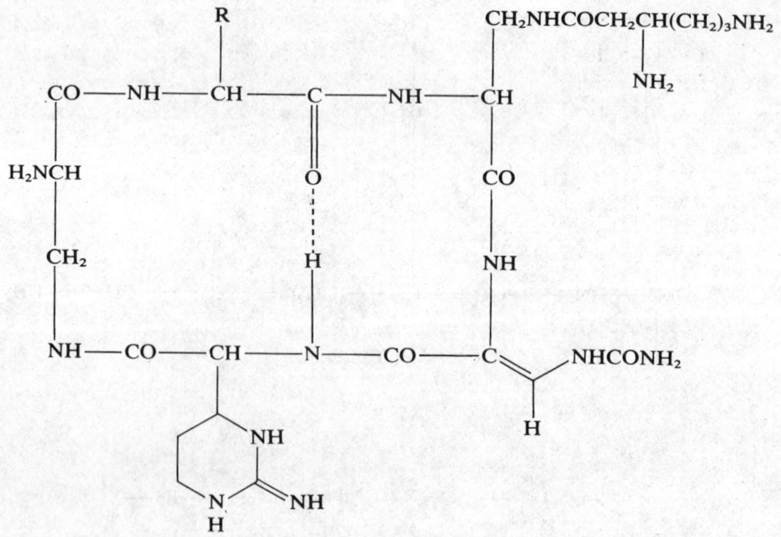

(**81**) Capreomycin IA, R = CH_2OH
Capreomycin IB, R = Me

The members of the viomycin,[120] tuberactinomycin,[121] and capreomycin[122] families have found limited clinical use for the treatment of tuberculosis, and are a particularly intriguing group, in that they are all cyclic pentapeptides with notable structural, conformational, and biosynthetic features. The families have a common pentapeptide skeleton which contains variants of the modified arginine residue, previously encountered in elastinal (26), a unique didehydroserine ureide unit, and two molecules of L-diaminopropionic acid. The ring system is homodetic, but includes an amide linkage involving a β-amino-group of one of the diaminopropionic acids. They differ in the point of attachment of the β-lysine side-chain; in the viomycin and tuberactinomycin group (**80**) it is to the α-amino-group of one of the diaminopropionic acids, and to the β-amino-group of the other diaminopropionic acid for the capreomycins (**81**).

Feeding studies have shown that the basic guanidine amino-acids and the β-lysine are derived from arginine and lysine, respectively.[123] β-Lysine occurs in only one other group of peptide antibiotics, the glycopeptides belonging to the streptothricin group, and is considered to be derived directly from lysine by an isomerase enzyme for which 5′-deoxyadenosylcobalamine (vitamin B_{12}) is a co-factor.

The peptides contain only L-amino-acids and crystallographic analyses of viomycin and tuberactinomycin O demonstrate that the amides all have the *trans* geometry. A further significant feature revealed by X-ray data, and subsequently corroborated by solution studies, is the hydrogen-bonded chelate ring, the conformation of which is similar to the β-turn structures common to many other cyclic peptides (see later). In the viomycin family, the corner positions of the β-turns are occupied by L-serine and the didehydroserine ureide unit.

The most fully authenticated series of cyclic hexapeptides are the iron-containing metabolites of the siderochrome class. The subject has been reviewed[124,125] and here only the general structural features and properties are outlined. Ferrichrome (**82**) was first isolated 20 years ago and shown to be a potent microbial growth factor. Related compounds were subsequently isolated and it was discovered that the metal-free peptides were produced in high yield when the organisms were grown in media lacking iron. This observation, together with other evidence, has led to the suggestion that these compounds act as cellular transport agents for iron in aerobic micro-organisms. The iron-binding site is furnished by three acylated *N*-hydroxyornithine units incorporated in a cyclic hexapeptide. The remaining amino-acids are either glycine or serine, and the acetyl group can be replaced by a *trans*-5-hydroxy-3-methylpent-2-enoyl group.[126] Removal of the metal ion removes the restrictions imposed on the side-chains of the ornithine residues, considerably augmenting the overall flexibility of the cyclic peptide.[127] The metal-free peptides are referred to as deferri-ferrichromes, and these are also capable of forming complexes with other metals, *e.g.* interaction with Al^{3+} or Ga^{3+} leads to the respective alumichromes or gallichromes.[128]

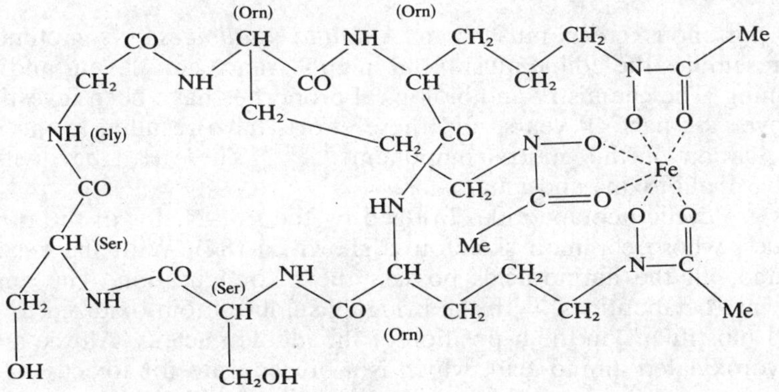

(82)

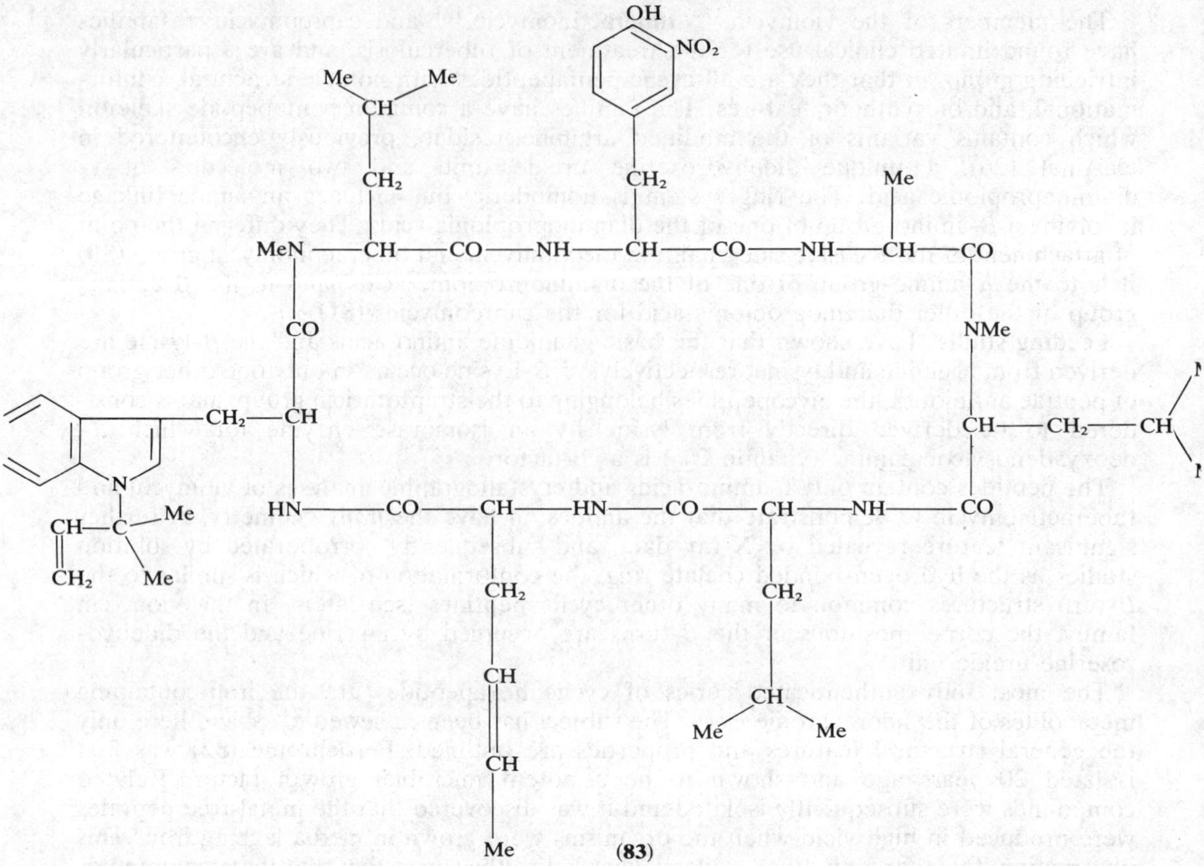

(**83**)

An X-ray analysis of the cyclic heptapeptide ilamycin B (**83**) accords with the structure originally proposed on the basis of chemical investigations. Isoprenylated tryptophan units had not previously been observed in large ring peptides, and the 1-(1′,1′-dimethylallyl)tryptophan unit is unique to this class of peptide antibiotic. The crystal conformation is also significant with *cis* amide linkages associated with both *N*-methyl amino-acid residues.[129] Another equally interesting structural peculiarity is that the carbonyl group of the nitrotyrosine residue simultaneously participates in two intramolecular hydrogen bonds. Spectral data suggest that these features are retained in solution.

The deadly poisonous green mushroom *Amanita phalloides* is sometimes called the 'deathcap'; it resembles the edible mushroom in appearance and flavour and is a common cause of poisoning. The chemistry and biological properties have been investigated almost continuously over the past 40 years, and these efforts have resulted in the isolation and structural elucidation of the major components.[130,131] These can be divided into two groups, namely, phallotoxins and amatoxins.

Phallotoxins are cyclic heptapeptides bridged by the side-chains of the tryptophan and cysteine residues, whose common skeleton is shown in (**84**). With the exception of the residue indicated, all the amino-acids possess the L-configuration. The amatoxins (**85**) have a larger ring (octapeptides) with the bridging sulphur atom oxidized to a sulphoxide and a hydroxyl substituent on the 6-position of the indole nucleus. All the amanita toxins contain a γ-hydroxylated amino-acid, which is a prerequisite for toxicity.

Despite the similar structures, the groups of toxins have different modes of action.

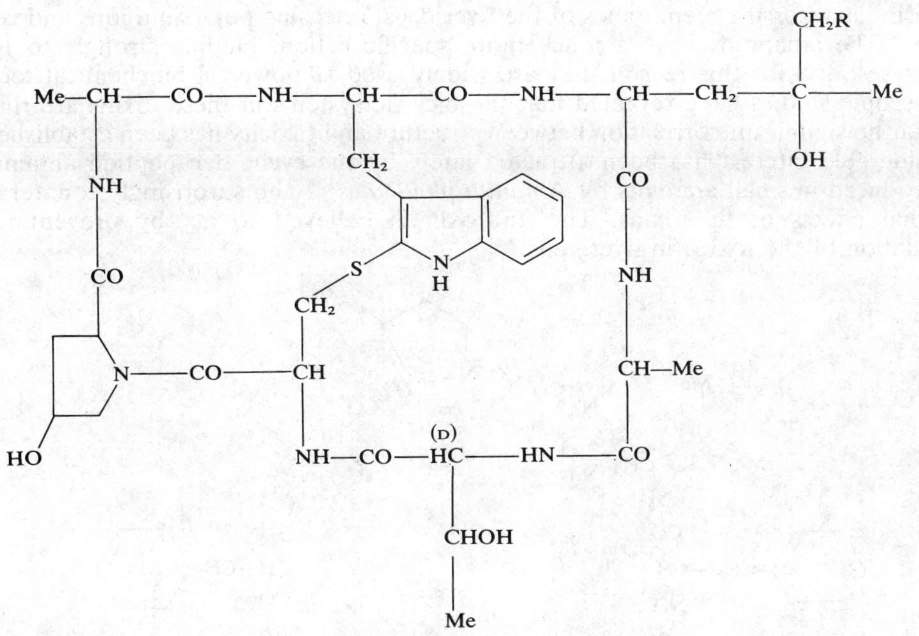

(**84**) Phalloidin, R = OH
Phalloin, R = H

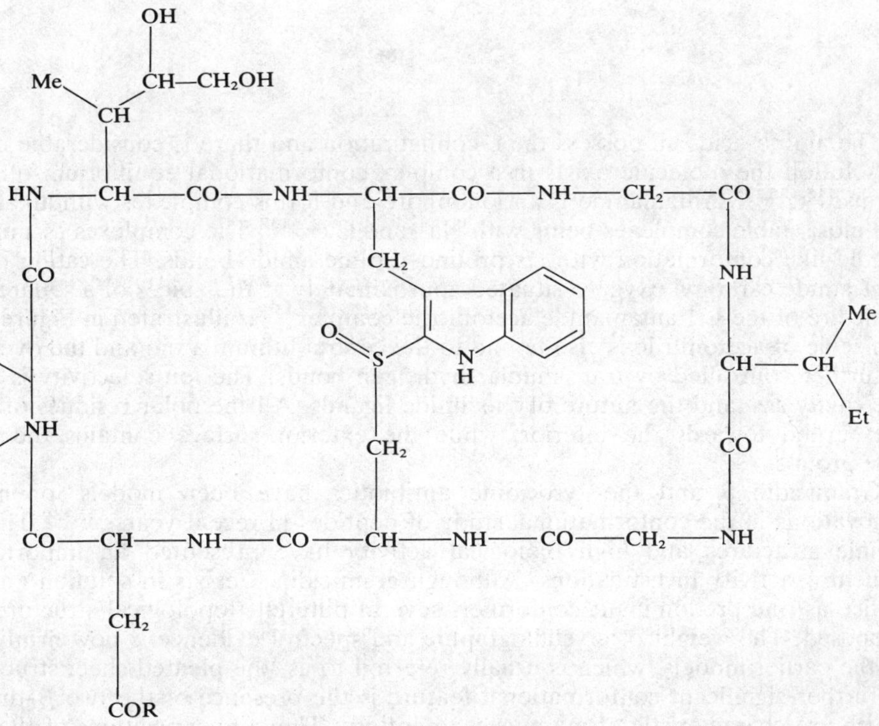

(**85**) α-Amanitin, R = NH_2
β-Amanitin, R = OH

Phalloidin damages the membranes of the liver cells, releasing potassium ions and certain enzymes. The amanitins have a much more specific action, binding strongly to RNA-polymerases, and for this reason they are widely used as powerful biochemical tools.[132] Spectroscopic studies have revealed that the bicyclic systems in these toxins are rigid,[133] but so far no significant correlation between structure and toxicity has been established.

Considerable interest has been attracted lately by the cyclic decapeptide antamanide (**86**), produced in small amounts by *Amanita phalloides*[134] and surprisingly counteracting the lethal effects of its toxins. The antitoxin is believed to act by preventing the accumulation of the toxins in liver cells.

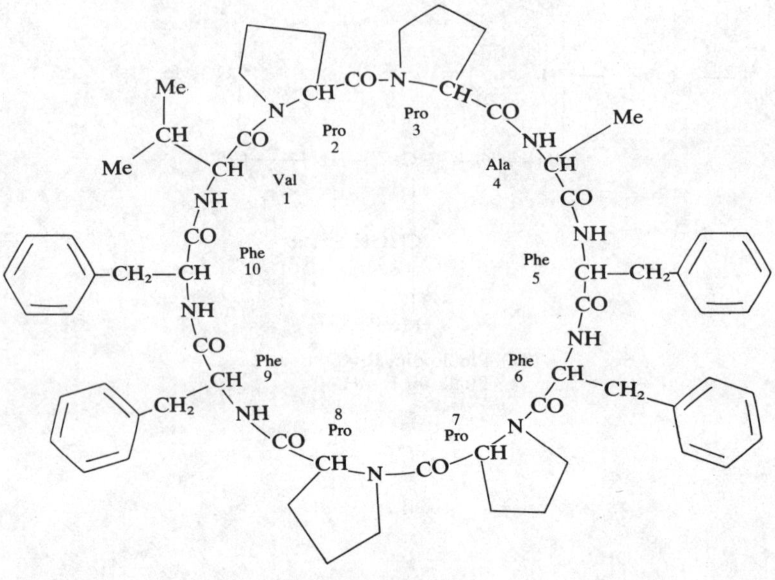

(**86**)

The amino-acids all possess the L-configuration and there is considerable evidence that in solution the molecule exists in a complex conformational equilibrium of at least four forms.[115,135,136] Antamanide is an ionophore and forms complexes with alkali metal ions, the most stable complexes being with Na^+ and Ca^{2+}.[137] The complexes assume a common saddle-like conformation with *cis* proline–proline amide bonds. The cation interacts with four amide carbonyl oxygens situated approximately at the apices of a square. The crystal structure of the Li^+ antamanide acetonitrile complex[138] is illustrated in Figure 1. A solvate molecule of acetonitrile is also bound to the central lithium atom and the overall topology is further controlled by transannular hydrogen bonds. The ion selectivity is attributed to the cavity size and the nature of the amide ligands. All the polar residues of the complex are turned towards the interior, while the exterior surface contains the hydrophobic side-groups.

Gramicidin S and the tyrocidine antibiotics have been models for many of the innovations in the conformational study of peptides in recent years.[115,139] Their relatively simple structures and high biological activity have presented an important area for structure–activity investigations. Although gramicidin S exists in solution and the crystal lattice as one predominant conformer, several different topological structures have been proposed. The weight of crystallographic and spectral evidence is now firmly behind one of the earlier models, which is usually referred to as 'the pleated sheet structure' (**87**).[140] A further significant conformational feature is the presence of the two β-turns or loops, which are characteristic of many cyclic peptides. The corner positions of the β-turns are occupied by the D-phenylalanine and L-proline residues and conformational energy

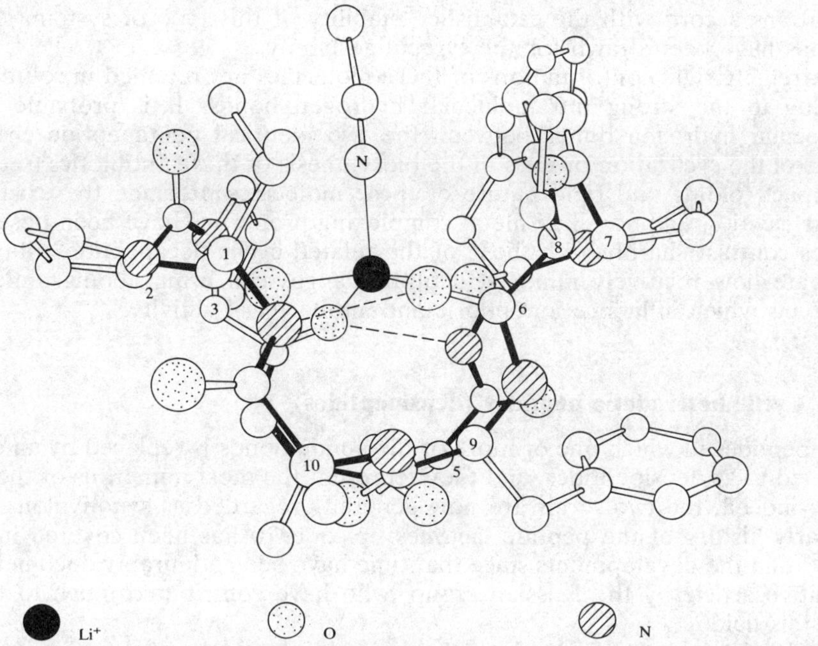

Figure 1 Li^+ antanamide–acetonitrile complex (the numbers refer to the amino-acid residues) (Reproduced with permission from Karle[138])

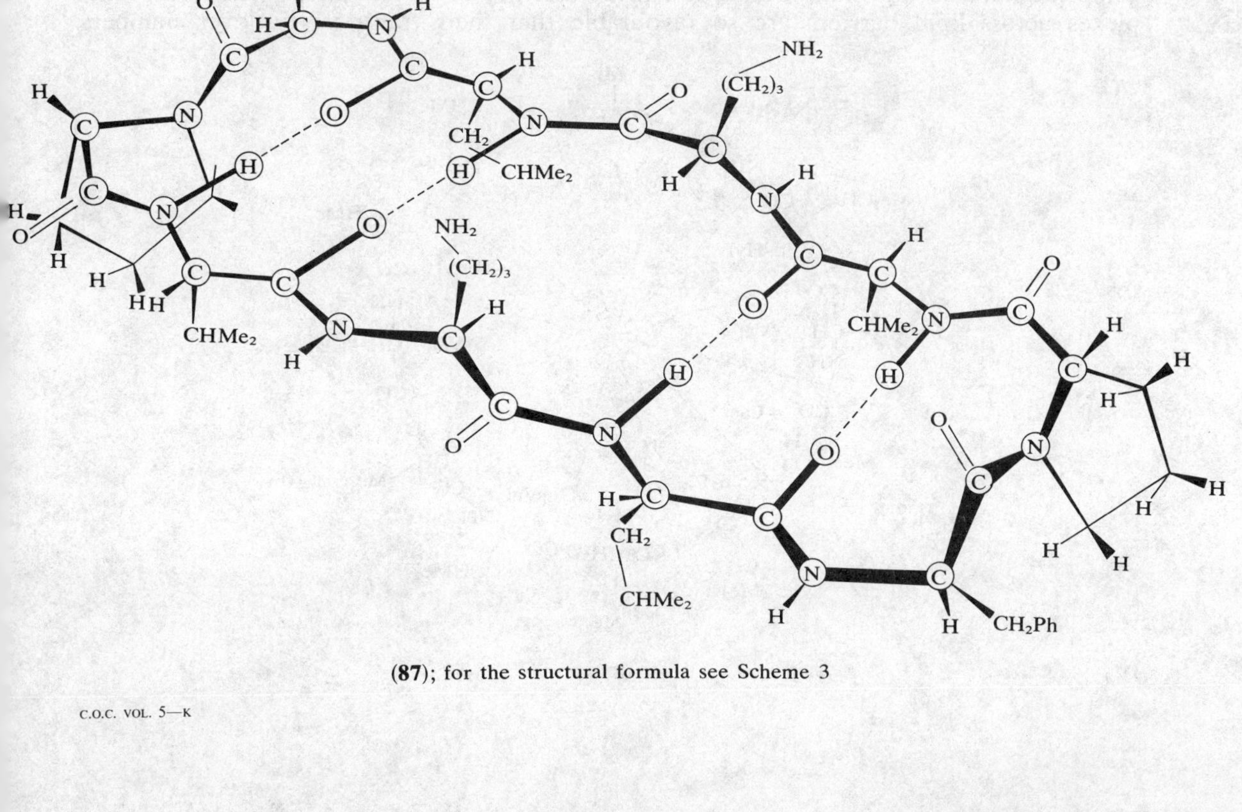

(**87**); for the structural formula see Scheme 3

considerations accord with the established stability of this type of system.[114–116] Similar conclusions have been drawn for the tyrocidine family.

The extremely rigid conformations of these molecules are retained in solution, presumably owing to the strong intramolecular hydrogen bonds. It is probable that similar intermolecular hydrogen bonds between the two identical pentapeptide chains may, in part, control the cyclization process in the biosynthesis of these antibiotics (see Scheme 3). The compact folding and rigid nature of these molecules preclude the existence of any molecular cavities, and no alkali metal complexing properties have been observed. These properties contrast sharply with those of the related cyclic decapeptide antamanide, and demonstrate how relatively minor structural changes can bring about subtle conformational effects which influence ionophoric and antibacterial activity.

23.4.2.4 Cyclic heterodetic peptides (depsipeptides)

Cyclic peptides in which one or more of the amide bonds is replaced by an ester linkage are referred to as depsipeptides, and they represent the most numerous of the heterodetic peptides; indeed, the two terms are now generally regarded as synonymous.

The early history of the peptide lactones up to 1970 has been covered in a thorough review,[141] and the developments since that time have been admirably documented[142] in an authoritative article by the Russian group, who have contributed much to the development of this field.

Cyclic depsipeptides may be broadly divided into two main groups, namely those possessing a regular alternating array of peptide and ester linkages, and those with irregular insertion of ester bonds. Valinomycin (**88**), the enniatins (**89**), and beauvericin (**90**), most of which were characterized 20 years ago, all belong in the former category. The observations in the mid-1960s that valinomycin and related compounds possessed unique selective ion-transporting properties led to a resurgence of interest in these compounds, which were subsequently referred to as ionophores. These peptides form lipid-soluble complexes with polar cations of which K^+, Na^+, Ca^{2+}, Mg^{2+}, and the biogenic amines are the most significant biologically. A variety of physical studies indicate that the complexation–decomplexation kinetics and diffusion rates of ionophores and their complexes across lipid barriers are so favourable that their transport turnover numbers

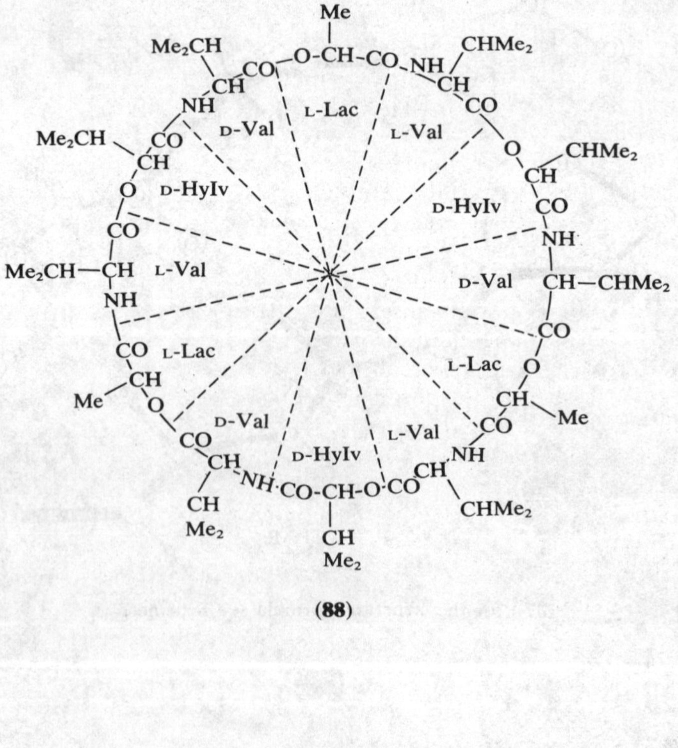

(**88**)

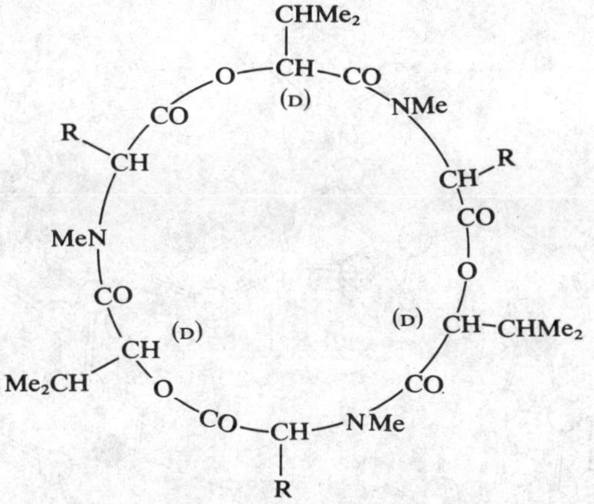

(89) R = CHMeEt (A); $CHMe_2$ (B); CH_2CHMe_2
(90) R = CH_2Ph

through biological and artificial membranes attain values in some cases exceeding those of enzyme systems. The biological applications of general ionophores, which also include the polyethers and synthetic compounds, have been comprehensively reviewed.[142,143]

Valinomycin with its 36-membered ring possesses a wealth of conformational possibilities. The spectral data have revealed that in solution valinomycin is an equilibrium of three major conformers. The relative amounts of the conformers depend on the solvent and temperature. Crystallographic data on two different modifications[144,145] showed little difference in the ring conformation, which resembles the proposed non-polar solution conformer.

Valinomycin is known to form more or less stable complexes with K^+, Rb^+, and Cs^+ in non-polar solvents. The K^+ complex is considerably more stable than the Na^+ complex and the potassium/sodium complexing selectivity is higher than for any other ionophore. The crystal conformations of valinomycin[145] and the potassium ion complex,[146] as well as a diagramatic representation of the general architecture of the potassium binding, are shown in Figure 2. The six arrows represent ester carbonyls oriented inward and binding the unsolvated cation in the cavity. The broken lines are the hydrogen bonds of the β-turns which stabilize the so-called bracelet backbone of the molecule in both the complexed and uncomplexed form. The lipophilic groups are all on the exterior of the complex and facilitate its solubility in non-polar solvents.

The valinomycin complexes with Rb^+ and Cs^+ are basically the same, and the increased size of the molecular cavity necessary to accommodate the ions is attained by small conformational changes. Similar changes cannot compensate for the smaller ionic size of Na^+, and this cation does not fully interact with complexing ester carbonyls. Equally extensive investigations on the enniatins and beauvericin and their complexes have been reported as part of a wide ranging programme of peptide ionophores.[115,142]

The most comprehensively studied cyclic depsipeptides belonging to the second category are undoubtedly the actinomycins. The general structure of the group (**91**) is well established and a considerable number have now been characterized, principally from *Streptomyces* species.[147,148] The most commonly isolated compound is actinomycin D (**92**), which possesses two structurally identical peptide rings. Metabolites of this series are referred to as the *iso*-series, and those with differing peptide rings as the *aniso*-series.

The considerable variation in the amino-acids within the peptide residues reflects the relatively low specificity in the biosynthesis. This can be exploited to produce higher yields of specific antibiotics by adding the appropriate amino-acids to the organism. In some cases this can be used to produce new antibiotics. For example, the actinomycins containing azetidine-2-carboxylic acid instead of proline or pipecolic acid are produced when the organism is supplied with an extraneous source of azetidine-2-carboxylic acid.[149]

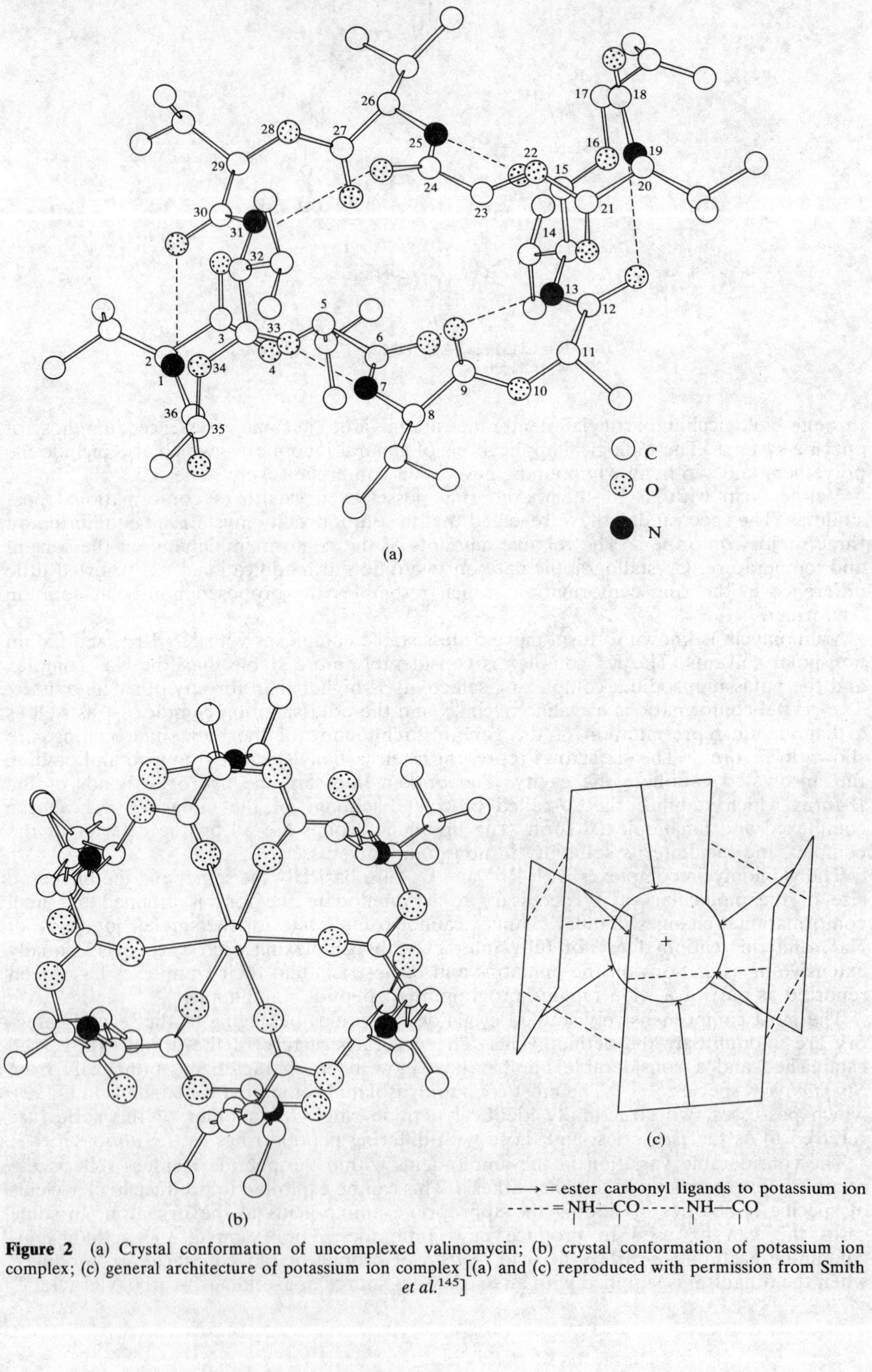

Figure 2 (a) Crystal conformation of uncomplexed valinomycin; (b) crystal conformation of potassium ion complex; (c) general architecture of potassium ion complex [(a) and (c) reproduced with permission from Smith *et al.*[145]]

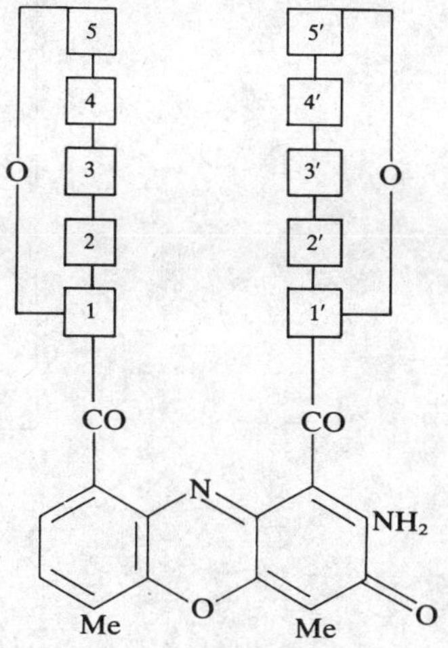

(91)

1,1′ = Thr or MeThr
2,2′ = D-Val or D-Leu
3,3′ = Pro, 4-oxo-5-methylproline, 4-*allo*-hydroxy-5-methylproline, *cis*-5-methylproline, pipecolic acid, 4-oxopipecolic acid, or 4-hydroxypipecolic acid.
4,4′ = Sar or Gly
5,5′ = MeVal, MeIle, or MeAla

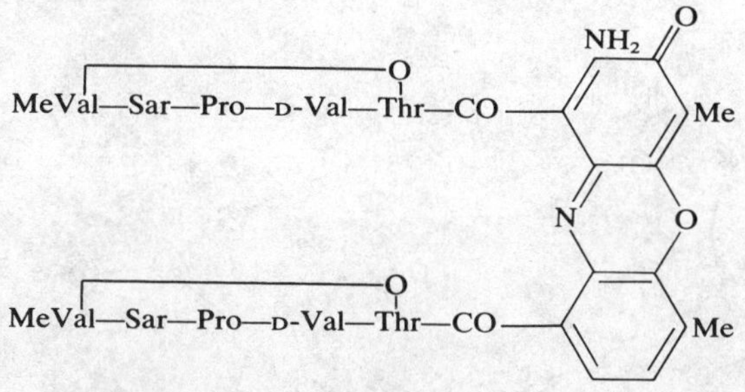

(92)

The degree to which this form of directed biosynthesis can be applied varies considerably and is dependent on the peptide and the organism. However, it does offer an interesting method of producing new antibiotics within a given series and has also been employed in cell-free systems (see Scheme 3). The proposed biosynthesis of the phenoxazone nucleus of the actinomycins is outlined in Scheme 7[41–43] (see also Section 30.3.5).

The actinomycins are potent inhibitors of DNA-dependent RNA synthesis, *i.e.* the transcription in protein biosynthesis (see Scheme 1), and have proved to be powerful biochemical tools. Actinomycin D has also found limited application in the clinical treatment of certain types of tumours. Their mode of action involves the formation of highly stable complexes with DNA, which precludes the latter from performing its biological function. Consequently, considerable effort has been directed to the investigation of the crystal and solution conformations of these molecules.[115,150] The current model for the interaction of the DNA double helix with actinomycin is based on the X-ray data obtained from a crystalline complex containing actinomycin and deoxyguanosine (Figure 3).[151] In this model the phenoxazone chromophore is intercalated between adjacent G–C base pairs of DNA, where the guanine residues are on opposite DNA strands and the two amino-groups of the guanines form specific hydrogen bonds with both cyclic peptides, which fit into a narrow groove. The model accords with data available and represents an important advance in molecular biology.

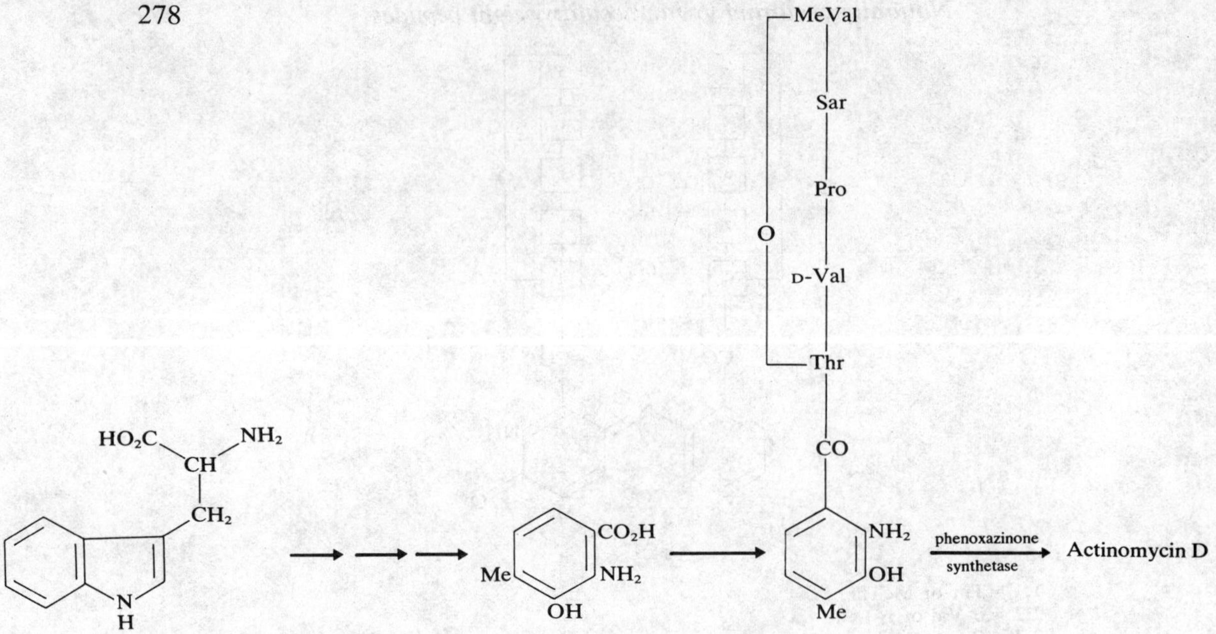

SCHEME 7 Outline of the proposed biosynthesis of the actinomycin chromophore

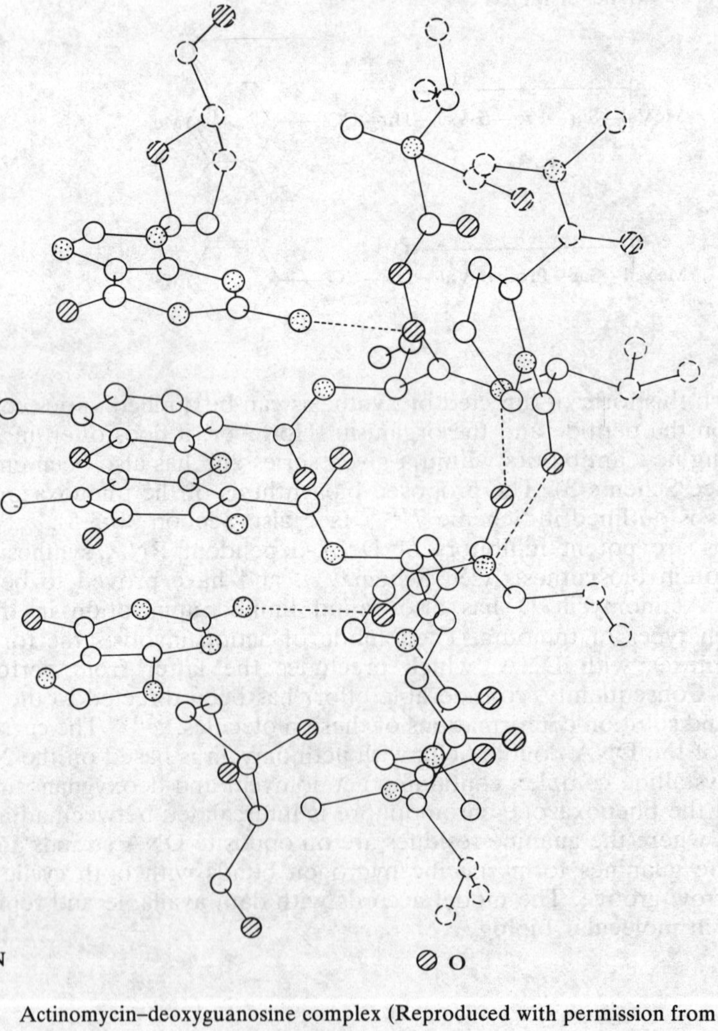

Figure 3 Actinomycin–deoxyguanosine complex (Reproduced with permission from Sobell *et al.*[151])

A wide variety of other cyclic depsipeptides with many of the atypical features already outlined in previous sections have been identified, and shown to exhibit an equally broad range of biological activity. Fortunately these compounds have been comprehensively reviewed[142] and only an illustration of structural types and properties will be given here. The destruxins (**93**) are a group of interesting insecticidal peptides isolated from *Aspergillus ochraceas.*[152] Surfactin (**94**) is a microbial anticoagulant containing all L-amino-acids,[153] and the antibiotic stendomycin (**95**) contains a number of D-amino-acid residues as well as a modified guanidine amino-acid,[154] *cf.* (**26**), (**80**) and (**81**). Recently interest has

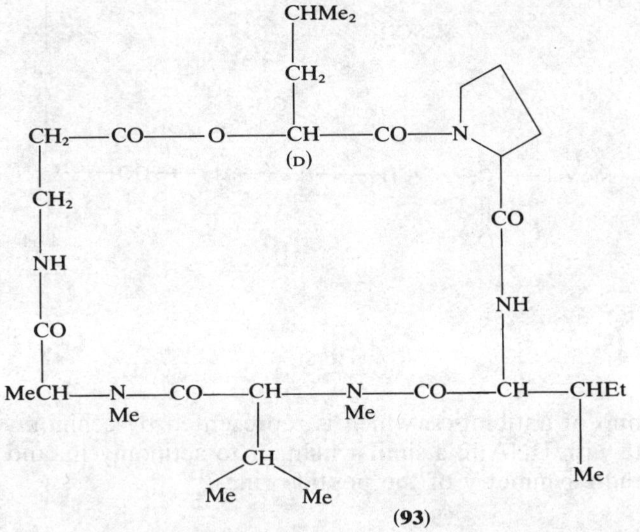

(**93**)

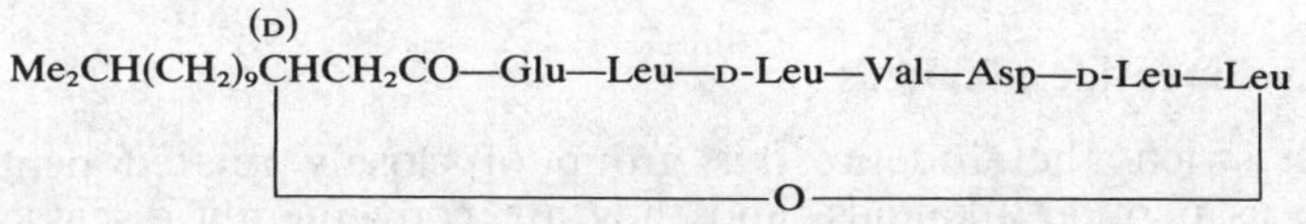

(**94**)

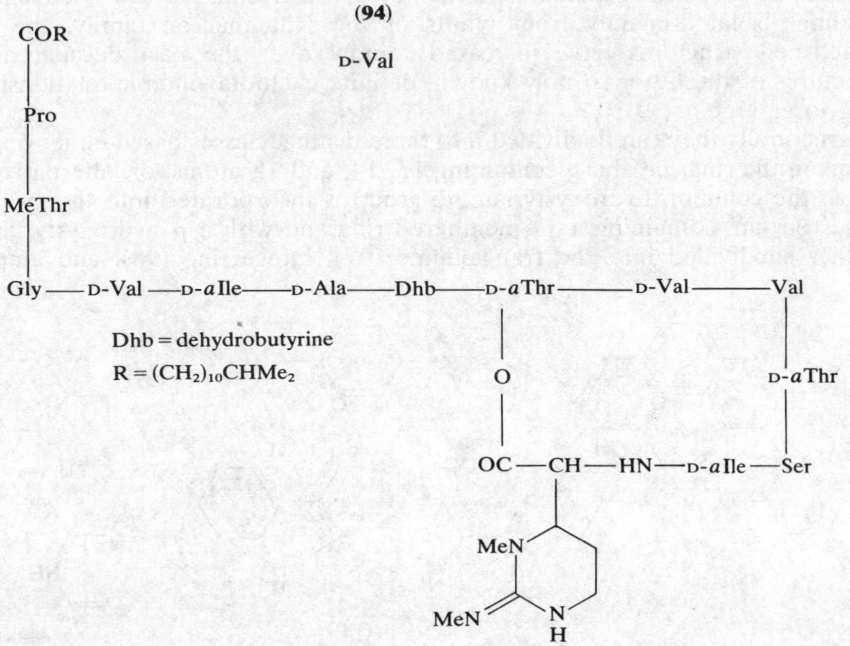

(**95**)

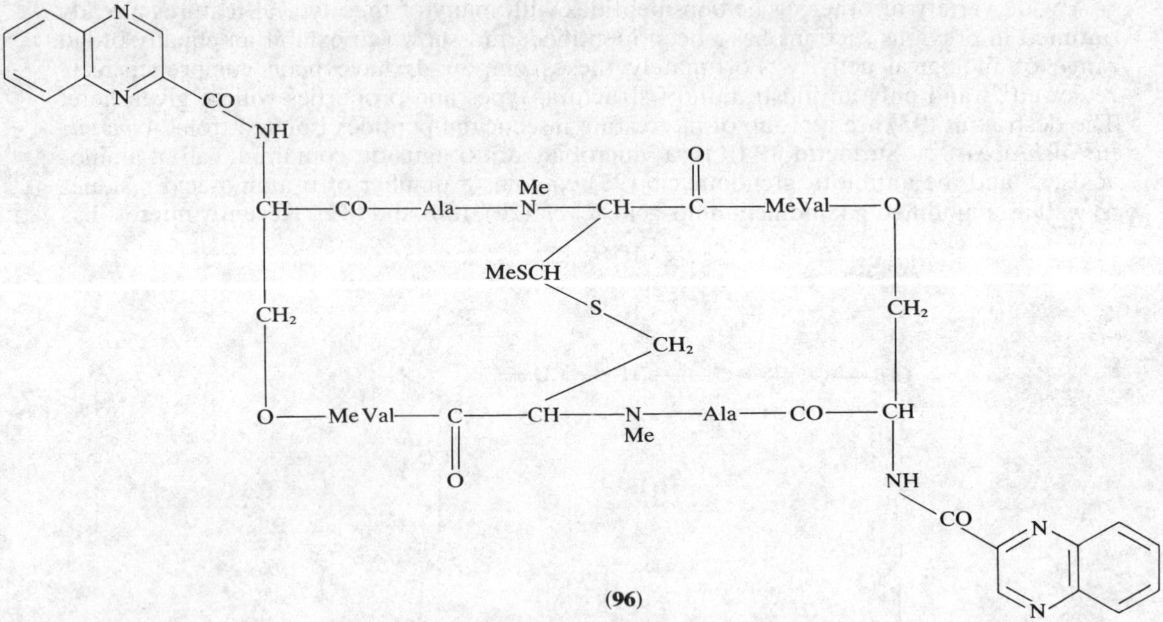

(96)

centred on a new group of antibiotics, which is represented by echinomycin (**96**). These compounds intercalate with DNA in a similar manner to actinomycin, and it is interesting to note the same pseudo-symmetry of the peptide rings.[155]

23.4.2.5 Cyclopeptide alkaloids

Although strictly heterodetic, this group of closely related peptides are invariably referred to as peptide alkaloids, and they are conveniently discussed under a separate heading. Isolated mainly from plants of the Rhamnaceae family, the numbers with established structures have increased rapidly over the past decade, and within the structures of the 60 or so now known, definite chemotaxonomic relationships are beginning to emerge.

Structurally they can be divided into three distinct classes based either on the number of atoms in the ring, *i.e.* those containing 13, 14, and 15 atoms, or, alternatively, the way in which the common hydroxystyrlamino-group is incorporated into the cyclic system. The largest group, containing a 14-membered ring and with a *p*-hydroxystyrlamino-group, is further subdivided into the frangalanine (**97**), integerrine (**98**), and amphibine-B (**99**)

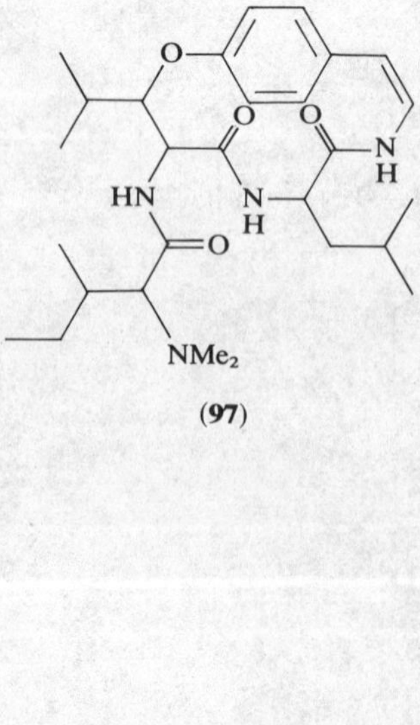

(97)

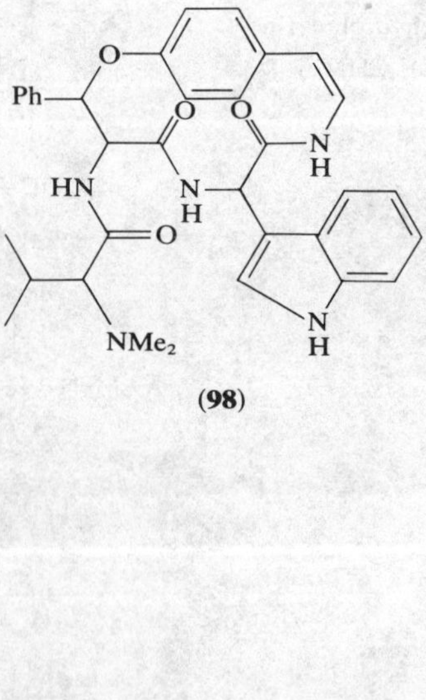

(98)

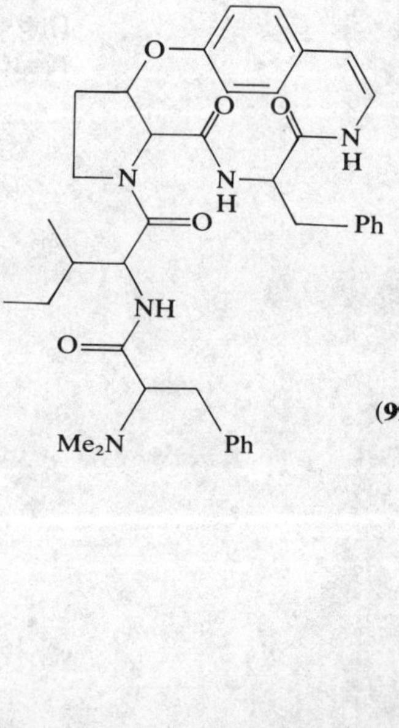

(99)

types. The smaller classes containing 13- and 15-membered rings are characterized by zizyphine-A (**100**) and mucronine-A (**101**). The chemistry, structural elucidation, and distribution of these plant products have been admirably covered in a recent review.[156]

As well as the unique hydroxystyrlamino-group, all the alkaloids possess a terminal *N*-dimethyl amino-acid residue, which is rarely observed in microbial peptides. In the main, the amino-acids of the cyclopeptides have the L-configuration, although scrutianine-E (**102**) appears to be an exception in that it contains a novel D-amino acid,

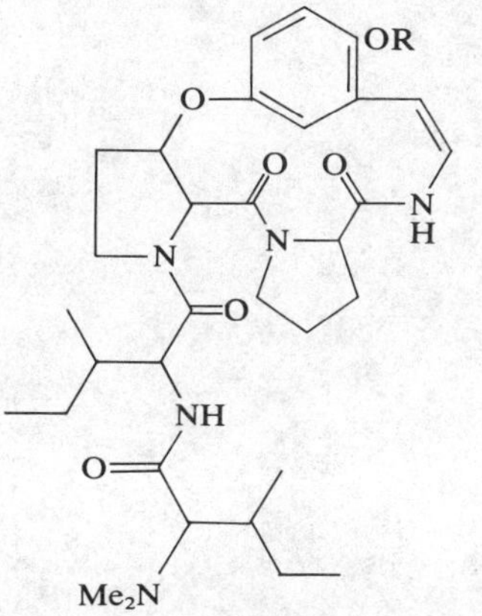

(**100**)

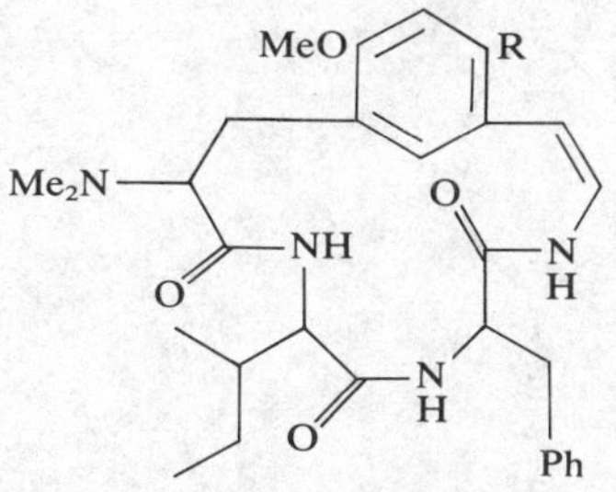

(**101**)

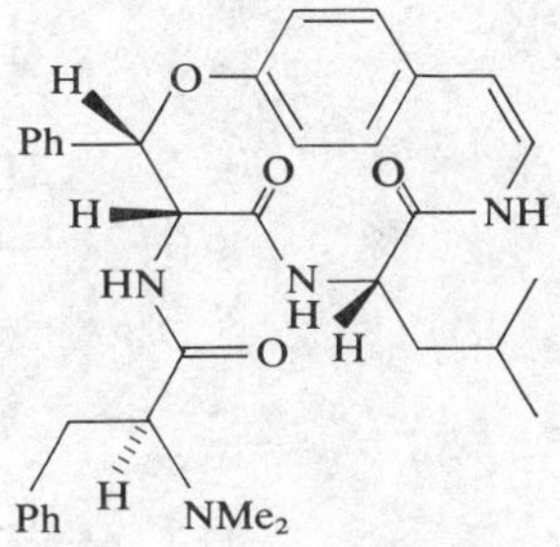

(**102**)

namely *threo*-β-phenylserine. As yet nothing is known about the biosynthesis of these peptides, and it will be of interest to await developments in this area, in order to compare the production of atypical peptides in plants and micro-organisms. In view of the considerable structural diversity of the microbial peptides, it is surprising that to date the cyclopeptide alkaloids represent the only major group of plant peptides. Since these compounds are basic and were originally isolated as alkaloids, it is possible that this represents a distorted picture of the distribution of atypical peptides in the plant kingdom, and that more careful investigation of plant material may lead to the isolation of other types of peptide.

23.4.2.6 Large modified peptides

The growing sophistication in isolation procedures and methods of structural elucidation has led to the characterization of modified microbial peptides, the size of which moves closer to the borderline between classical organic natural products and biomacromolecules. In general, these metabolites contain the same types of structural modifications as observed in the peptides described in the preceding sections, but their complexity warrants separate consideration.

The antibiotic subtilin from *Bacillus subtilis* has been assigned structure (**103**) on the basis of an extensive investigation.[157] The molecule is composed of 32 amino-acid residues, eight of which, *i.e.* four 3-methyl-lanthionines, two dehydroalanines, and one each of dehydro-α-aminobutyric acid and lanthionine, are not found in protein. The alanine residues of lanthionine and all four aminobutyric acid moieties of the 3-methyl-lanthionine residues possess the D-configuration. The novel thio-ether bridges probably result from intramolecular addition of cysteine thiol groups to dehydroalanine or dehydrobutyrine units, which in turn could be derived by the dehydration of serine and

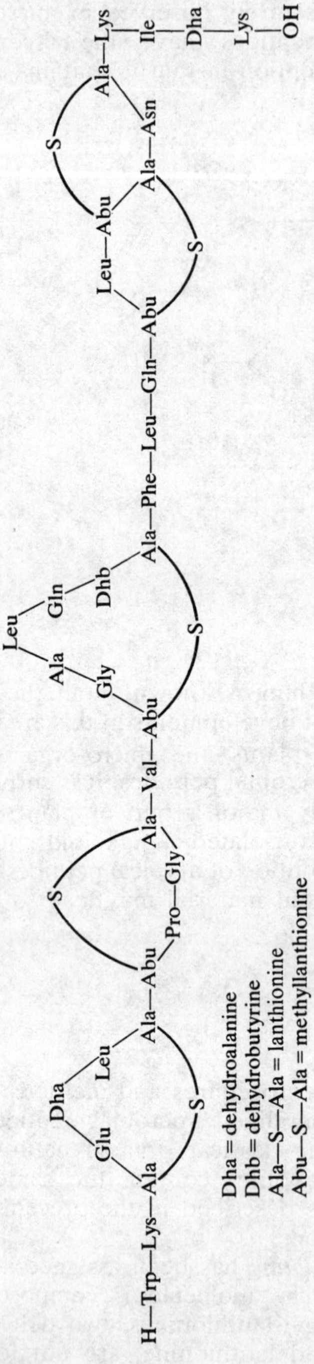

Dha = dehydroalanine
Dhb = dehydrobutyrine
Ala—S—Ala = lanthionine
Abu—S—Ala = methyllanthionine

(103)

Ac—Aib—Pro—Aib—Ala—Aib—Ala—Gln—Aib—Val—Aib—Gly—Leu—Aib—Pro—Val—Aib—Aib—Glu(Pheol)—Gln—OH

Aib = α-aminoisobutyric acid
Pheol = phenylalaninol

(104)

threonine, respectively. This leads to the interesting speculation that the precursor of subtilin could well be a linear peptide derived solely from primary amino-acids and constructed under ribosomal control. There is evidence that the biosynthesis of a closely related peptide, nisin,[158] may well be formed on a ribosomal template.[159] The possibility that more low molecular weight peptides with post-translation modifications will be isolated and characterized in the next decade is an exciting prospect.

The linear ion-transporting peptide alamethicin (**104**) is smaller than subtilin and is noteworthy because of the presence of a considerable number of α-methylalanine residues. In addition, a β-phenylalaninol residue is attached to the γ-carboxylglutamic acid.[160]

The complexity of thiostrepton (**105**) could only be unravelled by an X-ray analysis,

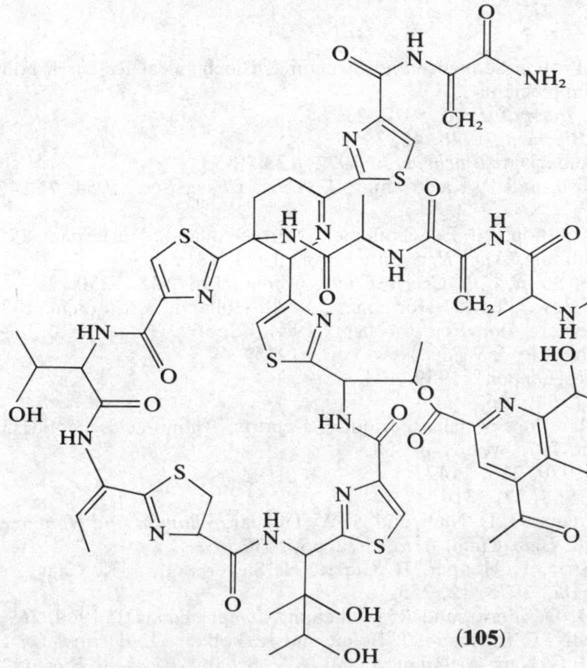

(**105**)

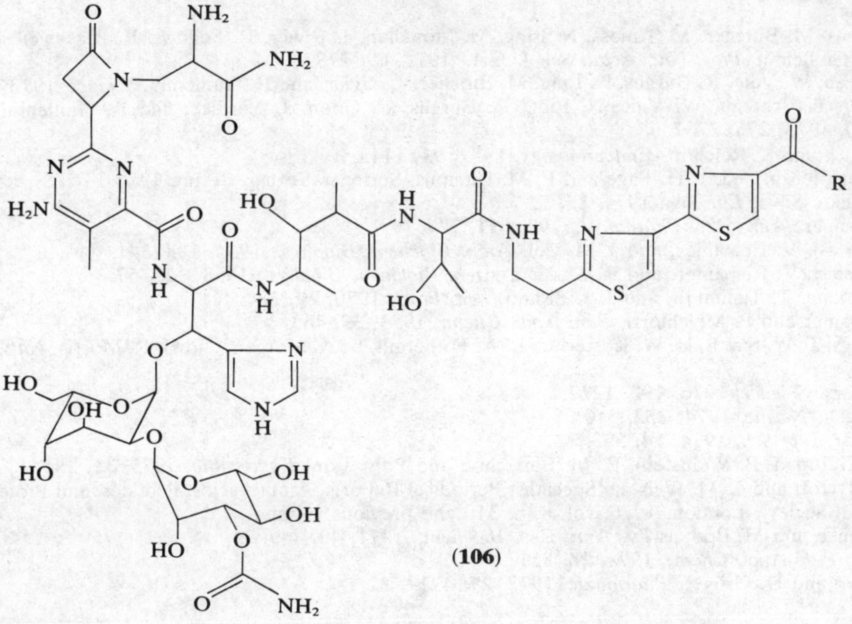

(**106**)

which has revealed a variety of modified units many of which have been observed in smaller metabolites.[161] Peptides related to thiostrepton which have recently been characterized include siomycin,[162] nosiheptide,[163] and berninamycin.[164]

Finally, it should be noted that a whole range of natural products derived from the interaction of peptides with products formed by other metabolic processes have been excluded from this review. These frequently possess novel structures and biological activity and contain many of the atypical features already described. Many of these observations are exemplified by the antitumour glycopeptides of the bleomycin family (**106**).[165]

References

1. M. O. Dayhoff, 'Atlas of Protein Sequence and Structure', Biochemical Research Foundation, Bethesda, 1972, vol. 5 and annual supplements.
2. W. Rüger, *Angew. Chem. Internat. Edn.*, 1972, **11,** 883.
3. H. F. Lodish, *Ann. Rev. Biochem.*, 1976, **45,** 39.
4. R. B. Loftfield and D. Vanderjagt, *Biochem. J.*, 1972, **128,** 1353.
5. V. du Vigneaud, D. T. Gish, and P. Katsoyannis, *J. Amer. Chem. Soc.*, 1954, **76,** 4751 and references cited therein.
6. R. Acher, in 'Molecular Evolution', ed. E. Schoffeniels, North-Holland, Amsterdam, 1971, vol II, p. 43.
7. W. H. Sawyer and M. Manning, *Ann. Rev. Pharmacol.*, 1973, **13,** 5.
8. M. Zaoral, J. Kolc, and F. Sorm, *Coll. Czech. Chem. Comm.*, 1967, **32,** 1250.
9. S. A. Berson and R. S. Yalow, 'Peptide Hormones', North-Holland, Amsterdam, 1973.
10. H. S. Tager and D. F. Steiner, *Ann. Rev. Biochem.*, 1974, **43,** 509.
11. W. R. Butt, 'Hormone Chemistry', Wiley, New York, 1975.
12. F. Sanger, *Chem. and Ind. (London)*, 1959, 104.
13. H. Zahn, *Naturwiss.*, 1967, **54,** 396.
14. D. J. Schafer and M. Szelke, in 'Specialist Periodical Reports, Amino-acids Peptides, and Proteins', The Chemical Society, London, 1976, vol. 8, p. 338.
15. M. I. Grossman, *Nature*, 1970, **228,** 1147.
16. J. S. Morley, *Fed. Proc.*, 1968, **27,** 1314.
17. J. T. Potts Jr., H. T. Keutmann, H. D. Niall, and G. W. Tregear, *Vitamins and Hormones*, 1971, **29,** 41.
18. P. Rivaille and G. Milhaud, *Helv. Chim. Acta*, 1972, **55,** 1617.
19. K. Folkers, N.-G. Johansson, F. Hooper, B. Currie, H. Sievertsson, J.-K. Chang, and C. Y. Bowers, *Angew. Chem. Internat. Edn.*, 1973, **12,** 255.
20. R. Burgus, T. F. Dunn, D. Desiderio, and R. Guillemin, *Compt. rend.(D)*, 1969, **269,** 1870.
21. C. Y. Bowers, A. V. Schally, F. Enzmann, J. Boler, and K. Folkers, *Endocrinology*, 1970, **86,** 1143.
22. H. Matsuo, Y. Baba, R. M. G. Nair, A. Arimura, and A. V. Schally, *Biochem. Biophys. Res. Comm.*, 1971, **43,** 1334.
23. R. Burgus, M. Butcher, M. Amoss, N. Ling, M. Monahan, J. Rivier, R. Fellows, R. Blackwell, W. Vale, and R. Guillemin, *Proc. Nat. Acad. Sci. U.S.A.*, 1972, **69,** 278.
24. P. Brazeau, W. Vale, R. Burgus, N. Ling, M. Butcher, J. Rivier, and R. Guillemin, *Science*, 1973, **179,** 77,
25. J. Rivier, P. Brazeau, W. Vale, N. Ling, R. Burgus, C. Gilon, J. Yardley, and R. Guillemin, *Compt. rend.(D)*, 1973, **276,** 2737.
26. M. Mituck and S. Reichlin, *Endocrinology*, 1972, **91,** 1145.
27. See 'Angiotensin', ed. I. H. Page and F. M. Bumpus, Springer-Verlag, Berlin, 1974.
28. M. Rocha e Silva, *Life Sci.*, 1974, **15,** 7.
29. V. Erspamer, *Ann. Rev. Pharmacol.*, 1971, **11,** 327.
30. A. Anastasi, V. Erspamer, and J. M. Cei, *Arch. Biochem. Biophys.*, 1964, **108,** 341.
31. A. Anastasi, V. Erspamer, and R. Endean, *Arch. Biochem. Biophys.*, 1968, **125,** 57.
32. T. Nakajima, T. Tanimura, and J. J. Pisano, *Fed. Proc.*, 1970, **29,** 284.
33. V. Erspamer and P. Melchiorri, *Pure Appl. Chem.*, 1973, **35,** 463.
34. J. Hughes, T. W. Smith, H. W. Kosterlitz, L. A. Fothergill, B. A. Morgan, and H. R. Morris, *Nature*, 1975, **258,** 577.
35. J. L. Marx, *Science*, 1976, **193,** 1227.
36. G. Metcalf, *Nature*, 1974, **252,** 310.
37. G. Ungar, *Life Sci.*, 1974, **14,** 595.
38. H. N. Guttman, B. Weinstein, R. M. Bartschot, and P. S. Tam, *Experientia*, 1975, **31,** 285.
39. B. W. Bycroft and C. M. Wels, in 'Specialist Periodical Reports, Amino-acids, Peptides, and Proteins', The Chemical Society, London, 1976, vol. 8, p. 310 and previous volumes.
40. D. Perlman and M. Bodanszky, *Ann. Rev. Biochem.*, 1971, **40,** 449.
41. E. Katz, *Pure Appl. Chem.*, 1971, **28,** 551.
42. W. Maier and D. Gröger, *Pharmazie*, 1972, **27,** 491.

43. K. Kurahashi, *Ann. Rev. Biochem.*, 1974, **43,** 445.
44. S. G. Laland and T. L. Zimmer, *Essays in Biochemistry*, 1973, **9,** 31.
45. F. Lipmann, *Accounts Chem. Res.*, 1973, **6,** 361.
46. F. Lipmann, *Science*, 1973, **173,** 875.
47. C. H. Tzeng, K. D. Thrasher, J. P. Montgomery, B. K. Hamilton, and D. I. C. Wang, *Biotechnol. Bioeng.*, 1975, **17,** 143.
48. See 'Glutathione', Proceedings of the 16th Conference of the German Society of Biological Chemistry, ed. L. Flohé, H. Ch. Benöhr, H. Sies, H. D. Waller, and A. Wendel, Tubingen, 1973.
49. A. Meister and S. S. Tate, *Ann. Rev. Biochem.*, 1976, **45,** 559.
50. A. Meister, *Science*, 1973, **180,** 33.
51. S. G. Waley, *Biochem. J.*, 1958, **68,** 189.
52. L. Fowden, *Ann. Rev. Biochem.*, 1964, **33,** 173.
53. S. G. Waley, *Adv. Protein Chem.*, 1966, **21,** 1.
54. H. Tabor and C. W. Tabor, *J. Biol. Chem.*, 1975, **250,** 2648.
55. H. Umezawa, *Pure Appl. Chem.*, 1973, **33,** 129.
56. D. L. Pruess and J. P. Scannell, *Adv. Appl. Microbiol.*, 1974, **17,** 19.
57. J. P. Scannell and D. L. Pruess, in 'Chemistry and Biochemistry of Amino-acids, Peptides, and Proteins', ed. B. Weinstein, Marcel Dekker, New York, 1974, vol. 3, p. 189.
58. K. Kawamura, S.-I. Kondo, K. Maeda, and H. Umezawa, *Chem. Pharm. Bull* (*Japan*), 1969, **17,** 1902.
59. S. Umezawa, K. Tatsuta, K. Fujimoto, T. Tsuchiya, H. Umezawa, and H. Naganawa, *J. Antibiotics*, 1972, **25,** 267.
60. T. Aoyagi, S. Kunimoto, H. Morishima, T. Takeuchi, and H. Umezawa, *J. Antibiotics*, 1971, **24,** 687.
61. A. Okura, H. Morishima, T. Takita, T. Aoyagi, T. Takeuchi, and H. Umezawa, *J. Antibiotics*, 1975, **28,** 337.
62. B. W. Bycroft, *Nature*, 1969, **224,** 595.
63. M. Bodanszky and D. Perlman, *Nature*, 1968, **218,** 291.
64. W. A. König, W. Loeffler, W. H. Meyer, and R. Uhmann, *Chem. Ber.*, 1973, **106,** 816.
65. K. H. Baggaley, B. Blessington, C. P. Falshaw, W. D. Ollis, L. Chaiet, and F. L. Wolf, *Chem. Comm.*, 1969, 101.
66. H. J. Klosterman, G. L. Lamoureux, and J. L. Parsons, *Biochemistry*, 1967, **6,** 170.
67. Y. Konda, Y. Suzuki, S. Omura, and M. Onda, *Chem. Pharm. Bull.* (*Japan*), 1976, **24,** 92.
68. R. F. Pittillo and D. E. Hunt, in 'Antibiotics', ed. D. Gottlieb and P. D. Shaw, Springer-Verlag, Berlin, 1966, vol. I, p. 481.
69. K. V. Rao, S. C. Brooks, M. Kugelman, and A. A. Romano, *Antibiot. Ann.*, 1960, 943.
70. E. Bayer, K. H. Gugel, K. Hagele, H. Hagenmaier, S. Jessipow, W. A. König, and H. Zahner, *Helv. Chim. Acta*, 1972, **55,** 224.
71. D. L. Pruess, J. P. Scannell, H. A. Ax, M. Kellett, F. Weiss, T. C. Demny, and A. Stempel, *J. Antibiotics*, 1973, **26,** 261.
72. R. E. Mitchell, *Nature*, 1976, **260,** 75.
73. W. W. Stewart, *Nature*, 1971, **229,** 174.
74. J. P. Scannell, D. L. Pruess, J. F. Blount, H. A. Ax, M. Kellett, F. Weiss, T. C. Demny, T. H. Williams, and A. Stempel, *J. Antibiotics*, 1975, **28,** 1.
75. T. Kamiya, in 'Recent Advances in the Chemistry of β-Lactam Antibiotics', ed. J. Elks, The Chemical Society, London, 1977, p. 281.
76. B. W. Bycroft and R. Pinchin, *J.C.S. Chem. Comm.*, 1975, 121.
77. H. Sakakibara, H. Naganawa, M. Ohno, K. Maeda, and H. Umezawa, *J. Antibiotics*, 1974, **27,** 897.
78. H. Maehr, *Pure Appl. Chem.*, 1971, **28,** 603; G. A. Snow, *Bacteriol. Rev.*, 1970, **34,** 99.
79. C. Ratledge and G. A. Snow, *Biochem. J.*, 1974, **139,** 407.
80. E. Hough and D. Rogers, *Biochem. Biophys. Res. Comm.*, 1974, **57,** 73.
81. P. G. Sammes, *Fortschr. Chem. org. Naturstoffe*, 1975, **32,** 51.
82. A. S. Kokhlov and G. B. Lokshin, *Tetrahedron Letters*, 1963, 1881.
83. M. O. Forster and W. B. Saville, *J. Chem. Soc.*, 1922, 816.
84. K. Kakinuma and K. L. Rinehart Jr., *J. Antibiotics*, 1974, **27,** 733.
85. A. J. Birch, R. A. Massey-Westropp, and R. W. Rickards, *J. Chem. Soc.*, 1956, 3717.
86. C. L. Atkin and J. B. Neilands, *Biochemistry*, 1968, **7,** 3734.
87. M. Ya. Karpeiskii, Yu. N. Breusov, R. M. Khomutov, E. S. Severin, and O. L. Polyanovskii, *Biokhimiya*, 1963, **28,** 342.
88. B. H. Arison and J. L. Beck, *Tetrahedron*, 1973, **29,** 2743.
89. J. C. MacDonald, *Canad. J. Chem.*, 1963, **41,** 165.
90. G. W. Kirby and S. Narayanaswami, *J.C.S. Perkin I*, 1976, 1564.
91. H. A. Akers, M. Llinas, and J. B. Neilands, *Biochemistry*, 1972, **11,** 2283.
92. J. C. MacDonald, *Biochem. J.*, 1965, **96,** 533.
93. P. S. Steyn, *Tetrahedron*, 1973, **29,** 107.
94. C. M. Allen, *J. Amer. Chem. Soc.*, 1973, **95,** 2386.
95. A. Dossena, R. Marchelli, and A. Pochini, *J.C.S. Chem. Comm.*, 1974, 771.
96. P. M. Scott, M. A. Merrien, and J. Polonsky, *Experientia*, 1976, **32,** 140.
97. A. J. Birch and J. J. Wright, *Tetrahedron*, 1970, **26,** 2329.

98. A. J. Birch, G. E. Blance, S. David, and H. Smith, *J. Chem. Soc.*, 1961, 3128.
99. G. Casnati, A. Pochini, and R. Ungaro. *Gazzetta*, 1973, **103,** 141.
100. N. Eickman, J. Clardy, R. J. Cole, and J. W. Kirksey, *Tetrahedron Letters*, 1975, 1051.
101. P. A. Stadler and P. Stütz, in 'The Alkaloids', ed. R. H. F. Manske, Academic Press, New York, 1975, vol. 15, p. 1.
102. H. G. Floss, *Tetrahedron*, 1976, **32,** 873.
103. P. Stütz, R. Brunner, and P. A. Stadler, *Experientia*, 1973, **29,** 936.
104. A. Taylor, in 'Microbial Toxins', ed. A. Ciegler, S. J. Ajl, and S. K. Adis, Academic Press, New York, vol. VII, chapter 10.
105. M. R. Bell, J. R. Johnson, B. S. Wildi, and R. B. Woodward, *J. Amer. Chem. Soc.*, 1958, **80,** 1001.
106. J. Fridrichsons and A. McL. Mathieson, *Acta Cryst.*, 1965, **18,** 1043.
107. E. Francis, R. Rahman, S. Safe, and A. Taylor, *J.C.S. Perkin I*, 1972, 470.
108. R. Nagarajan, L. L. Huckstep, D. H. Lively, D. C. DeLong, M. M. Marsh, and N. Neuss, *J. Amer. Chem. Soc.*, 1968, **90,** 2980.
109. H. Minato, M. Matsumoto, and T. Katayama, *J.C.S. Perkin I*, 1973, 1819.
110. A. G. McInnes, A. Taylor, and J. A. Walter, *J. Amer. Chem. Soc.*, 1976, **98,** 6741.
111. J. D. Bu'lock and C. Leigh, *J.C.S. Chem. Comm.*, 1975, 628.
112. N. Neuss, R. Nagarajan, B. B. Molloy, and L. L. Huckstep, *Tetrahedron Letters*, 1968, 4467.
113. T. Wieland and C. Birr, in 'International Review of Science, Organic Chemistry Series Two', ed. D. H. Hey, Butterworths, London, 1976, vol. 6, p. 183.
114. F. A. Bovey, A. I. Brewster, D. J. Patel, A. E. Tonelli, and D. A. Torchia, *Accounts Chem. Res.*, 1972, **5,** 193.
115. Yu. A. Ovchinnikov and V. T. Ivanov, *Tetrahedron*, 1974, **30,** 1871, and 1975, **31,** 2177.
116. C. M. Deber, V. Madison, and E. R. Blout, *Accounts Chem. Res.*, 1976, **9,** 106.
117. W. L. Meyer, G. E. Templeton, C. I. Grable, R. Jones, L. F. Kuyper, R. B. Lewis, C. W. Sigel, and S. H. Woodhead, *J. Amer. Chem. Soc.*, 1975, **97,** 3802.
118. J. L. Flippen and I. L. Karle, *Biopolymers*, 1976, **15,** 1081.
119. M. Bodanszky and G. L. Stahl, *Proc. Nat. Acad. Sci. U.S.A.*, 1974, **71,** 2791.
120. B. W. Bycroft, *J.C.S. Chem. Comm.*, 1972, 660; and in 'The Chemistry and Biology of Peptides', ed. J. Meienhofer, Ann Arbor Science, Ann Arbor, 1972, p. 665.
121. H. Yoshioka, T. Aoki, H. Goko, K. Nakatsu, T. Noda, H. Sakakibara, T. Take, A. Nagata, J. Abe, T. Wakamiya, T. Shiba, and T. Kaneko, *Tetrahedron Letters*, 1971, 2043.
122. T. Shiba, S. Nomoto, T. Teshima, and T. Wakamiya, *Tetrahedron Letters*, 1976, 3907.
123. J. H. Carter, R. H. Du Bus, J. R. Dyer, J. C. Floyd, K. C. Rice, and P. D. Shaw, *Biochemistry*, 1974, **13,** 1221 and 1227.
124. W. Keller-Schierlein, V. Prelog, and H. Zahner, *Fortschr. Chem. org. Naturstoffe*, 1964, **22,** 279.
125. J. B. Neilands, in 'Inorganic Biochemistry', ed. G. L. Eichhorn, Elsevier, New York, vol. I, 1973, p. 167.
126. R. Norrestam, B. Stensland, and C. I. Branden, *J. Mol. Biol.*, 1975, **99,** 501.
127. M. Llinas, D. M. Wilson, M. P. Klein, and J. B. Neilands, *J. Mol. Biol.*, 1976, **104,** 853.
128. M. Llinas, W. J. Horsley, and M. P. Klein, *J. Amer. Chem. Soc.*, 1976, **98,** 7554.
129. Y. Iitaka, H. Nakamura, K. Takada, and T. Takita, *Acta Cryst.*, 1974, **B30,** 2817.
130. Th. Wieland, *Fortschr. Chem. org. Naturstoffe*, 1967, **25,** 214.
131. Th. Wieland and O. P. Wieland, in 'Microbial Toxins', ed. A. Ciegler, S. J. Ajl, and S. K. Adis, Academic Press, New York, 1972, vol. VIII, p. 248.
132. See G. C. Strain, K. P. Mullinix, and L. Bogorad, *Proc. Nat. Acad. Sci. U.S.A.*, 1971, **68,** 2467; P. D. Thut and T. J. Lindell, *Mol. Pharmacol.*, 1974, **10,** 146; L. Fiume and G. Barbanti, *Experientia*, 1974, **30,** 76.
133. D. J. Patel, A. E. Tonelli, P. Pfaender, H. Faulstich, and Th. Wieland, *J. Mol. Biol.*, 1973, **79,** 185.
134. Th. Wieland, J. Faesel, and W. Konz, *Annalen*, 1969, **722,** 197.
135. Yu. A. Ovchinnikov, V. T. Ivanov, V. F. Bystrov, and A. I. Miroshnikov, in 'The Chemistry and Biology of Peptides', ed. J. Meienhofer, Ann Arbor Science, Ann Arbor, 1972, p. 111.
136. H. Faulstich and Th. Wieland, in 'The Peptides', ed. H. Hanson and H. D. Jakubke, North-Holland/Elsevier, Amsterdam, 1973, p. 312.
137. Th. Wieland, H. Faulstich, and W. Burgermeister, *Biochem. Biophys. Res. Comm.*, 1972, **47,** 984.
138. I. L. Karle, *J. Amer. Chem. Soc.*, 1974, **96,** 4000.
139. L. K. Ramachandran, *Biochem. Rev.*, 1975, **46,** 1.
140. D. C. Hodgkin and B. M. Oughton, *Biochem. J.*, 1957, **65,** 752.
141. A. Taylor, *Adv. Appl. Microbiol.*, 1970, **12,** 189.
142. Yu. A. Ovchinnikov and V. T. Ivanov in 'International Review of Science, Organic Chemistry Series Two', ed. D. H. Hey, Butterworths, London, 1976, vol. 6, p. 219.
143. B. C. Pressman, *Ann. Rev. Biochem.*, 1976, **45,** 501.
144. I. L. Karle, *J. Amer. Chem. Soc.*, 1975, **97,** 4379.
145. G. D. Smith, W. L. Duax, D. A. Langs, G. T. DeTitta, J. W. Edmonds, D. C. Rohrer, and C. M. Weeks, *J. Amer. Chem. Soc.*, 1975, **97,** 7242.
146. K. Neupert-Laves and M. Dobler, *Helv. Chim. Acta*, 1975, **58,** 432.
147. H. Brockmann, *Fortschr. Chem. org. Naturstoffe*, 1960, **18,** 1.
148. U. Hollstein, *Chem. Rev.*, 1974, **74,** 625.
149. J. V. Formica and M. A. Apple, *Antimicrob. Agents Chemother.*, 1976, **9,** 214.

150. H. Lackner, *Angew. Chem. Internat. Edn.*, 1975, **14,** 375.
151. H. M. Sobell, S. C. Jain, T. D. Sakore, and C. E. Nordman, *Nature New Biol.*, 1971, **231,** 200; *J. Mol. Biol.*, 1972, **68,** 1.
152. S. Tamura, S. Kuyama, Y. Kodaira, and S. Higashikawa, *Agric. Biol. Chem.* (*Japan*), 1964, **28,** 137.
153. A. Kakinuma, M. Hori, H. Sugino, I. Yoshida, M. Isono, G. Tamura, and K. Arima, *Agric. Biol. Chem.* (*Japan*), 1969, **33,** 1523.
154. M. Bodanszky, I. Izdebski, and I. Muramatsu, *J. Amer. Chem. Soc.*, 1969, **91,** 2351.
155. A. Dell, D. H. Williams, H. R. Morris, G. A. Smith, J. Feeney, and G. C. K. Roberts, *J. Amer. Chem. Soc.*, 1975, **97,** 2497.
156. R. Tschesche and E. U. Kaussmann, in 'The Alkaloids', ed. R. H. F. Manske, Academic Press, New York, 1975, vol. 15, p. 165.
157. E. Gross, H. H. Kiltz, and E. Nebelin, *Z. physiol. Chem.*, 1973, **354,** 810.
158. E. Gross and J. L. Morell, *J. Amer. Chem. Soc.*, 1971, **93,** 4634.
159. L. Ingram, *Biochim. Biophys. Acta*, 1970, **224,** 263.
160. D. R. Martin and R. J. P. Williams, *Biochem. J.*, 1976, **153,** 181.
161. B. Anderson, D. C. Hodgkin, and M. A. Viswamitra, *Nature*, 1970, **225,** 233.
162. K. Tori, K. Tokura, K. Okabe, M. Ebata, and H. Otsuka, *Tetrahedron Letters*, 1976, 185.
163. H. Depaire, J.-P. Thomas, A. Brun, A. Olesker, and G. Lukacs, *Tetrahedron Letters*, 1977, 1403.
164. J. M. Liesch and K. L. Rinehart Jr., *J. Amer. Chem. Soc.*, 1977, **99,** 1645.
165. H. Umezawa, *Pure Appl. Chem.*, 1971, **28,** 665.

23.5
β-Lactam Antibiotics

G. LOWE
University of Oxford

23.5.1 INTRODUCTION

The discovery and investigation of the β-lactam antibiotics is one of the most fascinating stories of modern science. The initial observation by Alexander Fleming that a contaminant mould of the *Penicillium* family produces a substance inhibitory to *Staphylococci*[1] was perhaps the single most important observation in chemotherapy, leading as it did to the successful search for antibacterial substances from microorganisms, many of which are now widely used in medicine. Penicillin, the inhibitory substance produced by *Penicillium notatum* and some other *Penicillium* species, proved difficult to isolate and it was not until Florey took up the investigation in Oxford that the active material was isolated from cultures of the mould. Fleming's original observations were confirmed and the therapeutic value of penicillin in combatting bacterial infection in man was demonstrated.[2]

The recognition of the potential value of penicillin in medicine led to the inauguration of a monumental Anglo-American research programme during World War II aimed at producing the antibiotic in sufficient quantity for widespread use.[3] Rapid developments in fermentation technology, particularly the use of submerged culture techniques and the development of high-yielding mutants of *Penicillium chrysogenum*, led to dramatic improvements in yield. The further discovery that corn-steep liquor added to the culture medium removed a rate-limiting step in fermentation production led to the recognition that side-chain precursors such as phenylacetic acid accelerated penicillin production.

Today, almost all penicillin is produced as penicillin G (**1**; R = $PhCH_2$) in yields approaching 20 grams per litre of culture fluid and at a cost of only a few pence per gram.

Following the demonstration of the chemotherapeutic properties of penicillin, a worldwide search for antibacterial substances from micro-organisms was undertaken. In Sardinia, Brotzu, supposing that the self-purification of water might be due in part to bacterial antagonism, examined the microbial flora of seawater near a sewage outlet at Cagliari. From this he isolated a fungus similar to *Cephalosporium acremonium* which, when grown on agar, secreted material that inhibited the growth of a variety of gram-positive and gram-negative bacteria.[4] A culture of the organism was sent to Florey at Oxford. Abraham and Newton isolated several antibacterial substances from the organism, the one principally responsible for the observations of Brotzu being penicillin N [**1**; R = D-$\bar{O}_2CCH(\overset{+}{N}H_3)(CH_2)_3$]. They noted, however, another substance with modest antibacterial activity against both gram-positive and gram-negative bacteria which was resistant to the penicillinases. This substance, named cephalosporin C [**2**; R = D-$\bar{O}_2CCH(\overset{+}{N}H_3)(CH_2)_3$] has become the precursor of a second important family of β-lactam antibiotics, the cephalosporins (**2**).[5]

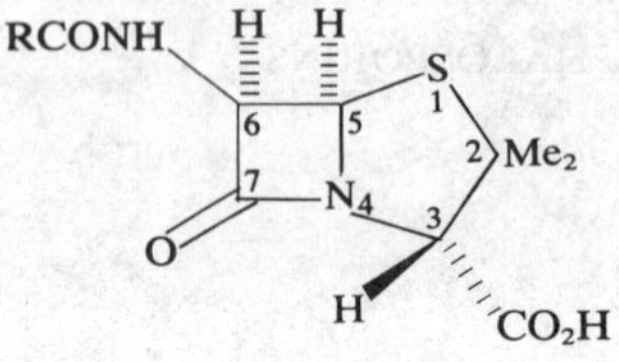

(1)

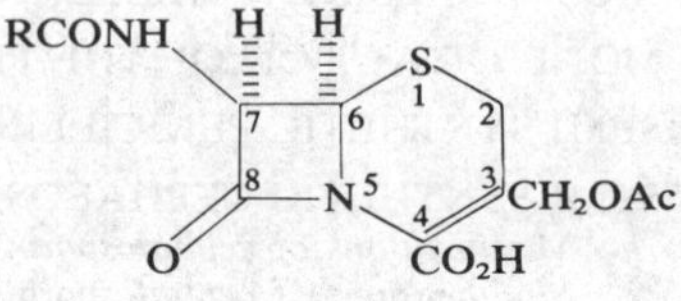

(2)

23.5.2 BIOSYNTHESIS OF THE PENICILLINS AND CEPHALOSPORIN C

In the absence of specific side-chain precursors, *Penicillium chrysogenum* produces a large number of penicillins with side-chains derived from natural carboxylic acids. One such penicillin has the amino-acid side chain of L-α-aminoadipic acid and is known as isopenicillin N [**1**; R = L-$\bar{O}_2CCH(\overset{+}{N}H_3)(CH_2)_3$]. The penicillin nucleus devoid of any side-chain, namely 6-aminopenicillanic acid, is also produced under these conditions.[6]

One of the initial reasons for interest in the Brotzu culture of the *Cephalosporium* species was that the principal antibiotic, namely penicillin N [**1**; R = D-$\bar{O}_2CCH(\overset{+}{N}H_3)$-$(CH_2)_3$], showed greater activity than penicillin G (**1**; R = $PhCH_2$) against gram-negative bacteria. Cephalosporin C, which was first discovered as a contaminant of penicillin N from *Cephalosporium acremonium*,[7] also has the D-α-aminoadipyl side-chain but with the β-lactam-dihydrothiazine nucleus of the cephalosporins.[8,9] 7-Aminocephalosporanic acid, *i.e.* cephalosporin C with the D-α-aminoadipyl side-chain removed, has not been found in fermentation media, however; nor have cephalosporins been isolated with side-chains derived from acids other than D-α-aminoadipic acid.

The structures of penicillin N and cephalosporin C can be seen to be formally derived from α-aminoadipic acid, cysteine, and valine. The isolation of the tripeptide δ-(α-aminoadipyl)-cysteinyl-valine from *Penicillium chrysogenum*[10] and a similar tripeptide from *Cephalosporium acremonium* which was shown to be δ-(L-α-aminoadipyl)-L-cysteinyl-D-valine[11] suggested that such tripeptides might well be biosynthetic intermediates of the β-lactam antibiotics. This hypothesis has been difficult to prove because micro-organisms do not take up the tripeptide. However, it has been shown recently that a cell-free extract from *Cephalosporium* protoplasts will take up and incorporate δ-(L-α-aminoadipyl)-L-cysteinyl-D-valine into penicillin N, but not δ-(L-α-aminoadipyl)-L-valine nor δ-(D-α-aminoadipyl)-L-cysteinyl-D-valine.[12,13]

Many suggestions have been made as to how the tripeptide might be converted into penicillin N or cephalosporin C, but no intermediates in the sequence of steps leading to the β-lactam-thiazolidine or β-lactam-dihydrothiazine nuclei have yet been obtained, probably because they are enzyme bound.[6] What has been shown is that [α-^{3}H]-L-cysteine is incorporated into penicillin G with the ^{3}H label as expected at C-6. α,β-Dehydrocysteine is not therefore formed at any point along the biosynthetic pathway.[14] Furthermore, using L-cysteine chirally labelled with ^{3}H at C-3, it has been shown that this carbon atom becomes C-5 of penicillin with overall retention of configuration.[15] Both D- and L-valine are incorporated into penicillins and experiments with [^{3}H]-labelled L-valine have shown that during its incorporation into the tripeptide δ-(L-α-aminoadipyl)-L-cysteinyl-D-valine, α,β-dehydrovaline is not formed since only the α-proton is lost.[16] It has also been shown that incorporation of δ-(L-α-aminoadipyl)-L-cysteinyl-D-valine into penicillin N by cell-free extracts of protoplasts of *Cephalosporium acremonium* occurs without loss of the α-D-valine proton.[13] This implies that the formation of the thiazolidine ring does not involve an α,β-dehydrovaline intermediate. When (2*S*,3*R*)-[4,4,4-$^2H_3$]valine and (2*S*,3*S*)-[4,4,4-^{2}H$_3$]valine are incorporated into penicillin N, the deuterium atoms are found in the β-methyl and α-methyl group of penicillin N, respectively. The formation of the thiazolidine ring has occurred therefore with retention of configuration at C-3 of valine, possibly by a radical or carbonium ion type of intermediate. Cephalosporin C which has incorporated the deuterium-labelled valines is labelled at C-2 and in the exocyclic methylene group at C-3, respectively (Scheme 1).[17] It has also been shown that penicillin N is converted into cephalosporin C by protoplasts of *Cephalosporium acremonium*. Since cyanide ions inhibit the conversion, an oxygenase is probably involved.[18]

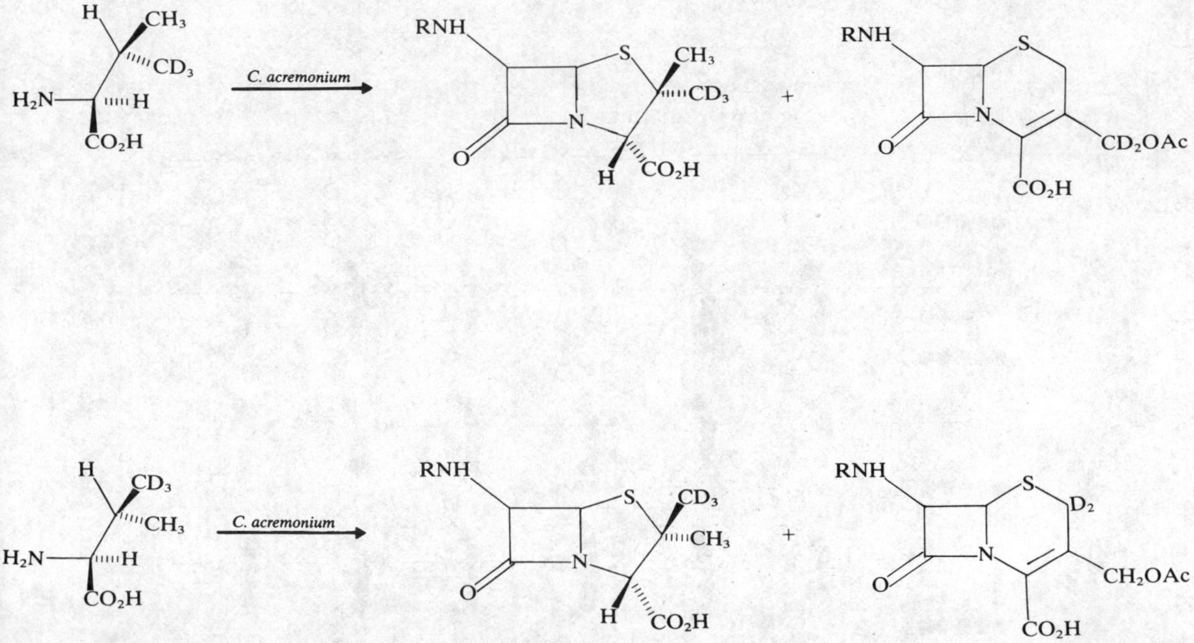

SCHEME 1

23.5.3 MODE OF ACTION OF THE PENICILLINS AND CEPHALOSPORINS

The antibacterial properties of the penicillins and cephalosporins stem from their ability to inhibit the enzyme responsible for the final step in bacterial cell wall biosynthesis. The bacterial cell wall is a macromolecular network that completely surrounds the cell and provides its structural integrity. In the presence of penicillins and cephalosporins the fine control of lytic and synthetic enzymes necessary for the proper development of the cell wall of a growing bacterium is perturbed and the cell wall so generated is defective and cannot support the fragile cytoplasmic membrane against the internal osmotic pressure. The cell fluid bursts through the cytoplasmic membrane and the organism is killed.[19,20]

The structure of the cell wall of *Staphylococcus aureus* is shown schematically in Figure 1. X and Y represent *N*-acetylglucosamine and *N*-acetylmuramic acid residues, respectively. The open circles represent amino-acids, of which there are five in the nascent peptidoglycan units, but on being cross-linked by the pentaglycine chains, represented by the filled-in circles, these lose their terminal amino-acid residue. This cross-linking step is shown in greater detail in Figure 2.

The cross-linking of the peptidoglycan chains is accompanied by the loss of the terminal D-alanine residue. The enzyme responsible for catalysing this step, a transpeptidase, is inhibited by penicillin. The terminal residues of the pentapeptide chain are seen to contain two D-alanine residues and Tipper and Strominger suggested that the conformation of these residues when bound to the enzyme was probably similar to the rigid conformation of the acyl-dipeptide moiety of the penicillins, as shown in Figure 3.[21] If this were so it would place the β-lactam at the site at which cleavage of the peptide bond of the natural substrate occurs, and in this way penicillin would irreversibly inhibit the enzyme by acylating an essential functional group. Evidence that penicillins are tightly bound to the transpeptidase but can be released by hydroxylamine or mercaptoethanol support the formation of a penicilloyl-transpeptidase. Recently, D-alanylcarboxypeptidases from a *Streptomyces* and *Bacillus stearothermophilus*, which are possibly solubilized forms of their transpeptidases, have been isolated and shown to be rapidly inhibited by penicillins, giving a penicilloyl-enzyme. The enzyme does, however, very slowly recover activity with the

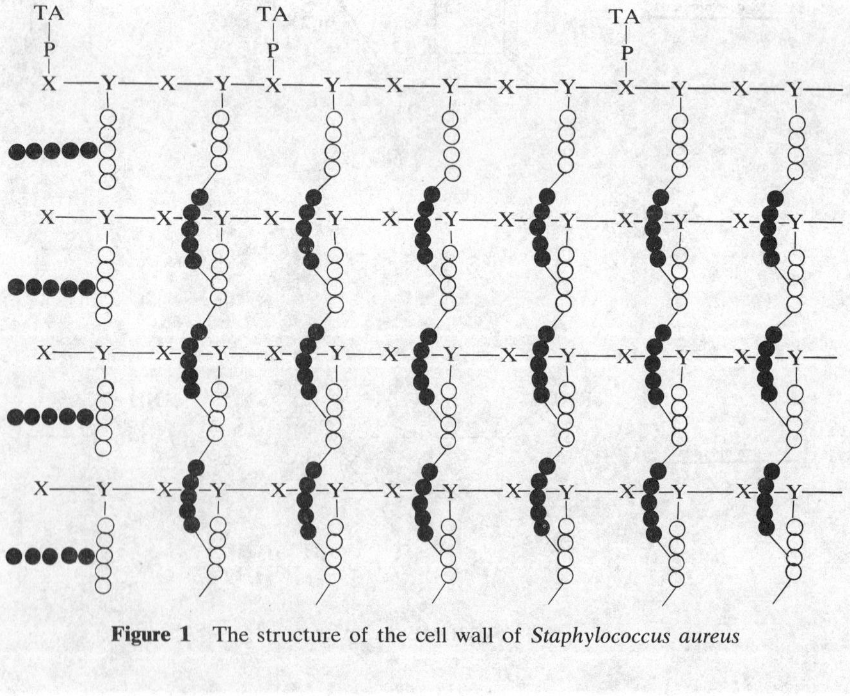

Figure 1 The structure of the cell wall of *Staphylococcus aureus*

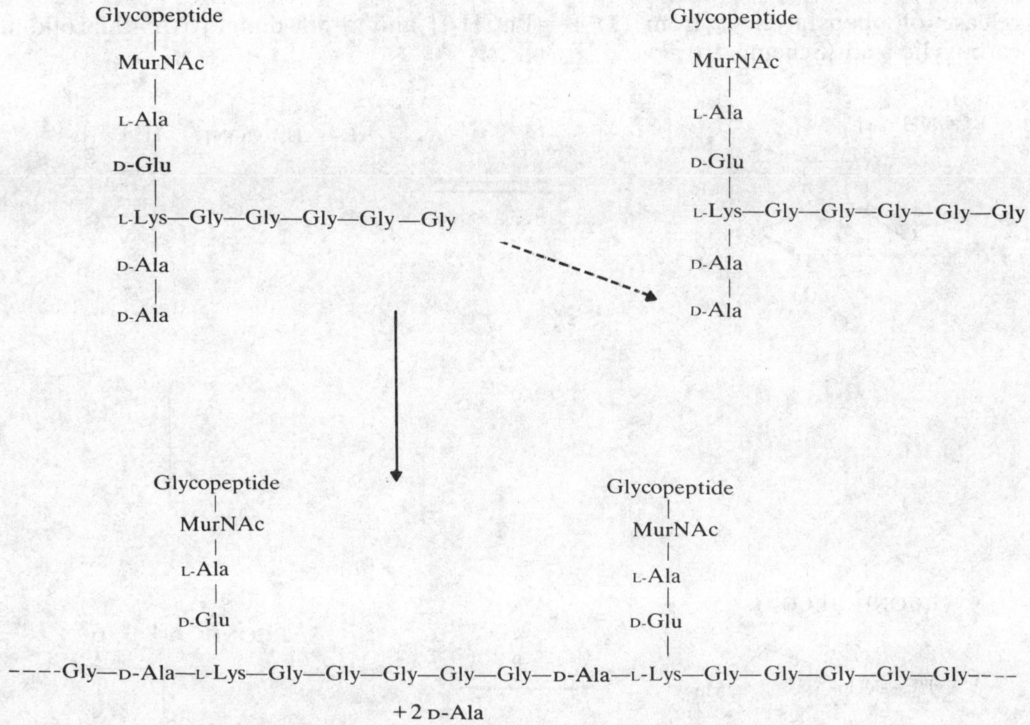

Figure 2 Transpeptidation of two peptidoglycan chains with formation of the interpeptide bridges and elimination of D-alanine

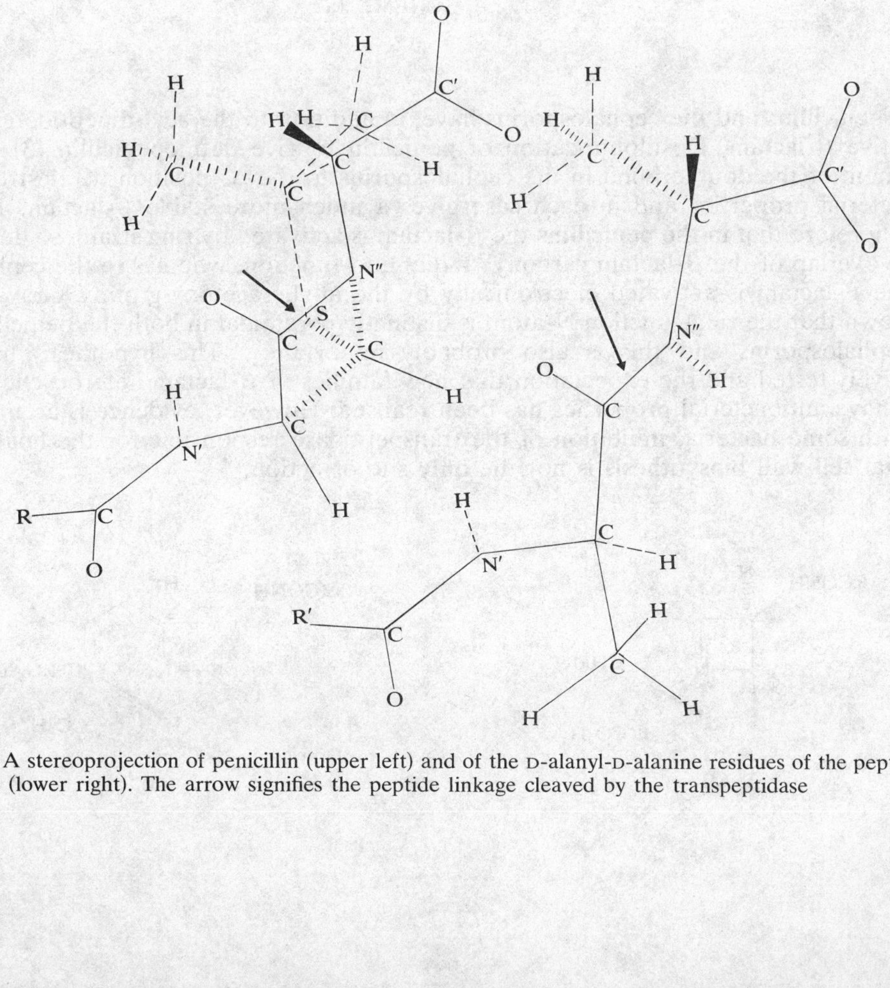

Figure 3 A stereoprojection of penicillin (upper left) and of the D-alanyl-D-alanine residues of the peptidoglycan chains (lower right). The arrow signifies the peptide linkage cleaved by the transpeptidase

release of phenylglycine [from (**1**; R = $PhCH_2$)] and D-5,5-dimethyl-Δ^2-thiazolidine-4-carboxylic acid (Scheme 1a).[21a]

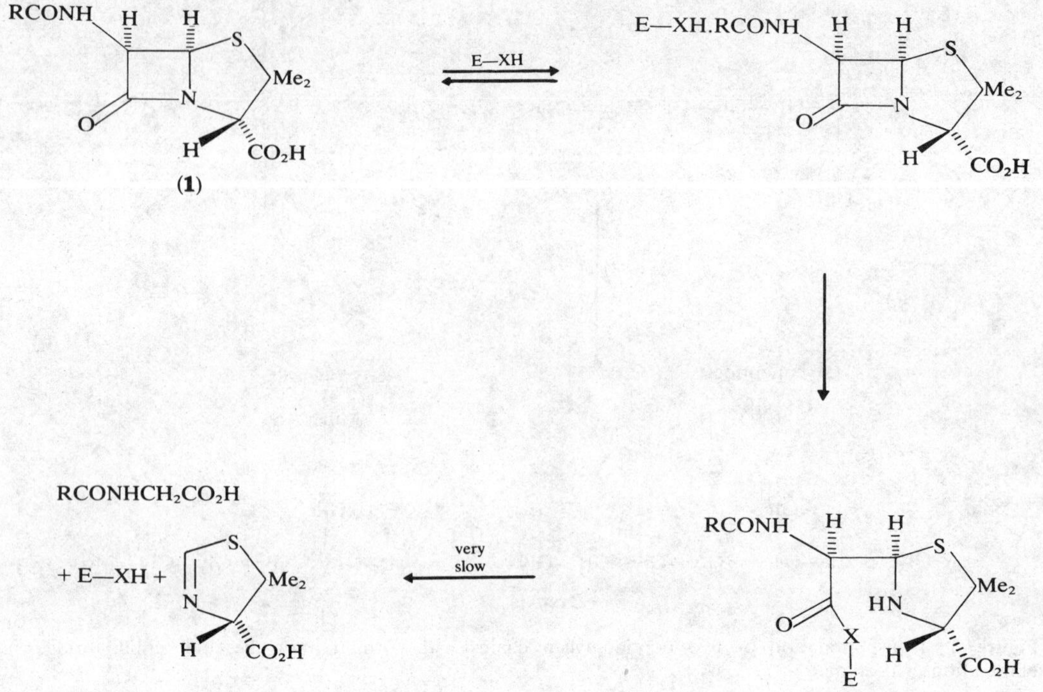

SCHEME 1a

The penicillins and the cephalosporins have, in addition to the acyl-dipeptide grouping, a reactive β-lactam. Desulphurization of penicillin to give dethiopenicillin (**3**) or rearrangement of the double bond in the cephalosporins to the Δ^2-position (**4**) destroys their antibacterial properties and in each case gives a much more stable β-lactam. It would seem therefore that in the penicillins the β-lactam is activated by ring strain, so decreasing orbital overlap of the β-lactam carbonyl and amino-function, whereas in the cephalosporins the β-lactam is activated electronically by the allylic acetoxy-group. X-ray analysis has shown that the ring-junction N-atom is distinctly pyramidal in both the penicillins and the cephalosporins and this is also probably important.[22] This hypothesis has been extensively tested and the expectation that new families of β-lactam heterocyclic systems might have antibacterial properties has been realised. However, evidence is accumulating that with some bacteria, inhibition of the transpeptidase responsible for the final step of bacterial cell wall biosynthesis is not the only site of action.[20]

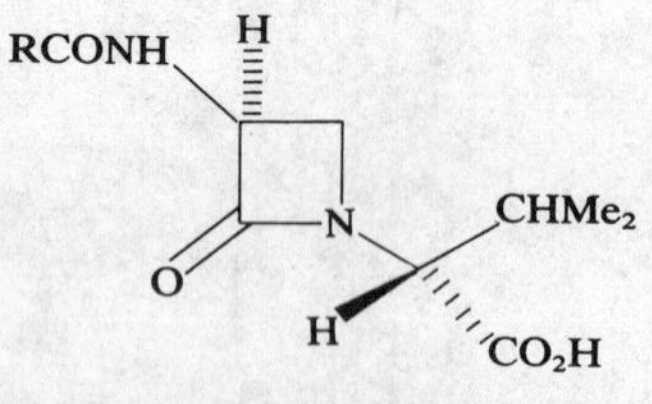

(3)

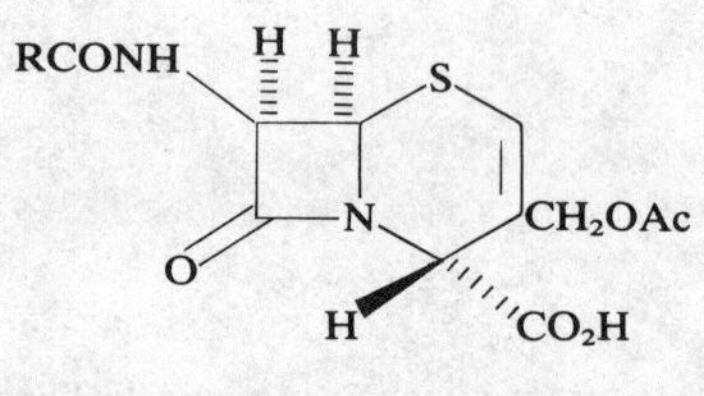

(4)

23.5.4 SEMI-SYNTHETIC PENICILLINS

In the late 1950s, penicillins were in widespread clinical use and a major problem was developing. Bacteria such as *Staphylococcus aureus* were becoming resistant to penicillins, particularly in hospitals. Strains of *staphylococci* survived which had the ability to produce an enzyme, penicillinase, which hydrolysed penicillins to penicilloic acids (**5**). It was indeed fortunate, therefore, that as this problem was becoming critical, 6-amino-penicillanic acid (**6**) was found in cultures of *Penicillium chrysogenum* when grown on media containing no side-chain precursors.

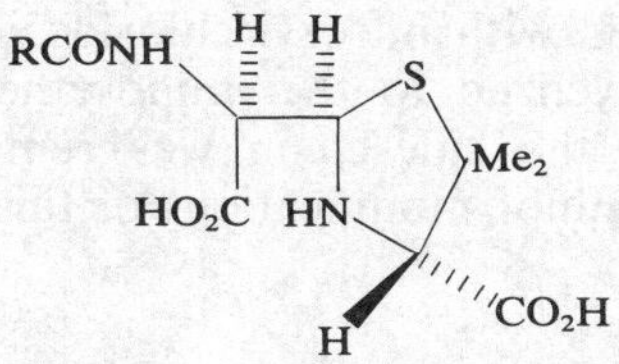

(5)

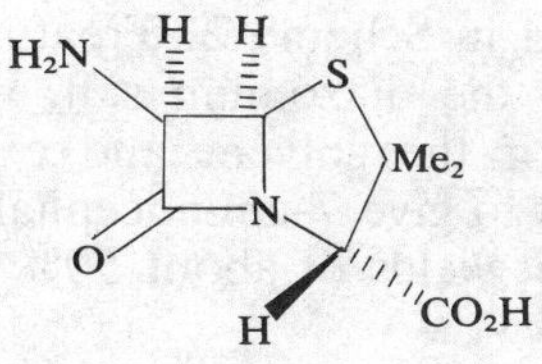

(6)

The isolation of 6-aminopenicillanic acid from fermentation media and its potential as an intermediate for the synthesis of novel penicillins led to a search for enzymes which would specifically remove the side-chain of penicillin G. Such enzymes were found to be quite widespread, those most specific for removing the phenylacetyl side-chain of penicillin G being in the genera *Escherichia* and *Alcaligenes.*[23] 6-Aminopenicillanic acid is now produced commercially by pumping a solution of penicillin G through a bed of sepharose to which the amidase from *Escherichia coli* has been covalently attached, until hydrolysis at a controlled pH is complete.

Several thousand new penicillins have been obtained by the acylation of 6-amino-penicillanic acid. One of the first clinically useful semi-synthetic penicillins made in this way was methicillin [**1**; R = 2,6-$(MeO)_2C_6H_3$], its virtue being that it was not hydrolysed by penicillinase and was therefore effective against the penicillin-resistant staphylococci. The two methoxy-groups are primarily responsible for protecting this penicillin from hydrolysis by the penicillinase.

The principal antibacterial activity of the fermentation derived penicillins is against gram-positive bacteria, except for penicillin N [**1**; R = D-$\bar{O}_2CCH(\overset{+}{N}H_3)(CH_2)_3$] which showed promise against gram-negative bacteria. This unusual antibacterial spectrum for a penicillin appeared to be associated with the presence of the free amino-function in the side-chain since when this group was acylated the gram-negative activity was greatly diminished and the gram-positive activity increased. Several amino-substituted penicillins have been synthesized from 6-aminopenicillanic acid and are amongst the most valuable penicillins available. Ampicillin [**1**; R = D-$PhCH(NH_2)$] has a broad antibacterial spectrum and can be orally administered.[24] Esters of ampicillin have the further advantage of being more readily absorbed by the intestinal wall and are hydrolysed on entering the blood stream. Amoxicillin [**1**; R = D-4-$HOC_6H_4CH(NH_2)$] is a more recently derived penicillin and is also effective against gram-negative bacteria and can be orally administered.

Carbenicillin [**1**; R = D-$PhCH(CO_2H)$], which has a carboxyl group in the side-chain, is also effective against gram-negative bacteria but has a different antibacterial spectrum, being particularly effective against *Proteus* and *Pseudomonas* species. The effectiveness of ampicillin and carbenicillin against gram-negative bacteria is probably associated with the more polar side-chain facilitating cell penetration.[25]

23.5.5 SEMI-SYNTHETIC CEPHALOSPORINS

23.5.5.1 Modifications of cephalosporins at C-7

The greater antibacterial activity of penicillin G (**1**; R = $PhCH_2$) compared with penicillin N [**1**; R = D-$\bar{O}_2CCH(\overset{+}{N}H_3)(CH_2)_3$] led to an attempt to convert cephalosporin C into 7-phenylacetamidocephalosporanic acid (**2**; R = $PhCH_2$). This was first achieved by mild acid hydrolysis to 7-aminocephalosporanic acid (albeit in very poor yield) followed by phenylacetylation.[26]

The search for an enzyme which would remove the D-2-aminoadipyl side-chain of cephalosporin C has been extensive, but no such acylase has been found. The side-chain, however, can be removed in high yield by chemical methods.[27] The first successful method is outlined in Scheme 2. Treatment of cephalosporin C with nitrosyl chloride in formic acid gave the diazonium salt, which spontaneously cyclized to the imino-ether. After removal of the solvent and excess volatile reagent, the side-chain was removed by hydrolysis to give 7-aminocephalosporanic acid. With minor modifications of this general procedure, yields of about 50% have been achieved.

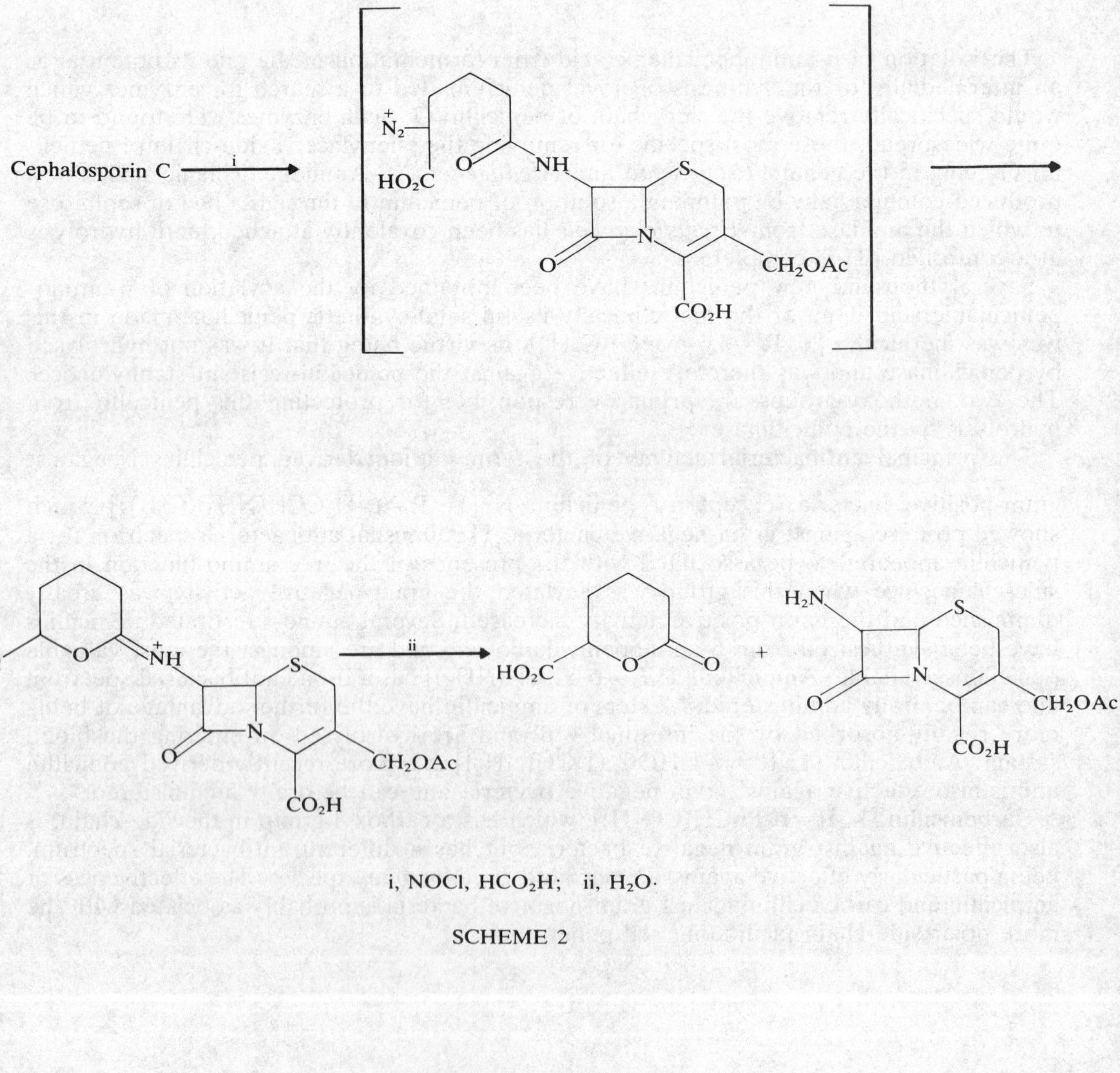

i, NOCl, HCO_2H; ii, H_2O.

SCHEME 2

The use of phosphorus pentachloride for the conversion of the 7-amido-group of cephalosporin C (with amino and carboxyl groups protected) into the imino-chloride, followed by treatment with an alcohol to form the imino-ether and then hydrolytic cleavage, proved to be the simplest and most efficient method. By using dimethyldichlorosilane to protect the amino and carboxyl functions, as outlined in Scheme 3,7-aminocephalosporanic acid was obtained in 92.5% yield. The advantage of the use of the silyl protecting group is its concomitant removal on treatment of the imino-ether with aqueous methanol.

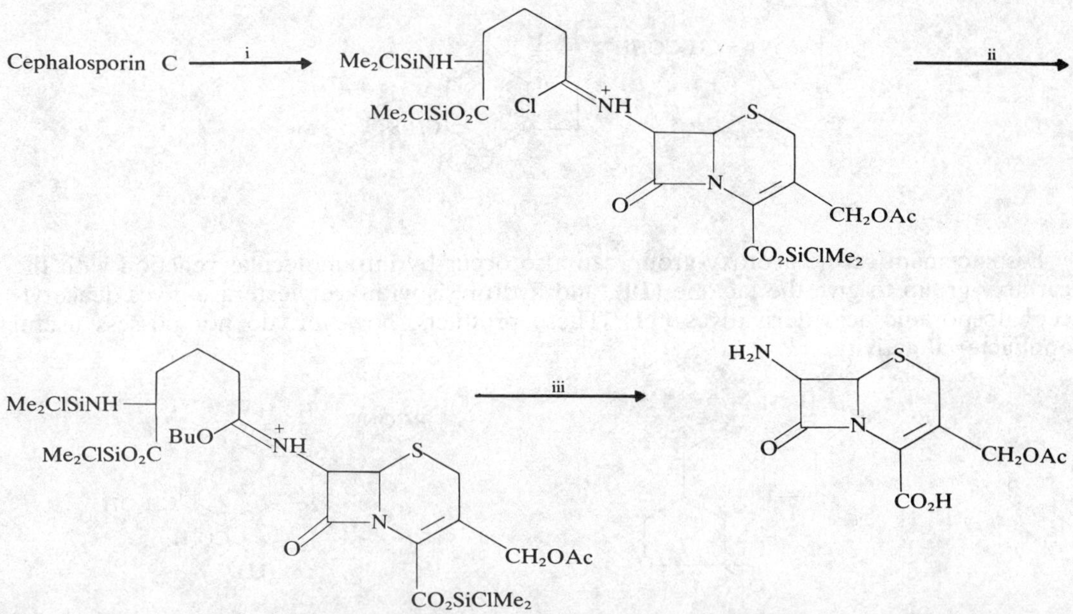

i, Me_2SiCl_2, $PhNMe_2$; PCl_5, −55 °C; ii, Bu^nOH; iii, MeOH, H_2O.

SCHEME 3

The availability of 7-aminocephalosporanic acid led to a spate of semi-synthetic cephalosporins, advantage being taken of the knowledge of the structure–activity relationship gained from semi-synthetic penicillins. Amongst the most clinically useful cephalosporins are cephalothin (**2**; R = 2-thienyl-CH_2) and cephaloglycin [**2**; R = D-PhCH(NH_2)]. Cephalothin exhibits bactericidal activity against both gram-positive and gram-negative bacteria and is effective against penicillinase-producing *Staphylococcus aureus.* Cephaloglycin can be administered orally and is particularly useful against urinary tract infections. Like ampicillin [**1**; R = D-PhCH(NH_2)], it shows high activity against gram-negative bacteria, but unlike ampicillin it is stable to penicillinase.

23.5.5.2 Modifications of cephalosporins at C-3

During the purification of cephalosporin C by electrophoresis in pyridine acetate buffer, a new antibacterial substance was obtained which has no net charge at pH 7 and was shown to be the pyridinium betaine (**7**). The remarkable ease with which nucleophilic displacement occurs is due to the stability of the carbonium ion generated on solvolysis of the acetoxy-group. Nucleophilic substitution at this centre has enabled hundreds of modified cephalosporins to be obtained. Two cephalosporins modified at both C-3 and C-7 which have valuable properties are cephaloridine (**8**) and cefazolin (**9**).[28]

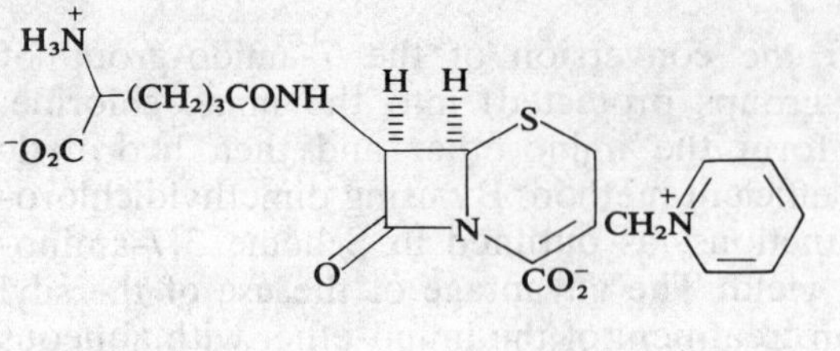

(7)

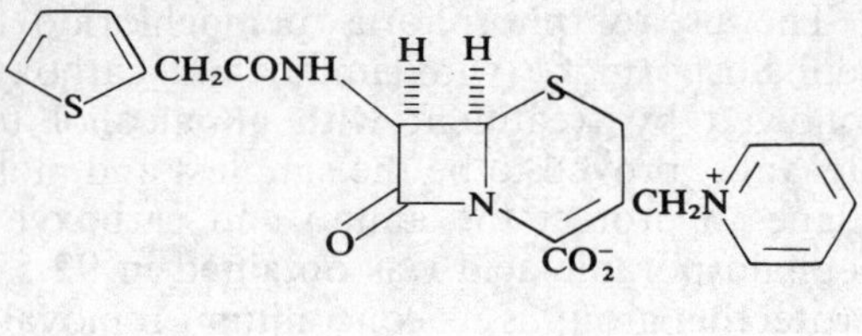

(8)

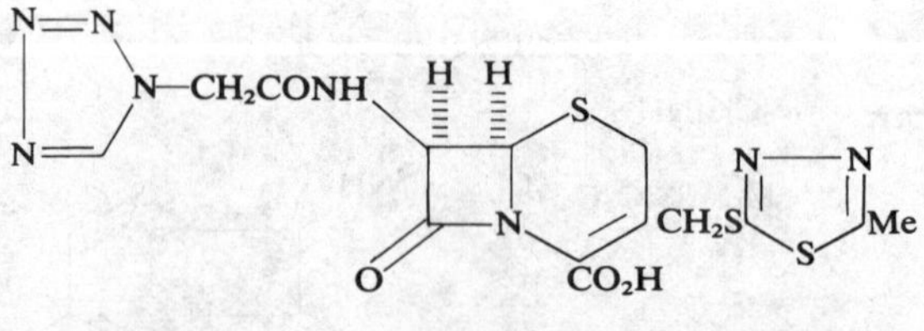

(9)

Displacement of the acetoxy-group can also occur by intramolecular reaction with the carboxy-group to give the lactone (**10**), and hydrolysis with acetylesterase gives deacetylcephalosporanic acid derivatives (**11**). These products, however, do not possess useful antibacterial activity.

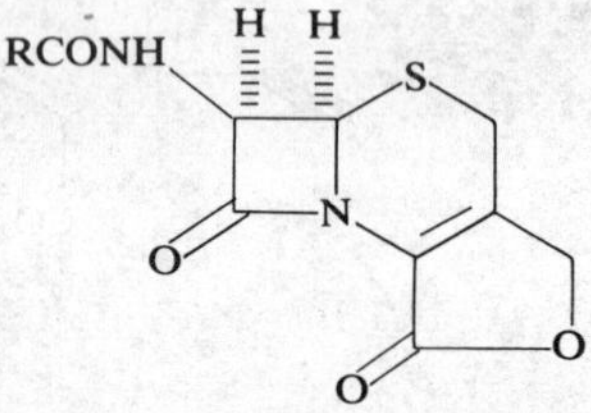

(10)

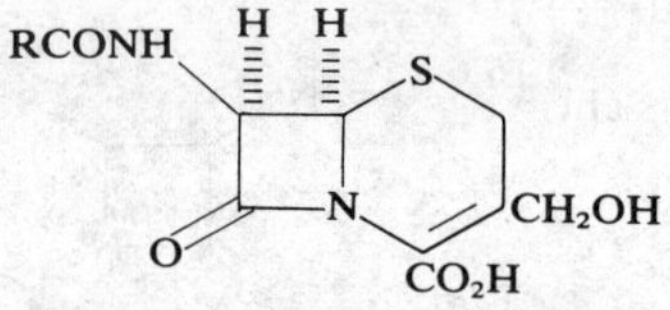

(11)

Deacetoxycephalosporanic acids (**12**) are particularly well absorbed through the intestinal wall and can be administered orally. Although they are intrinsically less active *in vitro* than the cephalosporins themselves, the high blood levels obtainable make them useful oral antibiotics.

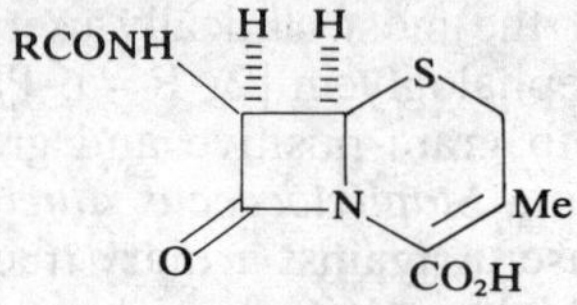

(12)

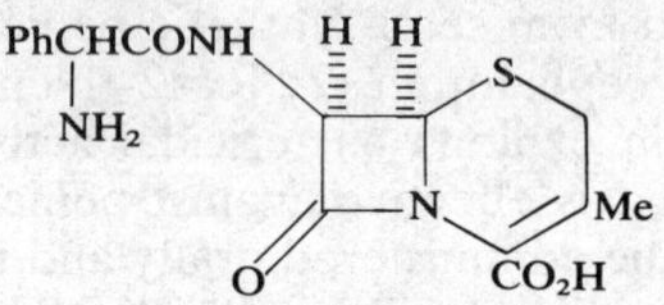

(13)

Deacetoxycephalosporanic acids were first obtained in rather poor yield by catalytic hydrogenolysis of cephalosporanic acids using a large amount of palladium catalyst. However, the transformation can be accomplished much more efficiently by chromium(II) salts,[29] or electrochemically,[30] to give 3-methylenecephams which can be isomerized quantitatively to deacetoxycephalosporanic acids as their silyl esters (Scheme 4). Cephalexin (**13**) can be prepared in this way and is one of the most valuable oral broad-spectrum antibiotics.

The 3-methylenecephems have also proved to be useful intermediates for the formation of 3-oxocephamoic acids, from which a further range of valuable therapeutic agents can be derived (Scheme 5).[31]

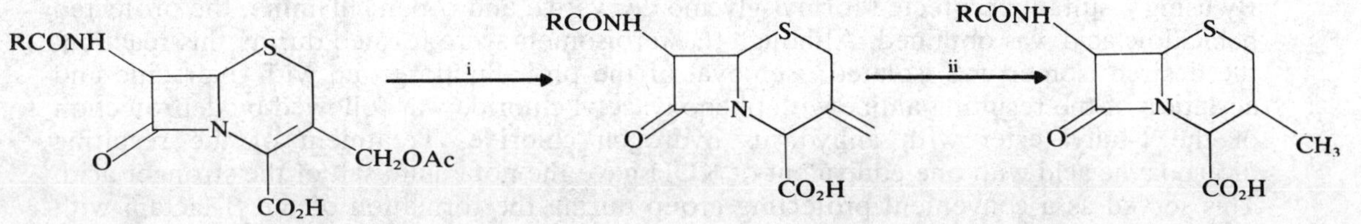

i, $Cr(OAc)_2$ or ε; ii, Me_3SiCl, py; H_2O.

SCHEME 4

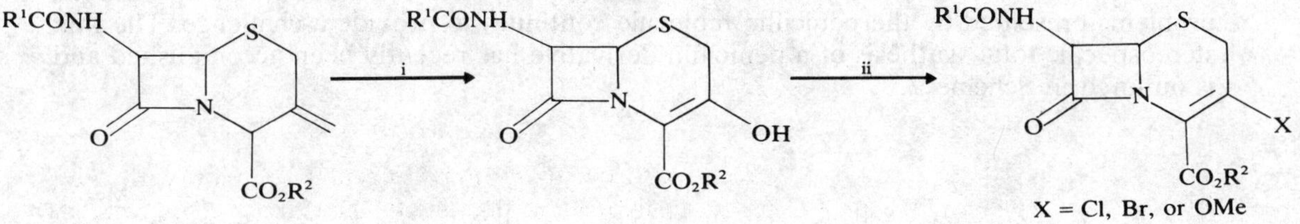

X = Cl, Br, or OMe

i, O_3, −80 °C; ii, PCl_3 or PBr_3 or CH_2N_2.

SCHEME

23.5.6 SYNTHESIS OF PENICILLINS

Since the structure of penicillin was elucidated, many attempts have been made to synthesize the molecule. It was not until the late 1950s, however, that the first synthesis of a penicillin in respectable yield was accomplished by Sheehan and Henery-Logan.[32] Their route is outlined in Scheme 6.

The condensation of aldehydes with 1,2-aminothiols gives thiazolidines in good yield.

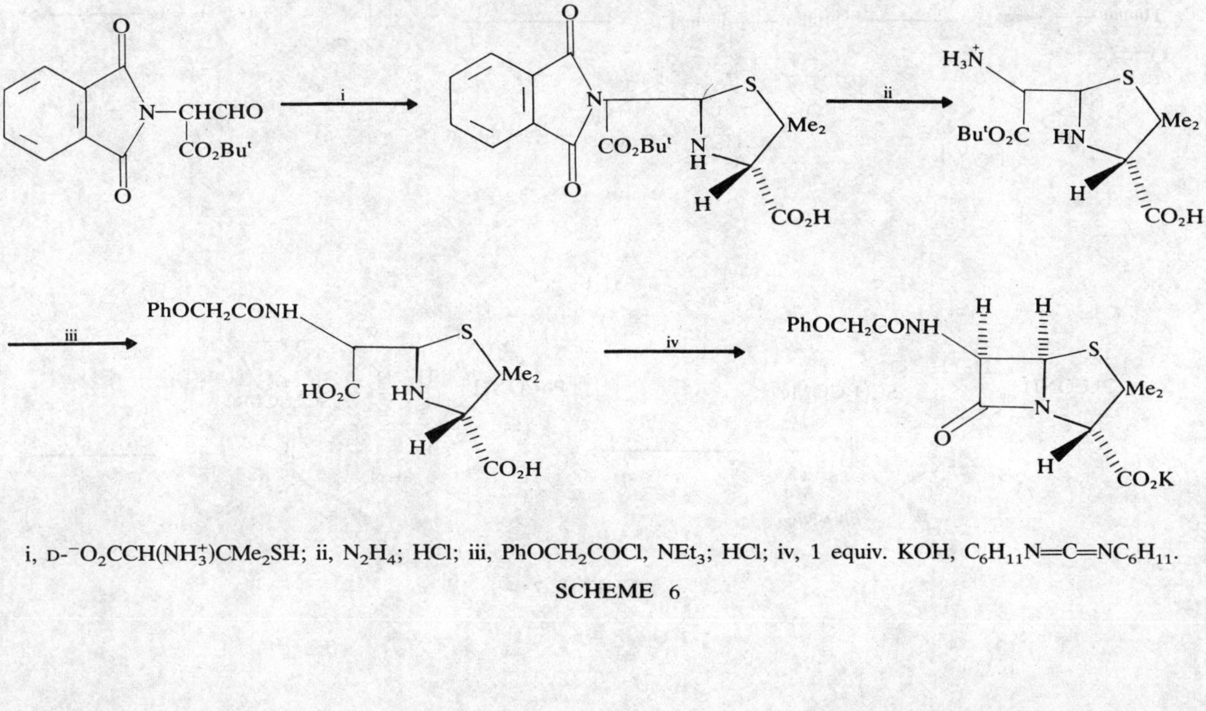

i, D-$^{-}O_2CCH(NH_3^+)CMe_2SH$; ii, N_2H_4; HCl; iii, $PhOCH_2COCl$, NEt_3; HCl; iv, 1 equiv. KOH, $C_6H_{11}N{=}C{=}NC_6H_{11}$.

SCHEME 6

By using a suitably protected formyl-glycine derivative and D-penicillamine, the protected penicilloic acid was obtained. Although diasteroisomers were formed during this reaction, the desired isomer was isolated. Removal of the phthalimido-group with hydrazine and acylation of the resulting amine with phenoxyacetyl chloride was followed by deprotection of the t-butyl ester with anhydrous hydrogen chloride. Treatment of the resulting dicarboxylic acid with one equivalent of KOH gave the potassium salt of the stronger acid. This served as a convenient protecting group during the formation of the β-lactam with dicyclohexylcarbodi-imide, a reagent which had just been introduced into peptide synthesis.

The total synthesis of penicillin was a great achievement and beautifully illustrated the skilful use of protecting groups in synthesis. However, the developments in fermentation technology which had taken place by this time made total synthesis of penicillins of no commercial interest. This is even more true today, but the fascination of the synthetic problems presented by the penicillin molecule continues to provide a challenge. The first stereospecific total synthesis of a penicillin derivative has recently been accomplished and is outlined in Scheme 7.[33]

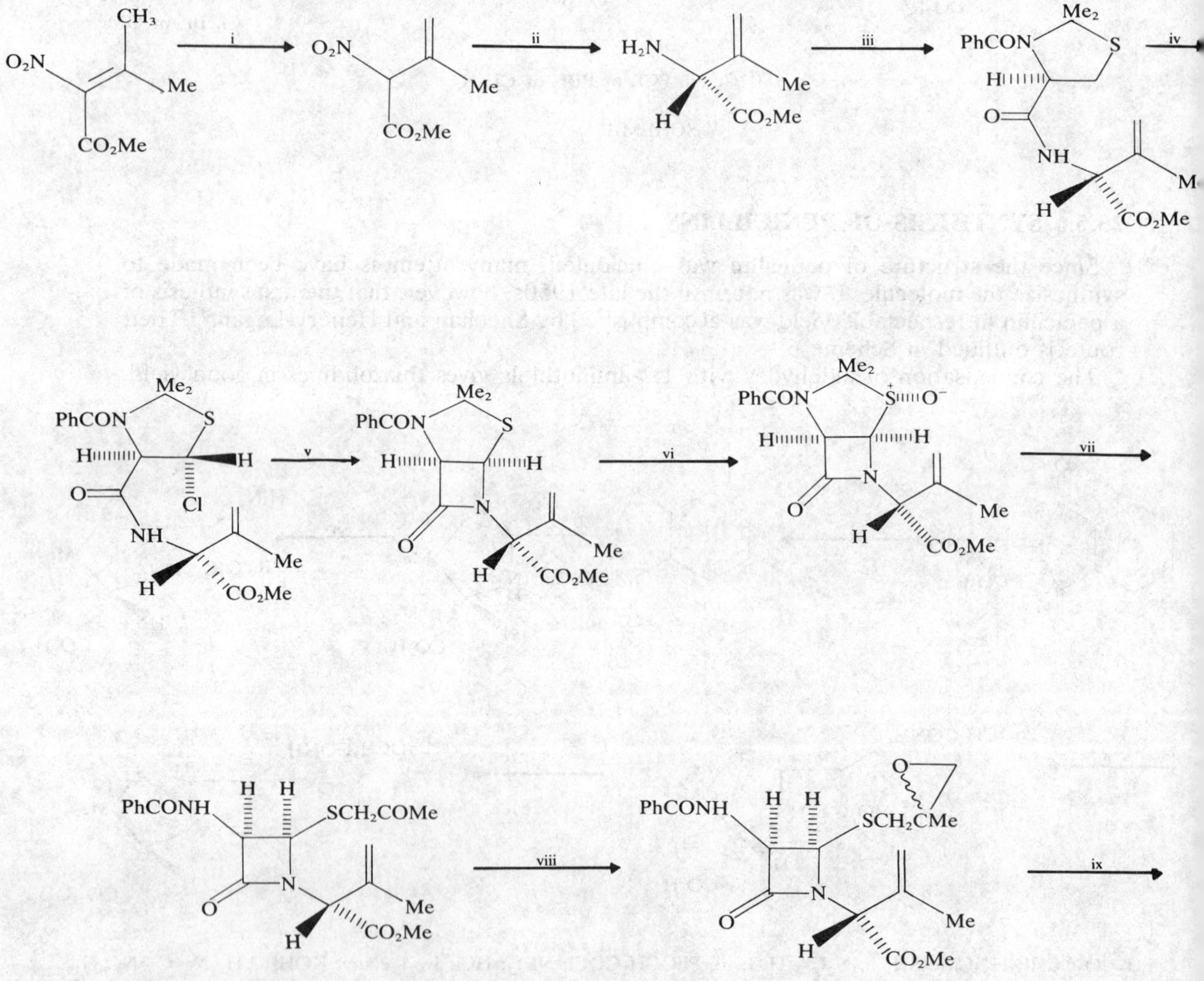

SCHEME 7

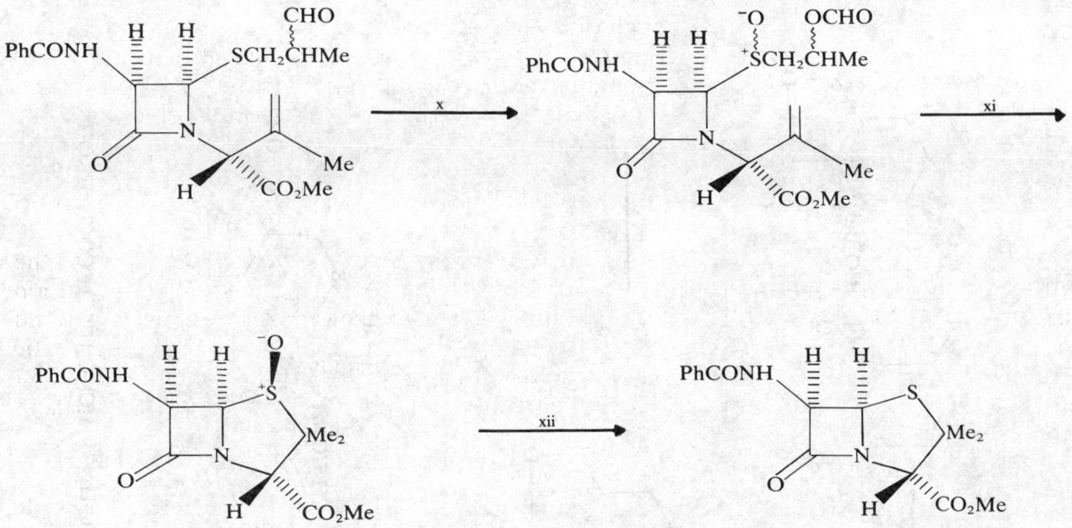

i, KH; HCl; ii, SnHCl; resolution as chloroacetyl derivative with hog acylase I;

iii, 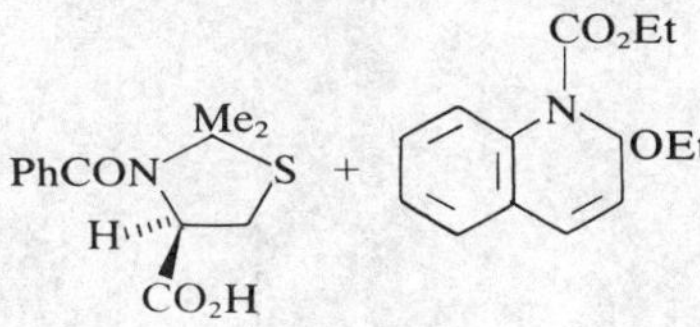

; iv, $(PhCOO)_2$; HCl; v, NaH;

vi, *m*-chloroperbenzoic acid; vii, heat, H_2SO_4; viii, CH_2N_2; ix, $BF_3.Et_2O$; x, *m*-chloroperbenzoic acid; xi, heat; xii, PBr_3.

SCHEME 7 (*continued*)

23.5.7 SYNTHESIS OF CEPHALOSPORINS

The general approach adopted by Sheehan and Henery-Logan for the synthesis of penicillin was successfully adapted for the synthesis of a cephalosporin, as outlined in Scheme 8. The Mannich base prepared from pyruvic acid, formaldehyde, and dimethylamine was converted to the thiolacetate by displacement of dimethylamine with thioacetic acid. Removal of the thiol protecting group under acidic conditions and condensation with the aminomethyleneglycine derivative gave both the *erythro*- and *threo*-isomers of the condensation product. Although the stereoisomers could be separated, after treatment with hydrazine (to remove the phthaloyl group) and acid (to remove the t-butyl group), both stereoisomers gave the desired *threo*-amino-acid. Tritylation of the amino-group greatly facilitates β-lactam formation, which was accomplished with dicyclohexylcarbodiimide. After removal of the trityl group by treatment with acid, the product was resolved with (+)-tartaric acid and then acylated with thienylacetyl chloride. The product was identical with a sample prepared from cephalothin. Conversion of the lactone to the hydroxy-acid had already been accomplished on material derived from natural cephalothin lactone.[34–36]

A very different synthetic approach was adopted by Woodward *et al.* in the synthesis of cephalosporins, first announced in his Nobel lecture in 1965 (Scheme 9).[37,38] Starting with

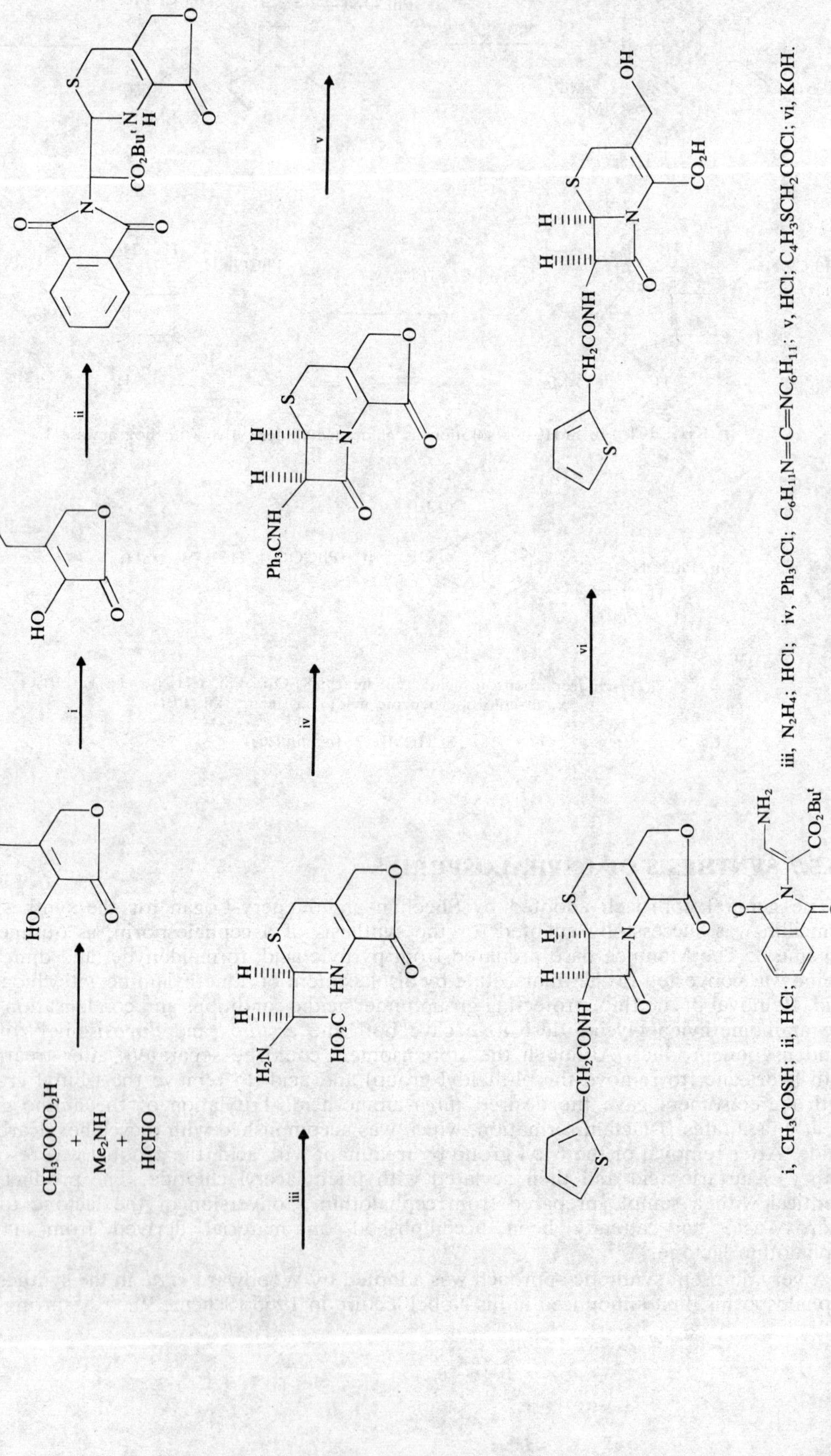

i, CH_3COSH; ii, HCl; iii, N_2H_4; HCl; iv, Ph_3CCl; $C_6H_{11}N{=}C{=}NC_6H_{11}$; v, HCl; $C_4H_3SCH_2COCl$; vi, KOH.

SCHEME 8

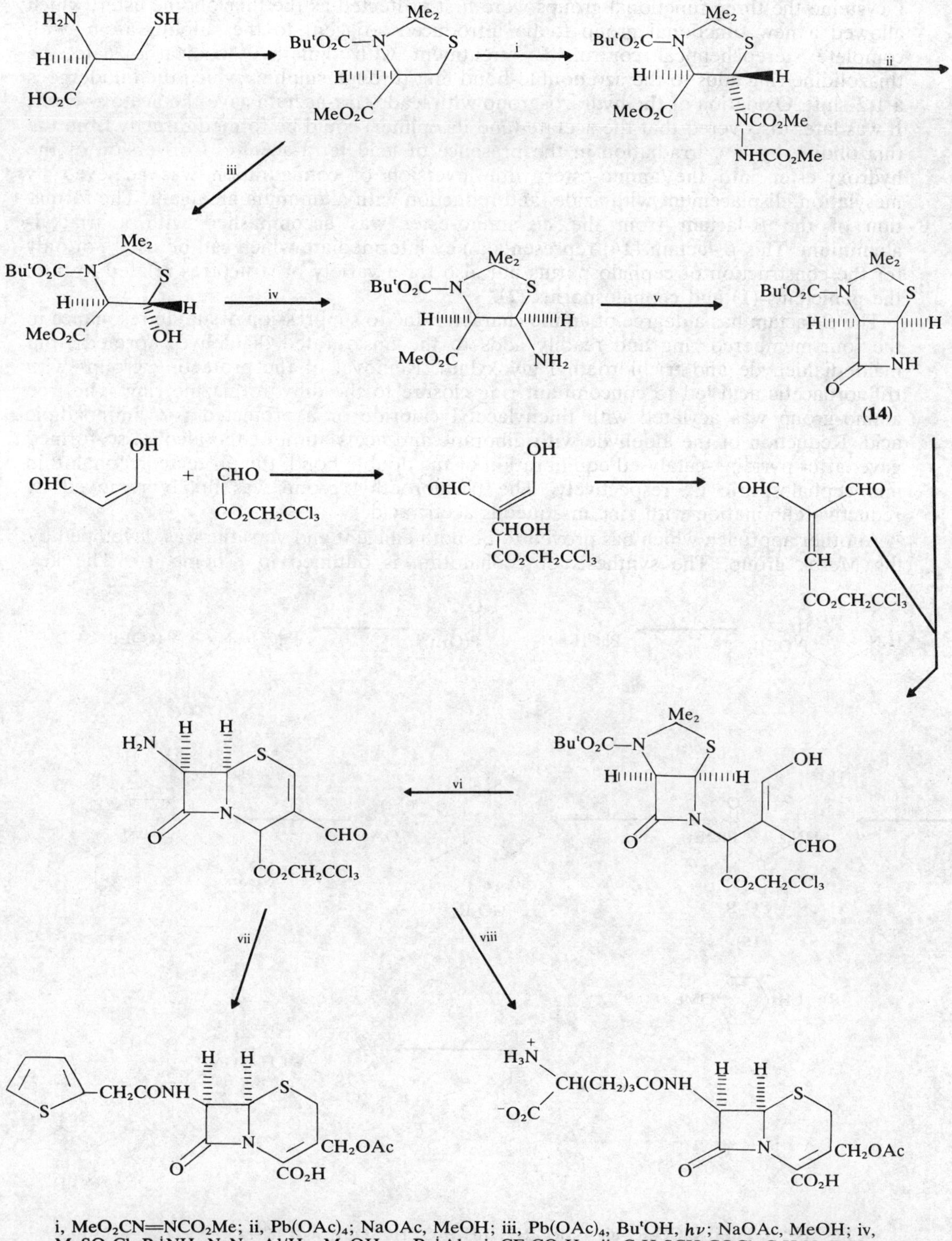

i, $MeO_2CN{=}NCO_2Me$; ii, $Pb(OAc)_4$; NaOAc, MeOH; iii, $Pb(OAc)_4$, Bu^tOH, $h\nu$; NaOAc, MeOH; iv, $MeSO_2Cl$, Pr^i_2NH; NaN_3; Al/Hg, MeOH; v, Bu^i_3Al; vi, CF_3CO_2H; vii, $C_4H_3SCH_2COCl$, C_5H_5N; B_2H_6; Ac_2O, C_5H_5N; Zn, aq. AcOH; viii, $Cl_3CCH_2O_2CNHCH(CO_2H)(CH_2)_3CO_2H + C_6H_{11}N{=}C{=}NC_6H_{11}$; Cl_3CCH_2OH, $C_6H_{11}N{=}C{=}NC_6H_{11}$, C_5H_5N; B_2H_6; Ac_2O, C_5H_5N; Zn, aq. AcOH.

SCHEME 9

L-cysteine the three functional groups were first protected as the thiazolidine ester, which allowed a new functional group to be introduced adjacent to the sulphur atom, with complete stereochemical control. By treatment with dimethyl azodicarboxylate the thiazolidine ring adds to the azo double-bond first through sulphur, which then undergoes a 1,2-shift. Oxidation of the hydrazo-group with lead tetra-acetate gave the acetoxy-ester. It was later discovered that the acetate (and its epimer) could be formed directly from the thiazolidine by u.v. irradiation in the presence of lead tetra-acetate. Conversion of the hydroxy-ester into the amino-ester with inversion of configuration was achieved by mesylation, displacement with azide, and reduction with aluminium amalgam. The formation of the β-lactam from the *cis*-amino-ester was accomplished with a trialkylaluminium. This β-lactam (**14**) represents a key intermediate which can be used not only for the construction of cephalosporins but also for a variety of structures related to both the penicillins (**1**) and cephalosporins (**2**).

The β-lactam has a degree of amine character due to suppression of amide resonance in the four-membered ring and readily adds to the unsaturated dialdehyde formed from malondialdehyde and trichloroethyl glyoxylate. Removal of the protecting groups with trifluoroacetic acid led to concomitant ring closure to the dihydrothiazine ring. The free amino-group was acylated with thienylacetyl chloride or a protected D-α-aminoadipic acid. Reduction of the aldehyde with diborane and acetylation of the alcohol so formed gave, after pyridine-catalysed equilibration of the double bond, the protected cephalothin and cephalosporin C, respectively. The trichloroethyl group was finally removed by reductive elimination with zinc in aqueous acetic acid.

Another approach which has proved to be both efficient and versatile was developed by the Merck group. The synthesis of cephalothin is outlined in Scheme 10. The key

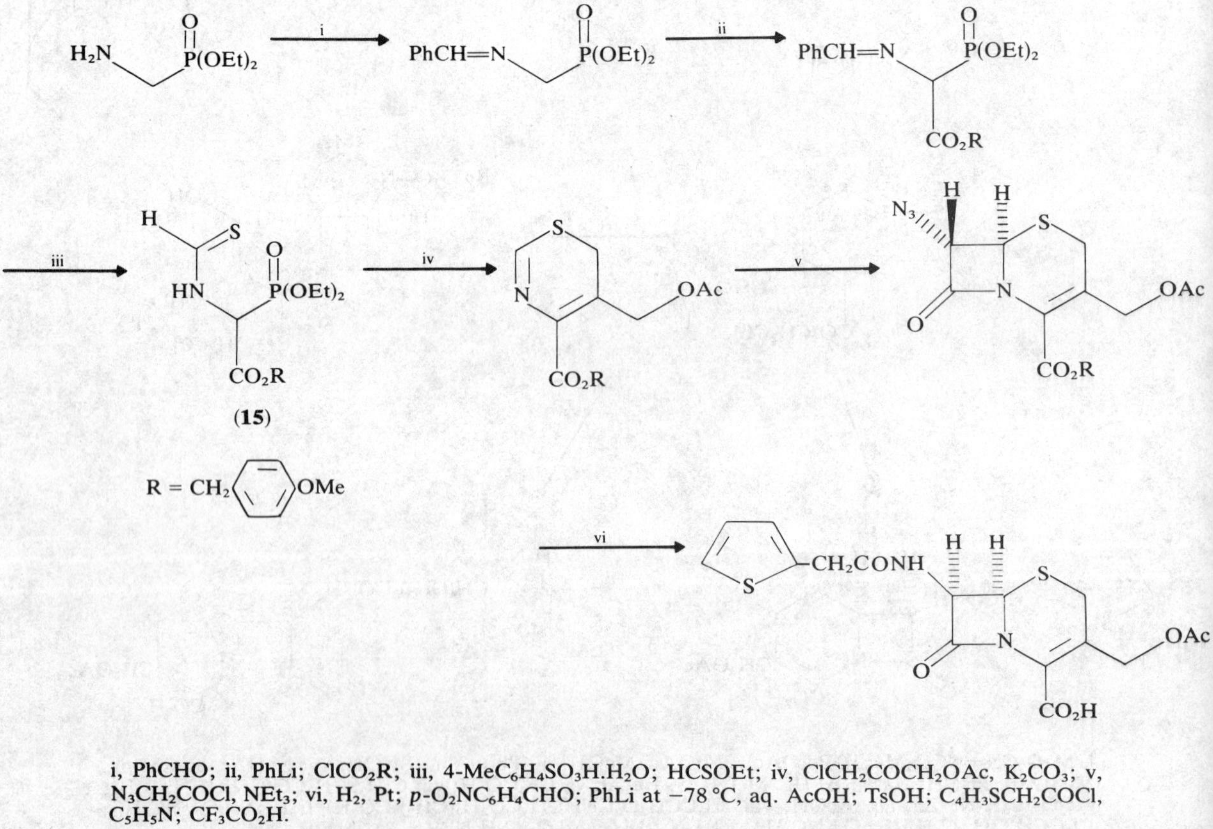

i, PhCHO; ii, PhLi; $ClCO_2R$; iii, 4-$MeC_6H_4SO_3H.H_2O$; HCSOEt; iv, $ClCH_2COCH_2OAc$, K_2CO_3; v, N_3CH_2COCl, NEt_3; vi, H_2, Pt; *p*-$O_2NC_6H_4CHO$; PhLi at −78 °C, aq. AcOH; TsOH; $C_4H_3SCH_2COCl$, C_5H_5N; CF_3CO_2H.

SCHEME 10

intermediate is the thioamide (**15**), from which the 6*H*-1,3-thiazine was prepared. A cycloaddition reaction between this 6*H*-1,3-thiazine and the keten obtained *in situ* from azidoacetyl chloride and triethylamine gave a mixture of diastereoisomeric azides which on hydrogenation gave the corresponding amines. Since the 7α-aminocephalosporanic ester can be isomerized to the 7β-aminocephalosporanic ester by way of its Schiff base, a good yield of the desired stereoisomer was obtained. Acylation of this with thienylacetyl chloride and removal of the ester protecting group gave racemic cephalothin.[39]

23.5.8 CONVERSION OF PENICILLINS TO CEPHALOSPORINS

In the total synthesis of cephalosporins by Woodward *et al.* the β-lactam (**14**) was a key intermediate.[37,38] This intermediate was subsequently synthesized from a penicillin, thereby achieving conversion of a penicillin to cephalosporins.[40]

Opening of the thiazolidine ring of penicillins without cleavage of the β-lactam can be accomplished in a number of ways (*e.g.* Scheme 11). By converting the acid of a penicillin to the acid azide, a Curtius rearrangement may be achieved.[41] If the isocyanate so formed is converted to the trichloroethylurethane with trichloroethanol, on removal of the trichloroethyl group with zinc and aqueous acetic acid the ring-opened aldehyde is obtained in good yield.[40]

Starting with penicillin (Scheme 12), the corresponding aldehyde (in equilibrium with the cyclic carbinolamine) on treatment with lead tetra-acetate gave the tertiary acetate which readily eliminated acetic acid to give the vinyl sulphide in good yield.

When the side-chain was that of t-butoxycarbonyl (Scheme 13), treatment of the corresponding vinyl sulphide with trifluoroacetic acid cleaved the t-butoxycarbonyl group and the resulting amine gave the thiazolidine ring. The t-butoxycarbonyl group was reintroduced to give the Woodward intermediate (**14**).

A much more direct and efficient conversion of penicillins to cephalosporins can be achieved by way of the penicillin sulphoxides. Since penicillins are very much cheaper to produce by fermentation than cephalosporin C, a great deal of attention has been given to this route.[42,43] It seems likely that as a result cephalosporins will in the future be prepared by chemical transformation of the penicillins.

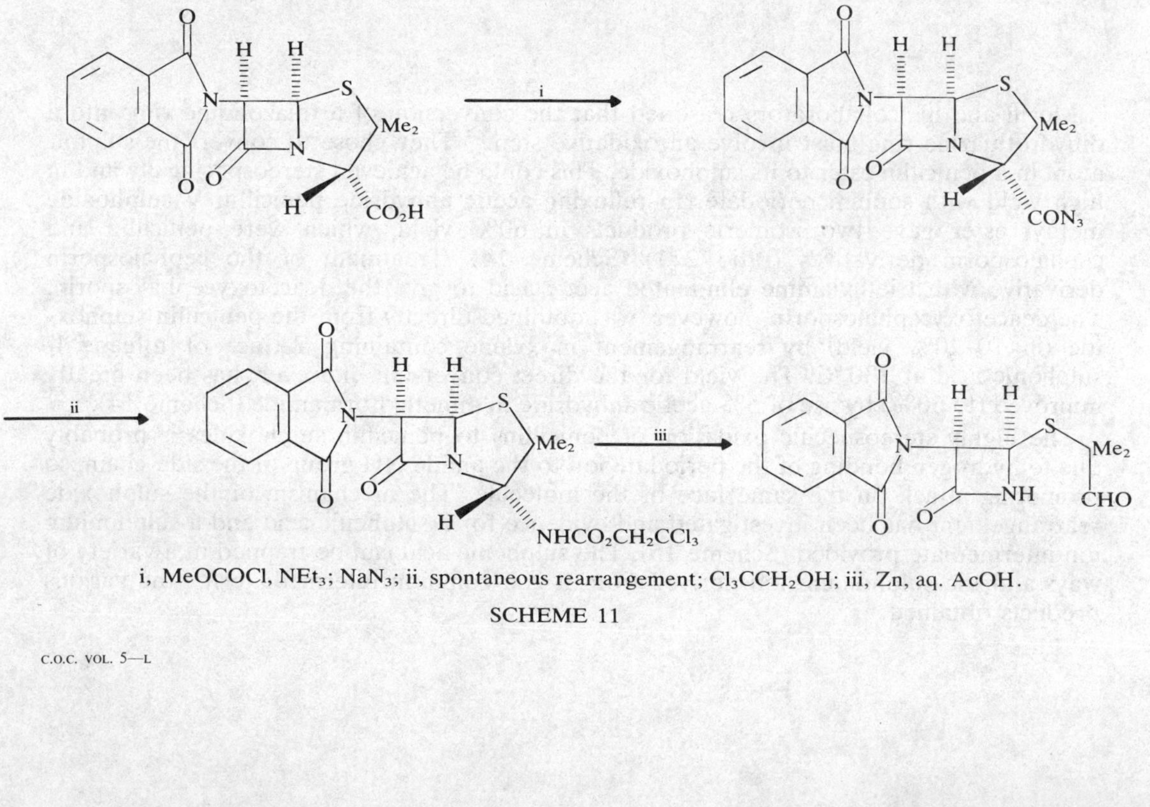

i, $MeOCOCl$, NEt_3; NaN_3; ii, spontaneous rearrangement; Cl_3CCH_2OH; iii, Zn, aq. AcOH.

SCHEME 11

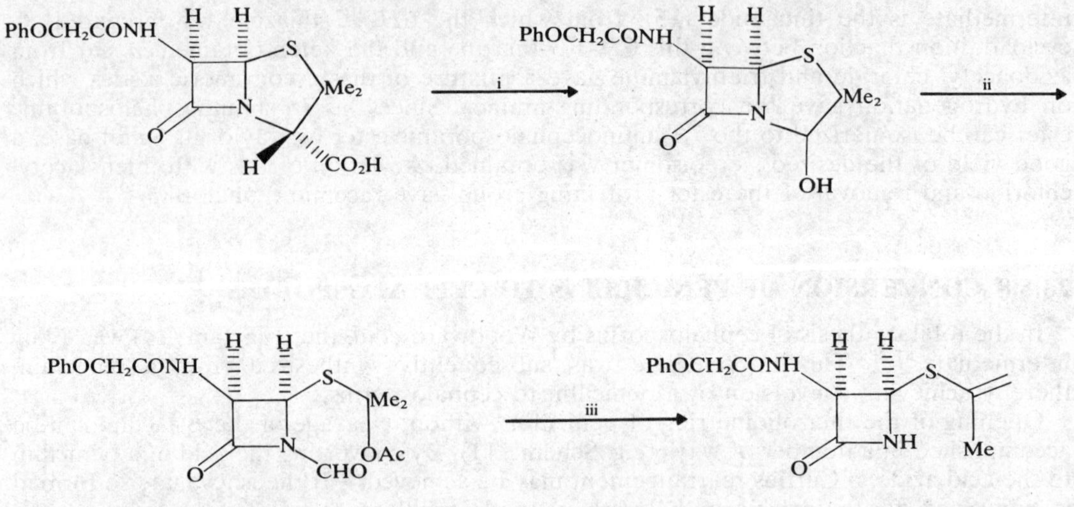

i, MeOCOCl, NEt_3; NaN_3; heat, Cl_3CCH_2OH; Zn, aq. AcOH; ii, $Pb(OAc)_4$; iii, NH_4OH

SCHEME 12

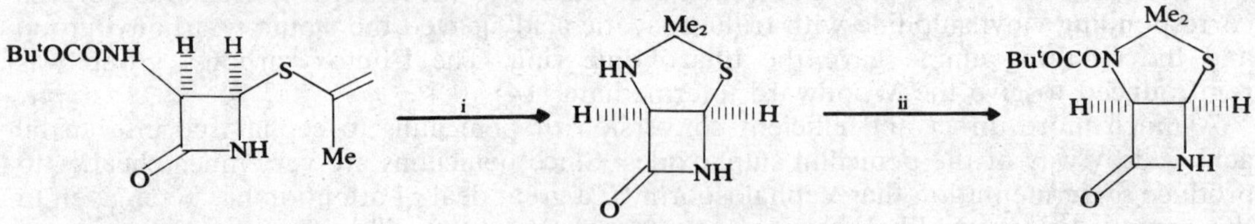

(14)

i, CF_3CO_2H; ii, (a) $COCl_2$, (b) Bu'OH.

SCHEME 13

Morin and his collaborators reasoned that the conversion of a thiazolidine ring into a dihydrothiazine ring must involve an oxidative step.[44] They chose to convert the sulphur atom in a penicillin ester to its sulphoxide. This could be achieved stereospecifically and in high yield with sodium periodate. In refluxing acetic anhydride penicillin V sulphoxide methyl ester gave two isomeric products in 60% yield, which were penicillin and cephalosporin derivatives (ratio 2:1) (Scheme 14). Treatment of the cephalosporin derivative with triethylamine eliminated acetic acid to give the deacetoxycephalosporin. The deacetoxycephalosporin, however, was obtained directly from the penicillin sulphoxide (in 10–20% yield) by rearrangement in xylene containing a trace of toluene-4-sulphonic acid at 130 °C. The yield for the direct conversion, however, has been greatly improved (to 60%) by use of 5% acetic anhydride in dimethylformamide (Scheme 14).[45]

The highly stereospecific oxidation of penicillins to penicillin sulphoxides is probably due to hydrogen bonding of the periodate ion to the amide NH group of the side-chain, so promoting attack on the same face of the molecule. The mechanism of the sulphoxide rearrangement has been investigated and evidence for a sulphenic acid and a sulphonium ion intermediate provided (Scheme 15). The sulphenic acid can be trapped in a variety of ways and the sulphonium ion derived from it is a common intermediate for the various products obtained.

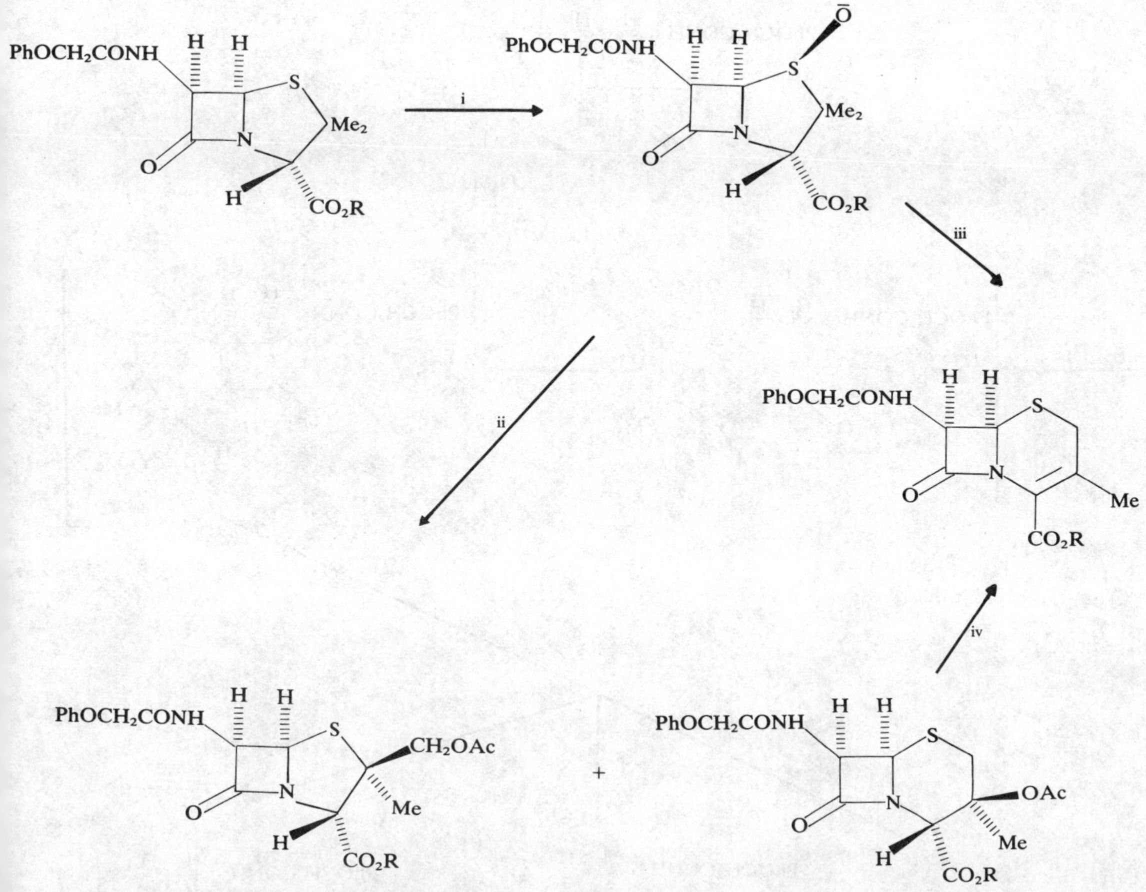

i, $NaIO_4$; ii, Ac_2O; iii, 5% Ac_2O–Me_2NCHO, 130 °C; iv, Et_3N.

SCHEME 14

Undoubtedly the most efficient conversion of a penicillin to a deacetoxycephalosporanic acid so far reported is the one-step conversion of penicillin G sulphoxide to the corresponding deacetoxycephalosporanic acid in greater than 80% yield with transient protection of the carboxylic acid as its trimethylsilyl ester using *N,O*-bis(trimethylsilyl)-acetamide.[45a]

Deacetoxycephalosporins which are now most efficiently produced from penicillins have valuable antibacterial properties in their own right, but can also be converted to cephalosporins. One of the best routes is outlined in Scheme 16.[43] By first converting the sulphur atom to the sulphoxide, it is protected from reaction itself with *N*-bromosuccinimide and directs the bromination to the methyl group. Displacement of the allylic bromide with acetate ion followed by reduction of the sulphoxide gave the cephalosporanic ester, which was readily deprotected to give the corresponding acid.

The recognition of 3-methylenecephams as valuable intermediates for the synthesis of the 3-halo- and 3-methoxy-cephems (Scheme 5) as well as cephalosporanic acids led to a search for a route from penicillins (Scheme 16a). Conversion of a penicillin sulphoxide ester into a sulphinyl halide with a *N*-halogenating agent (*e.g.* *N*-bromosuccinimide) when

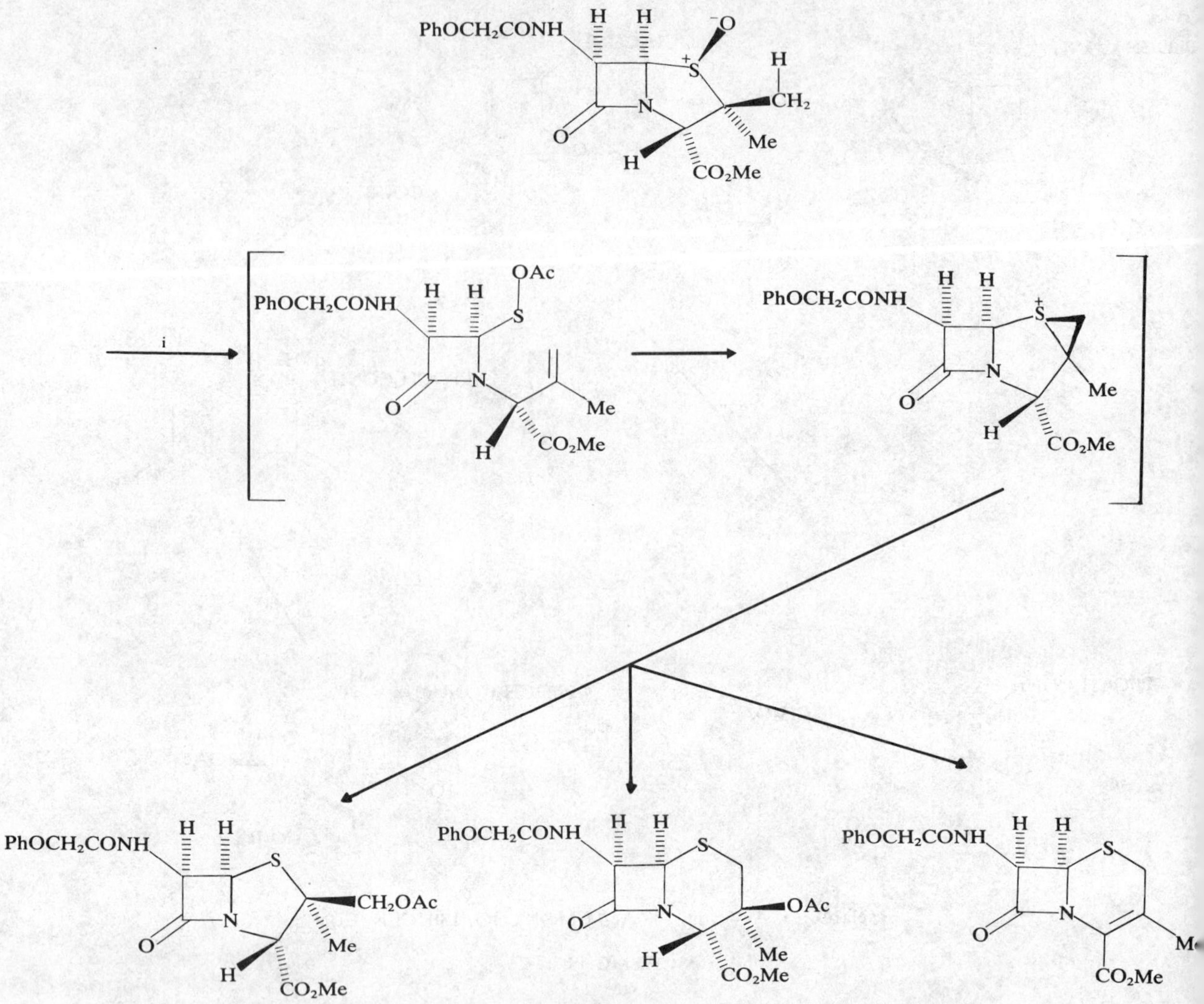

i, Ac_2O, Me_2NCHO.

SCHEME 15

followed by treatment with a Lewis acid (*e.g.* $TiCl_4$, $AlCl_3$, $ZnCl_2$) gave the 3-methylenecepham sulphoxide.[45b] Reduction of the sulphoxide with phosphorus trichloride followed by bromination in the presence of a strong organic base gave the 3-bromomethylcephem, which was converted to the cephalosporanic acid with silver acetate in acetic acid.[45c]

Conversion of the 3-methylenecepham sulphoxide to the 3-methoxy- and 3-halocephem sulphoxides was achieved by the methods outlined in Scheme 5, the reduction of the sulphoxide being finally accomplished with phosphorus trichloride.

There can be little doubt, however, that the most efficient route so far developed for the conversion of penicillins into the 3-heterosubstituted cephems is one recently reported by the Shionogi Research Laboratory (Scheme 17).[45d] Penicillin G, for example, is first converted into its sulphoxide *p*-nitrobenzyl ester and on treatment with trimethyl phosphite as previously described[45e] gave the fused thiazoline-β-lactam. Ozonolysis of the

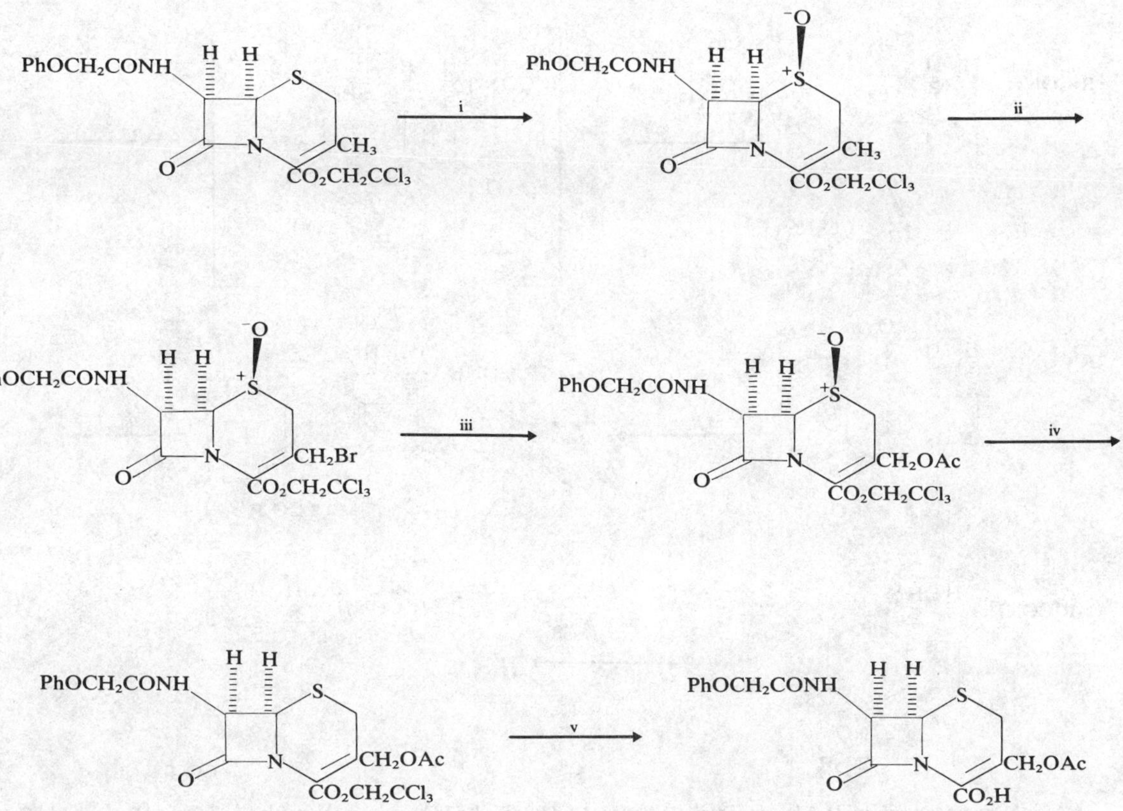

i, *m*-$ClC_6H_4CO_3H$; ii, *N*-bromosuccinimide, *hν*; iii, KOAc, Me_2NCHO; iv, PCl_3, Me_2NCHO; v, Zn, aq. AcOH.

SCHEME 16

olefin was then followed by a sequence of reactions which were all performed, without isolation, in a single reaction vessel. Initial mesylation of the enol was followed by conversion to the enamine with morpholine. Allylic bromination followed by hydrolysis of the enamine with hydrochloric acid gave the bromo-ketone, which spontaneously cyclized and tautomerized to the 3-hydroxycephem. The product crystallized from the reaction medium in 70% yield and undoubtedly provides a potentially valuable industrial process.[45e]

23.5.9 NUCLEAR ANALOGUES OF THE PENICILLINS AND CEPHALOSPORINS

From a consideration of the mode of action of the penicillins and cephalosporins, it appeared that the acyl-dipeptide moiety and the reactive β-lactam ring are essential features for antibacterial activity. There was, however, no compelling reason why the β-lactam ring should be fused to a thiazolidine or dihydrothiazine ring, and new families of β-lactam antibiotics containing different heterocycles have now been realised.

The intermediate used in Woodward's synthesis of cephalosporin C (**14**, Scheme 9), and which could also be obtained from penicillin, is a potential precursor of nuclear analogues

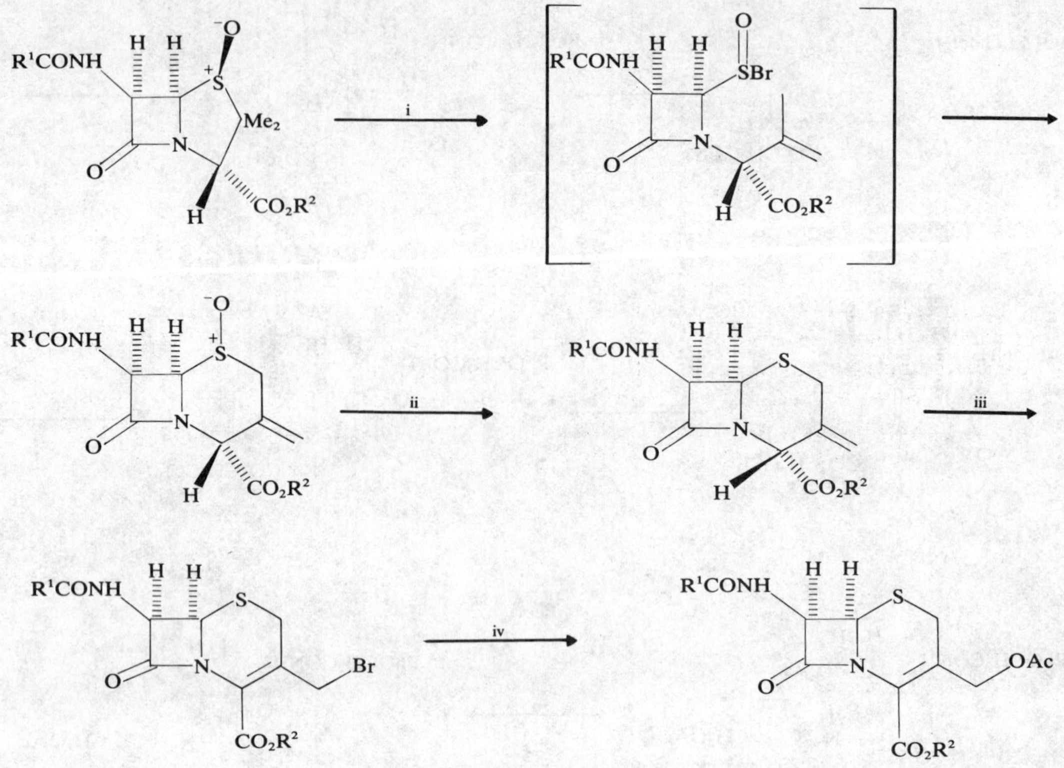

i, *N*-bromosuccinimide, Lewis acid; ii, PCl_3, Me_2NCHO; iii, 1,5-diazabicyclo[5,4,0]undec-5-ene, Br_2 at −80°C; iv, AgOAc, AcOH.

SCHEME 16a

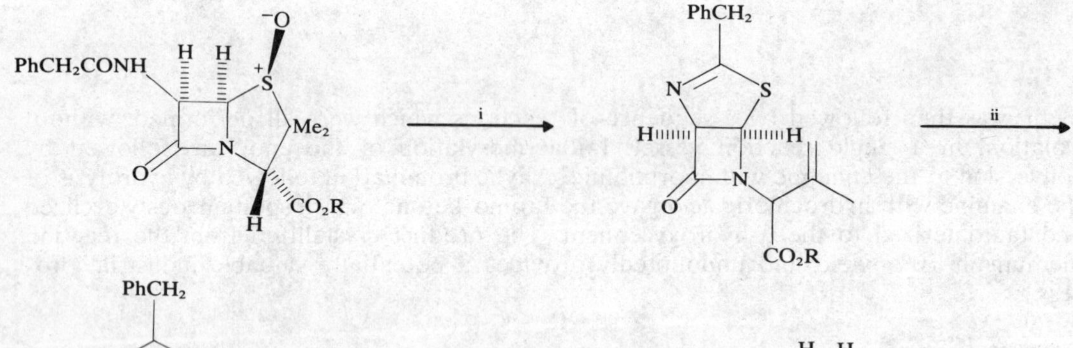

i, $(MeO)_3P$; ii, O_3 in CH_2Cl_2–MeOH at −20°C; ii, $MeSO_2Cl$, NEt_3; morpholine; Br_2, pyridine; 2N HCl, MeOH.

SCHEME 17

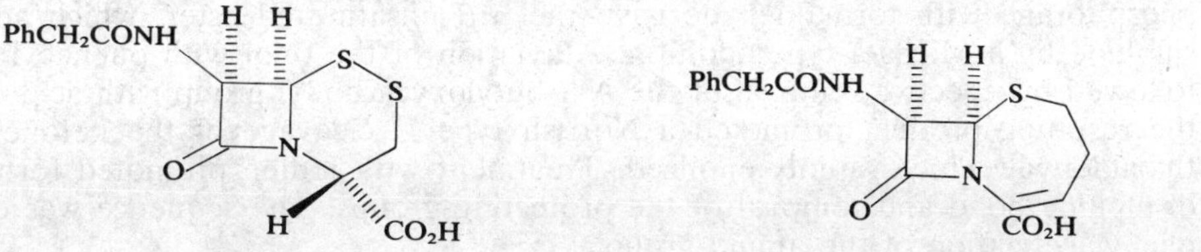

(16) (17)

of the penicillins and cephalosporins. This intermediate (and close variants of it) have been used to obtain several nuclear analogues of the cephalosporins, *e.g.* (**16**) and (**17**). Synthesis of the related analogue (**18**) is outlined in Scheme 18.[46] The β-lactam exhibits some amine-like character and readily adds to glyoxylic esters. The isomeric mixture of carbinols so formed are converted to chlorides with thionyl chloride, which on reaction with triphenylphosphine and base give a single stable phosphorane. Reaction of the

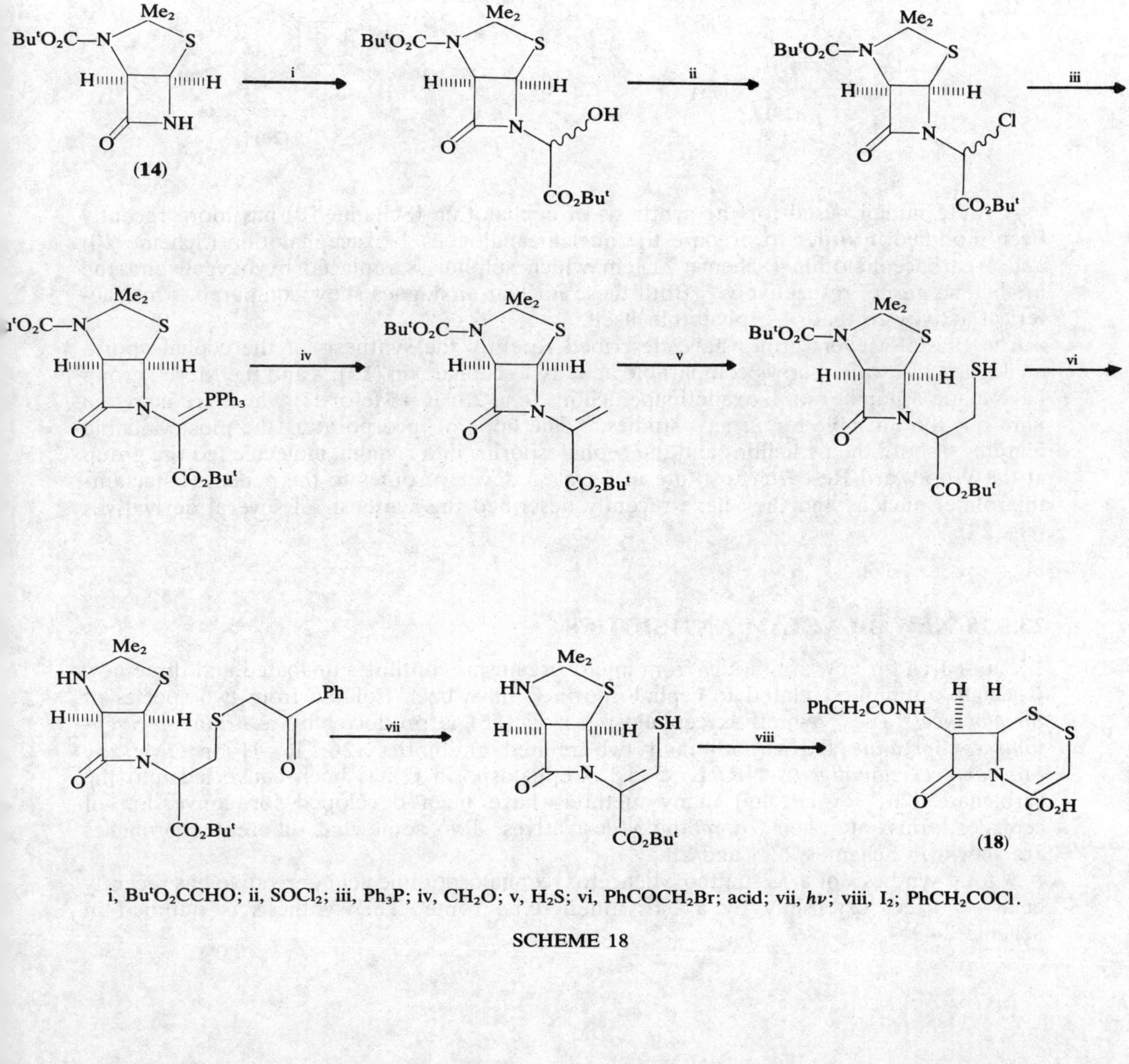

i, Bu^tO_2CCHO; ii, $SOCl_2$; iii, Ph_3P; iv, CH_2O; v, H_2S; vi, $PhCOCH_2Br$; acid; vii, *hν*; viii, I_2; $PhCH_2COCl$.

SCHEME 18

phosphorane with formaldehyde gave the α,β-unsaturated ester, which adds hydrogen sulphide by a Michael-type addition. Alkylation of the thiol with phenacyl bromide was followed by selective removal of the *N*-t-butyloxycarbonyl group with acid. Photolysis of the resulting product promoted a Norrish type II cleavage of the ketone, to give the thioaldehyde which rapidly enolized. Treatment with iodine promoted formation of the disulphide bond and removal of the protecting groups. The sequence was completed by phenylacetylation of the amino-group.

An important question concerning the structure–activity relationship of the penicillins and the cephalosporins was whether the sulphur atom could be replaced. A general method for synthesizing nuclear analogues of the penicillins and cephalosporins, *e.g.* (**19**) and (**20**), was developed. The synthesis of the analogue (**21**) is outlined in Scheme 19.[47] The key step in this reaction scheme is the photolysis of the diazo-amide to give regio- and stereo-specific carbene insertion into the heterocyclic ring. Only two stereoisomeric β-lactams were generated. The conversion of the carboxylic acid side-chain to the amino-function was achieved by Curtius rearrangement of the acid azide, the synthesis being completed by phenylacetylation and deprotection.

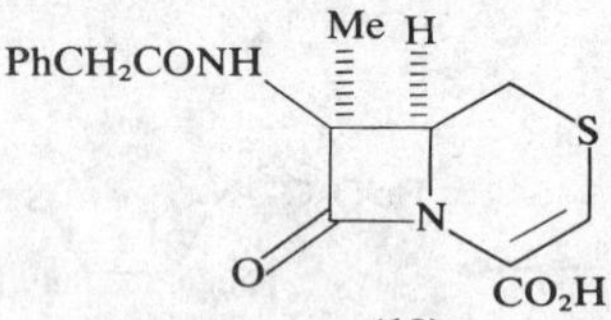

(**19**)

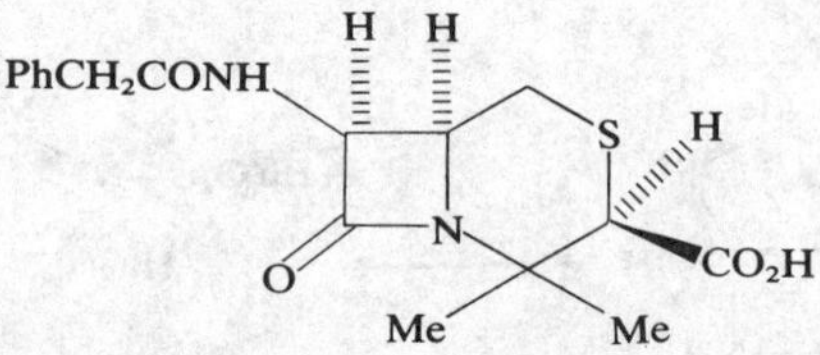

(**20**)

A route initially used for the synthesis of cephalothin (Scheme 10) has more recently been modified in order to prepare the nuclear analogues 1-oxacephalothin (Scheme 20) and 1-carbacephalothin (Scheme 21), in which sulphur is replaced by oxygen and the methylene group, respectively.[48] Both these nuclear analogues show comparable antibacterical activity to that of cephalothin itself.

The Bristol-Meyers group have described recently the synthesis of the cephalosporin analogue (**22**), which shows comparable activity to cephalexin (**13**),[49] and the Merck group have made a number of 1-oxadethiapenicillins (*e.g.* **23**; R = Me or H) which are active *in vitro* but too unstable for *in vivo* studies.[50] The hope of incorporating the most valuable features of both the penicillins and the cephalosporins into a single molecule led the group at the Woodward Research Institute in Basel to develop routes to the penem (β-lactam-thiazoline) nucleus and they have recently described the synthesis of several derivatives (*e.g.* **24**).[51]

23.5.10 NEW β-LACTAM ANTIBIOTICS

The search for new antibiotics from micro-organisms continues unabated and three new β-lactam antibiotics related to cephalosporin C have been isolated from two species of *Streptomyces*. The 7α-methoxycephalosporin C (**25**) is produced by a strain of *Streptomyces lipmanii* NRRL 3584 and two related antibiotics (**26**; R = H or OMe) by *Streptomyces clavuligerus* NRRL 3585.[52] Cephalosporin C has been converted into the carbamate (**26**; R = H) and many methods have been developed for conversion of cephalosporins into their 7α-methoxy-derivatives. Two somewhat different approaches are shown in Schemes 22[53] and 23.[54]

A total synthesis of a 7α-methoxydeacetoxycephalosporanic acid derivative has recently been reported, ostensibly by a biosynthetic-type route. The synthesis is outlined in Scheme 24.[55]

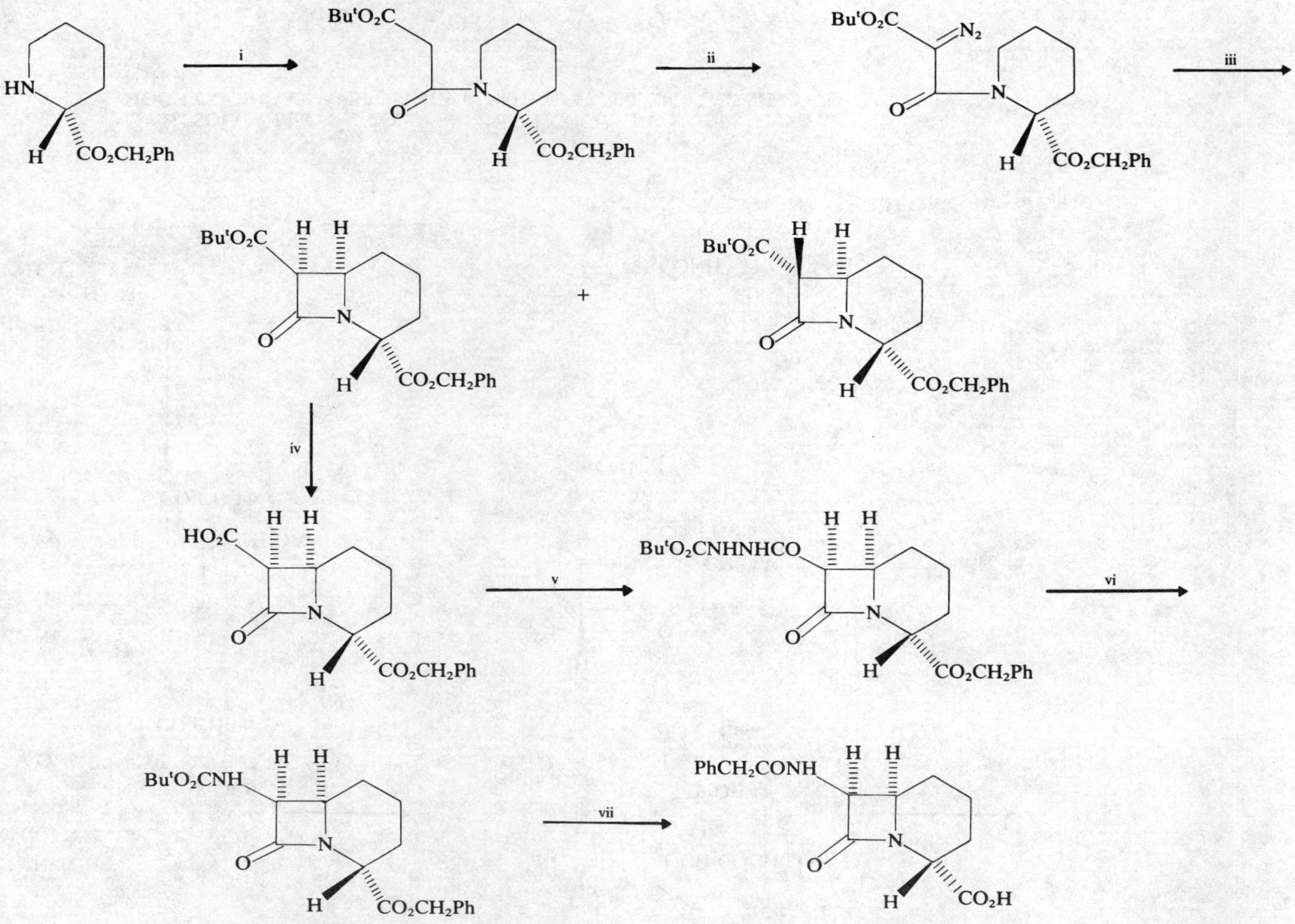

(21)

i, $Bu^tO_2CCH_2CO_2H$, $C_6H_{11}N{=}C{=}NC_6H_{11}$; ii, 4-$MeC_6H_4SO_2N_3$, NEt_3; iii, *hν*; iv, CF_3CO_2H; v, $Bu^tO_2CNHNH_2$, $C_6H_{11}N{=}C{=}NC_6H_{11}$; vi, CF_3CO_2H; $NaNO_2$, HCl; heat; Bu^tOH: vii, CF_3CO_2H; $PhCH_2COCl$, NEt_3; H_2, Pd.

SCHEME 19

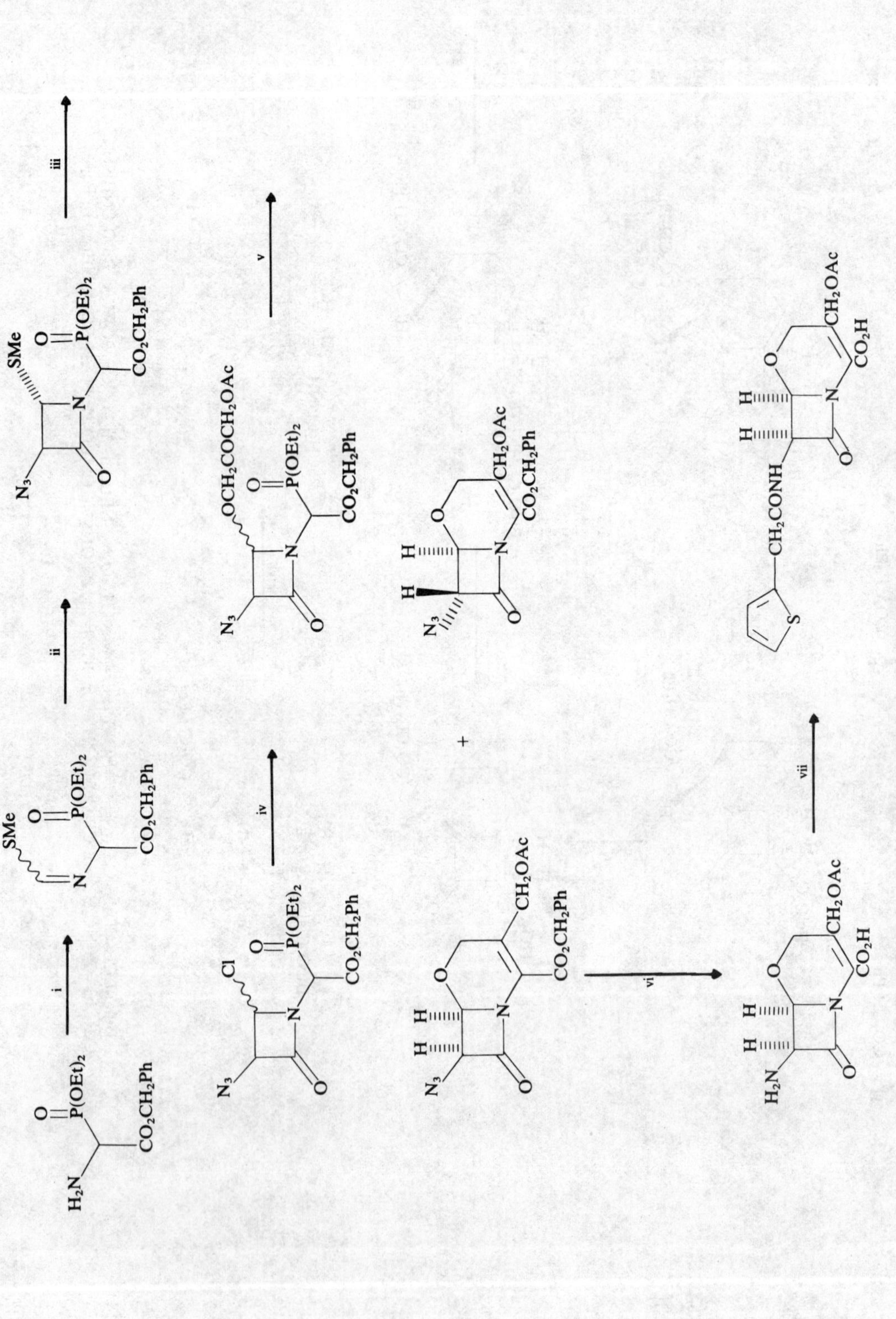

i, $HCSOEt$; MeI, K_2CO_3; ii, N_3CH_2COCl, NEt_3; iii, Cl_2; iv, $HOCH_2COCH_2OAc$, $AgBF_4$, Ag_2O; v, NaH; vi, H_2, Pd; vii, $C_4H_3SCH_2COCl$, $NaHCO_3$

SCHEME 20

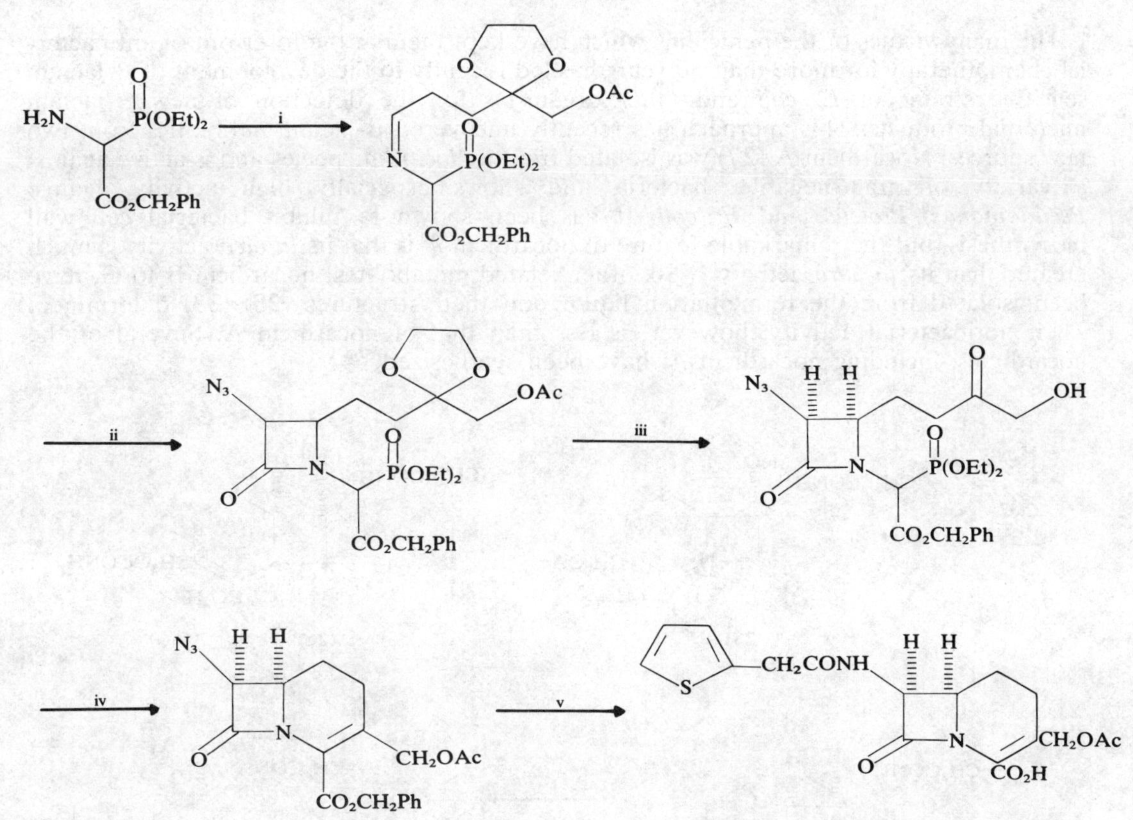

i, $OCH(CH_2)_2CCH_2OAc$; ii, N_3CH_2COCl, Et_3N, −78 °C; iii, 10% H_2SO_4–AcOH; iv, CH_3COCl, pyridine; NaH; v, H_2,Pd; $C_4H_3SCH_2COCl$, $NaHCO_3$.

SCHEME 21

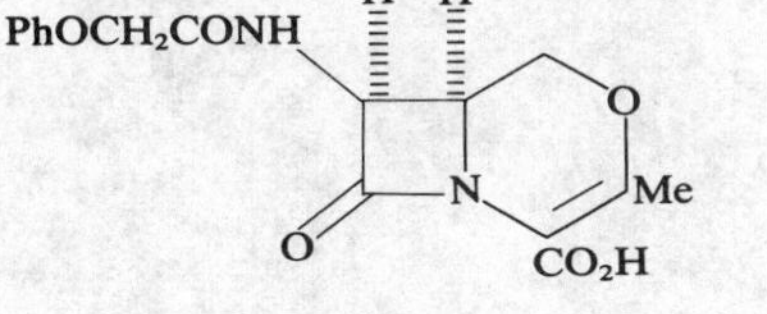

(22)

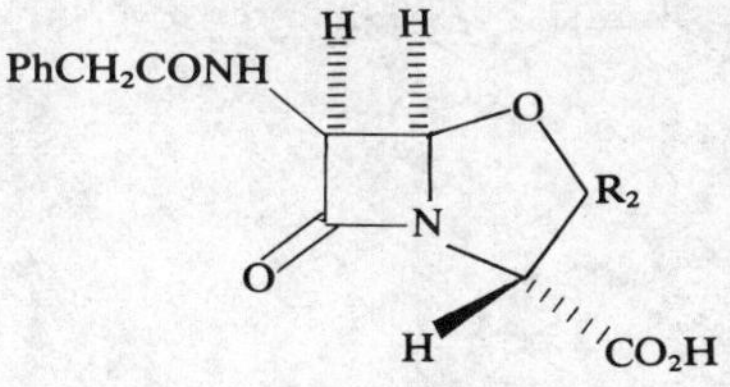

(23)

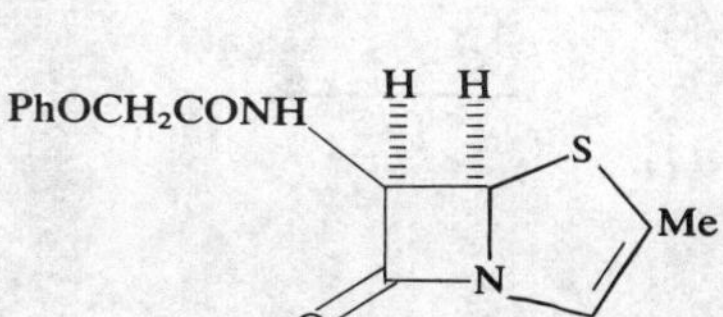

(24)

The many virtues of the penicillins which have kept them at the forefront of antibacterial chemotherapy for more than 30 years has led recently to the development of β-lactam sensitive strains of *E. coli* and other organisms for the detection of new β-lactam microbial products. This approach has recently uncovered β-lactam antibiotics from two new sources. Nocardicin A (**27**) was isolated from a *Nocardia* species and is active against a variety of gram-negative bacteria and shows especially high activity against *Pseudomonas*, *Proteus*, and *E. coli.* It has been shown to inhibit bacterial cell wall biosynthesis, but the remarkable feature of nocardicin A is that its *in vivo* activity is much greater than its *in vitro* activity.[56] Six other related metabolites, nocardicin B to G, have been isolated from the fermentation liquor and their structures (**28**)–(**33**) determined. Their antibacterial activity, however, is less than that of nocardicin A. Several of the nocardicins, including nocardicin A, have been synthesized.[56a]

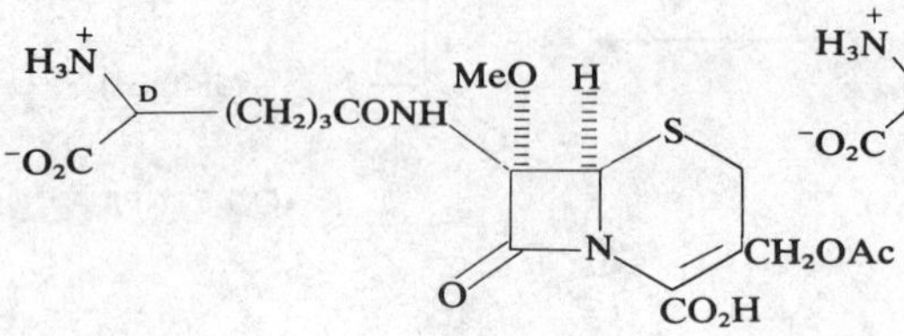

(25)

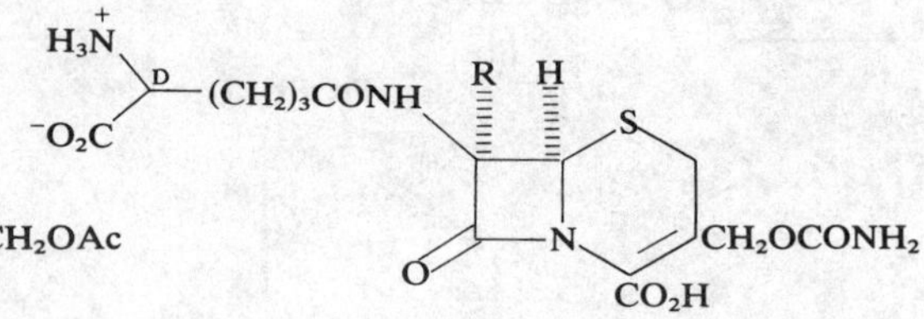

(26)

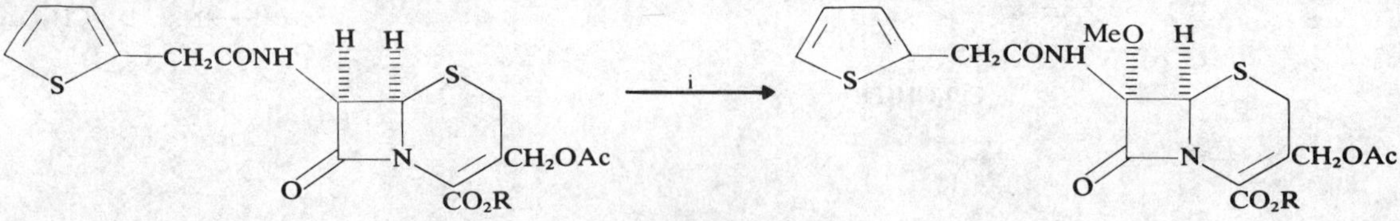

i, LiOMe, Bu^{t}OCl at −80 °C

SCHEME 22

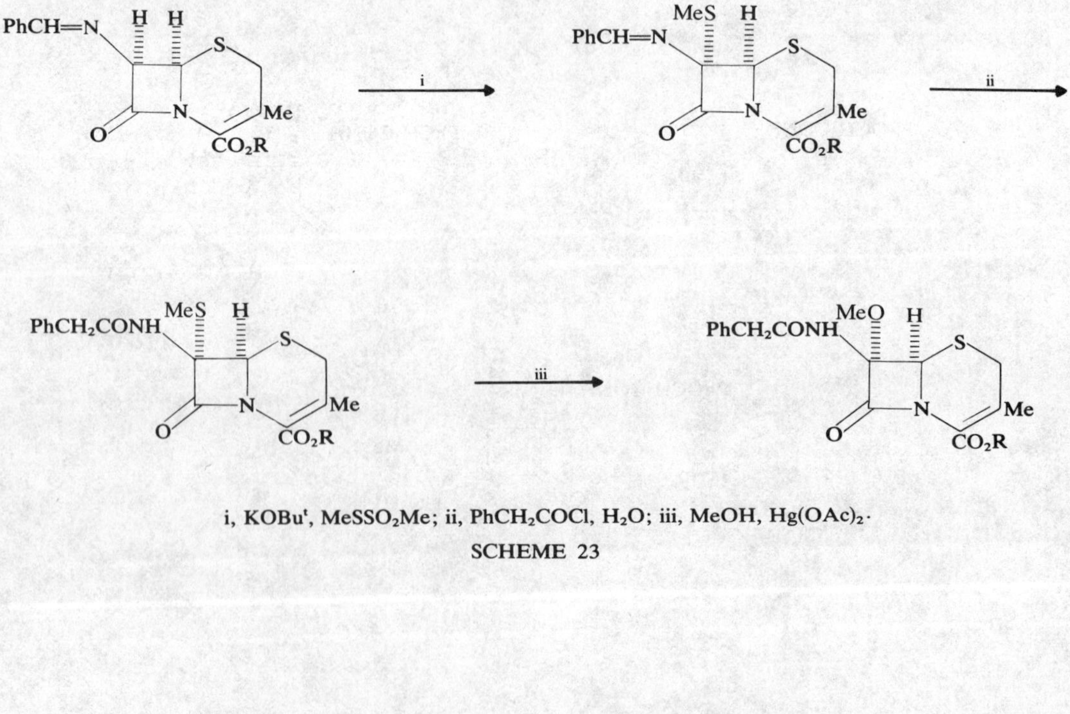

i, KOBu^t, $MeSSO_2Me$; ii, $PhCH_2COCl$, H_2O; iii, MeOH, $Hg(OAc)_2$.

SCHEME 23

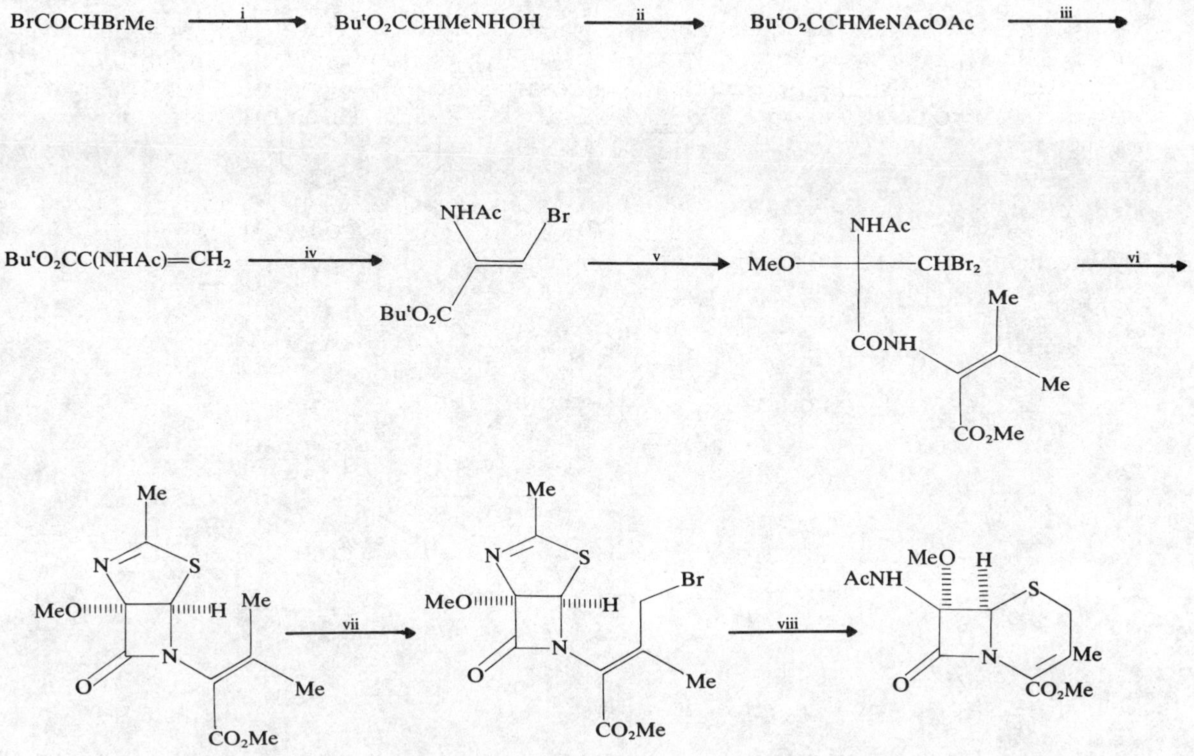

i, Bu^tOH, C_5H_5N; NH_2OH; ii, Ac_2O, 100 °C; iii, NEt_3; iv, Br_2; NEt_3; v, Br_2, MeOH; $C_6H_{11}N{=}C{=}NC_6H_{11}$, $MeO_2CCH(NH_2)C(SH)Me_2$; *p*-$MeC_6H_4SO_2Cl$, C_5H_5N; Et_3N; vi, P_2S_5; NaH; vii, *N*-bromosuccinimide; viii, room temp. for 3 days.

SCHEME 24

Thienamycin (**34**) was isolated from *Streptomyces cattleya* and is active against both gram-positive and gram-negative bacteria.[57] Its instability in aqueous solution, however, may mitigate against its clinical use.

Clavulanic acid (**35**), isolated from *Streptomyces clavuligerus*, was recognized as a β-lactamase inhibitor, and can be used as such in conjunction with penicillins.[58] It does show antibacterial activity itself even though it contains no side-chain at all.[59] These new β-lactam antibiotics indicate that our understanding of the structure–activity relationship in this area is not yet complete and that greater structural variation may be possible than has hitherto been considered likely.

β-Lactams are known to occur in several other natural products, including the antibiotics bleomycin and phleomycin, but the β-lactam is the only feature in common with the penicillins and cephalosporins and their site of action is almost certainly different.[60] For this reason they have not been included.

There seems little doubt that the β-lactam antibiotics will remain an active area of research for some time. Certainly a better understanding of the relationship between structure and antibacterial activity would greatly enhance the possibility of synthesizing powerful new antibiotics. Even though the armory of antibacterial agents may at the present time be impressive, bacteria are versatile creatures and their ability to become resistant yet again to existing antibiotics must not be overlooked.

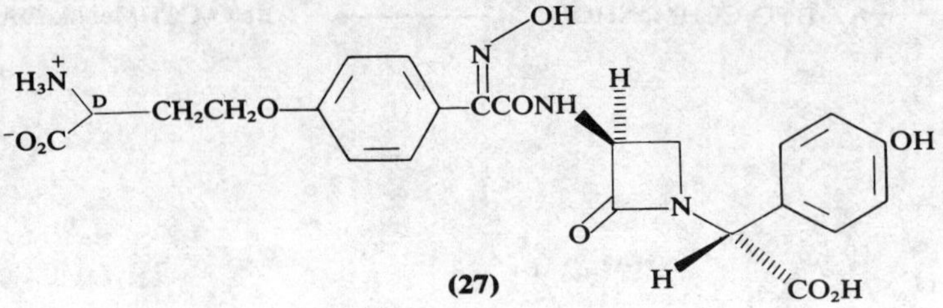

(27)

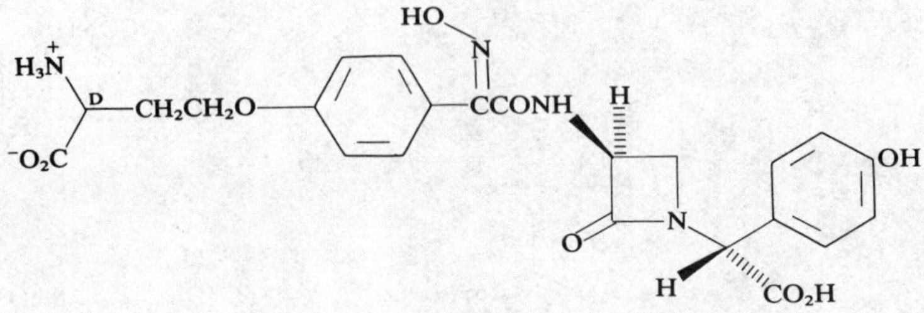

(28)

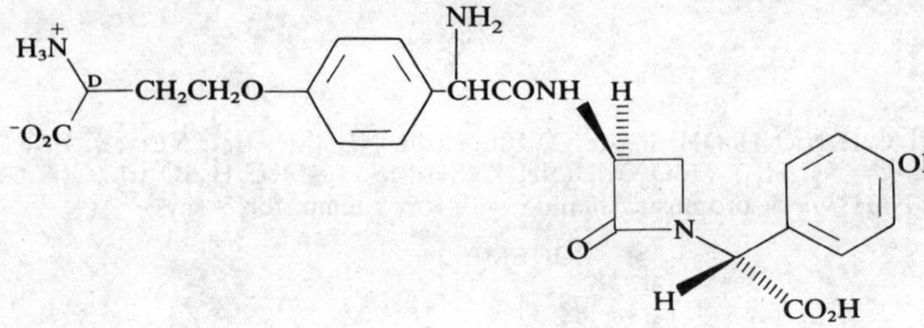

(29)

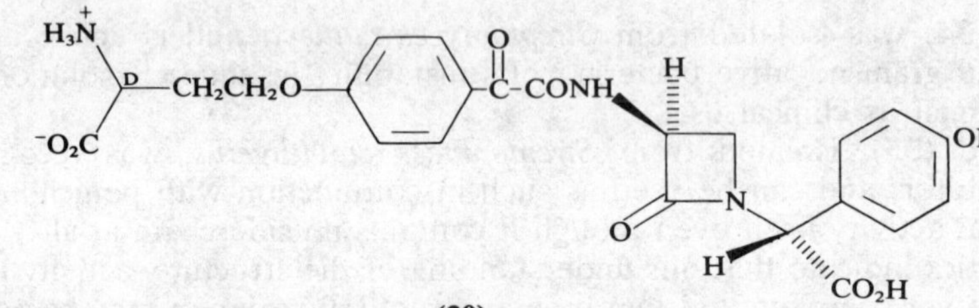

(30)

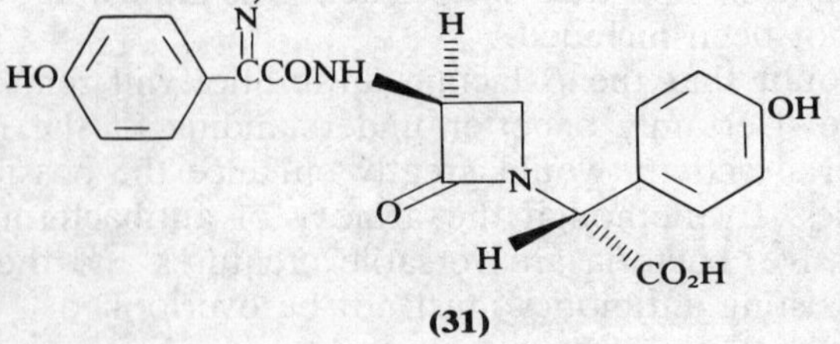

(31)

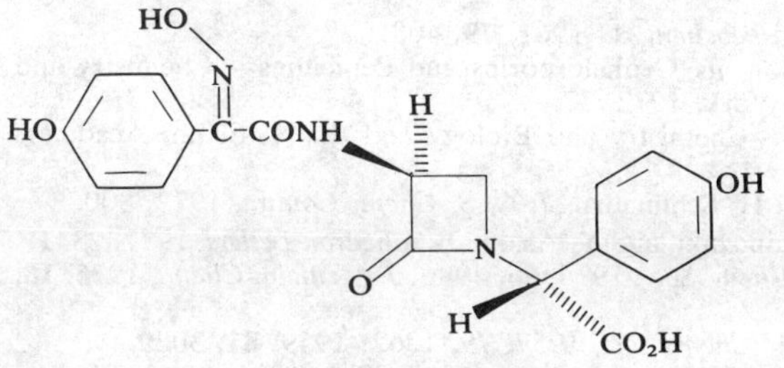

(32)

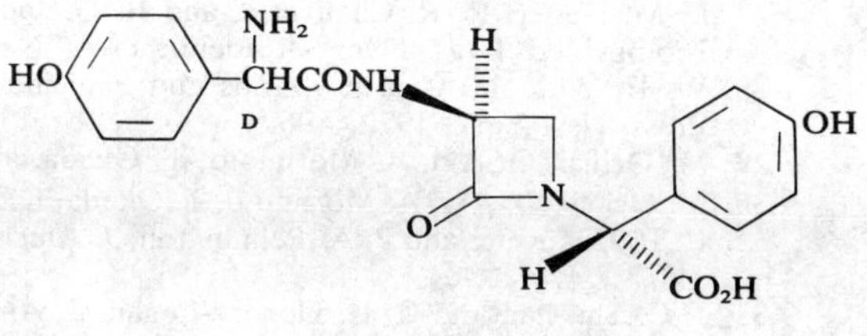

(33)

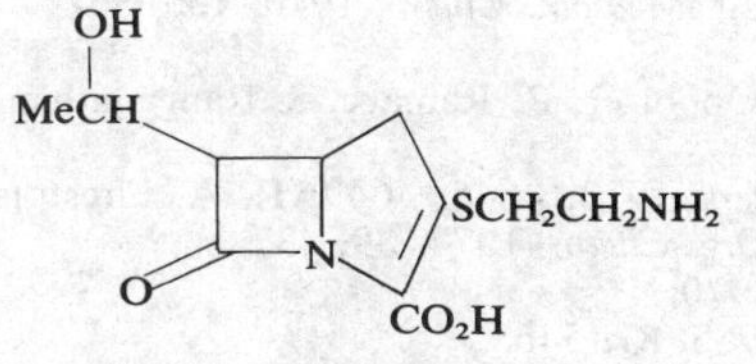

(34)

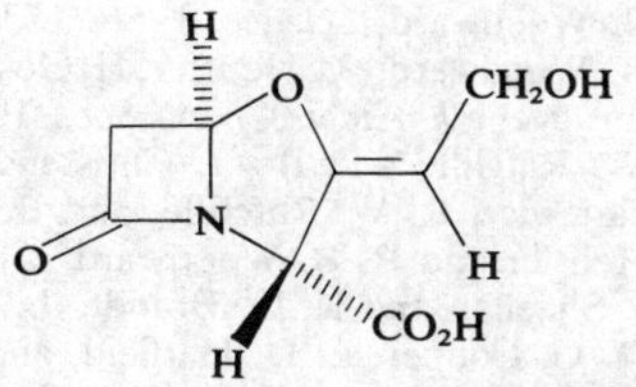

(35)

References

1. A. Fleming, *Brit. J. Exp. Pathol.*, 1929, **10,** 226.
2. H. W. Florey, E. Chain, N. G. Heatley, M. A. Jennings, A. G. Sanders, E. P. Abraham, and M. E. Florey, 'Antibiotics', Oxford Medical Publications, 1949, vol. I and II.
3. 'The Chemistry of Penicillin', ed. H. T. Clarke, J. R. Johnson, and R. Robinson, Princeton University Press, 1949.
4. G. Brotzu, Lavori dell'instituto D'lgiene di Cagliari, 1948.
5. E. P. Abraham and P. B. Loder, 'Cephalosporins and Penicillins — Chemistry and Biology', ed. E. H. Flynn, Academic Press, New York, 1972, 1.
6. P. A. Lemke and D. R. Brannon, 'Cephalosporins and Penicillins — Chemistry and Biology', ed. E. H. Flynn, Academic Press, New York, 1972, 370.
7. G. G. F. Newton and E. P. Abraham, *Biochem. J.*, 1954, **58,** 103; *Nature*, 1955, **175,** 548.
8. E. P. Abraham and G. G. F. Newton, *Biochem. J.*, 1961, **79,** 377.
9. D. C. Hodgkin and E. N. Maslen, *Biochem. J.*, 1961, **79,** 393.
10. H. R. V. Arnstein and D. Morris, *Biochem. J.*, 1960, **76,** 357.
11. P. B. Loder and E. P. Abraham, *Biochem. J.*, 1971, **123,** 471.
12. P. A. Fawcett, J. J. Usher, and E. P. Abraham, 'Proc. 2nd Internat. Symp. Genetics Industrial Microorganisms', ed. K. D. MacDonald, Academic Press, New York, 1976, 129.
13. P. A. Fawcett, J. J. Usher, J. A. Huddleston, R. C. Bleaney, J. J. Nisbet, and E. P. Abraham, *Biochem. J.*, 1976, **157,** 651.
14. B. W. Bycroft, C. M. Wels, K. Corbett, and D. A. Lowe, *J. C. S. Chem. Comm.*, 1975, 123.
15. D. J. Morecombe and D. W. Young, *J. C. S. Chem. Comm.*, 1975, 198.
16. F.-C. Huang, J. A. Chan, C. J. Sih, P. Fawcett, and E. P. Abraham, *J. Amer. Chem. Soc.*, 1975, **97,** 3858.
17. H. Kluender, F.-C. Huang, A. Fritzberg, H. Schnoes, C. J. Sih, P. Fawcett, and E. P. Abraham, *J. Amer. Chem. Soc.*, 1974, **96,** 4054.
18. M. Kohsaka and A. L. Demain, *Biochem. Biophys. Res. Comm.*, 1976, **70,** 465.
19. J. L. Strominger, *The Harvey Lectures*, 1970, **64,** 179.
20. J. L. Strominger, P. M. Blumberg, H. Suginaka, J. Umbreit, and G. G. Wickus, *Proc. Roy. Soc.*, 1971, **B179,** 369.
21. D. J. Tipper and J. L. Strominger, *Proc. Nat. Acad. Sci, U.S.A.*, 1965, **54,** 1133.

21a. J.-M. Frere, J-M. Ghuysen, J. Degelaen, A. Loffet, and H. R. Perkins, *Nature*, 1975, **258,** 168; *S.* Hammarstrom and J. L. Strominger, *Proc. Nat. Acad. Sci. U.S.A.*, 1975, **72,** 3463; *J. Biol. Chem.*, 1976, **251,** 7947.

22. R. M. Sweet, in 'Cephalosporins and Penicillins — Chemistry and Biology', ed. E. H. Flynn, Academic Press, New York, 1972, 280.
23. F. P. Doyle and J. H. C. Nayler, 'Advances in Drug Research', ed. N. J. Harper and A. B. Simmonds, Academic Press, New York, 1964, vol. 1, p. 1.
24. G. T. Stewart, 'The Penicillin Group of Drugs', Elsevier, Amsterdam, 1965.
25. M. Gorman and C. W. Ryan, in 'Cephalosporins and Penicillins — Chemistry and Biology', ed. E. H. Flynn, Academic Press, New York, 1972, 532.

26. B. Loder, G. G. F. Newton, and E. P. Abraham, *Biochem. J.*, 1961, **79,** 408.
27. F. M. Huber, R. R. Chauvette, and B. G. Jackson, in 'Cephalosporins and Penicillins — Chemistry and Biology', ed. E. H. Flynn, Academic Press, New York, 1972, 27.
28. W. E. Wick, in 'Cephalosporins and Penicillins — Chemistry and Biology', ed. E. H. Flynn, Academic Press, New York, 1972, 496.
29. M. Ochiai, O. Aki, A. Morimoto, T. Okada, and H. Schimadzu, *J. C. S. Chem. Comm.*, 1972, 800.
30. M. Ochiai, O. Aki, A. Morimoto, T. Okada, K. Shinozaki, and Y. Asahi, *Tetrahedron Letters*, 1972, 2341.
31. R. R. Chauvette and P. A. Pennington, *J. Amer. Chem. Soc.*, 1974, **96,** 4986; *J. Medicin. Chem.*, 1975, **18,** 496.
32. J. C. Sheehan and K. R. Henery-Logan, *J. Amer. Chem. Soc.*, 1957, **79,** 1262; 1959, **81,** 3089.
33. J. E. Baldwin, M. A. Christie, S. B. Haber, and L. I. Kruse, *J. Amer. Chem. Soc.*, 1976, **98,** 3045.
34. R. Heymes, G. Amiard, and G. Nomine, *Compt. rend.(C)*, 1966, **263,** 170.
35. G. Nomine, *Chim. Therapeut*, 1971, **6,** 53.
36. S. L. Neidleman, S. C. Pan, J. A. Last, and J. E. Dolfini, *J. Medicin. Chem.*, 1970, **13,** 386.
37. R. B. Woodward, *Science*, 1966, **153,** 487.
38. R. B. Woodward, K. Heusler, H. Gosteli, P. Naegeli, W. Oppolzer, R. Ramage, S. Ranganathan, and H. Vorbruggen, *J. Amer. Chem. Soc.*, 1966, **88,** 852.
39. R. W. Ratcliffe and B. G. Christensen, *Tetrahedron Letters*, 1973, 4645, 4649; R. A. Firestone, N. S. Maciejewicz, R. W. Ratcliffe, and B. G. Christensen, *J. Org. Chem.*, 1974, **39,** 437.
40. K. Heusler and R. B. Woodward, *Ger. Offer.* 1 935 607/1970.
41. J. C. Sheehan and K. G. Brandt, *J. Amer. Chem. Soc.*, 1965, **87,** 5468.
42. R. D. G. Cooper, L. D. Hatfield, and D. O. Spry, *Accounts Chem. Res.* 1973, **6,** 32.
43. R. D. G. Cooper and D. O. Spry, in 'Cephalosporins and Penicillins — Chemistry and Biology', ed. E. H. Flynn, Academic Press, New York, 1972, 183; R. D. G. Cooper, personal communication.
44. R. B. Morin, B. G. Jackson, R. A. Mueller, E. R. Lavagnino, W. B. Scanlon, and S. L. Andrews, *J. Amer. Chem. Soc.*, 1963, **85,** 1896.
45. R. R. Chauvette, P. A. Pennington, C. W. Ryan, R. D. G. Cooper, F. L. José, I. G. Wright, E. M. van Heyningen, and G. W. Huffman, *J. Org. Chem.*, 1971, **36,** 1259.
45a. J. J. de Koning, H. J. Kooreman, H. S. Tan, and J. Verweij, *J. Org. Chem.*, 1975, **40,** 1346.
45b. S. Kukolja, in 'Recent Advances in the Chemistry of β-Lactam Anitbiotics', ed. J. Elks, Special Publication No. 28, The Chemical Society, London, 1977, p. 181.
45c. G. A. Koppel, M. D. Kinnick, and L. J. Nummy, in Ref. 45b, p. 101.
45d. Y. Hamashima, K. Ishikura, H. Ishitobi, H. Itani, T. Kubota, K. Minami, M. Murakami, W. Nagata, M. Narisada, Y. Nishitani, T. Okada, H. Onoue, H. Satoh, Y. Sendo, T. Tsuji, and M. Yoshioka, in Ref. 45b, p. 243.
45e. R. D. G. Cooper and F. L José, *J. Amer. Chem. Soc.*, 1970, **92,** 2575.
46. K. Heusler, 'Cephalosporins and Pencillins — Chemistry and Biology', ed. E. H. Flynn, Academic Press, New York, 1972, p. 255.
47. G. Lowe, *Chem. and Ind. (London)*, 1975, 459.
48. L. D. Cama and B. G. Christensen, *J. Amer. Chem. Soc.*, 1974, **96,** 7582, 7584.
49. B. R. Belleau, T. W. Doyle, B. Y. Luh, and T. T. Conway, *Ger. Offen.* 2 600 039 (*Chem. Abs.*, 1977, **86,** 140 060).
50. B. G. Christensen and R. W. Ratcliffe, *Ger. Offen.*, 2 411 856/1974 (*Chem. Abs.*, 1975, **82,** 31 314); *Ann. Reports Medicin. Chem.*, 1976, **11,** 271.
51. R. B. Woodward, in 'Recent Advances in the Chemistry of β-Lactam Antibiotics', ed. J. Elks, Special Publication No. 28, The Chemical Society, London, 1977, p. 167.
52. R. Nagarajan, L. D. Boeck, M. Gorman, R. L. Hamill, C. E. Higgens, M. M. Hoehn, W. M. Stark, and J. G. Whitney, *J. Amer. Chem. Soc.*, 1971, **93,** 2308.
53. G. A. Koppel and R. E. Koehler, *J. Amer. Chem. Soc.*, 1973, **95,** 2403.
54. W. A. Slusarchyk, H. E. Applegate, P. Funke, W. Koster, M. S. Puer, M. Young, and J. E. Dolfini, *J. Org. Chem.*, 1973, **38,** 943.
55. S.-I. Nakatsuka, H. Tanino, and Y. Kishi, *J. Amer. Chem. Soc.*, 1975, **97,** 5008.
56. H. Aoki, H. Sakai, M. Kohsaka, T. Konomi, J. Hosoda, T. Kubochi, E. Iguchi, and H. Imanaka, *J. Antibiotics*, 1976, **29,** 492; M. Hashimoto, T.-A. Komori, and T. Kamiya, *J. Amer. Chem. Soc.*, 1976, **98,** 3028.
56a. T. Kamiya, in 'Recent Advances in the Chemistry of β-Lactam Anitbiotics', ed. J. Elks, Special Publication No. 28, The Chemical Society, London, 1977, p. 281.
57. J. S. Kahan, F. M. Kahan, E. O. Stapley, R. T. Georgeiman, and S. Hernandez, *U.S. Pat.* 3 950 357/1976.
58. T. T. Howarth, A. G. Brown, and T. J. King, *J. C. S. Chem. Comm.*, 1976, 266; A. G. Brown, J. Goodacre, J. B. Harbridge, T. T. Howarth, R. J. Ponsford, I. Stirling, and T. J. King, in Ref. 56a, p. 295.
59. I. D. Fleming, D. Noble, H. M. Noble, and W. F. Wall, *Ger. Offen.*, 2 604 697/1976.
60. H. Umezawa, 'Antibiotics', ed. J. W. Corcoran and F. E. Hahn, Springer-Verlag, Berlin, 1975, vol. III, p. 21.

23.6

Peptide Synthesis

R. C. SHEPPARD
MRC Laboratory of Molecular Biology, Cambridge

23.6.1 INTRODUCTION[1,2]

Formally, peptide synthesis consists simply of the elimination of water from two amino-acids (or peptides) (equation 1):

$$\overset{+}{N}H_3CHR^1CO_2^- + \overset{+}{N}H_3CHR^2CO_2^- \longrightarrow \overset{+}{N}H_3CHR^1CONHCHR^2CO_2^- + H_2O \quad (1)$$

Largely because of the dipolar nature of amino-acids, this reaction is thermodynamically unfavourable and would require unacceptably high temperatures. In any case, it would lead to complete ambiguity about the sequence of amino-acids in the product and to cyclization and polycondensation reactions. Rational peptide synthesis requires (a) destruction of the dipolar character of the reacting amino-acids; (b) differentiation of the *amino-component* (the amino-acid contributing its nitrogen atom to the newly formed peptide bond), and the *carboxy-component* which contributes its carbonyl group; and (c) further activation of the latter so that the coupling reaction can be achieved with high efficiency under very mild conditions.

Requirements (a) and (b) are usually met by the use of reversible protecting groups. The basic peptide synthesis (equation 2) is followed by removal of X (the *N*- or

$$\left.\begin{array}{l} \overset{+}{N}H_3CHR^1CO_2^- \longrightarrow XNHCHR^1CO_2H \\ \qquad\qquad\qquad\qquad \text{Carboxy-component} \\ \\ \overset{+}{N}H_3CHR^2CO_2^- \longrightarrow NH_2CHR^2CO_2Y \\ \qquad\qquad\qquad\qquad \text{Amino-component} \end{array}\right\} \longrightarrow XNHCHR^1CONHCHR^2CO_2Y \quad (2)$$

amino-protecting group, Section 23.6.2.1) and Y (the *C*- or carboxy-protecting group, Section 23.6.2.2). The amino- and carboxy-protecting groups should also be capable of cleavage independently of each other (and from any other protecting groups present in polyfunctional amino-acid side chains), enabling further lengthening of the peptide at either end. The amino-protecting group functions by suppressing the basicity and nucleophilicity of the nitrogen atom, preventing protonation and acylation and liberating the α-carboxy-group from its internal salt form for subsequent activation. Strictly speaking, a covalently linked carboxy-protecting group Y is not necessary, and metal or other salts of amino-acids (*e.g.* $NH_2CHRCO_2^-Na^+$) can sometimes be used as amino-components. However, solubility and isolation problems are reduced and peptide bond-forming reactions made more efficient by use of amino-acid ester or other only weakly reactive carboxylic acid derivatives as amino-components. This also allows greater flexibility in the choice of coupling procedures.

In principle, carboxy activation may be by many of the species commonly used in acylation reactions in general organic chemistry (acid chlorides, anhydrides, *etc.*). However, amino-acids and peptides are polyfunctional compounds, and unwanted reactivities are always present to some extent, even in fully 'protected' derivatives. These reactivities are often enhanced by the particular steric relationships between side-chain or terminal functional groups and intra-chain peptide bonds. β-Elimination and five-ring cyclization reactions, for example, may occur with unexpected facility. It is chiefly for this reason that peptide synthesis is such a demanding technical operation, for strict adherence to optimal reaction conditions is necessary in order to minimize the intervention of these side reactions. An important example of the latter is the formation of oxazolones (**1**) through reaction between activated terminal carboxy-groups and the adjacent peptide bond (equation 3).

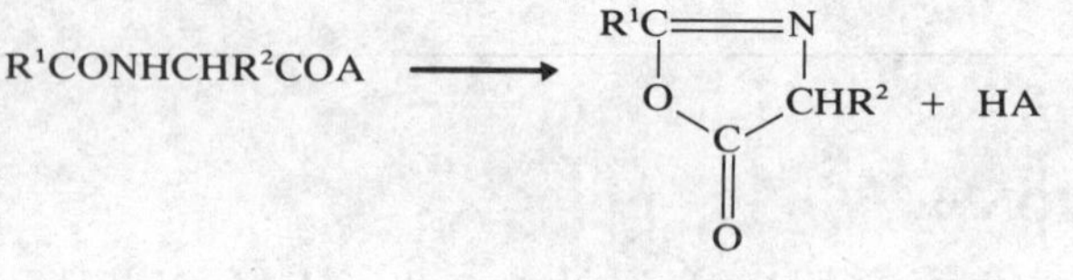

(3)

(**1**)

This is a particularly serious side-reaction because it usually results in racemization of the chiral α-centre. Racemization of individual amino-acid residues in polypeptides results in mixtures of diastereoisomers which often defy separation. Furthermore, the biological properties of peptides, the generation of which provides the most common motivation for peptide synthesis, are usually critically dependent upon the correct stereochemistry. Although there are a number of possible mechanisms for the racemization of α-amino-acid derivatives, it is probable that in the absence of special side-chain or *N*-substituent effects, oxazolone formation provides by far the most important pathway.[3] The fleeting optical activity of oxazolones (**1**) is usually ascribed to resonance stabilization

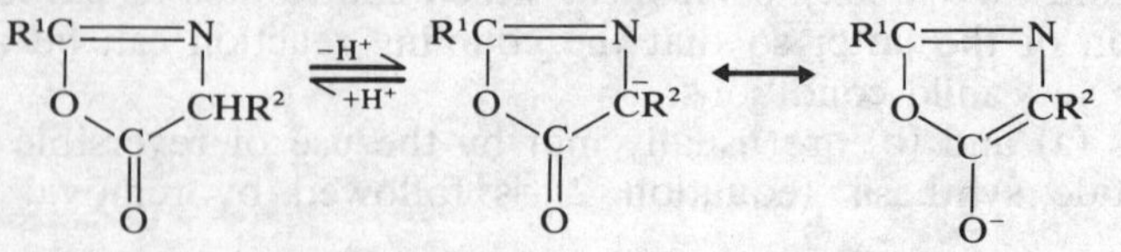

(**1**)

of the anion and activation procedures must therefore be chosen with great care to minimize the possibility of oxazolone formation. The formation of oxazolones is strongly influenced by the nature of the *N*-acyl substituent, as well as by solvent and base strength effects. All of these factors need to be taken into account in designing a practical peptide synthesis.

Activation of the amino-component for peptide bond formation is also conceptually possible. It is likely, however, that in procedures of this type, the activation is initially transferred to the carboxy-function. A straight-forward example[4] is the reaction between a carboxylic acid and an amino-component activated as its isocyanate derivative (Scheme 1). Formation of the peptide bond actually takes place by intramolecular rearrangement

$$\underset{\text{Amino-component}}{NH_2CHR^2CO_2Y} \xrightarrow{\text{i}} O{=}C{=}NCHR^2CO_2Y \xrightarrow{\text{ii}} \underset{N\text{-carboxy-anhydride}}{XNHCHR^1COOCONHCH_2CO_2Y}$$

$$\longrightarrow XNHCHR^1CO\underset{\displaystyle CO_2H}{\underset{|}{N}}CHR^2CO_2Y \longrightarrow XNHCHR^1CONHCHR^2CO_2Y$$

i, $COCl_2$; ii, carboxy-component, $XHNCHR^1CO_2H$.

SCHEME 1

of the intermiedate *N*-carboxy-anhydride.[5] Similar mechanisms may be written for other procedures in which the amino-component is initially reacted with an activating agent, for example in the reaction with phosphorus trichloride.[6]

23.6.2 PROTECTING GROUPS

23.6.2.1 Amino-protecting groups

Many factors are important in the choice or design of amino-protecting groups. They include ease of introduction into the individual amino-acids, adequate protection of the amino-function, stability under the conditions of peptide synthesis, protection of the adjacent chiral centre (in α-amino-acids other than glycine) from racemization, and ease of removal during, or at the completion, of the synthesis. Because amino-groups are also present in the side chains of certain polyfunctional amino-acids (lysine, ornithine), and because of the demands for coupling protected peptide derivatives as well as amino-acids, there is need for a range of protecting groups capable of selective removal. Most commonly this has been achieved through protecting groups of graded acid lability or by a combination of groups removed respectively by acidic and other (*e.g.* basic) reagents.

The amino-protecting group functions by reducing the nucleophilicity of the nitrogen atom. This can be achieved by conjugative electron withdrawal as in acyl derivatives, especially urethanes (**2**), by inductive electron withdrawal, as in *o*-nitrophenylsulphenyl derivatives (**3**), or by steric hindrance, as in triphenylmethyl (trityl) derivatives (**4**). Usually a combination of these factors is operative.

ROCONH—

(2)

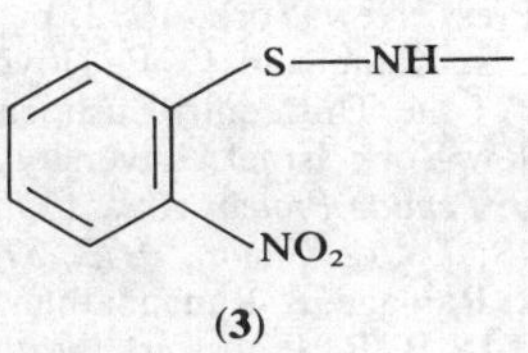

(3)

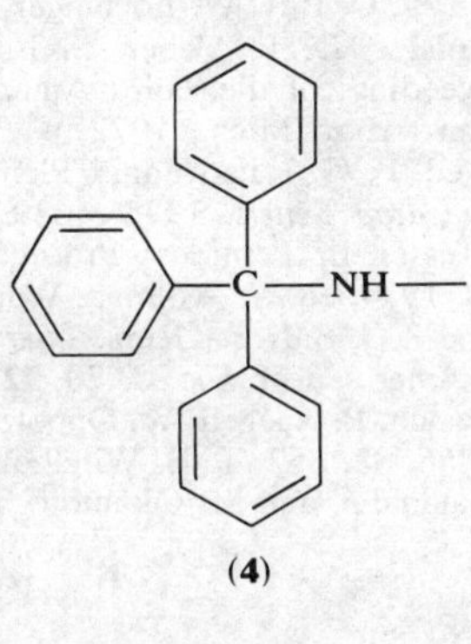

(4)

(i) Urethane protecting groups

Urethane-type amino-protecting groups constitute by far the most important class of amino-protecting groups; selected examples are collected in Table 1. The urethane structure in (**5**) provides a convenient and flexible way of protecting amino-functions as ester derivatives, since the corresponding carbamic acids (**6**) decarboxylate spontaneously under neutral or acidic conditions (equation 4).

TABLE 1
Urethane-type Protecting Groups

	Name	*Abbrev.*	*Typical cleavage reagents*	*Ref.*
1.	Benzyloxycarbonyl	Z	H_2, Pd; HBr, AcOH; HF	a
2.	*p*-Chlorobenzyloxycarbonyl	Z(Cl)	H_2, Pd; HBr, AcOH; HF	b
3.	*p*-Nitrobenzyloxycarbonyl	$Z(NO_2)$	H_2, Pd; HBr, AcOH; HF	c
4.	*p*-Phenylazobenzyloxycarbonyl	Pz	H_2, Pd; HBr, AcOH; HF	d
5.	*p*-Methoxybenzyloxycarbonyl	Z(OMe)	H_2, Pd; CF_3CO_2H; HCl, AcOH	e
6.	3,5-Dimethoxybenzyloxycarbonyl	$Z(OMe)_2$	H_2, Pd; CF_3CO_2H; photolysis	f
7.	t-Butoxycarbonyl	Boc	CF_3CO_2H; HCl, AcOH	g
8.	t-Amyloxycarbonyl	Aoc	CF_3CO_2H; HCl, AcOH	h
9.	Adamantyloxycarbonyl	Adoc	CF_3CO_2H	i
10.	Isobornyloxycarbonyl	Bornoc	CF_3CO_2H	j
11.	Biphenylisopropoxycarbonyl	Bpoc	AcOH	k
12.	3,5-Dimethoxy-α,α-dimethyl-benzyloxycarbonyl	Ddz	AcOH; photolysis	l
13.	Isonicotinyloxycarbonyl	iNoc	Zn, AcOH; H_2, Pd	m
14.	*p*-Dihydroxyborylbenzyloxycarbonyl	Dobz	H_2O_2, pH 9.5	n
15.	5-Benzisoxazolylmethoxycarbonyl	Bic	NEt_3	o
16.	β-Methylsulphonylethoxycarbonyl	Msc	NaOH	p
17.	9-Fluorenylmethoxycarbonyl	Fmoc	Morpholine; liq. NH_3	q
18.	β,β,β-Trichloroethoxycarbonyl	—	Zn, AcOH	r

[a] M. Bergmann and L. Zervas, *Ber.*, 1932, **65,** 1192. [b] R. A. Boissonnas and G. Preitner, *Helv. Chim. Acta*, 1953, **36,** 875; L. Kisfaludy and S. Bualszky, *Acta Chim. Acad. Sci. Hung.*, 1960, **24,** 301. [c] F. H. Carpenter and D. T. Gish, *J. Amer. Chem. Soc.*, 1952, **74,** 3818; D. T. Gish and F. H. Carpenter, *ibid.*, 1953, **75,** 950. [d] R. Schwyzer, P. Sieber, and K. Zatsko, *Helv. Chim. Acta*, 1958, **41,** 491. [e] F. C. McKay and N. F. Albertson, *J. Amer. Chem. Soc.*, 1957, **79,** 4686; F. Weygand and K. Hunger, *Chem. Ber.*, 1962, **95,** 1. [f] J. W. Chamberlin, *J. Org. Chem.*, 1966, **31,** 1658. [g] L. A. Carpino, *J. Amer. Chem. Soc.*, 1957, **79,** 98; F. C. McKay and N. F. Albertson, *ibid.*, 1957, **79,** 4686; G. W. Anderson and A. C. McGregor, *ibid.*, 1957, **79,** 6180. [h] S. Sakakibara, M. Shin, M. Fujino, Y. Shimonishi, S. Inouye, and N. Inukai, *Bull. Chem. Soc. Japan*, 1965, **38,** 1522; S. Sakakibara and M. Fujino, *ibid.*, 1966, **39,** 947; S. Sakakibara and M. Itoh, *ibid.*, 1967, **40,** 646; I. Honda, Y. Shimonishi, and S. Sakakibara, *ibid.*, 1967, **40,** 2415; S. Sakakibara, I. Honda, and K. Takada, *ibid.*, 1969, **42,** 809. [i] W. L. Haas, E. V. Krumkalns, and K. Gerzon, *J. Amer. Chem. Soc.*, 1966, **88,** 1988. [j] G. Jäger and R. Geiger, in 'Peptides', Proceedings of the Eleventh European Peptide Symposium, Vienna, 1971, ed. H. Nesvadba, North-Holland, Amsterdam, 1973, p. 78; G. Jäger and R. Geiger, *Annalen*, 1973, 1535; M. Fujino, S. Shinagawa, O. Nishimura, and T. Fukuda, *Chem. Pharm. Bull.* (*Japan*), 1972, **20,** 1017. [k] P. Sieber and B. Iselin, *Helv. Chim. Acta*, 1968, **51,** 614, 622; S. S. Wang and R. B. Merrifield, *Internat. J. Protein Res.*, 1969, **1,** 235; R. S. Feinberg and R. B. Merrifield, *Tetrahedron*, 1972, **28,** 5865; E. Schnabel, G. Schmidt, and E. Klauke, *Annalen.*, 1971, **743,** 69. [l] C. Birr, F. Flor, P. Fleckenstein, and Th. Wieland, in 'Peptides', Proceedings of the Eleventh European Peptide Symposium, Vienna, 1971, ed. H. Nesvadba, North-Holland, Amsterdam, 1973, p. 175; C. Birr, W. Lochinger, G. Stahnke, and P. Lang, *Annalen.*, 1972, **763,** 162; C. Birr, *Ibid.*, 1973, 1652. [m] D. F. Veber, S. F. Brady, and R. Hirschmann, in Chemistry and Biology of Peptides', Proceedings of the Third American Peptide Symposium, Boston, 1972, Ann Arbor Science Publishing, Ann Arbor, Mich., 1972, p. 315; R. Hirschmann and D. F. Veber, in 'The Chemistry of Polypeptides', ed. P. G. Katsoyannis, Plenum Press, New York, 1973, p. 125. [n] D. S. Kemp and D. C. Roberts, *Tetrahedron Letters*, 1975, 4629. [o] D. S. Kemp and C. F. Hoyng, *Tetrahedron Letters*, 1975, 4625. [p] G. I. Tesser, in 'Peptides', Proceedings of the Thirteenth European Peptide Symposium, Kiryat Anavim, Israel, 1974, ed. Y. Wolman, Wiley, New York–Israel University Press, Jerusalem, 1975, p. 53; G. I. Tesser and I. C. Balvert-Geers, *Internat. J. Peptide Protein Res.*, 1975, **7,** 295. [q] L. A. Carpino and G. Y. Han, *J. Amer. Chem. Soc.*, 1970, **92,** 5748; *J. Org. Chem.*, 1972, **37,** 3404. [r] R. B. Woodward, K. Heusler, H. Gosteli, P. Naegeli, W. Oppolzer, R. Ramage, S. Ranganathan, and H. Vorbruggen, *J. Amer. Chem. Soc.*, 1966, **88,** 852; T. B. Windholz and D. B. R. Johnston, *Tetrahedron Letters*, 1967, 2555; H. Yajima, H. Watanabe, and M. Okamoto, *Chem. Pharm. Bull.* (*Japan*), 1971, **19,** 2185.

$$\underset{(\mathbf{5})}{R^1OCONHCHR^2CO\text{—}} \longrightarrow \underset{(\mathbf{6})}{HOCONHCHR^2CO\text{—}} \longrightarrow CO_2 + NH_2CHR^2CO\text{—} \quad (4)$$

Thus the alcohol R^1OH from which the urethane is derived may be chosen from a wide variety of structural types, provided that the urethane ester bond is then capable of cleavage to the free acid under conditions which are sufficiently mild to leave peptide bonds and the various side-chain functions present in (protected) polyfunctional amino-acids and peptides unaffected. Alkaline hydrolysis is not a suitable way of cleaving these ester bonds because urethanes are markedly resistant to attack by hydroxide ion, and intramolecular cyclization to hydantoin derivatives (**8**) may intervene (equation 5).

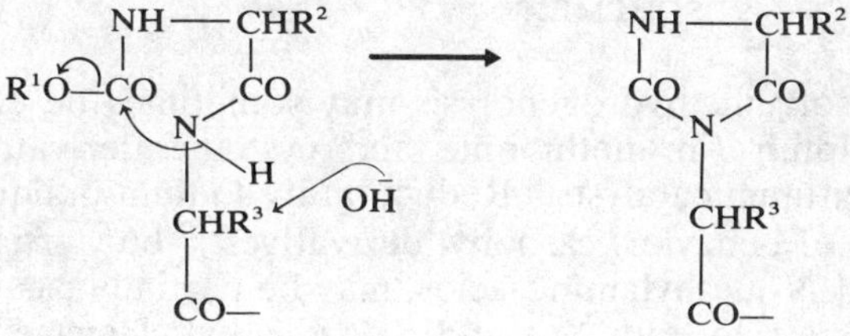

(**7**) (**8**) (5)

This side reaction is particularly facile when the penultimate amino-acid is glycine ($R^3 = H$ in **7**), otherwise urethane protecting groups with R = alkyl or aralkyl are generally unaffected by mildly basic reagents. *C*-Terminal peptide esters, for example, may often be saponified or converted to hydrazides in the presence of simple urethane amino-protecting groups.

Pre-eminent among urethane protecting groups are the benzyloxycarbonyl derivatives (**9**), the introduction of which by Bergmann and Zervas in 1932 initiated the present era of peptide synthesis.[7] They are readily prepared *via* the intermediate chloroformate (Scheme 2), and for most amino-acids are stable crystalline solids. A number of structural

$$PhCH_2OH \xrightarrow{\text{i}} PhCH_2OCOCl \xrightarrow{\text{ii}} \underset{(\mathbf{9})}{PhCH_2OCONHCHRCO_2H}$$

i, $COCl_2$; ii, $NH_2CHRCO_2^-Na^+$, controlled alkaline pH.

SCHEME 2

variants which may be employed to give more easily crystallized derivatives in difficult cases are included in Table 1.

One of the most important properties of the benzyloxycarbonyl group is the resistance it confers to the racemization of protected amino-acids. Thus benzyloxycarbonylamino-acids may be activated for peptide synthesis by any of a wide variety of methods (Section 23.6.3) with little risk of racemization of adjacent chiral centres. This property is apparently common to all urethane protected amino-acids (but not peptides) and is associated with a much reduced tendency to form oxazolones or oxazolinium salts of type (**10**) (Scheme 3). Although alkoxyoxazolones of this type are known, vigorous conditions are usually necessary for their formation and they then decompose to *N*-carboxyanhydrides. A typical example is the thermal decomposition of benzyloxy-carbonylamino-acid chlorides to cyclic *N*-carboxyanhydrides (**11**) (Scheme 3).[8]

Bergmann and Zervas originally used catalytic hydrogenolysis of the benzyl ester bond for removal of benzyloxycarbonyl groups, and this is still one of the mildest and most efficient deprotection procedures available to peptide chemists (equation 6).

$$PhCH_2OCONHCHRCONH\text{—} \xrightarrow{H_2,\,Pd} PhCH_3 + CO_2 + NH_2CHRCONH\text{—} \quad (6)$$

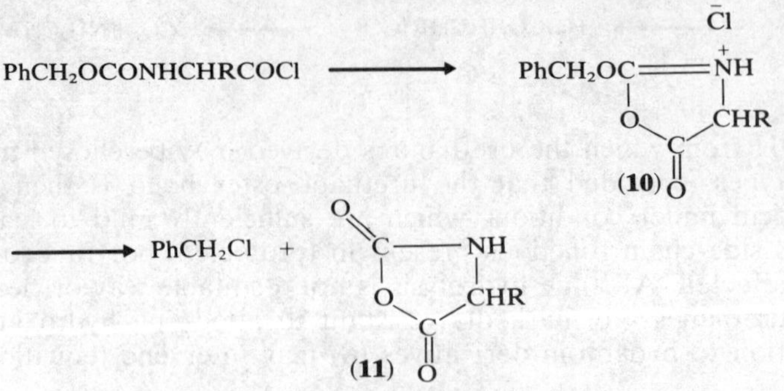

SCHEME 3

In liquid ammonia as solvent, hydrogenolysis may sometimes be effective even in the presence of divalent sulphur (in methionine and cysteine derivatives) which usually poisons palladium and platinum catalysts.[9] Reduction by sodium in liquid ammonia is also effective for the cleavage of benzyloxycarbonyl derivatives,[10] but tertiary amide bonds, as in peptides of proline and *N*-methylamino-acids, may be partially cleaved. Because of the susceptibility of benzyl esters to both S_N1 and S_N2 *O*-alkyl cleavage, benzyloxycarbonyl groups are also removed by acidolysis with strong anhydrous acids. Hydrogen bromide in acetic acid[11] and liquid hydrogen fluoride[12] are commonly used. Kinetic studies of the hydrogen bromide reaction show that both unimolecular (A) and bimolecular (B) decompositions of the *O*-protonated intermediate (**12**) are important (Scheme 4). There is the

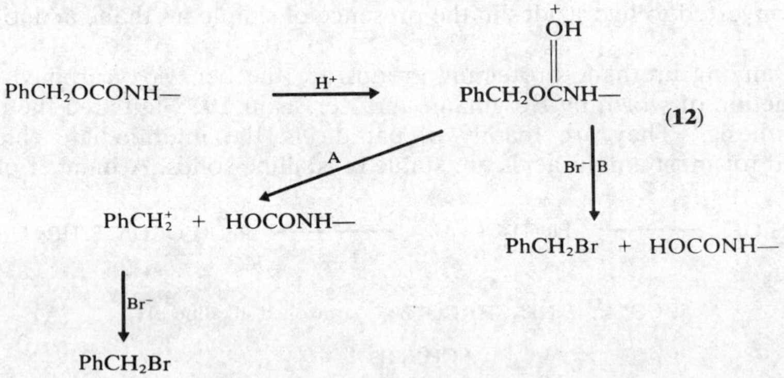

SCHEME 4

expected shift in mechanism towards the S_N1 process in the more acid labile *p*-methoxybenzyloxycarbonyl derivative, and towards the S_N2 process in the more stable *p*-nitro and chloro compounds. Pseudo-first-order rate constants for the cleavage of unsubstituted, *p*-methoxy- and *p*-nitro-benzyloxycarbonylglycines by 0.85M hydrogen bromide in acetic acid at 25 °C are 1.8×10^{-4}, 1.3×10^{-2}, and $2.6\times10^{-5}\ s^{-1}$, respectively.

p-Methoxybenzyloxycarbonyl derivatives are sufficiently labile to be cleaved by weaker acids, *e.g.* by trifluoroacetic acid. When cleavage reactions giving rise to carbonium ions are carried out with acidic reagents in which the anion is a relatively poor nucleophile, nucleophilic amino-acid side-chains (*e.g.* of tyrosine, tryptophan, and methionine) may compete for reaction with the liberated cation. Large excesses of 'scavenger' reagents (*e.g.* anisole, indole, or methyl ethyl sulphide) are commonly added to the reaction medium to avoid this complication.

The t-butoxycarbonyl amino-protecting group[13] is often used in a complementary manner to the benzyloxycarbonyl group. It is cleaved by trifluoroacetic acid and other acidic reagents to which benzyloxycarbonyl derivatives are stable, and is unaffected by catalytic hydrogenation (*cf.* *p*-methoxybenzyloxycarbonyl which has comparable acid lability). This enables benzyl derivatives to be used for side-chain protection and t-butyl derivatives for α-protection, and *vice versa.* It has come into special prominence because of its almost universal adoption in solid-phase peptide synthesis (Section 23.6.4). t-Butyl chloroformate is rather unstable and can be used for the preparation of t-butoxycarbonyl-amino-acids only at low temperatures. The corresponding azide (**13**), obtained by the action of nitrites on the carbazate (**14**), has most commonly been used[14] for the preparation of butoxycarbonyl derivatives, but a recent report indicates that this azide may be dangerously explosive.[15] Many other reagents [*e.g.* t-butyl *p*-nitrophenyl (**15**)[16] and t-butyl 2,4,5-trichlorophenyl (**16**)[17] carbonates, and t-butyl fluoroformate (**17**)[18] and the pyrocarbonate (**18**)][19] are also available. The last named is claimed to be an 'ideal' reagent.

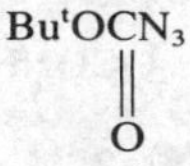

(**13**)

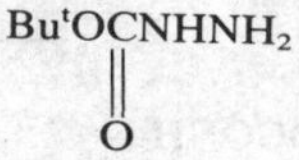

(**14**)

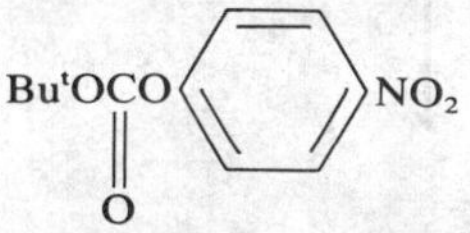

(**15**)

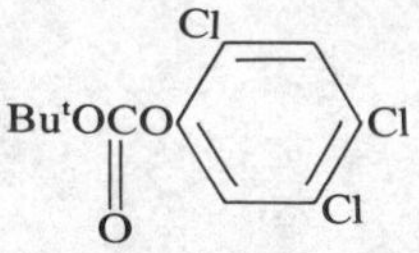

(**16**)

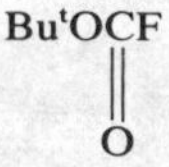

(**17**)

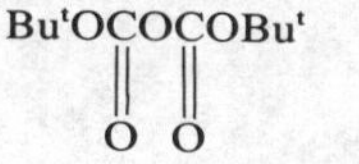

(**18**)

In the quest for milder and milder reaction conditions for de-protection reactions and also to provide increased selectivity in removal over benzyloxycarbonyl and t-butoxycarbonyl derivatives, urethane protecting groups of even greater acid lability have been developed. (See also trityl and *O*-nitrophenylsulphenyl derivatives, p. 329). Since t-butoxycarbonyl and (in part) benzyloxycarbonyl derivatives owe their acid lability to stabilization of the t-butyl and benzyl cations produced by S_N1 heterolysis, greater lability can be produced by combining the structural features of both. A range of urethane-type protecting groups of the general structure $ArCMe_2OCO$— have been devised.[20] They show quite extreme acid lability, since the cation which is initially produced on cleavage is both benzylic and tertiary. The *p*-phenyl substituted derivatives (**19**), for example, are cleaved by aqueous acetic acid at a rate some 10^4 times faster than the analogous t-butoxycarbonyl compounds. They can be introduced and removed selectively in the presence of both t-butyl- and benzyl-based protecting groups. The 3,5-dimethoxy-derivatives (**20**) have lability intermediate between (**19**) and t-butoxycarbonyl, but have the additional interesting property of cleavage by ultraviolet irradiation.[21]

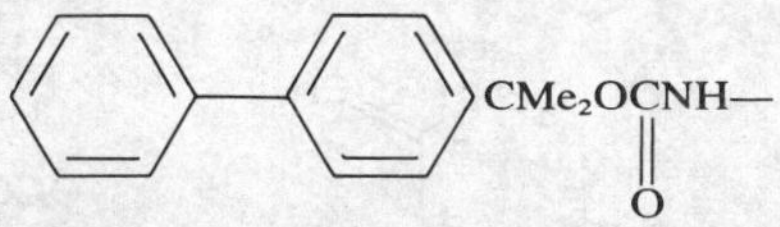

(**19**)

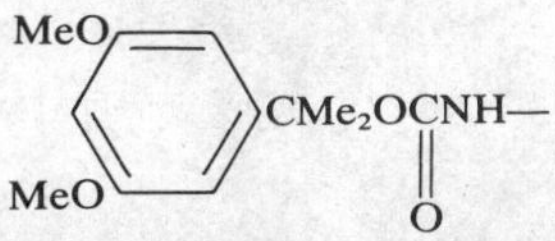

(**20**)

Considerable ingenuity has been used to devise urethane-type protecting groups which may be cleaved under specific, non-acidic, reaction conditions. Many of these are cleaved by base-catalysed elimination reactions. Simple examples are provided by methylsulphonylethoxycarbonyl (**21**)[22] and fluorenylmethoxycarbonyl derivatives (**23**).[23]

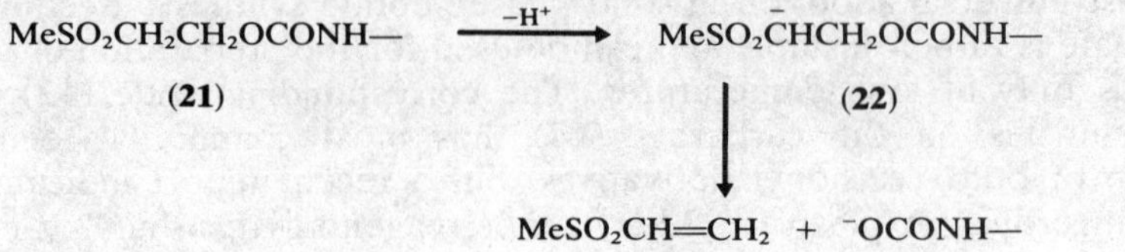

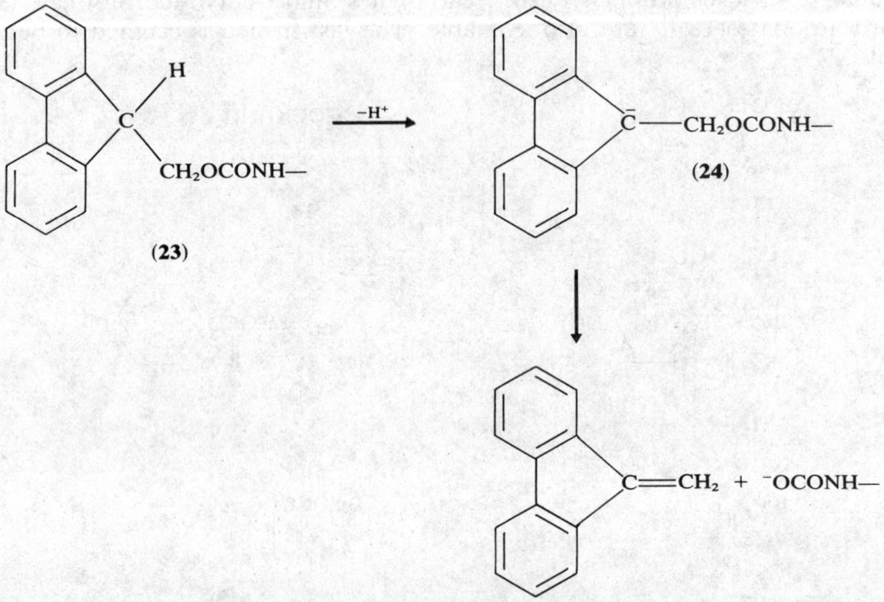

Derivatives of both types decompose by β-elimination processes, formation of the intermediate carbanions being facilitated by mesomeric and inductive factors in (**22**), and by delocalization of the negative charge over the whole aromatic dibenzocyclopentadienide system in (**24**). More complex elimination processes are operative in the recently introduced modified benzyloxycarbonyl derivatives (**25**)[24] and (**28**)[25]. The β-proton in benzisoxazole is particularly labile and is readily abstracted by tertiary base. Decomposition of (**25**) then takes place by way of the cyanophenolate anion (**26**) and quinone methide (**27**) (Scheme 5).[24]

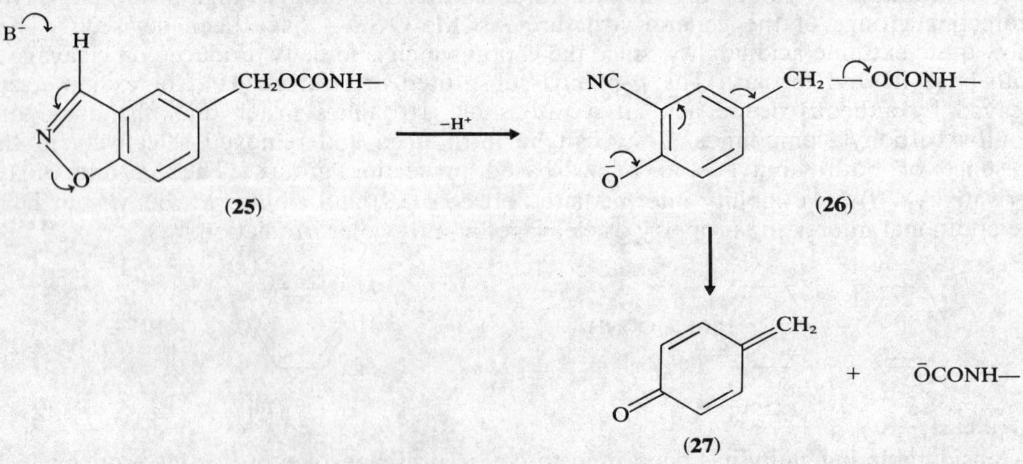

SCHEME 5

p-Dihydroxyborylbenzyloxycarbonyl derivatives (**28**) are cleaved oxidatively by hydrogen peroxide under mild conditions to the phenolate anion (**29**), and their decomposition presumably follows a similar pathway.[25]

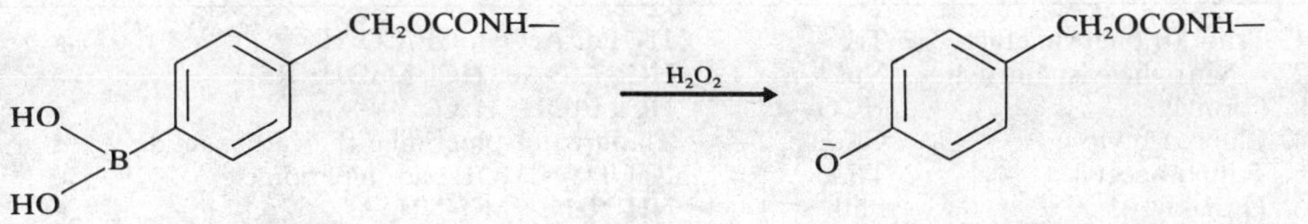

(**28**) (**29**)

A particularly interesting feature of this protecting group, and one that in general is likely to play an increasingly important role in peptide synthesis, is the special utility of the boronic acid substituent in manipulative and purification procedures. Thus peptide derivatives of the type (**28**) complex reversibly with diols, enabling for example, extraction into aqueous solution by chromotropic acid (**30**), or into non-polar organic media by lipophilic diols, *e.g.* *N*-octadecyldiethanolamine. The system also constitutes a potential ligand for affinity chromatography on diol-containing solid supports. The pyridine analogues (**31**) of benzyloxycarbonyl derivatives[26] and other benzyl-based protecting

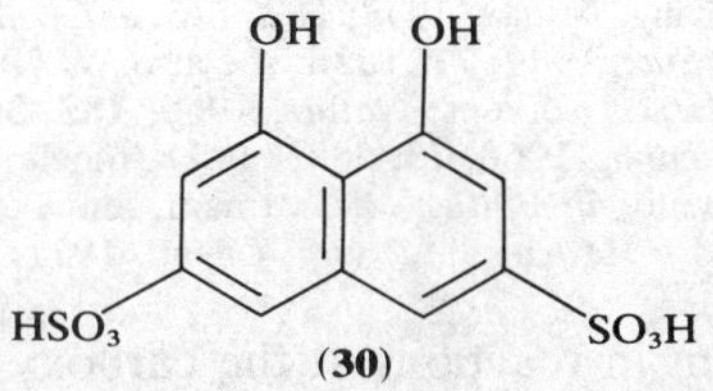

(**30**)

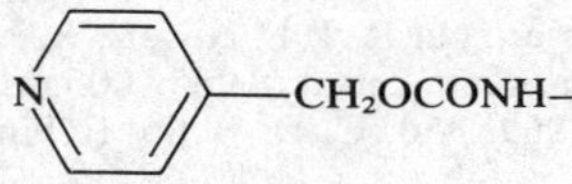

(**31**)

groups[27] also possess advantageous purification properties because of their weakly basic character. It should be noted that preferential protonation of the ring nitrogen effectively inhibits acid-catalysed cleavage of this protecting group. The susceptibility of all benzyloxycarbonyl-type derivatives to reductive cleavage, *e.g.* by catalytic hydrogenolysis, is, however, retained.

(*b*) *Other amino-protecting groups*

A number of protecting groups not involving urethane linkages have been devised and selected examples are shown in Table 2. The acid-labile trityl (triphenylmethyl) derivatives (**32**)[28] are prepared by way of trityl chloride. Their lability is due to the enhanced resonance stabilization of the triphenylmethyl cation as illustrated.

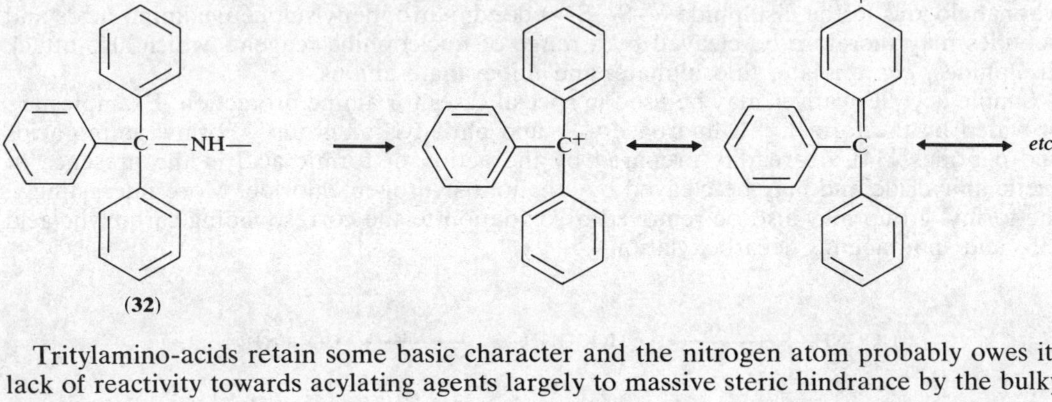

(**32**)

Tritylamino-acids retain some basic character and the nitrogen atom probably owes its lack of reactivity towards acylating agents largely to massive steric hindrance by the bulky

TABLE 2

Non-urethane Amino-protecting Groups

Name	*Abbrev.*	*Typical cleavage reagents*	*Ref.*
1. Trityl (triphenylmethyl)	Trt	H_2, Pd; AcOH; CF_3CO_2H	a
2. *o*-Nitrophenylsulphenyl	Nps	HCl, EtOAc; HCl, MeOH; RSH	b
3. Formyl	HCO—	HCl, EtOH; H_2O_2	c
4. Chloroacetyl	$ClCH_2CO$—	Thiourea; 1-piperidine thiocarboxamide	d
5. Trifluoroacetyl	TFA	NaOH; NH_4OH; aq. piperidine	e
6. Phthaloyl	Phth	NH_2NH_2; $PhNHNH_2$	f
7. Acetoacetyl	Acac	$PhNHNH_2$; NH_2OH	g

[a] A. Hillmann-Elies, G. Hillmann, and H. Jatzkewitz, *Z. Naturforsch.*, 1953, **8b,** 445; G. Amiard, R. Heymes, and L. Velluz, *Bull. Soc. chim. France*, 1955, 191, 201, 1464; *ibid.*, 1956, **97,** 698; L. Zervas and D. M. Theodoropoulos, *J. Amer. Chem. Soc.*, 1956, **78,** 1359. [b] L. Zervas, D. Borovas, and E. Gazis, *J. Amer. Chem. Soc.*, 1963, **85,** 3660; see also L. Zervas and C. Hamalidis, *ibid.*, 1965, **87,** 99; J. Goerdeler and A. Holst, *Angew. Chem.*, 1959, **71,** 775. [c] V. du Vigneaud, R. Dorfmann, and H. S. Loring, *J. Biol. Chem.*, 1932, **98,** 577; J. S. Fruton and H. T. Clarke, *ibid.*, 1934, **106,** 667; A. Stoll and Th. Petrzilka, *Helv. Chim. Acta*, 1952, **35,** 589; see also J. C. Sheehan, D. W. Chapman, and R. W. Roth, *J. Amer. Chem. Soc.*, 1952, **74,** 3822. [d] A. Fontana and E. Scoffone, *Gazzetta.*, 1968, **98,** 1261; C. Toniolo and A. Fontana, *ibid.*, 1969, **99,** 1017; M. Masaki, T. Kitahara, H. Kurita, and M. Ohta, *J. Amer. Chem. Soc.*, 1968, **90,** 4508. [e] F. Weygand and E. Csendes, *Angew. Chem.*, 1952, **64,** 136; F. Weygand, *Chem.-Ztg.*, 1954, **78,** 480. [f] D. A. A. Kidd and F. E. King, *Nature*, 1948, **162,** 776; *J. Chem. Soc.*, 1949, 3315; J. C. Sheehan and V. S. Frank, *J. Amer. Chem. Soc.*, 1949, **71,** 1856; see also W. Grassmann and E. Schulte-Uebbing, *Chem. Ber.*, 1950, **83,** 244; G. H. L. Nefkens, *Nature*, 1960, **185,** 309; G. H. L. Nefkens, G. I. Tesser, and R. J. F. Nivard, *Rec. Trav. chim.*, 1960, **79,** 688. [g] F. D'Angeli, F. Filira, and E. Scoffone, *Tetrahedron Letters*, 1965, 605; C. Di Bello, F. Filira, V. Giormani, and F. D'Angeli, *J. Chem. Soc.* (*C*), 1969, 350; C. Di Bello, F. Filira, and F. D'Angeli, *J. Org. Chem.*, 1971, **36,** 1818.

trityl group. This steric hindrance is also evident in reactions at the carboxy-group of *N*-tritylamino-acids, and for derivatives other than that of glycine, coupling reactions may be difficult. *o*-Nitrophenylsulphenyl derivatives (**33**)[29] are readily prepared by reaction between *o*-nitrophenylsulphenyl chloride and free amino-acids in alkaline media. This reaction is rapidly reversed by hydrogen chloride in anhydrous organic solvents, with cleavage of the protecting group.

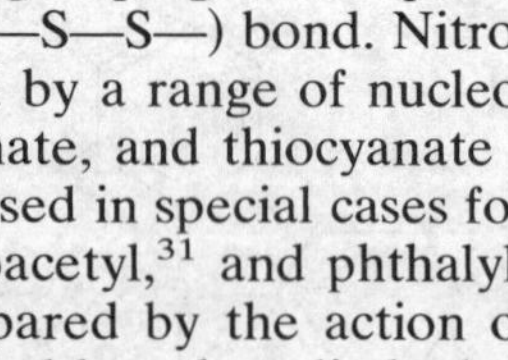

(**33**)

The sulphenamide (—S—NH—) grouping in nitrophenylsulphenyl derivatives is in some ways analogous to the disulphide (—S—S—) bond. Nitrophenylsulphenyl amino-acids and peptides may therefore be cleaved by a range of nucleophilic reagents which also attack disulphides, *e.g.* thiolate, thiosulphate, and thiocyanate anions.

Simple acyl derivatives may be used in special cases for amine protection. Examples are provided by the formyl,[30] trifluoroacetyl,[31] and phthalyl[32,33] groups. Formyl amino-acids and peptides (**34**) are readily prepared by the action of formic acid in the presence of acetic anhydride and may be cleaved by ethanolic hydrogen chloride. More interestingly, the formyl group may also be removed by oxidation to the corresponding carboxylic acid (**35**) and spontaneous decarboxylation.

$$\underset{(\mathbf{34})}{\text{HCONH—}} \xrightarrow{H_2O_2} \underset{(\mathbf{35})}{\text{HOCONH—}} \longrightarrow CO_2 + NH_2\text{—}$$

The trifluoroacetyl group (**36**) which is conveniently introduced into amino-acids by way of the thiol ester (**37**) owes its lability to the powerful electron-withdrawing effect of three fluorine atoms. Trifluoroacetylamino-acid peptide derivatives are readily cleaved by hydroxide ion and slowly by ethanolic hydrogen chloride. They are, however, stable to the anhydrous acidic conditions usually employed for the removal of t-butoxycarbonyl groups and may be used in combination with the latter. An important example of their use is for the protection of side-chain amino-groups, *e.g.* on lysine residues. The phthalyl protecting group (**38**) found early application in peptide chemistry. It may be conveniently introduced using the reagent *N*-carbethoxyphthalimide (**39**)[33] and is cleaved most commonly by the action of hydrazine or a substituted hydrazine. The powerful electron-attracting character of the phthalimide group favours direct ionization of the chiral α-hydrogen atom. Phthalylamino-acids are therefore very prone to racemization.

CF_3CONH—

(36)

CF_3COSEt

(37)

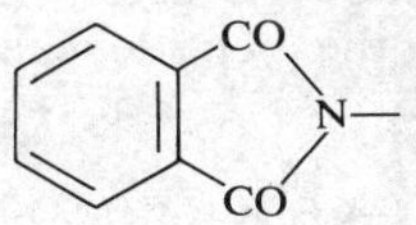

(38)

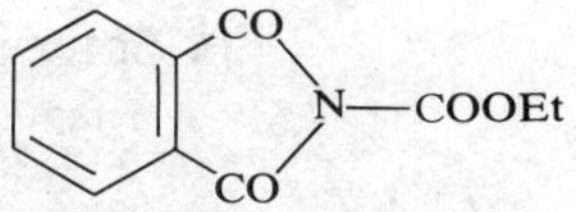

(39)

23.6.2.2 Carboxy-protecting groups

In so far as many of the commonly used amino-protecting groups are urethane derivatives, then there exists a corresponding set of simple esters which may function as carboxy-protecting groups. Thus benzyl esters (cleaved by hydrogenolysis or strong acids) and t-butyl esters (cleaved by milder acidic treatment) find wide application for protection of the *C*-terminal and side-chain carboxy-groups in amino-acid and peptide derivatives. Likewise, several ring-substituted benzyl and other esters analogous to urethanes listed in Table 1 have been used. Simple alkyl (methyl or ethyl) esters, with cleavage by saponification, find limited application for carboxy protection. Although *C*-terminal esters of peptides are intrinsically more electrophilic than normal aliphatic esters (owing to the electron-withdrawing character of the α-carboxamide substituent), the alkaline conditions necessary for their cleavage are too severe for all but the very simple peptide esters. They are likewise generally unsuitable for side-chain carboxy-group protection (p. 333) and the corresponding urethanes undergo intramolecular cyclization to hydantoin derivatives (p. 325) in preference to simple hydrolysis. Methyl and ethyl esters are, however, important intermediates in the preparation of *C*-terminal hydrazides for further peptide synthesis by the azide method (Section 23.6.3.4).

Phenyl esters are saponified under milder and less damaging alkaline conditions, although there is risk of racemization of optically active *C*-terminal residues presumably due to base-catalysed oxazolone formation. Their potential value has been much increased with the recognition that cleavage is enormously accelerated in the presence of hydroperoxide anion, which evidently functions as a very selective 'α-effect' nucleophile.[34] Phenyl esters have sufficient intrinsic reactivity towards amino nucleophiles to make the formation of diketo-piperazines (**40**) a serious side reaction in the conversion of dipeptide esters to tripeptide derivatives (*e.g.* equation 7). Synthetic routes utilising phenyl esters must be designed so as to avoid passing through the free dipeptide ester stage.

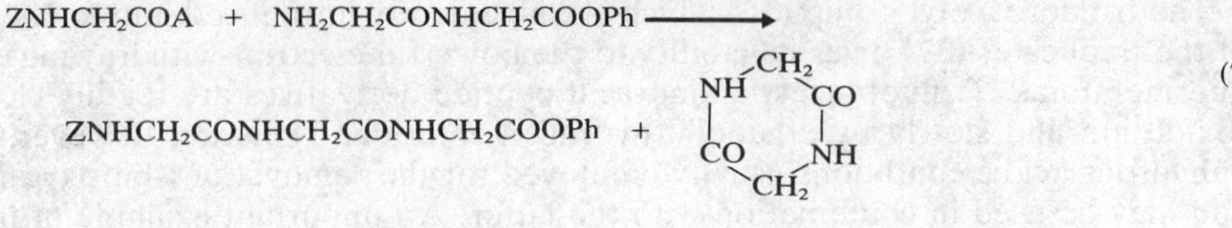

(7)

(40)

Esters which are cleaved by base-catalysed elimination reactions have been used for carboxy-group protection. Simple examples are provided by the sulphone (**41**) and sulphonium salt (**42**) derivatives, both of which are cleaved under mildly alkaline conditions by β-elimination (*cf.* **21**).[35] An interesting possibility is the formation of these esters at a late stage in the synthesis by oxidation or reaction with methyl iodide of the more stable methylthioethyl esters (**43**). In this way, basic conditions can be used during elaboration of the peptide chain with less risk of premature release of the terminal carboxy-group.

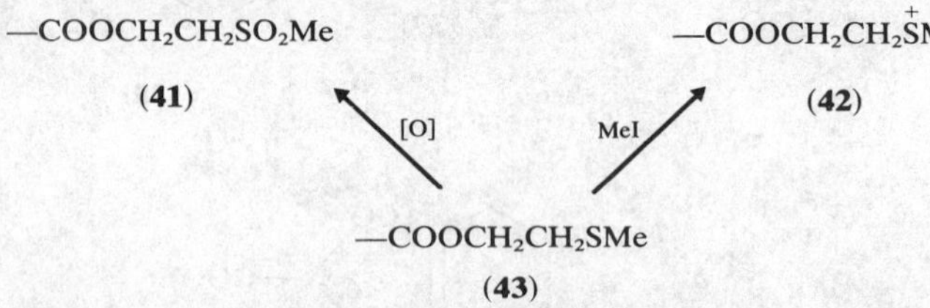

Esters of 2-hydroxymethyleneanthraquinol (**44**) undergo a more complex elimination process with liberation of the free carboxyl component (Scheme 6).[36] The fragmentation is initiated by reduction of the stable precursor anthraquinone ester (**45**) to the anthraquinol (**44**), which then undergoes elimination with tautomerism of the initially formed extended quinone methide (**46**) to the more stable 2-methylanthraquinone.

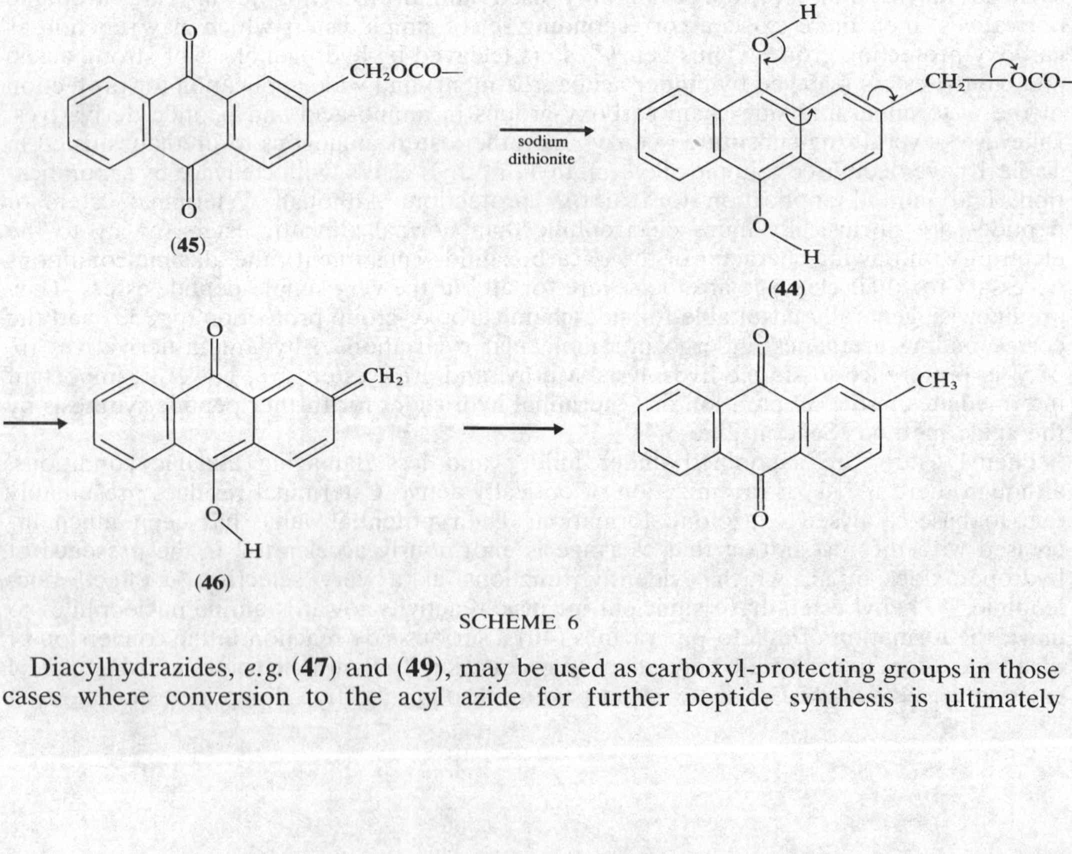

SCHEME 6

Diacylhydrazides, *e.g.* (**47**) and (**49**), may be used as carboxyl-protecting groups in those cases where conversion to the acyl azide for further peptide synthesis is ultimately required (p. 349). The monoacyl (peptide) hydrazide (**48**) is liberated under the usual conditions for cleavage of benzyloxycarbonyl[37] or t-butoxycarbonyl[38] groups from (**47**) and (**49**), respectively.

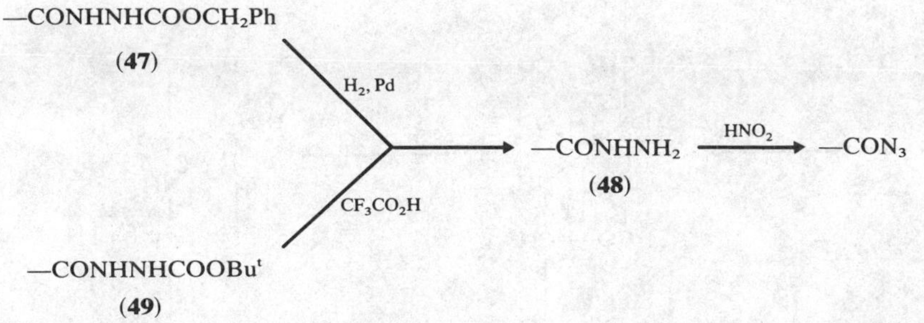

SCHEME 7

23.6.2.3 Side-chain protecting groups

The common or 'protein' amino-acids may be grouped according to the nature of their side chains. Those containing side-chain functionality, *e.g.* the acidic amino-acids aspartic and glutamic acid (carboxyl), the basic amino-acids lysine (amino), arginine (guanidino), and histidine (imidazole), as well as cysteine (thiol), and serine, threonine, and tyrosine (hydroxyl), may require protection depending on the coupling methods (Section 23.6.3) employed and the overall synthetic strategy (Section 23.6.5). Furthermore, for those amino-acids containing amino or carboxyl groups in their side chains, the synthetic problem requires efficient differentiation of these from the α-amino and carboxy groups in order that there should be no ambiguity about the backbone structure in the final product.

(*a*) *Lysine*

The ε-amino-group of lysine (α,ε-diaminohexanoic acid) may be easily brought into reaction selectively by way of the amino-acid copper salt (**50**). In the complex (**50**), the basicity of the α-amino-group is much reduced by coordination to the central copper atom. This enable specific acylation reactions at the free side-chain amino-group giving, for example, the ε-benzyloxycarbonyl derivative (**51**). After decomposition of the copper complex (**51**) with hydrogen sulphide or with a more powerful chelating agent (ethylenediaminetetra-acetic acid, 8-hydroxyquinoline) the free ε-benzyloxycarbonyl-lysine (**52**) may be further reacted, *e.g.* with t-butoxycarbonyl azide to give the versatile intermediate (**53**). The isomer (**54**) may be similarly prepared.

The benzyloxycarbonyl group is appreciably more acid-labile when attached to the strongly basic side-chain amino-function than when attached to the α-amino-group. It may therefore give inadequate protection to the side chain if, for example, α-t-butoxycarbonyl groups are to be removed repetitively in several steps of peptide synthesis. Ring-substituted benzyloxycarbonyl groups, *e.g.* *p*-nitro and various halogenated derivatives, may be used to increase the side-chain protecting-group stability. Alternatively, the more acid-stable trifluoroacetyl or phthaloyl groups may be used for side chain protection.

(*b*) *Aspartic and glutamic acids*

Unambiguous synthesis of aspartyl and glutamyl peptides linked exclusively through α-carboxy-groups requires protection of their respective β and γ side-chain carboxy-functions. Usually this is achieved through the side-chain benzyl or t-butyl esters. These monoesters are readily prepared, *e.g.* by direct acid-catalysed partial esterification or transesterification reactions, or by partial alkaline hydrolysis of diesters. Selectivity of attack at the more electrophilic α-carbonyl group is enhanced in this last method by the presence of cupric ion. Other routes are based on preferential ring-opening of cyclic

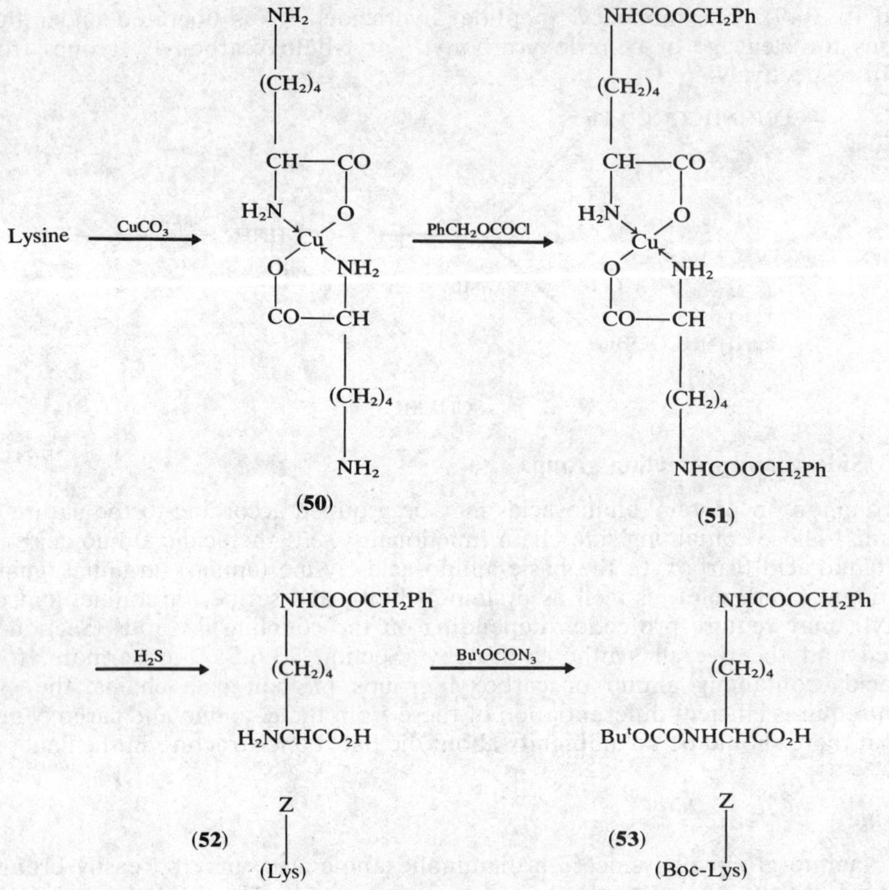

SCHEME 8

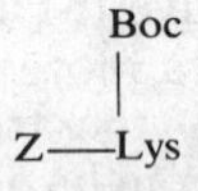

(54)

anhydrides of the dicarboxylic acid, *e.g.* benzyloxycarbonylglutamic anhydride (**55**) yields predominantly the α-benzyl ester (**56**) with benzyl alcohol in the presence of dicyclohexylamine.[39] Mixed diesters, *e.g.* the α-benzyl-γ-t-butyl (**57**) derivative and thence glutamic acid γ-t-butyl ester (**58**), may thus be obtained (Scheme 9).

An alternative route to the benzyloxycarbonyl derivative of (**58**) involves an oxazolidinone intermediate (Scheme 10).[40] This route is equally applicable to aspartic acid derivatives.

Simple alkyl esters of glutamic and aspartic acids are not generally applicable in peptide synthesis, since the final removal of the alkyl ester group by saponification may cause substantial $\alpha \rightarrow \gamma$ or $\alpha \rightarrow \beta$ transpeptidation. This reaction sequence (Scheme 11) is particularly serious with aspartyl peptides since cyclization to the five-ring imide intermediate (**59**) may occur with great ease. Its hydrolysis yields a mixture of α- and β-peptides. Formation of (**59**) is sequence dependent and is particularly favoured with -Asp-Gly- and -Asp-Ser- sequences. With the former, cyclization may occur during the cleavage of β-benzyl esters under acidic conditions, and this sequence is therefore particularly difficult to prepare.

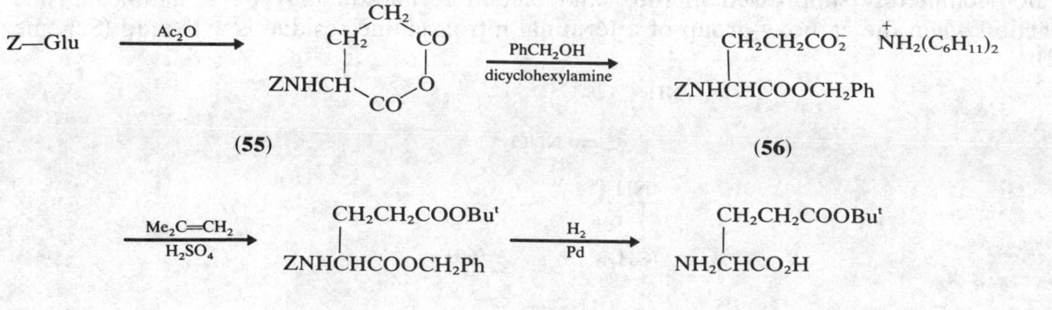

(55) (56) (57) (58)

SCHEME 9

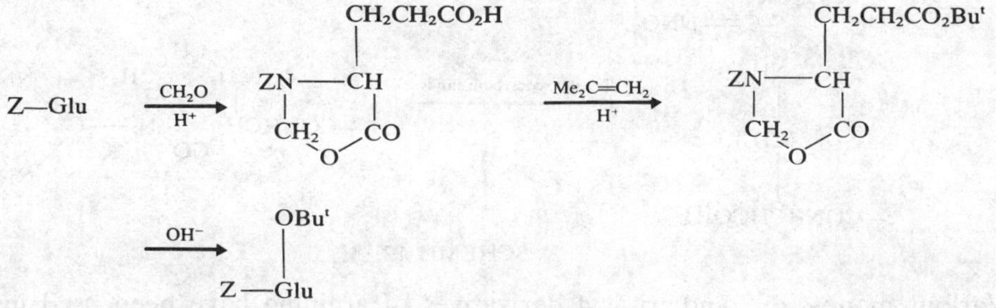

SCHEME 10

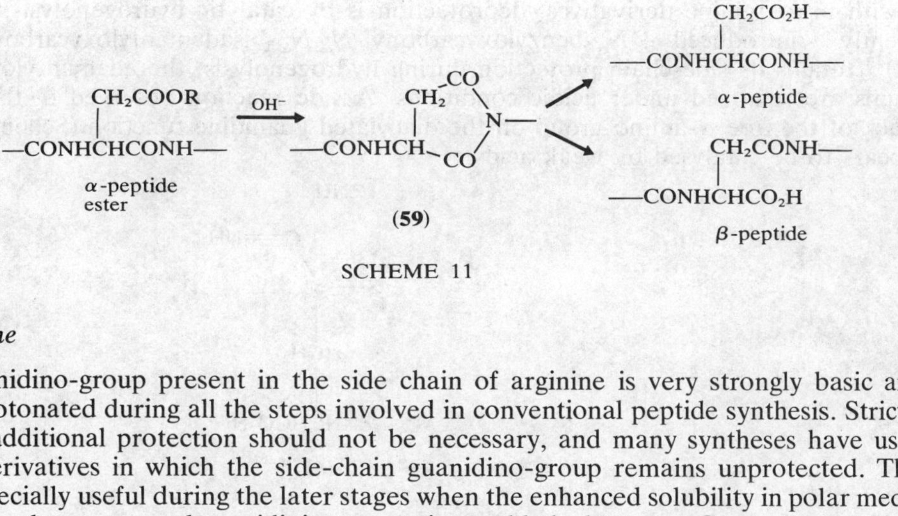

(59)

SCHEME 11

(*c*) *Arginine*

The guanidino-group present in the side chain of arginine is very strongly basic and remains protonated during all the steps involved in conventional peptide synthesis. Strictly speaking, additional protection should not be necessary, and many syntheses have used arginine derivatives in which the side-chain guanidino-group remains unprotected. This may be especially useful during the later stages when the enhanced solubility in polar media conferred by the protonated guanidinium group is an added advantage. Generally speaking, however, it is found advantageous to use side-chain protected forms of arginine in synthesis.

N_G-Nitroarginine (**60**)* is readily prepared from the free amino-acid by nitric–sulphuric acid mixtures, and the nitro-group serves to suppress the basicity of the guanidino-function.[41] It may be removed at the end of the synthesis by catalytic hydrogenation, along with all benzyl-based protecting groups. The nucleophilicity of the guanidino-group

* The position of the double bond in guanidine-substituted arginine derivatives, or of the relative locations of acyl groups within the guanidine system, should not be regarded as unequivocally established.

is not completely suppressed in (**60**), and lactam formation may be a significant side reaction when the carboxy-group of a terminal nitroarginine residue is activated (Scheme 12).

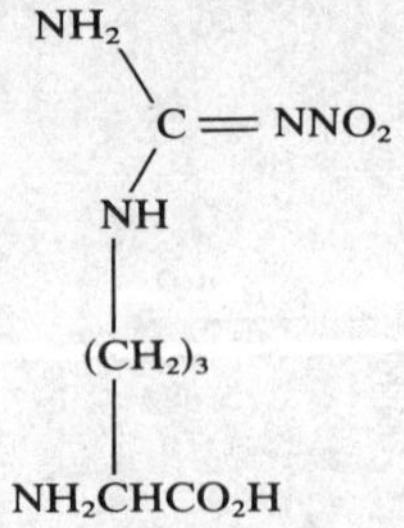

(**60**)

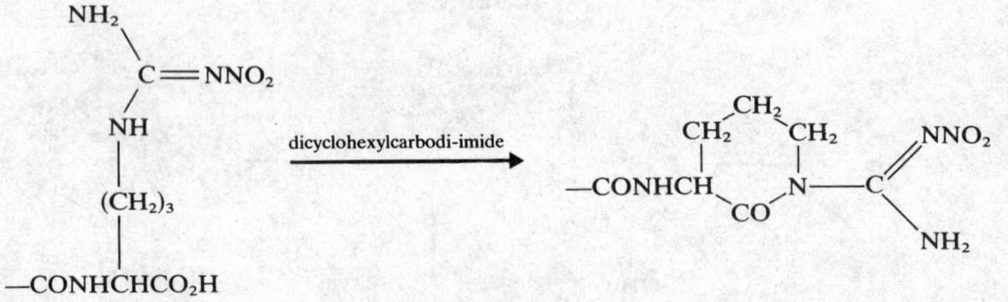

SCHEME 12

Various mono-, di-, and tri-acyl derivatives of arginine have been used in synthesis, though some tend to be unstable and decompose to give ornithine derivatives. N_{α},N_{ω},N_{ω}-Tribenzyloxycarbonylarginine (**61**)[42] can be prepared by reaction between arginine and benzyl chloroformate in strongly alkaline solution and has proved a useful intermediate; as with nitroarginine derivatives, deprotection is by catalytic hydrogenolysis. The more recently introduced N_{α}-benzyloxycarbonyl-N_{ω},N_{ω}-bisadamantyloxycarbonylarginine (**62**)[43] retains its side-chain protection during hydrogenolysis; the adamantyloxycarbonyl groups are removed under acidic conditions. A side reaction observed in this series is attack of the free α-amino-group on the diacylated guanidine function (Scheme 13).[43] It appears to be catalysed by weak acids.

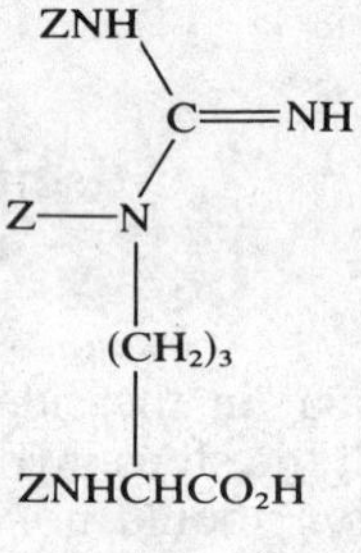

(**61**)

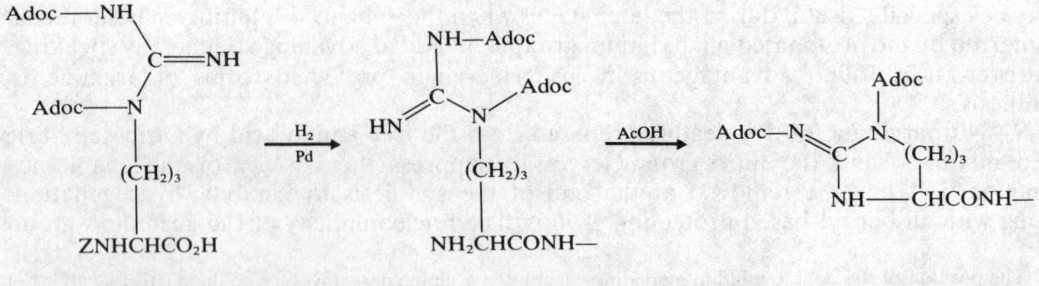

(**62**) SCHEME 13

The very stable N_a-toluene-*p*-sulphonyl derivative of arginine is finding increasing application, especially in solid-phase synthesis.[44,45] The toluene-*p*-sulphonyl group is removed only by very strong acids (liquid hydrogen fluoride)[46] or by sodium in liquid ammonia reduction.

An alternative approach to the preparation of arginine peptides is to use side-chain-protected ornithine derivatives throughout the synthesis, with conversion of the ornithine side-chain to that of arginine in the final step (Scheme 14). In this way, problems during synthesis associated with the guanidinine function are avoided.

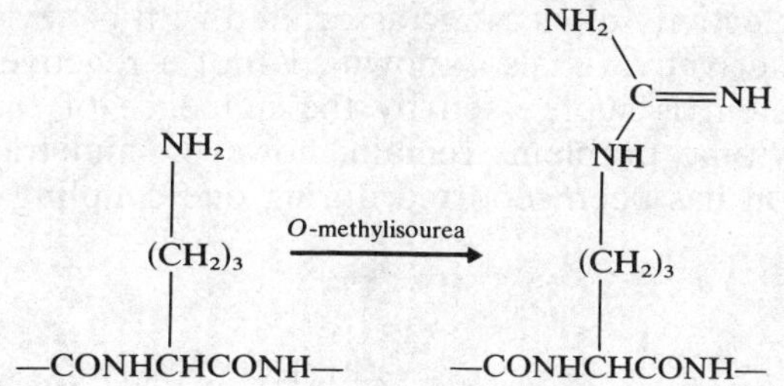

SCHEME 14

(*d*) *Histidine*

The efficient incorporation of the amino-acid histidine into synthetic peptides still presents formidable problems. This is because the imidazole ring possesses particularly awkward chemical properties. Free imidazole is an effective catalyst for ester and amide hydrolysis, as well as for racemization. Histidine derivatives themselves are particularly prone to racemization during peptide synthesis. If left unprotected, the imidazole ring may become acylated by activated carboxy-components, but the acyl derivatives are themselves reactive and may subsequently induce acyl transfer elsewhere in the molecule. For this reason, N_{im}-acyl derivatives of histidine are generally unsuitable as synthetic intermediates if the side-chain protection is to be retained during several synthetic steps. For stepwise synthesis, urethane-protected derivatives, *e.g.* the N_{α},N_{im}-bis-t-butoxycarbonyl derivative (**63**)*, may be used, both protecting groups being removed immediately after incorporation. The intermediate (**63**) has been used successfully in solid-phase synthesis.[47]

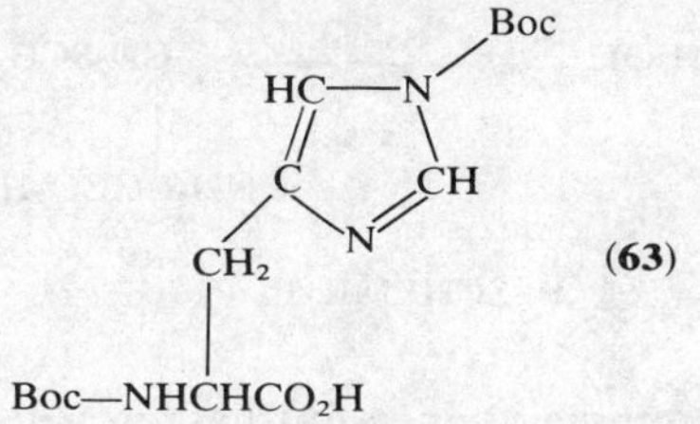

(**63**)

* In general, the position of N_{im} substituents specifically on one or other of the ring nitrogen atoms has not been established. The only case where the ring substituent has been definitely located is apparently the N_{im}-dinitrophenyl derivative (**64**), where the Dnp group is on the τ-nitrogen, as shown.[48]

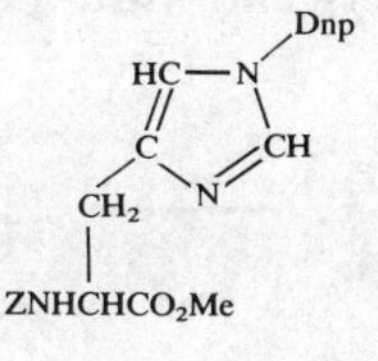

(**64**)

The *im*-benzyl derivative is resistant to acylation although the basic character of the imidazole ring is not suppressed. Its use in peptide synthesis has a long history.[49] Removal requires sodium in liquid ammonia reduction or prolonged hydrogenolysis. Other simple derivatives employed include N_{im}-dinitrophenylhistidine (**64**)[50] (removed by good nucleophiles, *e.g.* thiols), and the N_{im}-tosyl derivative.[51] The latter appears to have only marginal stability. It is removed by the hydroxybenzotriazole widely used in coupling procedures (p. 344), and by other nucleophiles. Transfer to the α-amino-group during further peptide chain elongation may be a serious risk.

The Ztf group (as in **65**) was devised, by Weygand *et al.*,[52] specifically for histidine protection. It has the selectivity of cleavage associated with benzyloxycarbonyl derivatives (the t-butoxycarbonyl derivative is also known), is not a reactive acylimidazole, yet the basicity of the ring system is suppressed by the influence of the electron-withdrawing trifluoromethyl group. Some problems remain, however, and transfer of the whole Ztf group to the N_α-position has been reported during one coupling procedure.

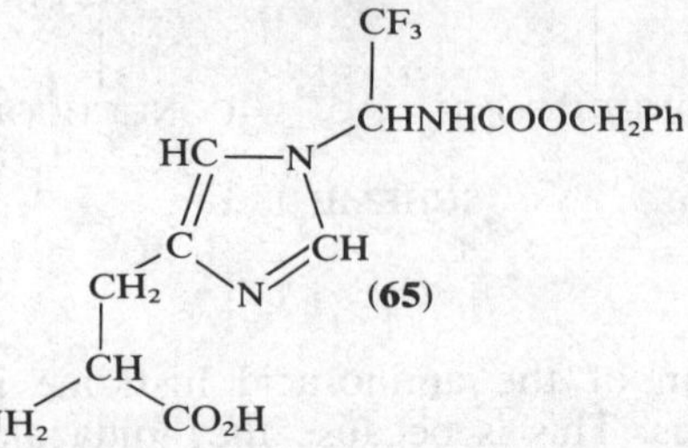

(**65**)

(*e*) *Cysteine*

A wide range of aralkyl groups has been used for blocking the reactive thiol group of cysteine. Benzyl[53] (and the more acid-labile *p*-methoxy[54] and *p*-methyl[55] substituted compounds), diphenylmethyl[56] and triphenylmethyl (trityl)[56,57] have all been employed for this purpose. The first of these provides a very stable protected derivative, requiring sodium/ammonia reduction or liquid HF acidolysis for its cleavage. At the other extreme, *S*-trityl derivatives are cleaved under relatively mild acid conditions ($HCl–CHCl_3$) by virtue of the greatly increased stability of the triphenylmethyl cation.

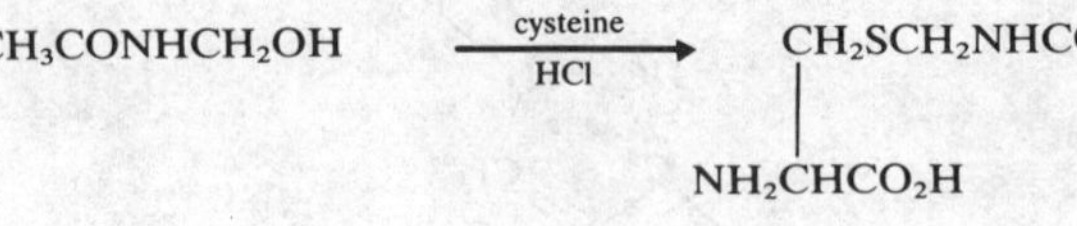

(**66**)

SCHEME 15

More recently, *S*-acetamidomethylcysteine (**66**) has proved a valuable intermediate in peptide synthesis.[58] It is readily prepared from acetamidomethanol (Scheme 15) and is stable to a wide range of acidic and basic conditions. The acetamidomethyl (Acm) group is smoothly removed at pH 4 in the presence of mercuric ion. Of particular interest is the oxidative removal with iodine with direct formation of disulphide-containing *cystine* peptides.[59] This may occur both inter- and intra-molecularly (Scheme 16).

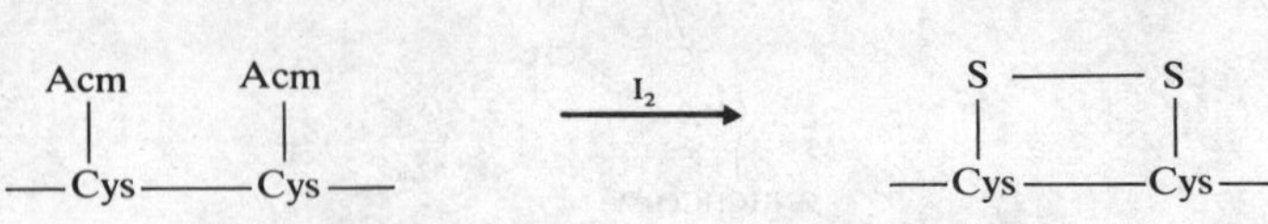

SCHEME 16

Cystine peptides may themselves be used in synthetic operations, partly obviating the need for thiol group protection.[60] Mixed disulphides, *e.g.* (**67**), have also been used successfully.

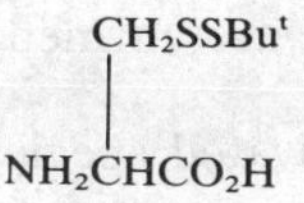

(**67**)

(*f*) *Serine, threonine, tyrosine*

The hydroxy-groups of these amino-acids may require protection according to the coupling methods to be employed. *O*-Benzyl or t-butyl ethers have been widely used, the former being removed by hydrogenolysis and the latter by mild acidic treatment. Removal of the *O*-benzyl group from tyrosine derivatives by treatment with strong acids (*e.g.* liquid HF) is less satisfactory since some rearrangement into the aromatic nucleus occurs[61] (equation 8). This, presumably intramolecular side reaction, is minimized by use of the 2,6-dichlorobenzyl ether derivative.

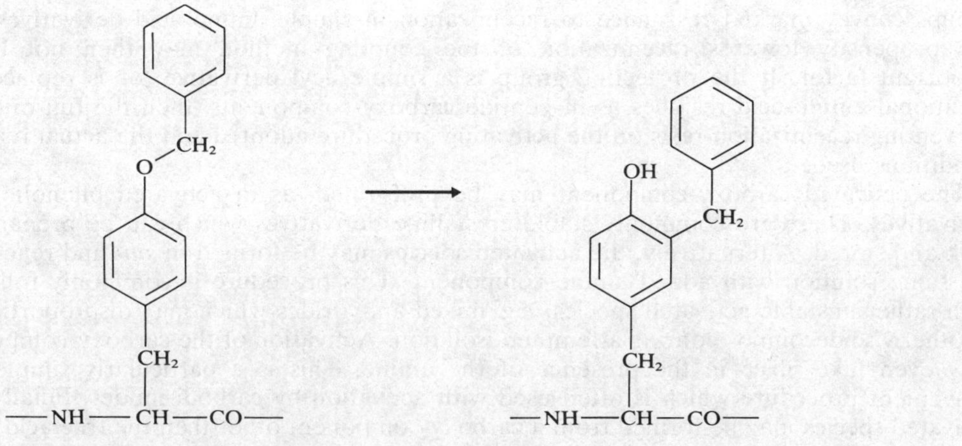

(8)

23.6.3 FORMATION OF THE PEPTIDE BOND

The usual process for the synthesis of peptide bonds is illustrated in Scheme 17. In the first step, an *N*-protected amino-acid or its carboxylate salt is reacted with an activating reagent to form the activated carboxy-component (**69**). The activating or leaving group (A in **69**) may take any of a very wide range of molecular forms. It may correspond to the anion of a simple inorganic acid (*e.g.* **69**; A = Cl, N_3, *etc.*), giving rise to the free anion of that acid on nucleophilic displacement by amines. Similarly, it may be derived from a

$$XNHCHR^1CO_2H \longrightarrow XNHCHR^1COA$$

(**68**) (**69**)

$$(\mathbf{69}) + NH_2CHR^2CO_2Y \longrightarrow XNHCHR^1CONHCHR^2CO_2Y$$

(**70**)

SCHEME 17

carboxylic acid, as in mixed and symmetrical anhydrides (**69**; A = $OCOR_3$), or a phenol, as in activated aryl esters (*e.g.* **69**; A = $OC_6H_4NO_2$). The alcohol component of other activated esters may be various nitrogenous derivatives, *e.g.* *N*-acylhydroxylamines, or be derived from enolic forms of carbonyl compounds. Particularly important examples of the last class are esters of isoureas, which are formed as intermediates in carbodi-imide activation reactions.

Although a very large number of activated species have been investigated for use in peptide synthesis, many are over-reactive (*e.g.* acid halides), or for other reasons give rise to undesirable side reactions. In practice, the great majority of complex peptide syntheses involve combination of only a few methods. The most important are described individually later.

The choice of coupling methods in each case is dictated by the overall synthetic strategies (see Section 23.6.5), the speed and efficiency of the coupling reaction, and by practical factors. Not the least important of these is the ease of isolation of the desired peptide from the inevitable co-product derived from the activating group. Thus activation by dicyclohexylcarbodi-imide (Section 23.6.3.1) gives rise to very insoluble dicyclohexylurea, whereas preformed esters of *N*-hydroxysuccinimide (Section 23.6.3.2) liberate water-soluble *N*-hydroxysuccinimide. Judicious choice of coupling reagent thus gives considerable flexibility in the work-up procedure of the reaction mixture. Choice of activating method is also dependent on the nature of the carboxy-component, and particularly on the amino-protecting group X (Scheme 17). Urethane-derived protecting groups convey marked resistance to racemization in simple amino-acid derivatives, and the propensity towards racemization of the coupling method may then not be an important factor. If the protecting group is a simple acyl derivative, or is replaced by additional amino-acid residues as in peptide carboxy-components, then the full onus for preventing racemization rests on the activating procedure adopted and the actual reaction conditions used.

The activated carboxy-component may be preformed, as in activated phenolic ester derivatives. These are commonly stable crystalline derivatives which can be prepared in bulk and stored. Alternatively, the activated species may be formed *in situ* and reacted in the same solution with added amino-component. This procedure is commonly followed with rather unstable activated species, *e.g.* mixed anhydrides which may disproportionate or otherwise decompose during attempted isolation. Activation of the carboxy-component may even take place in the presence of the amine. This is a particularly simple and convenient procedure which is often used with activation by carbodi-imides. Finally, the activated species may be formed from a carboxy-component other than the free acid or its salt, *e.g.* in the azide procedure (Section 23.6.3.4) from carboxylic esters by the action of hydrazine and then nitrous acid, or from protected hydrazide derivatives.

23.6.3.1 Carbodi-imides

The use of carbodi-imides for peptide synthesis by Sheehan and Hess[62] in 1955 has proved to be one of the most significant developments in peptide chemistry. The *N,N'*-dicyclohexyl derivative (**71**) is now by far the most widely used reagent for peptide bond formation. Its popularity can be ascribed to the readily availability of the reagent, the simplicity of its use, and, in appropriate solvents, the speed and general efficiency of coupling reactions. In the preparation of small, soluble peptide derivatives, the extreme insolubility of the co-product, *N,N'*-dicyclohexylurea, in most solvents other than the lower alcohols leads to relatively easy purification. The disadvantages of dicyclohexylcarbodi-imide are its toxicity, its ability to racemize the non-urethane protected amino-acids (including peptides), and the potential by-product formation through rearrangement reactions of activated intermediates. These last two unfavourable properties may be minimized by careful choice of reaction conditions and particularly by addition of certain hydroxylamine derivatives to reaction mixtures (see overleaf).

Carbodi-imides react only relatively slowly with amines so that activation of carboxy-components may be achieved in the presence of the amino-component. In practice, the reagent is usually simply added to a mixture of carboxy and amino derivatives dissolved in a suitable solvent. Relatively non-polar organic solvents such as methylene chloride are most suitable, but dimethylformamide and other polar media may be used if solubility problems intervene.

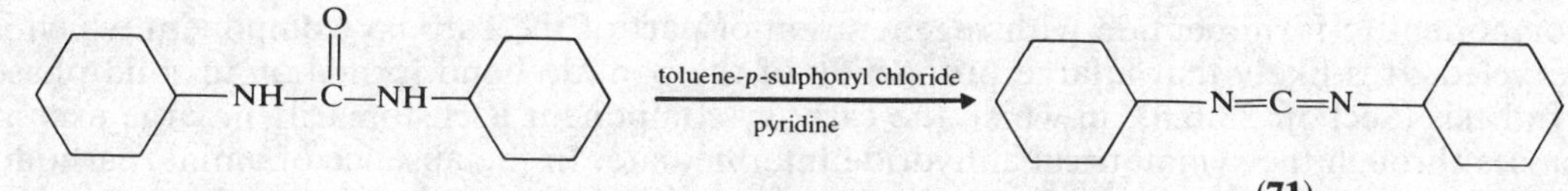

(71)

SCHEME 18

The common reagent, dicyclohexylcarbodi-imide (DCCI), is usually prepared by dehydration of the corresponding urea (Scheme 18). Presumably the *O*-sulphonylisourea is an intermediate in the dehydration reaction. An alternative earlier procedure involves removal of the elements of hydrogen sulphide from dicyclohexylthiourea using heavy metal salts. Other carbodi-imides may be prepared similarly, but the only examples which have had significant practical application are variants containing basic functions (*e.g.* **72**), which give rise to water- or acid-soluble ureas. These are sometimes of value in cases where separation of a sparingly soluble product from dicyclohexylurea is a problem.

$$EtN{=}C{=}N(CH_2)_3NMe_2$$

(72)

The mechanisms of carbodi-imide-mediated acylation reactions are moderately complex and the various pathways are summarized in Scheme 19.[63] In addition, there is the potential direct reaction between peptide amine and carbodi-imide, leading to guanidine derivatives (**76,** Scheme 20), but this does not seem to be significant under normal

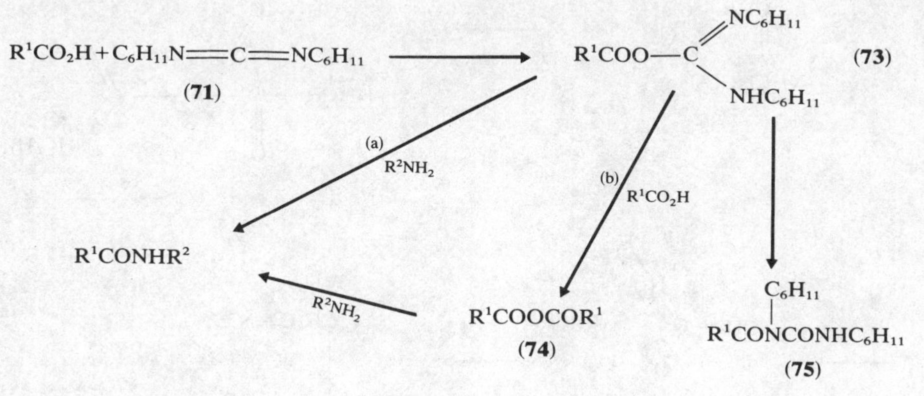

SCHEME 19

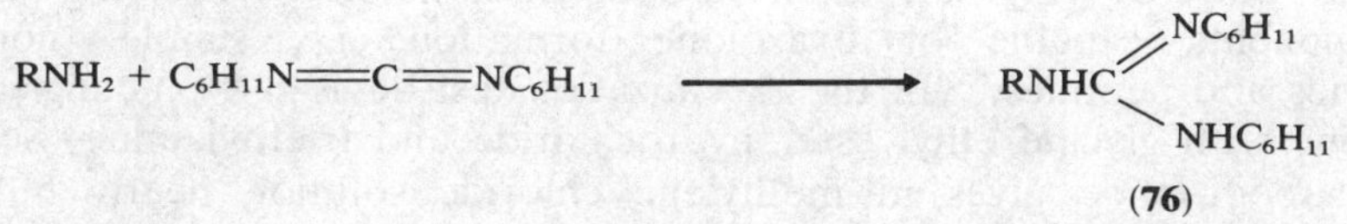

SCHEME 20

reaction conditions. In the first step, the carboxylic acid combines with carbodi-imide in an acid-catalysed step to yield the *O*-acylisourea (**73**). This adduct constitutes the primary activated form of the carboxy-component and may combine directly with peptide amine as in pathway (a). Mechanistic studies[63] have established that this is the major peptide bond-forming process in normal peptide synthesis in solution.

When the initial carboxy-component is in excess, the *O*-acylisourea (**73**) may react preferentially with carboxylic acid or carboxylate, leading to the symmetrical anhydride (**74**) (path b). This secondary activated species is in turn able to react with the amino-component to form peptide with regeneration of part of the carboxy-component which is recycled. It is likely that a large proportion of the peptide bond formation in solid-phase synthesis (Section 23.6.4), in which the carboxy-component is customarily in large excess, passes through the symmetrical anhydride intermediate. In the absence of amine, carbodi-imides may be used very effectively for the preparation of symmetrical anhydrides. Finally, again in the absence of amino-component or when reaction of (**73**) or (**74**) with amine is particularly sluggish, the stable *N*-acylurea (**75**) may be formed. This may arise either by intramolecular rearrangement of the *O*-acyl derivative (**73**) or by intermolecular acylation of dicyclohexylurea by (**74**) or (**73**). *N*-Acylureas are common by-products in carbodi-imide-mediated coupling reactions. The extent of their formation is also a function of carboxy-component structure and of reaction medium. *o*-Nitrophenylsulphenyl and phthaloyl amino-acids are particularly poor in this respect. *N*-Acylurea formation is minimized in solvents such as methylene chloride and acetonitrile.[64]

An amino-acid specific side-reaction of importance is the dehydration of side-chain amide groups. This occurs to a significant extent when derivatives of asparagine and glutamine are activated by dicyclohexylcarbodi-imide (Scheme 21). It does not occur when these amino-acid residues are already incorporated within the peptide chain and clearly the free α-carboxy-function is necessary. The probable mechanism is indicated in the scheme. The reaction is largely suppressed by addition of hydroxylamine derivatives to the reaction mixture (see below).

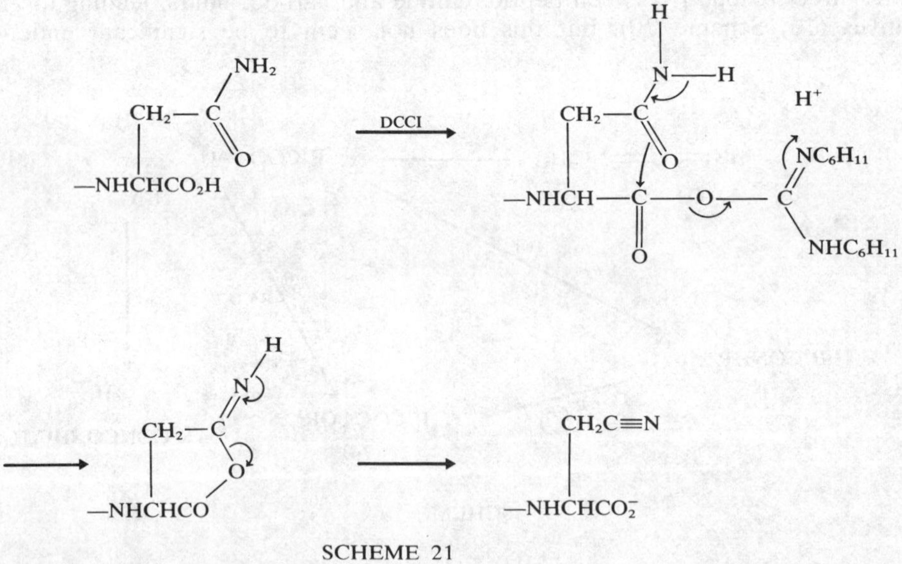

SCHEME 21

Racemization may be very serious in carbodi-imide-mediated coupling reactions for carboxy-components capable of oxazolone formation, *i.e.* simple (non-urethane) acylamino-acids and peptides.[65] In the racemization test devised by Young [coupling of benzoyl-leucine with glycine ethyl ester hydrochloride and triethylamine (Scheme 22)], dicyclohexylcarbodi-imide gives in methylene chloride solution nearly 50% racemic dipeptide derivative.[66] In the presence of excess triethylamine more than 80% racemate is

$$\text{PhCO-Leu-OH} + \overset{-}{\text{Cl}}\,\text{H}_2\overset{+}{\text{Gly}}\text{OEt} + \text{NEt}_3 \xrightarrow{\text{DCCI}} \text{PhCO-Leu-Gly-OEt} + \text{Et}_3\overset{+}{\text{N}}\text{H}\,\text{Cl}^-$$

SCHEME 22

formed. This is a very stringent test for racemizing ability since benzoylamino-acids form 2-phenyloxazolones with exceptional facility, but significant racemization has been observed in other more typical cases. Thus the coupling of t-butoxycarbonyl-L-leucyl-L-phenylalanine with L-valine t-butyl ester hydrochloride (Scheme 23) in a variety of

$$\text{Boc-L-Leu-L-Phe-OH} + \overset{-}{\text{Cl}}\,\text{H}_2\text{-L-}\overset{+}{\text{V}}\text{al-OBu}^t \xrightarrow[N\text{-methylmorpholine}]{\text{DCCI}} \text{Boc-L-Leu-(L+D)-Phe-L-Val-OBu}^t$$

SCHEME 23

solvents and using the weaker base *N*-methylmorpholine to liberate the free amino-ester gave up to 20% of the L-D-L diestereoisomer, equivalent to 40% racemization in the Young test.[67] This serious limitation on the use of carbodi-imide in coupling reactions of peptides (as opposed to urethane protected amino-acids) has been largely removed by the important observations of Weygand and Wünsch[68] that addition of *N*-hydroxysuccinimide to coupling reactions markedly reduced the extent of racemization (and of other side reactions). In the tripeptide synthesis mentioned above (Scheme 23), addition of *N*-hydroxysuccinimide to the coupling reaction in DMF reduced the formation of L-D-L tripeptide from 14% to 1%.

The use of preformed esters (**77**) of *N*-hydroxysuccinimide in peptide synthesis had been proposed earlier[69] (see Section 23.6.3.2) and favourable results obtained in racemization tests.[70] The beneficial effect of added *N*-hydroxysuccinimide in carbodi-imide coupling reactions is almost certainly due to the intermediacy of these esters. Hydroxysuccinimide is an 'α-effect' nucleophile and reacts very rapidly (Scheme 24) with *O*-acylisourea. Subsequent reaction of the resulting ester (**77**) with the amino-component is also rapid, probably because of the formation of the hydrogen bonded intermediate (**78**).

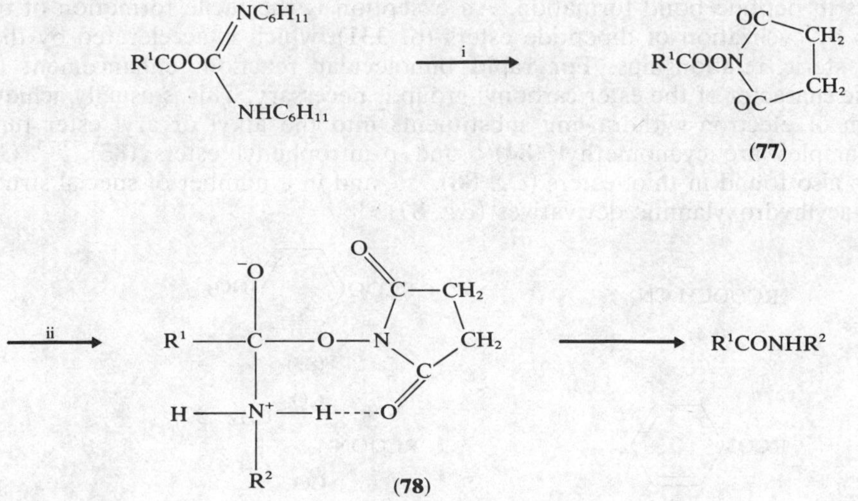

i, *N*-hydroxysuccinimide; ii, amino-component R^2NH_2.

SCHEME 24

Similar intermediates have been invoked to explain the exceptional reactivity with amine nucleophiles of other activated ester derivatives (Section 23.6.3.2). This enhanced reactivity would not be shown in oxazolone formation by reaction with the adjacent carbonyl group. It seems that rapid formation and aminolysis of (**77**) effectively precludes racemization by way of the oxazolone. Side-chain dehydration of asparagine and glutamine derivatives (Scheme 21, p. 342), which also involves attack of carbonyl oxygen on an activated intermediate, is similarly minimized by addition of *N*-hydroxysuccinimide to the reaction mixture. Several other hydroxylamine derivatives (**79**–**83**) have been shown to function similarly, and of these 1-hydroxybenzotriazole (**79**) now seems to be the additive of choice.[67]

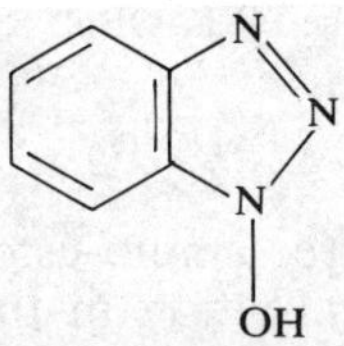

(79)

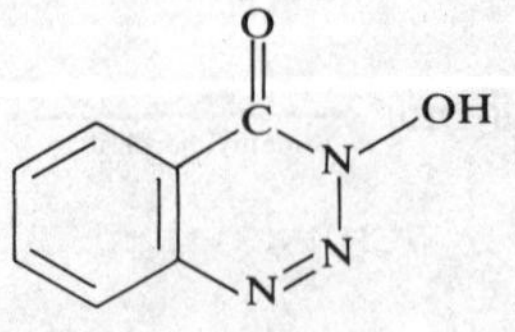

(80)

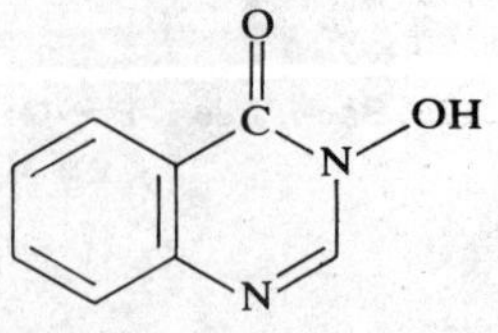

(81)

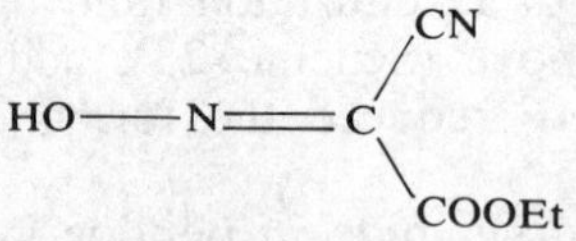

(82)

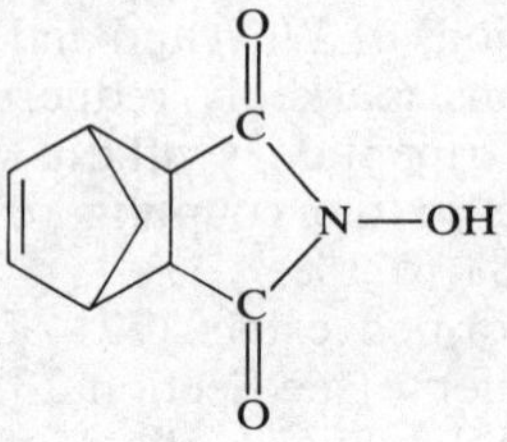

(83)

23.6.3.2 Activated esters

Simple alkyl and acyl esters aminolyse too slowly to function as activated carboxyl components in peptide bond formation. An exception is the facile formation of diketopiperazines by cyclization of dipeptide esters (p. 331), which is accelerated by the very favourable steric relationships. For rapid bimolecular reaction, enhancement of the electrophilic character of the ester carbonyl group is necessary. This is usually achieved by introduction of electron-withdrawing substituents into the alkyl or aryl ester function. Typical examples are cyanomethyl (**84**)[71] and *p*-nitrophenyl esters (**85**).[72,73] Greater reactivity is also found in thiol esters (*e.g.* **86**),[73,74] and in a number of special structures, notably *O*-acylhydroxylamine derivatives (*e.g.* **87**).[69]

$RCOOCH_2CN$

(84)

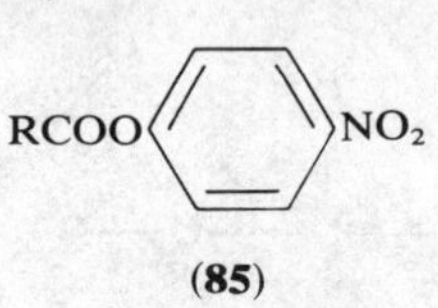

(85)

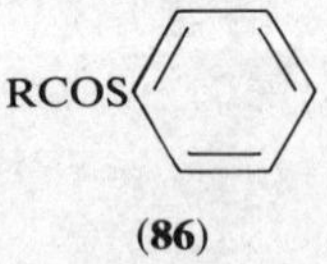

(86)

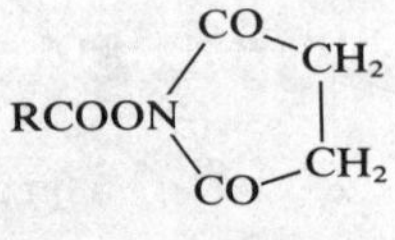

(87)

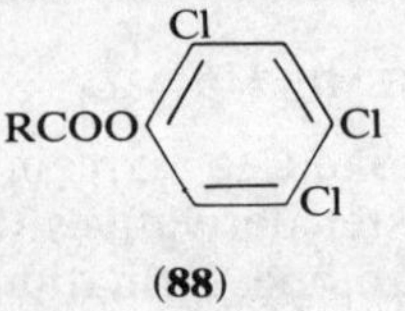

(88)

Modern peptide synthesis uses the activated ester method of coupling very extensively. Its advantages are simplicity, mildness of reaction conditions, and, because of the usually low degree of activation of the carboxy-component, general freedom from side reactions. On the other hand, separation of the peptide product from the alcohol or phenol liberated during the aminolysis reaction may present problems. These can be minimized by judicious initial choice of the particular activated ester derivative. A useful device for removing any excess of acylamino-acid active ester remaining at the end of the coupling reaction is addition of a highly reactive diamine, *e.g.* dimethylaminoethylamine (**89**). This reacts very rapidly with excess acylating agent to generate a basic, acid-soluble dimethylaminoethylamide (**90**) (equation 9).

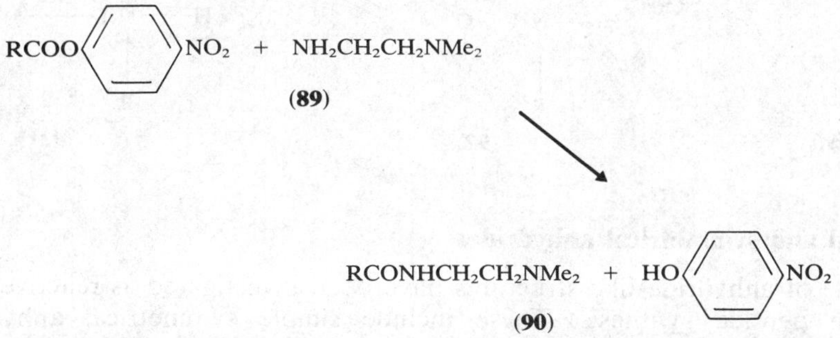

(9)

Preparation of the activated ester derivatives themselves involves an additional synthetic step, but this can be carried out on a large scale and the products stored until required. However, it should be noted that both active ester formation and aminolysis steps are in principle subject to racemization through oxazolone formation, and the activated ester method therefore finds its widest application in the stepwise coupling of optically stable urethane-protected (*e.g.* benzyloxycarbonyl and t-butoxycarbonyl) amino-acid derivatives.

A very wide range of activated esters have been investigated for use in peptide synthesis,[75] but only a few have achieved practical importance. Of these, the previously mentioned *p*-nitrophenyl esters introduced in 1957 are perhaps still the most important. They are generally easily prepared from urethane-protected amino acids and *p*-nitrophenol using dicyclohexylcarbodi-imide as the ester forming reagent.[76] An alternative preparation involves reaction of the carboxylic acid with bis-*p*-nitrophenyl sulphite[77] or tris-*p*-nitrophenyl phosphite[78] (*e.g.* equation 10).

$$ZNHCH_2CO_2H + (O_2NC_6H_4O)_2S{=}O \longrightarrow ZNHCH_2COOC_6H_4NO_2 + SO_2 + HOC_6H_4NO_2 \qquad (10)$$

Of the many other simple acyl esters that have been studied, probably only those derived from 2,4,5-trichlorophenol (**88**)[75] have found wide application. This phenol has pK_a 9.45, close to that of *p*-nitrophenol (pK_a 9.27), and generally the derived esters have similar reactivity in aminolysis. The more reactive pentachlorophenyl[79] and pentafluorophenyl esters[80] have also found significant application. Preferred use of halogen-substituted acyl esters as opposed to nitro-derivatives is indicated in syntheses involving hydrogenolysis of benzyloxycarbonyl groups, since otherwise traces of unremoved *p*-nitrophenol then give the very reactive *p*-aminophenol.

Esters of *N*-hydroxysuccinimide (**87**) are more reactive than those of *p*-nitro- and 2,4,5-trichloro-phenol and have the important additional feature of water solubility of the parent alcohol.[69] This can greatly simplify isolation procedures in peptide synthesis. As mentioned earlier, the exceptional reactivity and relative freedom from racemization of esters of this type is probably due to the formation of hydrogen-bonded intermediate complexes of type (**91**) with amino-components. Anchimeric acceleration has likewise been observed in esters of *o*-catechol[81] and 8-hydroxyquinoline,[82] which may form intermediates (**92**) and (**93**). As might be anticipated, coupling reactions using *p*-nitro- and 2,4,5-trichloro-phenol esters are accelerated by additives of the hydroxy-succinimide/hydroxybenzotriazole type, presumably through the intervention of hydroxy-succinimide and hydroxybenzotriazole esters.

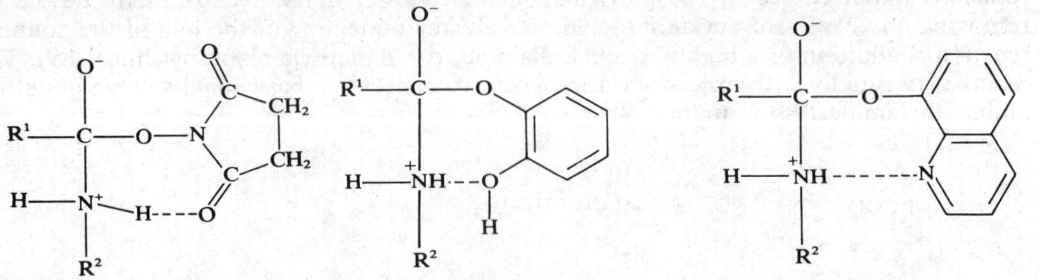

(**91**) (**92**) (**93**)

23.6.3.3 Mixed and symmetrical anhydrides

A wide range of anhydride-like structures have been investigated as reactive carboxy-components in peptide synthesis. These include simple symmetrical anhydrides of benzyloxycarbonyl- and t-butoxycarbonyl-amino-acids (*e.g.* **94**),[83] unsymmetrical anhydrides with other carboxylic acids (*e.g.* **95**)[84] or carbonic acid half esters (*e.g.* **96**),[85] anhydrides with various inorganic acids (*e.g.* **97**),[86] and cyclic *N*-carboxyanhydrides (*e.g.* **98**).[87] Internal anhydrides in which the second acid group derives from the side chain carboxy-groups of aspartic or glutamic acids (*e.g.* **99**) also found some early application in peptide synthesis.

$(\text{Boc-NHCH}_2\text{CO})_2\text{O}$
(**94**)

$\text{Boc-NHCH}_2\text{COOCOBu}^t$
(**95**)

$\text{Boc-NHCH}_2\text{COOCOOEt}$
(**96**)

$\text{ZNHCH}_2\text{COOSO}_3^-\ \text{Li}^+$
(**97**)

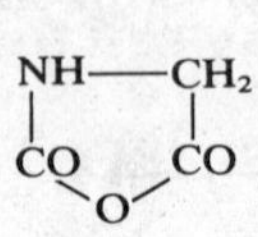

(**98**)

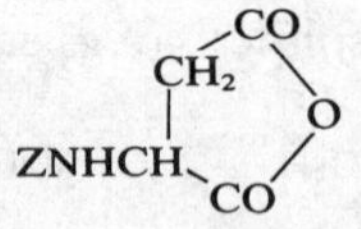

(**99**)

(*a*) *Symmetrical anhydrides*[83]

These are the simplest anhydride derivatives of acylamino-acids and are very easily prepared under mild conditions using, for example, dicyclohexylcarbodi-imide as dehydrating agent.[88] Nevertheless, their use in peptide synthesis has not been widespread. In part this is due to their uneconomic utilization of the often costly starting acylamino-acid, since only half is incorporated into the peptide (Scheme 25).

$$(\text{Boc-NHCHR}^1\text{CO})_2\text{O} + \text{NH}_2\text{R}^2 \longrightarrow \text{Boc-NHCHR}^1\text{CONHR}^2 + \text{Boc-NHCHR}^1\text{CO}_2\text{H}$$

SCHEME 25

The application of symmetrical anhydrides as carboxy-components has achieved most importance in solid-phase synthesis (*q.v.*), where freedom from any side reactions involving activating reagents is considered important.[88] The coupling method is essentially limited to urethane or other optically-stable protected amino-acid derivatives, since otherwise substantial or complete racemization may occur.

(b) Unsymmetrical carboxylic and carbonic anhydrides

These anhydrides, which are derived from an acylamino-acid salt and (usually) the acid chloride of a second carboxylic (*e.g.* Scheme 26)[84] or carbonic acid derivative (*e.g.* Scheme 27),[85] have achieved much wider importance. Anhydride formation is usually complete within a few minutes at 0 °C or below. This low temperature of formation and immediate use of the anhydrides as acylating agents without attempted isolation is important in order to avoid disproportionation to the mixture of two symmetrical anhydrides.

$$\text{ZNHCHRCO}_2^- \overset{+}{\text{N}}\text{HEt}_3 + \text{Bu}^t\text{COCl} \longrightarrow \text{ZNHCHRCOOCOBu}^t + \overset{+}{\text{N}}\text{HEt}_3\ \text{Cl}^-$$

SCHEME 26

$$\text{ZNHCHRCO}_2^- \overset{+}{\text{N}}\text{HEt}_3 + \text{EtOCOCl} \longrightarrow \text{ZNHCHRCOOCOOEt} + \overset{+}{\text{N}}\text{HEt}_3\ \text{Cl}^-$$

SCHEME 27

Because the mixed anhydride now contains two dissimilar carbonyl groups, nucleophilic attack by the amino-component may give rise to two products (Scheme 28). The second partner in the anhydride must therefore be chosen so as to minimize unwanted attack at its carbonyl group (Scheme 28, path b).

$$\text{ZNHCHR}^1\text{COOCOR}^2\ (\mathbf{100}) \xrightarrow[\text{(a)}]{\text{R}^3\text{NH}_2} \text{ZNHCHR}^1\text{CONHR}^3 + \text{R}^2\text{CO}_2\text{H}$$

$$\text{ZNHCHR}^1\text{COOCOR}^2\ (\mathbf{100}) \xrightarrow[\text{(b)}]{\text{R}^3\text{NH}_2} \text{ZNHCHR}^1\text{CO}_2\text{H} + \text{R}^2\text{CONHR}^3$$

SCHEME 28

The reaction of unsymmetrical carboxylic anhydrides with amino nucleophiles is influenced by steric and electronic effects in the anhydride, as well as by the nature of the solvent medium and amino-component. Steric hindrance is turned to advantage by using a bulky carboxylic acid as the second component. Anhydrides with pivalic (trimethylacetic) (**100**; $R^2 = Bu^t$)[84] and isovaleric acids (**100**; $R^2 = Bu^i$)[89] are attacked by amines in rather non-polar solvents very largely at the less-hindered acylamino-acid carbonyl group (Scheme 28, path a). Electronic effects, *i.e.* the relative electrophilicity of the two carbonyl groups in the anhydride, may be gauged approximately by the relative dissociation constants of the two parent acids. These reflect the electron-withdrawing or -donating character of their α-substituent. Nucleophilic attack is favoured at the carbonyl group derived from the stronger acid. Carbonic acid is particularly weak, and its half esters form nearly ideal partners in mixed anhydrides with the relatively strong acylamino-acids.[85] These anhydrides (*e.g.* **100**; $R^2 = COOEt$) react with amino-components under normal conditions almost exclusively at the acylamino-acid carbonyl yielding peptide derivatives (Scheme 28, path a).

An additional advantage in the use of mixed carboxylic–carbonic anhydrides is that only volatile coproducts (carbon dioxide and a lower alcohol) are produced. On the other hand, formation of these anhydrides from the corresponding chloroformate (Scheme 27) requires strictly anhydrous conditions, and racemization is serious in their aminolysis. The method is commonly used for coupling peptide fragments terminating in optically inactive (glycine) or optically stable (proline) residues, or with simple urethane-protected amino-acids. A particular application is the so-called REMA or repetitive excess mixed anhydride method in which single amino-acid residues are added sequentially to the amino-component without isolation of intermediate protected peptides.[90] More than one equivalent of the mixed anhydride is used at each stage to ensure rapid and near-quantitative acylation of the amino-component, the excess being destroyed by treatment with aqueous alkali. This accelerated method of synthesis is conceptually similar to the solid phase procedure of Merrifield (*q.v.*).

(*c*) *Anhydrides with inorganic acids*

A variety of inorganic acids have been proposed as partners in mixed anhydride derivatives for peptide synthesis, phosphorus being the inorganic element most commonly involved. Examples of this class include anhydrides prepared from acylamino-acids and diphenyl phosphoryl chloride (**101**),[91] dibenzyl phosphoryl chloride (**102**),[92] diethyl chlorophosphite (**103**), and *o*-phenylene chlorophosphite (**104**).[93] A recent addition to this list is diphenylphosphinyl chloride (**105**).[94] All form simple mixed anhydrides with carboxylate salts which react with amino-components seemingly exclusively at the carbonyl group. Anhydrides with sulphuric acid (*e.g.* **106**) constitute a special case since the

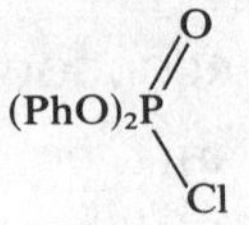

(**101**)

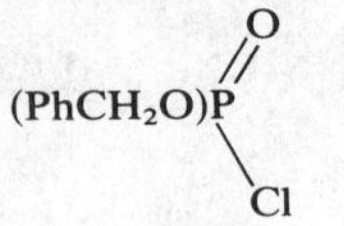

(**102**)

$(EtO)_2P{-}Cl$

(**103**)

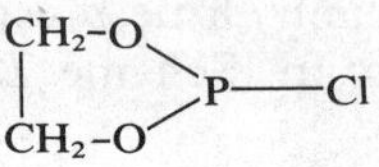

(**104**)

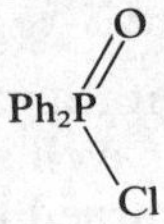

(**105**)

$ZNHCH_2COOSO_3^- \; Li^+$

(**106**)

$Me_2\overset{+}{N}{=}CHOSO_3^-$

(**107**)

salt of the dibasic acid anhydride confers water solubility, and these activated derivatives were considered especially suitable for acylation of free amino-acids and peptides in mildly alkaline aqueous solution. They are prepared[86] by reaction between the acylamino-acid or peptide carboxylate salt, and the complex (**107**) between sulphur trioxide and dimethylformamide. Racemization was not detected for peptide coupling reactions carried out in anhydrous media, and was low in aqueous solution.

(*d*) *Carboxyanhydrides*

Anhydrides of this class may be obtained by thermal elimination of benzyl chloride from benzyloxycarbonylamino-acid chlorides[95] (p. 325) or by direct reaction between amino-acids and phosgene (Scheme 29).[96] They react (Scheme 30) with amino nucleophiles predominantly at the amino-acid carbonyl group, the second anhydride carbonyl

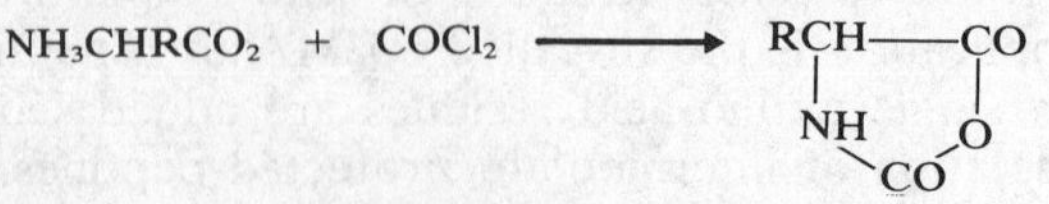

SCHEME 29

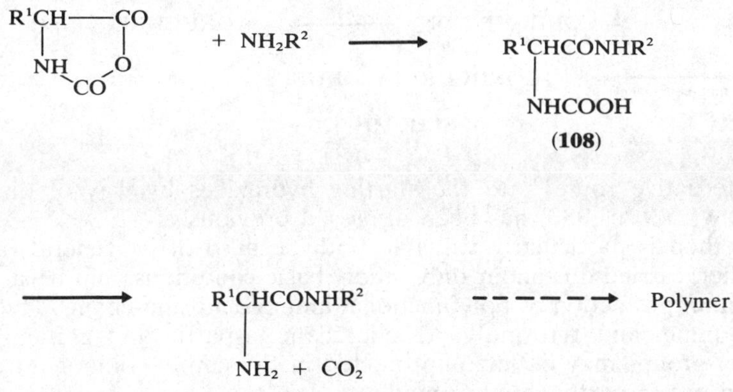

SCHEME 30

being deactivated to nucleophilic attack since it is part of a resonance-stabilized amide system. However, the resulting carbamic acids (**108**) are unstable and usually decompose during the reaction with liberation of a second amino-component. Further reaction with *N*-carboxyanhydride then gives rise to polymer. This polymerization of *N*-carboxyanhydrides is an important method for the preparation of high molecular weight poly(amino-acids).[97] When the initial reaction is carried out in the presence of excess carboxyanhydride and under conditions of alkaline pH and temperature such that decarboxylation of (**108**) does not occur, the method can be used for controlled peptide synthesis.[87]

23.6.3.4 Acid azides

For many years the procedure of choice for combining peptide fragments to form larger peptides was through acid azide derivatives.[98] This was primarily because of the method's long standing reputation as the only procedure completely free from the risk of racemization. This reputation is now know to be incompletely deserved, and racemization of azide-activated carboxy-components in the presence of tertiary base has been observed. Other more modern procedures (*e.g.* use of dicyclohexylcarbodi-imide in the presence of hydroxybenzotriazole) likely to be at least as free from racemization are now available. Nevertheless, the procedure retains a prominent place in current methodology.

In the original procedure of Curtius, acylamino-acid or peptide esters were converted by the action of hydrazine to the corresponding hydrazides, which were isolated and in turn reacted with nitrous acid to form acid azides (Scheme 31).

$$\text{—CONHCHRCO}_2\text{Me} \xrightarrow{\text{N}_2\text{H}_4} \text{—CONHCHRCONHNH}_2$$

$$\xrightarrow[\text{HCl}]{\text{NaNO}_2} \text{—CONHCHRCON}_3$$

SCHEME 31

This is still the most common method of preparing amino-acid and peptide hydrazides, but many improvements in technique have been introduced. Side reactions, particularly amide formation, are minimized when the azides are prepared at very low temperature (down to −25 °C) and using an organic nitrite ester in place of nitrous acid.[99] The low temperature also reduces risk of Curtius rearrangement of the azide. This is a potentially serious side reaction because the resulting isocyanate may subsequently react with the amino-component to form a urea derivative (Scheme 32), very similar to the desired

$$\text{—CONHCHR}^1\text{CON}_3 \longrightarrow \text{—CONHCHR}^1\text{N=C=O} \xrightarrow[\text{(amino-component)}]{\text{R}^2\text{NH}_2} \text{—CONHCHR}^1\text{NHCONHR}^2$$

SCHEME 32

peptide. An alternative route[100] to the starting hydrazides by way of diacylhydrazine derivatives (*e.g.* **47, 49,** p. 333) has been discussed previously.

The azide method is particularly valuable both because of its general freedom from racemization when carried out under only mildly basic conditions, and because it may be used with minimum protection of polyfunctional amino-acid side-chains. Hydroxy-groups (of serine, threonine, and tyrosine) and side-chain (aspartic and glutamic acids) and terminal carboxy-groups may be left unprotected in the amino-component and coupling reactions carried out in partly aqueous media.

23.6.3.5 Other coupling methods

The ingenuity of organic chemists has found free expression in devising new reactions and reagents for the formation of peptide bonds. *N*-Ethyl-5-phenylisoxazolium salts, *e.g.* the sulphonic acid derivative (**109**), react with acylamino-acid or peptide salts to form enol esters. These are sufficiently reactive to couple rapidly with amino-components to form peptides.[101] The method achieved early popularity because of its ingenuity, ease of product isolation from water-soluble sulphonic acid co-products, and because racemization tests were favourable. Subsequent experience showed that racemization may be significant. The proposed mechanism for the activation reaction is outlined in Scheme 33.

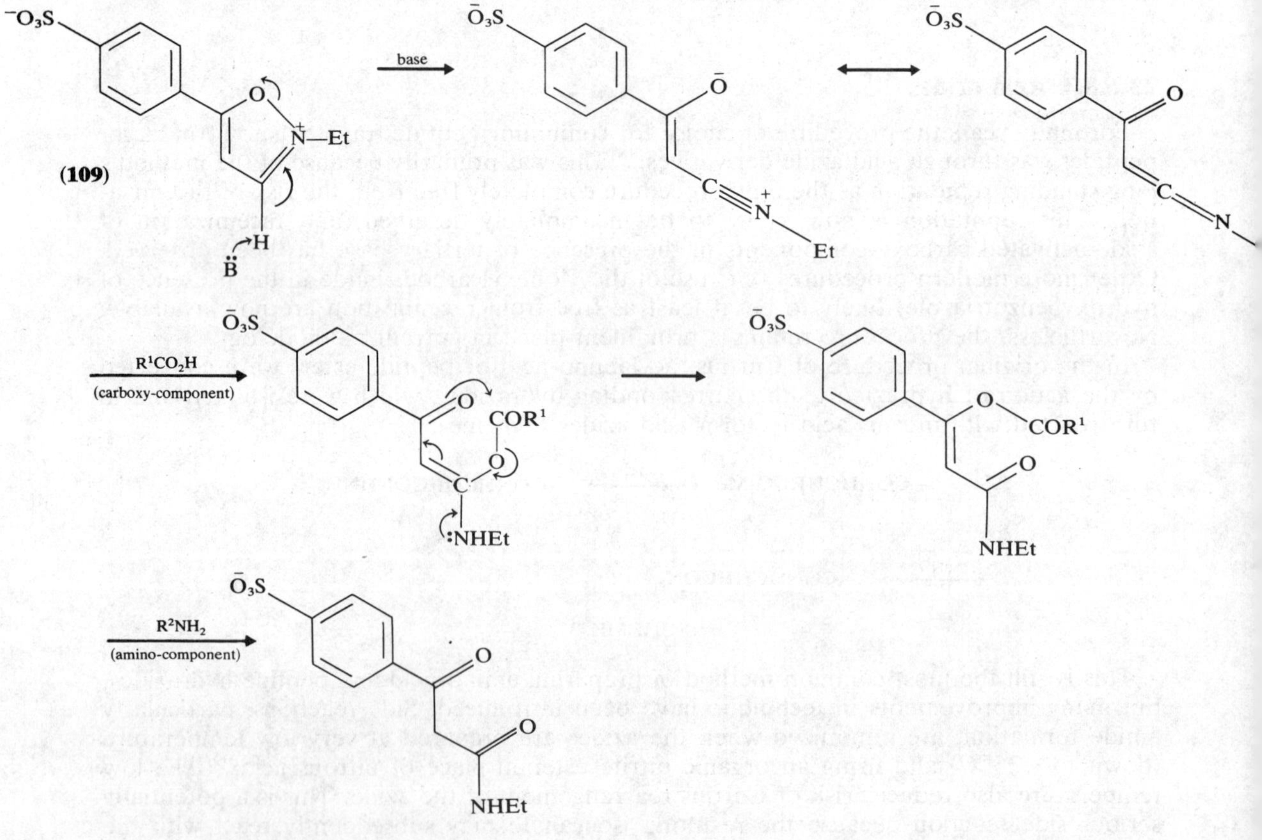

SCHEME 33

A number of variants on the original isoxazolium salt method have been proposed. An interesting example[102] is the 7-hydroxybenzoisoxazolium derivative (**110**). Reaction with the carboxy-component yields an intermediate enol ester (**111**) as before (Scheme 34), but

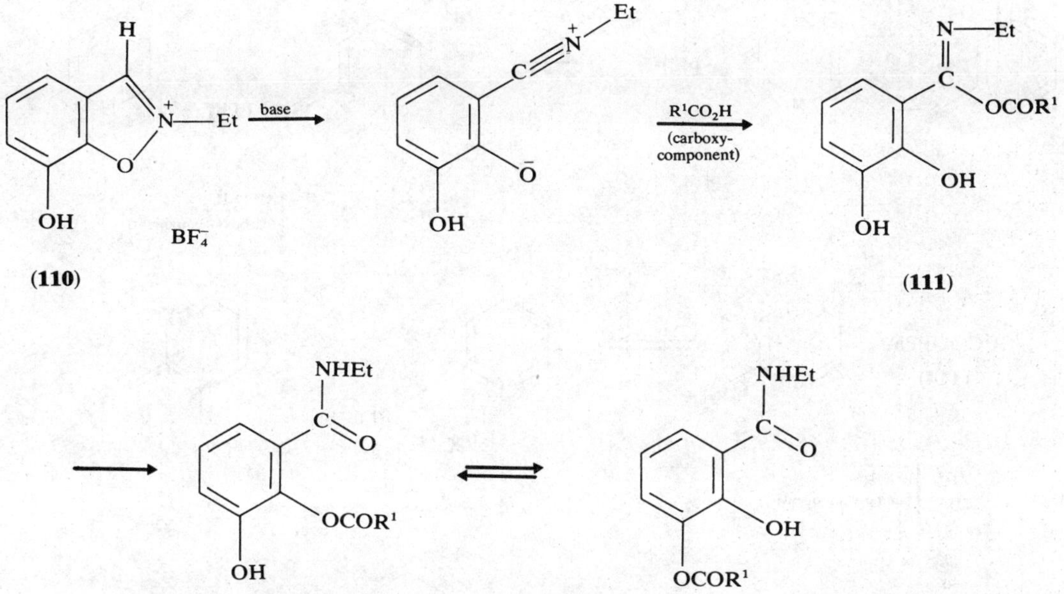

(**110**) (**111**)

SCHEME 34

this is now able to transfer its acyl group to one or other of the phenolic hydroxyls. The final activated species is thus a catechol ester, ammonolysis of which proceeds with anchimeric assistance from the adjacent free phenol or phenolate group (p. 345), and with marked resistance to racemization.

1-Ethoxycarbonyl-2-ethoxy-1,2-dihydroquinoline[103] (**112**) is another reagent which undergoes a multi-step reaction with carboxy-components to form an activated derivative (Scheme 35). In this case the activated species is probably the carboxylic–carbonic anhydride (**113**), identical to the intermediate in the ethyl chloroformate mixed anhydride procedure (p. 347).

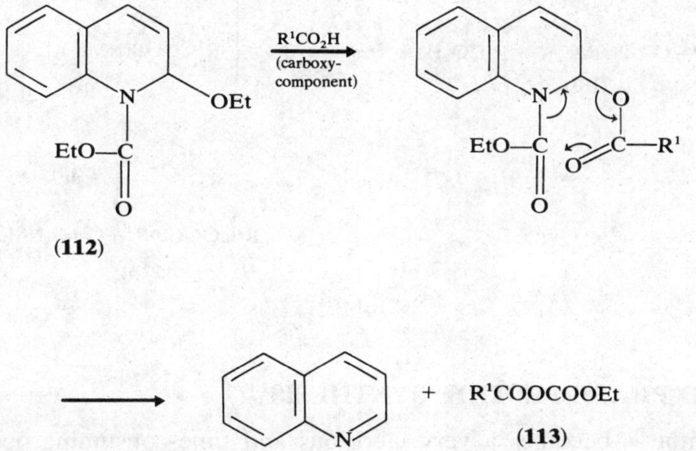

(**112**) (**113**)

SCHEME 35

Although formally a dehydration process, peptide bond formation may also be achieved by a combination of oxidizing and reducing agents.[104] One example of such a process is shown in Scheme 36.

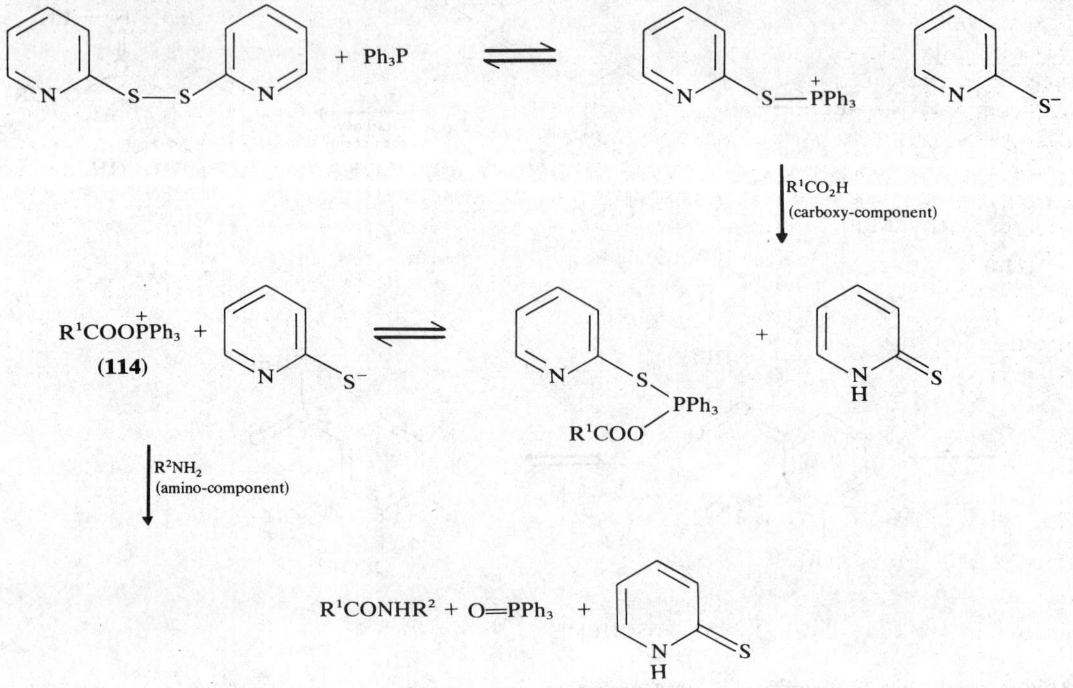

SCHEME 36

Acyloxyphosphonium salts of the type (**114**) have been postulated as intermediates in other peptide synthesis procedures, *e.g.* in the activation of carboxy-components by the bisphosphonium salt (**115**) (Scheme 37),[105] but evidence for their participation is lacking. In this last procedure, it seems likely that the symmetrical anhydride (**116**) is the principal acylating species involved.

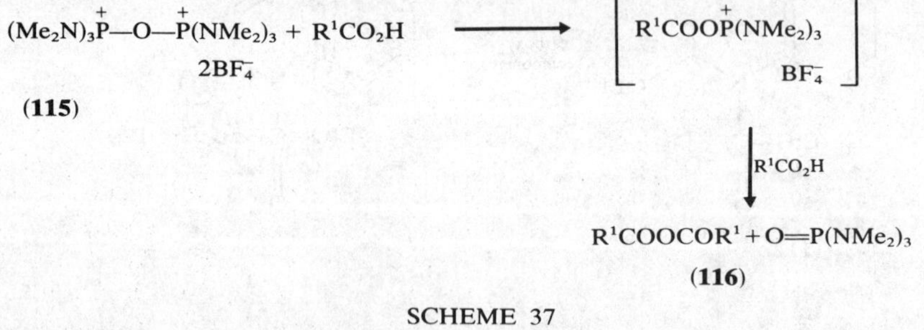

SCHEME 37

23.6.4 SOLID PHASE PEPTIDE SYNTHESIS[106]

Peptide synthesis becomes a very laborious and time-consuming operation when the synthetic objectives are large. Formation of each peptide bond requires the preparation of an appropriately protected amino-acid, coupling, and de-protection steps. Although the number of *consecutive* steps may be modified by choice of synthetic strategy (Section 23.6.5), the total number of discrete chemical operations is very great in the synthesis of the larger peptide hormones and small proteins. Peptide synthesis is also a technically very demanding process. Reaction conditions have to be investigated and optimized with great thoroughness to obtain adequate yields and to avoid side reactions. The isolation of the sensitive and often intractable reaction products requires considerable skill and experience. For these reasons, a number of schemes have been proposed for accelerated synthetic procedures which also simplify the many manipulative operations. Of these, the solid phase procedure introduced by Merrifield[107] has found wide application.

The general scheme of solid-phase synthesis is depicted in Scheme 38. Amino-acid sequences are assembled one residue at a time attached to an insoluble polymeric support. Each cycle of amino-addition involves coupling and deprotection steps. The synthesis is concluded by cleavage from the polymer support, usually with simultaneous removal of side chain and *N*-terminal protecting groups. During the synthesis the peptide is at all times attached to the insoluble support and the only physical operations involved are filtration and washing.

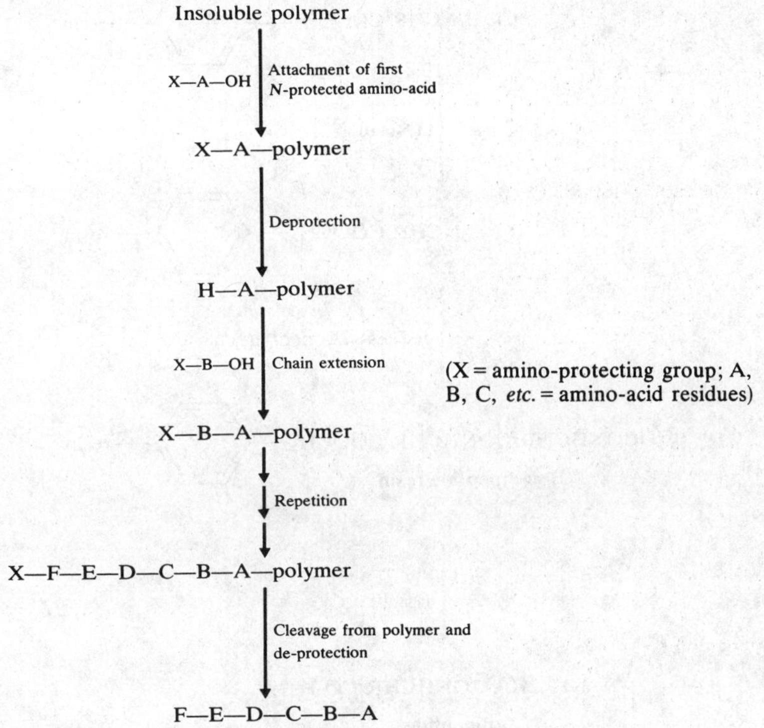

SCHEME 38 Merrifield solid-phase peptide synthesis

The most usual sequence of chemical operations used in practice for achieving solid-phase synthesis is shown in Scheme 39. t-Butoxycarbonylamino-acids and polystyrene supports have been used almost exclusively. The first amino-acid is attached to the resin as a substituted benzyl ester derivative by reaction with the partially chloromethylated polymer. Acid-catalysed removal of the butoxycarbonyl group followed by neutralization liberates the free terminal amino-group ready for coupling of the second amino-acid residue, usually by the dicyclohexylcarbodi-imide procedure. Excess (soluble) reagents are removed by thorough washing at each stage, and the deprotection, neutralization, and coupling sequence repeated until the assembly is complete. Removal from the resin involves cleavage of the benzyl ester linkage by very strong acid, usually anhydrous hydrogen fluoride. The essential features of solid-phase synthesis are therefore as follows.

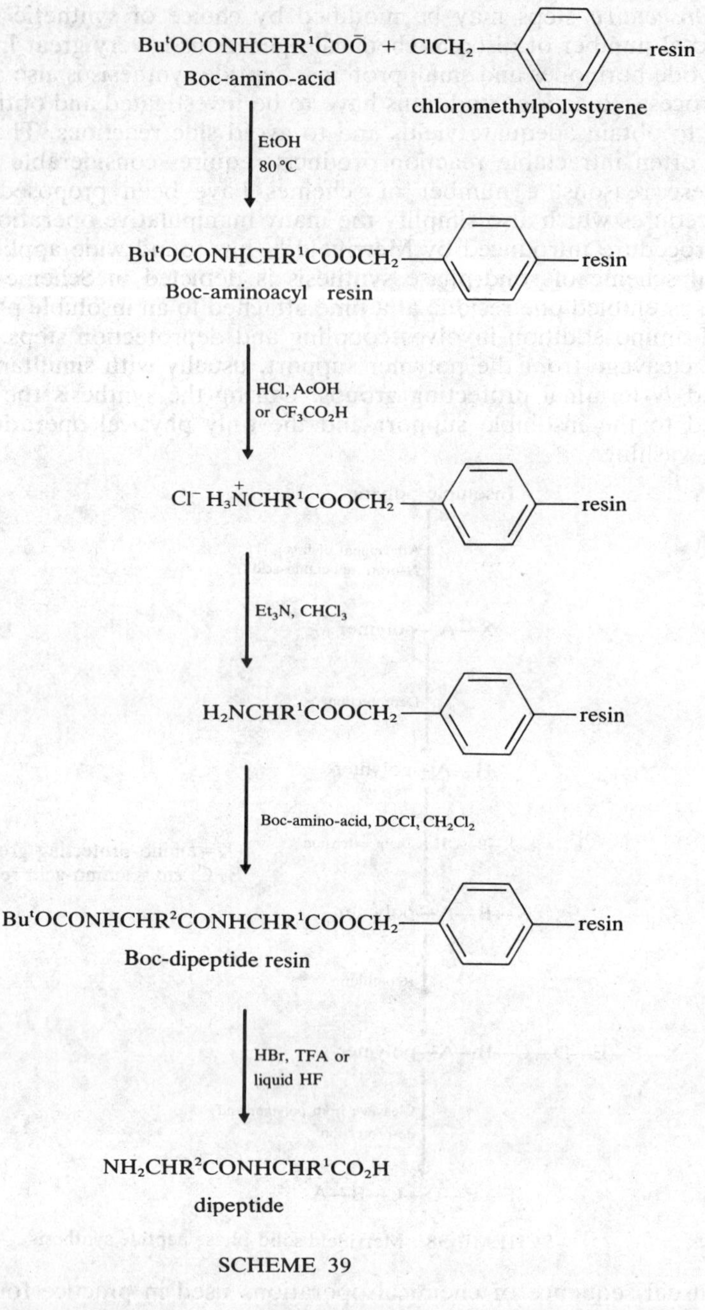

SCHEME 39

1. The synthesis is conducted in a semi-heterogeneous system by use of insoluble polymeric supports. The original method of Merrifield and the great majority of reported syntheses[106] have used styrene–divinylbenzene copolymer, but more polar support materials have recently been advocated.[108] The synthetic polymer is very lightly cross-linked (typically 1–2% in the case of polystyrene) and is therefore freely permeated by the reactants and solvent media employed. The polymer phase thus constitutes a gel system with considerable mobility of the polymer lattice and the attached growing peptides. Steric hindrance, which would be expected to be massive within a rigid polymer network, is thus minimized and reactions may proceed at rates not greatly dissimilar to those in free solution.

2. The insoluble nature of the polymer particles leads to great simplification of manipulation procedures, which may be machine-aided. The polymer is usually retained in the same reaction vessel throughout the synthesis, thus minimizing mechanical losses. The only physical operations involved are agitation, washing, and filtration.

3. Large excesses of chemical reagents can be used, speeding and making more efficient the coupling reactions. The excess soluble reagents are easily removed by filtration and thorough washing of the resin.

4. No purification of the intermediate polymer-bound peptides, apart from removal of soluble reactants and by-products, is feasible, and the losses entailed in more traditional purification procedures (recrystallization, chromatography, *etc.*) are eliminated. In principle, this is acceptable only if all the reactions in the synthetic sequence proceed quantitatively. Since this is unlikely to be achieved, the solid-phase method inevitably leads to a final crude product contaminated with shorter peptides lacking one or more amino-acids (failure sequences). Similarly, the intervention of side reactions may result in contamination by modified and partial (truncated) sequences. The success of solid-phase synthesis therefore ultimately depends upon whether the final product is capable of being adequately purified from closely related co-products. Very high coupling efficiencies are necessary in order to obtain products of even modest purity direct from the resin support. Thus average yields of about 98% for the formation of each peptide bond are necessary to obtain a resin-bound decapeptide in which 80% of the peptide chains are complete. For a 20-residue peptide, this figure rises to 99% average yield. In both cases, the remaining 20% of resin-bound product will consist of shorter peptide chains, mostly lacking just one or two amino-acid residues. The purification problems are usually therefore very formidable.

5. Because of the insoluble nature of the reaction product at all intermediate stages, quantitative analytical control is difficult and is usually not attempted. Rigorous characterization of the final purified product is thus mandatory.

The method has been applied with substantial success to the preparation of very many short (8–12 residue) biologically active peptides. At this level, failure sequences lacking one or more of the constituent amino-acids may often be separated. Solid-phase syntheses of very many larger molecules have also been carried out, and products with high biological activity obtained. It is very difficult, however, to establish complete homogeneity in such syntheses. The outstanding achievement thus far is the generation of enzymically active polypeptide by solid-phase assembly of the 124-residue sequence of ribonuclease A.[109]

23.6.5 STRATEGY AND TACTICS IN PEPTIDE SYNTHESIS[110]

In planning the synthesis of peptides of substantial size, particular attention has to be paid at the outset to the overall plan or strategy of synthesis and to the tactics whereby this plan might be most efficiently achieved. The main strategic decision is the manner in which the particular amino-acid sequence is to be assembled, whether in stepwise fashion, one residue at a time commencing at the amino or carboxy terminus, or by union of several part sequences (fragment condensation), whether in free solution or by solid phase synthesis, *etc.* Tactical considerations include the choice of protecting group combinations for the terminal amino and carboxy groups, and for the various amino-acid side chains. Some of these protecting groups will be 'permanent', in so far as they will be retained until the end of the synthesis; others will be 'temporary' and will be removed at intermediate stages to allow formation of a particular peptide bond or to change solubility properties, *etc.* The conditions for their removal need to be considered in terms of the amino-acid composition of the peptide. The choice of coupling procedures and solvent media also comes under the tactical heading, especially in relation to the avoidance of racemization.

23.6.5.1 Stepwise assembly and fragment condensation

The three basic approaches for elaboration of polypeptide chains are illustrated in Scheme 40, parts a–c. The first approach, stepwise elongation from the amino-terminus

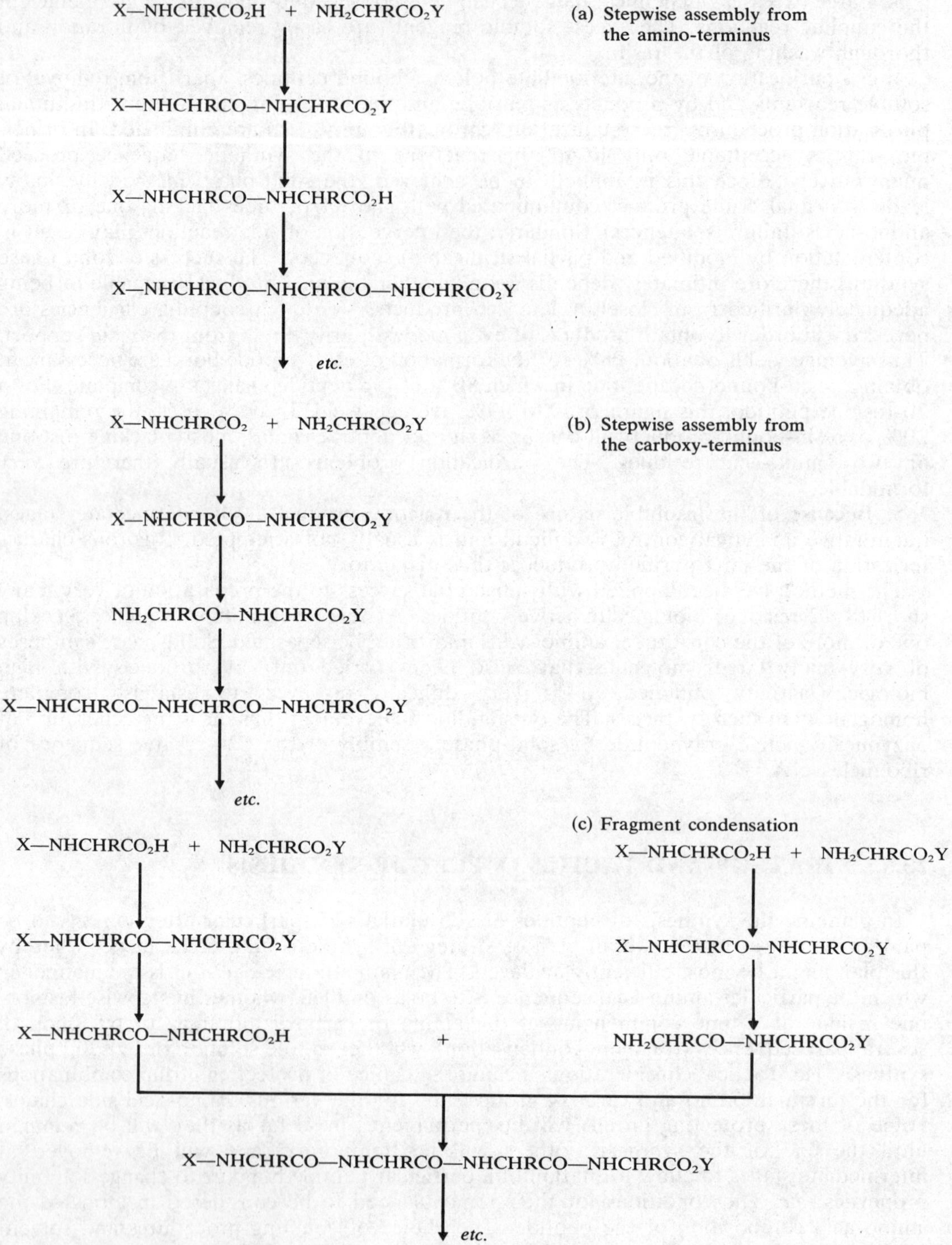

SCHEME 40

(Scheme 40, a) is rarely used, although it is directly analogous to that of protein biosynthesis. It involves removal of carboxy protecting groups and activation of the peptide carboxy-terminus. This approach therefore maximizes the opportunity for racemization and other potential side reactions involving the activated peptide. The reverse process, stepwise elongation from the carboxy-terminus (Scheme 40, b) is much more attractive. The activated species are now simple amino-acid derivatives which may be protected from racemization by the use of urethane-type or other carefully chosen amino-protecting groups. Furthermore, it is found in practice that yields in coupling reactions are in general considerably higher with activated acylamino-acids than with activated peptides. The former are of course much more readily accessible than larger peptides, and coupling yields can be further improved by use of moderate excesses of the acylating agent. This is especially appropriate in solid-phase synthesis, where excess soluble reagents are easily removed from the insoluble polymer phase. Despite the fact that the stepwise approach maximizes the number of consecutive coupling reactions, their relative efficiency makes this procedure often the method of choice for peptides of moderate length. It has been used successfully in solution for the synthesis of the entire sequence of a 27-residue hormone, secretin.[111] Biologically active preparations of even larger peptides and small proteins (including the enzyme ribonuclease, 124 amino-acid residues)[109] have been obtained using this approach in solid-phase synthesis.

Fragment condensation (Scheme 40, c) has the advantage of involving fewer consecutive coupling reactions, giving in principle at least the possibility of higher overall efficiency. The success of this approach may depend critically on the choice of division points of the target sequence. Glycine residues are obvious points for division into fragments since coupling of peptides with *C*-terminal glycine offers no possibility for racemization. Likewise proline is often placed at the *C*-terminus of peptide fragments since the formation of simple oxazolones is not possible with secondary amino-acids. Conversely, glutamine or γ-benzylglutamate residues should not be located at the amino-terminus of peptide fragments since there is risk of competitive cyclization reactions to pyrrolidonecarbonyl or pyroglutamyl derivatives (Scheme 41).

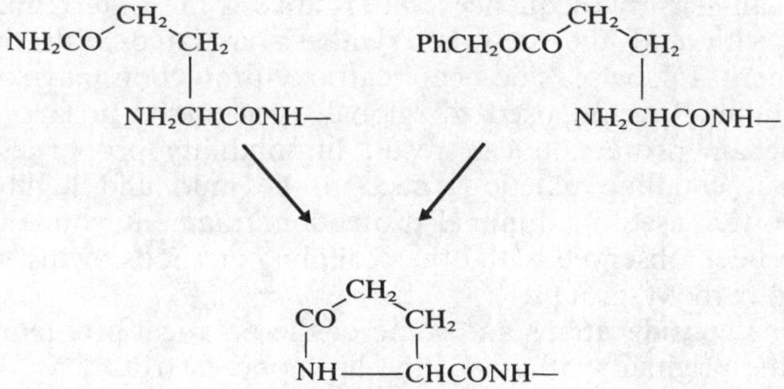

SCHEME 41

These reactions will obviously be favoured if the desired peptide bond-forming reaction is sluggish due to steric hindrance, and in general bulky amino-acid residues, particularly valine and isoleucine, should not be located at coupling points. In all cases involving coupling of peptide fragments with *C*-terminal residues other than glycine or proline, the coupling reagent must be chosen to minimize the risk of racemization.

23.6.5.2 Choice of protecting groups and activation procedures

As mentioned earlier, the protecting groups involved in peptide synthesis fall broadly into two categories — permanent and temporary. In the stepwise method of assembly (Scheme 40, b) for example, the various side-chain protecting groups and that masking the initial *C*-terminus will usually be retained until the end of the synthesis. On the other hand, the amino-protecting groups for the individual acylamino-acids will have only a short life and need to be selectively removable in the presence of the permanent groups. A common way of achieving this is by appropriate combination of benzyl- and t-butyl-based derivatives. Perhaps the mildest and cleanest de-protection method available to peptide chemists is catalytic hydrogenolysis. Thus a frequently used procedure is stepwise assembly of *N*-benzyloxycarbonylamino-acids bearing side-chain acid-labile t-butoxycarbonyl (lysine), t-butyl ester (aspartic acid, glutamic acid), or t-butyl ether (serine, threonine, tyrosine) groups. Usually activated ester derivatives are used in such stepwise syntheses since racemization is not a problem. Hydrogenolysis is generally inapplicable in the presence of divalent sulphur, so a different approach is necessary for peptides containing cysteine or methionine. Often it suffices to reverse the roles of the benzyl and t-butyl groups, and to assemble the sequence using *N*-t-butoxycarbonylamino-acids bearing side-chain benzyloxycarbonyl, benzyl ester, and benzyl ether groups. Repetitive removal of the amino-protecting groups is achieved by mild acidolysis, while the benzyl-based permanent groups are cleaved at the end of the synthesis by very strong acids.

Often the foregoing stepwise method is used for the preparation of partial sequences required for subsequent fragment condensation assembly. In these cases selective deprotection of the *C*-terminus may be required. In the absence of base-labile sequences (*cf.* p. 334), methyl esters may be used in combination with *N*-benzyloxycarbonyl and t-butyl based side-chain protection, cleavage of the methyl ester being by mild alkali or hydrazine treatment. The former deprotection procedure liberates the free peptide acid which may be activated for further condensation by, for example, dicyclohexylcarbodi-imide in the presence of hydroxybenzotriazole with minimal risk of racemization. When the *C*-terminal residue is glycine or proline, other methods of activation may be used; mixed anhydrides are often found effective and minimize isolation problems. Cleavage of the methyl ester protecting group by hydrazine enables further condensation reactions to be carried out by the azide procedure.

The more labile phenyl esters[112] may be used for carboxy-terminus protection in the presence of alkali-sensitive sequences. Alternatively, the *C*-terminal protecting group may be dispensed with and the stepwise synthesis initiated with the free amino-acid as amino-component. Likewise, side-chain carboxy-protection may sometimes be avoided, a strategy of 'minimal' as opposed to 'global' protection. In large molecules the use of minimal side-chain protection can result in solubility properties more favourable for purification, but coupling reactions need to be mild and highly selective. The azide procedure is often used in minimal-protection fragment condensation strategies. Side reactions have been observed with other coupling methods owing to transfer of activation to unprotected carboxy-groups.[113]

The foregoing considerations and some of the practical problems encountered are well illustrated in the original synthesis of the hormone gastrin.[114–117] This natural 17-residue peptide (**117**) is involved in the control of gastric acid secretion in the stomach. Its

Glu-Gly-Pro-Trp-Met-Glu-Glu-Glu-Glu-Glu-Ala-Tyr-Gly-Trp-Met-Asp-Phe-NH_2

1 2 3 4 5 6 7 8 9 10 11 12 13 14 15 16 17

(**117**)

amino-acid composition includes two residues of methionine (limiting use of hydrogenolysis during synthesis), two residues of acid-sensitive tryptophan, and six acidic amino-acid residues. The carboxy-terminus is blocked as a primary amide and the amino-terminus is a pyroglutamyl (cyclized glutamic acid) residue. The initial decision was to use a combination of stepwise and fragment condensation strategies with division of the chain at the 5–6 and 13–14 peptide bonds. The glycine residue at position-13 provided an obvious point for division since it placed a non-racemizable residue at the *C*-terminus of one fragment. Division at residue-5 was indicated because the presence of methionine at this position precluded further use of hydrogenolysis in elaboration of the long central sequence.

The central peptide comprising residues 6–13 was synthesized in a simple stepwise manner (Scheme 42).[115] Alanine methyl ester was used as the starting amino-component, benzyloxycarbonyl derivatives were used throughout for amino protection with removal

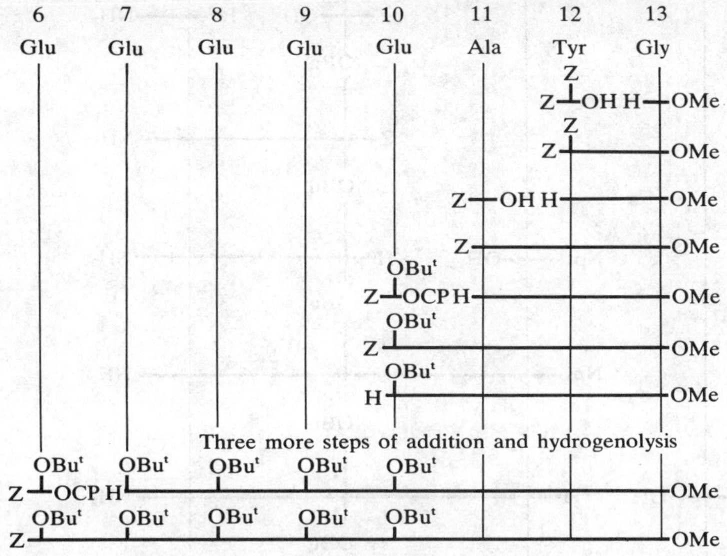

SCHEME 42

by hydrogenolysis, and t-butyl esters for side-chain carboxy-group protection. The *N,O*-bisbenzyloxycarbonyl derivative of tyrosine was used for the introduction of this amino-acid, but both protecting groups were removed at the first hydrogenolysis and thereafter the phenolic hydroxyl was left unprotected. The t-butyl ether of tyrosine would have provided a safer alternative, but its preparation is less easy. All five residues of γ-t-butyl glutamate were introduced by trichlorophenyl active-ester condensations; dicyclohexyl-carbodi-imide coupling proved adequate for the alanine and tyrosine derivatives, but in later syntheses active esters were used for these steps also.

The tetrapeptide residues 14–17 contain a difficult combination of amino-acids, and a range of protecting groups was necessary in its synthesis (Scheme 43).[114,117] The *C*-terminal dipeptide amide was readily prepared by dicyclohexylcarbodi-imide coupling of the two amino-acid derivatives and the benzyloxycarbonyl group was cleaved by hydrogenolysis. Further hydrogenolysis was precluded by the following methionine residue and the very acid-labile *o*-nitrophenylsulphenyl protecting group was adopted for this amino-acid. Acidolysis of this group was easily possible while leaving the aspartyl β-t-butyl ester intact, but attempts to add the next amino-acid (tryptophan) in the same manner were frustrated by intervention of an unexpected rearrangement reaction during attempted de-protection (Scheme 44).

No other suitable highly acid-labile *N*-protecting group had been devised at that time and recourse was therefore made to the t-butoxycarbonyl derivative of tryptophan. Removal of this protecting group from the completed tetrapeptide then resulted in simultaneous cleavage of the aspartyl side-chain t-butyl ester.

Alternative syntheses were subsequently devised[116,117] (*e.g.* scheme 45) using a simple 2+2 fragment condensation approach, again leading to the zwitterionic tetrapeptide amide. The later discovery of non-acidic methods for the cleavage of *o*-nitrophenylsulphenyl groups provided a simple solution to the original problem, and thiolysis of the tetrapeptide derivative (**118**) is a practical route to the side-chain protected tetrapeptide amide[117] (Scheme 46).

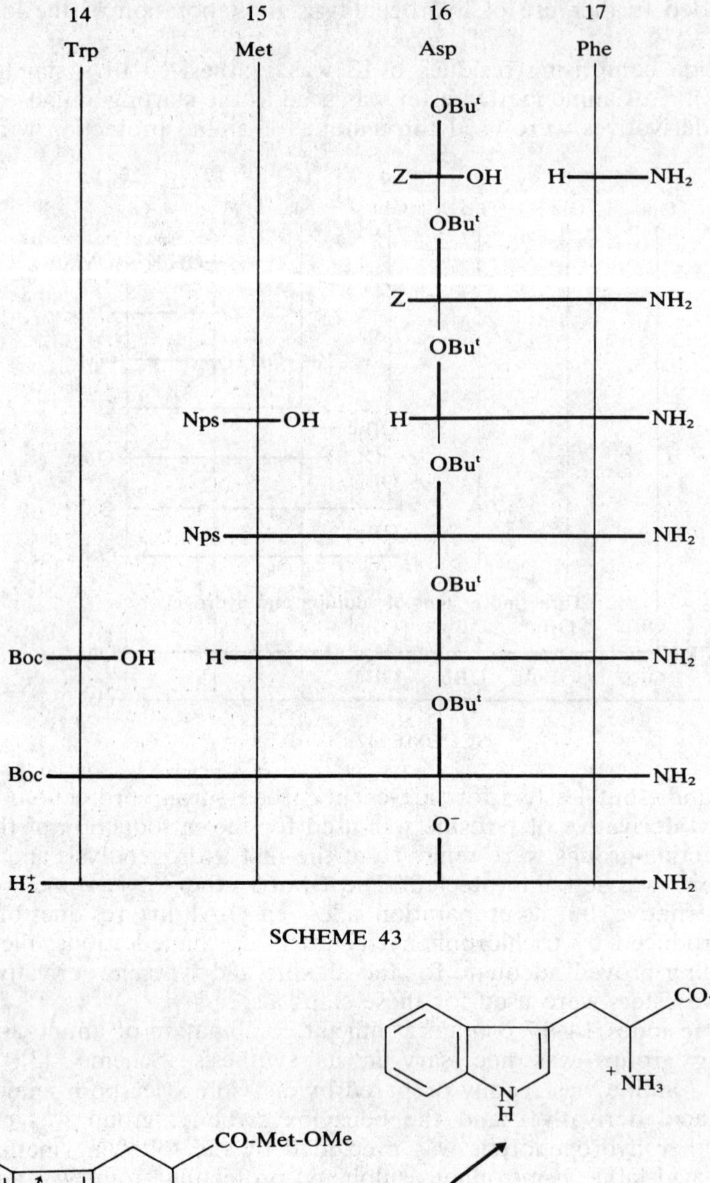

SCHEME 43

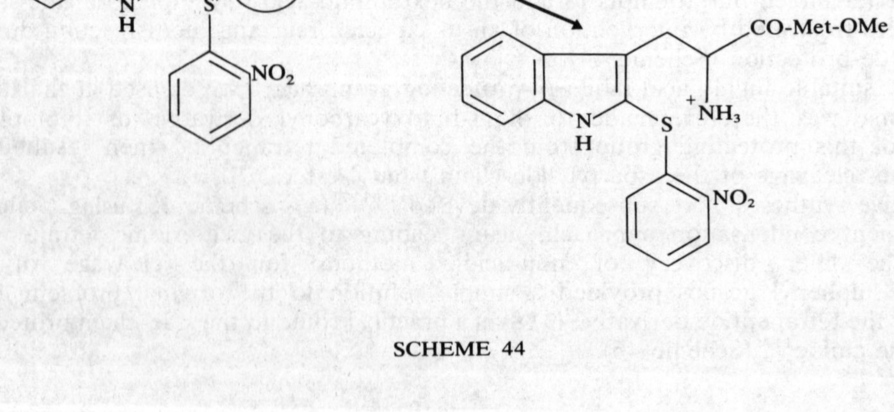

SCHEME 44

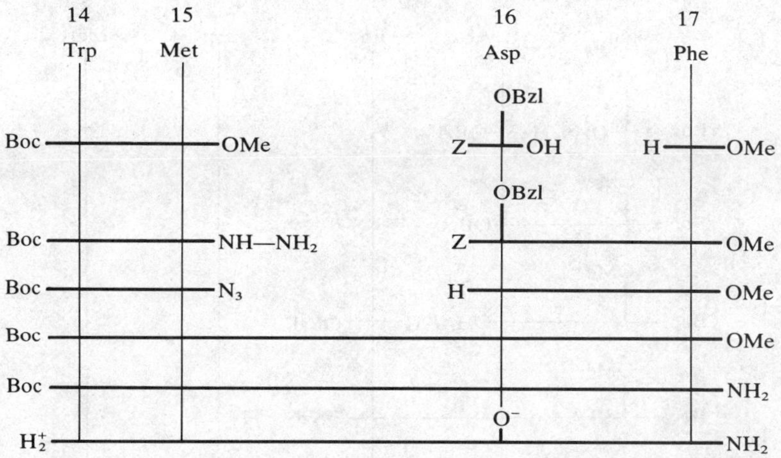

SCHEME 45

(**118**)

SCHEME 46

The presence of glycine and proline residues in the N-terminal fragment residues 1–5 allowed considerable flexibility in synthetic design, notwithstanding the more difficult Trp-Met sequence. The first synthesis[116] (Scheme 47) involved a straightforward 2+2

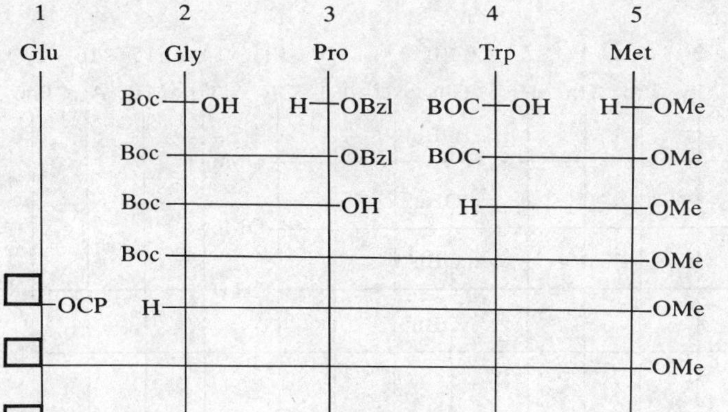

SCHEME 47

fragment condensation followed by addition of the N-terminal pyroglutamyl residue. Because the amino-group of this last residue is internally acylated, no protecting group was necessary. In an alternative synthesis (Scheme 48),[117] this residue was formed in the final step by cyclization of N-terminal γ-methylglutamate.

Several routes were explored for the final assembly of the N-terminal, central, and C-terminal fragments.[114,115] In the first (Scheme 49), the N-terminal pentapeptide methyl ester was converted into its hydrazide and thence the azide, and coupled with the central octapeptide. The reaction was slow (increasing hindrance of the N-terminus of the central

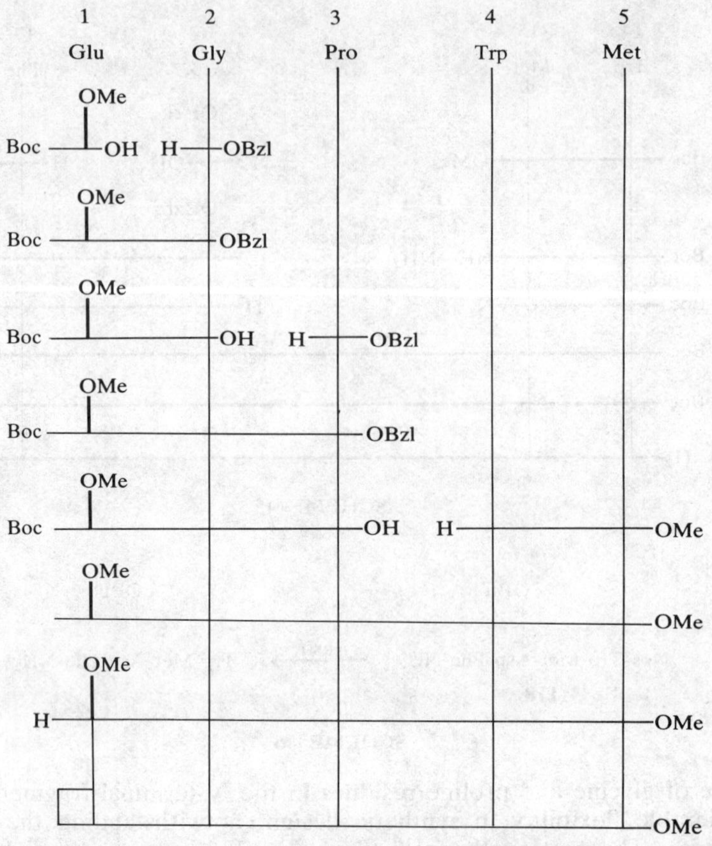

SCHEME 48

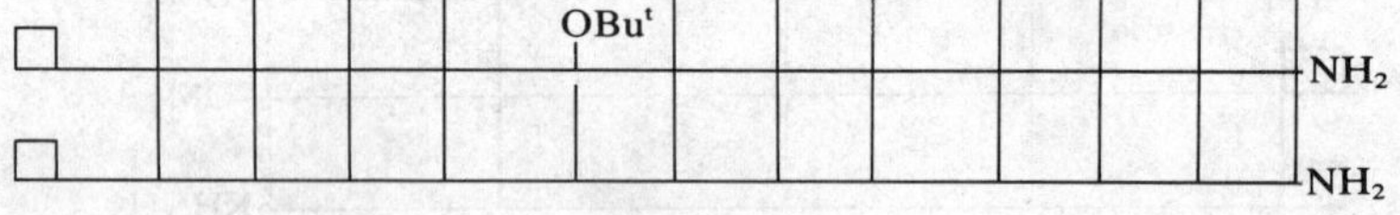

SCHEME 49

peptide t-butyl ester was noted during its stepwise preparation) and a substantial part of the azide component underwent a competitive unimolecular Curtius rearrangement (p. 349). An alternative procedure was therefore devised (Scheme 50) in which the *N*-t-butoxycarbonyloctapeptide was first saponified and combined (mixed anhydride method) with the *C*-terminal fragment. All the protecting groups were then stripped from the resulting tridecapeptide and coupling with the *N*-terminal pentapeptide azide then proceeded much more rapidly and smoothly with no evidence for Curtius rearrangement of the azide.

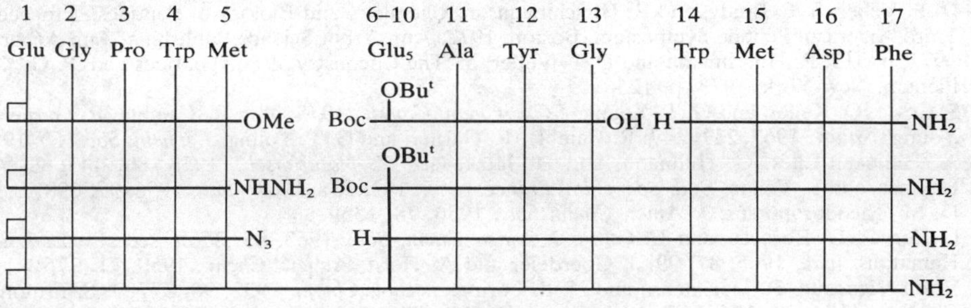

SCHEME 50

References

1. Abbreviations and nomenclature follow the recommendations of the I.U.P.A.C.–I.U.B. Commission on Biochemical Nomenclature published in 'Amino-Acids, Peptides, and Proteins', Specialist Periodical Reports of the Chemical Society, London, vols. 2, 4, 5, 8, and elsewhere.
2. For recent detailed reviews see (a) M. Bodanszky, Y. S. Klausner, and M. A. Ondetti, 'Peptide Synthesis' Wiley, New York, 2nd edn., 1976; (b) F. M. Finn and K. Hofmann, in 'The Proteins', ed. H. Neurath, R. L. Hill, and C.-L. Boeder, Academic Press, New York, 3rd edn., 1976, p. 106; (c) 'Methoden der Organischen Chemie (Houben-Weyl)', Vol. XXV/1,2, 'Synthese von Peptiden', ed. E. Wunsch, Georg Thieme Verlag, Stuttgart, 1974. Current developments are reviewed annually in (d) 'Amino-Acids, Peptides, and Proteins', Specialist Periodical Reports of the Chemical Society, London, which also includes lists of peptides synthesized and useful synthetic intermediates.
3. G. T. Young, in 'Peptides', Proceedings of the 8th European Peptide Symposium, Noordwijk, Holland, 1966, North Holland, Amsterdam, 1967, p. 55.
4. S. Goldschmidt, *Angew. Chem.*, 1950, **62,** 538; S. Goldschmidt and M. Wick, *Z. Naturforsch.*, 1950, **5b,** 170; *Ann. Chem.*, 1952, **575,** 217.
5. W. Dieckmann and F. Breest, *Ber.*, 1906, **39,** 3052; H. Staudinger, *Helv. Chim. Acta*, 1922, **5,** 87; C. Naegeli and A. Tyabji, *ibid.*, 1934, **17,** 931; *ibid.*, 1935, **18,** 142.
6. S. Goldschmidt and H. Lautenschlager, *Ann. Chem.*, 1953, **580,** 68.
7. M. Bergmann and L. Zervas, *Ber.*, 1932, **65,** 1192.
8. M. Bergmann, L. Zervas, and W. F. Ross, *J. Biol. Chem.*, 1935, **111,** 245; see also H. Leuchs, *Ber.*, 1906, **39,** 857; H. Leuchs and W. Manasse, *ibid.*, 1907, **40,** 3235; H. Leuchs and W. Geiger, *ibid.*, 1908, **41,** 1721.
9. J. Meienhofer and K. Kuromizu, *Tetrahedron Letters*, 1974, 3259; K. Kuromizu and J. Meienhofer, *J. Amer. Chem. Soc.*, 1974, **96,** 4978.
10. R. H. Sifferd and V. du Vigneaud, *J. Biol. Chem.*, 1935, **108,** 753; H. S. Loring and V. du Vigneaud, *ibid.*, 1935, **111,** 385.
11. D. Ben-Ishai and A. Berger, *J. Org. Chem.*, 1952, **17,** 1564; D. Ben-Ishai, *ibid.*, 1954, **19,** 62; G. W. Anderson, J. Blodinger, and A. D. Welcher, *J. Amer. Chem. Soc.*, 1952, **74,** 5309; R. A. Boissonnas and I. Schumann, *Helv. Chim. Acta*, 1952, **35,** 2229.
12. S. Sakakibara and Y. Shimonishi, *Bull. Chem. Soc. Japan*, 1965, **38,** 1412; S. Sakakibara, Y. Shimonishi, Y. Kishida, M. Okada, and H. Sugihara, *ibid.*, 1967, **40,** 2164.
13. L. A. Carpino, *J. Amer. Chem. Soc.*, 1957, **79,** 98; F. C. McKay and N. F. Albertson, *ibid.*, 1957, **79,** 4686; G. W. Anderson and A. C. McGregor, *ibid.*, 1957, **79,** 6180.
14. R. Schnabel, P. Sieber, and H. Kappeler, *Helv. Chim. Acta*, 1959, **42,** 2622.
15. *Chem. Eng. News*, 1976, **54,** 3.
16. G. W. Anderson and A. C. McGregor, *J. Amer. Chem. Soc.*, 1957, **79,** 6180.
17. W. Broadbent, J. S. Morley and B. E. Stone, *J. Chem. Soc.* (*C*), 1967, 2632.
18. E. Schnabel, H. Herzog, P. Hoffmann, E. Klauke, and I. Ugi, *Angew. Chem. Internat. Edn.*, 1968, **7,** 380; *Annalen*, 1968, **716,** 175; E. Schnabel, J. Stoltefuss, H. A. Offe, and E. Klauke, *ibid.*, 1971, **743,** 57.
19. L. Moroder, A. Hallett, E. Wunsch, O. Keller, and G. Wersin, *Z. physiol. Chem.*, 1976, **357,** 1651.
20. P. Sieber and B. Iselin, *Helv. Chim. Acta*, 1968, **51,** 614, 622.
21. C. Birr, F. Flor, P. Fleckenstein, and Th. Wieland, in 'Peptides 1971' Proceedings of the 11th European Peptide Symposium, Vienna, 1971, ed. H. Nesvadba, North-Holland, Amsterdam, 1973, p. 175; C. Birr, W. Lochinger, G. Stahnke, and P. Lang, *Annalen*, 1972, **763,** 162; C. Birr, *ibid.*, 1973, 1652.
22. G. I. Tesser, in 'Peptides', Proceedings of the 13th European Peptide Symposium, Kiryat Anavim, Israel, 1974, ed. Y. Wolman, Wiley, New York–Israel University Press, Jerusalem, 1975, p. 53; G. I. Tesser and I. C. Balvert-Geers, *Internat. J. Peptide Protein Res.*, 1975, **7,** 295.
23. L. A. Carpino and G. Y. Han, *J. Amer. Chem. Soc.*, 1970, **92,** 5748; *J. Org. Chem.*, 1972, **37,** 3404.
24. D. S. Kemp and D. C. Roberts, *Tetrahedron Letters*, 1975, 4629.
25. D. S. Kemp and C. F. Hoyng, *Tetrahedron Letters*, 1975, 4625.

26. D. F. Veber, S. F. Brady, and R. Hirschmann, in 'Chemistry and Biology of Peptides', Proceedings of the Third American Peptide Symposium, Boston, 1972, Ann Arbor Science Publishing, Ann Arbor, Michigan, 1972, p. 315; R. Hirschmann and D. F. Veber, in 'The Chemistry of Polypeptides', ed. P. G. Katsoyannis, Plenum, New York, 1973, p. 125.
27. S. Coyle, O. Keller, and G. T. Young, *J.C.S. Chem. Comm.*, 1975, 939; R. Camble, R. Garner, and G. T. Young, *Nature*, 1967, **217,** 247; R. Camble, R. Garner, and G. T. Young, *J. Chem. Soc.* (*C*), 1969, 1911.
28. A. Hillmann-Elies, G. Hillmann, and H. Jatzkewitz, *Z. Naturforsch.*, 1953, **8b,** 445; G. Amiard, R. Heymes, and L. Velluz, *Bull. Soc. chim. France*, 1955, 191, 201, 1464; *ibid.*, 1956, 97, 698; L. Zervas and D. M. Theodoropoulos, *J. Amer. Chem. Soc.*, 1956, **78,** 1359.
29. L. Zervas, D. Borovas, and E. Gazis, *J. Amer. Chem. Soc.*, 1963, **85,** 3660; see also L. Zervas and C. Hamalidis, *ibid.*, 1965, **87,** 99; J. Goerdeler and A. Holst, *Angew. Chem.*, 1959, **71,** 775.
30. V. du Vigneaud, R. Dorfmann, and H. S. Loring, *J. Biol. Chem.*, 1932, **98,** 577; J. S. Fruton and H. T. Clarke, *ibid.*, 1934, **106,** 667; A. Stoll and Th. Petrzilka, *Helv. Chim. Acta*, 1952, **35,** 589; see also J. C. Sheehan, D. W. Chapman, and R. W. Roth, *J. Amer. Chem. Soc.*, 1952, **74,** 3822.
31. F. Weygand and E. Csendes, *Angew. Chem.*, 1952, **64,** 136; F. Weygand, *Chem.-Ztg.*, 1954, **78,** 480.
32. D. A. A. Kidd and F. E. King, *Nature*, 1948, **162,** 776; *J. Chem. Soc.*, 1949, 3315; J. C. Sheehan and V. S. Frank, *J. Amer. Chem. Soc.*, 1949, **71,** 1856; see also W. Grassmann and E. Schulte-Uebbing, *Chem. Ber.*, 1950, **83,** 244.
33. G. H. L. Nefkens, *Nature*, 1960, **185,** 309; G. H. L. Nefkens, G. I. Tesser, and R. J. F. Nivard, *Rec. Trav. chim.*, 1960, **79,** 688.
34. G. W. Kenner and J. H. Seely, *J. Amer. Chem. Soc.*, 1972, **94,** 3259.
35. J. S. A. Amaral, G. C. Barrett, N. H. Rydon, and J. E. Willett, *J. Chem. Soc.* (*C*), 1966, 807; P. M. Hardy, H. N. Rydon, and R. C. Thompson, *Tetrahedron Letters*, 1968, 2525.
36. D. S. Kemp and J. Reczek, *Tetrahedron Letters*, 1977, 1031.
37. K. Hofmann, M. Z. Magee, and A. Lindenmann, *J. Amer. Chem. Soc.*, 1950, **72,** 2814; K. Hofmann, A. Lindenmann, M. Z. Magee, and N. H. Khan, *ibid.*, 1952, **74,** 470.
38. R. A. Boissonnas, St. Guttman, and P. A. Jaquenoud, *Helv. Chim. Acta*, 1960, **43,** 1349; R. Schwyzer, E. Surbeck-Wegman, and H. Dietrich, *Chimia* (*Aarau*), 1960, **14,** 366.
39. E. Klieger and H. Gibian, *Annalen*, 1962, **655,** 195.
40. M. Itoh, *Chem. Pharm. Bull.* (*Japan*), 1969, **17,** 1679.
41. K. Hofmann, A. Rheiner, and W. D. Peckham, *J. Amer. Chem. Soc.*, 1953, **75,** 6083; K. Hofmann, W. D. Peckham, and A. Rheiner, *ibid.*, 1956, **78,** 238; H. O. Van Orden and E. L. Smith, *J. Biol. Chem.*, 1954, **208,** 751.
42. L. Zervas, M. Winitz, and J. P. Greenstein, *Arch. Biochem. Biophys.*, 1956, **65,** 573; *J. Org. Chem.*, 1957, **22,** 1515; *J. Amer. Chem. Soc.*, 1961, **83,** 3300; see also G. Losse and C. Rueger, *Z. Chem.*, 1973, **13,** 344; E. Wunsch and G. Wendlberger, *Chem. Ber.*, 1967, **100,** 160.
43. G. Jager and R. Geiger, *Chem. Ber.*, 1970, **103,** 1727.
44. E. Schnabel and C. H. Li, *J. Amer. Chem. Soc.*, 1960, **82,** 457.
45. J. R. Ramachandran and C. H. Li, *J. Org. Chem.*, 1962, **27,** 4006.
46. D. Yamashiro, J. Blake, and C. H. Li, *J. Amer. Chem. Soc.*, 1972, **94,** 2855.
47. D. Yamashiro and C. H. Li, *J. Amer. Chem. Soc.*, 1973, **95,** 1310.
48. J. R. Bell and J. H. Jones, *J.C.S. Perkin I*, 1974, 2336.
49. V. du Vigneaud and O. K. Behrens, *J. Biol. Chem.*, 1937, **117,** 27.
50. S. Shaltiel and M. Fridkin, *Biochemistry*, 1970, **9,** 5122.
51. T. Fujii and S. Sakakibara, *Bull. Chem. Soc. Japan*, 1974, **47,** 3416.
52. F. Weygand, W. Steglich, and P. Pietta, *Chem., Ber.*, 1967, **100,** 3841.
53. R. H. Sifferd and V. du Vigneaud, *J. Biol. Chem.*, 1935, **108,** 753; H. S. Loring and V. du Vigneaud, *ibid.*, 1935, **111,** 385.
54. S. Akabori, S. Sakakibara, Y. Shimonishi, and Y. Nobuhara, *Bull. Chem. Soc. Japan*, 1964, **37,** 433.
55. D. Yamashiro, R. L. Noble, and C. H. Li, in 'Chemistry and Biology of Peptides', Proceedings of the Third American Peptide Symposium, Boston, 1972, Ann Arbor Science Publishing, Ann Arbor, Michigan, 1972, p. 197; see also D. Yamashiro and C. H. Li, *J. Amer. Chem. Soc.*, 1973, **95,** 1310; D. Yamashiro, R. L. Noble, and C. H. Li, *J. Org. Chem.*, 1973, **38,** 3561.
56. L. Zervas and I. Photaki, *Chimia*, 1960, **14,** 375; R. G. Hiskey and J. B. Adams, Jr., *J. Org. Chem.*, 1965, **30,** 1340; R. W. Hanson and H. D. Law, *J. Chem. Soc.*, 1965, 7285; I. Photaki, J. Taylor-Papadimitriou, C. Sakarellos, P. Mazarakis, and L. Zervas, *J. Chem. Soc.* (*C*), 1970, 2683.
57. G. Amiard, R. Heymes, and L. Velluz, *Bull. Soc. chim. France*, 1956, 698.
58. D. F. Veber, J. D. Milkowski, R. G. Denkewalter, and R. Hirschmann, *Tetrahedron Letters*, 1968, 3057; D. F. Veber, J. D. Milkowski, S. Varga, R. G. Denkewalter, and R. Hirschmann, *J. Amer. Chem. Soc.*, 1972, **94,** 5456.
59. P. Sieber, B. Kamber, A. Hartmann, A. Johl, B. Rimiker, and W. Rittel, *Helv. Chim. Acta*, 1977, **60,** 27.
60. H. Zahn and G. Schmidt, *Annalen*, 1970, **731,** 91, 101.
61. B. W. Erickson and R. B. Merrifield, *J. Amer. Chem. Soc.*, 1973, **95,** 3750.
62. J. C. Sheehan and G. P. Hess, *J. Amer. Chem. Soc.*, 1955, **77,** 1067.
63. J. Rebek, 'Peptides 1974', Proceedings of the 13th European Peptide Symposium, Israel, 1974, ed. Y. Wolman, Wiley, New York, 1975, p. 27.
64. M. C. Khosla, M. M. Hall, R. R. Smeby, and F. M. Bumpus, *J. Medicin. Chem.*, 1974, **17,** 431.
65. G. W. Anderson and F. Callahan, *J. Amer. Chem. Soc.*, 1958, **80,** 2902.

66. M. W. Williams and G. T. Young *J. Chem. Soc.*, 1963, 881.
67. W. Konig and R. Geiger, *Chem., Ber.*, 1970, **103,** 788, 2024, 2034, 2041.
68. F. Weygand, D. Hoffmann, and E. Wünsch, *Z. Naturforsch.*, 1966, **21b,** 426.
69. G. W. Anderson, J. E. Zimmerman, and F. M. Callahan, *J. Amer. Chem. Soc.*, 1964, **86,** 1839.
70. G. W. Anderson, J. E. Zimmerman, and F. M. Callahan, *J. Amer. Chem. Soc.*, 1967, **89,** 178.
71. R. Schwyzer, *Helv. Chim. Acta*, 1954, **37,** 647; R. Schwyzer, B. Iselin, and M. Feurer, *ibid.*, 1955, **38,** 69; R. Schwyzer, M. Feurer, B. Iselin, and H. Kagi, *ibid.*, 1955, **38,** 80.
72. M. Bodanszky, *Nature*, 1955, **175,** 685; *Acta Chim. Acad. Sci. Hung.*, 1957, **10,** 335; M. Bodanszky, M. Szelke, E. Tomorkeny, and E. Weisz, *Chem. Ind.* (*London*), 1955, 1517; *Acta Chim. Acad. Sci. Hung.*, 1957, **11,** 179; Th. Wieland and F. Jaenicke, *Annalen*, 1956, **599,** 125.
73. J. A. Farrington, G. W. Kenner, and J. M. Turner, *Chem. Ind.* (*London*), 1955, 601; J. A. Farrington, P. J. Hextall, G. W. Kenner, and J. M. Turner, *J. Chem. Soc.*, 1957, 1407; G. W. Kenner, P. J. Thomson, and J. M. Turner, *ibid.*, 1958, 4148.
74. Th. Wieland, W. Schafer, and E. Bokelmann, *Annalen*, 1951, **573,** 99.
75. See, for example, J. Pless and R. A. Boissonnas, *Helv. Chim. Acta*, 1963, **46,** 1609.
76. M. Rothe and F. W. Kunitz, *Annalen*, 1957, **609,** 88; D. F. Elliott and D. W. Russell, *Biochem. J.*, 1957, **66,** 49p.
77. B. Iselin, W. Rittel, P. Sieber, and R. Schwyzer, *Helv. Chim. Acta*, 1957, **40,** 373.
78. J. A. Farrington, G. W. Kenner, and J. M. Turner, *Chem. Ind.* (*London*), 1955, 601; J. A. Farrington, P. J. Hextall, G. W. Kenner, and J. M. Turner, *J. Chem. Soc.*, 1957, 1407; G. W. Kenner, P. J. Thomson, and J. M. Turner, *ibid.*, 1958, 4148.
79. G. Kupryszewski and M. Formela, *Roczniki Chem.*, 1963, **37,** 161; J. Kovacs and A. Kapoor, *J. Amer. Chem. Soc.*, 1965, **87,** 118; J. Kovacs, G. N. Schmit, and U. R. Ghatak, *Biopolymers*, 1968, **6,** 817; J. Kovacs, L. Kisfaludy, M. Q. Ceprini, and R. H. Johnson, *Tetrahedron*, 1969, **25,** 2555.
80. L. Kisfaludy, J. E. Roberts, R. H. Johnson, G. L. Mayers, and J. Kovacs, *J. Org. Chem.*, 1970, **35,** 3563; J. Kovacs, L. Kisfaludy, and M. Q. Ceprini, *J. Amer. Chem. Soc.*, 1967, **89,** 183; L. Kisfaludy, M. Low, O. Nyeki, T. Szirtes, and I. Schon, *Annalen*, 1973, 1421.
81. J. H. Jones and G. T. Young, *J. Chem. Soc.* (*C*), 1968, 436.
82. H. D. Jakubke, *Z. Naturforsch.*, 1965, **20b,** 273; H. D. Jakubke, *Z. Chem.*, 1965, **5,** 453.
83. F. Weygand, P. Huber, and K. Weiss, *Z. Naturforsch.*, 1967, **22b,** 1084; F. Weygand and C. DiBello, *ibid.*, 1969, **24b,** 314; Th. Wieland, C. Birr, and F. Flor, *Angew. Chem. Internat. Edn.*, 1971, **10,** 336; Th. Wieland, F. Flor, and C. Birr, *Annalen*, 1973, 1595; F. Flor, C. Birr, and Th. Wieland, *ibid.*, 1973, 1601.
84. M. Zaoral, *Coll. Czech. Chem. Comm.*, 1962, **27,** 1273.
85. Th. Wieland and H. Bernhard, *Annalen*, 1951, **572,** 190; R. A. Boissonnas, *Helv. Chim. Acta*, 1951, **34,** 874; J. R. Vaughan, Jr., and R. L. Osato, *J. Amer. Chem. Soc.*, 1952, **74,** 676.
86. G. W. Kenner and R. J. Stedman, *J. Chem. Soc.*, 1952, 2069; D. W. Clayton, J. A. Farrington, G. W. Kenner, and J. M. Turner, *ibid.*, 1957, 1398.
87. R. G. Denkewalter, H. Schwam, R. G. Strachan, T. E. Beesley, D. F. Veber, E. F. Schoenewaldt, H. Barkemeyer, W. J. Paleveda, Jr., T. A. Jacob, and R. Hirschmann, *J. Amer. Chem. Soc.*, 1966, **88,** 3163; R. Hirschmann, R. G. Strachan, H. Schwam, E. F. Schoenewaldt, H. Joshua, H. Barkemeyer, D. F. Veber, W. J. Paleveda, Jr., T. A. Jacob, T. E. Beesley, and R. G. Denkewalter, *J. Org. Chem.*, 1967, **32,** 3415; R. Hirschmann, *Intra-Sci. Chem. Reports*, 1971, **5,** 204.
88. H. Hagenmaier and H. Frank, *Z. physiol. Chem.*, 1972, **353,** 1973.
89. J. R. Vaughan, Jr., and R. L. Osato, *J. Amer. Chem. Soc.*, 1951, **73,** 5553.
90. M. A. Tilak, *Tetrahedron Letters*, 1970, 849; see also R. Sarges and B. Witkop, *J. Amer. Chem. Soc.*, 1965, **87,** 2020; A. van Zon and H. C. Beyerman, *Helv. Chim. Acta*, 1973, **56,** 1729.
91. H. Chantrenne, *Nature*, 1947, **160,** 603; 1949, **164,** 576; *Biochim. Biophys. Acta*, 1948, **2,** 286; 1950, **4,** 484.
92. J. C. Sheehan and V. S. Frank, *J. Amer. Chem. Soc.*, 1950, **72,** 1312.
93. G. W. Anderson, J. Blodinger, R. W. Young, and A. D. Welcher, *J. Amer. Chem. Soc.*, 1952, **74,** 5304; G. W. Anderson and R. W. Young, *ibid.*, 1952, **74,** 5307; G. W. Anderson, J. Blodinger, and A. D. Welcher, *ibid.*, 1952, **74,** 5309.
94. A. G. Jackson, G. W. Kenner, G. A. Moore, R. Ramage, and W. D. Thorpe, *Tetrahedron Letters*, 1976, 3627.
95. H. Leuchs, *Ber.*, 1906, **39,** 857; H. Leuchs, and W. Geiger, *ibid.*, 1908, **41,** 1721.
96. F. Fuchs, *Ber.*, 1922, **55,** 2943; C. J. Brown, D. Coleman, and A. C. Farthing, *Nature*, 1949, **163,** 834; V. Bruckner, T. Vajda, and J. Kovacs, *Naturwiss.*, 1954, **41,** 449.
97. E. Katchalski, *Adv. Protein Chem.*, 1951, **6,** 123.
98. For a review see Y. S. Klausner and M. Bodanszky, *Synthesis*, 1974, 549.
99. Y. S. Klausner and M. Bodanszky, *Synthesis*, 1974, 549.
100. K. Hofmann, M. Z. Magee, and A. Lindenmann, *J. Amer. Chem. Soc.*, 1950, **72,** 2814; K. Hofmann, A. Lindenmann, M. Z. Magee, and N. H. Khan, *ibid.*, 1952, **74,** 470; see also H. Boshagen and J. Ullrich, *Chem. Ber.*, 1959, **92,** 1478.
101. R. B. Woodward and R. A. Olofson, *J. Amer. Chem. Soc.*, 1961, **83,** 1007; R. B. Woodward and R. A. Olofson, *Tetrahedron suppl.*, 1966, No. 7, 415; R. B. Woodward, R. A. Olofson, and H. Mayer, *J. Amer. Chem. Soc.*, 1961, **83,** 1010; R. B. Woodward, R. A. Olofson, and H. Mayer, *Tetrahedron Suppl.*, 1966, No. 8, Part 1, 321.
102. D. S. Kemp and S. W. Chien, *J. Amer. Chem. Soc.*, 1967, **89,** 2743.

103. B. Belleau and G. Malek, *J. Amer. Chem. Soc.*, 1968, **90,** 1651.
104. For a review, see T. Mukaiyama, *Angew. Chem. Internat. Edn.*, 1976, **15,** 94.
105. A. J. Bates, I. J. Galpin, A. Hallett, D. Hudson, G. W. Kenner, R. Ramage, and R. C. Sheppard, *Helv. Chim. Acta*, 1975, **58,** 688.
106. For recent reviews, see B. W. Erickson and R. B. Merrifield, in 'The Proteins', ed. H. Neurath, R. L. Hill, and C.-L. Boeder, Academic Press, New York, 1976, vol. 3, p. 257; J. Meienhofer, in 'Hormonal Proteins and Peptides', ed. C. H. Li, Academic Press, New York, 1973, p. 46.
107. R. B. Merrifield, *J. Amer. Chem. Soc.*, 1963, **85,** 2149.
108. E. Atherton, D. L. J. Clive, and R. C. Sheppard, *J. Amer. Chem. Soc.*, 1975, **97,** 6584.
109. B. Gutte and R. B. Merrifield, *J. Biol. Chem.*, 1971, **246,** 1922.
110. For a full discussion, see M. Bodanszky, Y. S. Klausner, and M. A. Oudetti, in 'Peptide Synthesis', Wiley New York, 2nd edn., 1976, p. 177.
111. M. Bodanszky, M. A. Ondetti. S. D. Levine, and N. J. Williams, *J. Amer. Chem. Soc.*, 1967, **89,** 6753.
112. G. W. Kenner and J. H. Seely, *J. Amer. Chem. Soc.*, 1972, **94,** 3259.
113. S. Natarajan and M. Bodanszky, *J. Org. Chem.*, 1976, **41,** 1269.
114. J. C. Anderson, M. A. Barton, R. A. Gregory, P. M. Hardy, G. W. Kenner, J. K. MacLeod, J. Preston, R. C. Sheppard, and J. S. Morley, *Nature*, 1964, **204,** 933.
115. J. C. Anderson, G. W. Kenner, J. K. MacLeod, and R. C. Sheppard, *Tetrahedron Suppl.*, 1966, No. 8, 39.
116. J. C. Anderson, M. A. Barton, P. M. Hardy, G. W. Kenner, J. Preston, and R. C. Sheppard, *J. Chem. Soc. (C)*, 1967, 108.
117. J. M. Davey, A. H. Laird, and J. S. Morley, *J. Chem. Soc.* (*C*), 1966, 555.

23.7
Conformations of Polypeptides

G. C. BARRETT
Oxford Polytechnic

23.7.1 INTRODUCTION AND NOMENCLATURE

23.7.1.1 Introduction

A survey of the conformational properties of polypeptides, and the methods by which these properties are established, provides material of relevance to acyclic molecules more generally. Although an understanding of the conformational behaviour of this group of natural products is one essential factor when attempting to account for their biological functions, particularly oligopeptide hormone–receptor interactions and enzyme catalysis (Chapter 24.1), this topic occupies a rightful place in this volume because it illustrates the influence of organic functional groups in determining preferred conformations of acyclic molecules.

23.7.1.2 Primary structures and nomenclature

The general structure of a polypeptide (Figure 1) is most conveniently visualized as an array of linked amino-acid residues, since the chain is created by condensing amino-acids in sequence, and is broken down into amino-acids by hydrolysis. One alternative view of the repeating unit, less commonly used, is to visualize the chain as a sequence of peptide units (Figure 1). The remaining alternative view of the polypeptide chain, as a sequence of 'amide units' —CONHCHR^n— is not used in the literature; like the 'peptide unit' nomenclature it is untidy in the sense that there are groupings 'left over' at each end of the polypeptide chain. This nomenclature also involves a further possible source of confusion since the first 'amide unit', for example, incorporates the second side-chain. The 'amino-acid residue' nomenclature is therefore used throughout this section.

By convention, the polypeptide chain is arranged across the page with the —$\overset{+}{N}H_3$ group (the '*N*-terminus') to the left, and the —CO_2^- group (the '*C*-terminus') to the right. This display (Figure 1) is the fully-extended conformation.

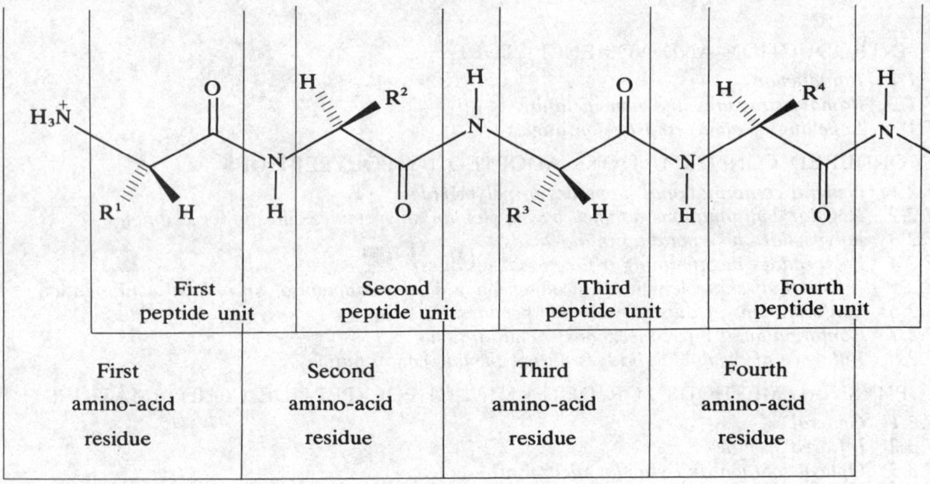

Figure 1 Polypeptide in the extended conformation (L-amino-acid residues)

23.7.1.3 Labelling of atoms and torsion angles

The perspective view of a polypeptide composed of L-amino-acid residues, labelled in terms of atoms and torsion angles, is shown in Figure 2. The atoms forming the polypeptide backbone are given subscript numbers corresponding to the number of the amino-acid residue in the polypeptide, numbering from the *N*-terminus. Side-chain atoms have the same subscript number, with carbon atoms labelled β, γ, δ, *etc.*, according to their relationship with the carbonyl carbon atom (C′).

Torsion angles are labelled ϕ and ψ for the sequences of atoms —C̰O—NH—CHR—C̰O— and —N̰H—C̰HR—C̰O—NH— respectively, and have the value 180° in the fully-extended conformation (Figure 2), for which the torsion angle ω describing rotation within the amide group is also 180° (*i.e. trans*-amide bond). The conformation of a polypeptide, as far as the run of the backbone through the molecule is concerned (the secondary structure), is therefore fully described by a set of torsion angles ϕ, ψ, and ω for each amino-acid residue. Corresponding data for side-chains (torsion angles χ_i, with appropriate superscript numbers determined by Sequence Rules[1] to specify the bond to which the angle refers) complete a full description of the tertiary structure of the polypeptide. Torsion angles are + for bonds with right-handed helicity, and *vice versa.*

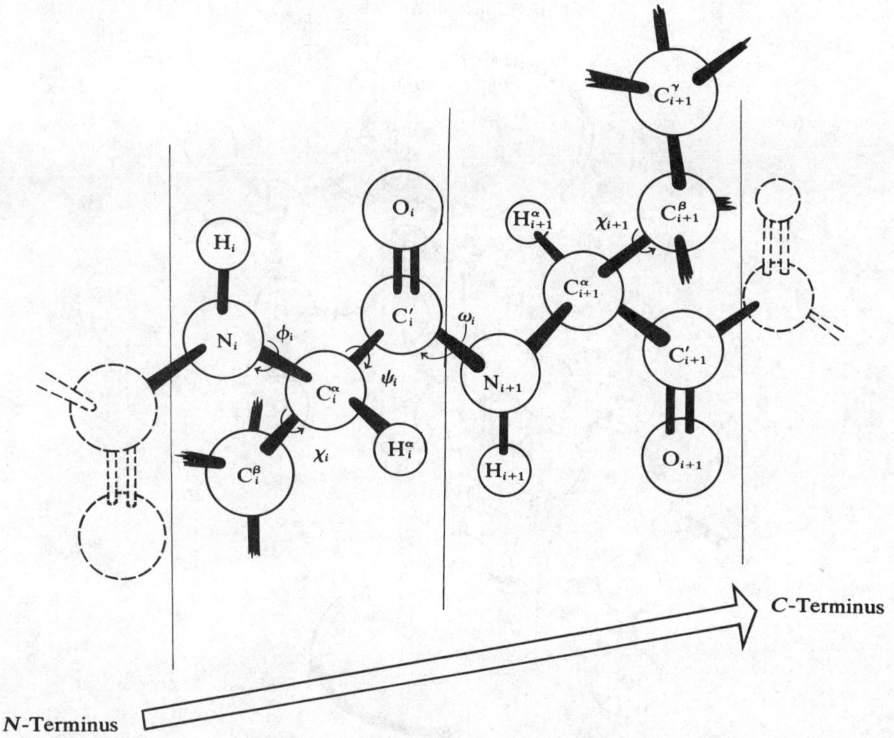

Figure 2 Two L-amino-acid residues of a polypeptide, showing conventions for labelling atoms and torsion angles

23.7.2 ORDERED CONFORMATIONS ADOPTED BY POLYPEPTIDES

23.7.2.1 General conformational properties of polypeptides

A feature of the simplest type of polypeptide (Figure 1) is the repetition of the same grouping —NHCHCO— throughout the backbone of the molecule. This repetitive feature fulfils one condition, encouraging a molecule to adopt an ordered conformation, and if two further conditions of structural regularity are also fulfilled by a polypeptide, namely (i) the same configuration at each chiral centre, and (ii) all side-chains identical or structurally similar (*e.g.* polarity) and of small size, then the polypeptide adopts one of two ordered conformations, a helix, or a sheet structure formed from aligned chains.

Although certain stereoregular polyalkenes adopt a helical conformation in the solid state,[2] the particular feature of the polypeptide which stabilizes an ordered conformation in the solid state as well as in solution is the hydrogen bonding which is possible between amide groups. In the right-handed α-helix (Figure 3), the hydrogen bond is formed between the amide N—H proton of the *i*th residue and the carbonyl group of the $(i-4)$th residue. An example of a structure fulfilling conditions for α-helix formation in solution is poly-(L-glutamic acid) in aqueous solution at pH 4.3.[3]

In the more general situation, for a polypeptide constructed by the stepwise condensation of different amino-acids, the effect of a diversity of side-chain structures is usually to overcome the tendency for ordered conformational behaviour imposed by the structural regularity of the backbone of the molecule. A 'random' or 'disordered' conformation is then commonly adopted in the general case. This situation applies in globular proteins,

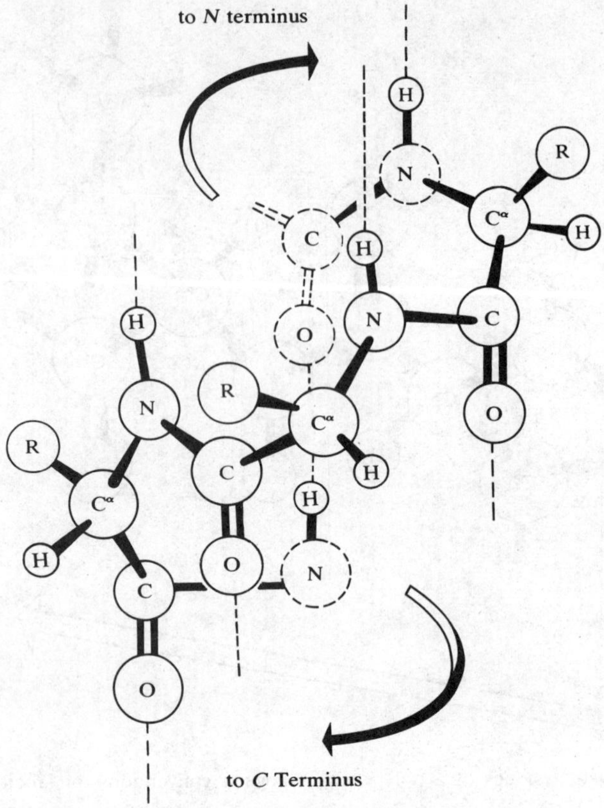

Figure 3 Right-handed α-helix (L-amino-acid residues)

including enzymes, where the molecule as a whole does not adopt an ordered conformation, but, nevertheless, regions of the polypeptide chain adopt ordered conformations where side-chains of complementary types are grouped together. As examples, regions of the insulin molecule are 'casually helical',[4] and regions of horse heart ferricytochrome *c* have been described similarly ('gently helical'),[5] while in other regions the polypeptide chain lies in a 'hairpin'-type conformation due to the formation of an element of the 'antiparallel sheet' or β-conformation (Figure 4).

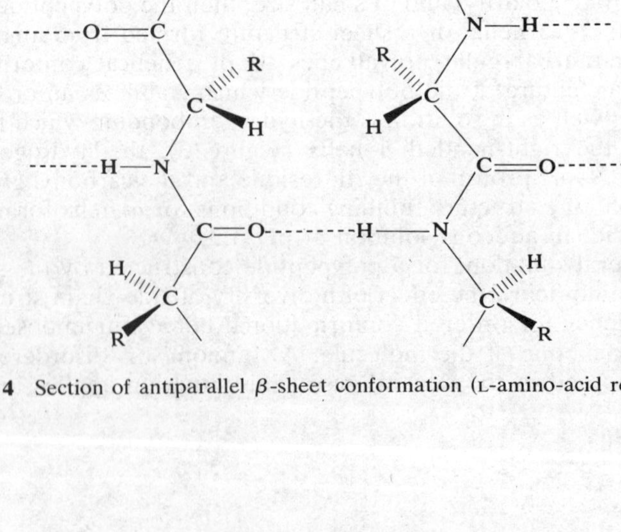

Figure 4 Section of antiparallel β-sheet conformation (L-amino-acid residues)

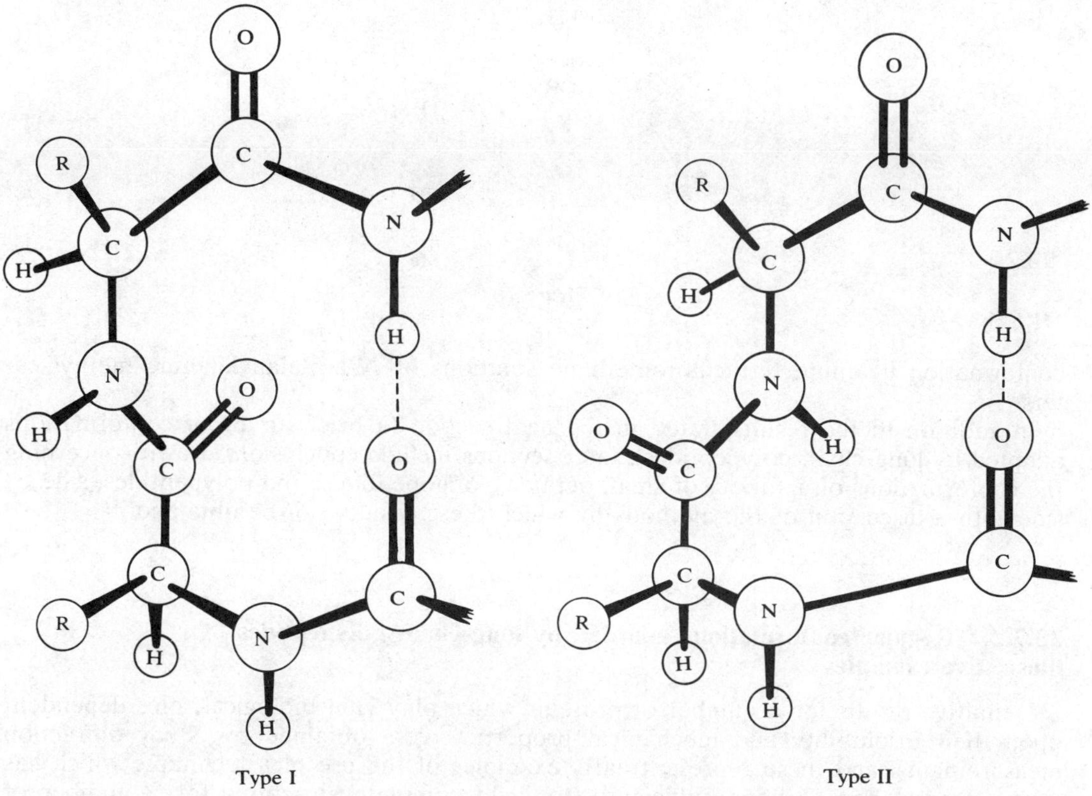

Figure 5 Type I and type II β-bends

There are other conformational features which have been recognized sufficiently often so that specific names have been suggested for them. One of these, a 'β-bend' or 'β-turn', or '3_{10}-bend', or '3_{10}-helix', is common and may adopt two forms, type I and type II (Figure 5), the latter being identical with type I except for opposite helicity for the bridging CHRNHCOCHR sequence. A γ-turn, and 11-membered hydrogen-bonded ring (Figure 6), has recently been demonstrated for the first time to be an independent conformational feature for some polypeptides.[6] Type II β-turns have been noted to occur in three localities in α-chymotrypsin,[7] and the 'casual helicity' of the carboxyl-terminal half of the insulin A-chain is in fact due to a succession of these features. Tetrapeptides containing glycyl, L-prolyl, and L-asparaginyl residues are likely to adopt this 10-membered hydrogen-bonded conformation,[8] rather than a seven-membered ring alternative form (Figure 7) which is present in equilibrium with a minor proportion of the random

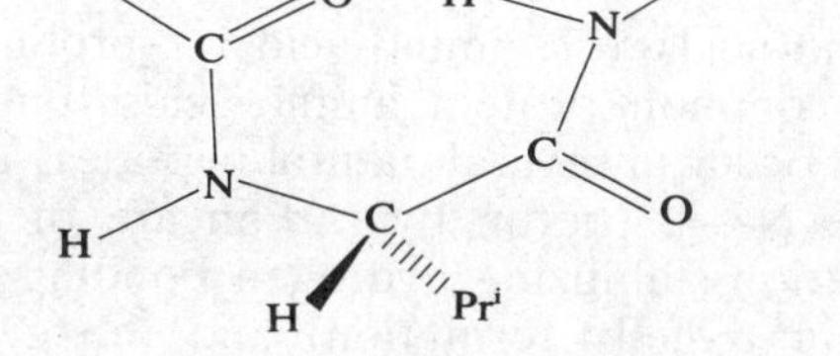

Figure 6 γ-Turn in Boc-Gly-L-Val-Gly-OMe

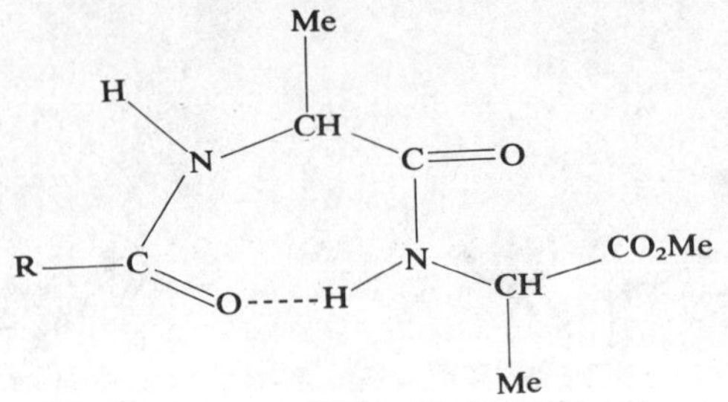

Figure 7

conformation in dilute tetrachloromethane solutions of *N*-acylalanylalanine methyl esters.[9]

In addition to the results stated in the next section (a brief survey of conformations adopted by long-chain polypeptides), later sections include conclusions drawn concerning the conformations of a variety of small peptides, oligopeptides, and polypeptides, side by side with a discussion of the methods by which these results can be obtained.

23.7.2.2 Regular conformations adopted by long-chain polypeptides: illustrative examples

Definitive results for a number of proteins which play vital biological roles dependent upon their insolubility and mechanical properties were obtained by X-ray diffraction measurements, and these represent early examples of the use of a technique which has more recently been refined sufficiently to yield complete structures for a number of crystalline globular proteins. The stretched or β-form of keratin illustrates the β-sheet, with both parallel and antiparallel arrangements of the peptide chains (Figure 4); in silk fibroin, only the parallel arrangement of peptide chains is found, with close to a planar sheet structure rather than the pleated sheet seen in keratins. The unstretched or α-form of keratin exemplifies the α-helix in its most compact form, in which five turns of the right-handed helix incorporate 18 amino-acid residues — hence it can be described as a 3.6-residue helical conformation. The right-handed helicity can be seen from molecular models to be preferred, since in comparison with the left-handed form a polypeptide composed of L-amino-acid residues disposes its side-chains outwards from the axis of the helix so that destabilizing repulsions, particularly involving carbonyl groups, are minimized (see Figure 3).

The structure of collagen, the major fibrous constituent of connective tissue, is that of a triple helix, with three peptide chains twisted in a right-handed sense about the same axis (approximately 3.3 amino-acid residues per turn of helix). The high proportion of glycine, proline, and hydroxyproline (see next section) accounts for this unusual and distinctive structure.

23.7.2.3 Polypeptides incorporating imino-acids

The secondary amino-acids (*i.e.* imino-acids; L-proline and hydroxy-L-proline) are numbered among the common protein amino-acids, but *N*-methyl derivatives of the simple amino-acids also occur in several natural peptides. The two major consequences of the replacement of the N—H proton by carbon are the inability of these residues to participate in conformation-stabilizing hydrogen bonding (making proline, for example, unable to participate in α-helix formation) and the greater relative stability of the *cis*-amide bond for tertiary amides.

23.7.2.4 Polypeptides incorporating trifunctional residues

Amino-acids carrying a functional group in the side-chain may introduce unusual conformational properties when incorporated into a polypeptide if the side-chain functional group competes with backbone amide groups for hydrogen-bonding opportunities. While poly-(L-lysine) [Figure 1; all side-chains —$(CH_2)_4NH_2$] exists in solutions at pH 11.75 as a stable helix, more stable than that formed by poly-(L-ornithine) (Figure 1; all side-chains —$CH_2CH_2CH_2NH_2$) under these conditions, changing the pH of the solutions to pH 7 leads to disordering ('denaturation').[10] Dipole–dipole interactions between the ester groups and backbone amide groups in poly-(β-benzyl-L-aspartate) encourage the formation of a left-handed helix, *i.e.* the opposite helicity to that adopted by other helix-forming poly-(L-amino-acids); a spectacular example of the dependence of solution conformation upon solvent is provided in this case, since change of solvent from chloroform to trimethyl phosphate reverses the sense of the helix (left to right).[11]

Studies with structurally homogeneous polypeptides, *i.e.* poly-(L-amino-acids), have also led to information about the effects of the steric size of side-chains on the propensity of particular amino-acid residues in polypeptides to participate in ordered conformational arrangements. Thus, valine and isoleucine, carrying the bulky isopropyl and s-butyl side-chains, respectively, discourage the formation of α-helical conformations in regions of a globular protein. Amino-acids carrying a heteroatom (*i.e.* serine, cysteine, or threonine) are similarly reluctant to become involved in helix formation. The most strongly bound helices are formed by poly-(L-alanine) and poly-(L-leucine) (side-chains Me and Bu^i, respectively), and by poly-(γ-alkyl-L-glutamates).[12]

Polyglycine and polyproline are exceptional cases, in that polyglycine helices involve interchain hydrogen bonds rather than the intramolecular hydrogen bonds of the α-helix,[13] while no hydrogen bonds are possible with proline residues but nevertheless the structural and stereochemical regularity is in itself sufficient to lead to left-handed helix formation. These are not α-helices, but possess threefold screw-axis symmetry with *trans*-amide bonds (a poly-L-proline with *cis*-amide bonds is known), and the polyglycine case is instructive in that α-helix formation is apparently dependent upon the presence of a side-chain larger than hydrogen.

Poly-(L-glutamic acid), like other poly(amino-acids) carrying ionizable side chains, is readily water-soluble, and adopts an α-helical conformation at pH values low enough to suppress ionization. An additional helix structure formed by the network of γ-carboxy-groups hydrogen-bonded to each other stabilizes the structure further; this 'super-helix' is strengthened by the addition of dioxan or hydroxylic solvents to the solution.

23.7.2.5 Influence of chain length of an oligopeptide on the adoption of an ordered conformation

Competition between ordered conformations involving intermolecular or intramolecular hydrogen-bonding is usually determined in favour of intramolecularly hydrogen-bonded forms for oligopeptides in solution, although a run of up to five, six, or seven residues is usually required before any ordering sets in.

An early study of poly-(γ-benzyl-L-glutamate) showed that a run of at least seven residues is required before a tendency towards α-helix formation can be established. Longer oligomers adopt mixtures of α-helix and β-sheet structures.[14] At least four residues are required in any case, before the first turn of the α-helix can be formed, and more recent studies[15,16] of structurally homogeneous oligopeptides Boc—(L-amino-*acid*)$_n$—OMe (where Boc = Bu^tOCO; L-amino-acid = Ala, Ile, Val, Leu, Met, β-OMeAsp, γ-OMeGlu, or Lys) show that in most cases a β-sheet conformation is preferred, especially at high concentrations and also when some water is present in the solution. The terminal acyl and ester groupings were included in the structure so that the solubility of the peptide in solvents such as chloroform could be increased. In summary, while seven residues of alanine, valine, or isoleucine favour the formation of the β-sheet conformation in this series[15–18] in trifluoroethanol solutions, and eight residues for the β-methylaspartate derivative (in $CHCl_3$), the other derivative show a preference for the α-helix (seven residues for leucine, and methionine, nine for the γ-methylglutamate derivative, and twelve for the lysine derivative). Generally, tendency to favour the α-helix conformation at longer than these minimum runs is shown throughout these series.[18]

Sequential oligopeptides H—(L-Tyr-L-Ala-L-Glu)$_n$ OH are unordered in aqueous solutions at pH 7.4 until the length of chain $n = 13$ is reached, when a significant amount of α-helix conformation is discernible.[19]

Oligo-glycines and -prolines are characteristically different from the oligo-(L-amino-acids) just described, with the onset of ordering occurring at three residues in the proline case but with no well-defined ordering in the glycine case being discernible with fewer than six residues (for aqueous solutions).[18]

23.7.2.6 Conformations of diastereoisomeric peptides

A dipeptide exists in an extended conformation in solution in the L,L-configuration. Although an intramolecular hydrogen bond leading to an eight-membered ring appears feasible, this involves substantial repulsive interactions between the two side-chains when the amide adopts the energetically favourable *trans*-configuration. However, the D,L-diastereoisomer adopts the more compact intramolecularly hydrogen-bonded structure (Figure 8a) in deuterium oxide.[20] An extension of these principles to longer oligopeptides shows that a D-residue in a run of L-amino-acid residues causes a change of direction for an extended conformation, or disrupts an ordered conformation.

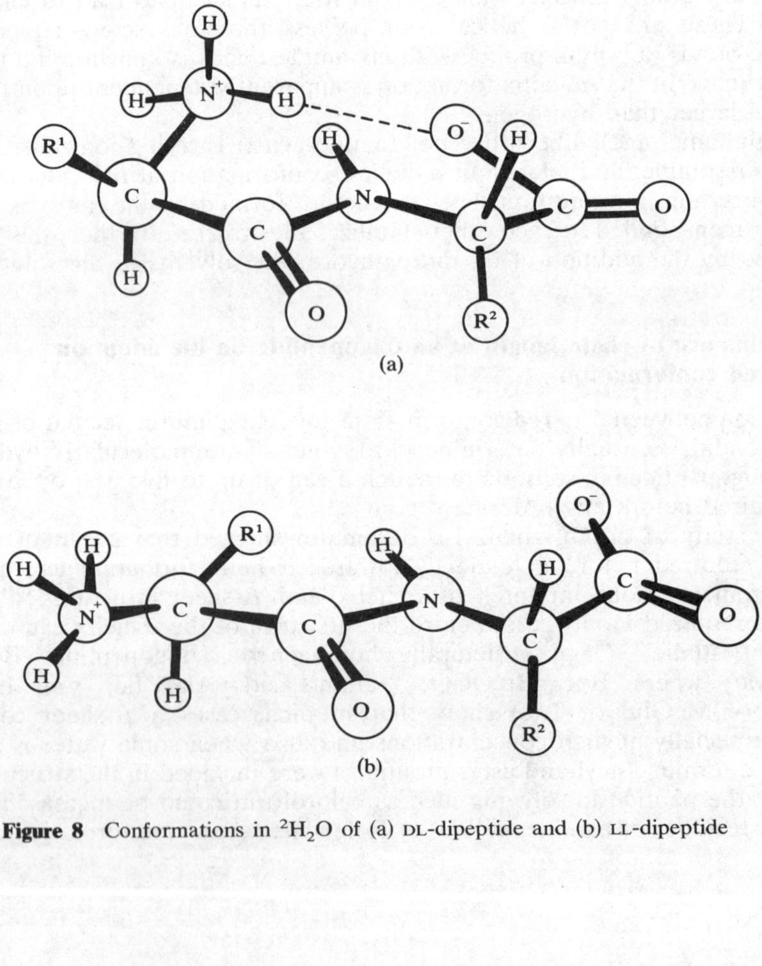

Figure 8 Conformations in ^2H_2O of (a) DL-dipeptide and (b) LL-dipeptide

23.7.2.7 Conformational interconversions: denaturation

Transformation of an α-helix conformation into the β-form in films of water-insoluble poly-(L-amino-acids) can be brought about by stretching in steam, and the process can be reversed by dissolving the film in chloroform (one of a range of 'helicogenic' solvents—those which do not compete significantly for the NH proton as hydrogen-bond acceptors).[21] The effect of moisture has been referred to above (Section 23.7.2.5) in promoting partial α-helix–β-sheet interconversion in water-soluble poly-(L-amino-acids). Two helical structures are adopted by poly-(L-proline), in one of which the amide bonds are *trans*, and in the other they are *cis*; the effect of high pressure applied to a solution of the more stable *trans*-isomer is to transform it into the more compact *cis*-form.[22]

The term 'denaturation' has been taken into the oligopeptide sphere from its original usage to describe disordering of globular protein conformations, but the more obvious term 'disordering', or phrases such as 'helix to random-coil transitions' are more commonly used. Acidic or basic side-chains stabilize α-helical conformations in solutions at pH values where the side-chain functional group is in its un-ionized form, but change of pH leads to disordering. Studies of poly-(L-amino-acids) in helicogenic solvents have provided precise information on the influence of added hydrogen-bond-breaking solvents (*e.g.* halogenoacetic acids), since changes from ordered to disordered conformations can be followed conveniently by spectroscopic methods (Section 23.7.3). Those poly-(L-amino-acids) which form strong helices [poly-(L-alanine) and poly-(L-leucine)] are still helical in dichloroacetic acid, and require high concentrations of trifluoroacetic acid to disorder them, while weak helices such as poly-(β-benzyl-L-aspartate) are disordered when 10% dichloroacetic acid is added to their solutions in chloroform. It is from results such as these that semiquantitative information on 'hydrophobic attraction' between hydrocarbon side-chains has been obtained.

High concentrations of urea or guanidine hydrochloride in aqueous solutions lead to total denaturation, which in some cases can be reversed by removal of the denaturant by dialysis. These particular denaturants continue to be widely used since very high concentrations can be achieved in aqueous solutions, but other salts promote disordering,[23] and so, to a limited extent, do detergents such as sodium dodecyl sulphate.[24] It is not possible to make generalizations in this area, since many unusual observations have been reported involving amino-acid residues carrying side-chain functional groups; for example,[25] poly-(1-benzyl-L-histidine) adopts unordered conformations in solutions containing small amounts of simple mineral acids, but a stoichiometric amount of perchloric acid leads to α-helix formation. Furthermore, it has been reported[26] that sodium dodecyl sulphate appears to increase the amount of ordering of the enzyme elastase from relatively low levels, and removal of the detergent by dialysis leaves the globular protein with enhanced β-structure.

23.7.2.8 Influence of disulphide bridges on polypeptide conformations

Intramolecular linking between side-chains of two cysteine residues creates a disulphide bridge which generally assists the maintenance of an ordered conformation. Many polypeptides with important biological functions have primary structures involving disulphide bridges between cysteine residues separated by several atoms along the polypeptide chain, so that many-membered rings are formed in this way. The effect of the disulphide bridge on the conformation of the polypeptide chain sited between the two cysteine residues is best seen in the increase in disordering which accompanies cleavage of the disulphide group. Lysozyme loses some 50% α-helical content after cleavage of its disulphide bonds,[27] but cleavage of the polypeptide chain at two points (at methionine residues) gives three peptide fragments connected by disulphide bridges which are devoid of α-helical regions. Results such as these reinforce the view that the primary structure (*i.e.* the amino-acid sequence) of the polypeptide is the dominant factor in determining its tertiary structure, and that additional covalent bonds such as disulphide bridges play only a supporting role. Cleavage of the disulphide loop in thyrocalcitonin, a polypeptide hormone containing 32 amino-acid residues, has little effect on the ordering of the polypeptide backbone, which is α-helical to only a small extent in water but to *ca.* 50% in 2-chloroethanol.[28]

23.7.3 PHYSICAL METHODS FOR DETERMINING POLYPEPTIDE CONFORMATIONS

23.7.3.1 General

The earliest data supporting the notion that certain proteins and poly(amino-acids) may adopt ordered conformations in the solid state were obtained using X-ray diffraction and infrared absorption methods. These techniques gave a substantial amount of important information in this area, at an early stage in the application of physical methods in structure determination, and using relatively unsophisticated apparatus.

From the 1960s to recent times, optical rotatory dispersion (o.r.d.) and circular dichroism spectroscopy (c.d.) were the most widely used techniques in this area, apart from the procession of definitive results from X-ray analysis of globular proteins (which, however, give little information on the dynamic properties of polypeptides in solution). O.r.d. and c.d. continue to provide useful information on the influence of physical changes in the environment of the polypeptide on its conformational behaviour, both in solution and, more recently, for membrane proteins in the solid state.

The volume of applications of nuclear magnetic resonance (n.m.r.) techniques for polypeptide conformational studies has approached or overtaken that of o.r.d. and c.d. in recent years, consistent not only with the wider availability of improved n.m.r. instrumentation, but also with the particular information on torsion angles which n.m.r. can provide.

Other physical methods providing information on polypeptide conformations are discussed in this section. These include conformational probing techniques, giving information on the conformational behaviour of selected regions of a polypeptide molecule.

23.7.3.2 Infrared spectra

Vibrational spectra of oriented films of polypeptides showing a peak at 1630 cm^{-1} can be diagnostic for the β-sheet structure; an additional peak at 1685–1700 cm^{-1} has been associated with the antiparallel packing mode. In contrast, the α-helix amide I absorption peak appears at 1650 cm^{-1} for the right-handed helix (at slightly higher frequencies for the left-handed form[29]), although unordered polypeptides show the corresponding band close by (1656 cm^{-1}). The amide I absorption frequency is identical for both *cis*- and *trans*-configurations, but there are differences for other deformations (*cis*: 1450, 1350 cm^{-1}, *trans*: 1550, 1250 cm^{-1} for the amide II and amide III regions, respectively). The amide V band is diagnostic of an extended conformation (700 cm^{-1}), the α-helix (620 cm^{-1}), or the disordered form (650 cm^{-1}), although this falls in a part of the i.r. range where many other absorption peaks arise.

Infrared studies of solution conformations of peptides have been less informative for polypeptides, and other spectroscopic methods have been found to be more suitable, but it has proved possible to identify intramolecular hydrogen-bonding in models for peptide units based on N—H stretching frequencies. Characteristic frequencies 3340 (hydrogen-bonded) and 3420 cm^{-1} (non hydrogen-bonded N—H) apply to the intramolecularly hydrogen-bonded conformation, and 3440 and 3460 cm^{-1} to the extended conformation, respectively, of *N*-acetyl amino-acid *N*-methylamides, according to Mizushima *et al.*,[30] and the appearance of all four peaks allows the analysis of conformer populations to be carried out for these compounds. Different assignments have been claimed for these peaks.[31]

The amide I and III bands for the β-sheet conformation mentioned above are also diagnostic of this conformation for solutions of oligopeptides, and conformational assignments have been made to oligo-(γ-ethyl-L-glutamates) on this basis.[32] Taken with circular dichroism data (Section 23.7.3.3), infrared spectra have provided information on the ordered structures adopted by protected L-alanine oligomers.[15] The use of Raman spectroscopy for deducing polypeptide conformations has been limited to poly-(L-amino-acids), but characteristic amide I frequencies can be exploited for the purpose, obtained for aqueous solutions using laser-Raman techniques.[33]

23.7.3.3 Optical rotation and circular dichroism

The variation of optical rotation with wavelength of the plane-polarized light used in its determination is referred to as optical rotatory dispersion (o.r.d.); an intimately-related technique, circular dichroism (c.d.), is the wavelength-dependence of the differential absorption of right- and left-circularly polarized light by a chiral molecule. Since c.d. is observed only in wavelength regions where absorption by the chiral molecule occurs, which for an amide is at wavelengths shorter than *ca.* 230 nm, application of this technique in the polypeptide area was precluded until instrumentation capable of providing differential absorption data into the 180–230 nm region became available. This was accomplished in the early 1960s, several years after a period of study which had established characteristic o.r.d. parameters by which the different ordered conformations could be recognized.

Optical rotations as a function of wavelength in the 350–600 nm range conform to a one-term Drude equation:

$$[\alpha] = \sum_i \left(\frac{b_0}{\lambda_i^2 - \lambda_0^2} \right)$$

$i = 1$ for a poly(amino-acid) in the disordered conformation in solution, while the data for a polypeptide in the α-helix form follow a two-term equation ($i = 2$). When the wavelength term λ_0 is 212 nm for a particular poly(amino-acid), then the constant b_0 has the value −630.[34] Using this value, the proportion of a polypeptide which exists in the right-handed α-helix can be calculated from a mathematical treatment of its o.r.d. spectrum measured in the visible wavelength range. A positive b_0 value corresponds to the rare cases of poly-(L-amino-acids) adopting the left-handed α-helix, if the value is near 630, but lower values between *ca.* +200 and −200 correspond to the β-sheet conformation.

Although these are clear rules for assigning the various conformations to polypeptides, they are of little help for determining the proportions in which mixed conformations are adopted by a globular protein, for example, in solution. Penetration into the far-u.v. region, through the absorption region of the amide chromophore, became possible with the more sophisticated spectropolarimeters of the 1960s, and characteristic superimposed Cotton effects were seen for each of the major conformations.

Circular dichroism spectra were shown to be more easily interpreted in general, since

TABLE 1
Circular Dichroism Features for Poly-(L-amino-acids)[36,70]

Conformation	*α-Helix*	*β-Sheet*	*Unordered*
Positive c.d. features	Strong positive c.d. (ε_{max} 191 nm)	Strong positive c.d. (ε_{max} 195 nm)	Weak positive c.d. (ε_{max} 195 nm)
Zero c.d.	180, 202, 250 nm	180, 207, 250 nm	211, 234, 250 nm
Negative c.d. features	Strong negative c.d. (ε_{max} 208, 222 nm)	Medium intensity c.d. (ε_{max} 217 nm)	Very weak negative c.d. (ε_{max} 238–240 nm)
	Medium intensity c.d. (ε_{max} *ca.* 165 nm)	Medium intensity c.d. (ε_{max} 150 nm)	Strong negative c.d. (ε_{max} 197 nm)

the various Cotton effects can be assigned to the respective electronic transitions involving the amide chromophore. Again, characteristic curves can be ascribed to the various ordered conformations, and the disordered form also has its own c.d. spectrum. A summary of the various c.d. spectra in the form of wavelengths of c.d. maxima and of crossover points between positive and negative Cotton effects is displayed in Table 1 relating to poly-(L-amino-acids) in different solvents. The use of these 'standard c.d. spectra' for computing the proportions of α-helix, β-sheet, and disordered conformations present in a globular protein, by attempting to simulate the experimentally determined c.d. spectrum of the protein by algebraic summation of various proportions of the spectra of the three 'pure' conformations, has been a major application of this spectroscopic technique in this field. The precision which can be achieved depends on the use of 'standards' of high conformational purity, and several poly-(L-amino-acids) have been advocated for the purpose. One particular uncertainty has been the conditions under which complete disordering of a poly(amino-acid) can be achieved. The c.d. spectrum of poly-(L-serine) at high salt concentrations has been proposed as a standard for the disordered conformation,[23] and c.d. spectra calculated for myoglobin, lysozyme, and ribonuclease, on the basis of tertiary structures established for these globular proteins by X-ray analysis, agree at all characteristic points with experimental c.d. spectra, using poly-(L-serine) as one of the standards.[35] Many examples of these curve-fitting procedures have been reported for native proteins and polypeptides, giving data on their degree of ordering in solution.[36] This, of course, does not give any indication which parts of the primary structure are α-helical, which are β-sheet structured, and which are disordered, but it is possible to hazard guesses on the localities of these conformations based on the amino-acid sequence of the polypeptide or protein. As discussed in Section 23.7.2.4, the protein amino-acids can be grouped into categories defining those which, when part of a polypeptide, promote the adoption of ordered structures, and those which do not. This information has been acquired by inspection of the library of three-dimensional structures of proteins established through X-ray analysis; the current literature includes many papers concerned with the prediction of polypeptide conformations using this knowledge, and also on the route taken to reach the biologically-active conformation as the protein is synthesized at the gene. Reviews of these two areas are given in references 37 and 38, respectively.

O.r.d. and c.d. provide a means of following conformational changes as a result of variations in temperature, ionic strength, pH, *etc.* Although n.m.r. methods are of greater precision in many ways, the n.m.r. signals would be entirely swamped by the various additives involved in many of the o.r.d. and c.d. conformational studies; for example, o.r.d. studies have shown that stabilization of the α-helical conformation for poly-(L-glutamic acid) occurs by binding acetic acid from aqueous solutions at pH values at which denaturation would occur if mineral acids were involved.[39]

Cotton effects arising from side-chain chromophores (*i.e.* groupings absorbing at longer wavelengths than the backbone amide chromophores) can be interpreted to show some local conformational features. The enhanced c.d. of ribonuclease in the 240–320 nm wavelength region has been ascribed to effects of tyrosine residue side-chains lying close together, buried within the tertiary structure.[40] Introduction of a chromophore at the *N*-terminus of an oligopeptide illustrates the same conformational probing approach; terminal *N*-thiobenzoyl peptides $PhC(=S)NHCHR^1CO\ldots$ show c.d. features near 390 nm due to perturbation of the thiobenzamide chromophore by the asymmetric centres in the *N*-terminal and adjacent amino-acid residues. Interpretation of the c.d. behaviour to reveal conformational relationships between the *N*-terminal residue and the penultimate *N*-terminal residue has been discussed.[41]

23.7.3.4 Nuclear magnetic resonance spectroscopy

Conformational studies of oligopeptides using nuclear magnetic resonance spectroscopy can take the form of interpretation of chemical shift data to reveal hydrogen bonding effects and the proximity of various groupings, or through Karplus-type relationships connecting ^{1}H coupling constants with NH—C$^\alpha$H torsion angles for amino-acid residues in peptides.[42] Rapid progress made in the late 1960s brought the technique to the point where every proton resonance in a relatively complex metal-free oligopeptide (*e.g.* the cyclic decapeptide antamanide[43]) can be assigned to particular nuclei, and where torsion angles can be determined in favourable cases. The conformation of antamanide in solution has been established by ^{1}H n.m.r. as presenting all amide carbonyl oxygen atoms outwards into the solution from one face of the molecule, and side-chains outwards from the opposite face.[43] Subsequent developments in ^{13}C n.m.r. spectroscopy based on spin–lattice relaxation times[44] and nuclear Overhauser effect studies[45] have widened the applications of n.m.r. in polypeptide conformational studies, providing information of a specific nature in a wide range of examples. The tripeptide amide L-prolyl-L-leucylglycinamide, exerting a hypothalamic function, has, for example, been shown to adopt a compact but flexible conformation in aqueous solution, based on ^{13}C n.m.r. spin–lattice relaxation time data,[44] and this conclusion is in line with conformational calculations[46] showing that the molecule adopts a type II β-turn conformation.

The n.m.r. technique is particularly successful for conformational assignments to small peptides, and current studies include a number of compounds possessing important physiological functions. These include mammalian peptide hormones and associated factors, as well as peptide antibiotics, and the advances being made will undoubtedly clarify the details of molecule–receptor interactions (see also Section 23.7.3.7).

The conformation of the antibiotic actinomycin has been the subject of study by several groups; it contains a macrocyclic pentapeptide lactone grouping

↳ L-Thr-D-Val-L-Pro-Sar-L-MeVal┘

in which the *N*-methylvalyl carboxy-group is esterified to the hydroxyl group of the threonine residue. N.m.r. studies leading to estimates of torsion angles χ for the C$^\alpha$—C$^\beta$ bonds of the valyl and *N*-methylvalyl residues have been described.[47] Previously similar studies established a stretched transannular hydrogen bond between the D-valyl NH proton and the sarcosyl oxygen atom, and torsion angles $\phi + 120°$ for the valyl residue and χ +90° for the threonine side-chain, and showed that all amide groups were in the *trans*-configuration.[48] However, some distortion from the all-*trans* amide structure is probably involved to avoid repulsions between the proline ring and the sarcosine *N*-methyl group.

Cyclic peptides such as those described in the preceding paragraphs have been particularly intensively studied both for their biological importance and as means of evaluating the scope of n.m.r. methods in conformational analysis. Related natural products, cyclic depsipeptides, are discussed together with cyclic oligopeptides in Chapter 23.4. Synthetic analogues whose conformations have been studied are cyclo(tri-L-prolyl),[49] cyclo-(L-Pro-L-Ser-Gly-L-Pro-L-Ser-Gly),[50] which exists in equilibrium between two structurally-similar β-conformations, and cyclo-(L-Ser-L-Pro-Gly-L-Ser-L-Pro-Gly),[51] which exists in equilibrium between a β-conformation and a more compact form (given the label Ω_c),[51] in which the proline amide groupings are in the *cis*-configuration.

Larger structures yield n.m.r. spectra of corresponding complexity, but information can be extracted from them in terms of conformation, especially the detection of conformational changes accompanying physical changes in the media in which polypeptides are studied. The 'sharpening' of the NH resonances of the three clusters of protonated arginine side-chains of lysozyme (seven arginine residues) before many other proton resonance peaks, when the solution of the enzyme is brought to pH 2.8, shows that disordering of the protein is a stepwise process involving the unfolding of the arginine clusters before other parts of the molecule.[52] A further part of this study was the demonstration, based on n.m.r. spectral changes, that α-lactalbumin is more easily denatured than lysozyme.[52]

The conformational properties of insoluble proteins can in some cases be established by X-ray diffraction studies, but in many important areas can only be approached indirectly. Elastin, a highly-crosslinked and insoluble protein of elastic fibres in connective tissue, has a soluble precursor (tropelastin) composed of a long-chain polypeptide with the repeating unit (-L-Val-L-Pro-Gly-L-Val-Gly-)$_n$. Model peptides related to this sequence have been studied by n.m.r. and show a pronounced tendency for the adoption of type II β-turns. Protected peptide models Boc-Gly-L-Val-Gly-OMe, HCO-(L-Val-L-Pro-Gly-Gly)$_n$-L-Val-OMe, and Boc-L-Val-L-Pro-Gly-L-Val-Gly-NH_2 used in these studies were shown to adopt a novel 11-membered hydrogen-bonded ring (a γ-turn; Figure 6),[6] a type II β-turn,[53] and the hydrogen-bonded network involving two 10-membered rings and one 7-membered ring, shown in Figure 9,[54] respectively.

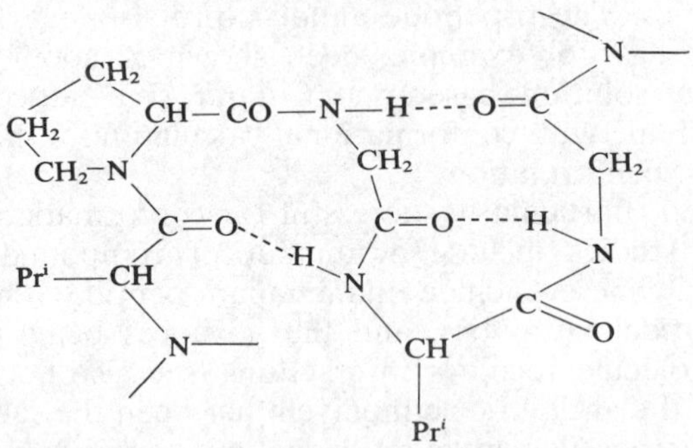

Figure 9 Conformation of L-Val-L-Pro-Gly-L-Val-Gly repeating unit of elastin

N.m.r. spectra of poly-(L-amino-acids) show a number of characteristic features for the various ordered conformations. Disordering of the α-helical conformation in a non-aqueous solvent causes a downfield shift of the $C^\alpha H$ resonances, and this fact can be used to infer structural information, *e.g.* from the n.m.r. behaviour of adrenocorticotrophic hormone, the information that the backbone conformation of this oligopeptide keeps the 1–10 and 11–24 residue regions apart from each other.[55] Similarly, the NH—C^αH coupling constants characteristic of the α-helix conformation are not shown by Gramicidin A′. This molecule is known to adopt a helical structure and on this basis must be a 'looser' helix (given the label $\pi_{L,D}$-helix).[56] ^{13}C N.m.r. spectra give clear indications of dissociation of the α-helical conformation of a poly-(L-amino-acid), with upfield shifts in resonances of C^α (by 3.0 p.p.m.) and C′ (by 2.7 p.p.m.) for poly-(γ-benzyl-L-glutamate) in chloroform accompanying denaturation by trifluoroacetic acid.[57] Features of n.m.r. spectra arising from side-chain protons can be interpreted in conformational terms where functional group interactions may be involved; thus, poly-(β-benzyl-L-aspartate) in chloroform adopts a left-handed α-helix conformation with no interaction between the benzyl groups and backbone amide groups, whereas the right-handed α-helix adopted by poly-(β-benzyl-L-glutamate) depends on stabilization by this type of interaction.[58]

23.7.3.5 Fluorescence spectroscopy

The natural fluorescence of polypeptides and proteins which contain tyrosine and tryptophan residues is sensitive to the environment of these residues. This can be exploited in a number of ways to provide information on the conformational mobility of the tyrosine and tryptophan side-chains in relation to nearby groupings within the polypeptide. One example is the interpretation of changes in the fluorescence of α-chymotrypsin accompanying variations in solvent composition to show that backbone —CO—NH— groups are close enough to tyrosine and tryptophan residues to allow charge-transfer interaction.[59] The accessibility of tyrosine residues in ribonuclease has been established by fluorescence studies of a more qualitative type;[60] three exposed tyrosine residues in the enzyme lose their fluorescence by *O*-acetylation, implying that the other three tyrosine residue side-chains are 'buried'. This conclusion is supported by the development of fluorescence when the tris-(*O*-acetyl)enzyme is denatured.[60]

Use of polarized light pulses and analysis of the intensities of the parallel and perpendicular components of the fluorescence as a function of time gives information on the conformational mobility of the fluorescent chromophore (the 'fluorophore').[61] The fluorescence anisotropy, relating the intensities of the parallel and perpendicular components of the fluorescence, is a measure of depolarization which occurs rapidly with an unhindered fluorophore, but slowly when the fluorophore is constrained. The polypeptide hormone glucagon includes a tryptophan residue amongst its 29 amino-acid residues, and fluorescence studies support o.r.d. evidence that the molecule exists to a greater extent in the α-helical conformation in aqueous solutions containing aliphatic quaternary ammonium salts, compared with pure water as solvent.[62]

An alternative application of fluorescence anisotropy measurements[63] is well illustrated by studies with the decapeptide hormone luliberin, 5-oxo-Pro-His-Trp-Ser-Tyr-Gly-Leu-Arg-Pro-Gly-NH_2. When studied in a viscous medium to reduce depolarization arising from thermal motion of the relatively free-moving aromatic side-chains, evidence of energy transfer from the phenol side-chain fluorophore of tyrosine to the tryptophan indole grouping indicates that the β-turn conformation suggested by c.d. studies involves the region of the polypeptide containing the tyrosine and tryptophan residues.[63]

23.7.3.6 Conformational probes: fluorescence spectroscopy and electron spin resonance

Examples of the use of natural fluorescence of polypeptides for deducing conformational behaviour are given in the preceding paragraphs. An extension of this approach, the introduction of covalently bound 'reporter' groups (either fluorophores,[64,65] or 'spin labels'[66] — groupings containing a localized unpaired electron) allows the neighbourhood of the reporter group to be explored through fluorescence spectroscopy and electron spin resonance spectroscopy, respectively. Reference has been made in Section 23.7.3.3 to the analogous study of 'tagged' peptides by c.d.

The introduction of the fluorescent 5-dimethylamino-1-naphthalenesulphonyl group into membrane proteins and analysis of the fluorescence spectrum indicate[63] that the reporter groups are strongly immobilized, and leads to the conclusion that cell membranes are not 'liquid-like' as previously supposed. Solubilization of some of the membrane protein brought about by the action of a detergent is accompanied by increased mobility of the tagging groups (see also Chapter 25.3).

The shape of the e.s.r. spectrum of polypeptides into which spin labels have been introduced can be interpreted to indicate the rotational mobility of the label.[65,67] Typical tagging groups for this purpose are *N*-(2,2,6,6-tetramethylpiperidin-1-oxy)maleimide[68] for covalent linkage to polypeptide thiol groups, and the related 2,2,6,6-tetramethyl-4-isocyanopiperidino-*N*-oxyl,[69] which has been used for specific bonding to the haem group in cytochrome P-450 in a search for connections between the function of the haem group and conformational characteristics of the neighbouring polypeptide.

23.7.3.7 Combined use of spectroscopic methods in the study of peptide conformations

One of the spectroscopic techniques, i.r. and Raman, o.r.d. and c.d., n.m.r., fluorescence, and e.s.r. spectroscopy, is often best suited for a particular conformational study, but the information which it provides is almost invariably usefully supplemented by data from one of the other techniques. Indications have been given in the preceding sections of a number of examples of the use of more than one spectroscopic method, and further examples dealing with oligopeptides are protected L-alanine oligomers (i.r. and c.d.),[15,70]

L-valine and L-norvaline analogues,[70] and Gramicidin A′ (n.m.r. and c.d.).[56] Hexafluoroisopropanol is a good solvent for oligopeptides, polypeptides, and proteins, and has been used for comparative i.r. and o.r.d. or c.d. studies.[71]

The more concentrated solutions used for n.m.r. spectroscopy in comparison with other spectroscopic methods may make comparative studies unreliable in certain cases,[72] since certain conformations (*e.g.* the β-sheet) are favoured at high concentrations. Reviews dealing with physical methods applied to polypeptide conformational analysis are available.[73–75]

23.7.3.8 Non-spectroscopic methods

Several physical parameters of peptides are conformation-dependent, and therefore capable of yielding structural information. The information which is most readily obtained, as for spectroscopic methods, is for the change from an ordered to a disordered conformation, or *vice versa.* This is revealed as a discontinuity in refractive index increments as a function of solvent composition, for example.[76] The other familiar physical property of a solution, its viscosity, is sensitive to the size and shape of solute molecules and gives clear information on conformational changes. Glucagon, for example (see also Section 23.7.3.5), exists in dilute aqueous solution in an overall spherical shape, but aggregates at pH 11 and gels in concentrated solutions through antiparallel β-sheet alignment of elongated chains.[77]

Potentiometric titration is less useful than c.d. for studying the effects of quaternary ammonium salts on the disordering of α-helical poly-(L-glutamic acid),[78] whereas the technique provides data for changes in free energy and enthalpy accompanying the ordering of poly-(L-lysine) into α-helical and β-sheet conformations[10,79] and supplements the information obtained for this poly(amino-acid) through o.r.d. and viscometry.

23.7.4 CONFORMATIONAL CALCULATIONS FOR PEPTIDES

23.7.4.1 General

A feature of recent work in the area of polypeptide conformational analysis has been the computation of preferred conformations, using three-dimensional structures of proteins established by X-ray analysis to check the effectiveness of the calculations. The results are of broader relevance since the uncertainties involved in deducing the minimum energy conformation of a molecule are well-illustrated with polypeptides.

23.7.4.2 Energy maps

Repulsive interactions between the atoms making up an amino-acid residue are conveniently portrayed by energy maps relating the potential energy of the residue to torsion angles ψ and ϕ. A simple example (Figure 10)[80] shows the torsion angle combinations (hatched areas) which bring the atoms of a glycine residue sufficiently close together to raise the potential energy of the molecule, of which the residue is a part, by an arbitrary 4 kcal mol^{-1} (~17 kJ mol^{-1}) above the minimum energy; the torsion angles corresponding to the minimum energy conformation and near-minimum energy conformation are sited in the white area of the map. Depending on the confidence limits adopted for van der Waals' radii of atoms, more contour lines may be drawn, spaced from 0.5 kcal mol^{-1} (~2 kJ mol^{-1}) apart, and with more complex amino-acid residues less symmetrical energy maps are obtained, showing energy wells representing the preferred backbone torsion angles for each residue. Energy–torsion angle χ relationships for

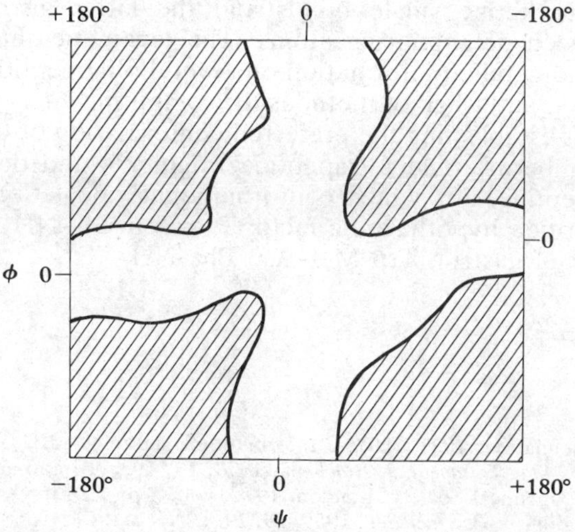

Figure 10 Energy map for a glycyl residue in a polypeptide

side-chain C—C bonds can also be represented graphically,[81] displaying energy barriers which must be overcome to permit rotation of side-chains relative to the peptide backbone.

The minimum energy structure for an isolated amino-acid residue may bear little relationship to the conformation imposed on that residue when part of a polypeptide. At first sight this may be thought to create a unique situation for each peptide sequence, but since many of the protein amino-acids are compatible with ordered backbone conformations (Section 23.7.2.2), there are several general cases which represent a substantial number of polypeptide structures. An example of molecular orbital calculations (using PCILO theory — perturbative configuration interactions using localized orbitals) for assessing the influence of an adjacent α-helix on the minimum energy conformation of an L-alanine residue uses *N*-acetyl-$(\text{L-Ala})_4$NHMe as a model, leading to an energy map for the *C*-terminal residue with torsion angles for the preceding residues held at the characteristic α-helix values (ϕ, −57°; ψ, −47°).[82]

23.7.4.3 Objectives of conformational calculations

Over a period of about 10 years, theoretical studies of peptide conformations have progressed from calculations for single amino-acid residues, through studies of amino-acid residues sited in oligopeptides and subjected to interactions with local backbone and side-chains, to current studies of oligopeptides with important biological functions. Several aspects of these studies have been reviewed.[83,84]

The total conformational energy of a polypeptide is the sum of the contributions of non-bonded interactions involving side-chain–side-chain and side-chain–backbone repulsions; one factor which simplifies the calculations is the effect of the rigid amide group in maintaining each amino-acid residue independent of its neighbours.[83] Although a unique minimum energy conformation can be arrived at for each polypeptide through an assessment of intramolecular interactions, one of the features of recent work has been the consideration of various conformations of near-minimum energy which might become favoured as a result of solvation effects and intramolecular interactions.

Molecular orbital calculations[83,85] avoid some of the limitations caused by uncertainties in van der Waals' radii used in conformational computations; extended Hückel methods and the CNDO/2 approach are complementary to some extent, the former giving reliable energy barriers for backbone single bonds and the latter giving good estimates for repulsion forces between neighbouring atoms. The techniques have been developed to sophisticated levels and detailed information on several oligopeptides has been computed. Full accounts have been published of methods for computing energy barriers for backbone single bonds,[86] and for establishing the preferred conformation of the luteinizing hormone releasing hormone luliberin[87] and tetrapeptide fragments and decapeptide analogues.[88] Examples of small peptides for which conformational calculations have been reported recently are the tripeptide hypothalamic factor Pro-Leu-Gly-NH_2,[46] and the *C*-terminal tetrapeptide fragment of gastrin, Trp-Met-Asp-Phe-NH_2.[89]

References

1. IUPAC–IUB Recommendations 1971 (IUPAC Information Bulletin No. 10, 1971); reproduced *inter alia* in *Biochemistry*, 1970, **9,** 3471; *European J. Biochem.*, 1971, **17,** 193; 'Amino-acids, Peptides, and Proteins', ed. G. T. Young, The Chemical Society, London, 1972, vol. 4, p. 455.
2. G. Natta and P. Corradini, *J. Polymer Sci.*, 1959, **39,** 29.
3. L. Velluz and M. Legrand, *Angew. Chem. Internat. Edn.*, 1965, **4,** 838.
4. D. C. Hodgkin, *Proc. Roy. Soc.*, 1974, **338,** 251; T. L. Blundell, J. F. Cutfield. S. M. Cutfield, E. J. Dodson, G. G. Dodson, D. C. Hodgkin, D. A. Mercola, and M. Vijayan, *Nature*, 1971, **231,** 506.
5. R. E. Dickerson, T. Takano, D. Eisenberg, O. B. Kallai, L. Samson, A. Cooper, and E. Margoliash, *J. Biol. Chem.*, 1971, **246,** 1511.
6. M. Abu Khaled, D. W. Urry, and K. Okamoto, *Biochem. Biophys. Res. Comm.*, 1976, **72,** 162.
7. J. J. Birktoft, D. M. Blow, R. Henderson, and T. A. Steitz, *Phil. Trans. Roy. Soc.* (*B*), 1970, **257,** 67.
8. K. D. Kopple and A. Go, *Biopolymers*, 1976, **15,** 1701.
9. S. L. Portnova, V. F. Bystrov, V. I. Tsetlin, V. T. Ivanov, and Y. A. Ovchinnikov, *Zhur. obshchei. Khim.*, 1968, **38,** 428.
10. P. Y. Chou and H. A. Scheraga, *Biopolymers*, 1971, **10,** 657.
11. V. Giancotti, F. Quadrifoglio, and V. Crescenzi, *J. Amer. Chem. Soc.*, 1972, **94,** 297.
12. E. R. Blout, in 'Polyamino-acids, Polypeptides, and Proteins'. ed. M. A. Stahmann, University of Wisconsin Press, Madison, 1962, p. 275.
13. F. H. C. Crick and A. Rich, *Nature*, 1955, **176,** 780.
14. E. R. Blout and A. Asadourian, *J. Amer. Chem. Soc.*, 1956, **78,** 955; J. C. Mitchell, A. E. Woodward, and P. Doty, *ibid.*, 1957, **79,** 3955.
15. C. Toniolo and G. M. Bonora, *J. Polymer Sci., Polymer Chem. Edn.*, 1976, **14,** 515.
16. C. Toniolo, *Biopolymers*, 1971, **10,** 1707; M. Goodman, F. Naider, and C. Toniolo, *ibid.*, p. 1719.
17. U. Widmer and G. P. Lorenzi, *Chimia*, 1971, **25,** 236.
18. R. T. Ingwall and M. Goodman, in 'International Review of Science, Organic Chemistry Series Two', ed. H. N. Rydon, Butterworths, London, 1975, vol. 6, p. 153.
19. B. Schechter, I. Schechter, J. Ramachandran, A. Conway-Jacobs, and M. Sela, *European J. Biochem.*, 1971, **20,** 301.
20. B. Weinstein, in 'Peptides: Chemistry and Biochemistry', ed. B. Weinstein and S. Lande, Dekker, New York, 1970, p. 371.
21. W. B. Gratzer, W. Rhodes, and G. D. Fasman, *Biopolymers*, 1963, **1,** 319.
22. J. M. Rifkind and J. Applequist, *J. Amer. Chem. Soc.*, 1968, **90,** 3650.
23. H. Rosenkranz and W. Scholtan, *Z. physiol. Chem.*, 1971, **352,** 896.
24. B. Jirgensons and S. Capetillo, *Biochim. Biophys. Acta*, 1970, **205,** 1.
25. A. Cosani, E. Peggion, M. Terbojevich, and M. Acampora, *Macromolecules*, 1971, **214,** 1.
26. L. Visser and E. R. Blout, *Biochemistry*, 1971, **10,** 743.
27. Y. Ohta, Y. Hibino, K. Asaba, K. Sugiura, and T. Samejima, *Biochim. Biophys. Acta*, 1971, **236,** 802.
28. H. B. Brewer and H. Edelhoch, *J. Biol. Chem.*, 1970, **245,** 2402.
29. E. M. Bradbury, B. G. Carpenter, and R. M. Stephens, *Macromolecules*, 1972, **5,** 8.
30. S. Mizishima, T. Shimanouchi, M. Tsuboi, T. Sugita, K. Kurosaki, N. Mataga, and R. Souda, *J. Amer. Chem. Soc.*, 1952, **74,** 4639.
31. M. Avignon, P. V. Huong, J. Lascombe, M. Marraud, and J. Neel, *Biopolymers*, 1969, **8,** 69; A. Burgess and H. A. Scheraga, *ibid.*, 1973, **12,** 2177.
32. M. Goodman, Y. Masuda, and A. S. Verdini, *Biopolymers*, 1971, **10,** 1031.
33. D. F. H. Wallach, J. M. Graham, and A. R. Oseroff, *FEBS Letters*, 1970, **7,** 330.
34. Reviews of mathematical treatment of o.r.d. data are given by G. D. Fasman, in 'Methods in Enzymology', Academic Press, New York, 1963, vol. 6, p. 928; see also Ref. 35.
35. Y. T. Yang, in 'Conformations of Biopolymers', Academic Press, New York, 1967, vol. 1, p. 173.
36. For leading references, see the review by G. C. Barrett, in 'MTP International Review of Science, Organic Chemistry Series One', ed. D. H. Hey and D. I. John, Butterworths, London, 1973, vol. 6, p. 77.

37. B. Robson and E. Suzuki, *J. Mol. Biol.*, 1976, **107,** 327.
38. B. Robson, *Trends Biochem. Sci.*, 1976, **1,** 55; D. B. Wetlaufer and S. Rostow, *Ann. Rev. Biochem.*, 1973, **42,** 135.
39. J. R. Cann, *Biochemistry*, 1971, **10,** 3707.
40. C. Sander and P.O.P.Ts'o, *Biochemistry*, 1971, **10,** 1953.
41. G. C. Barrett, *J. Chem. Soc.* (*C*), 1969, 1123.
42. A. E. Tonelli and F. A. Bovey, *Macromolecules*, 1970, **3,** 410.
43. A. E. Tonelli, D. J. Patel, M. Goodman, F. Naider, H. Faulstich, and T. Wieland, *Biochemistry*, 1971, **10,** 3211.
44. R. Deslauriers and I. C. P. Smith, *Topics ^{13}C N.M.R. Spectroscopy*, 1976, **2,** 1.
45. H. E. Bleich, J. D. Cutnell, and J. A. Glasel, *Biochemistry*, 1976, **15,** 2455.
46. S. Kang and R. Walter, *Proc. Nat. Acad. Sci. U.S.A.*, 1976, **73,** 1203.
47. P. De Santis, R. Rizzo, and G. Ughetto, *Tetrahedron Letters*, 1971, 4309.
48. H. Lackner, *Tetrahedron Letters*, 1970, 2807, 3189; 1971, 2221; *Chem. Ber.*, 1970, **103,** 2476.
49. C. M. Deber, D. A. Torchia, and E. R. Blout, *J. Amer. Chem. Soc.*, 1971, **93,** 4893.
50. D. A. Torchia, A. Di Corato, S. C. K. Wong, C. M. Deber, and E. R. Blout, *J. Amer. Chem. Soc.*, 1972, **94,** 609.
51. D. A. Torchia, S. C. K. Wong, C. M. Deber, and E. R. Blout, *J. Amer. Chem. Soc.*, 1972, **94,** 616.
52. J. H. Bradbury and N. L. R. King, *Austral. J. Chem.*, 1971, **24,** 1703.
53. D. W. Urry and M. M. Long, *Crit. Rev. Biochem.*, 1976, **4,** 1.
54. D. W. Urry, L. W. Mitchell, T. Ohnishi, and M. M. Long, *J. Mol. Biol.*, 1975, **96,** 101.
55. D. J. Patel, *Macromolecules*, 1971, **4,** 251.
56. J. D. Glickson, D. F. Mayers, J. M. Settine, and D. W. Urry, *Biochemistry*, 1972, **11,** 477.
57. L. Paolillo, T. Tancredi, P. A. Temussi, E. Trivellone, E. M. Bradbury, and C. Crane-Robinson, *J.C.S. Chem. Comm.*, 1972, 335.
58. D. N. Silverman, G. T. Taylor, and H. A. Scheraga, *Arch. Biochem. Biophys.*, 1971, **146,** 587.
59. R. W. Cowgill, *Biochim. Biophys. Acta*, 1970, **200,** 18.
60. R. W. Cowgill and N. K. Lang, *Biochim. Biophys. Acta*, 1970, **214,** 228.
61. P. Wahl, M. Kasai, and J.-P. Changeux, *European J. Biochem.*, 1971, **18,** 332.
62. H. Bornet and H. Edelhoch, *J. Biol. Chem.*, 1971, **246,** 1785.
63. P. Marche, T. Montenay-Garestier, C. Helene, and P. Fromageot, *Biochemistry*, 1976, **15,** 5730, 5738.
64. For reviews, see G. K. Radda, *Biochem. J.*, 1971, **122,** 385; also Ref. 65.
65. L. Stryer, in 'CIBA Foundation Symposium on Molecular Properties of Drug Receptors', ed. R. Porter, Churchill, London, 1970, p. 133.
66. For a review see H. M. McConnell and B. G. McFarland, *Quart. Rev. Biophys.*, 1970, **3,** 91.
67. V. P. Timofeev, O. L. Polianovsky, M. V. Volkenstein, and G. I. Lichtenstein, *Biochim. Biophys. Acta*, 1970, **220,** 357.
68. A. M. Gotto and H. Kon, *Biochemistry*, 1970, **9,** 4276.
69. L. M. Raikhman, B. Annaev, and E. G. Rozantsev, *Dokl. Akad. Nauk S.S.S.R.*, 1971, **200,** 387.
70. J. S. Balcerski, E. S. Pysh, G. M. Bonora, and C. Toniolo, *J. Amer. Chem. Soc.*, 1976, **98,** 3470.
71. J. R. Parrish and E. R. Blout, *Biopolymers*, 1971, **10,** 1491.
72. P. A. Temussi and M. Goodman, *Proc. Nat. Acad. Sci. U.S.A.*, 1971, **68,** 1767.
73. 'Physical Principles and Techniques of Protein Chemistry', ed. S. J. Leach, Academic Press, New York, 1970.
74. R. Wetzel, *Stud. Biophys.*, 1968, **8,** 127.
75. 'Amino-acids, Peptides, and Proteins', vols. 1 (1969) to 4 (1972), ed. G. T. Young, vols. 5 (1974) to 9 (1978), ed. R. C. Sheppard, The Chemical Society, London.
76. K. De Groot, J. Feyen, A. C. De Visser, G. Van de Ridder, and A. Bantjes, *Kolloid-Z.*, 1971, **246,** 578.
77. R. M. Epand, *Canad. J. Biochem.*, 1971, **49,** 166.
78. J. Steigman and A. Cosani, *Biopolymers*, 1971, **10,** 357.
79. T. V. Barskaya and O. B. Ptitsyn, *Biopolymers*, 1971, **10,** 2181; D. Pederson, D. Gabriel, and J. Hermans, *Biopolymers*, 1971, **10,** 2133.
80. D. A. Brant, W. G. Miller, and P. J. Flory, *J. Mol. Biol.*, 1967, **23,** 47.
81. *E.g.*, J. Caillet, B. Pullman, and B. Maigret, *Biopolymers*, 1971, **10,** 221.
82. G. Jeronimidis and A. Damiani, *Nature New Biol.* 1971, **229,** 150.
83. H. A. Scheraga, *Chem. Rev.*, 1971, **71,** 195; *Pure Appl. Chem.*, 1973, **36,** 1.
84. M. Goodman, A. S. Verdini, N. S. Choi, and Y. Masuda, *Topics Stereochem.*, 1970, **5,** 69.
85. H. A. Scheraga, F. A. Momany, R. F. McGuire, and J. F. Yan, *J. Phys. Chem.*, 1971, **75,** 2286.
86. A. T. Hagler, L. Leiserowitz, and M. Tuval, *J. Amer. Chem. Soc.*, 1976, **98,** 4600.
87. F. A. Momany, *J. Amer. Chem. Soc.*, 1976, **98,** 2990.
88. F. A. Momany, *J. Amer. Chem. Soc.*, 1976, **98,** 2996.
89. T. Yamada, H. Wako, N. Saito, Y. Isogai, and H. Watari, *Internat. J. Peptide Protein Res.*, 1976, **8,** 607.

PART 24

PROTEINS: ENZYME CATALYSIS AND FUNCTIONAL PROTEINS

24.1
Enzyme Catalysis

A. J. KIRBY
University of Cambridge

24.1.1 GENERAL FEATURES OF ENZYME CATALYSIS[1]

Enzymes are the most powerful and efficient catalysts known, yet at the same time the most subtle and most delicate. They catalyse almost all the reactions which go on in living systems, including many which the organic chemist finds difficult or impossible; and they do this under the mildest conditions of temperature and pH, in dilute solution in water.

Enzymes can also be put to work outside the living system. The food industry has traditionally used enzyme-catalysed reactions in such processes as brewing and baking, and the number of such applications is gradually increasing. Clinical biochemistry relies on enzymes for a large proportion of routine assays, and the synthetic chemist is turning more and more to enzymes for reactions where stereospecificity is important,[2] or the need for complex group-protection of multi-functional molecules makes conventional routes unattractive.[3] Last, and most important, we need to know more about the chemical machinery of the cell to be able to deal rationally with the system when it goes wrong. Some illnesses are known to involve specific biochemical disorders involving particular enzymes or enzyme systems, and more examples will undoubtedly emerge.

It is therefore of the greatest interest to understand how enzymes work. In this chapter we will approach enzyme catalysis from the standpoint of the organic chemist. The mechanism of action of a particular enzyme then becomes a problem in organic reaction mechanism, and can in principle be treated by the methods of physical organic chemistry. These presuppose that the structures of the starting materials and products of the reaction concerned are known, and it is the availability in recent years of detailed structural information for enzymes, from X-ray crystallography, that has made work of this sort a practical proposition.

24.1.1.1 Enzymes as proteins*

Enzymes are protein molecules, with molecular weights ranging from about ten thousand to several million. Many enzymes contain, or work in conjunction with, coenzymes or metal ions which are essential for catalytic activity. Many, too, are aggregates of one, or sometimes two, sorts of individual protein subunits. Still others are organized in groups of small numbers of different enzymes, either in solution (multi-enzyme complexes) or more or less firmly attached to specific subcellular structures.[4] These multi-enzyme systems can thus carry out sequences of several reactions, and contain one enzyme for each step of the sequence.

Since we are concerned here with the general chemical features of enzyme catalysis, we will mostly ignore such complications, and consider as far as possible only the simplest type of monomeric enzyme. This will be a large protein molecule, made up of a single polypeptide chain containing perhaps a few hundred amino-acids, which exists free in solution in the cell. Such a protein can be purified, using the powerful chromatographic and electrophoretic methods available for the separation of charged macromolecules in aqueous media,[5] to the point where it is homogeneous by several different criteria; at which point, given reasonable luck, it may crystallize. Thus many enzymes are available as well-defined organic compounds, which satisfy the usual standards of purity for organic compounds, and which can indeed be synthesized.[6] They may, too, be available in quite substantial amounts. All the enzymes which have been widely studied — trypsin, chymotrypsin, lysozyme, and ribonuclease, for example — are available in gram quantities, and a wide range of other enzymes can be obtained commercially in milligram amounts.

The characteristic properties of such a protein (**1**) are determined exclusively by the nature and sequence of its constituent amino-acids, *i.e.* by the arrangement and type of the side-chains R^i (a table of amino-acid structures is given in Section 23.2.1). Polyphenylalanine (**1**, $R^1 = R^2 = R^i = CH_2Ph$), for example, is extremely insoluble in water, whereas polymers of amino-acids with charged groups in the side-chain (lysine, glutamic acid, *etc.*)[7] are soluble. A given enzyme protein is made up of a unique, but apparently random,

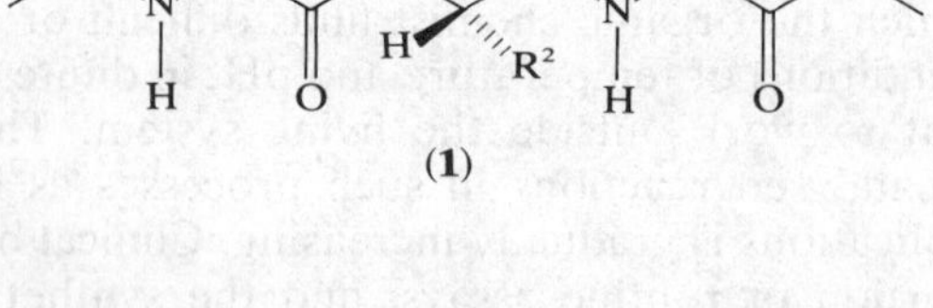

(**1**)

* See Chapter 23.1

sequence of all the 20 common amino-acids. Some parts of the polypeptide chain, bearing uncharged and particularly hydrocarbon side-chains (*e.g.* $R = CH_2Ph$), are therefore intrinsically hydrophobic, whereas the charged groups of functional side-chains are strongly solvated by water (hydrophilic). The result is a compromise: the polypeptide chain adopts a conformation which gives both types of groups, as far as possible, their preferred type of environment. The hydrophobic side-chains come together to form a group or groups near the centre of the final structure, surrounded and thus shielded from the solvent by those parts of the chain carrying hydrophilic side-chains, which make up the surface of the (roughly globular) protein. The surface is thus composed predominantly of hydrophilic groups, and the molecule as a whole is solvated efficiently enough to be soluble.[8]

This crude picture — similar in many respects to what is believed to be the structure of a micelle of detergent[9] — is sufficient to show how distant parts of the polypeptide chain are brought into close proximity in the most favourable conformation. Consider a portion of the primary sequence of an enzyme protein (Figure 1). The folding process which brings the hydrophobic groups into close proximity brings together the main polypeptide chain at these points, and thus also the side-chains of neighbouring amino-acid residues. If these happen to carry functional groups, it is easy to see how fairly precise arrangements of several functional groups in three dimensions can result at a particular point in the protein structure. This is how the active side of an enzyme is formed, when the linear polypeptide produced by the protein synthesis sequence folds to the characteristic stable

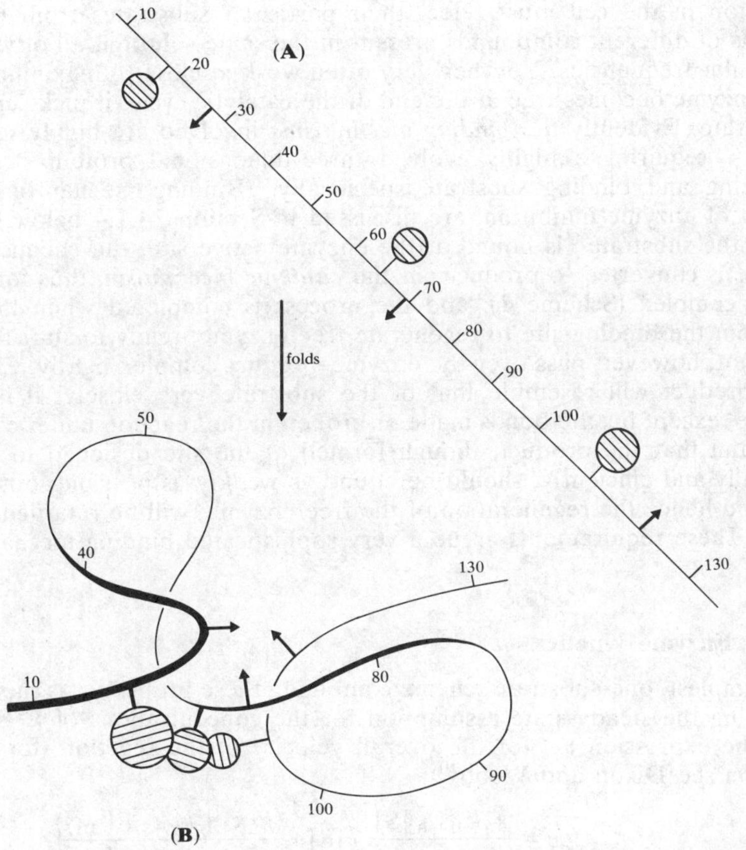

Figure 1 Formation of an active site. The linear polypeptide **A** produced by protein biosynthesis folds to adopt its most stable conformation (**B**) in water. Hydrophobic groups (shaded circles) come together, bringing nearby functional groups (arrows) into close proximity also. The stable structure is reinforced by hydrogen bonds, disulphide bridges, *etc.* (see text)

conformation of the enzyme protein. It is presumably also how the first enzymes developed, when particular arrangements of functional groups, formed accidentally in this way, turned out to have useful catalytic properties.

24.1.1.2 Enzymes as catalysts

Although we will generally discuss one enzyme-catalysed reaction at a time, and most mechanistic studies involve isolated enzymes, it is important to bear in mind that enzymes actually operate under conditions where many hundreds of chemical reactions are taking place simultaneously. Control of this enormous network of unrelated and interrelated reactions is achieved by mechanisms which control the concentrations and efficiency of individual enzymes. These *control mechanisms* (see below, Section 24.1.6) depend on high catalytic efficiency, so that small changes in concentration may have large effects on reaction flux, and on the high specificity characteristic of enzyme catalysis. Most enzymes catalyse one, and only one, reaction of one, and only one substrate, so that individual steps of particular pathways can be controlled by controlling the activity of individual enzymes.

The first step of any enzyme-catalysed reaction occurs when enzyme and substrate meet, usually as a result of a random collision in solution (and hence at a rate ($k \approx 10^8\ \mathrm{l\ mol^{-1}\ s^{-1}}$) appropriate for the diffusion together of a large and a small molecule in water) to form the (non-covalent) enzyme–substrate complex. Enzymes which act free in solution in the cell must select their particular substrates from the range of many hundreds of different compounds present in the same solution, all of which collide with it with similar frequencies. Yet they very often work at close to maximum capacity: as soon as the enzyme becomes free at the end of the catalytic cycle, it picks up another molecule of substrate. Evidently the *binding mechanisms* involved are highly selective — and very efficient — requiring a highly evolved three-dimensional protein structure, capable of recognizing and binding substrate specifically. (Binding mechanisms, and the related question of enzyme inhibition, are discussed in Section 24.1.4 below.)

Once the substrate* is bound at the enzyme active site, the chemistry can begin. The substrate is converted to product *via* the *catalytic mechanism*, thus forming the enzyme–product complex (Scheme 1), and the process is completed when the product diffuses away from the binding site to regenerate free enzyme, ready to start the cycle again. We should not, however, pass over the enzyme–product complex lightly. Clearly the structure of the product will resemble that of the substrate very closely: it is indeed the same molecule, except for the bonds made or broken in the reaction under consideration. Yet it is essential that the product, though formed at the site designed to bind the substrate specifically and efficiently, should be bound as *weakly* as possible; otherwise its diffusion away, and hence the regeneration of the free enzyme, will be retarded, and turnover will be lost. These requirements argue a very sophisticated binding mechanism.

24.1.1.3 Enzyme kinetics

The simplest, one-substrate scheme combining these properties is shown below (Scheme 1). Making the steady-state assumption for the concentrations of **E–S** and **E–P**, we can derive the expression (1) for the overall velocity of the reaction (for the details of this derivation see Dixon and Webb[10]):

$$v = \frac{k_3k_5[\mathrm{E_0}][\mathrm{S}]}{(k_3+k_4+k_5)} \Bigg/ [\mathrm{S}] + \frac{(k_2k_4+k_2k_5+k_3k_5)}{k_1(k_3+k_4+k_5)} \tag{1}$$

* Or substrates: most enzyme reactions are between substrate molecules, both bound to the enzyme. Here we limit discussion to the simplest possible case, for reasons which will become apparent!

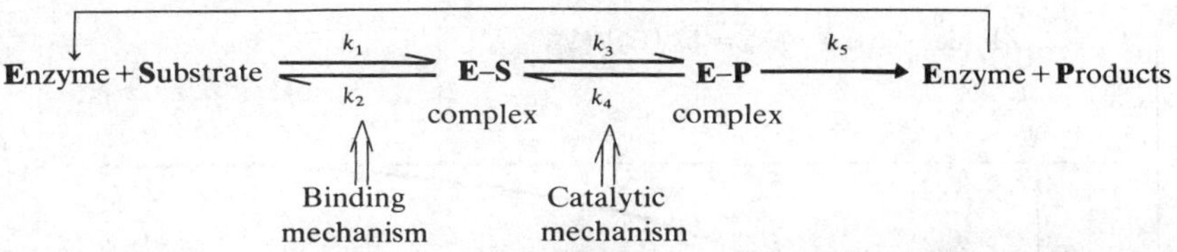

SCHEME 1

If we are interested specifically in the catalytic mechanism, we would like to disentangle k_3 from this expression. This can be done in many cases by a suitable choice of experimental conditions,* or by making assumptions (which may or may not turn out to be valid) about the relative values of the various rate constants. For example, if the product dissociates rapidly from the enzyme (*i.e.* k_5 is large compared with k_3 and k_4), equation (1) can be simplified to:

$$v = k_e[\mathrm{E_0}][\mathrm{S}] \Big/ \left([\mathrm{S}] + \frac{(k_2+k_3)}{k_1}\right) \quad (2)$$

More complex reaction schemes with more intermediates than Scheme 1 give more complex expressions, but still of the same general form:

$$v = k_{\mathrm{cat}}[\mathrm{E_0}][\mathrm{S}]/([\mathrm{S}] + K_m) \quad (3)$$

Here k_{cat} is the rate constant for the rate determining step, and $K_m = (k_2 + k_3)/k_1$ is the Michaelis constant, a measure of the affinity of the substrate for the enzyme. (In simple cases, where binding is a rapid pre-equilibrium process (*i.e.* $k_2 \gg k_3$), K_m becomes equal to the dissociation constant for the enzyme–substrate complex.) At high substrate concentrations, defined as $[\mathrm{S}] \gg K_m$, equation (3) reduces to

$$v \rightarrow V = k_{\mathrm{cat}}[\mathrm{E_0}] \quad (4)$$

V defines the maximum velocity, and is independent of substrate concentration. Combining equations (3) and (4) allows a practical definition of K_m, as the substrate concentration at which the rate is half the maximum velocity, V. At low substrate concentrations ($[\mathrm{S}] \ll K_m$) equation (3) becomes:

$$v = k_{\mathrm{cat}}[\mathrm{E_0}][\mathrm{S}]/K_m \quad (5)$$

Under these conditions the ratio k_{cat}/K_m is the apparent second-order rate constant for the reaction between enzyme and substrate.

The constants k_{cat} and K_m are readily determined, by measuring the initial rates of the enzyme-catalysed reaction as a function of substrate concentration (Figure 2), and are the most convenient indication of the efficiency of catalysis for a given substrate. A good substrate is one that binds strongly and reacts rapidly, and is therefore characterized by a small K_m and a large k_{cat}. However, for all but the very simplest cases, k_{cat} and K_m are composite constants [compare equations (3) and (1)], and detailed interpretation — for example, of k_{cat} values for a series of similar substrates as part of a mechanistic study — must be made with great care.

Our discussion of enzyme kinetics has thus far been limited to behaviour under steady-state conditions. Conveniently measurable rates of reaction are arranged, despite the very high rate constants associated with enzyme-catalysed reactions, by working with very low concentrations of enzyme (of the order of 10^{-7} to 10^{-10} mol l^{-1}). Measurements of this sort, readily performed with conventional apparatus, account for the great bulk of published work in enzyme kinetics. An area of growing importance, however, involves the use of modern fast-reaction techniques, which can give information about transient states.[11,12]

* We have already made one such choice implicitly in Scheme 1, by not writing a term, k_6, for the recombination of enzyme and product; this is valid in general if we measure initial rates of reaction, where the concentration of product is negligible.

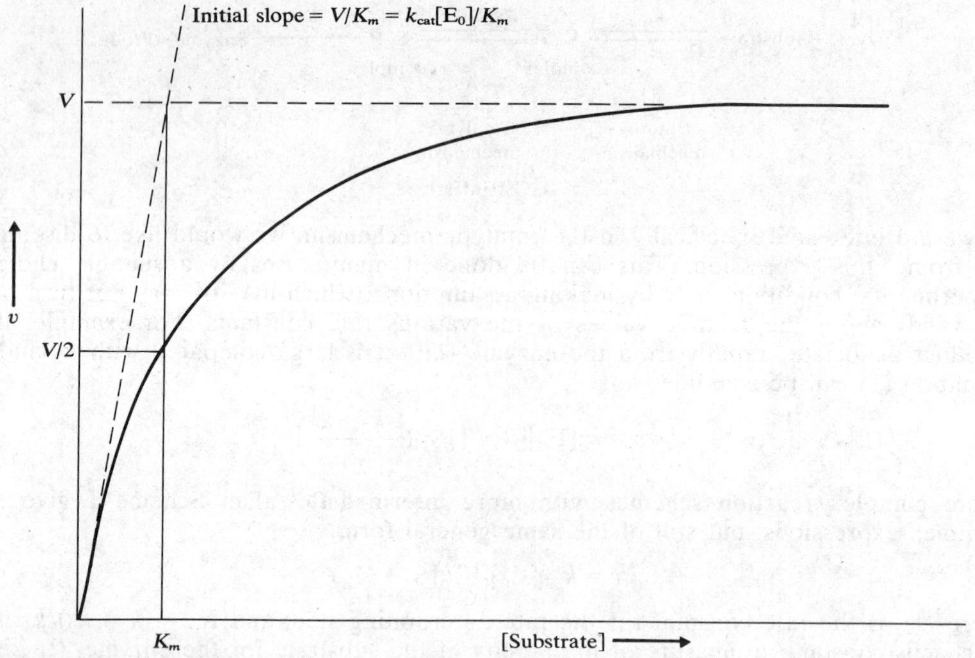

Figure 2 Plot of the initial rate (v) of an enzyme-catalysed reaction, against substrate concentration at a constant concentration of enzyme, E_0 (see text)

Stopped-flow techniques[13] can give information about processes with lifetimes of the order of milliseconds, limited by the mixing time of the apparatus. This allows higher enzyme concentrations (up to about 10^{-3} mol l^{-1}) to be used, and makes possible the study of the pre-steady state region, *i.e.* what happens when substrate meets enzyme for the first time. The binding step, and other processes which precede the rate-determining step, thus become accessible to study.

Faster still are relaxation methods, developed especially by Eigen[14] and his co-workers, which can measure reactions, such as proton transfers, with rate constants as large as 10^{10} l mol^{-1} s^{-1}. These methods involve following the return of the system to equilibrium (relaxation) after a sudden perturbation, and are limited largely by how fast the system can be perturbed. N.m.r.[15] and temperature-jump experiments extend the accessible time-constant region to 10^{-6} s. Electron spin resonance (e.s.r.) is faster still (10^{-9} s), but generally requires a spin-label to be introduced artificially into enzyme or substrate.

Each of these methods presents its own special problems of interpretation, particularly the identification of observed relaxation times with rate constants for specific steps. Nevertheless, the use of fast reaction techniques is now an essential part of the direct study of enzyme catalysis.

24.1.2 CATALYSIS IN AQUEOUS SOLUTION

Although the binding step and the catalytic process are in practice interdependent and inseparable, it is helpful to discuss them separately. In this section we begin to look at the sort of chemistry that goes on within the enzyme–substrate complex, as a problem in organic reaction mechanism. We start, as we must, with the assumption that the reactivity of the groups concerned, however remarkable, is normal under the conditions: that is to say, if we could somehow reproduce the conditions of the enzyme–substrate complex in the laboratory, we should be able to reproduce the high reactivity also.

The key questions then are: what are the mechanisms available for these reactions? Why are they so efficient? The first question is the subject of this section. The second involves the nature of the binding process as well as the mechanisms of catalysis, and is discussed in Section 24.1.4.

In contrast to the situation with coenzymes, where satisfactory mechanisms, backed up by working model systems, can be written in almost every case, no single enzyme-catalysed reaction is understood in detail. The objective of an investigation of organic reaction mechanism is an understanding of all the intermediate states along the reaction pathway — transition states as well as intermediates — in as much detail as we have of the stable initial and final states. At present we are not even able to identify all the intermediate states with any certainty for any enzyme-catalysed reaction.

We will consider first the functional groups available to enzymes; then the mechanisms by which these groups catalyse the reactions of typical substrate groups, as revealed by studies on model systems; finally we go on to discuss methods for identifying and studying enzyme reactions (Section 24.1.3).

24.1.1.2 Catalytic groups available to enzymes

Every protein molecule contains hundreds of functional groups. The most common are the amide groups of the main peptide chain, and of the amino-acids glutamine and asparagine. These have important structural roles, as hydrogen-bond donors and acceptors (see Chapter 23.7), and play an essential part in determining and stabilizing the conformation of the protein. They can play a similar role in stabilizing enzyme–substrate and enzyme–transition state complexes; but they are not thought to play an active part in catalysis, and are often assumed to be inert.

The functional groups involved in catalysis are carried on the side-chains of the amino-acid units. The guanidino-group of arginine* ($pK_a = 12.5$) is almost as weak an acid as the amide group, so the free base form can be neglected. It, too, is an important hydrogen-bond donor, particularly to dibasic groups like carboxylate and phosphate, and can be of considerable structural importance, but it does not normally play an active part in catalysis.

Just five functional groups account for the catalytic activity of most enzymes: these are listed in Table 1. The carboxyl group occurs in both acid and conjugate base forms, and is the most strongly acidic group available. The protonated imidazole group of histidine, with a pK_a close to physiological pH, can also act as a weak acid, but the conjugate base is the strongest base available under normal conditions, since the primary NH_2 group of lysine will normally be fully protonated near pH 7. Nevertheless, the proton of the NH_3^+

TABLE 1
The Catalytic Groups used by Enzymes

Group	pK_a[a]	*Amino-acid**
CO_2H, CO_2^-	4	Aspartic acid Glutamic acid
OH	14–15	Serine Threonine
SH	10	Cysteine
NH_3^+	10	Lysine
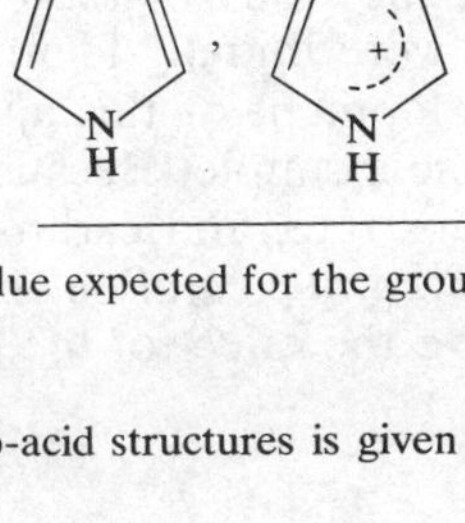	7	Histidine

[a] Approximate value expected for the group as a side-chain of a linear peptide.

* A table of amino-acid structures is given in Section 23.2.1.

group, like that of the SH and even the OH groups, is removed in the course of many enzyme reactions, and the main reactions of these groups are as nucleophiles.

There are two important additions to this list, which are used by many enzymes to extend the range of reactions catalysed by the five functional groups. These are the coenzymes, and metal ions. One obvious shortcoming of the list of groups in Table 1 is the lack of an efficient electron-sink. Many organic reactions, particularly those involving the cleavage of carbon–carbon bonds, require the delocalization of a pair of electrons, originating from the bond being broken, on to a suitable electronegative centre. A simple example is the decarboxylation of β-ketoacid (**2**):

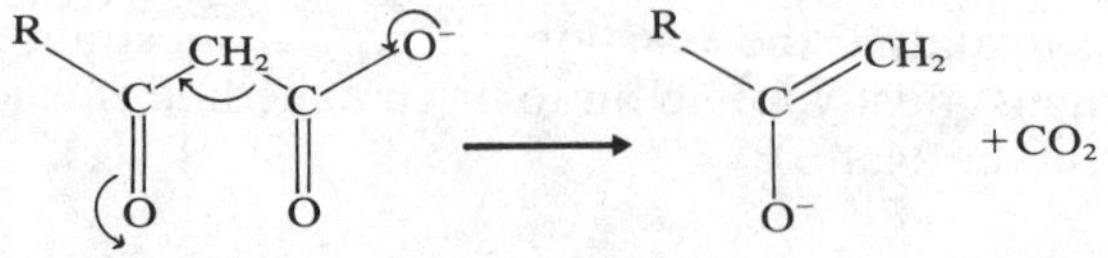

(2)

Enzymes catalysing the cleavage of carbon–carbon bonds often use coenzymes, such as thiamine pyrophosphate and pyridoxal, as portable electron-sinks; and other coenzymes play similarly specific roles which are not effectively fulfilled by the five primary functional groups.*The other missing function is any obvious reversible oxidation–reduction system, particularly for one-electron transfers. Here, too, coenzymes help fill the gap, as do metal ions in many systems. The various roles played by metal ions are discussed below (Section 24.1.2.5).

To the chemist the five functional groups of Table 1 do not look a very impressive array of reagents. There is no strong acid or base (as of course there cannot be near pH 7); and even the nucleophiles, which have perhaps the highest obvious intrinsic reactivity, are present mostly as the conjugate acids. Yet these are the groups responsible for the remarkable catalytic properties of enzymes, and it is to their chemistry we must look to begin to understand how this can be so.

24.1.2.2 Catalytic mechanisms available to enzymes

At the heart of every enzyme-catalysed reaction lies a reaction between a functional group of the substrate – ester, amide, acetal, phosphate, *etc.* – and one, perhaps two, sometimes three, catalytic groups on the enzyme. Any informed discussion of the mechanisms of enzyme catalysis must therefore be based on an understanding of how these groups can interact, and this understanding has come over the last 15–20 years from the study of such reactions in simple systems.[16–18]

Initially we can ask specific questions. How does an imidazole group interact with an amide group? Can a carboxyl group catalyse phosphate transfer? It has gradually become apparent that particular catalytic groups are well-suited for particular reactions; reactions of acetals, for example, always require acid catalysis, and only the carboxyl group, of the five groups available to the enzyme, seems to be a strong enough acid. On the other hand, amide hydrolysis, a reaction catalysed by a large number of enzymes, can be catalysed by at least four of the five. These conclusions come both from the identification of the catalytic groups on the relevant enzymes, and from the study of the chemistry of model systems. The basic mechanisms, on the other hand, were initially identified exclusively in simple systems, and we start by describing the development of this approach.

It is convenient to base the discussion on a specific example. Because the chemistry is familiar we choose the ester group as the model substrate, and carboxyl as the catalyst. The question in this case, then, is *how can the carboxyl group catalyse ester hydrolysis*? The experimental approach is simple. We need to observe the effect of the carboxyl group on

* The principal coenzymes are discussed in Chapter 24.3.

the hydrolysis of an ester: so we start by choosing the simplest possible compound containing a carboxyl group, say acetic acid; and a simple ester, for example ethyl acetate. The initial experiment is therefore to measure the rate of hydrolysis of ethyl acetate in water, or a predominantly aqueous solvent, perhaps at or near 37 °C, in the presence of varying concentrations of acetic acid. We know that ester hydrolysis is acid catalysed (Figure 3, below), so the pH must not vary. The simplest answer is to use an acetic acid–sodium acetate buffer to maintain constant pH, and to vary the concentrations of this buffer. This has the further advantage of allowing us to test the effect of both CO_2H and CO_2^- groups, and the result is almost invariably the same: there is no detectable catalysis of the hydrolysis of a simple substrate like ethyl acetate by acetic acid or acetate. Far from being able to say something about the mechansim, we have not even found the reaction!

Nevertheless, the reaction must be possible, if only because enzymes can do it. Evidently in our system it is too slow to detect. One answer might be to raise the temperature, and indeed many reactions of this sort have first been detected at high temperatures, of 100 °C or more. However, raising the temperature from 25 or 37 to 100 °C increases rates by factors of no more than about 10^3 for most of the reactions we are interested in. To find reactions which go at convenient rates under mild conditions, the simplest solution is to use more reactive substrates.

An ester can be made more reactive by putting electron-withdrawing substituents in either the acyl group or the leaving group. These effects are illustrated by the changing pH–rate profiles for hydrolysis shown in Figure 3. In either case the more electrophillic ester is slightly less reactive towards acid catalysis (which is no longer detectable above pH 0 for highly electrophilic esters like trifluoroacetates),[19] but much more readily hydrolysed by alkali. An entirely new feature is the pH-independent region near neutrality. This represents a spontaneous or water-catalysed reaction, evidently too slow, compared with the H_3O^+ and OH^- catalysed reactions, to show up in the hydrolysis of a simple ester like ethyl acetate, even at pH 5–7.

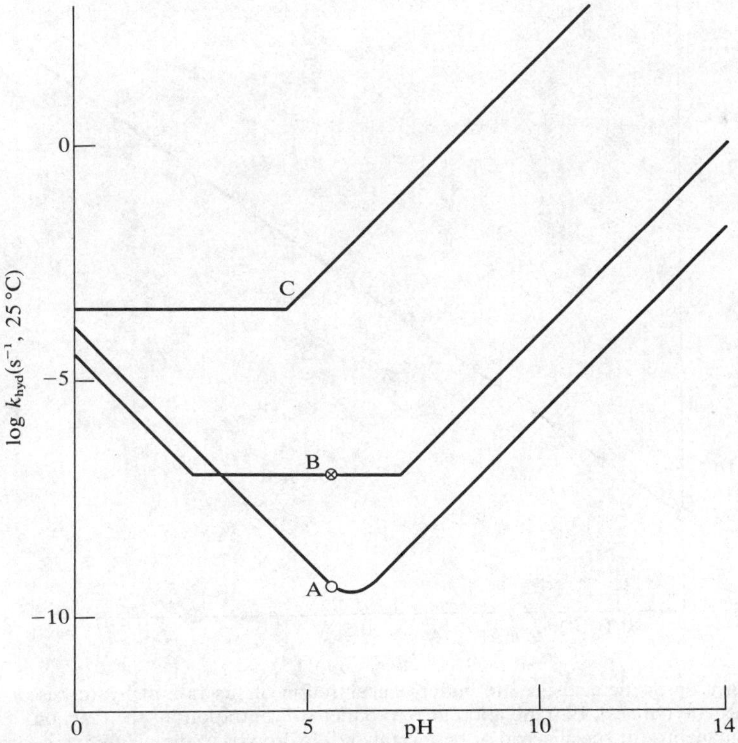

Figure 3 Approximate pH–rate profiles for the hydrolysis of ethyl acetate (A), phenyl acetate (B), and ethyl trifluoroacetate (C), at 25 °C. Curves B and C are based on extrapolations to zero buffer concentration (Figure 4)

Now water is not a powerful reagent. It is a weaker acid than acetic acid, and a weaker base, and nucleophile, than acetate ion. So if water can catalyse the hydrolysis of an ester like phenyl acetate, or ethyl trifluoroacetate, then we would expect acetate buffers to do so also. So we now repeat our first experiment with more confidence, using the more reactive substrate. Then we do indeed find that the rate of hydrolysis of the ester depends on the concentration of the buffer. The situation is summarized in Figure 4, where rate constants for the hydrolysis of phenyl acetate are plotted as a function of buffer concentration. The plot is linear in buffer concentration, so the reaction is first order in either acetate or acetic acid. Which is the active form is readily determined by changing the acetate : acetic acid ratio. In this case the buffers with more acetate are more active, so the active catalyst is acetate ion.

We have now found the sort of reaction we are looking for, and in these relatively simple cases it is possible to elucidate the mechanism in detail. Much careful work on systems of this sort has identified two mechanisms by which the carboxyl group can catalyse the hydrolysis of reactive esters, both involving the carboxylate anion.

The simpler mechanism is *nucleophilic catalysis* (Scheme 2). Nucleophilic attack by acetate on an aryl acetate first produces a tetrahedral addition intermediate, a highly unstable species, not produced in detectable amounts because it can break down very rapidly, either by loss of the acetate (k_2) to regenerate the starting materials, or by the elimination of aryl oxide (k_3). In the latter case the other product is acetic anhydride, which is rapidly hydrolysed by water.

The anhydride intermediate can be trapped by carrying out the reaction in the presence of low concentrations of a nucleophile, such as aniline, which does not react with the ester under the conditions of the reaction. Aniline is a more powerful nucleophile than water towards the carbonyl group, and the reaction is diverted to produce largely acetanilide,[20] though the rate of disapppearance of the ester is unchanged.

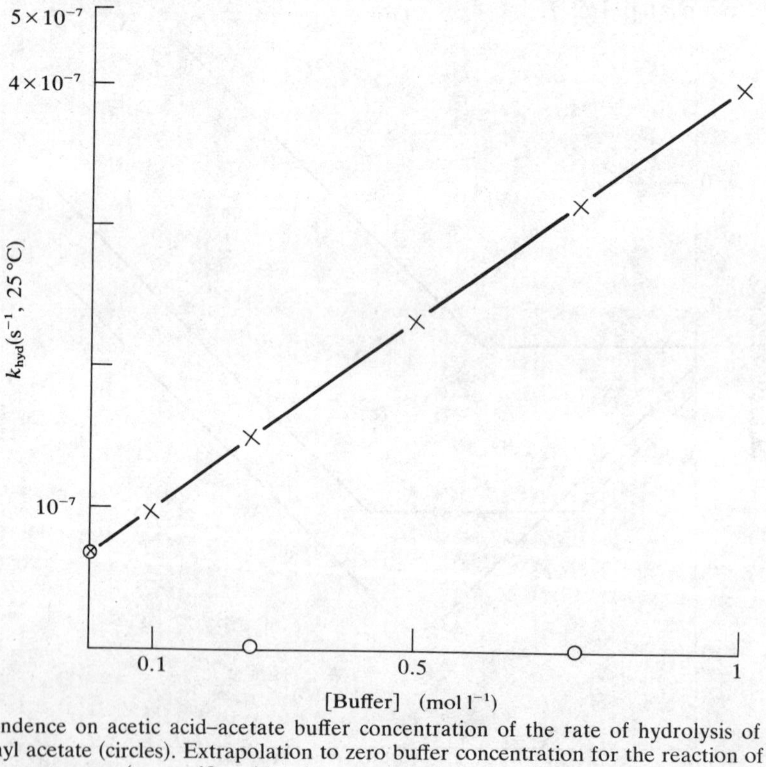

Figure 4 Dependence on acetic acid–acetate buffer concentration of the rate of hydrolysis of phenyl acetate (crosses) and ethyl acetate (circles). Extrapolation to zero buffer concentration for the reaction of phenyl acetate gives the rate of spontaneous (or specific acid or base catalysed) hydrolysis at the pH (5.5) concerned, and thus a point ⊗ on the pH–rate profile for hydrolysis (Figure 3). Since the hydrolysis of ethyl acetate is not buffer-catalysed, points (○) at any buffer concentration can be used

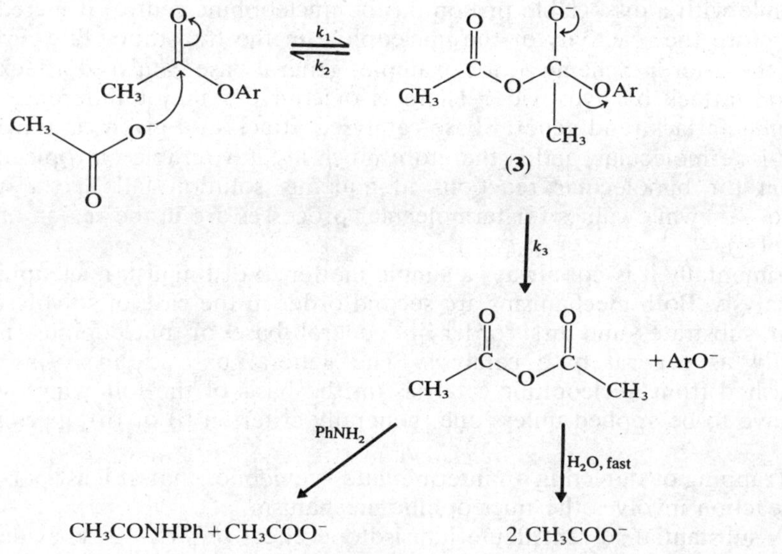

SCHEME 2

All five functional groups available at enzyme active sites can act as nucleophiles, and can be shown to act as nucleophilic catalysts in appropriate model systems. The main requirements are that the intermediate formed – the anhydride in Scheme 2 – should be more reactive than the starting materials, and that the nucleophile can readily displace the leaving group, *i.e.* in Scheme 2, that $k_3 \geqslant k_2$. Otherwise the tetrahedral intermediate (**3**) will simply revert to starting materials. This requirement is nicely demonstrated by the reaction shown in Scheme 2, the acetate-catalysed hydrolysis of aryl acetates. When the ester has a very good leaving group, as is the case for 2,4-dinitrophenyl acetate, the aniline trapping technique shows that the mechanism is nucleophilic catalysis, since the amount of acetanilide formed is sufficient to account completely for the observed reaction. On the other hand, as we have seen, acetate has no effect on the rate of hydrolysis of ethyl acetate, where the leaving group, ethoxide ion, is much poorer than acetate, and nucleophilic catalysis is ruled out.

The second mechanism is observed in reactive systems where nucleophilic catalysis is ruled out. For example, the hydrolysis of ethyl trifluoroacetate is catalysed by acetate, although it is quite clear that acetate cannot displace ethoxide ion under the conditions. This mechanism is shown in Scheme 3, and is known as *general base catalysis.*

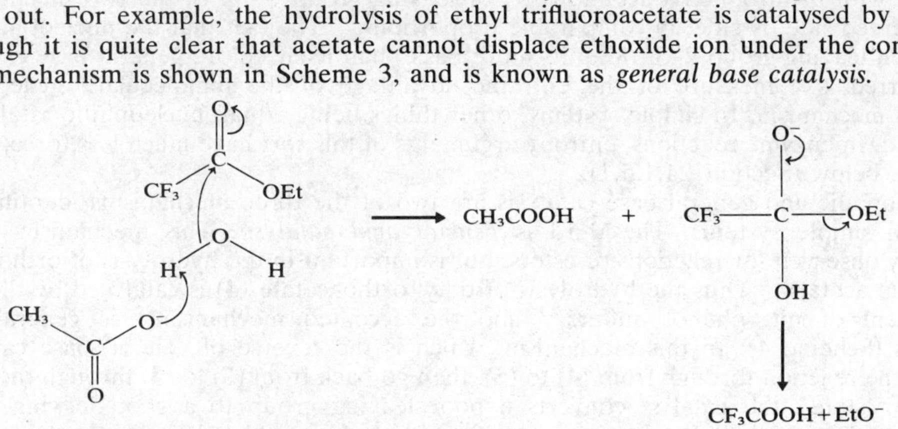

SCHEME 3

Here the *general base* is too poor a nucleophile to support nucleophilic catalysis, so it does the next best thing in assisting the attack of a water molecule on the electrophilic centre. By removing a proton from the attacking molecule (which can in principle be any nucleophile with a dissociable proton on the nucleophilic centre) it increases the basicity, and therefore the reactivity of the nucleophile in the transition state. From the point of view of the ester in Scheme 3, for example, general base catalysed attack of water differs little from attack by hydroxide. There is of course a major difference between simple nucleophilic attack, and general base catalysed attack at a given centre, in that the latter reaction is termolecular, and is therefore much less favourable entropically. (Entropies of activation for bimolecular reactions in aqueous solution fall in the vicinity of -80 $J\,K^{-1}\,mol^{-1}$,[21] while values for termolecular processes are in the region of -140 to -200 $J\,K^{-1}\,mol^{-1}$).[19]

Experimentally it is not always a simple matter to distinguish nucleophilic from general base catalysis. Both mechanisms are second order in the case of solvolysis reactions, first order in substrate, and first order in general base or nucleophile, *i.e.* both behave *kinetically* as general base catalysis. The general base catalysis *mechanism* can be distinguished from nucleophilic catalysis on the basis of the following five criteria, all of which have to be applied unless one [generally criterion (i) or (ii)] gives an unambiguous answer.

(i) Trapping or detecting an intermediate is evidence that at least part of the observed reaction involves the nucleophilic mechanism.

(ii) A substantial solvent deuterium isotope effect ($k_H/k_D \geqslant 2$) is evidence that a bond to hydrogen is broken in the rate-determining step, and thus that the mechanism is general base catalysis.

(iii) A large negative entropy of activation ($-140\ J\,K^{-1}\,mol^{-1}$ or less) is evidence for a termolecular transition state, and thus for general base catalysis.

(iv) If a Brønsted plot (of log k_{cat} *vs.* the pK_a of the catalyst) gives a good straight line, the mechanism is probably general base catalysis. Nucleophilic reactions are much more susceptible to steric effects, which show up very clearly as scatter on plots of this sort.

(v) If the pK_a of the conjugate acid of the leaving group is much higher than that of the nucleophile, it is unlikely to be displaced, and any observed catalysis is likely to involve the general base mechanism.

The useful generalizations which emerge from the study of these two mechanisms are that, for a given acyl group, the mechanism will change from nucleophilic to general base catalysis as the leaving group is varied. Thus catalysis by acetate (p$K_a = 4.76$) is 100% nucleophilic for reaction with 2,4-dinitrophenyl acetate (pK_a of leaving group $= 4.11$), but 100% general base for the reaction with phenyl acetate (the pK_a of phenol is 10.0). For reaction with *p*-nitrophenyl acetate (pK_a of leaving group $= 7.14$), the two mechanisms are observed side by side, in comparable proportions.[20] The fact that the nucleophile can displace a leaving group 2–3 pK units more basic than itself before general base catalysis is preferred is a measure of the entropic advantage of the bimolecular, nucleophilic catalysis mechanism. In simple systems, other things being equal, nucleophilic catalysis is preferred. In enzyme reactions, entropy arguments of this sort have much less force, as we shall see below (Section 24.1.5.1).

Nucleophilic and general base catalysis are two of the three mechanisms identified by work on simple systems. The third is *general acid catalysis.* This mechanism is not normally observed for reactions of esters, but is important in the hydrolysis of orthoesters and some acetals.[22] Thus the hydrolysis of ethyl orthoacetate (**4**) is catalysed by the acid component of nitrophenol buffers,[23] and the accepted mechanism[22] is general acid catalysis (Scheme 4). In this mechanism, which is the reverse of general base catalysis [follow the reaction through from (**4**) to (**5**), then go back from (**5**) to (**4**) through the same transition state], the catalyst converts a poor leaving group to a good leaving group by protonating it. The difference from specific acid catalysis, which depends on pH only and not on the concentration of the general acid (the phenol in this case), is that here the proton transfer and C—O bond-breaking steps are concerted.

Since the tetrahedral intermediates involved in ester hydrolysis are hemi-orthoesters, *e.g.* $MeC(OH)_2OEt$ in the hydrolysis of ethyl acetate, we would expect general acid catalysis

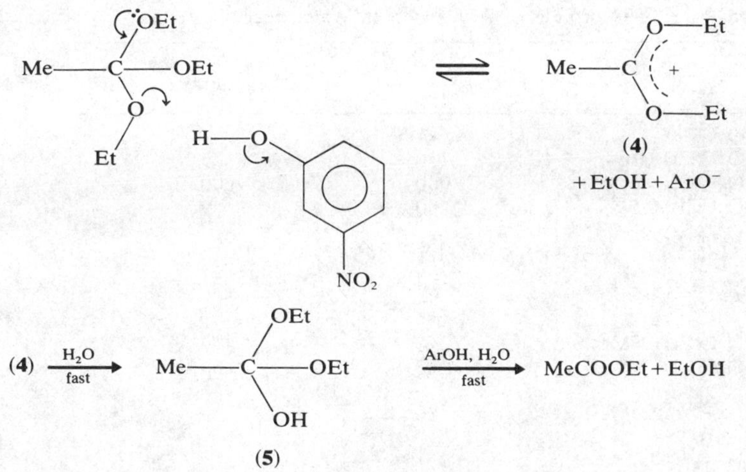

SCHEME 4

of the second stage of this and similar reactions, which involve the hydrolysis of intermediates of this type. Such catalysis is not normally observed simply because the first step in general base catalysed ester hydrolysis is securely rate determining, and the breakdown of the tetrahedral intermediate is too fast to be kinetically significant. In an enzyme-catalysed hydrolysis, on the other hand, the normal rate-determining step will necessarily be much accelerated, and other, normally fast steps, may well require catalysis if they are not to reduce the efficiency of the overall catalytic process.

These three mechanisms, general acid, general base, and nucleophilic catalysis, are found to account for the observed reactions of all five functional groups available to enzymes, with all the common substrate groups. So our initial conclusion is that these are the mechanisms involved when the reactions occur in the enzyme–substrate complex, and enzyme mechanisms are indeed discussed in terms of these three basic mechanisms. An important weakness in the argument, however, is that these conclusions are based almost exclusively on evidence from activated substrate systems. Clearly we should prefer evidence that the same mechanisms are involved in reactions with unactivated substrates in simple systems. This evidence comes from the study of intramolecular reactions.

24.1.2.3 Intramolecular catalysis[24,25]

When the catalytic and substrate groups are in the same molecule, dramatic increases in reactivity may result. Table 2 shows the effects of incorporating a carboxyl group into a series of aryl esters. The rates of hydrolysis of these compounds were measured at very low concentrations ($<10^{-4}$ mol l^{-1}), and the presence of such low concentrations of added carboxyl compound (acetate, for example) would have no detectable effect. It is clear that the hydrolysis of the ester group is catalysed by the neighbouring carboxylate group (the ionized form is active, as usual), which evidently has an 'effective molarity' much higher than its actual concentration in solution. This parameter (the final column of Table 2) is the best measure of the efficiency of intramolecular compared with intermolecular catalysis by a particular group. It is the ratio of a first-order and a second-order rate constant, so has the dimensions of molarity, and can be interpreted as the molarity of external catalyst (acetate in this case) which would be required to make the reaction of the standard substrate go as fast as that of the substrate with the catalyst built-in. All the effective molarities in Table 2 exceed the limit of solubility of acetate in water, so the concept is a hypothetical one. This parameter allows for the intrinsic reactivity of the catalytic group: in a given system the effective molarity of an efficient catalyst is likely to

TABLE 2

Relative Rates of Hydrolysis of Aryl Esters in Water near pH 7, at 25°

Ester	k^{a}_{rel}	*Effective molarity*[b] ($mol\,l^{-1}$)
1. PhO_2CMe	1	
2. $PhO_2C(CH)_3CO_2^-$ [c]	150	25
3. $PhO_2C(CH)_2CO_2^-$ [c]	23 000	4000
4. $PhO_2CCH_2CO_2^-$ [d]	50	9
5. 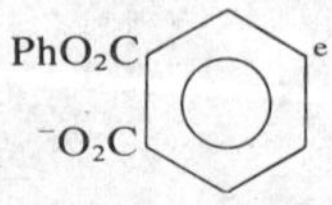 [e]	1.1×10^{6}	2×10^{5}
6. $PhO_2C(CH)_4NMe_2$ [f]	10^{6}	490
7. $PhO_2C(CH)_3NMe_2$ [f]	2.5×10^{6}	1260
8. 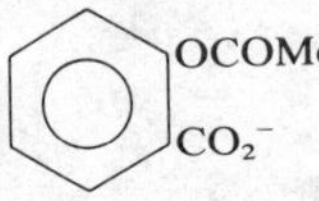	50	9

[a] k_{hyd}/k°_{hyd}, where k°_{hyd} is the first-order rate constant for the pH-independent (water-catalysed) hydrolysis of phenyl acetate ($6.6\times10^{-8}\,s^{-1}$ according to Ref. 20). [b] k_{hyd}/k_2, where k_2 is the second-order rate constant for catalysis by acetate (ref. 20) or trimethylamine (Ref. 28) of the hydrolysis of phenyl acetate. [c] E. Gaetjens and H. Morawetz, *J. Amer. Chem. Soc.*, 1960, **82,** 5328. [d] See Ref. 33. [e] See Ref. 33. [f] See Ref. 28. [g] See Ref. 29.

be similar to that of a poor one. (Compare the effective molarities of the Me_2N—* and —CO_2^- groups for entries 2 and 6, and 3 and 7, of Table 2.) So the alternative measure of catalytic efficiency, simply the ratio of the rates of the reaction in the presence and absence of the intramolecular catalyst, is sometimes more appropriate, since in this comparison the more efficient catalyst will produce a larger acceleration. (Compare k_{rel} for the same pairs of entries, 2 and 6, 3 and 7, of Table 2.)

All the intramolecular reactions shown in Table 2 involve nucleophilic catalysis (Scheme 5). This is expected for intramolecular catalysis by the dimethylamino-group (reactions 6 and 7), because (intermolecular) catalysis of the hydrolysis of phenyl acetate by the strongly basic trimethylamine involves this mechanism also. (Hammett's ρ value for

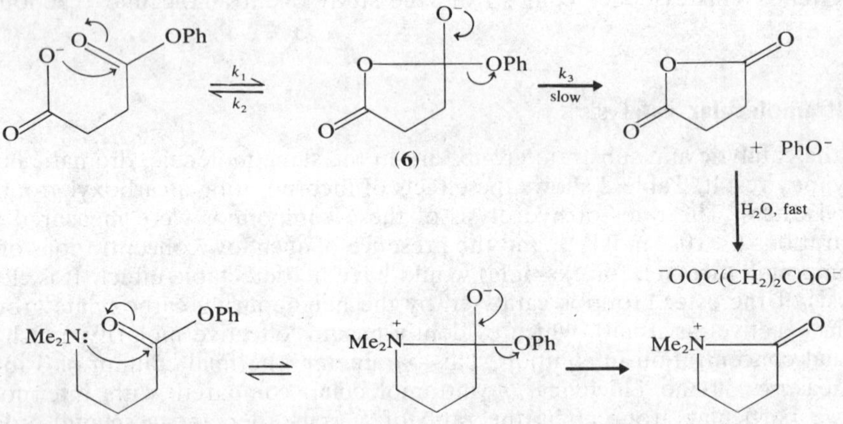

SCHEME 5

* Data for catalysis by imidazole would be more interesting than for the dimethylamino-group, but kinetic complications[26] make direct comparison difficult. The imidazole group, however, is known to be more reactive towards ester carbonyl.[26,27]

a series of substituted-phenol leaving groups is 2.2 for the intermolecular reaction with trimethylamine, very similar to the value (2.5) found for both intramolecular processes.)[28]

Catalysis of the hydrolysis of phenyl acetate by acetate ion, on the other hand, involves the acetate as a general base. Evidently making the reaction intramolecular favours the nucleophilic mechanism compared with general base catalysis. A possible explanation involves the tetrahedral intermediates involved, *e.g.* (**6**) (Scheme 5). The CO_2^- group is a better leaving group than PhO^- (*i.e.* $k_2 > k_3$), because of its lower basicity, so the breakdown of (**6**) to products (k_3) is the rate-determining step of the reaction. In the rate-determining step the molecule loses phenoxide ion, producing two molecules from one, a process which has a specially favourable entropy of activation. When the leaving group is not lost in this way, general base catalysis is observed for the intramolecular reaction also (Scheme 6). In these reactions of aspirin[24,29] (**7**) and the imidazole analogue (**8**),[30] nucleophilic catalysis is possible, but unfavourable for the same reasons that general base catalysis is preferred for the corresponding intermolecular reactions (see above, p. 399, and Ref. 24 for a fuller discussion).

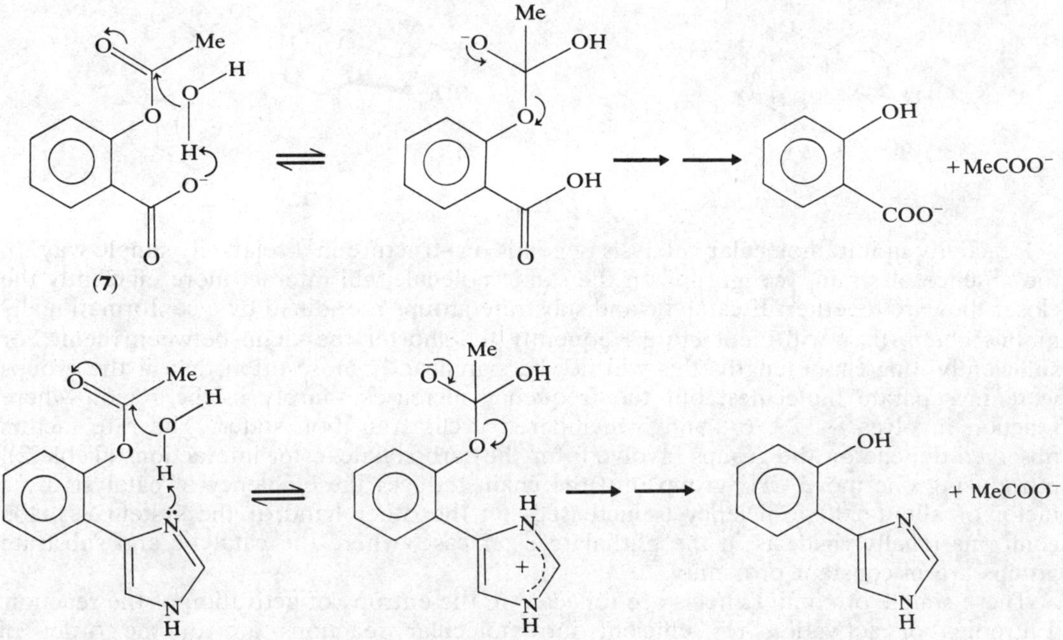

(**7**)

(**8**)

SCHEME 6

Apart from special cases of this sort, the general rule is that the mechanism of an intramolecular reaction will be the same as that of the corresponding intermolecular reaction between the same groups. Trivial exceptions are those where the groups are prevented from interacting at all, by the geometry of the molecule concerned. Obviously the carboxyl group of the acetate ester (**9**) of *p*-hydroxybenzoic acid cannot catalyse the hydrolysis of the ester group, because it cannot get close enough. In fact the hydrolysis of (**9**) makes a useful control for the hydrolysis of aspirin, because the electronic effects of the *para*-CO_2^- group are similar to those of *ortho*-carboxylate. A more subtle, but still readily identified, exception is exemplified by the hydrolysis of monophenyl malonate (entry 4 of Table 2). Here the nucleophilic mechanism would involve a four-membered ring anhydride. Compounds with four-membered rings cannot generally be made by reversible reactions, and any tetrahedral intermediate (**10**) will revert exclusively to starting materials. So the mechanism must be intramolecular general base catalysis, as is readily shown by inspection of the kinetic parameters for the reaction.[31]

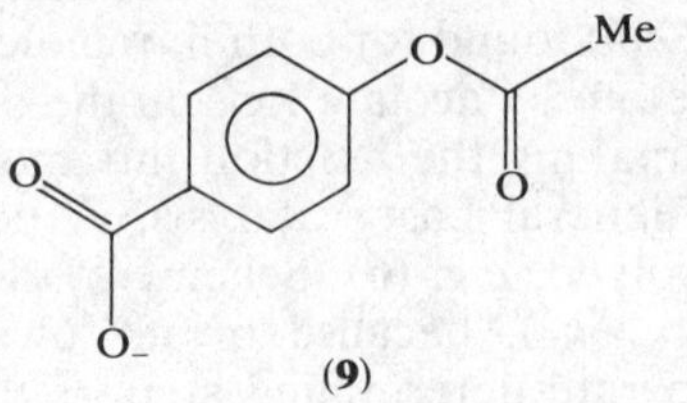

(9)

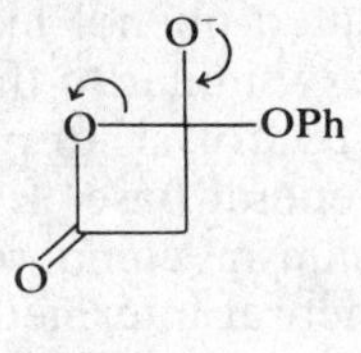

(10)

Intramolecular general acid catalysis is also readily demonstrated, particularly with acetals (**11**) derived from salicylic acid and aldehydes. These may be simple compounds like formaldehyde and benzaldehyde, or sugar aldehydes. The reactions of such compounds (**12**) are of interest in connection with the mechanism of action of enzymes which hydrolyse glycosides[24,32] (see below, Section 24.1.4.4).

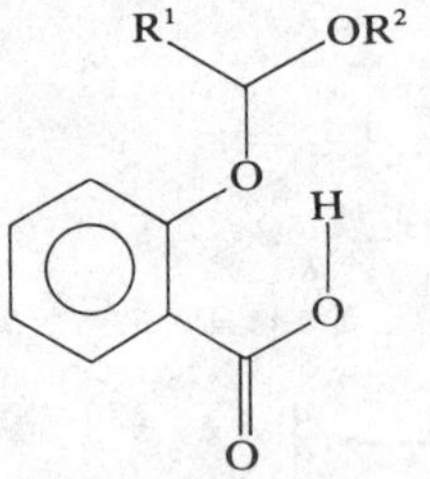

(11)

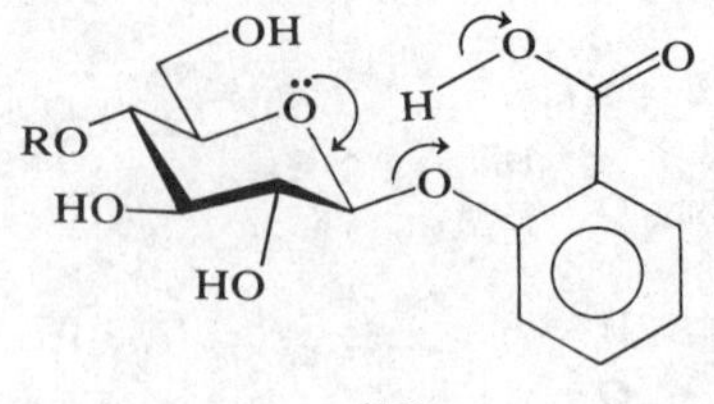

(12)

Reactivity in intramolecular catalysis depends on structure in a relatively simple way. In the absence of strain, two groups on the same molecule will interact more efficiently the closer they are together. If catalytic and substrate groups are joined by a conformationally mobile chain, they will meet more frequently the shorter the chain between them. For sufficiently long chain lengths this will not be significantly more often than if the groups were in separate molecules, but the frequency increases sharply in the region where reaction involves 8-, 7-, 6-, and 5-membered cyclic transition states. The rate factors observed depend on the groups involved: for the carboxylate–ester interactions (Table 2), introducing one more CH_2 group into the chain reduces the efficiency of catalysis by a factor of about 200. Efficiency is increased, on the other hand, if the system is made conformationally rigid, as in the phthalate ester case, where the catalytic and substrate groups are in constant proximity.

These simple proximity effects are reflected in the entropy of activation of the reaction. Entropies of activation for efficient intramolecular reactions are of the order of $80\ J\ mol^{-1}\ K^{-1}$ more positive than for the corresponding intermolecular processes, because incorporating the reacting groups into the same molecule reduces the molecularity of the reaction by one. The enthalpy of activation, on the other hand, is not necessarily significantly different, and many examples are known where it is the same within experimental error for the corresponding intra- and inter-molecular reaction.

The enthalpy of activation for the intramolecular process is changed, however, when either the ground or the transition state for the intramolecular reaction is strained. The wrong rigid conformation can hinder or even prevent reaction; bringing groups *too* close together can have the same effect, as in the case of monophenyl malonate (Table 2) discussed above. We will return to this subject in Section 24.1.5.

The intramolecular reactions we have seen so far all involve activated substrate groups, and are simply intramolecular versions of the reactions discussed in the previous section. The next step is to look for reactions of unactivated substrate groups. If we take the most reactive ester–carboxyl system of Table 2, the monoester of phthalic acid, and look at the hydrolysis not of the phenyl ester but of the monomethyl ester, catalysis is again found, though the reaction is very much slower. The hydrolysis of monophenyl phthalate anion has a half-life of about 30 min at 30 °C, while the anion of monomethyl phthalate is quite stable, even at 100 °C. Now, however, the *acid* form is hydrolysed, with a half-life of about one hour at 100 °C (equation 1).[33]

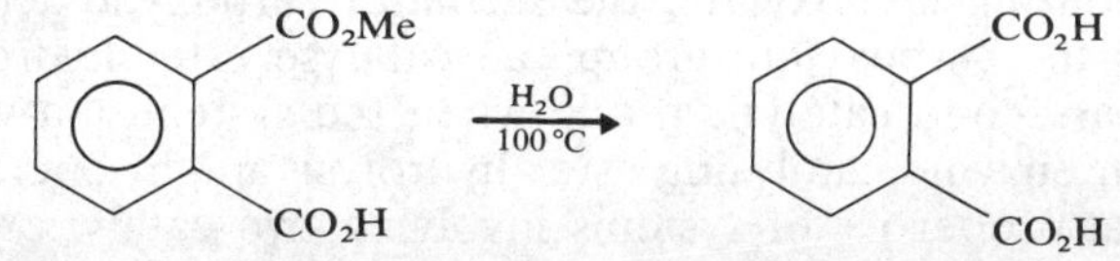

(1)

Kinetically this is intramolecular general acid catalysis by the carboxyl group, whatever the mechanism, since the group is involved in the rate law. This is the first time we have come across general acid catalysis of ester hydrolysis, and the first example of the hydrolysis of an unactivated ester. This illustrates an important principle: as we increase the reactivity of the systems under investigation, we may expect to see new mechanisms — or at least new rate-determining steps — appear, which could not be observed with less reactive compounds. Since we are ultimately interested in enzyme catalysis, where the chemical reactions are much faster than any we have seen so far, it is clearly important to study the most reactive simple systems possible.

In the case of ester hydrolysis catalysed by the carboxyl group, the most reactive system known is derived from dimethylmaleic acid. Methyl hydrogen dimethylmaleate (**13**) is hydrolysed with a half-life of just 30 s at 37 °C,[34] presumably by the same mechanism as methyl hydrogen phthalate: in this case the product is dimethylmaleic anhydride. However, the system is now so reactive that we see a reaction of the anion also. The ester (**13**) is hydrolysed at high pH with a half-life of less than two hours at 37 °C (Scheme 6a). This, in turn, is the first time we have seen intramolecular catalysis of an unactivated ester by the ionized carboxyl group. The mechanism turns out[34] to be very simple, namely, intramolecular nucleophilic catalysis by the carboxylate group (Scheme 6a), *via* a tetrahedral

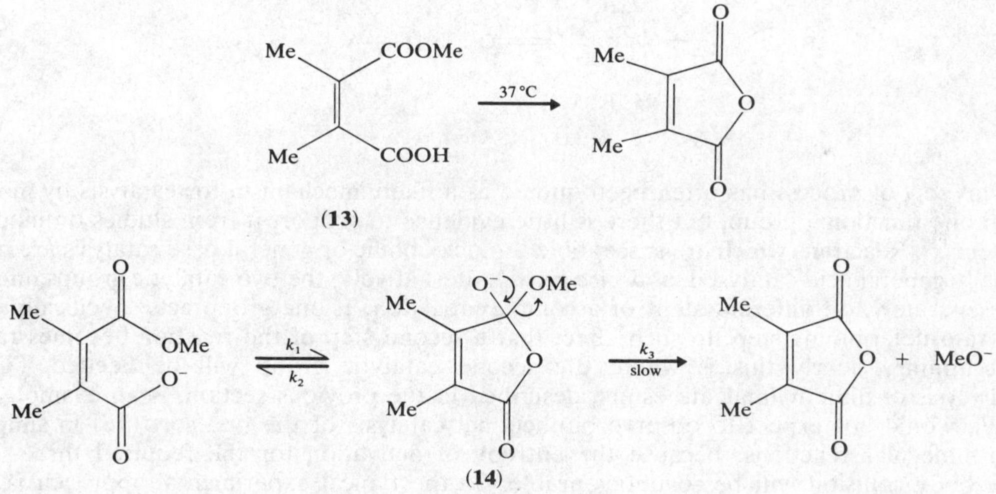

SCHEME 6a

intermediate (**14**) which is evidently much less reluctant to eliminate alkoxide than any we have previously encountered (see below). Nevertheless, there is no doubt that the loss of methoxide is the rate-determining step of the reaction. Now this is a reaction (rate-determining loss of a poor leaving group from a hemi-orthoester) which we could expect to show general acid catalysis (see p. 400), and the hydrolysis of the monomethyl dimethylmaleate anion is in fact general acid catalysed (by acetic acid*, for example).

* Also, therefore, by H_3O^+. Since the H_3O^+-catalysed reaction of the ester anion is kinetically indistinguishable from a reaction of the ester acid, this suggests that the mechanism of catalysis of the hydrolysis of this form (**13**) involves general acid catalysis of the breakdown of the tetrahedral intermediate (**14**) by H_3O^+.

The mechanism is thus general acid catalysis of intramolecular nucleophilic catalysis, and involves two catalytic groups acting simultaneously in the rate-determining step of the reaction, one the neighbouring carboxylate, the second a carboxylic acid group. Here a careful study of the way the carboxylate group can catalyse ester hydrolysis indicates a specific requirement for a second catalytic group when the system is a very reactive one (and suggests one way an enzyme catalysing ester hydrolysis might operate). We now go on to consider the special properties of systems involving two catalytic groups.

24.1.2.4 Bifunctional catalysis

Enzyme active sites commonly involve two or more catalytic groups. These may be brought to bear on a substrate group in two quite different ways. One involves nucleophilic (or general base) and general acid catalysis operating simultaneously, on the same transition state. A mechanism of this sort which might apply to ester hydrolysis is shown in (**15**).

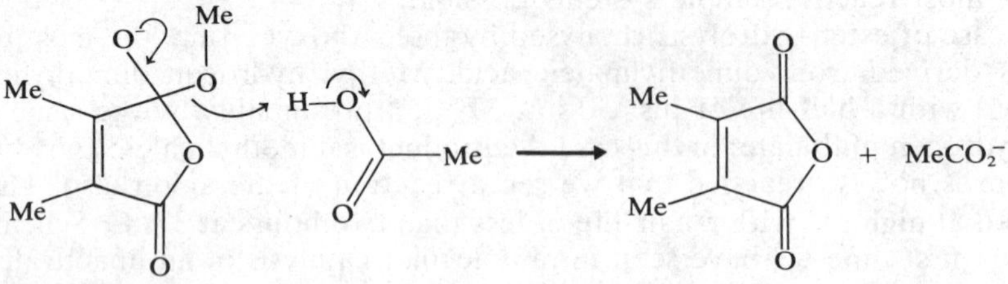

(14)

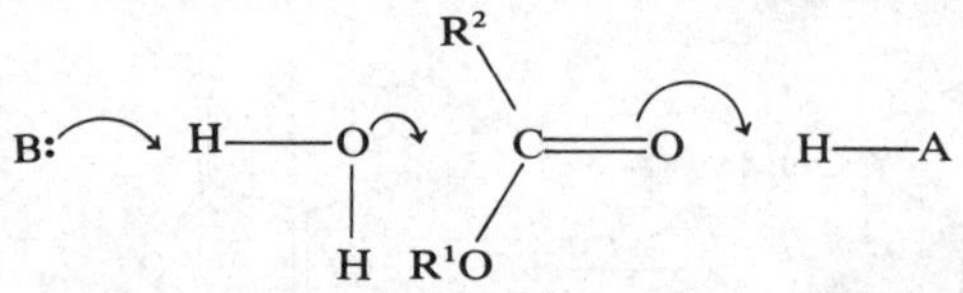

(15)

This sort of process has often been quoted as a likely mechanism for catalysis by more than one functional group, but there is little evidence to support it from studies on model systems.[32] Reactions which are susceptible to nucleophilic or general base catalysis are not usually general acid catalysed, and *vice versa*. Alternatively, the two catalytic groups might act separately, on different steps of a complex reaction. If one group acts specifically on the rate-determining step, to such effect that a second step of the reaction becomes rate determining, clearly this is where the second catalytic group will be needed. (The hydrolysis of dimethylmaleate esters, described in the previous section, is an example.)

We would not expect to observe bifunctional catalysis of the first sort (**15**) in simple intermolecular reactions, because the entropy of activation for the required three- or four-body collision will be so unfavourable. So the typical experimental approach is to observe the effect of introducing a second catalytic group on a known intramolecular reaction. The results are best illustrated by the changes observed in the pH–rate profile for the reaction.

If the hydrolysis of an ester is subject to catalysis by a neighbouring carboxyl group, we have seen that the active form of the intramolecular catalyst is CO_2^-. The ionized ester acid is therefore hydrolysed more rapidly than the undissociated compound, and the pH–rate profile for hydrolysis has the form shown in Figure 5. Specific acid- and base-catalysed reactions are observed at high and low pH, but the rate of reaction in the pH-independent region (A) is higher than expected for the compound lacking the carboxyl group. The magnitude of this acceleration (B in Figure 5) is a measure of the efficiency of the intramolecular catalysis. If the CO_2H group does not catalyse the

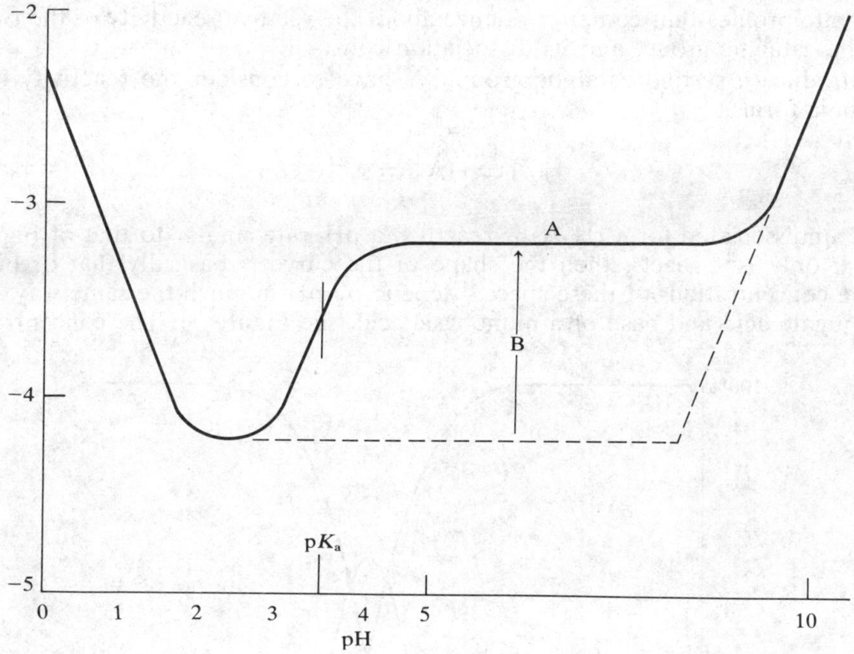

Figure 5 Catalysis by the CO_2^- group: pH–rate profile for the hydrolysis of aspirin (**7**)[29] at 39 °C

reaction, no such acceleration is apparent at low pH, so the rate of the reaction falls with decreasing pH in the region of the pK_a of the CO_2H group, as the proportion of reactive, ionized, compound falls. If, on the other hand, the CO_2H group is active, and the CO_2^- form unreactive, or less reactive, the rate is increased at lower pH (Figure 6). The shapes

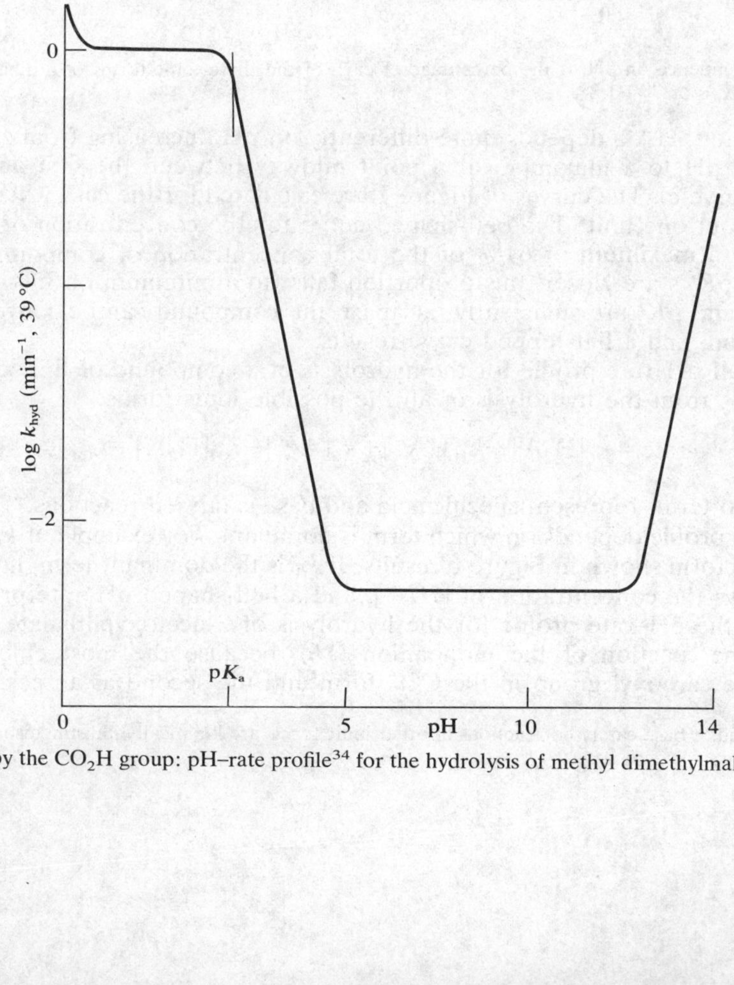

Figure 6 Catalysis by the CO_2H group: pH–rate profile[34] for the hydrolysis of methyl dimethylmaleate at 39 °C

of the pH–rate profiles thus contain evidence about the relative reactivity of the two ionic forms of the catalytic group, and its dissociation constant.

If we introduce a second catalytic group, we have to consider the reactivity of three different ionic forms:

$$H_2A \overset{K_1}{\rightleftharpoons} HA^- \overset{K_2}{\rightleftharpoons} A^{2-}$$

If only the undissociated form (H_2A) is reactive, a pH–rate similar to that of Figure 6 is observed; if only A^{2-} reacts, then the shape of the curve is basically that of Figure 5, because the concentrations of these species depend on pH in much the same way as those of the conjugate acid and base of a monobasic acid (see Figure 7). The concentration of

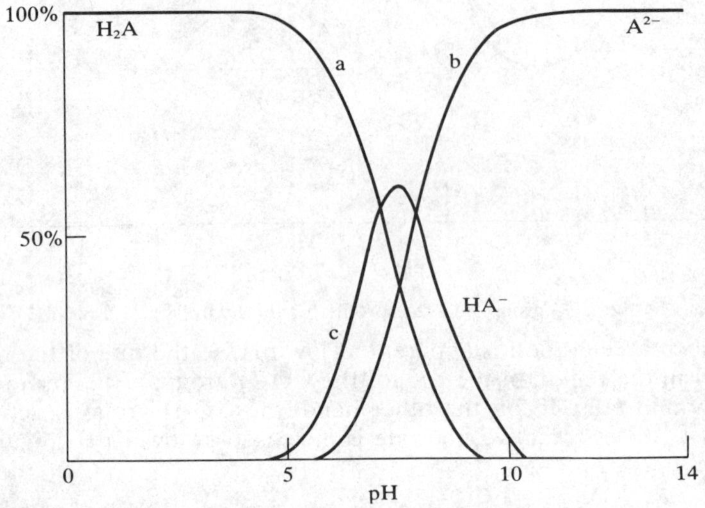

Figure 7 Dependence on pH of the percentage of each of the three ionic forms of a dibasic acid, H_2A, with approximate pK_a's of 7 and 8

the monoanion, HA^-, depends quite differently on pH, increasing from zero at very high or very low pH to a maximum at a point midway between the first and second pK_a's (Figure 7, curve c). The curves of Figure 7 are calculated for the case where the two pK_a's differ by about one unit. The bell-shaped curve for the concentration of the monoanion accounts for a maximum of 61% of the total concentration of compound added in this case. If the pK_a's are closer, this proportion falls (to a minimum of 33% when $K_1 = K_2$), and if pK_1 and pK_2 are sufficiently far apart, the compound can be converted completely to monoanion, and a flat-topped curve results.[35]

The overall pH–rate profile for the hydrolysis of a compound of this sort is the sum of contributions from the hydrolysis of all the possible ionic forms:

$$\text{rate} = k_0[H_2A] + k_1[HA^-] + k_2[A^{2-}] + k_H[H_2A] + k_{OH}[A^{2-}]$$

(The last two terms represent specific acid and base catalysed reactions.*) So the shape of the pH–rate profile depends on which term is dominant. For example, if $k_0 \gg k_1, k_2$, then a curve of the form shown in Figure 6 results. If k_1 is the dominant term, however, then the curve follows the concentration of $[HA^-]$, and a bell-shaped pH–rate profile results. An example is the pH–rate profile for the hydrolysis of 3-acetoxyphthalate, which is dominated by the reaction of the monoanion (**16**), because the most efficient mechanism requires one carboxyl group in the CO_2^- form and the second as an acid, CO_2H.[24,36]

* Specific acid or base catalysed reactions of other ionic forms are kinetically indistinguishable from one of the first three terms.

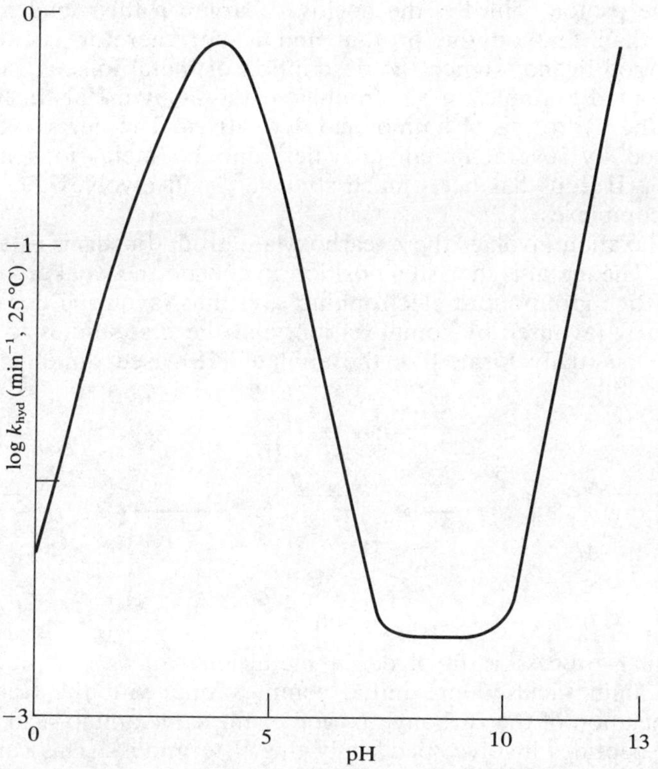

Figure 8 Catalysis by both CO_2H and CO_2^- groups: pH–rate profile for the hydrolysis of 3-acetoxyphthalate (**16**)[36]

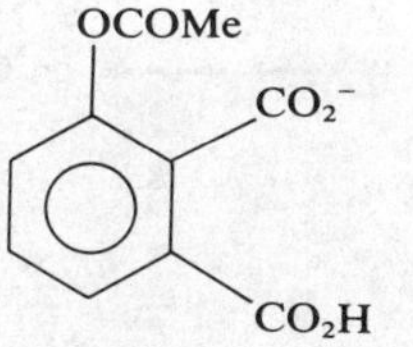

(**16**)

24.1.2.5 Catalysis by metals[37]

Many enzymes contain tightly bound metal (usually transition metal) ions, and many more are activated by specific metals. All the functional groups available to a protein are potential ligands, so a wide variety of complexes is possible, with the metal bound to oxygen, nitrogen, or sulphur donors. The properties of the metal depend critically on the ligands bound, but generally the metal acts as a Lewis acid or electrophilic catalyst*, accepting and thus withdrawing electrons from the donor, which itself becomes more electrophilic. This process is similar in some respects to general acid catalysis, but the right metal ion is considerably more versatile than the proton, and can be more effective for at least three reasons. Firstly, the concentration of the electrophile can be much higher, since the concentration of H_3O^+ is less than 10^{-7} mol l^{-1} above pH 7. Secondly, the metal ion can coordinate more than one ligand, so when it is bound close to the substrate group its effective concentration can be increased, like that of any other group, in a way not

* The cobamide enzyme reactions[38] provide (rare) examples of the opposite process, where the metal (Co) acts as a nucleophile.

available to the proton. Thirdly, the positive charge on the metal may be two or three times greater than that on the proton, and is not therefore neutralized by a single negatively charged ligand. Hence the description of metal ions as 'superacid' catalysts.

Well documented examples of electrophilic catalysis by metals include decarboxylation reactions and the hydrolysis of amino-acid derivatives. The decarboxylation of oxaloacetate is catalysed by several metalloenzymes, and by metal ions in aqueous solution. Catalysis by Cu(II) ions has been much studied,[39] and involves the formation of metal-oxaloacetate complexes (**17**).

Initial complexation involves the α-carboxylate group (if this is esterified the reaction is not observed). The metal is then in a position to chelate the weaker ketone oxygen donor, making the ketone group more electrophilic, and thus favouring decarboxylation. (Enolization is similarly favoured by complexation, and the first studies to demonstrate clearly that pyruvate is initially formed as the enolate (**18**) used dimethyloxaloacetate, which cannot enolize.[39a])

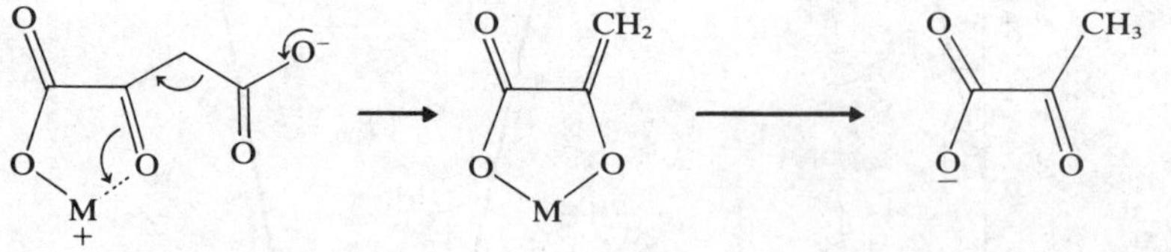

(**17**) (**18**)

A very similar process is involved in metal-ion catalysis of nucleophilic attack on derivatives of amino-acids. Here initial complexation with the free α-amino group is followed by chelation of the carbonyl oxygen of the same amino-acid residue (**19**) (so that reactions with peptides involve specifically the *N*-terminus). This carbonyl group is thus activated towards nucleophilic attack,[40] by factors greater than 10^6.*

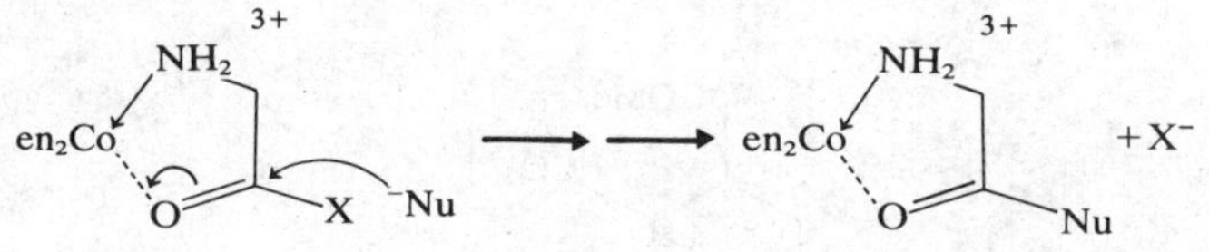

(**19**)

A different, also very efficient, mechanism for hydrolysis is observed when the amino-acid derivative is singly-bound, leaving a coordination position filled by H_2O or HO^-. Then intramolecular attack (**20**) by bound hydroxide can produce accelerations of the order of 10^{10} compared with bimolecular attack.[41]

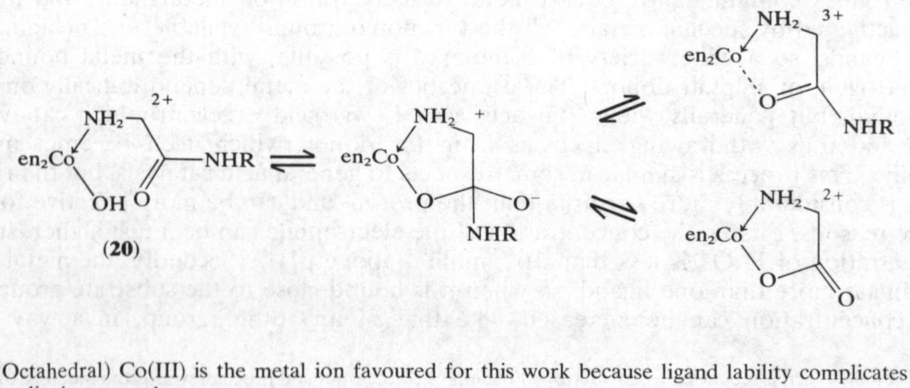

(**20**)

* (Octahedral) Co(III) is the metal ion favoured for this work because ligand lability complicates work with other, divalent, metals.

Both these mechanisms are likely to be of importance in enzyme-catalysed reactions, where the stability of complexes can be controlled by the three-dimensional structure of the protein (and to a certain extent *vice versa*, because bringing together ligands from different parts of the chain into the coordination sphere of a metal ion can play a structural role).

A final, very important role played by metal ions is as electron transfer agents,[37] particularly for one-electron transfers, where redox systems such as Fe(II)$\rightleftharpoons$Fe(III) and Cu(I)$\rightleftharpoons$Cu(II) are commonly involved. The redox potential is a sensitive function of ligand binding, and in many cases (haemoglobin, the cytochromes, chlorophyll, vitamin B_{12}) the metal is complexed not only by the protein, but also by macrocyclic tetradentate ligands (*e.g.* the porphyrin in haem), which leave free only one coordination position, with very special and closely controlled properties.[42]

24.1.3 CATALYTIC MECHANISMS USED BY ENZYMES

We began by resolving to approach enzyme mechanism like any other problem of organic reaction mechanism. The catalytic mechanisms used by enzymes can indeed be considered in terms of the interactions of a small number of functional groups within the enzyme–substrate complex. However, it is no small problem to discover just which groups on the enzyme are involved in catalysis. A protein has many functional groups, which will react indiscriminately with most reagents, and only reactions with substrates can be relied on to involve active site groups specifically. So the essential first step is a study of the specific reaction catalysed by the enzyme, designed to identify the functional groups involved in the catalytic mechanism.

24.1.3.1 Identification of catalytic groups

The simplest approach is to determine the pH–rate profile for the specific reaction. The rates of all enzyme-catalysed reactions depend on pH, and most show bell-shaped pH–rate profiles with well-defined pH-optima (see, for example, Figure 9). It is not surprising that a particular ionic form of an enzyme should be the most active, and since a protein has many ionizable groups it is reasonable that the activity should fall off at higher or lower pH. The activity, however, is generally determined by the state of ionization of individual groups at or close to the active site. We have seen already (Section 24.1.2.4) how a mechanism involving just two ionizable groups can account for a bell-shaped pH–rate profile, and how its shape depends on the pK_a's of the groups involved. It is possible to analyze the shapes of the pH–rate profiles obtained for enzyme-catalysed reactions,[43] on the assumption that they reflect specific functional group ionizations, and to deduce the pK_a's of the groups involved. Since the five catalytic groups available have characteristic pK_a's, the numerical values obtained might then be sufficient evidence to identify the catalytic groups involved. For example, an apparent pK_a in the region of 7 suggests the involvement of the imidazole group of a histidine residue at the active site.

In practice, matters are less simple. Apparent pK_a's in the region of 7, for example, are often observed, and include cases where histidine certainly is not involved at the active site. This can happen because group dissociation constants depend on environment, and may differ substantially, when the group is surrounded largely by protein, from the values found for the free amino-acids, or simple derivatives, in solution. Some relevant effects are illustrated in Table 3.

The usual effect of a less polar medium is to suppress the ionization, since charged species are less stable in a less polar medium. This accounts for the increased pK_a of acetic acid, and the decreased pK_a of ethylamine, in 90% aqueous ethanol. Note that both effects displace pK_a's towards neutrality. When the situation is complicated by adjacent protic groups which may be involved in internal solvation, as is the case with glycine, smaller but less easily predictable effects may be observed. When the neighbouring group

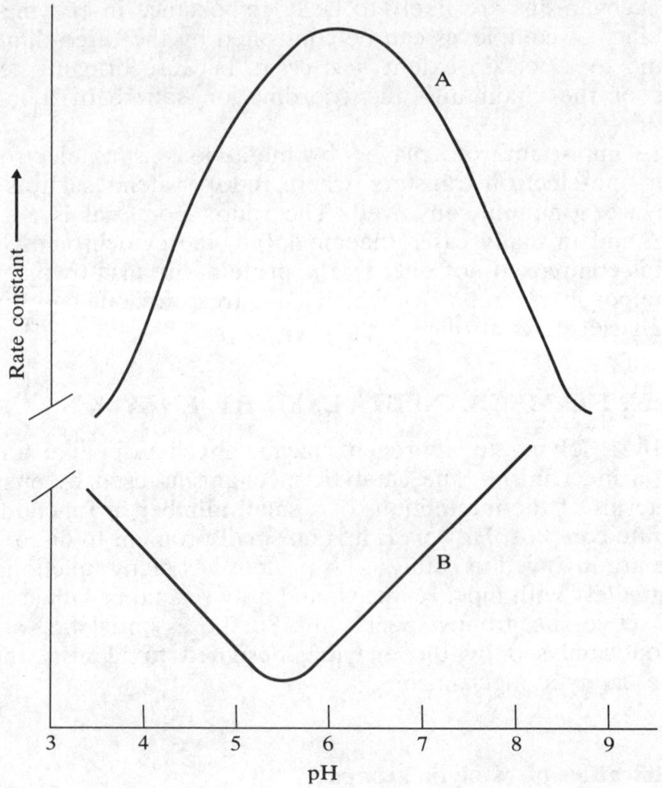

Figure 9 Bell-shaped pH–rate profile (curve A) for an enzyme-catalysed reaction (the hydrolysis of *N*-benzoylarginine ethyl ester, catalysed by papain, k_{cat}/K_m plotted). Curve B is the expected pH–rate profile for hydrolysis of this ester, and brings out the contrast between enzyme and specific acid–base catalysis. (The scales are arbitrary, but the enzyme-catalysed reaction is of course faster by many orders of magnitude)

can form a very strong hydrogen-bond to the basic form of one of the ionizable groups, as in racemic di-t-butylsuccinic acid, the monoacid-monobase form is selectively stabilized, and pK_1 and pK_2 are displaced away from neutrality. Strong local medium effects and H-bonding are to be expected in a protein, so it is clear that apparent dissociation constants alone do not allow reliable identification of catalytic groups.*

A celebrated example is the enzyme acetoacetate decarboxylase,[44] which catalyses the decarboxylation of acetoacetate, and depends on a basic group of apparent p$K_a = 5.9$. This group turns out to be the ε-NH_2 group of a lysine residue, which normally has a pK_a close to 10. The reaction is catalysed *in vitro* by secondary amines, and by aniline, *via* Schiff base (**21**) formation (Scheme 7), and a similar mechanism is thought to account for the enzyme-catalysed reaction. (Ketone carbonyl ^{18}O-enriched acetoacetate, for example, loses all its label during decarboxylation.) The amino-group at the enzyme active site which catalyses the reaction (RNH_2 in Scheme 7) is the terminal amino-group of a lysine residue, and the first step of Schiff-base formation requires this to attack the substrate carbonyl group. The amine must therefore be in the free base form. If the group had a normal solution pK_a of about 10 it would be 99.99% protonated at pH 6. With the pK_a shifted to 5.9 it exists mostly as the free base form at pH 6, and an increase in efficiency which could be as large as 10^4 is gained.

So there may be very good reasons why group pK_a's should be shifted in the special environment of the enzyme–substrate complex, and the apparent pK_a's obtained from the

* Thermodynamic parameters, particularly the heat of ionization, are less medium-dependent, and also characteristic of the ionizing groups, and such extra information may strengthen the original assignment.

TABLE 3
Effects of environment on group dissociation constants

1. *Medium effects*[a]	pK_a *in* *Water*	pK_a *in* 90% *Ethanol*
CH_3CO_2H	4.7	7.1
$CH_2(CO_2H)(NH_3^+)$: CO_2H	2.35	3.8
NH_3^+	9.8	10.0
$EtNH_3^+$	10.8	9.8
2. *Effects of H-bonding*[b]	pK_1	pK_2
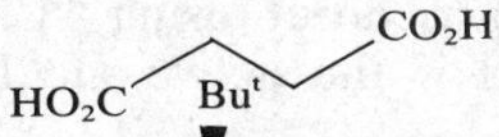	4.19	5.54
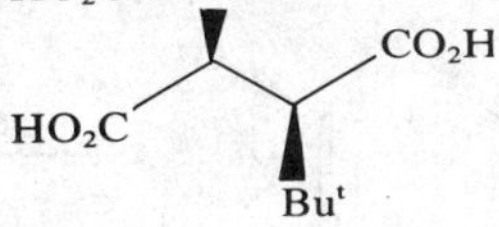	2.20	10.3
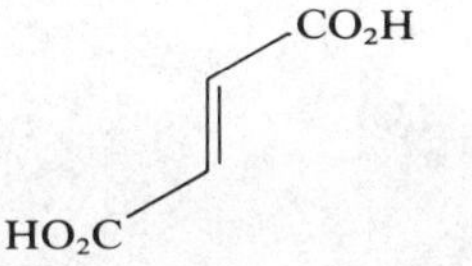	3.02	4.38
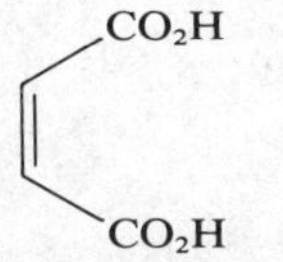	1.92	6.34
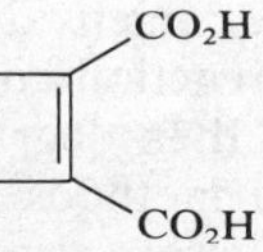	1.12	7.63

[a] J. E. Cohn and J. T. Edsall, 'Proteins, Amino-acids and Peptides as Ions and Dipolar ions', ACS Monograph 90, Reinhold, New York, 1943, p. 109. [b] L. Eberson, in 'The Chemistry of Carboxylic Acids and Amides', ed. S. Patai, Wiley-Interscience, New York, 1969, p. 274.

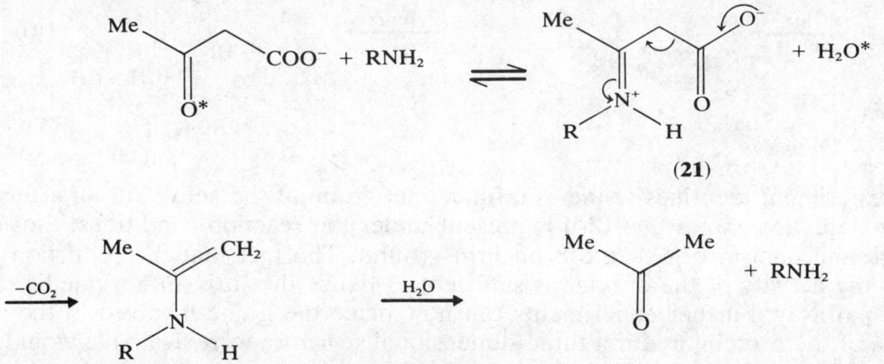

SCHEME 7

pH–rate profile must be regarded as just one piece of evidence which might be useful. This is a perfectly normal situation in mechanistic work, where most arguments are based on empirical comparisons, and evidence for a particular mechanism must be built up on the basis of results from as many independent tests as possible. Just occasionally, however, it is possible to do a single experiment which is definitive. In particular, it is sometimes possible to trap an intermediate and thus prove, on the basis of just one experiment, that the species is present under the reaction conditions. (Separate proof may be required that it lies on the main reaction pathway.) Because of the much greater complexity of the reaction, such an experiment would be invaluable in an enzyme-catalysed reaction, and the work on acetoacetate decarboxyls provides an example where this approach has been successful.

The mechanism involved (Scheme 7) is expanded in Scheme 8.[45] It is known that a Schiff base like (**22**) is readily reduced by hydride donors, and a suitable reagent which can be used in aqueous solution is sodium borohydride. It was hoped, therefore, to trap intermediate (**22**), reducing it to a stable secondary amine, by carrying out the reaction in the presence of $NaBH_4$. The experiment was carried out, using ^{14}C-labelled acetoacetate, and the enzyme was inactivated, as expected if the active-site NH_2 group had been

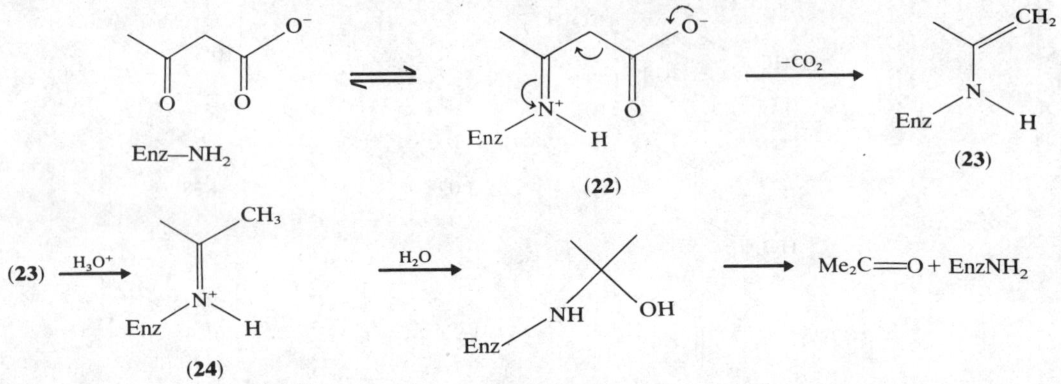

SCHEME 8

converted to a secondary amine. ($NaBH_4$ does not inactivate the enzyme in the absence of substrate.) The inert protein isolated from the reaction contained one equivalent of ^{14}C per molecule, as expected, and the protein was then degraded by conventional methods and the amino-acid composition analysed (these methods are discussed in Part 23). The amino-acid analysis showed that one lysine was missing, compared with the native enzyme, and that a new amino-acid had appeared. This was identified as ε-*N*-isopropyllysine (**25**): evidently the borohydride has reduced not (**22**) (Scheme 8) but the iminium intermediate (**24**) involved in the hydrolysis of (**23**) after decarboxylation.

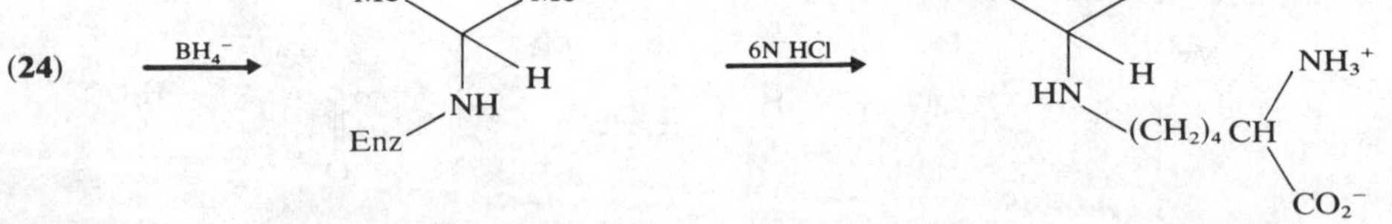

(**25**)

This experiment identifies lysine as a functional group at the active site of acetoacetate decarboxylase, and shows that (**24**) is present under the reaction conditions, thus putting the whole mechanism of Scheme 8 on firm ground. The fact that the reduction of (**24**) destroys the activity of the enzyme is sufficient evidence that it is an intermediate on the reaction pathway. Further experiments can now place the lysine involved in the primary sequence of the protein, and if a three-dimensional structure were available would lead to the identification of the active site.

A trapping experiment is therefore a very powerful tool for mechanistic work with enzymes, and borohydride has been used to trap iminium intermediates in a number of cases. However, we cannot expect that a non-specific reagent will generally be able to intervene in the chemistry of the enzyme–substrate complex. Borohydride is a special case, since it is a very small molecule, much the same size and shape as H_2O. Enzymes generally, of course, are designed to keep foreign molecules *out* of the active site. The best way to get a reagent in, in fact, is to 'disguise' it as a substrate: that is, to use a substrate analogue, which has the structural features necessary for binding at the active site, but carries also a functional group intended to react irreversibly with active site groups. (These reagents are therefore of most value for identifying functional groups at the active site, rather than trapping intermediates.) We will describe the substrate analogue approach in some detail, using some of the many examples available from work on chymotrypsin.

This cheap and readily available enzyme has been a proving ground for many approaches and reagents. Naturally an investigation of the active site done today, all in one laboratory, would be much briefer and more decisive, largely because of the availability of the three-dimensional structure of the enzyme from X-ray diffraction studies. However, this information is available for only a limited number of enzymes, and the substrate analogue approach is still one of the most useful ways of finding out about the active site.

24.1.3.2 The active site of chymotrypsin

This enzyme[46] catalyses the hydrolysis of peptide (amide) bonds, specifically those involving amino-acid residues, such as phenylalanine and tryptophan, with aromatic groups in the side-chain. This is because the enzyme has a specific binding site for such groups (see below, Section 24.1.3.3). The enzyme exhibits rather broad specificity, and will also catalyse the hydrolysis of amide (and ester) groups of many simpler compounds, including derivatives of *N*-toluene-*p*-sulphonyl (tosyl) phenylalanine. The reaction can be represented schematically as in (**26**): the aromatic residue is bound in such a way that the amide carbonyl group is held next to the catalytic group or groups in the active site.

A successful substrate analogue in this case is TPCK (tosylphenylalanine chloroketone, **27**) which is readily made from the tosyl amino-acid *via* the diazoketone. TPCK also

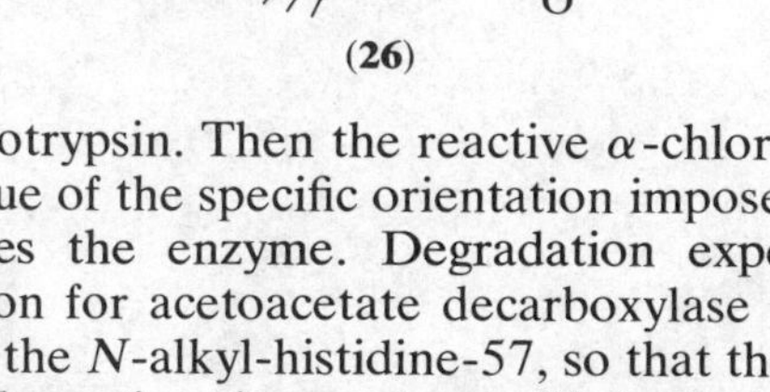

(**26**)

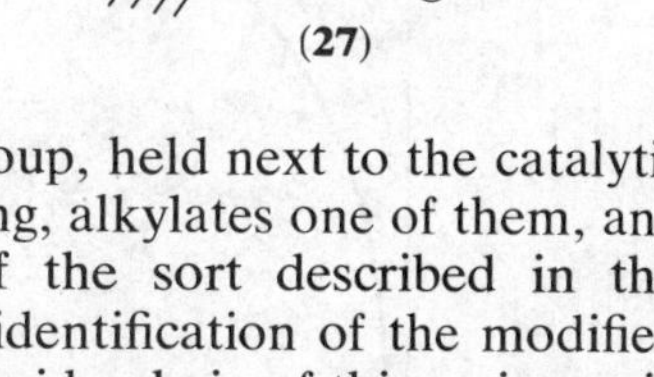

(**27**)

binds to chymotrypsin. Then the reactive α-chloroketone group, held next to the catalytic groups by virtue of the specific orientation imposed by binding, alkylates one of them, and thus inactivates the enzyme. Degradation experiments of the sort described in the previous section for acetoacetate decarboxylase led to the identification of the modified amino-acid as the *N*-alkyl-histidine-57, so that the imidazole side-chain of this amino-acid is present at the active site, and very likely involved in the catalytic mechanism.

Other evidence was already available that serine might also be present in the active site. The reagent di-isopropyl phosphorofluoridate (DFP; **28**), a nerve gas and a phosphorylating agent, was known to inactivate the enzyme in a 1:1 reaction, which was identified as the phosphorylation of serine by appropriate degradation experiments.

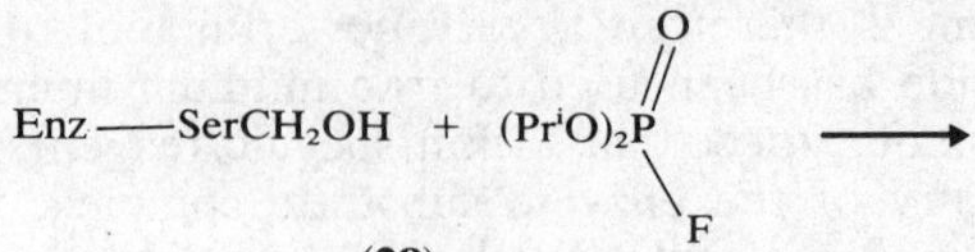

(**28**)

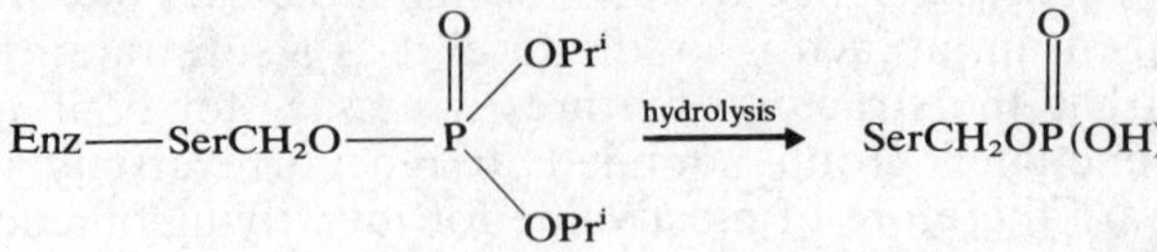

DFP is clearly not a specific substrate, but appears for some reason to behave like one, because other, more specific, acylating agents react similarly. Reaction with the very reactive *p*-nitrophenyl ester of *N*-acetyltryptophan (**29**), for example, can be shown to be

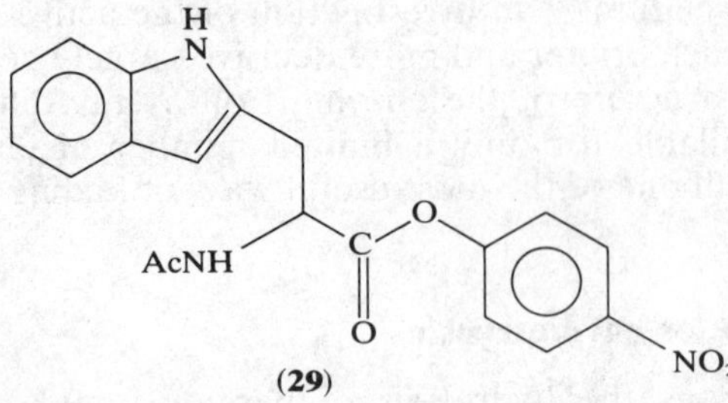

(**29**)

biphasic (Figure 10) if the first few milliseconds of the reaction are studied. (*p*-Nitrophenol is a favourite group from the experimental point of view because the anion absorbs strongly in the visible spectrum at 400 nm, and its release is readily followed spectroscopically.)

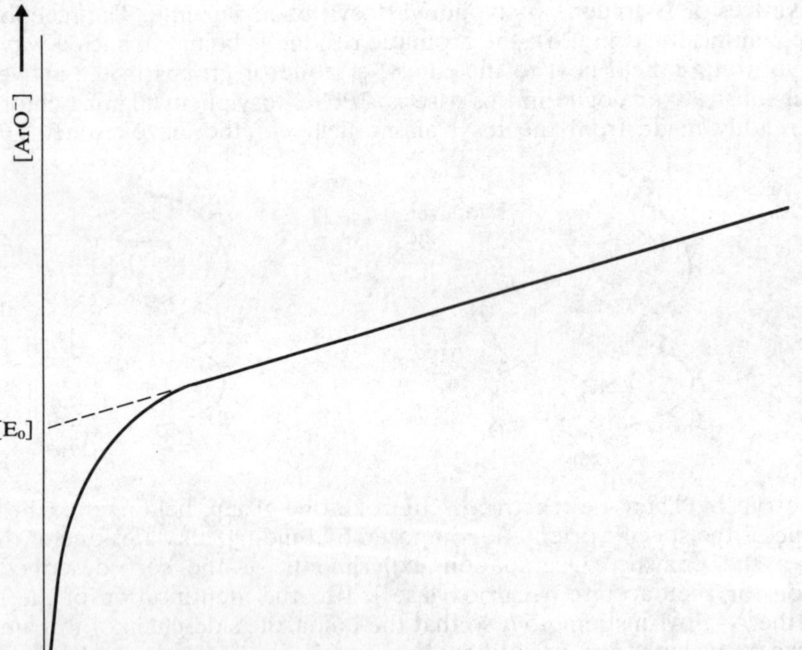

Figure 10 Time course for the release of *p*-nitrophenol from an ester RCO_2Ar, catalysed by chymotrypsin. See text

The explanation of the initial 'burst' is that the reaction involves acylation of the enzyme:

$$\text{Enz} + \text{RCO}_2\text{Ar} \longrightarrow \underset{\textbf{(30)}}{\text{Enz—COR}} + \text{ArO}^- \xrightarrow[\text{slow}]{\text{H}_2\text{O}} \text{Enz} + \text{RCO}_2^-$$

Acylation blocks the nucleophilic group of the active site, so that the enzyme is inactive until the acyl enzyme (**30**) produced is hydrolysed. If this second stage of the reaction is rate-determining, the initial attack of the free enzyme on the ester can lead to rapid displacement of ArO^-, but subsequent reaction goes at the slower rate. This mechanism predicts that the concentration of *p*-nitrophenoxide produced in the initial burst will be equal to the concentration of enzyme used, and this is found to be so.

The reaction thus appears to involve a two-stage mechanism, both stages involving nucleophilic attack at the carbonyl group of the substrate, and much evidence is available consistent with this conclusion. The two stages can be studied separately; pre-steady-state studies with (**29**), for example, give information about the formation of the acyl enzyme. A useful technique for following the hydrolysis of the acyl enzyme is to use a chromophoric acyl group. Cinnamoyl imidazole (**31**) rapidly acylates chymotypsin in a 1:1

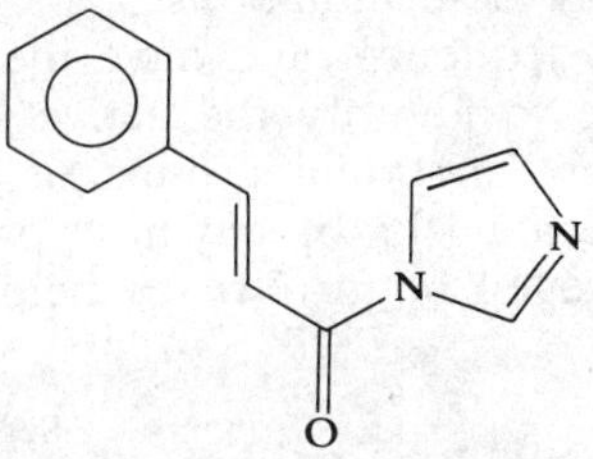

(**31**)

reaction, and the u.v. absorption of the cinnamoyl chromophore can be observed on the enzyme. This allows acyl enzyme to be prepared, and its hydrolysis followed independently of the acylation stage.

Although the evidence for the acyl enzyme mechanism is overwhelming, there are few specific indications that the site of acylation is the serine hydroxyl, as opposed, for example, to the imidazole nitrogen. Perhaps the clearest evidence for the obligatory involvement of serine comes from an experiment in which the OH group is eliminated. This is possible because the serine oxygen atom can be acylated, and thus converted to a good leaving group, by such non-specific reagents as DFP (above) and toluene-*p*-sulphonyl fluoride. The tosyl ester (**32**) prepared in this way loses toluene-*p*-sulphonic acid on treatment with N/10 NaOH for 4 hours at 0 °C, to produce anhydrochymotrypsin (**33**). This modified enzyme differs from native enzyme only in the absence of the serine hydroxyl group: it binds substrate fairly normally, for example. However, it is totally inactive, consistent with an essential role for the missing hydroxy-group in enzyme catalysis.

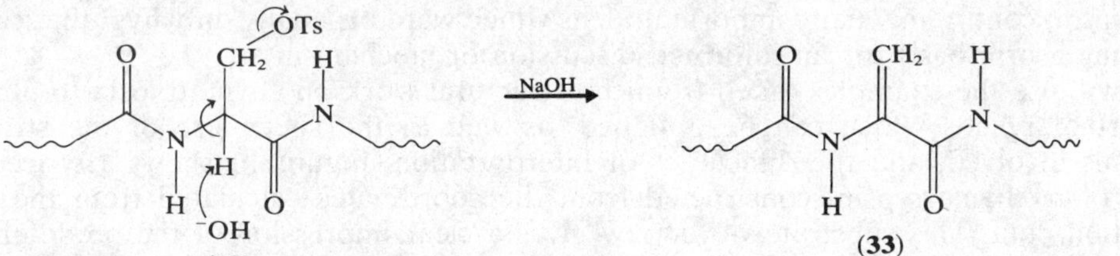

(**32**) (**33**)

Chemical evidence, however, is rarely as specific as this. What is really needed is detailed structural evidence about the identity and arrangement of functional groups at the active site. By far the most detailed and useful information of this sort is the three-dimensional structure of the enzyme as revealed by X-ray diffraction studies.

24.1.3.3 Evidence from X-ray diffraction[47]

To date (mid-1977) the structures of over 100 proteins, most of them enzymes, have been determined. The precision of such measurements is not as high as that available for small organic molecules, because all protein crystals exhibit a certain amount of disorder, limiting resolution generally to about 2 Å (200 pm). This means that side-chains with the same geometry cannot be distinguished (*e.g.* valine from threonine, or amide group from carboxyl in aspartate and glutamate residues), so that complete amino-acid sequences cannot be determined by the technique. The identity of such ambiguous amino-acids therefore has to be established by conventional sequencing methods (see Part 23). These, however, are relatively minor restrictions on a technique which provides structural information of unmatched precision and extent.

It is also important to be sure that the three-dimensional structure measured for the crystal does not differ significantly from the structure of the enzyme in solution. Two lines of evidence are reassuring on this point. A protein crystal contains a great deal of water of crystallization, and in some cases substrates can diffuse through the crystal, reacting apparently normally as they do so. Similarly, direct n.m.r. structural studies on proteins in solution where direct comparison is possible (for example with lysozyme[48]) give results in close agreement with X-ray structure determinations.

Thus if we know the amino-acid sequence, and have identified (or perhaps modified) an active-site residue, it is possible to identify the active site in the three-dimensional structure, and to 'see' which other functional groups, if any, are present there. For example, the hydroxy-group of serine-195 of chymotrypsin is actually linked to the imidazole of histidine-57 by a hydrogen bond (**34**; see below, Section 24.1.3.4).

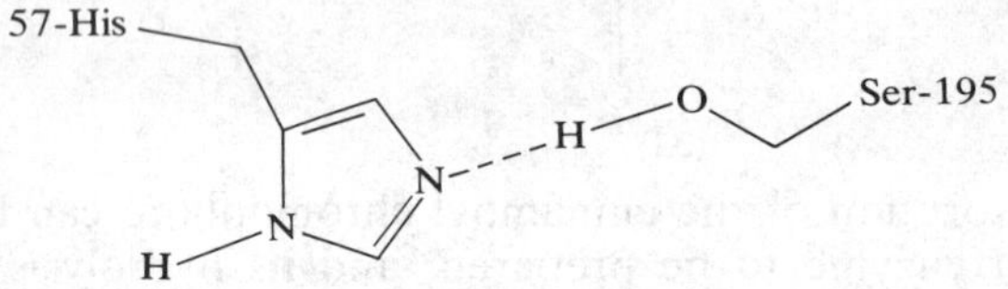

(**34**)

Clearly this sort of evidence about the active site is invaluable. Specifically, it tells us, almost for certain, which functional groups are likely to be involved in catalysis, thus defining the reaction. It may of course tell us more: in particular the three-dimensional arrangement of the functional groups may suggest something about the mechanism of catalysis. This is the case in (**34**), where the imidazole is ideally placed to act as a general base, to assist the attack of the serine-OH on an electrophilic centre. On the other hand, the three-dimensional structure of the active site may suggest nothing about the mechanism. All it necessarily tells us is the exact structure of (one of) the starting materials of the reaction, information we would always hope to have at the start of a mechanistic investigation. Those who expected the availability of three-dimensional structures of enzymes to solve most of the problems of enzyme mechanism have been disappointed. Nevertheless, this information does amount to a crucial advance in the study of enzyme mechanism, confirming many important ideas which were originally only hypothetical, and providing a firm basis for an informed discussion of mechanism.

We will use the examples taken from the structural work on chymotrypsin to illustrate the sort of results which can be obtained, as well as the sheer size of the structural problems involved, and the difficulties of interpretation. Figure 11 shows a space-filling model of α-chymotrypsin, constructed from the coordinates calculated from the X-ray diffraction data. This substrate's-eye view gives a clear impression of the predominantly polar character of the protein surface, which is very largely composed of oxygen and hydrogen (red and white) atoms. Useful landmarks are the yellow sulphur atoms of the disulphide bridge between residues 42 and 58, and, a little further to the right, that of methionine-192. The active site lies just below these sulphur atoms: even at this distance the cavity of the aromatic binding site can be discerned, below the sulphide sulphur of

ERRATA
page 419, Fig. 11 and page 420, Fig. 13
These figures have been printed upside down.

C.O.C. Volume 5

Figure 11 Space-filling model of α-chymotrypsin. Hydrogen atoms are white, oxygen red, carbon black, and nitrogen blue. Note the predominance of polar groups (mainly red, some blue, few black atoms) on the surface of the molecule, in contact with solvent. The active site is readily located by reference to the sulphur atoms visible near the centre of the molecule (see text, and Figure 13). Reproduced by permission of Professor D. M. Blow.

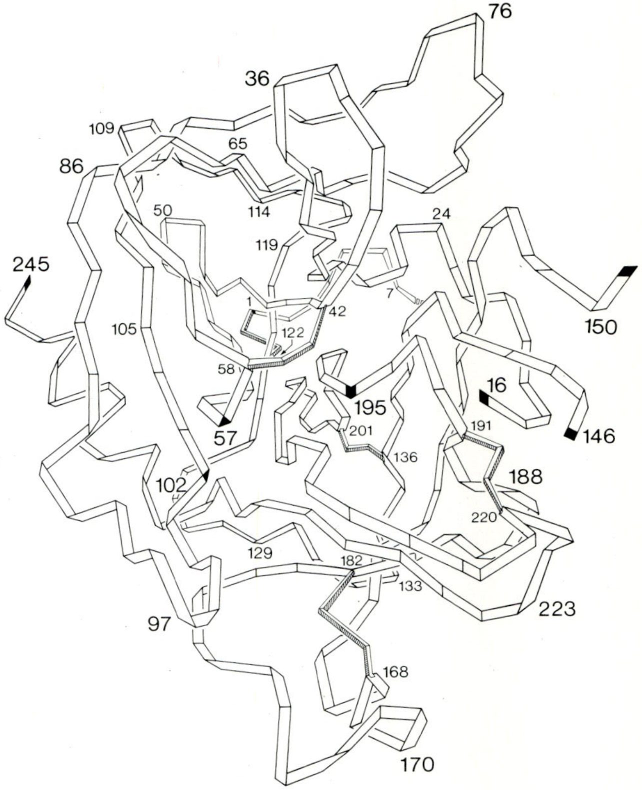

α - CHYMOTRYPSIN

Figure 12 Schematic representation of the primary chain of α-chymotrypsin, seen from the same viewpoint as Figure 11. Note the three segments of chain (residues 1–13, 16–146, and 149–245) joined by five disulphide bridges (*e.g.* 42–58). It is instructive to follow the sequence all the way from the *N*-terminus, 1 (near the centre), to the *C*-terminus, 245 (From Blow,[46a] by courtesy of Academic press).

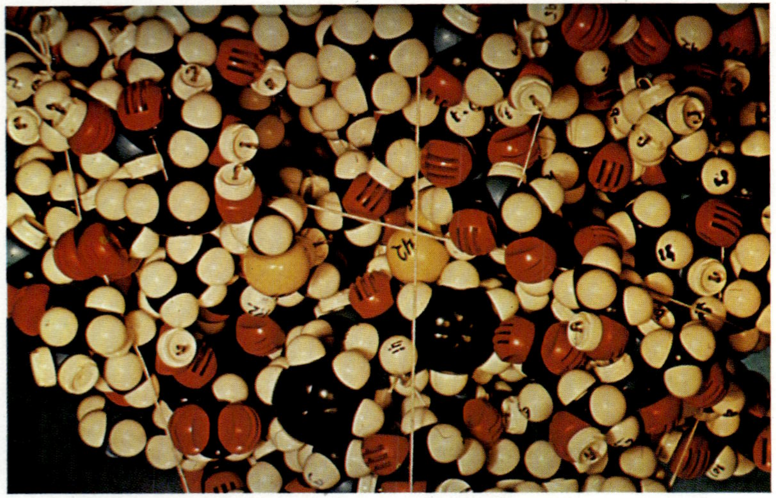

Figure 13 Close-up of the active-site region of the space-filling model of α-chymotrypsin (Figure 12). The hydrophobic pocket is clearly visible below the single (right hand) sulphur atom of methionine-192. The CH_2OH side-chain of serine-195 lies between the cavity and the sulphur atom of cysteine-42, with its hydrogen atom hydrogen-bonded to a nitrogen atom of the imidazole group of histidine-57 (see text). Reproduced by permission of Professor D. M. Blow

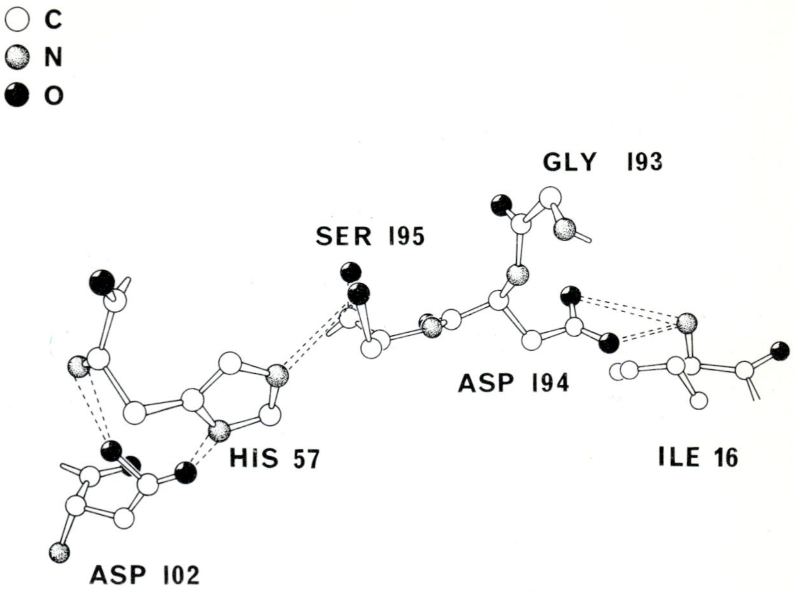

Figure 14 The arrangement of groups at the active site of α-chymotrypsin, as revealed by the X-ray structure (see text) (From Blow,[46a] by courtesy of Academic Press)

methionine-192. Figure 12 is a different representation, from almost the same point of view, designed to bring out the convolutions of the main chain. (Actually, chains: when α-chymotrypsin is formed from its precursor, chymotrypsinogen, residues 14 and 15, and 147 and 148, of the zymogen are lost. Thus the backbone of α-chymotrypsin is divided into three parts, chains A, B, and C, composed of residues 1–13, 16–146, and 149–245, respectively.)[49] Five disulphide bridges reinforce the binding of the primary chain, so that each chain segment is covalently linked to another, and disulphide bridges close to the active site residues 57 and 195 are clearly involved in determining the geometry of the catalytic centre.

Figure 13 is a closer view of the space-filling model, and represents nearly half the surface area visible in Figure 11. Using the same sulphur atoms as landmarks, the cavity of the aromatic binding site is clearly visible, and can be seen to be comparable in size with the two benzene rings at the top of the picture. On the lip of the cavity, in the direction of the sulphur atom of cysteine-42, is the OH group of sereine-195, hydrogen-bonded (white disc) to the nitrogen atom (just visible) of an imidazole ring turned almost at right angles to the surface. This is the side-chain of histidine-57. Out of sight, beneath this imidazole ring, and probably hydrogen-bonded to the NH group, is a carboxylate group, on the side-chain of aspartic acid-102. No other functional group lies close to the histidine–serine pair, so the catalytic mechanism presumably involves these three. The catalytic unit, often called the electron-relay system of chymotrypsin, as revealed by the X-ray structure, is shown in detail in Figure 14. Also included in this Figure is the amino-acid next to the active site serine, aspartic acid-194. A hydrogen bond between the carboxylate group of this amino-acid and the NH_3^+ group of the terminal isoleucine of the B-chain (absent in the zymogen) appears to act as a sort of tensioning device, and is an important factor in the maintenance of the correct stereochemistry at the active site.

We discuss the mechanism of catalysis by chymotrypsin below (Section 24.1.3.4), but first there is a point of very general importance to be made. Perhaps the single most important conclusion to be drawn from all the X-ray structural information available so far for enzymes is that the catalytic unit found in chymotrypsin occurs in other enzymes also. It has always been the hope, perhaps the reasonable expectation, that the many enzymes catalysing a given reaction will not all use different catalytic mechanisms. The many enzymes known to catalyse the hydrolysis of the amide bond fall into four main classes: metalloenzymes, of which the best known is carboxypeptidase; carboxyl enzymes, like pepsin; SH enzymes, like papain; and the serine proteinases. This last group includes chymotrypsin, trypsin, elastase, and thrombin. Amino-acid sequencing studies show that these four enzymes are very similar proteins, with obvious similarities of amino-acid sequence. Trypsin and chymotrypsin, for example, have over 40% of their amino-acid residues identical.[50] In particular, they all have the same sequence* (**35**) around the active site serine.

Gly——Asp——Ser——Gly——Pro

(**35**)

X-ray structures are now available for several of these serine proteinases (trypsin, chymotrypsin, elastase), and show that they also have the same three-dimensional structure at the active site, with charge-relay systems identical with that of chymotrypsin (Figure 14). This is perhaps not surprising, if, as seems likely, all these enzymes have evolved from a common ancestor. The logical sequence would be the development of an efficient catalytic mechanism, followed by the evolution of substrate specificity, so that enzymes further down the evolutionary chain would have the same catalytic unit, but different arrangements at the binding site, and perhaps elsewhere.

The striking discovery is that another serine proteinase, subtilisin, a quite different protein, with no sequence similarities to the chymotrypsin group, nevertheless uses exactly

* There is, course, no special reactivity associated with the serine in this short sequence. Syntheses of oligopeptide sequences from active sites produce compounds generally less active catalytically than simple compounds containing the side-chain groups involved.

the same catalytic unit. The X-ray structure of subtilisin reveals that it too has a hydrogen-bonded system, aspartic acid-32 . . . histidine-64 . . . serine-221, like that of chymotrypsin[51] (Figure 14). This must mean that the catalytic mechanisms used by these enzymes are also identical. The conclusion must surely be that two lines of protein evolution have arrived at the same solution to the problem of hydrolysing an amide bond. If this is true for the serine proteinases, it may also be true for proteinases based on other groups, and indeed for enzymes catalysing any given reaction. So there is some evidence to support our reasonable expectation that the very large number of enzymes involved in living systems may turn out to use a much smaller number of catalytic mechanisms.

24.1.3.4 The catalytic mechanism of the serine proteinases

The arrangement of the functional groups at the active sites of the serine proteinases (Figure 14) suggests very strongly that the serine hydroxyl group is the nucleophilic centre which attacks the substrate carbonyl group, assisted by the histidine imidazole group acting as a general base. All the evidence we have seen, and a great deal more that we have not, is consistent with a mechanism of this type, and it is generally accepted that this is the first step of the catalytic process. It would seem unreasonable not to assign a role also to the aspartate carboxyl group which is part of the charge-relay system of all the serine proteinases; here too the position of the group is ideal for a role as a general base, and the first step as generally written becomes that shown in Scheme 9.

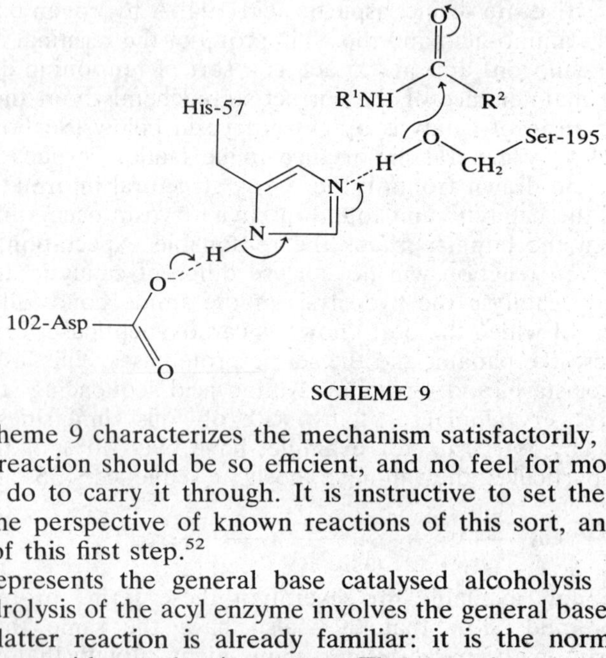

SCHEME 9

Although Scheme 9 characterizes the mechanism satisfactorily, it gives us no idea why this particular reaction should be so efficient, and no feel for most of the chemistry the enzyme has to do to carry it through. It is instructive to set the mechanism implied in Scheme 9 in the perspective of known reactions of this sort, and then to examine the consequences of this first step.[52]

Scheme 9 represents the general base catalysed alcoholysis of an amide, and the subsequent hydrolysis of the acyl enzyme involves the general base catalysed hydrolysis of an ester. The latter reaction is already familiar: it is the normal mechanism for the hydrolysis of esters with poor leaving groups. The example most relevant to the hydrolysis of an acyl chymotrypsin is the hydrolysis of *N,O*-diacetylserine amide (**36**). This reaction is very slow near pH 7, but general base catalysis by imidazole (Scheme 10) can be identified and characterized at 100 °C.[53]

Amides are even less reactive substrates, but intramolecular alcoholysis can be observed in favourable instances. The most reactive system studied is the hydroxyamide (**37**), which is converted to the lactone (**38**) in a reaction[54] not much slower than the corresponding reaction of chymotrypsin with substrate. Unlike slower reactions of this sort, phthalide formation is catalysed by imidazole. The effective concentration of the hydroxy-group of (**37**) appears to be comparable with that of the serine-hydroxyl of chymotrypsin in the enzyme–substrate complex; and the efficiency of catalysis by imidazole is such that an

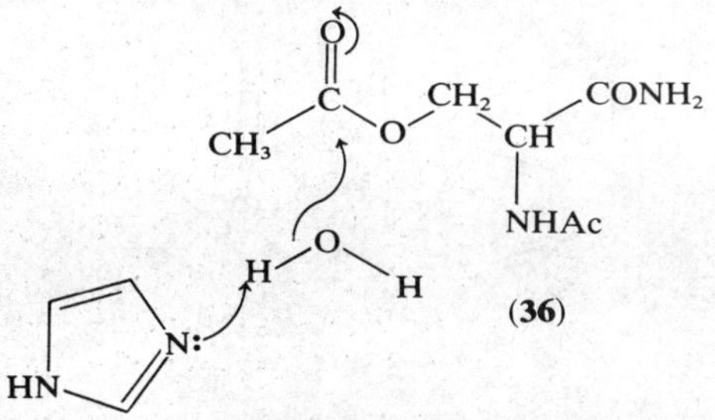

(36)

SCHEME 10

effective concentration of the side-chain group of histidine-57 of no more than about $20\ \mathrm{mol\ l^{-1}}$ in the enzyme–substrate complex would account for the observed rate of the chymotrypsin reaction with amide substrates.

The first step in the enzyme reaction (Scheme 9) involves the formation of a tetrahedral intermediate (**39**), which as initially formed would revert rapidly to starting materials (Scheme 11, arrows). A species like (**39**) is too reactive to be detected in ordinary systems,

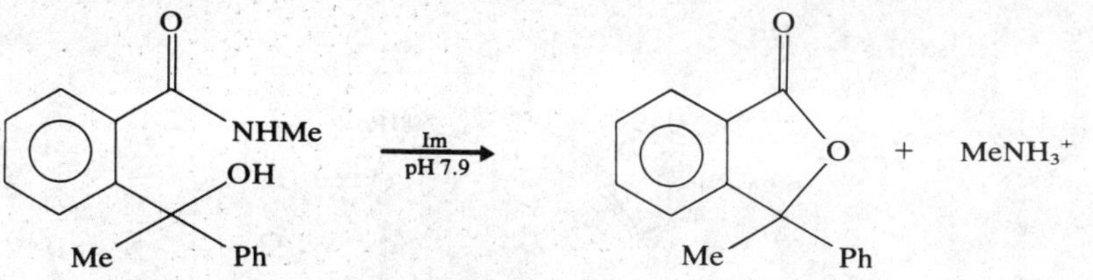

(**37**) (**38**)

(SCHEME 9→)

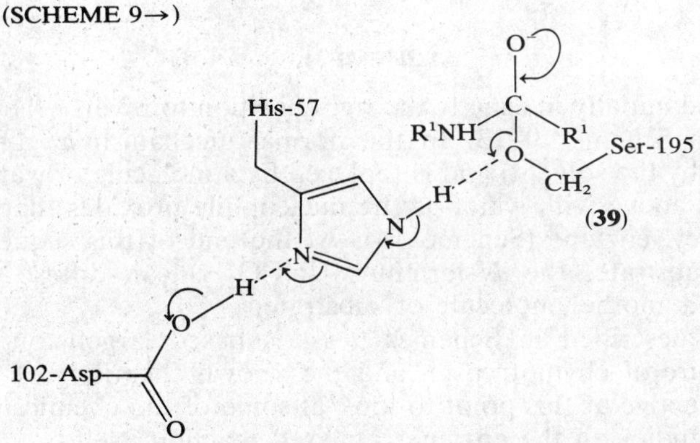

(**39**)

SCHEME 11

but structural evidence is available for this intermediate in the serine proteinase reaction. Naturally occurring protein inhibitors of trypsin form very stable complexes with the enzyme. X-ray structures of such complexes[55] show that an apparently normal peptide cleavage reaction has been frozen at about the position of the tetrahedral intermediate, as shown in Scheme 11. (R^2CONHR^1 is now the inhibitor, rather than the substrate.) Evidently the intermediate in these enzyme–inhibitor complexes can go neither back (Scheme 11) nor forward, presumably because the inhibitor is too tightly bound. The forward reaction, the breakdown of the tetrahedral intermediate, involves the cleavage of the C—N bond, and this will happen if the nitrogen atom is protonated, to convert it to a viable leaving group. A likely source of the necessary proton is the imidazole group of the charge-relay system. This can be made available if the conformation of the tetrahedral intermediate now changes away from that shown above (Scheme 11) (which has the important additional advantage of preventing the reversal of the first step) to that shown in Scheme 12.

(SCHEME 11→)

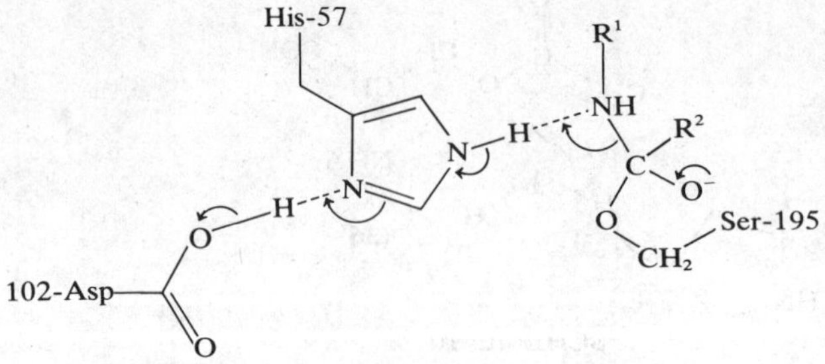

SCHEME 12

The tetrahedral intermediate can now break down (Scheme 12, arrows) to give the acyl enzyme (Scheme 13) and the free amine, RNH_2 (the *C*-terminal fragment of the original substrate), which is now released. This is the point where it becomes important that the leaving group should no longer be tightly bound to the enzyme, or efficiency will be lost.

(SCHEME 12→)

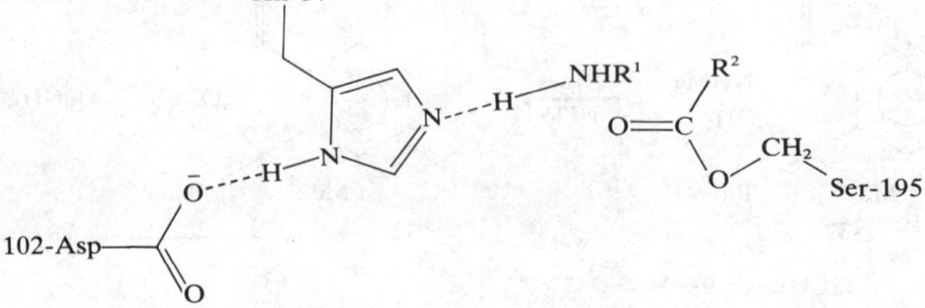

SCHEME 13

The amine is formed initially in exactly the right position to reverse the whole sequence of events described in Schemes 9–13. In the normal mechanism it diffuses away (to be rapidly protonated by the solvent) and is replaced by a molecule of water. Reversal of the sequence described above with water as the nucleophile provides the mechanism for the hydrolysis of the acyl enzyme (Scheme 14). At the end of this sequence the remaining fragment of the substrate, the *N*-terminus (R^2CO_2H) can diffuse away, leaving the enzyme free to bind another molecule of substrate.

The mechanism described in Schemes 9–14 is based largely on the known three-dimensional structure of chymotrypsin, and the known chemistry of the basic reactions involved. It is instructive at this point to look at some of the chemical evidence available from mechanistic studies on the enzyme-catalysed reaction itself.

The formation and hydrolysis of the acyl enzyme (acylation and deacylation of the enzyme) can be studied separately by using the appropriate conditions or substrates. Both steps depend on a basic group with pK_a near 7, which is taken to be the imidazole group of the charge-relay system. Acylation (only) depends also on an acidic group, with pK_a near 8.8, and thus this step, and the overall reaction, shows a bell-shaped pH–rate profile. The identification of this dissociation constant with a particular group was a source of controversy for many years, but the group involved is now thought to be the terminal NH_3^+ group of the B-chain, belonging to isoleucine-16. This is hydrogen-bonded to the carboxylate group of the side-chain of aspartic acid-194, next to the active site serine (Scheme 14), and the H-bond is broken when the NH_3^+ group loses a proton. This salt-bridge evidently plays an important part in controlling the stereochemistry of some part of the active site in the region around the first tetrahedral intermediate (Schemes 11, 12). (Here we see one potential pitfall of interpretation: the reaction depends on the dissociation of a group which takes no part in chemical catalysis, and is not even recognisably at the active site.)

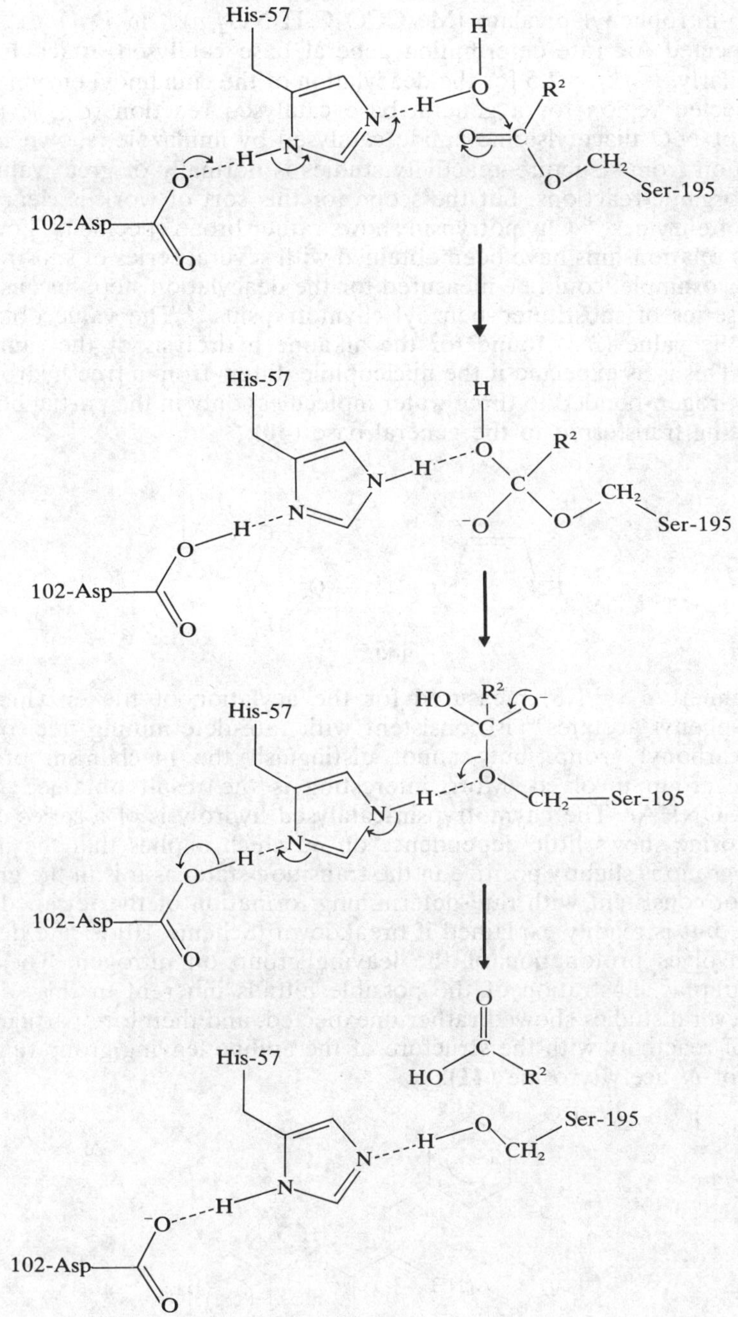

SCHEME 14

Another parameter which is easily measured (though not always easily interpreted) is the solvent deuterium isotope effect. Both acylation and deacylation steps involve general base catalysis, and would be expected to be 2–3 times slower in D_2O (see above, p. 400). In this case the evidence is clearly consistent with the proposed mechanism. By studying the acylation step with a reactive (nitrophenyl) ester substrate it is possible to isolate the formation of the tetrahedral intermediate (Scheme 9), since the breakdown (Scheme 12) involves the loss of a good (nitrophenolate) leaving group, which is rapid without general acid catalysis. The solvent deuterium isotope effect observed with the rather unreactive substrate *p*-nitrophenyl pivalate ($Me_3CCO_2C_6H_4NO_2$-*p*)[56] in D_2O *vs.* H_2O is k_H/k_D = 2.2, as expected for rate-determining general base catalysed attack by the serine-OD group. Similarly, $k_H/k_D = 2.5$ for the deacylation of the cinnamoyl enzyme in D_2O[57] is also in the expected region for a general base catalysed reaction (*e.g.* $k_H/k_D = 2.0$ for the hydrolysis of *N,O*-diacetylserine-amide catalysed by imidazole, shown in Scheme 10).

Information from structure–reactivity studies is normally of great value in mechanistic studies of organic reactions, but the scope for this sort of work is clearly limited by the specificity of enzymes.[58] Chymotrypsin shows rather broad specificity, however, and linear free energy relationships have been obtained with several series of substrates. A Hammett ρ-value, for example, could be measured for the deacylation step, because it is possible to prepare a series of substituted-benzoyl chymotrypsins.[59] The value obtained, $\rho = 2.1$, is similar to the value (2.3) found for the alkaline hydrolysis of the same series of alkyl benzoates. This is as expected if the nucleophile differs from a free hydroxide ion (*i.e.* one strongly hydrogen-bonded to three water molecules) only in the partial bond to the proton which is being transferred to the general base (**40**).

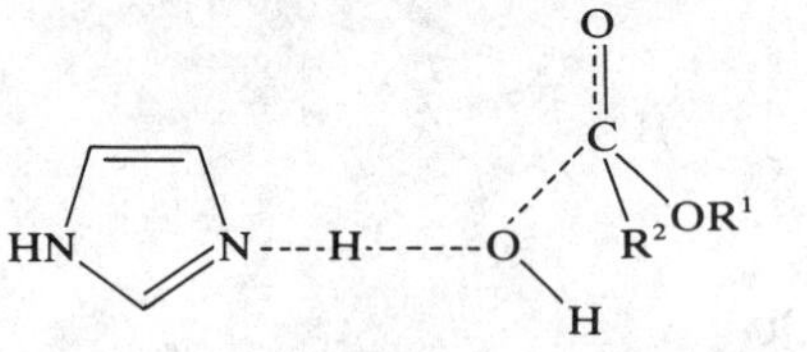

(40)

The Hammett ρ (= 1.8) measured for the acylation of the enzyme by a series of substituted-phenyl acetates[60] is consistent with rate-determining nucleophilic attack on the ester carbonyl group, but cannot distinguish the mechanism of attack, or the nucleophilic group involved. More interesting is the result obtained with a series of anilides, RCONHAr. The chymotrypsin-catalysed hydrolysis of a series of anilides of *N*-acetyl-L-tyrosine shows little dependence on σ, which implies that the nitrogen atom of the leaving group is slightly positive in the transition state, as it is in the ground state. This is clearly not consistent with rate-determining formation of the tetrahedral intermediate (Scheme 9), but is readily explained if breakdown (Scheme 10) is rate determining, since this step involves protonation of the leaving group on nitrogen. These investigations contain a graphic illustration of the possible pitfalls inherent in this sort of work with enzymes. Several studies showed rather unexpected, and therefore particularly interesting, variations of reactivity with the structure of the aniline leaving group (*e.g.* substituent X) of anilides of *N*-acetyltyrosine (**41**).

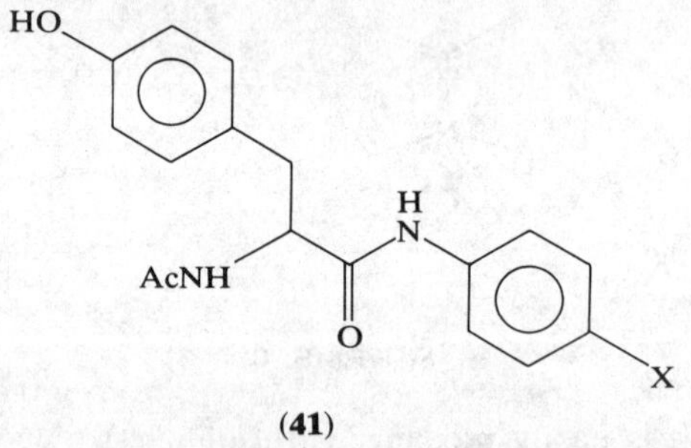

(41)

The explanation[61] turns out to be entirely unconnected with the chemical mechanism. With hydrophilic substituents (X = COMe, $^+NMe_3$), (**41**) is a normal substrate, but with hydrophobic groups (such as X = Cl) the aromatic ring of the leaving group, rather than that of the tyrosine, is bound in the aromatic binding site, a factor which can clearly lead to a large change in the efficiency of catalysis.

Finally, it is possible to use nucleophiles other than water to deacylate the acyl enzyme. Alcohols and primary amines react, for example, with furoyl chymotrypsin,[62] in a reaction where the nucleophile presumably takes the place of the water molecule in the normal hydrolysis mechanism (Scheme 14), and produces an amide or an ester (Scheme 15).

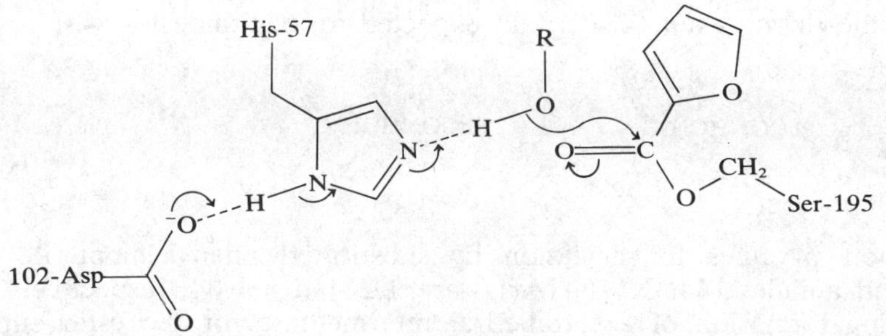

SCHEME 15

The reactivity of the nucleophile clearly depends on how well it binds, but is practically independent of its basicity (Bronsted coefficient $\beta \approx 0$).[62] This must be a result of the timing of the proton transfer in the general base catalysed reaction, and has the advantage for the enzyme reaction that the reactivity of a weak nucleophile such as water is little lower than that of more basic reagents, which are more strongly nucleophilic towards simple carbonyl compounds.

The mechanism of action of the serine proteinases is understood better than that of any other type of enzyme, and it illustrates several important points about enzyme catalysis. Amide hydrolysis may not appear a very complicated reaction to an organic chemist, but the very strict requirements for successful enzyme catalysis mean that steps which the chemist can happily ignore have to be specifically provided for, if they are not to hold up proceedings. Even the mechanism described in Schemes 9–14, in which we have distinguished nine separate steps, is certainly an oversimplification. (To take just one example, recent studies on chymotrypsin, using fast reaction techniques in aqueous dimethyl sulphoxide at −90 °C, have identified *four* separate processes preceding the formation of the tetrahedral intermediate (Scheme 9); the first of these is substrate binding, the rest presumably substrate-induced conformational changes necessary to set up the correct stereochemistry for catalysis.)[63] It is not difficult to see why specific provision would have to be made for catalysing the breakdown of such a high-energy species as a tetrahedral intermediate: such steps can, after all, be rate determining in quite simple reactions. However, in the context of the very high rates necessary for efficient enzyme catalysis, even such processes as proton transfers and conformational changes, which may have activation energies as small as 12–16 kJ mol^{-1}, are potential slow steps. In practice, enzymes do make specific provision for even the smallest details of the chemical mechanism of catalysis, with the result that the efficiency of catalysis is often limited not by the rate of any chemical step, but by physical events such as conformational changes, or, particularly, the diffusion of product away from the enzyme.[64]

24.1.3.5 Other proteolytic enzymes

The mechanism used by the serine proteinases is not nature's only solution to the problem of hydrolysing the amide bond. Three other major classes of proteolytic enzymes have been mentioned already, and each of the three different types of mechanism involved illustrates points of particular interest.

The cysteine proteinases[65] show many similarities to the serine enzymes. Most work has been done on papain, and the three-dimensional structure of this enzyme is known.[66] It consists of a single amino-acid chain, of much the same size (212 residues) as chymotrypsin, with a characteristic deep cleft where substrate binds. Papain is inactivated by iodoacetate, which is known to alkylate thiol groups, and since six of the seven cysteines are tied up in disulphide bonds, the single free SH group of cysteine-25 is involved. There is good evidence for an acyl enzyme intermediate: as with chymotrypsin, the cinnamoyl enzyme can be prepared, and its hydrolysis followed in the absence of complications from the acylation step.[67] Furthermore, the u.v. spectrum of the acyl enzyme obtained by reaction with the thiono-ester (**42**) is that expected for a dithioester (**43**).[68]

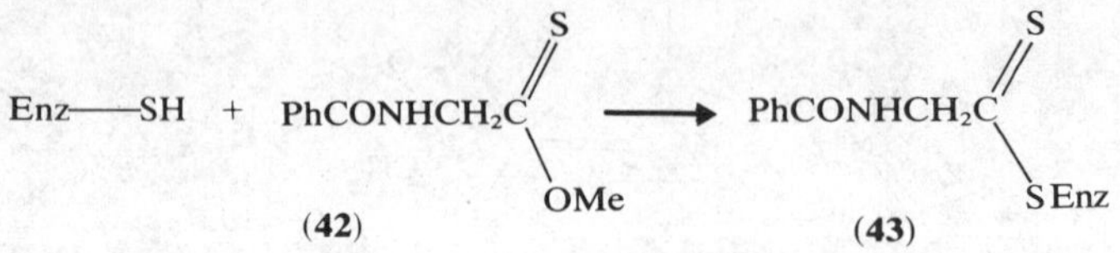

(**42**) (**43**)

(**44**)

The Hammett ρ-values for acylation by substituted phenyl hippurate esters (**44**; X = OAr)[69] and anilides (**44**; X = NHAr)[70] are 1.2 and -1.04, respectively, suggesting rate-determining formation of a tetrahedral intermediate for the ester substrate, and rate-determining breakdown, with general acid catalysis, for the amide, much as discussed above for similar reactions of chymotrypsin. Acylation, and also alkylation by iodoacetate, depends on apparent pK_a's of 4.2 and 8.2 (bell-shaped pH–rate profile), while deacylation depends only on a pK_a in the region of 4. The obvious conclusion appears, in this case, to be correct: that the higher pK_a is that of the SH group of the active site cysteine residue, which is acylated, and hence absent, in the acyl enzyme.

These properties are qualitatively very similar to those of the serine proteinases, and the mechanism of catalysis is very similar also. The X-ray structure shows the active site SH group in contact with the imidazole side-chain of a histidine residue on the opposite side of the cleft (histidine-159), which is hydrogen bonded through the distant ring nitrogen atom to the amide group of asparagine-175. Since an amide group does not act as a general base under mild conditions, the similarities do not extend to a full charge-relay system, but the accepted mechanism[71] (outlined briefly in Scheme 16 as far as the acyl enzyme) otherwise differs little from that discussed above for chymotrypsin (Schemes 9–14).

The basic group of pK_a about 4, on which both acylation and deacylation depend, is almost certainly the imidazole of histidine-159. The structural studies strongly support the involvement of this group as a general base in the first stage of the acylation process; and there is good evidence that deacylation is also general base catalysed. The hydrolysis of cinnamoyl papain,[67] and of the α-*N*-benzoyl-L-argininyl ester,[72] show large solvent deuterium isotope effects, $k_H/k_D \approx 3$. This is consistent with general base catalysis of the attack of water on the acyl enzyme, but does not conclusively identify the catalytic group concerned as an imidazole.

The two remaining classes of proteolytic enzymes use quite different mechanisms. Pepsin[73] is the major component of a group of proteolytic enzymes active at low pH in the stomach. Although it was the second enzyme to be obtained in crystalline form, in 1926, detailed three-dimensional structural data are only now becoming available. Porcine pepsin has a single polypeptide chain of 327 amino-acid residues, and the sequence is known.[73] It contains a large number (29) of side-chain carboxyl groups, and two or perhaps three of these are thought to be involved in catalysis. Thus the enzyme is inactivated by diazomethane at pH 5.5, although an excess of the reagent is required, and complete inactivation involves the esterification of as many as five carboxyl groups. Substrate analogue diazoketones are more specific inhibitors. Tosyl-L-phenylalanyldiazomethane (**45**), for example, gives rapid 1 : 1 inhibition.[74]

The pH–rate profile for hydrolysis of acetyl dipeptides shows a pH optimum near pH 2, and a dependence on groups of apparent pK_a's close to 1 and 4. These are assigned to active-site carboxyl groups, the one with the very low pK_a being perhaps involved in a strong hydrogen bond (compare the final dicarboxylic acid of Table 2).

Though details of the catalytic mechanism are still in doubt, it is well known that the hydrolysis of the amide group is subject to efficient nucleophilic catalysis by a neighbouring carboxyl. For example, the hydrolysis of phthalamic acid (**46**) is some 10^6 times faster than that of benzamide.[75] This reaction is very sensitive to structural variation, and with

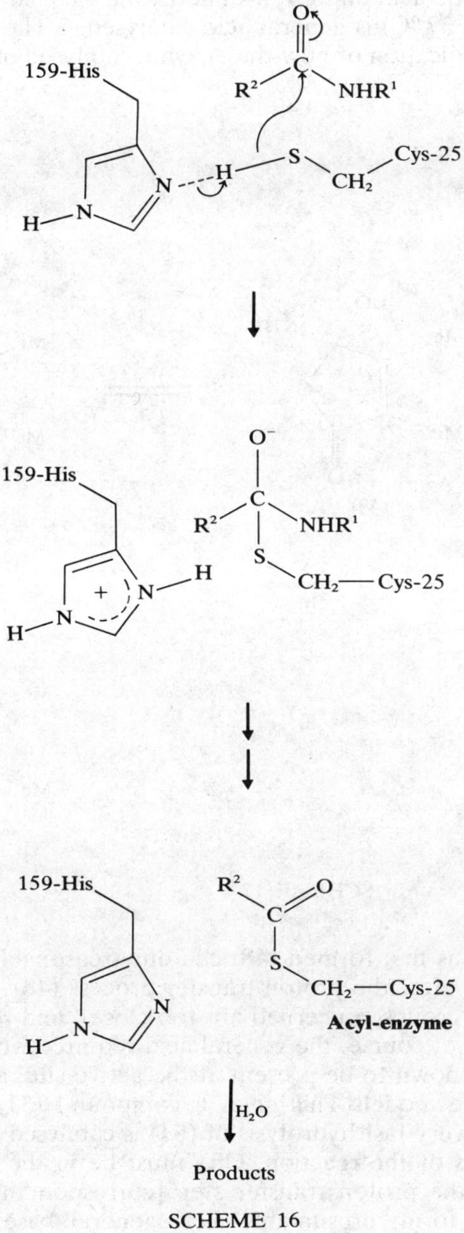
159-His
Cys-25
Acyl-enzyme
H_2O
Products

SCHEME 16

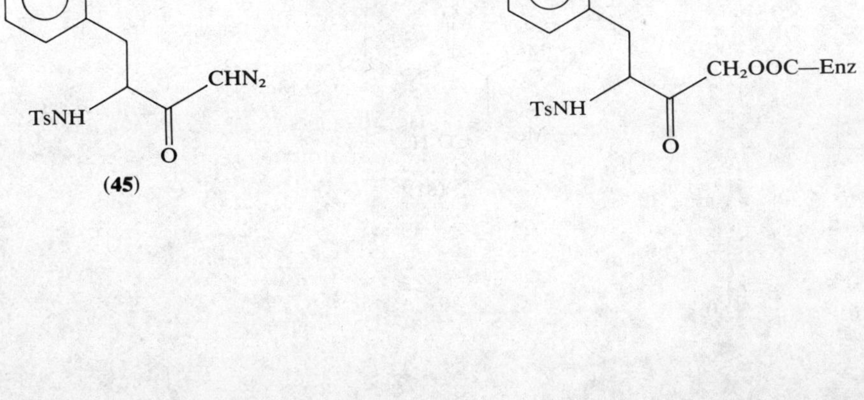
Enz—COOH
TsNH
CHN_2
CH_2OOC—Enz

(**45**)

the most reactive systems (*e.g.* the acid amides of dimethylmaleic acid (**47**), which have a half-life of less than a second at 37 °C) is general acid catalysed.[76] The mechanism of this reaction (Scheme 17) gives an indication of how the enzyme catalysed hydrolysis must go.

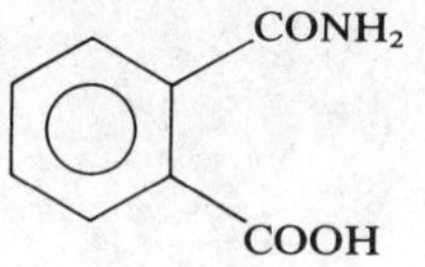

(**46**)

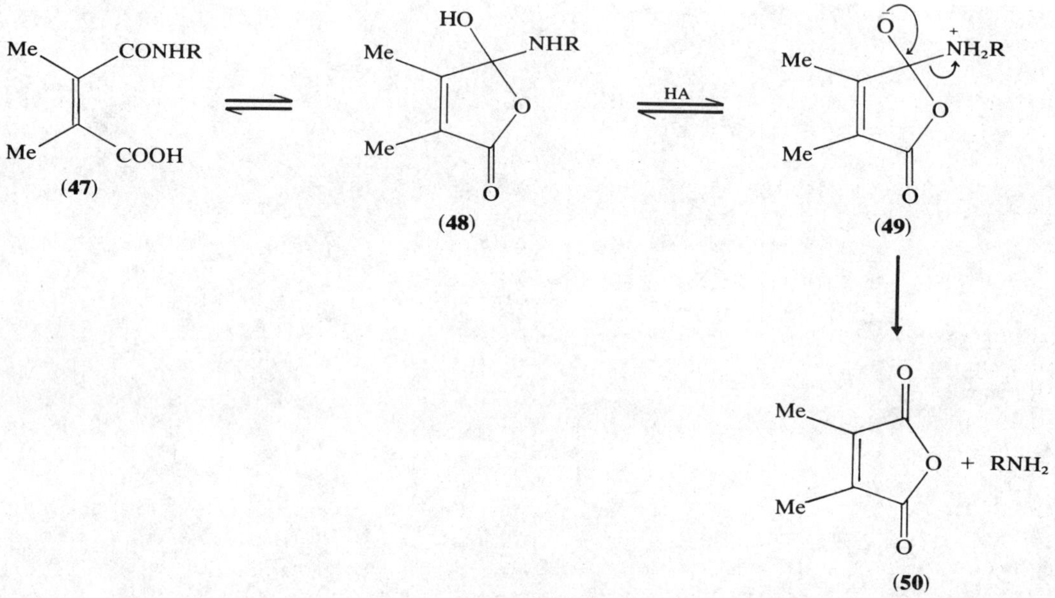

SCHEME 17

The tetrahedral intermediate as first formed (**48**) can only reasonably revert to starting materials, but it can lose RNH_2 after the proton transfer process (**48**)→(**49**). This proton transfer cannot go directly (the groups concerned are too close), and requires catalysis by a general acid.[76] In the enzyme, of course, the general acid required would be expected to be one of the carboxyl groups known to be present at the active site; and it is possible to build a carboxyl group into the model. This gives a compound (**51**), which no longer requires external catalysis. The very fast hydrolysis of (**51**) is catalysed by its two carboxyl groups, acting on separate steps of the reaction. One must be in the acid form, but the second, presumably catalysing the proton transfer step [corresponding to (**48**)→(**49**) of Scheme 17] acts in the CO_2^- form, presumably as a general base. As a result, the hydrolysis of (**51**) shows a bell-shaped pH–rate profile,[77] qualitatively similar to that observed for pepsin catalysis.

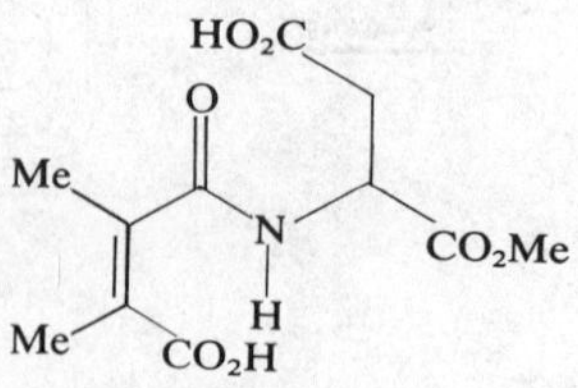

(**51**)

The best known of the metalloenzymes which catalyse the hydrolysis of the peptide linkage is carboxypeptidase. Here detailed structural information has been available for some time.[78] The enzyme is specific for the peptide bond to the *C*-terminal amino-acid residue of a chain, and requires a free terminal carboxylate group. The bovine pancreatic enzyme is a single polypeptide chain of 307 amino-acid residues, contains one atom of Zn, and shows bell-shaped pH–rate profiles for the hydrolysis of peptides, with pH optima near pH 7.5. The pH dependence is complex, and the derived apparent pK_a's of 6.7 ± 0.2 and 8.5 ± 1.0 have proved difficult to assign to particular active site groups.

A combination of X-ray structural and chemical evidence has identified the side-chain groups bound to the active site Zn. These are two imidazoles, of histidines-69 and -196, and the carboxylate group of glutamic acid-72. The Zn ion can be exchanged with other metals, and the new metalloenzymes formed have their own characteristic reactivities (or none) towards amide (and ester) substrates; but the apoenzyme (no metal present) is entirely inactive, as expected if the metal ion plays an essential part in catalysis. Current ideas about the mechanism are based partly on chemical evidence, but especially on the crystallographic work, which includes three-dimensional structures not only of the native enzyme, but also of the complex formed with glycyl-L-tyrosine when the dipeptide is allowed to diffuse into crystals of the enzyme.[78]

The substrate carboxylate group essential for binding is found to form a salt bridge with the guanidinium group of arginine-145, and this, and the preferred positions of binding of side-chain groups, bring the amide bond to be cleaved into close contact with the Zn atom. The only other functional groups now close to this amide linkage are the carboxyl of glutamic acid-270, which (like the arginine) has moved about 2 Å from its position in the native enzyme, and the phenolic hydroxyl of tyrosine-248. This last group is not one of the five thought to be commonly implicated in enzyme catalysis, but there is chemical evidence for an essential tyrosine in carboxypeptidase. The remarkable observation is that this group appears to be nowhere near the active site Zn in the native enzyme, but binding of glycyl-tyrosine induces an enormous conformational change, as part of which the phenol group of tyrosine-248 moves no less than 12 Å, from the surface of the protein to its position close to the substrate–peptide bond. This movement has the further effect of closing off the active site cavity, so that it appears no longer to be in equilibrium with the solvent.

In the proposed mechanism (Scheme 18) electrophilic catalysis (Section 24.1.2.5) by the

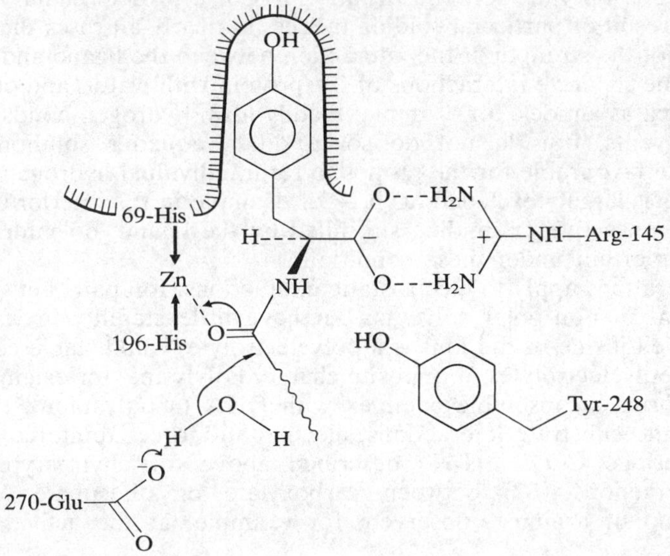

SCHEME 18

metal assists nucleophilic attack on the substrate amide group. The nucleophile is either the carboxylate group of glutamic acid-270, or alternatively a water molecule, as shown in the scheme. In the latter case the carboxylate acts as a general base to catalyse attack by water.[78,79] (There is no strong evidence for or against the anhydride intermediate which would be involved if the carboxylate group acted as a nucleophile.) The role assigned to the phenolic hydroxyl of tyrosine-248 is that of a general acid, protonating the leaving group nitrogen, presumably as part of the breakdown of the tetrahedral intermediate produced by the first step shown in Scheme 18. (The hydrolysis of some esters, with leaving groups which are good without protonation, do not depend on the phenolic hydroxyl group.[80])

24.1.4 BINDING MECHANISMS

We have managed so far with a very simple picture of a typical enzyme, as a single protein molecule in solution. In this and subsequent sections we will gradually have to give up this simple picture, because a protein in solution, and especially a protein in solution under physiological conditions, is rarely, if ever, alone in its solvation shell, as a simple organic molecule, for example, generally is.[81] The surface of a protein, although essentially hydrophilic, is of a sufficiently varied nature to bind all sorts of ligands, from small counterions to other macromolecules, more or less tightly. At any one time several such ligands may be arriving or departing from the surface; how long they stay depends on the strength of the binding interactions. These can vary over the complete range from negative binding (*i.e.* repulsion), to binding energies in the region of 80 kJ mol^{-1} (corresponding to dissociation constants as low as 10^{-14}–10^{-15}), as observed for the binding of lac repressor protein to DNA,[82] or of naturally occurring protein inhibitors to enzymes such as trypsin.[83] Enzyme–substrate binding is thus just one (albeit the most important one) of many ligand–protein interactions, of several different types, and we will first look briefly at the sorts of forces involved in such interactions.

24.1.4.1 Forces in aqueous solution[84]

The forces which operate between molecules in aqueous solution are, of course, the same as those which operate between distant parts of a protein chain which may come into contact as a result of particular folding modes. In nearly all cases the most important consideration is not the strength of the interaction between the ligand and the protein, but the strengths of the separate interactions of the protein with water, and of the ligand with water. Groups such as amides, for example, readily form hydrogen bonds with each other in non-polar solvents, but do not do so in dilute aqueous solution, because it is energetically more favourable for the groups to form individual hydrogen bonds to water. (Only when water is largely excluded, as it is for example in the interior of a protein, and as it might be in an active or binding site filled by the ligand, do interamide hydrogen bonds become important under these conditions.)

Similar considerations apply to electrostatic interactions. Ion pairs between monovalent ions are important in non-polar solvents, but have little stability in water. Significant effects are observed if one of the ions is a polyelectrolyte,[85] and stable complexes can be formed between polyelectrolytes of opposite charge. Polylysine, for example (a polycation at neutral pH) forms an insoluble complex with DNA (a polyanion).[86] In many intra-protein and enzyme–substrate interactions, electrostatic forces reinforce hydrogen bonds as in the salt bridge $CO_2^- \ldots H_3N^+$ described above for chymotrypsin, and in the bifunctional interactions (**52**) between carboxylate or phosphate anions and the guanidinium group of arginine, observed, for example, at the active site of creatine kinase.[87]

Probably more important than either hydrogen bonding or electrostatic attraction in

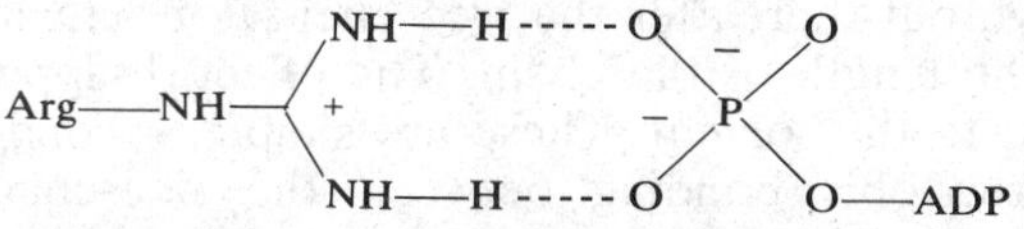

(52)

determining the strength of the interaction between two molecules in water is hydrophobic bonding.[8,84] Molecules, or parts of molecules, which are not effectively solvated by water disrupt the network of hydrogen bonds that makes up the structure of the solvent. The disruption is minimized if such molecules come together, to reduce the total area of non-polar surface in contact with water. Hydrocarbons, for example, form a second phase, while detergents, which are generally long-chain hydrocarbons with polar groups at one end, form micelles.[9] These are roughly spherical aggregates of molecules, with the charged end-groups at the surface, solvated by the water, and the hydrocarbon chains inside, in contact only with each other. Small non-polar regions, or cavities, on the surface of a protein are similarly poorly solvated by water, but do not control the state of aggregation of the molecule as a whole. They can, however, interact with hydrophobic molecules, or parts of molecules, of a similar size, coming together with them to reduce the total non-polar surface area in contact with water, as before. We have already seen one example of this mechanism in our discussion of the three-dimensional structure of chymotrypsin (p. 421). Close to the active site is a pocket, lined with hydrophobic groups,[46] of the correct size to accommodate the indole side-chain of tryptophan. Indole itself is tightly bound (binding energy about 60 kJ mol^{-1})[88] in this pocket, and the selectivity of chymotrypsin for particular peptide bonds is largely determined by how well the side-chain of the amino-acid residue concerned fits into this pocket.

A large part of the energy of hydrophobic binding thus comes from a reduction in the unfavourable energy of interaction of the molecules involved with water. Once they are in contact, however, other short-range forces come into play, which may add significantly to the total apparent energy of interaction. These are van der Waals, or London dispersion, forces between charges, dipoles, and induced dipoles, which give rise to a net attractive interaction.[84]

24.1.4.2 Association-prefaced catalysis

Although it is now possible to approach enzyme rates in simple intramolecular reactions, particularly in systems activated by ground-state strain (see Section 24.1.5.2), it has proved much more difficult to devise efficient models for the binding process, which would allow comparable reactivities in intermolecular processes. Several systems are known, however, in which catalytic efficiency is enhanced by prior association of catalyst and substrate molecules.

The simplest cases involve 1:1 interactions between molecules with hydrophobic side-chains. Straight-chain *N*-alkyl imidazoles (**53**) react up to to 200 times faster than *N*-methylimidazole with *p*-nitrophenyl decanoate (**54**).[89,90] No effect is observed with shorter

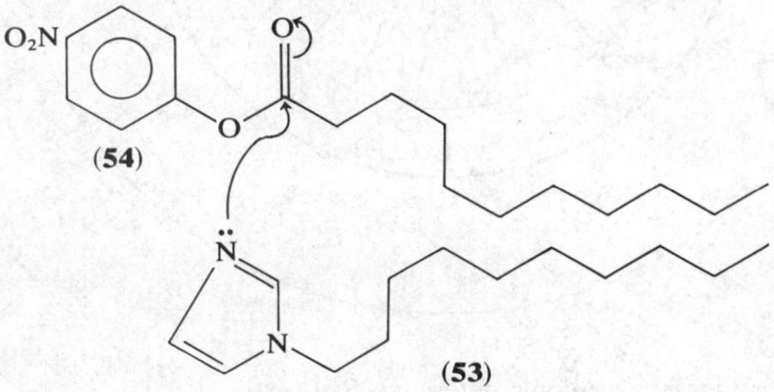

(54) **(53)**

alkyl chains, up to n-butyl, but thereafter the free energy of activation is lowered by an amount proportional to the length of the chain. The effect disappears completely when sufficient ethanol is added to the solvent. These are simple, second-order reactions, and presumably involve hydrophobic bonding between the side-chains, which will hold molecules of ester and imidazole (as well as ester and ester, *etc.*) together for long enough–if they are held in the correct orientation–to convert the bimolecular to an effectively intramolecular reaction.

As the concentration of ester (**54**) is increased the critical micelle concentration is reached, and micelles are formed (the CMC is typically between 10^{-2} and 10^{-4} mol l^{-1} for detergents of this sort with polar end groups).[8] This results in a further increase in catalytic efficiency, which must originate in a more effective binding interaction between the imidazole and the ester in the micelle. Figure 15 is a diagrammatic representation of a spherical micelle (of a charged detergent molecule). The hydrocarbon core has a diameter in the region of 20 Å, surrounded by the spherical Stern layer, a few ångstroms thick, which contains the charged or polar end-groups, tightly bound water of solvation, and some counter-ions. Most of the counter-ions, however, are found in a much thicker outer layer, where they are independently solvated, and exchange freely with other ions in the solvent.

The hydrocarbon core of the micelle is a more favourable environment than water for any hydrophobic molecule or part of a molecule, and the side-chain of an *N*-alkylimidazole (**53**), for example, will be bound to a micelle more effectively than to a single molecule of (**54**). In the favoured mode of binding the side-chain is absorbed into the core, and the polar end-group is thus held amongst the polar head-groups of the detergent, in this case the ester (**54**). The binding is thus not only stronger, but also involves a specific orientation of the 'substrate', which brings ester and imidazole groups into close proximity at the surface of the micelle.

Most known examples of micellar catalysis do not involve functional micelles, largely because the charged end-groups of the detergents which are readily available are the conjugate bases of strong acids, or tetra-alkylammonium cations, and are not reactive as

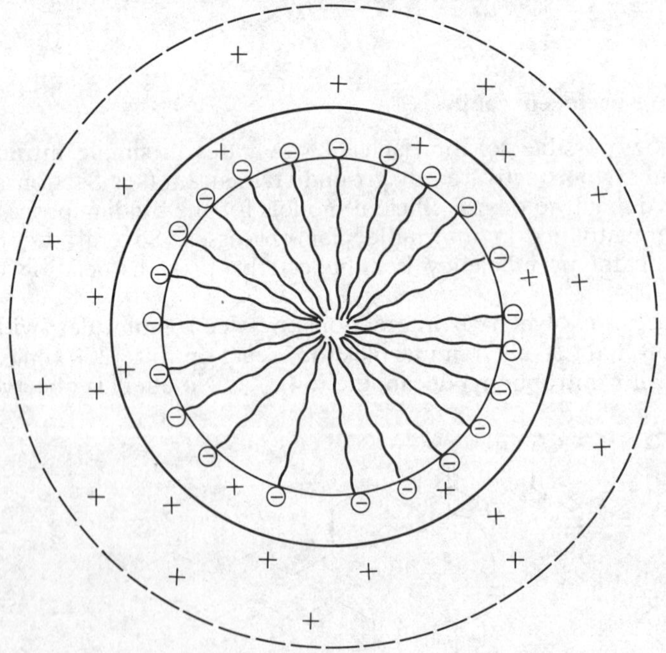

Figure 15 A micelle (schematic) of an anionic detergent. See text

nucleophiles or general bases. An anionic micelle (Figure 15) may catalyse specific acid-catalysed reactions, such as the hydrolysis of acetals or orthoesters with hydrophobic groups, by binding the substrate so that the polar functional group is held in the Stern layer, in a region where H_3O^+ ions are concentrated relative to the bulk solvent, and where the prevailing excess of negative charge might act to stabilize a cationic transition state, of the sort involved in acetal hydrolysis (p. 400). The anionic micelle may well *inhibit* the alkaline hydrolysis of an ester, because the Stern layer is not a likely place to find hydroxide ions; this reaction is catalysed by cationic micelles, where the counter-ions are hydroxide, and prevailing excess charge is positive. Both effects are reduced by increasing concentrations of salts, which replace most of the counter-ions by unreactive anions or cations.[8]

There are obvious similarities in the mechanisms of binding and catalysis by micelles to enzyme-catalysed reactions; and these extend to the observation of Michaelis–Menten kinetics in most cases, since it is clear that the micelle could become saturated with substrate. However, catalysis by aqueous micelles is generally modest, with rate enhancements rarely greater than 100, and even the degree of substrate specificity possible is relatively small. This is because the binding interactions involved are themselves fairly modest and non-specific, and an important objective of current research is to find simple systems which allow more specific binding.

The best known of these are the cyclodextrins.[91] These are cyclic oligomers of α-1,4-linked D-glucose, which are available from natural sources. α-Cyclodextrin, for example, contains six such glucose units, and its structure is known in detail from X-ray studies. It is a (ring) doughnut-shaped molecule (a torus), with the glucose units in the *C*1 conformation, which means that the secondary 2- and 3-hydroxy-groups of the sugar are on one face of the torus, and the primary 6-hydroxy-groups on the other (Figure 16). The torus encloses a cavity about 5 Å diameter (7 Å for the heptamer, β-cyclodextrin, and 10 Å for the octamer), which is lined with axial H atoms and the lone pairs of the glucosidic oxygens. The cavity is thus much less polar than water, and will accommodate a variety of mostly hydrophobic molecules of the correct size. This can lead to the formation of 1:1 inclusion complexes in aqueous solution, and, when an appropriate substrate molecule is bound, to intracomplex reactions with the hydroxy-groups of the cyclodextrin.

One such reaction is the hydrolysis of substituted phenyl acetates.[91] The most striking result involves the t-butyl substituent. The hydrolysis of *p*-t-butylphenyl acetate is 20%

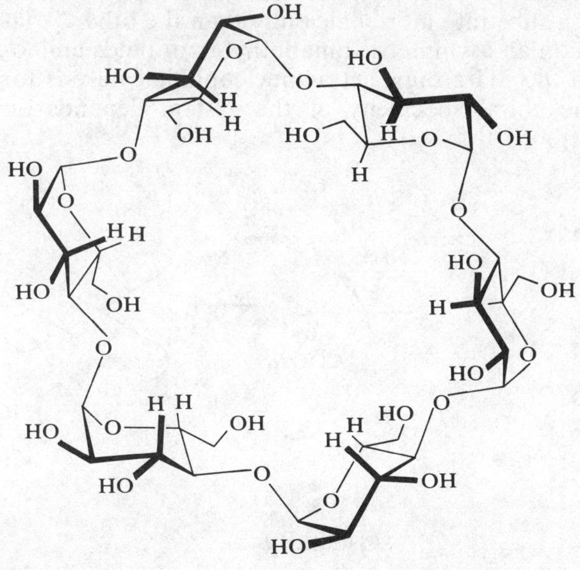

Figure 16 α-Cyclodextrin

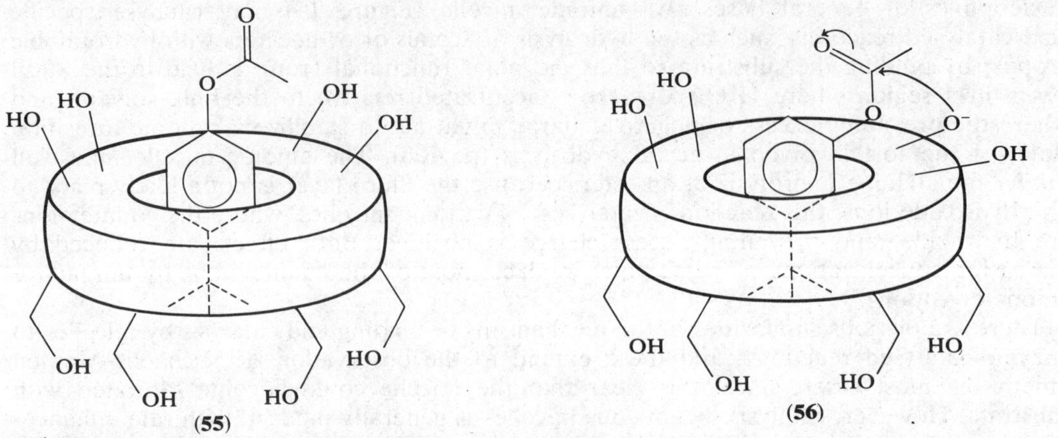

Figure 17 Binding of t-butylphenyl acetates by α-cyclodextrin. See text

faster in the presence of α-cyclodextrin at pH 10.6, but that of *m*-t-butylphenyl acetate is accelerated by a factor of 260. The explanation is illustrated in Figure 17.

In both cases the aromatic ring of the substrate is bound in the cavity, with its orientation controlled by the t-butyl group. This has little effect on the environment of the ester function of the *para*-compound (**55**), but the *meta-O*-acetyl group is held in close proximity to the 2- and 3-hydroxyl groups of the cyclodextrin. At the relatively high pH involved these will sometimes be ionized, and nucleophilic attack on the ester group by the alkoxide anion will readily displace a phenolate leaving group (**56**). The acyl group is thus transferred to the cyclodextrin, forming an intermediate (as with any other nucleophilic catalyst, from acetate to chymotrypsin) which has to be hydrolysed in a second, separate step. The intermediate in this case also can be identified if a chromophoric acyl group is used.

The cyclodextrins are thus very apposite models for enzyme–substrate reactions. Being formed from D(+)-glucose units they are in principle asymmetric catalysts, and do indeed show some small enantiomeric specificity.[92] However, this is very weak compared with the 100% enantiomeric specificity shown by an enzyme like chymotrypsin. Recent work has led to the synthesis of specially designed host molecules capable of binding, or reacting with, one enantiomer of a substrate more efficiently than the other.[93] The cyclic polyether (**57**), for example, based on an asymmetric binaphthyl unit, binds amino-acid ester cations in non-polar media, and the SH groups act as nucleophilic catalysts for the solvolysis of the bound ester*.[94] The chiral specificity of the system depends on the size of the

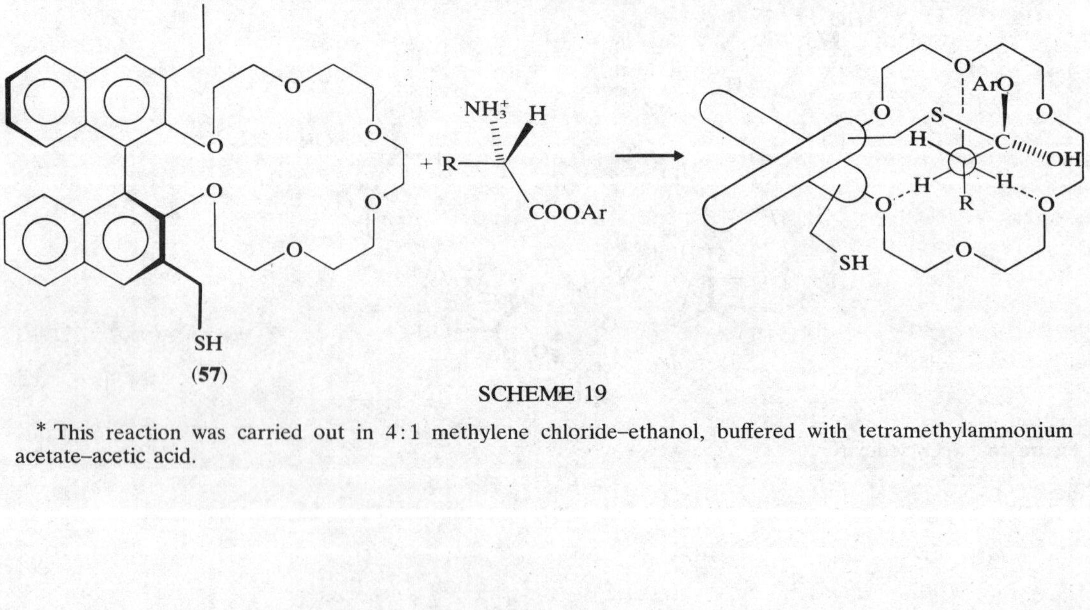

SCHEME 19

* This reaction was carried out in 4:1 methylene chloride–ethanol, buffered with tetramethylammonium acetate–acetic acid.

amino-acid side-chain R, but can be as high as 9–10 for R = isopropyl. (If the enantiomeric D-amino-acid ester were bound as in Scheme 19, R would have to be held close to the binaphthyl group.) This system is of particular interest because the binding interaction involves hydrogen-bonding rather than hydrophobic bonding, but here too the effect disappears if enough water is added to the solvent.

24.1.4.3 The binding of specific substrates by enzymes

The most important difference between enzyme catalysis and ordinary chemical catalysis is the binding step leading to the formation of the enzyme–substrate complex. As we have seen, this is the step most difficult to reproduce in attempts to approach enzyme rates and specificities in bimolecular reactions between simple compounds. Clearly the binding sites of enzymes must be highly developed structures, and as we might expect therefore, the most detailed evidence comes from X-ray crystallography.

The theory of substrate binding has been dominated by Fischer's idea, expressed as long ago as 1894[95] that the substrate fits the enzyme much as a key fits a lock. The fundamental idea is much older; Lucretius described the situation very aptly 2000 years ago:

> 'Things whose textures have such a mutual correspondence that cavities fit solids, the cavities of the first the solids of the second, the cavities of the second the solids of the first, form the closest union.'[96]

In terms of the forces in aqueous solution which control enzyme–substrate binding, we would now say that specificity is due to short-range interactions, especially van der Waals dispersion forces, which have an r^{-6} dependence. so that the net binding interaction depends critically on the size and shape of the molecules involved.

The study of enzyme–substrate interactions, especially over the last 20 years, has confirmed and extended these ideas. Enzymes which are truly specific, and accept just the substrate molecule and no other, are difficult to study systematically. However, many enzymes exist with sufficiently broad specificities that it is possible to measure how k_{cat} and K_m are affected by structural variation, under conditions which are optimal for the natural substrate. A useful alternative is to use compounds which are competitive inhibitors. These are substances which can bind at the same site as the substrate, generally because of their similarities of structure, and thus interfere with the enzyme–substrate reaction by occupying sites that would normally be binding substrate. This effect clearly depends on concentration, and a particular inhibitor can be characterized by measuring the effect of varying its concentration on the rate of the enzyme–substrate reaction. The efficiency is expressed as the inhibition constant, K_i, which, like K_m, becomes a dissociation constant under limiting conditions.

By varying the structure, either of a substrate or of an inhibitor, it is possible to determine which features of the substrate molecule are important for binding, and thus which groups are specifically bound by the enzyme. With enough data it becomes feasible to map the geometry of the binding site, assuming it to be complementary to the structures of the substrate or inhibitor molecule best bound: in other words, to deduce the geometry of the lock from the shape of the key. Very detailed investigations of this sort for chymotrypsin led to the characterization of requirements for all four positions around the tetrahedral α-carbon of the amino-acid residue attacked by the active site serine;[97,98] the substituents are designated as shown (**58**; Scheme 20).

It is found that a good substrate must have an α-hydrogen atom, an acylamino-group (R^1), a leaving group (R^3) in the correct position, as defined by (**58**), and a reasonably large (but not too large) hydrophobic group in the side-chain (R^2), which must not be branched at the first (β) carbon atom. These requirements lead to the conclusion that there are three binding sites, designated ρ_1, ρ_2, and ρ_3 (**59**) for the groups R^1, R^2, and R^3, respectively, and that their relative positions, and to some extent their shapes and sizes also, are complementary to the groups R^i of the best substrates.

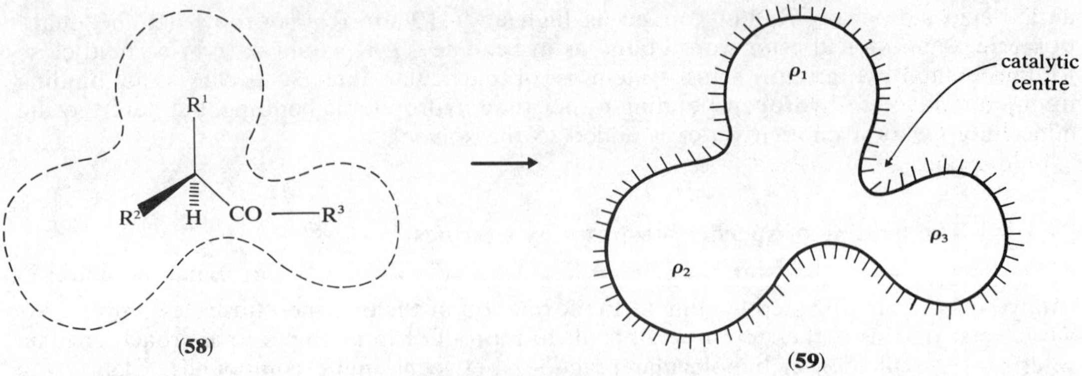

(58) (59)

SCHEME 20

This approach is time-consuming, but valuable because it gives information about binding under conditions where the enzyme is working normally. However, it cannot give detailed information about the group interactions involved in binding, of the sort which has become available in recent years from X-ray structural studies. Such studies are not generally possible on enzyme–substrate complexes, because their lifetimes are too short, and must necessarily be carried out on enzymes which are not working normally. Nevertheless, X-ray results for complexes of enzymes with inhibitors, or poor substrates, have given a great deal of evidence about the details of the binding of small and large molecules to enzymes, which in favourable cases is clearly relevant to substrate binding. The structures of complexes of chymotrypsin with *N*-formyl-L-tryptophan and -phenylalanine (**60** and **61**; X = OH, products of the hydrolysis of specific substrates) for example, are almost certainly close to those of the corresponding enzyme substrate complexes (with **60** and **61**; X = NHR), because the enzyme catalyses the exchange of ^{18}O from the carboxyl group of *N*-acyl derivatives of these amino-acids into solvent water.[99]

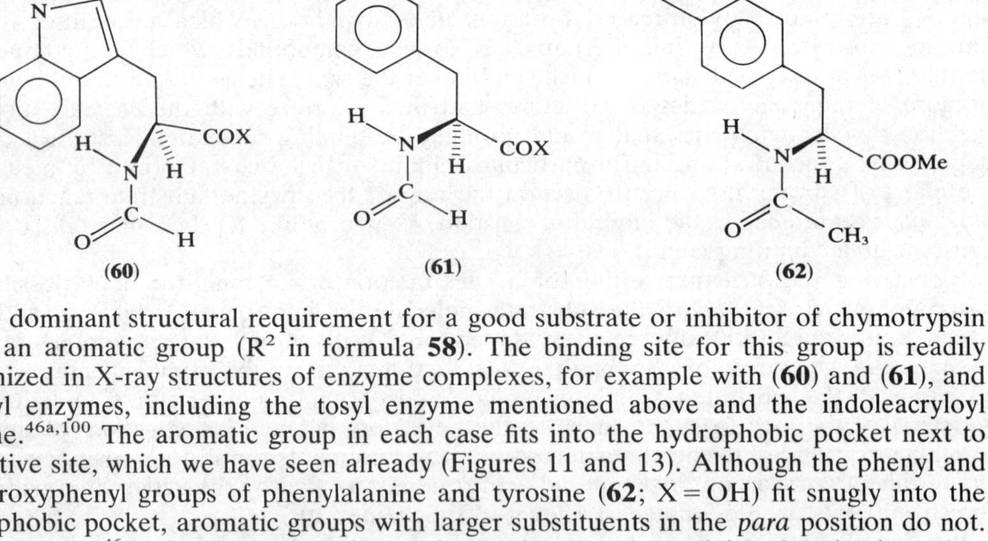

(60) (61) (62)

The dominant structural requirement for a good substrate or inhibitor of chymotrypsin is for an aromatic group (R^2 in formula **58**). The binding site for this group is readily recognized in X-ray structures of enzyme complexes, for example with (**60**) and (**61**), and of acyl enzymes, including the tosyl enzyme mentioned above and the indoleacryloyl enzyme.[46a,100] The aromatic group in each case fits into the hydrophobic pocket next to the active site, which we have seen already (Figures 11 and 13). Although the phenyl and *p*-hydroxyphenyl groups of phenylalanine and tyrosine (**62**; X = OH) fit snugly into the hydrophobic pocket, aromatic groups with larger substituents in the *para* position do not. This is thought[46a] to be the explanation of the much reduced reactivity of derivatives of *p*-iodo-L-phenylalanine and *O*-alkylated tyrosine (**62**; X = I, OMe, OEt, OPr^i).[101] These compounds are bound only slightly less efficiently than good substrates (**62**; X = H, OH), but k_{cat} is sharply depressed. Presumably the larger aromatic group fits as best it can into the hydrophobic pocket, but its tail, including the α-carbon atom in (**58**), projects too far for correct orientation of the other groups, particularly the peptide bond under attack, with respect to the catalytic groups of the active site.

The acylamino-group (**58**; R^1) is also involved in setting up the correct orientation for reaction.[102] *N*-Methyl compounds ($R^1 = R^4CONMe$), for example, are poor substrates.[101] In the crystal structure of the chymotrypsin–*N*-formyl-L-tryptophan complex the NH group forms a hydrogen bond to the (peptide) carbonyl oxygen atom of serine-214 (see **62a**).[103] This interaction fixes the position of the α-carbon atom, and thus of the substrate amide group (**58**; COR^3).[46a]

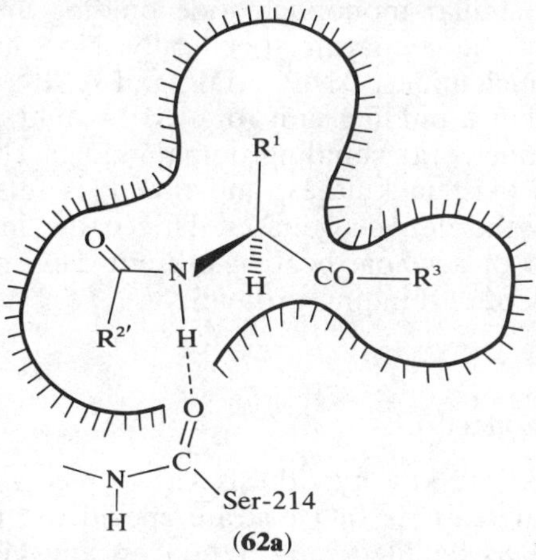

(**62a**)

We can now see how the chiral specificity of chymotrypsin may arise (see **62a**). If R^1 is bound tightly in the hydrophobic site ρ_1, and the acylamino-group held in its specificity site ρ_2, the positions of the remaining substituents H and COR^3 are fixed. Derivatives of L-amino acids are held in the reactive configuration shown in (**62a**), but in derivatives of D-amino-acids the groups H and COR^3 are interchanged, and the substrate carbonyl group is no longer in contact with the nucleophilic groups of the active site. Derivatives of D-amino-acids can be bound by the enzyme — they may indeed be bound more tightly than the natural substrates — but this 'non-productive' binding does not lead to catalysis.

Once the substrate binding site of chymotrypsin had been identified, it was of great interest to examine the binding mechanisms used by the related serine proteinases, trypsin and elastase. These enzymes use the same catalytic mechanism as chymotrypsin (Section 24.1.3.3), but have very different specificities. Whereas chymotrypsin attacks the carbonyl groups of amino-acid residues with aromatic side-chains, trypsin is specific for the positively charged amino-acids lysine and arginine,[104] while elastase, though markedly less narrowly specific, prefers aliphatic amino-acids, particularly alanine.[105] These different specificities are nicely explained by differences in the three-dimensional structures of the enzymes in the region of the binding pocket of chymotrypsin. This pocket is still readily identifiable in the other two enzymes, but it has been modified in each case. In the elastase structure[105] the hydrophobic side-chains of two amino-acids not present in chymotrypsin, valine-216 and threonine-226, restrict access to the hydrophobic pocket, preventing aromatic groups from being bound. Derivatives of L-alanine, on the other hand, fit well, in the same conformation as that deduced for the productive binding of substrates to chymotrypsin, with the alanine methyl group in van der Waals contact with side-chains at the mouth of the hydrophobic pocket. In the trypsin structure[55,106] the binding pocket is still apparent, but no longer hydrophobic. One of the amino-acid residues lining the pocket, serine-189 of chymotrypsin, has become aspartic acid-189 in trypsin. This puts a carboxylate group in the specificity pocket, and allows the formation, away from solvent, of the strong carboxylate–guanidinium salt bridge in (**52**) with the side-chain of an arginine residue on the substrate (and similarly with the NH_3^+ group of lysine).

These variations at the substrate binding site strongly suggest that this family of serine proteinases evolved relatively recently from a common ancestor. Much more extensive, and ancient, evolutionary relationships may account for the structural similarities found in a number of enzymes using the coenzyme nicotinamide-adenine dinucleotide* (NAD).[107] In several cases such enzymes (which include lactate, malate and glyceraldehyde 3-phosphate dehydrogenases) have remarkably similar three-dimensional structures in the regions which bind the coenzyme, though quite different structures elsewhere. This 'dinucleotide binding domain' occurs at widely varying points along the primary chain of those enzymes, and consists of two very similar mononucleotide binding units, one of which binds the adenine nucleotide half of the coenzyme specifically. Now many more enzymes catalyse reactions of the adenine nucleotides, AMP, ADP, and ATP*, than use NAD, and it might reasonably be expected that a binding unit for AMP would be even more evolutionarily likely than a common dinucleotide binding domain. In fact there is some evidence that such a binding unit does exist in kinases, and that it is related to the mononucleotide binding unit identified in the dehydrogenases. Hence the intriguing suggestion that the nucleotide binding region of a whole host of present-day enzymes is descended from a common 'primordial nucleotide binding protein'.[107]

24.1.4.4 The theory of induced fit

Although Fischer's lock-and-key hypothesis has proved immensely fruitful, and explains many of the general features of substrate specificity, its limitations have become more and more apparent as detailed information on enzyme–molecule interactions has accumulated in recent years. Taken literally the hypothesis presumes a rigid binding site, whereas there is a great deal of evidence for conformational mobility in proteins, both from X-ray structure determinations and from direct spectroscopic and kinetic measurements on enzymes in solution. We have seen how tyrosine-248 of carboxypeptidase apparently moves a distance of no less than 12 Å when glycyltyrosine is bound, clearly changing the nature and the shape of the binding site as well as the catalytic centre (Section 24.1.3.4). This may be an extreme example, but in every case where high-resolution X-ray diffraction data are available for complexes of an enzyme with a substrate analogue or an inhibitor, it is possible to identify significant conformational differences, compared with the structure of the native enzyme.

The theory of induced fit[108–110] suggests that such conformation changes, occurring when the substrate is bound to the enzyme, can play an important role in catalysis. This can happen if the conformation changes induced by binding the substrate affect the relative geometry of the catalytic groups at the active site, as appears to happen in the case of carboxypeptidase. Clearly the catalytic groups must be in their optimum positions in the reacting enzyme–substrate complex, so in such cases they evidently cannot be in their optimum positions in the native enzyme. The same considerations apply to the geometry of the binding site, which can also adjust itself to improve the fit of the substrate as part of the binding process.

There are many examples available of the dependence of catalysis and binding on conformational changes. The binding site of chymotrypsin depends critically on the salt bridge between aspartic acid-194 and the terminal NH_3^+ group of isoleucine-16 (Figure 14). In the inactive precursor, chymotrypsinogen, for example, the catalytic groups appear to be in much the same arrangement as in the native enzyme, but the hydrophobic pocket does not exist.[49] It is opened up when formation of the salt bridge induces changes in the conformations of aspartic acid-194 and the adjacent amino-acids, glycine-193 and methionine-192. Kinetic experiments on chymotrypsin suggest that a somewhat similar process occurs when enzyme with isoleucine-16 in the free base (NH_2) form is protonated. This high-pH form of the enzyme is inactive, because it cannot bind substrate. If the pH of a solution of the enzyme in this inactive form is rapidly reduced from 12 to 7, binding does occur, but only after a measurable delay (of less than a second) during which the

* These coenzymes are discussed, and their structures given, in Chapter 24.3.

enzyme takes up the active conformation.[111] In this case the conformation change must precede binding, and is clearly too slow to be part of the normal mechanism.

Cases where the geometry of the catalytic site is modified on binding are often easier to study. A simple, but suggestive, piece of evidence is the effect of inhibitors of chymotrypsin on the rate of hydrolysis of the acetyl-enzyme. The acetyl group is too small to reach from serine-195 to the hydrophobic pocket, so is not specifically bound. Where the acyl group is a substrate amino-acid, with the side-chain bound in the hydrophobic pocket, the acyl group is properly orientated for catalysis of hydrolysis: the *N*-acetyl-L-tyrosyl-enzyme, for example, is hydrolysed 12 000 times faster than acetylchymotrypsin.[112] This factor may not all be due to orientation of the ester group, because the rate of hydrolysis of acetylchymotrypsin is doubled in the presence of indole,[113] which is normally a competitive inhibitor of the enzyme. Indole apparently binds in the vacant hydrophobic pocket of the acetyl-enzyme, and clearly affects the interaction between the ester group of the acyl-enzyme and the general base groups of the active sites.

Induced fit is of particular importance for the large number of enzymes which transfer acyl or phosphoryl groups from one nucleophilic centre to another. Acylchymotrypsins exemplify what happens where there is little specificity for the nucleophile. Acetyl-[114] and furoyl-chymotrypsin[62] readily react with all sorts of alcohols, as well as with water, transferring the acyl group to ROH to form the ester. Indiscriminate acyl transfer of this sort is clearly unacceptable in cases where transfer of the acyl group to a specific receptor is the reason the enzyme exists, but since the specific alcohol will always be a molecule larger than water, there is no way that water can be excluded from the catalytic site. In this situation the enzyme must select between ROH and HOH by ensuring that HOH does not react, and induced fit provides a mechanism for specificity of this sort. Two very clear examples involve the transfer of a phosphoryl group to glucose, catalysed by phosphoglucomutase and hexokinase.

Phosphoglucomutase catalyses the interconversion of glucose 1- and 6-phosphates (**63** and **64**), by transferring a phosphoryl group from a phosphoryl-enzyme–serine residue to either sugar phosphate, to form glucose 1,6-diphosphate (**65**, Scheme 21). Phosphorylation of such a weak nucleophile as an alcohol normally requires a reactive phosphoryl

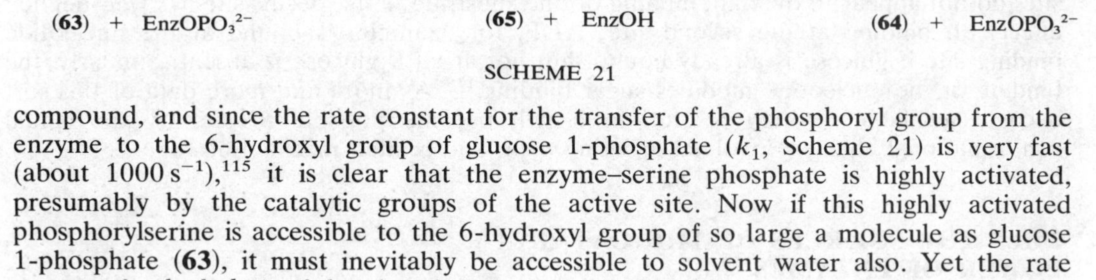

(**63**) + $EnzOPO_3^{2-}$ (**65**) + EnzOH (**64**) + $EnzOPO_3^{2-}$

SCHEME 21

compound, and since the rate constant for the transfer of the phosphoryl group from the enzyme to the 6-hydroxyl group of glucose 1-phosphate (k_1, Scheme 21) is very fast (about $1000\ s^{-1}$),[115] it is clear that the enzyme–serine phosphate is highly activated, presumably by the catalytic groups of the active site. Now if this highly activated phosphorylserine is accessible to the 6-hydroxyl group of so large a molecule as glucose 1-phosphate (**63**), it must inevitably be accessible to solvent water also. Yet the rate constant for *hydrolysis* of the phosphoryl-enzyme in the absence of substrate is only about $3\times10^{-8}\ s^{-1}$, smaller by a factor of over 10^{10}. Even this very slow rate is 300 times faster than expected for the hydrolysis of serine phosphate under the conditions (pH 7.5), so the enzyme–serine phosphate is slightly activated in the absence of substrate. However, it is clear that by far the major part of the high catalytic efficiency of this enzyme depends on having the substrate present: is probably caused, in fact, by substrate binding. In fact it has proved possible to assign substantial effects to the binding of specific parts of glucose 1-phosphate.[115]

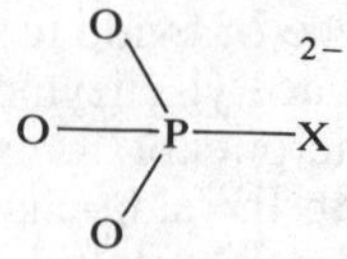

(66)

One group very likely to have a specific binding site is the terminal phosphate. Adding inorganic phosphate (**66**; X = OH), rather than the substrate, increases the rate of hydrolysis of the phosphoryl-enzyme 50-fold, while factors of several hundred are obtained if simple phosphate esters (**66**; X = OMe, OEt) or inorganic phosphite (**66**; X = H) are added. It seems certain that these effects, of binding a phosphoryl compound at a site several bond lengths away from the catalytic centre, must involve changes in the three-dimensional structure of the enzyme. Most of the remainder of the difference in reactivity of 10^{10}-fold between the enzyme phosphate and enzyme phosphate–substrate complex must be accounted for by the effects of binding the glucose residue. No significant increase in the rate of hydrolysis of the phosphoryl-enzyme is observed in the presence of 6-deoxyglucose 1-phosphate (**67**), no doubt because the binding of the substrate analogue prevents the approach of water to the active site. In the presence of α-D-xylose 1-phosphate (**68**), however, where water still seems unlikely to enjoy unrestricted access, the rate of hydrolysis of the phosphoryl-enzyme is greatly accelerated, by a factor of 2×10^5.[115]

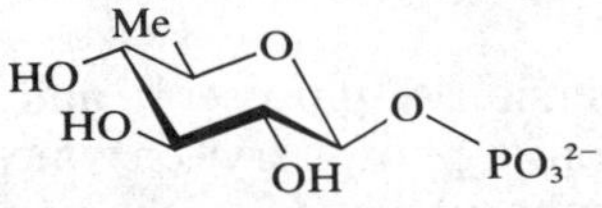

(67)

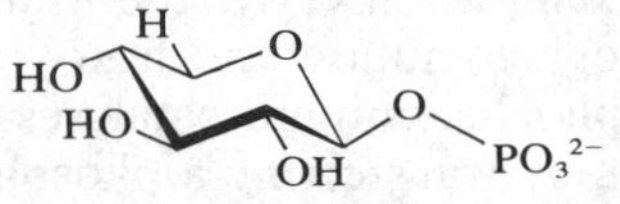

(68)

This is very strong evidence for induced fit. A similar case where specific structural evidence is available is the hexokinase reaction. Hexokinase[116] catalyses the transfer of the terminal phosphate group of bound ATP, also to the 6-hydroxyl group of D-glucose, and here too phosphoryl transfer to the substrate hydroxyl group is tens of thousands of times faster than to water. In this case X-ray structural information is available both for the enzyme itself[117] and for some enzyme–substrate and substrate analogue complexes.[118] The enzyme is made up of two identical subunits, unsymmetrically joined by non-covalent interactions. The nucleotide is bound in the region between the subunits, close to the receptor sugar molecule, which is itself bound in a deep cleft in the monomer. Though the sites do not appear to overlap, binding of one substrate in its specific site has well-defined effects on binding at the second site. ADP, for example, is bound at the nucleotide binding site if glucose is already bound, but not at all if glucose is absent. Similarly, the binding of the nucleotide modifies sugar binding.[118] As more and more data of this sort become available at high resolution, it will become possible to identify the actual conformational changes involved in induced fit, and related mechanisms.

24.1.5 THE SOURCES OF THE CATALYTIC POWER OF ENZYMES[119]

The evidence now available suggests that the catalytic mechanisms used by enzymes are normal reaction mechanisms, similar in detail to those studied by the physical organic chemist in simple systems; and thus that there is nothing special or unexpected about the reaction mechanisms concerned to account for the extraordinarily high rates of enzyme-catalysed reactions. For explanations of catalytic efficiency, therefore, we must look to the substrate binding step. We are only now beginning to understand just how complex the binding of a substrate to an enzyme can be, but it is clear that no single factor can account for the catalytic power of enzymes. Rather we have to consider several not entirely independent factors, and individual cases will depend differently on different factors.

The factors which have been discussed most often in this connection are entropy and strain, and we will consider these separately below. Strain is a vague term generally referring to distortion of substrates away from their most stable, ground state, geometries, but other types of destabilization may be equally important in individual cases. A particularly interesting possibility is desolvation. It is well known that polar molecules, and ions in particular, are strongly solvated in water. Solvation energies are large compared with the free energies of activation of reactions which go at a significant rate under mild conditions, so solvation is an important factor in stabilizing ground states. For example, before a nucleophile like RCO_2^- can attack an electrophilic centre, at least one of the lone-pair electrons of one of the oxygen atoms must become free, which means interrupting its solvation shell, and, specifically, breaking a hydrogen bond to a solvating water molecule (Scheme 22).

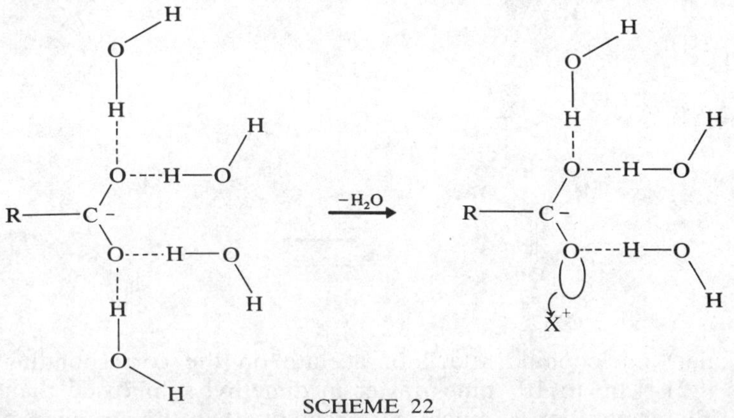

SCHEME 22

Anions in dipolar aprotic solvents, which are not specifically solvated, are many times more reactive than in water (by factors of over 10^6 for anions based on first- and second-row elements,[120] which include all those of major biochemical importance). So the nucleophilicity and basicity of nucleophiles could be greatly enhanced if they were effectively desolvated on binding to an enzyme. Raising the energy of the ground state in this way also accelerates unimolecular reactions of charged species which go *via* transition states in which the charge is less localized than in the ground state. Examples are decarboxylation and dephosphorylation, very similar processes which may lead to CO_2 or phosphoryl transfer if an appropriate nucleophile is present. The decarboxylation of benzisoxazole-3-carboxylic acid anions (**69**), for example, is up to 10^8 times faster in dipolar aprotic solvents than in water;[121] and the decomposition of a phosphate dianion with a good leaving group, such as (**70**), is similarly accelerated in dipolar aprotic solvents.[122] The effect in this case appears to be smaller (although the data are not directly comparable), no doubt because both ground state and transition state are doubly negatively charged in this case.

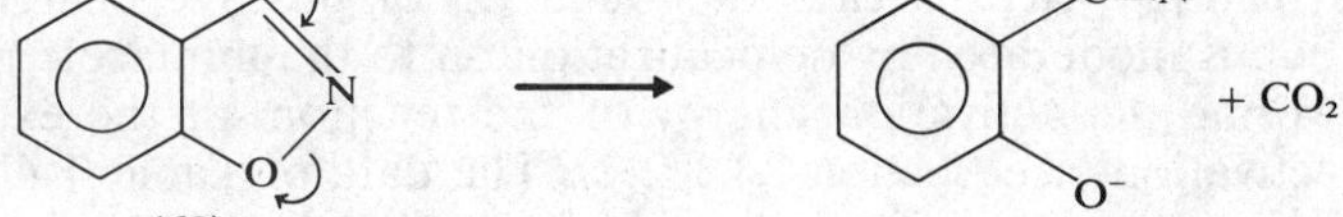

(**69**)

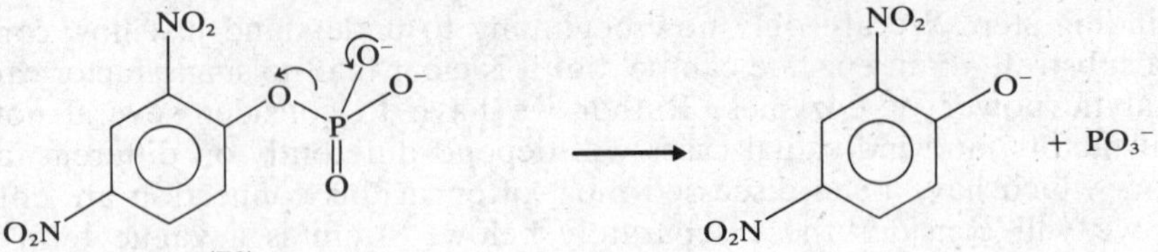

(70)

In contrast to these clear-cut results with unimolecular reactions, it has not proved possible to observe the expected effects of desolvation of the nucleophile on comparable intramolecular reactions in water. The possibility arises because, in such compounds as (**71**), models show that the carboxylate and ester groups must be in direct contact, because there simply is not room for a molecule of water between them.[123,124]

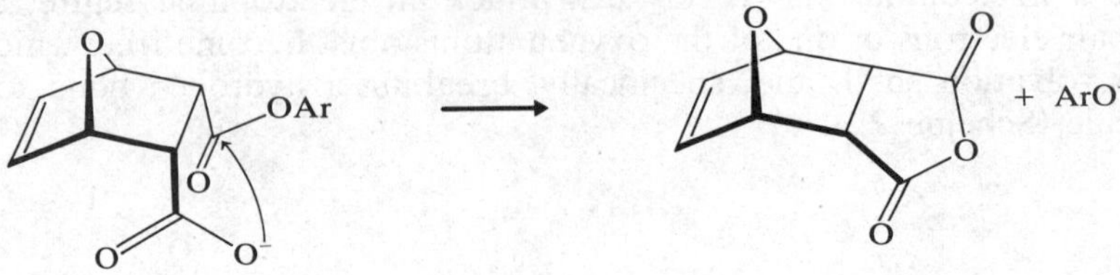

(71)

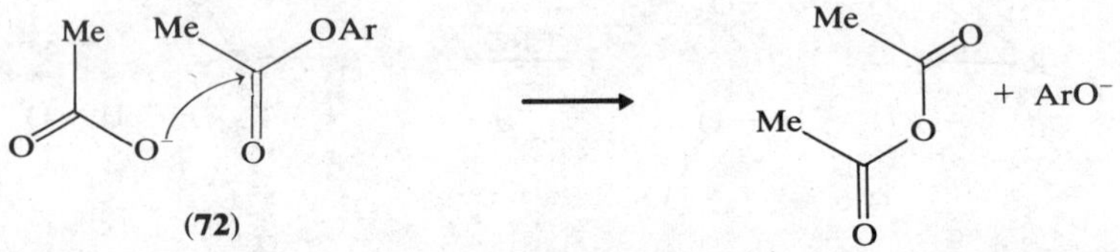

(72)

The (bimolecular) nucleophilic attack of acetate on the corresponding substituted-phenyl acetates (**72**) is up to 10^5 times faster in dimethyl sulphoxide than in water, as expected if the free acetate ion is much more reactive than the strongly solvated ion in aqueous solution. The intramolecular reaction of (**71**) is faster still (effective molarity of carboxylate up to 10^8 mol l^{-1} in water),[123] so a large part of the increased reactivity of the neighbouring CO_2^- group of (**71**) might be due to its enforced desolvation. This possibility was tested by measuring the rate of the intramolecular reaction in DMSO: if part of the high reactivity of (**71**) is really due to the partial desolvation of the carboxylate group, the rate enhancement on transfer to DMSO should be correspondingly reduced. In fact both intra- and inter-molecular reactions are accelerated to a similar extent on transfer from water to DMSO, so that ground-state desolvation is not an important feature of this, and other, similar systems tested.[123] Nevertheless, this mechanism for activation certainly should not be ruled out for enzyme–substrate reactions.

24.1.5.1 Entropy factors[119]

We saw in Section 24.1.3 how the catalytic mechanisms thought to be used by particular enzymes can in some cases be identified in simple systems. General base catalysis by imidazole, for example, of the hydrolysis of *N,O*-diacetylserine amide (**36**)[53] is a model for the reaction of chymotrypsin with an ester substrate. In an ionic reaction of this sort the transition state for the catalysed reaction is stabilized by the delocalization of charge over several centres. In this case the build up of positive charge on the nucleophilic hydroxyl group is moderated by delocalization on to the imidazole nitrogens. This has the effect of lowering the activation energy of the reaction, at the expense of an increased entropy of activation (see Section 24.1.2.2). The data of Table 4 illustrate this point: the unimolecular loss of 2,4-dinitrophenoxide from the phosphate monoester dianion has a

TABLE 4

Dependence of Entropy of Activation on Molecularity, in Phosphoryl Transfer Reactions[a]

	$\Delta H^{\ddagger}$ (kJ mol^{-1})	$\Delta S^{\ddagger}$ (J.K^{-1} mol^{-1})
Unimolecular[b]		
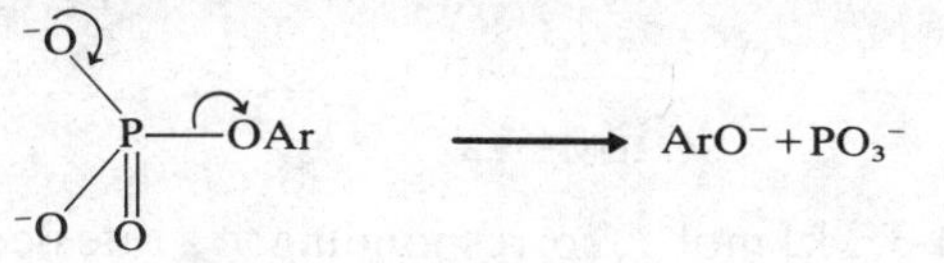	107.4	+27.6
Bimolecular[b,c]		
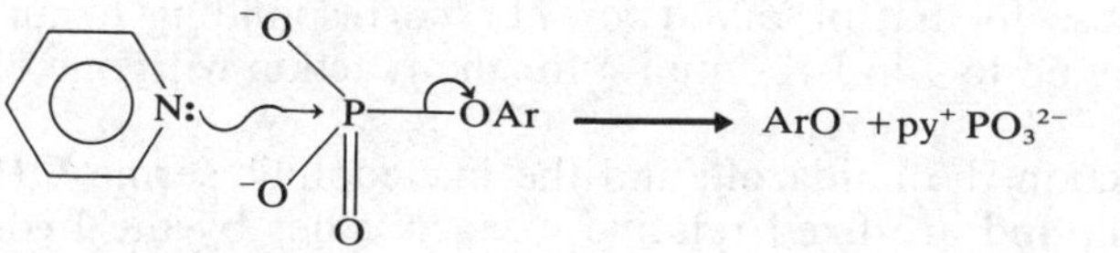	70.2	−81.1
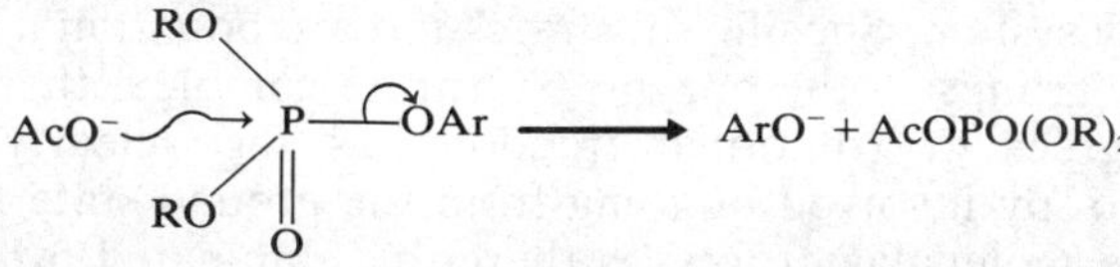	75.7	−72.3
Termolecular[b,c]		
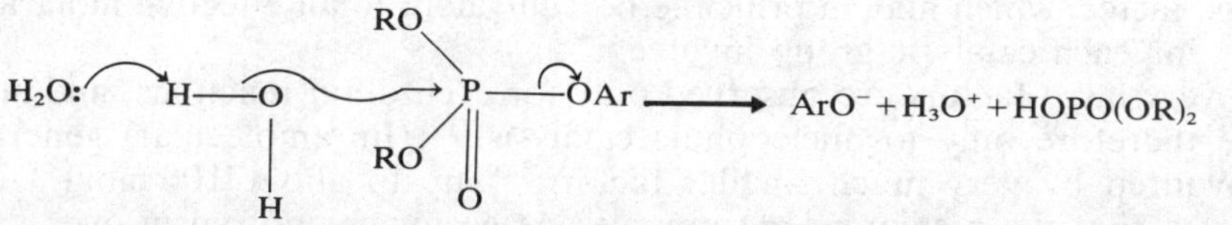	60.2	−148.8

[a] Data from A. J. Kirby and A. G. Varvoglis, *J. Chem. Soc. (B)*, 1968, 135; S. A. Kahn and A. J. Kirby, *ibid.*, 1970, 1172. [b] Ar = 2,4-dinitrophenyl. [c] $R_2 = (CH_2)_3$.

high enthalpy of activation, but is readily observed because of its very favourable entropy of activation. Nucleophilic catalysis of this reaction by pyridine is characterized by a much reduced activation enthalpy, since the pyridine nitrogen can accept positive charge in the transition state, and the high-energy metaphosphate intermediate [PO_3^-] is avoided; nevertheless, the involvement of the molecule of pyridine has to be paid for, in terms of a much less favourable entropy of activation. Similar parameters are observed for nucleophilic catalysis by acetate of the hydrolysis of the triester (**73**), which is also a bimolecular reaction. The neutral hydrolysis of (**73**), thought to involve the termolecular general base catalysis mechanism shown in Table 4, is a relatively slow reaction, but this is due entirely to the entropy term, which is much less favourable still. The enthalpy of activation is actually lower for the termolecular process, because delocalization over three molecules reduces still further the build up of charge at any single centre.

Similar considerations apply to the reaction of (**36**) (Scheme 23). The most obvious difference between this reaction and a corresponding reaction of chymotrypsin with an ester substrate is that the model reaction (Scheme 23) is slower by many orders of magnitude. This difference is due principally to the improbability of three separate molecules, each with its independent degrees of freedom of translation and rotation, coming together at the same time, in the same place, in the correct orientation for easy access to the transition state. It is statistical factors of this sort which determine the entropy of activation of a reaction. Bruice[124] finds an *average* (experimental) value of

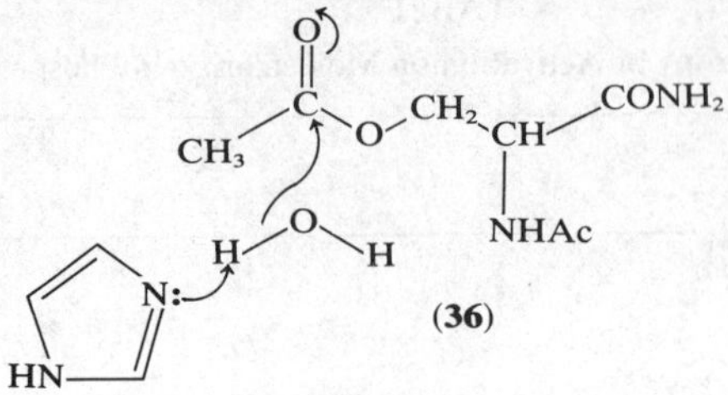

(36)

SCHEME 23

($-T\Delta S^{\neq}$/kinetic order) of 18.4±3.3 kJ mol^{-1}, corresponding to a rate decrease of 1.7±0.3×10^3 fold at 25 °C, for each extra species in the rate law, and hence in the transition state. Jencks[119] estimates a *maximum* effective molarity of the order of 10^8 mol l^{-1} for an intramolecular reaction (close to that observed for **71**), corresponding to an entropy of activation less favourable by up to 146 J K^{-1} mol^{-1} for the reaction with one extra species in the transition state.

In the chymotrypsin reaction the imidazole and the nucleophilic serine-OH group are parts of the same molecule, and are fixed relative to each other by the hydrogen-bond network of the charge-relay system. Specific substrates too are bound in a specific position and orientation. The three groups in the enzyme–substrate complex, therefore, may well have no independent degrees of rotational freedom, as they clearly have none of translation. The loss of entropy involved in going from the ground state to the transition state in a bimolecular or termolecular process is therefore eliminated, while the favourable enthalpy of activation is retained, making the enzyme–substrate reaction more favourable by a large factor, which may in principle be equivalent to an effective molarity of up to 10^8 mol l^{-1} for each catalytic group involved.

In practice these very large factors are observed only for cyclization reactions, and are probably applicable therefore only to nucleophilic catalysis.[119] (Intramolecular) general base catalysis is favoured by very much smaller factors,[119] up to about 100 mol l^{-1} in known examples,[125] so that the maximum entropy advantage for chymotrypsin over the reaction shown in Scheme 23 might be of the order of 10^{10} mol^2 l^{-2} (third-order compared with a first-order reaction).

Entropy factors can thus account for very large accelerations, and certainly represent the most generally important single factor involved in the high efficiency of enzyme catalysis. However, effective molarities greater than those expected from consideration of the entropy factor alone can be observed even in simple intramolecular processes. Such cases appear to involve strain in the ground state, and we see in the following section how this might be relevant to enzyme catalysis.

24.1.5.2 Strain

The fastest intramolecular reactions are cyclizations which relieve strong non-bonded interactions in the ground state. A simple example is the lactonization of 3,6-dimethyl-2-hydroxymethylbenzoic acid (**74**; R = Me). This is transformed to the phthalide (**75**; R = Me) over 300 times faster than the unsubstituted compound (**74**; R = H).[126]

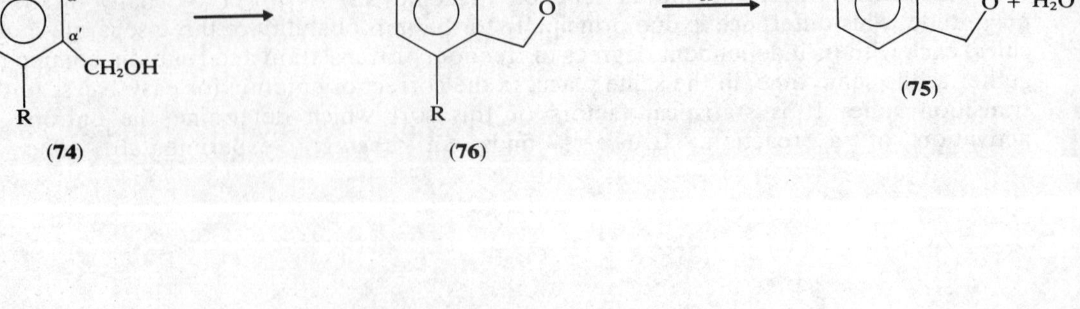

(74) **(76)** **(75)**

In the product, in the tetrahedral intermediate, and in the transition state for the reaction also, non-bonded interactions between the substituents on the 1,2,3,4-tetrasubstituted benzene (**74**; R=Me) are relieved, since the formation of the five-membered ring involves a decrease in the angles α,α'. The two inner substituents thus move inwards during the reaction, relieving the non-bonded interactions with the outer, methyl groups. Strain in the ground state is therefore part of the driving force for this reaction.

An extreme example is the similar reaction of the *o*-hydroxyphenylpropionic acid (**77**). In this case, lactonization gives a six-membered ring, and the introduction of the outer methyl groups (**78**) has only a small effect. When steric crowding is increased by making the α-carbon of the propionic acid side-chain tertiary (**79**), the rate of lactone formation is increased by a remarkable 10^{11}-fold.[127]

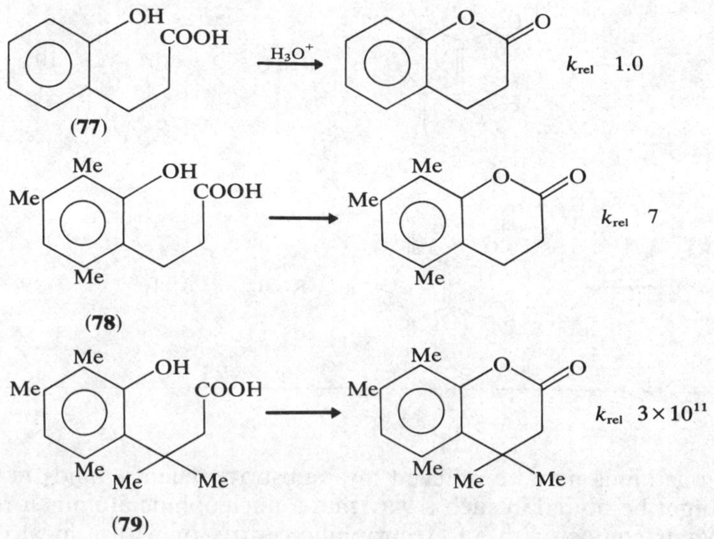

A comparison with the (estimated) rate of formation of phenyl acetate from phenol and acetic acid under the same conditions gives[127] an effective concentration of the neighbouring group of (**79**) of over 10^{15} mol l^{-1}, orders of magnitude greater than the largest rate acceleration expected from the more favourable entropy of activation of the intramolecular reaction. Here too the principal driving force is almost certainly the relief of ground-state strain.[128]

This effect would be expected to act on the enthalpy of activation. The relevant parameters have been measured for the hydrolysis of the maleamic acids shown in Table 5.[76,129] This reaction involves nucleophilic catalysis by the neighbouring carboxyl group, with formation of a five-membered ring anhydride by way of a tetrahedral intermediate. Formation of the five-membered ring from the tetrasubstituted ethylene (**82**) relieves non-bonded interactions between the four coplanar groups, much as in the reactions of the 1,2,3,4-tetrasubstituted benzenes already discussed. This factor disappears when the two alkyl substituents are absent: [the angles α,α' have opened up to near 130° in the ground state of (**81**) compared with near 120° in (**82**)].[130] Also, in the case of (**80**) the anhydride, and the transition state leading to it, are more strained than the ground state, because substituents on a double bond endocyclic to a five-membered ring naturally open out [α,α' are near 130° in (**80**)], and actually have to be forced together to make the second five-membered ring. The data in Table 5 show clearly how strain of this sort affects the enthalpy of activation. The entropies of activation change only a little in this series, and are if anything more favourable for the less reactive compounds.

TABLE 5
Effects of Ground-state Strain on the Enthalpy of Activation[a]

Reaction	Relative rate	$\Delta H^{\ddagger}$ (kJ mol^{-1})	$\Delta S^{\ddagger}$ (J K^{-1} mol^{-1})
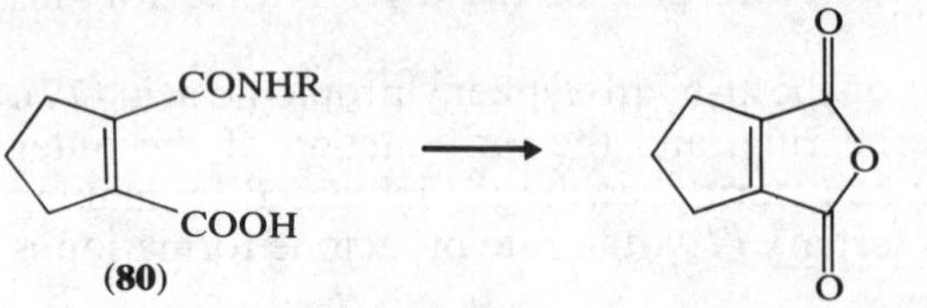 + RNH_2 (80)	1.0	142	+46
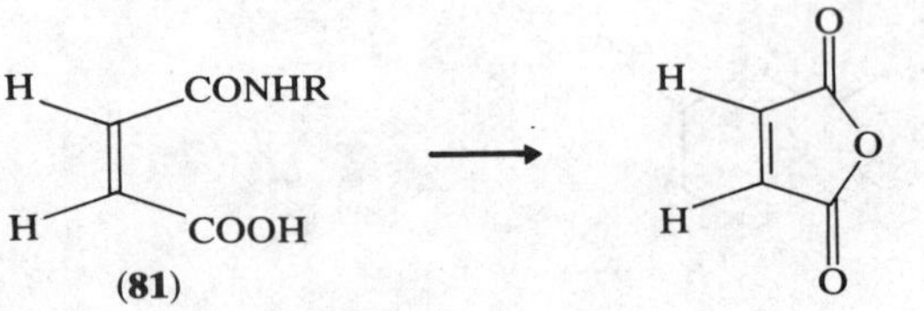 + RNH_2 (81)	2.4×10^{4}	105	+38
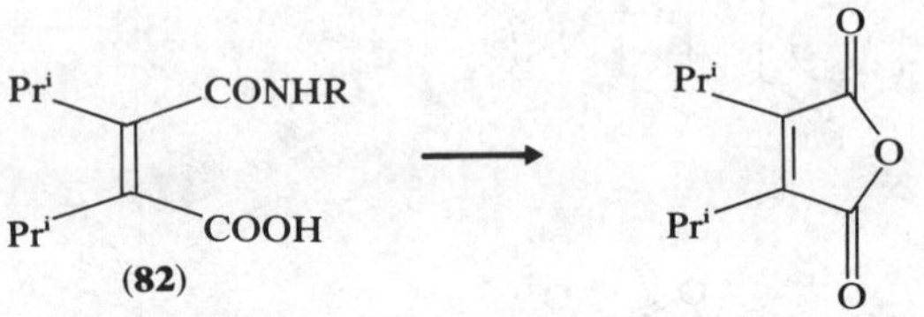 + RNH_2 (82)	6×10^{8}	79	+17

[a] See Ref. 76.

Strain of various kinds may be induced in a substrate when it binds to an enzyme.[119] The substrate might be bound in such a way that a nucleophilic atom is forced into closer than van der Waals contact with an electrophilic centre, or with a local deformation of bond-angles which raises the energy of the ground state. Binding in a less favourable conformation can have the same effect. Many lactones, for example, are 10^3–10^4 times more reactive towards alkaline hydrolysis than ordinary esters,[131] largely because they are fixed in the unfavourable *cis* conformation (*i.e.* **83**) while open-chain esters exist in the more stable *trans* conformation (**84**). In neither case is the delocalization energy of the ester group lost; clearly a much more substantial increase in ground-state energy would result if this happened, as it does, for example, in the lactam (**85**).[132] The cage structure of this compound fixes the lone pair electrons of nitrogen in the plane of the carbonyl group,

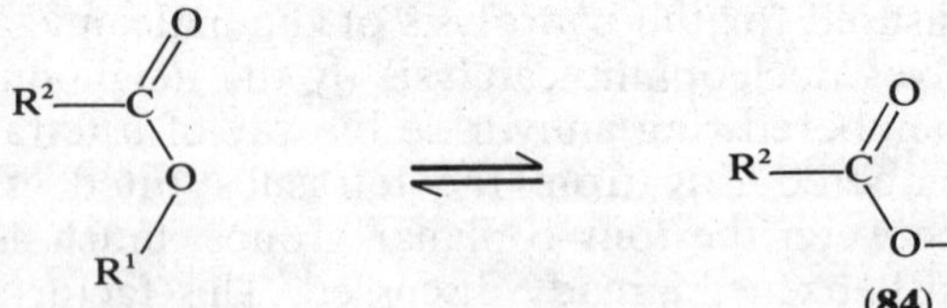

(**83**) (**84**)

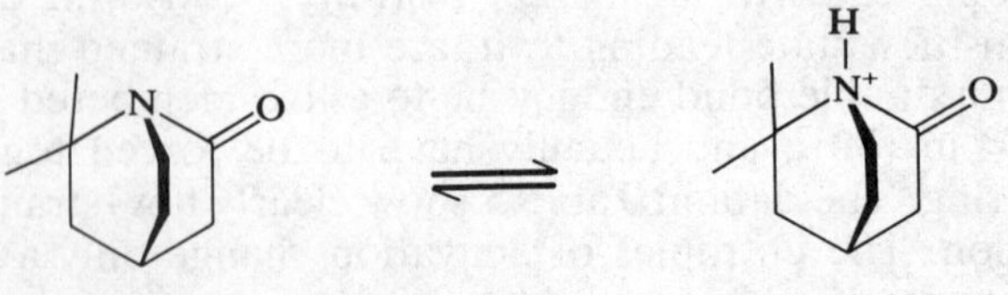

(**85**)

so that delocalization cannot occur. As a result, (**85**) behaves not as an amide, but as an amine with an adjacent electron-withdrawing group. For example, it shows a carbonyl absorption band in the infrared at 1762 cm^{-1}, and is much more basic than a true amide, with a pK_a of 5.33, so that a hydrochloride (ν_{max} 1799 cm^{-1}) is readily prepared. The carbonyl group of (**85**) is highly reactive, and the compound is hydrolysed in dilute acid with a half-life of about 15 min at 20 °C, conditions under which normal amides are stable indefinitely.

Thus the reactivity of an ester or amide substrate would be substantially increased on binding to an enzyme, if it were bound in a non-planar or even simply in a *cis* formation; and similar effects are possible for many other groups whose reactivity is sensitive to conformation or small changes of geometry. Such deformations would result, of course, from interactions between the substrate and the three-dimensional structure of the protein, and since the protein is not a rigid template, its structure will be deformed also. These deformations from the stable ground-state structures make such interactions energetically expensive, and reduce the total binding energy. They are thus only feasible where there is sufficient positive binding from interactions with other points on the substrate molecule (which would appear to rule out this mechanism for very small molecules). This means that the total binding energy is reduced; but as long as this remains high enough for efficient binding, so that the enzyme–substrate complex is fully formed at physiological levels of substrate concentration, the efficiency of catalysis is not impaired.

24.1.5.3 The binding of specific substrates by lysozyme[133]

Many of these effects are beautifully illustrated by the mechanisms of binding and catalysis used by the enzyme lysozyme. Lysozyme occupies a special place in the history of enzymology, as the first three-dimensional enzyme structure to be solved by the X-ray technique.[134] It is a small protein, consisting of a single chain of 129 amino-acid residues, which catalyses the hydrolysis of the glycosidic links of a carbohydrate component of bacterial cell walls (as part of the protective mechanism against bacterial infection). The natural substrate is an alternating copolymer (**86**) of *N*-acetyl-β-D-glucosamine (NAG) and *N*-acetyl-β-D-muramic acid (NAM), joined by β-1,4-glycosidic bonds, but most of the mechanistic work has been done with simpler substrates. Poly-*N*-acetylglucosamine, for example, is also hydrolysed, but the efficiency of this reaction depends critically on the size of the substrate, and the trisaccharide, $(NAG)_3$, is actually an inhibitor. Comparative X-ray structures of the enzyme and its complex with $(NAG)_3$ show the trisaccharide bound in a cleft in the enzyme, and allow a detailed examination of the way the three monosaccharide units are bound (at sites A, B, and C on the enzyme) by a combination of hydrophobic interactions and hydrogen bonds. (As we have seen in our discussion of the cyclodextrins (Section 24.1.4.2), glucose residues have hydrophobic faces, with the polar groups on the edges. The floor of the binding cleft of lysozyme is largely hydrophobic, while the polar side-chains which can form hydrogen bonds are closer to the surface of the enzyme.[135]) The cleft is filled by the trisaccharide for only part of its length, and model-building studies showed that up to a hexasaccharide could be reasonably accommodated. A likely conformation for a bound hexasaccharide $(NAG)_6$ was deduced, using the pattern of binding interactions observed for the sugar residues of the trisaccharide as a basis for assigning the positions (D, E, and F) of the groups on the extra three residues.

In this way it was possible to deduce a reasonable structure for an enzyme–hexasaccharide complex, which might be expected to be close to that of the productive-enzyme–substrate complex, since hexasaccharides are cleaved by the enzyme. In this model,[135] five of the six sugar residues (in sites A, B, C, E, and F) are bound in the stable chair conformation (as in **86**), but the fourth sugar from the non-reducing* end of the

* The reducing end of a sugar is the hemiacetal (aldehyde) group, *e.g.* the right hand end of (**86**).

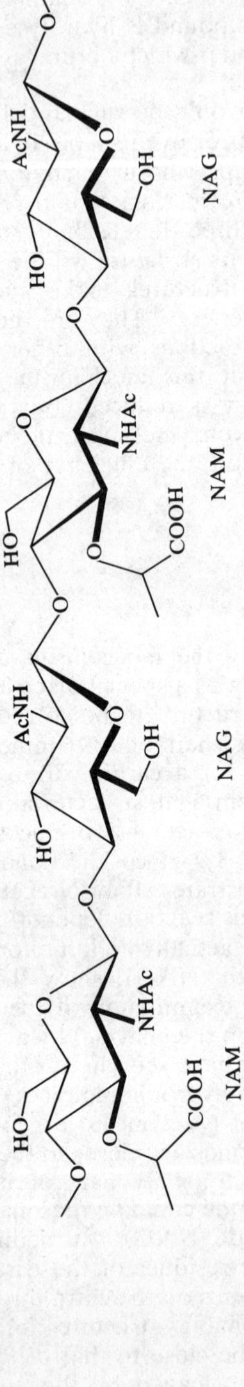
HO
AcNH
OH
NAG
HO
NHAc
COOH
NAM
HO
AcNH
OH
NAG
HO
NHAc
COOH
NAM

(86)

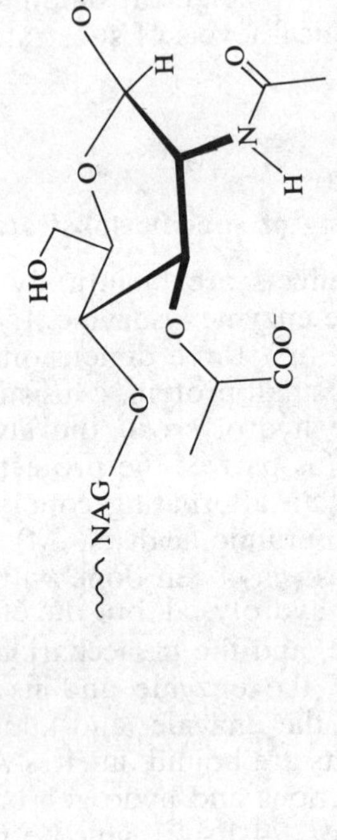
HO
NAG
NAG
COO⁻
H
N
H
O

(87)

hexasaccharide (at site D on the enzyme) has to be twisted into the half-chair conformation (**87**) to allow binding of residues five and six at sites E and F (see below). Furthermore, the fit around the 2-hydroxy-group of the sugar residue of $(NAG)_6$ in site C is too close to accommodate the larger lactic acid side-chain of a NAM residue. This suggests that the NAM units of the natural substrate are bound in sites B, D, and F, with NAG units in positions A, C, and E. Hydrolysis is known to produce a (reducing) terminal NAM unit, so the bond cleaved must be between the NAM and NAG residues in sites D and E (since the trisaccharide complex, at sites A–C, is stable); that is, it must be the glycosidic bond of the NAM residue in site D, with the distorted conformation (**87**).

Independent evidence for the distortion of the sugar residue in site D is available from binding studies. Oligomers, $(NAG)_n$ for example, are bound to the enzyme increasingly strongly, up to $(NAG)_3$,[136] but the tetramer, pentamer, and hexamer show no further increases in binding energy, and NAG-NAM-NAG-NAM is actually bound less tightly than NAG-NAM-NAG.[137] These deviations can be nicely rationalized in terms of the picture just described, derived from the structural work. Sites A, B, and C can each bind a sugar residue, so that the highest binding energy is observed when a trisaccharide is bound. The ground-state conformation of the fourth sugar unit of a tetrasaccharide is not bound to site D, however, and it is not until both ends of the oligosaccharide are tied down, by further binding of residues 5 and 6 at positions E and F, that the fourth sugar can be forced into site D (Figure 18). Any favourable binding energy to sites E and F is thus largely offset by the unfavourable term for site D, where energy must be expended to distort the NAM residue into the relatively high-energy conformation (**87**) in which it can bind. It is particularly interesting that the oxidation product of $(NAG)_4$, lactone (**88**), which has a planar C-1 and thus exists naturally in the half-chair conformation, is bound to lysozyme more tightly than the trisaccharide $(NAG)_3$.[138] This suggests that there is positive binding of the half-chair conformation of the natural substrate.*

These observations are directly relevant to the efficiency of catalysis, because known mechanisms of hydrolysis of acetals involve oxocarbonium ion intermediates. It is known that the glycosidic C—O bond is broken in the lysozyme reaction, so the expected intermediate in the lysozyme reaction is the cyclic oxocarbonium ion (**89**). This has a

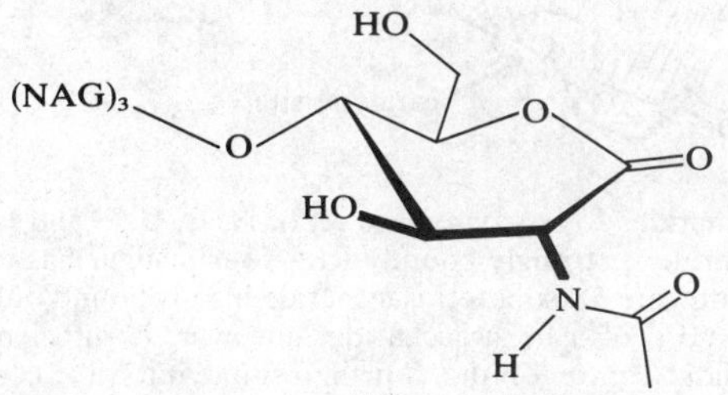

(**88**)

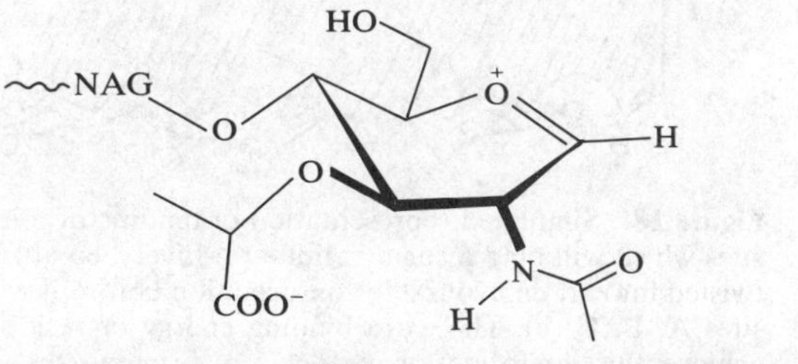

(**89**)

planar, sp^2-hybridized C-1, like (**88**), and will therefore also be expected to adopt a half-chair conformation. The result of the conformational distortion of the substrate on binding at site D is thus to force the reacting sugar residue into a conformation approaching that of the expected high-energy intermediate on the reaction pathway, and thus close to that of the transition state for the reaction. Part of the binding energy is thus used to raise the energy of the ground state, which will result in a reduced free energy of activation and thus faster reaction.†

Clues to the actual mechanism of catalysis emerge most clearly from the structural studies. In the model of the enzyme–$(NAG)_6$ complex described above, just two catalytic

* Very recent results[139] suggest that the conformational distortion of the sugar residue at site D is due largely to interactions involving the lactic acid side-chain of the NAm unit, and that a NAG residue can bind at this site. This could mean that the contribution of conformational distortion to catalysis may be smaller than was originally thought.

† More detailed discussions of this important point can be found in Refs. 119 and 140.

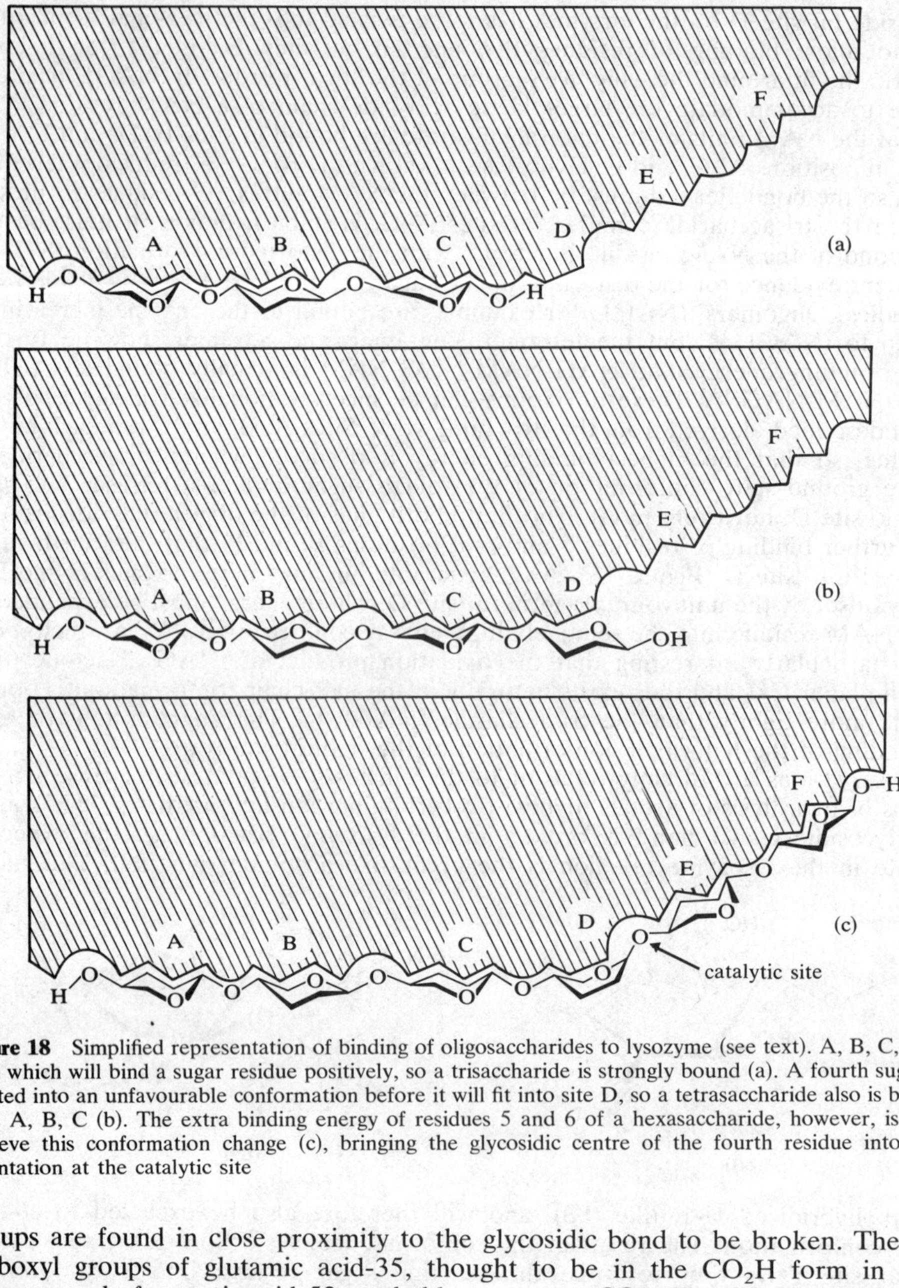

Figure 18 Simplified representation of binding of oligosaccharides to lysozyme (see text). A, B, C, E and F are sites which will bind a sugar residue positively, so a trisaccharide is strongly bound (a). A fourth sugar has to be twisted into an unfavourable conformation before it will fit into site D, so a tetrasaccharide also is bound only at sites A, B, C (b). The extra binding energy of residues 5 and 6 of a hexasaccharide, however, is sufficient to achieve this conformation change (c), bringing the glycosidic centre of the fourth residue into the correct orientation at the catalytic site

groups are found in close proximity to the glycosidic bond to be broken. These are the carboxyl groups of glutamic acid-35, thought to be in the CO_2H form in the active enzyme, and of aspartic acid-52, probably present as CO_2^-. The CO_2H group of glutamic acid-35 is positioned close to the leaving group oxygen, and is generally thought to act as a general acid, to assist the departure of the leaving group as OH rather than O^- (Scheme 24).[141,142] The role of aspartate-52, on the other hand, is still obscure. This group lies on the opposite side of the sugar residue, and one suggestion is that the electrostatic effect of the CO_2^- group stabilizes the developing oxocarbonium ion (**89**) (path a, Scheme 24).[133,143] Alternatively (path b), the carboxylate group might act as a nucleophile, producing a covalent intermediate (**90**), which would have to be hydrolysed at the acetal rather than the ester centre, since hydrolysis is known to proceed with retention at the glycosidic centre.[144] None of the evidence currently available either strongly supports or rules out either mechanism.[141]

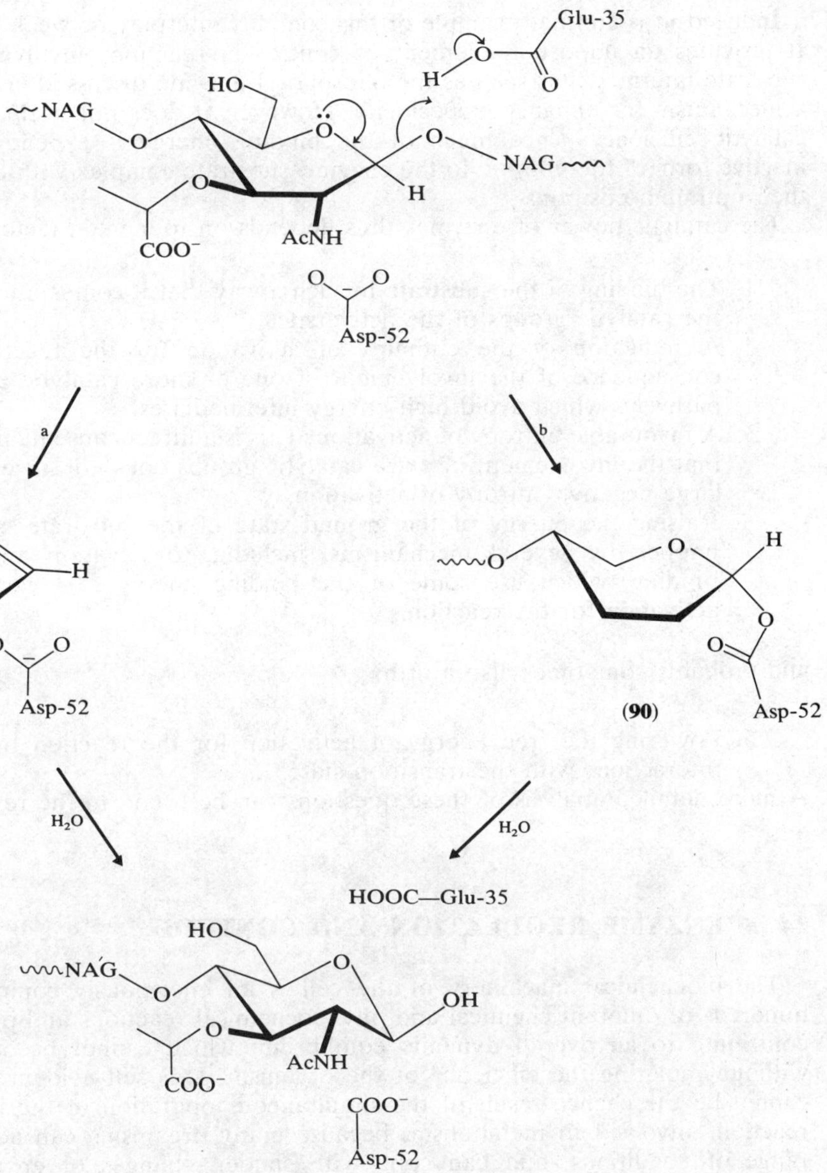

(**89**)

(**90**)

SCHEME 24

All the evidence we have discussed in this and the preceding section makes it plain that binding and catalysis may have a very complex mutual dependence. It is not true, for example, that the best substrates are those which are bound most tightly. A trisaccharide will bind very well to lysozyme, and a D-amino-acid derivative to chymotrypsin, but neither is a substrate, because the former is bound in the wrong place, and the latter in the wrong orientation. Furthermore, strain induced in the substrate on binding may increase the rate of the catalytic reaction, while at the same time reducing the efficiency of binding. Evidence of this kind has been cited to support the suggestion that the catalytic power of an enzyme is derived, at least in part, from its ability to bind the substrate in the transition state more strongly than in the ground state.[145] This may be because there are unfavourable enzyme–substrate interactions in the ground state, which are relieved in the transition state; this is the case for lysozyme. Alternatively, it may be because the transition state actually binds positively well. Only the latter situation necessarily leads to more efficient catalysis,[140] though both might do so under the right conditions.

Induced fit is another example of this complex interplay between binding and catalysis. It provides an important element of control in reactions involving covalent enzyme–substrate intermediates such as the phosphoryl-enzyme discussed in Section 24.1.4.4, and a mechanism for enhancing specificity. However, it does not itself contribute directly to catalytic efficiency, since much of the binding energy is expended in converting the inactive form of the enzyme to the enzyme–substrate complex with the catalytic groups in their optimum positions.

The catalytic power of enzymes thus depends on four main factors:

1. The binding of the substrate in such a way that it comes into correct contact with the catalytic groups of the active site.
2. A reduction of the enthalpy of activation for the reaction concerned, as a consequence of the involvement of one or more catalytic groups which provide pathways which avoid high-energy intermediates.
3. A favourable entropy of activation. This is a direct consequence of 1, which means that the involvement of extra catalytic groups does not have to be 'paid for' by a large negative entropy of activation.
4. Raising the energy of the ground state of the substrate on binding. This may happen by several mechanisms, including desolvation and strain induced on binding, which use some of the binding energy to lower the free energy of activation for the reaction.

and probably sometimes also a fifth:

5. Lowering the free energy of activation for the reaction by favourable binding interactions with the transition state.

A more detailed analysis of these questions can be found in the review by Jencks.[119]

24.1.6 ENZYME REGULATION AND CONTROL[146–148]

The biochemical machinery of the cell is an enormously complex system of many hundreds of different chemical and physicochemical reactions and processes, all of which contribute to an overall dynamic equilibrium which cannot be significantly disturbed without impairing the efficiency of the organism. It is self-evident that this equilibrium cannot be the chance result of the simultaneous operation of the hundreds of different reactions involved in metabolism, because many organisms can adapt to a fairly wide range of conditions, and can cope with sudden changes of environment. The great majority of these processes are enzyme-catalysed reactions, and it has become apparent in recent years that the activity of enzymes in the intact organism is subject to very close control, by a battery of mechanisms which are at least as complex as the mechanisms of binding and catalysis.

So far we have considered the enzyme and its substrate as an isolated system, insulated from other enzyme–substrate systems by the high specificity of the binding mechanisms involved. However, in the working cell the catalytic activity of any single enzyme depends on many more factors than the binding and catalytic mechanisms it employs, and the concentration of the substrate. The substrate itself will normally be the product from another — perhaps more than one — enzyme-catalysed reaction, and may also be a substrate for one or more other enzymes. The reaction product, similarly, will often be a substrate for one or more other enzymes, and so on. Clearly the concentration of the substrate can play an important role, but control is achieved mainly through the regulation of the concentrations and activities of enzymes, and these are the factors we will consider in this section.

24.1.6.1 The regulation of enzyme concentration

The concentration of a particular enzyme depends on the relative rates of its synthesis and its degradation. Under most conditions the concentration remains at a steady state, but it can clearly be adjusted by altering the rate of either process. The level of arginase in rat liver, for example, can vary by a factor of three with the protein content of the diet. Over 14 days on a protein-rich diet the concentration of the enzyme rises, specifically because of an increase in the rate of its biosynthesis. The concentration of arginase also rises during the first few days of starvation after a period of a low protein diet, but in this case because the degradition of the enzyme ceases, though synthesis continues.[149]

The biosynthesis of proteins is subject to genetic control. In bacteria at least this control is exercised at the level of the synthesis of messenger RNA, by the interaction of a particular ('regulatory') protein with a specific region of DNA (see Part 22, and Section 24.2.3); in animal tissues also, inhibitors of RNA synthesis interfere with the mechanisms which control enzyme levels.[149] Details of these control mechanisms are not important for this discussion. What matters is that mechanisms exist for the regulation of the concentrations of enzymes on particular metabolic pathways by the end-products of these pathways. Inducible enzymes, for example, are well-established in bacterial systems. These are enzymes which are not produced unless their substrates are present in the medium. Often the synthesis of the several enzymes of a particular pathway is induced by the presence of the substrate for the first enzyme of the pathway. Substrate induction thus provides a mechanism for increasing the concentration of an enzyme system when a job opportunity arises. The corresponding mechanism which reduces redundant enzyme concentration, when the enzyme or enzyme system is producing too much of a particular metabolite, is feedback repression. A classic example is the inhibition of the biosynthesis of histidine in *Salmonella typhimurium* by high concentrations of histidine. The concentrations of all ten enzymes on the biosynthetic pathway vary in exactly the same way in response to variations in the histidine concentration.[150]

Protein degradation appears to be much more random under normal conditions. The rate of degradation of a particular enzyme is characteristic, but there is an overall dependence on the size of the protein: larger proteins are degraded more rapidly, as they would be if the rate depended on the frequency of random collisions with proteolytic enzymes. (There is, for example, a general correlation between the rates of degradation of enzymes *in vivo*, and their rates of inactivation by the digestive enzymes, trypsin and chymotrypsin, *in vitro*.) As a consequence of this fairly constant rate of degradation, and a closely controlled rate of synthesis, the half-lives of different enzymes in animal tissues range from minutes to several days.

24.1.6.2 The regulation of enzyme activity[148]

Controlling enzyme concentration is clearly a very suitable way for the organism to adapt to relatively long-term changes in conditions, but it does not allow a rapid response to change. Stopping the synthesis of an enzyme with a half-life of several days, for example, has no immediate effect on the rates of removal of its substrate or of formation of its products. So it is not surprising that further mechanisms have been evolved, for the regulation of the *activity* of particular enzymes.

Perhaps the simplest of these involves the synthesis of the enzyme in the form of an inactive precursor. The best-known examples are the powerful proteolytic enzymes active in digestion. Such activity would clearly be undesirable in the cells producing these enzymes, so pepsin, trypsin, and chymotrypsin are all synthesized as inactive 'zymogens'. Pepsinogen, for example, is then secreted into the stomach, where the high acid concentration, and particularly the proteolytic action of the pepsin already present, combine to remove a 44-amino-acid peptide fragment and generate the active enzyme. Trypsinogen is activated, with removal of a hexapeptide, by the enzyme enterokinase, and autocatalytically by preformed trypsin; while chymotrypsin is formed from chymotrypsinogen by the proteolytic action of trypsin, which generates the terminal NH_3^+ group of isoleucine-16 essential for activity (see Section 24.1.3.4).

A more flexible mechanism involves reversible interconversion of the active and inactive forms. A notable example is the phosphorylase reaction. This enzyme catalyses the reversible addition of a molecule of glucose (as glucose 1-phosphate) to the polysaccharide glycogen, which is the molecular form in which carbohydrates, and thus readily available supplies of energy, are stored by animals. Phosphorylase thus holds the key to this energy store, and operates under a highly sophisticated control mechanism. The enzyme itself exists as two distinct forms, phosphorylase *a* and *b*. Phosphorylase *b* is inactive (at least in the absence of AMP—see below), and exists as a dimer. It is converted to phosphorylase *a** by the phosphorylation of a specific serine residue on each protein subunit, catalysed by the enzyme phosphorylase kinase. The reverse reaction is catalysed by a separate enzyme, phosphorylase phosphatase.

Without going into further details of a very complex process,[148] we can see how a system of this sort can provide a very sensitive mechanism for control of a particular process, in this case the mobilization of glucose from the glycogen bank. One molecule of phosphorylase *a* catalyses the release of thousands of molecules of glucose; and one molecule of phosphorylase kinase can activate thousands of molecules of phosphorylase.† If we add that phosphorylase kinase itself exists in active and inactive forms, and is activated by a phosphorylation catalysed by yet another enzyme, phosphorylase kinase kinase, which is itself regulated by a different mechanism altogether, then we can begin to see just how sophisticated and sensitive control mechanisms of this sort can be.

Logically the simplest mechanism of control, though far from simple in molecular terms, is feedback inhibition.[148] In many multienzyme systems the end product of the metabolic pathway may specifically inhibit an enzyme at or near the beginning of the sequence, so that the rate of formation of the product, which requires the complete sequence of reactions, is regulated by its concentration. Thus if product begins to accumulate, its rate of production is almost instantaneously reduced. An example is the biosynthetic pathway from threonine (**91**) to isoleucine (**92**, Scheme 25). Added L-isoleucine specifically inhibits threonine dehydratase, the enzyme catalysing the first of the five steps of the sequence, which must therefore have special features that make it an effective regulatory enzyme.

A typical regulatory enzyme is a protein of relatively high molecular weight, probably composed of an even number (often four) of subunits, which are identical individually or in pairs. Such enzymes are often highly specific for both substrate and inhibitor, which will

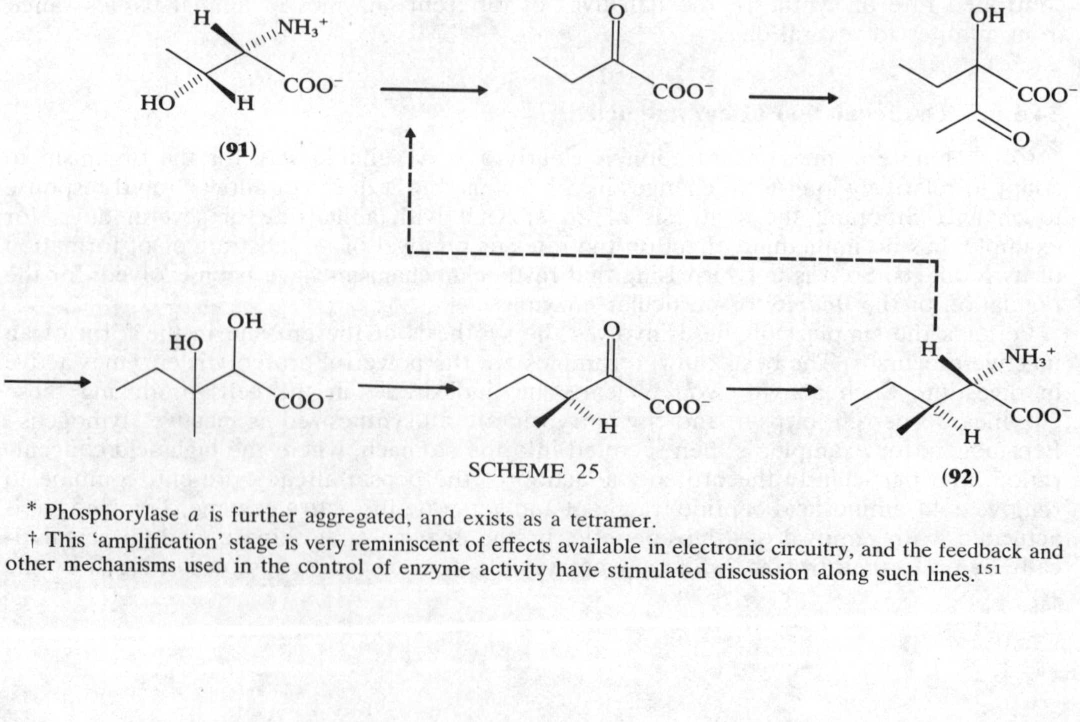

(91)

SCHEME 25

(92)

* Phosphorylase *a* is further aggregated, and exists as a tetramer.

† This 'amplification' stage is very reminiscent of effects available in electronic circuitry, and the feedback and other mechanisms used in the control of enzyme activity have stimulated discussion along such lines.[151]

generally have quite different three-dimensional structures (and are thus '*allosteric*'[152]). This suggests immediately that the inhibitor cannot bind at the active site, as classical (isosteric) inhibitors do. A great deal of evidence, including very specific X-ray structural evidence, confirms that regulatory enzymes have distinct catalytic (substrate binding) and allosteric (inhibitor binding) sites. The effects of inhibitor binding at the allosteric site are transmitted through the protein — they often have a readily detectable effect on subunit interaction, for example — *via* a series of conformation changes, which eventually affect the active site. In this way metabolites not directly involved in the reaction catalysed by a particular enzyme can regulate the activity of that enzyme,* modifying either substrate binding, or catalytic function, or both.[147,148]

The study of enzyme regulation and control is a young and rapidly expanding part of biochemistry, but already mechanisms of great sophistication have been identified. It is possible to distinguish mechanisms which are common to all enzymes, such as substrate specificity, optimum pH, and so on; mechanisms common to all organisms, which include feedback inhibition and repression; and mechanisms specific to higher organisms, where enzyme activity is further regulated, for example by the action of hormones. To take just one example, the whole complex network of reactions described above, which culminates in the phosphorylase-catalysed release of glucose from glycogen, can be triggered by the release of a few molecules of epinephrine.[148]

A picture as complex as this can only be put together by drawing on a great deal of painstaking work covering a whole range of disciplines, from physics on the one hand, through chemistry and biology to medicine. It is appropriate to end by noting that a great deal of contemporary scientific effort is regulated by enzymes.

References

1. All biochemistry text books have sections dealing with enzymes. For a particularly chemical treatment see H. R. Mahler and E. H. Corder, 'Biological Chemistry', Harper and Row, New York, 2nd edn., 1972. A valuable single-volume text devoted exclusively to the subject is M. Dixon and E. C. Webb, 'Enzymes', Longmans, London, 2nd edn., 1964 (new edition scheduled for 1978). The authoritative work is a multi-volume series 'The Enzymes', published by Academic Press. The current (third edition) runs to 13 volumes, published between 1970 and 1975, and many of these are cited individually below.
2. F. H. Westheimer, *Adv. Enzymol.*, 1962, **24,** 441. G. Popjak, in 'The Enzymes', ed. P. D. Boyer, Academic Press, New York, 3rd edn., vol. II, 1970, p. 115.
3. See, for example, A. Pollak, R. L. Baughn, and G. M. Whitesides, *J. Amer. Chem. Soc.*, 1977, **99,** 2366.
4. L. J. Reed and D. J. Cox, in 'The Enzymes', ed. P. D. Boyer, Academic Press, New York, 3rd edn., vol. I, 1970, p. 213.
5. 'Comprehensive Biochemistry', ed. M. Florkin and E. H. Stotz, Elsevier, Amsterdam, vol. IV, 1962; H. Neurath and R. L. Hill, 'The Proteins', Academic Press, New York, 1975.
6. (See Section 23.6, above.) B. Gutte and R. B. Merrifield, *J. Amer. Chem. Soc.*, 1969, **91,** 501; R. G. Denkewalter, D. F. Veber, F. W. Holly, and R. Hirschmann, *ibid.*, pp. 502, 503.
7. E. Katchalski, *Adv. Protein Chem.*, 1951, **6,** 123; E. Katchalski and M. Sela, *ibid.*, 1958, **13,** 243.
8. C. Tanford, 'The Hydrophobic Effect', Wiley, New York, 1973.
9. J. H. Fendler and E. J. Fendler, 'Catalysis in Micellar and Macromolecular Systems', Academic Press, New York, 1975.
10. M. Dixon and E. C. Webb, 'Enzymes', Longmans, London, 1958, p. 110.
11. D. N. Hague, 'Fast Reactions', Wiley, London, 1971.
12. G. G. Hammes and P. R. Schimmel, in 'The Enzymes', ed. P. D. Boyer, Academic Press, New York, 3rd edn., vol. II, 1970, p. 67.
13. H. Gutfreund, 'An Introduction to the Study of Enzymes', Blackwell, Oxford, 1965.
14. M. Eigen and L. de Maeyer, in 'Technique of Organic Chemistry', ed. S. L. Friess, E. S. Lewis, and A. Weissberger, Wiley, New York, 2nd edn., vol. VIII, 1963, p. 895.
15. A. F. Casy, 'PMR Spectroscopy in Medicinal and Biological Chemistry', Academic Press, New York, 1971. T. L. James, 'NMR in Biochemistry', Academic Press, New York, 1975.
16. T. C. Bruice and S. J. Benkovic, 'Bioorganic Mechanisms', Benjamin, New York, 1966.
17. W. P. Jencks, 'Catalysis in Chemistry and Enzymology', McGraw-Hill, New York, 1969.
18. (a) M. L. Bender, 'Mechanisms of Homogeneous Catalysis from Protons to Proteins', Wiley-Interscience, New York, 1971; (b) M. L. Bender and L. J. Brubacher, 'Catalysis and Enzyme Action', McGraw-Hill, New York, 1973.

* Either positively or negatively. The mechanism is just as suitable for activation as for inhibition.

19. A. J. Kirby, in 'Comprehensive Chemical Kinetics', ed. C. H. Bamford and C. F. H. Tipper, Elsevier, Amsterdam, 1972, vol. 10, chapter 2.
20. V. Gold, D. G. Oakenfull and T. Riley, *J. Chem. Soc. (B)*, 1968, 515.
21. L. L. Schaleger and F. A. Long, *Adv. Phys. Org. Chem.*, 1963, **1,** 1.
22. E. H. Cordes, *Progr. Phys. Org. Chem.*, 1967, **4,** 1; E. H. Cordes and H. G. Bull, *Chem. Rev.*, 1974, **74,** 581.
23. J. N. Bronsted and W. F. K. Wynne-Jones, *Trans. Faraday Soc.*, 1929, **25,** 59.
24. A. R. Fersht and A. J. Kirby, *Progr. Bioorganic Chem.*, 1971, **1,** 1.
25. 'Neighbouring Group Participation', ed. B. Capon, Plenum Press, New York, vol. I, 1976.
26. Ref. 16, p. 128.
27. W. P. Jencks and M. Gilchrist, *J. Amer. Chem. Soc.*, 1968, **90,** 2622.
28. T. C. Bruice and S. J. Benkovic, *J. Amer. Chem. Soc.*, 1963, **85,** 1.
29. A. R. Fersht and A. J. Kirby, *J. Amer. Chem. Soc.*, 1967, **89,** 4857.
30. G. A. Rogers and T. C. Bruice, *J. Amer. Chem. Soc.*, 1974, **96,** 2463.
31. A. J. Kirby and G. J. Lloyd, *J.C.S. Perkin II,* 1976, 1753.
32. T. H. Fife, *Adv. Phys. Org. Chem.*, 1975, **11,** 1.
33. J. W. Thanassi and T. C. Bruice, *J. Amer. Chem. Soc.*, 1966, **88,** 747.
34. M. F. Aldersley, A. J. Kirby and P. W. Lancaster, *J.C.S. Perkin II,* 1974, 1504.
35. M. Dixon and E. C. Webb, 'Enzymes', Longmans, London, 1958, p. 125.
36. A. R. Fersht and A. J. Kirby, *J. Amer. Chem. Soc.*, 1968, **90,** 5833.
37. M. L. Bender and L. J. Brubacher, ref. 18b; A. S. Mildvan, in 'The Enzymes', ed. P. D. Boyer, Academic Press, New York, 3rd edn., vol. II, 1970, p. 446.
38. B. M. Babior, *Accounts Chem. Res.*, 1975, **8,** 376; R. H. Abeles and D. Dolphin, *ibid.*, 1976, **9,** 114.
39. (a) R. Steinberger and F. H. Westheimer, *J. Amer. Chem. Soc.*, 1951, **73,** 429; (b) N. V. Raghavan and D. L. Leussing, *ibid.*, 1976, **98,** 723.
40. D. A. Buckingham, D. M. Foster, and A. M. Sargeson, *J. Amer. Chem. Soc.*, 1970, **92,** 5701; D. A. Buckingham, J. Dekkers, A. M. Sargeson, and M. Wein, *J. Amer. Chem. Soc.*, 1972, **94,** 4032.
41. D. A. Buckingham, F. R. Keene, and A. M. Sargeson, *J. Amer. Chem. Soc.*, 1974, **96,** 4981.
42. 'Porphyrins and Metalloporphyrins', ed. K. M. Smith, Elsevier, Amsterdam, 1975.
43. M. Dixon and E. C. Webb, Ref. 35, p. 133.
44. I. Fridovich, in 'The Enzymes', ed. P. D. Boyer, Academic Press, New York, 3rd edn., vol. VI, 1972, p. 255.
45. G. Hammons, F. H. Westheimer, K. Nakaoka, and R. Kluger, *J. Amer. Chem. Soc.*, 1975, **97,** 1568.
46. (a) D. M. Blow, in 'The Enzymes', ed. P. D. Boyer, Academic Press, New York, 3rd edn., vol. III, 1970, p.185; (b) G. P. Hess, *ibid.*, p. 213.
47. D. Eisenberg, in 'The Enzymes', ed. P. D. Boyer, Academic Press, New York, 3rd edn., vol. I, 1970, p. 1; D. M. Blow, *Ann. Rev. Biochem.*, 1970, **39,** 63.
48. R. A. Dwek, 'Nuclear Magnetic Resonance in Biochemistry: Application to Enzyme Systems', Oxford University Press, 1973, p. 101.
49. J. Kraut, in 'The Enzymes', ed. P. D. Boyer, Academic Press, New York, 3rd edn., vol. III, 1970, p. 165.
50. B. S. Hartley, *Phil. Trans. Roy. Soc. London*, 1970, **B257,** 77.
51. J. Kraut, in 'The Enzymes', ed. P. D. Boyer, Academic Press, New York, 3rd edn., vol. III, 1970, p. 547.
52. D. M. Blow, *Accounts Chem. Res.*, 1976, **9,** 145.
53. B. M. Anderson, E. H. Cordes, and W. P. Jencks, *J. Biol. Chem.*, 1961, **236,** 455.
54. K. N. G. Chiong, S. D. Lewis, and J. A. Shafer, *J. Amer. Chem. Soc.*, 1975, **97,** 418.
55. R. Huber, D. Kukla, W. Bode, P. Schwager, K. Bartels, J. Deisenhoffer, and W. Steigemann, *J. Mol. Biol.*, 1974, **89,** 73; R. M. Sweet, H. T. Wright, J. Janin, C. H. Chothia, and D. M. Blow, *Biochemistry*, 1974, **13,** 4212.
56. M. L. Bender and G. A. Hamilton, *J. Amer. Chem. Soc.*, 1962, **84,** 2570.
57. M. L. Bender, G. E. Clement, F. J. Kezdy, and H. d'A. Heck, *J. Amer. Chem. Soc.*, 1964, **86,** 3680.
58. J. F. Kirsch, in 'Advances in Linear Free Energy Relationships', ed. N. B. Chapman and J. Shorter, Plenum Press, London, 1972, p. 369.
59. M. Caplow and W. P. Jencks, *Biochemistry*, 1962, **1,** 883.
60. M. L. Bender and K. Nakamura, *J. Amer. Chem. Soc.*, 1962, **84,** 2577; A. Williams, *Biochemistry*, 1970, **9,** 3383.
61. J. Fastrez and A. R. Fersht, *Biochemistry*, 1973, **12,** 1067.
62. P. W. Inward and W. P. Jencks, *J. Biol. Chem.*, 1965, **240,** 1986.
63. A. L. Fink, *Biochemistry*, 1976, **15,** 1580.
64. W. W. Cleland, *Accounts Chem. Res.*, 1971, **8,** 145.
65. A. N. Glazer and E. L. Smith, in 'The Enzymes', ed. P. D. Boyer, Academic Press, New York, 3rd edn., vol. III, 1970, p. 502.
66. J. Drenth, J. N. Jansonius, R. Koekoek, and B. G. Wolthers, in 'The Enzymes', ed. P. D. Boyer, Academic Press, New York, 3rd edn., vol. III, 1970, p. 484.
67. L. J. Brubacher and M. L. Bender, *J. Amer. Chem. Soc.*, 1966, **88,** 5871.
68. G. Lowe and A. Williams, *Biochem. J.*, 1965, **96,** 189.
69. G. Lowe and A. Williams, *Biochem. J.*, 1965, **96,** 199.
70. G. Lowe and Y. Yuthavong, *Biochem. J.*, 1971, **124,** 117.

71. G. Lowe, *Tetrahedron*, 1976, **32,** 291.
72. J. R. Whitaker and M. L. Bender, *J. Amer. Chem. Soc.*, 1965, **87,** 2728.
73. J. S. Fruton, in 'The Enzymes', ed. P. D. Boyer, Academic Press, New York, 3rd edn., vol. III, 1970, p. 119; *Adv. Enzymology*, 1976, **44,** 1.
74. G. R. Delpierre and J. S. Fruton, *Proc. Nat. Acad. Sci. U.S.A.*, 1966, **56,** 1817.
75. M. L. Bender, Y.-L. Chow and F. Chloupek, *J. Amer. Chem. Soc.*, 1958, **80,** 5380.
76. M. F. Aldersley, A. J. Kirby, P. W. Lancaster, R. S. McDonald and C. R. Smith, *J.C.S. Perkin II*, 1974, 1487.
77. A. J. Kirby, R. S. McDonald and C. R. Smith, *J.C.S. Perkin II*, 1974, 1495.
78. J. A. Hartsuck and W. N. Lipscomb, in 'The Enzymes', ed. P. D. Boyer, Academic Press, New York, 3rd edn., vol. III, 1970, p. 1.
79. R. Breslow and D. Wernick, *J. Amer. Chem. Soc.*, 1976, **98,** 259.
80. B. L. Vallee, F. F. Riordan, and J. E. Coleman, *Proc. Nat. Acad. Sci. U.S.A.*, 1963, **49,** 109.
81. G. Weber, *Adv. Protein Chem.*, 1972, **26,** 243.
82. See Section 24.2.3.
83. J. P. Vincent and M. Lazdunski, *Biochemistry*, 1972, **11,** 2967.
84. Ref. 17, part II.
85. Ref. 17, p. 356.
86. M. Yarus, *Ann. Rev. Biochem.*, 1969, **38,** 841.
87. A. C. McLaughlin, J. S. Leigh, and M. Cohn, *J. Biol. Chem.*, 1976, **251,** 2777.
88. D. D. F. Shiao and J. M. Sturtevant, *Biochemistry*, 1969, **8,** 4910.
89. C. A. Blyth and J. R. Knowles, *J. Amer. Chem. Soc.*, 1971, **93,** 3021.
90. D. G. Oakenfull and D. E. Fenwick, *Austral. J. Chem.*, 1974, **27,** 2149.
91. Ref. 18a, p. 373.
92. F. Cramer and W. Dietsche, *Chem. Ber.*, 1959, **92,** 378, 1739.
93. D. J. Cram, R. C. Helgeson, L. R. Sousa, J. M. Timko, M. Newcomb, P. Moreau, F. de Jong, G. W. Gokel, D. H. Hofman, L. A. Domeier, S. C. Peacock, K. Madan, and L. Kaplan, *Pure Appl. Chem.*, 1975, **43,** 327.
94. Y. Chao and D. J. Cram, *J. Amer. Chem. Soc.*, 1976, **98,** 1015.
95. E. Fischer, *Ber.*, 1894, **27,** 2985.
96. Lucretius, 'De Rerum Natura', Book VI, lines 1084–1086. Translated by H. A. J. Munro, Bell, London, 1886.
97. L. Cunningham, *Comprehensive Biochemistry*, 1965, **16,** 85.
98. G. Hein and C. Niemann, *J. Amer. Chem. Soc.*, 1962, **84,** 4495.
99. M. L. Bender and K. C. Kemp, *J. Amer. Chem. Soc.*, 1957, **79,** 116.
100. R. Henderson, *J. Mol. Biol.*, 1970, **54,** 341.
101. R. L. Peterson, K. W. Hubele, and C. Niemann, *Biochemistry*, 1963, **2,** 942.
102. D. W. Ingles and J. R. Knowles, *Biochem. J.*, 1968, **108,** 561.
103. T. A. Steitz, R. Henderson, and D. M. Blow, *J. Mol. Biol.*, 1969, **46,** 337.
104. B. Keil, in 'The Enzymes', ed. P. D. Boyer, Academic Press, New York, 3rd edn., vol. III, 1970, p. 250.
105. B. S. Hartley and D. M. Shotton, in 'The Enzymes', ed. P. D. Boyer, Academic Press, New York, 3rd edn., vol. III, 1970, p. 322.
106. M. Krieger, L. M. Kay, and R. M. Stroud, *J. Mol. Biol.*, 1974, **83,** 209.
107. M. G. Rossmann, A. Liljas, C. I. Bränden, and L. J. Banaszak, in 'The Enzymes', ed. P. D. Boyer, Academic Press, New York, 3rd edn., vol. XI, 1975, p. 61.
108. D. E. Koshland, in 'The Enzymes', ed. P. D. Boyer, H. Lardy and K. Myrbäck, Academic Press, New York, 2nd edn., vol. I, 1959, p. 305; D. E. Koshland, *Proc. Nat. Acad. Sci. U.S.A.*, 1958, **44,** 98.
109. D. E. Koshland and K. E. Neet, *Ann. Rev. Biochem.*, 1968, **37,** 359.
110. N. Citri, *Adv. Enzymol.*, 1973, **37,** 397.
111. A. R. Fersht and Y. Requena, *J. Mol. Biol.*, 1971, **60,** 279.
112. T. Spencer and J. M. Sturtevant, *J. Amer. Chem. Soc.*, 1959, **81,** 1874.
113. E. Awad, Dissertation, University of Washington, Seattle, 1969; quoted in Ref. 97.
114. R. J. Foster, *J. Biol. Chem.*, 1961, **236,** 2461.
115. W. J. Ray and J. W. Long, *Biochemistry*, 1976, **15,** 3993; W. J. Ray, J. W. Long and J. D. Owens, *ibid.*, p. 4006.
116. S. P. Colowick, in 'The Enzymes', ed. P. D. Boyer, Academic Press, New York, 3rd edn., vol. IXB, 1973, p. 1.
117. T. A. Steitz, R. J. Fletterick, W. F. Anderson, and C. M. Anderson, *J. Mol. Biol.*, 1976, **104,** 197.
118. W. F. Anderson and T. A. Steitz, *J. Mol. Biol.*, 1975, **92,** 279.
119. W. P. Jencks, *Adv. Enzymol.*, 1975, **43,** 219.
120. C. D. Ritchie, in 'Solute-Solvent Interactions', ed. J. F. Coetzee and C. D. Ritchie, Dekker, New York, 1969, p. 219.
121. D. S. Kemp and K.G. Paul, *J. Amer. Chem. Soc.*, 1970, **92,** 2553; *ibid.*, 1975, **97,** 7305.
122. A. J. Kirby and A. G. Varvoglis, *J. Amer. Chem. Soc.*, 1967, **89,** 415.
123. T. C. Bruice and A Turner, *J. Amer. Chem. Soc.*, 1970, **92,** 3422.
124. T. C. Bruice, in 'The Enzymes', ed. P. D. Boyer, Academic Press, New York, 3rd edn., vol. II, 1970, p. 217.
125. A. J. Kirby and G. J. Lloyd, *J.C.S. Perkin II*, 1976, 1753.

126. J. F. Bunnett and C. F. Hauser, *J. Amer. Chem. Soc.*, 1965, **87,** 2214.
127. S. Milstien and L. A. Cohen, *J. Amer. Chem. Soc.*, 1972, **94,** 9158.
128. C. Danforth, A. W. Nicholson, J. C. James, and G. M. Loudon, *J. Amer. Chem. Soc.*, 1976, **98,** 4275; R. E. Winans and C. F. Wilcox, *ibid.*, p. 4281.
129. A. J. Kirby and P. W. Lancaster, *J.C.S. Perkin II*, 1972, 1206.
130. A. J. Kirby and G. J. Lloyd, *J.C.S. Perkin II*, 1976, 1753.
131. R. Huisgen and H. Ott, *Tetrahedron*, 1959, **6,** 253.
132. H. Pracejus, *Chem. Ber.*, 1959, **92,** 988.
133. T. Imoto, L. N. Johnson, A. C. T. North, D. C. Phillips, and J. A. Rupley in 'The Enzymes', ed. P. D. Boyer, Academic Press, New York, 3rd edn., vol. VII, 1972, p. 665.
134. C. C. F. Blake, D. F. Koenig, G. A. Mair, A. C. T. North, D. C. Phillips, and V. R. Sarma, *Nature*, 1965, **206,** 757; L. N. Johnson and D. C. Phillips, *ibid.*, 761.
135. C. C. F. Blake, L. N. Johnson, G. A. Mair, A. C. T. North, D. C. Phillips, and V. R. Sarma, *Proc. Roy. Soc.* (*B*), 1967, **167,** 378; D. C. Phillips, *Proc. Nat. Acad. Sci. U.S.A.*, 1967, **57,** 484.
136. J. A. Rupley, L. Butler, M. Gerring, F. J. Hartdegen, and R. Pecoraro, *Proc. Nat. Acad. Sci. U.S.A.*, 1967, **57,** 1088; D. M. Chipman, V. Grisaro, and N. Sharon, *J. Biol. Chem.*, 1967, **242,** 4388.
137. N. Sharon, *Proc. Roy. Soc* (*B*), 1967, **167,** 402.
138. I. I. Secemski, S. S. Lehrer, and G. E. Lienhard, *J. Biol. Chem.*, 1972, **247,** 4740.
139. M. Schindler, Y. Assaf, N. Sharon, and D. M. Chipman, *Biochemistry*, 1977, **16,** 423.
140. A. R. Fersht, *Proc. Roy. Soc.* (*B*), 1974, **187,** 397.
141. B. M. Dunn and T. C. Bruice, *Adv. Enzymol.*, 1973, **37,** 1.
142. C. A. Vernon, *Proc. Roy. Soc.* (*B*), 1967, **167,** 389.
143. T. H. Fife, *Adv. Phys. Org. Chem.*, 1975, **11,** 1.
144. M. A. Raftery and T. Rand-Meir, *Biochemistry*, 1968, **7,** 3281.
145. J. B. S. Haldane, 'Enzymes', Longmans, London, 1930.
146. A. B. Pardee, in 'The Enzymes', ed. P. D. Boyer, H. A. Lardy, and K. Myrbäck, Academic Press, New York, 2nd edn., vol. I, 1959, p. 681.
147. D. E. Koshland, in 'The Enzymes', ed. P. D. Boyer, Academic Press, New York, 3rd edn., vol. I, 1970, p. 342.
148. E. R. Stadtman, in 'The Enzymes', ed. P. D. Boyer, Academic Press, New York, 3rd edn., vol. I, 1970, p. 398.
149. R. T. Schimke, *Adv. Enzymol.*, 1973, **37,** 135.
150. B. N. Ames and R. G. Martin, *Ann. Rev. Biochem.*, 1964, **33,** 235.
151. D. E. Atkinson, in 'The Enzymes', ed. P. D. Boyer, Academic Press, New York, 3rd edn., vol. I, 1970, p. 461.
152. J. Monod, J. P. Changeux, and F. Jacob, *J. Mol. Biol.*, 1963, **6,** 306.

24.2
Chemistry of Other Proteins

D. T. ELMORE
The Queen's University of Belfast

24.2.1 POST-TRANSLATIONAL MODIFICATIONS OF PROTEINS

After completion of the translational process on ribosomes, proteins may undergo further modifications. These range in complexity from the oxidation of adjacent pairs of thiol groups to disulphide bonds, through the introduction of small groups such as hydroxyl, methyl, acetyl, carboxyl, and phosphate, to the attachment of sizable oligosaccharide units. In contrast, many proteins are biosynthesized and stored as biologically inactive molecules which undergo limited and controlled proteolytic degradation to give enzymes or polypeptide hormones.

24.2.1.1 Non-degradative modification of proteins

The reduced form of proteins that normally contain cystine probably have a conformation similar to the oxidized forms with pairs of thiol groups in juxtaposition so that correct pairing is favoured. Oxidized glutathione has been regarded as a candidate oxidant, but there is usually an excess of reduced glutathione in the cell owing to the presence of glutathione reductase. It is known that cysteamine (**1**) is oxidized to cystamine (**2**) by cysteamine oxidase in endoplasmic reticulum close to the cellular site of protein biosynthesis. Moreover, disulphide bonds are more abundant in extracellular than intracellular proteins and cysteamine oxidase is present in highest concentration in tissues producing extracellular proteins. It has been proposed,[1] therefore, that cystamine undergoes disulphide exchange reactions (*cf.* Section 23.3.8) with newly synthesized protein containing thiol groups as in equation (1). The resulting disulphide (**3**) then undergoes an internal exchange reaction (2). Both of the exchange reactions are believed to be enzyme catalysed.

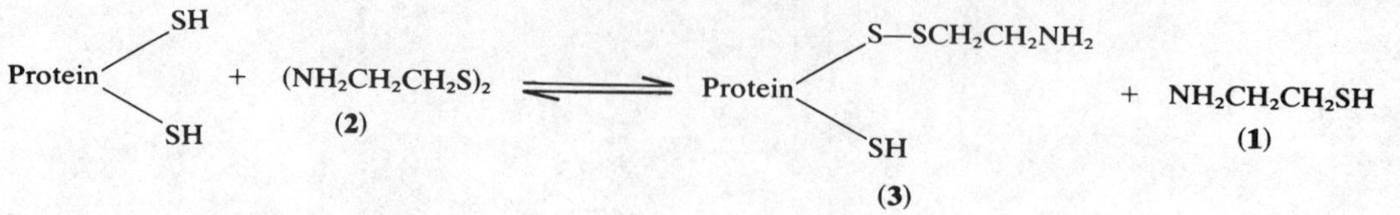
(1)

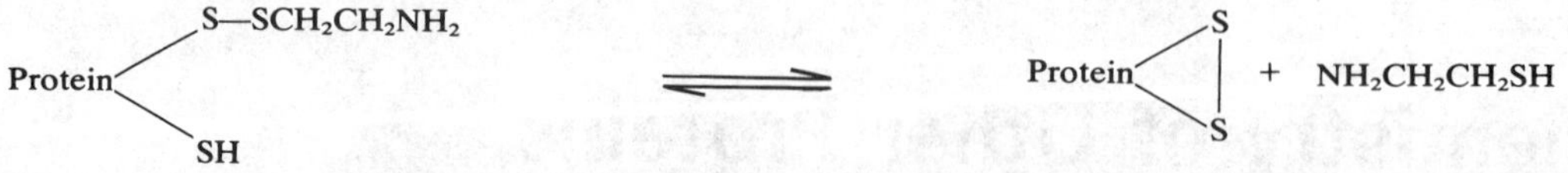
(2)

Post-translational hydroxylation involves proline and lysine. The former usually gives 4-hydroxyproline (**4**), but 3-hydroxyproline is also present to a smaller extent in collagen. Lysine gives rise to 5-hydroxylysine (**5**). Significantly it may be noted that both amino-acids have (*S,S*) configurations. Those proline residues which are hydroxylated occur in sequences of the type -Gly-X-Pro-Gly-. The enzyme responsible for this reaction, prolylhydroxylase, requires oxygen, Fe^{2+} ions, α-ketoglutarate, and ascorbic acid. The proposed mechanism[2] for the hydroxylation is given in Scheme 1. In collagen, but not elastin, some lysine residues in the triplets -Gly-X-Lys- are also hydroxylated. About 5–15 hydroxylysine residues are present per 1000 amino-acid residues, the richest source being bone collagen. A deficiency of lysylhydroxylase gives rise to a form of the Ehlers–Danlos syndrome in man in which the mechanical properties of the tissue are impaired.

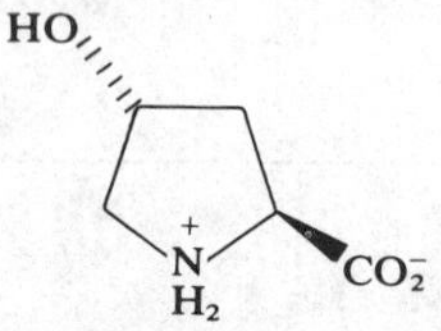
(**4**)

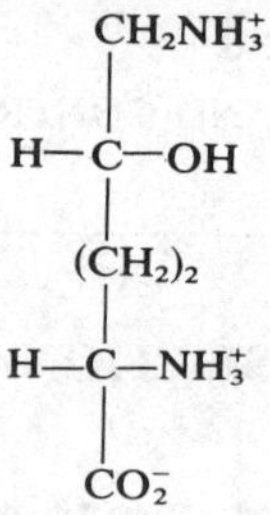
(**5**)

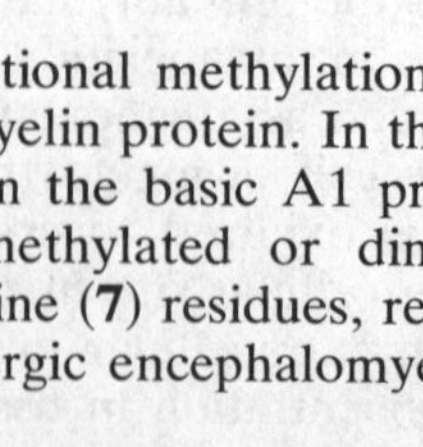
(**6**)

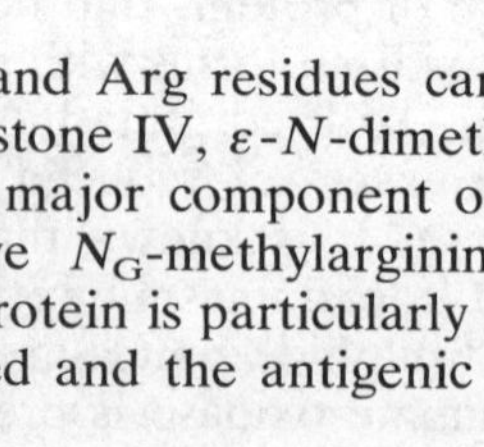
(**7**)

Post-translational methylation[3] of certain Lys and Arg residues can occur in histones and a basic myelin protein. In the arginine-rich histone IV, ε-*N*-dimethyl-lysine occurs at position-20. In the basic A1 protein, which is a major component of myelin, Arg^{107} is either monomethylated or dimethylated to give N_G-methylarginine (**6**) or N_G,N'_G-dimethylarginine (**7**) residues, respectively. This protein is particularly interesting because it induces allergic encephalomyelitis when injected and the antigenic determinant is the sequence:

114 115 116 117 118 119 120 121
Phe-Ser-Trp-Gly-Ala-Gly-Gln-Arg

Apparently, the presence of a methylated arginine residue is not directly responsible for the encephalitogenic behaviour. Separate enzymes bring about the methylation of lysine

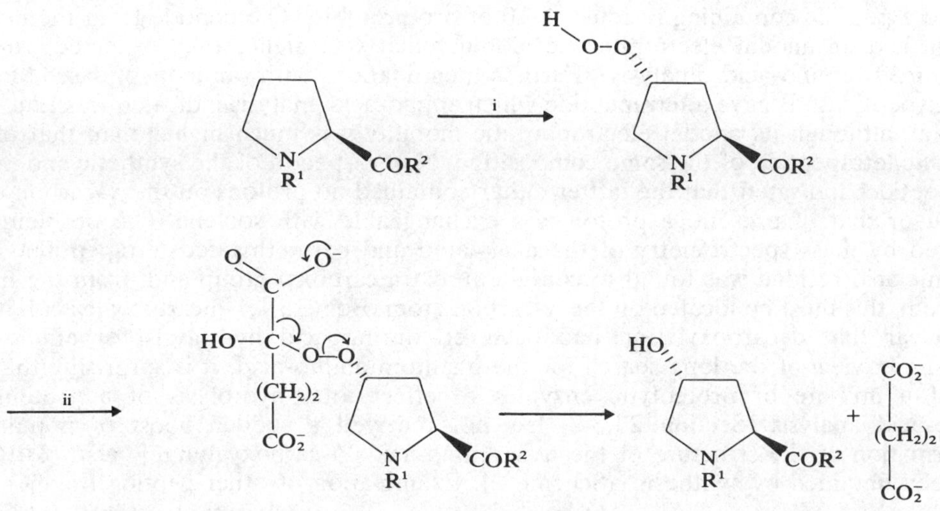

i, O_2, Fe^{2+}, prolylhydroxylase; ii, α-ketoglutarate (ascorbic acid is also required).

SCHEME 1

and arginine residues. Lysine residues are methylated by an enzyme in the nucleus, whereas the corresponding enzyme for arginine modification is found in the cytosol. Both enzymes, however, use *S*-adenosylmethionine as the source of methyl groups.

Acetylation of α-amino-groups and of ε-amino-groups of lysine residues occurs in some proteins. The substrates for acetylation range in size from α-melanocyte stimulating hormone:

Ac-Ser-Tyr-Ser-Met-Glu-His-Phe-Arg-Trp-Gly-Lys-Pro-Val-NH_2

to cytochrome *c* and histones. In histone IV from calf-thymus, the α-amino-group of the terminal serine residue and the side-chain of Lys^{16} are acetylated. Specific acetylases have been isolated and acetyl-CoA is the donor molecule. It should be noted that acetylation of the α-amino-group protects the protein from hydrolysis by aminopeptidases in the cell.

Another type of post-translational modification of proteins has been discovered recently. It has long been known that vitamin K deficiency disables the blood coagulation mechanism and various coumarin derivatives have been used as vitamin K antagonists to prolong the clotting time in patients with a history of coronary thrombosis. Clotting assays revealed that factor II (prothrombin) was one of the components of the blood coagulation system that was affected by vitamin K deficiency, but conventional amino-acid analysis of the normal and abnormal protein did not indicate any significant differences in composition. Prothrombin is a glycoprotein and vitamin K deficiency did not cause any obvious structural changes in the carbohydrate component. Immunochemical analysis demonstrated that the blood of vitamin K deficient animals contained normal levels of a protein which reacted with antibodies to prothrombin. Hence the transcriptional and translational processes were apparently unaffected. When activated by non-physiological catalysts such as trypsin or the venom from *Echis carinatus*, normal and abnormal prothrombin yielded identical amounts of thrombin with full clotting activity towards fibrinogen. One clear difference was established: normal prothrombin binds 10–12 Ca^{2+} ions whereas prothrombin from vitamin K deficient animals binds little or none. Calcium ions are known to be required for blood clotting and blood is usually collected in the presence of citrate or oxalate ions to prevent coagulation. In the course of determining the primary structure of prothrombin, a peptide containing the first 72 amino-acids from the *N*-terminus was obtained by cleavage with cyanogen bromide. This bound Ca^{2+} ions and the corresponding peptide from abnormal prothrombin did not, but the amino-acid compositions of the two peptides appeared to be identical. Further degradation by the action of trypsin yielded a peptide containing residues 4–10 of prothrombin. The peptide from the normal protein had an anodal electrophoretic mobility that was higher than expected from its determined amino-acid analysis. Further degradation with aminopeptidase M and carboxypeptidase B gave a tetrapeptide which amino-acid analysis indicated was Leu-Glu-Glu-Val, although its anodal electrophoretic mobility was much higher than that of the synthetic tetrapeptide of the same composition. N.m.r. spectra of the synthetic and native tetrapeptides indicated that the lattter either contained no protons on the γ-carbon atom of Glu or that, if present, a proton was exchangeable with solvent. The problem was resolved by mass spectrometry of the acetylated and permethylated tetrapeptide. Each glutamic acid residue was found to contain an extra carboxyl group and, from the n.m.r. spectrum, this must be located on the γ-carbon atom. Being a 1,1-dicarboxylic acid, it was now clear that decarboxylation had occurred during acid hydrolysis for amino-acid analysis. In view of the long search for the phantom amino-acid, it is surprising that the use of a mixture of proteolytic enzymes to effect total hydrolysis of a protein for amino-acid analysis (Section 23.3.2) has not received a sudden boost of popularity. Confirmation of the structure of the new amino-acid, γ-carboxyglutamic acid (**8**) (Gla), has been obtained by synthesis[4] (Scheme 2). Examination of other peptide fragments of

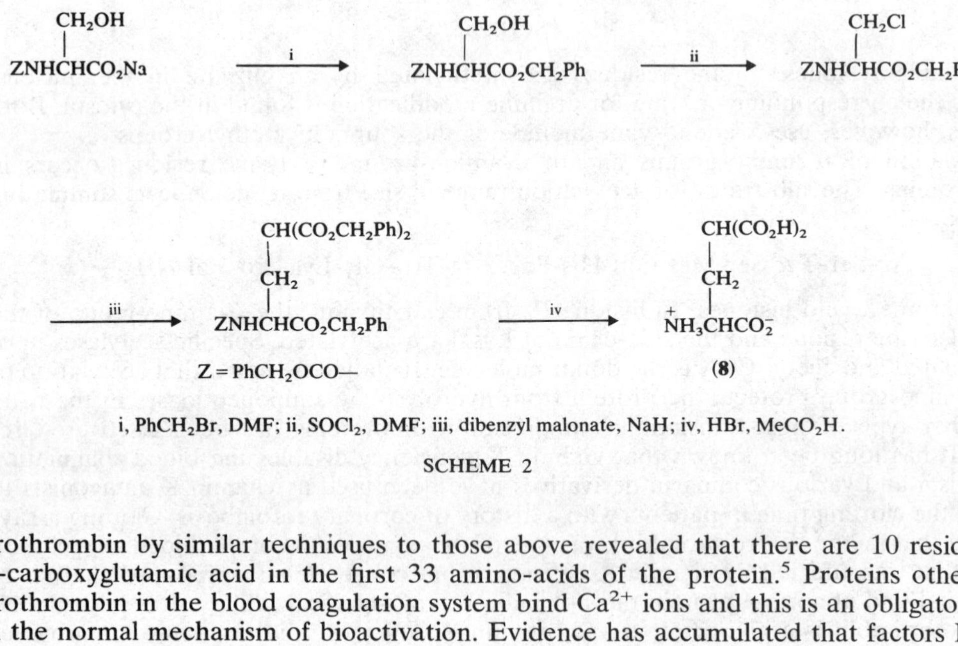

i, $PhCH_2Br$, DMF; ii, $SOCl_2$, DMF; iii, dibenzyl malonate, NaH; iv, HBr, $MeCO_2H$.

SCHEME 2

prothrombin by similar techniques to those above revealed that there are 10 residues of γ-carboxyglutamic acid in the first 33 amino-acids of the protein.[5] Proteins other than prothrombin in the blood coagulation system bind Ca^{2+} ions and this is an obligatory step in the normal mechanism of bioactivation. Evidence has accumulated that factors IX and X contain γ-carboxyglutamic acid in positions homologous to those in prothrombin (Figure 1).

Bovine prothrombin	**Ala-Asn-Lys- Gly-Phe-Leu-Gla-Gla- -Val-**
Factor IX	**Tyr-Asn-Ser- Gly-Lys- Leu-Gla-Gla-Phe-Val-**
Factor X	**Ala-Asn-Ser- -Phe-Leu-Gla-Gla- -Val-**
Bovine prothrombin	**-Arg-Lys-Gly-Asn-Leu-Gla-Arg-Gla-Cys-Leu-**
Factor IX	**-Arg- -Gly-Asn-Leu-Cys-**
Factor X	**-Lys-Gln-Gly-Asn-Leu-Gla-Arg-Gla-Cys-Leu-**
Bovine prothrombin	**-Gla-Gla-Pro-Cys-Ser-Arg-Gla-Gla-Ala-Phe-**
Factor X	**-Gla-Gla-Ala-Cys-Ser-Leu-Gla-Gla-Ala-Cys-**

Figure 1 *N*-Terminal sequences of prothrombin, factor IX, and factor X

The direct involvement of vitamin K in the biosynthesis of bioactivatable prothrombin was demonstrated[6] by measuring the incorporation of [^{14}C]bicarbonate into the protein contained in liver fractions from rats which were deficient in vitamin K. All incubations were carried out in the presence of cycloheximide to inhibit *de novo* protein biosynthesis. Isotopic incorporation was substantially increased in the presence of added vitamin K. Moreover, the presence of normal prothrombin in the incubation mixture was demonstrated by activation with factor Xa, factor V, phospholipid, and Ca^{2+} ions. Acid hydrolysis of the [^{14}C]-labelled prothrombin eliminated 50% of the radioactivity and amino-acid analysis showed that the remainder was located in glutamic acid, proving that vitamin K had participated in the incorporation of bicarbonate into γ-carboxyglutamic acid residues. Finally, it has been shown[7] that a mast cell protease rapidly and specifically cleaves a peptide containing the first 41 amino-acids and all the Gla residues from prothrombin, which is then inactive. A similar result was obtained with factor X.

The post-translational attachment of carbohydrate to proteins is of widespread occurrence. For example, all the known proteins in blood plasma are glycoproteins, with the exception of albumin and prealbumin. The glycoproteins cover a wide range of functions, including structural (*e.g.* collagen), enzymatic (*e.g.* thrombin), and hormonal (*e.g.* thyroglobulin). The carbohydrate content can range from about 0.5% in collagen to about 85% in blood-group substances. Only a relatively small number of carbohydrates, however, feature in glycoproteins and they include D-galactose, D-mannose, D-glucose, L-fucose, *N*-acetyl-D-glucosamine, *N*-acetyl-D-galactosamine, and sialic acids. L-Arabinose occurs in some plant glycoproteins.

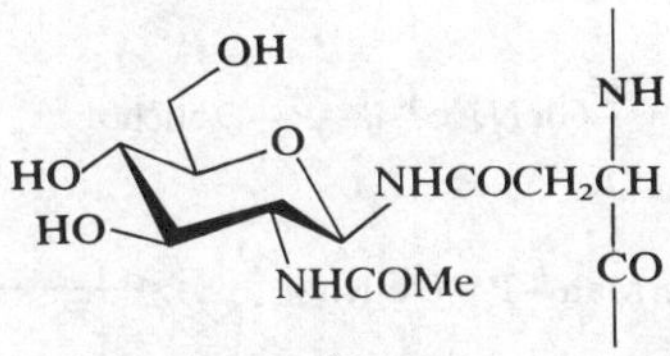

(9)

The most thoroughly studied glycoproteins contain carbohydrate attached to the side chains of asparagine, serine, threonine, and hydroxylysine, although cysteine and hydroxyproline are also known to be involved.[8] In the case of asparagine, the innermost carbohydrate unit is *N*-acetylglucosamine and attachment is by an *N*-β-glycosidic linkage to the amide group of asparagine (**9**). The asparagine always seems to occur in a sequence of the types Asn-X-Ser or Asn-X-Thr as in

-Val-Pro-Lys-Asn-Ile-Thr-Ser-	sheep pituitary luteinizing hormone
-Arg-Val-Glx-Asn-His-Thr-Glx-	sheep pituitary luteinizing hormone
-Glu-Pro-Ile-Asn-Ala-Thr-Leu-	sheep pituitary luteinizing hormone
-Ser-Asn-Ala-Thr-	bovine deoxyribonuclease

The carbohydrate component frequently contains only mannose and *N*-acetylglucosamine, but more complex examples are known which contain sialic acid, galactose, and fucose in addition to the above. For the simpler case, the general structure and the biosynthetic route[9] is given in Scheme 3. For most of the steps the hexose donor is either uridine diphosphate–*N*-acetylglucosamine (UDP–Glc*N*Ac) or guanosine diphosphate–mannose (GDP–Man), but the role of the phosphate ester of dolichol, a group of C_{80}–C_{110} polyprenols, should be noted.

No patterns are apparent relating either to the structures of oligosaccharides or the sequences around serine or threonine except that *O*-glycosides of these amino-acids tend

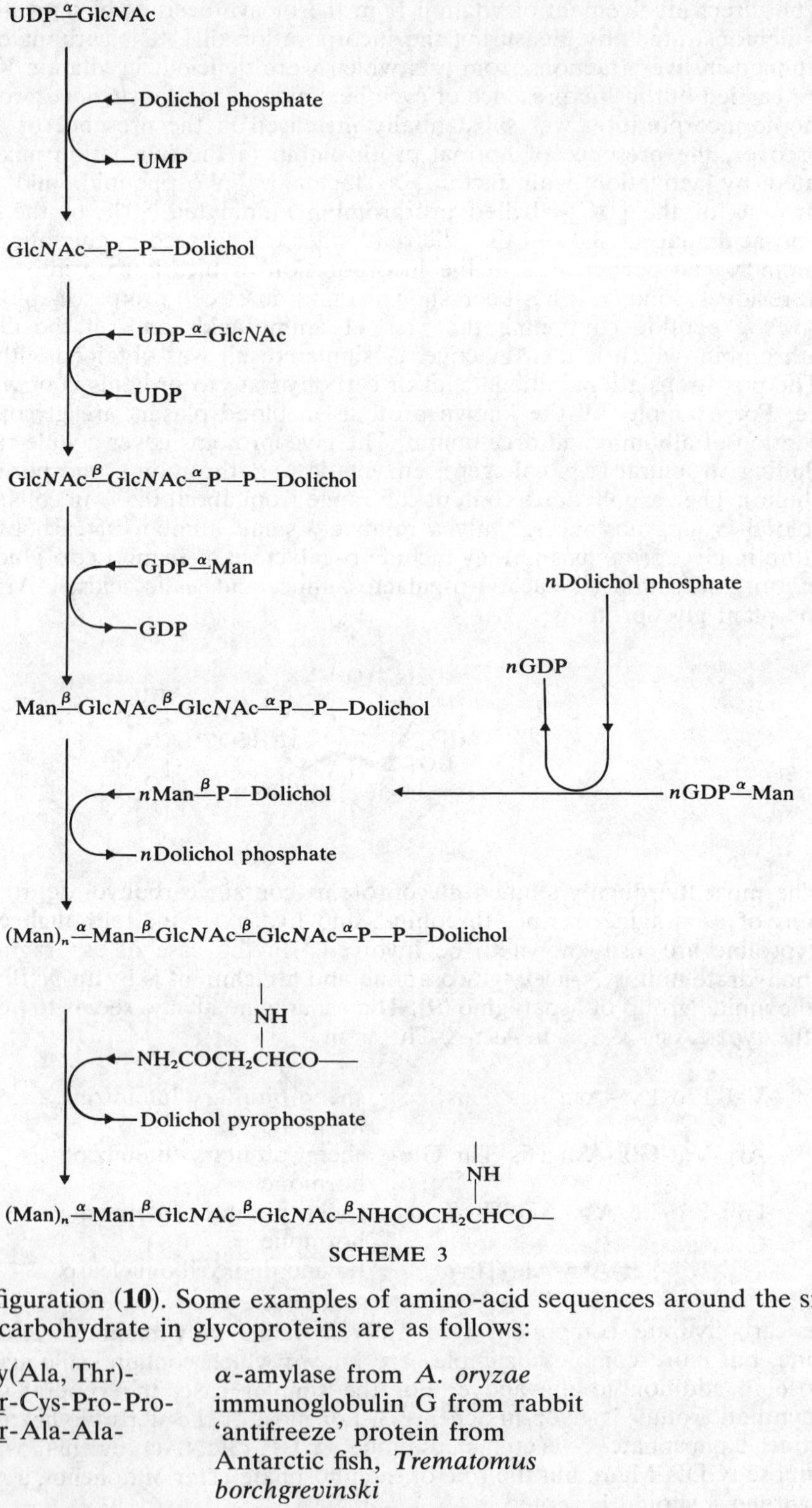

SCHEME 3

to have an α-configuration (**10**). Some examples of amino-acid sequences around the site of attachment of carbohydrate in glycoproteins are as follows:

-<u>Ser</u>- Glu-Asp-Gly(Ala, Thr)-	α-amylase from *A. oryzae*
-Ser- Lys-Pro- <u>Thr</u>-Cys-Pro-Pro-	immunoglobulin G from rabbit
-<u>Thr</u>-Ala-Ala- <u>Thr</u>-Ala-Ala-	'antifreeze' protein from Antarctic fish, *Trematomus borchgrevinski*

(The residue at which carbohydrate is attached is underlined.) *O*-Glycosides of serine and threonine are likely to undergo β-elimination under basic conditions.

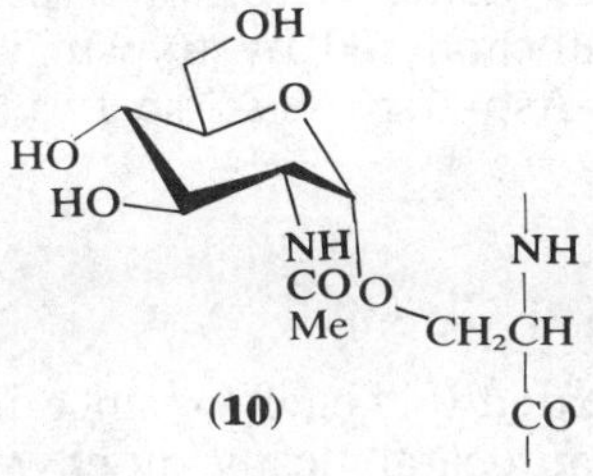

(10)

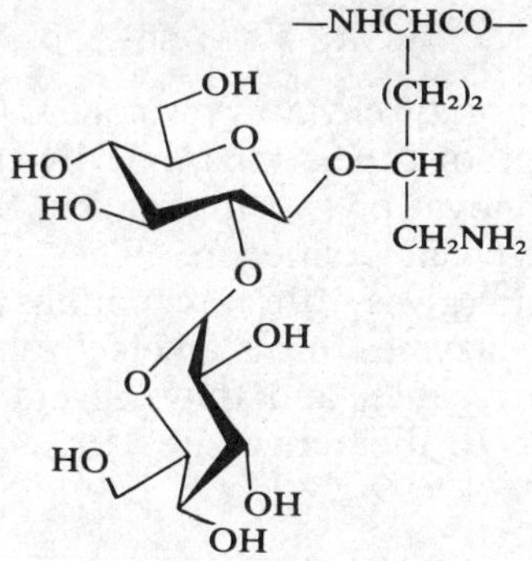

(11)

Attachment of carbohydrate to hydroxylysine occurs in collagen and a typical partial structure is given in (**11**).

The last type of non-degradative modification to be considered is phosphorylation. This usually involves the formation of a phosphomonoester on the side chain of serine or threonine catalysed by specific protein phosphokinases.[10] Glycogen phosphorylase, which catalyses the transfer of glucosyl units from glycogen to phosphate to form D-glucose-α-1-phosphate, exists in two forms. The less active phosphorylase *b* is converted into the more active phosphorylase *a* by the phosphorylation of a serine residue in the sequence

Arg Ile
Lys-Gln-Ile-Ser-Val-Arg

in a reaction catalysed by phosphorylase kinase. Phosphorylase kinase is itself converted from a less active to a more active form by phosphorylation in a reaction which requires ATP, cyclic AMP, and Mg^{2+} ions as well as the enzyme phosphorylase kinase kinase.

Phosphorylated enzymes also occur as intermediates in reactions catalysed by phosphatases and certain phosphate-transferring enzymes. Some phosphatases are phosphorylated by simply placing them in a phosphate buffer. The existence of a phosphorylated phosphatase as a true intermediate in the hydrolysis of phosphomonoesters demonstrates that the enzymatic mechanism is of the 'ping-pong' type. The sequences at the sites of phosphorylation of *E. coli* alkaline phosphatase and rabbit muscle phosphoglucomutase are respectively:

Val-Thr-Asp-<u>Ser</u>-Ala-Ala-Ser-

Thr-Ala-<u>Ser</u>-His-Pro-Gly-

24.2.1.2 Degradative modification of proteins

It is commonly found that biologically active proteins, especially those which are secreted by cells, such as enzymes and polypeptide hormones, are biosynthesized as inactive precursor molecules which become activated by specific hydrolytic removal of peptide fragments by proteolytic enzymes. This limited proteolysis effects a conformational change, bringing essential groups into the correct spatial relationship with each other, and occasionally the cleavage of a peptide bond may liberate an amino or carboxyl group that is essential for activity.

One of the simplest examples of limited proteolysis is the activation of trypsinogen to trypsin, which is catalysed by enteropeptidase and autocatalysed by trypsin itself. The process involves the cleavage of a hexapeptide Val-Asp-Asp-Asp-Asp-Lys from the *N*-terminus of trypsinogen:

$$\text{Val-Asp-Asp-Asp-Asp-Lys}\overset{\downarrow}{-}\text{Ile-Val-Gly-}\ldots$$

The activation of chymotrypsinogen A is more complicated (Scheme 4), since it involves the cleavage of four bonds, one by trypsin and three autocatalytically by chymotrypsin, and the removal of two dipeptides, Ser^{14}-Arg^{15} and Thr^{147}-Asn^{148}, from the interior of the single-chain zymogen. The existence of disulphide bonds Cys^{1} Cys^{122} and Cys^{136} Cys^{201} gives a three-chain enzyme. The protonated α-amino-group of *N*-terminal Ile in both enzymes forms an electrovalent bond with the carboxylate ion in the side chain of an Asp residue, and this triggers small conformational changes that make the final adjustments to the active site.

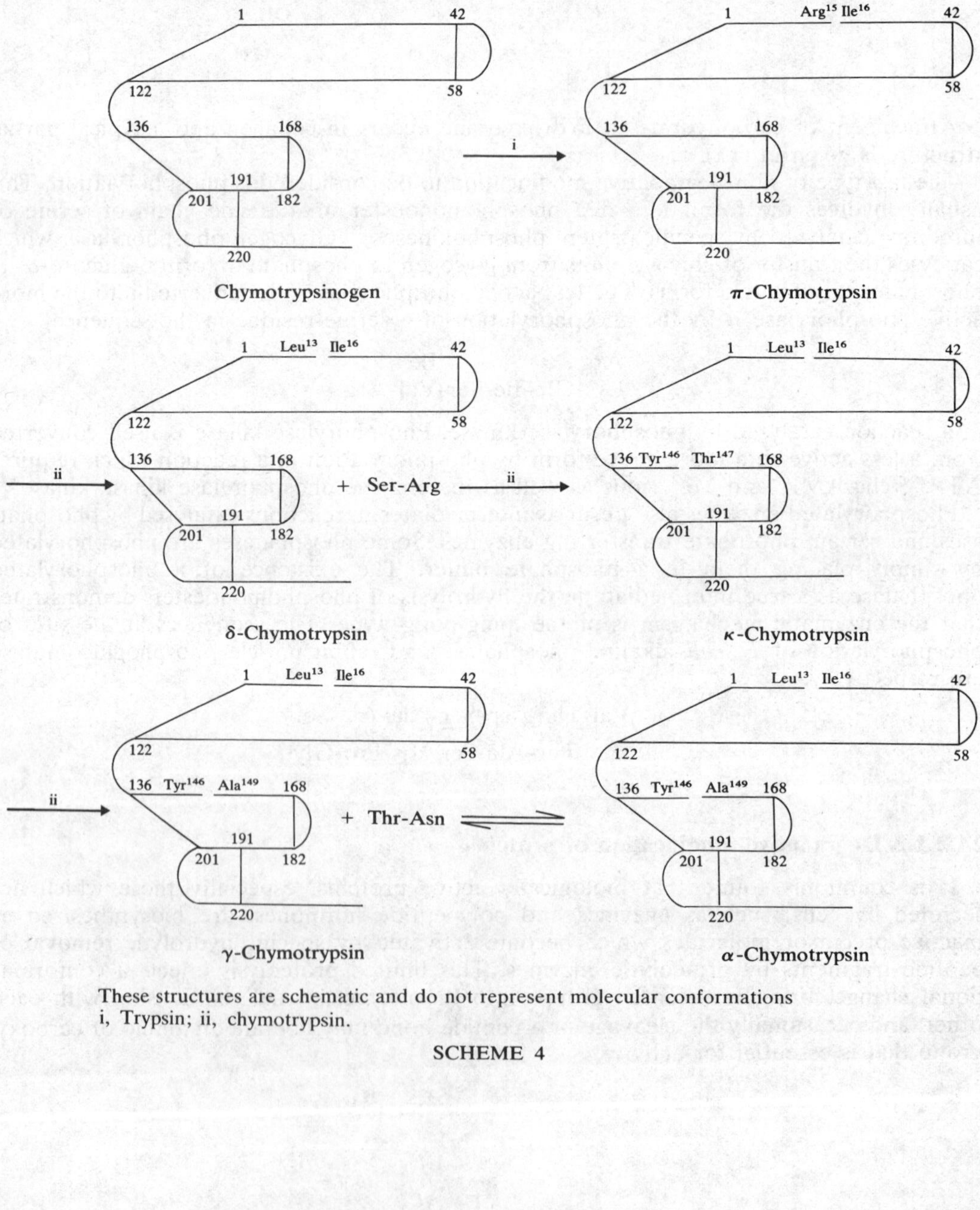

These structures are schematic and do not represent molecular conformations
i, Trypsin; ii, chymotrypsin.

SCHEME 4

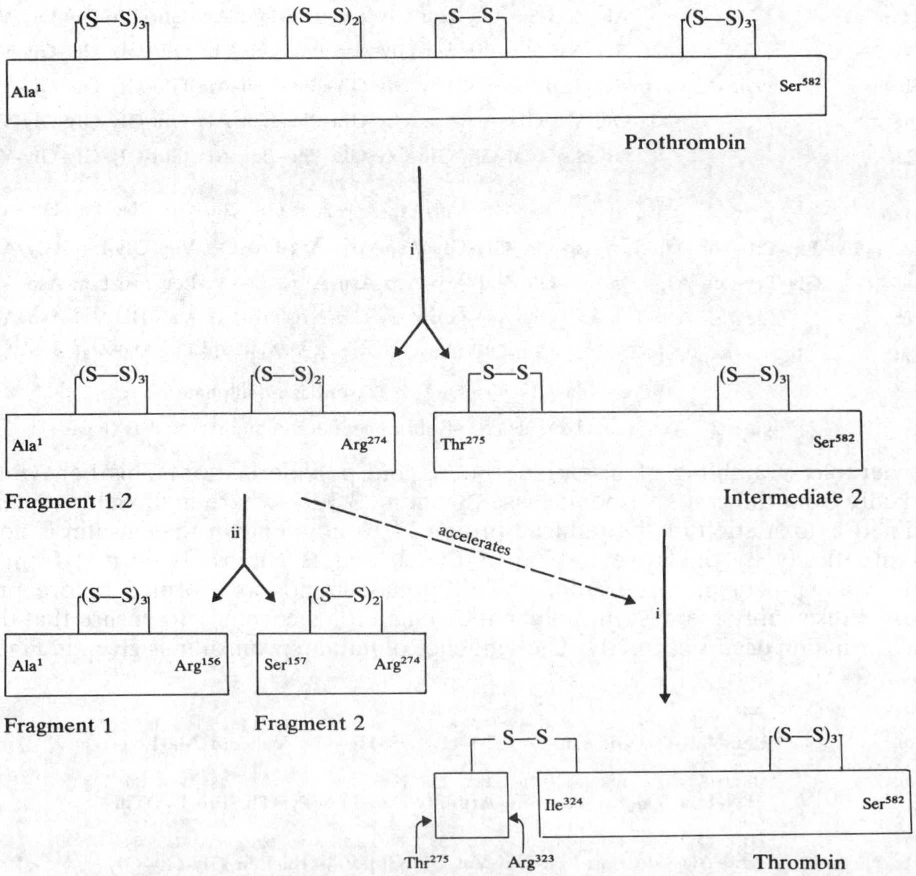

i, Factor Xa, factor V, phospholipid, Ca^{2+}; ii, thrombin.

SCHEME 5

Most of the components of the blood coagulation[11] and complement[12] systems are biosynthesized as inactive precursors for obvious biological reasons. Two examples will suffice to illustrate the processes. The importance of γ-carboxyglutamic acid residues near the *N*-terminus of prothrombin has been described above (Section 24.2.1.1). Prothrombin undergoes three cleavages (Scheme 5), one autocatalytically by thrombin and two by factor Xa, leaving a two-chain molecule linked by a single disulphide bond. Interestingly, one of the fragments removed in the activation process, fragment 1.2, accelerates the cleavage of the final Arg–Ile bond by factor Xa. The other example from the blood coagulation system concerns fibrinogen, the substrate for thrombin. This protein contains six polypeptide chains, two A, two B, and two C, held together by about 28 disulphide bonds. The conversion of fibrinogen into fibrin in the clotting process involves the cleavage of two peptides from the *N*-termini of the A and B chains. In each case the bond cleaved is Arg–Gly. It is interesting that these two peptides, fibrinopeptides A and B, differ in size and sequence between different species (Figure 2). This rapid evolutionary rate has focused attention on fibrinogen as a molecule suitable for constructing phylogenetic trees.[13]

Some smaller proteins and peptides are also liberated by limited proteolysis of larger molecules. Insulin consists of two chains linked through two disulphide bonds. The A chain contains 21 amino-acids and has an additional intramolecular disulphide bond, whereas the B chain has 30 amino-acids. Although both chains were chemically synthesized with protected cysteine side-chains in satisfactory yield, deprotection and oxidation to form disulphide bonds gave very poor yields of insulin. Presumably there was

Man	Ala-Asp-Ser*-Gly-Glu-Gly-Asp-Phe-Leu-Ala-Glu-Gly-Gly-Gly-Val-Arg
Ox	Glu-Asp-Gly-Ser-Asp-Pro-Pro-Ser-Gly-Asp-Phe-Leu-Thr-Glu-Gly-Gly-Gly-Val-Arg
Sheep	Ala-Asp-Asp-Ser-Asp-Pro-Val-Gly-Gly-Glu-Phe-Leu-Ala-Glu-Gly-Gly-Gly-Val-Arg
Pig	Ala-Glu-Val-Gln-Asp-Lys-Gly-Glu-Phe-Leu-Ala-Glu-Gly-Gly-Gly-Val-Arg
Cat	Gly-Asp-Val-Gln-Glu-Gly-Glu-Phe-Ile- Ala-Glu-Gly-Gly-Gly-Val-Arg
Man	Pyr-Gly-Val-Asn-Asp-Asn-Glu-Glu-Gly-Phe- Phe-Ser-Ala-Arg
Ox	Pyr-Phe-Pro-Thr-Asp-Tyr†-Asp-Glu-Gly-Glu-Asp-Asp- Arg-Pro-Lys-Val- Gly-Leu-Gly-Ala-Arg
Sheep	Gly-Tyr-Leu-Asp-Tyr†-Asp-Glu-Val-Asp-Asp-Asn-Arg-Ala-Lys-Leu-Pro-Leu-Asp-Ala-Arg
Pig	Ala-Ile -Asp-Tyr†-Asp-Glu-Asp-Glu-Asp- Gly-Arg-Pro-Lys-Val- His-Val-Asp-Ala-Arg
Cat	Ile-Ile-Asp-Tyr- Tyr-Asp-Glu-Gly-Glu-Glu-Asp-Arg-Asp-Val-Gly-Val-Val-Asp-Ala-Arg

* Present as *O*-phosphate † Present as *O*-sulphate

Figure 2 Amino-acid sequences of fibrinopeptides A (upper) and B (lower)

considerable scrambling of disulphide bonds and polymerization. This behaviour differs markedly from the case of ribonuclease (Section 23.3.9), where high yields of enzyme are obtained by oxidation of the reduced protein. It is now known that insulin is not formed biosynthetically by oxidative linking of the A and B chains. It derives from a larger single-chain precursor, proinsulin, and disulphide bonds are formed before proteolysis occurs. Presumably the nascent polypeptide chain folds correctly to ensure that disulphide bond formation occurs correctly. The sequence of human proinsulin is given in Figure 3 (see also p. 245).

1 2 3 4 5 6 7 8 9 10 11 12 13 14 15
Phe-Val-Asn-Gln-His-Leu-Cys-Gly-Ser-His-Leu-Val-Glu-Ala-Leu-

16 17 18 19 20 21 22 23 24 25 26 27 28 29 30
-Tyr-Leu-Val-Cys-Gly-Glu-Arg-Gly-Phe-Phe-Tyr-Thr-Pro-Lys-Thr-

31 32 33 34 35 36 37 38 39 40 41 42 43 44 45
-Glu-Ala-Glu-Asp-Leu-Gln-Val-Gly-Gln-Val-Glu-Leu-Gly-Gly-Gly-

46 47 48 49 50 51 52 53 54 55 56 57 58 59 60
-Pro-Gly-Ala-Gly-Ser-Leu-Gln-Pro-Leu-Ala-Leu-Glu-Gly-Ser-Leu-

61 62 63 64 65 66 67 68 69 70 71 72 73 74 75
-Gln-Gly-Ile-Val-Glu-Gln-Cys-Cys-Thr-Ser-Ile-Cys-Ser-Leu-Tyr-

76 77 78 79 80 81 82
-Gln-Leu-Glu-Asn-Tyr-Cys-Asn

Figure 3 Sequence of hyman proinsulin. Disulphide bonds are between Cys^{7} and Cys^{68}, Cys^{19} and Cys^{81}, Cys^{67} and Cys^{72}. Residues 1–30 comprise the B chain and residues 62–82 the A chain of human insulin

Angiotensin II is an octapeptide which is strongly hypertensive. The action of a specific proteinase, renin, in kidney liberates a decapeptide, angiotensin I, from a glycoprotein in blood plasma (Scheme 6). Treatment of the glycoprotein with trypsin or alkali under mild conditions liberates a tetradecapeptide and it has been suggested that the side chain of the *C*-terminal Ser residue of the tetradecapeptide participates in an ester bond which also involves the carboxyl group of arginine or lysine in the glycoprotein. The tetradecapeptide is also cleaved to give angiotensin I *in vitro*. Neither the tetradecapeptide or decapeptide are hypertensive. A specific converting enzyme, which is present at high levels in the lung but is widely distributed at lower concentrations in other tissues and blood plasma, cleaves the *C*-terminal dipeptide and forms the angiotensin II.

Bradykinin is a nonapeptide which is hypotensive, bronchoconstrictive, and produces pain. It has the sequence

Arg-Pro-Pro-Gly-Phe-Ser-Pro-Phe-Arg

and is produced by the action of plasma kallikrein, an enzyme which resembles trypsin in

Glycoprotein
|
Arg-Asp-Val-Tyr-Ile-His-Pro-Phe-His-Leu-Leu-Val-Tyr-Ser

↓ i

Arg-Asp-Val-Tyr-Ile-His-Pro-Phe-His-Leu   Angiotensin I

↓ ii

Arg-Asp-Val-Tyr-Ile-His-Pro-Phe   Angiotensin II

i, renin; ii, angiotensin converting enzyme.

SCHEME 6

specificity, on an acidic glycoprotein in blood plasma. Another kallikrein from submaxillary gland produces kallidin, which is a decapeptide and consists of bradykinin with an additional lysine residue at the *N*-terminus (see also p. 248).

24.2.2 PROTEINS WHICH INTERACT WITH SMALL MOLECULES

The two proteins myoglobin and haemoglobin play complementary roles in the transport and storage of oxygen. Myoglobin (Mb) has one polypeptide chain, which is about 78% helical, and comprises 153 amino-acids and one haem group (**12**). Haemoglobin

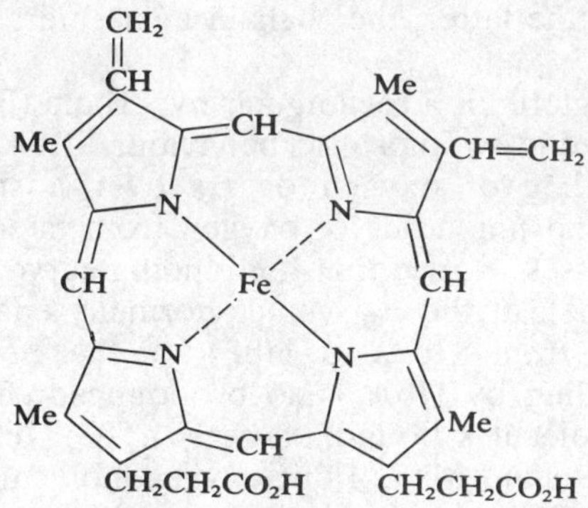

(**12**)

(HbA) contains four polypeptide chains, two α (141 amino-acids) and two β (146 amino-acids), and each is associated with a haem group. Fe(II) is normally hexa-coordinate and in the oxygenated forms of the haemoproteins the fifth and sixth positions are occupied by O_2 and a nitrogen atom in the imidazole ring (His^{93} in Mb, His^{87} in Hbα and His^{92} in Hbβ). The Fe(II) atom is diamagnetic and lies in the plane of the porphyrin ring. In deoxyhaemoglobin, however, Fe(II) is penta-coordinate, paramagnetic, and is displaced out of the plane of the porphyrin ring towards the imidazole nitrogen atom. As will be seen later, this small change triggers much greater conformational changes. In foetal haemoglobin (HbF), the two β chains are replaced by γ-chains of the same length. There is an additional minor constituent of adult blood, HbA_2, which contains two δ chains in place of the β chains. Analysis of the homology between the various chains suggests that

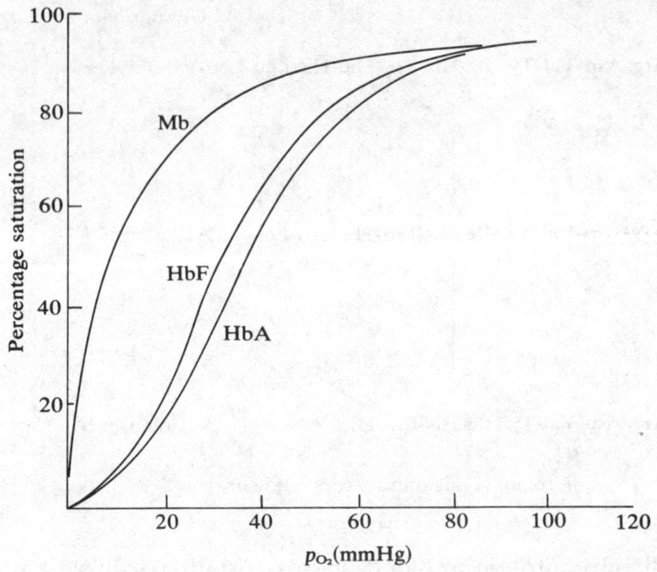

Figure 4 Oxygen saturation curves for myoglobin (Mb), adult haemoglobin (HbA), and foetal haemoglobin (HbF)

all have evolved from the same primaeval precursor gene. The single-chain myoglobin is regarded as the oldest protein and it is significant that the four-chain structure of haemoglobin did not appear, presumably by gene duplication, until after the lamprey in the evolutionary time scale. Comparison of α chains on the one hand with β, γ, and δ chains on the other shows 76–95 differences. Comparison of γ chains with β and δ chains shows 36–48 differences, while human β and δ chains differ at only 10 loci. This suggests that the subunits of the earliest haemoglobins consisted of an α chain and the precursor of the β, γ, and δ chains. From the latter, the γ chain next emerged and, most recently, the β and δ chains diverged.

The binding curve for myoglobin is a rectangular hyperbola (Figure 4), whereas those of haemoglobins are sigmoidal due to allosteric behaviour. This is a biological advantage because loading and unloading of oxygen occurs over a small range of p_{O_2}. Thus reoxygenation in the lungs and unloading of oxygen from haemoglobin to myoglobin in the tissues are efficient processes. Notice that the binding curve for HbF lies to the left of that for HbA. This means that, at the p_{O_2} values normally encountered in the placenta, oxygen will be transferred from HbA to HbF for the benefit of the foetus. The cooperativity of oxygen binding by HbA is so pronounced that if a molecule of HbA already has three oxygen molecules bound, there is a 70 : 1 chance that it will become saturated in preference to the acquisition of one oxygen molecule by deoxyhaemoglobin.

The binding curve for oxygen by haemoglobin is sensitive to pH; at a given p_{O_2} value, the affinity for oxygen decreases as the pH decreases (Bohr effect). Glycolysis is an anaerobic process and leads to the formation of lactate and carbon dioxide. Both tend to lower pH and facilitate the unloading of oxygen from oxyhaemoglobin where the need exists. Conversely, deoxyhaemoglobin has some slightly stronger basic groups (imidazole-N of His^{146} in the β chains and His^{122} in the α chains and the amino-group of Val^{1} in the α chains) than oxyhaemoglobin and so deoxyhaemoglobin binds protons as oxygen is released. This is important for transporting carbon dioxide back to the lungs. Formation of bicarbonate from carbon dioxide and water is catalysed by carbonic anhydrase in the erythrocytes and the bicarbonate ions can bind to protonated groups in the deoxyhaemoglobin. When the deoxyhaemoglobin is recharged with oxygen in the lungs the Bohr effect causes the unloading of bicarbonate and carbonic anhydrase catalyses the formation of carbon dioxide which is exhaled. Carbon dioxide is also transported by deoxyhaemoglobin by forming carbamic acid derivatives with amino-groups in the protein (equation 3).

Although oxyhaemoglobin also binds carbon dioxide, the capacity of deoxyhaemoglobin is greater because amino groups are more accessible.

$$\text{Hb}\sim\sim NH_2 + CO_2 \rightleftharpoons \text{Hb}\sim\sim NHCO_2H \qquad (3)$$

Deoxyhaemoglobin, but not oxyhaemoglobin, binds 2,3-diphosphoglycerate. This compound is produced from 1,3-diphosphoglycerate in erythrocytes in a shunt off the glycolysis sequence. Consequently, anaerobic conditions favour glycolysis and the production of 2,3-diphosphoglycerate. A higher concentration of this tends to shift the equilibrium towards deoxyhaemoglobin with unloading of oxygen.

It was mentioned above that binding of oxygen to the iron atom of a haem group converts the latter from a non-planar to a planar configuration. This triggers other conformational changes which result in cooperative interaction between subunits of haemoglobin so that once some oxygen is bound, the binding of additional molecules is favoured. We are now in a position to consider this in more detail. The penultimate residue in each of the α and β chains is a tyrosine residue and, in deoxyhaemoglobin, the aromatic side chains of Tyr^{140} and Tyr^{145} in the α and β chains respectively are buried in hydrophobic pockets. When the Fe(II) atom comes into the plane of the porphyrin ring system on binding oxygen, the side chain of the penultimate tyrosine residue in the same subunit is flipped out of the pocket and exposed to solvent. In deoxyhaemoglobin there is electrovalent bonding between the two α chains involving the NH_3^+ group of Val^1 in one chain and the carboxylate group of Arg^{141} in the other. There is also electrovalent bonding between α and β chains. The ε-NH_3^+ group of Lys^{40} in the α chain is adjacent to the carboxylate group of His^{146} in the β chain. As soon as a subunit has acquired an oxygen molecule the interchain electrovalent bonding in which it participated is broken. Consequently, adjacent subunits are thermodynamically influenced to assume the conformation of a subunit in the oxygenated form. The reason for the cooperative behaviour of haemoglobin in binding oxygen is therefore apparent. Perutz, who has been responsible for the elucidation of its three-dimensional structures and allosteric behaviour,[14] has aptly described haemoglobin as a molecular lung.[15]

Numerous abnormal human haemoglobins have been sequenced and some have been studied by X-ray crystallographic analysis so that it is possible to explain the pathological consequences of certain genetic errors at the molecular level. Sickle-cell anaemia, so called because the erythrocytes of patients collapse to a sickle-shape at low p_{O_2} values, is responsible for the deaths of about 80 000 children per annum. The abnormal haemoglobin, HbS, contains Val^6 in place of Glu^6 in the β-chains. The deoxygenated form of HbS appears to aggregate to a rather insoluble polymer. One proposed method of treatment involves administration of cyanate at low concentrations. This is believed to cause carbamoylation of the amino groups of Val^1 in the α chain and Val^1 in the β chains. The former are involved in intrachain bonding in deoxyhaemoglobin while the latter form electrovalent bonds with 2,3-diphosphoglycerate. Carbamoylation will prevent both interactions, thus favouring a shift towards the conformation of oxyhaemoglobin. The risk of aggregation is thereby diminished.

Some abnormal haemoglobins have either a low or a high affinity for oxygen. In oxy-HbA, the amide group of Asn^{102} in the β chain is hydrogen-bonded to the carboxyl group of Asp^{94} in the α-chain, but this hydrogen bond does not exist in deoxy-HbA. In Hb_{Kansas}, Asn^{102} in the β chain is replaced by Thr which does not form a hydrogen bond with Asp^{94} in the α chain. Hence the oxygenated form of Hb_{Kansas} is relatively unstable and has an abnormally low affinity for oxygen. In contrast, $Hb_{Kempsey}$ has an abnormally high affinity for oxygen and this is related to the substitution of Asn^{99} for Asp^{99} in the β chains. In deoxy-HbA, the carboxyl group of Asp^{99} is hydrogen-bonded to the phenolic hydroxyl group of Tyr^{42} in the α chains, but in oxy-HbA the side chain of Asp^{99} is unbonded. In deoxy-$Hb_{Kempsey}$, the amide group of Asn^{99} is unable to bond to Tyr^{42} in the α chain. Hence the deoxygenated form is relatively unstable.

Some abnormal haemoglobins are associated with methaemoglobinaemia in which Fe(II) is oxidized to Fe(III) and the haemoglobin no longer functions as an oxygen-transporting molecule. Mutations involving His^{87} in the α chain or His^{92} in the β chain, both of which are ligands for haem iron, are especially important in this connection. In Hb_{Iwate}, Tyr replaces His^{87} in the α chains whereas in $Hb_{Hyde\ Park}$, Tyr replaces His^{92} in the β chains. The phenolic hydroxyl of the tyrosine in either case probably forms an electrovalent bond with Fe^{3+}, thus stabilizing this state.

A complete contrast to haemoglobin is provided by avidin,[16] a protein that constitutes 0.05% of egg white. Although it exists as a tetramer like haemoglobin, it binds biotin (**13**)

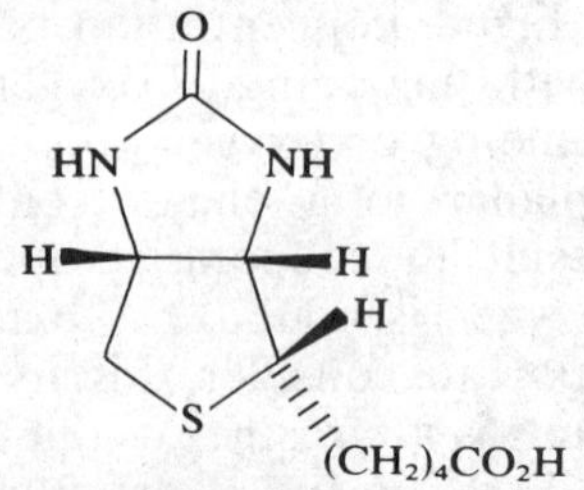

(**13**)

much more firmly ($K \approx 10^{-15}$ mol l^{-1}) than haemoglobin binds oxygen. Moreover, avidin does not exhibit cooperativity. The biological function of avidin is uncertain, although it may serve as an antibacterial agent. Interestingly, some species of *Streptomyces* produce a biotin-sequestering protein, streptavidin, which has antibacterial activity against gram-negative bacteria. The primary structure of avidin has been determined and it shows some resemblance to that of lysozyme, an enzyme from various sources including egg white that causes cell lysis of bacteria. Avidin contains 128 amino-acids and lysozyme contains 129. If lysozyme is regarded as having two extra residues at the *N*-terminus and one deletion near the *C*-terminus, there are 15 identities between the two proteins from hens' egg white and about as many conservative substitutions. Avidin, unlike lysozyme, is a glycoprotein with a carbohydrate of unknown structure attached to Asn.[17] The nature of the binding site of avidin is not certain, but modification with *N*-bromosuccinimide or mild oxidation with periodate as well as measurement of ultraviolet difference spectra suggest that tryptophan is implicated. The affinity of avidin for numerous analogues of biotin has been determined and, since quite simple models like urea and hexanoic acid are weakly bound, it is likely that almost every atom in biotin enters into some type of bonding with the protein.

Blood plasma albumin exhibits binding behaviour that differs from both haemoglobin and avidin. It is a monomeric protein which is able to bind a wide variety of cations, anions, and small neutral molecules. Not only normal components of plasma such as Ca^{2+} ions, fatty acids, and thyroxine are bound, but also numerous abnormal components such as barbiturates, sulphonamides, and penicillin. Albumin is the major component of plasma and apparently functions as a general transport system as well as helping to maintain the correct osmotic pressure of plasma for the maintenance of the cellular components of blood.

Numerous proteins bind metal ions. Some act as biological stores of metals whereas others serve to transport metals. Ferritin stores iron especially in liver and spleen as iron(III) oxyhydroxyphosphate with the approximate composition $(FeOOH)_8$,-FeO,PO_4H_2. This forms a core of 70 Å (7000 pm) diameter and is surrounded by 24 protein subunits, giving the spherical molecule a total diameter of about 120 Å. Transferrin, on the other hand, is a plasma protein which transports Fe^{3+} and Cu^{2+} ions. Available evidence implicates tyrosine and histidine at the binding site. For example, the absorption spectrum has a peak at 465 nm which is attributed to charge transfer between phenolate ion and Fe^{3+} ion.

In contrast to the two foregoing examples, many proteins bind metal ions either temporarily or throughout their existence in the organism. One example of temporary binding of Ca^{2+} ions has been given above in connection with the proteolytic activation of prothrombin and other components of the blood-coagulation system (Section 24.2.1.2). Another example is provided by alkaline phosphatases and phosphokinases. A divalent cation such as Mg^{2+} or Zn^{2+} is required presumably to shield the negative charges on the phosphate group in order to facilitate attack by a nucleophile at the phosphorus atom. More permanent binding of metal ions by proteins may serve one of several purposes. Trypsin is stabilized against autolysis by Ca^{2+} ions. Concanavalin A (*vide infra*) does not bind glucose derivatives unless it first contains one Ca^{2+} ion and one Mn^{2+} ion per subunit. Presumably the cations adjust the molecular conformation to form the glucose-binding site. Metal ions can also form part of the active sites of enzymes. Being Lewis acids they can effect powerful acid catalysis. Carbonic anhydrase, carboxypeptidase A, carboxypeptidase B, and thermolysin all contain one Zn^{2+} ion which is believed to function in this way. The side-chains of glutamic acid and histidine form three of the ligands in the above examples. The fourth is water or substrate. Transition elements by virtue of their different valency states can function in oxidoreductases. For example, in ascorbic acid oxidase, copper oscillates between the Cu(I) and Cu(II) states.

The final example of proteins which bind small molecules concerns lectins. These proteins, frequently but not exclusively from plants, bind carbohydrate derivatives with a considerable degree of stereospecificity. They first became interesting because of their ability to agglutinate erythrocytes by binding membrane glycoproteins. Some are specific for individual blood groups. Interest in lectins was heightened by the observation that some of them preferentially agglutinate malignant cells. By attaching lectins covalently to an insoluble support such as agarose, they can be used for the isolation of glycoproteins by affinity chromatography. Concanavalin A is the most thoroughly studied of the lectins. Its sequence of 238 amino-acids and the three-dimensional structure have been determined.[17] The protein has a remarkable conformation. Seven sections of the single chain are folded to form an antiparallel pleated sheet and six further sections form another antiparallel sheet perpendicular to the first. The Mn^{2+} ion is octahedrally coordinated to two water molecules and the side chains of His^{24}, Glu^{8}, Asp^{10}, and Asp^{14}. The Ca^{2+} ion situated 5 Å distant from Mn^{2+} shares the last two ligands and is also bonded to the carbonyl oxygen of Tyr^{12}, the side chain of Asn^{14}, and two water molecules in an octahedral configuration. Glucose and mannose residues are bound in a deep pocket $6 \times 7.5 \times 18$ Å surprisingly lined by hydrophobic residues.

24.2.3 PROTEINS WHICH INTERACT WITH OTHER PROTEINS

Polypeptide hormones circulate at extremely low concentrations and do not have to enter the cell to exert their regulatory effect on intracellular reactions. Specific receptors exist at cellular membranes and the hormonal signal is amplified, often through the enzyme adenyl cyclase, and relayed by a second messenger such as cAMP. Little is known of the chemistry of hormone receptors but they are probably proteins or conjugated proteins. In the case of glucagon (Figure 5), which stimulates glycogenolysis in liver, some information is available concerning the amino acids which are responsible for biological activity. De-His^{1}-glucagon, which lacks the *N*-terminal residue, binds to the liver receptors but does not activate adenyl cyclase. Fragments containing residues 1–21 and 20–29 failed to bind or to activate adenyl cyclase. His^{1} appears to be essential for biological activity and it is interesting that secretin (Figure 5), which also has *N*-terminal histidine, and glucagon stimulate the adenyl cyclase system in fat cells of rat but through different receptors. Some of the interactions between glucagon and the receptor appear to involve the hydrophobic residues near the *C*-terminus of the hormone. Unlike some other polypeptide hormones such as β-corticotropin and gastrin, however, in the case of glucagon the whole molecule appears to be necessary.

Another rather specific type of protein–protein interaction concerns the inhibition of trypsin by a small protein inhibitor from bovine pancreas. The latter protein contains 58 amino-acids in a rather compact structure containing three disulphide bonds. Accordingly, it is not very susceptible to proteolytic attack. The side chain of Lys^{15}, however, is quite

1 2 3 4 5 6 7 8 9 10 11 12 13 14 15
His-Ser-Gln-Gly-Thr-Phe-Thr-Ser-Asp-Tyr-Ser-Lys-Tyr-Leu-Asp-

16 17 18 19 20 21 22 23 24 25 26 27 28 29
-Ser-Arg-Arg-Ala-Gln-Asp-Phe-Val-Gln-Trp-Leu-Met-Asn-Thr

1 2 3 4 5 6 7 8 9 10 11 12 13 14 15
His-Ser-Asp-Gly-Thr-Phe-Thr-Ser-Glu-Leu-Ser-Arg-Leu-Arg-Asp-

16 17 18 19 20 21 22 23 24 25 26 27
-Ser-Ala-Arg-Leu-Gln-Arg-Leu-Leu-Gln-Gly-Leu-Val-NH_2

Figure 5 Amino-acid sequences of human and bovine glucagon (upper) and porcine secretin (lower)

exposed and this is the site for interaction with and inhibition of trypsin. The normal catalytic mechanism of 'serine proteinases', of which trypsin is an example, involves the formation of a non-covalent complex followed by the acylation of Ser^{195} in the enzyme by the carbonyl group of lysine or arginine and the liberation of the first product of reaction. Deacylation of the acyl-enzyme completes the process. When trypsin and the bovine pancreatic trypsin inhibitor interact, Lys^{15} in the inhibitor molecule is located in the specificity pocket of trypsin and the carbonyl-carbon atom of Lys^{15} is covalently bonded to the oxygen atom in the side chain of Ser^{195} in the enzyme. Nevertheless, the carbonyl-carbon atom of Lys^{15} in the inhibitor is still covalently linked to the amide-nitrogen of Ala^{16} in the inhibitor. Clearly a tetrahedral product is formed which is prevented from collapsing to the acyl-enzyme with an sp^2 carbon atom by the failure of the nitrogen atom of Ala^{16} to depart. There are several reasons for the failure of the reaction to proceed. One concerns the position of the imidazole ring of His^{57} which participates in the charge relay system at the active site of 'serine proteinases'. In tosylchymotrypsin and indoleacryloyl-chymotrypsin the N_τ atom of His^{57} points towards an oxygen atom in the acylating groups and to a water molecule that is hydrogen-bonded to it, whereas it points towards the oxygen atom in the side chain of Ser^{195} in the tetrahedral complex between trypsin and bovine pancreatic trypsin inhibitor. Moreover, there is no room for a water molecule between the protein surfaces in contact in the tetrahedral complex. Finally, the —NH— group of Ala^{16} in the inhibitor acts as a hydrogen-bond donor to the carbonyl-oxygen of Gly^{36} in the inhibitor, and this would have to be ruptured if the tetrahedral complex were to collapse to the acyl-enzyme.

The final example of protein–protein interaction to be considered concerns antibodies or immunoglobulins (IgG). These proteins are produced by B-lymphocytes when a 'foreign' macromolecule such as a protein or carbohydrate is introduced into the body. The 'foreign' macromolecule, known as an antigen, may enter the body as invading bacteria or viruses, through the skin (accidently through injury or deliberately through immunization), or through the gut in food allergy. If a protein is attached covalently to a small molecule (hapten) such as a drug and then introduced into the body, antibodies to the hapten are usually produced. This ready production by experimental animals of antibodies to a wide variety of molecules forms the basis of various immunochemical assays such as radioimmunoassays.

The basic structure (Figure 6) of IgG molecules consists of two identical light (L) chains of molecular weight 22 500 and two heavy (H) chains of molecular weight 50 000–75 000 linked by non-covalent bonds and disulphide bonds to form a macromolecule with twofold symmetry. Antibodies bind antigens very tightly in the region represented by the first 110–120 residues from the *N*-terminus of both the L and H chains. These regions, which constitute about half of the L chains and about a quarter of the H chains, have amino-acid sequences which vary from one antibody to another. They are denoted as V_L and V_H. Within the V_L and V_H regions, some sequences (24–34, 50–55, 89–96 in V_L and 31–35, 50–65, 95–102 in V_H) are hypervariable and provide the specificity required for binding antigens. It will be noticed that antibodies have two binding sites per molecule. The remainder of the light (C_L) and heavy (C_H) chains have constant sequences within the same class of immunoglobulins.

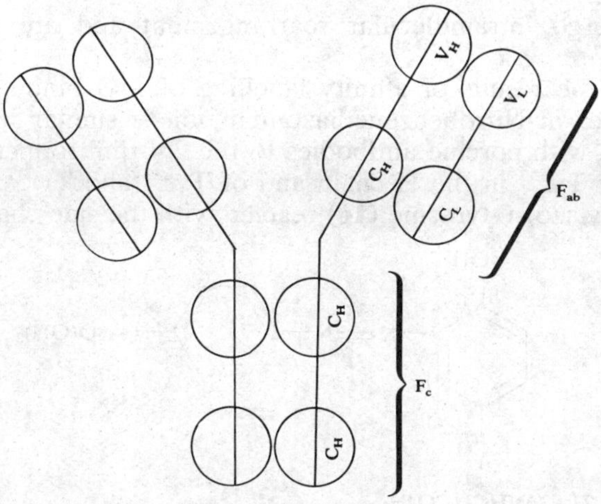

Figure 6 Schematic representation of IgG molecule

Structural studies of IgG molecules have been particularly facilitated by two observations. First, it is possible to obtain the F_{ab} and F_c fragments (Figure 6) with intact binding sites by degradation with papain. Secondly, patients with multiple myeloma excrete large quantities of light chains (Bence–Jones protein) in their urine. The sequences of the Bence–Jones proteins differ from patient to patient in the V_L regions. All three types of fragment have been crystallized and the three-dimensional structure has been determined by X-ray crystallographic analysis.[18]

In order to identify the amino-acids responsible for binding antigen, considerable use has been made of affinity labelling.[19] The technique involves the same principles as with enzymes (Section 23.3.10). A compound similar in structure to the antigen carries a reactive functional group which hopefully will form a covalent bond with the side chain of an amino-acid in the binding site. The technique has not been developed yet to the stage where the antigen is a polysaccharide or protein. This problem has another parameter which requires knowledge of the part of the antigen that is bound and specific attachment of a reactive probe to it. In view of these technical difficulties, investigations to date have concentrated on identifying the amino-acids involved in binding small haptens. The haptens used include derivatives of nitrobenzene, *m*-dinitrobenzene, phosphorylcholine, and lactose. The reactive groups used to form covalent bonds with the binding site include arenediazonium, bromoacetyl, and photochemically generated carbene or nitrene. As with affinity labelling of enzymes, the amount and validity of the information derived from such experiments depends crucially on the size and reactivity of the reactive group. If it is too large, it may react with amino-acids which are not directly involved in the binding process. If its reactivity is too selective, it may not react at all. The diazonium group reacts with the side chains of tyrosine, histidine, and lysine. The bromoacetyl group can alkylate all these together with the side chains of cysteine, methionine, aspartic acid, and glutamic acid. Photochemical conversion of a diazoketone into a carbene (Scheme 7) is potentially capable of labelling any amino-acid because it can insert into C—H bonds. The method has not completely fulfilled expectations, however, because rearrangement of the carbene (**14**) to the less reactive keten (**15**) competes strongly with free-radical insertion reactions.

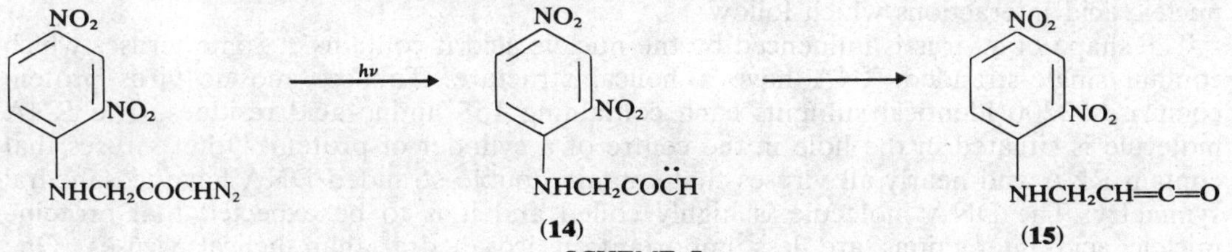

SCHEME 7

Nitrenes do not undergo intramolecular rearrangement and are more reactive than carbenes.[20]

A few examples of the results of affinity labelling of IgG binding sites will serve to illustrate the technique. *m*-Nitrobenzenediazonium ion is similar in size to *m*-dinitrobenzene and it reacted with porcine antibodies to the 2,4-dinitrophenyl hapten, coupling with the side chains of Tyr^{33} in the H chain and of Tyr^{33} and Tyr^{313} in the L chain. *N*-Bromoacetyl-(*m*-arsanylazo)-L-tyrosine (**16**) reacted with the side chain of Lys^{59} in the H

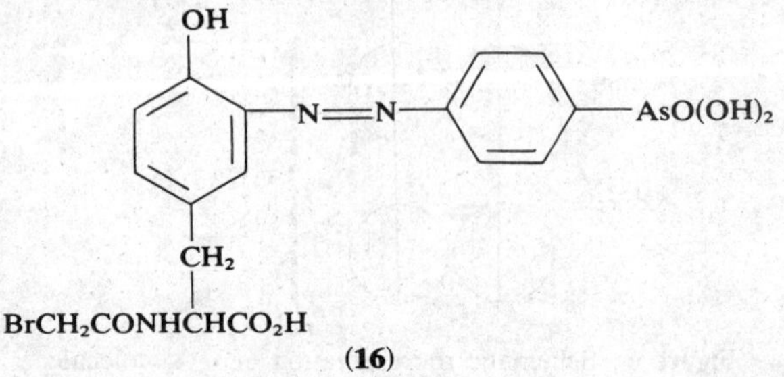

(**16**)

chain of guinea pig antiarsanyl antibodies. When *N*-dinitrophenylglycine diazoketone was allowed to bind to rabbit antibodies to bovine DNP-γ-globulin and then photolysed, the products (**14–15**) reacted predominantly with the H chains but the precise locus was not identified. Aryl nitrenes are generated by photolysis of aryl azides. Consequently, bovine γ-globulin was treated with 4-azido-2-nitrofluorobenzene and antibodies to the 4-azido-2-nitrophenyl hapten were raised in rabbits and isolated. Equilibration of the antibody with *N*-ε-(4-azido-2-nitrophenyl)-L-lysine followed by irradiation above 400 nm effected labelling of the H and, to a lesser extent, the L chain. Proteolytic degradation of the labelled antibody gave two peptides, X-Ala-Arg and Phe-Cys-Y-Arg. The sequence 91–94 in the H chain is -Phe-Cys-Ala-Arg- and it is apparent that the nitrene has reacted with Cys^{92} and Ala^{93}. This sequence is adjacent to one of the hypervariable regions in the H chain.

24.2.4 PROTEINS WHICH INTERACT WITH NUCLEIC ACIDS

A good example of a discrete isolatable system containing proteins and nucleic acid in close association is a virus. The simplest type of virus consists of RNA or DNA, either single- or double-stranded, surrounded by a protein coat consisting of subunits which may be identical or different but arranged in a symmetrical structure. More complex types of virus have an outer envelope consisting of lipid and glycoprotein. While there is a close relationship between nucleic acid and coat protein(s) in the sense that the genetic information for biosynthesis of the latter resides in the former and also because the protein coat protects the nucleic acid from the action of nucleases in the host cell, there is a closer physical relationship between the protein subunits. This has been demonstrated by disrupting tobacco mosaic virus and allowing the protein to reassemble spontaneously in the absence of nucleic acid. The empty shell or capsid, however, was less stable than reconstituted virus particles containing nucleic acid. The implication is that protein–nucleic acid interactions are important though perhaps less so than protein–protein interactions. Viruses, therefore, form a conceptual bridge between the previous section and histone–nucleic acid interactions which follow.

The shape of a virus is influenced by the nucleic acid it contains.[21] Some viruses which contain single-stranded RNA have a helical structure. Tobacco mosaic virus protein comprises 2200 identical subunits each containing 158 amino-acid residues. The RNA molecule is situated in the hole in the centre of a cylinder of protein. Other viruses that contain RNA and nearly all viruses that contain double-stranded DNA have icosahedral symmetry. The DNA molecule is highly coiled and it is to be expected that protein–nucleic acid interactions are less important in icosahedral than helical viruses. One important exception has been found in polio virus.[22] A small protein (VPg) of molecular weight <7000 is covalently attached to the 5′-terminal oligonucleotide of the polio RNA as VPg-pU-U-A-A-A-A-C-A-Gp, to nascent strands of polio replicative intermediate, and to the poly(U) sequence of the minus RNA strands. The nature of the covalent linkage has not been determined, but it has been suggested[22] that VPg is implicated in the initiation of RNA synthesis. It must be cleaved, however, before the RNA is translated to give viral protein.

Those viruses which contain dissimilar protein subunits apparently direct the host cell initially to synthesize a primary protein which contains the amino-acid sequences of the subunits in a single chain. During the process of virus self-assembly and maturation, this primary protein is cleaved selectively, possibly by host proteinases, to generate all the required subunits. It has been suggested,[21] for example, that picornaviruses such as enteroviruses are assembled and undergo maturation in a stepwise manner. The primary protein is cleaved initially into three fragments (VP0, VP1, VP3). These are believed to assemble through the intermediates VP0-VP1-VP3 and $(\text{VP0-VP1-VP3})_5$ to give the procapsid $(\text{VP0-VP1-VP3})_{60}$. Empty shells of this type have been isolated. RNA is then inserted into the procapsid and maturation is achieved by further proteolytic cleavage involving VP0. This gives rise to two subunits VP2 and VP4, so that the mature virus consists of a capsid $(\text{VP1-VP2-VP3-VP4})_{60}$ enclosing viral RNA. It is not known if the RNA plays some role in directing the maturation of the capsid proteins.

Protein–nucleic acid interaction is stronger in the chromosomes of eucaryotic cells. Here five classes of histones complex with DNA.[23,24] All the histones are very basic; H1 is very rich in lysine compared with arginine, H3 and H4 are arginine-rich, and H2A and H2B have lysine/arginine ratios intermediate between the others. The sequences of arginine-rich histones are highly conserved. The H4 histones from calf thymus and pea seedlings differ at only 2 loci in a chain of 102 amino-acids, and the H3 histones of calf thymus and carp display only one difference in 135 amino-acids. The *N*-terminal regions of H2A, H2B, H3, and H4 are much more basic than the *C*-terminal regions; the latter contain several amino-acids with apolar side chains. These four histones interact strongly with each other, possibly in the *C*-terminal regions, and it is believed that aggregates of the type $(\text{H2A-H2B-H3-H4})_2$ cluster, probably by interaction of the side chains of lysine and arginine, with phosphate groups in DNA. The clusters cover regions of about 200 base pairs each to form repeating structures called nucleosomes. Such structures have been observed by electron microscopy.

H1 histones are rather different from the others. They are larger (molecular weight approximately 23 000) and they are much more diverse in sequence even though about half the molecule consists of lysine and alanine. Within one species the H1 histone has been separated into several closely related proteins. They bind to DNA in a different manner from the other histones and seem to form cross-links between polynucleotide strands over a region of about 50 base pairs. As well as undergoing post-translational methylation and acetylation (Section 24.2.1.1), histones undergo phosphorylation in the side chains of particular residues of serine. This modification is especially interesting in the case of the H1 histones since phosphorylation is maximal at the time of cell division and this might be a mitotic trigger. The rate of phosphorylation of H1 histones is high in regeneration of liver after partial hepatectomy and correlates positively with the rate of tumour growth.

Another system involving strong interaction between protein and DNA concerns the regulation of an operon by a repressor protein. The most thoroughly studied case is the *lac* operon of *E. coli.*[25] The regulator gene codes for the synthesis of the *lac* repressor protein which then binds to the adjacent operator. A small inducer molecule such as isopropylthio-β-D-galactopyranoside binds to the repressor protein, causing it to dissociate from the operator site. Transcription of three adjacent genes in the operon proceeds and leads to the biosynthesis of three enzymes, β-galactosidase, galactose permease, and thiogalactoside transacetylase. The *lac* repressor protein is a tetramer containing identical subunits composed of 347 amino-acid residues. Its affinity for the operator DNA sequence is dependent on ionic strength, but the dissociation constant is probably less than 10^{-11} mol l^{-1} in the cell. The structure of the binding site for DNA in the repressor protein has not yet been elucidated, but removal of 59 residues from the *N*-terminus and 20 residues from the *C*-terminus by trypsin prevents binding. A little more is known of the binding site for the inducer. Fluorescence measurements indicate that tryptophan in the binding site moves into a less polar environment. Stopped-flow measurements of the fluorescence changes show that the binding is biphasic. The initial fast step follows second-order kinetics as expected, but the slower step is first order and probably represents a conformational change. This would be consistent with the postulate that the inducer produces a conformational change in the repressor protein that lowers its affinity for the operator DNA.

Other systems involving protein–nucleic acid interactions include the enzymes ribonucleases, deoxyribonucleases, DNA-dependent RNA polymerase, DNA methylases, and DNA-photoreactivating enzymes.

24.2.5 TOXIC PROTEINS

In view of the ubiquity and nutritional importance of proteins in living organisms, it is perhaps surprising that some proteins are exceedingly toxic. Since minute concentrations of polypeptide hormones, however, are capable of modulating cellular metabolism, it is not unreasonable that polypeptides of appropriate structure are capable of blocking cellular activity, for example by binding to membrane receptors.

Some of the most toxic proteins[26] known are produced by an anaerobic spore-forming bacillus *Clostridium botulinum*, which grows well on certain foodstuffs and especially on meat products. The toxins produced by the organism block release of acetylcholine and hence neural transmission at cholinergic synapses. The lethal dose for man is about 1 μg. Little is known about their structures.

Better understood are the neurotoxins from certain snakes, especially cobras (Elapidae family) and sea snakes (Hydrophiidae family). Toxins from a number of genuses and species of each family have been isolated and sequenced. Most snake neurotoxins so far studied fall into one of two classes. Type I contain 61–62 amino-acids (molecular weight 6700–7000) and type II contain 71–74 amino-acids (molecular weight 7800–8000). Type I contain four disulphide bonds and these occupy homologous positions in the sequences of both types I and II toxins. Type II toxins, however, contain an additional disulphide bond (Figure 7). When the sequences of 13 toxins were compared,[27] 18 amino-acids were constant and several conservative substitutions were apparent. Comparison of the sequences suggests that type II toxins have evolved by insertion of a segment containing a disulphide bond near the centre of the sequence and by addition of more amino-acids at the *C*-terminus. Numerous attempts have been made to identify the amino-acid residues which are essential for activity. For example, cobrotoxin (Figure 7) from *Naja naja atra*

	1 2 3 4 5 6 7 8 9 10 11 12 13 14 15 16 17 18 19 20
Naja naja atra	Leu-Glu-Cys-His-Asn-Gln-Gln-Ser-Ser-Gln-Thr-Pro-Thr-Thr-Thr-Gly-Cys-Ser-Gly-Gly-
Naja melanoleuca b	Ile-Arg-Cys-Phe -Ile-Thr-Pro-Asp-Val-Thr-Ser-Gln-Ile-Cys-Ala-Asp-Gly-
	21 22 23 24 25 26 27 28 29 30 31 32 33 34 35 36 37 38 39 40
Naja naja atra	-Glu-Thr-Asn-Cys-Tyr-Lys-Lys-Arg-Trp-Arg-Asp-His- -Arg-Gly-Tyr-Arg-
Naja melanoleuca b	-His-Val-Cys-Tyr-Thr-Lys-Thr-Trp-Cys-Asp-Asn-Phe-Cys-Ala-Ser-Arg-Gly-Lys-Arg-
	41 42 43 44 45 46 47 48 49 50 51 52 53 54 55 56 57 58 59 60
Naja naja atra	-Thr-Glu-Arg-Gly-Cys-Gly- -Cys-Pro- Ser-Bal-Lys-Asn-Gly- Ile- Glu-Ile-Asn-Cys-
Naja melanoleuca b	-Val-Asp-Leu-Gly-Cys-Ala-Ala-Thr-Cys-Pro-Thr-Bal-Lys- Pro-Gly-Bal-Asn-Ile- Lys-Cys-
	61 62 63 64 65 66 67 68 69 70
Naja naja atra	-Cys-Thr-Thr-Asp-Arg-Cys-Asn-Asn
Naja melanoleuca b	-Cys-Ser-Thr-Asp-Asn-Cys-Asn-Pro-Phe-Pro-Thr-Arg-Asn-Arg-Pro

Figure 7 Amino-acid sequences of the venoms of *Naja naja atra* and *Naja melanoleuca b*. Disulphide bonds link residues 3–24, 17–45, 49–60, 61–66; there is an additional disulphide bond in *Naja melanoleuca b* venom linking residues 30–34

(Formosan cobra) contains two tyrosine residues and one, Tyr^{35}, can be nitrated by tetranitromethane without affecting toxicity. The other, Tyr^{25}, is apparently buried but can be nitrated in guanidine hydrochloride with loss of activity. Since this residue is buried, it presumably does not interact with the cholinergic receptor but is essential for maintenance of the correct molecular conformation of the toxin. It is likely that several amino-acid side chains interact with or are adjacent to the receptor, and modification of individual residues may not throw much light on the mechanism of action of this molecule. Cobrotoxin has been synthesized by the solid-phase method[28] and it may be that the synthesis of analogues using natural or unnatural amino-acids to replace residues at selected loci would provide valuable information about the relationship between structure and toxicity.

24.2.6 STRUCTURAL AND CONTRACTILE PROTEINS

Proteins that do not have a structural function tend to have a high content of tertiary structure which is often easily disrupted by exposure to denaturing conditions such as shift of pH, addition of organic solvents, detergents, or certain salts at high ionic strength. In contrast, structural proteins tend to have a high content of secondary structure. This is favoured by a primary structure which has repeating patterns.

Keratins occupy a special place amongst structural proteins because they were the first proteins to be studied by X-ray diffraction methods by Astbury. Their insolubility and biochemical inertness, however, have not encouraged the level of research activity that they merit. Keratins form protective barriers against the environment in the form of horns, hooves, nails, hair, wool, and feathers. Feathers contain β-structures whereas hair and wool contain α-helical structures. The latter are composed of proteins with a low sulphur content, but the microfibrils are surrounded by a matrix of two other types. One has a high content of glycine and tyrosine while the other has a high sulphur content. During the synthesis of prekeratin in epithelial cells, the sulphur-rich proteins contain many thiol groups and these later form disulphide bonds which hardens the keratin. The loss of mechanical strength of hair by exposure to bleaching agents or reducing agents (permanent waving) can be explained in part by the cleavage of disulphide bonds. Reduction and carboxymethylation of disulphide bonds (Section 23.3.3) has made it possible to solubilize and fractionate some of the components of keratin for sequencing.[29] In one component containing 151 amino-acids, the sequence

$$\genfrac{}{}{0pt}{}{\text{Ser}}{\text{Thr}}\text{-Ser-Cys-Cys-}\genfrac{}{}{0pt}{}{\text{Arg}}{\text{Gln}}\,\genfrac{}{}{0pt}{}{\text{Ser}}{\text{Pro}}\text{-Thr-}\genfrac{}{}{0pt}{}{\text{Cys}}{\text{Ser}}\text{-}\begin{matrix}\text{Ser}\\ \text{Ile}\\ \text{Leu}\end{matrix}\text{-Gln}$$

is repeated three times. The α-amino group is acetylated.

Collagen[30] is a fibrous protein found in skin, tendon, cartilage, bone, and teeth. It is synthesized intracellularly in fibroblasts as a precursor, procollagen, with molecular weight 125 000–130 000 per chain. The extension at the *N*-terminus contains disulphide bonds and tryptophan, unlike collagen. Conversion of procollagen into collagen occurs extracellularly by procollagen peptidase, which removes the *N*-terminal segment containing cystine and tryptophan by cleaving X—Glu or X—Gln bonds leaving *N*-terminal residues of pyrrolid-2-one-5-carboxylic acid. Little is known about this proteinase except that it apparently contains an essential metal and is not a 'serine proteinase' or a 'thiol proteinase'. An hereditary disease in cattle, dermatosparaxis, is due to the failure to carry out this proteolytic step. Collagen consists of three polypeptide chains, each containing about 1000 amino-acid residues, and the chains form a triple right-handed helix with one intermolecular $>NH \cdots O{=}C<$ hydrogen bond every third residue. Non-helical telopeptide regions are present at the *N*- and *C*-termini. Adult collagen contains two types of chain, $\alpha 1$ and $\alpha 2$, forming $(\alpha 1)_2\alpha 2$ in the triple helix. On the other hand, foetal collagen and collagen laid down in the early stages of wound healing contains only one type of chain and is $(\alpha 1)_3$.

As described above (Section 24.2.1), post-translational modification of certain proline and lysine residues occurs within the fibroblast. Rather less than half the proline residues are converted into hydroxyproline and about 20% of the lysine residues are converted into hydroxylysine and carbohydrate is attached to one or more of these. Apart from the telopeptide regions, the remainder of the $\alpha 1$ and $\alpha 2$ chains consist of a regular sequence containing glycine at every third residue. The glycine is commonly followed by proline and preceded by hydroxyproline. Glycine, proline, and hydroxyproline constitute together about half the collagen molecule. Alanine is the next most common amino-acid, but there is very little histidine and tyrosine.

Collagen undergoes further covalent changes in a maturation process. This involves the formation of novel covalent cross-linkages which render the molecule insoluble. The modifications seem to be confined to the telopeptide regions and are initiated by the oxidative deamination of specific lysine or hydroxylysine residues by a copper-dependent lysyloxidase. The resulting aldehydes, allysine (**17**) or hydroxyallysine (**18**), form aldimines with lysine or hydroxylysine residues in an adjacent chain. The aldimines formed initially from hydroxyallysine (**19**) are stabilized by undergoing the Amadori rearrangement to give (**20**) (Scheme 8). In contrast, the aldimines from allysine (**21**) remain acid-labile and so these bonds can only be detected after reduction of the aldimine with potassium borohydride to give (**22**) (Scheme 8). There is evidence that some of these aldimines are reduced in the native state so that (**22**) can be detected in acid hydrolysates of collagen without prior reduction.

It is proposed that collagen assembles in filaments each containing five overlapping chains of triple helices (Figure 8). This would permit intermolecular cross-linking at overlap regions involving the telopeptides.

Elastin, unlike collagen, is found mainly in ligaments and large blood vessels, which are required to have elastic properties. Like collagen, it is rich in glycine, alanine, and proline and contains no cyst(e)ine. Unlike collagen, however, it contains no hydroxylysine. The chains of elastin are cross-linked intramolecularly and intermolecularly, but some of the chemical structures differ from those in collagen even though they are believed to arise from the oxidative deamination of lysine to allysine. Acid hydrolysates of mature elastin contain desmosine (**23**) and isodesmosine (**24**). These structures are believed to arise from the condensation of the side chains of three residues of allysine and one of lysine, but the exact mechanism has not been established. The expected initial products are dihydropyridine derivatives so that it must be presumed that an oxidative step is involved. Formation of desmosine and isodesmosine appears to occur in sequences such as -Lys-Ala-Ala-Lys- or -Lys-Ala-Ala-Ala-Lys-. One such sequence with cross-linking is shown in Figure 9. In addition to desmosine and isodesmosine, cross-linkages involving the reaction of a lysine side-chain with one (**21–22**) or two (**25–26**) allysine side-chains are formed. Copper deficiency in the diet lowers the activity of lysyloxidase and hence the number of cross-linkages. The consequent poor elasticity of large blood vessels gives rise to cardiovascular lesions such as dissecting aneurysm of the aorta. Cross-linkage of collagen is also affected and pathological skeletal defects are found. Marfan syndrome is an inherited defect in which cross-linkage of collagen and elastin is severely impaired. The genetic defect might involve a deficiency of lysyloxidase, a defect in transport of copper ions or abnormal sequences in the proteins.

The elastic properties of elastin have proved difficult to explain satisfactorily at a molecular level.[31] One hypothesis suggests that, between the regions which give rise to cross-linkages, sequences of the type -Pro-Gly-Gly-Val-Pro- occur frequently and that these form a wide left-handed helix ('oiled coil') in which the hydrophobic side-chains point inwards when the elastic fibre is relaxed. It is argued that extension of the elastic fibre would tend to expose these to water and so there would be an entropic driving force to restore the fibre to its original state. Elastic fibre contains at least one other protein

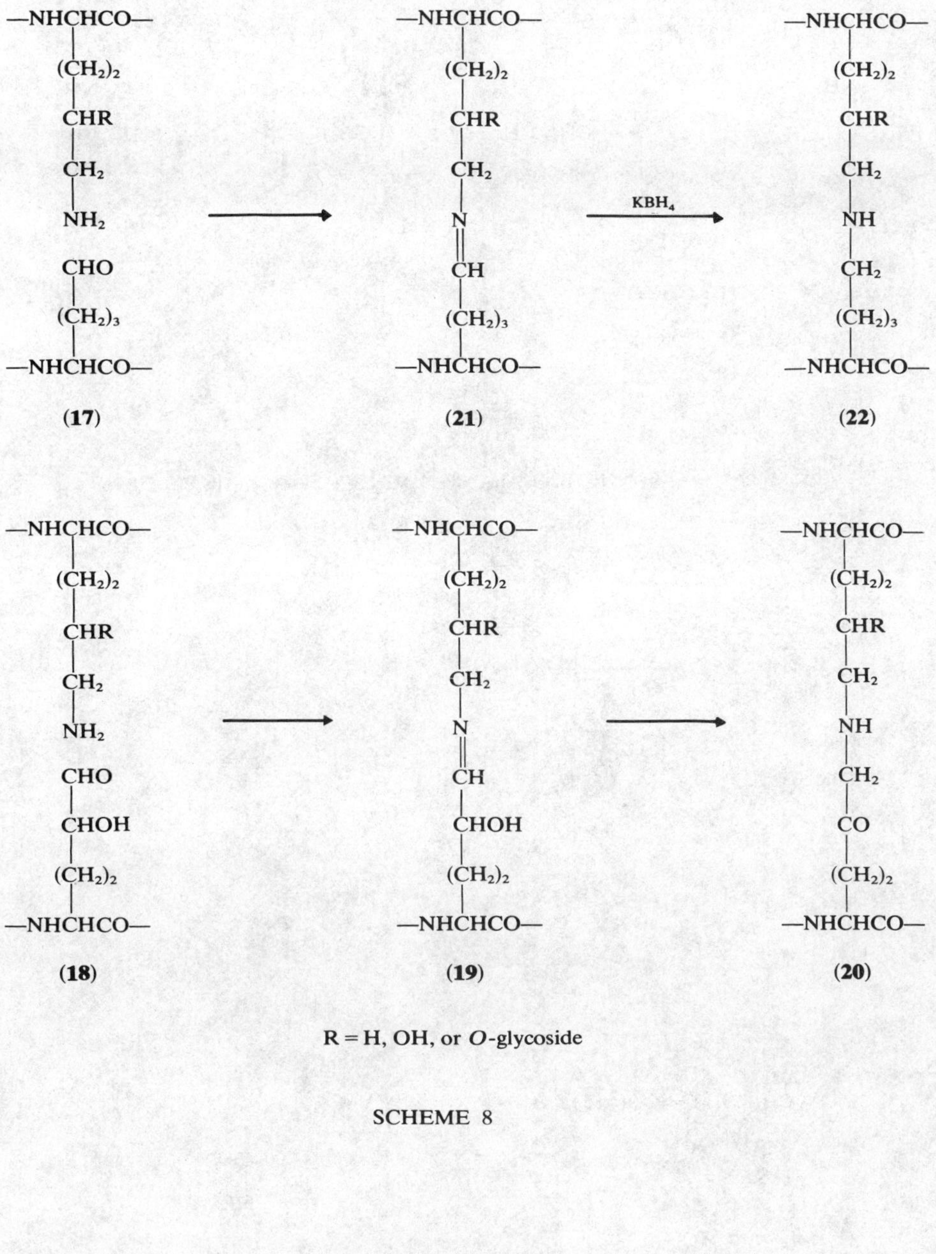

(**17**) (**21**) (**22**)

(**18**) (**19**) (**20**)

R = H, OH, or *O*-glycoside

SCHEME 8

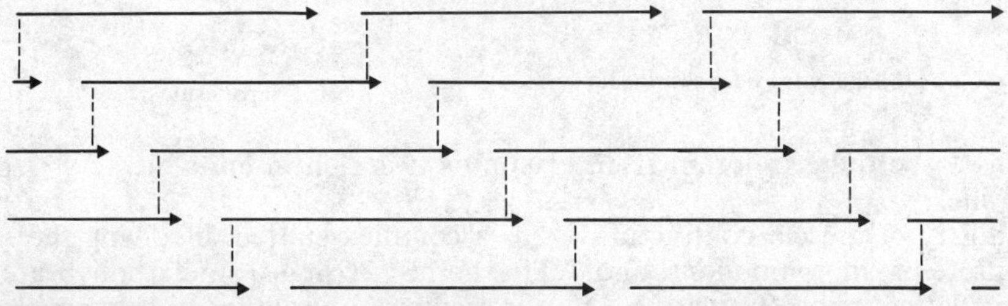

Figure 8 Schematic representation of staggered packing of five triple helices of collagen in a filament. Although drawn in two dimensions, the five triple helices pack around the circumference of a cylinder. Notice the sites of cross-linking in the telopeptide regions

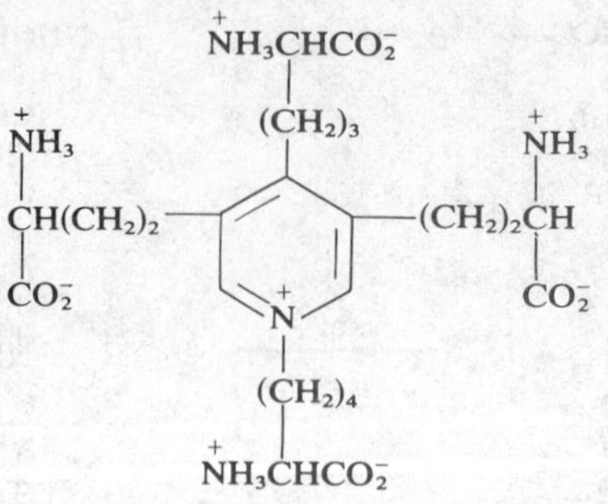

(**23**) Desmosine

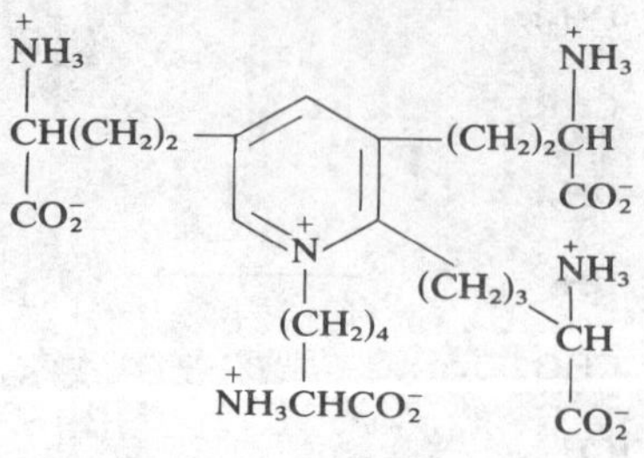

(**24**) Isodesmosine

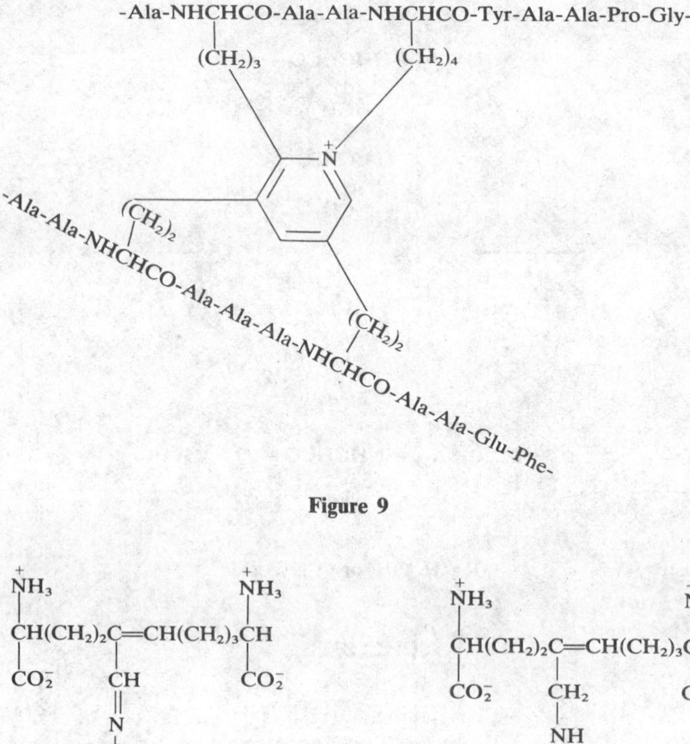

Figure 9

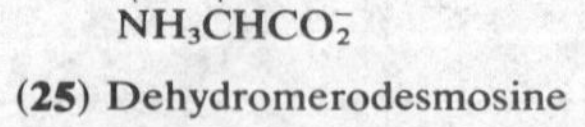

(**25**) Dehydromerodesmosine

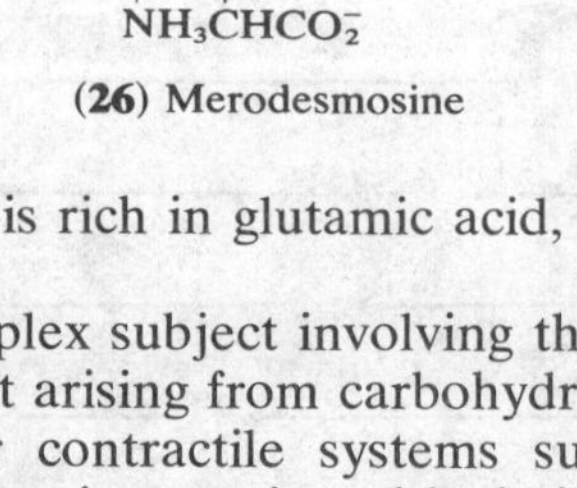

(**26**) Merodesmosine

which seems to be quite different from elastin as it is rich in glutamic acid, aspartic acid and, cyst(e)ine.

The chemistry of muscle contraction[32,33] is a complex subject involving the interaction of several proteins, inorganic ions, and ATP, the last arising from carbohydrate metabolism and oxidative phosphorylation. Non-muscular contractile systems such as those involved in phagocytosis, mitosis, and cytokinesis contain proteins with similar structures and properties.[34] Electron microscopic studies indicate that muscle fibres consist of bundles (myofibrils) of thick filaments (myosin) interspersed by thin filaments (actin) and

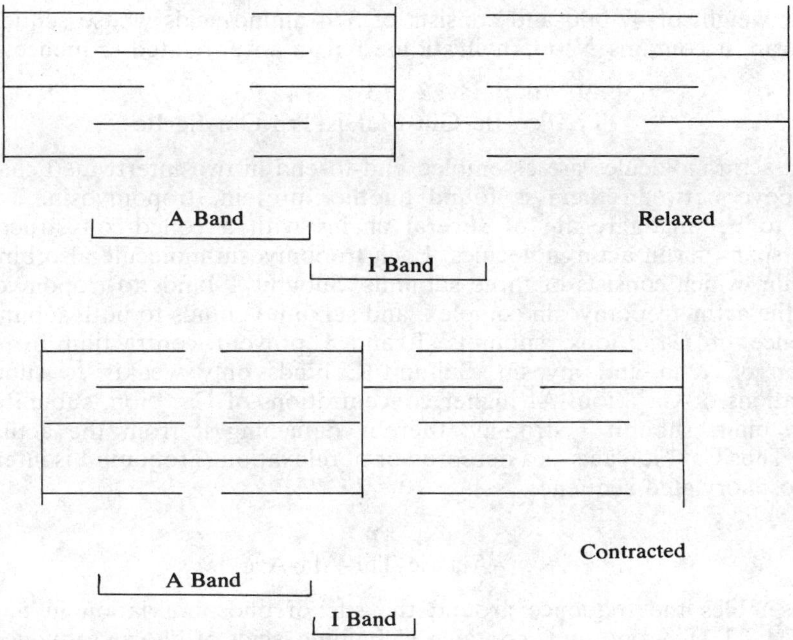

Figure 10 Schematic representation of skeletal muscle showing interrelationship of myosin and actin

surrounded by a fluid (sarcoplasm) and a sleeve-like system of vesicles (sarcoplasmic reticulum). Gaps (Figure 10) (I bands) between myosin filaments (A bands) are bridged by actin filaments. It was proposed by Huxley that when muscle contracts, the actin and myosin filaments slide past one another so that the gaps between the ends of myosin filaments shorten. It was further proposed that the relative movement of actin and myosin is accompanied by the rapid formation and cleavage of intermolecular covalent bonds. Contraction of muscle is triggered by the release of Ca^{2+} ions from sarcoplasmic reticulum into the sarcoplasm and this activates an adenosine triphosphatase in myosin. ATP diminishes interaction of myosin and actin and promotes relaxation. The action of adenosine triphosphatase lowers the concentration of ATP and increases the interaction between myosin and actin. This simple picture has been complicated considerably by the discovery of several other proteins which are involved in muscular contraction and relaxation, and the reader is referred to recent reviews[32,33] for a detailed discussion. A brief description of some of the components and their interrelationships is given here.

Myosin has a molecular weight of 460 000 and the filaments are 1600 Å long. The molecule contains two identical α-helical chains which are intertwined, but the *N*-terminal portion of each chain has a globular conformation and the adenosine triphosphatase is sited in this region. Myosin contains N_τ-methylhistidine, ε-*N*-methyl-lysine and ε-*N*-trimethyl-lysine. The N_τ-methylhistidine occurs in the sequence:

-Ile-Asp-Val-Asp-MeHis-Gln-Thr-Tyr-Lys-

Each chain has a molecular weight of 190 000 and the difference between the total molecular weight (460 000) and twice 190 000 is accounted for by the association of three small chains with the globular head of the molecule. One small chain can be removed without affecting the adenosine triphosphatase activity, but the other two are essential for activity. The two essential small chains are highly homologous but one is longer than the other. The extension at the *N*-terminus has a remarkable sequence containing alanine, proline, and lysine.

The actin referred to above is another complex assembly of proteins. Actin itself has a molecular weight of 42 000 and consists of 374 amino-acids whose sequence is known. Like myosin, it contains N_τ-methylhistidine but in an unrelated sequence:

69 70 71 72 73 74 75 76 77
-Tyr-Pro-Ile-Glu-MeHis-Trp-Glu-Ile-Ile-

α-Helical actin molecules are assembled end-to-end in two intertwined chains and in the spiral groove between them is found another protein, tropomyosin. Tropomyosin is believed to be an aggregate of several chains with a coiled-coil structure and each molecule spans seven actin molecules. Each tropomyosin molecule also binds a molecule of troponin, which consists of three subunits. Subunit T binds to tropomyosin, subunit I binds to the actin–tropomyosin complex, and subunit C binds to both subunits T and I. In the absence of Ca^{2+} ions, subunits T and I prevent contraction by inhibiting the interaction of actin and myosin. Subunit C binds only weakly to subunit I at low concentrations of Ca^{2+} ion. At higher concentrations of Ca^{2+} ion, subunit C binds Ca^{2+} and then binds subunit I strongly, thereby detaching it from the actin–tropomyosin complex. Thus Ca^{2+} ion acts as a derepressor of relaxation. Troponin I is interesting since it has a phosphorylated sequence

P
-Ala-Ile-Thr-Ala-Arg-

which resembles the sequence around the site of phosphorylation in phosphorylase *a* (Section 24.2.1.1). Troponin C contains 158 amino-acids of known sequence and has four binding sites for Ca^{2+} ions.

24.2.7 EPILOGUE

The proteins described in the foregoing sections are arranged to illustrate different aspects of structure and function. The classification is somewhat arbitrary, since the reader will appreciate that, for example, haemoglobin and muscle proteins could have been classified under protein–protein interactions, troponin C is an example of a protein which binds a metal ion, and myosin could have been cited as an example of a protein that undergoes post-translational methylation. Proteins may be studied from several angles including biosynthesis, structure, interactions, and biological role. Any attempt to classify them is likely to be only partially successful, but it provides an opportunity to highlight similarities and differences. The proteins selected cover a wide range and descriptions are necessarily brief. The reader is recommended to consult the reviews cited.

References

1. D. M. Zeigler and L. L. Poulsen, *Trends Biochem. Sci.*, 1977, **2,** 79.
2. G. J. Cardinale, R. E. Rhoads, and S. Udenfriend, *Biochem. Biophys. Res. Comm.*, 1971, **43,** 537.
3. G. L. Cantoni, *Ann. Rev. Biochem.*, 1975, **44,** 435.
4. H. R. Morris, M. R. Thompson, and A. Bell, *Biochem. Biophys. Res. Comm.*, 1975, **62,** 856.
5. S. Magnusson, T. E. Petersen, L. Sottrup-Jensen, and H. Claeys, in 'Proteases and Biological Control', ed. E. Reich, D. B. Rifkin, and E. Shaw, Cold Spring Harbor Laboratory, 1975, p. 123.
6. C. T. Esmon, J. A. Sadowski, and J. W. Suttie, *J. Biol. Chem.*, 1975, **250,** 4744.
7. A. R. G. Wylie, J. D. Lonsdale-Eccles, N. L. Blumsom, and D. T. Elmore, *Biochem. Soc. Trans.*, 1977, **5,** 1449.
8. R. Kornfeld and S. Kornfeld, *Ann. Rev. Biochem.*, 1976, **45,** 217.
9. C. J. Waechter and W. J. Lennarz, *Ann. Rev. Biochem.*, 1976, **45,** 95.
10. C. S. Rubin and O. M. Rosen, *Ann. Rev. Biochem.*, 1975, **44,** 831.
11. E. W. Davie and K. Fujikawa, *Ann. Rev. Biochem.*, 1975, **44,** 799.
12. H. J. Müller-Eberhard, *Ann. Rev. Biochem.*, 1975, **44,** 697.
13. R. F. Doolittle and B. Blombäck, *Nature*, 1964, **202,** 147.
14. M. F. Perutz and L. F. TenEyck, *Cold Spring Harbor Symp. Quant. Biol.*, 1972, **36,** 295.
15. M. F. Perutz, *New Scientist*, 1971, **50,** 676.
16. N. M. Green, *Adv. Protein Chem.*, 1975, **29,** 85.

17. G. M. Edelman, B. A. Cunningham, G. N. Reeke, J. W. Becker, M. J. Waxdal, and J. L. Wang, *Proc. Nat. Acad. Sci. U.S.A.*, 1972, **69,** 2580.
18. D. R. Davies, E. A. Padlan, and D. M. Segal, *Ann. Rev. Biochem.*, 1975, **44,** 639.
19. D. Givol, *Essays Biochem.*, 1974, **10,** 73.
20. J. R. Knowles, *Accounts Chem. Res.*, 1972, **5,** 155.
21. S. J. Martin, in 'The Biochemistry of Viruses', Cambridge University Press, Cambridge, 1978, chapter 5.
22. A. Nomoto, B. Detjen, R. Pozzatti, and E. Wimmer, *Nature*, 1977, **268,** 208.
23. B. Lewin, in 'Gene Expression', Wiley, New York, 1974, vol. 2, chapter 3.
24. S. C. R. Elgin and H. Weintraub, *Ann. Rev. Biochem.*, 1975, **44,** 725.
25. S. Bourgeois and M. Pfahl, *Adv. Protein Chem.*, 1976, **30,** 1.
26. D. A. Boroff and B. R. DasGupta, in 'Microbial Toxins', ed. S. Kadis, T. C. Montie, and S. J. Ajl, Academic, New York, vol. IIA, chapter 1.
27. D. J. Strydom, *J. Biol. Chem.*, 1972, **247,** 4029.
28. H. Aoyagi, H. Yonezawa, N. Takahashi, T. Kato, N. Izumiya, and C.-C. Yang, *Biochim. Biophys. Acta*, 1972, **263,** 823.
29. J. H. Bradbury, *Adv. Protein Chem.*, 1973, **27,** 111.
30. W. Traub and K. A. Piez, *Adv. Protein Chem.*, 1971, **25,** 243.
31. R. B. Rucker and D. Tinker, *Internat. Rev. Exp. Pathol.*, 1977, **17,** 1.
32. E. W. Taylor, *Ann. Rev. Biochem.*, 1972, **41,** 577.
33. S. Ebashi, *Essays Biochem.*, 1974, **10,** 1.
34. M. Clarke and J. A. Spudich, *Ann. Rev. Biochem.*, 1977, **46,** 797.

24.3 Coenzymes

H. C. S. WOOD
University of Strathclyde

24.3.1 INTRODUCTION

Many enzyme-catalysed reactions require a substance to be present in addition to the enzyme and the substrate in order that the reaction may proceed. Such substances are called coenzymes or cofactors and form an essential part of the catalytic mechanism. The intact enzyme system, or holoenzyme, is thus formed from a protein portion called the apoenzyme and a non-protein component referred to variously as a prosthetic group, a cofactor, and more commonly a coenzyme.

Coenzymes can frequently be dissociated from the enzyme protein and may thus be regarded on occasion as a special form of cosubstrate in the enzymic reaction. They function generally as acceptors or donors of functional groups or of atoms and frequently act by linking two enzymes to form an enzyme system.[1] Thus one enzyme transfers the group or atom from substrate to coenzyme, and the second enzyme transfers it from coenzyme to the second substrate. In most cases it is now possible to explain the transfer process in terms of organic reaction mechanisms.

The relationship of certain of the vitamins, particularly those in the B-group, to the coenzymes is a particularly close one and it is appropriate to examine this before proceeding further. In many cases it is helpful to think of a particular coenzyme as the metabolically active form of the corresponding vitamin. This conversion into a metabolically active form may take place in a variety of ways, such as by combination with adenosine triphosphate (ATP), by phosphorylation, or by reduction, *etc.*

The term vitamin is used to describe an organic compound which is necessary in addition to carbohydrate, fat, and protein for normal health and growth. Vitamins cannot be synthesized by the host and must therefore be obtained exclusively from the diet, in which they are normally present in small concentration. When absent from the diet, or not properly absorbed, the result is a specific deficiency disease leading ultimately to death of the organism. In practice the term vitamin is usually restricted to the requirements of mammalian species and sometimes to humans only. It is important to be clear on this point since species vary in their vitamin requirements. Many bacteria and other parasites are incapable of absorbing some vitamins from their environment and synthesize any coenzymes which are necessary for growth from simple precursors.

There is thus a fundamental difference in the way in which mammalian species, on the one hand, and many parasitic organisms, on the other, obtain the coenzymes necessary for continued growth and normal metabolic function. Exploitation of these differences is an approach to chemotherapy which has been studied extensively in recent years,[2] and which has led to a range of exceedingly useful antibacterial and antiprotozoal drugs. Now that the mechanism of action of many of the coenzymes can be explained in some detail, attempts are being made[3] to design compounds which will interfere specifically with the biological function of the coenzyme. It seems likely that these studies will also have far-reaching consequences in the treatment of disease and metabolic malfunction.

24.3.2 COENZYMES INVOLVED PRINCIPALLY IN OXIDATION–REDUCTION PROCESSES

Biological oxidations[4] are catalysed by enzymes, each of which functions in conjunction with a coenzyme which acts as electron carrier. There are numerous oxidative enzymes but relatively few coenzymes. It appears to be the latter which determine whether the electron transfer process involves single electrons, pairs of electrons, *etc.* Associated with the transfer of electrons is the transfer of hydrogen. One of the questions which is examined below is whether the redox reaction is best explained in terms of separate transfer of electrons and protons, or alternatively transfer of a species such as a hydrogen atom or a hydride ion.

24.3.2.1 Pyridine nucleotide coenzymes

Many biological redox reactions require nicotinamide adenine dinucleotide (NAD^+) (**1**) or its 2′-phosphate ($NADP^+$) (**2**) as coenzymes. In the older literature these compounds

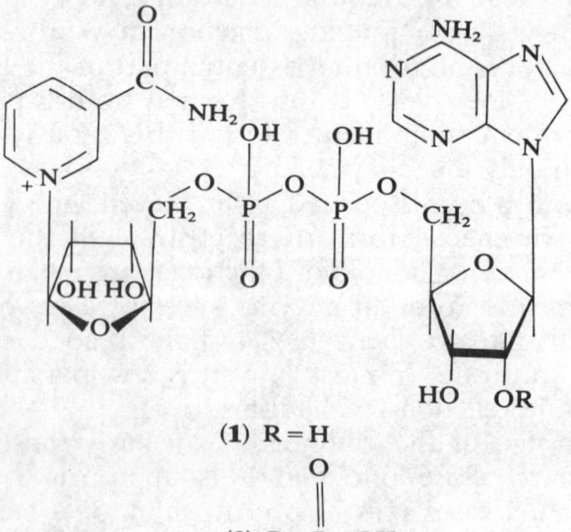

(**1**) $R = H$

(**2**) $R = P(=O)(OH)-OH$

are often referred to as diphosphopyridine nucleotide (DPN) and triphosphopyridine nucleotide (TPN), respectively. These names are clearly erroneous since DPN is not a nucleotide of 'diphosphopyridine'. They are thus not recommended, nor are the even older terms 'Coenzyme I' and 'Coenzyme II'.

These coenzymes are of interest for historical reasons since NAD^+ was the first coenzyme to be discovered. Harden and Young in 1906 demonstrated the existence of a heat-stable coenzyme involved in fermentation, but it was not until 1936 that NAD^+ was isolated in pure form. Meantime $NADP^+$ had been isolated as the coenzyme of glucose 6-phosphate dehydrogenase and shown to have properties similar to those of NAD^+. The early work is reviewed in detail by Singer and Kearney.[5]

The structure of NAD^+ has been established by a combination of degradation and synthesis. Hydrolysis with mineral acid gave adenine, nicotinamide, and two moles of D-ribose 5-phosphate. Alkaline hydrolysis gave adenosine diphosphate, the structure of which was well established, and enzymatic hydrolysis using a nucleotide pyrophosphatase gave quantitative conversion into adenosine 5′-phosphate and the 5′-phosphate of *N*-ribosylnicotinamide. The gross structure of NAD^+ as a P^1,P^2-diester of pyrophosphoric acid is thus established, the above evidence being summarized by Schlenk.[6] More recently,[7] the configuration of the glycosidic link in the ribosyl nicotinamide portion of NAD^+ has been established as β from studies of the ^1H n.m.r. spectrum. The configuration of the other glycosidic centre follows from the isolation of adenosine derivatives, which are known to be β-ribosides.

The synthesis of NAD^+ has been reported by Todd and his co-workers[8] and is summarized in Scheme 1. The β-anomer predominated in this synthesis. The final product was purified by selective enzymatic reduction of the β-anomer to the dihydro-form, destruction of the α-anomer with dilute alkali, and enzymatic reoxidation of the purified dihydro-β-anomer.

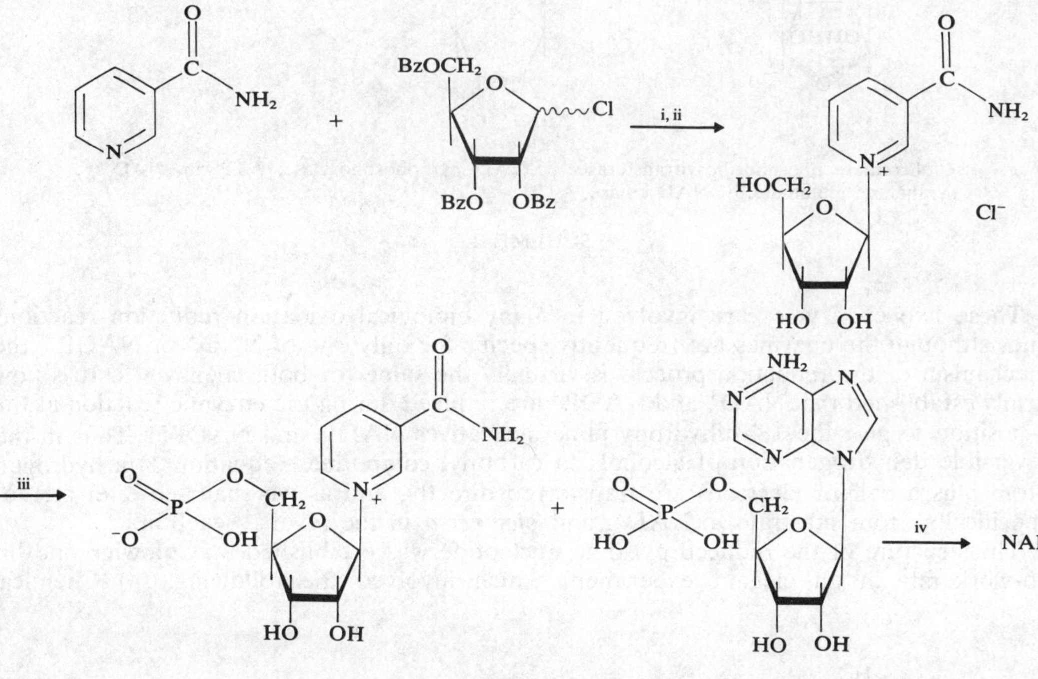

i, MeCN or $MeNO_2$, 0 °C; ii, MeOH, NH_3; iii, $POCl_3$, H_2O; iv, $C_6H_{11}N{=}C{=}NC_6H_{11}$.

SCHEME 1

The structure of NADP$^+$ was elucidated in similar fashion. The position of the third phosphate residue was established beyond doubt when adenosine 2′,5′-diphosphate was isolated[9] by enzymatic hydrolysis of NADP$^+$ using a nucleotide pyrophosphatase.

The relationship between the coenzymes NAD$^+$ and NADP$^+$ and the vitamins nicotinic acid and nicotinamide was clarified by studies[10] of the biosynthesis of the coenzymes. Deficiency of these vitamins leads, in man, to a nutritional disease termed pellagra. It seems likely that this condition is due to lack of NAD$^+$ and/or NADP$^+$, which are formed enzymatically from nicotinic acid in mammalian cells. It is perhaps surprising to discover that nicotinic acid, rather than the corresponding amide, is the preferred precursor for the biosynthetic sequence, but an enzyme exists in liver which can carry out the necessary hydrolysis. The presently accepted biosynthetic pathway is set out in Scheme 2.

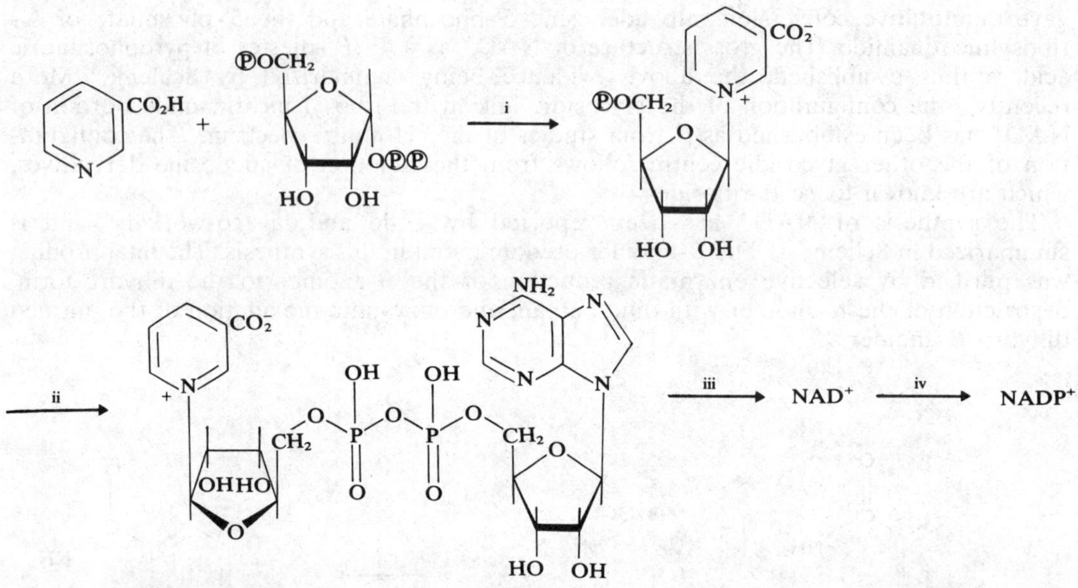

i, Nicotinate phosphoribosyltransferase; ii, NAD pyrophosphorylase, ATP; iii, NAD synthetase, glutamine; iv, NAD kinase, ATP.

SCHEME 2

These two coenzymes are involved in many biological oxidation–reduction reactions and, although the enzymes are frequently specific for only one of NAD$^+$ or NADP$^+$, the mechanism of the reduction process is virtually the same for both coenzymes. It is now firmly established that NAD$^+$ and NADP$^+$ are reduced during the enzymic reaction at the 4-position to give the 1,4-dihydropyridine derivatives NADH and NADPH. Thus in the reversible dehydrogenation of alcohols to carbonyl compounds (equation 1) a hydrogen atom plus a pair of electrons are transferred directly, and as we shall see later stereospecifically, from substrate to NAD$^+$, and *vice versa* in the reverse reaction.

The structure of the reduced pyridine nucleotide was established by Colowick and his co-workers[11] in an elegant experiment which involved the following. (a) Chemical

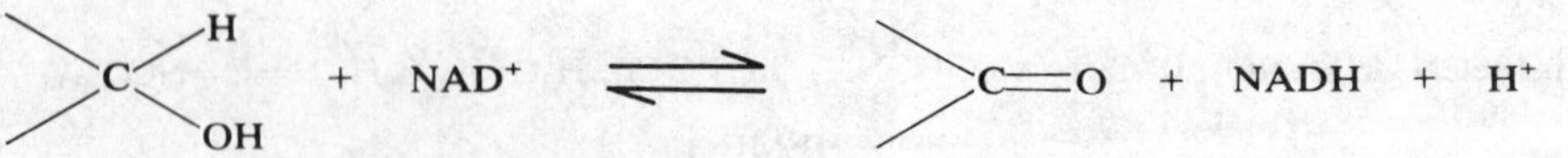

(1)

reduction of NAD^+ with sodium dithionite in deuterium oxide. This gave a deuterium-labelled NADH which was active enzymically. (b) Oxidation of this labelled NADH with yeast alcohol dehydrogenase and acetaldehyde gave the oxidized coenzyme NAD^+, which contained deuterium. The position of the deuterium was established by the degradative sequence set out in Scheme 3.

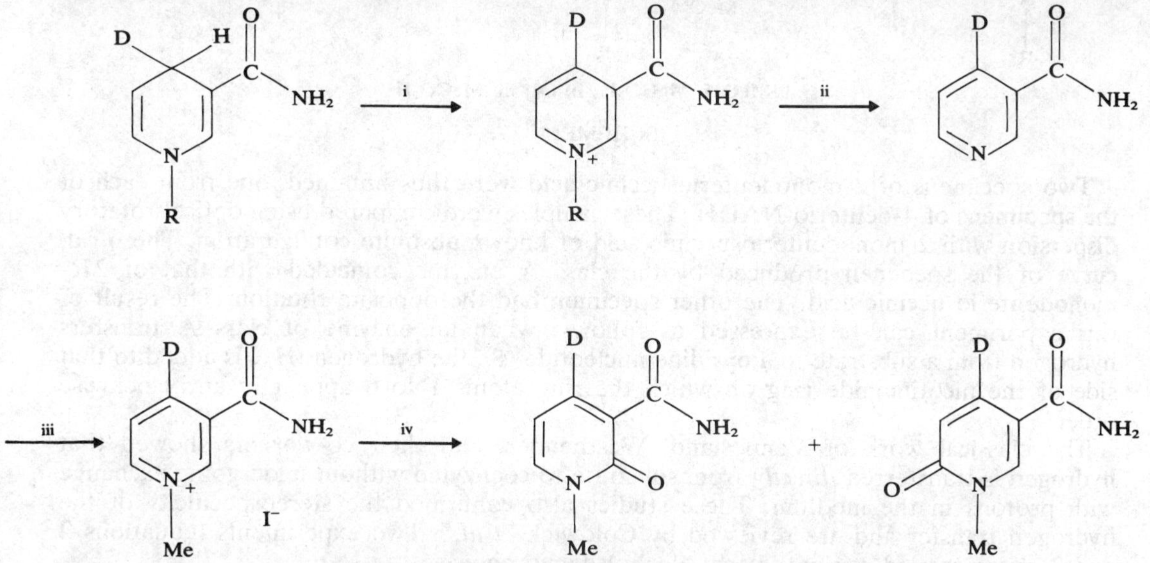

i, Yeast alcohol dehydrogenase, MeCHO; ii, NADase; iii, MeI; iv, $K_3Fe(CN)_6$, KOH.

SCHEME 3

The final oxidation yielded a mixture of approximately equal amounts of the 2- and 6-pyridones with *no loss of deuterium.* Thus the dithionite reduction did not give either the 1,2- or the 1,6-dihydropyridine derivative, and the enzymically active NADH must be the 1,4-dihydropyridine. Other observations have confirmed this assignment and direct proof for the presence of deuterium in the 4-position after dithionite reduction of an NAD^+ analogue in D_2O has been obtained from ^1H n.m.r. studies.[12]

The enzymic reduction is not only regiospecific as set out above but also stereospecific. Depending on the enzyme which is used, one or other of the isomers (**3**) and (**4**) is formed. The chemical reduction, of course, leads to both isomers being formed.

The enzymes can be classified[13] as of class A (*e.g.* yeast alcohol dehydrogenase) or of class B (*e.g.* β-hydroxy-steroid dehydrogenase) on the basis of their ability to transfer deuterium from either (**3**) or (**4**) respectively to a substrate. Elegant studies by Cornforth and his co-workers[14] have shown which of the two hydrogen atoms at C-4 of NADH is involved in the two classes of enzyme reaction. The assignment involved oxidative degradation (Scheme 4) applied to two specimens of 4-deuterio-NADH prepared by enzyme reactions of class A and class B, respectively.

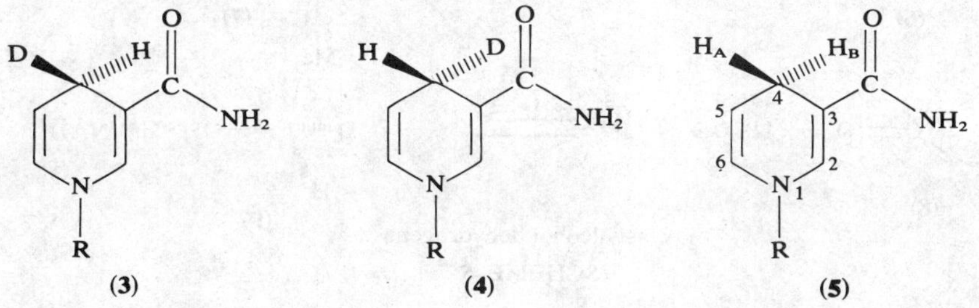

(**3**) (**4**) (**5**)

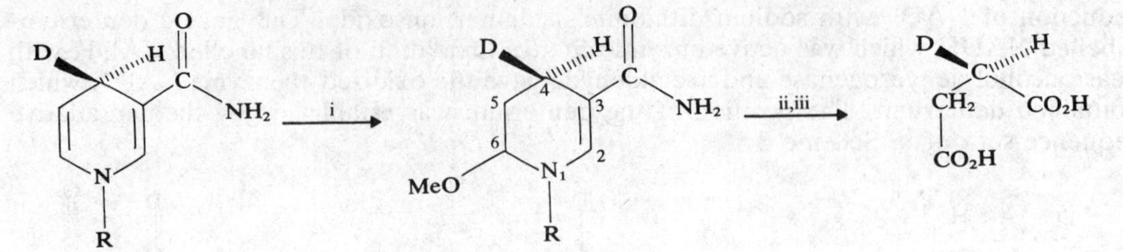

i, MeOH, $MeCO_2H$; ii, O_3; iii, $MeCO_2H$.

SCHEME 4

Two specimens of 2-monodeuteriosuccinic acid were thus obtained, one from each of the specimens of 4-deuterio-NADH. These samples were compared using optical rotatory dispersion with 2-monodeuteriosuccinic acid of known absolute configuration. The o.r.d. curve of the specimen produced by the class A enzyme coincided with that of 2*R*-monodeuteriosuccinic acid. The other specimen had the opposite rotation. The result of this experiment can be expressed as follows: when an enzyme of class A transfers hydrogen from a substrate to a pyridine nucleotide (**5**), the hydrogen (H_A) is added to that side of the nicotinamide ring on which the ring atoms 1 to 6 appear in anti-clockwise order.

The classical work of Vennesland, Westheimer, and their co-workers showed that hydrogen is transferred *directly* from substrate to coenzyme without undergoing exchange with protons in the medium. These studies also confirmed the stereospecificity of the hydrogen transfer and are reviewed by Colowick *et al.*[15] Two experiments (equations 2 and 3) were carried out with yeast alcohol dehydrogenase.

$$MeCD_2OH + NAD^+ \underset{}{\overset{D_2O}{\rightleftharpoons}} MeCHO + NADH + H^+ \qquad (2)$$

$$MeCH_2OH + NAD^+ \underset{}{\overset{D_2O}{\rightleftharpoons}} MeCDO + NADD + H^+ \qquad (3)$$

The NADH produced in reaction (2) does not contain deuterium, whereas that formed in reaction (3) contains 1 mole of deuterium per mole of reduced NAD^+, thus proving that a hydrogen atom has been transferred directly from the α-carbon of ethanol to the nicotinamide ring without exchange with the protons of the medium. Study of the reverse reactions confirms that alcohol dehydrogenase will remove hydrogen only from that side of the nicotinamide ring to which it had originally been added. Thus re-oxidation with acetaldehyde and the same enzyme of the enzymically produced NADD (equation 3) gave NAD^+ which contained no deuterium.

Yeast alcohol dehydrogenase is stereospecific not only so far as the NADH is concerned but also is stereospecific for the substrate ethanol or acetaldehyde.[16] This is illustrated in Scheme 5. 1-Deuterioacetaldehyde (**6**) when reduced with NADH in the presence of the

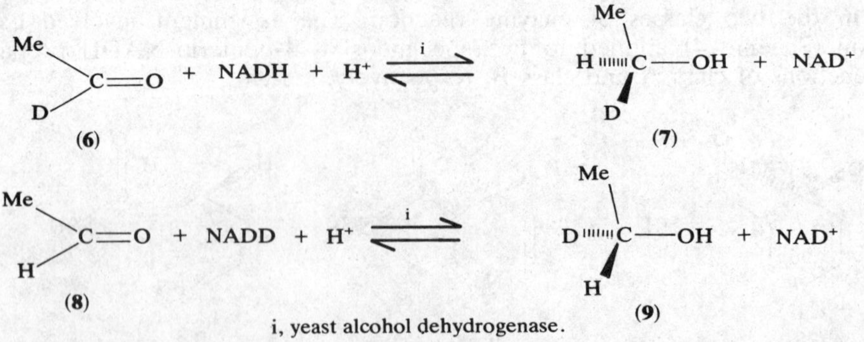

(6) (7) (8) (9)

i, yeast alcohol dehydrogenase.

SCHEME 5

enzyme gives 1*S*-deuterioethanol (**7**), whereas unlabelled acetaldehyde (**8**) with enzymically produced 4-deuterio-NADH gives the enantiomeric 1*R*-deuterioethanol (**9**). The reverse reactions give labelled and unlabelled acetaldehyde, respectively.

Absolute stereoselectivity in reduction has been demonstrated only with acetaldehyde as substrate, but high degrees of stereoselectivity have been reported[15] for other carbonyl derivatives and for alcohols in the reverse reaction.

A question which is still unresolved at the present time is the mechanism of hydrogen transfer between substrate and coenzyme. It has been tacitly assumed by many workers[17] that the reaction involves transfer of a hydride ion from substrate to coenzyme (equation 4). Other possibilities are the simultaneous transfer of two electrons followed by transfer of one proton (equation 5), successive transfer of two single electrons and one proton (equation 6), and separate transfer of one electron and one hydrogen atom (equation 7).

No free radicals have been detected[18] during the enzymic action of alcohol dehydrogenases and this would appear to rule out those routes described by equations (6) and (7). Model studies have not been particularly helpful perhaps because highly reactive abnormal substrates or untypical coenzyme models have been used. Such studies,[19] however, have led to claims that the mechanism involves hydride transfer (equation 4) or initial transfer of a single electron (equations 6 and 7).

$$NAD^+ \xrightarrow{H^-} NADH \tag{4}$$

$$NAD^+ \xrightarrow{2e^-} NAD^- \xrightarrow{H^+} NADH \tag{5}$$

$$NAD^+ \xrightarrow{e^-} NAD^{\cdot} \xrightarrow{e^-} NAD^- \xrightarrow{H^+} NADH \tag{6}$$

$$NAD^+ \xrightarrow{e^-} NAD^{\cdot} \xrightarrow{H^{\cdot}} NADH \tag{7}$$

Hamilton[20] has argued convincingly that a covalent compound is formed between NAD^+ and substrate, thus effecting the transfer of two electrons, and that hydrogen is transferred as a proton (*i.e.* a variant of equation 5). His proposals are set out in Scheme 6.

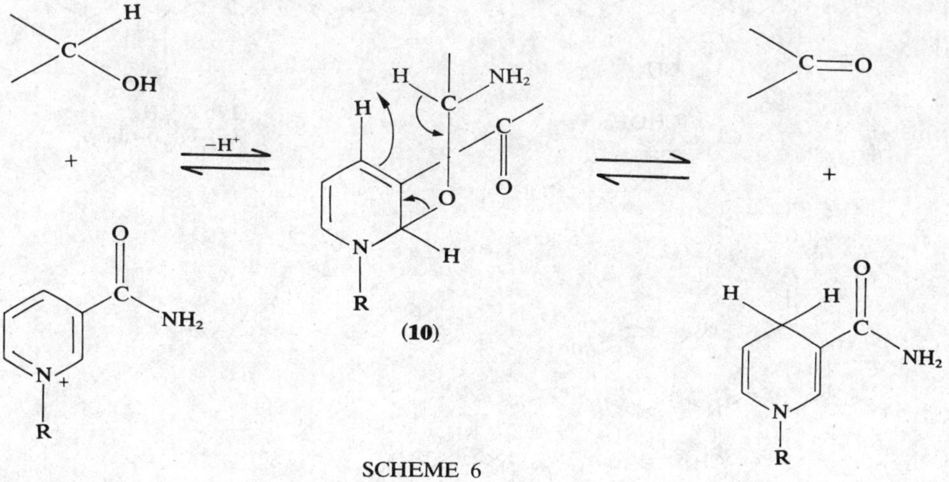

SCHEME 6

The key step in this proposal is the conversion of (**10**) into NADH and the carbonyl compound *via* a six-membered transition state in which the electrons can move as shown, formally the transfer of a proton, or in the opposite sense when hydrogen is transferred as a hydride ion. No definitive evidence has been produced in support of this hypothesis, but in the opinion of the author it remains an attractive possibility.

The ultraviolet absorption spectra of NAD^+ and NADH and their analogues are very characteristic and provide a convenient method for following the progress of enzymic reactions involving these coenzymes.[21] The technique is so simple that enzymes which do not require the coenzymes are often assayed by means of a coupled enzyme system in which the enzyme in question is coupled to another enzyme which does require NAD^+/NADH. The spectrum of NAD^+, unlike that of pyridine, is not affected by change of pH over the range 2–10 and shows λ_{max} 259 nm ($\varepsilon = 17.3 \times 10^3$). Enzymic (or chemical) reduction to NADH results in the appearance of a new maximum at 340 nm ($\varepsilon = 6.22 \times 10^3$) and enzyme assays are commonly carried out at this wavelength. The long wavelength absorption of NADH is due to contributions from the dipolar structure (**11**).

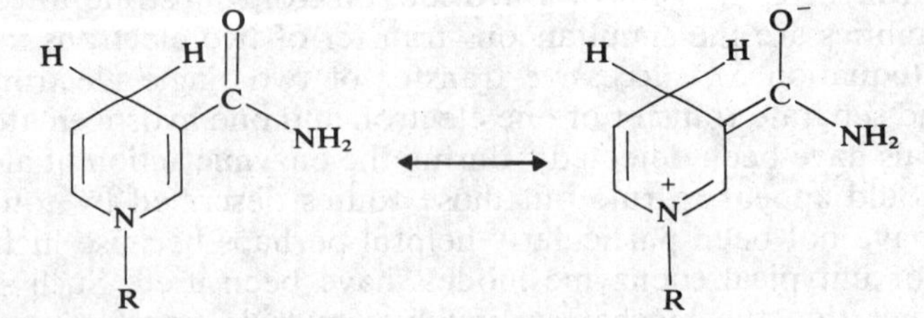

(11)

24.3.2.2 Flavin coenzymes

There are two flavin coenzymes which are usually referred to as flavin mononucleotide (FMN) (**12**) and flavin adenine dinucleotide (FAD) (**13**). This nomenclature is incorrect since FMN is not a nucleotide and FAD is not a dinucleotide. More appropriate names would be flavin monophosphate and flavin adenine diphosphate, but it seems wishful

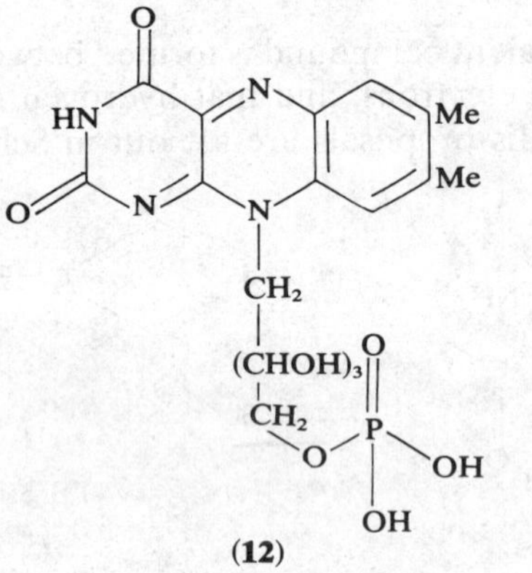

(12)

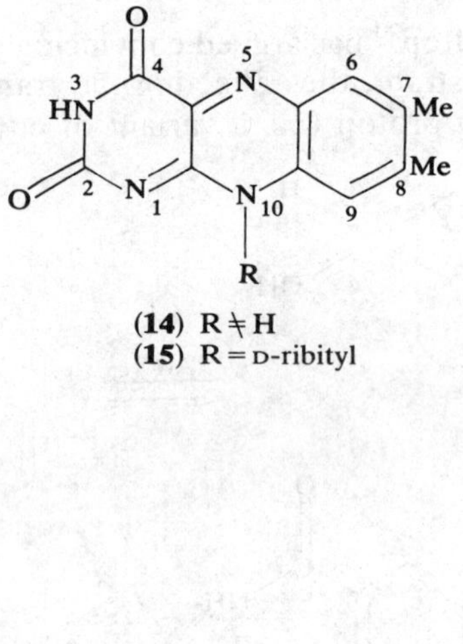

(14) R ≠ H
(15) R = D-ribityl

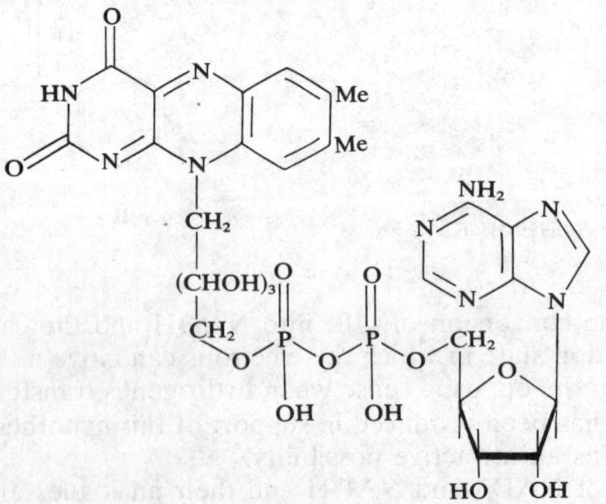

(13)

thinking to hope that the accepted abbreviations FMN and FAD will be changed in the near future. In these names the term 'flavin' is taken to be synonymous with riboflavin, but even here there are ambiguities since 'flavin' is used generally to describe the group of compounds (**14**) of which riboflavin is a member. Attention should also be paid to the numbering of the flavin ring system since many biochemists still use an old, and quite obsolete, German numbering system.

Information about the structure of FMN and FAD was obtained by study of the products of chemical and enzymatic hydrolysis, the most significant product thus obtained being riboflavin (**15**). The chemistry of riboflavin and its derivatives and numerous syntheses are reviewed by Wagner-Jauregg.[22] Confirmation of the structure of the coenzymes was obtained by synthesis. Scheme 7 summarizes the method of Kuhn and his co-workers[23] for the synthesis of FMN, although a more convenient method for the small-scale preparation of riboflavin 5′-phosphate is by direct phosphorylation of riboflavin (a number of different reagents have been used) to give the 4′,5′-cyclic phosphate which on acid hydrolysis gives FMN.[22]

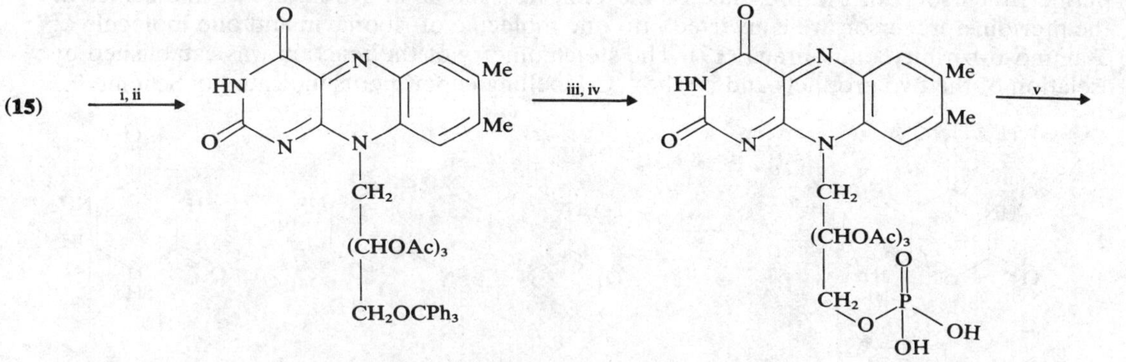

i, Ph_3CCl, py; ii, Ac_2O, py; iii, aq. HOAc; iv, $POCl_3$, py; v, aq. NaOH.

SCHEME 7

Similarly, the early synthesis of FAD by Todd and his co-workers,[24] which involved condensation of a protected adenosine benzylphosphorochloridate with the thallous salt of riboflavin 5′-phosphate and then removal of the protecting groups, has been modified and improved by several groups of workers. The most convenient method appears to be that of Moffatt and Khorana,[25] who condensed riboflavin 5′-phosphate with the phosphoramidate derived from adenosine 5′-phosphate (Scheme 8). These workers also report a convenient chromatographic procedure for the separation of flavin coenzymes using cellulose anion exchange resins.

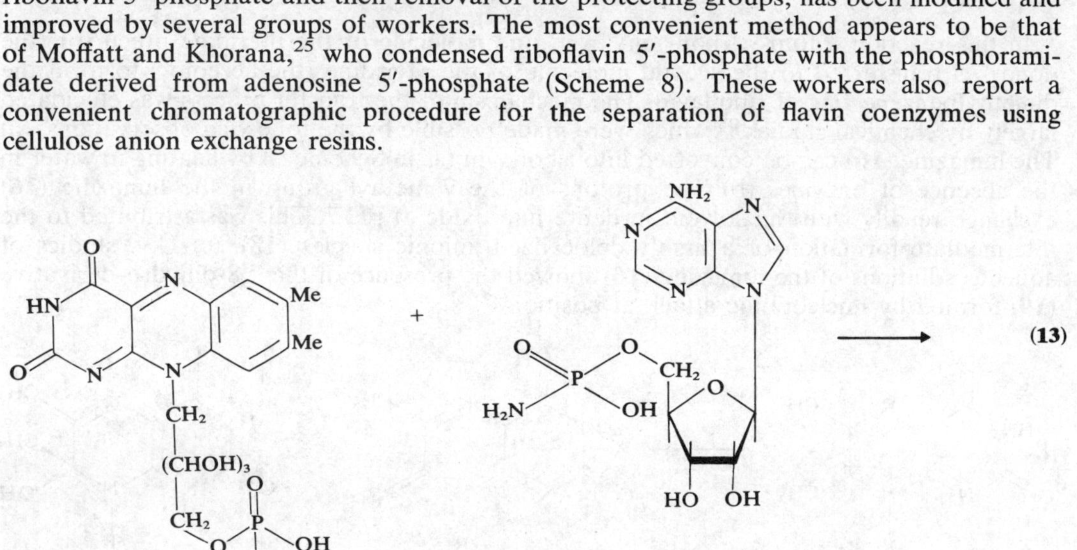

SCHEME 8

The relationship between FMN and FAD and the vitamin riboflavin is very obvious, and enzymes exist in mammalian cells which can carry out the necessary transformations.[26] Riboflavin kinase catalyses the conversion of riboflavin into FMN (equation 8) and FAD-pyrophosphorylase (FMN-adenyltransferase) the subsequent conversion into FAD (equation 9).

$$\text{Riboflavin} + \text{ATP} \longrightarrow \text{FMN} + \text{ADP} \quad (8)$$

$$\text{FMN} + \text{ATP} \longrightarrow \text{FAD} + \text{pyrophosphate} \quad (9)$$

The biosynthesis of riboflavin itself, a process peculiar to micro-organisms, has been studied extensively in recent years. Biochemical[27] and chemical[28] studies have combined to elucidate a most unusual biochemical pathway. The immediate precursor of riboflavin is a pteridine derivative, 6,7-dimethyl-8-D-ribityllumazine (**16**), which is itself formed from purine precursors. In the presence of the enzyme, riboflavin synthase, two molecules of the pteridine precursor are converted into one molecule of riboflavin and one molecule of 5-amino-6-D-ribitylaminouracil (**17**). The stoichiometry of the reaction was established by isolation of the two products and by the ^{14}C-labelling experiments indicated in Scheme 9.

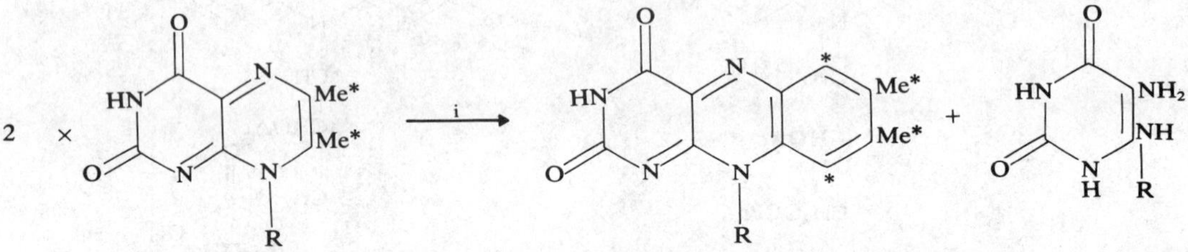

(**16**) R = D-ribityl (**15**) R = D-ribityl (**17**) R = D-ribityl

* = ^{14}C-label

i, Riboflavin synthase or H_2O, 100 °C.

SCHEME 9

In this reaction a four-carbon unit from one molecule of the pteridine precursor (the donor) is transferred to the second molecule of the pteridine (the acceptor) to form the dimethylbenzene ring of riboflavin. The mechanism of the transfer process was elucidated largely by chemical studies,[28] which were made possible by the following observations. (a) The lumazine (**16**) can be converted into riboflavin (as in Scheme 9) by heating in water in the absence of enzyme. (b) The protons of the 7-methyl group in the lumazine (**16**) exchange readily with the solvent in deuterium oxide at pH 7. This was attributed to the intermediate formation of a highly delocalized anionic species (**18**). (c) U.v. studies of aqueous solutions of the lumazine (**16**) showed the presence of the 7,8-dihydro-derivative (**19**) formed by nucleophilic attack at position-7.

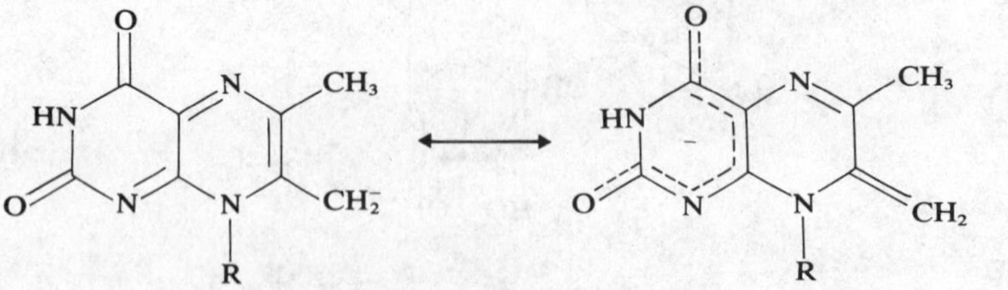

(**18**)

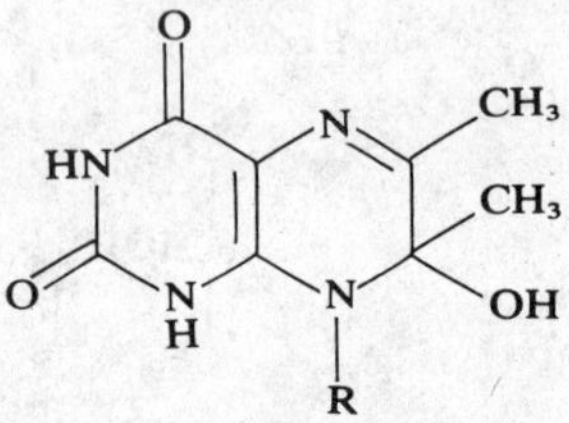

(**19**)

Repetition of the *chemical* experiment summarized in Scheme 9 using deuterium oxide as solvent gave a deuterium-labelled riboflavin (**20**), the position of the deuterium being elucidated by ^{1}H n.m.r. studies. Clearly the four-carbon unit is transferred in some very specific way and the pathway set out in Scheme 10 has been proposed for the chemical reaction. It has been confirmed[27] that the enzyme-catalysed reaction follows an essentially similar pathway, kinetic analysis showing that binding of the lumazine occurs with different affinities to the donor and acceptor sites on the enzyme.

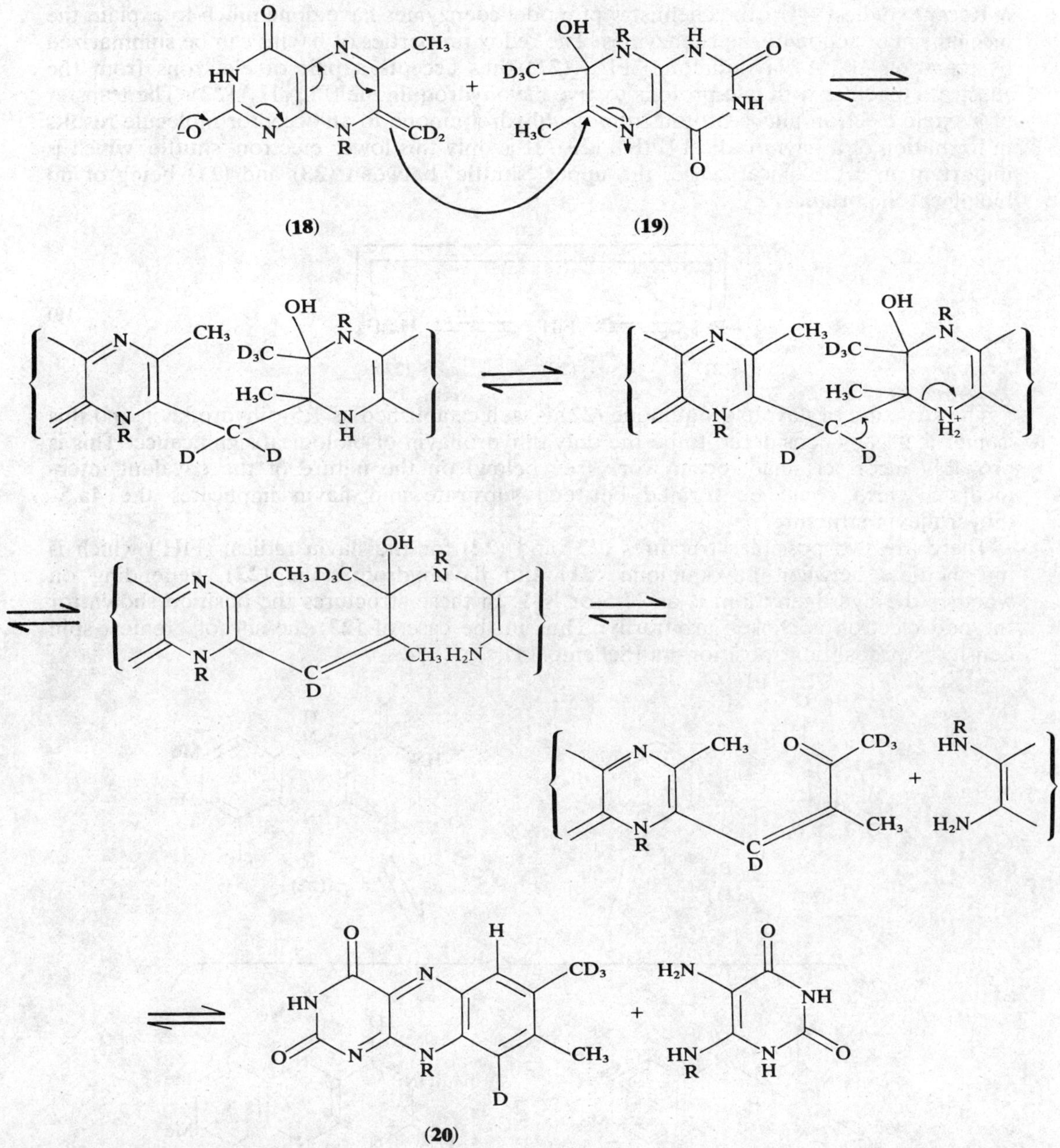

(**18**) (**19**)

(**20**)

SCHEME 10

The flavin coenzymes are involved in a wide variety of redox reactions. These include (a) a number of dehydrogenases, such as NADH-dehydrogenase, which are part of the respiratory chain, (b) flavin oxidases such as D-amino-acid oxidase, and (c) flavin oxygenases which are often involved in hydroxylation reactions. Especially significant is the role of flavin coenzymes in the respiratory chain. They appear to accept electrons from molecules such as NADH which are regarded as obligatory two-electron transfer agents, and donate these electrons one at a time to molecules such as ubiquinone and the cytochromes whose biological activity is strictly radical in nature.

Recent studies[29,30] of the chemistry of model coenzymes have done much to explain the mechanism of action of the coenzymes. The redox properties of flavins can be summarized by equation 10. A flavoquinone (Fl_{ox}) (**21**) thus accepts a pair of electrons from the substrate together with two protons to give flavohydroquinone ($Fl_{red}H_2$) (**22**). The transfer of a single electron plus a proton from flavohydroquinone to an acceptor molecule results in formation of a flavin radical ($FlH^{\bullet}$) (**23**). It is only this lower electron 'shuttle' which is important in a biological sense, the upper 'shuttle' between (**23**) and (**21**) being of no biological importance.

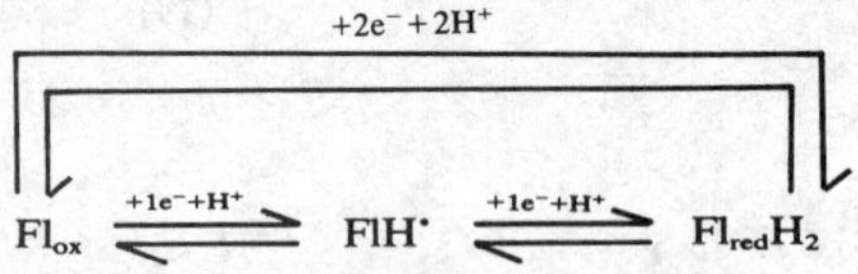

(10)

(**21**) (**23**) (**22**)

The structure of flavohydroquinone (**22**) is well established as 1,5-dihydroflavin and this isomer is usually considered to be the only dihydroflavin of biological significance. This is probably incorrect since recent work (see below) on the nature of the covalent intermediate which could be formed between substrate and flavin implicates the 4a,5-dihydroflavin structure.

There are two possible structures (**23**) and (**24**) for the flavin radical ($FlH^{\bullet}$) which is intermediate between flavoquinone (**21**) and flavohydroquinone (**22**), depending on whether the hydrogen atom is on N-5 or N-1. In these structures the position shown for the odd electron is chosen arbitrarily. Thus in the case of (**23**) the site of greatest spin density is probably at position-4a (Scheme 11).

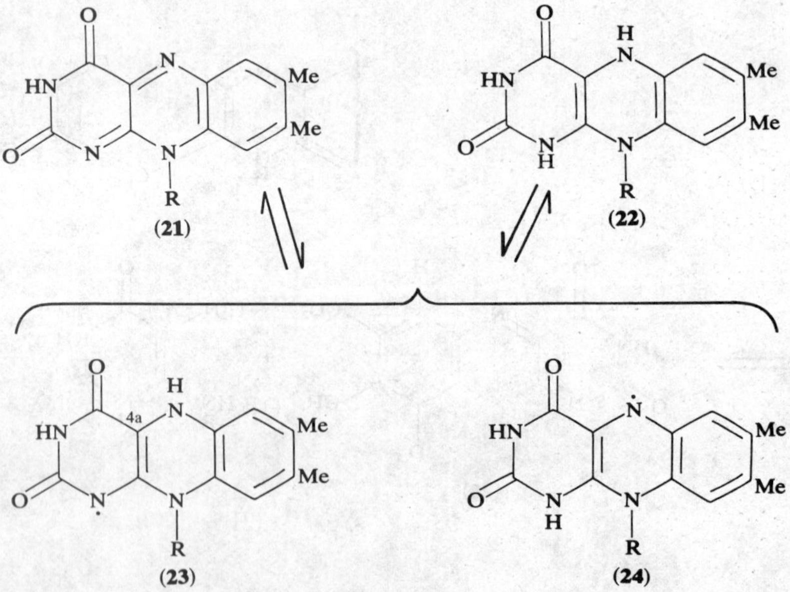

SCHEME 11

Studies using alkylated derivatives have shown that flavin radicals such as (**23**) are blue in colour ($\lambda_{max} \sim 580$ nm) and relatively stable, whereas radicals such as (**24**) are red in colour ($\lambda_{max} \sim 470$ nm) and unstable. It has been suggested[29] that one-electron and two-electron transfer can thus be switched on and off in flavoproteins by formation of strongly directed hydrogen bridges between enzyme protein and flavin coenzyme. If the bridge involves N-5 the radical is stabilized and two-electron transfer is suppressed, whereas a bridge to N-1 destabilizes the radical and suppresses one-electron transfer. Support for this hypothesis comes from the work of Massey *et al.*,[31] who were able to show that all flavoproteins (dehydrogenases) which were involved with one-electron transfer (either by input or output or both) gave blue radicals which were essential to the biological function, whereas flavoproteins (oxidases and oxygenases) which were solely concerned with two-electron transfer gave red radicals which were non-essential.

The nature of the interaction between flavin coenzyme and substrate has been the subject of intense interest in recent years.[20,29,30] Opinion now seems to favour formation of a covalent intermediate. The precise nature of such an intermediate is still an open question, but it seems likely that a carbanion derived from the substrate[32] attacks the flavoquinone (**21**) to give a 4a,5-dihydroflavin derivative (**25**). Transfer of a pair of electrons is thus achieved through the covalent intermediate, which then gives flavohydroquinone (**22**) and oxidized substrate. This is illustrated for a typical dehydrogenation in Scheme 12.

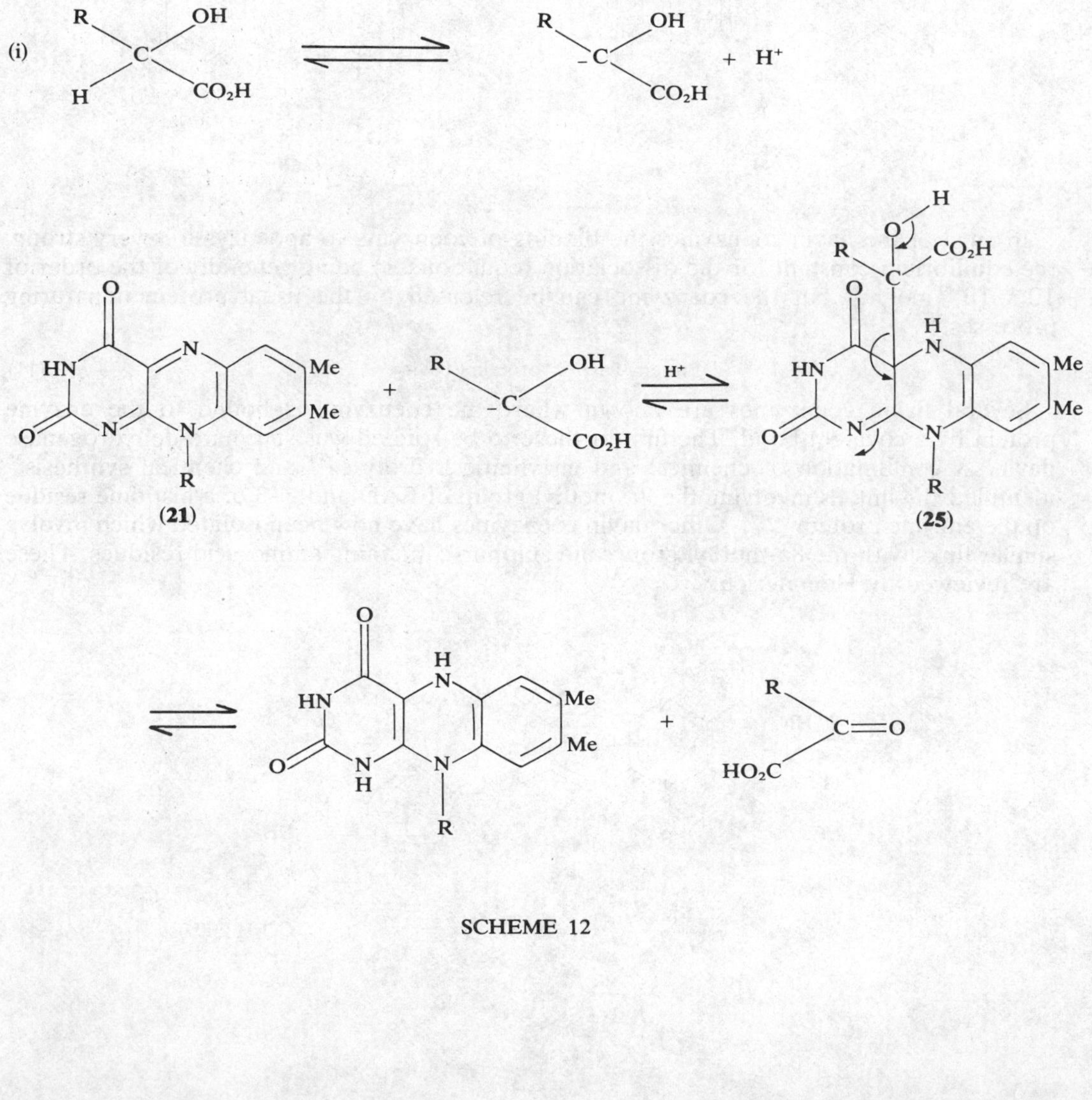

SCHEME 12

The method whereby electrons are transferred from flavohydroquinone (**22**) to an electron acceptor is not well understood. It has been suggested[29] that transfer of a single electron to an acceptor such as Fe(III) in a cytochrome molecule could involve a planar form of flavohydroquinone which normally adopts a 'butterfly' conformation. Such a planar structure would be formally 'anti-aromatic' and as such would be a potent one-electron donor.

The interaction of flavohydroquinone (**22**) with molecular oxygen which is necessary in the flavin oxidases and oxygenases must involve some type of dihydroflavin–oxygen complex. Two hydroperoxide structures (**26**) have been proposed.[31] There is some evidence[33] for the first of these being involved in flavin-dependent bacterial bioluminescence where, in the absence of substrate, the flavohydroquinone–enzyme complex reacts with oxygen to give a substance with $\lambda_{max} \sim 372$ nm. This is the absorption expected for a 4a,5-dihydroflavin. At the present time there is little more direct evidence to support these hypotheses although it may be that both, and possibly others,[29] are involved in different reactions. Theories[20,29,30] are also plentiful to explain how one or other of these hydroperoxide adducts can function as donor of an 'oxene' equivalent which is required to explain aromatic hydroxylation by flavin oxygenases (Chapter 30.3).

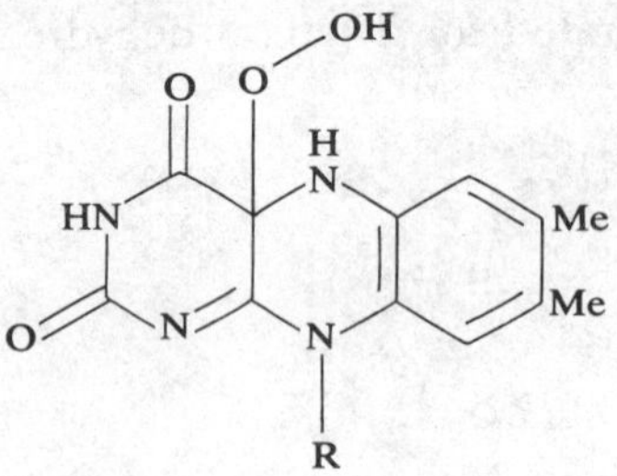

(**26a**)

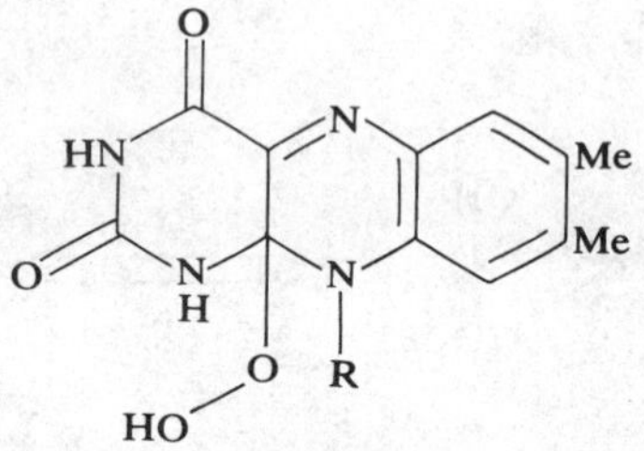

(**26b**)

In most of the flavin coenzymes the binding of coenzyme to apoenzyme is very strong, the equilibrium constant for the dissociation (equation 11) being generally of the order of 10^{-8}–10^{-9} mol l^{-1}, but the coenzyme can be released by the usual protein denaturing processes.

$$\text{Protein–flavin} \rightleftharpoons \text{protein} + \text{flavin} \qquad (11)$$

Several flavin coenzymes are known where the coenzyme is bound to the enzyme protein by a covalent bond. The first of these to be studied was succinate-dehydrogenase flavin. A combination of chemical and enzymatic hydrolyses[34] and chemical synthesis[35] identified the link as involving the 8a-methyl group of FAD and N-3 of a histidine residue on the enzyme protein (**27**). Other flavin coenzymes have now been isolated which involve similar links with the 8a-methyl group and sulphur-containing amino-acid residues. These are reviewed by Hemmerich.[29]

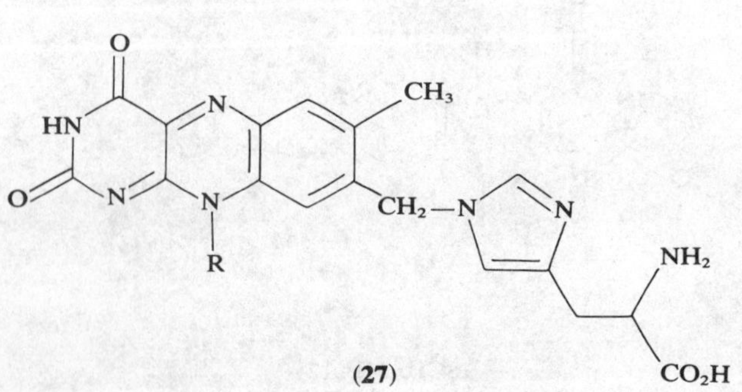

(**27**)

24.3.2.3 Ubiquinones

The ubiquinones, or coenzymes Q, are a group of benzoquinone derivatives which can be represented by the general formula (**28**). The number of isoprenoid units in the side-chain varies, in the naturally occurring compounds, from 6 to 10. Thus in the ubiquinones present in man and most mammals there are 10 such units, whereas in micro-organisms the number varies between 6 and 8. Nomenclature has been confused in the past and several systems have appeared in the literature. Ubiquinone, rather than coenzyme Q, is now preferred and the number of isoprenoid residues is indicated by the suffix 6–10. Thus ubiquinone-10 is that which occurs in man. Two other compounds related to ubiquinone-10 have been isolated from natural sources. One is ubichromenol (**29**), which is easily formed by cyclization of ubiquinone-10 and which may be an artefact. The other is a partially reduced derivative of ubiquinone-10 in which the double bond in the terminal isoprenoid unit is saturated.

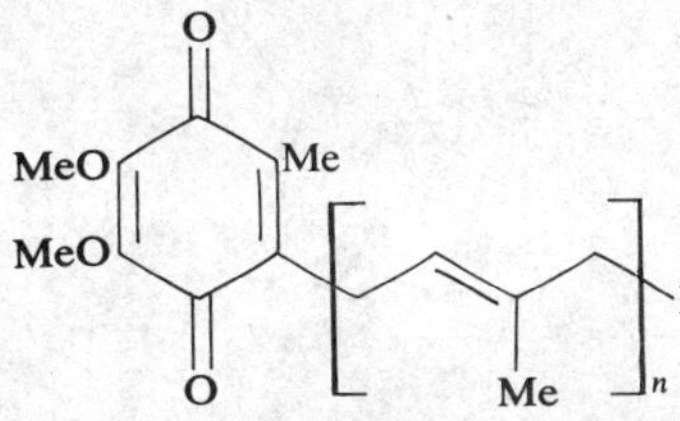

(**28**) $n = 6–10$

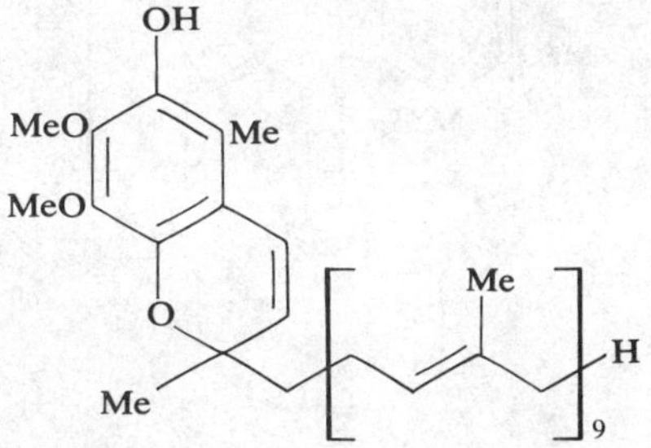

(**29**)

The structure of ubiquinone-10 was elucidated by a combination of degradative, spectroscopic, and synthetic studies and the evidence is reviewed by Hatefi.[36] The most important observations included elemental and functional group analysis, which established the molecular formula of $C_{59}H_{90}O_4$ and the presence of two methoxyl groups. The presence of the latter was confirmed by ^1H n.m.r. studies, which also showed the presence of an aromatic methyl group and a polyisoprenoid chain, and the absence of aromatic protons. Infrared studies[37] confirmed that the polyisoprenoid chain was all-*trans*. Recognition of the benzoquinone ring system followed from its reversible reduction under mild conditions to a quinol and the resultant change in the u.v. absorption maximum from 275 to 290 nm. The oxidative degradation summarized in Scheme 13 completed the structural assignment.

The synthesis[38] of ubiquinone-10 and its analogues involves condensation of 2,3-dimethoxy-5-methylhydroquinone (**30**) with the appropriate all-*trans* allylic alcohol (**31**) or alternatively the all-*trans* tertiary alcohol (**32**). The product of the condensation reaction is a hydroquinone derivative which is easily oxidized to the corresponding quinone (Scheme 14).

Elucidation of the biosynthetic pathway proved to be a more complicated study. The ubiquinones, unlike the structurally related compounds in the vitamin K group, can be biosynthesized by animal tissue and thus are not considered to be vitamins. The polyisoprenoid side-chain was shown[39] to be formed by the accepted biosynthetic pathways of terpene biosynthesis leading from mevalonate *via* allylic pyrophosphate esters. The aromatic ring appears to be formed from chorismic acid and the essential aromatic amino-acids, phenylalanine and tyrosine (**33**). The accepted biosynthetic pathway[40] leads *via* *p*-hydroxybenzoate (**34**), which is converted into a 3-polyprenyl-4-hydroxybenzoate (**35**) by alkylation with the appropriate allylic pyrophosphate. A sequence of decarboxylation, hydroxylation, and methylation gives 6-methoxy-2-polyprenylphenol (**36**), which is converted into 5-demethoxyubiquinone (**37**) by oxidation and methylation and thence to ubiquinone. The most significant work has been done using rat-liver homogenates or

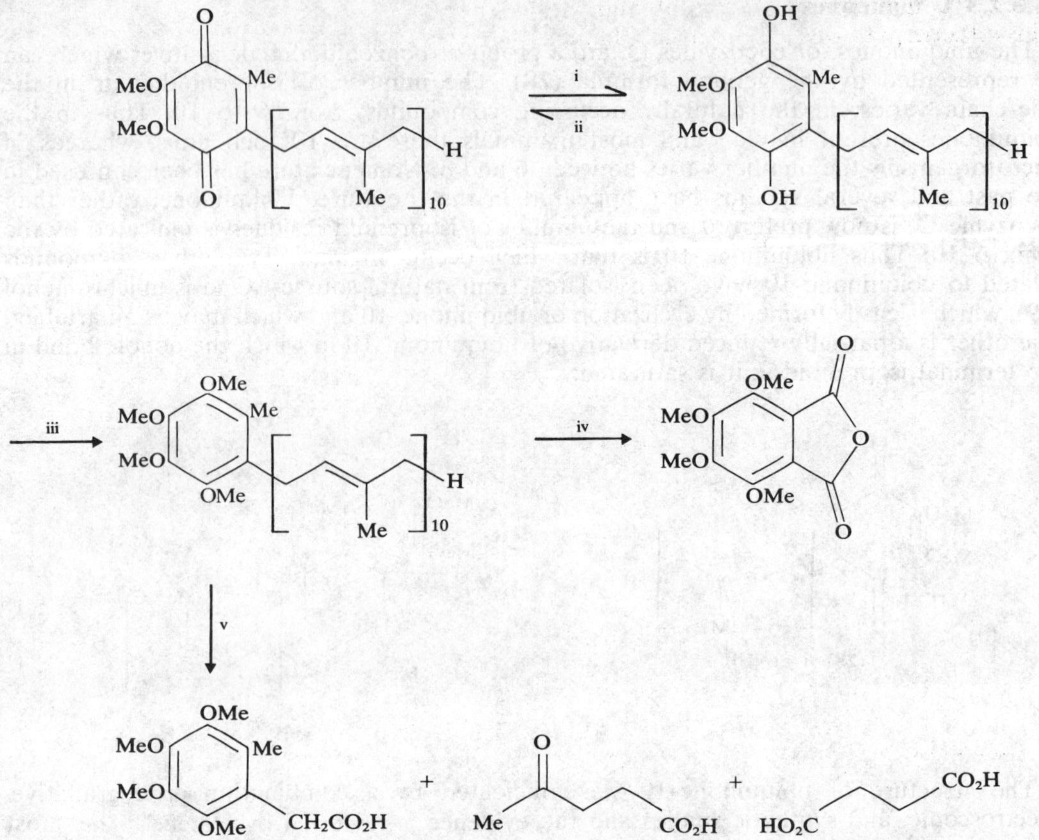

i, Zn, HOAc; ii, $FeCl_3$ or Ag_2O; iii, Me_2SO_4; iv, excess $KMnO_4$; v, 10 eq. $KMnO_4$ in Me_2CO.

SCHEME 13

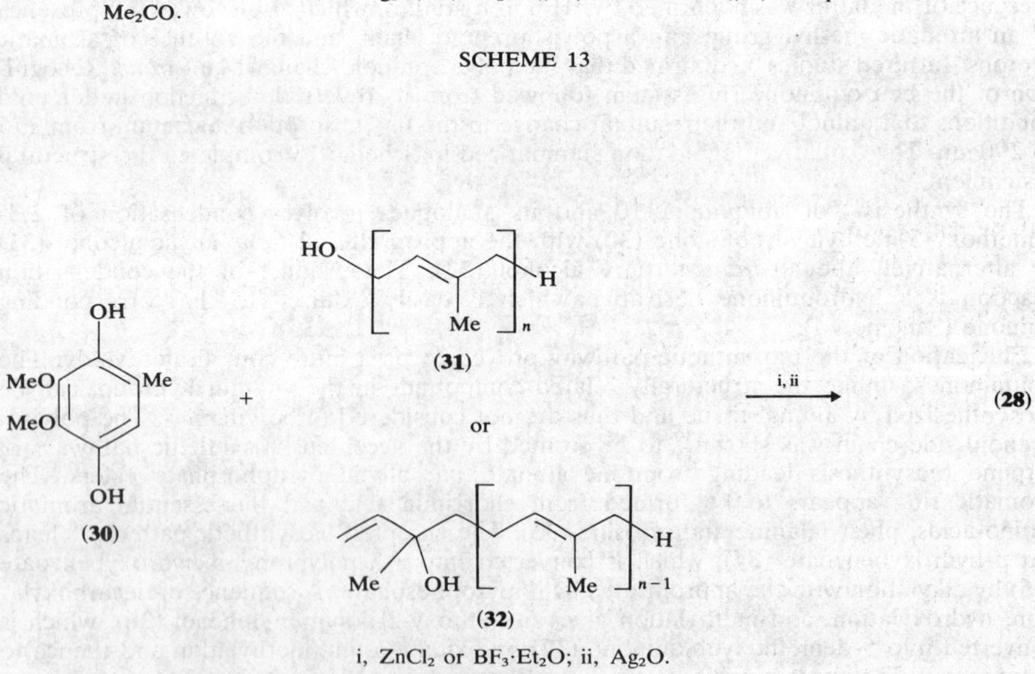

(30) (31) (32) (28)

i, $ZnCl_2$ or $BF_3{\cdot}Et_2O$; ii, Ag_2O.

SCHEME 14

slices, the final product being ubiquinone-9 (Scheme 15). Further observations on this biosynthetic pathway are given in Chapter 30.3.

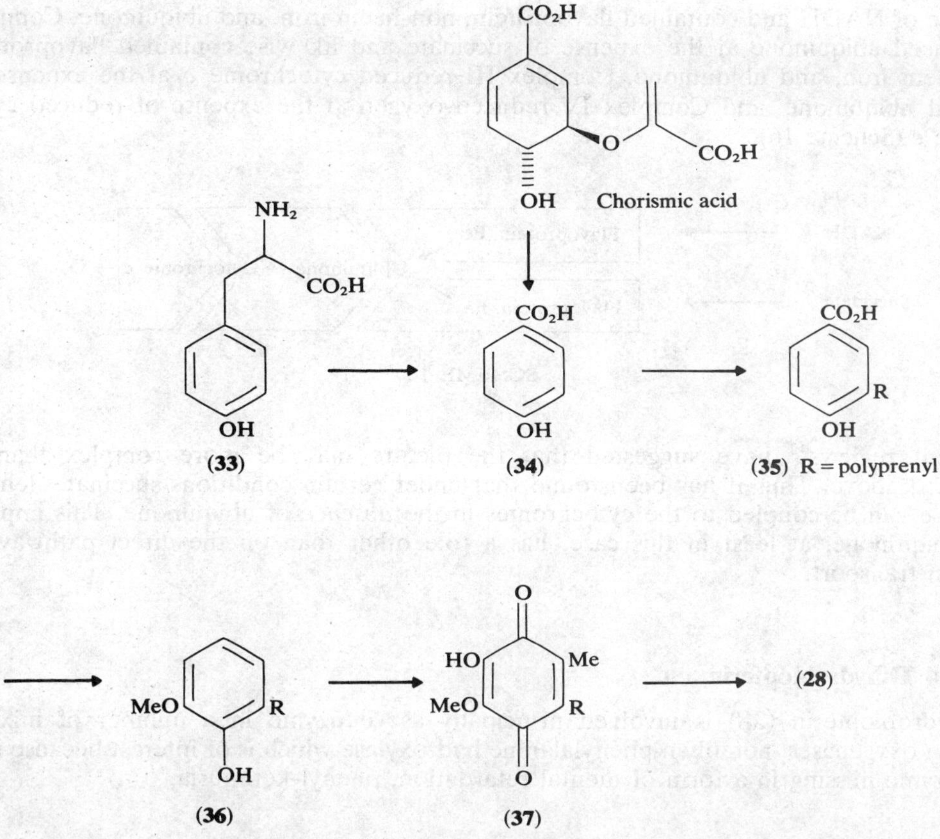

SCHEME 15

The biological function of the ubiquinones in redox reactions depends upon the reversible reduction to the corresponding hydroquinone. When the reducing agent is a one-electron donor the reaction can be viewed as a two-stage process involving an intermediate radical (equation 12). The ubiquinones thus function in electron transport in the respiratory chain. It is generally accepted that they are intermediate between the flavoproteins, which have a two-electron input and a one-electron output, and the cytochromes, which have a one-electron input corresponding to Fe(III) → Fe(II).

Support for this role for ubiquinone in respiratory electron transport comes from the fact that lyophilized mitochondria after extraction with pentane and resuspension in buffer were not able to conduct electron transport. Viability was restored, however, by addition

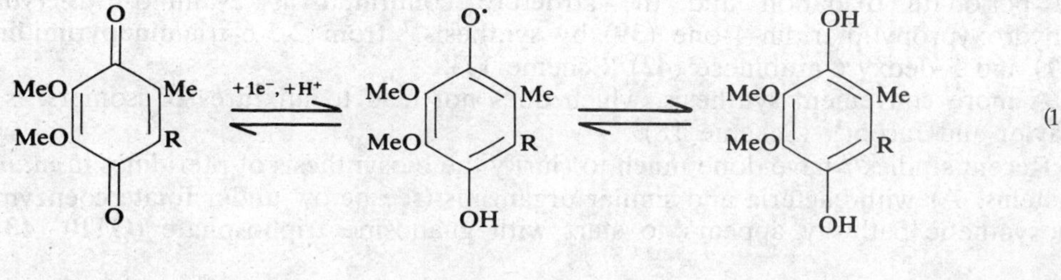

(12)

of ubiquinone-10 or a lower homologue.[41] Moreover, methods of mitochondrial fractionation have been developed which allow four complexes to be isolated which together are able to conduct electron transport in its entirety.[42] Complex I reduced ubiquinones at the expense of NADH and contained flavoprotein, non-haem iron, and ubiquinone. Complex II reduced ubiquinone at the expense of succinate and likewise contained flavoprotein, non-haem iron, and ubiquinone. Complex III reduced cytochrome *c* at the expense of reduced ubiquinone, and Complex IV reduced oxygen at the expense of reduced cytochrome *c* (Scheme 16).

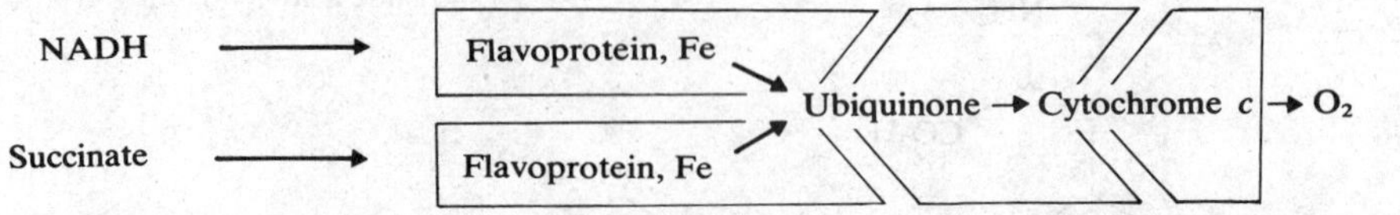

SCHEME 16

Recent reviews[43] have suggested that the picture may be more complex than is suggested above. Thus it has been found that under certain conditions succinate dehydrogenase can be coupled to the cytochromes in the *absence* of ubiquinone. This implies that ubiquinone, at least in this case, has a role other than on the direct pathway of electron transport.

24.3.2.4 Dihydrobiopterin

Dihydrobiopterin (**38**) is involved principally as coenzyme in a number of mixed-function oxygenases, notably L-phenylalanine hydroxylase which is of interest because it is the enzyme missing in a form of mental retardation, phenyl-ketonuria.[44]

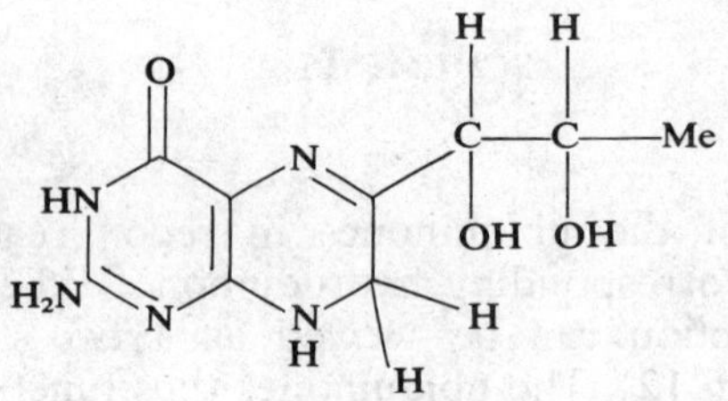

(**38**)

The coenzyme is the 7,8-dihydro-derivative of biopterin (**39**), a pteridine first isolated from human urine[45] and also from *Drosophila melanogaster*.[46] The structure of biopterin was elucidated by a combination of spectral studies, the u.v. spectrum being very similar to that of 2-aminopteridin-4-ones, and oxidation with hot alkaline permanganate which gave 2-amino-6-carboxypteridin-4-one (**40**). The nature of the side-chain was determined by periodate oxidation and the structure confirmed as 2-amino-6-(L-*erythro*-1,2-dihydroxypropyl)pteridin-4-one (**39**) by synthesis[47] from 2,5,6-triaminopyrimidin-4-one (**41**) and 5-deoxy-L-arabinose (**42**) (Scheme 17).

A more convenient synthesis, which does not lead to mixtures of isomers, is that of Taylor and Jacobi[48] (Scheme 18).

Recent studies[49] have done much to clarify the biosynthesis of pteridines in *mammalian* systems. As with bacteria and similar organisms (see below under folate coenzymes), the biosynthetic pathway appears to start with guanosine triphosphate (GTP) (**43**) which

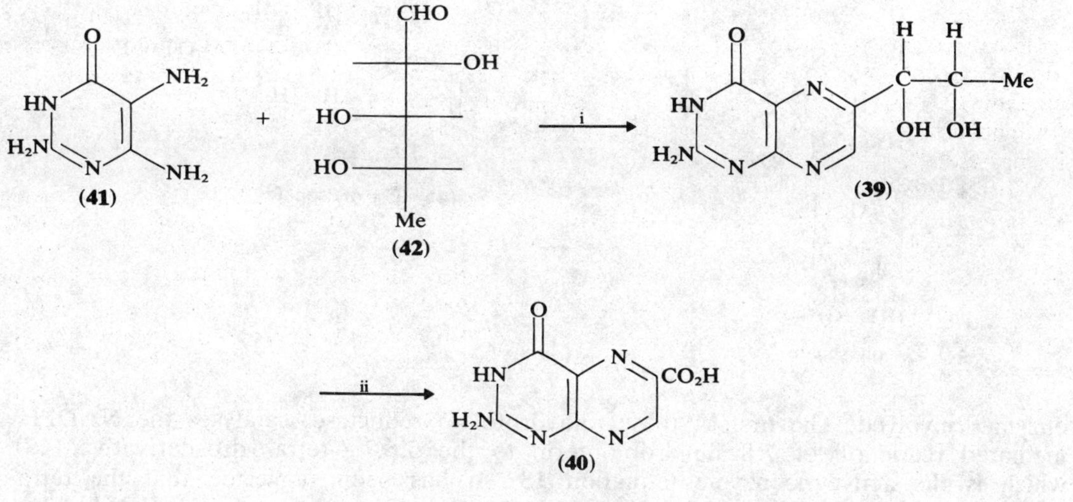

i, $N_2H_4 \cdot H_2O$; ii, alk. $KMnO_4$, 100 °C.

SCHEME 17

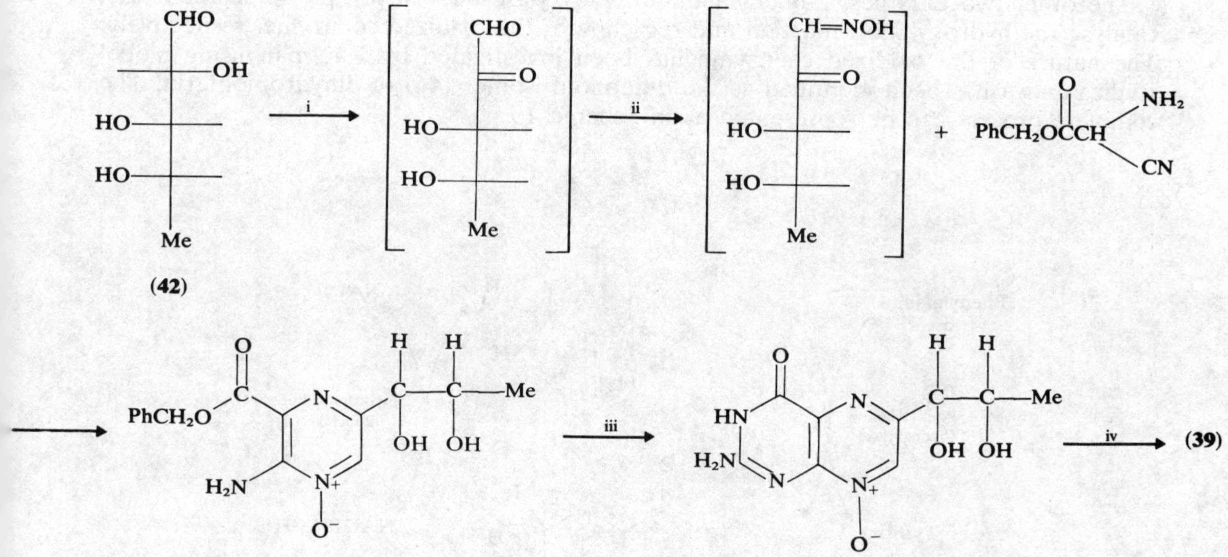

i, $Cu(OAc)_2$; ii, acetone oxime; iii, guanidine HCl, NaOMe; iv, $Na_2S_2O_4$.

SCHEME 18

undergoes ring-opening of the imidazole ring, loss of what was C-8 of the purine ring system, and ring-closure to give D-*erythro*-dihydroneopterin triphosphate (**44**), a sequence of reactions catalysed by GTP-cyclohydrolase. The way in which the side-chain of D-*erythro*-dihydroneopterin triphosphate is converted into that present in biopterin or 7,8-dihydrobiopterin is not yet known.

The role of dihydrobiopterin in the enzymatic hydroxylation of phenylalanine has been clarified by the elegant studies of Kaufman.[44] These show clearly that there are three

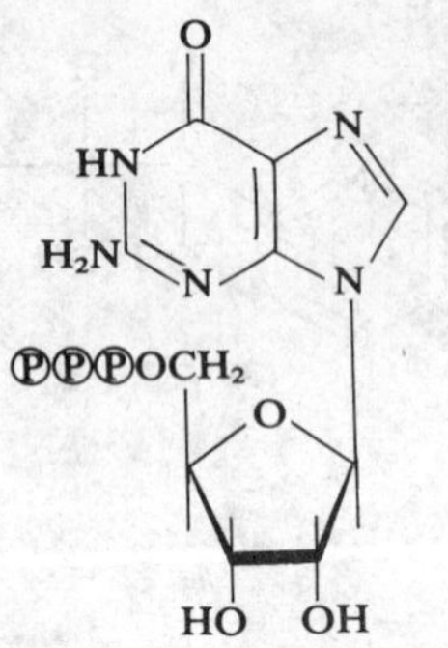

(43) Ⓟ = phosphate

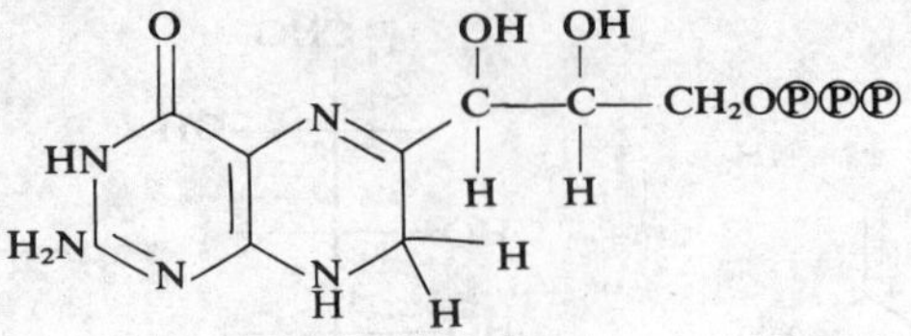

(44) Ⓟ = phosphate

enzymes involved. The first of these, dihydrofolate reductase, catalyses the NADPH-mediated reduction of 7,8-dihydrobiopterin to the 5,6,7,8-tetrahydro-derivative (**45**), which is the active coenzyme (equation 13). It has been suggested that the tetra-hydro-derivative (**45**) is probably the naturally occurring coenzyme, the 7,8-dihydro-compound being an artefact formed during isolation, and that the enzymic reduction (equation 13) is a salvage mechanism.

$$\text{7,8-Dihydrobiopterin} + \text{NADPH} + \text{H}^+ \rightarrow \text{Tetrahydrobiopterin} + \text{NADP}^+ \quad (13)$$

The other two enzymes, L-phenylalanine hydroxylase and dihydropteridine reductase, catalyse the hydroxylation reaction and recycling of the oxidized coenzyme, respectively. The nature of the oxidized coenzyme has been investigated by Kaufman using model studies and it has been identified as the quinonoid isomer (**46**) of dihydrobiopterin. The complete process can be represented as in Scheme 19.

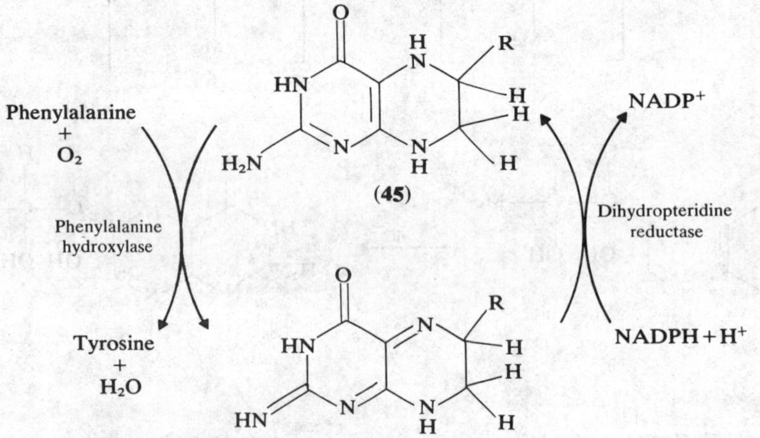

SCHEME 19

Most recent work has concentrated on the mechanism whereby tetrahydrobiopterin (**45**) reacts with molecular oxygen to give a species capable of carrying out the hydroxylation reaction. Evidence has been obtained[50] for such an intermediate species and the reaction catalysed by phenylalanine hydroxylase has now been formulated in more detail and involves the intermediate pteridine hydroperoxide (**47**) and hydrated pteridine derivative (**48**) (Scheme 20). An alternative proposal[51] involving the 8a-hydroperoxide (**49**) rather than the 4a-hydroperoxide has also been put forward, based largely on model studies.

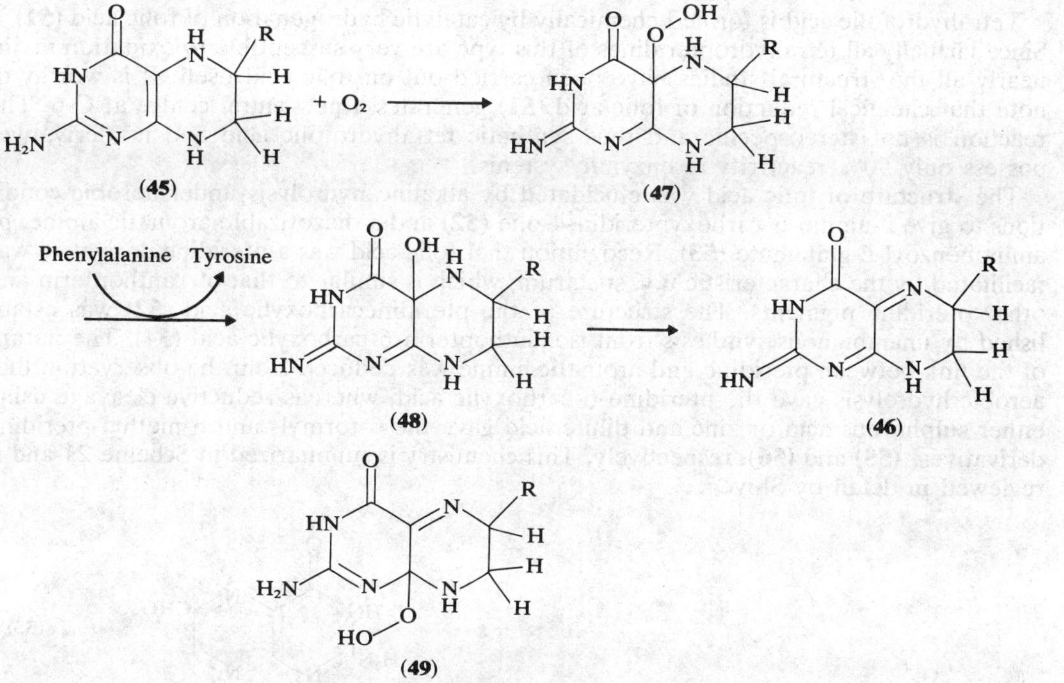

(45) (47) (48) (46) (49)

SCHEME 20

24.3.3 COENZYMES INVOLVED PRINCIPALLY IN GROUP TRANSFER REACTIONS

Enzymes frequently catalyse the transfer of groups of atoms from substrate to product. The function of the coenzyme is to accept the group from the substrate, usually by formation of a covalent derivative, and to donate this group to an acceptor molecule, thus forming the desired product. The nature of the covalent derivative formed between coenzyme and the group being transferred is such that (a) the group acquires the necessary reactivity to combine with the acceptor molecule, and (b) on occasion the nature of the group being transferred, for example its oxidation level, is modified while attached to the coenzyme.

24.3.3.1 Folate coenzymes

The folate coenzymes, which are involved in the transfer and interconversion of one-carbon atom units at the oxidation levels of formate, formaldehyde, and methanol, are all derivatives of 5,6,7,8-tetrahydrofolic acid (**50**).

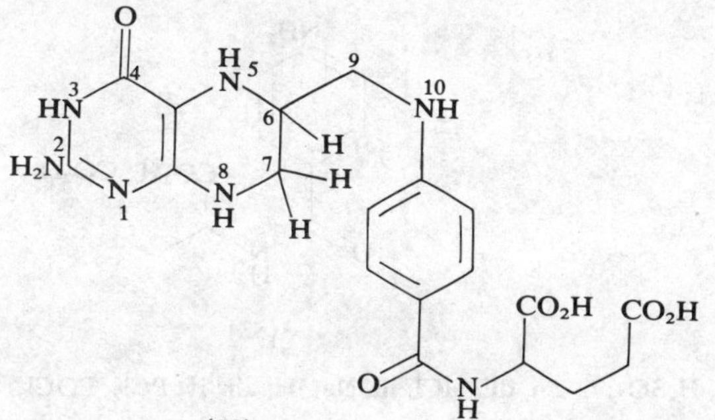

(**50**)

Tetrahydrofolic acid is formed chemically by catalytic hydrogenation of folic acid (**51**).[52] Since virtually all tetrahydropteridines of this type are very susceptible to oxidation in air, nearly all the structural studies have been carried out on folic acid itself. It is worthy of note that chemical reduction of folic acid (**51**) generates a new chiral centre at C-6. The reaction is not stereospecific and thus synthetic tetrahydrofolic acid and its derivatives possess only 50% reactivity in enzymic systems.

The structure of folic acid was elucidated by alkaline hydrolysis under aerobic conditions to give 2-amino-6-carboxypteridin-4-one (**52**) and a diazotizable aromatic amine, *p*-aminobenzoyl-L-glutamate (**53**). Recognition that folic acid was a pteridine derivative was facilitated by the characteristic u.v. spectrum, which is similar to that of xanthopterin and other pteridine pigments. The structure of the pteridinecarboxylic acid (**52**) was established by unambiguous synthesis from isoxanthopterin-6-carboxylic acid (**54**). The nature of the link between pteridine and aromatic amine was deduced from the observation that aerobic hydrolysis gave the pteridine-6-carboxylic acid, whereas reductive cleavage using either sulphurous acid or zinc and dilute acid gave the 6-formyl- and 6-methyl-pteridine derivatives, (**55**) and (**56**), respectively. This chemistry is summarized in Scheme 21 and is reviewed in detail by Shive.[53]

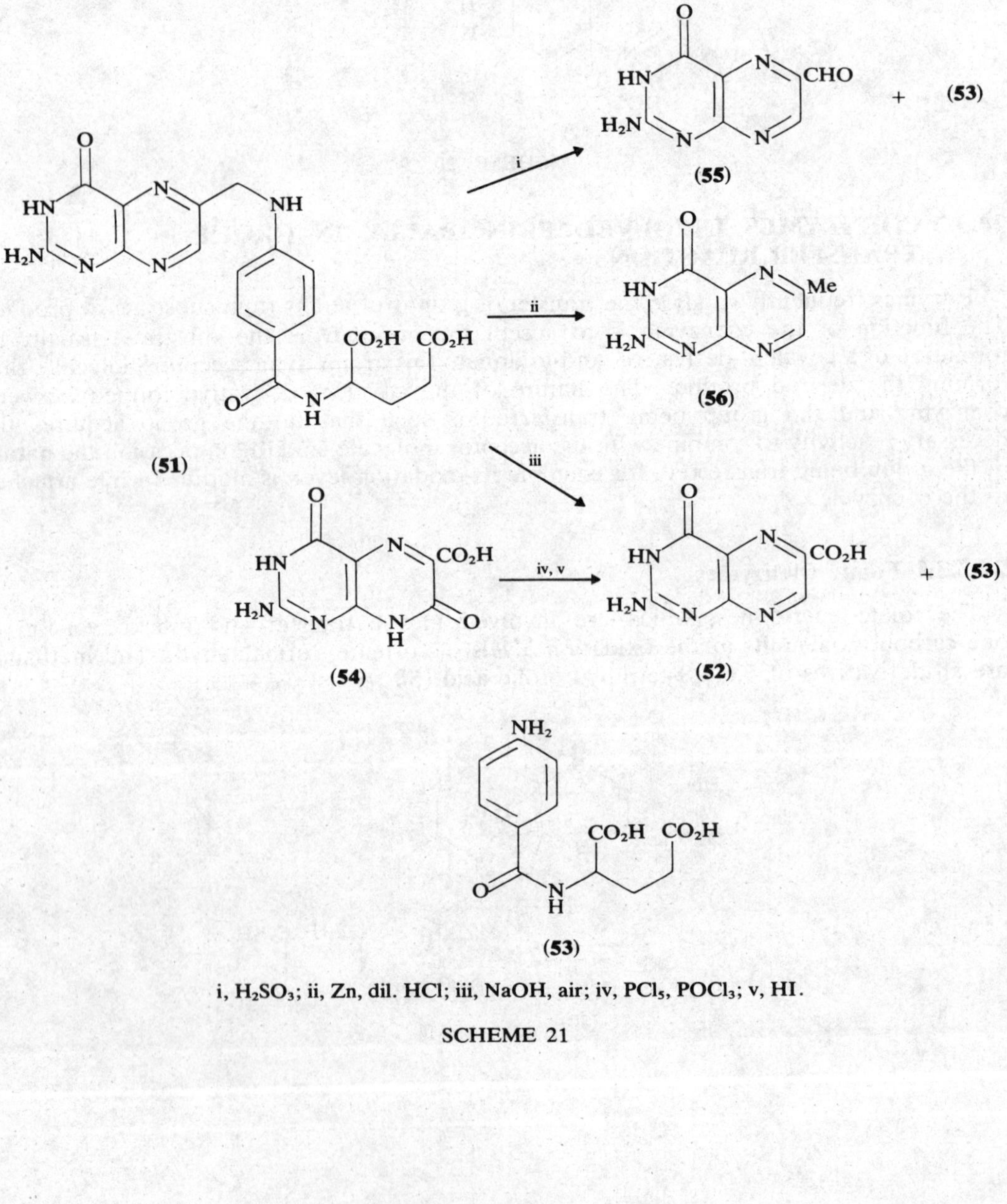

i, H_2SO_3; ii, Zn, dil. HCl; iii, NaOH, air; iv, PCl_5, $POCl_3$; v, HI.

SCHEME 21

Many syntheses of folic acid have been reported and these are of two general types. The first (Scheme 22) involves condensation of 2,5,6-triaminopyrimidin-4-one with a three-carbon compound such as 2,3-dibromopropanal in the presence of *p*-aminobenzoylglutamate. The reaction mixture contains about 25% of the desired compound and extensive purification is required.[53] A more effective type of synthesis (Scheme 23) starts from a preformed pteridine derivative, which on condensation with *p*-aminobenzoylglutamate gives folic acid in 60% yield in pure crystalline form.[54]

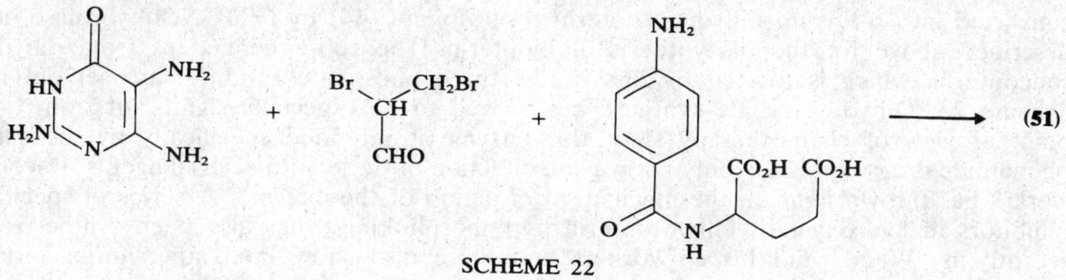

SCHEME 22

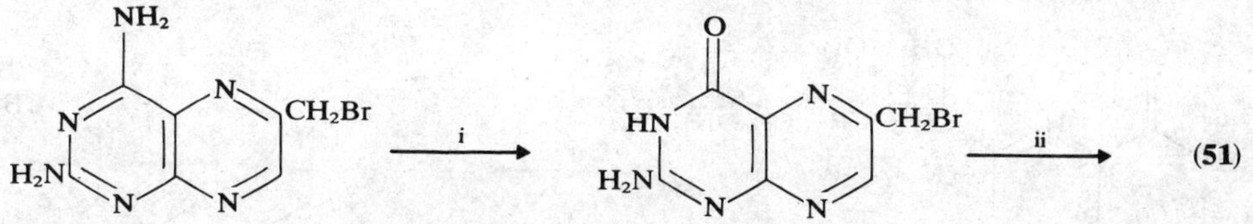

i, 48% HBr, 100 °C; ii, *p*-aminobenzoyl-L-glutamic acid.

SCHEME 23

The biosynthesis of tetrahydrofolic acid (**50**) is best considered in two stages. The first of these, the formation of 7,8-dihydrofolic acid (**57**) from purine nucleotide precursors, is a process peculiar to micro-organisms and other parasitic species, most of which have the necessary enzyme systems to carry out *de novo* synthesis. The second stage, the reduction of 7,8-dihydrofolic acid to give 5,6,7,8-tetrahydrofolic acid, is catalysed by an enzyme dihydrofolate reductase and is a process common to mammalian and bacterial systems. Specific inhibition of one or more of the enzymes involved in these biosynthetic pathways has proved to be a most valuable approach to the design of antibacterial and antiprotozoal agents, and it is thus worth considering both the approach[2] and the biosynthetic pathways in some detail.

The difference between host and parasite in relation to tetrahydrofolate biosynthesis is summarized in Scheme 24. Most parasitic organisms are incapable of absorbing folic acid from their environment and are dependent on that produced by *de novo* synthesis. Clearly

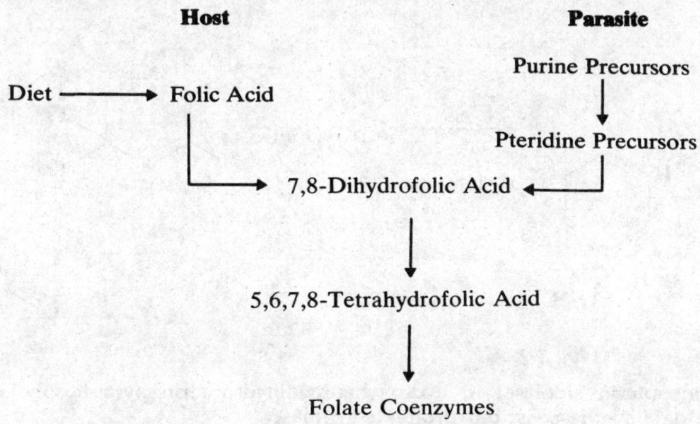

SCHEME 24

specific inhibition of one or more of the enzymes in the biosynthetic pathway (see below) leading to 7,8-dihydrofolic acid is likely to produce a selectively toxic agent. What is perhaps not so obvious is that specific inhibition of dihydrofolate reductase, an enzyme *common* to host and parasite, has also given rise to very valuable chemotherapeutic agents.

The biosynthesis of 7,8-dihydrofolic acid has been reviewed comprehensively by Shiota.[55] Guanosine triphosphate (GTP) (**43**) is the most effective precursor and is converted into D-*erythro*-dihydroneopterin triphosphate (**44**) by GTP cyclohydrolase, as described above for the biosynthesis of biopterin. The subsequent steps from dihydroneopterin, which is formed by loss of the three phosphate residues, are set out in Scheme 25. Dihydropteroate synthase has proved to be especially significant from the point of view of chemotherapy since this enzyme is inhibited specifically by the sulphonamide drugs and is the only known site of action of these antibacterial agents. Recent work[56] has thrown light on the mechanism of action of the enzyme. A series of specific inhibitors of hydroxymethyldihydropterin pyrophosphokinase has also been synthesized recently by Wood[2] and his co-workers and these also show interesting antibacterial activity.

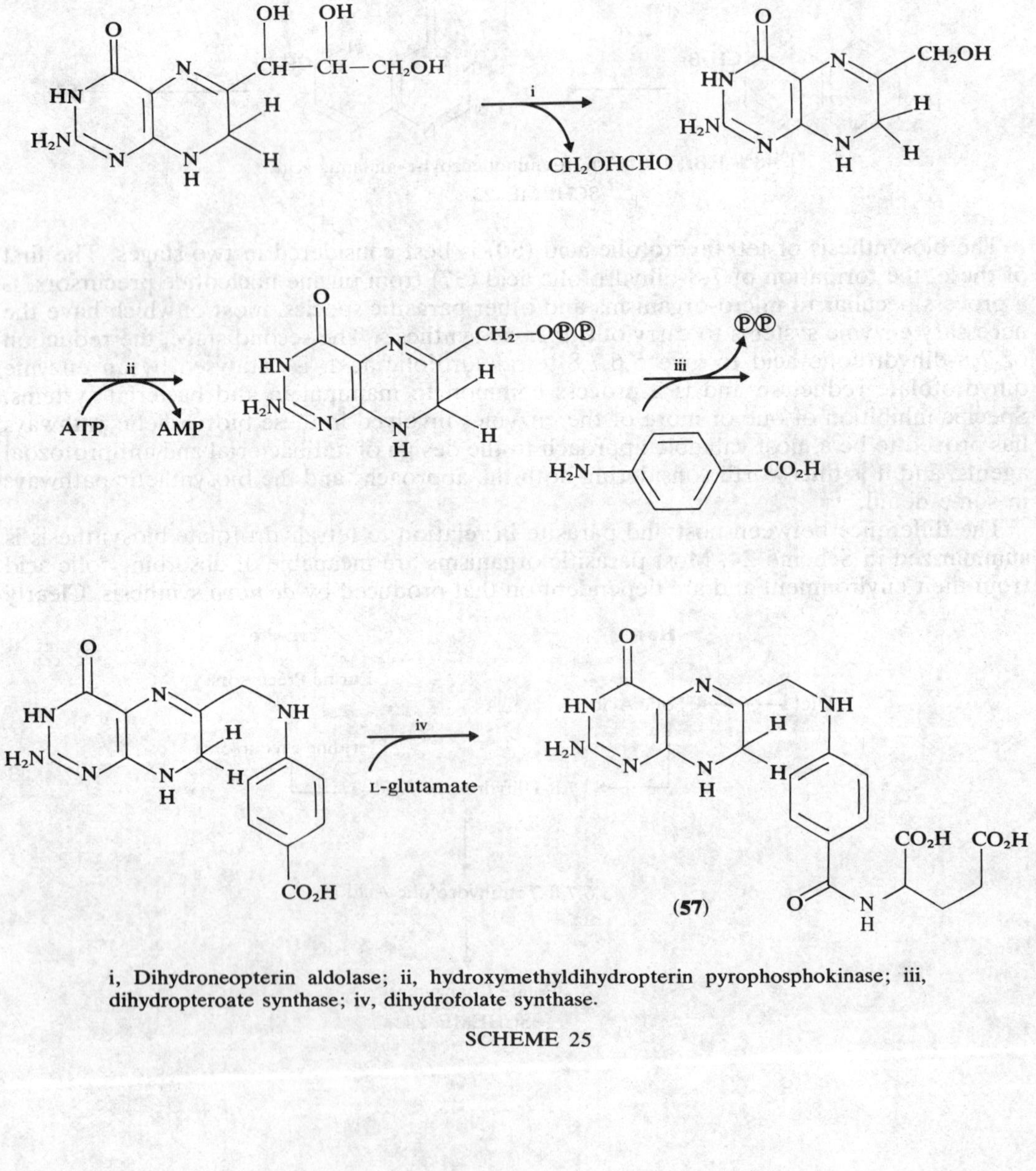

i, Dihydroneopterin aldolase; ii, hydroxymethyldihydropterin pyrophosphokinase; iii, dihydropteroate synthase; iv, dihydrofolate synthase.

SCHEME 25

The enzyme dihydrofolate reductase which catalyses the reduction of dihydrofolic acid (equation 14) has been studied extensively. Two types of inhibitor of this enzyme are known. The first type is exemplified by the isosteric analogue methotrexate (**58**). Such compounds are potent inhibitors of the enzyme ($K_i < 10^{-9}$ mol l^{-1}) and are used widely in cancer chemotherapy. They show little discrimination among enzymes from various sources and are not useful as antibacterial agents since, as with folic acid, bacteria lack the active-transport mechanism which enables such compounds to permeate the cell wall. Trimethoprim (**59**) is a good example of the second type of inhibitor. Compounds of this type are able to permeate bacterial cell walls by a passive diffusion mechanism. Moreover, it is possible to synthesize 2,4-diaminopyrimidine derivatives which will bind *selectively* to the dihydrofolate reductase of pathogenic species. Thus trimethoprim will cause 50% inhibition of the enzyme from a bacterial source at a concentration of 5×10^{-8} mol l^{-1} whereas a 10 000 fold increase is necessary to achieve the same inhibition of the mammalian enzyme. Such compounds are clearly effective antibacterial agents.[2]

$$\text{7,8-Dihydrofolic Acid} + \text{NAD(P)H} + \text{H}^+ \xrightarrow[\text{reductase}]{\text{Dihydrofolate}} \text{5,6,7,8-Tetrahydrofolic Acid} + \text{NAD(P)}^+ \quad (14)$$

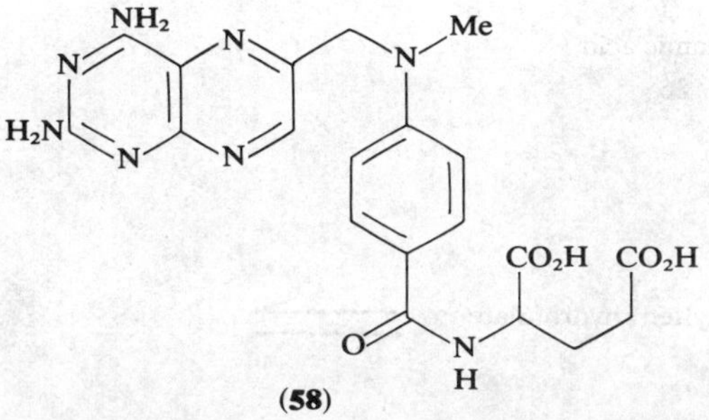

(**58**)

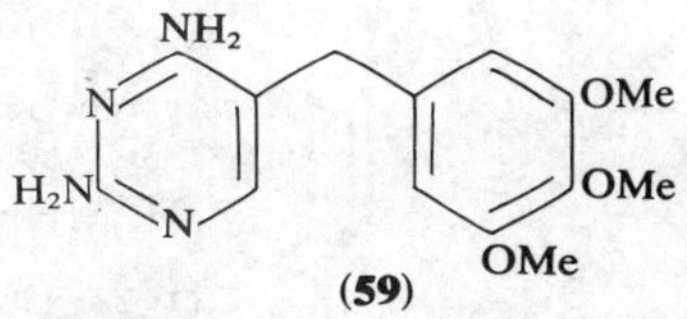

(**59**)

The mechanisms whereby one-carbon units at different oxidation levels are activated and interconverted in biochemical systems has been the subject of intensive study using both the natural coenzyme, tetrahydrofolate, and model systems. The results of these investigations are reviewed well by Bruice and Benkovic[57] and more recently by Benkovic and Bullard.[58] Enzymic aspects are emphasized in the review by Mudd and Cantoni.[59]

Enzymic activation of formate can be represented by equation 15. It is not known, however, whether the mechanism of the reaction involves formation of a ternary complex, or whether initial reaction occurs to give an intermediate such as a mixed anhydride of formic and phosphoric acids (formyl phosphate) or a phosphorylated derivative of tetrahydrofolate. In any event, the product of the reaction is N^{10}-formyltetrahydrofolate (**60**).

$$\text{Tetrahydrofolate} + \text{HCO}_2\text{H} + \text{ATP} \rightleftharpoons N^{10}\text{-Formyltetrahydrofolate} + \text{ADP} + \text{P}_i \quad (15)$$

The N^{10}-formyltetrahydrofolate (**60**) so formed is readily interconvertible, either enzymically or chemically, with 5,10-methenyltetrahydrofolate (**61**) and N^5-formyltetrahydrofolate (**62**) (Scheme 26).

Each of these tetrahydrofolate derivatives, and the closely related N^5-formiminotetrahydrofolate (**63**), is involved in a quite specific fashion in biological formyl transfer reactions. One example is the utilization of the N^{10}-formyl derivative (**60**) as the formyl donor in enzymic formylation of the aminoimidazolecarboxamide (**64**), an intermediate in the *de novo* biosynthesis of purine nucleotides (Scheme 27).

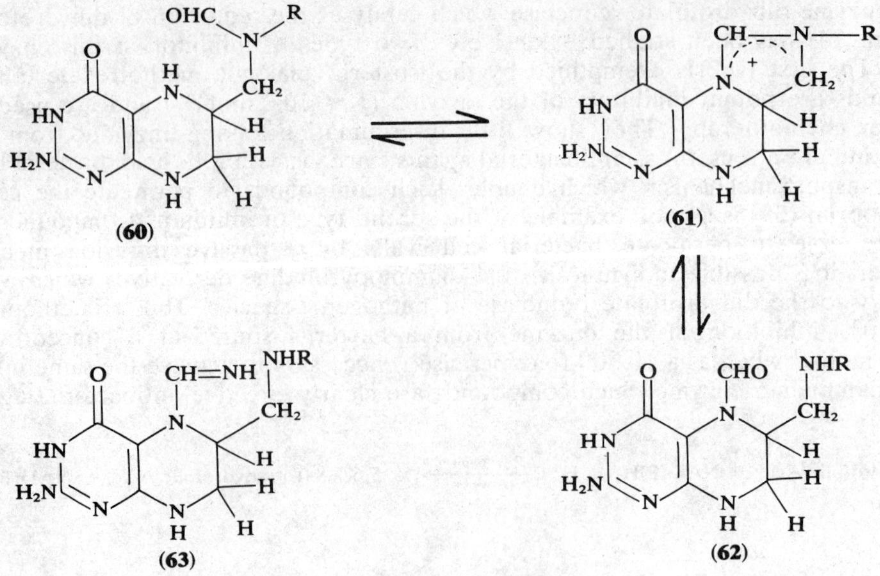

R = benzoylglutamic acid

SCHEME 26

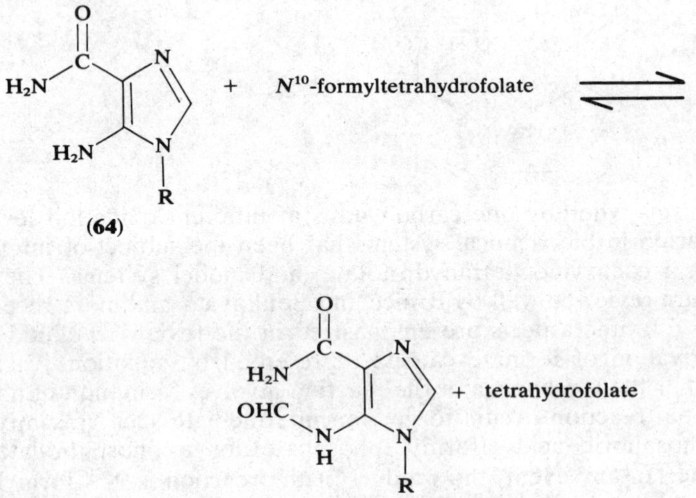

R = Ribosyl 5′-phosphate

SCHEME 27

Chemical reaction of formaldehyde with tetrahydrofolate to give 5,10-methylenetetrahydrofolate (**65**) occurs readily in the absence of enzyme. This compound can also be formed by (a) the enzymic reduction of 5,10-methenyltetrahydrofolate (**61**), a reaction which involves NADPH as hydrogen donor, and (b) the enzyme-catalysed reaction of serine (**66**) with tetrahydrofolate in the presence of pyridoxal phosphate. These interconversions are set out in Scheme 28. The resulting 5,10-methylenetetrahydrofolate (**65**) is the principal donor of hydroxymethyl groups in biological systems, particularly important being its involvement in the conversion of deoxyuridylate to thymidylic acid which is discussed below. It has been suggested that the Mannich-base form (**67**) might be the reactive species involved as 'hydroxymethyl donor'.

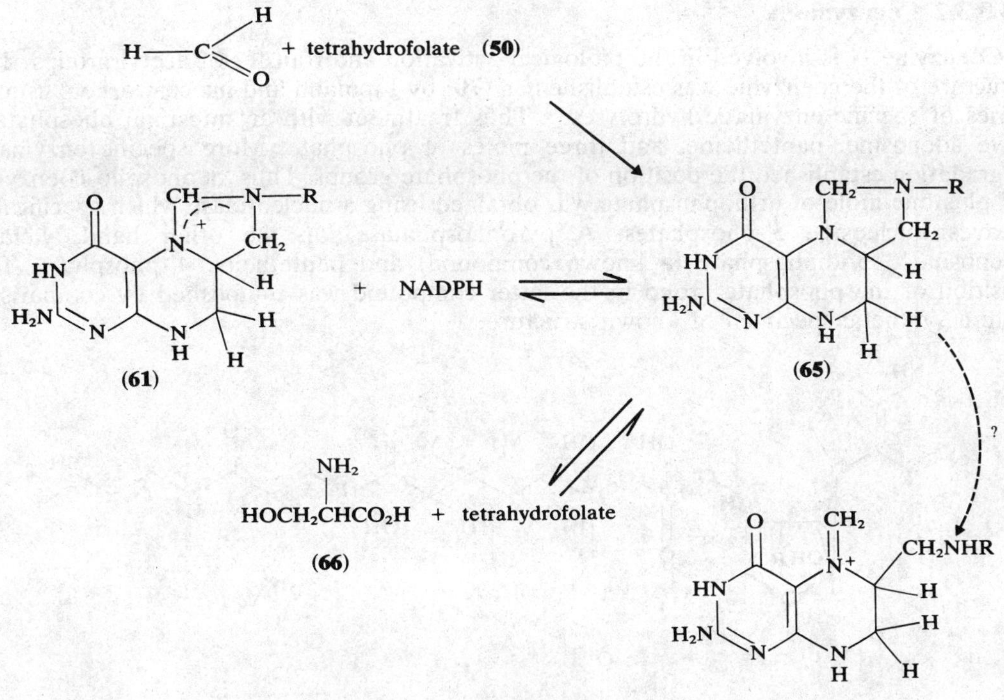

SCHEME 28

Methyl group transfer involving folate coenzymes is best exemplified by the conversion of deoxyuridylate (**68**) to thymidylic acid (**69**). In this reaction, 5,10-methylenetetrahydrofolate (**65**) functions as both the carbon donor and as reductant (Scheme 29). Various mechanisms have been proposed for this reaction but there is as yet little definitive evidence in favour of any one of these. More important is the fact that the coenzyme is converted into 7,8-dihydrofolate, which can be recycled to give tetrahydrofolate. This reduction is catalysed by dihydrofolate reductase, and it is probably at this point that inhibitors of the enzyme such as trimethoprim make their biggest impact. The stereochemistry of thymidylate synthesis has recently been elucidated.[60]

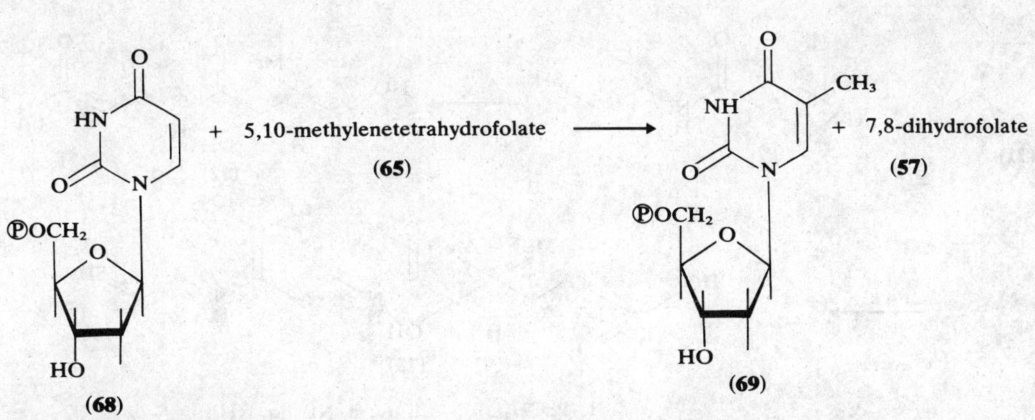

SCHEME 29

24.3.3.2 Coenzyme A

Coenzyme A is involved in the biological activation and transfer of acetyl groups. The structure of the coenzyme was established as (**70**) by Lipmann and his co-workers using a series of specific enzymatic hydrolyses.[61] Thus treatment with an intestinal phosphatase gave adenosine, pantetheine, and three moles of phosphate. More specific enzymatic degradation established the position of the phosphate groups. Thus 'dephospho-coenzyme A' plus one mole of orthophosphate was obtained using a nucleotidase which specifically cleaves nucleoside 3′-phosphates. A pyrophosphatase, on the other hand, yielded adenosine 3′,5′-diphosphate (a known compound) and pantetheine 4′-phosphate. The position of the phosphate group in the latter compound was established by comparison with a synthetic specimen of known structure.

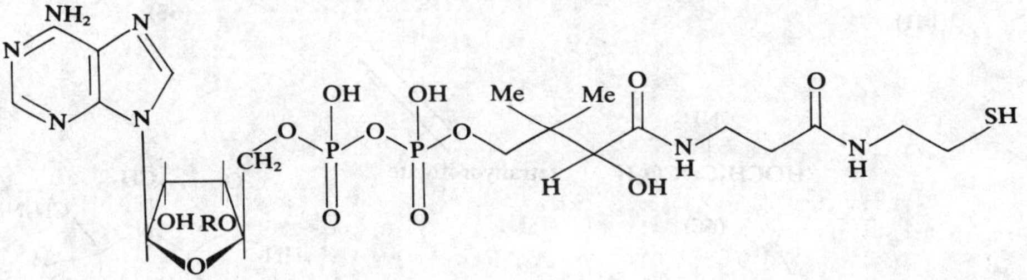

(**70**) R = P(=O)(OH)—OH

The synthesis of coenzyme A resolved itself into two component parts: firstly, the synthesis of pantetheine 4′-phosphate (**73**) and, secondly, coupling of this compound with a suitably activated adenosine phosphate. A typical example of the former synthesis is that reported by Baddiley and Thain[62] starting from D(−)-pantolactone (**71**) and summarized in Scheme 30. This particular synthesis, although convenient, is not unambiguous and the position of the phosphate group was established by a more complicated synthesis by the same workers. It is interesting to note, however, the preferential phosphorylation of the primary hydroxyl group of pantetheine (**72**).

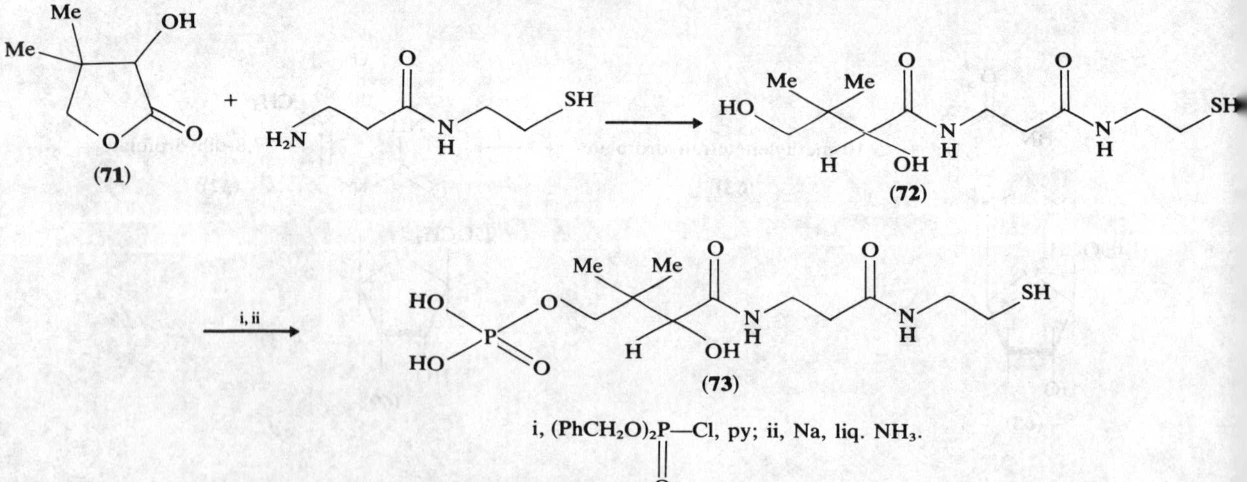

(**71**) (**72**) (**73**)

i, $(PhCH_2O)_2P(O)$—Cl, py; ii, Na, liq. NH_3.

SCHEME 30

Condensation of pantetheine 4′-phosphate (**73**) with the 5′-phosphoromorpholidate (**74**) of adenosine 2′,3′-cyclic phosphate gave a product which, when treated with dilute acid, gave a mixture of coenzyme A and iso-coenzyme A (the adenosine 2′-phosphate derivative). These were separated by chromatography on a cellulose ion-exchange resin. Adenosine 3′-phosphate 5′-phosphoromorpholidate could not be used in a more direct synthesis of the coenzyme because of competition for the phosphoromorpholidate between the primary phosphate residues at the 3′-position of adenosine and the 4′-position of pantetheine. The success of the method using the cyclic phosphate depends upon the preferential attack by a phosphomonoester rather than a phosphodiester on the phosphoromorpholidate.[63]

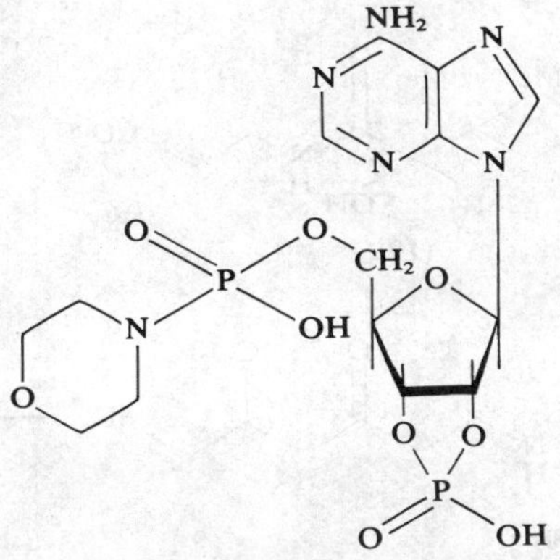

(**74**)

Studies of the biosynthesis of coenzyme A have established the relationship between the vitamin L-pantothenic acid (**78**) and the coenzyme. Pantothenic acid, one of the B-group of vitamins, cannot be synthesized by animals. Most micro-organisms, on the other hand, are able to make this vitamin from its precursors, pantoic acid and β-alanine. Both animals and bacteria can convert pantothenic acid into coenzyme A. An excellent review[64] by Brown summarizes both biosynthetic processes.

The enzymatic synthesis of pantothenic acid is outlined in Scheme 31. The precursor α-ketoisovaleric acid (**75**) is converted into ketopantoic acid (**76**) by reaction with

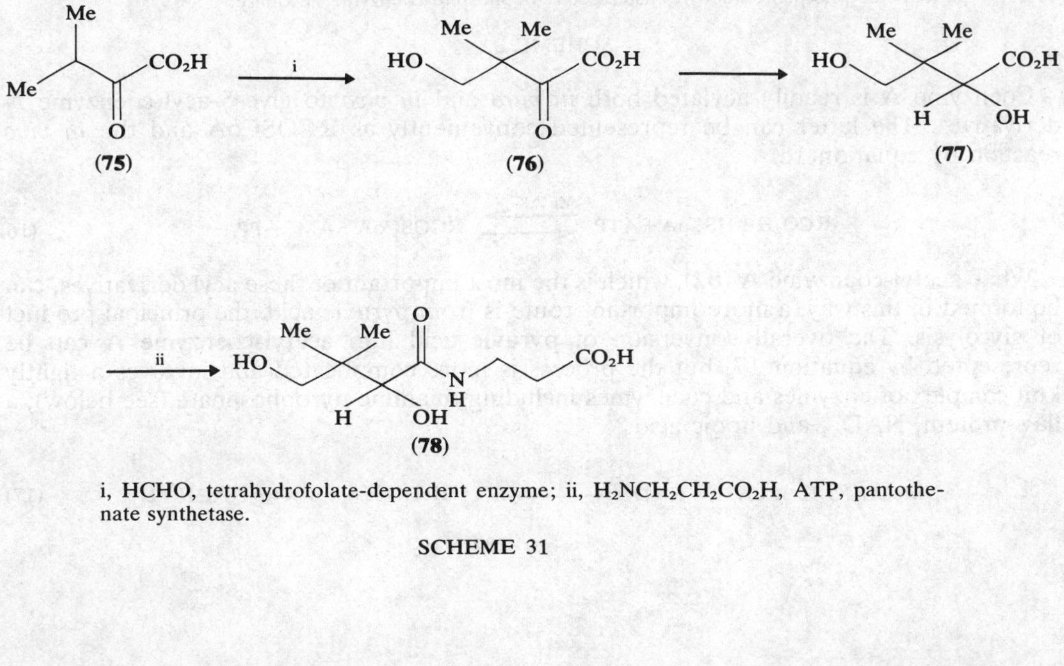

i, HCHO, tetrahydrofolate-dependent enzyme; ii, $H_2NCH_2CH_2CO_2H$, ATP, pantothenate synthetase.

SCHEME 31

formaldehyde in a tetrahydrofolate-dependent enzymic reaction. Subsequent reduction gives pantoic acid (**77**), but the nature of the reducing agent has not been established. The enzyme pantothenate synthetase catalyses the condensation of pantoic acid with β-alanine to give L-pantothenic acid (**78**).

Conversion of pantothenic acid into coenzyme A has been studied using enzyme preparations from both bacterial species and rat-liver.[65] Initial phosphorylation gives 4′-phosphopantothenate (**79**), which condenses with L-cysteine to give 4′-phosphopantothenoyl-L-cysteine (**80**). Subsequent decarboxylation to pantetheine 4′-phosphate (**73**), reaction with adenosine triphosphate to give dephospho-coenzyme A (**81**), and final selective phosphorylation yields coenzyme A (**70**) (Scheme 32). It now seems unlikely that an alternative pathway[66] involving initial condensation of pantothenic acid and cysteine is of biological significance.

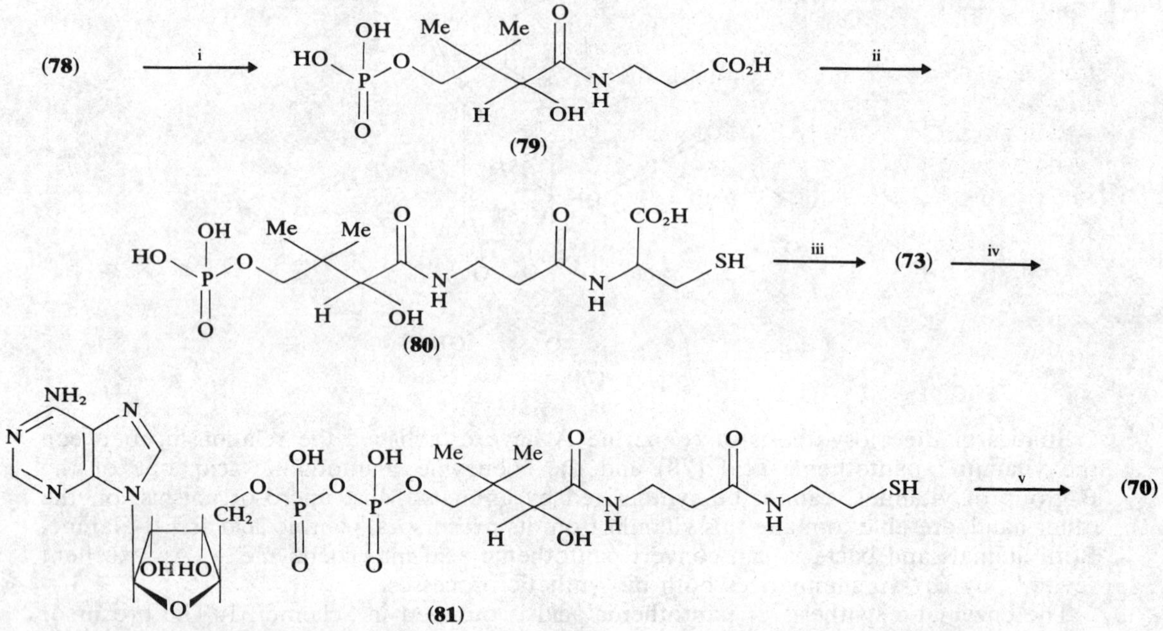

i, ATP, pantothenate kinase, ii, L-cysteine, synthetase; iii, decarboxylase; iv, ATP, pantetheine phosphate adenyltransferase; v, dephospho-coenzyme A kinase.

SCHEME 32

Coenzyme A is readily acylated both *in vitro* and *in vivo* to give 5-acyl-coenzyme A derivatives. The latter can be represented conveniently as RCOSCoA and the *in vivo* reaction by equation 16.

$$\mathrm{RCO_2H + HSCoA + ATP} \underset{}{\overset{\mathrm{Mg^{2+}, K^+}}{\rightleftharpoons}} \mathrm{RCOSCoA + AMP + PP_i} \qquad (16)$$

While acetyl-coenzyme A (**82**), which is the most important of these acyl derivatives, can be formed in this way, a more important route is from pyruvic acid, the principal product of glycolysis. The overall conversion of pyruvic acid into acetyl-coenzyme A can be represented by equation 17, but the process is more complicated and involves a tightly knit complex of enzymes and coenzymes including thiamine pyrophosphate (see below), a flavoprotein, NAD^+, and lipoic acid.[67]

$$\mathrm{MeCOCO_2H + CoASH + NAD^+} \longrightarrow \underset{(\mathbf{82})}{\mathrm{MeCOSCoA}} \mathrm{+ CO_2 + NADH + H^+} \qquad (17)$$

Coenzyme A is able to activate acetate in two ways so that the acetate residue can function either as an electrophile or as a nucleophile.[68] This can best be explained by considering the chemical properties of thiol esters. These esters can be represented by the canonical forms shown in Scheme 33.

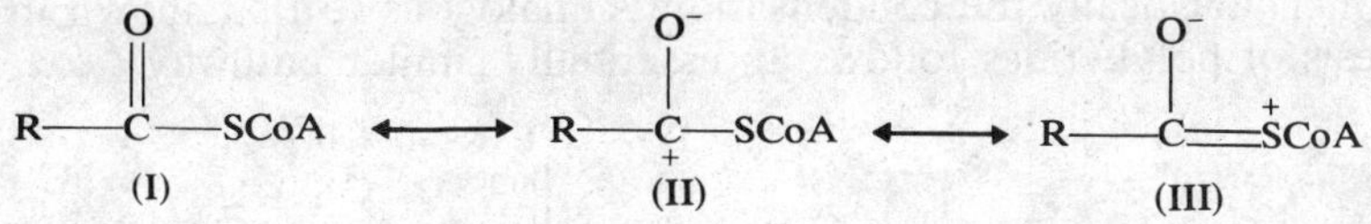

SCHEME 33

The contribution to the resonance hybrid from forms such as (III) is much less than is the case with ordinary esters and consequently the electrophilic character of the carbon atom of the carbonyl group is much enhanced. Thiol esters such as acetyl-coenzyme A are thus effective acylating agents, nucleophilic attack taking place at the electron-deficient carbon atom (Scheme 34). Typical nucleophiles in biological systems are amines, carboxylate ions, phosphate ions, and carbanions such as those discussed below.

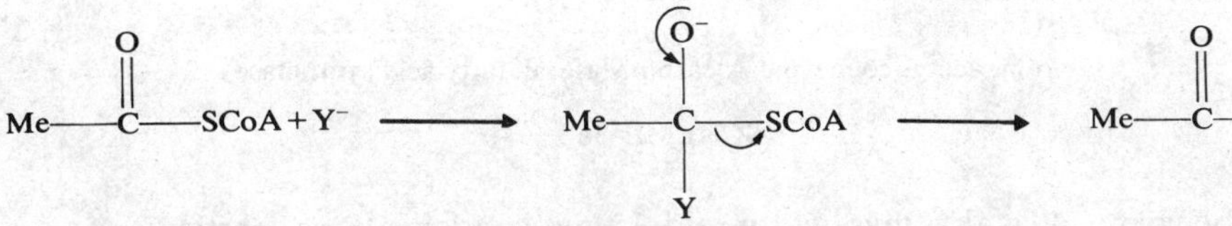

SCHEME 34

The carbanion formed by loss of a proton from the α-carbon atom of a thiol ester is also stabilized by effective delocalization of the negative charge, this delocalization being more effective than is the case with ordinary esters (Scheme 35). Acetyl-coenzyme A can thus behave as an effective nucleophile.

SCHEME 35

It is thus possible for acetyl-coenzyme A to undergo the biological equivalent of a Claisen condensation to give acetoacetyl-coenzyme A (equation 18). This reaction is important since it is apparently the first stage in the biosynthesis of mevalonate and hence of terpenes and steroids.[69] The equilibrium represented by equation 18 lies to the left and presumably the reaction is only effective by virtue of an irreversible step later in the biosynthetic sequence.

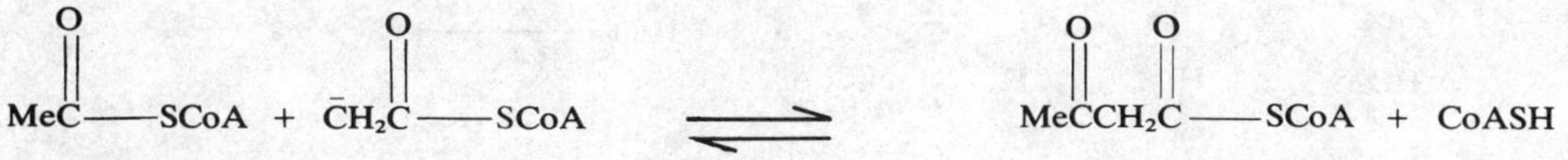 (18)

The biosynthesis of fatty acids by sequential addition of two-carbon atom units to an acetyl-coenzyme A molecule ('starter unit') could well be formulated in a similar way. However, it appears that in this case at least, a more effective nucleophilic species is required and the chain-extending agent is malonyl-coenzyme A (**83**).[70] This is formed from acetyl-coenzyme A in an ATP-dependent enzymic carboxylation. The driving force in the condensation reaction is the decarboxylation which shifts the equilibrium to the right with the resultant formation of the acetoacetyl derivative. It is probable that, prior to condensation, the acetyl and malonyl groups are transferred to a special carrier protein and thence to the enzyme (fatty acid synthetase). In each case, however, thiol esters are involved and mechanistically the condensation is analogous to that shown in Scheme 36. The biosynthesis of polyketides follows an essentially similar pathway.

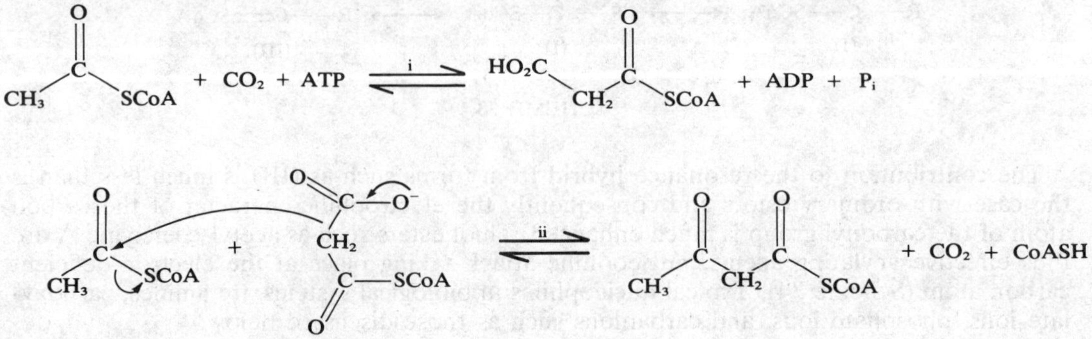

(83)

i, biotin, acetyl-coenzyme A carboxylase; ii, fatty acid synthetase.

SCHEME 36

Acetyl-coenzyme A is also involved in aldol-type condensations, particularly important being the reaction with oxalacetic acid to give citric acid (equation 19), a key step in the citric acid cycle. This reaction, which is catalysed by citrate synthetase, has been studied extensively.[71]

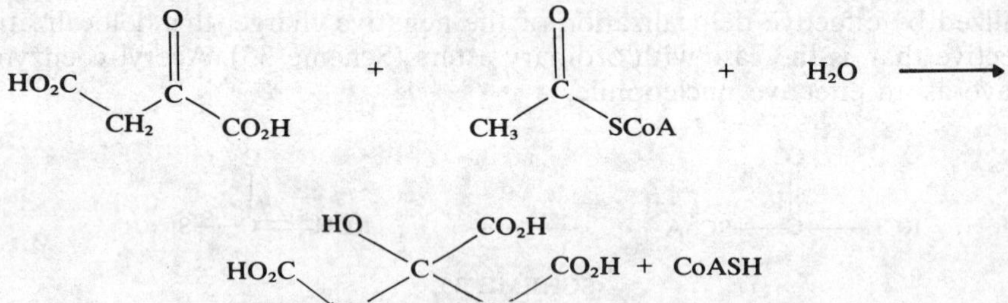

(19)

Similar reactions are those in which acetyl-coenzyme A reacts with glyoxylate and acetoacetyl-coenzyme A with the formation of malate and β-hydroxy-β-methylglutaryl-coenzyme A, respectively (equations 20 and 21). The latter reaction is a key step in the biosynthesis of mevalonate.

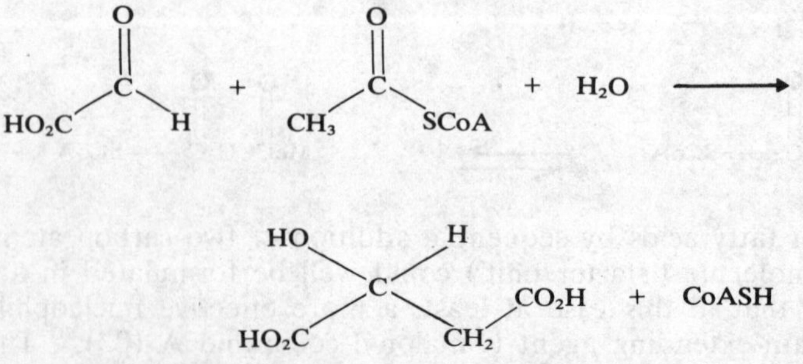

(20)

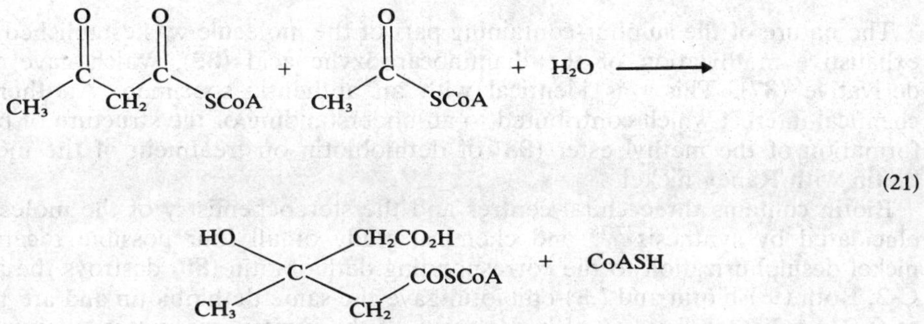

(21)

24.3.3.3 Biotin

The enzymatic role of biotin as the coenzyme responsible for the transfer of carbon dioxide in carboxylation reactions is now established unequivocally.[72] The structure of the coenzyme was determined largely by the work of du Vigneaud and his co-workers[73,74] and both structure and stereochemistry have been confirmed as (**84**) by synthesis[75] and by X-ray diffraction studies.[76]

The essential features of the structure elucidation are set out in Scheme 37. Biotin (**84**), $C_{10}H_{16}N_2O_3S$, formed a methyl ester with diazomethane and a sulphone on oxidation with hydrogen peroxide and acetic acid, thus confirming the presence of carboxyl and thioether groups. Hydrolysis yielded an optically active diaminocarboxylic acid (**85**), which could be reconverted into biotin by treatment with phosgene and which gave adipic acid on oxidation. The diamine (**85**) gave a quinoxaline derivative (**86**) on reaction with phenanthraquinone, thus confirming the 1,2-relationship of the amino-groups and the presence of the imidazolidone ring in biotin.

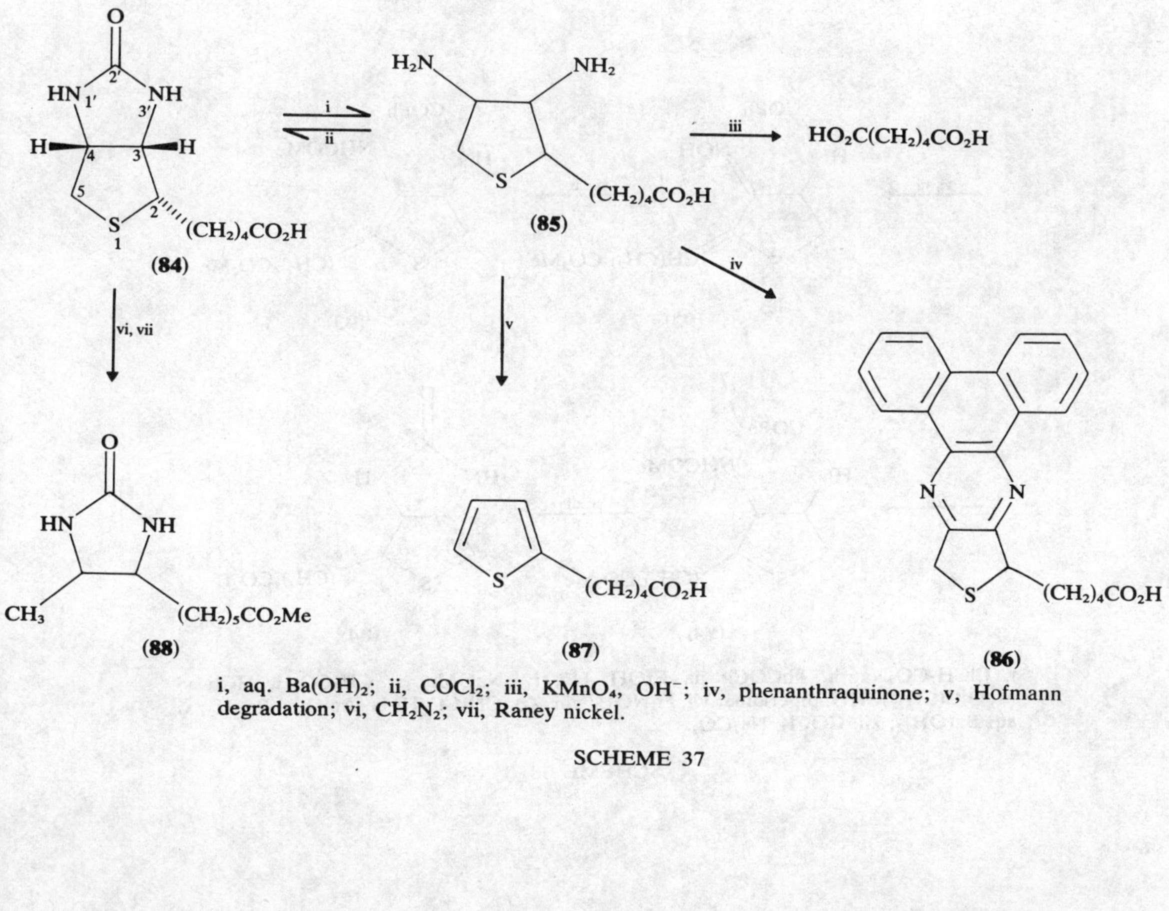

i, aq. $Ba(OH)_2$; ii, $COCl_2$; iii, $KMnO_4$, OH^-; iv, phenanthraquinone; v, Hofmann degradation; vi, CH_2N_2; vii, Raney nickel.

SCHEME 37

The nature of the sulphur-containing part of the molecule was established by Hofmann exhaustive methylation of the diaminocarboxylic acid (**85**), which gave the thiophen derivative (**87**). This was identical with an authentic specimen. Another reaction of chemical interest which contributed to an understanding of the structure of biotin was the formation of the methyl ester (**88**) of dethiobiotin on treatment of the methyl ester of biotin with Raney nickel.

Biotin contains three chiral centres and the stereochemistry of the molecule has been elucidated by synthesis[77,78] and chemical study of all four possible racemates. Raney nickel desulphurization to the corresponding dethiobiotin (**88**) destroys the asymmetry at C-2. Both (±)-biotin and (±)-epibiotin gave the same dethiobiotin and are thus epimeric at C-2. Studies of the rate of hydrolysis[78] of the various racemates to give the diaminocarboxylic acids (**85**) showed that this pair of racemates are hydrolysed more slowly than are the other pair. The conclusion was drawn that the rings are *cis*-fused in biotin and epibiotin. The stereochemistry of the third asymmetric carbon atom and the absolute configuration of (+)-biotin (the only biologically active stereoisomer) was established by X-ray analysis[79] of the bis-*p*-bromoanilide of $N^{1'}$-carboxybiotin.

Numerous syntheses of biotin have been reported and only two of these are selected for discussion here. The first is outlined in Scheme 38 and is that reported by Harris and his

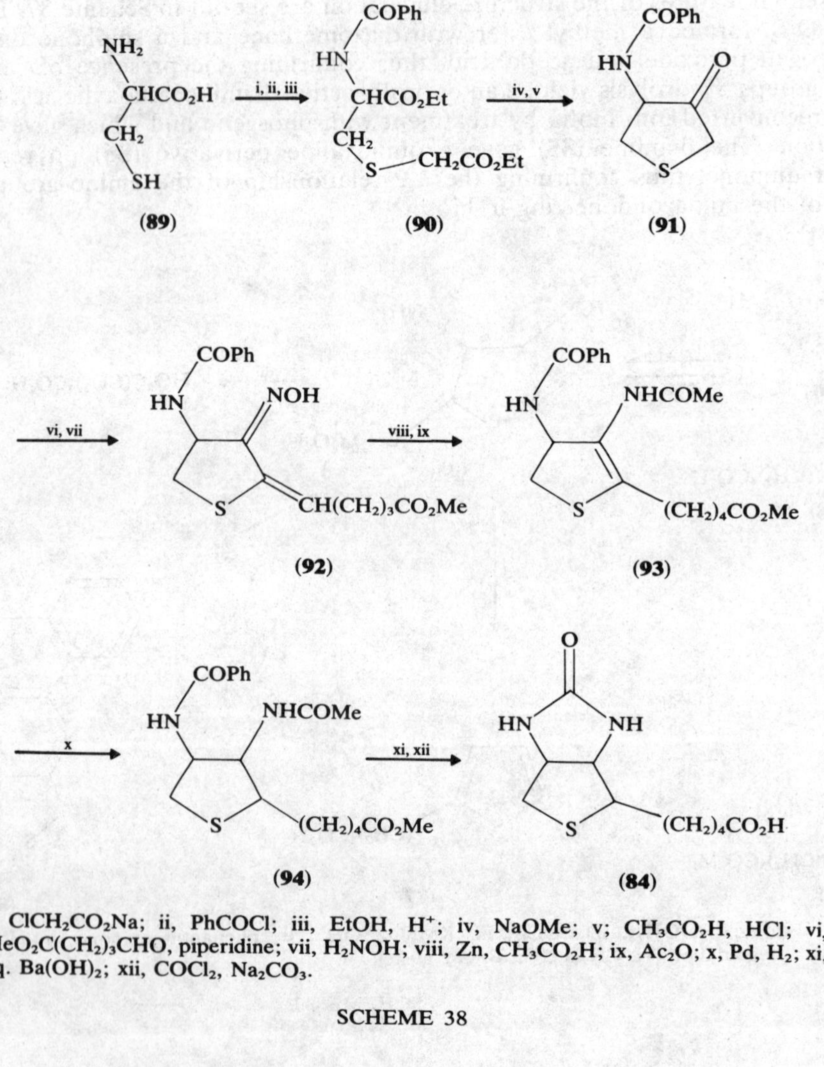

i, $ClCH_2CO_2Na$; ii, PhCOCl; iii, EtOH, H^+; iv, NaOMe; v; CH_3CO_2H, HCl; vi, $MeO_2C(CH_2)_3CHO$, piperidine; vii, H_2NOH; viii, Zn, CH_3CO_2H; ix, Ac_2O; x, Pd, H_2; xi, aq. $Ba(OH)_2$; xii, $COCl_2$, Na_2CO_3.

SCHEME 38

co-workers[75] which confirmed the gross structure of biotin. Treatment of the sodio-derivative of cysteine (**89**) successively with chloroacetic acid, benzoyl chloride, and ethanol in the presence of an acid catalyst gave the di-ester (**90**). Dieckmann cyclization, followed by hydrolysis and decarboxylation, gave the cyclic ketone (**91**). Condensation with methyl γ-formylbutanoate in the presence of base, and reaction with hydroxylamine gave the oxime (**92**). Reduction and acetylation gave the unsaturated derivative (**93**) and a structural isomer. Catalytic reduction of (**93**) gave two racemates of structure (**94**), which were separately hydrolysed to the corresponding diamines and then treated with phosgene. (±)-Biotin was obtained from one of these racemates and resolution, most effectively using L(+)-arginine,[80] gave (+)-biotin (**84**).

The second synthesis[81] differs in that (a) it is used on a commercial scale, (b) it is stereospecific, and (c) it involves initial formation of the imidazolidone ring rather than the tetrahydrothiophen ring. The synthesis is summarized in Scheme 39. Fumaric acid (**95**) was converted *via meso*-dibromosuccinic acid into *meso*-2,3-bisbenzylaminosuccinic acid

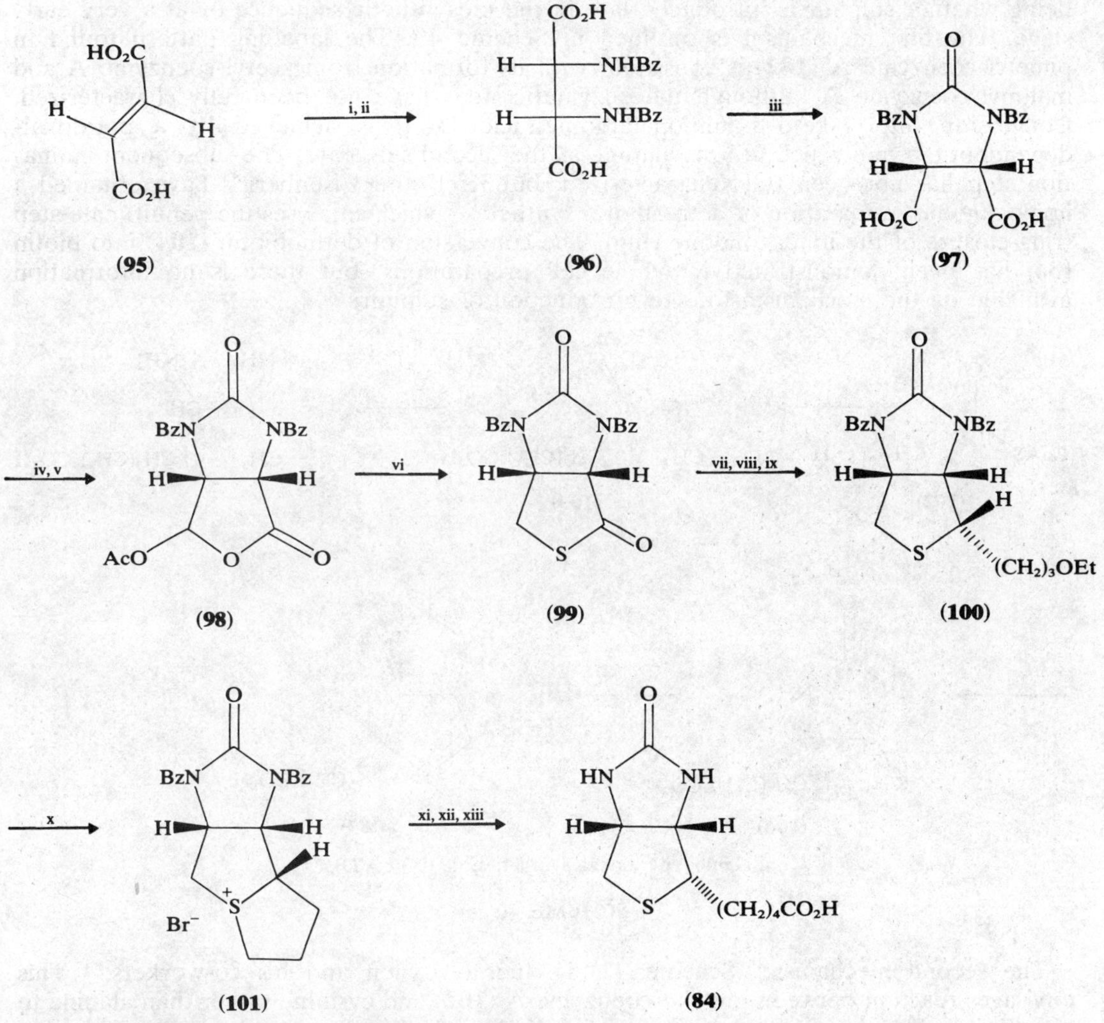

i, Br_2; ii, $PhCH_2Br$; iii, $COCl_2$; iv, Ac_2O; v, Zn, HOAc, Ac_2O; vi, H_2S, HCl; KSH; Zn, HOAc; vii, $EtO(CH_2)_3MgBr$; viii, HOAc, heat; ix, H_2, Pd; x, HBr, HOAc, xi, optical resolution; xii, $Na^+ \, ^-CH(CO_2Et)_2$; xiii, conc. HCl.

SCHEME 39

(**96**). Treatment with phosgene gave the imidazolidone derivative (**97**) in which the carboxyl groups are *cis*. The corresponding anhydride was reduced in the presence of acetic anhydride to give the cyclic aldehydo-acid derivative (**98**). Sulphur was introduced using hydrogen sulphide followed by reduction to give the thiolactone (**99**). Introduction of the side-chain was achieved by Grignard synthesis, dehydration, and catalytic reduction (which is also stereospecific) to give (**100**). The relative configuration of all three chiral centres is thus established. Treatment with hydrogen bromide in acetic acid gave the sulphonium salt (**101**), which was resolved using (+)-camphorsulphonic acid. The laevorotatory isomer was reacted with the sodium salt of diethyl malonate and thence by boiling with concentrated hydrochloric acid, which effected hydrolysis, decarboxylation, and debenzylation, (+)-biotin (**84**) was obtained.

Considerable progress has been made in the characterization of the enzymatic steps in the biosynthesis of biotin.[82] Biotin is a vitamin for man but it can be synthesized by intestinal bacteria and a true deficiency disease in man is virtually unknown. Two conflicting mechanisms of biotin biosynthesis have been proposed, the essential difference being whether sulphur is introduced late in the biosynthetic sequence or at a very early stage. The first mechanism is outlined in Scheme 40. The labelling pattern found in pimelyl-coenzyme A (**102**) is consistent with its formation from acetyl-coenzyme A and malonyl-coenzyme A, although the enzymatic steps have not been fully characterized. Conversion into 7-keto-8-aminopelargonic acid (**103**) is achieved by a pyridoxal-dependent enzyme which utilizes alanine as the second substrate. The subsequent amination step has not been fully characterized, but Krell and Eisenberg[83] have obtained a homogeneous preparation of dethiobiotin synthetase which catalyses the penultimate step (ring-closure of the imidazolidone ring). The conversion of dethiobiotin (**104**) into biotin (**84**) has been demonstrated in whole cell preparations, but there is no information available on the mechanism for the introduction of sulphur.

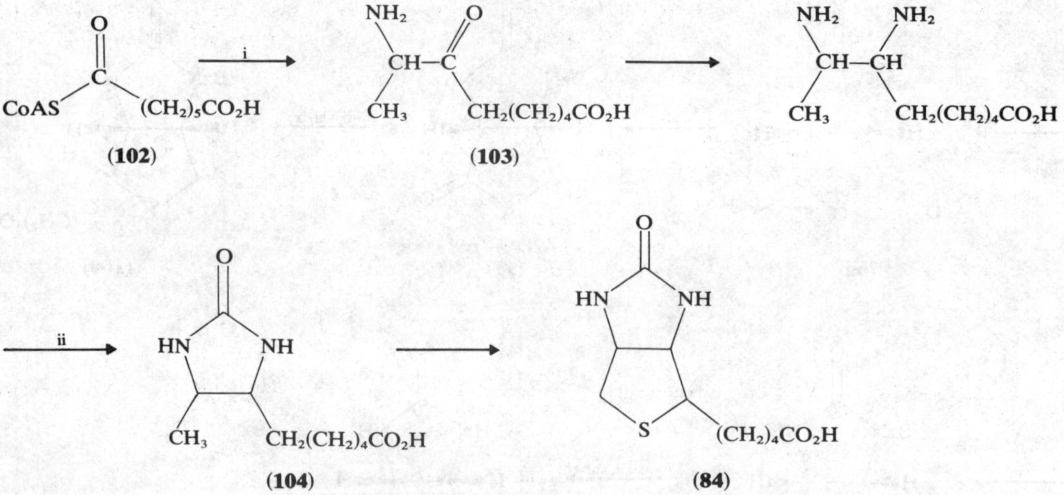

i, L-alanine, pyridoxal, synthetase; ii, CO_2, ATP.

SCHEME 40

The second mechanism (Scheme 41) is due to Lynen and his co-workers.[84] This envisages reaction between pimelyl-coenzyme A (**102**) and cysteine rather than alanine to give 7-keto-8-amino-9-thiopelargonic acid (**105**), which subsequently reacts with carbamyl phosphate as shown. It is easier to see the mechanism for ring-closure of the tetrahydrothiophen ring in this proposal, but the balance of biochemical evidence appears to favour the route shown in Scheme 40. Additional work, particularly on the mechanism for the introduction of sulphur, is clearly necessary.

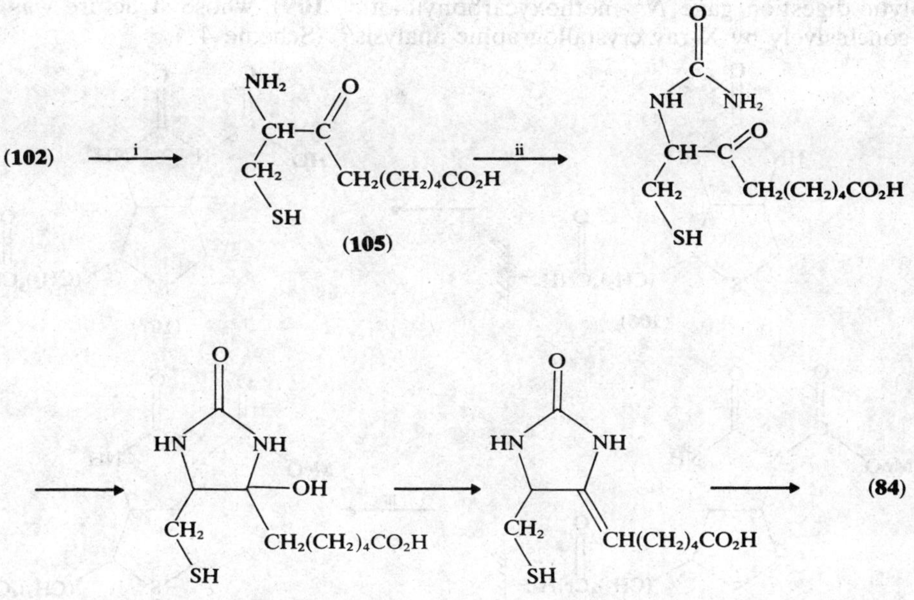

i, cysteine; ii, $H_2NCOOPO_3H_2$.

SCHEME 41

Before biotin can function in the enzymic transfer of carbon dioxide in carboxylation reactions it must become attached covalently to the apoenzyme to form the catalytically active holoenzyme.[72] Two enzymic reactions are involved (equations 22 and 23).

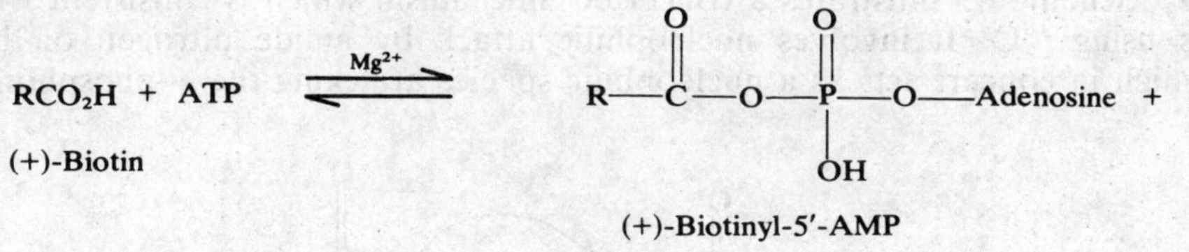

(22)

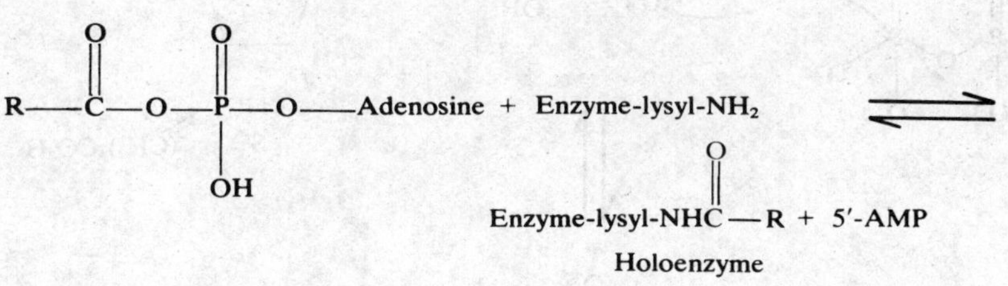

(23)

The net result of these reactions is to attach the biotin *via* an amide link to the ω-amino-group of a lysine residue in the apoenzyme. Support for this mechanism comes from the use[85] of chemically synthesized (+)-biotinyl-5′-AMP in equation (23), where it replaced completely (+)-biotin, ATP, and Mg^{2+}.

The nature of the compound formed between carbon dioxide and enzyme bound biotin (**106**) was established by the elegant work of Lynen, Knappe, and their colleagues.[86] The carboxylated product (**107**) was found to be relatively unstable, particularly at acid pH, and was therefore converted into the methyl ester (**108**) by treatment with diazomethane.

Proteolytic digestion gave $N^{1'}$-methoxycarbonylbiotin (**109**), whose structure was established conclusively by X-ray crystallographic analysis[79] (Scheme 42).

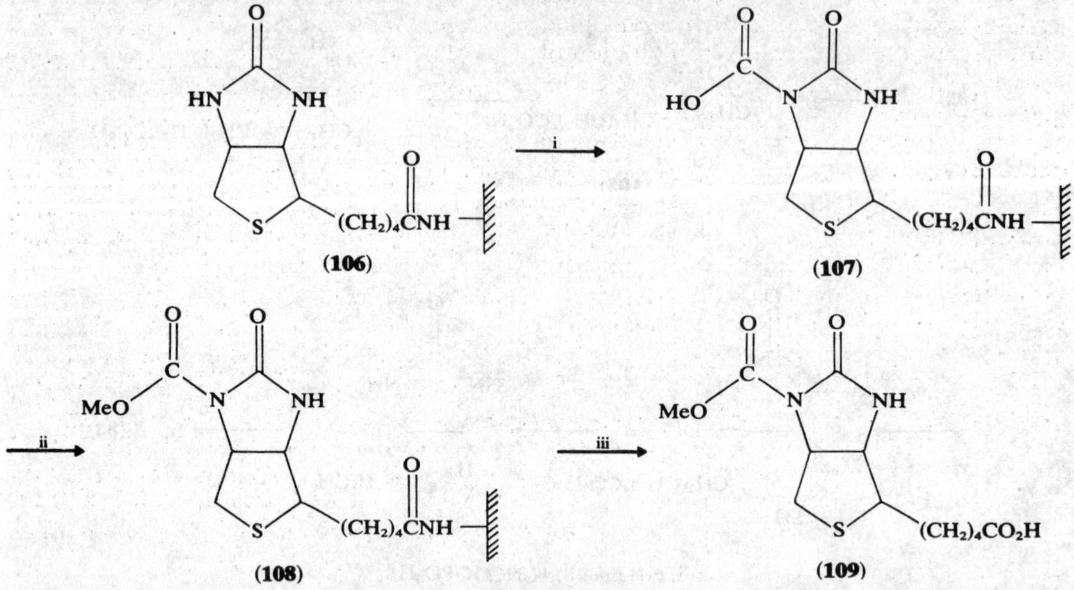

i, HCO_3^-, ATP; ii, CH_2N_2; iii, proteolytic enzymes.

SCHEME 42

It is generally accepted that bicarbonate ion, HCO_3^-, is the active form of carbon dioxide involved in the carboxylation reaction. Various mechanisms,[72] stepwise and concerted, have been put forward to explain the formation of carboxybiotin from biotin, ATP, and HCO_3^-. Scheme 43 illustrates a concerted mechanism which is consistent with labelling studies using ^{18}O. It involves nucleophilic attack by amide nitrogen on bicarbonate ion which in concert acts as a nucleophilic species attacking the γ-phosphorus atom of ATP.

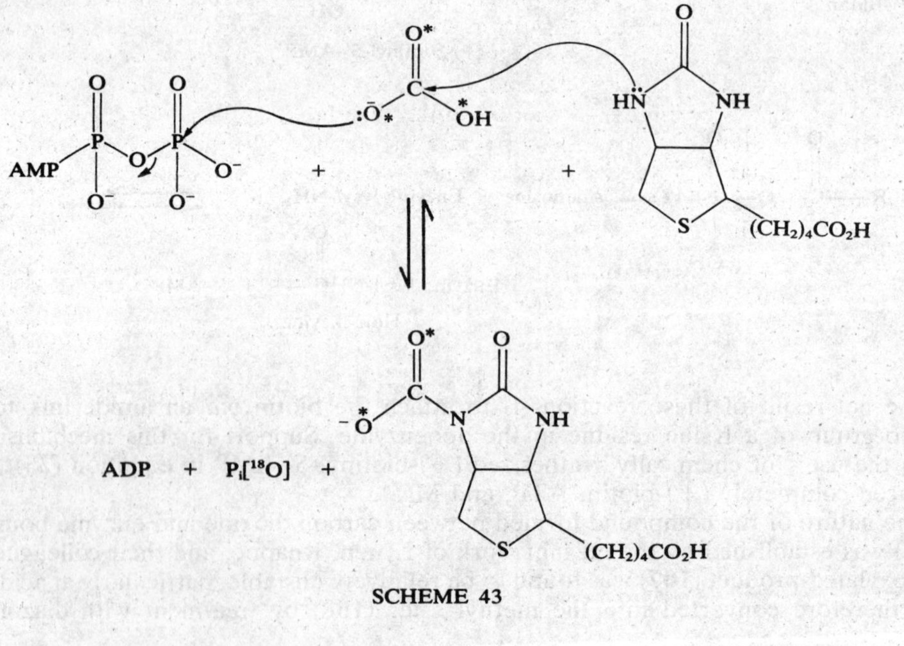

SCHEME 43

The transfer of carbon dioxide from carboxybiotin to an acceptor requires the carboxyl carbon atom to acquire electrophilic properties as well as enzyme-mediated activation of the acceptor molecule as a nucleophilic species. The latter has already been discussed for acyl-coenzyme A derivatives. It seems likely that general acid catalysis (such as that set out in Scheme 44) leading to protonation of the urea carbonyl oxygen atom could lead to decarboxylation (in the absence of a nucleophilic species) or to carboxyl transfer in the presence of an appropriate nucleophile.

These carboxylation reactions involving biotin are important in the enzymic synthesis of oxalacetic acid from pyruvate plus carbon dioxide. They are also involved in the enzymic conversion of acetyl-coenzyme A into the more nucleophilic species, malonyl-coenzyme A, required for fatty acid synthesis, and the related conversion of propionyl-coenzyme A to methylmalonyl-coenzyme A.

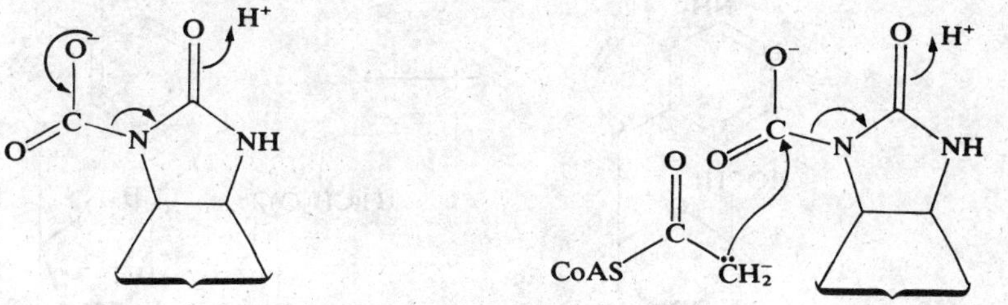

SCHEME 44

24.3.3.4 Adenosine diphosphate and triphosphate

Adenosine triphosphate is involved in a vast number of metabolic reactions and is a key intermediate in the transfer of energy in living organisms whether derived from oxidation, fermentation, or by photochemical means. In this section we are concerned with the role played by adenosine triphosphate (ATP) and adenosine diphosphate (ADP) in the transfer of phosphate groups.

The structure of ATP (and of ADP) was elucidated largely by hydrolysis experiments.[87] Thus hydrolysis with dilute alkali gave adenosine 5′-phosphate and inorganic pyrophosphate. Acidic hydrolysis, on the other hand, rapidly liberated two of the three phosphate residues as inorganic phosphate, the other products being adenine and ribose 5-phosphate. The position of attachment of the triphosphate group to adenosine was established by (a) deamination with nitrous acid which gave an analogous inosine 5′-triphosphate, indicating lack of substitution at the 6-amino-group, and (b) consumption of one molar equivalent of periodate which established that the 2′- and 3′-hydroxyl groups were not esterified. Titrimetric evidence confirmed the presence of three primary and one secondary phosphate dissociations. Enzymic removal of one phosphate group from ATP gave ADP. Similar degradation experiments and titration of acidic groups were carried out on this molecule. These experiments confirm the structures for adenosine triphosphate (**110**) and adenosine diphosphate (**111**).

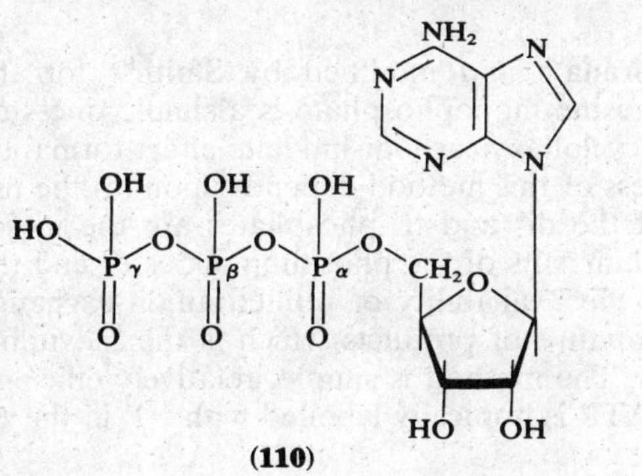

(110)

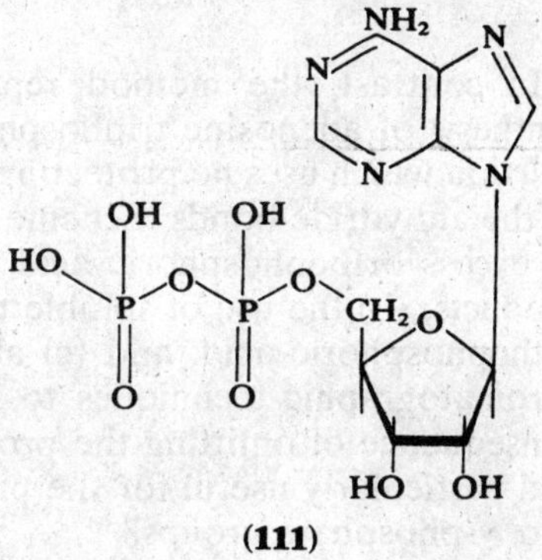

(111)

Numerous syntheses of ATP and ADP have been reported.[88] These are largely of academic interest at the present time since both coenzymes are readily available from commercial sources, having been obtained by extraction from biological materials. Two syntheses are outlined below to illustrate the general techniques used in this field. The pioneer synthetic work of Todd and his colleagues served to confirm the structure of ATP and ADP, and their synthesis[89] of the latter compound is set out in Scheme 45. It is characterized by the use of protecting groups (on both sugar and phosphate) to direct the synthesis in the appropriate way, and by application of the classical method for the preparation of mixed anhydrides, namely reaction of the salt of an acid with an acid chloride. Yields in this synthesis were low owing to the instability of the fully esterified phosphate intermediates.

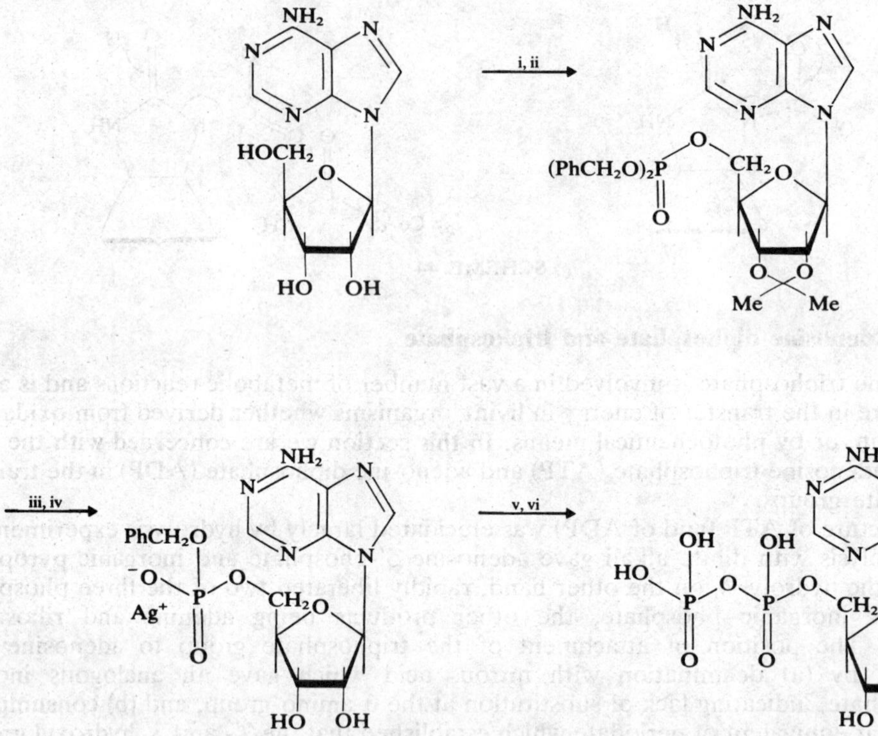

(111)

i, Me_2CO, H^+; ii, Cl—P(O)($OCH_2Ph)_2$, py; iii, 0.02N H_2SO_4; iv, $AgNO_3$; v, Cl—P(O)($OCH_2Ph)_2$, HOAc; vi, H_2, PdO.

SCHEME 45

In contrast, the method reported by Khorana[90] and modified by Smith[91] for the synthesis of adenosine triphosphate from adenosine monophosphate is a simple one-step reaction which uses no protecting groups and dicyclohexylcarbodi-imide to effect formation of the anhydride bonds (Scheme 46). The success of this method depends upon (a) the use of excess orthophosphoric acid to ensure that the di- and tri-phosphates are the major products, (b) the use of soluble trialkylammonium salts of the phosphomonoester and the orthophosphoric acid, and (c) above all upon the availability of refined anion-exchange chromatographic techniques to separate the mixture of products which is the inevitable consequence of omitting the protecting groups. The method is simple, relatively efficient, and particularly useful for the preparation of ATP isotopically labelled with ^{32}P in the β- and γ-phosphate groups.

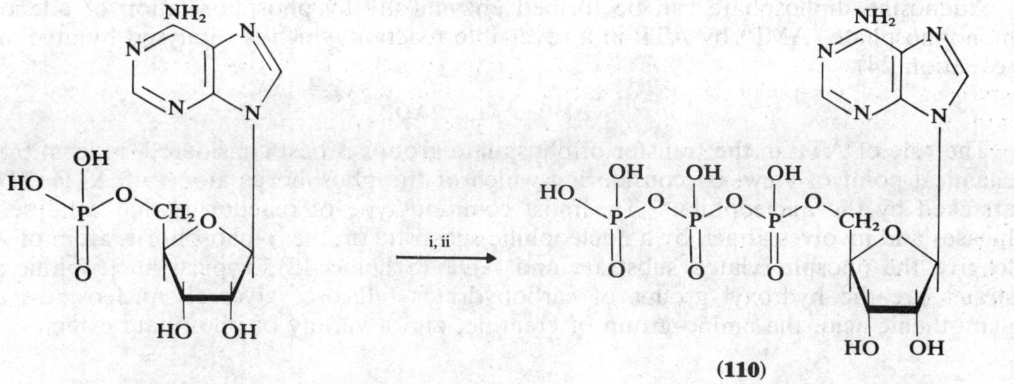

(**110**)

i, excess H_3PO_4, Bu^n_3N, $C_6H_{11}N{=}C{=}NC_6H_{11}$, py; ii, anion-exchange chromatography.

SCHEME 46

The biosynthetic processes leading to formation of ATP (in excess of that consumed) are numerous and include oxidative phosphorylation in the respiratory chain, oxidative phosphorylation at substrate level, and photosynthetic phosphorylation.[92] In each case the essential step involves phosphorylation of ADP to give ATP rather than phosphorylation of adenosine monophosphate. A typical reaction is that between ADP and a phosphorylating agent such as enolpyruvate 2-phosphate (Scheme 47).

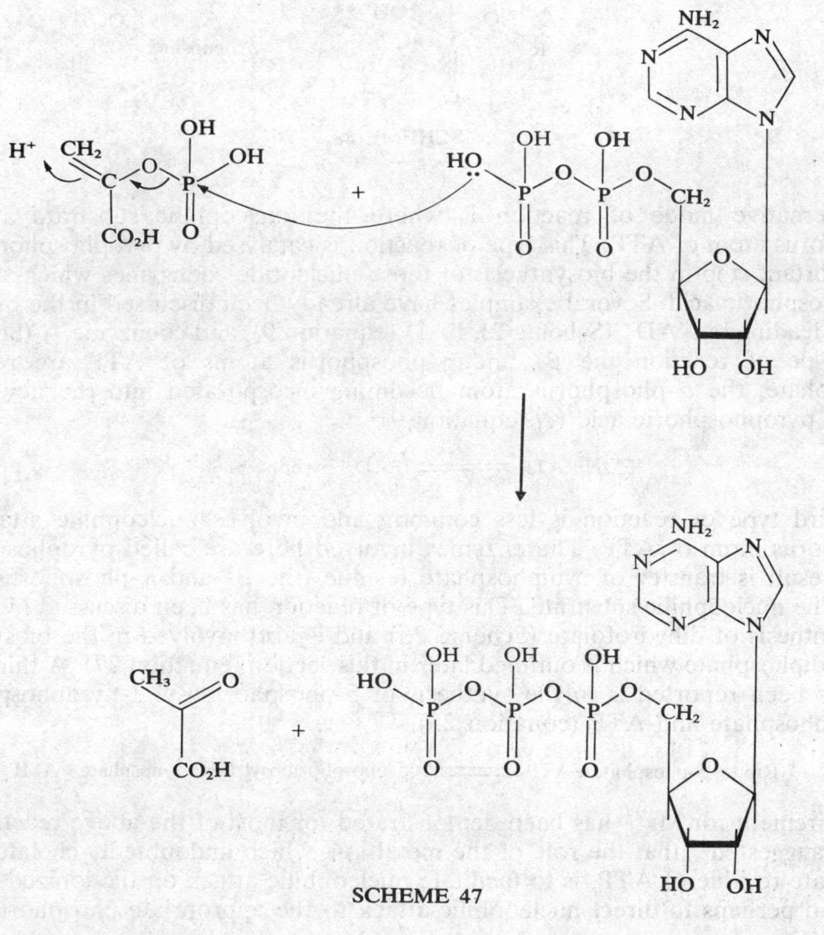

SCHEME 47

Adenosine diphosphate can be formed enzymically by phosphorylation of adenosine monophosphate (AMP) by ATP in a reversible reaction which is catalysed by myokinase (equation 24).

$$AMP + ATP \rightleftharpoons 2ADP \quad (24)$$

The role of ATP in the transfer of phosphate groups is best considered, at least from a chemical point of view, by considering which of the phosphorus atoms of ATP (**110**) is attacked by the nucleophile.[93] The most common type of reaction is that catalysed by kinases and involves attack by a nucleophilic substrate on the γ-phosphorus atom of ATP to give the phosphorylated substrate and ADP (Scheme 48). Typical nucleophilic substrates are the hydroxyl groups of carbohydrates (glucose, glycerol, nucleosides, *etc.*), pantothenic acid, the amino-group of creatine, and a variety of phosphate esters.

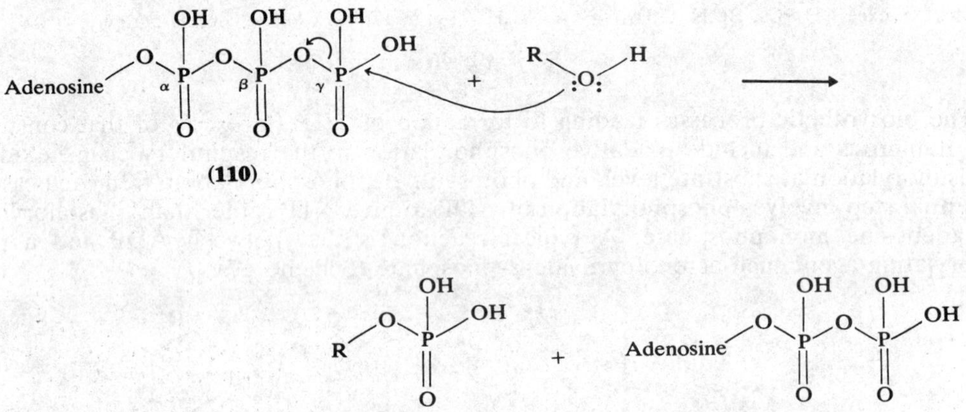

SCHEME 48

An alternative mode of reaction is where the nucleophilic substrate attacks the α-phosphorus atom of ATP. This type of reaction is catalysed by pyrophosphorylases and is an important step in the biosynthesis of those nucleotide coenzymes which are diesters of pyrophosphoric acid. Several examples have already been discussed in the biosynthetic pathways leading to NAD^+ (Scheme 2), FAD (equation 9), and coenzyme A (Scheme 32). In this type of reaction the β- and γ-phosphorus atoms of ATP are released as pyrophosphate, the α-phosphorus atom becoming incorporated into the newly formed diester of pyrophosphoric acid (*cf.* equation 9).

$$FMN + ATP \rightleftharpoons FAD + \text{pyrophosphate} \quad (9)$$

The third type of reaction is less common and involves nucleophilic attack on the β-phosphorus atom of ATP. The enzymes involved here are called pyrophosphokinases and the result is transfer of a diphosphate residue (the β- and γ-phosphorus atoms of ATP) to the nucleophilic substrate. This type of reaction has been discussed in relation to the biosynthesis of dihydrofolate (Scheme 25) and is also involved in the biosynthesis of thiamine diphosphate which is outlined later in this section (equation 27). A third example which has been reported is in the synthesis of 5-phosphoribosyl 1-pyrophosphate from ribose 5-phosphate and ATP (equation 25).

$$\text{Ribose 5-phosphate} + ATP \rightleftharpoons \text{5-phosphoribosyl 1-pyrophosphate} + AMP \quad (25)$$

A requirement for Mg^{2+} has been demonstrated for most of the above reactions and it has been suggested[94] that the role of the metal ion, which undoubtedly chelates with the triphosphate residue of ATP, is to facilitate nucleophilic attack on the ionized phosphate groups and perhaps to direct nucleophilic attack to the appropriate phosphorus atom.

24.3.4 COENZYMES INVOLVED PRINCIPALLY IN THE STABILIZATION OF CARBANIONS

The two coenzymes which are discussed in this section play rather a unique role in certain enzyme-catalysed reactions. While the chemical structure of these coenzymes is very different, consideration of the type of reaction in which they participate suggests that their biological function may be related, at least in so far as the mechanism by which they act is concerned.[95] Thus the role of these coenzymes can be considered to be the stabilization of a type of carbanion which might normally be regarded as an unstable, or even improbable, species. In order that they may carry out this function they must form covalent derivatives with the substrate in question, the nature of the derivative being such that the substrate acquires the necessary carbanionic character. The way in which this is achieved is discussed in the individual sections below.

24.3.4.1 Thiamine diphosphate

Thiamine diphosphate, which is sometimes referred to as cocarboxylase, is involved as coenzyme in the enzymic decarboxylation of α-keto-acids, in the oxidative decarboxylation of α-keto-acids, and in the formation of acetoin. Bruice and Benkovic[95] have summarized the mechanistic aspects of these reactions in the generalized equation (26), which involves an improbable acyl anion. The function of the coenzyme is to obviate the necessity for this acyl anion.

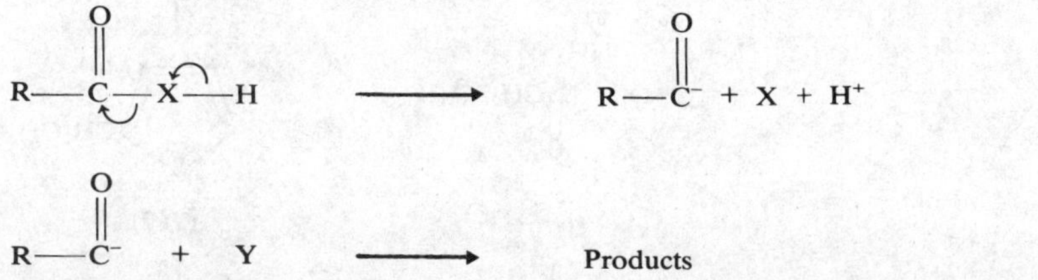

(26)

Thiamine diphosphate was first isolated in crystalline form from yeast by Lohmann and Schuster,[96] who also established the structure (**112**) for the coenzyme by titration and hydrolysis experiments. Thus titration established that the molecule contained one strongly acidic group and two more weakly acidic groups. Consideration of the pK_a values and the fact that enzymic hydrolysis of the coenzyme gave thiamine (**113**) and two equivalents of phosphoric acid suggested that the coenzyme was a pyrophosphate ester of thiamine. Alkaline hydrolysis and identification of inorganic pyrophosphate in the hydrolysis products confirmed this conclusion. That the diphosphate function was in the thiazole moiety was established by sodium sulphite cleavage of the coenzyme, which gave a phosphorus-free pyrimidine and a thiazole diphosphate. The position of attachment of the diphosphate residue was thus obvious and was later confirmed by synthesis.

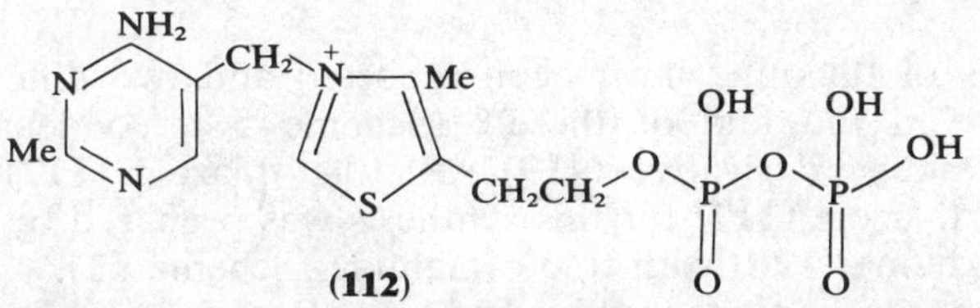

(**112**)

The structural relationship between thiamine diphosphate (**112**) and the vitamin thiamine (**113**) is thus clear and it is worth considering briefly the evidence[97] which led to elucidation of the structure of the latter compound (Scheme 49). The most significant reaction is cleavage of the vitamin into two components by aqueous sodium sulphite at room temperature. The first component was identified as the pyrimidine derivative (**114**) by (a) formation of sulphuric acid on treatment with water at 200 °C and sodium sulphite on fusing with alkali, (b) treatment with sodium in liquid ammonia to give pyrimidine (**115**), whose structure was confirmed by synthesis, and (c) acid hydrolysis which gave a third pyrimidine (**116**) whose structure was also confirmed by synthesis. The second component of the sulphite cleavage was identified as the thiazole (**117**). Oxidation of this compound with nitric acid resulted in loss of one carbon atom and formation of the known thiazolecarboxylic acid (**118**). The presence of a hydroxyl group in (**117**) was demonstrated by acylation and by replacement with chlorine on treatment with hydrochloric acid at 150 °C. The thiazole thus contained either an α- or β-hydroxyethyl substituent at the 5-position. The latter was preferred because (a) the vitamin was optically inactive, and (b) the thiazole (**117**) gave a negative iodoform test. The position of attachment of the thiazole nucleus to the pyrimidine was inferred from the position of the sulphonic acid residue in the pyrimidine (**114**) and was later confirmed by synthesis.

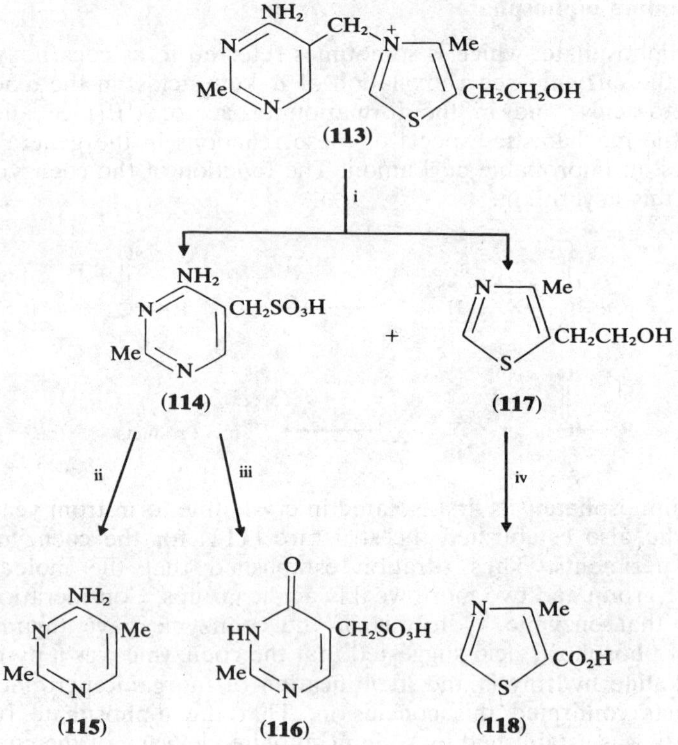

i, aq. Na_2SO_3; ii, Na, liq. NH_3; iii, H_3O^+; iv, HNO_3.

SCHEME 49

Several syntheses of thiamine have been reported and two different approaches are summarized below. In the first of these[98] (Scheme 50), condensation of a suitably functionalized pyrimidine derivative (**119**) with the thiazole (**117**) led directly to the vitamin (**113**). The thiazole (**117**) for this synthesis was prepared by condensation of the appropriate chloroketone (**120**) with thioformamide (Scheme 51).

The alternative approach, pioneered by Todd and Bergel,[99] involved synthesis of the 5-thioformamidomethylpyrimidine (**121**), which contained the thioformamide group as precursor of the thiazole ring. Condensation with the chloroketone (**120**) thus gave thiamine directly. The pyrimidine (**121**) is most conveniently prepared[100] by a sequence of reactions starting from acetamidine and ethoxymethylenemalononitrile (Scheme 52).

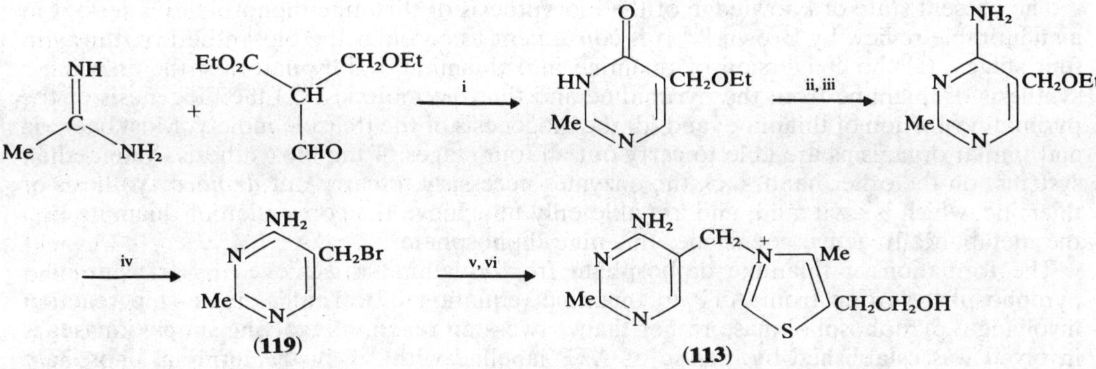

i, NaOEt, ii, $POCl_3$; iii, NH_3; iv, HBr; v, thiazole (**117**); vi, AgCl.

SCHEME 50

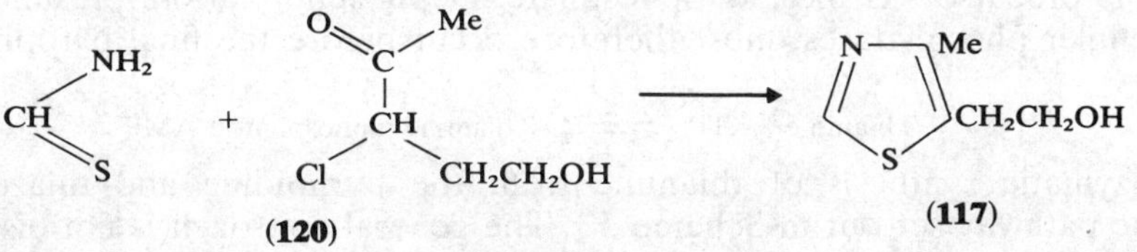

SCHEME 51

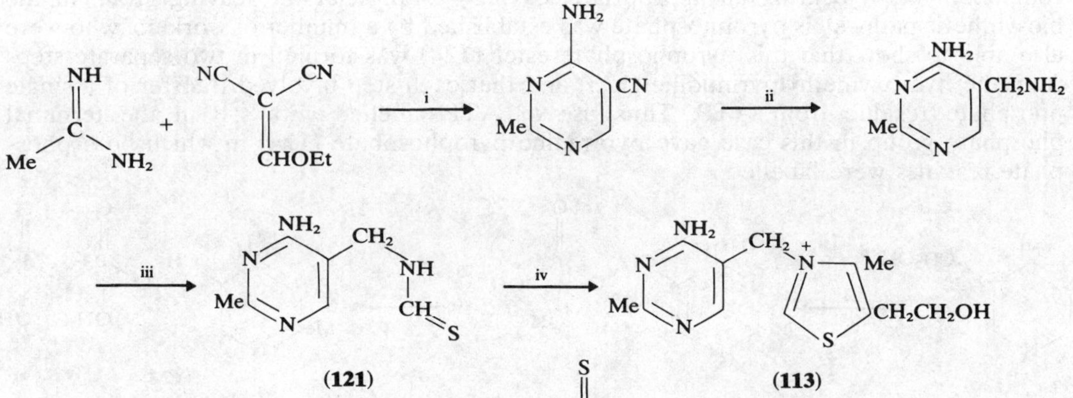

i, NaOEt; ii, Pd, H_2; iii, HC(=S)SK; iv, chloroketone (**120**).

SCHEME 52

Syntheses[101] of the coenzyme thiamine diphosphate (**112**) have usually involved phosphorylation of thiamine using reagents such as sodium pyrophosphate in orthophosphoric acid solution. These syntheses are not very efficient and it has been claimed[102] that treatment of the 5-(β-bromoethyl) analogue (**122**) of thiamine with silver pyrophosphate in phosphoric acid solution at 100 °C leads to better yields and a cleaner product.

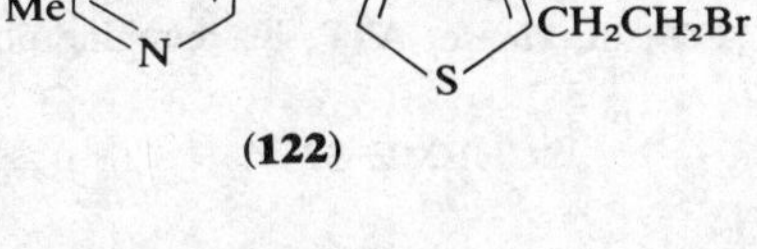

(**122**)

The present state of knowledge of the biosynthesis of thiamine diphosphate is set out in an admirable review by Brown.[103] It is convenient to consider the biosynthetic pathway in four stages: (a) the conversion of thiamine into thiamine diphosphate; (b) the enzymatic synthesis of thiamine from the pyrimidine and thiazole moieties; (c) the biogenesis of the pyrimidine portion of thiamine; and (d) the biogenesis of the thiazole moiety. Most bacteria and similar organisms are able to carry out all four stages of the biosynthesis. Mammalian systems, on the other hand, lack the enzymes necessary to carry out *de novo* synthesis of thiamine, which is a vitamin, and are able only to achieve the conversion of thiamine into the metabolically active coenzyme, thiamine diphosphate.

The formation of thiamine diphosphate from thiamine is achieved by transfer of a pyrophosphate group from ATP to thiamine (equation 27). That a single-step reaction involving a pyrophosphokinase rather than a two-step reaction involving simple kinases is involved was established by (a) use of ATP labelled with ^{32}P in the terminal phosphate group, which gave thiamine diphosphate labelled only in the terminal phosphate group, and (b) the observation that partially purified enzyme could use thiamine but not thiamine phosphate for the synthesis of the coenzyme. This last experiment is particularly relevant since it will be shown below that the primary product of *de novo* synthesis of thiamine from simple precursors is the monophosphate. Dephosphorylation, presumably catalysed by intracellular phosphatases, must therefore occur before the final pyrophosphorylation step.

$$\text{Thiamine} + \text{ATP} \rightleftharpoons \text{Thiamine diphosphate} + \text{AMP} \quad (27)$$

The enzymatic synthesis of thiamine from the pyrimidine and thiazole precursors follows the pathway set out in Scheme 53. The general approach is not dissimilar to that involved in many of the chemical syntheses (*cf.* Scheme 50) of thiamine in that the thiazole component (**126**) reacting as a nucleophilic species condenses with a pyrimidine component (**124**) bearing an appropriate leaving group. That the leaving group in the biosynthetic pathway is pyrophosphate was established by a number of workers, who were also able to show that this pyrophosphate ester (**124**) was formed in two separate steps from the hydroxymethylpyrimidine (**123**) and that each step involved transfer of a single phosphate residue from ATP. Thus use of ATP labelled with ^{32}P in the terminal phosphate group in this case gave pyrimidine pyrophosphate (**124**) in which both phosphate residues were labelled.

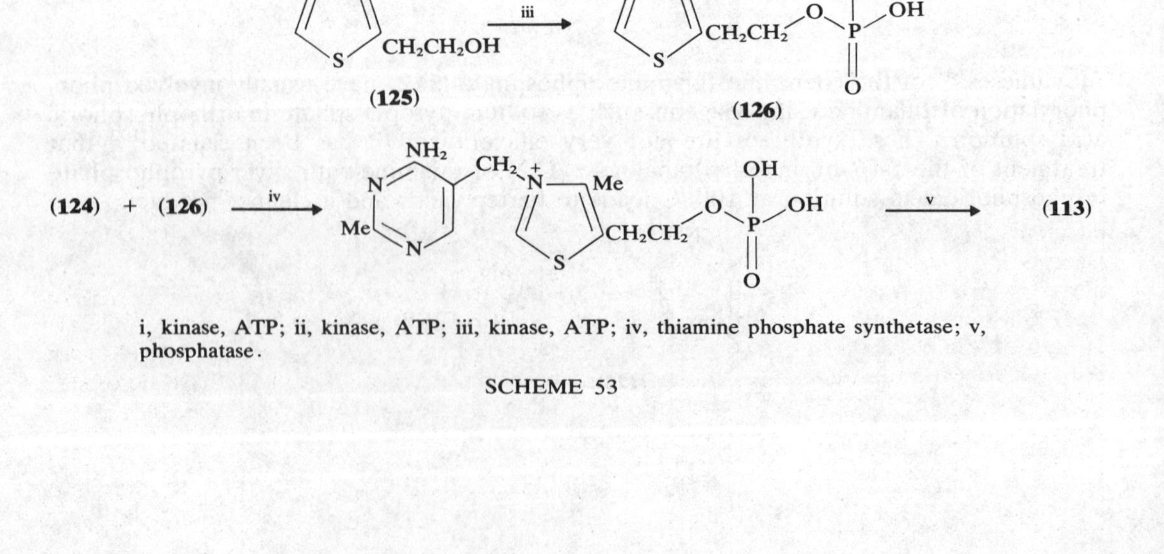

i, kinase, ATP; ii, kinase, ATP; iii, kinase, ATP; iv, thiamine phosphate synthetase; v, phosphatase.

SCHEME 53

Very little is known about the detailed steps involved in the biosynthesis of the pyrimidine (**123**) and thiazole (**125**) precursors. The most significant work on the biosynthesis of the pyrimidine has been carried out by Newell and Tucker.[104] They suggest that some of the steps in the *de novo* synthesis of purines also function in the biosynthesis of the pyrimidine component of thiamine. There is evidence that 5-aminoimidazole ribonucleotide (**127**) is a common intermediate. It has been further suggested by Diorio and Lewin[105] that the 5-aminomethyl- and 5-formyl-pyrimidine derivatives (**128**) and (**129**) are intermediates, although definite evidence is lacking, and no reasonable mechanisms for the transformations have been proposed. The partial biosynthetic sequence set out in Scheme 54 must therefore be regarded as speculative. There is even less information available on the biosynthesis of the thiazole component, investigations in this area having been restricted to attempts to identify possible precursors using radioactive incorporation experiments.

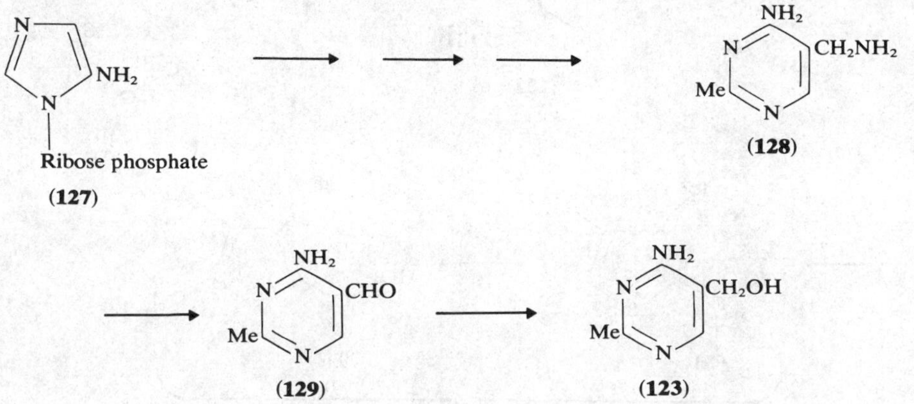

SCHEME 54

The key to the understanding of the mechanism of action of thiamine diphosphate as a coenzyme came from Breslow's observation[106] that the proton attached to the C-2 carbon atom of thiazolium salts exchanges rapidly with deuterium when the salt is dissolved in deuterium oxide. The dipolar ion or ylide (**130**), which is an intermediate in the exchange reaction, is formed particularly easily with *N*-benzylthiazolium salts and it is reasonable to suggest that the pyrimidine ring in the coenzyme has a similar function (equation 28).

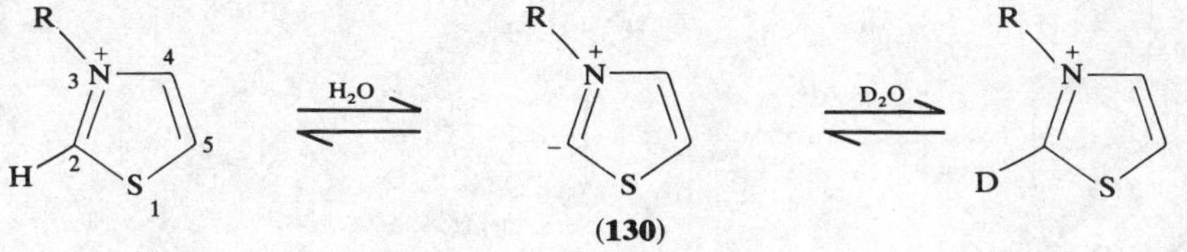

(28)

The ylide, abbreviated to (**131**), formed in this way from thiamine diphosphate is a nucleophilic species and can thus react with carbonyl derivatives such as pyruvate (**132**) to give an α-hydroxy-acid (**133**). Decarboxylation of this compound would be expected to take place as shown to give the stabilized acyl carbanion (**134**). Evidence has been obtained from deuterium exchange reactions which indicates that the ylide (**131**) is an intermediate in the decarboxylation of pyruvate *in vivo*. The carbanion (**134**) can readily undergo further reaction to give products expected from enzymic reactions which involve thiamine diphosphate as coenzyme. Thus protonation will lead to formation of acetaldehyde and regeneration of the ylide (**131**), as found in the enzymic decarboxylation of α-keto-acids. Reaction with an aldehyde such as benzaldehyde will lead to a 'crossed' acyloin condensation and formation of the acyloin (**135**). This product has been isolated from yeast fermentations to which benzaldehyde has been added. The role of the ylide (**131**) in the acyloin reaction is similar to that of cyanide in the familiar benzoin condensation which is (a) to act as an effective nucleophile towards the carbonyl group and (b) to stabilize the negative charge on carbon in the resulting adduct.[107] That the carbanion (**134**) is indeed an intermediate in the enzymic acyloin reaction was established by synthesis[108] of the protonated form, 2-(1′-hydroxyethyl)thiamine diphosphate, and showing that it would react with aldehydes in the acyloin reaction in the presence of a crude enzyme preparation (Scheme 55).

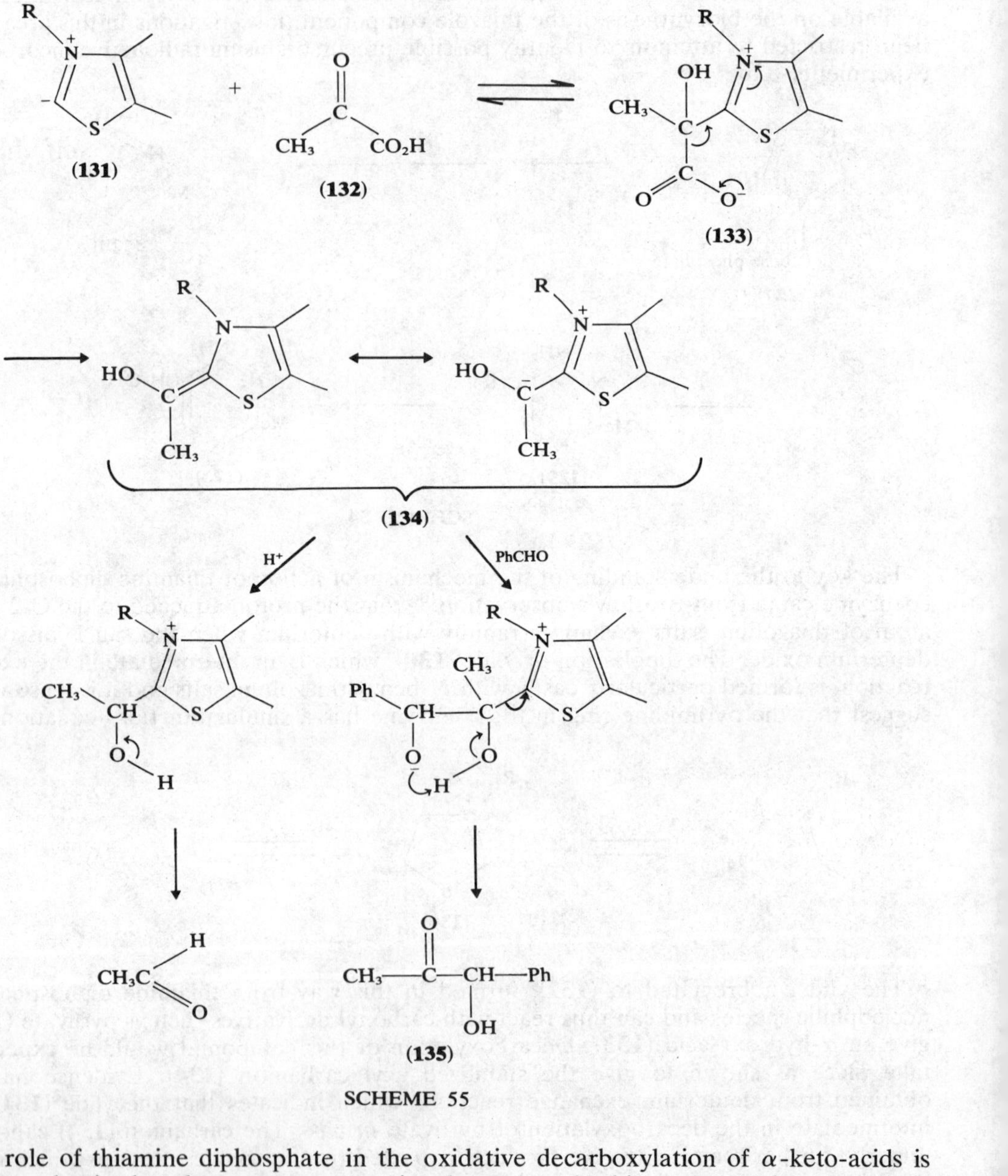

SCHEME 55

The role of thiamine diphosphate in the oxidative decarboxylation of α-keto-acids is more complex in that the enzyme system involves not only this coenzyme but also lipoic acid, coenzyme A, and NAD^+. The net result is that an 'acetaldehyde equivalent' is transferred from pyruvate and ends up as the acetyl group of coenzyme A. The necessary oxidation involves the carbanion (**134**), which is formed as in Scheme 55, and lipoic acid (abbreviated to **136**), and results in the formation of acetyldihydrolipoic acid (**137**) and the ylide (**131**) (Scheme 56). The acetyl group is subsequently transferred to coenzyme A, the function of the NAD^+ being to reoxidize the dihydrolipoic acid to lipoic acid.[109]

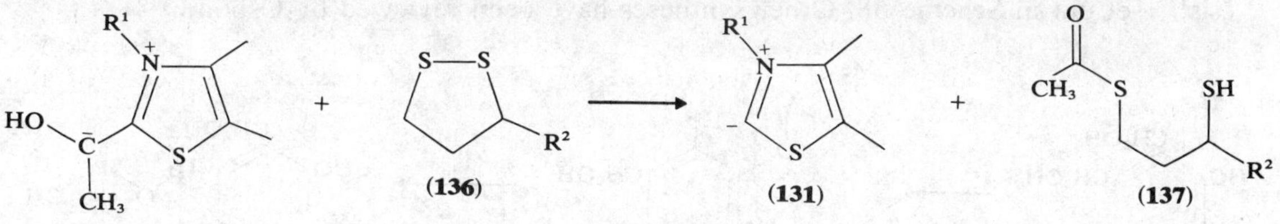

SCHEME 56

Further support for the involvement of carbanions such as (**134**) in thiamine diphosphate dependent enzymic reactions has come from an interesting paper by Gutowski and Lienhard.[110] These workers showed that the thiazolone (**138**), which they consider to be a transition state analogue[111] of the carbanion (**134**), binds very strongly to the pyruvate oxidase enzyme complex, the dissociation constant for the enzyme–inhibitor complex being at least 10^4 less than that for the enzyme–substrate complex under similar conditions. This observation suggests very strongly that the enamine form of the carbanion (**134**) is indeed present in the enzymic reaction (Scheme 57).

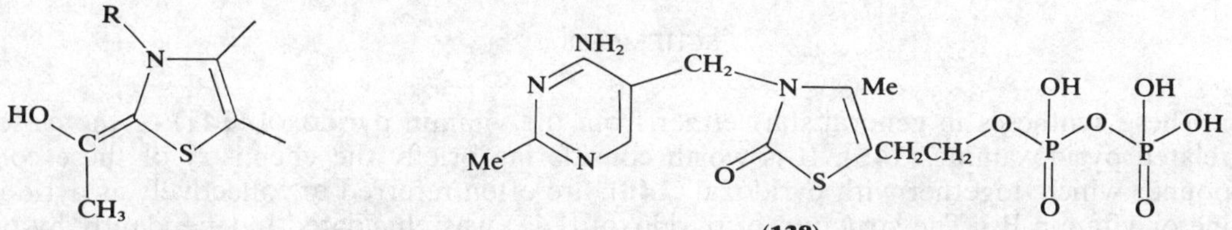

SCHEME 57

24.3.4.2 Pyridoxal phosphate

Pyridoxal phosphate (**139**) is the coenzyme concerned in a large number of reactions involved α-amino-acids, including racemization, decarboxylation, transamination, and elimination or replacement of substituents on the β- and γ-carbon atoms.[112] These reactions can all be classified in mechanistic terms as requiring the stabilization of a carbanion at the α- or β-position of an α-amino-acid. The pyridoxal phosphate fulfils this requirement by formation of an imine between the α-amino-group of the amino-acid and the aldehyde group of the coenzyme. As will be discussed later, effective delocalization of the negative charge on the carbanion then becomes possible.

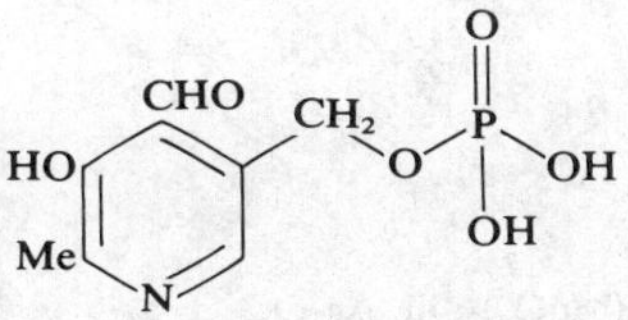

(139)

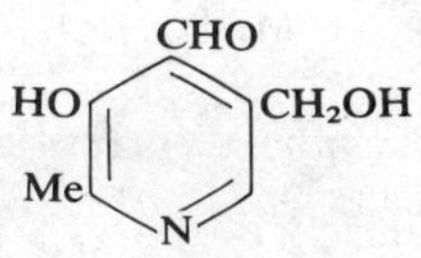

(140)

Pyridoxal phosphate, which is sometimes referred to as codecarboxylase, was first isolated[113] from yeast and shown to be identical with synthetic material prepared in low yield by phosphorylation of pyridoxal (**140**) using phosphoryl chloride in the presence of water.[114] The precise location of the phosphate residue was established by an elimination procedure and the structure (**139**) was eventually confirmed by the unambiguous synthesis[115] set out in Scheme 58. Other syntheses have been reviewed by Osbond.[116]

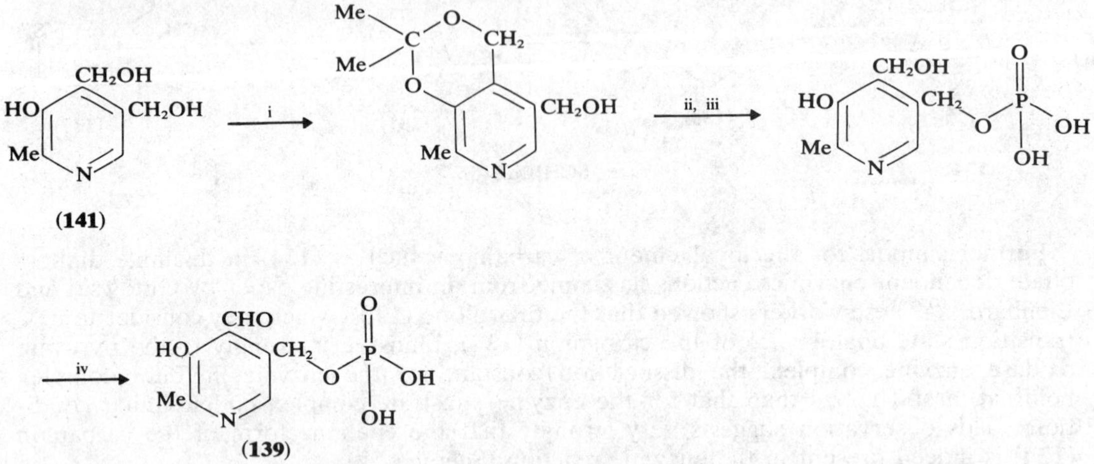

i, Me_2CO, H^+; ii, P_2O_5, H_3PO_4; iii, H_3O^+; iv, MnO_2, H_3O^+.

SCHEME 58

These syntheses in general start either from the vitamin pyridoxol (**141**) or the closely related pyridoxamine (**145**). It is worth considering briefly the chemistry of these compounds which, together with pyridoxal (**140**), are often referred to collectively as pyridoxine or vitamin B_6. The structure of pyridoxol (**141**) was elucidated independently by two groups of workers[117] and the significant reactions are set out in Scheme 59. The presence of a 3-hydroxypyridine nucleus was deduced from the characteristic u.v. spectrum, and the presence of three hydroxyl groups by the formation of a triacetate and tribenzoate. Only one of the hydroxyl groups could be methylated with diazomethane to give the methyl

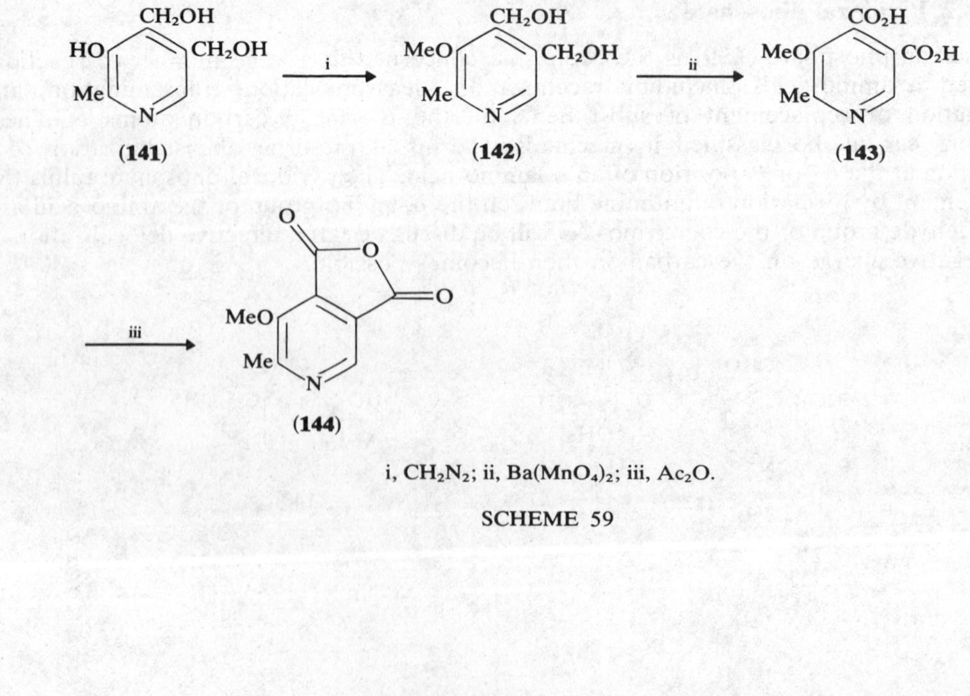

i, CH_2N_2; ii, $Ba(MnO_4)_2$; iii, Ac_2O.

SCHEME 59

ether (**142**). Careful oxidation with barium permanganate give the dicarboxylic acid (**143**), which contained all the carbon atoms of the original compound including the *C*-methyl group. The dicarboxylic acid readily gave an anhydride (**144**). The substitution pattern was deduced from the observation that pyridoxol (**141**), but not its methyl ether (**142**), gave a positive Gibb's phenol test which indicated that the position *para* to the phenolic hydroxyl group was unsubstituted.

Numerous syntheses of pyridoxol have been reported and are reviewed in detail by Osbond.[116] One particularly elegant synthesis[118] is outlined in Scheme 60.

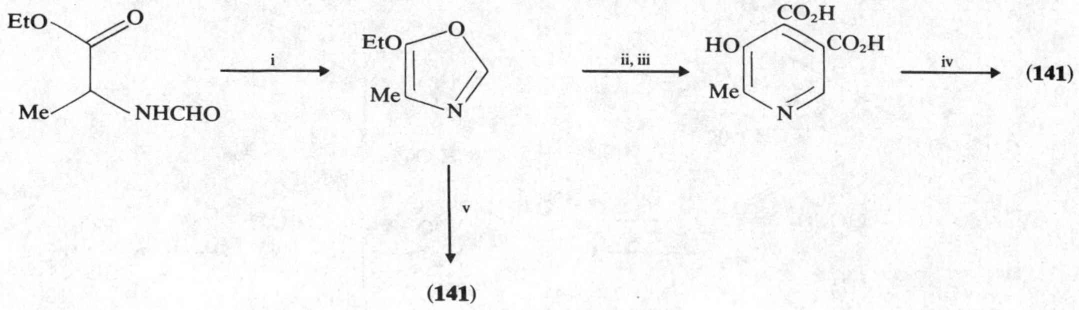

i, P_2O_5; ii, maleic anhydride; iii, EtOH, HCl; iv, $LiAlH_4$; v, $HOCH_2CH{=}CHCH_2OH$.

SCHEME 60

The inter-relationship between pyridoxol (**141**), pyridoxal (**140**), and pyridoxamine (**145**) was established[118] by the transformations set out in Scheme 61. Careful oxidation of pyridoxol with potassium permanganate gave pyridoxal isolated as its oxime (**146**). Treatment of the latter compound with nitrous acid gave pyridoxal, whereas catalytic reduction gave pyridoxamine. Alternative syntheses of these compounds have also been reported.[116]

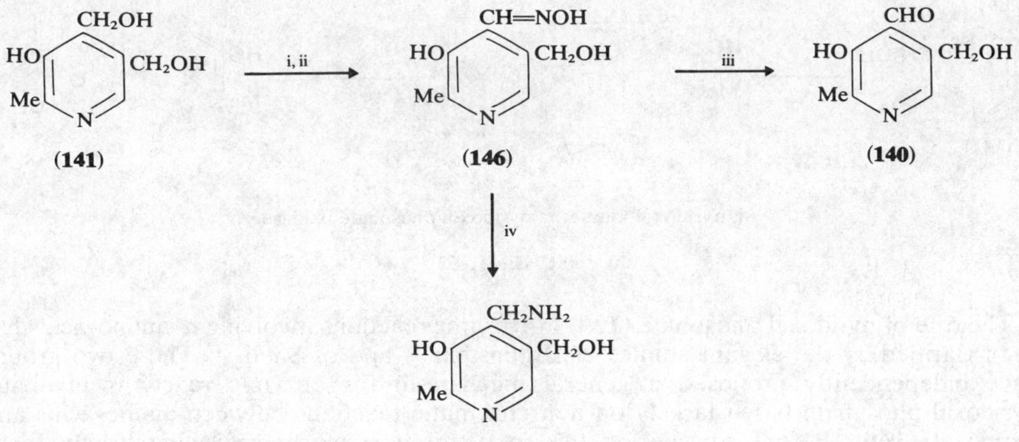

i, $KMnO_4$; ii, H_2NOH; iii, $NaNO_2$, HCl; iv, H_2, Pt.

SCHEME 61

Until recently, very little was known about the biosynthesis of pyridoxal phosphate, at least in so far as formation of the heterocyclic ring was concerned.[119] *De novo* biosynthesis of pyridoxine (vitamin B_6) is a process peculiar to micro-organisms and attempts to establish the nature of precursors, intermediates, and biosynthetic pathways were frustrated by the very small quantities of the vitamin produced by most organisms. Recent work[120] using *E. coli* mutants has elucidated the pathway to pyridoxol set out in Scheme 62. Some details of the transformation of glycerol into pyridoxol are still unexplained, but the essential features are now clear.

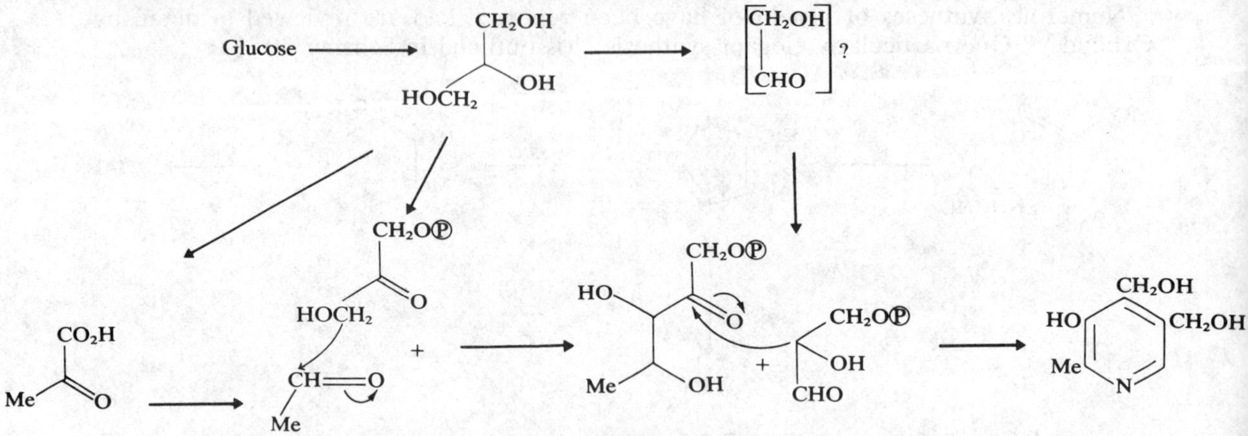

SCHEME 62

Considerable work has been done on the interconversion of the various forms of vitamin B_6 and their phosphorylated derivatives.[121] Most of these results relate to mammalian systems, but available evidence suggests comparable relationships for other organisms. It seems likely that the preferred route from pyridoxol (**141**) to pyridoxal phosphate (**139**) involves conversion of pyridoxol into the corresponding phosphate (**147**) by an appropriate intracellular kinase, followed by oxidation of the resulting phosphate to give pyridoxal phosphate (Scheme 63).

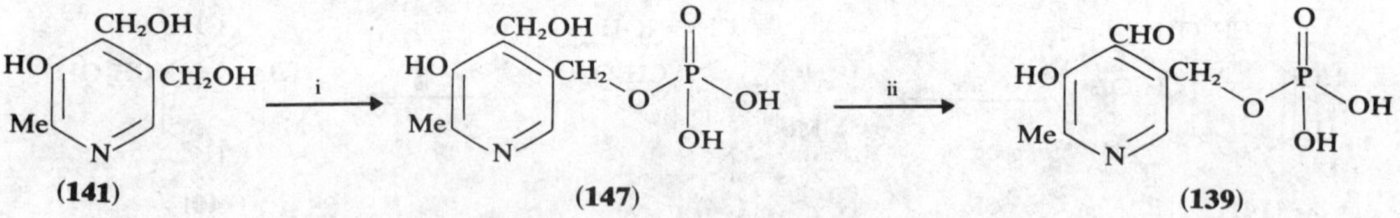

i, pyridoxol kinase; ii, pyridoxol phosphate oxidase.

SCHEME 63

The role of pyridoxal phosphate (**139**) in enzymic reactions involving α-amino-acids has been clarified by the elegant studies of Braunstein[112] and of Snell.[122] These two groups have independently proposed a general mechanism for enzymic reactions involving pyridoxal phosphate based largely on non-enzymatic reactions between amino-acids and pyridoxal (**140**). Before considering this in detail it is worth reviewing briefly those functional groups on the coenzyme which are necessary for bonding to the apoenzyme.

Non-phosphorylated analogues of the coenzyme are not bound to the enzyme protein, while phosphonate analogues which do bind strongly are not active catalytically. The aldehyde group is essential and is involved initially in formation of an imine with the ε-amino-group of a lysine residue. The u.v. spectrum of the complex is typical of a Schiff's base and sodium borohydride reduction of the complex leads to irreversible labelling of the lysine residue. The phenolic hydroxyl group is apparently not involved in binding since the corresponding methyl ether binds strongly, although the resulting holoenzyme is inactive. Finally, the methyl group is not involved in binding and indeed the 2-nor-pyridoxal phosphate complex is more active than that involving the normal coenzyme.

The model experiments referred to above led Braunstein and Snell to suggest that the essential feature of pyridoxal phosphate mediated reactions was the formation of an imine (Schiff's base) between the α-amino-group of the amino-acid and the aldehyde group of pyridoxal phosphate. This suggestion has been widely accepted. The model experiments typically included pyridoxal, a polyvalent metal ion (Ca^{2+}, Fe^{3+}, Al^{3+}), and the appropriate amino-acid substrate. A typical transamination reaction which could be effected in this way is illustrated in Scheme 64, a large excess of substrate being necessary to drive the reaction to completion.

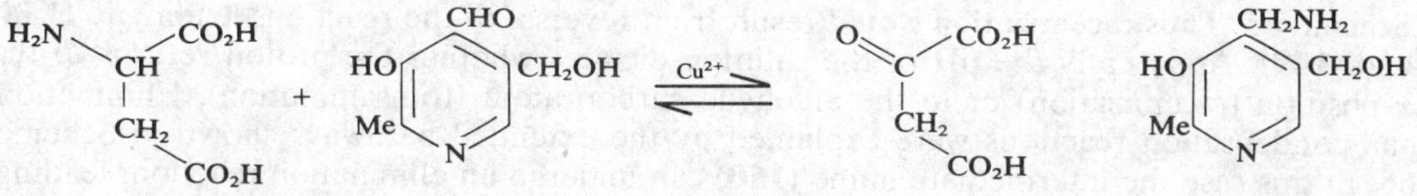

SCHEME 64

The mechanism put forward to explain such non-enzymic reactions is illustrated in Scheme 65.[123] The driving force for the reaction is the stabilization of the transition state for carbanion formation by delocalization of the charge through the conjugated system. The function of the metal ion in these model systems appears to be stabilization of the intermediate imine (**148**), provision of a planar conjugated system, and increasing the inductive effect on the α-carbon atom and thus facilitating release of the proton.

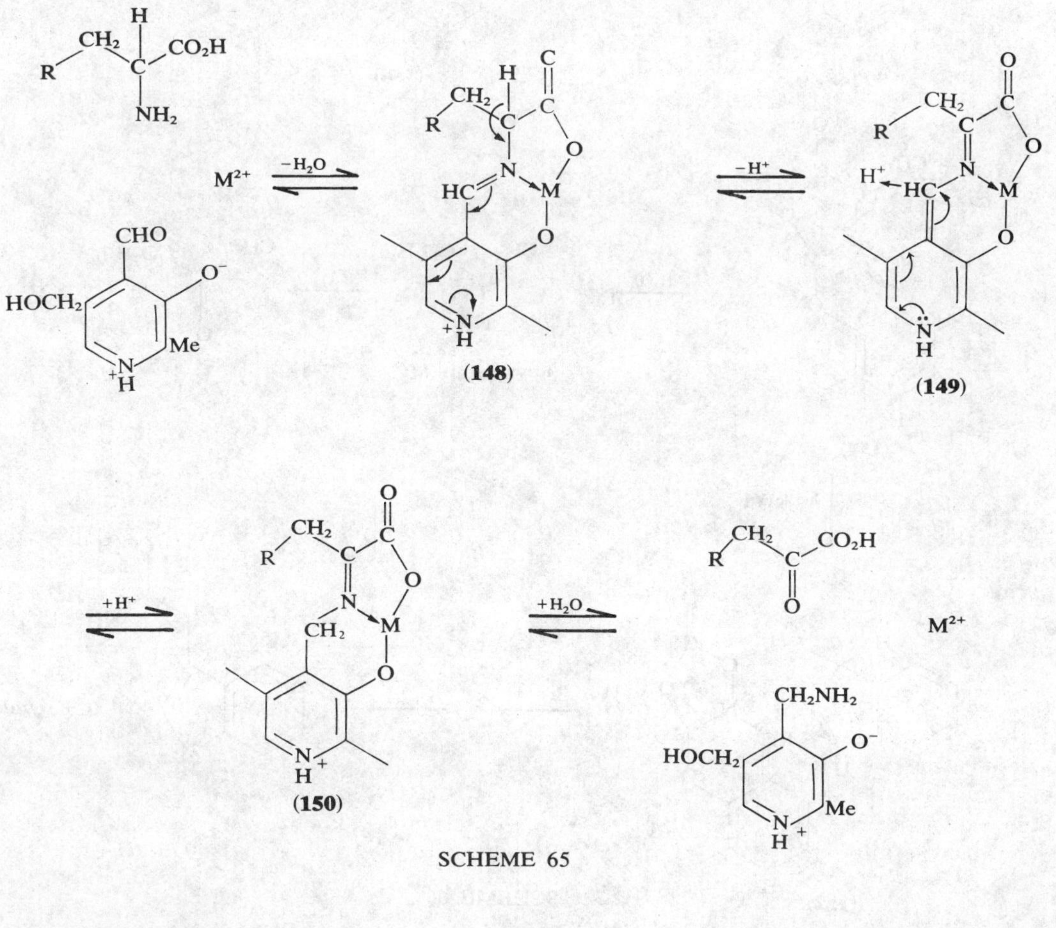

SCHEME 65

Similar model reactions have been carried out to demonstrate racemization, and elimination and condensation reactions. The experimental conditions and the nature of the substrates determine which type of reaction will occur in these model systems. Thus the transamination reaction outlined in Scheme 64 is favoured at lower pH values (*ca.* pH 5) and racemization at higher pH values (*ca.* pH 10). Use of an amino-acid such as serine, which has a leaving group in the β-position, in the presence of an appropriate nucleophile such as indole will lead to the non-enzymatic transformation of serine into tryptophan (equation 29).

$$\text{Serine} + \text{pyridoxal} + \text{indole} \underset{}{\overset{Al^{3+}}{\rightleftharpoons}} \text{Tryptophan} + \text{pyridoxal} \qquad (29)$$

These reactions were explained using mechanistic pathways similar to that set out in Scheme 65. Thus racemization would result from reversal of the reactions leading to (**148**) and (**149**). Apparently the pH of the solution dictates whether the proton returns to the α-position (racemization) or to the aldehyde carbon atom (transamination). Elimination and condensation reactions were explained by the extended pathways shown in Scheme 66. In this case the intermediate imine (**150**) can undergo an elimination reaction, leading

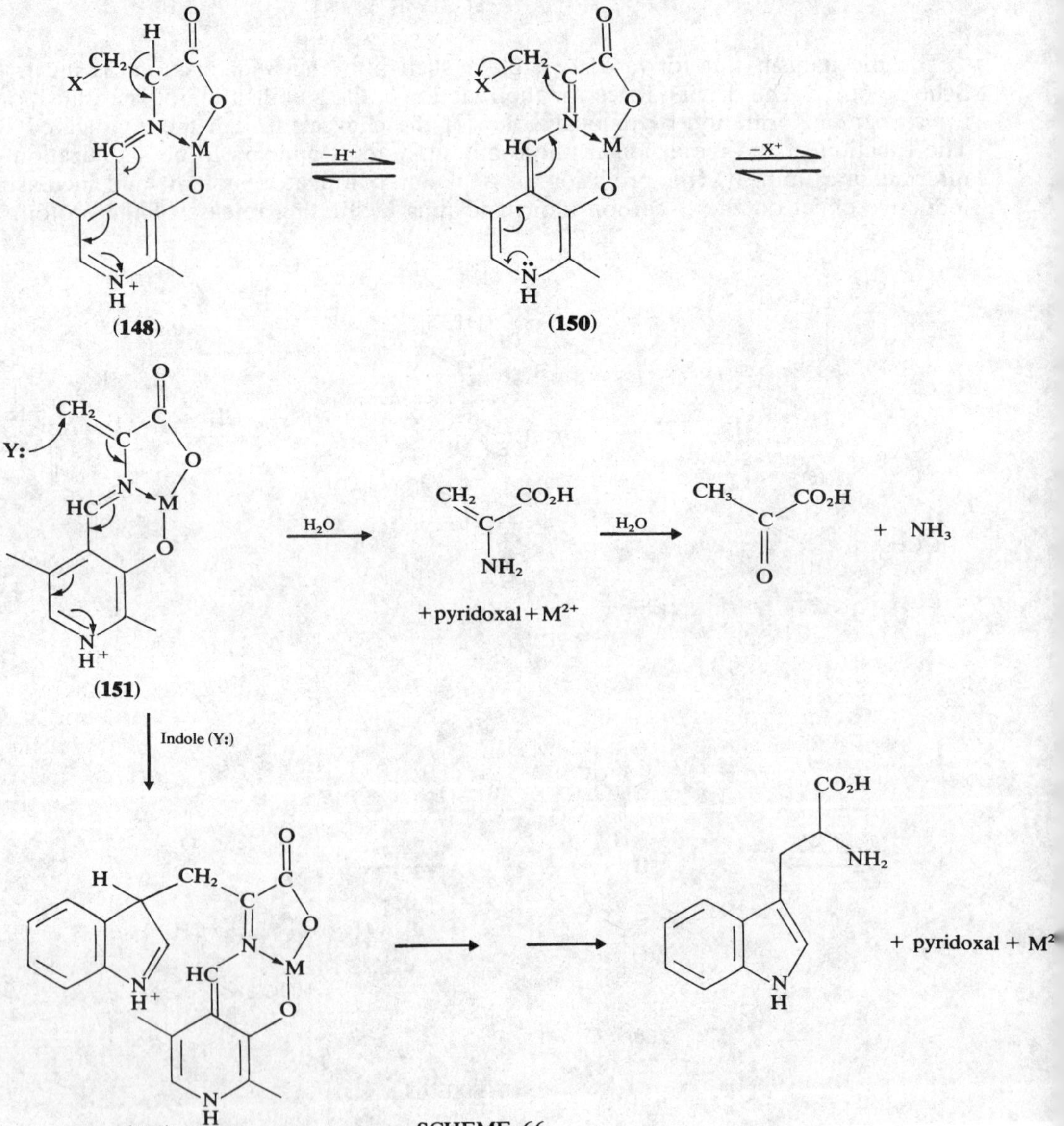

SCHEME 66

to the intermediate (**151**). Hydrolysis of this intermediate leads to pyruvate plus ammonia. Alternatively, reaction between the intermediate and indole will lead to intermediate (**152**) and thence to tryptophan.

Decarboxylation of α-amino-acids does not take place in these model systems and it has been shown that metal ions actually inhibit the decarboxylation reaction. This is not unexpected since the carboxyl group forms part of the chelate structure, the pair of electrons on the carboxyl anion being shared with the metal ion. These factors would be expected to decrease the tendency for carbon dioxide to be eliminated. The metal chelates would not be able to attain the preferred conformation for decarboxylation (see below), and this is probably a major factor in their reluctance to undergo decarboxylation.

The success of the model experiments using pyridoxal plus metal ions in duplicating many enzymic reactions of α-amino-acids tends to suggest that metal ions might play an important part in the corresponding enzymic reactions. This is probably not the case since highly purified enzyme systems have been prepared which require pyridoxal phosphate but which do not require metal ions for full activity.[124] The function of the metal ion in the model systems appears to be to maintain the correct geometry of the intermediate imines and thus facilitate charge delocalization. In the enzymatic reaction this role is filled by the enzyme protein. This factor apart, it seems that the role of pyridoxal phosphate in the enzymic reaction is very similar to that of pyridoxal in the model system. Since the initial reaction between coenzyme and apoenzyme is the formation of an imine between pyridoxal phosphate and the ε-amino-group of lysine to give the holoenzyme, an transimination reaction must take place in the presence of the amino-acid substrate to give the imine (**153**) involved in the enzymic reaction (Scheme 67).

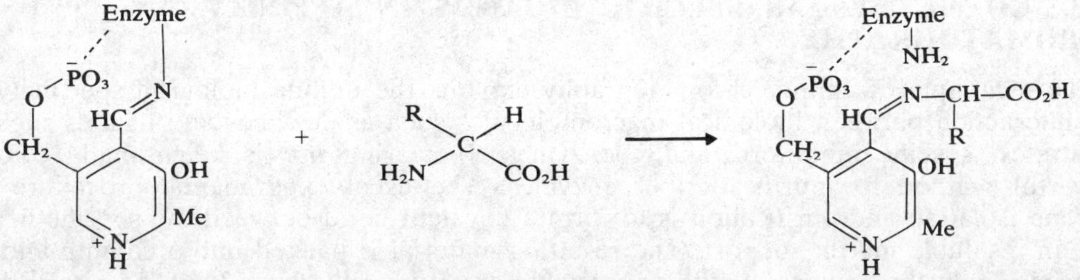

SCHEME 67 (**153**)

The enzymatic reactions catalysed by pyridoxal phosphate have been shown to be under strict stereochemical control.[125] An important factor in the 'activation' of σ-bonds by a π-system is the stereochemical relationship of the σ-bond to the adjacent π-orbitals. Thus it was argued by Dunathan[126] that the particular σ-bond to be broken should be in a plane perpendicular to the plane of the π-system in the coenzyme. This particular conformation would allow maximum σ–π overlap and minimize the energy of the transition state for bond breaking. He further suggested that the apoenzyme might control the conformation, possibly by binding of the carboxylate ion, and suggested that the imine (**153**) could adopt one of three possible conformations. These are represented in diagrammatic form in Scheme 68, where the rectangles indicate the plane of the pyridine ring and

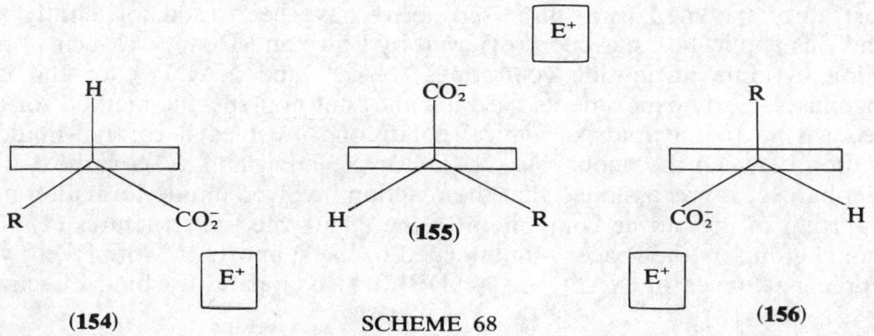

(**154**) (**155**) (**156**)

SCHEME 68

the conformations are viewed along the C^{α}—N bond. The labile σ-bond is that shown vertically in each case. Thus conformation (**154**) is that appropriate for transamination, *etc.*, conformation (**155**) for decarboxylation, and conformation (**156**) for removal of R as in serine hydroxymethylase.

Interesting confirmation of these ideas has come from the observation that if the configuration of the amino-acid is changed (from L to D), then conformation (**156**), which is that appropriate for loss of R, becomes conformation (**157**), which is that appropriate for loss of H and transamination (Scheme 69). Cross reactivity of this kind can be demonstrated and L-serine hydroxymethylase will catalyse transamination reactions of D-alanine.[125]

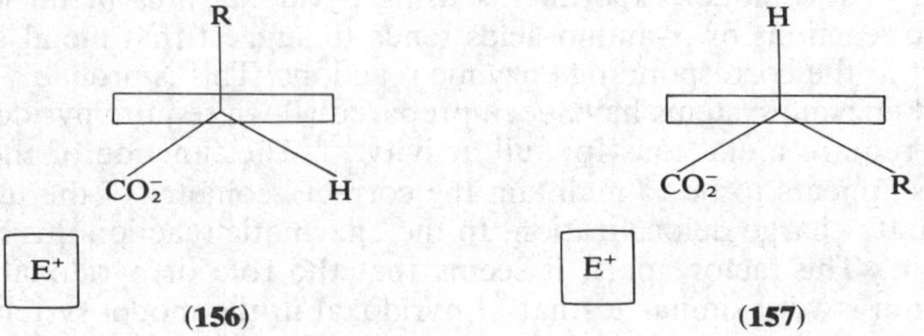

SCHEME 69

24.3.5 COENZYMES AS SPECIFIC LIGANDS IN AFFINITY CHROMATOGRAPHY

The technique of affinity chromatography exploits the unique biological specificity of the interaction between biological macromolecules such as enzymes and ligands such as substrates, specific inhibitors, and coenzymes. Increasing use is being made of this powerful method for purification of enzymes. The usual experimental procedure for enzyme isolation and purification is to form a covalent bond between the specific ligand and an insoluble matrix support. The resulting material is packed into a column and, in principle, only the enzyme(s) with appreciable affinity for the ligand will be retained on such a column, all other proteins passing through unretarded. Elution of the enzyme which has been specifically absorbed is achieved by altering the composition of the solvent so as to favour dissociation of the enzyme–ligand complex.[127]

Increasing use is being made in affinity chromatography of group specific adsorbents. Thus by using a ligand which is specific for a group of enzymes it is possible to circumvent some of the problems inherent in the design and synthesis of a more specific ligand. Some selectivity is lost in the adsorption phase of affinity chromatography when using such material, but may be regained in the elution phase by suitable choice of eluant conditions.[128] Coenzymes are almost ideal group-specific adsorbents since in general they function for a given group of enzymes. These enzymes have at least two specific binding sites, one for the coenzyme which will be common to all members of the group, and one or more for the particular substrate involved.

Almost all of the coenzymes discussed above have been used for affinity chromatography and this application has been reviewed by Lowe and Dean.[127] Much work has been done using pyridine nucleotide coenzymes (NAD^+ and $NADP^+$) for the isolation of dehydrogenases. Early experiments used a somewhat non-specific method for attachment of the coenzyme to the matrix. Typical conditions involved a carbodi-imide mediated condensation between the coenzyme (NAD^+ or $NADP^+$) and an ω-carboxyalkyl derivative of Sepharose. It was assumed that the reaction involved amide formation between the 6-amino-group of the purine component of the coenzyme [*cf.* structures (**1**) and (**2**)] and the carboxyl group on the spacer arm attached to the support.[129] More recent work[130] has used simple analogues of NAD^+ or $NADP^+$ which apparently bind effectively to the enzymes involved. Thus the adenosine 2′,5′-diphosphate derivative (**158**) has been linked to Sepharose *via* the terminal amino-group in the side-chain. The resulting material specifically absorbs NADP$^+$-dependent enzymes such as glucose 6-phosphate dehydrogenase and 6-phosphogluconate dehydrogenase, but NAD$^+$-dependent enzymes such as lactate dehydrogenase are not affected. Elution of the bound enzymes is effected by pulses of NADP$^+$.

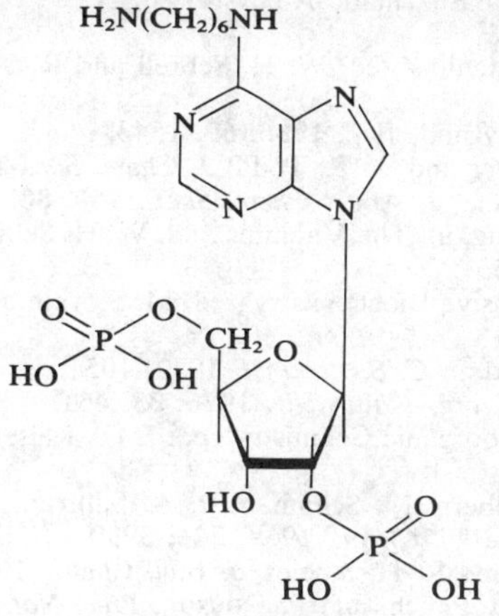

(**158**)

An alternative approach is to use specific inhibitors of the enzymes which are structurally related to the coenzyme. Thus methotrexate (**58**) is a potent inhibitor of dihydrofolate reductase and shows little discrimination among samples of the enzyme which have been isolated from various sources. This compound was covalently attached to 6-aminoethyl-Sepharose by a carbodi-imide condensation which presumably involves a carboxyl group of the methotrexate. The resulting material has been used for the large-scale isolation of dihydrofolate reductase from *E. coli.* Washing of the column with molar sodium chloride solution eluted less than 1% of the bound enzyme, and quantitative recovery was achieved by elution with dihydrofolate in molar sodium chloride solutions.[131]

References

1. M. Dixon and E. C. Webb, 'Enzymes', Longmans, London, 1958, chapter IX.
2. G. H. Hitchings and J. J. Burchall, *Adv. Enzymol.*, 1965, **27,** 417; H. C. S. Wood, in 'Chemistry and Biology of Pteridines', ed. W. Pfleiderer, de Gruyter, Berlin, 1976, p. 27.
3. R. H. Abeles and A. L. Maycock, *Accounts Chem. Res.*, 1976, **9,** 313.
4. A. White, P. Handler, and E. L. Smith, 'Principles of Biochemistry', 5th edn., McGraw-Hill, New York, 1973, chapters 14–16.
5. T. P. Singer and E. B. Kearney, *Adv. Enzymol.*, 1954, **15,** 79.
6. F. Schlenk, *J. Biol. Chem.*, 1942, **146,** 619.
7. R. U. Lemieux and J. W. Loun, *Canad. J. Chem.*, 1963, **41,** 889.
8. L. J. Haynes, N. A. Hughes, G. W. Kenner, and A. R. Todd, *J. Chem. Soc.*, 1957, 3727; N. A. Hughes, G. W. Kenner and A. R. Todd, *ibid.*, 1957, 3733.
9. A. Kornberg and W. E. Pricer, *J. Biol. Chem.*, 1950, **186,** 557.
10. J. Preiss and P. Handler, *J. Biol. Chem.*, 1958, **233,** 488 and 493.
11. M. E. Pullman, A. San Pietro, and S. P. Colowick, *J. Biol. Chem.*, 1954, **206,** 129.
12. H. E. Dubb, M. Saunders, and J. H. Wang, *J. Amer. Chem. Soc.*, 1958, **80,** 1767; R. F. Hutton and F. H. Westheimer, *Tetrahedron*, 1958, **3,** 73.
13. H. R. Levy, P. Talalay, and B. Vennesland, in 'Progress in Stereochemistry', ed. P. B. D. de la Mare and W. Klyne, Butterworths, London, 1962, vol. III, chapter 8.
14. J. W. Cornforth, G. Ryback, G. Popjak, C. Donninger, and G. Schroepfer, *Biochem. Biophys. Res. Comm.*, 1962, **9,** 371.
15. S. P. Colowick, J. van Eys, and J. H. Park in 'Comprehensive Biochemistry', ed. M. Florkin and E. H. Stotz, Elsevier, Amsterdam, 1966, vol. 14, chapter 1.

16. H. R. Levy, F. A. Loewus, and B. Vennesland, *J. Amer. Chem. Soc.*, 1957, **79,** 2949.
17. H. Sund, in 'Biological Oxidations', ed. T. P. Singer, Interscience, New York, 1968, p. 603.
18. H. R. Mahler, in 'Symposium on Free Radicals in Biological Systems', Stanford Biophysics Lab., California, 1960.
19. T. C. Bruice and S. J. Benkovic, 'Bioorganic Mechanisms', Benjamin, New York, 1966, vol. II, p. 343.
20. G. A. Hamilton, in 'Progress in Bioorganic Chemistry', ed. E. T. Kaiser and F. J. Kézdy, Wiley-Interscience, New York, 1971, vol. 1, p. 83.
21. M. Klingenberg, in 'Methods of Enzymatic Analysis', ed. H. U. Bergmeyer and K. Gawehn, Academic Press, 1974, vol. 4, p. 2045.
22. T. Wagner-Jauregg, in 'The Vitamins', ed. W. H. Sebrell and R. S. Harris, Academic Press, New York, 1972, vol. V, p. 3.
23. R. Kuhn, H. Rudy, and F. Weygand, *Ber.*, 1936, **69,** 1543.
24. S. M. H. Christie, G. W. Kenner and A. R. Todd, *J. Chem. Soc.*, 1954, 46.
25. J. G. Moffatt and H. G. Khorana, *J. Amer. Chem. Soc.*, 1958, **80,** 3756.
26. M. K. Horwitt and L. A. Witting, in 'The Vitamins', ed. W. H. Sebrell and R. S. Harris, Academic Press, New York, 1972, vol. V, p. 53.
27. G. W. E. Plaut, in 'Comprehensive Biochemistry', ed. M. Florkin and E. H. Stotz, Elsevier, Amsterdam, 1971, vol. 21, p. 11.
28. T. Paterson and H. C. S. Wood, *J. C. S. Perkin I*, 1972, 1051.
29. P. Hemmerich, *Fortschr. Chem. org. Naturstoffe*, 1976, **33,** 451.
30. T. C. Bruice, in 'Progress in Bioorganic Chemistry', ed. E. T. Kaiser and F. J. Kézdy, Wiley-Interscience, New York, 1976, vol. 4, p. 1.
31. V. Massey, F. Müller, R. Feldberg, M. Schuman, P. A. Sullivan, L. G. Howell, S. G. Mayhew, R. G. Matthews, and G. P. Foust, *J. Biol. Chem.*, 1969, **244,** 3999.
32. C. T. Walsh, A. Schonbrunn, and R. H. Abeles, *J. Biol. Chem.*, 1971, **246,** 6855.
33. J. W. Hastings, C. Balny, C. Le Peuch, and P. Douzou, *Proc. Nat. Acad. Sci. U.S.A.*, 1973, **70,** 3468.
34. W. C. Kenney, W. H. Walker, and T. P. Singer, *J. Biol. Chem.*, 1972, **247,** 4510.
35. W. H. Walker, T. P. Singer, S. Ghisla, and P. Hemmerich, *European J. Biochem.*, 1972, **26,** 279.
36. Y. Hatefi, *Adv. Enzymol.*, 1963, **25,** 275.
37. R. A. Morton, U. Gloor, O. Schindler, G. M. Wilson, L. H. Chopard-dit-Jean, F. W. Hemming, O. Isler, W. M. F. Leat, J. F. Pennock, R. Rüegg, U. Schwieter, and O. Wiss, *Helv. Chim. Acta*, 1958, **41,** 2343.
38. A. F. Wagner and K. Folkers, 'Vitamins and Coenzymes', Interscience, New York, 1964, p. 443.
39. U. Gloor, O. Schindler, and O. Wiss, *Helv. Chim. Acta*, 1960, **43,** 2089.
40. R. Bentley and I. M. Campbell, in 'The Chemistry of the Quinonoid Compounds', Interscience, London, 1974, pt. 2, p. 692.
41. L. Szarkowska, *Arch. Biochem. Biophys.*, 1966, **113,** 519.
42. Y. Hatefi, A. G. Haavik, L. R. Fowler, and D. E. Griffiths, *J. Biol. Chem.*, 1962, **237,** 2661.
43. E. C. Slater, *Quart. Rev. Biophys.*, 1971, **4,** 35; B. Chance, *FEBS Letters*, 1972, **23,** 3.
44. S. Kaufman and D. B. Fisher, in 'Molecular Mechanisms of Oxygen Activation', ed. O. Hayaishi, Academic Press, New York, 1974, p. 285.
45. E. L. Patterson, M. H. von Saltza, and E. L. R. Stokstad, *J. Amer. Chem. Soc.*, 1956, **78,** 5871.
46. H. S. Forrest and H. K. Mitchell, *J. Amer. Chem. Soc.*, 1955, **77,** 4865.
47. E. L. Patterson, R. Milstrey, and E. L. R. Stokstad, *J. Amer. Chem. Soc.*, 1956, **78,** 5868.
48. E. C. Taylor and P. A. Jacobi, *J. Amer. Chem. Soc.*, 1974, **96,** 6781.
49. K. Fukushima, I. Eto, T. Mayumi, W. Richter, S. Goodson, and T. Shiota, in 'Chemistry and Biology of Pteridines', ed. W. Pfleiderer, de Gruyter, Berlin, 1975, p. 247.
50. S. Kaufman, in 'Chemistry and Biology of Pteridines', ed. W. Pfleiderer, de Gruyter, Berlin, 1975, p. 291.
51. H. I. X. Mager, in 'Chemistry and Biology of Pteridines', ed. W. Pfleiderer, de Gruyter, Berlin, 1975, p. 753.
52. Y. Hatefi, P. T. Talbert, M. J. Osborn, and F. M. Huennekins, *Biochem. Preps.* 1960, **7,** 89.
53. W. Shive, in 'Comprehensive Biochemistry', ed. M. Florkin and E. H. Stotz, Elsevier, Amsterdam, 1963, vol. 11, p. 82.
54. J. A. Montgomery, J. D. Rose, C. Temple, and J. R. Piper, in 'Chemistry and Biology of Pteridines', ed. W. Pfleiderer, de Gruyter, Berlin, 1975, 485; J. R. Piper and J. A. Montgomery, *J. Org. Chem.*, 1977, **42,** 208.
55. T. Shiota, in 'Comprehensive Biochemistry', ed. M. Florkin and E. H. Stotz, Elsevier, Amsterdam, 1971, vol. 21, p. 111.
56. C. J. Suckling, J. R. Sweeney, and H. C. S. Wood, *J. C. S. Perkin I*, 1977, 439.
57. T. C. Bruice and S. J. Benkovic, 'Bioorganic Mechanisms', Benjamin, New York, 1966, vol. II, p. 350.
58. S. J. Benkovic and W. P. Bullard, in 'Progress in Bioorganic Chemistry', ed. E. T. Kaiser and F. J. Kézdy, Wiley-Interscience, New York, 1973, vol. 2, p. 133.
59. S. H. Mudd and G. L. Cantoni, in 'Comprehensive Biochemistry', ed. M. Florkin and E. H. Stotz, Elsevier, Amsterdam, 1964, vol. 15, p. 1.
60. C. Tatum, J. Vederas, E. Schleicher, S. J. Benkovic, and H. Floss, *J. C. S. Chem. Comm.*, 1977, 218.
61. L. Jaenicke and F. Lynen, in 'The Enzymes', ed. P. D. Boyer, H. Lardy, and K. Myrbäck, Academic Press, New York, 1960, vol. 3B, p. 3.
62. J. Baddiley and E. M. Thain, *J. Chem. Soc.*, 1952, 800; *ibid.*, 1953, 1610.
63. J. G. Moffatt and H. G. Khorana, *J. Amer. Chem. Soc.*, 1961, **83,** 663.

64. G. M. Brown, in 'Comprehensive Biochemistry', ed. M. Florkin and E. H. Stotz, Elsevier, Amsterdam, 1971, vol. 21, p. 73.
65. G. M. Brown, *J. Biol. Chem.*, 1959, **234,** 370; Y. Abiko, *J. Biochem.* (*Japan*), 1967, **61,** 290; *ibid.*, p. 300.
66. M. B. Hoagland and G. D. Novelli, *J. Biol. Chem.*, 1954, **207,** 767.
67. A. White, P. Handler, and E. L. Smith, 'Principles of Biochemistry', 5th edn., McGraw-Hill, New York, 1973, pp. 339–342.
68. P. Goldman and P. R. Vagelos, in 'Comprehensive Biochemistry', ed. M. Florkin and E. H. Stotz, Elsevier, Amsterdam, 1964, vol. 15, p. 71.
69. R. B. Clayton, *Quart. Rev.*, 1965, **19,** 168.
70. H. R. Mahler and E. H. Cordes, 'Biological Chemistry', 2nd edn., Harper and Row, New York, 1971, p. 714.
71. A. White, P. Handler, and E. L. Smith, 'Principles of Biochemistry', 5th edn., McGraw-Hill, New York, 1973, p. 343.
72. J. Moss and M. D. Lane, *Adv. Enzymol.*, 1971, **35,** 321.
73. V. du Vigneaud, D. B. Melville, K. Folkers, D. E. Wolf, R. Mozingo, J. C. Keresztesy, and S. A. Harris, *J. Biol. Chem.*, 1942, **146,** 475.
74. D. B. Melville, A. W. Moyer, K. Hofmann, and V. du Vigneaud, *J. Biol. Chem.*, 1942, **146,** 487.
75. S. A. Harris, D. E. Wolf, R. Mozingo, R. C. Anderson, G. E. Arth, N. R. Easton, D. Heyl, A. N. Wilson, and K. Folkers, *J. Amer. Chem. Soc.*, 1944, **66,** 1756.
76. W. Traub, *Nature*, 1956, **178,** 649.
77. A. Grüssner, J. P. Bourquin, and O. Schnider, *Helv. Chim. Acta*, 1945, **28,** 517.
78. B. R. Baker, W. L. McEwen, and W. N. Kinley, *J. Org. Chem.*, 1947, **12,** 322.
79. C. Bonnemere, J. A. Hamilton, L. K. Steinrauf, and J. Knappe, *Biochemistry*, 1965, **4,** 240; J. Trotter and J. A. Hamilton, *ibid.*, 1966, **5,** 713.
80. D. E. Wolf, R. Mozingo, S. A. Harris, R. C. Anderson, and K. Folkers, *J. Amer. Chem. Soc.*, 1945, **67,** 2100.
81. M. W. Goldberg and L. H. Sternback, *U.S. Pat.* 2 489 232, 2 489 235, and 2 489 238 (1949).
82. D. B. McCormick and L. D. Wright, in 'Comprehensive Biochemistry', ed. M. Florkin and E. H. Stotz, Elsevier, Amsterdam, 1971, vol. 21, p. 81.
83. K. Krell and M. A. Eisenberg, *J. Biol. Chem.*, 1970, **245,** 6558.
84. A. Lezius, E. Ringelmann, and F. Lynen, *Biochem. Z.*, 1963, **336,** 510.
85. M. D. Lane, K. L. Rominger, D. L. Young, and F. Lynen, *J. Biol. Chem.*, 1964, **239,** 2865.
86. M. D. Lane and F. Lynen, *Proc. Nat. Acad. Sci. U.S.A.*, 1963, **49,** 379.
87. J. Baddiley, in 'Chemistry of Carbon Compounds' ed. E. H. Rodd, Elsevier, 1960, vol. IVC, p. 1732.
88. H. G. Khorana, 'Some Recent Developments in the Chemistry of Phosphate Esters of Biological Interest', Wiley, New York, 1961, chapter 4; A. M. Michelson, 'The Chemistry of Nucleosides and Nucleotides', Academic Press, New York, 1963, chapter 4.
89. J. Baddiley and A. R. Todd, *J. Chem. Soc.*, 1947, 648.
90. H. G. Khorana, *J. Amer. Chem. Soc.*, 1954, **76,** 3517; M. Smith and H. G. Khorana, *ibid.*, 1958, **80,** 1141.
91. M. Smith, *Biochem. Prep* 1961, **8,** 1.
92. E. Racker, *Adv. Enzymol.*, 1961, **23,** 323.
93. S. Kit, in 'Metabolic Pathways', ed. D. M. Greenberg, Academic Press, New York, 1970, vol. IV, pp. 237–251.
94. T. C. Bruice and S. J. Benkovic, 'Bioorganic Mechanisms', Benjamin, New York, 1966, vol. II, pp. 167–176.
95. T. C. Bruice and S. J. Benkovic, 'Bioorganic Mechanisms', Benjamin, New York, 1966, vol. II, chapter 8.
96. K. Lohmann and P. Schuster, *Biochem. Z.*, 1937, **294,** 188.
97. E. P. Steyn-Parvé and C. H. Monfoort, in 'Comprehensive Biochemistry', ed. M. Florkin and E. H. Stotz, Elsevier, Amsterdam, 1963, vol. 11, p. 3.
98. R. R. Williams and J. K. Cline, *J. Amer. Chem. Soc.*, 1936, **58,** 1504.
99. A. R. Todd and F. Bergel, *J. Chem. Soc.*, 1937, 364.
100. A. Windaus, R. Tschesche, and R. Grewe, *Z. physiol. Chem.*, 1935, **237,** 98; R. Grewe, *ibid.*, 1936, **242,** 89.
101. J. Weijlard and H. Tauber, *J. Amer. Chem. Soc.*, 1938, **60,** 2263; P. Karrer and M. Viscontini, *Helv. Chim. Acta*, 1946, **29,** 711.
102. H. Weil-Malherbe, *Biochem. J.*, 1940, **34,** 980.
103. G. M. Brown, in 'Comprehensive Biochemistry', ed. M. Florkin and E. H. Stotz, Elsevier, Amsterdam, 1971, vol. 21, p. 1.
104. P. C. Newell and R. G. Tucker, *Biochem. J.*, 1968, **106,** 271, 279.
105. A. F. Diorio and L. M. Lewin, *J. Biol. Chem.*, 1968, **243,** 3999, 4006.
106. R. Breslow, *J. Amer. Chem. Soc.*, 1957, **79,** 1762; *ibid.*, 1958, **80,** 3719.
107. N. L. Allinger, M. P. Cava, D. C. de Jongh, C. R. Johnson, N. A. Lebel, and C. L. Stevens, 'Organic Chemistry', 2nd edn., Worth, New York, 1976, p. 464.
108. L. O. Krampitz, I. Suzuki, and G. Greull, *Fed. Proc.*, 1961, **20,** 971; L. O. Krampitz, G. Greull, C. S. Miller, J. B. Bicking, H. R. Skeggs, and J. M. Sprague, *J. Amer. Chem. Soc.*, 1958, **80,** 5893.
109. F. G. White and L. L. Ingraham, *J. Amer. Chem. Soc.*, 1962, **84,** 3109.
110. J. A. Gutowski and G. E. Lienhard, *J. Biol. Chem.*, 1976, **251,** 2863.

111. R. Wolfenden, *Nature*, 1969, **223,** 704; G. E. Lienhard, *Science*, 1973, **180,** 149.
112. A. E. Braunstein, in 'The Enzymes', ed. P. D. Boyer, H. Lardy, and K. Myrbäck, Academic Press, New York, 1960, vol. 2, chapter 6.
113. E. F. Gale and H. M. R. Epps, *Biochem. J.*, 1944, **38,** 250.
114. I. C. Gunsalus, W. W. Umbreit, W. D. Bellamy, and C. E. Foust, *J. Biol. Chem.*, 1945, **161,** 743.
115. J. Baddiley and A. P. Mathias, *J. Chem. Soc.*, 1952, 2583.
116. J. M. Osbond, *Vitamins and Hormones*, 1964, **22,** 367.
117. A. F. Wagner and K. Folkers, 'Vitamins and Coenzymes', Interscience, New York, 1964, p. 162.
118. S. A. Harris, D. Heyl, and K. Folkers, *J. Amer. Chem. Soc.*, 1944, **66,** 2088.
119. V. M. Rodwell, in 'Metabolic Pathways', ed. D. M. Greenberg, Academic Press, 1970, vol. IV, p. 4.
120. R. E. Hill, F. J. Rowell, R. N. Gupta, and I. D. Spenser, *J. Biol. Chem.*, 1972, **247,** 1869; R. E. Hill, P. Horsewood, I. D. Spenser, and Y. Tani, *J. C. S. Perkin I*, 1975, 1622; R. E. Hill, I. Miura, and I. D. Spenser, *J. Amer. Chem. Soc.*, 1977, **99,** 4179.
121. E. E. Snell and B. E. Haskell, in 'Comprehensive Biochemistry', ed. M. Florkin and E. H. Stotz, Elsevier, Ansterdam, 1971, vol. 21, p. 47.
122. D. E. Metzler, M. Ikawa, and E. E. Snell, *J. Amer. Chem. Soc.*, 1954, **76,** 648.
123. T. C. Bruice and S. J. Benkovic, 'Bioorganic Mechanisms', Benjamin, New York, 1966, vol. II, pp. 226–300.
124. B. M. Guirard and E. E. Snell, in 'Comprehensive Biochemistry', ed. M. Florkin and E. H. Stotz, Elsevier, Amsterdam, 1964, vol. 15, chapter V.
125. H. C. Dunathan, *Adv. Enzymol.*, 1971, **35,** 79.
126. H. C. Dunathan, *Proc. Nat. Acad. Sci. U.S.A.*, 1966, **55,** 712.
127. C. R. Lowe and P. D. G. Dean, 'Affinity Chromatography', Wiley, London, 1974, chapter I.
128. C. R. Lowe and P. D. G. Dean, 'Affinity Chromatography', Wiley, London, 1974, chapter III.
129. H. Guilford, *Chem. Soc. Rev.*, 1973, **2,** 249.
130. P. Brodelius, P. O. Larsson, and K. Mosbach, *European J. Biochem.*, 1974, **47,** 81.
131. M. Poe, N. J. Greenfield, J. M. Hirshfield, M. N. Williams, and K. Hoogsteen, *Biochemistry*, 1972, **11,** 1023.

24.4
Vitamin B12

B. T. GOLDING
University of Warwick

24.4.1 INTRODUCTION

24.4.1.1 Nomenclature[1]

Vitamin B_{12}(= cyanocobalamin; **1a**) belongs to a family of substances called corrinoids. They are derivatives of corrin (**2**), the core of the structure of cyanocobalamin. The numbering of the carbon and nitrogen atoms in (**2**) corresponds to that in the porphyrin nucleus, except that corrin lacks C-20 of the latter. As in porphyrins, the four five-membered rings of corrinoids are labelled A, B, C, and D, as shown in (**1**) and (**2**). The

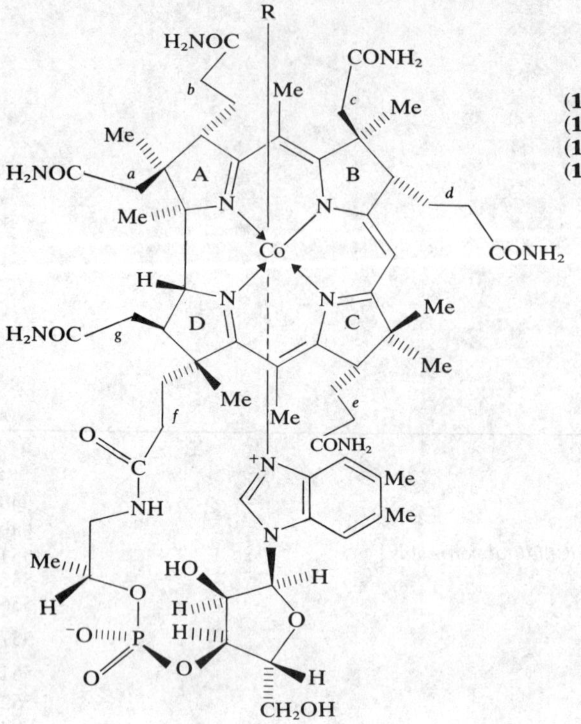

(**1a**) R = CN
(**1b**) R = OH
(**1c**) R = Adenosyl
(**1d**) R = Me

Adenosyl =

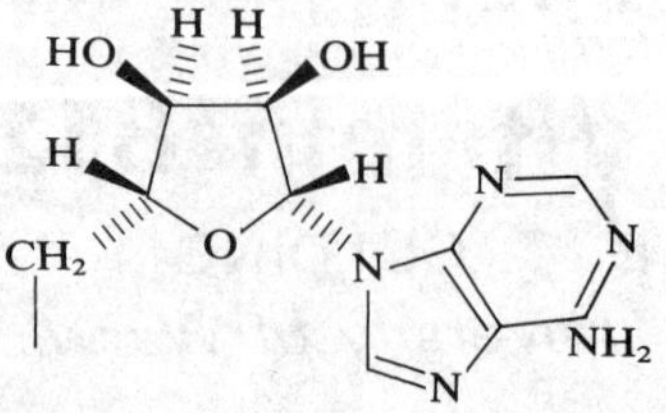

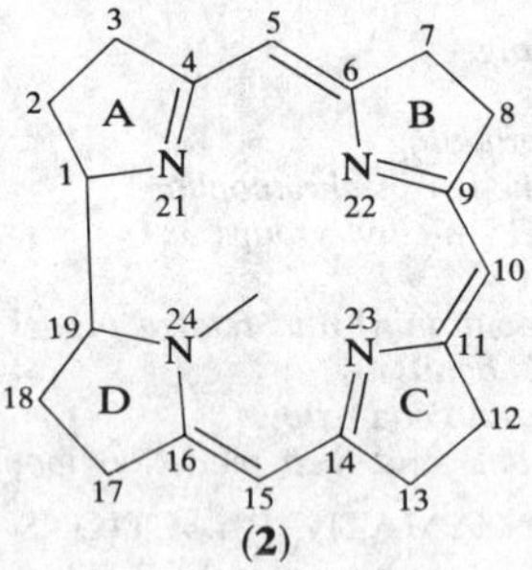

(**2**)

TABLE 1
Principal Corrinoids

Corrinoid	*Abbreviation*	*Side-chains*[a] *a–e, g*	f^{b}
Cobyrinic acid	—	CO_2H	CO_2H
Cobinic acid	—	CO_2H	$CONHCH_2CH(OH)Me$
Cobamic acid	—	CO_2H	$CONHCH_2CH(OR^1)Me$
Cobyric acid	Cby	$CONH_2$	CO_2H
Cobinamide	Cbi	$CONH_2$	$CONHCH_2CH(OH)Me$
Cobamide	Cba	$CONH_2$	$CONHCH_2CH(OR^2)Me$
Cobalamin	Cbl	$CONH_2$	$CONHCH_2CH(OR^3)Me$

[a] *cf.* formula (**1**).
[b] $R^1 = \alpha$-D-ribofuranose-3-phosphoryl; $R^2 = R^1$ + base joined to α-glycosidic bond; $R^3 = R^1$ + 5,6-dimethylbenzimidazole joined to α-glycosidic bond.

carboxyl side-chains (or derived amides) are given letters *a*, *b*, *c* . . . *g*, as in (**1**). The principal corrinoids (see Table 1) are given names based on the prefix cob, which denotes the presence of a cobalt ion. The lower and upper coordination sites of the cobalt ion are labelled α and β, respectively (*cf.* steroids). To name a corrinoid whose stereochemistry is known, its α-substituent is written first in brackets, followed by its β-substituent (not bracketed), followed by the name of the corrinoid. Since cobalamins, by definition, contain 5,6-dimethylbenzimidazole in the α-position, it is not necessary to specify this in their names, which reduce to X cobalamin (X-Cbl), where X is the ligand in the β-position. Examples are given in Table 2.

TABLE 2
Corrinoid Examples[a–d]

Name	*Abbreviation*
Cyanocobalamin (vitamin B_{12})	= CN-Cbl (**1a**)
Aquacobalamin (vitamin B_{12a}) / Hydroxocobalamin (vitamin B_{12b})	= OH-Cbl (**1b**)
Adenosylcobalamin	= AdoCbl (**1c**)
Methylcobalamin	= MeCbl (**1d**)
(Cyano)methylcobinamide	= (CN)MeCbi
Co^{α}-(5-Methoxybenzimidazolyl)-Co^{β}-cyanocobamide	= (5-MeOBza)CN-Cba

[a] It is sometimes necessary to specify the oxidation state of the cobalt ion as (I), (II), or (III), *e.g.** cob(I)alamin (Cbl^{I}) (formerly called B_{12s}) and cob(II)alamin (Cbl^{II}) (formerly called B_{12r}). [b] The terms coenzyme B_{12}, DBC, *etc.* for adenosylcobalamin should not be used. Likewise, the designation 'Factor' [*e.g.* Factor III_m for (5-MeOBza)CN-Cba] should be avoided. [c] Modification of an amide function is denoted by adding a specification of the alteration and its position to the basic name of the corrinoid, *e.g.* hydrolysing the *e*-amide group of CN-Cbl gives cyanocobalamin *a,b,c,d,g*-penta-amide-*e*-carboxylic acid [abbreviated as CN-Cbl-(*e*-OH)]. [d] Cobester is used as an abbreviation for dicyanocobyrinic acid *a,b,c,d,e,f,g*-heptamethyl ester.

Besides the abbreviations discussed above, the following representations will sometimes be used to denote 'base-on' (**3a**) [equivalent to (**1**)] and 'base-off' (**3b**) forms of cobalamins:

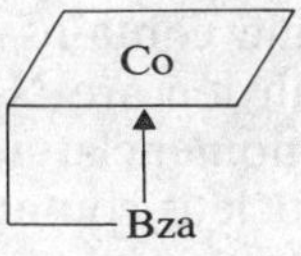

(**3a**)

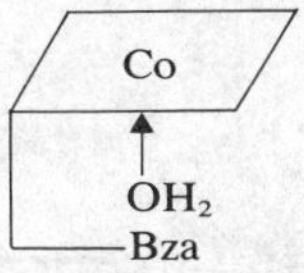

(**3b**)

24.4.1.2 Isolation and structure determination of natural corrinoids

Pernicious anaemia[2,3] was first clearly described in the early nineteenth century and is characterized by a diminution in the number of red blood cells, many of which lack their normal disc-like appearance. Some of the red cells ('megaloblasts') are much larger than usual. Further symptoms of the disease include a lemon-yellow complexion, redness and soreness of the tongue, an absence of hydrochloric acid in the stomach and, in the later stages, neural degeneration associated with gradual paralysis of limbs. The frequency of occurrence of pernicious anaemia in the United Kingdom is *ca.* 1 case per 1000 population, most patients being over 40 years old. Before 1922 the disease was without exception fatal, after a progressively worsening illness of about three years duration.

* This cobalamin is obtained by reduction, *e.g.* OH-Cbl (→ Cbl^{II} → Cbl^{I}) (see Section 24.4.4).

Following a lead from Whipple, the physicians Minot and Murphy reported in 1926 the beneficial effect on patients suffering from pernicious anaemia of a diet containing large amounts of raw liver. At this point, chemistry was faced with an exceptionally difficult problem in separating the active principle, or so-called anti-pernicious anaemia (APA) factor, which was present in liver at a concentration of only *ca.* 1 p.p.m.

Between 1926 and 1939, methods for isolating the APA factor were evolved principally by Cohn and his co-workers, Gänsllen, Dakin, and West, and LaLand and Klem. Their efforts rendered the treatment of pernicious anaemia a more pleasant affair. Meals of liver were replaced by periodic injections of an aqueous solution containing the partially purified active principle from liver. Paradoxically, this advance may have delayed further purification because sufferers from pernicious anaemia, on whom refined preparations could be tested, became scarcer.

Isolation of the APA factor was hastened by the intervention of the Glaxo Company in England (Smith, Parker, and their co-workers)[4] and Merck Laboratories in the United States (Folkers and his co-workers)[5]. Tons of liver were processed by these groups and, in April 1948, Folkers announced the isolation of a red, crystalline substance which he called vitamin B_{12} (= cyanocobalamin, CN-Cbl) and was shown to be extremely active against pernicious anaemia. The later stages of Folkers' work were aided by the development of a microbiological assay for cobalamins, which also enabled their detection in beef extract, powdered milk, and in the fermentation broth of several micro-organisms. Cobalamins are now produced commercially by fermentation processes.

The elucidation of the structure of CN-Cbl began in classical style. Degradative and synthetic studies by Folkers, Todd, Petrow, and their co-workers (reviewed by Folkers[5] and Smith[4]) identified the nucleotide portion of the molecule and hinted at the nature of the corrin nucleus. Owing to the involvement of haem and cobalamin in blood biochemistry, it was suspected that these natural products were structurally related. Indeed, fusion of CN-Cbl with alkali gave a pyrrole-containing distillate and cobalt was detected in the ash from its combustion. The cyanide ligand was identified by smelling hydrogen cyanide given off when CN-Cbl was treated with $KMnO_4$ in dil. H_2SO_4 at 0 °C! [confirmed by infrared spectroscopy of CN-Cbl ($\nu_{C\equiv N}$ 2130 cm^{-1})].

Todd's group, in collaboration with Smith, studied the products from acidic and alkaline degradations of CN-Cbl. Crystals of CN-Cbl and two of its degradation products were given to Hodgkin for an X-ray crystallographic study. The correct structure (**1a**) for CN-Cbl was proposed[6] in 1957. The molecule contains a hexacoordinated, diamagnetic cobalt(III) ion. The donor atoms to the cobalt ion are N-3 of the dimethylbenzimidazole and the four nitrogen atoms of the corrin (for nomenclature, see above); the sixth coordination site is occupied by cyanide. The corrin nucleus contains four partially reduced pyrrole rings (A–D), two of which are directly linked (the A/D link, C-1—C-19) whilst A/B, B/C, and C/D are joined in a porphyrin-like fashion.

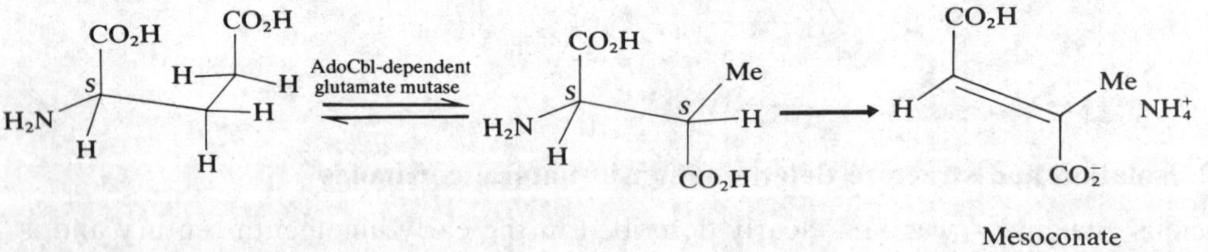

(1)

While the finishing strokes were being applied in the structural elucidation of CN-Cbl, Barker[7] was studying the two-step conversion of glutamate to ammonium ion and mesoconate *via* 3-methylaspartate catalysed by a cell-free extract of *Clostridium tetanomorphum* (equation 1). It was found that the first step could be blocked by treating the extract with charcoal. This removed an orange cofactor for the conversion of glutamate to 3-methylaspartate, whose isolation progressed after its instability to light had been recognized. A kinship with CN-Cbl was revealed on finding that cyanide ions converted the cofactor to a cyanocobamide. Further chemical studies nearly reached the correct structure of the cofactor, but missed a very important, unique feature. Crystals were once more sent to Hodgkin and she, with her collaborator, Lenhert, deduced[8] the remarkable structure (**1c**) (see also Figure 1). The unsuspected feature of (**1c**), its Co—C σ-bond, identified adenosylcobalamin (AdoCbl) (as the cofactor is now called) as the first naturally occurring organometallic compound. Subsequently, methylcobalamin (MeCbl; **1d**) joined this select company.

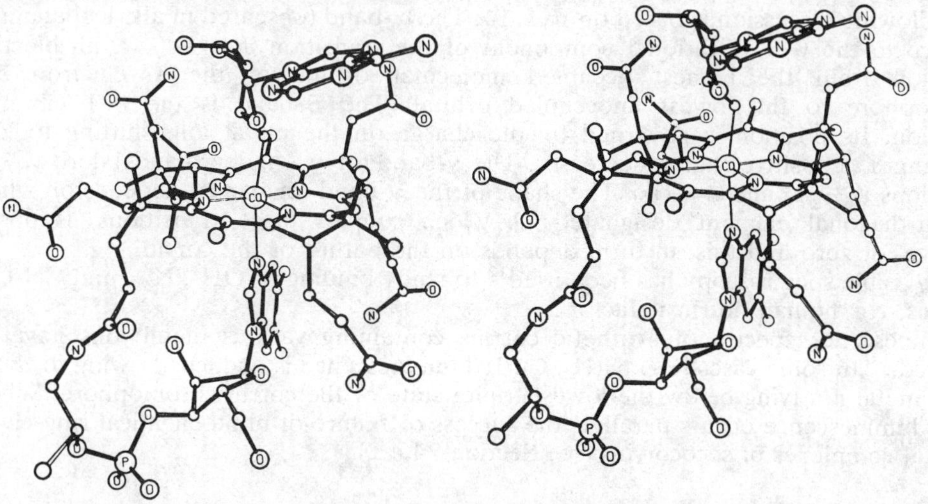

Figure 1 Stereoscopic view of the crystal structure of AdoCbl (From Lenhert,[8] by courtesy of the Royal Society)

It is now realised that CN-Cbl, as originally isolated, was an artefact of the isolation procedure, which converted AdoCbl to CN-Cbl. Unless strict precautions are taken to avoid exposure to light, AdoCbl and MeCbl are converted to hydroxocobalamin (OH-Cbl; **1b**), which reacts with cyanide ions (present in the charcoal columns originally used to isolate 'vitamin B_{12}') to give CN-Cbl.

Quantitative methods have been developed[9] to assay cobalamins in biological samples. The plasma of healthy, non-smoking adults contains MeCbl ($\sim$250 pg cm^{-3}), AdoCbl, and OH-Cbl (AdoCbl + OH-Cbl $\sim$ 125 pg cm^{-3} with AdoCbl > OH-Cbl). CN-Cbl is present ($\sim$2% of total cobalamin) in the blood of smokers (tobacco smoke contains HCN!) and individuals who partake of food (*e.g.* cassava) which releases cyanide ions. In erythrocytes, the chief cobalamin is AdoCbl (>50% of total cobalamin) followed by OH-Cbl ($\sim$25%), MeCbl (10–15%), and small amounts of CN-Cbl. Cobalamin-dependent enzymatic reactions in animals and micro-organisms involve either AdoCbl, OH-Cbl, or MeCbl as cofactor (see Section 24.4.5). There is no established role for CN-Cbl.

Analogues of AdoCbl in which the dimethylbenzimidazole is replaced by another base are found in sewage and are produced by certain micro-organisms, *e.g. Clostridium thermoaceticum* produces an analogue containing 5-methoxybenzimidazole; see below concerning its biosynthesis. The variety of naturally occurring corrinoids was further extended by the isolation of metal-free corrinoids from cells of certain photosynthetic bacteria, *e.g. Chromatium* strain D.[10]

24.4.1.3 Spectroscopy of corrinoids

The most useful spectroscopic techniques for corrinoids are electronic and n.m.r. spectroscopy and, in addition, for cobalt(II) corrinoids, e.s.r. spectroscopy. Among other spectroscopic techniques applied to corrinoids, circular dichroism[11] is useful in characterization and for studying their binding to proteins.[12]

(i) Electronic spectra

The u.v./visible spectra of corrinoids show several bands, the more prominent of which above 300 nm are labelled α, β, and γ, *e.g.* $\lambda_{max}(\varepsilon \times 10^{-5})$ for $(CN)_2$-Cbl in water: α, 584 nm (0.98 $m^2 mol^{-1}$); β, 543 (0.86); γ, 369 (3.04). These absorptions are assigned as transitions of the corrin chromophore rather than *d–d* or charge-transfer transitions of the metal ion, owing to their intensities and because similar bands appear in the spectra of metal-free corrinoids. Values of λ_{max} and ε for numerous corrinoids are given in Ref. 13. The following discussion is based on Ref. 14. The α-band (obscured in alkylcobalamins) is assigned to the 0–0 vibrational component of the transition $\psi_7 \rightarrow \psi_8$, *i.e.* an electronic transition from the highest occupied molecular orbital of the 14-electron corrin chromophore to the lowest unoccupied orbital. The β-band is the 0–1 vibrational transition. Its position is governed by the charge on the metal ion, shifting to longer wavelength as positive charge decreases. The γ-band (or γ_1- and γ_2-bands) derive(s) from transitions $\psi_6 \rightarrow \psi_8$ and $\psi_7 \rightarrow \psi_9$. The shape of the γ-band (or bands) depends on whether the off-diagonal element designated *Q*, which couples these transitions, is positive, negative, or zero and this, in turn, depends on the nature of the corrin.

U.v./visible spectroscopy has been used[12] to study binding of OH-, N_3-, and CN-Cbl to proteins, *e.g.* human 'intrinsic factor'.

Luminescence spectra of synthetic corrins containing various metal ions have been obtained.[15] In some cases [*e.g.* Ni(II), Cu(II)] luminescence is quenched owing to *d*-levels of the metal ion lying below the lowest triplet state of the corrin chromophore. Whether or not luminescence occurs parallels the success or failure of photochemical ring-closures of metal complexes of secocorrins (see Section 24.4.2).

(ii) ^1H and ^{13}C n.m.r. spectra

N.m.r. spectroscopy is a valuable aid in the characterization of cobalt(III) corrinoids in solution (*cf.* Refs. 16–18) and, in principle, can be used to determine their conformation. Some special features of the ^1H n.m.r. spectra of corrinoids are:

(a) Base-on cobalamins show a singlet at *ca.* δ 0.5 due to the C-1 Me group which lies over the dimethylbenzimidazole (*cf.* Figure 1) and consequently is shielded. In the spectra of cobinamides and base-off cobalamins the signal from the C-1 Me group appears at *ca.* δ 1.2.

(b) The C-5′ protons (Co-CH_2) of adenosylcorrinoids are diastereotopic and therefore have different chemical shifts, *e.g.* δ 0.38 and 0.63 for AdoCbi in ^2H_2O.

(c) Protons at the 2-position of the σ-alkyl group in alkylcorrinoids are shielded by the corrin chromophore and resonate at higher field, *e.g.* Pr^iCbi shows two signals at δ −0.70 and −0.94 due to the methyls of its isopropyl group.

(d) In the ^1H n.m.r. spectrum of AdoCbi in ^2H_2O, H-1′ (anomeric proton of the adenosyl group) resonates at δ 4.37, *ca.* 1 p.p.m. upfield from its position with free adenosine. To explain this observation it was suggested[18] that a conformational change takes place when adenosylcorrinoids are dissolved in water, whereby the adenine ring moves from an orientation over the corrin (in the crystal, *cf.* Figure 1) to an orientation perpendicular to the corrin. For such a far-reaching proposal the evidence appears to be rather tenuous.

^{13}C N.m.r. spectra of common corrinoids are illustrated and assigned in Ref. 19. This technique has been extremely useful in biosynthetic studies and for the detection of isomerization involving axial ligands to the cobalt ion. The isomers can be easily distinguished in spectra of selectively enriched corrinoids, *e.g.* aqueous methylcobinamide containing 90% ^{13}C in the methyl group shows signals at δ 128.4 and 130.3 p.p.m. (relative to benzene as external standard) of nearly equal intensity, due to (OH)MeCbi and (Me)OH-Cbi.[20] A study of the temperature-dependence of chemical shifts in the spectra of $^{13}CH_3$-Cbl, [1,2-$^{13}C_2$]EtCbl, and [5′-^{13}C]AdoCbl (all 90% enriched at the specified position) concluded that these cobalamins undergo a conformational change associated with the corrin system on varying the temperature.[21] It has earlier been erroneously suggested, on the basis of changes in electronic spectra and ^1H n.m.r. spectra observed on heating aquocobinamides and alkylcobalamins in solution, that many corrinoids are in equilibrium with five-coordinate forms (lacking the α-ligand) in aqueous solution.

(*iii*) *E.s.r. spectra*

This technique is applicable to the paramagnetic cobalt(II) corrinoids and has recently been used to characterize cobalt(IV) species derived from corrinoid model compounds (see below). Cobalt(II) corrinoids are five-coordinate, low-spin d^7 complexes with the unpaired electron occupying a d_{z^2} orbital. Typical e.s.r. spectra are shown in Figure 2.[16,22] The splittings arise from coupling of the unpaired electron to ^{59}Co ($I = \frac{7}{2}$) and ^{14}N ($I = 1$) of the dimethylbenzimidazole. For Cbl^{II} in pH 7.3 aq. buffer and 77 K, $a(Co) = 11.0$ mT, $a(N) = 1.8$ mT and $g = 2.006 \pm 0.007$. E.s.r. spectroscopy has been invaluable for detecting Cbl^{II} in enzymatic reactions (see Section 24.4.5).

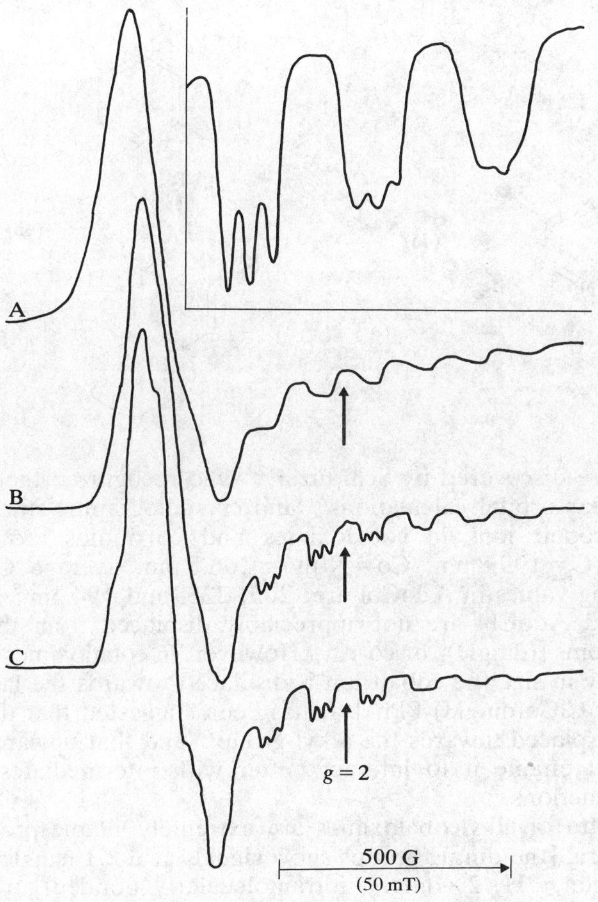

Figure 2 E.s.r. spectra of Cbl^{II}: (A) crystals at 298 K; (B) from photolysis of AdoCbl at 77 K in pH 7.3 aq. phosphate buffer containing dimethylglutarate; (C) in ethanol. Inset: high-field end of curve C recorded at high gain and low scanning rate to show resolution of the triplets of the last three hyperfine lines (From Bayston *et al.*,[22] by courtesy of the American Chemical Society)

Oxidation of alkyl(aquo)cobaloximes (see Section 24.4.4) with Ce(IV) in acidic aqueous methanol gives cobalt(IV) species [$RCo^{IV}(dmgH)_2OH_2$, R = Me, $PhCH_2$, *etc.*] stable for many hours at 195 K.[23] The e.s.r. spectra of these species at 77 K show eight relatively broad lines with *g* values in the range 2.012–2.033 and *a*(Co) 2.45–3.13 mT.

24.4.1.4 Model compounds for corrinoids

Thorough characterization of corrinoids is difficult; even combustion analyses are unreliable owing to variable hydration. It is also difficult to study the properties of the Co—C σ-bond in alkylcorrinoids because this structural feature is buried in a complex molecule. Well-defined model compounds have therefore been developed so that the properties of Co—C bonds in a corrinoid-like environment can be readily ascertained. Examples of useful model systems are alkyl-bis(dimethylglyoximato)cobalt complexes ('cobaloximes'[24]) [*e.g.* methyl(pyridine)cobaloxime (**4a**): $MeCo(dmgH)_2py$, where dmgH = monoanion of dimethylglyoxime and py = pyridine] and alkyl-*N,N'*-ethylenebis-(salicylideneiminato)cobalt complexes [RCo(salen); (**4b**)].[25] More work has been done with cobaloximes than any other model system. In this article, most of the references to models for corrinoids concern cobaloximes (for a review of all types of alkylcobalt compound, see Ref. 26).

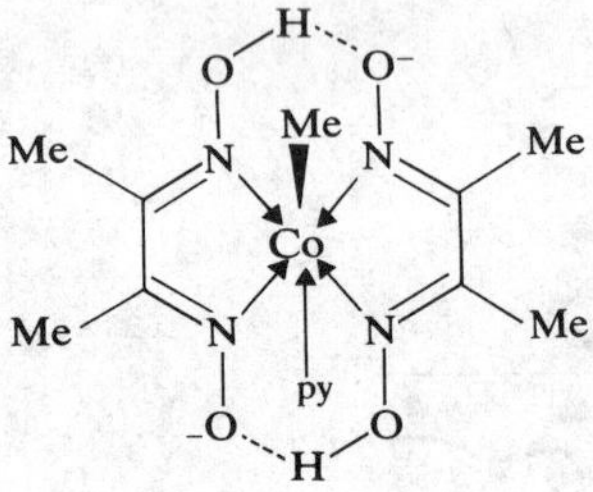

(**4a**)

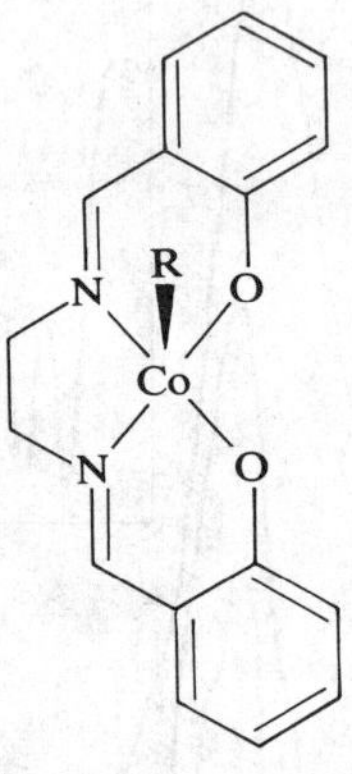

(**4b**)

Alkylcobaloximes were discovered by Schrauzer,[24] who recognized their similarities to alkylcorrinoids. Molecular orbital calculations[24] and crystallographic studies[27] show that the vicinities of the cobalt ions in cobaloximes and corrinoids are alike, *e.g.* for $MeCo(dmgH)_2py$: Co—C = 199.8 pm, Co—N(py) = 206.8 pm, average Co—N(dmgH) = 189.7 pm. Corresponding values in AdoCbl are: 205, 223, and 194 pm.[8] The cobalt ions of $MeCo(dmgH)_2py$ and AdoCbl are not appreciably displaced from the plane of the surrounding nitrogen atoms [$(dmgH)_2$ or corrin]. However, in cobaloximes where the axial groups differ appreciably in size the cobalt ion is displaced towards the larger group [*e.g.* displacement = 5 pm for $ClCo(dmgH)_2PPh_3$]. It has been suggested that the cobalt ion of s-alkylcobinamides is displaced towards the alkyl group[28] and that upwards displacement of the cobalt ion might enable it to interact better with intermediates in cobalamin-dependent enzymatic reactions.[29]

Since the n.m.r. spectra of alkylcobaloximes are extremely simple provided the alkyl group and base are achiral [the dmgH groups show signals at δ 2.1 (singlet, 12H, Me × 4) and *ca.* 18 (broad singlet, 2H, 2 × dmgH, intramolecularly bonded)], reactions of the σ-alkyl group can be conveniently monitored. The chemical properties of alkylcobaloximes, hydridocobaloximes, and cobaloxime(II) and cobaloxime(I) species show many parallels with the corresponding corrinoids, as described in Section 24.4.4. The mechanism of a reaction of a corrinoid is often best studied in depth with the corresponding cobaloxime [*cf.* reactions of alkylcobalt compounds with Hg(II): Section 24.4.4].

24.4.2 SYNTHESIS OF CORRINOIDS

In 1964, Bernhauer and his co-workers[30] reported the conversion of natural cobyric acid to CN-Cbl (*cf.* Scheme 1, steps 1 and 2). Efficient methods for converting CN-Cbl to MeCbl and AdoCbl were described soon afterwards (*cf.* Scheme 1, steps 3 and 4 for the synthesis of AdoCbl). Many analogues of the natural corrinoids, of interest for studies of structure–activity relationships, have been prepared by variants of Scheme 1.

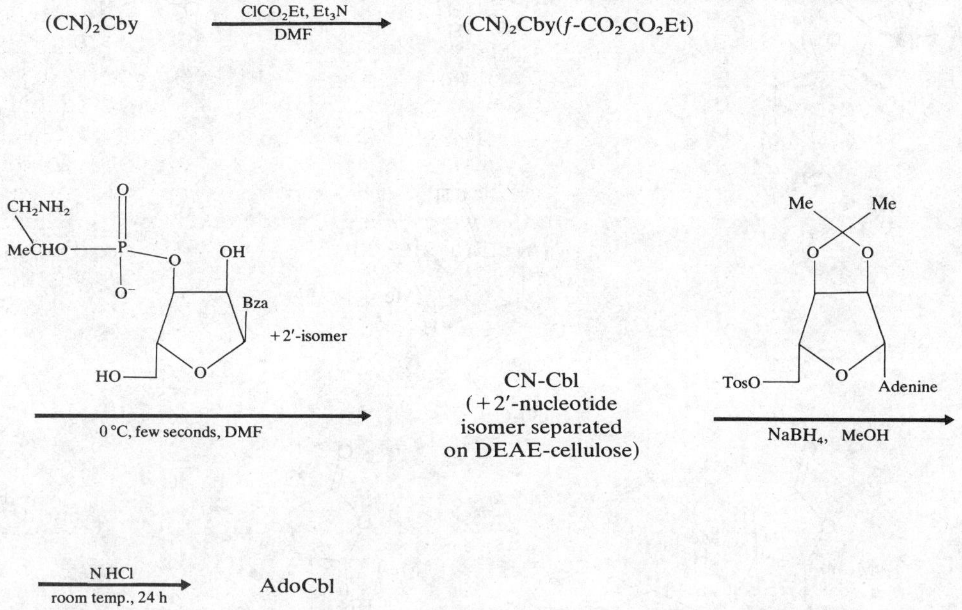

SCHEME 1

The total synthesis of 'vitamin B_{12}' requires the preparation of cobyric acid, a molecule with nine chiral centres, from simple precursors. A relatively easier problem is to synthesize the corrin chromophore or a model corrin. At Cambridge, England, research on the structure of CN-Cbl led to three synthetic endeavours: Clark and Todd chose nitrones as precursors of the corrin system and prepared model compounds corresponding to the AB and AD portions of CN-Cbl.[31] 'With one eye looking to the problem of vitamin B_{12} biosynthesis',[32] Johnson developed a simple route to corrins from pyrroles.[33] Cationic complexes of 1,19-dimethyl-2,3,7,8,12,13,17,18-octadehydrocorrin with nickel(II), cobalt(II), and cobalt(III) are very easy to prepare and can be hydrogenated to the corresponding metal complex of 1,19-dimethylcorrin. Cornforth ignored the easier path to model corrins and faced the challenge of cobyric acid from the outset. In 1962, he conceived a synthesis of cobyric acid which employed isoxazoles as latent functionalities. In the early stages, precursors of the five-membered rings of cobyric acid were synthesized. Cornforth's clever synthesis of a ring C precursor from D-(+)-camphor was subsequently used by Eschenmoser and Woodward in their synthesis of cobyric acid (see below). The Cornforth synthesis faltered when attempts to forge the A/D link by coupling radicals failed.[34] The use of isoxazoles as precursors of corrins was later independently conceived by Stevens,[35] who thereby synthesized a zinc(II) complex of 1,2,2,7,7,12,12,17,17-nonamethylcorrin and is progressing towards cobyric acid.

Eschenmoser (E. T. H. Zürich) and Woodward (Harvard) began their assault on the B_{12} citadel in 1960. Displaying his flair for combining mainly well-worn reactions in delectable sequences, Woodward[36] devised a *ca.* 37-stage synthesis of 'β-corrnorsterone', a precursor of compound (**5**) corresponding to the AD portion of cobyric acid. A blind alley in early attempts to synthesize (**5**) inspired Woodward and Hoffmanns' rationalization of pericyclic reactions.

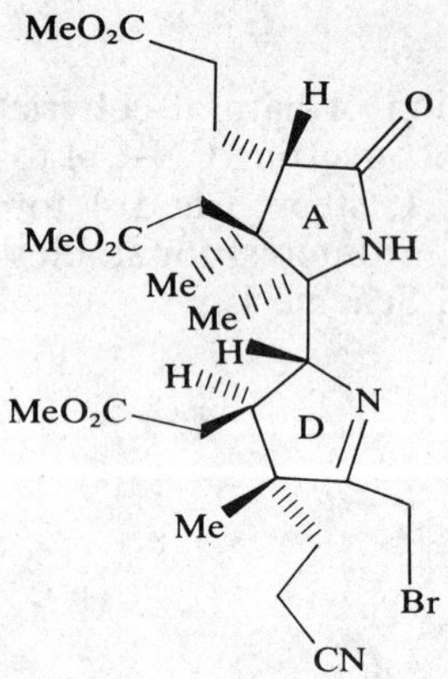

(**5**)

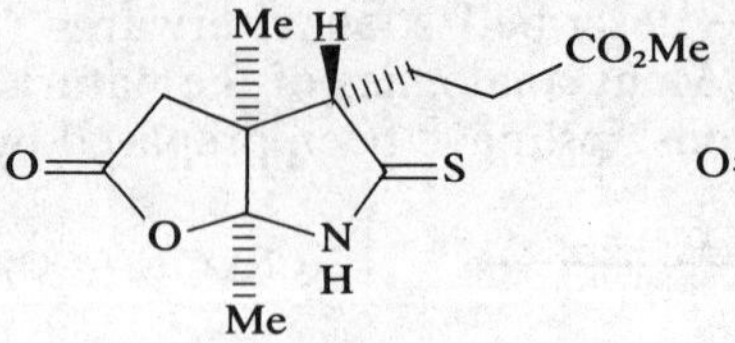

(**6**)

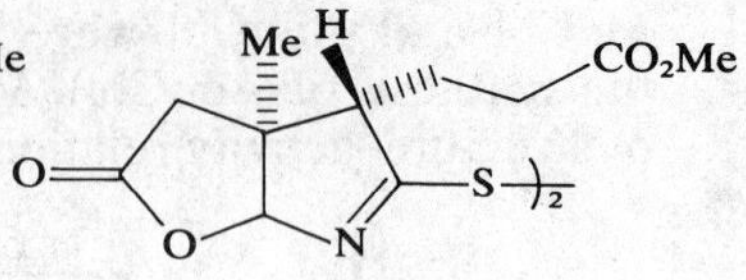

(**7**)

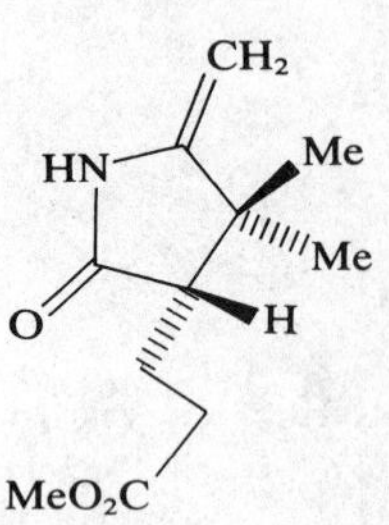

(**8**)

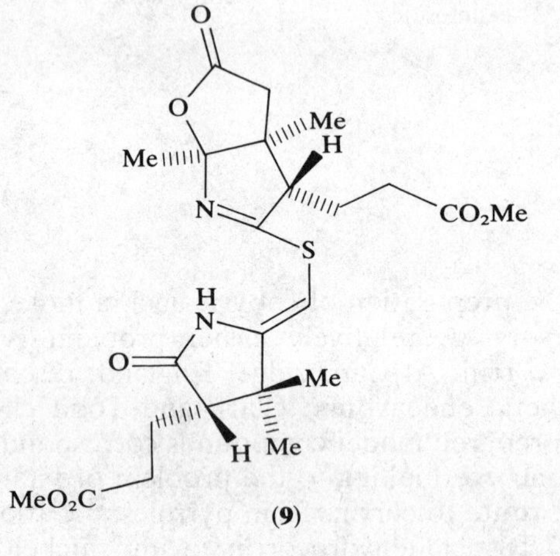

(**9**)

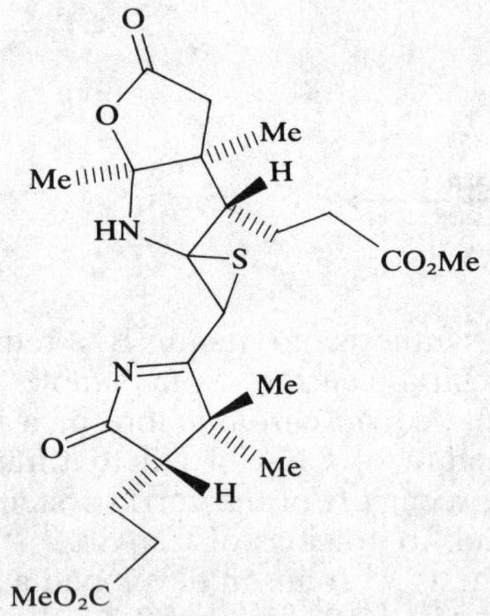

(**10**)

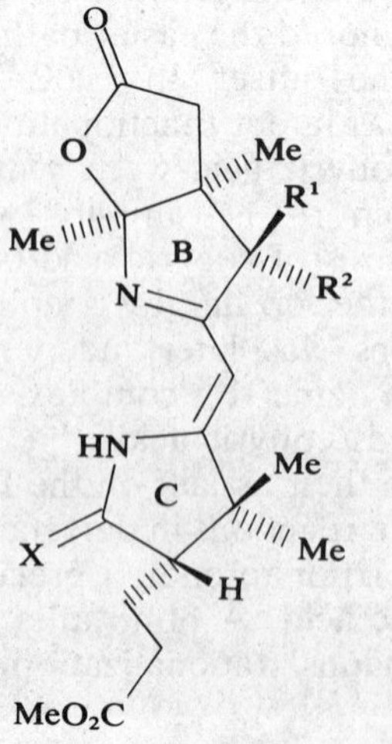

(**11**) $R^1 = CH_2CH_2CO_2Me$, $R^2 = H$, $X = O$
(**12**) $R^1 = H$, $R^2 = CH_2CH_2CO_2Me$, $X = O$
(**13**) R^1, R^2 as in (**11**), (**12**), $X = S$

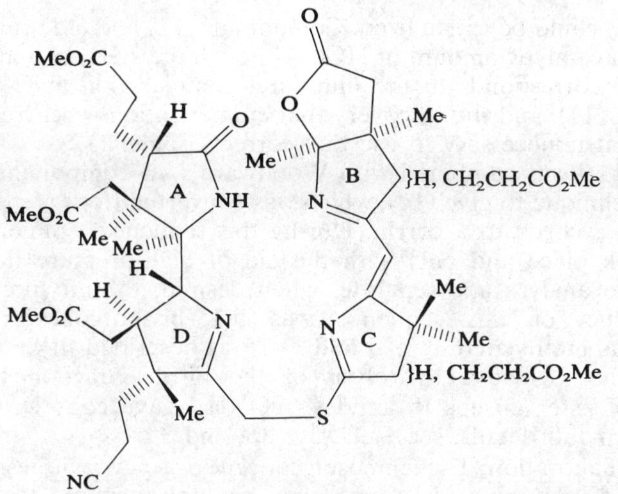

(14)

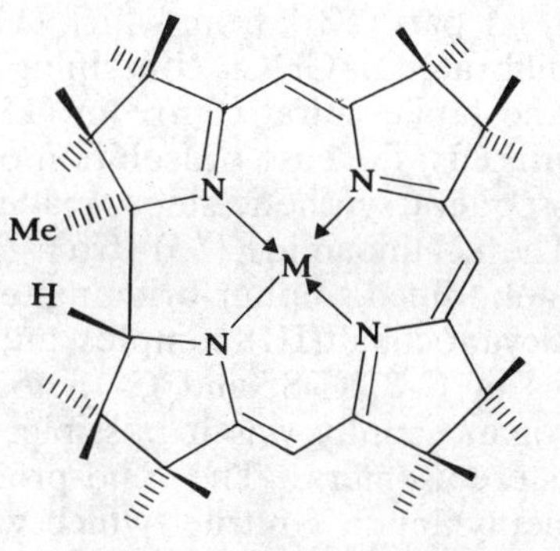

(16)

[Peripheral substituents as in (15)]

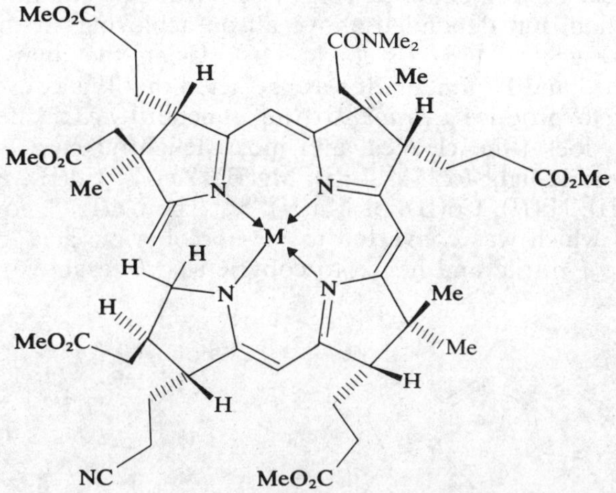

(15) M = metal ion

Eschenmoser conceived a quite different approach to the corrin system, involving condensations between enamides and iminoesters. This method served for the synthesis of model corrins[37] [*e.g.* dicyanocobalt(III)-1,8,8,13,13-pentamethylcorrin], but was inadequate for the real thing. Groups intended to become substituents around the corrinoid system of cobyric acid blocked the iminoester condensation. The solution to this problem depended on the creation of the 'sulphur-bridging method' for constructing the N═C—C═C—N units of the corrin chromophore.[38] In early 1966 it was found that the thioamide (**6**) could be oxidized quantitatively by benzoyl peroxide to the bis-imidoyl disulphide (**7**). In dichloromethane, containing traces of HCl, (**7**) reacted with the enamide (**8**) to give the products (**9**) and (**6**). For the efficient synthesis of a BC-precursor, 1 mol equiv. of thioamide (**6**) was oxidized with 0.5 mol equiv. of benzoyl peroxide in the presence of enamide (**8**) in dichloromethane containing catalytic HCl. The bis-imidoyl disulphide (**7**) was formed immediately and reacted with (**8**) to give (**9**) and thioamide (**6**). The latter was recycled by another 0.5 mol equiv. of benzoyl peroxide. It had been anticipated that sulphur could be expelled from (**9**), possibly *via* the tautomer (**10**) using a phosphorus(III) compound and, indeed, heating (**9**) with triethyl phosphite gave the BC-precursor (**11**). In this reaction there took place a subtle additional transformation of a nature which was to badger the remainder of the synthesis of cobyric acid ('the experimental battle against diastereoisomerism').[38] An equilibrium mixture of epimers was obtained [*ca.* 2 parts

(**11**) + 1 part (**12**)], from which (**11**) could be crystallized. Compounds (**11**) and (**12**) can be equilibrated in $CHCl_3$ containing a catalytic amount of HCl. Although the stereochemistry of the labile chiral centre in (**11**) corresponds to the unnatural configuration at C-8 in corrinoids, the ease of isolation of (**11**), and the discovery that epimerizations occurred in subsequent synthetic steps, made it unnecessary to use the 'correct' isomer (**12**).

The BC-thioamide (**13**) [from (**11**)] was combined with Woodward's AD-component (**5**) by a modified sulphur-bridging technique to give (**14**), which was converted (four steps) to a dicyanocobalt(III) complex and cyclized to a corrin. During this sequence, epimerizations at C-3, C-8, and C-13 took place and only with the aid of high-pressure liquid chromatography was it possible to analyse and separate, when desired, the mixtures of diastereoisomers. The end-product of this sequence was 5,15-bis-norcobester *c*-dimethylamide *f*-nitrile, which was methylated at C-5 and C-15 as described in Section 24.4.4 to afford cobester *f*-nitrile. Selective hydrolysis of the nitrile function to a carboxylic acid and conversion of ester groups to amides ($CONH_2$) gave cobyric acid, identical with natural material. For full details, see Refs. 36, 38, and 39.

During the Harvard–Zürich collaboration, Eschenmoser conceived a new synthesis of corrins which could be extended to cobyric acid. It embodied his long nurtured idea of using a single precursor to furnish each five-membered ring of cobyric acid. The new idea required the sulphur-bridging method, but depended above all on achieving, at the last stage, conversion of a secocorrin such as (**15**) to corrin (**16**). Using the analysis of pericyclic reactions due to Woodward and Hoffmann, it was predicted that this conversion would take place photochemically to produce a *trans* AD ring junction (*i.e.* the natural configuration of corrinoids). So it does ('the cleanest and most delightful step . . . ever encountered in our work')[32] with visible light for M = Li(I), Mg(II), Zn(II), Cd(II), Pd(II), or Pt(II), but not with M = H, Cu(II), Ni(II), Co(II), or Mn(II).[40,41] The Cd(II)Cl complex of (**15**) was photocyclized to (**16**), which was converted to bis-norcobyrinic acid *abdeg*-pentamethyl ester *c*-dimethylamide *f*-nitrile and hence to cobyric acid (see above).

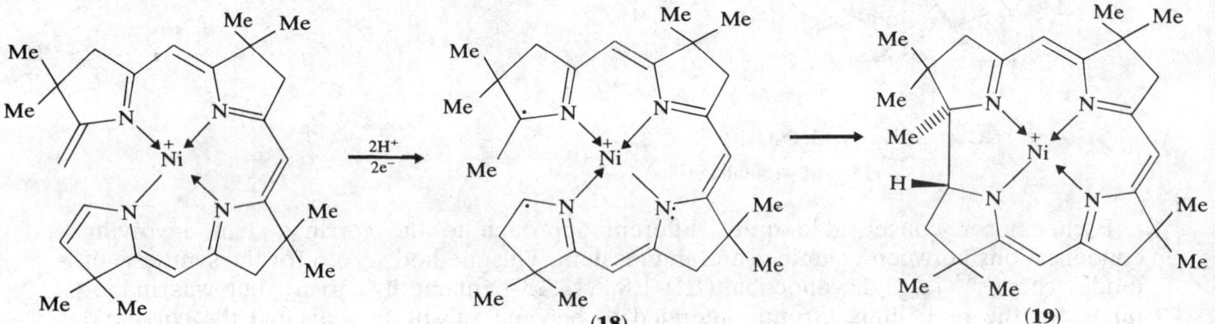

(**17**) (**18**) (**19**)

Electrolysis in acetonitrile/trifluoroacetic acid containing lithium perchlorate

Recently, Eschenmoser[32] has described a reductive, acid-catalysed cyclization of the Δ^{18}-dehydro-1,19-secocorrin complex (**17**) to corrin (**19**) (equation 2). The transformation of the putative intermediate (**18**) to (**19**) may be a model for one step in the biosynthesis of corrinoids.

In natural corrinoids, the corrin chromophore is embedded by the five-membered rings and their substituents and is blocked by methyl groups at the 5- and 15-positions. It is of interest to compare the properties of the chromophore in this shielded environment with the 'naked' corrin chromophore ('corromin';[42] for an alternative nomenclature, see Ref. 43) of compound (**20**). Metal complexes of two derivatives of (**20**) have been prepared by short routes [5,6,14,15-dibenzocorromin (**21**)[42] and 5,15-diphenylcorromin (**22**)[43]]. So far, detailed chemical investigations of these complexes have not been reported.

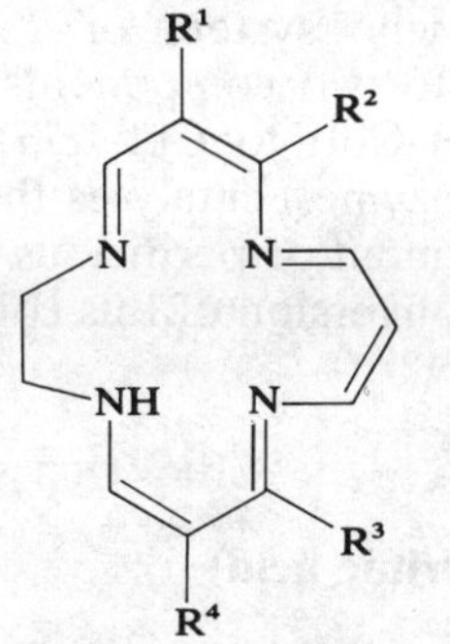

(20) $R^1 = R^2 = R^3 = R^4 = H$
(21) $R^1R^2 = R^3R^4 =$ benzo
(22) $R^1 = R^4 = Ph$, $R^2 = R^3 = H$

24.4.3 BIOSYNTHESIS OF CORRINOIDS

Cobalamins arise from several building blocks, which are shown in Scheme 2. Note that in the biosynthesis of adenosyl corrinoids, adenosylation can occur before OH-Cbl is formed (partially amidated corrinoids from *P. shermanii* contain a σ-adenosyl group). There is probably not a unique series of events in the conversion of cobyrinic acid to cobinamide, although amidation of *c*-OH always starts the sequence.[44] The production of AdoCbl from CN-Cbl (or OH-Cbl) has been investigated with *P. shermanii*, *C. tetanomorphum*, and

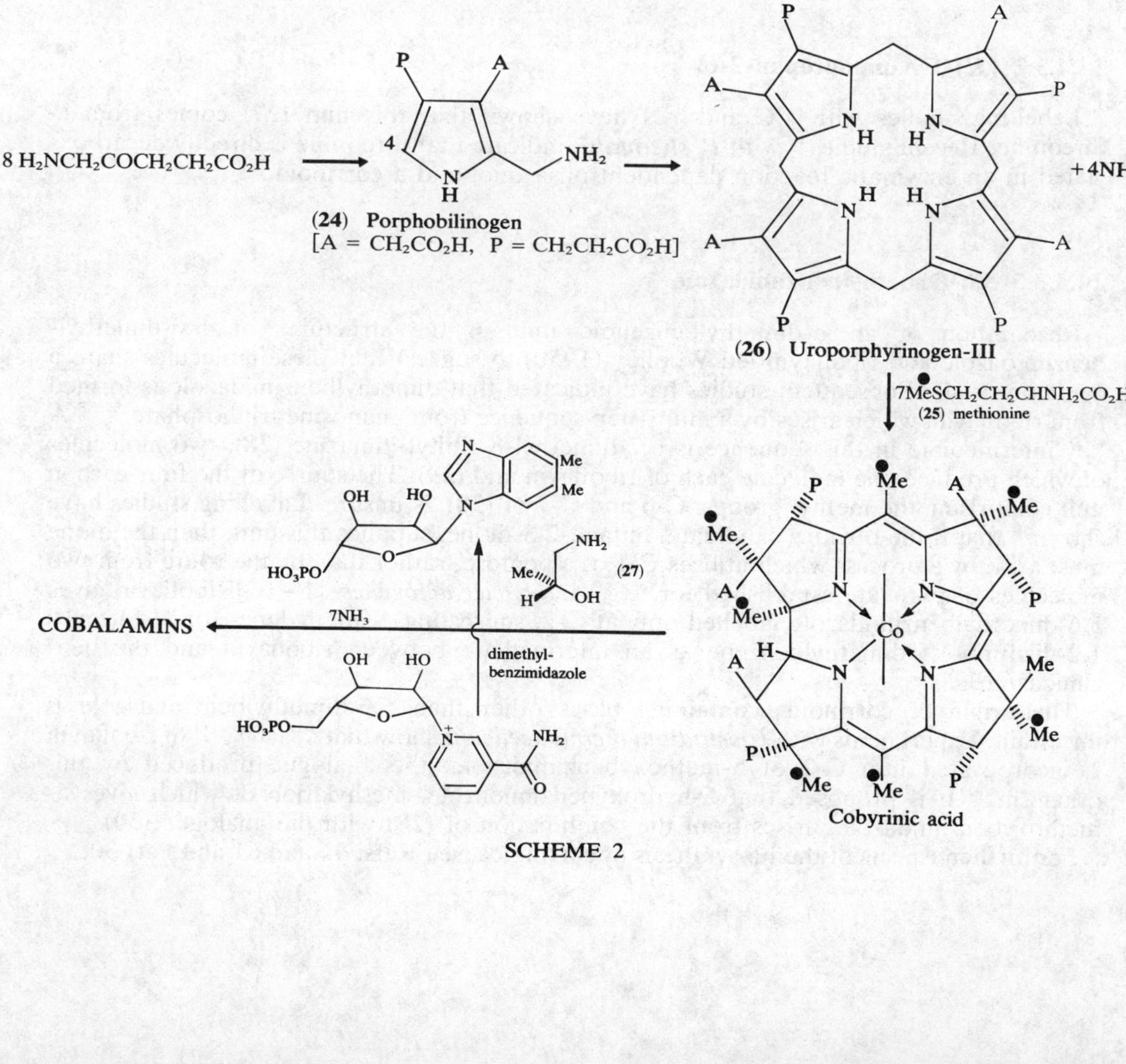

SCHEME 2

Lactobacillus leichmannii. The enzyme system of *L. leichmannii* is associated with ribosomes and has been resolved into two components.[45] One component ('vitamin B_{12} reductase') reduces CN-Cbl (or OH-Cbl) to Cbl^{I} *via* Cbl^{II} and is very unstable. The second component ('adenosylating enzyme') catalyses the reaction between Cbl^{I} and ATP to give AdoCbl. The successful treatment of pernicious anaemia by periodic injections of OH-Cbl depends on the metabolic conversion of this cobalamin to AdoCbl and MeCbl by endogeneous enzymic systems.

24.4.3.1 The corrinoid system (cobyrinic acid)

Pioneering studies[46,47] with ^{14}C-labelled δ-aminolaevulinic acid (**23**), porphobilinogen (**24**), and methionine (**25**) proved that these compounds are precursors of corrinoids. However, the information available from experiments with ^{14}C-labelled precursors was limited by experimental errors and by the small number of specific degradations known for CN-Cbl.

Recent studies with ^{13}C-labelled precursors[48–53] (in conjunction with n.m.r. spectroscopy) and the isolation of intermediates between uroporphyrinogen-III (**26**) and cobyrinic acid[54,55] have provided a detailed picture of corrinoid biosynthesis (see Chapter 30.2).

The timing and mechanism of the incorporation of cobalt ion into corrinoids is unknown. Once cobalt ion is incorporated into the corrin ring it is unlikely to be removed (there is no selective chemical method known for removing cobalt ion from corrinoids). Therefore, the discovery of metal-free corrinoids[10] suggests that cobalt ion does not play an important role in the biosynthesis of corrinoids.

24.4.3.2 (*R*)-1-Aminopropan-2-ol

Labelling studies with ^{14}C and ^{15}N have shown that this unit (**27**) comes from L-threonine. Recent studies[56] with *P. shermanii* indicate that threonine is directly decarboxylated in an enzymatic reaction dependent on a thiol and a corrinoid.

24.4.3.3 5,6-Dimethylbenzimidazole

Recognition of an *o*-dimethylbenzenoid unit in the structures of 5,6-dimethylbenzimidazole and riboflavin led Woolley (1950) to suggest that these molecules share a common origin. Subsequent studies have indicated that dimethylbenzimidazole is formed from riboflavin, which arises by a multi-step sequence from guanosine triphosphate.[57,58] A key intermediate in this sequence is 6,7-dimethyl-8-ribityl-lumazine (**28**), two molecules of which produce one molecule each of riboflavin and (**29**). The source of the four-carbon unit comprising the methyl groups, C-6 and C-7 of (**28**), is unsure. Labelling studies have shown[59] that if the obvious candidate, butane-2,3-dione, supplies this unit, then the dione must arise by a process which utilizes C-1 of a pentose, rather than by the route from two molecules of pyruvate established for, *e.g. Aerobacter aerogenes.* [1'-^{14}C]Riboflavin gives 5,6-dimethylbenzimidazole labelled only at C-2, suggesting *N*-methylene- (or *N*-formyl-) 1,2-diamino-4,5-dimethylbenzene as an intermediate between riboflavin and the benzimidazole.

The origin of corrinoids containing bases other than 5,6-dimethylbenzimidazole is uncertain. Experiments with *Clostridium thermoaceticum* show that ^{14}C at C-1' of riboflavin is incorporated into C-2 of 5-methoxybenzimidazole, the analogue produced by this organism.[60] It is proposed that 5-hydroxybenzimidazole, methylation of which gives 5-methoxybenzimidazole, arises from the combination of (**28**) with the analogue (**30**).

For further aspects of the biosynthesis of corrinoids, see Refs. 44 and 61 and Part 30.

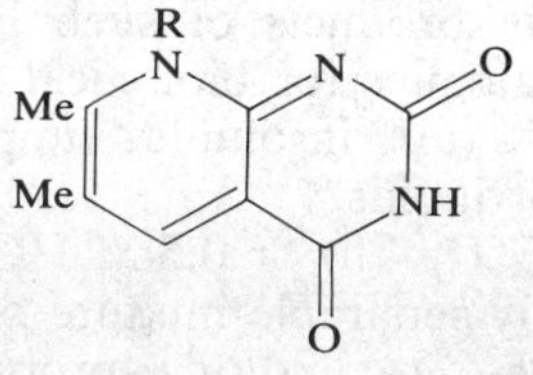

(28) R = ribityl

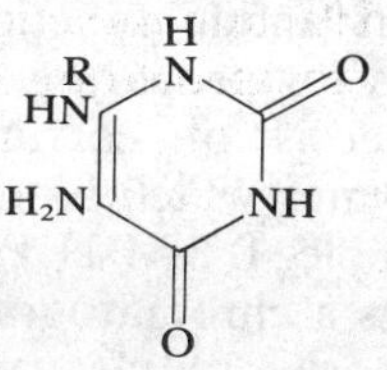

(29)

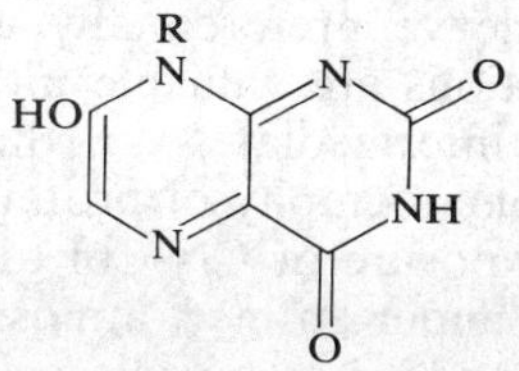

(30)

24.4.4 REACTIONS OF CORRINOIDS AND THEIR MODEL COMPOUNDS

24.4.4.1 Thermal stability

In the solid state or in solution, corrinoids are relatively stable compounds, *e.g.* at the optimum pH 4.5–5.0 an aqueous solution of CN-Cbl at room temperature retains all its biological activity for over two years.[5] The alkylcobalamins MeCbl and AdoCbl are recovered unchanged after heating in anaerobic aqueous solution at 94 °C for 5 h. However, EtCbl rapidly eliminates ethylene at 80 °C.[21] The order of thermal stability is primary > secondary > tertiary alkylcobalamins (the latter having not been prepared).

24.4.4.2 Protonation and ligand exchange

The pK_a of the protonation of OH-Cbl to aquacobalamin is 7.62 (0.1 M aq. KNO_3, 298 K).[62] At low pH, protonation of the dimethylbenzimidazole of cobalamins causes its removal from coordination, *e.g.* equation (3):

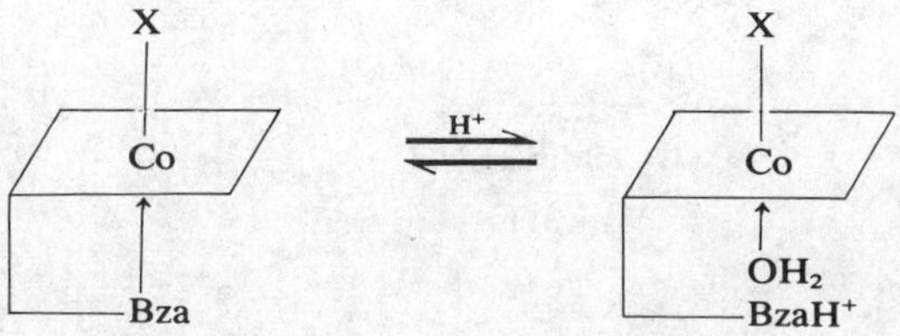

(3)

(for X = OH, $pK_a = -2.4$; for X = Me, $pK_a = 2.7$). The OH group of OH-Cbl is readily replaced by other ligands, *e.g.* equation (4):

$$\text{OH-Cbl} \xrightarrow{\text{aq. NaN}_3} \text{N}_3\text{-Cbl} \tag{4}$$

Cyanide ions displace dimethylbenzimidazole from cobalamins. With AdoCbl the adenosyl group is also removed (equation 5):

$$\text{AdoCbl} \xrightarrow{\text{excess CN}^-} (\text{CN})_2\text{-Cbl} + \text{CH}_2{=}\text{CH(CHOH)}_2\text{CH(OH)CN} + \text{adenine} \tag{5}$$

Ligand exchange in corrinoids is reviewed in Ref. 63 (see also Ref. 64 concerning related processes with cobaloximes).

The CN group of CN-Cbl (stability constant 10^{12} l mol^{-1}) is not easily replaced by other ligands. However, replacement can be achieved by photoaquation (→ OH-Cbl, $\phi \approx 10^{-4}$, independent of the wavelength of irradiation >300 nm).[65]

24.4.4.3 Hydrolysis of amide functions

Corrinoids contain several functional groups susceptible to acidic or basic hydrolysis. By exploiting possibilities for neighbouring group participation, methods have been devised to cleave preferentially a particular amide function. The products of such hydrolytic reactions are valuable intermediates for preparing: (i) analogues for biological testing;[66] (ii) intermediates in the biosynthesis of corrinoids;[67] (iii) insoluble supports for chromatography of proteins or enzymes which bind to corrinoids.[68]

Exposure of CN-Cbl to, for example, 0.8 M H_3PO_4 (75 °C, 4 h) or 0.5 M HCl (37 °C, 3 h) under an inert atmosphere gives a chromatographically separable mixture of mono-, di-, and tri-carboxylic acids, which arise by cleavage of *b*-, *d*-, and/or *e*-propionamide groups (relative rates of cleavage: $e > b \approx d$). The principal monocarboxylic acid was shown[69] to be CN-Cbl(*e*-OH) by X-ray and neutron diffraction. Hydrolysis of the *e*-propionamide is believed to be assisted by the nearby phosphate grouping. Standing an aqueous solution of CN-Cbl(*e*-OH) causes hydrolysis of its phosphate grouping [CN-Cbl and CN-Cbl(*b*-OH) are inert under these conditions].

Concentrated hydrochloric acid (65 °C, 18.5 min) selectively hydrolyses the *f*-amide group of CN-Cbl giving, after column chromatography (DEAE-cellulose) and preparative paper chromatography, a 10% yield of cobyric acid.[70] Preferential formation of cobyric acid depends on the isomerization of the *f*-amide to an *f*-ester (**31**), which undergoes acid-catalysed hydrolysis (*cf.* preferential cleavage of polypeptides at serine and threonine residues catalysed by strong acids). An alternative route[71] to cobyric acid due to Müller and Müller induces formation of the *f*-ester (**31**) using either anhydrous HF or methanolic $ZnCl_2$. After acetylating the amino-group of (**31**), the ester grouping can be cleaved by aqueous piperidine to give cobyric acid (yield up to 20%).

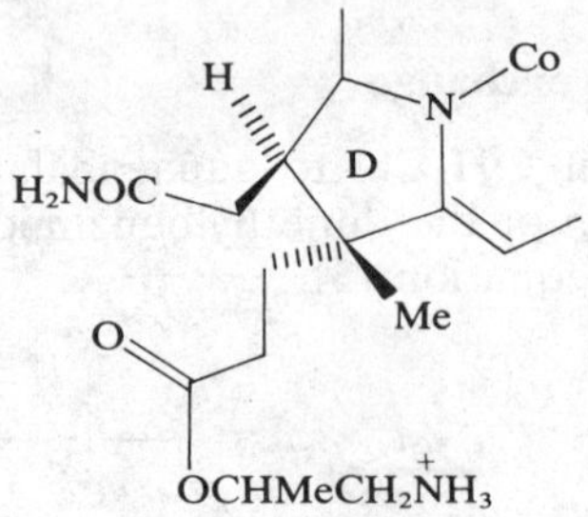

(**31**)

24.4.4.4 Hydrolysis of the phosphate grouping

The elegant method of Bernhauer and his co-workers (*cf.* Ref. 71) for converting cobalamins to cobinamides depends on the affinity of lanthanide ions for phosphate groupings, *e.g.* equation (6).

$$\text{CN-Cbl} \xrightarrow[\text{pH 8–9, 100 °C/50 min}]{\text{Ce(OH)}_3\text{, aq. HCN}} \sim 100\%\ \text{Cbi} + \alpha\text{-ribazole} + \text{insoluble phosphate} \qquad (6)$$

24.4.4.5 Electrophilic substitution on the corrin chromophore

Early investigators subjected corrinoids to standard electrophilic reagents (*e.g.* $^2H^+$, Cl^+, Br^+, NO_2^+) and obtained 10-substituted derivatives. With halogenating agents, depending on the conditions and type of corrin, formation of the *c*-lactone was shown to be an alternative reaction pathway, *e.g.* equation (7). (Note that *c*-lactones also arise during alkaline or acidic hydrolyses of corrinoids in air.[69])

Recently, Inhoffen and his co-workers[72] discovered an additional complication with halogenating reagents. Chlorination (15- to 20-fold excess of Cl_2 in CCl_4 at room temperature) of cobester gave within seconds the colourless adduct (**32**). Reduction of

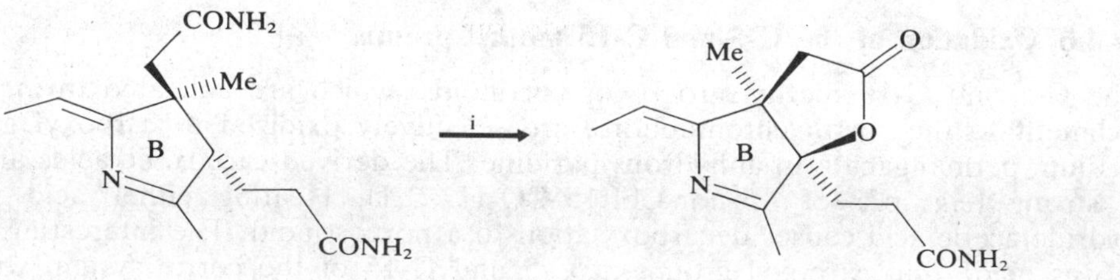

(7)

i, equimolar bromine in water (pH 4).

(**32**) with, for example, $[Rh(CO)_2Cl]_2$ in methanol at room temperature followed by treatment with aqueous KCN gave 10-chlorodicyanocobester.

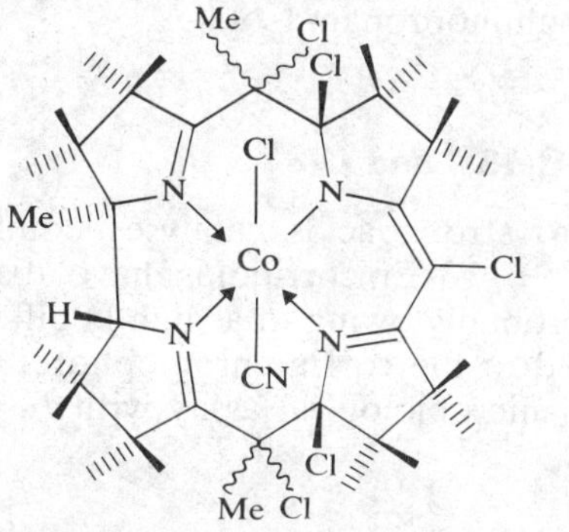

(**32**) (peripheral substituents as in **1a**)

Using simple synthetic norcorrins (*i.e.* corrins lacking substituents at C-5, C-10, and/or C-15 of the chromophore), Eschenmoser's students studied electrophilic substitutions in the corrin chromophore. As the biosynthesis of corrinoids may proceed through an intermediate 5,15-bis-norcorrin it was interesting to know whether or not enzymatic methylation (*i.e.* transfer of methyl groups from *S*-adenosylmethionine) of the corrin chromophore occurs at the positions of inherent greater reactivity. Eschenmoser needed to achieve regioselective substitutions in the corrin chromophore in order to convert synthetic 5,15-bis-norcobester to synthetic cobester. Molecular orbital calculations indicated that the reactivity of C-5 and C-15 towards electrophiles should be greater than C-10.[73] It was found[74] that racemic dicyano(7,7,12,12-tetramethylcorrin)cobalt(III) and racemic (7,7,12,12,19-pentamethylcorrin)nickel(II) chloride undergo preferential substitution at C-15 by $^2H^+$ or chlorosulphonyl isocyanate/DMF (→ 15-cyano-derivatives). This may be simply due to steric shielding of the C-5 and C-10 positions by *gem*-dimethyl groups (at C-7 and C-12, respectively).

Direct methylation of the corrin chromophore is not possible and it was necessary to devise reagents which introduce substituents at C-5 and C-15, afterwards convertible to methyl groups. Successful substitutions were only achieved with two reagents,[39] $(SCN)_2$ or $ClCH_2OCH_2Ph$, *e.g.* equation (8):

$$\text{Cobester} \xrightarrow[\text{in sulpholan, } 50\,^{\circ}\text{C}/12\text{ h}]{ClCH_2OCH_2Ph} \xrightarrow{\text{PHSH}} \begin{array}{c}\text{10-Phenylthio-}\\\text{methylcobester}\\(59\%)\end{array} \xrightarrow[\text{RaneyNi}]{H_2} \begin{array}{c}\text{10-Methyl-}\\\text{cobester}\\(75\%)\end{array} \quad (8)$$

A similar reaction to that in equation (8) was used in the synthesis of cobyric acid. 5,15-Bis-norcobester *c*-lactone-*f*-nitrile (mixture of C-13 epimers) with $PhCH_2OCH_2Cl$ (in MeCN–LiCl, 88 °C, 15 h) gave 47% of 5,15-bis-phenylthiomethylcobester *c*-lactone-*f*-nitrile (mixture of C-13 epimers), which was converted to synthetic cobyric acid (*cf.* Section 24.4.2). The use of 5,15-bis-norcobester *c*-lactone-*f*-nitrile (rather than 5,15-bis-norcobester or 5,15-bis-norcobester-*f*-nitrile) attenuated substitution at C-10 (less than 0.5% of C-10 substituted product was obtained). [Under the conditions of equation (8), cobester *c*-lactone was recovered unchanged.]

24.4.4.6 Oxidation of the C-5 and C-15 methyl groups[75]

The C-5 and C-15 methyl groups of corrinoids, which are activated owing to their attachment to the corrin chromophore, are selectively oxidized to carboxyl groups by potassium permanganate in anhydrous pyridine. The derived carboxylic acids are unusually strong (*e.g.* pK of 15-nor-Cbl-15-$CO_2H = 2.1$). Heating either acid in acetic anhydride/acetic acid causes decarboxylation to a norcorrinoid. The interesting observation was made that carboxyl groups at C-5 and C-15 of the corrin system will induce hydrolysis of nearby propionamide residues by Ce(III) hydroxide. Cbl(*e*-OH) was converted to its *e*-pentylamide and oxidized to a 15-norcarboxylic acid. Treating this acid with anhydrous hydrogen fluoride (to remove the nucleotide) followed by Ce(III) hydroxide (100 °C 3 h) released 2-aminopropa-1-nol and pentylamine. Similar treatment of Cbl(*b*-OH) released only 2-aminopropan-1-ol.

24.4.4.7 Epimerizations at C-13[76] and C-8[77]

Exposure of corrinoids to strong acids catalyses equilibration with the deeper red 13-epicorrinoids (equation 9). 13-Epicorrinoids have distinctive chiroptical properties compared with corrinoids, probably owing to a slightly differing skew about the C-12—C-13 bond which is transmitted to the corrin chromophore. CN-Cbl(13-epi) shows 14% of the activity of CN-Cbl in a microbiological assay with *E. coli.*

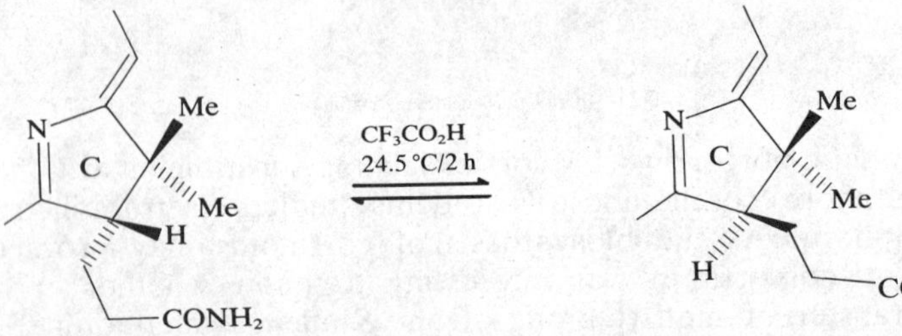

(9)

Reduction of the *c*-lactone of CN-Cbl with sodium borohydride gave Cbl-*abdeg*-penta-amide (50%), accompanied by its C-8 epimer (8%). Amidation of the latter gave CN-Cbl (8-epi). This epimer, like CN-Cbl (13-epi), also gave low growth rates with *E. coli* but, interestingly, was converted to CN-Cbl by growing cultures of *P. shermanii.*

24.4.4.8 Cbl^{II}, Cbl^{I}, H-Cbl (hydridocobalamin), and related model compounds

Corrinoids containing Co(I) or Co(II) are obtained by reducing, for example, OH-Cbl under anaerobic conditions (equations 10–12):[78,79]

$$\text{OH-Cbl} \xrightarrow[\text{0.2 M phosphate (pH 7.0)}]{\text{controlled potential reduction}} \text{Cbl}^{II} \qquad (10)$$

$$\text{OH-Cbl} \xrightarrow{\text{aq. NaBH}_4} \text{Cbl}^{I} \qquad (11)$$

$$\text{OH-Cbl} \xrightarrow{\text{Zn, HOAc}} \text{green soln. containing H-Cbl} \qquad (12)$$

For quantitative studies of the electrochemistry of cobalamins, see Ref. 80. Reduced model compounds are obtained by analogous methods. The following preparative method (equation 13)[81] *via* a cobaloxime(II) complex formed *in situ* is especially convenient:

$$CoCl_2{\cdot}6H_2O \xrightarrow[\text{py in MeOH}]{\text{2dmgH, 2NaOH}} Co^{II}(dmgH)_2py \xrightarrow{NaBH_4} {:}Co^{I}(dmgH)_2py \qquad (13)$$

When $ClCo(dmgH)_2PBu_3$ is reduced with $NaBH_4$ in aqueous MeOH buffered at pH 7, black, air-sensitive crystals of $HCo(dmgH)_2PBu_3$ are obtained.[79] This substance is soluble in non-polar solvents, in contrast to $Co^I(dmgH)_2L$. H-Cbl and hydridocobaloximes are the conjugate acids of Cbl^I and cobaloxime(I) nucleophiles, respectively, each of which bears a formal negative charge [for $HCo(dmgH)_2PBu_3$, $pK_a \approx 10.5$].

All of the above species are very sensitive to oxidation, *e.g.* dioxygen converts Cbl^I and Cbl^{II} eventually to OH-Cbl,[82] whilst N_2O selectively converts Cbl^I to Cbl^{II}.[83] At 174 K, Cbl^{II} reacts rapidly and reversibly with dioxygen to yield a complex of stoichiometry $CblO_2$ and probable structure (**33**).[84] E.s.r. spectroscopy of this species shows that its unpaired spin density resides mainly on the oxygen atoms. Steric factors prevent attack on (**33**) by another Cbl^{II} to give a peroxo-complex [Cbl-O—O-Cbl]. Cbl^I and H-Cbl rapidly decompose (to $Cbl^{II} + H_2$) in aqueous solution at low pH (*e.g.* half-life of Cbl^I at pH 8.01 = 67 min, at pH 6.98 = 22 min).[78]

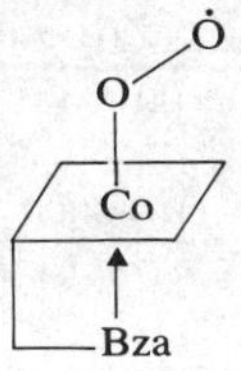

(**33**)

Cbl^I reacts rapidly with alkylating agents (alkyl halides and *O*-toluene-*p*-sulphonates, epoxides). This type of reaction provides an excellent general method for preparing alkylcobalamins, *e.g.* equation (14):[78]

$$Cbl^I + MeI \longrightarrow MeCbl + I^- \qquad (14)$$

The biosynthesis of AdoCbl involves the reaction between Cbl^I and ATP (see Section 24.4.3). The Co(I) derivatives of model compounds behave similarly to Cbl^I in their reactions with alkylating reagents. Schrauzer and Deutsch[85] found a second order rate law for the reactions between a variety of alkylating agents and either Cbl^I or (tributylphosphine)cobaloxime(I). The order of reactivity of alkyl halides was RI > RBr ≫ RCl and MeX > EtX > Pr^iX > Bu^tX. These reactions have the characteristics of S_N2 displacements and for several examples with cobaloxime(I) the stereochemical course has been shown to be inversion,[86,87] *e.g.* equation (15).

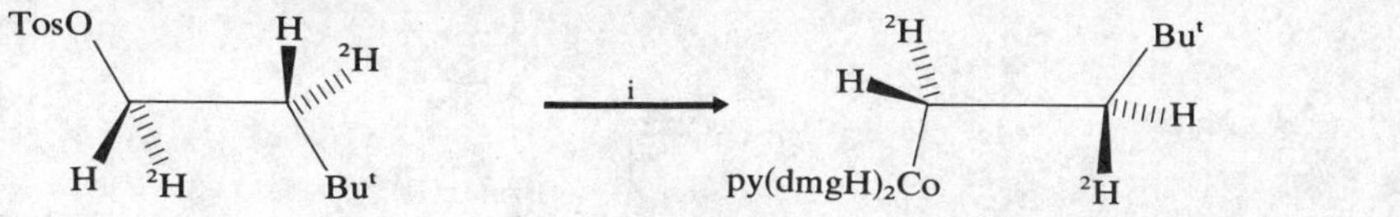

(15)

i, $:Co^I(dmgH)_2py$ in MeOH.

Nucleophilicities on the Pearson scale are as follows: 14.4 (Cbl^I), 14.3, 13.8, 13.6, 13.3, and 13.1[85] [(L)cobaloxime(I) with L = H_2O, py, Me_2S, Bu_3P, and $C_6H_{11}NC$, respectively]. Few nucleophiles can match these values (*cf.* CN^- 6.7), so Cbl^I and cobaloxime(I) species have been called 'supernucleophiles'.

An interesting exception to S_N2 behaviour has been observed by Breslow and Khanna,[88] who found that [1-^{2}H]cyclodecyl *O*-toluene-*p*-sulphonate did not react with Cbl^I (15 h, 50% aq. MeOH), whereas the corresponding iodide gave cyclodecylcobalamin within 15 min. However, this product contained deuterium at C-1 of its cyclodecyl group and *other* ring positions. Both cyclodecyl substrates are evidently too hindered to undergo S_N2 displacement by Cbl^I. However, with iodocyclodecane an alternative electron transfer pathway *via* the cyclodecyl radical may occur (equation 16):

$$Cbl^I + RI \rightleftharpoons Cbl^{II} + (RI)^{\overline{\cdot}} \longrightarrow I^- + R\cdot \xrightarrow{Cbl^{II}} R\text{-}Cbl \qquad (16)$$
(R = cyclodecyl)

Cobaloxime(II) complexes react with alkylating agents which are good electron acceptors, *e.g.* equation (17):

$$2Co^{II}(dmgH)_2py + PhCH_2Br \xrightarrow{\text{benzene}} BrCo(dmgH)_2py + PhCH_2Co(dmgH)_2py \quad (17)$$
$$[k \text{ (at 298 K)} = 3.0 \pm 0.2 \times 10^{-1}\ \text{l mol}^{-1}\ \text{s}^{-1}]$$

An electron transfer pathway analogous to equation (16) has been demonstrated for these reactions.[89]

A potentially useful property of H-Cbl and hydridocobaloximes is their ability to add to unactivated alkenes, *e.g.* equation (18):[79]

$$\text{H-Cbl} + C_2H_4 \xrightarrow{\text{acetic acid}} \text{EtCbl} \quad (18)$$

Note that Cbl^I does not react with ethylene. Both the hydrido-species and the Co(I) nucleophiles react with alkenes activated by an electron-withdrawing group, but their mode of addition is different (see equations 19 and 20):[79]

$$HCo(dmgH)_2PBu_3 \xrightarrow[\text{aq. MeOH}]{\text{excess } CH_2{=}CHCN} CH_3CHCNCo(dmgH)_2PBu_3 \quad (19)$$

$$:Co^I(dmgH)_2py \xrightarrow[pH \gg 7]{CH_2{=}CHCN} NCCH_2CH_2Co(dmgH)_2py \quad (20)$$

24.4.4.9 Electrophilic attack on Co—C bonds

The Co—C bond of alkylcorrinoids and their model compounds can be cleaved by halogenating agents (equation 21):

$$\text{AdoCbl} + I_2 \rightarrow \text{I-Cbl} + \text{AdoI} \quad (21)$$

The mechanistic interpretation of these reactions is confused.[90]

MeCbl undergoes a bimolecular electrophilic substitution (S_E2) with Hg(II) to give methylmercuric ion, MeHg(II).[91] The generation of toxic MeHg(II) from Hg(II) and endogeneous MeCbl can take place in the aqueous environment and within fishes and mammals. The dire consequences to an ecological system of exposure to excessive MeHg(II) and Hg(II) are illustrated by the events which occurred at Minamata, Japan.

Two competitive reactions are observed[91] spectrophotometrically when MeCbl is exposed to Hg(II) (aqueous acetate, pH 5.0, room temperature):

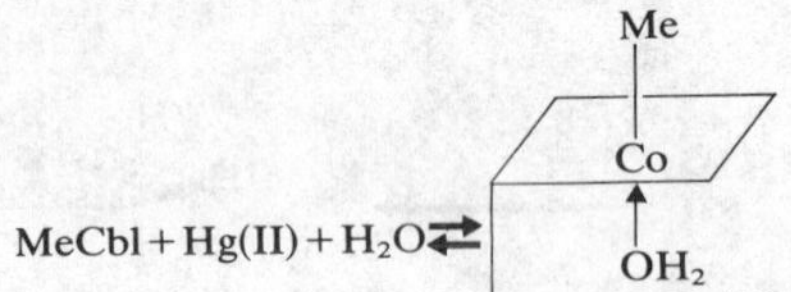

(22)

$$\text{MeCbl} + \text{Hg(II)} + H_2O \rightarrow \text{OH-Cbl} + \text{MeHg(II)} + H^+ \quad (23)$$

MeHg(II) reacts with MeCbl to give Me_2Hg and OH-Cbl, but this reaction is much slower than reaction (**23**). The base-off corrinoid from reaction (**22**) reacts with Hg(II), but with a rate constant *ca.* 10^{-3} of that for reaction (**23**). The contribution of reaction (**22**), therefore, is to reduce the concentration of the reactive species MeCbl. EtCbl behaves similarly to MeCbl with Hg(II), but the Co—C bond of AdoCbl is not cleaved. Studies of reactions between alkyl(aquo)cobaloximes (primary alkyl; secondary alkyls do not react) and Hg(II) indicate[92] a mechanism of dealkylation, similar to that which occurs with RCbl, but there was no analogous pre-equilibrium [*cf.* reaction (**22**) above]. The rate law for alkylcobaloximes reacting with Hg(II) in aqueous $HClO_4$ has been shown[87] to be:

$$\frac{d[\text{RHg(II)}]}{dt} = k[\text{RCo(dmgH)}_2H_2O][\text{Hg(II)}]$$

with k (at 298 K) $= 1.9 \times 10^{-3}$ l mol^{-1} s^{-1} for $R = Me_3CCH_2CH_2$. For one example, an inversion pathway for dealkylation has been clearly demonstrated (equation 24).[87] This result contrasts with the retention pathway observed in all other electrophilic substitutions at metal–carbon bonds brought about by Hg(II).

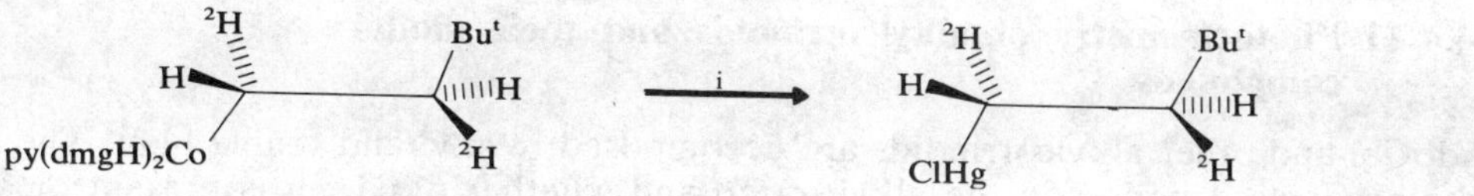

(24)

i, $Hg(ClO_4)_2$, aq. $HClO_4$, 75 °C, 17 h in darkness.

Many other metal ions [*e.g.* Cr(II), Ir(IV)Cl_6, Pd(II)Cl_4] react with alkyl cobalt compounds, effecting cleavage of their Co—C bond. Some of these reactions, *e.g.* MeCbl + Pd(II)Cl_4,[93] are mechanistically similar to the reactions with Hg(II) see above). Cr(II) effects a reductive cleavage of the Co—C bond of RCbl according to the following stoichiometry:[94]

$$\text{RCbl} + \text{Cr(II)aq.} \rightarrow \text{Cbl}^{II} + (\text{H}_2\text{O})_5\text{Cr(III)R} \xrightarrow{\text{H}^+} \text{RH} + \text{Cr(III)aq.} \quad (25)$$
(R = Me or Et)

The mechanism of this cleavage is either S_H2 (direct displacement) or redox S_E2 [*i.e.* carbanion (R^-) transfer from $RCbl^{II}$ to Cr(III)]. With Ir(IV)Cl_6 the metal ion oxidizes the cobalt chelate to an RCo(IV) species from which R is displaced by a chloride ion.[95] Such cobalt(IV) species have been generated electrochemically[96] and have been characterized spectroscopically[23] (see Section 24.4.1). Alkyl groups can also be removed from alkylcobalt compounds by cobalt(I), cobalt(II), and cobalt(III) complexes. The transfer of an alkyl group from an alkylcobaloxime to cobaloxime(I) or cobaloxime(II) species in methanol is a bimolecular displacement with inversion of configuration at the alkyl group.[97]

24.4.4.10 π-Complexes of cobalamins and cobaloximes

Evidence has been obtained that simple alkenes (*e.g.* ethylene) and alkenes carrying electron-releasing groups (*e.g.* vinyl ethers) form complexes with Co(III) (in cobalamins or cobaloximes),[98–101] whilst alkenes bearing electron-withdrawing groups (*e.g.* tetracyanoethylene) form complexes with Co(I) species.[102] Such complexes have been

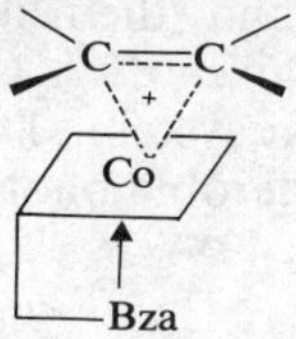

(34)

represented as π-complexes, *e.g.* (**34**). [For a discussion of their structure and bonding, see Refs. 98, 99, and 102]. π-Complexes of Co(III) are probably intermediates in solvolyses of 2-acetoxyalkyl-cobaloximes, *e.g.*[98]

$$\text{AcOCH}_2\text{CH}_2\text{Co(dmgH)}_2\text{py} \xrightarrow{\text{EtOH}} \text{EtOCH}_2\text{CH}_2\text{Co(dmgH)}_2\text{py} + \text{HOAc} \quad (26)$$
$$[k \text{ (at 298 K)} = 4.37 \times 10^{-6}\,\text{s}^{-1}]$$

in the acid-catalysed decompositions of 2-hydroxy- and 2-alkoxy-alkylcobalamins and cobaloximes, *e.g.*[103]

$$\text{HOCH}_2\text{CH}_2\text{-Cbl} \xrightarrow[100\,°\text{C, 2 min}]{0.01\,\text{M HCl}} \text{H}_2\text{O} + \text{CH}_2\text{=CH}_2 + \text{OH-Cbl} \quad (27)$$

and in the following type of synthesis of alkylcobalamins:[99]

$$\text{OH-Cbl} + \text{H}_2\text{C=CHOCH}_2\text{CH}_2\text{OH} \longrightarrow$$ $$\text{-CH}_2\text{-Cbl} + \text{H}_2\text{O} \quad (28)$$

They have also been suggested as intermediates in certain cobalamin-dependent enzymatic reactions[104] (see Section 24.4.5).

24.4.4.11 Photochemistry of alkylcorrinoids and their model compounds

AdoCbl and other alkylcorrinoids are decomposed by u.v. and visible light. The nature of the products depends on the alkyl group and whether dioxygen is present or absent. Aerobic photolysis of RCbl gives OH-Cbl and an aldehyde derived from R•, probably *via* $RO_2\cdot$. Alkylcobaloximes yield peroxy-compounds on irradiation (or sometimes thermolysis) in the presence of dioxygen,[105] *e.g.* equations (29a) and (29b).[106]

$$MeCH_2CHMeCo(dmgH)_2py \xrightarrow[O_2]{h\nu} MeCH_2CHMeOOCo(dmgH)_2py \quad (29a)$$

$$MeCbl \xrightarrow[\text{excess } O_2 \text{ in water}]{h\nu} OH\text{-}Cbl + CH_2O \quad (29b)$$

A detailed study[107] of the aerobic photodecomposition of several alkylcobalamins showed that quantum yields (ϕ) were dependent on the wavelength of the incident light (250–570 nm) and the pH of the solution (1 or 7). For MeCbl, $\phi_{570} = 0.12$, $\phi_{250} = 0.35$ (pH 7), and a similar wavelength dependence was found at pH 1 (for the base-off form). However, for EtCbl at pH 1 the quantum yield decreases from 0.38 (550 nm) to 0.28 (250 nm). MeCbl appears to resist anaerobic photolysis. Actually, its Co—C bond is cleaved very readily to Me• and Cbl^{II} (flash photolysis of MeCbl gave $k \approx 1.5 \times 10^9$ l mol^{-1} s^{-1} for this reaction),[108] but these species recombine unless scavengers (*e.g.* dioxygen, propan-2-ol) for Me• are present. EtCbl and AdoCbl are decomposed readily on anaerobic photolysis[21] because the organic radicals initially formed (they can be spin-trapped)[109] have reaction pathways available, which are not possible for Me•.

$$EtCbl \xrightarrow{h\nu} Cbl^{II} + \tfrac{1}{2}H_2 + C_2H_4 \quad (30)$$

$$AdoCbl \xrightarrow{h\nu} 8,5'\text{-cyclic-adenosine } (+\tfrac{1}{2}H_2\ ?) \quad (31)$$

Anaerobic irradiation of MeCbl in water with 150 W lamps gradually gave CH_4 and C_2H_6, the proportions of which varied with the pH of the solution (29/71 at pH 4, 57/43 at pH 12.5).[110] Addition of KCN (to 0.1 M) or 2-mercaptoethanol (to 1 M) gave a 100/0 ratio of methane/ethane. These additives are thought to coordinate to the cobalt ion (by displacing the dimethylbenzimidazole) and then displace Me^- from MeCbl. It was claimed[110] that irradiation of C^2H_3-Cbl gives 89% $C_2{}^2H_6$ and 11% $C_2H_3{}^2H_3$, the latter arising by Me abstraction from the corrin system. This interesting result requires confirmation. See Ref. 111 concerning the anaerobic photodecomposition of alkylcobaloximes in aqueous solution.

24.4.5 CORRINOID-DEPENDENT ENZYMATIC REACTIONS[112]

These reactions are listed in Table 3. Two of the reactions (catalysed by methylmalonyl-CoA mutase and N^5-methyltetrahydrofolate:homocysteine methyltransferase) are of importance in human physiology. Although humans possess a ribonucleotide reductase it belongs to the class of such enzymes which is *not* AdoCbl-dependent. Until recently, it was believed that plants do not contain corrinoids. However, the enzyme leucine 2,3-aminomutase, for which a dependence on AdoCbl has been demonstrated *in vitro*, has been detected in bean seedlings.[113] The remaining reactions of Table 3 are carried out by micro-organisms. Almost certainly, there are additional corrinoid-dependent enzymatic reactions awaiting discovery.

The reactions of Table 3 have been divided into two groups. Those in group (a) are synthetic reactions proceeding *via* MeCbl and are exemplified below by a discussion of

TABLE 3
Corrinoid-dependent enzymatic reactions

(a) Requiring MeCbl Reaction	
1. Synthesis of methionine (by N^5-methyltetrahydrofolate: homocysteine methyltransferase)	[see text and equation (33)]
2. Synthesis of methane	
3. Synthesis of acetate	
***(b) Requiring AdoCbl* Enzyme**	***Reaction***
4. Diol dehydrase	See text
5. Ethanolamine ammonia-lyase	*E.g.* $HOCH_2CH_2NH_2 \rightarrow CH_3CHO + NH_3$
6. Aminomutase utilizing either (i) (*S*)-3,6-diaminohexanoate (ii) (*R*)-2,6-diaminohexanoate (iii) (*R*)-2,5-diaminopentanoate (iv) α- or β-leucine	*E.g.* $H_2NCH_2CH_2CH_2\overset{S}{C}HNH_2CH_2CO_2H \rightleftarrows CH_3\overset{S}{C}HNH_2CH_2\overset{S}{C}HNH_2CH_2CO_2H$
7. Methylmalonyl-CoA mutase	See equations (34) and (35), and text
8. Glutamate mutase	See equation (1)
9. α-Methyleneglutarate mutase	$HO_2CCH_2CH_2C(CO_2H){=}CH_2 \rightleftarrows HO_2CCHMeC(CO_2H){=}CH_2$
10. Ribonucleotide reductase	See equation (37) and text

N^5-methyltetrahydrofolate:homocysteine methyltransferase. The reactions of group (b) require AdoCbl and have several common features. Reactions 4–9 and possibly 5 are molecular rearrangements in which a group X and a hydrogen atom exchange places:

$$\begin{array}{c} | \quad | \\ -C-C- \\ | \quad | \\ H \quad X \end{array} \;(\rightleftarrows)\; \begin{array}{c} | \quad | \\ -C-C- \\ | \quad | \\ X \quad H \end{array} \qquad (32)$$

In all reactions of this type a key step appears to be cleavage of the Co—C bond of AdoCbl, probably in a homolytic manner, to generate Cbl^{II} and an adenosyl radical (*cf.* Ref. 112b; for a contrary view, see Ref. 112c). The latter abstracts a hydrogen atom from substrate to produce a substrate-derived radical (S·) and 5′-deoxyadenosine (cleavage of the Co—C bond and transfer of an H atom from substrate may be a concerted process). S· rearranges to a product-related radical P·, which abstracts a hydrogen atoms from 5′-deoxyadenosine to give product PH and an adenosyl radical, which can recombine with Cbl^{II} or attack another substrate molecule. Evidence for this pathway and the mechanism of conversion of S· to P·, as well as possible alternative pathways, are discussed below for methylmalonyl-CoA mutase and diol dehydrase. Although it does not catalyse a molecular rearrangement, ribonucleotide reductase is included among group (b) reactions because it shows some mechanistic similarities to the other reactions of this group. This enzyme is also discussed in detail below. The corrinoid-dependent conversion of threonine to (*R*)-1-aminopropan-2-ol[56] seems to be of a different character from the reactions of Table 3. Until further information is available it is difficult even to speculate about its mechanism.

24.4.5.1 N^5-Methyltetrahydrofolate:homocysteine methyltransferase

In animal tissues and *E. coli*, methionine is synthesized from homocysteine in a cobalamin-dependent enzymatic pathway. The overall reaction is:

$$N\text{-Methyltetrahydrofolate} + \text{Homocysteine} \rightarrow \text{Tetrahydrofolate} + \text{Methionine} \qquad (33)$$

As isolated, the enzyme from *E. coli* is salmon-coloured due to a tightly-bound cob(III)alamin. The synthesis of methionine according to equation (33) also requires a reduced flavin and *S*-adenosylmethionine, which generates enzyme-bound MeCbl from the non-functional cob(III)alamin. Unlike free MeCbl, the enzyme-bound MeCbl is rather stable to light. Production of enzyme-bound MeCbl can be inhibited by iodopropane, which converts Cbl^I to PrCbl. Photolysis of the propylated enzyme restores activity. Taylor and Weissbach have proposed[114] a reaction mechanism (Scheme 3) for cobalamin-dependent methionine synthesis. Homocysteine displaces a methyl group from the enzyme-bound MeCbl to give methionine and Cbl^I, a reaction for which there are convincing models.[115] The least understood part of Scheme 3 concerns the reaction between Cbl^I and *N*-methyltetrahydrofolate.

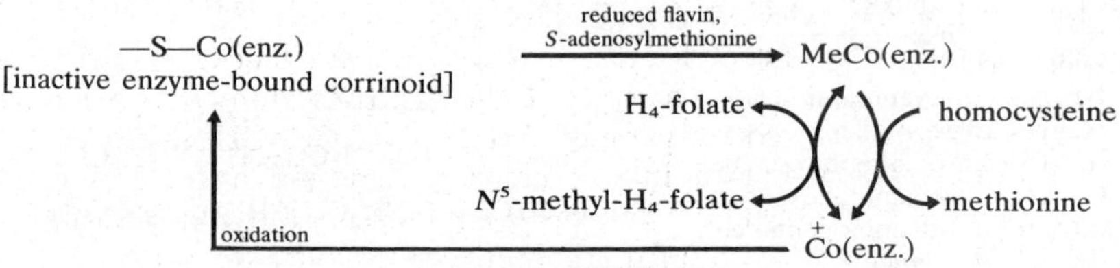

SCHEME 3

N-Methylamines are notoriously inert substrates in S_N2 reactions even when Cbl^I is the nucleophile (tetra-alkylammonium compounds react readily with cobalt(I) nucleophiles to give alkylcobalt compounds[116]). The enzyme might activate *N*-methyltetrahydrofolate by protonating N-5 and, indeed, it is reported[117] that at pH 4 Cbl^I reacts with *N*-methyltetrahydrofolate to give MeCbl (2.5%). However, the MeCbl was identified solely by a difference u.v./visible spectrum and so this report requires confirmation.

The significance of N^5-methyltetrahydrofolate:homocysteine methyltransferase to animals is a controversial question. Since methionine is a constituent of human diets, the role of the enzymatic route as a producer of methionine is believed to be unimportant. However, the methyltransferase may provide the only route for regenerating tetrahydrofolate from N^5-methyltetrahydrofolate. It has been proposed (the 'methylfolate trap' hypothesis)[118] that the clinical manifestations of pernicious anaemia, other than the neurological lesions for which there is no proven explanation yet, may be due to a 'conditioned folate deficiency' arising from cobalamin deficiency. The failure to convert N^5-methyltetrahydrofolate to tetrahydrofolate traps folate in a form which cannot accomplish all its biochemical functions. One consequence is a retardation of DNA synthesis (production of thymidine triphosphate by thymidylate synthetase requires 5,10-methylenetetrahydrofolate).

Certain tumour cells possess diminished activity of N^5-methyltetrahydrofolate:homocysteine methyltransferase and cannot produce sufficient methionine from homocysteine. In competition with normal cells in an environment containing homocysteine but not methionine, such cells die. This circumstance is being exploited in a new approach to cancer chemotherapy.[119]

24.4.5.2 Methylmalonyl-CoA mutase

This enzyme carries out one step in a remarkable series of reactions which are vital for the maintenance of health in humans. Propionyl-CoA is enzymatically carboxylated to (*S*)-methylmalonyl-CoA, which is epimerized to (*R*)-methylmalonyl-CoA. Only the (*R*)-isomer of methylmalonyl-CoA is a substrate for AdoCbl-dependent methylmalonyl-CoA mutase, which converts it to succinyl-CoA (equation 34). In this way, propionyl-CoA from, for example, the degradation of fats, is connected, *via* succinyl-CoA, to the Krebs cycle, and accumulation of toxic methylmalonic acid is prevented. In the rare genetic disease of children, methylmalonicaciduria,[120] proper functioning of methylmalonyl-CoA mutase is impaired. In one form of the disease ('B_{12}-responsive') there is insufficient of the

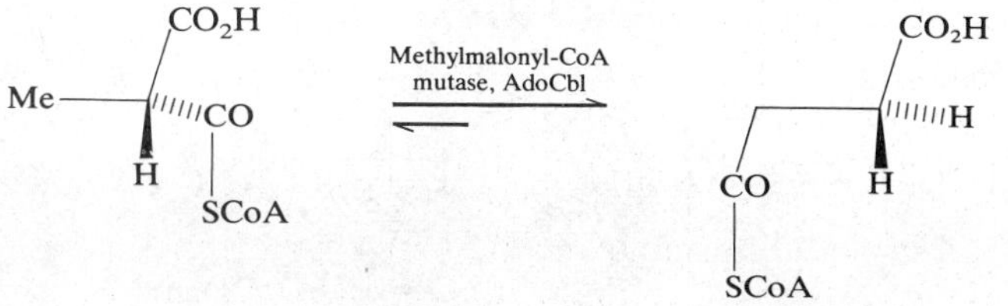
(34)

enzyme's cofactor owing, for example, to faulty biosynthesis of AdoCbl. It can be controlled by administering OH-Cbl (~1 mg per day). There is another form of the disease which does not respond to cobalamin treatment, owing to a structural defect in the enzyme.

Studies with methylmalonyl-CoA specifically labelled with ^{13}C and ^{14}C showed[121,122] that the rearrangement to succinyl-CoA involves an intramolecular 1,2-shift of the thioester grouping. Lynen *et al.*[121] compared the methylmalonyl-CoA mutase reaction with the 'Urry–Kharasch rearrangement' (*e.g.* $Me_2CPh\dot{C}H_2 \rightarrow Me_2\dot{C}CH_2Ph$) and suggested that radicals participated in the enzymic reaction also. However, the well-known rearrangements of β-arylalkyl radicals are fundamentally different from pathways *via* intermediate radicals conceivable for the methylmalonyl-CoA mutase reaction or any other AdoCbl-dependent rearrangement. With the β-arylalkyl radicals a bridged intermediate can arise in which the loss in resonance energy of the aromatic ring is partly offset by the increased stabilization of the radical centre. For the AdoCbl-dependent rearrangements it is necessary to consider the stability of intermediates of type (**35**) where X = OH, NH_2, or a carbon group (*e.g.* COSCoA). Since there are no *bona fide* examples of 1,2-shifts in radicals where the migrating group is a *saturated* group derived from a first-row element (C, N, or O), it appears that species (**35**) are energetically unfavourable, as expected on theoretical grounds.[123]

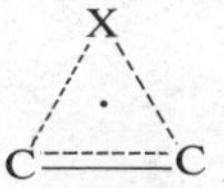
(**35**)

In 1964, Ingraham[124] christened the natural alkylcobalamins 'biolobical Grignard rearrangements' and proposed the mechanism of Scheme 4 for methylmalonyl-CoA mutase. It was envisaged that the Co—C bond of the intermediate organocobalamin derived from substrate and AdoCbl was polarized ($Co^{\delta+}$—$C^{\delta-}$) and this initiated the rearrangement. Subsequently, Lowe and Ingraham published[125] a model reaction in support of this mechanism. However, the very low yield (0.3%) of rearranged product makes this model unconvincing.

Meanwhile, labelling studies with ^2H and ^3H showed that the methylmalonyl-CoA mutase reaction possesses what can now be described as the typical features of AdoCbl-dependent reactions. For example, a mixture of unlabelled methylmalonyl-CoA and [*Me*-$^2H_3$]methylmalonyl-CoA was partially isomerized to succinyl-CoA. Analysis of the deuterium contents of product and recovered starting material led to the conclusion that a hydrogen atom removed from substrate is mixed with the 5′-CH_2 hydrogen atoms of AdoCbl in an intermediate containing a methyl group, and then a hydrogen is removed from this intermediate to become part of the product.[126] Retéy and Zagalak[127] extended prior studies of the stereochemistry of the methylmalonyl-CoA mutase reaction, by studying the action of the enzyme on ethylmalonyl-CoA. Using deuterium as a label, the stereochemical course shown in equation (35) was defined. In this case and with methylmalonyl-CoA, replacement of COSCoA by H takes place with retention of configuration [*cf.* the diol dehydrase reaction (see below) and glutamate mutase reaction (equation 1), where analogous processes occur with inversion of configuration].

Recent model studies have produced impressive results relevant to the mechanism of the methylmalonyl-CoA mutase reaction. Dowd and Shapiro[128] reacted Cbl^I with 3.1

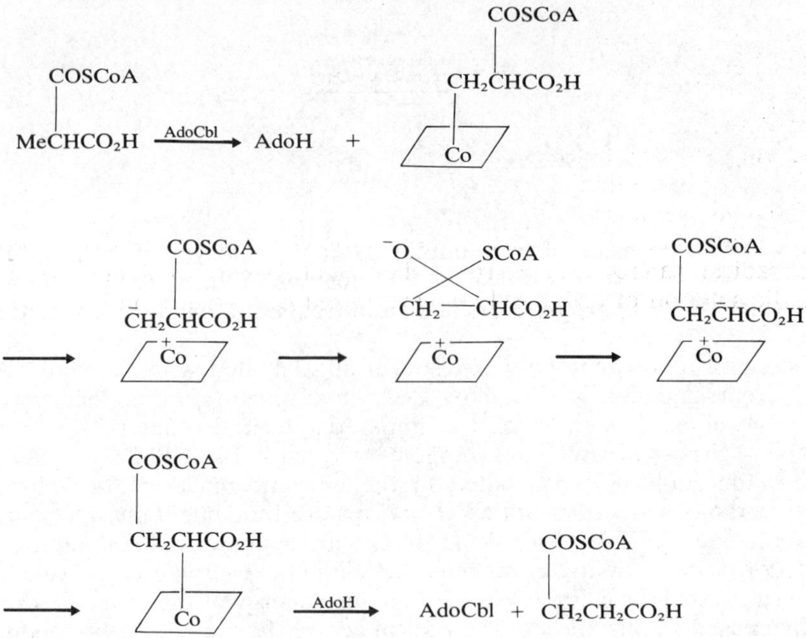

SCHEME 4

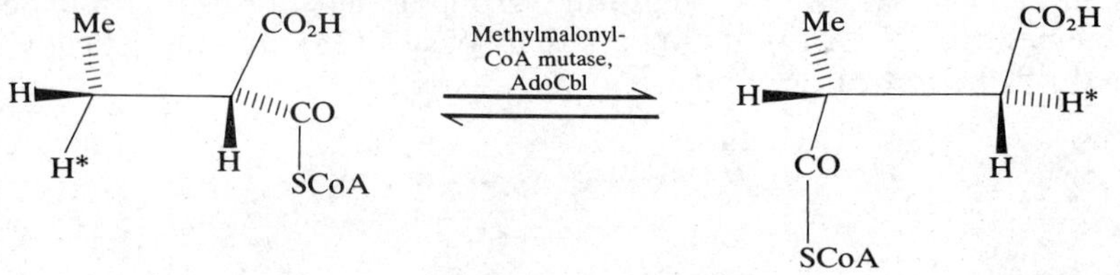

(35)

molar equivalents of dimethyl bromomethylmalonate in aqueous solution (pH 8–9). After 6 minutes, the electronic spectrum of a portion of the solution was consistent with formation of a cobalt–carbon bond [as in (**36**)] having occurred. However, (**36**) was too unstable to be isolated. The reaction mixture containing (**36**) was allowed to stand for 48 hours at room temperature, whereupon extraction with ether, treatment with alkali, and chromatography gave malonic acid, methylmalonic acid (18%), and succinic acid (3.7%) [yields based on OH-Cbl from which the Cbl^{I} was prepared (reduction with excess $NaBH_4$)]. The malonic acid probably arises from alkali-degradation of dimethyl bromomethylmalonate. A control experiment in which cobalt(II) nitrate was used instead of OH-Cbl did not give detectable succinic acid.

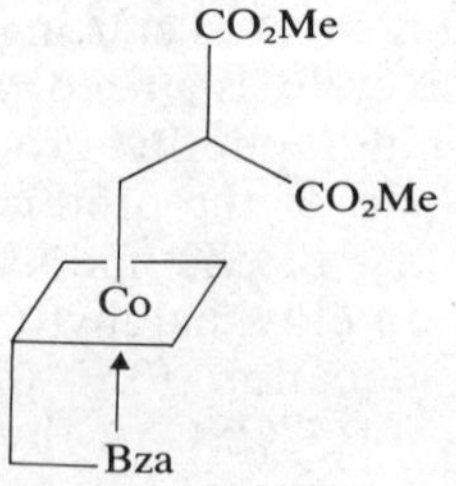

(36)

$(EtO_2C)_2CMeCH_2Co(dmgH)_2py$

(37)

Retéy *et al.*[129] irradiated the cobaloxime (**37**) in ethanol and obtained variable yields of diethyl dimethylmalonate and a rearrangement product, diethyl 2-methylsuccinate. The variability of yield in this experiment and the low yield in Dowd's experiment was ascribed to the situation that 'soon after homolytic cleavage of the cobalt–carbon bond the malonic ester substrate loses contact with the central cobalt atom and, thus deprived of that atom's catalytic effect, is no longer able to rearrange'.[130] A more sophisticated model compound, the 'bridged cobaloxime' (**38**) (structure proved by X-ray analysis) was therefore synthesized.[130] Now, after homolytic cleavage of the Co—C bond the derived organic radical cannot escape from the cobalt(II) atom. Anaerobic irradiation of a methanolic solution of (**38**) for 12 hours caused complete homolysis of its Co—C bond.

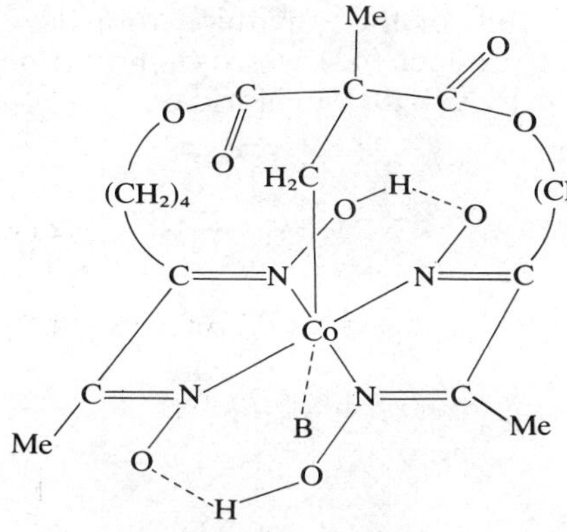

(**38**) B = methanol

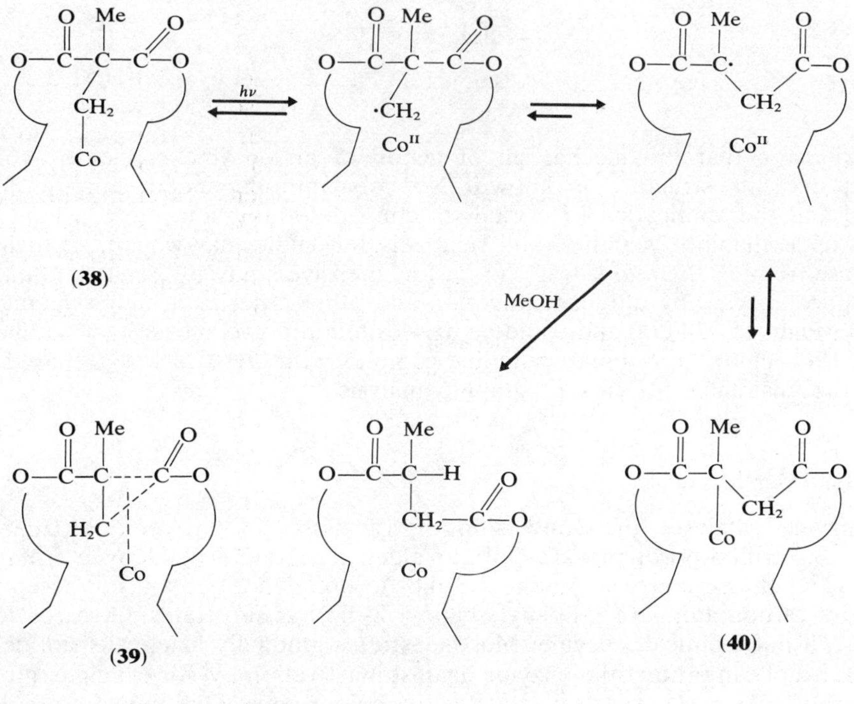

SCHEME 5

Following treatment of the crude product with ethanolic KOH to cleave ester bonds, 2-methylsuccinic acid was isolated in 83% yield. This brilliant experiment strongly suggests a role for Co(II) in the formation of the rearrangement product. Retéy *et al.*[130] suggest the sequence of Scheme 5, with a cobalt-stabilized intermediate (**39**) possibly connecting (**38**) and (**40**).

Scott and Kang[131] have developed Dowd's model system by incorporating a thioester grouping which has the opportunity to undergo a 1,2-shift in competition with an ester grouping. Cbl^{I} (from reducing OH-Cbl with excess $NaBH_4$) in aqueous solution (pH 8–9) was allowed to react with the bromide (**41**). After 30 minutes an aliquot of the solution showed an electronic spectrum like that of AdoCbl. After one day at room temperature the reaction mixture was extracted with ether, giving (**42**) and up to 70% (**43**) (yield based on OH-Cbl). It was suggested that an intermediate alkylcobalamin (**44**) is formed from which (**43**) is derived by a 1,2-shift of the ethylthiocarbonyl group. Control experiments showed that no rearrangement product (**43**) was obtained from reaction mixtures which omitted OH-Cbl or replaced it by cobalt(II) chloride.

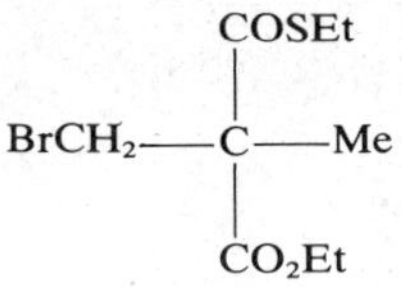

(**41**)

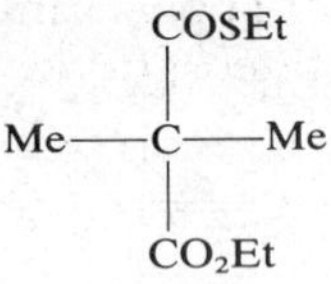

(**42**)

(**43**)

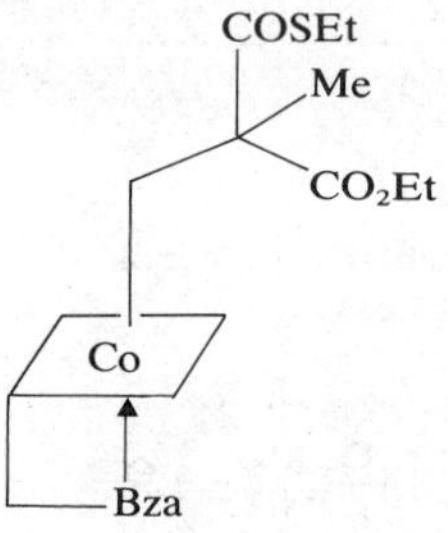

(**44**)

It is axiomatic that the mechanism of action of an enzyme cannot be completely elucidated until its structure is known. No cobalamin-dependent enzyme has been sequenced and the application of crystal-structure determination seems a long way off. Owing to its availability, stability, and relatively low molecular weight, methylmalonyl-CoA mutase from *P. shermanii* was chosen for intensive study by Zagalak *et al.*[132] They obtained pure enzyme by ultracentrifugation and other criteria. It had a weight average molecular weight of 124 000 and could be dissociated into two sub-units of similar size by guanidine hydrochloride. A pink crystalline complex with OH-Cbl was prepared, but the crystals were unsuitable for crystallographic analysis.

24.4.5.3 Diol dehydrase

This enzyme catalyses the conversion of 1,2-diols (C_2–C_4, *e.g.* propane-1,2-diol, glycerol, 3,3,3-trifluoropropane-1,2-diol) to aldehydes (*e.g.* acetaldehyde from ethane-1,2-diol).[133,134] It also converts butane-2,3-diol to butanone, but will not accept diols of five or more carbon atoms (*e.g.* phenylethane-1,2-diol) as substrates. Glycerol dehydrase is a similar, if not identical, enzyme. Most substrates gradually inactivate diol dehydrase. Propane-1,2-diol can protect the enzyme against inactivation by, for example, glycerol. In the absence of substrate, oxygen inactivates the holoenzyme. Diol dehydrase is inhibited by reagents which attack thiols.

Diol dehydrase is unusual in that it shows a lack of biological stereospecificity towards the enantiomers of its substrates. In competitive reactions, (*S*)-propane-1,2-diol is converted to propanal faster than (*R*)-propane-1,2-diol, but in reactions of the separate diols the (*R*)-isomer reacts more quickly![135] Evidently the (*S*)-isomer is bound more tightly by the enzyme, but reacts more slowly [for (*R*)-propane-1,2-diol, V_{max} (nmol min^{-1}) = 13.09 ± 0.60, K_m (mol l^{-1}) = $0.381 \pm 0.028 \times 10^{-4}$; for (*S*)-propane-1,2-diol, $V_{max} = 6.82 \pm 0.17$, $K_m = 0.123 \pm 0.07 \times 10^{-4}$].[134]

Isotopic labelling studies[112b,133,136] of the conversion of propane-1,2-diol to propanal catalysed by diol dehydrase show: (i) no incorporation of solvent H into propanal; (ii) ^2H (or ^3H) at C-1 of propane-1,2-diol is detected at C-2 of propanal and in the 5′-CH_2 of AdoCbl; (iii) ^3H in the 5′-CH_2 of AdoCbl is transferred to C-2 of propanal; (iv) with (*R*)-propane-1,2-diol, H_R is removed from C-1; with (*S*)-propane-1,1-diol H_S is removed from C-1; (v) there is inversion of stereochemistry at C-2, *i.e.* H is delivered to C-2 opposite to the departing OH group; (vi) *ca.* 1% of ^3H at C-1 of propane-1,1-diol is found at C-2 of the corresponding molecule of propanal; (vii) ^{18}O in the C-2 hydroxyl group of racemic propane-1,2-diol appears in propanal; therefore propane-1,1-diol is an intermediate, which undergoes stereospecific, enzymatic dehydration (loss of the *pro-R* hydroxyl group) to propanal.

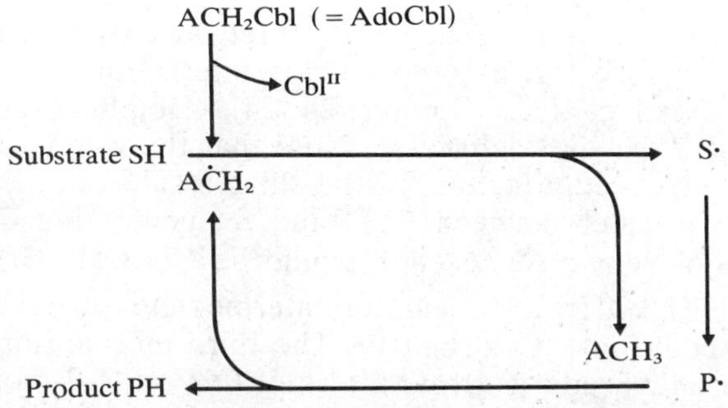

SCHEME 6

In 1972, Abeles[137] proposed the pathway of Scheme 6 for AdoCbl-dependent rearrangements. This scheme was derived mainly from his extensive work on diol dehydrase, with its extension to other reactions being presented as a possibility. Homolysis of the Co—C bond of AdoCbl and H-abstraction from a substrate molecule have been referred to in the introduction to this section. When propane-1,2-diol is substrate SH, S· is $CH_3CH(OH)\dot{C}HOH$, P· is $CH_3\dot{C}HCH(OH)_2$, and product PH is $CH_3CH_2CH(OH)_2$ ($\rightarrow CH_3CH_2CHO + H_2O$). Evidence that reactions catalysed by diol dehydrase involve Cbl^{II} and organic radicals has been obtained by e.s.r. spectroscopy. For example, a reaction mixture containing enzyme, AdoCbl, and propane-1,2-diol frozen to 77 K shows a broad signal at $g \approx 2.3$, corresponding to Cbl^{II}, and a doublet (splitting 15.4 mT, centred on *g* 2.0) ascribed to an organic radical.[138] The observed spectra can be accurately simulated by a model which places the organic radical ≥ 500 pm from the cobalt ion of Cbl^{II}.[138] The appearance of the signal assigned to an organic radical is kinetically competent, *i.e.* the rate of radical production monitored by this signal is faster than the overall rate of the enzymatic reaction. Its doublet character is explained as an electrostatic exchange interaction with Co(II).

The pathway of Scheme 6 includes 5′-deoxyadenosine as an intermediate. This compound has been isolated from experiments in which diol dehydrase is incubated with a substrate analogue (*e.g.* 2-hydroxyacetaldehyde) and then denatured. The most convincing evidence for its role as an intermediate comes from experiments with ethanolamine ammonia lyase.[139] This enzyme converts ethanolamine to acetaldehyde and ammonia and also 2-aminopropan-1-ol to propanal and ammonia. Incubating the enzyme for 15 s with [1-^3H]-2-aminopropan-1-ol followed by denaturation with trichloroacetic acid gave tritiated 5′-deoxyadenosine (>80% yield based on AdoCbl). However, if prior to treatment with trichloroacetic acid the enzyme–2-aminopropan-1-ol complex was incubated with ethanolamine for 90 s, then 80% AdoCbl is recovered and most of the tritium is found in acetaldehyde. Finally, incubating [1,1-$^{2}H_2$]-2-aminopropan-1-ol and [5′,5′-$^{2}H_2$]-AdoCbl with ethanolamine-ammonia lyase followed by denaturation gave 5′-deoxyadenosine, which was degraded to acetaldehyde *p*-nitrophenylhydrazone. This product contained mainly three atoms of deuterium in the methyl group (mass spectroscopic determination).

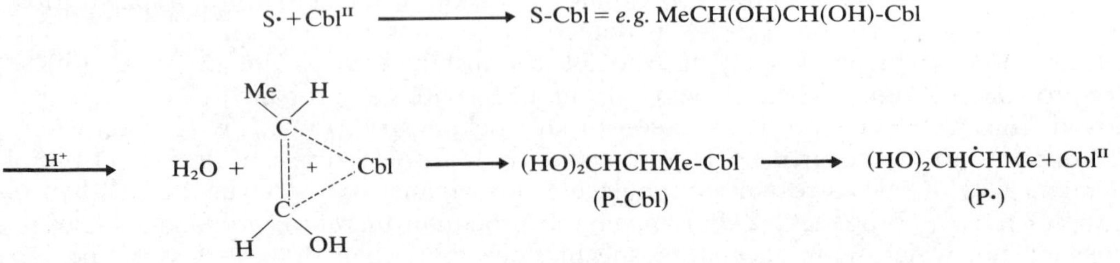

SCHEME 7

For the conversion of S· to P· (Scheme 6), three mechanisms have been proposed. Encouraged by the evidence for π-complexes as intermediates in certain reactions of cobalamins and cobaloximes (see Section 24.4.4), Dolphin *et al.*[104] proposed the mechanism of Scheme 7 for diol dehydrase. Note that the water molecule lost in step 1 must return in step 2, to account for the ^{18}O-labelling results of Arigoni *et al.*[136] A second possibility is electron transfer between Cbl^{II} and S·, generating Cbl^{I} and S^{+}. Such an intermediate in the diol dehydrase reaction would be $CH_3CH(OH)\overset{+}{C}HOH$, which could rearrange to $CH_3\overset{+}{C}HCH(OH)_2$ (P^{+}) *via* an intermediate protonated hydroxyepoxide. Finally, Cbl^{I} and P^{+} could give Cbl^{II} and P·. The third mechanism[123] was stimulated by reports that 2-hydroxyalkyl radicals (from attack of OH· on 1,2-diols) readily decompose to 1-formylalkyl radicals in the pH range 2.5–4, *e.g.* equation (36).

$$HOCH_2CH_2OH \xrightarrow{OH\cdot} H_2O + HOCH_2\dot{C}HOH \xrightarrow{H^+} H_2O + \dot{C}H_2CHO \qquad (36)$$

These reactions resemble the diol dehydrase reactions, and, moreover, proceed without involvement of a metal ion. Golding and his co-workers[140] carried out model studies designed to show that primary alkyl radicals could initiate reactions of the type in equation (**36**). They found that anaerobic photolysis of dihydroxyalkylcobaloximes (**45**; $n = 2, 3, 4$, or 9) yields exclusively pentanal when $n = 3$, a mixture of hexanal and hexan-2-one (ratio *ca.* 1:4) when $n = 4$, and no carbonyl-containing products when $n = 2$ or 9. A possible route to pentanal from (**45**; $n = 3$), is as follows: photolysis releases aquated

$$HOCH_2CHOH(CH_2)_nCo(dmgH)_2py$$

(45)

Co(II), dimethylglyoxime, and a 4,5-dihydroxypentyl radical, which rearranges by a favoured 1,5-H shift (supported by specific deuteration studies) to a 1,2-dihydroxypentyl radical; the latter decomposes to pentanal and water [*via* either $Pr\dot{C}HCHO$ or $Pr\dot{C}HCH(OH)_2$, which is neutralized by dimethylglyoxime]. Golding and Radom[123] considered the possibility that the interconversion of S· and P· (*cf.* Scheme 6) could go *via* a protonated bridged species (**46**) as in Scheme 8. This mechanism accounts for all the H- and O-labelling evidence obtained for diol dehydrase (the fate of labelled atoms is denoted by * in Scheme 8). *Ab initio* calculations indicated that the protonated bridged species was energetically comparable with the 'open' species (**47**) and (**48**), and therefore the activation barrier to the oxygen-shift should be relatively low. A possible sequence of events on the enzyme is shown in Scheme 9. (The required acidic catalyst could be a carboxyl group of the protein.)

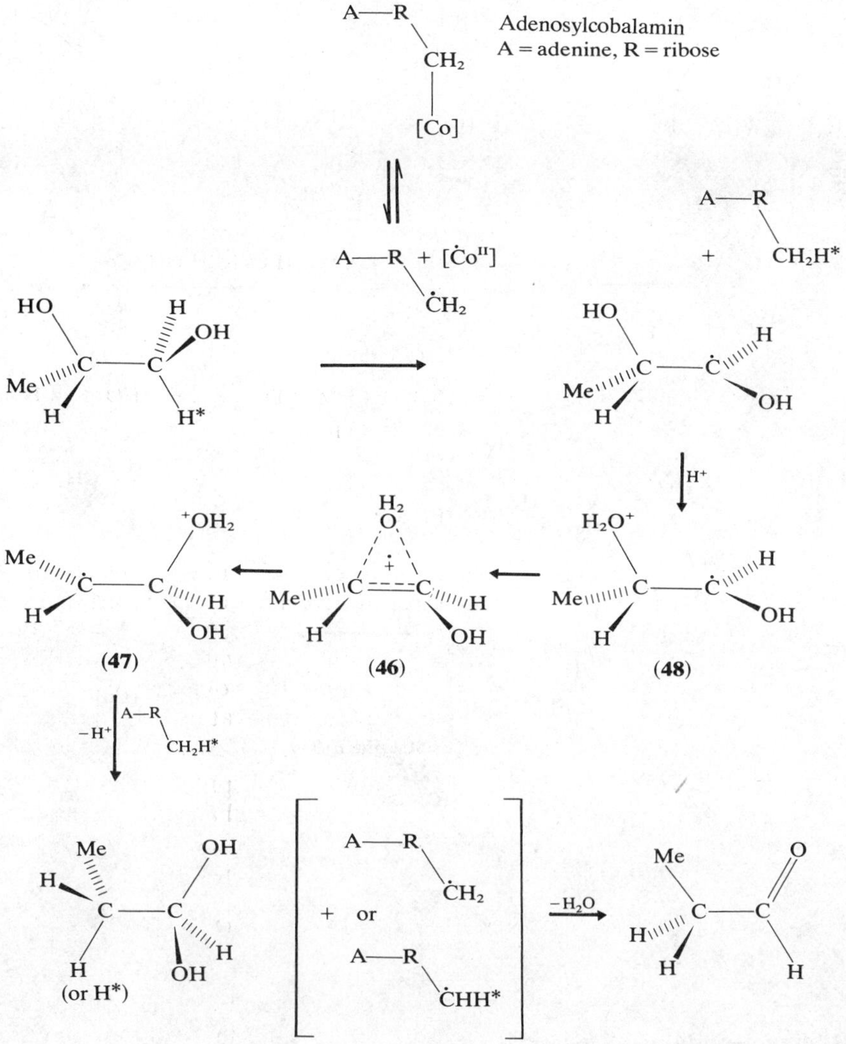

SCHEME 8

Three mechanisms have been proposed for diol dehydrase which do not invoke radical intermediates. After incorrectly asserting that organocobalt species are well founded intermediates in AdoCbl-dependent enzymatic reactions (there is no experimental evidence yet for such intermediates), Corey *et al.*[141] propose a multistep pathway for diol dehydrase, which features a host of alkyl and hydridocobalt intermediates. Salem *et al.*,[29] on the basis of molecular orbital calculations, have suggested a pathway for diol dehydrase which includes the interesting 'bicapped tetrahedron' transition state (**49**). Schrauzer[112c] has proposed mechanisms for diol dehydrase and ethanolamine ammonia lyase which depend on an initial heterolysis of the Co—C bond of AdoCbl to yield Cbl^I and 4′,5′-anhydroadenosine. The Cbl^I then attacks a substrate molecule to produce, for example, 2-hydroxypropylcobalamin from propane-1,2-diol. This intermediate is supposed to decompose to propanal and Cbl^I by an intramolecular 1,2-hydride shift. To explain the transfer of tritium from the 5′-CH_2 of the adenosyl group of AdoCbl to propanal observed by Abeles *et al.* (see above), Schrauzer suggests that this takes place *after* the formation of propanal by an exchange reaction involving 4′,5′-anhydroadenosine. However, the evidence put forward by Schrauzer for his hypothesis, including various model experiments,[142] is unconvincing.

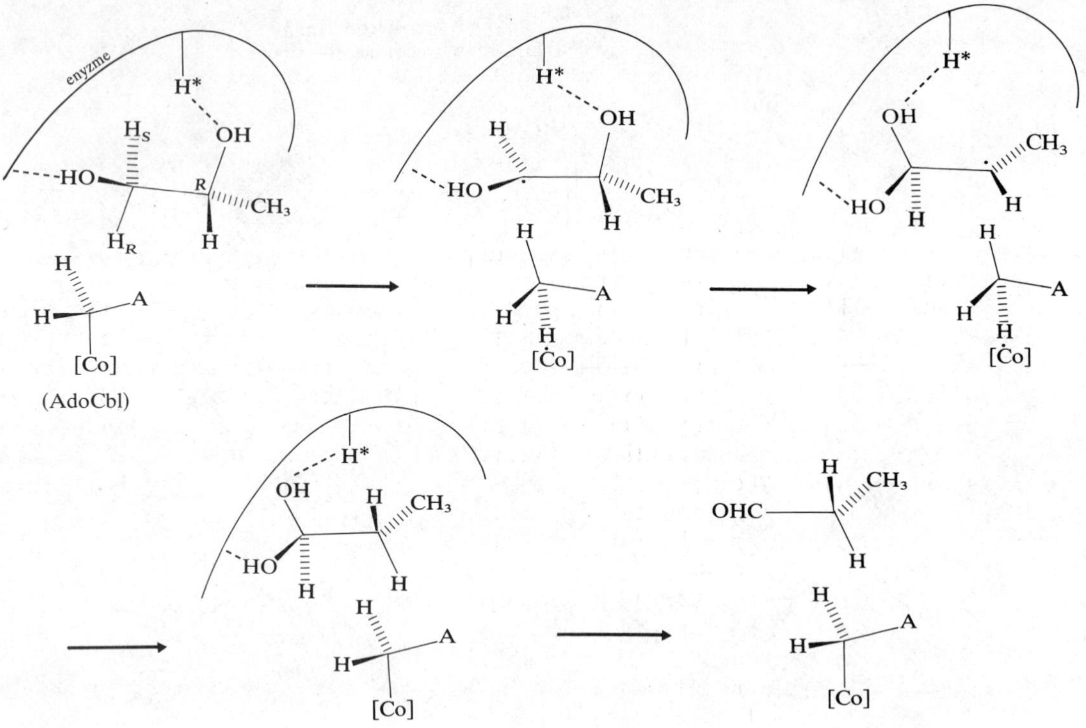

SCHEME 9

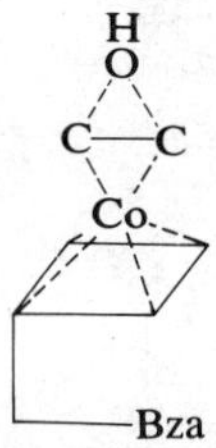

(49)

24.4.5.4 Ribonucleotide reductases[143]

These enzymes catalyse one of the steps (equation 37) in the biosynthesis of DNA and are therefore extremely important to cell growth. AdoCbl-dependent ribonucleotide reductases are produced only by micro-organisms, *e.g. L. leichmannii.* The enzyme from this micro-organism has been purified to homogeneity and found to be monomeric with a molecular weight of 76 000. Ribonucleotide reductases have an allosteric site (or sites) which allows the rate of reduction to be influenced by the product of reduction or by other deoxyribonucleotides (*e.g.* reduction of ATP by *L. leichmannii* enzyme is stimulated by dGTP). This property has been exploited for the large-scale separation of the enzyme from *L. leichmannii* on an affinity column containing dGTP attached to Sepharose.[144]

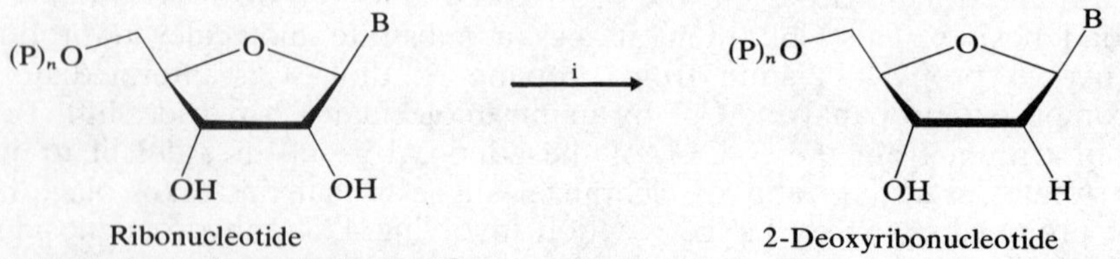

(37)

[B = base, $(P)_n$ = di- or tri-phosphate]

i, Ribonucleotide reductase, thioredoxin $(SH)_2$.

Ribonucleotide reductases catalyse the transfer of a hydrogen atom from dithiol $(SH)_2$ to C-2′ of the ribose ring. Running the reaction of ATP with *L. leichmannii* enzyme in ^2H_2O gave (2′*R*)-[2′-^{2}*H*]-*d*ATP (*i.e.* retention of configuration at C-2′). The dithiol reductant is of course rapidly deuterated ($SH \rightarrow S^2H$) under these conditions. Other labelling studies have demonstrated that during the reaction: (i) water/dithiol H is mixed with 5′-CH_2 protons of AdoCbl; and (ii) oxygen at C-2 is lost, whereas oxygen at C-3 is retained. The involvement of the 5′-CH_2 group of AdoCbl suggests that its Co—C bond is reversibly cleaved during the reaction, as in AdoCbl-dependent rearrangements. That a homolytic cleavage occurs with production of Cbl^{II} and an adenosyl radical is indicated by spectroscopic studies.[145] A mixture of enzyme from *L. leichmannii*, AdoCbl, dihydrolipoate (dithiol), and dGTP in buffer was incubated for 6–1000 ms at 310 K and then was frozen to 130 K. An e.s.r. spectrum taken at 78–85 K shows a complex signal centred on *g* 2.119 tentatively ascribed to Co(II) interacting with an adenosyl radical. When GTP is used instead of dGTP the signal appears after short incubations (10–40 ms) at 310 K, but is much diminished in intensity after longer times. Like dGTP, GTP also binds at an allosteric site and causes an essential conformational change, but, in addition, it reacts with the intermediate radical pair.

For the mechanism of the ribonucleotide reductase reaction, Schrauzer has proposed[112c] a scheme in which an intermediate enzyme-ribosyl cation is attacked by Cbl^I. However, this mechanism does not account for the following: (i) how the hydroxyl group is lost (OH^- is a very poor leaving group; there is evidence against its activation by, say, phosphorylation, although protonation is a possibility difficult to prove or disprove); (ii) the spectroscopic evidence of Blakley *et al.*[145] The author believes that the mechanistic question is still open and that non-enzymatic chemistry relevant to the ribonucleotide reductase reaction has yet to be discovered.

Finally, it is very interesting to note that an organic radical (located at the β-position of a tyrosine of the protein) has been shown to participate in reductions catalysed by the ribonucleotide reductase of *E. coli.* This enzyme is not AdoCbl-dependent, but contains two non-haem iron atoms in an 'antiferromagnetically coupled binuclear high-spin Fe(III) complex'.[146]

24.4.5.5 Intrinsic factor and transcobalamins[147]

Soon after the benefit of liver in the context of pernicious anaemia was discovered (see Section 24.4.1.2), Castle showed that a substance ('intrinsic factor') in gastric secretion, responsible for sequestering cobalamins in food and transporting them across the intestinal wall into the bloodstream, was absent in sufferers from the disease. Therefore, effective treatment of pernicious anaemia requires intramuscular injections (*e.g.* 1 mg dose of OH-Cbl, repeat five times at intervals of 2–3 days, then maintain with 1 mg every 2 months) rather than oral doses of the vitamin as these are not effectively transported to sites of action. Human intrinsic factor, which has recently been purified to homogeneity, is a glycoprotein of molecular weight *ca.* 44 000 and avidly binds cobalamins (for CN-Cbl, $K = 1{,}5 \times 10^{10}\ l\ mol^{-1}$). Transcobalamins (0, I, and II) are proteins which bind to cobalamins in plasma and are responsible for their transport into tissues.[148]

24.4.5.6 Cobalamins in nutrition[3]

Ruminants obtain cobalamins from bacterial flora in their gut. Their diet must contain Co(II) and cases of cobalamin deficiency have been observed, *e.g.* for sheep grazing on cobalt-deficient soil. Humans are totally dependent on receiving cobalamins in food, the daily requirement being *ca.* 3–7 μg (maintaining a cobalamin pool of *ca.* 5 mg). This amount is provided by most mixed diets, but may not be by a vegetarian diet. Cobalamins could not be detected in cereals, nuts, fruit, and vegetables, but are present in dairy products, fish, and meat [*e.g.* (figures in μg per 100 g) 0.36 in milk, 4.69 in salmon, 104 in lambs' liver].

References

1. *Cf.* (a) 'Cobalamin, Biochemistry and Pathophysiology', ed. B. M. Babior, Wiley, New York, 1975, appendix 1 (p. 453) or (b) *Biochemistry*, 1974, **13,** 1555.
2. For a more detailed history of pernicious anaemia and the isolation of cyanocobalamin, see W. B. Castle, 'Introduction' (p. 1) in Ref. 1a.
3. Clinical aspects: I. Chanarin, 'The Megaloblastic Anaemias', Blackwell, Oxford, 1969.
4. E. Lester Smith, 'Vitamin B_{12}', Methuen, London, 1965, chapters 3 and 4.
5. K. Folkers and D. E. Wolf, *Vitamins and Hormones*, 1954, **12,** 1.
6. R. Bonnett, J. R. Cannon, V. M. Clark, A. W. Johnson, L. F. J. Parker, E. Lester Smith, and Sir Alexander Todd, *J. Chem. Soc.*, 1957, 1158.
7. H. A. Barker, in 'Vitamin B_{12} und intrinsic Factor', (2nd European Symposium, Hamburg, 1961), ed. H. C. Heinrich, Ferdinand Enke Verlag, Stuttgart, 1962, p. 82.
8. P. G. Lenhert, *Proc. Roy. Soc.*, 1968, **A303,** 45.
9. J. C. Linnell, chapter 6 (p. 287) in Ref. 1a; E. V. Quadros, D. M. Mathews, I. J. Wise, and J. C. Linnell, *Biochim. Biophys. Acta*, 1976, **421,** 141.
10. (a) J. I. Toohey, in 'Methods in Enzymology', ed. D. B. McCormick and L. D. Wright, Academic Press, London, 1971, vol. XVIII, part C, p. 71; (b) V. B. Koppenhagen and J. J. Pfiffner, *J. Biol. Chem.*, 1971, **246,** 3075.
11. R. D. Fugate, C.-A. Chin and P.-S. Song, *Biochim. Biophys. Acta*, 1976, **421,** 1.
12. E. Nexø and H. Olesen, *Biochim. Biophys. Acta*, 1976, **446,** 143.
13. W. Friedrich, 'Vitamin B_{12} und verwandte Corrinoide', Georg Thieme Verlag, Stuttgart, 1975.
14. P. O'D. Offenhartz, B. H. Offenhartz, and M. M. Fung, *J. Amer. Chem. Soc.*, 1970, **92,** 2966.
15. M. Gardiner and A. J. Thomson, *J.C.S. Dalton*, 1974, 820.
16. H. A. O. Hill, J. M. Pratt, and R. J. P. Williams, in Ref. 10a, p. 5.
17. J. D. Brodie and M. Poe, *Biochemistry*, 1971, **10,** 914; *ibid.*, 1972, **11,** 2534.
18. P. Y. Law, D. G. Brown, E. L. Lien, B. M. Babior, and J. M. Wood, *Biochemistry*, 1971, **10,** 3428.
19. D. Doddrell and A. Allerhand, *Proc. Nat. Acad. Sci. U.S.A.*, 1971, **68,** 1083.
20. T. E. Needham, N. A. Matwiyoff, T. E. Walker, and H. P. C. Hogenkamp, *J. Amer. Chem. Soc.*, 1973, **95,** 5019.
21. H. P. C. Hogenkamp, P. J. Vergamini, and N. A. Matwiyoff, *J.C.S. Dalton*, 1975, 2628.
22. J. H. Bayston, F. D. Looney, J. R. Pilbrow, and M. E. Winfield, *Biochemistry*, 1970, **9,** 2164.
23. J. Halpern, M. S. Chan, J. Hanson, T. S. Roche, and J. A. Topich, *J. Amer. Chem. Soc.*, 1975, **97,** 106.
24. G. N. Schrauzer, *Accounts Chem. Res.*, 1968, **1,** 97; *Angew. Chem. Internat. Edn.*, 1976, **15,** 417.
25. G. Costa, *Pure Appl. Chem.*, 1972, **30,** 335.
26. D. Dodd and M. D. Johnson, *J. Organometallic Chem.*, 1973, **52,** 1.
27. A. Bigotto, E. Zangrando, and L. Randaccio, *J.C.S. Dalton*, 1976, 96.
28. J. D. Brodie, *Proc. Nat. Acad. Sci. U.S.A.*, 1969, **62,** 461.
29. L. Salem, O. Eisenstein, N. T. Anh, H. B. Burgi, A. Devaquet, G. Segal, and A. Veillard, *Nouveau J. Chim.*, 1977, **1,** 335.
30. K. Bernhauer, O. Müller, and F. Wagner, *Angew. Chem. Internat. Edn.*, 1964, **3,** 200.
31. See, for example, D. St. C. Black, V. M. Clark, B. G. Odell, and Lord Todd, *J.C.S. Perkin I*, 1976, 1944.
32. A. Eschenmoser, *Chem. Soc. Rev.*, 1976, **5,** 377.
33. A. W. Johnson, *Chem. Soc. Rev.*, 1975, **4,** 1; A. W. Johnson and W. R. Overend, *J.C.S. Perkin I*, 1972, 2681.
34. Unpublished work by A. L. Begbie, W. R. Bowman, V. M. Clark, J. W. Cornforth, B. T. Golding, and W. P. Watson (*cf.* W. P. Watson, Ph. D. Thesis, University of Warwick, 1974).
35. R. V. Stevens, *Tetrahedron*, 1976, **32,** 1599.
36. R. B. Woodward, *Pure Appl. Chem.*, 1973, **33,** 145.
37. A. Eschenmoser, R. Scheffold, E. Bertele, M. Pesaro, and H. Gschwend, *Proc. Roy. Soc.*, 1965, **288,** 306.
38. A. Eschenmoser, *Quart. Rev.*, 1970, **24,** 366; *Pure Appl. Chem. Suppl.*, 1971, **2,** 69; *Naturwiss.*, 1974, **61,** 513.
39. H. Maag, Diss. Nr. 5173, E.T.H. Zürich, 1973.
40. Y. Yamada, D. Miljkovic, P. Wehrli, B. T. Golding, P. Löliger, R. Keese, K. Müller, and A. Eschenmoser, *Angew. Chem. Internat. Edn.*, 1969, **8,** 343.
41. W. Führer, Diss. Nr. 5158, E.T.H. Zürich, 1973.
42. D. St. C. Black and A. J. Hartshorn, *Austral. J. Chem.*, 1976, **29,** 2271.
43. S. C. Tang and R. H. Holm, *J. Amer. Chem. Soc.*, 1975, **97,** 3359.
44. H. C. Friedmann and L. M. Cagen *Ann. Rev. Microbiol.*, 1970, **24,** 159.
45. H. Ohta and W. S. Beck, *Arch. Biochem. Biophys.*, 1976, **174,** 713.
46. R. C. Bray and D. Shemin, *J. Biol. Chem.*, 1963, **238,** 1501.
47. S. Schwartz, K. Ikeda, I. M. Miller, and C. J. Watson, *Science*, 1959, **129,** 40.
48. A. I. Scott, *Tetrahedron*, 1975, **31,** 2639.
49. C. E. Brown, D. Shemin, and J. J. Katz, *J. Biol. Chem.*, 1973, **248,** 8015.
50. A. R. Battersby, M. Ihara, E. McDonald, J. R. Redfern, and B. T. Golding, *J.C.S. Perkin I*, 1977, 158.
51. A. R. Battersby, E. McDonald, R. Hollenstein, M. Ihara, F. Satoh, and D. C. Williams, *J.C.S. Perkin I*, 1977, 166.

52. A. I. Scott, N. Georgopapadakou, K. S. Ho, S. Klioze, E. Lee, S. L. Lee, G. H. Temme, C. A. Townsend, and I. A. Armitage, *J. Amer. Chem. Soc.*, 1975, **97,** 2548.
53. M. Imfeld, C. A. Townsend, and D. Arigoni, *J.C.S. Chem. Comm.*, 1976, 541; A. I. Scott *et al.*, *ibid.*, p. 544.
54. R. Deeg, H.-P. Kriemler, K.-H. Bergmann, and G. Müller, *Z. physiol. Chem.*, 1977, **358,** 339.
55. A. R. Battersby, E. McDonald, H. R. Morris, M. Thompson, D. C. Williams, V. Ya. Byhovsky, N. I. Zaitseva, and V. N. Bukin, *Tetrahedron Letters*, 1977, 2217.
56. S. H. Ford and H. C. Friedmann, *Biochem. Biophys. Res. Comm.*, 1976, **72,** 1077.
57. G. W. E. Plaut, C. M. Smith, and W. L. Alworth, *Ann. Rev. Biochem.*, 1974, **43,** 899.
58. B. Mailänder and A. Bacher, *J. Biol. Chem.*, 1976, **251,** 3623.
59. W. L. Alworth, M. F. Dove, and H. N. Baker, *Biochemistry*, 1977, **16,** 526.
60. R. Wurm, R. Weyhenmeyer, and P. Renz, *European J. Biochem.*, 1975, **56,** 427.
61. H. C. Friedmann, chapter 2 (p. 75) in Ref. 1a.
62. W. J. Eilbeck, M. S. West, and Y. E. Owen, *J.C.S. Dalton*, 1974, 2205.
63. J. M. Pratt, 'Inorganic Chemistry of Vitamin B_{12}', Academic Press, London, 1972.
64. K. L. Brown, D. Lyles, M. Pencovici, and R. G. Kallen, *J. Amer. Chem. Soc.*, 1975, **97,** 7338.
65. A. Volger, R. Hirschmann, H. Otto, and H. Kunkely, *Ber. Bunsengesellschaft phys. Chem.*, 1976, **80,** 420.
66. H. C. Heinrich, W. Friedrich, and P. Riedel in Ref. 7, p. 244.
67. K. Bernhauer, F. Wagner, H. Michna, P. Rapp, and H. Vogelmann, *Z. phys. Chem.*, 1968, **349,** 1297.
68. R.-H. Yamada and H. P. C. Hogenkamp, *J. Biol. Chem.*, 1972, **247,** 6266; R. H. Allen and P. W. Majerus, *ibid.*, p. 7695.
69. K. Bernhauer, H. Vogelmann, and F. Wagner, *Z. phys. Chem.*, 1968, **349,** 1281.
70. R. Bonnett, J. M. Godfrey, and D. G. Redman, *J. Chem. Soc.* (*C*), 1969, 1163.
71. P. Renz, in Ref. 10a, p. 82.
72. A. Gossauer, K.-P. Heise, H. Laas, and H. H. Inhoffen, *Annalen*, 1976, 1150.
73. R. Keese, *Tetrahedron Letters*, 1969, 149.
74. D. Bormann, A. Fischli, R. Keese, and A. Eschenmoser, *Angew. Chem. Internat. Edn.*, 1967, **6,** 868.
75. D. Jauernig, P. Rapp, and G. Ruoff, *Z. phys. Chem.*, 1973, **354,** 957; P. Rapp, G. Bozler, and E. Fridrich, *ibid.*, p. 970.
76. R. Bonnett, J. M. Godfrey, V. B. Math, E. Edmond, H. Evans, and O. J. R. Hodder, *Nature*, 1971, **229,** 473.
77. P. Rapp and U. Oltersdorf, *Z. phys. Chem.*, 1973, **354,** 32.
78. D. Dolphin, in Ref. 10a, p. 34.
79. G. N. Schrauzer and R. J. Holland, *J. Amer. Chem. Soc.*, 1971, **93,** 1505; *ibid.*, p. 4060.
80. T. M. Kenyhercz, T. P. DeAngelis, B. J. Norris, W. R. Heineman, and H. B. Mark, *J. Amer. Chem. Soc.*, 1976, **98,** 2469; D. Lexa, J. M. Saveant, and J. Zickler, *ibid.*, 1977, **99,** 2786.
81. G. N. Schrauzer, *Inorg. Synth.*, 1968, **11,** 61.
82. T. M. Kenyhercz, A. M. Yacynych, and H. B. Mark, *Analyt. Letters*, 1976, **9,** 203.
83. R. Blackburn, M. Kyaw, and A. J. Swallow, *J.C.S. Faraday I*, 1977, **73,** 250.
84. J. H. Bayston, N. K. King, F. D. Looney, and M. E. Winfield, *J. Amer. Chem. Soc.*, 1969, **91,** 2775.
85. G. N. Schrauzer and E. Deutsch, *J. Amer. Chem. Soc.*, 1969, **91,** 3341.
86. F. R. Jensen, V. Madan, and D. H. Buchanan, *J. Amer. Chem. Soc.*, 1970, **92,** 1414.
87. H. L. Fritz, J. H. Espenson, D. A. Williams, and G. A. Molander, *J. Amer. Chem. Soc.*, 1974, **96,** 2378.
88. R. Breslow and P. L. Khanna, *J. Amer. Chem. Soc.*, 1976, **98,** 1297, 6765.
89. P. W. Schneider, P. F. Phelan, and J. Halpern, *J. Amer. Chem. Soc.*, 1969, **91,** 77.
90. R. Dreos, G. Tauzher, N. Marsich, and G. Costa, *J. Organometallic Chem.*, 1975, **92,** 227.
91. J. M. Wood, *Naturwiss.*, 1975, **62,** 357; R. E. DeSimone, M. W. Penley, L. Charbonneau, S. G. Smith, J. M. Wood, H. A. O. Hill, J. M. Pratt, S. Ridsdale, and R. J. P. Williams, *Biochim. Biophys. Acta*, 1973, **304,** 851.
92. J. H. Espenson, W. R. Bushey, and M. E. Chmielewski, *Inorg. Chem.*, 1975, **14,** 1302.
93. W. M. Scovell, *J. Amer. Chem. Soc.*, 1974, **96,** 3451.
94. J. H. Espenson and T. D. Sellers, *J. Amer. Chem. Soc.*, 1974, **96,** 94.
95. S. N. Anderson, D. H. Ballard, J. Z. Chrzastowski, D. Dodd, and M. D. Johnson *J.C.S. Chem. Comm.*, 1972, 685; P. Abley, E. R. Dockal, and J. Halpern, *J. Amer. Chem. Soc.*, 1972, **94,** 659.
96. I. Levitin, A. L. Sigan, and M. E. Vol'pin, *J.C.S. Chem. Comm.*, 1975, 469.
97. D. Dodd, M. D. Johnson, and B. L. Lockman, *J. Amer. Chem. Soc.*, 1977, **99,** 3664.
98. B. T. Golding, H. L. Holland, U. Horn, and S. Sakrikar, *Angew. Chem. Internat. Edn.*, 1970, **9,** 959.
99. R. B. Silverman and D. Dolphin, *J. Amer. Chem. Soc.*, 1976, **98,** 4626.
100. K. L. Brown, M. M. L. Chu, and L. L. Ingraham, *Biochemistry*, 1976, **15,** 1402.
101. E. A. Parfenov, T. G. Chervyakova, M. G. Edelev, and A. M. Yurkevich, *J. Gen. Chem. USSR*, 1974, **44,** 2319.
102. G. N. Schrauzer, J. H. Weber, and T. M. Beckham, *J. Amer. Chem. Soc.*, 1970, **92,** 7078.
103. H. P. C. Hogenkamp, J. E. Rush, and C. A. Swenson, *J. Biol. Chem.*, 1965, **240,** 3641.
104. B. M. Babior, *J. Biol. Chem.*, 1970, **245,** 6125; R. B. Silverman, D. Dolphin, and B. M. Babior, *J. Amer. Chem. Soc.*, 1972, **94,** 4028.
105. C. Bied-Charreton and A. Gaudemer, *J. Amer. Chem. Soc.*, 1976, **98,** 3997.

106. H. P. C. Hogenkamp, J. N. Ladd, and H. A. Barker, *J. Biol. Chem.*, 1962, **237,** 1950; H. P. C. Hogenkamp, *ibid.*, 1963, **238,** 477; H. P. C. Hogenkamp, *Biochemistry*, 1966, **5,** 417.
107. R. T. Taylor, L. Smucker, M. L. Hanna, and J. Gill, *Arch. Biochem. Biophys.*, 1973, **156,** 521.
108. J. F. Endicott and G. J. Ferraudi, *J. Amer. Chem. Soc.*, 1977, **99,** 243.
109. K. N. Joblin, A. W. Johnson, M. F. Lappert, and B. K. Nicholson, *J.C.S. Chem. Comm.*, 1975, 441.
110. G. N. Schrauzer, J. W. Sibert, and R. J. Windgassen, *J. Amer. Chem. Soc.*, 1968, **90,** 6681; G. N. Schrauzer, L. P. Lee, and J. W. Sibert, *ibid.*, 1970, **92,** 2997.
111. B. T. Golding, T. J. Kemp, P. J. Sellars, and E. Nocchi, *J.C.S. Dalton*, 1977, 1266.
112. Reviewed by (a) B. M. Babior, *Accounts Chem. Res.*, 1975, **8,** 376; (b) R. H. Abeles and D. Dolphin, *ibid.*, 1976, **9,** 114; (c) G. N. Schrauzer, *Angew. Chem. Internat. Edn.*, 1977, **16,** 233; (d) J. M. Poston and T. C. Stadtman, chapter 3 (p. 111) in Ref. 1a; (e) B. M. Babior, chapter 4 (p. 141) in Ref. 1a.
113. J. M. Poston, *Science*, 1977, **195,** 301.
114. R. T. Taylor and H. Weissbach, *Arch. Biochem. Biophys.*, 1969, **129,** 745; R. T. Taylor, *ibid.*, 1971, **144,** 352.
115. G. N. Schrauzer and R. J. Windgassen, *J. Amer. Chem. Soc.*, 1967, **89,** 3607.
116. G. Costa, A. Puxeddu, and E. Reisenhofer, *J.C.S. Dalton*, 1973, 2034.
117. H. Rüdiger, *European J. Biochem.*, 1971, **21,** 264.
118. V. Herbert and K. C. Das, *Vitamins and Hormones*, 1976, **34,** 1.
119. R. M. Halpern, B. C. Halpern, B. R. Clark, H. Ashe, D. N. Hardy, P. Y. Jenkinson, S.-C. Chou, and R. A. Smith, *Proc. Nat. Acad. Sci. U.S.A.*, 1975, **72,** 4018.
120. M. J. Mahoney and L. E. Rosenberg, chapter 8 (p. 369) of Ref. 1a; D. Gompertz in 'The Treatment of Inherited Metabolic Disease', ed. D. N. Raine, MTP, Lancaster, 1975, chapter 8.
121. H. Eggerer, E. R. Stadtmann, P. Overath, and F. Lynen, *Biochem. Z.*, 1960, **333,** 1.
122. R. W. Kellermeyer and H. G. Wood, *Biochemistry*, 1962, **1,** 1124; E. F. Phares, M. V. Long, and S. F. Carson, *Biochem. Biophys. Res. Comm.*, 1962, **8,** 142.
123. B. T. Golding and L. Radom, *J. Amer. Chem. Soc.*, 1976, **98,** 6331.
124. L. L. Ingraham, *Ann. New York Acad. Sci.*, 1964, **112,** 713.
125. J. N. Lowe and L. L. Ingraham, *J. Amer. Chem. Soc.*, 1971, **93,** 3801.
126. W. W. Miller and J. H. Richards, *J. Amer. Chem. Soc.*, 1969, **91,** 1498.
127. J. Retey and B. Zagalak, *Angew. Chem. Internat. Edn.*, 1973, **12,** 671.
128. P. Dowd and M. Shapiro, *J. Amer. Chem. Soc.*, 1976, **98,** 3724.
129. G. Bidlingmaier, U. Kempe, T. Krebs, J. Retey, and H. Flohr, *Angew. Chem. Internat. Edn.*, 1975, **14,** 822.
130. H. Flohr, W. Pannhorst, and J. Retey, *Angew. Chem. Internat. Edn.*, 1976, **15,** 561.
131. A. I. Scott and K. Kang, *J. Amer. Chem. Soc.*, 1977, **99,** 1997.
132. B. Zagalak, J. Retey, and H. Sund, *European J. Biochem.*, 1974, **44,** 529.
133. B. Zagalak, P. A. Frey, G. L. Karabatsos, and R. H. Abeles, *J. Biol. Chem.*, 1966, **241,** 3028.
134. W. W. Bachovchin, R. G. Eagar, K. W. Moore, and J. H. Richards, *Biochemistry*, 1977, **16,** 1082.
135. F. R. Jensen and R. A. Neese, *Biochem. Biophys. Res. Comm.*, 1975, **62,** 816.
136. J. Retey, A. Umani-Ronchi, and D. Arigoni, *Experientia*, 1966, **22,** 72; J. Retey, A. Umani-Ronchi, J. Seibl, and D. Arigoni, *ibid.*, p. 502.
137. T. H. Finlay, J. Valinsky, K. Sato, and R. H. Abeles, *J. Biol. Chem.*, 1972, **247,** 4197.
138. K. L. Schepler, W. R. Dunham, R. H. Sands, J. A. Fee, and R. H. Abeles, *Biochim. Biophys. Acta*, 1975, **397,** 510.
139. K. Sato, J. C. Orr, B. M. Babior, and R. H. Abeles, *J. Biol. Chem.*, 1976, **251,** 3734.
140. B. T. Golding, P. J. Sellars, and C. S. Sell, *J.C.S. Chem. Comm.*, 1976, 773.
141. E. J. Corey, N. J. Cooper, and M. L. H. Green, *Proc. Nat. Acad. Sci. U.S.A.*, 1977, **74,** 811.
142. G. N. Schrauzer and J. W. Sibert, *J. Amer. Chem. Soc.*, 1970, **92,** 1022.
143. Reviewed by H. P. C. Hogenkamp and G. N. Sando, *Structure and Bonding*, 1974, **20,** 23.
144. P. J. Hoffmann and R. L. Blakley, *Biochemistry*, 1975, **14,** 4804.
145. W. H. Orme-Johnson, H. Beinert, and R. L. Blakley, *J. Biol. Chem.*, 1974, **249,** 2338.
146. B.-M. Sjöberg, P. Reichard, A. Gräslund, and E. Ehrenberg, *J. Biol. Chem.*, 1977, **252,** 536.
147. L. Ellenbogen, chapter 5 (p. 215) in Ref. 1a.
148. J. M. England, M. C. Down, I. J. Wise, and J. C. Linnell, *Clinical Science and Molecular Medicine*, 1976, **51,** 47.

PART 25

LIPID CHEMISTRY AND BIOCHEMISTRY

25.1
Fatty Acids

F. D. GUNSTONE
University of St. Andrews

Although there is no generally agreed definition of the term 'lipid', this author prefers to consider lipids as natural derivatives of fatty acids and this structurally based definition is appropriate for a chemical treatise. It also suggests that a description of fatty acids should precede the account of the natural molecules in which they most frequently occur.

Lipids are found increasingly to be physiologically significant and chemically interesting. Renewed activity has arisen from the recognition of fatty acids as essential dietary requirements, from their link with the prostaglandins, and from their involvement in structural membranes. These discoveries followed principally from the development of improved chromatographic and spectroscopic techniques for studying these compounds.

25.1.1 INTRODUCTION

25.1.1.1 Nomenclature

Many fatty acids were given trivial names before their structures were fully known and these are so fully embedded in the literature that their use can hardly be avoided. The trivial names used in this article are listed alphabetically and defined in Table 1.

Systematic names based on IUPAC nomenclature indicate chain length and the position and nature of unsaturated centres and substituents (see, for example, linoleic acid, **1**).

Although not officially recognized, an abbreviated structural representation is widely used. A symbol such as 20:4 5*c*8*c*11*c*14*c* is used to designate the C_{20} acid with four unsaturated centres whose position and configuration is indicated by the symbols 5*c*, *etc.*; *cis* unsaturation (*c*), *trans* unsaturation (*t*), acetylenic unsaturation (*a*), and olefinic unsaturation of unknown configuration or where there is no isomerism (*e*) is indicated by the appropriate letter.

Natural polyene acids are almost entirely *cis* compounds with methylene-interrupted unsaturation, so that the designation of the position of one double bond is sufficient to indicate the remainder. This fact is exploited in an alternative system of numbering where the double bonds are related to the ω-methyl group rather than the carboxyl group and

TABLE 1
Trivial Names of Fatty Acids (in Alphabetical Order)

Trivial name	*Systematic name*	*Symbol*
Calendic	octadeca-8,10,12-trienoic	18:3 8*t*10*t*12*c*
Catalpic	octadeca-9,11,13-trienoic	18:3 9*t*11*t*13*c*
Crepenynic	octadec-9-en-12-ynoic	18:2 9*c*12*a*
Elaidic	octadec-9-enoic	18:1 9*t*
Eleostearic	octadeca-9,11,13-trienoic	18:3 9*c*11*t*13*t*
Erucic	docos-13-enoic	22:1 13*c*
Lauric	dodecanoic	12:0
Linelaidic	octadeca-9,12-dienoic	18:2 9*t*12*t*
Linoleic	octadeca-9,12-dienoic	18:2 9*c*12*c*
α-Linolenic	octadeca-9,12,15-trienoic	18:3 9*c*12*c*15*c*
γ-Linolenic	octadeca-6,9,12-trienoic	18:3 6*c*9*c*12*c*
Malvalic	[see (**14**) (p. 594)]	
Myristic	tetradecanoic	14:0
Oleic	octadec-9-enoic	18:1 9*c*
Palmitic	hexadecanoic	16:0
Parinaric	octadeca-9,11,13,15-tetraenoic	18:4 9*c*11*t*13*t*15*c*
Ricinoleic	12-hydroxyoctadec-9-enoic	12-OH 18:1 9*c*
Stearic	octadecanoic	18:0
Sterculic	[see (**14**) (p. 594)]	
Tariric	octadec-6-ynoic	18: 1 6*a*
Vaccenic	octadec-11-enoic	18:1 11*t*

designated by symbols such as ω6 or *n*-6. These alternative systems are illustrated for linoleic acid (**1**), which may also be represented as (**2**).

(**1**) $CH_3(CH_2)_4CH{=}CHCH_2CH{=}CH(CH_2)_7CO_2H$

(**2**)
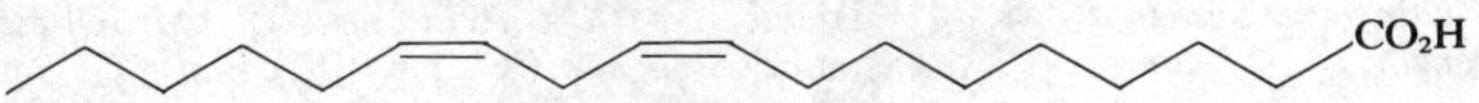

linoleic acid

18:2 9*c*12*c*

18:2ω6

octadeca-*cis*-9,*cis*-12-dienoic acid

25.1.1.2 Major literature sources

In addition to the normal range of chemical and biochemical journals, reports on fatty acids and lipids are to be found in specialized journals such as *J. Amer. Oil Chemists' Soc.*, *Lipids, J. Lipid Res., Biochim. Biophys. Acta* (*Lipids*), *Chem. Phys. Lipids, Fette Seifen Anstrichmittel*, and *Yukagaku* (*J. Japanese Oil Chem. Soc.*), and in the following review series: *Progress in the Chemistry of Fats and Other Lipids* (ed. R. T. Holman *et al.*), *Advances in Lipid Research* (ed. R. Paoletti and D. Kritchevsky), and *Topics in Liquid Chemistry* (ed. F. D. Gunstone) (three vols. only). A listing of books and reviews in the field has been published.[1]

25.1.2 NATURAL FATTY ACIDS: STRUCTURE AND OCCURRENCE

Over 500 natural fatty acids have been identified. A few of these are very common, much studied, and biologically important. On the basis of the 1964–1970 production of commercial oilseeds, Hitchcock and Nichols[2] estimated that the acids produced in largest amount were: oleic (34%), linoleic (34%), palmitic (11%), linolenic (5%), stearic (4%), lauric (4%), erucic (3%), and myristic (2%). A second group of acids, though less common, are well known and fit easily into the known biosynthetic pathways by which fatty acids are metabolized. Yet other acids occur only in restricted sources, have an unidentified biological significance, and for the most part have not been extensively studied. Nevertheless, many of these are of interest and taken together they serve to delineate the several features of fatty acid structure which can be combined in so many ways. Most newly discovered acids represent an unusual conjunction of familiar structural features; only rarely is a new structural unit discovered.

The following generalizations are based on the structure of the natural acids. To each statement, however, there are interesting and important exceptions.

(i) Most natural acids, whether saturated or unsaturated, are straight-chain compounds with an even number of carbon atoms; odd acids, branched-chain acids, and cyclic acids also exist.

(ii) Although the range of chain length is great (C_2 to $>C_{80}$), the most common chain-lengths are C_{16}, C_{18}, C_{20}, and C_{22}.

(iii) Mono-unsaturated acids usually contain a *cis* olefinic bond in a limited number of preferred positions in the carbon chain; *trans*-alkenoic and alkynoic acids are also known.

(iv) Most polyunsaturated acids have 2–6 *cis* double bonds in a methylene-interrupted pattern; acids with conjugated unsaturation also exist.

(v) Substituted acids are uncommon but hydroxy-, oxo-, epoxy-, and fluoro-containing acids are known. Only the branched-chain, cyclic, and substituted acids are potentially optically active.[3]

25.1.2.1 Saturated acids

Table 2 contains information on a range of saturated acids. The lower members (C_4–C_{10}) occur mainly in milk fats, those of intermediate chain length (C_8–C_{14}) are present

TABLE 2
n-Saturated Acids and their Methyl Esters

Chain length	*Systematic name*	*Trivial name*	*Acid*		*Methyl ester*	
			m.p. (°C)	*b.p.*[a] (°C)	*m.p.* (°C)	*b.p.*[a] (°C)
1	Methanoic	Formic	8.4	101	—	32
2	Ethanoic	Acetic	16.6	118	—	57
3	Propanoic	Propionic	−20.8	141	—	80
4	Butanoic	Butyric	−5.3	164	—	103
5	Pentanoic	Valeric	−34.5	186	−80.7	127
6	Hexanoic	Caproic	−3.2	206	−69.6	151
7	Heptanoic	Enanthic	−7.5	223	−55.7	174
8	Octanoic	Caprylic	16.5	240	−36.7	195
9	Nonanoic	Pelargonic	12.5	256	−34.3	214
10	Decanoic	Capric	31.6	271	−12.8	228
11	Hendecanoic	—	29.3	284	−11.3	250
12	Dodecanoic	Lauric	44.8	130[1]	+5.1	262
13	Tridecanoic	—	41.8	140[1]	+5.8	—
14	Tetradecanoic	Myristic	54.4	149[1]	19.1	114[1]
15	Pentadecanoic	—	52.5	158[1]	19.1	127[1]
16	Hexadecanoic	Palmitic	62.9	167[1]	30.7	136[1]
17	Heptadecanoic	Margaric	61.3	175[1]	29.7	148[1]
18	Octadecanoic	Stearic	70.1	184[1]	37.8	156[1]
19	Nonadecanoic	—	69.4	—	38.5	191[4]
20	Eicosanoic	Arachidic	76.1	204[1]	46.4	188[2]
21	Heneicosanoic	—	75.2	—	—	207[4]
22	Docosanoic	Behenic	80.0	—	51.8	206[2]
23	Tricosanoic	—	79.6	—	53.9	—
24	Tetracosanoic	Lignoceric	84.2	—	57.4	222[2]
25	Pentacosanoic	—	83.5	—	59.5	—
26	Hexacosanoic	Cerotic	87.8	—	63.5	237[2]
27	Heptacosanoic	—	87.6	—	64.6	—
28	Octacosanoic	Montanic	90.9	—	67.5	—
29	Nonacosanoic	—	90.4	—	68.8	—
30	Triacontanoic	Melissic	93.6	—	71.5	—

[a] B.p. at 760 mmHg or at 1, 2, or 4 mmHg, as indicated by superscript.

in the seed fats of the Lauraceae and Myristicaceae (hence lauric and myristic acid), whilst palmitic acid (the most common of all saturated acids) and stearic acid are present in most animal and vegetable fats.

25.1.2.2 Mono-unsaturated acids

Of more than 100 monoenoic acids, oleic acid (**3**) is the most common and has been extensively studied as the prototype of this class. It is conveniently isolated from oleic-rich sources such as olive oil.

(**3**) $CH_3(CH_2)_7CH\overset{cis}{=}CH(CH_2)_7CO_2H$ 18:1 9*c*

In the acids of this class the double bond usually has a *cis* configuration and occurs in certain preferred positions, reflecting their mode of biosynthesis. The most common desaturation process is effected by an oxygen-dependent Δ^9-desaturase so that the double bond is found in the Δ^9-position or in a related position following chain-extension or chain-shortening. Oleic acid, for example, is probably the precursor of the 20:1 11*c*, 22:1 13*c*, 24:1 15*c*, 26:1 17, 28:1 19*c*, and 30:1 21*c* monoene acids, in all of which the double bond is in a fixed position (ω9) relative to the methyl group. The Δ^9 C_{16}, C_{17}, and C_{19} acids can also act as precursors of other series of related monoene acids. There must also be a Δ^6-desaturase and a Δ^5-desaturase since monoenes with double bonds in these positions are well known.[4]

The following are a selection from the monoene acids (>100) now known (see also Sections 25.1.2.8 and 25.1.2.9): 14:1 9*c* (myristoleic); 16:1 3*t*, 6*t*, 9*c* (palmitoleic), 11*c*; 17:1 9*c*; 18:1 3*t*, 5*c*, 5*t*, 6*c* (petroselinic), 6*t* (petroselaidic), 9*c* (oleic), 9*t* (elaidic), 11*c* (*cis*-vaccenic), 11*t* (vaccenic); 19:1 9*c*; 20:1 5*c*, 9*c*, 11*c*; 22:1 5*c*, 11*c*, 13*c* (erucic); and 24:1 15*c* (nervonic).

25.1.2.3 Methylene-interrupted polyene acids

The common and important polyunsaturated acids have 2–6 *cis* double bonds arranged in a methylene-interrupted pattern and are typified by acids such as linoleic (**1**) and docosahexaenoic acid (**4**).

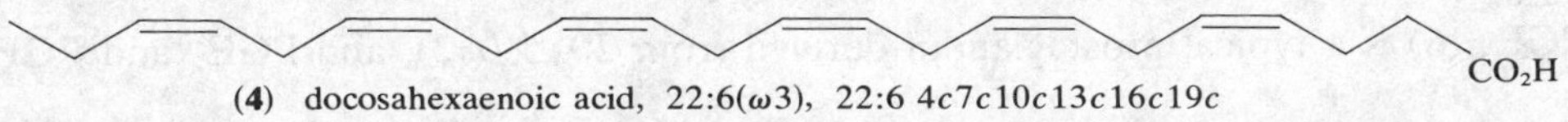

(**4**) docosahexaenoic acid, 22:6(ω3), 22:6 4*c*7*c*10*c*13*c*16*c*19*c*

The 100 or so acids of this type are mainly of even chain-length (especially C_{16}–C_{22}).[5] They are metabolically related by desaturation and chain-extension procedures (Section 25.1.5.4) and are grouped according to their ω classification. The most important are ω9 acids derived from oleic, ω6 acids from linoleic, ω3 acids from α-linolenic, and ω7 acids from hexadec-9-enoic (Table 3).

Linoleic and α-linolenic acids are present in vegetable oils and cannot be produced by animals. α-Linolenic acid is the major acid of linseed oil and linoleic is found in high proportions in seed oils such as sunflower, soya, corn, and safflower.[6] The C_{20} and C_{22} polyene acids are usually produced in animals from dietary sources of 16:1, 18:1, 18:2, and 18:3. In land animals the polyene acids are found mainly in the polar lipids; in fish they are present in neutral and polar lipids.

Some less common polyene acids, such as 18:3 3*t*9*c*12*c* and 20:3 5*c*11*c*14*c*, display the familiar methylene-interrupted pattern with an additional double bond in the Δ^3- or Δ^5-positions. This is not always in a methylene-interrupted arrangement and may be *cis* or *trans*. Acids with Δ^7-unsaturation (*e.g.* 22:2 7,11 and 22:2 7,13) are probably derived from Δ^5-acids by chain-extension.

TABLE 3
Typical Methylene-interrupted Polyene Acids (all Unsaturated Centres have the *cis* Configuration)

ω9 series	*ω3 series*
18:2 6,9	18:3 9,12,15[e]
20:2 8,11	18:4 6,9,12,15
20:3 5,8,11	20:4 8,11,14,17
22:3 7,10,13	20:5 5,8,11,14,17
22:4 4,7,10,13	22:5 7,10,13,16,19
	22:6 4,7,10,13,16,19
ω6 series	*ω7 series*
18:2 9,12[a]	16:2 6,9
18:3 6,9,12[b]	18:2 8,11
20:3 8,11,14[c]	18:3 5,8,11
20:4 5,8,11,14[d]	20:3 7,10,13
22:4 7,10,13,16	20:4 4,7,10,13
22:5 4,7,10,13,16	

Trivial names: [a] linoleic, [b] γ-linolenic, [c] dihomo-γ-linolenic, [d] arachidonic, [e] α-linolenic.

Long-chain acids discovered in mycobacteria and typified by the phleic acid (**5**) are unusual in that they display 1,5-unsaturation.

(**5**) $CH_3(CH_2)_{14}(CH{=}CHCH_2CH_2)_5CO_2H$ 36:5 4,8,12,16,20

25.1.2.4 Essential fatty acids, prostaglandins, thromboxanes

It has been known since 1929 (Burr and Burr) that linoleic acid (and possibly also α-linolenic acid) is a dietary requirement for the healthy growth of animals and such acids have been designated 'essential fatty acids'. The reason for dietary dependence is not fully known, but these C_{18} acids and/or the higher polyene acids produced from them by animals are involved in the production of structurally important membranes and of the prostaglandins and thromboxanes. First discovered in semen, the prostaglandins are widely distributed in animal tissues and exhibit a wide range of pharmacological activities.[7]

PGE_3 (**6**) is a typical prostaglandin derived from 20:5 (ω3), and PGE_2 and PGE_1 are

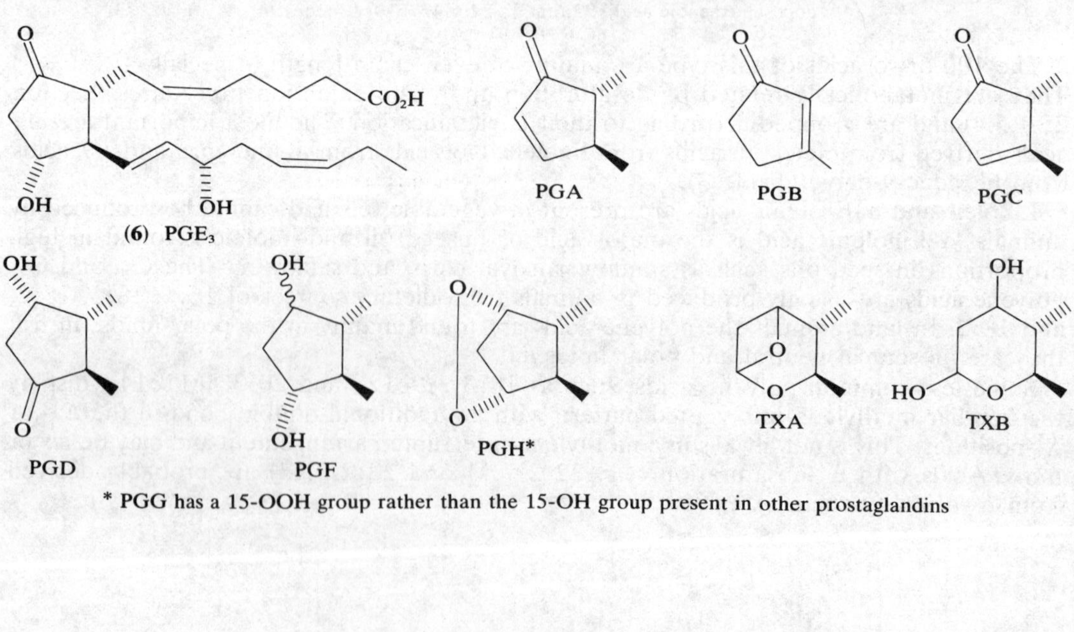

* PGG has a 15-OOH group rather than the 15-OH group present in other prostaglandins

similarly derived from 20:4 (ω6) and 20:3 (ω6), respectively. Other prostaglandins differ only in the nature of the five-membered carbocyclic ring and each of these exists in the 1, 2, and 3 series. The thromboxanes are a group of related compounds in which the cyclopentane ring is replaced by a tetrahydropyran system.

25.1.2.5 Conjugated polyenoic acids[8]

Some plants produce acids with conjugated unsaturation. These are mainly C_{18} acids with 2 to 4 double bonds, of which α-eleostearic acid (18:3 9*c*11*t*13*t*) is the best known since it is the major component in tung oil. Other C_{18} acids with conjugated unsaturation are the 8*c*10*t*12*c* (jacaric), 8*t*10*t*12*c* (calendic), 9*c*11*t*13*c* (punicic), and 9*t*11*t*13*c* (catalpic) isomers, and the 9*c*11*t*13*t*15*c* (α-parinaric) tetraene acid. Some acetylenic acids (Section 25.1.2.6) and some oxygenated acids (Section 25.1.2.7) also have conjugated unsaturation.

25.1.2.6 Acetylenic and allenic acids[3,8,9]

Excluding the oxygenated acids (Section 25.1.2.9), the acetylenic acids fall into four categories. (i) Monoacetylenic acids, such as 18:1 6*a* (tariric), are known but rare. (ii) Acetylenic analogues of the better known methylene-interrupted polyene acids such as 18:2 9*c*12*a* (crepenynic), 18:3 6*a*9*c*12*c*, 18:4 6*a*9*c*12*c*15*c*, and 20:3 8*a*11*c*14*c*. (iii) A range of highly unsaturated C_{18} and C_{17} acids with conjugated unsaturation, including 17:2 8*a*10*t*, 17:3 8*a*10*t*17*e*, 18:2 9*a*11*t* (ximenynic or santalbic), 18:3 9*a*11*a*13*c*, and 18:5 9*a*11*a*13*a*15*c*17*e*. (iv) A group of highly unsaturated C_9–C_{17} compounds, not always acids and seldom occurring in lipid combination, which are present in plants and in micro-organisms. Mycomycin (**7**) is a typical C_{13} member of this group

$$\text{(7)}\quad HC{\equiv}CC{\equiv}CCH{=}C{=}CHCH\overset{c}{=}CHCH\overset{t}{=}CHCH_2CO_2H$$

Among natural allenic acids such as 14:3 2*t*4*e*5*e*, 18:2 5*e*6*e*, and 18:3 5*e*6*e*16*t*, the C_8 hydroxy-acid (**8**) is an interesting member. It occurs in *Sapium sebiferum* glycerides associated with a C_{10} acid and the two together have possibly been produced by modification of a more typical C_{18} acid (9*c*12*c* or 6*c*9*c*12*c* or 6*a*9*c*12*c*).

$$\text{(8)}\quad CH_3(CH_2)_4CH\overset{c}{=}CHCH\overset{t}{=}CHCOOCH_2CH{=}C{=}CH(CH_2)_3COOCH_2CH(OCOR^1)CH_2OCOR^2$$

25.1.2.7 Branched-chain acids[10]

The important branched-chain acids contain one or more methyl groups and are generally optically active.[3] Polybranched isoprenoid compounds based on phytol are excluded from this review even though they occur in lipid combination.[11]

(i) Iso-acids [$CH_3CH(Me)(CH_2)_nCO_2H$] and anteiso-acids [$CH_3CH_2CH(Me)(CH_2)_nCO_2H$] are mainly even and odd, respectively, and the latter are generally *S*-(+). These are present in a number of animal fats, especially wool wax, and arise by a simple variant of normal fatty acid biosynthesis. They frequently occur also as monohydroxy-derivatives.

(ii) Tuberculostearic acid [*R*-(−)-10-methylstearic acid] from tubercle bacilli is typical of acids having a single mid-chain substituent. 7-Methylhexadec-7-enoic acid has been identified in sperm whale oil.

(iii) Polymethyl-substituted acids occur in the preen glands of birds (*e.g.* 2,6-dimethyldecanoic acid), in the fats of sheep fed on a barley-rich diet (*e.g.* 4,8-dimethyltetradecanoic acid), and in a number of bacterial lipids (*e.g.* 2,4,6,8-tetramethyloctacosanoic acid). In all cases the methyl group is believed to result from the replacement of one or more molecules of acetate (or malonate) with propionate (or methylmalonate) in the biosynthetic process.[12]

(iv) Mycolic and related acids (**9, 10**) are compounds of higher molecular weight, frequently containing one or more oxygenated functions and cyclopropane rings, and of bacterial origin.

(**9**) $CH_3(CH_2)_x\overset{CH_2}{\widehat{CHCH}}(CH_2)_y\overset{CH_2}{\widehat{CHCH}}(CH_2)_zCH(OH)CH(R)CO_2H$

mycolic acids $x = 17$, $y = 15$, $z = 16$, $R = C_{22}H_{45}$ or $C_{24}H_{49}$

(**10**) $CH_3(CH_2)_mCH(OH)CH(R)CO_2H$

corynomycolic acids $m = 14$, $R = C_{14}H_{29}$

nocardic acids $m = 30\text{–}34$, $R = C_8H_{15}\text{–}C_{12}H_{25}$

25.1.2.8 Cyclic acids

Natural fatty acids having a three-, five-, or six-membered alicyclic ring are known.

The cyclohexane derivatives (**11**), present in butter fat and in certain micro-organisms living in hot springs, are probably derived from a shikimic acid derivative by chain elongation.

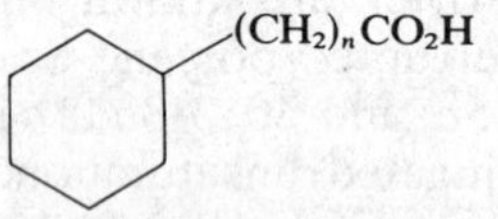

(**11**) $n = 10$ or 12

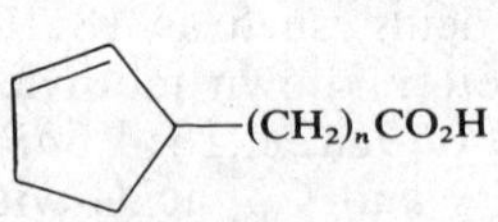

(**12**)

C_{16}: Δ^4, Δ^6, Δ^9
C_{18}: Δ^6 Δ^9
C_{20}: Δ^8 Δ^9

In addition to the prostaglandins (Section 25.1.2.4), which are disubstituted cyclopentanes, a series of monosubstituted cyclopentenes occur in the seed oils of the Flacourtiaceae. These exist as C_6–C_{20} acids (**12**; $n = 0\text{–}14$) and as C_{16}–C_{20} acids with additional unsaturation.

Natural cyclopropane acids are mainly C_{17} and C_{19} acids (**13**) present in bacterial lipids, where they accompany the C_{16} and C_{18} monoene acids from which they are formed. Many mycolic acids (Section 25.1.2.7) are also cyclopropane derivatives.[10,13]

$CH_3(CH_2)_5\overset{CH_2}{\widehat{CHCH}}(CH_2)_nCO_2H$ $\quad$ $CH_3(CH_2)_7\overset{CH_2}{\widehat{C{=}C}}(CH_2)_nCO_2H$

(**13**) $n = 7$ or 9 $\quad$ (**14**)

Cyclopropene acids are found in the seed fats of the Malvaceae, Bombacaceae, and Tiliaceae.[13] The most common are sterculic (**14**; $n = 7$) and malvalic (**14**; $n = 6$).

25.1.2.9 Oxygenated acids

The structure and chemistry of natural oxygenated acids have been extensively reviewed.[9,14,15,17,18]

(i) Hydroxy-acids are found in brain lipids, wool wax, milk lipids, cutins, plants, and micro-organisms. Thus brain lipids contain the odd and even C_{14}–C_{26} α-hydroxy acids and wool wax has α- and ω-hydroxy-acids of the n-, iso-, and anteiso-series. Hydroxy-acids of plant origin include ricinoleic acid (**15**; $n = 17$), lesquerolic acid (**15**; $n = 9$), dimorphecolic acid (**16**), and aleuritic acid (**17**).

(**15**) $CH_3(CH_2)_5CH(OH)CH_2CH\overset{c}{=}CH(CH_2)_nCO_2H$

(**16**) $CH_3(CH_2)_4CH{=}CHCH{=}CHCH(OH)(CH_2)_7CO_2H$

(**17**) $HOCH_2(CH_2)_5CH(OH)CH(OH)(CH_2)_7CO_2H$

(ii) Oxo-acids are relatively uncommon, though over 60 have been recognized as minor components of milk fat. Both eleostearic acid and parinaric acid also exist as 4-oxo-derivatives, and oxomycolic acids have been recognized.

(iii) Natural epoxy-acids of lipid origin are almost entirely confined to C_{18} acids derived from plants. The best known is vernolic acid (**18**).

(18) $CH_3(CH_2)_4\overset{O}{\widehat{CHCH}}CH_2CH{=}CH(CH_2)_7CO_2H$ (*cis*, *cis*)

(iv) Furan-containing acids (**19**) occur in fish lipids.

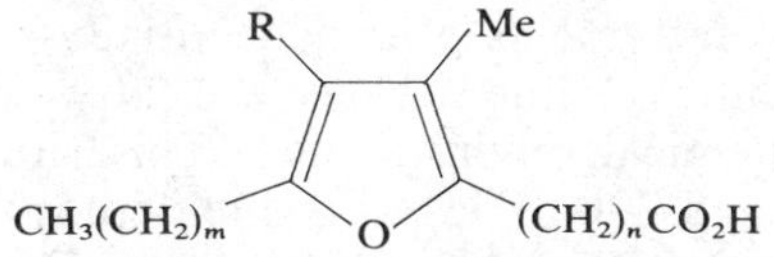

(**19**) R = H or Me, $m = 2$ or 4, $n = 8$, 10, or 12

25.1.3 STRUCTURE DETERMINATION

This section provides a brief account of the more important methods in current use for determining the structure of a fatty acid.[19]

Fatty acids occur as mixtures and the presence of a component of unusual structure is often apparent through unexpected behaviour during thin-layer or gas–liquid chromatographic analysis. Its chromatographic behaviour may indicate chain-length, degree of unsaturation, and the presence or absence of branching or of additional polar groups. The procedure thereafter depends on the amount of acid available, on the nature of the spectroscopic equipment which can be called on, and on the ability of the experimenter to interpret this information! Sometimes the structure can be completely recognized by spectroscopic techniques with a choice between ultraviolet, infrared, Raman, n.m.r., and mass spectrometry. On other occasions it is desirable to hydrogenate the acid first and to recognize the carbon skeleton of the perhydro-acid before tackling the problems related to unsaturation and other functional groups. Classical degradation procedures are now so refined that, with the assistance of g.l.c. to identify the products, these can be satisfactorily effected on a few milligrams of material or less, and can even be applied to mixtures of isomeric acids on a quantitative or semiquantitative basis. Special difficulties which may arise with polyunsaturated compounds can be solved using partial reaction techniques.

25.1.3.1 Gas–liquid chromatography[20]

The influence of unsaturated centres and other functional groups on g.l.c. behaviour and the secondary effects which depend on the position of the olefinic or acetylenic group have been so well documented that a good deal of structural information can be derived merely from retention data on one or more stationary phases. Natural mixtures containing only the more conventional acids are usually identified by this procedure alone, and whilst this may be satisfactory for major components most of the time, there is always the chance that compounds of unusual structure, present in trace amounts or in higher proportions, will be wrongly identified.

25.1.3.2 Ultraviolet, infrared, and Raman spectroscopy*

Ultraviolet spectroscopy is mainly of diagnostic value with acids having conjugated

* See also Sections 25.1.6.1 and 25.1.6.2.

unsaturation, since monoene and non-conjugated polyunsaturation show no useful absorption.[8]

Infrared spectra are used mainly to detect *trans* unsaturation in monoene and in non-conjugated and conjugated polyenes. Infrared spectra are of little value for studying *cis* olefinic or acetylenic unsaturation, but Raman spectroscopy can be used to detect *cis* and *trans* olefinic and acetylenic unsaturation.

25.1.3.3 Nuclear magnetic resonance spectroscopy

The ^{1}H n.m.r. spectra of fatty acids obtained initially with low-resolution instruments (40 and 60 MHz) were of limited value because the fatty acid molecules contain too many protons in very similar chemical environments. This situation has changed with more powerful instruments (100 and 220 MHz), with the deployment of shift reagents, and with the application of the more sensitive ^{13}C spectra (see Section 25.1.6.3).

25.1.3.4 Mass spectrometry

Mass spectrometry is increasingly important for the structural identification of fatty acids. Combined with the separating efficiency of gas chromatography, it provides one of the most powerful methods of identification since it requires only very small amounts of material. The tandem procedure is even more efficient when linked to a computer (see Section 25.1.6.4).

25.1.3.5 Oxidative fission coupled with partial hydrogenation

Alkyne unsaturation can usually be distinguished from alkene unsaturation by gas–liquid chromatography or by n.m.r. or Raman spectroscopy, and *cis* and *trans* alkenes are usually distinguished by infrared, Raman, or ^{13}C n.m.r. spectroscopy. The position of unsaturated centres, however, is most frequently determined by oxidative degradation followed by recognition of the fragments by g.l.c.

Oxidative degradation is effected by von Rudloff oxidation ($KMnO_4$–KIO_4) to short-chain acids or by ozonolysis. The latter procedure may give alcohols, aldehydes, or acids, but formation of aldehydes or acids is more common (Scheme 1).

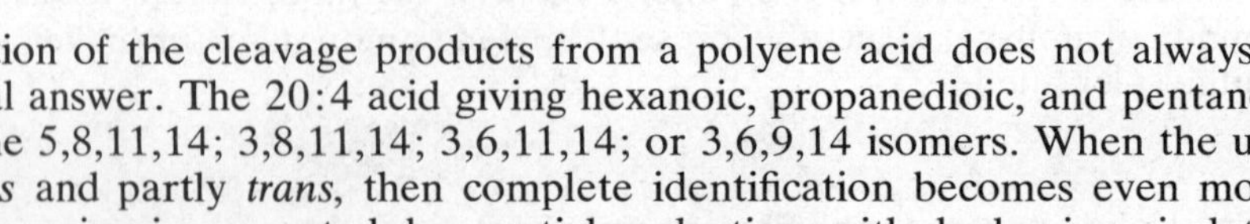

i, $LiAlH_4$ or H_2, Pd/C; ii, Ph_3P or $(NC)_2C{=}C(CN)_2$ or Me_2S or H_2, Lindlar's catalyst, or pyrolysis; iii, Ag_2O or $KMnO_4$ or H_2O_2.

SCHEME 1

Recognition of the cleavage products from a polyene acid does not always lead to an unequivocal answer. The 20:4 acid giving hexanoic, propanedioic, and pentanedioic acids could be the 5,8,11,14; 3,8,11,14; 3,6,11,14; or 3,6,9,14 isomers. When the unsaturation is partly *cis* and partly *trans*, then complete identification becomes even more difficult. This problem is circumvented by partial reduction with hydrazine, isolation of the monoene fraction (and if necessary the *cis* and *trans* monoenes separately) by silver ion chromatography, and oxidation. The following example (equation 1) indicates how this would lead to an unambiguous structure for the 20:4 5*t*8*c*11*c*14*c* acid:

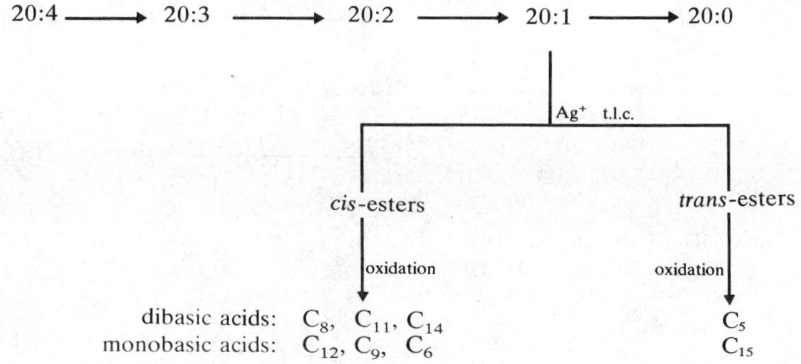

(1)

This problem has also been tackled by partial oxymercuration (Sections 25.1.6.4 and 25.1.11.2).

25.1.3.6 Identification of functional groups

Although not common, acids with branched methyl, cyclic, oxo-, hydroxy-, or epoxy-groups are known. Many of these functions can be recognized and placed by a combination of chromatographic and spectroscopic procedures, but chemical degradation is still useful on occasions.

Epoxides are cleaved by periodic acid to aldehydes. The position of a hydroxyl group can be determined by oxidation to the ketone, hydrogenation to the saturated oxo-ester (if necessary), preparation of its oxime, Beckmann rearrangement to two amides, and hydrolysis to an amine, an amino-acid, and a mono- and di-basic acid (equation 2), followed by identification of one or more of these products.

$$R^1COR^2 \longrightarrow R^1C(:NOH)R^2 \longrightarrow \begin{matrix} R^1CONHR^2 \\ R^1NHCOR^2 \end{matrix} \longrightarrow \begin{matrix} R^1CO_2H + H_2NR^2 \\ R^1NH_2 + HO_2CR^2 \end{matrix} \qquad (2)$$

$$R^1 = CH_3(CH_2)_n, \quad R^2 = (CH_2)_mCO_2Me \text{ or } (CH_2)_mCO_2H$$

25.1.4 FATTY ACID SYNTHESIS[21,22]

The availability of suitable synthetic procedures and a growing need for fatty acids not easily isolated from natural mixtures, or not known to occur naturally, and for labelled compounds required for biochemical studies, has led to extensive synthetic activity. The main demand is for unsaturated acids. The acetylenic route has been the most successful for a wide range of polyene acids, especially the important all-*cis* isomers, and the Wittig reaction has proved to be of value for acids with different types of unsaturation. No other method is widely used except that standard methods of chain-extension by one (*via* the nitrile), two (malonic ester synthesis), or a larger number of carbon atoms (enamine synthesis, anodic synthesis) are available for use when appropriate.

25.1.4.1 The acetylenic route to *cis* mono- and poly-enoic acids

The success of this approach depends on the ease with which acetylenic compounds can be alkylated and on the ability to reduce poly-ynes to the all-*cis* polyenes.

Mono-ynoic acids are prepared by alkylation of acetylene with an alkyl halide (or its equivalent) and an α,ω-chloroiodoalkane. The conversion of chloride to carboxyl *via* the cyanide or by malonic ester allows the introduction of a labelled carbon atom (Scheme 2).

$$HC{\equiv}CH \xrightarrow{i} RC{\equiv}CH \xrightarrow{ii} RC{\equiv}C(CH_2)_nCl$$

$$RC{\equiv}C(CH_2)_nCl \xrightarrow{iii} RC{\equiv}C(CH_2)_nCO_2Me$$

$$RC{\equiv}C(CH_2)_nCl \xrightarrow{iv} RC{\equiv}C(CH_2)_{n+1}CO_2Me$$

i, $NaNH_2$, RBr; ii, $NaNH_2$, $I(CH_2)_nCl$; iii, KCN; MeOH, HCl; iv, $CH_2(CO_2Et)_2$; MeOH, HCl.

SCHEME 2

The poly-ynoic acids required as intermediates to produce natural polyene acids result from condensation of an alkyne (as its Grignard complex) and a propargyl compound in the presence of cuprous salts (equation 3). Careful thought must be given to the selection

$$R^1C{\equiv}CCH_2X + BrMgC{\equiv}CR^2 \xrightarrow{Cu^I,\ THF} R^1C{\equiv}CCH_2C{\equiv}CR^2 \quad (3)$$

of those intermediates which give the best overall yield. Kunau claims that, under optimum conditions, pentaenes and heptaenes can be produced in about 20% and 10% yield, respectively. In one route to a pentaynoic acid (**24**) the alkynol (**20**) is combined first with hexadiynol (**21**) and then with the alkadiynoic acid (**22**) (Scheme 3).

(a) $HC{\equiv}CCH_2OH \xrightarrow{i} HC{\equiv}CCH_2Othp^* \xrightarrow{ii} CH_3(CH_2)_mC{\equiv}CCH_2OH$ (**20**)

$HC{\equiv}CCH_2Othp^* \xrightarrow{iii} HC{\equiv}CCH_2C{\equiv}CCH_2OH$ (**21**)

(b) $HC{\equiv}C(CH_2)_nCH_2OH \xrightarrow{i} HC{\equiv}C(CH_2)_nCH_2Othp^* \xrightarrow{iii} HC{\equiv}CCH_2C{\equiv}C(CH_2)_nCH_2OH$

$\xrightarrow{iv} HC{\equiv}CCH_2C{\equiv}C(CH_2)_nCO_2H$ (**22**)

(c) (**20**) $\longrightarrow$ (**20′**) $\xrightarrow{v} CH_3(CH_2)_mC{\equiv}CCH_2C{\equiv}CCH_2C{\equiv}CCH_2OH$ (**23**) $\longrightarrow$ (**23′**)

$\xrightarrow{vi} CH_3(CH_2)_mC{\equiv}CCH_2C{\equiv}CCH_2C{\equiv}CCH_2C{\equiv}CCH_2C{\equiv}C(CH_2)_nCO_2H$ (**24**)

i, dihydropyran, H^+; ii, EtMgBr, $CH_3(CH_2)_mBr$, H_3O^+; iii, EtMgBr, HC—CCH_2X (X= I, OMs, OTs), H_3O^+; iv, CrO_3, $COMe_2$; v, $BrMgC{\equiv}CCH_2C{\equiv}CCH_2Othp$ (*cf.* **21**), H_3O^+; vi, $BrMgC{\equiv}CCH_2C{\equiv}C(CH_2)_nCO_2MgBr$ (*cf.* **22**); (**20′**) and (**23′**) are the iodide, mesylate or tosylate of (**20**) and (**23**), respectively.

*thp = tetrahydropyranyl

SCHEME 3

The poly-yne acids are crystalline solids which can be preserved at low temperatures and can (must) be purified by crystallization prior to partial reduction with hydrogen and Lindlar's catalyst (palladium on calcium carbonate containing some lead) in the presence of quinoline to enhance selectively. The polyene is finally purified from compounds which have been over-reduced or contain *trans* double bonds by silver ion chromatography.

25.1.4.2 The Wittig reaction

The Wittig reaction (4) is a means of condensing aldehydes or ketones with a derivative of an alkyl halide to produce an alkene which usually has the *trans* configuration except that in the presence of Lewis bases such as Br^- or I^- almost pure *cis*-alkenes are

$$R^1CHO + BrCH_2R^2 \longrightarrow R^1CH{=}CHR^2 \qquad (4)$$

produced. The alkyl halide is the source of an alkylidenephosphorane ($R^2CH{=}PPh_3$) which can be replaced by phosphine oxides $R^2CH_2P(O)R^3_2$ or phosphonates $R^2CH_2PO(OEt)_2$, a modification which has been extensively employed in prostaglandin synthesis.

The more conventional Wittig reaction is illustrated in the synthesis of calendic and catalpic acids (Scheme 4) and of crepenynic and linoleic acids (Scheme 5). These reactions are easily adapted to give labelled compounds.

$$CH_3(CH_2)_mC{\equiv}CCH\overset{t}{=}CHCH_2OH \xrightarrow{\text{i}} CH_3(CH_2)_mC{\equiv}CCH{=}CHCH{=}PPh_3$$

$$\xrightarrow{\text{ii}} CH_3(CH_2)_mCH\overset{c}{=}CHCH\overset{t}{=}CHCH\overset{t}{=}CH(CH_2)_nCO_2Me$$

calendic acid $n = 6$, $m = 4$

catalpic acid $n = 7$, $m = 3$

i; PBr_3; Ph_3P; NaOMe; ii, $OHC(CH_2)_nCO_2Me$; H_2, Lindlar's catalyst.

SCHEME 4

$$CH_3(CH_2)_4C{\equiv}CCH_2CH_2Br \xrightarrow{\text{i}} CH_3(CH_2)_4C{\equiv}CCH_2CH{=}PPh_3 \xrightarrow{\text{ii}}$$

$$CH_3(CH_2)_4C{\equiv}CCH_2CH\overset{c}{=}CH(CH_2)_7CO_2Me \xrightarrow{\text{iii}} CH_3(CH_2)_4CH\overset{c}{=}CHCH_2CH\overset{c}{=}CH(CH_2)_7CO_2Me$$

i, Ph_3P; base; ii, $OHC(CH_2)_7CO_2Me$; iii, H_2, Lindlar's catalyst.

SCHEME 5

25.1.4.3 Prostaglandin synthesis

The potential pharmacological importance of the natural prostaglandins and of related unnatural compounds has given rise in recent years to a considerable synthetic programme. The Harvard synthesis of Corey and his colleagues (Scheme 6) has been the basis of many modified routes. In this synthesis, cyclopentadiene is converted to the bicyclic iodolactone (**25**) which has the required stereochemistry and two potential aldehyde functions for Wittig condensation with C_7 and C_5 compounds, furnishing the diene (**26**); this yields PGE_2 and $PGF_{2\alpha}$ and can be reduced to a monoene, yielding PGE_1 and $PGF_{1\alpha}$.

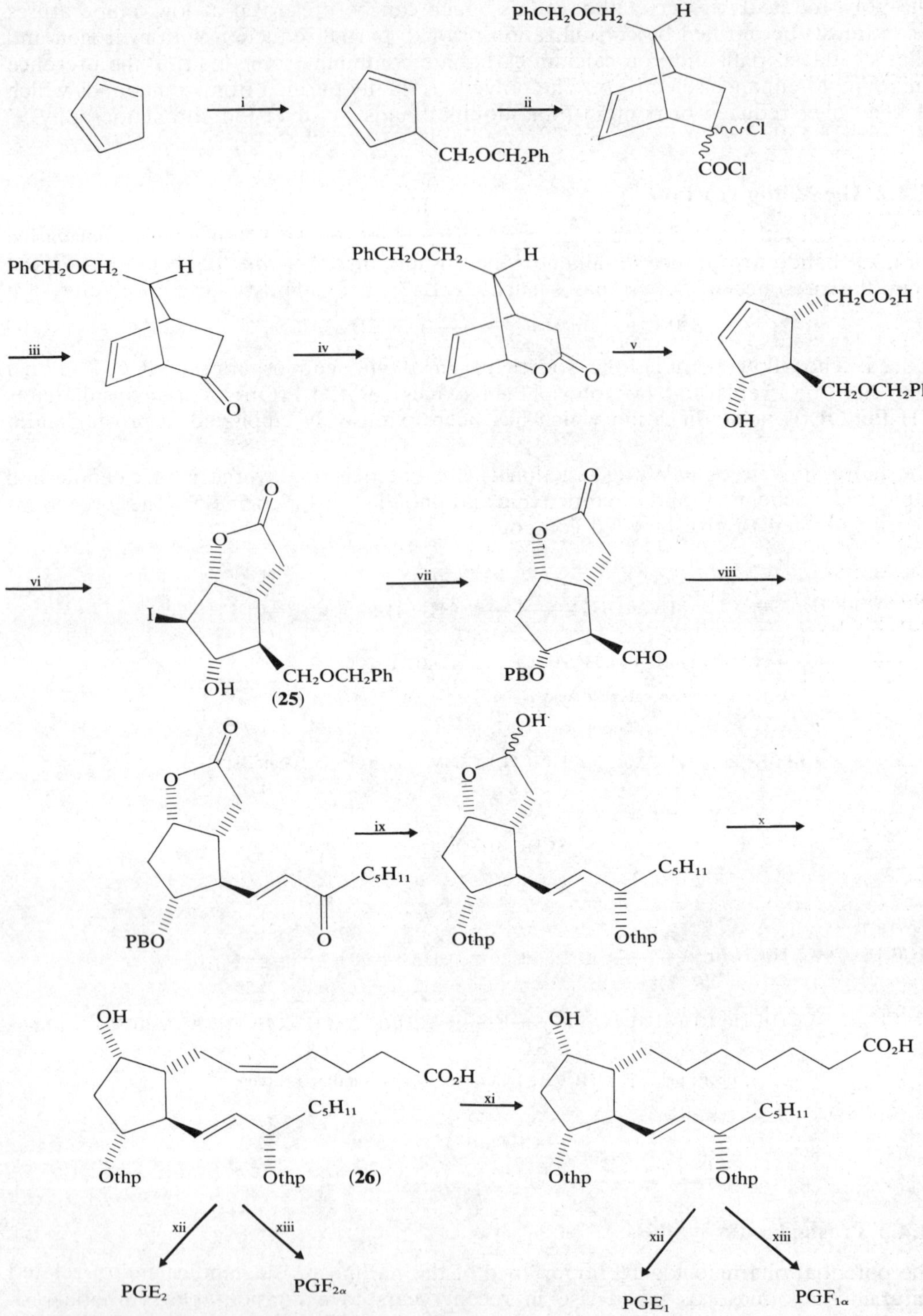

i, $Tl(OEt)_3$; $PhCH_2OCH_2Cl$; ii, $CH_2{=}C(Cl)COCl$; iii, NaN_3; heat, H^+; iv, m-$ClC_6H_4CO_3H$; v, NaOH; vi, KI, $NaHCO_3$; vii, p-$C_6H_5C_6H_4COCl$ (PBCl); Bu_3SnH; H_2, Pd/C, H^+; CrO_3, C_5H_5N; viii, Na^+ $(MeO)_2P(O)\bar{C}HCOC_5H_{11}$; ix, $LiBR_3H$; K_2CO_3; dihydropyran (thp = tetrahydropyranyl); Bu^i_2AlH; x, $Ph_3P{=}CH(CH_2)_3CO_2Na$; xi, H_2, Pd/C at −15 to −20 °C; xii, CrO_3; H^+; xiii, H^+.

SCHEME 6

25.1.5 FATTY ACID BIOSYNTHESIS

Because of the importance of fatty acid biosynthesis in plant and animal systems, extensive study has been undertaken and the basic biosynthetic procedures are now understood. Attention will be directed here mainly to the chemical nature of the various intermediates rather than to the enzymes. In some of the reactions to be discussed the true nature of the substrate is not fully known and reference to acids such as acetate or palmitate will imply the acid in an appropriate form, *i.e.* as a free acid or associated with a coenzyme or enzyme or even incorporated into a lipid. Most commonly the acids react as thiol esters and are frequently written as RCOSCoA or RCOSACP. These represent the acid RCO_2H attached to the thiol group of the phosphopantothiene moiety of coenzyme A or acyl carrier protein.

This account of fatty acid biosynthesis will be divided into the following stages: *de novo* synthesis of saturated acids, chain-elongation of saturated or unsaturated acids, formation of monoenes, and further desaturation to give polyenes. The subject has been extensively reviewed.[23–26]

25.1.5.1 *De novo* synthesis of palmitic acid

The first stage of fatty acid biosynthesis is usually the *de novo* synthesis of palmitic acid (16:0) from acetate (8 moles). Not all the acetate units are identical: one enters into the condensation reaction in this form but the remainder react in the chain-extension process as malonate. The additional carboxyl group is lost during condensation so that all 16 carbon atoms in palmitate are acetate-derived (equation 5).

$$\begin{array}{ccc} \text{acetate} + 7\ \text{malonate} & \longrightarrow & \text{palmitate} + 7CO_2 \\ \uparrow & & \\ 7\ \text{acetate} & & \end{array} \qquad (5)$$

The conversion of acetate to malonate is catalysed by a biotin-containing enzyme (acetyl-CoA carboxylase) and the remainder of the process is catalysed by a group of enzymes (fatty acid synthetase) which exist in animal tissues or yeast cells as a multienzyme complex or in bacterial plant tissues in a readily dissociable form. The synthetase contains (*a*) a transacylase to transfer acetyl and malonyl groups to appropriate thiol functions in acyl carrier protein (ACP); (*b*) a condensing enzyme to combine acetyl and malonyl units into 3-oxobutanoate (acetoacetate); (*c*) a reductase which, along with NADPH, furnishes 3-*R*-(−)-hydroxybutanoate; (*d*) a dehydrase furnishing but-*trans*-2-enoate; and (*e*) another reductase which with NADPH or NADH gives butanoate. At this stage the first cycle is complete (Scheme 7) and one acetate unit has been combined with a second acetate in the form of malonate to give butanoate. By repetition of this cycle the chain length is increased to the C_6, C_8, C_{10}, C_{12}, C_{14}, and C_{16} acids. Another enzyme then controls the removal of this acid so that the major product of *de novo* synthesis is palmitic acid. Sometimes this is accompanied by stearic acid and there must be plant systems where shorter-chain acids such as lauric (12:0) and myristic (14:0) are predominant.

This widely occurring biosynthetic pathway sometimes appears in a modified form. One unusual modification leads to monoenoic acids (Section 25.1.5.3). It is also possible for the primer molecule of acetate to be replaced by other acids which are chain-extended with malonate in the usual way to give different end-products. Thus propanoate (heptadecanoate), 2-methylpropanoate (16-methylheptadecanoate), and 2-methylbutanoate (14-methylhexadecanoate) give the products indicated in parenthesis. The acetate units which react as malonate can, in exceptional circumstances, be replaced by propanoate (or

(i) formation of malonyl-CoA

$$ATP + HCO_3^- + \text{biotin-enzyme} \xrightleftharpoons{Mn^{2+}} ADP + P_i + CO_2\text{-biotin-enzyme}$$

$$CO_2\text{-biotin-enzyme} + CH_3COSCoA \rightleftharpoons \text{biotin-enzyme} + HO_2CCH_2COSCoA$$

(ii) transfer to acyl carrier protein (ACPSH)

$$CH_3COSCoA + ACPSH \rightleftharpoons CH_3COSACP + CoASH$$

$$HO_2CCH_2COSCoA + ACPSH \rightleftharpoons HO_2CCH_2COSACP + CoASH$$

(iii) condensation

$$CH_3COSACP + HO_2CCH_2COSACP \rightleftharpoons CH_3COCH_2COSACP + CO_2 + ACPSH$$

(iv) reduction

$$CH_3COCH_2COSACP + NADPH + H^+ \rightleftharpoons CH_3CH(OH)CH_2COSACP + NADP$$

(v) dehydration

$$CH_3CH(OH)CH_2COSACP \rightleftharpoons CH_3CH\overset{t}{=}CHCOSACP + H_2O$$

(vi) hydrogenation

$$CH_3CH{=}CHCOSACP + NADPH + H^+ \rightleftharpoons CH_3CH_2CH_2COSACP + NADP$$

SCHEME 7 Biosynthesis of saturated acids by the *de novo* pathway

methylmalonate) and the branched-chain acids produced in lambs fed on a barley-rich diet may be produced in this way (Section 25.1.2.7; see also Section 25.1.5.2).

25.1.5.2 Chain-elongation

A different elongation process proceeding through similar intermediates converts the acid RCO_2H to its bis-homologue $RCH_2CH_2CO_2H$ and is effective on both saturated and unsaturated substrates (Sections 25.1.5.3 and 25.1.5.4). It may require acetate (mainly a mitochondrial process) or malonate (mainly a microsomal process) and involves coenzyme A derivatives throughout the whole cycle. The acetate or malonate can also be replaced by propanoate or methylmalonate and it is likely that some of the polymethyl branched acids are formed in this way (equation 6).

$$\begin{array}{l} 20{:}0 \longrightarrow \text{2-methyl } 22{:}0 \longrightarrow \text{2,4-dimethyl } 24{:}0 \longrightarrow \\ \qquad \text{2,4,6-trimethyl } 26{:}0 \longrightarrow \text{2,4,6,8-tetramethyl } 28{:}0 \end{array} \tag{6}$$

25.1.5.3 Monoene biosynthesis

Monoenes are produced in two ways. The common procedure is oxygen dependent; the less-common route is employed only by those lower forms of life that have an anaerobic existence.

Bacterial lipids rarely contain polyene acids. In addition to monoenes, they have cyclopropane acids derived from the monoenes. Bacteria living under anaerobic conditions produce their monoene acids by a modification of the *de novo* pathway. One step in the usual route to palmitic acid is the conversion of 3-hydroxydecanoic acid to its Δ^{2t}-analogue, which is reduced to decanoic acid. Anaerobic bacteria contain a special dehydrase operating best on the C_{10} hydroxy-acid and, less efficiently, on the C_8 and C_{12} acids. This produces the Δ^{2t}-monoene which equilibrates with the 3*c*-isomer and, while the former is reduced as usual and finally yields long-chain saturated acids, the 3*c*-isomer escapes reduction and the double bond remains to furnish 16:1 and 18:1 acids (equation 7).

$$\begin{array}{l} 2{:}0 \longrightarrow \longrightarrow \text{3-OH } 10{:}0 \longrightarrow 10{:}1\ 2t \longrightarrow \longrightarrow \longrightarrow \longrightarrow 16{:}0 \longrightarrow 18{:}0 \\ \qquad\qquad\qquad\qquad\qquad\qquad\quad \Updownarrow \\ \qquad\qquad\qquad\qquad\qquad\quad 10{:}1\ 3c \longrightarrow 12{:}1\ 5c \longrightarrow 14{:}1\ 7c \longrightarrow \\ \qquad\qquad\qquad\qquad\qquad\qquad\qquad 16{:}1\ 9c \longrightarrow 18{:}1\ 11c \end{array} \tag{7}$$

The more important route to monoenes occurring in plants, animals, and micro-organisms requires oxygen and NADH (or NADPH). The most common desaturase produces a *cis* double bond at the Δ^9-position by specific removal of the 9-*pro-R* and 10-*pro-R* hydrogen atoms. Reaction can occur with both CoA and ACP derivatives and possibly in some phospholipids. This desaturase produces the common Δ^9-acids of varying chain length, in addition to the acids derived from Δ^9-precursors by chain elongation (Section 25.1.5.2).

25.1.5.4 Polyene biosynthesis

The polyene acids present in plants and animals are important as precursors of prostaglandins and as constituents of membrane lipids. Plants convert monoenes to polyenes by further desaturation in the distal end of the molecule (between the existing double bond and the ω-methyl group) and only rarely in the proximal end (between the existing double bond and the carboxyl group). In contrast, animals insert additional double bonds in the proximal unit of monoenes and of plant-derived polyenes, but never in the distal unit (Scheme 7a).

$$\overbrace{CH_3CH_2CH_2CH_2CH_2CH_2CH_2CH_2}^{\text{distal}}CH{=}CH\overbrace{CH_2CH_2CH_2CH_2CH_2CH_2CH_2CO_2H}^{\text{proximal}}$$

	distal	proximal
plants:	commonly	rarely
animals:	never	commonly

SCHEME 7a

oleic 18:1 (9*c*)	—animals→	ω9 polyenes
↓ plants		
linoleic 18:2 (9*c*12*c*)	—animals→	ω6 polyenes
↓ plants		
linolenic 18:3 (9*c*12*c*15*c*)	—animals→	ω3 polyenes

SCHEME 7b

Plants convert oleate to linoleate and linolenate. These three C_{18} acids are the most common in the plant kingdom and serve as precursors for other unsaturated acids (Scheme 7b). An alternative route to linolenate (equation 8) operates in the chloroplast. The 12:3 acid will be the 3*c*6*c*9*c*-isomer.

$$12{:}0 \longrightarrow 12{:}1 \longrightarrow 12{:}2 \longrightarrow 12{:}3 \longrightarrow 14{:}3 \longrightarrow 16{:}3 \longrightarrow 18{:}3 \tag{8}$$

Animals, unable to insert double bonds in the distal end of the molecule, produce polyene acids (see Table 3) by desaturation and chain-elongation of oleate (ω9-acids) or hexadec-9-enoate (ω7-acids), which may be endogenous or exogenous, or of linoleate (ω6-acids) or linolenate (ω3-acids), which must be obtained from dietary plant sources.

In addition to the Δ^9-desaturase, which serves mainly to convert saturated acids to monoenes, there may be only three other common desaturases (Δ^6, Δ^5, Δ^4). The Δ^6-desaturase operates most effectively on Δ^9-substrates (these substrates may have additional double bonds in the distal unit). The formation of the Δ^6-acids is detailed in Scheme 8. Acids appearing in a horizontal line are linked by elongation and chain-shortening processes, but passage to the next horizontal line is limited to substrates which can be desaturated by a Δ^6-, Δ^5-, or Δ^4-desaturase. Linoleic acid is converted most often to arachidonic acid (20:4 ω6) and oleic (20:3 ω9) and linolenic acids (20:5 and 22:6 ω3) furnish mainly the polyene acids indicated in parenthesis. The reversal of the desaturation process is less fully studied, but the existence of a 4-enoylreductase required in the equilibria 22:5 (ω6)⇌22:4 (ω6)⇌20:4 (ω6) has been demonstrated.

18:2 (9, 12) ⇌ 20:2 (11, 14) ⇌ 22:2 (13, 16)

↓ i

18:3 (6, 9, 12) ⇌ 20:3 (8, 11, 14) ⇌ 22:3 (10, 13, 16)

↓ ii

20:4 (5, 8, 11, 14) ⇌ 22:4 (7, 10, 13, 16)

iv ↑↓ iii

22:5 (4, 7, 10, 13, 16)

i, Δ^6 desaturase; ii, Δ^5 desaturase; iii, Δ^4 desaturase; iv, 4-enoylreductase.

SCHEME 8 ω6 Polyene acids

25.1.5.5 Prostaglandins and thromboxanes

These pharmacologically significant compounds are C_{20} acids (Section 25.1.2.4) produced from natural polyene acids. The 20:3 (ω6), 20:4 (ω6), and 20:5 (ω3) acids furnish compounds of the PG_1, PG_2, and PG_3 series, respectively.

The minimum structural requirement appears to be an ω6,9,12 triene system, preferably in a C_{20} acid, and biosynthesis occurs through endoperoxides (Scheme 9) which are the precursors of both prostaglandins and thromboxanes.

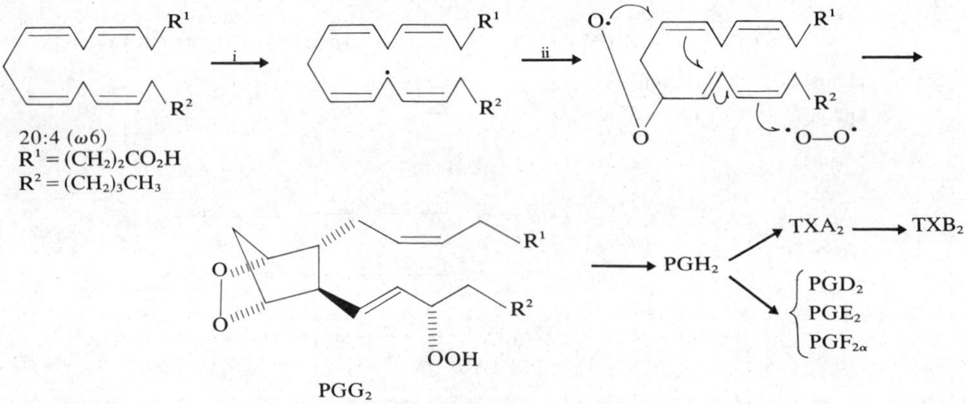

i, prostaglandin synthetase; ii, O_2.

SCHEME 9 Prostaglandin biosynthesis

25.1.6 SPECTROSCOPIC PROPERTIES

25.1.6.1 Ultraviolet spectroscopy[8,27]

Monoene and methylene-interrupted polyene acids absorb ultraviolet light at wavelengths too low for convenient study. Ultraviolet spectroscopy is, however, invaluable in the study of acids having conjugated unsaturation and of reactions producing such acids (Sections 25.1.8 and 25.1.11). Absorption data for some typical chromophores are given in Table 4.

TABLE 4
Ultraviolet Absorption of some Chromophores present in Natural Acids

Chromophore	*Acid*	*Solvent*	λ_{max} (nm) (log ε_{max})
Diene	18:2 9*c*11*c*	ethanol	235 (4.41)
	18:2 9*c*11*t*	ethanol	232
	18:2 9*t*11*t*	ethanol	230 (4.55)
Dienoic acid	10:2 2*t*4*c*	ethanol	260 (4.38)
Oxo-diene	9-oxo 18:2 10*t*12*t*	methanol	275 (4.44)
		cyclohexane	267 (4.44)
Enyne	18:2 9*a*11*c*		227 (4.15)
	18:3 9*c*12*a*14*c*	methanol	227, 235 (infl)
	18:2 9*a*11*t*	ethanol	229 (4.22) 240 (infl)
Triene	18:3 9*c*11*t*13*t*	cyclohexane	262, 272 (4.69), 283
	18:3 9*c*11*c*13*t*		261, 270, 280
	18:3 9*t*11*t*13*t*	ethanol	259, 268 (4.78), 279
	18:3 8*c*10*t*12*c*	cyclohexane	265, 275, 287
Tetraene	18:4 9*c*11*t*13*t*15*c*	ethanol	291.5 (4.67), 305 (4.86) 320 (4.83)
	18:4 9*t*11*t*13*t*15*t*	methanol	286 (4.77), 299 (4.97), 313 (4.94)
Dienyne	18:3 9*a*11*t*13*t*	ethanol	266.5 (4.59), 277 (4.49)

25.1.6.2 Infrared and Raman spectroscopy[8,27–30]

Solution spectra have been used to recognize functional groups generally, and *trans* double bonds in particular, while solid-state spectra have provided valuable information on structural questions and on matters of polymorphism, chain-packing, conformation, *etc.*

Oils and fats containing only the usual mixture of saturated and unsaturated acids have very similar infrared spectra with peaks at 1380 (CH_3), 1720 and 2915–2950 (CH_2), 1700–1715 (CO_2H), and 1180–1260 and 1740 cm^{-1} (CO_2R). In the presence of unusual structural features there may be additional absorption bands associated with hydroxyl (3450), keto (1725), cyclopropene (1850 and 1010), epoxide (850 and 825), allene (2220 and 1960), vinyl (990 and 910), and conjugated enyne (950 cm^{-1}).[31]

Infrared spectroscopy is most commonly used to examine acids suspected of containing *trans* unsaturation. One such bond produces a characteristic absorption at 968 cm^{-1}. The effect is roughly additive for non-conjugated polyene acids, so that linelaidic acid absorbs at the same position as elaidic acid but with increased intensity. Conjugated polyenes with one or more *trans* bonds show a shift in the position of absorption and, sometimes, additional peaks. Thus: *tt*, 998; *ct*, 985 and 950; *ttt*, 994; *ttc*, 993 and 965; *ctc*, 988 and 937; *tttt*, 997; and *cttc* 993 and 952 cm^{-1}.

In the crystalline state, i.r. spectra of long-chain acids show a series of absorption bands of uniform spacing and intensity between 1335 and 1175 cm^{-1} which result from interaction between the rocking and/or twisting vibrations of the CH_2 groups. The number of such bands is $n/2$ for even acids and $(n+1)/2$ for odd acids, where n is the number of carbon atoms in the acid.[30]

Although i.r. spectra are used to recognize *trans* unsaturation by the band at 968 cm^{-1} arising from out-of-plane bending of the olefinic hydrogens, the bands at 1657 (*cis*) and 1673 (*trans*) resulting from double bond vibration are hardly visible because of the local symmetry around the double bond. This does not apply to Raman spectra, which show strong absorption bands for *cis* (1656 ± 1), *trans* (1670 ± 1), and acetylenic unsaturation (2232 ± 1 and $2291 \pm 2\ cm^{-1}$). These values differ slightly when unsaturation is conjugated with the carboxyl group or is in a terminal position.[32]

25.1.6.3 Nuclear magnetic resonance spectroscopy[33–35]

The ^1H n.m.r. spectrum of a saturated ester such as methyl stearate shows signals for the ω-CH_3 group (δ 0.89), the ester methyl group (δ 3.65), the α-CH_2 function (δ 2.21), and the remaining CH_2 groups (δ 1.26). A more powerful instrument will also show a separate signal for the β-CH_2 group (δ 1.58), and signals for other CH_2 groups close to the ester function can be obtained with shift reagents.

In addition to these signals an olefinic ester such as methyl oleate will show a triplet for two olefinic protons (δ 5–6) and a signal for allylic protons (δ 1.99). *cis*- and *trans*-alkenes differ in the coupling constant for the vinyl protons (*cis* ~10 Hz and *trans* ~15 Hz) and in the chemical shift of the allylic protons (*cis*, δ 1.99; *trans*, δ 1.94). Methylene-interrupted polyene esters show a signal at δ 2.72 for the CH_2 group lying between two *cis* double bonds.

Smaller deshielding effects operate in an additive manner up to six atomic centres from the source of the deshielding influence, and small differences in the chemical shifts between isomeric compounds allow structural identification in many cases. Among mono-unsaturated C_{18} esters it is possible by 220 MHz spectroscopy to distinguish all the monoynoic isomers and all the *cis*- and *trans*-monoenoic esters except the Δ^{10}-, Δ^{11}-, and Δ^{12}-compounds, which can, however, be distinguished with the assistance of shift reagents.

Table 5 lists the chemical shifts of some of the more significant functional groups (when not influenced by any other group) and the deshielding effects of these groups along an alkyl chain. This information can be used to calculate the chemical shifts expected of protons in a wide range of fatty acids.

TABLE 5
^1H N.M.R. Spectra[a]

	Chemical shift[b]	Deshielding effect					
		α	β	γ	δ	ε	ζ
CO_2H	—	1.035	0.360	0.095	0.055	0.030	0.005
CO_2CH_3	3.65	0.955	0.320	0.060	0.035	0.020	0.005
CH_3	0.885	0.030	0.000	0.000	0.000	0.000	0.000
—CH=CH— (*c*)	5–6	0.730	0.065	0.020	0.015	0.005	0.000
—CH=CH— (*t*)	5–6	0.685	0.050	0.005	0.000	0.000	0.000
—C≡C—	—	0.820	0.160	0.130	0.040	0.025	0.015
—CH=CHC*H*$_2$CH=CH— (*c, c*)	2.72	0.770	0.075	0.025	0.025	0.010	0.000
—CHCH— (O bridged) (*c*)	2.70[c]	0.170	0.190	0.100	0.055	0.030	0.015
—CHCH— (O bridged) (*t*)	2.45[d]	0.170	0.150	0.075	0.035	0.020	0.010
—CHCH— (CH_2 bridged) (*t*)	0.2[e]	−0.100	0.075	0.020	0.000	0.000	0.000

[a] Chemical shifts (δ, p.p.m. downfield from $SiMe_4$) for the major functional groups in fatty acids; deshielding effects of functional groups along an alkyl chain (in p.p.m. in CCl_4 solution). Basic chemical shift for CH_2 group is 1.255. Slightly different deshielding effects are observed for a CH_3 group. D. J. Frost and F. D. Gunstone, *Chem. Phys. Lipids*, 1975, **15,** 53; D. J. Frost, Ph.D. Thesis, University of Amsterdam, cited in part by F. D. Gunstone and H. R. Schuler, *Chem. Phys. Lipids*, 1975, **15,** 189.

[b] Also —C*H*(OH)—, 3.42; —C*H*(OOH)—, 3.6; and $RCOOCH_2CH(OCOR)CH_2OCOR$, —C*H*$_2$O—, 4.2; >C*H*O—, 5.2.

[c] $J = 2.5$–5.0. [d] $J = 0.5$–2.5. [e] *cis* isomer, −0.3 (1H) and +0.6 (3H).

The greater sensitivity of ^{13}C n.m.r. spectra to chemical environment make these more informative than ^{1}H spectra despite the low abundance of ^{13}C in unenriched samples. A saturated methyl ester shows seven clearly separated signals along with a complex signal in the 29–30 p.p.m. range in which it is possible to distinguish five more signals. These arise from the deshielding influence of the CO_2CH_3 and CH_3 groups and have been allocated as follows:

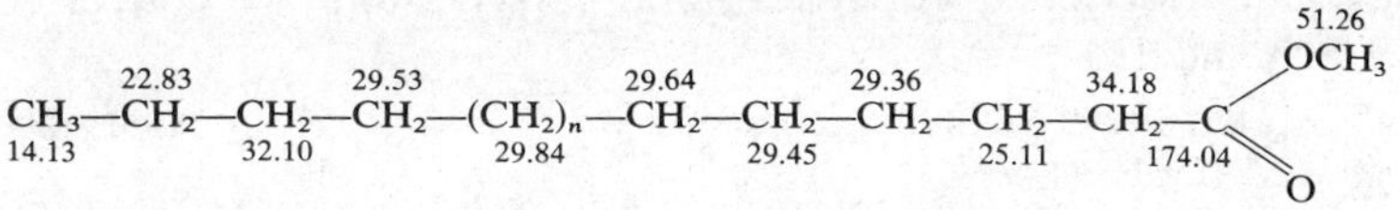

The presence of unsaturated centres or other functional groups in the alkyl chain provides further signals associated with the new group and causes shifts in the signals observed in saturated esters of the same chain length. These are usually sufficient to indicate both the nature and the position of the functional group.

Unsaturated carbon atoms unaffected by any other deshielding influence give signals at 129.90 (*cis*-alkenes), 130.40 (*trans*-alkenes), and 80.19 (alkynes), which frequently appear as two signals under the influence of the ω-CH_3 or CO_2Me function. The extent of this splitting is often sufficient to place the unsaturated centre in the saturated chain. The deshielding influence of one unsaturated centre on another has also been resolved and it is possible to allocate all the signals observed for a polyene ester (Table 6). The assignment of the 16 signals observed for methyl arachidonate is as follows:

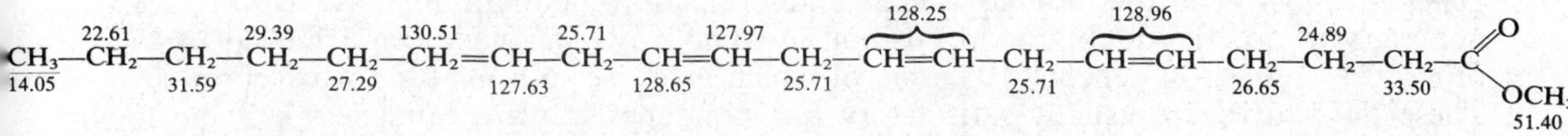

TABLE 6

Changes in Chemical Shift of CH_2 Groups[a]

	Carbon of functional group	*Deshielding effect* α	β	γ	δ	ε	ζ
—CO_2H	180.60	+4.43	−5.00	−0.58	−0.42	−0.25	0.00
—CO_2CH_3	{174.04, 51.26	+4.34	−4.73	−0.48	−0.39	−0.20	0.00
—CH_3	14.13	−7.01	+2.26	−0.31	0.00	0.00	0.00
—CH=CH— (*c*)	129.90	−2.50	−0.05	−0.45	−0.20	−0.10	0.00
—CH=CH—(*t*)	130.40	+2.85	−0.10	−0.55	−0.20	−0.10	0.00
—C≡C—	80.19	−10.96	−0.56	−0.84	−0.53	−0.16	−0.11
—CH(OH)—	71.85	+7.80	−4.00	+0.06	−0.06	−0.09	−0.05
—CH(OAc)—	74.35	+4.40	−4.40	−0.20	−0.20	−0.10	−0.05
—CO—	181.10	+13.10	−5.75	−0.40	−0.25	−0.20	−0.08

[a] Acids 29.80, esters 29.84,[b,c] 29.75,[d] and 29.65.[e] [b] F. D. Gunstone, M. R. Pollard, C. M. Scrimgeour, N. W. Gilman, and B. C. Holland, *Chem. Phys. Lipids*, 1976, **17,** 1. [c] F. D. Gunstone, M. R. Pollard, C. M. Scrimgeour, and H. S. Vedanayagam, *Chem. Phys. Lipids*, 1977 **18,** 115. [d] A. P. Tulloch and M. Mazurek, *Lipids*, 1976, **11,** 228. [e] J. Bus, I. Sies, and M. S. F. Lie Ken Jie, *Chem. Phys. Lipids*, 1976, **17,** 501; *ibid.*, 1977, **18,** 130.

25.1.6.4 Mass spectrometry[36,37]

The mass spectrum of a saturated ester such as methyl stearate gives a molecular ion peak (M^+) and peaks at 74 $[CH_2{=}C(OH)OMe]^+$, $M-31$, $59+14n$ from $(CH_2)_nCO_2Me$, and $15+14n$ from $CH_3(CH_2)_n$. Although there is some selectivity, fission occurs between each pair of CH_2 groups.

In the presence of a branched methyl group (**27**) or an oxygen-containing substituent (**28–31**), α-cleavage becomes more dominant and the major fragment ions usually indicate the nature and position of the additional functional group. α-Cleavage of ketones is accompanied by β-cleavage with McLafferty rearrangement. Fragments containing the ester function frequently undergo further loss of 32 mass units (CH_3OH). Oxo-esters are usually examined as such; hydroxy-esters are converted to their trimethylsilyl esters (which also have superior g.l.c. properties), and epoxy-esters are converted to ethers of some kind (see Scheme 10).

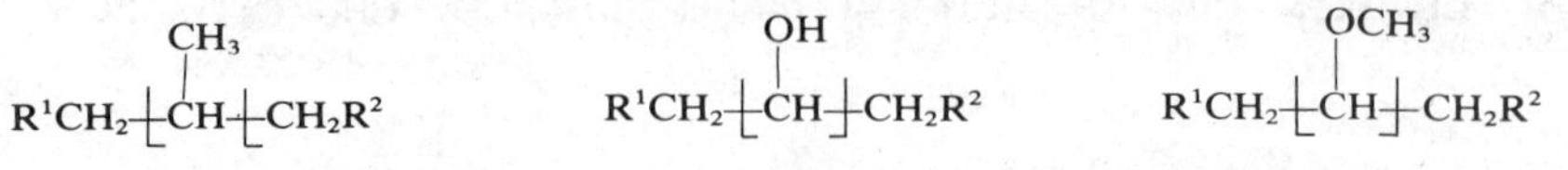

(**27**) (**28**) (**29**)

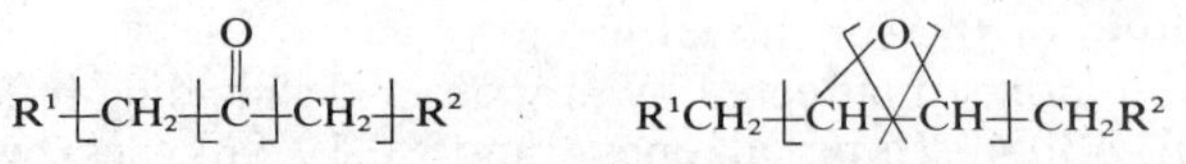

(**30**) (**31**)

$R^1 = CH_3(CH_2)_n \quad R^2 = (CH_2)_mCO_2CH_3$

Although methyl oleate, linoleate, and linolenate give distinctive mass spectra, they are not readily distinguished from isomeric monoene, diene, and triene esters because of the general mobility of the double bonds under electron bombardment. It is therefore necessary to 'fix' the double bond position in some way. This has been achieved by the procedures set out in Scheme 10, some of which give rise to a mixture of two products. These derivatives are usually satisfactory for monoenes but are often less suitable for polyene esters since the more complex fragmentation patterns may not provide a unique solution. For polyene acids, perhydroxylation with osmium tetroxide followed by formation of the poly(trimethylsilyl) esters is usually recommended.

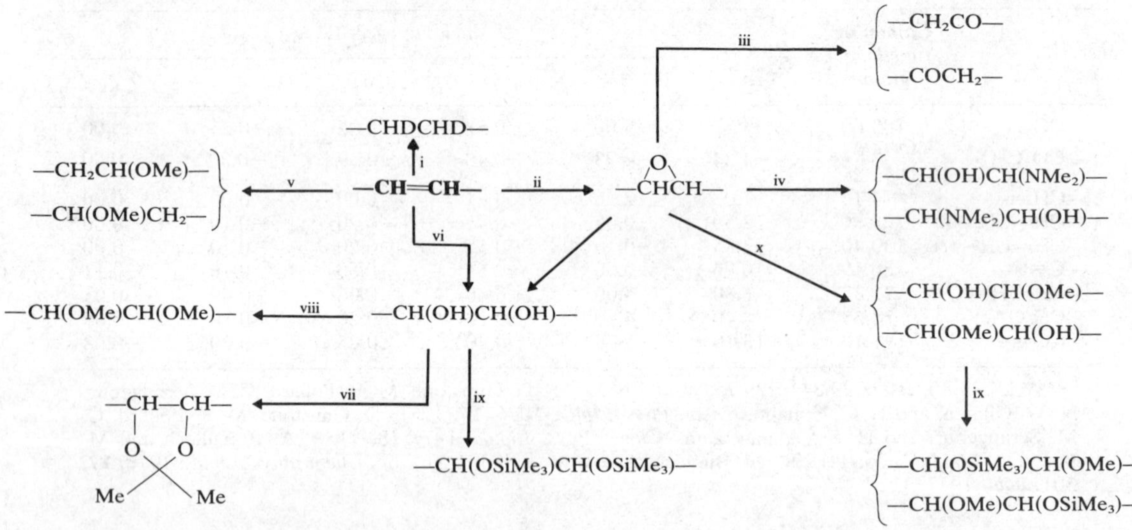

i, D_2, catalyst; ii, RCO_3H; iii, NaI; iv, Me_2NH; v, $Hg(OAc)_2$, MeOH; $NaBH_4$; vi, OsO_4; vii, $COMe_2$; viii, MeI; ix, $(Me_3Si)_2NH$ and Me_3SiCl or $(Me_3Si)_2NAc$; x, BF_3, MeOH.

SCHEME 10

This last problem has been overcome in the oxymercuration–demercuration process (Scheme 10, reaction v) by confining reaction to one double bond and hydrogenating the remaining unsaturated centres. Methyl linolenate is thereby converted to a mixture of six methoxystearates (9, 10, 12, 13, 15, and 16). This gives a mass spectrum which can be interpreted in these terms and so indicates the position of the original double bonds.[38]

It has also been claimed that the double bond migration observed with the methyl esters is much reduced with the *N*-acylpyrrolidides. The spectra of the octadecenylpyrrolidides contain clusters of peaks corresponding to fragments with 18, 17, 16, *etc.* carbon atoms attached to the pyrrolidide moiety. The major peaks in each cluster differ by 14 mass units except around the double bond, where one major peak differs by only 12 mass units from its nearest neighbour. This easily-observed change can be related to the double bond position.[39,40]

25.1.7 CATALYTIC HYDROGENATION AND CHEMICAL REDUCTION

25.1.7.1 Catalytic hydrogenation[41,42]

Natural glycerides are solid or liquid at room temperature. Solid fats are in greater demand in most temperate climates, but liquid fats of vegetable or fish origin are more readily available and thus hydrogenation is an important process whereby natural liquid fats can be modified to the desired melting range whilst retaining high nutritional value. At the same time the product may be more stable to atmospheric oxidation and therefore less likely to develop unwanted flavours. Several million tons of oil — mainly soybean, other vegetable oils, and some fish oil — are hydrogenated annually.

The aim of industrial fat hydrogenation is to produce a solid which melts in the mouth and has the required plastic properties. Such products result from partial hydrogenation and the incomplete catalytic process gives more complex products than might be expected.

Hydrogenation is effected with a range of heterogeneous and homogeneous metal catalysts (Pt, Pd, Ni, Cu, Co) of which only nickel is extensively employed industrially. An understanding of the processes involved in the partial hydrogenation of natural unsaturated glycerides has developed from careful study of the hydrogenation and deuteration of pure esters such as methyl oleate, linoleate, and linolenate.

Important features of partial hydrogenation are selectivity (the differential hydrogenation of monoenes, dienes, and trienes) and the fact that reaction is accompanied by double bond migration and stereomutation. This last produces *trans*-compounds, which are higher melting than their *cis*-isomers. This term selectivity may refer to the preferential reduction of triene to diene or of diene to monoene, or merely to the production of the high-melting saturated and *trans*-monoene esters compared with *cis*-monoenes.

Complete hydrogenation of methyl oleate gives only methyl stearate, but partial reaction gives stearate, unchanged oleate, and iso-oleate (a mixture of isomers having the double bond in other positions and largely in the *trans* configuration) (Scheme 11). Depending on the experimental conditions, the double bond in iso-oleate may occupy almost every position in the carbon chain. A study of the hydrogenation of methyl oleate (18:1 9*c*) and methyl elaidate (18:1 9*t*) showed that although these two esters are hydrogenated at the same rate, double bond migration occurs quicker in the 9*t*-ester.

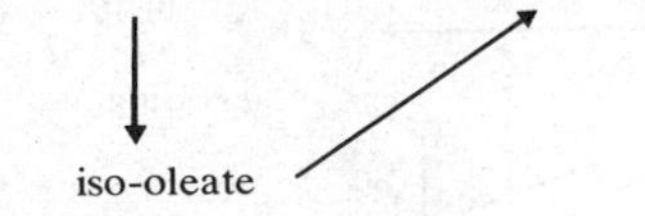

SCHEME 11

The possible isomerization and reduction reactions occurring with methyl linoleate are outlined in Scheme 12. The simplest process would be reduction to 18:1 (9*c* and 12*c*) esters and thereafter to stearate, but isomerization reactions compete with the hydrogenation process to produce several dienes (conjugated and non-conjugated) and monoenes. Isomers result from double bond migration and stereomutation. A partially reduced product is thus quite complex and its composition depends on the catalyst, temperature, pressure, and other factors affecting the availability of hydrogen on the catalyst surface. Also of significance are the adsorption–desorption properties of the various esters on the catalyst and their rate of hydrogenation. With a mixture of esters the relative ease of adsorption may be more significant than the rate of hydrogenation of an adsorbed species. The availability of hydrogen may decide whether an adsorbed species is hydrogenated or isomerized.

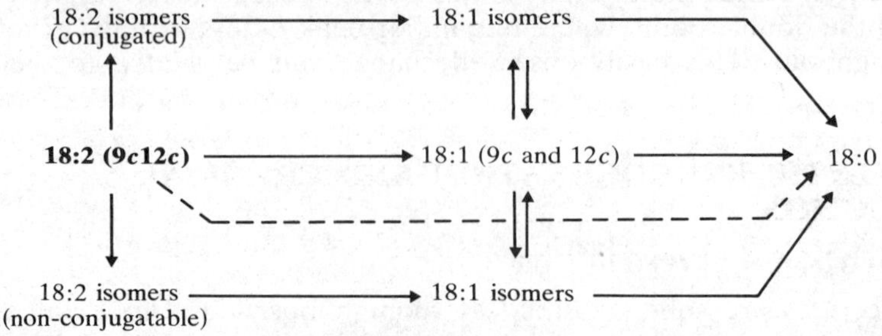

SCHEME 12

The formation of conjugated dienes (mainly 9*c*11*t* and 10*t*12*c*) is particularly significant, since with catalysts such as copper chromite hydrogenation is confined to conjugated dienes produced from methylene-interrupted isomers; non-conjugatable dienes and monoenes are not reduced by this catalyst. With nickel catalysts this is a significant reaction, but not the only one. The quick reaction occurring *via* conjugated isomers is thought to account for the enhanced reactivity of methylene-interrupted polyenes over monoenes and polyenes which cannot furnish conjugated isomers. Some diene molecules are reduced directly to stearate, *i.e.* they undergo hydrogenation twice without desorption between each step.

The reaction with methyl linolenate is even more complex (Scheme 13). Direct hydrogenation furnishes, in turn, a number of methylene-interrupted (conjugatable) dienes, then monoenes, and finally stearate. Conjugation produces trienes having diene conjugation and, after more extensive double bond migration, a conjugated triene. The conjugated compounds are usually reduced more quickly than their non-conjugated isomers. The intermediate dienes are of three kinds: those with conjugated unsaturation, those with methylene-interrupted double bonds which readily yield conjugated dienes, and those with two double bonds so well separated (*e.g.* 18:2 9*c*15*c*) that conjugation is unlikely. The first two diene types are readily hydrogenated but the latter — like the monoenes — react more slowly and tend to accumulate during partial hydrogenation. All unsaturated intermediates may undergo double bond migration and/or stereomutation before or instead of hydrogenation, so that many isomers other than those designated in Scheme 13 are actually present. Some molecules may be reduced two or even three times on a single occasion of adsorption. The relative importance of these several pathways depends on the choice of catalyst and on other experimental factors.

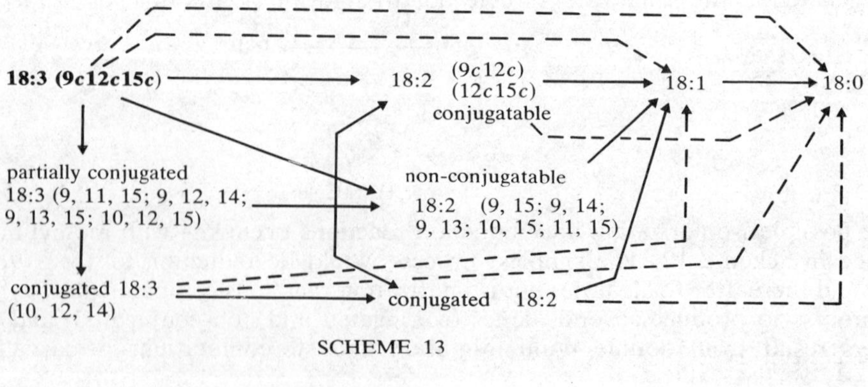

SCHEME 13

Because linolenic acid (present in Soybean oil) with its $\omega 3$ double bond furnishes undesirable flavours after oxidation, and because linoleic acid has a desirable dietary value, considerable effort has been made to produce catalysts with high linolenic/linoleic selectivity. Limited success has been achieved with copper chromite, which effects hydrogenation *only* after conjugation.

The competing processes of double bond migration and hydrogenation (or deuteriation) have been discussed in terms of half-hydrogenated species which may lose hydrogen to regenerate the original unsaturated species or an isomer or may react with a second hydrogen to produce the reduced product (Scheme 14). Alternatively π-allyl complexes might be involved. These ideas are outlined for the simple reaction of a monoene with deuterium, which gives the saturated (dideuterio) compound along with the original alkene and a number of (monodeuterio) isomers. All the monoenes can re-enter a similar reaction cycle, leading to extensive double bond movement (and deuterium incorporation). Up to 30 atoms of deuterium may be introduced into methyl stearate during deuteriation of oleate. (See Ref. 43 for a recent discussion of this problem.)

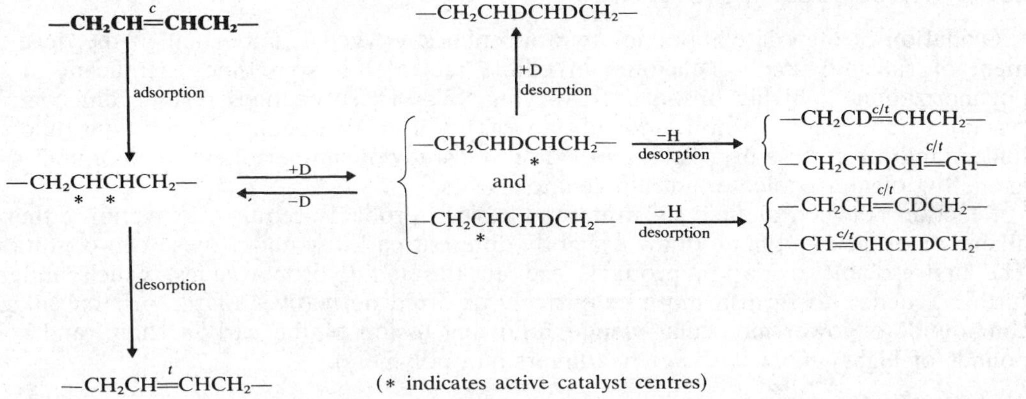

SCHEME 14

25.1.7.2 Chemical reduction

The conversion of an unsaturated acid or ester to its perhydro-derivative — sometimes a significant step in the identification of new fatty acids — is readily achieved by catalytic hydrogenation using a platinum, palladium, or nickel catalyst. The partial and stereospecific reduction of alkynes to *cis*-alkenes is most commonly effected with Lindlar's catalyst (Section 25.1.4.1).

Non-catalytic reduction of alkenes is conveniently carried out with hydrazine. Oxygen is a necessary reagent and the reaction probably involves di-imine (N_2H_2). The reaction is stereospecific and occurs without double bond migration or stereomutation. Partial reduction of a polyene therefore gives a simpler product than catalytic processes. A typical experiment with linolenic acid is summarized in equation (9) and shows the range of products after incomplete reduction. This *cis* addition procedure provides the best route to *vic*-dideuterio-acids of known stereochemistry (equation 10).

18:3	→ 18:2	→ 18:1	→ 18:0 (9)
9*c*12*c*15*c* (26%)	9*c*12*c* (15%)	9*c* (9%)	(5%)
	12*c*15*c* (15%)	12*c* (9%)	
	9*c*15*c* (13%)	15*c* (8%)	

$$\underset{cis\ \text{or}\ trans}{\text{—CH=CH—}} \xrightarrow{N_2D_4} \underset{erythro\ \text{or}\ threo}{\text{—CHDCHD—}} \qquad (10)$$

25.1.7.3 Biohydrogenation

Micro-organisms in the rumen have long been known to hydrogenate dietary polyunsaturated 18:2 and 18:3 acids to stearic acid and to unsaturated acids having double bonds in unusual positions and mainly with the *trans* configuration.[44] One species of rumen bacteria (*Butyrivibrio fibrosolvens*), for example, contains an isomerase and a reductase which convert linoleic acid to vaccenic acid (equation 11). Another rumen bacterium promotes the reduction of oleic and linoleic acid to stearic acid and of linolenic acid to 18:1 (15*c*).

$$18{:}2\ (9c12c) \longrightarrow 18{:}2\ (9c11t) \longrightarrow 18{:}1\ (11t) \qquad (11)$$

Rumen hydrogenation can be circumvented by a special feeding regimen in which unsaturated acids are protected during their passage through the rumen.

25.1.8 OXIDATION BY OXYGEN

Oxidation of olefinic compounds by atmospheric oxygen is important in the development of rancidity and off-flavours in edible fats and is sometimes significant in the polymerization of highly unsaturated (drying) oils. These changes result from complex reactions occurring in complicated substances, often under undefined conditions, and our understanding of these processes is based on the study of simpler olefinic compounds such as methyl oleate, linoleate, and linolenate.

Reaction between olefinic substrate and oxygen probably requires activation either of alkene or oxygen and may follow a slightly different pathway under these two conditions. The first isolable oxidation products are unsaturated hydroperoxides, which undergo further reaction to furnish more extensively oxidized derivatives of the original alkene, compounds of lower molecular weight following fission of the carbon chain, and compounds of higher molecular weight (dimers and polymers).

25.1.8.1 Autoxidation[45]

The major non-enzymic oxidation process is a radical chain reaction involving initiation, propagation, and termination steps (Scheme 15).

initiation	production of $R\cdot$ or $RO_2\cdot$ radicals
propagation	$R\cdot + O_2 \longrightarrow RO_2\cdot$
	$RO_2\cdot + RH \longrightarrow RO_2H + R\cdot$
termination	interaction of radicals to produce non-initiating and non-propagating products
	(RH represents alkene substrate)

SCHEME 15

The nature of the initiation reaction is still uncertain, although it is known that hydroperoxides once formed furnish additional initiating radicals and that the reaction of alkenes with singlet oxygen (Section 25.1.8.2) to produce hydroperoxides may play a key role in the initiation of autoxidation. The propagation sequence involves the production of a radical $R\cdot$ from the alkene RH and its subsequent reaction with oxygen. The radical, produced by reaction at the allylic position, is resonance-stabilized and this affects the structure of the reaction product. The termination reactions have not been extensively studied.

The prevention (or reduction) of autoxidation is possible through the addition of phenolic compounds (antioxidants), which react with the propagating radicals to give non-propagating species.

Autoxidation of pure methyl oleate at room temperature is a slow reaction, occurring only after a long induction period. The reaction can be accelerated by addition of a radical source, by irradiation, or by raising the temperature. Samples of oleate containing linoleate also have shorter induction periods because the more readily formed products of linoleate oxidation can initiate oleate oxidation. The hydroperoxides produced from methyl oleate are a mixture of the *cis*- and *trans*-isomers of 8-hydroperoxy Δ^9-, 9-hydroperoxy Δ^{10}-, 10-hydroperoxy Δ^8-, and 11-hydroperoxy Δ^9-octadecenoates. The formation of these compounds is explained in terms of the propagation sequence which occurs *via* two resonance-stabilized allyl radicals (Scheme 16).

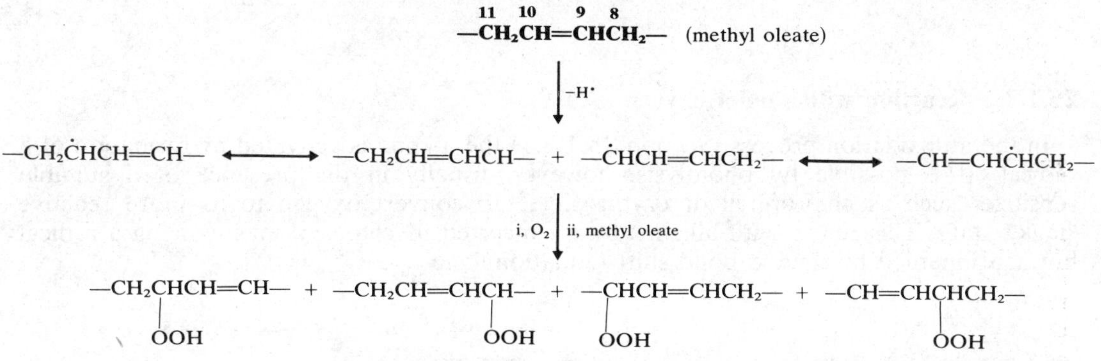

SCHEME 16

Methyl linoleate reacts 10–40 times more quickly because of the enhanced reactivity of the C-11 methylene group lying between two double bonds. The reaction product is mainly (or entirely) a mixture of the 9- and 13-hydroperoxyoctadecadienoates, and there is no firm evidence for the formation of a 11-hydroperoxydiene. The *cis,trans* conjugated dienes produced at 0 °C isomerize readily to *trans,trans*-isomers at room temperature and above (Scheme 17).

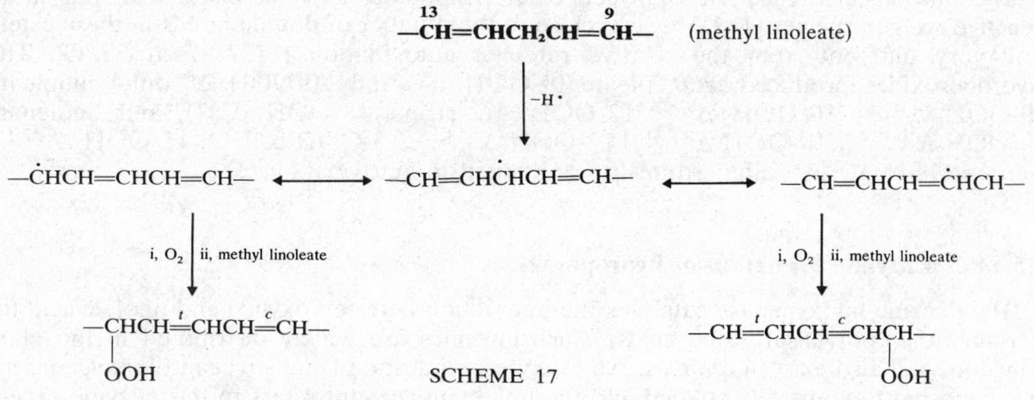

SCHEME 17

Methyl linolenate undergoes similar oxidation at the 9,12- or 12,15-diene system to yield initially four major hydroperoxides with three unsaturated centres, including a conjugated diene (Scheme 18). In addition, there is evidence that, in methylene-interrupted polyenes containing three or more double bonds, intramolecular reaction

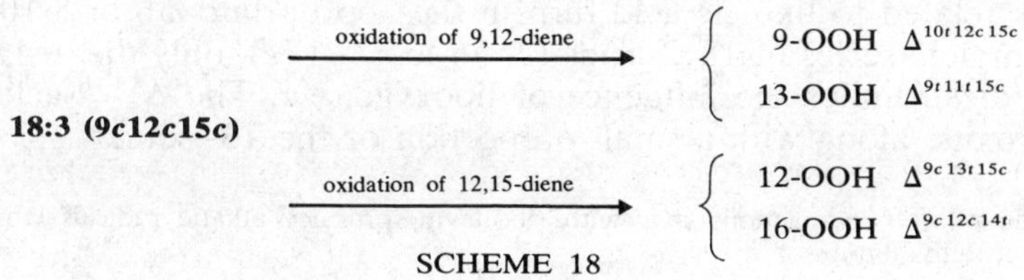

SCHEME 18

between peroxyl radicals and double bonds occur, producing endoperoxides such as (**32**) and (**33**).

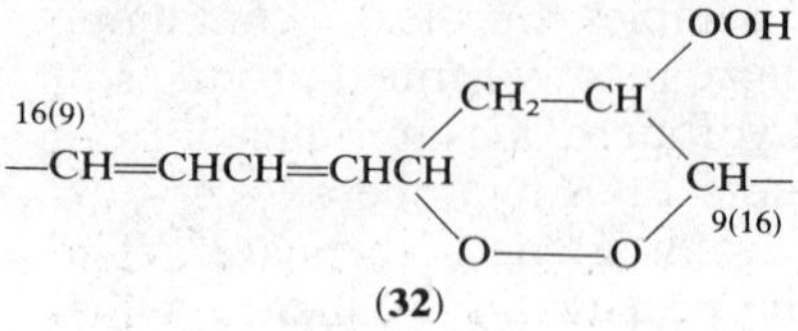

(**32**)

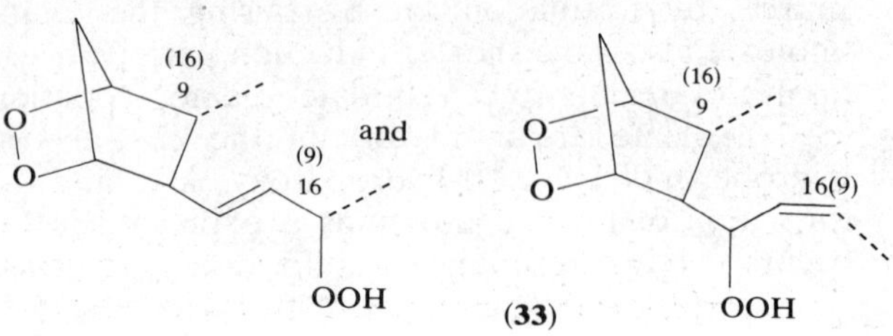

(**33**)

25.1.8.2 Reaction with singlet oxygen

In the autoxidation process (Section 25.1.8.1) the alkene is activated by formation of a radical. It is possible by photolysis, however, usually in the presence of a suitable sensitizer such as chlorophyll or erythrosine,* to convert oxygen to its more reactive singlet state. This reacts with alkenes by a concerted mechanism not involving a radical but accompanied by double bond shift (equation 12).

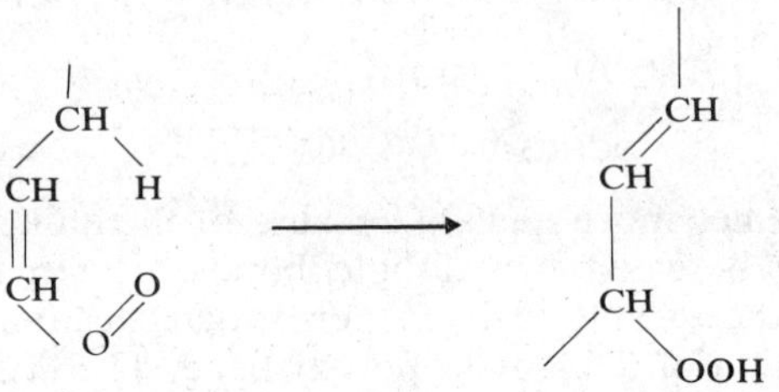

(12)

The difference in reactivity between oleate, linoleate, and linolenate with this more reactive oxygen species (1:1.3:2.3) is close to the number of double bonds in these esters and very different from the relative rates of autoxidation (1:27:77 at 37 °C). The hydroperoxides produced from oleate (9-OOH Δ^{10} and 10-OOH Δ^{8} only), linoleate (9-OOH $\Delta^{10,12}$; 10-OOH $\Delta^{8,12}$; 12-OOH $\Delta^{9,13}$; and 13-OOH $\Delta^{9,11}$), and linolenate (9-OOH $\Delta^{10,12,15}$; 10-OOH $\Delta^{8,12,15}$; 12-OOH $\Delta^{9,13,15}$; 13-OOH $\Delta^{9,11,15}$; 15-OOH $\Delta^{9,12,16}$; and 16-OOH $\Delta^{9,12,14}$) differ from those obtained in autoxidation.

25.1.8.3 Enzymic formation of hydroperoxides[46]

The enzyme lipoxygenase catalyses the interaction between oxygen and linoleic acid (or certain other polyunsaturated acids). Such enzymes are widely distributed in the plant kingdom and also exist in animals. An enzyme preparation from soybean (lipoxygenase I) has been most extensively studied and the following account refers to this enzyme except where otherwise indicated.

With this lipoxygenase, linoleic acid is oxidized to the (13*S*)-hydroperoxy-Δ^{9c11t} acid and other enzyme preparations furnish the (9*R*)-hydroperoxy-Δ^{10t12t} acid or mixtures of the two optically active hydroperoxides. In contrast, autoxidation produces racemic products. Acids related to linoleic acid furnish the appropriate $\omega 6$ or $\omega 10$ hydroperoxy-acid. Of the complete series of 18:2 acids (2*c*5*c* to 14*c*17*e*), only the 9*c*12*c* and 13*c*16*c* isomers are oxidized under the influence of lipoxygenase. The $\Delta^{13,16}$-acid furnishes the (17*S*)-hydroperoxide along with a small proportion of the 13-isomer.

* Other sensitized photolytic reactions (*e.g.* with riboflavin) produce alkene radicals which give the same products as in direct autoxidation.

The enzyme, containing one atom of iron in its molecule, exists in three states described as the colourless (native) enzyme, the yellow enzyme, and the purple enzyme. It promotes both aerobic and anaerobic reactions.

The aerobic reaction starts by removal of a hydrogen from C-11. Removal of the 11*S* atom is followed by oxygen insertion at C-13 and removal of 11*R* hydrogen is followed by oxygen insertion at C-9. The reaction is believed to involve the sequence (13); see also Scheme 19.

$$-\dot{C}H- \xrightarrow{O_2} -CH(O\dot{O})- \xrightarrow{Fe^{2+}} -CH(O\bar{O})- \longrightarrow -CH(OOH)- \qquad (13)$$

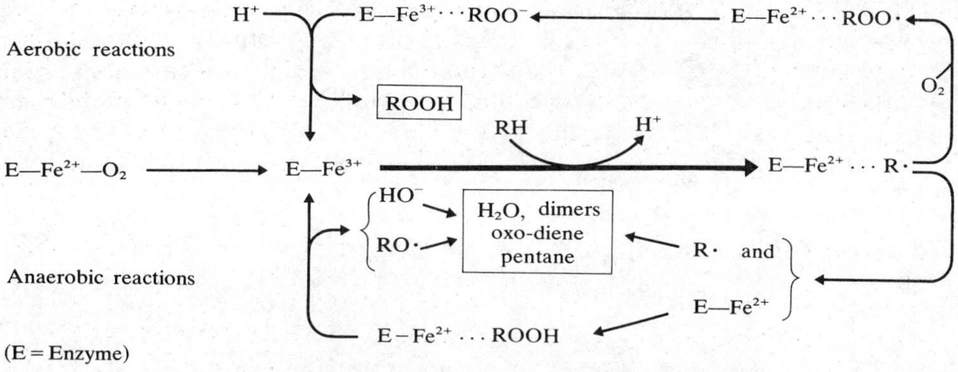

SCHEME 19 Lipoxygenase-catalysed oxidation of linoleic acid under aerobic and anaerobic reaction conditions

The reaction follows a different course when there is deficiency of oxygen (Scheme 19). Linoleic acid and the 13-hydroperoxide already formed interact to give a complex mixture of products including pentane, the C_{13} aldehydes (**34**), a C_{18} oxo-diene (**35**), and dimers involving two molecules of linoleic acid (**36**) or one of linoleic acid and one of the hydroperoxide (**37a**, **b**). These reaction products can be explained by the formation and coupling of carbon radicals from linoleic acid and alkoxyl radicals from the hydroperoxide.

$OHCCH\overset{t}{=}CHCH\overset{c/t}{=}CH(CH_2)_7CO_2H$

(**34**)

$-COCH\overset{t}{=}CHCH\overset{t}{=}CH-$

13-oxo Δ^{9t11t}

(**35**)

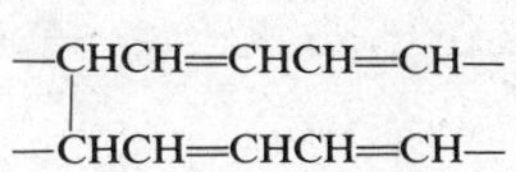

C(13)—C(13) dimer, also C(13)—C(11), C(13)—C(9) and C(11)—C(9) dimers

(**36**)

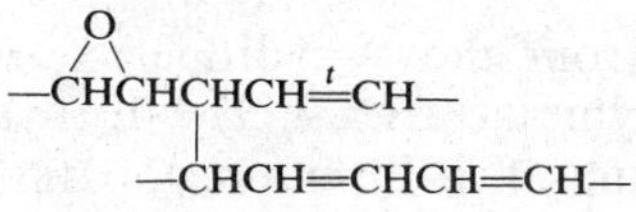

C(11)—C(13) dimer

(**37a**)

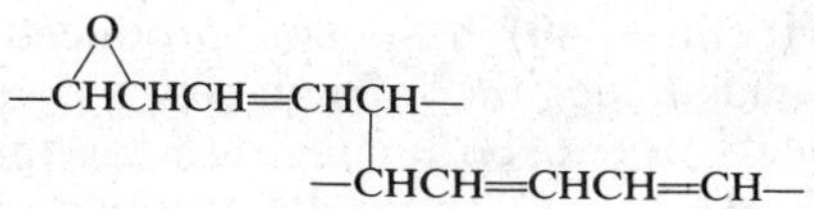

C(9)—C(13) dimer

(**37b**)

The products of primary oxidation are subject to further reaction, as indicated in the following section.

25.1.8.4 Reactions of hydroperoxides

Unsaturated hydroperoxides readily undergo further reaction under both enzymic and non-enzymic conditions.[47] The secondary products result from 13-OOH Δ^{9c11t} or 9-OOH Δ^{10t12c} or from a mixture of the two and are usually complex mixtures of oxygenated acids. Compounds (**38**)–(**54**) have been recognized. Each partial structure represents the C-8 to C-14 fragment, with isomeric compounds being produced from the two substrates so that (**38**) represents the 13-hydroxy-Δ^{9c11t} and 9-hydroxy-Δ^{10t12c} acids. Stereochemical details have not been confirmed in all cases.

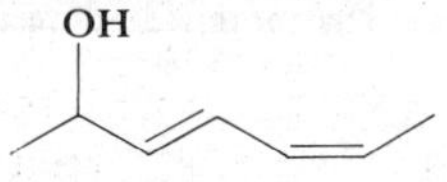
(**38**)

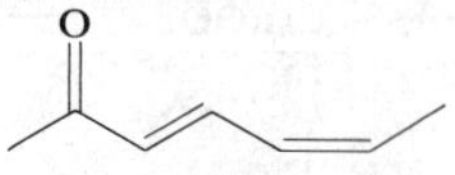
(**39**)
ct and *tt* isomers

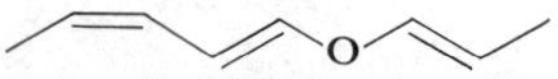
(**40**) [C(6)—C(13)]

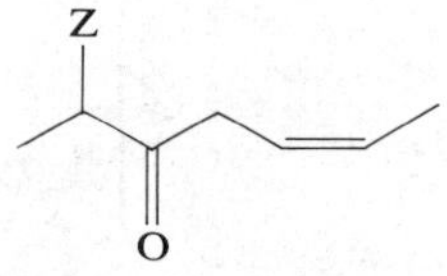
(**41**) (α-ketol)

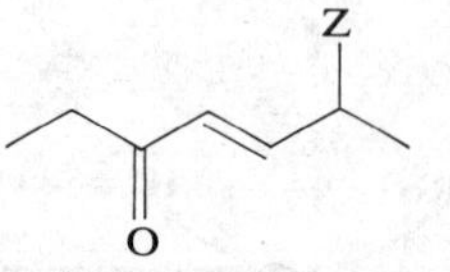
(**42**) (γ-ketol)

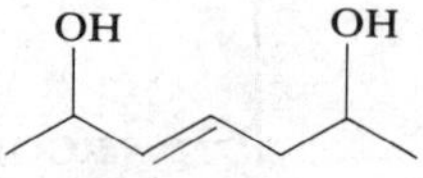
(**43**)

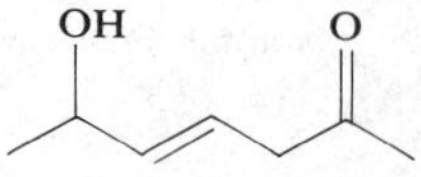
(**44**)

(Z = OH, OMe, SEt, OCOR)

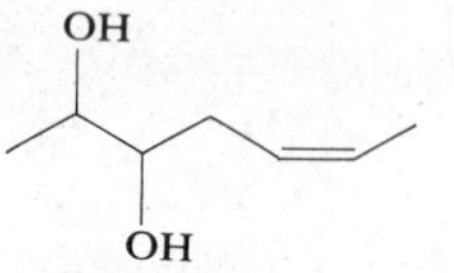
(**45**)

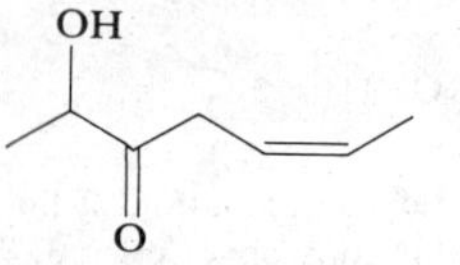
(**46**)

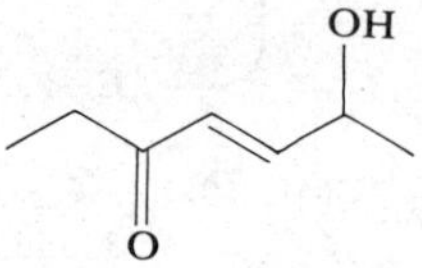
(**47**)

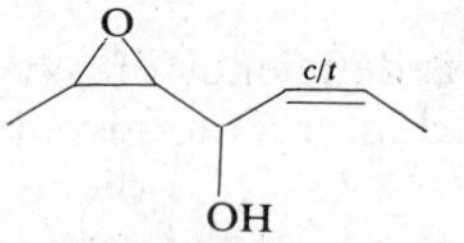
(**48**)

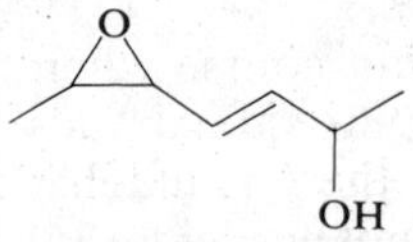
(**49**)

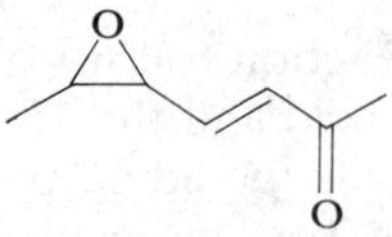
(**50**)

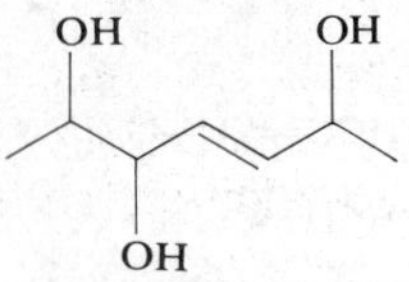
(**51**)

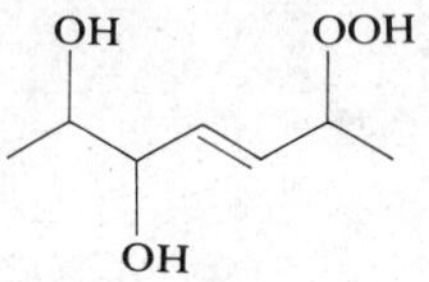
(**52**)

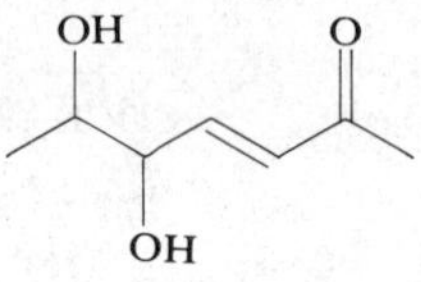
(**53**)

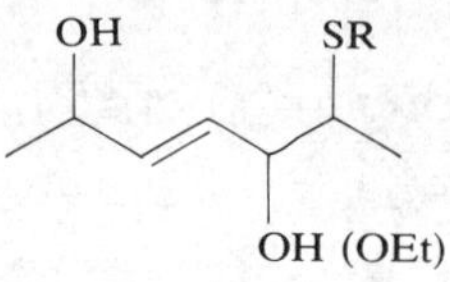
(**54**)

RSH = *N*-acetylcysteine

The vinyl ether (**40**) has been produced only from the 9-hydroperoxy-Δ^{10t12c} and -$\Delta^{10t12c15c}$ acids under the influence of a potato enzyme at pH > 7. The 13-hydroperoxy-isomer appears not to be a substrate for this enzyme. The formation of the α- and/or γ-ketol (**41**, **42**; Z = OH) results through the operation of an enzyme in an aqueous environment. Modified products are formed in the presence of other nucleophiles such as methanol, ethanethiol, or oleic or linoleic acid. The carbonyl oxygen is derived from the hydroperoxide group and the other oxygen atom from the solvent.

Short-chain products resulting from fission of the C_{18} chain include hydrocarbons, alcohols, aldehydes, ketones, acids, esters, and lactones, all of which may be saturated or unsaturated.[48] Typical aldehydes include alkanals, alkenals (Δ^{2t}, ^{3c}, ^{4c}), alkadienals (Δ^{2t4c}, ^{2t5c}, ^{2t6c}), and alkatrienals (Δ^{2t4c7c}), all of which may affect flavour in a desirable or undesirable way. In the complex question of flavour, human response varies with the concentration of a particular component and on the presence of other flavour-producing compounds. The threshold of organoleptic observation of many of these compounds is measured in parts per million or per billion, so that exceedingly small amounts may have significant consequences on flavour. Although the formation of many of these molecules is easily rationalized in terms of the structure of the original hydroperoxide and its major pathways of breakdown, some are not and it is uncertain whether these arise from unexpected substrates or from unrecognized breakdown pathways. Hydroperoxides of unexpected structure could result from conventional acids by an unusual oxidation route or by the standard oxidation of a minor fatty acid not yet recognized in the substrate. Fish oils and vegetable oils — even before the complications produced during partial hydrogenation (Section 25.1.7) — may contain traces of unidentified acids with double bonds in unexpected positions.

$$R^1CH{=}CHCH(OOH)R^2 \begin{cases} \xrightarrow{-\dot{O}H} R^1CH{=}CHCH(O\cdot)R^2 \longrightarrow R^1CH{=}CHCHO \text{ or } R^2CHO \\ \xrightarrow[-H_2O]{H^+} R^1CH{=}CHO\overset{+}{C}HR^2 \xrightarrow[H_2O]{} R^1CH{=}CHOCH(\overset{+}{O}H_2)R^2 \\ \qquad\qquad\qquad\qquad\qquad\qquad \longrightarrow R^1CH_2CHO + R^2CHO \end{cases} \quad (14)$$

The decomposition of hydroperoxides to short-chain aldehydes gives different products by homolytic and heterolytic processes (equation 14). Homolytic cleavage of the 13- and 9-hydroperoxides of linoleic ester would explain the presence of the 6:0 and 10:2 2*t*4*c* aldehydes. The latter conveys a 'deep fried' flavour at a threshold concentration of 5×10^{-10}. Other aldehydes may be produced from 10-, 11-, and 12-hydroperoxy-compounds. The four hydroperoxides from oleic acid (Scheme 16) give the 8:0, 9:0, 10:1 2*t*, and 11:1 2*t* aldehydes. The situation is more complicated with linolenic acid and with partially hydrogenated fats.

25.1.9 OTHER OXIDATION REACTIONS

25.1.9.1 Epoxidation

Epoxidation of alkenes is conveniently effected by a wide range of peracids (equation 15). Unless carefully buffered, the lower aliphatic members (RCO_3H, $R = H$, CH_3, CF_3) give acylated diols through ring-opening of the epoxide, but perlauric, monopersuccinic, monoperphthalic, and perbenzoic acid smoothly effect epoxidation and the commercially available *m*-chloroperbenzoic acid is widely employed. This reaction is a *cis* addition, so that *cis* and *trans* epoxides result from *cis*- and *trans*-alkenes, respectively. Peracids react much less readily with acetylenic centres and it is easy to carry out a selective epoxidation of enynes (equation 16).

$$-CH{=}CH- \xrightarrow{RCO_3H} -\overset{O}{\widehat{CHCH}}- \xrightarrow{RCO_2H} -CH(OH)CH(OCOR)- \quad (15)$$

$$R^1C{\equiv}CCH_2CH{=}CHR^2 \xrightarrow{ArCO_3H} R^1C{\equiv}CCH_2\overset{O}{\widehat{CHCH}}R^2 \quad (16)$$

18:2 (9*c*12*a*)

Monoenoic esters are readily converted to epoxides with the stereochemistry indicated above, and all the isomeric epoxyoctadecanoates have been prepared from the corresponding octadecenoates. These monoepoxides contain two chiral centres; the synthetic products are racemic but natural epoxy-acids (Section 25.1.2.9) are usually enantiomeric.

The melting points of the isomeric *cis*-epoxystearic acids show alternation, but this is not apparent with the *trans*-isomers so that the higher-melting acid is sometimes the *cis*- and sometimes the *trans*-epoxy-acid.

With polyene acids and esters the number of isomeric epoxides is greater. Methyl linoleate furnishes two isomeric *cis,cis*-bisepoxides and the corresponding acids, melting at 78 and 41 °C, have been separated by crystallization.

The epoxides are reactive compounds which serve as useful intermediates (Sections 25.1.6.4 and 25.1.9.2).[49] Structural information can be obtained by mass spectrometry or from cleavage of epoxides to aldehydes by periodic acid. Alkenes can be regenerated from epoxides (equation 17). Partially epoxidized fatty oils or esters are produced on an industrial scale for use as plasticizers.

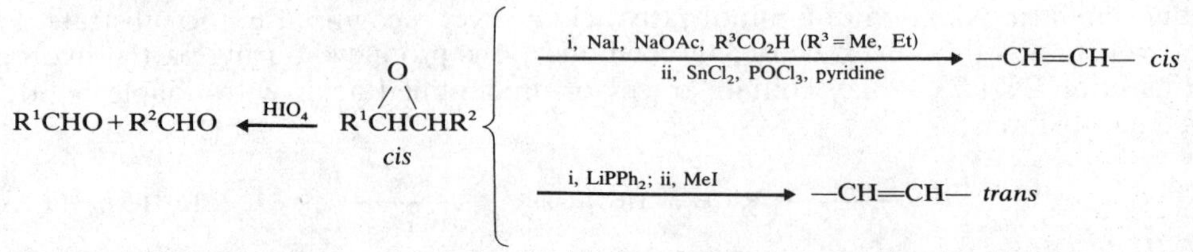

(17)

The compounds described above are the familiar 1,2-epoxides. Larger ring (1,4- and 1,5-) epoxides have been prepared from diene or hydroxy esters by a variety of procedures, of which reactions (18) and (19) are typical.[50]

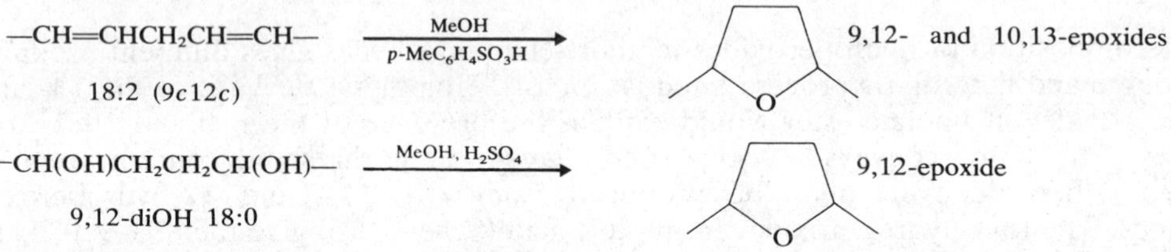

(18)

(19)

25.1.9.2 Hydroxylation[51]

The conversion of an alkene to its diol (hydroxylation) can be carried out by several reagents which effect *cis*-hydroxylation or *trans*-hydroxylation to give *threo*- and *erythro*-diols, as shown in equation (20).

—CH(OH)CH(OH)—	$\xleftarrow[\text{hydroxylation}]{\textit{trans}}$	—CH=CH—	$\xrightarrow[\text{hydroxylation}]{\textit{cis}}$	—CH(OH)CH(OH)—	(20)
threo-diol		*cis*-alkene		*erythro*-diol	
erythro-diol		*trans*-alkane		*threo*-diol	

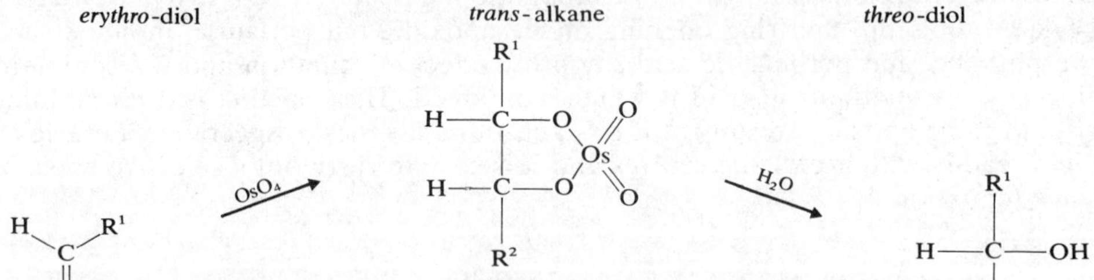

(55)

(56)

(21)

Potassium permanganate and osmium tetroxide effect *cis* hydroxylation and the two reactions are believed to occur through similar cyclic intermediates, (**55**) and (**56**). The osmate ester is subsequently decomposed by sodium sulphate, formaldehyde, hydrogen sulphide, or an excess of mannitol. This reaction occurs in high yield and is simple to perform, but suffers from the disadvantage that osmium tetroxide is expensive and toxic. Modified reaction procedures employ osmium tetroxide only as a catalyst in the presence of an oxidizing agent to regenerate continually the tetroxide. Metal chlorates, hydrogen peroxide–t-butanol, t-butyl hydroperoxide, and *N*-methylmorpholine *N*-oxide have been successfully used for this purpose.

Halogens and metal salts are used in different ways for *cis* and *trans* hydroxylation. *cis* Hydroxylation (Woodword procedure) occurs through reaction of an alkene with iodine and silver acetate in wet acetic acid (equation 22), although better results are claimed with cupric or potassium salts in place of silver acetate. The Prévost reaction, using iodine and silver benzoate under anhydrous conditions, leads to *trans* hydroxylation (equation 22).

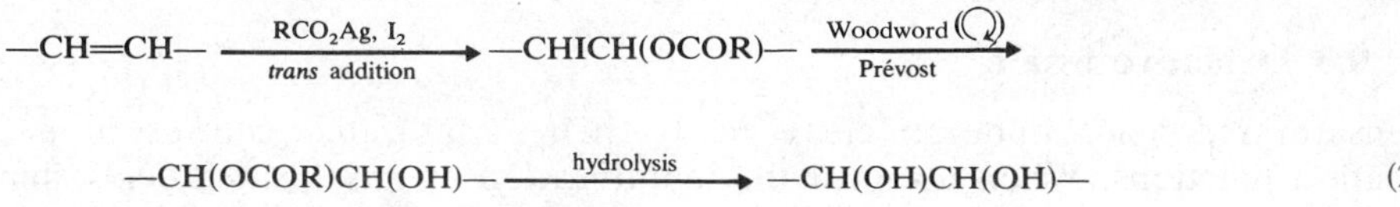

(22)

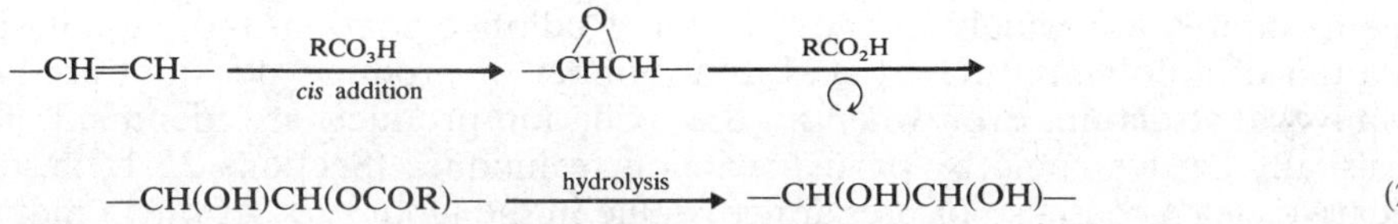

(23)

trans Hydroxylation is more usually carried out with peracids. These furnish epoxides (Section 25.1.9.1) which can be subjected to acid-catalysed hydration, a reaction which occurs with inversion (equation 23). This is conveniently achieved in a one-pot reaction with the lower aliphatic peracids such as performic or peracetic, prepared when *in situ* by mixing the carboxylic acid and hydrogen peroxide along with an acidic catalyst. As in the epoxidation process itself, the reaction can be applied selectively to the double bond in an enyne system. Whilst monoepoxy-compounds are smoothly hydrated to *vic*-diols, the reaction of diepoxy-compounds can be more complex with 1,4- or 1,5-epoxides being formed in preference to the desired tetrol, possibly through an intromolecular reaction of the type shown in equation (24). This unwanted product can be minimized if water is avoided during the ring-opening stage.

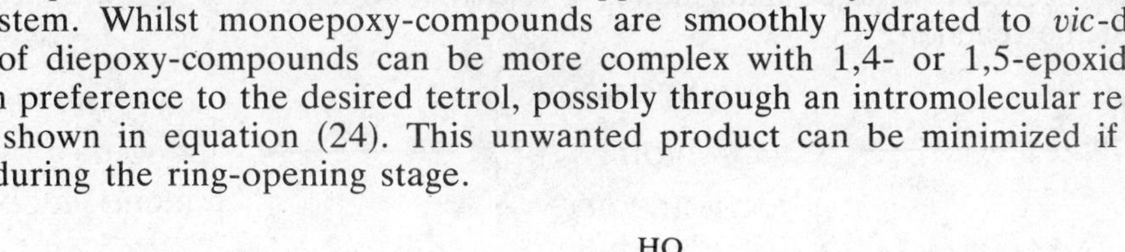

(24)

methyl 9,10;12,13-diepoxystearate   methyl 10,13-epoxy-9,12-dihydroxystearate

Simple dihydroxy-esters contain two chiral centres and the products of the hydroxylation reactions described above are usually racemic products with the *threo* or *erythro* configuration (see equation 20). The *erythro*-isomer is usually — but not always — the higher melting. Linoleic acid furnishes 9,10,12,13-tetrahydroxystearic acid, which can exist in eight racemic forms. Two (m.p. 174 and 164 °C) are produced by *cis*-hydroxylation of linoleic acid and a further pair (m.p. 148 and 126 °C) by *trans*-hydroxylation. The same four acids may be prepared from linelaidic acid (18:2 9*t*12*t*) and the remainder would have to be prepared from the 9*c*12*t* or 9*t*12*c* isomers. Eight of the 16 possible enantiomeric forms have been prepared from the natural 12*S*,13*R*-epoxyoleic acid. Ricinoleic acid (12-hydroxyoleic acid) with its C-12 chiral group yields four enantiomeric 9,10,12-trihydroxystearic acids, all of which have been isolated and characterized.

The *vic*-diols are cleaved by periodic acid (to aldehydes) or permanganate (to acids). After suitable derivatization they can be submitted to gas chromatographic separation and identified by mass spectrometry (Section 25.1.6.4).

Alkenes can be regenerated from diols by reactions (25) and (26).

$$\underset{\textit{erythro}}{\text{—CH(OH)CH(OH)—}} \xrightarrow[\text{pyridine}]{\text{MsCl}} \text{—CH(OMs)CH(OMs)—} \xrightarrow[\text{DMF}]{\text{NaI, Zn}} \underset{\textit{cis}}{\text{—CH=CH—}} \qquad (25)$$

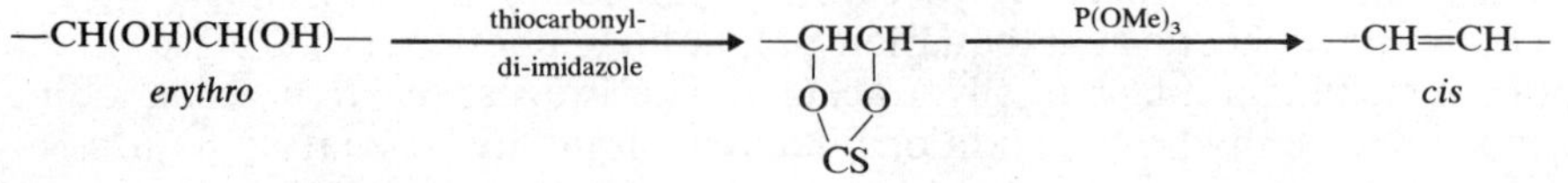

(26)

25.1.9.3 Oxidative fission[52]

Unsaturated acids undergo cleavage at their unsaturated centres in a number of oxidation reactions. When used for the identification of unsaturated acids, simple procedures giving unequivocal results with the minimum quantity of material are required. Only two procedures are widely employed: von Rudloff oxidation and ozonolysis. Complete oxidation of a polyunsaturated acid to a number of products does not always lead to an unequivocal structure, even when all the oxidation products are identified. This difficulty can usually be overcome by partial reaction techniques (Sections 25.1.3.5 and 25.1.6.4). Oxidative cleavage is also of preparative value in the laboratory or on an industrial scale.

Vigorous oxidation by potassium permanganate results in over-oxidation, but this problem is overcome in the von Rudloff oxidation, which employs as oxidant a 1:39 mixture of potassium permanganate and sodium metaperiodate under conditions where the permanganate is little more than a catalyst. The reaction proceeds in aqueous alkaline or aqueous t-butanol solution at room temperature during 6–24 hours through the stages of diol or ketol formation, cleavage, and oxidation (equation 27). The excess of periodate continually regenerates the permanganate, so that the latter persists at low concentration throughout the reaction.

$$R^1CH{=}CHR^2 \xrightarrow{KMnO_4} \left\{\begin{array}{c} R^1CH(OH)CH(OH)R^2 \\ R^1COCH(OH)R^2 \\ R^1CH(OH)COR^2 \end{array}\right\} \xrightarrow{NaIO_4} \left\{\begin{array}{c} R^1CHO + R^2CHO \\ R^1CO_2H + R^2CHO \\ R^1CHO + R^2CO_2H \end{array}\right\} \qquad (27)$$

$$\xrightarrow{KMnO_4} R^1CO_2H + R^2CO_2H$$

Ozonolysis[53–55] is now the more commonly used procedure for both analytical and preparative purposes. Reaction probably occurs *via* a molozonide, which is thought to have the structure (**57**) and to result from a 1,3-dipolar cycloaddition. This breaks down to a carbonyl compound and the Criegee zwitterion (**58**); the latter reacts with the carbonyl compound to produce an ozonide (**59**) or with alcohols or acids used as solvents to give alkoxyhydroperoxides (**60**) or acyloxyhydroperoxides (**61**).

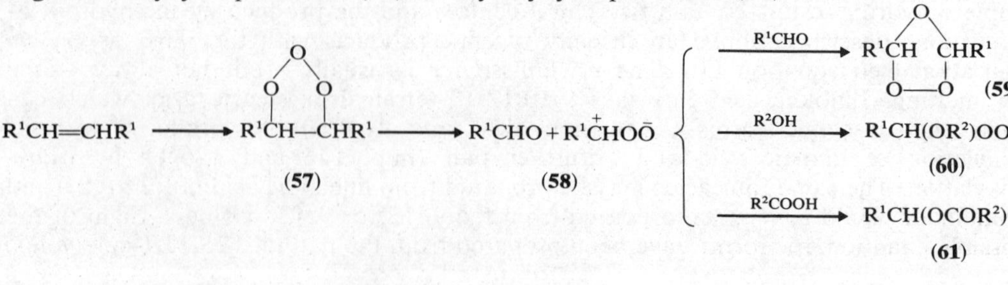

This appreciation of the reaction pathway suggests that not only can the cyclic ozonide exist in *cis-* and *trans*-forms but, because of the possibility of cross-coupling during ozonolysis, an unsymmetrical alkene should yield six different ozonides. This has been demonstrated with methyl oleate (equation 28).

Of greater practical importance are the fission products resulting from further reaction of the ozonide. These may be alcohols, aldehydes, ketones, acids, or even amines.

$$R^1CH{=}CHR^1 \longrightarrow \begin{matrix} R^1\overset{+}{C}HO\bar{O} + R^2CHO \\ R^1CHO + R^2\overset{+}{C}HOO^- \end{matrix} \longrightarrow$$

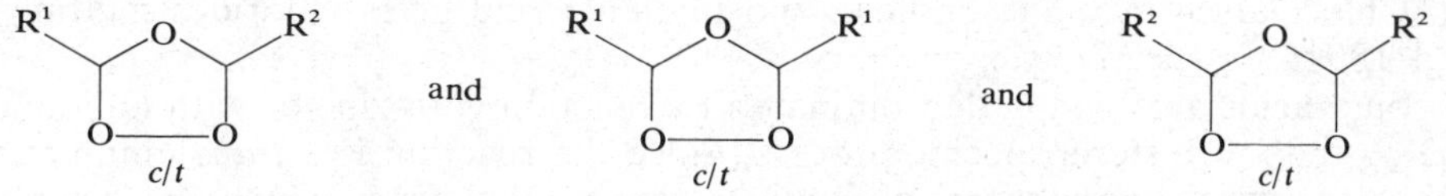

(28)

Alcohols are formed by reduction of the ozonide with metal hydrides (lithium aluminium hydride or sodium borohydride) or by catalytic hydrogenation with nickel or platinum catalysts (equation 29). Aldehydes result under milder reducing conditions, and zinc–acid, triphenylphosphine, dimethyl sulphide, or Lindlar's catalyst are commonly employed for the purpose (equation 30). Alkynes give carboxylic acids under conditions which convert the alkenes to aldehydes. Amines are produced by reduction of ozonides over Raney nickel in the presence of ammonia (equation 31) or are prepared from the aldehydes by reduction of their oximes. Acids are formed with a wide range of oxidizing agents, including peracids or silver oxide (equation 32).

$R^1CH{=}CHR^2 \longrightarrow$ 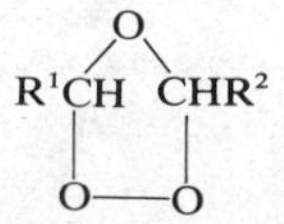

$$\longrightarrow R^1CH_2OH + R^2CH_2OH \quad (29)$$
$$\longrightarrow R^1CHO + R^2CHO \quad (30)$$
$$\longrightarrow R^1CH_2NH_2 + R^2CH_2NH_2 \quad (31)$$
$$\longrightarrow R^1CO_2H + R^2CO_2H \quad (32)$$

In one ozonolysis procedure the olefin, dissolved in pentane at −65 to −75 °C, is mixed with a pentane solution of ozone (0.03 M) at the same temperature. Reaction occurs in a few minutes and the ozonide is decomposed with hydrogen at 0 °C in the presence of Lindlar's catalyst. In another process, ozonization is effected at 0 °C in methanol solution, and the alkoxyhydroperoxide is reduced with zinc and acetic acid or with hydrogen and palladium/charcoal at 20–25 °C. These methods can be used on a macro- and micro-scale and are usually applied to acids or simple esters.

The ozonolysis of oleic acid to furnish azelaic (nonanedioic) acid is effected on an industrial scale and ozonolysis of other unsaturated acids or glycerides has furnished interesting difunctional compounds, including hydroxy-esters, aldehydo-esters, and unsaturated dibasic acids, as in equations (33)–(35).

$$RCH{=}CH(CH_2)_nCO_2H \longrightarrow HO_2C(CH_2)_nCO_2H \quad (33)$$

18:1 9*c*; 11:1 10*e*; 22:1 13*c* — $n = 7, 8, 10, 11, 13$

18:1 (9*c* and 12*c* and 15*c*)

$$R^1CH{=}CH(CH_2)_nCO_2Me \longrightarrow R^2(CH_2)_nCO_2Me \quad (34)$$

$(R^2 = OHC \text{ or } HOCH_2)$

$$CH_3(CH_2)_{10}CH{=}CH(CH_2)_4CO_2H \longrightarrow CH_3(CH_2)_{10}CH_2NH_2 + H_2NCH_2(CH_2)_4CO_2H \quad (35)$$

25.1.10 OTHER ADDITION REACTIONS

25.1.10.1 Halogenation

Halogenation is a polar *trans* addition so that, for example, bromination of oleic and elaidic acid give the *threo* (m.p. 28.5–29 °C) and *erythro* (m.p. 29.5–30 °C) isomers of 9,10-dibromostearic acid, respectively. Bromination of linoleic acid should give two *threo,threo*-tetrabromostearic acids and the crystalline product (m.p. 115 °C) is believed to be the *R*/*S*9,*R*/*S*10,*R*/*S*12,*R*/*S*13-racemate and the liquid isomer is the *R*/*S*9,*R*/*S*10,*S*/*R*12,*S*/*R*13-isomer. Linelaidic acid (18:2 9*t*12*t*) gives a tetrabromide (m.p. 78 °C) and linolenic acid has furnished only one crystalline hexabromide (m.p. 185 °C). Chlorination of oleic, linoleic, and linolenic acid gives chlorides melting at 37, 123, and 189 °C, respectively.

Polybromo-acids are readily debrominated by reaction with zinc or with iodide ion, the latter being the more stereospecific process. Since the reaction is a *trans* elimination, the alkene has the same stereochemical configuration as that from which the bromide was produced, so that the bromination–debromination procedure provides a method of protecting double bonds.

Dehydrohalogenation of a *vic*-dihalide with base leads to an alkyne only if the correct isomer is used. The reaction is a *trans* elimination procedure and the first step occurs easily with both the *threo*- and *erythro*-dihalides to furnish the *Z*- and *E*-vinyl halides, respectively. Only the *Z*-isomer reacts readily in the second elimination to furnish the alkyne (equation 36); the more resistant *E*-isomer gives a mixture of allenes and isomeric alkynes produced wholly or partly by allene rearrangement (equation 37). The *Z*- and *E*-isomers of 9(10)-bromo-octadec-9-enoic acid melt at 37 and 8 °C, respectively.

$$\underset{cis}{\text{—CH=CH—}} \longrightarrow \underset{threo}{\text{—CHXCHX—}} \longrightarrow \text{—CH=CX—} \longrightarrow \text{—C≡C—} \qquad (36)$$

$$\underset{trans}{\text{—}CH_2\text{CH=CH—}} \longrightarrow \underset{erythro}{\text{—}CH_2\text{CHXCHX—}} \longrightarrow \underset{E}{\text{—}CH_2\text{CX=CH—}} \longrightarrow \text{—CH=C=CH—} \qquad (37)$$

25.1.10.2 Oxymercuration

Alkenes react with mercuric acetate and methanol to give a methoxy mercuriacetate adduct by *trans* addition. This intermediate regenerates the alkene in its original stereoisomeric form on treatment with hydrochloric acid, and the mercury group is replaceable by halogen through reaction with bromine or by hydrogen in a radical reduction with sodium borohydride (equation 38).

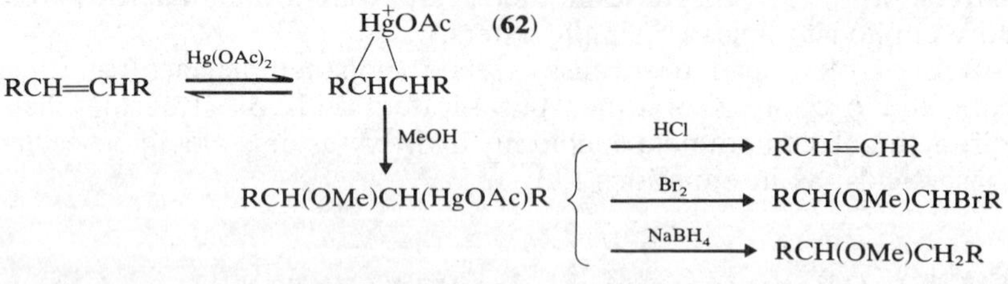

(38)

These processes occur under mild conditions in high yield and can be exploited through several useful modifications. Other mercury salts (nitrate, trifluoroacetate) sometimes give better results than the acetate. The reaction is believed to occur *via* the rapid and reversible formation of the intermediate (**62**), and methanol may be replaced by a wide range of other reagents which give the substituent indicated: EtOH (OEt), AcOH, (OAc), H_2O (OH), H_2O_2 (OOH), Bu^{t}OOH (OOBu^t), and MeCN (NHCOMe). The oxymercuration reaction has been recommended as a method of locating double bonds by mass spectrometry (Section 25.1.6.4).

With appropriately substituted alkenes the reactive intermediate (**62**) may undergo a competitive intramolecular reaction in preference to the intermolecular processes discussed above. This has been observed when hydroxyalkenes are used as substrates (*e.g.* arachidonyl alcohol, equation 39) or during the hydration of methyl linoleate.

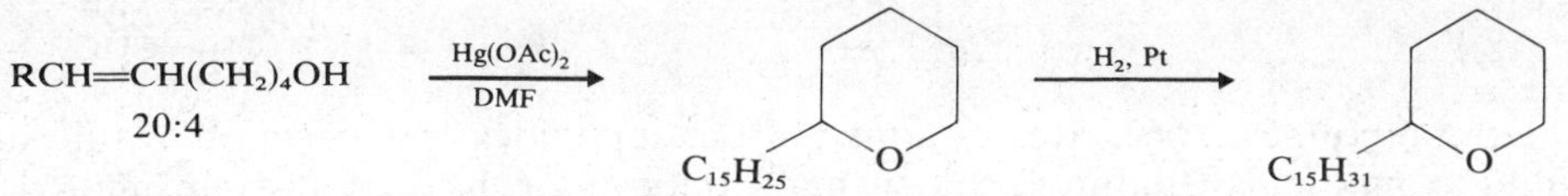

(39)

Acetylenic compounds also react with mercuric acetate and methanol. The organomercuri-compound gives a ketone when treated with acid and an alcohol results from reaction with borohydride.

25.1.10.3 Nitrogen and sulphur addition compounds[56]

This section is concerned with diamines (**63**) aziridines or epimino-compounds (**64**), thiols or mercapto-compounds (**65**), and thiirans or epithio-compounds (**66**).

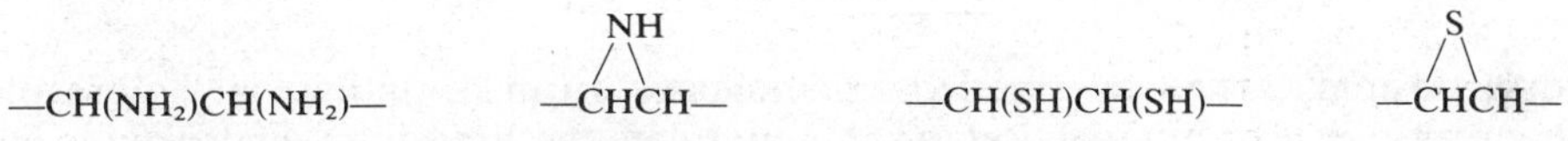

(**63**) (**64**) (**65**) (**66**)

Aziridines may be prepared from alkenes by addition of iodine isocyanate, iodine azide, or *N,N*-dichlorourethane (equation 40). Aziridines, like epoxides, are reactive compounds

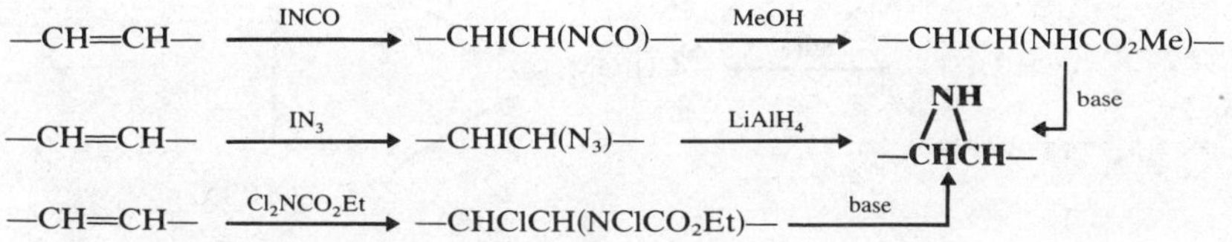

(40)

furnishing a wide range of derivatives (equation 41), including the *vic*-diamines (**63**) which have also been prepared from epoxides. The *threo*- and *erythro*-9,10-diaminostearic acids melt at 98 and 123 °C, respectively.

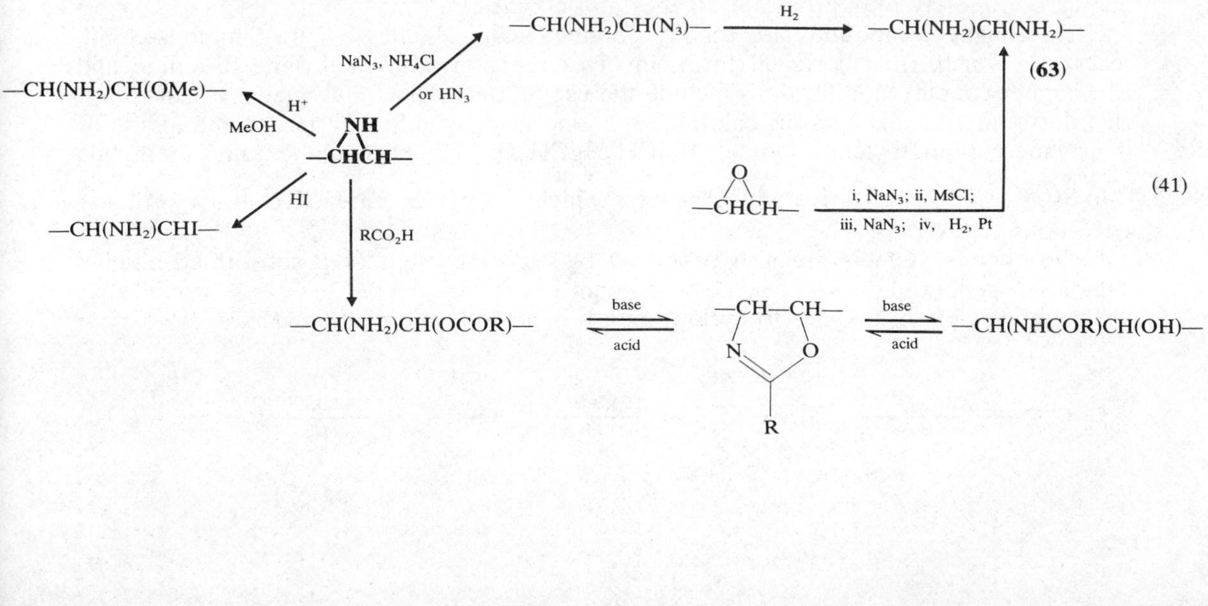

(41)

In the Ritter reaction, alkenes are converted to amido-derivatives by reaction with a nitrile in the presence of strong acid (equation 42). Apart from the formyl derivative, the amides are resistant to hydrolysis. Initial protonation of the alkene gives a carbonium ion, which may give rise to an extensive mixture of products through rearrangement.

$$\text{—CH=CH—} \xrightarrow{H^+} [\text{—CH}_2\overset{+}{\text{C}}\text{H—}] \xrightarrow{\text{RCN}} \text{—CH}_2\text{CH(NHCOR)—} \qquad (42)$$
$$(\text{R} = \text{H, CH}_3\text{,CH}_2\text{=CHCH}_2\text{—})$$

Epithio-compounds (**66**) are obtained from epoxides by reaction with sulphur-containing compounds (thiourea, thiocyanate, thiocyanogen, or potassium methyl xanthate) followed by base or by reaction with 3-methylbenzothiazol-2-one and trifluoroacetic acid. 9,10-Epithiostearic acid melts at 58 (*cis*) and 64 °C (*trans*).

Many thiols undergo radical addition to a double bond in the presence of light and/or a radical initiator, and substrates such as methyl oleate thereby produce a mixture of the 9- and 10-substituted compounds (equation 43). Acetylmercapto-derivatives are easily hydrolysed to the thiols under acidic or basic conditions. Hydrogen sulphide gives a better yield of thiol in the polar reaction with boron trifluoride at −70 °C.

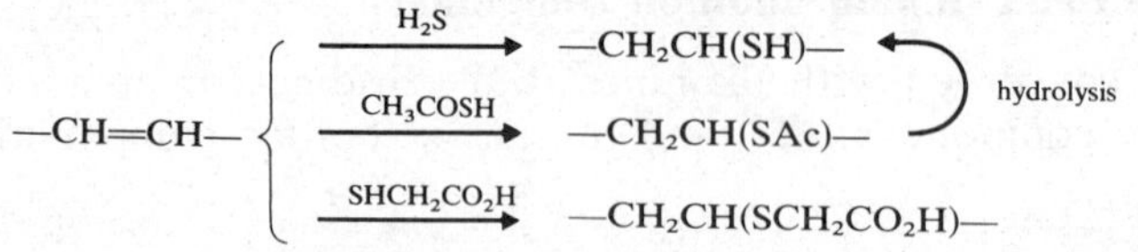

(43)

Hydroxy-esters such as methyl ricinoleate furnish individual mercapto- and acetylmercapto-esters by reaction of the mesylate with sodium hydrogen sulphide or potassium thioacetate in DMF solution (equation 44). With appropriate substrates, 1,4-epithio-, 1,3-epidithio-, and 1,4-epidithio-compounds have been obtained.

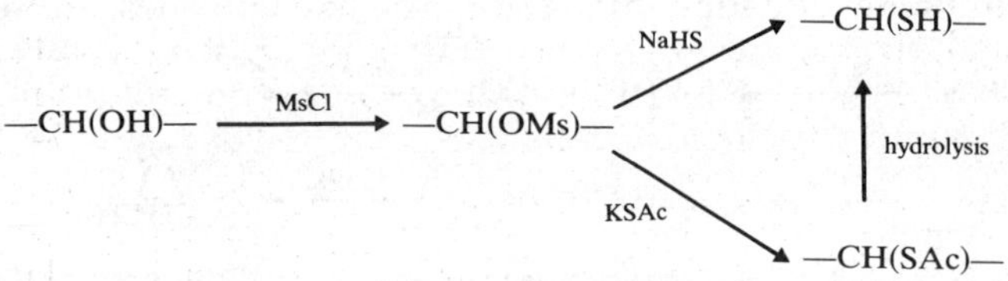

(44)

25.1.10.4 Conversion of alkenes (alkynes) to cyclopropanes (cyclopropenes)

The natural occurrence of both cyclopropane- and cyclopropene-containing fatty acids has led to an interest in chemical methods of preparing these compounds by addition of a one-carbon moiety to appropriate alkenes and alkynes.[13]

Cyclopropane compounds are readily obtained from alkenes by the Simmons–Smith reaction[57] which, in its original form, involves reaction with methylene di-iodide and zinc/copper couple. Modifications include the use of diethylzinc, magnesium or zinc films, zinc dust and aqueous cuprous chloride, and zinc/silver couple with methylene iodide or (benzylmercury)methylene iodide ($PhCH_2HgCH_2I$). Dimethylsulphonium methylide $[Me_2\overset{+}{S}(O)\text{—}\overset{-}{C}H_2]$ is effective with Δ^2-esters which are rather unreactive under ordinary conditions.

Cyclopropane systems are also produced by rearrangement of β-substituted alkenes capable of generating a homoallylic carbonium ion.[50]

The conversion of alkynes to cyclopropenes is possible by reaction (45).[58]

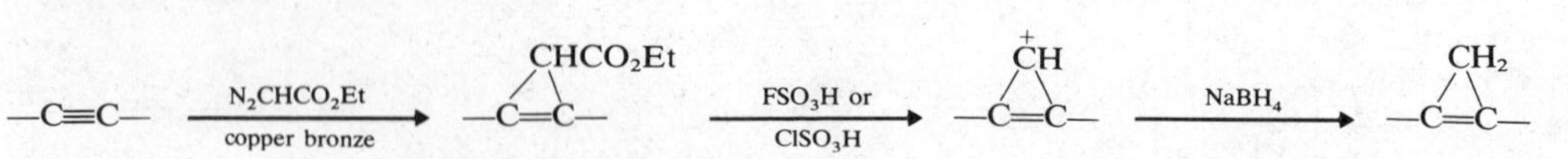

(45)

25.1.10.5 Metathesis[59,60]

The olefin metathesis reaction (46) may be promoted by both homogeneous (a transition complex with an organometallic derivative or Lewis acid) and heterogeneous catalysts (molybedenum or tungsten oxide on alumina or silica). Using tungsten hexachloride and tetramethyltin as catalyst, methyl oleate gives a hydrocarbon and a diester (equation 47), whilst methyl linolenate furnishes cyclohexa-1,4-diene and a 12:1 ester (equation 48). Glycerides give novel compounds and cross-coupling between an unsaturated ester and a hydrocarbon gives products of varying chain length (equation 49). The new unsaturated centres are *cis* and *trans*.

$$R^1CH{=}CHR^1 + R^2CH{=}CHR^2 \rightleftharpoons 2R^1CH{=}CHR^2 \qquad (46)$$

$$CH_3(CH_2)_7CH{=}CH(CH_2)_7CO_2Me \rightleftharpoons CH_3(CH_2)_7CH{=}CH(CH_2)_7CH_3 + MeO_2C(CH_2)_7CH{=}CH(CH_2)_7CO_2Me \qquad (47)$$

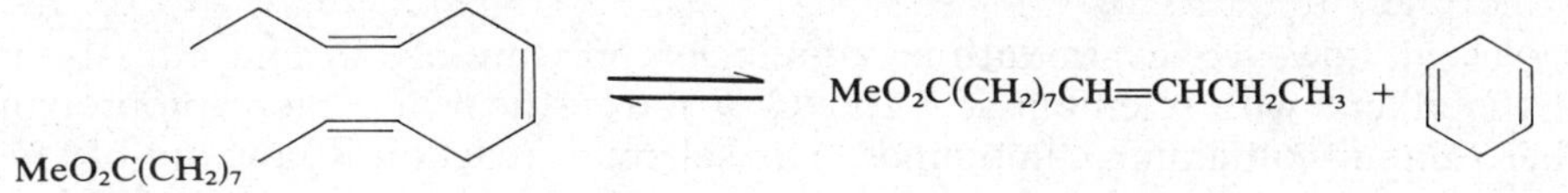

(48)

$$CH_3(CH_2)_7CH{=}CH(CH_2)_7CO_2Me + EtCH{=}CHEt \longrightarrow \text{hydrocarbons (18:1 and 12:1)} + \text{diester (18:1)} + \text{monoester (12:1 } \Delta^9) \qquad (49)$$

25.1.10.6 Other double bond reactions

In acid-catalysed addition reactions the new substituent is seldom confined to the unsaturated carbon atoms and reaction is frequently accompanied or preceded by double bond migration and stereomutation. Thus the simple Friedel–Crafts reaction between oleic acid (or its ester, alcohol, or nitrile) and benzene (or other aromatic compound) in the presence of aluminium trichloride gives a mixture of the 5- to 17-phenyloctadecanoic acids.

Reactions involving addition of carbon monoxide to alkenes yield a compound with an additional formyl or related group. Best known is the hydroformylation (oxo) reaction between a compound like oleic acid, synthesis gas (H_2+CO) under pressure (500–2000 psi), and cobalt carbonyl. Reaction (50) occurs at ~100 °C to furnish a mixture of formylstearic acids accompanied by some hydroxymethylstearic acids. With a rhodium catalyst and triphenylphosphine, isomerization is insignificant and the product is almost entirely 9(10)-formylstearic acid. The reaction may proceed as shown in equation (51).

$$-CH{=}CH- \xrightarrow{\text{hydroformylation}} -CH_2CH(CHO)- \qquad (50)$$

$$Co_2(CO)_8 \overset{H_2}{\rightleftharpoons} HCo(CO)_4 \overset{\text{alkene}}{\rightleftharpoons} \underset{HCo(CO)_3}{-CH{=}CH-} \overset{CO}{\rightleftharpoons} -\underset{Co(CO)_4}{CH}-CH_2-$$

$$\overset{CO}{\rightleftharpoons} -\underset{COCo(CO)_4}{CH}-CH_2- \xrightarrow{H_2} -\underset{CHO}{CH}CH_2- + HCo(CO)_4 \qquad (51)$$

Methyl oleate is converted directly to carboxy-stearates by reaction with carbon monoxide and water using a range of catalysts which include conc. sulphuric acid (Koch reaction), hydrogen fluoride, nickel carbonyl, or a mixture of palladium chloride and triphenylphosphine.

25.1.11 STEREOMUTATION, DOUBLE BOND MIGRATION AND CYCLIZATION, DIENE ADDITION, AND DIMERIZATION

25.1.11.1 Stereomutation

cis- and *trans*-isomers can be interconverted by a series of stereospecific reactions which add up to a stereomutation. A single pure product is usually obtained (equation 52; see also Ref. 61).

$$\underset{cis}{\text{—CH=CH—}} \xrightarrow{ArCO_3H} \underset{cis}{\overset{\text{O}}{\text{CHCH}}} \xrightarrow{LiPPh_2} \underset{threo}{\text{—CH(OH)CH(OPPh}_2\text{)—}} \xrightarrow{MeI} \underset{trans}{\text{—CH=CH—}} \qquad (52)$$

It is more usual, however, to produce an equilibrium mixture of *cis*- and *trans*-isomers by treating the alkene with selenium at ~200 °C or nitrous acid or 3-mercaptopropanoic acid or other sulphur-containing compounds. The selenium reaction is now out of favour because it is known to promote double bond migration and hydrogen transfer, leading to more saturated compounds.

In monoenes and non-conjugated polyenes, each *cis* double bond is converted to its *trans*-form to the extent of 75–80%. Among monoenoic acids the *trans*-isomer is usually the less soluble and can be purified by crystallization. The mixture resulting from a polyene is more complex and whilst the all-*trans* isomer can sometimes be crystallized it is more usual to use silver ion chromatography.

Conjugated polyenes isomerize more readily and give a higher proportion of the all-*trans* isomer. This change is usually effected by exposure to light and iodine.[8]

25.1.11.2 Double bond migration and cyclization

When treated with strong base a double bond may migrate through the sequence of events shown in reaction (53). With methylene-interrupted polyenes reaction occurs under milder conditions and leads to compounds with conjugated unsaturation. Linoleic acid, for example, gives the 9*c*11*t* and 10*t*12*c* octadecadienoic acids, although other compounds may be produced after prolonged reaction. Linolenic acid gives, initially, a mixture of triene acids with diene conjugation and with triene conjugation. Individual acids can sometimes be crystallized from the reaction mixtures and silver ion chromatography is also useful in simplifying complex mixtures of isomers.

$$\text{—CH=CHCH}_2\text{—} \underset{H^+}{\overset{\bar{O}H}{\rightleftharpoons}} \text{—CH=CH}\bar{\text{C}}\text{H—} \longleftrightarrow \text{—}\bar{\text{C}}\text{HCH=CH—} \underset{\bar{O}H}{\overset{H^+}{\rightleftharpoons}} \text{—CH}_2\text{CH=CH—} \qquad (53)$$

Alkali-isomerization of linolenic acid and other triene acids also furnishes mono- and bi-cyclic C_{18} acids as a result of more extensive change. Linolenic acid and linseed oil, which contains appreciable amounts of this acid, treated with alkali in a nitrogen atmosphere at 220–295 °C, furnish compounds with a cyclohexane or indane nucleus such as those shown in equation (54). The indane derivative probably results from an intramolecular diene synthesis requiring a 1,3,8-triene unit. Both types of cyclic acids are readily reduced to perhydro-acids or dehydrogenated to aromatic compounds (equation 54).

Similar compounds are produced from tung oil, which already contains a conjugated trienoic acid (18:3 9*c*11*t*13*t*). Alkali is not required here but the thermal reaction can be facilitated by sulphur, which presumably assists *cis–trans* isomerization. The bicyclic acids such as (**67**) present in tall oil are probably produced from the $\Delta^{5,9,12}$ acid, *via* the 5,10,12-isomer, under the alkaline conditions of wood pulping.

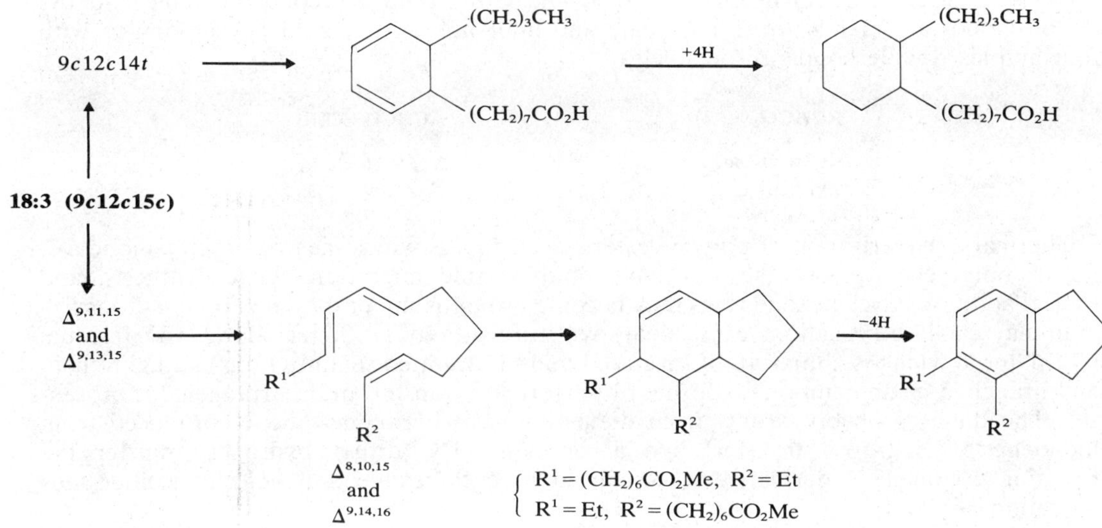

(54)

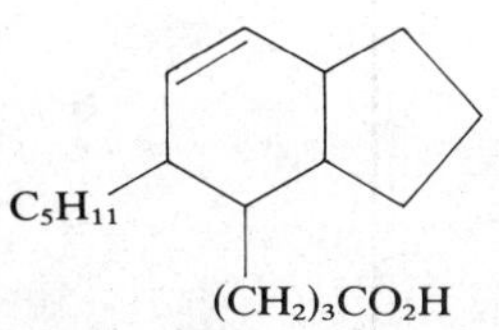

(**67**)

25.1.11.3 Diene addition[62]

The classical diene addition process (Diels–Alder reaction) has had a limited application to unsaturated fatty acids. Diene substrates include natural acids with conjugated unsaturation, alkali-isomerized polyene acids, and dehydrated castor oil. Since there is a requirement for a *trans,trans* conjugated diene, the reaction can be used to determine the configuration of conjugated polyenes. For example, in α-eleostearic acid (9*c*11*t*13*t*) only the all-*trans* 11*t*13*t* system forms an adduct (equation 55).

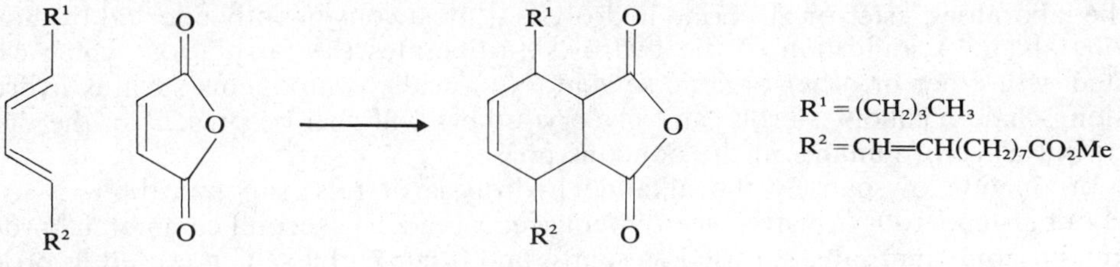

(55)

18:3 (9*c*11*t*13*t*)

Dienophiles include maleic anhydride, tetracyanoethylene, acrylic acid, propiolic acid, and acetylenedicarboxylate. The reaction with acrylic acid, for example, gives a cyclic dibasic acid with interesting properties. Under more vigorous conditions, ethene, propene, and but-1-ene will also serve as dienophiles.

25.1.11.4 Dimerization

Derivatives of dimeric acids remain liquid at very low temperatures and have been used as surface-active agents, corrosion inhibitors, lubricants, and as a source of improved alkyd resins. Radical sources, such as di-t-butyl peroxide, convert esters to acyclic dimers retaining all the original double bonds. Stearate and other saturated esters furnish the 2,2′-dimer (**68**), whilst methyl oleate gives mixed dimers of the type (**69**) in which the new bond is between C(8–11) in one molecule and C(8–11) in a second molecule and the double bonds are mainly *trans*. Linoleate and linolenate give similar acyclic dimers with four and six double bonds, respectively.

$RCHCO_2Me$ | $RCHCO_2Me$

(**68**) $R = C_{16}H_{33}$

ACHCH═CHB | XCHCH═CHY

(**69**)

Thermal polymerization of polyene esters gives dimers which may be acyclic, monocyclic, or polycyclic by processes involving double bond migration, diene synthesis, and radical addition. Cyclohexene rings may become aromatic under the reaction conditions.

In clay-catalysed reactions even oleate will furnish a cyclic dimer, along with residual C_{18} material which is a mixture of saturated and unsaturated (mainly *trans*) straight-chain and branched-chain compounds formed by hydrogen transfer and rearrangement processes. The dimer probably results from diene synthesis between a diene (produced from monoene by hydrogen transfer) and a monoene. By further hydrogen transfer the resulting cyclohexene derivative is converted to cyclohexane and benzene compounds (equation 56).

$$R^1CH{=}CHR^2 \xrightarrow{-2H} \left.\begin{matrix} R^3CH{=}CHCH{=}CHR^4 \\ +R^1CH{=}CHR^2 \end{matrix}\right\} \longrightarrow$$

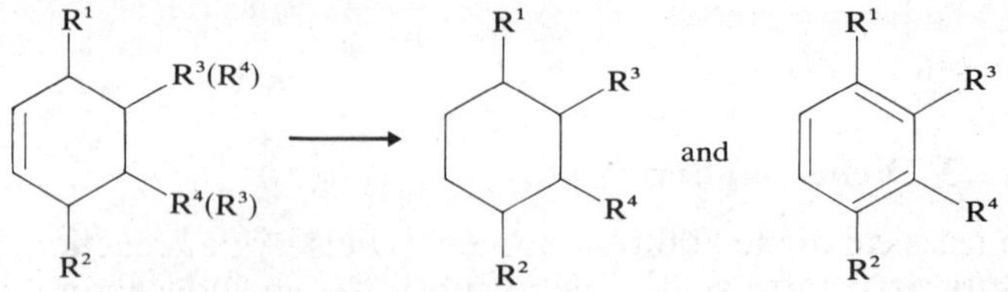

(56)

25.1.12 REACTIONS OF THE CARBOXYL GROUP

25.1.12.1 Hydrolysis

In the laboratory, ester or glyceride hydrolysis is most conveniently effected by aqueous ethanolic alkali. Acidification of the hydrolysate liberates the fatty acids, which can be extracted with ether or other organic solvent. Non-acidic components such as hydrocarbons, long-chain alcohols, sterols, and glyceryl ethers will also be present in the organic extract, but glycerol remains in the aqueous phase.

The production of soap by the alkaline hydrolysis of fats (saponification) is usually carried out around 100 °C, with glycerol being recovered as a second commercial product. Sodium and potassium salts are used as soaps, and other metal salts are used to promote polymerization of drying oils, in the manufacture of grease and lubricants, and as ingredients in plastics formulations.[63]

Fats can also be hydrolysed to free acids by water in what is considered to be a homogeneous reaction between fat and water dissolved in the oil phase (fat splitting). This process occurs in the presence of sulphonated long-chain alkylbenzenes (24–28 h, ~100 °C) or of the oxides of zinc or magnesium or calcium (2–3 h, 250 °C, 700 psi).

25.1.12.2 Esterification, alcoholysis, acidolysis, and interesterification[64,65]

Esters can be prepared by interaction of a carboxylic acid or other acyl compound with an alcohol or equivalent substance (esterification) or by reaction of an ester with an alcohol (alcoholysis), an acid (acidolysis), or another ester (inter-esterification). Each process usually requires an acidic or basic catalyst.

(*i*) *Esterification**

Methyl esters are made by reaction with methanol containing sulphuric acid (1–2%), hydrogen chloride (5%), or boron trifluoride (12–14%) as catalyst, or by reaction with diazomethane. The acid-catalysed procedures are not suitable for acids containing cyclopropane, cyclopropene, or epoxide groups, nor for conjugated polyenes or acids having a hydroxyl group adjacent to a double bond system. Such acids, however, can be safely esterified with diazomethane or the ester can be obtained by base-catalysed trans-esterification procedures.

(*ii*) *Alcoholysis*

Alcoholysis (equation 57) is widely employed to convert a lipid directly to a methyl ester without going through the free acid and to prepare partial glycerides through interaction of a triglyceride with glycerol.

$$RCOOR^1 + R^2OH \xrightleftharpoons{\text{catalyst}} RCOOR^2 + R^1OH \tag{57}$$

For methanolysis the glyceride is dissolved in an excess of methanol containing an acidic (sulphuric acid, hydrogen chloride, boron trifluoride) (equation 58) or basic catalyst (0.5 M sodium methoxide) (equation 59), usually with benzene or dichloromethane as a co-solvent.

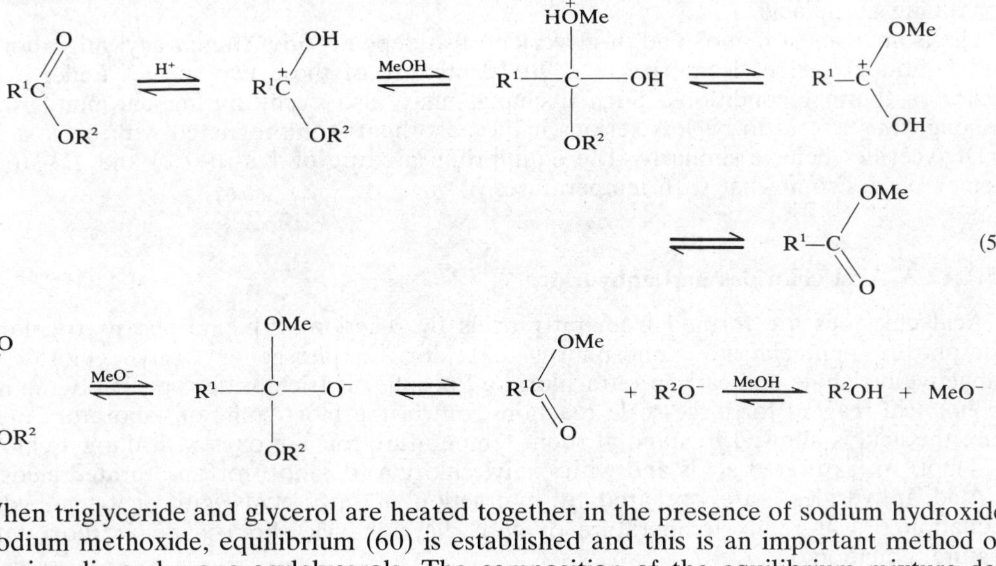

(58)

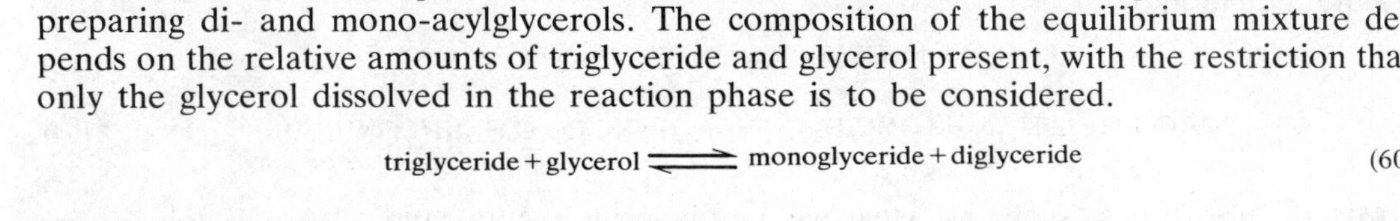

(59)

When triglyceride and glycerol are heated together in the presence of sodium hydroxide or sodium methoxide, equilibrium (60) is established and this is an important method of preparing di- and mono-acylglycerols. The composition of the equilibrium mixture depends on the relative amounts of triglyceride and glycerol present, with the restriction that only the glycerol dissolved in the reaction phase is to be considered.

$$\text{triglyceride} + \text{glycerol} \rightleftharpoons \text{monoglyceride} + \text{diglyceride} \tag{60}$$

(*iii*) *Acidolysis*

This involves the interaction of an ester with an acid in the presence of sulphuric acid or

* The acylation of glycerol to give glycerides or phosphoglycerides is discussed in Section 25.2.3.

zinc oxide or calcium oxide (equation 61). Applied to natural glycerides and lauric acid, for example, C_{16} and C_{18} acids are displaced by the C_{12} acid.

$$R^1COOR + R^2COOH \rightleftharpoons R^2COOR + R^1COOH \quad (61)$$

(*iv*) *Inter-esterification*[66]

A natural fat or oil is a mixture of triacylglycerols in which the fatty acids present are distributed in a selective manner (Section 25.2.2.4). Under the influence of a basic catalyst (sodium hydroxide or sodium methoxide or a sodium potassium alloy), at ~80 °C the acyl groups are redistributed until a random distribution is obtained with a consequent change in the physical properties of the mixture. Thus the melting point of soybean oil can be raised from −7 to +6 °C and of cottonseed oil from 10 to 34 °C.

Applied to a mixture of oils and fats, the fatty acids in each are randomly distributed through the whole mixture. This provides a method of transferring saturated fatty acids to predominantly unsaturated glycerides and *vice versa.*

A slightly different result is obtained if the inter-esterification process is carried out at such a temperature (10–40 °C) that fully saturated glycerides (S_3) crystallize from the reaction mixture (directed inter-esterification). Crystallization disturbs the equilibrium (62), so that more S_3 is produced and again separates, producing finally a mixture with more S_3, more fully unsaturated glyceride (U_3), and less of the glycerides with both saturated and unsaturated acyl groups. This causes an increase in the melting range of the fat.

$$U_3 \rightleftharpoons U_2S \rightleftharpoons US_2 \rightleftharpoons S_3 \quad (62)$$

Such procedures are used industrially to improve the physical properties of lard, to produce cocoa butter substitute from more readily available oils, and to produce fats containing acetic acid.

The isomerism of mono- and di-glycerides also depends on a similar acyl migration. 1- and 2-monoglycerides isomerize to a 90:10 mixture of these two species under acidic, basic, or thermal conditions. Such a change may also occur during chromatography, although this is said to be less serious if the adsorbent is impregnated with boric acid.

Diglycerides behave similarly. The equilibrium mixture of 1,2- (40%) and 1,3- (60%) isomers varies somewhat with temperature.

25.1.12.3 Acid chlorides and anhydrides

Acid chlorides are formed from fatty acids by reaction with phosphorus trichloride, phosphorus pentachloride, phosphorus oxychloride, phosgene, oxalyl chloride, or triphenylphosphine and carbon tetrachloride. Phosphorus trichloride is probably the most economical reagent for large-scale reactions, but in the most common laboratory procedure the acid is allowed to stand at room temperature for 3–5 days with thionyl chloride (1.4 mol) for saturated acids and with oxalyl chloride (0.8 mol) for unsaturated acids.

Acid anhydrides[67] are prepared by interaction of fatty acids with acetic anhydride (equation 63) at reflux temperature or with dicyclohexylcarbodi-imide at room temperature (equation 64).

$$2RCO_2H + (MeCO)_2O \rightleftharpoons (RCO)_2O + 2MeCO_2H \quad (63)$$

$$2RCO_2H + C_6H_{11}N{=}C{=}NC_6H_{11} \longrightarrow (RCO)_2O + C_6H_{11}NHCONHC_6H_{11} \quad (64)$$

Mixed anhydrides such as stearoyl methanesulphonate and stearoyl toluene-*p*-sulphonate are said to be particularly effective acylating agents (equations 65 and 66).

$$RCOOC(CH_3){=}CH_2 + MeSO_3H \longrightarrow RCOOSO_2Me + (CH_3)_2CO \quad (65)$$

$$RCOCl + AgOSO_2C_6H_4Me \longrightarrow RCOOSO_2C_6H_4Me + AgCl \quad (66)$$

25.1.12.4 Nitrogen-containing derivatives[68,69]

This section is concerned with fatty acid-derived long-chain nitrogen compounds used in plastics formulations and as lubricants, detergents, and froth-flotation agents.

Amides ($RCONH_2$) are obtained industrially by reaction of fatty acids with ammonia (or another amine) at ~200 °C.

Nitriles (RCN), used mainly to produce amines, are themselves made either by thermal decomposition of amides or by direct reaction of acids and ammonia in the presence of a dehydrating catalyst such as alumina.

Primary amines (RCH_2NH_2) are most commonly prepared on an industrial scale by catalytic reduction (Raney nickel) of nitriles in the presence of ammonia. Under other conditions the secondary amine $(RCH_2)_2NH$ is the major product.

25.1.12.5 Sulphur compounds[70,71]

Sulphuric acid, sulphur trioxide, or sulphamic acid form sulphates of long-chain alcohols. The related methanesulphonates (or toluene-*p*-sulphonates), often more conveniently prepared and more reactive than alkyl halides, are readily converted to hydrocarbons, alkyl halides, nitriles, malonates, esters, ethers, hydroperoxides, and aldehydes.

25.1.12.6 The preparations and reactions of α-anions of carboxylic acids[72]

Saturated and monoene carboxylic acids react with lithium di-isopropylamide to form dianions ($R\bar{C}HCO_2^-$) which are useful intermediates for the preparation of α-alkylated, α-hydroxy-, α-hydroperoxy-, and α-iodo-acids and for aldehydes and nitro-compounds which are formed with concommitant decarboxylation.

References

1. F. D. Gunstone, *Prog. Chem. Fats Lipids*, 1977, **15,** 75.
2. C. Hitchcock and B. W. Nichols, 'Plant Lipid Biochemistry', Academic Press, London, 1971, p. 3.
3. C. R. Smith, Jr., *Topics Lipid Chem.*, 1970, **1,** 277.
4. L. A. Witting in 'Modification of Lipid Metabolism', ed. E. G. Perkins and L. A. Witting, Academic Press, New York, 1975, p. 19.
5. H. Schlenk, *Prog. Chem. Fats Lipids*, 1970, **9,** 587.
6. G. F. Spencer, S. F. Herb, and P. J. Gormisky, *J. Amer. Oil Chemists' Soc.*, 1976, **53,** 94.
7. E. W. Horton, *Chem. Soc. Rev.*, 1975, **4,** 589.
8. C. Y. Hopkins, *Topics Lipid Chem.*, 1972, **3,** 37.
9. C. R. Smith Jr., *Prog. Chem. Fats Lipids*, 1970, **11,** 137.
10. N. Polgar, *Topics Lipid Chem.*, 1971, **2,** 207.
11. A. K. Lough, *Prog. Chem. Fats Lipids*, 1975, **14,** 1.
12. J. S. Buckner and P. E. Kolattukudy, in 'Chemistry and Biochemistry of Natural Waxes', ed. P. E. Kolattukudy, Elsevier, Amsterdam, 1976, p. 148.
13. W. W. Christie, *Topics Lipid Chem.*, 1970, **1,** 1.
14. D. T. Downing, *Rev. Pure App. Chem.*, 1961, **11,** 196.
15. C. Hitchcock and B. W. Nichols, Ref. 2, p. 22.
16. 'Recent Advances in the Chemistry and Biochemistry of Plant Lipids', ed. T. Galliard and E. I. Mercer, Academic Press, London, 1975.
17. C. Hitchcock, Ref. 16, p. 1.
18. P. E. Kolattukudy, Ref. 16, p. 203; P. E. Kolattukudy and T. J. Walton, *Prog. Chem. Fats Lipids*, 1972, **13,** 119.
19. F. D. Gunstone, Ref. 16, p. 21.
20. G. R. Jamieson, *Topics Lipid Chem.*, 1970, **1,** 107; *J. Chromatog. Sci.*, 1975, **13,** 491.
21. J. M. Osbond, *Prog. Chem. Fats Lipids*, 1966, **9,** 119.
22. W. H. Kunau, *Chem. Phys. Lipids*, 1973, **11,** 254.
23. C. Hitchcock and B. W. Nichols, Ref. 2, p. 96.

24. S. J. Wakil, in 'Lipid Metabolism', ed. S. J. Wakil, Academic Press, London, 1970, p. 1.
25. J. L. Harwood, Ref. 16, p. 44.
26. P. K. Stumpf, Ref. 16, p. 95.
27. D. Chapman, 'The Structure of Lipids by Spectroscopic and X-Ray Techniques', Methuen, London, 1965.
28. W. G. de Ruig, 'Infrared Spectra of Monoacid Triglycerides', Agricultural Research Report 759, Wageningen, 1971.
29. J. S. Showell, *Prog. Chem. Fats Lipids*, 1965, **8,** 253.
30. I. Fischmeister, *Prog. Chem. Fats Lipids*, 1975, **14,** 91; R. N. Jones, *Canad. J. Chem.*, 1962, **40,** 301.
31. I. A. Wolff and T. K. Miwa, *J. Amer. Oil Chemists' Soc.*, 1965, **42,** 208.
32. J. E. D. Davies *et al.*, *J.C.S. Perkin II*, 1972, 1557; *Chem. Phys. Lipids*, 1975, **15,** 48, 157.
33. C. Y. Hopkins, *Prog. Chem. Fats Lipids*, 1965, **8,** 213.
34. D. J. Frost and F. D. Gunstone, see references in Table 5.
35. See references in Table 6.
36. J. A. McCloskey, *Topics Lipid Chem.*, 1970, **1,** 369.
37. A. Zeman and H. Scharman, *Fette Seifen Anstrichm*, 1972, **74,** 509; 1973, **75,** 32.
38. R. D. Plattner, G. F. Spencer, and R. Kleiman, *Lipids*, 1976, **11,** 222.
39. B. Å. Andersson and R. T. Holman, *Lipids*, 1974, **9,** 185; B. Å. Andersson. W. H. Heimermann, and R. T. Holman, *ibid.*, 1974, **9,** 443.
40. B. Å. Andersson, W. W. Christie, and R. T. Holman, *Lipids*, 1975, **10,** 215; B. Å. Andersson, and R. T. Holman, *ibid.*, 1975, **10,** 716.
41. H. J. Dutton, *Prog. Chem. Fats Lipids*, 1968, **9,** 349.
42. E. N. Frankel and H. J. Dutton, *Topics Lipid Chem.*, 1970, **1,** 161.
43. P. van der Plank and H. J. van Oosten, *J. Catalysis*, 1975, **38,** 223.
44. R. Viviani, *Adv. Lipid Res.*, 1970, **8,** 267.
45. W. O. Lundberg and P. Järvi, *Prog. Chem. Fats Lipids*, 1968, **9,** 377.
46. G. A. Veldink, J. F. G. Vliegenthart, and J. Boldingh, *Prog. Chem. Fats Lipids*, 1975, **15,** 131.
47. H. W. Gardner, *J. Agric. Food Chem.*, 1975, **23,** 129.
48. D. A. Forss, *Prog. Chem. Fats Lipids*, 1972, **13,** 177.
49. D. Swern, *J. Amer. Oil Chemists' Soc.*, 1970, **47,** 424.
50. F. D. Gunstone, *Accounts Chem. Res.*, 1976, **9,** 34.
51. F. D. Gunstone, *Adv. Org. Chem.*, 1960, **1,** 103.
52. O. S. Privett, *Prog. Chem. Fats Lipids*, 1966, **9,** 91.
53. R. G. Kadesch, *Prog. Chem. Fats Lipids*, 1963, **6,** 291.
54. E. H. Pryde and J. C. Cowan, *Topics Lipid Chem.*, 1971, **2,** 1.
55. R. Criegee, *Angew. Chem. Internat. Edn.*, 1975, **14,** 745.
56. G. Maerker, *Topics Lipid Chem.*, 1971, **2,** 159.
57. H. E. Simmons, T. L. Cairns, S. A. Vladuchick, and C. M. Hoiness, *Org. Reactions*, 1973, **20,** 1.
58. J. L. Williams, and D. S. Sgoutas, *J. Org. Chem.*, 1971, **36,** 3064; *Chem. Phys. Lipids*, 1972, **9,** 295.
59. R. J. Haines and G. J. Leigh, *Chem. Soc. Rev.*, 1975, **4,** 155.
60. P. B. van Dam, M. C. Mittelmeijer, and C. Boelhouwer, *J. Amer. Oil Chemists' Soc.*, 1974, **51,** 389.
61. P. E. Sonnet and J. E. Oliver, *J. Org. Chem.*, 1976, **41,** 3279, 3284.
62. H. P. Kaufmann, *Fette Seifen Anstrichm.*, 1962, **64,** 1115.
63. N. Pilpel, *Chem. Rev.*, 1963, **63,** 221.
64. W. W. Christie, *Topics Lipid Chem.*, 1972, **3,** 171.
65. W. W. Christie, 'Lipid Analysis', Pergamon, Oxford, 1973.
66. H. H. Hustedt, *J. Amer. Oil Chemists' Soc.*, 1976, **53,** 390.
67. N. O. V. Sonntag, J. R. Trowbridge, and I. J. Krems, *J. Amer. Oil Chemists' Soc.*, 1954, **31,** 151.
68. N. O. V. Sonntag, in 'Fatty Acids', ed. K. S. Markley, 2nd edn. vol. 3, p. 1551.
69. S. H. Shapiro, in 'Fatty Acids and their Industrial Applications', ed. E. S. Pattison, 2nd edn., Arnold, London, 1968, p. 77.
70. K. S. Markley, in 'Fatty Acids, ed. K. S. Markley, 2nd edn., vol. 3, p. 1717.
71. F. Spener, *Chem. Phys. Lipids*, 1973, **11,** 229.
72. F. D. Gunstone, *Rev. franc. Corps Gras*, 1976, **23,** 539.

25.2

Lipids

F. D. GUNSTONE
University of St. Andrews

25.2.1 LIPID STRUCTURE

25.2.1.1 Introduction

A simple chemical division can be made between lipids in which fatty acids occur as *esters* usually, but not always, of glycerol (propane-1,2,3-triol) and those in which the fatty acids are present as *amides* of long-chain amines. Another classification into simple or neutral lipids (triacylglycerols, some ether lipids, cholesterol esters, wax esters) and complex or polar lipids (phosphoglycerides, glycosyldiacylglycerols, sphingolipids) is based mainly on chromatographic behaviour. Any lipid which contains a sugar residue is a *glycolipid*, any lipid based on phosphoric acid is a *phospholipid*, and any lipid containing a sulphate group is a *sulpholipid*.

Appropriate derivatives of glycerol contain a chiral centre and may exist in enantiomeric forms which can be described in D/L or R/S terminology. Both these systems, however, suffer from certain disadvantages and a new system is now generally employed.[1] When glycerol is represented by the Fischer projection (**1**) the three carbon atoms are numbered 1–3 reading from top to bottom and the prefix *sn*- (*stereospecific numbering*) is used to indicate that this system is being employed. The compounds (**2**)–(**4**) are examples of this system. Symbols α and β are sometimes used to designate groups attached to the primary and secondary hydroxyl groups, respectively.

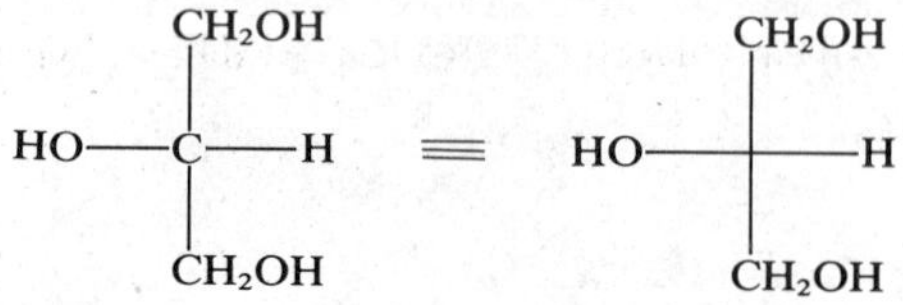

(**1**)

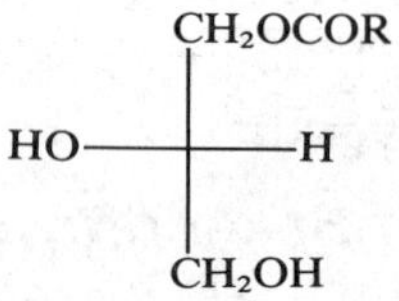

(**2**) 1-Acyl-*sn*-glycerol

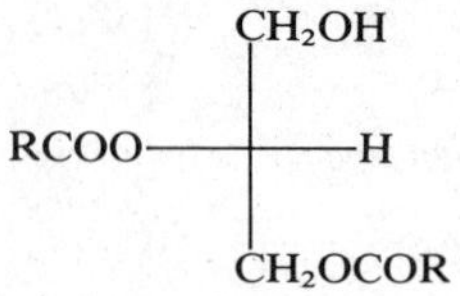

(**3**) 2,3-Diacyl-*sn*-glycerol

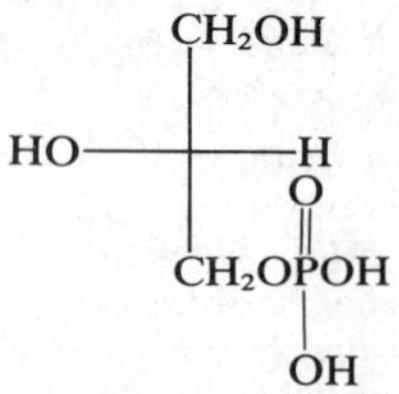

(**4**) *sn*-Glycerol 3-phosphate (more correctly *sn*-glycero-3-dihydrogen phosphate)

25.2.1.2 Acylglycerols

The most familiar lipids are the acyl esters of glycerol known as *monoacylglycerols* (monoglycerides), *diacylglycerols* (diglycerides), or *triacylglycerols* (triglycerides). The last group are commonly described as fats when solid at ambient temperature and as oils when liquid.

Monoacylglycerols exist as 1-acyl (**5**) or 2-acyl (**6**) isomers which may be designated α- and β-monoglycerides, respectively. The 1- and 3-acyl isomers are enantiomers which together comprise the racemic α-monoglyceride.

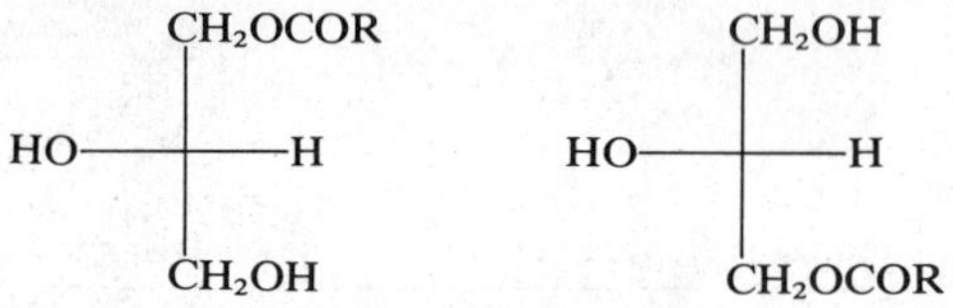

(**5**) 1(3)-Acylglycerol

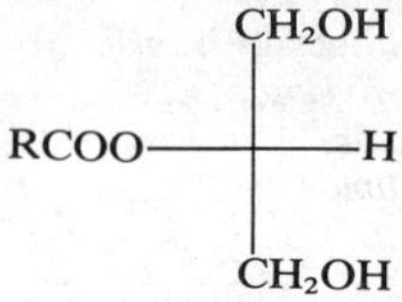

(**6**) 2-Acylglycerol

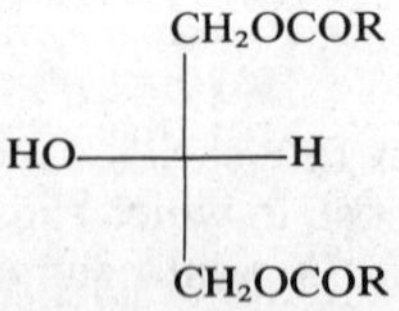

(**7**) 1,3-Diacylglycerol

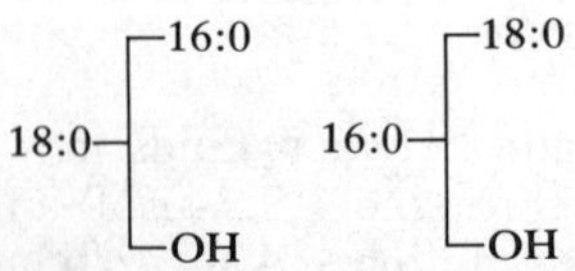

(**8***) 1-Palmitoyl-2-stearoyl-*sn*-glycerol and its isomer

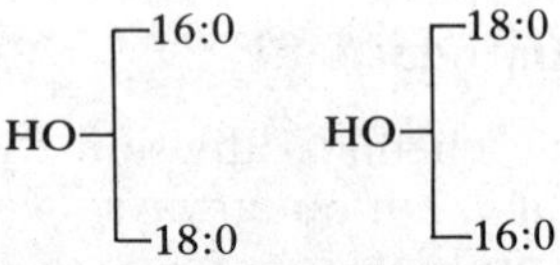

(**9***) 1-Palmitoyl-3-stearoyl-*sn*-glycerol and its enantiomer

* These symbols indicate the glycerol structure with attached hydroxy- or acyloxy-groups.

Diacylglycerols exist as α,β (1,2- and 2,3-diacyl esters, **3**) and as α,α' (1,3-diacylglycerols, **7**).

Natural oils and fats are mainly triacylglycerols, this being the most significant class of storage lipids in plants and in most animals, although not necessarily in marine lipids. Natural triacylglycerols usually contain two or three different acyl groups selected from all those present in the source. The number of possible compounds rises rapidly with the number of available acyl groups and is even greater if allowance is made for enantiomeric possibilities (see Section 25.2.2.1). There are eight triacylglycerols containing only palmitic acid and/or oleic acid, *viz.* PPP, PPO, POP, OPP, POO, OPO, OOP, and OOO, where P and O represent palmitoyl and oleoyl groups attached to C-1, C-2, and C-3 of glycerol, respectively; PPO and OPP are an enantiomeric pair, as are POO and OOP. With three different acyl groups there are 27 different compounds, including nine enantiomeric pairs.

On hydrolysis with aqueous alkali or acid, acylglycerols furnish glycerol and the acids present in the glyceride. Enzymic hydrolysis is more selective. Lipases present in animals, higher plants, and micro-organisms usually remove acyl groups only from the primary hydroxyl groups, leaving a 2-monoacylglycerol which is deacylated only after isomerization to an α-monoacylglycerol. This selectivity is important in the digestion of dietary fat and is exploited in a method of investigating the distribution of acyl groups in triacylglycerols (Section 25.2.2.4).

25.2.1.3 Acyl derivatives of other alcohols (waxes, cutins and suberins, diol lipids, cyanolipids, sterol esters)

The term wax[2,3] is sometimes used in a physical sense to describe a range of long-chain compounds which share the property of giving the surface they cover a characteristic sheen and having the ability to repel water. This latter property is important in conserving water within the organism and in providing a barrier against the environment. The word wax is also used in a restricted chemical sense to describe only those compounds (*wax esters*) which are esters of long-chain acids and long-chain alcohols and is so employed in the following discussions.

Ester waxes occur widely as a protective coating on plant leaves, fruit skins, insects, and on bird feathers[4] and sometimes as internal storage lipids, particularly in marine organisms. Waxes are usually esters of long-chain monohydric alcohols and long-chain monobasic acids (total chain-lengths C_{30}–C_{60}) or diesters from long-chain alkane-1,2- and -1,3-diols or from hydroxy-acids linked with an acid and an alcohol. The acids and the alcohols may be saturated or unsaturated and are frequently branched or multibranched.

Carnauba wax is a typical vegetable wax containing about 36% of ester waxes. These are C_{46}–C_{64} compounds, mainly C_{54} (13%), C_{56} (27%), C_{58} (16%), and C_{60} (17%) esters derived from C_{30} (16%), C_{32} (60%), and C_{34} (17%) alcohols and appropriate fatty acids.

Cutins and *suberins* serve as protective layers on leaves and other plant tissues and are mainly polymers derived from hydroxy-acids. The cutins are produced from C_{16} and C_{18} hydroxy-acids (see Section 25.1.2.9), whilst the suberins are derived from C_{16}–C_{22} mono- and poly-hydroxy-acids and also contain phenolic compounds as a major component.[5]

Although most lipids are based on glycerol, lipids from a wide range of sources including yeast, seed oils, egg, and liver, contain minor, and occasionally major, components which are acylated derivatives of short-chain diols such as ethane-, propane-, and butane-diols. The *diol lipids* exist as analogues of glycerides and phosphoglycerides. For example, so-called triacylglycerols isolated from the starfish (*Distolasterias nipon*) contain up to 35% of ethanediol derivatives, including the saturated and unsaturated ether lipids (**10**) and (**11**).[6]

Another unusual class of lipids are mono- and di-acyl derivatives of the cyano C_5 alcohols (**12**)–(**15**).

Most samples of extracted lipid contain some free sterol (cholesterol from animal sources, stigmasterol, β-sitosterol, and ergosterol from plant sources) along with their acylated derivatives.

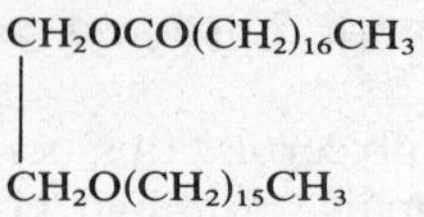

(10)

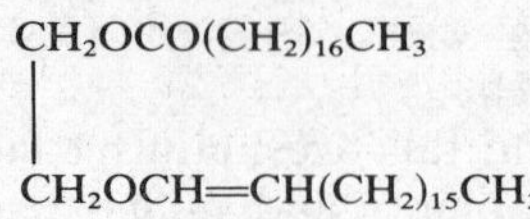

(11)

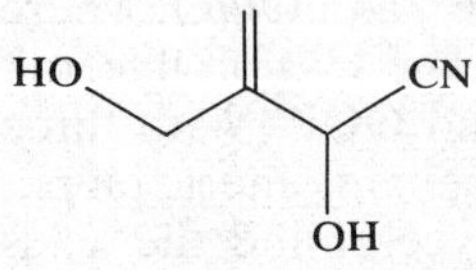

(12)

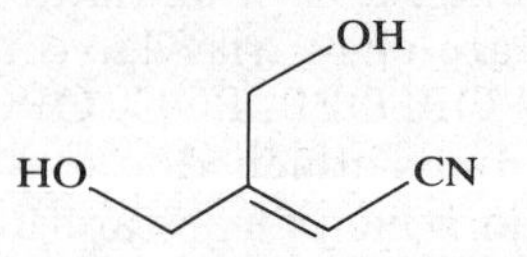

(13)

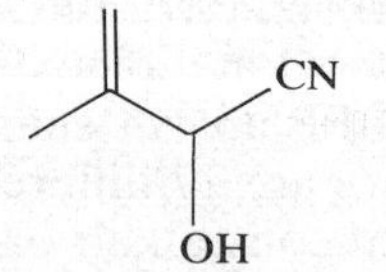

(14)

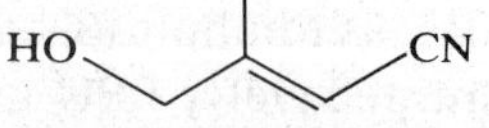

(15)

25.2.1.4 Glycosyldiacylglycerols[7,8]

These glycolipids consist of glycerol acylated at the C-1 and C-2 positions and linked glycosidically at C-3 to a mono-, di-, or tri-saccharide. In higher plants and algae, and particularly in the chloroplast, the sugar residue is gal (**16**), gal-gal, or 6-sulphoquinovose (**17**),[9] and the associated fatty acids are highly unsaturated.

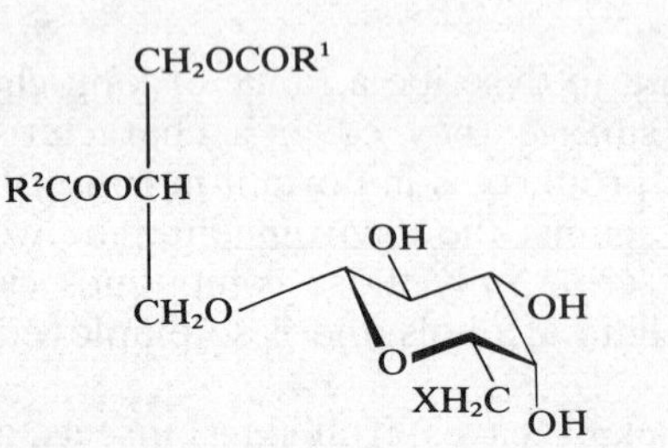

Monogalactosyldiglyceride (MGDG) 1,2-diacyl-[β-D-galactopyranosyl-(1′→3)]-*sn*-glycerol

(**16**) X = OH

Plant sulpholipid (sulphoquinovosyldiglyceride) 1,2-diacyl-[6-sulpho-α-D-quinovopyranosyl-(1′→3)]-*sn*-glycerol

(**17**) X = SO_3H

In bacteria, branched-chain fatty acids become significant and galactose may be accompanied or replaced by glucose or mannose or di- and tri-saccharides built up from these.

25.2.1.5 Phosphoglycerides

Phosphoglycerides are derivatives of glycerophosphoric acid (**18**) which occur widely throughout the animal and vegetable kingdoms, being particularly associated with biological membranes. These not only provide a physical barrier separating cells and subcellular organelles from their environment, but they are intimately involved in essential life processes. They facilitate and control the transport of metabolites between the environments they separate and are involved in many cell functions (see Section 25.3.1).

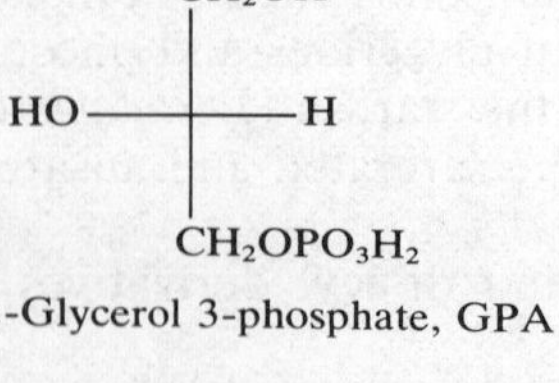

(**18**) *sn*-Glycerol 3-phosphate, GPA

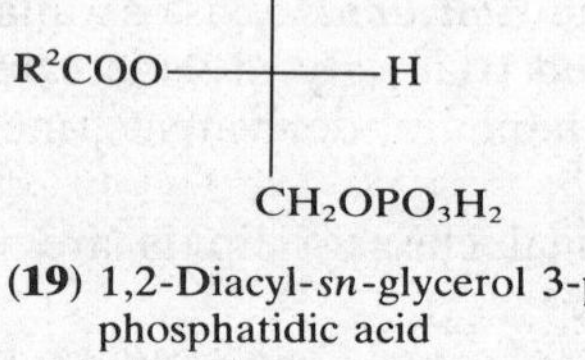

(**19**) 1,2-Diacyl-*sn*-glycerol 3-phosphate, phosphatidic acid

The *phosphatidic acids* (**19**), which are diacylated derivatives of *sn*-glycerol 3-phosphate (**18**), are important as structural units of other compounds and as biosynthetic intermediates. The ester linkage between glycerol and phosphoric acid is more resistant to hydrolysis — especially by alkali — than the glycerol–fatty acid linkages, so that mild hydrolysis gives fatty acids and a mixture of the 3- and 2- (α and β) isomers of glycerol phosphate. When phosphatidic acid is combined with other hydroxyl-containing compounds it is designated by the term phosphatidyl, which includes the two acyl groups attached to the glycerol chain.

Other lipids which on complete hydrolysis furnish only glycerol, phosphoric acid, and fatty acids include *phosphatidylglycerol* (**20**), an important component of photosynthetic tissue where it is particularly associated with the 16:1 3*t* acid, and *diphosphatidylglycerol* (**21**, cardiolipin) which is usually rich in linoleic acid and is a major lipid component of mitochondria. These compounds exist in partially acylated forms which are metabolically important.

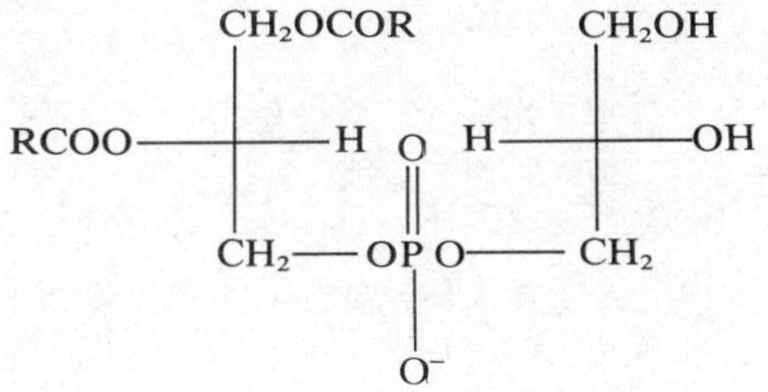

(**20**) 3-*sn*-Phosphatidyl-1′-*sn*-glycerol, PG

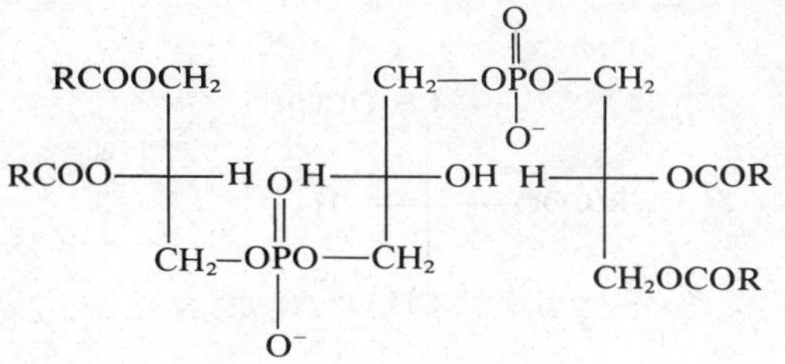

(**21**) 1′,3′-Di-*O*-(3-*sn*-phosphatidyl)-*sn*-glycerol, cardiolipin

The most important phosphoglycerides are the *phosphatidyl esters*[7] in which phosphatidic acid (**19**) is esterified with another alcohol to produce a diester of phosphoric acid (a monohydrogen phosphate). The hydroxy-compounds most frequently incorporated include ethanolamine, choline, serine, and inositol, furnishing the phosphatidyl esters (**22**)–(**25**). Other alcohols which may be incorporated include *N*-methyl- and *N,N*-dimethyl-ethanolamine and carnitine (3-hydroxy-4-trimethylaminobutanoic acid). Phosphatidylcholine was formerly known as lecithin and the term cephalin described a mixture of phosphatidylethanolamines and phosphatidylserines. The correct names of these esters are given along with their structures and accepted abbreviations (**22**)–(**25**). In these

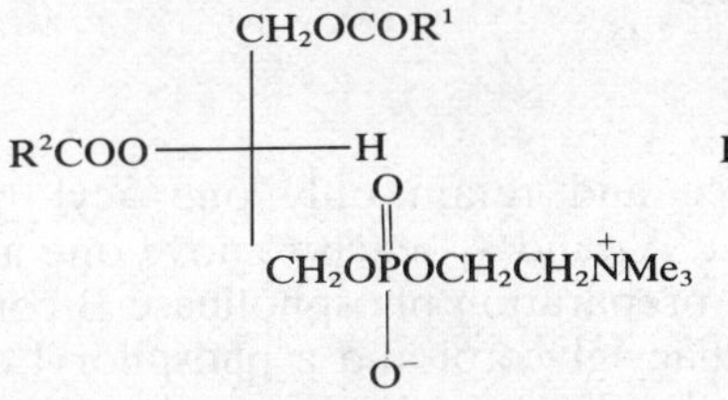

(**22**) 3-*sn*-Phosphatidylcholine, PC

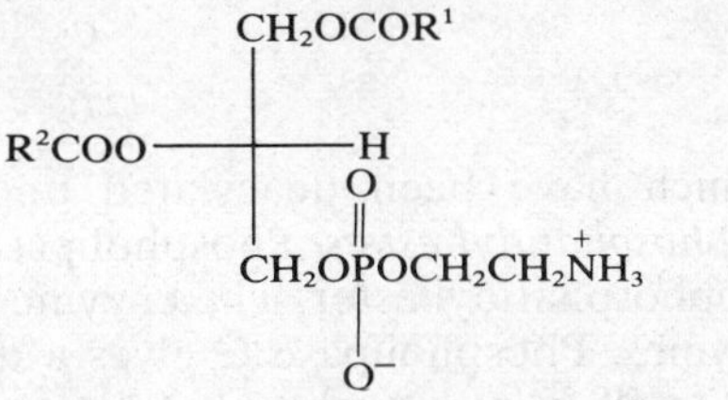

(**23**) 3-*sn*-Phosphatidylethanolamine, PE

phospholipids, unsaturated acids are usually concentrated at C-2 and saturated acids at C-1. Phosphatidylinositol also exists in forms in which other hydroxyl groups in the inositol segment are linked to additional phosphoric acid groups or to sugar units. All the natural phosphatidyl esters are based on *sn*-glycerol 3-phosphate.

Other structural variations include ether derivatives of the phosphoglycerides (Section 25.2.1.6) and phosphono-lipids (such as **26**) which are derivatives of the hypothetical dibasic phosphonic acid [$HPO(OH)_2$] rather than the tribasic phosphoric acid [$PO(OH)_3$]. These have been discovered as ethanolamine derivatives in marine invertebrates and in protozoa.

Complete hydrolysis of the phosphatidyl esters is easily effected in acidic media. With alkali the acyl groups are hydrolysed quickly and the remaining compound, such as glycerophosphorylcholine, is only slowly converted to choline and the isomeric mixture of

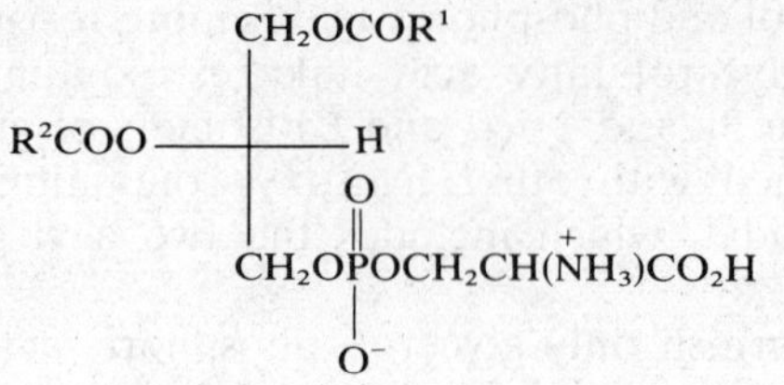

(**24**) 3-*sn*-Phosphatidylserine, PS

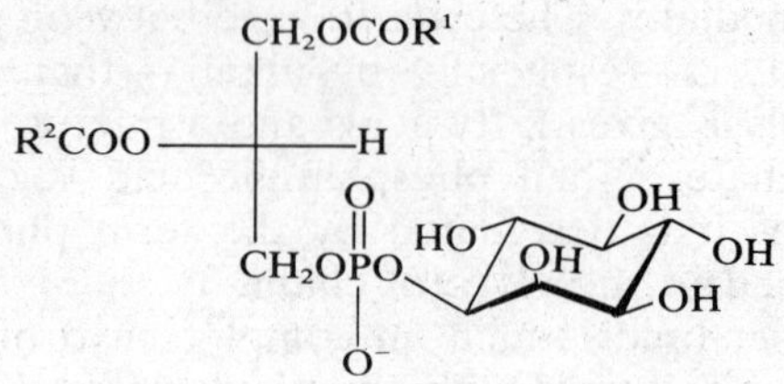

(**25**) 3-*sn*-Phosphatidyl-1′-myoinositol, PI (also occurs as the 4′-phosphate and the 4′,5′-diphosphate)

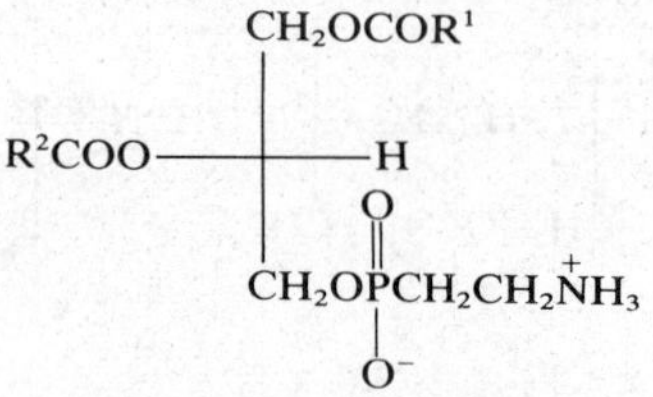

(**26**) a phosphono lipid

α- and β-glycerophosphoric acids. Enzymes operate specifically on most of the phosphoglycerides, effecting hydrolysis selectively at one or other of the four ester linkages (**27**).

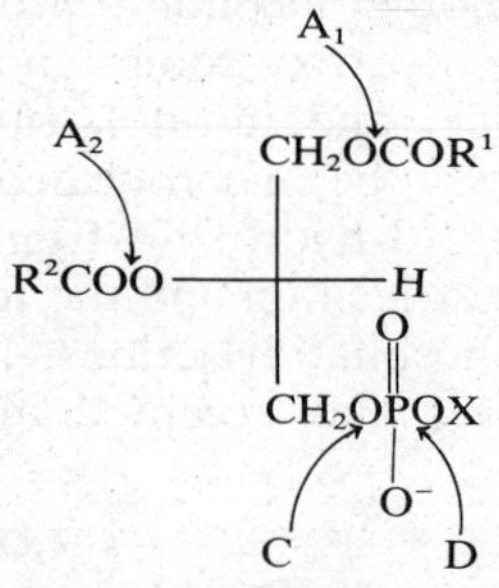

(**27**)

Compounds which have been deacylated once and retain only one acyl group are designated *lysophosphatidyl esters.* Phospholipase A_1 and A_2 each remove one acyl group and leave a lysophosphatidyl ester. The enzyme preparation phospholipase B contains the A_1 and A_2 enzymes. Phospholipase C gives a diacylglycerol and a phosphoryl ester, and phospholipase D will furnish a phosphatidic acid and the appropriate alcohol.

25.2.1.6 Ether lipids[10,11]

Ether lipids are glycerides in which the C-1 acyl group is replaced by an alkyl group, thus creating an ether bond in place of the usual ester bond. Ether lipids (**28**)–(**31**) are based on triacylglycerols or phosphoglycerides and may contain alkyl or alk-*cis*-1′-enyl chains.

When the alkyl group is saturated or has unsaturation in some position (commonly $\Delta^{9'}$) other than $\Delta^{1'}$, the compound behaves as a typical ether. It is resistant to alkaline hydrolysis and to reduction with lithium aluminium hydride and is cleaved only under vigorous acidic conditions. Alkyldiacylglycerols occur in marine lipids and hexadecylglycerol (chimyl alcohol), octadecylglycerol (batyl alcohol), and octadec-9-enylglycerol (**32**, selachyl alcohol) are typical hydrolysis products of such oils.

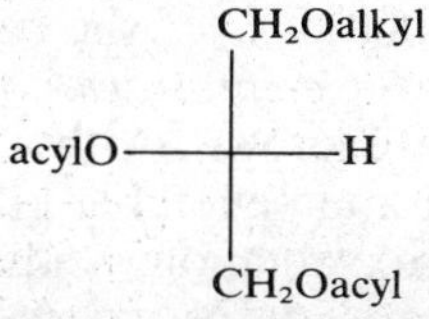

(28)

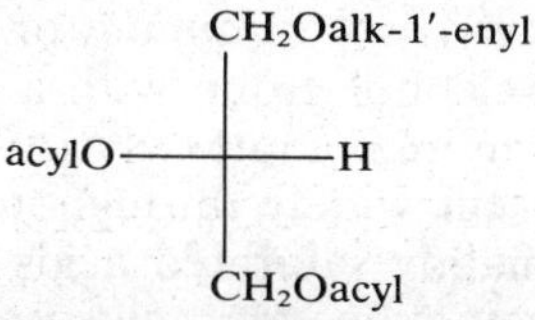

(29)

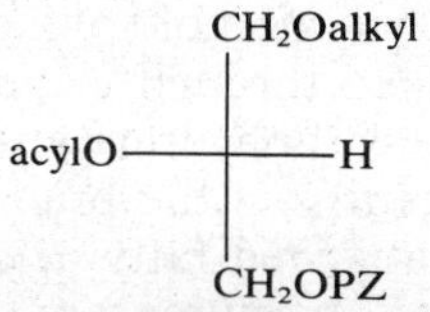

(30)

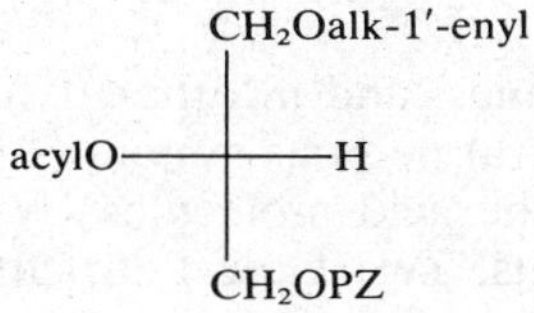

(31)

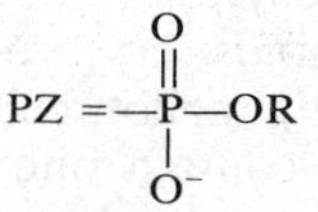

where R is ethanolamine, choline, *etc.*

(32) $CH_3(CH_2)_7CH{=}CH(CH_2)_8OCH_2CH(OH)CH_2OH$

Related to these, but showing novel properties, are lipids containing an alk-*cis*-1′-enyl group and choline and ethanolamine, phosphoglyceride ether lipids of this type, occur widely in plants and animals.[12] Although stable to alkali and to lithium aluminium hydride reduction, these vinyl ethers are readily hydrolysed by dilute acid, furnishing aldehydes in addition to the other expected products (equation 1).

$$R^1CH{=}CHOCH_2R^2 + H_2O \xrightarrow{H^+} R^1CH_2CHO \qquad (1)$$

25.2.1.7 Sphingolipids[13]

The sphingolipids are amides with fatty acids acylating a long-chain amine such as sphingenine. The acylated sphingenines (*ceramides*, **33**) occur mainly as derivatives in which the primary hydroxyl function is associated with a sugar unit or with the kind of phosphate ester already encountered in the phospholipids. The sphingolipids resemble the phosphoglycerides in having two long hydrophobic chains and a hydrophilic head group.

The natural long-chain bases are C_{12}–C_{22} compounds of two types.[14,15] The unsaturated molecules with three functional groups (**34**) are mainly of animal origin and the saturated molecules with four functional groups (**35**) are mainly of plant origin. All the chiral

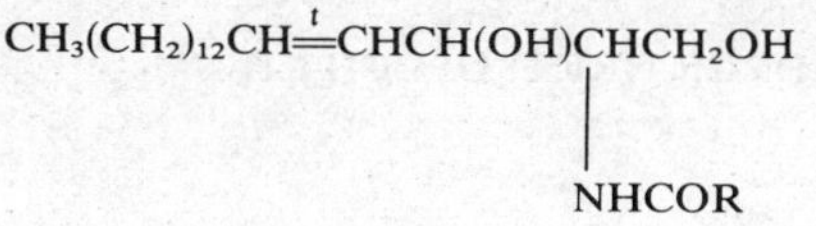

(33)

$RCH\overset{t}{=}CHCH(OH)CH(NH_2)CH_2OH$ **(34)**

$RCH_2CH(OH)CH(OH)CH(NH_2)CH_2OH$ **(35)**

centres in both kinds of molecules have the D-configuration. The C_{18} bases (**34**) and (**35**) were called sphingosine and phytosphingosine, respectively, but a new semisystematic nomenclature is based on the allocation of the name sphinganine to the dihydro-derivative of the C_{18} molecule (**34**). This name incorporates the absolute configuration of the chiral centres. On this basis sphingosine is sphing-*trans*-4-enine and the C_{20} compound (**35**) is hydroxyeicosasphinganine.

In one kind of sphingolipid (*glycosyl-ceramides* or glycosphingolipids)[16–19] the ceramide is linked through the primary alcohol group with a sugar moiety. *Cerebrosides* contain a single carbohydrate group, but more complex glycosylceramides are known. Although first isolated from the brain they occur widely throughout the animal and vegetable kingdom. The attached fatty acids are mainly saturated acids or 2-D-hydroxyalkanoic acids. Some typical structures are (Cer = ceramide, glc = glucose, gal = galatose, gal*N*Ac = *N*-acetyl-galatosamine):

monoglycosylceramides (cerebrosides)
- Cer-gal and Cer-glc
- Cer-gal 3-sulphate (cerebroside sulphates or sulphatides)[9]

diglycosylceramides
- Cer-glc(4←1)-gal (lactosylceramide cytolpin H)
- Cer-gal(4←1)-gal (digalactosylceramide)

Gangliosides, present in brain lipids, erythrocyte membranes, and in other tissues, are found in high concentration in the ganglion cells of the central nervous system.[20–23] They are glycosylceramides which also contain one or more sialic acid molecules (*N*-acetyl-neuraminic acid, NANA, **36**) and attached to galactose units. Two typical structures are given in (**37**) and (**38**).

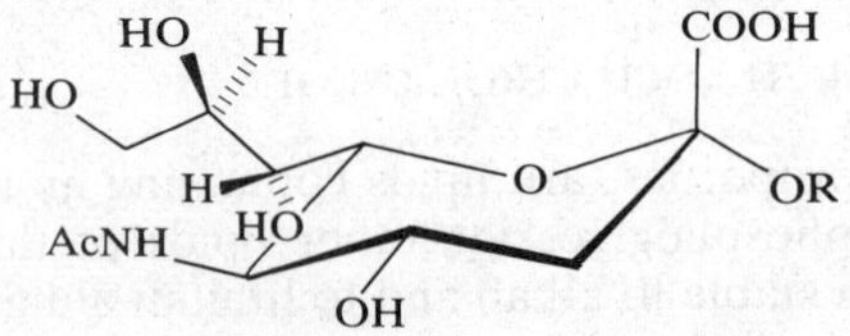

(**36**) α-Ketoside

Cer-glc(4 ←β— 1)-gal(4 ←β— 1)-gal*N*Ac(3 ←β— 1)-gal (**37**) Monosialoganglioside
(3 ↑ 2) on gal
NANA

Cer-glc(4 ←β— 1)-gal(4 ←β— 1)-gal*N*Ac(3 ←β— 1)-gal (**38**) Disialoganglioside
(3 ↑ 2) on first gal and (3 ↑ 2) on terminal gal
NANA NANA

Another kind of sphingolipid (*phosphosphingolipids* or *sphingomyelin*) contains phosphoric acid as phosphorylcholine or phosphorylethanolamine or, less commonly, as a phosphonate. Typical animal (**39**) and plant (**40**) phosphosphingolipids are formulated.

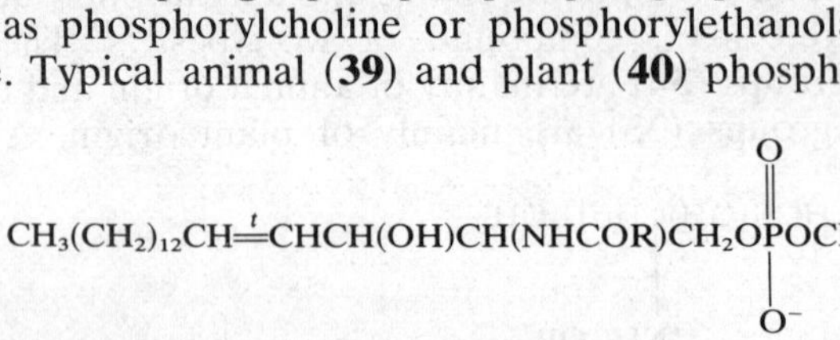

(**39**)

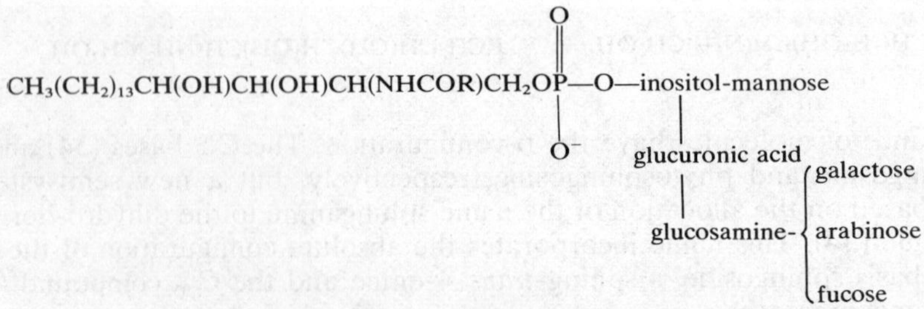

(**40**)

25.2.2 LIPID ANALYSIS

25.2.2.1 The nature of the problem

At one time it was possible to measure only the average properties of a lipid sample, such as its mean unsaturation (iodine value), the average molecular weight of its component acids (saponification equivalent), or the overall content of an element such as nitrogen, phosphorus, or sulphur. Lipid analysis is now conducted at a much more sophisticated level, mainly through the exploitation of various forms of chromatography and of selective deacylation procedures.

The study of a lipid sample may be conducted at several analytical levels. It is possible to determine, quantitatively and qualitatively, the nature of the fatty acids (or alcohols or aldehydes) associated with the whole sample under investigation, or to separate the lipid extract into its several types before determining the component acids (*etc.*) from each type. Furthermore, the fatty acids attached to each position in a triacylglycerol or phosphoglyceride can be distinguished by enzymatic processes and, finally, by a combination of chromatography and enzymic deacylation, individual molecular species can sometimes be identified.

The complexity of the problem arises from the large number of molecular species which are possible. For example, a natural wax ester containing as few as six alcohols and six acids could be a mixture of 36 different esters. Phosphoglycerides contain two acyl groups in each molecule and if the lipid contains ten acids there may be up to 100 different lipid molecules. With triacylglycerols the situation is more complex and the number of possible compounds from n fatty acids is n^3 if all isomers, including enantiomers, are distinguished, $(n^3+n^2)/2$ if optical isomers are not distinguished, and still $(n^3+3n^2+2n)/6$ if no isomers are distinguished. These numbers soon become quite large so that if there are five (125, 75, 35), ten (1000, 550, 220), or 20 component acids (8000, 4200, 1540) they have the values indicated in parenthesis. Most seed oils contain 5–10 different acids and animal fats are often more complex, with 10–40 acids.

The results that have been obtained show interesting patterns of association and distribution. They reveal, for example, that fatty acids are not distributed at random but that there is a marked selectivity in the acyl chains associated with each hydroxyl group.

25.2.2.2 Component acids, alcohols, and aldehydes[24,26]

Natural mixtures of long-chain acids, alcohols, or aldehydes differ mainly in chain-length and degree of unsaturation, but may also contain branched-chains, cyclic systems, and additional functional groups. After appropriate derivatization the mixtures are analysed quantitatively by gas chromatography. With very complex mixtures, or if a more refined analysis is required, gas chromatography is carried out on more than one phase or is combined with other separation processes such as silver ion chromatography or partition chromatography.

For gas chromatography, acids are usually examined as methyl esters; alcohols as acetates, trifluoroacetates, or trimethylsilyl ethers; 1,2-diols as bis(trimethylsilyl) ethers or isopropylidene derivatives; 1,2- and 1,3-diols as n-butyl boronates; aldehydes as such, as dimethyl acetals, or after conversion to alcohols or acids; and amines as trimethylsilyl derivatives or as aldehydes after oxidation with sodium metaperiodate (equation 2).

$$RCH{=}CHCH(OH)CH(NH_2)CH_2OH \longrightarrow RCH{=}CHCHO \qquad (2)$$

25.2.2.3 Component lipids[27–29]

The recognition of lipid types in a mixture and their quantitative determination is achieved mainly by chromatographic separation.

Phospholipids can be separated from neutral lipids by taking advantage of the virtual insolubility of the former, particularly in the salt form, in cold acetone containing a little magnesium chloride. The insoluble portion contains 95% of the lipid phosphorus and only traces of neutral lipid.

The most widely used procedures, however, are based on adsorption chromatography using silica or acid-washed florisil. Column or thin-layer techniques are employed and quantitative analysis is achieved by weighing, by densitometry or fluorimetry on thin layer systems only, or by gas chromatography of component acids after addition of an internal standard. Neutral lipids are eluted from a column with chloroform, glycolipids with acetone, and phospholipids with methanol.

Neutral lipids are then separated on a second silica column: hydrocarbons are eluted with hexane alone; cholesterol esters with 2% ether in hexane; triacylglycerols with 5% ether; cholesterol and diacylglycerols with 15% ether; and monoacylglycerols with ether alone: Similar separations are obtained on thin layers with mixtures of hexane (90–70%) and ether (10–30%), usually with added formic or acetic acid (1–2%).

Complex lipids are further separated on columns with chloroform containing increasing quantities of methanol or on thin layers with solvent systems such as chloroform–methanol–water (65 : 35 : 5 or 25 : 10 : 1) or, for plant lipids, di-isobutyl ketone–acetic acid–water (40 : 25 : 3.7). With the chloroform–methanol systems, acidic and non-acidic complex lipids elute in the order shown (the two groups are eluted in constant order but the region of overlap may vary):

acidic lipids: (i) cardiolipin, (ii) phosphatidic acid, (iii) cerebroside sulphate, (iv) phosphatidylglycerol, (v) phosphatidylserine, (vi) phosphatidylinositol, and (vii) di- and tri-phosphoinositides.

non-acidic complex lipids: (i) ceramide, (ii) ceramide monohexoside and glycosyldiglycerides, (iii) phosphatidylethanolamine, (iv) phosphatidylcholine, lypsophosphatidylethanolamine, and ceramide dihexoside, (v) sphingomyelin, and (vi) lysophosphatidylcholine.

Developments in gas chromatography make it increasingly possible to separate the more volatile lipid classes from one another after suitable derivatization (Section 25.2.2.5).

25.2.2.4 Enzymatic deacylation of lipids[30–32]

(i) Hydrolysis with pancreatic lipase[33]

Pig pancreatic lipase promotes the hydrolysis of acyl groups attached to the two primary glycerol hydroxyl functions, so that the final product is a 2-monoacylglycerol and the fatty acids originally attached to the C-1 and C-3 positions (equation 3). 2-Monoacylglycerols are only hydrolysed by the enzyme after rearrangement to the 1- (or 3-) monacyl isomer. Esters of fatty acids with chain-branching or unsaturation close to the carboxyl group (*e.g.* 16 : 1 3*t* and the C_{20} and C_{22} polyene acids with unsaturation starting at Δ^4 or Δ^5) are hydrolysed more slowly than the normal range of acids, and esters of short-chain acids are hydrolysed more quickly. The method is thus reliable for almost all natural triacylglycerols except the fish oils rich in C_{20-22} polyene acids and milk fats rich in short-chain acids.

$$\text{Triacylglycerols} \rightleftharpoons \begin{matrix}\text{1,2- and 2,3-diacyl-}\\ \text{glycerols + fatty acids}\end{matrix} \rightleftharpoons \begin{matrix}\text{2-acylglycerols +}\\ \text{fatty acids}\end{matrix} \qquad (3)$$

Hydrolysis to the 50–60% level is effected in a buffered solution (pH 8.5) at 37 °C in the presence of bile salts and calcium ions and the 2-monoacylglycerols are isolated chromatographically and their component acids determined by gas chromatography. The nature of the acyl groups attached at C-1 and C-3 is derived by comparison of the composition of the original triacylglycerols and the derived 2-monoacylglycerols.

This enzymatic procedure was the first to be generally applied to triacylglycerol analysis and revealed that acyl groups attached to the secondary hydroxyl group differ markedly from those attached to the primary groups. It was not then possible to distinguish acyl groups attached to C-1 and C-3.

In vegetable fats, C-2 is acylated entirely by unsaturated C_{18} acids, while saturated acids and long-chain unsaturated acids (20:1, 22:1) appear at C-1/3 and rarely at C-2 (Table 1).

TABLE 1

The Component Acids of Some Vegetable and Animal Fats and of the 2-Monoacylglycerols derived from them by Lipolysis[a]

		16:0	16:1	18:0	18:1	18:2
Almond	TG	7		2	70	21
	MG	1		0	64	34
Poppy	TG	10		2	11	76
	MG	1		0	9	89
Palm	TG	44		6	39	9
	MG	11		2	65	22
Rape[b]	TG	3		2	22	15
	MG	1		0	37	37
Pig	TG	28	3	15	42	9
	MG	72	4	4	12	3
Sheep	TG	27	3	27	35	2
	MG	14	5	9	58	6

[a] Adapted from F. D. Gustone, 'An Introduction to the Chemistry and Biochemistry of Fatty Acids and their Glycerides', 2nd edn., Chapman and Hall, London, 1967, p. 165.
[b] Also 18:3 (14, 23), 20:1 (15, 0), and 22:1 (28, 1). The two figures in parenthesis indicate the percentage of acid in triacylglycerol and 2-monoacylglycerol, respectively.

No clear pattern has emerged from the study of animal fats and fish oils. The pig and wild boar are unusual in that palmitic acid predominates in the 2-position. In other animals, oleic and hexadecenoic acid are enriched at C-2, while stearic acid consistently concentrates at C-1/3. In fish oils, the polyenoic (22:6, 20:5) and lower saturated (14:0, 16:0) acids are concentrated in the 2-position; monoene acids, particularly 20:1 and 22:1, are concentrated at the 1/3-positions.

(*ii*) *Regio- and stereo-specific analysis of triacylglycerols*[34]

A number of methods of analysing the acyl chains at each of the three glycerol positions have been pioneered by Brockerhoff (Scheme 1).

The triacylglycerol (**41**) is treated for one minute with ethylmagnesium bromide, which reacts in a random manner with the ester linkages. α,β-Diacyl-glycerols (**42**), separated by thin-layer chromatography from α,α'-diacylglycerols and tertiary alcohols, are treated with phenyl dichlorophosphate to give phosphatidylphenols of two types (**43**). Only one of these is hydrolysed by phospholipase A, so that the reaction product comprises unreacted phosphatidylphenol (**46**) containing the acyl groups originally attached to C-2 and C-3, free acid from C-2 (**45**), and lysophosphatidylphenol (**44**) containing the C-1 acyl groups. The C-2 acyl groups are also determined by pancreatic lipase hydrolysis. Although the procedure requires two synthetic steps, two enzymatic hydrolyses, two thin-layer separations, five transesterifications, and five gas chromatographic analyses, consistent and reliable results can be obtained. The fatty acids at C-1 are obtained from a study of the lysophosphatidylphenol (**44**), those at C-2 from the 2-monoacylglycerol (**47**), and those at C-3 are calculated from the other results in two ways:

C-3 acids $= 3\times(\mathbf{41})-(\mathbf{44})-(\mathbf{47})$ or $2\times(\mathbf{46})-(\mathbf{47})$ where (**41**), (**44**), (**46**), and (**47**), represent the acids present in these compounds (Scheme 1)

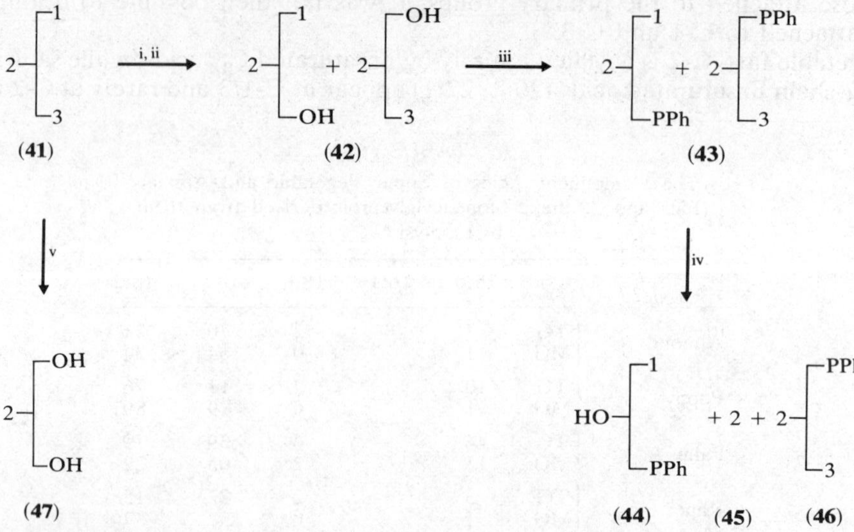

Brockerhoff's stereospecific analysis of triacylglycerols. The symbols indicate the glycerol backbone and the acyl groups attached to the *sn*-1, -2, and -3 positions; PPh = —O—P(=O)(O⁻)—OPh i, EtMgBr; ii, thin layer chromatography; iii, phenyl dichlorophosphate, pyridine; iv, phospholipase A; v, pancreatic lipase.

SCHEME 1

The results obtained by this and similar procedures showed that in most cases, though not in all, natural triacylglycerol mixtures have different fatty acids associated with each hydroxyl group. Even the information obtained by this procedure does not, of itself, indicate the individual triacylglycerols present and these can be calculated only on the *assumption* that each set of acyl groups is distributed at random.

The general conclusions derived from such studies have been summarized by Brockerhoff [34] and by Litchfield.[31] Some typical results are given in Table 2 and the following generalizations have been made — some of them on rather scanty evidence. (i) In all fats, acids outside the usual range are most likely to be attached to C-3. (ii) In plants, saturated acids and long-chain unsaturated acids concentrate at C-1 and C-3 and unsaturated acids at C-2; unusual acids (acetic, estolides), when present, are at C-3. (iii) In many animals, saturated acids concentrate at C-1, short-chain acids and unsaturated acids at C-2, and long-chain unsaturated acids at C-3. In pigs and related animals and in fish, however, palmitic acid concentrates at C-2. In birds the C-1 and C-3 positions seem to contain the same acids, while the short-chain acids in ruminant milk fats and the $C_{20/22}$ polyene acids in mammals concentrate at C-3.

(*iii*) *Stereospecific analysis of phosphoglycerides*

Phospholipase A_2, from snake venom, selectively deacylates practically all phosphoglycerides except the polyphosphoinositides. The natural derivatives of *sn*-3-phosphatidic acid furnish a lysophosphoglyceride with release of the fatty acids attached at C-2 (equation 4), and the analysis of both the liberated fatty acids and the lysophosphatide provides information about the acyl groups attached to C-1 and C-2. Results show fairly consistently that saturated acids predominate at C-1 and unsaturated acids at C-2, although disaturated and diunsaturated phosphoglycerides are known. In bacterial lipids the saturated cyclopropane acids, like their monoene precursors, predominate in the

TABLE 2

Stereospecific Analysis of some Natural Fats[30]

		16:0	18:0	18:1	18:2	*Other acids*[a]
Vegetable fats						
Soybean	1	14	6	23	48	18:3 (9, 7, 8)
	2	1	tr	22	70	
	3	13	6	28	45	
Cocoa butter	1	34	50	12	1	20:0 (1, 0, 2)
	2	2	2	87	9	
	3	37	53	9	tr	
Rape	1	4	2	23	11	18:3 (6, 20, 3); 20:1 (16, 2, 17); 22:1 (35, 4, 51)
	2	1	0	37	36	
	3	4	3	17	4	
Animal fats						
Man	1	39	10	33	3	14:0 (4, 11, 1); 16:1 (5, 11, 4)
	2	10	2	50	9	
	3	25	9	51	5	
Pig	1	16	21	44	12	14:0 (2, 4, tr); 16:1 (3, 4, 3)
	2	59	3	17	8	
	3	2	10	65	24	
Ox	1	41	17	20	4	14:0 (4, 9, 1); 16:1 (6, 6, 6)
	2	17	9	41	5	
	3	22	24	37	5	
Cow (milk)	1	36	15	21	1	4:0 (5, 3, 43); 6:0 (3, 5, 11) 8:0 (1, 2, 2): 10:0 (3, 6, 4) 12:0 (3, 6, 4); 14:0 (11, 20, 7); 16:1 (3, 2, 1)
	2	33	6	14	3	
	3	10	4	15	0	
Chicken	1	25	6	33	14	16:1 (12, 7, 12)
	2	15	4	43	23	
	3	24	6	35	14	
Herring gull	1	22	13	41	7	16:1 (4, 3, 5); 20:1 (7, 6, 7); 22:1 (3, 4, 5)
	2	15	9	48	11	
	3	17	7	46	9	
Cod	1	15	6	28	2	14:0 (6, 8, 4); 16:1 (14, 12, 14); 20:1 (12, 7, 17); 22:1 (6, 5, 7); 20:5 (2, 12, 13); 22:5 (1, 3, 1); 22:6 (1, 20, 6).
	2	16	1	9	2	
	3	7	1	23	2	

[a] The three figures in parenthesis indicate the percentage of acid at the 1, 2, and 3 positions, respectively.

TABLE 3

The Distribution of Major Component Acids in some Phosphatidylcholines[a]

	16:0		18:0		18:1		18:2		20:4		*Other acids*[b]
	1	2	1	2	1	2	1	2	1	2	
Salmon	37	10	3	0	8	23	—	—	—	—	20:5 (14, 17); 22:6 (33, 46)
Egg	61	5	25	2	10	59	2	26	0	6	
Bovine liver	19	1	46	0	19	17	3	16	2	18	22:1 (2, 15); 22:4 (1, 7); 22:5 (2, 10); 22:6 (0, 6)
Bovine milk	47	35	20	3	23	32	3	11	—	—	14:0 (6, 13)
Human serum	66	3	26	0	6	22	1	47	0	12	20:3 (0, 5); 22:6 (0, 6)

[a] Adapted from F. D. Gunstone, 'An Introduction to the Chemistry and Biochemistry of Fatty Acids and their Glycerides', 2nd edn., Chapman and Hall, London, 1967, p. 173.

[b] The two figures in parenthesis indicate the percentage of acid at the 1 and 2 positions respectively.

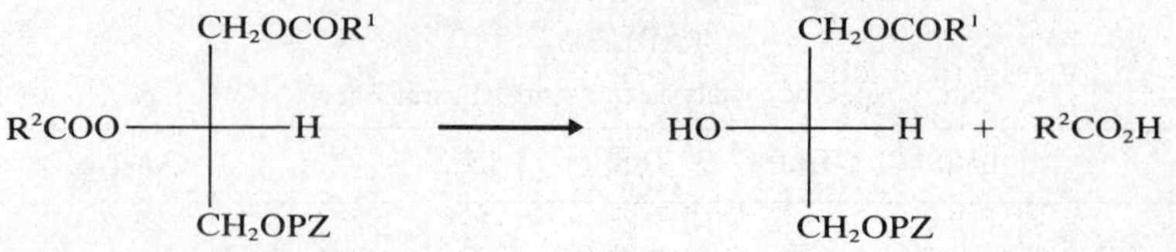

(4)

2-position. It is possible to calculate component phosphoglycerides from phospholipase hydrolysis results only if it is *assumed* that the two sets of acyl groups are distributed statistically, and there is little or no evidence in favour of this view. Some typical results are given in Table 3.

25.2.2.5 Attempts to isolate individual lipid species[30,31,35–38]

In this section, reference is made to attempts to isolate individual lipids or to obtain mixtures simple enough to be analysed unambiguously. The procedures which have given useful information involve the separation of triacylglycerols by silver ion thin-layer chromatography, thin-layer partition chromatography, and gas chromatography. These methods can also be applied to ester waxes, mono- and di-acylglycerols, and phosphoglycerides after appropriate derivatization.

Chromatography on silica impregnated with silver nitrate, which permits separation according to the degree of unsaturation, has been applied successfully to triacylglycerols and to diacylglycerols after acetylation. Reverse-phase partition chromatography separates triacylglycerols on the basis of their partition coefficients. An additional double bond exerts approximately the same influence as subtracting two carbon atoms so that, for example, tripalmitoylglycerol and trioleoylglycerol behave very similarly. Both paper and thin-layer techniques have been employed and typical systems involve a hydrocarbon or silicone as a non-polar stationary phase and mixtures of acetone with acetonitrile or methanol or acetic acid as the mobile phase.

Glycerides with up to ~90 carbon atoms can be separated by gas chromatography on a short column (15 cm upwards) coated at a low level (1–3%) with a non-polar stationary phase such as the methylsiloxanes SE-30, JXR, or OV-1, usually with temperature-programmed elution around 250–350 °C. Unsaturated glycerides are not separated from their saturated analogues under these conditions, but it is not difficult to separate glycerides differing by two carbon atoms. With additional care it is even possible to separate glycerides differing by only one carbon atom. New siloxane liquid phases with increased polarity and moderate temperature stability permit the resolution of neutral lipid mixtures according to molecular weight *and* degree of unsaturation.

These three techniques can be applied satisfactorily to cholesterol esters, mono- and di-acylglycerols after suitable derivatization, triacylglycerols, wax esters, and to phosphoglycerides, glycosyldiglycerides, and sphingolipids after appropriate modification. It is advisable to separate α- and β-monoacylglycerols and the α,β- and α,α'-diacylglycerols on silica impregnated with boric acid before the other separation procedures are applied.

Partial glycerides must be derivatized to reduce polarity and to increase stability before being examined by the above separation procedures. Acetates are preferred for the thin-layer systems and trimethylsilyl ethers for gas chromatography. The elution order of acetyldiacylglycerols on silver ion chromatography is given by the sequence 00 > 01 > 11 > 02 > 12 > 22 > 03 > 13 > 04 > 14 > 24 > 05 > 06, in which two digits represent the number of double bonds in the two long acyl chains.

Triacylacylcerols containing unusual acids, such as milk fats with short-chain acids or seed oils with the less-common oxygenated acids, can sometimes be separated on thin layers of silica, but silver ion chromatography is more useful and has been successfully applied to unsaturated vegetable oils with glycerides having 0–9 double bonds and to fish oils with acyl groups having 0–6 double bonds. The elution order for seed oils (containing mainly C_{18} acids) is given by the following sequence, where the three digits represent the number of double bonds in each acyl chain: 000>001>011>002>111>012>112>022>003>122>013>113>222>023>123>223>033>133>233>333.

Table 4 includes results obtained by silver ion chromatography followed by gas chromatographic examination of the component esters of each fraction, while Table 5 lists data available from a combination of silver ion chromatography and gas chromatography of triacylglycerols.

TABLE 4

Component Glycerides of some Vegetable Oils by Silver Ion Chromatography[a]

	322[b]	321	320	222	221	220	211	210	200	111	110
Soybean[c]	7	5	4	15	16	13	8	12	4	2	5
Peanut[d]	—	—	—	—	5	—	23	14	4	26	23

[a] F. D. Gunstone and M. I. Qureshi, *J. Amer. Oil Chemists' Soc.*, 1965, **42,** 957, 961. [b] The symbol 322 refers to all the triacylglycerols having one linolenic (3) and two linoleic (2) chains. The other three-figure arrays are to be interpreted similarly. [c] Also 332, 1; 330, 1; 311, 2; 310, 3; 300, 2. [d] Also 100, 5.

TABLE 5

Distribution by Carbon Number[a] of Triacylglycerols of differing Unsaturation in Sheep Depot Fat[30]

Carbon number[a]	*Saturate*	*Monoene*	*Diene*	*Triene*	*Tetraene*	*Total*
44	1.7	0.3	—	—	—	0.4
46	6.7	1.3	0.2	0.3	1.0	2.3
48	17.6	6.5	1.3	2.3	2.9	7.5
50	29.5	21.3	8.7	7.8	11.2	18.5
52	29.8	38.7	41.3	30.5	33.1	37.5
54	14.3	31.1	47.9	57.7	49.9	32.9
56	0.2	0.6	0.6	1.4	1.9	0.8
Total	31.3	45.8	19.4	2.6	0.8	100.0

[a] The carbon number is the total number of carbon atoms in the three acyl chains and does not include the three glycerol carbon atoms.

Before they are analysed by the separation procedures under consideration, phosphoglycerides are modified by conversion to acetyldiacylglycerols or to dimethyl phosphatidic acids (Scheme 2). Illustrative results are given in Table 6.

There are no enzymes readily available to convert glycosyldiacylglycerols to diacylglycerols. For silver ion chromatography they are examined in their native form or after acetylation, and their trimethylsilyl ethers are suitable for gas chromatography.

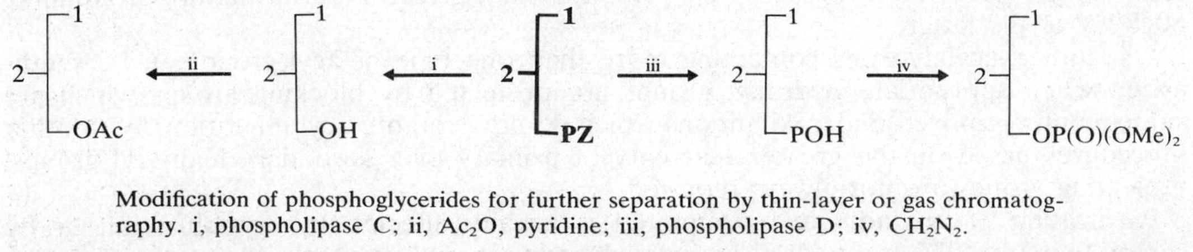

Modification of phosphoglycerides for further separation by thin-layer or gas chromatography. i, phospholipase C; ii, Ac_2O, pyridine; iii, phospholipase D; iv, CH_2N_2.

SCHEME 2

TABLE 6
Comparison of the Major Phosphatidyl-cholmes and -ethanolamines of Rat Liver[30]

Component acids C-1	C-2	PC	PE
16:0	18:1	6.4	0.9
16:0	18:2	15.7	7.2
18:0	18:2	9.0	5.3
16:0	20:4	18.9	14.2
18:0	20:4	20.9	35.0
18:1	20:4	3.6	4.1
16:0	22:6	4.8	12.4
18:0	22:6	5.6	8.8

Sphingomyelins containing 2-hydroxy-acids are easily separated from those without hydroxy-acids by chromatography on silica. Phospholipase C is used to convert sphingolipids to ceramides, which are further separated on silica impregnated with sodium arsenite depending on whether the molecules contain two, three, or four hydroxyl groups. Ceramides can also be separated on silver ion plates after acetylation or by gas chromatography after conversion to acetates or trimethylsilyl ethers.

25.2.3 LIPID SYNTHESIS

Most natural lipids are mixtures and isolation of pure individual compounds is difficult. Pure lipids are therefore frequently obtained by synthetic methods.

25.2.3.1 Mono-, di-, and tri-acylglycerols[39,40]

(*i*) *Introduction*

The position of acyl groups attached to glycerol will be shown by the numbers 1, 2, and 3 and the compounds will be racemic unless otherwise indicated by the prefix *sn* (Section 25.2.1.1).

A major difficulty in the synthesis of pure acylglycerols is the ease with which acyl groups in mono- and di-acylglycerols migrate from the oxygen to which they are attached to an adjacent free hydroxyl group by a transesterification process catalysed by acid or base or heat. Both 1- and 2-monoacylglycerols readily furnish an equilibrium mixture containing 10% of the latter, and 1,2- and 1,3-diacylglycerols form mixtures containing 60–80% of the latter.

Di- and tri-acylglycerols containing more than one type of acyl group can be synthesized when appropriate hydroxyl groups are protected by blocking groups which are subsequently removed under conditions which do not promote acyl migration. With other procedures based on the greater reactivity of primary over secondary hydroxyl groups, protecting groups are not always required.

Purification of an individual compound is usually effected by crystallization or by chromatography. The purity of a synthetic glyceride should be checked by gas chromatography of its component acids, by thin-layer chromatography, and by the enzymatic procedures described in Section 25.2.2.4.

(*ii*) *Acylation procedures*

Acylation is generally effected in chloroform solution by reaction with a slight excess of acyl halide and an equivalent amount of pyridine or other tertiary base. The acid chlorides are prepared by reaction with excess of thionyl chloride (for saturated acids) or oxalyl chloride (for unsaturated acids).

(*iii*) *Protecting groups*

With an acid catalyst (toluene-*p*-sulphonic acid), glycerol reacts with acetone to form a 1,2-acetal and with benzaldehyde to give a mixture of 1,2- and 1,3-acetals which can be separated by crystallization (equation 5). These protecting groups can be removed by aqueous acid which may, however, promote acyl migration. The benzylidene group can also be removed by hydrogenolyses, but this is inappropriate when an unsaturated acyl group has been introduced. These difficulties are overcome by the milder reaction with boric acid in trimethyl or triethyl borate to give a borate ester which can be hydrolysed by water at room temperature (equation 6).

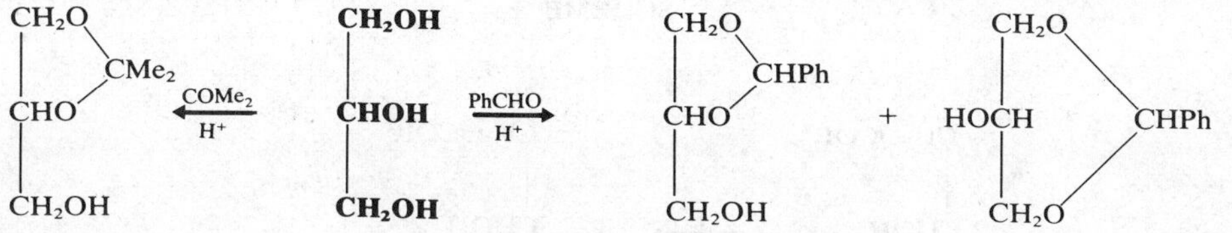

(5)

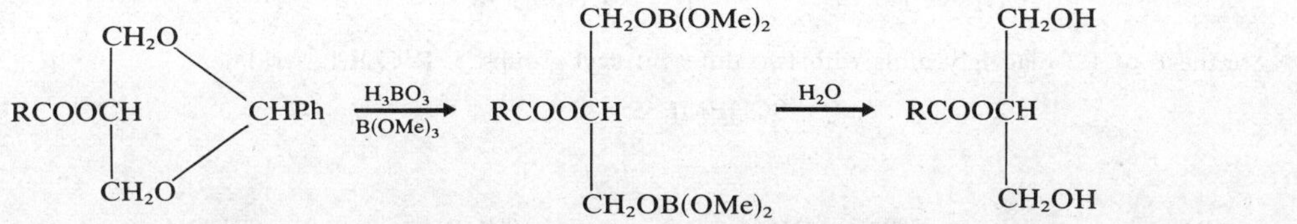

(6)

The 1- and 2-benzylglycerols are prepared from 1,2-isopropylideneglycerol and 1,3-benzylideneglycerol, respectively, by benzylation followed by decomposition of the acetal. The benzyl group can be removed subsequently by hydrogenolysis.

Glycerol, trityl chloride, and pyridine furnish 1-trityl and 1,3-ditrityl derivatives when reacted together at room temperature. The trityl group can be removed by (i) reaction with hydrogen bromide in acetic acid or hydrogen chloride in ether or petrol, (ii) hydrogenolysis, or (iii) percolation through a column of fresh silica.

Glycerol carbonate, made by reaction with phosgene, can be decomposed by hydrolysis with mild alkali and the 2,2,2-trichloroethoxycarbonyl protecting group is removed with zinc and acetic acid.

(*iv*) 1- *and* 2-*monoacylglycerols*

These are prepared from 1,2-isopropylideneglycerol (Scheme 3) and 1,3-benzylideneglycerol (Scheme 4), respectively.

(*v*) 1,3-*Diacylglycerols*

These compounds are prepared by direct acylation of the appropriate 1-monoacylglycerol, from dihydroxyacetone, or by direct acylation of glycerol (Schemes 5

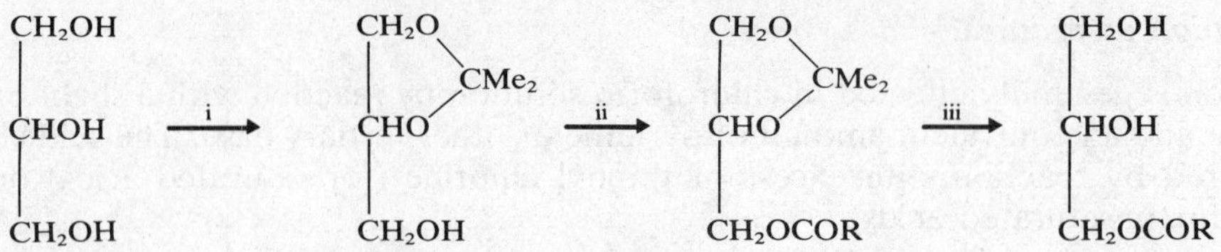

Preparation of 1-monoacylglycerols. i, $COMe_2$, C_6H_6, p-$MeC_6H_4SO_3H$; ii, RCO_2H, p-$MeC_6H_4SO_3H$; iii, H_3BO_3, $B(OMe)_3$; H_2O.

SCHEME 3

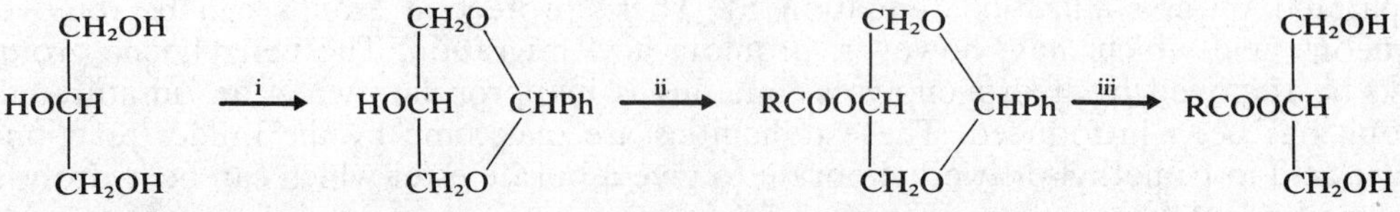

Preparation of 2-monoacylglycerols. i, PhCHO, p-$MeC_6H_4SO_3H$; ii, RCOCl, pyridine; iii, H_3BO_3, $B(OMe)_3$; H_2O.

SCHEME 4

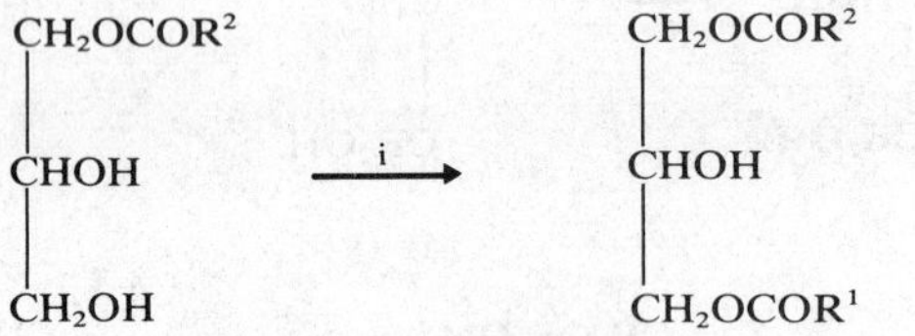

Synthesis of 1,3-diacylglycerols with two different acyl groups. i, R^1COCl, pyridine.

SCHEME 5

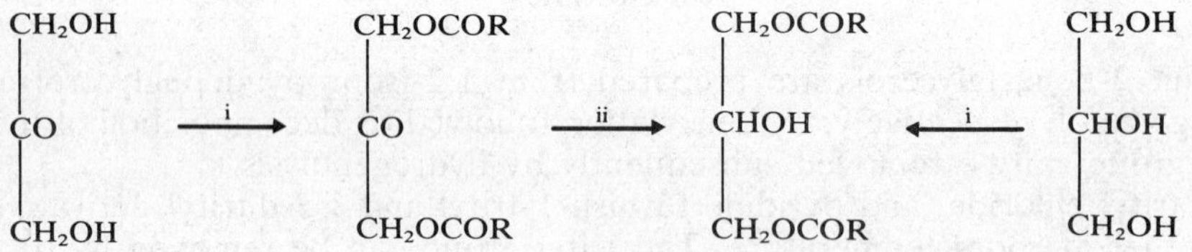

Synthesis of 1,3-diacylglycerols with two identical acyl groups. i, RCOCl, pyridine; ii, $NaBH_4$

SCHEME 6

and 6). The last method depends on the greater reactivity of the primary hydroxyl groups and on the ease with which the product can be purified by crystallization.

(*vi*) 1,2-*Diacylglycerols*

These are more difficult to prepare than their 1,3-isomers because of the readiness with which they undergo acyl migration and because they are not so easily crystallized.

The sequence (Scheme 7) proceeding through 1-benzylglycerol is suitable for saturated compounds. Alternative routes employ the tetrahydropyranyl ether of glycerol prepared from allyl alcohol (Scheme 8) or glycerol-1,2-carbonate (Scheme 9). These normally give monoacid diesters, but the 1-acyl group can be removed by pancreatic lipase and replaced by the required acyl group (see Scheme 9). Other methods of preparing 1,2-diacylglycerols are described in (viii) below.

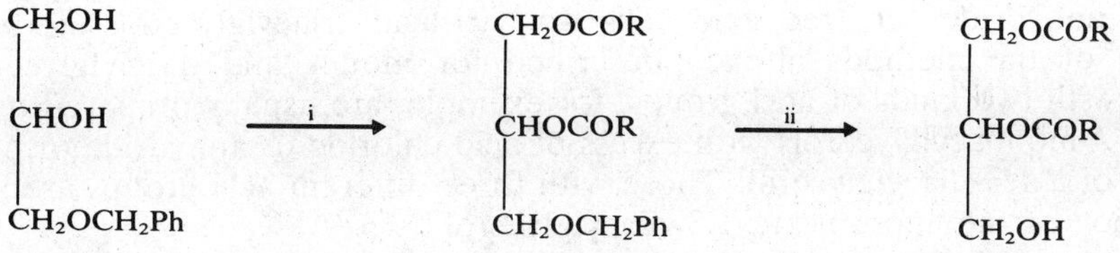

Synthesis of 1,2-diacylglycerols from 1-benzylglycerol. i, RCOCl, pyridine; ii, Ni, H_2.

SCHEME 7

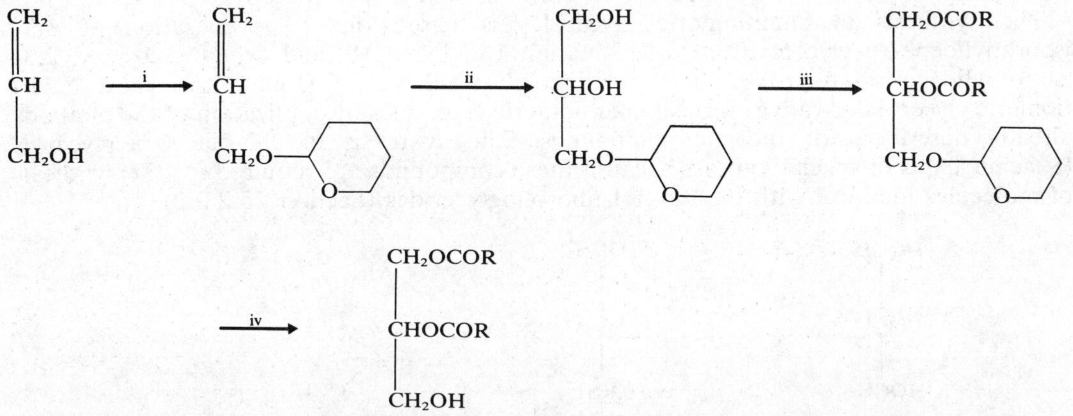

Synthesis of 1,2-diacylglycerols from allyl alcohol. i, dihydropyran, H^+; ii, $KMnO_4$; iii, RCOCl, pyridine; iv, HCl or H_3BO_3.

SCHEME 8

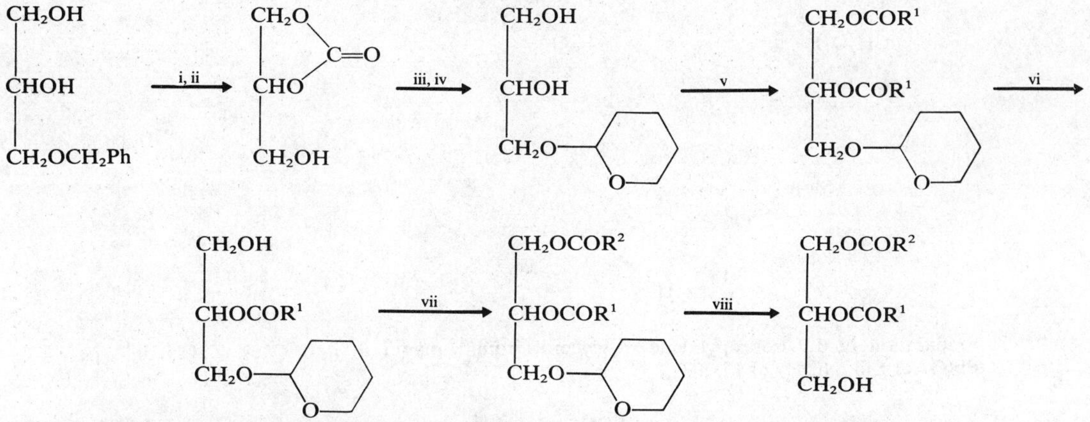

Synthesis of a diacid 1,2-diacylglycerol from 1-benzylglycerol. i, $(EtO)_2CO$, $NaHCO_3$; ii, Pd, H_2; iii, dihydropyran, H^+; iv, KOH; v, R^1COCl, pyridine; vi, pancreatic lipase; vii, R^2COCl, pyridine; viii, H_3BO_3.

SCHEME 9

(*vii*) *triacylglycerols*

Monoacid triacylglycerols are easily made by direct acylation of glycerol using acid chloride, anhydride, or free acid. Di- and tri-acid triacylglycerols are produced by extension of the methods already described for mono- and di-acylglycerols. Triacylglycerols with two kinds of acyl groups, for example, are usually prepared by acylation of the 1- or 2-monoacylglycerols with excess of acid chloride or anhydride or, if possible, by acylation of a 1,3-diacylglycerol. Those with three different acyl groups are also prepared by acylation of the appropriate 1,3-diacylglycerol.

(*viii*) *Optically active glycerides*

The methods so far discussed produce only racemic glycerides and special procedures are required when enantiomeric glycerides are to be prepared. Reactions which furnish enantiomeric triacylglycerols with three long-chain acyl groups usually give products with no measurable rotation and their stereochemical integrity must be assessed in other ways, such as enzymatic hydrolysis (Section 25.2.2.4) or n.m.r. spectroscopy.

The synthesis of enantiomeric glycerides is based on the preparation of 1,2-isopropylidene-*sn*-glycerol from D-(+)-mannitol (Scheme 10) and, when necessary, 2,3-isopropylidene-*sn*-glycerol from the less readily available L-(−)-mannitol. These enantiomeric glycerol derivatives furnish enantiomeric glycerides by application of the methods already described for racemic compounds. The synthesis of 1,2-diacyl-*sn*-glycerols (Scheme 11) is of special interest because these compounds are required for the synthesis of molecules identical with the natural phosphoglycerides (Section 25.2.3.2).

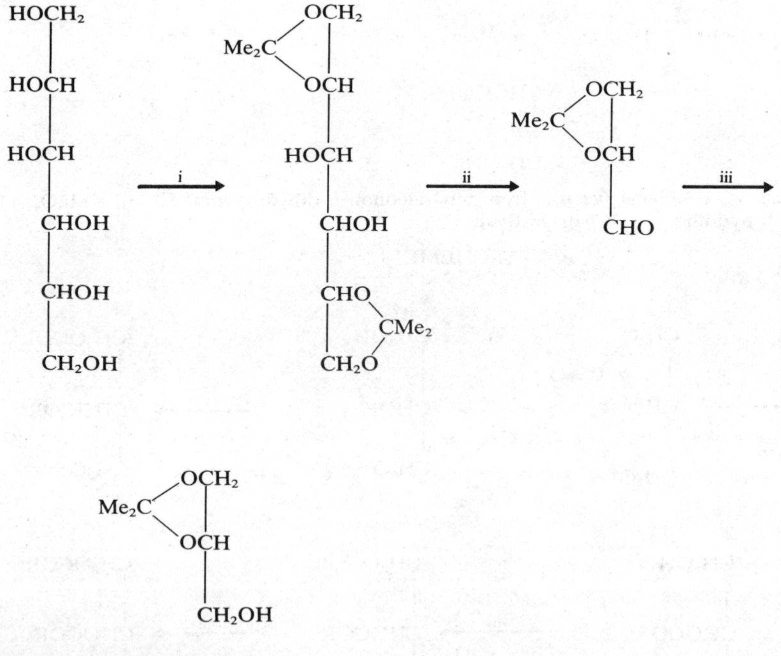

Preparation of 1,2-isopropylidene-*sn*-glycerol from D-mannitol. i, $COMe_2$, $ZnCl_2$; ii, $Pb(OAc)_4$; iii, Ni, H_2 or $LiAlH_4$.

SCHEME 10

A new approach to chiral glycerides, illustrated in Scheme 12, is based on the preparation of D- and L-glycidol (2,3-epoxypropanol) from L- and D-serine (3-hydroxy-2-aminopropanoic acid), and has the advantage that both isomers are readily available.[41]

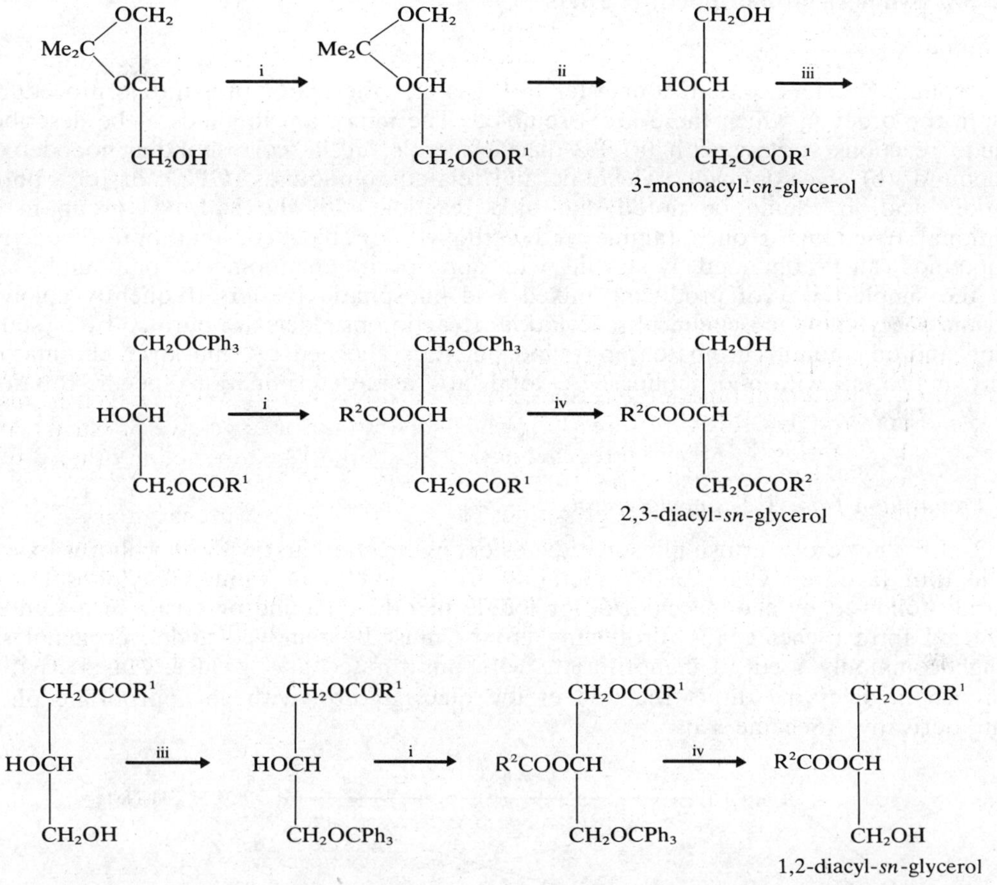

Preparation of 1,2- and 2,3-diacyl-*sn*-glycerols with two different acyl groups (including polyunsaturates). The 1-acyl-*sn*-glycerol can be prepared easily from 2,3-isopropylidene-*sn*-glycerol or, by a more involved reaction sequence, from 1,2-isopropylidene-*sn*-glycerol. i, acid chloride, pyridine; ii, HCl; iii, Ph_3CCl; iv, H_3BO_3 on silica.

SCHEME 11

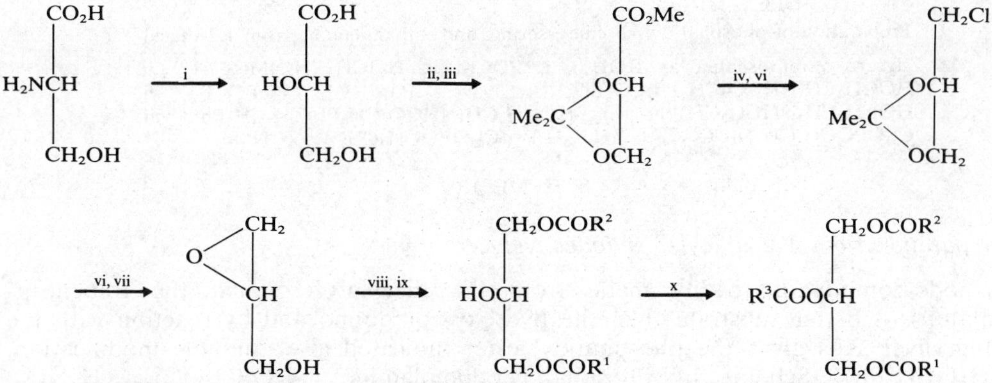

Preparation of 1,3-diacyl- and 1,2,3-triacyl-*sn*-glycerols from L-serine. i, $NaNO_2$, HCl; ii, MeOH, $Me_2C(OMe)_2$; iii, $COMe_2$, $Me_2C(OMe)_2$; iv, $LiAlH_4$; v, Ph_3P, CCl_4; vi, aq. AcOH; vii, Na, Et_2O; viii, R^1COCl; ix, R^2CO_2H, $Et_4\overset{+}{N}\,\bar{B}r$; x, R^3COCl.

SCHEME 12

25.2.3.2 Synthesis of phosphatidyl esters[42–44]

(*i*) *Introduction*

Phosphatidyl esters contain four ester linkages and the major preparative procedures vary in the order in which these are assembled. The important methods to be described include reactions starting with (a) 1,2-diacylglycerols or the corresponding iododeoxy-compound, (b) glycerophosphorylcholine and related compounds (GPZ), or (c) a phosphatidic acid. It should be noted that in a reaction with the hydroxyl group in Z, additional functional groups (amine and carboxyl) are blocked, that optically active compounds can be obtained by starting with appropriate enantiomeric compounds, and that the simplest way of producing mixed acid phosphatidyl esters frequently involves enzymic deacylation and chemical reacylation. Reaction products are purified by crystallization and/or column chromatography and purity is checked by thin-layer chromatography, hydrolysis with phospholipase A, total fatty acid determination, specific rotation, and P/N ratio.

(*ii*) *Preparation from* 1,2-*diacylglycerols*

1,2-Diacylglycerols furnish phosphatidyl esters by reaction first with phosphorus oxychloride (not favoured when better methods are available) or monophenylphosphoroyl chloride followed by choline chloride or iodide or ethanolamine or serine in a suitably protected form (Scheme 13). Protecting groups must be removed and hydrogenolysis, though commonly used, is inappropriate with unsaturated acyl groups. An alternative approach involves the direct reaction of the diacylglycerol with an appropriate phosphate derivative (Scheme 13).

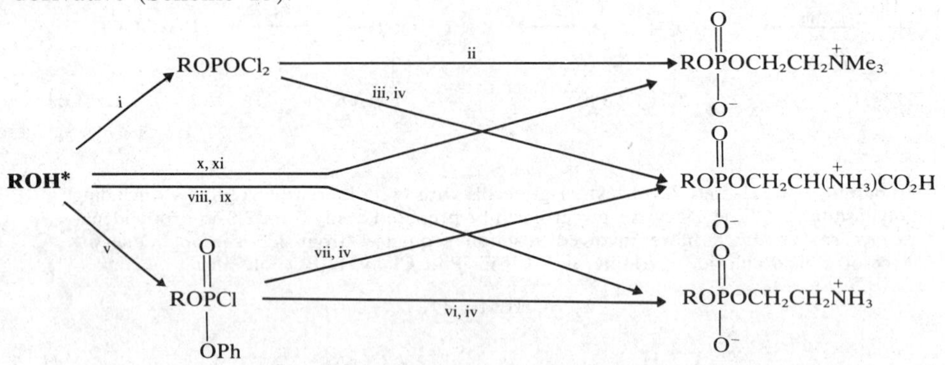

Preparation of phosphatidyl-choline, -serine, and -ethanolamine from 1,2-diacylglycerols (represented as ROH*). i, $POCl_3$; ii, $HOCH_2CH_2\overset{+}{N}Me_3\ Cl^-$; iii, $HOCH_2CH(NHCO_2CH_2Ph)CO_2Bu^t$; iv, H_2, Pd; v, $PhOPOCl_2$; vi, $HOCH_2CH_2NHCO_2CH_2Ph$; vii, $HOCH_2CH(NHCO_2CH_2Ph)CO_2CH_2Ph$; viii, $Cl_2OPOCH_2CH_2N(CO)_2C_6H_4$; ix, N_2H_4; x, $Cl_2OPOCH_2CH_2Br$; xi, NMe_3.

SCHEME 13

(*iii*) *Preparation from* 1,2-*diacyl*-3-*iododeoxyglycerol*

This iodo-compound, readily made from the diacylglycerol *via* the toluene-*p*-sulphonate, is a better substrate than the hydroxy-compound and by reaction with the following silver salts gives the phosphatidyl ester indicated after suitable modification: $AgOPO(OCH_2Ph)_2$ (Scheme 14, phosphatidylethanolamine), $AgOPO(OCH_2C_6H_4NO_2$-*p*$)OCH_2CH_2Cl$ (phosphatidylcholine), $AgOP(OY)CH_2CH_2NHX$ (phosphatidylethanolamine; $Y = Bu^t$, Ph, or $PhCH_2$; $X = (CO)_2C_6H_4$, CO_2CH_2Ph, or CPh_3), $AgOPOCH_2CH(NHCO_2Bu^t)CO_2Bu^t$ (phosphatidylserine), or silver 1,3,4,5,6-pentaacetylmyoinositol phenyl phosphate (phosphatidylinositol).

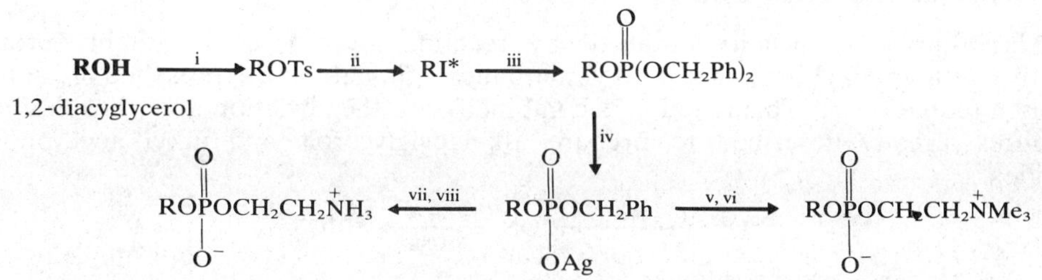

Preparation of phosphatidylcholine and phosphatidylethanolamine from 1,2-diacyl-3-iododeoxyglycerol (represented as RI*). i, TsCl; ii, NaI; iii, $AgOPO(OCH_2Ph)_2$; iv, NaI, $AgNO_3$; v, $Br(CH_2)_2\overset{+}{N}Me_3$ picrate$^-$; vi, NaI; vii, $Br(CH_2)_2N(CH_2Ph)_2$; viii, H_2, Pd.

SCHEME 14

(*iv*) *Acylation of glycerylphosphorylcholine and related compounds*

Glycerylphosphorylcholine can be obtained from pure phosphatidylcholine isolated from egg yolk. Deacylation with tetrabutylammonium hydroxide is followed by acylation of the glycerolphosphorylcholine or its cadmium chloride complex, usually by reaction at 80 °C for 4 hours with a mixture of the appropriate anhydride and its potassium salt.

Although less readily available from natural sources, glycerylphosphorylethanolamine has been isolated from the phospholipids of soybean and of egg yolk. Before acylation the amino-group is protected by the phthaloyl or trityl group. Protected forms of glycerylphosphoryl-ethanolamine and -serine have been synthesized from 1,2-isopropylidene-*sn*-glycerol by reaction with phosphorus oxychloride followed by the protected alcohol (as in Scheme 13).

(*v*) *Preparation of phosphatidic acids and their subsequent use to prepare phosphatidyl esters*

Phosphatidic acids, prepared from diacylglycerols or iododeoxy-compounds (Scheme 15), are converted to phosphatidyl esters by reaction with the following alcohols in the presence of 2,4,6-tri-isopropylbenzenesulphonyl chloride: $HOCH_2CH_2\overset{+}{N}Me_3Ac\bar{O}$ (phosphatidylcholine), $HOCH_2CH_2NHCPh_3$ (phosphatidylethanolamine), $HOCH_2CH_2N(Me)$-CO_2CH_2Ph (phosphatidyl-*N*-methylethanolamine), $HOCH_2CH_2NMe_2$ (phosphatidyl-*N,N*-dimethylethanolamine), and $HOCH_2CH(NHCO_2CH_2Ph)CO_2CH_2Ph$ (phosphatidylserine). When required, the protecting group is removed by hydrogenolysis.

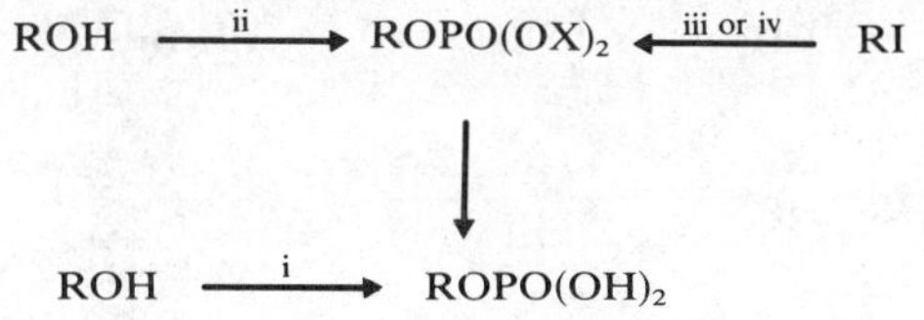

Synthesis of phosphatidic acids from diacylglycerols (ROH) or their iododeoxy-derivatives (RI). i, $POCl_3$, H_2O; ii, $(PhO)_2POCl$ followed by H_2, Pd; iii, $AgOPO(OCH_2Ph)_2$ followed by BaI_2; iv, $AgOPO(OBu^t)_2$ followed by HCl.

SCHEME 15

(*vi*) *The preparation of phosphonate analogues*

The preparation of phosphonates is carried out by similar procedures, employing appropriate derivatives of phosphonic acid such as $ClP(O)CH_2CH_2N(CO)_2C_6H_4$.

25.2.3.3 Synthesis of ether lipids[45]

O-Alkyl lipids are usually obtained by reacting glycerol or a suitably protected derivative with an alkyl bromide or alkyl methanesulphonate in the presence of sodium or potassium (equation 7). This reaction is combined with the acylation and phosphorylation procedures already described to prepare alkylacylglycerols and alkyl acyl phosphoglycerides.

$$\underset{\text{glycerol derivative}}{R^1OH} \xrightarrow{Na} R^1ONa \xrightarrow[R^2OMs]{R^2Br\text{ or}} R^1OR^2 \qquad (7)$$

1-*O*-Alkylglycerol is prepared from 1,2-isopropylideneglycerol. The readily-available 1,2-isopropylidene-*sn*-glycerol will furnish a 3-*O*-alkyl-*sn*-glycerol. Its more-interesting isomer 1-*O*-alkyl-*sn*-glycerol can be prepared from the less readily-available acetal or *via* a Walden inversion by the reaction sequence outlined in Scheme 16. 2-*O*-Alkylglycerols are prepared from 1,3-benzylideneglycerol.

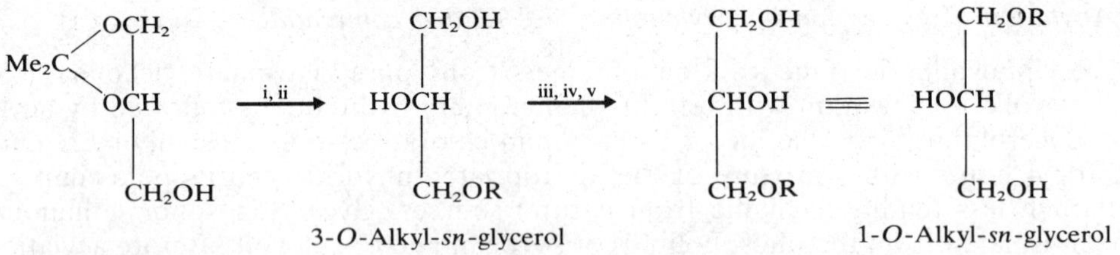

Synthesis of 3-*O*- and 1-*O*-alkyl-*sn*-glycerol. i, ROMs, KOH; ii, HCl; iii, TsCl; iv, AcOK; v, KOH.

SCHEME 16

The production of alk-1′-enyl ethers generally involves the elimination of a fragment HX (X = OEt, OTs, Cl, I) from a 1′- or 2′-substituted ether, as in equation (8). The

$$R^1OCHXCH_2R^2 \longrightarrow R^1OCH{=}CHR^2 \longleftarrow R^1OCH_2CHXR^2 \qquad (8)$$

product is a mixture of *cis*- and *trans*-isomers which are separated by silver ion chromatography. The natural compounds have the *cis* configuration. A typical synthesis is outlined in Scheme 17.

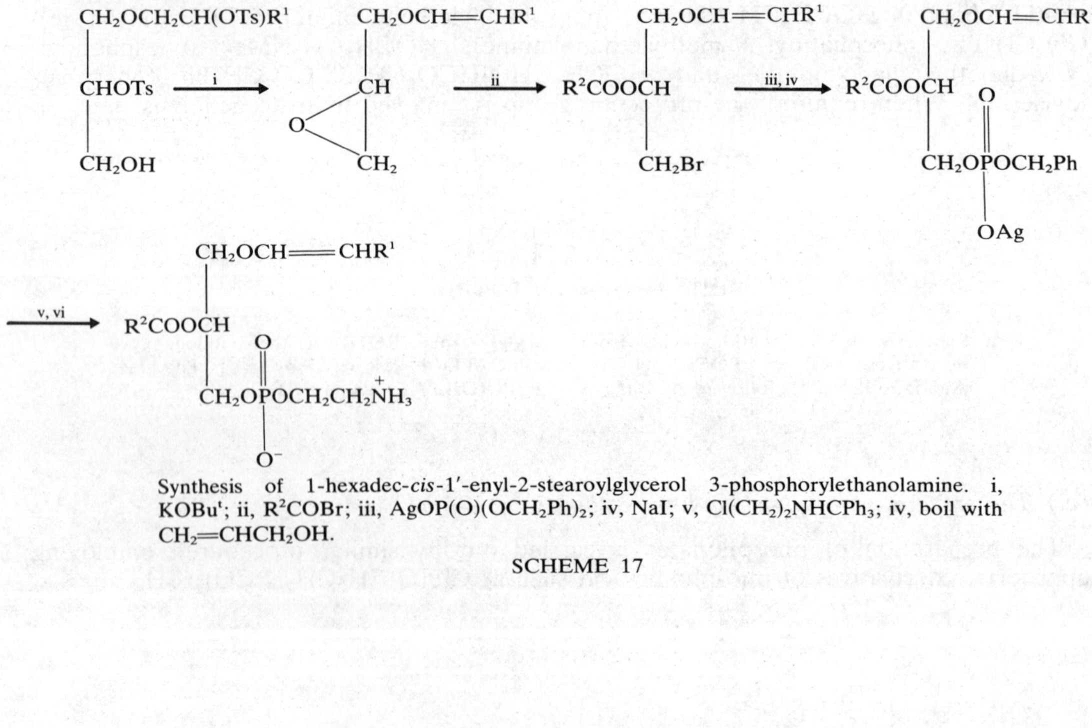

Synthesis of 1-hexadec-*cis*-1′-enyl-2-stearoylglycerol 3-phosphorylethanolamine. i, $KOBu^t$; ii, R^2COBr; iii, $AgOP(O)(OCH_2Ph)_2$; iv, NaI; v, $Cl(CH_2)_2NHCPh_3$; iv, boil with $CH_2{=}CHCH_2OH$.

SCHEME 17

25.2.3.4 Synthesis of sphingolipids[46]

The complete synthesis of a sphingolipid involves the preparation of a long-chain base, its *N*-acylation, and linking of the primary hydroxyl group with the necessary sugar or phosphate residue to give a glyco- or phospho-sphingolipid.

(*i*) *Sphing*-trans-4-*enine and related compounds*

Sphinganine and its 4*c* and 4*t* derivatives have been prepared starting with a sugar derivative of appropriate stereochemistry (Scheme 18). L-Serine is also a convenient substrate for the synthesis of D-*threo*-sphinganine and D-*erythro*-sphing-*trans*-4-enine (Scheme 19). The success of these reactions depends on the stereospecific reaction of the ketone (**50**) with $LiA1H(OBu^t)_3$ and of the aldehyde (**49**) with a di-isobutylalane.

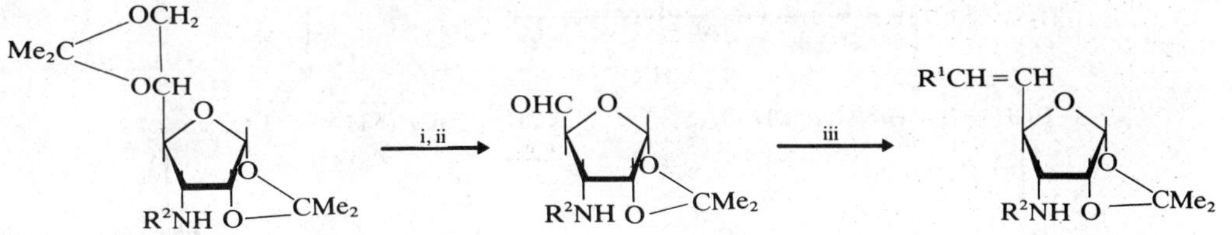

R^2 = Ac or $PhCH_2OCO$

$\xrightarrow{\text{i, ii}}$ $R^1CH{=}CHCH(OH)CH(NHR^2)CHO$ $\xrightarrow{\text{iv, v}}$ $R^1CH{=}CHCH(OH)CH(NH_2)CH_2OH$

$\xrightarrow{\text{iv, vi}}$ $R^1CH_2CH_2CH(OH)CH(NH_2)CH_2OH$

Synthesis of sphinganine and sphing-*cis*-4- and *trans*-4-enine. i, aq. AcOH; ii, $NaIO_4$; iii, $R^1CH_2\overset{+}{P}Ph_3\ \overset{-}{Br}$ ($R^1 = C_{13}H_{27}$), PhLi; iv, $NaBH_4$; v, $Ba(OH)_2$; vi, Pd/C, H_2, AcOH. (The unsaturated compound is a mixture of *cis* and *trans* isomers which can be separated.)

SCHEME 18

$HO_2CCH(NH_2)CH_2OH$ $\xrightarrow{\text{i, ii}}$ $HO_2CCH(NZ)CH_2OAc$ (**48**) $Z = (CO)_2C_6H_4$ $\xrightarrow{\text{iii, iv}}$ $OHCCH(NZ)CH_2OAc$ (**49**)

$\xrightarrow{\text{v}}$ $R^1CH{=}CHCH(OH)CH(NZ)CH_2OAc$ $\xrightarrow{\text{vi, vii}}$ $R^1CH{=}CHCH(OH)CH(NH_2)CH_2OH$
D-*erythro*-Sphing-*trans*-4-enine

(**48**) $\xrightarrow{\text{iii, viii}}$ $N_2CHCOCH(NZ)CH_2OAc$ $\xrightarrow{\text{ix}}$ $R^2CH_2COCH(NZ)CH_2OAc$ (**50**)

$\xrightarrow{\text{x, vi, vii}}$ $R^2CH_2CH(OH)CH(NH_2)CH_2OH$
D-*threo*-Sphinganine

Synthesis of sphing-*trans*-4-enine and D-*threo*-sphinganine from L-serine. i, $C_6H_4(CO)_2$-NCO_2Et; ii, Ac_2O; iii, $SOCl_2$; iv, H_2, Pd; v, $Me(CH_2)_{12}CH{=}CHAlBu^i_2$ (*trans*); vi, MeOH, H^+; vii, N_2H_4; viii, CH_2N_2; ix, R^2_3B, $R^2 = CH_3(CH_2)_{13}$; x, $LiAlH(OBu^t)_3$.

SCHEME 19

(*ii*) 4-*Hydroxysphinganine*

The synthesis of the racemates of the four isomers of 4-hydroxysphingenine from $C_{14}H_{29}C{\equiv}CCH(NHAc)CH_2OH$ (Scheme 20) is based on the stereospecific hydroxylation of the *cis*- and *trans*-alkenes, which result from partial reduction of the acetylenic bond, with suitable reagents. In each case the product should be a mixture of two racemates, but only one may be obtained as a consequence of steric factors during hydroxylation or of solubility effects during isolation.

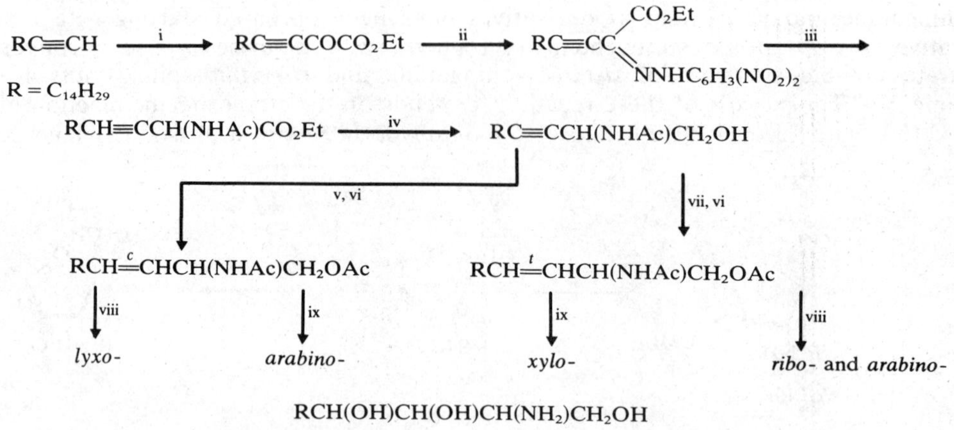

Synthesis of the four racemic 4-hydroxysphinganines. i, EtMgBr, $(CO_2Et)_2$; ii, 2,4-$(NO_2)_2C_6H_3NHNH_2$: iii, Zn; iv, $LiAlH_4$; v, H_2, Pd; vi, Ac_2O; vii, Na, NH_3; viii, HCO_3H, KOH; ix, I_2, AgOAc; KOH.

SCHEME 20

(*iii*) *Glyco- and phospho-sphingolipids*

The synthesis of a glycosphingolipid (Scheme 21) has been achieved starting with a natural 4-hydroxysphinganine. With the secondary hydroxyl groups blocked, the primary hydroxyl function is linked to a suitable galactose derivative to furnish finally a galactose ceramide based on 4-hydroxysphinganine.

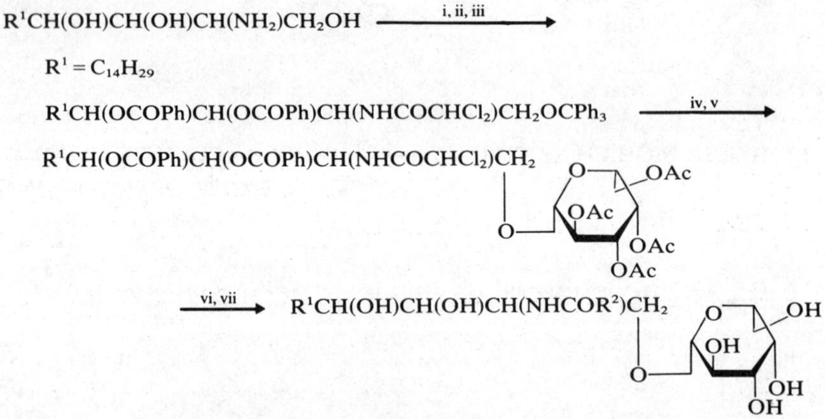

Synthesis of galactosylceramides based on 4-hydroxysphinganine. i, Cl_2CHCO_2Me; ii, Ph_3CCl; iii, PhCOCl; iv, AcOH; v,

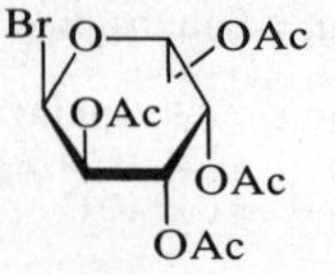

; $Hg(CN)_2$; vi, KOH; vii, $R^2CO_2C_6H_4NO_2$-*p* or $R^2CH(OAc)CO_2C_6H_4NO_2$-*p* for 2-hydroxy-acids.

SCHEME 21

Phosphosphingolipids are prepared from ceramides (with their secondary hydroxyl(s) blocked) and an appropriate phosphorylating agent (Scheme 22).

$$R^2CH{=}CHCH(OCOPh)CH(NHCOR^1)CH_2OH \xrightarrow{\text{i, ii}} R^2CH{=}CHCH(OH)CH(NHCOR^1)CH_2O\overset{\overset{\displaystyle O}{\|}}{\underset{\underset{\displaystyle O^-}{|}}{P}}O(CH_2)_2\overset{+}{N}Me_3$$

Synthesis of ceramidephosphorylcholines. ($R^2 = C_{13}H_{27}$, obtained from synthetic racemic material. R^1CO_2H represents a fatty acid or a 2-acetoxy fatty acid). i, $Cl_2OPO(CH_2)_2Br$, H_2O; ii, NMe_3, ^-OH.

SCHEME 22

25.2.4 LIPID BIOSYNTHESIS

25.2.4.1 Introduction[47–49]

Several pathways of lipid biosynthesis have been demonstrated in different biological systems. These are summarized in Scheme 23 and some general points are made before discussing the various lipid classes.

(i) The glycerol carbon atoms, present in all lipids except waxes and sphingolipids, are derived from glycerol (1,2,3-trihydroxypropane), glyceraldehyde (2,3-dihydroxypropanal), or dihydroxyacetone (1,3-dihydroxypropan-2-one) and all these result from the metabolism of carbohydrate.

(ii) Most lipids can be produced by more than one pathway and in any biological system one or more of these may be operative.

(iii) Apart from the ether lipids, all glycerolipids are made from phosphatidic acids or diacylglycerols, which are themselves interconvertible.

(iv) Structural units to be attached to the phosphatidic acids or diacylglycerols are made available in a modified form associated with a nucleotide such as cytidine or uridine phosphate.

(v) Acylation is effected by a fatty acid attached to coenzyme A (CoASH) or to some other enzyme system:

$$RCO_2H + CoASH + ATP \longrightarrow RCOSCoA + ADP + H_3PO_4 \qquad (9)$$

(vi) In most lipids, different acyl groups are associated with each hydroxyl group (Section 25.2.2.4) and the various lipids from a single source are frequently quite different in their component acids. This may arise in one or more of the following ways: (a) selectivity of the acid involved in acylation or of the substrate being acylated imposed by the enzyme; (b) selectivity of the acid entering an acylation reaction because of the composition of the fatty acid pool; (c) modification of an acyl group in a preformed lipid by a chemical reaction which may be specific in terms of the acyl group undergoing reaction, its point of attachment to the lipid molecule, and the nature of the polar head group (the best known example of this is the conversion of alkenyl groups to cyclopropane systems effected by *S*-adenosylmethionine which only occurs on certain alkyl chains in the 1-position of a phosphatidylethanolamine); (d) modification of the fatty acids in a preformed lipid by transacylation reactions which may be specific in terms of the entering acyl group, the departing acyl group, the site of the exchange, and the nature of the lipid head group.

25.2.4.2 Phosphatidic acids and diacylglycerols

The phosphatidic acids and 1,2-diacyl-*sn*-glycerols are intermediates in the synthesis of all glycerolipids except the ether lipids, and these two compounds are readily interconverted by phosphorylation and dephosphorylation. The diacylglycerols are made by acylation of 2-acyl-*sn*-glycerols available by metabolism of dietary triacylglycerols, and

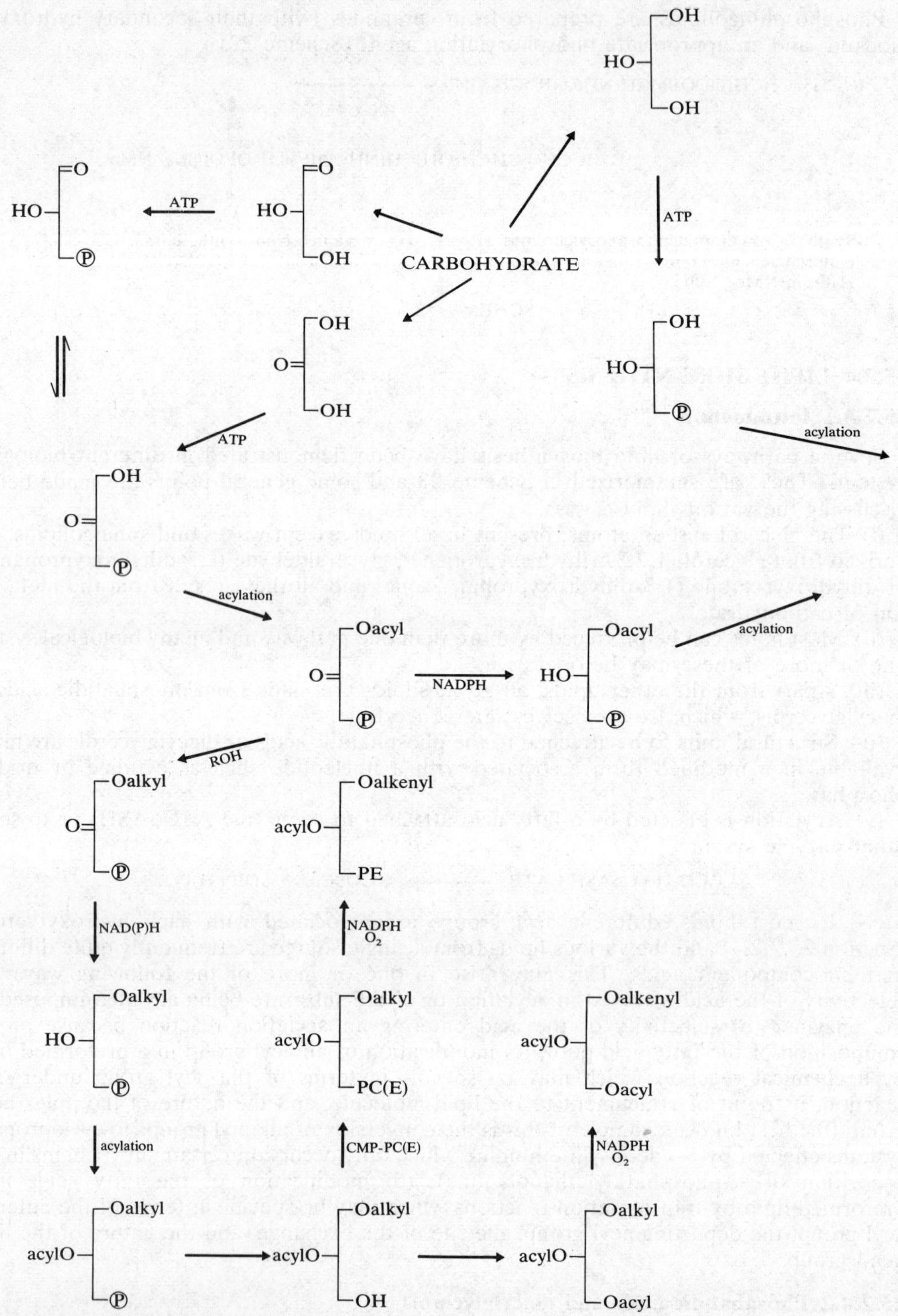
ATP
ATP
CARBOHYDRATE
ATP
acylation
acylation
acylation
NADPH
ROH
NAD(P)H
NADPH
O2
acylation
CMP-PC(E)
NADPH
O2
Oacyl
Oalkyl
Oalkenyl
acylO
PE
PC(E)

4

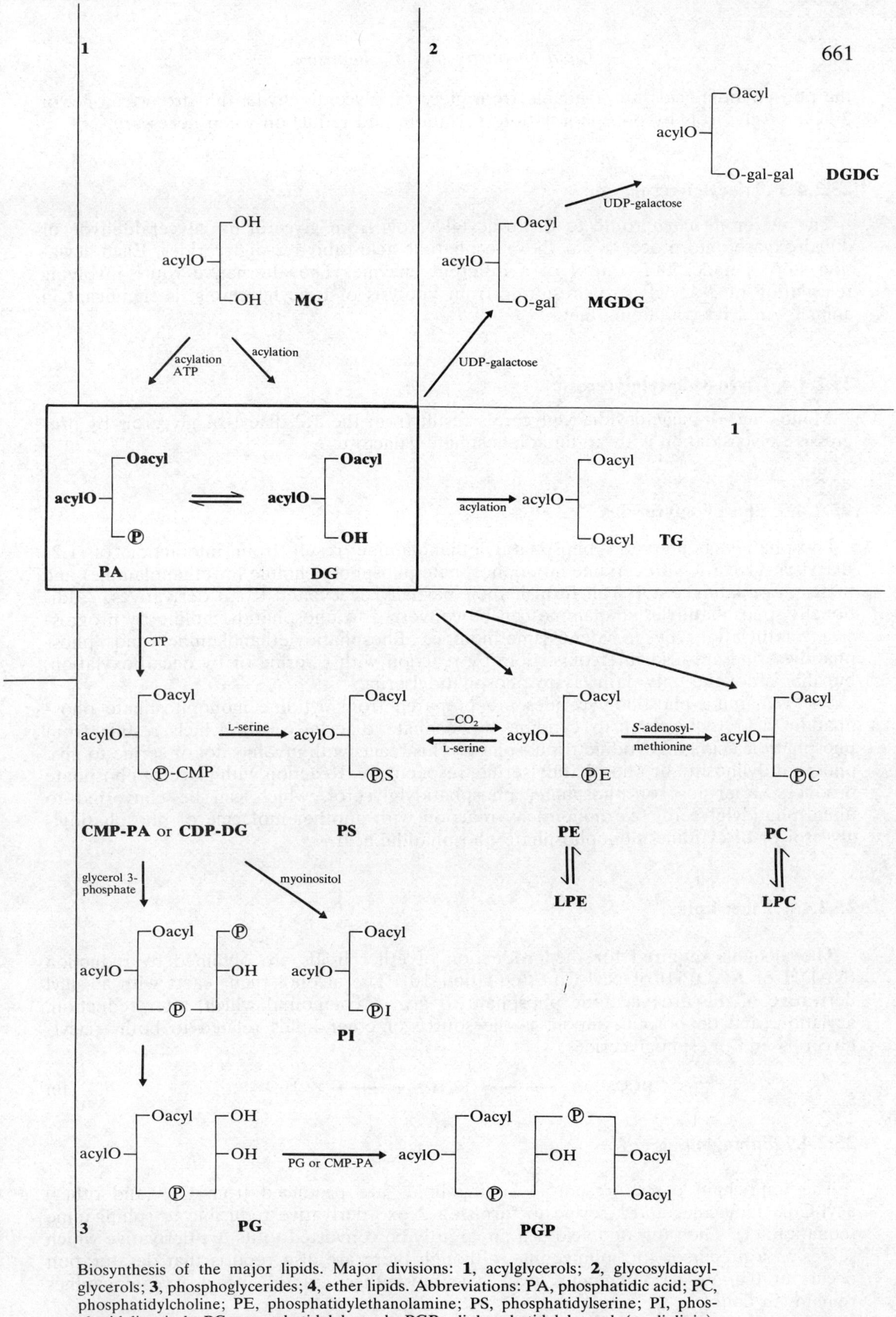

Biosynthesis of the major lipids. Major divisions: **1**, acylglycerols; **2**, glycosyldiacylglycerols; **3**, phosphoglycerides; **4**, ether lipids. Abbreviations: PA, phosphatidic acid; PC, phosphatidylcholine; PE, phosphatidylethanolamine; PS, phosphatidylserine; PI, phosphatidylinositol; PG, pnosphatidylglycerol; PGP, diphosphatidylglycerol (cardiolipin); MG, monoacylglycerol; DG, diacylglycerol; TG, triacylglycerol; MGDG (DGDG), mono- and di-galactosyldiacylglycerol; acyl^n, acylation; ATP, adenosine triphosphate; CMP, cytidine monophosphate; CTP, cytidine triphosphate; UDP-gal, uridine diphosphate galactose

SCHEME 23

the phosphatidic acids are available from glycerol, glyceraldehyde, dihydroxyacetone, or 2-acyl-*sn*-glycerols by phosphorylation, acylation, and reduction when necessary.

25.2.4.3 Triacylglycerols

The major *de novo* route to the triacylglycerols from glycerol (or glyceraldehyde or dihydroxyacetone) proceeds *via* the phosphatidic acids and 1,2-diglycerides. Each acylation step is under the control of a different enzyme. The alternative route involving reacylation of 2-acylglycerol, resulting from lipolysis of triacylglycerols, is significant in animals on a fat-containing diet.

25.2.4.4 Glycosyldiacylglycerols[50]

Mono- and di-galactosyldiacylglycerols result from the 1,2-diacyl-*sn*-glycerols by progressive glycosidation with uridine diphosphate galactose.

25.2.4.5 Phosphoglycerides

Phosphatidylcholine and phosphatidylethanolamine result from interaction of 1,2-diacyl-*sn*-glycerol with cytidine monophosphate phosphorylcholine (or ethanolamine) and these phosphatidyl esters can furnish their partially deacylated (lyso) derivatives. Additionally, phosphatidylethanolamine can be converted to phosphatidylcholine by progressive methylation by *S*-adenosylmethionine. Phosphatidylethanolamine and phosphatidylserine are also interconvertible by reaction with L-serine or by decarboxylation, but this is not the only pathway to phosphatidylserine.

The remaining phosphoglycerides are prepared from cytidine monophosphate phosphatidic acid (equivalent to cytidine diphosphate diacylglycerol), which results from phosphatidic acid and cytidine triphosphate. This reacts with myoinositol or serine to give phosphatidylinositol or phosphatidylserine, respectively. Reaction with glycerol phosphate produces, after loss of phosphate, phosphatidylglycerol, which can be converted to diphosphatidylglycerol (cardiolipin) by reaction with another molecule of phosphatidylglycerol or of cytidine monophosphate phosphatidic acid.

25.2.4.6 Ether lipids[51]

The alcohols required for the production of ether lipids are obtained by reduction (NADH or NADPH) of acyl-CoA (equation 10). The alcohol then reacts with an acyl derivative of dihydroxyacetone phosphate to give a compound which, after reduction, acylation, and dephosphorylation, is the source of ether lipids related to both triacylglycerols and phosphoglycerides.

$$\text{RCOSCoA} \longrightarrow \text{RCHO} \longrightarrow \text{RCH}_2\text{OH} \qquad (10)$$

25.2.4.7 Sphingolipids[52,53]

The long-chain bases present in sphingolipids are produced from C_{16} (and other) acyl-CoA molecules and L-serine to furnish a 3-oxo-derivative reducible to sphinganine (equation 11). The saturated acid can apparently be converted to its Δ^2t derivative which serves as a precursor for sphingenine, although there are also reports that desaturation occurs at the *N*-acylsphinganine level. The biosynthesis of the 4-hydroxysphinganines remains in doubt.

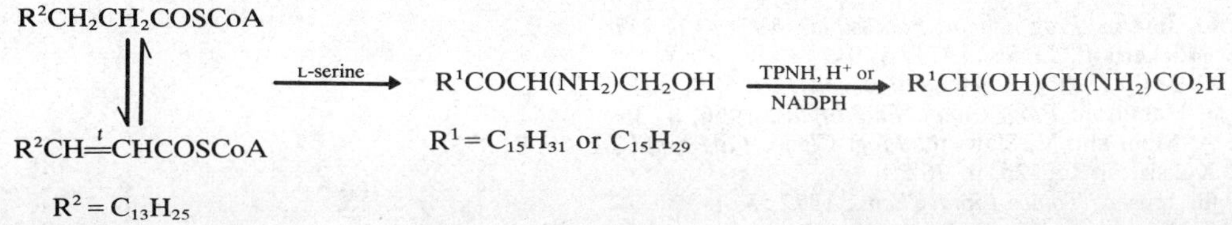

(11)

The long-chain bases are converted to glycosphingolipids and phosphosphingolipids as shown in Scheme 24.

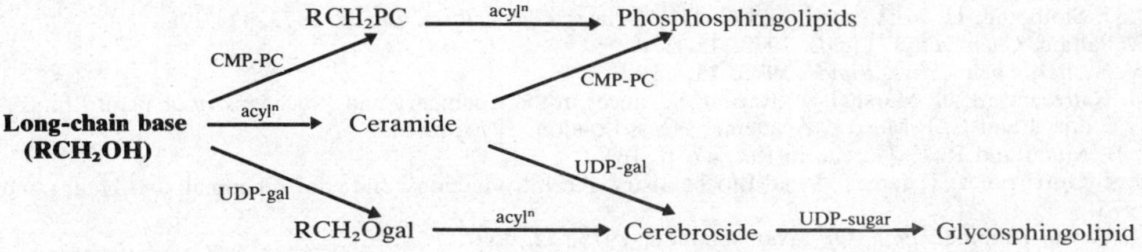

Biosynthesis of phospho- and glyco-sphingolipids (see Scheme 23 for information about abbreviations)

SCHEME 24

References

1. Anon., *European J. Biochem.*, 1967, **2,** 127; *Biochem. Biophys. Acta*, 1968, **152,** 1; *Chem. Phys. Lipids*, 1968, **2,** 156.
2. S. Hamilton and R. J. Hamilton, *Topics Lipid Chem.*, 1972, **3,** 199.
3. 'Chemistry and Biochemistry of Natural Waxes', ed. P. E. Kolattukudy, Elsevier, Amsterdam, 1976.
4. G. Odham and E. Stenhagen, *Accounts Chem. Res.*, 1971, **4,** 121.
5. P. E. Kolattukudy and T. J. Walton, *Prog. Chem. Fats Lipids*, 1972, **13,** 119; in 'Recent Advances in the Chemistry and Biochemistry of Plant Lipids', ed. T. Galliard and E. I. Mercer, Academic Press, London, 1975, p. 203.
6. L. D. Bergelson, *Prog. Chem. Fats Lipids*, 1969, **10,** 239; *Fette Seifen Anstrichm.*, 1973, **75,** 89.
7. M. Kates, *Adv. Lipid Res.*, 1970, **8,** 225.
8. P. S. Sastry, *Adv. Lipid Res.*, 1974, **12,** 251.
9. T. H. Haines, *Prog. Chem. Fats Lipids*, 1971, **11,** 297.
10. F. Snyder, *Prog. Chem. Fats Lipids*, 1969, **10,** 287.
11. F. Snyder, 'Ether Lipids: Chemistry and Biology', Academic Press, New York, 1972.
12. E. Klenk and H. Debuch, *Prog. Chem. Fats Lipids*, 1963, **6,** 1.
13. D. Shapiro, 'Chemistry of Sphingolipids', Hermann, Paris, 1969.
14. K. A. Karlsson, *Lipids*, 1970, **5,** 878; *Chem. Phys. Lipids*, 1970, **5,** 6.
15. W. Stoffel, *Chem. Phys. Lipids*, 1973, **11,** 318.
16. E. Mårtensson, *Prog. Chem. Fats Lipids*, 1970, **10,** 365.
17. N. S. Radin, *Chem. Phys. Lipids*, 1970, **5,** 178.
18. H. Wiegandt, *Adv. Lipid Res.*, 1971, **9,** 249.
19. J. Kiss, 'Carbohydrate Chemistry and Biochemistry', ed. L. Wolfram, R. S. Tipson, and D. Horton, Academic Press, New York, 1970, vol. 24.
20. H. Wiegandt, *Angew. Chem. Internat. Edn.*, 1968, **7,** 87.
21. D. Shapiro, *Chem. Phys. Lipids*, 1970, **5,** 80.
22. R. Ledeen, *Chem. Phys. Lipids*, 1970, **5,** 205.
23. R. H. McCluer, *Chem. Phys. Lipids*, 1970, **5,** 220.
24. R. G. Ackman, *Prog. Chem. Fats Lipids*, 1972, **12,** 165.
25. 'Analysis of Lipids and Lipoproteins', ed. E. G. Perkins, American Oil Chemists' Society, 1975.
26. N. Pelick and V. Mahadevan, in Ref. 25, p. 23.
27. M. Kates, 'Techniques in Lipidology: Isolation, Analysis, and Identification of Lipids', North Holland, Amsterdam, 1972.
28. W. W. Christie, 'Lipid Analysis', Pergamon, Oxford, 1973.
29. M. L. Blank and F. Snyder, in Ref. 25, p. 68; G. J. Nelson, in Ref. 25, p. 70; L. A. Witting, in Ref. 25, p. 90.
30. A. Kuksis, *Prog. Chem. Fats Lipids*, 1972, **12,** 1.
31. C. Litchfield, 'Analysis of Triglycerides', Academic Press, New York, 1972.
32. B. L. Walker, in Ref. 25, p. 108.

33. R. G. Jensen, *Prog. Chem. Fats Lipids*, 1971, **11,** 347.
34. H. Brockerhoff, *Lipids*, 1971, **6,** 942.
35. D. C. Malins, *Prog. Chem. Fats Lipids*, 1966, **8,** 301.
36. J. G. Hamilton, *Prog. Chem. Fats Lipids*, 1966, **8,** 359.
37. R. A. Stein and V. Slawson, *Prog. Chem. Fats Lipids*, 1966, **8,** 373.
38. A. Kuksis, in Ref. 25, p. 36.
39. R. G. Jensen, *Topics Lipid Chem.*, 1972, **3,** 1.
40. R. G. Jensen, and R. E. Pitas, *Adv. Lipid Res.*, 1976, **14,** 213.
41. C. M. Lok, J. P. Ward, and D. A. van Dorp, *Chem. Phys. Lipids*, 1976, **16,** 115.
42. A. J. Slotboom and P. P. M. Bonsen, *Chem. Phys. Lipids*, 1970, **5,** 301.
43. R. G. Jensen and D. T. Gordon, *Lipids*, 1972, **7,** 611.
44. A. J. Slotboom, H. M. Verheij, and G. H. de Haas, *Chem. Phys. Lipids*, 1973, **11,** 295.
45. F. Paltauf, *Chem. Phys. Lipids*, 1973, **11,** 270.
46. W. Stoffel, *Chem. Phys. Lipids*, 1973, **11,** 318.
47. M. Kates and M. O. Marshall in 'Recent Advances in the Chemistry and Biochemistry of Plant Lipids', ed. T. Galliard and E. I. Mercer, Academic Press, London, 1975, p. 115.
48. J. B. Mudd and R. E. Garcia, in Ref. 47, p. 161.
49. M. I. Gurr and A. T. James, 'Lipid Biochemistry, An Introduction', 2nd edn., Chapman and Hall, London, 1975.
50. H. C. van Hummel, *Prog. Org. Nat. Products*, 1975, **32,** 267.
51. F. Snyder, *Adv. Lipid Res.*, 1972, **10,** 233.
52. E. E. Snell, S. J. Dimari, and R. N. Brady, *Chem. Phys. Lipids*, 1970, **5,** 116.
53. W. Stoffel, *Chem. Phys. Lipids*, 1970, **5,** 139.

25.3

Membranes and Lipoproteins

P. F. KNOWLES
University of Leeds

25.3.1 THE BIOLOGICAL ROLE OF MEMBRANES AND THEIR STRUCTURAL COMPLEXITY

The traditional concept of a membrane is that it provides compartmentation either between cells or within a cell. This role is indeed true of myelin, which surrounds nerve cells and insulates them from interactions with their neighbours. The lipid components of membranes are well suited to a compartmentational role and in fact myelin has a high proportion of lipid relative to the other major membrane component, protein (see Table 1).

Most other biological membranes have much higher proportions of protein (50–60%) as well as carbohydrate (0–10%). One of several consequences of the higher protein content is that, whilst the advantages of the membrane in providing compartmentation are maintained, permeability to different metabolites is enhanced. The permeability of biological membranes is selective to particular metabolites and, for certain membranes, is able to proceed against a concentration gradient (this is termed 'active' as opposed to the normal 'passive' transport). A wide diversity of proteins is required to provide the membrane with its special permeability properties.

The majority of proteins found in membranes are enzymes.[1] Apart from the enzymes involved in permeability processes, the membrane location of other enzymes favours their catalytic role; thus enzymes catalysing oxidation processes or the assembly of certain macromolecules benefit from the apolar environment inside the membrane. The membrane also provides a favourable spatial relationship between enzymes whereby the product of one enzyme-catalysed reaction becomes the substrate for a laterally adjacent enzyme. In addition, the product(s) of one enzyme-catalysed reaction may be displaced to the opposite side of the membrane to the site of substrate attachment, provided that the

TABLE 1
Composition of Typical Biological Membranes

Component	*Human myelin*	*Human erythrocyte*	*Bovine liver endoplasmic reticulum*	*Guinea pig liver inner mitochondrial*	*E. coli (inner)*	*B. megaterium (inner)*	*Spinach chloroplast*	*Human serum high density lipoprotein 2*
Protein	18	49	55	74	68	75	50	41
Carbohydrate[a]	3	8	—	2	—	—	—	—
Lipid	79	43	45	24	32	25	50	59

[a] 'Carbohydrate' content refers to oligosaccharide in glycoproteins; glycolipids are included under 'Lipid'.

enzyme spans the membrane; a good example of this phenomenon is found in cytochrome oxidase.[2] It follows from the special properties of such enzymes that membranes must be asymmetric. Not only are proteins arranged asymmetrically in membranes but also the lipids themselves which compose the membrane are asymmetrically disposed; membrane asymmetry and its biological importance will be discussed in more detail in Section 25.3.3.3.

Cellular organization requires that the activity of enzymes should be subject to control. In common with most enzymes, the activity of enzymes located in membranes can be regulated through conformation changes in the protein induced by direct interaction with effectors. Additionally, the activity of enzymes in membranes can be regulated through interactions which involve the membrane lipids; very sophisticated control of enzyme activity results from this dual mechanism.

At the intercellular level, hormones are of great importance in regulating metabolic processes. The binding of hormones to their receptors is a process similar to the reaction between an enzyme and its substrate. An important, if not the only, group of hormone receptors are known to be glycoproteins located in membranes. It has been speculated[3] that the location of receptors in membranes might help hormones to reach their target receptor through two-dimensional diffusion along the membrane rather than by a less efficient three-dimensional diffusion process.

Enzyme–substrate and hormone–receptor formation both imply molecular recognition. At a higher level of organization, cells possess the ability to recognize each other. Thus the leucocytes in the blood stream recognize and destroy foreign cells, for example bacteria, whilst leaving unattacked cells which are native to the blood stream. Another aspect of cellular recognition is that cells from higher organisms exhibit contact inhibition; when cells of a specific type (*e.g.* muscle cells) are grown in a nutrient medium, they continue to divide until they come into contact with other cells at which point growth stops. Cancer cells under similar conditions continue to divide. Both these examples of cellular recognition, which have obvious medical importance, involve surface antigens. The uniqueness of a specific type of cell indicates that the diversity of surface antigens must be high and thus adds further to the complexity of biological membranes. Cellular recognition processes depend on the mobility of components in the membrane, which is probably controlled by microtubules* in the cytoplasm.[4]

From the above brief survey it can perhaps be appreciated that biological membranes have a wide diversity of functions. This diversity is achieved with a range of membrane components and a complexity of organization which consequently present substantial problems for the investigator.

25.3.2 THE COMPOSITION OF BIOLOGICAL MEMBRANES

From Table 1 it can be seen that myelin has a higher lipid/protein ratio than the other examples of membranes given and, as discussed above, this correlates with the limited functional role of myelin. By contrast, the highly efficient oxidative processes occurring in the inner mitochondrial membrane require a variety of enzymes and the lipid/protein ratio is correspondingly low. It can further be seen from Table 1 that the erythrocyte membrane has, in addition, a relatively high proportion of carbohydrate. The major glycoprotein in erythrocyte membranes, glycophorin, has been shown[6] to be oriented at the membrane surface so that the *N*-terminal part of the polypeptide chain, which bears all the covalently bound carbohydrate, extends into the external medium; the exposed oligosaccharide provides some of the blood group antigens and receptors, including the receptor for influenza virus. A schematic representation showing how proteins, lipids, and carbohydrates might be arranged in a biological membrane is given in Figure 1, which is

* Microtubules have been implicated in cell division and other biological processes;[5] they probably provide structure within the cytoplasm which allows cellular information to be transferred between organelles.

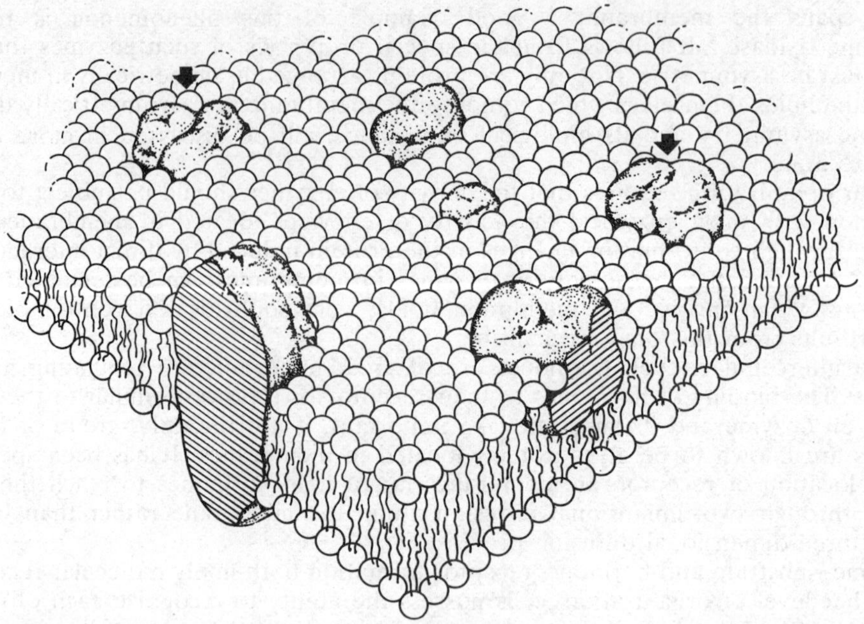

Figure 1 A schematic representation of the 'fluid mosaic' model for biological membranes. (From Singer,[7] by courtesy of the New York Academy of Science)

based on the 'fluid mosaic' model of Singer and Nicholson.[7] The polar lipid molecules form a bilayer, evidence for which is summarized in Section 25.3.3, whilst the protein can be either surface bound (termed 'extrinsic proteins') or intercalated into the bilayer (termed 'intrinsic' or 'integral' proteins); in certain cases, the protein can span the bilayer. The fluid mosaic model, which has gained general acceptance, requires that the membrane is fluid rather than static under physiological conditions. Thus, the lipid and protein components in isolated biological membranes undergo translational diffusion in the plane of the membrane at rates determined in part by the effective viscosity of the lipid matrix.

The lipid composition of various biological membranes is given in Table 2. It is clear that the relative proportions of phospholipid to glycolipid varies from species to species and even between cell types within a single species; there is also variation in the cholesterol content and in the relative proportions of different classes of phospholipid. These differences must relate to the functional demands on the membrane, a point which will be considered further in Section 25.3.3.3.

The percentage fatty acid composition of the major lipids in membranes of human erythrocytes is given in Table 3. In part, the complexity of the fatty acid composition amongst the different classes of phospholipid is necessary to maintain the fluidity of the membrane; fluidity depends on all lipid bilayers being lyotropic liquid crystals. At a temperature characteristic of a particular phospholipid, a phase transition from the rigid gel phase to the fluid liquid crystal phase occurs. The topic of fluidity and phase transitions will be considered later in more detail (see Section 25.3.3.1).

25.3.3 THE LIPID COMPONENTS OF MEMBRANES

The first two sections outlined the structural complexity of membranes and possible reasons (in terms of function) for this complexity. Our present knowledge of biological membranes is based upon studies with defined lipid and protein systems, which will be the subject of this section and Section 25.3.4.

TABLE 2
Lipid Composition of Typical Biological Membranes

	Human myelin	*Human erythrocyte*	*Bovine liver endoplasmic reticulum*	*Guinea pig liver inner mitochondrial*	*E. coli (inner)*	*B. magaterium (inner)*	*Spinach chloroplast*	*Human serum high density lipoprotein 2*
Phospholipids	38	72	82	100	89	48	12	60
Phosphatidyl-choline	11	23	55	48	—			
Phosphatidyl-ethanolamine	14	20	18	28	82			
Phosphatidyl-serine	7	11	9	2	—			
Phosphatidyl-glycerol	—	—	—	—	7			
Sphingomyelin	6	18	—	2	—			
Others	—	—	—	20[a]	20			
Glycolipids	37	3	—	—	—	52	80	—
Triglycerides	—	—	—	—	—	—	—	8
Cholesterol and cholesterol esters	25	25	6	—	—	—	—	32
Others	—	—	13	—	—	—	8	—

[a] Mainly diphosphatidylglycerol ('cardiolipin').

TABLE 3
Percentage Fatty Acid Composition of the Major Lipids in Human Erythrocyte Membranes[a]

	Total	*PC*	*PE*	*Sph*	*PS*
C16:0	25	34	29	28	14
C18:0	19	13	9	7	36
C18:1	16	22	22	6	15
C18:2	11	18	6	2	7
C20:4	15	6	18	—	21
Remainder	14	7	16	—	7

[a] PC = phosphatidylcholine; PE = phosphatidylethanolamine; Sph = sphingomyelin; PS = phosphatidylserine.

25.3.3.1 Phospholipids

(*i*) *Formation of bilayer structures*

The majority of investigations have been made on phospholipids rather than glyco- or sulpho-lipids, although the limited number of studies made with the latter lipids suggests that they behave similarly. Phospholipids spontaneously form bilayers when dispersed in water, with their polar head groups pointing out into the aqueous medium and the apolar chains pointing towards the centre of the bilayer (see Figure 1). The evidence for bilayers in these dispersions is based on experimental data obtained from several different lines of investigation.

(*a*) *X-ray diffraction studies.*[8]. The electron-density profile from oriented samples of dipalmitoyl phosphatidylcholine (DPPC) is characterized by peaks separated by approximately 50 Å and with a deep central well. It has been shown that the peaks could reasonably be ascribed to the polar regions of the lipid whilst the central well would be predicted by a model in which the terminal methyl groups of the alkyl chains are localized near the centre of the bilayer. The gel–liquid-crystal phase transition of DPPC is 42 °C; thus the sample above would be in the gel phase at the temperature of examination (23 °C). Investigations with phosphatidylcholine from egg yolk, which is in its liquid-crystalline phase at 23 °C, gave an electron density profile qualitatively similar to that from DPPC; however, the spacing of the peaks was now 36 Å and the central trough was broadened. These results are consistent with the bilayer thinning through the phase transition and the terminal methyl groups of the alkyl chain becoming more disordered through increased molecular motion. Similar electron-density profiles have been found for biological membranes, showing that bilayers of lipid are the basis for their structure.

The fluidity of phospholipid bilayers has hindered progress towards obtaining X-ray crystallographic information on phospholipid assemblies. However, Hitchcock *et al.*[9] have succeeded in crystallizing dilaurylphosphatidylethanolamine and have determined its structure; on the basis of this structure, a quantitative interpretation of the X-ray diffraction data from bilayers of phosphatidylethanolamine may be made (see Figure 2).

(*b*) *Electron microscopic studies.* Electron microscopy on ultra-thin sections of myelin shows the characteristic 'tramline' structure (Figure 3). More detailed investigations have shown[10] that the fundamental radial unit in myelin is formed by two adjacent bimolecular layers of lipid approximately 60 Å in thickness. Similar multibilayer structures to those from myelin are observed when pure phospholipids are dispersed in water (Figure 4a) and this clearly establishes that physical rather than biological factors are primarily responsible for bilayer formation. Sonication of phospholipid dispersions produces single

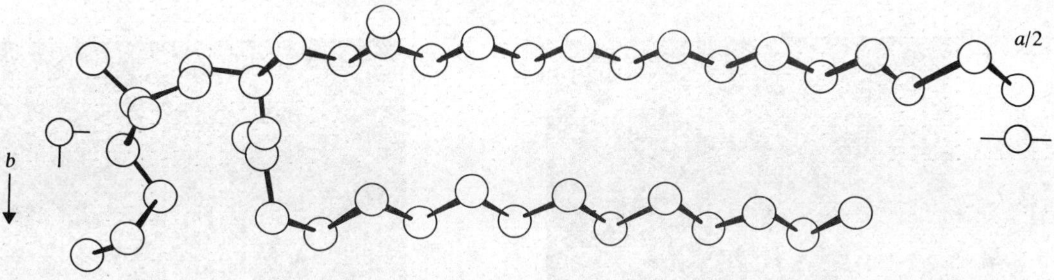

Figure 2 Projected molecular structure within a bilayer of 1,2-dimyristoyl-D,L-phosphatidylethanolamine. (From Hitchcock *et al.*,[9] by courtesy of *J. Mol. Biol.*)

bilayer vesicles (Figure 4b) which are particularly well suited to studies of permeability to various solute molecules.

The technique of freeze-fracture electron microscopy provides confirmation that the structures seen in the ultra-thin section studies are not the result of the chemical processes used in sample preparation (Figures 4c and 4d).

(c) *Spectroscopic studies.* Electron spin resonance (e.s.r) spectroscopy has provided elegant support for the bilayer structure. Oriented bilayers of phospholipid incorporating low levels (approximately 1 mole per cent) of a paramagnetic spin label probe whose structure closely resembles that of the host lipid (see Figure 5) can readily be prepared on glass surfaces. The e.s.r. spectrum is anisotropic, *i.e.* it depends on the direction along which the magnetic field is applied; thus the spectrum obtained when the field is applied in the plane of the glass surface differs from that when the field is applied perpendicular to the surface. From these results, it can readily be deduced that the spin labels and hence the host phospholipid molecules are aligned with their long axes perpendicular to the surface. Similar results have been obtained with other spectroscopic techniques[11] and provide further convincing evidence for bilayers in model and biological membranes.

(*ii*) *Model* vs. *biological membranes*

From the preceding discussion, it can be seen that X-ray diffraction, electron microscopy, and spectroscopic studies indicate that there are close parallels between the bilayer

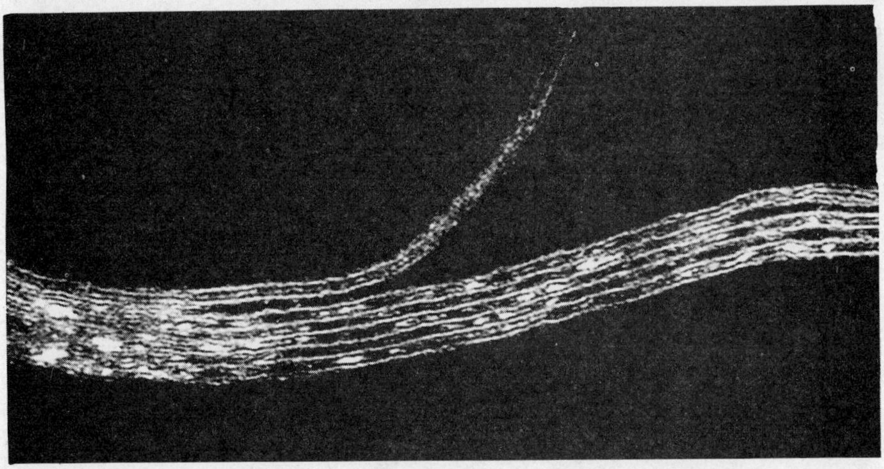

Figure 3 Electron micrograph of myelin sheath. (Reproduced by courtesy of Dr. S. G. R. Aparicio, Leeds University)

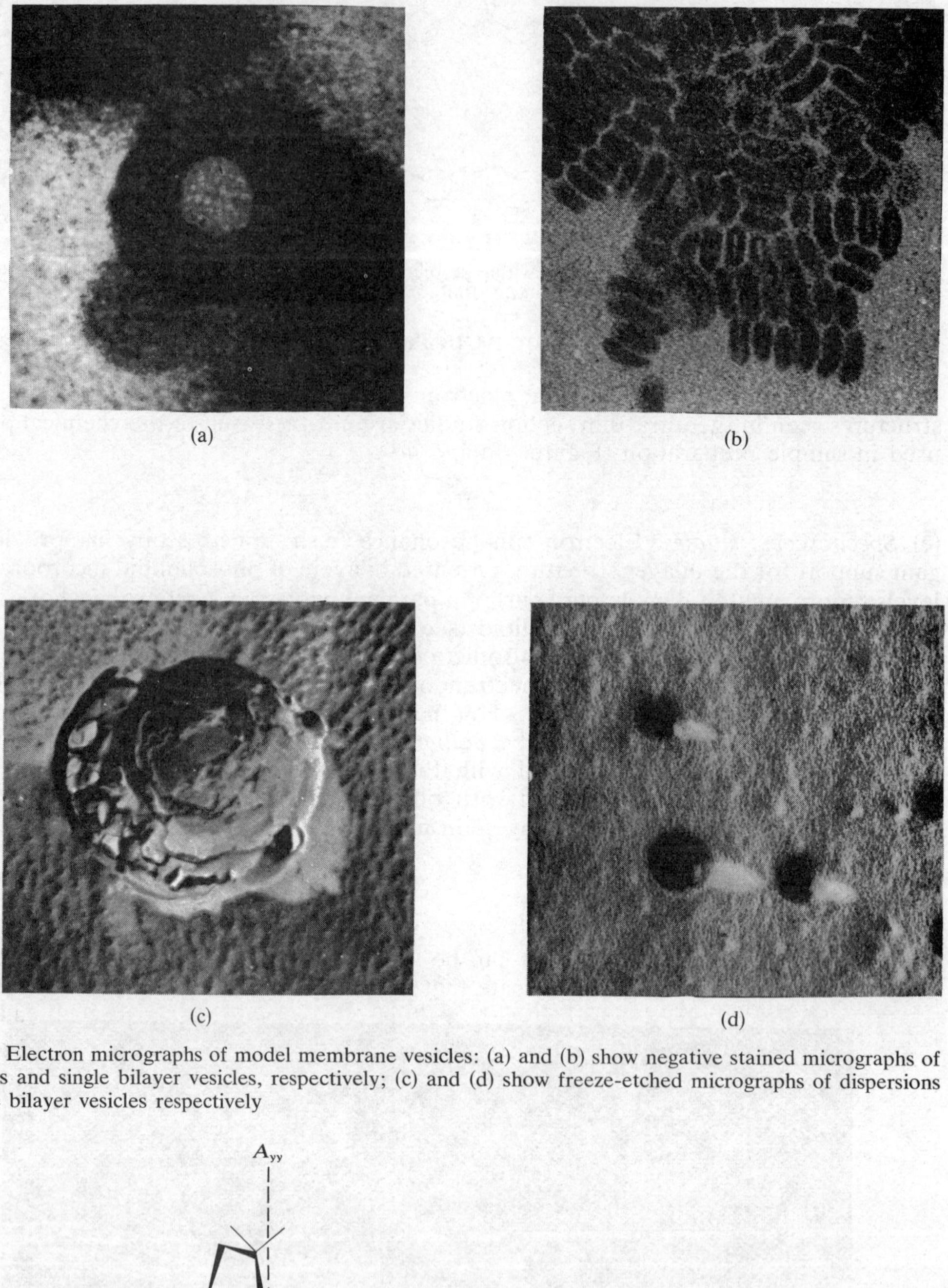

Figure 4 Electron micrographs of model membrane vesicles: (a) and (b) show negative stained micrographs of dispersions and single bilayer vesicles, respectively; (c) and (d) show freeze-etched micrographs of dispersions and single bilayer vesicles respectively

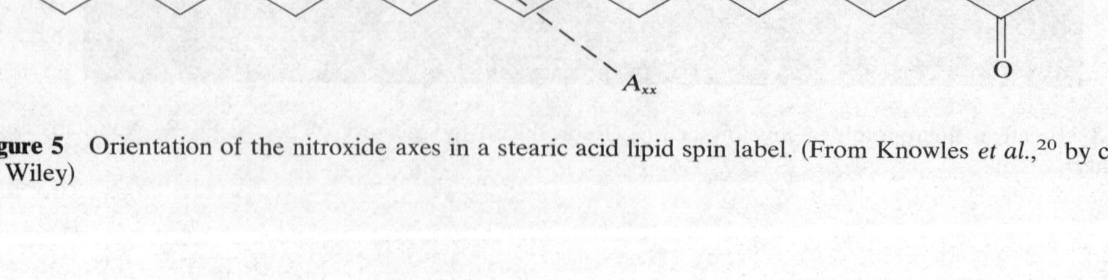

Figure 5 Orientation of the nitroxide axes in a stearic acid lipid spin label. (From Knowles *et al.*,[20] by courtesy of Wiley)

structures in biological membranes and phospholipid dispersions ('model membranes'). As seen in Table 4, these parallels extend to other physical characteristics of the two systems and justify continued study of model membranes as a basis for the understanding of the structure of biological membranes.

The major difference noted in Table 4 between model and biological membranes is in permeability. A strong body of evidence suggests that the permeability of bilayers to small molecules depends on the presence in the bilayer of other components, *e.g.* proteins, peptides or antibiotics.[12]

TABLE 4

Comparative Physical Characteristics of Natural and Biomolecular Lipid Membranes

Property	*Natural membranes*	*Bilayer membranes*
Thickness (Å)		
electron microscopy	40–130	60–90
X-ray diffraction (myelin)	40–85	
optical methods		46
Resistance (ohm cm^{-1})	10^2–10^5	10^3–10^9
Capacitance ($\mu F\ cm^{-2}$)	0.5–1.3	0.3–1.3
Resting potential difference (mV)	10–88	0–140
Breakdown voltage (mV)	100	100–550
Interfacial tension (10^{-7} J cm^2)	0.03–3.0	0.2–6.0
Permeability to water (10^{-4} cm s^{-1})	25–28	2.3–24

(*iii*) *Gel–liquid-crystal phase transitions*

Bilayers composed of a single phospholipid species undergo a well defined phase transition at a characteristic temperature from the relatively rigid gel phase to the fluid liquid-crystalline phase. In addition to the spectroscopic and diffraction methods discussed above, calorimetry has also been used to monitor this phase transition.

The temperature of the transition depends on the chain length, degree of unsaturation of the alkyl chains, and the nature of the head group (see Table 5). Calorimetric, spectroscopic, and diffraction studies in addition reveal a pre-transition at temperatures a few degrees below the tabulated values; the underlying structural and dynamic processes are not understood, although diffraction evidence suggests that one structural change at the pre-transition stage could involve tilted alkyl chains becoming perpendicular to the bilayer surface.[13]

TABLE 5

Phase Transition Temperatures (°C) of Phospholipid Bilayers: Variation with Head Group, Alkyl Chain Length, and Unsaturation

Phospholipid	*Alkyl chain composition* C14	C16	C18	C18:1
Phosphatidylcholine	24	41	58	~ -10
Phosphatidylethanolamine	50	63	—	<10

^{31}P and ^{2}H n.m.r. studies clearly establish differences in the conformation and mobility of polar head groups through the lipid phase transition (see references in Kohler and Klein[14]). The different classes of phospholipid can be distinguished in their behaviour in this respect. Of particular interest are the studies on phospholipids having a net negative charge in their head groups, *e.g.* phosphatidylserine, phosphatidic acid, phosphatidylglycerol, diphosphatidylglycerol and phosphatidylinositol, since it has been shown[15] that in these systems *isothermal* phase transitions can be produced by changes in pH and ionic strength; undoubtedly these changes have physiological significance.[15,16]

(*iv*) *Dynamic properties of lipid bilayers*

The nature of dynamic processes in the liquid crystalline state of bilayers may now be discussed. In Section 25.3.3.1(i) the use of spin-label probes to provide evidence for bilayer structures was briefly described. It is evident from the e.s.r. spectra that the probe spins rapidly about its long axis since this averages out the anisotropy in the x and y directions (see Figure 5). Similar results showing rapid rotational motion of lipid molecules in the bilayer have been obtained using fluorescent probes (Figure 6), since

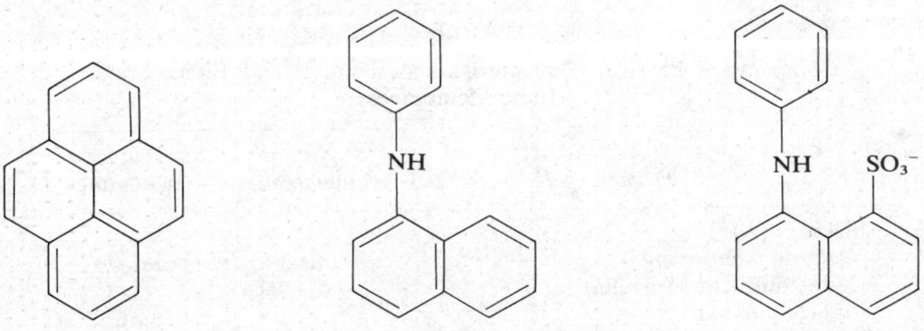

(1) **(2)** **(3)**

Figure 6 Fluorescent probes used in membrane studies: (**1**), pyrene; (**2**), n-phenyl-l-naphthylamine; (**3**), l-anilinonaphthalene-8-sulphonate

fluorescence polarization is also averaged out by this motion. Thus the results indicate that the interior of membranes is in a highly fluid condition with a viscosity comparable to a light oil. The next question might be, is the fluidity the same at all places inside the bilayer? E.s.r., n.m.r., and fluorescence methods can provide information bearing on this question.

A series of spin-labelled phospholipids have been synthesized bearing the nitroxide grouping at different positions down one of the alkyl chains of phosphatidylcholine (Figure 7). Incorporation of these spin labels into bilayers allows different parts of the

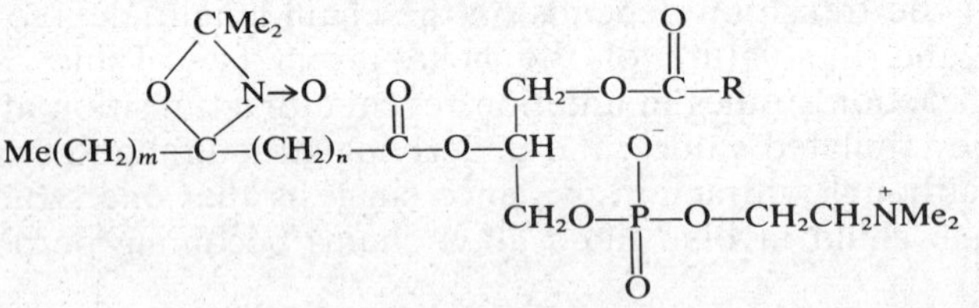

Figure 7 Phospholipid spin labels

bilayer to be monitored by e.s.r. From the spectra, two other types of motion in addition to the rapid long axis rotation can be distinguished.[17] Firstly, the whole lipid molecule precesses as a 'rigid stick' about the perpendicular to the bilayer surface with an anchor point at the surface. The amplitude of this motion depends on whether the lipid composing the bilayer is in its gel or liquid-crystalline state and also on the presence of other membrane components; thus, for example, the presence of cholesterol decreases the amplitude of precessional motion.[11] Secondly, there is a segmental motion within the fatty acid chains through rapid isomerizations between *gauche* and *trans* conformations about carbon–carbon single bonds; this motion would be cumulative down the chain, *i.e.* the terminal methyl at the centre of the bilayer would have greater mobility than the methylene groups close to the glycerol backbone, and is termed a 'fluidity gradient'. It may be noted that the e.s.r. results using the probes shown in Figure 7 further indicate that the polarity of the bilayer decreases towards the centre and hence indicate the existence of a hydrophobic barrier to permeability by charged solute molecules.

N.m.r. studies using ^{1}H, ^{2}H, and ^{13}C isotopes have also revealed the presence of flexibility gradients;[11,18] n.m.r. has the advantage over e.s.r. studies based on paramagnetic probes that the measurements are made on the phospholipid molecules themselves. Thus ^{13}C n.m.r. at natural abundance levels shows that the rates of molecular motion increase from the carbon atoms of the glycerol backbone to the terminal methyl groups of the alkyl chains. More detailed information on chain motion can be obtained by ^{2}H n.m.r. using phospholipids selectively deuteriated at different positions along the alkyl chains; it has been shown that the conformations of the two alkyl chains are distinct (Ref. 13; see also Figure 2) and that the introduction of a single *cis* double bond into one of the alkyl chains causes a local rigidity in the vicinity of the double bond yet increases the overall mobility of both chains.[19]

From the above discussion, it is overwhelmingly clear that phospholipid bilayers, in their liquid-crystalline state, are fluid. One consequence of this property is that they are able to diffuse laterally within the plane of the membrane ('lateral diffusion'). Several methods have been developed to measure lateral diffusion coefficients (D_{diff}) in membranes (see Refs. 11 and 20). For model bilayers and diverse isolated biological membranes, the values of D_{diff} lie close to 10^{-7} cm^2 s^{-1}. Such a value implies that a lipid molecule could, if motion were in a straight line, travel the length of a bacterial cell (a distance of 10^{-4} cm) in 2 seconds. From these values for D_{diff} of lipid molecules, corresponding values for D_{diff} of intercalated proteins can be estimated. The value of D_{diff} for a protein of molecular weight 100 000 daltons is approximately 3×10^{-10} cm^2 s^{-1}, which agrees reasonably with the value measured for the mixing of surface antigens during the formation of hybrid cells by virus-induced fusion.[21] It should be noted, however, that D_{diff} for certain membrane components *in vivo* might well be slower than these values; indeed, lateral diffusion in biological membranes is probably a highly regulated phenomenon. Two instances may be cited; firstly, it has been suggested that high lateral mobility of membrane components distinguishes unregulated cancer cells from their normal counterparts;[22] and secondly, the phenomenon of 'capping', whereby complexes of surface antigens with their antibodies segregate at the end of the cell remote from the nucleus, supports the proposal that, in normal cells, membrane motion is controlled.[4]

Yet another dynamic process is the transverse motion of phospholipid molecules between the inside and outside halves of the bilayer.[11] In model membranes this is slow ($t_{\frac{1}{2}} \sim 6.5$ hours), although the presence of other components in biological membranes can enhance this rate dramatically; thus the value for $t_{\frac{1}{2}}$ in the electroplax cell membranes of the electric eel is 5 minutes, which is fast enough to suggest a functional significance.

25.3.3.2 Other classes of lipid

(*i*) *Polar lipids*

There have been relatively few studies on the structure and mobility of polar lipids other than phospholipids, but it is clear that they too exhibit gel–liquid-crystal phase behaviour.[8] Glycolipids form bilayers whose thickness and area per molecule are similar to the values discussed for phospholipids. It is of interest to note that the phase transition temperature of extracted beef brain cerebrosides is approximately 70 °C owing to the predominance of 24:0 and 24:1 alkyl chains; the physiological value of this high phase transition temperature is not thoroughly understood. By contrast, the phase transition temperatures of mono- and di-galactosyl diglycerides from chloroplasts are below 0 °C and these lipids are therefore in the liquid-crystalline phase at physiological temperatures. The variety of residues which occur in the polar head region of glycolipids clearly must contribute to the nature of cell surfaces; for example, blood group specificities are associated both with glycoproteins *and* glycolipids in the erythrocyte membrane.

(ii) Neutral lipids

As is discussed in Section 25.3.3.3, cholesterol has a considerable effect on the properties of phospholipid bilayers. The sterols stigmasterol and ergosterol possibly exert similar effects in the membranes of plants and eukaryotic micro-organisms (*e.g.* yeasts), respectively.

Although the neutral lipids (for example, triglycerides and sterol esters) are not present to any significant extent in biological membranes, they are important components of the soluble lipoprotein complexes. This topic will be discussed further in Section 25.3.4.

25.3.3.3 Mixtures of lipids

Biological membranes are complex mixtures of lipids. Towards an understanding of principles important in biological membranes, studies have been made on the properties of defined mixtures of lipids.

(i) Lateral phase separations

Lipid bilayers composed of mixtures of phospholipids with different phase transition temperatures do not undergo a single phase transition; there is a much broader transition in which different proportions of fluid and solid lipid coexist in equilibrium over a range of temperatures. The detailed phase diagrams can be constructed from calorimetric or spectroscopic data.[11] It has further been shown that Ca^{2+} induces 'lateral phase separations' in membranes composed of mixtures of phosphatidylcholine and phosphatidylserine: this result compares with the effect of ionic strength in inducing isothermal phase transitions to bilayers composed of phosphatidylserine.

Biological membranes composed of complex mixtures of lipid classes and alkyl chain compositions are probably in a condition of lateral phase separation at physiological temperatures. The high lateral compressibility which results from having fluid and solid phases present at the same time could modulate changes in the activity of intrinsic enzymes, allow new components to be introduced into the membrane, and also affect transport processes. Studies have been made[23] on the properties of membranes from mutants of the bacterium *E. coli* which require fatty acids for growth; the composition of the inner membrane can be enriched with respect to a particular alkyl chain component by addition of the appropriate fatty acid to the growth medium. Figure 8a shows how the fluidity of the inner membrane varies with temperature when linoleic acid is the supplement; the breaks in the Arrhenius plot correspond to the extremes of lateral phase separation. Figure 8b shows the variation with temperature in the rate of β-glucoside transport in this same membrane; again there are changes in the activation energies for transport which correlate approximately with the limits for lateral phase separation. A striking observation is that when elaidic acid is used as the supplement, the rate of transport *increases* by a factor of two as the temperature is *decreased* by 1 °C at the upper extrema of lateral phase separation; this can be rationalized on the basis that transport is favoured when the membrane is in the condition of lateral phase separation.

As a general statement, membranes endeavour to regulate their lipid composition so that they are in the condition of lateral phase separation at physiological temperatures. For micro-organisms, this is achieved by alterations in the composition of the lipid alkyl chains, whilst in animals, variations in concentration of cholesterol fulfil a similar role.

(ii) Lipid asymmetry

When aqueous dispersions of mixtures of different lipid classes are subjected to ultrasonic radiation, single bilayer vesicles (see Figure 4) are formed. The relative

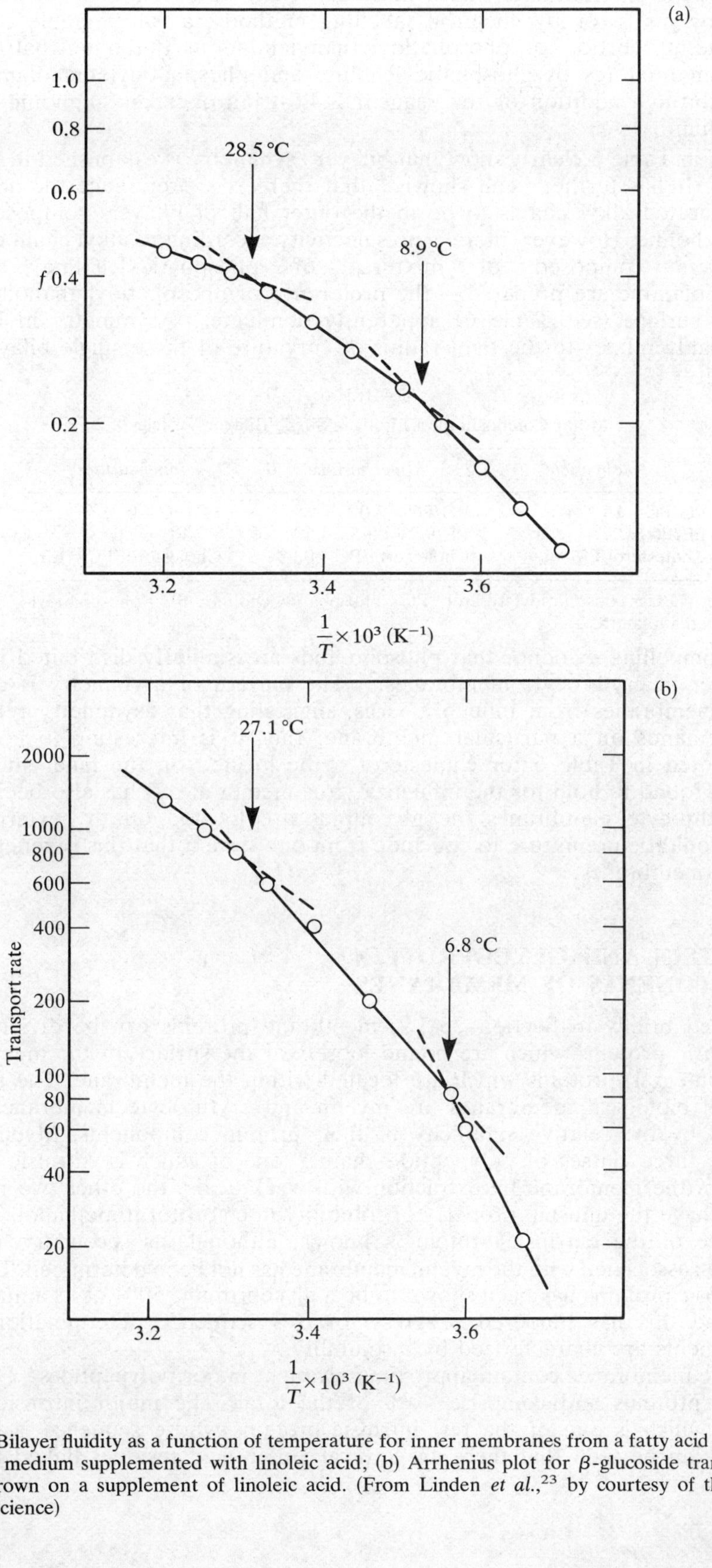

Figure 8 (a) Bilayer fluidity as a function of temperature for inner membranes from a fatty acid auxotroph of *E. coli* grown on medium supplemented with linoleic acid; (b) Arrhenius plot for β-glucoside transport in the *E. coli* mutant grown on a supplement of linoleic acid. (From Linden *et al.*,[23] by courtesy of the US National Academy of Science)

proportions of the different lipid classes in the inner and outer halves of the bilayer can be determined by n.m.r. or by chemical labelling methods; as an example of the latter approach, the proportion of phosphatidylethanolamine in the outer half of vesicles prepared from mixtures of phosphatidylcholine and phosphatidylethanolamine can be determined through addition of the reagent 2,4,6-trinitrobenzenesulphonic acid to the external medium.

The results in Table 6 clearly show that bilayer asymmetry is established in these mixed lipid vesicles. It has further been shown[24] that there is a preference for phospholipids having unsaturated alkyl chains to be in the outer half of bilayers composed solely of phosphatidylcholine. However, there is no selectivity according to alkyl chain composition when bilayers composed of mixtures of phosphatidylcholine and phosphatidylethanolamine are prepared — the preference of phosphatidylethanolamine to be in the inner surface (see Table 6) apparently dominates. Asymmetry in binary lipid systems probably relates to the tight radius of curvature of these single bilayer vesicles.

TABLE 6

Molar Composition of Lipids in Single Bilayer Vesicles[a]

Total vesicle	*Outer surface*	*Inner surface*
PG/PC = 1	PG/PC = 2.0	PG/PC = 0.33
PE/PC = 1	PE/PC = 0.65	PE/PC = 3.0
Cholesterol/PC = 1	Cholesterol/PC = 0.89	Cholesterol/PC = 1.3

[a] PG = phosphatidylglycerol; PC = phosphatidylcholine; PE = phosphatidylethanolamine.

There is compelling evidence that phospholipids are similarly distributed in an asymmetric manner in erythrocyte membranes.[25] The pattern of asymmetry is qualitatively different in membranes from influenza virus, suggesting that asymmetry relates to the functional demands on a particular membrane; thus it is interesting to note that the preference noted in Table 6 for cholesterol to be located on the inner surface of the bilayer is not found to hold for the influenza virus membrane. It has also been observed[6] that for erythrocyte membranes the glycolipids are located on the external surface, although it would be premature to conclude from one system that this is generally true of all biological membranes.

25.3.4 PROTEIN AND GLYCOPROTEIN COMPONENTS OF MEMBRANES

As discussed briefly in Section 25.3.2, membrane proteins can be divided into two classes: *extrinsic* proteins which are bound loosely at the surface of the membrane, and *intrinsic* (or integral) proteins which are located within the membrane. The most extensively studied biological membranes are myelin and erythrocyte membranes; both are characterized by the relative simplicity of their protein components. Myelin probably contains only three classes of polypeptide chain,[26] one of which is extrinsic and can be removed from the membrane by extraction with weak acids; the other two proteins are intrinsic and have the unusual property of solubility in chloroform/methanol. The amino-acid sequence of the extrinsic protein is known, although its secondary and tertiary structure when associated with the myelin membrane has not been determined. The major of the two intrinsic proteins has been shown to be a glycoprotein; 50% of its amino-acids are non-polar and this has hindered progress towards sequence determination since the peptide fragments are characterized by insolubility.

Erythrocyte membranes contain approximately eight major polypeptides.[6] Five of these are extrinsic proteins and comprise 40% of the total. The major intrinsic protein is glycophorin, which is one of the few intrinsic proteins whose sequence is known (see Figure 9). It can be seen that there are several amino-acid residues linked to oligosac-

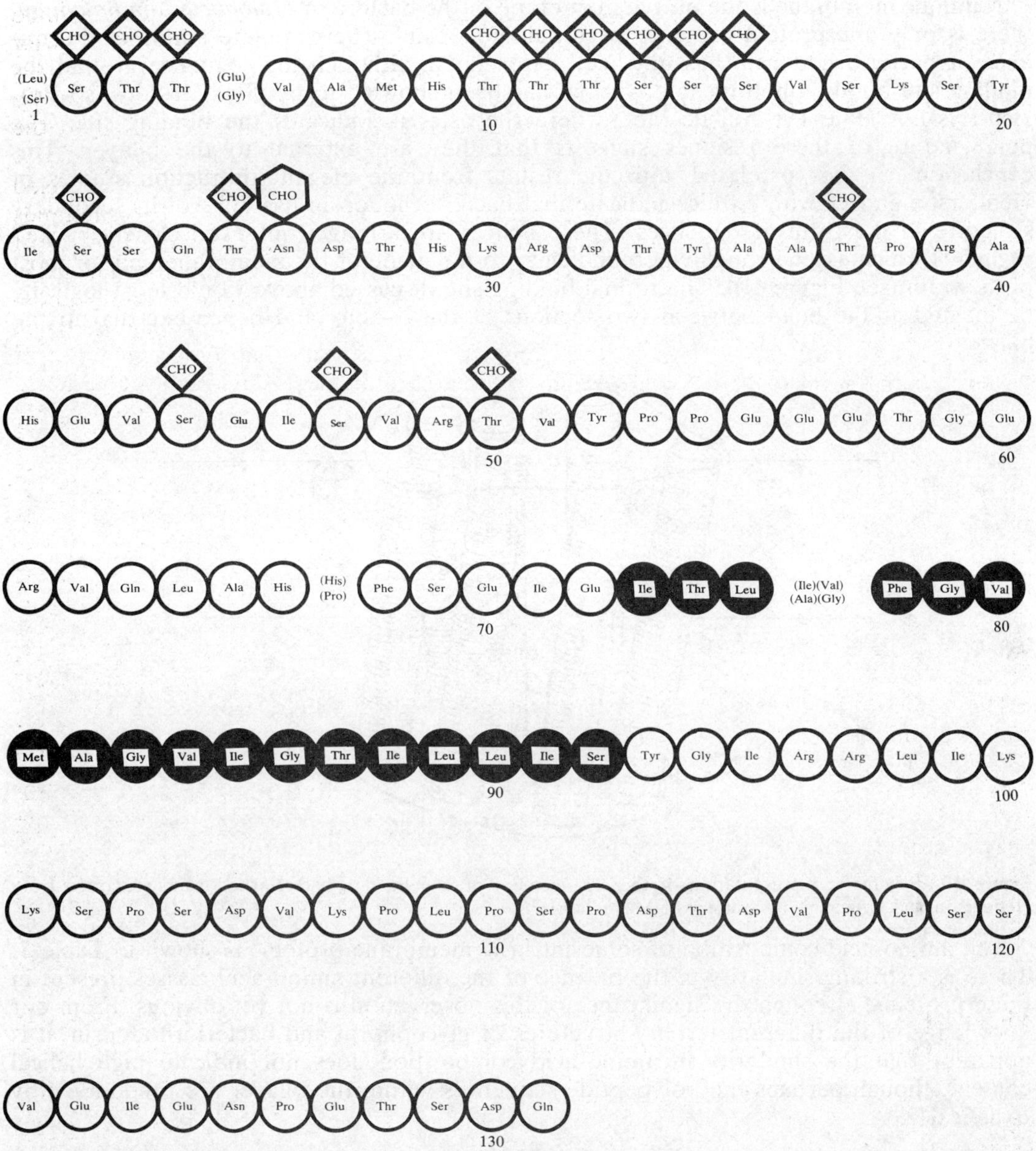

Figure 9 Amino-acid sequence of glycophorin A from erythrocyte membranes. The convention of numbering from the amino to the carboxylate terminus is followed. The hydrophobic part of the molecule, which is thought to extend through the bilayer, is indicated by filled circles. Carbohydrate residues are represented as 'CHO'

charide moieties and it is these residues which largely determine the antigen and receptor binding properties of erythrocytes; the oligosaccharides are located exclusively in the *N*-terminal one third of the sequence and are on the outside surface of the membrane. The high concentration of acidic amino-acids in the *C*-terminal sequence is also noteworthy. However, most interest has been aroused by the amino-acid residues (approximately 20) which link the *N*- and *C*-terminal sequences; from Figure 9 it can be seen that they are all non-polar. A peptide fragment containing this non-polar sequence has been isolated from glycophorin and shown to adopt an α-helical conformation in trifluorethanol and hence possibly also in the erythrocyte membrane. An α-helix composed of 20 amino-acids would have a length of approximately 30 Å, which is almost sufficient to span the bilayer.

A unique membrane is the purple membrane of the bacterium *Halobacterium halobium;* there is only one protein present, 'bacterial rhodopsin'. The complete amino-acid sequence of bacteriorhodopsin has not been determined, although the sequence around the binding site for the photoreceptor retinal has been shown[27] to be Gly-Val-Ser-Asp-Pro-Asp-Lys-Lys*-Phe-Tyr-Ala-Ile-Met, where the asterisk indicates the binding site. The polar nature of these residues suggests that they are external to the bilayer. This conclusion can be correlated with the results from the elegant diffraction studies of Henderson and Unwin,[28] which indicate that bacteriorhodopsin has 70% of its secondary structure in the form of α-helices. These helices are arranged in seven closely packed segments which extend roughly perpendicular to the plane of the membrane and for most of its width (see Figure 10). The retinal binding site discussed above would most logically be located at the bend between two sections of the α-helix and hence external to the bilayer.

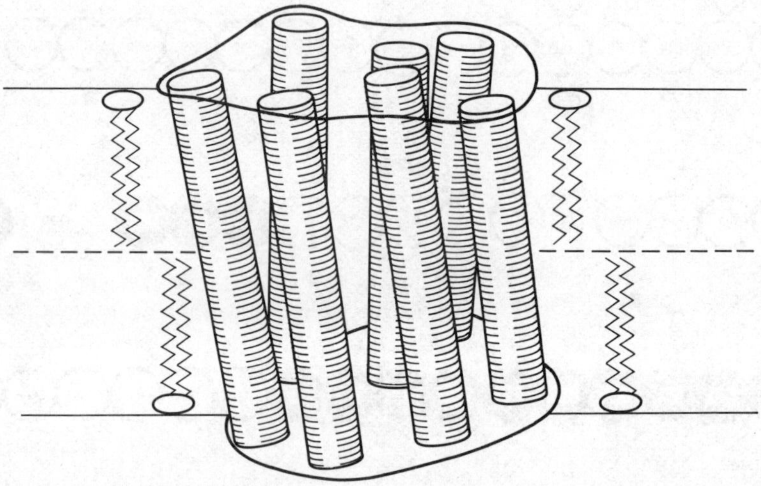

Figure 10 Schematic representation of the arrangement of α-helices in bacteriorhodopsin. (Adapted from Henderson and Unwin,[28] by courtesy of Macmillan)

The amino-acid composition of some intrinsic membrane proteins is shown in Table 7. There is a striking similarity in the balance of the different amino-acid classes present in these proteins, although the significance of this observation is not yet obvious. From our knowledge of the different tertiary structures of glycophorin and bacteriorhodopsin, it is apparent that the similarity in amino-acid composition does not indicate high helical content, though perhaps the polypeptide sequences within the bilayer are predominantly α-helical.

TABLE 7

Amino-acid class	*Human glycophorin*	*Bovine rhodopsin*	*Bacterial rhodopsin*	*Myelin proteolipid*	*Sarcoplasmic reticulum ATPase*
Basic	11.46	8.51	7.44	9.34	11.00
Acidic and polar	32.01	32.33	30.99	29.55	34.19
Hydrophobic	49.06	52.34	51.66	50.67	47.34

Although they are not strictly *membrane* proteins, the important group of serum lipoproteins may conveniently be discussed in this section. These protein complexes allow lipid to be transported between sites in the body as water soluble packets. The composition of one serum lipoprotein is given in Table 1 and can be seen to contain cholesterol esters and triglycerides in addition to phospholipids and proteins; several of the apoproteins have been sequenced.[29] It is generally agreed that serum lipoproteins have a micellar structure, but the detailed arrangement of protein and the various classes of lipid within this structure is not fully established.

25.3.5 BIOLOGICAL MEMBRANES

25.3.5.1 Location of proteins in membranes

The asymmetry of different lipid classes in model and biological membranes has been discussed in Section 25.3.3.1. There is convincing evidence[24] to suggest that the carbohydrate and protein components in biological membranes are distributed asymmetrically. Thus, for example, oligosaccharides bound to intrinsic proteins have always been found to be extracellular; this relates to the site of assembly of the glycoprotein complexes and to their function as centres for cellular recognition.[29,30]

A variety of labelling[24,30] and cross-linking methods[31-34] have been developed to examine the vectorial arrangement of proteins in membranes and also the arrangement of oligomers in multi-subunit enzymes. The cross-linking of proteins proximal in the biological membrane or the linking of a protein to an adjacent lipid gives scope for considerable ingenuity on the part of biochemists. Wang and Richards[30] have cross-linked proteins in the erythrocyte membrane through oxidation of their intrinsic SH groups; following isolation of the complex, the linkages can be split by reductive cleavage to allow identification of the component proteins. An alternative approach used by Khorana and co-workers[32] and by Stoffel and co-workers[33] has been to incorporate biosynthetically fatty acids bearing a photosensitive group into biological membranes; cross-linking between the fatty acid derivative and an adjacent protein can be induced by photolysis. Similar methods have been used[34] to cross-link proteins in the erythrocyte membrane.

25.3.5.2 Protein–lipid interactions in membranes

Spectroscopic techniques, particularly e.s.r., n.m.r., and fluorescence, are being applied increasingly to the study of protein–lipid interactions in membranes. Intrinsic membrane proteins can be extracted from their parent membrane using organic solvents or (better) detergents and purified. There have been several convincing demonstrations that the protein biologically functional state can be reconstituted by integration into a membrane of defined lipid composition.

As an example of the part that e.s.r. can play in the study of protein–lipid interactions in such reconstituted systems, the pioneer work of Griffith and co-workers[35] can be cited. Cytochrome oxidase was purified and depleted of its associated lipid by solvent extraction. By titrating back lipid bearing spin-label probe (see Figure 5), the existence of a layer of lipid tightly bound to the protein could be demonstrated (see Figure 11). This boundary

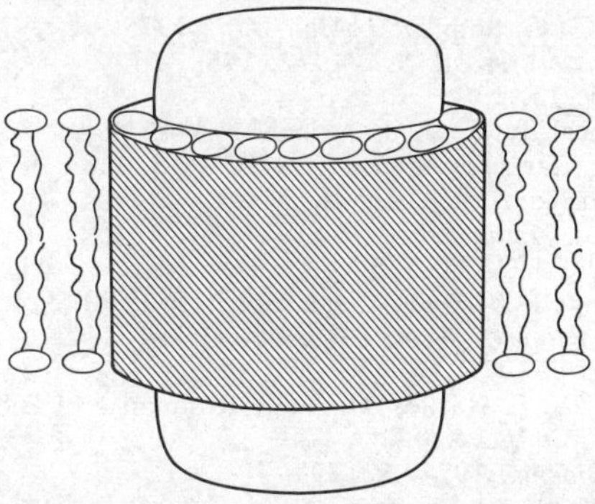

Figure 11 Diagrammatic representation of cytochrome oxidase and associated phospholipid (From Jost *et al.*,[35] by courtesy of the US National Academy of Science)

layer, composed of approximately 50 lipid molecules per molecule of cytochrome oxidase, was additionally shown to be a requirement for enzyme activity.

Biochemical studies have demonstrated that a boundary layer of lipid also controls the activity of the calcium-dependent ATPase from sarcoplasmic reticulum; again spin labels have been used to show that the boundary layer is immobilized.[35] The significant conclusions from these studies are firstly that the boundary layer reduces the level of perturbation to the bilayer by the intercalated protein and, secondly, that the boundary layer acts as a 'mediator' between the bilayer and the protein through which the functional role of the protein can respond to phase transitions and phase separations in the bilayer lipid.

Brulet and McConnell[37] have used n.m.r. to study the interaction between the protein glycophorin from erythrocyte membranes and dipalmitoyl(phosphatidyl)choline labelled with ^{13}C in the methyl groups of the choline head group. At temperatures below the phase transition of the phospholipid, the ^{13}C n.m.r. spectrum showed two components, one sharp and one broad. The sharp component was assigned to the choline head group, which is suggested to be more mobile in the immediate vicinity of the protein; this conclusion does not exclude the possibility that the alkyl chains of these boundary lipids are immobilized and thus is not necessarily in contradiction to the results of the studies on boundary lipid in cytochrome oxidase and the calcium-dependent ATPase which have been discussed above.

In the examples of protein–lipid interactions considered so far, enzyme activity is enhanced by a more fluid bilayer environment. Van Deenan and co-workers[38] have shown, however, that the activity of phospholipase A2 in catalysing the hydrolysis of phospholipids is optimal at the phase transition of the phospholipid. This result can be understood through a consideration of the special properties of lipid which is interfacial between the ordered and fluid domains present at the phase transition.[39] The evidence therefore suggests that the activities of proteins in membranes are dependent on both the boundary layer of lipid associated with the protein and interfaces between the various lipid domains.

References

1. R. Coleman, *Biochim. Biophys. Acta*, 1973, **300,** 1.
2. G. D. Eytan, R. C. Carroll, G. Schatz, and E. Racker, *J. Biol. Chem.*, 1975, **250,** 8598.
3. G. Adam and M. Delbrück, in 'Structural Chemistry and Molecular Biology', ed. N. Davidson and A. Rich, Freeman, San Francisco, 1968, p. 198.
4. M. S. Bretscher and M. C. Raff, *Nature*, 1975, **258,** 43.
5. 'The Biology of Cytoplasmic Microtubules', in *Ann. New York Acad. Sci.*, 1975, **253,** 1–848.
6. V. T. Marchesi, H. Furthmayr, and M. Tomita, *Ann. Rev. Biochem.*, 1976, **45,** 667.
7. S. J. Singer, *Ann. New York Acad. Sci.*, 1972, **195,** 16.
8. G. G. Shipley, in 'Biological Membranes' ed. D. Chapman and D. H. F. Wallach, Academic Press, New York, 1973, p. 1.
9. P. B. Hitchcock, R. Mason, and G. G. Shipley, *J. Mol. Biol.*, 1975, **94,** 297.
10. H. Fernandez-Moran, *Ann. New York Acad. Sci.*, 1972, **195,** 376.
11. D. Marsh, *Essays Biochem.*, 1975, **11,** 139.
12. 'Carriers and Channels in Biological Systems', in *Ann. New York Acad. Sci.*, 1975, **264,** 1–485.
13. R. P. Rand, D. Chapman, and K. Larsson, *Biophys. J.*, 1975, **15,** 1117.
14. S. J. Kohler and M. P. Klein, *Biochemistry*, 1977, **16,** 519.
15. H. Trauble and H. Eibl, *Proc. Nat. Acad. Sci. USA*, 1974, **71,** 214.
16. D. Chapman, *Quart. Rev. Biophys.*, 1975, **8,** 185.
17. W. L. Hubbell and H. M. McConnell, *J. Amer. Chem. Soc.*, 1971, **93,** 314.
18. A Seelig and J. Seelig, *Biochem. Biophys. Acta*, 1975, **406,** 1.
19. A Seelig and J. Seelig, *Biochemistry*, 1977, **16,** 45.
20. P. F. Knowles, D. Marsh, and H. W. E. Rattle, 'Magnetic Resonance of Biomolecules', Wiley, New York, 1976.
21. M. Edidin, *Ann. Rev. Biophys. Bioeng.*, 1974, **3,** 179.
22. J. L. Marx, *Science*, 1974, **183,** 1279.
23. C. D. Linden, K. L. Wright, H. M. McConnell, and C. F. Fox, *Proc. Nat. Acad. Sci. USA*, 1973, **70,** 2271.
24. P. L. Yeagle, W. C. Hutton, R. B. Martin, B. Sears, and C.-H. Huang, *J. Biol. Chem.*, 1976, **251,** 2110.
25. J. E. Rothman and J. Lenard, *Science*, 1977, **195,** 743.

26. G. Guidotti, *Ann. Rev. Biochem.*, 1972, **41,** 731.
27. J. Bridgen and I. D. Walker, *Biochemistry*, 1976, **15,** 792.
28. R. Henderson and P. N. T. Unwin, *Nature*, 1975, **257,** 28.
29. J. D. Morrisett, R. L. Jackson, and A. M. Gotto, *Ann. Rev. Biochem.*, 1975, **44,** 183.
30. R. C. Hughes, *Essays Biochem.*, 1975, **11,** 1.
31. K. Wang and F. M. Richards, *J. Biol. Chem.*, 1974, **249,** 8005.
32. C. R. Greenberg, P. Chakrabarti, and H. G. Khorana, *Proc. Nat. Acad. Sci. USA*, 1976, **73,** 86.
33. W. Stoffel, K. Salm, and U. Körkemeier, *Z. Physiol. Chem.*, 1976, **357,** 917.
34. R. B. Mikkelsen and D. F. H. Wallach, *J. Biol. Chem.*, 1976, **251,** 7413.
35. P. C. Jost, O. H. Griffith, R. A. Capaldi, and G. Vanderkooi, *Proc. Nat. Acad. Sci. USA*, 1973, **70,** 480.
36. T. R. Hesketh, G. A. Smith, M. D. Houslay, K. A. McGill, N. J. M. Birdsall, J. C. Metcalfe, and G. B. Warren, *Biochemistry*, 1976, **15,** 4145.
37. P. Brulet and H. M. McConnell, *Biochem. Biophys. Res. Comm.*, 1976, **68,** 363.
38. J. A. F. Op den Kampf, M. Th. Kauerz, and L. L. M. van Deemen, *Biochim. Biophys. Acta*, 1975, **406,** 169.
39. D. Marsh, A. Watts, and P. F. Knowles, *Biochemistry*, 1976, **15,** 3570.

PART 26

CARBOHYDRATE CHEMISTRY

26.1
Monosaccharide Chemistry

L. HOUGH and A. C. RICHARDSON
Queen Elizabeth College, University of London

26.1.1 INTRODUCTION

The carbohydrates constitute a highly significant class of natural products. To the chemist they offer a complete series of diastereoisomers, readily obtainable in a state of high optical purity, and in the course of investigations of their chemistry they have afforded considerable information on mechanistic and stereochemical problems. Significant contemporary developments have involved their use as chiral synthons and templates for the stereospecific synthesis of compounds such as prostaglandins, amino-acids, heterocyclics, lipids, *etc.* For the biologist the significance of carbohydrates lies in the dominant role they perform in living systems, and the intriguing complexity of their functions, the molecular basis of which is slowly and painstakingly evolving. The carbohydrate species involved in the majority of biological processes are macromolecular, although free mono- and di-saccharides occur in many biological fluids, and most plants contain free glucose, fructose, and sucrose. Plants alone are responsible for the primary synthesis of carbohydrates *via* photosynthesis, in which atmospheric carbon dioxide is converted into carbohydrates in the presence of light as the energy source (see Chapter 28.2). Through this process, enormous amounts of the homo-polysaccharides cellulose (a structural material) and starch (a food storage product) are laid down, and this photosynthetic process alone maintains the balance between 'living' and 'non-living' carbon. Certain plants, notably sugar cane and beet, accumulate relatively high concentrations of the unique disaccharide sucrose (α-D-glucopyranosyl β-D-fructofuranoside), which is extracted and easily crystallized, and has long been a product of commerce and trade. It is now produced on a substantial scale at a rate of 82×10^6 tonnes per annum and as a chemical is the cheapest *pure* organic substance available and, in contrast to petroleum-derived products, is replenishable. D-Glucose (Gk. γλυκυσ, meaning sweet) has been known for several centuries because of its crystallization from granulated honey and wine musts. It is also produced commercially by the hydrolysis of starch, and this process is now carried out by a continuous process using enzymes on a solid polymer support.

26.1.2 THE STRUCTURES OF MONOSACCHARIDES

The proof of structure and the stereochemical relationships between the eight D-hexoses and the four D-pentoses (Scheme 1) were founded on the brilliant deductive work of Emil Fischer, for which he was honoured by the award of the Nobel Prize for chemistry in 1901.[1] Fischer's deductions were based on the observation that (a) monosaccharides, epimeric at C-2, gave the same phenylosazone on reaction with phenylhydrazine; (b) reduction of the aldose to the alditol (Section 26.1.5) or oxidation to the dicarboxylic acids (glycaric acids) (Section 26.1.6.3) gave a product which was either optically active or inactive, depending upon configuration; and (c) ascent of the series (Section 26.1.4.2) allowed the inter-relationship of pentoses with hexoses, *etc.* to be established. Thus, D-glucose and D-mannose were epimeric at C-2, since they both afforded the same phenylosazone and must therefore have identical configurations at C-3, C-4, and C-5. In addition, the enantiomers of these two hexoses arose from the application of methods for 'ascent of the series' (Section 26.1.4.2) to the pentose arabinose (from gum arabic). Since neither glucose, mannose, nor arabinose formed an optically inactive alditol or glycaric acid, the configuration of arabinose was defined (Scheme 1). The stereochemistries of mannose and glucose were differentiated by the fact that glucitol (from glucose) was enantiomeric with gulitol (from gulose) (Section 26.1.5). Fischer arbitrarily assigned the absolute configuration to the penultimate carbon of glucose, later defined by the Cahn-Ingold-Prelog convention as *R*, and called this the D-series (Scheme 1). The mirror image series was termed the L-series. Fortunately, the configuration at C-5 of D-glucose arbitrarily assigned in this way was later confirmed by Bijvoet *et al.*[2] to be the correct absolute configuration.

The acyclic structures of aldoses shown in Scheme 1, however, did not adequately represent all the chemical properties of these compounds. For example, D-glucose did not give a Schiff's test for an aldehyde and, depending upon conditions, D-glucose crystallized in two forms (α and β) which showed different initial optical rotations, $[\alpha]_D$ +111° and +19°, but which progressively equilibrated to the same value (+53°). In addition, attempted formation of dimethyl acetals by acid-catalysed reaction with methanol resulted in the introduction of not two but only one *O*-methyl group. Significantly, two isomeric monomethyl derivatives, termed methyl glycosides, were formed ($[\alpha]_D$ +159° and −34°, respectively). These observations were rationalized by postulating that aldoses existed as cyclic hemiacetals, in which one of the hydroxyl groups in the chain has reacted with the

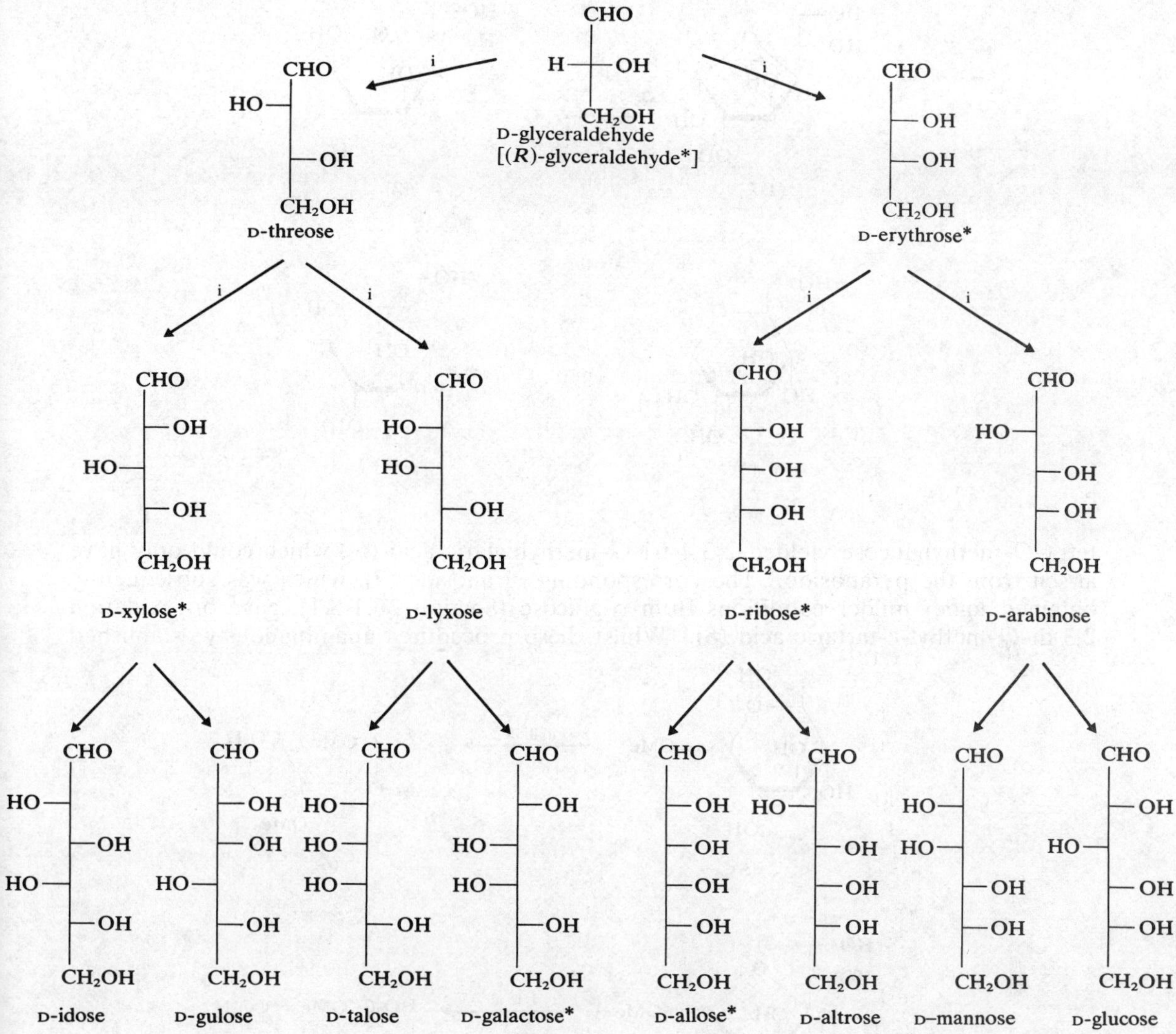

*yield optically inactive alditol and glycaric acid on reduction and oxidation, respectively
i, ascent of series, followed by separation of diastereoisomers (see p. 000).

SCHEME 1 Relationship of the D-hexoses, D-pentoses, and D-tetroses to D-glyceraldehyde [(*R*)-glyceraldehyde]

carbonyl group at C-1 as in (**1**)–(**4**). These structures explained why D-glucose existed in two forms, since intramolecular hemiacetal formation resulted in the formation of a new chiral centre. The methyl glycosides retained the hemiacetal ring structure of the sugar, but acetal formation had been completed by replacement of the hydroxy groups at C-1 by methoxy groups.

Until about 1923 the five-membered furanose ring was assumed to be the predominant species, largely on circumstantial evidence. However, later work by Haworth and Hirst established that aldoses favoured the six-membered pyranose rings rather than the five-membered furanoses.[3] For example, methylation of the two thermodynamically favoured methyl glycosides from D-glucose, followed by acid hydrolysis, gave the same tetra-*O*-methylglucose, indicating a common ring structure.[4] Vigorous oxidation of this

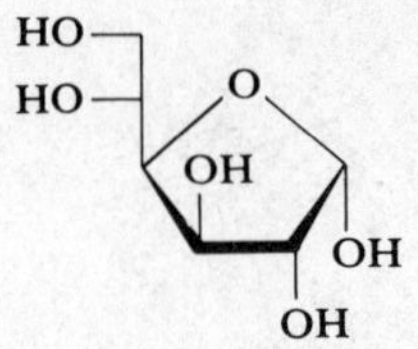

(1)

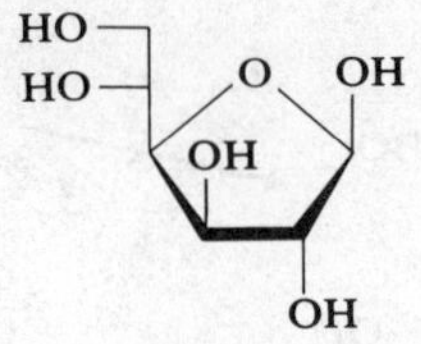

(2)

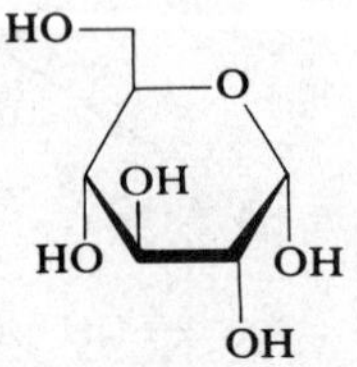

(3)

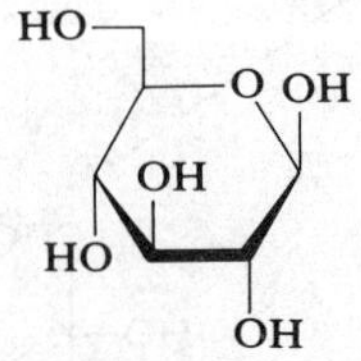

(4)

tetra-*O*-methylglucose yielded 2,3,4-tri-*O*-methylxylaric acid (**6**), which could only have arisen from the pyranoside.[5] The corresponding furanoside (**7**), which was subsequently obtained under milder conditions from D-glucose (Section 26.1.8.1), gave on oxidation 2,3-di-*O*-methyl-L-tartaric acid (**8**).[6] Whilst these procedures unambiguously established

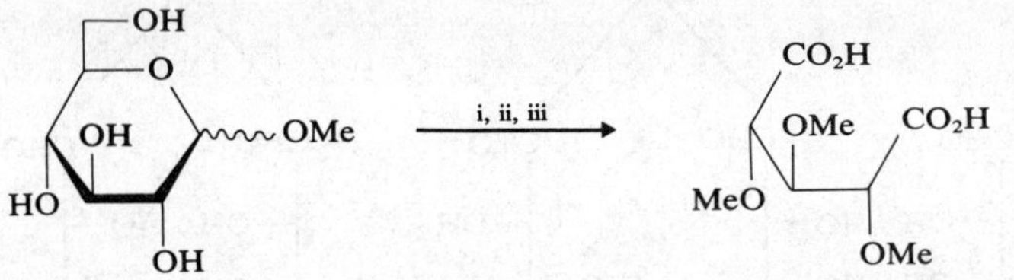

(5) (6)

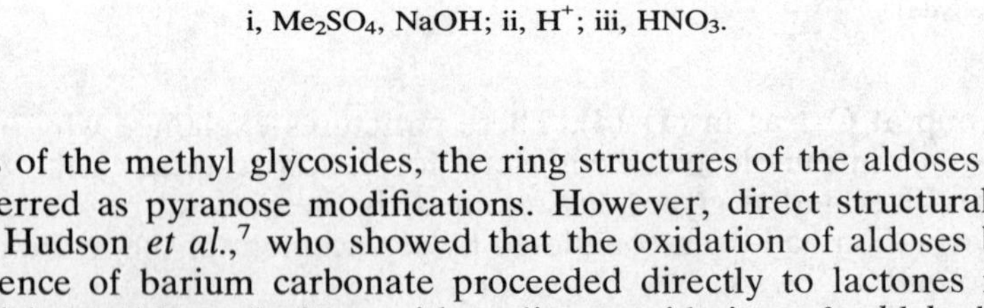

(7) (8)

i, Me_2SO_4, NaOH; ii, H^+; iii, HNO_3.

the ring structures of the methyl glycosides, the ring structures of the aldoses themselves could only be inferred as pyranose modifications. However, direct structural proof was later provided by Hudson *et al.*,[7] who showed that the oxidation of aldoses by bromine water in the presence of barium carbonate proceeded directly to lactones rather than aldonic acids, which would result from either direct oxidation of *aldehydo* forms or ring-opening of the lactones by hydrolysis.[8] Thus application of this method of oxidation to fresh solutions of crystalline α- and β-D-glucose afforded, in both cases, the same 1,5-lactone, showing that D-glucose exists in the pyranose forms (**3**) and (**4**). Both α- and β-D-galacturonic acids (**9**) afforded an optically active mixture of the 1,5-lactone (**10**) and the 1,4-lactone,[9] whereas had the lactones been formed *via* the acyclic galactaric (mucic)

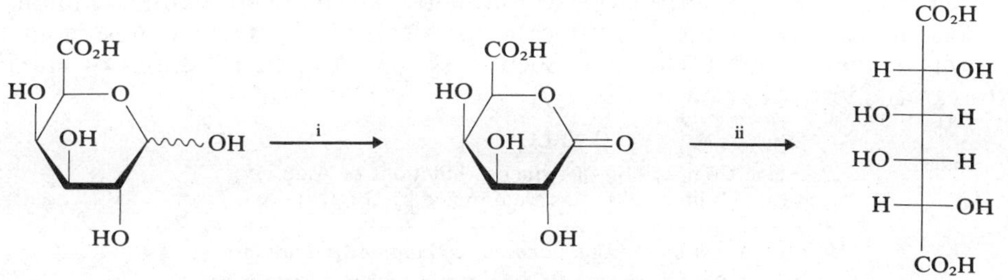

(9) (10) (11)

i, Br_2, H_2O, $BaCO_3$; ii, H_2O.

acid, which is a *meso* compound, then they would have been racemic. This result clearly showed that the oxidation is direct and does not proceed *via* open-chain intermediates, and, since both 1,4- and 1,5-lactones are formed, that D-galacturonic acid exists as a mixture of furanose and pyranose forms.

Oxidation with glycol cleaving agents (lead tetra-acetate, periodic acid, *etc.*) has proved to be a powerful and easily applicable method for the establishment of ring structures of carbohydrates (Section 26.1.6.5).[10] Thus, in 1934 Herrissey *et al.*[11] showed that methyl α- and β-D-glucopyranoside (**5**) both consumed two moles of oxidant and liberated one mole of formic acid, which is consistent with the pyranoside ring structure, whereas the furanoside consumed initially two moles of oxidant, but liberated one mole of formaldehyde. Application of periodate oxidation to aqueous solutions of reducing sugars likewise indicated that they adopted pyranose forms. Thus oxidation of D-glucose at pH 3.7, at which pH the intermediary 2-*O*-formylglyceraldehyde (**12**) is stable, resulted in the

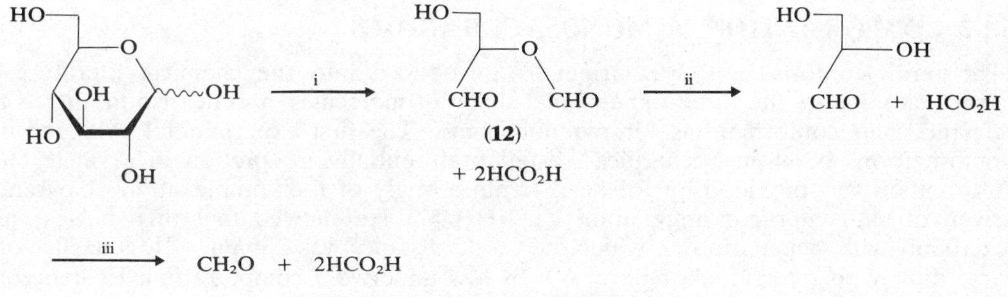

(12)

i, $3NaIO_4$; ii, pH>7; iii, $2NaIO_4$.

rapid consumption of 3 moles of oxidant and the liberation of 2 moles of formic acid. The formyl ester (**12**) underwent slow hydrolysis (at this pH) to give glyceraldehyde, which then consumed a further two moles of oxidant and liberated a further two moles of formic acid and one mole of formaldehyde.[12] Had the furanose form been present, formaldehyde would have been liberated during the first stage of the reaction owing to the cleavage of the 5,6-bond.

The above chemical methods of structure determination have now largely been supplanted by physical methods, in particular n.m.r. spectroscopy (^1H and ^{13}C),[13] mass spectrometry,[14] and X-ray crystallography,[15] which not only give information on primary structure, but also the molecular shape or conformation.

Monosaccharides in solution are regarded as equilibrium mixtures of all the possible forms, including the acyclic *aldehydo* forms and possibly the seven-membered septanose form. The major form is dictated by the thermodynamic stabilities of the various forms but in most cases the predominant species are pyranoses, although this may vary according to the solvent. The percentage composition of equilibrated aqueous solutions of monosaccharides has been determined by g.l.c. after trimethylsilylation,[16] which shows that only in the cases of ribose, altrose, idose, and talose are there substantial amounts of furanoses present, and this is probably due to the destabilizing factors present in the pyranose forms of these monosaccharides (Section 26.1.3). The compositions of mutarotated mixtures of aldoses is given in Table 1.

TABLE 1
Percentage Composition of Aqueous Solutions of Aldoses at Equilibrium at 40 °C as Determined by G.L.C.[16]

Aldose	*α-Pyranose*	*β-Pyranose*	*α-Furanose*	*β-Furanose*
Ribose	20	56	6	18
Arabinose	63	34	—— 3 ——	
Xylose	33	67	—— <1 ——	
Lyxose	71	29	—— <1 ——	
Allose	18	70	5	7
Altrose	27	40	20	13
Glucose	36	64	—— <1 ——	
Mannose	67	33	—— <1 ——	
Gulose	<22	>78	—— <1 ——	
Idose[a]	31	37	16	16
Galactose	27	73	—— <1 ——	
Talose	40	29	20	11

[a] At 60 °C.

However, in spite of the fact that acyclic, septanose and furnaose forms are present in only small amounts at equilibrium, reactions of monosaccharides frequently proceed by way of one of these forms, giving products corresponding to these modifications.

26.1.3 CONFORMATION OF MONOSACCHARIDES

The term 'conformation' was originally introduced into the chemical literature by Haworth[3] to denote the three-dimensional shape of molecules, and he also predicted the preferred chair conformations for pyranose rings. The first experimental evidence that pyranose forms of monosaccharides existed preferentially in solution in a single chair conformation was provided by Reeves,[17] from a study of the complexation of pyranoid derivatives with cuprammonium ion, $[Cu(NH_3)_4]^{2+}$. He showed that this species complexed only with vicinal diols, in which the O—O distance was 286 pm or less. Hence only vicinal diols with a torsional angle of 60° or less underwent complexation. Evidence for complex formation was provided by two parameters: (a) an increase in specific conductivity of the solution, and (b) a change in specific rotation for chiral compounds. The first parameter was an indication of the strength of the complex and the second was related to the handedness of the diol grouping.[17] For example, if the HO/HO torsional angle was positive (anticlockwise) there was a positive increase in rotation and *vice versa* (Figure 1).

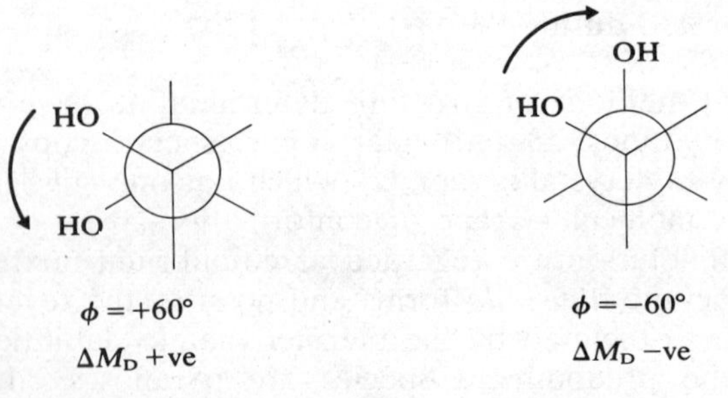

Figure 1

In the case of methyl α-D-glucopyranoside, complexation with cuprammonium took place without significant change in the molecular rotation because, in the favoured $^4C_1$ conformation (**13**), the 2,3- and 3,4-diol groups would lead to negative (*laevo*) and positive (*dextro*) changes in rotation respectively, and therefore compensate each other. This conclusion is supported by the observation that the 4-*O*-methyl ether forms a *laevo* complex, whilst the 2-*O*-methyl ether gives a *dextro* complex and the 3-*O*-methyl ether does not form a complex at all. Reeves suggested that methyl α-D-glucopyranoside and its simple derivatives exist in the $^4C_1$ conformation (**13**) since the alternative $^1C_4$ conformation (**14**), with all hydroxyl groups in axial positions should not permit complex formation with the cuprammonium ion.

By the use of the cuprammonium complexation method, Reeves was able to show that glycosides of the common hexoses glucose, mannose, galactose, and many others existed in the $^4C_1$ conformation, in which most of the groups were equatorial, but α-D-idosides (**15**) and α-D-altrosides (**16**) showed conformational instability and existed in a mixture of the two possible conformations. This exceptional behaviour was later confirmed by n.m.r. studies, which revealed that α-D-idopyranose[18] and α-D-altropyranose[19] derivatives marginally prefer the $^4C_1$ conformation.

(**13**) (**14**)

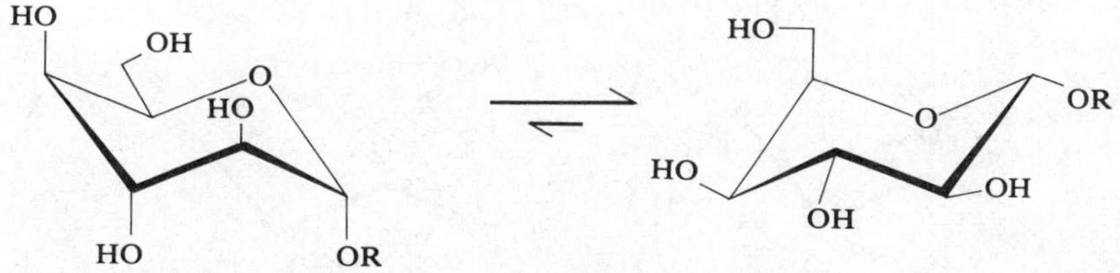

(**15**)

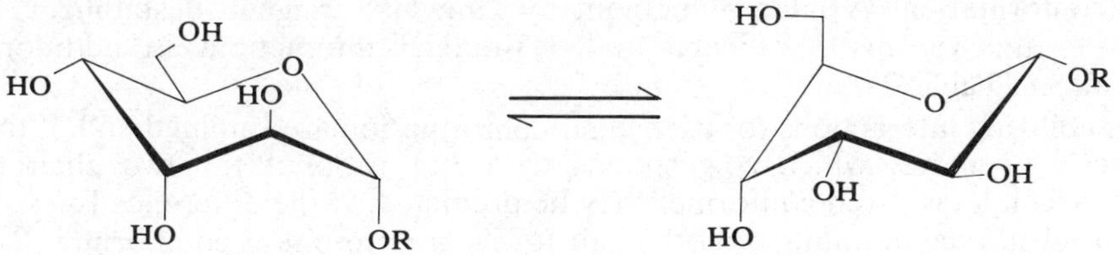

(**16**)

Edward[20] noted that the aglycone group at C-1 (the anomeric carbon) of a pyranoid ring usually preferred the axial configuration, particularly in less-polar solvents. This preference is known as the *anomeric effect* and is due to stereo-electronic factors arising from the more favourable dipolar interactions between the dipole associated with the ring-oxygen and the dipole associated with the substituent at C-1 in the axial position.[21] Thus it can be seen in the Newman projection along the C-1—O-5 bond of the β-anomer

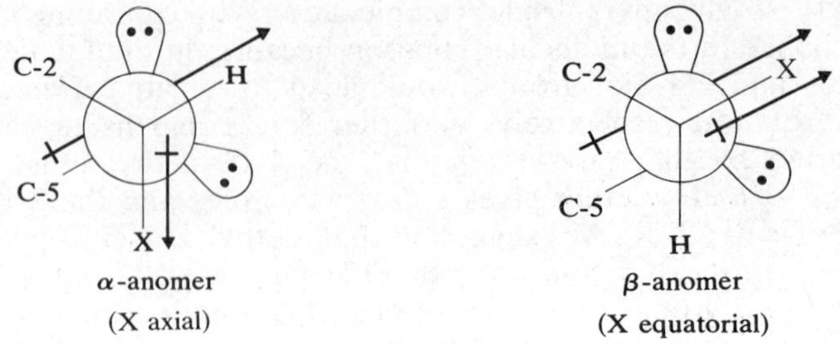

Figure 2

that the dipole of the C-1—X bond is parallel and opposed to the dipole of the ring-oxygen, whereas in the α-anomer they are no longer in opposition (Figure 2). The extent of the anomeric effect depends upon the nature of X and the solvent. For example, when X = OH in aqueous solution the anomeric effect is weakened by solvation, which reduces the net dipole moments. Consequently, at equilibrium D-glucose contains only 36% of the α-anomer. However, equilibration of either methyl α-D-glucopyranoside (**13**) or its β-anomer, in acidified methanol at 35 °C, favours the α-anomer to the extent of 66%. The more electronegative the anomeric substituent the greater will be the effect. For example, when X = F or Cl, then the requirement for this substituent to become axial is so great that the conformation can be forced from an all-equatorial to an all-axial conformation. Thus n.m.r. studies have shown that 2,3,4-tri-*O*-acetyl-β-D-xylopyranosyl chloride in deuteriochloroform exists in the seemingly unfavourable $^1C_4$ conformation (**17**).[22]

The corollary of the anomeric effect is the *reverse-anomeric effect.*[23] When the anomeric substituent is more electropositive than carbon, then the situation is reversed and the equatorial β-anomer becomes the more favoured. For example, *N*-glycosylpyridinium salts such as (**18**) exist in the apparently less-stable $^1C_4$ conformation.[23]

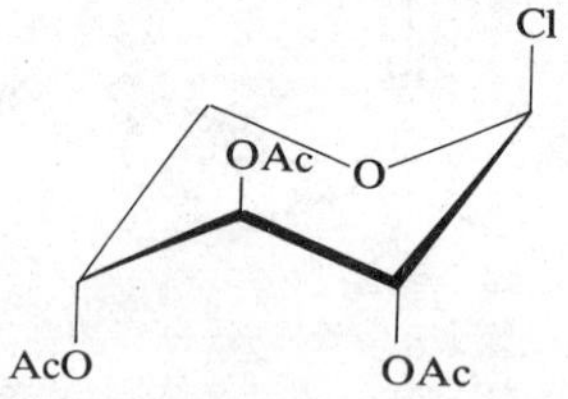

(17)

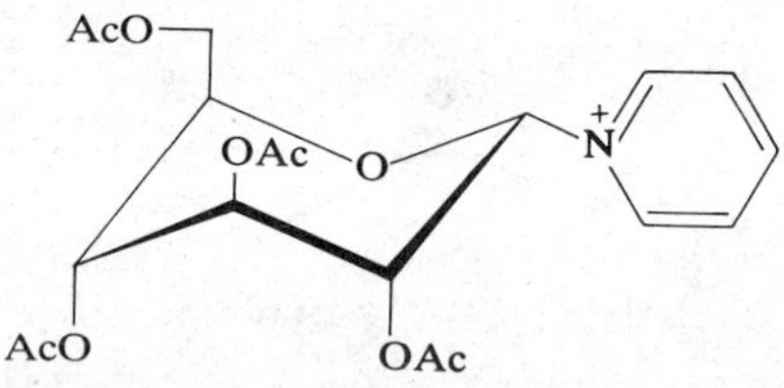

(18)

Reeves[17] originally introduced the concept of 'instability factors', which could be summed for each possible conformation, that with the lowest instability rating being the preferred conformation. A later refinement by Angyal[24] assigned destabilizing energies (kJ mol^{-1}) to the various 1,2-vicinal and 1,3-diaxial interactions in addition to the anomeric effect (Table 2).

The destabilizing interactions for each chair conformation are summed and, if they differ by 3 kJ mol^{-1} or more, which corresponds to a 3 : 1 ratio of the two chair forms at equilibrium, then a favoured conformer may be predicted. If the difference is less than this value, then substantial amounts of both chair forms are present at equilibrium. By the use of this empirical result, which generally gives good agreement with experimental observations, it is possible to predict not only the favoured conformer, but also the favoured anomer. For example, the calculated energy difference between the $^4C_1$ conformers of β- and α-D-glucopyranose, (**4**) and (**3**) respectively, is 1.46 kJ mol^{-1}, corresponding to 36% of the α-anomer. This agrees well with the figure of 36.2% calculated from the optical rotation of the equilibrium mixture of mutarotated D-glucose. Similarly, the calculated amount of the α-anomer of D-mannose is 68%, whereas the value from the specific rotation is 68.8%.

TABLE 2
Destabilizing Effects for Pyranoses in Aqueous Solution[24]

	Destabilizing energy (kJ mol^{-1})
Diaxial interactions	
O_a/H_a*	1.88
C_a/H_a	3.77
O_a/O_a	6.28
O_a/C_a	10.46
Vicinal interactions	
O_a/O_e or O_e/O_e	1.46
C_a/O_e or C_e/O_e	1.88
Anomeric effect (*include if* 1-OH equatorial)	
If 2-OH equatorial	2.30
If 2-OH axial	4.18
If 2-OH and 3-OH axial	3.56
If 2- and 3-OH's absent	3.56

* a and e designate axial and equatorial, respectively.

The conformations of furanose rings are less easily defined, but it is known that they are buckled, and exist in either envelope (*E*; one carbon out of plane) or twist (*T*; two vicinal carbons out of plane) conformations, (**19**) and (**20**), respectively.[13] The energy differences

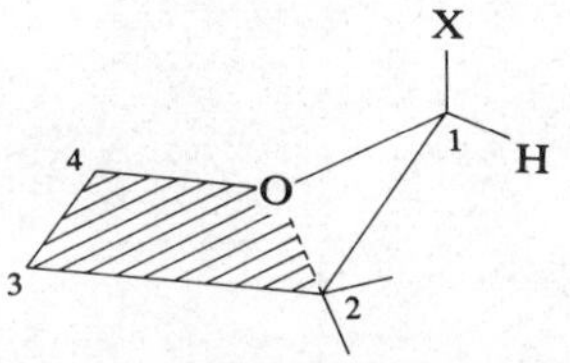

(**19**) E^1

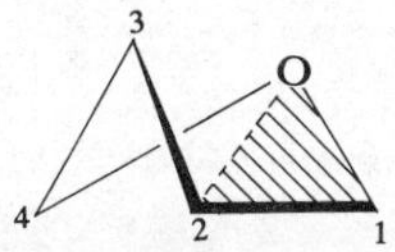

(**20**) $^3T_4$

between the various envelopes and twists is probably not large, but large substituents attached prefer the *quasi*-equatorial positions. Alternatively, electronegative substituents at C-1 prefer to be *quasi*-axial because of the anomeric effect, as in (**19**). In general vicinal *cis* arrangements are avoided, so that in furanoses the 1,2-*trans* anomer is the favoured one (Table 1).

26.1.4 SYNTHESIS OF MONOSACCHARIDES[25]

Monosaccharides which are readily available from natural sources (*e.g.* D-glucose, D-galactose, D-mannose, *etc.*) are usually employed as starting materials for the less readily available sugars. The total synthesis of many sugars has been achieved from non-carbohydrate materials, but this mode of synthesis has the disadvantage that an optical resolution is usually required at some stage in the synthesis.

26.1.4.1 Descent of series

The removal of one of the terminal carbon atoms of a monosaccharide can be achieved by a variety of methods, to give the lower aldose. One of the classical methods is the *Ruff degradation*, in which oxidation of the derived aldonic acid (Section 26.1.6.1) with hydrogen peroxide in the presence of ferric ions affords the lower aldose *via* the

2-ketoaldonic acid (Scheme 2). This method usually results in low yields although the use of ion exchange resins can improve yields considerably. For example, the yield of D-lyxose from D-galactose was improved in this way from 17% to 41%.[26]

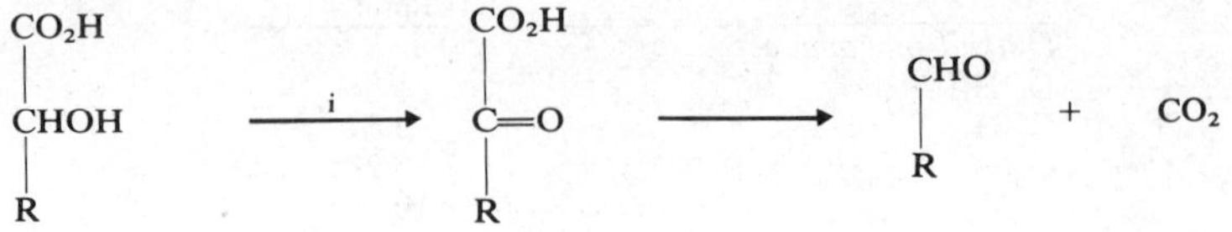

i, H_2O_2, Fe^{3+}.

SCHEME 2

The *Wohl degradation* (Scheme 3) involves the dehydration of the aldose oxime to give the *O*-acetylated nitrile, followed by ammonia-induced loss of hydrogen cyanide and *O*-deacetylation to give the 1,1-bisacetamido derivative (formed from the aminated aldehyde by O→N acetyl migration), from which the lower aldose is generated by mild acid hydrolysis.

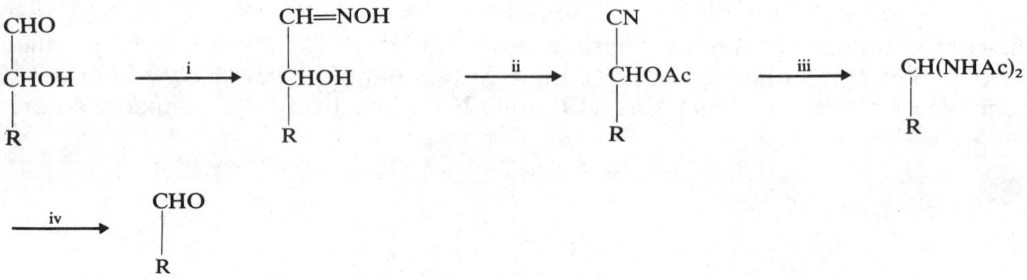

i, H_2NOH; ii, Ac_2O; iii, NH_3; iv, H^+.

SCHEME 3

The mildest and highest yielding chain shortening procedure is the *disulphone degradation*, introduced originally by Fischer,[27] and further developed by Hough *et al.*[28] The aldose is first converted to the dialkyl dithioacetal (mercaptal) by reaction with the appropriate alkanethiol, followed by oxidation with perpropionic acid to give the disulphone. These disulphones exist either as the acyclic form depicted (Scheme 4), or they undergo dehydration to give cyclic pyranoid or furanoid forms.[28] Regardless of their structure, they undergo cleavage of the C-1—C-2 bond in dilute ammonia, giving bisalkylsulphonylmethane and the next lower aldose. The high yield and purity of the final product makes this a valuable and superior method for the descent of the aldose series.

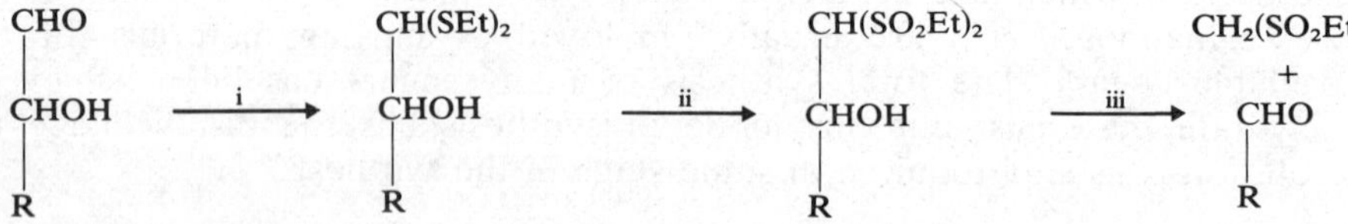

i, EtSH, HCl; ii, $EtCO_3H$; iii, NH_4OH.

SCHEME 4

Where suitably blocked derivatives are available with a terminal diol group present, glycol-cleaving agents can accomplish a descent of the series in a single step. For example, 2,4-*O*-benzylidene-D-glucitol (**21**) undergoes cleavage of the 5,6-diol grouping by lead tetra-acetate to give formaldehyde and an aldopentose acetal, which upon mild acid hydrolysis gives L-xylose (**22**).[29] Similarly, 3-*O*-benzyl-D-glucose reacts preferentially in the pyranose form to give, after hydrolysis of the formyl ester, 2-*O*-benzyl-D-arabinose.[30]

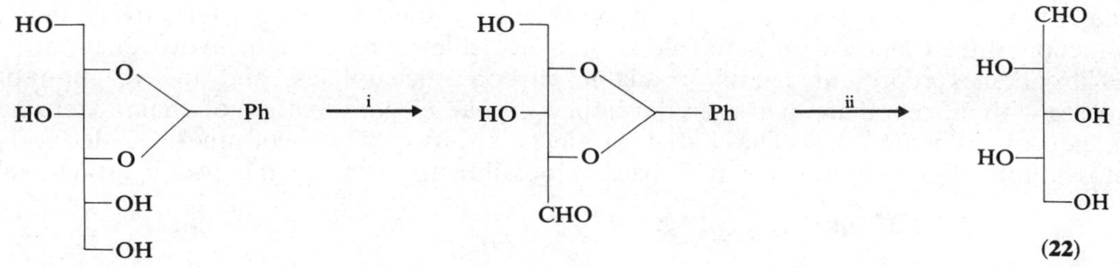

(**21**) (**22**)

i, $Pb(OAc)_4$; ii, H^+.

26.1.4.2 Ascent of series

Two very widely used methods are employed for proceeding from one aldose to a mixture of two epimeric higher aldoses, by the addition of a 'carbon nucleophile' to the potential aldehyde group of the monosaccharide. The classical method, the *Fischer–Kiliani cyanohydrin synthesis*,[31] involves the addition of hydrogen cyanide across the carbonyl double bond, thereby creating a new chiral centre at C-2, followed by hydrolysis of the nitriles and, after lactonization of the resulting acids, reduction to the aldoses (Scheme 5). Separation of the two epimeric products is usually conducted at the aldonic

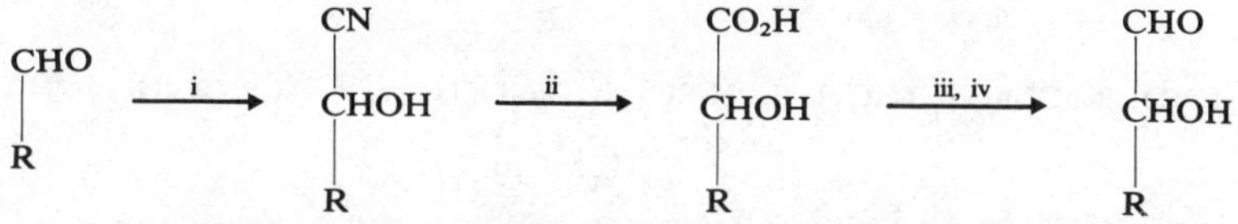

i, HCN or NaCN; ii, NaOH; iii, heat to lactonize; iv, Na/Hg.

SCHEME 5

acid or aldono-lactone stages. Stereoselectivity in the initial addition of hydrogen cyanide to the carbonyl group is often noted, and this is frequently seen to depend upon reaction conditions. For example, the reaction of L-arabinose with alkali metal cyanides at pH 9.0–9.1 goes rapidly to completion to give mainly L-gluconic acid,[32] whereas at acid pH the reaction is incomplete but gives predominantly L-mannonic acid. Cram's rule[33] predicts that L-gluconic acid should predominate, and this suggests that the reaction at alkaline pH may be kinetically controlled whilst that at acid pH may be under thermodynamic control, and thus favour the 2,4-*threo* configuration, in which the 2,4-hydroxyl groups are as far apart as possible (Maltby's rule). In the zig-zag conformation of acyclic compounds, 1,3-*erythro* interactions are destabilizing and therefore avoided.[34] L-Mannononitrile would be particularly favourable in this conformation (**23**) since there are no 1,3-*erythro* arrangements and this probably accounts for its preponderance under acid conditions which favour thermodynamic equilibration.

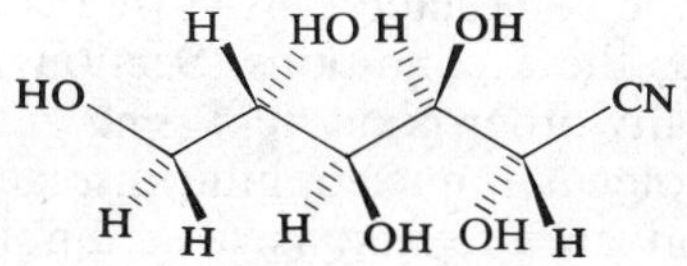

(**23**)

The conditions for the reduction of aldono-lactones to aldoses are crucial if over-reduction to the alditol is to be avoided. The original method employed sodium amalgam with rigorous maintenance of pH between 3.0 and 3.5, but reduction by sodium borohydride at 0 °C and pH 3–4 was later found to achieve the same result.[35] Alternatively, direct reduction of the cyanohydrin to the aldose may be achieved by catalytic hydrogenation.[36]

Nitroalkanes represent readily available carbon nucleophiles, and the reaction of aldoses with nitromethane has been developed into an elegant method of chain extension of aldoses (Scheme 6).[37] The initial products are two diastereoisomeric 1-deoxy-1-nitroalditols, and at this stage it is usually possible to separate the two by fractional

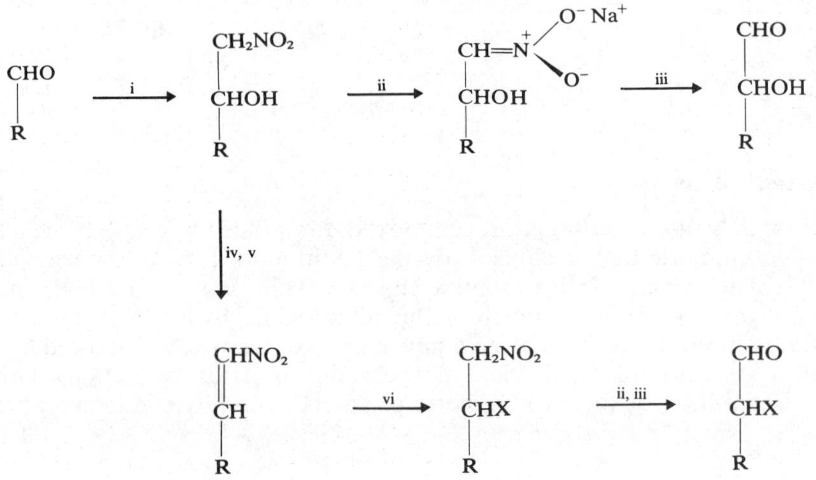

i, CH_3NO_2, NaOMe; ii, NaOH; iii, H_2SO_4; iv, Ac_2O, H^+; v, $NaHCO_3$, C_6H_6; vi, HX.

SCHEME 6

crystallization. Since the reaction is undoubtedly under thermodynamic control, the nitroalditol with the 2,4-*threo* arrangement (Maltby rule) usually predominates. The nitroalditols undergo the *Nef reaction*, by successive treatment with base and acid, to give the aldose (Scheme 6). The scope of this reaction may be extended by conversion of the nitroalditol to the 1-nitro-1-ene, which undergoes nucleophilic addition across the double bond to give a variety of 2-substituted 1-nitroalditols, the latter serving as precursors of the corresponding aldoses (Scheme 6).

Other 'carbon nucleophiles' used in the ascent of the series are diazomethane,[38] Grignard reagents,[39] malonate esters,[40] Wittig reagents,[41] *etc.*

26.1.4.3 Changes of configuration

The preparation of some rare sugars is readily achieved by inversion of configuration at one or more chiral centres of a readily available stereoisomer. For example, D-galactose is readily available but its 2-epimer, D-talose, is not. Epimerization at C-2 has been accomplished by base-catalysed epimerization of the derived D-galactonic acid to give D-talonic acid which, after lactonization, was reduced to D-talose. This process gives low yields, however, and can only be applied if the starting material is abundant. The epimerization cannot normally be conducted on the free sugars, since they often undergo extensive rearrangements under these conditions (Section 26.1.7).

Sulphonate ester groups, or any other good departing group, if suitably placed within a monosaccharide, undergo bimolecular nucleophilic displacement (S_N2) by oxygen nucleophiles such as benzoate and acetate anions in a dipolar aprotic solvent (*e.g.* *N,N*-dimethylformamide).[42] Inversion of configuration occurs at the site of the substitution reaction, providing it is chiral. The reactivity in these S_N2 displacement reactions varies considerably with the monosaccharide and the position of substitution. Richardson[43] has outlined rules based on stereoelectronic factors in the transition state, whereby the relative reactivity of sulphonyloxy groups attached to pyranose rings may be predicted.

Sulphonyloxy groups attached to *exo*-cyclic carbons generally undergo nucleophilic displacement with ease. Thus, the 5,6-ditosylate (**24**) of a glucofuranose on treatment with sodium benzoate in *N,N*-dimethylformamide gives 1,2-*O*-isopropylidene-β-L-idofuranose as a triester (**25**) with inversion of chirality at C-5, thereby providing a convenient access to L-idose.[44]

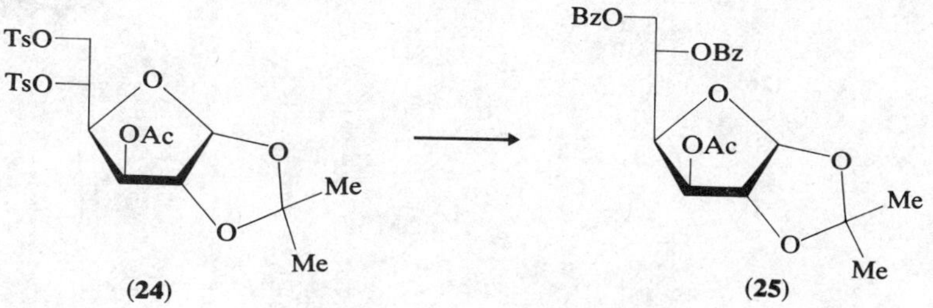

(**24**) (**25**)

The S_N2 transition state involves the formation of two transitory dipoles, and the ease of development of these will be conditioned by their interactions with the permanent dipoles existing within the molecule. Thus dipoles associated with substituents adjacent to the site of substitution will exert a marked influence, and on most occasions they oppose the establishment of the transitory dipoles and therefore impede the reaction (Figure 3).

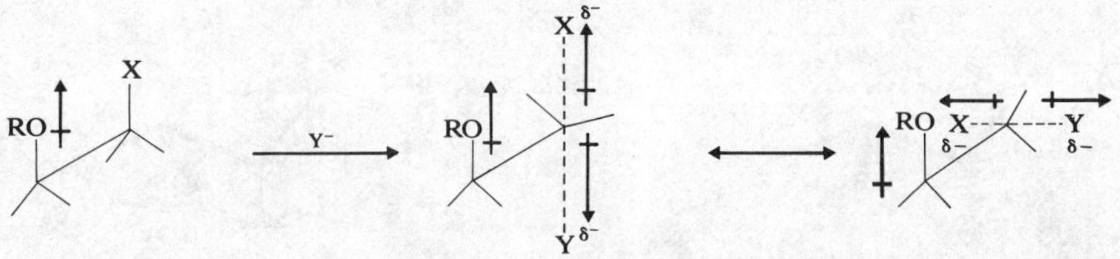

Figure 3

In addition, there will be a dependence on the torsional angle between the opposing dipoles, which will be maximal when parallel and minimal when perpendicular (crossed).

In cases where there are three or more vicinal permanent dipoles, nucleophilic displacement is impeded so severely that reaction is usually impossible when anions are the nucleophilic species. Hence attempted displacement at secondary positions adjacent to the anomeric carbon is rarely successful with charged nucleophiles because of the opposing three permanent dipoles, two from C-1 substituents and the other from the C-3 substituent. On the other hand, displacements have been achieved using neutral nucleophilic species, such as ammonia and hydrazine, because these reverse the polarity of one of the transitory dipoles and give rise to an attractive dipolar interaction. As one might expect, the removal of the polar 3-substituent, as in the 3-deoxy-2-tosylate (**26**), results in an easier displacement giving, in the case of azide anion, the 2-azide (**27**).[45] However, a related 2-axial tosylate (**28**) preferentially undergoes elimination to give the 2-ene,[46] probably because the 2-tosyloxy group is situated antiperiplanar to one of the C-3 hydrogens, which greatly favours an *E*2 elimination.

Displacements at other positions in pyranoid rings are governed by the orientation of vicinal substituents, and by the disposition of the substituents at β-positions. If there is a group, other than hydrogen, bearing a β-*trans*-axial relationship to a departing group, then nucleophilic displacement will be severely hindered owing to steric hindrance of the approaching nucleophile.[42] Thus, reaction of methyl 2,4,6-tri-*O*-benzoyl-3-*O*-tosyl-α-D-glucopyranoside (**29**) with benzoate anion proceeded only very sluggishly compared with the β-anomer (**30**).[47] In each case the corresponding allopyranosides, (**31**) and (**32**), were

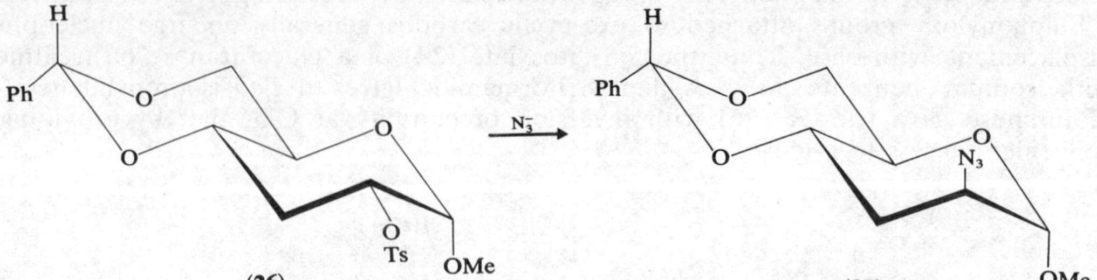

(**26**) (**27**)

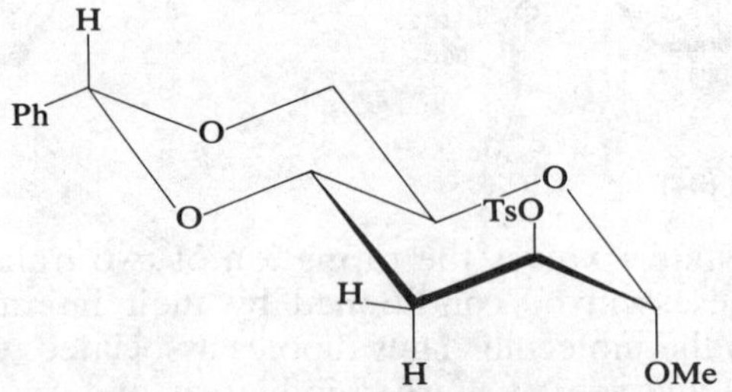

(**28**)

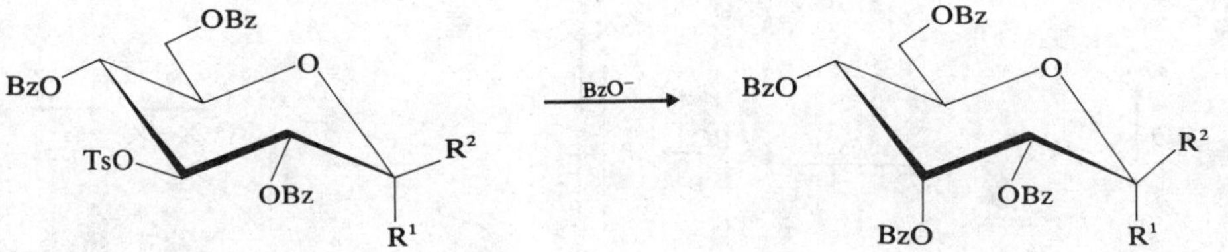

(**29**) $R^1 = OMe$, $R^2 = H$
(**30**) $R^1 = H$, $R^2 = OMe$

(**31**) $R^1 = OMe$, $R^2 = H$
(**32**) $R^1 = H$, $R^2 = OMe$

obtained, and this constitutes a convenient synthesis of methyl allopyranosides. Similarly, methyl 2,3-di-*O*-isopropylidene-4-*O*-tosyl-α-D-mannopyranoside does not undergo substantial direct nucleophilic displacement, but usually undergoes ring contraction in preference.[48]

The presence of a vicinal axial substituent also hinders nucleophilic displacement owing to unfavourable dipolar interactions in the transition state. For example, 3-sulphonates of glucopyranoid derivatives undergo ready nucleophilic displacement reactions providing the anomeric linkage is β as in (**30**), but the corresponding galactopyranoid derivatives, *e.g.* (**33**), do not undergo displacement owing to the axial group at C-4.[43] This effect is best visualized by a Newman projection along the C-3—C-4 bond of both the ground state (**34**) and the transition state (**35**). It can be seen that one of the transitory dipoles of the transition state becomes almost aligned with the permanent dipole associated with the C-4 substituent, resulting in maximal dipolar repulsion. Indeed, Miljkovic *et al.*[49] have shown that displacement by benzoate anion at C-2 of methyl 3,4,6-tri-*O*-methyl-2-*O*-tosyl-β-D-glucopyranoside (1-OMe equatorial) can be achieved in good yield by prolonged reaction, whereas the α-anomer (1-OMe axial) does not undergo displacement because of the vicinal axial polar substituent at C-1.

By nucleophilic substitution it is possible to achieve ready displacement of the 4-sulphonate group in α- and β-*gluco*-, *galacto*-, *xylo*-, *arabino*- and, as a result of a

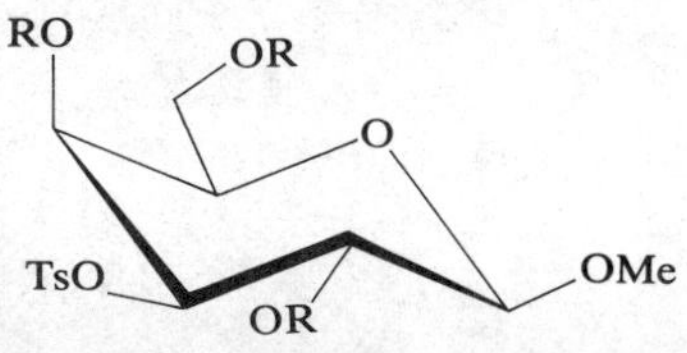

(33)

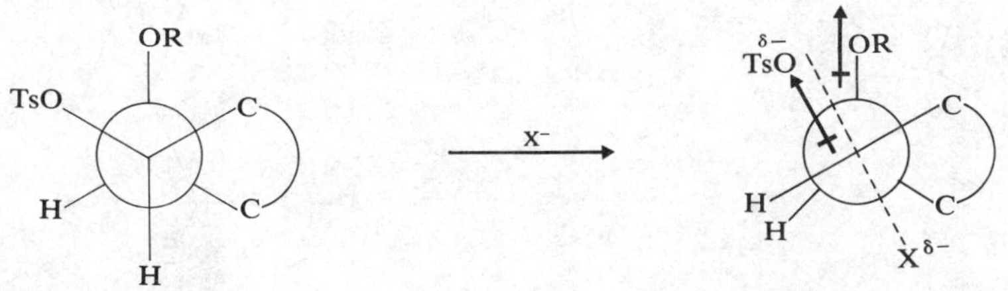

(34) (35)

conformational change, -*altro*-pyranosides, and the 3-sulphonate group in β-*gluco*-, β-*xylo*-, and, probably, β-*allo*-pyranosides.

A further use of sulphonate esters is their conversion to epoxides, when a vicinal hydroxyl group is *trans*.[42,50] The reaction occurs in the presence of base and, usually, in high yield. For example, the 2- and 3-tosylate esters, (**36**) and (**37**) respectively, of methyl 4,6-*O*-benzylidene-α-D-glucopyranoside give the 2,3-anhydrides with the *manno* and *allo* configurations, (**38**) and (**39**) respectively. Reaction of these epoxides with nucleophiles under basic or neutral conditions favours the formation of products with the antiperiplanar arrangement of the 2- and 3-substituents, as in (**40**) and (**41**). Thus, the reaction of either epoxide with sodium hydroxide gives the altropyranoside (**40**; X = OH) in high yield, from which D-altrose is formed by deblocking.

Epoxides which are situated at a terminal position undergo preferential ring-opening by nucleophilic attack at the primary position, in agreement with an S_N2 mechanism. Epoxides which are not attached to a rigid fused ring structure may give mixtures of products owing to reaction *via* two transition states relating to the two chair conformations of the pyranoid ring (Section 26.1.8.2).

With the advent of selective and highly effective oxidizing agents for the conversion of hydroxyl groups to carbonyl functions[51] (*e.g.* methyl sulphoxide in combination with a variety of acid anhydrides and carbodi-imides, *etc.*; ruthenium tetroxide, *etc.*) it is possible to oxidize sequentially an isolated hydroxyl (Section 26.1.6.4) and then reduce the ketone back to an hydroxyl group with overall inversion of configuration. Since sodium borohydride reductions occur by approach of the reagent from the least-hindered side of the carbonyl group, it is usually possible to predict inversion of configuration or, if overall retention of configuration is predicted, then alternative means of reduction may give the required inversion. Thus, sequential oxidation and reduction of 1,2:5,6-di-*O*-isopropylidene-α-D-glucofuranose (**42**) gives the corresponding allofuranose (**44**) because underside attack of the 3-ketone (**43**) requires an *endo* approach, whereas topside attack takes place from the less hindered *exo* direction.[52] The intermediate keto derivatives such as (**43**) often offer scope for inversion at neighbouring chiral centres. For example, acetylation of (**43**) to give the enol acetate (**45**) followed by catalytic reduction gives 1,2:5,6-di-*O*-isopropylidene-α-D-gulofuranose (**46**).[53] Reduction of 2-keto derivatives of pyranosides is governed by the anomeric configuration. The α-anomer affords the glucoside since the axial C-1 substituent hinders equatorial approach, whereas in the β-anomer, equatorial attack is preferred giving the mannopyranoside.[54]

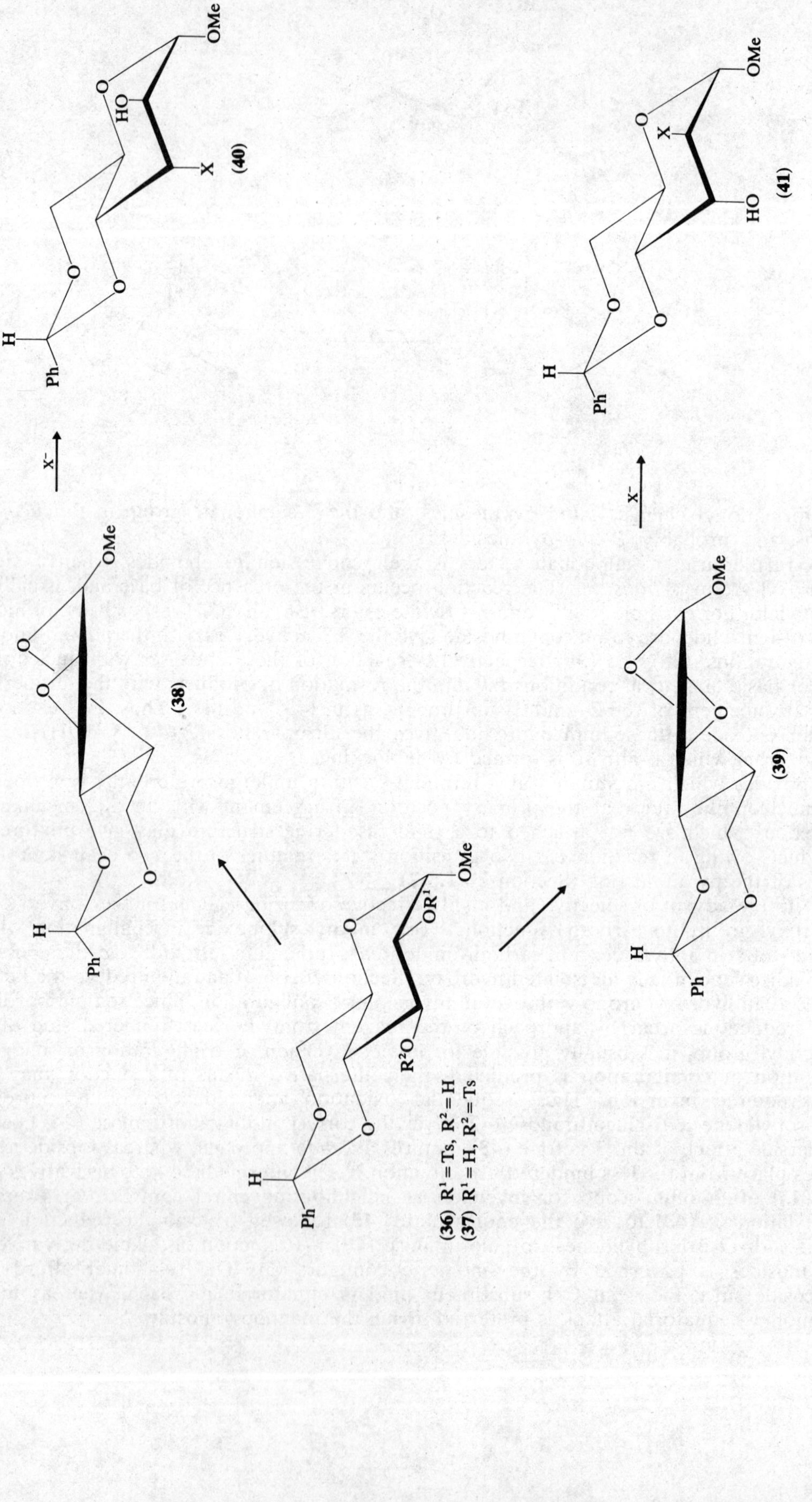
(36) R¹ = Ts, R² = H
(37) R¹ = H, R² = Ts
(38)
(39)
(40)
(41)
X⁻
X⁻
Ph
H
OMe
HO
X
OR¹
R²O

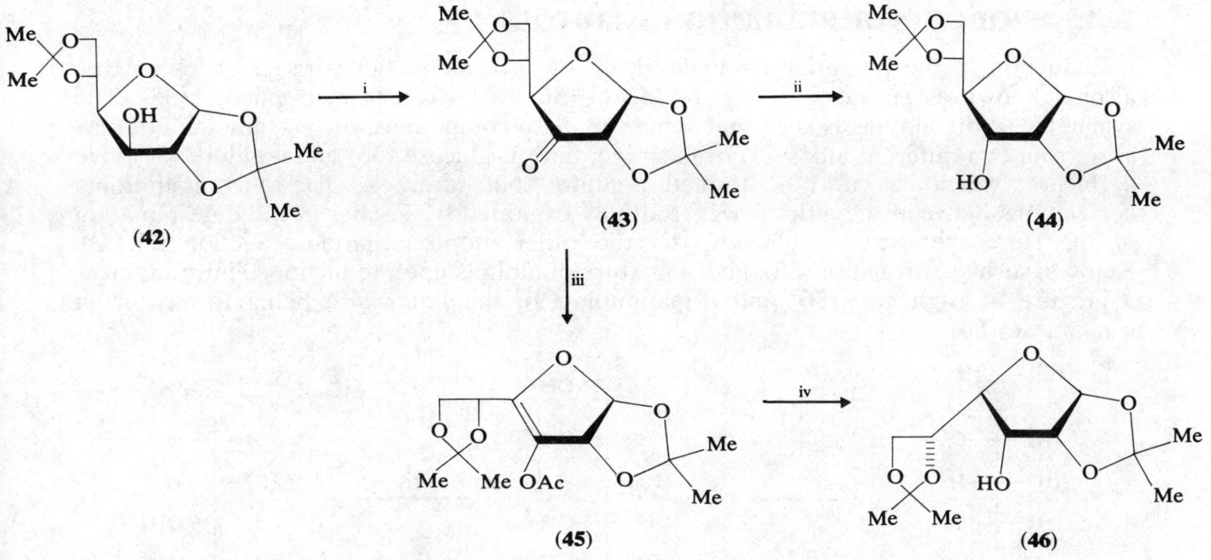

i, $Ru(OH)_2$, $NaIO_4$; ii, $NaBH_4$; iii, Ac_2O, py; iv, [H].

The action on per-acetylated sugars or glycosyl halides of acidic but weakly nucleophilic reagents such as anhydrous hydrogen fluoride or antimony pentachloride[55] often results in inversion of at least one, and often several, chiral centres. Particularly noteworthy is the antimony pentachloride induced conversion of 2,3,4,6-tetra-*O*-acetyl-α-D-glucopyranosyl chloride (**47**) directly into a mixture of α-D-idopyranose tetra-acetates, (**52**) and (**53**).[56] The reaction mixture is in effect an equilibrium mixture of the acetoxonium ions (**48**)–(**51**) in which each acetoxonium ring is successively ruptured as a result of participation by the neighbouring *trans*-acetoxyl group, with the simultaneous formation of a new acetoxonium group. The corresponding *ido*-ion (**51**) is the least favoured thermodynamically but fortuitously the least soluble in dichloromethane, so that it crystallizes out of the reaction mixture in good yield when this solvent is used. Decomposition of the ion (**51**) with water gives the mixture of idose tetra-acetates (**52**) and (**53**), from which α-D-idopyranose penta-acetate (**54**) and D-idose are easily available.

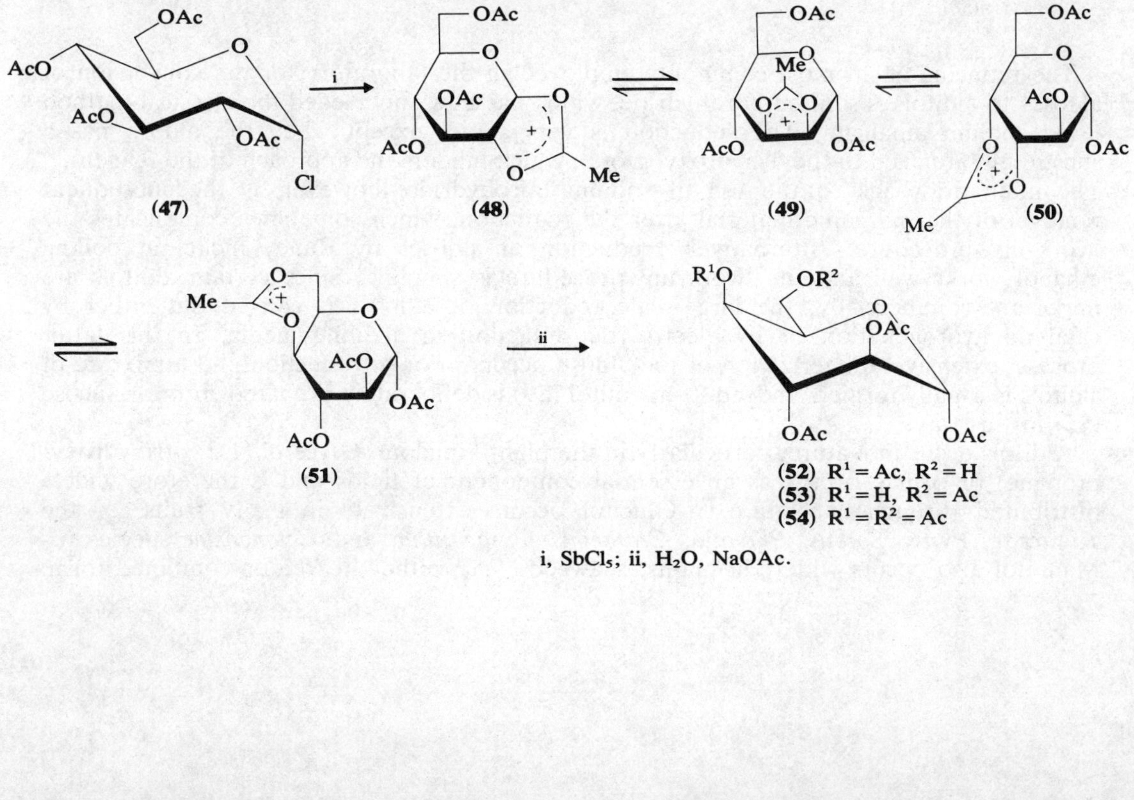

i, $SbCl_5$; ii, H_2O, NaOAc.

26.1.5 PRODUCTS OF REDUCTION: ALDITOLS

Reduction of the potential aldehyde group of each aldose gives rise to a polyhydric alcohol known as an alditol,[57] *e.g.* D-glucitol (**56**) is derived from D-glucose (**55**). The symmetry of the alditols is such that some are *meso* compounds, or the same alditol may arise from two different aldoses. For example, both D-glucose (**55**) and L-gulose (**57**) give D-glucitol, which can also be termed L-gulitol, but rarely is. These configurational correlations between the alditols were skilfully exploited by Fischer in the determination of the stereochemistry of glucose and the other monosaccharides (Section 26.1.2). Ketoses, such as D-fructose (**58**), give a mixture of alditols upon reduction. Thus (**58**) gives a mixture of D-glucitol (**56**) and D-mannitol (**59**), the latter also being formed from D-mannose (**60**).

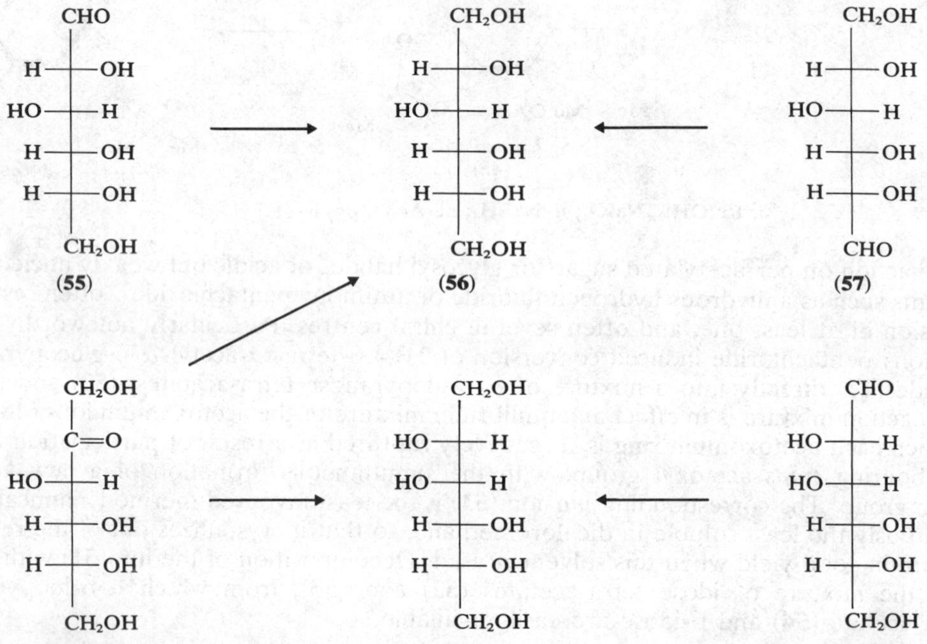

(55) (56) (57) (58) (59) (60)

The reducing agent most commonly employed in the laboratory for the conversion of aldoses to alditols is sodium borohydride, which has now superseded the classical method using sodium amalgam. The reduction is very rapid, except when the aldose has a substituent attached to the 3-hydroxyl group, which hinders the approach of the reagent.[58] The main drawback to the use of sodium borohydride, however, is the subsequent removal of the inorganic material after the reduction, which sometimes complicates the 'work-up' procedure. Alternatively, reduction of aldoses by Raney nickel in boiling ethanol works well and the 'work-up' procedure is simple.[59] Since certain alditols are important commercially, the large-scale reduction of aldoses is carried out either by catalytic hydrogenation or by electrolytic reduction in alkaline media. In this latter process, extensive epimerization of the aldose occurs prior to reduction and a mixture of alditols is usually formed. Indeed, D-mannitol (**59**) is deliberately prepared from D-glucose (**55**) in this way.

Alditols occur in Nature, particularly in the plant kingdom. Glycerol (1,2,3-trihydroxypropane) of course occurs as an essential component of lipids and is therefore widely distributed throughout Nature. D-Glucitol occurs extensively in many fruits, of the *Rosaceae, Pyrus, Sorbus, Photinia, Crataegus, Pyracantha,* and *Cotoneaster* species. D-Mannitol also occurs widely in plants, seaweeds, *etc.*, either in free or combined form.

Several mannas contain D-mannitol in high concentrations (30–90%) and its highly crystalline nature makes it easily isolated in a pure form.

D-Glucitol (**56**) also termed sorbitol, is non-toxic, slightly sweet, and mildly hygroscopic, properties which make it useful as a humectant in cosmetic and pharmaceutical products.

26.1.6 PRODUCTS OF OXIDATION

Aldoses, as one would expect from their multifunctional nature, undergo several modes of oxidation with a variety of oxidizing agents.[51] They can cause oxidation at either the reducing end or both ends of the chain, at individual hydroxyl groups, or cause cleavage of the carbon–carbon chain.

26.1.6.1 Aldonic acids

These acids are produced from aldoses under mild oxidative conditions; thus halogens (chlorine, bromine, or iodine) at pH 5 oxidize aldoses directly to the corresponding lactones, which then hydrolyse slowly to the free acids. Whilst the mechanism of bromine oxidation of aldoses is obscure, the observation that at pH 5 β-D-glucopyranose is oxidized 250 times faster than the α-anomer indicates that the initial attack may involve the 1-hydroxyl group, since the equatorial hydroxyl in the β-anomer would be the more reactive.[8] Esters of aldonic acids are available directly by oxidation of alkyl β-glycosides with ozone; the α-anomers are, however, unreactive.[60]

The free aldonic acids are sometimes difficult to isolate because they show a marked tendency to dehydrate to give a mixture of lactones, and consequently they are usually isolated as salts, amides, or phenylhydrazides. Evaporation of aqueous solutions of an aldonic acid usually gives the five-membered 1→4 γ-lactone in preference to the six-membered 1→5 δ-lactone, and this contrasts with the behaviour of the corresponding aldoses in cyclic hemi-acetal formation. The most likely reason for this preference is the polar interaction between the dipole associated with the ring oxygen and that of the carbonyl group, which are parallel and opposed in the 1→5 lactone, whereas the 1→4 lactone is able to adopt an energetically favoured conformation in which this interaction is lessened.

26.1.6.2 Alduronic acids

The term alduronic acid refers to aldoses possessing a terminal carboxylic acid grouping, in addition to the reducing group, at the other end of the chain. They occur widely in Nature, and D-glucuronic acid plays an important role in animal metabolism by aiding the excretion of phenols, steroids, and aromatic carboxylic acids, which are eliminated in conjugate form as glucuronides.[61] D-Galacturonic acid is a component of fruit pectin from which it is readily isolated after enzymic hydrolysis. D-Mannuronic and L-guluronic acids occur in combined form in various seaweeds. L-Iduronic acid is a component of heparin and other polysaccharides. They are not easily isolated from natural sources, since they undergo easy decarboxylation and elimination under the vigorous conditions required for the cleavage of glucuronides. However, Capon and Ghosh[62] have noted that the un-ionized carboxylic acid group lends greater stability to the glycosidic bond, and that the ionized carboxylic acid group has the opposite effect. Hence, in M HCl, 2-naphthyl glucuronide is hydrolysed 45 times slower than methyl β-D-glucopyranoside, whereas at pH 4.79, when the acid group occurs to a much greater extent in the ionized form, the glucuronide is hydrolysed 35 times more rapidly than the glucoside.

Alduronic acids are synthesized most directly by selective catalytic oxidation of aldoses or their glycosides using platinum as catalyst.[63] For example, catalytic oxidation of 1,2-*O*-isopropylidene-α-D-glucofuranose (**61**), followed by acid hydrolysis, gives D-glucuronic acid (**63**) in good yield, *via* its 1,2-*O*-isopropylidene derivative (**62**).[64] Where suitably blocked carbohydrate derivatives are available with only the primary hydroxyl group free, then oxidation with potassium permanganate can often be employed with advantage.

Alduronic acids usually exist in the form of lactones; thus D-glucuronic acid exists as the 6→3 lactone of the furanose form, termed glucurone (**64**), whereas in the case of D-galacturonic acid this furanose form would require a *trans* junction between the two five-membered rings, and hence the 6→3 lactone adopts the pyranose form (**65**).

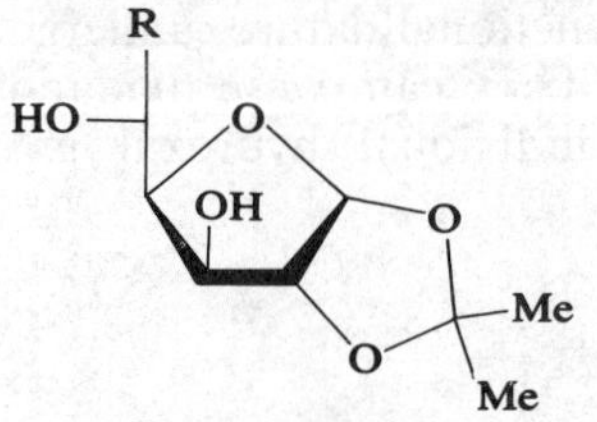

(61) R = CH_2OH
(62) R = CO_2H

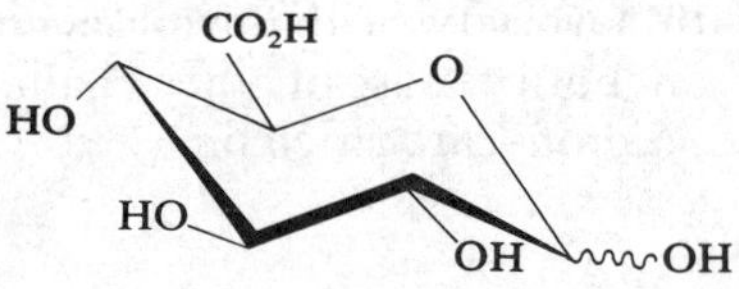

(63)

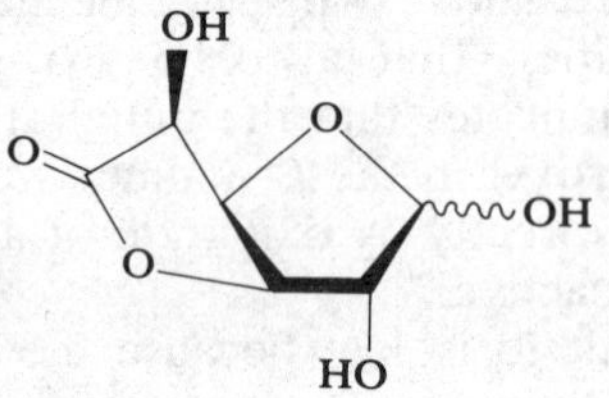

(64)

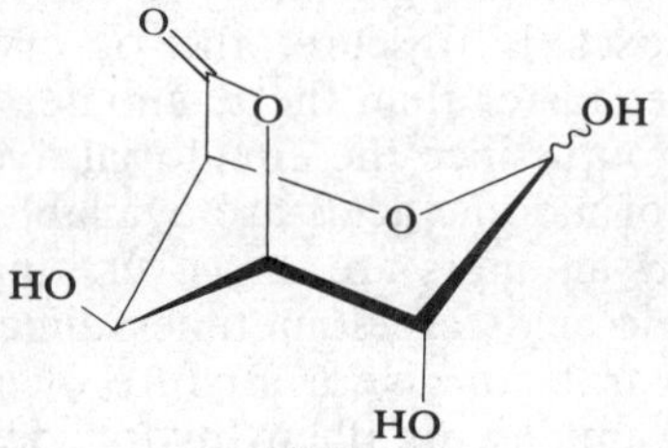

(65)

26.1.6.3 Aldaric acids

The dicarboxylic acids derived from aldoses by oxidation of both terminal sites are termed aldaric acids. They are formed by oxidation of aldoses with nitric acid. Galactaric acid (mucic acid), a *meso* compound, is the most readily isolated because of its limited solubility in water (0.3% w/v), and this forms the basis of a gravimetric method of analysis for galactose. Ketoses are oxidized under the same conditions with cleavage of the 1,2-bond. Thus, D-fructose gives D-arabinaric acid, which is also derived from D-arabinose or D-lyxose. Similarly, 6-deoxyhexoses oxidize by cleavage of the 5,6-bond, so that 6-deoxy-L-galactose (L-fucose) gives D-arabinaric acid.

26.1.6.4 Keto derivatives (aldosuloses)

Oxidation at secondary hydroxyl groups cannot usually be accomplished in good yield directly from the free sugars, but is usually applied with success to suitably blocked derivatives. The free aldosulose is rarely liberated from blocked derivatives, but these are usually employed as synthetic intermediates. One other complication is that the free aldosulose is usually not crystalline owing to the wide variety of cyclic structures which it can adopt. Whilst the aldo-2-uloses (osones) have been known for many years and are available from either phenylosazones or by direct oxidation of aldoses with cupric ions,[65] they are particularly difficult to purify and yields are low.

The oxidation of secondary hydroxyl groups of blocked carbohydrates was revolutionized by the discovery of the methyl sulphoxide-based oxidizing agents originally described by Moffat and Pfitzner.[66] Until then, chromium trioxide in pyridine had been used, but this suffered from the need for repeated application before a reasonable degree of reaction could be achieved.[67] Various oxidizing agents based on methyl sulphoxide (DMSO) have been described;[51] thus in addition to the original Pfitzner–Moffatt reagent,[66] which is composed of DMSO, dicyclohexylcarbodi-imide and phosphoric acid, DMSO–acetic anhydride,[68] DMSO–phosphorus pentoxide,[69] and DMSO–sulphur trioxide–triethylamine[70] have been used. In many cases, however, side products have been encountered which reduce yields and complicate the isolation of the required product. In the case of DMSO–Ac_2O, the acetate ester is formed, sometimes in substantial quantities, and in all cases the methylthiomethyl ether ($ROCH_2SCH_3$) is formed in varying amounts. Ruthenium tetroxide[71] and the chromium trioxide–bipyridyl complex[72] are claimed to be superior to DMSO-based oxidations. The former has been widely exploited, usually gives good yields, and is often carried out using catalytic amounts of hydrated ruthenium dioxide with sodium periodate, which serves to generate the tetroxide from the dioxide.[73]

Oxidation of secondary or primary hydroxyl groups which are located next to a chiral centre may also result in inversion of configuration at that position. For example, oxidation of the 2-acetamidoaltroside (**66**) gives the 3-ulose (**67**), in which inversion of configuration has occurred at C-2, to give the thermodynamically more favoured 2-equatorial isomer (**67**).[74] It is also possible to use these keto derivatives to specifically deuteriate or tritiate certain adjacent positions.[75] For example, treatment of the 2-azido-3-ulose (**68**) with deuterium oxide under mildly acidic conditions gave specifically the 2-deuterio derivative of (**68**).

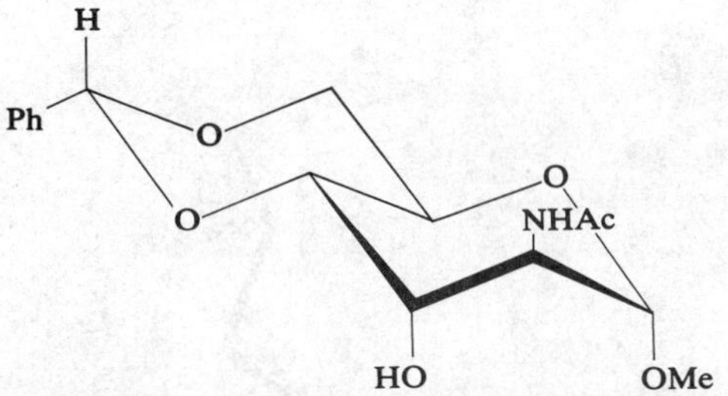

(**66**)

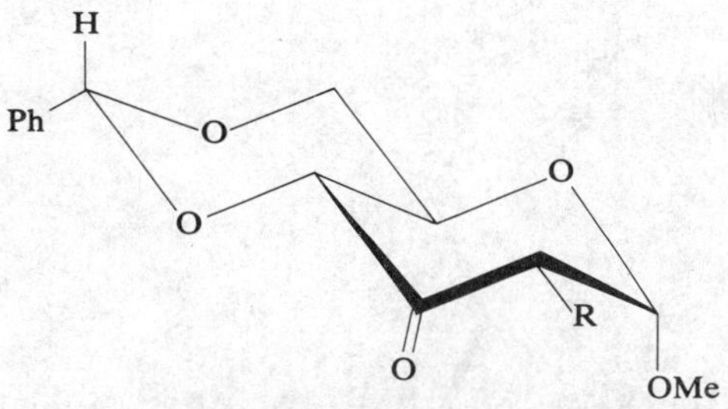

(**67**) R = NHAc
(**68**) R = N_3

Angyal[76] has described the oxidation of acetals using chromium trioxide in acetic acid, which affords keto derivatives in high yields. Thus, oxidation of 3,4-*O*-ethylidene-D-mannitol tetra-acetate (**69**) gives the penta-acetate (**70**) of D-*arabino*-3-hexulose in good

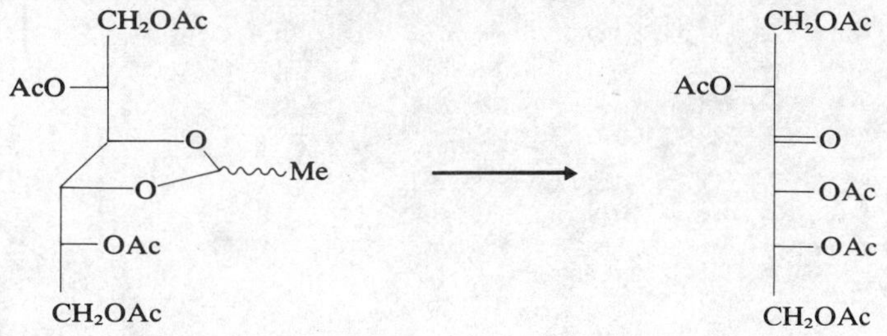

(**69**) (**70**)

yield; this method constitutes an excellent synthesis of the otherwise difficultly accessible 3-hexuloses. Methyl glycosides, which are cyclic hemiacetals, similarly undergo oxidation, although the anomeric configuration is a critical factor. In the case of β-glycosides, such as methyl β-D-glucopyranoside tetra-acetate (**71**), oxidation proceeds *via* the C-1 radical (**72**) and the C-1 Cr^{IV} ester (**73**) to give the 5-ketoaldonic acid ester (**74**). On the other hand, the corresponding α-anomer (**75**) reacts *via* the *O*-methylene radical (**76**) which, by a similar process, gives the 1-formyl ester (**77**). Hence the two glycosides react by abstraction of different hydrogen atoms.[77]

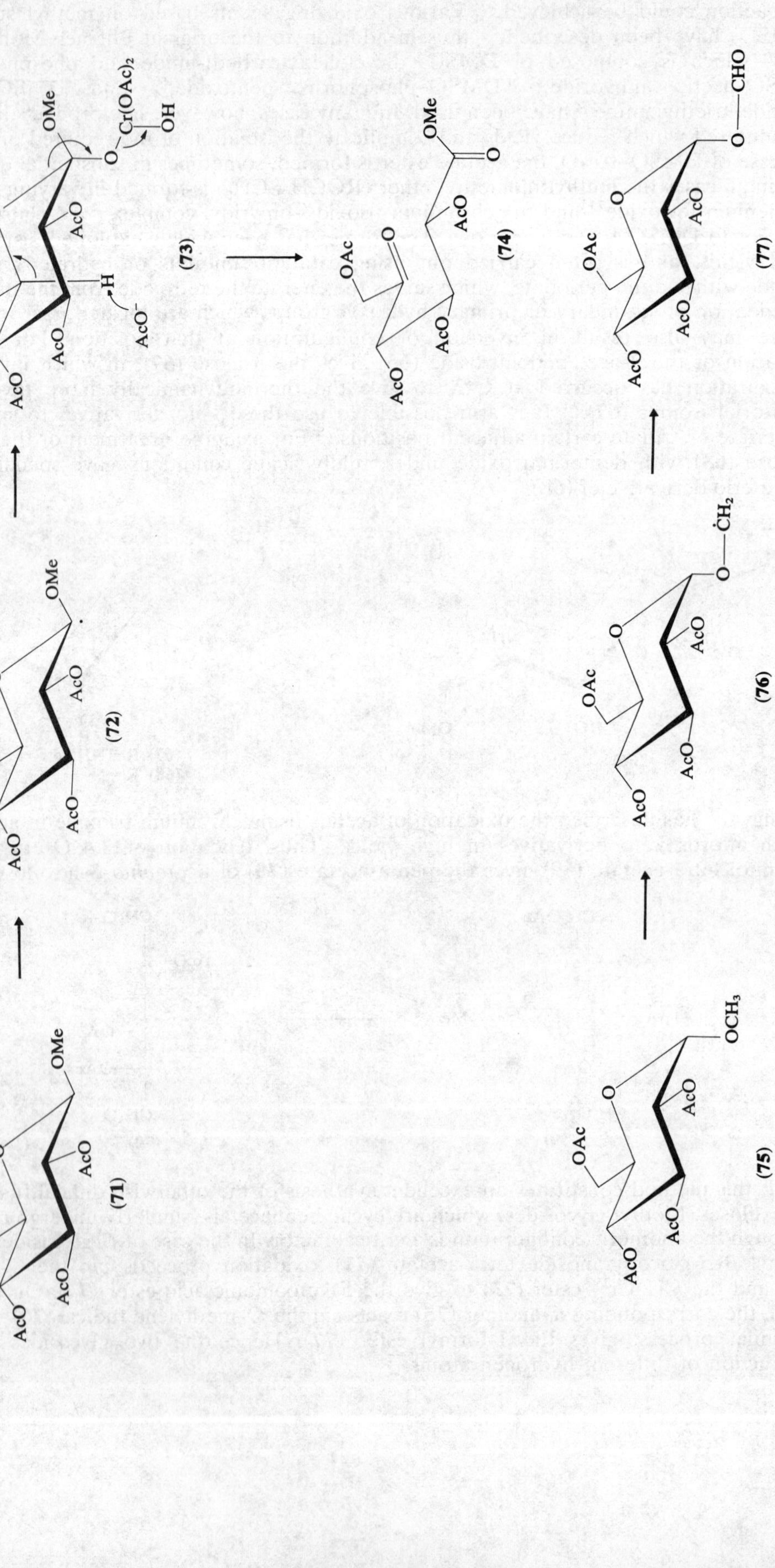

(71) (72) (73) (74) (75) (76) (77)

26.1.6.5 Oxidation with glycol-cleaving reagents

These reagents, notably periodic acid and its salts and lead tetra-acetate, have been of inestimable value in carbohydrate chemistry.[10,78] They have been employed not only as tools in structural studies (Section 26.1.2) but also in synthetic work (Section 26.1.4.1). These reagents selectively oxidize α,β-diol groups with cleavage of the intervening carbon–carbon bond. The intermediates are believed to be cyclic diesters (Scheme 7).

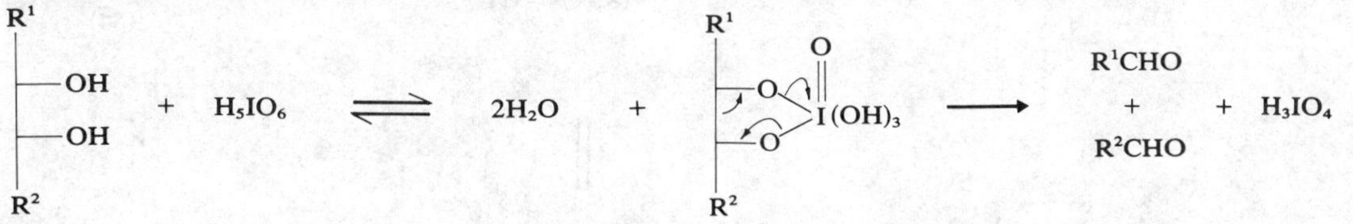

SCHEME 7

Because of the steric requirements for the formation of such cyclic esters, the steric relationship between the two hydroxyl groups will markedly affect the rate of oxidation. Thus, *cis*-diol groups in cyclic compounds will undergo oxidation more rapidly than *trans*-diols. In pyranoid systems, diols which are antiperiplanar (diaxial), as in methyl 4,6-*O*-benzylidene-α-D-altropyranoside, are oxidized only very slowly, if at all. Consequently, although the oxidation of a compound by a glycol-cleaving agent is normally indicative of a diol grouping, there are exceptions. In addition, oxidation may be inhibited by the presence of a large vicinal group. For example, methyl 4,6-*O*-benzylidene-β-D-glucopyranoside (**78**) slowly consumes one mole of oxidant, the phenyl glycoside (**79**) does so extremely slowly, and the bulky theophyllyl derivative (**80**) is not oxidized at all.[79]

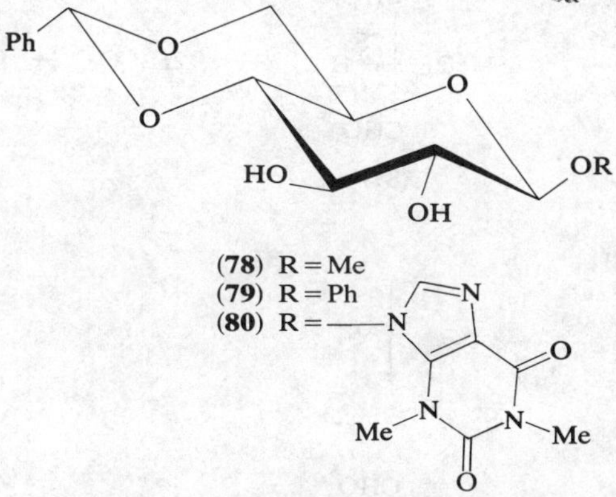

One further application of periodate oxidation is the correlation of anomeric configuration between glycosides. For example, methyl β-D-glucopyranoside (**81**) and methyl β-D-ribofuranose (**84**) both give the same dialdehyde (**82**), proving that they both have the same chirality at C-1. The dialdehydes, such as (**82**), usually exist as mixtures of hemialdals, *e.g.* (**85**), and are rarely crystalline so they are often oxidized further by bromine water to the corresponding dicarboxylic acid (**83**) and isolated as a crystalline salt.[78]

Anomalous results are obtained in certain cases, through non-Malapradian oxidation. When a malondialdehyde (**87**), or related species, is generated in a periodate oxidation, then it undergoes oxidation to give the hydroxymalondialdehyde (**88**) which then consumes a further two moles of oxidant with the formation of two moles of formic acid and one mole of carbon dioxide. This behaviour is observed during the periodate oxidation of 3-*O*-methylglucitol (**86**) at pH 1, in which 7 moles of oxidant are consumed with the liberation of two moles of formaldehyde, four moles of formic acid, and one mole of carbon dioxide.[80]

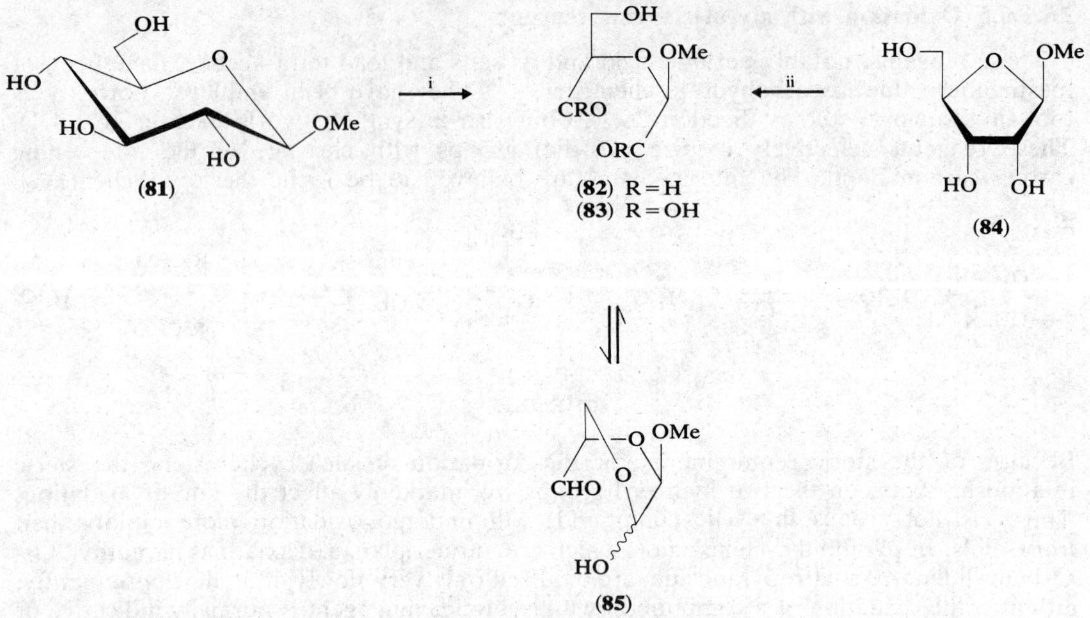

i, 2 $NaIO_4$; ii, 1 $NaIO_4$.

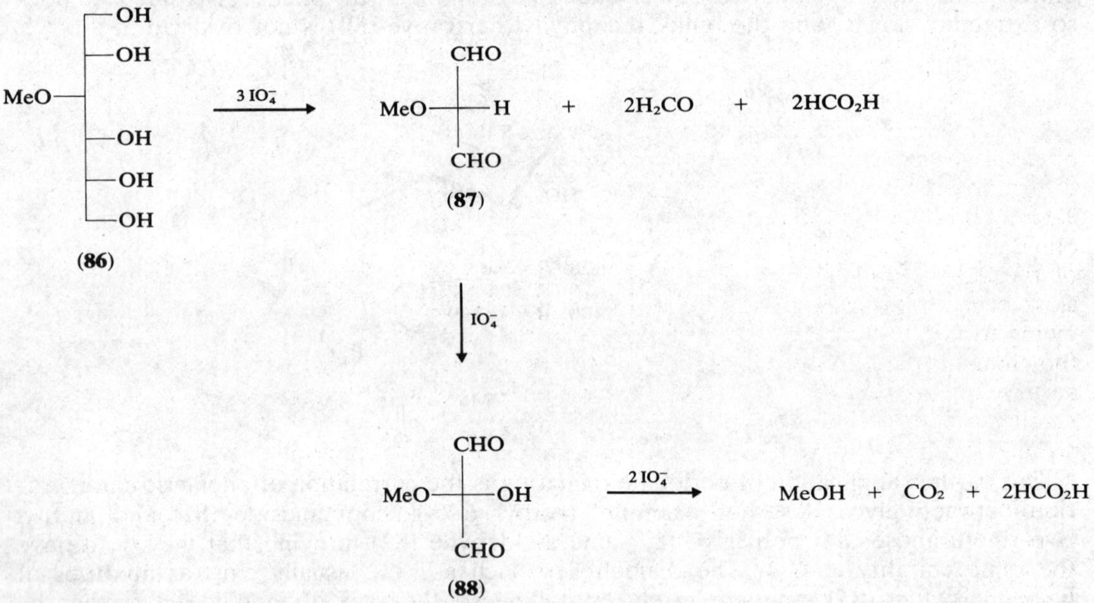

26.1.7 ACTION OF ACIDS AND BASES

Aldoses and ketoses undergo isomerization and degradation with acids and bases to give a variety of products. It can be assumed that the masked carbonyl group is responsible for this instability since alditols are remarkably stable under these conditions.

Under mildly basic conditions (organic bases such as pyridine are often used), epimerization at the carbon adjacent to the carbonyl group is observed together with an aldose⇌ketose isomerization. The reaction proceeds *via* enediols (Scheme 8).[81] Thus

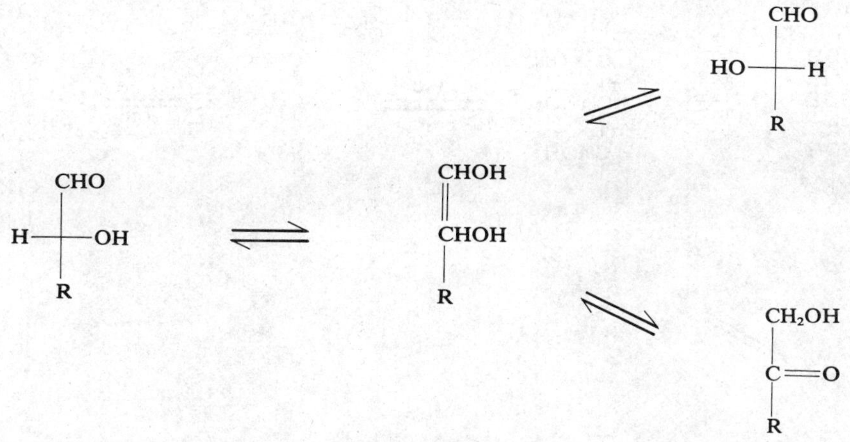

SCHEME 8

D-glucose in 8×10^{-3} M NaOH at 35 °C for 4 days gives a mixture containing D-fructose (28%) and D-mannose (3%) in addition to unchanged D-glucose. In cases where the 2-hydroxy-group is substituted, isomerization to the ketose is precluded and only epimerization at C-2 is observed. Thus, 2,4,6-tri-*O*-methyl-D-mannose is a convenient precursor of 2,4,6-tri-*O*-methyl-D-glucose by base-catalysed epimerization.[82]

Under more strongly basic conditions, rearrangements of a more profound nature occur with the formation of saccharinic acids[83] (Scheme 9). Thus, D-glucose and its derivatives on treatment with lime water (*ca.* 0.15 M) undergo formation of (a) 3-deoxyhexonic acids (metasaccharinic acids), (b) 2-*C*-methylpentonic acids (saccharinic acids), and (c) 3-deoxy-2-*C*-hydroxymethylpentonic acids (isosaccharinic acids). These are formed by (a) β-elimination from the free aldose followed by benzilic acid rearrangement of the α-dicarbonyl compound; (b) formation of the 2,3-enediol, followed by β-elimination of the 1-OH and benzilic acid rearrangement of the 2,3-dicarbonyl derivative; and (c) β-elimination of the 4-OH from the 2,3-enediol followed by a benzilic acid rearrangement of the 2,3-dicarbonyl derivative. In the case of D-glucose these reactions are relatively slow because the elimination of hydroxyl groups under alkaline conditions is sluggish owing to the existence of some hydroxyls in the form of their conjugate base. However, the elimination of alkoxy-groups is much more favoured and consequently the presence of an alkyloxy-group at C-3 will favour the formation of metasaccharinic acids, whereas 4-*O*-alkyl derivatives will favour the formation of the isosaccharinic acid. Saccharinic acids are best prepared from 1-*O*-alkylketoses, or the ketoses themselves.

The formation of these degradation products is not usually stereospecific and isosaccharinic and metasaccharinic acids are generally formed as equimolar mixtures of the two diastereoisomers. However, the formation of saccharinic acids appears to be highly stereospecific and D-fructose and its 1-*O*-benzyl ether give only 2-*C*-methyl-D-*ribo*-pentonic acid (**91**).[84] In the case of saccharinic acid formation, the benzilic acid rearrangement occurs at a carbon adjacent to a chiral centre, and might therefore be expected to be influenced by asymmetric induction, whereas in the other two cases the site of rearrangement is more remote from a chiral centre. The highly stereospecific nature of saccharinic acid formation suggests that the intermediate prior to the benzilic acid rearrangement might be held rigidly as a chelate with the cation so that the rearrangement of the methyl group can occur from one direction only. For example, the 4-OH of the diulose (**89**) might exert a directive influence on the attack by hydroxide anion to give an intermediate (**90**) in which hydroxide has added on the side opposite to the 4-OH. The alkoxide could then be held rigidly as the chelate (**90**) in which rearrangement of the methyl group must occur at the rear of the molecule and so give the *ribo*-isomer (**91**).

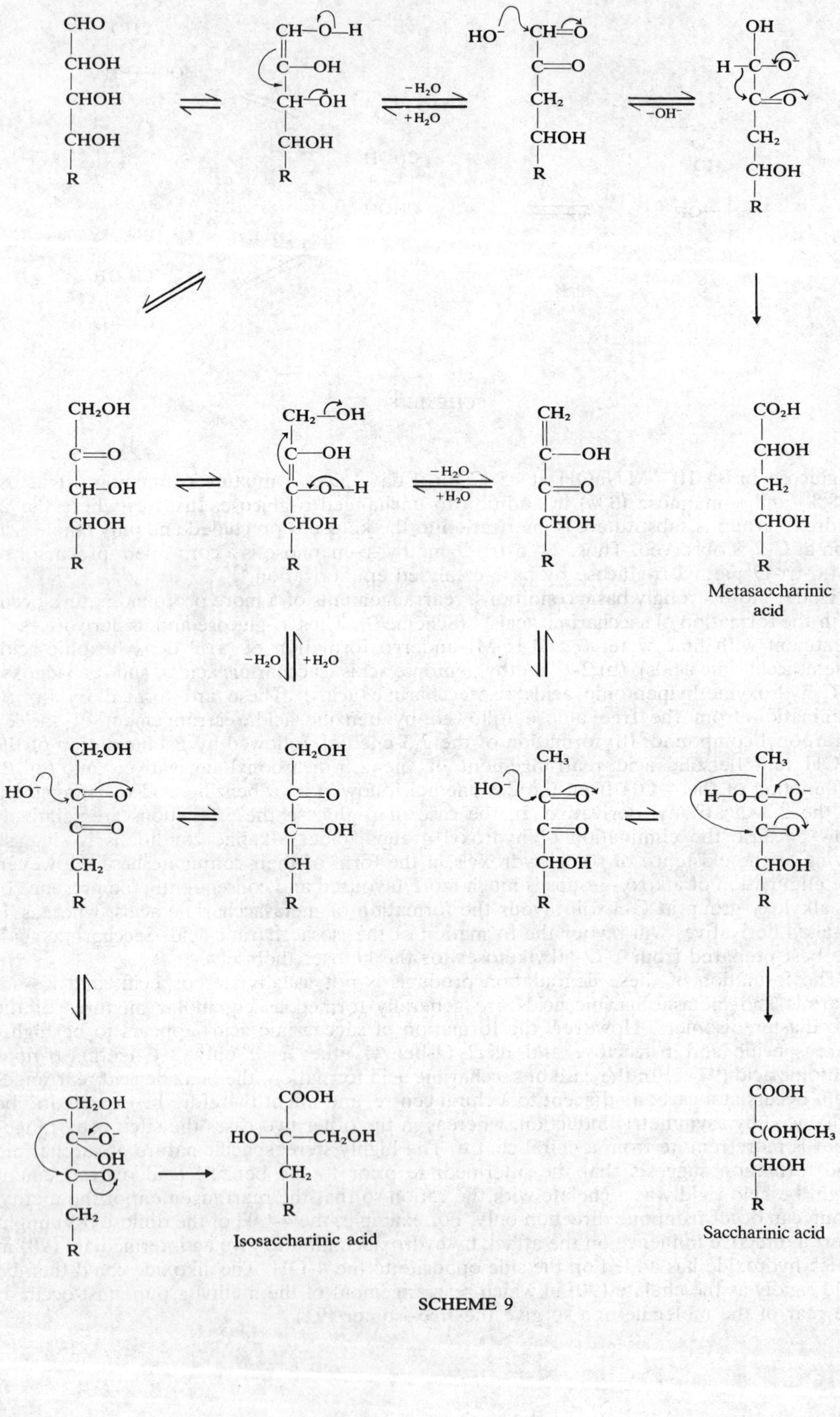
Metasaccharinic acid
Isosaccharinic acid
Saccharinic acid

SCHEME 9

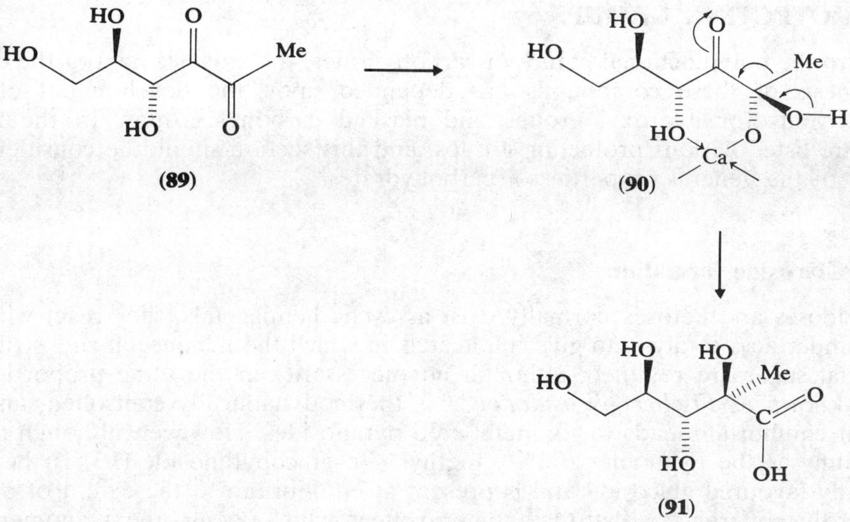

(89) (90) (91)

More strongly basic conditions result in complex reverse-aldol reactions, which give three-carbon fragments such as 2-hydroxypropionaldehyde, pyruvic acid, lactic acid, *etc.*

Whilst aldoses are invariably more stable towards acid than they are towards alkali, they do undergo dehydration, the extent of which depends upon the conditions. Evaporation of solutions of aldoses in dilute mineral acid (10^2–10^{-4} M) causes intermolecular condensation reactions akin to glycoside formation (Section 26.1.8.1) which is termed '*reversion*', to give small amounts of di-, tri-, and higher oligo-saccharides. Hexoses and higher analogues, in which the energy difference between the two chair conformations is small, undergo facile intramolecular dehydration to give 1,6-anhydro-β-pyranoses. The reaction is thermodynamically controlled and the amount of anhydride formed is dependent upon the stability of the aldose in its ${}^1C_4$ conformation (Section 26.1.8.2). Under more drastic conditions, aldoses and ketoses undergo more extensive dehydration to give furan derivatives (Scheme 10).[85] In the case of hexoses and hexuloses, the product is 5-hydroxymethylfurfuraldehyde (**92**), which under more drastic conditions undergoes ring-opening to give laevulinic acid (**93**) and formic acid. The colorimetric assay of sugars is based upon transformation into furfuraldehydes under carefully controlled conditions, followed by reaction with various phenols and aromatic amines. In some instances this reaction permits differentiation of different types of sugars.[86]

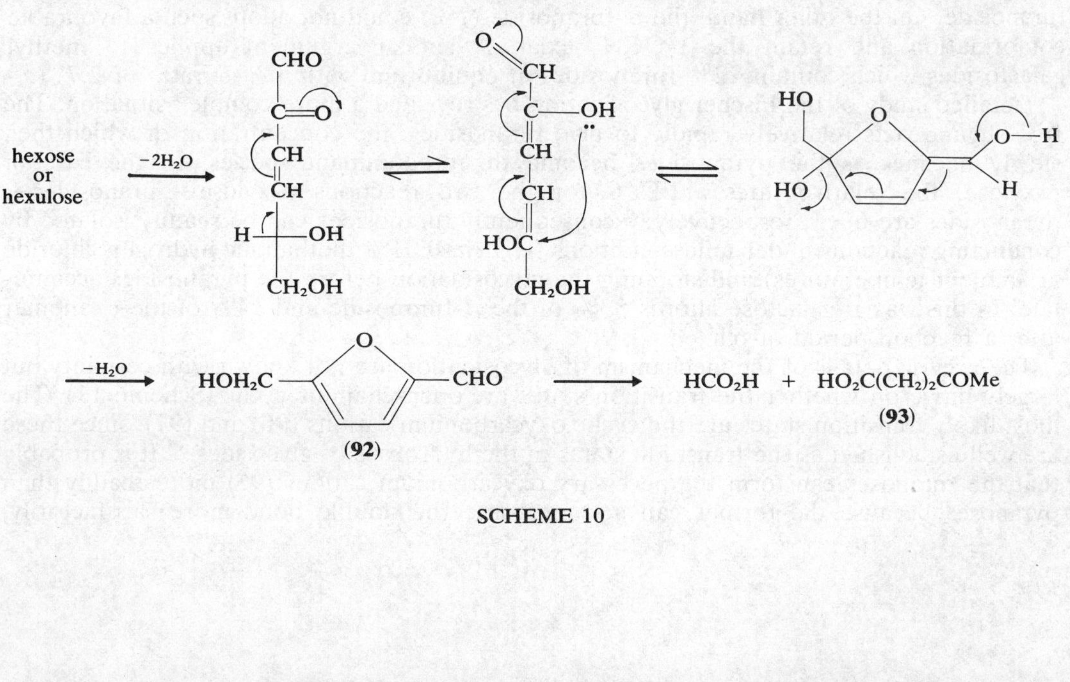

(92) (93)

SCHEME 10

26.1.8 PROTECTING GROUPS

Owing to the polyfunctional nature of carbohydrates, the advancement of the chemistry and synthesis of these compounds has depended upon the development of specific blocking agents for hydroxyl groups and masked carbonyl groups. In the following sections are listed various protecting groups, and this section should be considered as an extension of the general properties of carbohydrates.

26.1.8.1 Glycoside formation

Since aldoses and ketoses normally exist as cyclic hemiacetals, they react with simple alcohols under acid catalysis to give full acetals in which the hemiacetal ring structures of the original sugar are retained, although not necessarily in the same proportions. This reaction, known as *Fischer glycosidation*,[87] is thermodynamically controlled, and consequently at equilibrium leads to six-membered pyranosides. However, although D-glucose exists mainly as the β-anomer (64%), methyl α-D-glucopyranoside (**13**) is the thermodynamically favoured glycoside and is present at equilibrium to the extent of 66%. The reason for this difference is that the anomeric effect, which favours the α-anomer (Section 26.1.3), is more pronounced in less-polar solvents such as methanol. Furthermore, although D-arabinose exists only to a minor extent (3%) in furanose forms, the methyl furanosides are markedly more abundant (29%) at equilibrium in acidified methanol (Table 1). The reason for this difference is obscure, but it may be due to the operation of an anomeric effect in the furanosides in the less-polar methanol which, as in the case of D-glucose, is less significant in water. The β-furanoside, which predominates (22%), could exist in the E^0 conformation (**94**), or related twist, in which the 1-OCH_3 would be

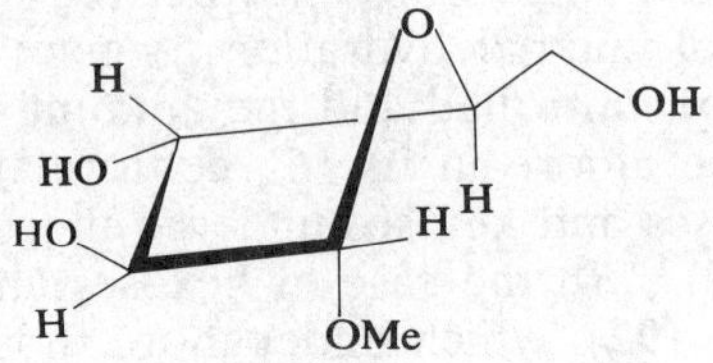

(94)

quasi-axial (favourable anomeric effect) and the bulky hydroxymethyl group would be *quasi*-equatorial. In addition, there would be no vicinal-*cis* interactions in the β-furanoside. On the other hand, the α-furanoside (7%) could not adopt such a favourable conformation and retain the 1-OCH_3 axial. A similar argument applies to methyl galactosides which contain 22% furanosides at equilibrium with a $\beta:\alpha$ ratio of 2.7:1.[88]

Detailed study of the Fischer glycosidation has revealed a more complex situation. The free sugar reacts relatively rapidly to give furanosides, the concentration of which then slowly declines as the pyranosides become the predominant species. In the case of D-xylose, the relative rates at 25 °C for the two reactions D-xylose→furanosides→pyranosides are 60:1, respectively;[89] consequently furanosides can be readily isolated by conducting reactions under mild conditions (*e.g. ca.* 0.01% methanolic hydrogen chloride at ambient temperatures) and stopping the glycosidation before the pyranosides accumulate. In this way D-galactose affords 53% of the β-furanoside and 14% of the α-anomer after a reaction period of 6 h.

The precise details of the mechanism of glycosidation are not known with certainty but largely hinge on whether the transition states are open-chain or cyclic (Scheme 11). The most likely transition states are the cyclic oxycarbonium cations (**95**) and (**97**), since these are well established as the transition states in the hydrolysis of glycosides.[89] It is probable that the furanoses can form the necessary oxycarbonium cations (**95**) more readily than pyranoses because the former can accommodate the double bond more satisfactorily, thereby increasing the stabilization of the C-1 carbonium ion. However, if the acyclic hemiacetal (**96**) or the acyclic oxycarbonium ion (**98**) were involved, they would be expected to undergo preferential ring-closure to the five-membered ring, a more kinetically favoured process, than closure to the six-membered ring.[89] The ring-expansion of furanosides into pyranosides must involve acyclic oxycarbonium ions, such as (**98**), whereas it is well established that anomerization of methyl pyranosides in acidified CD_3OH proceeds *via* the cyclic oxycarbonium ion (**97**) since the aglycone in the initially formed products arises from solvent.

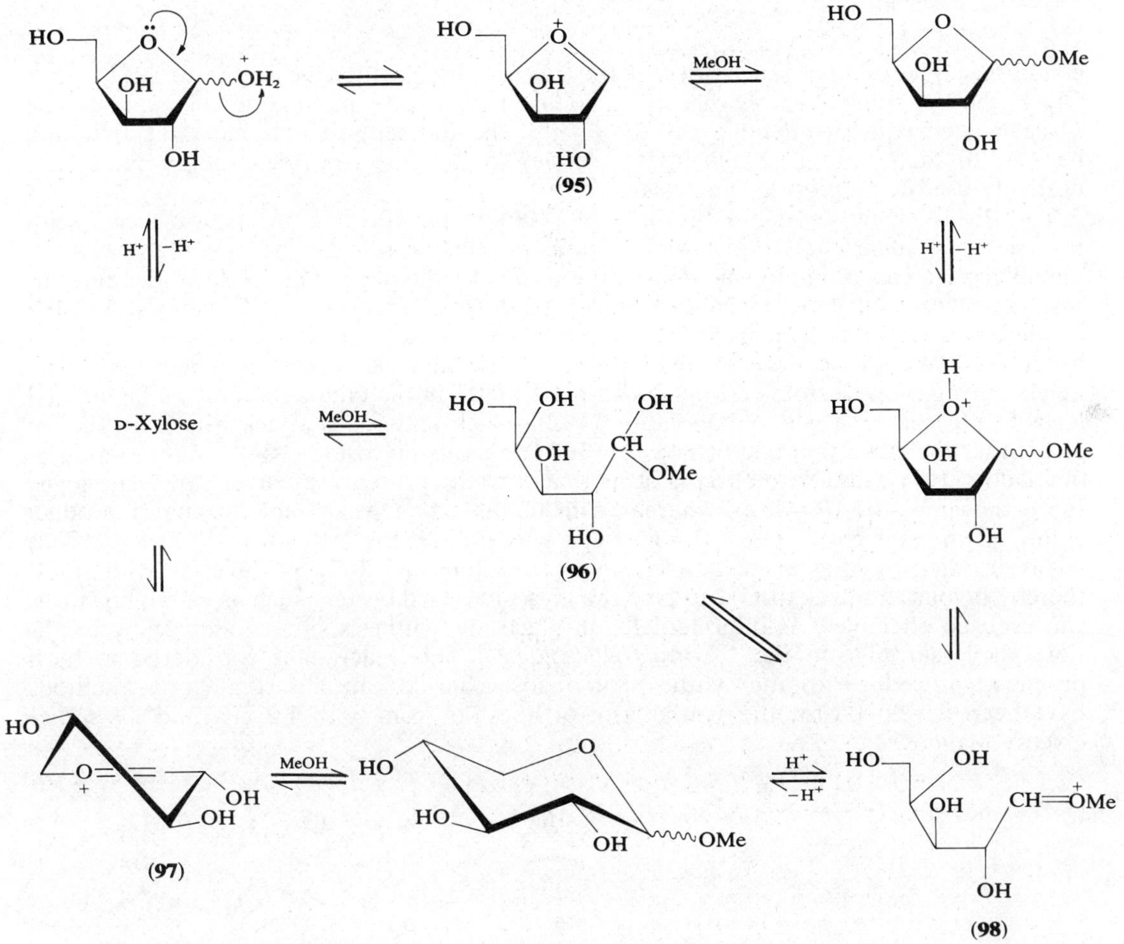

SCHEME 11

The Fischer glycosidation procedure is suitable only for simple alcohols which are volatile and, since it is necessary to use a large excess, they should also be preferably abundantly available. The synthesis of those glycosides that are not available directly is usually accomplished by the *Koenigs–Knorr* synthesis in which the appropriate glycosyl halide is reacted with the alcohol in the presence of an acid acceptor such as silver oxide.[90]

Glycosyl halides are prepared by reaction of the appropriate hydrogen halide with the per-*O*-acylaldose, in which the 1-*O*-acyl substituent is replaced by halogen. Thus, reaction of either α- or β-D-glucopyranose penta-acetate (**99**) with hydrogen bromide in acetic acid gives a high yield of 2,3,4,6-tetra-*O*-acetyl-α-D-glucopyranosyl bromide (**100**). Under these conditions the α-anomer always predominates because of the powerful anomeric effect (Section 26.1.3), although the β-anomers can be prepared under special

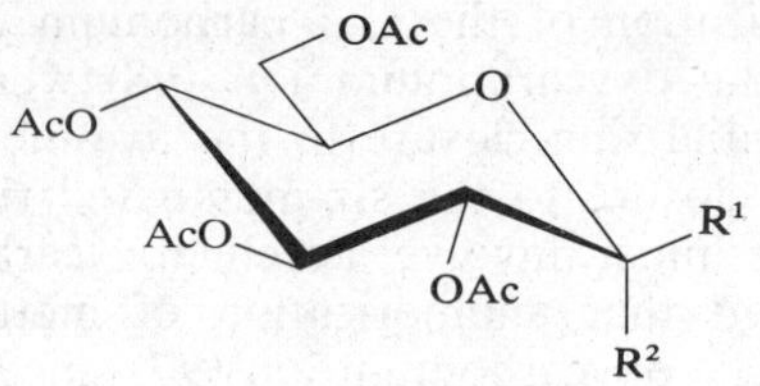

(**99**) $R^1 = OAc$, $R^2 = H$
(**100**) $R^1 = H$, $R^2 = Br$

conditions. The relative reactivities of the glycosyl halides decline in the order I > Br > Cl > F, the last being so stable that the halide is unaffected by methoxide anion and can be *O*-deacylated without affecting the fluoride.[91] The iodides are extremely reactive and because of their instability difficult to handle, so that the chlorides and bromides are normally used for glycoside synthesis.

The displacement of the α-bromide in (**100**) by an alcohol or phenol occurs with inversion of configuration to give the 1,2-*trans* product despite the fact that the reaction is unimolecular. This is due to shielding of the α-side by the departing halide anion, and this results is always observed in halides having a 1,2-*cis* configuration.[92] In cases where the halide has a 1,2-*trans* configuration, such as 2,3,4,6-tetra-*O*-acetyl-α-D-mannopyranosyl bromide (**101**), the course of the reaction may vary according to conditions. The neighbouring *trans*-acetoxyl group at C-2 participates in the elimination of the halide and leads to a cyclic 1,2-acyloxonium cation (**102**), which undergoes attack by the alcohol to give either the corresponding orthoester (**103**) or the α-glycoside (**104**). Thus reaction of the halide (**101**) with a large excess of methanol in the presence of silver carbonate gives the orthoester (**103**, R = Me), whereas with diethyl ether as diluent the major product is the 1,2-*trans*-glycoside (**104**; R = Me) and very little of the orthoester is formed.[93] This has been interpreted as protection of the acetoxonium ion (**102**) by solvation with ether, thereby favouring attack at C-1 to give the glycoside. Orthoesters such as (**103**), however, can be used effectively as intermediates in glycoside synthesis since under acidic conditions they rearrange to the 1,2-*trans*-glycosides.[94] This reaction is considered to be a preferred procedure for the synthesis of oligosaccharides. In this connection, methods have been developed for the synthesis of orthoesters from both 1,2-*cis*- and 1,2-*trans*-glycosyl halides.[95]

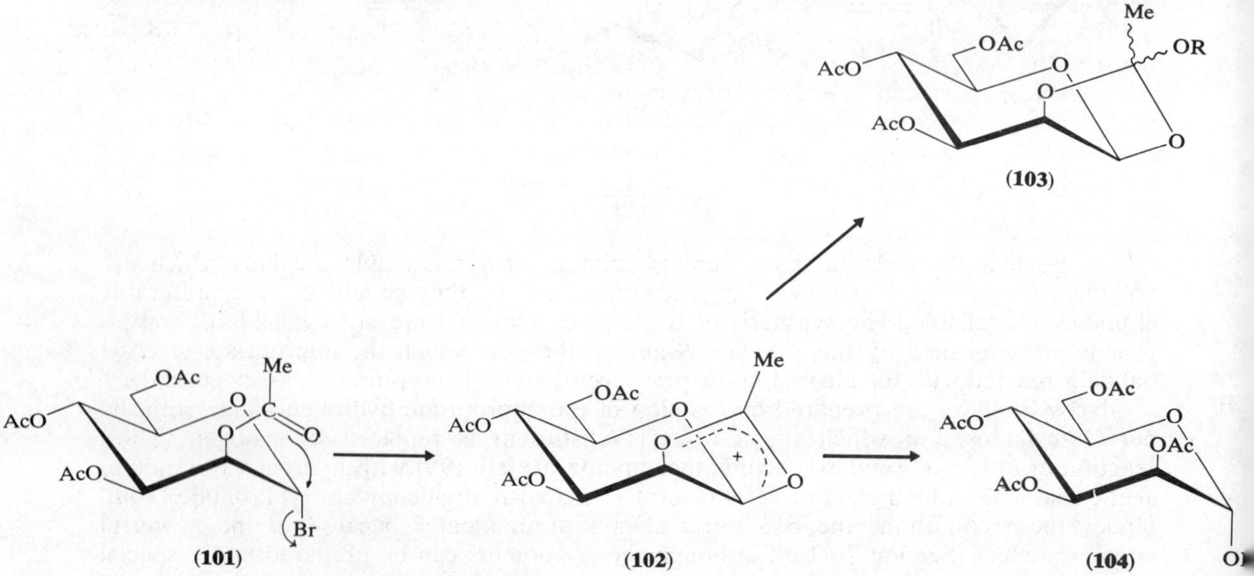

For many years, efficient syntheses of 1,2-*cis*-glycosides were elusive and represented a major challenge. However, as a result of recent work, the establishment of these α-linkages to give α-D-glucopyranosides, α-D-galactopyranosides, *etc.* is feasible on a routine basis. Providing that a non-participating group is present at C-2, *i.e.* benzyloxy, *etc.*, the β-D-glycopyranosyl halide affords a good yield of the α-glycoside when allowed to react with an alcohol, silver perchlorate and *sym*-collidine in diethyl ether as solvent.[96] Since the β-glycopyranosyl chlorides are less readily available than the α-anomers, Lemieux has devised a more convenient method in which the α-anomer may be used as starting material and anomerized *in situ* to the β-anomer. Thus reaction of the appropriate α-D-glycopyranosyl bromide with the alcohol is conducted in the presence of a large excess of tetraethylammonium bromide, which ensures that the α→β anomerization is more rapid than the displacement of the less reactive α-bromide by the alcohol. Providing the reaction is conducted in the absence of metal ions or highly polar solvents, a high degree of selectivity is obtained. The reaction is usually carried out in *N,N*-dimethylformamide with a hindered base, such as ethyldi-isopropylamine, as acid acceptor.[97] This latter method has been used in the elegant synthesis of several oligosaccharides which form part of the antigenic determinant of various blood group substances, which were used in the preparation of semisynthetic antigens.[98]

Phenolic glycosides[99] are prepared either by the action of the phenoxide anion on the glycosyl halide, which usually gives the 1,2-*trans* product when a participating group is present at C-2, or by fusion of the fully acetylated aldose with the phenol in the presence of an acid catalyst. The latter process usually gives a mixture of α,β-glycosides. Phenolic glycosides are of interest since they occur in a wide variety of natural products and are also used for the assay of several enzymes. For example, 4-nitrophenyl β-D-glucopyranoside is hydrolysed by β-glucosidase with the release of *p*-nitrophenol which can be monitored by its absorption at *ca.* 400 nm after basification; the extent of phenol liberation is directly related to the amount of enzyme present. The 7-methyl umbelliferyl glycosides are superior for this purpose, since the released aglycone can be determined at much lower concentrations by its fluorescence.[100]

In certain cases, phenolic glucuronides are available by feeding the appropriate phenol to small mammals (*e.g.* rabbits) and harvesting the glucuronide from their urine. For example, various nitrophenyl glucuronides are readily available by this technique.[101]

The mechanism of acid hydrolysis of glycopyranosides is in essence the reverse of their formation by the Fischer procedure, involving protonation of the glycosidic oxygen, fission of the glycosidic C—O linkage to give the oxycarbonium cation (**97**) which is then attacked by water. However, other pathways involving acyclic intermediates cannot be completely discounted and, in the case of t-butyl glycosides, fission of the oxygen–aglycone linkage is the preferred route because of the greater stability of the t-butyl carbonium ion and leads to a large (>500) enhancement in rate of hydrolysis. Predictably, electronic effects, particularly those arising from the substituents at C-2 and C-5, have a marked effect on rates of hydrolysis. Electron-withdrawing groups at C-2 and C-5 result in a reduction of the rate of hydrolysis. Hence, methyl pyranosides (**105**) are hydrolysed in the order Y,X = H,H ≫ H,OH > H,NHAc > H,$\overset{+}{N}H_3$ > Cl,Cl > F,F and R = H > CH_3 > CH_2OH > CO_2H (Scheme 12). The reason for this is two-fold: electron-withdrawing groups at these positions tend to decrease the amount of the conjugate acid (**106**) formed

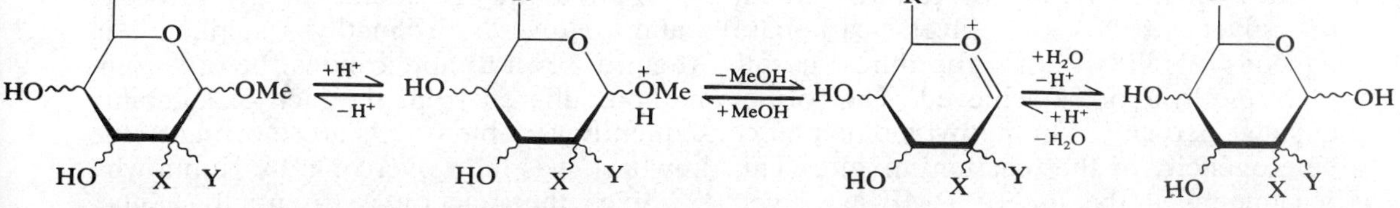

(**105**) (**106**) (**107**)

SCHEME 12

and to destabilize the oxycarbonium ion (**107**)[89,102] Methyl 2-acetamido-2-deoxy-β-D-glucopyranoside, in which the *N*-acetyl group is *trans* to the aglycone, is hydrolysed abnormally rapidly due to anchimeric assistance of the acetamido-group in the departure of the aglycone residue. In general, β-glycopyranosides are more rapidly hydrolysed than their α-anomers owing to a greater degree of protonation at the sterically more accessible equatorial aglycone. In addition, the protonated aglycone exerts a powerful reverse-anomeric effect (Section 26.1.3) which stabilizes the equatorial anomer.

In contrast to the pyranosides, where the entropies of hydrolysis are all strongly positive, the furanosides are not only more readily hydrolysed but this occurs with negative entropies. This is probably indicative of different mechanisms of hydrolysis and furanosides are believed to react either by a ring-opening mechanism or by direct nucleophilic displacement of the protonated aglycone by water.[89]

Phenolic and enolic glycosides are also sensitive to alkali,[103] and in some cases 1,6-anhydrides are formed in high yields. Thus aryl β-glucopyranosides (**108**) readily afford 1,6-anhydro-β-D-glucose (**110**) whereas the α-anomers do not. The observation that the 2-*O*-methyl derivative of (**108**) does not form the 1,6-anhydride, whereas the 3-*O*-methyl derivative does, suggests a reaction pathway proceeding *via* the 1,2-anhydride (**109**). There appear to be several modes of reaction of phenolic glycosides with base,

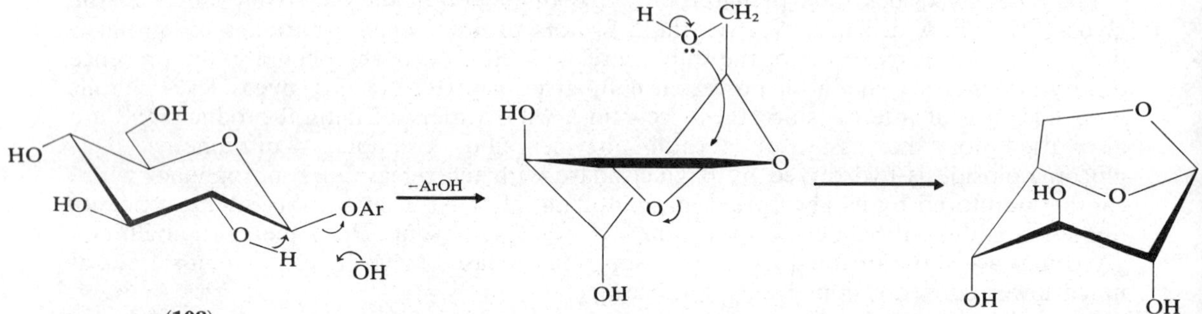

(**108**) (**109**) (**110**)

which vary with sugar configuration. For example, phenyl α-D-galactopyranoside gives 85% of the 1,6-anhydride after prolonged reaction, in spite of a 1,2-*cis* arrangement, probably by direct displacement of the aglycone by the ionized 6-OH. Furthermore, the β-mannopyranoside, also with a 1,2-*cis* arrangement, gives the anhydride in 57% yield, probably *via* the 1,4-anhydride. The mode of reaction is possibly a reflection of the relative configurations and acidities of the various hydroxyl groups.

26.1.8.2 Ethers including anhydrides

Methyl ethers were a major feature in classical carbohydrate chemistry since they were used in the determination of the ring structures of sugars (Section 26.1.2) and in the structural elucidation of many oligo- and poly-saccharides. Their main advantage as blocking groups is their stability under a variety of conditions, but for synthetic objectives this is a disadvantage since there is a lack of reliable deblocking procedures.

Classical methylating procedures are due to *Purdie* (methyl iodide in the presence of silver oxide or silver carbonate) and *Haworth* (dimethyl sulphate in aqueous alkali). Both procedures usually require several applications before complete methylation is achieved. The former method suffered from the lack of solubility of sugar derivatives in methyl iodide and consequently variable results arose owing to the heterogeneity of the reaction mixture. This drawback was later overcome by Kuhn, who recommended the use of DMF as solvent.[104] Since the reaction is essentially a nucleophilic displacement reaction, with the alcohol as nucleophilic species, the choice was fortuitous since this solvent is known to have a profound effect upon such reactions.

Hence, methylations conducted in DMF are usually accomplished rapidly and efficiently, without the need for repeated applications, and barium or strontium oxides, or hydroxides, can be used in place of the expensive silver salts. A particularly effective alternative procedure is the prior formation of the alkoxide using sodium hydride in either DMF or DMSO, followed by reaction at ambient temperature with methyl iodide or dimethyl sulphate.[105] All these procedures employ basic conditions under which acyl groups may undergo migration or even cleavage. However, diazomethane in the presence of boron trifluoride permits methylation to be accomplished under conditions where *O*-acyl migration does not occur.[106] On the other hand, methylation of methyl 2,3,4-tri-*O*-acetyl-α-D-glucopyranoside by the Purdie procedure gives the triacetate of methyl 2-*O*-methyl-α-D-glucopyranoside owing to acyl migration.[107]

The removal of methyl and presumably other alkyl groups is best accomplished by reaction with boron trichloride, although this is far from selective and in addition cleaves acetal and glycoside bonds. Many other reagents have been described for *O*-demethylation, but most have disadvantages.

A greater degree of lability can be built into alkyl ether groups by placing suitable substituents at the α-carbon. For example, benzyl ethers,[108] which are prepared similarly to methyl ethers using benzyl halides, can be removed by hydrogenolysis and are consequently more widely used than methyl ethers as blocking groups. Allyl ethers, which are also conveniently introduced *via* allyl bromide, are stable towards acids and can be removed by base-induced isomerization to the prop-1-enyl ether and this, being a vinyl ether, is readily removed by mild acid.[109] The triphenylmethyl (trityl) ether has a special place in carbohydrate chemistry. As a result of its bulk, it can usually be introduced selectively at primary positions by reaction of trityl chloride with the carbohydrate in the presence of pyridine. Secondary hydroxyls are only substituted with difficulty under forcing conditions. Furthermore, because of the stability of the triphenylmethyl carbonium ion, cleavage of the O–trityl linkage occurs readily under mildly acidic conditions, so that the trityl group can be removed in the presence of ester, ether, and glycosidic linkages and, if hydrogenolysis is used, in the presence of acetals also.

Trimethylsilyl ethers are available by reaction of sugars with trimethylsilyl chloride or hexamethyldisilazane [$(Me_3Si)_2NH$] in pyridine. The reaction is rapid and quantitative and the ethers are suitable for examination by g.l.c.[16] or for vacuum distillation. Since the etherification is rapid compared with the rate of mutarotation, analysis of a mutarotated mixture or a mixture of sugars can be accomplished by trimethylsilylation followed by g.l.c. (Section 26.1.2). These ethers are, however, hydrolytically unstable and are not suitable as blocking groups in synthesis. By contrast, the t-butyldimethylsilyl ethers are hydrolytically stable towards acid and alkali and can, with ease, be introduced selectively at primary positions. Thus reaction of sucrose with t-butyldimethylsilyl chloride in pyridine gave the 1′,6,6′-triether in good yield. Removal of the ether group is accomplished by treatment with tetrabutylammonium fluoride.[110]

Tetrahydropyran-2-yl 'ethers', and related blocking groups, provide a useful base-stable, acid-sensitive group which has been employed fairly widely. It is introduced by addition of the alcohol across the double bond of dihydropyran, but since the reaction generates a new chiral centre, a mixture of diastereoisomers is formed. This drawback has been overcome[111] by the use of 4-methoxydihydro-1*H*-pyran and related compounds which react with alcohols without the generation of a new chiral centre. These ethers are misnomers, since they are strictly mixed acetals.

Anhydro derivatives of aldoses are intramolecular ethers, and are important compounds in synthesis.[112] Some anhydrides are formed merely by heating aldoses and alditols, and alditols show a marked tendency to dehydrate to form anhydrides with the stable tetrahydro-furan or -pyran rings. For example, D-glucitol, D-mannitol, and L-iditol, and presumably their enantiomorphs, irreversibly form 1,4:3,6-dianhydrides (**111**) on heating with acid.[113] Aldoses show a lesser tendency to form anhydrides upon heating, because imposition of a fused ring on to the cyclic aldose is, in most cases, kinetically and thermodynamically unfavourable. However, hexoses and higher sugars which display conformational instability form 1,6-anhydrides upon heating in acid solution.[114] The compounds are intramolecular glycosides and the dehydration is reversible. The product distribution is therefore under thermodynamic control, so that the proportion of anhydride depends upon the number of equatorial groups present in the anhydride. Thus, D-glucose equilibrates with only *ca.* 0.2% of its 1,6-anhydride whereas D-idose forms about 86% of its 1,6-anhydride (**112**), in which all three hydroxyl groups are equatorial.

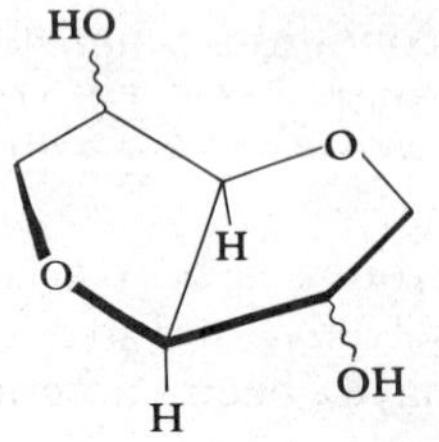

(**111**)

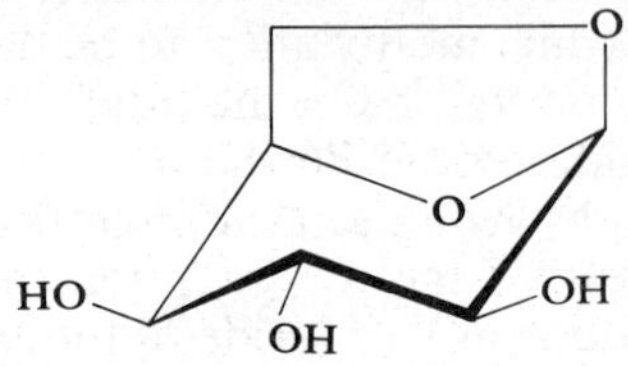

(**112**)

The most important destabilizing interaction for 1,6-anhydride formation appears to be the 1,3-diaxial interaction between an axial 3-OH and the anhydro bridge, since 1,6-anhydro-D-talose, in which the 3-hydroxyl is the only one which is axial, is formed to the extent of only 3% at equilibrium.[115] 1,6-Anhydrides, particularly the *gluco-*, *galacto-*, and *manno*-isomers, are prepared either by alkaline hydrolysis of the appropriate phenyl glycoside (Section 26.1.8.1), or from the thermal depolymerization of an appropriate polysaccharide.[114]

1,2-Anhydrides have been made, but are very reactive; 1,3-anhydropyranoses are unknown, but 1,4-anhydrides are available from 4-*O*-sulphonyl esters. For example, 1-*O*-acetyl-2,3,6-tri-*O*-benzoyl-4-*O*-mesyl-*α*-D-glucopyranose (**113**) gives the 1,4-anhydride (**116**) on treatment with nucleophilic anions (*e.g.* N_3^-, OBz^-) in DMF.[116] The reaction must proceed by initial cleavage of the 1-acetyl group to generate the 1-alkoxide (**114**), which anomerizes to (**115**) and displaces the 4-sulphonyloxy group to give the anhydride (**116**).

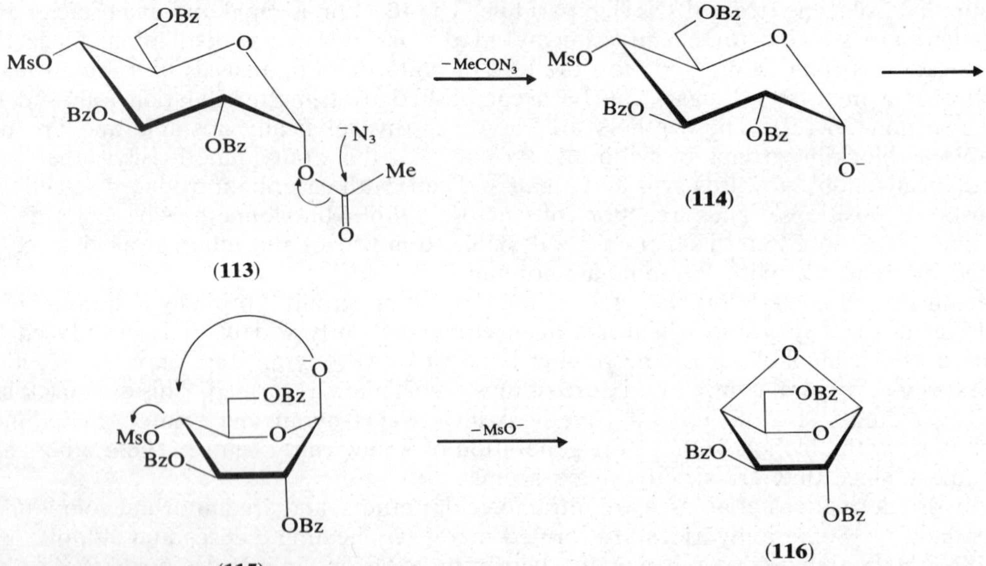

(**113**) (**114**) (**115**) (**116**)

The reaction appears to be of general application, and gives good yields, and is independent of the relative configuration of the substituents at C-1 and C-4.

Anhydrides in which the reducing group is not involved are formed from the appropriate sulphonate ester, which undergoes intramolecular nucleophilic displacement by suitably placed hydroxyl groups to give three-, four-, five-, and perhaps six-membered anhydrides with ease, in spite of the apparent steric and angular strain required for their formation. For example, methyl 6-*O*-tosyl-α-D-glucopyranoside (**117**) gives a high yield of the 3,6-anhydride (**118**) under basic conditions in spite of the unfavourable change from the $^4C_1$ to the $^1C_4$ conformation required for reaction.[117] The tetrahydrofuran ring formed in this way is very stable and the process cannot usually be reversed. However, the inherent strain of the dioxa[1,2,3]bicyclo-octane ring system causes a rapid acid-catalysed rearrangement of (**118**) to the more favourable furanose (**120**) with a dioxa[0,3,3]bicyclo-octane ring system, and there is evidence to suggest that the anomeric configuration is unaltered in the rearrangement, suggesting a concerted process as in (**119**).[118]

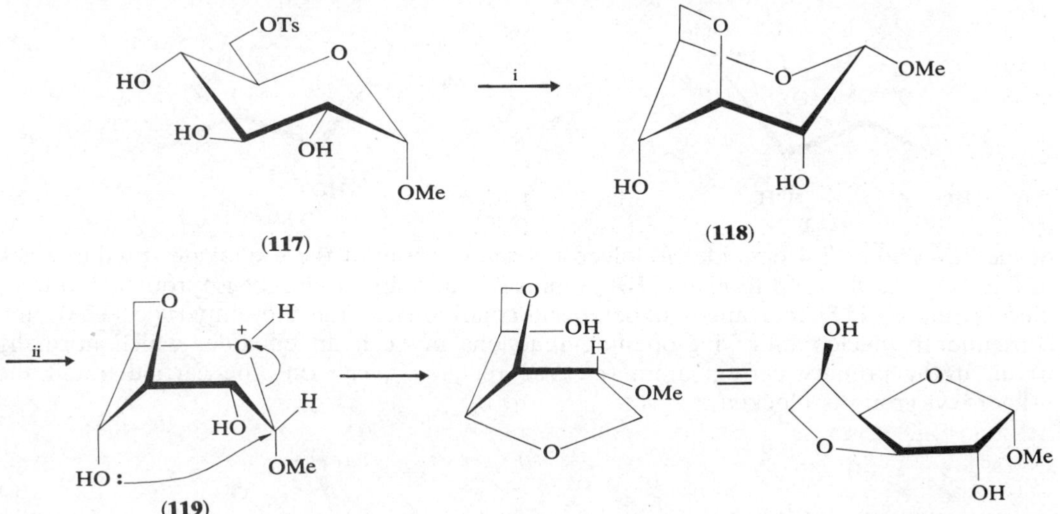

(**117**) (**118**) (**119**) (**120**)

i, base; ii, acid.

Oxiran (epoxide) derivatives of carbohydrates are extremely useful synthetic intermediates, being easily prepared by the action of base on vicinal *trans*-diols in which one or both of the hydroxyl groups bears a sulphonyl ester group.[119] For example, selective tosylation of methyl 4,6-*O*-benzylidene-α-D-glucopyranoside gives the 2-ester (**36**) in high yield, which in turn is converted almost quantitatively into the 2,3-epoxide (**38**) with the *manno*-configuration. The 3-tosylate (**37**) or the 2,3-ditosylate afford the 2,3-epoxide (**39**) with the *allo*-configuration. The value of epoxides is that the oxiran ring readily undergoes nucleophilic opening, leading to a variety of sugar derivatives. The reactions of epoxides in which the conformation of the molecule is rigid occur stereoselectively, since the incoming and outgoing groups must achieve in the transition state a co-planar relationship approximating to the S_N2 transition state (Furst–Plattner rule).[119] This requirement leads therefore to the 2,3-*trans*-diaxial altropyranosides (**40**) and (**41**), where $X = N_3$, NR_2, OR, SR, CN, *etc.*

Prediction of the mode of ring-opening of epoxides with non-rigid structures is often difficult because the assessment of the relative stabilities of the transition states arising from the various ground state conformations is not always possible with accuracy. For example, methyl 2,3-anhydro-6-deoxy-α-L-talopyranoside reacts with methoxide anion *via* the HC_0^5 conformation (**121**) to give the L-idoside (**123**), but its 4-*O*-methyl ether must react *via* the alternative half-chair conformation (**122**), since it gives the galactoside (**124**). It has been suggested that the 4-hydroxy-group may be hydrogen-bonded to the ring oxygen, thus stabilizing the HC_0^5 conformation (**121**), whereas in its 4-*O*-methyl derivative, the larger group shows a preference for an equatorial situation.[120]

Epoxides which contain a suitably placed free hydroxyl group may undergo rearrangement of the anhydro ring with some ease. Thus attempted formation of an epoxide from methyl 3-*O*-tosyl-α-D-glucopyranoside gave the 3,6-anhydride (**118**) by initial formation

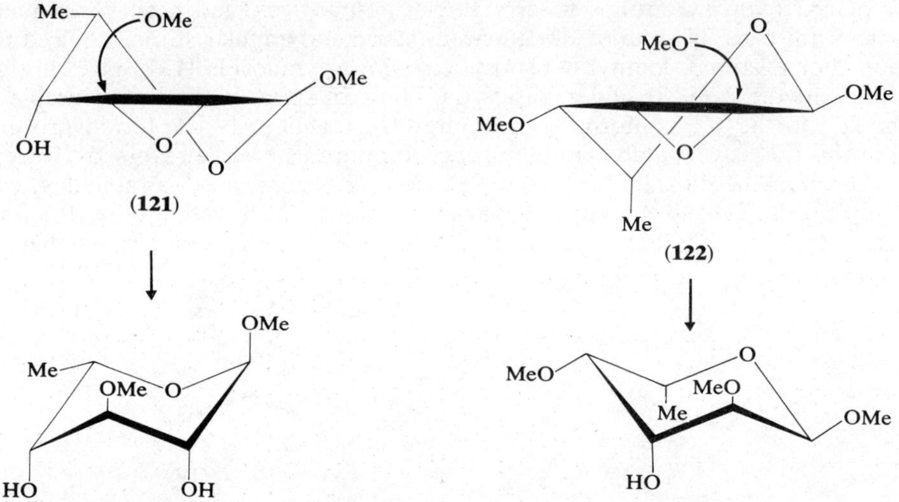

(121) (122) (123) (124)

of the 2,3- and/or 3,4-epoxides followed by ring opening of these epoxides by the attack of the 6-hydroxyl group to give (**118**). Similarly, 5,6-anhydro-1,2-*O*-isopropylidene-*α*-D-glucofuranose (**125**) rearranges under basic conditions to the 3,6-anhydride (**126**), and consequently nucleophilic ring-opening reactions of such an epoxide, which normally occur at the primary carbon atom (*cf.* S_N2 reactions), can only be carried out if the 3-hydroxyl group is blocked.

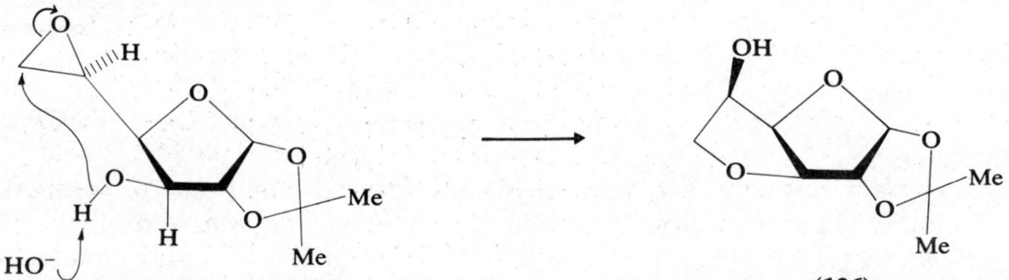

(125) (126)

The stereospecificity of epoxide ring-opening can be completely altered by participation of an *O*- or *N*-acyl group. Thus methyl 3,4-anhydro-6-*O*-trityl-α-D-altropyranoside upon acid-catalysed ring-opening with aqueous acetic acid underwent predominant *trans*-diaxial ring-opening of the favoured HC_5^0 (see **127**). On the other hand, its 2-*O*-acetyl derivative underwent a more rapid reaction by participation of the acetoxy-group to give exclusively the mannoside (**128**).[121] *N*-Acyl groups are even more powerfully participating and

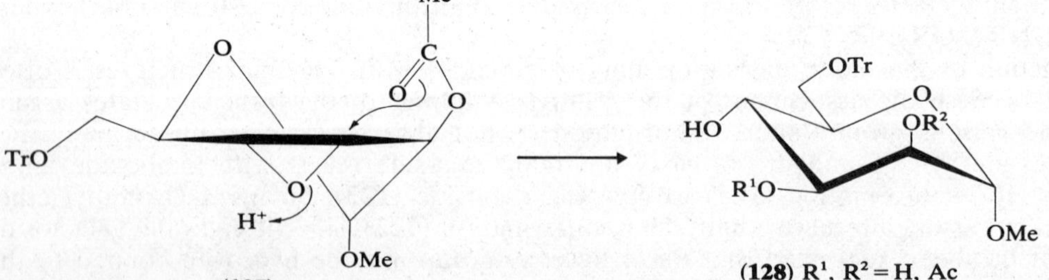

(127) (128) R¹, R² = H, Ac

attempts to prepare methyl 3.4-anhydro-2-benzamido-2-deoxy-*α*-D-galactoside (**129**) from a 4-sulphonate gave an almost quantitative yield of an oxazoline (**130**) with the *gulo*-configuration, which arose from the intermediate epoxide by attack of the *N*-benzoyl group.[122]

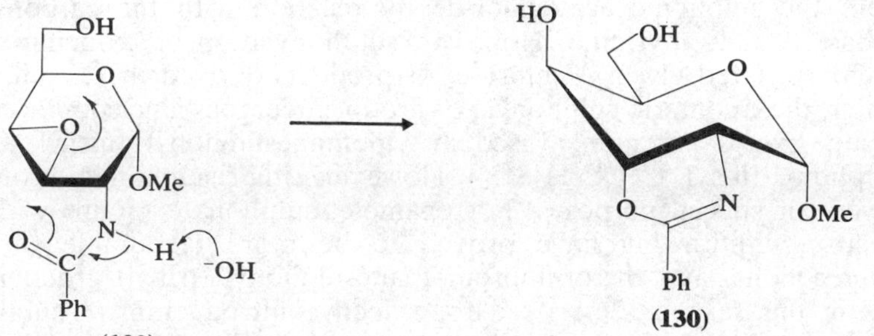

(129) (130)

26.1.8.3 Esters

Carboxylic esters of carbohydrates were initially used for characterization purposes since they were often highly crystalline, particularly acetates and benzoates. They have also found wide applicability as blocking groups and are easily introduced by reaction with the appropriate acid chloride or acid anhydride in pyridine, or other catalyst. Acetylation of D-glucose gives one of two products depending on the catalyst. Using acetic anhydride–sulphuric acid the thermodynamically more stable α-anomer is formed since under acid conditions anomerization of the α- and β-penta-acetates occurs. However, in cold pyridine, acylation is faster than mutarotation of unreacted sugar and consequently the proportion of the two anomers reflects the composition of the glucose employed. On the other hand, in hot acetic anhydride with sodium acetate as catalyst, mutarotation of the sugar is very rapid compared with acetylation and since the equatorial-anomer is more rapidly acetylated than the axial-anomer, the final product contains mainly the β-penta-acetate.

The acetate groups are stable to mildly acidic conditions and removed under basic conditions, usually by ester exchange using methanolic sodium methoxide. However, they do show a tendency to migrate to free hydroxyl groups under mildly basic conditions, *e.g.* in methylations using silver oxide. Benzoates, on the other hand, are less labile towards mildly basic conditions than acetates, and show little tendency to migrate. They will also withstand acid conditions better than the corresponding acetates. This behaviour must be due to the decreased electrophilic character of the ester carbonyl group in benzoates owing to resonance with the benzene ring. As would be expected, chloroacetyl and trifluoroacetyl groups are more labile, and hence very easily and selectively removed than *O*-acetyl groups.

Selective introduction of acyl and sulphonyl groups into carbohydrates is feasible in many cases, and the selectivity of the hydroxyl groups of a carbohydrate to different acylating agents does vary.[123] Thus, selective monoacetylation of a series of methyl 3-acetamido-3,6-dideoxyglucopyranosides produced mainly 2-esters when acetyl chloride was used and mainly 4-esters when the anhydride was employed.[124] Selective benzoylation using benzoyl chloride in pyridine has been widely exploited and good yields have often been encountered of partially substituted compounds. For example, in the case of methyl α-D-glucopyranoside (**13**) the order of reaction of hydroxyls is $6>2>3>4$, whereas in the case of the α-mannopyranoside, it is $6>3>2>4$. However, the galactoside shows $6>3\approx2>4$.[125] There appears to be a relationship between reactivity of a hydroxyl group and the presence of a vicinal 'OR' group. For example, selective benzoylation or tosylation of methyl 4,6-*O*-benzylidene-α-D-glucopyranoside (**136**) with the appropriate acid chloride gives the 2-ester in each case in high yield because of the *cis*-OMe at C-1, whereas the β-anomer (**78**) shows little selectivity between the 2- and 3-hydroxyl groups.

In contrast to carboxylic esters, sulphonate esters[42,126] are more versatile because they undergo alkyl–oxygen fission rather than O—S fission, and are used as intermediates for the synthesis of a variety of carbohydrate derivatives (see Section 26.1.9). They are introduced *via* the sulphonic acid chloride by reaction with the carbohydrate in the presence of base, usually pyridine. However, sulphonylation of reducing sugars usually affords the sulphonylated glycosyl chloride, or products derived therefrom, owing to the ease with which the anomeric sulphonyloxy group undergoes nucleophilic displacement. The sulphonate groups commonly used are methanesulphonyl (mesyl, CH_3SO_2) and toluene-*p*-sulphonyl (tosyl, $CH_3C_6H_4SO_2$). However, other sulphonate groups are sometimes employed for special purposes. For example, sulphonate groups with substituents that are electron-withdrawing have proved to be more favourable for nucleophilic displacement reactions, and the trifluoromethanesulphonyl (triflyl) group is particularly advantageous in this respect. The need for selective introduction of sulphonyl groups, particularly at primary postitions, for subsequent nucleophilic displacement has led to the use of 'bulky' sulphonyl chlorides such as mesitylenesulphonyl chloride and *sym*-tri-isopropylbenzenesulphonyl chloride. Thus, reaction of sucrose, which contains three primary hydroxyl groups, with tosyl chloride afforded the 1′,6,6′-tritosylate in low yield after extensive chromatography. However, the use of mesitylenesulphonyl chloride gave the 1′,6,6′-trisulphonate crystalline in good yield without the need for chromatography. Mesitylenesulphonylation at secondary hydroxyl groups is also very selective, because the steric hindrance is such that once one group is established, reaction at vicinal hydroxyl groups is effectively blocked. Hence, treatment of methyl 4,6-*O*-benzylidene-α-D-glucopyranoside (**136**) with even an excess of mesitylenesulphonyl chloride gives only the 2-ester.[127] Care must be exercised in the preparation of sulphonate esters because the primary sulphonyloxy groups undergo displacement by chloride and, possibly, pyridine (to give a pyridinium salt). This latter reaction takes place with great ease in the case of the trifluoromethanesulphonates[128] and has been postulated in other sulphonylations to be the cause of lower than expected yields, since the derived pyridinium salts would be lost in the water extracts during the 'work-up' of reaction mixtures.

The use of sulphonyl esters as blocking groups is limited by their reactivity towards S_N2 displacements, and the difficulty in achieving the necessary alkyl–oxygen fission for their removal.[126] The attempted removal of a sulphonate ester by alkali in the presence of a suitably placed hydroxyl group (actual or potential) results in anhydride formation, but if such hydroxyl groups are protected by a base-stable group, such as an ether group, then oxygen–sulphur fission will occur smoothly under basic conditions to give the alcohol. Sulphonates are also reconverted back to the alcohol by treatment with either Raney nickel or sodium amalgam, but again intramolecular anhydride formation may interfere. Two efficient but mild methods of effecting alkyl–oxygen fission have been reported: firstly, photolytic cleavage of tosylates in the presence of a stoichiometric amount of sodium methoxide results in conversion to the alcohol without anhydride formation;[129] and, secondly, reaction with sodium naphthalene[130] is reported to cause selective oxygen–sulphur cleavage without side reactions.

The cleavage of sulphonate esters with lithium aluminium hydride occurs by either nucleophilic displacement of the sulphonyloxy group by 'hydride anion' to give deoxy-derivatives (Section 26.1.9.3) or by attack at sulphur to give the original alcohol. The former mode of action is observed only in the case of primary sulphonyloxy-groups, but reaction conditions must be carefully controlled. With secondary sulphonates, attack always occurs, albeit slowly, at sulphur unless anhydride formation is possible. If an epoxide is formed by intramolecular displacement, then this reacts further with the reducing agent to give the deoxy-compound.

26.1.8.4 Cyclic acetals

The polyhydroxylated nature of carbohydrates means that pairs of hydroxyl groups, if suitably placed, can form cyclic acetals with aldehydes and ketones in the presence of acid catalysts.[131] These cyclic acetals, being easily introduced under mild conditions, are stable to base and readily removed by mild acid hydrolysis. They are of the greatest value for the selective protection of pairs of hydroxyl groups (Scheme 13).

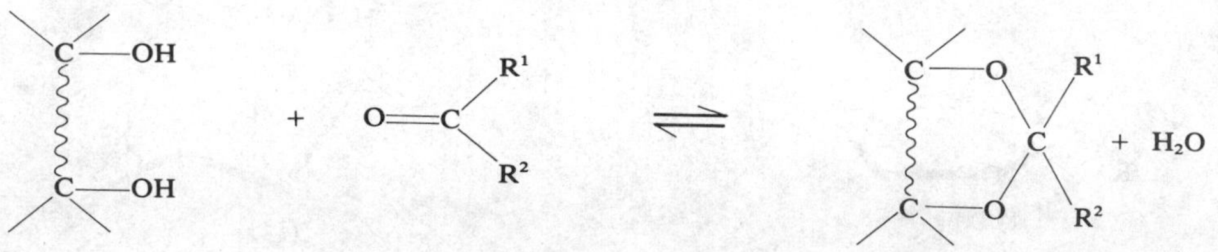

SCHEME 13

Acetaldehyde, benzaldehyde, and acetone, which give ethylidine, benzylidene, and isopropylidene derivatives respectively, are the most commonly used carbonyl compounds, but many other carbonyl compounds have been used. In most cases the carbonyl compound is also utilized as the solvent, and the most favoured catalysts are sulphuric acid (0.1–5%), hydrogen chloride (0.2–1%), and Lewis acids such as zinc chloride and boron trifluoride. In the case of isopropylidene formation from acetone, anhydrous copper sulphate has been incorporated to maintain anhydrous conditions. More recently the hydrogen form of cation exchange resins have proved to be effective acid catalysts in the presence of anhydrous calcium sulphate.

Since acetal formation takes place by a reversible process, akin to glycoside formation, the products are formed by thermodynamic control, and since an enormous excess of the carbonyl compound is normally present, the product is usually as highly substituted as is sterically feasible. Ketones normally show a preference for the formation of five-membered rings to give 1,3-dioxalans (**131**) and therefore react preferentially with vicinal diols. If the diol grouping is attached to a five- or six-membered ring, then a *cis* stereochemistry is required. On the other hand, aldehydes show a preference for six-membered 1,3-dioxan rings (**132**), providing that bridged ring systems are avoided. The reason for these preferences lies in the conformational stability of the acetal ring. When ketones form six-membered 1,3-dioxans, one of the alkyl groups must occupy an axial position which destabilizes it relative to the 1,3-dioxolan ring. Under acidic conditions, aldehydes condense with diols to give 1,3-dioxans with the large substituent equatorial, but in cases where they are restricted to the formation of 1,3-dioxolans, two diastereoisomers are encountered. Thus, D-glucose reacts with acetone in the presence of acid to give the 1,2:5,6-di-*O*-isopropylidene-α-D-glucofuranose (**133**) in high yield, whereas with benzaldehyde (with zinc chloride as catalyst) it gives mostly 4,6-*O*-benzylidene-α-D-glucopyranose (**135**) together with lesser amounts of the 1,2:4,6-diacetal. The 4,6-acetal (**135**) is particularly favoured in hexoses because it possesses a *trans*- or *cis*-decalin type of ring system. In the case of the 1,2:5,6-diacetal (**133**), the furanose modification is preferred because it can accommodate two acetal groups, whereas in the α-pyranose form only the 1,2-diol grouping would be suited for acetal formation. In the case of D-galactose and L-arabinose the α-pyranoses have two vicinal *cis*-diol groups and therefore these react with acetone to give 1,2:3,4-diacetals (**137**) and (**138**), respectively. In the case of galactose the 1,2:5,6-diacetal is also present as a minor product, but it is more readily available by inversion of configuration at C-4 of the glucofuranose (**133**) (Section 26.1.4.3). All other hexoses and pentoses react with acetone mainly in the furanose modification to give either the 1,2:5,6- or the 2,3:5,6-diacetal, although D-altrose gives a mixture of 1,2:5,6- (furanose) and 1,2:3,4- (pyranose) diacetals. In the case of D-xylose, the 1,2:3,5-diacetal (**139**) is formed, which possesses a six-membered acetal ring. The greater strain in this six-membered ring is revealed by the ease with which (**139**) is selectively hydrolysed to the 1,2-acetal. In some cases, incorporation of methanol into the reaction mixture results in the acetal of the glycoside. For example, L-gulose affords methyl 2,3:5,6-di-*O*-isopropylidene-β-L-gulofuranoside (**140**) in methanolic acetone, whereas in the absence of the alcohol it gives the 1,2:5,6-diacetal.[132]

Aldehydes condense with methyl hexopyranosides to give mainly the 4,6-acetal, and, if the 2,3-diol grouping is *cis*, then further reaction occurs to give the 2,3:4,6-diacetal, usually as a mixture of diastereoisomers differing in configuration at the 2,3-acetal carbon.

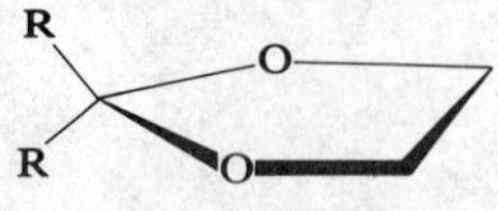

(131)

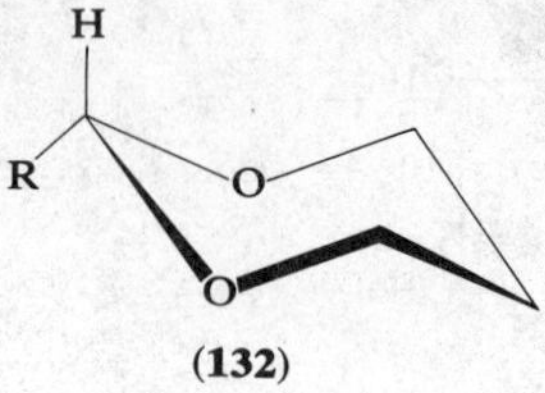

(132)

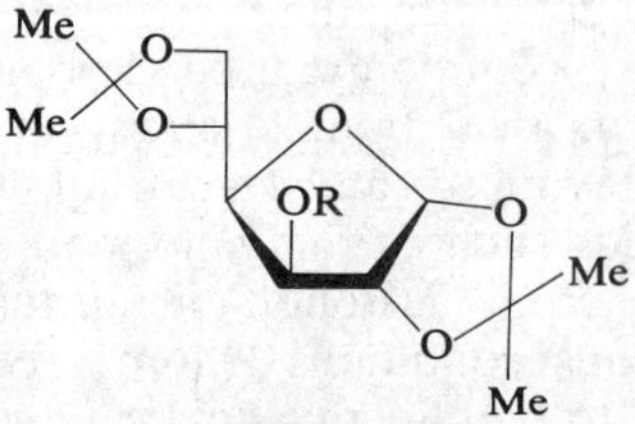

(133) R = H
(134) R = Ts

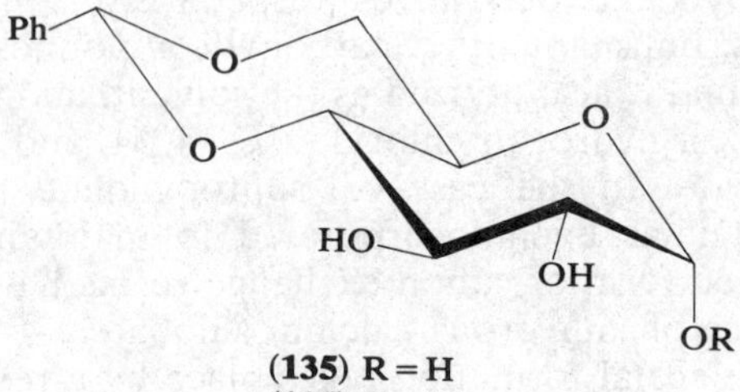

(135) R = H
(136) R = Me

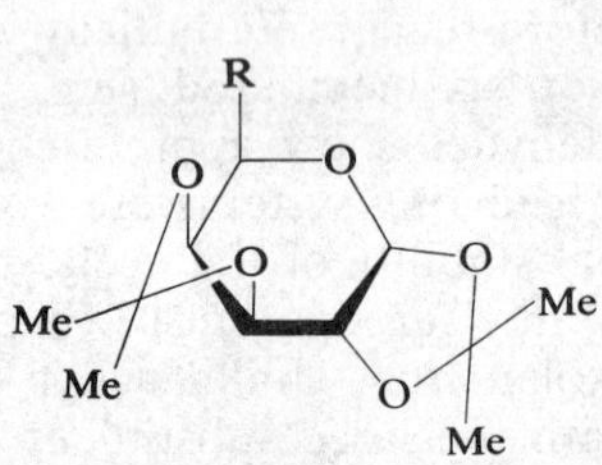

(137) R = CH_2OH
(138) R = H

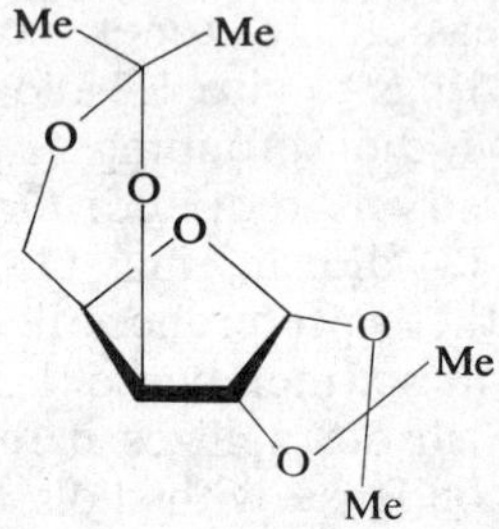

(139)

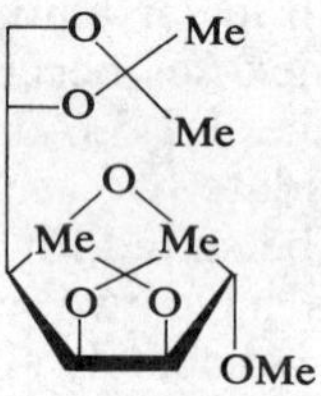

(140)

Thus, methyl α-D-glucopyranoside (**13**) gives the 4,6-acetal (**136**), whereas the mannoside gives the 2,3:4,6-diacetal. The 4,6-*O*-benzylidene derivatives, such as (**136**), are of interest because of their rigid conformations, and they have been used widely in synthesis and in studies of the stereochemical requirements for a variety of reactions occurring at the 2- and 3-positions. 4,6-*O*-Benzylidene derivatives have also been made under kinetic control using benzal bromide in the presence of a suitable base (potassium t-butoxide, pyridine, *etc.*), and under these conditions mixtures of the two diastereoisomers are produced, which have been separated.[133] As anticipated, the isomer with the axial phenyl group rapidly isomerizes to the equatorial isomer in mildly acidic media.

With ketones, pyranosides give five-membered acetals involving pairs of *cis*-diols where possible. Thus galactosides and arabinosides give 3,4-acetals, whereas mannosides and lyxosides give 2,3-acetals. Allopyranosides and ribopyranosides contain two pairs of *cis*-diol groupings and give mixtures of 2,3 and 3,4-acetals. Methyl α-D-glucopyranoside (**13**) contains no *cis*-diols and reacts only to a small extent to give a 4,6-acetal. However, under conditions of acetal exchange it is possible not only to place an isopropylidene group at the 4,6-positions, but also to span vicinal *trans*-diols attached to six-membered rings. Thus, methyl α-D-glucopyranoside (**13**) on treatment with 2,2-dimethoxypropane in DMF in the presence of a catalytic amount of toluene-*p*-sulphonic acid afforded mainly the 4,6-acetal together with about 2% of the 2,3:4,6-diacetal.[134] When 1,1-dimethoxycyclohexane was used, the corresponding diacetal was the major product.[135] The conditions of acetal exchange are particularly suited to the acetalation of acid-sensitive compounds. For example, prior to 1974 numerous attempts had been made to prepare cyclic acetals of sucrose, but the glycosidic bond would not withstand the acid environment necessary for acetalation. However, application of acetal exchange afforded a high yield of the 4,6-acetal along with smaller amounts of the 1′,2:4,6-diacetal, which is unique in that one acetal group spans the two sugar residues of the disaccharide.[136] Similar reactions take place when vinyl ethers are used, such as 2-methoxypropene. The reason for the differences in products formed by acetal exchange and direct acetalation are not clear, but acetal exchange often takes place with stoichiometric amounts of the acetal, whereas direct acetalation seems to require an enormous excess of the carbonyl compound. This suggests that acetal exchange may be kinetically controlled, whereas direct acetalation is under thermodynamic control.

Acetal formation with acyclic compounds such as alditols is governed by somewhat different factors, but reaction with aldehydes preferentially results in 1,3-dioxan rings and ketones show a preference for 1,3-dioxalan rings. However, when D-glucitol (**141**) reacts with acetaldehyde it has been shown that the first-formed kinetic product is the 2,3-acetal, which subsequently rearranges to the thermodynamically favoured 2,4-acetal (**142**). An inspection of the conformation (**142**) of the 2,4-acetal reveals that it is the most stable structure since the side chains C-1 and C-5,6 are held equatorially on the six-membered ring.[137] Further reaction of the 2,4-acetal (**142**) occurs to give the 1,3:2,4-diacetal (**143**) and subsequently the 1,3:2,4:5,6-triacetal (**144**).

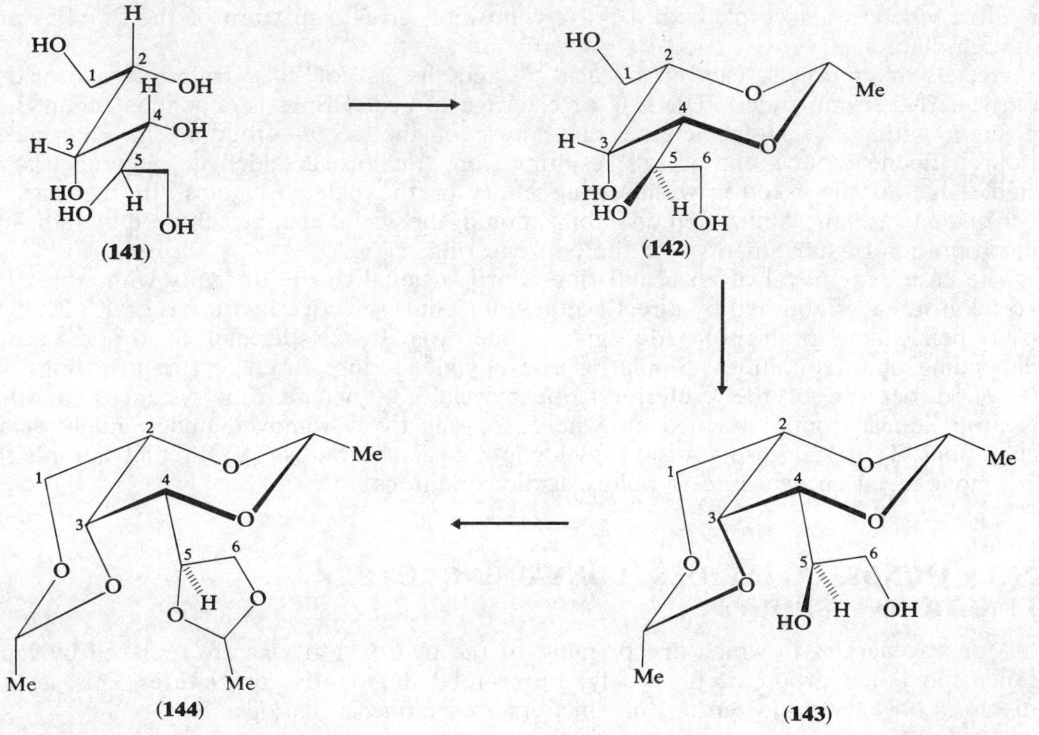

These results, and others, have led to the following generalizations for the formation of acetals of acyclic compounds from aldehydes. (a) Six-membered acetal rings formed from two secondary hydroxyl groups with an *erythro* relationship are preferred to six-membered rings involving a primary hydroxyl group. (b) Less favoured are five-membered rings formed from either terminal or *threo*-diols, and six- or seven-membered rings from *threo*-diols. The relative stabilities of these ring systems are difficult to predict, and may vary according to conditions and the aldehyde used. (c) All other acetal ring systems are highly unfavoured.

Hence, D-mannitol and galactitol preferentially form 1,3:4,6-diacetals; allitol and D-iditol form 2,4:3,5-diacetals. However, L-arabinitol (**146**) appears to have the choice of two similar monoacetals, (**145**) and (**147**), formed from terminal diols. However, the 1,3-acetal (**145**) is formed preferentially, probably due to the ability of the axial 2-hydroxyl group in (**145**) being able to hydrogen bond to the ring-oxygens.[138]

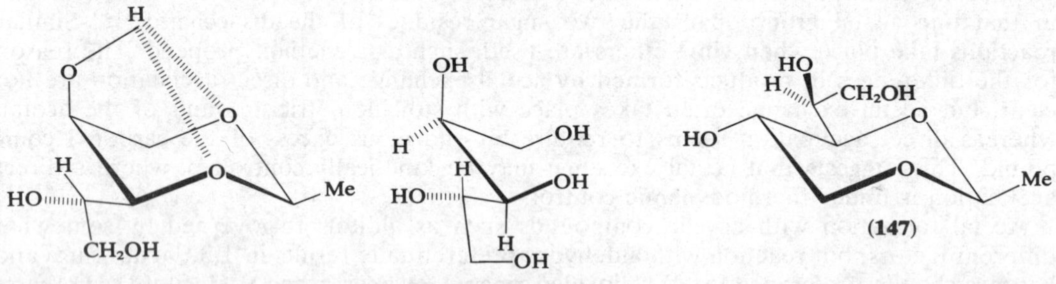

(**145**) (**146**) (**147**)

Reaction of alditols and other acyclic derivatives with ketones gives mainly acetals resulting from reaction of vicinal *threo*-diols or terminal diols. Formation of an acetal from an α-*threo*-diol gives a 1,3-dioxalan with the residues of the carbon chain in a *trans* relationship, whereas an α-*erythro*-diol would give an acetal with these substituents *cis*. Thus, D-glucitol (**141**) and D-mannitol with 3,4-*threo*-diols give 1,2:3,4:5,6-triacetals readily, whereas galactitol, with a 3,4-*erythro*-diol, gives a mixture of the 2,3:4,5- and 2,3:5,6-diacetals.[137]

Acetals are normally cleaved by acid,[139] and the use of 90% trifluoroacetic acid is particularly recommended. The rate of cleavage of acetal rings is dependent upon their position within the molecule and the nature of the acetal group itself. Fluorinated isopropylidene groups, and acetals resulting from trichloroacetaldehyde, are remarkably stable due to the electron-withdrawing effect of the halogen atoms. In the case of benzylidene acetals, labilization or stabilization of the acetal may be achieved by judicious incorporation of substituents into the benzene ring.

The ease of removal of an acetal ring is proportional to the difficulty with which the acetal group is established by direct acetalation. Thus, selective hydrolysis of 1,3:2,4:5,6-tri-*O*-benzylidene-D-glucitol (**144**) gives either the 1,3:2,4-diacetal or the 2,4-acetal depending upon conditions. Similarly, isopropylidene rings fused to furanose rings are favoured because of the preferred trioxabicyclo[3,3,0]octane ring system, and other isopropylidene groups attached elsewhere are selectively removed under mildly acidic conditions. Thus 1,2:5,6-di-*O*-isopropylidene-α-D-glucofuranose (**133**) readily affords the 1,2-monoacetal on exposure to mildly acidic conditions.

26.1.9 MONOSACCHARIDES CONTAINING OTHER FUNCTIONAL GROUPS

Monosaccharides in which one or more of the hydroxyl groups are replaced by some other functional group occur widely, but rarely abundantly, in Nature. This section discusses only the major groups of functionalized monosaccharides.

26.1.9.1 Amino- and nitro-sugars

2-Amino-2-deoxy-D-glucose (**148**) (glucosamine) is the most abundant amino-sugar, and occurs widely as its *N*-acetyl derivative (**149**) in polysaccharides and glycoproteins. Chitin, a widely distributed structural polysaccharide of crustaceans, is a homopolysaccharide, analogous to cellulose, consisting of 2-acetamido-2-deoxy-β-D-glucopyranose (**149**) units. 2-Acetamido-2-deoxy-D-galactose, -α-D-mannose, and -D-talose occur also in glycoproteins, but to a more limited extent. Amino-sugars[140] rarely occur free, or as simple derivatives, but almost always as components of polysaccharides, oligosaccharides, or as glycosides. Two notable exceptions are 5-amino-5-deoxy-D-glucose (**151**) (nojirimycin), and *N*-carbamyl-*N'*-methyl-*N'*-nitroso-D-glucosamine (**150**) (streptozoticin), antibiotics elaborated by several strains of *Streptomyces.* In the former compound the amino-group is involved in ring formation with the reducing group to give a piperidinose ring structure. A particularly abundant source of a variety of amino-sugars is the *Streptomyces* family of moulds which elaborate a variety of antibiotics containing amino-sugars.[141] Representative examples are kanamycin B (**152**), which contains 2,6-diamino-2,6-dideoxy-D-glucose and 3-amino-3-deoxy-D-glucose, and methymycin (**153**), which contains 3,4,6-trideoxy-3-dimethylamino-D-*xylo*-hexose (desosamine).

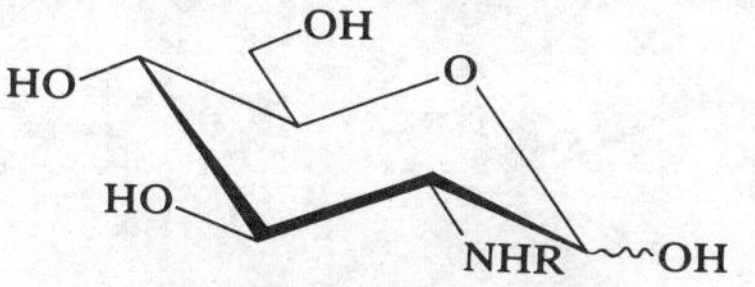

(**148**) R = H
(**149**) R = Ac
(**150**) R = CON(NO)Me

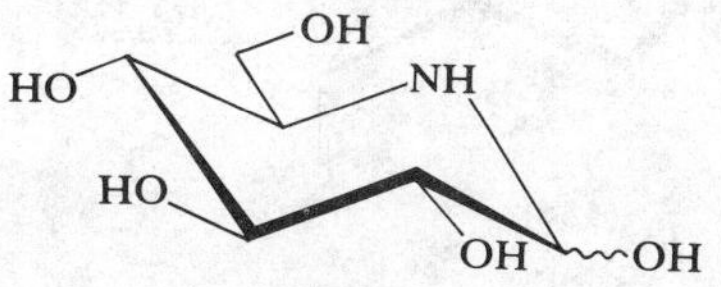

(**151**)

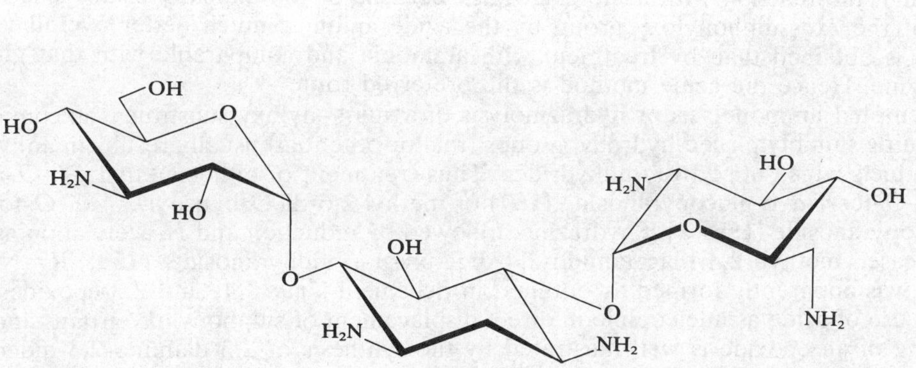

(**152**)

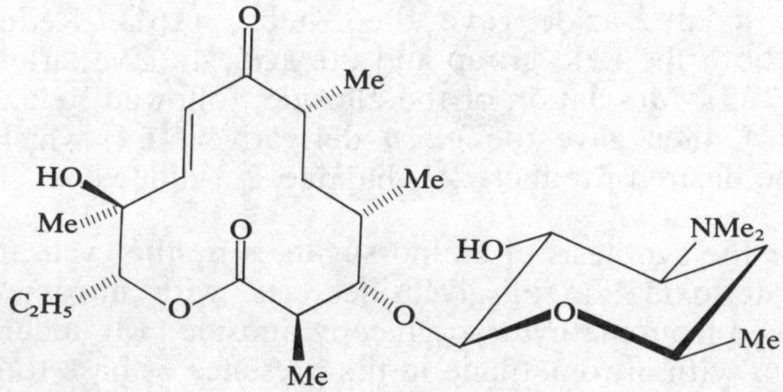

(**153**)

Sulphonate esters (Section 26.1.8.3) are the most common intermediates in synthesis of amino-sugars.[142] The sulphonyloxy group is displaced either directly by ammonia, hydrazine, or azide anion if feasible (see Section 26.1.4.3) or by initial conversion to an epoxide, followed by ring-opening by ammonia or azide anion. The use of ammonia has the disadvantage that the first-formed amine is sometimes a better nucleophile than ammonia itself and therefore competes for unreacted starting material, forming secondary amines as side products. This problem is not observed with hydrazine, which can sometimes be used with advantage as a bidentate nucleophile. Thus, methyl 2,3-di-*O*-benzyl-4,6-*O*-mesyl-α-D-glucopyranoside (**154**) reacts with hydrazine to give the cyclic 4,6-hydrazide (**155**), from which the 4,6-diamine is available by reductive fission of the hydrazide.[143,144] Displacements with azide are subject to the limitations proposed by Richardson[43] for S_N2 displacements (Section 26.1.4.3), but the use of this reagent is usually preferred. The reduction of the intermediary azide is accomplished by catalytic hydrogenation, Raney nickel–hydrazine, sodium borohydride, lithium aluminium hydride, *etc.* 3-Amino-3-deoxy-1,2:5,-di-*O*-isopropyllidene-α-D-allofurnaose (**156**) has

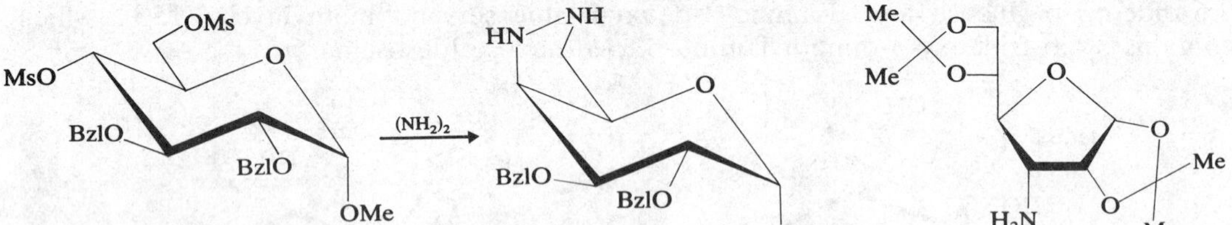

(154) **(155)** **(156)**

been synthesized from the corresponding 3-tosylglucofuranoside (**134**) using either ammonia, hydrazine, or azide (in hexamethylphosphoric triamide) as the nucleophilic species. Although the reaction with azide is sluggish because of the difficulty of approach to the rear of the *exo*-sulphonyloxy group by the azide anion, a much better overall yield of amine is obtained than by treatment with ammonia and comparable with that given by hydrazine. Hence the azide method is the preferred route.[142]

Attempted ammonolysis or hydrazinolysis of a sulphonyloxy substituent in compounds containing suitably placed hydroxyl groups (real or potential) usually results in anhydrides or products which arise from anhydrides. Thus treatment of either methyl 4,6-*O*-acetyl-2,3-di-*O*-tosyl-α-D-glucopyranoside (**157**) or methyl 2,6-di-*O*-benzoyl-3,4-di-*O*-tosyl-α-D-glucopyranoside (**158**) with hydrazine, followed by reduction and *N*-acetylation gave, in each case, methyl 2,4-diacetamido-2,4-dideoxy-α-D-idopyranoside (**159**; R = NHAc), which was apparently formed as outlined in Scheme 14 *via* 3,4- and 2,3-epoxides.

The use of azide as nucleophile in direct displacement of sulphonyloxy groups and ring-opening of an epoxide is well illustrated by the synthesis of 2,3-diamino-2,3-dideoxy-D-glucose from methyl 2,3-anhydro-4,6-*O*-benzylidene-α-D-allopyranoside (**39**). Reaction of the latter with sodium azide in ethanol using ammonium chloride as buffer gave, by *trans*-diaxial opening, the 2-azidoaltroside (**41**; $X = N_3$) in high yield. Oxidation of this azide with DMSO and acetic anhydride resulted in formation of the 3-ulose (**160**), and epimerization of the axial 2-azide gave the 3-ulose (**161**). Reduction with sodium borohydride reduced both the keto-group and the azide to give, after *N*-acetylation, the 2-acetamidoalloside (**162**). Mesylation of the alloside, followed by azide displacement of the 3-sulphonate (**163**), then gave the *gluco* derivative (**164**) which on reduction and hydrolysis afforded the desired product, 2,3-diamino-2,3-dideoxy-D-glucose (**165**) in high overall yield.[145]

A novel method for the synthesis of amino-sugars is by the cyclization of dialdehydes, derived from periodate oxidation of glycosides, *etc.* with nitromethane.[146] Thus the dialdehyde (**166**) derived from methyl α-D-glucopyranoside (**13**) undergoes two successive aldol addition reactions with nitromethane in the presence of base to give a reconstituted pyranoside ring with a 3-nitro substituent (Scheme 15). Since three new chiral centres are

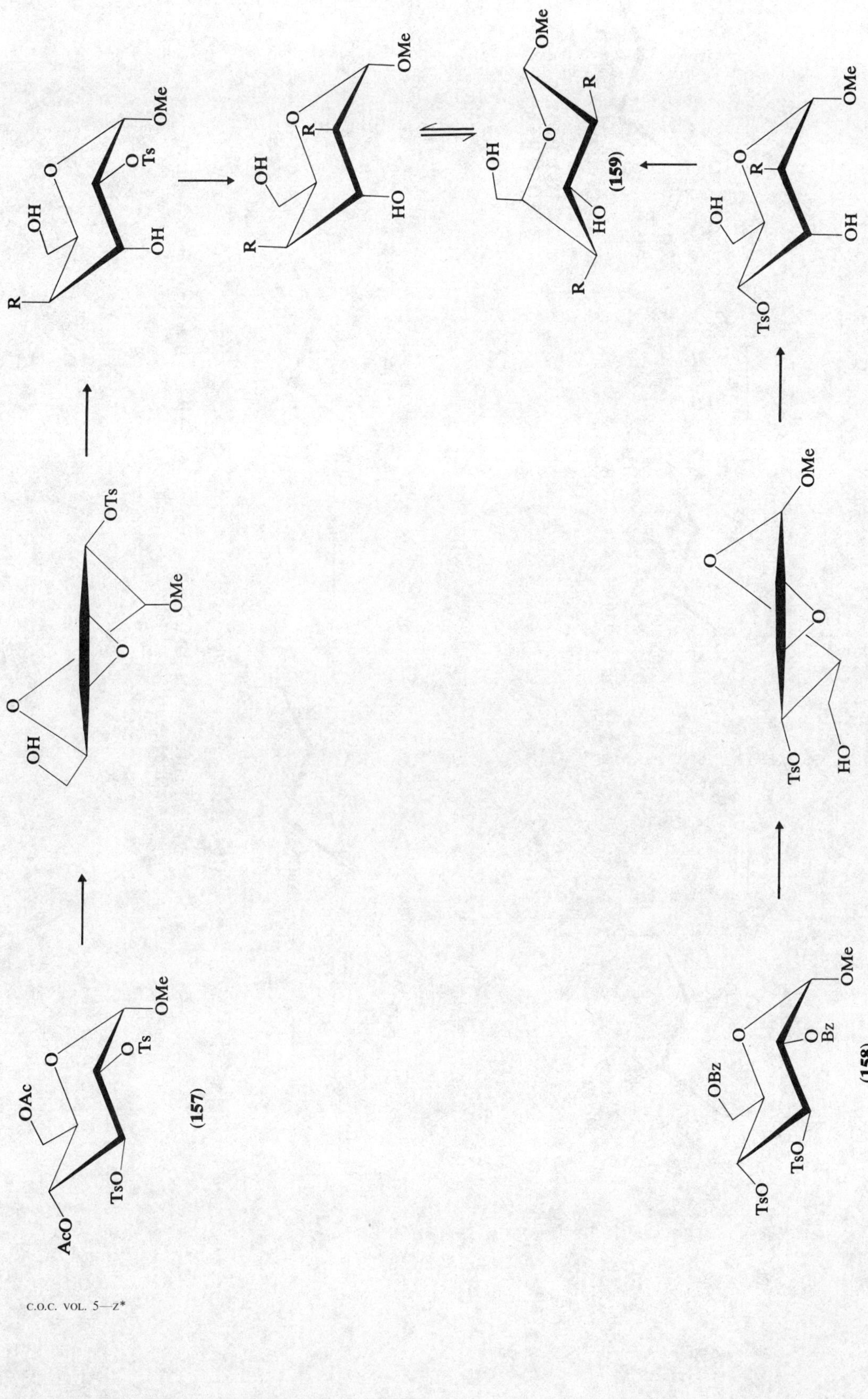

(157)

(158)

(159)

$R = H_2NH_2 \longrightarrow NHAc$

SCHEME 14

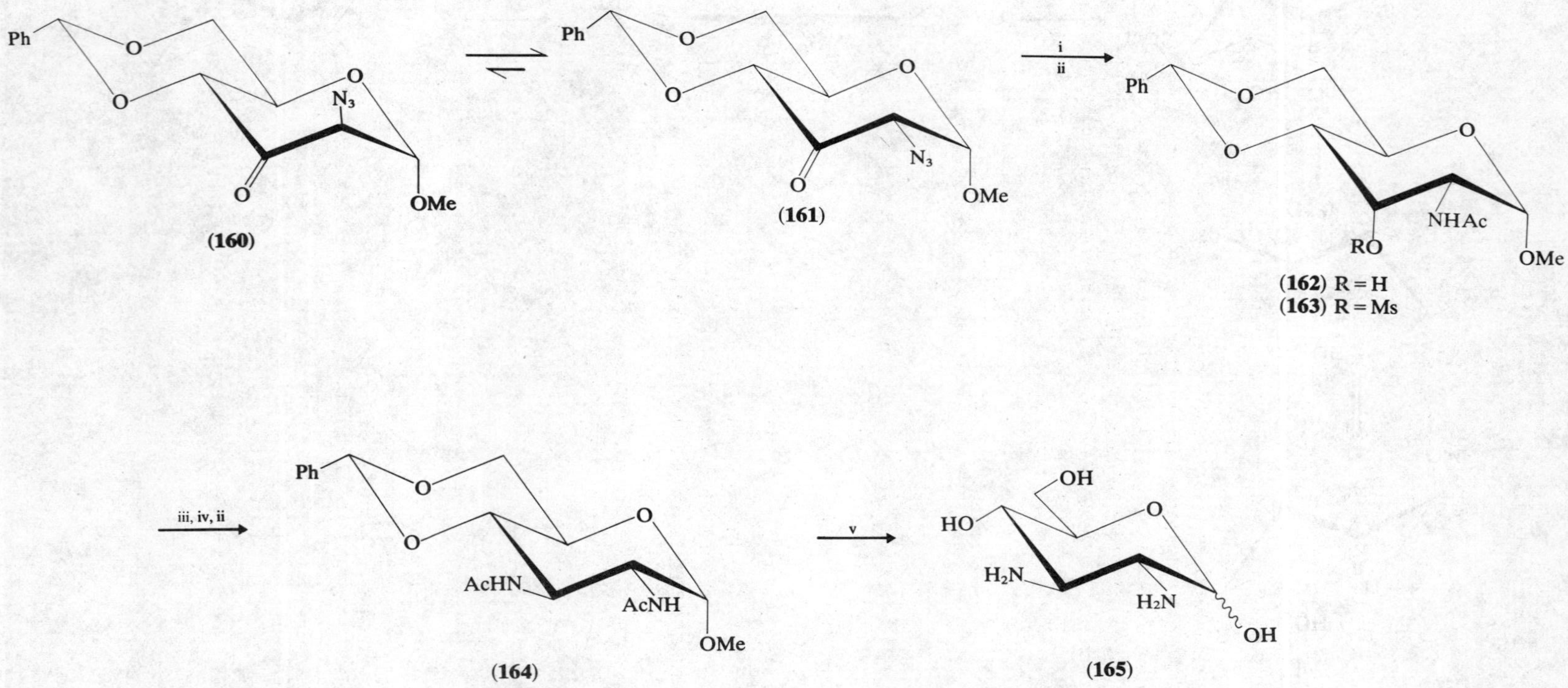

i, $NaBH_4$, MeOH, DMF (→R = H); MsCl, py (→R = Ms); ii, Ac_2O; iii, NaN_3, DMF; iv, Ni, H_2; v, H^+.

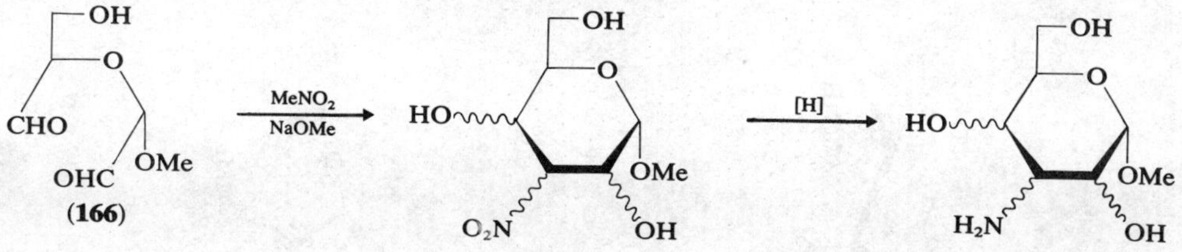

SCHEME 15

established, eight disastereoisomers are possible. However, the nitro-group always adopts an equatorial orientation, and consequently only the *gluco*-, *manno*-, *galacto*-, and *talo*-isomers are formed. In the early stages of the reaction the ratio of the four is 41:42:9:8, but upon standing the reaction products become subject to thermodynamic control and the *talo*-isomer becomes the predominant species. At first sight this appears to be surprising since the *talo*-isomer (**169**) has two axial hydroxyl groups, but since the *aci*-nitro forms of the nitropyranosides are the species undergoing equilibration, the steric clash between the *aci*-nitro oxygens and neighbouring equatorial substituents ($\Delta^{1,3}$ effect) destabilizes the '*gluco*' isomer (**167**) and favours the diaxial '*talo*' isomer (**168**). Therefore,

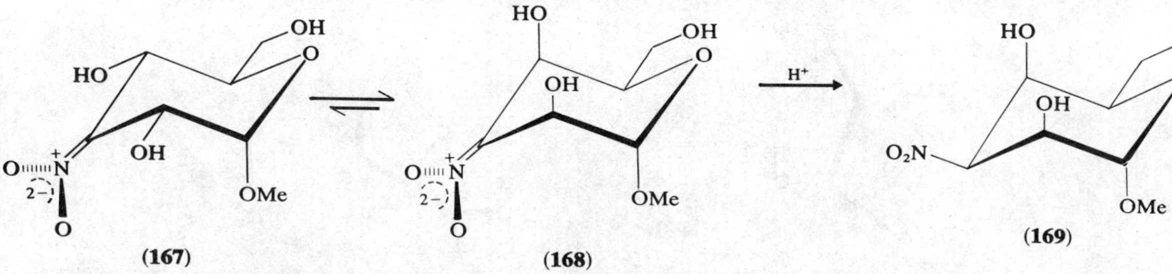

using a variety of glycosides, nucleosides, 1,6-anhydro-pyranoses, *etc.* as starting materials, the synthesis of a variety of 3-nitro derivatives is possible, and from these the corresponding amino-pyranoses are available by reduction.

The 3-nitro-pyranosides such as (**169**) also serve as versatile synthetic intermediates for a range of amino-sugars.[147] The 4,6-*O*-benzylidene derivative (**170**) of methyl 3-deoxy-3-nitro-α-D-glucopyranoside on sequential acetylation and basification gives the 3-nitro-2-ene (**171**) (Schmidt–Rutz reaction). The 2-ene (**171**) undergoes a Michael-type addition of

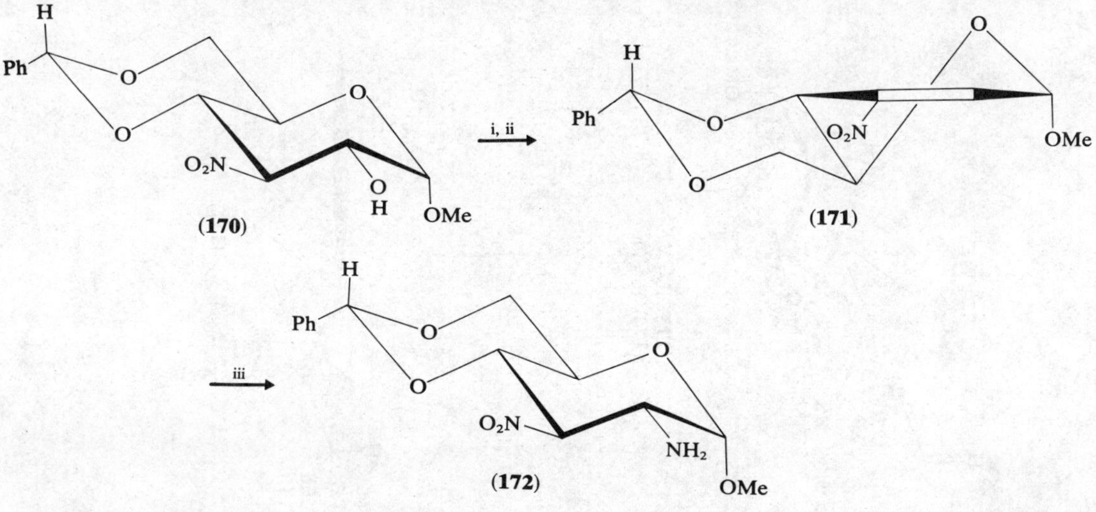

i, Ac_2O, $BF_3.Et_2O$; ii, $NaHCO_3$; iii, NH_3.

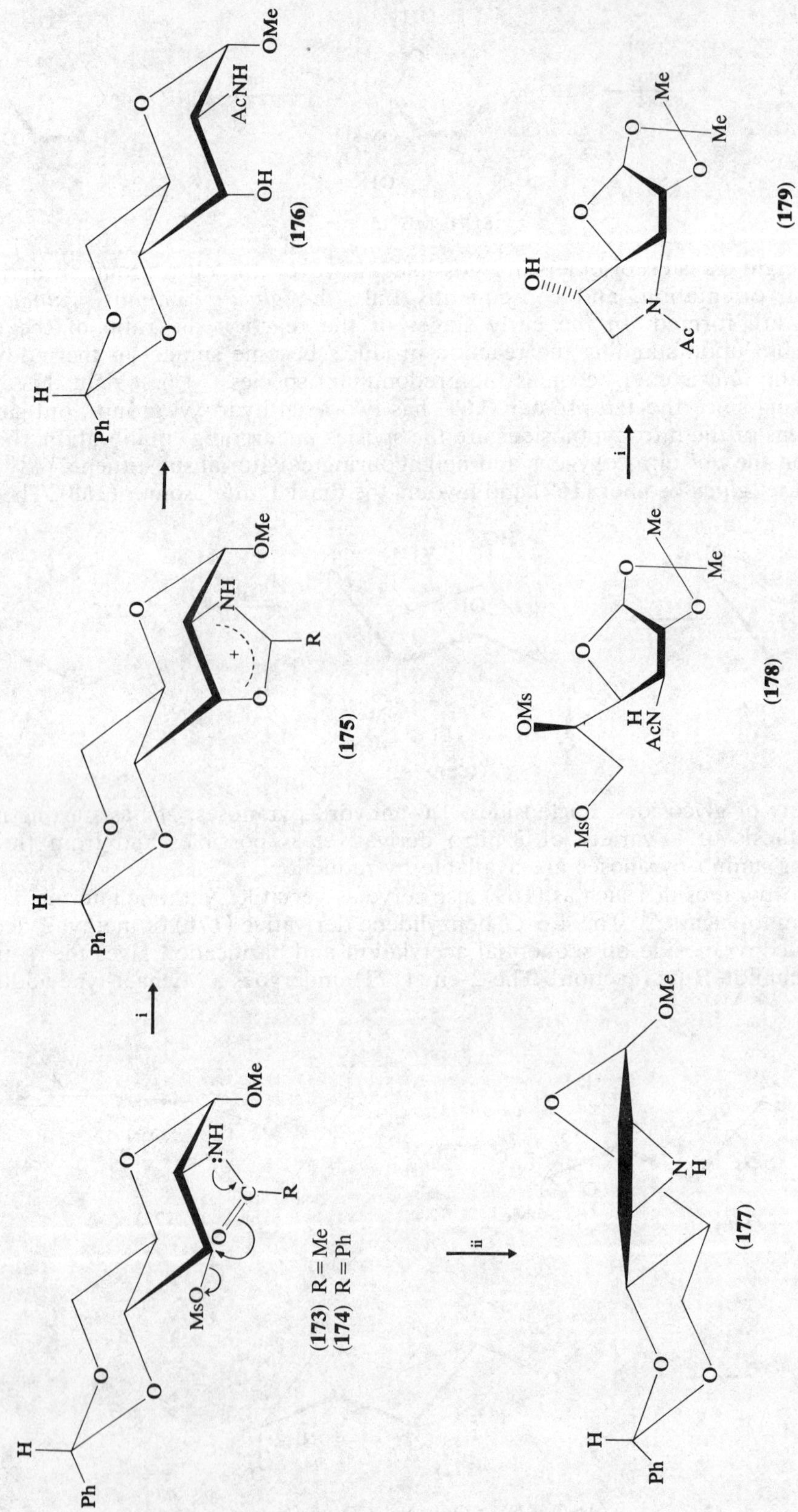

i, NaOAc; ii, (R = Ph), $Me_2CHO^- Na^+$.

a variety of nucleophilic reagents, giving 2-substituted 3-nitropyranosides. Thus the 2-amino-3-nitroglucoside (**172**) is formed from (**171**) with ammonia, from which 2,3-diamino-2,3-dideoxy-D-glucose may be obtained by reduction and hydrolysis. These addition reactions, which are generally performed under basic conditions, always lead to 2,3-diequatoral products. However, under mildly acidic or neutral conditions the 2-axial kinetic product is formed in high yield with reagents such as HCN, HN_3, *etc.*[148] Other deoxynitro derivatives of sugars are available by oxidation of oximes and amines by peracids.

Transformation of readily available amino-sugars, such as glucosamine (**148**), is sometimes employed for the synthesis of other amino-sugars; of particular interest is the ability of acylamido-groups to participate in reactions at neighbouring positions.[149] Thus inversion of configuration of a *trans*-hydroxyl group adjacent to an acetamido function is accomplished by *O*-mesylation followed by treatment with sodium acetate in aqueous 2-methoxyethanol. By this method methyl 2-acetamido-4,6-*O*-benzylidene-2-deoxy-3-*O*-mesyl-α-D-glucopyranoside (**173**) gives the allopyranoside (**176**) *via* the oxazolinium ion (**175**) intermediate.[150] In contrast, treatment of the corresponding *N*-benzoyl derivative (**174**) with strong base results in participation by the nitrogen atom to give the three-membered epimine ring.[151] Because of the decreased resonance between the nitrogen of the epimine ring and the *N*-benzoyl group, the *N*-benzoyl group is extremely sensitive to base and is rapidly hydrolysed so that the parent epimine (**177**) is obtained. These epimines are unlike epoxides (Section 26.1.8.2) in that they are very stable towards basic reagents, and moderately stable to acids, enabling the 4,6-*O*-benzylidene group to be removed from (**177**) without rupture of the epimine ring. Participation by acylamido-groups may occur over several carbon atoms. For example, solvolysis of 3-acetamido-3-deoxy-5,6-di-*O*-mesyl-1,2-*O*-isopropylidene-α-D-glucofuranose (**178**) results in the elimination of the 5-sulphonyloxy group by amide 'oxygen' attack, and the 6-sulphonate by amide 'nitrogen' attack to give the 3,6-imino derivative (**179**).[152]

26.1.9.2 Halogenated sugars

Although only one halogeno-derivative of a carbohydrate has been isolated from natural sources (the antibiotic nucleocidin), there has been a great deal of biological interest in halogenated sugars[153] because in other areas of natural products chemistry (*e.g.* steroids) the introduction of halogen has often led to interesting biological properties. This interest has been centred on fluoro- and chloro-derivatives, and some fluoro-derivatives have shown competitive inhibition of certain enzymes.

The most common way of introducing halogen into a carbohydrate is either by nucleophilic displacement of sulphonate esters by halide anion, or by nucleophilic opening of epoxide rings. In the former the relative ease of displacement is I > Br > Cl > F and the reaction proceeds only where the polar and steric effects favour an S_N2 reaction (Section 26.1.4.3), so that the introduction of halogen at primary positions is usually quite easy, but less so at secondary positions. The introduction of fluorine at secondary positions by displacement of sulphonate ester groups is difficult owing to the poor nucleophilicity and high basicity of the fluoride anion, which often favours elimination reactions. The use of tetrabutylammonium fluoride in acetonitrile is the most effective reagent for these displacements. Thus the synthesis of 3-deoxy-3-fluoro-D-glucose from the 3-tosylate of 1,2:5,6-isopropylidene-α-D-allofuranose (**44**) is accomplished by treatment with the above reagent, followed by acid hydrolysis.[154] Displacement reactions with fluoride occur with inversion of configuration, but with the other halogens further nucleophilic displacement of the halogen atom may occur to give either the product with overall retention of configuration, or a mixture of the two epimers. There are several related procedures by which the hydroxyl group is directly replaced by halogen, without the need to isolate the intermediate ester, or other species. The oldest method is that due to Helferich, in which the sugar is reacted with sulphuryl chloride. The first-formed chlorosulphate ester then undergoes displacement by the liberated chloride anion, providing the steric and polar factors are opportune for an S_N2 reaction. The reaction occurs under very mild conditions and those hydroxyls for which an S_N2 reaction is unfavourable remain either as the chlorosulphate ester, or may form a cyclic sulphate ester with a vicinal hydroxyl group. Thus, reaction of methyl α-D-glucopyranoside (**13**) with sulphuryl chloride gives the 4,6-dichlorogalactoside either as the bischlorosulphate (**180**) or the cyclic sulphate (**181**),

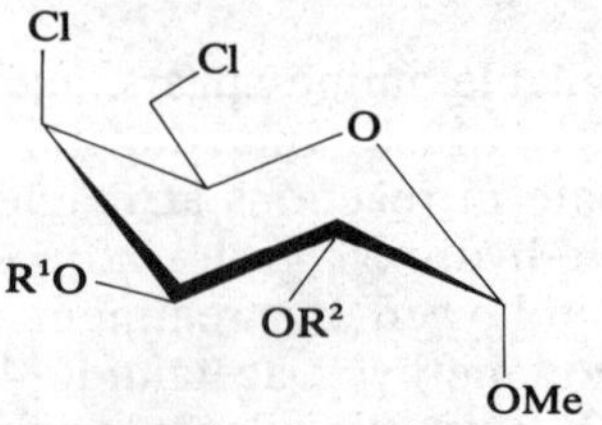

(**180**) $R^1 = R^2 = SO_2Cl$
(**181**) $R^1,R^2 = —SO_2—$
(**182**) $R^1 = R^2 = H$

depending upon the conditions. The subsequent removal of chlorosulphate groups is facile, requiring treatment with sodium iodide in methanol, whereas cyclic sulphate groups are removed less readily by sequential treatment with alkali and acid. Unfortunately this reaction cannot be applied for the preparation of fluoro-, bromo-, and iodo-sugars. Other reagents used for the direct replacement of hydroxyl groups by halogen include DMF–mesyl chloride, which at ambient temperatures gives only primary chlorination, but at elevated temperatures gives extensive chlorination at secondary centres;[155] the intermediate ester is believed to be $ROCH{=}\overset{+}{N}Me_2$, which is also formed on reaction of the carbohydrate with chloro- and bromo-methylenedimethylammonium chloride ($Me_2\overset{+}{N}{=}CHX\,Cl^-$). Similarly, triphenylphosphine–carbon tetrahalide and triphenylphosphine–*N*-halosuccinimide cause halogenation (not fluorination) *via* the species $R—O\overset{+}{P}(C_6H_5)_3$, and triphenyl phosphite methiodide (Rydon reagent) or the corresponding dihalide proceeds *via* the ester $RO\overset{+}{P}(CH_3)(OC_6H_5)_2$.

Ring-opening of epoxides by halide anions or by hydrogen halides gives the corresponding *trans*-halo-alcohol. Reaction with halide ions must be conducted under buffered conditions, since the alkoxide generated gives rise to basic conditions causing side reactions and low yields. Hence, the reaction is usually performed in the presence of acetic acid–sodium acetate. The reaction obeys the Furst–Plattner rule for anhydropyranosides and gives the *trans*-diaxial isomer preferentially (Section 26.1.8.2), so that methyl 4,6-*O*-benzylidene-α-D-mannopyranoside (**38**) and -allopyranoside (**39**) give mainly the 3-halo- (**40**) and 2-halo-altropyranosides (**41**), respectively. Hydrogen halides cause more rapid ring-opening of epoxides, and usually give the same *trans*-diaxial products. In the case of fluorides, potassium hydrogen difluoride in ethylene glycol is the preferred reagent. For example, reaction of methyl 2,3-anhydro-6-*O*-benzyl-α-D-lyxopyranoside (**183**) with the above reagent affords the 3-fluoride (**184**) in 30% yield.

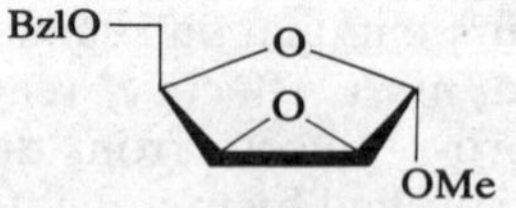

(**183**)

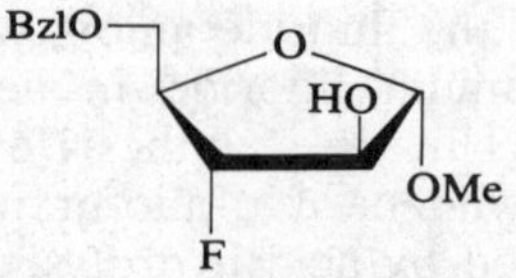

(**184**)

A particulary valuable way of making 6-bromo-6-deoxyhexosides is by reaction of the 4,6-*O*-benzylidene acetal with *N*-bromosuccinimide in carbon tetrachloride. The reaction proceeds by free-radical bromination at the benzylic position followed by ionization to give the benzoxonium salt, followed by bromide attack at C-6. Thus the 4,6-acetal (**136**) gives methyl 4-*O*-benzoyl-6-bromo-6-deoxy-α-D-glucopyranoside in quantitative yield.

On the other hand, the corresponding β-galactopyranoside gives a 3-bromo-gulopyranoside in addition to the expected 6-bromide. In this case it appears that migration of the 4,6-benzoxonium ion to the 3,4-positions is significant and leads to bromide attack at C-3. In the case of benzylidene acetals attached to two secondary positions, a mixture of bromides is usually obtained.

The addition of bromine, chlorine, and pseudohalogens (IN_3, IF, *etc.*) to unsaturated sugars (Section 26.1.9.6) affords a variety of halogenated derivatives, and the addition of the pseudohalogen trifluoromethoxy fluoride (F_3COF) to glycals, such as (**185**), constitutes the only convenient method of placing fluorine at C-2 of an aldose. Thus reaction of (**185**) with the reagent gave a mixture of four products, namely trifluoromethyl 3,4,6-tri-*O*-acetyl-2-deoxy-2-fluoro-α-D-glucopyranoside (**186**) and the corresponding glycosyl fluoride in a combined yield of 60%, together with the related β-*manno*-isomers in a combined yield of 14%.[154] What is particularly remarkable about the reagent is that it functions as $CF_3O^- F^+$ (*i.e.* electrophilic fluorine) and therefore fluorine attaches to the more nucleophilic end of the double bond. However, the 1,2-*cis* stereochemistry of the products suggests that the addition must be a concerted process by attack of CF_3OF to one side or the other of the olefin. The origin of the 1-fluoro-group is obscure, but it has been suggested that it may arise by decomposition of the trifluoromethyl glycoside with loss of F_2CO, or it could arise *via* the five-centre transition state (**187**).

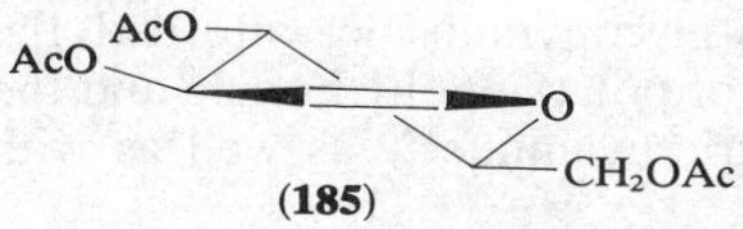

(**185**)

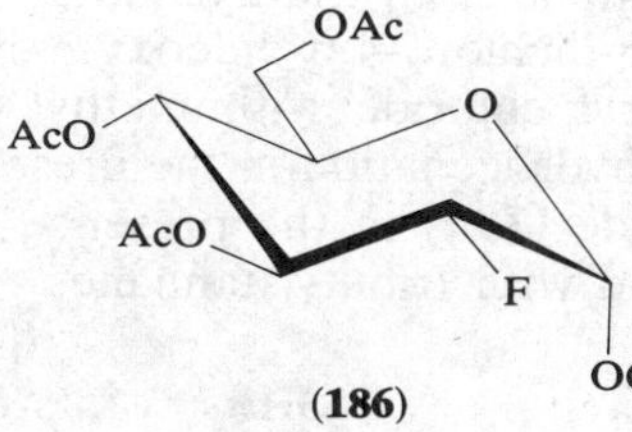

(**186**)

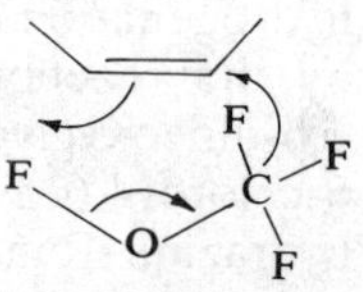

(**187**)

It is noteworthy that replacement of the 7-hydroxyl group in lincomycin (**188**) by chlorine to give (**189**) resulted in a remarkably improved antibiotic known as clindamycin. Even more surprising is the report that replacement of various hydroxyl groups of sucrose by chlorine results in considerably enhanced sweetness.[156] The 1′,4,6′-trichloride (**190**) is 2000 times more sweet than sucrose itself.

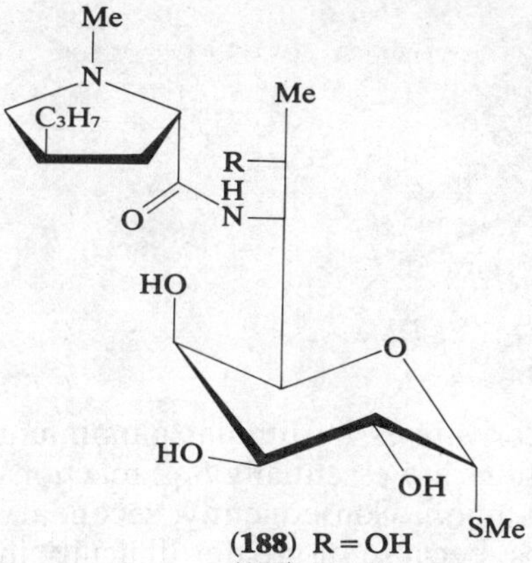

(**188**) R=OH
(**189**) R=Cl

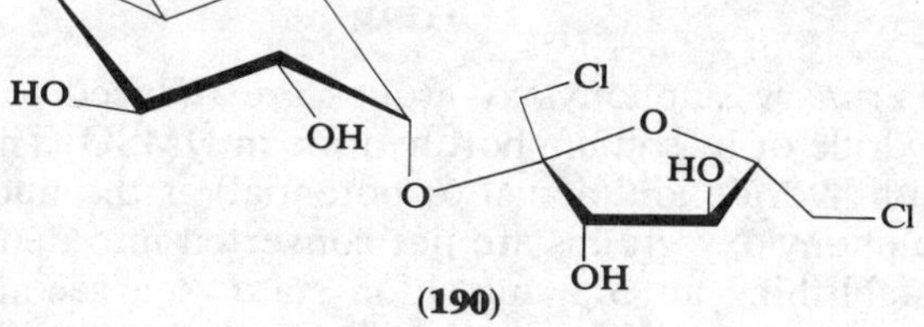

(**190**)

26.1.9.3 Deoxy-sugars

Aldoses in which one or more hydroxyl groups are replaced by hydrogen occur widely throughout Nature,[157] particularly those bearing a terminal deoxy-group such as L-rhamnose (6-deoxy-L-mannose) (**191**) and L-fucose (6-deoxy-L-galactose) (**192**), which are found as components of many polysaccharides, glycoproteins, and plant glycosides. 2-Deoxy-D-*erythro*-pentose (2-deoxy-D-ribose) (**193**) occurs throughout Nature in DNA, and many deoxy- and dideoxy-sugars occur as components of antibiotics.[141]

In many respects the synthesis of deoxy-sugars is dependent upon the availability of the corresponding halogen derivative (Section 26.1.9.2), from which the deoxy-sugar can be derived by reductive dehalogenation by a variety of methods. The ease with which reduction is accomplished is in the order I>Br>Cl, with fluorine being inert to all reducing agents. The reduction of iodo- and bromo-derivatives can be accomplished by catalytic hydrogenation in the presence of an acid acceptor (*e.g.* sodium acetate), although secondary halo-substituents can sometimes be stubborn. The use of hydrazine–Raney nickel is particularly effective and has the advantage that hydrogenation equipment is not required, and its usefulness is enhanced by its selectivity in reducing secondary chloro-substituents whilst leaving those at primary carbons intact. This same selectivity is also observed with Raney nickel in the presence of triethylamine, whereas in the presence of potassium hydroxide both secondary and primary chloro-groups are reduced.[158] Reduction of chloro-substituents by tributylstannane also shows a similar selectivity for secondary chlorine atoms[159] and it is well established that this reagent reacts by a free-radical mechanism. Because of this similarity in selectivity, it is possible that hydrogenation with Raney nickel–triethylamine must proceed *via* a radical process, perhaps initiated by abstraction of a hydrogen atoms from triethylamine by Raney nickel. Thus Raney nickel hydrogenation of methyl 4,6-dichloro-4,6-dideoxy-α-D-galactopyranoside (**182**), prepared by the reaction of sulphuryl chloride with methyl α-D-glycopyranoside, afforded the 4,6-dideoxyglycoside upon hydrogenation in the presence of potassium hydroxide and the 6-chloro-4,6-dideoxyglycoside (**194**) in the presence of triethylamine[158] as well as with hydrazine–Raney nickel, and with tributylstannane.

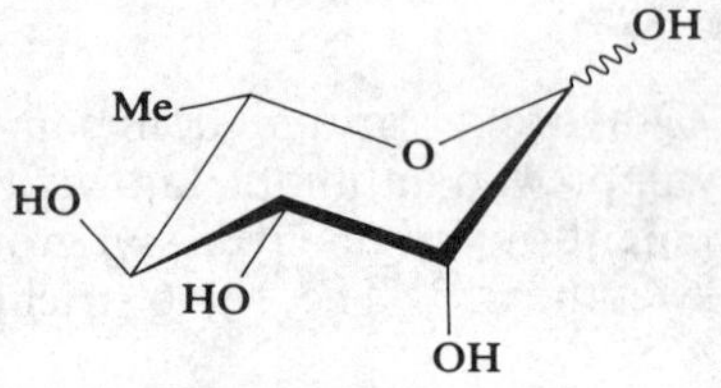

(**191**)

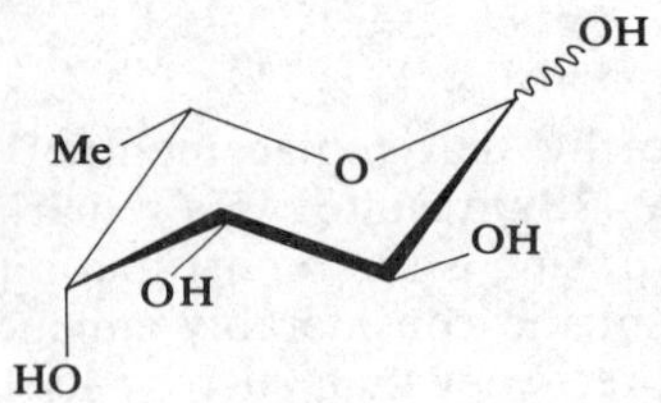

(**192**)

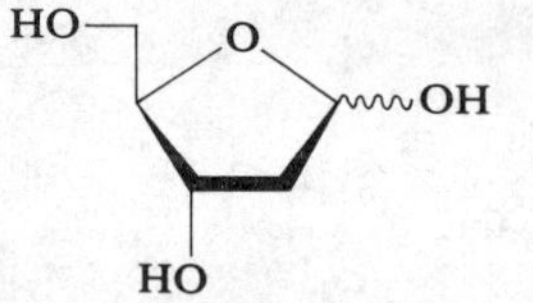

(**193**)

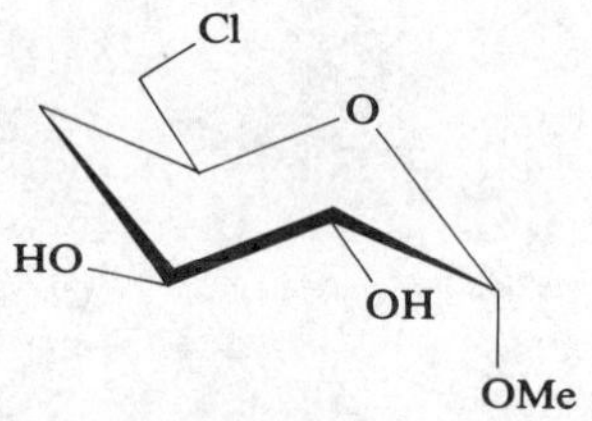

(**194**)

Primary sulphonyloxy-groups are reduced to deoxy-derivatives by lithium aluminium hydride or by sodium borohydride in DMSO. These reactions are essentially S_N2 reactions with hydride ion (actual or potential) as the nucleophilic anion. Consequently, secondary sulphonyloxy-groups are not converted into deoxy-groups, because of greater difficulty in establishing an S_N2 transition state at a secondary carbon. In some cases the hydride attacks at sulphur, resulting in O—S cleavage and generation of the parent alcohol.

Reduction of epoxides with lithium aluminium hydride occurs with great ease to give the 'diaxial product' as predicted by the Furst–Plattner rule (Section 26.1.8.2). Reduction of methyl 2,3-anhydro-4,6-*O*-benzylidene-α-allopyranoside (**39**) gave the corresponding 2-deoxyhexopyranoside (**41**; X = H), whereas the corresponding mannopyranoside (**38**) on reduction gave the 3-deoxypyranoside (**40**; X = H) in high yield.

Other methods for the preparation of deoxy-sugars include the Raney nickel induced desulphurization of thio-sugars (Section 26.1.9.4) and the reduction of unsaturated derivatives (Section 26.1.9.6).

26.1.9.4 Thio-sugars

The replacement of the constituent oxygen atoms of sugars by sulphur gives rise to thio-sugars.[160] The only thio-sugars to occur naturally are the group of thioglycosides, an example of which is sinigrin (a component of black mustard) and the unique 5′-methylthioadenosine (vitamin L_2) from yeast. In the thioglycosides the glycosidic oxygen is replaced by sulphur; over 50 compounds of this type are known to occur naturally and are important flavour constituents known as the 'mustard-oil' glycosides.[160]

The greatly enhanced nucleophilicity of sulphur compared with oxygen facilitates its introduction into sugar residues by nucleophilic displacements of halides and sulphonate esters and by ring-opening of epoxides. A further result of the greater nucleophilicity of sulphur is that it competes successfully with hydroxyl groups for the carbonyl group of sugars if the formation of a thiapyranose or thiafuranose ring is possible. Thus 4-thioglucose exists only in the furanose modification.[161]

The sulphur nucleophiles used for synthesis include potassium thiocyanate, sodium thiosulphate, alkali metal salts of thioacetic and thiobenzoic acids, and sodium benzyl-thiolate. For example, reaction of 1,2-*O*-isopropylidene-5-*O*-tosyl-α-D-xylofuranose (**195**) with any one of these reagents gives the corresponding 5-thio-analogue (**196**), where R = CN, SO_3Na, COMe, COPh, or CH_2Ph. Removal of the *S*-substituent is accomplished by sodium sulphide, sodium borohydride, sodium methoxide, sodium methoxide, or sodium in liquid ammonia, respectively, to give the 1,2-isopropylidene-5-thio-α-D-xylofuranose (**196**; R = H).[162] The xylothiapyranose (**197**) is obtained from the latter by acid hydrolysis.

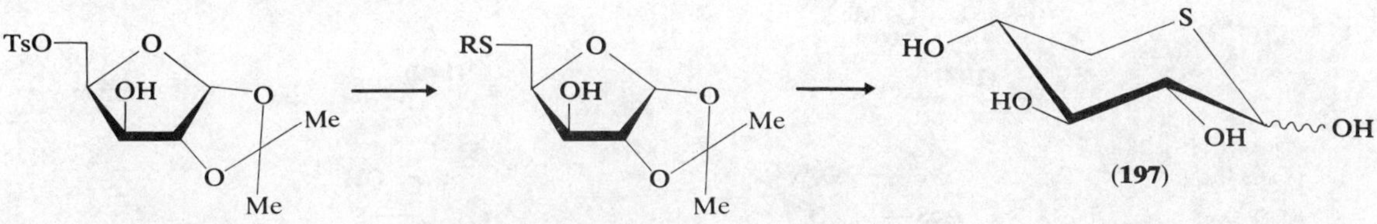

(**195**) (**196**) (**197**)

Reaction of epoxides with thiocyanates and with thiourea follow a surprising course giving, finally, the corresponding episulphide with inversion of configuration (Scheme 16). Thus reaction of 5,6-anhydro-3-*O*-benzyl-1,2-*O*-isopropylidene-β-L-idofuranose (**198**) with thiourea gave the 5,6-episulphide (**199**) with the *gluco* configuration. Subsequent nucleophilic ring-opening of the episulphide with potassium acetate in acetic anhydride afforded the acetylated 5-thiol (**200**), from which 5-thio-D-glucopyranose (**201**) was liberated.[163] Since the mechanism outlined in Scheme 16 requires free rotation about the C—C bond, it does not work well for epoxides fused to other rings. However, the 2,3-episulphides corresponding to the epoxides (**38**) and (**39**) are obtained indirectly by treatment of the appropriate epoxide with potassium thiocyanate, followed by mesylation and treatment of the product with base.

The powerful participating capacity of sulphur results in several diverse rearrangements, which have found use in synthesis. For example, solvolysis of 5-*O*-tosyl-D-xylose diethyl dithioacetal (**202**) gives 5-*S*-ethyl-1,5-dithio-L-arabinofuranosides (**203**) by 'S' participation in the displacement of the sulphonyloxy-group (Scheme 17). A considerably more profound rearrangement occurs upon ethanethiolysis of 3-*O*-benzoyl-1,2:5,6-di-*O*-isopropylidene-α-D-glucofuranoside, from which ethyl 4-*O*-benzoyl-2,3,6-tri-*S*-ethyl-1,2,3,6-tetrathio-α-D-mannopyranoside (**206**) is isolated in 40% yield. The *S*-ethyl groups

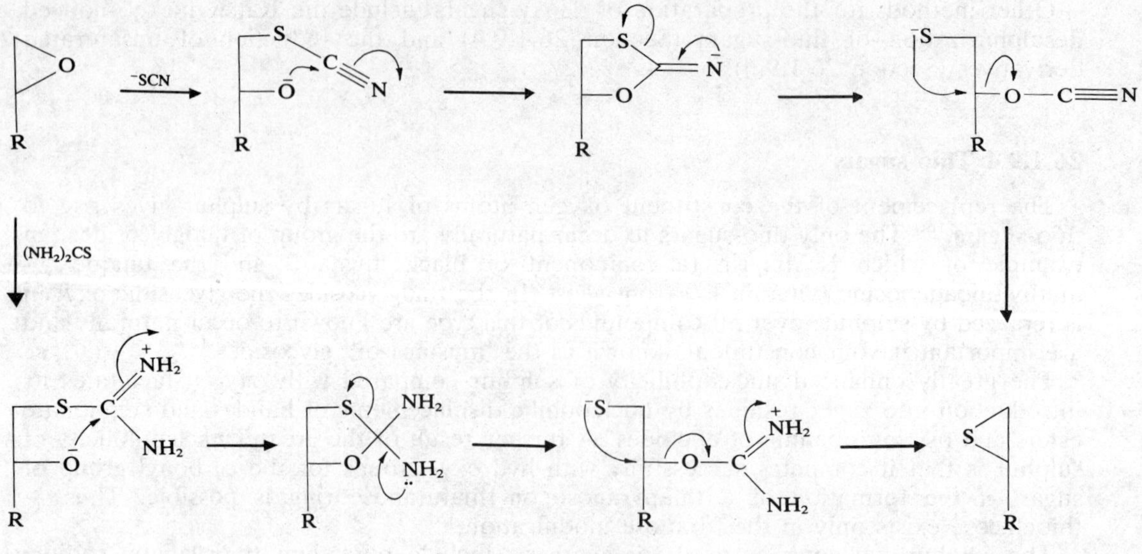

SCHEME 16

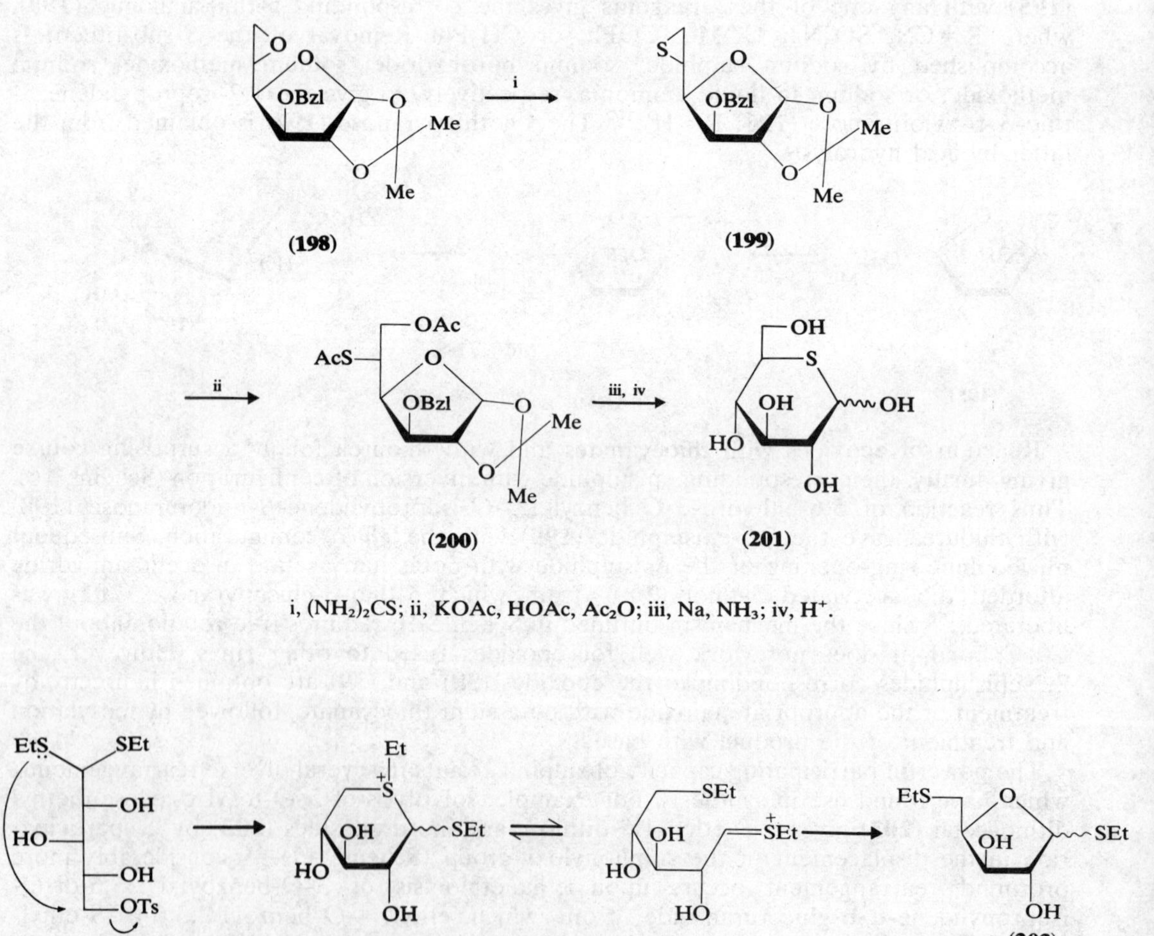

(198) (199) (200) (201)

i, $(NH_2)_2CS$; ii, KOAc, HOAc, Ac_2O; iii, Na, NH_3; iv, H^+.

(202) (203)

SCHEME 17

of the first-formed dithioacetal (**204**) migrate 1→2, probably *via* the 2,3-benzoxonium ion (**205**). The successive feeding of 'SEt' from C-1 proceeds in the sequence 1→2→3→6 *via* 2,3-, and 3,4-, and 4,6-benzonium ions, with the formation of the thiopyranoside (**206**) as the end product.[164] Many similar rearrangements are known.

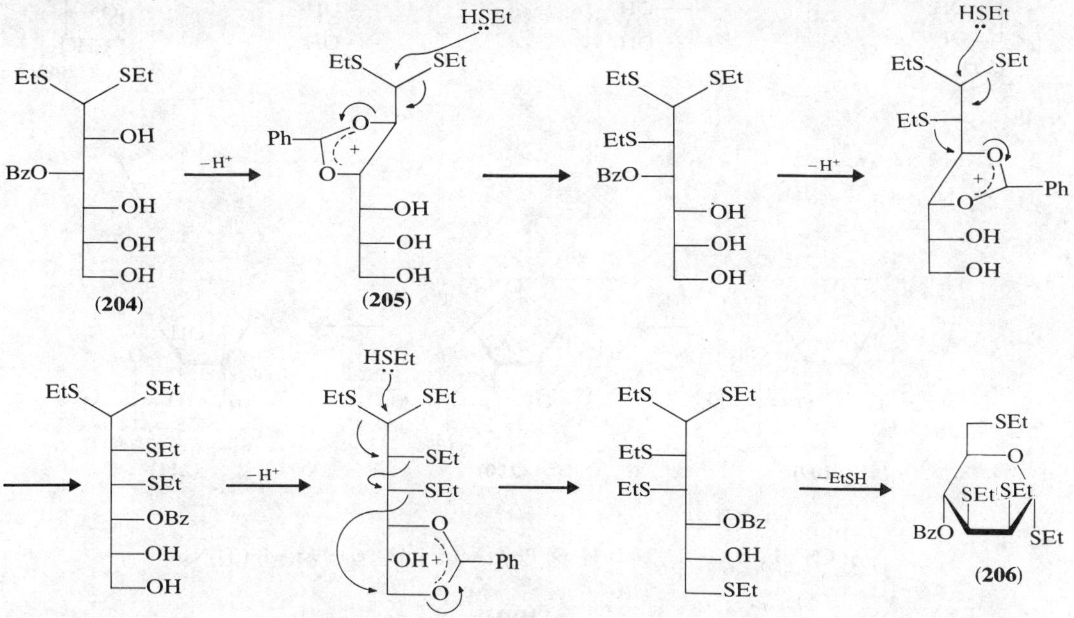

SCHEME 18

26.1.9.5 Branched-chain sugars

This class of compound is unique in that it does not possess a linear carbon backbone.[165] Branched-chain sugars occur in antibiotics, *e.g.* streptose (**216**) from streptomycin, and also in several plants. For example, 3-*C*-hydroxymethyl-D-*glycero*-tetrose[166] (apiose) (**208**) occurs as a glycoside in parsley and 2-*C*-hydroxymethyl-D-ribose (hamamelose) occurs as a digalloyl ester in the bark of witch-hazel.

The introduction of a branch chain into an aldose is carried out by way of the appropriate *keto* derivative, using various 'carbon nucleophiles' such as Grignard reagents, cyanide anion, diazomethane, nitromethane, Wittig reagents, *etc.* Thus apiose (**208**) has been synthesized (Scheme 19) from 3-*O*-benzyl-D-fructose (**207**) by reaction with cyanide, hydrolysis of the cyanohydrin to the acid, followed by lactonization, reduction of the lactone with sodium borohydride, and cleavage of the 4,5- and 5,6-bonds with lead tetra-acetate to give the *O*-benzylapiose, from which apiose was obtained by catalytic hydrogenation. Alternatively, reaction of the ketotetrose (**209**) with diazomethane is stereospecific and the nucleophile approaches from an *exo*-position leading to spiro-epoxide (**210**), hydrolysis of which leads to apiose.[166]

The importance of steric control in the introduction of functionalized side chains has led to the development of the use of 2-lithio-1,3-dithians, such as (**213**).[167] The bulk of these reagents usually ensures a good degree of stereoselectivity and the dithian ring can be desulphurized to give the alkyl chain, or cleaved to give the keto (or aldehydo) side chain. For example, the synthesis of streptose (**216**) from the 3-ketone (**212**) is illustrated in Scheme 20, in which the dithian is inserted *exo* with respect to the fused bicyclic ring system. The introduction of the carboxaldehyde branch in streptose has also been accomplished from (**212**) using vinylmagnesium bromide, followed by ozonolysis.

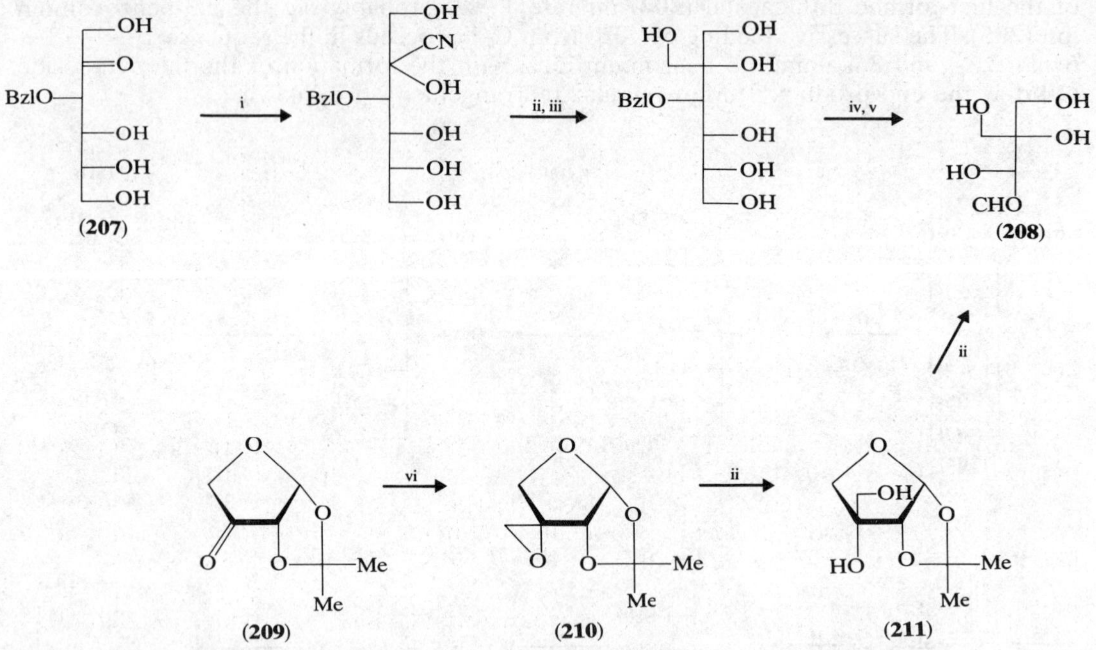

i, CN^-; ii, H_3O^+; iii, $NaBH_4$; iv, $Pb(OAc)_4$; v, H_2, catalyst; vi, CH_2N_2.

SCHEME 19

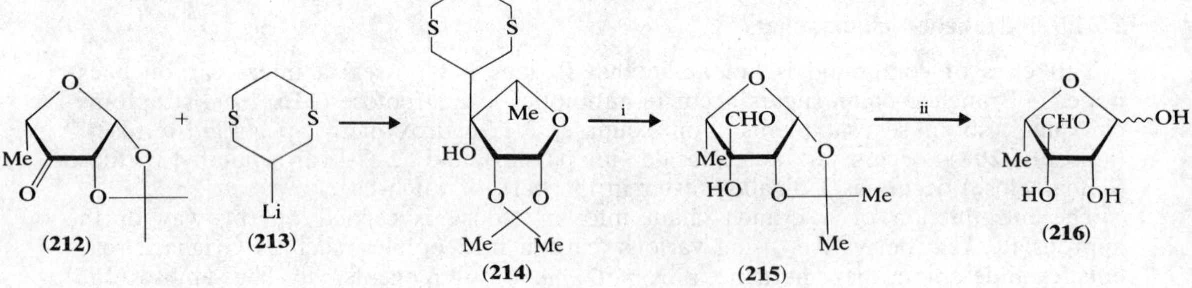

i, HgO, BF_3; ii, H_3O^+.

SCHEME 20

The aforementioned procedures give rise to branched-chain derivatives with tertiary hydroxyl groups at branching-points. However, many branched-chain sugars do not possess a hydroxyl group at this carbon and are synthesized either by ring-opening of epoxides by 'carbon nucleophiles', or by the Wittig reaction. For example, reaction of methyl 2,3-anhydro-4,6-*O*-benzylidene-α-D-mannopyranoside (**38**) with Grignard agents, 2-lithio-1,3-dithians, *etc.*, gives the branched-chain derivative (**40**; X = branch chain). Alternatively, reaction of Wittig agents, such as $(MeO)_2P(O)CHCO_2Et$, with keto derivatives such as the 3-ulose (**43**) gives the branched-chain analogue (**217**), which undergoes stereospecific hydrogenation from the less-hindered face of the molecule to give the *allo*-isomer (**218**), which is capable of further functionalization.

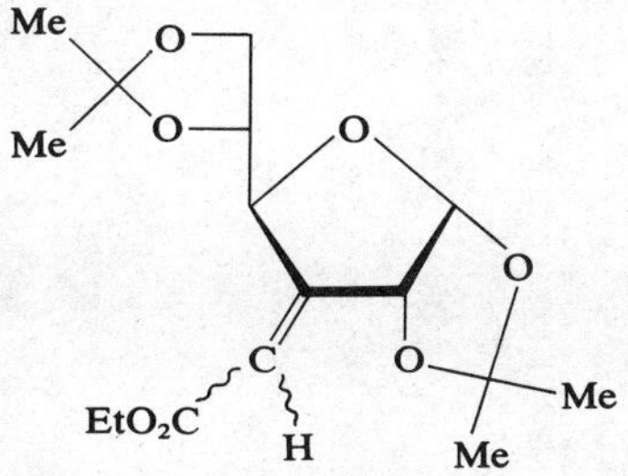

(217)

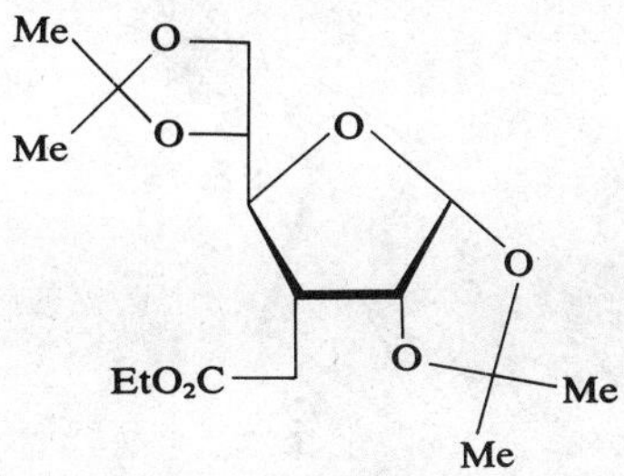

(218)

26.1.9.6 Unsaturated sugars[168]

Sugars containing a carbon–carbon double bond are widely used as synthetic intermediates, although the availability of many of these compounds is quite recent. The classic pyran-1-enoses, trivially termed glycals (*e.g.* **185**), are readily prepared from the corresponding glycosyl bromide (*e.g.* **101**) by reduction with zinc in acetic acid. The related 2-acetoxyglycal is available either by sequential treatment of the glycosyl bromide with iodide anion and triethylamine or directly using 1,5-diazabicyclo[5,4,0]undec-5-ene (DBU).[169]

Addition reactions of glycals have been extensively studied.[168] The mode of addition of ionic reagents is governed by the mesomeric interaction of the ring oxygen with the double bond, so that C-2 being the most nucleophilic, accepts the incoming electrophile (Scheme 21). These addition reactions lead to four possible diastereoisomers, but in

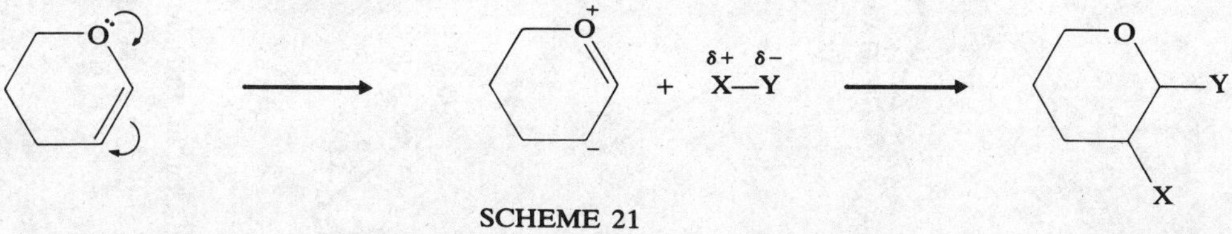

SCHEME 21

practice substantial stereoselectivity is observed which varies with reagents. Thus, iodo(methoxyl)ation of tri-*O*-acetylglucal (**185**) affords the two 1,2-*trans*-2-iodoglycosides with the *manno* and *gluco* configurations in a 2:1 ratio. The *trans* stereochemistry is a result of formation of the 1,2-iodonium ion which undergoes rear-face attack. In contrast, bromination gives the α-D-gluco (1,2-*cis*) and the α-D-manno (1,2-*trans*) isomers in a 2:1 ratio. The *gluco*-isomer in this case is formed from the 1,2-bromonium ion as the β-anomer which, because of the unfavourable anomeric effect, rearranges under the reaction conditions to the α-anomer.

The addition of the pseudohalogen nitrosyl chloride (NOCl) to glycals is of great significance because it forms the basis of a highly selective synthesis of complex α-glycosides.[170] For example, reaction with glucal (**185**) afforded 3,4,6-tri-*O*-acetyl-2-deoxy-2-nitroso-α-D-glucopyranosyl chloride as the dimer (**219**). Reaction of (**219**) with alcohols and phenols gives the corresponding α-glycoside stereospecifically during which the nitroso-group tautomerizes to the oxime (**220**). The sequence is then completed either by deoximation and reduction of the ketone by borohydride, or by reduction of the oxime itself, which gives the 2-amino-2-deoxy glycoside (Scheme 22).

The addition of water, alcohols, phenols, *etc.* to glycals occurs in variable yields under the influence of acid catalysis to give 2-deoxy-sugar derivatives. By contrast, in the absence of acid catalysis, rearrangement occurs. Thus allylic rearrangement of triacetylglucal (**185**) occurs in boiling water to give 4,6-di-*O*-acetyl-2,3-dideoxy-D-*erythro*-hex-2-enose (**221**).

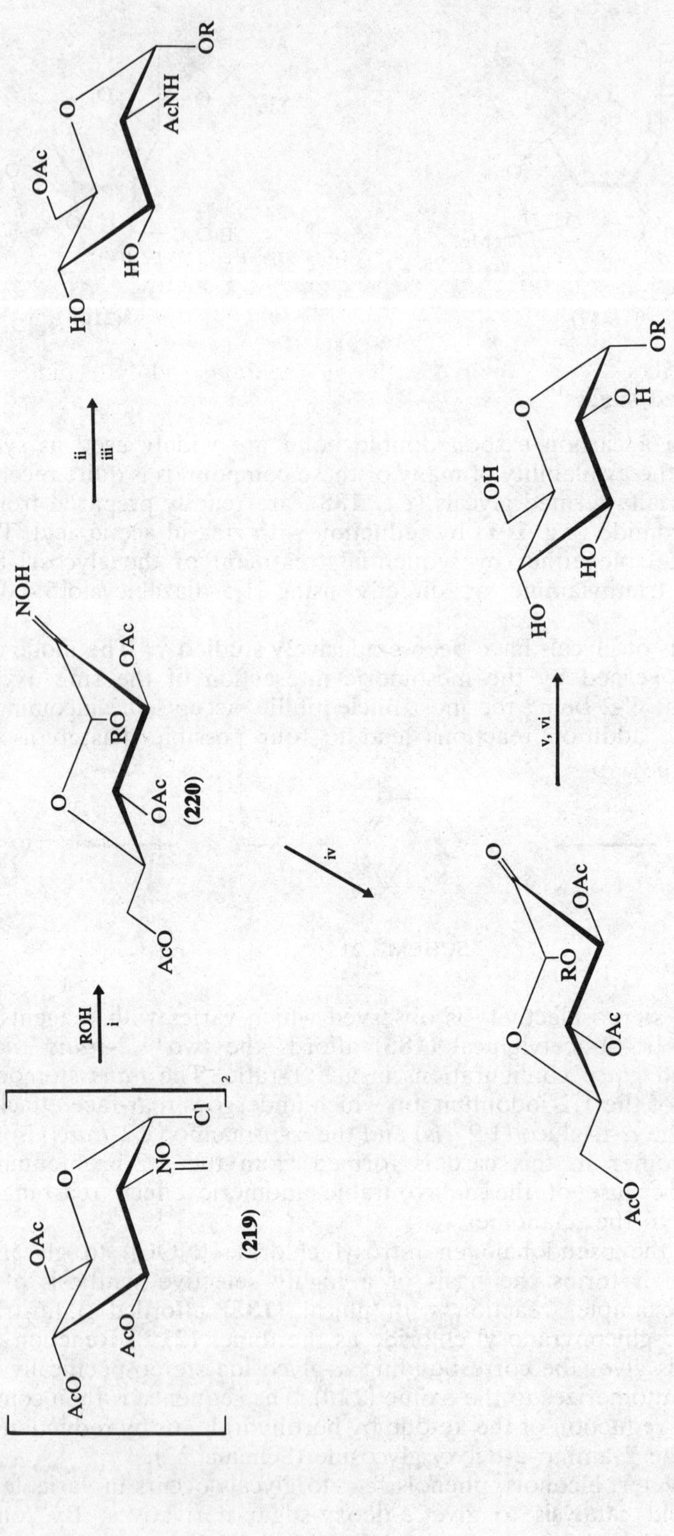

i, ROH, base; ii, H_2, Pd; iii, Et_3N, MeOH; iv, Me_2CO; v, $NaBH_4$; vi, NaOMe.

SCHEME 22

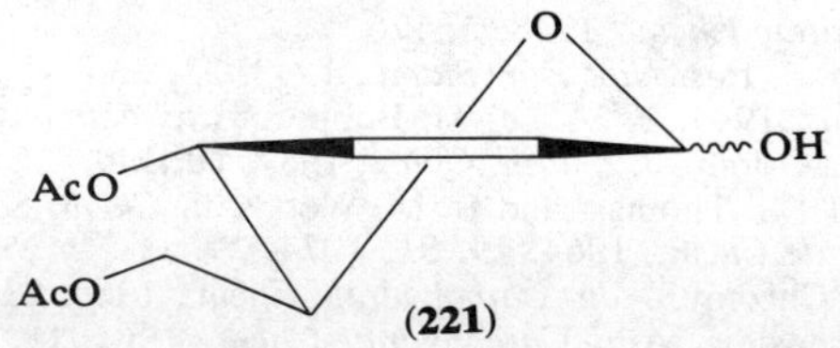

(**221**)

The introduction of unsaturation at other positions in a pyranoid or furanoid sugar is illustrated in Scheme 23 with reference to the synthesis of methyl 4,6-*O*-benzylidene-2,3-dideoxy-α-D-*erythro*-hex-2-enoside (**222**). One of the most widely applied methods for the introduction of unsaturation into carbohydrates is the Tipson–Cohen procedure[171] in which an α,β-ditosylate ester is heated with zinc dust and sodium iodide in DMF.

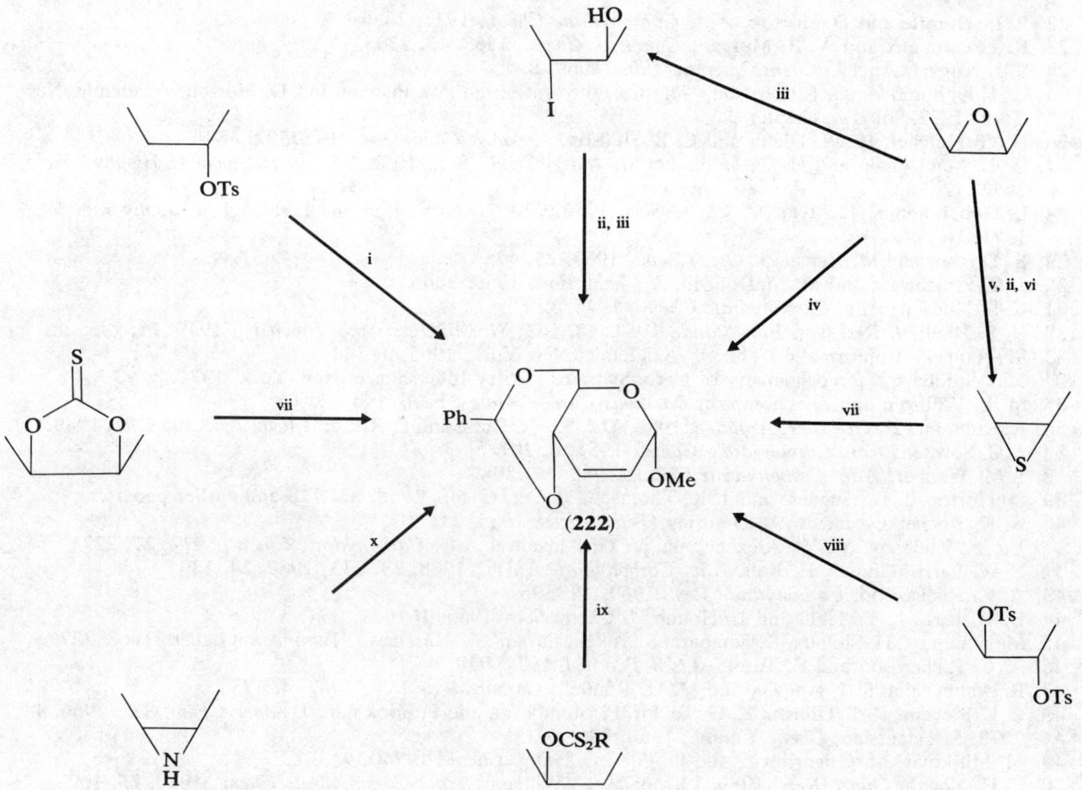

i, base (*e.g.* soda lime); ii, MsCl, py; iii, NaI; iv, $KOCS_2Et$; v, KSCN; vi, NaOMe; vii, $(EtO)_3P$; viii, NaI, Zn, DMF; ix, heat; x, HNO_2.

SCHEME 23

References

1. K. Freudenberg, 'Emil Fischer and his contribution to carbohydrate chemistry', *Adv. Carbohydrate Chem.*, 1966, **21,** 1.
2. J. M. Bijvoet, A. F. Peerdeman, and A. J. van Bommel, *Nature*, 1951, **168,** 271.
3. W. N. Haworth, 'Constitution of the Sugars', Arnold, London, 1929.
4. T. Purdie and J. C. Irvine, *J. Chem. Soc.*, 1903, 1021.
5. E. L. Hirst, *J. Chem. Soc.*, 1926, 350.
6. W. N. Haworth and E. Hirst, *J. Chem. Soc.*, 1926, 1858; 1927, 1237, 2436.
7. H. S. Isbell and C. S. Hudson, *J. Res. Nat. Bur. Stand.*, 1932, **8,** 327, 615; 1933, **10,** 337.

8. J. W. Green, *Adv. Carbohydrate Chem.*, 1948, **3,** 176.
9. H. S. Isbell and H. L. Frush, *J. Res. Nat. Bur. Stand.*, 1943, **31,** 33.
10. E. L. Jackson, *Org. Reactions*, 1944, **2,** 341; J. M. Bobbitt, *Adv. Carbohydrate Chem.*, 1956, **11,** 1.
11. H. Herrissey, P. Fleury, and M. Jolly, *J. Pharm. Chim.*, 1934, **20,** 149.
12. L. Hough, T. J. Taylor, G. H. S. Thomas, and B. M. Woods, *J. Chem. Soc.*, 1958, 1212.
13. L. D. Hall, *Adv. Carbohydrate Chem.*, 1964, **19,** 51; 1974, **29,** 11.
14. N. K. Kochetkov and O. S. Chizhov, *Adv. Carbohydrate Chem.*, 1966, **21,** 39.
15. G. A. Jeffrey and R. D. Rosenstein, *Adv. Carbohydrate Chem.*, 1964, **19,** 7 (For updates see *ibid.*, 1970, **25,** 53; 1974, **30,** 445; 1975, **31,** 347; 1976, **32,** 353; 1976, **33,** 387; 1977, **34,** 345).
16. T. E. Acree, R. S. Shallenberger, C. Y. Lee, and J. W. Einset, *Carbohydrate Res.*, 1969, **10,** 355; A. S. Hill and R. S. Shallenberger, *ibid.*, 1969, **11,** 541.
17. R. E. Reeves, *Adv. Carbohydrate Chem.*, 1951, **6,** 107.
18. N. S. Bhacca, D. Horton, and H. Paulsen, *J. Org. Chem.*, 1968, **33,** 2484.
19. B. Coxon, *Carbohydrate Res.*, 1966, **1,** 357.
20. J. T. Edward, *Chem. and Ind.* (*London*), 1955, 1102.
21. R. U. Lemieux, in 'Molecular Rearrangements Part II', ed. P. De Mayo, Interscience, New York, 1963, p. 735.
22. P. L. Durette and D. Horton, *Adv. Carbohydrate Chem.*, 1971, **26,** 49.
23. R. U. Lemieux and A. R. Morgan, *Canad. J. Chem.*, 1965, **43,** 2205.
24. S. J. Angyal, *Angew. Chem. Internat. Edn.*, 1969, **8,** 157.
25. L. Hough and A. C. Richardson, 'The Carbohydrates', ed. W. Pigman and D. Horton, Academic, New York, 1972, vol. 1A, chapter 3.
26. H. G. Fletcher, H. W. Diehl, and C. S. Hudson, *J. Amer. Chem. Soc.*, 1950, **72,** 4546.
27. D. L. MacDonald and H. O. L. Fischer, *J. Amer. Chem. Soc.*, 1952, **74,** 2087; *Biochem. Biophys. Acta*, 1953, **12,** 203.
28. L. Hough and T. J. Taylor, *J. Chem. Soc.*, 1956, 970; A. Farrington and L. Hough, *Carbohydrate Res.*, 1971, **16,** 59.
29. E. Dimant and M. Banay, *J. Org. Chem.*, 1960, **25,** 475.
30. J. C. P. Schwarz and M. McDougall, *J. Chem. Soc.*, 1956, 3065.
31. C. S. Hudson, *Adv. Carbohydrate Chem.*, 1945, **1,** 1.
32. H. S. Isbell, *J. Res. Nat. Bur. Stand.*, 1952, **48,** 163; W. Militzer, *Arch. Biochem.*, 1949, **21,** 143.
33. M. Hanack, 'Conformation Theory' Academic, New York, 1965, p. 344.
34. J. F. Stoddart, 'Stereochemistry of Carbohydrates', Wiley-Interscience, New York, 1971, p. 93.
35. M. L. Wolfrom and A. Thompson, *Methods Carbohydrate Chem.*, 1963, **2,** 65.
36. R. Kuhn and H. Grassner, *Annalen*, 1958, **612,** 55; R. Kuhn and P. Klesse, *Chem. Ber.*, 1958, **91,** 1989.
37. J. C. Sowden, *Adv. Carbohydrate Chem.*, 1951, **6,** 291.
38. J. M. Webber, *Adv. Carbohydrate Chem.*, 1962, **17,** 20.
39. D. Horton, J. B. Hughes, and J. K. Thomson, *J. Org. Chem.*, 1968, **33,** 728 and earlier papers.
40. N. K. Kochetkov and B. A. Dmitriev, *Tetrahedron*, 1965, **21,** 803.
41. Yu. A. Zhdanov, Yu. E. Alexeev, and V. G. Alexeeva, *Adv. Carbohydrate Chem.*, 1972, **27,** 227.
42. F. W. Parrish and D. H. Ball, *Adv. Carbohydrate Chem.*, 1968, **23,** 233; 1969, **24,** 139.
43. A. C. Richardson, *Carbohydrate Res.*, 1969, **10,** 395.
44. D. H. Buss, L. D. Hall, and L. Hough, *J. Chem. Soc.*, 1965, 1616.
45. M. Nakajima, H. Shibata, K. Kitahara, S. Takahashi, and A. Hasegawa, *Tetrahedron Letters*, 1968, 2271.
46. A. C. Richardson and E. Tarelli, *J.C.S. Perkin I*, 1972, 949.
47. R. Ahluwahlia, S. J. Angyal, and M. H. Randall, *Carbohydrate Res.*, 1967, **4,** 478.
48. C. L. Stevens, R. P. Glinski, K. G. Taylor, P. Blumbergs, and F. Sirokman, *J. Amer. Chem. Soc.*, 1966, **88,** 2073; S. Hanessian, *Chem. Comm.*, 1966, 796.
49. M. Miljkovic, M. Gligorijevic, and D. Glesin, *J. Org. Chem.*, 1974, **39,** 3223.
50. F. H. Newth, *Quart. Rev.*, 1959, **13,** 30; N. R. Williams, *Adv. Carbohydrate Chem.*, 1970, **25,** 109.
51. R. F. Butterworth and S. Hanessian, *Synthesis*, 1971, **49,** 2755.
52. D. C. Baker, D. Horton and C. G. Tindall, Jr., *Carbohydrate Res*, 1972, **24,** 192.
53. K. N. Slessor and A. S. Tracey, *Canad. J. Chem.*, 1970, **48,** 2900; R. U. Lemieux and R. V. Stick, *Austral. J. Chem.*, 1975, **28,** 1799.
54. R. U. Lemieux, T. L. Nagabhushan, and K. James, *Canad. J. Chem.*, 1973, **51,** 1; R. U. Lemieux, T. L. Nagabhushan, K. J. Clemetson, and L. C. N. Tucker, *ibid.*, 1973, **51,** 53.
56. H. Paulsen and C. P. Herold, *Chem. Ber.*, 1970, **103,** 2450.
57. L. Hough and A. C. Richardson, in 'Rodd's Chemistry of Carbon Compounds', Elsevier, Amsterdam, 1967, vol. 1F, chapter 22.
58. P. D. Bragg and L. Hough. *J. Chem. Soc.*, 1957, 4347.
59. M. L. Wolfrom and J. N. Schumacher, *J. Amer. Chem. Soc.*, 1955, **77,** 3318.
60. P. Deslongchamps and C. Moreau, *Canad. J. Chem.*, 1971, **49,** 2465.
61. R. S. Teague, *Adv. Carbohydrate Chem.*, 1954, **9,** 185.
62. B. Capon and B. C. Ghosh, *Chem. Comm.*, 1965, 586.
63. K. Heyns and H. Paulsen, *Adv. Carbohydrate Chem.*, 1962, **17,** 169.
64. C. L. Mehltretter, *Adv. Carbohydrate Chem.*, 1953, **8,** 231.
65. S. Bayne and J. A. Fewster, *Adv. Carbohydrate Chem.*, 1956, **11,** 43.

66. K. E. Pfitzner and J. G. Moffatt, *J. Amer. Chem. Soc.*, 1965, **87,** 5661, 5670.
67. O. Theander, *Adv. Carbohydrate Chem.*, 1962, **17,** 264.
68. J. D. Albright and L. Goldman, *J. Amer. Chem. Soc.*, 1967, **89,** 2416.
69. K. Onodera, S. Hirano, and N. Kashimura, *Carbohydrate Res.*, 1968, **6,** 276.
70. J. R. Parikh and W. von E. Doering, *J. Amer. Chem. Soc.*, 1967, **89,** 5505.
71. P. J. Beynon, P. M. Collins, and W. G. Overend, *Proc. Chem. Soc.*, 1964, 342.
72. R. E. Arrick, D. C. Baker, and D. Horton, *Carbohydrate Res.*, 1973, **26,** 441.
73. V. M. Parikh and J. K. N. Jones, *Canad. J. Chem.*, 1965, **43,** 3452.
74. B. R. Baker and D. H. Buss, *J. Org. Chem.*, 1965, **30,** 2304, 2308.
75. D. Horton, J. S. Jewell, E. K. Just, and J. D. Wander, *Carbohydrate Res.*, 1971, **18,** 49.
76. S. J. Angyal and K. James, *Chem. Comm.*, 1970, 320.
77. S. J. Angyal and K. James, *Austral. J. Chem.*, 1970, **23,** 1209.
78. R. D. Guthrie, *Adv. Carbohydrate Chem.*, 1961, **16,** 105.
79. W. E. Harvey, J. J. Michalski, and A. R. Todd, *J. Chem. Soc.*, 1951, 2271.
80. M. Cantley, L. Hough, and A. O. Pittet, *J. Chem. Soc.*, 1963, 2527.
81. J. E. Hodge, *Adv. Carbohydrate Chem.*, 1955, **10,** 169.
82. N. Prentice, L. S. Cuendet, and F. Smith, *J. Amer. Chem. Soc.*, 1956, **78,** 4439.
83. J. C. Sowden, *Adv. Carbohydrate Chem.*, 1957, **12,** 35; J. D. Crum, *ibid.*, 1958, **13,** 169.
84. J. C. Sowden and D. R. Strobach, *J. Amer. Chem. Soc.*, 1960, **82,** 3707.
85. E. F. L. J. Anet, *Adv. Carbohydrate Chem.*, 1964, **19,** 181.
86. Z. Dische, *Methods Carbohydrate Chem.*, 1962, **1,** 477.
87. W. G. Overend, in 'The Carbohydrates', ed. W. Pigman and D. Horton, Academic, New York, 1972, vol. 1A, chapter 9.
88. C. T. Bishop and F. P. Cooper, *Canad. J. Chem.*, 1962, **40,** 224; 1963, **41,** 2743.
89. B. Capon, G. W. Loveday, and W. G. Overend, *Chem. and Ind.* (*London*), 1962, 1537; B. Capon, *Chem. Rev.*, 1969, **69,** 407.
90. G. Wagner and P. Nahn, *Die Pharmazie*, 1966, **21,** 205, 261.
91. F. Micheel and A. Klemer, *Adv. Carbohydrate Chem.*, 1961, **16,** 85.
92. B. Capon, *Chem. Rev.*, 1969, **69,** 462.
93. H. S. Isbell and H. L. Frush, *J. Res. Nat. Bur. Stand.*, 1949, **43,** 161.
94. N. K. Kochetkov and A. F. Bochkov, *Recent Develop. Chem. Natural Carbon Compounds*, 1971, **4,** 75.
95. R. U. Lemieux and A. R. Morgan, *Canad. J. Chem.*, 1965, **43,** 2199.
96. K. Igarashi, J. Irisawa, and T. Homma, *Carbohydrate Res.*, 1975, **39,** 213, 341.
97. R. U. Lemieux, K. B. Hendricks, R. V. Stick, and K. James, *J. Amer. Chem. Soc.*, 1975, **97,** 4057.
98. R. U. Lemieux and H. Driguez, *J. Amer. Chem. Soc.*, 1975, **97,** 4063, 4069; R. U. Lemieux, D. R. Bundle, and D. A. Baker, *ibid.*, 1975, **97,** 4076.
99. J. Conchie, G. A. Levvy, and C. A. Marsh, *Adv. Carbohydrate Chem.*, 1957, **12,** 157.
100. D. H. Leaback and P. G. Walker, *Biochem. J.*, 1961, **78,** 151.
101. D. Robinson, J. N. Smith, and R. T. Williams, *Biochem. J.*, 1952, **50,** 221.
102. J. N. BeMiller, *Adv. Carbohydrate Chem.*, 1967, **22,** 25.
103. C. E. Ballou, *Adv. Carbohydrate Chem.*, 1954, **9,** 59.
104. R. Kuhn, I. Low, and H. Trischmann, *Chem. Ber.*, 1957, **90,** 203.
105. J. S. Brimacombe, B. D. Jones, M. Stacey, and J. J. Willard, *Carbohydrate Res.*, 1966, **2,** 167; *Methods Carbohydrate Chem.*, 1972, **6,** 376.
106. J. O. Deferrari, E. G. Gros, and I. M. E. Thiel, *Methods Carbohydrate Chem.*, 1972, **6,** 365.
107. J. M. Sugihara, *Adv. Carbohydrate Chem.*, 1953, **8,** 1.
108. C. M. McCloskey, *Adv. Carbohydrate Chem.*, 1957, **12,** 137.
109. R. Gigg and C. D. Warren, *J. Chem. Soc.*, 1965, 2205.
110. F. Franke and R. D. Guthrie, *Austral. J. Chem.*, 1977, **30,** 639.
111. C. B. Reese, R. Saffhill, and J. E. Sulston, *Tetrahedron*, 1970, **26,** 1023.
112. S. Soltzberg, *Adv. Carbohydrate Chem.*, 1970, **25,** 229.
113. L. F. Wiggins, *Adv. Carbohydrate Chem.*, 1950, **5,** 191.
114. M. Cerny and J. Stanek, *Fortschr. Chem. Forsch.*, 1970, **14,** 526.
115. S. J. Angyal and K. Dawes, *Austral. J. Chem.*, 1968, **21,** 2747.
116. C. Bullock, L. Hough, and A. C. Richardson, *Chem. Comm.*, 1971, 1276; J. S. Brimacombe, J. Minshall, and L. C. N. Tucker, *J.C.S. Perkin I*, 1973, 2691.
117. S. Peat, *Adv. Carbohydrate Chem.*, 1946, **2,** 37.
118. G. Birch, C. K. Lee, and A. C. Richardson, *Carbohydrate Res.*, 1971, **19,** 119.
119. N. R. Williams, *Adv. Carbohydrate Chem.*, 1970, **25,** 109.
120. G. Charalambous and E. Percival, *J. Chem. Soc.*, 1954, 2443.
121. J. G. Buchanan and R. Fletcher, *J. Chem. Soc.*, 1965, 6316.
122. M. Horner, L. Hough and A. C. Richardson, *J. Chem. Soc.* (*C*), 1971, 99.
123. A. H. Haines, *Adv. Carbohydrate Chem.*, 1976, **33,** 11.
124. K. Capek, J. Capkova-Steffkova, and J. Jary, *Coll. Czech. Chem. Comm.*, 1966, **31,** 1854.
125. J. M. Williams and A. C. Richardson, *Tetrahedron*, 1967, **23,** 1369, 1641.
126. R. S. Tipson, *Adv. Carbohydrate Chem.*, 1953, **8,** 107.
127. S. E. Creasey and R. D. Guthrie, *J.C.S. Perkin I*, 1974, 1373.

128. L. D. Hall and D. C. Miller, *Carbohydrate Res.*, 1975, **40,** Cl.
129. S. Zen, S. Tashima, and S. Koto, *Bull. Chem. Soc. Japan*, 1968, **41,** 3025; A. D. Barford, A. B. Foster, and J. H. Westwood, *Carbohydrate Res.*, 1970, **13,** 189.
130. H. C. Jarrell, R. G. S. Ritchie, W. A. Szarek, and J. K. N. Jones, *Canad. J. Chem.*, 1973, **51,** 1767.
131. A. N. de Belder, *Adv. Carbohydrate Chem.*, 1965, **20,** 219; A. B. Foster, *Ann. Rev. Biochem.*, 1961, **30,** 45.
132. M. E. Evans and F. W. Parrish, *Carbohydrate Res.*, 1973, **28,** 359.
133. P. J. Garegg and C. G. Swahn, *Acta Chem. Scand.*, 1972, **26,** 3895; N. Baggett, J. M. Duxbury, A. B. Foster, and J. M. Webber, *Carbohydrate Res.*, 1965, **1,** 22.
134. M. E. Evans, F. W. Parrish, and L. Long, *Carbohydrate Res.*, 1967, **3,** 453.
135. F. H. Bissett, M. E. Evans, and F. W. Parrish, *Carbohydrate Res.*, 1967, **5,** 184.
136. R. Khan, *Carbohydrate Res.*, 1974, **32,** 375; 1975, **43,** 247.
137. S. A. Barker and E. J. Bourne, *Adv. Carbohydrate Chem.*, 1952, **7,** 137; J. A. Mills, *ibid.*, 1955, **10,** 1; R. J. Ferrier and W. G. Overend, *Quart. Rev.*, 1959, **13,** 265.
138. J. S. Brimacombe, A. B. Foster, and M. Stacey, *Chem. and Ind.* (*London*), 1958, 1228.
139. T. H. Fife, *Accounts Current Res.*, 1972, **5,** 264.
140. 'The Amino Sugars', ed. R. W. Jeanloz and E. E. Balazs, Academic, New York, 1969, vol. 1A.
141. S. Umezawa, in 'Internat. Rev. Sci., Org. Chem. Ser. 2', Butterworths, London, 1976, vol. 7, chapter 5.
142. A. C. Richardson, in 'Internat. Rev. Sci., Org. Chem. Ser. 1', Butterworths, London, 1973, vol. 7, chapter 4.
143. H. Paulsen and D. Stoye, *Chem. Ber.*, 1969, **102,** 3833.
144. T. Suami and T. Shoji, *Bull. Soc. Chem. Japan*, 1970, **43,** 2948.
145. Y. Ali and A. C. Richardson, *J. Chem. Soc.* (*C*), 1968, 1764; W. Meyer zu Reckendorf, *Chem. Ber.*, 1964, **97,** 1275.
146. F. W. Lichtenthaler, *Fortschr. Chem. Forsch.*, 1970, **14,** 556.
147. H. H. Baer, *Adv. Carbohydrate Chem.*, 1969, **24,** 67.
148. T. Sakakibara and R. Sudoh, *J.C.S. Chem. Comm.*, 1974, 69.
149. L. Goodman, *Adv. Carbohydrate Chem.*, 1967, **22,** 109.
150. R. W. Jeanloz, *J. Amer. Chem. Soc.*, 1957, **79,** 2591.
151. Y. Ali, A. C. Richardson, C. F. Gibbs, and L. Hough, *Carbohydrate Res.*, 1968, **7,** 255.
152. J. S. Brimacombe and J. G. H. Bryan, *Carbohydrate Res.*, 1968, **6,** 423.
153. J. E. G. Barnett, *Adv. Carbohydrate Chem.*, 1967, **22,** 177; S. Hanessian, *Adv. Chem. Ser.*, 1968, **74,** 159; W. A. Szarek, *Adv. Carbohydrate Chem.*, 1973, **28,** 225.
154. A. B. Foster and J. H. Westwood, *Pure Appl. Chem.*, 1973, **35,** 147.
155. R. G. Edwards, L. Hough, A. C. Richardson, and E. Tarelli, *Carbohydrate Res.*, 1974, **35,** 111; R. S. Bhatt, L. Hough, and A. C. Richardson, *ibid.*, 1976, **49,** 103.
156. L. Hough and S. Phadnis, *Nature*, 1976, **263,** 800.
157. S. Hanessian, *Adv. Carbohydrate Chem.*, 1966, **21,** 143.
158. B. T. Lawton, W. A. Szarek, and J. K. N. Jones, *Carbohydrate Res.*, 1970, **14,** 255; 1970, **15,** 397.
159. H. Arita, K. Fukukawa, and Y. Matsushima, *Bull. Chem. Soc. Japan*, 1972, **45,** 3611.
160. D. Horton and D. H. Hutson, *Adv. Carbohydrate Chem.*, 1963, **18,** 115.
161. H. Paulsen and K. Todt, *Adv. Carbohydrate Chem.*, 1968, **23,** 115.
162. J. C. P. Schwarz and K. C. Yule, *Proc. Chem. Soc.*, 1961, 417, T. J. Adley and L. N. Owen, *J. Chem. Soc.* (*C*), 1966, 1287; D. L. Ingles and R. L. Whistler, *J. Org. Chem.*, 1962, **27,** 3896.
163. R. L. Whistler and W. C. Lake, *Methods Carbohydrate Chem.*, 1972, **6,** 286.
164. G. S. Bethell and R. J. Ferrier, *J.C.S. Perkin I*, 1972, 2873; 1973, 1400.
165. F. Shafizadeh, *Adv. Carbohydrate Chem.*, 1956, **11,** 263.
166. R. R. Watson and N. S. Orenstein, *Adv. Carbohydrate Chem.*, 1975, **31,** 135; H. Paulsen, *Stärke*, 1973, **12,** 389.
167. A. M. Sepulchre, A. Gateau-Olesker, G. Lukacs, G. Vass, and S. D. Gero, *Tetrahedron Letters*, 1972, 3945; H. Paulsen, V. Sinnwell, and P. Stadler, *Chem. Ber.*, 1972, **105,** 1978.
168. R. J. Ferrier, *Adv. Carbohydrate Chem.*, 1965, **20,** 67; 1969, **24,** 199; in 'Internat. Rev. Sci., Org. Chem. Ser. 2', Butterworths, London, 1976, vol. 7, chapter 2; B. Fraser-Reid, *Accounts Chem. Res.*, 1975, **8,** 192.
169. R. J. Ferrier and G. H. Sankey, *J. Chem. Soc.* (*C*), 1966, 2339; N. A. Hughes, *Carbohydrate Res.*, 1972, **25,** 242; D. R. Rao and L. M. Lerner, *ibid.*, 1972, **22,** 345.
170. R. U. Lemieux, T. L. Nagabhushan, and I. K. O'Neill. *Canad. J. Chem.*, 1968, **46,** 413.
171. R. S. Tipson and A. Cohen, *Carbohydrate Res.*, 1965, **1,** 338; S. umezawa, Y. Okazaki, and T. Tsuchiya, *Bull. Chem. Soc. Japan*, 1972, **45,** 3619.

26.2
Oligosaccharide Chemistry

L. HOUGH and A. C. RICHARDSON
Queen Elizabeth College, University of London

26.2.1 INTRODUCTION

Linkage of one monosaccharide to another by a glycosidic linkage with the formal elimination of 1 mole of water gives what is known as a disaccharide. The process can be repeated giving tri-, tetra-, penta-saccharides and higher saccharides to the macromolecular species known as polysaccharides. Although not rigorously defined, oligosaccharides are usually considered to be of less than seven or eight monosaccharide units. However, most of the known and readily available members of this class are disaccharides. The most abundant is sucrose,[1] α-D-glucopyranosyl β-D-fructofuranoside (**1**), a non-reducing disaccharide which is ubiquitous in plants. The disaccharide lactose,[2] 4-*O*-β-D-galactopyranosyl-D-glucose (**3**), is an animal product, a constituent of mammalian milk, and is a major waste product of the diary industry. The only other naturally occurring oligosaccharide which is abundantly available is trehalose,[3] α-D-glucopyranosyl α-D-glucopyranoside (**4**), a non-reducing disaccharide that occurs in substantial amounts in dried yeast. Trehalose plays an essential role in the insect kingdom as a storage carbohydrate, from which glucose may be obtained as required. Raffinose (**5**), the only readily available natural trisaccharide occurs, together with sucrose, in sugar beet, but has only 1/10th the sweetness of sucrose. Some oligosaccharides are fairly readily available from the partial hydrolysis of polysaccharides, for example maltose (**6**) from starch and cellobiose (**2**) from cellulose.

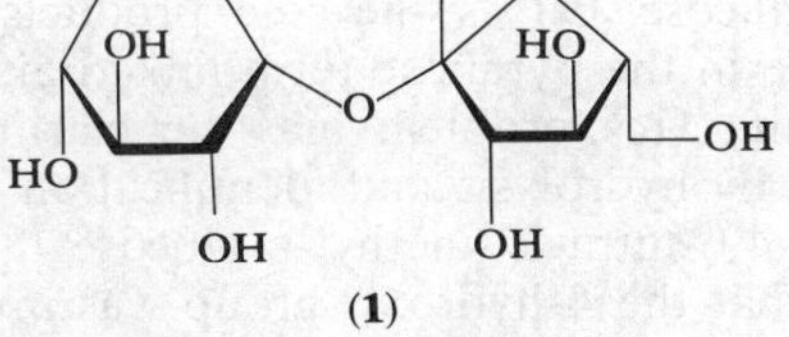

(**1**)

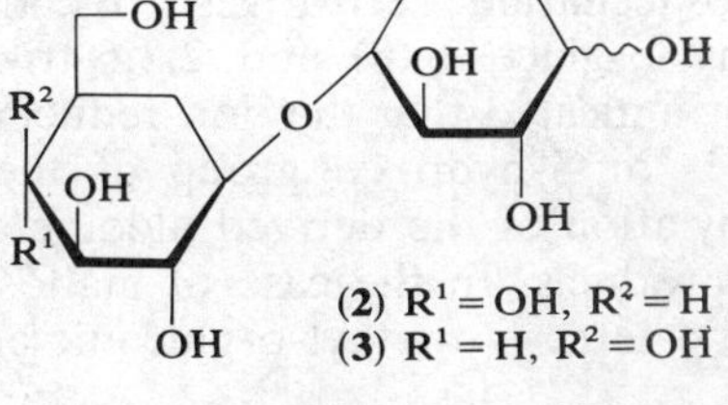

(**2**) $R^1 = OH$, $R^2 = H$
(**3**) $R^1 = H$, $R^2 = OH$

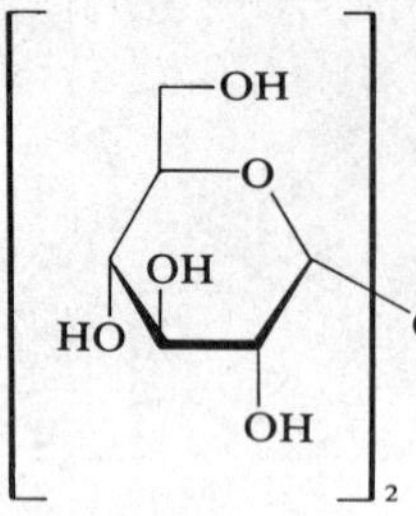

(4)

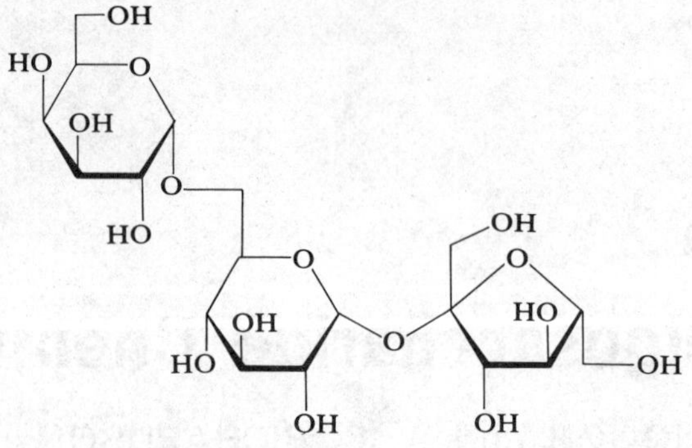

(5)

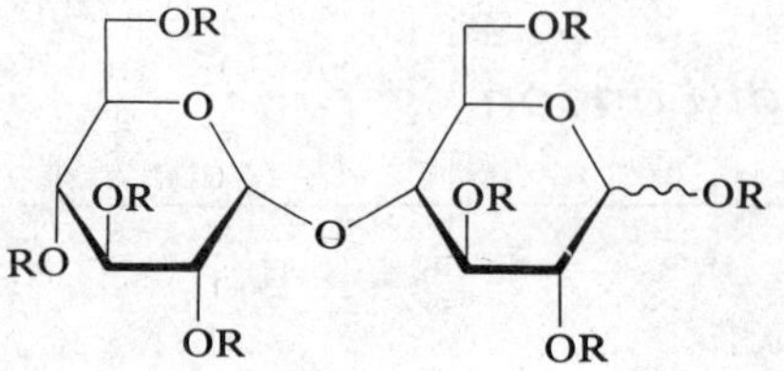

(6) R = H
(7) R = Me

26.2.2 STRUCTURAL DETERMINATION[4]

The first stage in the determination of the structure of oligosaccharides is the identification of the constituent monosaccharides which are formed upon hydrolysis with either dilute acid or enzymes. The specificity of the latter often yields further information regarding the α- or β-configuration of the linkage. The liberated sugars may then be identified chromatographically and the application of g.l.c. is of particular importance[5] in this respect. Thus methanolysis of oligosaccharides produces a possible four glycosides from each common aldose, two pyranosides and two furanosides. After trimethylsilylation and subsequent g.l.c., the retention times and intensity of each component are absolutely definitive for that particular sugar and if, in place of methanol, a chiral alcohol is used such as (−)-butan-2-ol, then the D- and L-enantiomers of an aldose are distinguishable.[6]

If the oligosaccharide possesses a reducing terminus, then this may be identified by either reduction to the alditol or oxidation to the aldonic acid, followed by acid hydrolysis and identification of the appropriate 'modified' aldose. Lactose on oxidation with bromine water affords lactonic acid, hydrolysis of which gives D-galactose and D-gluconic acid. Thus D-glucose occupies the reducing unit of the disaccharide.

The major problem in structure determination of oligosaccharides is the establishment of ring sizes and the positions involved in the glycosidic bonds. The original method developed by Haworth, involving methylation and hydrolysis of the oligosaccharide, has been considerably refined in recent years[7] and may now be carried out with minimal quantities of material, but the basic method remains unchanged. The free hydroxyls in the oligosaccharide are converted into methyl ethers by methylation with dimethyl sulphate–caustic soda or, better, with methyl iodide–methylsulphinyl cation (Hakomori procedure), followed by either acid hydrolysis or methanolysis and identification of the resultant monosaccharide derivatives.[7] By this procedure maltose (**6**) yielded 2,3,4,6-tetra-*O*-methyl-D-glucose (**8**) and 2,3,6-tri-*O*-methyl-D-glucose (**9**) as the sole products. This result indicated that the non-reducing unit must be in the pyranose form linked either to the 4- or 5-hydroxyl group of the reducing unit. This problem may be clarified by methylation of the derived aldonic acid, followed by hydrolysis and identification of the two products. In the case of maltose it gives 2,3,4,6-tetra-*O*-methyl-D-glucose (**8**), and 2,3,5,6-tetra-*O*-methyl-D-gluconic acid, showing that the 4-hydroxyl group was involved in the interglycosidic linkage. The position of linkage, however, is more readily determined by contemporary methods, such as mass spectrometry on microgram quantities.[8,9]

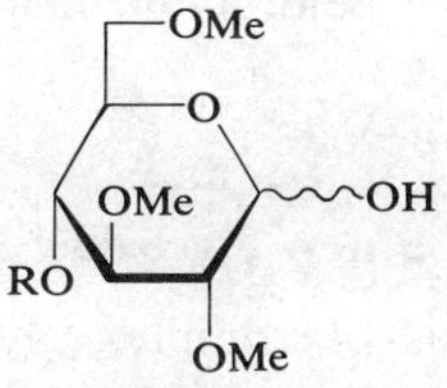

(8) R = Me
(9) R = H

The determination of the anomeric configuration of the interglycosidic bond has been based mainly on enzymic studies, which illustrate the value of enzymology in configurational analysis of oligo- and poly-saccharides. For example, sucrose was recognized as an α-D-glucopyranoside by its hydrolysis with maltase, an enzyme which hydrolyses only α-D-glucopyranosides. Sucrose is also hydrolysed by an enzyme specific for β-D-fructofuranosides and consequently it was assigned as a α-glucosyl β-fructofuranoside. Although enzymes are still widely used in structural work of higher saccharides, ^{1}H and ^{13}C n m.r. spectroscopy[10,11] has largely superseded their uses for the lower oligosaccharides. For example, the ^{1}H n.m.r. spectrum of sucrose shows a low-field doublet ($J_{1,2}$ 3.5 Hz) at about δ 5.5, whereas a β-linkage as in cellobiose would have exhibited a high-field doublet with a larger splitting ($J_{1,2}$ 8.0 Hz). In the case of raffinose (**5**) the ^{1}H n.m.r. spectrum shows two anomeric proton doublet resonances with splittings of *ca.* 3–4 Hz, indicating that both hexose units are attached by α-linkages.

26.2.3 CHEMICAL SYNTHESIS OF OLIGOSACCHARIDES

In principle there are two basic approaches to the synthesis of oligosaccharides: either conversion of one oligosaccharide into another without the need to establish new glycosidic linkages, or construction of the oligosaccharide from simpler units by formation of glycosidic bonds.

26.2.3.1 Interconversions of oligosaccharides

This method is severely limited by the fact that the set of available oligosaccharides for modification is limited to sucrose, lactose, cellobiose, trehalose, raffinose, and maltose [(**1**)–(**6**)]. The reducing disaccharides are easily able to be modified by ascent (Section 26.1.4.2) or descent (Section 26.1.4.1) of the series. Difficulties associated with the synthesis of sucrose (Section 26.2.3.2) and to a lesser extent trehalose has meant that these non-reducing disaccharides are the natural starting materials for most analogues. For example, α-D-galactopyranosyl β-D-fructofuranoside is available from sucrose (**1**) by sequential trimolar tritylation, acetylation, detritylation using conditions under which the 4-*O*-acetyl group migrates to the primary position C-6, tritylation at the remaining two primary positions, and mesylation of the 4-hydroxyl group. Benzoate anion displacement of the sulphonyloxy-group gave the 4-benzoate from which 'galacto' sucrose, a tasteless compound, was available by deblocking.[12]

The symmetrical nature of trehalose (**4**) greatly facilitates its conversion into symmetrical analogues, and in its reactions it resembles those of methyl α-D-glucopyranoside; the corresponding 'allo', 'galacto', 'altro' trehaloses have been synthesized along with many variously substituted derivatives.[13] Non-symmetrical analogues of trehalose are available by interrupting the reactions of trehalose at an intermediary point. For example, first-order reactions of each sugar residue of trehalose may be treated as consecutive reactions, in which the probability of the second residue reacting is only half that of the first. Thus there is always a build-up of non-symmetrical products during a reaction to a maximum of 50%. Therefore hydrolysis of 4,6:4′,6′-di-*O*-benzylidenetrehalose tetrabenzoate gives the monobenzylidene derivative in 43% yield,[14] and from this a variety of non-symmetrical trehaloses have been prepared.

26.2.3.2 Synthesis by formation of a new glycosidic linkage

The procedures for the synthesis of complex glycosides have been summarized in Section 26.1.8.1, and these methods apply to the synthesis of oligosaccharides. Hence, the synthesis of a particular disaccharide requires that the monosaccharide which is to become the reducing part of the final product be suitably protected so that the hydroxyl group which is to be involved in glycoside formation remains unsubstituted. It is then reacted with the appropriate glycosyl halide or related compound. The 'oxazoline method' is one of the most efficient and general methods for the synthesis of 1,2-*trans*-glycosides (Section 26.1.8.1). Thus reaction of 1,2,3,4-tetra-*O*-acetyl-β-D-glucopyranose (**10**) with the 1,2-orthobenzoate (**11**) gave the disaccharide ester (**12**) in 93% yield.

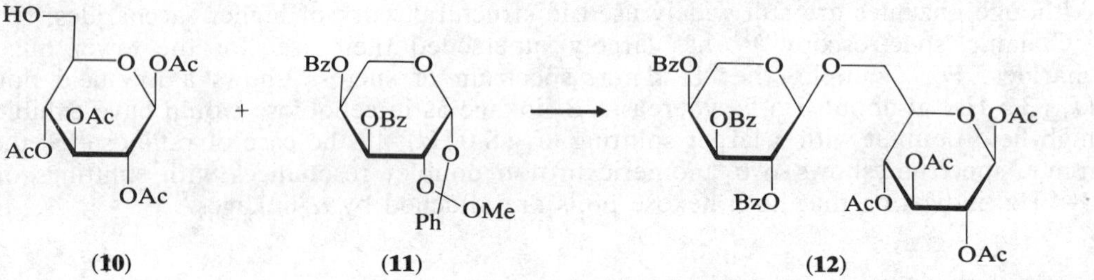

(**10**) (**11**) (**12**)

The synthesis of 1,2-*cis*-glycosides requires special considerations which are discussed in Section 26.1.8.1. In particular, non-participating *O*-blocking groups must be employed and the 1,2-*trans*-glycosyl halide must be the glycosidating species. For example, the disaccharide derivative (**13**) was glycosidated with 2,3,4-tri-*O*-benzyl-6-deoxy-α-L-galactopyranosyl bromide in the presence of tetraethylammonium bromide (to anomerize the α-bromide to the β-anomer) and ethyldi-isopropylamine (a hindered base) in dichloromethane–DMF to give the trisaccharide (**14**).[15] The synthesis of 1,2-*cis*-linked oligosaccharides of 2-amino-2-deoxyhexoses, important in the total synthesis of aminoglycoside antibiotics, *etc.*, has been achieved using the non-participating azide group at C-2, which was then subsequently converted into the amine at the last stage.[16] An alternative procedure for the synthesis of similar glycosides is discussed in Section 26.1.9.6.

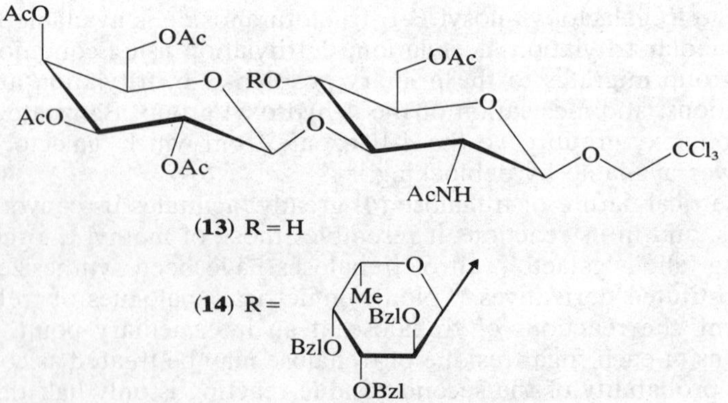

(**13**) R = H

(**14**) R =

References

1. R. Khan, *Adv. Carbohydrate Chem.*, 1976, **33,** 236.
2. J. R. Clamp, L. Hough, J. L. Hickson, and R. L. Whistler, *Adv. Carbohydrate Chem.*, 1961, **16,** 159.
3. G. G. Birch, *Adv. Carbohydrate Chem.*, 1963, **18,** 201.
4. J. Stanek, M. Cerny, and J. Pacak. 'The Oligosaccharides', Academic, New York, 1965.
5. G. G. S. Dutton, *Adv. Carbohydrate Chem.*, 1973, **28,** 11; 1974, **30,** 9.
6. J. P. Kamerling. G. J. Gerwig, and J. F. G. Vliegenthart, *Carbohydrate Res.*, 1978, in press.
7. S.-I. Hakomori, *J. Biochem.* (*Japan*), 1964, **55,** 205.
8. J. Lonngren and S. Svensson, *Adv. Carbohydrate Chem.*, 1974, **29,** 41.
9. N. K. Kochetkov and O. S. Chizhov, *Adv. Carbohydrate Chem.*, 1966, **21,** 39.
10. L. D. Hall, *Adv. Carbohydrate Chem.*, 1964, **19,** 51; 1974, **29,** 11.
11. A. S. Perlin, in 'Internat. Rev. Sci., Org. Chem. Ser. 2', Butterworths, London, 1976, vol. 7, chapter 1.
12. P. H. Fairclough, L. Hough, and A. C. Richardson, *Carbohydrate Res.*, 1975, **40,** 285.
13. A. F. Hadfield, L. Hough, and A. C. Richardson, *Carbohydrate Res.*, 1978, **63,** 51 and earlier papers.
14. A. C. Richardson and E. Tarelli, *J. Chem. Soc.* (*C*), 1971, 3733.
15. J. M. Sugihara, *Adv. Carbohydrate Chem.*, 1953, **8,** 1.
16. H. Paulsen and W. Stenzel, *Angew. Chem. Internat. Edn.*, 1975, **14,** 558.

26.3

Polysaccharides

J. F. KENNEDY
University of Birmingham

AND

C. A. WHITE
Lea Castle Hospital, Kidderminster

26.3.1 NATURE, OCCURRENCE, AND CLASSIFICATION OF POLYSACCHARIDES

Polysaccharides are high molecular weight carbohydrates formed as a result of condensation reactions in which normal neutral monosaccharides (or monosaccharides such as 2-amino-2-deoxyhexoses or hexuronic acids derived from them) are glycosidically linked with elimination of water between the hydroxyl group at C-1 of a monosaccharide unit and an available hydroxyl group of another unit. These linkages are the same as those in oligosaccharides, the difference between the two molecular types being only in molecular weight. Since there are no easily defined upper or lower limits for molecular weight, the division between oligosaccharides and polysaccharides is arbitrarily set at 10 monosaccharide units, but this division is made easier since carbohydrates containing 5–15 units rarely occur in nature; the majority which are found contain between 80 and 100 units, with only a few in the range 25–75 units. There are some condensation polymers which contain more than 100 units, *e.g.* native cellulose contains an average of 3000 units. More accurately, these polysaccharides exists as a homologous series of polymers with distributions of molecular weights about a mean value, not as discrete macromolecules of identical molecular weight.

Polysaccharides are natural macromolecules occurring in almost all living organisms and are one of the largest groups of all natural compounds classified thus far. They function either as an energy source or as structural units in the morphology of the living material to which they are endogenous. For examples of structural functions, cellulose, a polymer of D-glucose, is the most abundant organic substance in nature and is the structural material of plants, whilst chitin, a polymer of 2-acetamido-2-deoxy-D-glucose, is the major organic component of the exoskeleton of crabs, lobsters, *etc.* As one of the main sources of energy for living organisms, certain polysaccharides form part of the central pathway of energy in most cells. The starches and glycogens, long-chain polymers of D-glucose, are the media for energy storage in plants and animals, respectively. Polysaccharides also perform more specific roles, such as being responsible for the type specificity of the *Pneumococci.* Other natural macromolecules, which are not composed entirely of sugar units, contain blocks of monosaccharide units as part of the molecular structure and contribute extensively to the production and maintenance of living tissue of animals. The blood group substances, for example, are a group of glycoproteins in which the arrangements of monosaccharide residues in the carbohydrates sub-units contribute towards the blood group specificity of the overall molecule.

Trivial names of the polysaccharides usually reflect their origin; examples include cellulose, the principle component of cell walls in plants, and dermatan, a polysaccharide normally occurring in its sulphated form and originally found in the dermal layer of skin, *derma.* The trivial names can also reflect some property of the isolated polymer, *e.g.* starch, a name derived from *stercan,* meaning to stiffen. The origin of a polysaccharide leads to differences within a polysaccharide type. Thus since, for example, starches from various plant sources are readily distinguished chemically, it is necessary to specify the origin in naming the starch definitively, *e.g.* maize starch. The traditional names of long standing, such as cellulose, glycogen, and amylose, are still retained, but with the increase in knowledge of the structure of these compounds, nomenclature and classification are

TABLE 1
Structures of the Common Monosaccharides[a]

Pentoses

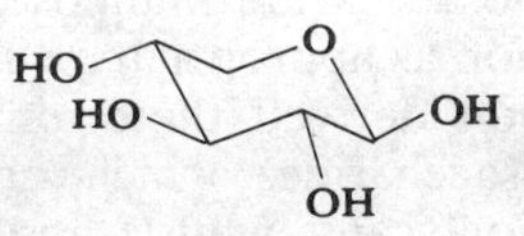
D-Xylopyranose

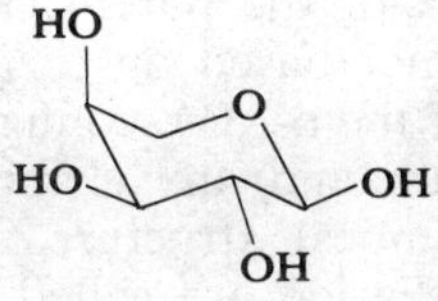
L-Arabinopyranose

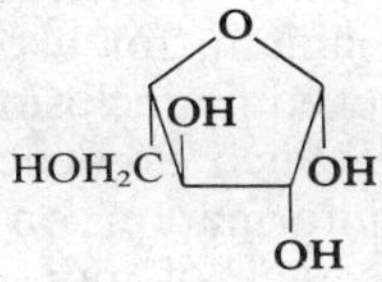
L-Arabinofuranose

Hexoses

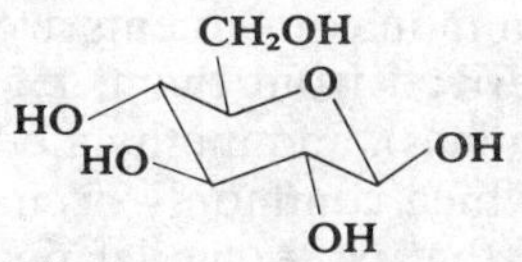
D-Glucopyranose

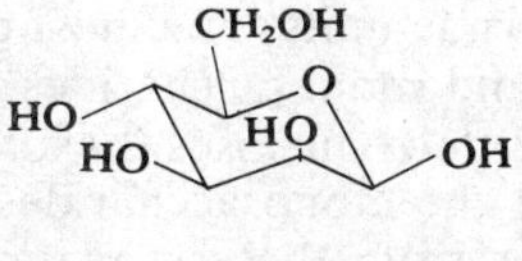
D-Mannopyranose

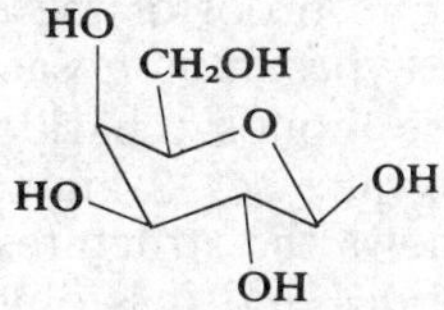
D-Galactopyranose

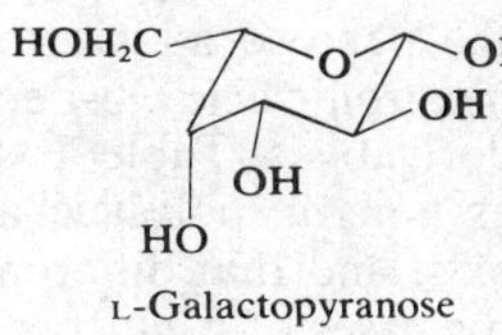
L-Galactopyranose

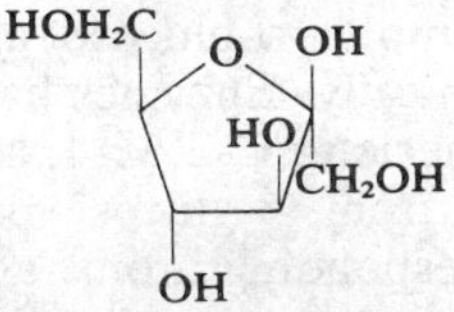
D-Fructofuranose

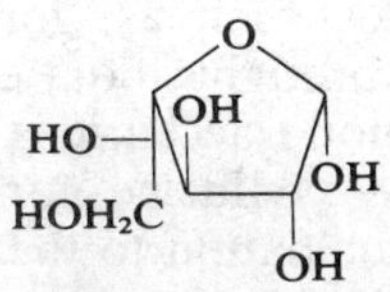
D-Galactofuranose

Hexuronic acids

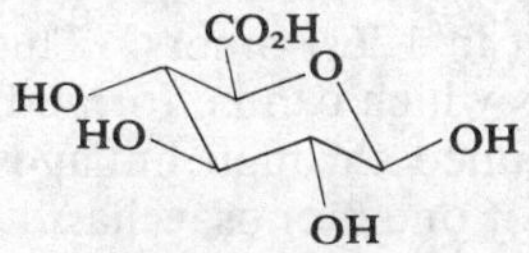
D-Glucopyranuronic acid

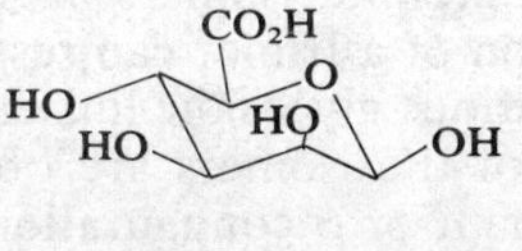
D-Mannopyranuronic acid

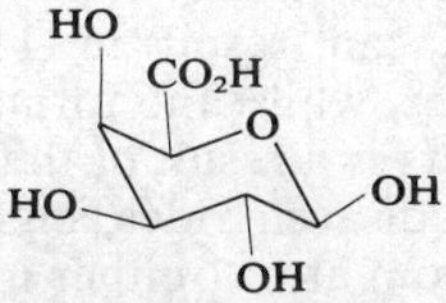
D-Galactopyranuronic acid

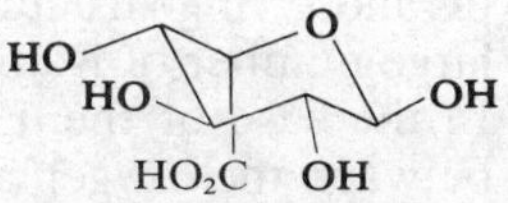
L-Idopyranuronic acid

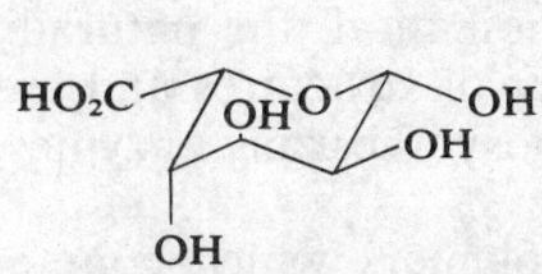
L-Gulopyranuronic acid

2-Amino-2-deoxyhexoses (hexosamines)

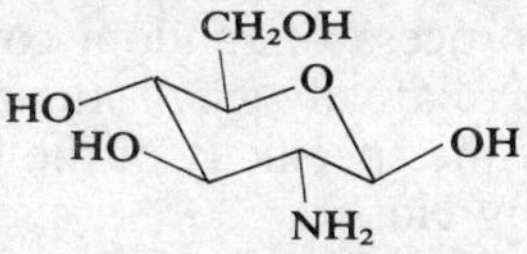
2-Amino-2-deoxy-D-glucopyranose (D-Glucosamine)

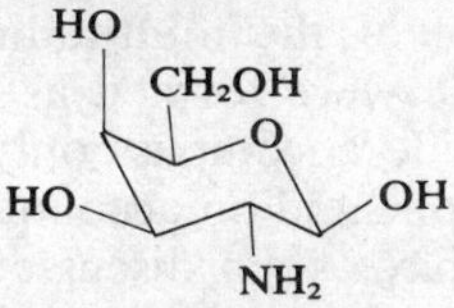
2-Amino-2-deoxy-D-galactopyranoside (D-Galactosamine)

Esters

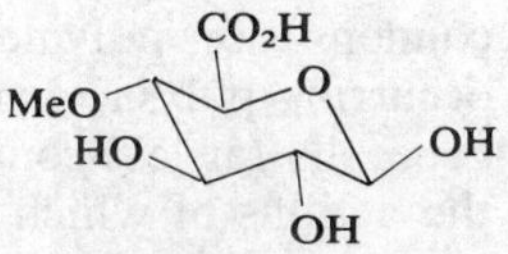
4-*O*-Methyl-D-glucopyranuronic acid

Sialic acids

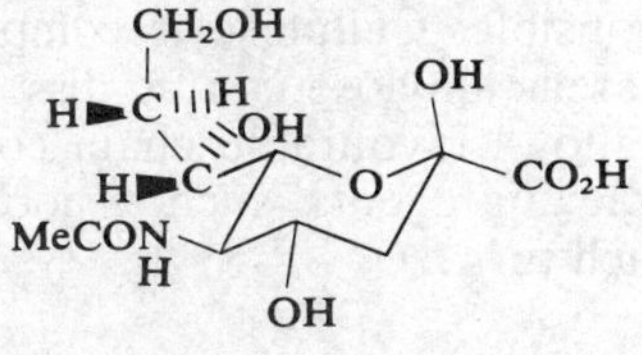
5-Acetamido-3,5-dideoxy-D-*glycero*-D-*galacto*-2-nonulosonic acid

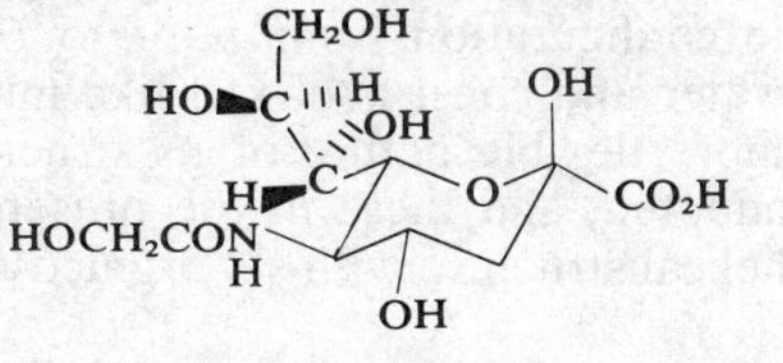
5-Glycolylamido-3,5-dideoxy-D-*glycero*-D-*galacto*-2-nonulosonic acid

[a] Each shown in one anomeric form only.

now being made primarily in terms of structure, and in the interests of systematization all new discoveries should be named in this way. The term *glycan*, derived from glycose, meaning a simple sugar, is another word for polysaccharide, but a more specific term is obtained by using the configurational prefix of the parent sugar with the suffix *-an* to signify a polymer, *e.g.* mannan, for a polymer based on the monosaccharide mannose. Further specificity is achieved by inclusion of the D- or L-configuration as appropriate, *e.g.* D-glycan from D-glucose. Such nomenclature and any classification derived therefrom should ideally include information on chemical structure. Polysaccharides which on hydrolysis yield only one type of monosaccharide are called homoglycans, whilst those which hydrolyse to yield more than one type of monosaccharide are called heteroglycans, with designatory prefixes of di- or tri- for the number of monosaccharide types involved.

There is, at present, no proof of the existence of polysaccharides which contain more than about six different types of monosaccharide units. The most common constituents are the pentose and hexose monosaccharides and monosaccharides derived from them, *e.g.* hexuronic acids, 6-deoxyhexoses, 2-amino-2-deoxyhexoses (hexosamines), and methylated derivatives. Table 1 shows the structures of the monosaccharides which commonly occur as units in polysaccharides and it is obvious from these structures that close similarities exist and that differences largely exist on account of differences in stereochemistry, as in the cases of D-glucose and D-galactose, where the only difference is the stereochemistry of one asymmetric carbon atom.

Interconversion of pairs of monosaccharides such as, for example, D-glucose and D-mannose, which are epimeric at C-2, can be brought about enzymically. Enzymes have been isolated which can interconvert pairs of monosaccharides epimeric at C-2, C-4, and C-5. The pentoses D-xylose and D-arabinose could be formed from D-glucose and D-galactose, respectively, through a process of oxidation to the corresponding uronic acid and subsequent decarboxylation, reactions which also have been found to occur naturally.

Formation of a disaccharide by a condensation reaction between two identical monosaccharide units can result in 11 different isomers if one considers only the pyranose ring structures, whilst the formation of a trimer can result in 176 isomers.[1] This large number of isomers is a result of the various glycosidic linkages which can be formed. In the case of the 11 disaccharides, eight of the isomers are obtained through linkages between the oxygen atom at C-1 either in the α or β configuration of one monosaccharide to carbon atoms C-2, C-3, C-4, or C-6 [termed α-(1 → 2)-, β-(1 → 3)-linkages, *etc.*] of another monosaccharide. The other three isomers are obtained by bond formation between both C-1 carbon atoms *via* the glycosidic oxygen atom in $\alpha\alpha$, $\alpha\beta$, or $\beta\beta$ configuration. In spite of the multitudinous possibilities which could therefore exist if condensation polymers were formed as a process of random combination, the naturally occurring polysaccharide encompass only a relatively small number of structures and are generally far less complicated on account of the specificity of the biosynthesizing enzymes, the actions of which have been discussed by Hassid.[2]

The simplest types of homopolysaccharides are the linear polymers which contain glycosidic linkages of one type only, *e.g.* amylose (**1**) and cellulose (**2**), which only contain α-(1 → 4)- and β-(1 → 4)-linked D-glucose units, respectively. The differences in properties between these two polysaccharides is very marked and the apparently simple change in configuration from α-D- to β-D- is responsible. Cellulose is completely insoluble in water and consists of rod-like molecules whereas amylose can be dissolved, owing to the more flexible nature of its structure which allows favourable chain conformations to be adopted, and these in the presence of complexing agents, such as iodine, can exist as a helical structure with six D-glucose units in each turn.

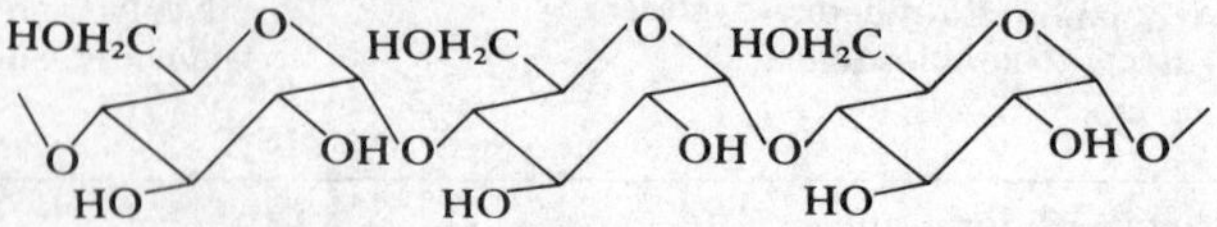

(**1**)

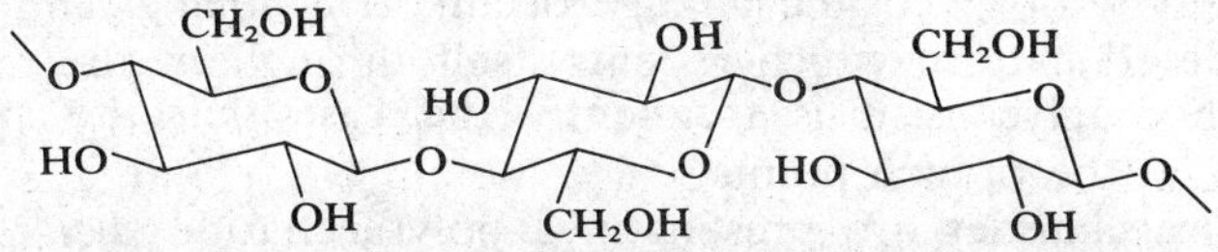

(**2**)

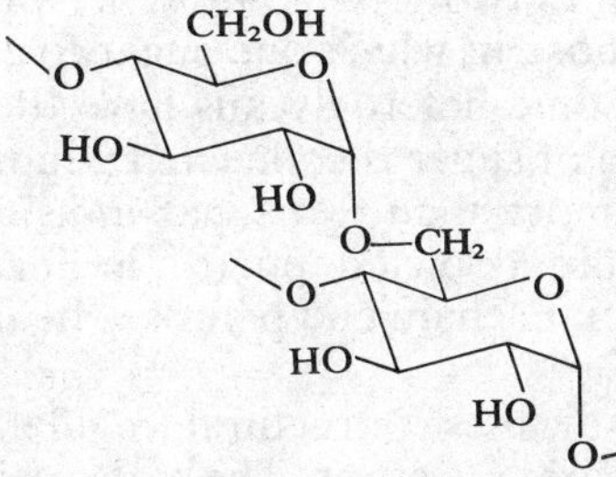

(**3**)

The degree of complexity increases when the hydroxyl group at C-1 from two monosaccharide units both form glycosidic linkages with carbon atoms, other than at C-1, of a third monosaccharide unit, forming a branch point (**3**). Figure 1 shows the way in which the complexity increases with the degree of branching, from a linear molecule depicted as (a) to a highly branched tree like molecule (d). In no known case has a three dimensional cage structure been found to exist.

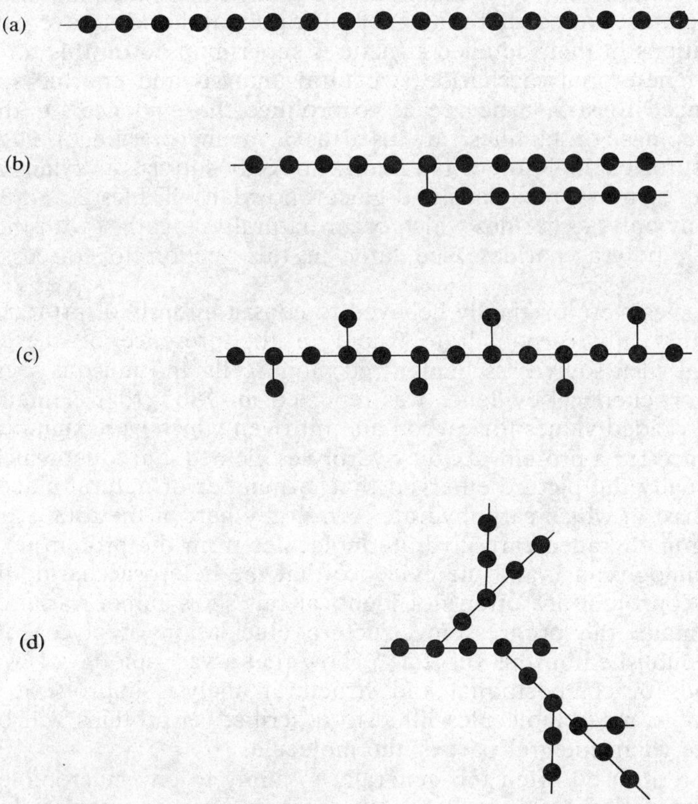

Figure 1 Possible arrangements of monosaccharides in polysaccharides.

When more than one type of glycosidic linkage occurs in a *homoglycan* the linkages tend to occur in an ordered pattern which repeats itself throughout the molecule. For example, in branched homoglycans it is frequently the case that the same type of glycosidic linkage occurs at the branch points.

If two or more monosaccharides are present in a polysaccharide (*heteroglycan*), the complexity of the situation is greatly increased, but it is generally the case that the polysaccharide consists of an ordered arrangement. The glycans which consist of two types of monosaccharides can be divided into two groups: those in which both monosaccharides exist in the same linear chain, and those in which one sugar forms the main chain and the other is present in the side chains. Some heteroglycans have the structure depicted by (c) in Figure 1, in which the second type of sugar is connected pendantly to the main chain by identical glycosidic linkages. When more than two sugar units are present the usual type of structure is the tree-like structure depicted by (d) in Figure 1, with hexoses and hexuronic acids predominant in the main chain and pentoses being predominant in the side branches.

The naming of heteroglycans uses these structural regularities with configurational prefixes for each type of monosaccharide present. The last prefix is usually that denoting the monosaccharide which dominates the main chain of a linear chain or forms the main chain of a branched glycan, with the first prefix(es) denoting the monosaccharide(s) attaching to this main chain. For example, a branched structure consisting of D-galactose units which are attached to a main chain of D-mannose units would be described as a D-galacto-D-mannan rather than a D-manno-D-galactan. In more complex polysaccharides which have less defined arrangements the prefixes are often placed in alphabetical order, but to avoid over-complex and cumbersome names it is often the case that the names of only one or two of the more important constitutional monosaccharides are used as prefixes. Alternatively, complex polysaccharides are named after their natural sources.

It has recently been discovered that many complex polysaccharides have common root structures with variations in their detailed structures superimposed on this root structure. The development of heteropolysaccharide structural analysis and enzyme susceptibility chemistry has advanced to such a degree as to produce the evidence for the common features. These groupings, or families, are used as a means of categorizing the many varied heteroglycans, for example the xylan family not only consists of xylans in the strict sense of the term but also arabinoxylans and glucuronoxylans. Tables 2 and 3 give some indication of the many polysaccharides which occur naturally, together with the structures and sources of such polysaccharides (see later in this section for discussion of the glycosaminoglycans).

Many macromolecules were originally believed to consist entirely of protein and it was formerly believed that any carbohydrate found in the presence of these biological macromolecules from such sources as human red blood cells and mucous secretions was an impurity. However, chemical evidence was reported, in 1865, that elemental analysis of a purified mucin[3] yielded values for carbon and nitrogen which were significantly lower than would be obtained for a protein. Acidic hydrolyses yielded a product which appeared to be glucose. Gradually the picture emerged that a number of natural macromolecules (*i.e.* glycoproteins) exist in which carbohydrate forms only part of the total structure. The difficulty in isolating undegraded carbohydrate molecules from the protein (except in the case of the glycosaminoglycans), and the evidence that the heterosaccharide present in a single species of glycoprotein are often not identical but show minor variations in their composition, have made the progress in structure elucidation of glycoproteins much slower, but reviews published on this subject[4–6] show that a vast amount of work is being pursued. As methods of compositional and structural analysis improve it is virtually certain that many more macromolecules hitherto described as proteins will be found to contain carbohydrate as an integral part of the molecule.

Glycoprotein is a term used often too-generally to apply to any macromolecule which contains carbohydrate and protein, and in such areas of use the term really applies to molecules which if properly classified come under the headings of glycoproteins, proteoglycans, and carbohydrate–protein complexes. Glycoproteins contain a protein chain which

TABLE 2
Structure and Source of the Common Homopolysaccharides

Linkage	*Source*	*Common name*
L-Arabinans		
α-1,3-, α-1,5- branched	Plant pectic substances	
D-Fructans		
β-1,2- linear	Dandelions, dahlias, Jerusalem artichokes	Inulin
β-2,6- linear	Various grasses	Levans
β-2,6-, β-2,1- branched	Plants and bacteria	Levans
L-Fucans		
α-1,2-, α-1,4- branched	Brown seaweeds	Fucoidan
D-Galactans		
β-1,3-, α-1,4-linear	Red seaweeds	Carrageenan
β-1,3-, β-1,6- branched	Beef lung	
β-1,4- linear	Plant pectic substances	
β-1,5- linear	Penicillin mould	Galactocarolose
D-Galacturonans		
α-1,4- linear	Plant pectic substances	Pectic acid
D-Glucans		
β-1,2-	Agrobacteria	
α-1,3-, α-1,4-	*Aspergillus niger*, Iceland Moss	Nigeran, isolichenan
β-1,3-	Brown seaweeds, plants, algae, fungi and yeasts	Laminaran, callose
β-1,3-, β-1,4-	Iceland Moss, cereal grains	Lichenan
α-1,4-	Plant starches	Amylase
α-1,4-, α-1,6- linear	Fungi	Pullulan
α-1,4-, α-1,6- branched	Plant starches, animals, and micro-organisms	Glycogen, amylopectin
β-1,4- linear	Plant cell walls	Cellulose
α-1,6-, α-1,3- branched	Bacteria	Dextran
β-1,6- linear	Lichen	Pustulan
2-Amino-2-deoxy-D-glucans		
β-1,4- linear	Crab and lobster shells, fungi	Chitin
D-Mannans		
α-1,2-, α-1,6- branched	Yeasts	
β-1,4- linear	Seaweeds, plants	
D-Xylans		
β-1,3- linear	Green seaweed	Rhodymenan
β-1,3-, β-1,4- linear	Red seaweed	
β-1,4- linear	Plant cell walls	

TABLE 3
Structure and Source of some Heteropolysaccharides

Constituent monosaccharides and chain type	*Source*	*Common Name*
L-Arabinose, D-galactose, branched	Coniferous woods	
L-Arabinose, D-xylose, branched	Plant cell walls	
DL-Galactose, linear	Red seaweeds	Agarose, porphyran
branched	Snails	
D-Galactose, 2-amino-2-deoxy-D-glucose, linear	Cornea	Keratan sulphate
D-Galactose, D-mannose, branched	Leguminous seeds, fungi	
2-Amino-2-deoxy-D-galactose, D-glucuronic acid, linear	Cornea, cartilage	Chondroitin, chondroitin sulphates
2-Amino-2-deoxy-D-galactose, L-iduronic acid, linear	Skin	Dermatan sulphate
D-Glucose, D-mannose, linear	Coniferous woods, seeds, bulbs	
2-Amino-2-deoxy-D-glucose, D-glucuronic acid, linear	Animal and mammalian tissues	Hyaluronic acid
D-Glucuronic acid, D-xylose, branched	Plant cell walls	
L-Guluronic acid, D-mannuronic acid, linear	Bacteria, brown seaweeds	Alginic acid

consists of about 300 or so amino-acid units which are any of the 20 naturally occurring L-α-amino-acids. Covalently attached to this protein backbone and pendant to it is the carbohydrate part of the molecule, which consists of hetero-oligosaccharide chains. These are usually branched (see Figure 2) and can contain neutral monosaccharides (D-glucose, D-galactose, D-mannose, or L-fucose), basic monosaccharides (2-amino-2-deoxy-D-glucose or 2-amino-2-deoxy-D-galactose), and acidic monosaccharides (5-amino-3,5-dideoxy-D-*glycero*-D-*galacto*-2-nonulosonic acid). The basic carbohydrates may be *N*-acetylated and the acidic units may be *N*-glycolylated or *N*- and, in some cases, *O*-acetylated, which therefore renders the oligosaccharide chain mildly acidic.

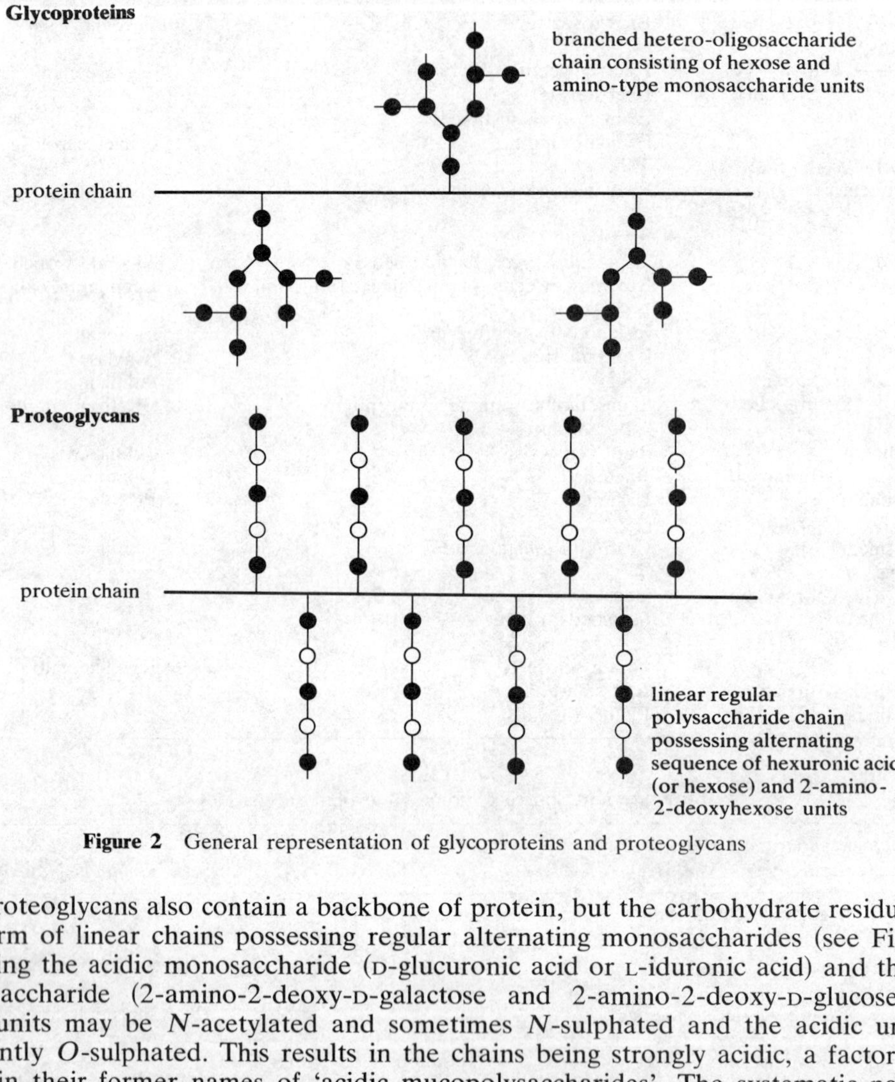

Figure 2 General representation of glycoproteins and proteoglycans

Proteoglycans also contain a backbone of protein, but the carbohydrate residue takes the form of linear chains possessing regular alternating monosaccharides (see Figure 2) involving the acidic monosaccharide (D-glucuronic acid or L-iduronic acid) and the basic monosaccharide (2-amino-2-deoxy-D-galactose and 2-amino-2-deoxy-D-glucose). The basic units may be *N*-acetylated and sometimes *N*-sulphated and the acidic units are frequently *O*-sulphated. This results in the chains being strongly acidic, a factor recognized in their former names of 'acidic mucopolysaccharides'. The systematic name for these chains is 'glycosaminoglycans'.

A factor distinguishing glycoproteins from proteoglycans, as the names suggest, is the number of carbohydrates units per unit length (or molecular weight) of protein backbone, with protein being predominant in glycoprotein and carbohydrates predominating in proteoglycans. The term carbohydrate–protein complex is used to describe molecules which contain protein and carbohydrates and which are linked by non-covalent (usually ionic) bonds. A full discussion of the nomenclature of glycosaminoglycans and glycoproteins has been published.[7]

26.3.2 STRUCTURAL ANALYSIS

The first problem of structural analysis is one which is common to the analysis of other groups of macromolecules, and is that of isolating the material in a pure form. The definition of purity is not as clear-cut as was originally thought, because microheterogeneity, the phenomenon of minor variations within a single species of compound, is now well recognized. The ensuing discussion on the separation of the carbohydrate species from various impurities, including inorganic salts and low molecular weight materials, and also macromolecular species such as proteins and lignins, is written from the general viewpoint, but it must be borne in mind that each polysaccharide will have its own peculiarities.

Wherever possible, the first stage involves solubilization in an aqueous or an aprotic solvent such as ethylene glycol or dimethyl sulphoxide, but care must be taken to ensure that the method used and solvents chosen do not modify or degrade the structure of the macromolecule. This eliminates the use of acids, alkalis, or enzymes. The removal of low molecular weight impurities is readily performed by dialysis (distinction on basis of molecular size), ion exchange chromatography (distinction on basis of molecular charge), or gel filtration (distinction on basis of molecular size) (see Section 26.3.2.6). The last two techniques are also used extensively for the separation of the desired material from contaminant macromolecules. Removal of the macromolecule from solution can be achieved by precipitation with solvents such as ethanol or acetone, or by complexation with, for example, metal ions, and, for acidic polysaccharides, quaternary ammonium salts.

Purification stages must proceed until a material of constant composition is obtained. In the past the criteria for estimation of constant composition have relied on physical and chemical measurements such as functional group analysis, specific rotation, and carbohydrate composition after hydrolysis. More recently ultracentrifugation, which gives a measure of sedimentation in a high force field, electrophoresis, which measures the mobility of the polysaccharide in an applied electric field, and the aforementioned chromatographies have all been used to investigate the purity of a preparation. The best definition of homogeneity is one based on assessment by two or more methods which rely on different criteria.

Once a pure sample of the polysaccharide has been obtained the first step in elucidating its structure is to identify and estimate the component monosaccharides. The molecule is broken down by acid hydrolysis and the hydrolysate analysed by chromatographic techniques,[8] which can be fully automated.[9–12] The conditions for hydrolysis must be controlled such that complete hydrolysis is achieved with little or no degradation of the monosaccharide units — use of more than one set of hydrolysis conditions may be necessary. The ease of hydrolysis of different linkages and the stabilities of the various monosaccharides mean that the optimum conditions for each polysaccharide have to be determined. Polysaccharides containing furanose or sialic acid units, and 2-deoxy-hexoses or -pentoses, are more readily hydrolysed than those containing hexuronic acids or 2-amino-2-deoxyhexoses, with hexose-containing polysaccharides being intermediate. Conditions which have been found appropriate for hexose-containing polysaccharides are 1M sulphuric acid at 100 °C for four hours, with the use of 0.25M sulphuric acid at 70 °C for pentose-containing polysaccharides. Degradation frequently occurs in direct hydrolysis whatever conditions are used, *e.g.* in the case of glycosaminoglycans, 4M hydrochloric acid at 100 °C for nine hours is necessary to liberate all the 2-amino-2-deoxyhexose units,[13] but under such conditions the majority of the hexuronic acid units are decomposed. Partial hydrolysis (under appropriately milder conditions) to give a small number of oligosaccharides can be useful for structural analysis, but care must be taken in interpreting the results since monocaccharides can recombine, under certain hydrolysis conditions, to give oligosaccharides linked in a manner different from that in the original polysaccharide. This process is called reversion. Certain functional, *etc.*, groups are lost on hydrolysis and specific methods are required for their estimation. These groups include acetyl, carboxyl, carbonyl, and ether, and methods for their estimation have been reviewed.[14]

No one method available for structural analysis will provide sufficient data to allow the structure of a polysaccharide to be defined in terms of its component monosaccharide units, the inter-unit linkages, and the sequence in which the units are linked. This information can only be obtained using a number of different techniques in conjunction with one another.

26.3.2.1 Methylation

Once the individual monosaccharide components have been estimated, the manner in which they are linked to each other, and the sequence, has to be determined. If all the free hydroxyl groups in the polysaccharide can be reacted to form derivatives which are stable to acid hydrolysis, the hydroxyl groups produced by hydrolysis of the glycosidic linkages will indicate where linkage points were formerly located on each monosaccharide unit. Haworth[15] originally developed a technique of methylating the free hydroxyl groups by reaction of the polysaccharide with dimethyl sulphate and sodium hydroxide in aqueous solution. This method was modified[16] to improve the degree of methylation, as it is essential that all free hydroxyl groups are reacted. This modification uses methyl iodide and silver oxide in an organic solvent, but as polysaccharides are insoluble in organic solvents this method is usually used after a certain degree of methylation has been achieved by Haworth's method. The successor to these methods — the Hakomori method[17,18] — which uses the dimethylsulphinyl anion in dimethyl sulphoxide, with subsequent treatment with methyl iodide, has led to increased effectiveness. Another recent method[19] involves dissolving the polysaccharide in dimethyl sulphoxide or dimethylformamide and treatment with barium oxide and methyl iodide. Polysaccharides containing hexuronic acids can be methylated only with considerable difficulty using the thallium salt of the hexuronic acid and reacting it with methyl iodide and thallium hydroxide,[20] but this reaction must be carefully controlled to avoid degradation and demethylation of the polysaccharide. Recently the Hakomori method has been applied to hexuronic acid-containing polysaccharides to achieve complete methylation of hydroxyl and carboxyl groups in one step.[21]

The fully methylated polysaccharide is hydrolysed to its constituent methylated monosaccharides using sulphuric acid and trifluoroacetic acid. The hydrolysis mixture can be fractioned by partition chromatography on cellulose or silica gel,[22] by adsorption chromatography, or, best of all, by gas–liquid chromatography of volatile derivatives such as the methyl per-*O*-methyl glycosides,[23] partially methylated alditol acetates,[24] or partially methylated *O*-trimethylsilyl ethers.[25] An important extension to gas–liquid chromatography for the further identification of these volatile derivatives is the use of mass spectrometry linked to gas–liquid chromatography[26,27] (see below, Section 26.3.2.8).). Reviews have been published giving the characteristic data for known, standard, partially methylated compounds.[28]

Methylation analysis is not without its problems, the most common being the presence of hexuronic acid units which, as already mentioned, are difficult to hydrolyse completely, but this can be overcome by (indirect) reduction of the acid to an alcohol with sodium borohydride. Methylation analysis on its own will not give structural sequence data, but does identify the monosaccharide components of the polysaccharide and the types of intermonosaccharide linkages involved.

26.3.2.2 Partial hydrolysis

If the hydrolysis of the polysaccharide is stopped before the reaction goes to completion, fragments of intermediate molecular weight can be isolated and fractionated using a number of chromatographic techniques such as gel filtration, ion-exchange chromatography, and partition chromatography. Determination of the structure of these simpler oligosaccharides is generally easier than determinations carried out on the parent polysaccharide. If the glycosidic linkages in the polysaccharide are all hydrolysed at the same rate as in, for example, the linear *homopolysaccharides*, the product of partial hydrolysis will consist, in the case of amylose, of a range of oligosaccharides such as glucose, maltose, maltotriose, and maltotetraose. In *heteropolysaccharides* there are a number of types of glycosidic linkages and their respective rates of hydrolysis will differ, giving a degree of selectivity to the reaction. In general terms furanosides are hydrolysed at a greater rate than pyranosides by factors between 10 and 1000 which will result in removal of, for example, arabinofuranoside residues attached to xylanopyranoside residues in arabinoxylans. The conditions of the hydrolysis will also affect the specificity of the degradation. In mineral acids, $(1 \rightarrow 6)$-linkages are more stable than $(1 \rightarrow 4)$-linkages, but if this reaction were to be carried out in acidified acetic anhydride (containing approx. 5% sulphuric acid) the $(1 \rightarrow 6)$-linkages are less stable. The use of both these methods of hydrolysis will lead to different fragments which will give overlapping data to provide a better picture of the complete polysaccharide. The concentration of carbohydrate material must be kept below about 0.5% to prevent acid-catalysed polymerization of the fragments (acid reversion),[29] which leads to artifacts in structural analysis. Some glycoside bonds in polysaccharides can be cleaved specifically by enzymes to give oligosaccharides in a controlled manner. This method will be discussed later (Section 26.3.2.11).

26.3.2.3 Periodate oxidation

Oxidation of monosaccharides by glycol cleavage is a widespread method of analysis. The use of lead tetra-acetate[30] has found little application to polysaccharide chemistry owing to the lack of suitable solvents for the carbohydrates in which the reagent does not decompose, although the use of lead tetra-acetate in pyridine has been found to oxidize rigidly held diaxial diols which do not react readily with periodic acid. In a much more commonly employed method, periodic acid[31] and its salts are used in aqueous solutions of pH 3–5 to avoid acid hydrolysis and the non-selective oxidations which can occur at higher pH values. The reagent reacts with vicinal hydroxyl groups to cleave the linkage between them with the consumption of one mole of periodate per diol. The products of the reaction depend on the linkages between the monosaccharide units. Oxidation of a primary hydroxyl group adjacent to a secondary hydroxyl group, as in the case of a furanose ring structure, leads to the formation of formaldehyde, whilst vicinal triol groups (see Scheme 1d) yield formic acid. Reaction of polysaccharides with periodic acid is followed by measurement of the amount of reagent consumed and of the formic acid and, less frequently, of the formaldehyde produced. This allows the distinction between 1,3-linked (Scheme 1b) and 1,6-linked (Scheme 1d) units and allows a measurement of the total amount of 1,2- and 1,4-linked units (Schemes 1a and 1c).

The dialdehyde-type products are unstable in water and it is therefore desirable to reduce them, usually with sodium borohydride, to alcohols before acid hydrolysis (to split the oxidized material into the component units) is carried out. Analysis of these component products is essential because it provides a means of distinguishing between 1,2- and 1,4-linked units and determines whether periodate resistant residues are 1,3-linked units or 1,2,4-linked branched points. The products of hydrolysis, such as glycerol, glycol aldehyde, glyceraldehyde, tetritols (such as erythritol), and free monosaccharides (resulting from periodate-resistant residues) are usually determined by gas–liquid chromatography as their trimethylsilyl ethers. This method again does not give complete linkage sequences, but gives information which is used in conjunction with other methods.

26.3.2.4 Smith degradation

An important modification to the above reaction sequence is the controlled hydrolysis of the polyalcohol known as the Smith[32] degradation, which utilizes dilute mineral acid at

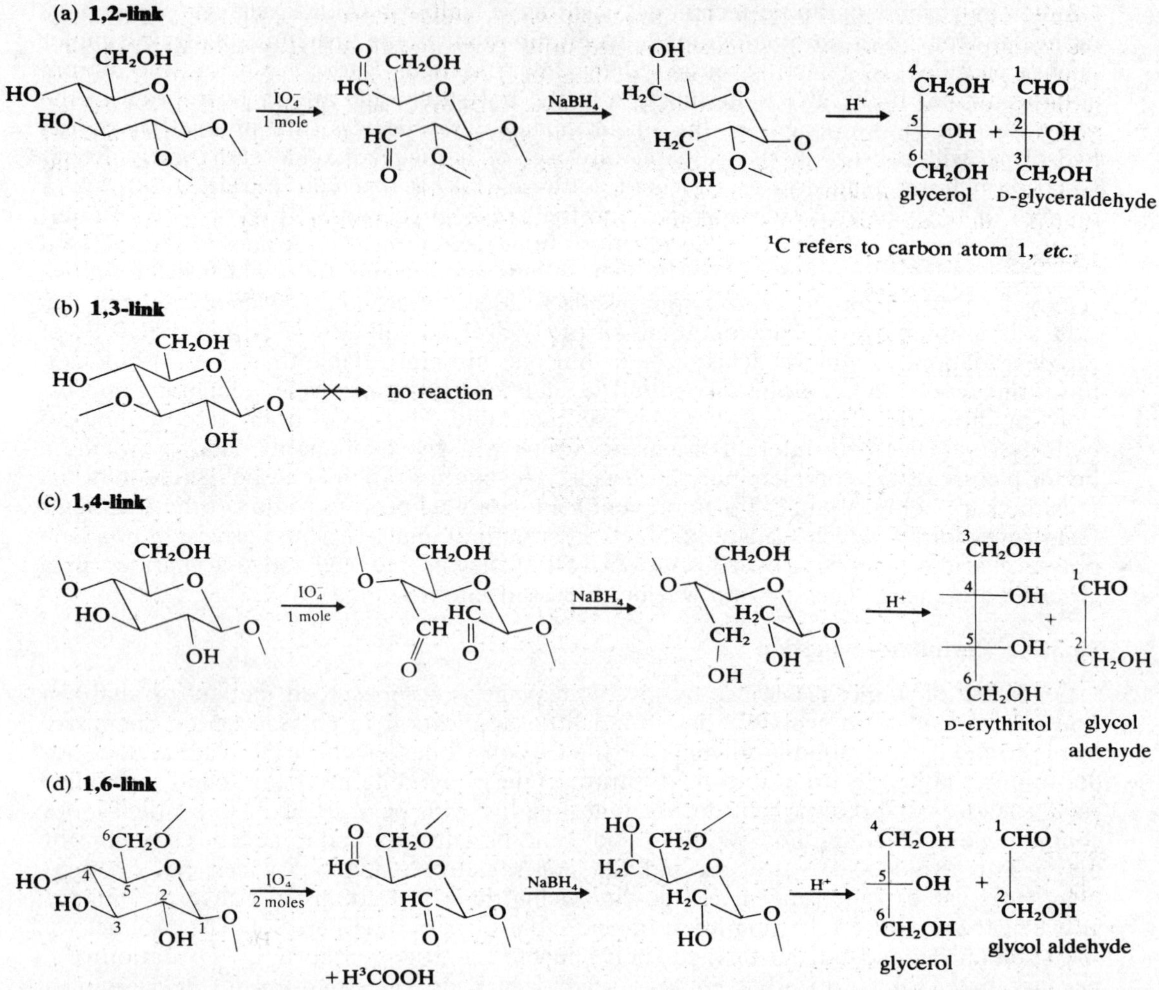

SCHEME 1

room temperature to degrade the polyalcohol partially, to give specific glycosides of oligosaccharides characteristic of the original polysaccharide. This is due to the comparative stability of the glycosidic linkage of the sugar unit which was resistant to periodate oxidation. For example, Smith degradation of a glucan consisting of alternate 1,3- and 1,4-linkages would result in the production of 3-*O*-glucosylerythritol (**4**).

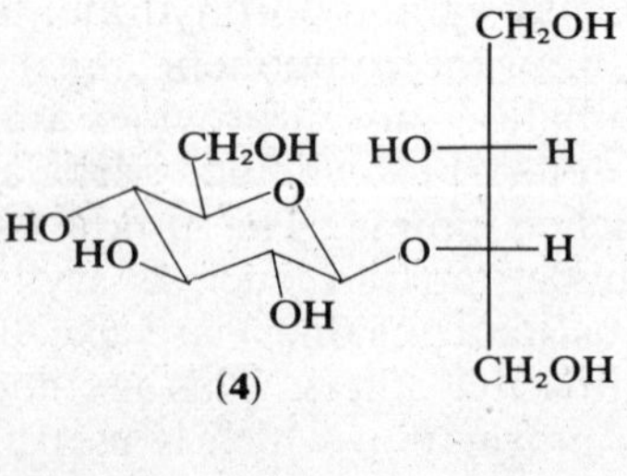

(**4**)

26.3.2.5 Alkaline degradation

This method[33] of analysis of polysaccharides provides little new information about the overall structure, but since the use is often made of alkali in the isolation of a purified sample the type of reactions which occur should be understood. The most common reactions are the hydrolysis of ester groups attached *via* hydroxyl or carboxylic acid groups of the monosaccharide. The reaction which yields most structural information is the progressive erosion of monosaccharide units from the reducing end of the polysaccharide, the so-called 'peeling' reaction. This peeling of monosaccharide units from the reducing end of the molecule is frequently more definitive than methods such as enzyme hydrolysis (which degrade from the non-reducing end of the molecule), when branched structures have to be analysed owing to there being only one reducing end to the molecule. The reaction sequences for the degradation of 1,3- and 1,4-linked monosaccharide units are shown in Scheme 2, from which it is obvious that analysis of the

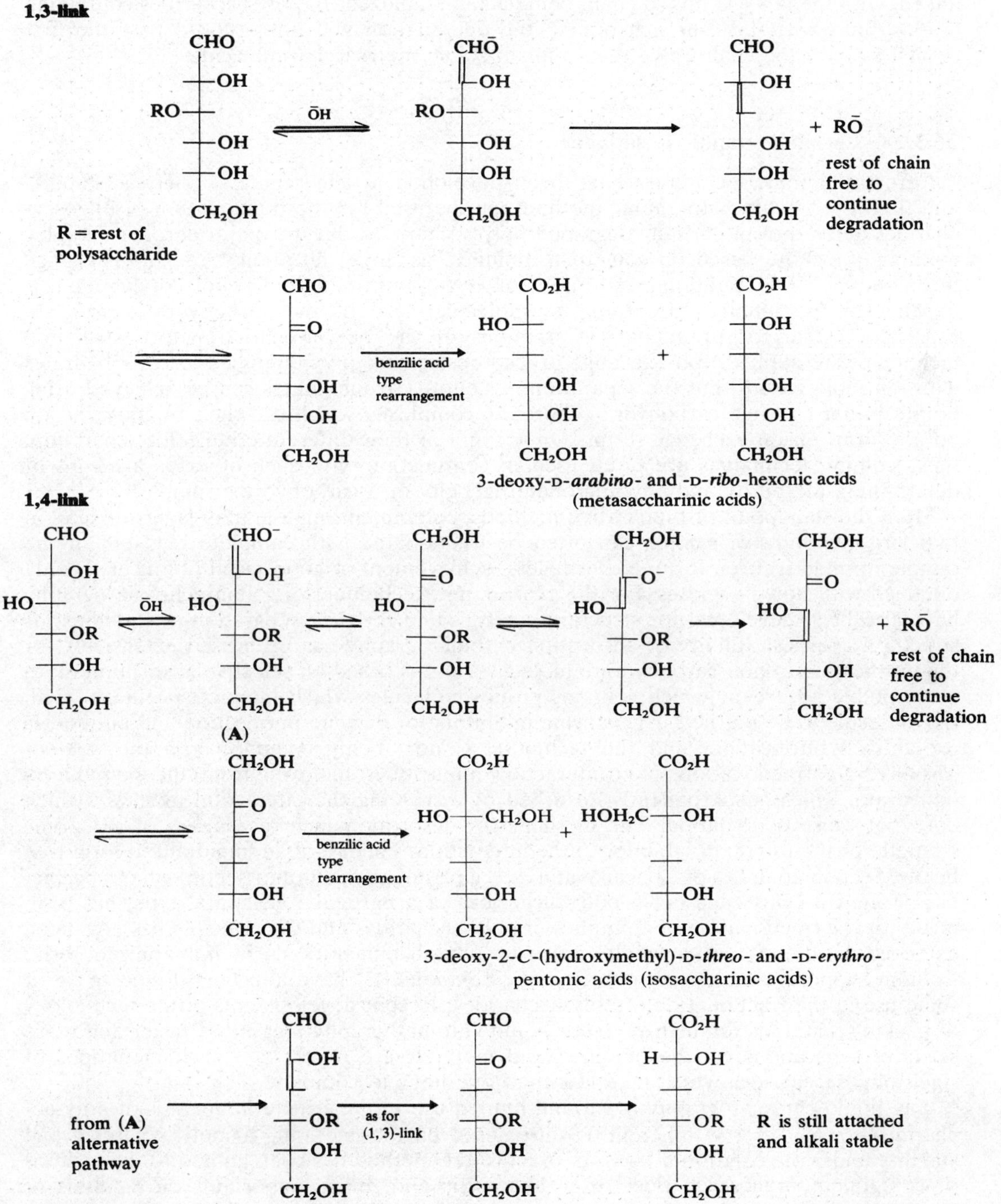

SCHEME 2

saccharinic acid produced will give the position of the original link. The (1 → 4)-linked polysaccharides are not degraded completely, owing to the competing reaction which produces alkali-stable polysaccharides. Another problem in the method is the relative rates of reaction; the (1 → 3)-links are degraded up to ten times faster than (1 → 4)-linkages. This means that as soon as a (1 → 4)-link is degraded the next unit, if it is (1 → 3)-linked, is also immediately degraded, thus making sequence studies difficult.

At branch points the degradation only proceeds along one branch. In a polysaccharide containing 1,3,4-branch points the peeling proceeds along the chain which is (1 → 3)-linked, with the (1 → 4)-linked chain being alkali-stabilized. In polysaccharides containing (1 → 4)-links with 1,4,6-branch points, the degradation will not proceed past the first branch as both the chains give rise to alkali-stable metasaccharinic acids.

26.3.2.6 Chromatographic techniques

Chromatographic techniques have been developed to aid structural analysis. Paper[34] and thin-layer[35] chromatographic methods can be used for the fractionation of oligosaccharides derived from partially degraded polysaccharides, but the major chromatographic methods are those based on column techniques, using gel filtration[36] and ion exchange techniques.[37,38] Gel filtration procedures using cross-linked dextran or polyacrylamide gels depend on fractionation according to molecular size of the carbohydrates and are especially useful for preliminary separations of the polysaccharide before structural techniques are applied. Ion-exchange procedures can readily be applied to carbohydrates with ionizable groups, but the separation of neutral carbohydrates can be achieved using borate buffers to give carbohydrate–borate[39] complexes which are then fractionated on borate forms of ion-exchange resins. On account of their different criteria for separation, these column techniques are often used in conjunction with each other as a means of determining the purity and any microheterogeneity of a carbohydrate sample.

From the standpoint of preparative methods, column chromatography is most useful in that large amounts of sample can often be fractionated with complete recovery of the sample in underivatized form. Nevertheless, achievement of good resolution is often only obtained with slow flow rates. For this reason, new techniques of column chromatography have also been developed for structural analysis and for the preparation of fractions for structural analysis. Affinity or adsorption chromatography has been used extensively for the purification of non-carbohydrate molecules and is based on selective adsorption on to an insoluble adsorbent which contains groups/molecules which interact specifically with the molecule to be purified, *e.g.* enzyme inhibitors for enzyme purification and antibodies for antigen purification, and the technique is now being extended into the field of carbohydrate fractionation. Non-interacting impurity is removed from the bound carbohydrate, which is subsequently desorbed by disrupting the interaction in a way which does not cause degradation. The methods for desorption include changes of pH, ionic strength, or the use of an inhibitor to the interaction. The use of an immobilized form (see below, Section 26.3.7.6) of Concanavalin A,[40] a phytohaemaglutinin (lectin) which specifically reacts with branched-chain polysaccharides of a particular structural type, has been made for the fractionation of a number of polysaccharides and the principle has now been extended to the use of a whole series of phytohaemaglutinins in immobilized form. Column supports coated with polyaromatic compounds[41] have also been found to be of some use in the fractionation of polysaccharides. Recent developments in the technology of packing materials for high-pressure liquid chromatography mean that faster and more selective separations can be obtained and methods are reported[42] of fractionations of small oligosaccharides which take less than one hour to complete.

Gas–liquid chromatography has found limited use in the structural analysis of polysaccharides, owing to the inherent requirements of the method, namely volatility and stability under the conditions used. In practice the method has been limited to the analysis of component monosaccharides after hydrolysis and, more important, the analysis of partially methylated sugars produced in methylation analysis (Section 26.3.2.1). Disaccharides can be made sufficiently volatile for gas–liquid chromatography. The method has the advantage of rapid analysis compared with column techniques, but the method will also, at the same time, separate isomers of the monosaccharides, thus making the number of peaks on the chromatograms larger than the number of component monosaccharides. The need for volatile compounds for analysis has led to a number of methods of derivatization, methyl ethers,[23] alditol acetates,[24] and trimethylsilyl ethers[25] being used commonly, but other methods based on volatile products have also been employed, including the use of isopropylidene derivatives.[43] The use of trimethylsilyl ethers of carbohydrates is preferred, owing to the ease of preparation of the derivatives at room temperature in a few minutes. This method has been extended to include derivatives of acidic monosaccharides[44–46] and basic monosaccharides.[47] More recent developments that increase the value of gas–liquid chromatography in the structural analysis include the use of specific detectors and direct coupling of the gas–liquid chromatograph to radioactive counters (gas–liquid radiochromatography) and to mass spectrometers (Section 26.3.2.7).

26.3.2.7 Mass spectrometry

Mass spectrometry plays a large role in the structural analysis of polysaccharides not only in the identification of compounds derived from methylation analysis (Section 26.3.2.1), but also in the analysis of oligosaccharides directly after preparation of one of the volatile derivatives mentioned earlier[23–25,44–47] (Section 26.3.2.6). The molecular weight of small oligosaccharides can be measured and the sequence of monosaccharide units and position of glycosidic bonds have been determined, although some information on the nature of the residues present is also usually needed[48,49]. The direct mass spectrometric identification of oligosaccharides containing more than four residues with the use of trimethylsilyl derivatives is difficult, but characteristic fragmentation patterns of peracetylated glycoside derivatives of pentasaccharides have been obtained[50] and, more recently, a method has been described for the detection of D-fructose units in permethylated oligosaccharides which also gave information on the ratio of pyranose to furanose units and the positions of the glycosidic linkages.[51]

Methods involving chemical ionization[52] rather than electron impact are more sensitive than conventional methods of analysis,[53,54] and such methods have been used in the analysis of oligopeptides, low molecular weight fragments obtained from hydrolysis of glycoproteins. Not only was the amino-acid sequence obtained but the carbohydrate–peptide linkages could be determined by comparing the fragmentation patterns with those obtained for the various monosaccharide amino-acid derivatives.[55]

26.3.2.8 Nuclear magnetic resonance spectroscopy

A method which will give information of the anomeric linkages in polysaccharides, provided the monosaccharide components and substitution positions are known, has been developed using n.m.r. spectroscopy. The hydroxyl groups of the sugar units are preferably converted to derivatives such as their methyl or trimethylsilyl ethers to eliminate from the spectrum the peak due to hydroxyl groups. The protons of the anomeric carbon atom occur at lower field than protons on the other carbon atoms, with those in the equatorial position showing larger chemical shifts than those in axial positions. Complete structural analyses of polysaccharides have been obtained with ^{1}H n.m.r. of methylated monosaccharides, and simpler polysaccharides such as glucogens.[56] The use of ^{13}C, ^{19}F, and ^{31}P n.m.r. may also be useful in determination of the position of substitution of monosaccharide by another, the last two methods using derivatives of the polysaccharides such as [^{19}F]trifluoroacetates.

26.3.2.9 Electrophoretic techniques

Electrophoresis is not a substitute for chromatography, but provides very useful complementary information because it utilizes different criteria for separation, namely molecular charge, size, and shape. The use of high-voltage paper electrophoresis has been applied to the separation not only of monosaccharides but also oligosaccharides. The method is not restricted to carbohydrate derivatives which possess an electric charge of their own, such as hexuronic acids, amino-monosaccharides, and monosaccharide sulphates and phosphates, but has been extended to include neutral compounds which can form electrically charged complexes with electrolytes such as sodium borate, arsenite, or molybdate. The relative mobilities of the carbohydrates can be varied[57] by changing the complexing agent used, when steric factors often determine the formation of different complexes. Choice of electrolytes will often lead to identification of the carbohydrate and its structure and bonding. Separations of acidic polysaccharides[58] have been obtained directly, using high-voltage paper electrophoresis, but separation of neutral polysaccharides requires pre-conversion to their borate derivatives.[59]

The development of better supporting media, such as cellulose acetate strips and polyacrylamide gels in the form of rods or slabs, for electrophoretic purposes has meant that purer chromatographic materials which are of homogeneous character and possess minimal adsorption properties are available, thus reducing tailing and resulting in quicker separations on a small scale. Methods are reported for the separation of acidic polysaccharides on cellulose acetate[60] and polyacrylamide gels,[61] and the application of the latter method to molecular weight determination has been discussed.

26.3.2.10 Immunochemical techniques

Polysaccharides have been found to be determinants of the immunological specificites of many types of micro-organisms. The specific interaction depends on the association of multiple reactive groups in both the polysaccharide antigen and protein antibody, and so a method based on this type of interaction is usually specific for the structure of a polysaccharide. If the appropriate antisera can be prepared against polysaccharides of known structure, it can be used to indicate structural similarities in unknown polysaccharides. An example of this specificity was shown in the discovery of the heterogeneity of a beef lung galactan. The precipitate formed with anti-*Pneumococcus* type XIV sera contained proportions of D-galactose and D-glucuronic acid different from those in the original preparation.[62]

26.3.2.11 The use of enzymes in structural analysis

Hydrolysis by enzymes provides an alternative method for the controlled hydrolysis of polysaccharides. The information obtained is not limited to that obtainable by analysis of the hydrolysis fragments because the specificity of enzyme action, a specificity based on type of monosaccharide and type of linkage, leads to significant data being obtained, by a process of elimination, from enzyme-resistant structures and partially hydrolysed structures. The enzymes which hydrolyse polysaccharides are, for convenience, divided into two groups, *endo*- and *exo*-hydrolases. *endo*-Hydrolases are specific for linkage and monosaccharide unit, and cause random fragmentation of homopolysaccharides to give a homologous series of oligosaccharides. Examples of this type of enzyme include α-amylase, which gives a random series of D-glucose oligomers on reaction with amylose. *exo*-Hydrolases are specific for monosaccharide unit and stereochemistry at C-1 but do not differentiate between the units attached glycosidically at C-1. They cleave polysaccharides by sequential removal of residues from one end of the molecule, usually the non-reducing end which is the opposite end to that from which alkaline degradation starts

(Section 26.3.2.5). Examples of *exo*-hydrolases include β-amylase, which removes maltose units sequentially from amylose, producing an almost quantitative amount of maltose if the reaction goes to completion.

The first uses of enzyme analysis was in the determination of chain-length and degrees of branching in highly branched polysaccharides such as glycogen and amylopectin. Traditional methods for this analysis required the estimation of the non-reducing end group by chemical methods. The use of enzymes allows smaller quantities of material to be used and increases the speed of the determination.[63,64] The method used by Lee and Whelan[63] was based on the use of two enzymes, one of which (pullulanase) specifically hydrolyses the α-(1 → 6)-links at the branch points to give linear chains of α-(1 → 4)-linked D-glucose units which are degraded by β-amylase to give maltose and D-glucose units, the latter arising from degradation of chains with an odd number of D-glucose units. Analysis of the hydrolysis mixture for D-glucose gives a measure of the chain length because one D-glucose unit is produced from one chain containing an odd number of D-glucose units and, using the assumption that there is an equal number of odd and even chains, one D-glucose unit is produced from two unit chains.

More recently, a number of *exo*-glycoside glycohydrolases have been produced in sufficiently pure form to allow the development of a method of determination of monosaccharide sequences based on these enzymes. These enzymes will remove specific monosaccharide units linked by specific linkages from the non-reducing end of a polysaccharide. For example, β-D-galactosidase will remove D-galactose units linked β-glycosidically to the polysaccharide. Table 4 gives a listing of the major polysaccharide degrading enzymes, and reviews of the sources and methods of purification of these enzymes are available.[65] The method used can rely on the use of the enzymes sequentially or together, as in the case of enzymic degradation of keratan sulphate with β-D-galactosidase, β-D-2-acetamido-2-deoxyglucosidase, and a sulphatase,[66] a method which showed that some D-galactose and 2-acetamido-2-deoxy-D-glucose units at the non-reducing end of the molecule are non-sulphated. The use of sequential enzyme hydrolyses is a well-established technique,[67] particularly for the analysis of the carbohydrate residues of macromolecules.

A number of *endo*-glycosidases (glycopeptidases) which will certainly prove of great value in the analysis of the carbohydrate moiety of glycoproteins have recently been isolated. These enzymes cleave the carbohydrate glycosidic bond adjacent to the amino-acid residue. An example of this type of enzyme is 4-L-aspartyl-β-D-glucosylamine aminohydrolase, which cleaves the bond between the 2-acetamido-2-deoxy-D-glucose units of glycoproteins containing the sequence D-mannose linked to di-(*N*-acetyl)chitobiose linked in turn to an L-asparagine residue (**5**).[68]

β-D-Man*p*-(1 → 4)-β-D-Glc*p*NAc-(1 → 4)-β-D-Glc*p*NAc-(1 →)-L-Asn

(**5**)

The hydrolysis of glycosidic linkages by enzymes involves scission of the glycosyl–oxygen bond, but a number of enzymes known as eliminases or lyases, usually of bacterial origin, react by a different mechanism and cause cleavage of the oxygen–aglycone bond (see Figure 3) in acidic polysaccharides (such as pectins), producing unsaturated hexuronic acid units.

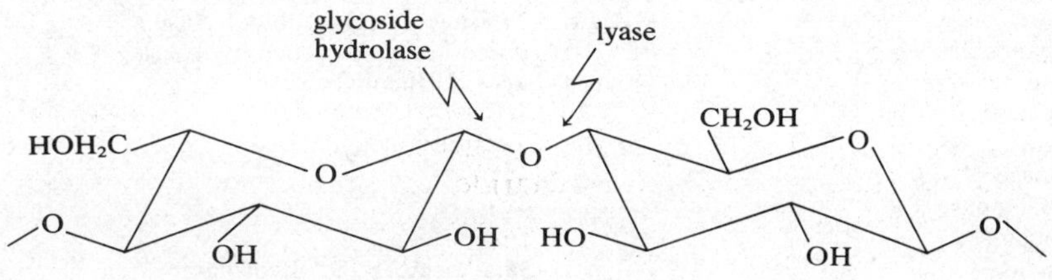

Figure 3 Position of action of glycosidases and lyases

TABLE 4

Enzymes which may be used for the Structural Analysis of Polysaccharides

Trivial name	*Systematic name*	*E.C. No.*[a,b]
2-Acetamido-2-deoxy-D-galactose 4-sulphate sulphatase		
α-D-2-Acetamido-2-deoxygalactosidase	2-Acetamido-2-deoxy-α-D-galactoside acetamidodeoxygalactohydrolase	3.2.1.49
β-D-2-Acetamido-2-deoxygalactosidase	2-Acetamido-2-deoxy-β-D-galactoside acetamidodeoxygalactohydrolase	3.2.1.53
exo-β-D-Acetamidodeoxyglucanase		
2-Acetamido-2-deoxy-D-glucose 6-sulphate sulphatase		
α-D-2-Acetamido-2-deoxyglucosidase	2-Acetamido-2-deoxy-α-D-glucoside acetamidodeoxyglucohydrolase	3.2.1.50
β-D-2-Acetamido-2-deoxyglucosidase	2-Acetamido-2-deoxy-β-D-glucoside acetamidodeoxyglucohydrolase	3.2.1.30
αβ-D-2-Acetamido-2-deoxyglucosidase		
endo-β-D-2-Acetamido-2-deoxyglucanase	*see* Chitinase	3.2.1.14
exo-β-D-2-Acetamido-2-deoxyglucosidase		
exo-β-D-2-Acetamido-2-deoxyglycanase		
β-D-2-Acetamido-2-deoxyhexosidase	2-Acetamido-2-deoxy-β-hexoside acetamidodeoxyhexohydrolase	3.2.1.52
exo-β-D-2-Acetamido-2-deoxyhexosidase		
N-Acetylmuramyl-L-alanine amidase	Mucopeptide amidohydrolase	3.5.1.28
α-Amylase	1,4-α-D-Glucan glucanohydrolase	3.2.1.1
β-Amylase	1,4-α-D-Glucan maltohydrolase	3.2.1.2
Arabinanase		
α-L-Arabinofuranosidase	α-L-Arabinofuranoside arabinohydrolase	3.2.1.55
Arylsulphatase	Aryl sulphate sulphohydrolase	3.1.6.1
L-Ascorbic acid 2-sulphate sulphatase		
4-L-Aspartyl-β-D-glucosylamine amidohydrolase	2-Acetamido-1-*N*-(4-L-aspartyl)-2-deoxy-β-D-glucosylamine glucosylamidohydrolase	3.5.1.37
Bacterial proteinase		
Carbohydrate isomerases		
Carbohydrate oxidases		
Carbohydrate transferases		
Cellulase	1,4-(1,3; 1,4)-β-D-Glucan 4-glucanohydrolase	3.2.1.4
Cerebroside sulphatase	Cerebroside 3-sulphate 3-sulphohydrolase	3.1.6.8
Chitinase	Poly-[1,4-β-(2-acetamido-2-deoxy-D-glucoside)]glycanohydrolase	3.2.1.14
Chitosanase		
Chondroitin sulphate lyase ABC	Chondroitin ABC lyase	4.2.2.4
Chondroitin sulphate lyase AC	Chondroitin AC lyase	4.2.2.5
Chondroitin 4-sulphate lyase		
Chondroitin 6-sulphate lyase		
Chondroitin 4-sulphate sulphatase	$\Delta^{4,5}$-β-D-Glucuronosyl-(1,4)-2-acetamido-2-deoxy-D-galactose 4-sulphate 4-sulphohydrolase	3.1.6.9
Chondroitin 6-sulphate sulphatase	$\Delta^{4,5}$-β-D-Glucuronosyl-(1,4)-2-acetamido-2-deoxy-D-galactose 6-sulphate 6-sulphohydrolase	3.1.6.10
Dextranase	1,6-α-D-Glucan 6-glucanohydrolase	3.2.1.11
β-D-Fructofuranosidase	β-D-Fructofuranoside fructohydrolase	3.2.1.26
α-L-Fucosidase	α-L-Fucoside fucohydrolase	3.2.1.51
α-1,2-L-Fucosidase		
β-D-Fucosidase	β-D-Fucoside fucohydrolase	3.2.1.38
endo-Galactanase		
endo-β-D-Galactanase		
exo-β-D-Galactanase		
α-D-Galactosidase	α-D-Galactoside galactohydrolase	3.2.1.22
β-D-Galactosidase	β-D-Galactoside galactohydrolase	3.2.1.23

Table 4 (*cont'd*)

Trivial name	*Systematic name*	*E.C. No.*
β-D-Galactosidases (glycosphingolipid-specific)		
β-D-Galactosphingosidase		
β-D-Galactosylceramidase	D-Galactosyl-*N*-acylsphingosine galactohydrolase	3.2.1.46
β-D-Galactosyl-D-glucosylceramidase		
endo-β-1,3-D-Glucanase	1,3-β-D-Glucan glucanohydrolase	3.2.1.39
endo-β-1,6-D-Glucanase	1,6-β-D-Glucan glucanohydrolase	3.2.1.75
exo-D-Glucanase		
exo-β-1,3-D-Glucanase	1,3-β-D-Glucan glucohydrolase	3.2.1.58
exo-β-1,4-D-Glucanase	1,4-β-D-Glucan glucohydrolase	3.2.1.74
exo-α-1,6-D-Glucanase	1,6-α-D-Glucan glucohydrolase	3.2.1.70
D-Glucanases (miscellaneous)		
Glucoamylase	1,4-α-D-Glucan glucohydrolase	3.2.1.3
α-D-Glucosidase	α-D-Glucoside glucohydrolase	3.2.1.20
β-D-Glucosidase	β-D-Glucoside glucohydrolase	3.2.1.21
exo-β-1,3-D-Glucosidase	1,3-β-D-Glucan glucohydrolase	3.2.1.58
exo-β-1,4-D-Glucosidase	1,4-β-D-Glucan glucohydrolase	3.2.1.74
exo-α-1,6-D-Glucosidase	1,6-α-D-Glucan glucohydrolase	3.2.1.70
β-D-Glucosylcerebrosidase		
β-D-Glucuronidase	β-D-Glucuronide glucuronohydrolase	3.2.1.31
Glycanase		
Glycolipid sulphate sulphatase		
Glycopeptidases		
Glycopeptide-linkage hydrolases		
β-D-2-Glycylamido-2-deoxyglucosidase		
Heparin sulphatase		
Hyaluronidase	Hyaluronate 4-glycanohydrolase	3.2.1.35
α-L-Iduronidase	α-L-Iduronide iduronohydrolase	3.2.1.76
L-Iduronic acid 2-sulphate sulphatase		
Isoamylase	Glycogen 6-glucanohydrolase	3.2.1.68
Isomaltase	*see* α-D-Glucosidase	
Keratan sulphate hydrolase		
Lactose sulphate sulphatase		
Laminarinase	1,3-(1,3; 1,4)-β-D-Glucan 3(4)-glucanohydrolase	3.2.1.6
Lichenase	1,3-1,4-β-D-Glucan 4-glucanohydrolase	3.2.1.73
Lysozyme	Mucopeptide *N*-acetylmuramylhydrolase	3.2.1.17
'Maltase'	*see* α-D-Glucosidase	
D-Mannanases (miscellaneous)		
α-D-Mannosidase	α-D-Mannoside mannohydrolase	3.2.1.24
β-D-Mannosidase	β-D-Mannoside mannohydrolase	3.2.1.25
Neuraminidase	Acylneuraminyl hydrolase	3.2.1.18
Oligo-1,6-D-glucosidase	Dextrin 6-α-glucanohydrolase	3.2.1.10
Pectate hydrolases		
Pectate lysase	Poly-(1,4-α-D-galacturonide) lyase	4.2.2.2
Pectinesterase	Pectin pectylhydrolase	3.1.1.11
Pectin lyase	Poly(methoxygalacturonide) lyase	4.2.2.10
β-D-Phosphogalactosidase		
β-D-Phosphoglucosidase		
Polygalacturonase	Poly-(1,4-α-D-galacturonide) glycanohydrolase	3.2.1.15
exo-Polygalacturonase	Poly-(1,4-α-D-galacturonide) galacturonohydrolase	3.2.1.67
exo-Poly-α-galacturonosidase	Poly-(1,4-α-D-galactosiduronate) digalacturonohydrolase	3.2.1.82
Polysaccharide deacetylase		
Pullulanase	Pullulan 6-glucanohydrolase	3.2.1.41
Sucrose α-D-glucohydrolase	Sucrose α-D-glucohydrolase	3.2.1.48
Sulphamidase		3.10.1.1
Sulphatases (miscellaneous)		
D-Thioglucosidase	Thioglucoside glucohydrolase	3.2.3.1
αα-Trehalase	αα-Trehalose glucohydrolase	3.2.1.28

Table 4 (*cont'd*)

Trivial name	*Systematic name*	*E.C. No.*
Xylanase (miscellaneous)		
D-Xylose isomerase	D-Xylose ketol-isomerase	5.3.1.5
β-D-Xylosidase	β-D-Xyloside xylohydrolase	3.2.1.37

[a] Enzyme Nomenclature: Recommendations (1972) of the International Union of Pure and Applied Chemistry and the International Union of Biochemistry. [b] Supplement 1. Corrections and Additions (1975), *Biochim. Biophys. Acta*, 1976, **429,** 1.

Interpretation of the results from enzyme analysis must be carried out with caution. The mode of action of the *exo*-glycosidases is such that it is not possible to determine from which branch(es) the terminal residue(s) have been removed; this is in contrast to the alkaline degradation method (Section 26.3.2.5). Microhetereogeneity of chains within the same molecule will also make interpretation uncertain. It is essential that the enzyme used is highly purified, as other glycosidases present in the enzyme can also lead to ambiguities and incorrect assumptions. It was originally thought, for example, that all D-mannose residues in glycoproteins were α-linked, but the use of α-D-mannosidase, purified to remove all traces of β-D-mannosidase activity, disproved this.

26.3.2.12 Molecular size and shape

A complete description of a polysaccharide involves an estimate of its molecular size and shape. Some of the methods described above will give a measure of the size and shape as part of the analysis (*e.g.* gel filtration, Section 26.3.2.1, and non-reducing end group analysis by methylation, Section 26.3.2.5, or periodate oxidation, Section 26.3.2.3), but specific methods are available for characterization of the polysaccharides to give data on molecular weight, size, and distribution in any given sample.

The use of electron microscopy has been limited by the small size of the molecules which are being dealt with. They are too small to scatter electrons themselves, but the technique of casting a metal shadow on the molecules has led to single molecular patterns being obtained, as in the case of *O*-(hydroxyethyl)cellulose,[69] but the method more frequently gives information only on molecular aggregates and conformational shape.[70]

X-Ray diffraction[71,71a] is another method which provides information and polysaccharides which form fibres give satisfactory diffraction diagrams. These are usually linear molecules but the attachment of single unit side chains, if not too frequent, does not interfere with the formation of crystals and hence with the method. In highly branched polysaccharides crystallinity is only found if the high degree of substitution shows a regular pattern, but in the majority of polysaccharides crystallinity is only partial, resulting in dislocations in the crystal lattice and large areas of amorphous material, which makes the interpretation of results more difficult. This method of analysis has shown, for example, how the repeating units in glycosaminoglycans which consist of a disaccharide unit[72] are arranged in chains, *etc.*[73]

Colligative property measurements have been used for molecular weight determination. Below a molecular weight of 20 000 the method involving measurement of differences in vapour pressure or boiling points of pure solvent and solutions can be used, but these are limited by the sensitivity of the techniques available for measuring the small differences. Above a molecular weight of 20 000 the only method which can satisfactorily be used is the measurement of osmotic pressure. The limitation on this method is the sensitivity of measurement of the pressure differences for high molecular weight substances, the upper limit of molecular weight being of the order of 500 000. At the opposite end of the molecular weight range the nature of the semipermeable membrane dictates the limitation of the method. The newer techniques of vapour phase osmometry[74] and dynamic osmometry[75] have been used successfully.

Light scattering by dilute solutions of the polysaccharide provides an absolute method for the detrmination of molecular weights. Solutions of the polysaccharides are illuminated with monochromatic polarized or unpolarized light, and the scattered light intensity is measured as a function of the scattering angle. From these data the shape of the molecules and the molecular weight can be obtained.[76]

Molecular weights and shapes can also be obtained from studies of sedimentation velocities of a solution of the polysaccharides under the influence of a high force field by following the changes in refractive index gradients. The rate of sedimentation obtained from these ultracentrifugation studies provides a measure of molecular weight and, on comparison with the calculated behaviour for molecular models, provides a basis for assessment of molecular shape. Laurent *et al.*[77] used this method for the characterization of hyaluronic acid. Ultracentrifugation has not been used as a means of purification, but is commonly used as a test of homogeneity of a purified sample.

26.3.3 PLANT AND ALGAL POLYSACCHARIDES

26.3.3.1 Starch

The principle food-reserve polysaccharide in the plant kingdom is starch. It forms the major source of carbohydrates in the human diet and is therefore of great economic importance, being isolated on an industrial scale from many sources. Starch has been found in some protozoa, bacteria, and algae, but by far the major source is plants, where it occurs in the seeds, fruits, leaves, tubers, and bulbs in varying amounts from a few percent to over 75% in the case of cereal grains. Starch occurs in granular form, the shape of the granules being characteristic of the source of the starch. Isolation of these granules from the plant tissues can be achieved without degradation because they are insoluble in cold water, whereas many of the contaminants are soluble. The granules swell reversibly in cold water, a phenomenon used in the industrial extraction of starch to loosen the granules in the matrix,[78] but as the temperature is raised the process becomes irreversible and eventually the granule bursts to form a starch paste. Not all starch granules in a sample burst at the same temperature, but the range of temperature of gelatinization is characteristic for a particular starch.

The starch granule can be separated into two distinctly different components (a phenomenon first discovered in the early 1940s, although its heterogeneity was indicated earlier). The two components, amylose and amylopectin, vary in amount among the different sources from less than 2% of amylose in waxy maize to about 80% of amylose in amylomaize, but the majority of starches contain between 15% and 35%. It has been proposed that the ratio of amylose : amylopectin is a function of the ratio of starch synthetases present in the plant, one synthetase being responsible for the production of amylose whilst the other, a complex branching synthetase, is responsible for amylopectin production.[79] The determination of this characteristic ratio for a given starch sample is based on the binding capacity of the starch for iodine. It can be seen from Figure 4 that amylose binds iodine, giving a blue colouration, up to a limiting value which is determined partly by the experimental conditions, but in this case is about 20% of iodine weight. Amylopectin, on the other hand, binds little iodine and in so doing gives a red colouration and starch, as a mixture, is of intermediate binding ability. Extrapolation of the linear portions of the binding curve to zero free iodine concentration, as shown, will give the iodine binding capacity of the starch. The amylose content is obtained by expressing the starch binding capacity as a percentage of the amylose binding capacity. Potentiometric measurements[80] have traditionally been used for such determinations, but it is now claimed that the use of spectrophotometric techniques at two wavelengths is a much more accurate method.[81]

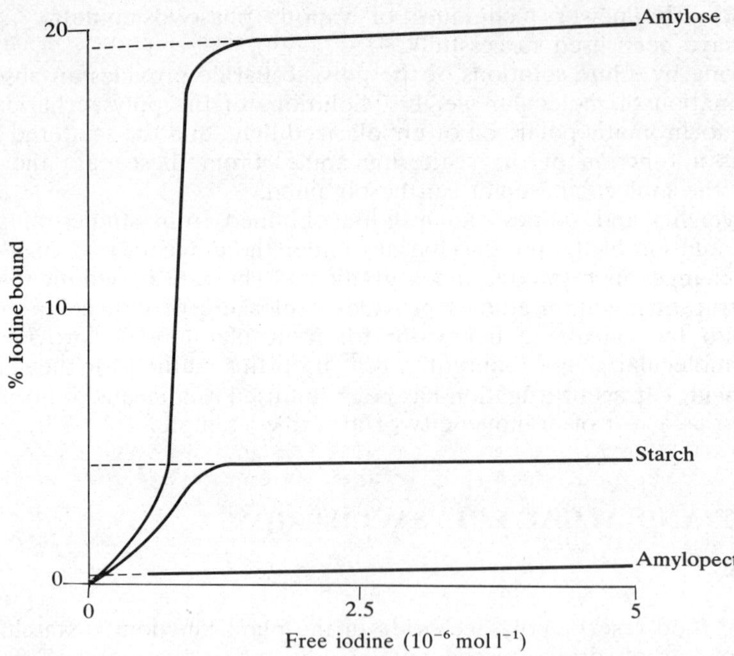

Figure 4 Adsorption of iodine by starch, amylose, and amylopectin

The most widely used method of fractionating starch[82] into its components is to add a polar, organic substance to an aqueous dispersion of the starch granules, which causes the amylose to form an insoluble complex. The amylose can be purified by reprecipitation, usually with a different organic substance. The most suitable precipitants have been found to be thymol for the initial precipitation and butanol for the subsequent purification.[83] The amylopectin fraction is removed from the supernatant liquors that remain after the removal of the amylose complex. To prevent degradation of the fractions it is necessary to carry out the purification in the absence of oxygen, and it is frequently necessary to pretreat the granule with dimethyl sulphoxide to ensure complete dispersion.

The characteristic structural features of amylose and amylopectin are α-D-(1→4)-linked (**1**) glucans with, in the case of amylopectin, α-D-(1,4,6)-linked branch points (**3**) and these have been known for many years as a result of methylation and hydrolysis analysis. Recently, significant proportions of D-glucose 6-phosphate were detected from acid hydrolysis of waxy maize and waxy rice starches.[84] Subsequent analysis showed there to be an average of one mole of phosphate to six D-glucose units in the amylopectins. Methylation analysis of amylose yields 2,3,6-tri-*O*-methyl-D-glucose as the major product, with less than 0.4% of 2,3,4,6-tetra-*O*-methyl-D-glucose, resulting from the non-reducing end of the molecule, and shows that the molecule is linear with a unit chain of about 200–350 D-glucose units. Osmotically determined molecular weights have agreed with this chain length,[85] but analysis of an unbranched structure is difficult. Not only is it difficult to measure the small amount of end groups in relation to the chain-forming units, but degradation has a considerable effect—the rupture of one bond can halve the unit chain length. Physical methods of chain length determination, provided that independent methods for homogeneity are used, have shown values for the size of the amylose molecules to be greater than values obtained by chemical methods. Light scattering and ultracentrifugation analysis show that chain lengths of up to 6000 units are frequently found. Enzyme analysis of amylose with β-amylase originally showed the molecule to be linear, producing maltose as its sole degradation product, but contamination of the enzyme with another α-amylolytic enzyme would remove any other α-linked D-glucose unit. A study of the action of pullulanase and other amylolytic enzymes on various amyloses has indicated that there are some α-(1→6)-linkages as branch points in the molecule.[63,64] This, plus the hydrodynamic behaviour of amylose fractions, has led to the acceptance of there being a limited degree of branching in amylose.

Many of the properties of amylose can be explained in terms of its ability to adopt different molecular conformations in solution. In neutral aqueous solutions the normal conformation is that of a random coil. If there are complexing agents in the solution, amylose will form a helical structure consisting of about six D-glucose units per helical turn. This is the conformation that gives the characteristic blue colouration of amylose–iodine complexes and is responsible for the formation of complexes with fats and polar organic solvents, with the complexing agent at the centre of the helix. The various forms of retrograded amyloses are due to variations in conformation of the amylose in solution. Retrogradation of amylose is the autodeposition of the polysaccharide in an insoluble form from solution, a phenomenon which rarely occurs with amylopectin. X-Ray patterns of the retrograded amyloses have shown that the size and type of amylose and the concentration, temperature, and pH of solution all contribute to the structure, but two distinct X-ray patterns can be observed. Type A is characteristic of cereal starches and amyloses produced by retrogradation above 50 °C and Type B of tuber starches and retrogradation of amyloses at room temperature. The slow formation of these forms of amylose will allow the formation and alignment of linear chains and through the formation of hydrogen bonds eventually results in the insoluble particles. No clear picture of the conformations which exist in these forms can be given at the moment, but it is thought that the B-form may comprise intertwined double helices.

A third crystalline form of amylose, of which more information is available, is the so-called V-form and is obtained as a result of retrogradation in the presence of complexing agents. X-Ray patterns of this form indicate a flexible helical arrangement with six or seven D-glucose units per turn, depending on the size of the complexing agent. The nature of the B-form has been related to the V-form through a mechanism involving the extension of the helices and changes in hydrogen bonding.[86] Comparisons of X-ray diffraction patterns of the V-form have been made with those from a series of cyclic oligossacharides which are prepared by the action of an amylase from *Bacillus macerans* on starch. These cyclic oligosaccharides (cycloamyloses, Shardinger dextrins)[87] are oligosaccharides of α-(1→4)-linked D-glucose units in a closed-loop structure, and cyclohexamylose gave an X-ray pattern similar to that of the V-form and is considered to be analogous to one turn of the helix.

Amylopectin, on methylation analysis, yield 2,3,6-tri-*O*-methyl-D-glucose as its main product, but the amount of the 2,3,4,6-tetra-*O*-methyl ether of about 4% shows that the unit chain length is smaller than in amylose. Isolation of 2,3-di-*O*-methyl-D-glucose as an additional product indicates the presence of (1,4,6)-linked branch points. Measurement of the unit chain length was traditionally carried out by estimating the formic acid liberated by periodate oxidation of the non-reducing end units, but is more accurately determined by enzymic methods (Section 26.3.2.11).[63,64] Values for the unit chain length are usually within the range 17–26 units. The arrangement of these unit chains to give a branched amylopectin could be in a variety of ways. Figure 5 shows three possible arrangements, the laminated structure (a) proposed by Haworth,[88] the herringbone structure (b) proposed by Staudinger and Husemann,[89] and the branched tree-like structure (c) proposed by Meyer and Bernfeld.[90] There are three types of chain present in these structures: A-chains are side chains linked only through their reducing ends to the rest of the molecules, B-chains are those to which A-chains are attached, and the C-chain which carries the reducing group (there can only be one C-chain per molecule). Enzymic studies,[91] based on the degradation with β-amylase and a debranching enzyme, favour a multiple-branched structure, but results obtained from degradation with isoamylose, phosphorylase, and β-amylase have indicated that the A- and B-chains are not arranged in the regularly branched structure (Figure 5c).[92]

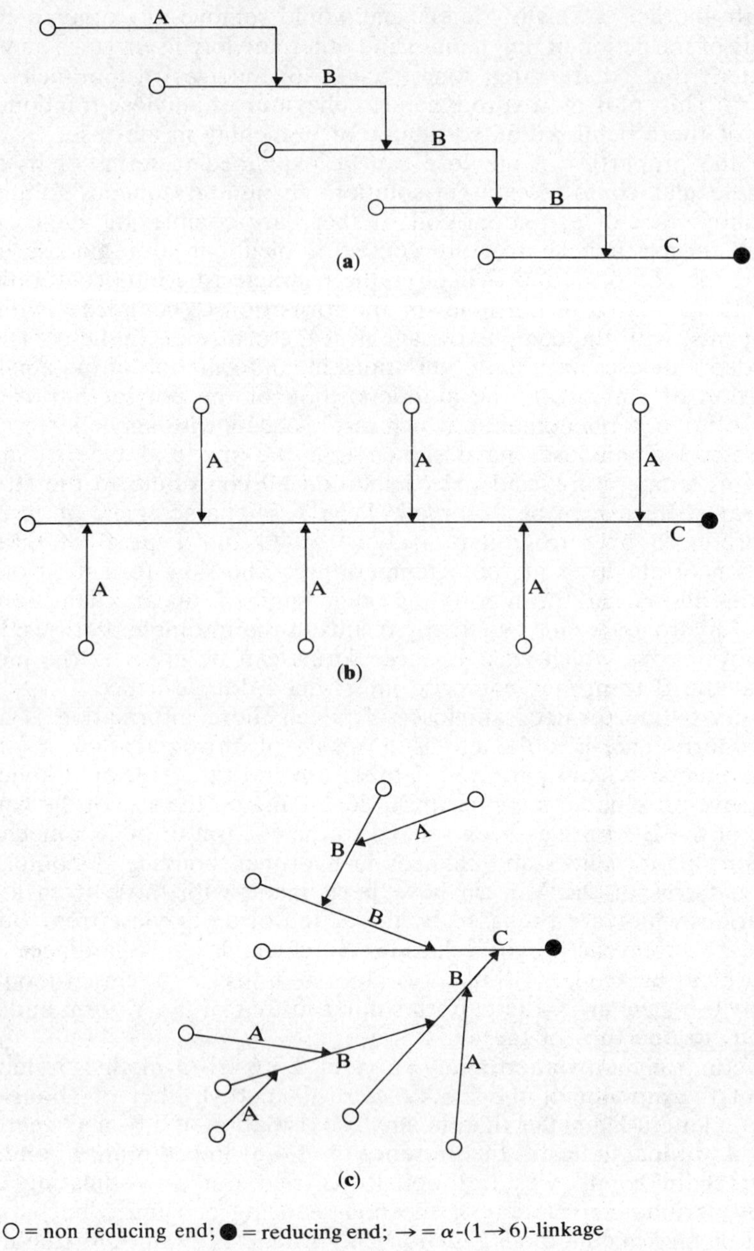

Figure 5 Schematic representation of the proposed arrangements of α-D-(1 → 4)-linked glucose polymer chains of *ca.* 20 units

The molecular size of amylopectin is almost too large to determine accurately. However, light-scattering results indicate that there are of the order of 10^6 D-glucose units per molecule, making amylopectin one of the largest naturally occurring molecules. The size of the molecule has been shown to increase with increasing maturity of the parent starch in the plant.[93]

The inability of amylopectin to bind much iodine, thereby producing the characteristic red colouration, is attributed to the large number of branch points in the molecule which disrupt any possible helix formation.

26.3.3.2 Cellulose

Cellulose is the most abundant organic substance found in nature. It is the principal constitutent of cell walls in higher plants, forming the main structural element. It occurs in an almost pure form (98%) in cotton fibres and to a lesser extent in flax (80%), jute (60–70%), and wood (40–50%), and has also been found as a constituent of some algae and as a product of bacterial synthesis. Isolation of cellulose from most of its sources is difficult owing to its insolubility in most solvents, and involves rather the solubilization of the contaminating compounds such as hemicelluloses (Section 26.3.3.6.) and lignin. The usual methods are based on alkaline pulping to remove the non-cellulosic polysaccharides, but it is difficult to remove all the contaminating monosaccharides, and this has led to reports of the presence of monosaccharides other than D-glucose in trace quantities.

The primary structure of cellulose was determined in the 1930s and little modification has had to be made to it since. Methylation analysis yields over 90% of 2,3,6-tri-*O*-methyl-D-glucose, showing that the molecule is essentially linear and, since partial hydrolysis yields cellobiose (**6**), it consists of β-(1$\rightarrow$4)-linkages. The β configuration of the inter D-glucose linkages was verified by enzymic studies. Determination of chain length by end-group analysis is inaccurate for the essentially linear molecule and has led to very low values of the order of 200 units, owing to degradation during the analysis. Physical methods of chain length determination have shown that chains of up to 10 000 D-glucose units exist. Kinetic studies of the hydrolysis of cellulose has shown that over 99% of the linkages are the same type[94] and those which appear different do so as a result of physical effects. There has been no evidence for the presence of any other type of linkage.

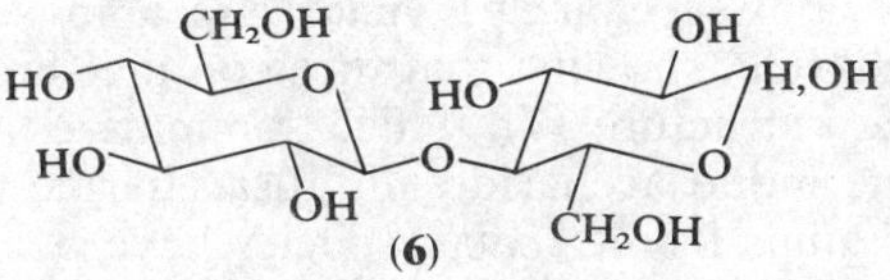

(**6**)

The characteristic properties of cellulose are due to the tendency of the individual chains to form microfibrils through inter- and intra-molecular hydrogen bonding to give a highly ordered structure. The microfibrils associate in a similar manner to give fibres, but in this case the axis of the fibres is at an angle to the axis of the microfibril whereas the individual molecules lie parallel to the microfibril axis. This regular arrangement of molecules is sufficient to allow X-ray diffraction patterns to be obtained and these indicate that cellulose from almost every natural source has an essentially similar pattern,[95] but which is different from that regenerated from derivatives or from solution. Cellulose I is the crystalline arrangement found in nature and cellulose II is that of regenerated cellulose. Variants in this classification have been noted and four other forms of cellulose have been distinguished from each other by their X-ray patterns.[96] Cellulose III_I and cellulose IV_I had similarities in the X-ray pattern with cellulose I, whilst cellulose III_{II} and cellulose IV_{II} had similarities with cellulose II. Whereas all the polymers had different molecular conformations in the unit cell, the cellulose I family (I, III_I, and IV_I) had 'bent' chain conformations and the cellulose II family (II, III_{II}, and IV_{II}) had 'bent-twisted' chain conformations.[97] The term 'bent' chain refers to the cellobiose units which correspond to the repeat distance of the fibres. The original proposed structures were based on 'straight' chain conformations[98] which had all the glucose residues in the same plane. The adoption of 'bent' chain chain structures with alternate D-glucose units in the same plane is a more satisfactory conformation because it eliminates the disadvantages of overlap of the vàn der Waals radii of the O-2 and C-6 atoms and of the repulsions between the hydrogen atoms at C-1 and C-4 and retains the intermolecular hydrogen bonding which is consistent with polarized infrared studies.[99] Cellulose II appears to be thermodynamically more stable than cellulose I and transformation of cellulose I into cellulose II can take place in sodium hydroxide, or preferably in a mixture of sodium hydroxide and sodium sulphide.[100]

26.3.3.3 Gums

The plant gums or exudate gums are essentially polysaccharides, containing hexuronic acid units as salts and a number of neutral monosaccharides units which are often esterified in highly branched structures. These gums, which may be formed spontaneously or may be induced by deliberately cutting the bark or fruit, are exuded as viscous liquids which become hard nodules on dehydration and protect the site of the injury from micro-organisms. Many of these gums find industrial application as thickening agents or emulsion stabilizers. Determination of the structure of these gums utilizes the differing rates of hydrolysis of the various glycosidic linkages to produce, by selected degradation, oligosaccharides whose structures can be determined with greater ease and certainty. The most common conditions for hydrolysis include autohydrolysis (heating the acidic gum in aqueous solution) and mild hydrolysis with 0.01 M acid to cleave L-arabinofuranose units, acid hydrolysis with 0.1–0.5 M acid or extended autohydrolysis to rupture the more labile hexopyranosidic linkages, and strong acid hydrolysis with 0.5 M sulphuric acid to hydrolyse all but hexuronic acid linkages and which results in the isolation of acidic oligosaccharides, particularly of disaccharides containing one acidic unit. There are a number of reviews[101–103] giving details of the extraction and occurrence of these gums and in many cases partial structures are advanced. In some cases the structures appear to be related to those of other less-complex plant polysaccharides.[102]

Gum arabic, from various species of *Acacia*, is probably the best known example of a plant gum and is typical of a number of gums which contain interior chains of β-(1→3)-linked-D-galactose units to which chains comprised of L-arabinofuranose, L-rhamnopyranose, and D-glucopyranuronic acid units are attached. Autohydrolysis of arabic acid, the salt free polysaccharide, yields L-arabinose, L-rhamnose, and a D-galacto-L-arabinose disaccharide, showing that these entities are linked to the main chains as shown in the generalized structure (**7**), which summarizes the structural features but does not place uniquely the monosaccharide and disaccharide units attached to the outer chain. A number of other gums from *Acacia* species have been shown to have the same general D-galactan core with variations in the degree of branching[104] and in the nature and attachment of the peripheral L-arabinose and L-rhamnose units. Some gums, such as the gum from *A. karroo*,[105] also contain α-D-glucopyranuronic acid units (1→4)-linked to

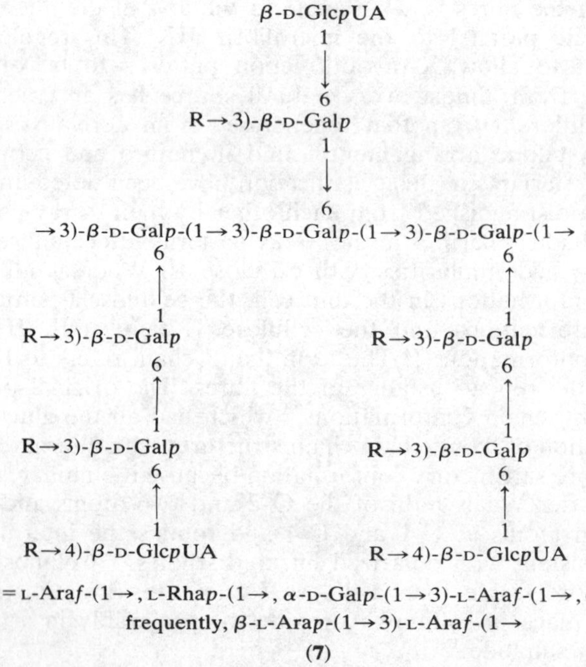

R = L-Ara*f*-(1→, L-Rha*p*-(1→, α-D-Gal*p*-(1→3)-L-Ara*f*-(1→, or, less frequently, β-L-Ara*p*-(1→3)-L-Ara*f*-(1→

(**7**)

D-galactose units. More recently an *Acacia* gum from species of a subseries Juli florae has been shown to contain more acidic groups, more methoxy-groups and a higher molecular weight, but with lower proportions of L-rhamnose and L-arabinose units.[106]

Gum ghatti from *Anogeissus latifolia* has interior chains of alternate D-glucuronic acid and D-mannose units with a high proportion of L-arabinofuranose end groups. A small proportion of L-arabinopyranose units are present in the more acid stable part of the polysaccharide. The structure (**8**) has been formulated to show the major structural features.

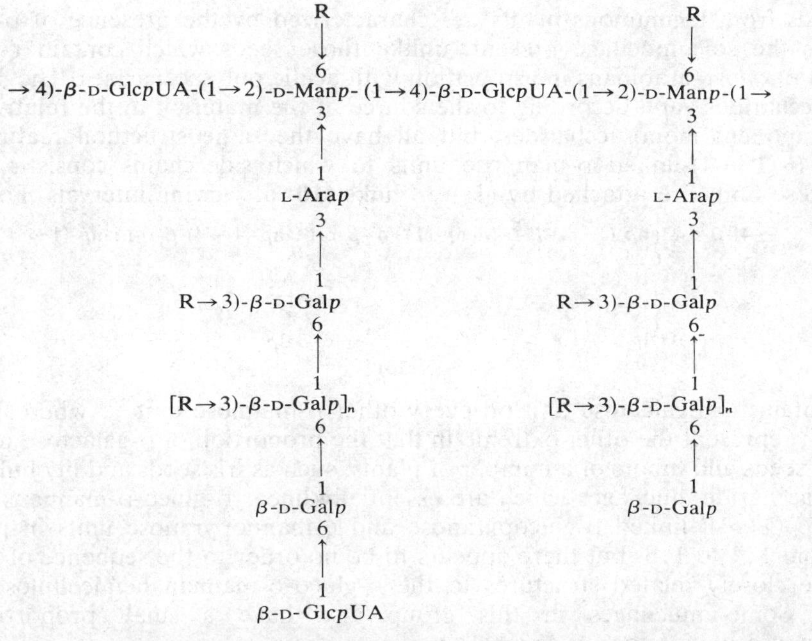

[R = L-Ara*f*-(1→, or less frequently L-Ara*f*-(1→2)-L-Ara*f*-(1→, L-Ara*f*-(1→3)-L-Ara*f*-(1→, or L-Ara*f*-(1→5)-L-Ara*f*-(1→]

(**8**)

Gum tragacanth has interior chains of 4-*O*-substituted α-D-galacturonic acid units and provides an example of a gum which is structurally related to the pectic acids (Section 26.3.3.5). The gums from *Sterculia* and *Khaya* species also contain a D-galacturonic acid interior chain, but with varying amounts 2-*O*-substituted L-rhamnose units interspersed and a variety of constituent monosaccharide (including D-glucuronic acid) units, which occur only as non-reducing end groups, unlike the D-galacturonic acid units which are present mainly in the inner chains. Gum tragacanth is only partially soluble in water but its major component, tragacanthic acid, has been purified and found to contain D-galacturonic acid, D-xylose, L-fucose, and D-galactose, with small amounts of D-glucuronic acid, L-rhamnose, and L-arabinose; the important structural features are summarized in (**9**). A minor component of the gum has been found to be a highly branched L-arabino-D-galactan.[107]

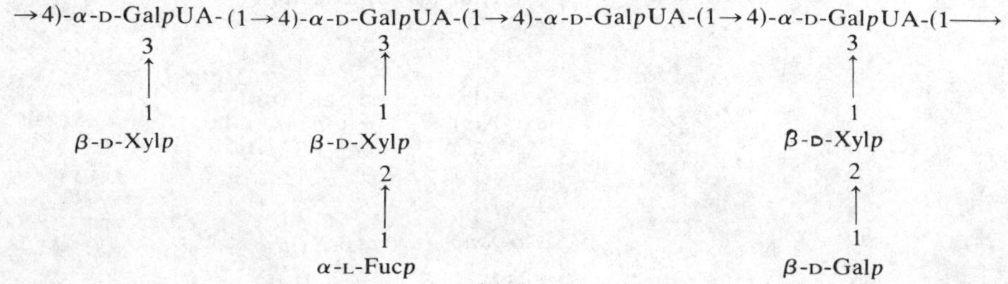

(**9**)

A polysaccharide containing the unusual L-glucose unit has been found in mesquite gum (see also Section 26.3.3.7.).

26.3.3.4 Mucilages

Mucilages derived from the bark, seeds, roots, and leaves of plants fall into several structural types and can be divided into two groups: acidic and neutral polysaccharides. Their role is probably to act as reservoirs for water retention to protect the seeds from desiccation.

The seeds from leguminous plants are characterized by the presence of D-galacto-D-mannan as the sole mucilage present, unlike those seeds which contain L-*arabino*-D-xylans and D-*xylo*-L-arabinans in association with acidic polysaccharides. The structure of the polysaccharides vary, according to the source of the material, in the relative amounts of the component monosaccharides, but all have the same structural feature, namely chains of β-(1 → 4)-linked D-mannose units to which side-chains consisting of single α-D-galactose units are attached by (1 → 6) links (**10**) at varying intervals. For example,

→4)-β-D-Man*p*-(1→4)-β-D-Man*p*-(1→4)-β-D-Man*p*-(1→4)-β-D-Man*p*-(1→

(first and third Man*p* units each carry, at O-6: 6 ↑ 1 α-D-Gal*p*)

(**10**)

guaran contains a D-galactose unit on every other D-mannose unit,[108] whereas ivory nut mannans[109] represent the other extreme in that the proportion of D-galactose units is very small. The seeds and tubers of a number of plants, such as iris seeds and lily bulbs, contain only polysaccharide mucilages which are essentially linear D-gluco-D-mannans possessing chains of β-(1 → 4)-linked D-glucopyranose and D-mannopyranose units in proportions ranging from 1:1 to 1:3, but there appears to be no order to the sequence of the chains, which have closely related structures to the D-gluco-D-mannan hemicelluloses (Section 26.3.3.6.). Some mucilages in this group also have a small proportion of D-galactopyranose units attached as single-unit side-chains.

Another group of neutral mucilages which occur in association with acidic polysaccharides come from cereal gums and consist of highly branched L-*arabino*-D-xylans. The D-xylan chains of these polysaccharides are β-(1 → 4)-linked (**11**), and to them are

→4)-β-D-Xyl*p*-(1→4)-β-D-Xyl*p*-(1→4)-β-D-Xyl*p*-(1→4)-β-D-Xyl*p*-(1→

(second and fourth Xyl*p* units each carry, at O-3: 3 ↑ 1 α-L-Ara*f*)

(**11**)

attached in an irregular manner single α-L-arabinofuranose units *via* α-(1 → 3) links. The mucilage of cress seed has been found to contain a neutral polysaccharide comprising of α-(1 → 5)-linked L-arabinofuranose chains to which are attached L-arabinose and D-xylose units (**12**).

→5)-α-L-Ara*f*-(1→5)-α-L-Ara*f*-(1→5)-α-LAra*f*-(1→

(second and third Ara*f* units each carry, at O-3: 3 ↑ 1 α-L-Ara*f*, which carries at O-3: 3 ↑ 1 α-D-Xyl*p*)

(**12**)

The acidic mucilages have in general been less well characterized but have been shown to contain chains of alternating D-galacturonic acid units and L-rhamnose units to which are attached side-chains consisting of 3-*O*-methylated D-galactose, and free D-galactose, L-rhamnose, D-xylose, and 4-*O*-methyl-D-glucuronic acid. Slippery elm mucilage (**13**)[110] and the acidic mucilage (**14**) from cress seed[111] are characteristic of this type of compound.

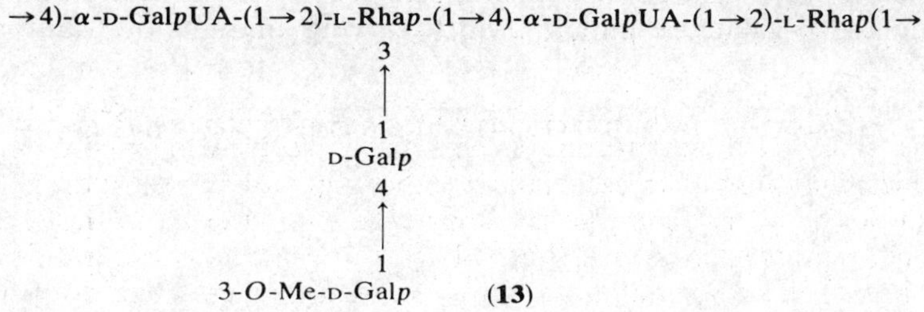

(**13**)

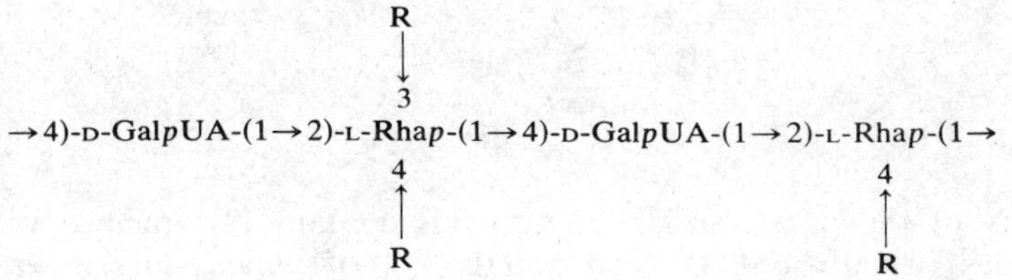

[R = L-Rha*p*-(1→, D-Gal*p*-(1→, 4-*O*-Me-D-Glc*p*UA-(1→4)-D-Gal*p*-(1→, or D-Xyl*p*-(1→4)-D-Gal*p*(1→]

(**14**)

26.3.3.5 Pectins

The pectins are a group of substances found in the primary cell walls and intercellular layers in land plants. They make up 15% of apples and 30% of the rinds of citrus fruits, but occur only in small proportions in woody tissues. Polysaccharides in which a proportion of D-galacturonic acid units are present as methyl esters are termed pectinic acids, whilst those in which no ester groups occur are termed pectic acids. Pectinic acids are very easily extracted with water and possess considerable gelling powers which are used commercially for gelation of fruit juices to give jellies. In contrast, the pectic acids frequently occur as calcium salts of their hexuronic acid units, are less soluble, and require the use of reagents such as bis[di(carboxymethyl)amino]ethane (H_4edta, ethylenediaminetetra-acetic acid) or sodium hexametaphosphate for their extraction. Among the pectic substances are those types of homopolysaccharides, D-galactans, L-arabinans, and D-galacturonans, but the most common pectic substances are heteropolysaccharides containing both acidic and neutral sugars.

Structural studies carried out on these homopolysaccharidic pectins established the salient features of the more complex pectins. The increased recognition of the susceptibility of pectins to acidic and basic hydrolysis indicates that these homopolysaccharides may have arisen from degradation of more complex nature polysaccharides. D-Galactans, containing β-(1→4)-linked D-galactose units in linear chains, have been isolated from white lupin seeds[112] and from red spruce compression wood, and a more complex acidic D-galactan, which has the same major structural features, has been isolated from Norwegian spruce.[113] Highly branched L-arabinofurans (**15**), containing no other monosaccharide units, have been isolated from mustard seeds,[114] and a D-galacturonan has been

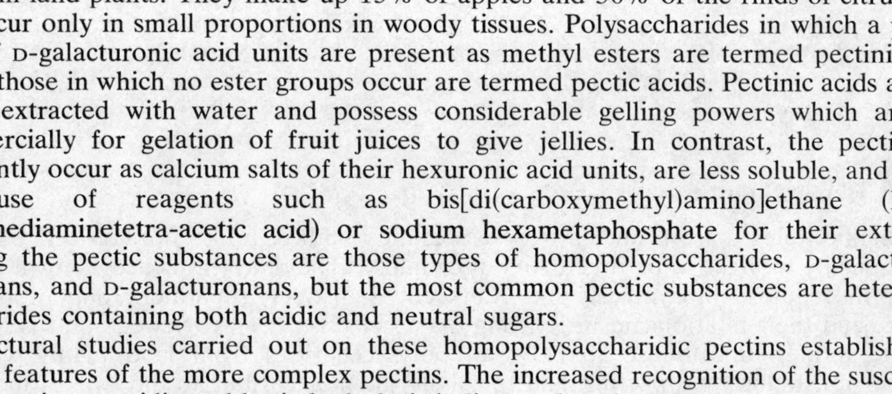

(**15**)

isolated from sunflower heads,[115] although this type of acidic homopolysaccharide is of infrequent occurrence, D-galacturonic acid being more frequently found in heteropolysaccharidic pectins. The structure of this D-galacturonan has linear chains of α-(1 → 4)-linked D-galacturonic acid units.

A pectin (**16**) from soybean is characteristic of the L-arabino-D-galactans which are the only neutral heteropolysaccharidic pectins known to possess structures containing chains of β-(1 → 4)-linked D-galactose units to which varying amounts of L-arabinose units are (1 → 3)-linked.

→4)-β-D-Gal*p*-(1→4)-β-D-Gal*p*-(1→4)-β-D-Gal*p*-(1→4)-β-D-Gal*p*-(1→
3
↑
1
L-Ara*f*
5
↑
1
L-Ara*f*

(**16**)

The majority of the pectic and pectinic acids contain varying proportions of neutral monosaccharides (usually between 10 and 25%), of which only L-rhamnose is found interrupting the D-galacturonic acid chains, the others being attached as side chains (**17**). A pectic acid isolated from soybean was the first example to show the nature of the D-xylose side-chains[116] and their similarity to those in tragacanthic acid (Section 26.3.3.3).

→4)-α-D-GalUA*p*-(1→2)-L-Rha*p*-(1→4)-(D-GalUA*p*)$_n$-(1→4)-α-D-GalUA*p*-(1→
↑ (R¹, to GalUA*p* or Rha*p*) ... 3 ← R²

R¹ = (D-Gal*p*)$_n$-(1→
or (L-Ara*f*)$_n$-(1→

R² = β-D-Xyl*p*-(1→
α-L-Fuc*p*-(1→2)-D-Xyl*p*-(1→
or β-D-Gal*p*-(1→2)-D-Xyl*p*-(1→

(**17**)

26.3.3.6 Hemicelluloses

The hemicelluloses are found in close association with cellulose in plant cell walls and were originally thought to be precursors of cellulose. They have now been shown to have no part in cellulose biosynthesis but represent a distinct, separate group of polysaccharides, and their relationship in growing plant cells has been reviewed.[117] It is usual to limit the term hemicellulose to cell-wall polysaccharides from land plants with the exclusion of cellulose and pectins, and to classify them according to the type of monosaccharide unit present. The majority of the hemicelluloses are relatively small molecules consisting of between 50 and 200 monosaccharide units, whilst those from hardwoods are larger molecules (150–200 units). Some of these compounds have crystalline structures and it has been found that the backbone of the molecule of the D-xylans[118] consists of D-xylose units with a rotation of 120° between successive units.

D-Xylans are the most common of the hemicelluloses, occurring in all parts of all land plants in quantities varying between 7 and 30%. The backbone of the molecule is an essentially linear chain of β-(1 → 4)-linked D-xylose units. Homoglycans are neither common or abundant, but a typical example is the D-xylan from Esparto grass, which has been found to contain β-(1 → 4)-linked D-xylose chains which, as indicated by the isolation of small amounts of 2,3,4-tri-*O*-methyl-D-xylose from methylation analysis, contain a single branch point located at C-3 of a D-xylose residue. The most common hemicellulose in soft woods is an (*O*-acetyl-L-arabino)-(4-*O*-methyl-D-glucurono)-D-xylan, but the most common side-chain is an α-(1 → 2)-linked or, less frequently, an α-(1 → 3)-linked 4-*O*-methyl-D-glucuronic acid, although annual plants have been shown to contain unmethylated units. The number of D-glucuronic acid side-chains varies considerably, with the hard wood D-xylans having, on average, one acidic unit for every ten D-xylose units. Another common side-chain in D-xylans is α-(1 → 3)-linked L-arabinose, but these units do not always occur as non-reducing end units. A typical example of non-terminal L-arabinose units occurs in barley husk xylan (**18**).[119]

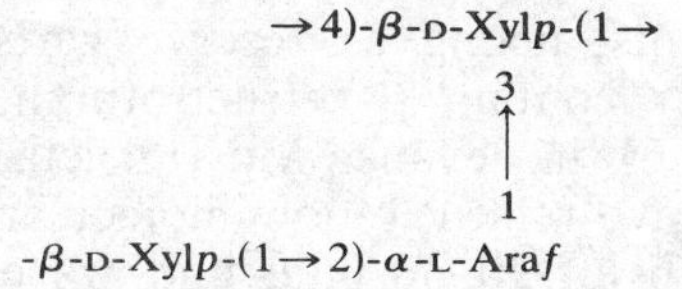

(**18**)

D-Mannans, containing not less than 95% D-mannose units, occur in ivory nuts, green coffee beans, and a number of other plant sources and have a common carbohydrate structure[120] consisting of linear chains of β-(1 → 4)-linked D-mannose units, which differ in chain length depending on the source. D-*gluco*-D-Mannans have been isolated from hard woods in which the ratio of D-glucose to D-mannose is 1:2. Partial hydrolysis has shown the existence of disaccharides of D-glucose and D-mannose together with cellobiose, mannobiose, and mannotriose, which suggests that these hemicelluloses have a random distribution of β-(1 → 4)-linked D-glucose and D-mannose units in linear chains. Similar compounds have been isolated from seeds of certain iris, orchid, and lily bulbs, but they contain different proportions of D-glucose to D-mannose, varying between 1:1 and 1:4. D-*galacto*-D-Gluco-D-mannans occur together with D-*gluco*-D-mannans in coniferous woods. Structural studies have shown that these have the same backbone as the D-gluco-D-mannans with α-(1 → 6) linked D-galactose units as single-unit side-chains. There is also evidence that D-glucose and D-mannose occur as non-reducing end groups. These side-chains are responsible for the greater solubility of these compounds compared with that of the D-*gluco*-D-mannans. This solubility is probably brought about by the side-chains preventing the macromolecules from forming strong, intermolecular hydrogen bonds in the same way as the side-chains of the other hemiculluloses such as 4-*O*-methyl-D-glucuronic acid, L-arabinose, and *O*-acetylated units.

Another group of hemicelluloses is the water-soluble L-*arabino*-D-galactans which are highly branched molecules with β-(1 → 3)-linked D-galactose chains to which are attached side-chains of L-arabinofuranose and L-arabinopyranose as well as D-galactose and D-glucuronic acid units. A typical example is the L-arabino-D-galactan from maritime pine (**19**).

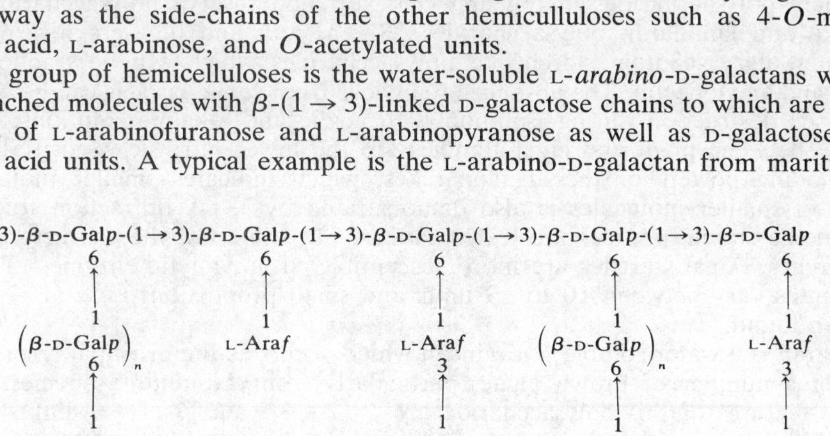

(**19**)

26.3.3.7 Miscellaneous plant polysaccharides

Lichenan is a D-glucan from Iceland moss and contains a random mixture of β-(1 → 4)- and β-(1 → 3)-linked D-glucose units in linear chains, with a higher proportion (about 70%) of β-(1 → 4) links. Iceland moss also contains a polysaccharide which is the

α-analogue of lichenan. This so-called isolichenan contains α-(1 → 4)- and α-(1 → 3)-linked D-glucose units with a higher proportion of the latter linkage. It also has a lower chain length of about 40–45 units compared with 60–360 for lichenan. These polysaccharides have been important in the understanding of the mode of action of glycanases.[121]

There are a number of glycans which occur in plants. Nigeran is an α-D-glucan with approximately equal proportions of alternating α-(1 → 3) and α-(1 → 4) linkages. The β-D-glucans found in plants include pustulan, which contains β-(1 → 6) linkages. Fructans also commonly occur in plants, acting as reserve carbohydrates either alone, or in conjunction with starch. They contain β-D-fructrofuranose residues which are linked (2 → 1) in inulins and (2 → 6) in levans. Most fructans contain D-glucose units as non-reducing end units and have relatively low molecular weights of the order of 8000. Grass levans have chain lengths of 20–30 units and are essentially linear.

Amyloids are a group of water-soluble polysaccharides found in the seeds of a number of plants, including tamarind and nasturtium, which are so-called because of their staining reactions with iodine. They contain β-(1 → 4)-linked D-glucose units to which are linked single D-xylose units and 2-*O*-β-D-galacto-α-D-xylose residues as side chains *via* α-(1 → 6) links.[122]

The unusual monosaccharide, L-glucose, has been reported to be among the hydrolysis products of the leaves of jute and *Grindelia* species.[123]

26.3.3.8 Algal polysaccharides

It is difficult to define algae because they encompass not only microscopic unicellular organisms but also the seaweeds, the arms of which can extend to over 150 feet in length. They have no roots, stems, or leaves, as have the higher land plants, and they are most frequently found in fresh or salt water, occasionally free floating in the water as in the case of the brown seaweeds. The polysaccharides obtained from those sources can be grouped into three classes: food reserve, structural, and sulphated polysaccharides. Whilst many compounds are known, their structures still have to be determined and reviews on these compounds reflect the increasing knowledge of their structures.[103,124]

The main polysaccharides in the first class, of food reserve polysaccharides, include starch-like and laminaran polysaccharides. Green, red, and blue-green seaweeds, and fresh water algae, contain starch-like polysaccharides which can be fractionated into amylose and amylopectin. The absence of amylose from some extracts can be explained in terms of its destruction during isolation with acidic and alkaline solutions. The major difference between plant and algal starches are the lower intrinsic viscosities and lower iodine binding powers of the algal starches, which indicates smaller molecules. The presence of smaller molecules is also demonstrated by X-ray diffraction studies, which show that the starch granules are less organized but still show the characteristics of the plant starches. Algal starches are more susceptible to amylolytic enzymes. The average chain lengths vary between 10 to 19 units and small proportions of α-(1 → 3) linkages have been found.[125]

Laminaran is a water soluble β-D-glucan which occurs as the principal reserve polysaccharide in a number of brown algae, particularly from *Laminaran* species. The main structural features of this polysaccharide are β-(1→3)- and β-(1→6)-linked D-glucose units forming chains with some β-(1 → 6) linkages as branch points which are terminated by reducing D-glucose or non-reducing D-mannitol units. The ratio of these so-called G- and M-chains varies, but is about 50% for the majority of samples. The average chain lengths vary between 7 and 11, with molecular weights corresponding to about 30 units for soluble forms of laminaran, which indicates an average of 2–3 branch points per molecule, whilst insoluble forms contain 16–24 units with average chain lengths of 15–19 units and indicate essentially linear structures. Chains terminating in D-mannitol have not been found in the D-glucans isolated from members of the Chrysophycaae and Bacillariophycaae, but they resemble laminaran in other respects.

Brown, red, and green algae have, as a structural polysaccharide, the second class of algal polysaccharides, a cellulose which is essentially similar to that of land plants and comprises about 10% of the dry weight. In very close association with this cellulose are a number of hemicelluloses and similar compounds which include D-mannans, D-xylans, and a lichenan type of polysaccharide (Section 26.3.3.7) consisting of linear chains of β-(1 → 4)- and β-(1 → 3)-linked D-glucose units with a lower proportion of β-(1 → 3) linkages than is found in plant lichenan. The D-mannans isolated from red seaweed resemble the ivory nut D-mannan in that they contain essentially linear chains of about 16 β-(1 → 4)-linked D-mannose units. The D-xylans found in seaweeds differ from those in land plants by being predominantly true D-xylans containing, in the case of the red seaweed *Rhodymenia palmata*, chains of β-(1 → 3) and β-(1 → 4) linkages in an irregular distribution,[126] whereas the D-xylan from green seaweed, *Caulerpa filiformis*, contains only β-(1 → 3)-linked D-xylose units.[127]

Alginic acid is a commercially important component of brown seaweeds and is the most important mucilaginous polysaccharide which prevents desiccation of the seaweed when exposed to the air, *e.g.* at low tide. Industrially it is important as a thickening agent and emulsion stabilizer. It has been shown that the molecule readily produces fibres which according to X-ray diffraction are essentially linear. It was originally thought to be a β-(1 → 4)-linked D-mannuronan, but the composition of alginic acid actually consists of D-mannuronic and L-guluronic acid units. Partial hydrolysis gives degraded polysaccharides containing entirely one or the other acid unit and a number of oligosaccharides containing both units, and it has been suggested that alginic acid contains crystalline regions of β-D-mannuronic acid and of α-L-guluronic acid and amorphous regions containing a mixture of both residues, all (1 → 4)-linked.[128]

The third group of polysaccharides isolated from algae, the sulphated polysaccharides, include carrageenan and agar. Carrageenan can be fractionated into a number of polysaccharides, proportions of which vary from species to species and with the season and environment. The major structural features of this group of polysaccharides are an alternating sequence of β-(1 → 3)-and α-(1 → 4)-linked D-galactose residues containing various degrees and sites of sulphation. κ-Carrageenan, for instance, is an insoluble galactan containing alternating 3-*O*-substituted β-D-galactopyranose 4-sulphate and 4-*O*-substituted 3,6-anhydro-α-D-galactopyranose units and small proportions of (1 → 4)-linked D-galactose 6-sulphate units,[129] whereas μ-carrageenan is soluble and contains only α-(1 → 3)-linked D-galactose 4-sulphate and β-(1 → 4) linked D-galactose 6-sulphate. ι-Carrageenan has a repeating structure of α-(1 → 3)-linked D-galactose 4-sulphate and β-(1 → 4)-linked 3,6-anhydro-D-galactose 2-sulphate with approximately 1 in 10 of these units being replaced by a β-(1 → 4)-linked D-galactose 2,6-disulphate.[130] ξ-Carrageenan is a polymer of D-galactose 2-sulphate with β-(1 → 4) and α-(1 → 3) linkages.[131] λ- And χ-carrageenans are similar, each containing α-(1 → 3)-linked D-galactose 2-sulphate and β-(1 → 4)-linked D-galactose 2,6-disulphate and 3,6-anhydrogalactose 2-sulphate units; χ-carrageenan is rich in 3,6-anhydro-D-galactose compared with λ-carrageenan.

Agar is thought to consist of chains having alternating α-(1 → 3) and β-(1 → 4) linkages with three extremes of structure: neutral agarose, which contains β-(1 → 3) D-galactose

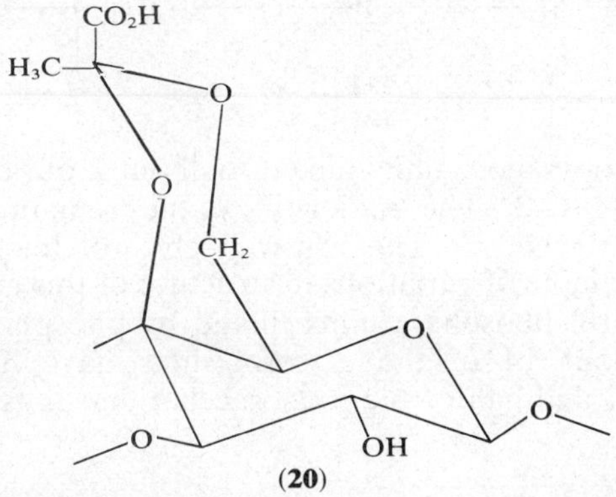

(20)

and α-(1 → 4) 3,6-anhydro-L-galactose units, a pyruvated agarose in which the D-galactose residues are substituted with pyruvic acid in acetal linkages (**20**), and a sulphated galactan with few, or no, 3,6-anhydro-L-galactose or pyruvated D-galactose residues.[132]

There are also a number of algal mucilages which have similar properties and whose structures have not yet been determined. They frequently contain L-rhamnose, D-xylose, D-glucuronic acid, D- and L-galactose, and D-mannose and are typified by the mucilage from fresh water red algae,[133] which may contain a D-galactose and D-glucuronic acid repeating sequence, together with the above neutral units and methylated L-rhamnose and D-galactose units; and by a mucilage from brown seaweed[134] which contains L-fucose, D-xylose, and D-galactose as non-reducing terminal residues with D-mannose and D-galactose occupying branch positions.

26.3.4 MICROBIAL POLYSACCHARIDES

Microbes give rise to many polysaccharides and also to macromolecules to which carbohydrate chains are attached, such as glycopeptides, glycoproteins, and glycolipids.[135] Many of these compounds are unique to microbes and are not found in any other areas of the plant and animal kingdoms. The polysaccharidic material can occur as an integral part of the cell wall, and as capsules surrounding the cell, or can be elaborated in culture media (extracellular polysaccharides).

26.3.4.1 Teichoic acids

These unusual phosphate polymers comprise up to 50% of the dry weight of cell walls of certain Gram-positive bacteria. They also occur as membrane or intracellular components of bacteria. They are firmly attached to the cell wall and extraction requires an agent such as trichloroacetic acid. Their occurrence, structure, and properties have been reviewed.[136] Two types of teichoic acids are known, one containing chains of D-glycerol and the other chains of D-ribitol, both types of polyol being joined by phosphate diester linkages. Glycosidically bound carbohydrate units are present in the ribitol teichoic acids but only in some of the glycerol teichoic acids.

The simplest teichoic acids are the glycerol polymers, such as the membrane teichoic acid from a strain of *Lactobacillus casei* which contains D-alanine residues linked to the glycerol phosphate backbone (**21**). The membrane teichoic acid from *Lactobacillus*

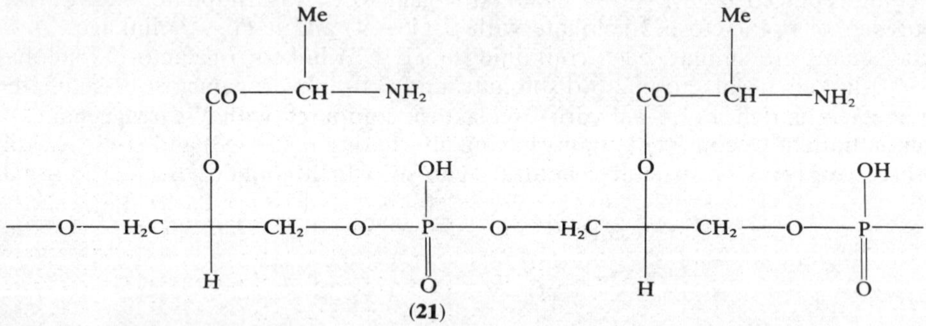

(21)

arabinosus contains D-glucopyranose units linked to about 2 out of 18 glycerol phosphate units through the glycerol C-2. The majority of the remaining glycerol units have D-alanine ester residues attached.[137] The cell-wall glycerol teichoic acid from *Bacillus stearothermophilus* is an example of variations in structure of this group. It contains a chain of approximately 18 glycerol phosphate units linked by phosphodiester linkages to C-2 and C-3 of glycerol. About 14 of the glycerol units have α-D-glucopyranose units glycosidically linked to C-1 and half of the monosaccharides units are esterified at C-6 by D-alanine (**22**).[138]

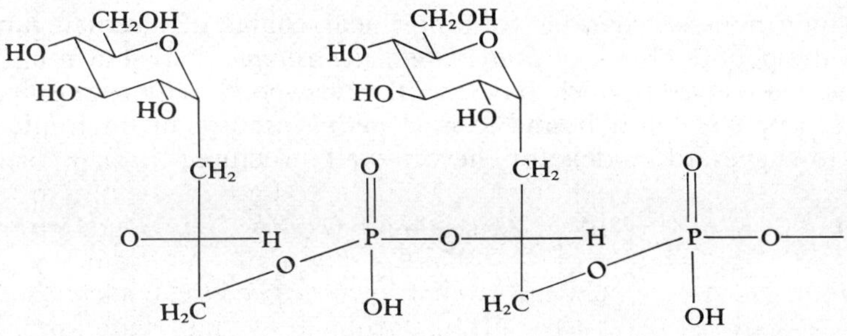

(**22**)

The ribitol teichoic acids, which only occur in cell walls, contain chains of ribitol phosphate units linked through C-1 and C-5 with monosaccharide units glycosidically linked at C-4, or possibly at C-4 and C-3, and D-alanine ester linkages at C-2 or C-3 of the ribitol units.[139] Examples of this group of compounds include the cell-wall teichoic acid from *Bacillus subtilis* (**23**)[140] and from *Staphylococcus aureus*,[141] in which the monosaccharide units present are 2-acetamido-2-deoxy-D-glucose.

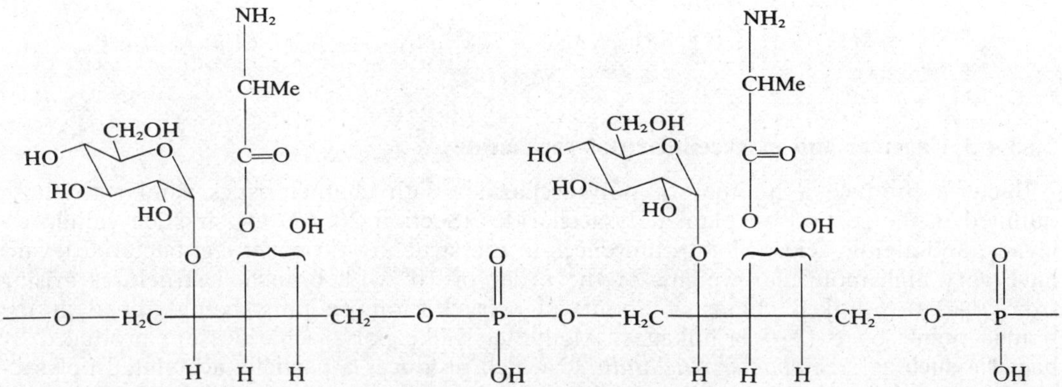

(**23**)

26.3.4.2 Cell-wall peptidoglycans (mureins)

These highly branched complex macromolecules occurring in bacterial cell walls consist of polysaccharide chains, in which individual residues carry an amino-group, cross linked by peptide bridges. The characteristic features of this unique type of structural polymer are the amino-type monosaccharides, 2-acetamido-2-deoxy-D-glucose and muramic acid [2-acetamido-3-*O*-(2-carboxyethyl)-2-deoxy-D-glucose] (**24**), and the amino-acids D- and L-alanine, D-glutamic acid, and L-lysine and other amino-acids depending on the bacterium,

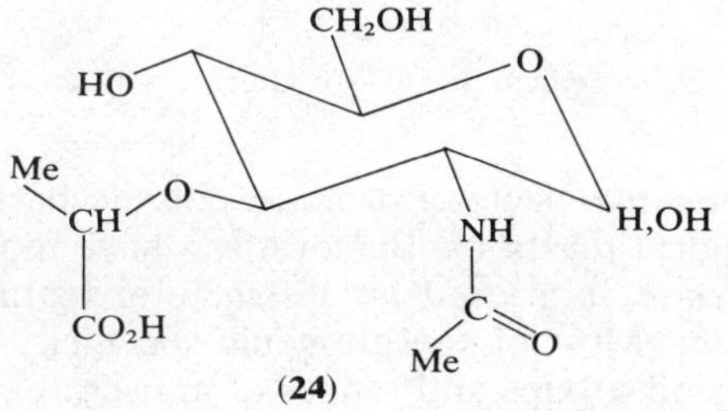

(**24**)

with the foregoing monosaccharides forming linear chains of alternate units. A typical fragment from the peptidoglycan of *Spirochaeta stenostrepta*[142] is shown in (**25**) by way of example of the many types which have been reviewed.[143] Hydrolysis has shown that teichoic acids can be covalently bound to mureins *via* phosphate–muramic acid linkages, which has led to suggested models for the cell-wall structure.[144]

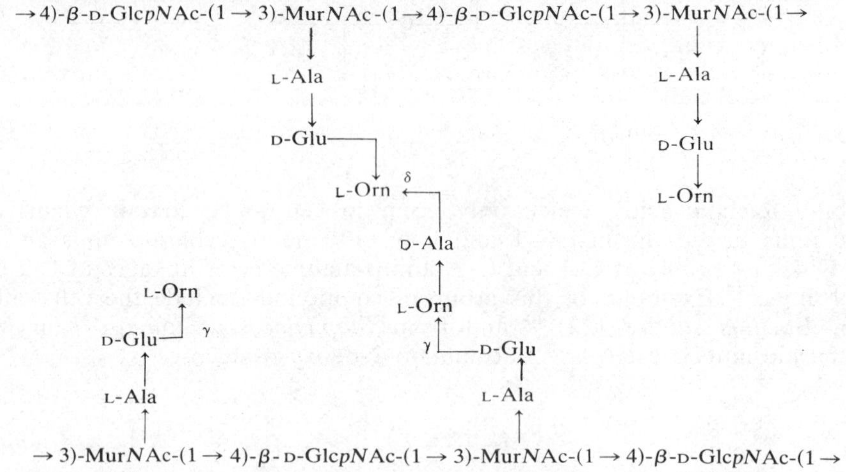

(**25**)

26.3.4.3 Capsular and extracellular polysaccharides

Bacteria produce a number of polysaccharides with characteristics similar to those outlined in the section on plant polysaccharides (Section 26.3.3) and include celluloses, levans, and alginic acids. Slight differences in the structure do exist, *e.g.* bacterial levans have very high molecular weights of the order of 10^6 with branched structures arising from β-(2→6)-linked chains of about 10 D-fructofuranose units being joined at the branch points by β-(2→1) linkages. Alginic acid-like polysaccharides are produced by bacteria such as *Azotobacter vinelandii*,[145] which produces a partially acetylated polysaccharide containing mainly D-mannuronic acid units with a small proportion of L-guluronic acid units.

More important are the dextrans, which find uses[146] as plasma substitutes and are used as molecular sieves in a modified form (Section 26.3.7.5). The principal linkage in the D-glucopyranose chain is α-(1→6), but branching does occur to very varying degrees through α-(1→3) and α-(1→4) branch points. An example of the different degrees of branching can be found in the dextrans from *Leuconostoc mesenteroides*[147,148] and *Betacoccus arabinosaceous*.[149] The dextran from the former is an essentially linear molecule with (1→3)-linked side-chains consisting of mono- or di-saccharide residues, whereas the latter is a highly branched structure (**26**) with a unit chain length of 6 or 7 units.

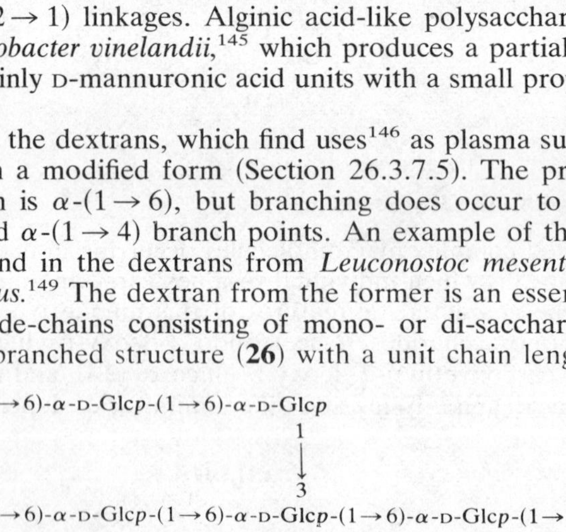

(**26**)

Many of the Gram-negative bacteria produce polysaccharides which are responsible for type specific immunological reactions. Bacteria can have more than one antigenic characteristic, many of which are sponsored by a structural feature of a/the capsular polysaccharide of the bacterium. Most of the antigenic specificity of the polysaccharides arises from the non-reducing end-groups and ranges of antisera have been used in the structural analysis of such polysaccharides. The structural basis for these serological reactions can be shown by the following example. The acidic polysaccharides from *Klebsiella aerogenes* and *Enterobacter 349* cross react to the extent of about 50% with the heterologous antibodies (*i.e.* from the antiserum to the opposite polysaccharide). Structural analysis has shown that these polysaccharides both contain 1,3-linked D-galactose and 1,3- and 1,4-linked D-mannose units, but that the *Klebsiella aerogenes* polysaccharide contains D-mannuronic acid units forming disaccharide repeating units with the D-mannose units, whereas the *Enterobacter* polysaccharide contains disaccharide repeating units of D-mannose attached to D-galacturonic acid and to D-glucuronic acid. A number of less common monosaccharides are found in the polysaccharides from Gram-negative bacteria, including 3,6-dideoxy-derivatives of D-glucose, D-galactose, and D-mannose and a number of aldoheptoses, the structures of which are given in Table 5. A unique homopolysaccharide from Gram-negative bacteria is a polymer of 5-acetamido-3,5-dideoxy-D-*glycero*-D-galacto-2-nonolosonic acid, colominic acid (**27**),[150] which contains (2 → 8) linkages and also, in some samples, (1 → 9) internal ester linkages between adjacent residues.

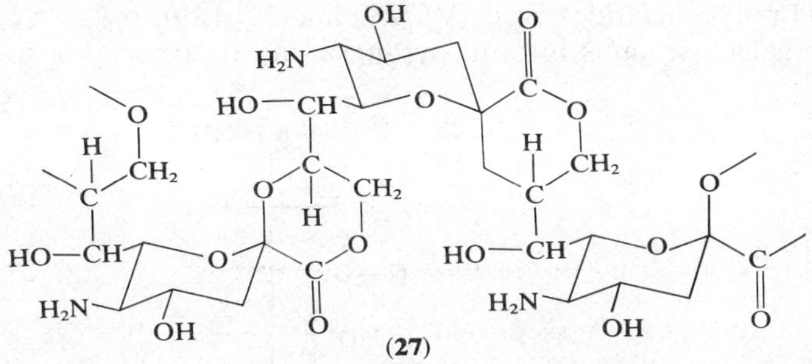

(27)

TABLE 5
Structures of Less Common Monosaccharides

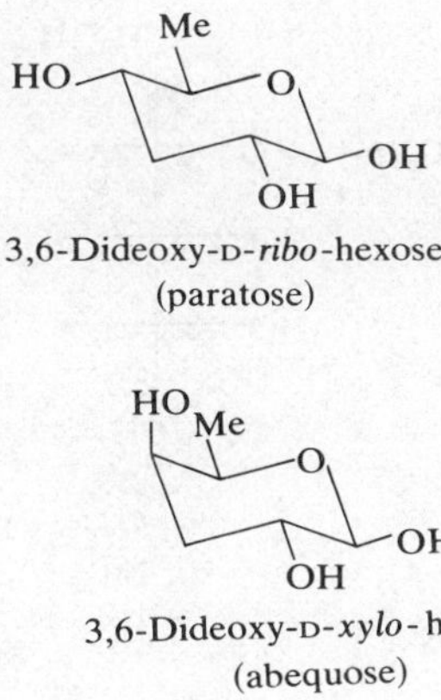

3,6-Dideoxy-D-*ribo*-hexose
(paratose)

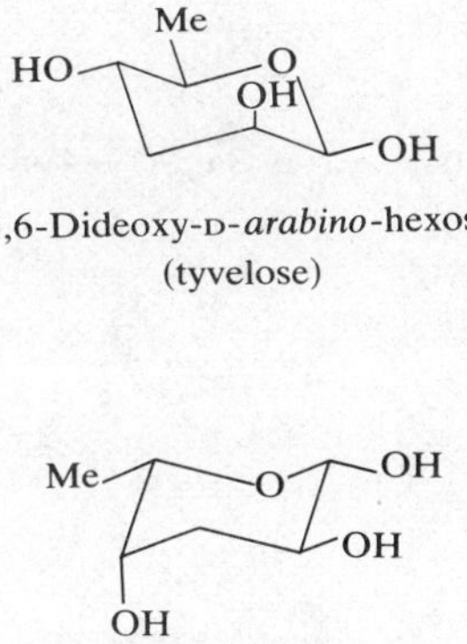

3,6-Dideoxy-D-*arabino*-hexose
(tyvelose)

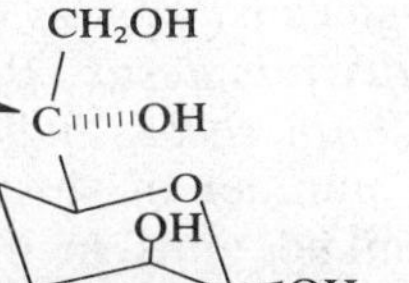

3,6-Dideoxy-D-*xylo*-hexose
(abequose)

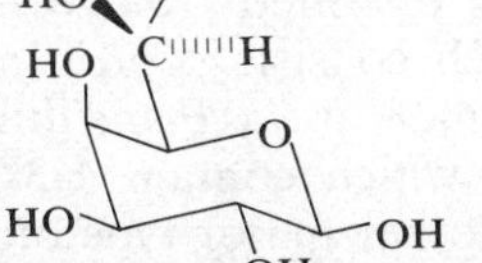

3,6-Dideoxy-L-*xylo*-hexose
(colitose)

L-*glycero*-D-*manno*-Heptose

D-*glycero*-D-*galacto*-Heptose

The Gram-positive bacteria produced type-specific polysaccharides which have complex structures and frequently contain amino-type monosaccharide units as well as neutral and acidic monosaccharides. The structures of a number of these polysaccharides have been determined and have been found to consist of a variety of polysaccharide types,[103,151,152,152a] Pneumococcal polysaccharides include: type II, which has 1,6-linked D-glucans to which are attached, *via* (1→4) linkages, chains of 1,3-linked L-rhamnose units with D-glucose and D-glucuronic acid units also attached to the D-glucan chain; type III, which has alternating 1,3-linked D-glucuronic acid and 1,4-linked D-glucose units; and type VIII, which consists of D-glucuronic acid, D-glucose, and D-galactose units in the molar ratio 1:2:1 and most probably contains a repeating sequence (**28**). There are also a number of

→4)-β-D-GlcpUA-(1→4)-β-D-Glcp-(1→4)-α-D-Glcp-(1→4)-D-Gal*p*-(1→

(**28**)

polysaccharides which show structural resemblances to the teichoic acids, as typified by the Pneumococcal polysaccharides types VI (**29**) and X_A (**30**), the latter being unusual in that it contains D-galactose units in both pyranose and furanose ring forms.

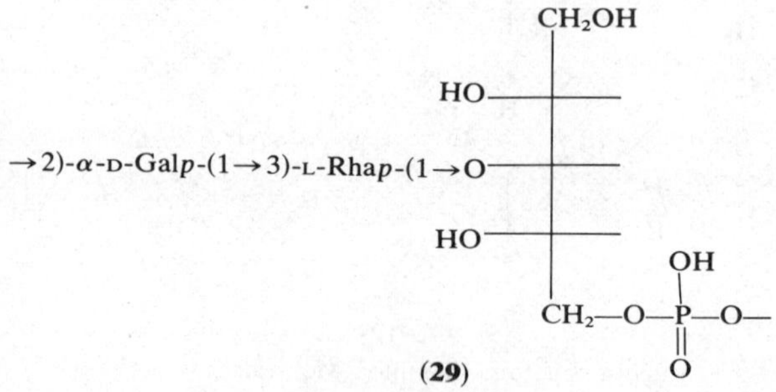

(**29**)

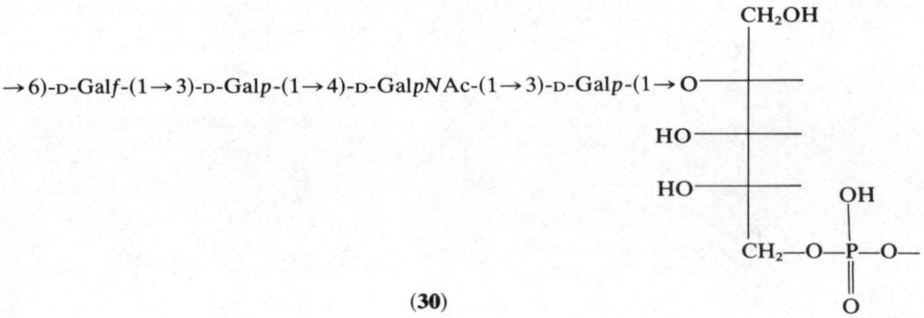

(**30**)

26.3.4.4 Fungal polysaccharides

Polysaccharides elaborated by fungi have been the subject of a number of reviews.[103,153,154] Those of particular interest are: α-D-glucans, such as one containing only 1,3-linked units obtained from cell walls of *Polyporus tumulosus*; β-D-glucans, such as pachyman, which consists of 1,3-linked units from *Porin cocos*, and luteose, which is a 1,6-linked polymer, from *Penicillium luteum*; and a number of storage products of the laminarin type which contain 1,3-, 1,4-, and 1,6-linked units in linear and branched structures. The other major type of fungal polysaccharides are the D-mannans, typified by the 1,6-linked D-mannan from *Saccharomyces rouxii* and the 1,2-linked D-mannan from *Saccharomyces cerevisiae.* The latter D-mannan has also been shown to contain phosphorylated side-chains[155] and has the structure (**31**) arising from the addition of 1,3-linked

α-D-Manp-(1→3)-α-D-Manp-(1→2)-α-D-Manp-(1→2)-α-D-Manp-(1→
|
α-D-Manp-(1→3)-α-D-Manp→Ⓟ

→Ⓟ = orthophosphate diester linkage

(31)

D-mannose units to the D-mannosylphosphorylmannotriose chains. A D-galactan, galactocarolose, from *Penicillium charlesii*, is a low molecular weight polysaccharide consisting of unbranched chains of approximately ten β-(1→5)-linked D-galactofuranose units.

A number of heteropolysaccharides have also been isolated and characterized, including the D-galacto-D-mannan from *Cladosporium herbarum*[156] which contains D-galactose units in both pyranose and furanose forms, and a D-xylo-D-mannan from *Cryptococcus leuerentii*.[157]

26.3.5 ANIMAL POLYSACCHARIDES

26.3.5.1 Glycogen

The principal reserve polysaccharide in the animal kingdom is glycogen. It is found in most tissues and the most convenient source for the purpose of extraction is usually liver or muscle. Human liver contains glycogen as up to 10% of its dry weight. Unlike starch, isolation and purification of glycogen is not simple. The classical method was to use strongly alkaline solutions at 100 °C for about 3 hours to dissolve the tissue and then to precipate the glycogen with ethanol, but with the development of understanding of alkaline degradation (Section 26.3.2.5) the use of milder techniques had to be sought. The use of cold dilute trichloroacetic acid for the extraction procedure results in a product the molecular size of which is some 10 times larger[158] than that obtained by the traditional method. Methods are now available which avoid more completely destruction during isolation[159] so that it is possible to investigate realistically molecular weights of the isolated polysaccharide. It has been found that, for example, glycogen from the liver in cases of general glycogen storage disease has a lower molecular weight than normal.

Classical analytical methods such as methylation have shown the structure of glyogen to be chains of α-(1,4)-linked D-glucose residues with α-(1,4,6) branch units. The use of amylolytic enzymes for determination of the fine structure has indicated a tree-like structure (Figure 5c) with a unit chain length of 12 D-glucose units. This short chain length in a molecule which can possess a molecular weight in the range 10^7–10^8 necessarily results in a highly branched structure, a consequence of which is the extremely limited uptake of iodine by the molecular compared even with that of amylopectin. Regions of dense branching that are resistant to the action of α-amylase are randomly distributed throughout the molecule.[160] With the availability of paracrystalline glycogen it should be possible to use physical methods to examine the structure in greater detail.[161] The occurrence, isolation, structure, and enzymic degradation of glycogen have been reviewed.[162–164]

26.3.5.2 Chitin

The most abundant of polysaccharides containing amino-type monosaccharides is chitin, which occurs as a structural polysaccharide in the shells of Crustacea. It also occurs in fungi and some green algae. The chemical evidence for its structure has been based on the isolation and characterization of oligosaccharides obtained as a result of partial hydrolysis. This shows that chitin is a homopolymer of 2-acetamido-2-deoxy-D-glucose, each unit being β-(1,4)-linked to form linear chains (**32**). The polysaccharide may be considered as an analogue of cellulose, the hydroxyl groups at C-2 positions of cellulose being replaced by acetamido-groups.

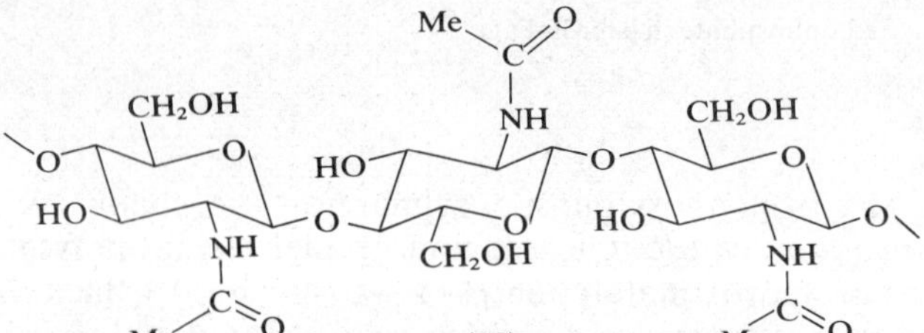

(**32**)

Chitin rarely occurs alone in nature and is usually found complexed or covalently bound to protein[165] in the shells of crabs and lobster. This property may be attributed to the more recently discovered fact that not all the amino-groups of the majority of chitins are *N*-acetylated. Accordingly, they can operate as basic groups and thereby complex with other molecules of a suitable ionic disposition. Chitin is insoluble in water and many organic solvents. This has made its structure determination difficult, its insolubility being reflected, for example, in its low reactivity towards methylation. The majority of samples obtained as a result of treatment with mineral acid have a degree of de-*N*-acetylation and are also of lower molecular weight than chitin in its native state. X-Ray diffraction studies of crystalline chitin have shown that the unit cell contains two chains with bent conformations with inter- and intra-molecular hydrogen bonding in a manner similar to cellulose (Section 26.3.3.2).

26.3.5.3 Glycosaminoglycans

The structures of this group of related carbohydrate substances, with the exception of keratan sulphate, consist of alternating units of hexuronic acids and 2-amino-2-deoxyhexose, the latter being *N*-acetylated, frequently *O*-sulphated, and sometimes *N*-sulphated. Historically these compounds have been known by a number of different names, including mucopolysaccharides[166] and acidic mucopolysaccharides. This terminology arose before it was known that these polysaccharides are attached covalently to protein. The modern name for these compounds is glycosaminoglycans for the carbohydrate residues and proteoglycans for the complete carbohydrates linked to protein compounds.[7]

The proteoglycans are important in the maintenance of tissue structure, *etc.*, being one of the essential building blocks of the macromolecular framework of connective and other tissues. A number of diseases which involve skeletal ot tissue changes have been shown to implicate proteoglycans. Hyaluronic acid appears to act, on account of its viscosity in solution, as a lubricant and shock-absorbing gel in limb joints. The size and charged nature of the proteoglycans may allow them to operate as filtration media for the exclusion of foreign matter from the body. Heparin is unique among the glycosaminoglycans in that it possesses biological activity as a blood anticoagulant. A group of diseases (hyperglycosaminoglycanuria, mucopolysaccharidoses) which has been studied widely is characterized by the excretion of excessive amounts of the various glycosaminoglycans attached to peptide. Much of the recent work in this field has gone into microtechniques for the demonstration of glycosaminoglycan disorders in various tissue conditions[13,167] and monitoring correction of the disorder during treatment,[168–170] but research is now being directed into identification on a basis of chemical structure[171] and screening tests for the diseases.[60,172,173] The role of proteoglycans and glycosaminoglycans in health and disease has been the subject of a number of reviews.[7,174–177]

The glycosaminoglycans can be distinguished by their compositions and primary structures, since the repeating disaccharide structures, which have been discovered as general features of all of them, are individually characteristic.[5] Hyaluronic acid has the simplest of the glycosaminoglycan structures with a repeating unit composed of the monosaccharides D-glucuronic acid and 2-acetamino-2-deoxy-D-glucose linked such that the linkage between the D-glucuronic acid and the 2-acetamido-2-deoxy-D-glucose is β-D-(1 → 3), *i.e.* the uronic acid has the β-D-anomeric configuration at C-1 from which it is linked glycosidically to position-3 of the 2-acetamido-2-deoxy-D-glucose unit. Between the 2-acetamido-2-deoxy-D-glucose and the D-glucuronic acid the linkage is β-D-(1 → 4), which results in the structure (**33**). Chondroitin is the only other non-sulphated glycosaminoglycan which occurs naturally and is isomeric with hyaluronic acid. The repeating unit (**34**) is

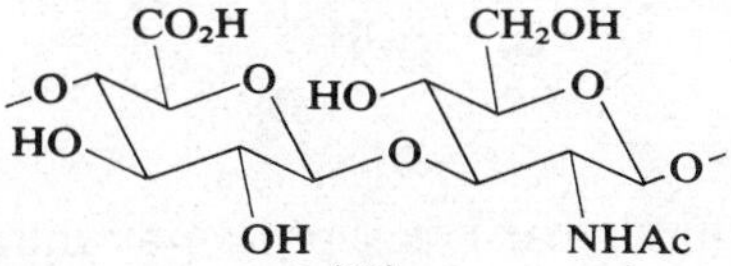

(**33**)

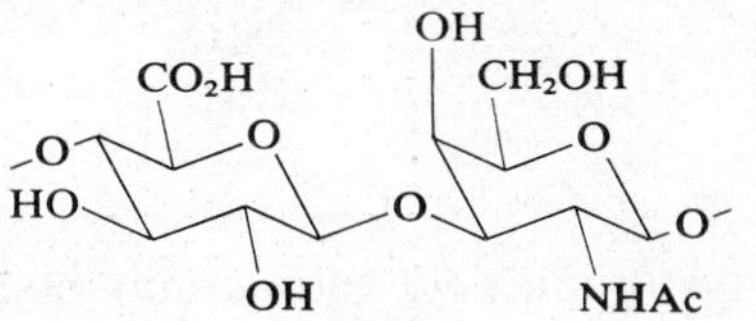

(**34**)

composed of D-glucuronic acid units linked *via* β-D-(1 → 3) links to 2-acetamido-2-deoxy-D-galactose units, which are in turn linked *via* β-D-(1 → 4) links to D-glucuronic acid units. This makes the only difference between hyaluronic acid and chondroitin the orientation of one hydroxyl group on every other monosaccharide unit along the chain.

Chondroitin 4-sulphate and chondroitin 6-sulphate are sulphated varieties of chondroitin, the sulphate group being located on the 4- and 6-positions of the 2-acetamido-2-deoxy-D-galactose units, resulting in the repeating structures (**35**) and (**36**), respectively.

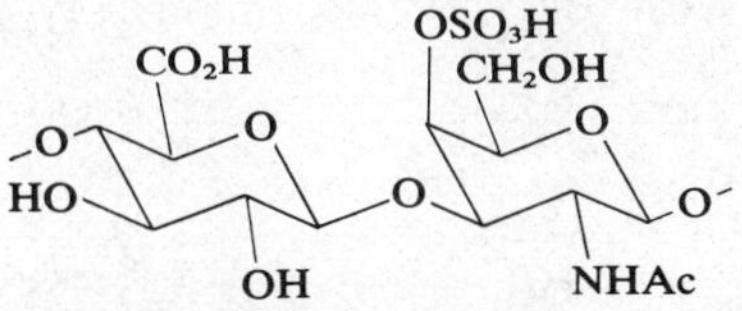

(**35**)

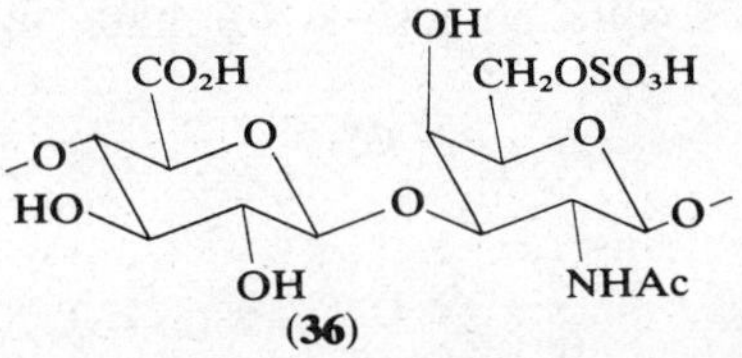

(**36**)

Variants of chondroitin sulphate with sulphate contents greater than 1 mole per mole of disaccharide unit are known but have no specific name. The so-called chondroitin sulphate D, obtained from shark cartilage, is in fact an over-sulphated chondroitin 6-sulphate.

Dermatan sulphate is an isomer of chondroitin 4-sulphate with L-iduronic acid units replacing D-glucuronic acid units but with the absolute configurations of the linkages remaining the same. This results in a repeating structure (**37**) which differs from chondroitin 4-sulphate only in the orientation of the carboxyl group at C-5 on every other

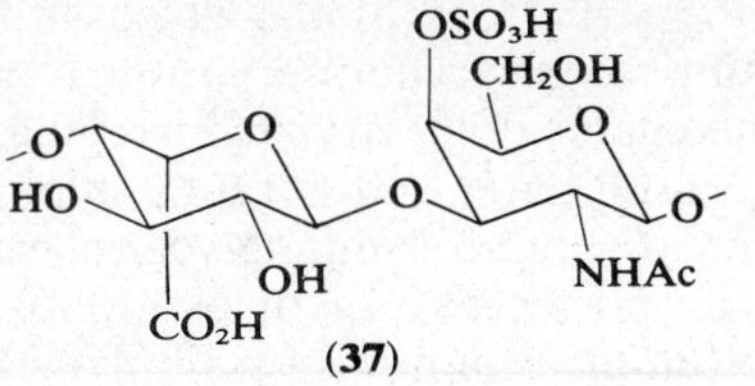

(**37**)

residue. Irregularities in the structure are known to exist, with one or two randomly spaced D-glucuronic acid units replacing L-iduronic acid units[178] and the repeating disaccharide may be di- or non-sulphated.[179] The non-sulphated form of dermatan sulphate, *i.e.* isomeric chondroitin and hyaluronic acid, and dermatan 6-sulphate, isomeric with chondroitin 6-sulphate, have not been detected in nature, but this does not preclude their existence. Dermatan can be prepared by chemical desulphation of dermatan sulphate.

The repeating unit in heparin was much more difficult to define and, in spite of extended efforts, it was not until the late 1960s that the location of the sulphate groups could be assigned with confidence. It is now recognized that the composition of heparin is a mixture of two disaccharide repeating units, comprising L-iduronic acid 2-sulphate and 2-deoxy-2-sulphamido-D-glucose 6-sulphate (**38**) and D-glucuronic acid and 2-deoxy-2-sulphamido-D-glucose 6-sulphate (**39**). Although the monosaccharide units are analogous

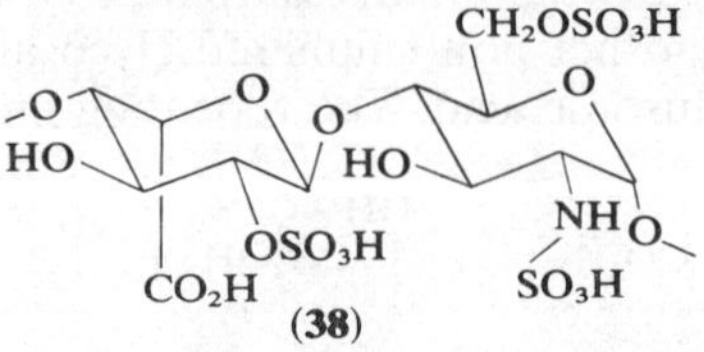

(**38**)

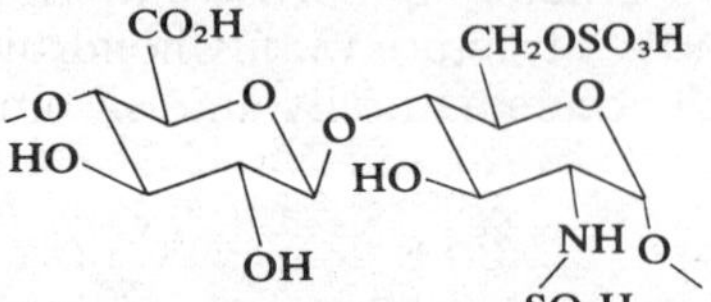

(**39**)

to those in hyaluronic acid and dermatan sulphate, heparin is significantly different owing to the α-anomeric configuration of the D-glucose and 2-amido-2-deoxy-D-glucose units. The overall sulphate content is in the range 2–3 moles sulphate per mole disaccharide repeating unit. Heparan sulphate is not, as the name might suggest, a further sulphated version of heparin. The backbone carbohydrate structure of heparan sulphate is similar to that of heparin but the 2-amido-2-deoxy-D-glucose units are less frequently *O*-sulphated and may be *N*-acetylated. Enzymic hydrolysis[180] and nitrous acid degradation[181] studies have led to suggestions of some repeating monosaccharide sequences.

Keratan sulphate is unlike the other glycosaminoglycans in that it contains no hexuronic acid units. It has a repeating disaccharide unit (**40**) consisting of D-galactose and 2-acetamido-2-deoxy-D-glucose 6-sulphate units with the reverse linkage types of the

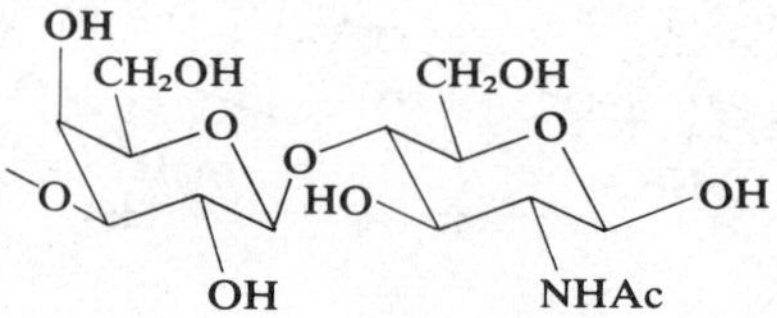

(**40**)

glycosaminoglyuronans, being β-D-galactose-(1 → 4)-2-acetamido-2-deoxy-D-glucose 6-sulphate (1 → 3). Variants of keratan sulphate are known in which other neutral sugars such as D-mannose, L-fucose, and sialic acid[182] occur and in which there is more than 1 mole sulphate per mole disaccharide repeating unit, additional sulphation occurring at C-6 of the neutral monosaccharide units.[183] Occasionally the neutral monosaccharides are non-sulphated.[184] Whereas the structures of the glycosaminoglycans are presented here largely as based on repeating oligosaccharide units, it must be borne in mind that infrequent irregularities in the chains do exist; such irregularities are most apparent in the structures of heparin, heparan sulphate, and keratan sulphate.[7]

In the overall proteoglycan, the glycosaminoglycans are linked to the protein *via* a glycopeptide linkage which, in most cases, involves monosaccharide units different from those of the main polysaccharide chain. These linkage monosaccharides are linked glycosidically to one another, with the reducing terminal group being linked to a side chain of an amino-acid residue. The chemical aspects of these linkages have been reviewed.[185] One of the first linkage regions to be studied was that occurring in chondroitin 4-sulphate proteoglycan, and the structure (**41**) has ultimately been assigned to this linkage region. The glycopeptide linkage is between the reducing group of D-xylose units and the side chain of L-serine units, with the other three monosaccharide units comprising the linkage region being one D-glucuronic acid and two D-galactose units. The actual linkage of chondroitin 6-sulphate to protein and of dermatan sulphate to protein has been found to be identical in the amino-acid and tetrasaccharide units (**42**). It is not

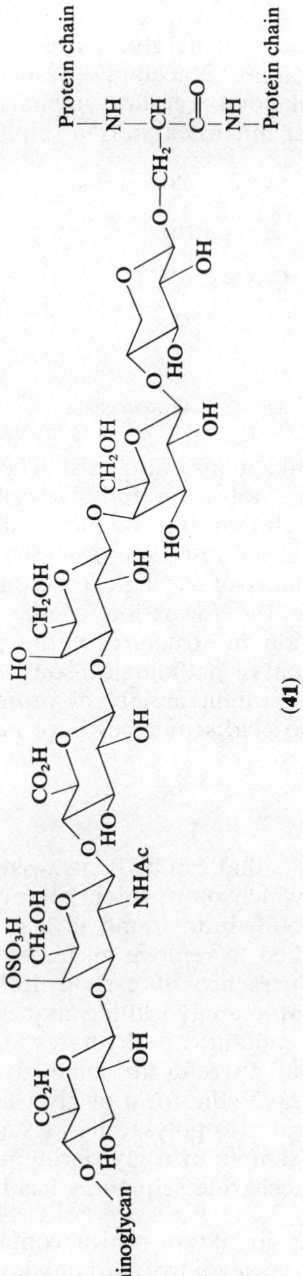
Protein chain
Protein chain
Glycosaminoglycan chain
NHAc

(41)

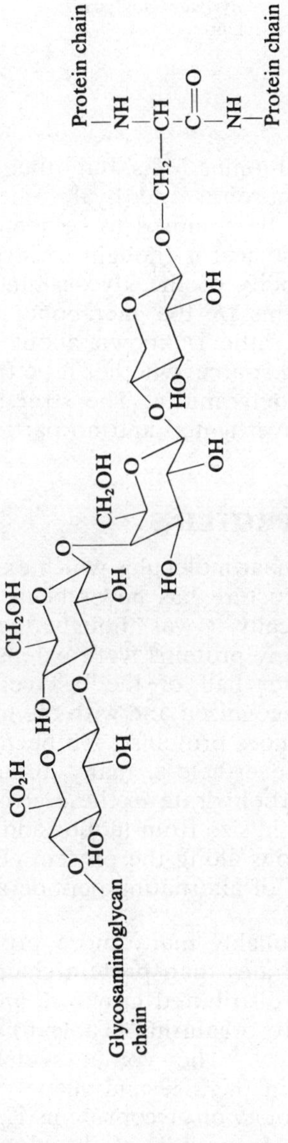
Protein chain
Protein chain
Glycosaminoglycan chain

(42)

clear whether heparin occurs as a proteoglycan, but all heparin samples contain amino-acids and the linkage region has the same structure as the foregoing glycopeptide linkages (**42**).

Corneal keratan sulphate is apparently unique in that its glycopeptide linkages have similarities with that in many glycoproteins, namely 2-acetamido-2-deoxy-D-glucose linked to L-aspargine units (**43**). In contrast, skeletal keratan sulphate involves 2-acetamido-2-deoxy-D-glucose, which does not occur in corneal kertan sulphate, linked to

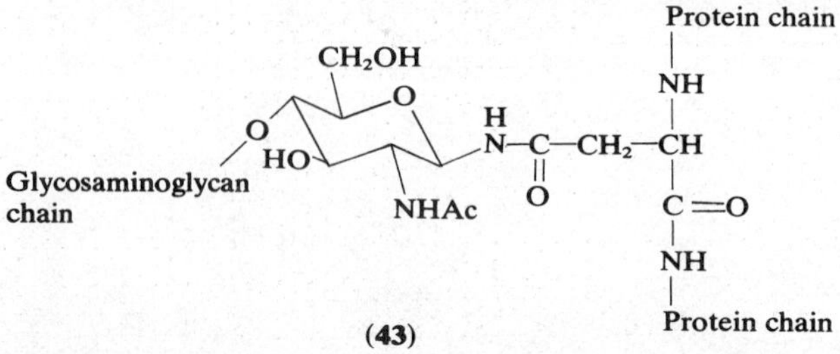

(**43**)

L-serine and L-threonine units, but other types of linkage may also exist. The glycopeptide linkages of chondroitin and hyaluronic acid have not been fully described. That of chondroitin may be assumed to be identical with chondroitin 4- and 6-sulphate, whilst that of hyaluronic acid is thought to involve neutral monosaccharide units.

The heterogeneity of the glycosaminoglycan chains of a single proteoglycan can be explained in terms of the metabolic process for the formation of the proteoglycan molecule,[177] but little is known about the variation in structure of the proteoglycans depending on the source, whether it be from a normal or pathological source, *e.g.* cases of hyperglycosaminoglycanuria. The structure of the protein moiety of proteoglycans has not been fully investigated and no particular amino-acid sequences have been observed.

26.3.6 GLYCOPROTEINS

Proteins are linear molecules which exist as strands that can be bent, twisted, or folded, but a branch structure has never been found in which one protein strand is joined to another. Historically it was thought that any carbohydrate found in a protein was an impurity and many proteins were extensively purified to remove all traces of carbohydrate. In the latter half of the last century the presence of carbohydrate in salivary excretions was recognized and with the improvement in analytical techniques over the last 20 years many more proteins have been found to contain carbohydrates and are in fact glycoproteins. Nevertheless, many authors continue to refer to some glycoproteins as proteins. The carbohydrate in these glycoproteins takes the form of glycosidically linked units which vary in size from mono- and di-saccharides to polysaccharides and are linked at various positions along the protein chain. No example of a glycoprotein existing as a block copolymer of alternating peptide and oligosaccharide sequences has been found in nature.

There are probably many more protein chains in nature which contain covalently bound carbohydrates than protein chains which are devoid of carbohydrate. Glycoproteins are widely distributed in nature and are found distributed amongst higher animals, plants, and micro-organisms (Table 6) and have been the subject of many reviews in recent years.[4,6,186,187] They contain widely differing carbohydrate contents (Figure 6), and have variations in the size and shape of the carbohydrate side chains. The role of the carbohydrate moiety of glycoproteins is not fully understood. As will be shown later, the terminal carbohydrate units of the blood group substance glycoproteins are responsible for their specificity, whilst the degradation of sialic acid units in the antifreeze glycoproteins, with neuraminidase, removes their antifreeze properties. There are a number of mysteries, one of which is the enzymic activities of ribonuclease A and B. Ribonuclease B,

TABLE 6

Distribution and Function of some Glycoproteins

Presumed function	*Name*
Structural	Collagen
	Bacterial cell wall
	Extension (plant cell wall)
Food reserve	Caesin
	Ovalbumin
	Endosperm glycoproteins
	Pollen allergens
Enzyme	Porcine ribonuclease B
	Porcine deoxyribonuclease
	Porcine α-amylase
	Acetyl cholinesterase
	Ficin
	Bromelain
	Taka-amylase (a fungal α-amylase)
	Yeast invertase (a β-D-fructofuranosidase)
Transport	Ceruloplasmin
	Haptoglobin
	Transferrin
Hormone	Thyroglobulin
	Human chorionic gonadotrophin
	Luteinizing hormone
	Follicle stimulating hormone
	Eyrthropoietin
Protective	Fibrinogen
	Interferon
	Immunoglobulin
	Mucins
Plasma and body fluids	α-, β-, and γ-Glycoproteins
Toxins	Ricin
	Fungal phytotoxins
Unknown	Blood group substances
	Avidin (egg white)

a glycoprotein, exhibits exactly the same activity as ribonuclease A which has no sugar component, showing that carbohydrate does not affect the molecule's enzymic activity. A number of proposals have been put forward to explain the presence of carbohydrate, ranging from providing components for intercellular communication and handles for transportation of proteins from one part of a cell to another, to rendering proteins resistant to enzymatic degradation but allowing them to be recognized by certain receptor sites.

As many of the glycoproteins contain only a small proportion of carbohydrate, structural investigations are facilitated by the use of proteolytic enzymes (*e.g.* pronase) to give glycopeptides which contain a small number of amino-acid residues to which are attached the intact carbohydrate units. These glycopeptides are analysed, not without problems,[188] by the classical methods of periodate oxidation[189] and methylation, or by sequential enzyme hydrolysis methods (Section 26.3.2.11) to characterize the component monosaccharide units and ideally give a single amino-acid residue linked to a monosaccharide unit. Only two types of linkage have been found to occur, namely *O*-glycosidic linkages to serine, threonine, hydroxyproline, and hydroxylysine, and *N*-glycosylamide linkages to asparagine. Only five monosaccharide units have been shown to be involved in these linkages: L-arabinose, D-xylose, D-galactose, 2-acetamido-2-deoxy-D-glucose, and 2-acetamido-2-deoxy-D-galactose. By way of example, a number of glycoproteins will be discussed to show the structure of their carbohydrate residues, since no classification can readily be made in terms of structural features.

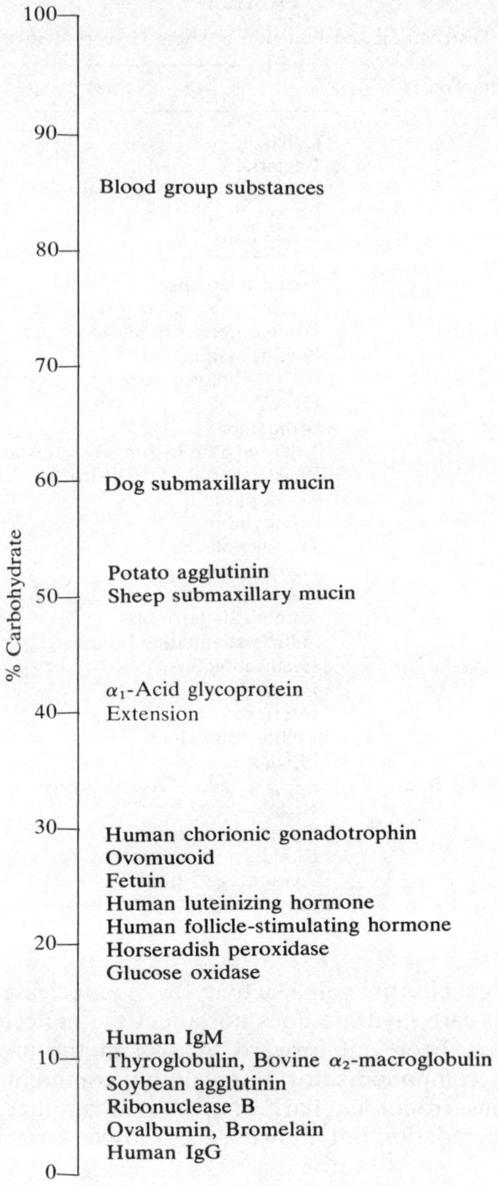

Figure 6 Carbohydrate content of some glycoproteins

26.3.6.1 Hormonal glycoproteins

This group of glycoproteins includes follicle-stimulating hormone, luteinizing hormone, human chorionic gonadotrophin, human menopausal gonadotrophin, pregnant mare serum gonadotrophin, and thyroid-stimulating hormone. The determination of the primary structures of the carbohydrates in these glycoproteins is still in its infancy, owing to the problems of isolating quantities of a pure hormone from closely similar (in chemical and physical nature) hormones and other macromolecules, including glycoproteins present in the media of origin of the sample. A recent two-volume review on experimental methods in hormone research has been published.[190] The purification of a particular hormone involves a complex series of purification stages in which advantage is taken of the properties peculiar to the hormone molecule, namely acidic and basic groups from certain amino-acid residues and acidic groups from 5-acetamido-3,5-dideoxy-D-*glycero*-D-*galacto*-nonulosonic acid. It is essential to monitor purification stages by determination of activity because no chemical method is hormone specific, and loss of a few residues can remove all biological activity; thus conditions which do not deviate markedly from the physiological have to be used. As the purification proceeds, it becomes more difficult and an example of a typical purification process is that applied to human follicle-stimulating hormone, which involves a number of chromatographic separations[191–193] on calcium phosphate, ion exchange resins, and gel filtration media, but which has achieved a purification of 5000-fold.[191] Immunoadsorption methods of separation are showing great promise as the purification can theoretically take place in one stage[194] if the hormone can be isolated in very high purity initially in order to allow the specific antibody to be raised.

Immunological reactions of the hormonal glycoproteins have shown that considerable similarities exist between those hormones; for example, the antiserum to follicle-stimulating hormone cross-reacts with luteinizing hormone, human chorionic gonadotrophin, and thyroid-stimulating hormone, indicating the presence of a number of of common antigenic groups in these hormones as well as their specific antigenic groups. These hormones have been shown to contain subunits,[195,196] which can be formed by the action of trypsin, 1M propionic acid, 8M urea, or sodium dodecyl sulphate. Although not all the human glycoproteins have been tested, a pattern is now emerging and it seems quite certain that the β-subunits are hormone specific whereas the α-subunits are interchangeable. Such a finding is in keeping with the similarities of the amino-acid sequences of the α-subunits and the unique characters of the β-subunits. Recombination of the subunits can be effected by incubation together under physiological conditions, the resulting biological activity being greater than the sum of the biological activities of the separate subunits. It has also been found that the subunits of a particular hormone from various species are interchangeable so far as activity is concerned. More important, however, is the fact that hybrid molecules can be produced by combining subunits from different hormonal glycoproteins,[196,197] the type of hormonal activity of the hybrid glycoprotein being designated by the activity in which the β-subunit used was originally involved.

Some success in structural analysis has been achieved.[175] The application of glycoside hydrolases to the intact human follicle-stimulating hormone for the purpose of carbohydrate sequence determination has proved disappointing owing to the inhibition of release of the monosaccharide units by adjacent parts of the molecule, but it has identified the non-reducing terminal units as 5-acetamido-3,5-dideoxy-D-*glycero*-D-*galacto*-nonulosonic acid which are adjacent to D-galactose units.[198] Methylation and identification of the hydrolysis products by gas–liquid chromatography and mass spectrometry demonstrated that L-fucose units occupy non-reducing terminal positions, D-galactose units are linked in the 1- and 2-positions, the D-mannose residues are present as non-reducing terminal units, 1,6-linked units, and 1,3,4-linked branch points, with the 2-acetamido-2-deoxy-D-glucose units 1,6-linked, all the sugars being in the pyranose form.[199]

The use of glycoside hydrolases and chemical methods in the analysis of human chorionic gonadotrophin has been more successful with some studies on the linkage[200] and sequence[201,202] of the carbohydrate units on the intact molecule. Studies on the pure α subunit[203] revealed the presence of two glycopeptides (**44**) and (**45**).

26.3.6.2 Serum and plasma glycoproteins

Many of the proteins in plasma and serum have been found to be glycoproteins; only serum albumin and prealbumin have no carbohydrate. The structure and functions of many of these have been reviewed,[204] including the role of the carbohydrate in regulating the survival time of plasma glycoprotein in sera.[205] Examples of these glycoproteins include α_1-acid glycoprotein (orosomucoid), transferrin, and fetuin; aspects of their carbohydrate structures are discussed by way of examples.

α-NeuNAc or α-NeuNG}(2→6)-β-D-Gal*p*-(1→6)-β-D-Glc*p*NAc-(1→2)-α-D-Man*p*-(1→6)-α-D-Man*p*-(1→6)-α-D-Man*p*-(1→6)-β-D-Glc*p*NAc-(1→)-L-Asn—L-Val—L-Thr—L-Ser—L-Glx—L-Ser—L-Thr—L-Cys—L-Cys—L-Val—L-Ala—L-Lys—

(on the third α-D-Man*p*: 2 ← 1 β-D-Glc*p*NAc 6 ← 1 α-D-Man*p*)

(44)

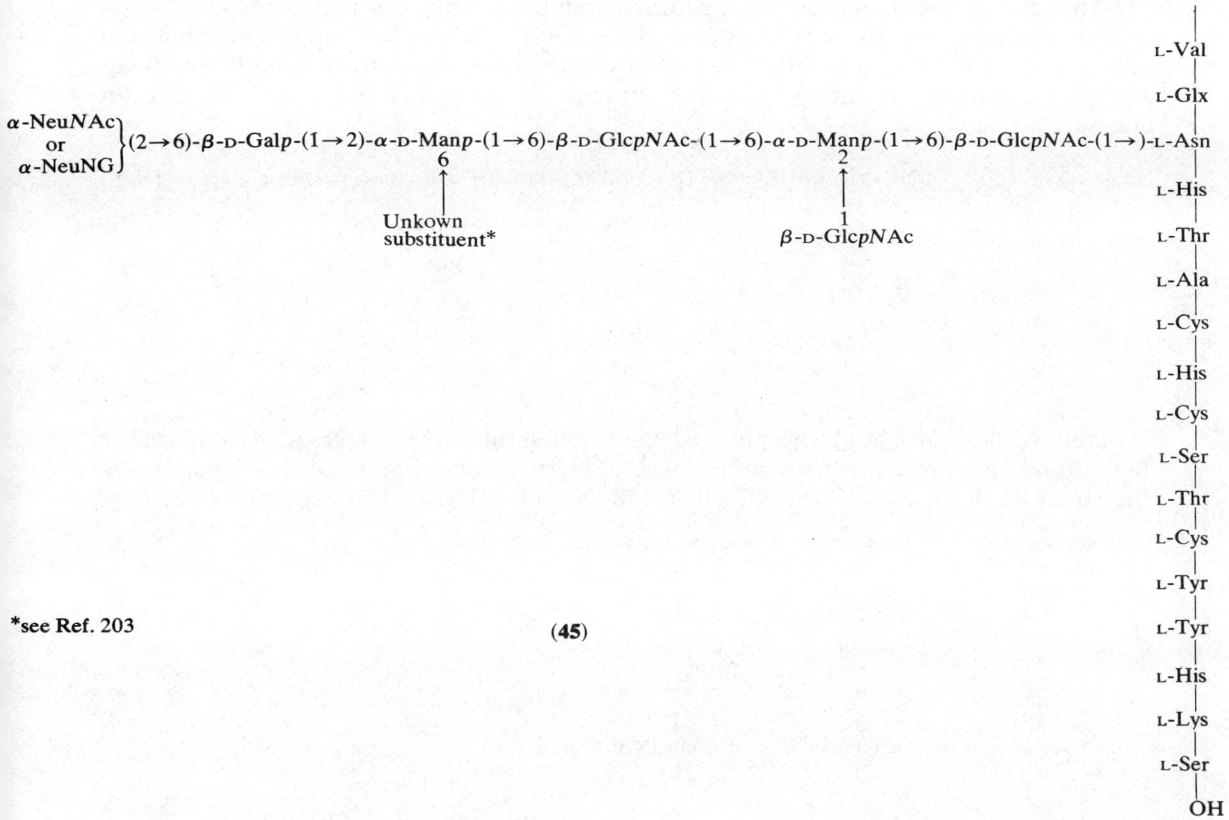

*see Ref. 203

(**45**)

α_1-Acid glycoprotein is the serum glycoprotein having the highest carbohydrate content (*ca.* 40%) and much work has been carried out on its structure since variations in its content in serum occurs in some diseases. Carbohydrate analyses have shown the components to be 2-acetamido-2-deoxy-D-glucose (12.2–15.3%), D-mannose (4.7–6.5%), D-galactose (6.5–12.2%), L-fucose (0.7–1.5%), and 5-acetamido-3,5-dideoxy-D-*glycero*-D-*galacto*-nonulosonic acid (10.8–14.7%) in which L-fucose and 5-acetamido-3,5-dideoxy-D-*glycero*-D-*galacto*-nonulosonic acid units are the non-reducing terminal units linked principally to D-galactose units. A number of possible structures have been postulated for the desialysed glycoprotein which, although similar to (**46**), were thought to possess the

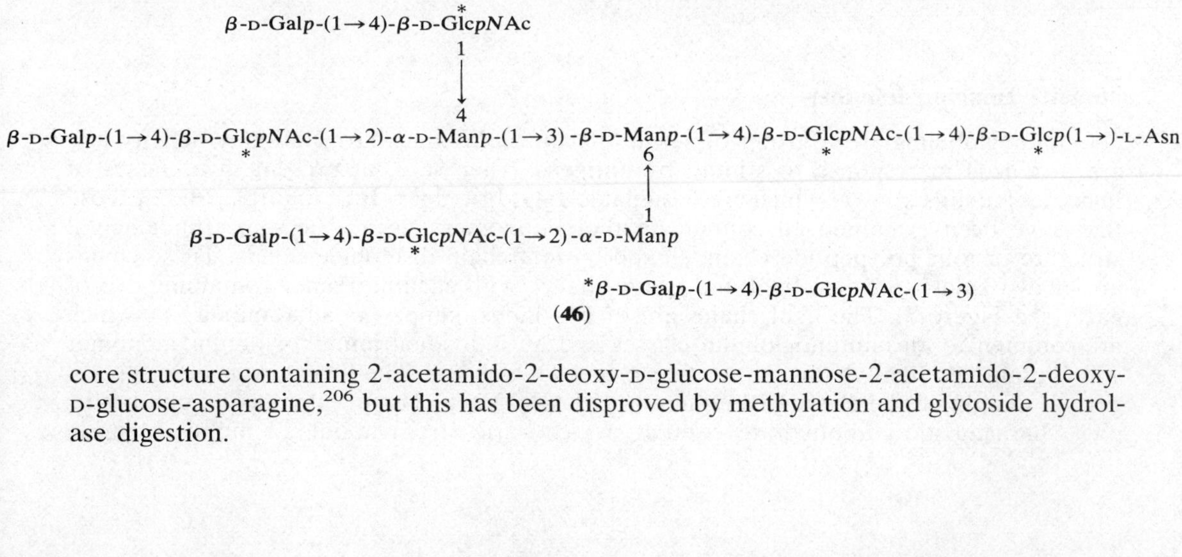

(**46**)

core structure containing 2-acetamido-2-deoxy-D-glucose-mannose-2-acetamido-2-deoxy-D-glucose-asparagine,[206] but this has been disproved by methylation and glycoside hydrolase digestion.

Transferrin is a glycoprotein which forms complexes with iron and is responsible for transporting iron from the storage form in tissues, especially in liver, to the metabotically functioning iron in haemoglobin. The carbohydrate chain was characterized by periodate oxidation, methylation, and glycosidase digestion[207] to give a structure (**47**) for the oligosaccharide chains.

NeuNAc-(2→6)-D-Gal-(1→3 or 4)-D-GlcNAc-(1→3)-D-Man-(1→2 or 4)-D-Man-(1→2 or 4)-D-Man-(1→2, 4, or 6)-

D-Man-(1→3 or 4)-D-GlcNAc-(1→)-L-Asn
3
↑
1
NeuNAc-(2→6)-D-Gal-(1→3 or 4)-D-GlcNAc-(1→3 or 4)-D-GlcNAc

(**47**)

Fetuin is the principle glycoprotein in foetal calf serum, which is gradually replaced by the normal glycoproteins. It resembles α_1-acid glycoprotein in many of its properties and the structure of an oligosaccharide chain (**48**) has been shown to have similarities in its

α-NeuNAc α-NeuNAc α-NeuNAc
2 2 2
↓ ↓ ↓
3 3 3
β-D-Galp β-D-Galp β-D-Galp
1 1 1
↓ ↓ ↓
4 4 4
β-D-GlcpNAc β-D-GlcpNAc β-D-GlcpNAc
1 1 1
↓ ↓ ↓
2 3 or 4 2, 4, or 6
α-D-Manp-(1→2 or 6)-α-D-Manp-(1→3)-β-D-Manp-(1→4)-β-D-GlcpNAc-(1→4)-β-D-GlcpNAc-(1→)-L-Asn

(**48**)

core structure (the carbohydrate units nearest the amino-acid unit).[208] However, oligosaccharides have been found which are linked glycosidically to serine and threonine (**49**), some of which lack the (2→6)-linked 5-acetamido-3,5-dideoxy-D-*glycero*-D-*galacto*-nonulosonic acid unit,[209] showing the presence of a different type of oligosaccharide chain attached to the glycoprotein.

α-NeuNAc-(2→3)-β-D-Galp-(1→3)-α-D-GalpNAc-(1→)-L-Ser (or L-Thr)

(**49**)

26.3.6.3 Immunoglobulins

Immunoglobulins are a group of serum glycoproteins that have antibody activity and are produced in response to stimuli by antigens. There are, at present, five classes of immunoglobulins known, which are designated IgG, IgA, IgM, IgD, and IgE, and of those that have been examined all contain, in their monomeric form, the same fundamental structure of four polypeptide chains linked by interchain disulphide bonds. These chains are of two kinds, so-called light and heavy chains, with each monomer containing two of each (see Figure 7). The light chains are of two kinds, kappa (κ) and lambda (λ), which are common to all immunoglobulin classes and an individual immunoglobulin monomer will contain two light chains, either two κ or two λ chains. The heavy chains are specific to, and determine the class of, the immunoglobulin. Each class of immunoglobulin has its own characteristic carbohydrate content, which varies from about 22 monosaccharide units in IgG to about 82 units in monomeric IgM. Polymeric forms of immunoglobulin have been reported and include dimeric IgA and pentameric IgM. Macromolecular IgM is thought to contain five monomer units joined in the form of a ring from which radiate the five arms.

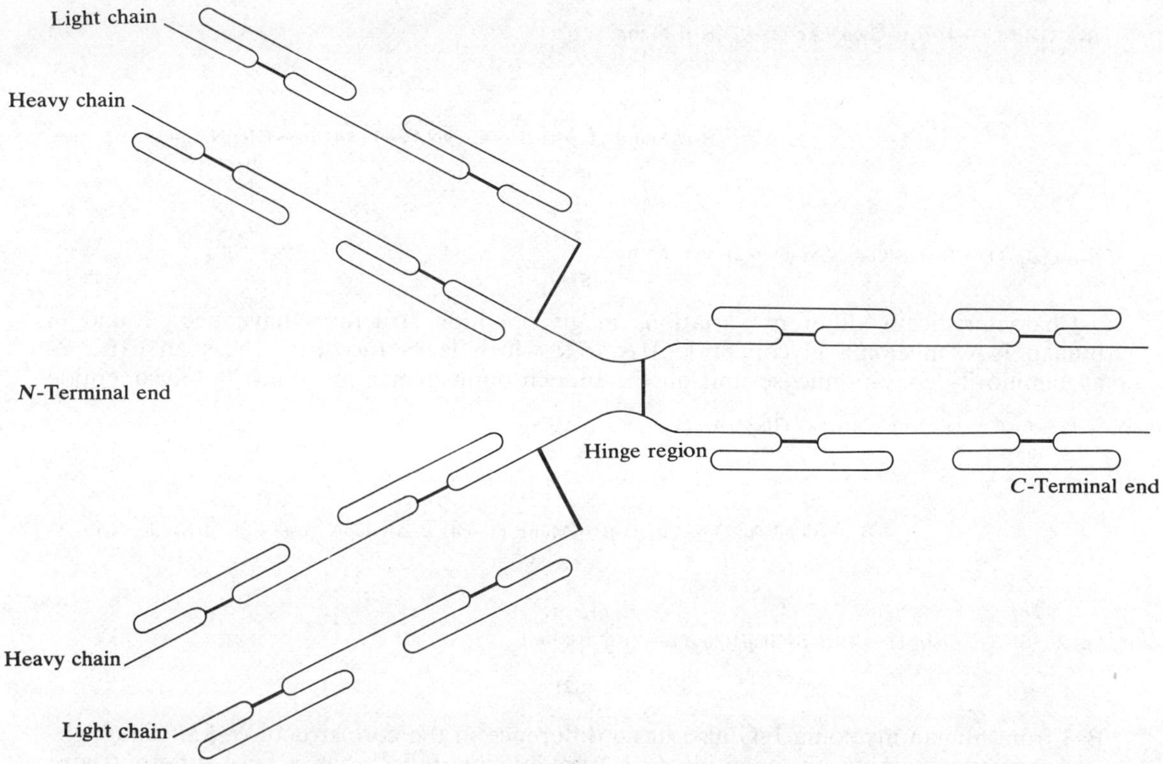

Figure 7 Schematic representation of an immunoglobulin molecule

The first studies on the oligosaccharide structure were carried out by Porter[210] and by Smith and his co-workers.[211] Normal IgG was shown to contain three glycopeptides representing two types of oligosaccharide units.[211] Periodate oxidation has shown that not all the 2-acetamido-2-deoxy-D-glucose and D-mannose units were destroyed and other results based on mild acid hydrolysis suggest that L-fucose and sialic acid units are present at non-reducing terminal positions, linked to D-galactose with D-mannose units substituted at C-3 or occurring at branch points, and some 2-acetamido-2-deoxy-D-glucose units, those oxidized by periodate, linked *via* C-6 or in a terminal position. These results, plus information obtained from glycoside hydrolase degradation studies, led to the proposal of structure (**50**) for human IgG.[212] Similar structures have been found for human IgE[213] and

α-NeuNAc-(2→6)-β-D-Gal*p*-(1→4)-β-D-Glc*p*NAc-(1→2)-α-D-Man*p*
1
↓
3
β-D-Man*p*-(1→4)-β-D-Glc*p*NAc-(1→4)-β-D-Glc*p*NAc-(1→)-L-Asn
6 6
↑ ↑
1 1
α-NeuNAc-(2→6)-β-D-Gal*p*-(1→4)-β-D-Glc*p*NAc-(1→2)-α-D-Man*p* α-D-Fuc*p*

(**50**)

IgA,[214] all having varying amounts of non-reducing terminal sialic acids and D-galactose units, indicating varying stages of completion of the outer chains or microheterogeneity. Bovine IgG[215] contains a glycopeptide (**51**) which is identical to human IgG minus sialic acid residues.

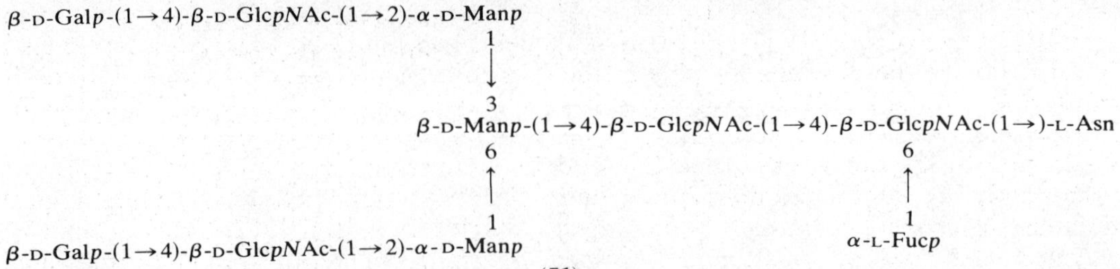

(**51**)

Characteristically different variations in glycopeptide structures have been found in human IgA, myeloma glycopeptide IIA (**52**) which lacks fucose and has an extra 2-acetamido-2-deoxy-D-glucose unit on the branch point D-mannose unit.[214] Glycopeptide

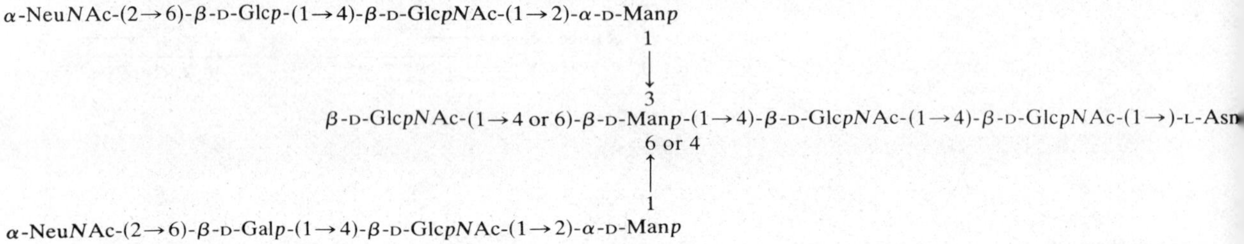

(**52**)

B-3 from human myeloma IgG also has a difference in the core structure (**53**). The high-D-mannose containing glycopeptide C-1 from human IgE[216] has a very different core

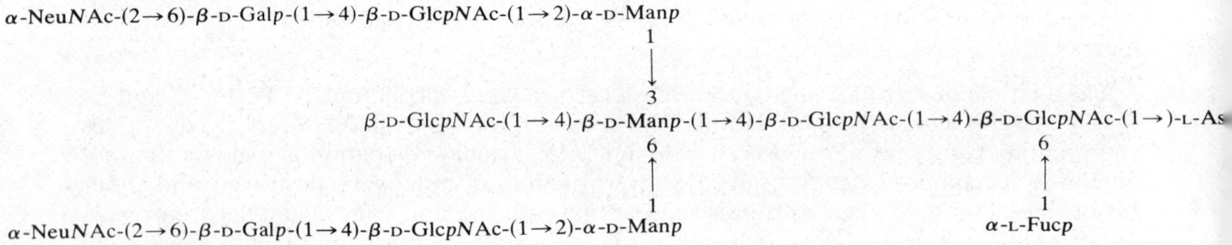

(**53**)

structure (**54**) containing alternating D-mannose and 2-acetamido-2-deoxy-D-glucose units, one of which is uniquely α-linked. These changes in the core structure are to be distinguished from the microheterogeneity resulting from incomplete outer chains, in that changes in the core structure can reflect genetic differences (*e.g.* myeloma).

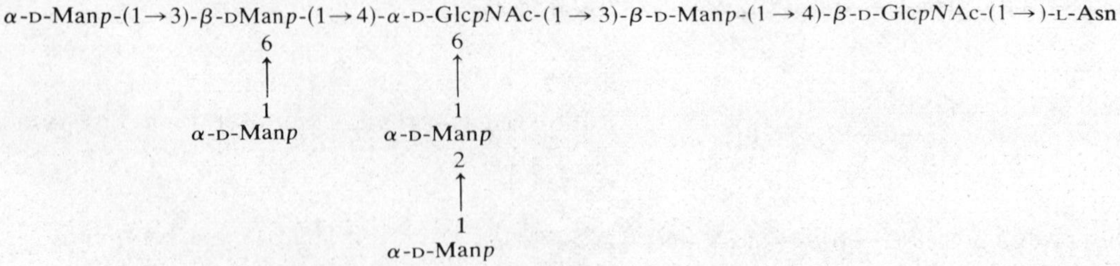

(**54**)

26.3.6.4 Blood group substances

Blood group substances are a group of structurally similar glycoproteins which occur on the surface of red blood cells and are responsible for the determination of particular blood-group types. The importance of the oligosaccharide moiety of the glycoprotein in this specificity has been demonstrated by enzymic removal of terminal units of α-linked 2-acetamido-2-deoxy-D-galactose from type A erythrocytes or of α-linked D-galactose from type B erythrocytes, which resulted in both being converted to type O erythrocytes.[6] Confirmation of these results is found in the determination of specific enzymes (glycosyltransferases) in the blood. Type A blood has an enzyme which will transfer 2-acetamido-2-deoxy-D-galactose to a core whilst type B blood contains an enzyme to transfer D-galactose to the same core. Type O blood has neither enzyme.

The structure of the carbohydrate chains has been analysed using immunological methods to determine changes in serological activity on hydrolysis by acid or specific enzymes. This gives information on the specific terminal carbohydrate units responsible for the immunological specificity. Alkaline degradation has shown that linkage of the oligosaccharide chain to the protein molecule is *via* glycosidic linkages to serine or threonine.[217,218] The structure of these glycoproteins with respect to their immunological properties have been reviewed[219,220] and a composite structure has been proposed[221] to indicate the residues which confer blood group specificity for A, B, H(O), Le^a, and Le^b (**55**).

26.3.7 SYNTHETIC DERIVATIVES OF POLYSACCHARIDES

The commercial importance of some derivatives of polysaccharides has been known and exploited for some time (for example, cellulose xanthate for rayon, and cellulose nitrate for explosives) but, with the advent of water-insoluble reagents for immunoadsorbents and for insolubilization of enzymes and antibodies, a new range of derivatives of polysaccharides has been developed. The use of polysaccharides for the purpose of insolubilizing other molecules is particularly advantageous owing to the residual hydroxyl groups of the polysaccharide providing a stabilizing hydrophilic environment for the attached macromolecule in the solid state. These derivatives have been included in many reviews for a number of polysaccharides, including cellulose,[222,223] starch,[224,225] and others[226] and are the subject of reviews in their own right.[65,227] In order to facilitate reference to methods of preparation, these derivatives will be classified in terms of the type of substituent and not the parent polysaccharide.

26.3.7.1 Ethers

Mention has already been made to the methods of preparation of methylated polysaccharides[15–19] and their use in structural analysis (Section 26.3.2.1). Partially methylated derivatives of cross-linked dextrans (*e.g.* Sephadex) have enabled separations by gel filtration and ion exchange to be carried out in non-aqueous solutions. Reaction of alkyl ethers with monochloroacetic acid[228] introduces carboxy-alkyl groups into polysaccharides. These derivatives are used as ion exchange materials (*e.g.* carboxymethylcellulose) and also as a means of activating the hydroxyl groups of polysaccharides to facilitate reaction with, for example, thionyl chloride to give reactive acid chlorides[229] or with ethyl chloroformate to produce, ultimately, an oxide,[230] these derivatives being useful for insolubilization of biological macromolecules. Alkylamino and alkylaminoalkyl ethers have been prepared and are used for ion exchange resins (*e.g.* diethylaminoethyl-Sephadex) and electrophoresis and immunoelectrophoresis (diethylaminoethyl-agarose gels). Treatment of these tertiary amino-groups containing polysaccharides with

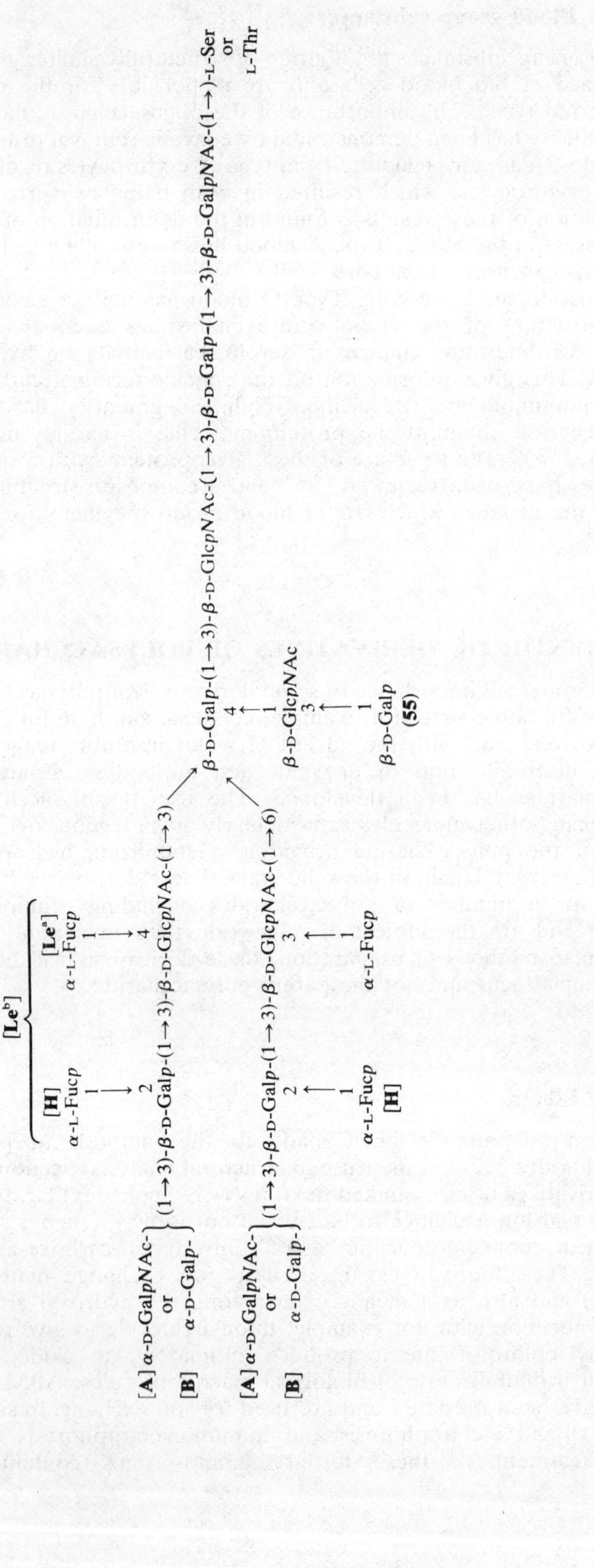
[A] α-D-GalpNAc-
or
[B] α-D-Galp-
(1 → 3)-β-D-Galp-(1 → 3)-β-D-GlcpNAc-(1 → 3)
[H]
α-L-Fucp
[Leᵃ]
α-L-Fucp
[Leᵇ]
(1 → 3)-β-D-Galp-(1 → 3)-β-D-GlcpNAc-(1 → 6)
β-D-Galp-(1 → 3)-β-D-GlcpNAc-(1 → 3)-β-D-Galp-(1 → 3)-β-D-GalpNAc-(1 → 3)-L-Ser
or
L-Thr
β-D-GlcpNAc
β-D-Galp

(55)

epichlorohydrin[231] produces a product with reactive epoxide side-chains which are used to form cross-linked polysaccharides, or insolubilized biologically active macromolecules.[232] Complex ethers of cellulose containing aromatic amino-groups such as, for example, (**56**) have been prepared for enzyme immobilization.[233]

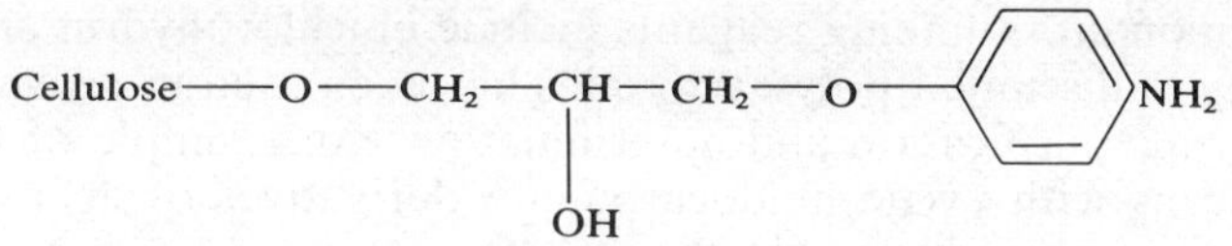

(**56**)

26.3.7.2 Esters

The preparation of starch esters and their use in the food industries have been reviewed.[224,225] Anhydrides and chlorides of aliphatic and aromatic carboxylic acids have been used[234–236] to prepare polysaccharide esters for media for chromatographic separations. Phosphate esters, prepared using tetrapolyphosphoric acid,[237] have been obtained for a number of polysaccharides. These esters can be used as cross-linking agents; phosphate cross-linked starches are used in the food industry. Sulphates[238] of many polysaccharides have been prepared; some have anticoagulant and anti-inflammatory properties similar to heparin (Section 26.3.5.3). Esters of sulphonic acid and its derivatives, particularly toluene-*p*-sulphonic acid, are used as protecting groups, rendering glycosidic linkages in polysaccharides more acid resistant.

26.3.7.3 Carbonates

The formation of carbonate derivatives of polysaccharides can be brought about with ethyl chloroformate.[239] Careful choice of conditions can lead to production of cyclic carbonate derivatives[240] which have been use as supports for the insolubilization of enzymes.[241–243]

26.3.7.4 Dye derivatives

A number of coloured derivatives of polysaccharides have been formed as a result of depositing insoluble dye on the carbohydrate fibres, or by hydrogen bonding between the carbohydrate and dye molecules, but the most important derivatives are those in which the dye molecule is covalently bound to the polysaccharide. The diazine and triazine types of dye which have a colour-producing group attached to, for example, 2,4,6-trichloro-*sym*-triazine, react with polysaccharides[244,245] to give a product (**57**). Many polysaccharides

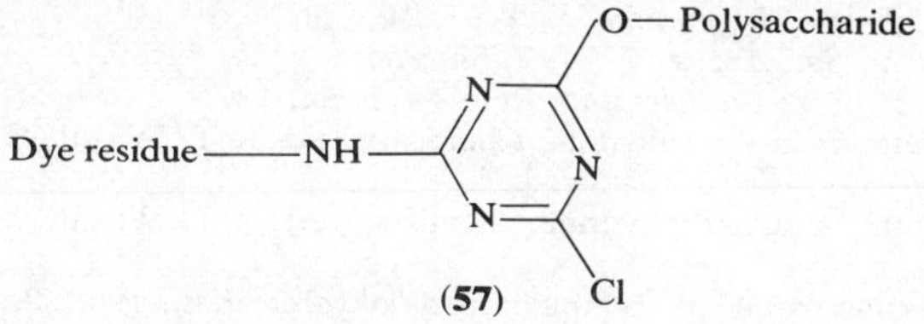

(**57**)

have been dyed with Cibachron Blue, a monochlorotriazinyl dye, and used as substrates for the determination of a number of glycanases,[246,247] the amount of material of small molecular weight released by the enzyme being determined spectrophotometrically. The dyeing of polysaccharides has been advocated as an aid to their identification in chromatographic and electrophoretic separations, but the characteristics of the polysaccharide under investigation are altered by the presence of the dye molecule.

26.3.7.5 Cross-linked polysaccharides

The use of cross-linking agents to vary the molecular parameters of polysaccharides without causing major alterations in their chemical properties has found many applications in industry. Those of particular interest are cross-linked celluloses, starches, dextrans (Sephadex), and agaroses (Sepharose), which have been used as chromatographic supports, and the most common cross-linking reagents include epichlorohydrin and formaldehyde.

The cross-linking of dissimilar polysaccharides has been used to provide water-insoluble substrates for enzymic purification and determination. An example of this is the treating of glycosaminoglycans with cyclic imidocarbonate derivatives of agarose[248] to give substrates for the assay of glycosaminoglycanases.

26.3.7.6 Water-insoluble macromolecules

Water-insoluble enzymes have been produced using attachment to insoluble polysaccharide supports. The literature on the preparation and uses of these products is vast and many reviews have been published.[65,227,249] Insolubilized enzymes are principally used to effect the reaction catalysed by the free enzyme, but in a simplified form because the enzyme, which is insoluble, can be removed very easily from the reaction mixture by filtration or centrifugation, whereas the removal of soluble enzymes would require laborious separation techniques.

Antibodies have been insolubilized on polysaccharide supports and used for the purification of antigens, usually by column chromatography (immunoadsorption) in which a solution of the impure antigen is passed through a bed of insolubilized antibody. The specific antigen is adsorbed to the antibody whilst impurities are washed through the column. Subsequently the antigen can be desorbed from the column in pure form.[250]

A comprehensive review[227] of methodology and descriptions of the insoluble products and the uses to which they are put has been published and greater details of the above examples and many others can be found therein.

References

1. I. Danishefsky, R. L. Whistler, and F. A. Bettelheim, in 'Carbohydrates, Chemistry and Biochemistry', ed. W. Pigman and D. Horton, Academic Press, New York, 2nd edn., 1970, vol. IIA, chap. 35.
2. W. Z. Hassid, in 'Carbohydrates, Chemistry and Biochemistry', ed. W. Pigman and D. Horton, Academic Press, New York, 2nd edn., 1970, vol. IIA, chap. 34.
3. E. Eichwald, *Ann. Chem. Pharm.*, 1865, **134,** 177.
4. 'Glycoproteins', ed. A. Gottschalk, Elsevier, Amsterdam, 2nd edn., 1972.
5. J. S. Brimacombe and J. Webber, 'Mucopolysaccharides', Elsevier, Amsterdam, 1964.
6. N. Sharon, 'Complex Carbohydrates, Their Chemistry and Biochemistry', Addison-Wesley, Massachusetts, 1975.
7. J. F. Kennedy, in 'Proteoglycans—Biological and Chemical Aspects in Human Life', Elsevier, Amsterdam, 1978.
8. O. Samuelson and L. Thede, *J. Chromatog.*, 1967, **30,** 556.
9. J. F. Kennedy, *Biochem. Soc. Trans.*, 1974, **2,** 54.
10. J. F. Kennedy and J. E. Fox, *Carbohydrate Res.*, 1977, **54,** 13.
11. J. F. Kennedy, in 'Methods in Carbohydrate Chemistry', ed. R. L. Whistler, Academic Press, New York, vol. VIII, in press.
12. J. F. Kennedy, in 'Methods in Carbohydrate Chemistry', ed. R. L. Whistler, Academic Press, New York, vol. VIII, in press.
13. S. A. Barker, J. F. Kennedy, and P. J. Somers, *Carbohydrate Res.*, 1968, **8,** 482.
14. D. Aminoff, W. W. Binkley, R. Schaffer, and R. W. Mowry, in 'The Carbohydrates, Chemistry and Biochemistry', ed. W. Pigman and D. Horton, Academic Press, New York, 2nd edn., 1970, vol. IIB, chap. 45.
15. W. N. Haworth, *J. Chem. Soc.*, 1915, **107,** 8.
16. T. Purdie and J. C. Irvine, *J. Chem. Soc.*, 1903, **83,** 1021.
17. S.-I. Hakomori, *J. Biochem.* (*Japan*), 1964, **55,** 205.
18. D. M. W. Anderson, G. M. Cree, J. J. Marshall, and S. Rahman, *Carbohydrate Res.*, 1966, **2,** 63.
19. K. Wallenfels, G. Bechtler, R. Kuhn, H. Trischmann, and H. Egge, *Angew. Chem. Internat. Edn.*, 1963, **2,** 515.

20. C. M. Fear and R. C. Menzies, *J. Chem. Soc.*, 1926, 937.
21. B. Lindberg and J. Lönngren, in 'Methods in Carbohydrate Chemistry', ed. R. L. Whistler and J. N. BeMiller, Academic Press, New York, 1976, vol. VII, p. 142.
22. W. W. Binkley, *Adv. Carbohydrate Chem.*, 1955, **10,** 55.
23. C. T. Bishop, *Adv. Carbohydrate Chem.*, 1964, **19,** 95.
24. H. Björndal, B. Lindberg, and S. Svensson, *Carbohydrate Res.*, 1967, **5,** 433.
25. C. C. Sweeley, W. W. Wells, and R. Bently, in 'Methods in Enzymology', ed. E. F. Neufeld and V. Ginsburg, Academic Press, New York, 1966, vol. 8, p. 95.
26. N. K. Kochetkov and O. S. Chizhov, *Adv. Carbohydrate Chem.*, 1966, **21,** 39.
27. J. Lönngren and S. Svensson, *Adv. Carbohydrate Chem. Biochem.*, 1974, **29,** 41.
28. H. Björndal, C. G. Hellerqvist, B. Lindberg, and S. Svensson, *Angew. Chem. Internat. Edn.*, 1970, **9,** 610.
29. A. Thompson, M. L. Wolfrom, and E. J. Quinn, *J. Amer. Chem. Soc.*, 1953, **75,** 3003.
30. A. S. Perlin, *Adv. Carbohydrate Chem.*, 1959, **14,** 9.
31. J. M. Bobbitt, *Adv. Carbohydrate Chem.*, 1956, **11,** 1.
32. J. K. Hamilton, G. W. Huffman, and F. Smith, *J. Amer. Chem. Soc.*, 1959, **81,** 2176.
33. R. L. Whistler, and J. N. BeMiller, *Adv. Carbohydrate Chem.*, 1958, **13,** 289.
34. T. Yamada, M. Hisamatsu and M. Taki, *J. Chromatog.*, 1975, **103,** 390.
35. C. E. Weill and P. Hanke, *Analyt. Chem.*, 1962, **34,** 1736.
36. M. John, G. Trénel and H. Dellweg, *J. Chromatog.*, 1969, **42,** 476.
37. S. A. Barker, B. W. Hatt, J. F. Kennedy, and P. J. Somers, *Carbohydrate Res.*, 1968, **9,** 327.
38. K. Mopper and E. T. Degens, *Analyt. Biochem.*, 1972, **45,** 147.
39. S. A. Barker, E. J. Bourne, and O. Theander, *J. Chem. Soc.*, 1955, 4276.
40. J. F. Kennedy and A. Rosevear, *J. C. S. Perkin I*, 1973, 2041.
41. J. F. Kennedy, S. A. Barker, and C. A. White, *Carbohydrate Res.*, 1974, **38,** 13.
42. F. M. Rabel, A. G. Caputo, and E. T. Butts, *J. Chromatog.*, 1976, **126,** 731.
43. H. W. Kircher, in 'Methods in Carbohydrate Chemistry', ed. R. L. Whistler and M. L. Wolfrom, Academic Press, New York, 1962, vol. 1, p. 13.
44. J. F. Kennedy, S. M. Robertson, and M. Stacey, *Carbohydrate Res.*, 1976, **49,** 243.
45. J. F. Kennedy, S. M. Robertson, and M. Stacey, *Carbohydrate Res.*, 1977, **57,** 205.
46. J. F. Kennedy and S. M. Robertson, *Carbohydrate Res.*, in press.
47. M. B. Perry, *Canad. J. Biochem.*, 1964, **42,** 451.
48. N. K. Kochetkov and O. S. Chizhov, *Adv. Carbohydrate Chem.*, 1966, **21,** 39.
49. O. S. Chizhov, N. V. Molodtsov, and N. K. Kochetkov, *Carbohydrate Res.*, 1967, **4,** 273.
50. G. S. Johnson, W. S. Ruliffson, and R. G. Cooks, *Chem. Comm.*, 1970, 587.
51. J. Moor and E. S. Waight, *Biomed. Mass. Spectrometry*, 1975, **2,** 36.
52. M. S. B. Munson and F. H. Field, *J. Amer. Chem. Soc.*, 1966, **88,** 2621.
53. W. R. Gray and U. E. Del Valle, *Biochemistry*, 1970, **9,** 2134.
54. W. R. Gray, L. H. Wojcik, and J. H. Futrell, *Biochem. Biophys. Res. Comm.*, 1970, **41,** 1111.
55. V. N. Reinhold, J. Biller, and K. Biemann, 18*th Ann. Conf. Mass Spectrometry and Allied Topics*, 1970, **18,** 1379.
56. T. Terui, T. Yadomae, H. Yamada, O. Hayashi, and T. Miyazaki, *Chem. and Pharm. Bull. (Japan)*, 1974, **22,** 2476.
57. B. Lindberg and B. Swan, *Acta Chem. Scand.*, 1960, **14,** 1043.
58. S. Gardell, A. H. Gordon, and S. Aqvist, *Acta Chem. Scand.*, 1950, **4,** 907.
59. D. H. Northcote, in 'Methods in Carbohydrate Chemistry', ed. R. L. Whistler, Academic Press, New York, 1965, vol. V, p. 49.
60. P. W. Lewis, D. N. Raine, and J. F. Kennedy, *Ann. Clin. Biochem.*, 1974, **11,** 67.
61. M. B. Mathews, in 'Methods in Carbohydrate Chemistry', ed. R. L. Whistler and J. N. BeMiller, Academic Press, New York, 1976, vol. VII, p. 116.
62. M. Heidelberger, Z. Dische, W. B. Neely, and M. L. Wolfrom, *J. Amer. Chem. Soc.*, 1955, **77,** 3511.
63. E. Y. C. Lee and W. J. Whelan, *Arch. Biochem. Biophys.*, 1966, **116,** 162.
64. W. Banks and C. T. Greenwood, *Carbohydrate Res.*, 1968, **6,** 177.
65. J. F. Kennedy, in 'Carbohydrate Chemistry — Specialist Periodical Reports' ed. J. S. Brimacombe, The Chemical Society, London, 1971–77, vols. 4–9, part II.
66. M. Nishida-Fukuda and F. Egami, *Biochem. J.*, 1970, **119,** 39.
67. Y. T. Li and S.-C. Li, in 'Methods in Carbohydrate Chemistry', ed. R. L. Whistler and J. N. BeMiller, Academic Press, New York, 1976, vol. VII, p. 221.
68. A. L. Tarentino, T. H. Plummer, jun., and F. Maley, *J. Biol. Chem.*, 1974, **249,** 818.
69. N. Gellerstedt, *Arkiv Kemi*, 1963, **20,** 147.
70. F. A. Bettelheim and D. E. Philpott, *Biochim. Biophys. Acta*, 1959, **34,** 124.
71. R. H. Marchessault and A. Sarko, *Adv. Carbohydrate Chem.*, 1967, **22,** 421.
71a. R. H. Marchessault and P. R. Sundararajan, *Adv. Carbohydrate Chem. Biochem.*, 1976, **33,** 387.
72. F. A. Bettelheim, *Biochim. Biophys. Acta*, 1964, **83,** 350.
73. J. M. Guss, D. W. L. Hukins, P. J. C. Smith, W. T. Winter, S. Arnott, R. Moorhouse, and D. A. Rees, *J. Mol. Biol.*, 1975, **95,** 359.
74. D. E. Burge, *J. Phys. Chem.*, 1963, **67,** 2590.
75. I. C. McNeill, *Polymer*, 1963, **4,** 247.
76. F. A. Bettelheim, Y. Hashimoto, and W. Pigman, *Biochim. Biophys. Acta*, 1962, **63,** 235.

77. T. C. Laurent, M. Ryan, and A. Pietruszkiewicz, *Biochim. Biophys. Acta*, 1960, **42,** 476.
78. G. K. Adkins and C. T. Greenwood, *Stärke*, 1966, **18,** 213.
79. S. Schiefer, E. Y. C. Lee, and W. J. Whelan, *F. E. B. S. Letters*, 1973, **30,** 129.
80. F. L. Bates, D. French, and R. E. Rundle, *J. Amer. Chem. Soc.*, 1943, **65,** 142.
81. M. Sanyal, V. S. Rao, and K. B. De, *Z. analyt. Chem.*, 1974, **271,** 208.
82. T. J. Schoch, *Adv. Carbohydrate Chem.*, 1945, **1,** 247.
83. C. T. Greenwood and J. Thomson, *J. Chem. Soc.*, 1962, 222.
84. S. Tabata, K. Nagata, and S. Hizukuri, *Stärke*, 1975, **27,** 333.
85. K. H. Meyer, M. Wertheim, and P. Bernfeld., *Helv. Chim. Acta*, 1940, **23,** 865; 1941, **24,** 378.
86. J. J. Cael, J. L. Koenig, and J. Blackwell, *Carbohydrate Res.*, 1973, **29,** 123; *Biopolymers*, 1975, **14,** 1885.
87. D. French. *Adv. Carbohydrate Res.*, 1957, **12,** 189.
88. W. N. Haworth, E. L. Hirst, and F. A. Isherwood, *J. Chem. Soc.*, 1937, 577.
89. H. Staudinger and E. Husemann, *Annalen*, 1937, **527,** 195.
90. K. H. Meyer and P. Bernfeld, *Helv. Chim. Acta*, 1940, **23,** 875.
91. S. Peat, W. J. Whelan, and G. J. Thomas, *J. Chem. Soc.*, 1952, 4546.
92. Z. Gunja-Smith, J. J. Marshall, C. Mercier, E. E. Smith, and W. J. Whelan, *F.E.B.S. Letters*, 1970, **12,** 101.
93. R. Geddes, C. T. Greenwood, and S. MacKenzie, *Carbohydrate Res.*, 1965, **1,** 71.
94. K. Freudenberg and G. Blomqvist, *Berzelius*, 1935, **68,** 2070.
95. R. H. Marchessault and A. Sarko, *Adv. Carbohydrate Chem.*, 1967, **22,** 421.
96. J. Hayashi, A. Sueoka, J. Ookita, and S. Watanabe, *J. Chem. Soc. Japan*, 1973, 146.
97. J. Hayashi, A. Sueoka, and S. Watanabe, *J. Chem. Soc. Japan*, 1973, 153.
98. K. H. Meyer and L. Misch, *Helv. Chim. Acta*, 1937, **20,** 232.
99. C. Y. Liang and R. H. Marchessault, *J. Polymer Sci.*, 1959, **37,** 385.
100. I. Y. Levdik, N. M. Birbrover, N. A. Dobrynin, T. V. Mlechko, and V. N. Nikitin, *Cellulose Chem. Technol.*, 1974, **8,** 141.
101. F. Smith and R. Montgomery, 'The Chemistry of Plant Gums and Mucilages', Reinhold, New York, 1959.
102. G. O. Aspinall, *Adv. Carbohydrate Chem.*, 1969, **24,** 333.
103. R. J. Sturgeon, in 'Carbohydrate Chemistry — Specialist Periodical Reports', ed. J. S. Brimacombe, The Chemical Society, London, 1972–77, vols. 5–9, part II.
104. D. M. W. Anderson, E. L. Hirst, and J. F. Stoddart, *J. Chem. Soc.* (*C*), 1967, 1476; J. F. Stoddart and J. K. N. Jones, *Carbohydrate Res.*, 1968, **8,** 29.
105. A. J. Charlson, J. R. Nunn, and A. M. Stephen, *J. Chem. Soc.*, 1955, 1428.
106. D. M. W. Anderson and M. C. L. Gill, *Phytochemistry*, 1975, **14,** 739.
107. G. O. Aspinall and J. Baillie, *J. Chem. Soc.*, 1963, 1714.
108. Z. F. Ahmed and R. L. Whistler, *J. Amer. Chem. Soc.*, 1950, **72,** 2524.
109. G. O. Aspinall, R. B. Rashbrook, and G. Kessler, *J. Chem. Soc.*, 1958, 215.
110. R. J. Beveridge, J. K. N. Jones, R. W. Lowe, and W. A. Szarek, *J. Polymer Sci., Part C, Polymer Symposia*, 1971, **23,** 461.
111. J. M. Tyler, *J. Chem. Soc.*, 1965, 5300.
112. E. L. Hirst, J. K. N. Jones, and W. O. Walder, *J. Chem. Soc.*, 1947, 1225.
113. H. O. Bouveng and H. Meier, *Acta Chem. Scand.*, 1959, **13,** 1884.
114. D. A. Rees and N. G. Richardson, *Biochemistry*, 1966, **5,** 3099.
115. V. Zitko and C. T. Bishop, *Canad. J. Chem.*, 1966, **44,** 1275.
116. G. O. Aspinall, K. Hunt, and I. M. Morrison, *J. Chem. Soc. (C)*, 1967, 1080.
117. P. Albersheim, *Sci. Amer.*, 1975, **232(4),** 80.
118. R. H. Marchessault and C. Y. Liang, *J. Polymer Sci.*, 1959, **37,** 385.
119. G. O. Aspinall and K. M. Ross, *J. Chem. Soc.*, 1963, 1681.
120. H. Pringsheim and A. Genin, *Z. physiol. Chem.*, 1964, **140,** 229.
121. A. S. Perlin, in 'Enzymatic Hydrolysis of Cellulose and Related Materials', ed. E. T. Reese, Macmillan, New York, 1963, p. 185.
122. J. E. Courtois and P. le Dizet, *Compt. rend.* (*D*), 1973, **277,** 1957.
123. H. Saha and K. N. Choudhury, *J. Chem. Soc.*, 1922, **121,** 1044.
124. E. Percival and R. H. McDowell, 'Chemistry and Enzymology of Marine Algal Polysaccharides', Academic Press, New York, 1967.
125. S. Peat, J. R. Turvey, and J. M. Evans, *J. Chem. Soc.*, 1959, 3223, 3341.
126. H. Björndal, K.-E. Eriksson, P. J. Garegg, B. Lindberg, and B. Swan, *Acta Chem. Scand.*, 1965, **19,** 2309.
127. I. M. Mackie and E. Percival, *J. Chem. Soc.*, 1959, 1151.
128. A. Haug, B. Larsen, and O. Smidsrød, *Carbohydrate Res.*, 1974, **32,** 217.
129. A. Penman and D. A. Rees, *J. C. S. Perkin I*, 1973, 2191.
130. N. S. Anderson, T. C. S. Dolan, and D. A. Rees, *J. C. S. Perkin I*, 1973, 2173.
131. A. Penman and D. A. Rees, *J. C. S. Perkin I*, 1973, 2188.
132. M. Duckworth and W. Yaphe, *Carbohydrate Res.*, 1971, **16,** 189.
133. J. R. Turvey and L. M. Griffiths, *Phytochemistry* 1973, **12,** 2901.
134. A. F. Pavlenko, A. V. Kurika, V. A. Khomenko, and Yu. S. Ovodov, *Khim. prirod. Soedinenii*, 1974, 172.
135. L. Glaser, *Ann. Rev. Biochem.*, 1973, **42,** 91.
136. K. W. Knox and A. J. Wicken, *Bacteriol. Rev.*, 1973, **37,** 215.
137. P. Critchley, A. R. Archibald, and J. Baddiley, *Biochem. J.*, 1962, **85,** 420.

138. A. J. Wicken, *Biochem. J.*, 1966, **99,** 108.
139. J. Baddiley, *Fed. Proc.*, 1962, **21,** 1084.
140. J. J. Armstrong, J. Baddiley, and J. G. Buchanan, *Biochem. J.*, 1961, **80,** 254.
141. A. R. Sanderson, J. L. Strominger, and S. G. Nathenson, *J. Biol. Chem.*, 1962, **237,** 3603.
142. K. H. Schleifer and R. Joseph, *F.E.B.S. Letters*, 1973, **36,** 83.
143. K. H. Schleifer and O. Kandler, *Bacteriol. Rev.*, 1972, **36,** 407.
144. P. A. J. Gorin and J. F. T. Spencer, *Canad. J. Chem.*, 1966, **44,** 993.
145. A. R. Archibald, J. Baddiley, and J. E. Heckels, *Nature New Biol.*, 1973, **241,** 29.
146. R. L. Sidebotham, *Adv. Carbohydrate Chem. Biochem.*, 1974, **30,** 371.
147. B. Lindberg and S. Svensson, *Acta Chem. Scand.*, 1968, **22,** 1907.
148. E. J. Bourne, R. L. Sidebotham, and H. Weigel, *Carbohydrate Res.*, 1974, **34,** 279.
149. S. A. Barker, E. J. Bourne, G. T. Bruce, W. B. Neely, and M. Stacey, *J. Chem. Soc.*, 1954, 2395.
150. G. T. Barry and W. F. Goebel, *Nature*, 1957, **179,** 206.
151. M. Stacey and S. A. Barker, 'Polysaccharides of Microorganisms', Oxford University Press, 1960.
152. M. J. How, J. S. Brimacombe, and M. Stacey, *Adv. Carbohydrate Chem.*, 1964, **19,** 303.
152a. O. Larm and B. Lindberg, *Adv. Carbohydrate Chem. Biochem.*, 1976, **33,** 295.
153. R. J. Sturgeon, in 'Plant Carbohydrate Biochemistry', ed. J. B. Pridham, Phytochemical Society Symposium, Academic Press, New York, 1974, no. 10, p. 219.
154. C. E. Ballou, *Adv. Enzymol.*, 1974, **40,** 239.
155. L. Rosenfeld and C. E. Ballou, *J. Biol. Chem.*, 1974, **249,** 2319.
156. T. Miyazaki and Y. Naoi, *Chem. and Pharm. Bull. (Japan)*, 1974, **22,** 1360.
157. J. S. Schutzbach, M. K. Raizada, and H. Ankel, *J. Biol. Chem.* 1974, **249,** 2953.
158. M. R. Stetten, H. M. Katzen, and D. Stetten, *J. Biol. Chem.*, 1956, **222,** 587.
159. R. D. Edstrom, *Arch. Biochem. Biophys.*, 1970, **137,** 293.
160. G. L. Brammer, M. A. Rougvie, and D. French, *Carbohydrate Res.*, 1972, **24,** 343.
161. H. De Wulf and H. G. Hers, in 'Biosynthesis of the Glycosidic Linkage', ed. R. Piras and H. G. Pontis, Academic Press, New York, 1972, p. 399.
162. C. T. Greenwood, in 'The Carbohydrates: Chemistry and Biochemistry', ed. W. Pigman and D. Horton, Academic Press, New York, 2nd edn., 1970, vol. IIB, chap. 38.
163. R. D. Marshall, in 'Carbohydrate Chemistry — Specialist Periodical Reports', ed. J. S. Brimacombe, The Chemical Society, London, 1975–76, vols. 7 and 8, part II.
164. B. J. Catley, in 'Carbohydrate Chemistry — Specialist Periodical Reports', ed. J. S. Brimacombe, The Chemical Society, London, 1977, vol. 9, part II.
165. K. M. Rudall, *Adv. Insect Physiol.*, 1963, **1,** 257.
166. K. Meyer, *Cold Spring Harbor Symp. Quant. Biol.*, 1938, **6,** 91.
167. S. A. Barker, J. F. Kennedy, and C. N. D. Cruickshank, *Carbohydrate Res.*, 1969, **10,** 65.
168. S. A. Barker and J. F. Kennedy, *Life Sci.*, 1969, **8,** part II, 989.
169. E. J. Moynahan and J. F. Kennedy, *Proc. Roy. Soc. Med.*, 1966, **59,** 1125.
170. E. J. Moynahan and J. F. Kennedy, 'XIII Congressus Internationalis Dermatologiae — München, 1967', Springer-Verlag, Berlin, 1968, p. 1543.
171. J. F. Kennedy, C. H. Sinette, and J. B. Familusi, *Clin. Chim. Acta*, 1970, **29,** 37.
172. P. W. Lewis, J. F. Kennedy, and D. N. Raine, *Biochem. Soc. Trans.*, 1973, **1,** 844.
173. C. A. White, J. F. Kennedy, and D. N. Raine, unpublished results.
174. J. F. Kennedy, *Biochem. Soc. Trans.*, 1973, **1,** 807.
175. J. F. Kennedy, "The Meldola Medal Lecture", *Chem. Soc. Rev.*, 1973, **2,** 355.
176. J. F. Kennedy and M. Stacey, *Egypt. J. Chem. Spec. Spec. Issue 'Tourky'*, 1973, 223.
177. J. F. Kennedy, *Adv. Clin. Chem.*, 1976, **18,** 1.
178. L.-Å. Fransson and L. Rodén, *J. Biol. Chem.*, 1967, **242,** 4161, 4170.
179. K. Murata, T. Harada, T. Fujiwara, and T. Furuhashi, *Biochim. Biophys. Acta*, 1971, **230,** 583.
180. A. Linker and P. Hovingh, *Biochim. Biophys. Acta*, 1968, **165,** 89.
181. J. A. Cifonelli, *Carbohydrate Res.*, 1968, **8,** 233.
182. N. Toda and N. Seno, *Biochim. Biophys. Acta*, 1970, **208,** 227.
183. V. P. Bhavanandan and K. Meyer, *J. Biol. Chem.*, 1968, **243,** 1052.
184. S. Hirano and K. Meyer, *Biochem. Biophys. Res. Comm.*, 1971, **44,** 1371.
185. U. Lindahl and L. Rodén, in 'Glycoproteins', ed. A. Gottschalk, Elsevier, Amsterdam, 2nd edn., 1972, p. 491.
186. R. D. Marshall and A. Neuberger, *Adv. Carbohydrate Chem. Biochem.*, 1970, **25,** 407.
187. R. Kornfeld and S. Kornfeld, *Ann. Rev. Biochem.*, 1976, **45,** 217.
188. J. F. Kennedy, in 'Structure–Activity Relationships of Protein and Polypeptide Hormones', ed. M. Margoulies and F. C. Greenwood, International Congress Series 241, Exerpta Medica, Amsterdam, 1972, part 2, p. 360.
189. J. F. Kennedy and W. R. Butt, *Biochem. J.*, 1969, **115,** 225.
190. 'Methods in Enzymology, Hormone Action', ed. B. W. O'Malley and J. G. Hardman, Academic Press, New York, 1975, vols. 36 and 37.
191. W. R. Butt, S. S. Lynch, and J. F. Kennedy, in 'Structure–Activity Relationships of Protein and Polypeptide Hormones', ed. M. Margoulies and F. C. Greenwood, International Congress Series 241, Exerpta Medica, Amsterdam 1972, part 2, p. 355.

192. A. S. Hartree, *Biochem. J.*, 1966, **100,** 754.
193. S. A. Barker, C. J. Gray, J. F. Kennedy, and W. R. Butt, *J. Endocrinol.*, 1969, **45,** 275.
194. D. Gospodarowicz, *J. Biol. Chem.*, 1972, **247,** 6491.
195. W. R. Butt and J. F. Kennedy, in 'Structure–Activity Relationships of Protein and Polypeptide Hormones', ed. M. Margoulies and F. C. Greenwood, International Congress Series 241, Exerpta Medica, Amsterdam, 1971, part 1, p. 115.
196. J. F. Kennedy, *Endocrinol. Experimentalis*, 1973, **7,** 5.
197. L. E. Reichert, M. A. Rasco, D. N. Ward, G. D. Niswender, and A. R. Midgley, *J. Biol. Chem.*, 1969, **244,** 5110.
198. M. F. Chaplin, L. J. Gray, and J. F. Kennedy, in 'Gonadotrophins and Ovarian Development', ed. W. R. Butt, A. C. Cooke, and M. Ryle, Livingstone, Edinburgh, 1970, p. 77.
199. J. F. Kennedy and M. F. Chaplin, *Biochem. J.*, 1972, **130,** 417.
200. J. F. Kennedy, M. F. Chaplin, and M. Stacey, *Carbohydrate Res.*, 1974, **36,** 369.
201. O. P. Bahl, *J. Biol. Chem.*, 1969, **244,** 575.
202. O. P. Bahl, in 'Structure–Activity Relationships of Protein and Polypeptide Hormones', ed. M. Margoulies and F. C. Greenwood, International Congress Series 241, Exerpta Medica, Amsterdam, 1971, part 1, p. 99.
203. J. F. Kennedy and M. F. Chaplin *Biochem. J.*, 1976, **155,** 303.
204. H. G. Schwick, *Naturwiss.*, 1974, **61,** 484.
205. G. Ashwell and A. G. Morell, *Adv. Enzymol.*, 1974, **41,** 99.
206. P. V. Wagh, I. Bornstein, and R. J. Winzler, *J. Biol. Chem.*, 1969, **244,** 658.
207. G. A. Jamieson, M. Jett, and S. L. DeBernardo, *J. Biol. Chem.*, 1971, **246,** 3686.
208. R. G. Spiro, *Adv. Protein Chem.*, 1973, **27,** 349.
209. R. G. Spiro and V. D. Bhoyroo, *J. Biol. Chem.*, 1974, **249,** 5704.
210. R. R. Porter, *Biochem. J.*, 1959, **73,** 119.
211. J. W. Rosevear and E. L. Smith, *J. Biol. Chem.*, 1961, **236,** 425.
212. R. Kornfeld, J. Keller, J. Baenziger, and S. Kornfeld, *J. Biol. Chem.*, 1971, **246,** 3259.
213. J. Baenziger, S. Kornfeld, and S. Kochwa, *J. Biol. Chem.*, 1974, **249,** 1897.
214. J. Baenziger and S. Kornfeld, *J. Biol. Chem.*, 1974, **249,** 7260.
215. T. Tai, S. Ito, K. Yamashita, T. Muramatsu, and A. Kobata, *Biochem. Biophys. Res. Comm.*, 1975, **65,** 968.
216. J. Baenziger, S. Kornfeld, and S. Kochwa, *J. Biol. Chem.*, 1974, **249,** 1889.
217. B. Anderson, N. Seno, P. Sampson, J. G. Riley, P. Hoffman, and K. Meyer, *J. Biol. Chem.*, 1964, **239,** PC2716.
218. B. Anderson, P. Hoffman, and K. Meyer, *J. Biol. Chem.*, 1965, **240,** 156.
219. V. Ginsburg, *Adv. Enzymol.*, 1972, **36,** 131.
220. S.-I. Hakomori and A. Kobata, in 'The Antigens', ed. M. Sela, Academic Press, New York and London, 1974, vol. 2, p. 80.
221. T. Feizi, E. A. Kabat, G. Vicari, B. Anderson, and W. L. Marsh, *J. Immunol.*, 1971, **106,** 1578.
222. 'Cellulose', in 'Methods in Carbohydrate Chemistry', ed. R. L. Whistler, Academic Press, New York, 1963, vol. III.
223. 'Cellulose and Cellulose Derivatives', 'High Polymers', ed. N. M. Bikales and L. Segal, Wiley-Interscience, New York, 2nd edn., 1971, vol. 5, parts IV and V.
224. J. A. Radley, 'Starch and its Derivatives', Chapman and Hall, London, 4th edn., 1968.
225. 'Starch, Chemistry and Technology', ed. R. L. Whistler and E. F. Paschall, Academic Press, New York, 1965, vol. I, and 1967, vol. II.
226. 'General Polysaccharides', in 'Methods in Carbohydrate Chemistry', ed. R. L. Whistler, Academic Press, New York, 1965, vol. V.
227. J. F. Kennedy, *Adv. Carbohydrate Chem. Biochem.*, 1974, **29,** 305.
228. H. Vink, *Makromol. Chem.*, 1969, **122,** 271.
229. C. Arsenis and D. B. McCormick, *J. Biol. Chem.*, 1964, **239,** 3093.
230. M. A. Mitz and L. J. Summaria, *Nature*, 1961, **189,** 576.
231. D. M. Soignet, R. J. Berni, and R. R. Benerito, *Textile Res. J.*, 1966, **36,** 978.
232. J. Porath and N. Fornstedt, *J. Chromatog.*, 1970, **51,** 479.
233. S. A. Barker, P. J. Somers, and R. Epton, *Carbohydrate Res.*, 1968, **8,** 491.
234. K. Graves, *Tappi*, 1972, **55,** 263.
235. J. W. Sedat, R. B. Kelly, and R. L. Sinsheimer, *J. Mol. Biol.*, 1967, **26,** 537.
236. I. Gillam, S. Millward, D. Blew, M. von Tigerstrom, E. Wimmer, and G. M. Tener, *Biochemistry*, 1967, **6,** 3043.
237. G. A. Towle and R. L. Whistler, *Methods Carbohydrate Chem.*, 1972, **6,** 408.
238. R. L. Whistler and W. W. Spencer, *Analyt. Chim. Acta*, 1971, **55,** 448.
239. W. M. Doane, B. S. Shasha, E. I. Stout, C. R. Russell, and C. E. Rist, *Carbohydrate Res.*, 1967, **4,** 445.
240. S. A. Barker, H. Cho Tun, S. H. Doss, C. J. Gray, and J. F. Kennedy, *Carbohydrate Res.*, 1971, **17,** 471.
241. J. F. Kennedy, S. A. Barker, and A. Rosevear, *J. C. S. Perkin I*, 1973, 2293.
242. J. F. Kennedy and A. Rosevear, *J. C. S. Perkin I*, 1974, 757.
243. J. F. Kennedy and A. Zamir, *Carbohydrate Res.*, 1973, **29,** 497.
244. H. N. Fernley, *Biochem. J.*, 1963, **87,** 90.

245. R. H. Hackman and M. Goldberg, *Analyt. Biochem.*, 1964, **8,** 397.
246. J. F. Kennedy, *Stärke*, 1976, **28,** 196.
247. J. F. Kennedy, *Stärke*, 1977, **29,** 114.
248. P.-H. Iverius, *J. Biol. Chem.*, 1972, **247,** 2607.
249. O. R. Zaborsky, 'Immobilised Enzymes', C. R. C. Press, Cleveland, 1973.
250. J. F. Kennedy, P. A. Keep, and D. Catty, *Biochem. Soc. Trans.*, 1976, **4,** 135.

26.4 Polysaccharides: Conformational Properties in Solution

D. A. REES
Unilever Research, Bedford

26.4.1 INTRODUCTION

The methods and concepts of organic chemistry are most powerful when applied to problems of equilibrium, reactivity, and biological function in highly solvated states and thus the emphasis in this chapter is on the conformational properties of polysaccharides in solution rather than in the solid state. Nevertheless, some significant recent developments in the characterization of polysaccharide conformation by X-ray diffraction are noted in passing.

Up-to-date reviews which cover aspects of polysaccharide conformation in more detail include accounts of the binding behaviour of carbohydrate chains in relation to their stereochemistry,[1] methods of prediction of the parameters for disordered polysaccharide shapes in solution,[2] and the secondary and tertiary structure of polysaccharide shapes in solutions and gels.[3] An earlier review outlined the basic principles of polysaccharide conformation in detail.[4] A short text has recently become available on the subject of 'Polysaccharide Shapes'.[5] Since these published reviews between them give a fairly complete account of the experimental results, this chapter will concentrate mainly on the background theory and the experimental approaches.

26.4.2 CONFORMATIONS CHARACTERIZED BY X-RAY DIFFRACTION ANALYSIS

Many polysaccharides, especially the industrially valuable starch and cellulose, occur naturally in a crystalline or semi-crystalline state and the course of modification and substitution reactions is therefore determined by the shapes and packing of the chains in this state. Another reason why organic chemists need to note the nature of conformations characterized by X-ray diffraction is that without such information as background, the methods for examining conformations in solution would be totally incapable of arriving at any degree of useful detail.

Knowledge of conformations in the solid state was arrested for many years because detailed characterization was too difficult for the methods of X-ray diffraction analysis then available. With the development of computer methods for the systematic generation and refinement of trial models, notably by Arnott and his co-workers,[6] it seems that these difficult problems have at last been resolved. Thus the balance of evidence is now accumulating that the natural form of cellulose (cellulose I) has all chains running parallel rather than anti-parallel (that is, within a given crystallite, all non-reducing termini occur at one end and all reducing termini at the other),[7] whereas the regenerated or mercerized form of cellulose (cellulose II), to which the natural form is often converted in industrial processing, has anti-parallel chains.[8] Likewise, there is now firm evidence from X-ray diffraction studies that the so called V-form of amylose — the hollow helix conformation adopted by the amylose component of starch when it forms the familiar inclusion complex with iodine/tri-iodide to give the intense blue-black colouration (used as an indicator in volumetric analysis and in histochemical stains) — is left- rather than right-handed.[9] Finally, the natural conformation of starch chains in cereal grains and vegetable tubers, for so long a subject of disagreement and mystery, has been very recently shown to be a double helix[10] — as had been proposed, but could not be proved, a number of years before.[11]

The important family of glycosaminoglycans include hyaluronate, chondroitin sulphates, and keratan sulphate. They have been shown to be capable of existence in oriented films in a wide range of interconvertible conformations which are strongly dependent on the cations present.[12] These conformations represent a family of left-handed helices which seem always to pack in anti-parallel fashion, and which differ mainly in the degree of extension. The most compressed form is one of the hyaluronate conformations which twists around a partner to occur as a double helix.[13] All other known forms pack in side-to-side fashion. In some cases it has been possible to carry the structure refinement to a level of detail which is quite unusual for fibrous (rather than truly crystalline) materials; it is then possible even to visualize the position of water molecules and the coordination geometry around cations.[14] Two other X-ray structure determinations which will probably prove important reference states for understanding the conformational principles for carbohydrate chains in general are (i) the first example of an ordered conformation of a branched polysaccharide (the extracellular polysaccharide from a strain of *Escherichia coli*), which suggests an important function of chain branching in folding down the side chain against the backbone to stabilize the polysaccharide conformation through non-bonded interactions;[15] and (ii) the first glycoprotein crystal structure to show the carbohydrate part in an ordered conformation.[16] At the time of writing this second structure determination — which is for the F_c fragment of immunoglobulin IgG molecule — is still incomplete but it already suggests some important conclusions to which we shall return below.

26.4.3 PRINCIPLES OF POLYSACCHARIDE CONFORMATION IN SOLUTION

The shapes of the individual sugar rings in carbohydrate chains are usually predicted readily from first principles and confirmed by analogy with determinations for relevant models of low molecular weight. The only exceptions are for some carbohydrate chains in the disordered state (when it may be necessary to take into account the minor proportion of higher-energy ring conformations that must exist at equilibrium[2]) and for ordered states in the relatively rare circumstances in which one of the sugar residues is conformationally unstable (such as for L-iduronate residues in several glycosaminoglycans when the alternative possibilities are $^4C_1$, $^1C_4$, and a twist-boat form[17]). For the more straightforward and fortunately more common situations, conformational analysis is greatly simplified because the chain can be regarded as made up of a series of rigid elements connected by a limited number of bonds which allow free rotation; when the glycosidic linkage is to a secondary hydroxy-group there are two such bonds (**1**) and when it is to a primary hydroxy-group there are three (**2**).

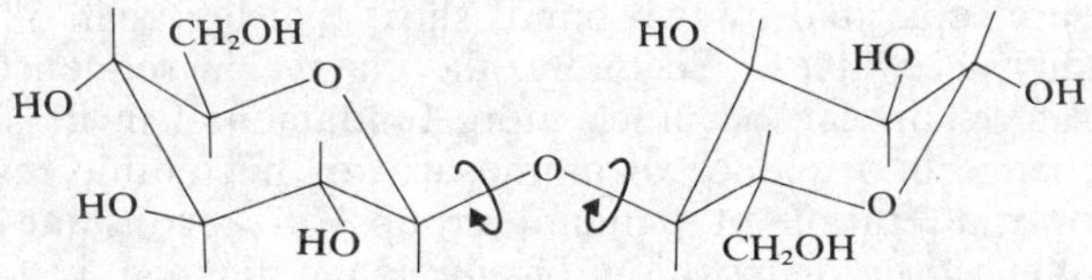

(1)

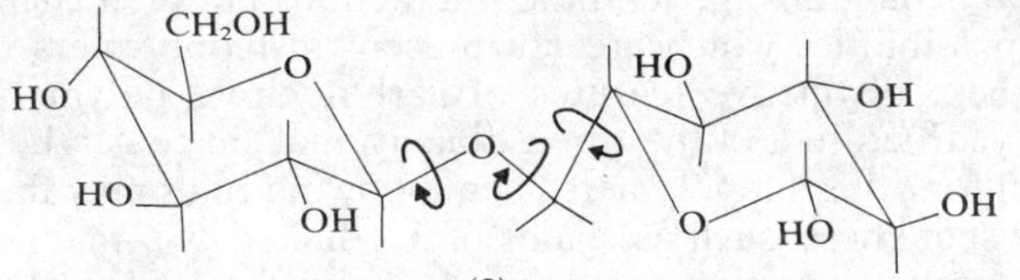

(2)

It is of course a simple matter to analyse the alternative, staggered conformational states that arise from the bond rotations (**1** and **2**) and by consideration of non-bonded interactions rank them in the order in which they would be energetically preferred. Unfortunately this leads to a picture of the overall conformation which is inadequate, whether the polymer chain is ordered or disordered. This is because, as shown by physical models as well as mathematical model building, the overall shape and dimensions are sensitive to minor adjustments within each energy minimum. Therefore, if predictive methods are to lead beyond vague generalities, they must point not only to a preference for a particular energy minimum but also whereabouts in that minimum. For a full understanding of disordered conformations there is an additional requirement for the estimation of how the molecule will partition quantitatively within and between the alternatives at a given temperature. The extent to which all this is possible with current methods is discussed elsewhere.[2,4,5]

A major issue for theoretical prediction or experimental determination is whether a particular carbohydrate chain is ordered or disordered in conformation. In the ordered state the bond rotation angles (**1** or **2**) have fixed values that are likely to be identical between all pairs of covalently equivalent residues along the chain — in contrast to the disordered state in which continuous rotation and oscillation occurs about these bonds, and it is improbable that an equivalent stereochemical relationship would exist between any pair of residues at any instant. As explained in more detail elsewhere,[4] the existence of an ordered state requires *cooperative interactions* within or between chains; this means that the minimization of interaction energy between one pair of residues simultaneously sets up the minimization for a neighbouring pair and so on throughout the structure. The disordered state is favoured, on the other hand, by *conformational entropy*; this derives from the probability that the chain will in time sample all conformational states as a result of thermal motion, and implies that the greater the conformational flexibility within a carbohydrate chain the more likely is the disordered state. Prediction of whether a chain would be ordered or disordered would require computation of the free energy corresponding to each of these two possibilities. At present this cannot be done reliably and the definitive answer must therefore be found experimentally. The predictive methods are useful, however, for pointing to likely properties of the ordered and disordered states separately.

26.4.3.1 Ordered states

Inspection of space-filling models, and more especially the result of model-building calculations in the computer, suggests immediately some characteristics of the ordered conformations of carbohydrate chains that can be distinguished from those of the other important biopolymers — proteins and nucleic acids. Firstly, the carbohydrate chains are much stiffer; hence a given chain is more restricted by mere steric exclusion in the range of shapes that it can adopt. Hard-sphere calculations for chains in which successive residues are separated by two bonds show that less than 5% of the conformational space is readily accessible.[18] Secondly, the change in sequence of sugar residues within a carbohydrate chain can be much more fundamental in its stereochemical consequences than a change in sequence of amino-acid or nucleotide residues, because the latter may cause only alteration of pendant groups on a constant backbone whereas change in the configuration or position of glycosidic linkage is fundamental to the backbone itself. Thirdly, carbohydrate chains are frequently branched structures with a variety of types of branching linkage, and the interactions between chains held together in this way introduce possibilities which are not present in biopolymers that are linear.

It follows from these distinctive features of carbohydrate polymers that correlations between the primary structure and the three-dimensional shape can be more obvious and more consistent than for other biopolymers. The following rules may then be formulated.

(i) Homoperiodic sequences, such as chains of 1,4-linked β-D-mannopyranose residues (**3**) or 1,3-linked α-D-glucopyranose residues (**4**), for which the dihedral angle between the aglycone and glycosidic bonds projected across each residue is close to 180°, are only likely to exist in ordered conformations which are extended and ribbon-like.

(ii) Homoperiodic sequences, such as 1,3-linked β-D-xylopyranose (**5**) or 1,4-linked α-D-glucopyranose (**6**), for which this projected angle is close to 0°, are only likely to exist in hollow helix-like conformations. Just as sequences of type (i) above are likely to pack in dense, fibrous aggregates, these chains typically form inclusion complexes or combine with like chains in double or triple helices.

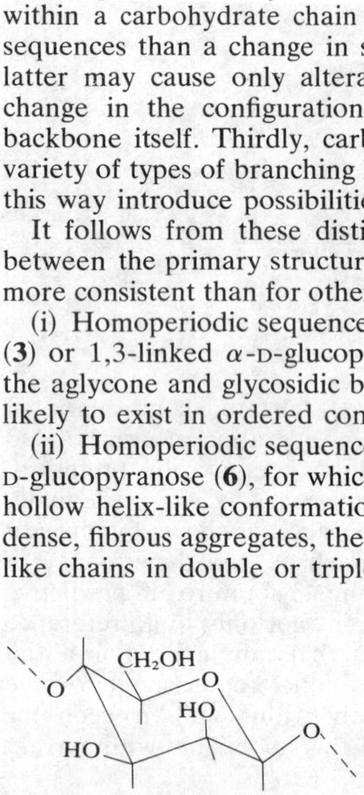

(3)

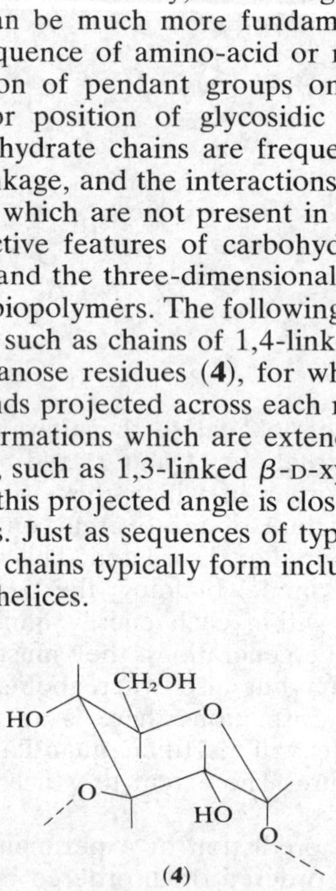

(4)

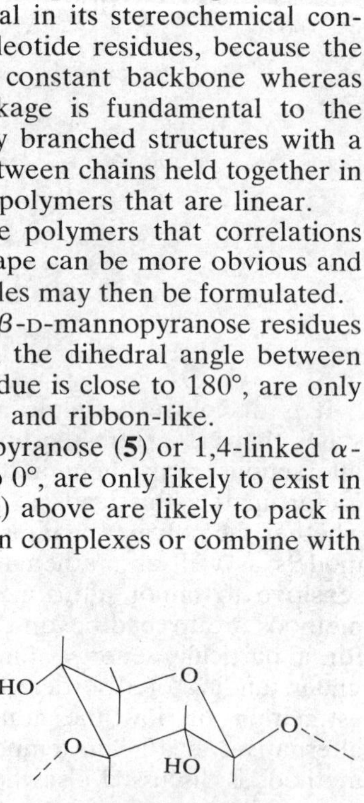

(5)

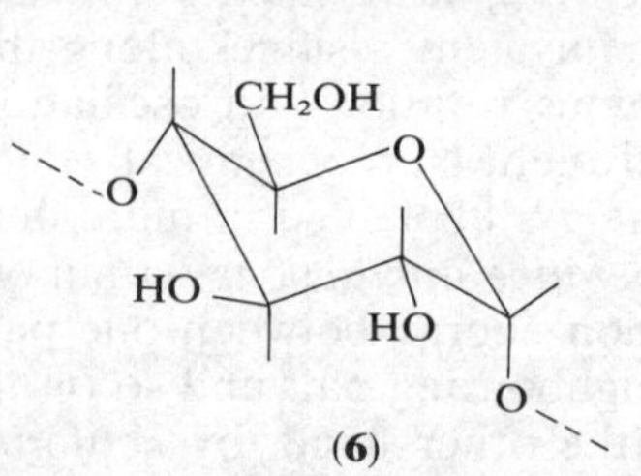

(6)

(iii) Heteroperiodic sequences of residues will likewise form periodic ribbon-like or helical conformations, but prediction is more difficult by mere inspection. Model building can, however, still be very informative and the steric restriction within the chain makes analogies possible between related covalent sequences. For example, the rather extensive group of alternating copolysaccharides (**7–11**) form a family of extended hollow helix conformations, some of which exist as double-stranded structures.[1,10,13]

(iv) When periodic sequences are interrupted by 'foreign' residues, conformational interruption very often results. This seems to represent a biological device for terminating chain-associations which depend on the interlocking of regular conformations and hence building up cross-linked network structures or gels. Some types of conformational ordering that have been shown to be responsible for such cross linking are shown in Figure 1, with corresponding covalent sequences as well as sequences that terminate the ordered conformations.

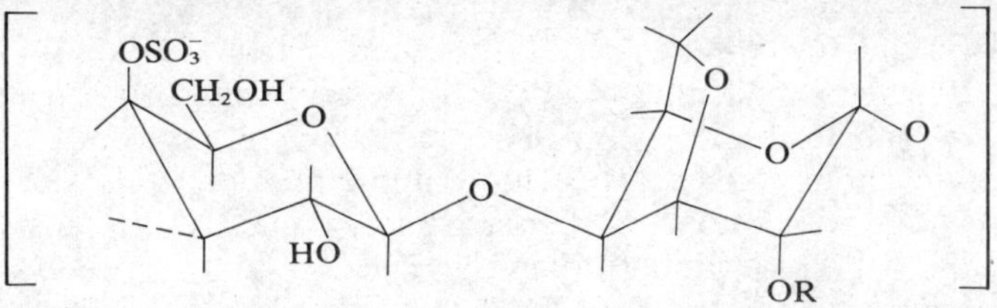

(7)

R = H, κ-Carrageenan

R = SO_3^-, ι-Carrageenan

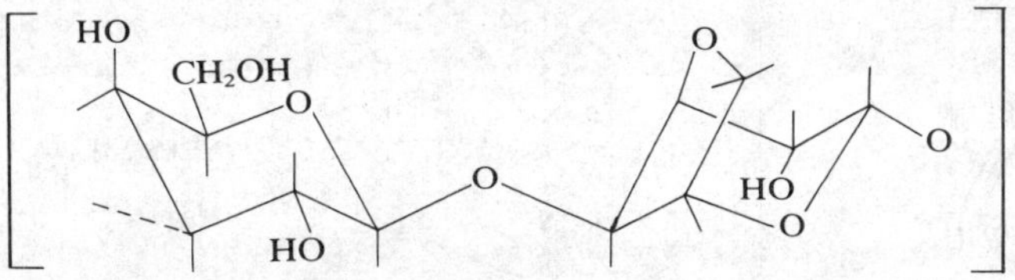

(8) Agarose

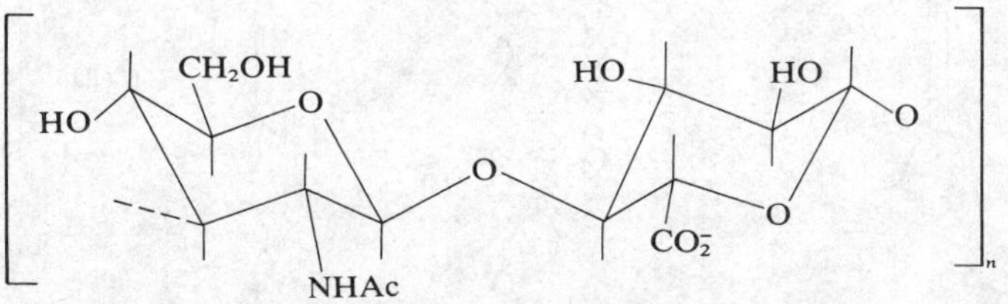

(9) Hyaluronate

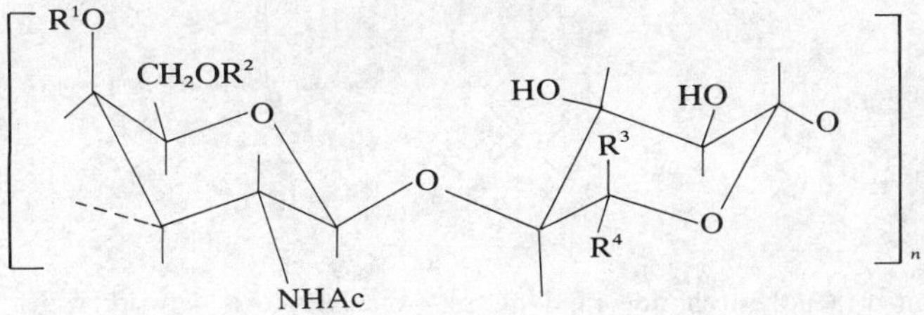

(10)

$R^1 = SO_3^-$, $R^2 = R^3 = H$, $R^4 = CO_2^-$, Chondroitin 4-sulphate

$R^2 = SO_3^-$, $R^1 = R^3 = H$, $R^4 = CO_2^-$, Chondroitin 6-sulphate

$R^1 = SO_3^-$, $R^2 = R^4 = H$, $R^3 = CO_2^-$, Dermatan sulphate

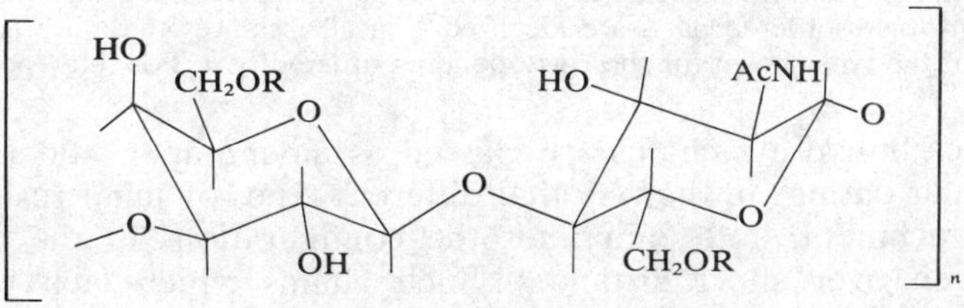

(11)

R = H or SO_3^-, Keratan sulphate

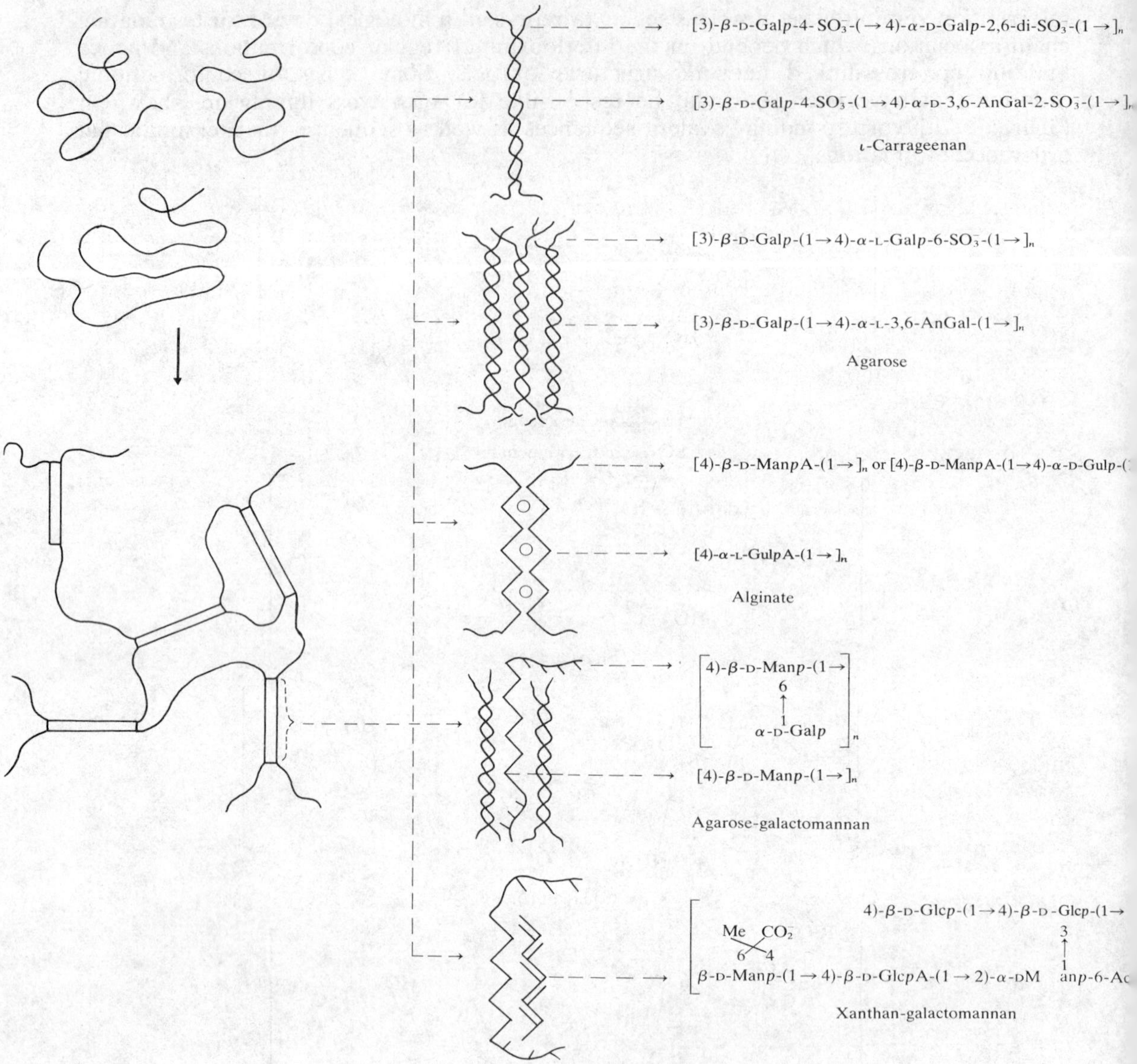

Figure 1 Conversion of a polysaccharide random coil (top, left) to a network which is cross-linked by the conformational ordering of chain segments and their consequent association (bottom, left). On the left, the ordered entities are shown as generalized rectangles; their actual shapes are shown schematically on the right with the covalent sequences in the ordered and disordered parts, respectively, of the network. These are: the carrageenan double helix, the agarose bundle of double helices, the alginate egg-box (circles represent complexed Ca^{2+} ions), the cooperative association of unlike polysaccharides as between the agarose bundles and galactomannan ribbons, and the association between the xanthan rods and galactomannan ribbons. Note that the sequences shown for disordered parts are idealized — in all instances these are actually likely to incorporate some residues of the type shown for the corresponding ordered part. For references to the literature, see text

(v) Many carbohydrate chains are classed as having aperiodic sequences because they have no regular pattern in the way that different types of sugar residues, their positions of glycosidic attachment, and sometimes the configurations at the glycosidic linkages are arranged, even over short distances. Such chains which lack periodicity in covalent sequence cannot adopt periodic ordered conformations such as the helices or ribbons in (i) above. A further feature of these structures is that very frequently they are branched. As shown by model-building calculations,[18] this can introduce dramatic stiffening of the chain and further restriction of possible conformations around the branching unit, especially when branching is through adjacent secondary positions on the ring. In at least some instances, branching may introduce the possibility of chain folding with alignment of the side chain against the backbone to provide mutual conformational stabilization through favourable non-bonded interactions. The best characterized example is in a branched extracellular polysaccharide from a strain of *Escherichia coli.*[15] Another, in which the ordered state persists into solution, is the extracellular polysaccharide from bacterial plant pathogens of the genus *Xanthomonas.*[19] Finally, recent evidence on the solid-state conformation of components of starch[10] would appear to support the possibility[11] of alignment of α-1,4 linked chains in amylopectin and glycogen which are joined together by an α-1,6 branch linkage — in this instance to combine in double helix formation.

Carbohydrate chains of the aperiodic branched type are very common components of glycoproteins and a number of glycolipids. From the principles outlined above, any ordered conformations are likely to be compact and globular rather than extended and periodic. This raises exciting possibilities for structure–function relationships, for example at cell surfaces in cellular recognition and hormone–receptor interactions. The first crystal structure of a glycoprotein[16] has indeed suggested that alignment and conformational stabilization can occur between carbohydrate and peptide chains.

26.4.3.2 Properties of disordered carbohydrate chains

The disordered state, or *random coil,* exists because the polymer chain has a large number of bonds which allow internal rotation with relatively little change of energy. Such rotations cause the chain to take up many alternative shapes, and the random coil is that state in which the chain is fluctuating continuously between the various possibilities. The greater the internal freedom, the greater will be the number of accessible transient forms and consequently the more difficult it will be to overcome thermal motion to settle in the form in which the potential energy is minimized. Thus the disordered state is favoured by the existence of three-bond connections between residues (compare **1** and **2**). For two-bond connections, model-building calculations show that chain flexibility is restricted with increasing size of equatorial groups adjacent to the linkage and also with the number of axial bonds to the glycosidic oxygen. Thus linkage (**12**) is less flexible than (**1**), which is less flexible than (**13**). It has already been pointed out that still further restriction is introduced by chain branching, especially when this involves adjacent secondary positions on the ring.

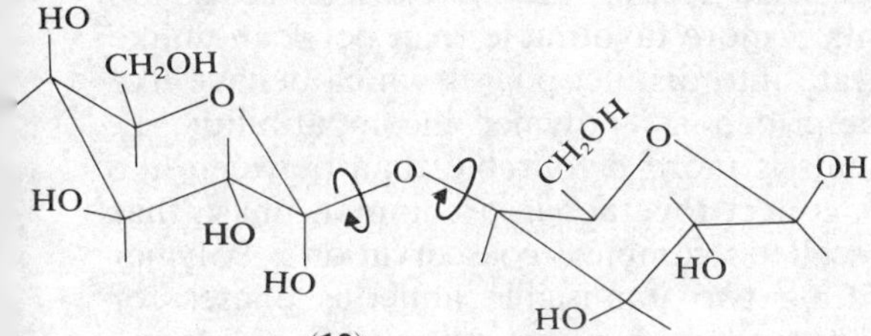

(**12**)

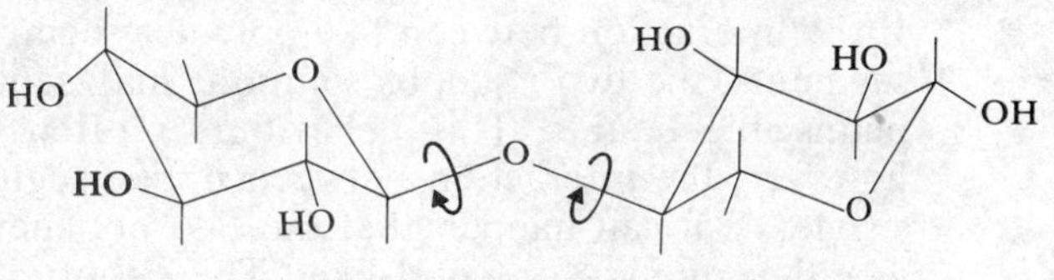

(**13**)

It is useful to characterize the disordered state in terms of its average *overall* dimensions rather than the average *local* conformation. This is because such properties as bulk viscosity and water binding behaviour are determined by the total solution space swept out by the mobile chain. It can be shown mathematically that the problem of calculating the average overall dimensions reduces to the problem of defining the average orientation of one sugar residue with respect to its neighbour, and in principle this can be done by an extension of computer model-building methods.[2] All possible mutual orientations are surveyed to calculate the corresponding interaction energy at each stage for averaging according to the Boltzmann distribution, and the mean square end-to-end length of the chain may then be obtained through expressions that apply to all polymers. The results can be compared with parameters derived by experiment, particularly from light-scattering measurements. It emerges that the two main families of periodic homopolysaccharides, which can be recognized from their distinct types of ordered conformation (see above), also differ in their general properties as random coils. Members of the ribbon family are correctly predicted[20] to sweep out a larger volume of solution space (the characteristic ratio, C_∞, is typically of the order of 100), compared with the hollow helix type (C_∞ is typically of the order of 10).

The predictions cannot, of course, take into account any variations in degree of interaction between the chain and the solvent — when, for example, the chain tends to be pulled out to become more thoroughly solvated or to contract in withdrawal of chain elements from solution. The results therefore correspond to those conditions (known as the 'θ-conditions') under which the polymer chain is said to be 'unperturbed', when such tendencies to expand and contract the chain are exactly balanced. Unfortunately, this does not necessarily correspond to the conditions which are most relevant to biological function or properties of technological interest. It must also be said that the functions available for estimating interaction energies within the chain are still rather imprecise, and therefore it is reasonable only to expect to be successful in predicting general trends. Some success has, however, been achieved in this direction.[21] The most interesting general properties of carbohydrate chains in their disordered states are their ability to bind water and ions and to exclude or attract other polymer chains into their domain.[1] Water binding arises because the driving force (conformational entropy) that favours the random coil must also favour a large aqueous domain to allow the shape to fluctuate. The binding of cations is usually fleeting and by chance collision without necessarily having reproducible geometry; this leads to an ion atmosphere around the polyelectrolyte chain analogous to the atmosphere of gases that may surround a planet. They are retained by the need for overall electrical neutrality but entropically driven to mix to the maximum extent with solvent; this enhances the internal osmotic pressure to imbibe water into the polymer domain.[1]

Polymer chains in expanded, disordered conformations must make many segment–segment contacts and this gives rise to certain typical interaction properties. The entropy of mixing of solutions of two different polymers depends to a first approximation on the number of molecules involved in the mixing and therefore is independent of molecular weight. In contrast, the interaction energy between two mixed polymers depends on the number of segment–segment contacts and must, for a given number of molecules, increase with molecular weight. Thus the entropy term becomes increasingly unimportant with molecular weight, and the behaviour of mixed polymers is determined by interaction energies even when the segment–segment contacts are fleeting and the energies are small. If the interaction between like polymer segments is more favourable than between unlike segments, the two aqueous solutions may separate into distinct phases which behave like immiscible liquids. This behaviour is often described as 'polymer incompatibility'. If, however, the interaction between unlike segments is more favourable than between like segments, it can happen that the two polymers collect together in a common phase that is either liquid- or solid-like. This is often called 'complex coascervation'. Polymer incompatibility is useful, for example, in creating two immiscible aqueous phases for biochemical separations, as in a well-known procedure for the separation of plasma membranes in which the polysaccharide dextran is used for one of these phases.[22] Complex coascervates of polysaccharides and proteins having opposite charges — especially gum arabic and gelatin — form the basis of modern microencapsulation technology.

26.4.4 STRATEGY FOR CHARACTERIZATION OF SHAPES IN SOLUTION

It will be apparent from the foregoing that the first question to answer, in any attempt to characterize a polysaccharide conformation in solution, is whether it is ordered or disordered. Relevant information can be obtained from classical methods of analysis of size, shape, and dissymmetry of polymer molecules, such as those based on light scattering, small-angle X-ray scattering, viscosity measurements (including, for polyelectrolytes, the response to ionic strength[23]), and sedimentation. All these are generally applicable to polymers of many types and details of the theoretical principles and experimental methods are given in the standard works on polymer solutions.[24,25] These methods suffer from the limitation that the information they give is about average properties of the polymer molecule as a whole. Hence, if part of the molecule is ordered whereas the remainder is disordered, the evidence will show the average shape to be disordered. This is inconvenient because any specificity in physical or biological properties is likely to arise from the ordered domain and therefore it is important to detect and characterize this part. A further drawback is inherent in the definition of the ordered conformation as one that is constrained close to a potential energy minimum by cooperative interactions. The existence of such cooperative interactions can usually only be inferred rather than demonstrated directly by these methods of polymer characterization.

In recent years, two other approaches have been found to be particularly useful for giving direct evidence for regions that are cooperatively stabilized. Because of their special value, they are discussed in detail below.

26.4.4.1 Nuclear magnetic relaxation

The relevant n.m.r. parameter here is the spin–spin relaxation time, T_2, measured either directly as the time constant of the exponential decay of magnetization induced in a pulse spectrometer, or from the linewidth ($\Delta\nu_{\frac{1}{2}}$) in the high-resolution spectrometer. For a Lorentzian lineshape:

$$\Delta\nu_{\frac{1}{2}} = (\pi T_2)^{-1}$$

The time measures the rate of dephasing of nuclear spins, a process which is slowed down by thermal motions. It follows that rapid motions lead to narrow spectral lines while slow motions may lead to lines so broad that they are undetectable in a conventional high-resolution spectral display. Thus it is possible to make a direct distinction between random coil molecules in which, by definition, thermal motion rapidly interconverts each local segment between various energetically accessible stages, and conformations in which local chain segments are so constrained by cooperative interactions that the polymer merely undergoes thermal motion as a whole. In consequence, polysaccharides in the disordered state show high-resolution ^{1}H n.m.r. spectra that are reasonably well resolved and average values for T_2 are of the order of 10 ms at laboratory temperature, whereas for polysaccharides in cooperatively ordered conformations the lines are too broad to be detected and T_2 may be of the order of 100 μs.[26] A good example of these effects is provided by the order–disorder transition for xanthan, the extracellular polysaccharide of the bacterium *Xanthomonas campestris.* This structure (shown at the bottom of Figure 1) contains two *C*-methyl groups which show well-resolved peaks in the ^{1}H n.m.r. spectrum at high temperatures, which collapse and disappear completely on cooling as the polysaccharide chain converts to an ordered state (Figure 2).[19]

In using the n.m.r. method, the high-resolution approach is usually more convenient because suitable instrumentation is more generally available. There are, however, two serious limitations to be noted. If different regions of a chain are in different types of conformation, the existence of a rapidly relaxing component must be inferred by difference. Accurate methods of peak integration are available involving the use of external standards in which the peak shapes have been adjusted to be as similar as possible as those to be measured,[27] but indirect evidence of this sort is always unsatisfactory. Secondly, the actual relaxation time is not measured for the rapidly relaxing component because the width of the invisible peak cannot of course be measured; therefore this valuable parameter for characterizing the degree of rigidity is not available. Both problems can in principle be overcome by direct measurement of the decay of magnetization, but this is still instrumentally difficult and short T_2 values may have to be obtained as averages over chemically different nuclei. Thus it is best to use the two methods together and regard them as complementary.

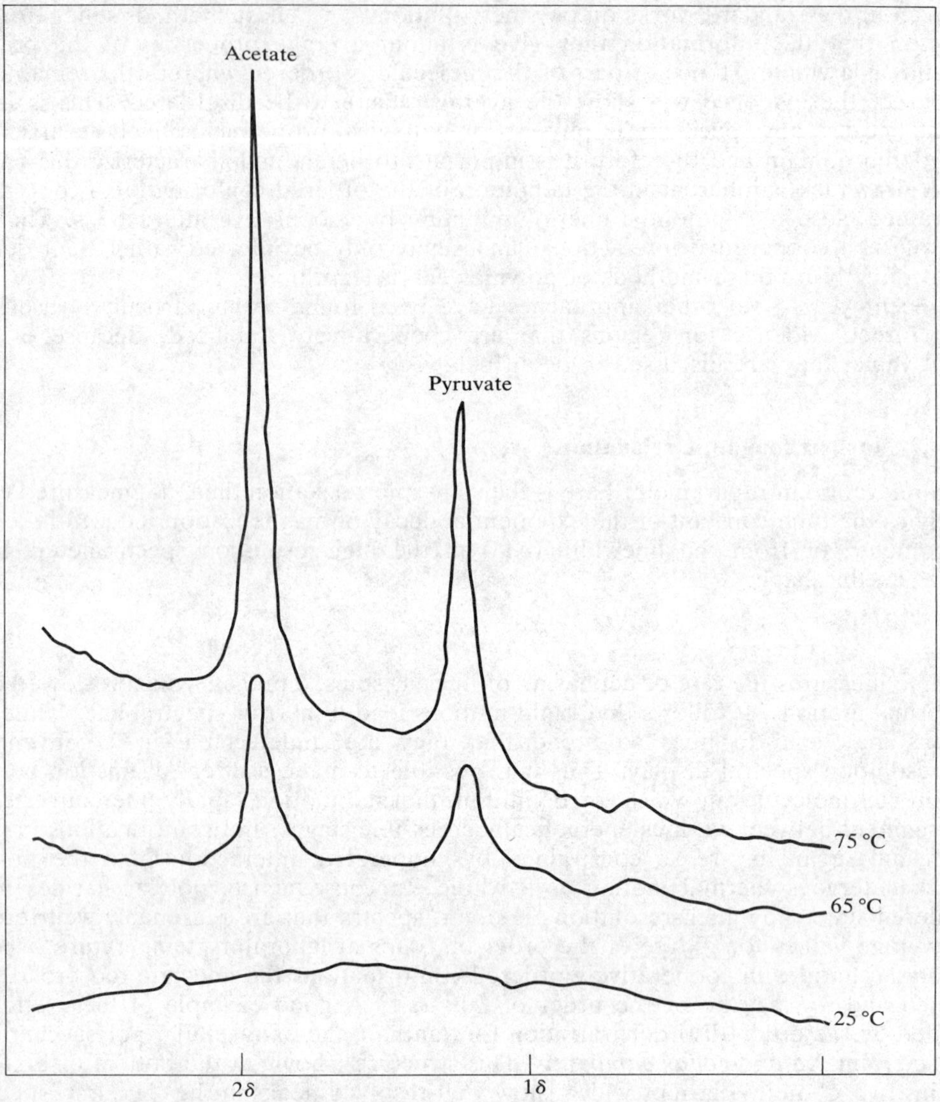

Figure 2 Part of the high-resolution ^1H n.m.r. spectrum of xanthan, showing the effect of temperature

26.4.4.2 Order–disorder transitions

An important characteristic of conformations stabilized by cooperative interactions is that their interconversion with the disordered state occurs rather suddenly, whether induced by change in temperature, solvent, ionic strength, or another variable. Often this approximates to an 'all or none' event which is quite different from the gradual shift of conformational equilibrium in small molecules. These sharp transitions can be detected by measurement of any physical parameter that is sensitive to overall conformation. The characteristic sigmoidal curve is illustrated by the example of the xanthan transition monitored by viscosity, monochromatic optical rotation or the area of detectable n.m.r. signal (Figure 3), as well as by the c.d. amplitude at appropriate wavelength, and other methods.

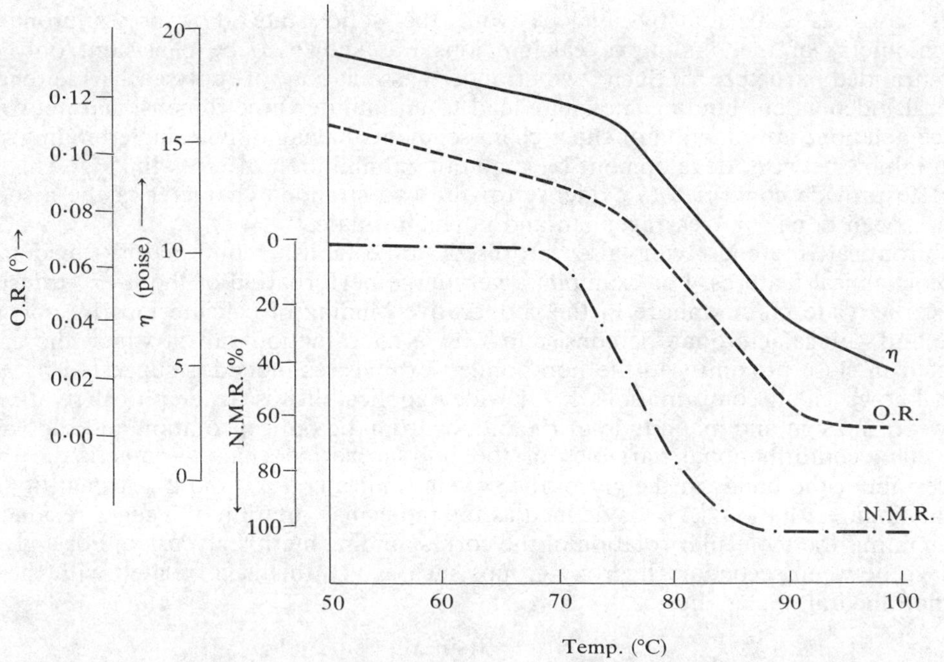

Figure 3 Influence of temperature on optical rotation (o.r.), low shear viscosity (η), and integrated area of the observable n.m.r. signal in high-resolution instrument for xanthan solution

In principle, the existence of a stepwise change in any such physical measurement is clear evidence for cooperativity in the interactions which stabilize at least one of the species in equilibrium. Although this might arise from interactions within the solvent, between solvent and cosolvent or solvent and cosolute, it is usually possible at least in the case of temperature-induced transitions to trace the origin of the effect confidently to perturbation of the macromolecule itself. When the change is induced by variation of such other properties as solvent composition, concentration of cosolute, or pH, alternative interpretations may be more difficult to eliminate. For this reason and because absolute thermodynamic relationships are more readily applied, it is preferable whenever possible to use temperature as the independent variable for inducing conformation change.

Cooperativity can also be recognized from the variation of the properties of the series of homologous oligosaccharides with chain lengths that lie on either side of the critical threshold for the ordered conformation. This has been elegantly shown by a study of the calcium-binding properties of oligogalacturonates and oligoguluronates;[28] both series show a sharp sigmoidal increase in affinity for calcium ions at an appropriate chain length, above and below which the binding properties change only gradually.

26.4.4.3 Characterization of ordered conformations in solution

If it has been established that a particular polysaccharide is partially or completely ordered in its solution conformation, the next problem is to characterize the geometry in as much detail as possible. All the approaches available at present rely heavily on comparison with reference conformations that have been established by X-ray diffraction analysis in the solid state. Comparison of some general features is possible from analysis of the stoichiometry of the order–disorder transition; an example is whether the ordered conformation is a single, double, or multi-stranded structure. For example, the concentration-dependence of the transition showed that xanthan forms a conformation which is ordered intramolecularly,[19] whereas ι-carrageenan forms an ordered dimer,[29] in both cases as expected by analogy with the solid state. For polyguluronate, the stoichiometry of the binding of calcium ions was shown to be consistent only with a two-stranded structure.[30] Such two-stranded association of polysaccharide chains at several independent binding sites may lead to an infinite three-dimensional network and hence gelation; introduction of short-chain segments having only a single binding site can then inhibit network development by competitive inhibition of cross-linking. This can be used to provide confirmatory evidence for the two-stranded character of the association, as has been done for ι-carrageenan and polyguluronate.[31]

Chiroptical methods can also be used to establish confirmatory evidence for stereochemical features. For example, a very large perturbation of the n–π^* transition of the carboxylate chromophore in the cooperative binding of calcium ions by polyguluronate and polygalacturonate is consistent with a coordination site in which the cation is located in close proximity to the non-bonded orbitals, as indeed is suggested by analogy with known chain conformations.[32] Of wider applicability is an empirical relationship[33] between the sign and magnitude of the monochromatic optical rotation and the values of the main conformational variables of the polysaccharide chain — namely the dihedral angles about the bonds in the glycosidic system (indicated in **1** and **2**). A quantity known as the linkage rotation, $[\Lambda]_D$, is defined as the molecular rotation of a sugar residue in the chain minus the molecular rotation of the corresponding methyl glycoside. For a glycosidic linkage between secondary hydroxy-groups such as (**1**), this is correlated with the values of the dihedral angles thus:

$$[\Lambda]_D = A - B(\sin \Delta\phi + \sin \Delta\psi)$$

where A and B are constants whose values are known and $\Delta\phi$ and $\Delta\psi$ are the dihedral angles. An analogous expression can be written for the linkage to the primary position (*e.g.* **2**). The relationship has been extensively checked against model compounds and shown to be successful in correlating the solution and/or gel conformation of carrageenan,[34] agar,[35] and certain arabinoxylans,[36] and a polymer of 3-*O*-methyl-D-glucose from *Mycobacterium smegmatum*,[37] with reference conformations established in the solid state. Unfortunately this analysis is complicated if chromophores are present in the accessible ultraviolet which are optically active and sensitive to the overall conformation, since they can have a swamping effect on the monochromatic optical rotation.

26.4.4.4 Characterization of disordered conformations in solution

In addition to well-established methods for the determination of overall dimensions of a random coil that have general applicability to polymer solutions,[24,25] several special methods are available for characterizing the local conformations at the glycosidic linkages in carbohydrate chains. In dimethyl sulphoxide solution it is possible to observe that the resonances from hydroxyl protons engaged in inter-residue hydrogen bonds are shifted downfield. The extent of this shift is related to the strength of the hydrogen bond.[38] Analysis of this spectrum can assist in identifying the inter-residue hydrogen bonds.[38,39] Analysis of ^{13}C–H coupling constants also offers promise for the estimation of these dihedral angles.[40]

The relationship between monochromatic optical rotation and dihedral angles at the glycosidic linkage (see above) is also applicable to fluctuating disordered conformations, but here the result corresponds to a weighted average over all species in equilibrium. For some linkages, such as in cellobiose and its oligomers and lactose, the evidence suggests that the residues in solution fluctuate in the neighbourhood of the conformation that is captured in the crystal.[33,41] For maltose and its oligomers and polymers, however, the observed optical rotation is explained in this way only for dimethyl sulphoxide solutions, but not for those in water. The leads to the prediction[33,41] that a separate, distinct energy minimum must be populated to some extent in aqueous solution — which has very recently been supported by the discovery of precisely this form in the crystal structure of a glycoside.[42] Another prediction, that cyclohexa-amylose exists in symmetrical form with six-fold rotation symmetry in dimethyl sulphoxide or when complexed with a hydrophobic guest molecule in aqueous solution, but not when in uncomplexed form in aqueous solution, has received some confirmation from crystal structures determined subsequently.[43]

26.4.5 EXAMPLES OF ORDERED POLYSACCHARIDE SHAPES IN SOLUTIONS AND GELS

Many of the simpler periodic polysaccharides are insoluble in their ordered states — examples are cellulose and starch. This happens firstly because the adoption of an ordered conformation generally facilitates the packing of chains with favourable non-bonded interactions between them to minimize the overall potential energy, and secondly because the ordering of the chain means a stiffening and loss of conformational entropy which is an important part of the driving force to solution. Therefore chain aggregation and precipitation are favoured even when polymer–polymer contacts are only marginally more favourable than polymer–solvent contacts and even when the packing is disordered and glass-like rather than ordered and crystalline. Ordered conformations for periodic polysaccharides therefore persist into solution only when additional factors are favourable. For example:

(i) Polysaccharide polyelectrolytes may be soluble in ordered conformations because the backbone charge together with the tendency for counterions to mix with solvent favour solution more than aggregation.
(ii) Chains that are awkward to pack — such as the five-fold xanthan conformation with its knobbly branches and some polyelectrolyte character — may be more stable in solution or in a weakly aggregated state rather than totally separated from solution.
(iii) Chains may have ordered conformations and hence have aggregated states that are only weakly cooperative, so that equilibrium always exists with the soluble random coil.

A further biological device by which polysaccharide chains are maintained in an ordered conformation under conditions of high hydration is seen in polysaccharides of the interrupted periodic type (see Section 26.4.3.1). In these structures, blocks of periodic sequence having a propensity for conformational ordering are interrupted by modified sequences which are conformationally disordered and therefore soluble. The tendency for separation of ordered segments from solution, either because of their facility to pack or because of the entropic factors mentioned above, is opposed by the tendency of disordered segments to remain in contact with solvent. If the ordered state is two or more stranded, this results in a cross-linked network in which there are also topological constraints on the aggregation of ordered domains. These networks represent the 'gel state' which is an important and natural state for polysaccharides. Some types of conformational ordering that have been shown to be responsible for the cross-linking in these structures are shown in Figure 1, with the corresponding covalent sequences as well as the sequences that terminate ordered conformations. For further details of the ordered conformations of polysaccharides that have been characterized in solutions and gels, reference may be made to recent reviews.[3,5]

References

1. D. A. Rees, *M.T.P. Internat. Rev. Sci., Biochem. Series One*, 1975, **5**, 1.
2. D. A. Brant, *Quart. Rev. Biophys.*, 1976, **9**, 527.

3. D. A. Rees and E. J. Welsh, *Angew. Chem. Internat. Edn.*, 1977, **16,** 214.
4. D. A. Rees, *M.T.P. Internat. Rev. Sci., Org. Chem. Series One*, 1973, **7,** 251.
5. D. A. Rees, 'Polysaccharide Shapes', Chapman and Hall, London, 1977.
6. S. Arnott, *Trans. Amer. Crystallogr. Assoc.*, 1973, **9,** 31.
7. K. H. Gardner and J. Blackwell, *Biopolymers*, 1974, **13,** 1975.
8. F. J. Kolpak and J. Blackwell, *Macromolecules*, 1976, **9,** 273.
9. V. G. Murphy, B. Zaslow, and A. D. French, *Biopolymers*, 1975, **14,** 1487.
10. H.-C. H. Wu and A. Sarko, *Carbohydrate Res.*, 1978, **61,** 7; *ibid.*, p. 27.
11. K. Kainuma and D. French, *Biopolymers*, 1972, **11,** 2241.
12. S. Arnott and W. T. Winter, *Fed. Proc.*, 1977, **36,** 73; E. D. T. Atkins, D. H. Isaac, I. A. Nieduszynski, C. F. Phelps, and J. K. Sheehan, *Polymer*, 1974, **15,** 263.
13. J. K. Sheehan, K. H. Gardner, and E. D. T. Atkins, *J. Mol. Biol.*, 1977, **117,** 113.
14. W. T. Winter, P. J. C. Smith, and S. Arnott, *J. Mol. Biol.*, 1975, **99,** 219.
15. R. Moorhouse, W. T. Winter, S. Arnott, and M. E. Bayer, *J. Mol. Biol.*, 1977, **109,** 373.
16. R. Huber, J. Deisenhofer, P. M. Colman, M. Matsushima, and W. Palm, *Nature*, 1976, **264,** 415.
17. A. S. Perlin, B. Casu, G. R. Sanderson, and J. Tse, *Carbohydrate Res.*, 1972, **21,** 123.
18. D. A. Rees and W. E. Scott, *J. Chem. Soc.* (*B*), 1971, 469.
19. E. R. Morris, D. A. Rees, G. Young, M. D. Walkinshaw, and A. Darke, *J. Mol. Biol.*, 1977, **110,** 1.
20. S. G. Whittington and R. M. Glover, *Macromolecules*, 1972, **5,** 55.
21. E. R. Morris, D. A. Rees, E. J. Welsh, L. G. Dunfield, and S. G. Whittington, *J.C.S. Perkin II*, 1978, in the press.
22. D. M. Brunette and J. E. Till, *J. Membrane Biol.*, 1971, **5,** 215.
23. O. Smidsrød and A. Haug, *Biopolymers*, 1971, **10,** 1227.
24. C. Tanford, 'Physical Chemistry of Macromolecules', Wiley, New York, 1961.
25. H. Morawetz, 'Macromolecules in Solution', 2nd edn., Wiley, New York, 1975.
26. S. Ablett, E. R. Morris, D. A. Rees and E. J. Welsh, to be published.
27. E. G. Finer, R. Henry, R. B. Leslie, and R. N. Robertson, *Biochim. Biophys. Acta*, 1975, **380,** 320.
28. R. Kohn and O. Luknar, *Coll. Czech. Chem. Comm.*, 1977, **42,** 731 and earlier papers.
29. T. A. Bryce, D. A. Rees, D. S. Reid, and A. H. Clark, in preparation.
30. J. Boyd, E. R. Morris, D. A. Rees, D. Thom, and J. R. Turvey, in preparation.
31. E. R. Morris, D. A. Rees, and G. Young, in preparation.
32. G. T. Grant, E. R. Morris, D. A. Rees, P. J. C. Smith, and D. Thom, *F.E.B.S. Letters*, 1973, **32,** 195.
33. D. A. Rees, *J. Chem. Soc.* (*B*), 1970, 877.
34. D. A. Rees, W. E. Scott, and F. B. Williamson, *Nature*, 1970, **227,** 390.
35. I. C. M. Dea, A. A. McKinnon, and D. A. Rees, *J. Mol. Biol.*, 1972, **68,** 153.
36. I. C. M. Dea, D. A. Rees, R. J. Beveridge, and G. N. Richards, *Carbohydrate Res.*, 1973, **29,** 363.
37. D. A. Rees, unpublished work.
38. B. Casu, M. Reggiani, G. G. Gallo, and A. Vigevani, *Tetrahedron*, 1966, **22,** 3061.
39. M. St.-Jacques, P. R. Sundararajan, K. J. Taylor, and R. H. Marchessault, *J. Amer. Chem. Soc.*, 1976, **98,** 4386.
40. R. U. Lemieux and S. Koto, *Tetrahedron*, 1974, **30,** 1933.
41. D. A. Rees and D. Thom, *J.C.S. Perkin II*, 1977, 191.
42. J. Tanaka, A. Tanaka, T. Ashida, and M. Kakudo, *Acta Cryst.*, 1976, **32B,** 155.
43. P. C. Manor and W. Saenger, *J. Amer. Chem. Soc.*, 1974, **96,** 3630.

PART 27

SYNTHESIS OF ORGANIC MACROMOLECULES AND THEIR USES IN ORGANIC CHEMISTRY

27

Synthesis of Organic Macromolecules and their Uses in Organic Chemistry

P. HODGE
University of Lancaster

This chapter is concerned with synthetic organic macromolecules and, in view of their close relationship to these substances, with natural rubber and gutta-percha. The chemistry of the other major natural polymers, *viz.* nucleic acids, proteins, and polysaccharides, are considered elsewhere in this volume. Many articles on synthetic organic macromolecules emphasize their uses as materials. In the present chapter, these aspects are only of relatively minor concern. Organic macromolecules are considered simply as another type of organic compound and the chapter is concerned with their synthesis, reactions, and applications in organic chemistry. The last two areas, in particular, are of considerable and increasing interest. It is hoped that the discussion on applications might encourage organic chemists to take more interest in this area. It is one to which they can contribute a great deal, but it tends to have become the preserve of physical chemists and materials scientists.

27.1 SYNTHESIS OF ORGANIC MACROMOLECULES

The account of the synthesis of organic macromolecules given here is necessarily very brief. The topic has been considered extensively elsewhere and the references given in this section of the chapter are mainly to books and reviews. For more detailed discussion of the theoretical aspects, the books by Odian[1] and by Jenkins and Ledwith[2] are particularly useful. For practical details, the books by Braun *et al.*[3] and by Sorenson and Campbell[4] are recommended. The 'Encyclopaedia of Polymer Science and Technology' gives much information on all aspects of the subject.[5]

Polymerization processes can be grouped on a mechanistic basis into two main types: chain-reaction polymerizations and step-reaction polymerizations. The range of macromolecules available may be extended by chemically modifying the polymerization products.

27.1.1 Chain-reaction polymerization

Chain-reaction polymerization, also known as addition polymerization, involves joining together unsaturated monomers to give macromolecules. In many instances the monomer is a vinyl compound and the process is known as vinyl polymerization.[6] Polymerization may be initiated in various ways and these are considered separately in the following sections.

(i) *Free-radical chain-reaction polymerization*[7]

In this case an initiator is used which generates free radicals. Addition of a free radical to a molecule of the unsaturated monomer generates another free radical. This may then add to another molecule to generate a larger radical, and so on. The chain continues to grow until the process is terminated by destruction of the radical. This usually occurs by two radicals combining or disproportionating. The reactions are outlined in Scheme 1 using $CH_2{=}CHX$ as the monomer. Chain-reaction polymerizations generally have the following features which differ from those of step-reaction polymerizations: (i) growth occurs by the rapid addition of monomer to a small number of active centres; (ii) the rate of polymerization rises very quickly to a maximum and remains more or less unchanged until the initiator is consumed; (iii) monomer concentration decreases steadily throughout the reaction; and (iv) high polymer is present even at low conversion.

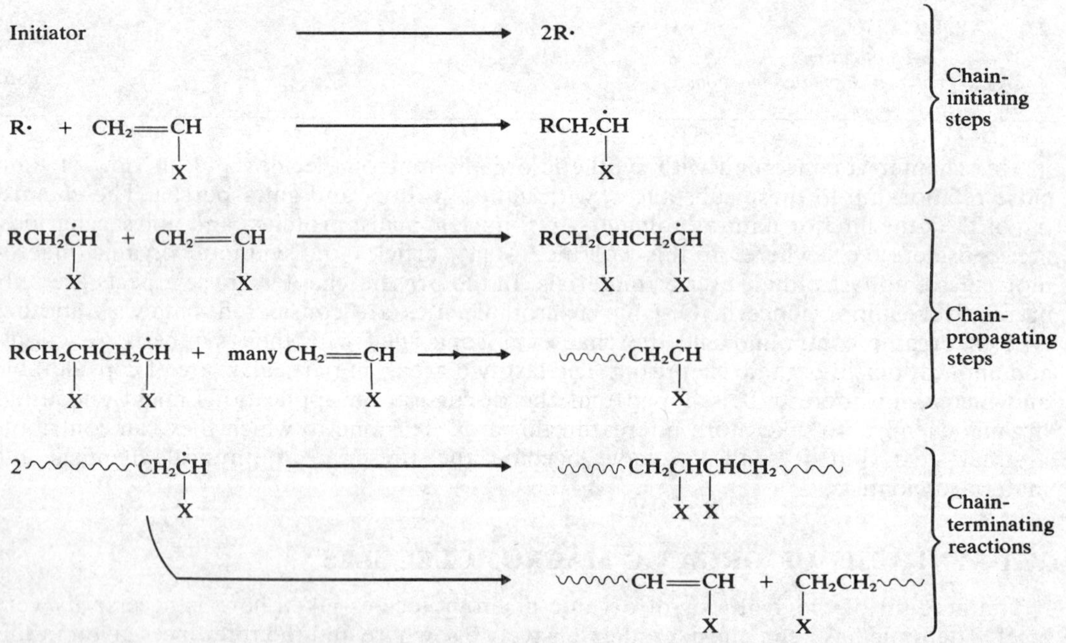

SCHEME 1

Many monosubstituted and 1,1-disubstituted derivatives of ethylene can be polymerized using free-radical chain-reactions. For example, vinyl chloride, vinyl acetate, styrene, 1,1-dichloroethylene, buta-1,3-diene, acrylonitrile, acrylamide, methyl acrylate, and methyl methacrylate. The substitutuents stabilize the radicals formed in the propagation reactions, such as reaction (1). This ensures that all, or essentially all, the monomer units are incorporated into the growing chain in this manner rather than by reactions such as reaction (2). For steric reasons, 1,2-di-, tri-, and tetra-substituted ethylenes are generally only polymerized with difficulty, if at all. Notable exceptions are tetrafluoroethylene and 1,2-disubstituted olefins, where the substituents form part of a strained ring system.

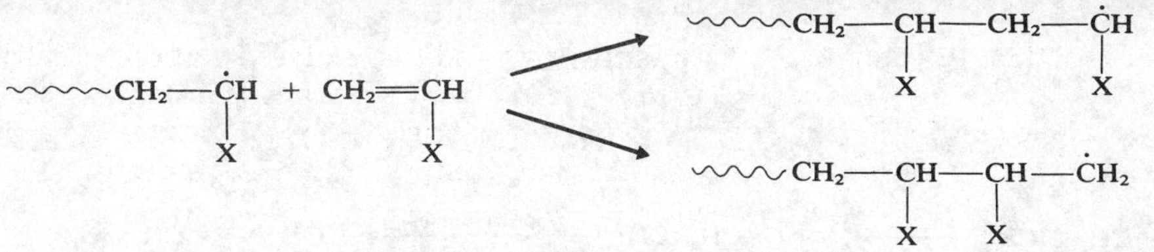

(1)

(2)

Vinyl polymerization is usually a strongly exothermic process and this generally precludes the possibility of polymerizing the olefins in bulk. The problem of heat dissipation may be overcome by employing solution, suspension, or emulsion procedures. Many initiators have been used, perhaps the most common being benzoyl peroxide and azobisisobutyronitrile (AZBN). On heating these decompose as shown in equations (3) and (4), respectively. When aqueous emulsion polymerizations are carried out, water-soluble initiators such as potassium persulphate or a wide variety of redox systems such as hydrogen peroxide and ferrous ion may be used.

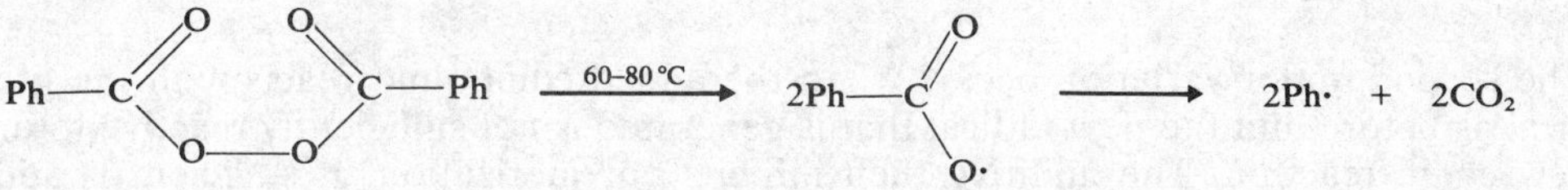

(3)

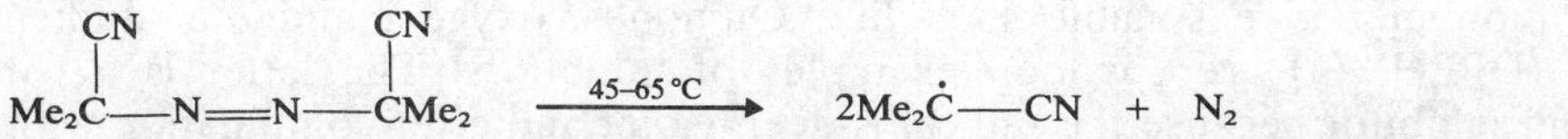

(4)

The polymerization process, whilst generally being exothermic, involves a substantial decrease in entropy. The free energy of polymerization will, therefore, only be negative, and polymerization will only be possible below a certain temperature. This is called the ceiling temperature, T_c. The precise value depends on the reaction conditions (solvent, monomer concentrations, *etc.*). Knowledge of the ceiling temperature becomes particularly important when the reaction is not strongly exothermic because T_c may well then be below 100 °C. The familiar example is α-methylstyrene, which at a concentration of 0.76 mol l^{-1} in tetrahydrofuran has a T_c of 0 °C and in bulk a T_c of 61 °C.

Free radicals are very reactive species and the presence in the reaction mixture of small amounts of chemicals other than the initiator and monomer can drastically modify the polymerization process. Very pure monomers must be used in order to obtain high molecular weight polymers. There are two major variations to Scheme 1 which may be caused by additives. An example of the first of these occurs in the polymerization of styrene in the presence of small amounts of carbon tetrachloride. The polymerization proceeds at the same rate as before but the polystyrene produced has a lower average molecular weight and contains traces of chlorine. This is due to chain transfer: termination of the original chain reaction being followed by the initiation of another (see Scheme 2). The number of growing chains and, therefore, the rate of polymerization is unaltered, but the number of chain-propagation steps before termination is reduced. A particularly important example of chain transfer is that which occurs when the macromolecules themselves act as a transfer agent. This leads to branching (Scheme 3a) and the branches may be very long. Short branches will be formed if the growing radical 'bites back' into its own chain (Scheme 3b). Common chain-transfer agents are carbon tetrachloride, toluene, and thiols.

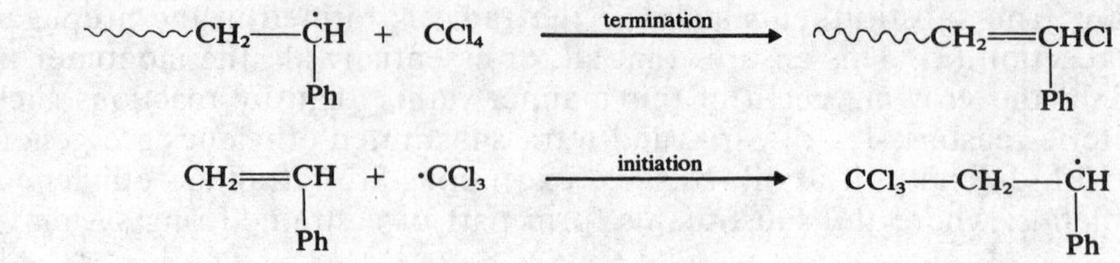

SCHEME 2

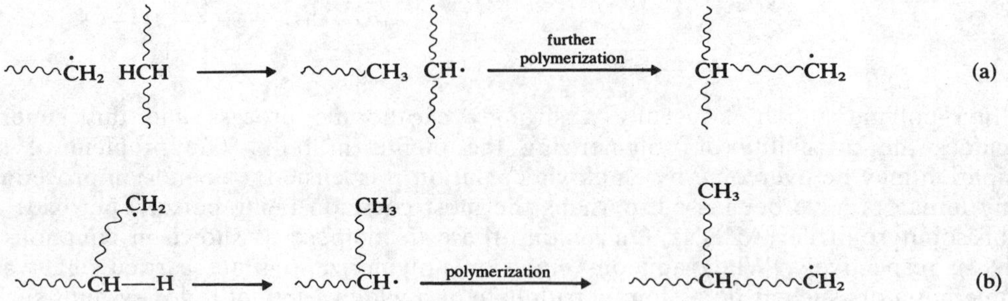

SCHEME 3

The second major variation occurs when the added compound reacts with the growing radical as before, but the new radical that is generated is not sufficiently reactive to initiate a new chain reaction. The additive then inhibits polymerization. If sufficient is added it will prevent polymerization. Inhibitors are often added to reactive monomers in order to give the monomer a reasonable shelf life. Quinones, oxygen, and α,α'-diphenylpicrylhydrazyl (DPPH) (**1**) are common examples of inhibitors. The latter is a particularly convenient inhibitor because the radical is deep violet and after combination with another radical the product is yellow or colourless.

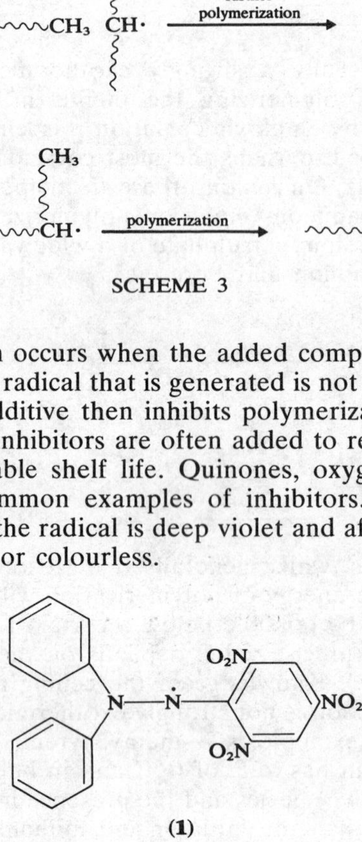

(**1**)

If instead of using just one type of monomer in a polymerization (homopolymerization) a mixture of monomers is used, copolymers may be obtained (copolymerization), each macromolecule containing more than one kind of monomer unit.[8] In the simplest case of a copolymerization involving just two monomers (M_1 and M_2), the reaction can proceed *via* either of two types of growing radical: those ending in an M_1 unit and those ending in an M_2 unit. If each type shows the same relative reactivity to the two monomers, the only factors affecting the probability of a given monomer being incorporated into the chain are its relative concentration (which can be controlled by adjusting the feed ratio) and its relative reactivity with respect to the other monomer. In this situation a random copolymer is formed:

$$M_1M_2M_2M_1M_2M_2M_1M_1M_2M_1M_1M_2 \qquad \text{Random copolymer}$$

An example of such a copolymerization is that between styrene and buta-1,3-diene. If, however, a growing radical ending in one monomer unit reacts only with the opposite monomer, an alternating copolymer will result:

$$M_1M_2M_1M_2M_1M_2M_1M_2M_1M_2M_1M_2 \qquad \text{Alternating copolymer}$$

A copolymerization of this type is that between stilbene and maleic anhydride. The two examples chosen are extreme cases. Most copolymerizations show some tendency to alternation, but it is not complete. The tendency to alternate has been attributed to polar factors. Essentially, it occurs when one olefin has electron-releasing substituents and the other has electron-attracting substituents. One very important feature of copolymerization is that monomers such as stilbene and maleic anhydride which homopolymerize with great difficulty, or not at all, often copolymerize readily.

(*ii*) *Cationic chain-reaction polymerization*[9]

Cationic chain-reaction polymerizations generally proceed by a mechanism that is analogous to that of the free-radical chain-reaction polymerizations discussed above except that the reactive chain-carrying species are carbonium ions or other positively charged entities. Depending on the nature of the growing cation and its counter ion, the solvating and dissociating ability of the medium employed, and the temperature, both free cations and various ion-pair species may contribute to the reaction.

A wide variety of initiators have been used. These include sulphuric acid, aluminium trichloride, boron trifluoride plus a trace of water, trialkyloxonium salts, and combinations which react to furnish carbonium ions, such as acid or alkyl chlorides plus a Lewis acid.

Polymerization occurs most readily if the monomer reacts to give a carbonium ion which is stabilized. Such monomers are isobutene, vinyl ethers, styrene, α-methylstyrene, and butadiene, but not, for example, acrylamide. The widely different reactivities of the monomers makes cationic copolymerization difficult.

The usual chain-termination reactions are transfer of a proton to a monomer or combination of the carbonium ion with an anion. Compounds such as amines, ethers, and sulphides which react with carbonium ions to give more stable onium ions are, of course, inhibitors. Not all cationic polymerizations are vinyl polymerizations. For example, the polymerizations shown in reactions (5) and (6) are possible. In ring-opening polymerization, the driving force for reaction is largely the relief of ring strain.

$$HCHO \xrightarrow{BF_3} \sim\sim OCH_2OCH_2OCH_2O \sim\sim \qquad (5)$$

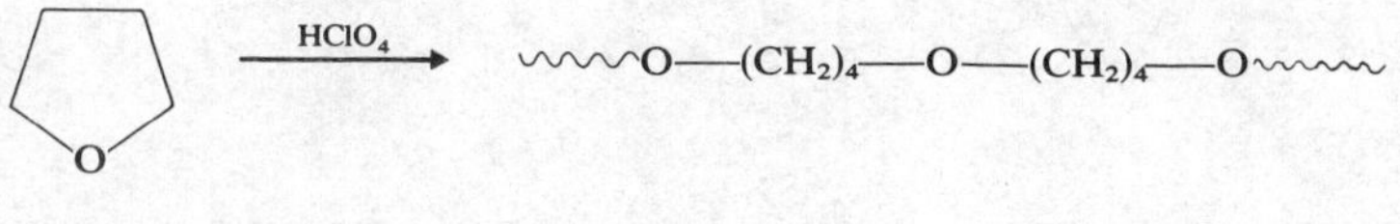

(6)

(*iii*) *Anionic chain-reaction polymerization*[10]

Anionic chain-reaction polymerizations proceed by a mechanism analogous to the free-radical initiated polymerizations discussed above, but in this case the chain-carrying species are anions. There is, however, a very important difference. This is that if the monomer and initiator are pure and the solvent is inert, there is usually no effective termination reaction. Hence, once the polymerization is initiated, it continues until all the monomer is consumed. The mixture then contains a 'living' polymer. If more monomer is added, polymerization continues and the macromolecules increase in size. This provides an opportunity to make block copolymers:[11,12]

$$M_1M_1M_1M_1M_1M_1M_2M_2M_2M_2M_2M_2M_2M_1M_1 \qquad \text{Block copolymer}$$

This is achieved by polymerizing one monomer, say styrene, and then when all the styrene is consumed the 'living' polymer is reacted with a second monomer, say buta-1,3-diene.

Anionic polymerizations are initiated by strong bases such as the alkali metals, potassium amide, or n-butyl-lithium. They occur most readily with monomers that react to give stabilized anions, *i.e.* monomers such as vinyl chloride, styrene, buta-1,3-diene, acrylonitrile, and methyl methacrylate. When polymerization is complete, the 'living' polymer is quenched using an acid or any other compound that will react with the carbanions, such as epoxides or alkyl halides. One interesting application of this is to quench 'living' polystyrene by allowing the anionic centres to react with the ester groupings of poly(methyl methacrylate). This produces

$$M_1M_1M_1M_1M_1M_1M_1M_1M_1M_1M_1M_1M_1M_1$$

$$\begin{array}{cccc} M_2 & M_2 & M_2 & M_2 \\ M_2 & M_2 & M_2 & M_2 \\ M_2 & M_2 & M_2 & M_2 \\ M_2 & M_2 & M_2 & M_2 \\ M_2 & & M_2 & M_2 \end{array}$$

M_1 = methyl methacrylate units

M_2 = polystyrene units

which is termed a comb polymer. Another way to produce this type of copolymer is to react a copolymer of *p*-chlorostyrene and styrene with sodium and to use the organometallic derivative obtained to serve as an initiator for the anionic polymerization of styrene. The product is also an example of a graft copolymer.[12,13]

One interesting initiator system is sodium metal and naphthalene. When used to initiate the polymerization of styrene, this system generates, *via* the naphthalene radical-anion, the styrene radical-anion which dimerizes as shown in reaction (7). Anionic polymerization then proceeds at both ends of the chain.

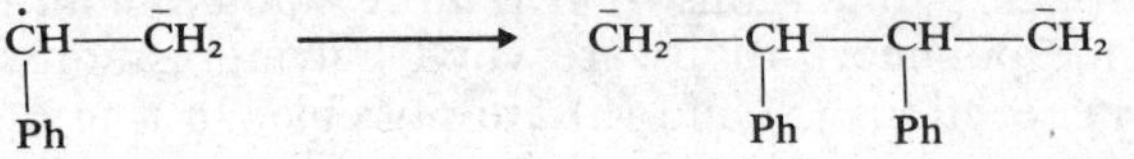

(7)

Anionic polymerization is not limited to vinyl polymerization. Ethylene oxide, for example, is converted by a small amount of base into a high molecular weight polyether as shown in Scheme 4.

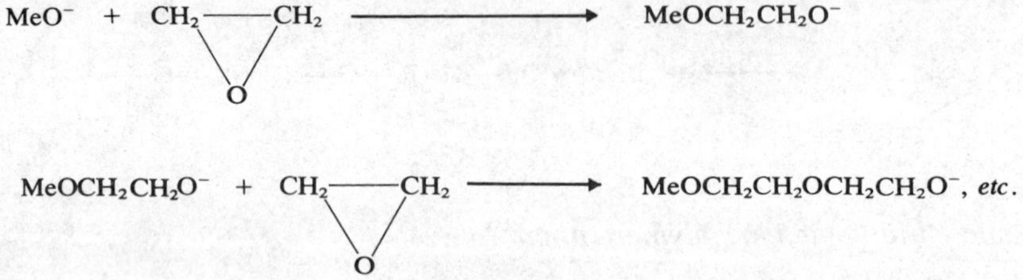

SCHEME 4

(*iv*) *Coordinative chain-reaction polymerization*[14,15]

Coordinative chain-reaction polymerization is the term given to polymerization brought about under the influence of Ziegler–Natta catalysts. These are complexes of transition metal halides with organometallic compounds, typically titanium tetrachloride–triethylaluminium and vanadium tetrachloride–diethylaluminium chloride, but there are many others. Some other catalysts such as dicobalt octacarbonyl and certain π-alkylnickel halides appear to operate in a similar manner. The identity of the reactive intermediates in these systems is the subject of debate, but the polymerizations appear to involve insertion of the vinyl monomers into transition metal–carbon bonds:

$$M—R + CH_2{=}CH_2 \longrightarrow M—CH_2CH_2R \longrightarrow M—(CH_2CH_2)_n—R$$

Important monomers which undergo coordination polymerization are ethylene, propene, buta-1,3-diene, and isoprene.

Coordination polymerization has two important advantages over free-radical polymerization. The first is that it produces macromolecules relatively free from branching. Thus, polyethylene prepared using titanium tetrachloride–aluminium triethyl as a catalyst has one side-chain per 200 or more carbon atoms in the main chain, whereas when a free-radical initiator is used at high pressure, there is one side-chain for every 30–50 carbon atoms in the main chain. The second advantage is that it permits some stereochemical control.[16] Polymerization of propene, for example, could give three different stereochemical arrangements: isotatic (**2**), with all the methyl groups on one side of an extended chain; syndiotatic (**3**), with methyl groups alternating regularly from side to side; and atactic (**4**), with methyl groups distributed at random. By a suitable choice of reaction conditions, each of these isomeric types has been made. Stereochemical control about carbon–carbon double bonds is also possible using coordination catalysts. Thus, isoprene has been polymerized to give a material virtually identical with natural rubber, *i.e. cis*-1,4-polyisoprene.

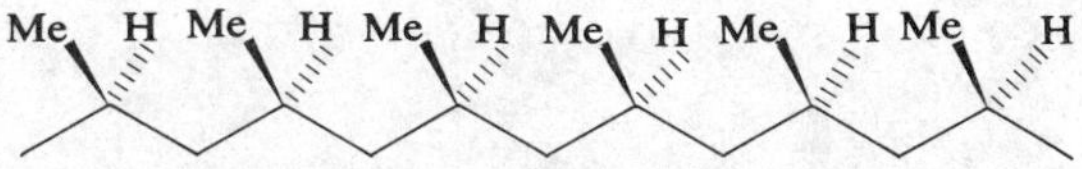

(**2**)

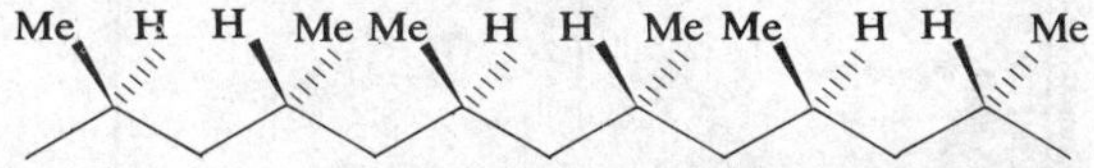

(**3**)

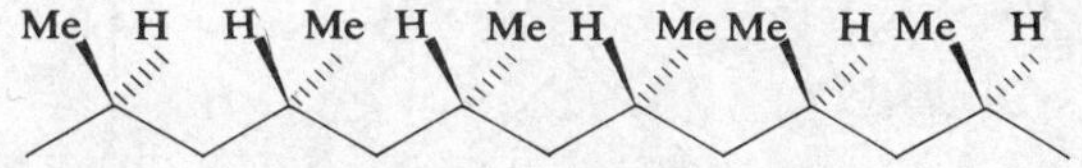

(**4**)

27.1.2 Step-reaction polymerization[17]

In this type of polymerization, difunctional monomers react together, often with the formation of ester or amide bonds, to give polymers. Important examples are shown in Scheme 5. When polymers are formed by the loss of water, methanol, or other small molecules, the polymerization process is often termed condensation polymerization and the products condensation polymers.

In step-reaction polymerization, the mechanism of each reaction step is usually the same as in simple monomeric systems. All the molecules present are capable of reacting at any time. Thus, monomer reacts initially to give oligomers and, when all the monomer is consumed, oligomers react to give higher polymers, and so on. To obtain a polymer of high molecular weight requires very high reaction yields. This means there must be no side reactions and the monomers (and the solvent if one is used) must be very pure. Step-reaction polymerizations differ in several respects from chain reaction polymerizations. Thus (i) growth occurs by the coupling of any two species present in the system; (ii) the rate of polymerization is a maximum at the start and decreases continuously during the reaction: (iii) monomer concentration decreases rapidly before any high polymer is formed; and (iv) high polymer is present only at very high conversions.

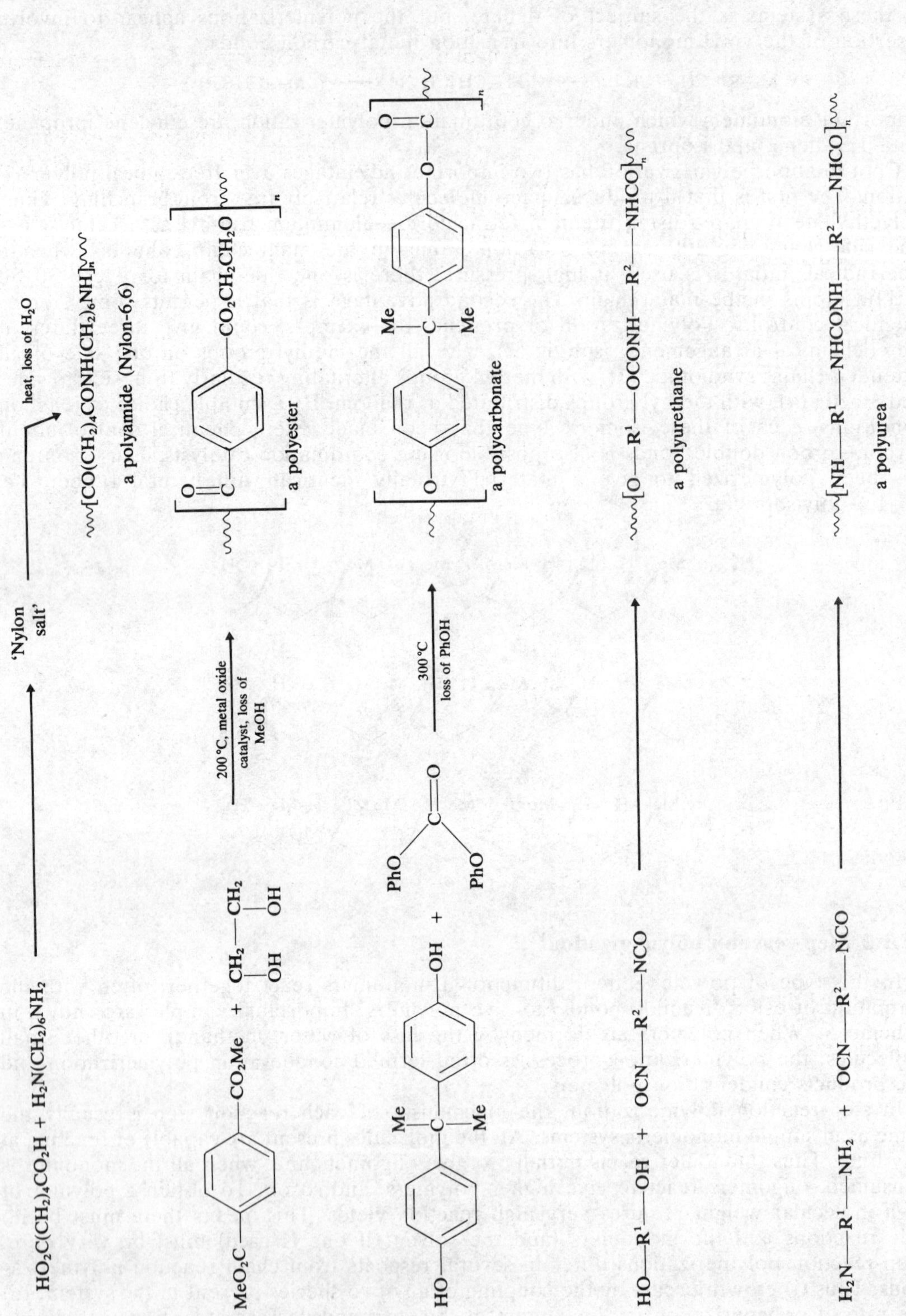
'Nylon salt'
heat, loss of H_2O
a polyamide (Nylon-6,6)
200 °C, metal oxide catalyst, loss of MeOH
a polyester
300 °C loss of PhOH
a polycarbonate
a polyurethane
a polyurea

SCHEME 5

27.1.3 Three-dimensional polymer networks

Although some polymer chemists use the term resin to describe all solid or semi-solid polymers, it will be convenient in this chapter to reserve the term to describe polymers which have a three-dimensional network.

Resins may be prepared directly using polymerization processes similar to to those discussed in the preceding sections. Perhaps the best known resin is cross-linked polystyrene. This is prepared by copolymeriztion of styrene with a small amount (usually less than 10%) of *p*-divinylbenzene (DVB). The DVB molecules become incorporated into two polymer chains; the chains, therefore, are cross-linked, and a three-dimensional network is obtained. When such copolymerizations are carried out, the reaction mixture appears much the same as it does in the preparation of the corresponding linear polymer until, at a certain point, insoluble polymer appears. This is called the gel point and the reaction mixture then consists of an insoluble polymer fraction (the gel) and a soluble polymer fraction (the sol). As the reaction proceeds further, the gel fraction increases at the expense of the sol fraction. For many applications, polystyrene resins prepared with 1–2% of DVB are used, but for some purposes, macroporous resins are more suitable. These may be prepared by copolymerizing styrene with DVB either in the presence of a substance (*e.g.* linear polymer) which can subsequently be washed away, or in the presence of a substance (*e.g.* heptane) which is a good solvent for the monomers but a poor swelling agent for the polymer.[18] The macroporous resins obtained have relatively rigid structures and large pores that remain even in the absence of solvent. Other difunctional compounds that have been used as co-monomers to bring about cross-linking include the diester (**5**) and the diamide (**6**).

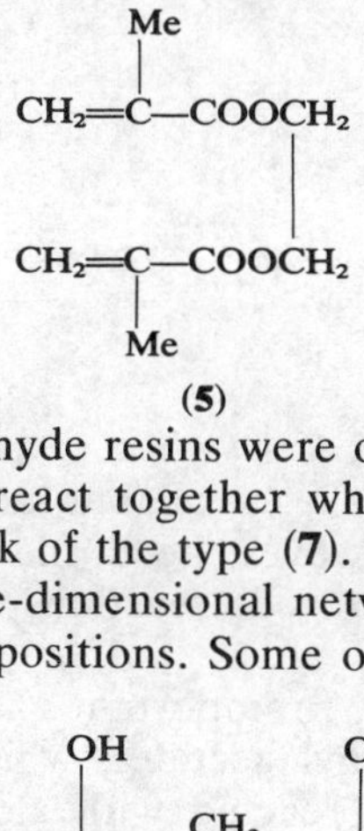

(5)

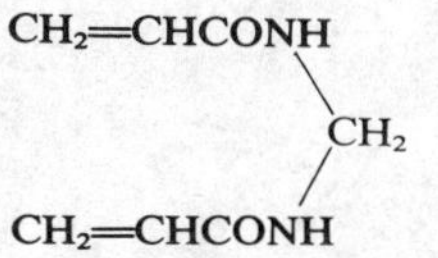

(6)

Phenol–formaldehyde resins were one of the first types of resin to be studied.[19] Phenol and formaldehyde react together when heated in the presence of base to give a three-dimensional network of the type (**7**). The reactions proceed *via o*- and *p*-hydroxymethylphenols and a three-dimensional network is obtained because each phenol molecule can react at three ring positions. Some other phenols and aldehydes react similarly.

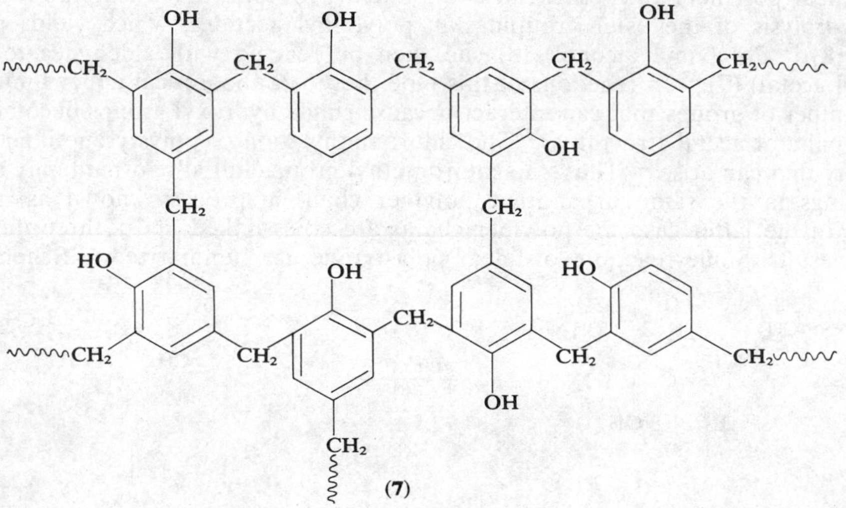

(7)

Another important group of resins are the urea–formaldehyde resins.[20] These are formed by reaction sequences of the type shown in Scheme 6.

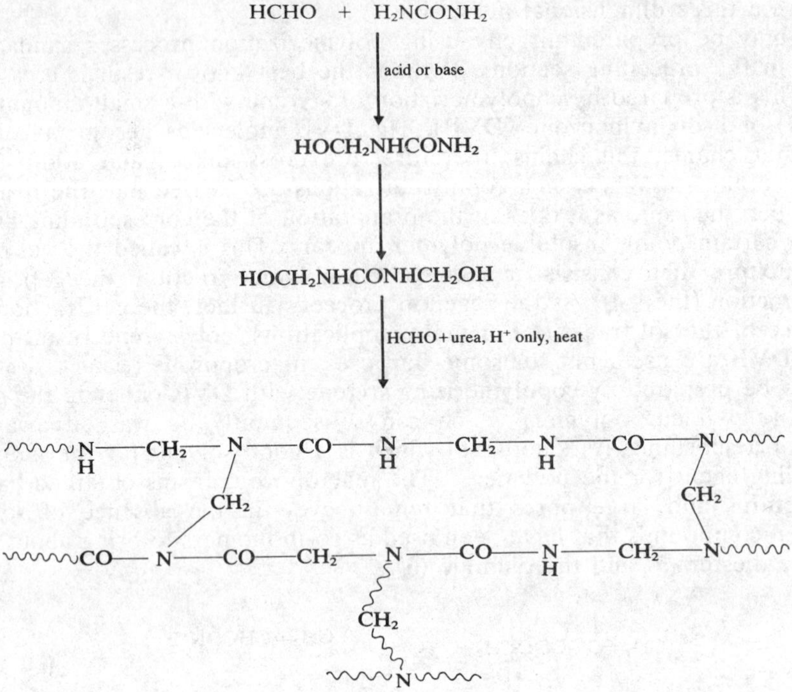

SCHEME 6

27.1.4 Modification of preformed macromolecules

Chemical modification of preformed polymers makes available many macromolecules not otherwise readily obtainable.[21,22] It should, however, be borne in mind that if the reactions do not proceed to completion, or if there are side reactions, the macromolecular chain will carry a range of functional groups.

Both linear polymers and resins can be modified. An important example of the former is the hydrolysis of the ester groupings in poly(vinyl acetate), which yields poly(vinyl alcohol) (**8**).[23] Poly(vinyl alcohol) (**8**) can then be reacted with aldehydes to give the poly(vinyl acetal) (**9**).[23] In reactions of this type there is a theoretical upper limit (86.5%) to the number of groups that can interact because single hydroxyl groups become isolated between pairs of acetal structures.[24] The chloromethylation of polystyrene illustrates one complexity that can arise.[25] Thus, the chloromethyl groups initially formed may react with phenyl rings in the same or another polymer chain, a process known as methylene bridging. In the latter case the polymer chains are cross-linked and a three-dimensional network results. Some reactions of linear polystyrene are summarized in Scheme 7.

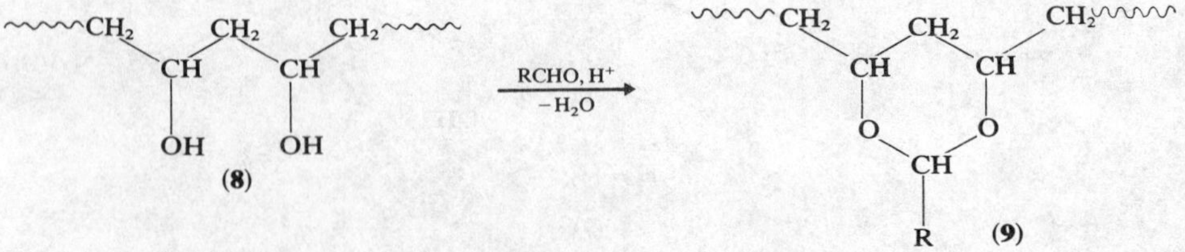

(**8**) (**9**)

In some cases cross-linking may be the aim of the reaction. Thus, cross-linked products are obtained by reacting chloromethylated linear polystyrene with ammonia or diamines.[25] The vulcanization of natural rubber with sulphur is also a cross-linking process (see Section 27.2). Cross-linking of a linear polymer (*e.g.* polyethylene) may be achieved by γ-irradiation.[26] The irradiation generates a few radicals on the polymer chain and combination of the radicals on two separate chains results in cross-linking.

The chemical modification of resins is very important, but it is not experimentally as easy as the modification of linear polymers because the resins are insoluble in all solvents. With conventional resins the reaction solvent must, therefore, be a solvent which will swell the resin so that the reagent can gain access to the interior of the network. Appreciable swelling is generally only possible if the resin is lightly cross-linked (<10%) and even then the range of solvents that can be used is limited. To reduce these problems macroporous resins are sometimes used because a greater range of reagents can penetrate such resins. Some important modifications of cross-linked polystyrenes are summarized in Scheme 7.

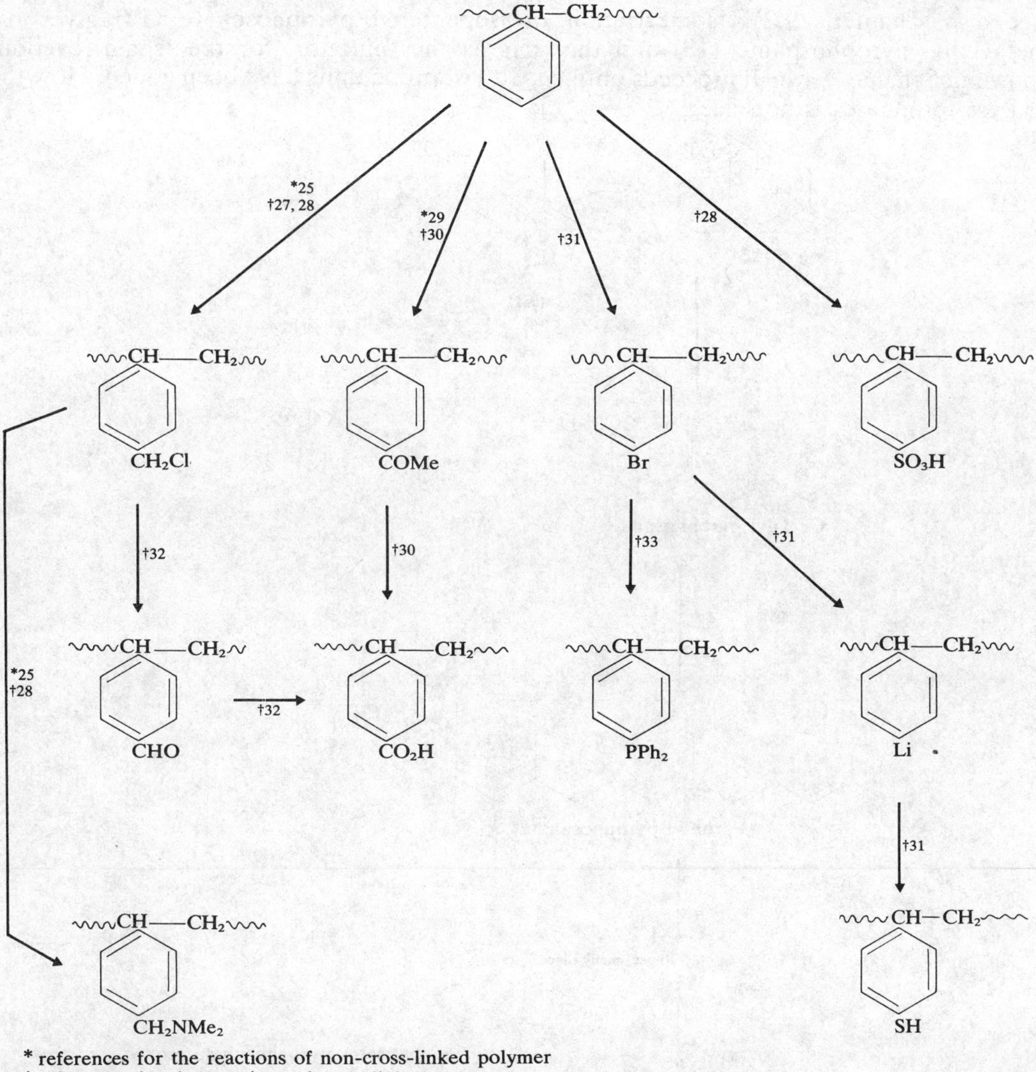

* references for the reactions of non-cross-linked polymer
† references for the reactions of cross-linked polymer

SCHEME 7

27.2 BIOLOGICAL SYNTHESIS OF RUBBER AND GUTTA-PERCHA

Numerous higher plants, and in particular the rubber tree (*Hevea brasiliensis*), produce natural rubber (**10**). A smaller number, notably *Palaquium gutta*, produce gutta-percha (**11**). Both these natural products are polyisoprenoids, the former containing exclusively

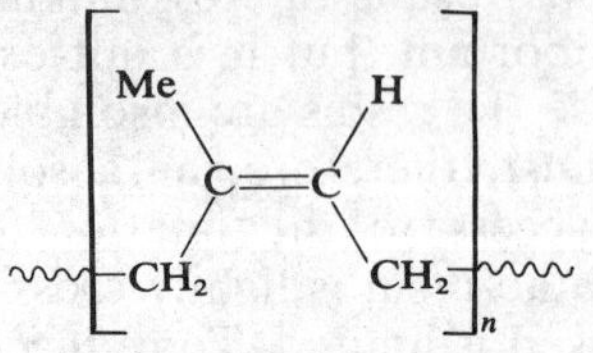

(10)

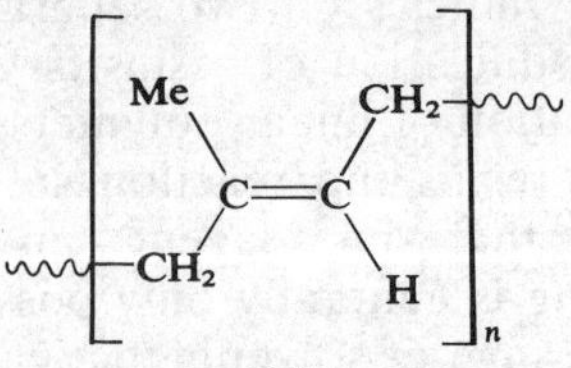

(11)

cis-double bonds and the latter exclusively *trans*-double bonds. Gutta-percha appears to be biosynthesized from isopentenyl pyrophosphate (**12**) (formed from acetate *via* mevalonate) in essentially the same way as the terpenes geraniol, farnesol, and geranylgeraniol[34] (see also Chapter 29.2). Isomerization of isopentenyl pyrophosphate (**12**) gives dimethylallyl pyrophosphate (**13**) and this acts as the 'initiator' for the 'chain-reaction' shown in Scheme 8, which proceeds until *ca.* 100 isoprene units have been added. How the process terminates is not clear.

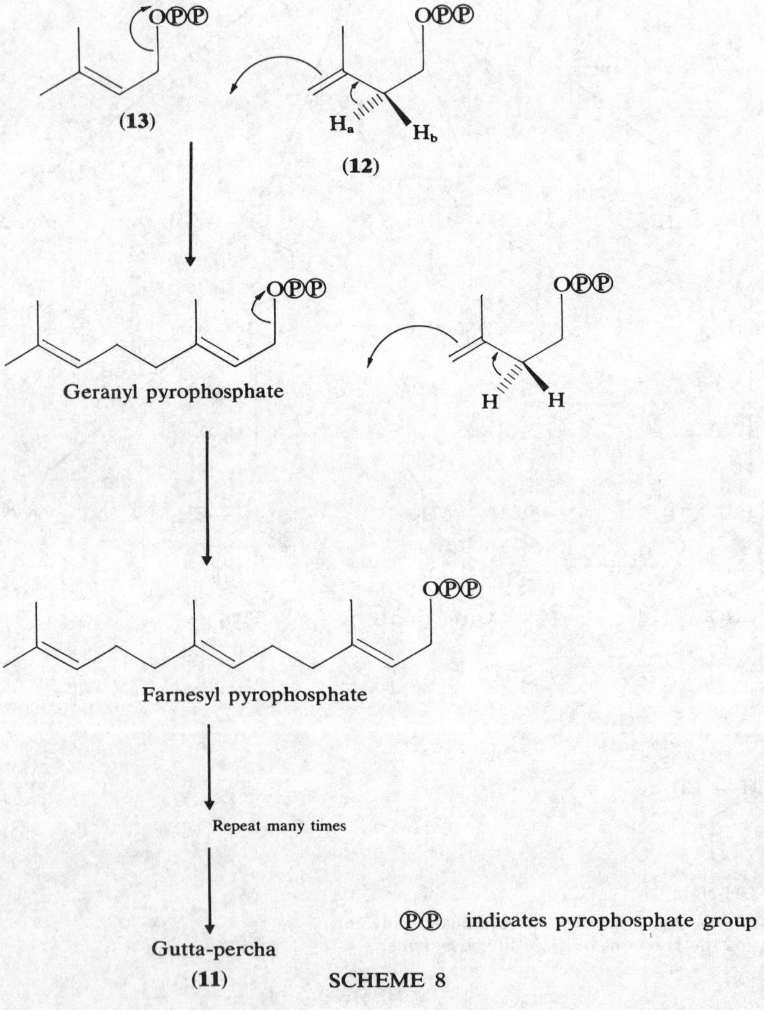

SCHEME 8

Tracer experiments indicate that natural rubber is formed by a similar polymerization process, dimethylallyl pyrophosphate (**13**) probably again being the initiator.[34] However, in this case H_b of the isopentenyl pyrophosphate (**12**) is specifically lost under enzymic control and this results in the formation of a *cis*-double bond, and the process continues until between 500 and 5000 isoprene units have been added.[35] The formation of a *cis*-double bond, which is exceptional in terpene biosynthesis, does not occur *via* isomerization of the more usual *trans*-double bond.[35]

For commercial use the properties of natural rubber are greatly improved (harder, stronger, less tacky) by vulcanization. This involves cross-linking the chains of the macromolecule and is commonly affected by heating the natural rubber with sulphur.[36] Reaction occurs at the reactive allylic positions (see Scheme 9).

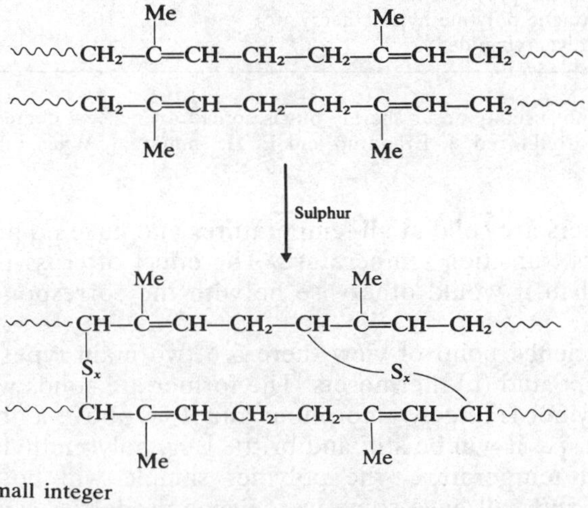

SCHEME 9

27.3 PROPERTIES

The chemical properties of macromolecules are similar to those of analogous small molecules. Numerous reactions of macromolecules are discussed in Sections 27.1.4 and 27.4.2 and this section is, therefore, mainly concerned with the physical properties of macromolecules both in the solid form and in solution.[37]

Pure non-cross-linked polymers of high molecular weight can show a range of physical properties depending on the temperature. When a sample of liquid polymer is slowly cooled, it becomes increasingly viscous and eventually viscoelastic. On cooling further, it becomes leathery and then solidifies. The temperature at which this occurs is called the melting point, T_m. Table 1 lists some values for common polymers. It is less easy to order macromolecules than it is to order smaller molecules and proper crystals are not obtained. Some regions of the sample become ordered but the remainder stays amorphous. The ordered parts are called crystallites. The degree of crystallinity is the proportion of the sample, usually on a weight basis, which is crystalline, and it can be as high as 90%. It tends to be greater with macromolecules which have (i) repeat units capable of close packing (as in polyethylene and polytetrafluoroethylene); (ii) stereoregularity (isotactic polymers are more crystalline than the corresponding atactic polymers); and (iii) strong interchain forces (as, for example, hydrogen bonding in polyamides). On cooling below the T_m, the next important physical change occurs when rotation about the inter-unit bonds in the amorphous regions of the solid polymer become frozen. This is called the glass transition temperature, T_g, and as with other transition temperatures of macromolecules, it is not sharp. Below T_g the amorphous polymer behaves as a glass; above T_g it is elastic. Table 1 lists T_g values of some common polymers.

TABLE 1
The Glass Transition Temperatures and Melting Points of Various Polymers

Polymer	T_g (°C)[a]	T_m (°C)[a]
cis-Polyisoprene	−72	28
Poly(ethylene adipate)	−63	*ca.* 50
Poly(methyl acrylate)	5	—
Poly(ethylene terephthalate)	69	*ca.* 270
Atactic polystyrene	100	240
Atactic poly(methyl methacrylate)	105	—
Polyacrylamide	195	—

[a] The transition does not usually occur sharply but is spread over a few degrees. Values taken from 'Polymer Handbook', 2nd edn., ed. J. Brandrup and E. H. Immergut, Wiley-Interscience, New York, 1975.

Cross-linked polymers are solid at all temperatures and have no melting point. They do, however, show a glass transition temperature. The effect of cross-linking is to make this temperature higher than it would otherwise be with the corresponding non-cross-linked polymer.

From a materials science point of view there are two main types of solid polymer: (a) thermoplastic polymers, and (b) thermosets. The former are solids which soften reversibly on heating. If the polymer is largely amorphous and T_g is above ambient temperature, the sample will be a glass, *i.e.* it will be stiff and brittle [*e.g.* poly(methyl methacrylate)], but if T_g is below ambient temperature, the polymer sample will be a flexible solid (*e.g.* *cis*-1,4-polyisoprene) and will have some viscoelastic properties, *i.e.* if a stress is applied briefly, the sample will deform reversibly. The latter corresponds to the reversible stretching of the polymer chains. If the stress is not immediately removed, segments of the macromolecules slowly slide past one another and the process is not then reversible. This can be prevented by lightly cross-linking the polymer chains. The classical example of such an elastomer is vulcanized natural rubber (see Scheme 9). If the polymer sample is largely crystalline, it may be suitable for making fibres. Fibres tend to have the macromolecules lined up (lining up is often assisted by the process of drawing) and held together by strong inter-chain forces (*e.g.* polyamides and polyesters). The second major type of solid polymer, the thermosets, contain non-cross-linked or lightly cross-linked chains which, on heating, irreversibly and extensively cross-link to give hard rigid solids (*e.g.* resols, which are intermediates in the formation of phenol–formaldehyde resins).[19]

Non-cross-linked macromolecules can usually be dissolved in one or more solvents. In solution a flexible macromolecule tends to be in the form of a loose coil which, as a result of bond rotation, is constantly changing shape. The tightness of the coil depends to what extent the solvent interacts with the chain.

Many measurements can be made on a solution of a polymer sample, including determination of the molecular weight. It is clear from the method of preparation that not all the molecules present in a sample will have the same molecular weight. The number-average molecular weight, $\bar{M}_n$, of a sample is the arithmetic mean of the molecular weights of the macromolecules present. That is

$$\bar{M}_n = \frac{\sum^{i} n_i M_i}{\sum^{i} n_i} \quad \text{where } n_i \text{ is the number of molecules of molecular weight } M_i$$

The value of $\bar{M}_n$ can be determined by colligative methods, osmometry being the most useful. It is also of interest to know the distribution of the molecular weights. One method of determining this is fractional precipitation followed by a molecular weight measurement on each fraction. A less tedious and more modern method is gel permeation chromatography (see Section 27.4.1(i)).

Colligative methods are not the only ones which can be used to determine molecular weights. Two other commonly used methods depend on light-scattering and viscosity measurements. The former method is based on the fact that the direction and intensity of the light scattered by a solution of a polymer is a function of the physical size and shape of the scattering particles. This method method allows the determination of the weight-average molecular weight, $\bar{M}_w$. The value of $\bar{M}_w$ is always greater than $\bar{M}_n$ and the spread of molecular weights in the distribution curve can be characterized by the ratio $\bar{M}_w/\bar{M}_n$.

$$\bar{M}_w = \frac{\sum^{i} n_i M_i^2}{\sum^{i} n_i M_i} \qquad \text{where } n_i \text{ is the number of molecules of molecular weight } M_i$$

The narrower the molecular weight distribution the closer the ratio is to 1. Viscosity measurements of polymer solutions allow the determination of the viscosity-average molecular weight, $\bar{M}_v$. The value of $\bar{M}_v$ is intermediate between $\bar{M}_w$ and $\bar{M}_n$ and depends upon the solvent chosen for the measurement.

Having a solution of the polymer sample also enables infrared, ultraviolet and ^{1}H and ^{13}C n.m.r. spectra to be measured. These naturally give much information about the composition of the sample.[38] The ^{1}H n.m.r. spectra of copolymers are of particular interest because detailed analysis of the spectrum often gives valuable information about the sequence of the various monomer units.

Polymers which have three-dimensional networks (resins) are generally insoluble in all solvents. They may nevertheless interact with the solvent and swell, possibly up to as much as ten times the original volume. The extent of swelling of a given polymer depends on the solvent used and on the extent of cross-linking. The lower the latter the more the sample will swell. The insolubility of resins, whilst being an advantage for some puposes, makes it impossible to measure ultraviolet spectra of the samples and very difficult to measure ^{1}H or ^{13}C n.m.r. spectra. Infrared spectra can, however, be measured.

27.4 APPLICATIONS

There are two main areas where organic chemists have used synthetic macromolecules in the laboratory. One is chromatography, the other is polymer-supported reactions. The two areas are not un-related. Ion exchange resins, for instance, may be used in chromatography or as polymer-supported catalysts and certain materials used in affinity chromatography may also be used as polymer-supported reagents. Most applications use resins because of their insolubility in all solvents. Non-cross-linked polymers are occasionally used. These have the advantage that the reactions or processes occurring on the polymer can be studied without diffusion and the insolubility of the resin raising problems. In both cases the polymers may be prepared by modification of preformed polymers (see Section 27.1.4) or directly by copolymerization as discussed above. The most widely used synthetic macromolecules are based on cross-linked polystyrene, polyacrylates, polymethacrylates, and polyacrylamides. Various natural polysacchardies (*e.g.* agarose) or modified natural polysaccharides (*e.g.* sephadex) are also used, especially when a hydrophilic material is required.

27.4.1 Chromatography

Chromatographic separations are achieved by the distribution of the components of a mixture between a stationary and a mobile phase. The present section is concerned with forms of chromatography where synthetic resins are used as the stationary phase. Usually the mixture to be separated is applied to the top of a column of the stationary phase. The mobile phase is then passed down the column. Each component of the mixture migrates down the column. Separation is achieved because different substances have different affinities for the stationary phase and for the mobile phase and they, therefore, migrate at different rates. Basically, the various forms of chromatography only differ in the nature of the interaction between the substances to be separated and the stationary phase. There are many possibilities and four of these are discussed briefly in the following sections.

(*i*) *Gel chromatography*

Perhaps the simplest type of chromatography is that which uses porous inert stationary phases and where separation is achieved on the basis of size and shape. In gel permeation chromatography[39] the stationary phase is usually a cross-linked polystyrene resin and the mobile phase an organic solvent which, for practical reasons, must be passed down the column under high pressure. Gel filtration[40] is a very closely related technique where the stationary phase is a cross-linked polyacrylamide (or other hydrophilic material) and the mobile phase is an aqueous solvent which is passed down the column under low pressure. A more general name covering both techniques is gel chromatography.

Separation is achieved because the stationary phase has pores of various sizes. The smaller molecules can pass into more pores than the larger molecules and so their passage down the column gets delayed with respect to the latter, which are eluted first. Molecules greater than a certain size will be unable to enter any pores of a given support (the exclusion limit) and will pass straight through the column unresolved. The length of time a molecule takes to pass down the column is closely related to its molecular size and weight and if the column is calibrated using samples of known molecular weight, the technique can be used to determine molecular weights.

The most important application of gel permeation chromatography is determining the molecular weight distribution of synthetic polymer samples. From the data, the number-average molecular weight, $\bar{M}_n$, and weight-average molecular weight, $\bar{M}_w$, can be calculated.

Gel filtration is used to separate proteins and to estimate their molecular weights.

(*ii*) *Ion exchange chromatography*

In ion exchange chromatography[41,42] the stationary phase is a water-insoluble material with attached ionic groups. The counter ions associated with these groups can exchange with ions of similar charge in the surrounding medium. Different ions have different affinities for the ionic groups on the resin and so separations can be achieved.

Originally, natural inorganic ion exchangers such as zeolites were used, but for many purposes these have been displaced by synthetic organic ion exchange resins. The resins may be classified according to the nature of the ionic groups. Strongly acidic cation exchangers contain residues such as sulphonic acid groups, which will be ionized to yield anionic centres over all the normal working pH range. Weakly acidic cation exchangers contain residues, such as carboxylic acid groups, which will only yield anionic centres at pH values greater than *ca.* pH 6. Strongly basic anionic exchangers contain cationic centres such as quaternary ammonium groups which will be present over the whole pH range. Weakly basic anionic exchangers contain basic groups such as *N,N*-dialkylbenzylamine residues which will react with acids to give cationic centres. The most important supports are cross-linked polystyrene and cross-linked poly(methacrylic acid). Phenol–formaldehyde resins have also been used, but they are not so chemically inert. Table 2 lists some common ion exchange resins. A somewhat similar family of ion exchange materials prepared from modified natural polysaccharides are also available.

TABLE 2
Some Common Ion Exchange Resins

Functional groups	*Polymer matrix*	*Type of exchange resin*	pH *at which ionic form is present*
Aromatic sulphonic acid	Crosslinked polystyrene	Strongly acidic cation exchange	all pH
Aliphatic sulphonic acid	Phenolic		
Aliphatic carboxylic acid	Polyacrylic acid or polymethacrylic acid	Weakly acidic cation exchange	$>$ about pH 6
N,N-Dialkylbenzylamine	Crosslinked polystyrene	Weakly basic anion exchange	$<$ about pH 8
Benzyltrimethylammonium	Crosslinked polystyrene	Strongly basic anion exchange	all pH
N-Methylpyridinium	Crosslinked polystyrene		

The exchange process is essentially a diffusion process and the rate-determining step can be either diffusion of ions across the liquid film immediately adjacent to the resin particle, or diffusion of ions within the pores of the resin particle. The former is usually the more important with dilute solutions. The latter becomes more important at high concentrations and with large ions. The affinity of ions for a particular resin depends on many factors, including the valency and size of the ion, concentration, temperature, and solvent. With aqueous solvents at low concentrations ($<$N/10) the affinity for the resin increases with the valency of the ion and, for ions of the same valency, decreasing diameter of the hydrated ion. The precise order of affinities for a series of ions will depend greatly on the particular resin. The lists below show the order of affinities for some typical commerical products used under normal conditions:

Cation exchange

strong acid $Fe^{3+} > Cu^{2+} > Ag^{+} > NH_4^{+} > K^{+} > Na^{+} > H^{+} > Li^{+}$

weak acid $H^{+} \gg Ag^{+} > K^{+} > Na^{+} > Li^{+}; H^{+} > Fe^{2+}$

Anion exchange

strong base $NO_3^- > {}^-CN > HSO_3^- > NO_2^- > Cl^- > AcO^- > OH^- > F^-$

weak base $HNO_3 > HCl > HF > HOAc$

Ion exchange chromatography has, of course, been widely used to separate inorganic species. In organic chemistry it can be used to separate mixtures of acids or bases, a classical application being the separation of the mixture of amino-acids produced by hydrolyses of peptides and proteins.[43] Peptides, proteins, and enzymes which carry acidic and/or basic groups can also be separated by ion exchange chromatography. One interesting application of ion exchange resins involves using a strongly basic resin in the bisulphite form.[44] When mixtures of aldehydes and ketones are chromatographed on this resin, they reversibly form bisulphite complexes on the resin and this allows separation to be achieved.

Apart from these chromatograhpic uses, ion exchange resins can be used to completely replace one ion by another. For example, if a sodium salt of a carboxylic acid is passed down a column of a strongly acidic resin in its acidic form, all the sodium ions displace protons and the eluant contains only the carboxylic acid. This is a useful way of obtaining certain water-soluble acids from their salts and could be regarded as an example of the use of a polymer-supported reagent. Similarly, if a solution of a quaternary ammonium chloride is passed down a column of a strongly basic resin in the hydroxide form, the chloride ions displace the hydroxide ions. The eluant, therefore, contains just the quaternary ammonium hydroxide and this is a useful way of preparing such compounds.

For further information on the separations that can be achieved, details of elution procedures, and practical details the reader is referred to standard texts.[42] Further uses of ion exchange resins are discussed in Section 27.4.2(ii).

(*iii*) *Complexing resins*

Resins have been prepared which contain groups that can complex with certain species in solution.[45] A common difficulty with these resins is that the exchange processes tend to be slow. One of the most intensively studied is a cross-linked polystyrene containing the functional group (**14**), the precise species present depending on the pH value of the eluant. The resin has a large affinity for many divalent ions, particlarly Cu^{2+}.[46]

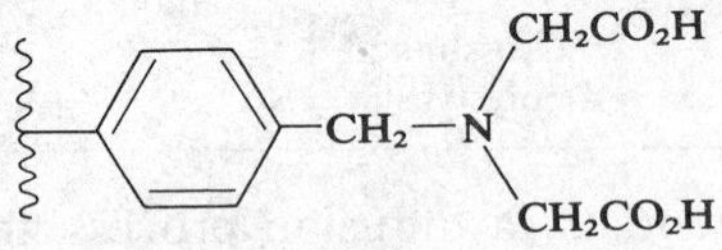

(14)

The groups attached to the stationary phase need not be ionic and a particularly interesting example of a non-ionic type is a resin recently described by Cram.[47] A chiral crown-ether complex was bound to cross-linked polystyrene to give the resin (**15**). This proved to be a suitable stationary phase for the separation of the optical isomers of amino-acid and amino-ester perchlorate salts (**16**). In most instances the (*R*)-stereoisomer was more tightly bound to the crown-ether site than the (*S*)-stereoisomer.

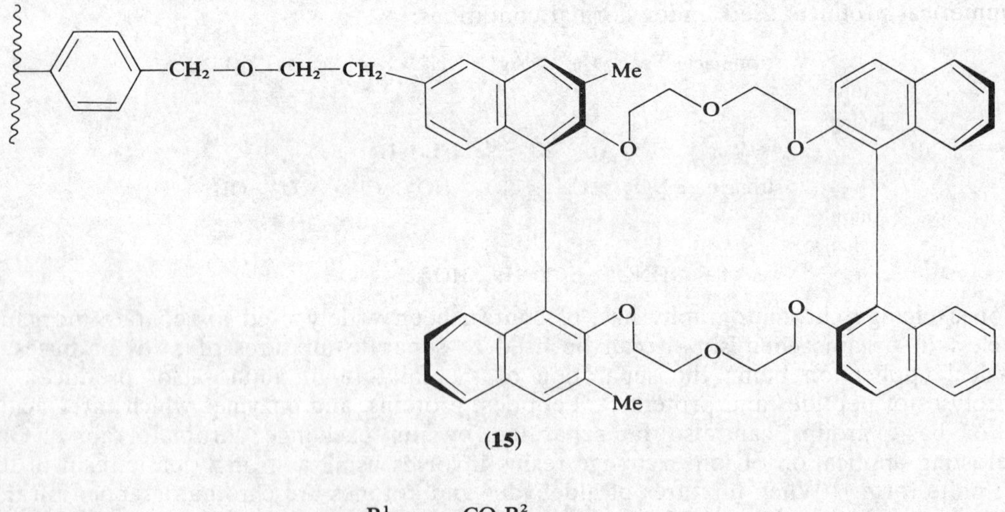

(15)

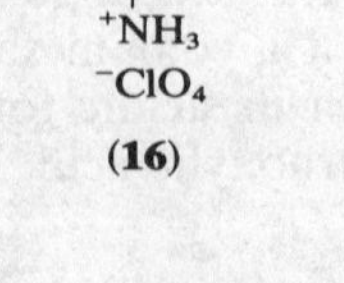

(16)

R^1 = usual natural protein amino-acid side-chains
R^2 = H or Me

(*iv*) *Affinity chromatography*

Chromatograhpic methods which achieve separations on the basis of the sizes, shapes, or overall charges of the molecules in a mixture are often not sufficiently powerful to separate mixtures of proteins. In such instances, affinity chromatography may be useful.[48] This depends on finding a substance that specifically interacts with the protein to be purified. For an enzyme this might be a competitive inhibitor of the enzyme-catalysed reaction, and for a hormone receptor the corresponding hormone. The substance is bound to a suitable insoluble hydrophilic support and this is used as the stationary phase in chromatography. The substance chosen will often itself be a large natural macromolecule and the technique used for binding it to supports are similar to some of those used in the preparation of immobilized enzymes (see Section 27.4.2(v)). The reactions in Scheme 10

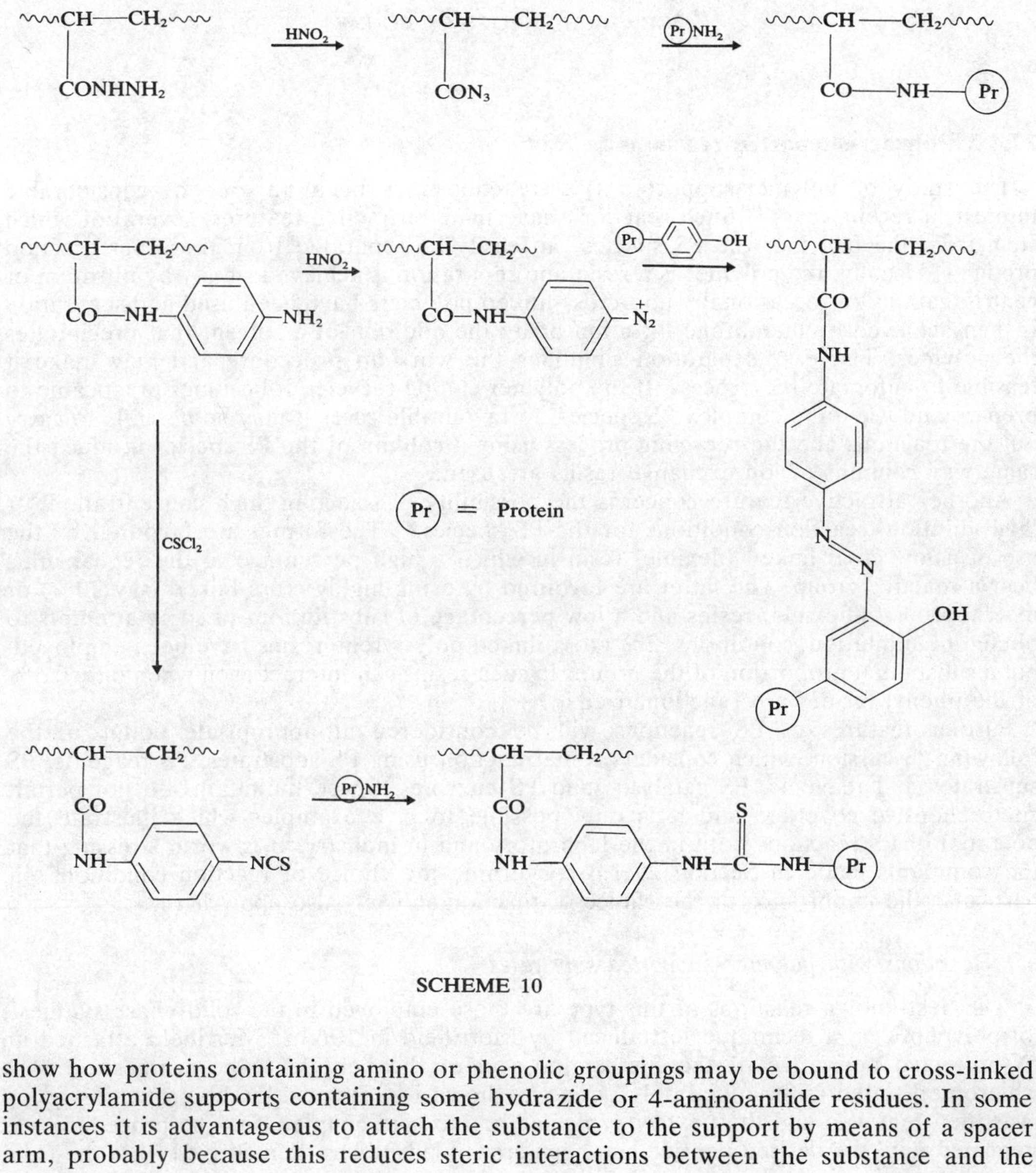

SCHEME 10

show how proteins containing amino or phenolic groupings may be bound to cross-linked polyacrylamide supports containing some hydrazide or 4-aminoanilide residues. In some instances it is advantageous to attach the substance to the support by means of a spacer arm, probably because this reduces steric interactions between the substance and the support. The following is an example:

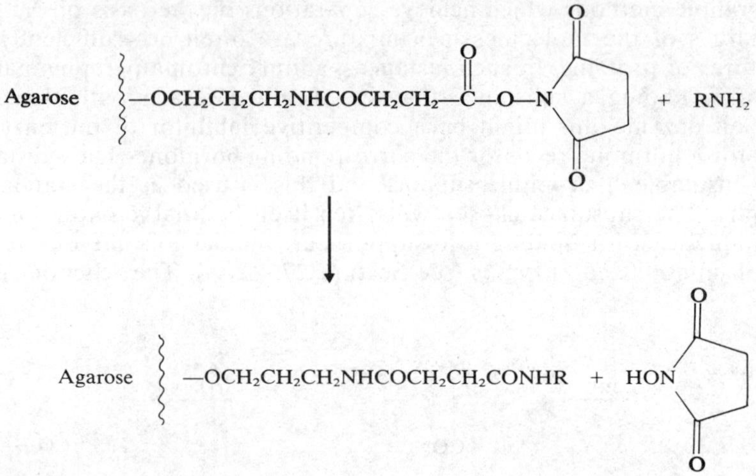

27.4.2 Polymer-supported reactions

The study of polymer-supported (PS) reactions has been an area of considerable interest in recent years.[49] Such reactions have many attractive features, several of which stem from the fact that the PS species can easily be separated from the other reaction products. Usually, the polymer is a resin and separation is achieved simply by filtration or centrifugation, but occasionally non-cross-linked polymers have been used and separation is then achieved by membrane filtration or by the addition of a solvent that precipitates the polymer. The easy separation simplifies the work-up procedure and may make it feasible to automate the process. If the polymer can be recycled it becomes practicable to prepare and use quite complex PS species. In favourable cases it may be possible to carry out the reactions and the recycling process using a column of the PS species in much the same way columns of ion exchange resins are used.

Another attractive feature concerns the possibility of selecting 'high concentration' or 'high dilution' reaction conditions for the PS species.[50] The former are favoured by the use of a low cross-linked (flexible) resin in which a high percentage of the repeat units bear a reactive group. The latter are favoured by using highly cross-linked (say 20%) or macroporous (inflexible) resins and a low percentage of substitution. In some attempts to obtain 'high dilution' conditions, 2% cross-linked polystyrene resins have been employed, but a substantial proportion of the groups in such resins can interact even when only 0.5% of the phenyl resides are functionalized.[51]

Further features of PS reactions will be considered at appropriate points in the following discussion, which considers some reaction using PS substrates, PS reagents, PS substrates and reagents, PS catalysts, and PS enzymes. Space limitations do not permit comprehensive coverage and it is only possible to give examples which illustrate the potential of PS reactions both in the laboratory and in industry. It is worth stressing that the comments made in Section 27.1.4 concerning the choice of reaction conditions, in particular the importance of the choice of reaction solvent, also apply here.

(*i*) *Reactions with polymer-supported substrates*

The best-known reactions of this type are those employed in the solid-phase synthesis of polypeptides, a technique introduced by Merrifield in 1963.[52] Merrifield attached an *N*-protected derivative of the first amino-acid of the peptide to be synthesized to a chloromethylated cross-linked polystyrene as shown in Scheme 11. The protecting group was then removed and the next *N*-protected amino-acid coupled on. The procedure was repeated until the desired peptide had been assembled. It was then cleaved from the resin

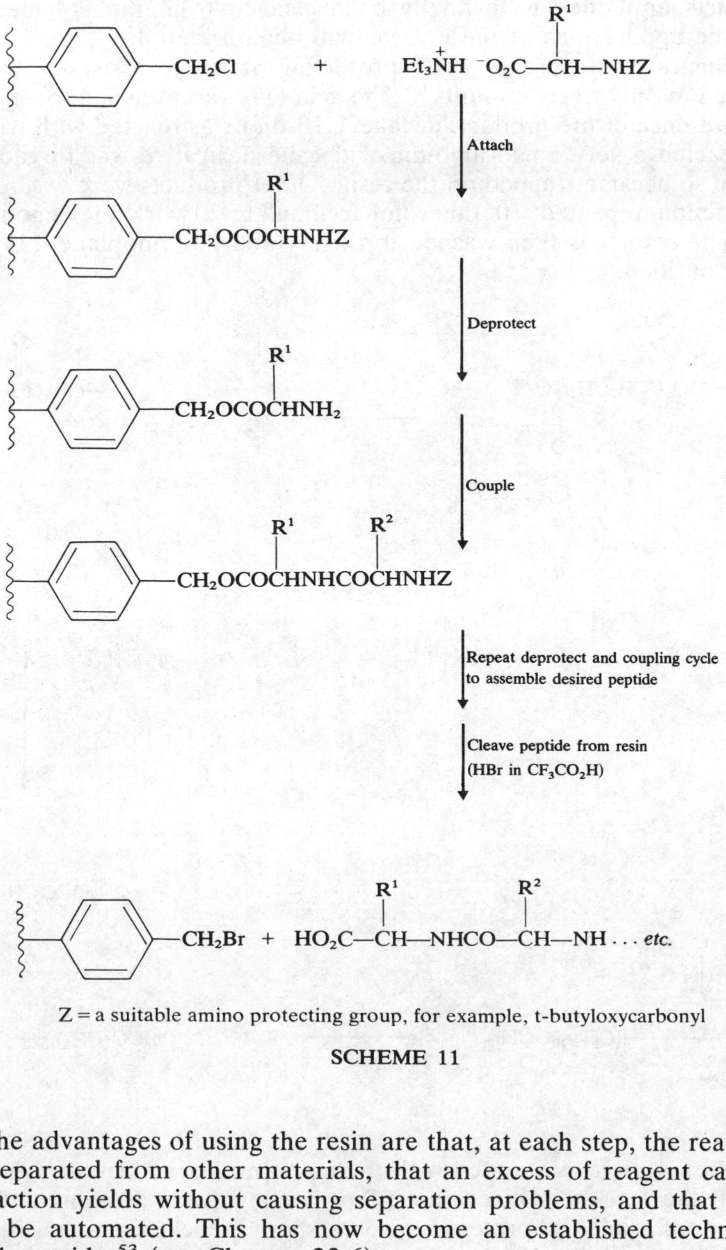

Z = a suitable amino protecting group, for example, t-butyloxycarbonyl

SCHEME 11

and purified. The advantages of using the resin are that, at each step, the reaction product can be easily separated from other materials, that an excess of reagent can be used to obtain high reaction yields without causing separation problems, and that the synthetic procedure can be automated. This has now become an established technique for the synthesis of polypeptides[53] (see Chapter 23.6).

Efforts to develop similar procedures for the synthesis of oligosaccharides[54] and oligonucleotides[55] have been less successful (see Chapter 22.4). The synthesis of such compounds, however, is inherently more difficult than peptides synthesis, partly because more protecting groups must be used and, in oligosaccharide synthesis, because coupling of the monosaccharide residues can give epimeric products. There is clearly a need in these areas for the development of more satisfactory polymer supports.

The above methods can be regarded as examples of the use of PS protecting groups and the PS peptide synthesis illustrates well the merits of using such protecting groups. PS protecting groups can be used with advantage in other areas, the main limitation being that each reaction must proceed in good yield without the formation of side products because the latter and any unreacted starting material cannot be separated from the desired products until they have finally been cleaved from the polymer support. Two examples of the applications of these areas will be considered.

The first illustrates the use of a PS protecting group to assist in the isolation of a product formed in only trace amounts.[56] The acid (**17**) was attached to chloromethyl resin (**18**). In the presence of the product, decane-1,10-diol was reacted with trityl chloride. At the time it reacted a very amall amount of decane derivative was threaded through the macrocyle and so became trapped on the resin. Other products were washed away and the tritylation reaction repeated 70 times to accumulate a workable amount of threaded macrocycle. The resin was then washed throughly and the 'hooplane' (**19**) released from the resin and purified.

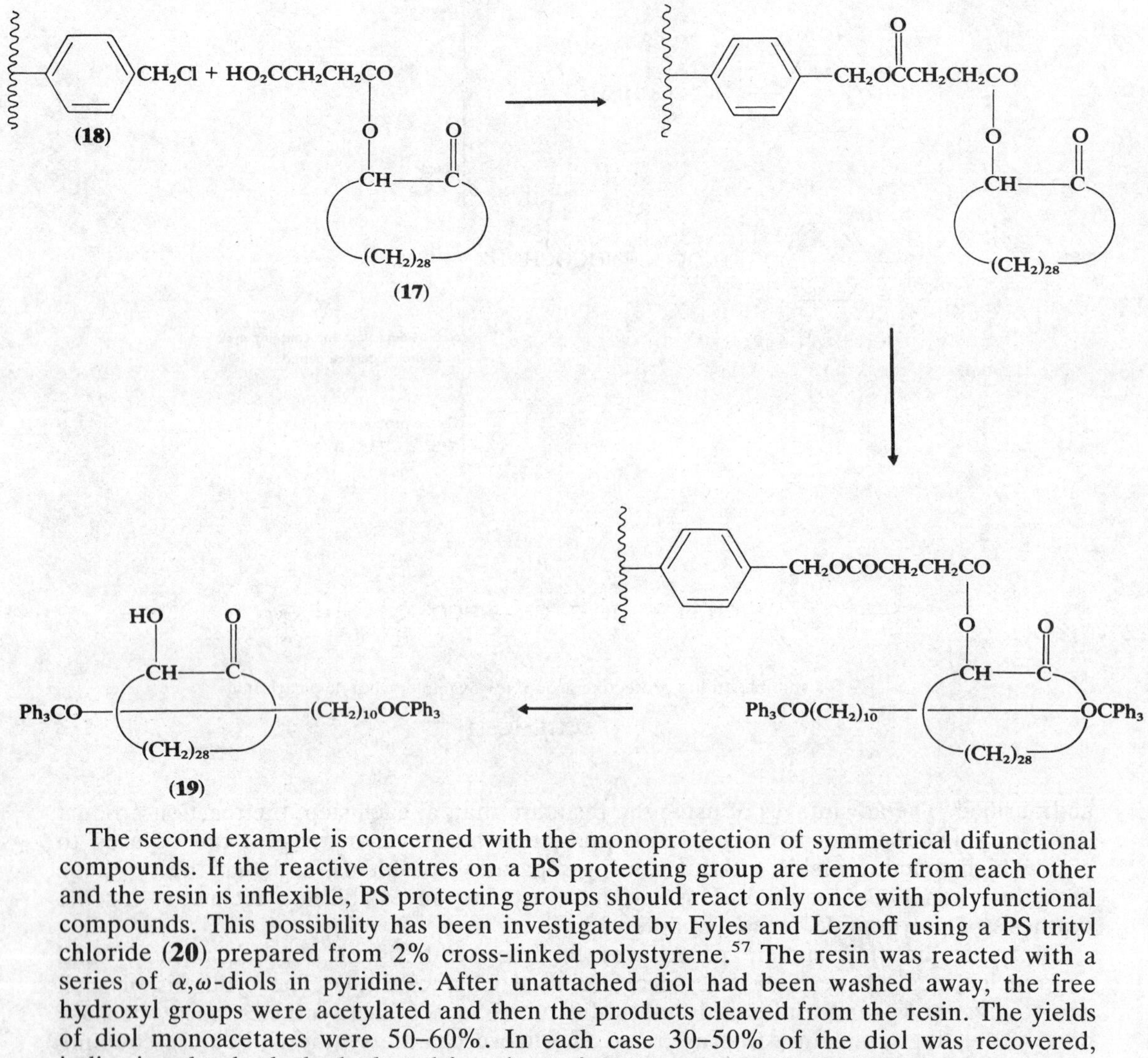

(**18**) (**17**) (**19**)

The second example is concerned with the monoprotection of symmetrical difunctional compounds. If the reactive centres on a PS protecting group are remote from each other and the resin is inflexible, PS protecting groups should react only once with polyfunctional compounds. This possibility has been investigated by Fyles and Leznoff using a PS trityl chloride (**20**) prepared from 2% cross-linked polystyrene.[57] The resin was reacted with a series of α,ω-diols in pyridine. After unattached diol had been washed away, the free hydroxyl groups were acetylated and then the products cleaved from the resin. The yields of diol monoacetates were 50–60%. In each case 30–50% of the diol was recovered, indicating that both the hydroxyl functions of these molecules had reacted with the resin. When the PS trityl chloride resin (**21**) was used, diol monoacetate was obtained in 90% yield and no diol was recovered. The latter resin was probably more rigid and the reactive groups more remote from each other than in the former resins. Leznoff and Fyles[58] have used this method to prepare the acetylenic alcohols (**22a–c**). These were subsequently converted to the insect sex attractants *cis*-dodec-7-enyl acetate, *cis*-tetradec-9-enyl acetate and *cis*-tetradec-11-enyl acetate, respectively.

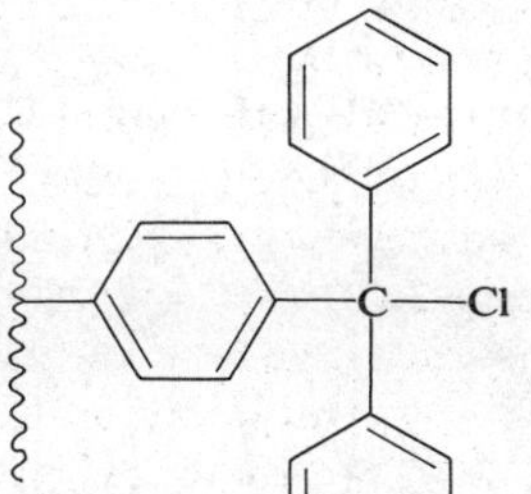

(20)

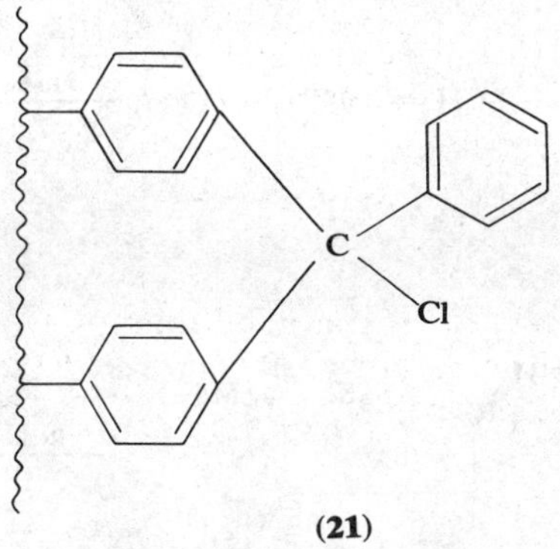

(21)

$$HO{-}(CH_2)_n{-}OH \longrightarrow CH_3{-}(CH_2)_m{-}C{\equiv}C{-}(CH_2)_n{-}OH$$

(22) a; $n=3$, $m=6$
b; $n=3$, $m=8$
c; $n=1$, $m=10$

(*ii*) *Reactions with polymer-supported reagents*

The most common advantage of PS reagents is the easy separation of the excess and spent reagent from the reaction mixture. This simplifies the reaction work-up and may avoid the need to expose the reaction product to water or to carry out chromatography. Also, excess of reagent may be used to obtain high reaction yields. As many PS reagents have been described it is only feasible to give examples which illustrate these and other advantages.

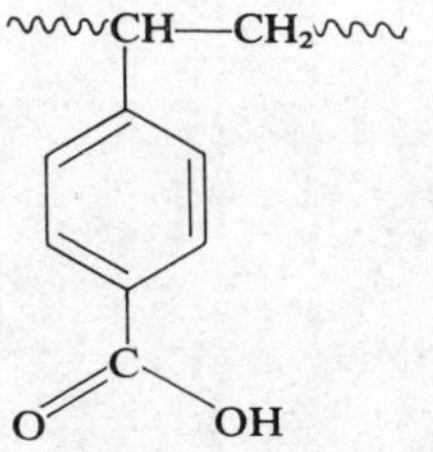

(23)

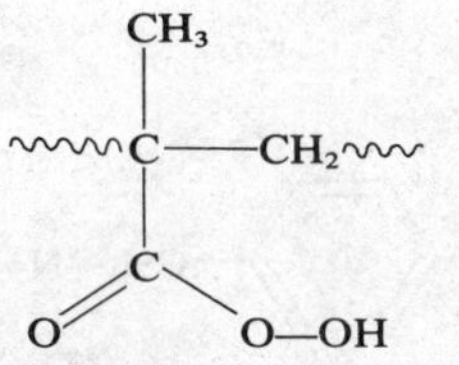

(24)

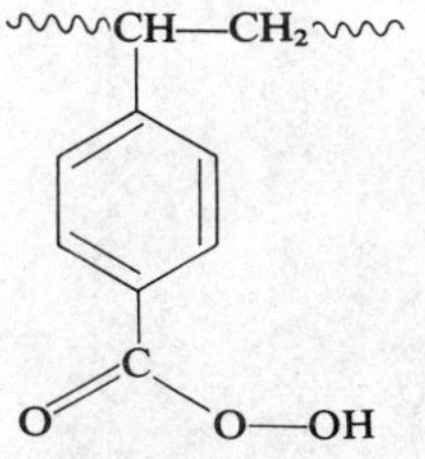

(25)

Treatment of cross-linked poly(methacrylic acid) or the resin carboxylic acid (**23**) with hydrogen peroxide and acid yields resins containing peroxyacid residues. The products prepared from the former contain aliphatic peroxyacid residues (**24**) and tend to be explosive.[59] The resins prepared from the latter contain aromatic peroxyacid residues (**25**) and are quite stable.[60,61] Tri- and di-substituted olefins are generally epoxidized in good yield by these resins[59,60] and sulphides, including penicillins and deacetoxycephalosporins, are efficiently oxidized to sulphoxides and/or sulphones.[59,62] Some examples are given in Scheme 12. Penicillin G (**26**) has been oxidized in 91% yield by passing a solution in acetone down a column of resin (**25**).[62] One advantage of using these peroxyacid resins is that they can be used with equal facility with acidic substrates, for there is no difficulty in separating the acidic product from the spent peroxyacid as there can be when monomeric peroxyacids are used. The spent resin can be recycled.

The Moffat oxidation (carbodi-imide, DMSO, H^+ catalyst) converts alcohols to aldehydes under mild conditions, but separating the aldehydes from the ureas which are also formed can present problems. These can be overcome by using resin (**27**).[63] Some examples of reaction carried out using the resin are shown in Scheme 13. The spent resin can be recycled, but with some loss of activity.

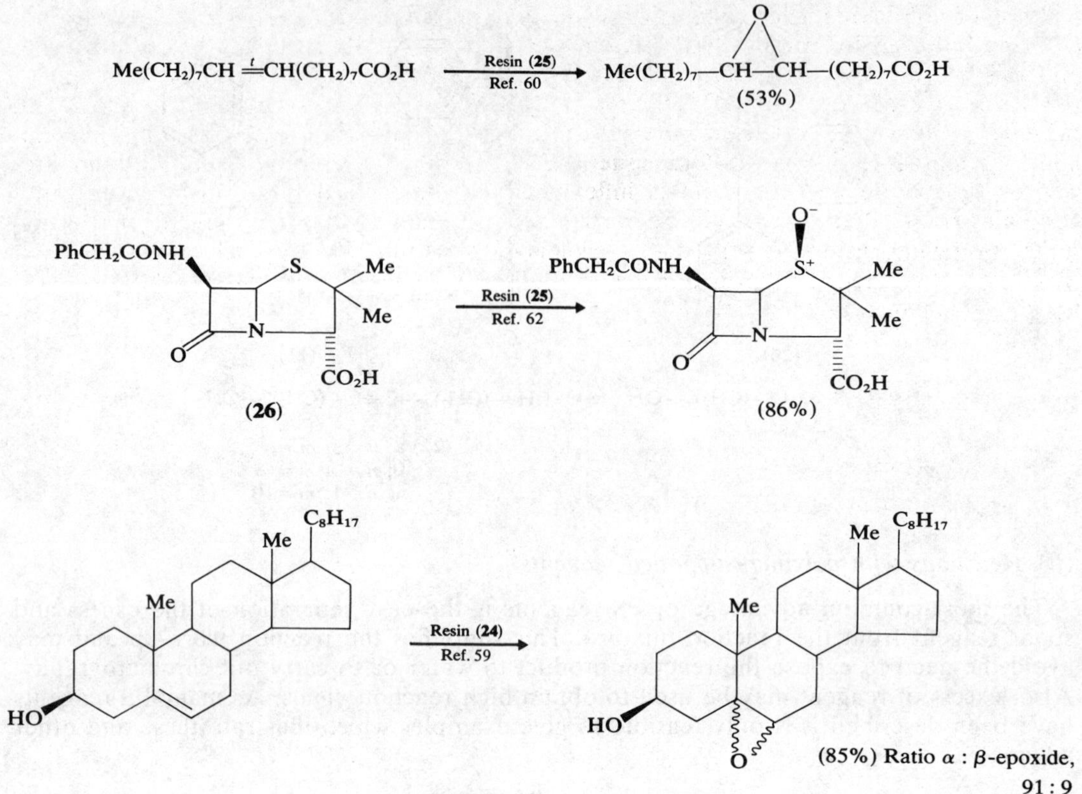

SCHEME 12

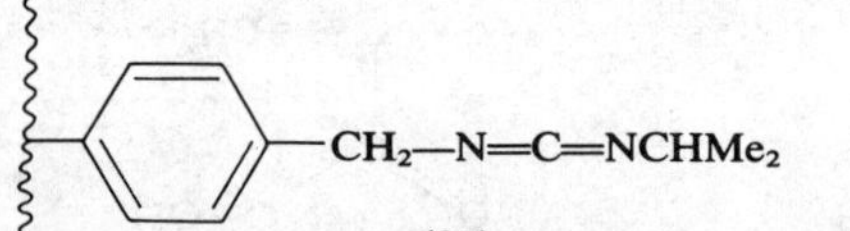

(27)

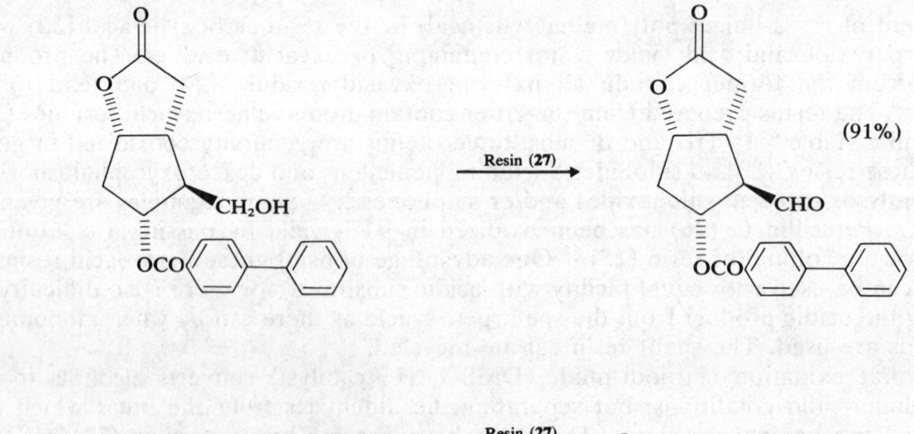

SCHEME 13

Another mild oxidant for converting alcohols to aldehydes or ketones is thioanisole and chlorine followed by triethylamine. Resin (**28**) can be employed in place of thioanisole and this has the advantage that the resin is odourless and at the end of the reaction is readily recovered for re-use.[64] The reactions involved in the oxidation are shown in Scheme 14. It will be noted that the substrate becomes attached to the resin in one step and is released in the next. Consequently, if the sulphide residues on the resin are sufficiently remote and the resin is inflexible, only one hydroxyl group of a substrate containing more than one should be oxidized. This possibility has been investigated using a macroreticular resin with heptane-1,7-diol as the substrate. The best selectively achieved was a hydroxyaldehyde to diaıdehyde ratio of 23:1, but, unfortunately, the yield of hydroxyaldehyde was then only 50%.

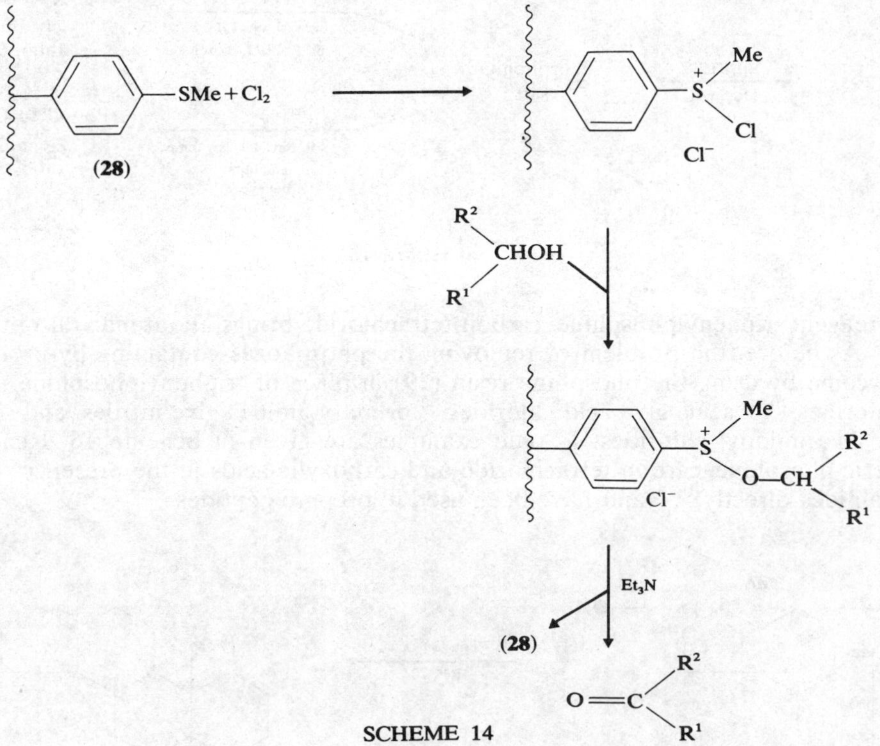

SCHEME 14

A difficulty sometimes encounted in Wittig reactions is separating the triphenylphosphine oxide produced from the olefin. This difficulty can be overcome by using a cross-linked polymer containing triphenylphosphine residues in place of triphenylphosphine and several research groups have investigated PS Wittig reactions.[65–68] The polymer (**29**) can be prepared from cross-linked polystyrene as outlined in Scheme 7 or

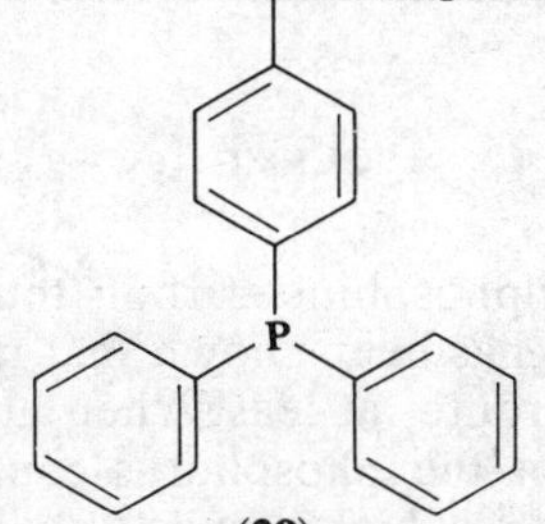

(**29**)

directly by copolymerization of styrene, divinylbenzene and 4-diphenylphosphinylstyrene.[65] Examples of the reactions studied are shown in Scheme 15. These include procedures designed to give mainly *cis*- or mainly *trans*-olefin.[68]

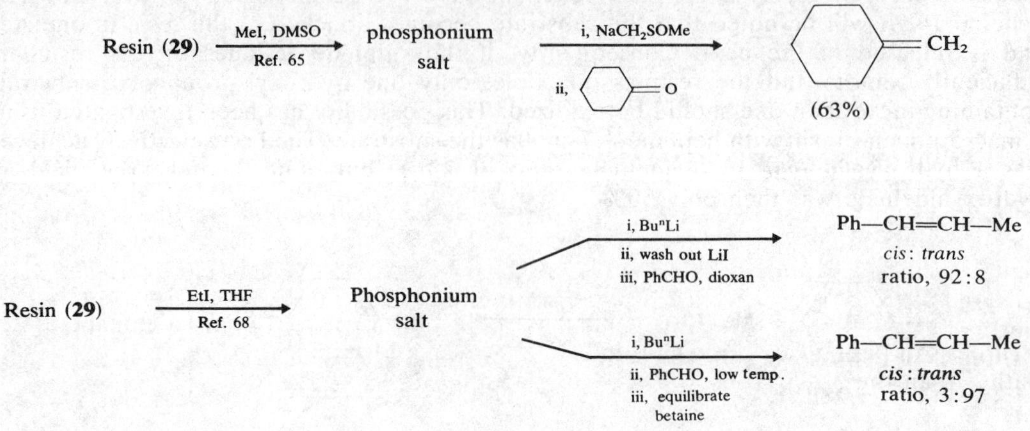

SCHEME 15

The reagent triphenylphosphine–carbon tetrachloride brings about many useful conversions.[69] As before, the problem of removing the phosphorus-containing by-products can be overcome by using the phosphine resin (**29**) in place of triphenylphosphine. Alcohols give chlorides,[70,71] acids give acid chlorides,[70] primary amides give nitriles, and secondary amides give imidoyl chlorides.[72] Some examples are given in Scheme 16. Reactions of polymeric phosphine–carbon tetrachloride and carboxylic acids in the presence of amines yields amides directly[70,73] and have been used to prepare peptides.

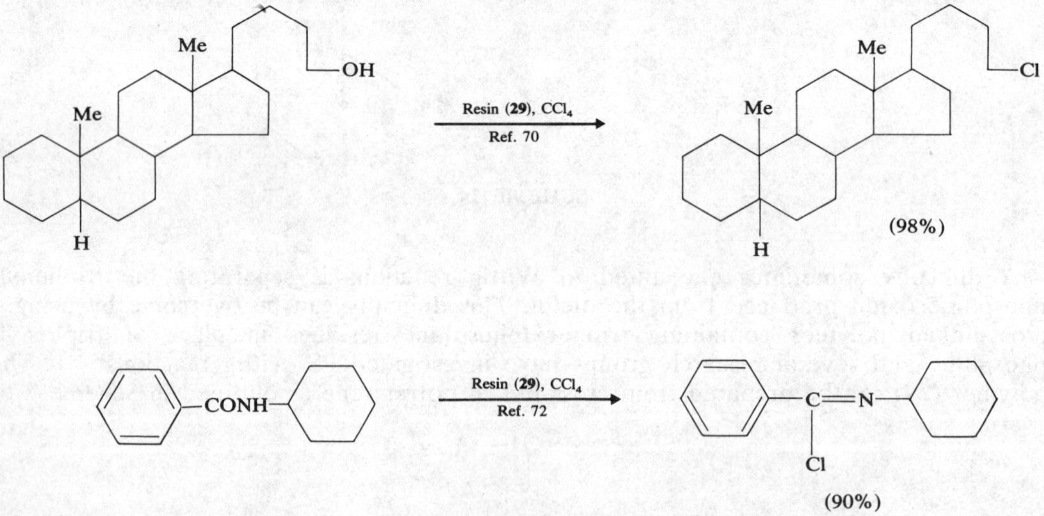

SCHEME 16

The reactions between triphenylphosphine–carbon tetrachloride and alcohols proceed by two pathways, (A) and (B).[69] It has been shown that in the resin reaction the analogue of the latter pathway is the main route, at least when alcohols are the substrates.[74] This requires that a high proportion of the phosphorus-containing residues can reach each other. This is probably easier with a highly substituted flexible resin than it is in the monomeric reaction, and may partly explain why the polymeric reactions proceed somewhat faster than those using triphenylphosphine (*ca.* 15 times) or 4-diphenylphosphinousisopropylbenzene (*ca.* 3 times) (a better analogue of the polymeric phosphine).[74]

(A) $$Ph_3P + CCl_4 + ROH \longrightarrow Ph_3P{=}O + CHCl_3 + RCl$$

(B) $$2Ph_3P + CCl_4 + ROH \longrightarrow Ph_3P{=}O + Ph_3\overset{+}{P}{-}CHCl_2 + RCl \quad Cl^-$$

Many PS acylating agents have been prepared, mainly with a view to use in peptide synthesis.[75] One of the most useful resins in this connection is the nitrophenol resin (**30**). Fridkin *et al.*[76] reacted this with *N*-protected amino-acids in the presence of dicyclohexylcarbodi-imide to obtain acylating agents. These reacted smoothly with various amino-acid and peptide derivatives. By using the acylating agent in considerable excess, high yields (>85%) of acylated product were obtained and these were easily isolated and purified. Fridkin has used this technique to synthesize bradykinin, a nonapeptide.[76] Synthesis of peptides in this way is in effect a 'reverse Merrifield' procedure as the reagent rather than the substrate is attached to the polymer.

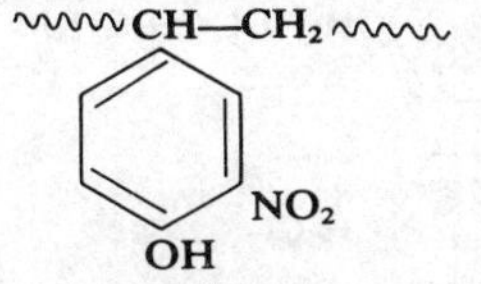

(**30**)

The reactive groups is PS reagents need not be covalently attached. Anion exchange resins with anions such as halides,[77] carboxylate,[78] cyanide,[74] periodate,[74] or anions from β-dicarbonyl compounds[79] can be used as reagents. Apart from the usual advantages of polymer-supported reagents, these have the interesting feature that, if macroporous anion exchange resins are used, they are a means of making anions available for reaction in non-polar solvents such as cyclohexane, benzene, and methylene chloride. They are, therefore, an alternative to phase-transfer catalysts and crown-ether catalysts. Examples of reactions using PS anions in non-polar solvents are given in Scheme 17.

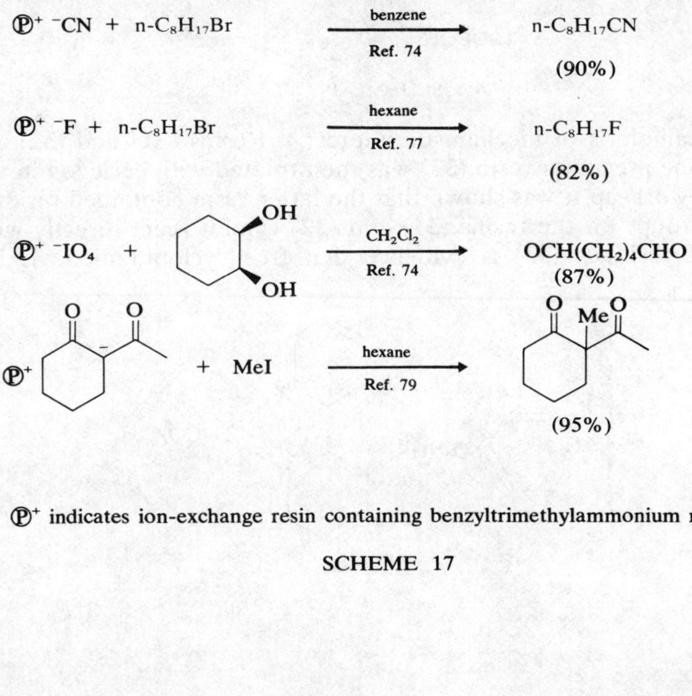

Ⓟ⁺ indicates ion-exchange resin containing benzyltrimethylammonium residues

SCHEME 17

(iii) Reactions with two polymer-supported reactants

There are obviously two situations possible: the reactants may be on the same resin or on separate resins. An example of each is given.

The first is concerned with the synthesis of unsymmetrical ketones.[80] An enolizable acid (*e.g.* $PhCH_2CH_2CO_2H$) was attached to a small proportion of the sites (equivalent to $0.1\ mmol\ g^{-1}$) on a chloromethylated polystyrene resin. A non-enolizable acid (*e.g.* $PhCO_2H$) was then attached to a large proportion of the remaining sites (equivalent to $1.0\ mmol\ g^{-1}$). The effect was that when base was added and the enolate anion generated, the only available groups for reaction were the non-enolizable ester groupings (see Scheme 18). Hence, after cleavage from the resin and decarboxylation of the β-keto-acid group, the only ketone obtained was the desired unsymmetrical ketone.

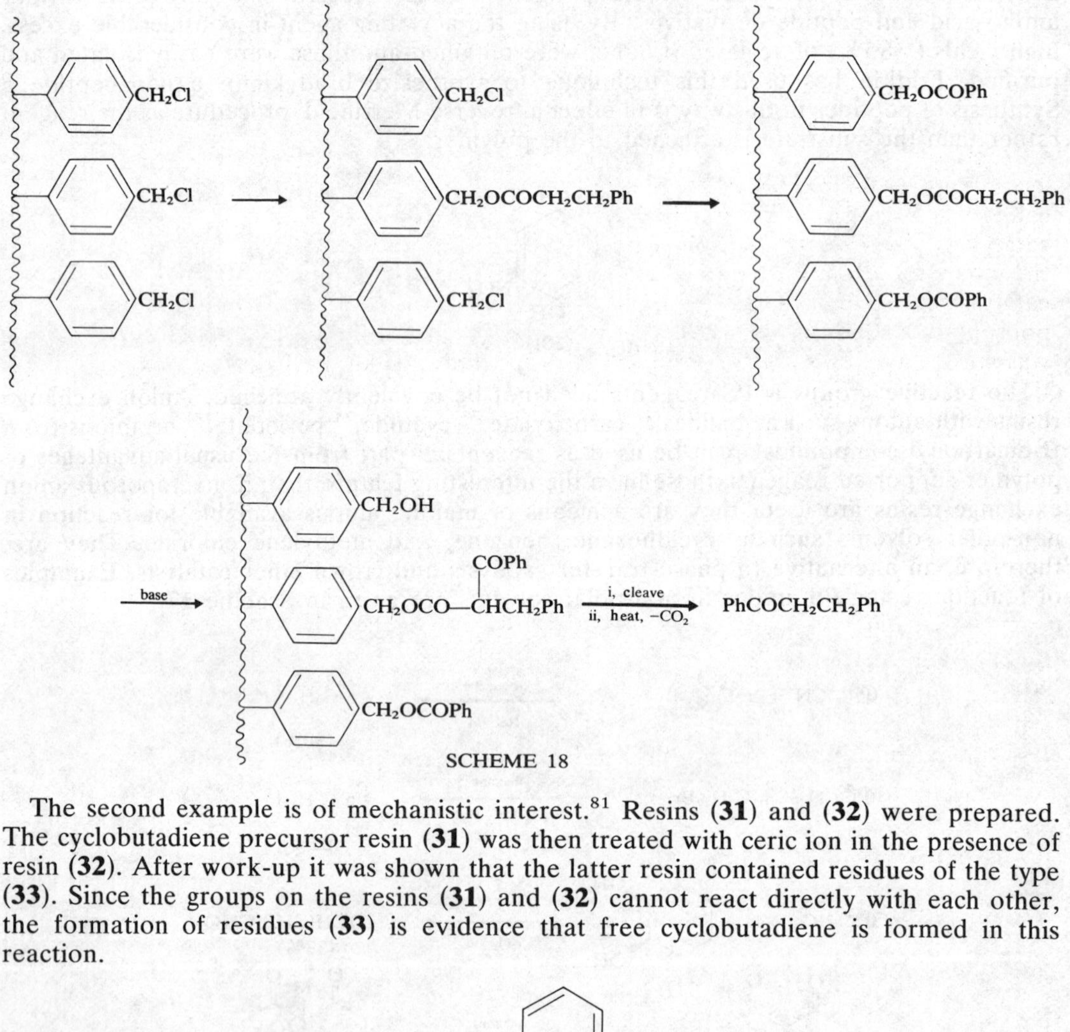

SCHEME 18

The second example is of mechanistic interest.[81] Resins (**31**) and (**32**) were prepared. The cyclobutadiene precursor resin (**31**) was then treated with ceric ion in the presence of resin (**32**). After work-up it was shown that the latter resin contained residues of the type (**33**). Since the groups on the resins (**31**) and (**32**) cannot react directly with each other, the formation of residues (**33**) is evidence that free cyclobutadiene is formed in this reaction.

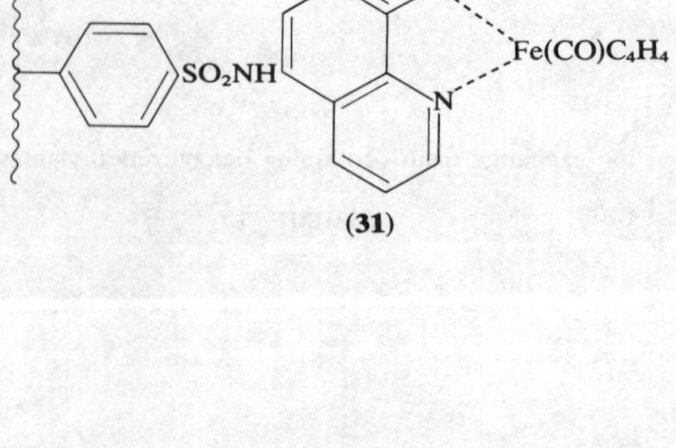

(**31**)

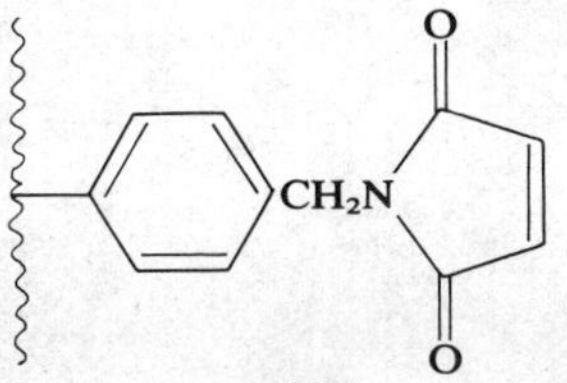

(32)

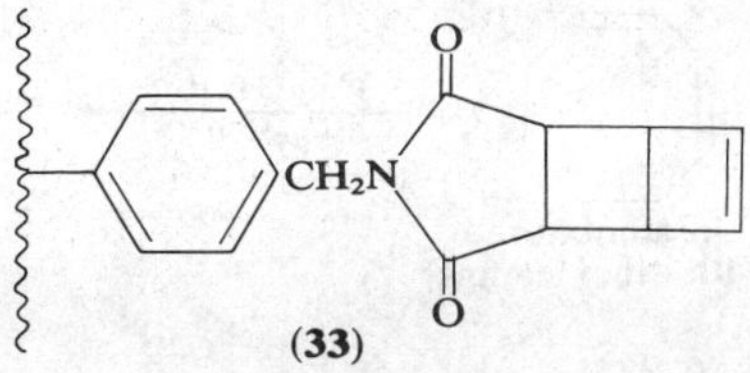

(33)

(*iv*) *Reactions using polymer-supported catalysts*

The best known reactions of this type are those involving acidic or basic ion-exchange resins.[82] Acidic ion-exchange resins can, for example, be used to catalyse esterification, ester hydrolyses, alcohol dehydration, and the inversion of sucrose. Basic ion-exchange resins can be used to catalyse Knoevenagel-type condensations, and aldol and benzoin condensations. The advantage of these catalysts over soluble catalysts is that they are easily removed at the end of the reaction, and they tend to give cleaner products.

Numerous transition metal complexes have been described in recent years which are useful catalysts for hydrogenation, hydroformylation, isomerization, *etc.* Many of these catalysts have now been attached to resins. This has the advantage that the valuable catalyst is easily removed from the reaction mixture and is available for re-use. Many of the monomeric catalysts contain triphenylphosphine ligands and the PS analogues of these catalysts have generally been prepared by replacing one or more of these ligands with phosphine residues of resin (**29**). For example, when a suspension of resin (**29**) in toluene was stirred with $(Ph_3P)_3RhCl$, the PS hydrogenation catalyst (**34**) was obtained.[83] Reaction of resin (**29**) with $(Ph_3P)_3RhH(CO)$ and with $(Ph_3P)_2Ni(CO)_2$ similarly gave the hydroformylation catalyst (**35**) and cyclo-oligermerization catalyst (**36**), respectively.[83] Some reactions using these catalysts are given in Scheme 19.

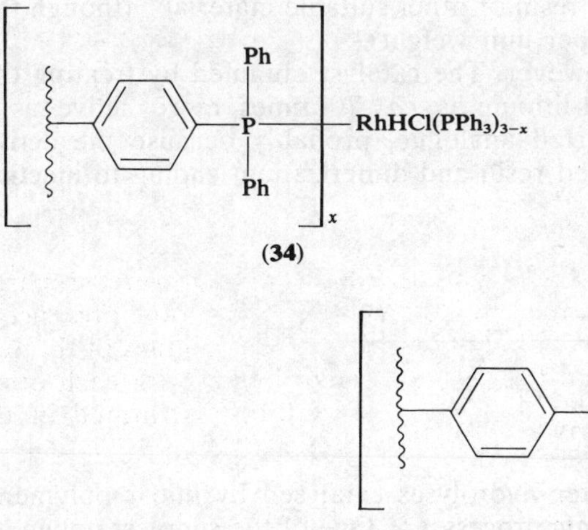

(34)

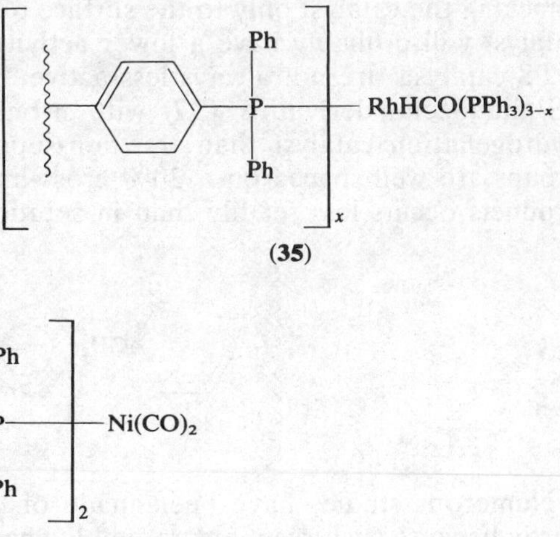

(35)

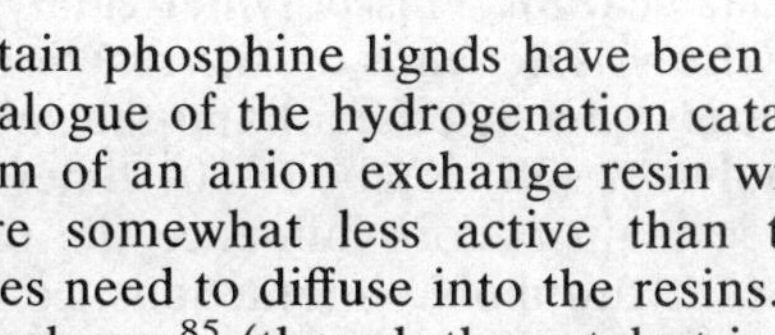

(36)

Catalysts which do not contain phosphine lignds have been attached to resins by other means. For example, a PS analogue of the hydrogenation catalyst K_2PdCl_4 was prepared by treating the hydroxide form of an anion exchange resin with this salt.[84]

Often the PS catalysts are somewhat less active than their soluble counterparts, probably because the substrates need to diffuse into the resins. The diffusion problem can be avoided by using a linear polymer[85] (though the catalyst is not easily recovered) or by

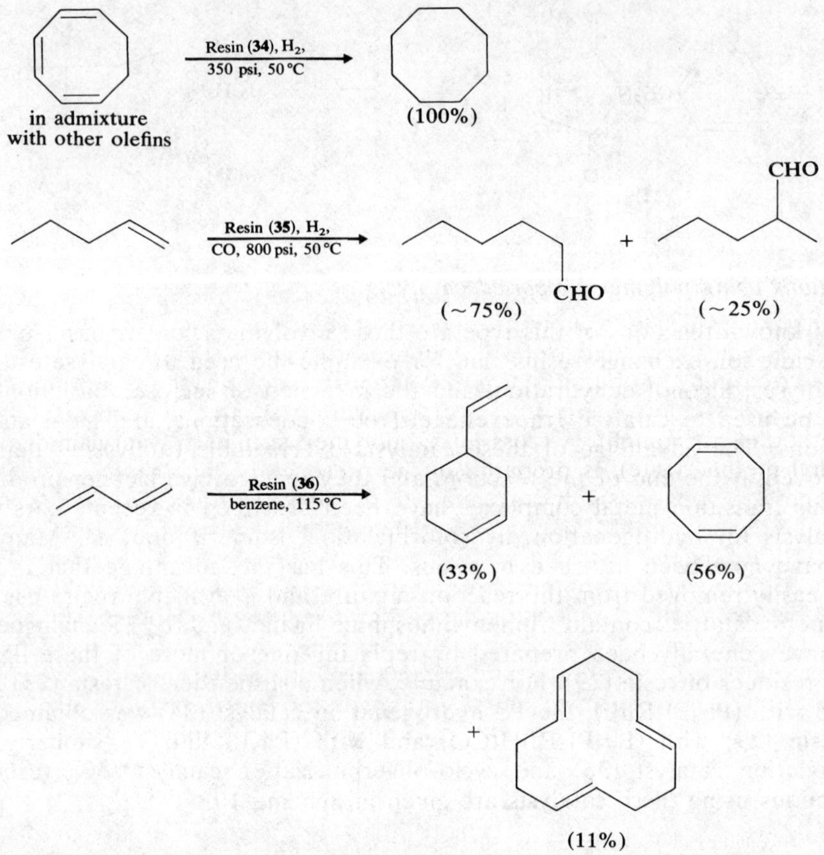

SCHEME 19

attaching the catalyst only to the surface of a resin or other suitable material[86] (though the catalyst will probably have a lower activity per unit weight).

PS catalysts are not always less active, however. The catalyst obtained by treating the PS titanocene derivative (**37**) with n-butyl-lithium is *ca.* 70 times more active as a hydrogenation catalyst than the non-supported analogue, probably because the active groups are well spaced on a 20% cross-linked resin and dimerization leading to inactive products occurs less readily than in solution.[87]

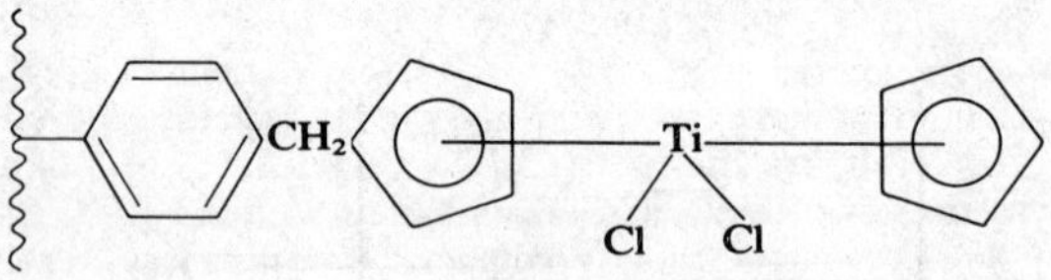

(**37**)

Numerous studies have been made of ester hydrolyses catalysed by linear polymers, partly because such reactions resemble enzymic processes.[88] One of the simplest polymers investigated is polyvinylimidazole (**38**). This polymer catalyses the hydrolysis of esters of the type (**39**). The rate of hydrolysis depends on the reaction solvent and the length of the acyl chain. When the solvent is aqueous ethanol and the substrate is an ester (**39**; $n = 11$) the reaction catalysed by polymer (**38**) is about 10^3 times faster than that catalysed by imidazole.[89] In general, three main factors influence the rate of hydrolysis catalysed by synthetic polymers. These are hydrophobic interactions, co-operative effects, and electrostatic interactions.

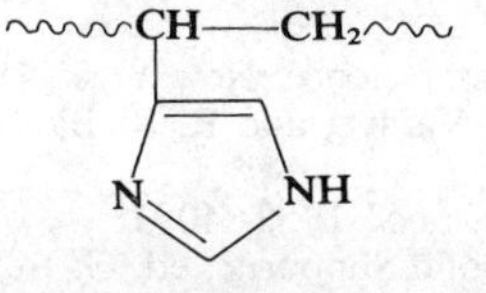

(38)

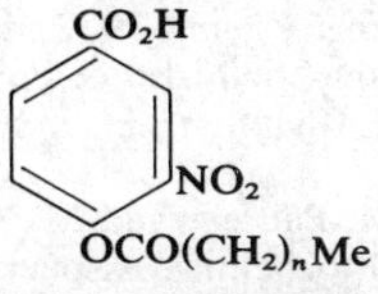

(39)

(*v*) *Polymer-supported enzymes*

Many enzymes have been immobilized on polymer supports.[90] This has the advantage that the enzyme, which is usually an expensive chemical, can easily be recovered and re-used. PS enzymes also tend to have enhanced storage stability.

Enzymes are usually bound to hydrophilic supports by one of three methods. The first is covalent attachment and some examples of this were presented in Section 27.4.1(iv). The second method is entrapment. A cross-linked polymer, such as polyacrylamide or poly-(2-hydroxyethyl methacrylate), is prepared in the presence of the enzyme. With a suitable degree of cross-linking a significant number of enzyme molecules become physically trapped in the resin. The third method is by physical adsorption of the enzyme on to inert supports or ion exchange resins.

The enzyme may well be denatured in the process of attaching it to the support. When an enzyme is successfully bound, its kinetic behaviour may be modified by a number of effects which include (a) microenvironmental change in the vicinity of the active site, (b) steric interactions between the enzyme, the substrate, and the support, (c) diffusion, and (d) effects resulting from chemical modifications of the enzyme molecule.

References

1. G. Odian, 'Principles of Polymerization', McGraw-Hill, New York, 1970.
2. 'Reactivity, Mechanism and Structure in Polymer Chemistry', ed. A. D. Jenkins and A. Ledwith, Wiley, London, 1974.
3. D. Braun, H. Cherdron, and W. Kern, 'Techniques of Polymer Synthesis and Characterisation', Wiley-Interscience, New York, 1971.
4. W. R. Sorenson and T. W. Campbell, 'Preparative Methods of Polymer Chemistry', 2nd edn., Interscience, New York, 1968.
5. 'Encyclopaedia of Polymer Science and Technology', ed. H. F. Mark, N. G. Gaylord, and N. M. Bikales, vols. 1–15 and a supplement to vol. 1, Interscience, New York, 1964–1976.
6. 'Kinetics and Mechanisms of Polymerization: vol. 1, Vinyl Polymerisation', ed. G. E. Ham. Edward Arnold, London, 1967.
7. C. H. Bamford and C. F. H. Tipper, 'Comprehensive Chemical Kinetics, vol. 14A, Free-radical Polymerisation', Elsevier, Amsterdam, 1976.
8. 'Copolymerization', ed. G. E. Ham, Interscience, New York, 1964.
9. 'The Chemistry of Cationic Polymerisation', ed. P. H. Plesch, Pergamon Press, Oxford, 1963.
10. M. Szwarc, 'Carbanions, Living Polymers and Electron-Transfer Processes', Interscience, New York, 1968.
11. 'Block Copolymers', ed. D. C. Allport and W. H. Janes, Applied Science, London, 1973.
12. 'Block and Graft Copolymerisation', ed. R. J. Ceresa, Wiley, London, 1973.
13. H. A. J. Battaerd and G. W. Tregear, 'Graft Copolymers', Interscience, New York, 1967.
14. 'Coordination Polymerisation', ed. J. C. W. Chien, Academic Press, New York, 1975.
15. L. Reich and A. Schindler, 'Polymerisation by Organometallic Compounds', Interscience, New York, 1966.
16. C. E. H. Bawn and A. Ledwith, *Quart. Rev.*, 1962, **16,** 361.
17. 'Kinetics and Mechanisms of Polymerisation: vol. 3, Step-Growth Polymerisation', ed. D. H. Solomon, Dekker, New York, 1972.
18. Ref. 5, vol. 7, p. 701; K. A. Kun and R. Kunin, *J. Polymer Sci., Polymer Chem.*, 1968, **6,** 2689.
19. A. A. K. Whitehouse, E. G. K. Pritchett, and G. Barnett, 'Phenolic Resins', Iliffe, London, 1967.
20. C. P. Vale and W. G. K. Taylor, 'Aminoplastics', Iliffe, London, 1964.
21. 'Chemical Reactions of Polymers', ed. E. M. Fettes, Interscience, New York, 1964.
22. 'Reactions on Polymers', ed. J. A. Moore, Reidel, Dordrecht, 1973.
23. 'Properties and Applications of Polyvinyl Alcohol', Monograph No. 30, Society of Chemical Industry, London, 1968; Ref. 3, pp. 255, 256.
24. P. J. Flory, *J. Amer. Chem. Soc.*, 1939, **61,** 1518; 1942, **64,** 177.

25. G. D. Jones, *Ind. Eng. Chem.*, 1952, **44,** 2686.
26. A Chapiro, 'Radiation Chemistry of Polymeric Systems', Interscience, New York, 1962, chapters 9 and 10; D. A. Laufer, T. M. Chapman, D. I. Marlborough, V. M. Vaidya, and E. R. Blout, *J. Amer. Chem. Soc.*, 1968, **90,** 2696.
27. K. W. Pepper, H. M. Paisley, and M. A. Young, *J. Chem. Soc.*, 1953, 4097.
28. J. A. Patterson, in 'Biochemical Aspects of Reactions on Solid Supports', ed. G. R. Stark, Academic Press, New York, 1971, chapter 5.
29. Ref. 3, p. 257.
30. R. L. Letsinger, M. J. Kornet, V. Mahadevan, and D. M. Jerina, *J. Amer. Chem. Soc.*, 1964, **86,** 5163.
31. M. J. Farrall and J. M. J. Frechet, *J. Org. Chem.*, 1976, **41,** 3877.
32. C. R. Harrison, P. Hodge, J. Kemp, and G. M. Perry, *Makromol. Chem.*, 1975, **176,** 267.
33. H. M. Relles and R. W. Schluenz, *J. Amer. Chem. Soc.*, 1974, **96,** 6469.
34. J. Bonner, in 'Biogenesis of Natural Compounds', ed. P. Bernfeld, Pergamon Press, Oxford, 1963, chapter 16.
35. B. L. Archer, D. Barnard, E. G. Cockbain, J. W. Cornforth, R. H. Cornforth, and G. Popjak, *Proc. Roy. Soc.* (*B*), 1966, **163,** 519.
36. D. Craig in Ref. 21, chapter IXC.
37. P. Meares, 'Polymers: Structure and Bulk Properties', Van Nostrand, London, 1965.
38. F. A. Bovey, 'High Resolution NMR of Macromolecules', Academic Press, New York, 1972; 'Structural Studies of Macromolecules by Spectroscopic Methods', ed. K. J. Ivin, Wiley, London, 1976.
39. 'Gel Permeation Chromatography', ed. K. H. Altgelt and L. Segal, Dekker, New York, 1971.
40. J. Reiland, in 'Methods in Enzymology', ed. W. B. Jakoby, Academic Press, New York, 1971, vol. 22, p. 287.
41. F. Helfferich, 'Ion Exchange', McGraw-Hill, New York, 1962.
42. J. Inczedy, 'Analytical Applications of Ion Exchangers', Pergamon Press, Oxford, 1966.
43. D. H. Spackman, W. H. Stein, and S. Moore, *Analyt. Chem.*, 1958, **30,** 1190.
44. G. Gabrielson and O. Samuelson, *Svensk kem. Tidskr.* 1950, **62,** 214; 1952, **64,** 150 (*Chem. Abs.*, 1951, **45,** 4168; 1952, **46,** 9018).
45. Ref. 42, chapter 10; G. Schmuckler, *Talanta*, 1965, **12,** 281; A. Patchornik and M. A. Kraus, Ref. 5, supplement to vol. 1, p. 471.
46. D. K. Hale, *Research*, 1956, **9,** 104.
47. G. Dotsevi, Y. Sogah, and D. J. Cram, *J. Amer. Chem. Soc.*, 1976, **98,** 3038.
48. H. Guilford, *Chem. Soc. Rev.*, 1973, **2,** 249.
49. C. G. Overberger and K. N. Sannes, *Angew. Chem. Internat. Edn.*, 1973, **13,** 99; C. C. Leznoff, *Chem. Soc. Rev.*, 1974, **3,** 65; L. P. Ellinger, *Ann. Reports* (*B*), 1973, 322; A. Patchornik and M. A. Kraus, in Ref. 5, supplement to vol. 1, p. 468.
50. J. I. Crowley and H. Rapoport, *Accounts Chem. Res.*, 1976, **9,** 135.
51. J. I. Crowley, T. B. Harvey, and H. Rapoport, *J. Macromol. Sci. Chem.*, 1973, **A7,** 1117.
52. R. B. Merrifield, *J. Amer. Chem. Soc.*, 1963, **85,** 2149.
53. G. R. Marshall and R. B. Merrifield, in Ref. 28, chapter 3.
54. J. M. J. Frechet and C. Schuerch, *J. Amer. Chem. Soc.*, 1972, **94,** 604; R. D. Guthrie, A. D. Jenkins, and J. Stehlicek, *J. Chem. Soc.* (*C*), 1971, 2690.
55. H. Köster and F. Cramer, *Annalen*, 1974, 946.
56. I. T. Harrison and S. Harrison, *J. Amer. Chem. Soc.*, 1967, **89,** 5723.
57. T. M. Fyles and C. C. Leznoff, *Canad. J. Chem.*, 1976, **54,** 935.
58. C. C. Leznoff, T. M. Fyles, and J. Weatherston, *Canad. J. Chem.*, 1977, **55,** 1143.
59. T. Takagi, *J. Appl. Polymer Sci.*, 1975, **19,** 1649.
60. C. R. Harrison and P. Hodge, *J.C.S. Perkin I*, 1976, 605.
61. J. M. J. Frechet and K. E. Haque, *Macromolecules*, 1975, **8,** 130.
62. C. R. Harrison and P. Hodge, *J.C.S. Perkin I*, 1976, 2252.
63. N. M. Weinshenker and C. M. Shen, *Tetrahedron Letters*, 1972, 3281, 3285.
64. G. A. Crosby, N. M. Weinshenker, and H.-S. Uh, *J. Amer. Chem. Soc.*, 1975, **97,** 2232.
65. S. V. McKinley and J. W. Rakshys, *J.C.S. Chem. Comm.*, 1972, 134.
66. W. Heitz and R. Michels, *Angew. Chem. Internat. Edn.*, 1972, **11,** 298.
67. F. Camps, J. Castells, J. Font, and F. Vela, *Tetrahedron Letters*, 1971, 1715.
68. W. Heitz and R. Michels, *Annalen*, 1973, 227.
69. R. Appel, *Angew. Chem. Internat. Edn.*, 1975, **14,** 801.
70. P. Hodge and G. Richardson, *J.C.S. Chem. Comm.*, 1975, 622.
71. S. L. Regen and D. P. Lee, *J. Org. Chem.*, 1975, **40,** 1669; D. C. Sherrington, D. J. Craig, J. Dagleish, G. Domin, J. Taylor, and G. V. Meehan, *European Polymer J.*, 1977, **13,** 73.
72. C. R. Harrison, P. Hodge, and W. J. Rogers, *Synthesis*, 1977, 41.
73. R. Appel, W. Strüver, and L. Willms, *Tetrahedron Letters*, 1976, 905.
74. C. R. Harrison and P. Hodge, unpublished work.
75. A. Patchornik and M. A. Kraus, in Ref. 5, supplement to vol. 1, p. 471.
76. M. Fridkin, A. Patchornik, and E. Katchalski, *J. Amer. Chem. Soc.*, 1968, **90,** 2953.
77. G. Cainelli, F. Manescalchi, and M. Panunzio, *Synthesis*, 1976, 472.
78. G. Cainelli, and F. Manescalchi, *Synthesis*, 1975, 723.
79. G. Gelbard and S. Colonna, *Synthesis*, 1977, 113.

80. M. A. Kraus and A. Patchornik, *J. Amer. Chem. Soc.*, 1971, **93,** 7325.
81. J. Rebek and F. Gavina, *J. Amer. Chem. Soc.*, 1975, **97,** 3453.
82. F. Helfferrich, *Angew. Chem.*, 1954, **66,** 241, 327.
83. C. U. Pittman, L. R. Smith, and R. M. Hanes, *J. Amer. Chem. Soc.*, 1975, **97,** 1742.
84. R. L. Lazcana and J.-E. Germain, *Bull. Soc. chim. France*, 1971, 1869.
85. E. Bayer and V. Schurig, *Angew. Chem. Internat. Edn.*, 1975, **14,** 493.
86. K. G. Allum, R. D. Hancock, S. McKenzie, and R. C. Pitkethly, in 'Proceedings of 5th International Congress on Catalysis, Miami Beach, 1972', North-Holland, Amsterdam, 1973, p. 477.
87. W. D. Bonds, C. H. Brubaker, E. S. Chandrasekaran, C. Gibbons, R. H. Grubbs, and L. C. Kroll, *J. Amer. Chem. Soc.*, 1975, **97,** 2128.
88. J. H. Fendler and E. J. Fendler, 'Catalysis in Micellar and Macromolecular Systems', Academic Press, New York, 1975, p. 451; C. G. Overberger and J. C. Salamone, *Accounts Chem. Res.*, 1969, **2,** 217.
89. C. G. Overberger, M. Morimoto, I. Cho, and J. C. Salamone, *Macromolecules*, 1969, **2,** 553.
90. R. Goldman, L. Goldstein, and E. Katchalski, in Ref. 28, chapter 1; K. Mosbach, *Sci. Amer.*, 1971, **224,** (3), 26.

PART 28

BIO-ORGANIC CHEMISTRY

28.1
Biosynthesis

R. THOMAS
University of Surrey

28.1.1 INTRODUCTION

The purpose of the present article is to introduce the topic of biosynthesis and, as a consequence, it will be primarily concerned with the more general aspects of the subject. Our present understanding of the frequently straightforward, but sometimes interestingly more subtle, origins and interrelationships of natural products, which have emerged from this comparatively new discipline, will be illustrated through reference to selected miscellaneous metabolites. In subsequent chapters, the biosynthetic pathways leading to the individual major classes of natural products will be described in more detail.

The historical background to the development of biosynthetic studies will be briefly reviewed, as will the experimental approaches, which up to the present have been predominantly based on the use of isotopically-labelled intermediates. Over the past quarter of a century, biosynthetic studies have been particularly successful in establishing many general precursor–product relationships. This has been valuable in facilitating the formulation of a simple biosynthetic classification of natural products, which in some respects parallels and yet in others greatly contrasts with the earlier traditional structure-based system.

In some early *in vivo* biochemical applications, radioactively-labelled precursors were utilized as a means of detecting metabolic intermediates through autoradiographic procedures, as in the study of chlorophyll-dependent pathways[1] involving short-term exposure of plants to $^{14}CO_2$. In contrast to this approach, the majority of natural product investigations have so far been designed to test the specific incorporation of more advanced intermediates. The choice of candidate precursors is normally based on hypothetical considerations of potential biosynthetic schemes. These schemes are generally arrived at following preliminary structural analysis of a natural product in terms of possible simple primary precursors, and also by comparison with other known products of related structure. Some examples of this speculative approach will be briefly outlined.

The successful elucidation of many pathways, sometimes in considerable detail, inevitably attracted organic chemists to devise non-enzyme catalysed laboratory analogies, so-called biomimetic syntheses. These chemical transformations of hypothetical biosynthetic intermediates have in a number of instances succeeded in producing moderate yields of the expected natural products. However, in the absence of the corresponding enzyme-controlled conversions, this apparent biogenetic evidence in support of a particular speculative scheme is at best circumstantial, and can on occasion be misleading.

28.1.2 THE HISTORICAL DEVELOPMENT OF BIOSYNTHETIC STUDIES

The term 'natural product' is a general name for any organic substance occurring in living organisms, whether plants, animals or micro-organisms. Several thousand substances have been isolated and structurally characterized over the past century, the vast majority of which are of plant and microbial origin. Natural products are the result of predominantly enzyme-catalysed metabolic processes, as distinct from the much larger number of 'unnatural products', as one might by way of contrast label the fruits of the synthetic endeavours of generations of organic chemists.

Interest in the biosynthetic origin of natural products developed in parallel with the accumulation of structural data. The realization that these compounds were formed by basically unexceptional chemical processes, and that they were attainable targets for laboratory syntheses, stemmed from early findings such as the classic observation of Wöhler in 1828 that the well-known animal metabolite urea could be obtained by the pyrolytic rearrangement of ammonium cyanate.

This important finding served to remove the mystique of natural products, which until then had been regarded rather as supernatural products, the formation of which required an unknown vital force. The vitalistic concept was finally demolished by Pasteur, who in the middle of the last century demonstrated that micro-organisms, such as yeasts and bacteria, were not the outcome of a process of spontaneous generation, and that they were in fact the source of known fermentation products such as ethanol, acetic, and butyric acids.

Since these early studies, new natural products, often with highly complex structures, have been characterized at an increasing rate, although the time taken for their complete structural elucidation has sometimes been extremely prolonged. For example, the structure of the well-known poisonous plant alkaloid strychnine, which was isolated in 1818, was not finally established until 1946. The fact that this complex heptacyclic product with multiple chiral centres was subsequently completely synthesized[2] in 1954 and that, in the past decade, its biosynthetic origin has also been largely clarified, is an indication of the recent rapid advances in the relevant chemical technology.

The period since 1940 has witnessed substantial progress in a number of important areas. There have been great improvements in the development of new fractionation techniques, which can separate even micro-quantities of chemically similar substances. Thus the various forms of chromatography, together with the use of ion exchange resins and molecular sieves, have made possible the isolation of minor components of natural product extracts. Coupled with the remarkable progress in spectroscopic aids to structure determination, such as ultraviolet, infra-red, nuclear magnetic resonance, and mass spectroscopy, and also X-ray analysis, the number of known natural product structures has increased at an impressive rate.

From the biosynthetic viewpoint, it is particularly fortunate that these advances also coincided with the development of nuclear technology, which led to the commercial availability of isotopic tracers. This has been so successfully applied to the investigation of metabolic pathways that biosynthesis can now be considered as an integral part of natural product studies.

In addition, advances in enzymic and mutant techniques, which were originally applied to the more traditional areas of biochemistry and genetics, have now added these valuable biosynthetic probes to the armamentarium of the natural product investigator.

Despite this recent considerable progress, the actual detailed knowledge of the pathways leading to the major groups of natural products is still at present fragmentary. However, in the light of the numerous developments in experimental techniques, further rapid enhancement of our understanding of the more intimate details of biosynthetic pathways is only limited by the funding of the necessary interdisciplinary collaboration.

28.1.3 BIOSYNTHETIC CLASSIFICATION OF NATURAL PRODUCTS

The carbon skeletons of the majority of natural products are assembled by characteristic sequences of enzyme-catalysed reactions, starting from glucose, which is itself produced by reduction of CO_2 in plants and autotrophic micro-organisms (Chapter 28.2). These interrelated metabolic sequences provide the basis for a biosynthetic classification in which it is convenient to divide natural products, somewhat arbitrarily, into two groups, namely primary and secondary metabolites (Table 1).

TABLE 1
Characteristics of Primary and Secondary Metabolites

Primary metabolites
(a) Products of general metabolism
(b) Broad distribution in plants, animals and micro-organisms, *e.g.* amino-acids, acetyl-CoA, monosaccharides, mevalonic acid, nucleotides
Secondary metabolites
(a) Products of specialized pathways
(b) Biosynthesized from primary metabolites
(c) Restricted distribution found mostly in plants and micro-organisms (often characteristic of individual genera, species or strains), *e.g.* alkaloids, terpenes, phenols, oligosaccharides, flavanoids, antibiotics

The classification criteria are mainly biosynthetic and to a lesser extent functional. Thus the first group comprises those substances of widespread distribution (*i.e.* general metabolites), which are recognized as being metabolically essential. For example, sugars, amino-acids, nucleotide bases, and general intermediary metabolites, together with a variety of biopolymers such as the common polysaccharides, proteins, and nucleic acids.

A small group of these primary metabolites function as precursors of all the other substances which do not fall into this category. These substances, which include the vast majority of natural products, are accordingly classed as secondary metabolites (Table 2).

This secondary status does not do justice to the often highly complex nature of their diverse molecular structures, whose origins cut right across the traditional groupings of the earlier purely chemical classification. A common feature of the majority of metabolites in this second group is that we are generally ignorant of the nature of their biochemical significance. However, this is presumably a temporary situation, which has no permanent

TABLE 2
Biosynthetic Classification of Secondary Metabolites

Primary precursors	*Secondary metabolites*
Amino-acids	Alkaloids, Peptide derivatives (*e.g.* oligopeptides, diketopiperazines and penicillins) Non-nitrogenous products (*e.g.* cinnamic acid derivatives)
Mevalonic acid $(HOCH_2CH_2—C(Me)(OH)—CH_2COOH)$	Terpenes, sterols, carotenoids and general polyisoprenoids
Glucose	Glycosides, modified sugars (*e.g.* kojic acid) and oligosaccharides (*e.g.* streptomycin)
Acetyl-CoA and propionyl-CoA	Polyketides: fatty acid derivatives, polyacetylenes, macrolides and numerous phenols
C_1 units (*ex* formate and methionine)	—OMe, —SMe, >NMe, >CMe, >CCH$_2$OH groups, *etc.*

value as a classification characteristic. Nonetheless, at this stage of development it is not unreasonable, as noted by one author,[3] to categorize primary and secondary metabolites respectively as "biochemists' compounds" and "chemists' compounds", since this duly reflects the different motivational interests of the corresponding groups of investigators concerned with these two classes of natural products.

Perhaps the most significant characteristic of many secondary metabolites is their comparatively restricted distribution, which is sometimes limited to single species or sub-species. These products, since they are by definition non-general metabolites, are consequently a manifestation of the individuality of the host plant. Their production may, in some instances, be a direct cause rather than an indirect consequence of those differences which presumably confer on individual species some survival advantage.

A variety of functions have, in fact, been attributed to a number of so-called secondary substances, relating to their effects on different species sharing the same environment. In this respect they serve as allelochemic agents,[4] which can profoundly influence the growth, health, behaviour, and population biology of other living organisms.

It is not unlikely that, while some natural products may indeed be non-essential or just simply evolutionary relics, the majority could possess metabolic, ecological, or other biological significance, the nature of which will only become apparent following detailed investigation of the 'biochemistry of individuality'.

28.1.4 LIMITATIONS OF NATURAL PRODUCT STUDIES

Natural product investigations typically follow the sequence of detection, isolation, and structure determination. An indication of the growing recognition of the importance of biosynthetic considerations is apparent from the current natural product literature, where new structures are rarely described without some discussion of their origin or interrelationships.

Since the criteria of detection, (colour, antibiotic activity, *etc.*) and the extraction procedures both profoundly influence the general types of structures which are isolated, it is inherently unlikely that the existing known groups of natural products, which have resulted from these limited, highly selective, and often arbitrary exploratory procedures, are representative of the total range of biosynthetic activities of living organisms. Thus, entirely new classes of structures may yet await discovery, although in view of the scale of the investigations over the past century, *a priori* it would appear more than likely that any remaining new groups of natural products will be found as quantitatively minor constituents.

A prerequisite of any systematic search for such fundamentally novel types of compounds must necessarily include a broad non-discriminatory detection procedure. Fortunately, such a technique is available, namely that of autoradiography (or, as it is alternatively known, radioautography). The scope of this simple and yet extremely valuable approach will be described later.

Several thousand natural products are detailed in the existing literature and assembled as structurally-related or taxonomically-related groups in a variety of compendia. It is difficult to estimate the precise number with any accuracy, since at present there is no single compilation of structures, nor any provision for the systematic storage of new data which are now accumulating at a rapid rate.

In view of the increasing industrial, medicinal, and agricultural importance of these substances, there is clearly a growing need for such a comprehensive compendium, as well as the means of searching systematically for useful correlations, for example structure–activity relationships of pharmacologically active metabolites, and structurally common denominators of biosynthetic significance, such as the presence of primary precursor units within individual carbon skeletons (*cf.* Table 2).

While the compilation of the necessary data would be a major undertaking, the essential information is already largely documented in the existing literature and its systematic collection and presentation on demand would be a relatively straightforward task. This would involve the creation of an unexceptional computerized data storage system, with an appropriate cross-referenced information-retrieval programme.

As an aid to biosynthetic studies, such a system would be of particular value in providing a potential means of carrying out structural analyses of natural products for the presence of primary precursor units, and also chemotaxonomically by facilitating correlations of individual products with similarly-derived metabolites of different origins. Considerations of this type currently form the basis of speculations leading to hypothetical biosynthetic schemes, which can then be tested, usually with the aid of isotopically-labelled candidate precursors. At present such structural analyses and potential interrelationships can only be formulated by protracted searches of multiple sources of information which are seldom collectively available to individual investigators.

28.1.5 THE METHODOLOGY OF BIOSYNTHESIS

As noted previously, the detailed investigation of biosynthetic pathways is necessarily a multi-disciplinary exercise. A reasonably full description of the origin and interrelationships of any natural product would require an understanding of both the molecular structures of its metabolic precursors and the mechanisms of their interconversions. In addition, it would be necessary to define such fundamental details as the structure, mode of action, and control of each of the individual enzymes, which would include genetic mapping of the relevant genes responsible for the biosynthesis and regulation of all of these enzymes.

The required collaboration of a team of chemists, enzymologists, geneticists, physiologists, taxonomists, and other appropriate specialists is as yet an unrealized academic ideal, particularly in the area of secondary metabolites. This is in part due to the unknown metabolic significance, if any, of the majority of these compounds, which have to date primarily excited the interests of organic chemists and those concerned with their potential medicinal applications.

It is consequently quite remarkable that by the intelligent use of virtually a single probe, namely isotopically-labelled precursors, in conjunction with simple biological incorporation procedures, it has proved possible to establish the general pathways leading from a small number of primary precursors to the known major groups of natural products. This has been made possible by the application of sophisticated chemical and physical techniques to determine the locations of both stable and radioactive isotopes (predominantly ^{2}H, ^{3}H, ^{13}C, ^{14}C, ^{15}N, ^{18}O, and ^{35}S) in natural products derived from specifically-labelled precursors.

Incorporation of these precursors[5] is normally effected in plants by feeding solutions to the appropriate tissue, or, for micro-organisms, by their addition to the culture medium during the phase of maximum production of the required metabolite. The optimum efficiency of incorporation is determined following exploratory feeding experiments. This is a particularly important preliminary when using stable isotopes with an appreciable normal abundance, such as ^{13}C (1.1%), where the minimum dilution of the isotopic label is required. With radioactively-labelled precursors this aspect is not always as critical, since the resulting metabolite is often sufficiently active to allow subsequent dilution with unlabelled (cold) product, prior to systematic chemical degradation to determine the distribution of labelling. The standard assay procedures employ either a sensitive Geiger–Müller end window counter or, more commonly, a scintillation counting system. This is the original and currently still the most commonly-utilized exploratory procedure.

More recently, non-destructive methods of locating labelled atoms have been devised,[6] based on nuclear magnetic resonance spectroscopic recognition of suitably-enriched isotopes. This direct approach, which has met with considerable success, was applied initially to ^{13}C, and subsequently to ^{3}H and ^{15}N, and even ^{2}H (which can be observed[7] by modified resonance techniques). Such procedures are even capable of directly demonstrating stereospecific labelling at selected prochiral centres where the two chemically identical substituents are magnetically non-equivalent.

The use of ^{13}C is now comparatively commonplace in studies of microbial metabolite biosynthesis, where the rapid growth of micro-organisms facilitates the efficient incorporation of labelled precursors. This is the limiting factor which determines the applicability of the n.m.r. approach, which is dependent upon obtaining a significant isotopic enrichment relative to the normal abundance of ^{13}C. For this reason, its application to plant products has thus far been more restricted.

Although the corresponding utilization of ^{3}H n.m.r. is a relatively recent development, it is more convenient to illustrate the direct n.m.r. approach using this isotope since by comparing the spectrum of [^{3}H]penicillic acid[8,9] (shown in Figure 1) with the normal proton spectrum, the triton locations are immediately apparent. The spectrum indicates some of the advantages of this particular isotope, namely the closely predictable chemical shifts and coupling constants and the proportional relative intensities of the integrals. It also clearly demonstrates the stereospecific labelling of the C-5 methylene group, where, in addition, the reduced integral shows partial loss of tritium relative to the other tritiated centres, which, in turn, has interesting mechanistic implications.[8]

A subsequent development of the direct ^{13}C n.m.r. approach has been the exploitation of ^{13}C–^{13}C spin coupling. Thus the presence of this splitting in the spectrum of a metabolite derived from a precursor containing ^{13}C at two adjacent carbon atoms in the same molecule, is a direct indication of the specific incorporation of that precursor without cleavage of the original carbon–carbon bond. A particularly subtle application of this coupling phenomenon is its use to distinguish between alternative modes of cyclization of a polyketide chain, as in the instance of the fungal phenalenone deoxyherqueinone[10] (Figure 2).

Unexpected ^{13}C–^{13}C coupling has been occasionally observed in feeding experiments where the precursor is only singly labelled. This can occur when high feeding levels increase the frequency of biosynthesis of molecules incorporating more than one labelled precursor unit. Hanson and his colleagues have ingeniously utilized this finding to define the folding pattern of the farnesyl pyrophosphate precursor of dihydrobotrydial prior to its cyclization and rearrangement.[11]

The spin coupling criterion is also potentially applicable to other suitable isotopic combinations, for example ^{13}C and ^{3}H. This may be useful as a direct probe for hydride shift mechanisms, as in certain terpenoid biosynthetic pathways where it could be studied using appropriately labelled mevalonic acid or even ^{13}C,^{3}H-labelled acetate.

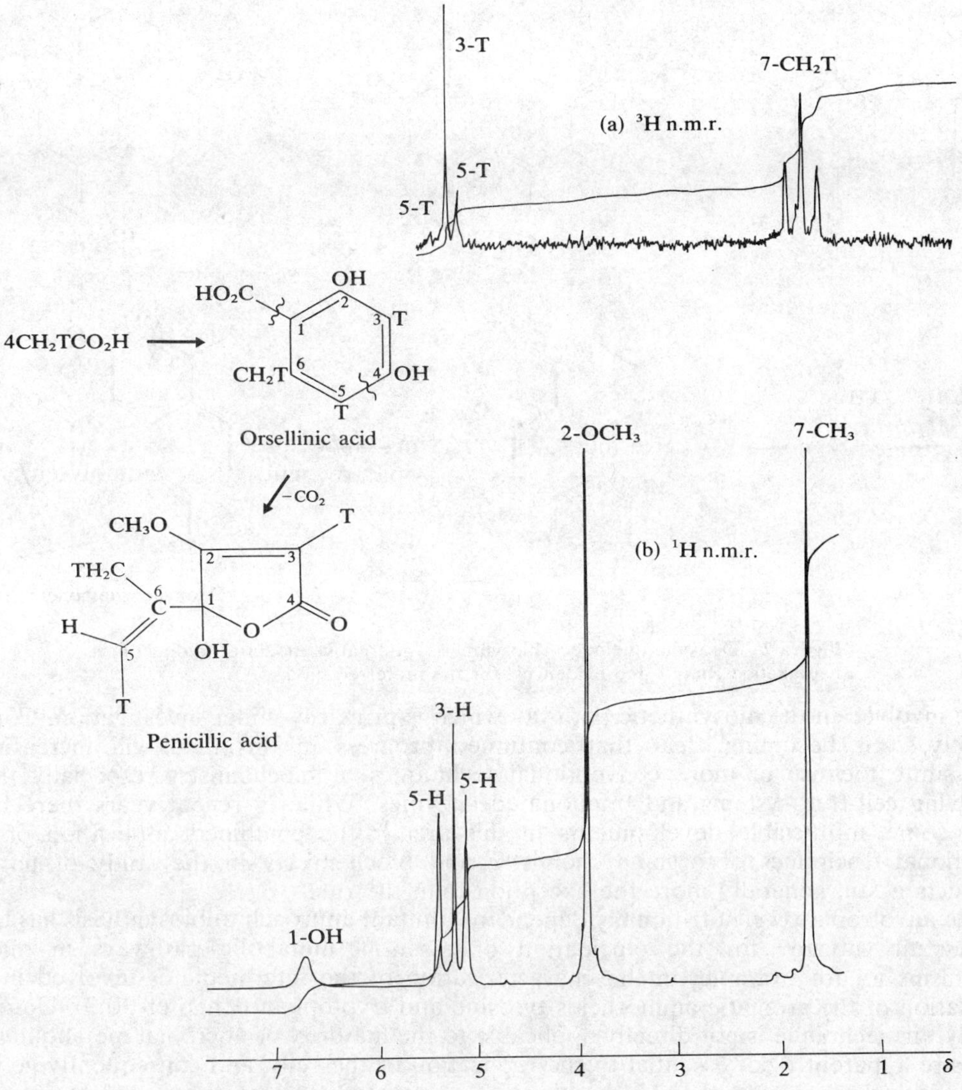

Figure 1 Penicillic acid triton (^{3}H n.m.r.) and proton (^{1}H n.m.r.) spectra

Double-labelling techniques involving conventional assay procedures have been widely used in biosynthetic investigations. This well-established approach requires the use of two different isotopes, the ratio of which, in the initial precursor, can be compared directly with that in the resulting metabolite without prior degradation to separate the different isotopic species. Examples include the use of [^{14}C, $^2H_3$]methionine to explore details of methylation pathways,[12] the corresponding investigations of the mechanisms of incorporation of [^{14}C, ^{15}N]-labelled amino-acids into alkaloids[13] and of [^{14}C, ^{18}O]acetate into the polyketide orsellinic acid.[14]

The majority of studies of natural product biosynthesis have utilized whole plants and intact microbial cells. This approach has been generally applicable to a range of simple precursors which are not normally excluded by cell-wall barrier limitations, which can sometimes be a serious restriction, particularly with certain types of advanced intermediates, for example oligopeptides. Even where permeability is not the limiting factor, complex intermediates once inside the cell can be metabolized by enzymes other than

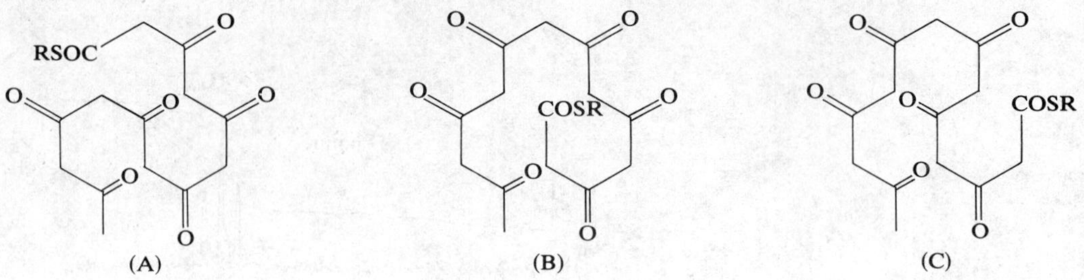

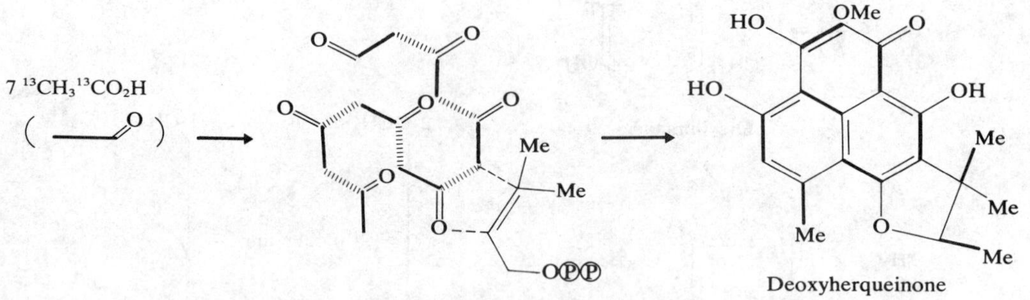

Figure 2 Deoxyherqueinone biosynthesis: alternative modes of folding of the heptaketide chain[36] and evidence[10] for the preferred mode (A)

those involved in the biosynthetic pathway which is primarily under investigation. Consequently, it is becoming clear that continued progress in this field will increasingly necessitate the use of more conventional techniques of biochemistry, especially those involving cell-free systems and fractionated enzymes. While in recent years there have been some appreciable developments in this area,[15] the combined application of the traditional disciplines of organic chemistry and biochemistry in the study of natural products is still generally more the exception than the rule.

The auxotrophic (*i.e.* nutritionally dependent) mutant approach to biosynthesis has been successfully utilized for the elucidation of essential metabolic pathways in micro-organisms as, for example, in the characterization of the intermediates involved in the formation of the aromatic amino-acids tyrosine and tryptophan (Chapter 30.3). Unfortunately this technique is not directly applicable to the majority of microbial metabolites, as they are apparently not essential to the replication of the cells, and consequently genetic blocks which interrupt their biosynthesis are not lethal. However, useful microbial mutants can certainly be prepared in the normal way, although their selection and isolation is more tedious than the more direct auxotrophic procedure.

It was observed in studies with the tetracycline antibiotics of *Streptomyces aureofaciens*[16] that certain artificial mutants, which were individually selected for their inability to produce active tetracyclines, were collectively able to do so when grown in mixed culture. This co-synthetic behaviour was considered to be a consequence of an overlap of genetic information in the individual mutants, which permitted the synthesis of a diffusible normal intermediate in one strain, which lay beyond the blocked step in the tetracycline pathway of the second mutant strain. Using this approach, McCormick and his colleagues,[16] by preparing large numbers of mutants, were able to isolate many new metabolites, some of which were direct tetracycline precursors, while others were found to be shunt metabolites. The resulting sequence (abbreviated in Figure 3) is one of the most detailed pathways yet established for a complex natural product.

Interestingly, no pre-tetracyclic intermediates were discovered, which is consistent with the hypothesis that the primary polyketide precursors (in this instance probably nine malonate units) are assembled into a long polyketide chain, which then undergoes cyclization without the involvement of enzyme-free intermediates.

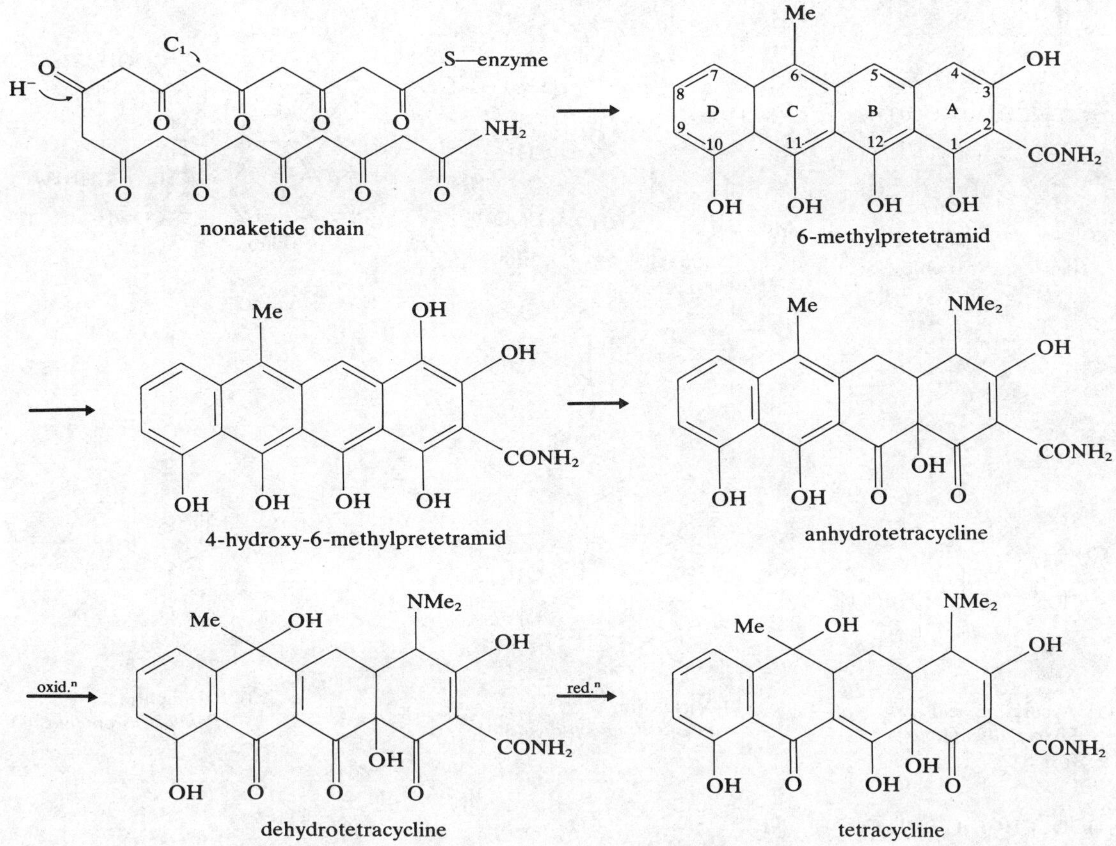

Figure 3 Key intermediates in the biosynthesis of tetracycline[16]

28.1.6 STRUCTURAL ANALYSIS AND NATURAL PRODUCT INTERRELATIONSHIPS

Long before the initiation of experimental investigations of the biosynthesis of natural products, a sufficient number of metabolites of plants and micro-organisms had been identified to allow a classification based primarily on chemical structure. The principal categories of products (terpenes, sterols, alkaloids, phenols, carbohydrates, pigments *etc.*) were formulated according to the characteristic groupings present in their molecular skeletons.

This led to the recognition of certain fundamental structural interrelationships and often successful speculation regarding the nature of the primary building blocks of precursors. Among the distinguished early contributors to biogenetic theory, based on the recognition of apparently common structural components and also chemically-feasible mechanisms of assembly under physiological conditions, must be included the names of Ruzicka, who drew attention to the polyisoprenoid character of terpenes and sterols,[17] and Sir Robert Robinson, who contributed to very many fields of natural products but perhaps especially to that of the alkaloids.[18]

In the biosynthetic context, so-called structural analysis is simply a paper-chemistry exercise, the object of which is the recognition of components of secondary product structures likely to arise from the known primary precursors (as summarized in Table 2). The examples in Figures 4–8 depict its application principally to a variety of microbial and plant metabolite structures, the implicit origins of most of which have subsequently been established by incorporation of the relevant isotopically-labelled precursors.

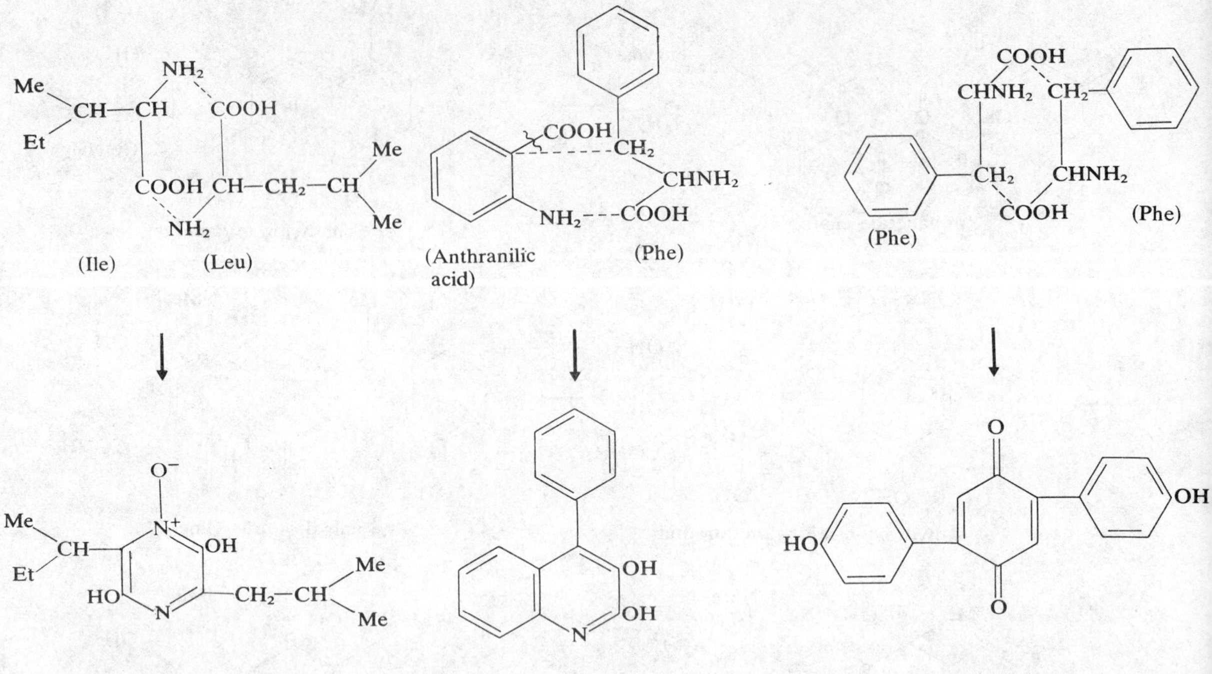

(1) Aspergillic acid (*Aspergillus flavus*)

(2) Viridicatin (*Penicillium viridicatum*)

(3) Atromentin (*Paxillus atromentosus*)

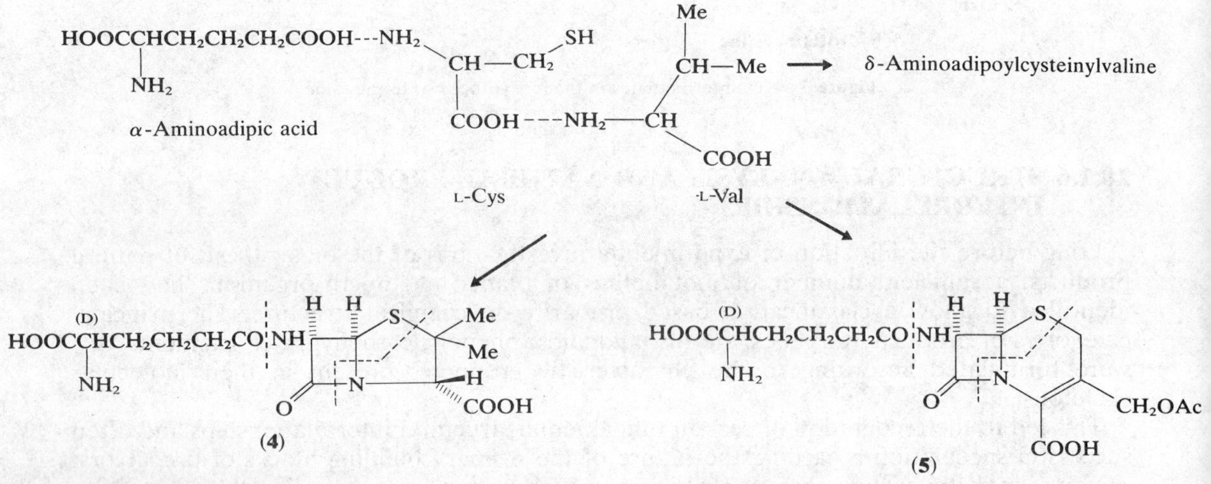

(4)

Penicillin N (D-δ-aminoadipoyl-6-APA)
Cephalosporium spp. and
Penicillium chrysogenum

(5)

Cephalosporin C (D-δ-aminoadipoyl-7-ACA)
Cephalosporium spp.

Other *P. chrysogenum* penicillins:[49]

	Side chain	*Precursor*
Penicillin F	$MeCH{=}CHCH_2CH_2CO{-}$	Acetyl coenzyme A
Penicillin G	$PhCH_2CO{-}$	Phenylacetic acid
	$MeSCH_2CH_2CO{-}$	$MeSCH_2CH_2CHNH_2COOH$
	$EtSCH_2CO{-}$	$EtSCH_2CHNH_2COOH$

Figure 4 Amino-acid-derived metabolites

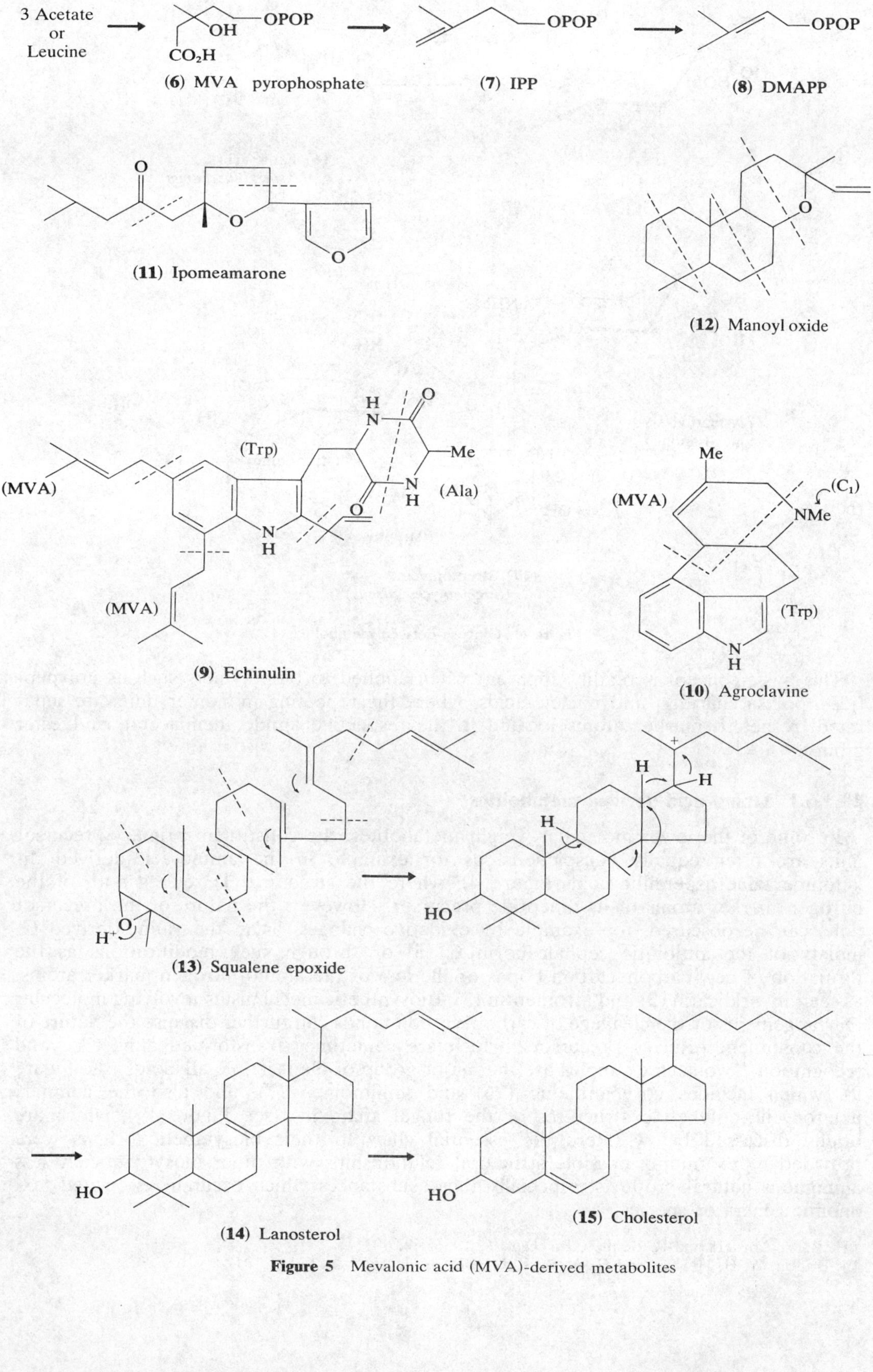

Figure 5 Mevalonic acid (MVA)-derived metabolites

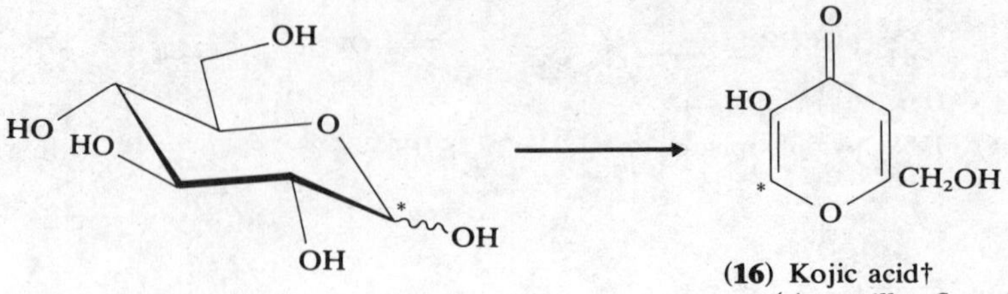

D-Glucose

(16) Kojic acid†
(*Aspergillus flavus*)

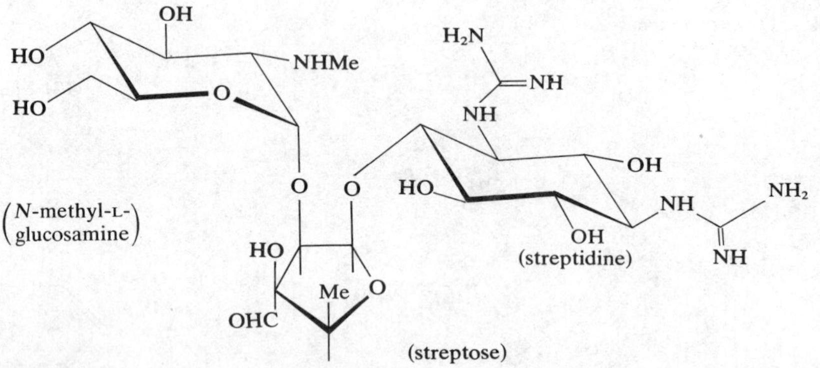

(17) Streptomycin‡
(*Streptomyces griseus*)

Figure 6 Glucose-derived metabolites

This basic concept is readily apparent when applied to biopolymers such as polypeptides, polysaccharides, and nucleic acids, where the repeating monomer units are separated by hetero marker atoms located in the respective amide, hemiacetal, and ester groups.

28.1.6.1 Amino-acid-derived metabolites

In some of the lower molecular weight metabolites, the constituent primary precursor units are often equally conspicuous, as for example in the amino-acid-derived diketopiperazine aspergillic acid (Figure 4), where the structure (**1**) retains both of the nitrogen marker atoms of its dipeptide precursor. However, the nature of the precursor units can be obscured, for example by oxidative changes, as in the valine-derived C_5 moiety of the antibiotic cephalosporin C (**5**) or through such modifications as the formation of new carbon–carbon bonds or the loss of the amino-nitrogen marker atoms, as seen in viridicatin (**2**) and atromentin (**3**). Biosynthetic mechanisms involving molecular rearrangements or the cleavage of carbon–carbon bonds can further disguise the nature of the constituent primary precursors, which are sometimes transformed almost beyond recognition. Two such examples are the major group of plant indole alkaloids (see Figure 9), which includes corynantheine (**26**) and yohimbine (**27**), and also the complex heterocyclic polyketide structures of the fungal aflatoxins (see Figure 19), which are briefly discussed below. Here the essential clues to their biosynthetic origins were provided by examining possible structural relationships with other biosynthetically less ambiguous natural products, especially those substances which occur in associated taxonomic genera or species.

† H. R. V. Arnstein and R. Bentley, *Biochem. J.*, 1956, **62,** 403.
‡ J. Bruton, W. H. Horner, and G. A. Russ, *J. Biol. Chem.*, 1967, **242,** 813.

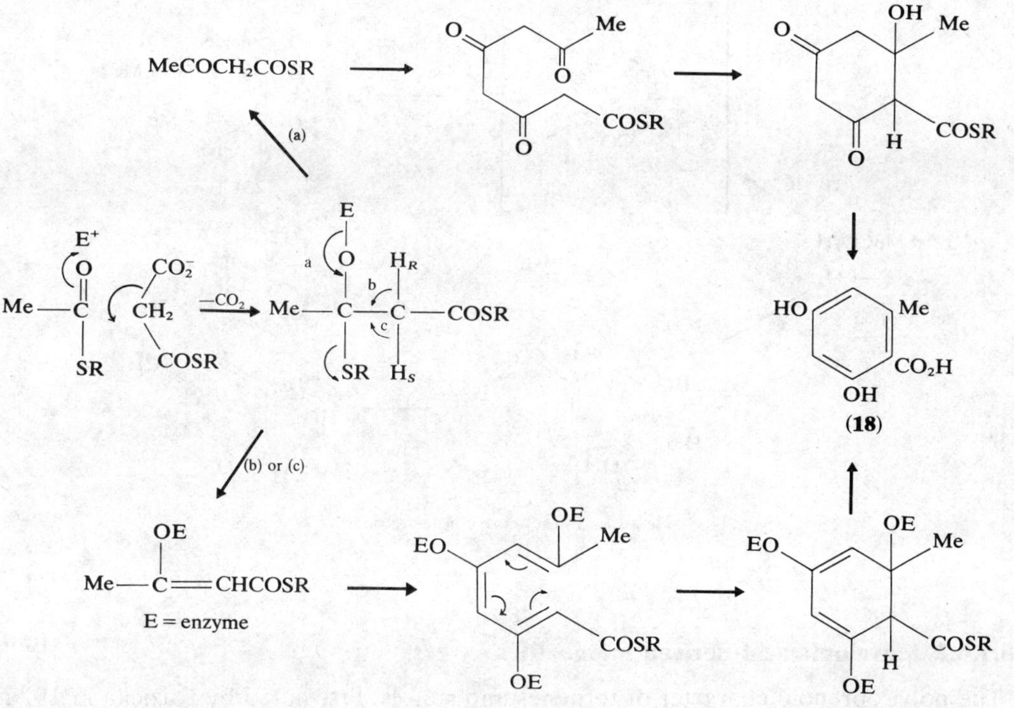

(18)

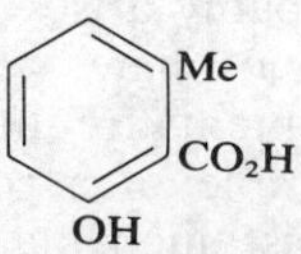

(19)

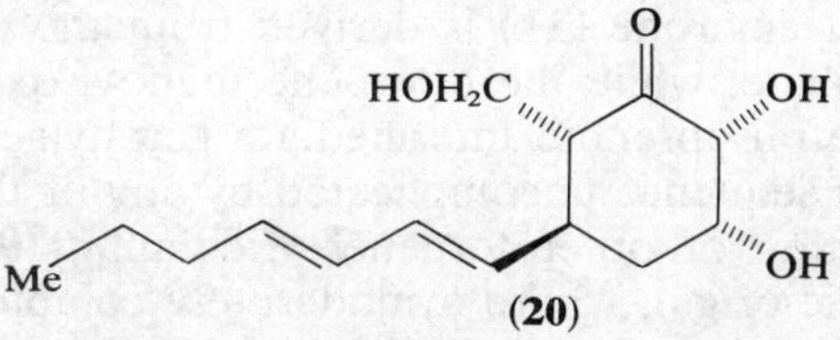

(20)

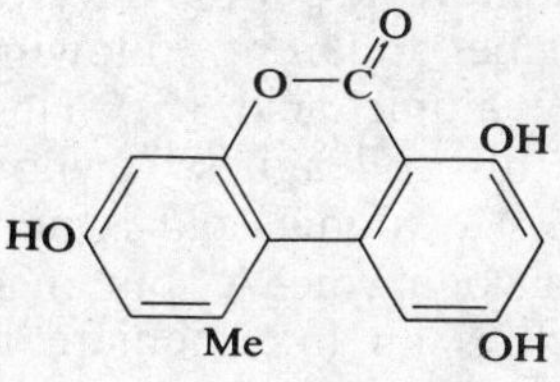

(21)

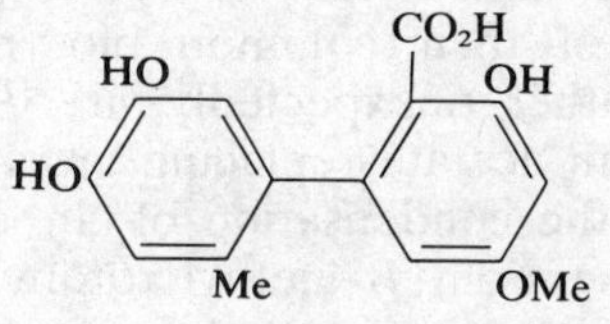

(22)

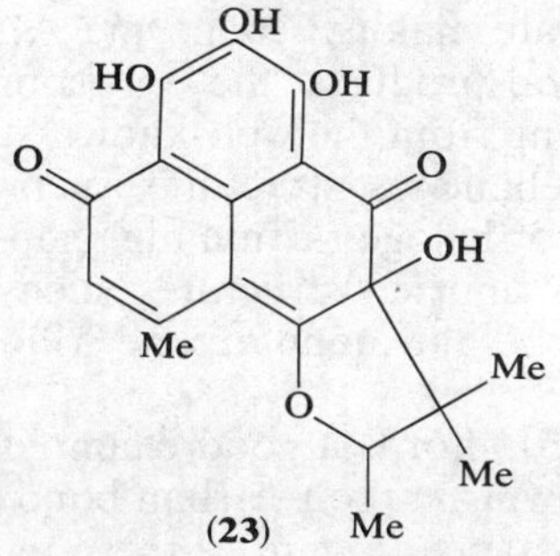

(23)

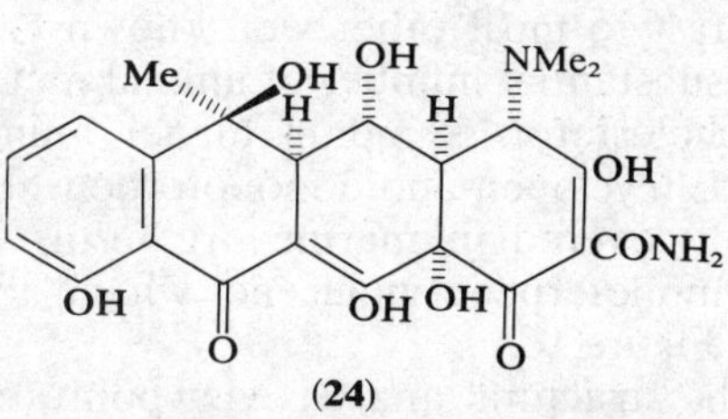

(24)

Figure 7 Polyketides derived from acetate

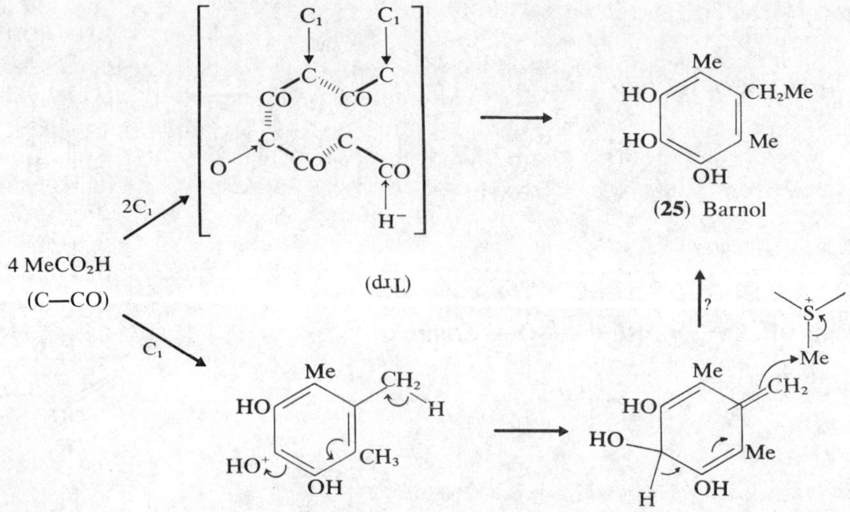

Figure 8 Barnol biosynthesis (tetraketide + 2C₁ units)

28.1.6.2 Mevalonic acid-derived metabolites

The polyisoprenoid character of terpenes and sterols, first noted by Ruzicka in 1922, is illustrated in Figure 5 by three different types of fungal isoprenoids. The simple sesquiterpene ipomeamarone (**11**) is derived from the common tri-isoprenoid precursor farnesyl pyrophosphate, while the diterpene manoyl oxide (**12**) probably arises from the corresponding tetraisoprenoid intermediate geranylgeraniol pyrophosphate, which undergoes a cyclization sequence uncomplicated by any of the rearrangements frequently observed in the formation of other terpenes. Echinulin (**9**) is an example of a product of mixed biosynthetic origin, where a moderately complex structure is amenable to a straightforward structural analysis which clearly indicates its derivation from tryptophan, alanine, and three isoprene units.

The nature of the actual monoisoprenoid intermediates, namely isopentenyl pyrophosphate (**7**) and dimethylallyl pyrophosphate (**8**), only became apparent following the identification of their common biosynthetic precursor, mevalonic acid (**6**). This was discovered, rather unexpectedly, in 1956 by Folkers and his colleagues,[19] who were investigating an acetate-replacing growth factor of bacteria. Mevalonic acid is normally formed from the condensation of three molecules of acetyl coenzyme A, but it is also known to arise from β-methylcrotonyl coenzyme A, which is an intermediate in the metabolism of the amino-acid leucine. Its conversion to (**6**) involves the uptake of a molecule of CO_2, in a process which is analogous to the biotin-dependent carboxylation of acetyl coenzyme A, leading to the polyketide intermediate malonyl coenzyme A.

In contrast to most other well-known groups of natural products, the polyisoprenoids include a substantial number of animal metabolites ranging from the well-known steroids, such as cholesterol, found in higher animals, to miscellaneous structures such as the ant iridoids (cyclopentanoid monoterpenes).[20] A variety of halogen-containing substances recently discovered in marine organisms include some unusual structures such as the apparent monoterpene violacene, which, together with the ant monoterpene iridodial, is shown in Figure 9.

From the structural analysis viewpoint, cholesterol (**15**) affords a good example of the previously mentioned complications which can arise following carbon–carbon bond formation combined with molecular rearrangements. In this instance (Figure 5) these modifications occur during the cyclization of the hexaisoprenoid intermediate squalene epoxide

(**13**) to lanosterol (**14**), which then undergoes the oxidative loss of three methyl substituents. The exploration of this fascinating pathway also provides an excellent illustration of the successful application of tracer techniques, initially using deuterium, but subsequently tritium and ^{14}C isotopes. The numerous subtle details were elucidated in a series of elegant labelling experiments, principally by the three groups of Bloch,[21] Cornforth and Popjak,[22] and Lynen.[23] The proposal by Robinson[24] of the intermediate role of the hydrocarbon squalene as an advanced acyclic precursor of cholesterol, made as long ago as 1934, preceded the determination of the constitution of the triterpenes such as lanosterol. This is a good example of the value of imaginative structural analysis considerations as an aid to identifying biosynthetic interrelationships. In this instance, although the hypothetical cyclization originally envisaged required subsequent minor modification, the suggested precursor role of squalene preceded its experimental verification by 20 years.

28.1.6.3 Metabolites of mixed precursor origin

Biosynthetic pathways, requiring mixed precursors, are quite common, as already exemplified by the fungal alkaloids echinulin (**9**) and agroclavine (**10**),[25] where the structural analyses are fairly self-evident. Less obvious is the established involvement of tryptophan and mevalonic acid in the biosynthesis of a large, and often structurally complex, class of plant indole alkaloids. Representative of two of the major sub-groups are yohimbine, which has a pentacyclic structure (**27**), and the closely-related tetracyclic corynantheine (**26**), in which the carbocyclic ring E is absent (Figure 9). The structure of the quinoline alkaloid emetine (**28**) illustrates a close relationship between the indicated C_9 moiety and the corresponding C_{10} unit of corynantheine.

Although the detailed pathway leading to these alkaloids is considerably more complicated than that of either echinulin or agroclavine, the common role of tryptophan as a precursor was correctly predicted by applying the normal structural analysis criteria.[26] The origin of the unusual C_{10} units corresponding to the non-tryptamine moieties (**29**) and (**30**) was suggested by their skeletal relationships with several monoterpenes present in closely-allied plant families, and also in a taxonomically totally unrelated source, namely *Iridomyrmex*, a genus of ants. Structural comparisons of this cyclopentanoid monoterpene series (iridoids), with their ring-cleaved analogues (seco-iridoids) as typified by the plant lactones swertiamarin (**31**) and erythrocentaurin (**32**), led to the hypothesis[27,28] outlined in Figure 9.

This concept was supported by the subsequent identification of the strychnine-congener loganin (**33**), which was seen as a probable intermediate, possessing precisely the required cyclopentanoid monoterpene carbon skeleton.[29] The general hypothesis was later confirmed, following a sophisticated series of alkaloid-labelling experiments, by the separate groups of Arigoni, Battersby, and Scott.[30] These included the demonstration of the incorporation of loganin, and also secologanin (**34**), into indole alkaloids of *Vinca rosea*.

It is interesting to compare the cyclization sequence (**29**)→(**30**) of the cyclopentanoid monoterpene hypothesis with the reverse sequence, which formed the basis of a previous highly original proposal. One of the earliest schemes was the suggestion of Barger[31] and Hahn,[32] that the non-tryptamine C_{10} unit of yohimbine (including ring E) originated from a disubstituted phenylalanine unit. Following the structural elucidation of strychnine, a most ingenious biogenetic scheme was proposed by Woodward,[33] requiring the plausible oxidative ring-fission of a corresponding intermediate, possibly related to 3,4-dihydroxyphenylalanine (DOPA). This fascinating concept was of direct value to Robinson in his derivation of the structure of emetine,[34] which was considered to provide strong support for the Woodward fission hypothesis, as did the subsequent characterization of many alkaloids apparently corresponding to a 'seco-ring E' series of the corynantheine type. From a retrospective viewpoint, the outstanding merit of this earlier hypothesis was

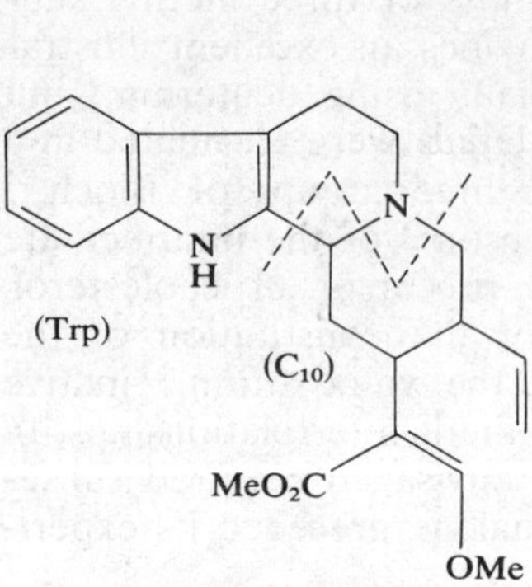

(**26**) Corynantheine

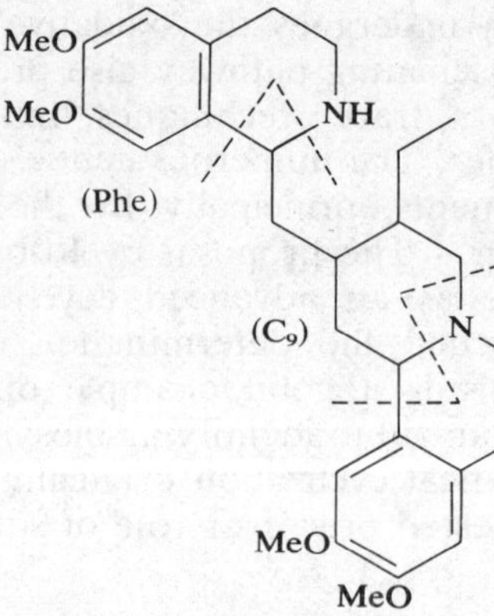

(**28**) Emetine

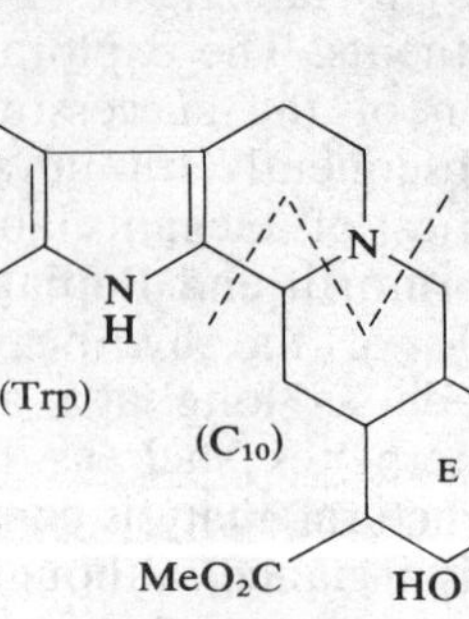

(**27**) Yohimbine

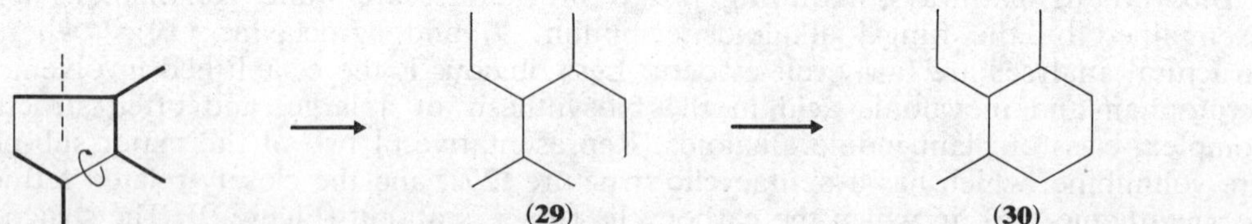

(**29**) (**30**)

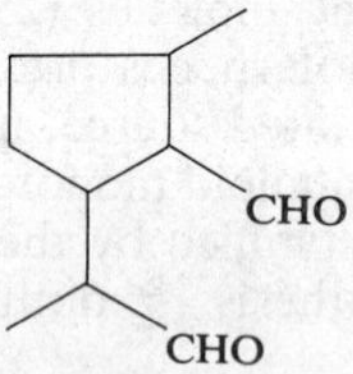

Iridodial[20]

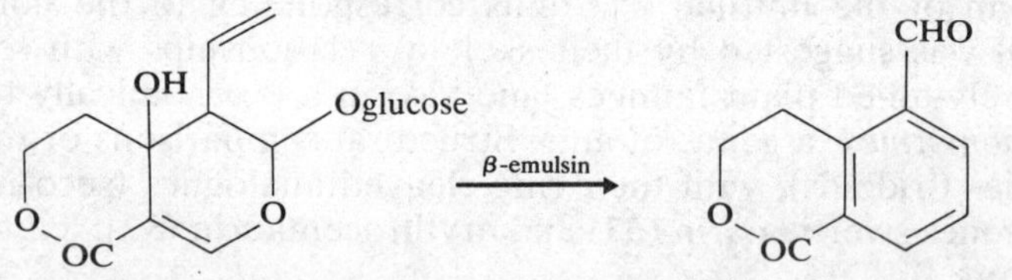

(**31**) Swerteamarin (**32**) Erythrocentaurin

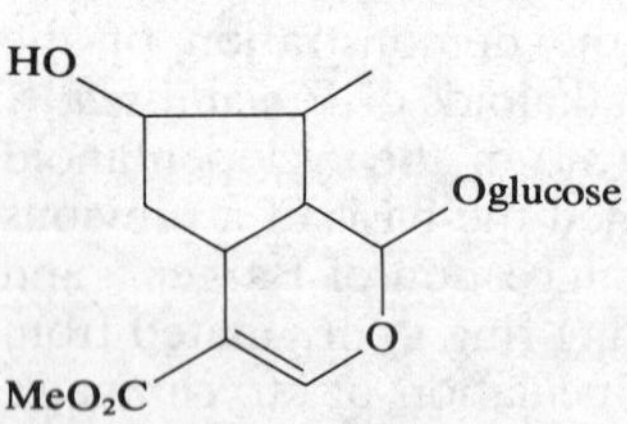

(**33**) Loganin

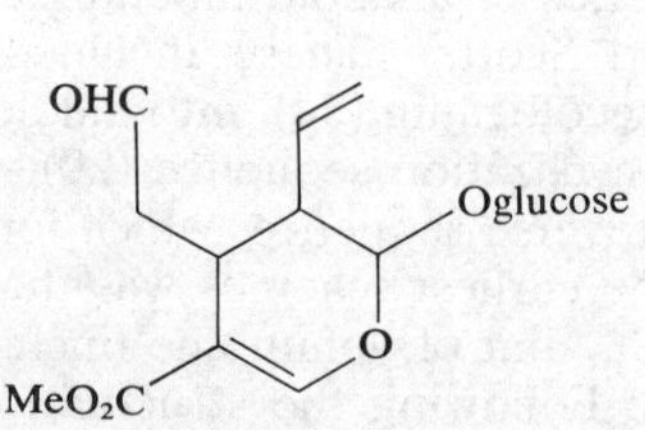

(**34**) Secologanin

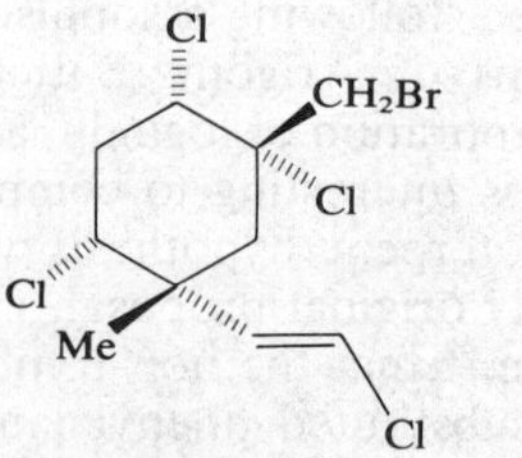

Violacene*

Figure 9 Monoterpenoid derivatives

* D. Van Engen, J. Clardy, E. Kho-Wiseman, P. Crews, M. D. Higgs, and D. J. Faulkner, *Tetrahedron Letters*, 1978, 29.

the recognition of a potential precursor–product relationship between the corynantheine and yohimbine groups of alkaloids.

The numerically predominant 'seco' series is now considered biosynthetically to precede the formation of the carbocyclic ring E group, the mechanism of cyclization possibly parallelling that illustrated in Figure 9 for the conversion of swertiamarin (**31**) to erythrocentaurin (**32**), following enzymic hydrolysis of the β-glucoside with emulsin. This cyclization step may involve an intermediate acyclic diene grouping, also present in corynantheine (**26**) and secologanin (**34**). However, the actual mechanism of formation of the alicyclic ring E of yohimbine still remains to be determined. The specific interrelationships of the various groups of indole alkaloids have to a considerable extent been clarified by appropriate labelling studies, and these are reviewed in a subsequent chapter and elsewhere[30] (Chapter 30.1).

28.1.6.4 Glucose-derived metabolites

The ubiquitous carbohydrate glucose, as previously noted, can serve indirectly as the source of all primary and hence all secondary metabolites. It is, however, also a direct precursor of a variety of products, for example numerous glucosides, and, after appropriate modifications, the unusual glycosides frequently encountered in *Streptomyces* antibiotics, such as erythromycin (**35**)[48] (Figure 10). The role of glucose as a direct precursor has been established for the *Aspergillus oryzae* metabolite kojic acid (**16**) (Figure 6), which is apparently formed without cleavage of any of the original glucose carbon–carbon bonds. This direct utilization of glucose may similarly take place in the biosynthesis of certain oligosaccharides, such as the antibiotics streptomycin (**17**)[48] and neomycin,[35] although clearly its conversion to the branched-chain residue of streptose would require a rearrangement of the glucose skeleton.

28.1.6.5 Acetate- and propionate-derived polyketides

The two-carbon primary precursor acetic acid, which is normally utilized in the form of its coenzyme A thioester, has been shown to be involved in the assembly of a remarkable range of natural products. Thus, labelling studies have shown many types of structures (*e.g.* fatty acids, certain phenols, and even some alkaloids) to be derived from polyacetates, currently referred to as polyketides, after Collie's original term,[39] used to describe the synthetic products which he prepared from β-diketones and β-keto esters.

This pathway, which is the subject of a later section (Chapter 29.1) is illustrated in Figure 7. The archetypal tetraketide orsellinic acid (**18**), first described in 1959, may be formed by sequential Claisen condensations of an acetate primer unit with three malonate units, leading through enzyme-bound thioesters of acetoacetic and triacetic acids to tetracetic acid thioester. This is then considered to undergo cyclization *via* an aldol condensation followed by the elimination of water. The hypothetical poly-β-ketonic intermediates are reflected in the resulting alternating oxygen substituents of the predominantly resorcinol-based structures, which are a common feature of many polyketide-derived phenolic metabolites, as typified by alternariol (**21**). This characteristic distribution of oxygen atoms is often masked by oxidative and reductive processes, which lead, for example, to variants of orsellinic acid and alternariol, such as 6-methylsalicyclic acid (**19**) and altenusin (**22**), the polyketide origin of which is less immediately apparent.

An alternative hypothetical mechanism for the incorporation of acetate and malonate units into orsellinic acid is also shown in Figure 7.[36] This pathway is devoid of ketonic intermediates, which are replaced by enzyme-stabilized polyenolates and which, given the appropriate sequence of conjugated *cis* and *trans* double bonds, could form orsellinic acid by a direct electrocyclic rearrangement. In higher polyketides, leading to polycarbocyclic structures, individual types of ring systems would be predetermined by the specific geometrical sequences of *cis* and *trans* double bonds.

Although the early synthetic work of Collie, at the turn of the century, brilliantly anticipated the isolation of the numerous polyketide phenolic and heterocyclic products which are now known to exist, very few of these had been characterized at the time of his investigations, and the potential significance of this valuable biogenetic contribution was largely overlooked. Interestingly, Raistrick and Clark, who in 1919 confirmed Wehmer's earlier observation of the formation of citric acid in fungi, actually quoted Collie's polyketide concept in support of their far-sighted prediction[37] of the basic mechanism of biosynthesis of citric acid, involving an aldol condensation of oxaloacetic and acetic acid units.

Raistrick's pioneering investigations of fungal metabolites over the next four decades led to the isolation of an impressive number of mono-, di-, and tri-cyclic phenols, many of which were subsequently shown to be polyketides, although the possible relevance of Collie's work was not considered in his 1949 Bakerian Lecture, reviewing this field.[38] As it happens, it was biosynthetic investigations of some of these fungal phenols which, not long afterwards, provided the experimental verification of the acetate hypothesis, as elaborated by Birch and Donovan[39] in 1953, to account for the origin of this major class of natural products. Robinson, in the course of a subsequent lecture series,[40] recalled Collie's early publication, and correctly postulated that the tetracycline antibiotics also fall into this biosynthetic category. The specific incorporation of ^{14}C-labelled acetate into a polyketide was first demonstrated by Birch and his colleagues,[41] who applied this technique to 6-methylsalicylic acid (**19**). This was soon confirmed in numerous investigations of a variety of mould products both by Birch's group and also by others.[42] A selection of differing structural types of cyclic microbial polyketides is shown in Figure 7, namely palitantin (**20**),[43] alternariol (**21**),[44] norherqueinone (**23**),[45] and the antibiotic oxytetracycline (**24**).[46]

The latter two products contain, in addition to their polyketide-derived ring systems, isopentenyl and methyl substituents respectively. These are further examples of mixed precursor metabolites and, in norherqueinone, this was shown to involve a typical MVA-derived isoprenoid unit.[45] This is attached to the phenalenone ring by its branched carbon, which is the less usual mode of linkage, previously exemplified in Figure 5, by one of the three isopentenyl substituents of the fungal indole echinulin (**9**). The *C*-methyl and *N*-dimethyl substituents of oxytetracycline provide examples of the utilization of C_1 units (*cf.* Table 2) derived from the *S*-methyl group of methionine or formate. Substitution of these units into polyketides with the formation of carbon–carbon bonds occurs at the potentially anionic acetate-methyl derived carbon, consistent with an enzymic mechanism involving a nucleophilic displacement of the methyl group attached to the sulphonium atom of an *S*-adenosylmethionine donor molecule. There is good experimental evidence, based on a variety of studies, which supports the view that, where this type of C_1 substitution takes place, it does so prior to cyclization of the polyketide chain. Thus, 6-methylpretetramid (Figure 3), the fully aromatized tetracyclic precursor of oxytetracycline, cannot be replaced by pretetramid itself, which is a known intermediate leading to the 6-demethyltetracycline series.[16]

While C_1 substitution of linear acetate-derived polyketides is also commonplace, it is not apparently utilized as a direct metabolic route leading to the formation of propionate from acetate or malonate, although such a reaction would not be at all mechanistically exceptional. This might almost be considered to be a good opportunity wasted, in view of the direct access of propionate to the tricarboxylic acid cycle *via* succinate. In fact, the C_1 substitution of the primary chain-initiating acetate unit does appear to take place in the interesting fungal phenol barnol (**25**)[47] (Figure 8), the symmetry of which, incidentally, subtly complicates the normally facile structural analysis of most monobenzenoid polyketides. Conceivably, this methylation could occur after aromatization and oxidative insertion of the *para*-hydroxyl group as shown, although recent evidence favours the introduction of both C_1 units at the more usual pre-aromatic stage.[50]

In marked contrast to the fungi, the actinomycetes, in addition to carrying out the above C_1 substitution of polyketides, as seen in tetracycline biosynthesis, also frequently utilize propionate as a direct polyketide precursor unit. A good example is seen in the biosynthesis of the *Streptomyces* antibiotic erythromycin (Figure 10).[48]

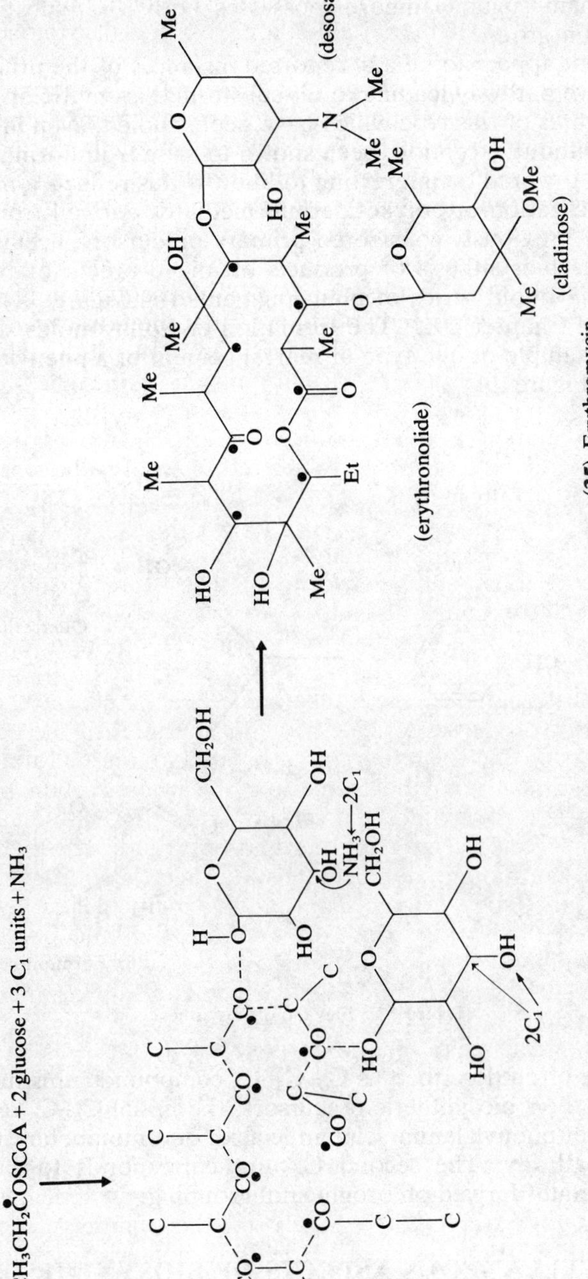

(35) Erythromycin
(*Streptomyces erythreus*)

Figure 10 Polyketides derived from propionate

The biosynthetic 'shorthand', used to indicate the primary precursors, is self-explanatory and illustrates a number of interesting biosynthetic features mentioned previously. For example, the heptaketide aglycone is totally derived from seven propionate units and is glycosidically linked to two glucose-derived monosaccharide units. One of these, the amino sugar desosamine, contains a C_1-substituted nitrogen, while the other, the C_7 branched-chain sugar cladinose, possesses both a C_1-substituted oxygen and a C_1-substituted carbon group.

As yet there do not appear to be any reported instances of the utilization of propionate units as an alternative pathway leading to C_1-substituted aromatic or alicyclic products. In the numerous structures of this type which have been studied, even in *Streptomyces*, the C_1 substituents have, without exception, been shown to arise from formate or methionine but not propionate, and it would be interesting to know if this reflects a mechanistic limitation in respect of the cyclization of polyketide intermediates derived from propionate.

As with the other previously considered primary precursors, acetate and also propionate contribute to the biosynthesis of products of mixed-precursor origin. A particularly large group is the flavonoid series of plant pigments (*cf.* Figure 11), the biosynthesis of which is discussed in Chapter 29.1. The formation of isoflavonoids such as formononetin (**36**)[48] provides an example of one type of rearrangement of a phenylalanine unit, which is again referred to in Figure 24.

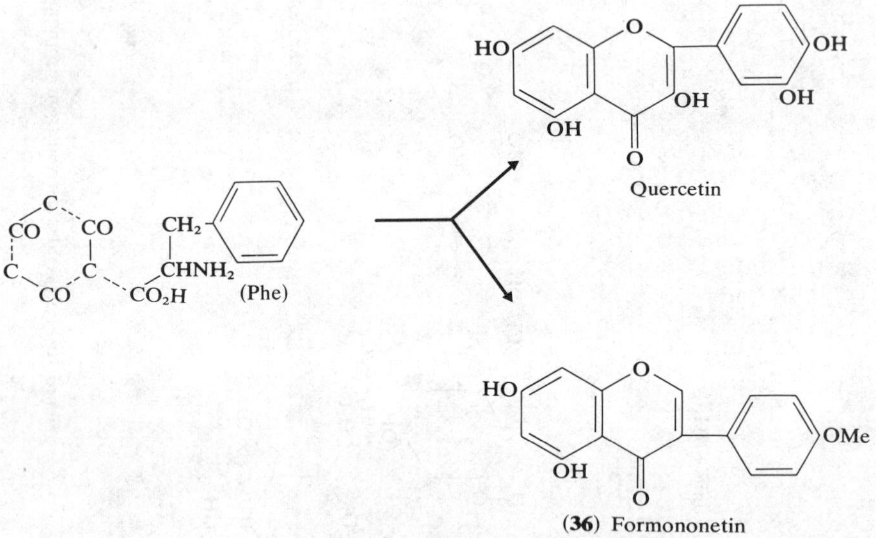

(**36**) Formononetin

Figure 11 Flavanoid biosynthesis

The flavonoids are often described as C_6–C_3–C_6 compounds, this numerical representation itself indicating their biosynthetic precursors, the initial C_6–C_3 unit corresponding to the carbon skeleton of phenylalanine, cinnamic acid, or a metabolic equivalent, formed by the shikimic acid pathway. The second C_6 unit corresponds to a triketide, or, more specifically, to an acetate-derived phloroglucinol grouping.

28.1.7 SOME MISCELLANEOUS ASPECTS OF BIOSYNTHESIS

Undoubtedly, purely speculative exercises, such as structural analysis and the recognition of potential biosynthetically-significant molecular interrelationships, have their own intrinsic fascination for the natural product chemist. Interesting though this aspect may be, the ultimate object of the exercise is, of course, to utilize these hypothetical considerations as a basis for the design of meaningful experimental explorations of metabolic pathways.

28.1.7.1 Autoradiographic applications

As previously indicated, our present considerable knowledge of the many classes of natural product structures is the outcome of more than a century of sustained, yet basically empirical experimental effort. In harvesting these varied fruits, inevitably the best, or at least the most conspicuous, 'plums' were the first to be picked. The search is now becoming more subtle, but certainly no less satisfying, and the need to systematize future explorations is increasingly apparent. One simple, yet relatively inexpensive, approach is based on the autoradiographic examination of extracts of plants, animals, and micro-organisms, grown in the presence of selected radioactively-labelled primary precursors.

This method has many advantages, predominant among which is the sensitive detection criterion which is independent of either chemical structure or special properties, other than those which determine their characteristic mobilities in various chromatographic or zone electrophoretic systems. In addition, their isolation, at least initially in minor amounts, is reasonably assured, as their detection is dependent upon the ability of these chromatographic or electrophoretic techniques to effect the resolution of mixtures into their individual constituents. Finally, the knowledge of their primary precursors automatically places the resulting metabolites in a particular biosynthetic class (*cf.* Table 2), which thus provides potentially useful structural information.

This approach is also likely to be of considerable taxonomic value, which can be considered at two levels. The first does not involve the structural characterization of new metabolites, but simply fingerprinting of the characteristic patterns of components labelled from selected precursors. This has the potential of providing a new set of taxonomic characteristics, which could be applied on a facile and inexpensive routine basis to the large groups of micro-organisms maintained in existing type-culture collections. It can also be applied to the direct examination of plants by short-term incubation of selected plant tissues at various times of growth.

The simplicity and small scale of operation of the experimental system ideally lends itself to adaptation as a comparatively inexpensive general screening procedure. Thus, satisfactory autoradiograms are readily obtained, using less than one gram (wet weight) of freshly grown microbial cells shaken overnight in 2 ml or less of buffer containing low levels (*e.g.* 10 μCi) of ^{14}C-labelled primary precursors, such as acetate, mevalonate, or amino-acids. This technique selectively detects metabolites of individual precursors and has been used for a range of microbial products, including penicillins,[49] tetracyclines,[54] alternariol,[51] and the fungal phenalenones.[52]

The same procedure has been applied to plants on a test-tube scale, for example in a preliminary examination of three species of the genus *Haemodorum* (an Australian monocotyledon) to determine suitable conditions for producing labelled haemocorin.[53] Freshly sliced rhizomes of *H. corymbosum*, *H. planifolium*, and *H. spicatum* were incubated overnight with solutions of ^{14}C-labelled precursors (acetate, phenylalanine, and tyrosine) and methanolic extracts subsequently chromatographed on paper using 70% propanol. The resulting autoradiograms showed multiple components, some of which (*e.g.* haemocorin and other unknown pigments) were common to all three species. These products, which incorporated each of the above precursors but with differing efficiencies, were therefore characteristic of the genus *Haemodorum*. Other components were representative of the individual species, and were therefore taxonomically useful species indicators.

A corresponding simple experiment,[54] in which a fresh slice of bark of *Strychnos lucida* was incubated with [1-^{14}C]acetate, also yielded multiple components, one of which was characterized by chromatographic comparison and subsequent isolation as the alkaloid brucine.

This biosynthetic approach to chemotaxonomic classification is easily adaptable to general screening procedures, using standard ^{14}C-labelled precursors in conjunction with a selected range of chromatographic systems or zone electrophoresis conditions. Since the metabolites are directly characterized as a function of their chromatographic or electrophoretic mobilities, they may be immediately classified on a numerical basis (*e.g.* in terms of R_f values), which is ideally suited to data collection and retrieval.

The second level at which the biosynthetic and taxonomic potential of autoradiography may be utilized is in the search for specific types of metabolites. A simple example is seen in the investigation of the utilization of the amino-acid methionine in the biosynthesis of penicillins by *Penicillium chrysogenum*.[49] The chromatographically-resolvable penicillin components of the fungus can be identified using a specific colorimetric spray reagent,[55] together with a standard bioautographic procedure for detecting paper chromatographic zones showing antibacterial activity. These zones are also seen in autoradiograms of extracts following incubation of the mould with [^{35}S]sulphate, which serves as a source of the ring sulphur of the penicillin thiazolidine nucleus. Growth in the presence of methionine was found to produce a new penicillin, which could also be detected by autoradiography, following small-scale feeding experiments with the labelled amino-acid. Furthermore, the chromatographic zone corresponding to the new penicillin was found to be radioactive when the mycelium was incubated with [^{14}C]methionine labelled at C-2 and in the methyl group, but not when the label was at C-1. This provided a significant structural clue, namely that all of the methionine skeleton between C-2 and the terminal methyl group was incorporated, whereas the carboxyl group was not. These conclusions were supported by the subsequent finding that substitution of methionine by its oxidative decarboxylation product, namely *S*-methylmercaptopropionic acid (**37**), yielded a chromatographically indistinguishable penicillin (Figure 12). The methionine isomer *S*-ethylcysteine (**38**) was similarly converted to a specific penicillin, which was isolated and structurally characterized as sodium 6-ethylmercaptoacetamidopenicillanate (**39**).

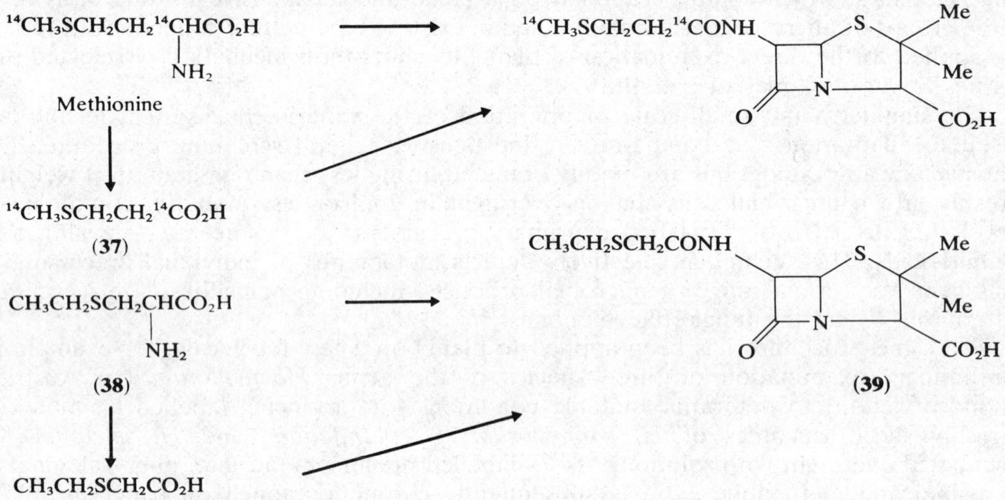

Figure 12 Utilization of methionine as a penicillin side-chain precursor[49]

In the course of these and associated studies of the biosynthesis of new penicillins and cephalosporins, [^{14}C]valine, [^{35}S]methionine (as a source of the penam and cephem ring sulphur atoms), and [^{14}C]acetate (a fungal source of α-aminoadipic acid) were mutually incorporated, not only into penicillin N and cephalosporin C, but also into unknown, possibly tripeptide, derivatives. These observations illustrate the value of this simple means of detecting specific types of precursor-defined intermediates.

Another example of the useful application of autoradiographic techniques is seen in the direct search for new *P. herquei* phenalenones or biosynthetic intermediates. Prior to this study, three structurally-related compounds were known, namely atrovenetin (**40**), herqueinone (**41**), and norherqueinone (**23**). The latter pigment had previously been shown to be an isoprenoid-substituted polyketide.[45]

Following an autoradiographic study along the lines already indicated, a radioactive zone, corresponding to an unidentified metabolite, was detected among the t.l.c.-resolved components of extracts labelled both from [^{14}C]acetate and from [^{14}C]mevalonic acid, consistent with an isoprenoid-polyketide structure. A known congener, namely the polyketide anthraquinone physcion (**43**) (emodin-7-methyl ether), predictably incorporated both ^{14}C-labelled acetate and formate, but not [^{14}C]mevalonic acid. It was therefore possible to exclude this and similarly labelled components from the search for advanced herqueinone precursors. This is a further example of the selective scope of autoradiography when applied to metabolites of mixed precursor origin. The new mevalonate- and acetate-derived pigment was subsequently isolated, by means of preparative t.l.c., and identified as deoxyherqueinone (**42**).

As indicated in Figure 13, deoxyherqueinone (**42**) proved to be the most likely

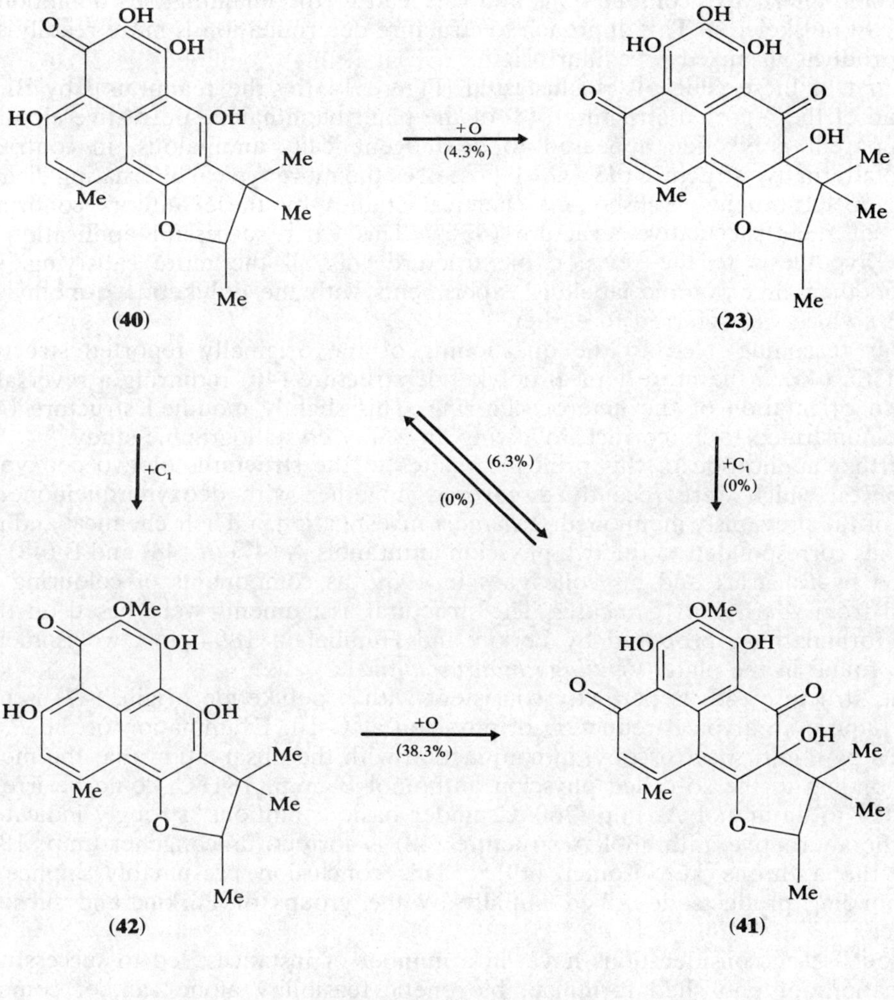

Figure 13 Interconversion of fungal phenalenones

metabolic intermediate between atrovenetin (**40**) and herqueinone (**41**). This was established by determining the efficiency of interconversion of all four ^{14}C-labelled phenalenones.[52] Norherqueinone, which on structural grounds had previously appeared to be the obvious intermediate, in the light of these results is currently viewed as an irreversibly-formed shunt metabolite of atrovenetin.

28.1.7.2 Biosynthetic approach to structure determination

The success with which structural analysis criteria, based on the predictable assembling of a limited number of primary precursor molecules, can account for the formation of the majority of known natural products (*cf.* Table 2), has inevitably led chemists to question the validity of any structures which do not conform to the most straightforward biogenetic interpretation. In a number of instances, such considerations have led to the revision of erroneously assigned structures. This approach has been particularly useful in studies of polyisoprenoids and polyketides.

A second structural application requires the initial biosynthetic classification of an unknown product by means of isotopic labelling experiments designed to identify its primary precursor units. The labelled metabolite is then selectively degraded to yield overlapping molecular fragments, the correlation of the specific activities of which limit the possible alternative combinations and can lead to the unambiguous definition of the parent carbon skeleton. This approach to structure determination is more readily applicable to products of mixed precursor origin.

The first of these concepts is illustrated (Figure 14) by the reappraisal by Birch and Donovan of the reported structure (**44**) of the plant naphthalene derivative eleutherinol. This apparent polyketide appeared to be biogenetically anomalous, in contrast to a possible alternative structure (**45**) which possesses the more typical alternating distribution of oxygen substituents. Subsequent chemical studies by these authors confirmed the validity of their speculative structure (**45**).[56] This early successful application of the general hypothesis to the revision of structure was all the more satisfying, since it preceded their first isotopic labelling experiments with the polyketide 6-methylsalicylic acid (**19**) which was referred to earlier.[41]

Similar reasoning[45] led to the questioning of the originally reported structure for atrovenetin (**46**),[57] the more typical polyketide structure (**40**) requiring a reversal of the proposed orientation of the heterocyclic ring. This slightly modified structure (**40**) was later demonstrated to be correct, following an X-ray crystallographic study.[58]

A further application of this principle relates to the structures of two deoxyphyscion metabolites, which were recently discovered, together with deoxyherqueinone, in the course of the previously mentioned *P. herquei* investigation.[52] Their chemical and physical properties corresponded to the two physcion anthranols A (**47**) or (**48**) and B (**48**) or (**47**), reported by Raistrick and his colleagues in 1939, as components of colouring matters isolated from *Aspergillus glaucus*.[59] The structural assignments were based on the even earlier formulations, proposed by Perkin and Humbel in 1894, for two isomeric substances found in the plant *Ventilago madraspatana*.[60]

While structure (**48**) is perfectly consistent with a polyketide origin, (**47**) is not, and would require an atypical reduction of physcion at C-10. Examination of the *P. herquei* products by n.m.r. spectroscopy, in conjunction with the observation that the metabolite corresponding to the so-called physcion anthranol B, m.p. 181 °C, could be irreversibly converted to anthranol A, m.p. 260 °C, under basic conditions, strongly indicated that, while the alternative anthranol A structure (**48**) is correct, its congener (m.p. 181 °C) is simply the anthrone keto-isomer (**49**).[61] This conclusion presumably applies to the corresponding products described initially by the groups of Perkin, and subsequently Raistrick.

Although such considerations have, in a number of instances, led to successful minor modifications of published formulae, biogenetic feasibility alone can, of course, only provide circumstantial support for the validity of any structure. Thus, in the polyketide category, alternariol (**21**), the first *Alternaria alternata* polyhydric phenol to be characterized,[62] is a classical heptaketide, since no rearrangement, oxidation, or reduction steps are required to account for its biosynthesis from acetate. However, had its structure determination been preceded by that of its congener altenusin (**22**),[63,64] where the typical polyketide resorcinol character of ring A has been replaced by a catechol grouping, then, on biogenetic grounds, the *ortho*-dihydroxy substitution of this ring might have been

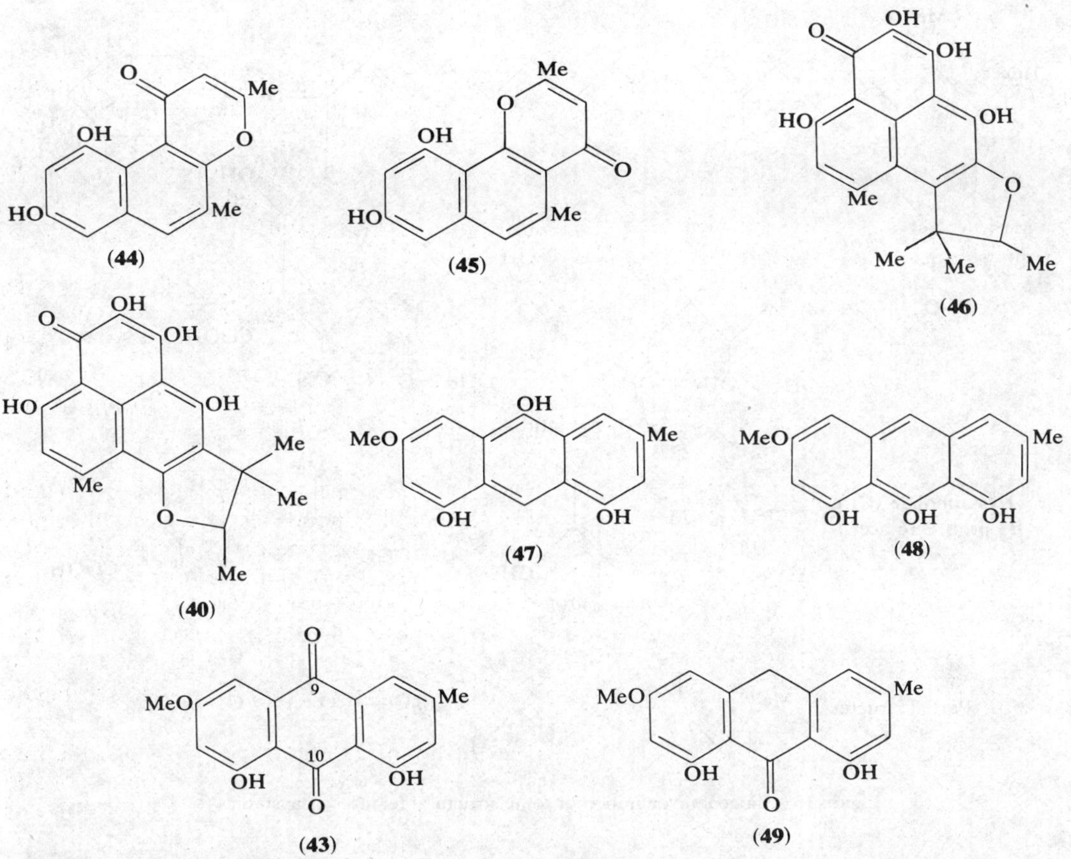

Figure 14 Application of biogenetic criteria as an aid to structure determination

wrongly questioned. Incidentally, if, as seems probable, altenusin is sequentially derived from alternariol, it represents a novel reaction involving in effect the enzyme-catalysed isomerization of a typical polyketide resorcinol nucleus to a catechol derivative.

The second structural application of biogenetic criteria, as mentioned above, involves comparison of the specific activities of a natural product derived from known primary precursors, with those of its chemically-defined degradation products. This is illustrated in Figure 15 by its use in the biosynthetic definition of some structural features of the antifungal *Streptomyces noursei* metabolite nystatin (**50**).[65]

The aglycone nystatinolide, at the time of initiating this investigation, had been considered to be a polyisoprenoid, and this was consistent with the facile incorporation of [^{14}C]acetate, but not the repeated failure to incorporate [^{14}C]mevalonate. In fact, as far as this author is aware, the successful incorporation of this characteristic isoprenoid precursor into any *Streptomyces* metabolite apparently remains to be demonstrated.

The C_3 precursor propionate was subsequently shown to account for the origin of the non-acetate derived carbon atoms, thereby establishing the total polyketide character of the aglycone. Comparison of the specific activities of nystatin with those of its degradation products permitted the calculation that the C_{41} aglycone was derived from 16 acetate and 3 propionate units and furthermore that the carbon skeleton of the terminal 22 atoms was defined by the linear sequence Ac-Pr_2-Ac_7 (where Ac and Pr refer to acetate and propionate units, respectively). This left one partially oxidized propionate to be located among the remaining eight acetate units. The final constitution (**50**) which was subsequently determined by conventional chemical procedures,[66] confirmed the structural conclusions based on the above biosynthetic approach.

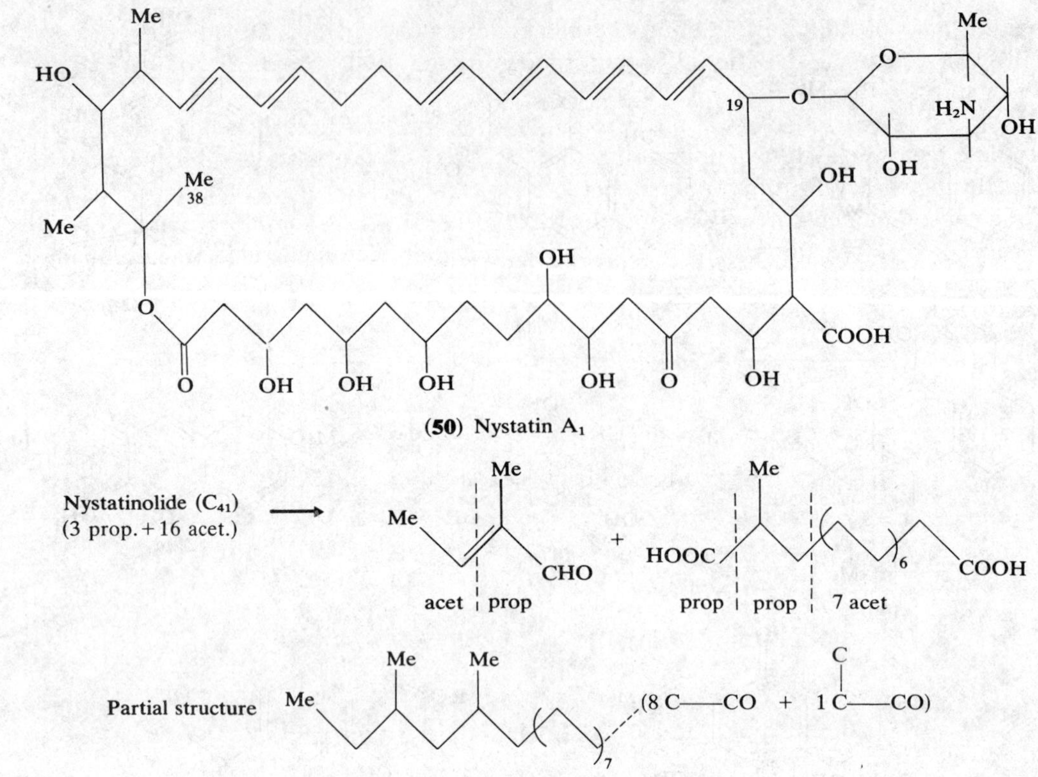

Figure 15 Biogenetic definition of some structural features of nystatin

28.1.7.3 Ring cleavages and molecular rearrangements

The scope of biosynthetic pathways is greatly enhanced by reactions involving either cleavage or rearrangement of carbon–carbon bonds. Where the cleavage involves a ring structure, there is obviously no direct elimination of any fragment of the carbon skeleton. These two processes can lead, *via* comparatively few steps, to highly complex molecules, which often bear little resemblance to the structures of their intermediates.

Both processes are well established in various primary metabolic pathways described in general biochemistry texts. An example of ring cleavage is seen in the degradation of tryptophan *via* anthranilic acid and catechol to *cis, cis*-muconic acid, involving sequential oxidative cleavages of the indole and benzene rings (Figure 16).

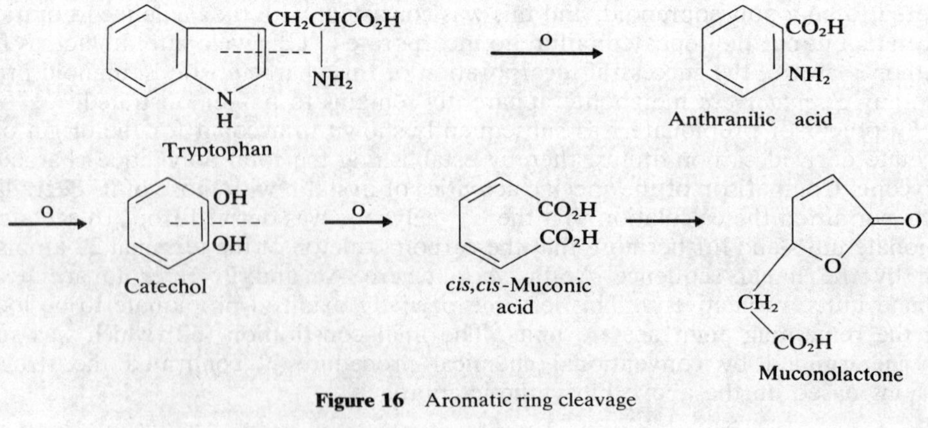

Figure 16 Aromatic ring cleavage

An instance of the cleavage of an alicyclic cyclopentane ring has already been discussed in the conversion of the iridoid monoterpenes to the seco-iridoids, *e.g.* loganin (**33**)→ secologanin (**34**) (Figure 9).

Instances of rearrangements are also fairly common, particularly among the polyisoprenoids, a now classic example being the 1,2-shifts of two methyl substituents, which occur during the cyclization of squalene epoxide (**13**) to lanosterol (**14**) (Figure 5). These are also accompanied by two less obvious 1,2-hydride shifts. A simpler example is seen in the pinacol-type rearrangement of the valine precursor acetolactic acid (Figure 17).[67]

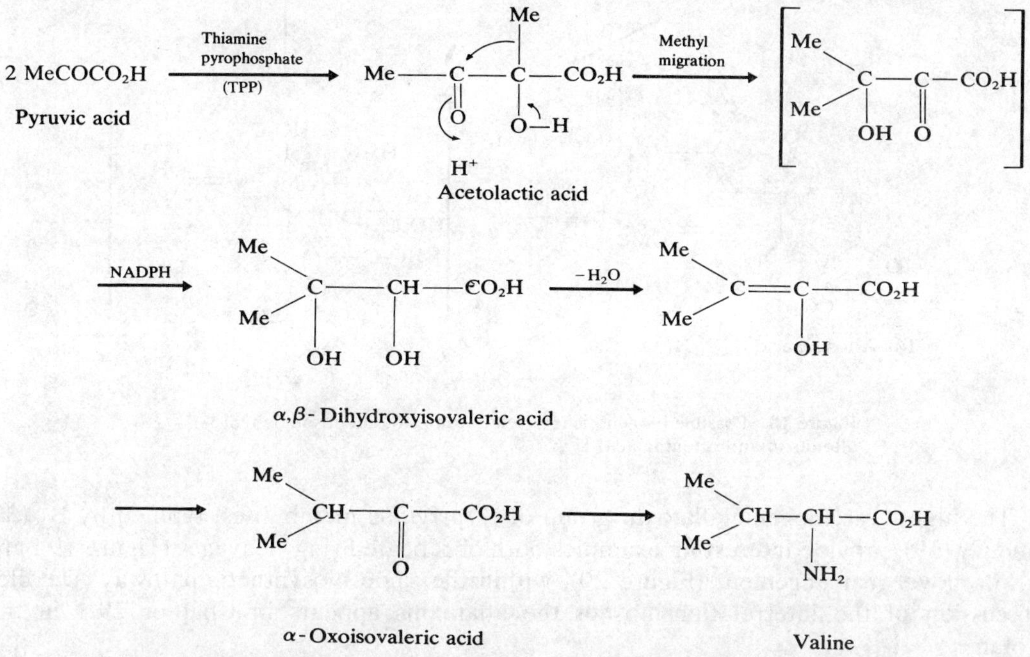

Figure 17 The pinacol-type rearrangement of acetolactic acid in the biosynthesis of valine

Various aspects of the role of ring cleavages in biosynthetic pathways are discussed in a previous review[29] which covered a range of alicyclic and aromatic natural products. In the present article, a simple but unusual type of fission of the phenolic polyketide orsellinic acid (**18**), leading to penicillic acid, has already been noted (Figure 1). In contrast to the majority of known cleavages of carbon–carbon bonds, where the oxidation yields two carbonyl groups (or carboxyl groups as in the above catechol cleavage), the formation of penicillic acid does not involve such an advanced level of oxidation. While in theory penicillic acid could arise *via* cleavage of either the C-1—C-2 or the C-4—C-5 bonds of orsellinic acid, the predominating operation of the latter pathway has been demonstrated by three contrasting isotopic studies. Thus the mode of incorporation of specifically ^{14}C-labelled orsellinic acid was followed by standard degradation procedures,[68] that of [^3H]acetate by tritium n.m.r.,[8,9] and the uptake of [1,2-^{13}C]acetate by the now common ^{13}C–^{13}C spin-coupling technique.[69]

The catechol–muconic acid type oxidative cleavage is paralleled in secondary metabolism by the relationship of the *Alternaria* metabolites altenusin (**22**)[63,64] and altenuic acid II (**24**),[70] which probably arises spontaneously from the substituted muconic acid intermediate (**23**) (Figure 18), possibly *via* dehydroaltenusin (**25**)[63] (*cf.* muconic acid → muconolactone, Figure 16). The role of the different types of oxygenase enzymes involved in oxidative ring cleavages is discussed in a recent comprehensive review.[71]

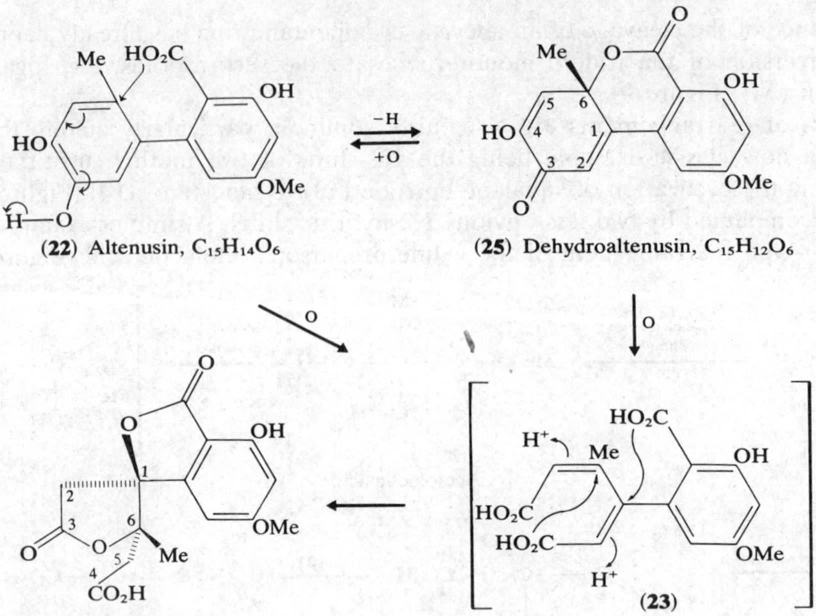

(22) Altenusin, $C_{15}H_{14}O_6$

(25) Dehydroaltenusin, $C_{15}H_{12}O_6$

(24) Altenuic acid II, $C_{15}H_{14}O_8$

Figure 18 Possible biogenetic interrelationships of altenusin, dehydroaltenusin, and altenuic acid II

The highly carcinogenic aflatoxin group of *Aspergillus* metabolites, typified by B_1 (**58**) and G_1 (**59**), provide interesting examples both of repeated ring cleavages (Figure 19), and also a novel rearrangement (Figure 20), within the same biosynthetic pathway. Detailed discussion of the interrelationships of the aflatoxins appears in Chapter 29.1 in this volume.

Labelling experiments[72] supported the predicted derivation of the non-bisfuranoid moiety by a process which involves ring fissions of a bisfuranoanthraquinone (**51**).[29] This hypothetical intermediate was subsequently identified as a metabolite of *Aspergillus versicolor*, namely 6-deoxyversicolorin-A, which occurs together with the xanthone demethylsterigmatocystin (**55**).[73] The biogenetic scheme was originally suggested by the presence of the unusual bisfuranoid group in both the aflatoxins and in the xanthone sterigmatocystin (**54**), which, after an initial negative incorporation experiment,[74] was later confirmed as a precursor of aflatoxin B_1 (**58**).[75] The semi-synthetic 5-hydroxydihydrosterigmatocystin (**56**) was similarly incorporated into the dihydro-series of aflatoxins (B_2 and G_2).[73]

The bisfuranoid substituent was considered to result from a novel oxidative rearrangement of a linear C_6 side-chain of the type present in averufin (**60**).[76] The hypothetic role of an averufin-type intermediate has received support from feeding experiments which have demonstrated the utilization of ^{14}C-labelled averufin as an aflatoxin precursor.[77,78]

It would appear that, in the course of the conversion of averufin to the aflatoxins, one of the hydroxyl groups (at C-6 on the resorcinol ring A) must be reductively removed, as it is not present in the established intermediate sterigmatocystin (**54**)[75] or its 5-hydroxy-derivative (**56**), which also serves as an aflatoxin precursor.[73] Comparison of the hydroxyl substitution patterns of related anthraquinones, for example versicolorin C (**52**) with its typical polyketide alternating oxygen distribution, 6-deoxyversicolorin-A (**51**), and dothistromin (**53**), raises the possibility of interconversions involving the direct removal of

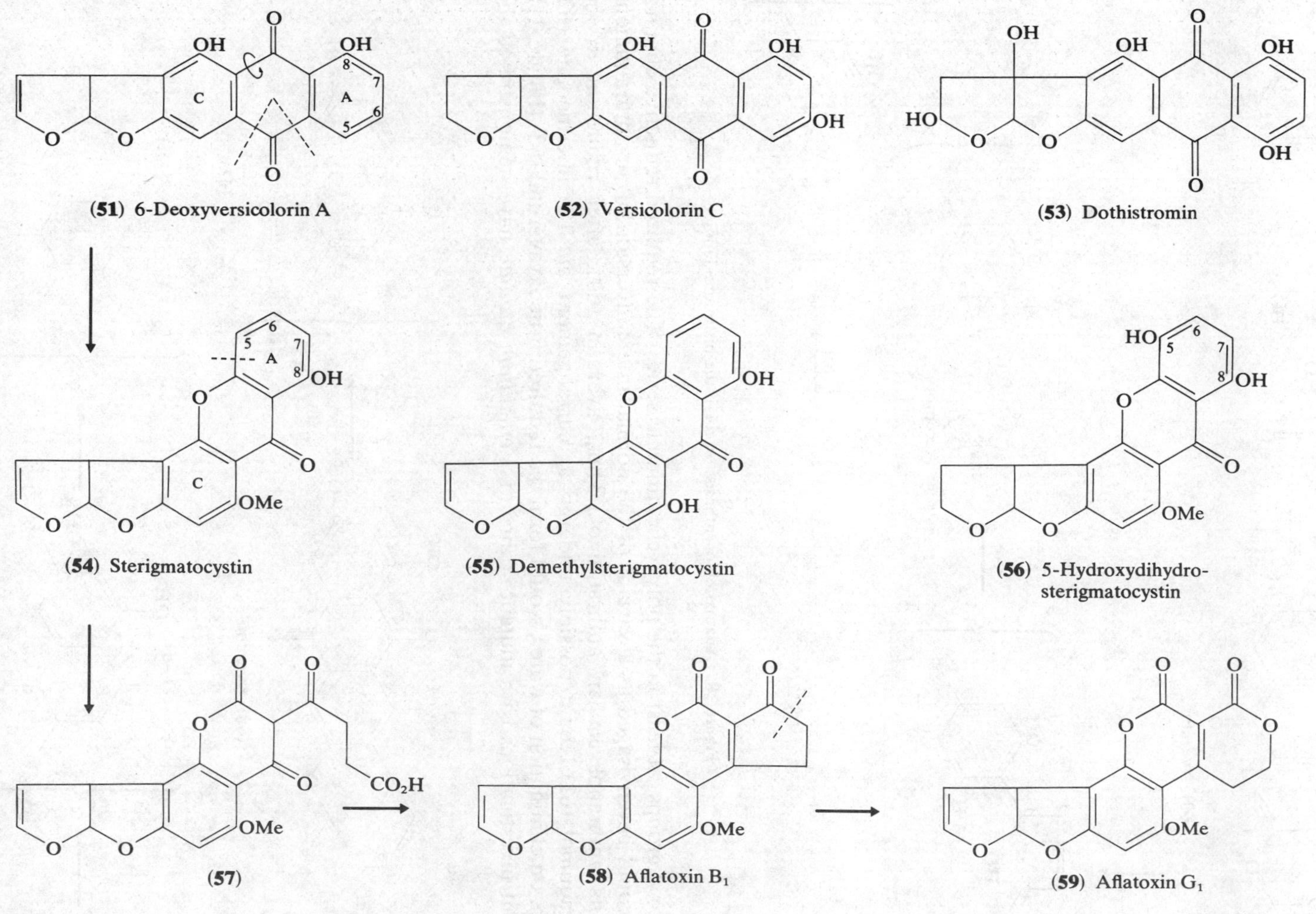

(**51**) 6-Deoxyversicolorin A

(**52**) Versicolorin C

(**53**) Dothistromin

(**54**) Sterigmatocystin

(**55**) Demethylsterigmatocystin

(**56**) 5-Hydroxydihydro-sterigmatocystin

(**57**)

(**58**) Aflatoxin B_1

(**59**) Aflatoxin G_1

Figure 19 Aflatoxin biosynthesis: origin of the non-bisfuranoid moiety

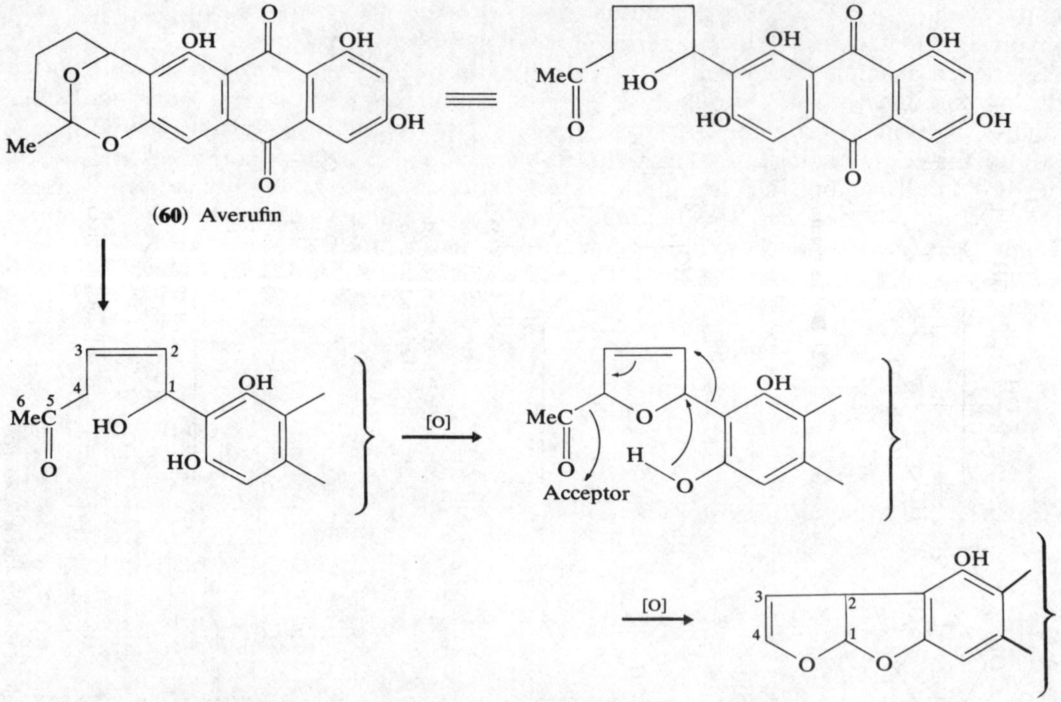

Figure 20 Aflatoxin biosynthesis: origin of the bisfuranoid moiety

hydroxyl groups, at least in the polycyclic quinone series. The required reductive elimination could be envisaged as a variant of the normal interconversion of an anthraquinone and its anthranol *via* an anthraquinol. Thus, averufin (**60**), after reduction to the anthraquinol (**61**), could possibly undergo a rearrangement *via* the intermediate (**62**) which, on elimination of water, would form the required 6-deoxyaverufin (**63**) (Figure 21). Recent investigations with mutant strains of *Verticillium dahliae* have demonstrated the

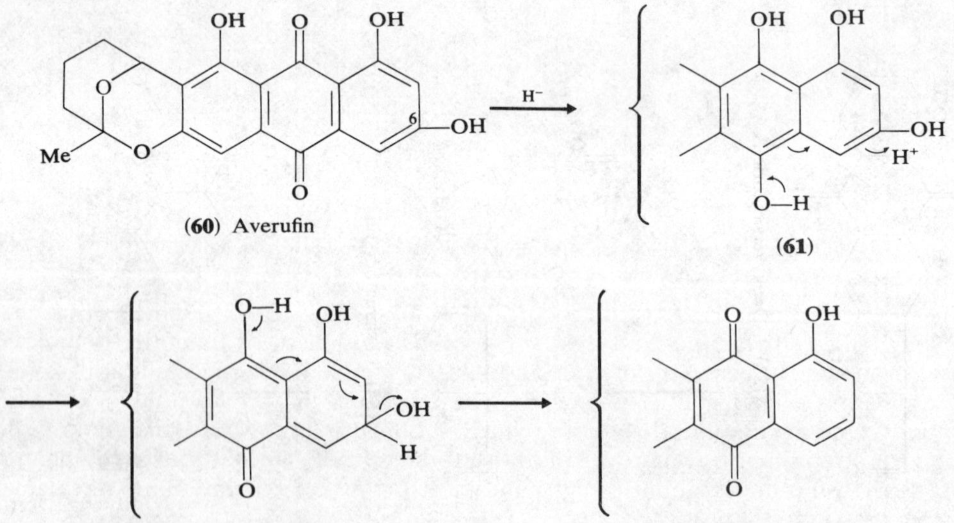

(**60**) Averufin

(**61**)

(**62**)

(**63**) 6-Deoxyaverufin

Figure 21 Hypothetical scheme for the 6-deoxygenation of averufin conversion of 1,3,8-trihydroxynaphthalene to 1,8-dihydroxynaphthalene by a pathway involving, in effect, the reductive removal of a phenolic hydroxyl group.[79]

A recent example of a novel type of trytophan ring cleavage has been shown to occur during the biosynthesis of the *Streptomyces* antibiotic streptonigrin (**64**).[80] Labelling studies in conjunction with ^{13}C n.m.r. supported the following pathway (Figure 22 path 'a') for the origin of rings C and D. With the exception of carbon-6′, both rings are entirely derived from tryptophan following a non-oxidative cleavage of the indole ring between the pyrrole nitrogen and the benzene ring. The *C*-methyl substituent and the three *O*-methyl groups were shown to originate from methionine.

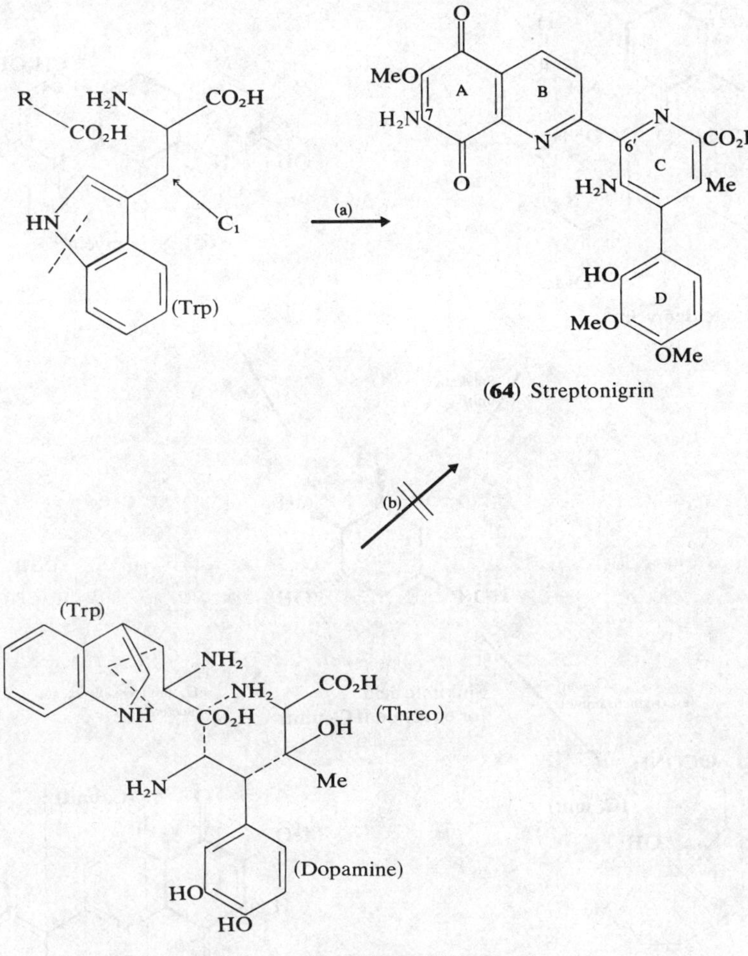

Figure 22 Origin of the 4-phenylpipecolic acid moiety of streptonigrin

The labelled tryptophan was not incorporated into rings A and B, which was somewhat surprising, since this amino-acid is a well-established source of quinoline rings in other natural products.[34] This result provides a further example of the necessity for experimental examination of biogenetic schemes before accepting the implications of even apparently unexceptional structural analyses, which, in this instance, would suggest a reasonable scheme for streptonigrin (Figure 22 path 'b'), based on the amino-acids tryptophan (rings A and B), tyrosine or dopamine (ring D and part of ring C), and threonine (the remaining nitrogen-substituted crotonyl moiety).

Another quinoline derivative with a second nitrogen substituent at C-7 is the unusual 1,8-diaza-anthracene antibiotic nybomycin (**65**), the tricyclic nucleus of which has been shown to be derived jointly from acetate and a shikimate-type precursor, as indicated in Figure 23.[81] The fully substituted benzoquinone ring A of streptonigrin may alternatively reflect an even more direct carbohydrate precursor, since the 1,3-diamino and four oxygen substituents correspond to the skeleton of the streptidine residue of streptomycin (**17**), and also the similar deoxystreptamine unit of the neomycins.[35]

The aminobenzoquinone component of streptonigrin also bears a close structural relationship to the corresponding moiety of the ansamycin antibiotics, as seen in geldamycin (Figure 23).

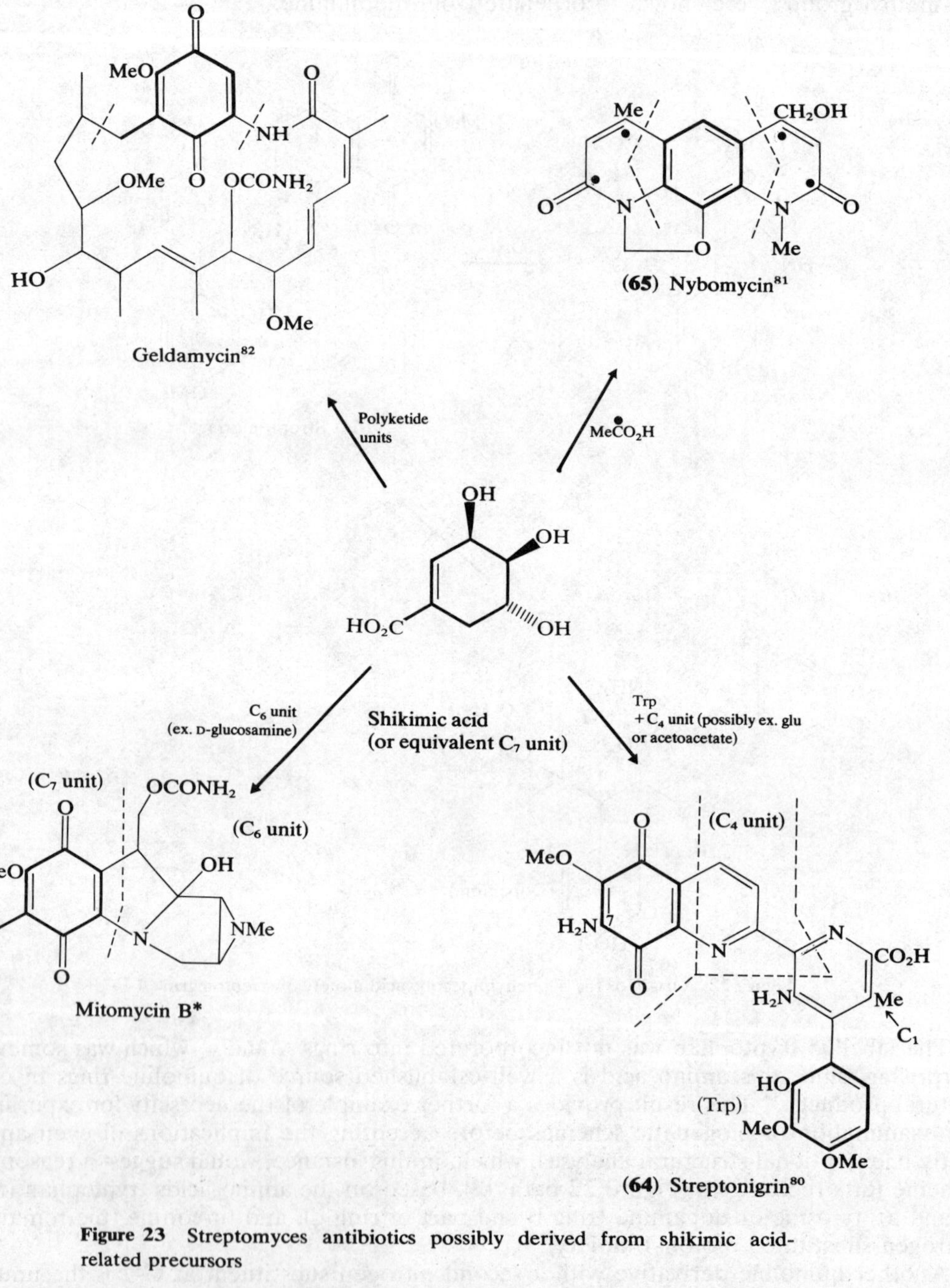

Figure 23 Streptomyces antibiotics possibly derived from shikimic acid-related precursors

* U. Hornemann, J. P. Kehrer, and J. H. Eggert, *J.C.S. Chem. Comm.*, 1974, 1045.

The mode of formation of the ansamycins appears to involve a novel biosynthetic pathway. Extensive labelling studies of geldamycin[82] and the rifamycins[83] indicate that the C_7 quinonoid unit may possess a shikimate-type origin, although the incorporation of shikimic acid itself has not been demonstrated. This is linked through a polyketide chain (mainly polypropionate) to form a variety of macrocyclic structures.

Another *Streptomyces* antibiotic, mitomycin B, is a substituted toluquinone, the origin of which may be related to that of the C_7 unit of the ansamycins. The remaining C_6 substituent appears to be derived directly from D-glucosamine, as indicated.[84]

If the benzoquinonoid ring A of streptonigrin is similarly formed from the cyclohexane ring of a shikimate-type precursor, it would be joined to the tryptophan-derived component of rings C and D, through one nitrogen and four carbon atoms (Figure 23). These could be derived from an amino-acid such as glutamate with loss of CO_2, or alternatively a polyketide (acetoacetate) unit, which would further parallel the ansamycin analogy (*cf.* rifamycin[83]). The diverse biosynthetic origins of quinones, including ansamycins, have been extensively reviewed by Bentley.[84]

Molecular rearrangements, in addition to the above examples among polyisoprenoids and polyketides, occur during the biosynthesis of many other types of natural products. These include carbohydrates, such as the previously mentioned streptomycin component streptose (Figure 6),[48] isoflavonoids, and also alkaloids, which in the specific instances of formononetin (**36**)[48] and the tropic acid (**66**)[85] moiety of atropine interestingly represent two alternative rearrangements of the C_6–C_3 skeleton of their common phenylalanine precursor (Figure 24).

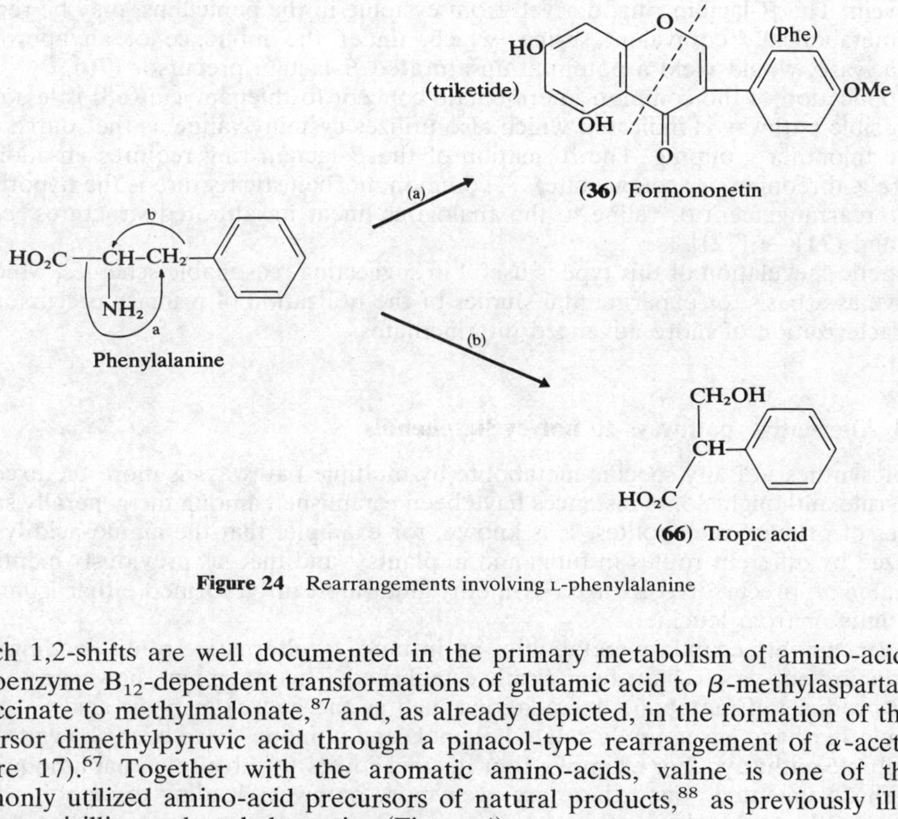

(**36**) Formononetin

(**66**) Tropic acid

Figure 24 Rearrangements involving L-phenylalanine

Such 1,2-shifts are well documented in the primary metabolism of amino-acids, as in the coenzyme B_{12}-dependent transformations of glutamic acid to β-methylaspartate[86] and of succinate to methylmalonate,[87] and, as already depicted, in the formation of the valine precursor dimethylpyruvic acid through a pinacol-type rearrangement of α-acetolactate (Figure 17).[67] Together with the aromatic amino-acids, valine is one of the most commonly utilized amino-acid precursors of natural products,[88] as previously illustrated by the penicillins and cephalosporins (Figure 4).

As a general biogenetic principle, where taxonomically close species are found to produce structures possessing a major common skeletal feature, prior consideration should always be given to the possibility of their formation from a common advanced intermediate. This represents the application of the Occam's Razor principle* to biosynthesis, and is clearly seen in the relationship of the penam and cephem ring systems. As it happens, these structures are atypically common both to species of fungi, and also the taxonomically remote actinomycetes.

The characteristic penam β-lactam ring has recently been discovered in two novel types of *Streptomyces* metabolites, namely clavulanic acid (**67**)[89] and the olivanic acids, *e.g.* thienamycin (**68**).[90] While structural analysis considerations alone do not immediately indicate any common biosynthetic interrelationship with the cysteinylvaline-derived penam nucleus, it is instructive to apply the above biogenetic principle. Thus, in addition to their analogous bicyclic systems, which represent a compelling common structural feature, the mutual derivation from a related dipeptide of both 6-aminopenicillanic acid (6-APA) and clavulanic acid would be consistent with the presence of C_3 and C_5 units in their respective four-and five-membered heterocyclic rings. The main difference between the C_5 units is that in penicillanic acid it possesses the same isopentane skeleton as its valine precursor, whereas in clavulanic acid this unit is linear. Its formation from valine would therefore require a rearrangement analogous to the known 1,2-shift involved in the prior formation of this amino-acid from acetolactate (Figure 17).[67]

A hypothetical scheme, based on this prerequisite, is outlined in Figure 25, in which the valine is initially oxidized to the same level as in the corresponding 7-aminocephalosporanic acid (7-ACA) moiety of cephalosporin C. This would provide for the formation of a suitable intermediate (**69**), which could directly yield the required linear C_5 skeleton of clavulanic acid (**67**) *via* a further pinacol-type rearrangement. This scheme depicts a possible common origin of 6-APA, 7-ACA, clavulanic acid, and thienamycin. The β-lactam ring, derived from cysteine in the penicillins, may be replaced by the metabolically equivalent serine, which, under the influence of an appropriate ammonia lyase, would yield a potential unsaturated β-lactam precursor (**70**).

The application of the common intermediate concept to thienamycin (**68**) is less direct, but a possible pathway is indicated which also utilizes cysteinylvaline as the source of the pyrroline thioether grouping. The formation of the β-lactam ring requires an additional C_4 unit (*e.g.* threonine or acetoacetate). A common mechanistic feature is the hypothetical oxidative rearrangement of valine to the analogous linear unsaturated structures [*cf.* (**69**) → (**67**) and (**71**) → (**72**)].

Biogenetic speculation of this type is useful in suggesting reasonable schemes, which can then serve as a basis for experimental studies of the utilization of primary precursors and the characterization of more advanced intermediates.

28.1.7.4 Alternative pathways to polycyclic phenols

The biosynthesis of any specific metabolite by multiple pathways is more the exception than the rule, althought some instances have been established among the generally smaller molecules of primary metabolites. It is known, for example, that the amino-acid lysine is synthesized by different routes in fungi and in plants[91] and that, as previously mentioned, the mevalonate precursor β-hydroxy-β-methylglutarate can be formed either from three acetate units or from leucine.

However, among secondary metabolites no instance of alternative routes leading to the same structure appears to have been firmly established. This statement, however, excludes terminal products of metabolic degradations such as the hydroxybenzoic acids.

On the other hand, structurally-related phenols and quinones have been shown to result from multiple pathways. For example, four distinct routes to substituted naphthoquinones have been documented,[92] and there are even more pathways leading to monobenzenoid derivatives, although the biosyntheses of the majority of such compounds are based on the predominating shikimate (Chapter 30.3) and polyketide pathways (Chapter 29.1).

The anthraquinones emodin (**73**) and alizarin (**74**) both occur in plants, and labelling

* 'It is vain to do with more, what can be done with fewer.'

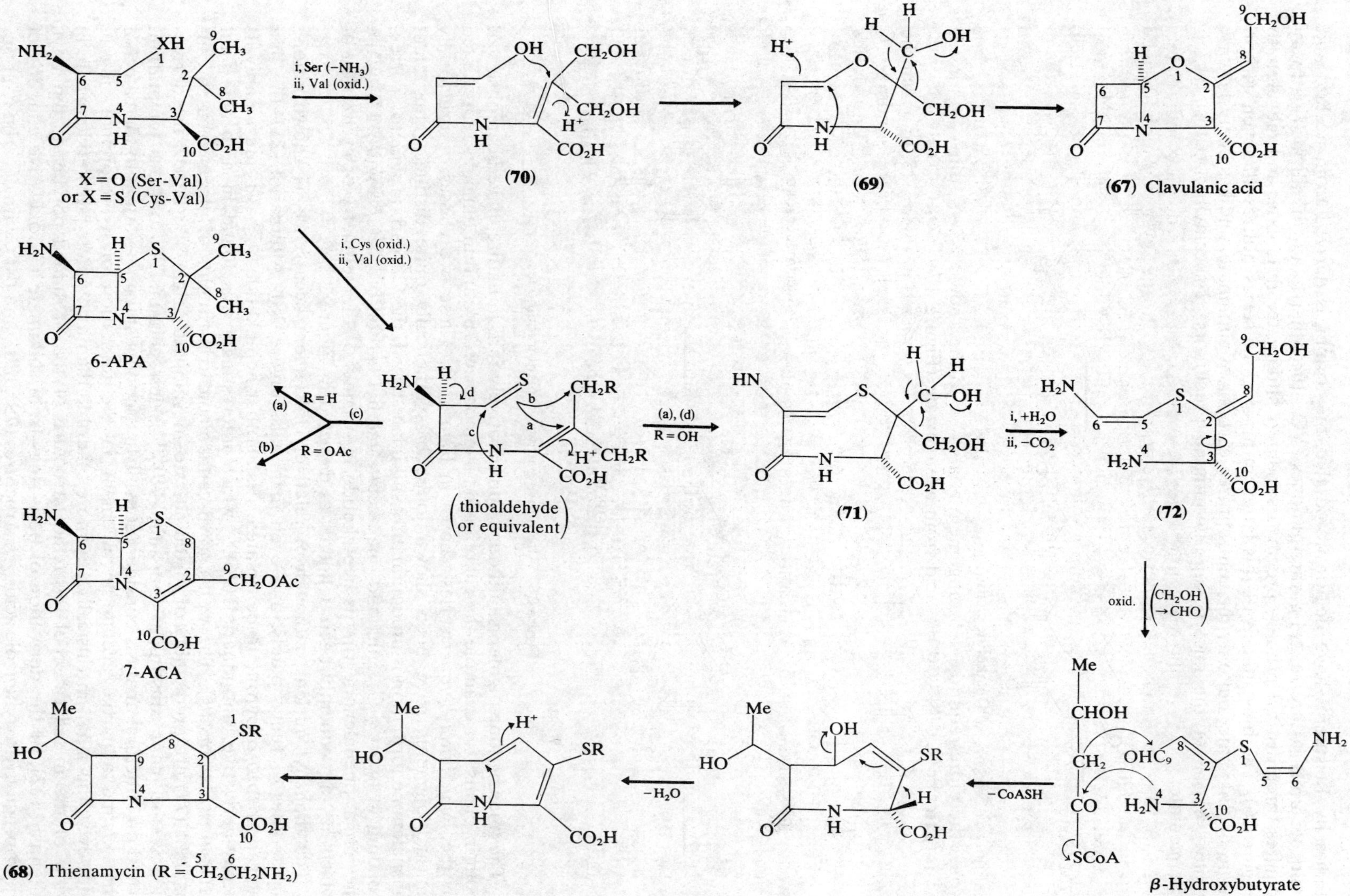

FIGURE 25 Hypothetical biogenetic interrelationship of 6-APA, 7-ACA, clavulanic acid, and thienamycin

studies by Leistner[93] have demonstrated that the former is derived from acetate and malonate units, as are fungal anthraquinones.[94] On the other hand, the alizarin pathway requires three distinct primary precursors, namely shikimate, glutamate, and mevalonate (Figure 26). The biosynthesis of this latter anthraquinone series utilizes all seven carbon atoms of shikimate, but only the central three carbons of glutamate, and has been shown[95] to involve one of the known naphthoquinone precursors, 2-succinylbenzoate,[84] as an intermediate.

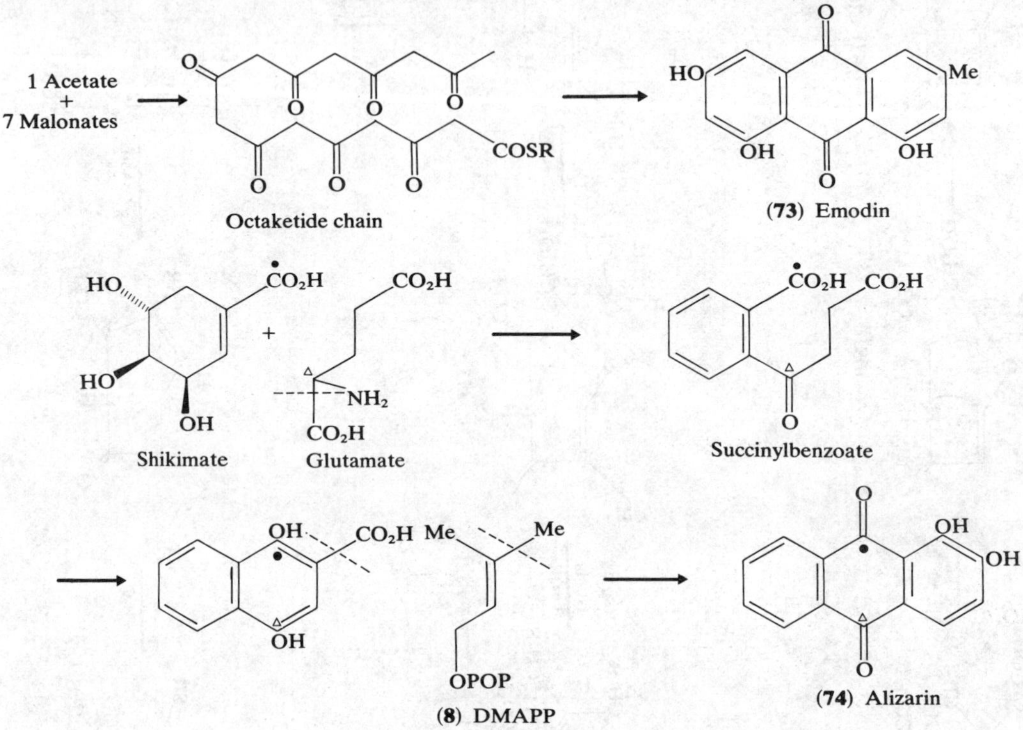

Figure 26 Alternative pathways to anthraquinones

The phenalenone pigments, although comparatively rare in nature, have been observed both in plants, for example the cellobioside haemocorin (**75**), and in fungi, as represented by the *Penicillium herquei* series, shown in Figure 13. These two classes of structurally unusual pigments have been shown to be biosynthetically quite distinct. Thus plant phenalenone formation is predominantly shikimate-based, the C_{19} carbon skeleton corresponding to two C_6–C_3 units and one acetate-methyl carbon[36,53,96] (*cf.* Figure 27), while the fungal phenalenone ring is heptaketide in origin,[36,45] the geometry of cyclization of which has subsequently been defined[10] as depicted in Figure 2.

The diphenyl nucleus, although of relatively infrequent occurrence, is known among metabolites of fungi, plants, and even mammals (Figure 28). Alternariol (**21**), the first fungal dibenzo-α-pyrone to be isolated,[97] is another heptaketide,[44,98] as is presumably its deoxy-derivative, autumnariol (**76**),[99] which interestingly was found in a higher plant (*Eucomis autumnalis*). The well-known cannabinoid constituents of plants such as cannabinol (**77**) are predictably of mixed acetate–mevalonate origin,[100] whereas the derivation of the plant constituent sappanin (**78**)[101] is less certain. The obvious temptation, based on structural analysis considerations, is to ascribe a shikimate origin to the catechol ring A, and a partial acetate involvement in the formation of its resorcinol ring B. However, this is not intrinsically any more plausible than a complete polyketide origin, in view of the mixed catechol–resorcinol structure of the presumed polyketide diphenyl, altenusin (**22**), or the derivation of both rings from shikimate *via* oxidative coupling of appropriate hydroxybenzoic acids, *cf.* structures (**79**), (**80**), and (**81**), with decarboxylation.

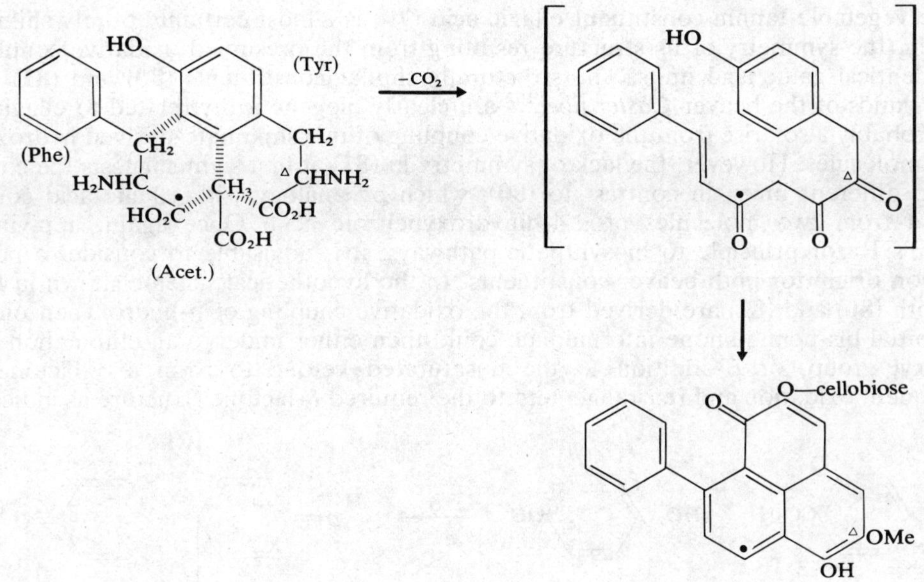

(75) Haemocorin

Figure 27 Biosynthesis of the plant phenalenone haemocorin

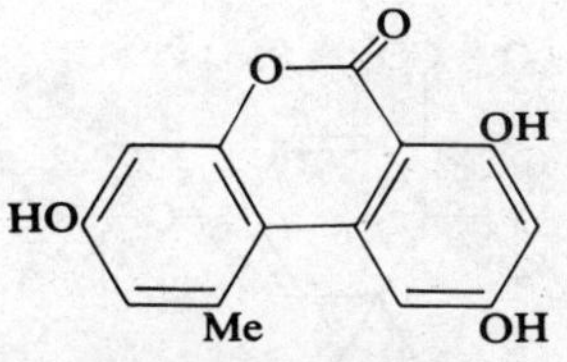

(21) Alternariol

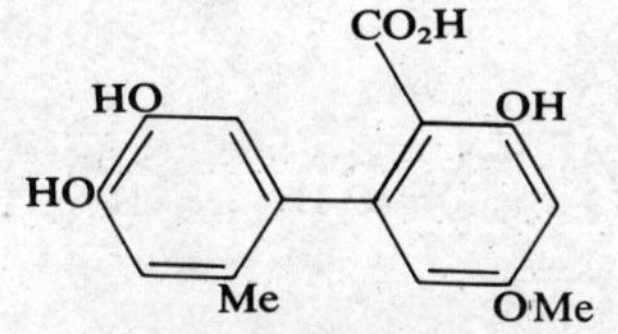

(22) Altenusin

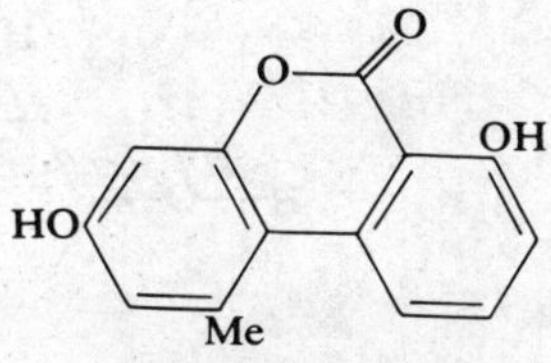

(76) Autumnariol

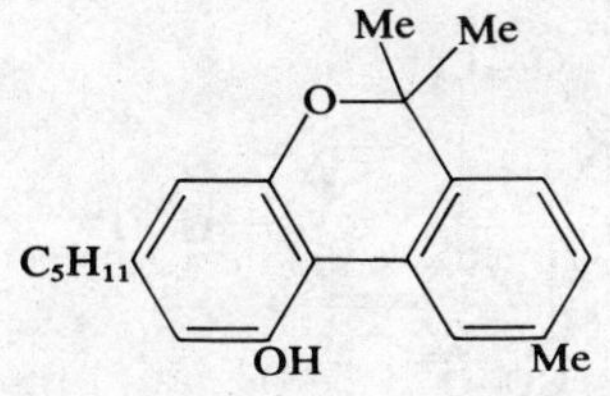

(77) Cannabinol

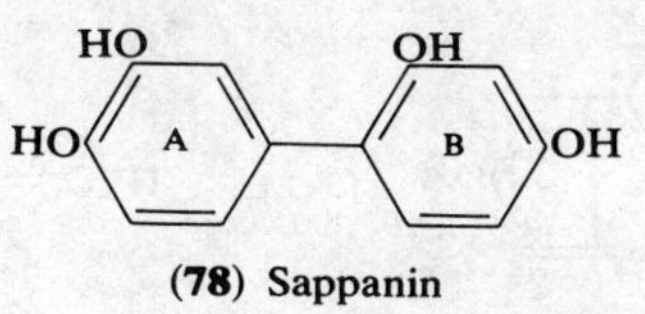

(78) Sappanin

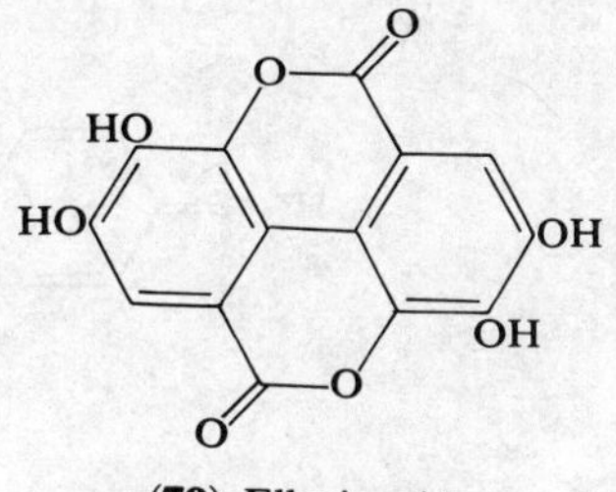

(79) Ellagic acid

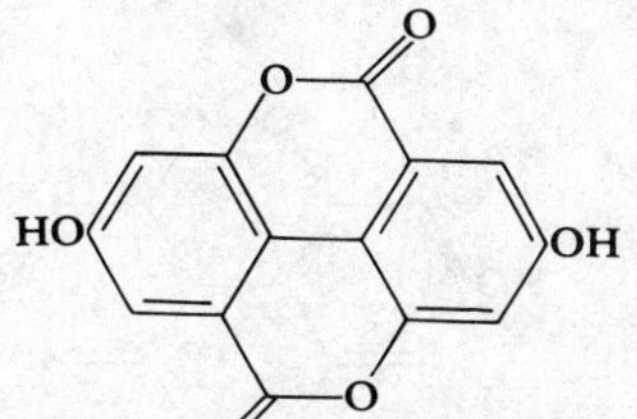

(80)

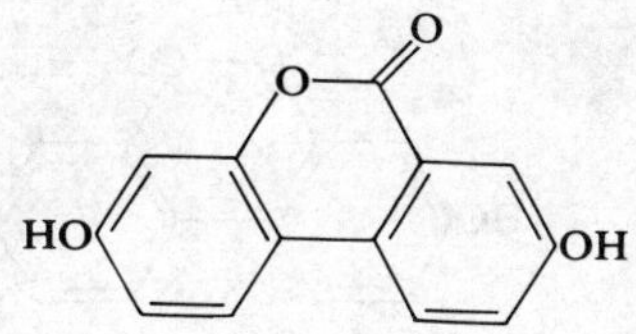

(81)

Figure 28 Naturally-occurring diphenyl derivatives

The vegetable tannin constituent ellagic acid (**79**) is almost certainly purely shikimate-derived, the symmetry of its structure resulting from the presumed oxidative coupling of two identical gallic acid units. The structurally similar constituents (**80**) and (**81**) of the scent glands of the beaver *Castor fiber*[102] are clearly biogenetically related to ellagic acid, and probably also arise from the oxidative coupling of two shikimate-derived hydroxybenzoate molecules. However, the lack of symmetry in (**81**) at first sight suggests the coupling of two different units, in contrast to (**80**), which by analogy with ellagic acid could be formed from two molecules of 2,4-dihydroxybenzoic acid. Once again, applying the Occam's Razor principle to biosynthetic pathways, it is advisable to consider a possible common origin for both beaver constituents. In the hypothetical scheme shown in Figure 29, both (**80**) and (**81**) are derived from the oxidative coupling of *p*-hydroxybenzoic acid. The initial bis-hemiquinone intermediate could then either undergo an elimination of the carboxyl group, or β-addition to the unsaturated ketone to form a γ-lactone with subsequent oxidation and rearrangement to the required δ-lactone structure as indicated.

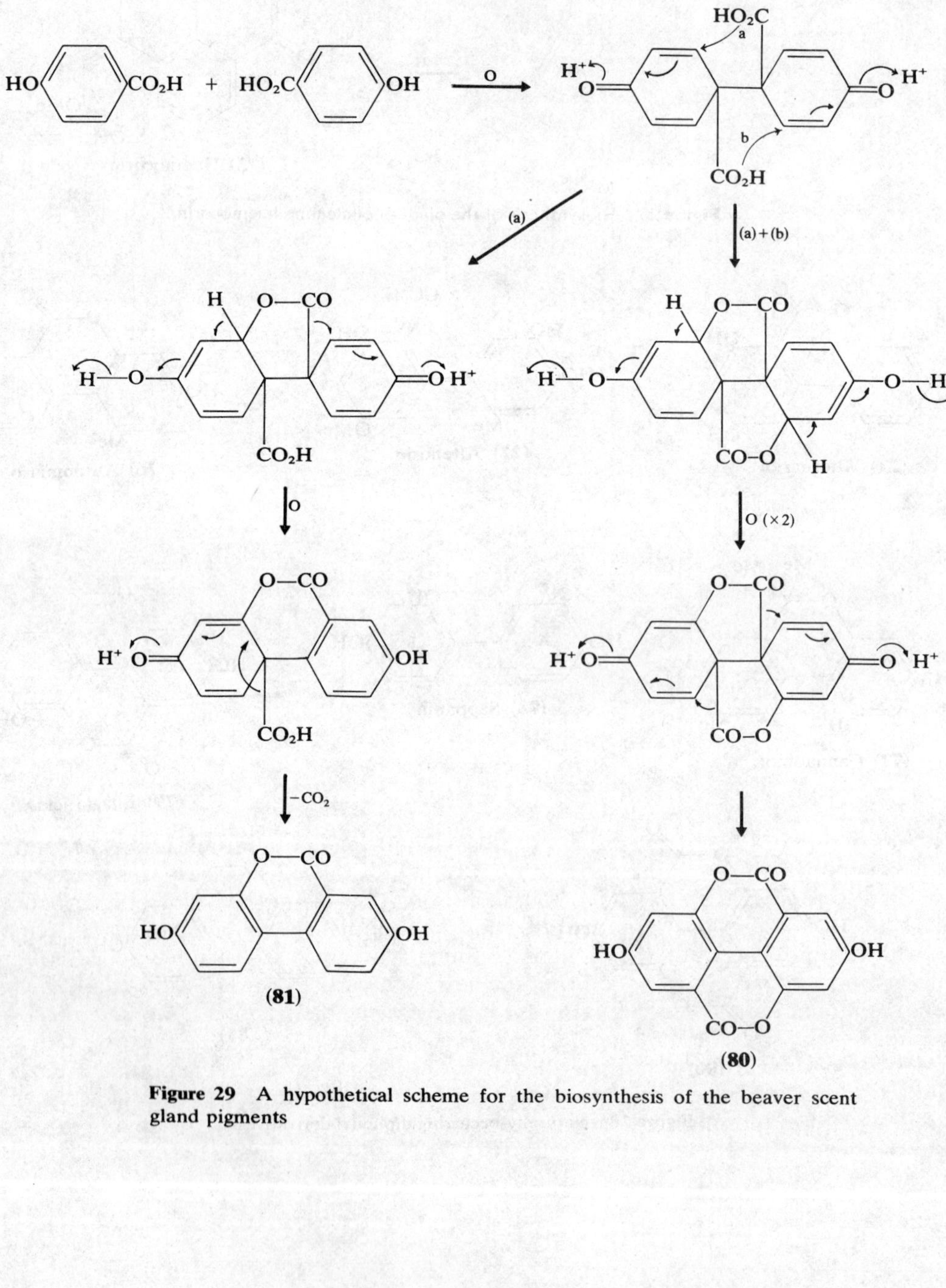

(**81**) (**80**)

Figure 29 A hypothetical scheme for the biosynthesis of the beaver scent gland pigments

The required 1,2-shift of the carboxyl group finds an analogy in the previously mentioned rearrangement of phenylalanine, as in the formation of tropic acid (**66**) (*cf.* Figure 24).

28.1.7.5 Racemic phenolic metabolites

The *Alternaria* metabolites, (±)-dehydroaltenusin (**25**) and (±)-alternuene (**82**), represent two modified diphenyl products, almost certainly derived from alternariol, in which the aromaticity of ring A has been lost through oxidation and reduction respectively (Figure 30). An interesting feature of both products is their occurrence as racemic mixtures (despite the presence in alternuene of three chiral centres), since the formation of both enantiomers is more often an indication of the operation of a spontaneous (*i.e.* non-enzymic) process in the generation of a chiral centre.

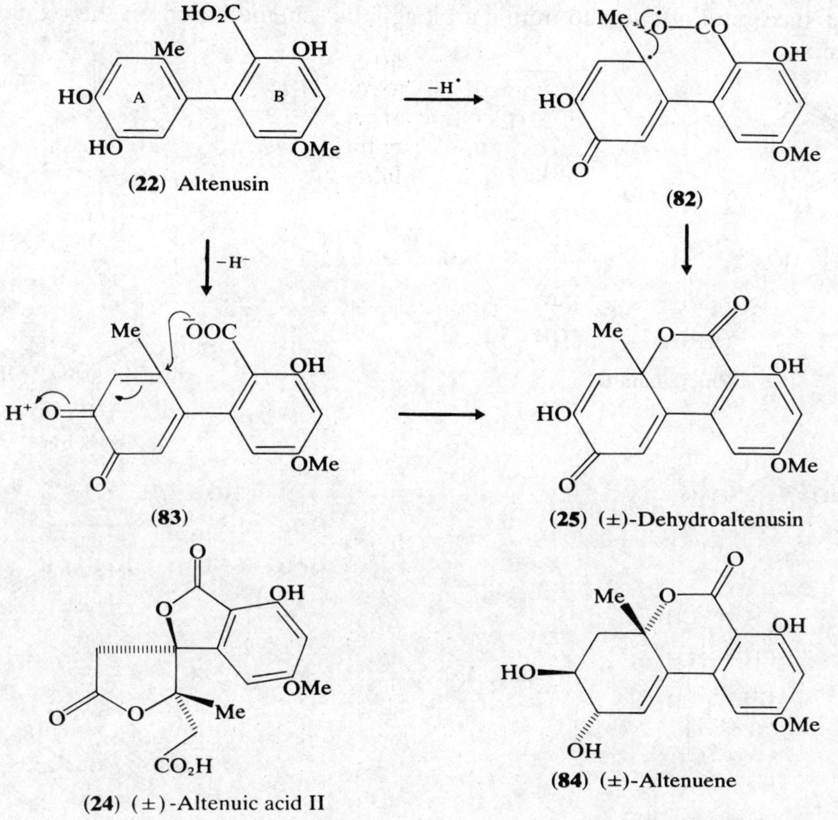

Figure 30 Racemic phenolic metabolites of *Alternaria alternata*

The formation of (**25**) by oxidation of the catechol ring of altenusin (**22**) is known to be a facile step, which is readily effected by aqueous ferric chloride. This is normally considered to represent an oxidative coupling process involving free-radical intermediates such as (**82**). A possible alternative enzymic intermediate would be the *ortho*-quinone (**83**), which then forms both enantiomers of the cross-conjugated dienone dehydroaltenusin, following a spontaneous β-addition of the ring B carboxyl group. This would involve an unexceptional ionic oxidative pathway, in contrast to the frequently non-stereospecific free-radical oxidative coupling mechanism discussed later (Section 28.1.7.6).

If dehydroaltenusin is the normal precursor of alternuene, then a subsequent enzyme-catalysed stereospecific reduction would be required to introduce the two adjacent chiral centres, with the appropriate relative configurations shown in (**84**). If this enzyme can utilize as alternative substrates both antipodal forms of (±)-dehydroaltenusin (**25**), then it would account for the observed production of both enantiomers of (±)-alternuene.

Altenusin or dehydroaltenusin probably serve as precursors of the three co-existing optically inactive isomeric altenuic acids I, II, and III,[103] one of which, (±)-altenuic acid II (**24**),[70] has been shown to contain two chiral centres. As previously indicated (Figure 18), it appears likely that the actual enzymically-formed metabolite is the corresponding muconic acid (**23**), which then undergoes spontaneous β-additions of the carboxyl groups, to form all three optically inactive altenuic acid isomers.

28.1.7.6 Oxidative coupling and biomimetic synthesis

Many natural products occur in dimeric forms ranging from bisflavonoids and bisanthraquinones to lignans, bisterpenoids, and dimeric alkaloids (Figure 31). The implicit general precursor role of the monomer unit of usnic acid was utilized by Barton and colleagues in the first successful attempt to mimic a biogenetic scheme based on the oxidative coupling of phenols.

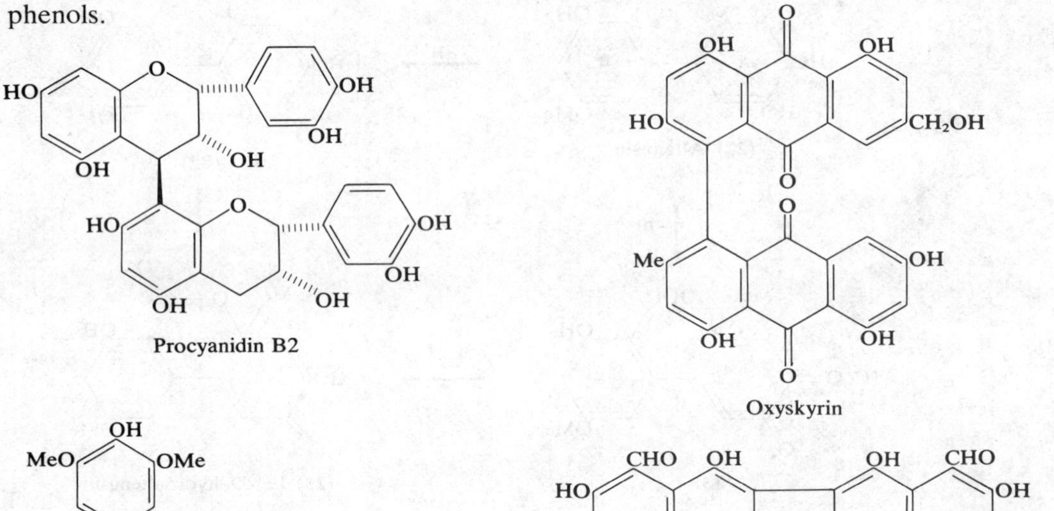

Figure 31 Dimeric natural products formed by oxidative coupling

Usnic acid (**85**), which accounts for up to 20% of the dry weight of some lichens, was prepared in an elegant two-step synthesis[104] involving the direct oxidation of methylphloracetophenone with ferricyanide. The proposed mechanism (Figure 32), involving single-electron oxidation followed by coupling of two of the resulting free-radical intermediates, is thought to parallel the natural process. This is supported by the subsequent enzymic synthesis of usnic acid by a peroxidase-catalysed oxidation of the same polyketide precursor.[105]

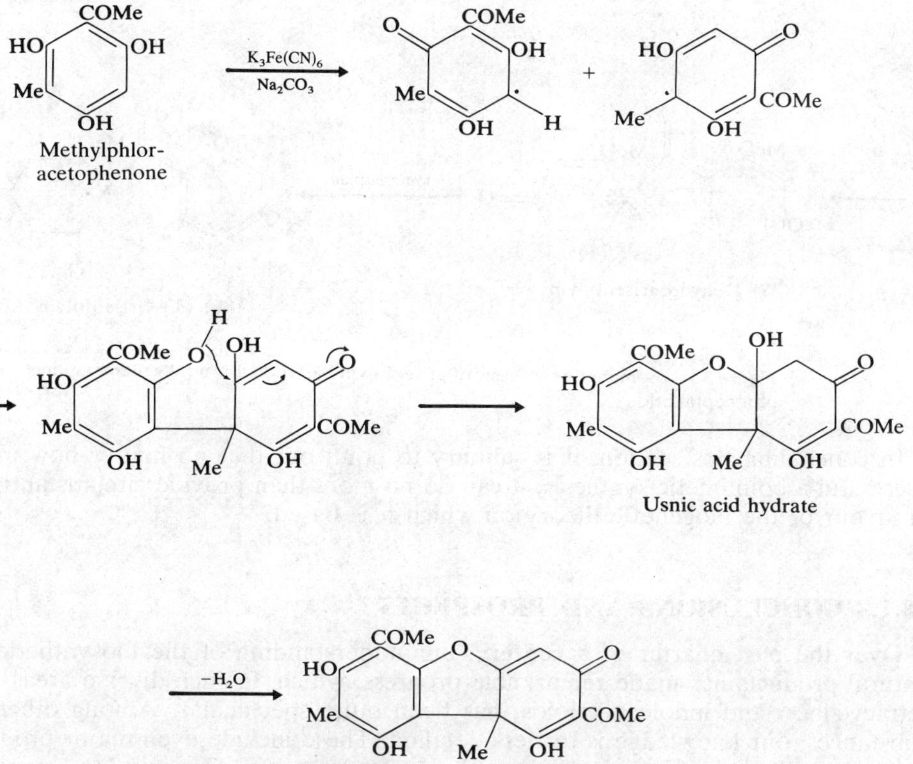

(**85**) (±)-Usnic acid

Figure 32 Biomimetic synthesis of usnic acid based on intermolecular coupling of tetraketide units

In keeping with other *in vitro* peroxidase-induced oxidations, the resulting usnic acid was racemic, whereas the natural product has been isolated both in optically active and racemic forms.

Intramolecular oxidative coupling has been applied to the synthesis of the antifungal metabolite griseofulvin (**88**),[106] based on the same general approach starting from griseophenone A (**86**) as shown in Figure 33.

The intermediate dehydrogriseofulvin (**87**) contains a similar cross-conjugated dienone structure to that present in dehydroaltenusin (**25**), which, as previously mentioned, is readily formed by oxidation of the catechol ring of altenusin by ferric ions (Figure 30).

In addition to syntheses based on oxidative coupling, a variety of mechanisms involved in general biosynthesis and metabolism have been successfully imitated. These have led to often efficient laboratory syntheses of a range of alkaloids,[107] terpenes,[108] and polyketides.[109] In fact, the success of the organic chemist in this area has led to the general adoption of the term 'biomimetic synthesis' to describe those natural product syntheses which are designed to parallel the processes taking place in nature.

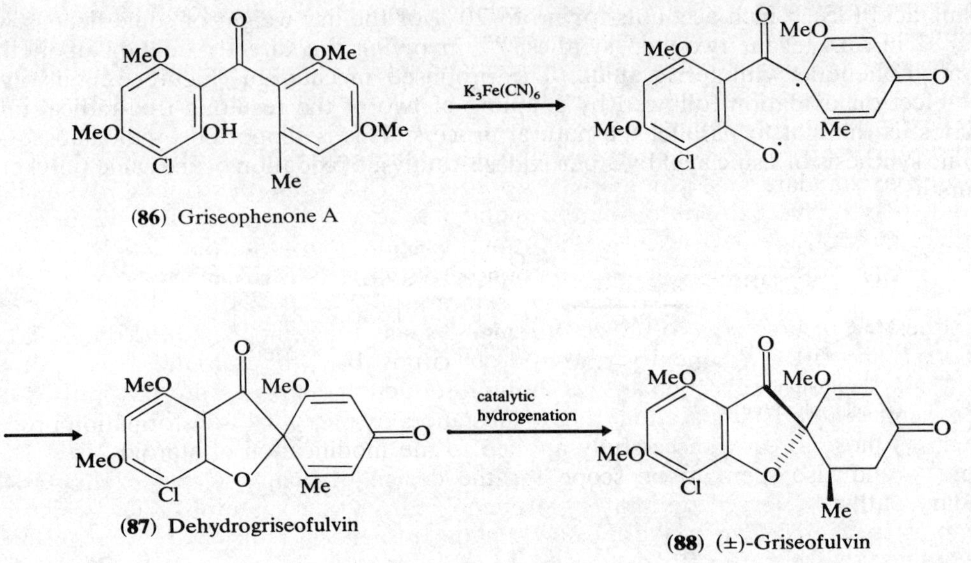

Figure 33 Biomimetic synthesis of griseofulvin based on intramolecular coupling of a heptaketide

In concluding this section, it is salutary to point out that no matter how ingenious or successful a boimimetic synthesis, it can do no more than provide circumstantial evidence in favour of the biogenetic theory on which it is based.

28.1.8 CONCLUSIONS AND PROSPECTS

Over the past quarter of a century, our understanding of the biosynthetic origins of natural products has made remarkable progress, which, in such diverse areas as steroids, tetracyclines, and indole alkaloids, has been quite spectacular. Among other groups of substances, our knowledge is far less detailed. These include even major products with a comparatively long history of biosynthetic investigation. For example, concerning the origin of the penicillin nucleus, we have as yet little detailed understanding of the mechanism of cyclization of its tripeptide precursor. However, recent developments, such as the determination of the stereochemistry of incorporation of the prochiral β-carbons of cysteine[110,111] and valine,[112,113] as well as the application of protoplast and cell-free enzyme techniques, offer encouraging prospects of early progress in this area.[114,116] Investigations of other classes of secondary products at the enzymic level are also meeting with increasing success.[115]

It is now possible to define in a general way 'what' types of secondary structures are produced in nature, and, at least in reasonable mechanistic terms, 'how' they are formed from a very limited range of primary precursors. The predominant question for the future is surely 'why' are they produced? Do secondary metabolites have any useful function, or, if not, and they are consequently only evolutionary relics or other metabolic nonentities, why have they or their host organisms survived so long in a highly competitive environment, where inefficient utilization of resources is normally lethal? A possible answer to this key question may be provided by considerations of the biochemical nature and prerequisites of the individuality of taxonomically-related species.

Natural product studies to date have been exceedingly prolific, and yet, because of the comparatively arbitrary basis of the procedures applied to their detection and isolation, our present knowledge is almost inevitably fragmentary. Thus, the currently available information is unlikely to be representative, either qualitatively, in terms of structural diversity, or quantitatively, as regards the relative amounts of individual metabolites and their intermediates at any given stage of growth. In order to obtain a better perspective, it will clearly be necessary to plan a more systematic search for new products. One possible approach which is considered (Section 28.1.7.1) involves the general application of a simple experimental technique, based on the incorporation of radioactively-labelled primary precursors into secondary metabolites, the formation of which can then be monitored by standard autoradiographic procedures. This would permit both the qualitative and quantitative exploration of the metabolic activities of individual organisms in depth, in contrast to the present rather opportunistic approach, where normally only major products with specific chemical or biological properties are selected for investigation.

The question of how best to utilize our rapidly increasing understanding of secondary metabolism should be frequently reviewed. Is it possible, for example, to control the quantity and diversity of secondary metabolite production, and what are the limits to the efficient semi-synthetic preparation of new substances by microbial transformation processes, such as those already successfully applied to the modification of steroids?

There would also seem to be scope for the design of highly selective inhibitors of secondary pathways, based on strategically modified advanced intermediates, which can function as antimetabolites, as do the sulphonamide folic acid antagonists. Thus, preliminary studies[117] have indicated the feasibility of inhibiting penicillic acid formation with 5-substituted orsellinic acids at concentrations which do not significantly inhibit growth or general metabolism. These inhibitors appear to block effectively the normal enzymic cleavage of the 4,5-bond of orsellinic acid (*cf.* Figure 1), leading to the accumulation of a previously reported penicillic acid intermediate.[118] This is analogous to the situation frequently observed in studies of blocked pathways of specific enzyme-deficient mutants. If generally applicable, this approach could be used for the selective inhibition of biosynthesis of undesirable metabolites, such as microbial toxins (phytotoxins, aflatoxins, *etc.*). It could also conceivably be applied to increasing the yields of antibiotics or specific intermediates, either by blocking alternative pathways which compete for the available pools of essential precursors, or by inhibiting other pathways involved in the further metabolism of the desired intermediate. The design of the required selective inhibitor of a specific pathway inevitably requires an understanding of the structure of at least one essential advanced intermediate, as in the use of 5-chloro-orsellinic acid to block the conversion of orsellinic acid to penicillic acid.

Further clarification of the nature of individual biosynthetic pathways and their potential regulation will undoubtedly require a deeper knowledge of the enzymes involved and their genetic control. This important area is currently receiving considerable attention, both in industrial and academic laboratories.

In developing our understanding, over the past two or three decades, of the fruits of more than two billion years of continuous biosynthetic evolution, it is important to continually assess our perspective, and to document our findings in such a way as to facilitate the recognition of significant correlations. With this object in mind, it has been argued, in this chapter, that a comprehensive compendium with appropriate data storage and retrieval capabilities is a prerequisite of efficient long-term progress in this area.

There can be little doubt that the future prospects of this comparatively young discipline of biosynthesis will more than match the highly successful and exciting achievements of its brief past.

References

1. M. Calvin and J. A. Bassham, 'The Photosynthesis of Carbon Compounds', Benjamin, New York, 1962.
2. R. B. Woodward, M. P. Cava, W. D. Ollis, A. H. Hunger, H. U. Daeniker, and K. Schenker, *J. Amer. Chem. Soc.*, 1954, **76**, 4749.
3. J. D. Bu'Lock, 'The Biosynthesis of Natural Products', McGraw-Hill, London, 1965, p. 2.
4. R. H. Whittaker and P. P. Feeny, *Science*, 1971, **171**, 757.

5. For a general review, see S. A. Brown, in 'Biosynthesis', ed. T. A. Geissman, Specialist Periodical Reports, The Chemical Society, London. 1972, vol. 1, p. 1.
6. For a general review, see M. Tanabe, in 'Biosynthesis', ed. T. A. Geissman, Specialist Periodical Reports, The Chemical Society, London, 1973, vol. 2, p. 241.
7. Y. Sato and T. Oda, *J.C.S. Chem. Comm.*, 1977, 415; for a review see H. H. Mantsch, H. Saito, and I. C. P. Smith, *Prog. N.M.R. Spectroscopy*, 1977, **11,** 211.
8. J. M. A. Al-Rawi, J. A. Elvidge, D. K. Jaiswal, J. R. Jones, and R. Thomas, *J.C.S. Chem. Comm.*, 1974, 220.
9. J. A. Elvidge, D. K. Jaiswal, J. R. Jones, and R. Thomas, *J.C.S. Perkin I*, 1977, 1080.
10. T. J. Simpson, *J.C.S. Chem. Comm.*, 1976, 258.
11. A. P. Bradshaw, J. R. Hanson, and M. Siverns, *J.C.S. Chem. Comm.*, 1977, 819.
12. L. J. Goad and T. W. Goodwin, *Prog. Phytochem.*, 1972, **3,** 113.
13. N. Castagnoli, K. Corbett, E. B. Chain, and R. Thomas, *Biochem. J.*, 1970, **117,** 450.
14. S. Gatenbeck and K. Mosbach, *Acta Chem. Scand.* 1959, **13,** 1561.
15. P. Dimroth, H. Walter, and F. Lynen, *European J. Biochem.*, 1970, **13,** 98.
16. J. R. D. McCormick, in 'Biogenesis of Antibiotics' ed. Z. Vaněk and Z. Hǒstálek, Czech. Acad. Sci., Prague, 1965, p. 73.
17. L. Ruzicka, A. Eschenmoser, and H. Heusser, *Experientia*, 1953, **9,** 357.
18. R. Robinson, 'Structural Relations of Natural Products', Clarendon, Oxford, 1955.
19. K. Folkers, C. H. Shunk, B. O. Linn, F. M. Robinson, P. E. Wittreich, J. W. Huff, J. L. Gilfillan, and H. R. Skeggs, 'CIBA Foundation Symposium: Biosynthesis of Terpenes and Sterols', ed. G. E. W. Wolstenholme and M. O'Connor, Little, Brown, Boston, 1959, p. 20.
20. G. W. K. Cavill, *Rev. Pure Appl. Chem.*, 1960, **10,** 169.
21. K. Bloch, 'CIBA Foundation Symposium: Biosynthesis of Terpenes and Sterols', ed. G. E. W. Wolstenholme and M. O'Connor, Little, Brown, Boston, 1959, p. 4.
22. J. W. Cornforth and G. Popjak, *Adv. Enzymol.*, 1960, **22,** 281.
23. F. H. Lynen, H. Eggerer, U. Henning, I. Kessel, and E. Ringelman, 'CIBA Foundation Symposium: Biosynthesis of Terpenes and Sterols', ed. G. E. W. Wolstenholme and M. O'Connor, Little, Brown, Boston, 1959, p. 95.
24. R. Robinson, *Chem. and Ind.* (*London*), 1934, **53,** 1062.
25. R. Thomas and R. A. Bassett, *Prog. Phytochem.* 1972, **3,** 47.
26. E. Leete, *Chem. and Ind.* (*London*), 1960, 692.
27. R. Thomas, *Tetrahedron Letters*, 1961, 544.
28. E. Wenkert, *J. Amer. Chem. Soc.*, 1962, **84,** 98.
29. R. Thomas, in 'Biogenesis of Antibiotics' ed. Z. Vaněk and Z. Hǒstálek, Czech. Acad. Sci., Prague, 1965, p. 155.
30. A. I. Scott, *Accounts Chem. Res.*, 1970, **3,** 151.
31. G. Barger and C. Scholz, *Helv. Chim. Acta*, 1933, **16,** 1343.
32. G. Hahn and H. Werner, *Annalen*, 1935, **520,** 123.
33. R. B. Woodward, *Nature*, 1948, **162,** 155.
34. R. Robinson, *Nature*, 1948, **162,** 524.
35. K. L. Rinehart, J. M. Malik, R. S. Nystrom, R. M. Stroshane, S. T. Truitt, M. Taniguchi, J. P. Rolls, W. J. Haak, and B. A. Ruff, *J. Amer. Chem. Soc.*, 1974, **96,** 2263.
36. R. Thomas, *Pure Appl. Chem.*, 1973, **34,** 515.
37. H. Raistrick and A. B. Clark, *Biochem. J.*, 1919, **13,** 329.
38. H. Raistrick, *Proc. Roy. Soc.*, 1949, **A199,** 141.
39. A. J. Birch and F. W. Donovan, *Austral. J. Chem.*, 1953, **6,** 360.
40. R. Robinson, 'Structural Relations of Natural Products', Clarendon, Oxford, 1955, p. 7.
41. A. J. Birch, R. A. Massy-Westropp, and C. J. Moye, *Chem. and Ind.* (*London*), 1955, 683.
42. A. J. Birch, *Proc. Chem. Soc.*, 1962, 3.
43. P. Chaplen and R. Thomas, *Biochem. J.*, 1960, **77,** 91.
44. R. Thomas, *Biochem. J.*, 1961, **78,** 748.
45. R. Thomas, *Biochem. J.*, 1961, **78,** 807.
46. S. Gatenbeck, *Biochem. Biophys. Res. Comm.*, 1962, **6,** 422.
47. I. Ljungcrantz and K. Mosbach, *Biochem. Biophys. Acta*, 1964, **86,** 203.
48. H. Grisebach, in 'Biosynthetic Patterns in Microorganisms and Higher Plants', Wiley, New York, 1967.
49. E. Albu and R. Thomas, *Biochem. J.*, 1963, **87,** 648.
50. J. Better and S. Gatenbeck, *Acta Chem. Scand.*, 1977, **B31,** 391.
51. J. W. Sime, *Ph.D. Thesis*, University of London, 1969, p. 98.
52. A. Kreigler and R. Thomas, *Chem. Comm.*, 1971, 738.
53. R. Thomas, *Chem. Comm.*, 1971, 739.
54. R. Thomas, unpublished observations.
55. R. Thomas, *Nature*, 1961, **191,** 1161.
56. A. J. Birch and F. W. Donovan, *Austral. J. Chem.*, 1953, **6,** 373.
57. D. H. R. Barton, P. de Mayo, G. A. Morrison, W. H. Schaeppi, and H. Raistrick, *Chem. and Ind.* (*London*), 1956, 552.
58. I. C. Paul and G. A. Sim, *J. Chem. Soc.*, 1965, 1097.

59. J. N. Ashley and H. Raistrick, *Biochem. J.*, 1939, **33,** 1291.
60. A. G. Perkin and J. J. Hummel, *J. Chem. Soc.*, 1894, **65,** 923.
61. A. Kreigler and R. Thomas, unpublished observations.
62. H. Raistrick, C. E. Stickings, and R. Thomas, *Biochem. J.*, 1953, **55,** 421.
63. D. Rogers, D. J. Williams, and R. Thomas, *Chem. Comm.*, 1971, 393.
64. R. G. Coombe, J. J. Jacobs, and T. R. Watson, *Austral. J. Chem.*, 1970, **23,** 2343.
65. A. J. Birch, C. W. Holzapfel, R. W. Rickards, C. Djerassi, M. Suzuki, J. Westley, J. Dutcher, and R. Thomas, *Tetrahedron Letters*, 1964, 1485.
66. E. Borowski, J. Zielinski, L. Falkowski, T. Ziminski, J. Golik, P. Kolodziejczyk, E. Jereczek, M. Gdulewicz, Yu. D. Shenin, and T. V. Kotienko, *Tetrahedron Letters*, 1971, 685.
67. H. R. Mahler and E. H. Cordes, 'Biological Chemistry', Harper and Row, New York, 1966, p. 677.
68. K. Mosbach, *Acta Chem. Scand.*, 1960, **14,** 457.
69. H. Seto, L. W. Carey, and M. Tanabe, *J. Antibiotics*, 1974, **27,** 558.
70. D. J. Williams and R. Thomas, *Tetrahedron Letters*, 1973, 639.
71. T. Matsuura, *Tetrahedron*, 1977, **33,** 2869.
72. M. Bollaz, G. Büchi, and G. Milne, *J. Amer. Chem. Soc.*, 1970, **92,** 1035.
73. G. C. Elsworthy, J. S. E. Holker, J. M. McKeown, J. B. Robinson, and L. J. Mulheirn, *Chem. Comm.*, 1970, 1069.
74. J. S. E. Holker and J. G. Underwood, *Chem. and Ind. (London)*, 1964, 1865.
75. D. P. H. Hsieh, M. T. Lin, and R. C. Yao, *Biochem. Biophys. Res. Comm.*, 1973, **52,** 922.
76. R. Thomas, personal communication to M. O. Moss, in 'Phytochemical Ecology', ed. J. B. Harborne, Academic, London, 1972, p. 140.
77. M. T. Lin, D. P. H. Hsieh, R. C. Yao, and J. A. Donkersloot, *Biochemistry*, 1973, **12,** 5167.
78. D. P. H. Hsieh, R. C. Yao, D. L. Fitzell, and C. A. Reece, *J. Amer. Chem. Soc.*, 1976, **98,** 1020.
79. R. D. Stipanovic and A. A. Bell, *J. Org. Chem.*, 1976, **41,** 2468.
80. S. J. Gould and C. C. Chang, *J. Amer. Chem. Soc.*, 1977, **99,** 5496.
81. W. M. J. Knöll, R. J. Huxtable, and K. Rinehart, *J. Amer. Chem. Soc.*, 1973, **95,** 2703; A. M. Nadzan and K. L. Rinehart, *ibid.*, 1976, **98,** 5102.
82. R. D. Johnson, A. Haber, and K. L. Rinehart, *J. Amer. Chem. Soc.*, 1974, **96,** 3316.
83. R. J. White and E. Martinelli, *FEBS Letters*, 1974, **49,** 233.
84. R. Bentley, in 'Biosynthesis', ed. T. A. Geissman, Specialist Periodical Reports, The Chemical Society, London, 1974, vol. 3, p. 181.
85. E. Leete, in 'Biosynthesis', ed. T. A. Geissman, Specialist Periodical Reports, The Chemical Society, London, 1973, vol. 2, p. 106.
86. A. A. Iodice and H. A. Barker, *J. Biol. Chem.*, 1963, **238,** 2094.
87. P. Overath, G. M. Kellerman, and F. Lynen, *Biochem. Z.*, 1962, **335,** 500.
88. R. Thomas and R. A. Bassett, *Prog. Phytochem.*, 1972, **3,** 91.
89. T. T. Howarth, A. G. Brown, and T. J. King, *J.C.S. Chem. Comm.*, 1976, 266.
90. A. G. Brown, D. F. Corbett, A. J. Eglington, and T. T. Howarth, *J.C.S. Chem. Comm.*, 1977, 523.
91. H. R. Mahler and E. H. Cordes, 'Biological Chemistry', Harper and Row, New York, 1966, p. 675.
92. J. B. Harborne, in 'Biosynthesis', ed. T. A. Geissman, Specialist Periodical Reports, The Chemical Society, London, 1972, vol. 1, p. 119.
93. E. Leistner, *Phytochemistry*, 1971, **10,** 3015.
94. S. Gatenbeck, *Acta Chem. Scand.*, 1958, **12,** 1211.
95. E. Leistner, *Phytochemistry*, 1973, **12,** 337.
96. A. D. Harmon, J. M. Edwards, and R. J. Highet, *Tetrahedron Letters*, 1977, 4471.
97. H. Raistrick, C. E. Stickings, and R. Thomas, *Biochem. J.*, 1953, **55,** 421.
98. R. Thomas, *Proc. Chem. Soc.*, 1959, 88.
99. W. T. L. Sidwell, H. Fritz, and Ch. Tamm., *Helv. Chim. Acta*, 1971, **54,** 207.
100. O. E. Schultz and G. Haffner, *Arch. Pharm.*, 1960, **293,** 1.
101. E. Spath and K. Gibian, *Monatsh.*, 1930, **55,** 342.
102. E. Lederer, *Nature*, 1946, **157,** 231.
103. T. Rosett, R. H. Sankhala, C. E. Stickings, M. E. U. Taylor, and R. Thomas, *Biochem. J.*, 1957, **67,** 390.
104. D. H. R. Barton, A. M. Deflorin, and O. E. Edwards, *J. Chem. Soc.*, 1956, 530.
105. A. Penttila and H. M. Fales, *Chem. Comm.*, 1966, 656.
106. A. I. Scott, in 'Oxidative Coupling of Phenols', ed. W. I. Taylor and A. R. Battersby, Dekker, New York, 1967, p. 95.
107. D. H. R. Barton, G. W. Kirby, W. Steglich, and G. M. Thomas, *Proc. Chem. Soc.*, 1963, 203.
108. D. Goldsmith, *Fortschr. Chem. org. Naturstoffe*, 1971, **29,** 363.
109. T. M. Harris and C. M. Harris, *Tetrahedron*, 1977, **33,** 2159.
110. D. J. Morecombe and D. W. Young, *J.C.S. Chem. Comm.*, 1975, 198.
111. D. J. Aberhart, J. Y. R. Chu, and L. J. Lin, *J.C.S. Perkin I*, 1975, 2517.
112. N. Neuss, C. H. Nash, J. E. Baldwin, P. A. Lemke, and J. B. Grutzner, *J. Amer. Chem. Soc.*, 1973, **95,** 3797.
113. H. Kluender, C. H. Bradley, C. J. Sih, P. Fawcett, and E. P. Abraham, *J. Amer. Chem. Soc.*, 1973, **95,** 6149.

114. P. A. Fawcett and E. P. Abraham, in 'Biosynthesis', ed. J. D. Bu'lock, Specialist Periodical Reports, The Chemical Society, London, 1976, vol. 4, p. 248.
115. E. Leete, in 'Biosynthesis', ed. J. D. Bu'Lock, Specialist Periodical Reports, The Chemical Society, London, 1977, vol. 5, p. 136.
116. D. J. Aberhart, *Tetrahedron*, 1977, **33,** 1545.
117. E. Babaie, J. Lari, and R. Thomas, unpublished observations.
118. J. Better and S. Gatenbeck, *Acta Chem. Scand.*, 1976, **B30,** 368.

28.2
Photosynthesis, Nitrogen Fixation, and Intermediary Metabolism

E. HASLAM
University of Sheffield

28.2.1 INTRODUCTION

Intermediary metabolism consists of a complex network of chemical reactions which degrade, interconvert, and synthesize the organic molecules of nature. The purposes of this intense chemical activity are twofold: to release energy to carry out useful work, and to create new cellular material. A useful distinction which is frequently employed is to separate formally these chemical reactions into two types, namely those which produce energy, and those which lead to the synthesis of new metabolites. Studies of this latter area are generally referred to as studies of biosynthesis and the reactions themselves as biosynthetic reactions. This useful working distinction should, nevertheless, not be allowed to obscure the fact that many of the intermediates involved in the classical energy-releasing degradation pathways (*e.g.* glycolysis and the Krebs cycle) are themselves the branch points from which purely biosynthetic routes to other cellular constituents arise.

In a somewhat analogous way, chemists and biochemists have over the past few decades developed separate interests in the molecular aspects of the natural world. The dynamics of the living cell, its separate functions, and their control have in the main become the province of the biochemist. Correspondingly the interests of organic chemists have focused upon those metabolites which accumulate in cells — the primary metabolites such as carbohydrates, proteins, nucleic acids, lipids, and steroids, and the plethora of secondary metabolites such as alkaloids, terpenes, phenols, quinones, and the many microbial antibiotics. This division of interests should similarly not obscure the common interests and common goals which the two groups of scientists pursue. Although therefore the major emphasis in the discussions of biosynthesis both in the chapters which follow and elsewhere in the text concerns topics of particular interest to the chemist, it is considered important initially to place these developments in the context of the broad picture of intermediary metabolism and of the two fundamental biosynthetic processes of photosynthesis and nitrogen fixation. These provide the starting points and the basis for the biosynthetic discussions which follow.

28.2.2 PHOTOSYNTHESIS

Biosynthesis begins with photosynthesis.[1] All life on earth depends on the ability of certain organisms — green plants, algae, and photosynthetic bacteria — which contain the characteristic photosynthetic pigments to utilize the sun's radiant energy to synthesize organic molecules from inorganic substrates such as carbon dioxide, nitrogen, and sulphur. The products of photosynthesis then provide not only the substrate materials but also the chemical energy for all subsequent biosynthetic reactions. It is customary in some texts to describe photosynthesis as a process for the formation of *only* carbohydrates, and whilst in some cases starch, cellulose, and sucrose constitute the major products of photosynthesis, in others carbohydrate synthesis may account for perhaps only a third of the total carbon fixed and reduced during photosynthesis. When the light is on, no clear distinction can be made between the products of photosynthesis and other types of biosynthetic activity in the cell and for which intermediates in the photosynthetic carbon reduction cycle may be utilized.

The energy of the incident light is adsorbed by the photosynthetic pigments in organelles called photoreceptors (chloroplasts in higher plants, plastids in algae, and chromatophores in photosynthetic bacteria). The predominant pigment is chlorophyll and each organism which is capable of performing photosynthesis contains at least one species of chlorophyll. Green plants contain, for example, chlorophyll *a* (**1a**) and chlorophyll *b* (**1b**) and two other types of chlorophyll, P700 and P670, which are thought to be of considerable importance in photosynthesis. Other pigments may be components of the photosynthetic apparatus and they play a secondary role in the photosynthetic act. Typical of these are the yellow-bronze carotenoids in higher plants and, in the photosynthetic algae, the blue-red phycobilins.

Photosynthesis in green plants may be formulated in terms of the overall reaction shown in equation (1), which accommodates the well-known fact that in plants water is essential for photosynthesis to occur and oxygen (from the water) is evolved as a by-product of the reaction. In photosynthetic bacteria, oxygen is not produced and other hydrogen donors [H_2X—such as H_2S and lactate, $CH_3CH(OH)CO_2^-$] are utilized—equation (2). Hill (in 1937) and Arnon (in 1954) showed that the production of respectively the NADPH and of the ATP required to fix carbon dioxide is independent of their utilization in the photosynthetic carbon reduction cycle. These observations permitted the formal separation of the photosynthetic reaction into a light reaction (NADPH and ATP production) and a dark reaction in which carbon dioxide is converted to carbohydrate.

$$2H_2O + CO_2 \longrightarrow [CH_2O] + H_2O + O_2 \qquad (1)$$

$$2H_2X + CO_2 \longrightarrow [CH_2O] + H_2O + 2X \qquad (2)$$

Although many of the physico-chemical details of the light reaction remain to be fully clarified, the overall sequence can be briefly summarized. The primary reaction is the excitation of a chlorophyll molecule by electron capture to a higher energy level. Ultimately this electronic energy of excitation reaches a reaction centre where it may be used either to synthesize ATP from ADP or to reduce NADP to NADPH (~3 molecules of ATP and 2 molecules of NADPH are required for each molecule of carbon dioxide which is reduced). Water (or H_2X) participates in this overall reaction and oxygen (or 2X) is released.

The sequence of steps in which carbon dioxide is fixed in photosynthesis was proposed in the 1950s by Calvin and is frequently referred to as the Calvin cycle or the photosynthetic carbon reduction cycle (see Scheme 2). Unlike the light reaction, which is peculiar to photosynthetic tissues, the synthesis of carbohydrate from carbon dioxide has much in common with the steps that are employed for the synthesis of carbohydrate in non-photosynthetic organisms. The scale of the operation in green plants nevertheless remains impressive, and conservative estimates place the amount of carbon fixed by plants at around 35×10^{15} kg per annum, and for each gram of carbon fixed a plant must process more than 6250 litres of air. Although normally 99% of the carbon dioxide that is extracted by plants from the air is fixed by photosynthetic reactions in the light, there are dark carboxylations[2] which in certain plants, particularly succulents (members of the Crassulaceae), are rapid and make significant contributions to the total carbon which is fixed.

The key initial step is that in which carbon dioxide enters the photosynthetic carbon reduction cycle by carboxylation of ribulose 1,5-diphosphate (**2**) to give the putative intermediate (**4**), which is then hydrolysed to give two molecules of 3-phosphoglyceric acid (**5**; PGA). The enzyme which catalyses this reaction, presumably on the enolic form (**3**) as substrate—ribulose 1,5-diphosphate carboxylase (E.C. 4.1.1.39)—has been isolated and purified.[3] The two molecules of 3-phosphoglyceric acid (**5**) are then reduced to D-glyceraldehyde 3-phosphate (**6**) in a two-stage process which requires NADPH and ATP (Scheme 1). This carboxylation and reduction sequence may be considered to occur three times for each turn of the cycle, so that three molecules of ribulose 1,5-diphosphate (**2**) are carboxylated by three molecules of carbon dioxide to produce finally six molecules of phosphorylated triose (**6**) (Scheme 2). Five of these triose molecules then undergo a series of condensations and intramolecular transfer reactions, the net effect of which is to lead to the formation of three molecules of ribulose 1,5-diphosphate (**2**), the receptor molecule for the carbon dioxide. There is thus a net gain of one triose molecule for each turn of the cycle and this may be withdrawn from the cycle either as PGA or as any of the sugar phosphates for storage purposes or for other synthetic activities.

28.2.3 NITROGEN FIXATION

Nitrogen is essential to all forms of life but, in the form which it exists in the atmosphere, nitrogen is an inert gas to all but a comparatively few organisms which have the ability to convert the molecule to a combined form (ammonia). The latest estimates show that biological nitrogen fixation accounts for about two-thirds of all fixed nitrogen and amounts to approximately 178×10^9 kg per annum. About half that quantity is also fixed by physical and chemical means. Ionizing radiation, fuel combustion, and lightning give rise to fixed nitrogen as oxides of nitrogen, and the Haber process of nitrogen fixation to nitrogen as ammonia. Indeed, of all man's recent interventions in the cycles of nature, the industrial fixation of nitrogen for agricultural practice far exceeds all others in its magnitude.

Biological nitrogen fixation is brought about by certain procaryotic organisms—bacteria and blue-green algae. The nitrogen-fixing bacteria may be further sub-divided into the free living and those which fix nitrogen whilst living in a symbiotic association with plants. Particularly significant in this latter category are the genus Rhizobium, which associate with the agriculturally important legumes (soyabean, clover, and lucerne) to form nodules, able to fix nitrogen, on the roots of these plants. Both plants and bacteria show specificity in the association, although the biochemical basis of the symbiotic interaction is not understood. However, it is believed that the bacterium itself contains all the genetic information which is necessary for the synthesis of the enzyme nitrogenase, which catalyses the process of nitrogen fixation. After the roots of the host plant have been invaded by the rhizobial bacterium, the latter develops into enlarged non-reproductive cells (bacteroids) which, enclosed by a membrane, live within the cytoplasm of the host cell.

A major breakthrough in the understanding of the biochemistry and chemistry of nitrogen fixation came in 1960 when the first cell-free extract of the enzyme nitrogenase was prepared from the free-living anaerobic bacterium *Clostridium pasteurianum*.[4] The enzyme has now also been isolated from at least 16 other organisms and a great deal of data relating to nitrogenase has been accumulated.[5,6] However, the exact mode of

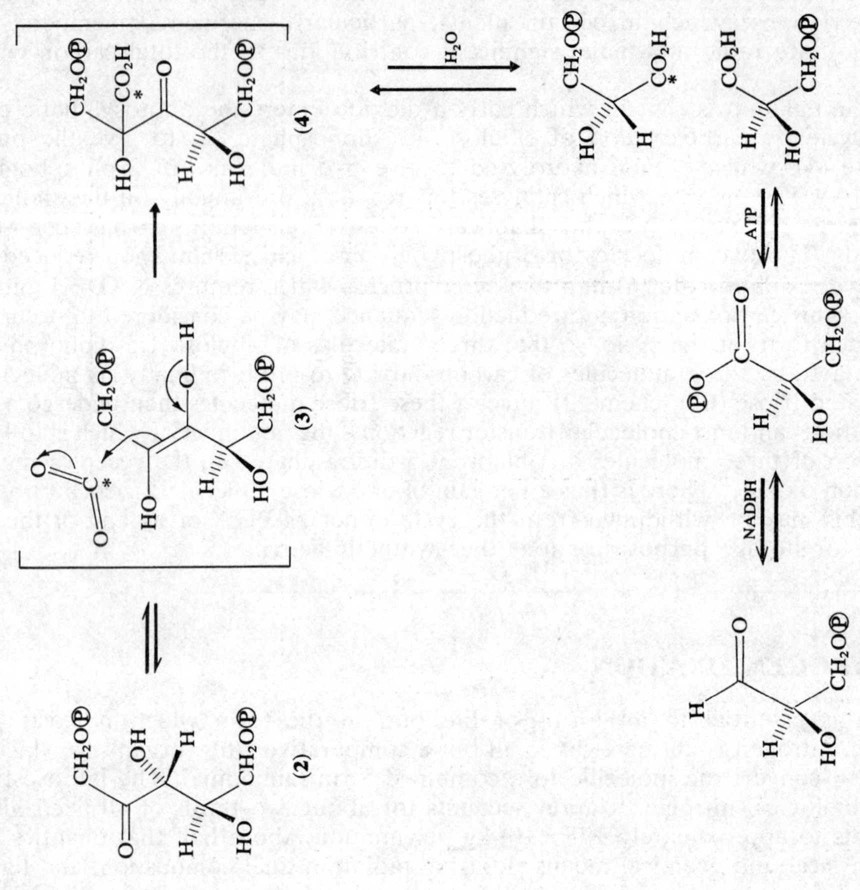

SCHEME 1 Carboxylation of ribulose 1,5-diphosphate and reduction to give glyceraldehyde 3-phosphate

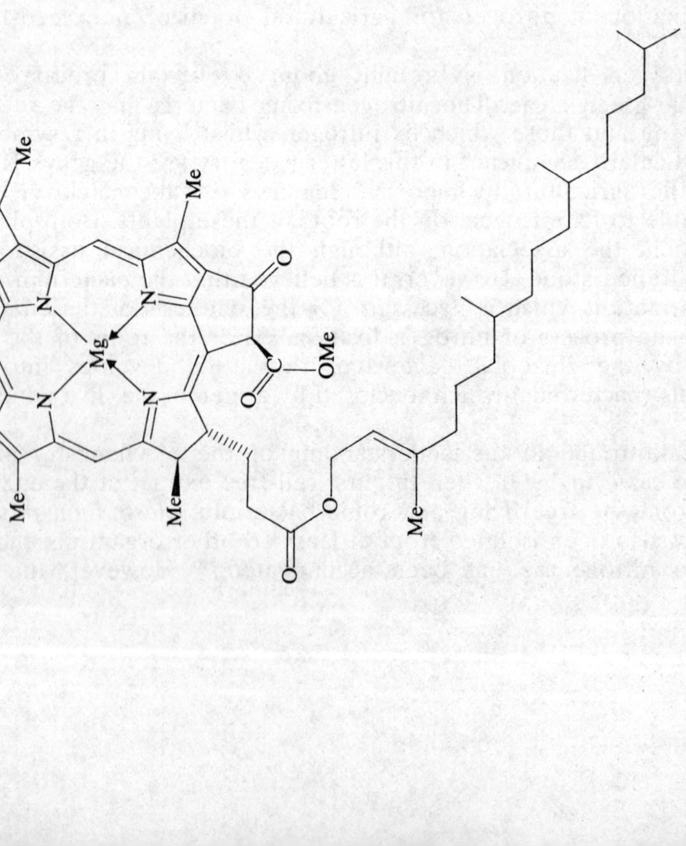

(1a) Chlorophyll a, R = Me
(1b) Chlorophyll b, R = CHO

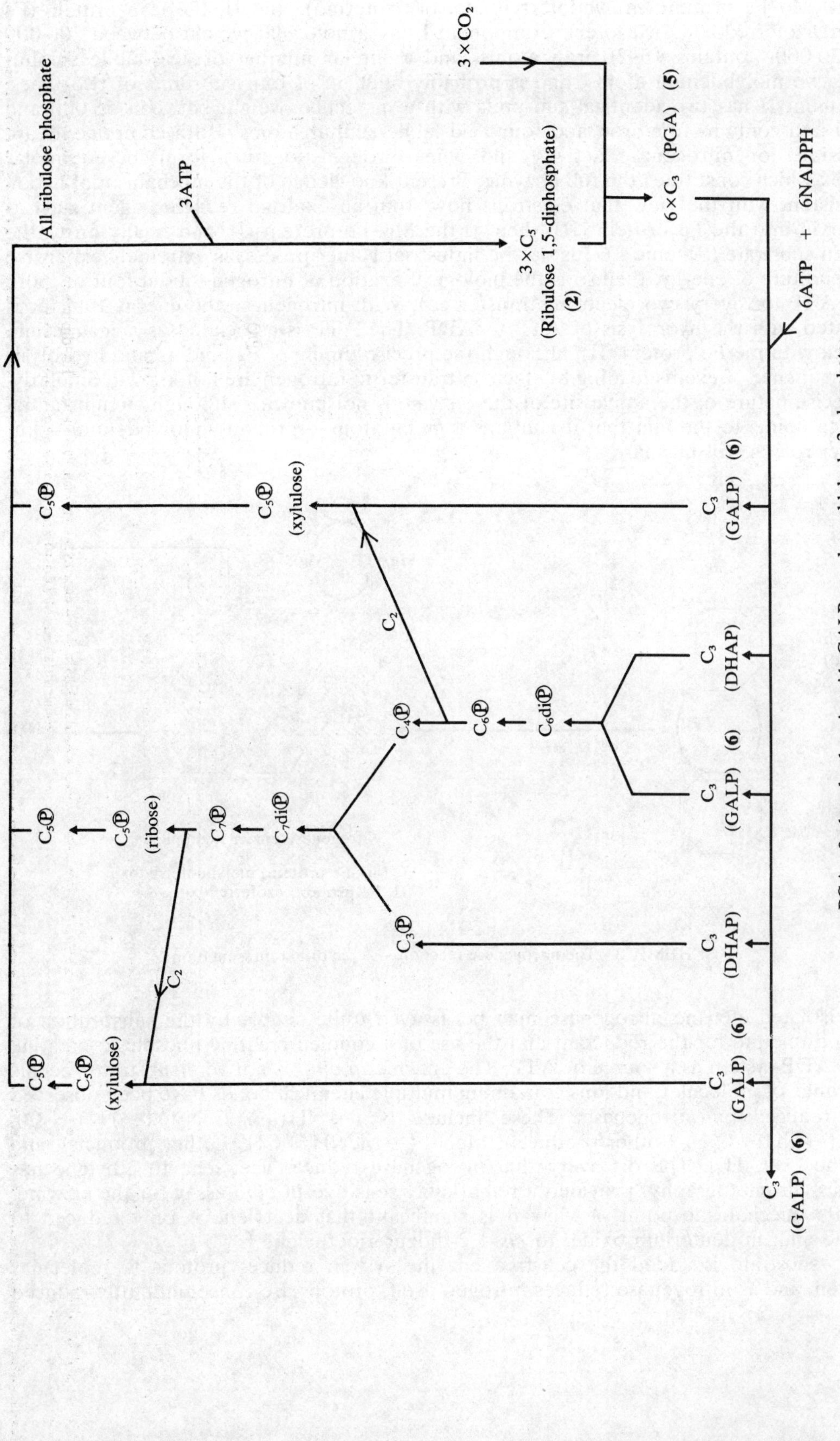

PGA = 3-phosphoglyceric acid; GALP = D-glyceraldehyde 3-phosphate;
DHAP = dihydroxyacetone phosphate

SCHEME 2 Calvin photosynthetic carbon reduction cycle

nitrogen reduction and fixation still remains a mystery. Nitrogenase is extremely sensitive to oxygen and the enzyme complex consists of two components: I, the molybdenum–iron protein (Mo–Fe protein, molybdoferredoxin, or azofermo); and II, the iron protein (Fe protein, azoferredoxin, or azofer). Component I has a molecular weight between 200 000 and 230 000, contains 15–30 iron atoms, and a similar number of acid-labile sulphur atoms, two molybdenum atoms, and is probably built up of four sub-units of two types. Component II has two identical sub-units with a molecular weight between 55 000 and 60 000 and contains four iron and four acid-labile suphur atoms. Both components are responsible for nitrogenase activity and when brought together form an equimolar complex which constitutes the full enzyme. Present knowledge of the mechanism of action is consistent with the view that electrons flow from an external reducing agent such as ferredoxin into the Fe protein (II), then to the Mo–Fe protein (I), and finally on to the nitrogen substrate (Scheme 3). Just as the industrial Haber process is extremely expensive in expenditure of energy, then so is the biological fixation of nitrogen (about four or more moles ATP for every two electrons transferred). With nitrogenase the energy transfer is associated with the hydrolysis of ATP to ADP. The ATP is associated as a magnesium complex with the Fe protein (II), although the precise timing of its binding and hydrolysis in the sequence of events leading to electron transfer to nitrogen are not known. Similarly, the precise nature of the active site of the enzyme is not known, although circumstantial evidence points to the fact that it contains a metal atom — probably molybdenum — but its exact role is still unclear.

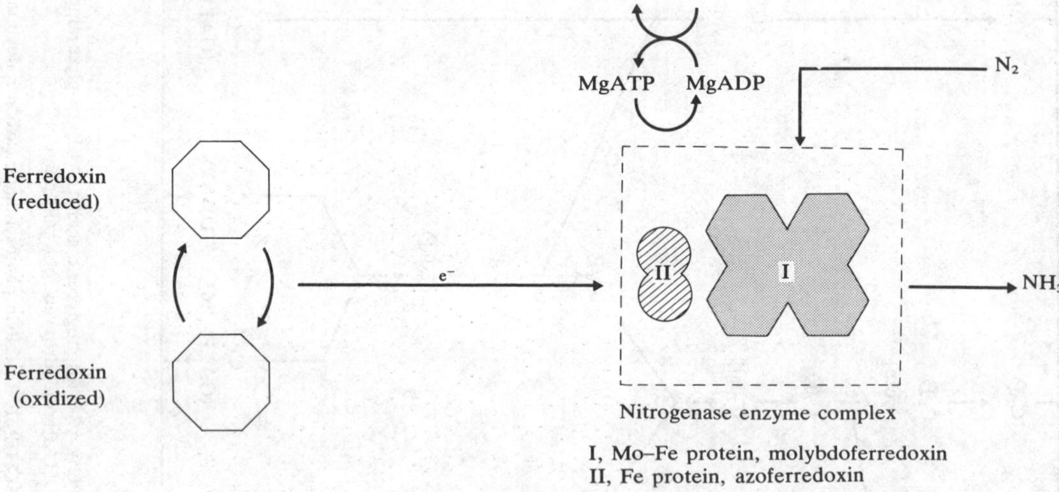

SCHEME 3 The nitrogenase reaction — a pictorial representation

The isolated enzyme nitrogenase may be assayed quite simply by the substitution of sodium dithionite for the reductant and the use of a coupled creatine phosphate–creatine kinase–ADP system as a source of ATP. The enzyme *in vitro* is not very substrate-specific and a range of molecules and ions containing multiple chemical bonds have been observed to be reduced by nitrogenase. These include N_3^- ($\rightarrow NH_3+N_2$), N_2O ($\rightarrow N_2+H_2O$), MeCN ($\rightarrow NH_3+C_2H_6$+other products), MeNC ($\rightarrow MeNH_2+CH_4$+other products), and acetylene ($\rightarrow C_2H_4$). The discovery that nitrogenase reduces acetylene to ethylene has (with gas chromatography) provided a remarkably sensitive *in vitro* assay for the enzyme. From the mechanistic point of view it is significant that acetylene is only reduced to ethylene and, in deuterium oxide, to *cis*-1,2-dideuterioethylene.[7]

If no substrate is added the cell-free enzyme system reduces protons to molecular hydrogen, and as nitrogenase reduces nitrogen itself, protons are concommitantly reduced to give hydrogen. Surveys[8] indeed indicate that hydrogen evolution is a general phenomenon associated with nitrogen fixation in nodulated nitrogen-fixing symbionts, and this energy loss in terms of electron transfer to nitrogen, *via* nitrogenase, may seriously effect the efficiency of nitrogen fixation in many legumes. Significantly, carbon monoxide inhibits all nitrogenase reactions except this production of hydrogen.

The mechanism of nitrogen fixation has long been a challenging chemical and biochemical problem, not least because of the characteristic chemical inertness of the nitrogen molecule. The oldest and most widely accepted proposal, due to Wieland (1922), is that the nitrogen molecule is reduced in three stages (equation 3).

$$N_2 \xrightarrow[2H^+]{2e^-} NH{=}NH \xrightarrow[2H^+]{2e^-} H_2N{-}NH_2 \xrightarrow[2H^+]{2e^-} 2NH_3 \qquad (3)$$

However, neither of these postulated intermediates (di-imine and hydrazine) has been detected during nitrogen reduction. Moreover, di-imine is not reduced by the enzyme, although recent reports indicate that the second postulated intermediate (hydrazine) is converted to ammonia by nitrogenase. Recent chemical investigations have sought to throw light on this problem. Chatt and his colleagues[9] have shown that metal–dinitrogen complexes of the type $M(N_2)_2(PR_3)_4$ (where M = Mo or W) yield up to 90% of ammonia when treated with sulphuric acid in methanol. By using different phosphine ligands and acids they were able to derive tungsten and molybdenum complexes containing nitrogen ligands (N_2H, N_2H_2, and N_2H_3) which represent various stages of nitrogen reduction. Van Tamelen and Brulet[10] similarly found that the molybdenum complex (**7**) gave 0.36 mole of ammonia per mole of complex when treated with hydrobromic acid in *N*-methylpyrrolidone.

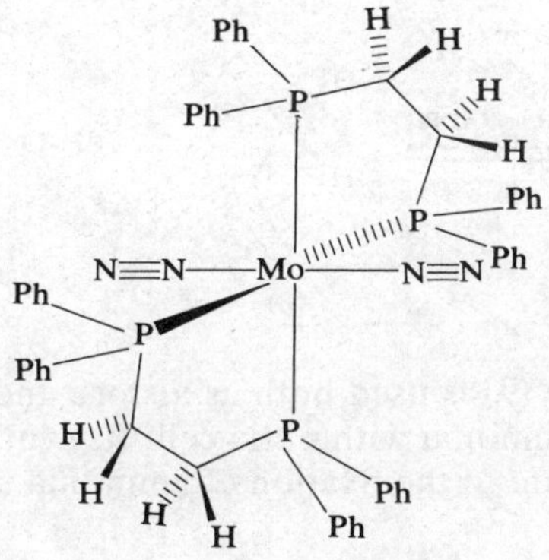

(**7**)

Schrauzer and his group[11,12] have developed alternative model systems for nitrogenase which operate in aqueous media. In a system composed of alkali molybdate, a sulphur ligand such as the amino-acid L-cysteine, ferredoxin models of the type $[Fe_4S_4(SR)_4]^{2-}$, and a reducing agent ($Na_2S_2O_4$ or $NaBH_4$), they have been able to duplicate the known *in vitro* reactions of the isolated enzyme system. In a hypothesis for nitrogenase action based on these models, Schrauzer has suggested that in the *in vivo* situation the nitrogen molecule forms an unusual labile 'side on' complex with molybdenum in the enzyme active site. The nitrogen molecule is then reduced in the *cis* mode (analogous to acetylene reduction) to di-imine. Subsequently the di-imine dissociates and disproportionates to yield nitrogen, hydrogen, and hydrazine; the latter is then finally reduced to ammonia by the enzyme.

Although ammonia is the product of biological nitrogen fixation, it is not accumulated by organisms which are able to carry out this process. If the nitrogen-fixing organism is associated with a higher plant, the ammonia may be stored as the amino-acids asparagine and glutamine. In other cases excess nitrogen fixed (as ammonia) may be excreted into the environment of the nitrogen-fixing organism. Here it may be directly utilized by other organisms incapable of nitrogen fixation or it may be subjected to the process of nitrification in which the ammonia is oxidized to nitrate ion by nitrifying bacteria (equation 4).

$$NH_3 \xrightarrow{\text{Nitrosomonas}} NO_2^- \xrightarrow{\text{Nitrobacter}} NO_3^- \qquad (4)$$

Nitrate is the most abundant form of nitrogen in the soil and plants and soil organisms generally assimilate this species by reduction back to ammonia, using the enzymes nitrate reductase (i), nitrite reductase (ii), hyponitrite reductase (iii), and hydroxylamine reductase (iv) (equation 5).

$$NO_3^- \xrightarrow{\text{(i)}} NO_2^- \xrightarrow{\text{(ii)}} N_2O_2^{2-} \xrightarrow{\text{(iii)}} NH_2OH \xrightarrow{\text{(iv)}} NH_3 \qquad (5)$$

Some micro-organisms, including for example *Escherichia coli* and *Bacillus subtilis*, use nitrate ion as a terminal electron acceptor instead of oxygen and reduce nitrate to ammonia in a process termed nitrate respiration. Certain bacteria indeed carry out nitrate respiration to the ultimate point of producing nitrogen, which is returned to the atmosphere, instead of ammonia.

There are three principal reactions by which ammonia is incorporated into organic molecules in nature. The reductive amination of α-ketoglutarate to give L-glutamic acid (**8**) (equation 6) is in many organisms the sole reaction in which inorganic nitrogen can be fixed to give the α-amino-group of an amino-acid. The amino-group of L-glutamic acid can then be utilized to form the amino-group of other amino-acids by the transamination reaction, which involves pyridoxal phosphate as a co-factor.

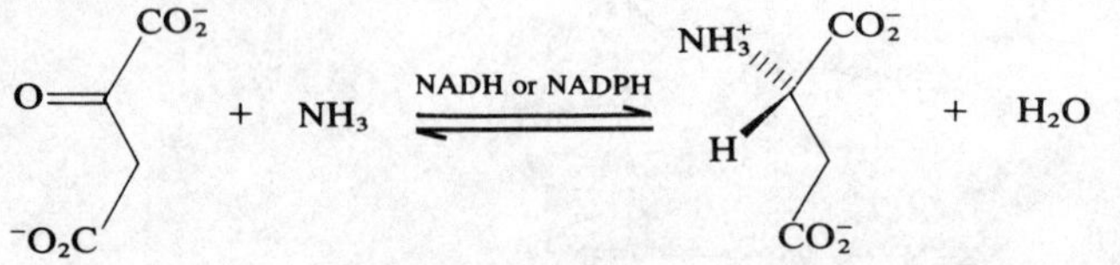

(6)

(**8**) Glu

The amino-acid L-glutamine (**9**) is used both as a store and a donor of the amino-group and as a means to transport ammonia within the cell. Its synthesis from L-glutamic acid (**8**) constitutes a second major route to the fixation of ammonia in organic molecules (equation 7).

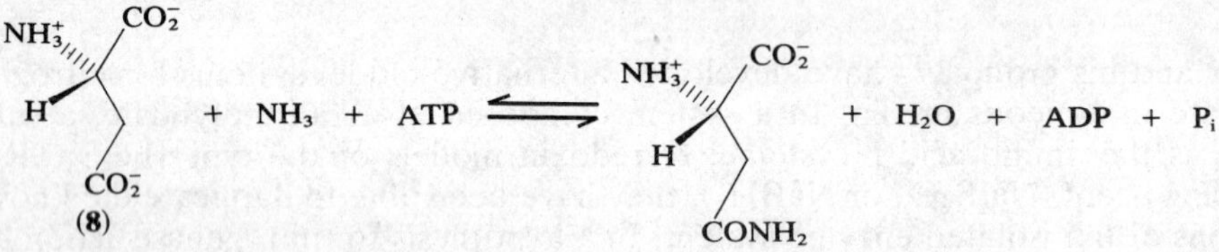

(**8**) (7)

(**9**) Gln

The third reaction utilized for the fixation of ammonia is that in which carbamoyl phosphate (**10**) is synthesized from carbon dioxide, ammonia, and ATP (equation 8). Carbamoyl phosphate is an intermediate in the synthesis of urea and pyrimidines.

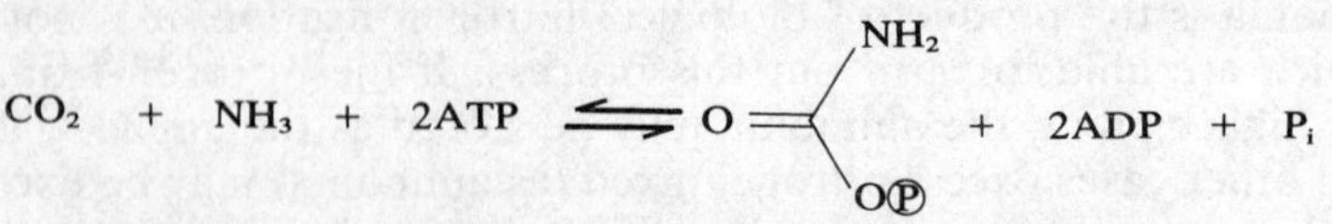

(8)

(**10**)

These three synthetic reactions constitute the principal means by which the requirements for nitrogen in biosynthetic reactions in various organisms are met.

28.2.4 THE SULPHUR CYCLE

Sulphur occurs in many important compounds of a biological origin and it is therefore pertinent to discuss briefly the way in which this element is incorporated into general metabolism. The sulphur atom occurs in a variety of inorganic forms of differing oxidation states. It must first be reduced to the oxidation state of sulphide (S^{2-}) before it can enter into organic molecules. Many micro-organisms and higher plants are able to utilize sulphate ion as their source of sulphur and the anion is reduced to sulphide in a series of reactions (equation 9) analogous to those employed in nitrate assimilation (equation 5). Sulphate can also serve as a terminal oxidant for certain anaerobic bacteria, and in this case electron transfer also produces a stepwise reduction to sulphide.

$$SO_4^{2-} \longrightarrow SO_3^{2-} \longrightarrow S^{2-} \qquad (9)$$

When the sulphur is released from organic molecules, as for example in decaying organic matter, it can be oxidized by soil micro-organisms back to its highest oxidation level of the sulphate anion.

The principal route whereby sulphur is assimilated into organic compounds is *via* the synthesis of the amino-acids cysteine (**11**), homocysteine, and methionine. The acceptor for sulphide is *O*-acetylserine (**12**), and plants and micro-organisms fix sulphide by reaction with (**12**) to give cysteine (equation 10).

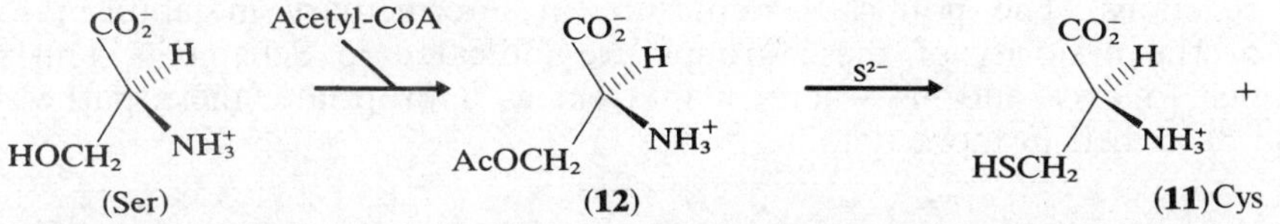

(10)

28.2.5 INTERMEDIARY METABOLISM AND BIOSYNTHESIS

The carbohydrates of the pentose phosphate cycle, the triose phosphates (*e.g.* **6**), and 3-phosphoglyceric acid (**5**; PGA) are not only important intermediates in the photosynthetic fixation of carbon dioxide but they are also key compounds in the intermediary metabolism of carbohydrates in all organisms. Although it should be noted that not all organisms can carry out all the reactions depicted, some of the broad biochemical interrelationships which exist at the primary and secondary metabolic levels are shown in Scheme 4. Glucose and other sugars are metabolized *via* the well-known glycolytic pathway and the tricarboxylic cycle and this route accounts for the bulk of carbohydrate metabolism in most organisms. Although quantitatively of less importance than glycolysis and the tricarboxylic acid cycle, an alternative pathway for the oxidation of carbohydrate is the pentose phosphate pathway. The chief importance to the organism of this route is that it provides a source of NADPH and the pentose phosphates. The key step in this pathway involves the oxidation of D-glucose 6-phosphate *via* D-gluconolactone 6-phosphate to D-ribulose 5-phosphate with the generation of NADPH. The pentose phosphate is then metabolized to other sugar phosphates such as D-fructose 6-phosphate and D-glyceraldehyde 3-phosphate, which may be further metabolized by the glycolytic pathway, and D-erythrose 4-phosphate, which along with phosphoenolpyruvate (PEP) is a starting point for aromatic amino-acid synthesis *via* the shikimate pathway.

These routes of metabolism not only provide means whereby carbon compounds are degraded to simpler products with the release of energy, but they also have a twin function in that they provide intermediates which are used as branch points for activities of a purely biosynthetic nature. The tricarboxylic acid cycle is thus used for the final oxidation to carbon dioxide of all carbon sources, whether carbohydrate, protein, or lipids. In addition, many of the compounds that are intermediates in the cycle of oxidation are the precursors of other primary and secondary cell constituents. Thus, for example, oxaloacetate and α-ketoglutarate are respectively essential for the synthesis of the amino-acids aspartic acid (Asp) and glutamic acid (Glu) and those other amino-acids which in turn are derived from them. Succinyl-coenzyme A is similarly an intermediate in the synthesis of the porphyrin and corrin ring systems. Analogously, in the glycolytic pathway, 3-phosphoglyceric acid (PGA), pyruvic acid, and phosphoenol pyruvate (PEP) are starting points for amino-acid and other biosynthetic activities. However, as an intermediate in or starting point for primary and secondary biosynthetic activity, acetyl-coenzyme A must hold the pre-eminent position amongst the intermediates of intermediary metabolism.

The more intimate details of the various biosynthetic pathways which lead to many of the primary metabolites such as amino-acids, purines, and pyrimidines are generally more than adequately described in textbooks of biochemistry. The primary purpose of the subsequent discussion in this text is to describe the knowledge which has been garnered over the past quarter of a century to elucidate the pathways of biosynthesis of some of the more complex natural molecules such as the steroids, haem, chlorophyll, and vitamin B_{12} whose biological functions are partially, if not fully, understood. In addition, its purpose is to outline the biosynthetic routes which nature has chosen to that plethora of secondary metabolites such as the polyketides, alkaloids, phenols, quinones, and the various microbial antibiotics. The organic chemist has made substantial efforts to unravel the intricate details of many of these pathways, not only in tracing the biosynthetic details but also at the enzymic level in delineating many of the more subtle stereochemical features of these biosynthetic reactions. The points of departure in intermediary metabolism for the biosynthesis of the majority of these groups are indicated in Scheme 4, and in the discussion which follows, and elsewhere in the text as appropriate, these pathways are amplified and described in more detail.

References

1. M. Calvin and J. A. Bassham, 'The Photosynthesis of Carbon Compounds', Benjamin, New York, 1962.
2. D. A. Walker, *Endeavour,* 1966, **25,** 21.
3. J. R. Quayle, R. C. Fuller, A. A. Benson, and M. Calvin, *J. Amer. Chem. Soc.*, 1954, **76,** 3610; A. Weissbach, G. L. Horecker, and J. Hurwitz, *J. Biol. Chem.*, 1956, **218,** 795; J. Mayaudon, A. A. Benson, and M. Calvin, *Biochim. Biophys. Acta,* 1957, **23,** 342.
4. J. E. Carnaham, L. E. Mortenson, H. F. Mower, and J. E. Castle, *Biochim. Biophys. Acta,* 1960, **38,** 188.
5. J. R. Postgate, 'The Chemistry and Biochemistry of Nitrogen Fixation', Plenum, London, 1971.
6. D. Kleimer, *Angew. Chem. Internat. Edn.*, 1975, **14,** 80; H. C. Winter and R. H. Burris, *Ann. Rev. Biochem.*, 1975, **45,** 409.
7. M. J. Dilworth, *Biochim. Biophys. Acta,* 1966, **127,** 285.
8. K. R. Schubert and H. J. Evans, *Proc. Nat. Acad. Sci. U.S.A.*, 1976, **73,** 1207.
9. J. Chatt, A. J. Pearman, and R. L. Richards, *Nature,* 1975, **253,** 39.
10. E. E. van Tamelen and C. R. Brulet, *J. Amer. Chem. Soc.*, 1975, **97,** 911.
11. G. N. Schrauzer, *Angew. Chem. Internat. Edn.*, 1975, **14,** 514.
12. E. L. Moorehead, P. R. Robinson, T. M. Vickrey, and G. N. Schrauzer, *J. Amer. Chem. Soc.*, 1976, **98,** 6555; 1975, **97,** 7069.

PART 29

BIOSYNTHETIC PATHWAYS FROM ACETATE

29.1
Polyketide Biosynthesis

J. D. BU'LOCK
University of Manchester

29.1.1 INTRODUCTION

The polyketides are a group of natural products whose diversity even exceeds that of the isoprenoids. Like the isoprenoids, they were first defined as a result of comparative analysis of natural product structures,[1] but with the significant difference that this was done at a time when basic experimental confirmation of the resultant hypotheses could also be attained rather quickly. A good historical account, from the school to which primary credit is due, is that of Birch.[2] Because of its general importance, the topic of polyketide biosynthesis is covered in most general accounts of natural product biosynthesis,[3–7] and its progress is regularly reviewed, both for the topic as a whole[8] and for individual groups of polyketides.[9]

29.1.1.1 Definition, examples, and occurrence

The category of polyketides is described in biosynthetic rather than structural terms. Broadly it comprises structures derived wholly or partly, formally or strictly, from poly-β-ketomethylene chains, —[CHRCO]$_n$—, where n may be as high as 19–20 (in macrolide antibiotics) but is most often 4–8, and R is often, but by no means invariably, hydrogen. These sub-units are often referred to as derived from the corresponding simple aliphatic acids:

—CH_2CO—	'acetate'
—CHMeCO—	'propionate'
—CHEtCO—	'butyrate' *etc.*

or alternatively from the malonic acids, which are closer to the actual biosynthetic intermediates (malonate, methylmalonate, *etc.*) (see below). The terms 'acetate-derived', or the cacologism 'acetogenin', are neither adequate nor equivalent. The basic assembly process by acylation reactions adequately distinguishes the polyketides from the isoprenoids, even though the fundamental precursor of the latter is acetyl-CoA. The derivation of the overall skeletal structures from the ketomethylene chain includes, particularly, cyclizations of the chain by aldol- or Claisen-type condensations, rearrangements, and ring-cleavages, combined with a series of ancillary processes in which some carbons may be removed, others added, and new functionalities introduced.

Many natural polyketides can legitimately be described as 'wholly acetate-derived'. In these, the poly-β-ketomethylene chain is in fact built up (as discussed more fully in Section 29.1.2) from one molecule of acetyl-coenzyme A, the 'starter' unit, and several (usually 3 to 7) molecules of malonyl-CoA, which is also derivable from acetyl-CoA (see Section 29.1.1.3), which are added to the chain by stepwise condensations with concomitant decarboxylations (equation 1).

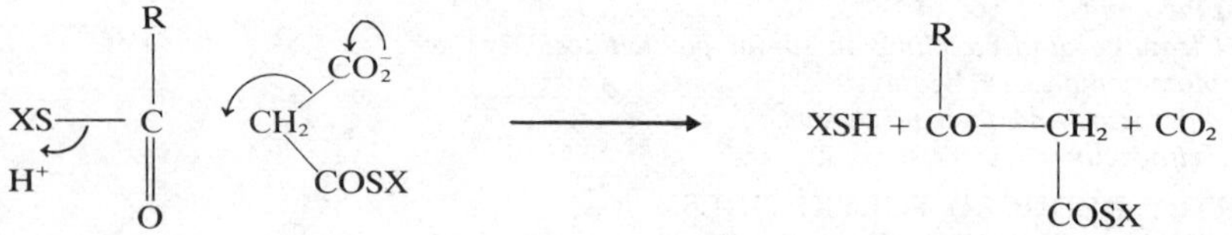

(1)

More generally, however, the starter acyl unit may be contributed by any one of a considerable variety of acyl-CoA — fatty acids up to C_{18}, benzoic, cinnamic, nicotinic, *etc.* The range of 'malonyl' units is less varied, but methylmalonyl units ('propionate') are quite common in some natural product series and 'butyrate' units, presumed to come from ethylmalonyl-CoA, are also known.

Where all the successive ketomethylene units are fully reduced — normally during the process of chain assembly — the resultant polyketide is of course a fatty acid. There are so many examples of natural products of intermediate types, either with part-reduction of many individual units or complete reduction of just some of them, that the distinction of fatty acids in particular from polyketides in general is not very logical. Thus, although the subject of fatty acid synthesis fully merits separate and detailed discussion (given here in Section 25.1.5 by F. D. Gunstone), it will be essential to reconsider the topic here.

Besides reduction, ancillary transformations of the ketomethylene system which can be integral with the main biosynthetic process can include alkylations, particularly *C*-methylation and -prenylation, and other (electrophilic) substitutions, and the resultant assembly, besides being subject to various modes of cyclization, may undergo more or less extensive subsequent transformations. Polyketide biosynthesis therefore leads to a great variety of natural product structures. The scope of this overall variety can be illustrated even for quite small polyketides, *e.g.* by the series of structures (**1**)–(**13**), all of which are tetraketides, *i.e.* are derived from four acyl units. Several of the examples cited here are discussed in greater detail in subsequent sections.

Both the acetophenone (**1**) and orsellinic acid (**2**) are formed from one acetyl-CoA and three malonyl-CoA units, assembled linearly (the linear chain is shown, conventionally, as an unbroken heavy line); they result by different cyclizations at the triketo-octanoyl stage, as indicated in formulae (**1a**) and (**2a**), and a useful terminology is to distinguish the two patterns of cyclization as 'Claisen' (involving the terminal acyl unit) and 'aldol', respectively. Among the naturally-formed homologues of orsellinic acid, compound (**3**) results when the acetyl-CoA starter is replaced by propionyl-CoA, whereas (**4**) results from *C*-methylation prior to cyclization. Since such *C*-methylations occur by electrophilic attack [of *S*-adenosylmethionine (see Section 29.1.2.2)], they involve precisely the same carbon atoms of an acetate-derived chain as those which would carry methyl groups when a methylmalonyl-CoA ('propionate') unit is incorporated. Consequently the two biosynthetic routes cannot be distinguished *a priori*, by simple inspection of the structure, but only experimentally, *i.e.* by testing for the incorporation of the alternative precursors.

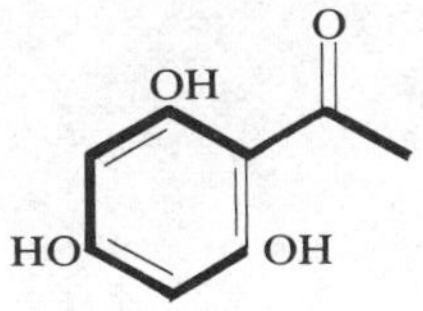

(1)

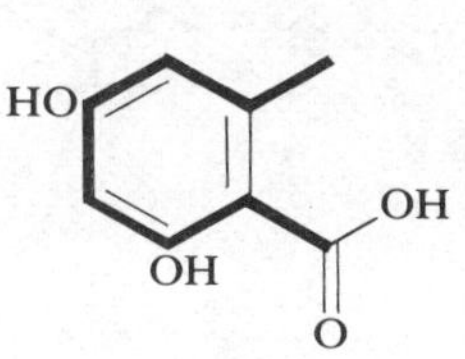

(2)

Heavy line shows polyketide chain

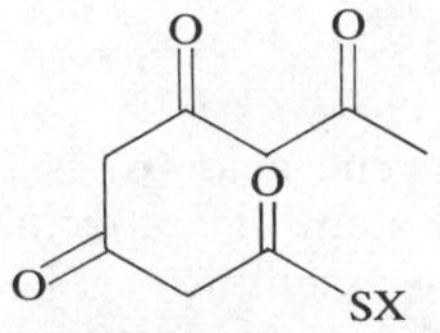

(1a)

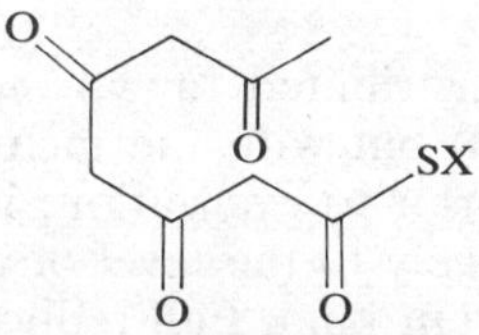

(2a)

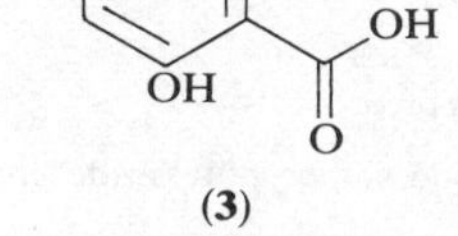

(3)

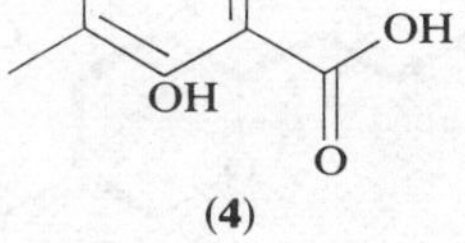

(4)

More apparent features are exemplified in the flavanone (**5**), which is the parent compound of a very numerous series of plant phenolics and which results from the combination of a *p*-coumaryl-CoA starter with three malonyl-CoA, followed by 'Claisen' cyclization.

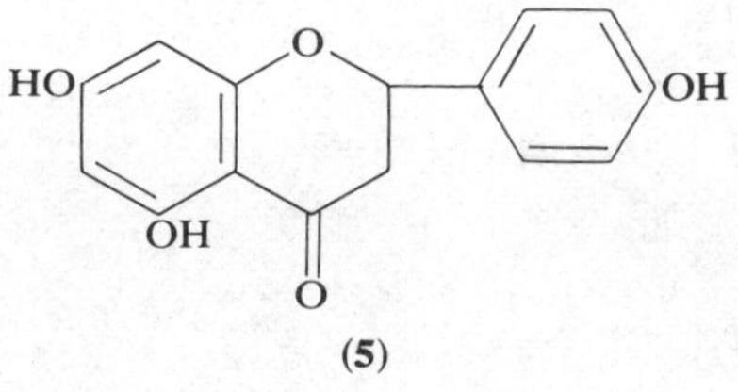

(5)

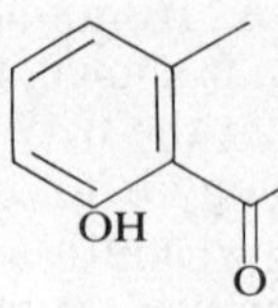

(6)

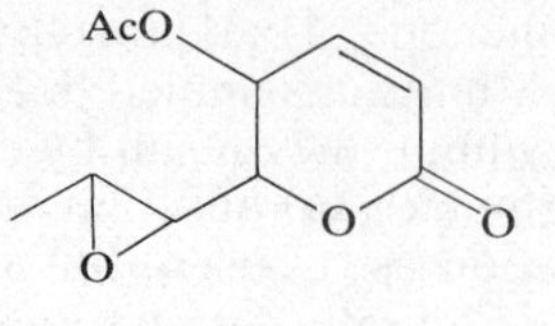

(7)

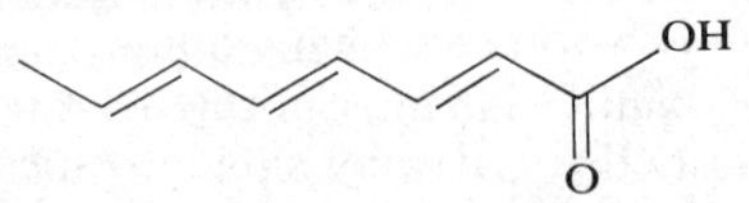

(8)

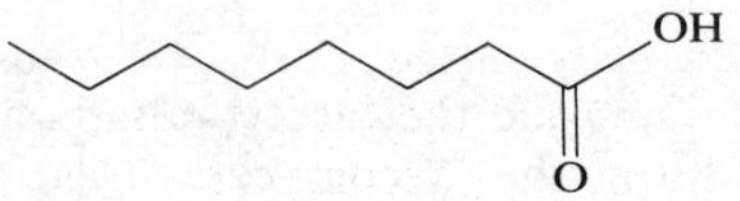

(9)

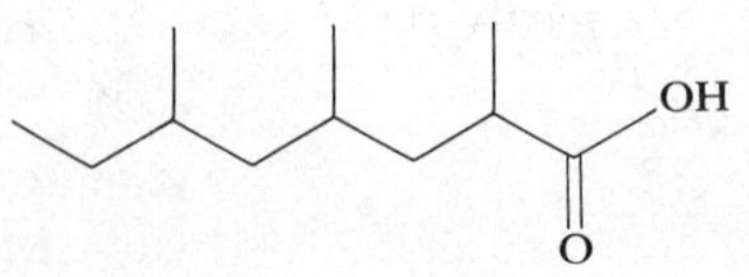

(10)

The widely-distributed fungal metabolite 6-methylsalicyclic acid (**6**) is formed in the same way as (**2**) but with the inclusion of one reduction step, at a specific point in the assembly and prior to cyclization; in asperlin (**7**) the same assembly has been more fully reduced, effectively to the level of octatrienoic acid (**8**), and its cyclization is correspondingly of a different kind. Full reduction, of course, gives octanoic acid (**9**), while with the homologous assembly from three methylmalonyl-CoA it affords 2,4,6-trimethyloctanoic acid (**10**). In radicinin (**11**), only part of the molecule (shown by a heavy line) is the C_8 tetraketide assembly, and the remainder is an attached acetoacetyl residue. The tropolone stipitatic acid (**12**) is a transformation product of the orsellinic homologue (**4**), and the enol-lactone patulin (**13**) is similarly a special metabolite of 6-methylsalicylic acid (**6**); these two examples illustrate the scope of skeletal transformations in polyketides.

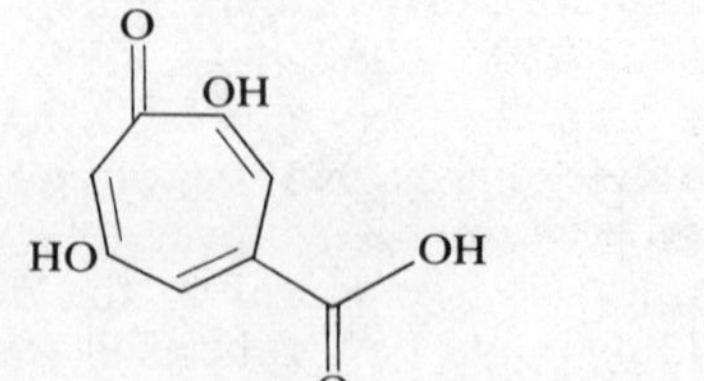

(11)

Heavy line shows polyketide chain

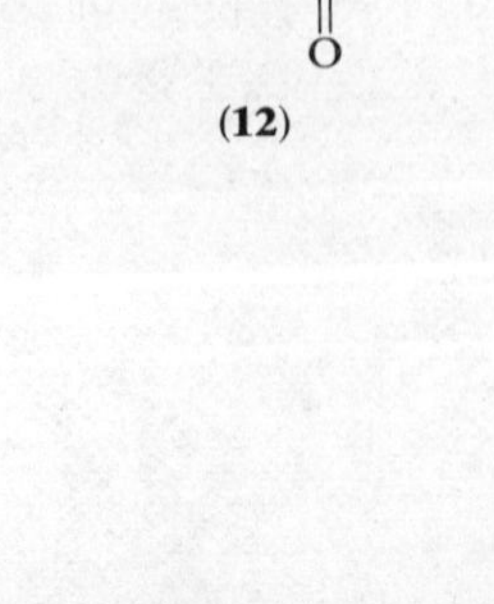

(12)

(13)

The distribution of polyketides in nature is extensive, and more usefully considered in relation to particular sub-categories. Fatty acids of the more common types occur almost universally as essential cell components and storage compounds, but specific fatty acids and such specialized derivatives as the polyacetylenes (see Section 25.1.2.6) can also accumulate as 'secondary metabolites' with more selective distributions in specific families of plants and micro-organisms. Apart from these, the remaining polyketides nearly all belong to the category of secondary metabolites, and as such their enormous structural variety is normally confined with relatively restricted occurrence. They are rare in vertebrates, but are found sporadically in insects, notably as fatty-acid-related pheromones in considerable variety and as a series of pigments based on polycyclic aromatic polyketides.

The distribution of flavonoids and related cinnamoyl tetraketides covers higher plants (universally), lichens, and even a few fungi, and in the higher plants the taxonomic groupings are characterized in terms of particular details of the overall pattern of flavonoid biosynthesis rather than by major biosynthetic classes; on the other hand, other types of polyketides from plants do have quite narrow species distributions. The starter units in plant polyketides are conspicuously varied, perhaps because of the quite general occurrence of the 'cinnamoyl' series alongside the 'acetyl' allows a wider variety to have been evolved. In fungi, and especially in the Ascomycetes (and the related 'imperfect' fungi), the classic acetate-derived polyketides occur in their greatest number and variety.[4] In the filamentous actinomycete bacteria there is also a considerable variety of longer-chain polyketides, particularly those with partly-reduced chains incorporating propionyl- and methylmalonyl-CoA units, as in the various macrolide antibiotics (see Section 29.1.4.2). In other, 'true', bacteria polyketides are much less common.

In general, the relative importance of polyketides in different types of organism partly reflects the relative importance of the corresponding acyl-CoA species in their general metabolism. For example, the widespread variety of aryl polyketides in higher plants reflects the importance in these organisms of aromatic acid biosynthesis as a connecting link between photosynthesis and lignification; in fungi the range of acetate-derived polyketides reflects the regulatory importance of acetyl-CoA in their metabolic responses to changing environment; the predominance of 'propionate' polyketides in actinomycetes probably relates to similar specific features in their — largely unexplored — intermediary metabolism. Similarly, polyketide production often reflects the extent to which organisms utilize secondary metabolism as part of their regulatory equipment. On the other hand, various 'uses' for these secondary products have been evolved, under selection pressure, throughout the biological world. Not only the insect polyketides, already mentioned, but the full panoply of plant flavonoids and other phenolics have major roles in the general physiology of the producing organisms and in the complex interactions between plants and insects. Individual microbial polyketides have considerable importance as antibiotics, toxins, *etc.*, and though the evolutionary 'usefulness' of these properties is certainly questionable, the employment of polyketides as structural or light-absorbing components in the reproductive structures, which in micro-organisms are developed out of vegetative cells, is well authenticated.

29.1.1.2 Historical

The acetate-chain origin of fatty acids was first postulated because of their predominantly even-numbered chain length and, particularly, in the light of early work on their metabolic degradation by the process of β-oxidation. In chemical terms this process involves oxidation of the acids to β-ketoacyl derivatives, from which 'acetate' is cleaved by a reverse Claisen reaction, leaving a shorter-chain acid for repetition of the process. Chemically, too, the sequence could be seen as potentially reversible and hence as a possible route of biosynthesis. In fact, biosynthesis is only rarely the direct reverse of catabolism. Nevertheless, the conclusion that fats might be synthesized from 'acetate' was a fruitful *non sequitur*, the overall truth of which was eventually proved by some of the earliest experiments employing isotope labelling of precursors as a test system,[10] though the distinction between the degradative and the synthetic processes was not appreciated until considerably later.

The existence of a wider class of polyketide natural products was first postulated, and the term introduced, by Collie (1907), from considerations of *in vitro* reactions in which poly-β-ketoacyl compounds afforded aromatic products, notably orsellinic acid (**2**). However, few authentic natural product structures were available at that time, so that critical comparative data were not accessible, and the idea was not further developed in any fruitful sense until it was independently formulated by Birch in 1952–53.[1,2] Comparisons of the structures of a large number, and significant variety, of natural phenolic compounds were now possible, and these showed that in a great many of them there was a prevailing pattern of carbon skeletons which could be written as folded linear chains of C_2 units and of oxygenation on alternate carbon atoms of those chains. Explained in the context of what was then known about fatty acid biosynthesis, this led to the formulation of the 'acetate hypothesis' in a form which could be tested, namely that these structures were formed from a linear chain of 'acetate' units, like that in a fatty acid but with the oxygens of the initial Claisen condensation products still present and available for effective cyclizations, to give alternately-oxygenated phenolic products.

The first 'tests' were of the predictive value of the acetate hypothesis, for example in amending the then proposed structure of eleutherinol (**14**) to the correct (**15**), since only the latter can be drawn out as a single unbranched and alternately-oxygenated chain as the hypothesis would require. The tests by labelled-precursor incorporation experiments (see Section 29.1.5.2) which soon followed[11] led to a much wider recognition, and indeed the validation of the acetate hypothesis also played a major part in establishing the study of natural product biosynthesis as an accessible, legitimate, and significant extension of experimental organic chemistry.

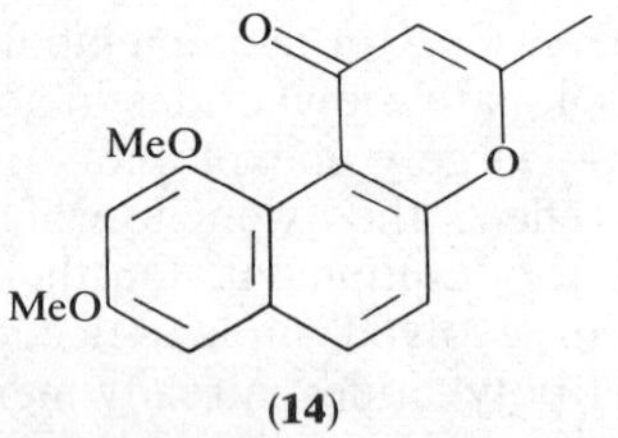

(14)

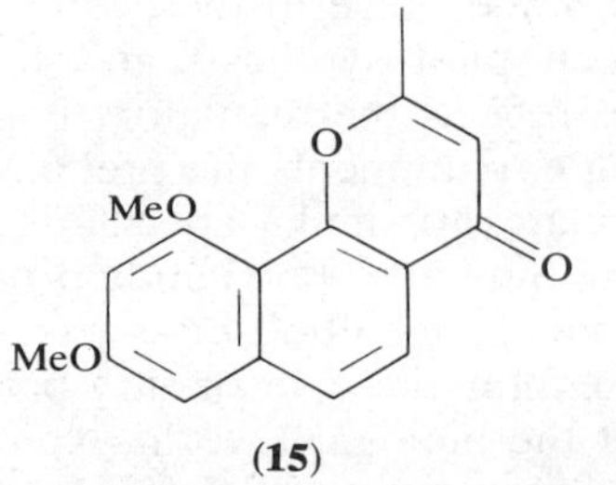

(15)

Since then, biosynthetic studies by this technique, mostly but not exclusively with micro-organisms, have provided the main body of evidence on polyketide biosynthesis. Almost all of the examples cited in the present account have been studied experimentally, and, increasingly, by the use of appropriate labelling methods and experimental design (see Section 29.1.5), considerable detail can now be elucidated. Other evidence has come from the detailed study of co-metabolites and, to a limited extent, of the kinetics of their interconversions. However, except for the rather special case of fatty acid synthesis, it is only recently that the biochemical approach, by appropriate enzymological methods, has begun to contribute significantly, and only in a very limited number of instances.

29.1.1.3 Polyketide units

The biochemical identities and origins of the natural precursors which are utilized in polyketide biosynthesis are important both as background information and also for an understanding of the ways in which much of the available evidence has been obtained, since most labelled-precursor experiments depend upon (and often assume) efficient conversion of the administered precursor into the metabolically active precursor.

Acetyl-CoA (**16**; R = 3′-phosphoadenosyl-phosphatyl) (see Chapter 24.3) is a key metabolic intermediate in most organisms (see Section 28.2.4). Some of the important routes leading to and from acetyl-CoA are represented in Scheme 1. It results from the

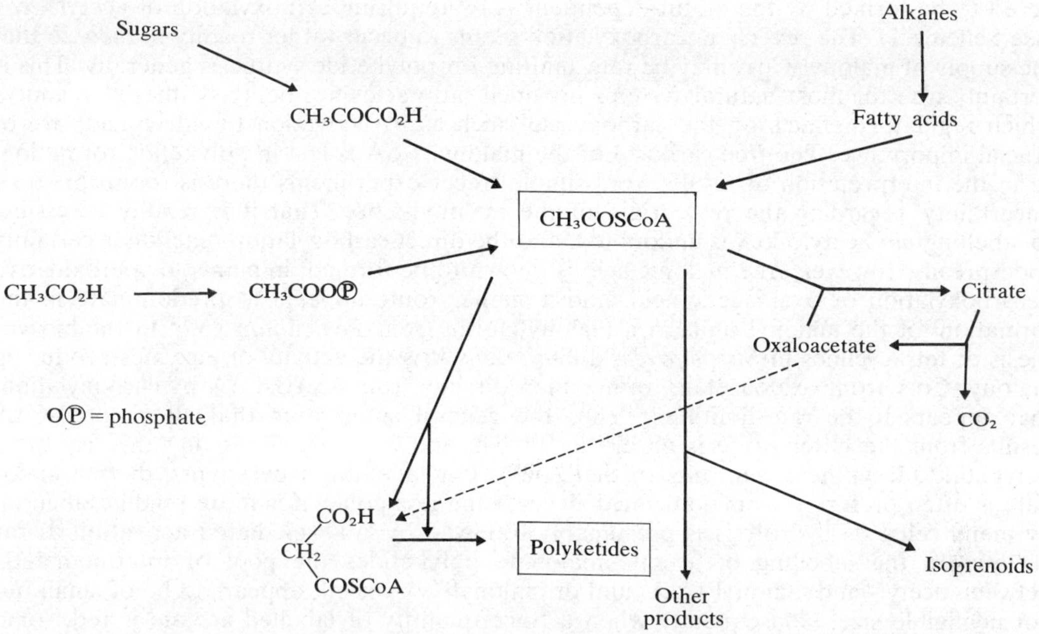

SCHEME 1 Major pathways of intermediary metabolism involving acetyl-CoA

metabolism of sugars by way of the oxidative decarboxylation of pyruvic acid, and it is also the end-product of β-oxidation when hydrocarbons or fatty acids are used as the carbon and energy source. In labelling experiments with intact organisms, acetyl-CoA is usually formed without difficulty from exogenous acetate by ATP-dependent 'activation'. In aerobic metabolism the major fate of acetyl-CoA is condensation with oxaloacetate, giving citric acid, and thence by oxidation through the tricarboxylic acid cycle (and the glyoxalate 'bypass'). Because of the detailed mechanisms of the tricarboxylic acid cycle, added [2-^{14}C]acetate usually also labels, to a significant extent, carbons derived from C-1 of acetate, and its conversion into CO_2 is correspondingly slower. In contrast, this does not happen to [1-^{14}C]acetate, which in the normal cycle is either retained as C-1 labelling or lost as CO_2, and this is accordingly a more specific precursor though overall somewhat less 'efficient'. The effects of these metabolic processes are reduced when a large quantity of 'label' (under such circumstances the term is not strictly applicable) is added, as is often the case in experiments with heavy-isotope labelling (see Section 29.1.5).

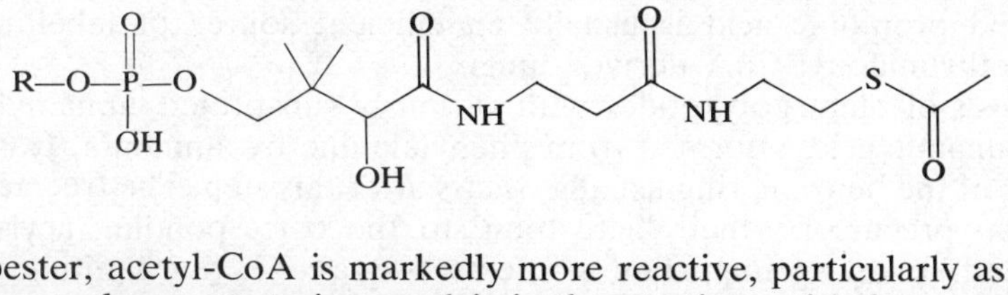

(**16**)

As a thioester, acetyl-CoA is markedly more reactive, particularly as an (electrophilic) acylating agent, than acetate ion, and it is the starting point for a number of different biosynthetic routes, including that to terpenoids (see Section 28.2.4), additional to those leading to different polyketides. In the actual assembly reactions of polyketide synthesis, coenzyme A is replaced by its 'macromolecular equivalent', an acyl carrier protein (ACP) [acetyl-ACP (**16**; R = protein, attached through a serine residue)]; in some enzymological studies, coenzyme A can conveniently be replaced by simpler thiols like cysteinamine (β-aminoethanethiol) or its *N*-acetyl derivative.

Malonyl-CoA, the other precursor for 'acetate-derived' polyketides, is usually considered to be formed by the biotin-dependent ATP-requiring carboxylation of acetyl-CoA (see Scheme 1). The reverse decarboxylation seems to occur rather readily *in vivo*, so that the supply of malonyl-CoA may be rate-limiting for polyketide synthesis generally. This is certainly true for most natural systems in which fatty acids are being synthesized, and in which regulatory effects on the carboxylase (such as its activation by citric acid) are of crucial importance. The free carboxyl of the malonyl-CoA is lost in polyketide formation, so in the interpretation of results from simple tracer experiments there is room for some uncertainty regarding the real origin of the malonyl-CoA. That it is readily accessible to labelling *via* acetyl-CoA is undoubted, and the direct carboxylation reaction is certainly widespread. However, free malonic acid is known to be formed, in plants, by peroxidative decarboxylation of oxaloacetic acid, and a similar route appears to predominate in the formation of the malonyl units in a malonylglucan from *Penicillium* sp.[12] In the biosynthesis of tetracyclines in *Streptomyces aureofaciens* it is the activity of enzymes producing malonyl-CoA from oxaloacetate, rather than directly from acetyl-CoA by carboxylation, that appears to be rate-limiting.[13] Thus the general assumption that all malonyl-CoA results from the latter route is hardly justifiable, and we must admit that this aspect of polyketide biosynthesis warrants further study. For labelling experiments, diethyl malonate is often preferred to malonic acid since, being less polar, it is more readily taken up by many cells; its hydrolysis is presumably followed by ATP-mediated activation as for acetate. In the labelling of acetate–malonate polyketides the pool of intermediate(s) between acetyl- and malonyl-CoA, and of malonyl-CoA itself, appears to be of small but not negligible size. Consequently when a trace quantity of labelled acetate is fed, some preponderance of label in the acetyl-CoA starter unit is often detectable. In the converse experiment, labelled malonate usually contributes significant label to the starter unit through decarboxylation, and in unfavourable circumstances the labelling may even be uniform. Taking one set of data from the author's own laboratory as an example, [1-^{14}C]acetate added to cultures of *Penicillium urticae* contributed *ca.* 5% excess labelling to the acetyl-CoA-derived starter unit in 6-methylsalicylic acid (**6**), while under comparable conditions [2-^{14}C]malonate gave a starter residue carrying *ca.* 25% of the activity found in each of the malonyl-CoA-derived units.

The origins of propionyl-CoA and methylmalonyl-CoA, involved in the biosynthesis of other polyketides, have been relatively less explored, because they are mostly important in actinomycete bacteria, whose intermediary metabolism has been surprisingly little investigated. In some cases, particularly in anaerobic bacteria, propionic acid may be formed from sugars by way of pyruvate and lactate, but the route of most general importance is by way of the B_{12}-dependent isomerization of succinyl-CoA, an intermediate in the tricarboxylic acid cycle. This gives methylmalonyl-CoA directly, and this can afford propionyl-CoA by decarboxylation. Thus, although propionyl-CoA can also be carboxylated to methylmalonyl-CoA, the normal situation is just the reverse of that which occurs with acetyl- and malonyl-CoA, so far as their order of formation is concerned. Despite this, added propionic acid is usually an efficient source of labelling for both propionyl- and methylmalonyl-CoA-derived units.

In the biosynthesis of plant polyketides with variously-substituted cinnamyl-CoA starters, the parent cinnamic acid is formed from phenylalanine by ammonia (lyase) elimination. Oxygenation of the benzene ring usually occurs after this step. The free acids are thus the normal *in vivo* precursors; their 'activation' to the corresponding acyl-CoA is an obligatory step and in many important contexts it may be rate-limiting. The other acyl-CoA species which are occasionally involved in polyketide synthesis are usually formed from the corresponding acids, but otherwise they are too miscellaneous an assortment for their biochemistry to be usefully summarized here.

29.1.2 THE BASIC MECHANISMS OF POLYKETIDE BIOSYNTHESIS

For most polyketide natural products our evidence about their biosynthesis predominantly comes from studying the incorporation of labelled precursors in experiments using intact cells. From such data certain general principles emerge, which can usefully be summarized using the tetraketides already exemplified, (**1**)–(**13**), as illustrative instances:

(a) Biosynthesis is divided between a basic assembly process, in which the component acyl and malonyl units are brought together and a stable small molecule produced [*e.g.* one acetyl-CoA and three malonyl-CoA to give orsellinic acid (**2**)] and subsequent modification processes [*e.g.* (**4**) → (**12**) or (**6**) → (**13**)].

(b) In labelling experiments the product of the basic assembly step is accessible to label from the precursor acyl and malonyl units but not from 'free' molecules equivalent to assemblies of intermediate size. For example, added acetoacetyl-CoA is not incorporated into (**2**), nor is butyryl-CoA into (**9**). In contrast the products of the later modifying processes are so accessible; for instance, added (**6**) is tranformed into (**13**).

(c) The dilutions of added label are normally the same in 'successive' malonyl units of the assembly, *i.e.* there are no significant pools of the partly-assembled intermediates. More strictly, the pool size is fixed by the concentration of the enzyme molecules in the system and not by that of the substrates.

(d) Certain variations in biosynthesis are integral with the basic assembly process; thus the different cyclization patterns in (**1**) and (**2**) are intrinsic to the corresponding assembly systems; moreover, the 'missing' OH group in (**6**) and the 'extra' CH_3 group in (**4**) each result from reactions taking place during the assembly process, and not from transformations of pre-formed (**2**).

From such generalizations we can deduce that the assembly of the acyl and malonyl units takes place in an 'all-or-nothing' enzyme system which accepts the precursor coenzyme A derivatives and does not release them until it produces a stabilized polyketide. The intermediates are apparently all enzyme-bound, yet the overall process must involve an organized multiplicity of enzymes, comprising catalysts for the assembly process and for the additional processes associated with it [above, (d)] and macromolecular carriers for the intermediates. These are general conclusions for the whole range of polyketide natural products. Their enzymological basis has only been directly investigated for the particular case of fatty acid synthesis and for a very small number of other polyketides; the work of Lynen and his collaborators is particularly noteworthy in this area. The biochemical results provide a striking validation of the more general chemical reasoning as well as a detailed picture for particular examples. However, the chemical reasoning still extends further than the enzymological evidence would itself allow. For example, the structures of some of the macrolide polyketides from actinomycetes (see Section 29.1.4.3) show, when a sufficient number are compared, that their biosynthesis involves enzyme systems capable of assembling a sequence of different acyl precursors in a specific order, to part-assemblies that somehow correspond to 'block templates' for major portions of the eventual molecular product. The only enzymological data for systems with any such properties actually relate to oligopeptide biosynthesis and not to polyketides at all. Doubtless in due course some relevant data for a few typical systems will be obtained more directly, but in the meantime we must continue to rely upon chemical insight.

29.1.2.1 Fatty acid synthetase

The most fully-explored type of polyketide synthetase is certainly the multi-enzyme complex system for fatty acid synthesis [*cf.* Section 25.1.5]. This centres upon the acyl carrier protein (ACP), which is a macromolecular equivalent of coenzyme A (**16**); its operation is outlined in Scheme 2.[14] Note that the multi-enzyme complex, represented in the scheme by a letter E, has two distinct thiol sites; one, marked (*a*) in the scheme, is

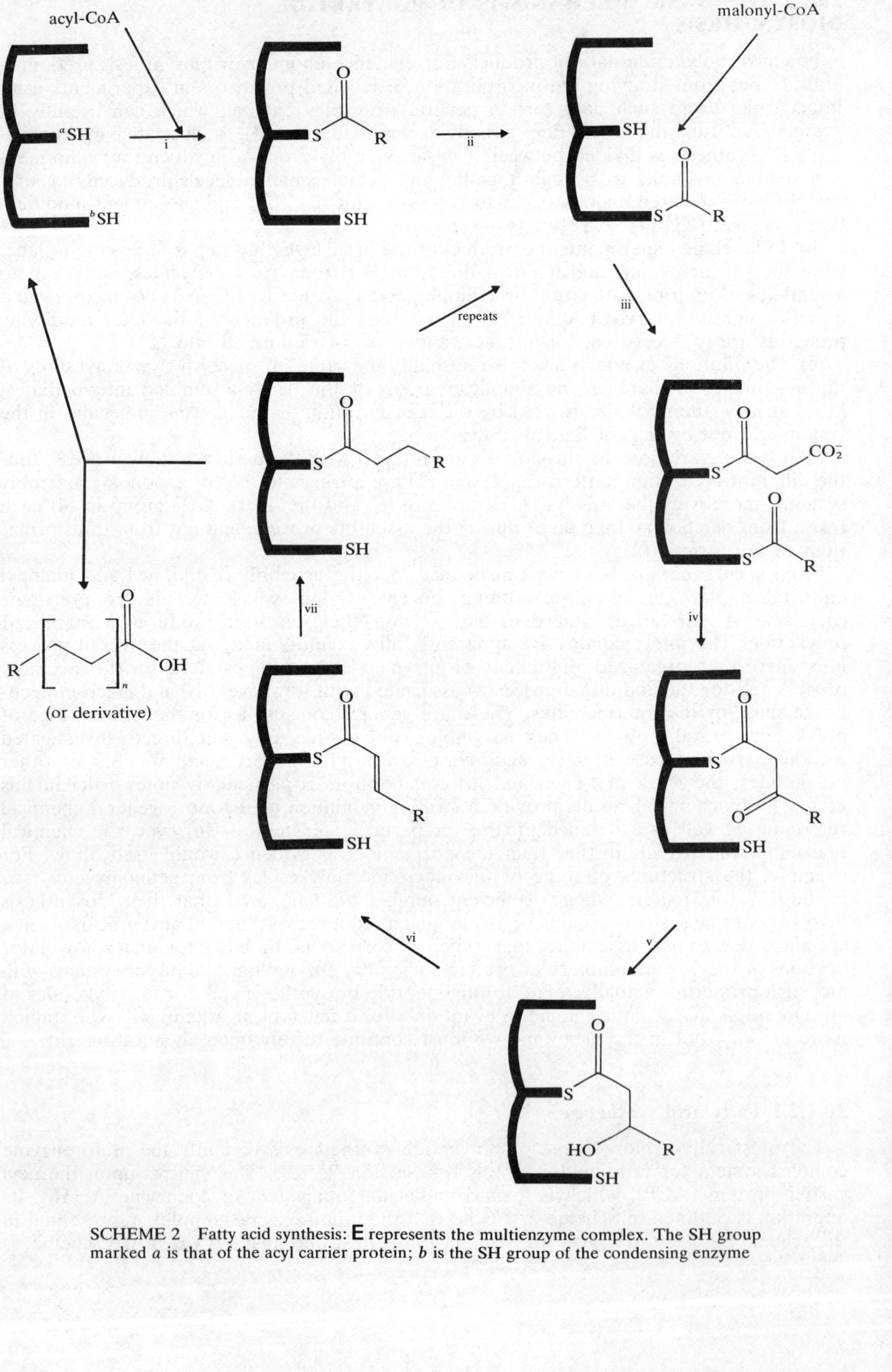

SCHEME 2 Fatty acid synthesis: **E** represents the multienzyme complex. The SH group marked *a* is that of the acyl carrier protein; *b* is the SH group of the condensing enzyme

that of the acyl carrier protein, and the second belongs to the enzyme which effects the acylation of malonyl-ACP.

In Scheme 2, the first step is acylation of the ACP by an acyl-CoA. This is catalyzed by a specific transacylase. It is the specificity of step (i), coupled with the availability of various acyl-CoA species, that determines what the 'starter' group can be. In step (ii) the acyl group — whether the one introduced in step (i) or one elaborated in subsequent steps — is transferred to the thiol site on the condensing enzyme. This leaves the carrier protein available for acylation by malonyl-CoA in step (iii), which is catalysed by a second transacylase. The two thiol sites are so located that acylation of the malonyl-ACP can now occur [step (iv)], giving a β-ketoacyl-ACP. This is next reduced [steps (v)–(vii)], through the β-hydroxy- and α,β-unsaturated acyl-ACP stages, to give an acyl-ACP. This eventual acyl-ACP may now undergo either transfer to the condensing enzyme, a repeat of step (ii), which can then be followed by a further cycle of acylation, *etc.*, or it may be removed from the system by a third transacylase [step (vii)]. Depending upon the system, this last step may give either a free fatty acid or some particular acylated acceptor species, including acyl-CoA.

It is not absolutely essential for a reaction series of this kind to be carried out by enzymes physically contiguous in a structured aggregate. The necessary degree of organization can be conferred by the specificities of the enzymes themselves, and indeed in typical bacteria the enzymes of fatty acid synthesis and the acyl carrier protein itself are normally found in solution. The situation in higher plant cells may be similar. However, in fungi and in animal cells the enzymes are combined as a multi-enzyme complex, the 'fatty acid synthetase' whose structure favours, rather precisely, the successive macromolecular interactions which Scheme 2 requires. Such a structure will allow the reaction cycle to occur without involving the diffusion or displacement of macromolecules and will permit much more favourable kinetics. The yeast complex, with a molecular weight of 2.3×10^6, contains three sets of the enzymes, 21 in all, plus three molecules of ACP. Formation of these multi-enzyme complexes reinforces the 'all-or-nothing' characteristics of the system. For example, no new starter acyl-CoA can come into the system until the completed acyl group has been transferred off the ACP site. Such complexes are of special interest since they also offer the best model for polyketide synthesis generally.

In the normal biosynthesis of fatty acids the reduction steps are absolutely essential for the progress of the reaction sequence, and in the absence of the requisite hydride donor (NADPH) the reaction stops, typically at the stages of acetoacetyl- and 3,5-dioxohexanoyl-ACP.[15] Normally only malonyl-CoA is accepted by the system (step iii, Scheme 2). On the other hand, there may be considerable elasticity in the specificities of steps (i) and (viii). Thus these two transacylase steps effectively control what spectrum of fatty acids can be produced. For example, the usual spectrum of normal even-numbered, normal odd-numbered, *iso* and *anteiso* fatty acids in bacteria is determined by the competition between corresponding acyl-CoA starters for reaction in step (i). Whereas in *Escherichia coli* the acceptor transacylase is fairly specific for acetyl-CoA, in other bacteria it will accept straight- or branched-chain acyl-CoA from C_4 to C_6,[16] and in others again, cyclic acids from cyclobutylacetic to cycloheptanecarboxylic.[17] On the other hand, the spectrum of chain lengths eventually released from the synthetase sequence in these same bacteria depends upon the specificity of the transacylases for step (viii) and not upon the size of the starter or the number of malonyl units which have been added to it.

29.1.2.2 Polyketide synthetases

With respect to more general polyketide biosynthesis, and in particular the biosynthesis of aromatic polyketides, only that of 6-methylsalicyclic acid (**6**) has been investigated in comparable detail.[18,19] In *Penicillium patulum* and related fungi, (**6**) is synthesized by a multi-enzyme complex having several similarities to fatty acid synthetases and provides us with a good basis for more general understanding of polyketide formation. The complex contains two different types of thiol group (analogous to those of the ACP carrier and the condensing-enzyme sites in the fatty acid synthetase), and it has similar 'all-or-nothing' characteristics. The molecular weight has been variously determined as *ca.* 3.7×10^5 and *ca.* 1.3×10^6, possibly referring to monomer and trimer forms of the complex. The action of the synthetase, as elucidated by Lynen and others,[19] is set out in Scheme 3 in a form comparable with the representation of fatty acid biosynthesis.

The most obvious difference between this scheme and the mechanism of fatty acid synthesis is the omission of nearly all the reduction steps. This is very general in one class of polyketides and it leads to the diagnostic pattern of alternate oxygenation on an aromatic ring. A less obvious, but more deeply significant, feature is that in the three successive acylation steps, marked (ii) on Scheme 3, the acylating species are all chemically rather different — acetyl, acetoacetyl, and 5-ketohex-3-enoyl — whereas in fatty acid synthesis successive acylations are by a homologous series of saturated acyl species.

In more complex polyketides (see, for example, Section 29.1.4.3) the malonyl species undergoing acylation may also vary, in a specified sequence. In general, then, the special feature of polyketide synthetases is their ability to effect successive acylations of a specified sequence of malonyl or homologous residues with a series of chemically-differentiated residues, whose nature is in turn determined by the inclusion, or omission, of steps between the successive acylations. All of this is combined with a means of stabilizing the growing polyketide chain against adventitious cyclizations and eventually of imposing a specific cyclization at the end of the sequence.

Thus in 6-methylsalicylate synthetase there is one reduction step, effected specifically at the C_6 stage. If the requisite NADPH is not provided, the enzyme-bound 3,5-dioxohexanoyl group can only be split off as triacetic lactone (**17**), as happens with fatty acid synthetase in similar circumstances.

For comparison, the biosynthesis of alternariol [the lactone form of structure (**18**)] is carried out by a synthetase complex which has also been studied in a cell-free system.[20] In this case the acylating step must accommodate successive acylations of malonyl groups by acyl species from C_2 to C_{12}. There are no reduction steps, and the need to prevent 'premature' cyclizations is apparent. Similarly, in forming the tetracycline precursor (**20**) (see Section 29.1.3.6), at least the first seven units of the presumed intermediate (**19**) marked out in the structure must be assembled into this form (or an enolized equivalent) before even one cyclization can occur. On the other hand, in this same example we can also note the apparent influence of the one reduction/dehydration step in this biosynthetic sequence, which by giving the *cis* double bond marked in (**19**) 'turns' the growing assembly and thus helps to determine the eventual arrangement for cyclization.

Another kind of reaction that is interpolated into the polyketide assembly process is *C*-methylation, with *S*-adenosylmethionine as the methyl donor and a specific trans-methylase as part of the multi-enzyme complex. The reaction always proceeds by electrophilic attack at the enolizable 'methylene' carbons [a good example is seen in (**19**)] and all three of the methyl hydrogens are retained (equation 2):[21]

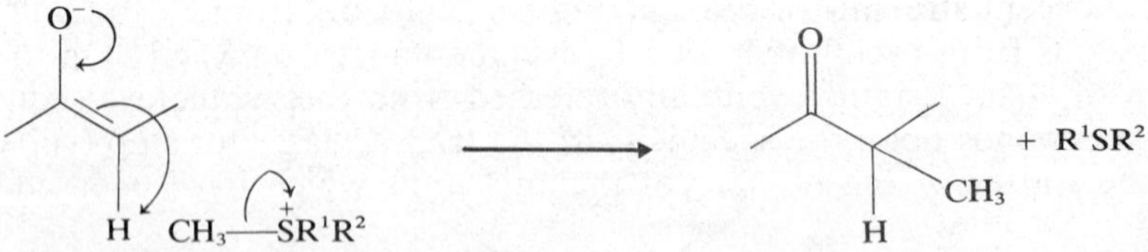

(2)

The stage of polyketide synthesis at which such reactions occur is a matter of interest and importance. Direct evidence that a *C*-methylation involves the part-assembled enzyme-bound polyketide came[22] from work on *Aspergillus flavipes*, which normally produces 5-methylorsellinic acid (**22**) and its further metabolite flavipin (**23**). A cell-free protein preparation was obtained which would incorporate label from [$^{14}CH_3$]-*S*-adenosylmethionine into (**23**), without the addition of any polyketide acceptor. Clearly a methylation substrate was present in this preparation, and this substance was identified as

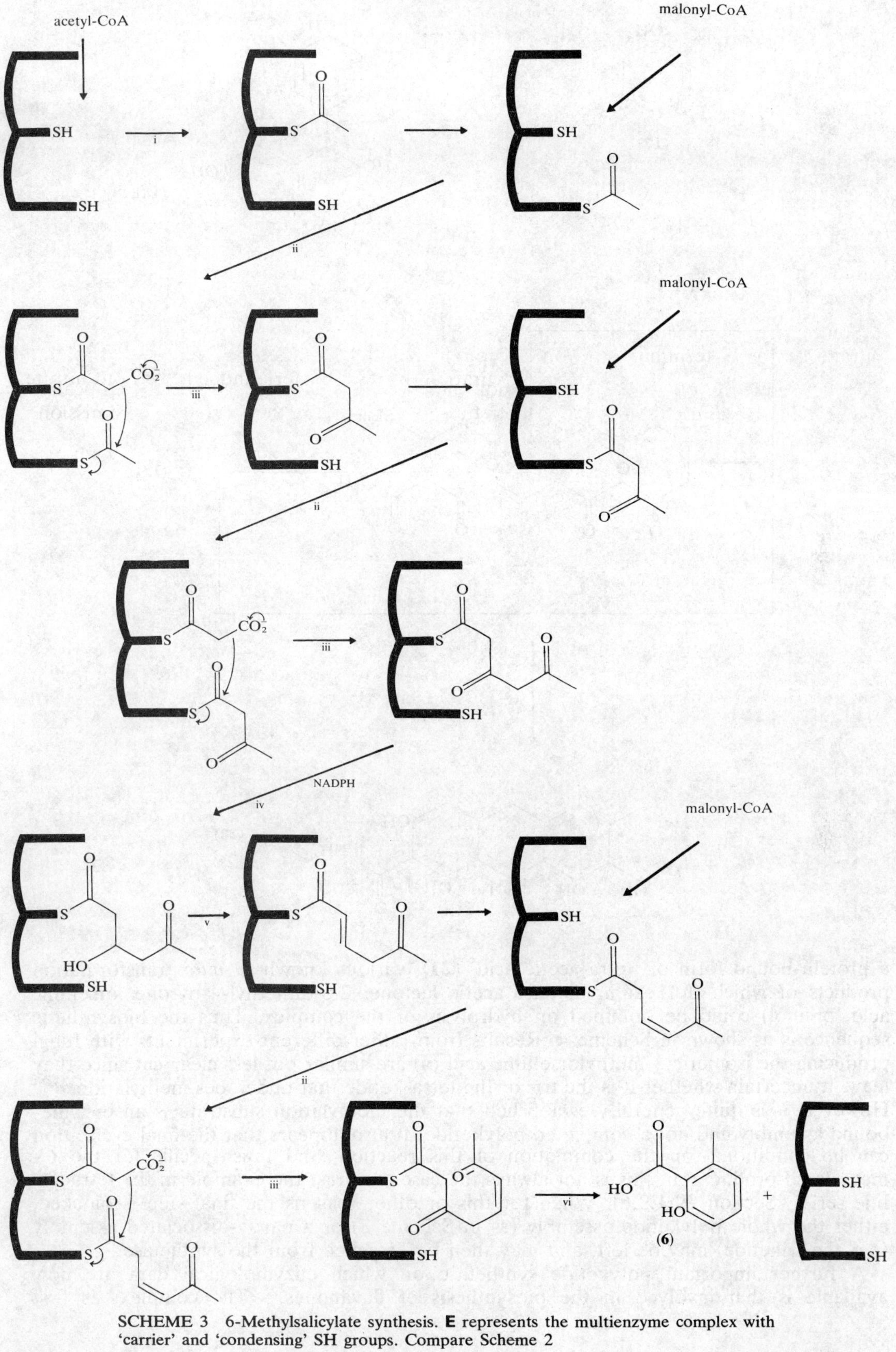

SCHEME 3 6-Methylsalicylate synthesis. **E** represents the multienzyme complex with 'carrier' and 'condensing' SH groups. Compare Scheme 2

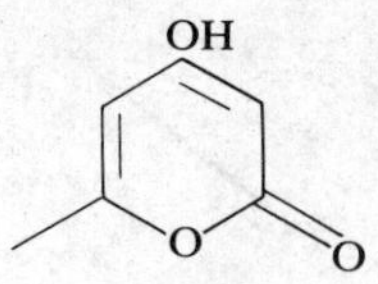

(17)

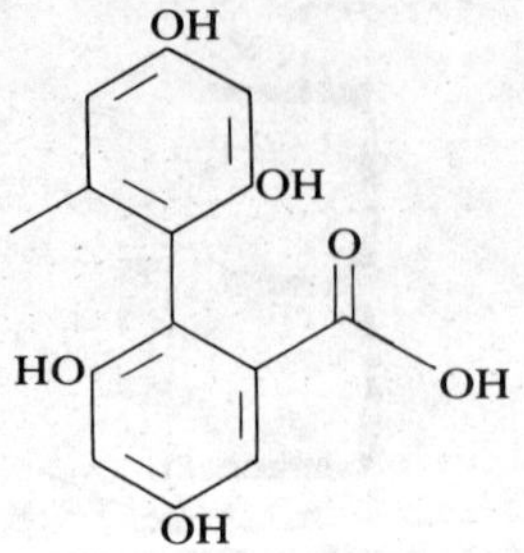

(18)

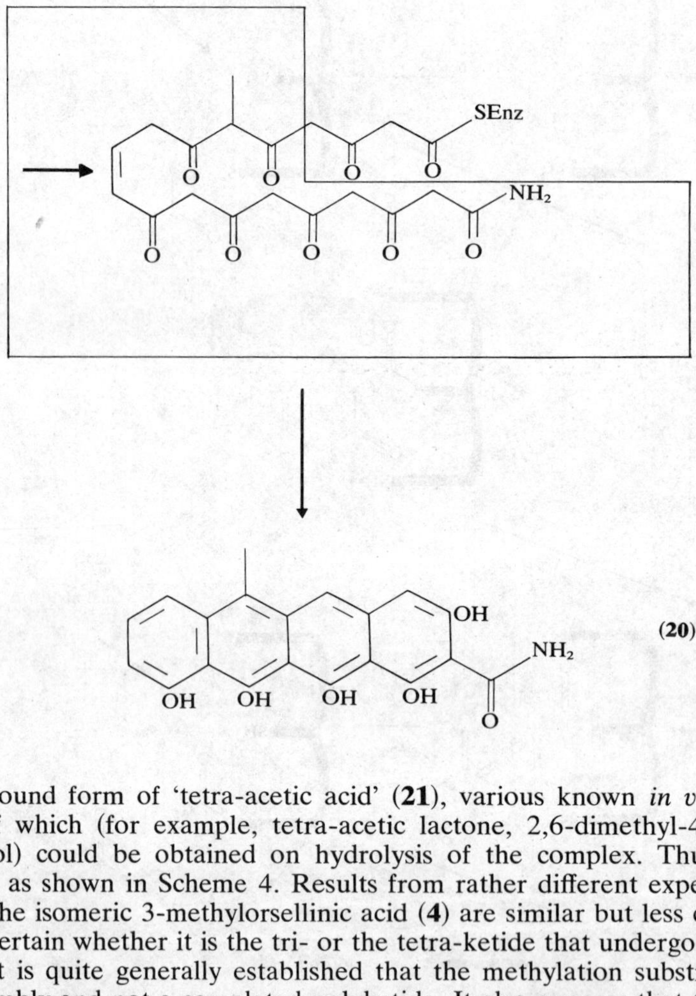

(19)

(20)

a protein-bound form of 'tetra-acetic acid' (**21**), various known *in vitro* transformation products of which (for example, tetra-acetic lactone, 2,6-dimethyl-4-pyrone, orsellinic acid, orcinol) could be obtained on hydrolysis of the complex. Thus the biosynthetic sequence is as shown in Scheme 4. Results from rather different experiments with fungi producing the isomeric 3-methylorsellinic acid (**4**) are similar but less clear-cut since they leave it uncertain whether it is the tri- or the tetra-ketide that undergoes methylation.[23,24] However, it is quite generally established that the methylation substrate is an enzyme-bound assembly and not a completed polyketide. It also appears that the final cyclization can be conditional on the completion of this reaction, and thus specific for the *C*-methylated product, but this is not always the case (contrast the example in the tetracycline series, Section 29.1.3.6). When for this or other reasons the final step is blocked, either the whole polyketide assembly (as in Scheme 3) or a partly-dissociated assembly (*e.g.* the triketide) may be left, and may then be detached from the synthetase.

A further important polyketide synthetase on which enzymological data are now available is that involved in the biosynthesis of flavanones.[25] The complex, as first

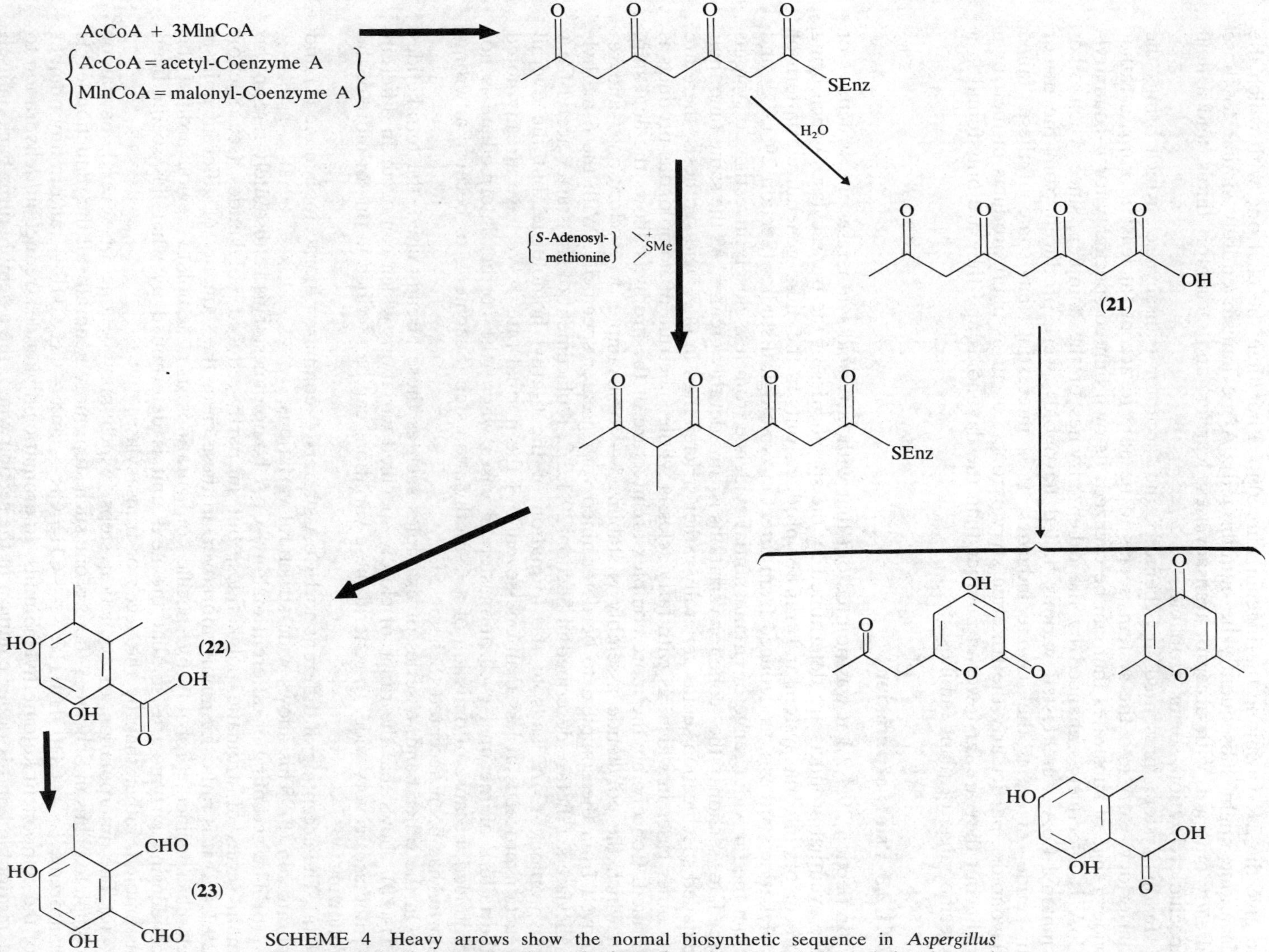

SCHEME 4 Heavy arrows show the normal biosynthetic sequence in *Aspergillus flavipes* and the lighter arrow shows the formation of products by hydrolytic release of the parent polyketide when *C*-methylation does not take place

obtained from cell cultures of parsley by Grisebach and co-workers, accepts *p*-coumaryl-CoA as the 'starter' and three successive malonyl-CoA thereafter, and it has been shown to give the 2*R*-flavanone (naringenin) directly (rather than the chalcone with which the flavanone might subsequently be equilibrated). A mechanism entirely similar to those already outlined, with acceptor transacylases, 'carrier' and 'acceptor' thiols, and a final specific cyclization, seems probable.

In this system, presented in Scheme 5, the 'acceptor transacylase', which brings the starter acyl-CoA into the system, seems to be able to accept (at least *in vitro*) a rather wider range of substrates that can be cyclized; its own limited specificity for *p*-coumaryl-CoA is effectively enhanced by the other enzymes of the complex, since when other cinnamyl-CoA were tested various 'failed flavonoids' were produced from the di- or tri-ketide stages, as the scheme indicates. It is interesting that some of these 'failed flavonoids', particularly the triketide styrylpyrones, occur as plant products in their own right, but they are far fewer in number than the flavonoids and they are only found in a specific range of plant families.

29.1.2.3 Points of variation

In Section 29.1.2.1 it was pointed out that with fatty acid synthetases the structures of the eventual products are determined by the specificities of the transacylases which bring precursors into the system and transfer products out. In the more generalized pattern of polyketide synthesis, as deduced from direct enzymological studies (Section 29.1.2.2) and from the far wider range of circumstantial evidence from less closely-investigated systems (see later sections), the scope for variations is considerably greater. At the same time their detailed specification seems generally rather stricter than in fatty acid synthesis; there are no other cases in which, as with fatty acids, a whole spectrum of homologous products is formed. Conceivably the more complex architecture of the products allows the final step, in which the polyketide assembly is stabilized — for example, by cyclization — and removed from the complex, to be determined with greater specificity. With the 6-methylsalicylate synthetase described in Section 29.1.2.2, replacement of the starter acetyl-CoA by propionyl-CoA leads to the formation of the 6-ethyl homologue, but the overall reaction proceeds at less than one-seventh of the normal rate.[26] We have already noted the failure of the final cyclization step in cases where the correct intermediate is not produced, because of omission of a *C*-methylation step (Scheme 4) or because a 'wrong' starter has been accepted (Scheme 5).

On the other hand, a series of examples such as those already cited, the tetraketides (**1**)–(**10**), shows that the range of closely-specified variations which can occur in different synthetases is very wide. In general these variations can be ascribed to one or more of the following:

(a) The accepting of different acyl-CoA starters — compare phloracetophenone (**1**) and naringenin (**5**). This allows wide structural variations.

(b) The accepting of different malonyl-CoA homologues. This is absolutely rare in the main series of aromatic polyketides, but in partly-reduced polyketides (see Section 29.1.4.2) it is rather common; moreover, in these cases the sequence of different units is closely specified, probably by specific configurations in a 'template' region around the condensing enzyme site so that the next unit is determined by what has so far been assembled, rather than by the acceptor transacylase.

(c) The interpolation of reduction steps, which may be one or several, partial or complete. Thus in Scheme 3 there is one such step; more generally all the reduction levels can occur, *i.e.* —CH(OH)CHR—, —CH=CR—, and —CH_2CHR—, and in any number of the component ketomethylene units. In aromatic polyketides, reduction/dehydration to a *cis* double bond is rather common in the section where the assembly turns back on itself [for example, see (**19**)→(**20**)]; in partly-reduced polyketides the reduction level of successive units in the sequence is probably determined by a similar mechanism to that suggested in (b), above.

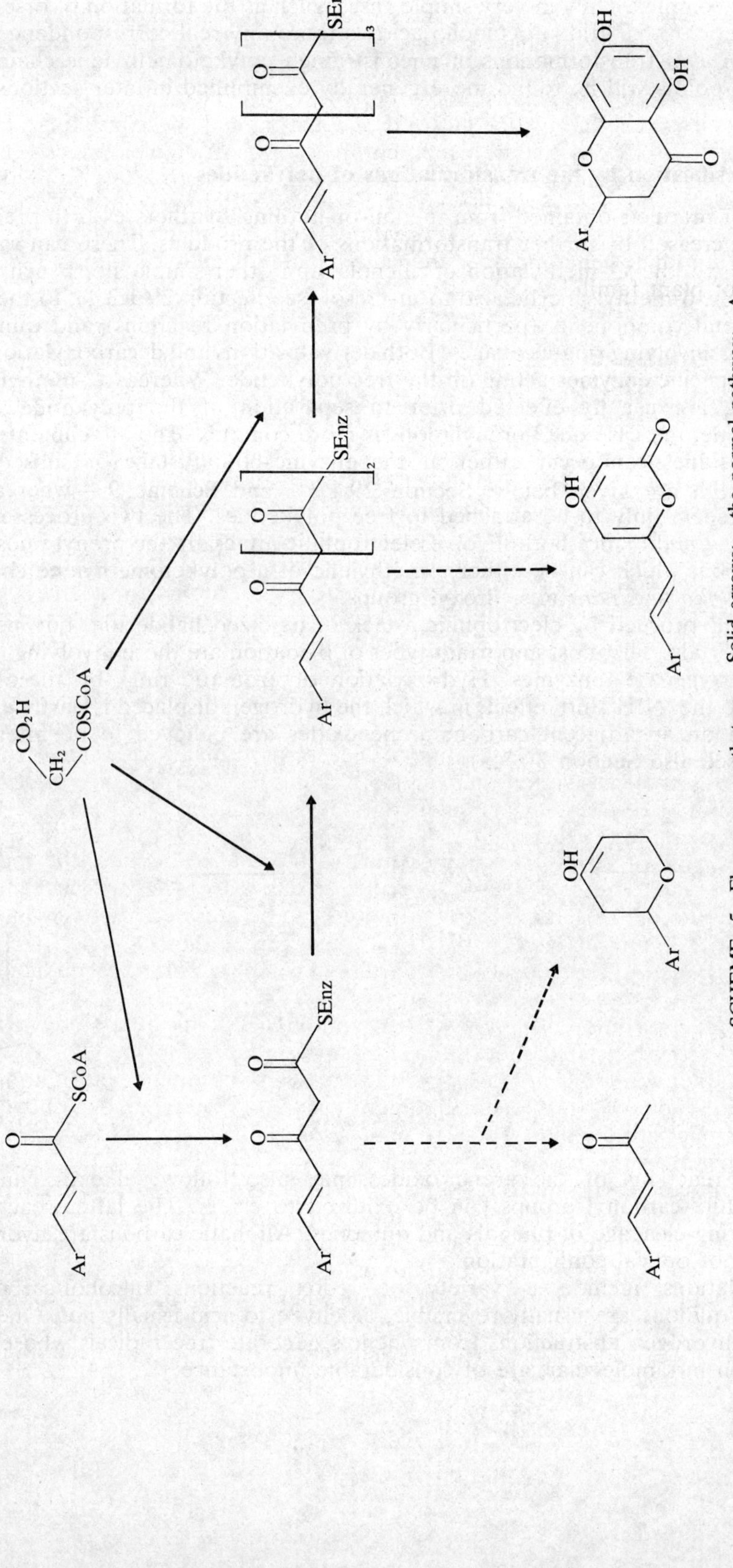

SCHEME 5 Flavanone synthetase. Solid arrows: the normal pathway; Ar = *p*-hydroxyphenyl; broken arrows: formation of 'failed flavonoids', Ar = 3,4-dihydroxyphenyl or 4-hydroxy-3-methoxyphenyl

(d) The interpolation of *C*-methylations as already discussed (compare Scheme 4). These are usually closely specified; some *C*-prenylations may have a similar status.

(e) Alternative cyclizations — phloracetophenone (**1**) and orsellinic acid (**2**) provide the simplest example — which are determined by the configuration imposed on the assembly by the enzyme complex. Only in very simple cases, such as the formation of orsellinic acid from 'tetra-acetic' (3,5,7-trioxo-octanoic) acid, is there any real correspondence with the spontaneous *in vitro* transformations of unconstrained polyketomethylene chains.

All of these points will be found more generally exemplified in later sections.

29.1.2.4 Diversification by the transformations of polyketides

The range of products obtained from the 'all-or-nothing' synthetases is in practice very considerably increased by further transformations of the products. These can range from derivatization, *e.g.* the *O*-methylation of phenols, and other simple steps such as decarboxylation,[27] *e.g.* 6-methylsalicylic acid to *m*-cresol (see Section 29.1.3.1), to the addition of major skeletal components (particularly by prenylation reactions) and quite radical transformations involving ring-cleavages. Both derivatizations and decarboxylations usually involve fairly specific enzymes acting on the free polyketide. Whereas *C*-methylations, as already noted, are usually effected prior to separation of the polyketide from the assembly enzyme, the case of *C*-prenylations is more complex. The attachment of simple isopentenyl residues can occur either at the enzyme-bound stage or subsequently — examples of both are given below, Section 29.1.3.1 and Scheme 9 — whereas larger prenyl groups seem only to be attached to free polyketides. The two processes are not distinguishable *a priori* since both involve electrophilic attack by the prenyl phosphate on a reactive carbon, which can be either a methylene of a polyketomethylene chain or an aromatic CH *ortho* and *para* to hydroxyl groups.

Halogenations proceed by electrophilic attack of oxidized halide ion, but mechanistic details are uncertain. The most important types of oxidation are those involving molecular oxygen and 'oxygenase' enzymes. Hydroxylation of aromatic rings by these enzymes usually leads to the 'NIH shift' effect, in which the hydrogen displaced is partly exchanged with hydrogen on an adjacent carbon; arene oxides are believed to be intermediates (equation 3) (see also Section 30.3.5).

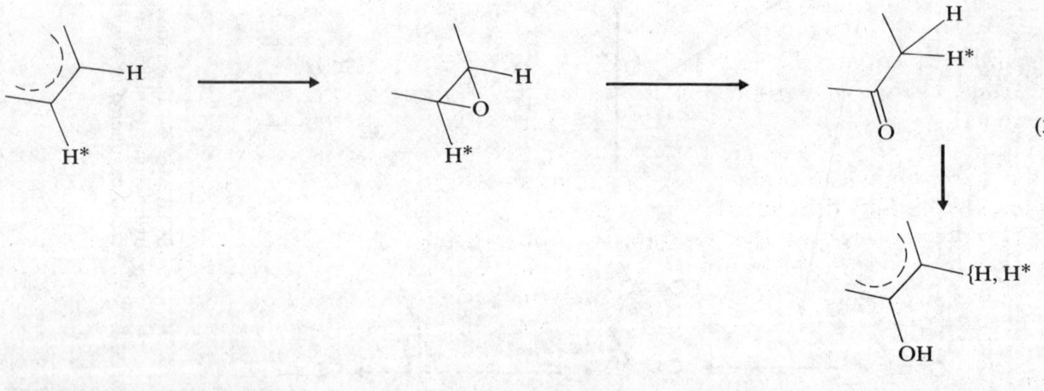

(3)

Alternative reactions of the arene oxides may also follow; alkenes can also be epoxidized, while carbonyl groups can be oxidized to esters. The latter reaction leads effectively to ring-cleavage of phenols and quinones. Aliphatic carbons are hydroxylated, with retention of optical configuration.

Other oxidations include a variety of redox reactions: alcohol/carbonyl and hydroquinone/quinone are usually reversible, aldehyde to acid usually not. One-electron oxidations or hydrogen abstractions from phenols generate free radicals whose coupling reactions, often intramolecular, are of considerable importance.

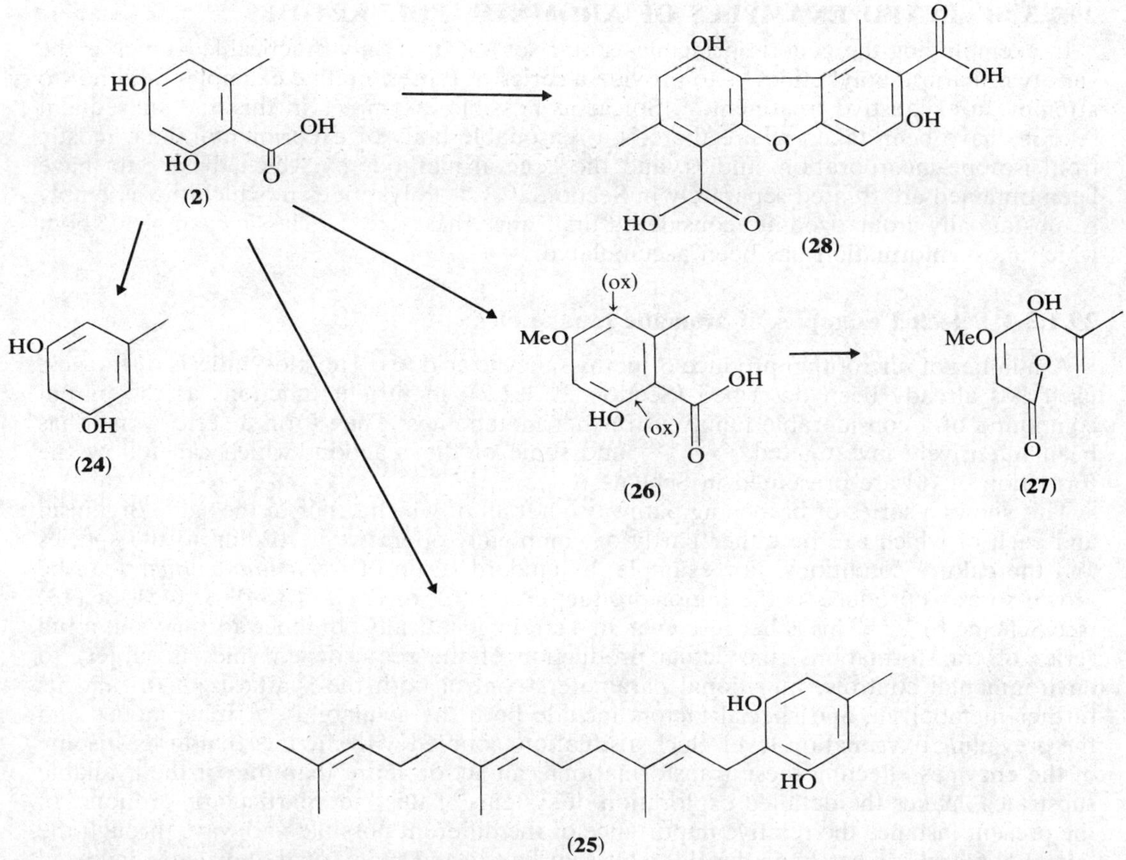

A few examples of the scope of these further transformation reactions are illustrated, structures (**24**)–(**28**), all of which are fungal metabolites of the simple tetraketide orsellinic acid (**2**). Orcinol (**24**) is its decarboxylation product, produced by a specific decarboxylase,[27] while grifolin (**25**) is one of a series[28] of farnesyl-substituted metabolites of (**2**) [compare mycophenolic acid, Section 29.1.3.1, (**45**); the electrophilic alkylation of aromatic rings by prenyl pyrophosphates occurs in a variety of natural products and is not confined to polyketides]. The lichen product (**28**) clearly results from oxidative 3,5-coupling of (**2**), whereas penicillic acid (**27**) results from oxygenase attack on the methyl ether (**26**), known to occur at the positions shown.[29] A fuller variety of such transformations is exemplified in subsequent sections.

The prevalence of extensive transformations in the range of natural polyketides most frequently encountered is often such that the real 'parent' products of many polyketide synthetase products are only found as quite minor metabolites, accompanying more or less complex mixtures of their transformation products. The transforming enzymes often lack absolute specificity for their substrates, and this adds to the complexity of the product mixture. This 'disappearing' tendency of the parent polyketides is so general that it may be functional, perhaps avoiding feed-back effects on the synthetases, and perhaps also in spreading secondary metabolism over a greater range of products in each particular organism. For reasons which are not really understood, metabolic diversity both within and between species seems to have been an evolutionary target, which is rather economically attained by branching systems of diversifying transformations. Certainly the chemical distinctions between, for example, different lichen species[30] seem to lie more in the pattern of these diversifying processes than in the range of basic polyketide synthesis which provide their substrates.

29.1.3 SELECTED EXAMPLES OF AROMATIC POLYKETIDES

In exemplifying the general principles so far set out, it is only practicable — such is the variety of natural polyketides — to provide a series of representative examples, and not to attempt an exhaustive treatment.[4–8] So far as possible, examples in this and subsequent sections have been taken where there is a reasonable body of experimental data, mostly from isotope-incorporation studies, and the general methods by which these data have been obtained are treated separately in Section 29.1.5. Polyketides in which the assembly is substantially aromatized are considered first, since these are the classic examples, about which most information has been accumulated.

29.1.3.1 Selected examples of aromatic tetraketides

A number of microfungi produce 6-methylsalicylic acid (**6**). The biosynthesis of the acid itself has already been described (Section 29.1.2.2); in turn it functions as the parent compound of a considerable family of further metabolites. They form a series which has been intensively investigated,[4,8,11,31–35] and some of the reactions which can follow the formation of (**6**) are presented in Scheme 6.

This shows a series of branching pathways, not all of which occur in the same organism and each of which can be either partly or completely operative according to the species and the culture conditions. For example, a standard strain of *Penicillium urticae* can be grown so as to produce as the major product either (**6**), or (**31**), or (**34**), or (**43**), or (**13**) (see Scheme 6).[31,32] This is because even in a strain genetically equipped to carry out a full series of transformations, the actual production of the requisite enzymes is subject to environmental controls; nutritional parameters control both the synthesis of (**6**) and its further metabolism, and special factors include both the availability of trace metals and the prevailing oxygenation level. Such a situation, coupled with the fact that at least some of the enzymes effecting these transformations can act on more than one of the available substrates, makes the detailed exploration of systems of this kind particularly arduous. In the present instance the relative importance of the different possible pathways through the 'metabolic grid'[3,31] has been considerably clarified,[33] and many of the enzymes involved have been studied in isolation, whereas in other comparable sets of compounds the detailed picture remains obscure.

Of the different pathways in Scheme 6, that from (**6**) *via* (**29**) to 3-hydroxyphthalic acid (**30**) is of minor importance, in most species that produce (**6**), compared with the decarboxylation to *m*-cresol (**31**). Two different O_2-requiring hydroxylases[34] convert (**31**) into either toluquinol (**32**) or *m*-hydroxybenzyl alcohol (**33**). The first of these will also convert (**33**) into gentisyl alcohol (**34**), but the second is more specific and will not effect the corresponding conversion of (**32**) into (**34**). Where toluquinol (**32**) is the preferred product, it is in redox equilibrium with the corresponding quinone (**37**), and it can also be further metabolized to the epoxides (**35**) and (**36**). Exactly parallel transformations of (**34**) will correspondingly give (**38**) or (**39**).

A different range of products is formed by the dehydrogenation of the alcohol (**33**) to *m*-hydroxybenzaldehyde (**40**). This dehydrogenase is relatively specific,[33] and it is uncertain whether gentisaldehyde (**41**) is formed by the hydroxylation of (**40**) or by the parallel dehydrogenation of (**34**). The equilibrium between (**33**) and (**40**) favours the former, but the product can be oxidized, irreversibly, to *m*-hydroxybenzoic acid (**42**); similarly, gentisic acid (**43**) is an end-product from the oxidation of (**41**).

In many species the eventual product of 6-methylsalicylate metabolism is the antibiotic and mycotoxin patulin (**13**), and the investigation of its biosynthesis has provided a major research objective for a number of years. The ring-cleavage reaction by which this is formed has been plausibly formulated as proceeding from the aldehyde (**41**) by way of a Baeyer–Villiger type of oxidation (route *a*, in Scheme 6). However, an alternative possibility starts from an epoxide such as (**39**) (or its equivalent in the aldehyde series), and this is shown as route *b* in Scheme 6. The reaction involves yet another oxygenase,[35] and recent work suggests that a route such as *b* may actually operate.[36]

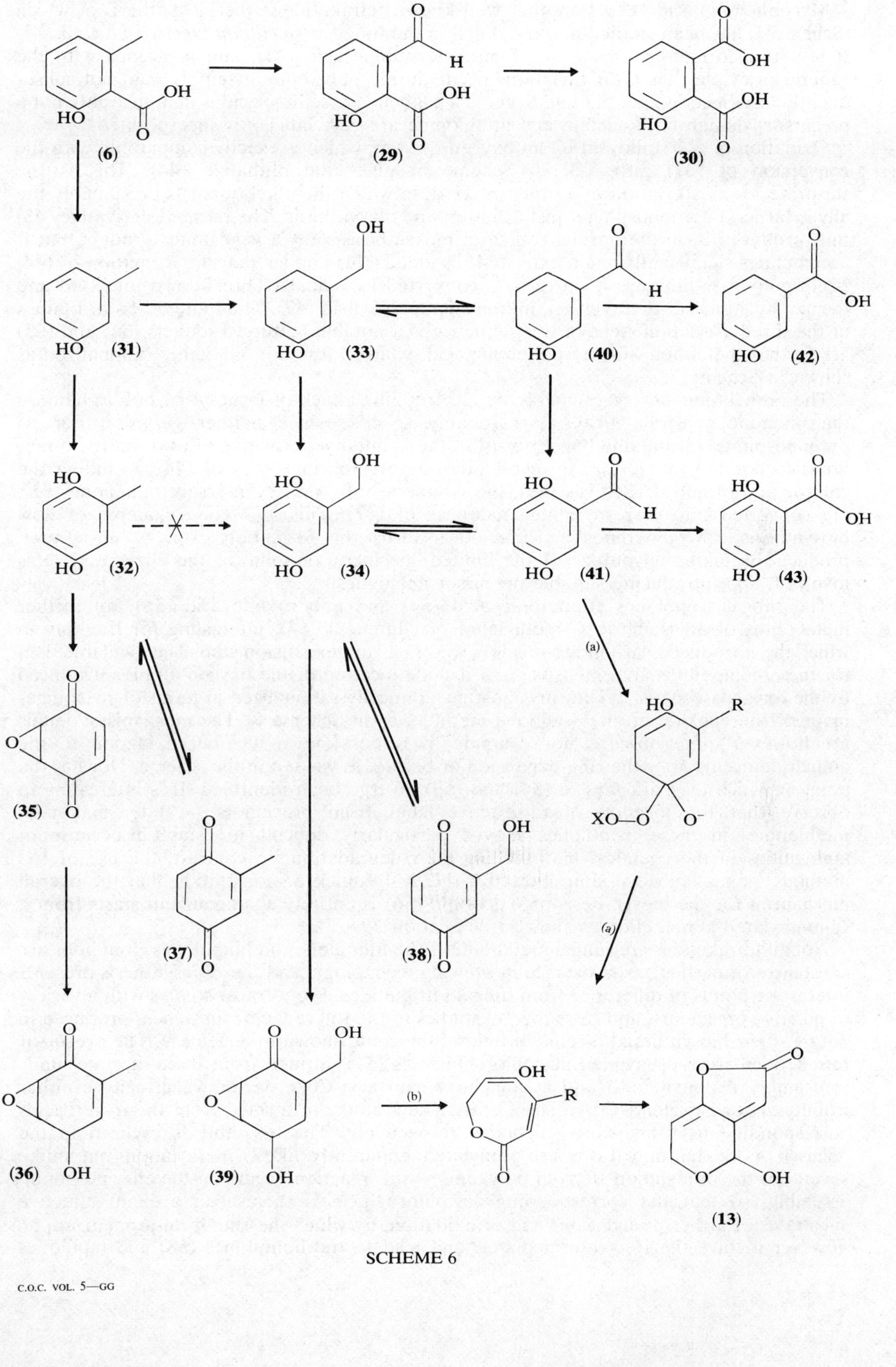

SCHEME 6

Mycophenolic acid (**47**) is another well-known tetraketide,[37] the biosynthesis of which (Scheme 7) has been studied in some detail in the fungus *Penicillium brevicompactum*.[37–41] It is a transformation product of 5-methylorsellinic acid (**22**), and in accord with the general principle that *C*-methylations occur during polyketide assembly, and not subsequently (see Section 29.1.2.2 and Scheme 4], labelled orsellinic acid is demonstrably not a precursor, though both acetate and methionine are very efficiently incorporated.

Oxidation of (**22**) followed by an oxygenase step, which is exactly comparable with the conversion of (**31**) into (**33**) in Scheme 6, gives the phthalide (**44**). This is the substrate for *C*-alkylation of a different kind, in which the alkyl donor is presumably the allylic farnesyl pyrophosphate, just as in terpene biosynthesis. The farnesyl derivative (**45**) thus produced is, in the normal course of biosynthesis, only a very minor product, but it accumulates significantly when extra (**44**) is added; this implies that the formation of (**44**) is normally rate-limiting. It, in turn, is converted by oxidative breakdown into (**46**) and thence, by a final *O*-methylation, into mycophenolic acid (**47**). The quite different timings of the three alkylation steps in this sequence is a notable feature. Products (**52**) and (**53**) are further metabolites of mycophenolic acid, whose formation is readily rationalized as shown in Scheme 7.

The prenylation step, presumably by electrophilic attack of farnesyl pyrophosphate on the phenolic substrate (**44**), is reasonably specific, since neither geraniol (nor its pyrophosphate) on the one hand, nor (**48**), the *O*-methyl derivative of (**44**), on the other, are incorporated into (**47**). However, other resorcinol analogues of (**44**), including the chloro- and bromo-derivatives (**49**), the indanone (**50**), and even resacetophenone (**51**), are converted into corresponding analogues of (**47**). This is a good example of how biosynthesis can sometimes be directed towards the formation of new 'un-natural' products by taking advantage of the limited specificity of some of the enzyme systems involved, and introducing suitable precursor analogues.

The fungal tropolones stipitatonic acid (**54**) and puberulonic acid (**55**) are further metabolites of an isomeric *C*-methylated orsellinic acid (**4**), interesting for the way in which the introduced carbon atom is involved in a ring-expansion step. The available data for these compounds are extensive, and include location of the oxygen atoms introduced by the oxygenase steps.[42] They are most economically rationalized in terms of rearrangement of the epoxide of an *o*-quinone methide, as in Scheme 8. The mechanistic details are however quite obscure; for example, it is not known whether oxidation to the anhydride occurs after the ring-expansion or before, as written in the scheme. Nor has the point at which the pathways to (**54**) and (**55**) diverge been identified. It is interesting to observe that the location of radioactive label from precursors (acetate, malonate, methionine) in these tropolones proved particularly difficult to establish because of ambiguities in the chemical methods for their degradation; in contrast, the use of ^{13}C methods[43] on a later occasion allowed a relatively quick demonstration that the overall mechanism for the biosynthesis of sepedonin (**56**) is entirely analogous but starts from a *C*-methylated pentaketide as shown (see Section 29.1.3.3).

Not all tetraketides are fungal metabolites. The humulones and lupulones from hops are substances of particular importance in brewing technology, and their biosynthesis presents interesting points of difference from fungal tetraketides. From tracer studies with a variety of possible precursors, and from careful studies of the full range of substances produced in nature, their biosynthesis[44] seems to follows the route shown in Scheme 9. The precursor tetraketides are 3-prenylated acylphloroglucinols (**57**), formed from three 'acetate' units (presumbly malonyl-CoA) and a branched-chain acyl-CoA starter which can be either isobutyryl- or isovaleryl-CoA. These starter acyl units are formed from the α-ketoacids corresponding to valine and isoleucine, respectively. The assembly is cyclized in the 'Claisen' mode, but only if it is also prenylated; apparently, like *C*-methylation (but unlike several of the prenylation steps in polyketides) this reaction occurs on the enzyme-bound assembly, so that the corresponding acylphloroglucinols themselves are not effective substrates. Further prenylations and/or oxidation, by which the ring is de-aromatized, do however involve the free intermediates, and lead to the humulones (**58**) and lupulones

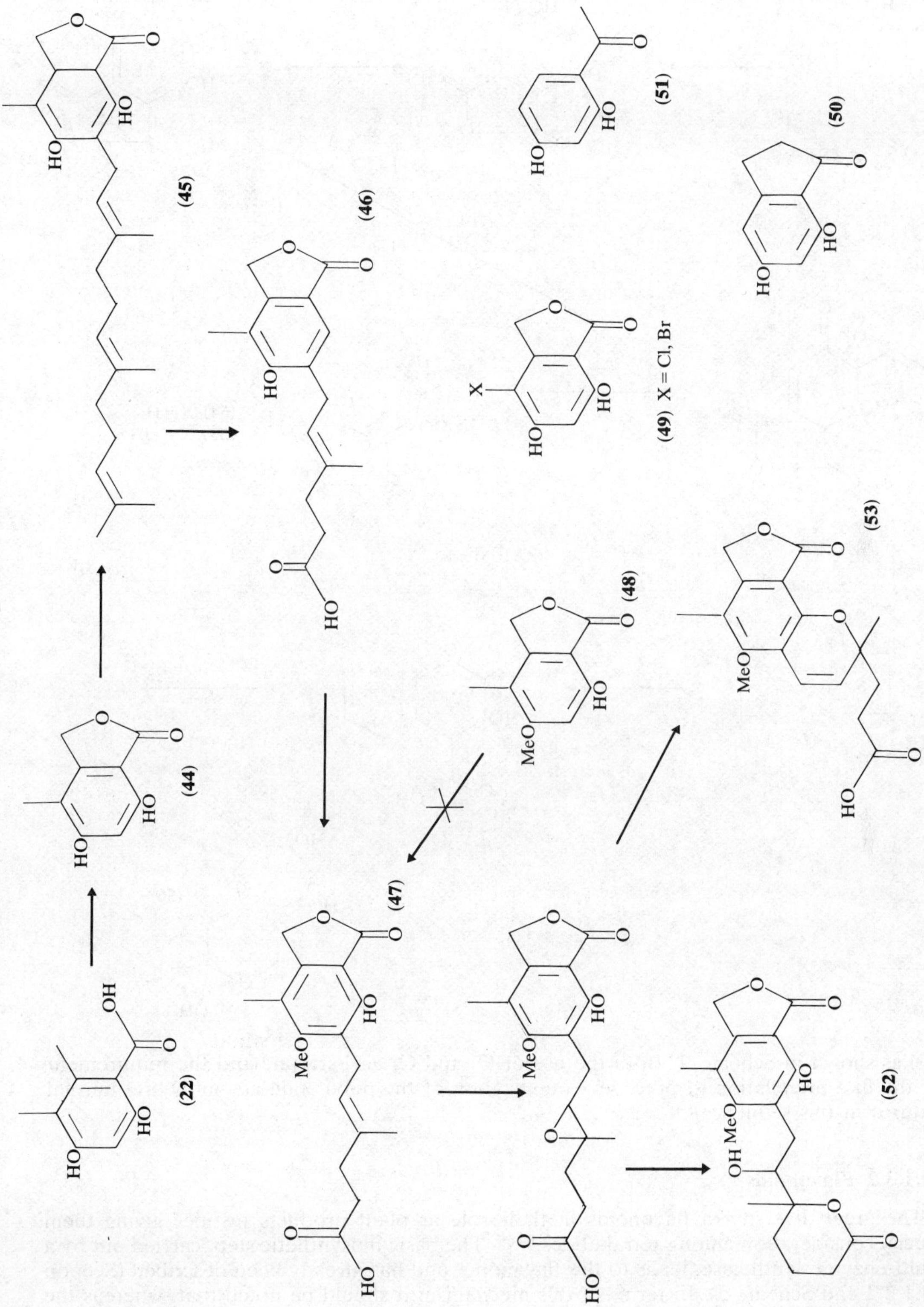

SCHEME 7

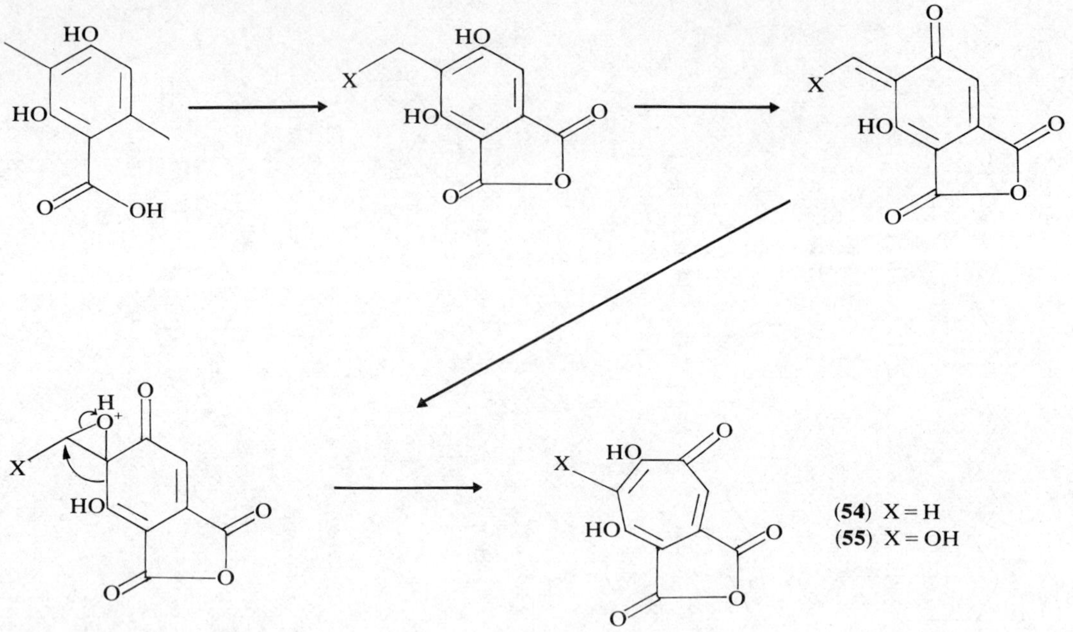

(**54**) X = H
(**55**) X = OH

SCHEME 8

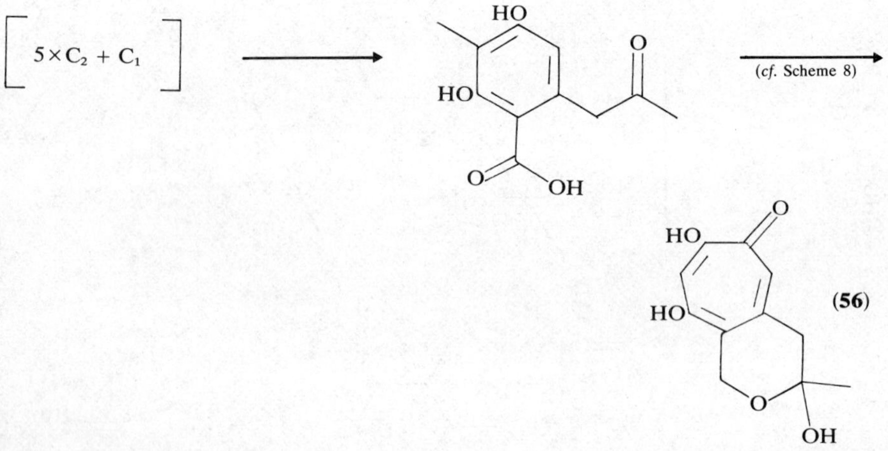

(**56**)

(**59**) as shown in Scheme 9. Both the use of C_4 and C_5 acyl starters and the requirement for the first prenylation to precede aromatization of the polyketide assembly are unusual features in this sequence.

29.1.3.2 Flavonoids

The great diversity of flavonoids in their role as plant products justifies giving them special consideration among tetraketides.[45–47] The basic biosynthetic step, carried out by a multi-enzyme synthetase, leads to the flavanones and has already been described (Section 29.1.2.2 and Scheme 5). In terms of this mechanism it should be noted that, whereas the initial oxygenation pattern of ring C is determined by the specificity of the starter

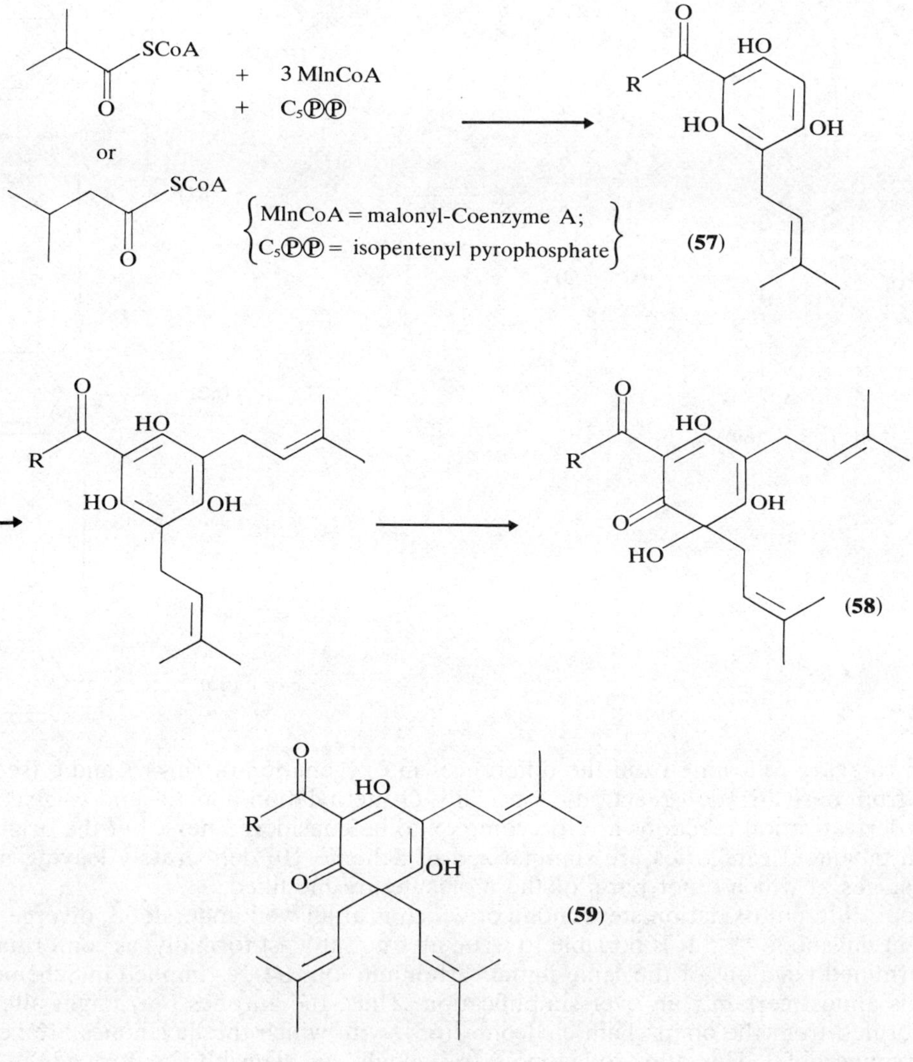

SCHEME 9

transacylase, that of ring A [in such structures as (**60**) and (**61**)] is determined by the interpolation of reduction steps [*e.g.* to give (**60**)] or their absence [*cf.* (**61**)] in the tetraketide assembly sequence. However, additional oxygenations, *O*-methylations, *etc.*, can also be effected in the completed flavonoid [*e.g.* converting (**61**) to (**62**) and (**63**)]. Thus the co-occurrence of flavonoids with different oxygenation patterns might imply either the co-existence of differently-constituted synthetase complexes or the action of modifying enzymes subsequent to flavanone formation. Current views of this complex problem[47] seem overall to favour the first explanation for differences in ring A, and the second for differences in ring C, at least so far as the most common patterns [exemplified in (**60**)–(**63**)] are concerned. In other words, incorporation of a *p*-coumaryl-CoA starter leads to flavonoids like (**61**), which are subsequently converted to those like (**62**) and (**63**). However, the existence of synthetases with a different transacylase specificity, accepting other substrates from the range of natural cinnamyl-CoA derivatives, is not excluded by the evidence so far available.

The other categories of natural flavonoids[45] arise from flavanones by an important series of transformations affecting the heterocyclic B ring,[46,47] further variety from within

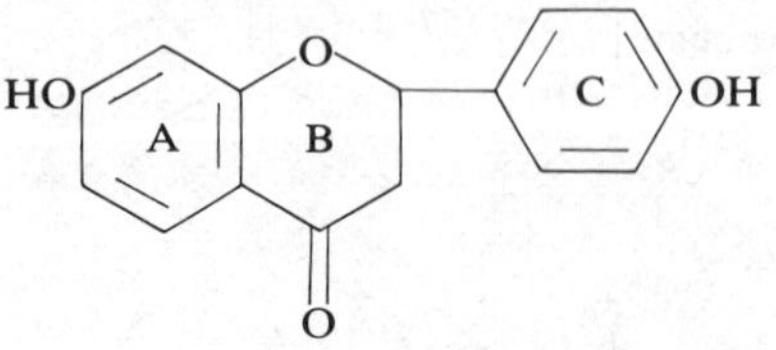

(60)

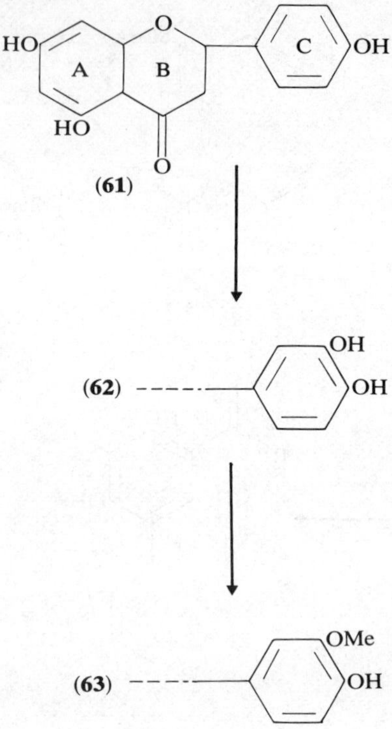

(61)

(62)

(63)

each category following from the differences in oxygenation of rings A and C (see above), and from derivatization reactions, especially *O*-methylation and *O*- and *C*-glycosylation. The derivatization reactions are too complex to be considered here, but the origins of the main flavonoid categories are summarized in Scheme 10, deliberately leaving undefined the stages at which other parts of the molecule are modified.

Four different oxidation steps, none of which is at all well understood, diverge from the parent flavanone.[46–48] It is possible to write all four, at least formally, as conformationally-determined reactions of the same initial carbonium ion (**64**), as implied in Scheme 10, but this is almost certainly an over-simplification. Thus, the aurones (**66**) might alternatively be formed from the open-chain chalcones (**65**), with which the flavanones are very freely interconvertible. The dihydroflavonols (**67**) could as plausibly be formed by a direct oxygenation of the flavanone. The spirodienones, written as a possible intermediate for the aryl migration in formation of the isoflavones (**68**), could equally well collapse to give the flavones (**69**) without rearrangement. In isoflavone formation there are suggestions that electrophilic *O*-methylation on ring C may be an alternative to protonation in promoting the collapse of the spirodienone, thus giving the known series of 4′-methoxyisoflavones directly.[49]

Further elaboration of the dihydroflavonols (**67**) seems to be a particularly important source of yet further flavonoid series,[47,48] and some possibilities are represented in Scheme 11. Direct dehydrogenation leads to flavonols (**70**) and direct reduction to the flavandiols (**71**); an important variant of the reduction pathway leads by way of flav-3-en-3-ols to the catechins and epicatechins (**72**), but the carbonium ion suggested as an intermediate in this last step can also, apparently, react by electrophilic substitution into (**72**), generating the oligomeric 'proanthocyanidins', of which (**73**) is a simple example.[50,51] Note that in these conversions the original *R*-configuration at C-2 persists, but there is no fixed configuration at the other centres in ring B; thus both the catechin (3*R*) and epicatechin (3*S*) configurations in (**72**) occur. However, in forming the proanthocyanidins the group at C-4 is always attacked *trans* to the hydroxyl at C-3. The true anthocyanidins (**74**) are also formed from dihydroflavonols, possibly also by way of the flav-3-en-3-ols.

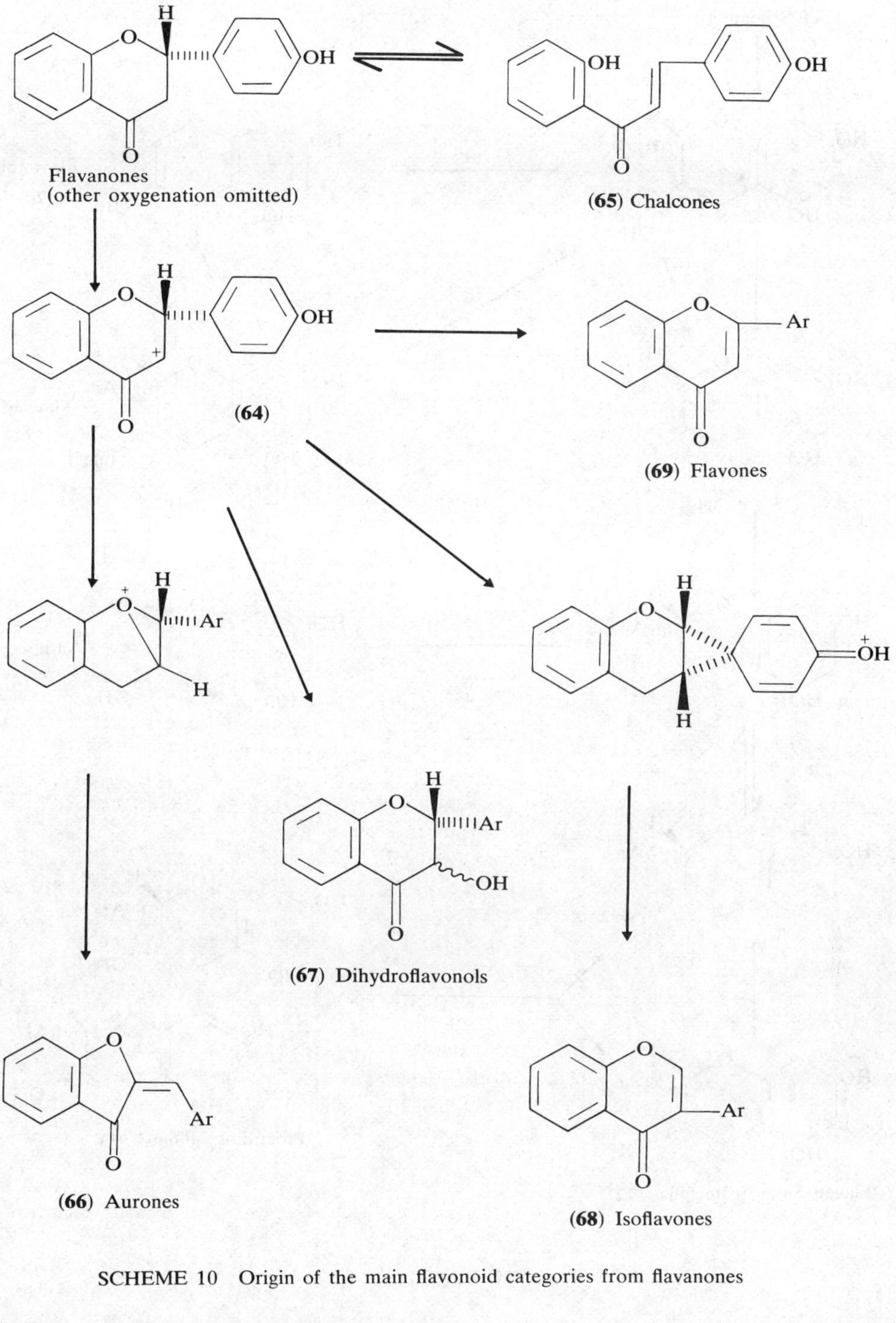

SCHEME 10 Origin of the main flavonoid categories from flavanones

One of several interesting continuations of the isoflavone sequence which has been studied is that to the insecticidal rotenoids,[49] a well-defined group of substances of rather more restricted distribution in nature than the main flavonoid categories. From precursor incorporation data the main further steps leading to rotenoids can be set out in the order shown in Scheme 12. As noted above, the 4′-methoxyisoflavone (**75**) is apparently formed by concerted oxidation, methylation, and rearrangement of the flavanone precursor.

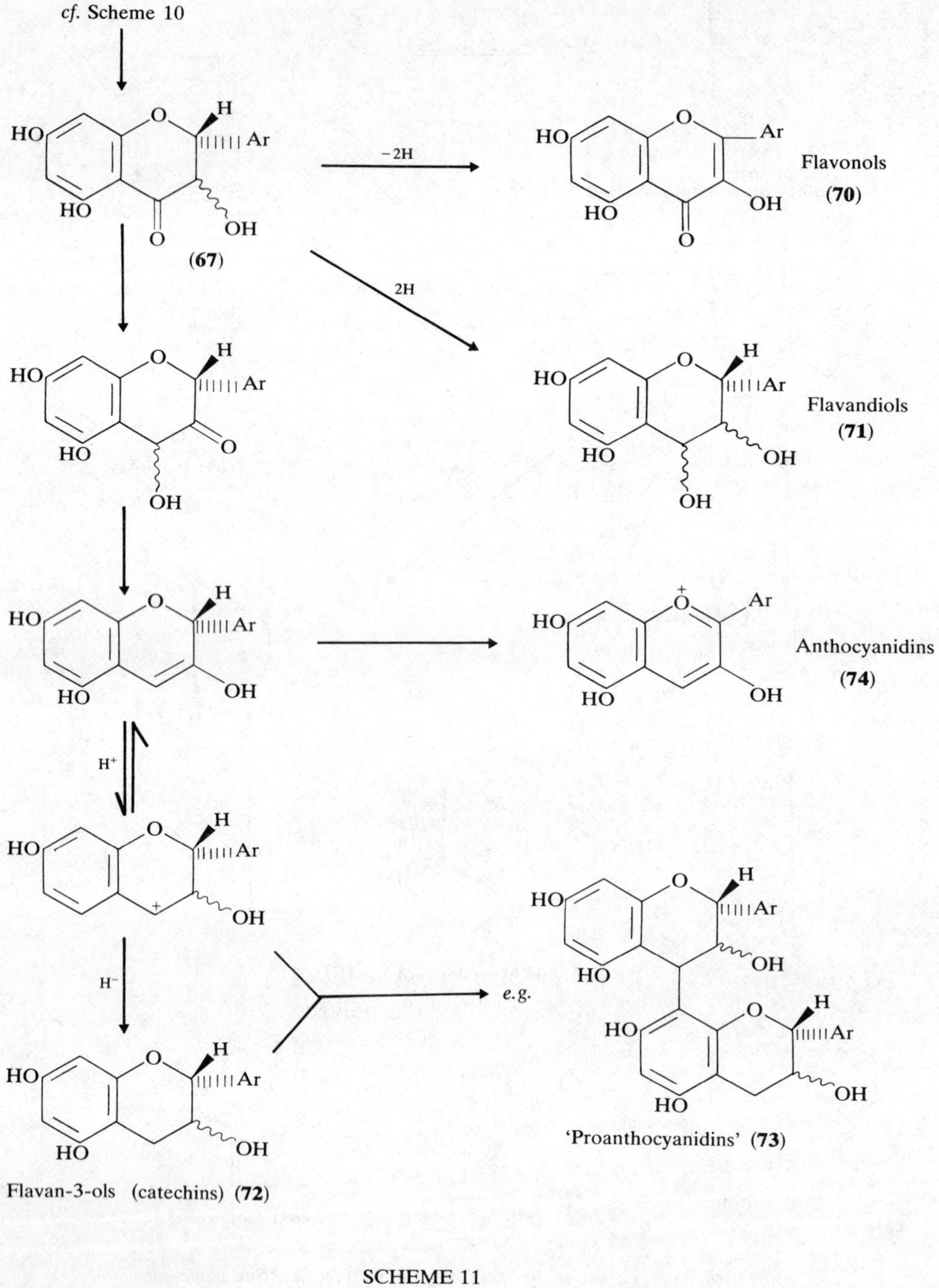

SCHEME 11

Further hydroxylations and *O*-methylations give (**76**), and direct oxidation of the *O*-methyl of (**76**) followed by stereospecific cyclization generates the second pyran ring of (**77**). This step is followed, as shown in the scheme, by successive prenylation, epoxidation, and cyclization, and final dehydration to rotenone (**78**). Other rotenoids are derived either from rotenone itself or from its precursors in this sequence by further transforming reactions.

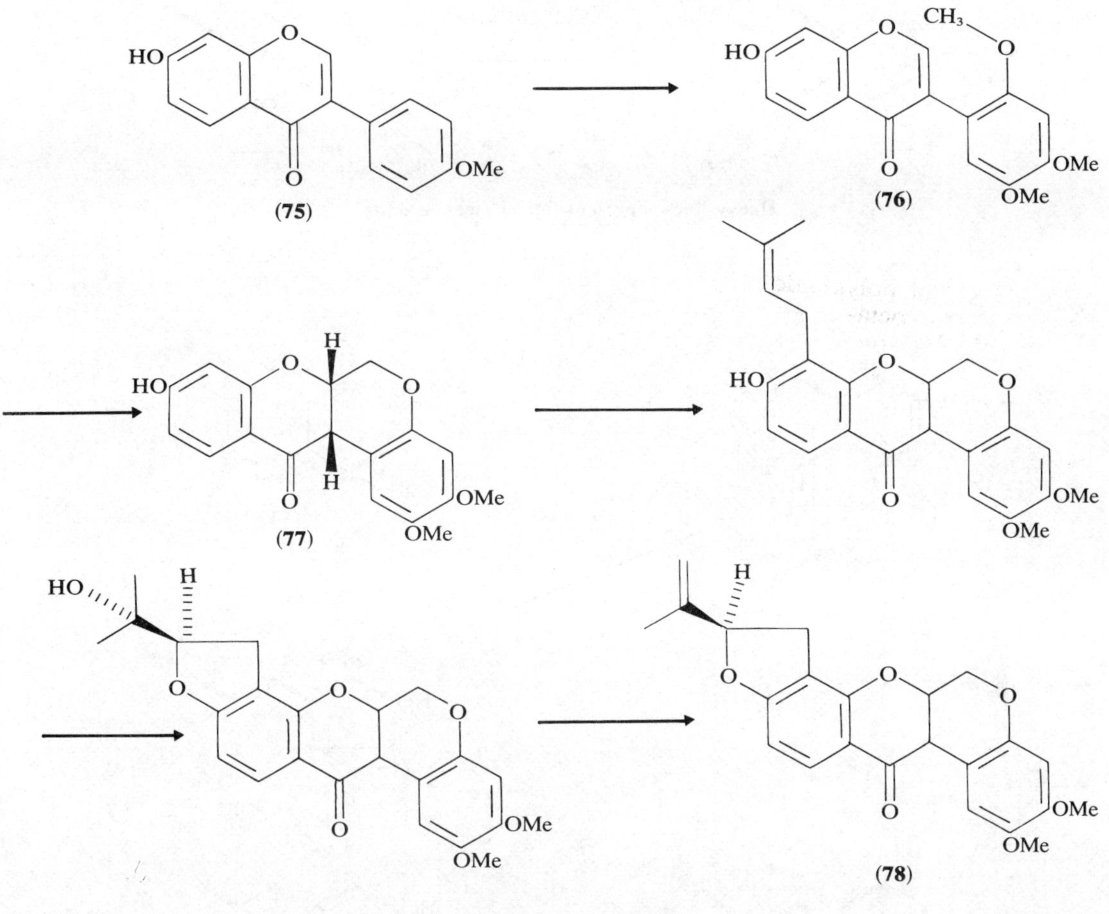

SCHEME 12

29.1.3.3 Cleavages and couplings in simple polyketides

The examples of patulin [(**13**); Scheme 6], penicillic acid (**27**), and the tropolones [(**54**)–(**56**); Scheme 8] illustrate how a relatively simple aromatic polyketide is sometimes quite radically transformed by oxidative ring-cleavage and associated skeletal rearrangements. All three of these cases have been confirmed by direct demonstrations of the inferred transformations and identification of the un-rearranged precursors. Other examples must be inferred from the patterns of labelled-precursor incorporation into the final product. The method employing [$^{13}C_2$]acetate as a precursor, in which n.m.r. measurements on the product allow 'intact' and 'cleaved' C_2 units to be located in its structure (see Section 29.1.5.4), has proved particularly powerful in this respect, since it provides an element of information about intermediate stages in the biosynthetic sequence.

Thus the isocoumarin (**79**), which is a typical pentaketide analogue of orsellinic acid, accompanies the cyclopentanone terrein (**80**) in some strains of *Aspergillus terreus*. Labelling from [$^{13}C_2$]acetate suggests that (**80**) arises from (**79**) by the overall mechanism shown, with decarboxylation and a ring contraction. It was demonstrated that each of the hydroxyl-bearing carbons in (**80**) is the residue of a separate two-carbon unit, and the inferred mechanism was confirmed by the incorporation of ^{14}C-labelled (**79**) into (**80**).[52]

Similar studies[53] on another pair of pentaketide co-metabolites, (**81**) and (**82**), suggested a rather different cleavage and ring-contraction mechanism. Without knowing

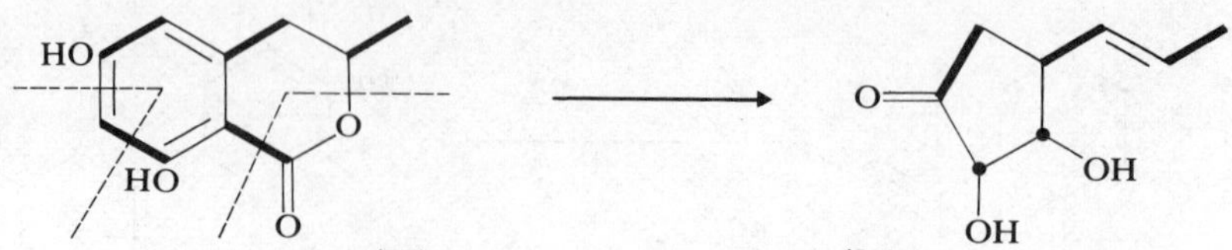

(79) (80)

Heavy lines represent 'intact' acetate units

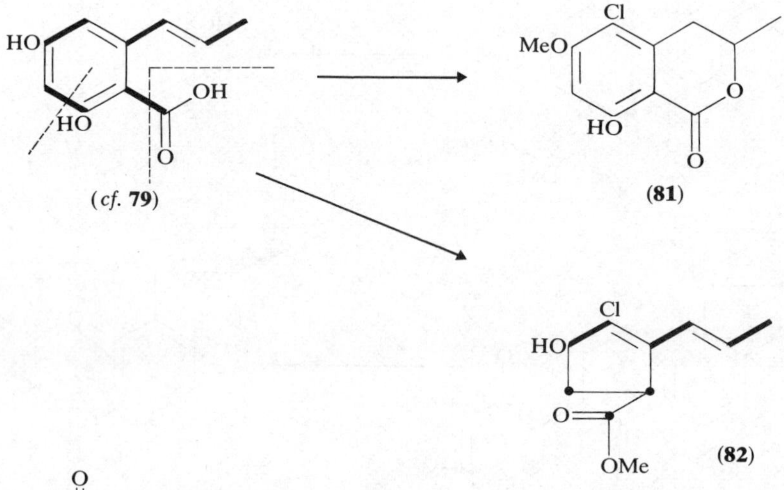

(*cf.* 79) (81) (82)

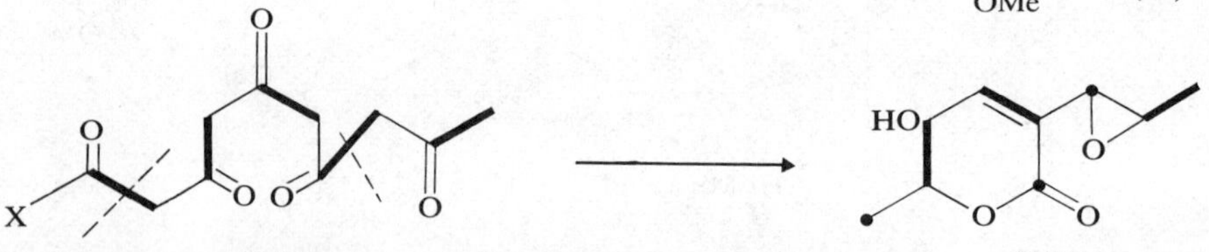

(83) (84)

details of the remaining steps, including *O*-methylation and electrophilic halogenation, the common precursor of both products is deduced as the free acid corresponding to (**79**), as shown here. A third rearrangement through unidentified intermediates is similarly evidenced[54] as leading from the pentaketide assembly (**83**) to the pyrone (**84**). Here the oxidation level of the product appears to be that of an epoxidized polyene rather than an aromatic polyketide [compare the rather similar regular polyketide (**7**)], but only the overall origin of the carbon skeleton is actually known. On the other hand, the $^{13}C_2$ data do establish which 'end' of the assembly is intact and which has lost a carbon atom. Some important examples of ring-cleavage and rearrangement of large polyketides are considered below (Sections 29.1.3.3 and 29.1.3.5).

The oxidative coupling of phenols is known as an important route to more complex natural products in several different biosynthetic groups, for example in alkaloid biosynthesis (see Chapter 30.1). A relatively simple and historically important example[55] of an oxidatively-coupled polyketide product is the lichen metabolite usnic acid (**86**). Here the parent phenol, 3-methylphloracetophenone (**85**), is a *C*-methylated tetraketide [compare the 3-prenyl analogues (**57**)], and it undergoes oxidative coupling as shown in Scheme 13. As in other cases noted previously, pre-formed phloracetophenone is not a substrate for the *C*-methylation step and this must occur at the polyketide assembly stage.[56] The

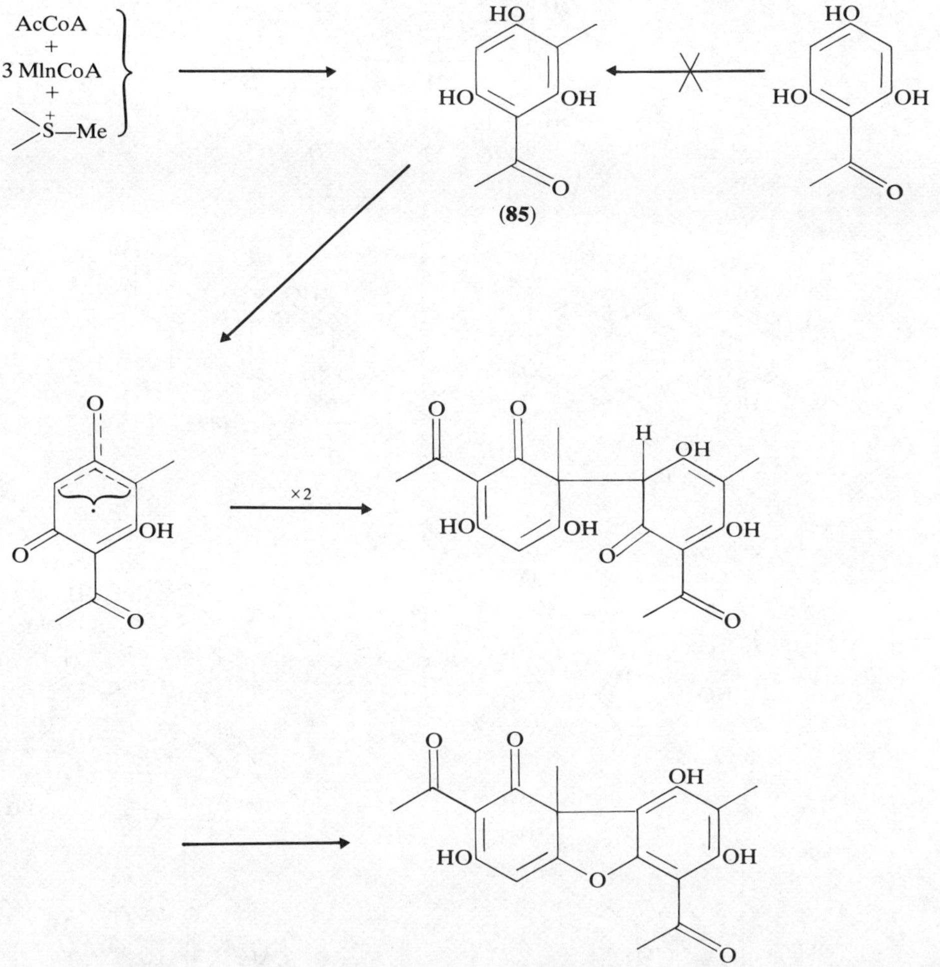

(85)

(86)

SCHEME 13

coupling itself, though fully analogous to reactions which can be effected *in vitro* with a variety of one-electron oxidizing agents, must take place under enzymic control since the product is optically active. As in other cases, representation of the coupling process as the union of two free radicals is purely formal, since it might equally involve radical substitution into the parent phenol followed by a second one-electron transfer. Further examples of coupling reactions, to which similar observations equally apply, will be found in later sections. An ionic mechanism for oxidative coupling has been illustrated for proanthocyanidin biosynthesis (Section 29.1.3.2 and Scheme 11) and illustrates an alternative concept which has attracted rather less consideration than the one-electron process.

29.1.3.4 More complex cyclizations

With larger polyketide assemblies we would expect to find greater variety in cyclization patterns, which might involve all the assembly or only parts of it, and also to find scope for a greater range of subsequent modification reactions. Such is in fact the case. For example, the compounds (**87**)–(**93**) are all regular heptaketide metabolites from fungi; in structures (**87**)–(**92**) the 'methyl' and carboxyl' ends of the assembly are marked * and •, respectively.

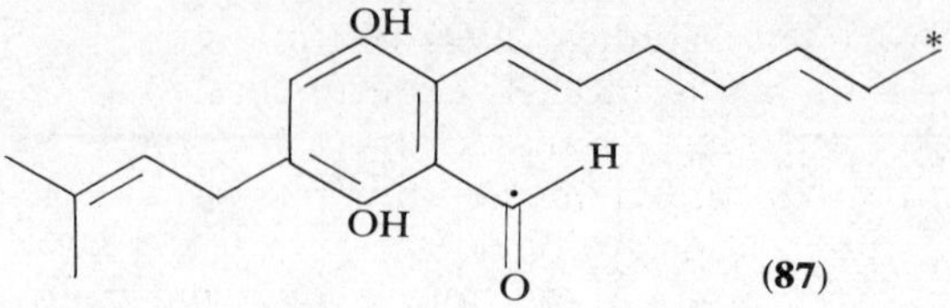

(**87**)

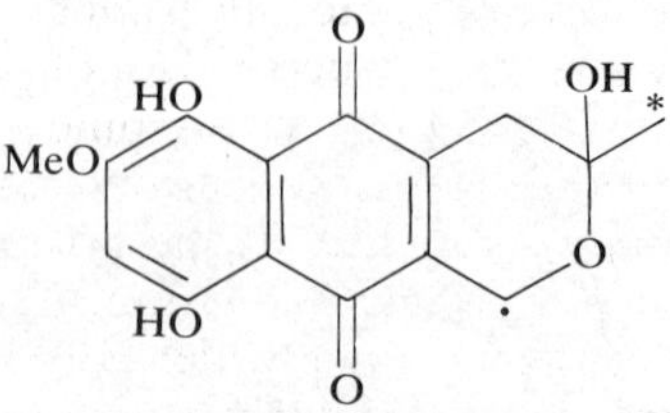

(**88**)

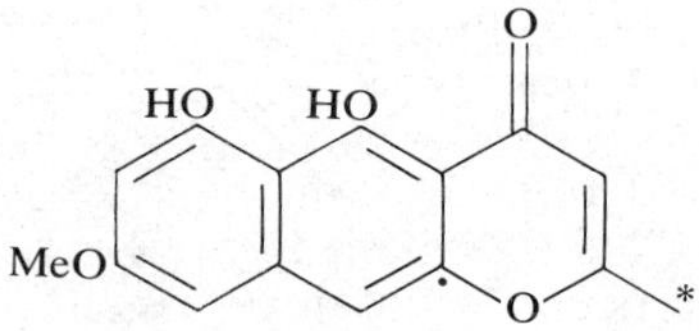

(**89**)

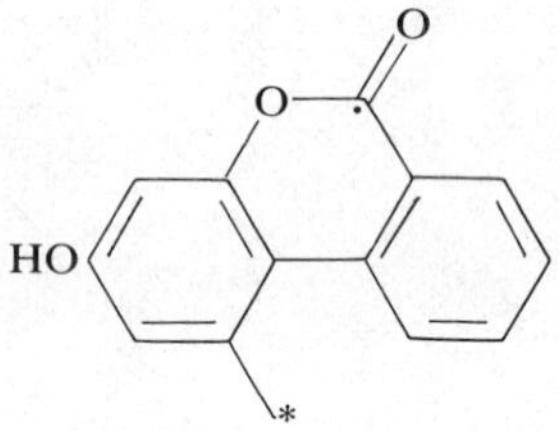

(**90**)

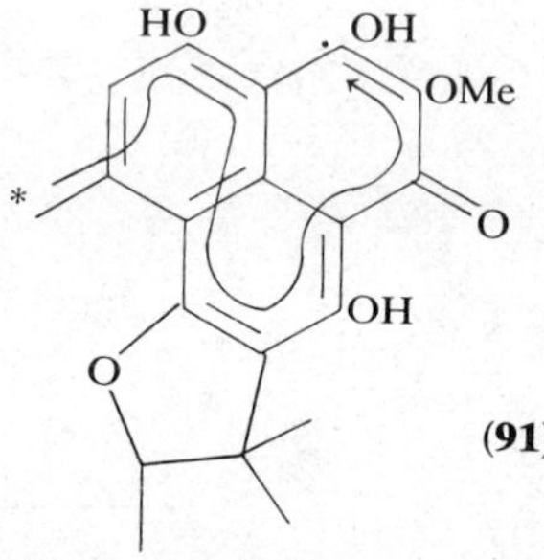

(**91**)

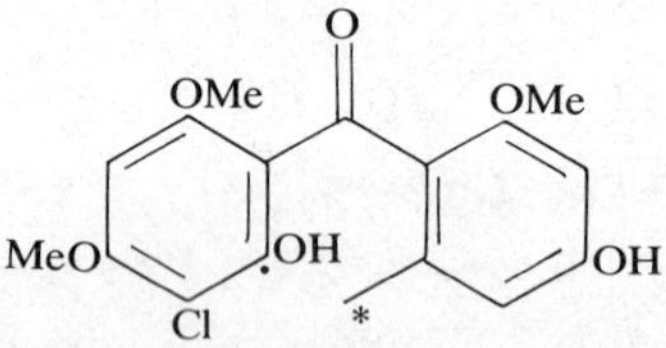

(**92**)

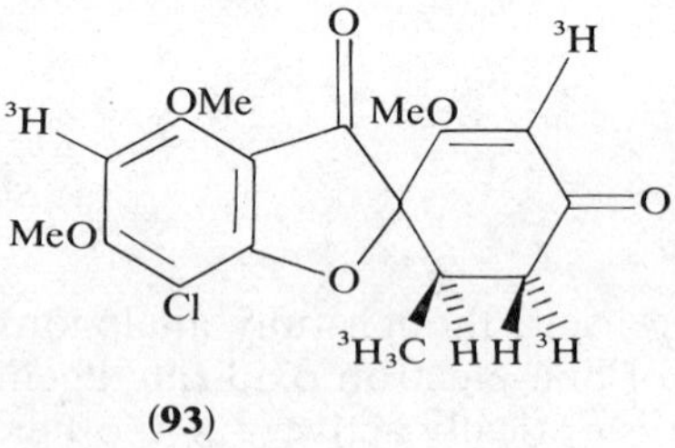

(**93**)

Auroglaucin (**87**), which also carries an introduced prenyl residue, illustrates how part of a polyketide chain may be reduced and part aromatized. Three ketomethylene groups are reduced to the polyene level; in a co-metabolite, flavoglaucin, this side-chain is fully saturated but nothing is known about the biosynthetic relationship between the two products. The reduction of the carboxyl terminal seen in (**87**) is a less common feature.

Fusarubin (**88**) and rubrofusarin (**89**) are products from related *Fusarium* species, and each also typifies a further series of related metabolites. The naphthalene elements in (**88**) and (**89**) are the products of 'aldol' and 'Claisen'-type cyclizations, respectively.

An entirely different, and rather rare, type of cyclization is seen in alternariol (**90**) and some related diphenyl derivatives. The heptaketide framework in deoxyherqueinone (**91**) and other fungal phenalenones could arise by several different foldings, but that shown has been established, at least for this particular compound, by the $[^{13}C_2]$-labelling method.[57] The 'retro'-attached prenyl residue which appears in (**91**) is probably formed by a Fries rearrangement of a regular 3,3-dimethylallyl ether at the adjacent oxygen.

The case of the important antifungal antibiotic griseofulvin (**93**) has been an important one for the progress of biosynthetic studies. Demonstration of its overall heptaketide origin by [^{14}C]acetate incorporation was an early experimental vindication of the 'acetate hypothesis',[58] depending upon a perceptive dissection of the known structure, while the implied conversion of an initial benzophenone assembly such as (**92**) (one of a series of very minor co-metabolites) into an optically active spiro-dienone system provided a striking example in elaborating views on the role of enzyme-controlled one-electron coupling processes in natural product biosynthesis.[59] Introduction of the chlorine substituent apparently occurs by electrophilic attack on the parent assembly, the normal (but largely unknown) mechanism in such cases, but there is some uncertainty regarding the timing of the three *O*-methylation steps in the overall sequence.[60] Doubts as to whether griseofulvin really is a 'regular' polyketide, and not the product of some more complex cleavage and re-cyclization process, have been voiced, but are surely resolved by the observation[61] that tritium from [2-^{3}H,2-^{14}C]acetate is retained at the positions shown in (**93**), in particular with three labelled atoms on the *C*-methyl group. This result also established the stereospecificity of the reduction of the dienone, which is probably the last step in the formation of (**93**), by the introduction of two hydrogens which are mutually *trans* and both axial.

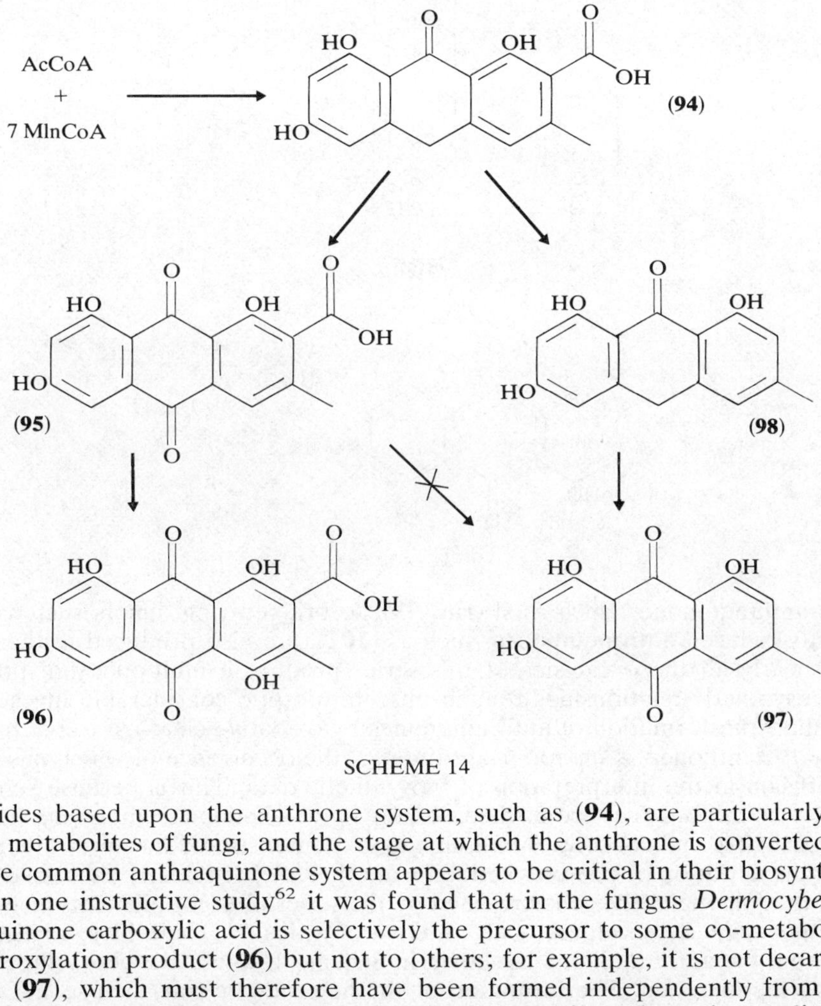

SCHEME 14

Octaketides based upon the anthrone system, such as (**94**), are particularly frequent among the metabolites of fungi, and the stage at which the anthrone is converted (if at all) to the more common anthraquinone system appears to be critical in their biosynthesis. For example, in one instructive study[62] it was found that in the fungus *Dermocybe* endocrin (**95**) the quinone carboxylic acid is selectively the precursor to some co-metabolites such as the hydroxylation product (**96**) but not to others; for example, it is not decarboxylated to emodin (**97**), which must therefore have been formed independently from (**94**) (see Scheme 14). The most probable intermediate in the emodin pathway is emodin anthrone (**98**), and in several *Penicillium* species this figures as the parent substance of a particularly complex series of anthraquinone derivatives, including dimeric products which range from

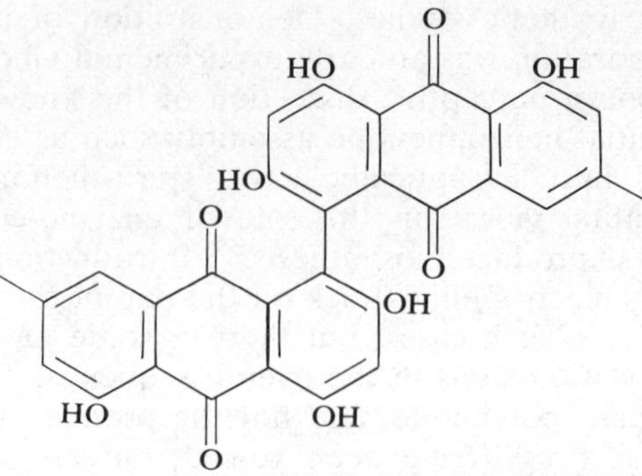

(99)

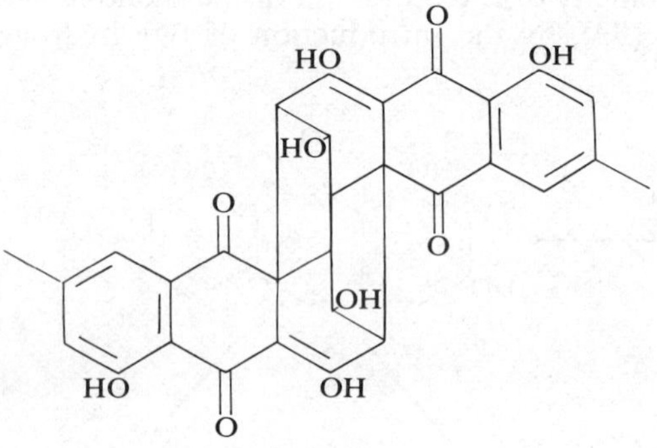

(100)

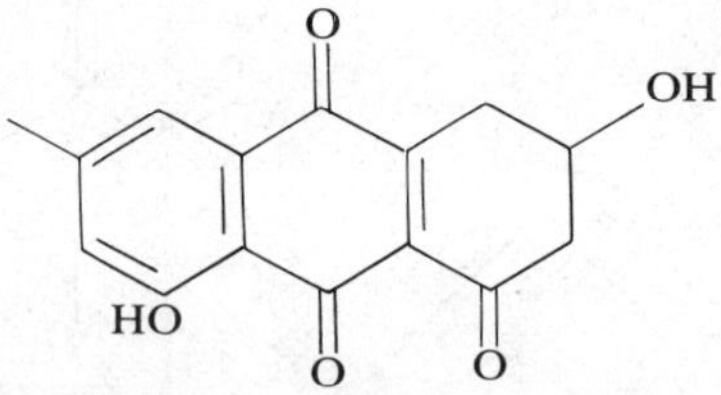

(101)

simple bis-anthraquinones, such as skyrin (**99**), to bridged-ring dimers such as rugulosin (**100**). Partly-reduced anthraquinones, such as (**101**), are also produced in these systems, and since nearly all the organisms of this series produce a multiplicity of products the detailed biosynthetic relationships remain obscure, despite considerable investigation.[63]

The further transformation of anthraquinones by oxidative cleavage to benzophenones, and thence to xanthones, is an important process, the occurrence of which has sometimes led to confusion in the interpretation of biosynthetic data. This is because some superficially similar benzophenones are formed directly by polyketide assembly [*cf.* the griseofulvin precursor (**92**)], and conceivably (though in no demonstrated instance) others could be formed by the acylation of one polyketide by another. A particularly clear instance of this type of anthraquinone cleavage is provided by the series of ergot pigments (ergochromes), of which the epimeric secalonic acids (**102**) are typical.[64] Their biosynthesis probably proceeds from the anthraquinone as shown in Scheme 15, with Baeyer–Villiger oxidation in ring B as the first step. The formation of the heterocyclic ring then seems to be most plausibly formulated as proceeding by 1,4-addition of a ring A phenol to a ring C quinone, as shown. In this scheme it is the eventual reduction of the resulting intermediate which generates the different naturally-occurring epimers of (**102**). Oxidative coupling to the

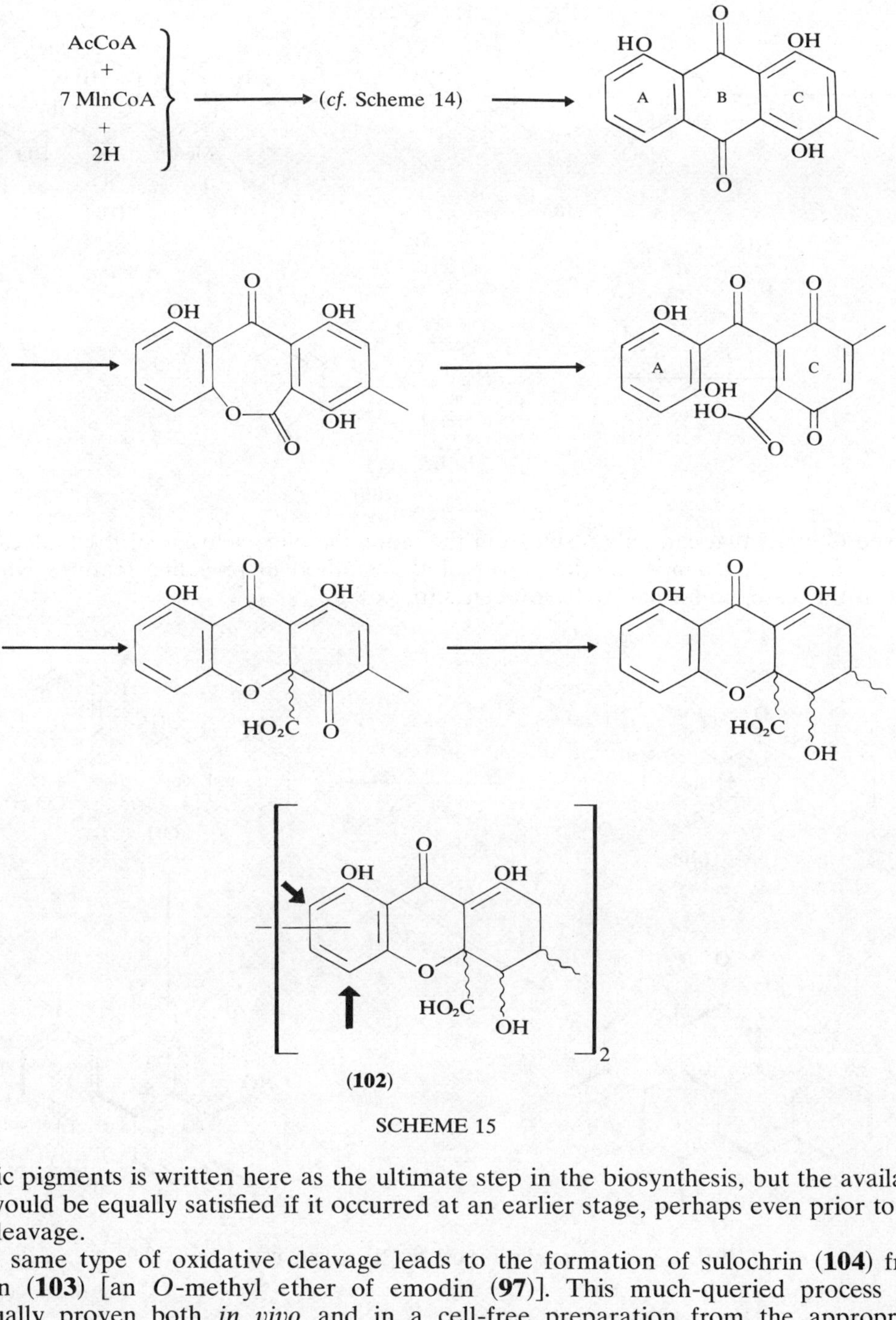

(**102**)

SCHEME 15

dimeric pigments is written here as the ultimate step in the biosynthesis, but the available data would be equally satisfied if it occurred at an earlier stage, perhaps even prior to the ring cleavage.

The same type of oxidative cleavage leads to the formation of sulochrin (**104**) from questin (**103**) [an *O*-methyl ether of emodin (**97**)]. This much-queried process was eventually proven both *in vivo* and in a cell-free preparation from the appropriate fungus.[65,66] In this case, subsequent oxidative coupling of the benzophenone leads on to the superficially griseofulvin-like spirodienone (**105**). For further important cases of the ring-cleavage process, see Section 29.1.3.5; an example of their sometimes surprising consequences is provided by the actinomycete antibiotic chartreusin. The aglycone (**108**) is clearly a cleaved polyketide and from simple inspection of the structure it was quite plausibly thought to result from a single cleavage of the decaketide assembly (**106**). However, investigation by the [$^{13}C_2$]acetate method revealed[67] a pattern of intact and

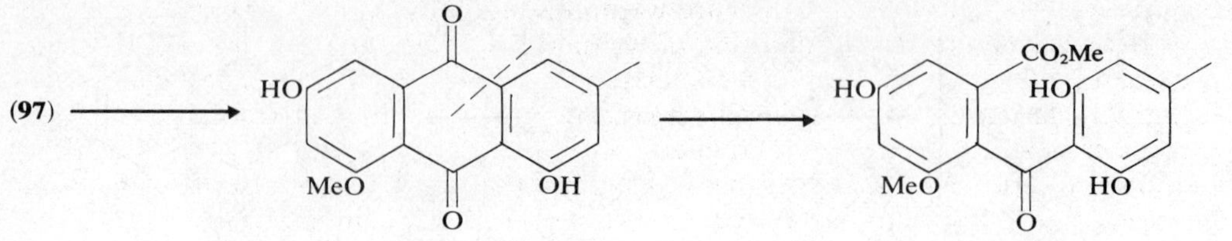

(103) (104)

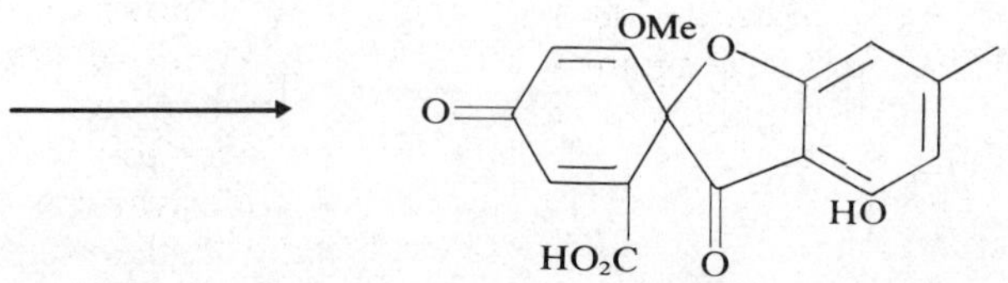

(105)

cleaved C_2 units that can only result from the more complex cleavage of the undecaketide (**107**) — a further example of the power of this method in revealing features which, at least in this case, could hardly have been anticipated.

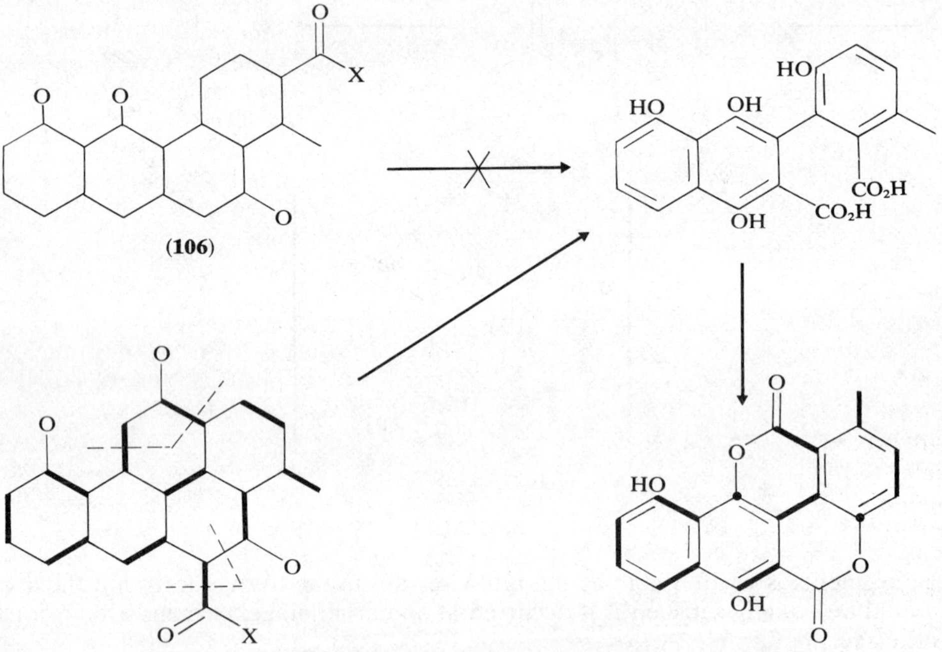

(107) Heavy lines show 'intact' acetate units (108)

29.1.3.5 Aflatoxins and their precursors

The biosynthesis of the aflatoxins by *Aspergillus flavus* and related fungi is clearly an important problem, because of the economic significance of these toxins, and also a difficult one, because the structures of the aflatoxins are highly modified versions of their precursors, while the experimental handling of these potent carcinogens is not a particularly attractive proposition. The study of the problem has been marked both by elegance and accuracy on the one hand and by error and confusion on the other. A pathway which accommodates at least the more securely-established findings is set out in Schemes 16 and 17. Relevant information (at the time of writing) comprises: (a) the considerable range of co-metabolites which has been characterized in various strains of *Aspergillus flavus* and *Aspergillus versicolor*, together with some data on their interconvertibilities;[68,69] (b) the patterns of labelling from singly-labelled and (particularly) from [$^{13}C_2$]acetate which have been meticulously established for some key metabolites;[70] and (c) fruitful suggestions by Thomas[71] and Holker[72], supported by parallel instances in other fungal systems (*cf.* Scheme 15), as to feasible transformation mechanisms.

The parent compound is an anthraquinone with a six-carbon side-chain (**110**), formed from a decaketide assembly known to be folded as shown in (**109**);[70] averufin (**111**) is typical of a series of products which are known to be formed in this way. One of this series, unidentified but possibly with a side-chain of the type shown in (**112**), plausibly undergoes (stereospecifically) a rearrangement in which the point of attachment of the side-chain shifts from the α- to the β-carbon and a two-carbon unit is eliminated. A possible mechanism[71] is illustrated, which leads from (**112**) either to versicolorin A (**113**) or its deoxy-derivative (**114**). A parallel series of dihydro-derivatives, as in (**115**), is known, implying that reduction of the furan ring A can occur at this stage. In a blocked mutant, labelled averufin (**111**) is an effective precursor of (**113**).[69] Further transformations, leading successively to the sterigmatocystins and the aflatoxins, might plausibly involve compounds of the deoxy-series, as (**114**). This would normally imply that the deoxygenation has occurred at the polyketide stage, *cf.* (**109**), so that the pathways *via* (**113**) and (**114**) would be separate throughout. However, the alternative route, by some unknown mechanism of deoxygenation at an anthraquinone stage, seems to be followed in this series; in particular, labelled versicolorin A is an efficient precursor of aflatoxins.[69]

The ensuing cleavage reaction is most convincingly formulated[72] if it is preceded by the introduction of a *para*-hydroxyl group into ring E, as shown in (**116**). This allows the ring-cleavage to be followed by ring-closure, to form the xanthones, by way of 1,4-addition to a quinone, (**117**) $\rightarrow$ (**118**), just as in the formation of the ergochromes (Scheme 15). From the [$^{13}C_2$]acetate data,[70] the orientation of ring E is known to remain as shown [*e.g.* in the postulated intermediate (**118**)], so that the xanthone ring-closure cannot be formulated as one which involves the hydroxyl group originally available on ring E. On this mechanism there could still be a choice between two oxygens on ring C, each potentially available for forming the xanthone, and this 'choice' may be forced, either by the stereochemistry of an enzyme site or by *O*-methylation, conceivably concerted with fragmentation of the Baeyer–Villiger cleavage intermediate. Certainly (as indicated in Scheme 16) the *O*-methyl derivatives figure as genuine intermediates from that point onwards. The immediate product of ring-closure would then be (**118**), and this can, apparently, aromatize either by reduction followed by elimination/decarboxylation, giving sterigmatocystin (**119**), or by direct decarboxylation of the enol, giving 5-hydroxysterigmatocystin (**120**). Labelled sterigmatocystin is a very efficient precursor of the aflatoxins.[69]

Only the 5-*O*-methyl derivative of (**120**) has actually been isolated from the fungi, but experimentally the (ring A) dihydro-analogue of (**120**) is the only *selective* precursor of the corresponding series of aflatoxins so far identified.[68] Clearly, however, the formation of (**120**) by oxygenation of (**119**) is not ruled out as a route. In either case, formation of ring E of the aflatoxins (Scheme 17) is then best formulated[71] as proceeding from (**120**) by a second oxidative cleavage, on this occasion followed by a Claisen condensation. This leads to the B series of aflatoxins (**121**), from which yet a third Baeyer–Villiger oxidation would generate the aflatoxins of the G-series (**122**).

29.1.3.6 Tetracyclines

The range of tetracycline antibiotics from streptomycete bacteria provides further important instances of the sequential transformations of polyketide precursors, particularly interesting for the manner in which parallel transformations of different substrates

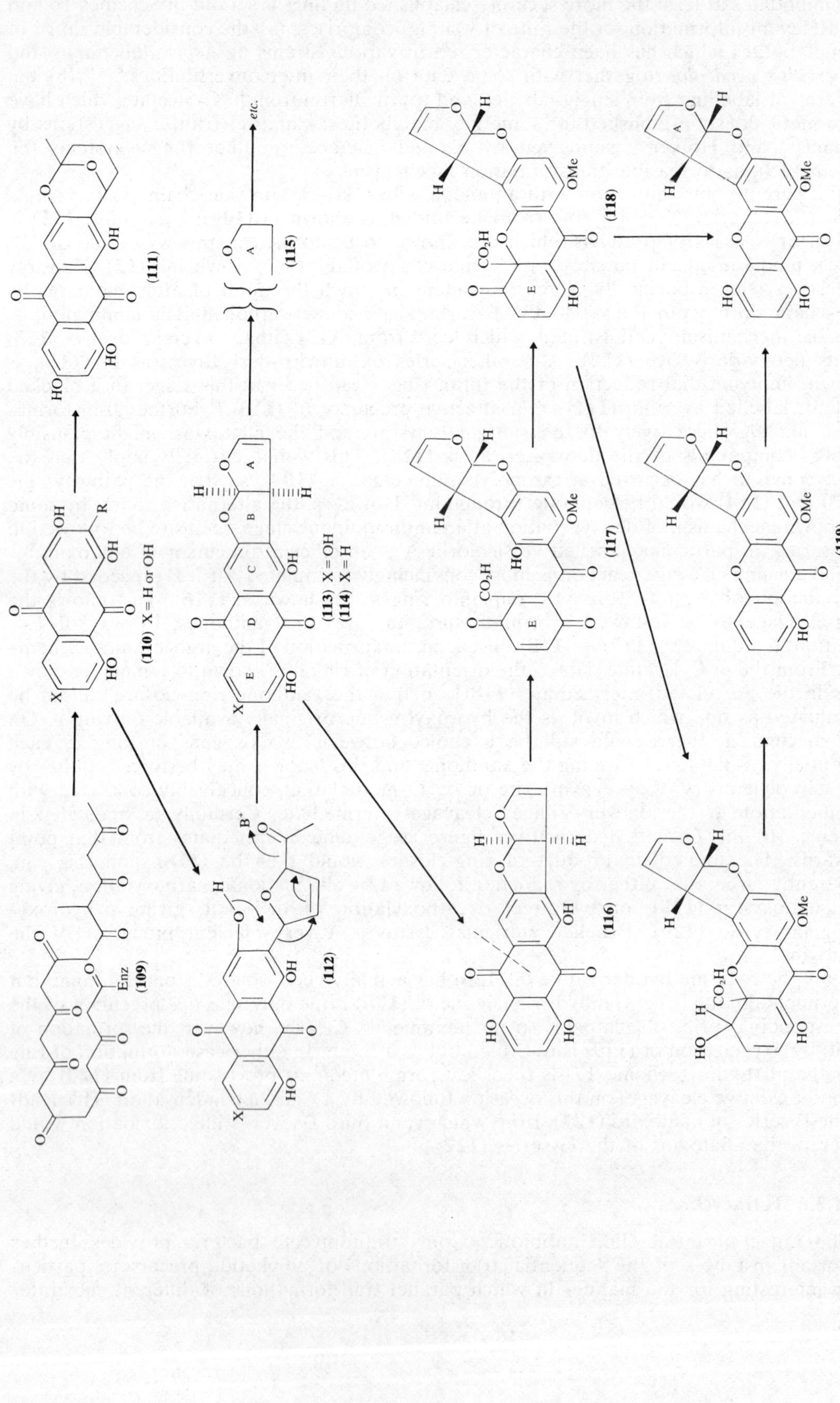

SCHEME 16 Biosynthesis of sterigmatocystins (aflatoxin precursors)

(119)

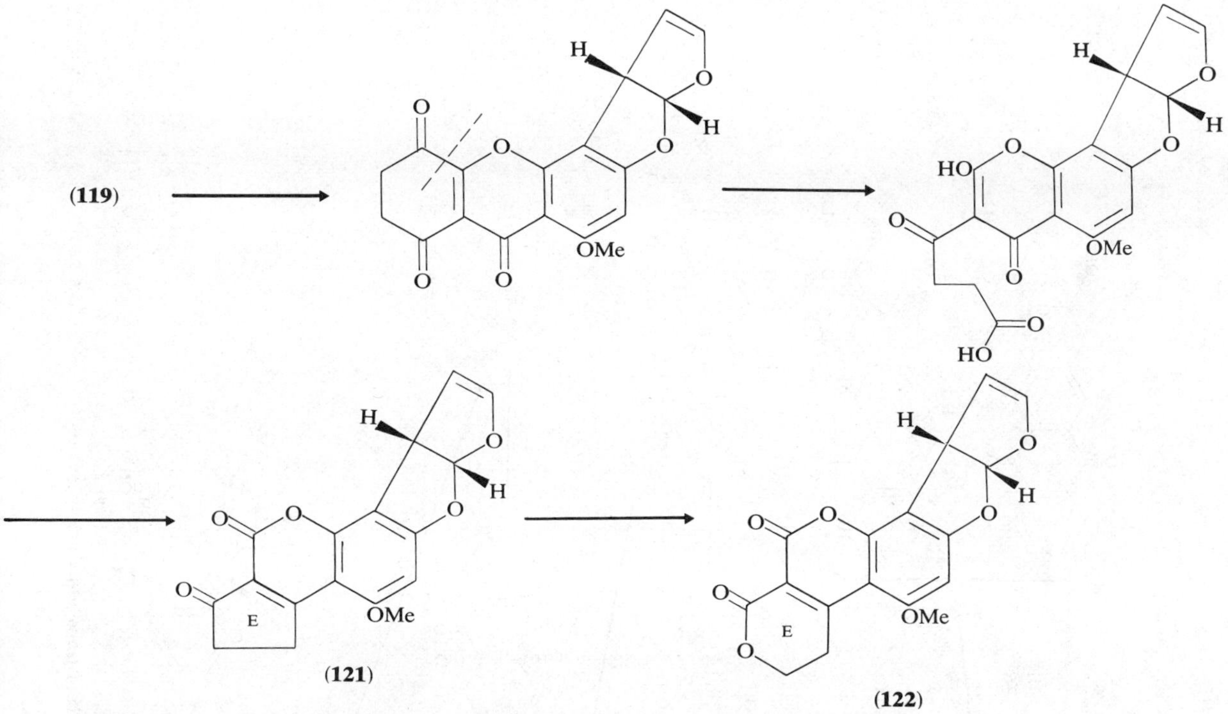

SCHEME 17 Aflatoxin biosynthesis from sterigmatocystins

can proceed. These transformations have been investigated in considerable detail,[73,74] and the 'central' pathway, to 7-chlorotetracycline, is known to be as illustrated in Scheme 18.

In the parent nonaketide precursor the starter unit is a malonamyl residue (see Section 29.1.1.3); the nonaketide is deoxygenated and *C*-methylated during assembly, giving 6-methylpretetramid (**123**). Mutants in which the process of assembly and cyclization is defective are known, and some of these accumulate tricyclic nonaketides such as (**124**) and (**125**). These products are entirely comparable with the products of 'incomplete' polyketide synthesis noted in describing simpler systems.

Extensive experiments, in which the products from differently blocked mutants were identified and tested for conversion into tetracyclines by other 'non-producing' strains, showed that (**123**) is next hydroxylated to give (**126**); it is presumably the corresponding quinone which then undergoes the stereospecific addition of water to give (**127**), which is known to be the substrate for the next step, electrophilic chlorination. The amino-group is apparently introduced reductively (and stereospecifically), followed by two *N*-methylations to give (**128**); two further stereospecific steps, of hydroxylation and, finally, reduction, lead to chlorotetracycline (**129**). Thus the specific stereochemistry of the eventual product is introduced by a stepwise series of transformations.

The parallel biosynthetic sequences mentioned above are of several different kinds. Some occur as the principal pathways in different strains of the producing streptomycetes. Altered specificity for the starter acyl group (*e.g.* acetoacetyl replacing malonyl) gives rise to a wholly independent series of 2-acetyl analogues, while altered specificity in the assembly complex which allows omission of the *C*-methylation step leads to a full series of 6-demethylated tetracyclines. The chlorination step can likewise be omitted without blocking the subsequent transformations, whereas introduction of an additional hydroxyl at C-5 is an optional extra step. Thus the very high degree of reaction specificity shown by the enzymes of tetracycline biosynthesis is combined with relatively loose substrate specificity, and in this respect the tetracycline system provides a well-studied example of what is, apparently, a not uncommon situation in natural product biosynthesis.

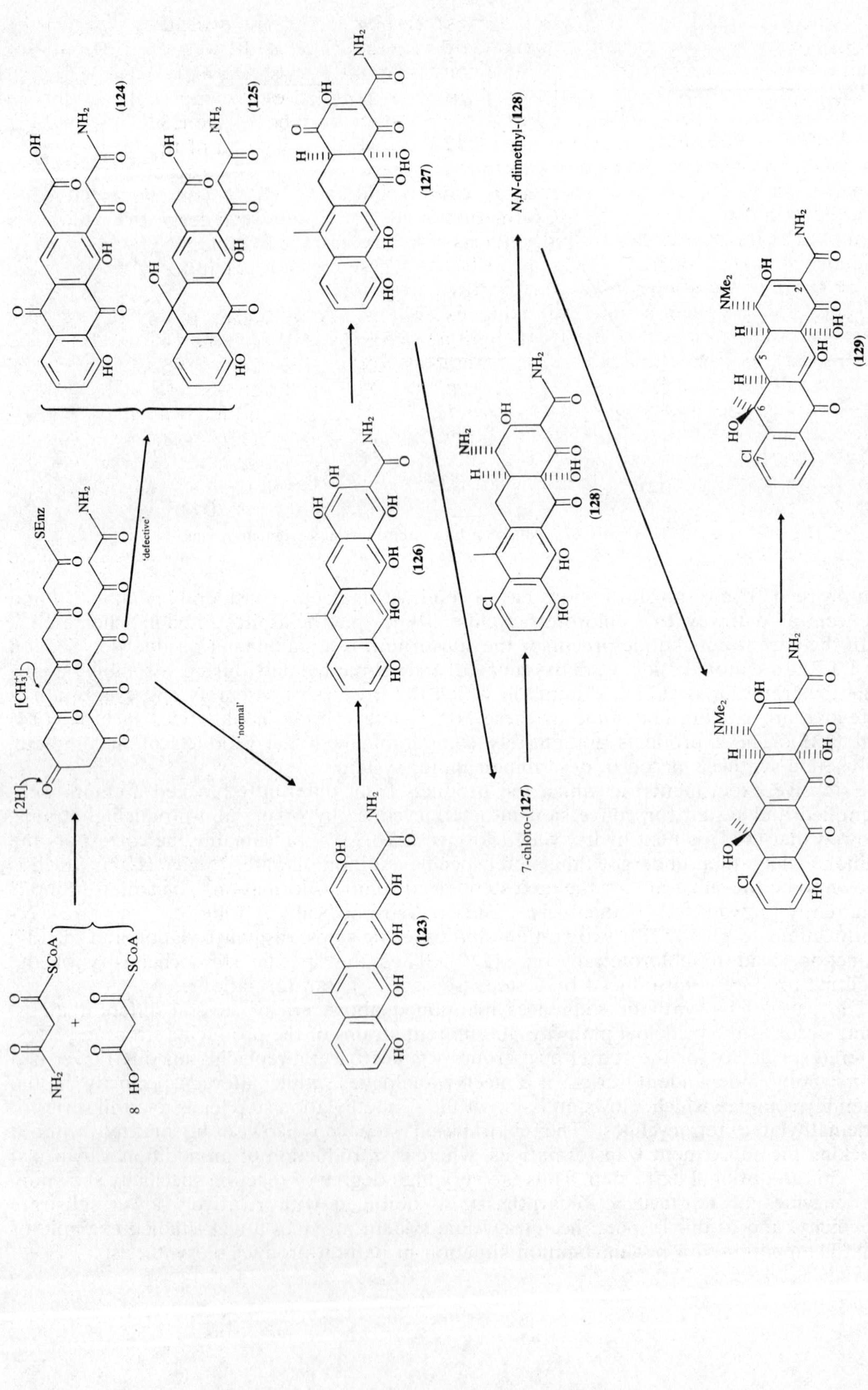

SCHEME 18 Biosynthesis of tetracyclines

29.1.4 PARTLY REDUCED POLYKETIDES

29.1.4.1 Types of partial reduction

As already noted, and as illustrated by several examples, the overall mechanisms of polyketide biosynthesis readily accommodate cases intermediate between the extremes of complete reduction of nearly all the acyl units, as in fatty acids, and little or no reduction at all, giving the alternately-oxygenated aromatic polyketides. Examples of assemblies which are partly of one kind and partly of the other have been given, for instance the decaketides of the aflatoxin series, (**110**)–(**122**), in which only seven of the acyl units are involved in aromatization, or auroglaucin (**87**), where only four of the seven units are not reduced. In either of these cases it is conceivable that the biosynthesis could have proceeded in two stages, *i.e.* by the formation of a C_6 or C_8 fatty acid derivative which was then used as the acyl starter for biosynthesis of an aromatic polyketide, but the balance of evidence is against such explanations, either for these examples or indeed for the much wider range of variations which are actually encountered.

There are, however, some interesting exceptions. A few higher plants such as the gingko and the cashew produce long-chain alkyl phenols and phenol acids, of the types (**130**)–(**133**), in which the R groups are unbranched odd-numbered C_{13} to C_{19} alkyl chains with up to three double bonds. Their biosynthesis by the polyketide route with common C_{14}–C_{20} fatty acids as starter units is *a priori* most plausible,[75] and indeed added labelled malonate is selectively incorporated into the aromatic part of (**130**).[76] Such a two-step biosynthesis is in several ways similar to the mechanisms found in animal systems for the conversion of common C_{16}–C_{18} fatty acids into more specialized C_{22}–C_{24} homologues (*cf.* Chapter 25.2).

(**130**) 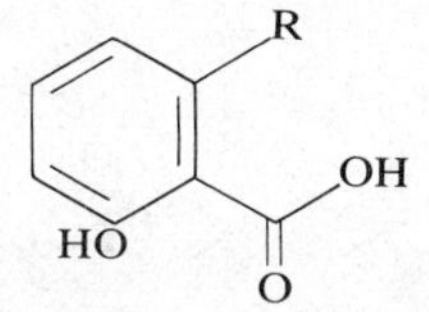

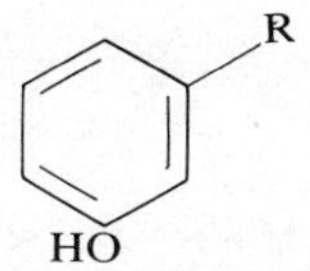 (**131**)

(**132**) 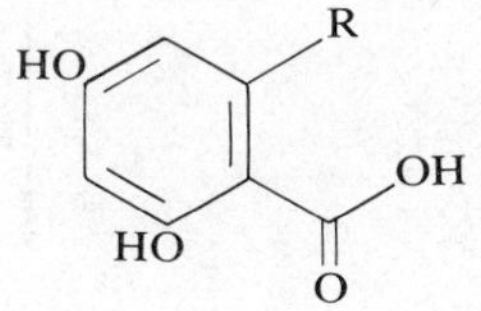

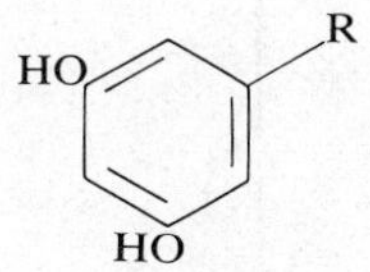 (**133**)

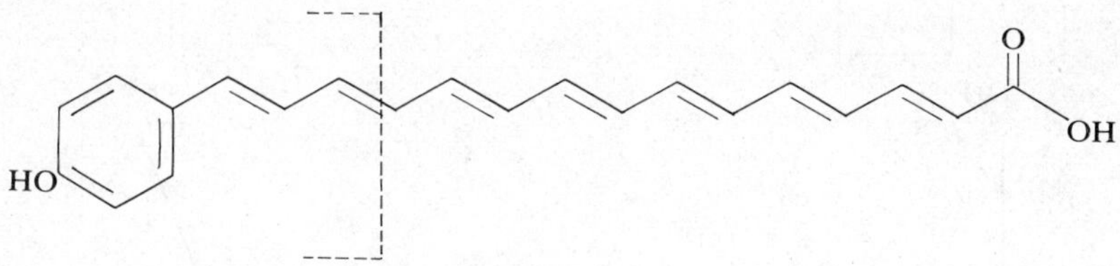

(**134**)

More generally, however, there are important groups of natural polyketides in which the normal assembly sequence appears to be acylation followed by just one reduction per acyl unit, thus giving either 1,3-polyol or conjugated polyene structures over substantial stretches of the assembly:

$$—[CHR—CO]_n— \rightarrow —[CHR—CHOH]_n— \rightarrow —[CR{=}CH]_n—$$

A simple example of this pattern is exemplified in the fungal pigment cortisalin (**134**), which is almost certainly formed by the condensation of *p*-coumaryl-CoA with six malonyl-CoA, accompanied by reduction and dehydration at each step. As with other polyketides, departures from a prevailing reduction pattern can be expected, so that the overall pattern of biosynthesis ends as one of some complexity. This is all the more true since the category of 'partly-reduced polyketides' thus loosely defined includes several major groups of natural products, particularly from actinomycetes, in which a variety of malonyl-CoA analogues are also selectively employed in the assembly sequence. Unfortunately, hardly any mechanistic details for the biosynthesis of these products are yet available, and our own subdivision of the examples given is necessarily somewhat arbitrary.

29.1.4.2 Some fungal products

Variations in the pattern of predominant reduction are well exemplified in erythroskyrine (**135**), which is built up from a decaketide assembly and an *N*-methylated valine residue.[77] Scheme 19 is an attempt to present its biosynthesis formally, but it must be emphasized that the order in which steps occur is really unknown. The parent decaketide is written as a β-ketoacyl derivative with eight residues reduced to the polyene/polyol level; this allows formal derivation of the furanofuran system from a bis-epoxide at one end of the chain (sequence *a*) and of the lactam from the β-ketoacyl end (sequence *b*). The second type of process is one which can be exemplified in several series of natural products, including the corresponding oxygen analogues, the enol-lactones and tetronic acids.[4]

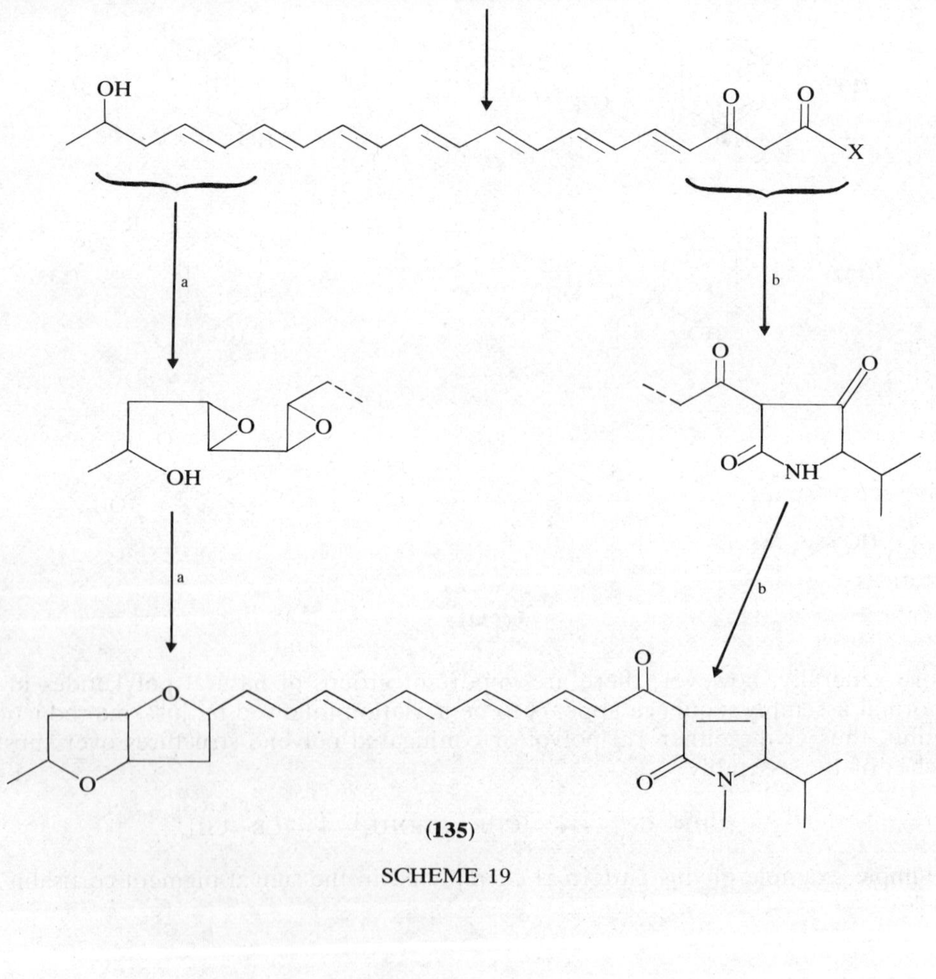

(135)

SCHEME 19

A variant, again in combination with a partly-reduced polyketide assembly, is found in bassianin (**136**). The structure and the biosynthesis of this compound were elucidated simultaneously and interdependently, mostly by ^{13}C and ^{15}N n.m.r. methods.[78] The parent assembly is a hexaketide, twice *C*-methylated, with one fully-reduced and three part-reduced units, and this is combined with a rearranged phenylalanine as shown (Scheme 20); again, the rearrangement is one which is found in other phenylalanine-derived plant and fungal products.

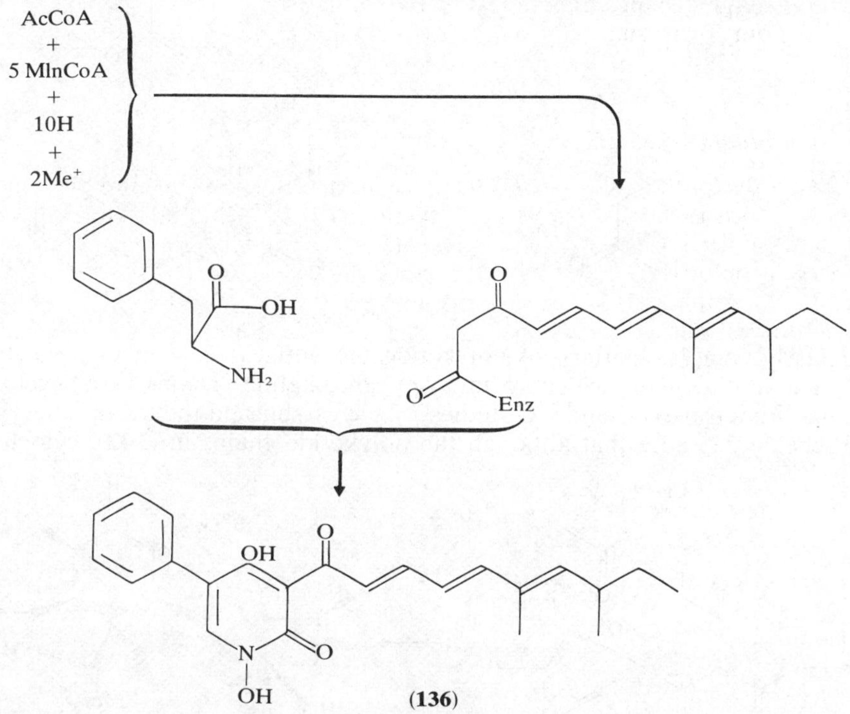

(**136**)

SCHEME 20

In other types of fungal polyketide the general level of reduction is closer to that of the fatty acids, but with sufficient departure from complete reduction to allow further structural elaboration. The antibiotic lactone brefeldin A (**137**) was once, erroneously, thought to derive from palmitic acid, but it now appears to be a specifically-assembled polyketide,[79] even to the extent that both lactone oxygens (*i.e.* those attached at C-1 and C-15) are derived from the oxygen of acetyl-CoA.[80] The oxygens at C-4 and C-7, however, are separately derived from two oxygen molecules. Significant ideas as to how the five-membered ring is produced, by linking two carbonyl-derived carbons, are lacking, but the formation of the macrocyclic lactone ring may well be a direct outcome of the assembly mechanism. It has been remarked that in the great majority of aromatic polyketides the 'methyl' and 'carboxyl' ends of the assembly are in close proximity in the final product, and the formation of macrocyclic lactones may still be a counterpart to this effect in cases where the polyketide chain is reduced and aromatization is limited or absent. The implication is that in both cases the assembly is built up as a growing loop between two more or less fixed sites, and that it can be split off by a reaction between the groups at these sites when the intervening 'template' region occupied by the loop is 'full'. Further macrocyclic lactones illustrating this feature in polyketides from fungi are the octa- and nona-ketides curvularin (**138**),[81] zearalenone (**139**),[82] and monorden (**140**). At first inspection the reduction levels of successive units along the assembly in these three structures are so various that the availability of three or four adjacent ketomethylenes for aromatization seems almost fortuitous, but in fact the assemblies corresponding to (**138**) and (**140**) could each have cyclized in a different way, so specific cyclization mechanisms are still implied.

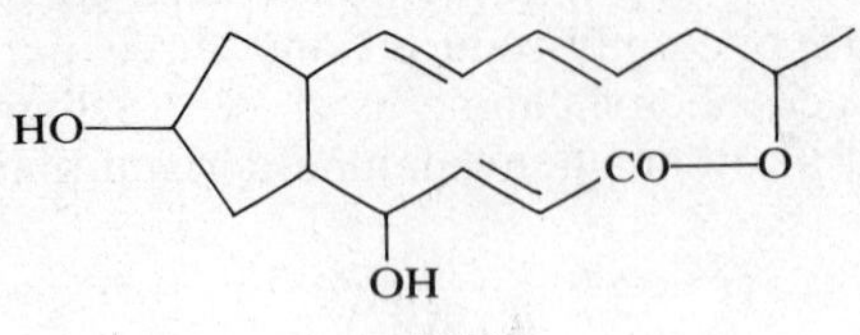

(**137**)

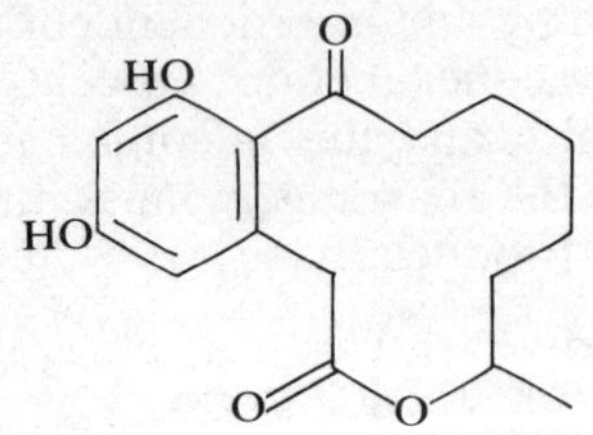

(**138**)

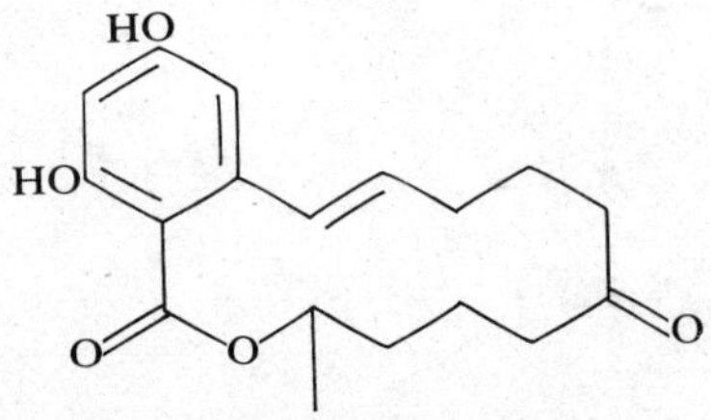

(**139**)

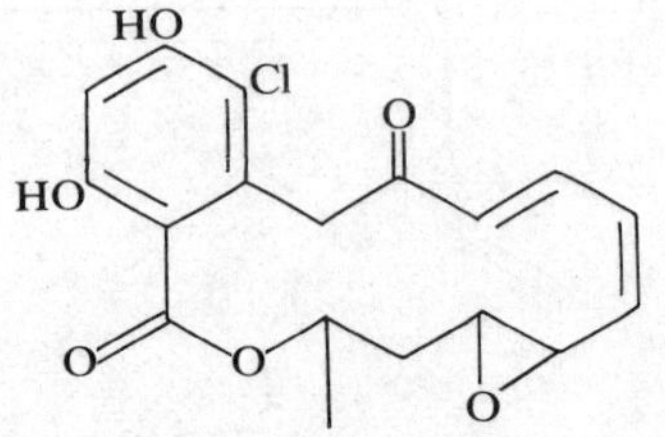

(**140**)

A particularly complex pattern of polyketide biosynthesis is seen in the cytochalasin series of metabolites, in which octa- or nona-ketide chains are cyclized with phenylalanine. Extensive ^{14}C and ^{13}C studies[83] have established the overall biosynthesis as shown in Scheme 21; note that although the polyketide chains in (**142**), cytochalasin D,

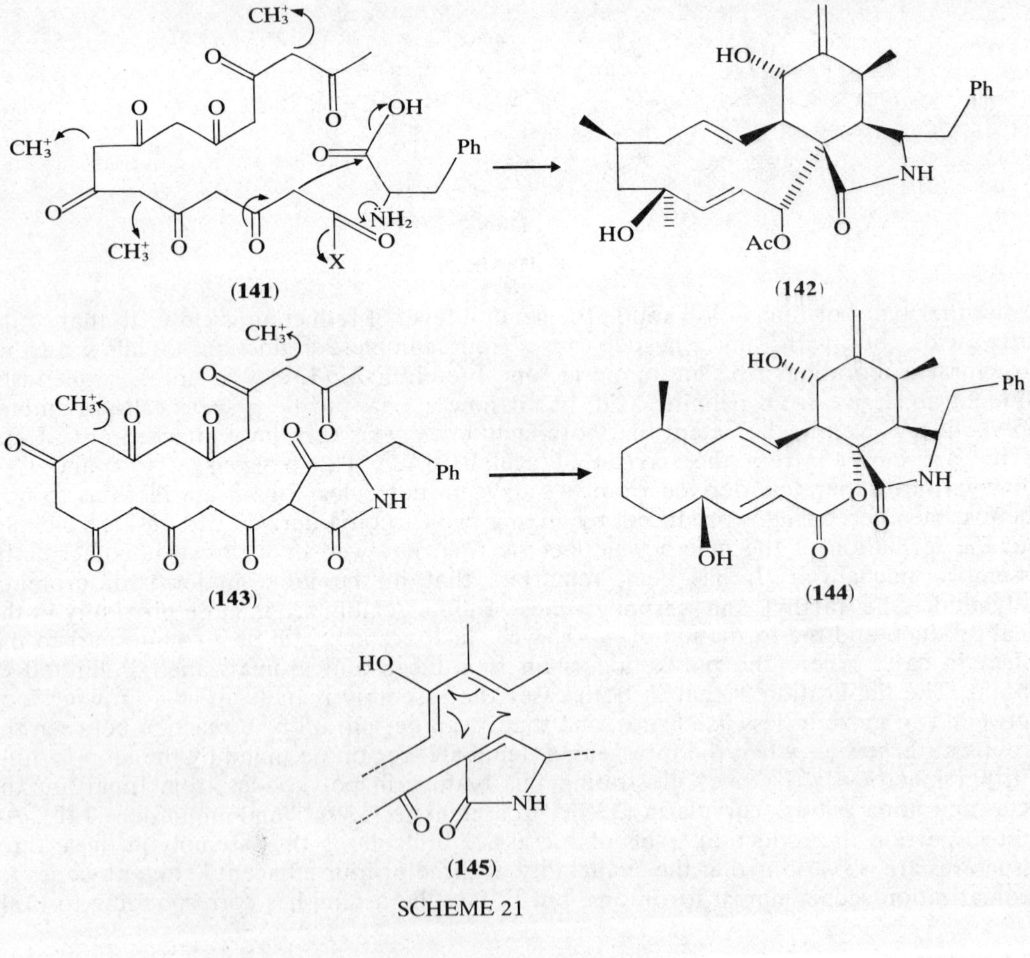

SCHEME 21

and (**144**), phomin, are of different lengths, the patterns of their further elaboration— oxidation level, *C*-methylation, cyclization—are quite similar at each end, so that the 'extra' C_2 unit in (**144**) appears, in effect, in mid-chain. Such comparisons, which are further exemplified below, provide glimpses of unknown mechanisms of high specificity. The initial attachment of phenylalanine is probably to a β-ketoacyl intermediate as shown in (**141**) and (**143**) and can be compared with the formation of (**135**) and (**136**) from the same amino-acid. It has been suggested[84] that the formation of the cyclohexane ring, which cannot be formulated in terms of normal mechanisms in the polyketide chain itself, proceeds by an electrocyclic process, perhaps as indicated in (**145**); as well as being economical, this hypothesis would also explain the stereochemistry that is generated in the final products. Finally, in the case of phomin (**144**) there is a clear implication that the lactone oxygen is inserted into a carbocyclic analogue of (**142**) by a Baeyer–Villiger type of reaction, at a late stage in the biosynthesis and (as expected) with retention of configuration.

29.1.4.3 Macrolides and related antibiotics from Streptomycetes

The fullest exploitation of the flexibility of polyketide biosynthesis seems to have been developed amongst the streptomycete bacteria, where it is employed in the production of a seemingly inexhaustible range of structures, all of which have been discovered in the course of searching for new antibiotics. These products illustrate in full variety the assembly of specific sequences of different acyl precursors, with different reduction levels selected following each assembly step, with specific foldings of the completed assembly, and with controlled further transformations. There is a marked tendency towards the formation of macrocyclic products along the lines discussed in the previous section.

Examples of 'acetate-derived' and largely aromatized polyketide antibiotics from Streptomycetes have already been given [*e.g.* (**108**) and (**129**)], but there is an even greater variety amongst the products where the *typical* reduction level is the intermediate polyene/polyol type (see Section 29.1.4.1).

Although in many cases the component acyl units have been identified experimentally by precursor-incorporation studies, virtually nothing is known about the mechanisms through which structural specificity is attained in the assembly of these precursors. For example, the aglycone of nystatin (**146**) is built up from 16 'acetate' units (presumably 15 from malonyl-CoA) and 3 'propionate' or methylmalonyl units [heavy lines in (**146**)],[85] arranged in the sequence:

$$(\text{acetate}) \rightarrow (\text{propionate})_2 \rightarrow (\text{acetate})_8 \rightarrow (\text{propionate}) \rightarrow (\text{acetate})_7$$

The reduction level is predominantly that of the polyene/polyol, but not exclusively so, and the polyene and polyol patterns are both present but in segregated sequences. Comparisons with other products bring out further remarkable features. For instance, the heptaene amphotericin B (**147**) is obviously very closely related to (**146**), but we have no inkling as to what mechanisms underly the fact that the differences between (**146**) and (**147**) are confined to the central regions of both the polyene and the polyol segments.

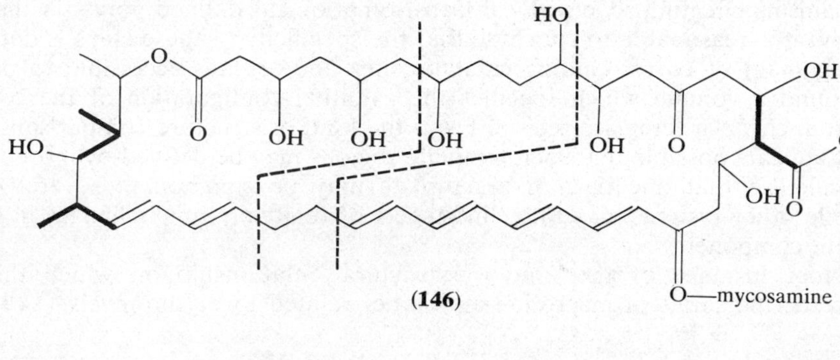

(**146**)

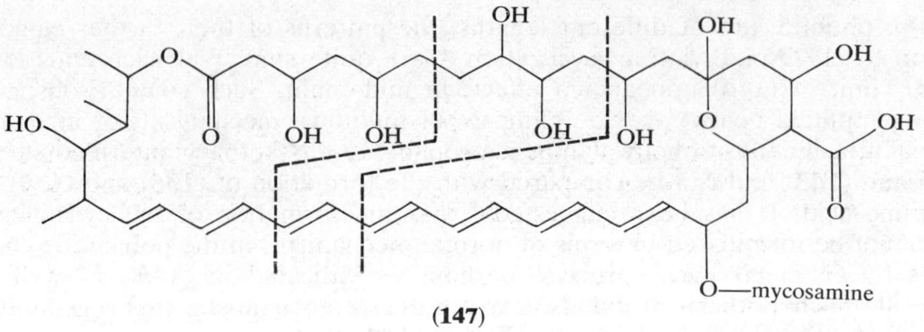

(**147**)

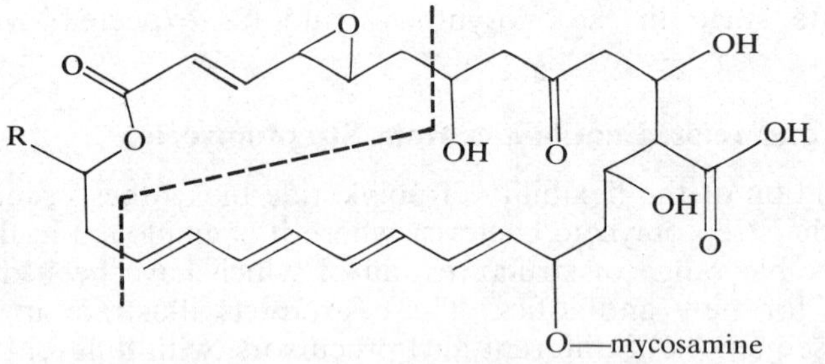

(**148**) pimaricin, R = Me; lucensomycin, R = Bu^n

That this is not a mere coincidence is emphasized by the fact that in the structures of two further polyene macrolides, pimaricin (**148**; R = Me) and lucensomycin (**148**; R = Bu^n), which are assembled from the quite different precursor sequence:

$$(\text{acetate or n-valerate}) \rightarrow (\text{acetate})_6 \rightarrow (\text{propionate}) \rightarrow (\text{acetate})_5$$

A large part of the eventual structure is the same as one of the elements common to (**146**) and (**147**), as indicated in the structures.

Similar patterns of regularity can be picked out in quite different groups of polyene and other macrolide antibiotics when their structures[86] are compared in these terms, and further examples will be given, but in the absence of direct experimental information on their biosynthetic basis only the most general of interpretations can be offered. In the biosynthesis of oligopeptide antibiotics in bacteria, a topic which has been more extensively investigated,[87] the basic assembly mechanism is very similar to that in polyketide synthesis in that it involves a multi-enzyme complex with acceptor sites, a carrier element, and a condensation mechanism based on the reactivity of thioesters. The oligopeptide synthetases show a fairly high, but not absolute, specificity for the substrate amino-acids and these are combined in specific sequences, which it is believed are established by the spatial arrangement of acceptor sites on the complex. Some such mechanism might equally explain the specification of precursor unit sequences in the more complex polyketides. In addition, however, the polyketide synthetases must accommodate additional mechanisms effecting specific local transformations at defined points in the overall assembly, and it is reasonable to conclude that the specificity of these steps is determined by an arrangement of corresponding catalytic sites and/or sites accessible to particular reagents around a volume which specifies the eventual configuration of the completed assembly product, *i.e.* a template region. From the kind of structure-comparisons already indicated, it appears possible that such template regions may be defined by more than one macromolecule, so that one part of a template may be common to several different systems, while other systems may differ in respect of relatively small changes in one part of a template component.

As a further instance of the kind of structural relationship on which the above discussion rests, the series of macrolide antibiotics related to erythromycin (**149**) can be illustrative. The aglycone of (**149**) is assembled from one propionyl-CoA and six methylmalonyl-CoA, and is thus wholly 'propionate-derived'.[88] The component units are indicated by heavy lines in the structure, and numbered (i)–(vii) from the starter end. The aglycone of narbomycin (**150**) is almost identical with (**149**) except that unit (iii) is here an 'acetate' residue. However, in tylomycin (**151**) the changes are more considerable. The third unit is replaced by two: (iiia) a 'propionate' and (iiib) an 'acetate' residue; unit (v) is now a butyrate-derived C_4 unit, and unit (vii) an 'acetate'. Nevertheless, the overall pattern of the molecule is retained to a marked extent. Conversely, methymycin (**152**) remains closely analogous to (**150**) although unit (vi) is omitted altogether. Again, platenomycin (**153**) is obviously very closely related to (**151**) even though it is assembled from an entirely different precursor sequence:

$$(\text{acetate})_4 \rightarrow (\text{propionate}) \rightarrow (\text{butyrate}) \rightarrow (\text{X}) \rightarrow (\text{acetate})$$

where X is an unidentified precursor of the 'methoxyacetate' unit.[89] As with the polyene macrolides, this list could be very considerably extended in terms of known structures,[86] and the close relationships between them extend to features such as stereochemistry and

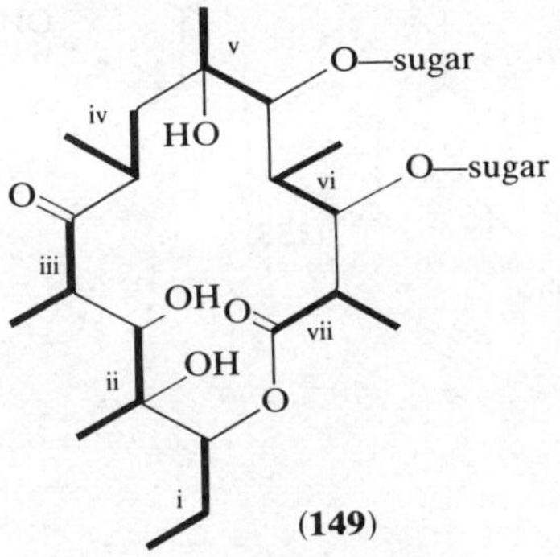

(**149**)

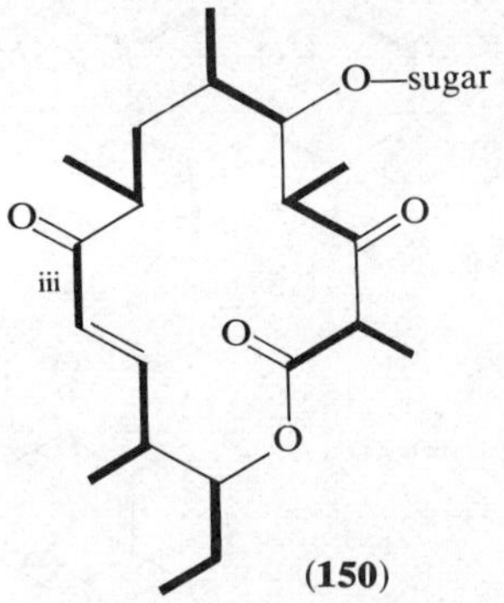

(**150**)

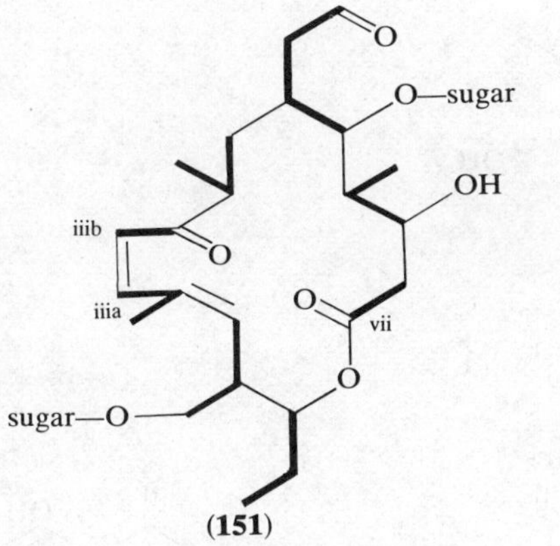

(**151**)

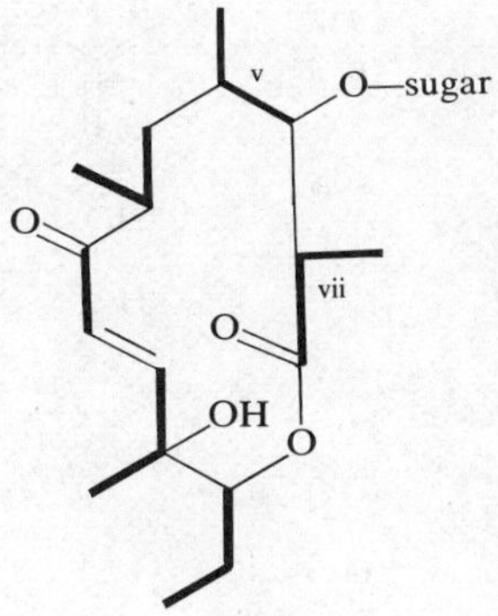

(**152**)

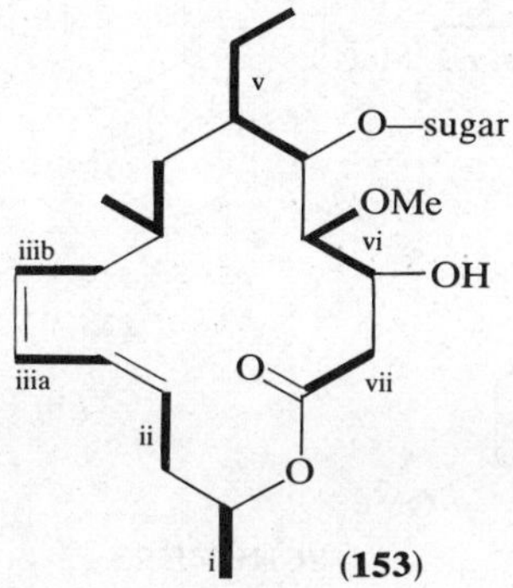

(**153**)

detailed glycosylation, not shown in (**149**)–(**153**). The extent to which the organisms producing these different products are biologically related, and implicitly the corresponding extent to which the macromolecules involved in their biosynthesis might be the same or closely similar, is obscure (and indeed controversial), but on chemical grounds alone we can see evidence for close similarities between what we have termed the 'template' functions, and their partial independence of the functions determining the precursor sequence.

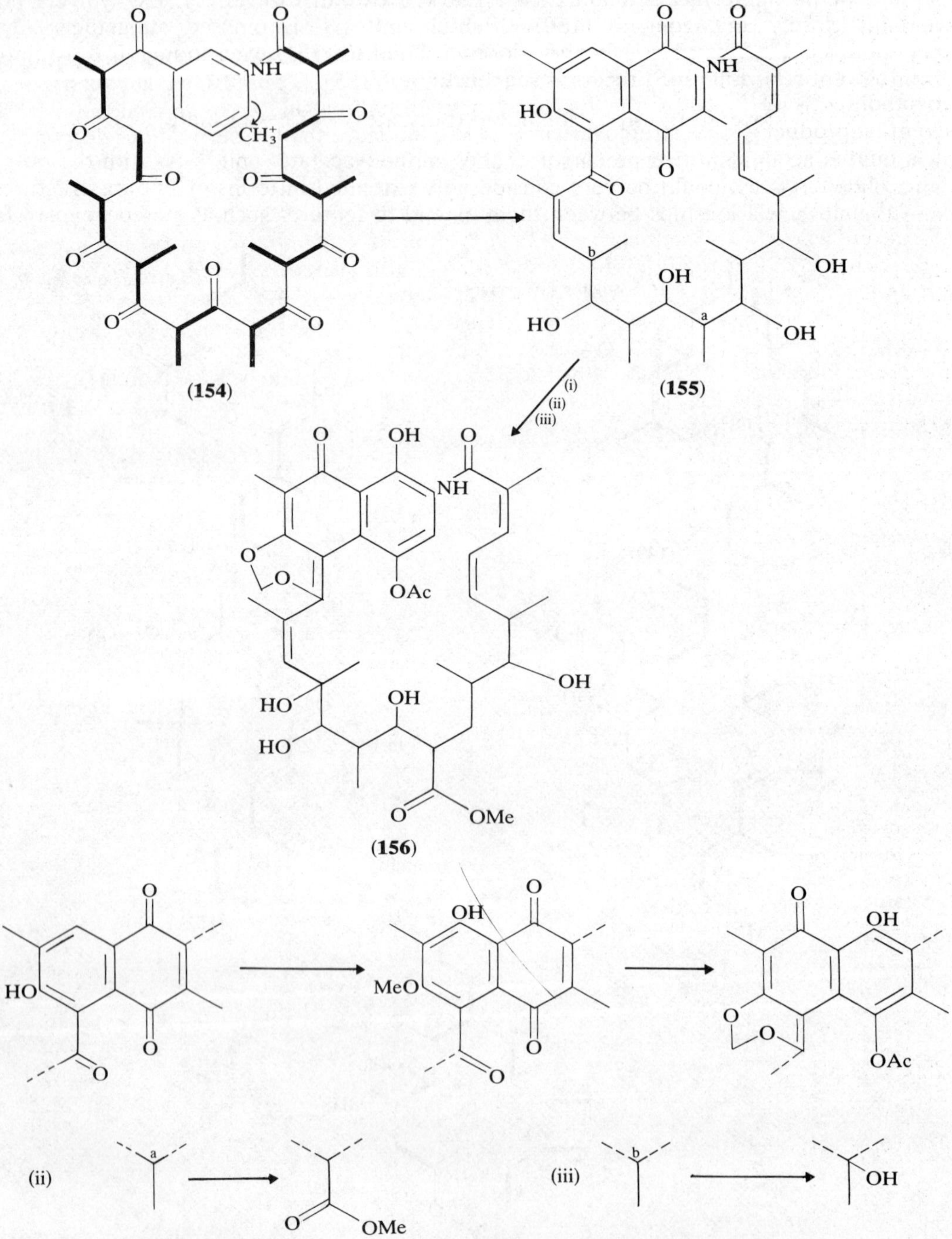

SCHEME 22

The 'ansa' group of actinomycete antibiotics, so-called because of their macrocyclic bridged-aromatic structures, are in several respects similar to other macrolides, and though (again) little is known about their basic assembly mechanism, rather more has been ascertained about their interrelationships.

The starter acyl species in these antibiotics has been identified by tracer experiments as a sugar-derived substance with the *m*-aminobenzoic skeleton,[90] as in (**154**), which in one series, the streptovaricins, is also *C*-methylated as shown. The incorporation of other ^{13}C-labelled precursors has been extensively studied and indicates the overall assembly shown in (**154**), with eight propionate-derived and two acetyl-derived units, the whole chain terminating by acylation of the amino-group of the starter.[91] The earliest product of this process so far identified is protostreptovaricin I (**155**). From this, a complex of further metabolites is formed, by the localized transformations indicated in Scheme 22. The eventual product here is streptovaricin C (**156**), but since the transformations do not occur in a unique sequence, a series of different 'intermediates' is generated.[92]

In the closely-related rifamycin series, which has also been extensively studied by similar methods, the sugar-derived starter unit lacks the *C*-methyl group but otherwise the parent assembly is identical with (**154**).[93] However, its transformations differ considerably (Scheme 23). Already in the earliest identified product, rifamycin W (**157**), a methyl group has been hydroxylated. Labelled (**157**) is demonstrably[94] a precursor of rifamycin S (**158**), by an oxygen insertion reaction which perhaps proceeds as shown [Scheme 23, reaction (i)] in which ring-opening of an intermediate epoxide leads to loss of an oxidized methyl group as CO_2. Minor transformations elsewhere in the chain, reaction (ii), are also involved. A further transformation, clearly by an oxidative ring contraction, converts (**158**) into the pyrone analogue (**159**).

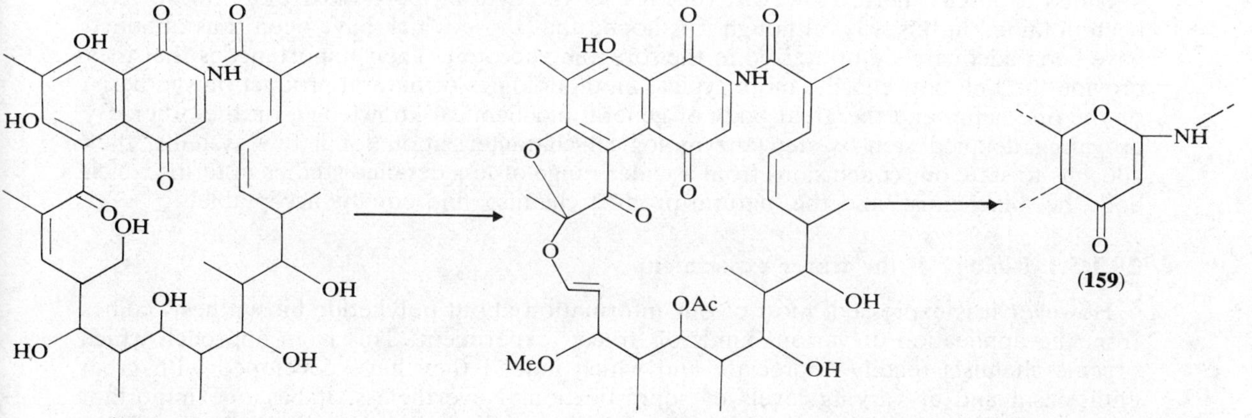

(**157**) *cf.* (**154**) (**158**) (**159**)

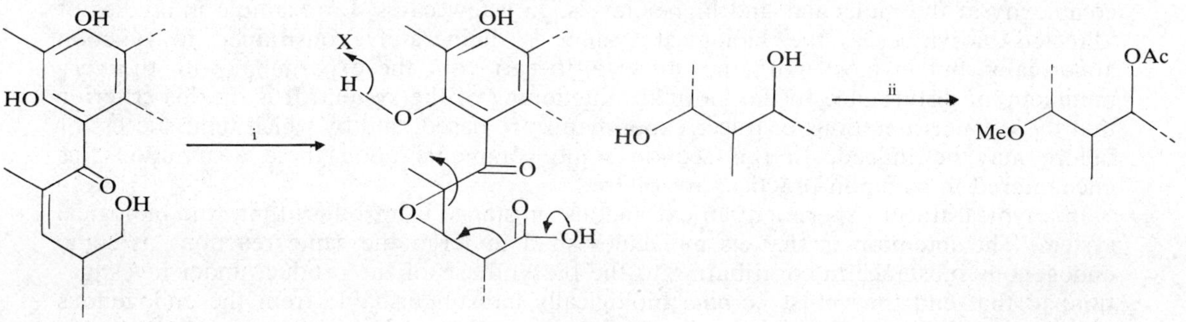

SCHEME 23

Geldanamycin (**161**) represents a third type of ansa-antibiotic. Investigation by ^{13}C methods[95] establishes the parent assembly as (**160**), which is clearly homologous with (**154**) although three units shorter.

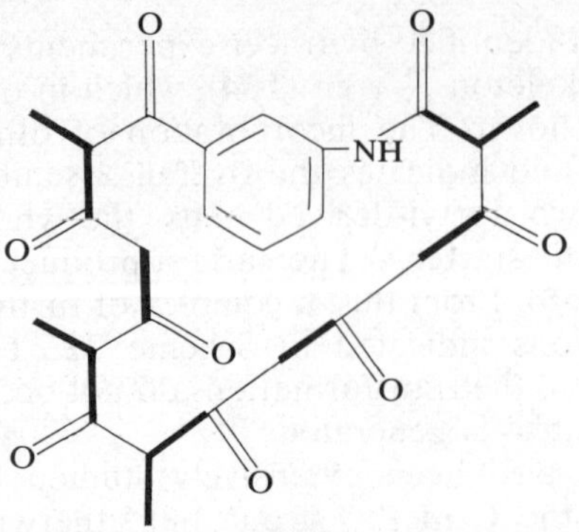

(**160**)

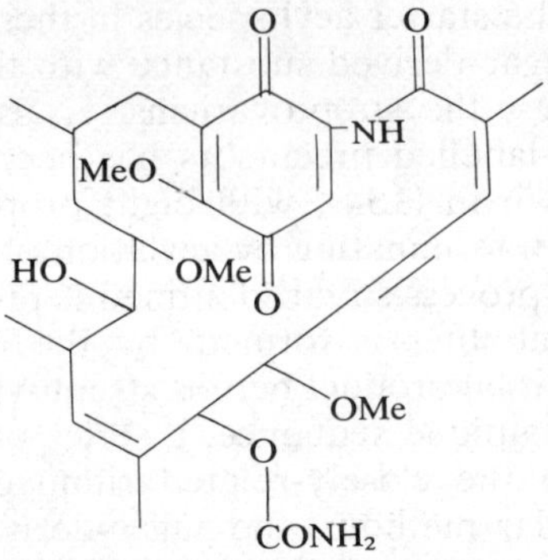

(**161**)

29.1.5 INVESTIGATIVE METHODS

Our knowledge of biosynthesis results from experimental studies; this is reflected both in its known and its unknown aspects. The classic method of studying chemical reactions in biological systems is by the methods of biochemical enzymology, which it is not intended to discuss here. However, very few of the data on polyketide biosynthesis have been obtained in this way, although it is hoped that those which have been thus obtained have been adequately emphasized in the foregoing account. Their importance is that they provide the link between the more typical methodologies of natural product biosynthesis, on the one hand, and the great body of general biochemical knowledge, on the other. By providing detailed step-by-step enzymological characterization for a few systems, they allow us to state our conclusions from a wider range of less detailed studies in terms which both the 'biochemist' and the 'natural product chemist' find equally acceptable.

29.1.5.1 Biology of the tracer experiment

However it is expressed, most of our information about polyketide biosynthesis comes from the application of various kinds of 'tracer' experiment. This is an approach which organic chemists readily appreciate and which indeed they have developed with great enthusiasm and at varying levels of sophistication. Nevertheless, it has one important feature which distinguishes it from most other branches of modern organic chemistry, namely that it is an *observational* science. In a biosynthetic experiment, the controlled variables can only be manipulated within limits which are imposed by the biological system under investigation — typically, a more or less intact organism of enormous complexity at the molecular and higher levels. In a few cases, for example in studies of 'directed biosynthesis', the biological system is deliberately constrained to function abnormally, but in most work the object is to carry out the experiment with the very minimum of disturbance to the 'normal' functioning of the system. It is on this criterion that the usual assumptions of tracer experiments are based, and by which their success or failure may be judged. In this section some comments upon these assumptions, as encountered in common practice, are offered.

In a typical tracer experiment an extraneous substance is introduced into the biological system. The intention is that its molecules shall undergo the same reactions as some endogenous substance in contributing to the biosynthesis of the product under investigation; to that end they must *become* biologically indistinguishable from the endogenous substrate, and this requirement applies both to the *nature* and to the *amount* of the tracer.

For example, a tracer quantity of sodium acetate is added to a culture of a mould; the acetate ion (or acetic acid) is to be taken up by the cells, without significantly disturbing the energy-linked transport mechanisms in the cell membranes, and within the cells is to be converted into acetyl-coenzyme A without significantly changing the concentrations of the substrates required for these steps (ATP, coenzyme A) or of the products (ADP, acetyl-coenzyme A). Finally, the acetyl-coenzyme A thus produced is to be completely mixed with the pools of acetyl-coenzyme A formed by several quite different routes within the cells so that it is representatively taken into the process of polyketide biosynthesis. On the other hand, at the conclusion of the experiment it must be possible by some appropriate technique to distinguish the tracer component from the endogenous contributions to the biosynthetic product, for example by measuring the level of radioactivity if the original exogenous acetate contained a proportion of ^{14}C or ^{3}H. In the last analysis, the two sets of requirements are mutually contradictory; it is only by extrapolation from the properties of the perturbed system to its unperturbed state, and to the extent that the observer's powers of discrimination are greater than the discriminatory properties of the biological system, that the experiment can be made to work.

Sometimes very crude molecular distinctions can be sufficient. For instance, in many but not all organisms, propionic or even isobutyric acid could be used to 'trace' the contribution of the starter acetyl-CoA to fatty acid synthesis, by giving a proportion of n-C_{2n+1} or iso-C_{2n} fatty acid in the resultant fatty acid mixture. However, this method would not 'trace' the contribution of acetyl-CoA to the remainder of the fatty acid molecules, because the enzyme converting acetyl-CoA into malonyl-CoA is more specific, *i.e.* too discriminatory. This latter situation is generally more common, so that the experimenter's discriminants must be more subtle; hence the development of the range of isotope-labelling techniques which is summarized in later sections. It is the biological criteria that we shall consider here; from this aspect, experimental failure can arise from three types of factors.

First is general experimental design. The added precursor must be able to enter the system more or less freely and to become metabolically equivalent to the endogenous substrate it is intended to label. For example, this is generally true for added acetate ion, but less strictly true for malonate ion (and frequently quite untrue for added mevalonate ion). The organism must of course be producing the desired product from the endogenous substrate during the experiment (and not, for example, from some accumulated later intermediate). The experiment must also allow a distinction between the 'direct' route of incorporation which is being tested and any unsuspected and often highly indirect routes. For example, the structures of many polyketides are such that by simple degradative steps the labelled polyketide may be quite a good source of specifically-labelled acetyl-CoA which can then be incorporated into some quite different product. Again, such diverse and superficially unrelated amino-acids as glycine, serine, and tryptophan can all be effective precursors of a *C*-methyl group; a quantitative comparison with labelled methionine would reveal the latter as a much better precursor, but only some knowledge of intermediary metabolism would indicate the correct interpretation of the results given by the other amino-acids. Correct biological design may also be advantageous in enabling the most economic or the most sensitive experiment to be carried out. As discussed in later sections, the different types of isotope labelling now in use give best results at different degrees of incorporation and this must be taken into account, for example by carrying out preliminary studies to optimize conditions for precursor incorporation. The quantitative kinetics of precursor incorporation can be exceedingly complex, and according to the degree to which they are understood they can present either an unsuspected difficulty, or an obstacle successfully avoided, or even an additional investigative method. The topic has been very adequately reviewed in simple terms,[1,96,97] while the rigorous uses of a detailed mathematical analysis, perhaps of less general applicability but of fundamental significance nonetheless, have been described.[98,99]

A second biological factor to be taken into account in the experimental design has to do with perturbations of the normal state of the system by the amount of added tracer. As explained below, experiments using radioactive tracers have the advantage of a discriminatory detector system that can be extremely sensitive, so that the absolute amounts of labelled precursor used can be correspondingly small. For most precursors they will also be small in relation to the metabolic generation of the corresponding endogenous substrate, so that the requirement for minimal quantitative disturbance can be at least approximately met. However, exceptions will arise on the one hand when the endogenous supply of the substrate is itself small, and on the other when the radioisotope content of the exogenous tracer is inconveniently low. Unfortunately, these two circumstances often coincide; when a rather complex precursor or presumed intermediate is being tested, its endogenous level may be very low indeed (*e.g.* if it is the product of a rate-limiting step), while the need to synthesize the more complex isotopically-labelled precursor may have tempted the chemist to eke out his yields with too much 'cold' material. With heavy-isotope labelling and the less sensitive detector systems (see Section 29.1.5.2) such a situation is much more likely, particularly since the emphasis then given to securing minimal *dilution* of the added label (as distinct from maximal *incorporation*) implies making the proportion of 'tracer' to endogenous substrate as *high* as possible.

The perturbations caused by the resultant over-feeding may be quite profound, because the normal biological system is closely regulated by far-reaching mechanisms with particularly important effects. In brief, the excess of a so-called tracer may be metabolized by pathways which are not normally operative, either because some pre-existing enzyme only acts on that substrate when its concentration is high, or because a new enzyme is produced in the system as a response. Equally the excess 'tracer' may inhibit the synthesis of, and/or promote competing pathways for the removal of, the endogenous substrate, through normal regulatory responses. Such effects may well repay exploration in their own right, and once a biosynthetic system is reasonably well understood they can be taken into account and even exploited, but in the general case they can be seriously misleading. They can be minimized by careful experimental design, notably by adding the precursor (and recovering the product) at truly optimal times, and if necessary by adding the precursor gradually throughout the whole experimental period instead of in a single dose.

A third type of difficulty arises when the isotopically labelled material is itself biologically distinguishable, *i.e.* when there is an 'isotope effect'. Fortunately, biological isotope effects have the same basis as, and obey the same rules as, the effects found in chemical systems, and their consideration should present the chemist with few serious problems. In particular, the effects are seldom significant, or even detectable, except with the hydrogen isotopes. Here it is essential to bear in mind that with radioisotopes under normal circumstances only a minority of the 'labelled positions' are actually occupied by the isotope. For example, in a sample of [1-^{14}C,2-^{3}H]acetate, most of the molecules will contain neither isotope, hardly any will contain both, and virtually none will contain more than one tritium atom; if the sample is converted, whether chemically or biologically, into (say) a product $CHCl_2COR$, we would *not* expect two-thirds of the tritium content to be lost, and the precise result to be expected would depend on detailed mechanistic considerations. The case is quite different when all the possible positions are actually occupied by isotopic atoms, a situation which usually obtains with heavy isotopes, *e.g.* with a sample of [2-^{2}H$_3$]acetate. Thus in order to determine how many hydrogens are transferred with the carbon in a *C*-methylation it is standard practice to use [methyl-^{2}H$_3$]methionine (which would lead to an eventual analysis by mass spectrometry). The case is also different for stereospecific labelling, for example with a proportion of ^{3}H in a prochiral —CH_2— group, a method which has been widely used to explore the stereochemistry of biosynthetic processes. In both cases, however, it is still essential to remember that the labelled species may not react at the same *rates* as the unlabelled, and to interpret results with corresponding caution; in general, an experiment that gives a 'yes/no' answer is preferable to one which can only be interpreted in terms of quantitatively uncertain isotope effects.

Finally we must note that the eventual interpretation of the results of tracer experiments will be expressed in terms of the definition of a metabolic pathway. Nearly always the definition is incomplete and the result is an 'overall' scheme in which details of mechanism and even the sequence of steps are entirely lacking. In biological fact there may or may not be a unique sequence; the degree of definition that can be attained by strict logic is usually quite limited and in nearly all cases it must be eked out by reference to a large body of similar data and to a small group of well-explored examples. Because biological systems are built along certain common lines, which at the biochemical level are increasingly well understood, this extension of the conclusions is really a matter of biological 'common sense'.

29.1.5.2 Tracer methods: radioisotopes

Until quite recently, virtually all the experimental data on polyketide biosynthesis were obtained by the use of radioisotope tracers,[97] sometimes using ^{3}H but most commonly with ^{14}C. This is, of course, paralleled by the very general utility of radiotracers in biochemistry, which has been very thoroughly monographed.[100–103] Both isotopes are soft β-emitters with half-lives long enough to permit the commercial distribution of labelled compounds and to avoid the need to correct for decay during the duration of an experiment (^{3}H has the shorter half-life, but it is radiation-induced chemical decomposition which limits the shelf-life of the labelled compounds). With modern equipment, detection of both isotopes is convenient, quantitatively accurate (weighing the sample is often less accurate than counting its radioactivity), and sensitive — products with only a few hundred-thousandths of the activity fed are regularly and usefully recovered. The same sample can be used for simultaneous ^{3}H and ^{14}C measurements, making the powerful technique of double labelling especially convenient; less accurate counting methods can be applied directly to a variety of chromatographic separations and provide a valuable auxiliary method of working.

Results are expressed in several ways. Most procedures give the total number of counts (per unit time) from a sample, which is multiplied by the 'efficiency' of the counter to give *total radioactivity*, *e.g.* in disintegrations per minute. Divided by the sample weight, this gives the *specific activity*; when required, for example in degradative sequences, the molar specific activity can be divided by the number of labelled positions in the molecule. The results of an incorporation experiment can then be expressed either as *incorporation* (total activity in product as a fraction of the total activity fed) or as *dilution* (specific activity of precursor/specific activity of product) or as *specific incorporation* (1/dilution). The specific activity method is often preferred when for experimental reasons the actual recovery of products is low. All methods of expressing the result contain — or conceal — a number of complexities concerning the pool size of endogenous precursors and intermediates, and the rate of the process being studied relative to general metabolic rates.[96–98] In feedings of labelled acetate, incorporation into typical polyketides is usually 1–10% in micro-organisms and less by one or two orders of magnitude in plants, while the dilution is extremely variable ($\times 100$–$\times 10\,000$). When more complex substrates are being fed, dilution is perhaps a more significant measure and is normally low ($\times 1$–$\times 100$). Note that very low dilutions indicate that there is a substantial departure from 'tracer' conditions, as discussed in Section 29.1.5.1.

Isotope effects are insignificant with ^{14}C and labelling is seldom lost by exchange reactions; a wide range of labelled substances is available commercially, either for direct use or as starting materials for synthesis. However, the synthesis of labelled compounds with specific activities high enough to be used as tracers in biosynthetic work will often call for special skills,[104,105] to minimize losses from handling small quantities and to avoid unacceptable dilution. Biosynthetic incorporation of ^{14}C from a simple precursor has often been used to produce a more complex labelled metabolite for subsequent re-incorporation. However, this procedure inevitably means that the tracer is diluted twice, and losses are also bound to be large. In addition, there is the danger that many natural polyketides labelled in such a manner are quite liable to afford significant yields of specifically-labelled acetate by biological degradation. The detection of radioactivity is so sensitive that indirect incorporation by such a route can be mistaken for direct incorporation if internal checks (such as testing for simultaneous incorporation into some quite unrelated product, such as a fatty acid) are not carried out. The results of such experiments must therefore be very unambiguous if they are to be at all acceptable.

The situation with ^3H is somewhat different. As already noted, isotope effects here may be very significant, and in addition many hydrogen atoms in organic molecules will exchange, more or less rapidly, with water. The possibilities of chemical exchange, as a way of introducing label as well as of losing it, are usually obvious and indeed when label thus introduced, perhaps under rather drastic conditions, is subsequently stable, this offers an important method for preparing ^3H-labelled compounds; on the other hand, there are several quite common enzyme-catalysed processes which lead to hydrogen exchanges *in vivo* and which might not be expected under mild chemical conditions. The range of ^3H-labelled compounds available commercially is less wide than for ^{14}C, but some—including ^3H_2O and $^3H_2$ for exchange reactions—are available at very high specific activity. Again this is convenient, but in practice it can also be misleading because indirect incorporation is so easily detected; an additional problem is that at high specific activities of ^3H the radiation decomposition of a stored substance can be considerable and the identity of what has eventually been fed may be correspondingly dubious.

One of the most fruitful applications of ^3H is in double-labelling, usually with ^{14}C (since a common detector system can then be used). This method can be used to test details of a proposed biosynthetic mechanism, particularly stereochemical details (where stereospecifically ^3H-labelled substrates can be prepared)[106,107] and also to test for the 'intact' incorporation of a complex intermediate which might alternatively be incorporated by an indirect route (in which the proportions of ^3H and ^{14}C originally present would almost certainly be very considerably altered). Although not without pitfalls, this method circumvents many of the uncertainties already alluded to. At high specific activities (about 1 millicurie of the ^3H compound is required), ^3H can also be detected—and located within the molecule—by n.m.r. methods; the technique has been elegantly applied in biosynthetic studies,[108] but its usefulness compared with alternative techniques seems rather limited.

In some experiments, the bare information that label from a [^{14}C]- or [^3H]-labelled precursor is incorporated into a product with such-and-such an efficiency is a sufficient conclusion in itself. More generally, however, it is desirable, or even essential, to locate the distribution of labelling in the product. For example, location of the label from [1- or 2-^{14}C]acetate at alternate positions in a molecule has been the classic demonstration of the entire polyketide concept.[2,11] It is the comparative difficulty and inconvenience of this requirement which most seriously limit the usefulness of the radioisotope method. The location of label at specific positions in a molecule requires its degradation, by chemical reactions whose specificity is assured and whose yield is high, to progressively smaller components, and ideally, to one-carbon products (such as CO_2 or methylamine).[109] Purity of reactants and products is essential. The reactions are necessarily carried out sequentially, and even in skilled hands the amount of material required (to say nothing of patience) is often considerably more than a biosynthetic experiment conveniently affords.

A typical sequence from the early work, carried out by Birch and co-workers on griseofulvin labelled from [1-^{14}C]acetate,[110] is shown in Scheme 24; the labelled carbons are marked, and the figures are molar specific activities (on an un-calibrated scale). Typically, this particular sequence falls short of complete degradation to specific one-carbon products, but is quite sufficient to show the alternate labelling pattern and also to show that the labelling is uniform to a first approximation. The need to simplify sequences of this kind, or to abandon them half-way when the material 'runs out', is obvious, and yet short-cuts can be misleading. The chemical reactions used may have unsuspected features which could cause problems; to cite only one example, Kuhn–Roth oxidation offers a useful route to the isolation of specific carbon atoms as acetic acid, but in a molecule containing several *C*-methyl groups the acetic acid—normally produced in less than

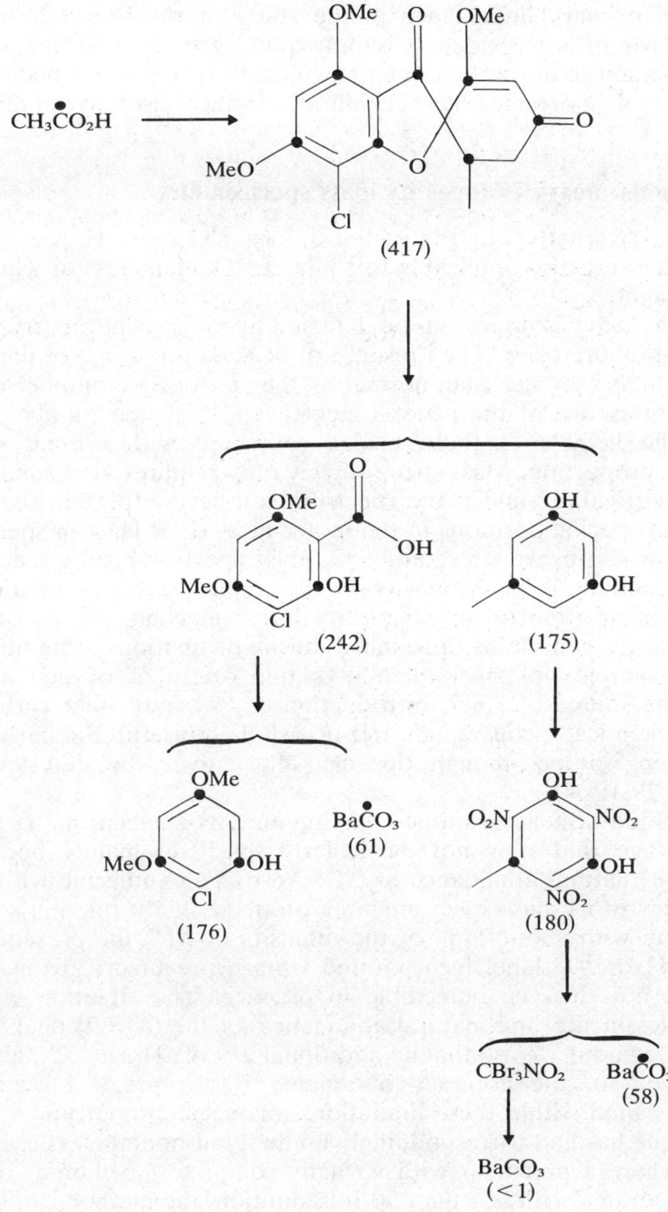

SCHEME 24

theoretical yield — often 'represents' one *C*-methyl group more than the rest, while the CO_2 which is simultaneously formed may be quite unrepresentative of the remaining carbon atoms from which, in principle, it derives.

In recent years this problem has become even more acute; many new natural products have only been obtained in quite small amounts, sufficient for their characterization by spectroscopic means alone, and often in a rather impure condition. As a result, even if the labelled product were then made available in sufficient quantity for degradations to be carried out, the necessary preliminary knowledge of the chemistry of the product is often lacking. Then only the most rudimentary degradative schemes can be safely attempted. Nevertheless, though noting these difficulties we must also firmly acknowledge that the method of radioisotope tracer incorporation, followed by chemical degradation for isotope location, has been of primary importance in the study of the biosynthesis of natural products and particularly of polyketides. It is only quite recently, with the availability of specialized instrumentation in more than one or two centres, that it has been relegated to second place in favour of 'more modern' techniques — which also have their drawbacks, as discussed below.

29.1.5.3 Tracer methods: heavy isotopes by mass spectrometry

The heavy isotopes alternative to the radioisotopes ^{14}C and ^{3}H are ^{13}C and ^{2}H; heavy-isotope techniques are also applicable to ^{15}N and ^{18}O, elements for which no useful radioisotopes are available.

In principle, all these heavy isotopes can be detected by mass spectrometry, a technique now accessible to most laboratories. The presence of peaks, both in molecular ions and in fragments, at mass numbers greater than normal by the appropriate number of units, will be diagnostic for the presence of the isotopic nuclei, and for their number, and within instrumental limits the heights of those peaks, compared with normal spectra, will measure their relative proportion. Mass spectrometry only requires very small samples of the product under investigation, and if the spectrum can be interpreted, useful tests for locating the isotopes at specific positions in the molecule — or at least in specific regions, which may be sufficient — are available. For ^{18}O, mass spectrometry is indeed the only useful investigative technique. It has been successfully employed in a number of cases and for specific purposes, most often to investigate oxidation mechanisms, by observing the number and location of oxygen atoms introduced during incubations in an atmosphere of $^{18}O_2$, but also, for example, to check the polyketide origin of oxygen atoms, using [$^{18}O_2$]acetic acid as the source of label. Introduction of ^{18}O into other carboxyl groups (from $C^{18}O_2$) is also chemically convenient and occasionally useful. Similarly, ^{15}N is the only available nitrogen isotope, though this can alternatively be detected by n.m.r. methods (see Section 29.1.5.4).

However, the technique of heavy-isotope labelling and mass spectrometry suffers from the general disadvantage that it is not particularly sensitive, mainly because of the relatively high (*ca.* 1%) natural abundance of ^{13}C. An organic molecule with ten carbon atoms will already show in its mass spectrum an 'isotope peak', at one mass unit greater than the molecular ion, with about 11% of the intensity of M^+; the presence of 2% of molecules with one ^{2}H or ^{13}C label incorporated from a precursor, giving an $(M+1)^+$ peak of 13%, would not then be detectable in practice. The situation with multiple labelling is of course less acute: the 'natural abundance' of the $(M+2)$ peak in the same spectrum would only be about 1%, so that an additional 2% of $^{2}H_2$ or $^{13}C_2$ label would be readily observed. Even so, the accuracy obtainable is not great. Provided that an experiment can be designed within these limitations, mass spectrometry is a very convenient tool; the technique has had correspondingly limited, but nonetheless useful, applications. For example, where a precursor with virtually complete labelling at one or more positions can be incorporated with less than 50-fold dilution, the method can be used with sufficient accuracy to show quite unambiguously how many labelled atoms have been incorporated into the product. This is a particularly appropriate technique for use with ^{2}H, where the requirement for complete labelling in the precursor is often attainable and may in any case be desirable to avoid isotope effects; the widespread use of [*methyl*-$^{2}H_3$]methionine to study *C*-methylation processes is a clear example of this situation.

29.1.5.4 Tracer methods: heavy isotopes by n.m.r.

The technique of nuclear magnetic resonance spectroscopy can in principle be used to detect, locate, and quantify ^{2}H (and ^{3}H), ^{13}C, and ^{15}N, and also linked labels such as ^{13}C–^{13}C, ^{13}C–^{15}N, *etc.* The observations can be made without chemical manipulations of the labelled product, and even without isolating it in a particularly pure form, and they can be interpreted in terms of basic data obtained from unlabelled material by the same techniques. With increasing availability of the requisite instrumentation — particularly for ^{13}C — the method is being increasingly applied. Not only can it replace the radioisotope method, it can also yield results of a kind not obtainable by any other means. Consequently its advent has led to an advance in biosynthetic studies second only to that which followed the introduction of the ^{14}C method. Nevertheless, its use is not free from limitations and disadvantages.

Some of the first biosynthetic work with ^{13}C was carried out by observing the 'satellite bands' due to ^{13}C–^{1}H couplings (~100 Hz) in the ^{1}H n.m.r., but these can only be observed in favourable spectra (and not at all for fully-substituted carbon atoms); direct ^{13}C n.m.r. is obviously preferred, given the availability of Fourier-transform spectrometers fully capable of measuring ^{13}C spectra at natural abundance. This is the basic technique for ^{13}C biosynthetic studies, which have been very adequately reviewed.[111,112] For biosynthetic work, it is an essential pre-requisite that the natural-abundance spectrum can be assigned;[113–115] although obvious, this requirement is neither insignificant nor trivial, and the work of Steyn *et al.* on the aflatoxin problem[70] illustrates its importance particularly well. However, since the necessity of assigning the spectrum replaces the even more tedious task of chemical degradation to locate labelling, chemists will undertake the task cheerfully!

In the simplest type of ^{13}C-labelling experiment, equivalent in design to conventional ^{14}C-labelling work, the appreciable natural abundance of ^{13}C is a very significant factor. In the first place, of course, it is this which has brought the whole ^{13}C n.m.r. technique into use and which makes assignment of the spectra possible. However, it also limits the sensitivity of the n.m.r. method as a detector for label incorporation. Supposing a precursor carrying 100% of ^{13}C at a particular position is fed, corresponding signals in the ^{13}C n.m.r. of the product will be more intense than at natural abundance, which is *ca.* 1%. For the effect to be unambiguously detectable, the maximum permissible dilution of the added precursor is only about ×100, and for it to be at all accurately measured it should be appreciably less. Moreover, such an experiment ideally requires the quantitative comparison of the incorporation into several labelled positions, and here the Fourier-transform n.m.r. method introduces its own problems. First, the signal intensities in a ^{13}C spectrum are markedly affected by relaxation effects which differ from one carbon atom to another; secondly, they are seldom very reproducible from one spectrum to another. These problems can be overcome,[70] by using paramagnetic 'relaxation reagents' (such as tris-acetylacetonato-Cr^{III}),[116,117] special decoupling modes,[118] and an increased number of data points for the Fourier transform.[119] Additionally, wholly non-labelled signals (if necessary, from groups introduced chemically, *e.g.* by acetylation) can be used to calibrate one spectrum against another. Unless such measures are taken, the quantitative significance of ^{13}C incorporation studies by n.m.r. is highly suspect. Even when they are applied, the inherent insensitivity of the method remains. For simple polyketide synthesis in micro-organisms the sensitivity is usually adequate when a precursor like 95%-enriched [^{13}C]acetate is used, except when (as sometimes happens) the yield of the metabolite is low compared with the metabolic turnover of its precursors. In such circumstances there is a natural tendency to add larger quantities of precursor, and the resulting perturbations of normal biochemistry may be very apparent (see Section 29.1.5.1). In other systems, *e.g.* in higher plants, the requirement for rather low dilutions effectively restricts the simple ^{13}C method to the study of more complex precursors, whose metabolic dilution is normally much less; the same cautions apply.

The detection of ^{15}N by direct n.m.r. methods involves similar considerations, except that fewer signals are usually involved and the natural abundance is somewhat lower, and presents no additional problems where the appropriate equipment is available. In principle, ^{2}H can be observed in the same manner, by working in the appropriate frequency range and with selective or noise proton decoupling. However, both ^{15}N and ^{2}H are more sensitively, and more usefully, detected in combination with ^{13}C.

Whereas the use of single ^{13}C labelling simply provides a convenient — but not unlimited — alternative to the use of ^{14}C in biosynthetic studies, the introduction of

$^{13}C_2$-labelled substrates, particularly (but by no means exclusively) [$^{13}C_2$]acetate, introduces a new dimension, by providing a type of information not previously attainable; an early and particularly striking application was the simultaneous elucidation of both the structure and the outline biosynthesis of the mould metabolites tenellin and bassianin (**136**) (see Section 29.1.4.2) by Vining *et al.*,[78] using both single- and double-^{13}C (and also ^{13}C–^{15}N) methods. The $^{13}C_2$ method depends upon the observation of ^{13}C–^{13}C couplings in the n.m.r. spectrum. At natural abundance these are not seen, since the natural abundance of *pairs* of ^{13}C atoms is 1% of 1%, *i.e.* 0.01% only. Consequently if pairs of ^{13}C atoms are incorporated from a precursor, their presence is evidenced by the appearance of doublet signals, spaced equally on either side of the natural-abundance singlet in the proton-decoupled ^{13}C n.m.r. With adequate instrumentation and spectroscopic technique, the method is considerably more sensitive than single-labelling (so that its use for studies of biosynthesis in higher plants becomes feasible); additionally, by providing information about the fate of individual pairs of atoms during biosynthesis it yields unique data, particularly in relation to cleavage and rearrangement processes. Several examples of this have been given in previous sections, especially 29.1.3.3 and 29.1.3.4. By feeding the correct precursor, information can also be obtained about 'pairs' formed during a biosynthesis by bonding between originally non-adjacent atoms. An extension of the method to include ^{13}C–^{15}N and ^{13}C–^{2}H 'pairs' is entirely feasible, though additional spectrometer facilities may be needed.

The main drawback of the $^{13}C_2$ method is the need to secure a correct degree of dilution in the labelled product. If the dilution is too great, the 'satellite' signals will be correspondingly weak compared with the natural abundance, necessitating long accumulation-times for the spectra and becoming indistinguishable from background noise and impurities, particularly in crowded regions of the ^{13}C spectrum. On the other hand, if the dilution effect is not great enough, so that the total ^{13}C content of the product is fairly high, further splittings due to adjacent ^{13}C atoms not derived from a $^{13}C_2$ 'pair' will become apparent. This will occur either when the total dilution is low or, more insidiously, when a significant *proportion* of the product is formed rather rapidly from the added precursor, and hence is at low dilution, and the rest formed at much greater dilution from endogenous precursor. Since other techniques, such as the ^{14}C incorporation measurements often made in preliminary investigations, measure only the overall dilution they do not provide a warning of this situation. The resultant complexities in the ^{13}C n.m.r. data can be handled in various ways, for example by combining selective ^{13}C–^{13}C decoupling with ^{1}H noise-decoupling,[120] but it would seem preferable — where possible — to secure the right degree of dilution, first of all by using a suitably-diluted precursor and second by adding it in successive small aliquots.

Until now, most of the studies using the $^{13}C_2$ method have been carried out with the simplest precursor, [$^{13}C_2$]acetate, which like the singly-labelled species is commercially available. Some significant work has been done with more complex precursors, requiring their laboratory synthesis. The techniques needed are marginally less demanding than those used with radioisotopes, if only because rather larger quantities of material are usually handled, and with the general success of the method a greater range of precursors will certainly come into use. One by-product of the somewhat critical dilution requirements for a successful experiment will certainly be an increased knowledge of the kinetic aspects of biosynthesis, and increased attention to this aspect by the workers involved.

References

1. A. J. Birch and F. W. Donovan, *Austral. J. Chem.*, 1953, **6,** 360.
2. A. J. Birch, *Science*, 1967, **156,** 202.
3. J. D. Bu'Lock, 'Biosynthesis of Natural Products', McGraw-Hill, London, 1965, and in preparation.
4. W. B. Turner, 'Fungal Metabolites', Academic Press, London, 1971.
5. N. M. Packter, 'Biosynthesis of Acetate-Derived Compounds', Wiley, London, 1973.

6. M. Luckner, 'Secondary Metabolism in Plants and Animals', Chapman and Hall, London, 1972.
7. J. Mann, 'Secondary Metabolism', Oxford University Press, 1977.
8. *E.g.* by successive contributors (J. B. Harborne, T. Money, J. Simpson) in 'Specialist Periodical Reports: Biosynthesis', ed. T. A. Geissman and J. D. Bu'Lock, The Chemical Society, London, from 1972, onwards.
9. *E.g.* in *Fortschr. Chem. org. Naturstoffe, passim.*
10. D. Rittenberg and K. Bloch, *J. Biol. Chem.*, 1944, **154,** 311.
11. A. J. Birch, R. A. Massy-Westropp, and C. J. Moye, *Chem. and Ind.* (*London*), 1955, 683; *Austral. J. Chem.*, 1955, **8,** 529.
12. S. Gatenbeck and A. K. Mahlen, *Acta Chem. Scand.*, 1968, **22,** 1696.
13. V. Behal and Z. Vanek, *Folia Microbiol.*, 1970, **15,** 354.
14. F. Lynen, *Biochem. J.*, 1967, **102,** 381.
15. M. Yalpani, K. Willecke, and F. Lynen, *European J. Biochem.*, 1969, **8,** 495.
16. P. H. W. Butterworth and K. Bloch, *European J. Biochem.*, 1970, **12,** 496.
17. M. de Rosa, A. Gambacorta, and J. D. Bu'Lock, *Phytochemistry*, 1974, **13,** 905.
18. R. J. Light and L. P. Hager, *Arch. Biochem. Biophys.*, 1968, **125,** 326.
19. P. Dimroth, H. Walter, and F. Lynen, *European J. Biochem.*, 1970, **13,** 98.
20. S. Sjoland and S. Gatenbeck, *Acta Chem. Scand.*, 1966, **20,** 1053.
21. G. Jaureguiberry, M. Lenfant, B. C. Das, and E. Lederer, *Tetrahedron*, 1966, suppl. 8, part 1, 27.
22. S. Gatenbeck, P. O. Eriksson, and Y. Hansson, *Acta Chem. Scand.*, 1969, **23,** 699.
23. S. W. Tanenbaum, S. Nakajima, and G. Marx, *Biotechnol. Bioeng.*, 1969, **11,** 1135.
24. R. Bentley and P. M. Zwitkowits, *J. Amer. Chem. Soc.*, 1967, **89,** 676, 681.
25. G. Hrazdina, F. Kreuzaler, K. Hahlbrock, and H. Grisebach, *Arch. Biochem. Biophys.*, 1976, **175,** 392.
26. P. Dimroth, E. Ringelmann, and F. Lynen, *European J. Biochem.*, 1976, **68,** 591.
27. R. Bentley, in 'Biogenesis of Antibiotic Substances', ed. Z. Vanek and Z. Hostalek, Academic Press, New York, 1965, p. 241.
28. K. T. Suzuki and S. Nozoe, *Bioorg. Chem.*, 1974, **3,** 72.
29. K. Mosbach, *Acta Chem. Scand.*, 1960, **14,** 457.
30. C. F. Culbertson, 'Chemical and Botanical Guide to Lichen Products', University of North Carolina Press, Chapel Hill, 1969.
31. J. D. Bu'Lock, D. Hamilton, M. A. Hulme, A. J. Powell, H. M. Smalley, D. Shepherd, and G. N. Smith, *Canad. J. Microbiol.*, 1965, **11,** 765.
32. J. D. Bu'Lock, D. Shepherd, and D. J. Winstanley, *Canad. J. Microbiol.*, 1969, **15,** 279.
33. P. I. Forrester and G. M. Gaucher, *Biochemistry*, 1972, **11,** 1102.
34. G. Murphy and F. Lynen, *European J. Biochem.*, 1975, **58,** 467.
35. A. I. Scott and L. Beadling, *Bioorg. Chem.*, 1974, **3,** 281.
36. G. M. Gaucher, conference communication, 1977.
37. A. J. Birch, R. J. English, R. A. Massy-Westropp, and H. Smith, *J. Chem. Soc.*, 1958, 369.
38. I. M. Campbell, C. H. Calzadilla, and N. J. McCorkindale, *Tetrahedron Letters*, 1966, 5107.
39. L. Canonica, W. Kroszczynski, B. M. Ranzi, B. Rindone, E. Santaniello, and C. Scolastico, *J.C.S. Perkin I*, 1972, 2639.
40. L. Canonica, B. Rindone, C. Scolastico, F. Aragozzini, and R. Craveri, *J.C.S. Chem. Comm.*, 1973, 222.
41. F. Aragozzini, P. Toppino, R. Craveri, M. G. Beretta, B. Rindone, and C. Scolastico, *Bioorg. Chem.*, 1975, **4,** 127.
42. A. I. Scott and K. J. Wiesner, *J.C.S. Chem. Comm.*, 1972, 1075.
43. J. Wright, D. G. Smith, A. G. McInnes, L. C. Vining, and D. W. S. Westlake, *Canad. J. Biochem.*, 1969, **47,** 945; *Chem. Comm.*, 1971, 325.
44. F. Drawert and J. Beier, *Phytochemistry*, 1976, **15,** 1695.
45. J. B. Harborne, 'Comparative Biochemistry of the Flavonoids', Academic Press, London, 1967.
46. H. Grisebach and K. Hahlbrock, in 'Recent Advances in Phytochemistry, ed. V. Runeckles and E. Conn, Academic Press, New York, 1976, vol. 8.
47. J. B. Harborne, in 'Specialist Periodical Reports: Biosynthesis', ed. J. D. Bu'Lock, The Chemical Society, London, 1977, vol. 5.
48. A. J. Birch, *Ann. Rev. Plant Physiol.*, 1968, **19,** 321.
49. L. Crombie, P. M. Dewick and D. A. Whiting, *Chem. Comm.*, 1970, 1469; 1971, 1182, 1183.
50. E. Haslam, *J.C.S. Chem. Comm.*, 1974, 594.
51. E. Haslam, C. T. Opie, and L. J. Porter, *Phytochemistry*, 1977, **16,** 99.
52. R. A. Hill, R. H. Carter, and J. Staunton, *J.C.S. Chem. Comm.*, 1975, 380.
53. J. S. E. Holker and K. Young, *J.C.S. Chem. Comm.*, 1975, 525.
54. T. J. Simpson and J. S. E. Holker, *Tetrahedron Letters*, 1975, 4693.
55. D. H. R. Barton, A. M. Deflorin, and O. E. Edwards, *J. Chem. Soc.*, 1956, 530.
56. H. Taguchi, U. Sankawa, and S. Shibata, *Tetrahedron Letters*, 1966, 5211.
57. T. J. Simpson, *J.C.S. Chem. Comm.*, 1976, 258.
58. A. J. Birch, R. A. Massy-Westropp, R. W. Rickards, and H. Smith, *J. Chem. Soc.*, 1958, 360.
59. D. H. R. Barton and I. Cohen, 'Festschrift A. Stoll', Birkhauser, Basel, 1956, p. 117.
60. A. Rhodes, G. A. Somerfield, and M. P. McGonagle, *Biochem. J.*, 1963, **88,** 349.
61. Y. Sato, T. Machida, and T. Oda, *Tetrahedron Letters*, 1975, 4571.
62. W. Steglich, R. Arnold, W. Lösel, and W. Reininger, *J.C.S. Chem. Comm.*, 1972, 102.

63. N. Takeda, S. Seo, Y. Ogihara, U. Sankawa, I. Iitaka, I. Kitagawa, and S. Shibata, *Tetrahedron*, 1973, **29,** 3703.
64. B. Franck and H. Flasch, *Prog. Chem. Org. Nat. Prod.*, 1973, **30,** 151.
65. A. Mahmoodian and C. E. Stickings, *Biochem. J.*, 1964, **92,** 369.
66. S. Gatenbeck and L. Malmstrom, *Acta Chem. Scand.*, 1969, **23,** 3493.
67. P. Canham, L. C. Vining, A. G. McInnes, J. A. Walker, and J. L. C. Wright, *J.C.S. Chem. Comm.*, 1976, 319.
68. G. C. Elsworthy, J. S. E. Holker, J. M. McKeown, J. B. Robinson, and L. J. Mulheirn, *Chem. Comm.*, 1970, 1069; J. C. Roberts, *Prog. Chem. Org. Nat. Prod.*, 1974, **31,** 119.
69. R. Singh and D. P. H. Hsieh, *Arch. Biochem. Biophys.*, 1977, **178,** 285.
70. P. S. Steyn, R. Vleggaar, P. L. Wessels, and D. B. Scott, *J.C.S. Chem. Comm.*, 1975, 193; C. P. Gorst-Allman, K. G. R. Pachler, P. S. Steyn, P. L. Wessels, and D. B. Scott, *ibid.*, 1976, 916; *J.C.S. Perkin I*, 1976, 1182.
71. R. Thomas, in 'Biogenesis of Antibiotic Substances', ed. Z. Vanek and Z. Hostalek, Academic Press, New York, 1965, p. 155.
72. J. S. C. Holker, personal communication.
73. J. R. D. McCormick, in 'Antibiotics, Vol. II, Biosynthesis', ed. D. Gottlieb and P. D. Shaw, Springer, New York, 1967, p. 113.
74. J. R. D. McCormick, E. R. Jensen, N. H. Arnold, H. S. Corey, U. H. Joachim, S. Johnson, P. A. Miller, and N. O. Sjolander, *J. Amer. Chem. Soc.*, 1968, **90,** 7127.
75. A. J. Birch, *Prog. Chem. Org. Nat. Prod.*, 1957, **14,** 186.
76. J. L. Gellerman, W. H. Anderson, and H. Schlenk, *Biochim. Biophys. Acta*, 1976, **431,** 16.
77. S. Shibata, U. Sankawa, K. Yamasaki and H. Taguchi, *Chem. Pharm. Bull.* (*Japan*), 1966, **14,** 474.
78. A. G. McInnes, D. G. Smith, J. A. Walter, L. C. Vining and J. L. C. Wright, *J.C.S. Chem. Comm.*, 1974, 281, 283; *Tetrahedron Letters*, 1975, 4103.
79. B. E. Cross and P. Hendley, *J.C.S. Chem. Comm.*, 1975, 124; G. N. Smith, unpublished results.
80. C. T. Mabuni, L. Garlaschelli, R. A. Ellison, and C. R. Hutchinson, *J. Amer. Chem. Soc.*, in press.
81. A. J. Birch, O. C. Musgrave, R. W. Rickards, and H. Smith, *J. Chem. Soc.* 1959, 3146.
82. J. A. Steele, J. R. Lieberman, and C. J. Mirocha, *Canad. J. Microbiol.*, 1974, **20,** 531.
83. W. Graf, J. L. Robert, J. C. Vederas, C. Tamm, P. H. Solomon, I. Miura, and K. Nakanishi, *Helv. Chim. Acta*, 1974, **57,** 1801.
84. W. B. Turner, unpublished results.
85. A. J. Birch, C. W. Holzapfel, R. W. Rickards, C. Djerassi, M. Suzuki, J. Westley, J. D. Dutcher, and R. Thomas, *Tetrahedron Letters*, 1964, 1485.
86. W. Keller-Schierlein, *Forschr. Chem. org. Naturstoffe*, 1973, **30,** 313.
87. L. C. Vining and J. L. C. Wright, in 'Specialist Periodical Reports: Biosynthesis', ed. J. D. Bu'Lock, The Chemical Society, London, 1977, vol. 5, p. 240.
88. H. Grisebach, H. Archenbach, and W. Hofheinz, *Z. Naturforsch.*, 1960, **15b,** 560; 1962, **17b,** 64.
89. S. Omura, A. Nakagawa, H. Takeshima, K. Atsumi, J. Miyazawa, F. Piriou, and G. Lukacs, *J. Amer. Chem. Soc.*, 1975, **97,** 6600; *Tetrahedron Letters*, 1975, 4503.
90. R. J. White and E. Martinelli, *FEBS Letters*, 1974, **49,** 233.
91. B. Milavetz, K. Kakinuma, K. L. Rinehart, J. P. Rolls, and W. J. Haak, *J. Amer. Chem. Soc.*, 1973, **95,** 5793.
92. P. V. Desmukh, K. Kakinuma, J. J. Ameel, K. L. Rinehart, P. F. Wiley, and L. H. Li, *J. Amer. Chem. Soc.*, 1976, **98,** 870.
93. E. Martinelli, R. J. White, G. G. Gallo, and P. J. Beynon, *Tetrahedron Letters*, 1974, 1367.
94. R. J. White, E. Martinelli, and G. Lancini, *Proc. Nat. Acad. Sci. USA*, 1974, **71,** 3260.
95. R. D. Johnson, A. Haber, and K. L. Rinehart, *J. Amer. Chem. Soc.*, 1974, **96,** 3316.
96. T. Swain, in 'Biosynthetic Pathways in Higher Plants', ed. J. B. Pridham and T. Swain, Academic Press, London, 1965, p. 9.
97. S. A. Brown, in 'Specialist Periodical Reports: Biosynthesis', ed. T. A. Geissman, The Chemical Society, London, 1971, vol. 1, p. 1.
98. I. M. Campbell, *Phytochemistry*, 1975, **14,** 683.
99. I. M. Campbell, *Phytochemistry*, 1976, **15,** 1367.
100. G. D. Chase and J. L. Rabinowitz, 'Principles of Radioisotope Methodology', Burgess, Minneapolis, 1967.
101. M. D. Kamen, 'Isotopic Tracers in Biology', Academic Press, New York, 1957.
102. J. R. Catch, 'Carbon-14 Compounds', Butterworth, London, 1961.
103. E. A. Evans, 'Tritium and its Compounds', Van Nostrand, London, 1966.
104. M. Bubner and L. Schmidt, 'Die Synthese Kohlenstoff-14-markierter organischer Verbindungen', Thieme, Leipzig, 1966.
105. A. Murray and D. L. Williams, 'Organic Syntheses with Isotopes', Interscience, New York, 1958.
106. J. W. Cornforth, *Quart. Rev.*, 1969, **23,** 125.
107. J. W. Cornforth, *Chem. Soc. Rev.*, 1973, **2,** 1.
108. J. M. A. Al-Rawi, J. A. Elvidge, D. K. Jaiswal, J. R. Jones, and R. Thomas, *J.C.S. Chem. Comm.*, 1974, 220.
109. H. Simon and H. G. Floss, 'Bestimmung der Isotopenverteilung in markierten Verbindungen', Springer, Berlin, 1967.

110. A. J. Birch, R. A. Massy-Westropp, R. W. Rickards and H. Smith, *Proc. Chem. Soc.*, 1957, 98.
111. T. J. Simpson, *Chem. Soc. Rev.*, 1975, **4,** 497.
112. M. Tanabe, in 'Specialist Periodical Reports: Biosynthesis', ed. T. A. Geissman and J. D. Bu'Lock, The Chemical Society, London, 1973, vol. 2, p. 241; 1975, vol. 3, p. 247; 1976, vol. 4, p. 204.
113. G. C. Levy and G. L. Nelson, 'Carbon-13 Nuclear Magnetic Resonance for Organic Chemists', Wiley, New York, 1972.
114. J. B. Stothers, 'Carbon-13 N.M.R. Spectroscopy', Academic Press, New York, 1972.
115. L. F. Johnson and W. C. Jankowski, 'Carbon-13 Nuclear Magnetic Resonance Spectroscopy', Wiley, New York, 1972.
116. R. Freeman, K. G. R. Pachler, and G. N. La Mar, *J. Chem. Phys.*, 1971, **55,** 4586.
117. M. Tanabe, K. Suzuki, and W. C. Jankowski, *Tetrahedron Letters*, 1973, 4723.
118. L. Cattel, J. F. Grove, and D. Shaw, *J.C.S. Perkin I*, 1973, 2626.
119. H. M. Pickett and H. L. Strass, *Analyt. Chem.*, 1972, **44,** 265.
120. A. G. McInnes, D. G. Smith, J. A. Walter, L. C. Vining, and J. L. C. Wright, *J.C.S. Chem. Comm.*, 1975, 66.

29.2
Terpenoid Biosynthesis

J. R. HANSON
University of Sussex

29.2.1 INTRODUCTION

Terpenoids and steroids are spread throughout nature. When considering their biosynthesis, the widespread integrity of the isoprene unit, coupled with the variety of terpenoid carbon skeleta, has attracted natural product chemists for almost a century. The structural similarity of a number of monoterpenoids and the formation of isoprene by thermal decomposition suggested as early as 1884 to Tilden that a common five-carbon unit existed as a structural fragment in the monoterpenoids. This structural unity was eventually expressed by Ruzicka in 1921 in terms of the isoprene rule that the carbon skeleta of terpenoids are composed of isoprene units linked in a regular arrangement.[1,2] This rule served as a significant aid to the structural elucidation of sesquiterpenoids and diterpenoids during the following decades. However, the accumulation of a number of apparent exceptions led to a restatement of the rule as the 'biogenetic isoprene rule'[3] which introduced the idea of various 'permitted' rearrangements during biosynthesis. In terms of the biogenetic isoprene rule, terpenoids are compounds formed by the combination of isoprene units to afford aliphatic substances such as geraniol, farnesol, geranylgeraniol, squalene, and others of a similar kind. Terpenoids can then be derived from these aliphatic precursors by accepted cyclization and, in certain cases, cyclization and rearrangement reactions.

There were many speculations on the origin of the 'active isoprene unit', involving compounds such as leucine and senecioic acid. For reasons of their biological importance, many of the early studies concentrated on the steroids. The first studies were recorded not long after the elucidation of the structure of these compounds. In one of the first applications in 1937[4] of isotopic methods to biosynthetic problems, Sonderhoff fed trideuterioacetic acid to yeast cells and found that the yeast sterols contained a large amount of deuterium. Studies beginning in 1946 by Bloch[5] and by Cornforth and Popjak[6] on the utilization of methyl and carboxyl ^{14}C-labelled acetic acid (**1**) were very informative in elucidating the pathway for the conversion of acetic acid into the steroids. These experiments showed that the methyl and carboxyl carbon atoms alternate around the carbon skeleton of cholesterol (**7**), whilst the pendant carbon atoms arose from the methyl group of acetic acid.

(**1**)

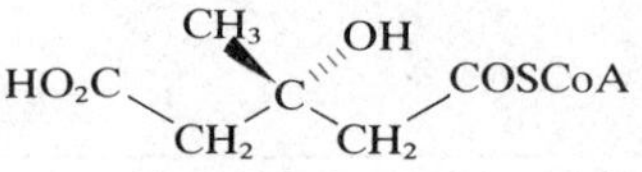

(**2**)

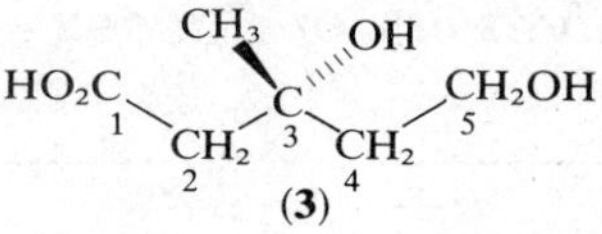

(**3**)

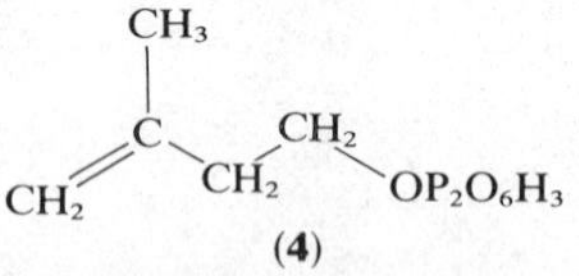

(**4**)

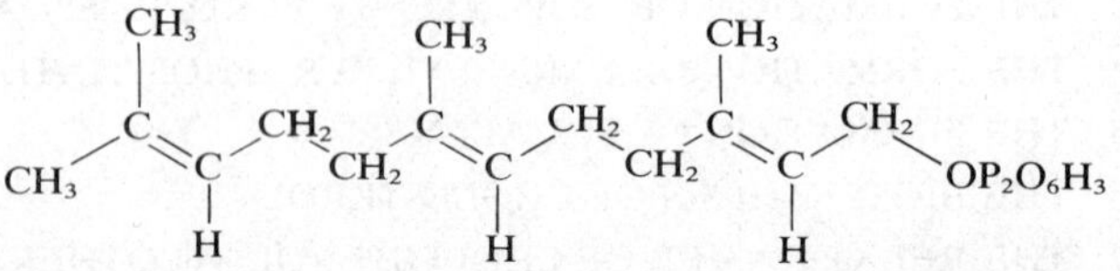

(**5**)

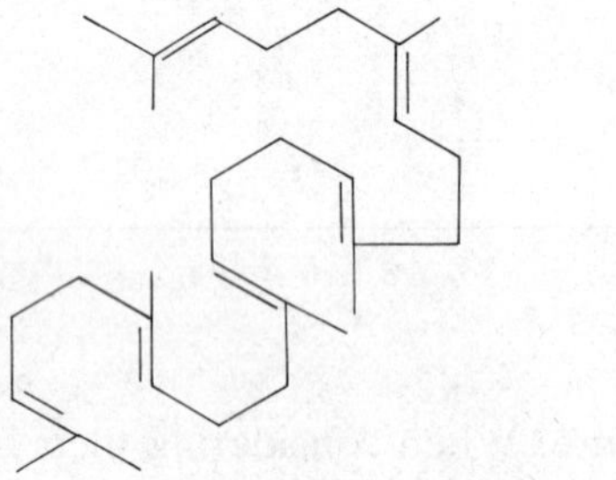

(**6**)

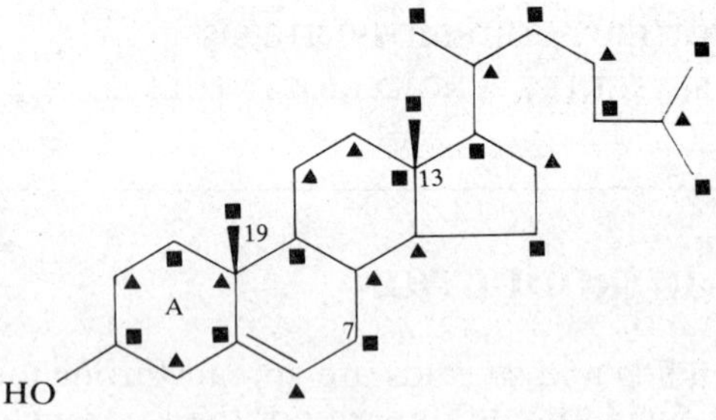

(**7**)

In 1951 the so-called 'active acetate' was identified[7] as the acetic acid thiol ester of coenzyme A, acetyl-CoA. This in turn led[8] to the implication of the six-carbon unit 3-hydroxy-3-methylglutaryl-CoA (HMGCoA) (**2**) in isoprenoid biosynthesis. The loss of an acetate carboxyl as carbon dioxide has been observed in sterol biosynthesis, in accord with a six-carbon fragment derived from three acetate units acting as a precursor of the fundamental five-carbon isoprene unit. The identification of the next important intermediate arose from experiments which were not directed at the problem of sterol biosynthesis. In 1956, workers at the laboratories of Merck, Sharp, and Dohme found[9] that 3,5-dihydroxy-3-methylvaleric acid, mevalonic acid (**3**), could replace acetate as an essential growth factor for a mutant of *Lactobacillus acidophilus.* It was shown by Tavormina, Gibbs, and Huff that[10] cell-free extracts of liver were able to incorporate radioactively labelled mevalonic acid into cholesterol (**7**). This work was rapidly followed by the demonstration that mevalonic acid was a general precursor in terpenoid biosynthesis.[11]

Squalene (**6**) had been proposed[12] by Channon in 1926 as a possible intermediate in sterol biosynthesis. In 1934 Robinson suggested[13] a folding (**8**) of squalene for its conversion into cholesterol. By 1952 Bloch had shown that the carbon atoms in the side chain of cholesterol (**7**) originated from acetate in accordance with the isoprene rule, whilst Cornforth and Popjak had established the origin of the carbon atoms of ring A and of C-19. However, in 1953 Bloch showed that the carbon atoms C-13 and C-7 did not originate from the carboxyl group of acetic acid, as required by folding (**8**), but instead came from the methyl group. This, coupled with structural studies on lanosterol (**9**), led Woodward and Bloch[14] in 1953 to propose a different folding of squalene (**6**). Experimental support for the role of squalene and lanosterol was soon forthcoming. Thus labelled acetate administered to rats yielded labelled squalene and the latter gave labelled cholesterol.

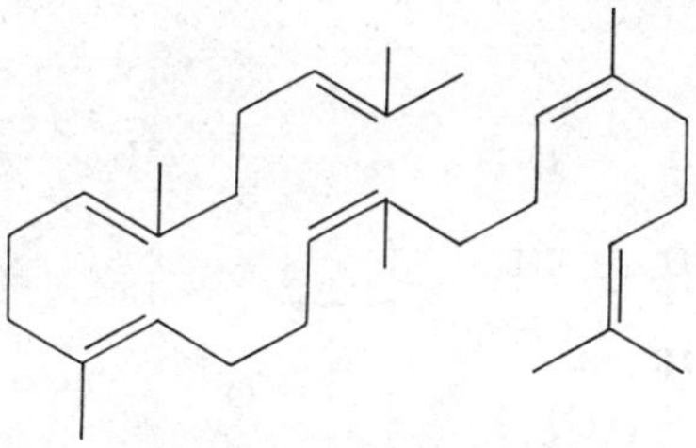

(**8**)

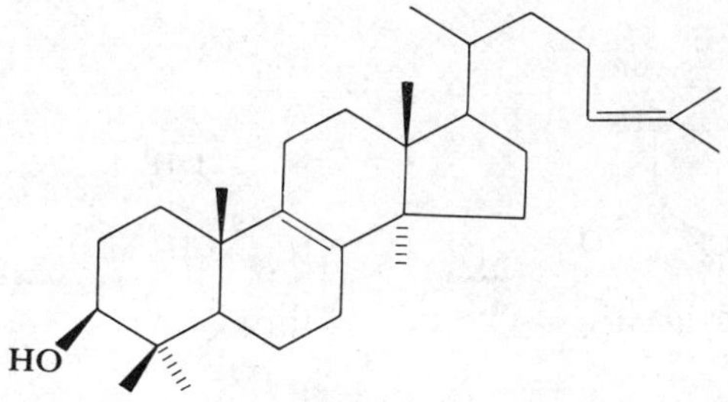

(**9**)

Cell-free extracts from yeast and liver were obtained that were capable of synthesizing squalene from mevalonic acid. These required, as co-factors, ATP, Mg^{2+} ions, and NADPH. In the absence of NADPH, farnesyl pyrophosphate (**5**) accumulated. This could in turn be converted into squalene (**6**). In the presence of an inhibitor (M/200 iodoacetamide) a further intermediate, isopentenyl pyrophosphate (**4**), was detected. This five-carbon compound has the key role of the 'active isoprene' unit. Thus by 1960 the major stages of acetate, mevalonate, isopentenyl pyrophosphate, farnesyl pyrophosphate, squalene, and lanosterol in cholesterol biosynthesis had been defined.

At this point it is appropriate to dissect the study of terpenoid biosynthesis into four areas and discuss these in more detail. These are (i) the formation of isopentenyl pyrophosphate (IPP): (ii) the oligomerization of IPP to the prenyl pyrophosphates, squalene, and rubber; (iii) the cyclization stages in steroid and terpenoid biosynthesis; (iv) the formation of individual terpenoids by hydroxylations and rearrangements, *etc.* In these studies, mevalonic acids, isotopically labelled at specific centres, have played an important role and it is therefore useful to describe the syntheses of these prior to a discussion of the biosynthetic studies.

29.2.2 PREPARATION OF LABELLED MEVALONATES[15,16]

Natural mevalonic acid is optically active and it is the 3(*R*)-isomer (**3**) which is utilized in terpenoid biosynthesis. Mevalonate kinase has been shown to phosphorylate only this enantiomer. The role of mevalonic acid as an irreversible precursor of terpenoid biosynthesis has led to a large number of syntheses of variously labelled mevalonates, which have been designed to incorporate labels at specific sites in terpenoid end-products.

The earliest studies of mevalonate utilization were based on [2-^{14}C]-labelled material which was prepared (Scheme 1) by a Reformatski reaction between methyl bromoacetate and 1-acetoxybutan-3-one or 1,1-dimethoxybutan-3-one. An alternative method utilized 1-chlorobutan-3-one as the four-carbon component.

In more recent syntheses, the base-catalysed condensation of labelled ethyl acetate or of the dilithio anion of acetic acid with a 1-substituted butan-3-one has been employed[16]

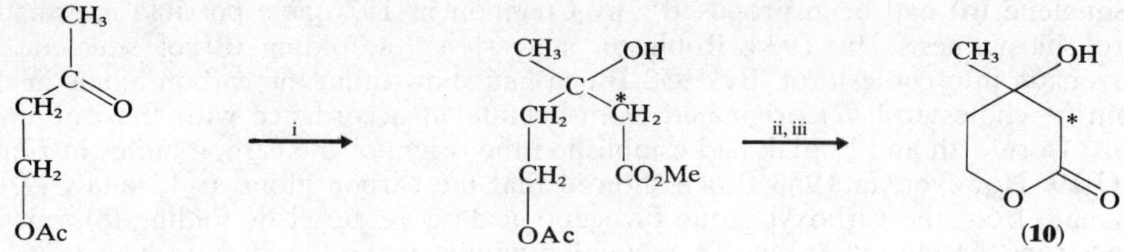

i, Zn, $BrCH_2CO_2Me$; ii, OH^-; iii, H^+. * = ^{14}C.

SCHEME 1

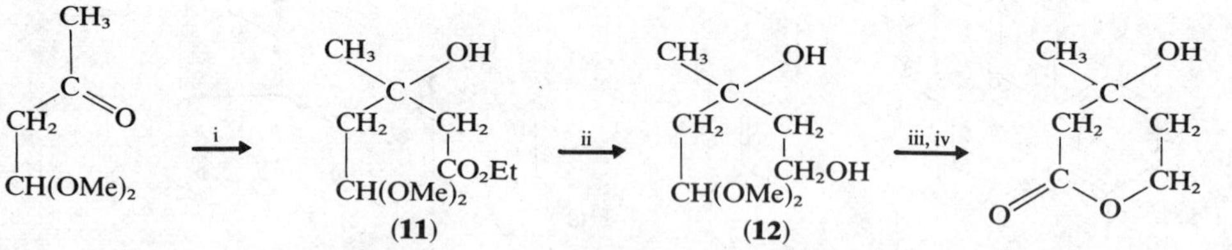

i, $LiCH_2CO_2Et$; ii, $LiAlH_4$; iii, H^+; iv, Br_2, H_2O.

SCHEME 2

to prepare carbon-13 labelled material. By reduction of the ester (**11**) to the alcohol and oxidative hydrolysis of the dimethoxyacetal (**12**), these syntheses can be adapted (Scheme 2) to provide C-4 and C-5 labelled material.

The four-carbon component, 1-acetoxybutan-3-one, labelled with carbon-13 at positions-2 and -4 has been prepared (Scheme 3) from [2-^{13}C]acetic acid *via* diketen. An

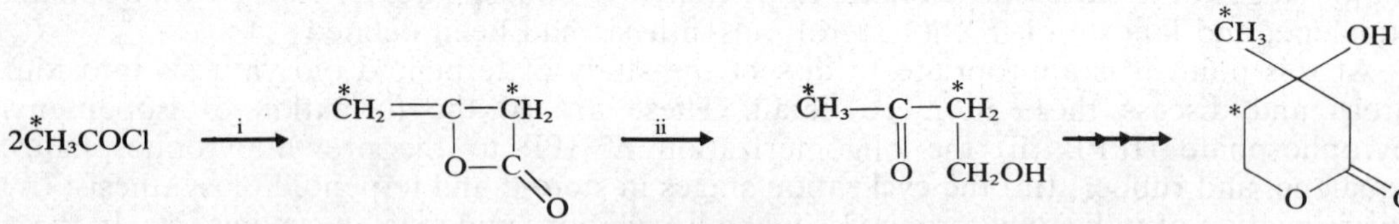

i, Et_3N; ii, $LiAlH_4$. * = ^{13}C.

SCHEME 3

alternative synthesis from acetyl chloride and the monoethyl ester of malonic acid has also been published. Some of these methods have been extended to the synthesis of non-stereospecifically labelled tritiated mevalonates. Thus methyl [2-$^3H_2$]bromoacetate affords [2-$^3H_2$]mevalonate by the Reformatski method. The preparation of [5-$^3H_2$]mevalonate has been effected by the reduction of the ester of 5,5-dimethoxy-3-methyl-3-hydroxypentan-1-oic acid with lithium [^{3}H]borohydride followed by oxidative hydrolysis, or by the reduction of the aldehyde, mevaldic acid (**28**), with sodium [^{3}H]borohydride.

The preparation of stereospecifically labelled mevalonates has been solved[17] by several ingenious methods, which are summarized in Scheme 4. The starting material for the synthesis of stereospecifically labelled 4(*R*)- and 4(*S*)-[4-^{3}H]mevalonic acid was the acetoxy-ester (**13**), which on mild hydrolysis gave a separable mixture of a lactone (**14**) (from the *cis*-isomer) and the *trans*-hydroxy-acid (**15**). Each isomer was separately converted to its epoxide (*e.g.* **16** and **17**) and these were reduced with lithium [^{2}H]- (or [^{3}H]-) borohydride to afford the labelled mevalonic acids (**18**) and (**19**). The opening of an epoxide in this manner leads to a *trans* relationship between the hydroxyl group and the

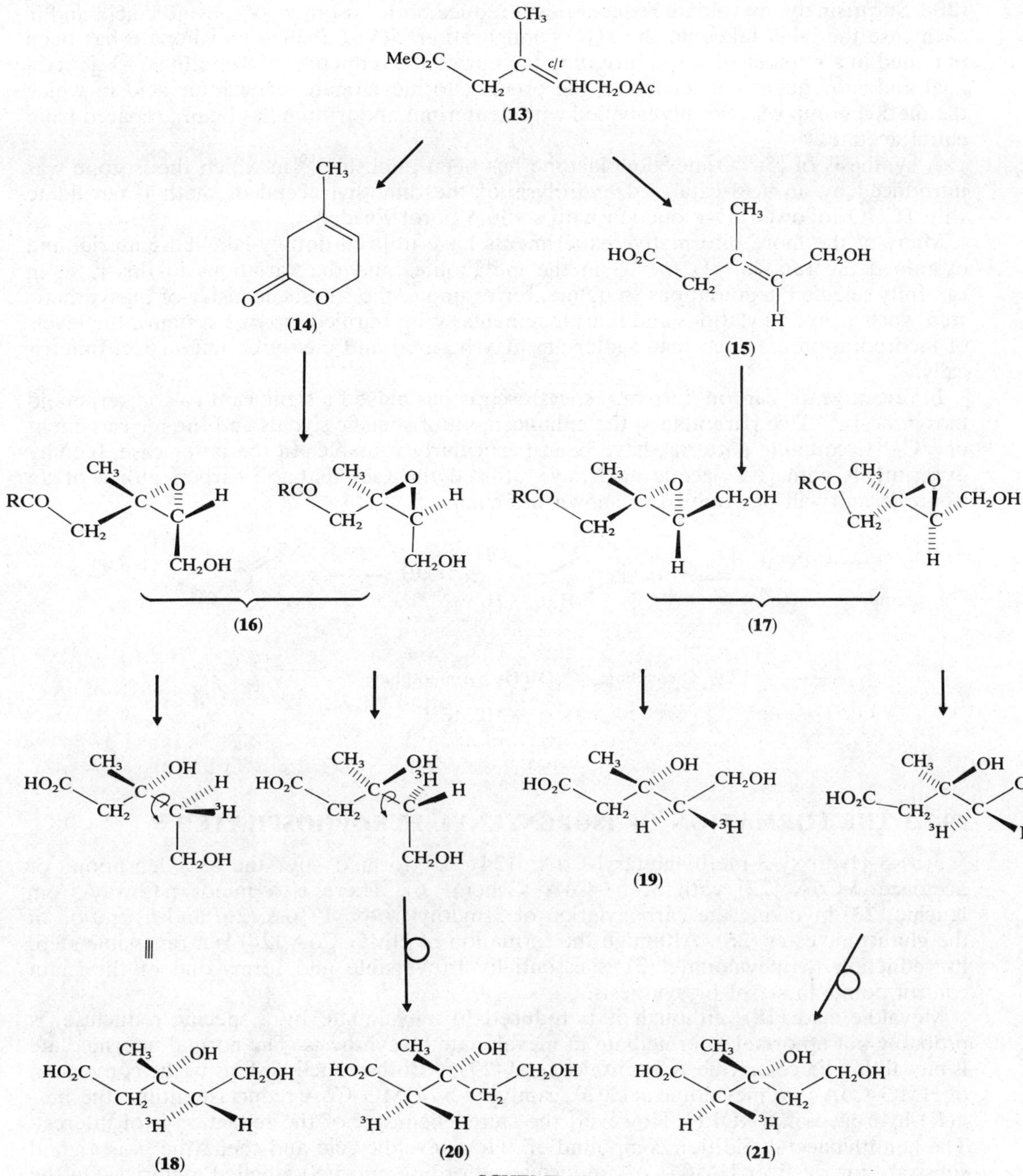

SCHEME 4

incoming nucleophilic hydride ion. Since only the 3(*R*) isomer (*e.g.* **18**) serves as a substrate for mevalonate kinase, this defines the chirality at C-4 in the mevalonate utilized for biosynthetic purposes. By interconversion of the alcohol and carboxyl functions, the chirality which was introduced at the C-4 position was transferred to the C-2 position to afford 2(*S*) and 2(*R*) (*e.g.* **20** and **21**) labelled material. Mevalonate bearing a stereospecific label at C-5 was prepared by an enzymatic method. Mevaldate reductase from pig or rat liver is an 'A' type enzyme transferring a label (^{2}H or ^{3}H) from the 'A' face of NADH [the 4(*R*) position] to generate a 5(*R*) labelled mevalonate from the mevaldic acid

(**28**). Surprisingly, mevaldate reductase will reduce both 3-isomers of mevaldic acid and in each case the label takes up the 5(*R*) configuration. 5(*S*)-Labelled mevalonate has been obtained in a number of ways, including the enzymatic reduction of 3-methyl[1-^{3}H]but-3-enal and subsequent conversion of the product to mevalonate. Mevalonic acid in which the methyl group was chirally labelled with deuterium and tritium has been prepared from chiral acetate.[18]

A synthesis of [5-^{18}O]mevalonolactone has been published[19] in which the isotope was introduced by an acid-catalysed hydrolysis of the dimethyl acetal of methyl mevaldate with $H_2{}^{18}O$ followed by reduction with sodium borohydride.

Many of the more informative experiments have utilized doubly-labelled material and examined the ratio of ^{3}H to ^{14}C in the metabolites and the variations in this ratio in carefully selected degradations to define, for example, the stereochemistry of biosynthetic steps such as hydroxylations and rearrangements. With purified enzyme systems, the levels of incorporation are such that deuterium may be used and measured mass spectrometrically.

In recent years, carbon-13 n.m.r. spectroscopy has played a significant role in terpenoid biosynthesis.[20] Two parameters, the enhancement of specific signals and the measurement of ^{13}C–^{13}C coupling patterns, have been particularly valuable. In the latter case, feeding experiments with [$^{13}C_2$]acetic acid have utilized the fact that the carbon atoms of an isoprene unit will be coupled as shown in Scheme 5.

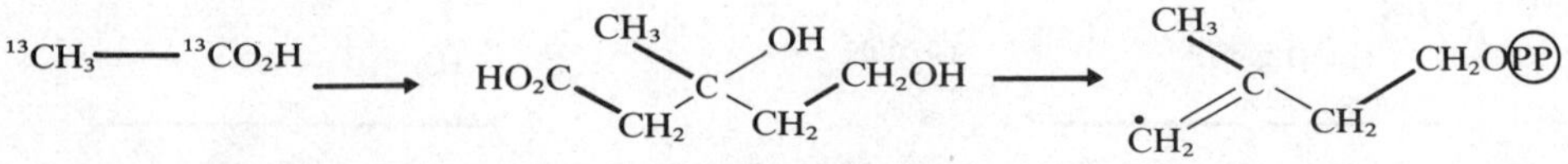

—— = ^{13}C–^{13}C coupling; O(PP) = pyrophosphate

SCHEME 5

29.2.3 THE FORMATION OF ISOPENTENYL PYROPHOSPHATE[21–23]

3(*S*)-3-Hydroxy-3-methylglutaryl-CoA (**24**) is formed by the condensation of acetoacetyl-CoA (**22**) with acetyl-CoA (Scheme 6). There is a minor pathway from leucine (**23**) involving the carboxylation of 3-methylcrotonyl-CoA (**26**) and hydration of the glutaconic ester (**25**). Although the formation of HMG-CoA (**24**) is a reversible step, its reduction to mevalonate (**3**) is essentially irreversible and forms one of the main control points in sterol biosynthesis.

Mevaldic acid (**28**), although it is reduced to mevalonate by a specific reductase, is probably not a normal intermediate in mevalonate biosynthesis. The normal intermediate is mevaldic acid coenzyme A hemithioacetal (**27**).[24] Both reductive steps in the conversion of HMG-CoA into mevalonic acid (**3**), catalysed by HMG-CoA reductase, utilize the *pro*-4(*R*)-hydrogen of NADPH. However, the stereochemistry of the reduction is of interest. The hemithioacetal addition compound of 3(*R*)-mevaldic acid and coenzyme A is a good substrate for rat-liver HMG-CoA reductase, affording *pro*-5(*S*)-labelled material. On the other hand, reduction of 3(*R*)-mevaldic acid by mevaldate reductase from rat liver affords *pro*-5(*R*)-labelled material.

The pattern of labelling of squalene and cholesterol biosynthesized from [2-^{14}C]-mevalonic acid conforms to its direct utilization.[25] Furthermore, labelled mevalonic acid has been trapped when unlabelled material was added to a liver enzyme system during the biosynthesis of squalene from [^{14}C]acetate.[26] The 3(*R*)-enantiomorph is utilized with the concomitant loss of the carboxyl group. The configuration of the biologically active form of mevalonic acid at C-3 was established through direct correlation of its antipode with quinic acid and by its synthesis from (*S*)-linalool.[27]

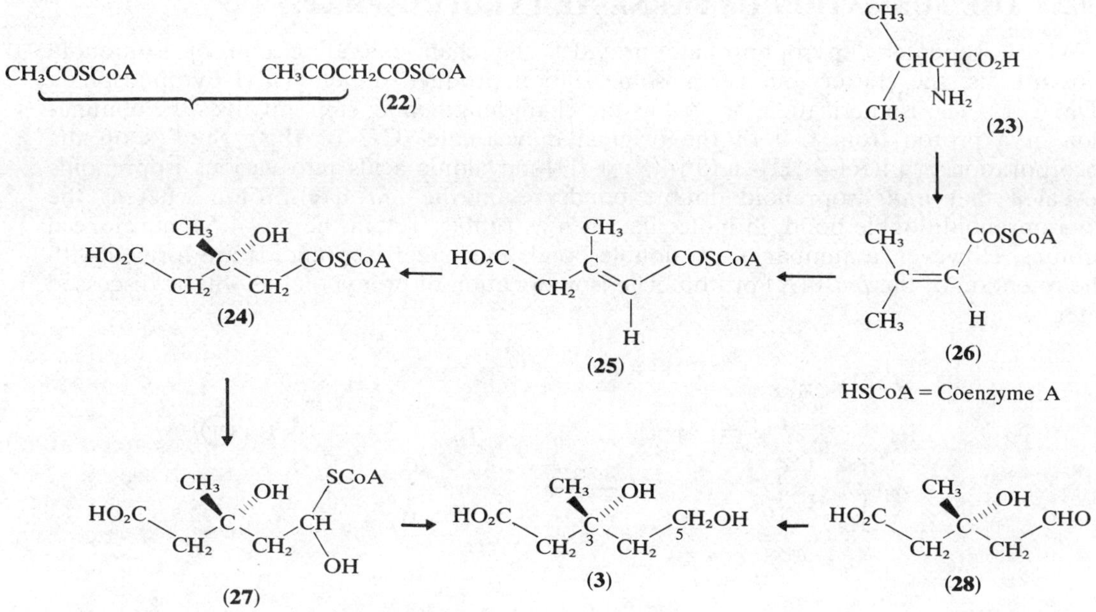

SCHEME 6

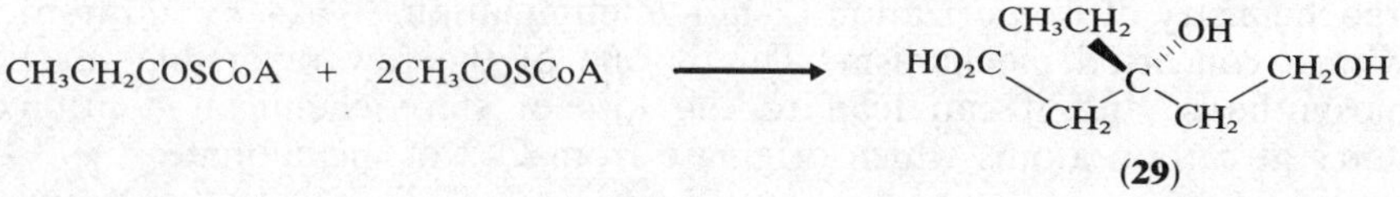

SCHEME 7

3-Homomevalonic acid (**29**) has been implicated[28] in the biosynthesis of the Insect Juvenile Hormone. It is formed (Scheme 7) from propionyl-coenzyme A and two moles of acetyl-coenzyme A.

The conversion of mevalonate into isopentenyl pyrophosphate requires ATP and Mg^{2+}. The formation of a monophosphorylated derivative of mevalonic acid, which served as an efficient precursor of squalene, was shown to be catalysed by a mevalonate kinase. Preparations of mevalonate kinase from several plants have been shown to be inhibited by geranyl and other prenyl pyrophosphates, and hence this enzyme system may serve as a further control point in isoprenoid biosynthesis. Subsequently it was shown that the monophosphate of mevalonic acid underwent a second ATP-dependent phosphorylation to form the C-5 pyrophosphate. This was followed by a concerted dehydration and decarboxylation, which required a third molecule of ATP.

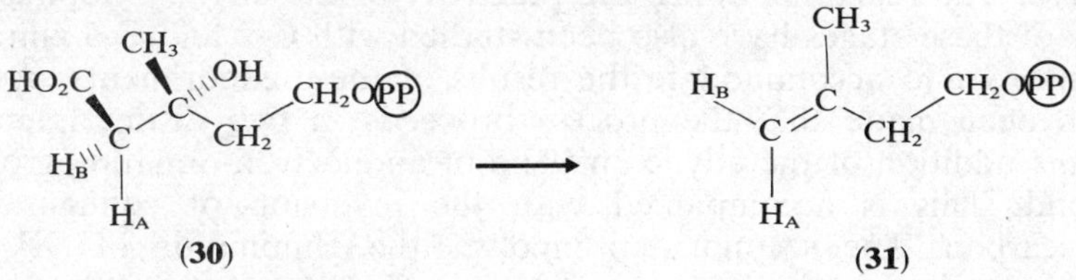

The stereochemistry of this decarboxylation and dehydration (**30**→**31**) has been studied[17] utilizing [2(*S*)-2-^{2}H]- and [2(*R*)-2-^{2}H]-labelled material. The stereochemistry of the deuterium labels on the double bond of the resultant isopentenyl pyrophosphates reflected the stereochemistry of the reaction and showed that the process was a *trans* elimination.

29.2.4 THE FORMATION OF FARNESYL PYROPHOSPHATE[29]

Whilst isopentenyl pyrophosphate provides the chain-extending unit of isoprenoid biosynthesis, the starter unit is an isomerization product, dimethylallyl pyrophosphate (DMAPP). The isomerization, as well as the chain-lengthening step, involves the elimination of a proton from C-4 of the original mevalonate (C-2 of IPP). Studies on the incorporation of [4(*R*)-4-^{3}H]- and [4(*S*)-4-^{3}H]-mevalonic acids into various isoprenoids revealed that *trans*-isoprenoid double bonds retain the *pro*-4(*R*)-proton whereas the *cis*-isoprenoid double bond, in molecules such as rubber, retain the *pro*-4(*S*) mevalonoid protons. However, a number of *cis* double bonds in smaller isoprenoids are formed with the retention of the *pro*-4(*R*) proton. The isomerization of prenyl olefins will be discussed later.

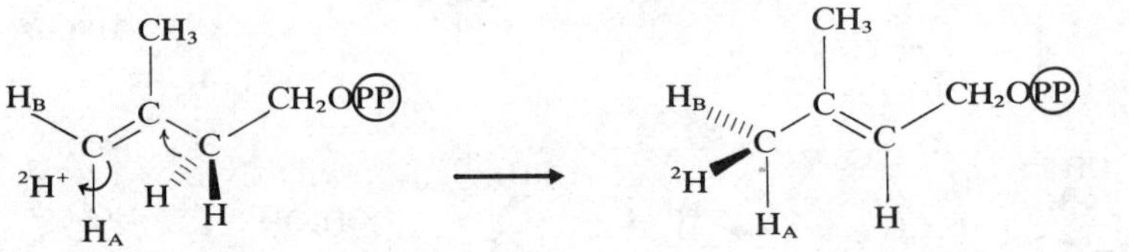

(32)

Studies utilizing the formation of chiral acetate have shown[30] that the addition of hydrogen to the double bond of isopentenyl pyrophosphate is from the 3-*re*,4-*re* face, so that the stereochemistry of isomerization (2-*pro*-*R* elimination, 3-*re*,4-*re* addition) (**32**) is consistent with a concerted mechanism. This is one of the few reversible reactions in terpenoid biosynthesis, and it can lead to the loss of stereochemical identity of the hydrogen atoms at carbon atoms which originate from C-2 of mevalonate.

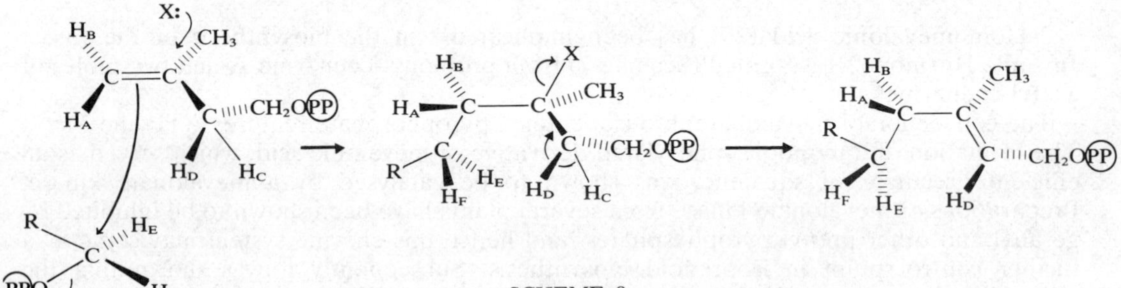

SCHEME 8

The olefin alkylation stages which lead[31] to the prenyl pyrophosphates are catalysed by prenyl transferase. The reactions utilize the reactivity of the allylic pyrophosphates. The stereochemistry of these stages have also been studied with C-2 and C-5 chirally labelled mevalonates. In order to accommodate the results of these experiments, the suggestion (Scheme 8) has been made that the process proceeds in two distinct steps. The first involves the *trans* addition of the allylic unit and of an electron-donating group X across the double bond. This is accompanied with the inversion of configuration at the pyrophosphate carbon. The second step involves the elimination of 'X' and of the hydrogen atom from the carbon atom originating in C-4 of mevalonate.

Farnesyl pyrophosphate (**34**) and later geranyl pyrophosphate (**33**) have been identified as precursors of squalene. Subsequently these prenyl pyrophosphates together with geranylgeranyl pyrophosphate (**35**) and farnesylgeranyl pyrophosphate (**36**) have been identified as the parents of the mono-, sesqui-, di-, and sester-terpenoids and of the carotenoids.

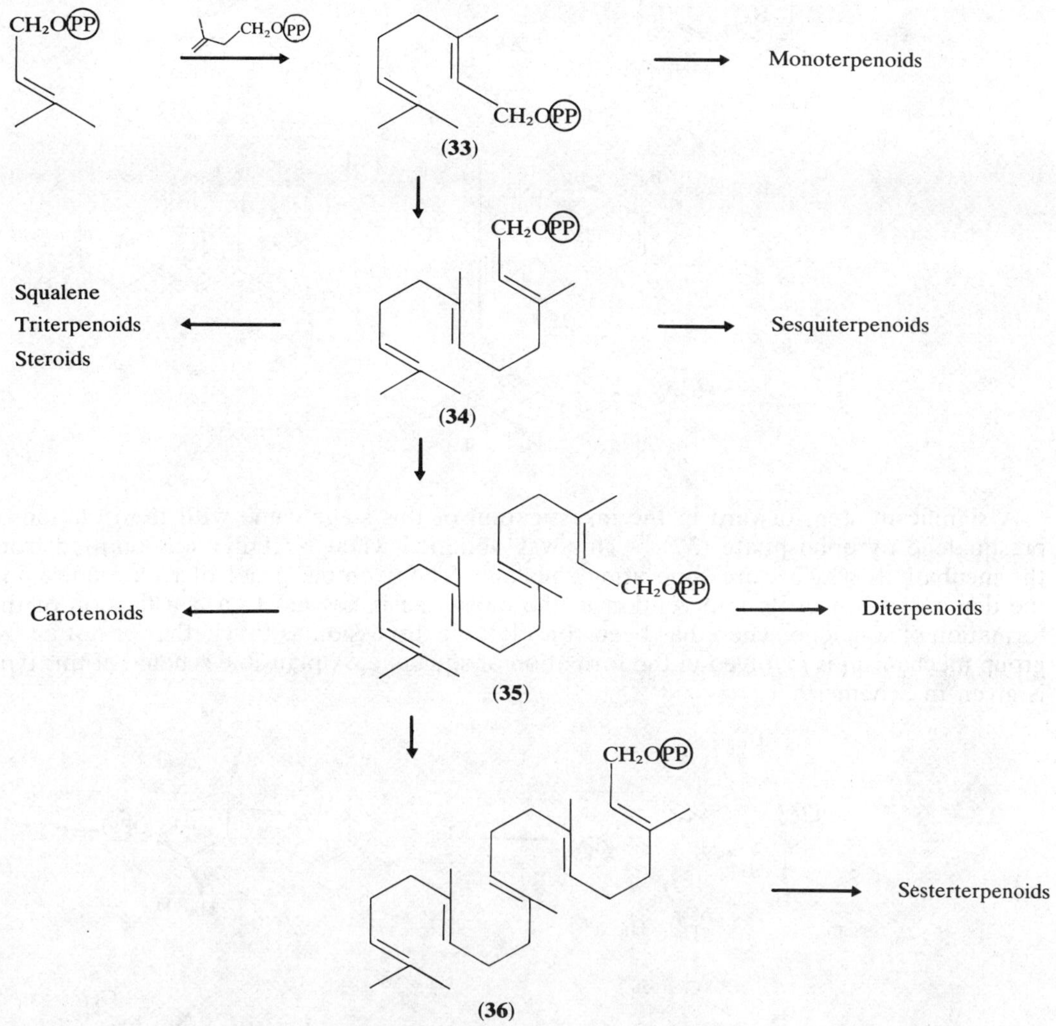

29.2.5 THE FORMATION OF SQUALENE[22,29,32]

The mechanism of the head-to-head condensation of two molecules of farnesyl pyrophosphate (**34**) to form squalene (**6**) has attracted considerable interest. When squalene was biosynthesized from [5-$^2H_2$]mevalonate by a rat-liver enzyme system, 11 of the maximum possible 12 deuterium atoms were retained. One label was lost from the central two carbon atoms of squalene and replaced stereospecifically by the *pro*-4(*S*) hydrogen atom from the 'B' face of NADPH. Ozonolysis of the squalene produced biosynthetically from [5-$^2H_2$]mevalonate gave a trideuteriosuccinic acid with an *S* configuration. This led to the prediction that the tritium introduced into squalene from [^{3}H]NADPH would ultimately appear at the C-11α and C-12β positions in the steroid nucleus. By utilizing 1(*R*)-[1,5,9-$^2H_3$]farnesyl pyrophosphate formed from [5(*R*)-5-^{2}H]mevalonate, it was shown that there was no loss of this label in the biosynthesis of squalene and that both the central carbon atoms retained their label. Thus a *pro*-1(*S*) hydrogen atom of farnesyl pyrophosphate is lost in the formation of squalene. Furthermore, the dideuteriosuccinic acid obtained by ozonolysis of the squalene was an optically inactive *meso*-(*RS*) isomer. Thus the overall process of the union of two farnesyl pyrophosphate molecules (**34**) to yield squalene (**6**) results in the retention of configuration at C-1 of one molecule and inversion at the other.

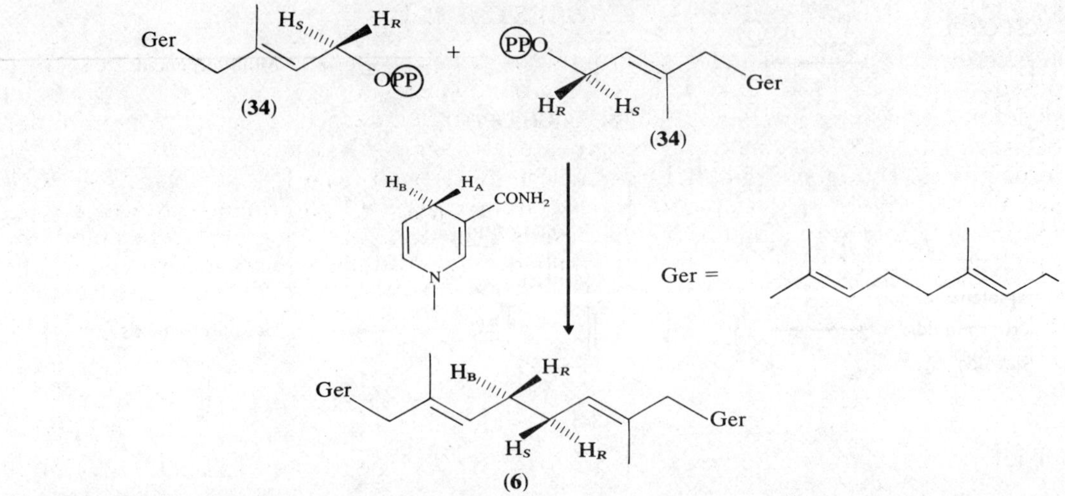

(34) (34) (6)

A significant step forward in the investigation of this stage came with the isolation of presqualene pyrophosphate (**37**).[33] This was obtained when NADPH was omitted from the incubations which were generating squalene. The eventual proof of its structure and the demonstration of its intervention in the biosynthesis has led to a clarification of the formation of squalene. There has been considerable discussion as to whether or not an 'X' group mechanism is involved in the formation of squalene. A plausible scheme of this type is given in Scheme 9.

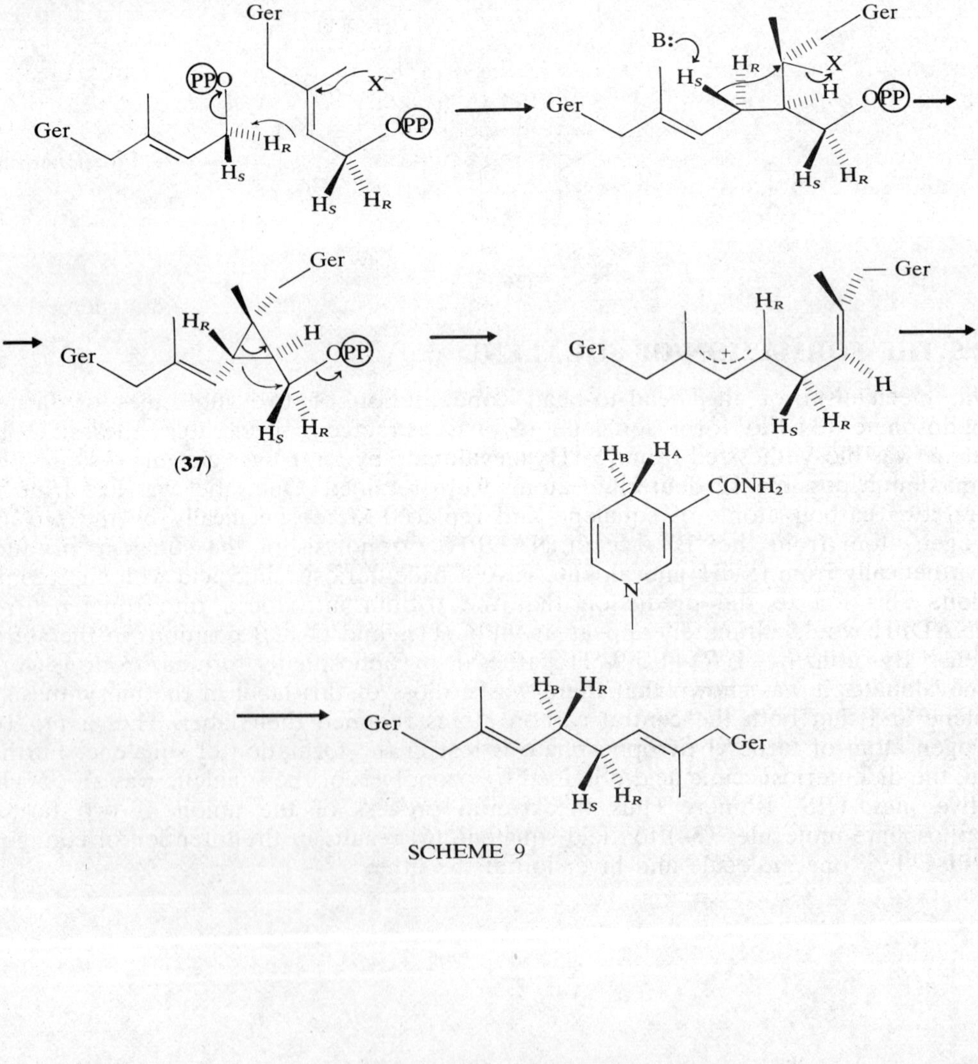

(37)

SCHEME 9

29.2.6 THE BIOSYNTHESIS OF CHOLESTEROL[22,34,35]

Although a biogenetic relationship between squalene and cholesterol had been proposed as early as 1926, the correct folding of squalene in the generation of the carbon skeleton of cholesterol was not defined until 1953 after the acetate-labelling pattern had been established. The acetate and mevalonate labelling patterns are shown in (**7**) and (**38**), respectively. The labelling pattern of cholesterol derived from [5-^{13}C]- and [3′,4-$^{13}C_2$]-mevalonate has recently been determined[36] using carbon-13 n.m.r. methods. The stereospecificity of the cyclization of squalene results in the *trans*-methyl group at the end of the squalene molecule, a group which is labelled by [2-^{14}C]mevalonic acid, generating the C-4α methyl group of lanosterol.

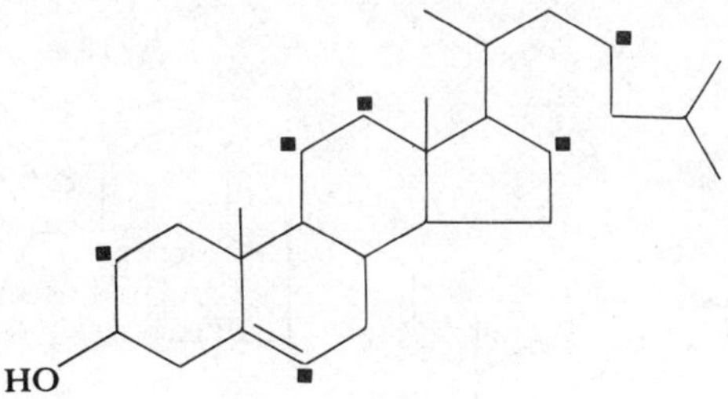

(**38**)

■ = label from [5-^{13}C]MVA

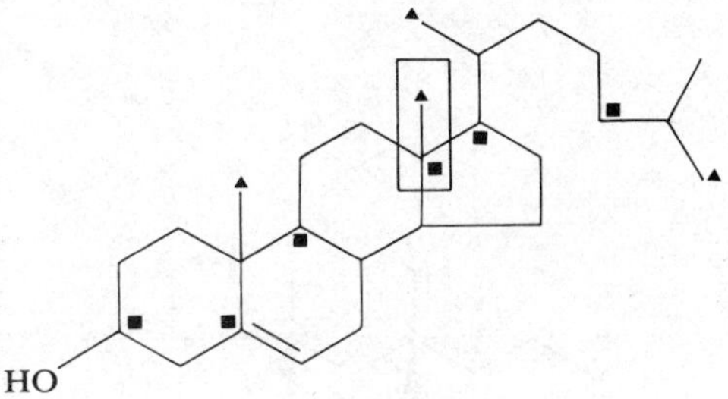

(**39**)

▲ = label from [3′-^{13}C]MVA
■ = label from [4-^{13}C]MVA

Originally the biological cyclization of squalene (**40**) to form lanosterol (**44**) was thought to be initiated by the attack of a formal cation, OH^+. Tchen and Bloch observed[37] that the cyclization of squalene was aerobic and in the presence of ^2H_2O or $H_2{}^{18}O$ the lanosterol which was formed was devoid of deuterium or oxygen-18. However, in the presence of $^{18}O_2$, isotopically labelled material was obtained. Thus the process does not involve free partially cyclized intermediates requiring reprotonation for further cyclization. In 1966 Corey and van Tamelen showed[38] that a new intermediate, 2,3-epoxysqualene (**41**), was involved in the biosynthesis. This compound has been shown to play a principal role in the formation of many triterpenes. In the presence of a carrier, radioactivity from squalene has been trapped in the 2,3-epoxide. This epoxide has been detected in a number of plants and its absolute stereochemistry has been established. 2,3-Iminosqualene acts as a competitive inhibitor of the cyclization and the 2,3-epoxide accumulates, whilst if the substrate is 10,11-dihydrosqualene this is oxidized to the epoxide but not further metabolized. Hence there are two enzyme systems involved, an epoxidase and a cyclase.

The formation of lanosterol and its subsequent conversion to cholesterol has been the subject of considerable study. It has been proposed (see Scheme 10) that the cyclization proceeds to form a protolanosterol carbonium ion (**42**) which may be stabilized on the enzyme surface. This is followed by a series of hydrogen and methyl group migrations which terminate in the loss of the C-9 hydrogen atom and the formation of lanosterol (**44**). Squalene, containing six *pro*-4(*R*) mevalonoid hydrogen atoms, gave[39] lanosterol retaining only five of these labels, in accord with the loss of the C-9 hydrogen atom. Degradation of cholesterol biosynthesized from this material showed that the protons at C-17α and C-20β originated from C-13 and C-17 of the protolanosterol. Furthermore, Barton has shown[40] that lanosterol obtained from 2,3-oxido[11,14-^{3}H]squalene had 94% of its remaining tritium label at C-17. The C-11 ^{3}H-label of squalene epoxide becomes the C-9 ^{3}H-label of protolanosterol and is therefore lost in the formation of the lanosterol. The use of [3′,4-$^{13}C_2$]mevalonate in an elegant experiment by Cornforth revealed[41] that the rearrangement of the methyl groups from C-8 to C-14 and from C-14 to C-13 involved two 1,2 migrations rather than a single 1,3 (*i.e.* C-8 to C-13) migration. The migration within a single isoprene unit of a [^{13}C]-labelled methyl group from C-14 to

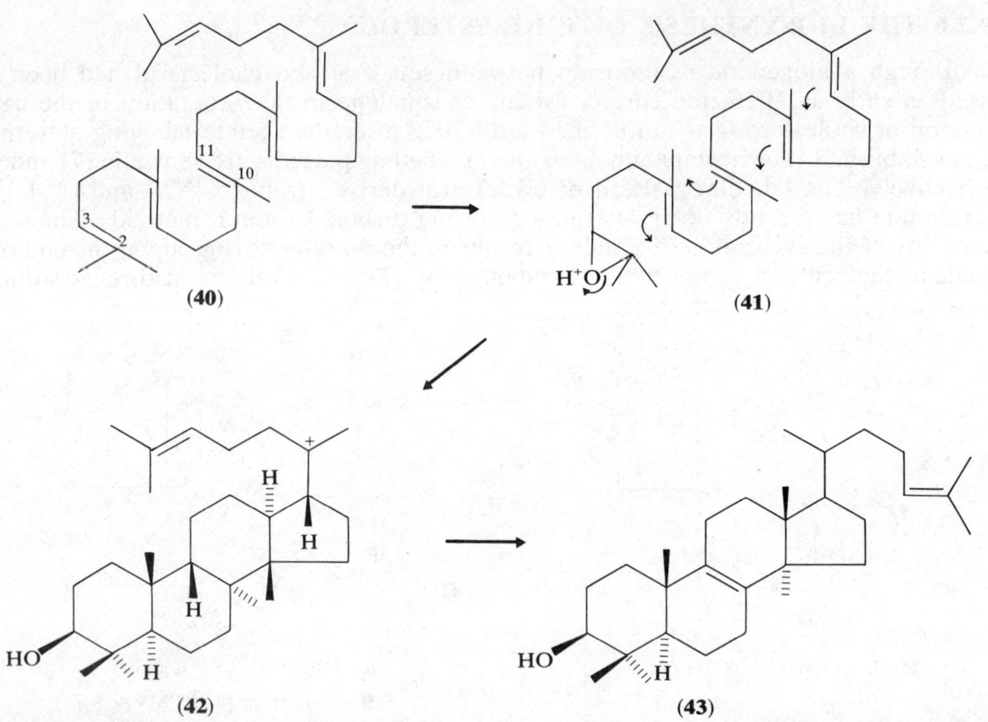

SCHEME 10

C-13, which was itself labelled by carbon-13 from the C-4 position of mevalonate, generated a species from which [$^{13}C_2$]acetic acid could be obtained on Kuhn–Roth degradation. By using mevalonic acid chirally labelled in the methyl group, it has been shown[18] that the steroidal C-18 methyl group retains its configuration during this rearrangement.

The conversion of lanosterol to cholesterol involves the loss of three methyl groups, the saturation of the side-chain double bond and the migration of the $\Delta^{8,9}$ double bond to the $\Delta^{5,6}$ position. During the conversion of lanosterol into cholesterol the methyl groups at C-4 are removed as carbon dioxide[42] and that at C-14 as formic acid.[43] 4,4′-Dimethylcholesta-8,24-dien-3β-ol has been detected in mammalian systems and shown to be readily converted into cholesterol, and thus the first methyl group to be eliminated is that at C-14. Labelling experiments with [3(*R*),2(*S*)-2-^3H]mevalonate showed[44] that the loss of a C-15α hydrogen atom is associated with the loss of the C-14α methyl group. The resultant 8,14-diene is then reduced in a *trans* manner with the addition of 14α and 15β hydrogen atoms. The removal of the two methyl groups at C-4 takes place in a stepwise fashion in which the C-4α methyl group is removed first. A C-3α label is lost during this process whilst the methyl groups eventually appear as carbon dioxide. Using 4,4-dimethyl-5α-cholest-7-en-3β-ol as a substrate, Gaylor showed[45] that this oxidative de-methylation stage could be differentiated into several steps. The first step, which required NADPH and oxygen, involved the hydroxylation of the methyl group. The second step, which proceeded under anaerobic conditions and utilized the oxidized pyridine nucleotide for a dehydrogenation reaction, involved the conversion of the alcohol to an aldehyde. The loss of a C-3α hydrogen atom is consistent with the formation of a 3-ketone, which by generating a β-dicarbonyl compound facilitates the loss of a C-4 function. However, the 4-methyl-3-keto steroids are not substrates for the hydroxylase and the 3β-hydroxy-4-methyl sterols are formed prior to further hydroxylation and removal of the second C-4 methyl group.

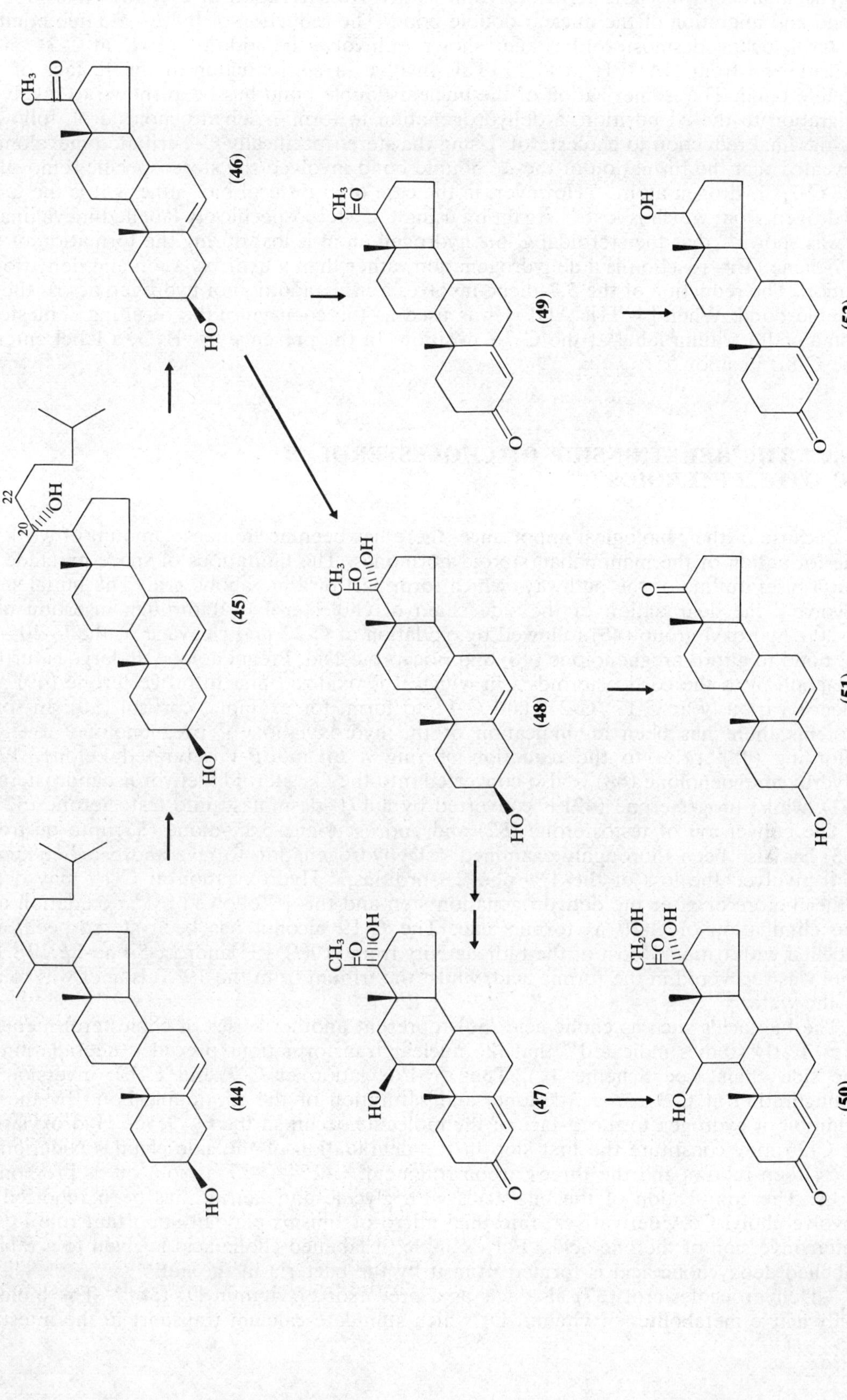

(44)
(45)
(46)
(47)
(48)
(49)
(50)
(51)
(52)
CH₃
CH₂OH
OH
HO
O
20
22

The final steps in cholesterol biosynthesis involve saturation of the side-chain double bond and migration of the nuclear double bond. The reduction of the Δ^{24} double bond of a sterol such as desmosterol has been shown to involve the addition of H^+ at C-24 and a hydride ion from NADPH at C-25. This involves a *cis* reduction at the *re* face of the double bond. The isomerization of the nuclear double bond has been shown to involve a migration to the Δ^7 position, a dehydrogenation to form 7-dehydrocholesterol, followed by the final reduction to cholesterol. Using the stereospecifically C-2-tritiated mevalonates revealed that the formation of the Δ^7 double bond involved the stereospecific removal of the C-7β hydrogen atom.[44] However, in the case of ergosterol biosynthesis it is the C-7α hydrogen atom which is lost.[45] Again by using C-5 stereospecifically labelled mevalonates it was shown[46] that the steroidal C-6α hydrogen atom is lost during the formation of the 5,7-diene. This reaction is a dehydrogenation rather than a hydroxylation and dehydrogenation. The reduction of the 5,7-diene involves a *trans*-addition of hydrogen across the Δ^7 double bond. When [4-^{3}H]NADPH was used as the coenzyme, the resulting cholesterol contained a tritium label at the C-7α position. In the presence of ^3H_2O, a label entered the C-8β position.[47]

29.2.7 THE RELATIONSHIP OF CHOLESTEROL TO OTHER STEROIDS[48,49]

Because of their biological importance, there has been an immense amount of work on the formation of the mammalian steroid hormones. The limitations of space preclude all but a brief outline of the pathways which form part of a metabolic grid. The initial steps involve[50] the degradation of the side chain of cholesterol (**44**) through insertion of a C-20α hydroxyl group (**45**) followed by oxidation at C-22 and cleavage of the C-20—C-22 bond to afford pregnenolone (**46**) and isocaproic acid. Pregnenolone undergoes further metabolism to the corticosteroids[51] in which it is oxidized first to progesterone (**49**) and thence variously at C-17, C-21, and C-11 to form, for example, cortisol (**50**). In some systems there has been an indication of the hydroxylation of pregnenolone at C-17, affording (**48**), prior to the oxidation of ring A to an $\alpha\beta$-unsaturated ketone. 17α-Hydroxypregnenolone (**48**) is also converted into the C_{19} steroid, dehydroisoandrosterone (**51**), whilst progesterone (**49**) is converted by a 17β-desmolase into testosterone (**52**).

The conversion of testosterone (**52**) and androst-4-ene-3,17-dione (**53**) into oestrone (**55**) has also been thoroughly examined.[48] Dehydrogenation to give androstadienedione (**54**) involved the loss of the 1β- and 2β-protons.[52] Hydroxylation at C-19 may occur either before or after the dehydrogenation step and this is followed by an oxidation and the elimination of C-19 as formic acid. The C-19 alcohol has been stereospecifically labelled with tritium. Most of the radioactivity from [19(*R*)-^{3}H]androst-5-ene-3β,17β,19-triol was recovered in the formic acid, whilst the tritium from the 19(*S*) isomer was found in the water.[53]

The bile acids such as cholic acid (**56**) represent another series of cholesterol metabolites. Early studies indicated[48] that the nuclear transformations preceded degradation of the side chain (see Scheme 11). Thus hydroxylation at C-7 and C-12, inversion of configuration at C-3 *via* a 3-ketone, and saturation of the Δ^5 double bond by the *cis* addition of hydrogen to the β-face of the molecule occurs at the C_{27} level. Hydroxylation at C-26 may constitute the first step in the degradation of the side chain. Oxidation at C-24 then follows and the three-carbon fragment, C-25—C-27, is split off as propionyl-CoA. The conjugation of the bile acids with glycine and taurine has been reported to involve cholyl-CoA derivatives. Intestinal micro-organisms play an important role in the interconversion of the bile acids. For example, if labelled cholic acid is given to a rabbit, labelled deoxycholic acid is formed from it by the bacteria in the gut.

7-Dehydrocholesterol (**57**) also acts as a precursor of vitamin D_3 (**58**).[54] The biologically active metabolites of vitamin D_3 which stimulate calcium transport in the intestine

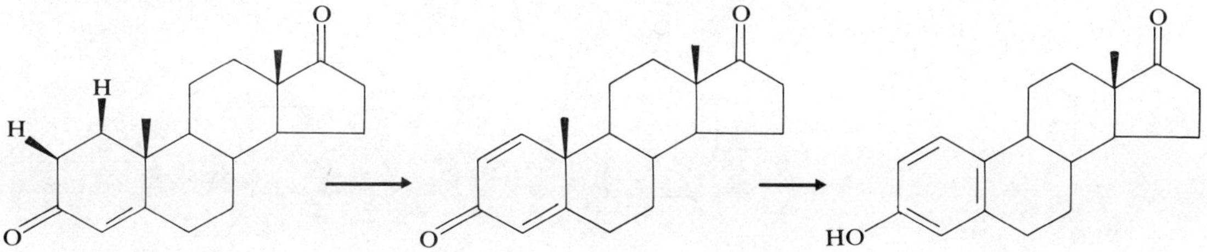

(53) (54) (55)

are 25-hydroxy- (**59**) and 1α,25-dihydroxy-vitamin D_3 (**60**) derivatives. The 1α,25-dihydroxyvitamin D_3 is produced in the kidney and has many of the properties of a hormonal compound. Its biological activity is linked with the formation of a peptide, the parathyroid hormone. Another metabolite is the 1α,24(*R*),25-trihydroxy-derivative.

Insects rely on a dietary source of sterols for the formation of the insect moulting hormones of the ecdysone series.[55] Again nuclear transformation precedes side-chain modification. Cholesterol is converted to 7-dehydrocholesterol by the loss of the C-7β and C-8β hydrogen atoms prior to oxidation to afford the characteristic 14α-hydroxy-7-en-6-one system. 3β,14α-Dihydroxy-5β-cholest-7-en-6-one (**61**) has been shown by labelling studies to be hydroxylated first at C-25 and then at C-22 and C-21 in the biosynthesis of ecdysterone (**62**). Phytophagous insects are capable of dealkylating the side chain of phytosterols in the formation of ecdysone. A surprising aspect of compounds of the ecdysone type is that they occur in substantial quantities in some plants, and their biosynthesis has also been studied in these systems.

29.2.8 THE BIOSYNTHESIS OF PLANT STEROLS[56,57]

The biosynthesis of the plant sterols differs in details from that of the animal steroids. Whereas lanosterol is of limited occurrence in higher plants, the related cyclopropyl triterpenes, cycloartenol (**63**) and 24-methylenecycloartanol, are widespread. Thus when the standard terpenoid precursors such as [2-^{14}C]mevalonate were fed to the leaves of *Zea mays* or *Pisum sativum*, these compounds rather than lanosterol were labelled. Furthermore, when cycloartenol and 24-methylenecycloartanol were biosynthesized from [4(*R*)-4-^{3}H,2-^{14}C]mevalonate, they retained six labelled atoms as opposed to the five which would be expected if lanosterol were implicated in their biosynthesis. The formation of squalene 2,3-epoxide has been demonstrated in tobacco tissue cultures and this system incorporated radioactivity from [^{14}C]squalene 2,3-epoxide into cycloartenol and 24-methylenecycloartanol. A number of plant systems have been obtained which will convert these triterpenes into plant sterols such as β-sitosterol. Thus cycloartenol (**63**) is the initial product of cyclization of squalene 2,3-epoxide in plants and is the precursor of the phytosterols.

Alkylation of the side chain[58] by *S*-adenosylmethionine to form the 24-methylene sterols can take place either at this stage, prior to the removal of the nuclear methyl groups, or at a much later stage, as in the biosynthesis of ergosterol. In those molecules in which there is an ethyl group in the side chain, both carbon atoms come from methionine. The alkylation reaction is accompanied by the migration of a hydrogen atom from C-24 to C-25.

A number of sterols found in higher plants possess a Δ^{22} *trans*-double bond. The stereochemistry of the hydrogen elimination from C-22 and C-23 in poriferasterol has been examined using stereospecifically C-2- and C-5-labelled mevalonates. The overall process involves the *cis*-removal of the 22-*pro*-(*R*) and 23-*pro*-(*R*) hydrogen atoms. However, in the case of ergosterol biosynthesis it is the *pro*-(*S*) atom at C-22 which is removed. As far as the sequence of nuclear demethylation is concerned, it is probable that one of the C-4 methyl groups is lost prior to demethylation at C-14 [*cf.* the occurrence of

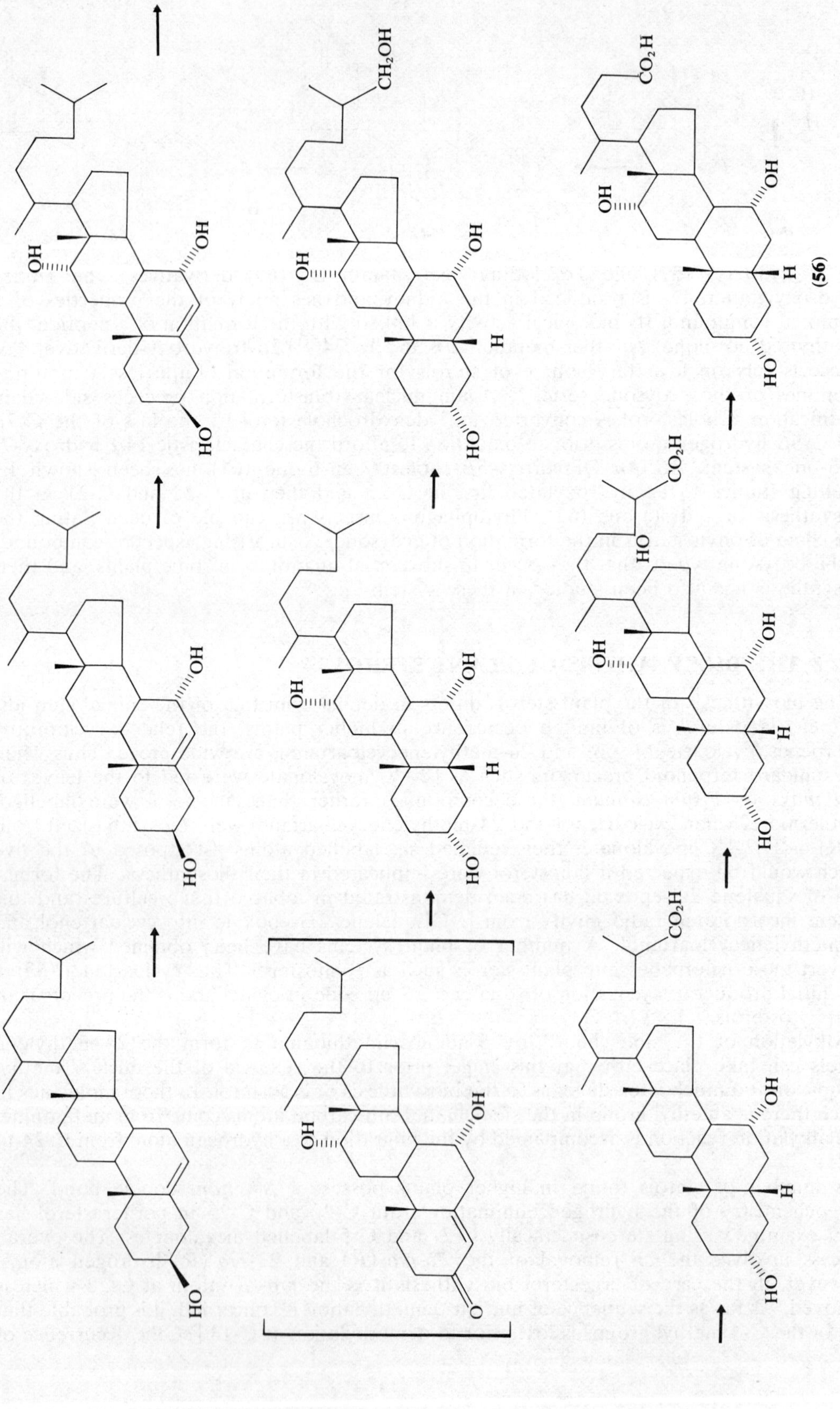

(56)

SCHEME 11

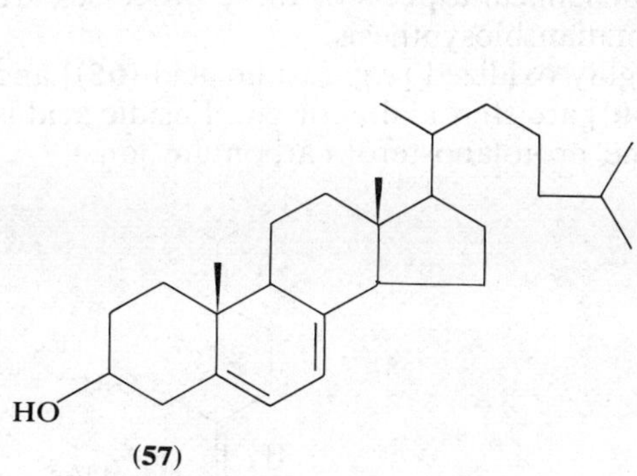

(**57**)

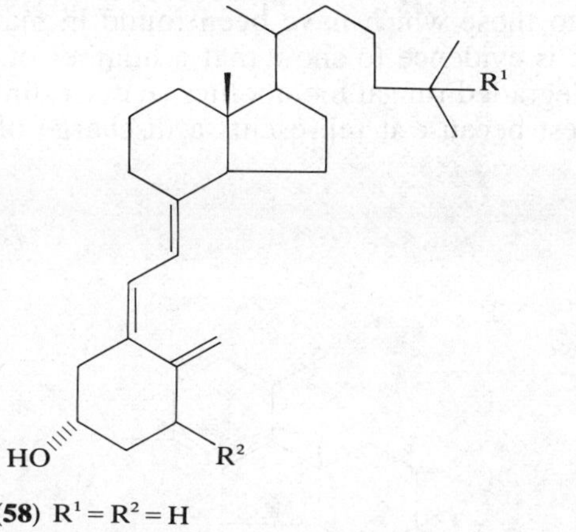

(**58**) $R^1 = R^2 = H$
(**59**) $R^1 = OH,\ R^2 = H$
(**60**) $R^1 = R^2 = OH$

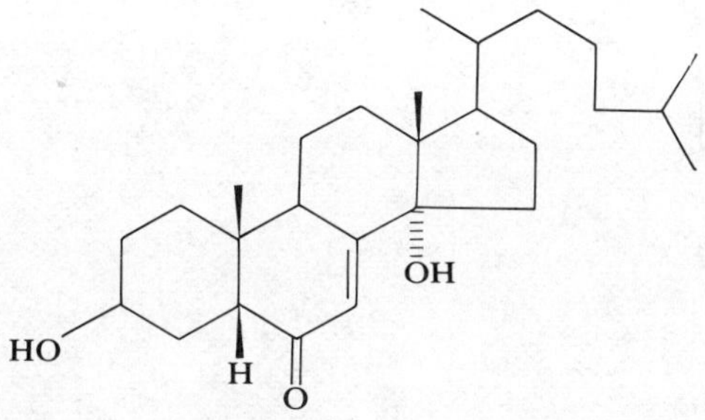

(**61**)

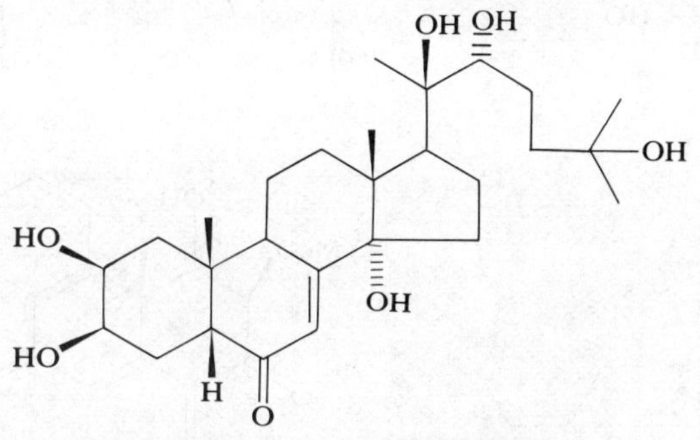

(**62**)

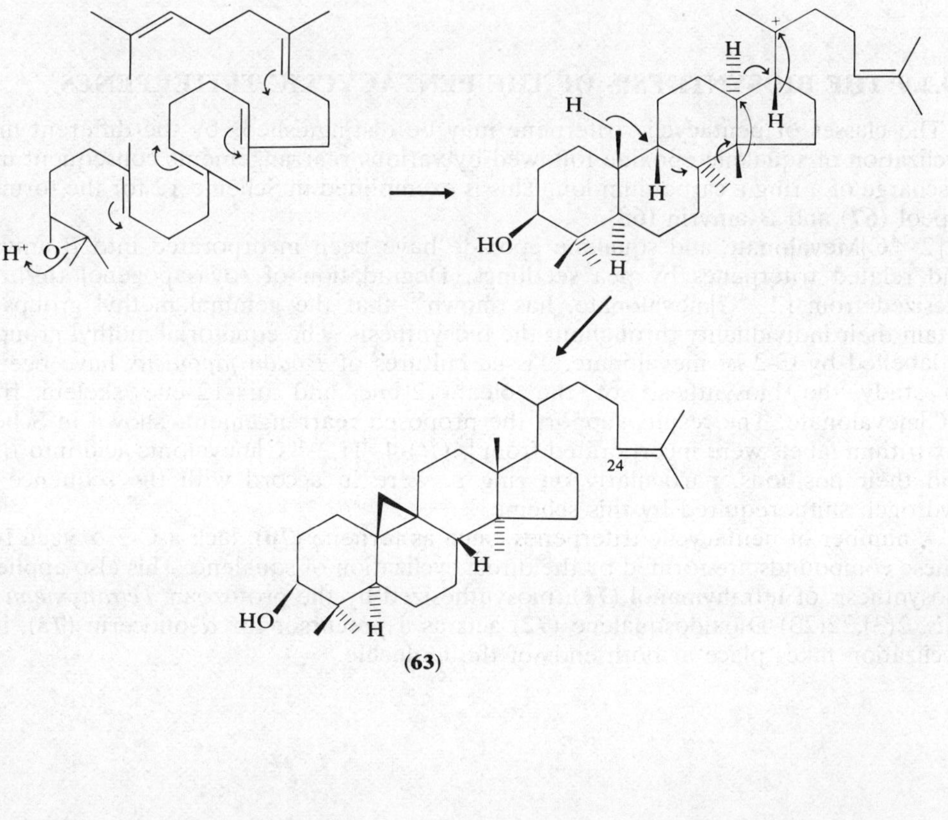

(**63**)

cycloeucalenol (**64**)]. However, many other stereochemical aspects of these processes are similar to those which have been found in mammalian biosynthesis.

There is evidence to show that a number of highly oxidized [*e.g.* fusidic acid (**65**)] and highly degraded fungal metabolites [*e.g.* viridin (**66**)] are steroidal in origin. Fusidic acid is of interest because it represents a discharge of the protolanosterol carbonium ion.

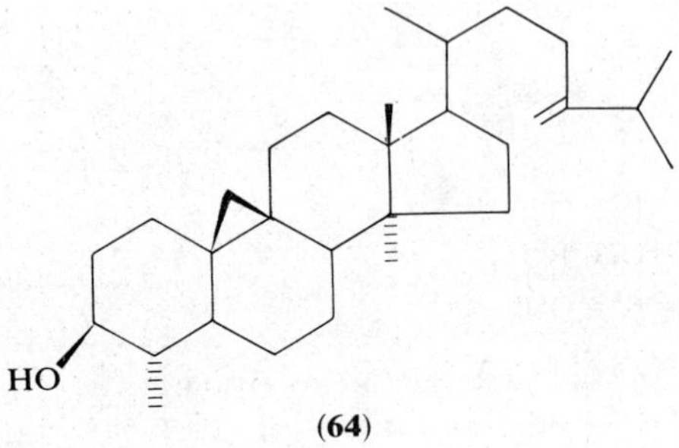

(**64**)

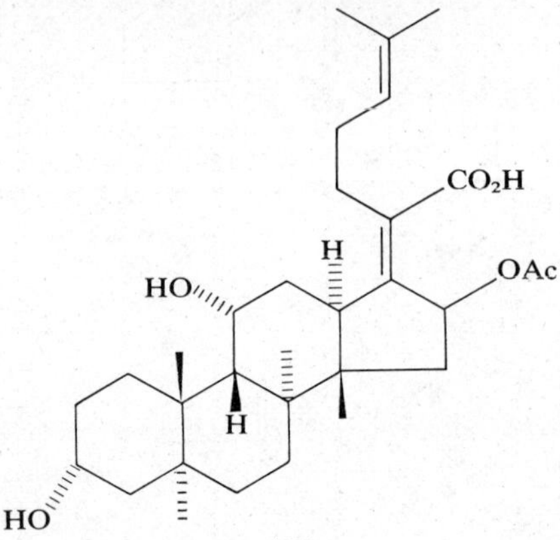

(**65**)

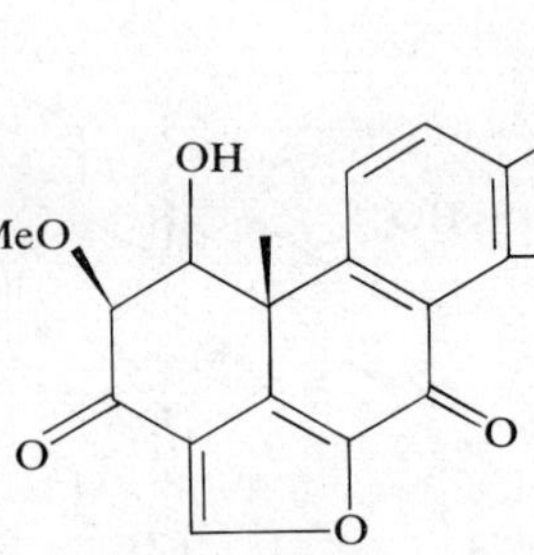

(**66**)

29.2.9 THE BIOSYNTHESIS OF THE PENTACYCLIC TRITERPENES[2]

The classes of pentacyclic triterpene may be distinguished[2] by the different modes of cyclization of squalene epoxide followed by various rearrangements consequent upon the discharge of a ring E carbonium ion. This is exemplified in Scheme 12 for the formation of lupeol (**67**) and β-amyrin (**68**).

[2-^{14}C]Mevalonate and squalene epoxide have been incorporated into β-amyrin (**67**) and related triterpenes by pea seedlings. Degradation of soyasapogenol (**69**), biosynthesized from [2-^{14}C]mevalonate, has shown[59] that the geminal methyl groups at C-4 retain their individuality throughout the biosynthesis. The equatorial methyl group at C-4 is labelled by C-2 of mevalonate. Tissue cultures of *Isodon japonicus* have been used[60] to study the biosynthesis of the olean-12-ene and urs-12-ene skeleta from [4-^{13}C]mevalonate. The results support the proposed rearrangements shown in Scheme 12. Six tritium labels were incorporated from [4(*R*)-4-^{3}H,2-^{14}C]mevalonic acid into β-amyrin and their positions, particularly on ring E, were in accord with the sequence of 1,2-hydrogen shifts required by this scheme.

A number of pentacyclic triterpenes, such as fernene (**70**), lack a C-3 oxygen function. These compounds are formed by the direct cyclization of squalene. This also applies to the biosynthesis of tetrahymanol (**71**), biosynthesized by the protozoan *Tetrahymena pyriformis.* 2(3),22(23)-Dioxidosqualene (**72**) acts as a precursor for α-onocerin (**73**), in which cyclization takes place at both ends of the molecule.

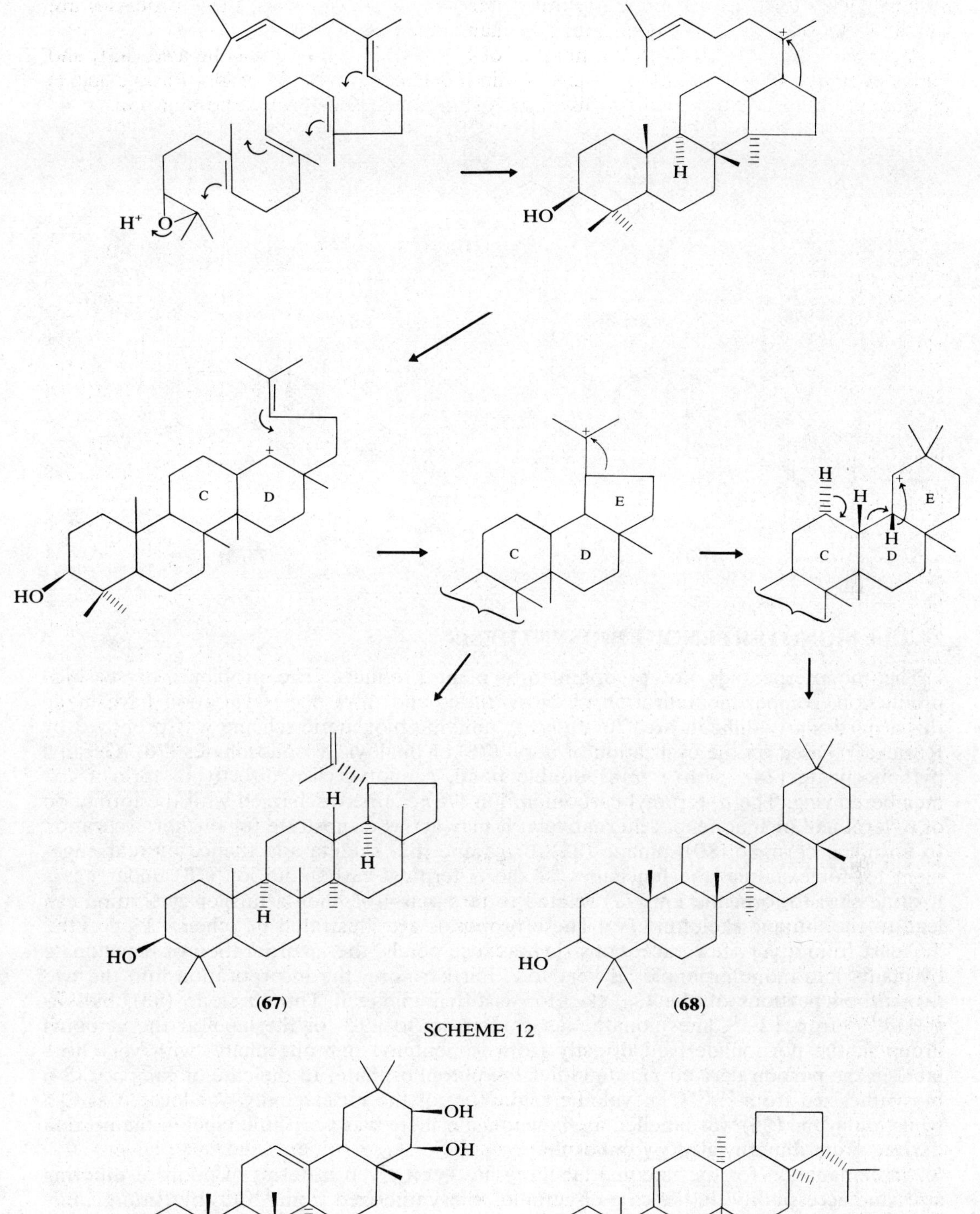

(67) **(68)**

SCHEME 12

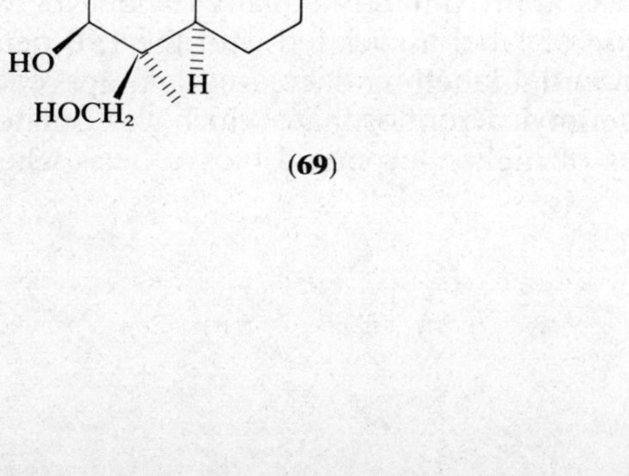

(69)

(70)

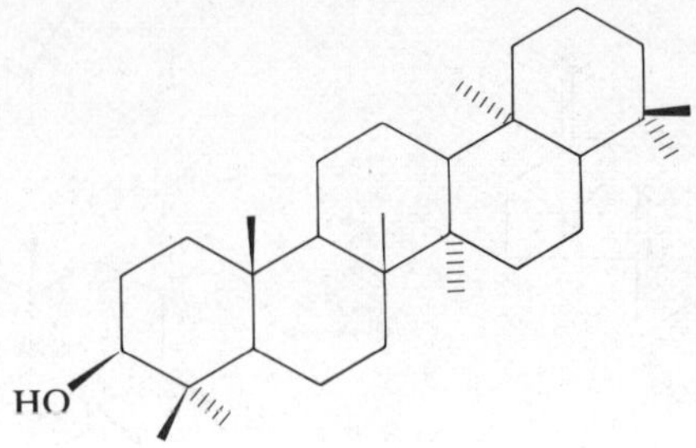

(71)

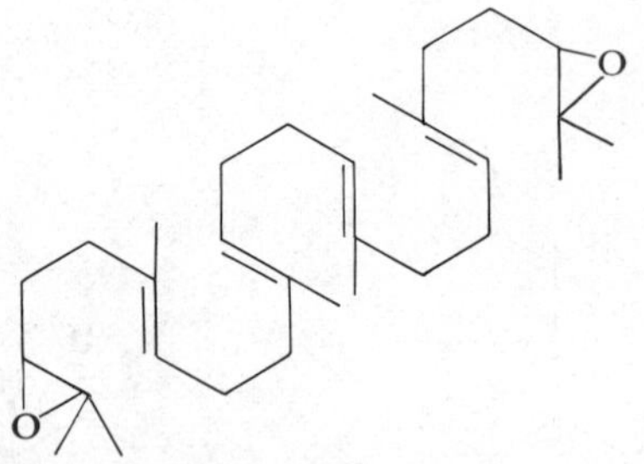

(72)

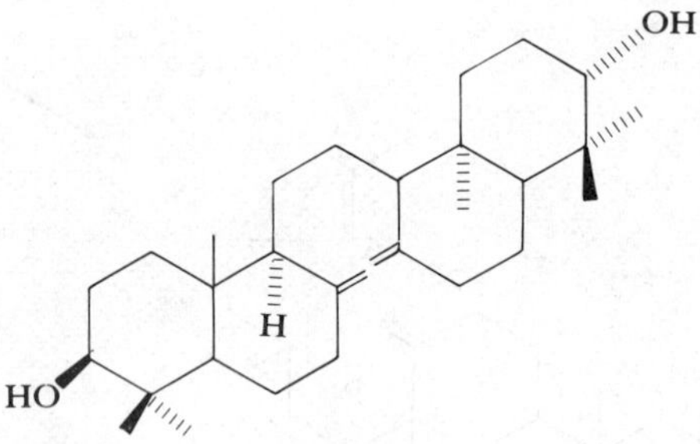

(73)

29.2.10 MONOTERPENOID BIOSYNTHESIS[61]

The monoterpenoids are predominantly plant products. The problems of seasonal production, compartmentalization of biosynthesis and infraspecies variation have made this a particularly difficult area to study. A unifying biogenetic scheme was proposed by Ruzicka[2,3] based on the cyclization of neryl (**75**) or linaloyl pyrophosphates (**76**). Geranyl pyrophosphate (**74**), with a *trans* double bond, cannot cyclize directly to form a six-membered ring. The α-terpinyl carbonium ion (**78**) can be discharged with the formation of α-terpineol or limonene. Alternatively, it may act as a substrate for further cyclization to form the bornane (**80**), pinane (**81**), or carane (**82**) skeleta and thence by rearrangement to, for example, the fenchanes. If the α-terpinyl carbonium ion (**78**) undergoes a hydride shift to form the ion (**77**) related to terpinen-4-ol, then a further cyclization can lead to the thujane skeleton (**79**). These proposals are illustrated in Scheme 13.

Apart from a very few exceptions such as rose petals, the incorporation of mevalonate by plants into monoterpenoids is very low. Furthermore, the incorporation into the two five-carbon portions of the C_{10} skeleton is often unequal. Thus thujone (**83**), biosynthesized[62] from [2-^{14}C]mevalonate, can contain up to 99% of the label at the carbonyl group in the portion derived directly from isopentenyl pyrophosphate, with very little label in the portion derived *via* dimethylallyl pyrophosphate. In the case of camphor (**84**) biosynthesized from [2-^{14}C]mevalonic acid, most of the radioactivity was located at C-6 whilst pulegone (**85**) was labelled as shown; again there was very little label in the portion derived from dimethylallyl pyrophosphate.

An explanation for the unequal labelling has been given in terms of pools of differing sizes and accessibility. In the case of geraniol, biosynthesized from $^{14}CO_2$ by *Perlargonium graveolens,* degradation showed that there was a greater turn-over of label in the portion derived directly from IPP. At first the label was approximately equally divided between the two portions, but after 12 hours the portion of label associated with the IPP-derived half increased to 78%. Subsequently the preferential labelling decreased and approached an equal distribution. This suggested that the geranyl pyrophosphate which was isolated at first arose as a result of leakage from the sites of higher terpenoid biosynthesis whereas

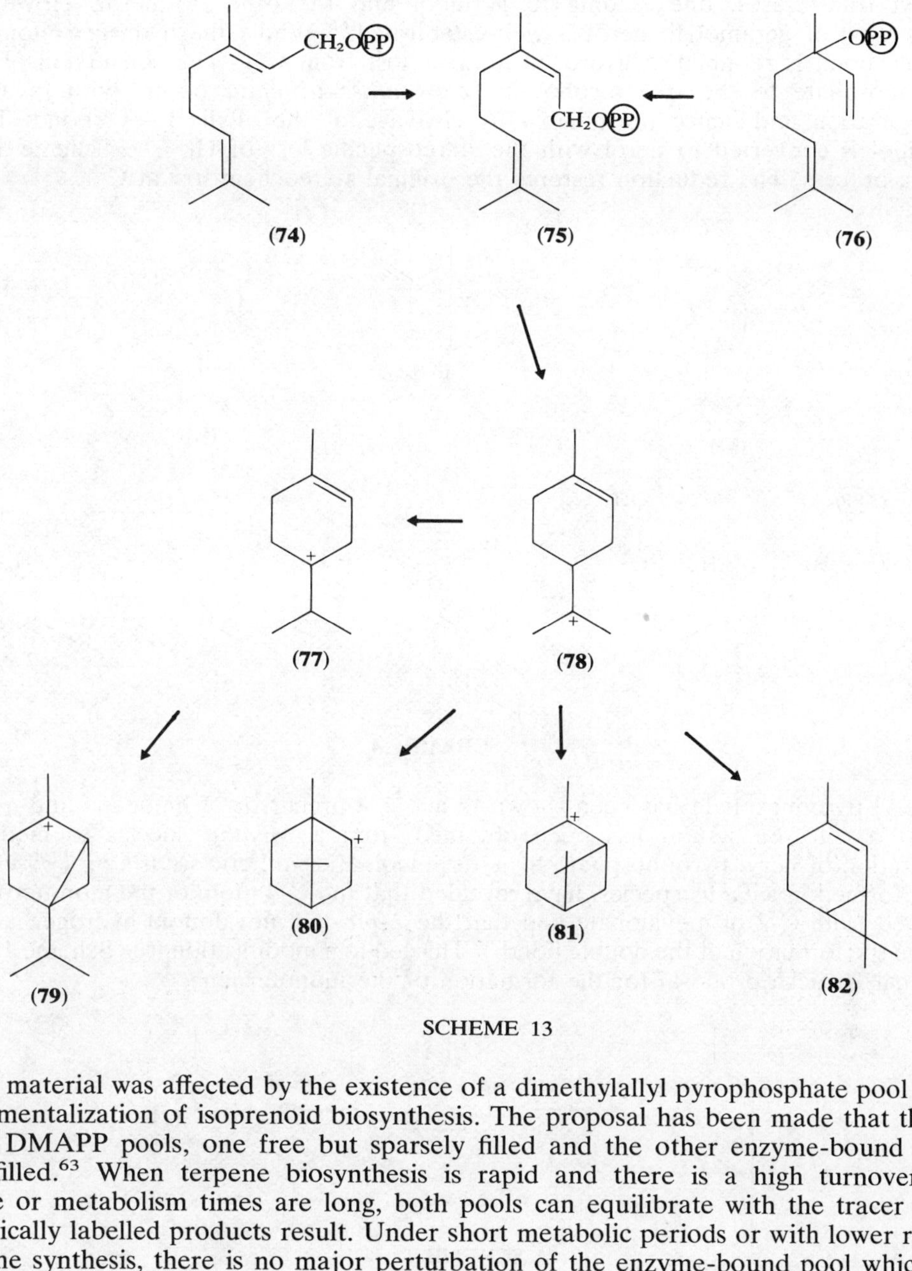

SCHEME 13

the later material was affected by the existence of a dimethylallyl pyrophosphate pool and compartmentalization of isoprenoid biosynthesis. The proposal has been made that there are two DMAPP pools, one free but sparsely filled and the other enzyme-bound and largely filled.[63] When terpene biosynthesis is rapid and there is a high turnover of substrate or metabolism times are long, both pools can equilibrate with the tracer and symmetrically labelled products result. Under short metabolic periods or with lower rates of terpene synthesis, there is no major perturbation of the enzyme-bound pool which is utilized for terpenoid synthesis and hence asymmetric labelling occurs. Low incorporations of amino-acids such as leucine and valine have been reported for geraniol, citronellol, and linalool. Partial degradation established the presence of the major portion of the label in the DMAPP portion of the monoterpenoids, but it was not established how much of the carbon skeleton of the amino-acids entered the DMAPP pool intact.

Geranyl pyrophosphate has been shown[64] to act as the parent of a number of cyclic monoterpenoids, such as cineole and the iridoids. In order to form the cyclohexane monoterpenoids and their bicyclic relatives, the *trans* double bond of geranyl pyrophosphate has to adopt the *cis* geometry of nerol. This isomerization and the relationship between geraniol and nerol has, as in the comparable problem in the sesquiterpenoid series, attracted considerable attention. There is some evidence for the presence of two prenyl transferases, one leading to geraniol and the other to nerol. However, the conversion of geraniol to nerol is well-established.[65] Whilst the hydrogen atom on the double bond is retained, a hydrogen atom is lost from C-1. The conversion of geranyl pyrophosphate to the free alcohol in *Menyanthes trifoliata* occurs with inversion of configuration and hence presumably by cleavage of the allylic C—O bond. The free geraniol is converted to nerol with the stereospecific loss of H_B (see Scheme 14) by a redox process. The reduction restores the original stereochemistry at C-1.

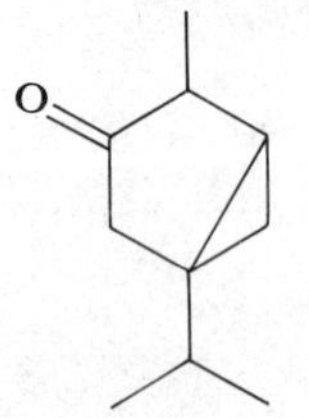

(83)

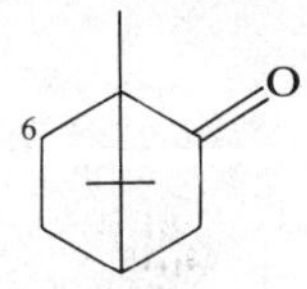

(84)

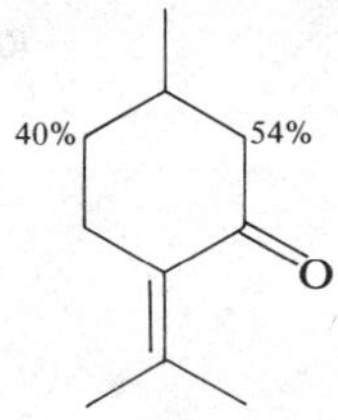

(85)

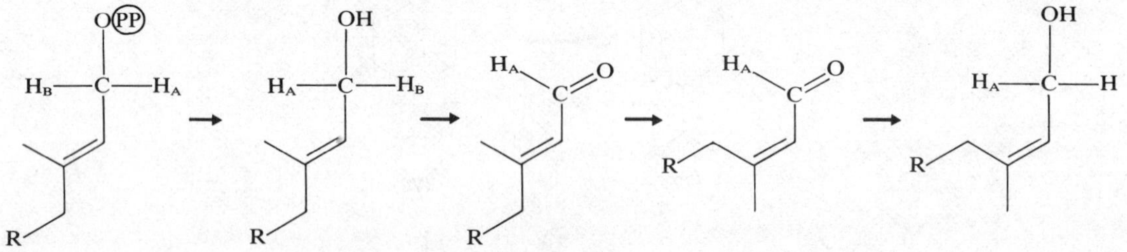

SCHEME 14

Neryl pyrophosphate has been shown to act as a precursor of limonene and α-pinene, whilst a cell-free system has been obtained from a *Mentha* species to mediate the cyclization of neryl pyrophosphate to α-terpineol. Biosynthetic studies on (+)-car-3-ene (**86**), formed in a *Pinus* species, have revealed that the C-4 atom of the monoterpene was derived from C-2 of mevalonate and that the 2-*pro*-(*S*) mevalonoid hydrogen atom was lost in the formation of the double bond.[66] This led to a modification (see Scheme 15) of the original Ruzicka proposal for the formation of the monoterpene.

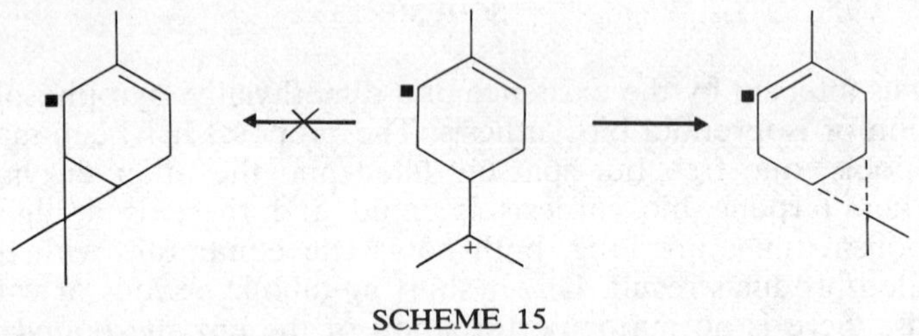

SCHEME 15

Biosynthetic studies on *Mentha* species have established a relationship between piperitone (**87**), pulegone (**88**), menthone (**89**), and menthol (**90**), on the one hand, and between piperitone (**91**), isomenthone (**92**), and isomenthol (**93**), on the other.

A number of monoterpenoids such as artemisia ketone (**94**), chrysanthemic acid (**95**), and lavandulol (**96**) possess an abnormal isoprenoid skeleton and consequently their biosynthesis has attracted considerable speculation.[67—69] Current biogenetic proposals suggest that these compounds are formed by routes not involving the normal C_{10} intermediates of terpenoid biosynthesis. [2-^{14}C]Mevalonic acid and isopentenyl pyrophosphate have been incorporated into artemisia ketone (**94**) such that the five-carbon fragment not containing the carbonyl group was preferentially labelled. Geranyl

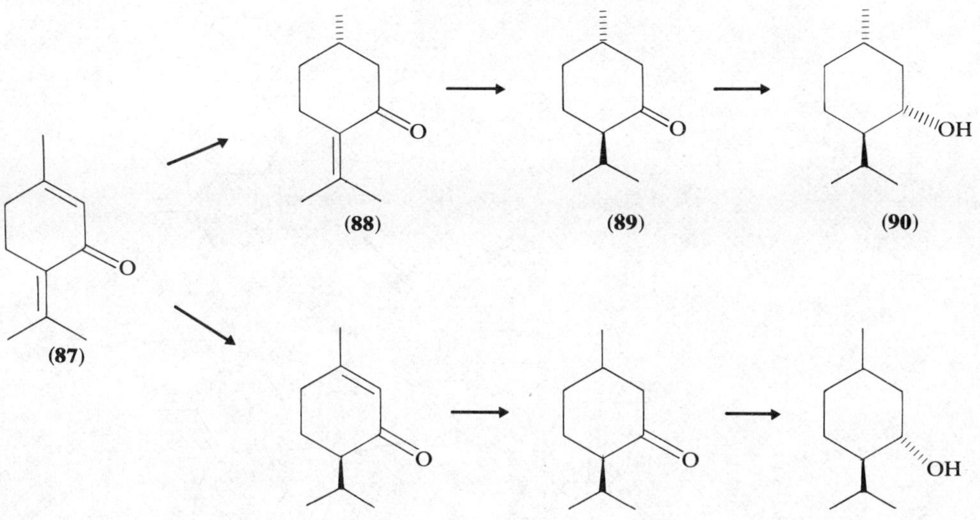

(87) (88) (89) (90) (91) (92) (93)

pyrophosphate did not appear to be an intermediate. Chrysanthemic acid (**95**) has also been shown to incorporate [2-^{14}C]mevalonic acid. Feeding and trapping experiments in *Chrysanthemum cinerariaefolium* have established that chrysanthemyl alcohol is an intermediate in the biosynthesis of the chrysanthemic acid, whereas lavandulol and artemisyl alcohol, from which the cyclopropane ring might have been generated, were not incorporated.

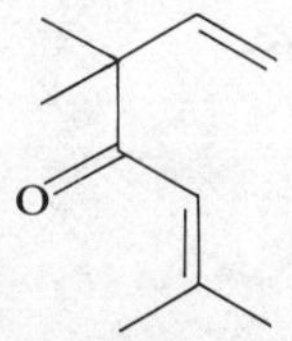

(94)

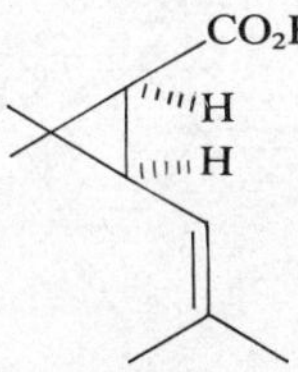

(95)

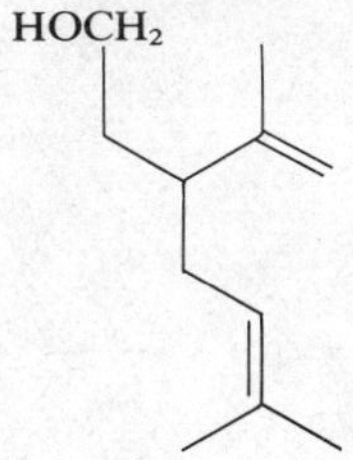

(96)

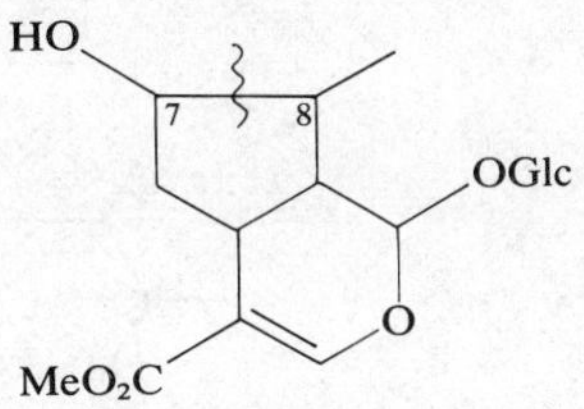

(97)

The cyclopentane monoterpenoids, based on the carbon skeleton of loganin (**97**) and its 7,8-seco derivatives, form an interesting group of monoterpenoids whose biosynthetic relationship with the indole alkaloids has been extensively explored (see Chapter 30.1).

29.2.11 SESQUITERPENOID BIOSYNTHESIS[70,71]

Although there are many different sesquiterpenoid skeleta, these may be grouped in terms of a few primary cyclizations of farnesyl pyrophosphate. The major cyclization

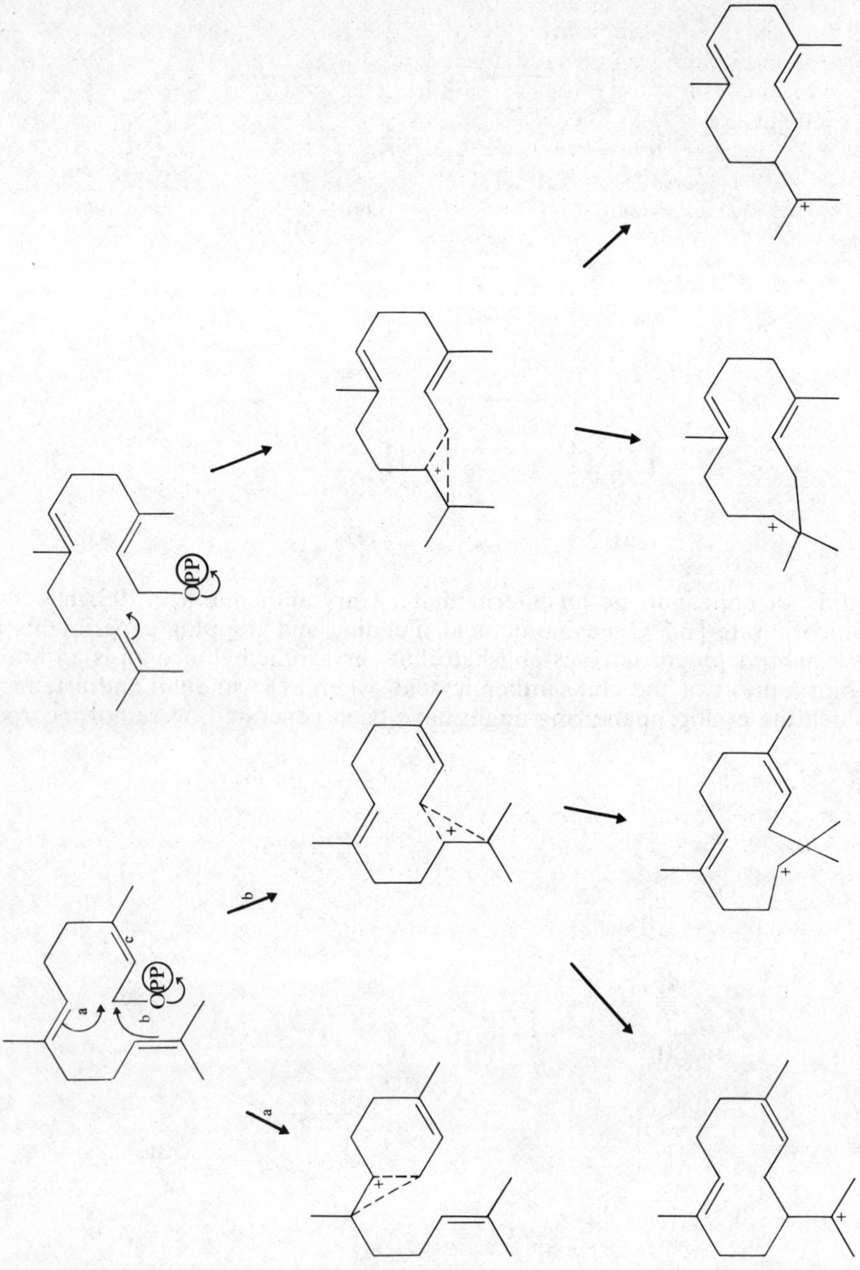

SCHEME 16

reactions involve the pyrophosphate attacking the central or the distal double bond. In order to attack the central double bond, the C-2 double bond must adopt the *cis* geometry. This geometrical requirement is also necessary for further secondary cyclizations which generate the polycyclic sesquiterpenoids containing five-, six-, or seven-membered rings (Scheme 16).

There are a few substances such as drimenol (**98**) whose origin may involve a cyclization typical of the triterpenoids. A few others such as cyclonerodiol (**99**) may be related to a derivative of nerolidol (**100**).

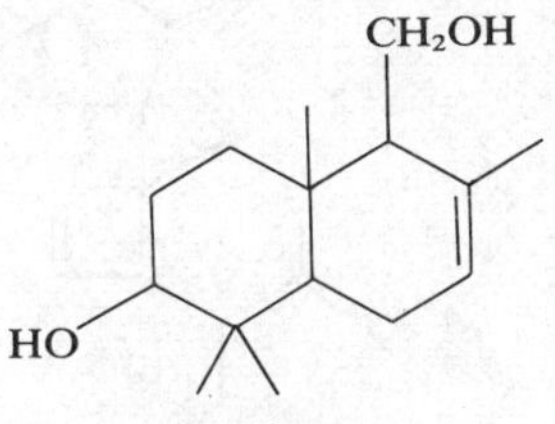

(98)

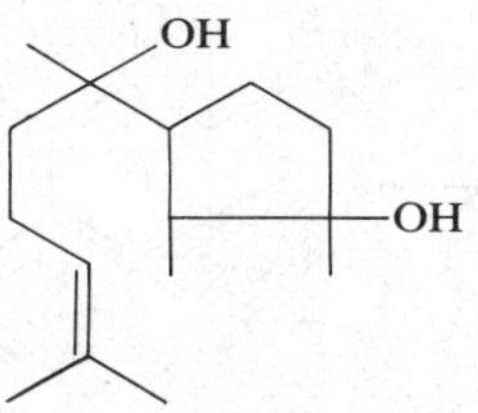

(99)

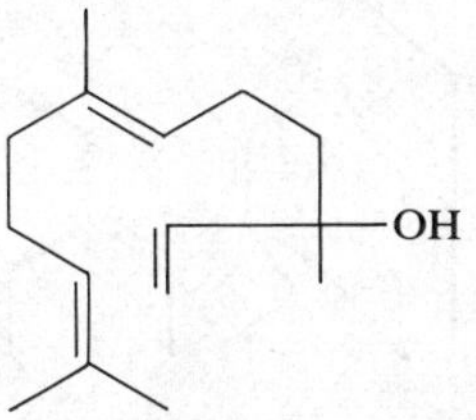

(100)

The isomerization of all-*trans*-farnesyl pyrophosphate and the parent alcohol to the 2-*cis*-isomer may follow a number of paths. One pathway which is found in the fungus, *Helminthosporium sativum*, involves[72] a redox relationship with the corresponding aldehyde. In this oxidation a C-1 *pro*-(*R*) hydrogen atom is lost. A second pathway, studied[73] in tissue cultures from *Andrographis paniculata*, involved the stereospecific loss of the C-1 *pro*-(*S*) hydrogen in the forward reaction from 2-*trans*-farnesyl pyrophosphate to the 2-*cis*-isomer. The reverse isomerization of the 2-*cis*-isomer into the 2-*trans*-isomer involved the loss of the C-1 *pro*(*R*) hydrogen atom. The isomerization of all-*trans*-farnesyl pyrophosphate by *Trichothecium roseum* and the formation of trichodiene also involved[74] the loss of the C-1 *pro*(*S*) hydrogen atom and its ultimate replacement with the C-4 *pro*-(*S*) hydrogen atom of NADPH in the formation of trichodiene. In another example studied[65] in the biosynthesis of longifolene, both hydrogen atoms were retained but with inversion of configuration at C-1. An explanation has been provided in terms of a suprafacial anionotropic rearrangement of the *trans*-pyrophosphate (**101**) to nerolidyl pyrophosphate (**102**), a rotation (**102** → **103**), and reversal of the anionotropic rearrangement to afford the *cis*-isomer (**104**).

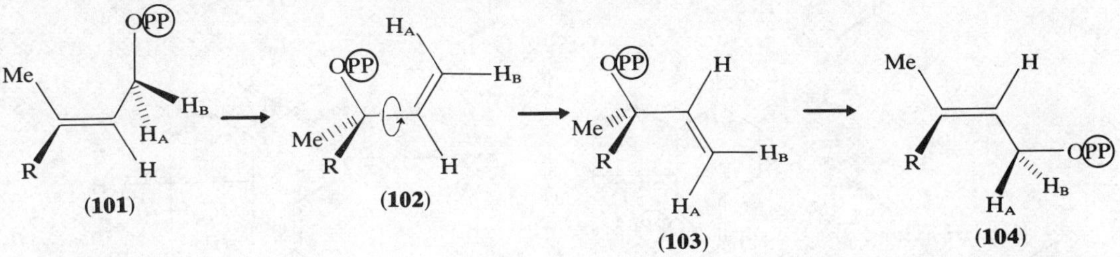

(101) **(102)** **(103)** **(104)**

Recently, carbon-13 n.m.r. methods have become widely used in defining the manner of folding of farnesyl pyrophosphate in the formation of cyclic sesquiterpenoids. These have included[20,73,75,76] both enrichment studies and coupling constant measurements. The labelling and coupling patterns of some sesquiterpenoids biosynthesized from labelled acetate and mevalonate are given in Scheme 17. In several instances these studies augment earlier carbon-14 experiments.

During the course of these cyclizations a number of hydride shifts and other rearrangements occur. In the case of trichothecin (**105**) and helicobasidin (**106**) there is a hydride shift from the central isoprene unit to the starter unit. In the case of the trichothecenes[77] this serves to initiate the methyl group migrations (see Scheme 18).

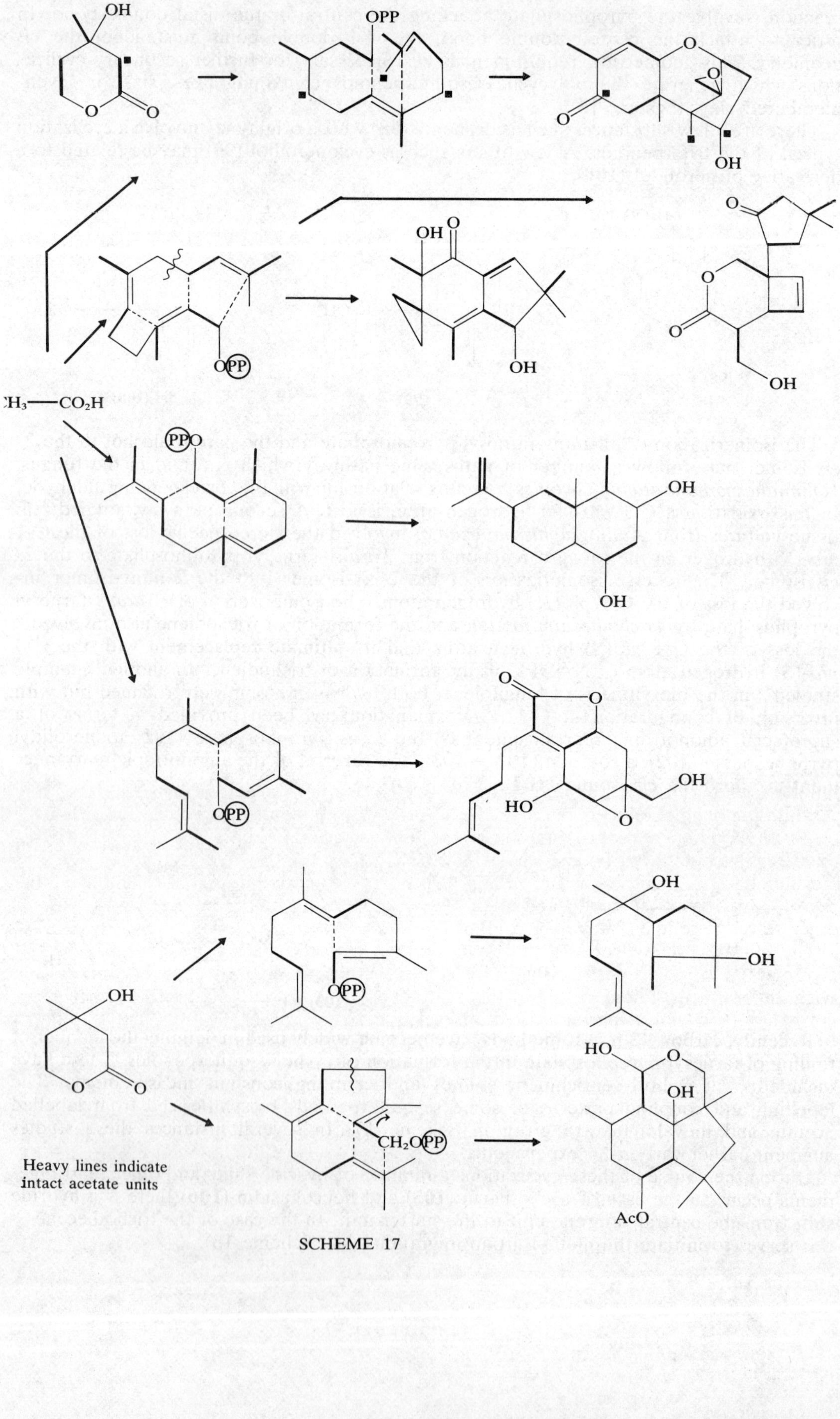

Heavy lines indicate intact acetate units

SCHEME 17

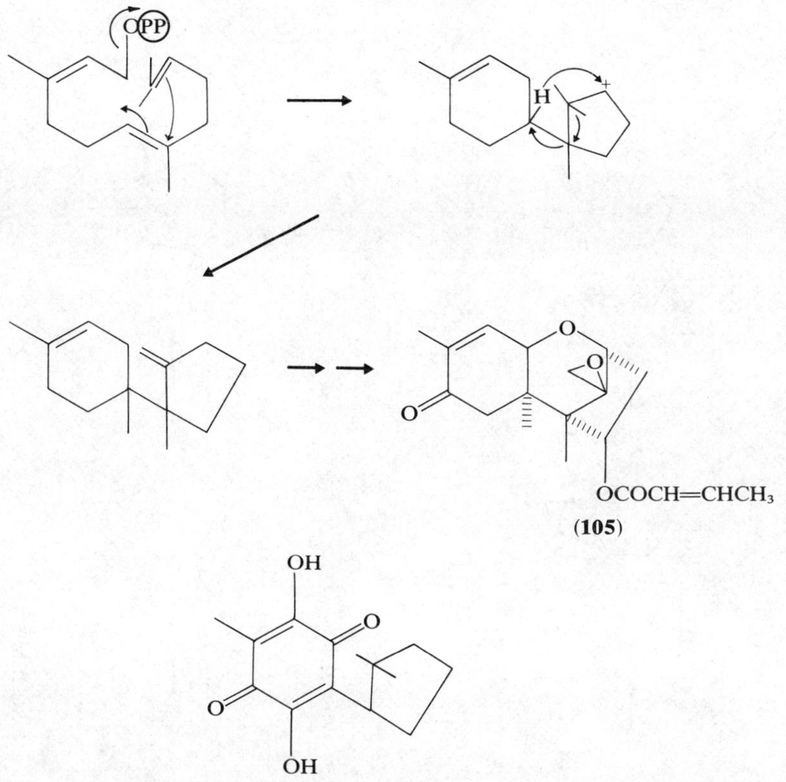

(**105**)

(**106**)

SCHEME 18

A number of polycyclic sesquiterpenoids arise through the initial attack of the C-1 pyrophosphate on the distal double bond followed by a further cyclization. In these cases a number of 1,3-hydrogen shifts have been detected. These serve to transfer a cationic centre from C-10 or C-11 at the distal end of farnesyl pyrophosphate back to C-1 to initiate further isomerization and secondary cyclizations of the 10- or 11-membered ring. A route has been presented[65] (Scheme 19) for the formation of the tricyclic hydrocarbons (−)-sativene (**117**) and ent-longifolene (**108**) *via* 10-membered and 11-membered intermediates, respectively. In the case of (−)-sativene, the *pro*-5(*R*) hydrogen (H_a) of mevalonate was involved in the migration whereas in the case of ent-longifolene, the *pro*-5(*S*) hydrogen was involved. Similar rearrangements have been detected in the case of other sesquiterpenoids, including culmorin, avocettin, and dendrobine.

The mevalonoid hydrogen-labelling pattern of several groups of sesquiterpenoid, including the trichothecenes, illudins, and coriolins, culmorin, avocettin, and sativene, have been determined. These results indicate that processes such as hydroxylation follow normal stereochemical patterns. Several biosynthetic sequences, such as that leading to trichothecin *via* trichodiene and trichodermol, have been determined.

Abscisic acid (**109**) is a widespread plant-growth regulator whose biological importance has led to extensive studies on its biosynthesis.[78] A cell-free system has been developed from avocado fruit for this purpose. Although it might be regarded as a degraded carotenoid, when [^{14}C]phytoene was applied to avocado fruit, radioactivity was incorporated into carotenoids but not into abscisic acid. However, abscisic acid acquired three labels from [2-^{14}C]mevalonate. The terminal *cis* double bond retains a label from C-4 *pro*-(*R*) mevalonate and hence it is produced at some stage from a *trans* double bond. Labelling studies with C-2 and C-5 stereospecifically labelled mevalonates have shown that there is an overall *trans* elimination from C-4 and C-5 of an abscisic acid progenitor

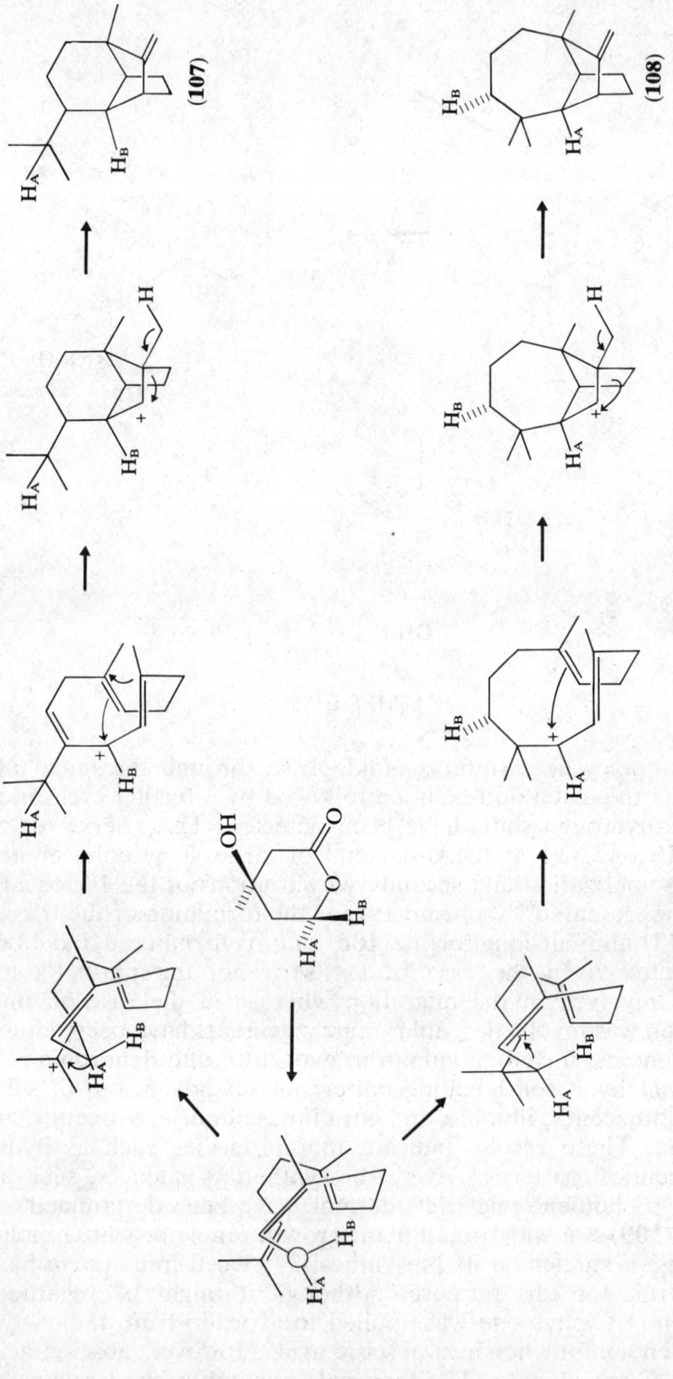

(107)

(108)

SCHEME 19

in the formation of the double bond, with the loss of a 2-*pro*-(*S*) and 5-*pro*-(*R*)-mevalonoid hydrogen atom. The hydroxy-epoxide, (+)-2-*cis*-xanthoxin acid (**110**), has been shown to be the precursor of the γ-hydroxy-αβ-unsaturated ketone of abscisic acid. A 2-*pro*-(*S*) hydrogen of mevalonate is lost from C-3′ in the formation of the double bond and a 4-*pro*-(*R*) mevalonoid hydrogen from C-1′ on insertion of the oxygen function. The methyl groups are labelled as shown in (**109**). The metabolism of abscisic acid, including its conversion into phaseic acid (**111**), has been studied in the bean *Phaseolus vulgaris*.

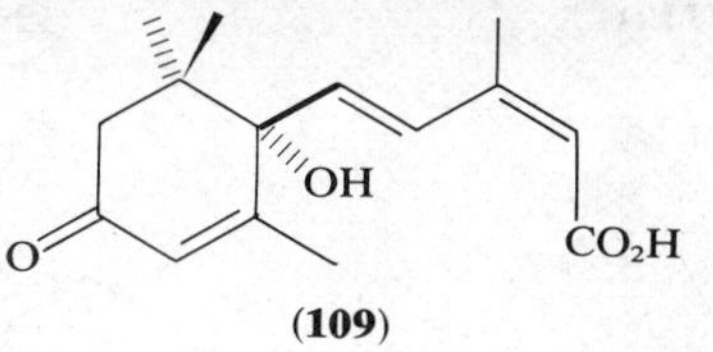

(**109**)

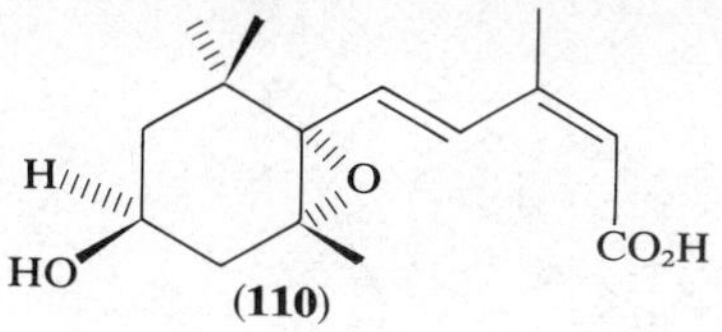

(**110**)

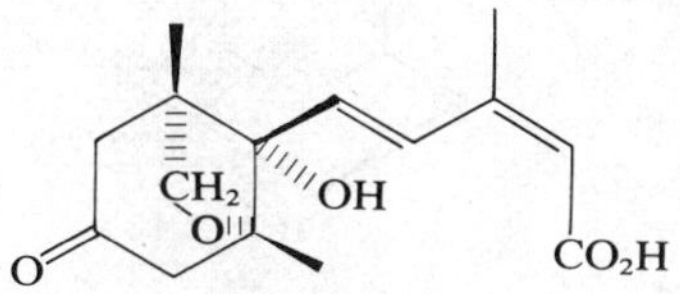

(**111**)

29.2.12 DITERPENOID BIOSYNTHESIS[79]

The diterpenoids are derived from geranylgeranyl pyrophosphate and contain representatives of the two main modes of terpenoid cyclization. On the one hand there are compounds which are formed by a proton-initiated reaction at the double bond of the distal isoprene unit and having a structure reminiscent of the higher terpenoids. However, unlike the majority of the triterpenoids, it is the olefin rather than the epoxide which undergoes cyclization. There are also a substantial number of diterpenoids which are probably formed by the initial attack of the carbonium ion derived from the terminal pyrophosphate on the distal double bond. The macrocyclic compounds which are so formed may then undergo further cyclization. A number of discrete phases are involved in the first mode of cyclization. Initially, geranylgeranyl pyrophosphate (**112**) cyclizes to form a bicyclic labdadienol pyrophosphate (**113**). Immediately one of the first characteristics of the diterpenoids becomes apparent; the formation of normal (steroid-like) and antipodal ring junctions. The constant absolute sterochemistry, which is a feature of the triterpenoids and steroids, is not observed in the diterpenoids. Subsequent modification and cyclization of the labdadienol pyrophosphate may lead to tri- and tetra-cyclic diterpenoids (*e.g.* **114** and **115**). In 1955, Wenkert suggested[80] that the tetracyclic diterpenoids might arise through the cyclization of suitable oriented pimaradienes (**114**) involving a non-classical carbonium ion (**116**), which could then collapse in a variety of ways leading to compounds now identified as possessing the phyllocladene-kaurene (**115**), atisine (**117**) and beyerene (stachene) (**118**) skeleta.

A number of diterpenoids are of fungal origin and this has facilitated the study of their biosynthesis. The labelling pattern from [^{14}C]acetate and [2-^{14}C]mevalonate has been determined for a number of these diterpenoids, including rosenonolactone (**119**), gibberellic acid (**120**), and pleuromutilin (**121**). More recently this has been supplemented with carbon-13 n.m.r. studies on these compounds and on the virescenosides (**122**). These studies have defined the manner of folding of geranylgeranyl pyrophosphate in these diterpenoids.

The biosynthesis of the tricyclic fungal metabolite rosenonolactone (**124**), which is produced by the fungus *Trichothecium roseum*, has been studied in some detail. Biogenetic speculation had suggested that[81] the rearranged skeleton might arise during the cyclization of the bicyclic labdadienol pyrophosphate (**123**). This compound has been shown to be a specific precursor of the rosanes. The use of multiply-labelled mevalonates enabled[82]

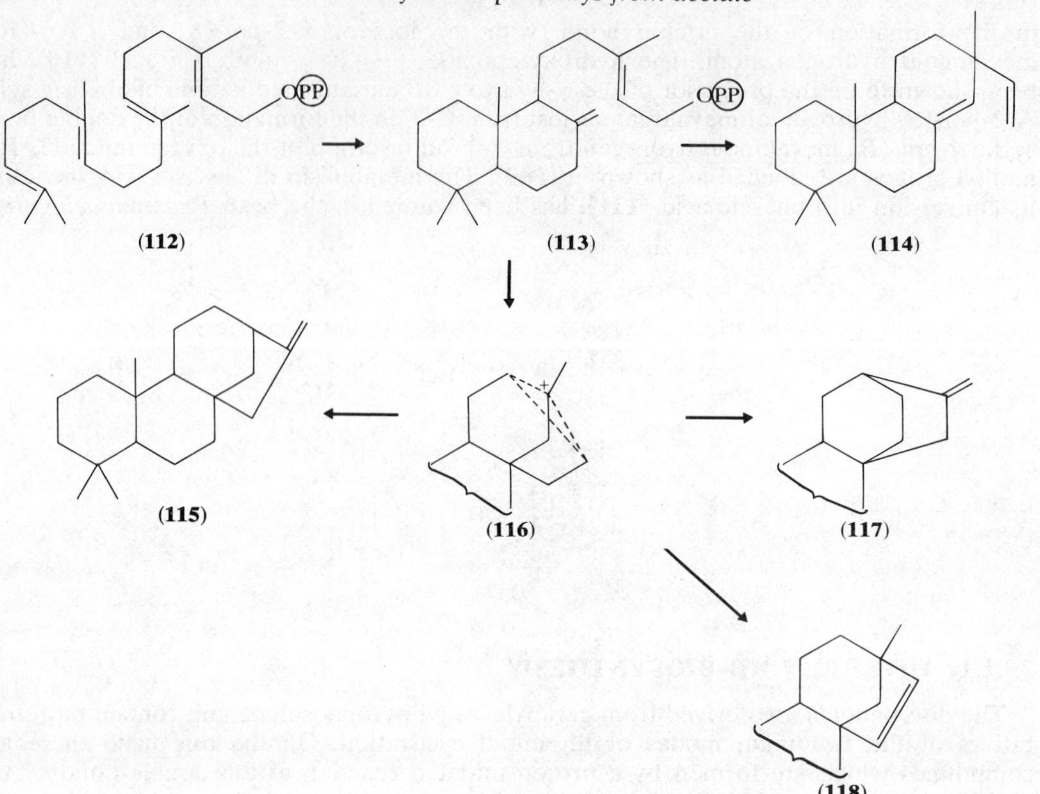

SCHEME 20

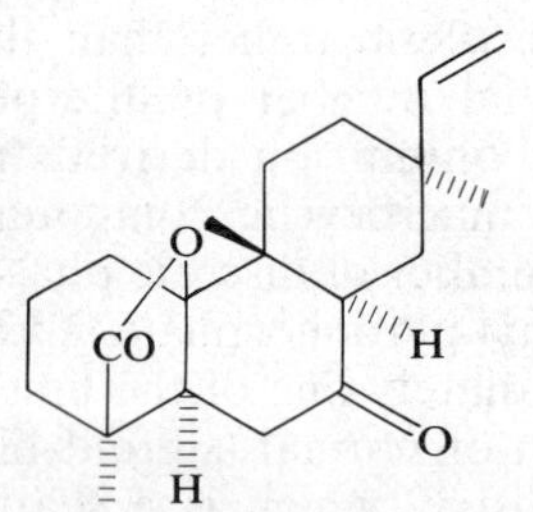

(119)

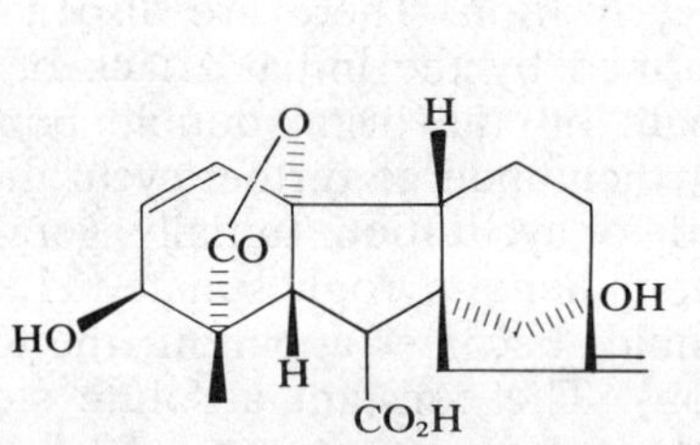

(120)

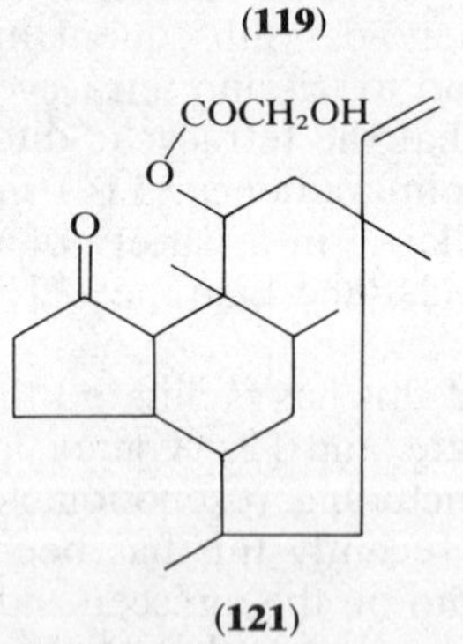

(121)

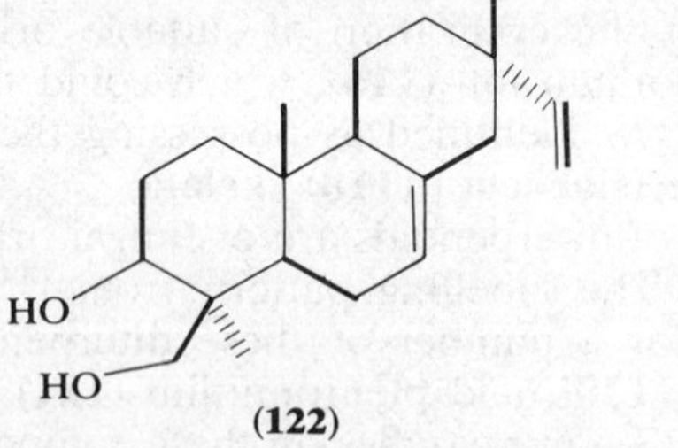

(122)

information to be obtained on these cyclization stages. If the postulated hydride shift occurs then a C-9 hydrogen atom will be retained at C-8. This hydrogen atom would be expected to arise from the C-4 position of mevalonate and indeed a 4-*pro*-(*R*) mevalonoid hydrogen atom was located at C-8. Furthermore, both C-1 hydrogens, the C-5 hydrogen, and the C-6 hydrogens were labelled by the respective mevalonates. These results confirmed the postulated hydride shift and excluded $\Delta^{1(10)}$, $\Delta^{5(10)}$, and $\Delta^{5,6}$ intermediates from the biosynthesis. Hence the C-10 carbonium ion may be stabilized by interaction with a nucleophile prior to lactonization. The sequence deoxyrosenonolactone, 7β-hydroxyrosenonolactone, rosenonolactone, has been established in this biosynthesis.

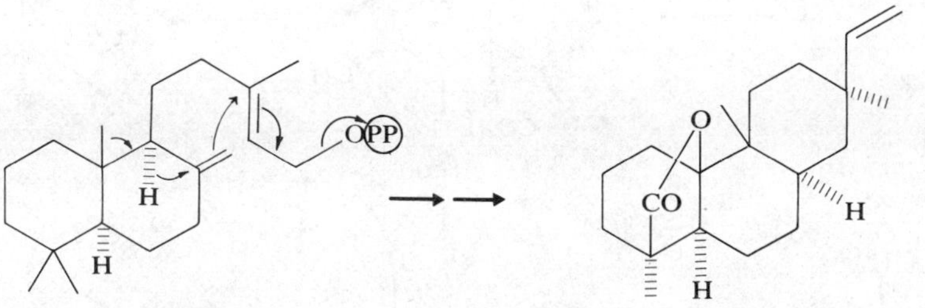

(**123**) (**124**)

Pleuromutilin (**121**)[83] represents another example of a tricyclic diterpenoid in which a skeletal rearrangement has occurred. During the biosynthesis a 4-*pro*-(*R*) mevalonoid hydrogen was shown to migrate from C-9 to C-8. Since this hydrogen atom is *cis* to the migrating methyl group, a C-9 enzyme-bound intermediate was proposed with the further migrations and ring contraction of ring A as the second step. An unexpected feature of this biosynthesis was the discharge of the carbonium ion formed during the last cyclization, by a hydride shift from C-11 to C-4 involving a C-5 *pro*-(*S*) mevalonoid hydrogen atom.

Because of their importance as plant-growth hormones, the biosynthesis of the gibberellins has received considerable attention.[84,85] The first experimental evidence for the diterpenoid nature of gibberellic acid (**120**) came[81] from a study of the incorporation of [1-^{14}C]acetic acid and [2-^{14}C]mevalonic acid. The ring A methyl group and the carboxyl carbon atom were shown to originate from C-2 of mevalonate. A careful study of the metabolites which co-occur with gibberellic acid revealed the presence of a number of kauranoid diterpenoids, including the hydrocarbon ent-kaurene (**125**). This hydrocarbon was shown to be a specific precursor of gibberellic acid.[86] Thus the study of gibberellin biosynthesis has been divided into a number of phases. The first involves the cyclization of geranylgeranyl pyrophosphate leading to ent-kaurene; the second involves the hydroxylation of ent-kaurene; the third involves the ring contraction; and the fourth involves the relationship between the various gibberellins, including the conversion of the C_{20} to the C_{19} compounds.

A soluble enzyme system has been obtained from plant and fungal sources which is capable of converting geranylgeranyl pyrophosphate and ent-labdadienol pyrophosphate (copalol pyrophosphate) into ent-kaurene. The incorporation of mevalonoid hydrogen into ent-kaurene (**125**) has shown[87] that a free tricyclic pimaradiene is not involved in this cyclization. Ent-kaurene has been shown to act as the parent hydrocarbon of the gibberellins, the kaurenolides (*e.g.* **127**), and steviol. In all three cases the next stage in the biosynthesis involves the stepwise oxidation of the C-19 methyl group through the alcohol and aldehyde to a carboxylic acid. Ent-7α-hydroxykaurenoic acid (**126**) is a substrate for ring contraction. This has been shown[88] to proceed by the abstraction of the equatorial 6β-hydrogen atom. The corresponding 6β,7β-diol was not an intermediate in the biosynthesis. The gibbane aldehyde (**129**) formed a divergent point in the biosynthesis.[89] Oxidation of the aldehyde to the carboxylic acid gibberellin A_{12} (**130**) afforded a series of gibberellins (A_{15}–A_9) (**133**) lacking a ring A hydroxyl group. Alternatively, prior hydroxylation of ring A led, *via* gibberellin A_{14} aldehyde (**128**), to the ring A hydroxylated series. The final stages of gibberellin biosynthesis involve the loss of the C-20 carbon atom, probably *via* a carboxyl group, and the conversion of gibberellin A_4 (**131**) *via* the Δ^1-olefin, gibberellin A_7, to gibberellic acid (**132**). Studies utilizing stereospecifically labelled mevalonates have shown that the hydroxylation of ring A occurs with retention of configuration and that the formation of the ring A double bond involves the *cis* loss of hydrogen from the α-face of the molecule. The metabolism of a number of gibberellins, such as gibberellin A_1 and A_4, has been studied in higher plants.

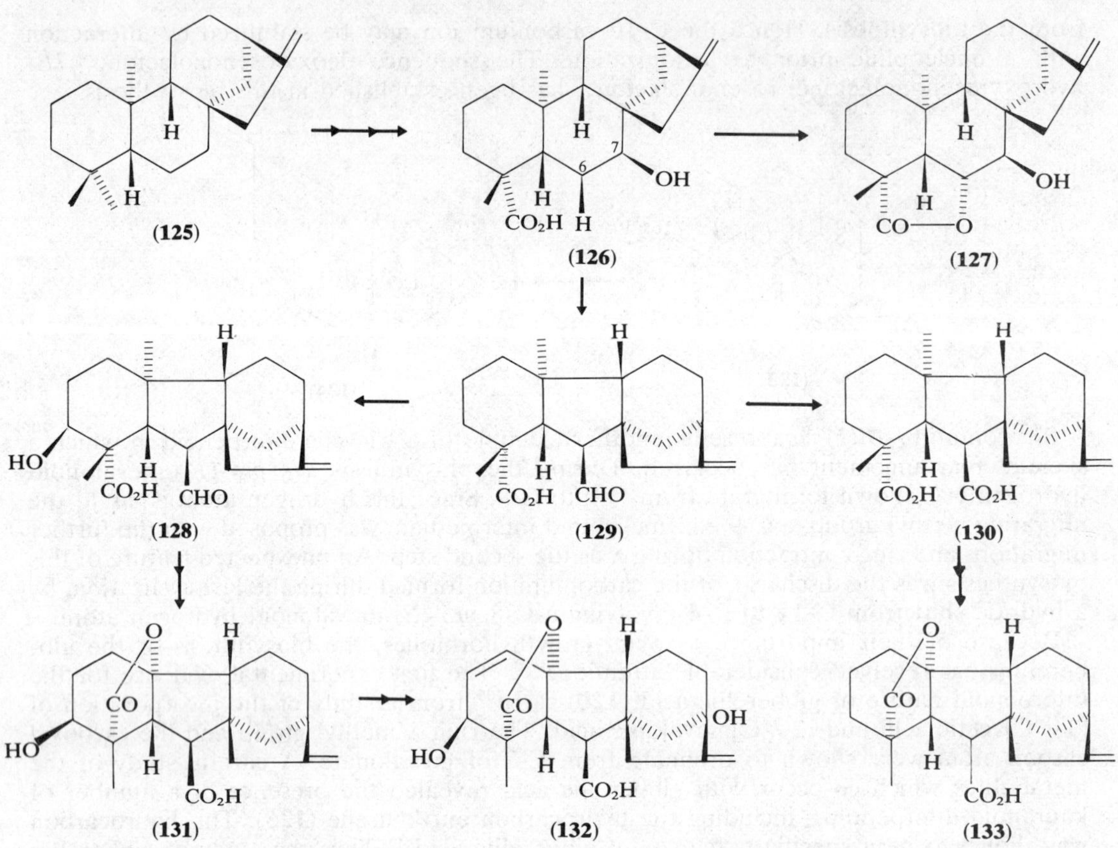

(125) (126) (127) (128) (129) (130) (131) (132) (133)

The fusicoccins (*e.g.* **134**) are metabolites of *Fusiccocum amygdali*. Whilst they have carbon skeleta reminiscent of the sesterterpenoids, they have been shown to be diterpenoids.[90] The labelling patterns of fusicoccins biosynthesized from [1-^{13}C]- and [2-^{13}C]-acetate and from [3-^{13}C]- and [4(*R*)-4-^{3}H]-mevalonic acid are consistent with a biosynthesis from geranylgeranyl pyrophosphate, as set out in Scheme 21.

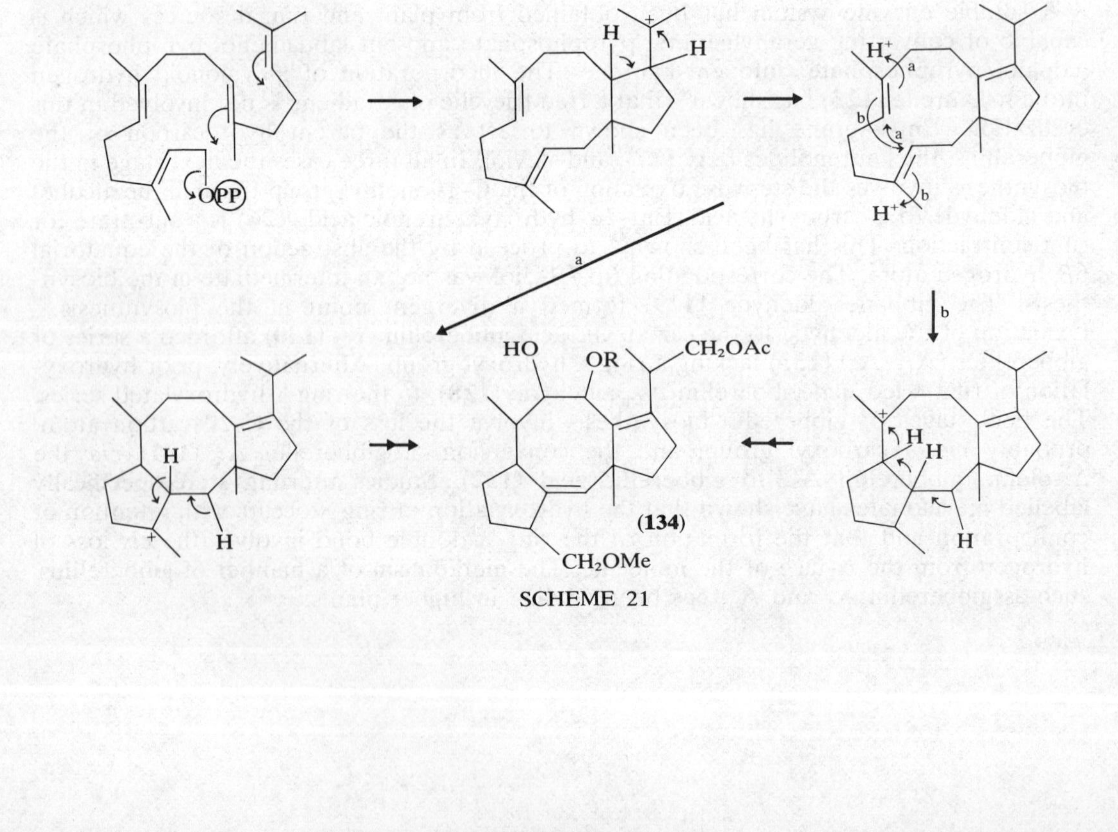

(134)

SCHEME 21

29.2.13 SESTERTERPENOID BIOSYNTHESIS[91]

The sesterterpenoids are a relatively small class of C_{25} terpenoid. The biosynthesis of the ophiobolins, which are the phytotoxic metabolites of some *Helminthosporium* and *Cochliobolus* species, has been studied. Geranylfarnesyl pyrophosphate (**135**) has been shown to be a precursor of ophiobolin F (**137**). During the biosynthesis a C-2 *pro*-(*R*) mevalonoid hydrogen atom migrates in a stereospecific 1,5-hydrogen shift (**136**) from position-8 to position-15. The proposals for the formation of sesterterpenoids possessing the ophiobolin skeleton are set out in Scheme 22.

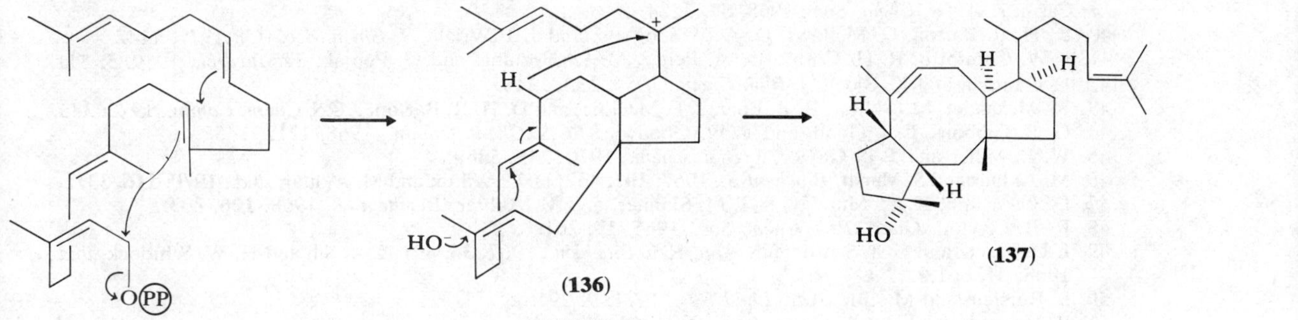

(**135**) (**136**) (**137**)

SCHEME 22

References

1. L. Ruzicka and J. Meyer, *Helv. Chim. Acta*, 1921, **4,** 505.
2. L. Ruzicka, *Proc. Chem. Soc.*, 1959, 341.
3. L. Ruzicka, A. Eschenmoser, and H. Heusser, *Experientia*, 1953, **9,** 357.
4. R. Sonderhoff and H. Thomas, *Annalen*, 1937, **530,** 195.
5. K. Bloch, *The Harvey Lectures*, 1952/3, series 48, p. 68.
6. J. W. Cornforth, G. D. Hunter, and G. Popjak, *Biochem. J.*, 1952, **54,** 597.
7. F. Lynen, L. Wessely, O. Wieland, L. Rueff, and E. Reichert, *Angew. Chem.*, 1952, **64,** 687.
8. H. Rudney and J. Ferguson, *J. Amer. Chem. Soc.*, 1957, **79,** 5580; *J. Biol. Chem.*, 1957, **227,** 363; G. Popjak, *Ann. Rev. Biochem.*, 1958, **27,** 533.
9. D. E. Wolf, C. H. Hoffman, P. E. Aldrich, H. R. Skeggs, L. D. Wright, and K. Folkers, *J. Amer. Chem. Soc.*, 1956, **78,** 4499; 1957, **79,** 1486.
10. P. A. Tavormina, M. H. Gibbs, and J. W. Huff, *J. Amer. Chem. Soc.*, 1956, **78,** 4498.
11. CIBA Foundation Symposium on the 'Biosynthesis of Terpenes and Sterols', ed. G. Wolstenholme and M. O'Connor, Churchill, London, 1959.
12. H. J. Channon, *Biochem. J.*, 1926, **20,** 400.
13. R. Robinson, *Chem. and Ind.* (*London*), 1934, **53,** 1062.
14. R. B. Woodward and K. Bloch, *J. Amer. Chem. Soc.*, 1953, **75,** 2023.
15. J. W. Cornforth and R. H. Cornforth, in 'Biochem. Soc. Symposium 29,' Academic Press, London, 1970, p. 5.
16. J. S. Lawson, W. T. Colwell, J. I. DeGraw, R. H. Peters, R. L. Dehn, and M. Tanabe, *Synthesis*, 1975, 729.
17. J. W. Cornforth, R. H. Cornforth, C. Donninger, and G. Popjak, *Proc. Roy. Soc.* (*B*), 1965, **163,** 492; C. Donninger and G. Popjak, *ibid.*, **163,** 465.
18. K. H. Clifford and G. T. Phillips, *European J. Biochem.*, 1976, **61,** 271.
19. J. W. Cornforth and R. T. Gray, *Tetrahedron*, 1975, **31,** 1509.
20. T. J. Simpson, *Chem. Soc. Rev.*, 1975, **4,** 497.
21. F. Lynen and U. Henning, *Angew. Chem.*, 1960, **72,** 820.
22. R. B. Clayton, *Quart. Rev. Chem. Soc.*, 1965, **19,** 168.
23. T. W. Goodwin, *Essays Biochem.*, 1973, **9,** 103.
24. J. Retey, E. von Stetten, U. Coy, and F. Lynen, *European J. Biochem.*, 1970, **15,** 72.
25. J. W. Cornforth, R. H. Cornforth, G. Popjak, and I. Y. Gore, *Biochem. J.*, 1958, **69,** 146.
26. H. J. Knauss, J. W. Porter, and G. Wasson, *J. Biol. Chem.*, 1959, **234,** 2835.
27. M. Eberle and D. Arigoni, *Helv. Chim. Acta*, 1960, **43,** 1508; R. H. Cornforth, J. W. Cornforth, and G. Popjak, *Tetrahedron*, 1962, **18,** 1351.
28. R. C. Jennings, K. J. Judy and D. A. Schooley, *J.C.S. Chem. Comm.*, 1975, 21.
29. G. Popjak and J. W. Cornforth, *Biochem. J.*, 1966, **101,** 553.
30. K. H. Clifford, J. W. Cornforth, R. Mallaby, and G. T. Phillips, *J.C.S. Chem. Comm.*, 1971, 1599; J. W. Cornforth, *Chem. Brit.*, 1970, **6,** 431.

31. J. W. Cornforth, *Angew. Chem. Internat. Edn.*, 1968, **7,** 903.
32. G. Popjak, in 'Biochem. Soc. Symposium 29', Academic Press, London, 1970, p. 17.
33. H. C. Rilling, *J. Biol. Chem.*, 1966, **241,** 3233; W. W. Epstein and H. C. Rilling, *ibid.*, 1970, **245,** 4597; H. C. Rilling, L. Kuehl, F. Muscio, and J. P. Carlson, *ibid.*, 1974, **249,** 2746.
34. L. J. Goad, in 'Biochem. Soc. Symposium 29', Academic Press, London, 1970, p. 45.
35. L. J. Mulheirn and P. J. Ramm, *Chem. Soc. Rev.*, 1972, **1,** 259.
36. G. Popjak, J. Edmond, F. A. L. Anet, and N. R. Easton, *J. Amer. Chem. Soc.*, 1977, **99,** 931.
37. T. T. Tchen and K. Bloch, *J. Amer. Chem. Soc.*, 1956, **78,** 1516.
38. E. J. Corey, W. E. Russey, and P. R. Ortiz de Montellano, *J. Amer. Chem. Soc.*, 1966, **88,** 4750; E. E. van Tamelen, J. D. Willet, R. B. Clayton and K. E. Lord, *ibid.*, 1966, **88,** 4752.
39. J. W. Cornforth, R. H. Cornforth, C. Donninger, G. Popjak, Y. Shimizu, S. Ichii, E. Forchielli, and E. Caspi, *J. Amer. Chem. Soc.*, 1965, **87,** 3224.
40. D. H. R. Barton, G. Mellows, D. A. Widdowson, and J. J. Wright, *J. Chem. Soc.* (*C*), 1971, 1142.
41. J. W. Cornforth, R. H. Cornforth, A. Pelter, M. G. Horning, and G. Popjak, *Tetrahedron*, 1959, **5,** 311.
42. F. Gautschi and K. Bloch, *J. Biol. Chem.*, 1958, **233,** 1343.
43. K. Alexander, M. Akhtar, R. B. Boar, J. F. McGhie, and D. H. R. Barton, *J.C.S. Chem. Comm.*, 1972, 383.
44. G. F. Gibbons, L. J. Goad, and T. W. Goodwin, *J.C.S. Chem. Comm.*, 1968, 1212, 1458.
45. W. L. Miller and J. L. Gaylor, *J. Biol. Chem.*, 1970, **245,** 5369.
46. M. Akhtar and S. Marsh, *Biochem. J.*, 1967, **102,** 462; D. C. Wilton and M. Akhtar, *ibid.*, 1970, **116,** 337.
47. D. C. Wilton, K. A. Munday, S. J. M. Skinner, and M. Akhtar, *Biochem. J.*, 1968, **106,** 803.
48. R. B. Clayton, *Quart. Rev. Chem. Soc.*, 1965, **19,** 201.
49. I. D. Frantz and G. J. Schroepfer, *Ann. Rev. Biochem.*, 1967, **36,** 691; C. J. Sih and H. W. Whitlock, *ibid.*, 1968, **37,** 661.
50. S. Burstein and M. Gut, *Adv. Lipid Res.*, 1971, **9,** 291.
51. P. Morand and J. Lyall, *Chem. Rev.*, 1968, **68,** 85.
52. T. Nambara, T. Anjyo, and H. Hosoda, *Chem. Pharm. Bull.* (*Japan*), 1972, **20,** 853.
53. D. Arigoni, R. Battaglia, M. Akhtar, and T. Smith, *J.C.S. Chem. Comm.*, 1975, 185.
54. H. F. de Luca and H. K. Schnoes, *Ann. Rev. Biochem.*, 1976, **45,** 631; P. E. Georghiou, *Chem. Soc. Rev.*, 1977, **6,** 83.
55. H. H. Rees and T. W. Goodwin, *Biochem. Soc. Trans.*, 1974, **2,** 1027.
56. L. J. Goad and T. W. Goodwin, *Prog. Phytochem.*, 1972, **3,** 113.
57. B. A. Knight, *Chem. Brit.*, 1973, **9,** 106.
58. E. Lederer, *Quart. Rev. Chem. Soc.*, 1969, **23,** 453.
59. D. Arigoni, *Experientia*, 1958, **14,** 153.
60. S. Seo, Y. Tomita, and K. Tori, *J.C.S. Chem. Comm.*, 1975, 270.
61. D. V. Banthorpe, B. V. Charlwood, and M. J. O. Francis, *Chem. Rev.*, 1972, **72,** 115.
62. D. V. Banthorpe, J. Mann, and K. W. Turnbull, *J. Chem. Soc.* (*C*), 1970, 2689.
63. K. G. Allen, D. V. Banthorpe, B. V. Charlwood, O. Ekundayo, and J. Mann, *Phytochemistry*, 1976, **15,** 101.
64. B. Achilladelis and J. R. Hanson, *Phytochemistry*, 1968, **7,** 1317.
65. D. Arigoni, *Pure Appl. Chem.*, 1975, **41,** 219.
66. D. V. Banthorpe and O. Ekundayo, *Phytochemistry*, 1976, **15,** 109.
67. W. Epstein and C. D. Poulter, *Phytochemistry*, 1973, **12,** 737.
68. G. Pattenden, C. R. Popplestone, and R. Storer, *J.C.S. Chem. Comm.*, 1975, 290.
69. K. G. Allen, D. V. Banthorpe, B. V. Charlwood, and C. M. Voller, *Phytochemistry*, 1977, **16,** 79; D. V. Banthorpe, S. Doonan, and J. A. Gutowski, *ibid.*, 1977, **16,** 85.
70. W. Parker, J. S. Roberts, and R. Ramage, *Quart. Rev. Chem. Soc.*, 1967, **21,** 331.
71. G. A. Cordell, *Chem. Rev.*, 1976, **76,** 425.
72. K. Imai and S. Marumo, *Tetrahedron Letters*, 1974, 4401.
73. K. H. Overton and D. J. Picken, *J.C.S. Chem. Comm.*, 1976, 105.
74. R. Evans and J. R. Hanson, *J.C.S. Perkin I*, 1976, 326.
75. J. R. Hanson, T. Marten, and M. Siverns, *J.C.S. Perkin I*, 1974, 1033.
76. M. Tanabe and K. T. Suzuki, *Tetrahedron Letters*, 1974, 4417; D. E. Cane and R. H. Levin, *J. Amer. Chem. Soc.*, 1975, **97,** 1282.
77. B. A. Achilladelis, P. M. Adams, and J. R. Hanson, *Chem. Comm.*, 1970, 511; *J.C.S. Perkin I*, 1972, 1425; S. Nozoe, M. Morisaki, and H. Matsumoto, *Chem. Comm.*, 1970, 926; D. Arigoni, D. E. Cane, B. Muller, and C. Tamm, *Helv. Chim. Acta*, 1973, **56,** 2946.
78. B. V. Milborrow, *Phytochemistry*, 1975, **14,** 123, 2403.
79. J. R. Hanson, *Forsch. Chem. org. Naturstoffe*, 1971, **29,** 395.
80. E. Wenkert, *Chem. Ind.* (*London*), 1955, 282.
81. A. J. Birch, R. W. Rickards, H. Smith, A. Harris, and W. B. Whalley, *Tetrahedron*, 1959, **7,** 241.
82. B. Achilladelis and J. R. Hanson, *J. Chem. Soc.* (*C*), 1969, 2010.
83. A. J. Birch, C. W. Holzapfel, and R. W. Rickards, *Tetrahedron*, 1966, Suppl. 8, 359; D. Arigoni, *Pure Appl. Chem.*, 1968, **17,** 331.
84. B. E. Cross, *Prog. Phytochem.*, 1968, **1,** 195.
85. J. MacMillan, 'Chemistry and Biochemistry of Plant Hormones', ed. V. C. Runeckles, E. Sondheimer, and D. C. Walton, Academic Press, New York, 1974.
86. B. E. Cross, R. H. B. Galt, and J. R. Hanson, *J. Chem. Soc.*, 1964, 295.

87. R. Evans and J. R. Hanson, *J.C.S. Perkin I*, 1972, 2382.
88. J. E. Graebe, P. Hedden, and J. MacMillan, *J.C.S. Chem. Comm.*, 1975, 161.
89. J. R. Bearder, J. MacMillan, and B. O. Phinney, *J.C.S. Perkin I*, 1975, 721; J. R. Hanson and R. Evans, *ibid.*, 1975, 663.
90. K. D. Barrow, R. B. Jones, P. W. Pemberton, and L. Phillips, *J.C.S. Perkin I*, 1975, 1405; A. Banerji, R. B. Jones, G. Mellows, L. Phillips, and Keng-Yeow Sim, *ibid.*, 1976, 2221.
91. G. A. Cordell, *Phytochemistry*, 1974, **13,** 2343.

29.3

Carotenoid Biosynthesis and Vitamin A

G. BRITTON

University of Liverpool

29.3.1 CAROTENOID BIOSYNTHESIS[1–18]

29.3.1.1 Introduction

Most natural carotenoid pigments are tetraterpenes, biosynthesized from two C_{20} units (geranylgeranyl pyrophosphate). Some recently described bacterial C_{30} carotenoids are formed in an analogous way from two C_{15} units (farnesyl pyrophosphate). The small number of natural C_{45} and C_{50} carotenoids are produced by addition of C_5 units to a C_{40} precursor.

The pathway of carotenoid biosynthesis involves a small number of basic reactions, *e.g.* production of phytoene, the first C_{40} intermediate, desaturation, cyclization and related processes, that are common to the formation of almost all carotenoids. The individual structures of the 400–500 naturally occurring carotenoids arise from a wide variety of final modifications.

29.3.1.2 Formation of phytoene[1,2]

The C_{20} carotenoid precursor geranylgeranyl pyrophosphate (**1**) is formed by the general terpenoid biosynthetic pathway (see Sections 29.2.1–3). Two molecules of geranylgeranyl pyrophosphate are used to form the first C_{40} carotenoid intermediate, which is phytoene (7,8,11,12,7′,8′,11′,12′-octahydro-ψ,ψ-carotene, **4**) and not as first thought lycopersene (7,8,11,12,15,7′,8′,11′,12′,15′-decahydro-ψ,ψ-carotene, **5**), the C_{40} analogue of squalene. The mechanism of formation of phytoene, however, has much in common with that of squalene biosynthesis. In particular, the reaction involves pre-phytoene pyrophosphate (**2**), a C_{40} cyclopropane intermediate analogous to presqualene pyrophosphate, and probably produced by an analogous reaction (see Section 29.2.5). The

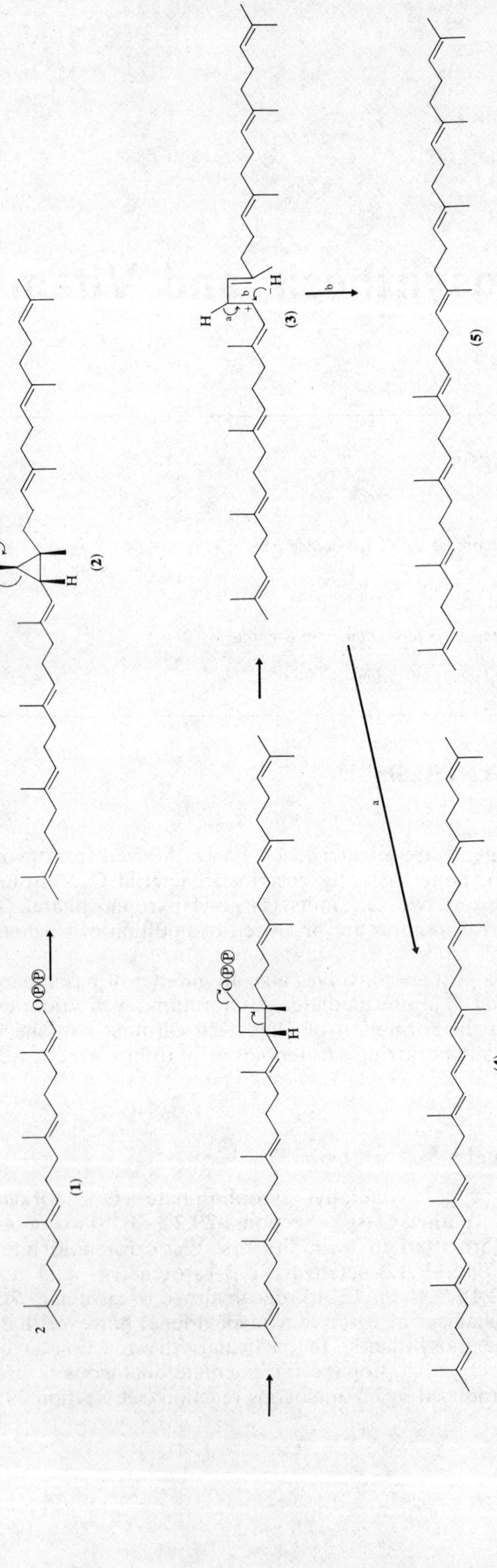

SCHEME 1

formation of phytoene from prephytoene pyrophosphate is also similar to the conversion of presqualene pyrophosphate into squalene, differing only in the last step. The final 'carbonium ion' intermediate (**3**) is stabilized by proton loss to give phytoene rather than by formal addition of H^- from NADPH, which would give lycopersene by analogy with squalene production (Scheme 1).

The phytoene produced by most carotenogenic systems appears to be the 15,15′-*cis*-isomer. Formation of this isomer involves loss of the 1-*pro-S* hydrogen atom of each molecule of geranylgeranyl pyrophosphate (originally the 5-*pro-S* hydrogen atom of mevalonate). In some bacterial systems all-*trans*-phytoene is produced by loss of the 1-*pro-S* hydrogen atom of one molecule of geranylgeranyl pyrophosphate and the 1-*pro-R* hydrogen atom from the other.[9] Details, including stereochemistry, of the likely mechanism of formation of all-*trans*- and 15-*cis*-phytoene from prephytoene pyrophosphate are given in Scheme 2. It is thought that the biosynthesis of the C_{30} carotenoids involves dehydrosqualene formed by an analogous reaction from presqualene pyrophosphate.

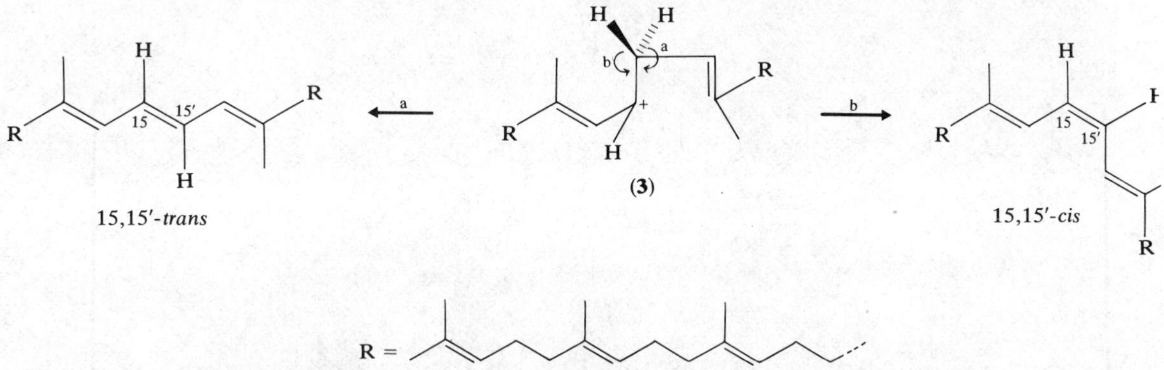

SCHEME 2

29.3.1.3 Desaturation[1,2]

The formation of coloured carotenoids from phytoene first involves a series of four desaturations, each step resulting in the introduction of a double bond and consequent extension of the chromophore by two conjugated double bonds (Scheme 3). The intermediates in this sequence are phytofluene (7,8,11,12,7′,8′-hexahydro-ψ,ψ-carotene, **6**), ζ-carotene, two possible isomers with the heptaene chromophore located either symmetrically (7,8,7′,8′-tetrahydro-ψ,ψ-carotene, **7**) or unsymmetrically (7,8,11,12-tetrahydro-ψ,ψ-carotene, **8**), and neurosporene (7,8-dihydro-ψ,ψ-carotene, **9**), and the final product of the desaturations is lycopene (ψ,ψ-carotene, **10**).

Lycopene and almost all natural carotenoids have the all-*trans* configuration, so in those tissues which produce 15-*cis*-phytoene an isomerization must be involved at some stage in the desaturation sequence. There is evidence that this isomerization may take place at different stages, *e.g.* phytoene, phytofluene, ζ-carotene, in different systems.

The mechanism of the desaturation reaction has not been established, but experiments with mevalonate labelled stereospecifically with ^3H at C-2 or C-5 have shown that the introduction of each double bond occurs by *trans* elimination of hydrogen (Scheme 4).

29.3.1.4 Cyclization and other reactions at the C-1,2 double bond[3]

In most carotenogenic systems lycopene is not the end product but is merely an intermediate in the biosynthesis of the normal main carotenoids present. In particular, lycopene may undergo various reactions at the C-1,2 double bond to give rise to the series of acyclic carotenoids characteristic of many photosynthetic bacteria, or to the more familiar monocyclic and bicyclic carotenoids that are typical of plants.

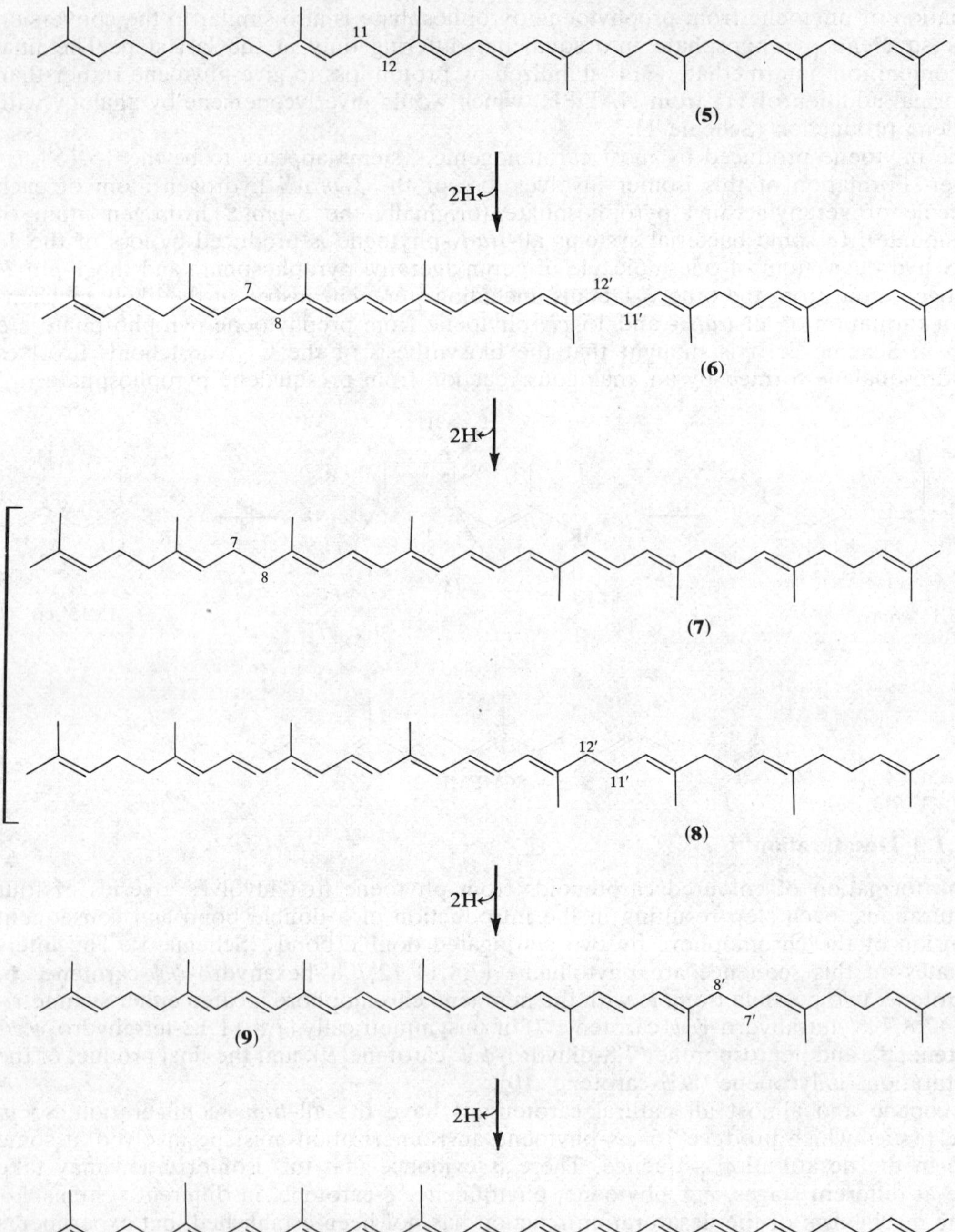

SCHEME 3

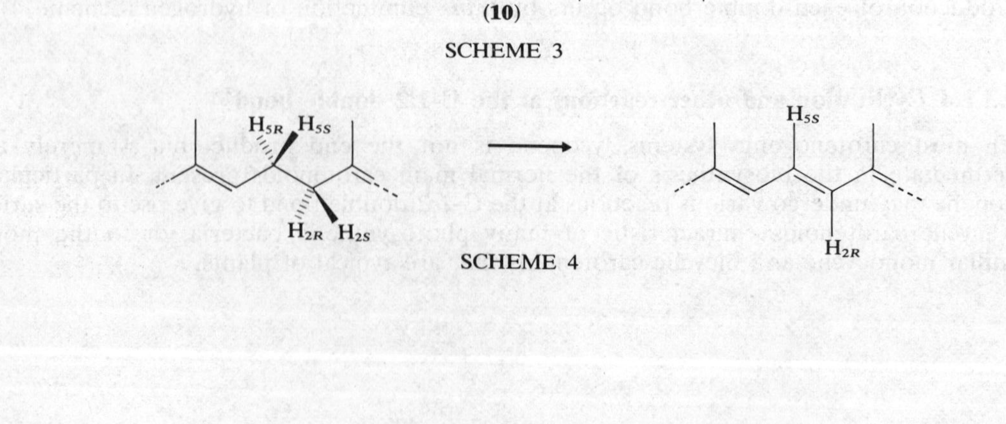

SCHEME 4

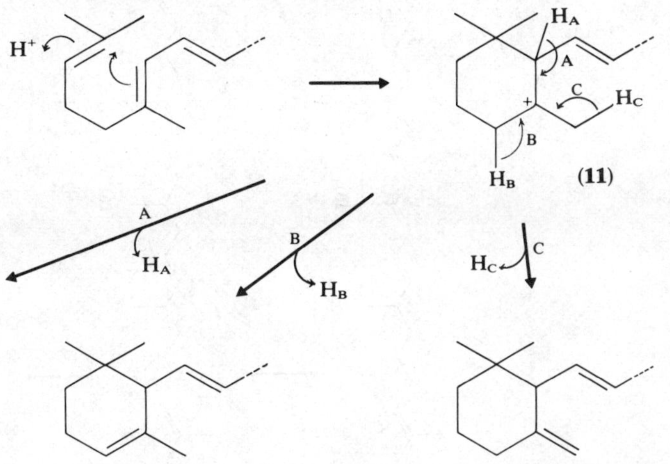

β-ring (12) ε-ring (13) γ-ring (14)

SCHEME 5

The rigidity of the conjugated polyene system of the carotenoid intermediates precludes extensive cyclizations of the kind that occur in the di- and tri-terpenoid series. In carotenoids, cyclization is limited to the formation of a single six-membered ring at one or both ends of the acyclic precursor molecule. It is generally accepted that the cyclization of carotenoid intermediates is an addition process initiated by proton attack at C-2 of the terminal double bond. Cyclization occurs as illustrated in Scheme 5 to give a 'carbonium ion' (**11**) which can be stabilized by proton loss from either C-6, C-4, or C-18 to give the β-ring (**12**), ε-ring (**13**) or less common γ-ring (**14**), respectively. The different ring types are not interconverted. Thus lycopene can be converted *via* γ-carotene (β,ψ-carotene, **16**) into β-carotene (β,β-carotene, **17**) or α-carotene (β,ε-carotene, **18**), and alternative pathways have been proposed for β-carotene synthesis from neurosporene *via* β-zeacarotene (7′,8′-dihydro-β,ψ-carotene, **15**) (Scheme 6). Similarly, if the first cyclization produces an ε-ring, neurosporene and lycopene give α-zeacarotene (7′,8′-dihydro-ε,ψ-carotene, **19**) and δ-carotene (ε,ψ-carotene, **20**) respectively as intermediates in alternative pathways for the formation of α-carotene and ε-carotene (ε,ε-carotene, **21**) (Scheme 7). In all cases, however, cyclization always occurs in a carotenoid 'half-molecule' that has reached the lycopene level of desaturation.

In certain non-photosynthetic bacteria the cyclization can be initiated by an electrophilic C_5 species (Scheme 8), resulting in C_{45} and C_{50} carotenoids with C_5 substituents at C-2 of the β-, ε-, or γ-ring, *e.g.* 'C.p. 450' [2-(4-hydroxy-3-hydroxymethylbut-2-enyl)-2′-(3-methylbut-2-enyl)-β,β-carotene, **22**], decaprenoxanthin [2,2′-bis-(4-hydroxy-3-methylbut-2-enyl)-ε,ε-carotene, **23**], and sarcinaxanthin [2,2′-bis-(4-hydroxy-3-methylbut-2-enyl)-γ,γ-carotene, **24**].

The stereochemical behaviour of the C-1 methyl substituents during cyclization to the β-ring and the stereochemistry of the initial H^+ attack on the C-1,2 double bond have been elucidated. [2-^{13}C]Mevalonate labels the C-16 methyl group of the acyclic precursor lycopene, *i.e.* the methyl group that is *trans* to the bulk of the molecule. In zeaxanthin (3*R*,3′*R*-β,β-carotene-3,3′-diol, **25**), the 1α-methyl substituent carries the ^{13}C label. Also, cyclization of lycopene in bacterial cells suspended in deuterium oxide results in the incorporation of deuterium into the 2β-position of zeaxanthin. The results of these experiments indicate the stereochemistry of folding, hydrogen attack, and cyclization illustrated in Scheme 9.

The stereochemistry of formation of the ε-ring involves two additional chiral or prochiral centres, C-6 and C-4. In all (unsubstituted) ε-ring carotenoids so far studied, the absolute configuration at C-6 is *R*, and in the one example yet examined, formation of the

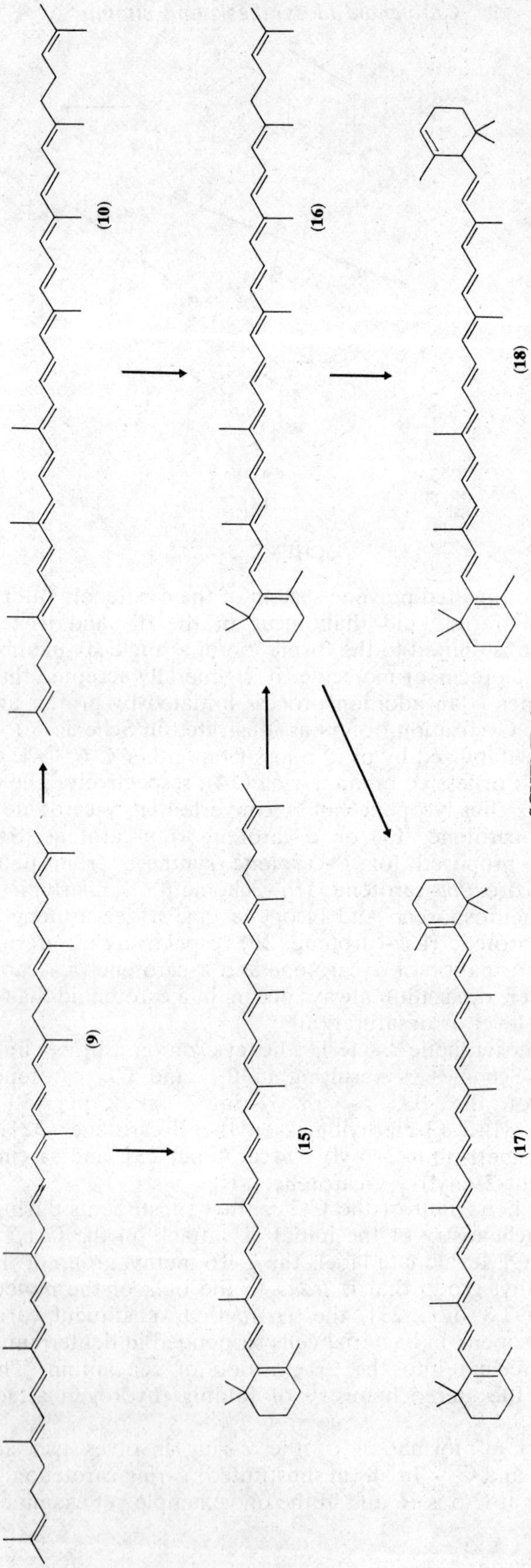

SCHEME 6

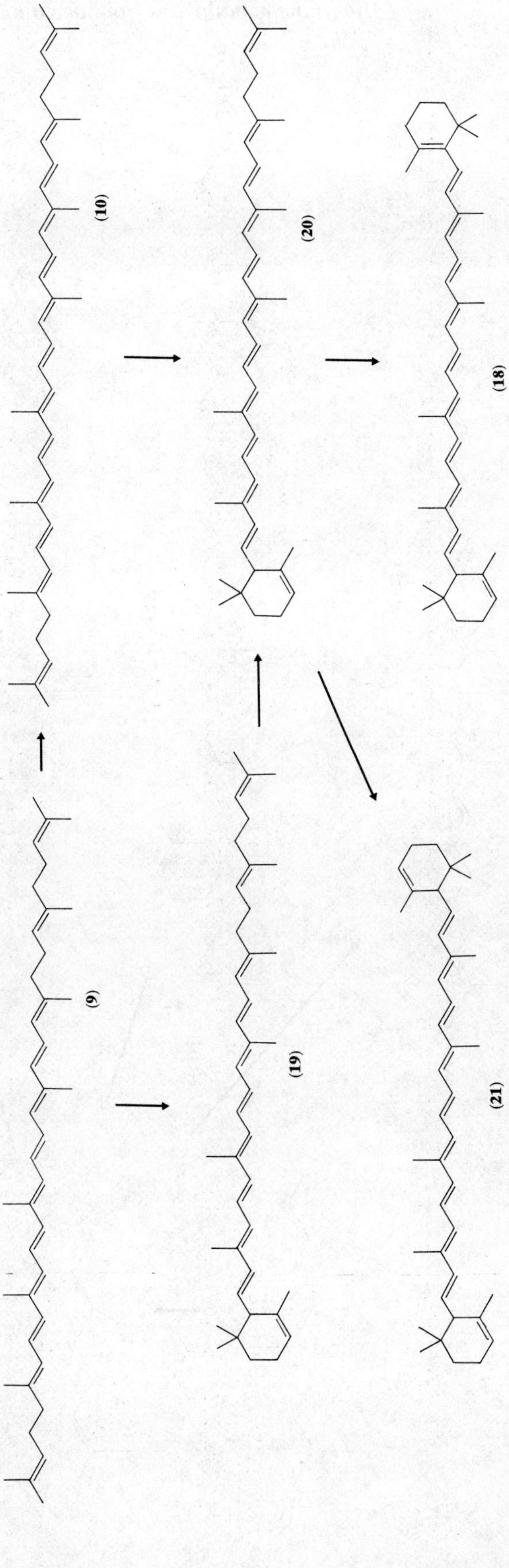

SCHEME 7

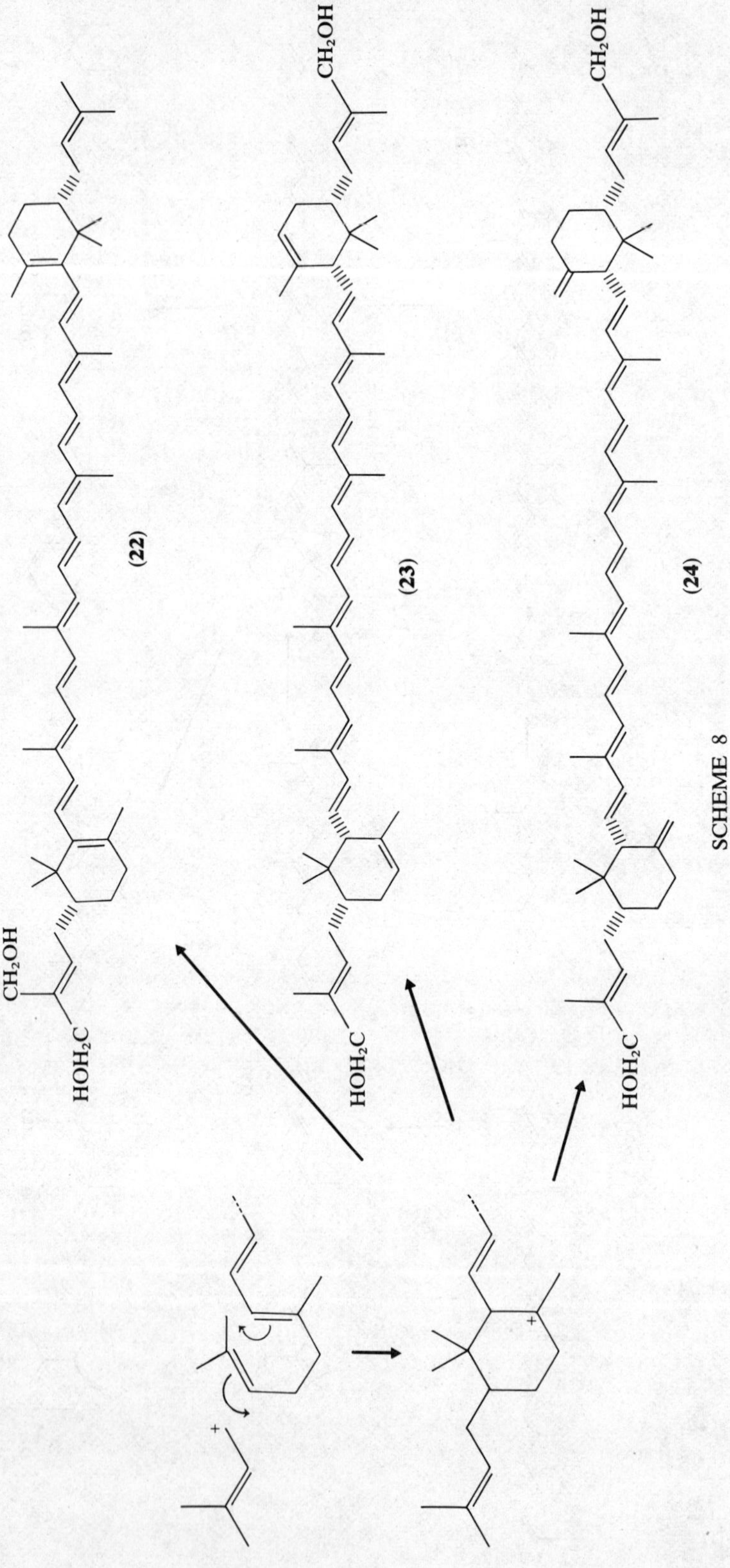

(22)

(23)

(24)

SCHEME 8

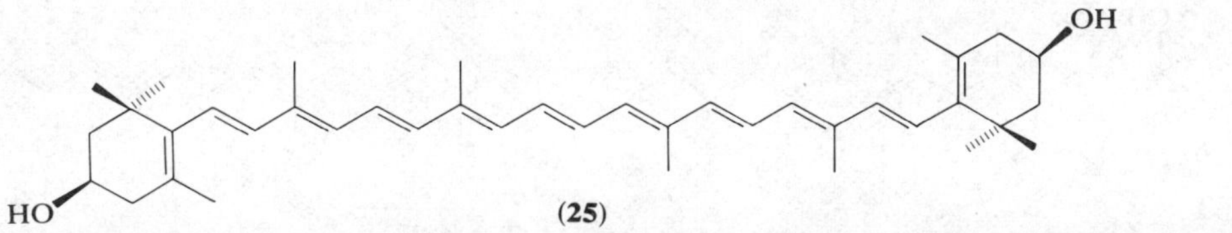

(**25**)

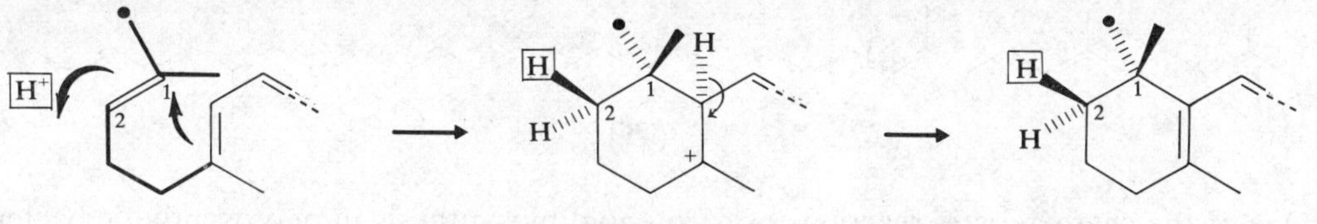

SCHEME 9

ε-ring from the presumed intermediate carbonium ion involves the loss from C-4 of the hydrogen atom which was originally the 2-*pro-S* hydrogen atom of mevalonate. The stereochemistry of the initial hydrogen attack and the stereochemical behaviour of the C-1 methyl groups during cyclization to the ε-ring have not been determined, but if they are the same as for the β-ring then cyclization to produce the ε-ring occurs as illustrated in Scheme 10.

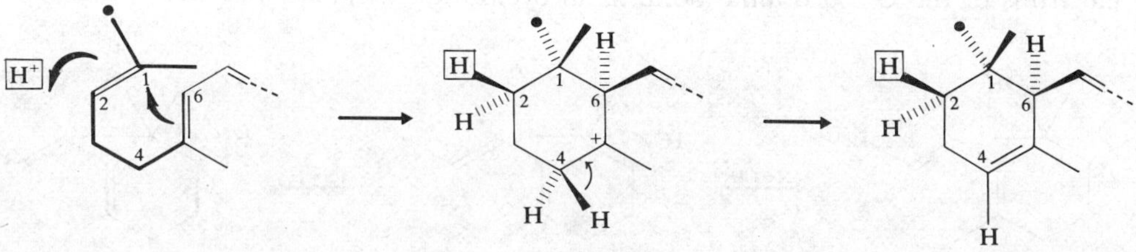

SCHEME 10

In the C_{50} carotenoids the C_5 substituent at C-2 occupies the 2α-position, *i.e.* opposite to that taken up by the hydrogen introduced during formation of the unsubstituted ring. Also, in the C_{50} carotenoid decaprenoxanthin the chirality at C-6 (**26**) is opposite to that (**27**) in the unsubstituted ε-ring carotenoids, *e.g.* α-carotene. These findings may indicate that electrophilic attack by the C_5 species occurs on the opposite face of the C-1,2 double bond and that the entire stereochemistry of folding may be different from that in the biosynthesis of the unsubstituted rings.

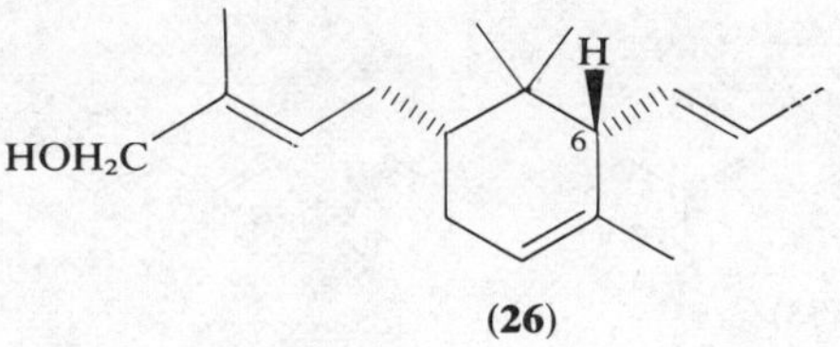

(**26**)

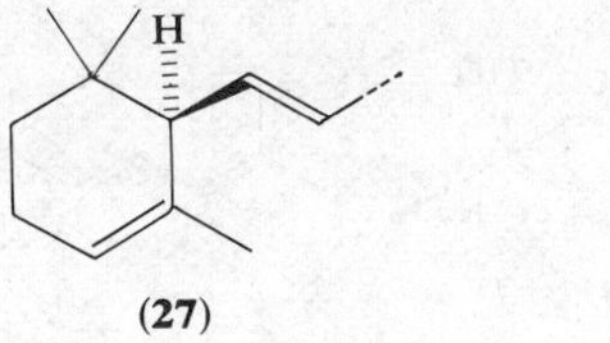

(**27**)

Besides cyclization, other, simpler additions to the carotenoid C-1,2 double bond occur. Thus 1,2-epoxides of acyclic carotenes, *e.g.* lycopene 1,2-epoxide (1,2-epoxy-1,2-dihydro-ψ,ψ-carotene, **28**), have been found in certain higher plants. Their biosynthesis may be similar to the formation of squalene 2,3-epoxide (see Section 29.2.6).

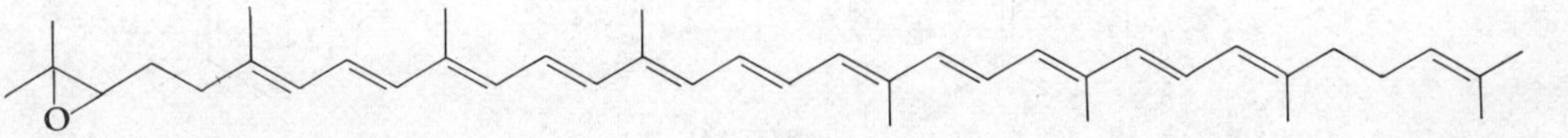

(**28**)

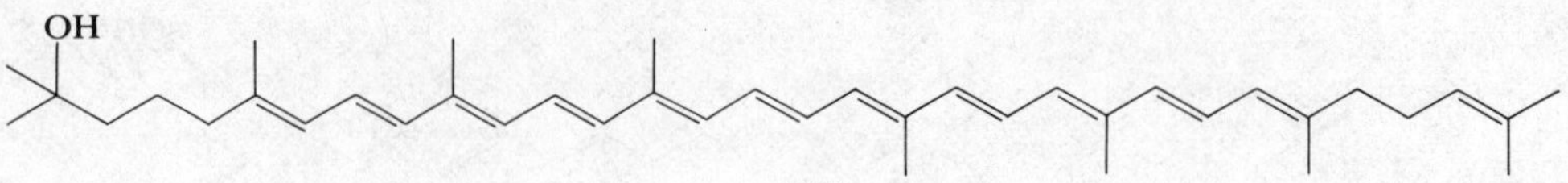

(29)

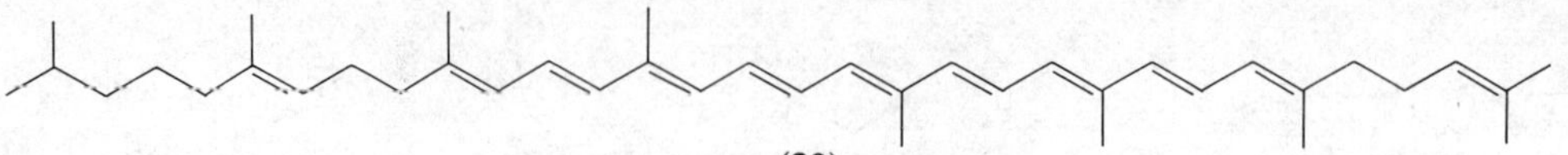

(30)

One of the characteristic reactions of carotenoid biosynthesis in photosynthetic bacteria is the addition of water across the C-1,2 double bond to give the 1-hydroxy-1,2-dihydro end group (**31**), as in rhodopin (1,2-dihydro-ψ,ψ-caroten-1-ol, **29**). In one case it is hydrogen that is added, to give a series of carotenes with a 1,2-dihydro end group (**32**), *e.g.* 1,2-dihydroneurosporene (1,2,7,8-tetrahydro-ψ,ψ-carotene, **30**). These reactions are inhibited by the same substances (*e.g.* nicotine) as inhibit the carotenoid cyclization, and it is probable that they too occur as addition reactions initiated by proton attack at C-2. The electron deficiency thus produced at C-1 is then neutralized by addition of HO^- or H^-, not by electrons of the C-5,6 double bond as in cyclization (Scheme 11).

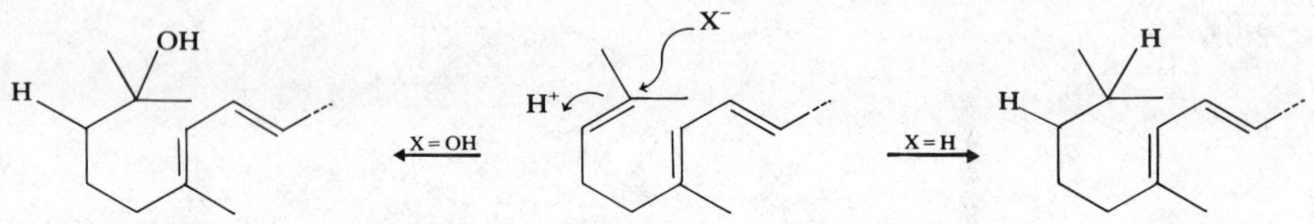

SCHEME 11

As was the case with cyclization, a reaction analogous to the C-1,2 hydration can be initiated by an electrophilic C_5 species in place of H^+ (Scheme 12), resulting in the formation of acyclic C_{45} and C_{50} carotenoids such as bacterioruberin [2,2′-bis-(3-hydroxy-3-methylbutyl)-3,4,3′,4′-tetradehydro-1,2,1′,2′-tetrahydro-ψ,ψ-carotene-1,1′-diol, **33**]. The absolute configuration at C-2 in bacterioruberin is the same as in the cyclic C_{50} carotenoids, perhaps reflecting the similarity of the two biosynthetic reactions.

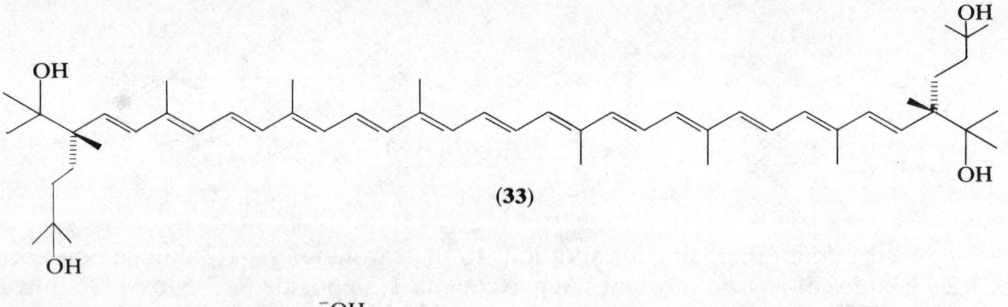

(33)

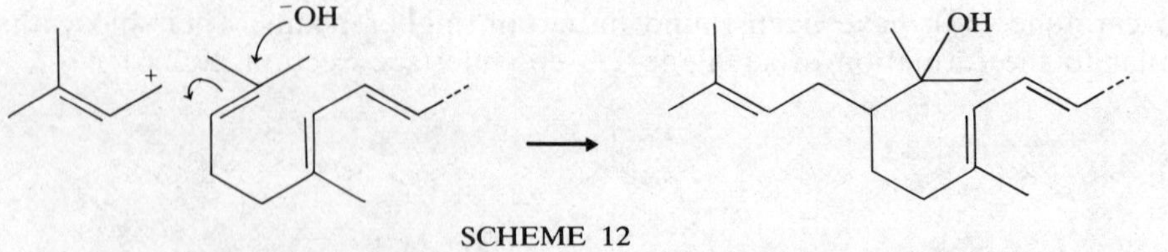

SCHEME 12

29.3.1.5 Final modifications[2,3]

The preceding sections describe the main reactions by which the basic acyclic and cyclic carotenoid structures are formed. There is then a wide variety of further transformations by which the individual carotenoid structures can be elaborated. Some of these processes occur commonly, notably introduction of oxygen functions. Other modifications appear to be unique to the formation of a single carotenoid in one species or group of species. The range of structural modifications in cyclic carotenoids is wider than in the acyclic series.

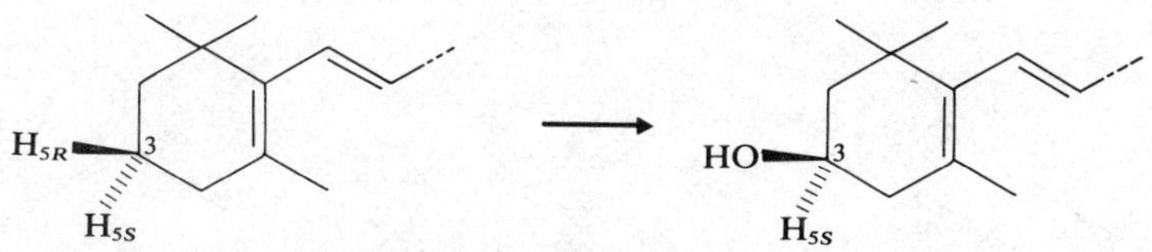

SCHEME 13

The most common structural feature present in cyclic carotenoids is the hydroxy-group. Hydroxylation occurs most frequently at C-3, but 2-hydroxy- and 4-hydroxy-carotenoids are also common (the latter often being oxidized further to 4-oxocarotenoids), and hydroxy-groups in other positions in the molecule (*e.g.* C-19) are sometimes found. Of these only the introduction of the C-3 hydroxy-group has been studied. In plants and bacteria, (3*R*,3′*R*)-zeaxanthin (**25**) is formed by hydroxylation of β-carotene. The hydroxy-group arises from molecular O_2, and the reaction is a mixed-function oxidase process in which the hydroxy-group directly replaces the C-3 hydrogen atom that was originally the 5-*pro-R* hydrogen atom of mevalonate (Scheme 13). The most abundant leaf xanthophyll lutein (β,ε-carotene-3,3′-diol, **34**) is probably formed similarly by hydroxylation of α-carotene. The chirality at C-3′, however, is opposite to that at C-3, and to that in zeaxanthin, so the stereochemistry of hydroxylation is obviously different.

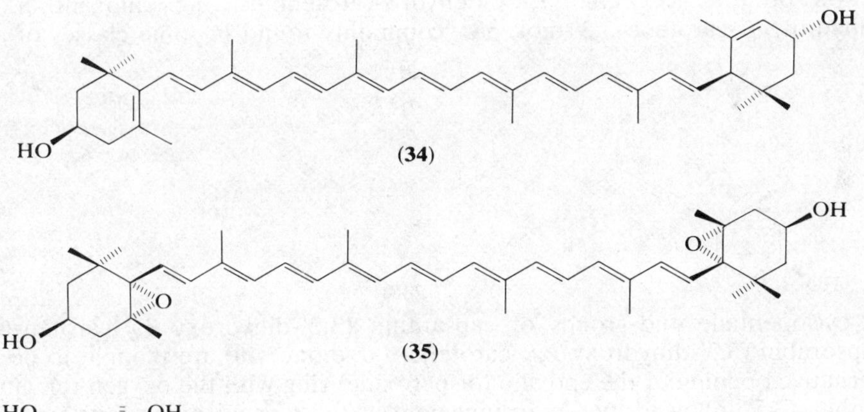

(34)

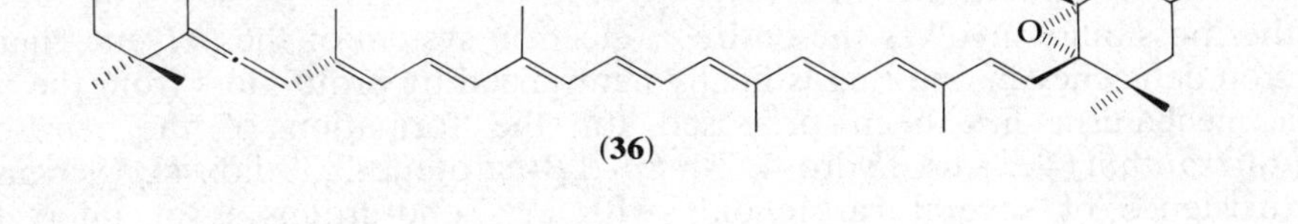

(35)

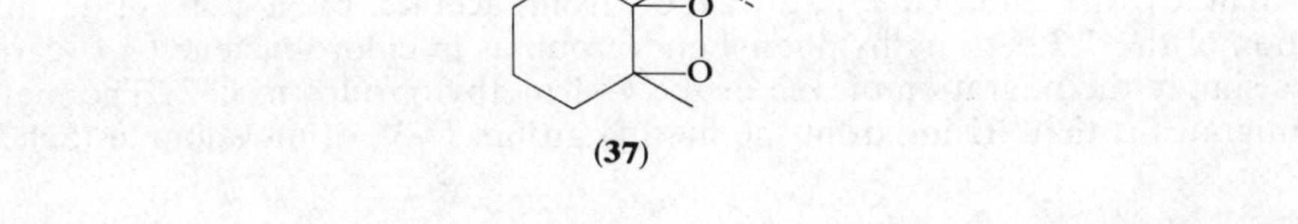

(36)

(37)

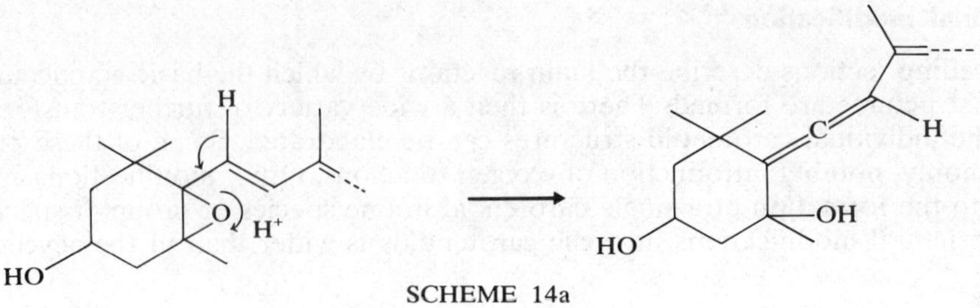

SCHEME 14a

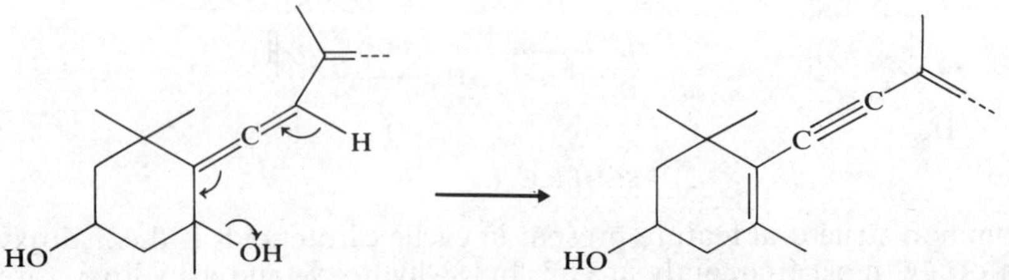

SCHEME 14b

The chloroplast xanthophylls violaxanthin (5,6,5′,6′-diepoxy-5,6,5′,6′-tetrahydro-β,β-carotene-3,3′-diol, **35**) and neoxanthin (5′,6′-epoxy-6,7-didehydro-5,6,5′,6′-tetrahydro-β,β-carotene-3,5,3′-triol, **36**) have the 5,6-epoxy-group. The enzymic epoxidation of zeaxanthin has been described. Such epoxides (or related peroxides, **37**) may also be important intermediates in several interesting modifications, especially of the carotenoid end-group. Although the mechanisms proposed for these reactions seem very reasonable, there is little or no supporting evidence. Thus opening of the epoxide (or peroxide) ring followed by proton loss from C-7 (Scheme 14a) would give the allenic end-group of neoxanthin. Rearrangement of such an allenic group (Scheme 14b) could account for the production of the acetylenic 7,8-didehydro-carotenoids, *e.g.* alloxanthin (7,8,7′,8′-tetradehydro-β,β-carotene-3,3′-diol, **38**), commonly found in some classes of algae.

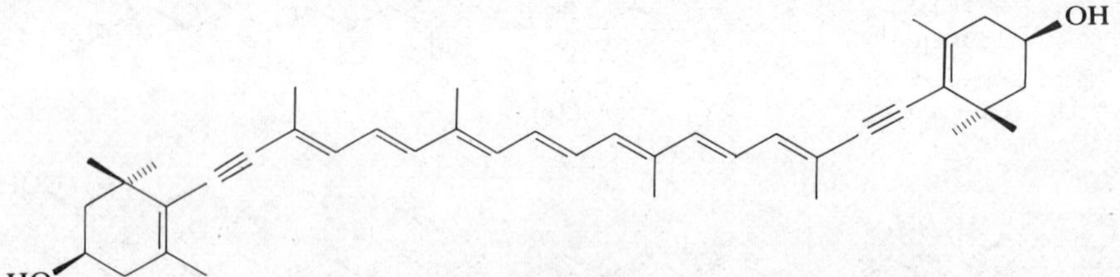

(38)

The cyclopentane end-groups of capsanthin (3,3′-dihydroxy-β,κ-caroten-6′-one, **39**) and capsorubin (3,3′-dihydroxy-κ,κ-carotene-6,6′-dione, **40**) are thought to be formed by an alternative opening of the epoxide (or peroxide) ring with the oxygen remaining at C-6 rather than C-5, followed by a rearrangement of the pinacol–pinacolone type to give the cyclopentane ring system (Scheme 15).

Another possibility involves the entire π-electron system of the polyene chain, so that an electron deficiency in one ring is finally neutralized by proton loss from the other ring. Such a mechanism has been proposed for the formation of the *retro*-carotenoid eschscholtzxanthin (4′,5′-didehydro-4,5′-*retro*-β,β-carotene-3,3′-diol, **41**) (Scheme 16).

The existence of several carotenoids with aryl end-groups is of interest since it illustrates a novel pathway for biosynthesis of the aromatic ring system from mevalonate rather than by the shikimate pathway or from acetate by a polyketide mechanism. Formation of the 1,2,5-trimethylphenyl end-group as in chlorobactene (ϕ,ψ-carotene, **42**) involves simply the migration of one of the C-1 methyl groups to C-2. The methyl group which migrates is that arising from the methyl group, C-3′, of mevalonate (Scheme 17).[10]

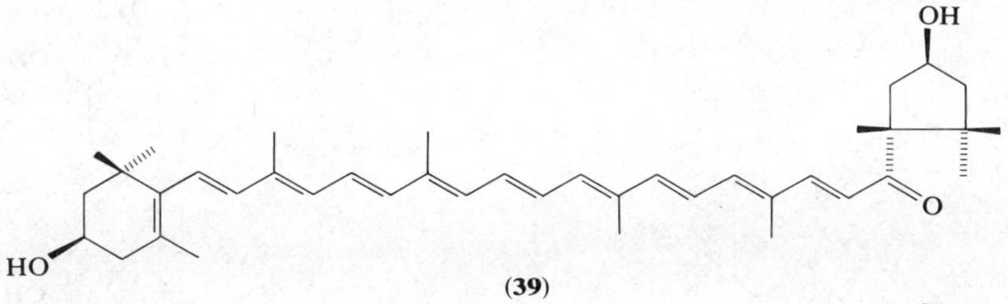

(39)

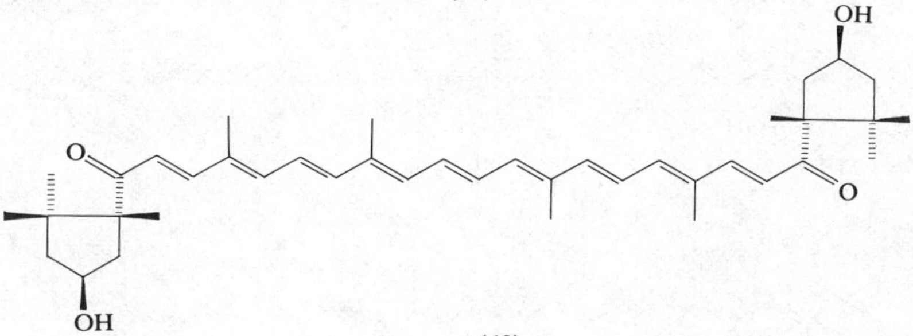

(40)

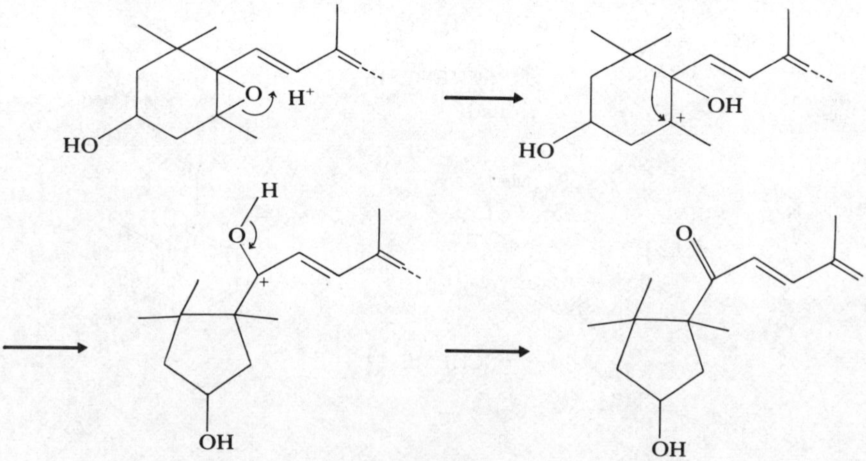

SCHEME 15

A much more complex rearrangement is involved in the biosynthesis of the 1,2,3-trimethylphenyl end-group present in okenone (1′-methoxy-1′,2′-dihydro-χ,ψ-caroten-4′-one, **43**). Although no supporting evidence is yet available, it has been suggested that an intermediate Ladenburg prism structure (Scheme 18) may be involved.[11]

The range of further modifications in the acyclic carotenoid series is much smaller. In the photosynthetic bacteria, hydration of the C-1,2 double bond is commonly followed by *O*-methylation (with *S*-adenosylmethionine) of the tertiary hydroxy-group thus introduced and desaturation of the C-3,4 bond to give carotenoids such as spirilloxanthin (1,1′-dimethoxy-3,4,3′,4′-tetradehydro-1,2,1′,2′-tetrahydro-ψ,ψ-carotene, **44**) (Scheme 19). It is not known whether the stereochemistry and mechanism of the C-3,4 desaturation are similar to those of the phytoene–lycopene desaturation series.

Carbonyl functions can be introduced at various positions in these acyclic carotenoids. The best-known of these processes is the conversion of spheroidene (1-methoxy-3,4-didehydro-1,2,7′,8′-tetrahydro-ψ,ψ-carotene, **45**) into spheroidenone (1-methoxy-3,4-didehydro-1,2,7′,8′-tetrahydro-ψ,ψ-caroten-2-one, **46**), the ketone group introduced at C-2 arising from molecular O_2. In other photosynthetic bacteria oxidation to a keto-group at C-4 or oxidation of the C-20 methyl group to an aldehyde occur under anaerobic conditions, probably *via* a hydroxy-intermediate (Scheme 20).

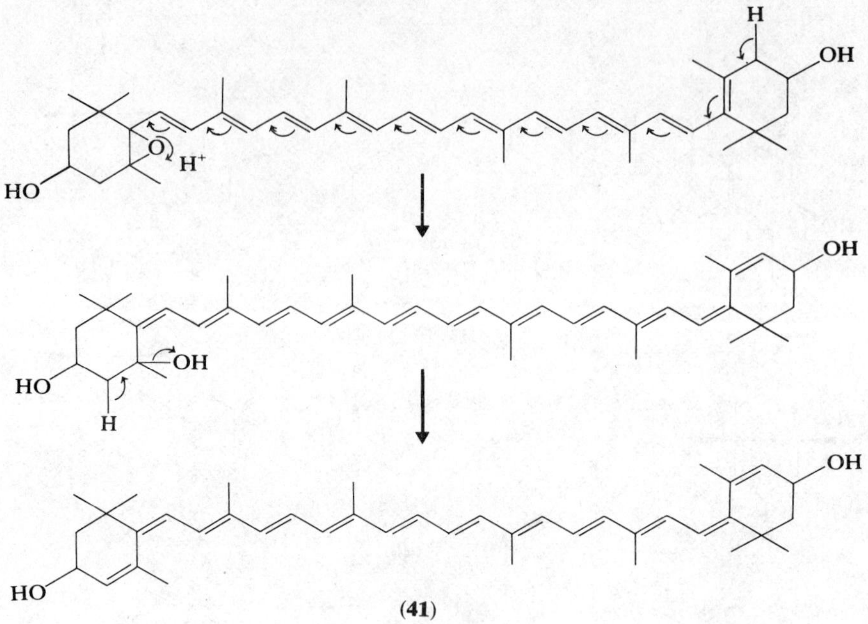

(**41**)

SCHEME 16

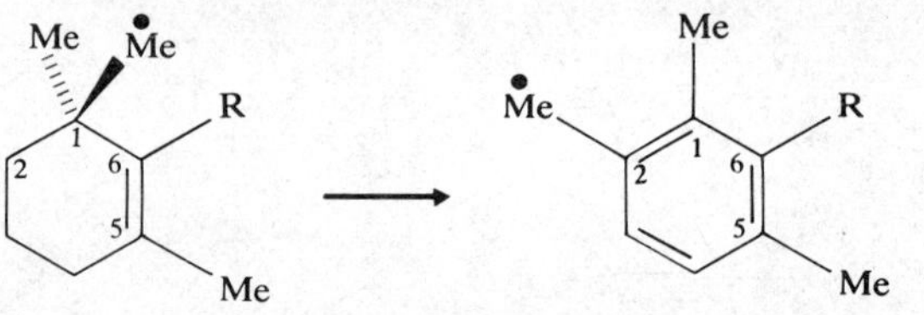

SCHEME 17

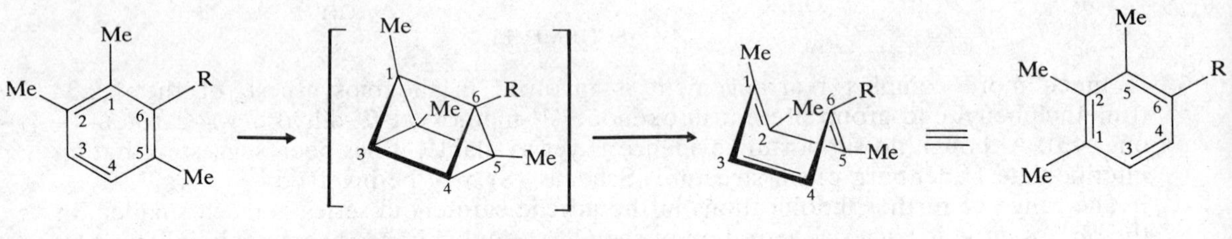

SCHEME 18

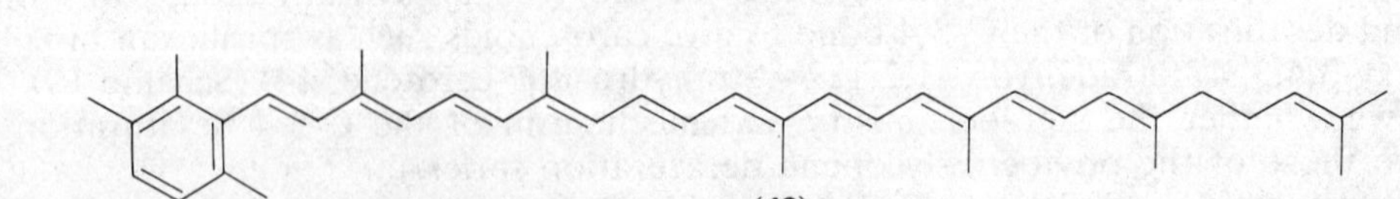

(**42**)

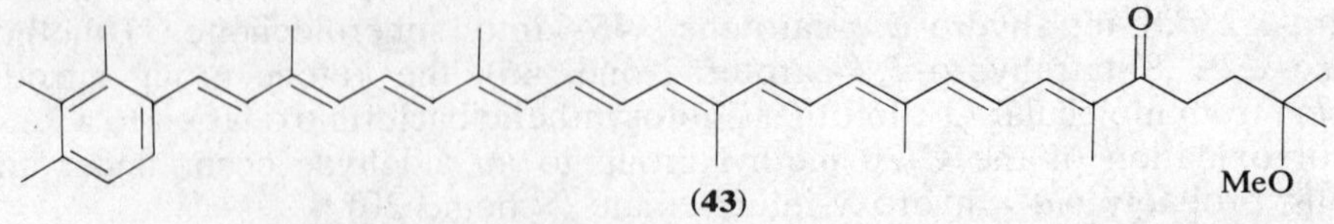

(**43**)

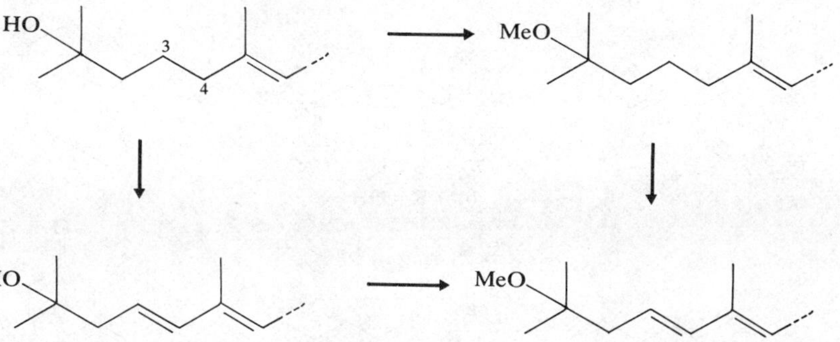

SCHEME 19

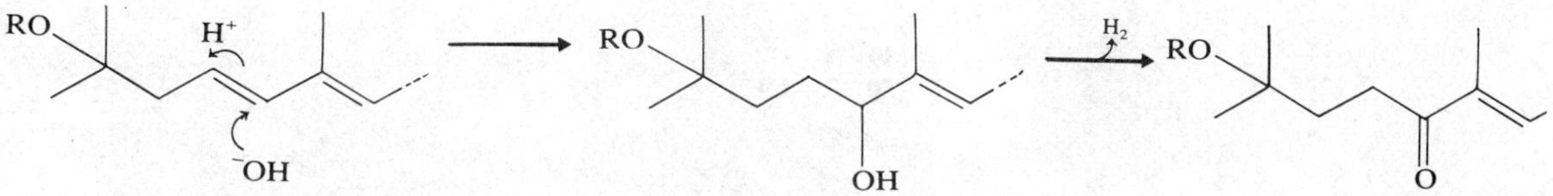

SCHEME 20

29.3.1.6 Structural modifications by animals[12]

It is believed that carotenoids cannot be synthesized by animals but have to be obtained from the diet or from microbial symbionts. Many animals (fish, birds, insects, and other invertebrates) can, however, modify carotenoid structures, particularly by introducing oxo-groups at C-4, *e.g.* by converting β-carotene (**47**) and zeaxanthin (**48**) into canthaxanthin (β,β-carotene-4,4′-dione, **49**) and astaxanthin (3,3′-dihydroxy-β,β-carotene-4,4′-dione, **50**), respectively.

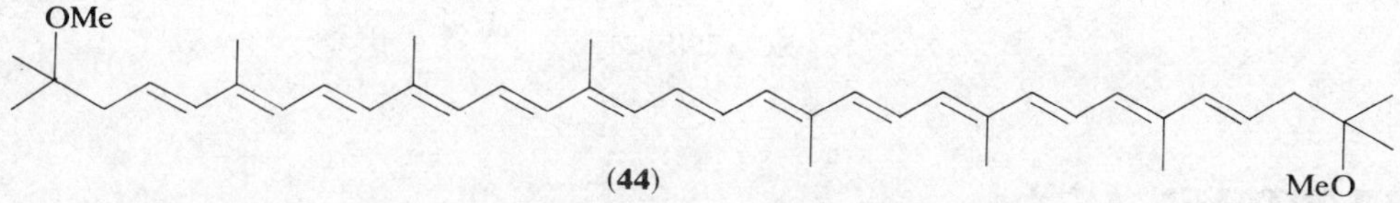

(44)

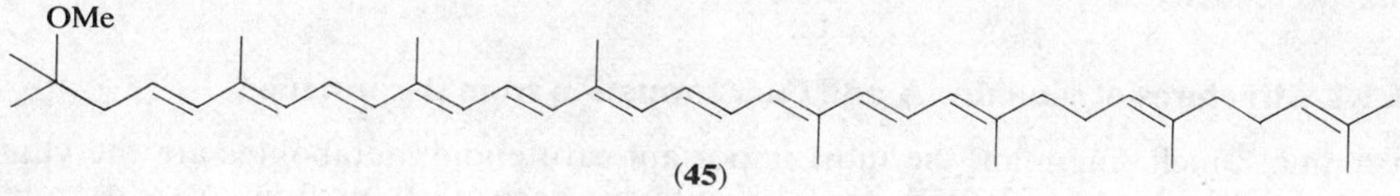

(45)

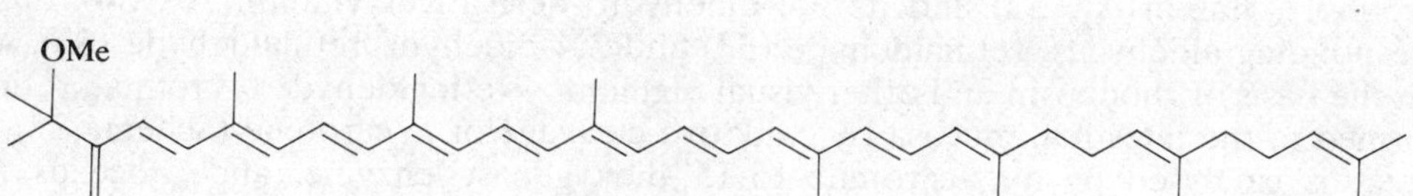

(46)

An interesting modification is the formation of the nor-carotenoid actinioerythrin (3,3′-dihydroxy-2,2′-dinor-β,β-carotene-4,4′-dione diester, **51**) by a process involving contraction from a six-membered to a five-membered ring. The postulated mechanism[11] (Scheme 21) includes a benzylic acid rearrangement of a triketo-intermediate.

Oxocarotenoids, especially astaxanthin, are the characteristic pigments of many marine invertebrates, where they commonly occur in stoichiometric association with protein, as water-soluble carotenoprotein complexes.[13,14] The binding is not covalent, but is usually specific and results in a large bathochromic shift in light absorption maxima. The mode of binding has not been elucidated, but almost certainly the oxo-groups of the carotenoid are involved.

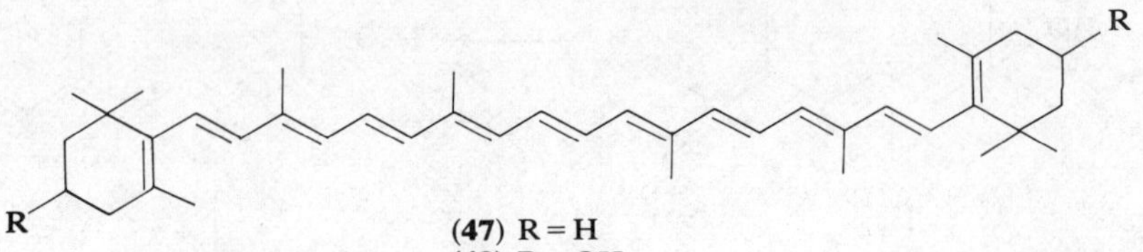

(**47**) R = H
(**48**) R = OH

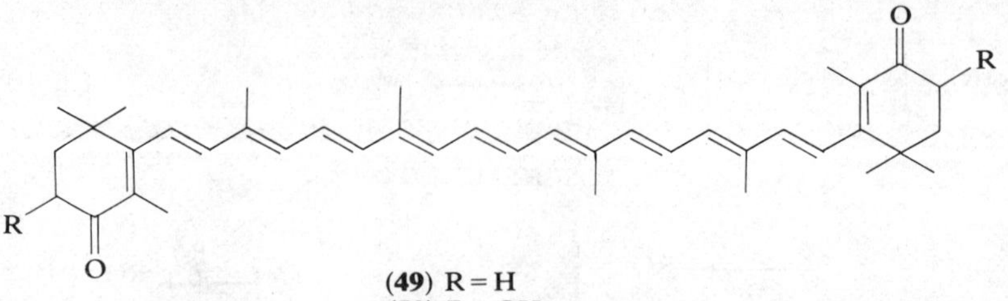

(**49**) R = H
(**50**) R = OH

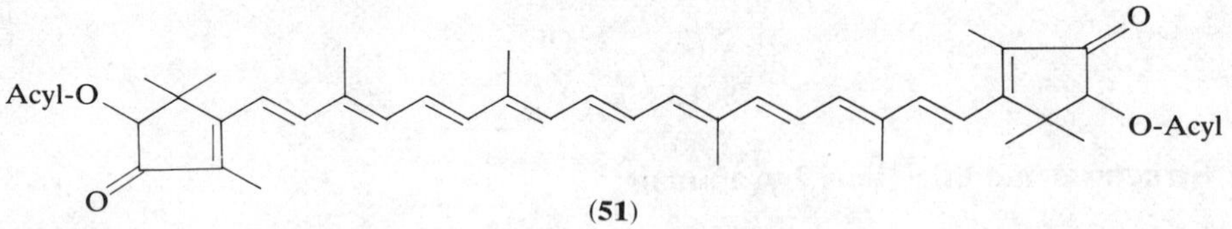

(**51**)

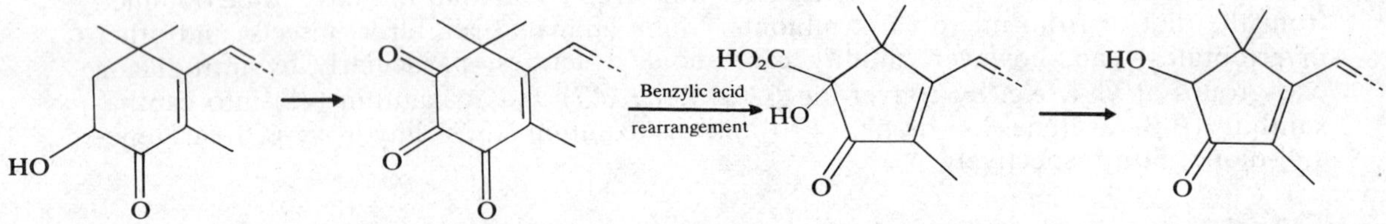

SCHEME 21

29.3.2 VITAMIN A[19–21]

29.3.2.1 Structures of vitamins A and their formation from β-carotene

In animals, including man, the most important carotenoid metabolites are the vitamins A, retinol (vitamin A_1, **52**) and its 3,4-didehydro-derivative (vitamin A_2, **54**), and the corresponding aldehydes, retinaldehyde (**53**) and 3,4-didehydroretinaldehyde (**55**), which form the basis of rhodopsin and other visual pigments. Retinaldehyde (≡ retinal, retinene) is formed in the intestinal mucosa by oxidative cleavage of β-carotene (Scheme 22). This process is catalysed by a β-carotene-15,15′-dioxygenase enzyme, and proceeds *via* a transient peroxide intermediate (**56**). Retinaldehyde and retinol are readily interconverted in the presence of NAD(H) or NADP(H) by dehydrogenase enzymes present in various tissues, notably the liver and the retina. For its transport in the blood, vitamin A is complexed with a lipoprotein, retinol binding protein, and stores of retinyl esters (predominantly palmitate) are maintained in the liver.

29.3.2.2 Visual pigments

The processes of vision depend upon a group of photosensitive pigments located in the retina of the eye. The most extensively studied example of these visual pigments is rhodopsin which in mammals, including man, is the photoreceptor of the retinal rod cells

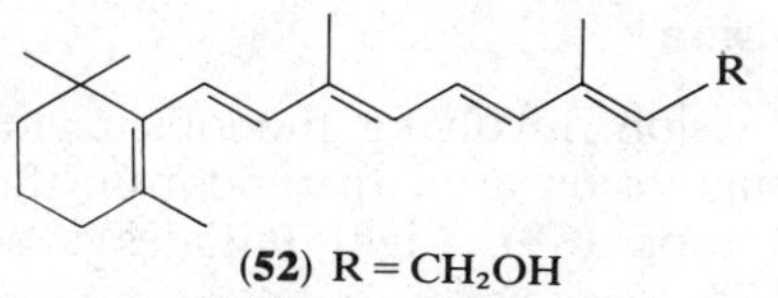

(52) R = CH_2OH
(53) R = CHO

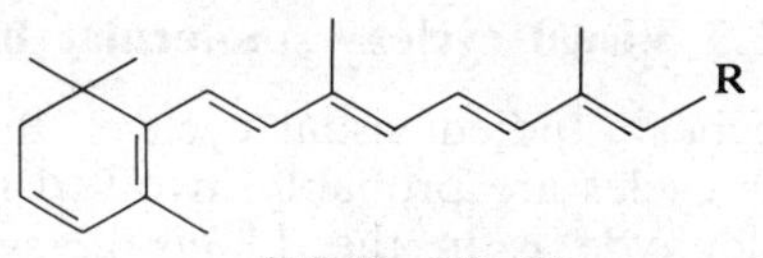

(54) R = CH_2OH
(55) R = CHO

that are responsible for vision in dim light. Rhodopsin is a complex between a glycoprotein, opsin, and 11-*cis*-retinaldehyde. The binding occurs through the formation of a Schiff base (**57**) between the aldehyde group of retinaldehyde and the ε-amino-group of a lysine residue in the opsin molecule. Although retinaldehyde is itself virtually colourless [λ_{max} (ethanol) 383 nm], formation of the protonated Schiff base (**58**) results in a large

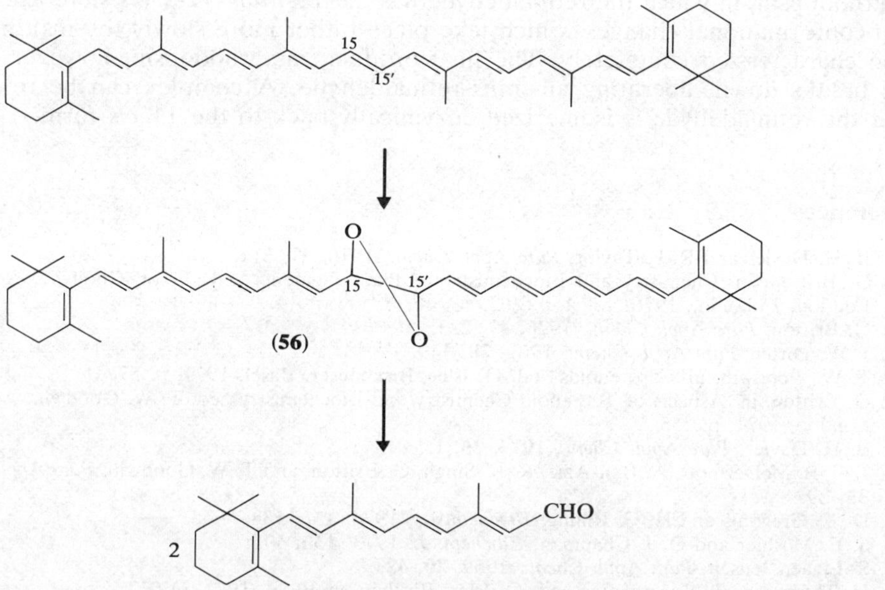

(**56**)

2

SCHEME 22

bathochromic shift, and consequently rhodopsin absorbs light of visible wavelength (λ_{max} 500 nm). Related complexes between retinaldehyde or 3,4-didehydroretinaldehyde and different cone opsins function in the retinal cone cells as the receptor pigments in colour vision. Several of these visual pigments have been recognized which have different absorption maxima and hence are sensitive to light of different wavelengths. They include iodopsin and cyanopsin formed from cone opsins with retinaldehyde and 3,4-didehydroretinaldehyde, respectively.

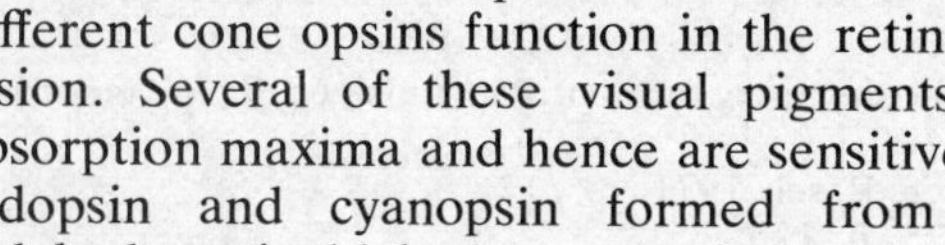

(**57**)

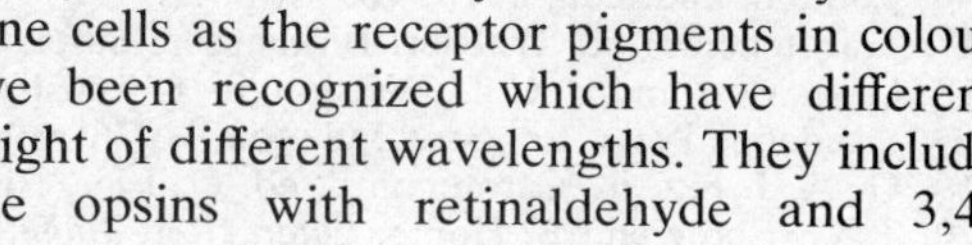

(**58**)

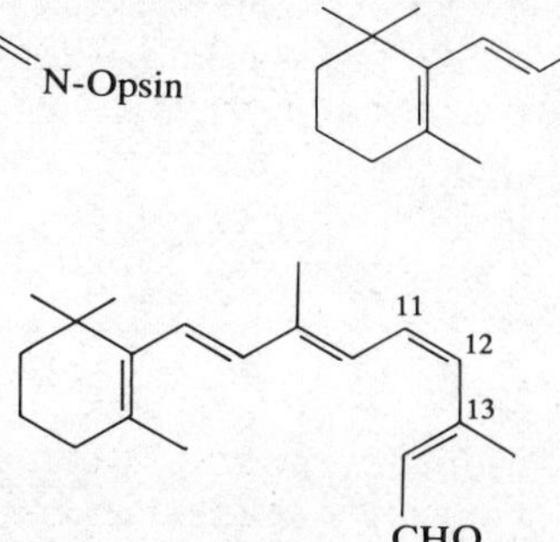

(**59**)

29.3.2.3 Visual cycles — molecular biology of vision

The best studied visual cycle is that of night vision, involving rhodopsin. Somewhat similar cycles are probably involved in day (colour) vision also. In rhodopsin the bound retinaldehyde is in the 11-*cis*-12-*s*-*cis* conformation (**59**). Light initiates a series of configurational and conformational changes thus:

Rhodopsin (500 nm) → Prelumirhodopsin (543 nm) →
Lumirhodopsin (497 nm) → Metarhodopsin I (478 nm) →
Metarhodopsin II (380 nm) → Metarhodopsin III (465 nm)
→ all-*trans*-Retinaldehyde + Opsin

The primary photoreaction is the extremely rapid (6 ps) isomerization to prelumirhodopsin, in which the retinaldehyde has the 11-*trans*-12-*s*-*cis* stereochemistry. The later conformational changes, which take place rather more slowly (ns–ms), have not yet been characterized completely. The final product, metarhodopsin, however, is unstable and breaks down, liberating all-*trans*-retinaldehyde. A complex can be reformed only after the retinaldehyde is isomerized enzymically back to the 11-*cis*-form.

References

1. B. H. Davies and R. F. Taylor, *Pure Appl. Chem.*, 1976, **47,** 211.
2. G. Britton, in 'Chemistry and Biochemistry of Plant Pigments', ed. T. W. Goodwin, Academic Press, London, 2nd. edn., 1976, vol. I, p. 262.
3. G. Britton, *Pure Appl. Chem.*, 1976, **47,** 223.
4. J. W. Porter, *Pure Appl. Chem.*, 1969, **20,** 449.
5. T. W. Goodwin, in 'Carotenoids', ed. O. Isler, Birkhäuser, Basel, 1971, p. 577.
6. G. Britton, in 'Aspects of Terpenoid Chemistry and Biochemistry', ed. T. W. Goodwin, Academic Press, London, 1971, p. 255.
7. B. H. Davies, *Pure Appl. Chem.*, 1973, **35,** 1.
8. J. C. B. McDermott, A. Ben-Aziz, R. K. Singh, G. Britton, and T. W. Goodwin, *Pure Appl. Chem.*, 1973, **35,** 29.
9. D. E. Gregonis and H. C. Rilling, *Biochemistry*, 1974, **13,** 1538.
10. S. E. Moshier and D. J. Chapman, *Biochem. J.*, 1973, **136,** 395.
11. S. Liaaen-Jensen, *Pure Appl. Chem.*, 1969, **20,** 421.
12. H. Thommen, in 'Carotenoids', ed. O. Isler, Birkhäuser, Basel, 1971, p. 637.
13. D. F. Cheesman, W. L. Lee, and P. F. Zagalsky, *Biol. Rev.*, 1967, **42,** 132.
14. P. F. Zagalsky, *Pure Appl. Chem.*, 1976, **47,** 103.
15. G. Britton and T. W. Goodwin, *Methods Enzymol.*, 1971, **18C,** 654.
16. B. H. Davies, in 'Chemistry and Biochemistry of Plant Pigments', ed. T. W. Goodwin, Academic Press, London, 2nd. edn, 1976, vol. II, p. 38.
17. 'Carotenoids', ed. O. Isler, Birkhäuser, Basel, 1971.
18. G. P. Moss and B. C. L. Weedon, in 'Chemistry and Biochemistry of Plant Pigments', ed. T. W. Goodwin, Academic Press, London, 2nd. edn., 1976, vol. I, p. 149.
19. G. A. J. Pitt, in 'Carotenoids', ed. O. Isler, Birkhäuser, Basel, 1971, p. 717.
20. T. G. Ebrey and B. Honig, *Quart. Rev. Biophys.*, 1975, **8,** 129.
21. R. Hubbard, P. K. Brown, and D. Bownds, *Methods Enzymol.*, 1971, **18C,** 615.

PART 30

BIOSYNTHESIS: A GENERAL SURVEY

30.1
Alkaloid Biosynthesis

R. B. HERBERT
University of Leeds

30.1.1 INTRODUCTION

The number of plant alkaloids is legion, the structural diversity is enormous and yet, as was recognized quite early, structural relationships can be discerned between individual alkaloids,[1] particularly between those from the same species or family. Moreover, consideration of likely relationships can be extended to quite simple molecules which are intermediates in primary metabolism, thus suggesting possible routes of biogenesis. Tropine (**2**) (which occurs naturally as various esters) and hygrine (**1**) may be taken as examples of this thinking. They are formed in plants of the same species, they are manifestly of related structure, and they have a plausible ancestry which can be traced to the primary metabolites, ornithine and acetic acid (Scheme 1). These alkaloids can also be used to illustrate a further important aspect of structural relationships, that of the possible sequence of formation in the plant; examination of the two structures suggests that (**1**) is an intermediate in the formation of (**2**).

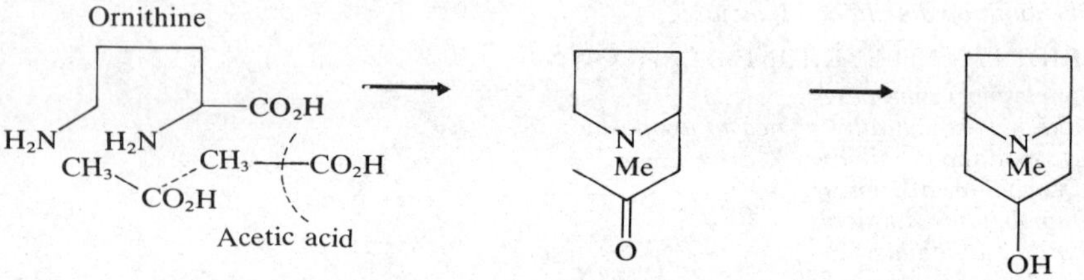

(**1**) (**2**)

SCHEME 1

Hand in hand with speculation of this kind has gone a consideration of plausible mechanisms by which alkaloids could arise from primary metabolites. The most pervasively useful mechanistic hypothesis for predicting and understanding biosynthetic pathways involves condensations with imines [as (**3**)] (which may be extended to include

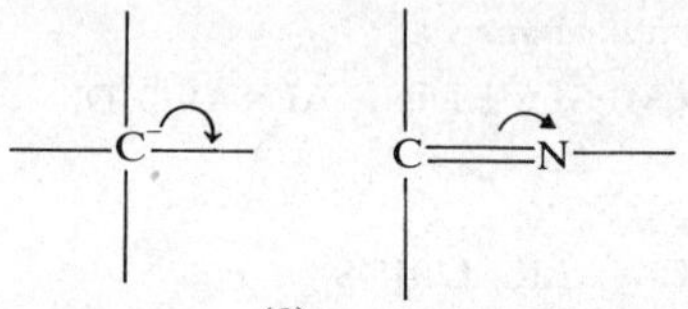

(**3**)

enamines). An example of this reaction is proposed (see Scheme 2) for hygrine biosynthesis and its feasibility is supported by successful laboratory examples (see Section 30.1.2.1).

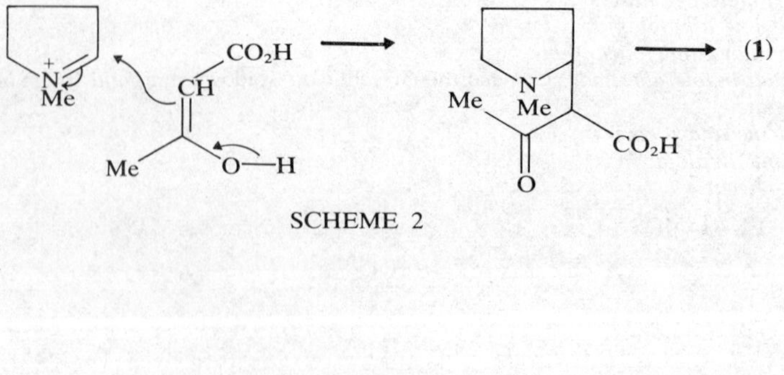

SCHEME 2

A further mechanistic hypothesis of great general importance, and particular application in understanding the biosynthesis of alkaloids with aromatic rings, is that of oxidative phenol coupling.[2] In essence the hypothesis is that new carbon–carbon, and carbon–oxygen, bonds can be formed by the coupling of mesomeric radicals arising from the one-electron oxidation of each of a pair of phenols. Chemical conversion of *p*-cresol into the ketone (**5**) provides an example in which coupling occurs between the position *ortho* to one phenolic hydroxy-group and the *para* position of the other (alternatives involve coupling between two *ortho* or two *para* positions, or coupling in which C—O bond formation occurs). Aromaticity is regained by proton loss from the coupling site. In the case of (**5**) only one ring can become aromatic in this way; the methyl group in the other secures it in the dienone form (**4**). Such dienones are reactive in the way shown (see also Section 30.1.4.4), but more importantly may also undergo rearrangement leading to aromatic products.

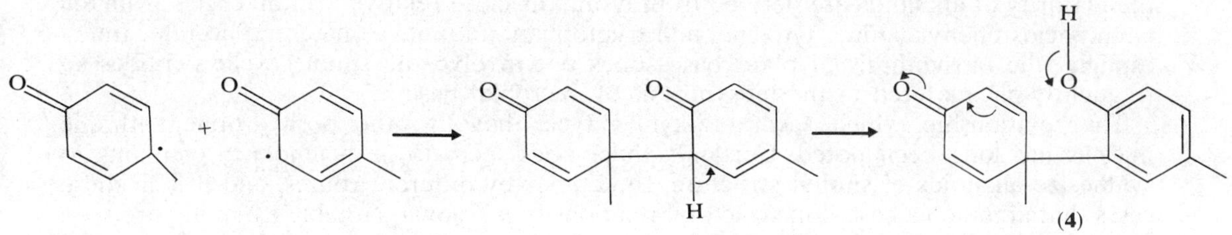

(4)

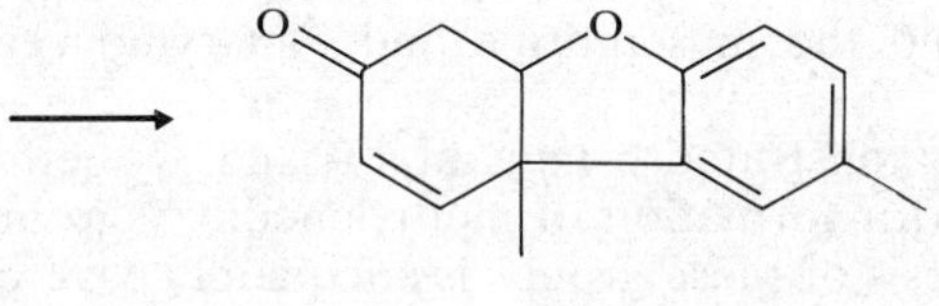

(5)

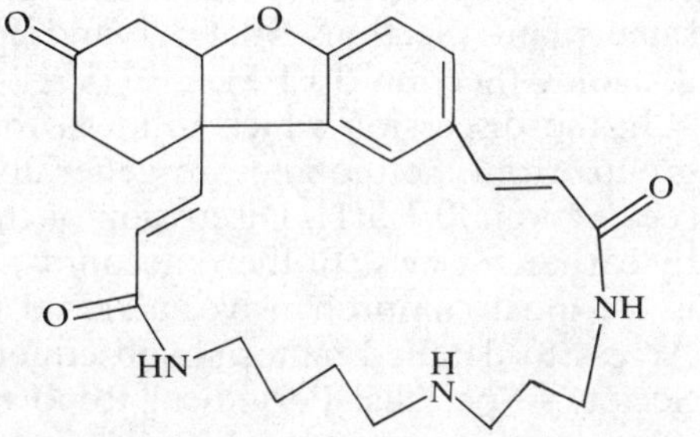

$H_2N(CH_2)_3NH(CH_2)_4NH_2$

(7)

(6)

These ideas have had a seminal influence in studies of aromatic alkaloid biosynthesis. Rigorous and extensive study has demonstrated most convincingly the correct applicability of the hypothesis in this area (see Sections 30.1.4 and 30.1.5). In every relevant case which has been examined, a new bond between aromatic rings is only forged *ortho* or *para* to phenolic hydroxy-groups; no reaction occurs in the plant with precursors where, for example, these hydroxy-groups are alkylated. The biosynthesis of lunarine (**6**) may be taken as a simple illustration. Most strikingly it is a strict analogue of (**5**) with one less double bond, and reasonably arises by a similar biological mechanism from *p*-hydroxycinnamic acid. This is supported by the observation that phenylalanine, the probable precursor for this acid, together with spermidine (**7**), is a precursor for the alkaloid.[3]

So far we have dealt almost exclusively with hypothesis and until radioactive isotopes became available this was the limit to progress in this area. Then in a great blossoming of research which continues, these hypotheses were tested in plants by experiments with compounds labelled with ^{14}C and ^{3}H. (For a useful review on techniques, see Ref. 4.) Some hypotheses were refuted but many were proved correct, and important detail was added in defining the actual biosynthetic pathways. Further definition has come with increasingly sophisticated experiments, including work with enzymes. Attention has turned recently to an exploration of the ability of plants to make unnatural analogues of the alkaloids they normally produce.[5,6] Apart from the production of potentially interesting compounds, some insight can be gained in this way into the enzymatic reactions of alkaloid biosynthesis.

It is quite remarkable that so much of the speculation on alkaloid biosynthesis proved correct. Nonetheless, it is also remarkable that in secondary metabolism in general (of which alkaloid biosynthesis is part) in contrast to primary metabolism, the reactions involved are simple, almost repetitive. Certainly these reactions are easily within the compass of conventional organic chemistry, which no doubt improves the chances of correct speculation on biosynthetic pathways. A further reflection of this is seen perhaps in the ease with which the pathway for the biosynthesis of an alkaloid can be used for its chemical synthesis.

A further notable feature of alkaloid biosynthesis is that there are very few branch points from primary metabolism which lead to the vast pantheon of different alkaloids; the majority of alkaloids are derived from lysine, its close relative ornithine, the aromatic amino-acids phenylalanine, tyrosine, and tryptophan, and mevalonate and acetate. Interestingly, the biosynthesis of plant bases does not involve the simple cyclic peptides so frequently encountered in the biosynthesis of microbial bases.

The relationship which several alkaloid types have to the perhydrophenanthrene nucleus has long been noted. Certainly there does seem to be a tendency of plants to synthesize alkaloids of similar structure, sometimes by different routes, and it is in these cases that arguments based on structural relations break down. Notable examples are seen in the quite different pathways which lead to the structurally related alkaloids coniine and *N*-methylpelletierine (Section 30.1.3.1), and anabasine and anatabine, which occur in the same plant (Sections 30.1.3.1 and 30.1.3.2), and the mesembrine and Amaryllidaceae alkaloids (Section 30.1.5).

In the discussion which follows, the amply demonstrated origins of *O*- and *N*-methyl groups from methionine are generally assumed (for formation of methylenedioxy-groups see Section 30.1.5.1). Discussion of the biosynthesis of these groups is adequately covered by earlier reviews. In these accounts,[7–10] of which the one[8] edited by Mothes and Schütte is the most comprehensive, alkaloid biosynthesis is surveyed up to approximately 1967. Access to detailed material subsequent to this date may be gained through the Chemical Society's Specialist Periodical Reports.[11,12]

30.1.2 PYRROLIDINE ALKALOIDS

30.1.2.1 Tropane and related alkaloids

The classical synthesis[13] of tropinone (**12**) from succindialdehyde, methylamine, and acetonedicarboxylic acid under 'physiological conditions' is quite spectacular, but the biosynthesis of these alkaloids follows another course, one which begins with the α-amino-acid ornithine. Specific incorporation of [2-^{14}C]ornithine into various tropane alkaloids, *e.g.* hyoscyamine (**15**), has been observed,[14,15] as has the incorporation of its decarboxylation product, putrescine; label from [1,4-$^{14}C_2$]putrescine was confined to the bridgehead carbon atoms.[16] An ingenious degradation sequence,[14] which allowed distinction between the bridgehead carbons in hyoscyamine, led to the deduction that label from C-2 of ornithine is confined to only one bridgehead carbon atom. It follows that the incorporation of ornithine into hyoscyamine (**15**) is unsymmetrical and cannot involve the symmetrical base, putrescine, as an intermediate. As a result of corroborating tracer experiments[18,19] obtained for tropane bases, and the related alkaloid, cuscohygrine (**13**), it is currently believed that ornithine asymmetry is maintained *via* δ-*N*-methylornithine (**8**), which is incorporated into the tropane nucleus through *N*-methylputrescine (**9**) (which may also be derived from the proven precursor putrescine[17]). This pathway (Scheme 3) differs from that deduced for piperidine alkaloids, where non-symmetrical incorporation of the amino-acid lysine is accounted for more simply (Section 30.1.3.1) and is a point taken up in the discussion of nicotine biosynthesis (Section 30.1.2.2).

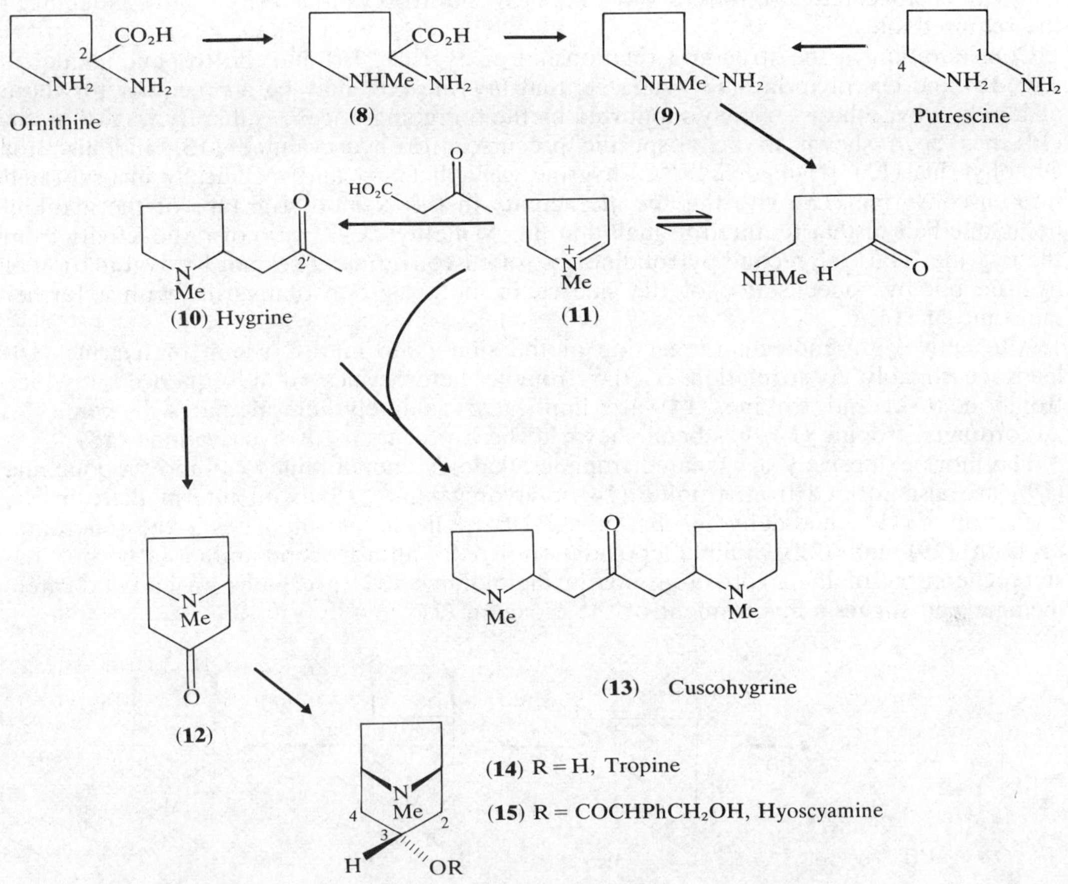

SCHEME 3

Cocaine (**16**) is an example of a tropane-like base with a vestigial carboxy-group, which co-occurs with hygrine and tropine esters. Unfortunately its biosynthesis is still obscure, for only minute incorporations of the expected precursors have been recorded. On the other hand, the betaine stachydrine (**18**) has been shown[20] to arise directly from proline through hygric acid (**17**) and thus to have a biosynthesis different from the tropane alkaloids.

N-Methyl-Δ^1-pyrroline (**11**), formed plausibly by oxidation of *N*-methylputrescine, is a likely if untested intermediate of importance in tropane alkaloid biosynthesis. It has been shown to be a nicotine precursor (see below) and reacts *in vitro* with acetoacetic acid to give hygrine (**10**)[21] [*cf.* Scheme 2; cocaine (**16**) may also arise by this pathway, but without

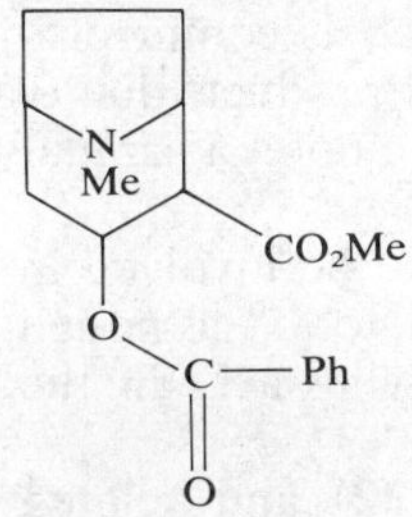

(**16**) Cocaine

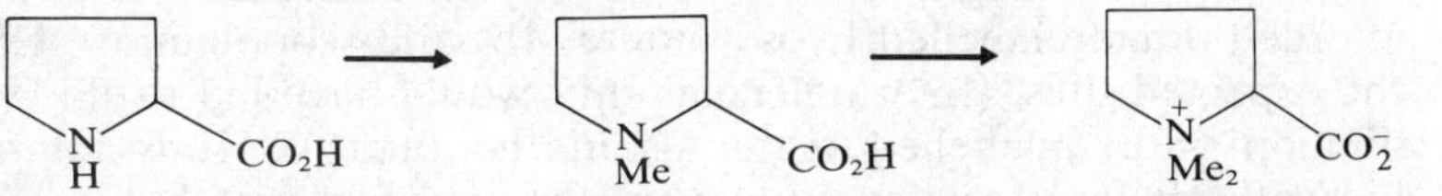

Proline (**17**) (**18**) Stachydrine

loss of the carboxy-group]. In accord with this hypothesis, which may be extended to that shown in Scheme 3, acetate has been shown to be a specific precursor for hygrine (**10**), and with acetoacetate also for cuscohygrine (**13**) and hyoscyamine (**15**), to give labelling of the required sites.[19,22,23]

Consideration of the structural relationship of hygrine (**10**) and the tropane alkaloids, *e.g.* (**15**) and cuscohygrine (**13**), suggests that hygrine (**10**) may be a precursor for these alkaloids, a hypothesis strongly supported by the foregoing tracer results. By way of proof, (**10**) has been shown to be a specific precursor for hyoscyamine (**15**) and also for cuscohygrine (**13**). [N-*methyl*,2′-$^{14}C_2$]hygrine was efficicntly and specifically incorporated into cuscohygrine (**13**) with the *specific* activity in the *N*-methyl groups of the alkaloid being one half of that required to maintain the *N*-methyl : C-2′ ratio of radioactivity from the hygrine. Thus the second pyrrolidine ring of cuscohygrine arises not by degradation of hygrine but by condensation of the side-chain methyl group of hygrine with a further molecule of (**11**).[22]

Alternative, intramolecular, reaction at the side-chain methyl group of hygrine (**10**) leads reasonably to formation of the tropane heterocycles in a sequence in which tropinone (**12**) and tropine (**14**) are implicated as likely intermediates (Scheme 3). Accordingly, tropine (**14**) has been shown to be a precursor for hyoscyamine (**15**).[24]

The more extensively oxygenated tropane alkaloids, meteloidine (**22**) and scopolamine (**19**), are also formed from tropine (**14**).[24] Hyoscyamine (**15**) is an intermediate in the formation of (**19**)[25] and dehydro-derivatives (**20**) are likely intermediates in the generation of both (**19**) and (**22**). Simple epoxidation would afford scopolamine (**19**), but the stereochemistry of the hydroxy-groups in meteloidine (**22**) precludes an epoxide intermediate and suggests involvement of the dioxetan (**21**).

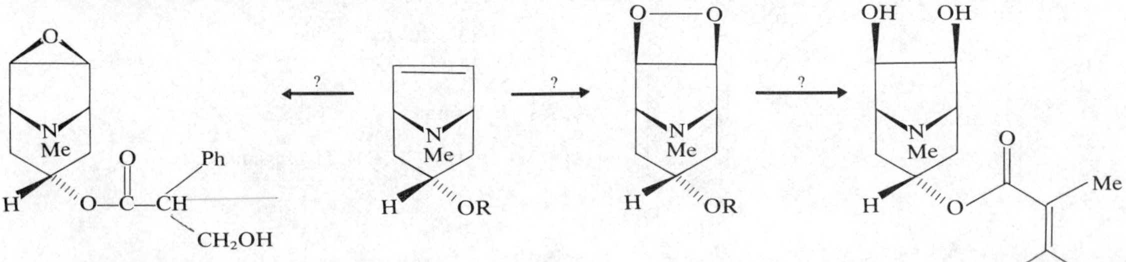

(**19**) Scopolamine (**20**) (**21**) (**22**) Meteloidine

The tropane alkaloids commonly occur as esters of tropic [as (**15**)] and tiglic [as (**22**)] acids. Extensive research (see Refs. 26 and 27) on the biosynthesis of the former acid (**23**) has shown that it arises preferentially from phenylalanine (other origins deduced as phenylacetic acid and tryptophan are less secure[27]). The entire carbon skeleton of phenylalanine is involved and from the results of ^{14}C-labelling studies, shown in Scheme 4, it is clear that in the rearrangement reaction which occurs during the biosynthesis it is the carboxy-group which migrates. Proof of an intramolecular shift for this group was obtained by showing that [1,3-$^{13}C_2$]phenylalanine (majority of molecules doubly labelled) afforded doubly labelled hyoscyamine (**15**) and scopolamine (**19**), with labels confined to the expected sites. (Intermolecular shift would have led to singly labelled products due to dilution with unlabelled materials in the plant.)[26] It is interesting to note a similar biosynthesis for a related system in the microbial metabolite tenellin.[28]

Both phenylpyruvic acid (**24**) and phenyl-lactic acid (**25**) are thought to be involved in tropic acid biosynthesis[29] and recent evidence implicates cinnamic acid too,[30] although earlier results were negative. (Phenyl-lactic acid appears as the esterifying acid in the alkaloid littorine and its origins are proven to be in phenylalanine.[31])

The biosynthesis of the tigloyl [as (**27**)] residues of meteloidine (**22**) and related alkaloids has been found to be similar to that leading to tiglic acid in animals. Isoleucine is

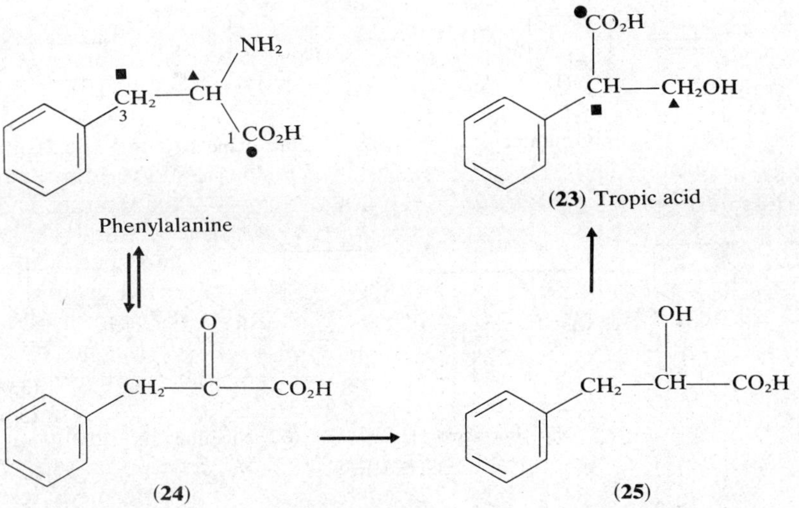

SCHEME 4

incorporated, with loss of the carboxy-group by way of 2-methylbutyric acid (**26**).[32,33] L-Isoleucine is also a precursor for the 2-methylbutanoyl residue in (**28**), which significantly has the same absolute configuration at the relevant centre.[34] That introduction of the double bond in tiglic acid is a dehydrogenative rather than dehydrative step is indicated by the failure of 2-hydroxy-2-methylbutyric acid to act as a precursor.[33]

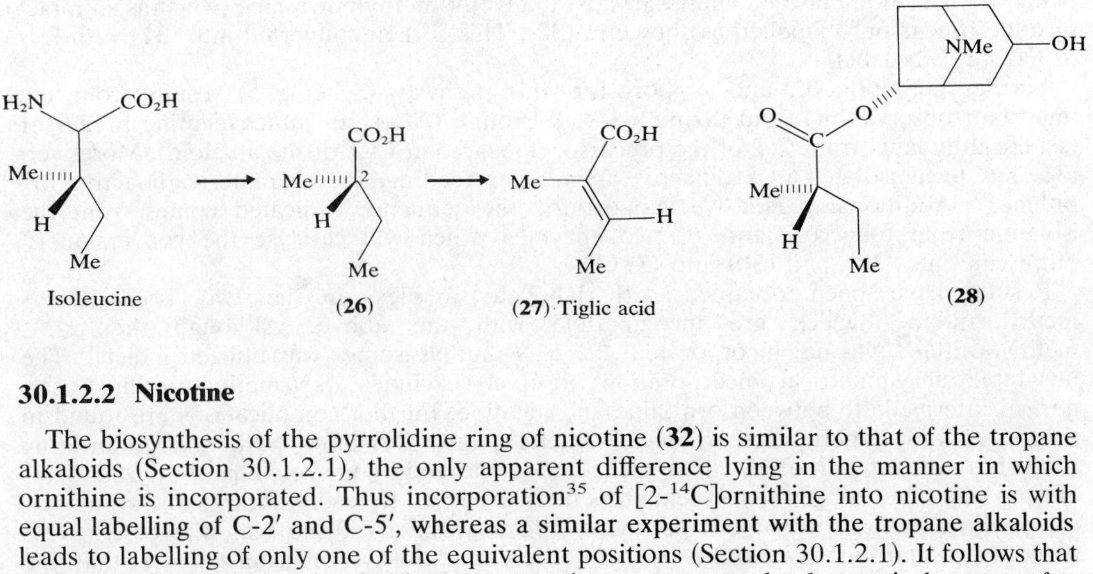

Isoleucine (**26**) (**27**) Tiglic acid (**28**)

30.1.2.2 Nicotine

The biosynthesis of the pyrrolidine ring of nicotine (**32**) is similar to that of the tropane alkaloids (Section 30.1.2.1), the only apparent difference lying in the manner in which ornithine is incorporated. Thus incorporation[35] of [2-^{14}C]ornithine into nicotine is with equal labelling of C-2′ and C-5′, whereas a similar experiment with the tropane alkaloids leads to labelling of only one of the equivalent positions (Section 30.1.2.1). It follows that utilization of ornithine in the former case, in contrast to the latter, is by way of a symmetrical intermediate, for which the diamine, putrescine, is an obvious candidate (Scheme 5). In agreement, [1,4-$^{14}C_2$]putrescine was incorporated and labelled C-2′ and C-5′ of nicotine.[36]

Other experiments support the role of ornithine as a precursor formed from glutamic acid (**33**),[36,37] but important definition of ornithine incorporation comes from experiments[38] with [2-^{14}C,α-^{15}N]- and [2-^{14}C,δ-^{15}N]-ornithine. Firstly, ^{15}N was found to be incorporated into the pyrrolidine ring of nicotine from the latter, but not the former, precursor. This eliminates as intermediates glutamic semialdehyde (**34**), which would result in loss of the δ-amino-group, and free putrescine arising by direct decarboxylation

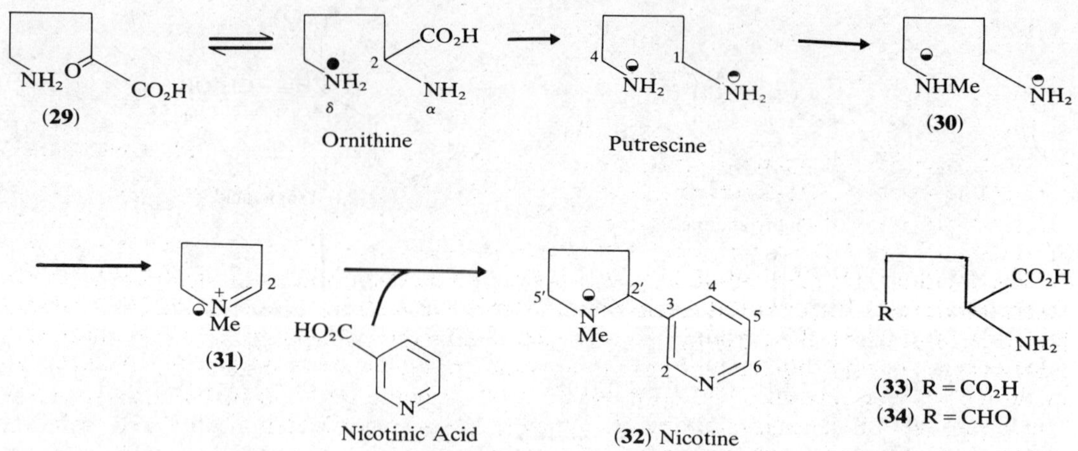

SCHEME 5

of ornithine (or alternative known pathway[39] *via* arginine) as this would result in incorporation of label from both the α- and δ-amino-groups. Putrescine may still be implicated as an intermediate, however, if the α-amino-group is lost by rapid equilibration of ornithine with δ-amino-α-ketovaleric acid (**29**) before conversion into putrescine (Scheme 5). Secondly, in the experiments with ^{15}N-labelled ornithine, the [2-^{14}C,δ-^{15}N]ornithine was found to give nicotine with loss of half the ^{15}N label, consistent with a plausible pathway from the amino-acid *via* putrescine, methylation of which occurs (without distinction between amino-groups) to give *N*-methylputrescine (which is an intact nicotine precursor[40]) labelled as shown in (**30**). This is then converted into (**31**) with loss of half the ^{15}N label.

Further definition of, and support for, this pathway (Scheme 5) results from the important observation[41] that *N*-methyl-Δ^1-pyrroline (**31**) is an intact nicotine precursor (significantly label from C-2 of the precursor appeared at C-2′ of the alkaloid). Moreover, (**31**) has been isolated in radioactive form after feeding, for example, radioactive ornithine.[42] Additional persuasive support for the sequence indicated comes from the isolation from tobacco plants of two enzymes which will catalyse the conversion of putrescine into (**30**), and (**30**) into (**31**).[43]

Results have been obtained with labelled samples of the two isomeric *N*-methylornithines which are incompatible with the above. Although the α-*N*-methylornithine was not incorporated, the δ-*N*-methyl isomer was utilized intact.[44] The inevitable unsymmetrical incorporation observed excludes δ-*N*-methylornithine as a normal intermediate between ornithine and nicotine. Further complications are found in on-going $^{14}CO_2$ experiments, where the results suggest formation of the pyrrolidine ring through a symmetrical intermediate on the one hand and by unsymmetrical intermediates only on the other. More rigorous experimentation has led to results which show symmetrical labelling patterns in some experiments and unsymmetrical patterns in others.[45] This suggests elaboration of nicotine by two different pathways or, alternatively, that the normal biosynthesis of nicotine, and perhaps that of the related tropane alkaloids too, resembles (in part) the route to piperidine alkaloids (Section 30.1.3), *i.e.* variable/incomplete desorption and readsorption of enzyme-bound putrescine (which arises by enzyme-catalysed decarboxylation of ornithine), leading to nicotine formed in part unsymmetrically (from enzyme bound putrescine) and in part symmetrically (from putrescine which had been desorbed from, and readsorbed to, the enzyme surface prior to further reaction). To accommodate the tropane alkaloids within this scheme (in spite apparently of tracer evidence to the contrary) requires simply that no normal equilibration occurs of bound with unbound putrescine; and accordingly the alkaloids will be formed in unsymmetrical fashion.

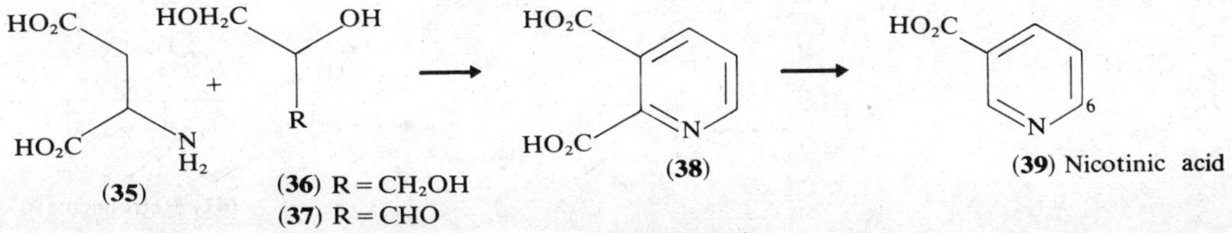

SCHEME 6

The pyridine ring of nicotine has been proved to have its genesis in nicotinic acid (**39**). Incorporation of this acid is such that the pyrrolidine ring becomes attached to the position from which the carboxy-group is lost. Analogous incorporation of quinolinic acid (**38**) occurs and the sequence of biosynthesis may be defined as (**38**) → (**39**), which reacts with (**31**) to give nicotine.[46] The answer to the intriguing question of the mechanism of this last reaction is suggested by the results obtained with deuteriated and tritiated nicotinic acid samples, where hydrogen isotope appears to be lost only from C-6. This is not the result of hydroxylation at this site for 6-hydroxynicotinic acid is not a nicotine precursor. Instead a dihydronicotinic acid intermediate (**40**) may be proposed which leads as shown (Scheme 7) to nicotine (**32**).[47]

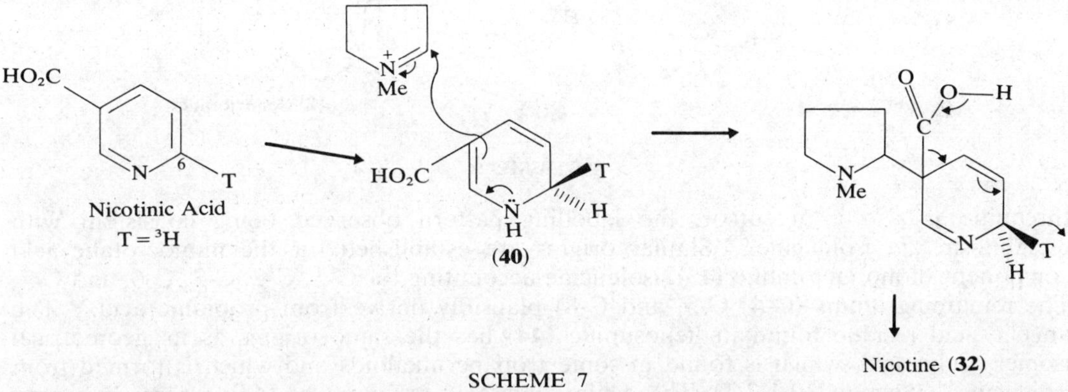

SCHEME 7

The biosynthesis of nicotinic acid in animals and some micro-organisms is well established to be by degradation of tryptophan, whereas in plants nicotinic acid has its origins in aspartic acid and glycerol. Such also are the origins of the pyridine ring of nicotine (**32**). Glycerol (**36**), and more immediately, glyceraldehyde (**37**), provides carbons 4, 5, and 6. Carbon atoms 2 and 3, on the other hand, have their genesis in a precursor derived from aspartic acid (**35**) (Scheme 6).[48]

30.1.2.3 Pyrrolizidine alkaloids

The pyrrolizidine alkaloids may be exemplified by senecionine (**42**). The basic fragment retronecine (**41**) seen in this alkaloid derives from ornithine and putrescine (and also their precursor arginine[49]), and two molecules are involved (Scheme 8). That much is agreed by separate investigators. However, one set of results indicates formation of alkaloid by unsymmetrical intermediates[50] from ornithine, and the other through a symmetrical intermediate for one ring at least[51] (*cf.* the discussion on nicotine biosynthesis above—the explanation may be similar). Additional work is clearly required both to clarify and further to define retronecine biosynthesis.

Available evidence on the biosynthesis of the necic acids associated with retronecine (**41**) in senecionine (**42**) and its congeners points to the derivation of these acids universally from branched-chain α-amino-acid precursors, but as yet the mechanism of

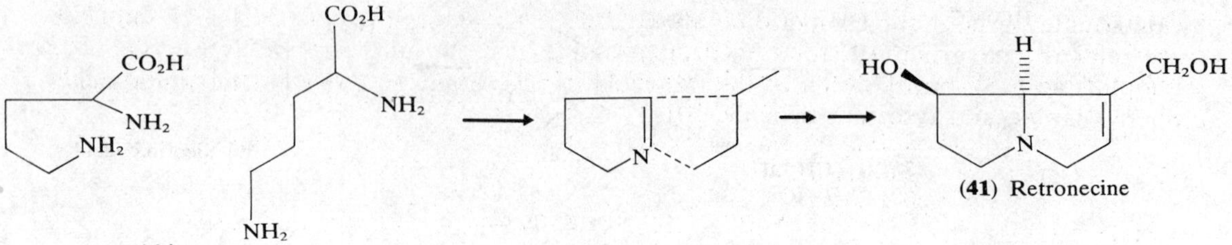

Ornithine

(**41**) Retronecine

SCHEME 8

their elaboration is unknown. The C_{10} necic acids of the senecic acid type, which appear in alkaloids like senecionine (**42**), are formed from two units of L-isoleucine (Scheme 9);

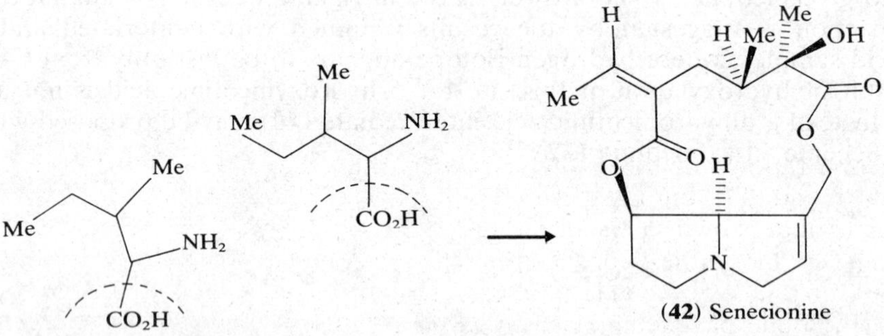

Isoleucine

(**42**) Senecionine

SCHEME 9

threonine is also a precursor, the labelling pattern observed being consistent with metabolism *via* isoleucine.[52] Similar origins are established for the monocrotalic acid component of monocrotaline (**43**), isoleucine accounting for C-1, C-2, C-3, C-6, and C-7. The remaining atoms (C-4, C-5, and C-8) plausibly derive from propionic acid.[53] The angelic acid residue found in heliosupine (**44**) has the same origins as its geometrical isomer, tiglic acid, which is found in some tropane alkaloids and which is formed from isoleucine[54] (Section 30.1.2.1). The echimidinic acid fragment of (**44**) is formed in part from another branched-chain amino-acid, valine.[54] This amino-acid appears to provide all but C-5 and C-6.

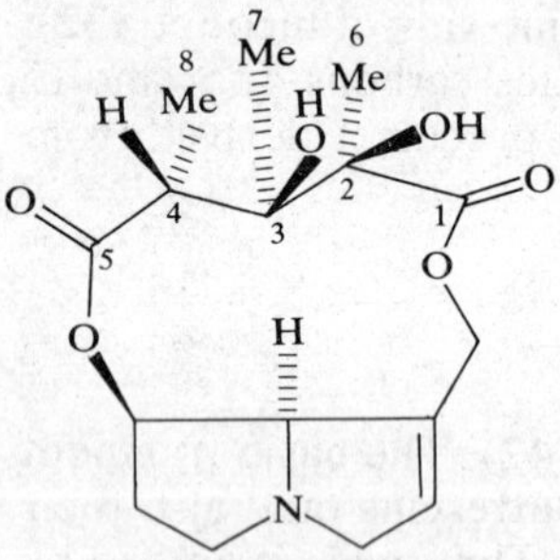

(**43**) Monocrotaline

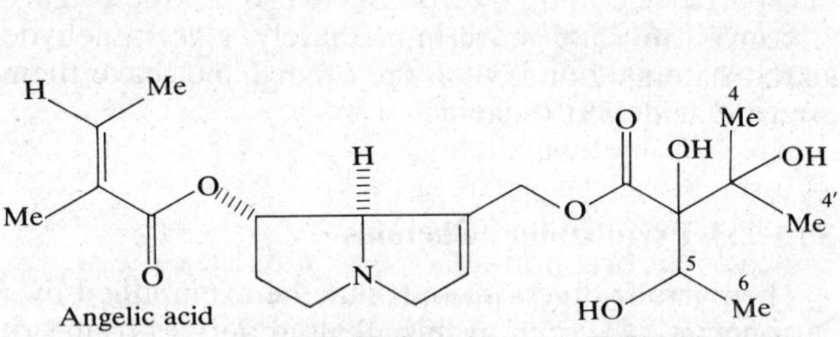

Angelic acid

(**44**) Heliosupine

Echimidinic acid

30.1.2.4 Phenanthroindolizidine alkaloids

It is the intermediacy of the 2-phenacylpyrrolidines, *e.g.* (**48**), in the biosynthesis of phenanthroindolizidine alkaloids, *e.g.* tylophorinine (**53**), which suggests their inclusion in the discussion on pyrrolidine alkaloids, since these pyrrolidines are the five-membered ring counterparts of some piperidine alkaloids, *e.g.* sedamine (Section 30.1.3.1) and they are related to hygrine (**10**) too. Accordingly, ornithine accounts apparently for ring E of tylophorine (**45**), but the bulk of the molecule has its origins in the aromatic amino-acids phenylalanine and tyrosine (Scheme 10).[55]

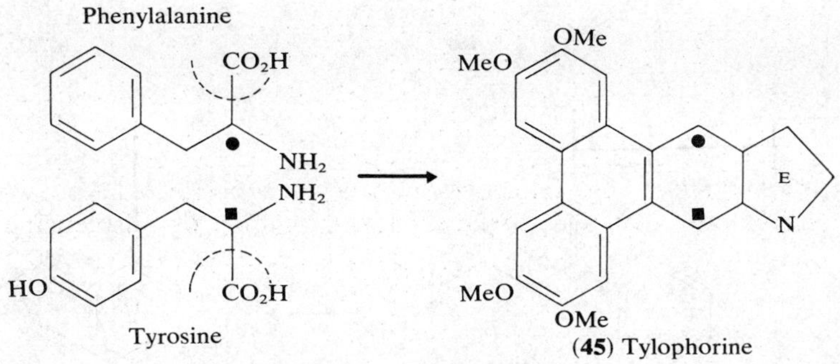

(**45**) Tylophorine

SCHEME 10

Intact incorporation[56] of the 2-phenacylpyrrolidines (**48**), (**49**), and (**50**) as well as the keto-acids (**46**) and (**47**), but not (**54**), into tylophorinine (**53**), define the part sequence of biosynthesis as that shown in the early part of Scheme 11.

Later stages of biosynthesis are suggested by the natural occurrence of septicine (**51**) and the variation in oxygenation pattern found in bases related to tylophorinine (**53**), against a background of reactions centred around enamines and phenol coupling (Section 30.1.1) as the likely means of converting the septicine type into tylophorinine type (Scheme 11).

The biosynthesis of the quinolizidine alkaloids, *e.g.* cryptopleurine (**55**), has not yet been studied but it is likely to be similar to that of the phenanthroindolizidine bases since their structures are obviously similar and six-membered ring analogues of (**50**) and (**51**) occur naturally.

30.1.3 PIPERIDINE AND PYRIDINE ALKALOIDS

30.1.3.1 Simple piperidine bases

Simple piperidine bases may be exemplified by *N*-methylpelletierine (**66**) and the hemlock alkaloid, coniine (**61**). The very close similarity in structure of these bases suggests a similar biogenesis, but this is deceptive for tracer experiments have shown that whereas *N*-methylpelletierine is elaborated from lysine and acetate, the coniine skeleton is formed exclusively from acetate.

Labelling of the alternate atoms C-2′, C-2, C-4, and C-6 of coniine by [1-^{14}C]acetate indicates an origin in a C_8 polyketide or equivalent.[57] In such a pathway, 5-oxo-octanoic acid (**57**) is a likely intermediate, and indeed tracer experiments have indicated that it and the corresponding aldehyde (**58**) are involved in coniine biosynthesis.[58] In the course of these experiments serendipity played a part when, for no good reason, labelled octanoic acid (**56**) was tested as a coniine precursor with positive results. It follows that coniine may be derived oxidatively from a C_8 fatty acid (octanoic acid) rather than reductively from a C_8 polyketide. If this is correct, then this base is unique in the field of secondary metabolism since, apart from the polyacetylenes, no other acetate-derived metabolite so far investigated is known to arise through a fatty acid.

The next intermediate in coniine biosynthesis after (**58**) is likely to be, *a priori*, the transamination product (**59**) $\leftrightharpoons$ γ-coniceine (**60**) (which occurs also in hemlock). This is strongly supported, as is the intermediacy of (**58**), by the isolation of an enzyme from

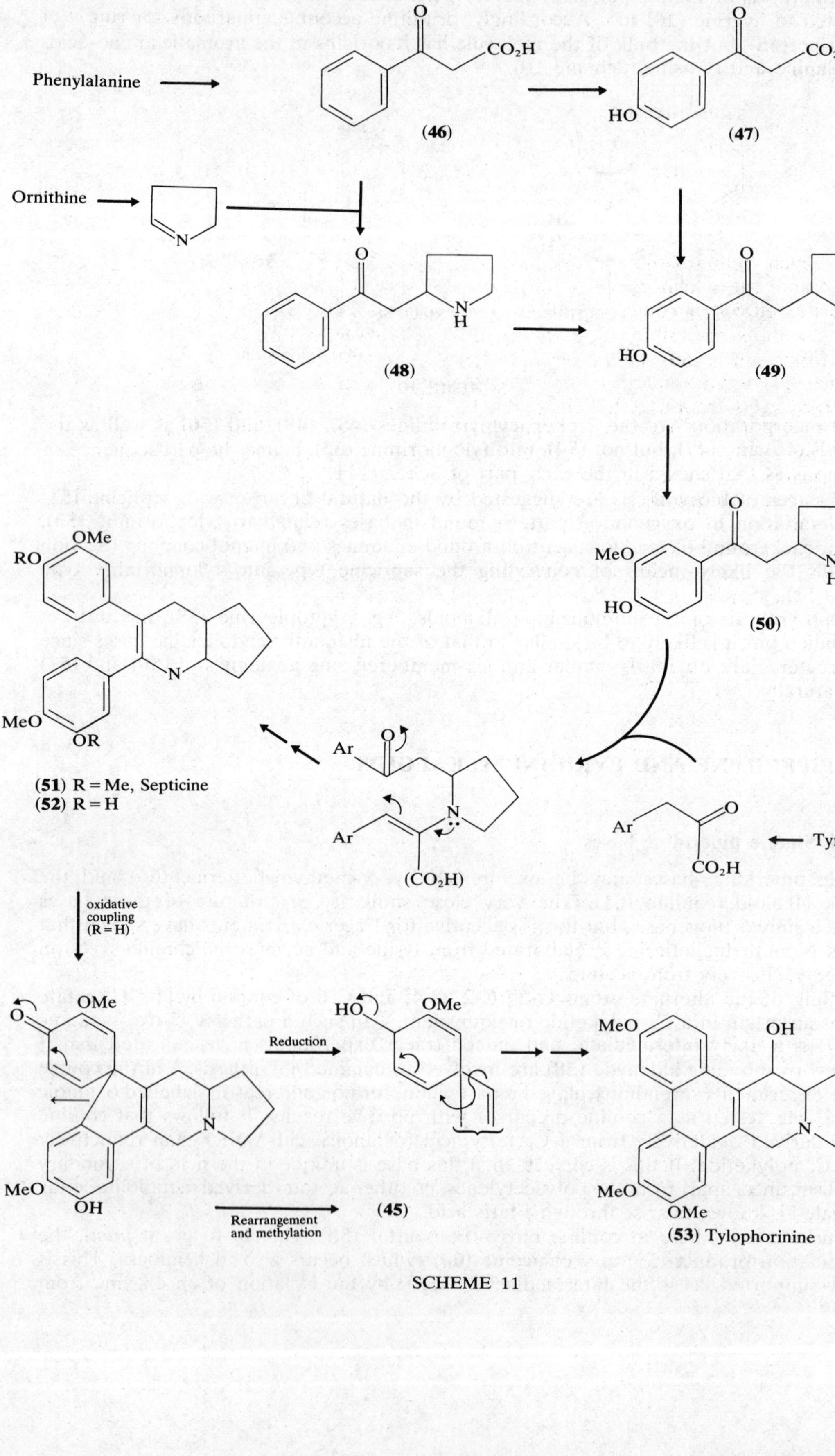
Phenylalanine
(46)
(47)
Ornithine
(48)
(49)
(50)
(51) R = Me, Septicine
(52) R = H
Tyrosine
oxidative coupling (R = H)
Reduction
Rearrangement and methylation
(45)
(53) Tylophorinine

SCHEME 11

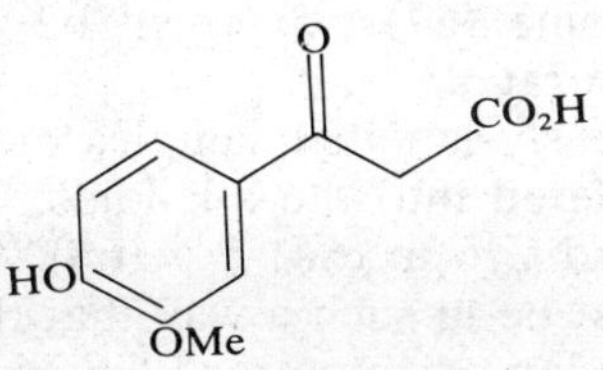

(54)

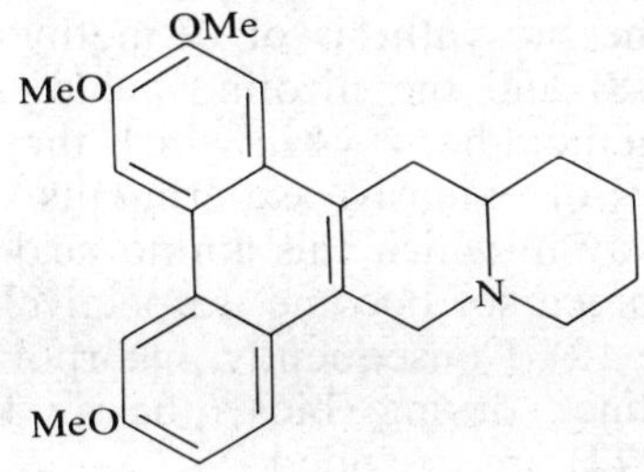

(55)

hemlock which catalyses the conversion of (**58**) into γ-coniceine (**60**) in the presence of alanine,[59] and one which converts (**60**) into coniine (**61**).[60] Moreover, tracer evidence indicates not only that coniceine (**60**) is converted into coniine (**61**), but also that the reaction is rapidly reversible.[58,61] Experiments with $^{14}CO_2$ confirm that the sequence of alkaloid formation is (**60**)→(**61**)→*N*-methylconiine.[62]

Pinidine (**62**) is an alkaloid which resembles coniine in structure, and it is similarly derived exclusively from acetate,[63] assembly of these C_2 units occurring in a simple linear manner: [1-^{14}C]acetate labels C-2, C-4, C-6, and C-9. In this case, decision between potential precursors is not quite as straightforward as for coniine since formation of pinidine clearly involves loss of a C_1 unit (carboxy-group) from either end of a C_{10} chain. Thus both (**63**) and (**64**) are potential precursors. However, insignificant incorporations have been recorded for (**63**) and the decarboxylation product of (**64**) as well as for decanoic acid. Nevertheless, a derivative of (**63**) but not (**64**) may still be implicated since inactive sodium acetate fed at the same time as diethyl [1-^{14}C]malonate diluted activity from C-2, which must thus be the 'starter' acetate unit.[64]

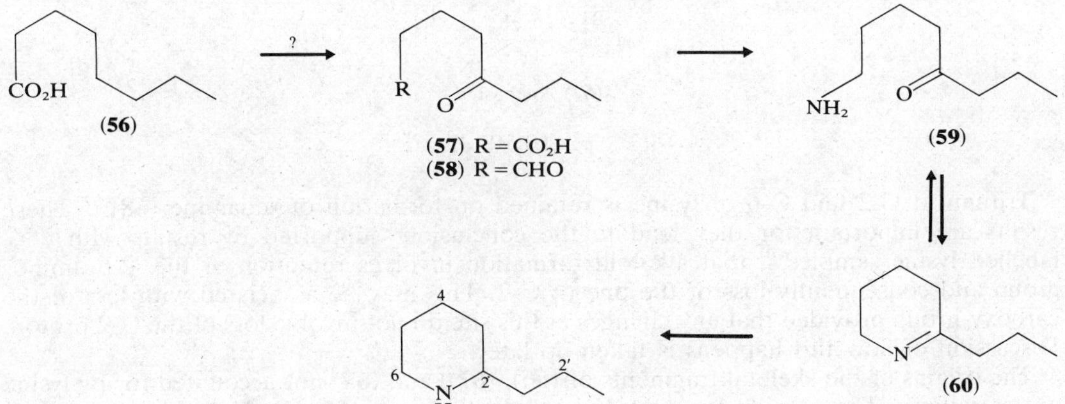

(56) **(57)** R = CO_2H **(58)** R = CHO **(59)** **(60)** **(61)** Coniine

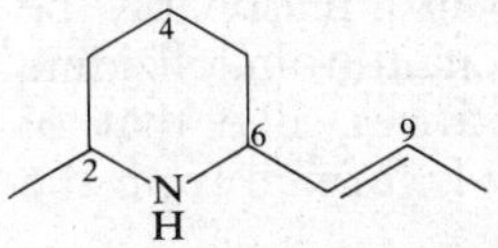

(62) Pinidine

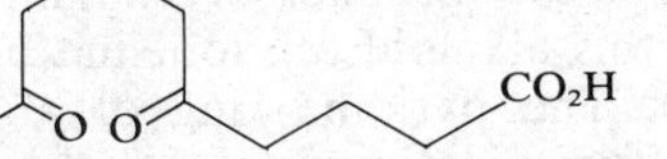

(63)

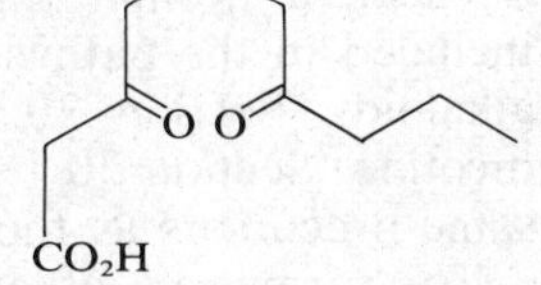

(64)

The pathways deduced for coniine and pinidine are exceptional ones. That deduced for *N*-methylpelletierine (**66**) can be considered much more typical of piperidine alkaloids in that the amino-acid lysine plays an important role. Because of interlocking evidence it is clear that the biosynthesis of *N*-methylpelletierine is generally very similar to that of sedamine (**68**) and the nicotine analogue, anabasine (**67**), so it will be convenient to consider together the ways in which they are elaborated.

The results of extensive experiments with variously labelled samples of lysine closely define the way in which this amino-acid is assimilated into the alkaloids. Thus C-2 and C-6 of the precursor become, respectively, C-2 and C-6 in (**66**),[65,66] (**68**),[67,68] and (**67**)[69] (see Scheme 12). Consequently, incorporation must be in such a way that these positions remain distinct during biosynthesis. In particular, symmetrical intermediates like cadaverine (**72**) are excluded.

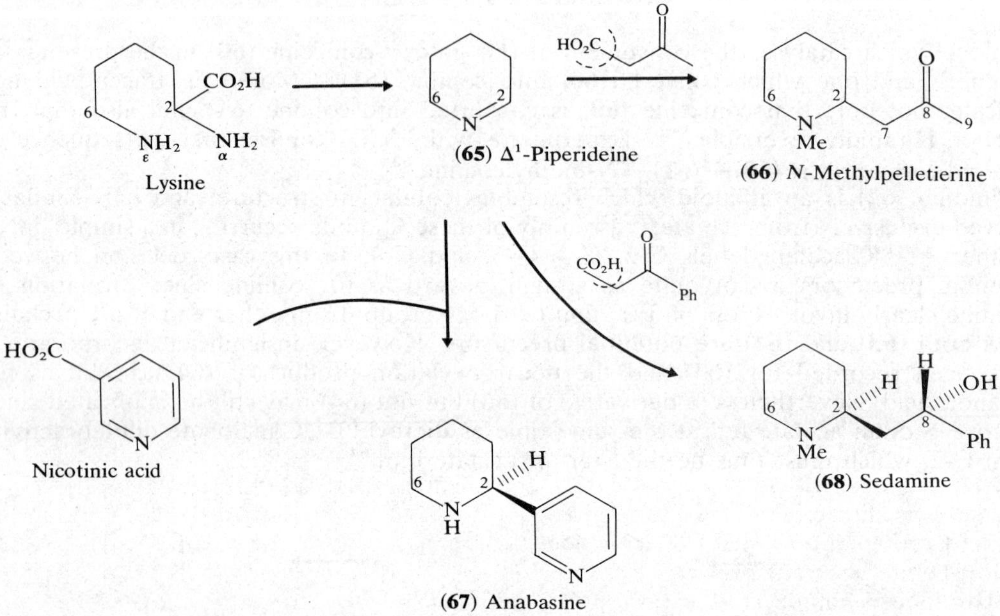

SCHEME 12

Tritium at C-2 and C-6 of lysine is retained on formation of sedamine (**68**).[67] These results are important for they lead to the conclusion, supported by results with ^{15}N-labelled lysine samples,[70] that alkaloid formation involves retention of the C-6 amino-group and consequently loss of the one at C-2. This may be associated with loss of the carboxy-group provided that any changes at this site do not involve loss of the C-2 proton. Discussion of how this happens is taken up later.

The origins of the skeletal fragments of (**66**), (**67**), and (**68**) not accounted for by lysine are as follows. The C_3 side-chain of *N*-methylpelletierine (**66**) has its origins in acetate, which is incorporated plausibly through acetoacetate (C-8 is labelled by [1-^{14}C]acetate whilst C-7 and C-9 are labelled by [2-^{14}C]acetate).[66] On the other hand, the side chain of sedamine is derived from phenylalanine, probably by way of cinnamic acid.[68] Benzoylacetic acid, a normal transformation product of cinnamic acid, may also reasonably be included in the pathway to this alkaloid (*cf.* lobeline below, and phenanthroindolizidine alkaloids, Section 30.1.2.4). The pyridine ring of anabasine (**67**) arises, like that of nicotine (Section 30.1.2.2), from nicotinic acid, which quite naturally is formed from the same precursors as those for nicotine.[71]

The formation of anabasine (**67**) from a lysine-derived unit and nicotinic acid can reasonably be thought of as being similar to that deduced for nicotine (Section 30.1.2.2) with substitution of Δ^1-piperideine (**65**) for *N*-methyl-Δ^1-pyrroline (**31**) in the condensation with dihydronicotinic acid (*cf.* Scheme 7). The specific incorporation of [6-^{14}C]-Δ^1-piperideine into anabasine (**67**) (with location of the label exclusively at C-6) gives substance to this view.[72] Alternative condensation of Δ^1-piperideine (**65**) with β-keto-acids, as shown in Scheme 12, leads to sedamine (**68**) and *N*-methylpelletierine (**66**).

Cadaverine (**72**) has been shown, like lysine, to be a precursor for the piperidine rings of anabasine, sedamine, *N*-methylpelletierine, and pseudopelletierine (**69**).[73] Such a symmetrical molecule cannot be an intermediate in the conversion of lysine into these piperidine alkaloids, for otherwise the identities of C-2 and C-6 of the amino acid would be lost, in contradiction of the tracer evidence. It was suggested that these identities could be retained by *N*-methylation, *i.e.* with ε-*N*-methyl-lysine and *N*-methylcadaverine (**70**) as intermediates following lysine, in parallel with pyrrolidine alkaloid biosynthesis. This was not, however, supported by the results of subsequent experiments.

(**69**) Pseudopelletierine

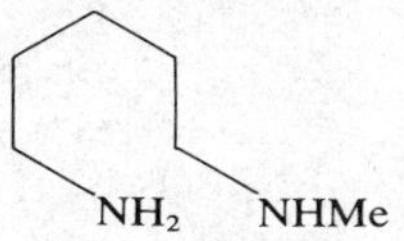

(**70**)

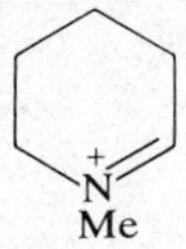

(**71**)

Firstly, although ε-*N*-methyl-lysine is a natural constituent of *Sedum acre* [a source of sedamine (**68**)], it was formed from [6-^{3}H]lysine and [*Me*-^{14}C]methionine with an isotope ratio which differed from that of sedamine formed in the same experiment, clearly excluding a precursor–product relationship between the two compounds.[73a,74] In addition, attempts to demonstrate the presence of *N*-methylcadaverine in *Sedum* species and *Nicotiana glauca* (a source of anabasine) were unsuccessful.[73a]

Secondly, if ε-*N*-methyl-lysine were to be an alkaloid precursor, then so also would *N*-methyl-Δ^1-piperideine (**71**). However, although labelled (**71**) gave anabasine (**67**) as well as *N*-methylanabasine, the latter is clearly not a natural alkaloid since it was formed with the same specific activity as the precursor and it was also not labelled by [2-^{14}C]lysine in another experiment. It may be concluded, therefore, that transformation of (**71**) into alkaloids is an aberrant process.[75]

A brilliantly simple and satisfying solution[73a] to the observations on lysine and cadaverine incorporation has been proposed which is consistent with all the evidence, in particular the observed incorporation of lysine with distinction maintained between C-2 and C-6, and which at the same time allows normal incorporation of cadaverine (**72**). Central to the proposal is enzyme-catalysed decarboxylation of lysine which proceeds *via* orthodox pyridoxal-linked intermediates as shown in Scheme 13. The proposal is more than mechanistically attractive because L-lysine decarboxylase and diamine oxidase, the two enzymes whose participation in the conversion of lysine into Δ^1-piperideine is likely, both require pyridoxal phosphate as coenzyme. Intermediate (**73**) can be seen to be derivable not only from lysine but also from cadaverine (**72**) and convertible into it. This then can satisfactorily explain why cadaverine as well as lysine is incorporated into the piperidine alkaloids (**66**), (**67**), and (**68**). However, it is an important point that if the cadaverine formed from lysine remains enzyme-bound, the necessary distinction between C-2 and C-6 is maintained through into Δ^1-piperideine (**65**) and hence to the alkaloids. Equilibration of bound cadaverine with unbound material would lead to loss of this identity, so it is postulated not to occur in the course of the normal biosynthesis of these alkaloids. However, as we shall see later, there are some piperidine alkaloids which are derived from lysine in such a way that C-2 and C-6 of the amino-acid become equivalent. This observation is simply explained in terms of the model above by allowing extensive equilibration of (**73**) with unbound cadaverine. (As indicated in Section 30.1.2, it is possible to apply this model in modified form to the genesis of the pyrrolidine alkaloids.)

On the basis of the above hypothesis, cadaverine (**72**) assumes importance as a normal precursor for piperidine alkaloids, and it has been shown that its incorporation into *N*-methylpelletierine involves stereospecific loss of one of the C-1 protons (the *pro-S* hydrogen atom) and this is accommodated within the model sequence (Scheme 13).[73a,76]

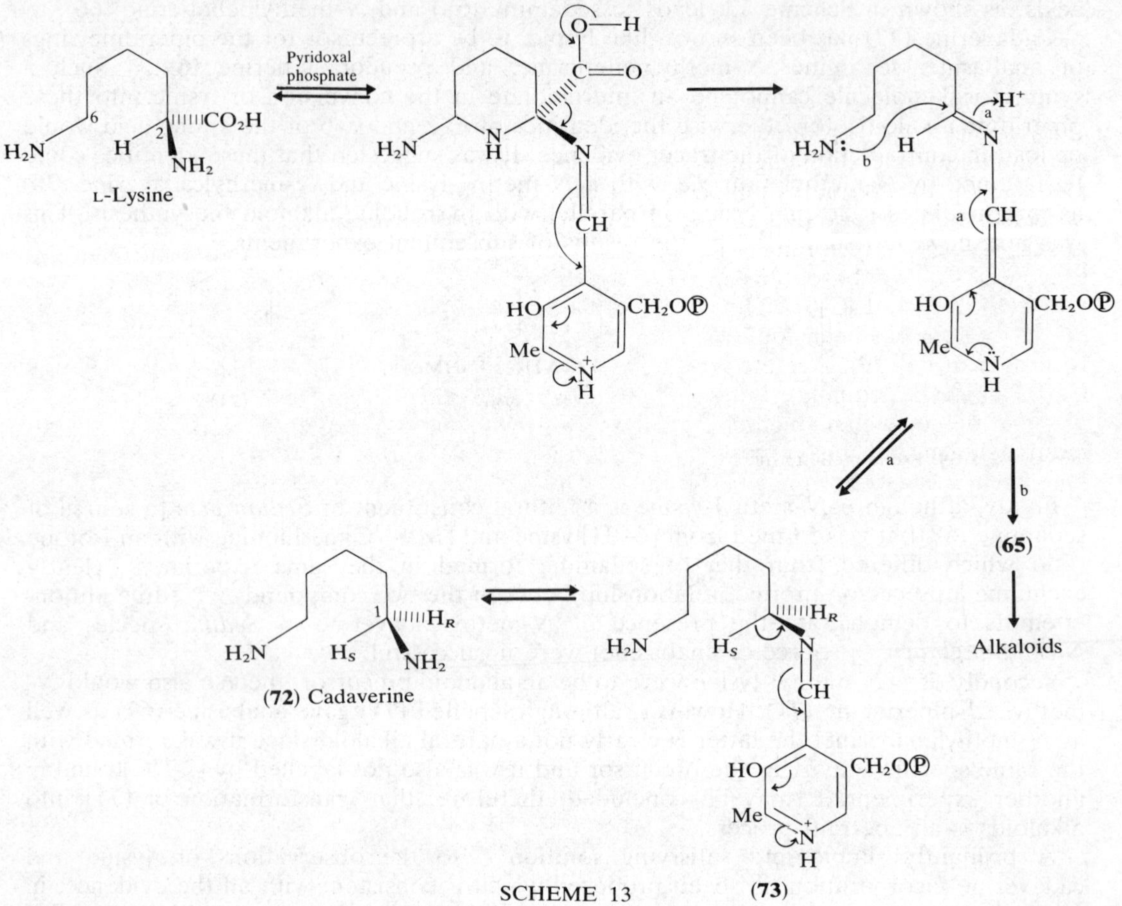

SCHEME 13

Further stereospecificity in this sequence is seen in the conversion of lysine into piperidine alkaloids since the L-isomer is a much better precursor than D-lysine for sedamine (**68**), (**67**), (**66**), *N*-methylallosedridine (**74**) (notably of opposite configuration

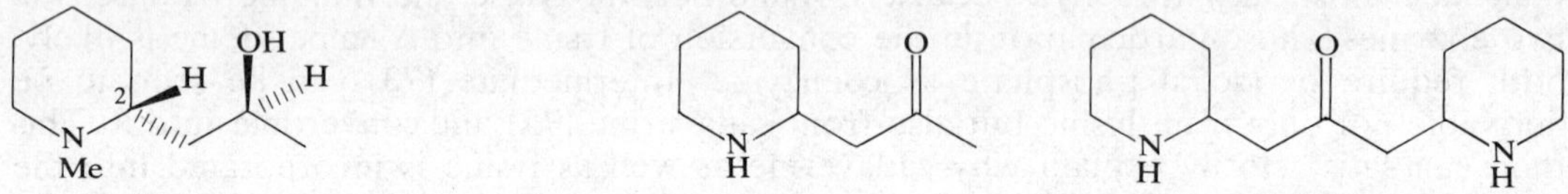

(**74**) (**75**) Pelletierine (**76**) Anaferine

to sedamine at C-2), and lycopodine (**94**)[77,78] (which is formed in symmetrical fashion from lysine; see below). In contrast, pipecolic acid (**77**) which is found widely in plants, animals, and micro-organisms is derived preferentially from D-lysine.[77,78] Significantly, the biosynthetic pathway[67,79] to pipecolic acid (**77**), illustrated in Scheme 14, is different to that which leads to piperidine alkaloids (Scheme 12).

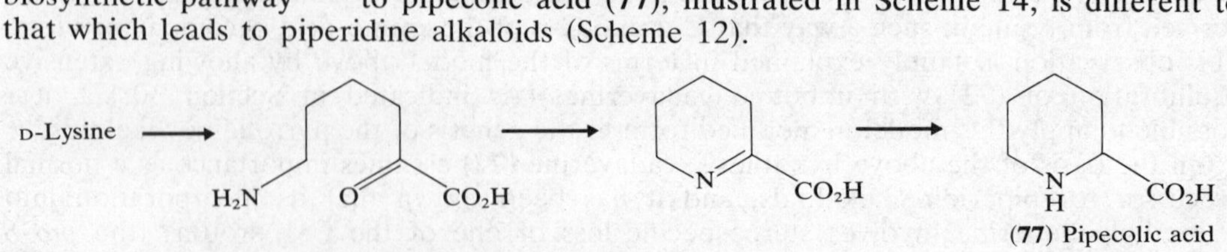

SCHEME 14

Although many examples exist in studies in micro-organisms, it is worth noting that this is only the second time attention has been paid to the chirality at the α centre of the amino-acid in plant alkaloid biosynthesis. The other study in plants was with Amaryllidaceae alkaloids, where no distinction was observed with D- and L-tyrosine incorporation into norpluviine and lycorine.[80]

By analogy with tropane alkaloid biosynthesis (Section 30.1.2.1), anaferine (**76**) and pseudopelletierine (**69**) might be expected to have a related genesis through an intermediate of the *N*-methylpelletierine type. In accord, *N*-methylpelletierine (**66**) is a precursor for (**69**), and pelletierine (**75**) is incorporated into anaferine (**76**). The manner in which lysine and acetate are incorporated is, moreover, consistent with such a biogenesis.[66] The base (**78**) also derives from lysine (with the customary distinction between C-2 and C-6). The C_5 side-chain should derive from acetate, but so far [2-^{14}C]acetate has been found not to label it.[81] Difficulties have also been experienced[82] with dioscorine (**79**). Acetate was incorporated as a C_7 unit ([1-^{14}C]acetate labelled C-5, C-10, and C-12 equally), but only negative results have been obtained with lysine and Δ^1-piperideine, which ought to be used for the construction of the remaining carbon atoms (shown with heavy bonding). Inspection of the dioscorine (**79**) structure suggests that pelletierine (**75**) might be implicated as a biosynthetic intermediate, as it is in the pathway to many other piperidine alkaloids.

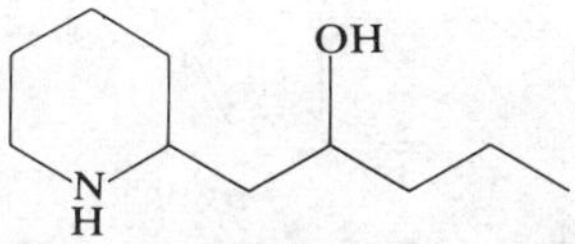

(**78**)

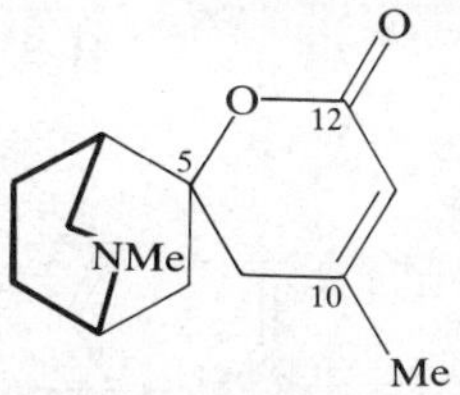

(**79**) Dioscorine

8-Phenyl-lobelol-I which differs from sedamine (**68**) only in the configuration at C-8, occurs together with lobeline (**80**) in *Lobelia inflata*. A biogenesis for these bases along the lines already discussed might be expected and indeed lysine and Δ^1-piperideine (**65**) are lobeline precursors.[83,84] However, the incorporation of lysine into (**80**), unlike sedamine (**68**), is such that C-2 and C-6 of the amino-acid become equivalent, indicating at least one symmetrical intermediate in the course of biosynthesis.[83] The almost symmetrical structure of lobeline suggests that this symmetrization occurs at a late stage. This is supported by the highly efficient incorporation of the 'symmetrical' lobelanine (**81**)[85] and the incorporation of cadaverine (**72**) at a much lower level than lysine.[84]

It is an orthodox assumption that lobeline (**80**) [and lobelanine (**81**)] may arise by

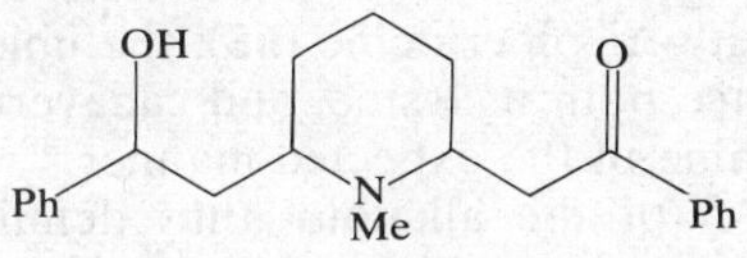

(**80**) Lobeline

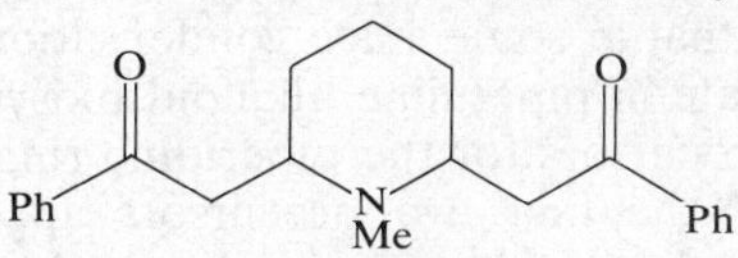

(**81**) Lobelanine

successive additions of benzoylacetic acid to Δ^1-piperideine (**65**), and this is substantiated by the specific incorporation into the side chains of the alkaloid of logical precursors for this keto-acid (phenylalanine, cinnamic acid, and 3-hydroxy-3-phenylpropionic acid).[83,84]

There are some *Lobelia* alkaloids with C_4 rather than the C_3 side-chains seen in alkaloids like *N*-methylpelletierine (**66**). The labelling patterns observed after feeding lysine and acetate indicate that the biosynthesis of these bases may be accounted for in terms of a pathway in which truncation occurs of, for example, (**82**) leading to (**83**).[84]

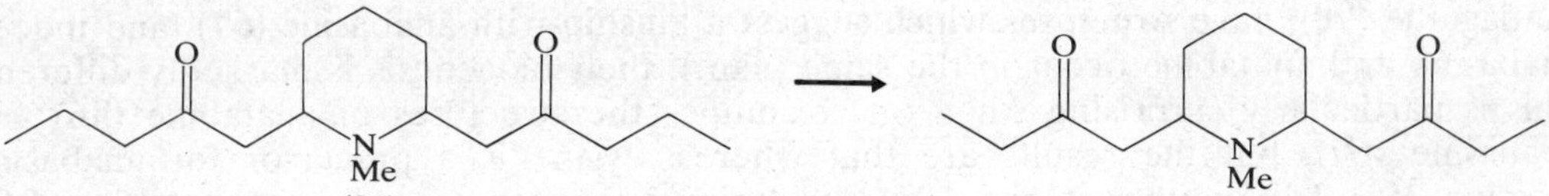

(**82**) (**83**)

A more complex *Lobelia* alkaloid is lobinaline (**84**), but the complexity is deceptive, for dissection along the dotted line as shown indicates a simple biogenesis from two units of the sedamine (**68**) type. In accord with this hypothesis, phenylalanine, cinnamic acid, and lysine have been found to be specific precursors for both 'halves' of the alkaloid. Label appeared at the expected sites (lysine was incorporated with distinction maintained between C-2 and C-6) and with apparently equal distribution of radioactivity between the two 'halves'. Incorporation of [4-^{14}C]-2-phenacylpiperidine [as (**83a**)] with equal labelling of C-2 and C-6′ of (**84**) completes the definition of the pathway to the alkaloid and proves the dimer hypothesis to be correct.[86]

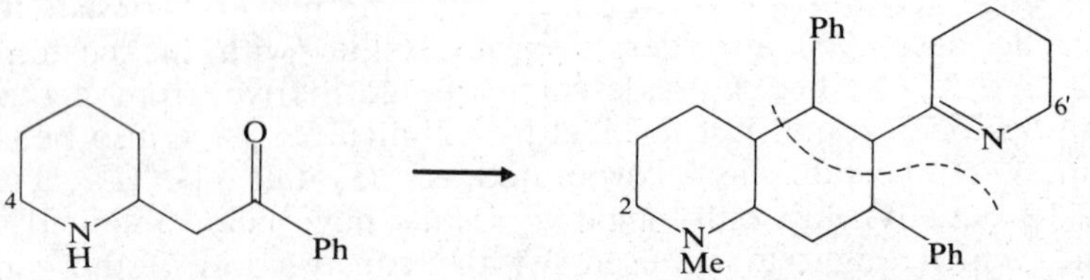

(**83a**) (**84**) Lobinaline

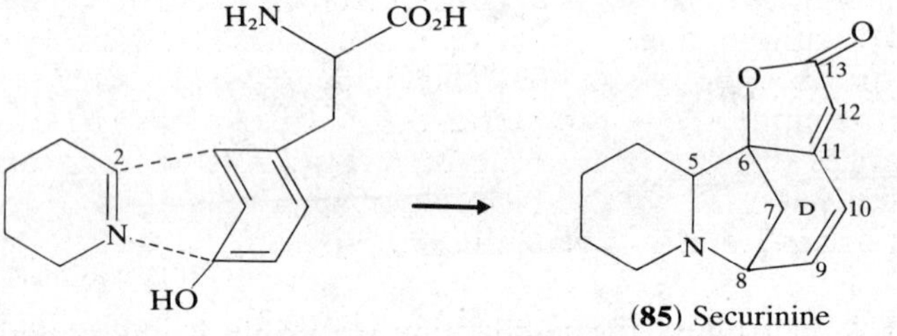

(**85**) Securinine

Δ^1-Piperideine Tyrosine

Many of the piperidine alkaloids are derived, as we have seen, along pathways common to which is the condensation of a β-keto-acid with Δ^1-piperideine to form compounds of the type (**75**) and (**83a**), which are either alkaloids or biosynthetic intermediates. There are exceptions of course. Anabasine (**67**) is one and several of the alkaloids discussed below provide further examples. Securinine (**85**) is another such exception, as perhaps befits its unique tetracyclic structure. Carbons 6–13 of securinine (**85**) are known to be derived from tyrosine,[87,88] but the way in which the aromatic amino-acid is modified in the course of biosynthesis is not yet known, no analogy even being to hand. It is clear, however, that at some stage condensation occurs with Δ^1-piperideine (**65**), the ubiquitous intermediate in piperidine alkaloid biosynthesis, since both it, lysine, and cadaverine are specific precursors for the piperidine ring of securinine in the expected manner;[88,89] label at C-2 of the first two precursors appears at C-5 of the alkaloid, thus defining the orientation of the Δ^1-piperideine double bond as that shown.[89] It is interesting to note from the results that incorporation of lysine avoids a symmetrical intermediate and it is so far the only base with a nitrogen atom common to two rings which does so[89] (*cf.* the discussion below).

30.1.3.2 Alkaloids structurally related to anabasine

Although anatabine (**86**) and the tetrahydroanabasine alkaloids, adenocarpine (**88**) and santiaguine (**90**), have structures which suggest a kinship with anabasine (**67**) (and indeed anabasine and anatabine occur in the same plant), their biogenesis is markedly different. This is particularly surprising when one examines the structures of anatabine (**86**) and anabasine (**67**), but the results are that whereas lysine is a precursor for anabasine (Section 30.1.3.1), neither it nor 4-hydroxylysine are incorporated into anatabine (**86**), and results of $^{14}CO_2$ experiments indicate that anabasine is not a precursor for anatabine. On the other hand,[90] [6-^{14}C]nicotinic acid gave radioactive anatabine (**86**), but surprisingly the activity was found to be equally divided between C-6 and C-6′. It follows that nicotinic acid is the source of both rings of the alkaloid, in contrast to anabasine (**67**) where it acts as the precursor for only one. The equal distribution of label indicates that the two units from which the anatabine (**86**) is formed are closely related, if not identical. The pathway suggested is illustrated in Scheme 15.

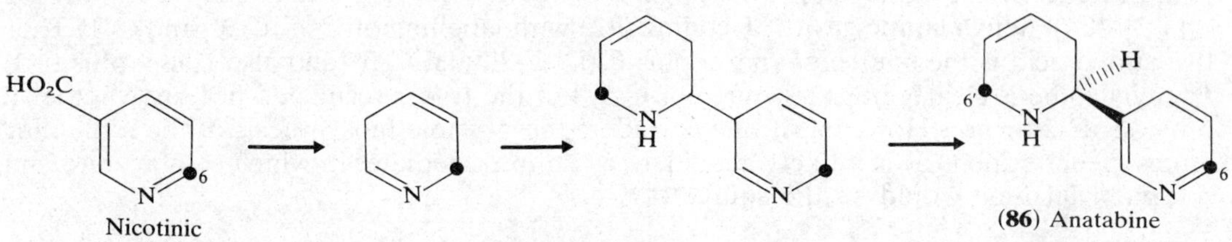

SCHEME 15

The tetrahydroanabasine units of adenocarpine (**88**) and santiaguine (**90**) are again constituted quite differently. Lysine is incorporated into all four piperidine rings of santiaguine [see (**90**)] without intervention of a symmetrical intermediate.[91] More immediate precursors are Δ^1-piperideine (**65**)[91] and adenocarpine (**88**).[92] The piperideine dimer (**87**) may also be implicated as an intermediate as it is a precursor for adenocarpine (**88**).[93] Santiaguine arises then by the $2_\pi + 2_\pi$ dimerization of adenocarpine, although an alternative, if apparently less important, route involves combination of truxillic acid (**89**) with (**87**); (**89**) itself arises from phenylalanine *via* cinnamic acid.[91,92]

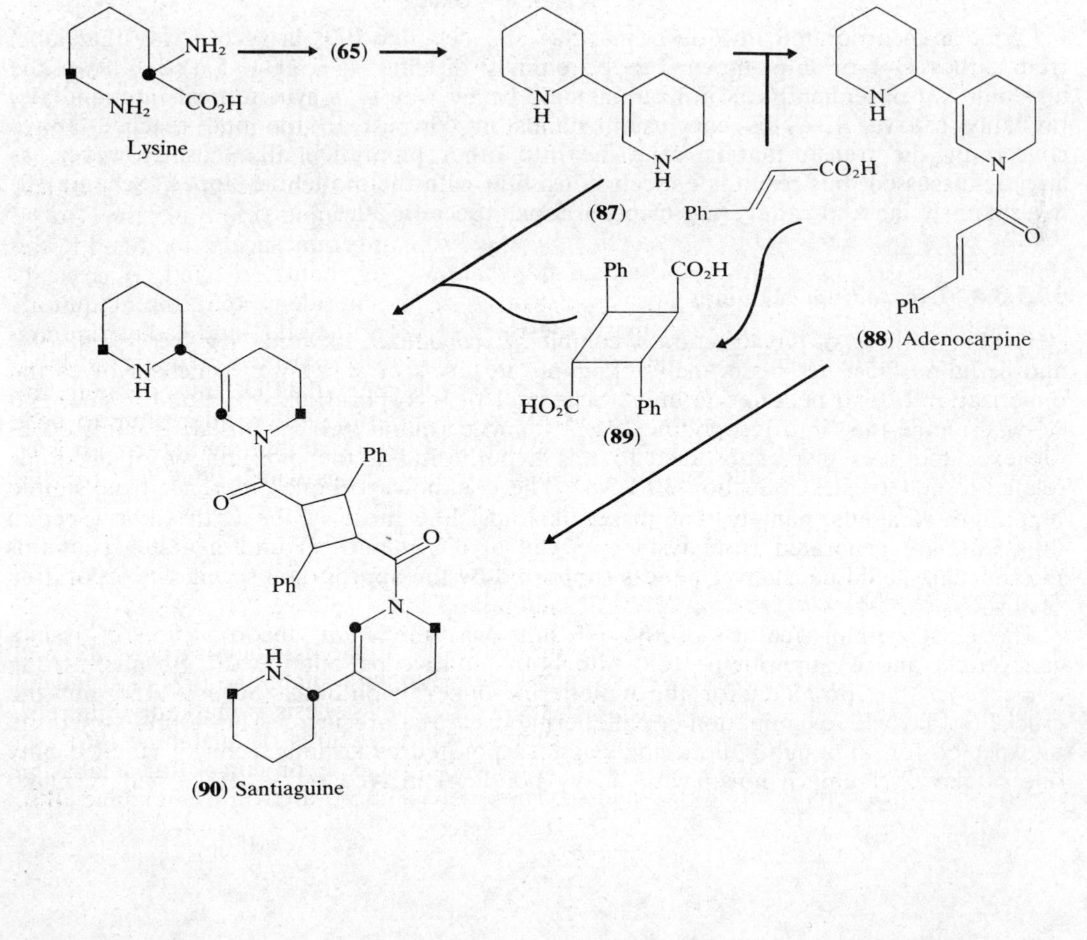

(**90**) Santiaguine

30.1.3.3 Lythraceae alkaloids

A C_6–C_3 unit which may be hypothetically derived from 4-hydroxycinnamic acid is manifest in the Lythraceae alkaloids, cryogenine (**91**) and decodine (**92**). Although this unit [ring D and side chain in (**91**)] can, with little doubt, be linked by oxidative coupling (Section 30.1.1) to a second oxygenated aromatic unit [ring C of (**91**)], it is not clear by inspection how many carbon atoms of the side chain of this latter unit are used in the biosynthesis of the quinolizidine ring of these alkaloids. Attendant on this is the question of the origins of the other atoms of the ring system.

[1,3-$^{14}C_2$]Phenylalanine gave[94] decodine (**92**) with labelling of C-1′, C-3′, and C-1. Thus this amino-acid is the source of ring D plus C-1′, C-2′, and C-3′, and also ring C plus C-1. Phenylalanine is clearly not the source of C-3, but the tracer results do not say whether it provides C-2 or not. However, if one considers the possible biosynthesis of the remaining atoms, pelletierine (**75**) is a likely candidate as an intermediate, in which case acetate and not phenylalanine would be the source of C-2.

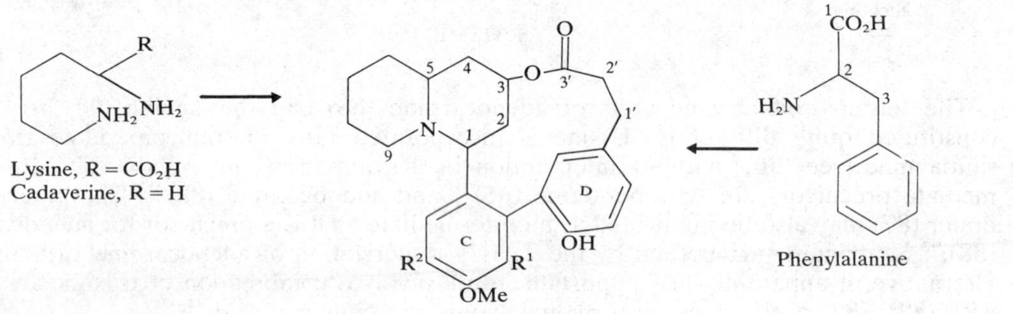

(**91**) $R^1 = H$, $R^2 = OMe$, $\Delta^{1'}$
(**92**) $R^1 = OH$, $R^2 = H$
(**93**) $R^1 = H$, $R^2 = OMe$

Lysine is incorporated into decodine (**92**) and decinine (**93**) in such a way that label from either C-2 or C-6 appears to be equally distributed over C-5 and C-9 in the alkaloids.[95] Consequently its utilization must be by way of a symmetrical intermediate, probably cadaverine. This conclusion stands in contrast to the one reached above concerning the transformation of lysine into other piperidine alkaloids. However, as already discussed, this result is entirely consistent with the model developed (Scheme 13) to explain lysine and cadaverine incorporations (Section 30.1.3.1).

30.1.3.4 *Lycopodium* alkaloids

The biosynthesis of the structurally complex *Lycopodium* alkaloids, *e.g.* lycopodine (**94**) and cernuine (**95**), has been analysed simply in terms of a pathway which involves the dimerization of two pelletierine units (Scheme 16). Incorporations of acetate, lysine, and Δ^1-piperideine (**65**) into lycopodine (**94**)[96,97] and cernuine (**95**)[98] with labels confined to the expected sites gives substance to this hypothesis (*cf.* the biosynthesis of alkaloids related to pelletierine in Section 30.1.3.1). There is, however, one difference from simple piperidine alkaloids, namely that these alkaloids, like those of the Lythraceae (Section 30.1.3.3), are generated from lysine by way of a symmetrical intermediate. That this intermediate could be cadaverine was supported by the appropriate specific incorporation of this base.[96,98,99]

The most striking feature of these results was the *equal* incorporation of lysine, cadaverine, and Δ^1-piperideine (**65**) into both putative pelletierine (C_8N) units. Strong evidence is thus provided for the pelletierine-dimer hypothesis (Scheme 16), but the crucial test lay with examination of pelletierine itself as a precursor. The results were quite unexpected for, although pelletierine was incorporated as an intact unit, it labelled only one of the C_8N units (shown with heavy bonding) in lycopodine (**94**)[97] and cernuine

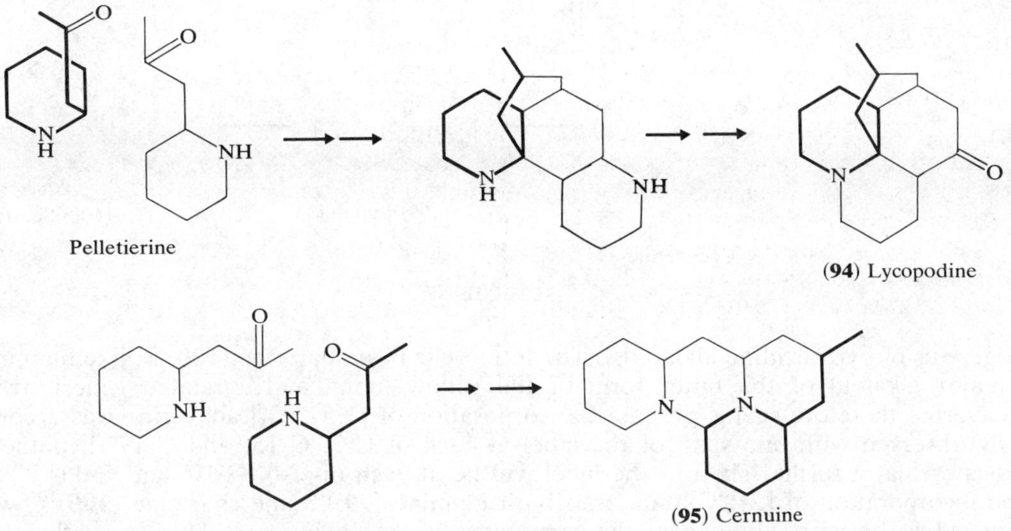

Pelletierine

(**94**) Lycopodine

(**95**) Cernuine

SCHEME 16

(**95**).[98] Pelletierine was subsequently shown, by an isotope dilution experiment, to be a normal constituent of the plant used to study lycopodine biosynthesis and also to be formed from lycopodine precursors. Further, it was found that unlabelled pelletierine very efficiently diluted activity from the one C_8N unit (heavy bonding) of lycopodine (**94**), for which it is a precursor, when fed together with radioactive cadaverine or Δ^1-piperideine.[99] It is clear from this evidence that pelletierine is an intermediate in the biosynthesis of lycopodine (and cernuine), but unlike acetate, lysine, and Δ^1-piperideine, pelletierine is involved in the biosynthesis of only one unit. Since lysine and Δ^1-piperideine label both C_8N units equally, (**94**) and (**95**) must be formed from a second C_8N intermediate which is very closely related to pelletierine (significant differences in structure would be associated with differing pathways and hence unequal incorporation of label into the two C_8N units). The results argue thus for the pelletierine-dimer route, but in a modified form. It is not easy, however, to propose modifications to the pathway which are entirely consonant with the results. Hypotheses surrounding (**96**) and (**97**) as intermediates have not been supported by subsequent experimental results.[78,99] So the puzzle remains, but there is a candidate for the second C_8N unit which has not so far been examined. That is compound (**98**), formed

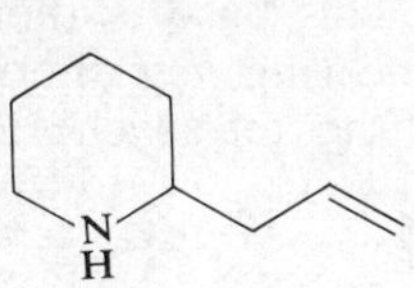

(**96**)

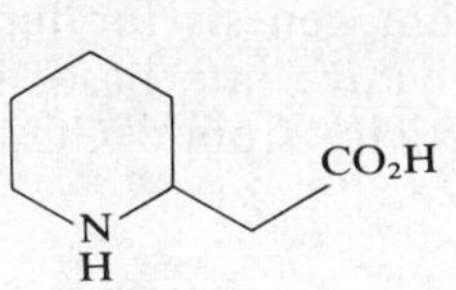

(**97**)

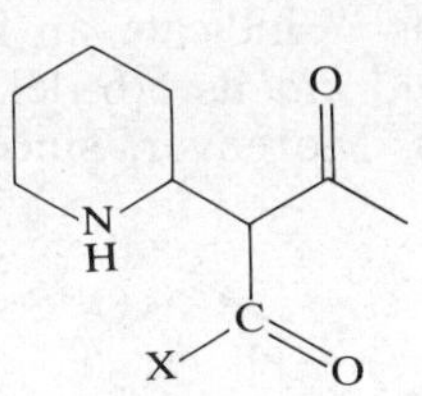

(**98**)

by condensation of (**65**) with acetoacetate, and which is logically the immediate precursor for pelletierine (*cf.* Scheme 2).

30.1.3.5 Quinolizidine alkaloids

Lupinine (**99**) is structurally the simplest of the quinolizidine alkaloids. Specific incorporation[100] of [1,5-$^{14}C_2$]cadaverine in the manner shown (Scheme 17) demonstrates that this molecule is derived from two of cadaverine (*cf.* Section 30.1.2.3 for a possibly related

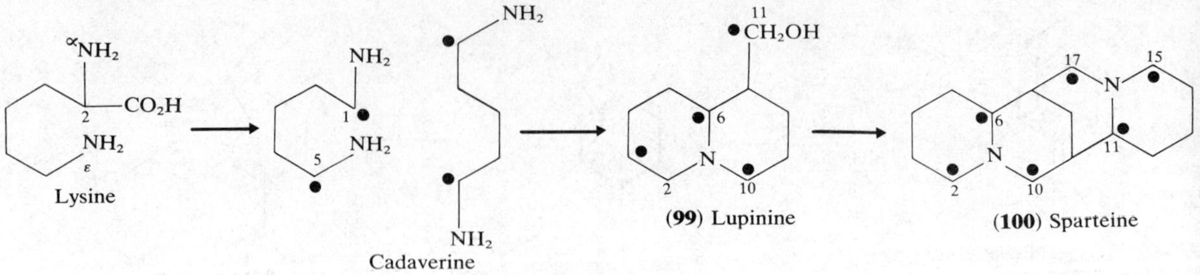

(**99**) Lupinine
(**100**) Sparteine

SCHEME 17

biogenesis of pyrrolizidine alkaloids). The tetracyclic base sparteine (**100**) is, like lupinine, a major alkaloid of the bitter form of the yellow lupin, and a related genesis from cadaverine therefore seems probable. Incorporation of [1,5-$^{14}C_2$]cadaverine was accordingly observed with one sixth of the label at each of C-2, C-15, and C-17 (hypothesis indicates that a further sixth of the label will be at each of C-6, C-10, and C-11).[101,102] The incorporation of [2-^{14}C]lysine into both lupinine (**99**)[103] and sparteine (**100**)[101] was deduced to be with labelling of the same sites as the cadaverine. Thus, as with other piperidine alkaloids having nitrogen common to two rings, excepting securinine (see above), lysine is incorporated *via* a symmetrical intermediate, most probably cadaverine (see Section 30.1.3.1 for a fuller discussion of the likely mechanism of incorporation of these precursors).

Samples of [2-^{14}C,ε-^{15}N]- and [2-^{14}C,α-^{15}N]-lysine both gave sparteine (**100**) with three times as much ^{14}C as ^{15}N incorporation. This is consistent with a biosynthetic pathway in which lysine incorporation occurs through cadaverine; taking into account the necessary incorporation of three lysine/cadaverine carbon chains but only two nitrogen atoms, this is the ^{14}C:^{15}N ratio expected.[101,104] Further evidence supporting the role of cadaverine as a biosynthetic intermediate in the formation of alkaloids of this type is the isolation of radioactive cadaverine from lupin plants after feeding radioactive lysine,[105] and the depression of radioactive lysine incorporation [into thermopsine (**103**)] by inactive cadaverine.[106]

The implied relationship between lupinine (**99**) and sparteine (**100**) is given substance by the observation that lupinine is a precursor for the tetracyclic base (**100**).[107]

Although lupanine (**101**) is a precursor[107,108] for its 11-hydroxy-derivative (**102**) (both of which arise from three molecules of lysine *via* cadaverine),[109] its reported derivation from sparteine (**100**)[108] is suspect, since, in experiments[110] with $^{14}CO_2$, there was no correlation in the appearance of radioactivity in (**100**) and (**101**). Instead the $^{14}CO_2$ results[110] indicate an independent genesis for lupanine (**101**). This base is then transformed *via* its 5,6-dehydro-derivative into bases with a pyridone ring, *e.g.* thermopsine (**103**). Moreover, since radioactivity from $^{14}CO_2$ appeared in the tetracyclic alkaloids

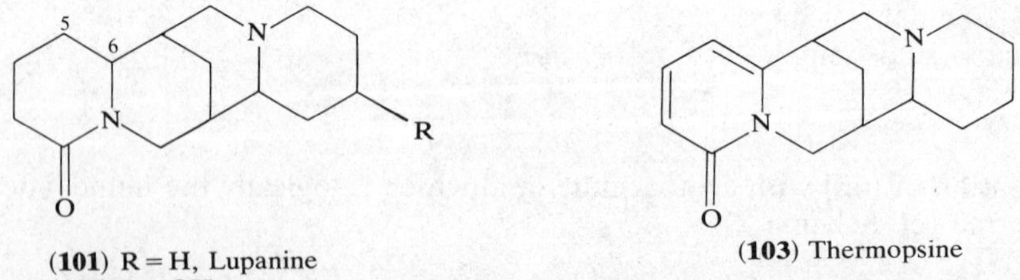

(**101**) R = H, Lupanine
(**102**) R = OH

(**103**) Thermopsine

before the tricyclic ones, *e.g.* (**104**) and (**105**), it follows that the tricyclic bases are formed by fragmentation of the tetracyclic skeleton. Further, rhombifoline (**104**) appears to play a primary role in the formation of cytisine (**105**) and *N*-methylcytisine. (These conclusions are supported by a less extensive study[106] with [2-^{14}C]lysine.)

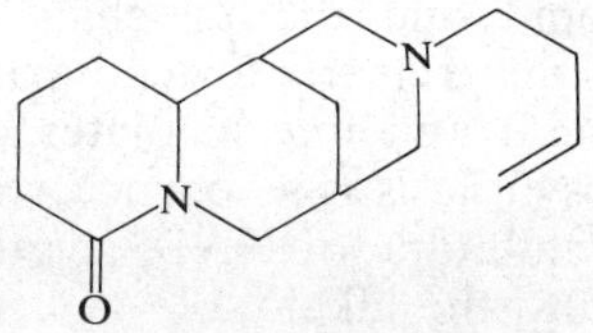
(**104**) Rhombifoline

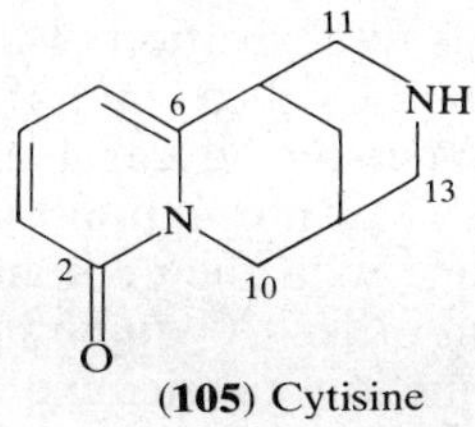
(**105**) Cytisine

In accord with the conclusion that cytisine (**105**) is derived from tetracyclic precursors (by way of rhombifoline), the alkaloid is labelled specifically at C-2 and C-11 with one fifth of the activity from [2-^{14}C]lysine and [1,5-$^{14}C_2$]cadaverine (remaining activity is assumed to reside at C-6, C-10, and C-13).[111]

An alternative tetracyclic skeleton to the one found in sparteine (**100**) is to be seen in matrine (**106**), and a similar derivation from three molecules of lysine *via* cadaverine has been demonstrated (● = ^{14}C labels as in Scheme 17).[112,113] A little is understood[114] about interrelationships of alkaloids of this type but, as in the case of the more extensively studied alkaloids derived from lupinine, nothing is known of the intermediates between cadaverine and the first recognized alkaloids. This is a problem of some significance and so far almost all the evidence is negative. The compounds (**107**),[109a] (**108**),[105] and (**87**)[115]

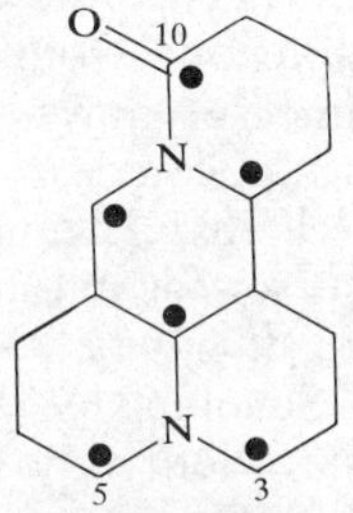
(**106**) Matrine

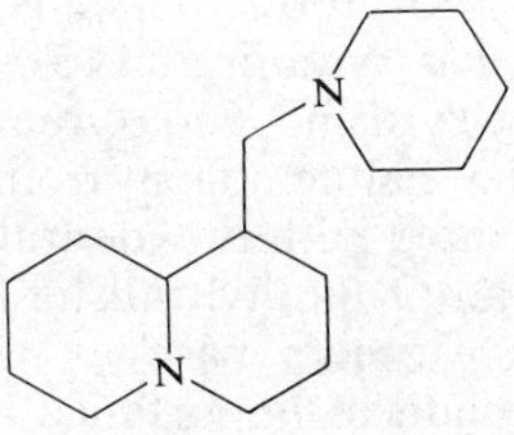
(**107**)

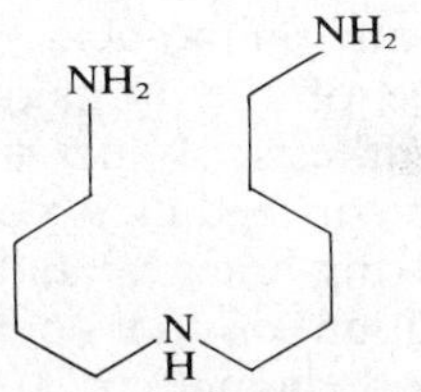
(**108**)

are poor precursors for various alkaloids. However, a specific incorporation of [2-^{14}C]-Δ^1-piperideine [as (**65**)] into matrine has been recorded,[113] although activity appeared to be very unequally distributed between C-10 (10%) and C-3+C-5 (90%).

30.1.3.6 Pyridine alkaloids

The biosynthesis of the pyridine alkaloids nicotine (**32**), anabasine (**67**), and anatabine (**86**) has already been discussed (Sections 30.1.2.2, 30.1.3.1, and 30.1.3.2, respectively). The pyridine ring of each of these alkaloids is known to arise from nicotinic acid, which also provides the tetrahydropyridine ring of anatabine (**86**). Such a derivation for the latter ring in anatabine has been noted and stands in marked contrast to that deduced for the related fragment of anabasine which is known to originate from lysine.

Contrariness is also apparently found in the biosynthesis of mimosine (**109**), where the

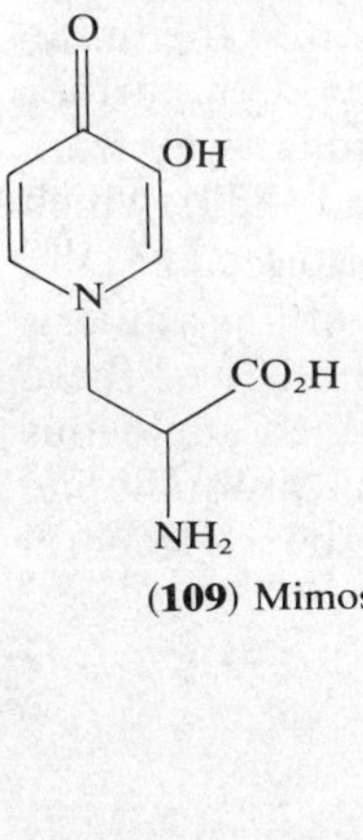
(**109**) Mimosine

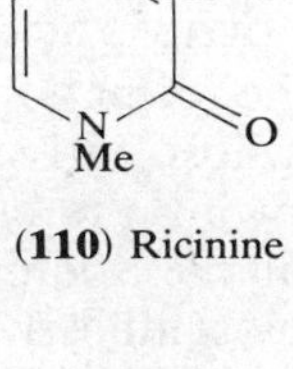
(**110**) Ricinine

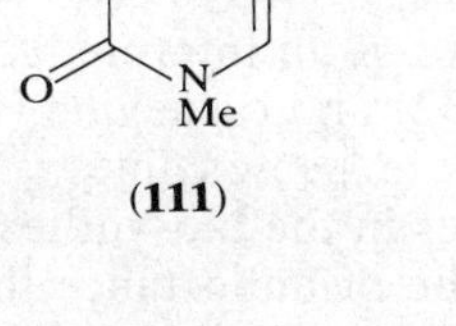
(**111**)

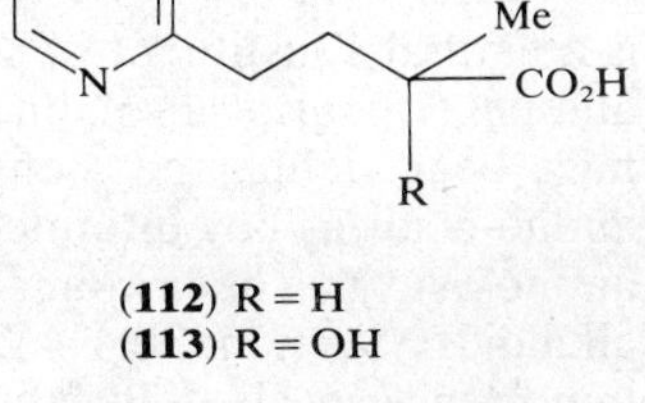
(**112**) R = H
(**113**) R = OH

evidence[116] points to derivation of the alkaloid from lysine (the side chain appears to arise from alanine). On the other hand, ricinine (**110**) is quite clearly formed from nicotinic acid (**39**) (and its precursors glycerol and aspartic acid) and intermediates of the pyridine nucleotide cycle.[117] Confirmation that ricinine biosynthesis is associated with operation of this cycle has come with the demonstration that inhibitors for NAD^+ synthetase decrease the incorporation of [6-^{14}C]quinolinic acid (**38**) into the alkaloid.[118]

The nitrile function of ricinine arises by amide dehydration, as evidenced by the incorporation of [$^{15}NH_2$]nicotinamide and the *N*-methyl group arises from methionine.[117]

Nicotinic acid and nicotinamide have been found to be progenitors of the pyridone (**111**)[119] and also[120] of wilfordic acid (**112**) and hydroxywilfordic acid (**113**), which are hydrolysis products of the complex ester alkaloids of *Tripterygium wilfordii.*

30.1.4 ISQUINOLINE AND RELATED ALKALOIDS

30.1.4.1 Phenethylamines and simple isoquinolines

The simplest of the phenethylamine alkaloids are *N*-methyltyramine (**116**) and hordenine (**117**), which are found in barley roots. Their genesis is a simple one from the aromatic amino-acid tyrosine (**114**) *via* tyramine (**115**), with methionine serving as the source of the *N*-methyl groups.[121] Tyrosine and tyramine are also implicated in the biosynthesis of mescaline (**126**), the hallucinogen from the peyote cactus. Extensive precursor feeding experiments, associated with the identification of precursors as naturally occurring bases,[122] and a study with an *O*-methyltransferase from peyote,[123] have allowed definition of the dominant pathway to this base as that shown in Scheme 18. The phenethylamine (**122**) is an intermediate of importance in this sequence. Methylation of the hydroxy-group at C-3 leads to mescaline (**126**), whereas methylation of the alternative hydroxy-function at C-4 leads in peyote to the isoquinoline alkaloids anhalamine (**127**) and anhalonidine (**128**).

The derivation of (**127**) and (**128**) from (**122**) *via* (**121**) accounts for all but one and two, respectively, of the carbon atoms in the two alkaloids. It was the origins of these almost trivial if distinguishing units which was to prove, for a time, quite elusive. Although [1-^{14}C]- and [2-^{14}C]-acetate was incorporated into pellotine (**129**), the label was scattered over the C_2 unit.[124] Moreover, *N*-acetyl- and *N*-formyl-phenethylamines were found not to be precursors for isoquinoline alkaloids[125–127] (*cf.* Section 30.1.7.2 for different results in the apparently related case of β-carboline alkaloids). On the other hand, pyruvic acid was incorporated [into (**128**)] with less scatter of label than that observed with acetate,[128] which was more encouraging. The solution to the problem, however, lay hidden in the old literature.[129] The biosynthesis of, for example, anhalonidine was there proposed to be by condensation of a phenethylamine with pyruvic acid to give an amino-acid [as (**123**)] which could afford the alkaloid upon decarboxylation. Specific incorporation of the key putative intermediate (**123**) into anhalonidine (**128**), demonstration of its presence as a natural constituent of peyote, and its decarboxylation by fresh peyote slices into the imine (**125**) establish the essential correctness of the hypothesis with the addition of further detail in the intermediacy of (**125**).[125] Anhalamine (**127**) is formed in a similar way from the amino-acid (**124**), derived from glyoxylic rather than pyruvic acid,[125] and (**130**), which is a natural constituent of peyote, is an intermediate in the formation of salsoline (**131**)[130] and possibly also salsolidine (**132**) and carnegine (**133**), whose early stages of biosynthesis have been delineated along the expected lines.[131] Following the establishment of these amino-acids as key intermediates in the biosynthesis of simple isoquinolines, an analogous amino-acid has been found to be of importance in the biosynthesis of benzylisoquinoline alkaloids (Section 30.1.4.2), and it is likely that such amino-acids will be of general importance in alkaloid biosynthesis.

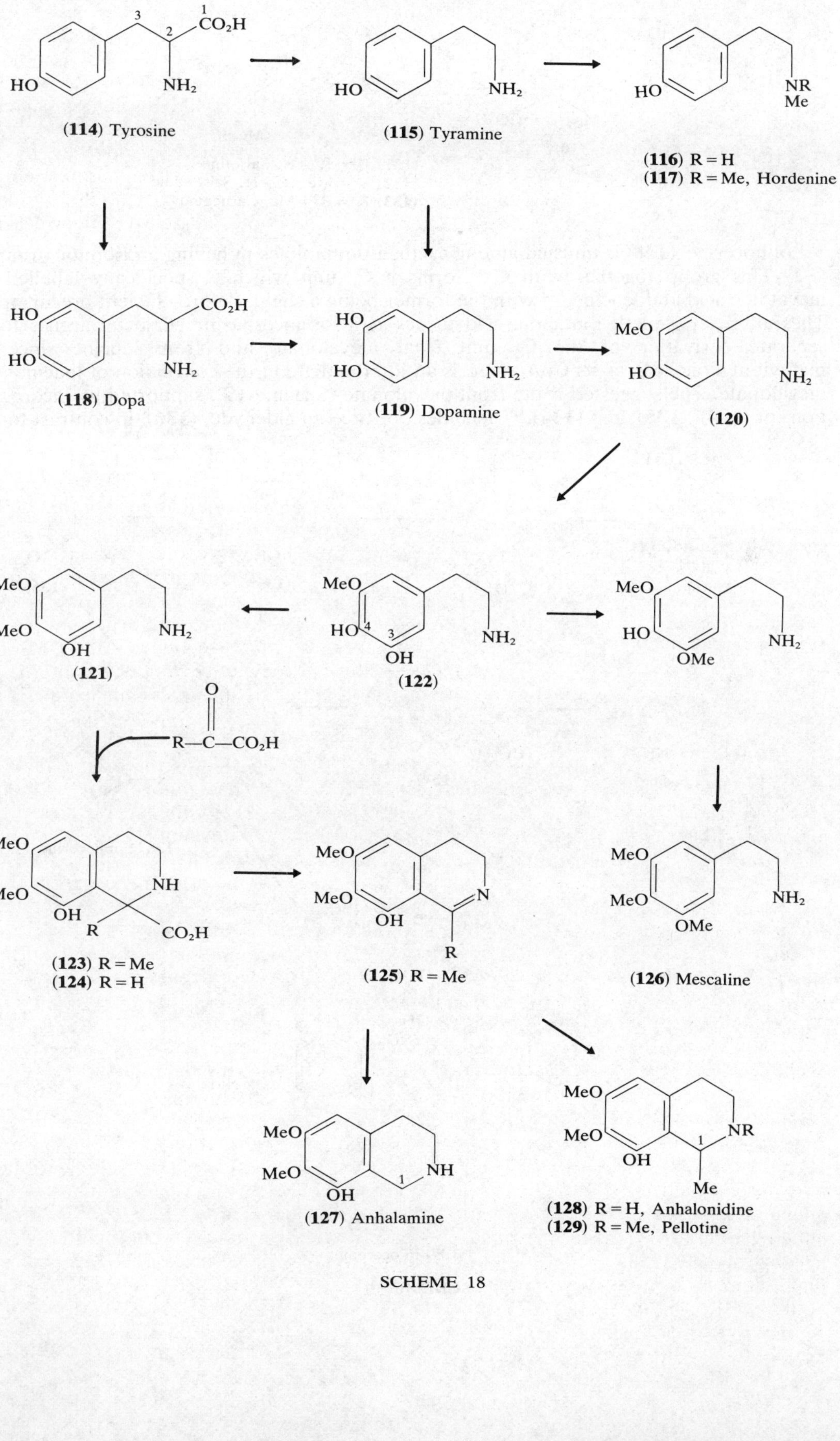
(114) Tyrosine
(115) Tyramine
(116) R = H
(117) R = Me, Hordenine
(118) Dopa
(119) Dopamine
(120)
(121)
(122)
(123) R = Me
(124) R = H
(125) R = Me
(126) Mescaline
(127) Anhalamine
(128) R = H, Anhalonidine
(129) R = Me, Pellotine

SCHEME 18

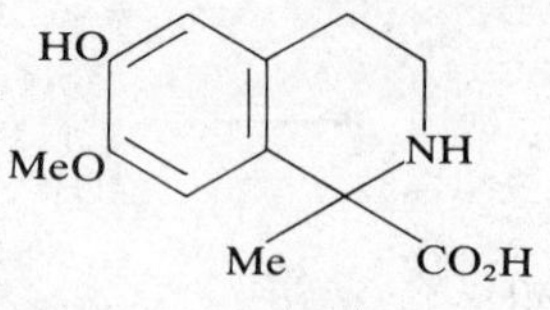

(130)

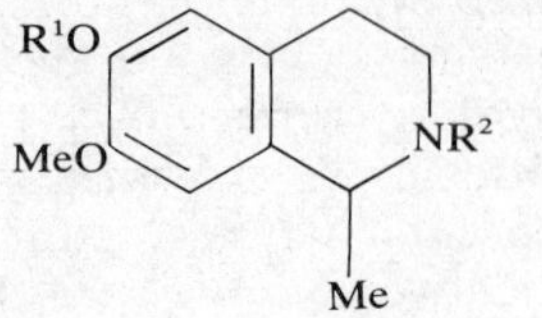

(131) $R^1 = R^2 = H$, Salsoline
(132) $R^1 = Me$, $R^2 = H$, Salsolidine
(133) $R^1 = R^2 = Me$, Carnegine

Lophocerine (**138**) is unusual among cactus isoquinolines in having an isobutyl group at C-1. This group together with C-1 forms a C_5 unit which is specifically labelled by mevalonic acid and leucine,[132] with the former being a slightly more efficient precursor.[133] The failure of 3-methylbutanoic acid to act as a lophocerine precursor[133] argues for a separate derivation of the C_5 unit from mevalonate and from leucine since 3-methylbutanoic acid, as its CoA ester, is an intermediate in the conversion of leucine into mevalonate. The suggested route from mevalonate (Scheme 19), supported by incorporations of (**134**), (**135**), and (**136**),[133] includes perforce an aldehyde, (**136**), in contrast to the

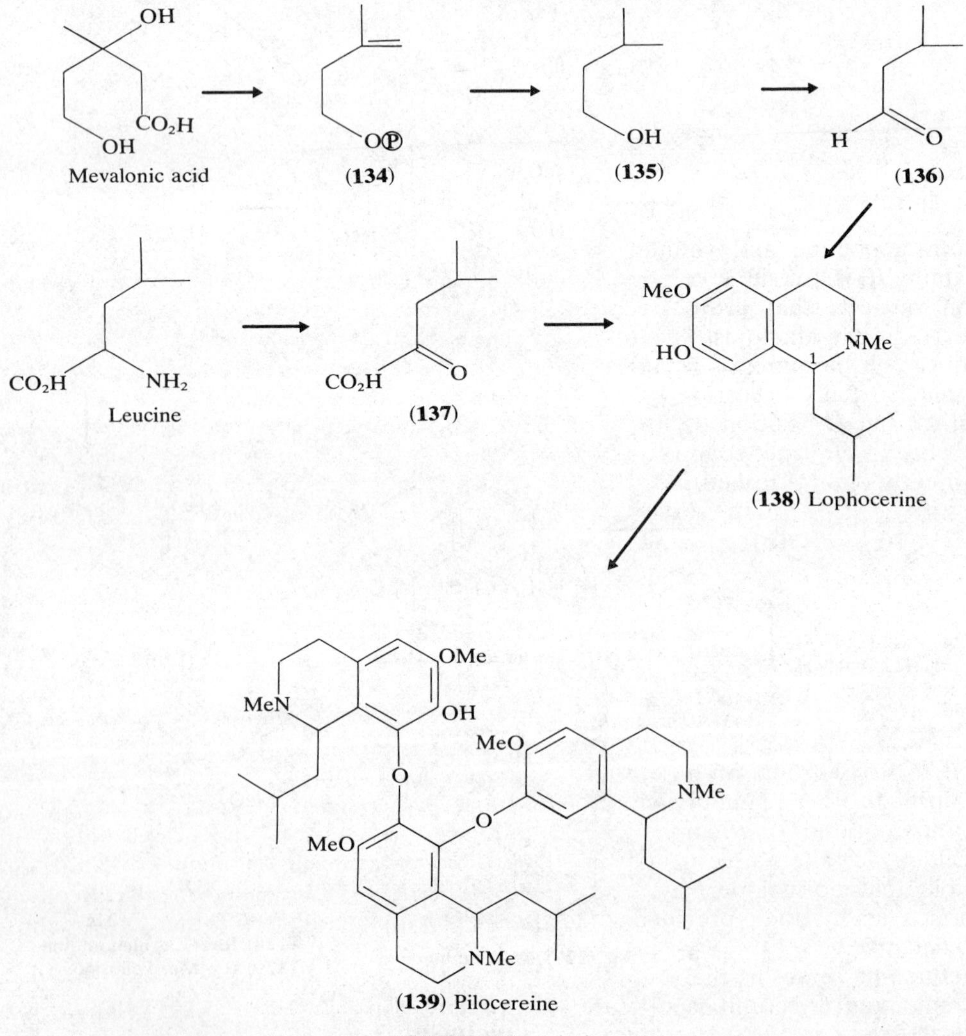

SCHEME 19

α-carbonyl acids utilized for the biosynthesis of the other isoquinoline alkaloids discussed above (but see cryptostyline below). On the other hand, leucine may be implicated *via* the keto-acid (**137**) simply derived from it by transamination. The phenethylamine residue in (**138**) is labelled by tyrosine as expected.[132]

Pilocereine (**139**) occurs naturally with lophocerine (**138**) and is transparently the product of phenol oxidative coupling (see Section 30.1.1) of three lophocerine units, and indeed, by way of confirmation of this hypothesis, a specific incorporation of [N-*methyl*-^{14}C]lophocerine into the alkaloid has been recorded.[134]

A unique 1-phenyl substituted isoquinoline structure is found in the cryptostyline alkaloids, *e.g.* cryptostyline-I (**142**), which makes them a group of alkaloids of particular biosynthetic interest. The biosynthesis of the skeleton, apart from C-1 and the attached phenyl group, is accounted for in terms of an evidently orthodox pathway from tyrosine (**114**), tyramine (**115**), and dopa (**118**) through dopamine (**119**) and the phenethylamine (**140**) [and not (**120**)].[135]

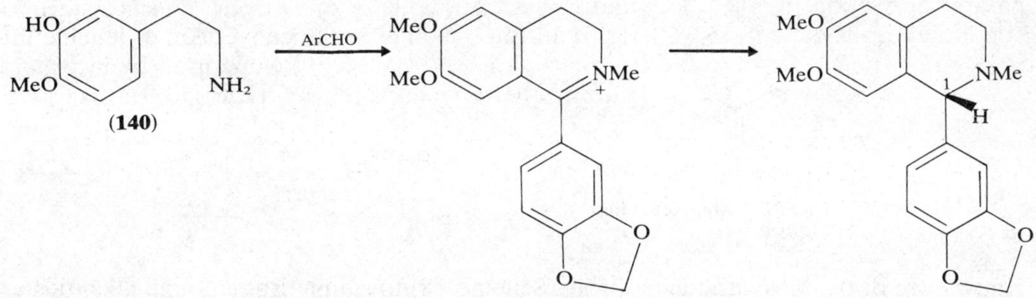

(**140**) (**141**) (**142**) Cryptostyline-I

Both dopamine and vanillin (but not isovanillin) can serve as precursors for the remaining C_6–C_1 residue in cryptostyline-I (**142**),[136] but the low level of vanillin incorporation suggests that protocatechualdehyde, the precursor for the C_6–C_1 unit in the Amaryllidaceae alkaloids (Section 30.1.5.1), may be a more important precursor. Further detail of the biosynthesis is that (**141**) is a highly efficient precursor for cryptostyline-I.

Some naturally occurring phenethylamines, *e.g.* macromerine (**143**), bear a hydroxy-group in the β-position of the side chain. This does not affect their origins, however, which are in tyrosine. It appears further that the distinguishing β-hydroxylation step may be an early one.[137] Ephedrine (**144**) has a *C*-methyl group in its side-chain in addition to

(**143**) Macromerine (**144**) Ephedrine

the β-hydroxy-group of bases like macromerine. Curiously, in spite of its structural kinship with other phenethylamines, the genesis of ephedrine is different. Phenylalanine was incorporated (*via* cinnamic acid) but only as a C_6–C_1 unit. Benzoic acid and benzaldehyde (alternative C_6–C_1 units) were incorporated at a higher level and it seems possible that normal derivation of the C_6-C_1 fragment in ephedrine is directly from shikimic acid (the precursor of aromatic amino-acids) rather than through phenylalanine.[138]

Methionine serves as the source of the *N*-methyl group of ephedrine, but the origin of the remaining C_2N unit is obscure, except that aspartate or a close relative may be implicated.

30.1.4.2 Simple benzylisoquinolines

The 1-benzyltetrahydroisoquinoline skeleton is of singular importance in the elaboration of alkaloids by plants. Apart from the terpenoid indole skeleton, no other appears in such diversely modified form, and as a substrate for modification the base reticuline (**145**) is pre-eminent.

Studies on various modified isoquinolines have provided consistent information about the way in which the benzylisoquinoline skeleton is constructed, and more recently the biosynthesis of reticuline (**145**) itself has been examined.[139] Affirmation has been obtained that its mode of genesis is from two molecules of tyrosine (**114**). Dopa (**118**) was incorporated essentially only *via* dopamine (**119**), which is the precursor for ring A plus attached C-2 unit (see Ref. 140).

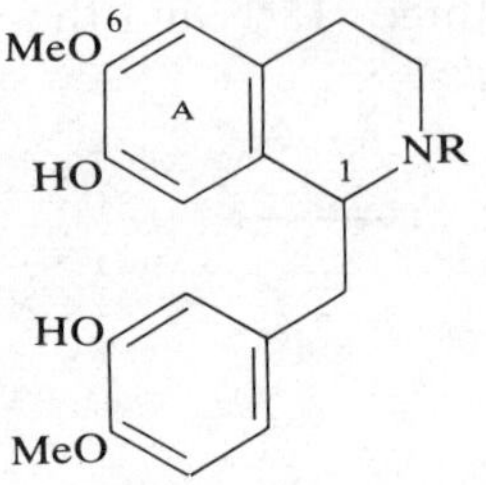

(**145**) R = Me, Reticuline
(**146**) R = H

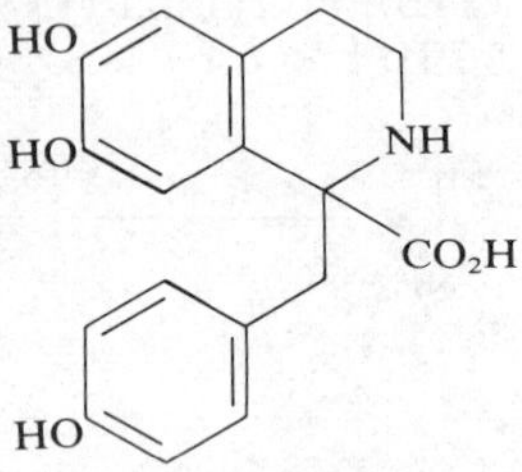

(**147**)

Knowledge of the biosynthetic pathway (Scheme 18) to simple isoquinoline alkaloids *via* amino-acids of the type (**123**) suggests a related pathway to benzylisoquinolines involving condensation of a dopamine derivative with a keto-acid derived from tyrosine to give (**148**) (Scheme 20). The correctness of this idea, in particular the intermediacy of (**148**) in biosynthesis, is supported by experimental results. This compound has been isolated in radioactive form after feeding labelled tyramine and dopa,[140] and was found to be a specific precursor for norlaudanosoline (**150**)[140] and morphine (**190**).[141] The trihydroxy-analogue (**147**) was a poor precursor for morphine[141] and reticuline,[139] which indicates that both aromatic building blocks must be dihydroxylated before they are joined together. (One of these building blocks is dopamine and it seems that little if any of the required α-keto-acid arises from dopa.) Chemical decarboxylation of (**148**) occurs with facility to give (**149**), which was also found to be a morphine precursor[141] and was itself labelled by radioactive dopa.[140] Norlaudanosoline (**150**) and reticuline (**145**) are also specific precursors for morphine (see Section 30.1.4.4) and so the evidence[139—141] is interlocking and there can be little doubt that the pathway which the results indicate (Scheme 20) is general for the biosynthesis of all benzylisoquinoline alkaloids.

Perhaps the simplest 1-benzyltetrahydroisoquinoline derivative is papaverine (**151**). A very early hypothesis was that this alkaloid was biosynthesized by way of norlaudanosoline (**150**) from two molecules of dopa.[142] The specific incorporation of two molecules of tyrosine and of norlaudanosoline (**150**) prove the essential correctness of this hypothesis.[143,144] Further detail[145] is that tetrahydropapaverine (**152**) is an important intermediate, probably formed from nor-reticuline (**146**), or (**153**). There appears to be some selectivity for benzylisoquinolines with particular methylation patterns and also stereoselectivity [the stereoisomers (−)-(*R*)-nor-reticuline and (−)-(*R*)-tetrahydropapaverine are preferred] in papaverine biosynthesis.

Simple intermolecular oxidative coupling between two phenolic benzylisoquinoline residues can be visualized as the way in which the complex bisbenzylisoquinoline alkaloids are formed. Coupling may occur with formation of either C—C or C—O bonds. In the case of epistephanine (**154**), the only alkaloid of this kind so far examined with tracers, there are two C—O linkages formed between benzylisoquinolines of the coclaurine (**155**) type. Both *N*-methylcoclaurine and coclaurine (**155**) were incorporated.[146] The former

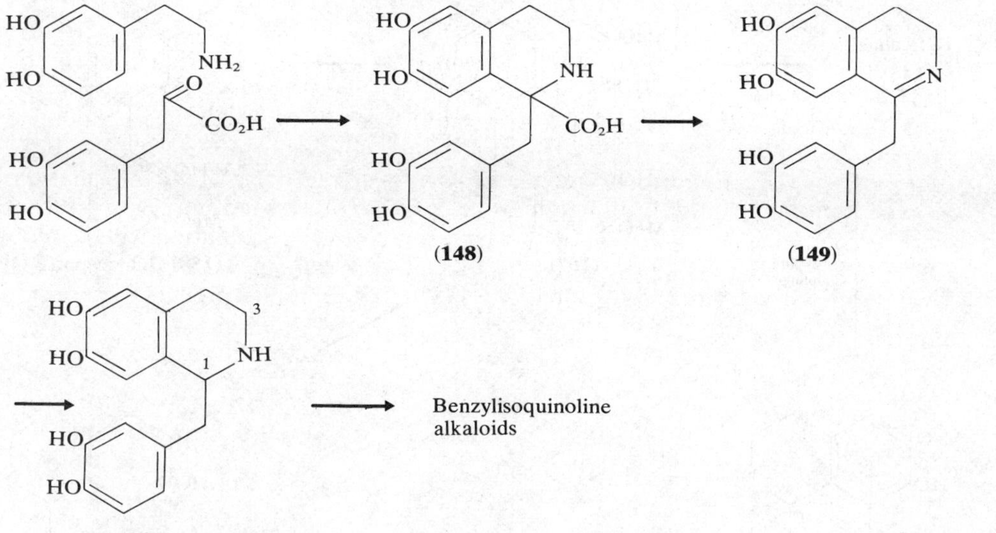

(**148**) (**149**)

(**150**) Norlaudanosoline

SCHEME 20

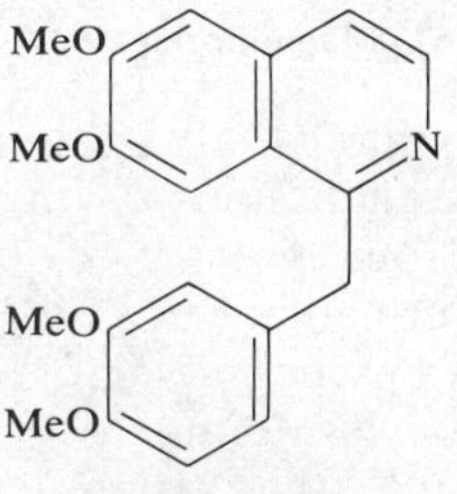

(**151**) Papaverine

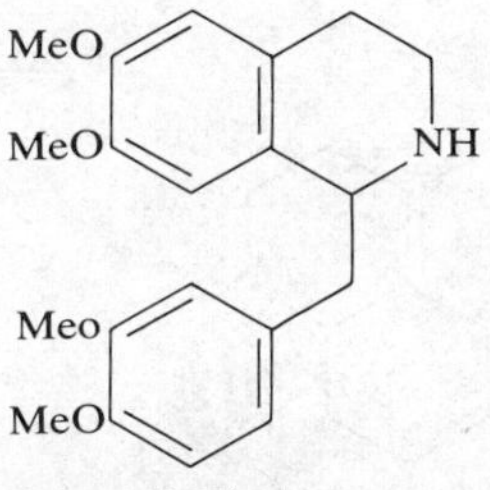

(**152**)

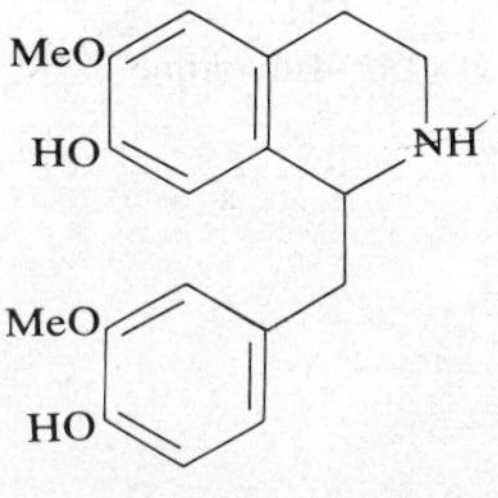

(**153**)

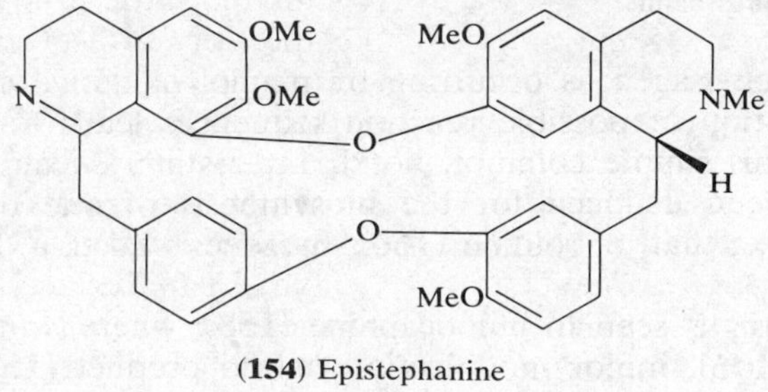

(**154**) Epistephanine

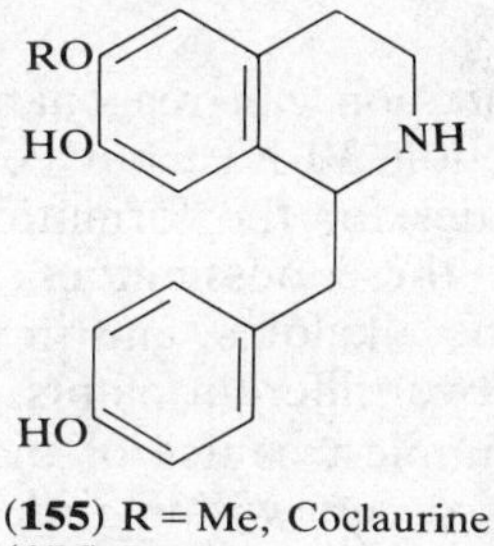

(**155**) R = Me, Coclaurine
(**156**) R = H

was assimilated without loss of the *N*-methyl group and only into the *N*-methylated half of epistephanine; the (−)-isomer [same configuration as (**154**)] was the preferred one. As expected, [2-^{14}C]tyrosine was an epistephanine precursor and, interestingly, was equally incorporated into both halves of the molecule.

30.1.4.3 Aporphines

Forging of a single new bond is all that is required for the transformation of the benzylisoquinoline skeleton into that of the aporphines, *e.g.* bulbocapnine (**158**). The

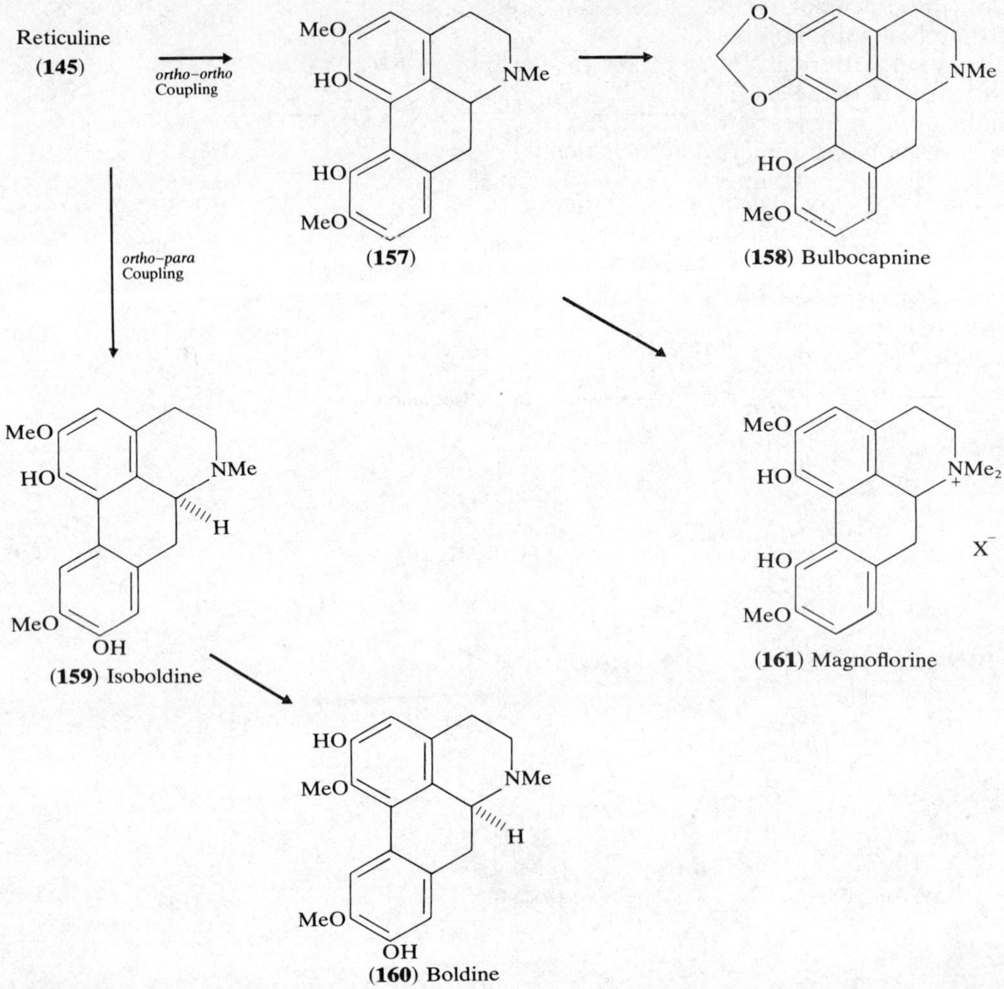

transformation can reasonably be envisaged as occurring by phenol oxidative coupling (see Section 30.1.1), but consideration of possible reaction sequences leads to several possibilities for the formation of this simple common bond. Interestingly, examples of most of these possibilities have been deduced for the biosynthesis of one or more aporphine alkaloids, and in one case, that of boldine (**160**), biogenesis takes a different path in two different plants.

The simplest course of biosynthesis is seen in bulbocapnine (**158**), where *ortho–ortho* coupling occurs within reticuline (**145**), minor modification of the product (**157**) then affording the aporphine (**158**).[147] The base (**157**) is also probably an intermediate in magnoflorine (**161**) biosynthesis since reticuline (**145**) [and not norprotosinomenine (**166**)] was incorporated.[148] Reticuline is also a precursor for isoboldine (**159**), a minor alkaloid of *Papaver somniferum* [reticuline is likewise the precursor of morphine (**190**) in this plant; see Section 30.1.4.4]. Orientaline (**162**), with an alternative methylation pattern which is *a priori* a possible precursor by way of the dienone (**163**), was not incorporated, nor was (**166**). Thus a unique pathway *via* reticuline (**145**) is indicated, a pathway which involves simple *ortho–para* coupling.[149]

Rigorous experimental distinction has been made[150] in *Litsea glutinosa* between possible benzylisoquinoline precursors for boldine (**160**), and it has been shown that the alkaloid is derived from reticuline (**145**) *via* isoboldine (**159**); (+)-reticuline with the same absolute configuration as boldine and not its isomer is involved. Notably, norprotosinomenine

(**166**) (the precursor for bases including boldine in *Dicentra eximia*; see below) was not incorporated into boldine (**160**) and it is clear from the combined evidence that the methylation pattern in boldine (**160**) is misleading as far as biosynthesis in this plant is concerned. It is the methylation pattern in isoboldine (**159**) which reflects the mode of coupling and it appears that the conversion of (**159**) into boldine (**160**) largely involves demethylation and remethylation rather than methyl transfer since boldine derived from [6-$O^{14}CH_3$,1-^{3}H]reticuline [as (**145**)] showed a 64% loss of ^{14}C label (similar results have been obtained in the related case of crotonosine; see below).

Thebaine (**185**) and isothebaine (**165**) are found in *Papaver orientale.* Although they are both elaborations of the norlaudanosoline (**150**) skeleton, thebaine is derived from reticuline (Section 30.1.4.4) whereas isothebaine is formed from orientaline (**162**).[151] This provides an excellent example of the methylation pattern coding the course of phenol coupling.

The clue to the biosynthesis of isothebaine (**165**) is the absence of an oxygen function at C-10 in the alkaloid. This can only plausibly have happened as a result of biosynthesis along the route shown in Scheme 21 from orientaline (**162**), where loss of this oxygen

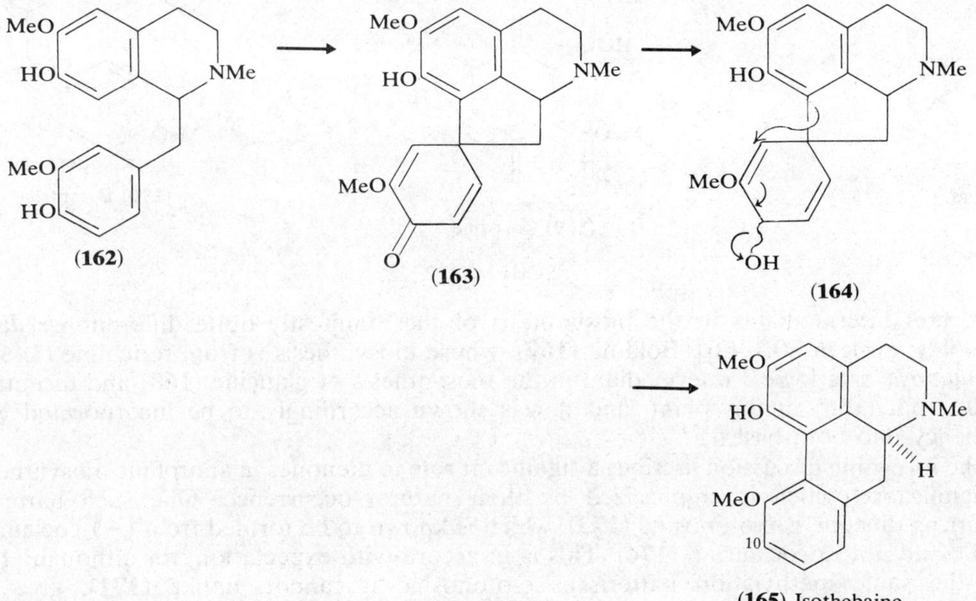

SCHEME 21

function may occur in the course of a 'dienol–benzene' rearrangement of (**164**). Tracer experiments have established the correctness of the route. Thus the key dienone (**163**) was shown to be a specific precursor for isothebaine (**165**) and a natural constituent of *P. orientale.*[152] One of the dienols (**164**) obtained on reduction of (**163**) was an efficient precursor for isothebaine, the other dienol being utilized with low efficiency, apparently *via* (**163**) and the first dienol.[153]

Inspection of the structures of the *Dicentra eximia* bases, glaucine (**168**), corydine (**169**), and dicentrine (**170**), suggests a biogenesis along the lines already discussed. Any confidence in this view is dispelled by the observation that neither reticuline (**145**) nor orientaline (**162**) were incorporated, although the unmethylated base, norlaudanosoline (**150**), was.[154] Painstaking, step-by-step introduction of methyl groups into various positions on this skeleton and testing in tracer experiments then allowed definition of the key intermediate for coupling as norprotosinomenine (**166**), the methylation pattern of which indicates the unexpected pathway shown in Scheme 22; the point of *N*-methylation is unknown [it is of considerable interest that norprotosinomenine and the dienone (**167**) are

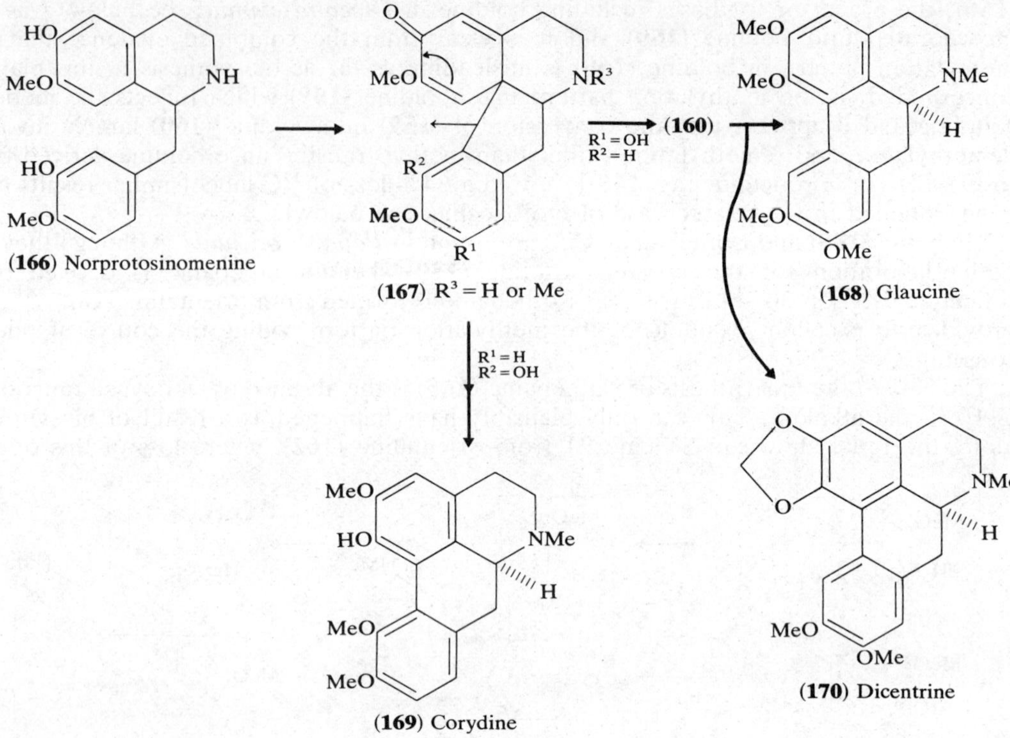

SCHEME 22

important intermediates in the biosynthesis of the manifestly quite different *Erythrina* alkaloids; Section 30.1.4.6]. Boldine (**160**), whose biosynthesis is from reticuline (**145**) in *L. glutinosa*, is a logical intermediate in the biosynthesis of glaucine (**168**) and dicentrine (**170**) from (**166**) in this plant, and it was shown accordingly to be incorporated with efficiency into both bases.[154]

The foregoing discussion ascribes a significant role to dienones in aporphine biosynthesis, a significance which is emphasized by their natural occurrence. One such naturally occurring dienone is crotonosine (**173**), which is known to be formed from (+)-coclaurine (**171**), and not isococlaurine (**174**). This is in accord with expectation, for although (**174**) has the same methylation pattern as crotonosine it cannot, unlike (**171**), give the crotonosine skeleton by phenol coupling.[155] Demethylation and remethylation to give (**173**) thus occurs in the course of biosynthesis, and since norcoclaurine (**156**) is a less efficient precursor than coclaurine, this sequence must take place after coupling.

An extension of the coclaurine/crotonosine pathway is found[156] in the biosynthesis of roemerine (**177**). Here it is (+)-methylcoclaurine (**172**) which is the latest isoquinoline precursor. The next known intermediate is mecambrine (**175**), which may afford roemerine *via* the dienol (**176**) (*cf.* isothebaine biosynthesis above). The related base, anonaine (**178**), arises from coclaurine (**155**), along what must be a similar pathway,[156] in which crotsparine (**179**) is a likely intermediate. This is a naturally occurring compound and, along with crotsparinine (**180**) and sparsiflorine (**181**) in the same plant, has been shown to arise from coclaurine (**155**).[157] It follows that crotsparine is a probable intermediate in the biosynthesis of (**181**) too.

30.1.4.4 Morphine and related alkaloids

The idea that morphine (**190**) was a modified benzylisoquinoline alkaloid helped with the assignment of the correct structure to this alkaloid; twisting of the benzylisoquinoline

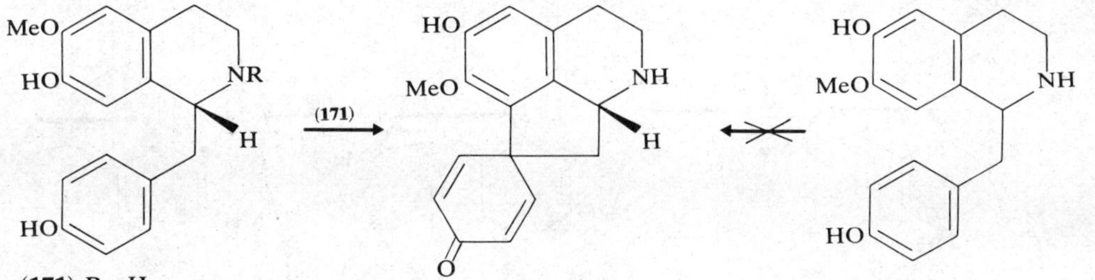

(171) R = H
(172) R = Me, enantiomer
(173) Crotonosine
(174)

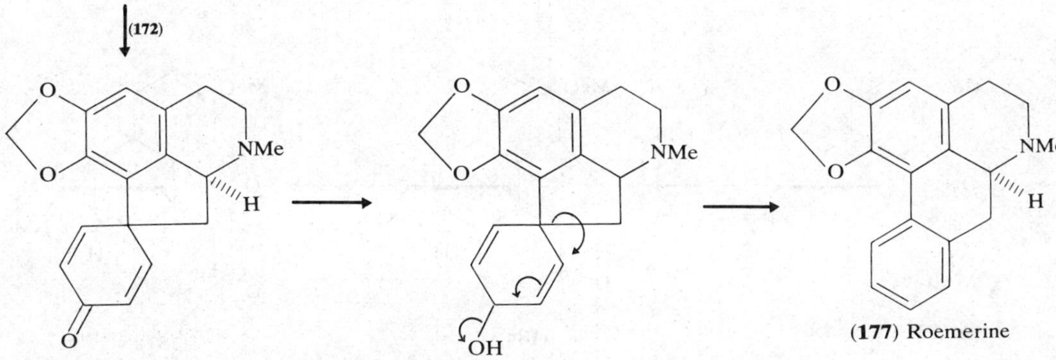

(175)
(176)
(177) Roemerine

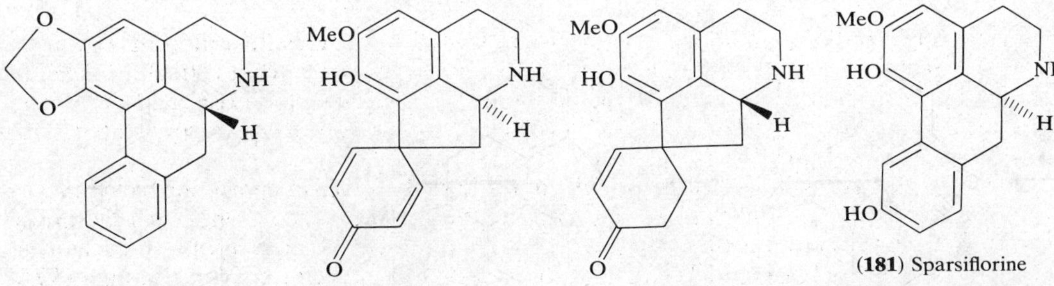

(178) Anonaine
(179) Crotsparine
(180)
(181) Sparsiflorine

skeleton into that shown for reticuline in (**182**) illustrates this relationship and suggests a possible biogenesis.[1] Study of the biosynthesis of morphine and the related hydrophenanthrene alkaloids of *Papaver somniferum*, thebaine (**185**), and codeine (**188**) has become one of the classics of alkaloid biosynthesis, one in which the correctness of the benzylisoquinoline hypothesis has been proved and extended.

Incorporation of [2-^{14}C]tyrosine into C-9 and C-16 of (**190**) provided preliminary support for the hypothesis[158] ([1-^{14}C]dopamine only labelled C-16),[159,160] but the crucial test lay with examination of a benzylisoquinoline as a precursor. In the first experiments ever to be carried out with complex plant-alkaloid precursors, [1-^{14}C]- and [3-^{14}C]-norlaudanosoline [as (**150**)] were fed, and specific labelling of the hydrophenanthrene alkaloids at C-9 [in (**190**)] and C-16 [in (**188**)], respectively, was recorded, thus establishing the genesis of these alkaloids as being by modification of the benzylisoquinoline skeleton.[144,161]

Examination of various benzylisoquinolines as precursors for morphine and its congeners has allowed definition of reticuline [as (**182**)] as the key intermediate after norlaudanosoline.[144,162,163] In an experiment of particular interest in defining the uniqueness

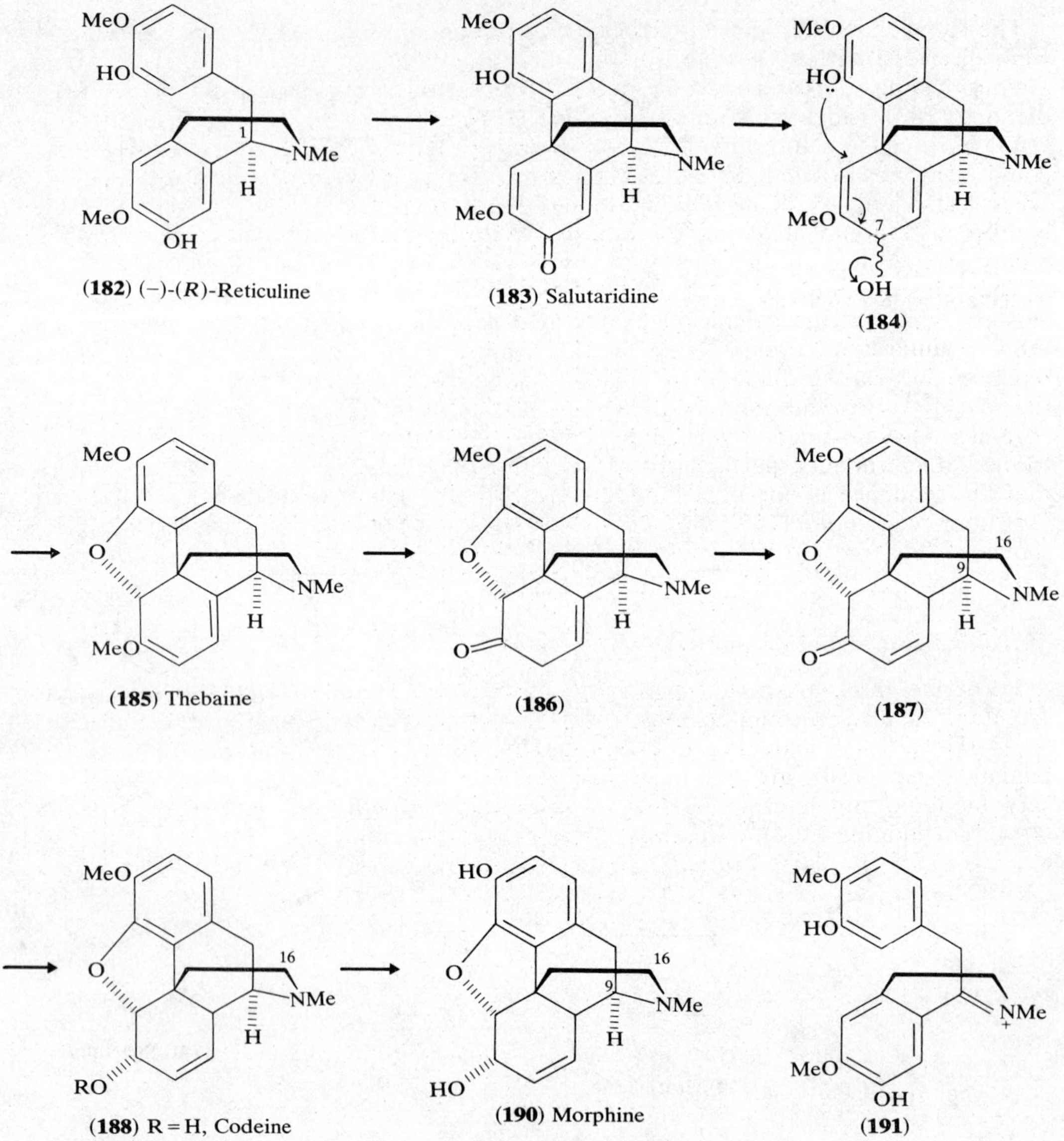

(**182**) (−)-(*R*)-Reticuline; (**183**) Salutaridine; (**184**); (**185**) Thebaine; (**186**); (**187**); (**188**) R = H, Codeine; (**189**) R = Me; (**190**) Morphine; (**191**)

SCHEME 23

of reticuline as a hydrophenanthrene precursor, it was shown that reticuline [as (**182**)] was incorporated into thebaine (**185**) and not isothebaine (**165**) in the same plant, whereas orientaline (**162**) was assimilated into isothebaine but not thebaine.[151] Both (*R*)-reticuline (**182**) and its (*S*)-isomer were similarly well incorporated into the alkaloids of *P. somniferum*, with loss of tritium label from C-1.[163] This suggests interconversion of the reticuline enantiomers by way of (**191**), before utilization for further biosynthesis, and in support this compound was found to be an efficient precursor for morphine.[163] Moreover, in the case of (*R*)-reticuline, the isomer with same configuration as, for example, morphine (**190**), loss of tritium from C-1 was incomplete, which indicates incomplete conversion into (**191**) before further biosynthesis. The demonstration that reticuline was present in *P. somniferum*[164] and was formed in radioactive form from $^{14}CO_2$ before the appearance of radioactive thebaine[165] further establishes its role as an intermediate in the biosynthesis of hydrophenanthrene alkaloids.

The steps which lie between tyrosine and norlaudanosoline (**150**) and reticuline (**182**) were discussed earlier (Section 30.1.4.2) and attention is now turned to those which lie after reticuline. *Ortho–para* coupling of this diphenol can be conceived of as giving the dienone (**183**), which could afford thebaine (**185**) as shown. In support of this hypothesis (**183**) (now called salutaridine) was found to be a natural constituent of another plant. It could also be isolated in radioactive form from *P. somniferum* after feeding [2-^{14}C]tyrosine and [3-^{14}C]norlaudanosoline. Proof that it is actually an intermediate in hydrophenanthrene alkaloid biosynthesis results from the observation that (**183**) is a highly efficient and specific precursor for the alkaloids.[162]

The transformation of salutaridine (**183**) into thebaine (**184**) requires a further ring-closure, which occurs chemically under mild conditions when the two epimeric dienols (**184**) (Salutaridinol-I and -II) are treated with acid. In contrast to the purely chemical reaction, only salutaridinol-I was efficiently converted into thebaine *in vivo*, indicating that the reaction is enzyme mediated and thus part of normal biosynthesis.

It has been assumed so far that thebaine is the first hydrophenanthrene base to be formed. Experiments, particularly with $^{14}CO_2$, have established that this is correct and that the sequence is one of demethylation: thebaine (**185**) → codeine (**188**) → morphine (**190**); and codeinone (**187**) and neopinone (**186**), but not (**189**), are to be included in the sequence (Scheme 23).[166]

30.1.4.5 Protoberberine and related alkaloids

Berberine (**194**) exemplifies a group of alkaloids based on the benzylisoquinoline skeleton, the distinguishing feature of which is an extra carbon atom appearing as a bridge at C-8: the so-called 'berberine bridge'. Closer to benzylisoquinolines, like norlaudanosoline (**150**), are the protoberberine alkaloids, *e.g.* stylopine [as (**199**)], which serve as important intermediates in the elaboration of alkaloids as diverse as berberine (**194**), chelidonine (**204**), narcotine (**209**), and corydaline (**213**).

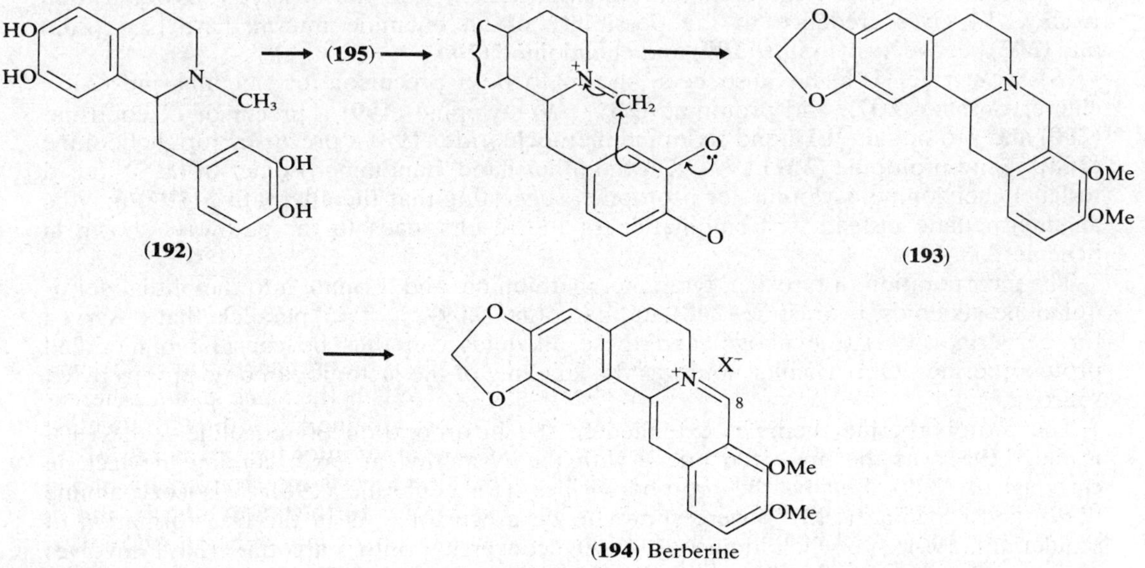

SCHEME 24

Results of experiments on the biosynthesis of berberine (**194**) showed that it was derived from two molecules of tyrosine, one of which was incorporated *via* dopamine.[167] In short, a pathway expected of a modified benzylisoquinoline. Both radioactive methionine and formate labelled the methylenedioxy- and methoxy-groups of berberine and, more significantly, C-8 as well.[168] The question of how this carbon atom arises from C_1 precursors like methionine is answered with brilliant simplicity in terms of an oxidative cyclization of the *N*-methyl group of a benzylisoquinoline precursor in a manner (Scheme 24) analogous to the formation of a methylenedioxy-group (see Section 30.1.5.1); validation was obtained with the specific and intact incorporation of label from the *N*-methyl groups of laudanosoline (**192**) and reticuline [as (**195**)] into C-8 of berberine [of several possible norlaudanosoline derivatives, reticuline (and nor-reticuline) was the only precursor — (*S*)-isomer (**195**) preferred over (*R*) — along with the protoberberine, canadine (**193**)].[169,170]

Robinson suspected[1] that chelidonine (**204**) was to be included in the group of modified protoberberine bases, an idea supported by the natural co-occurrence of stylopine (**199**) and chelidonine (**204**). Conversion of the one skeleton into the other can be envisaged as involving cleavage of the C-6—N bond of, say, (**199**) followed by reclosure at C-13. The way in which [2-^{14}C]tyrosine was incorporated (labelling of C-6 and C-13 of chelidonine) is consistent with this view — [1-^{14}C]dopamine gave the alkaloid with activity confined to C-6.[171] More importantly, stylopine (**199**) was shown to be a precursor for chelidonine,[172,173] and scoulerine (**196**) and the ubiquitous benzylisoquinoline, reticuline (**195**), are also implicated in the biosynthesis of this alkaloid.[170,173] The sequence thus indicated, (**195**) → (**196**) → (**199**) → chelidonine (**204**), is also one which involves the (*S*)-isomer in each case.

Tritium-labelling results establish that C-1 and C-9 of reticuline (**195**) are unaffected by transformation through scoulerine (**196**) to stylopine (**199**). The conversion of the *N*-methyl group of reticuline into the 'berberine-bridge' atom (C-8) of stylopine (**199**), which necessarily involves loss of one hydrogen/tritium atom, was found to occur with very high tritium retention, consonant with an expected tritium isotope effect.[170] Subsequent transformation of stylopine to chelidonine occurs with retention of tritium at C-8 and C-5 but loss of tritium from C-14[173] and loss of the 13-*pro-S* hydrogen atom (retention of *pro-R* proton).[174] The last observation is consistent with enzyme-mediated proton removal from C-13 at a stage after stylopine, and taken with the observed loss of tritium from C-14 gives credence to the possibility of an enamine intermediate [see (**202**) and (**203**)] between stylopine (**199**) and chelidonine (**204**).

(*S*)-Scoulerine (**196**) has also been shown to be a precursor for sanguinarine (**206**), chlereryanthine (**207**), and protopine (**201**), (*S*)-stylopine (**199**) a precursor of coptisine (**200**) and protopine (**201**), and stylopine methochloride (**198**) a precursor for chelidonine (**204**)[173] and protopine (**201**).[173,175] On the other hand, nandinine [isomer of (**197**)] failed to label chelidonine, stylopine, or protopine, suggesting that the alternative (**197**) may be an intermediate instead.[173] Combination of the results leads to the pathways shown in Scheme 25.

The incorporation of tyrosine, tyramine, methionine, and formate into the phthalideisoquinoline alkaloids, hydrastine (**208**) and narcotine (**209**),[167,168,176] parallels that observed for berberine (**194**) (see above) and these alkaloids can thus be considered modified protoberberines. Of particular note was the labelling of the lactonic carbonyl groups by C_1 sources.

The protoberberine theme is extended in the incorporation of reticuline — the (+)-isomer (**195**) was the preferred one — with the *N*-methyl group becoming the lactone carbonyl of (**209**). Finally, the protoberberines, (*S*)-scoulerine (**196**),[176] isocorypalmine (**210**), and canadine (**193**)[173] were shown to be precursors. As in the transformation of scoulerine (**196**) into chelidonine (**204**), its conversion into narcotine (**209**) involves removal of the 13-*pro-S* hydrogen atom.[174] Introduction of the oxygen atom at C-13 thus proceeds with orthodox retention of configuration[177] (see also refs. cited in Ref. 174).

Alkaloids of the alpinigenine (**212**) type are plausibly modified protoberberines. This is supported by the incorporation into (**212**) of [3-^{14}C]tyrosine [labelling of (**212**): •], methionine [labelling of (**212**): *], tetrahydropalmatine (**211**), and its *N*-methyl derivative. The latter was incorporated without loss of its *N*-methyl group and a role for the

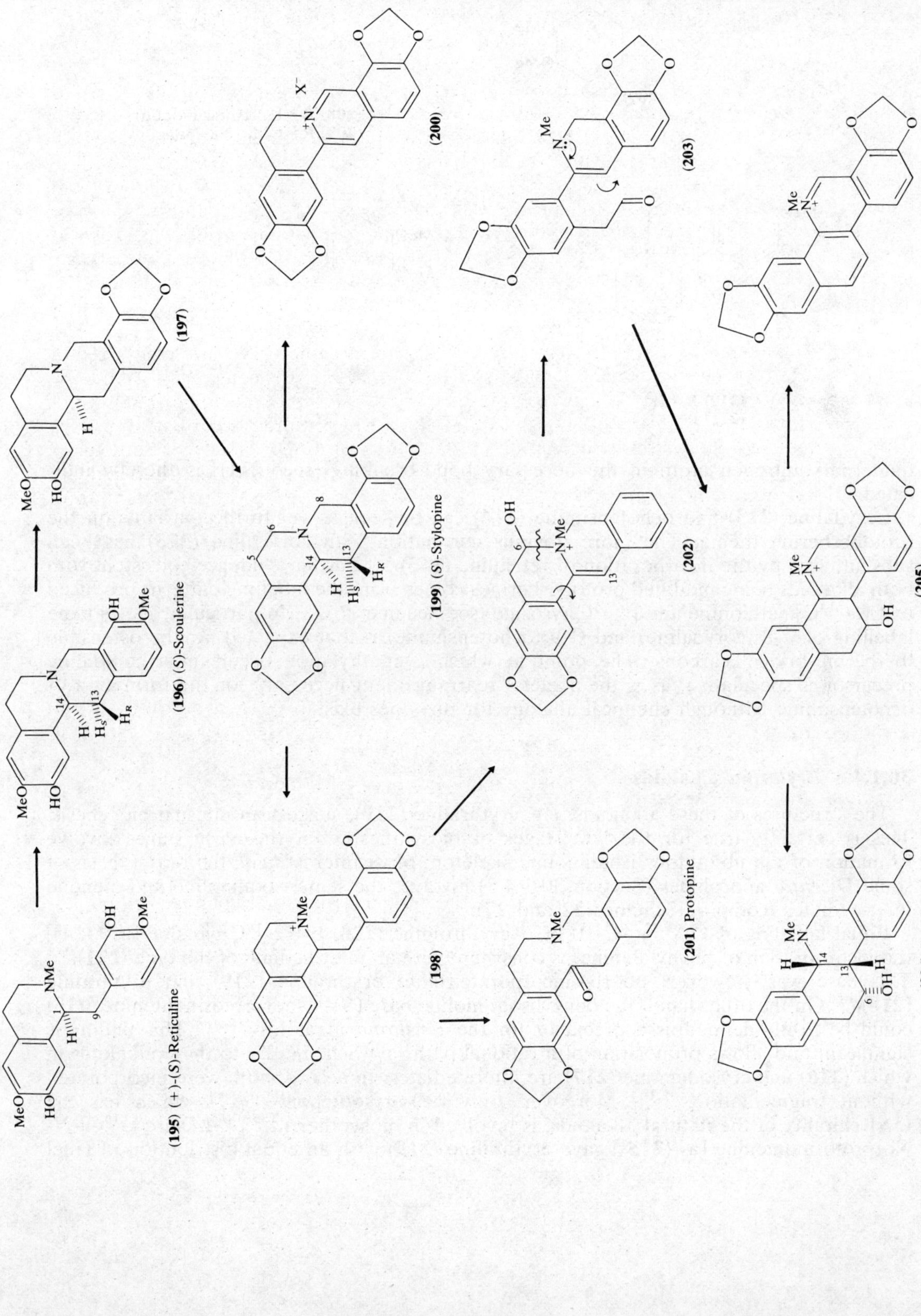
(195) (+)-(S)-Reticuline
(196) (S)-Scoulerine
(197)
(198)
(199) (S)-Stylopine
(200)
(201) Protopine
(202)
(203)
(204) Chelidonine
(205)
(206)

SCHEME 25

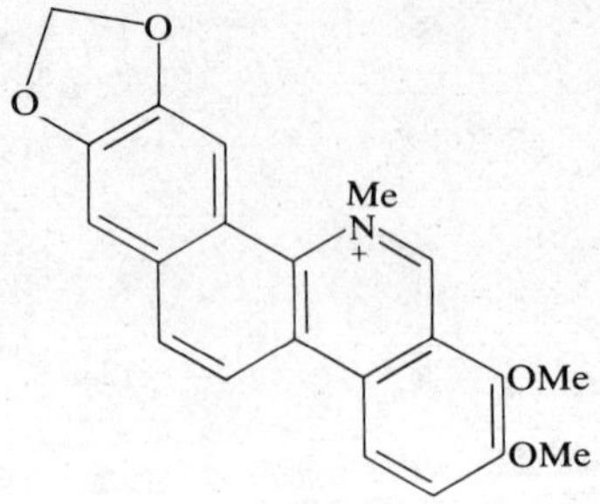

(**207**)

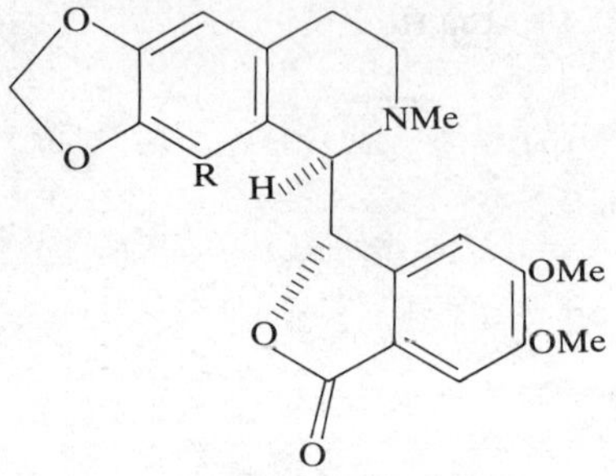

(**208**) R = H, Hydrastine
(**209**) R = OMe, Narcotine

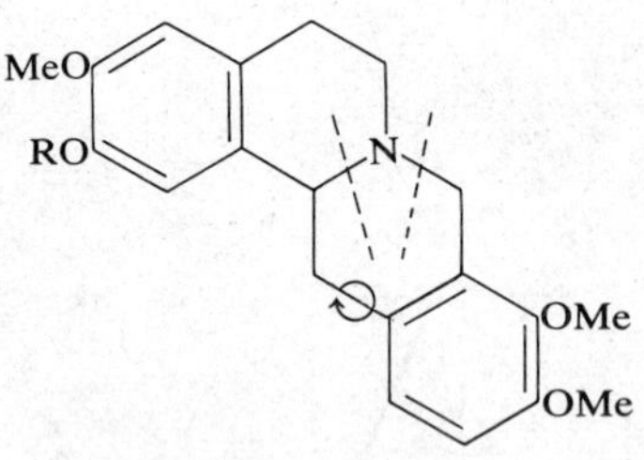

(**210**) R = H
(**211**) R = Me

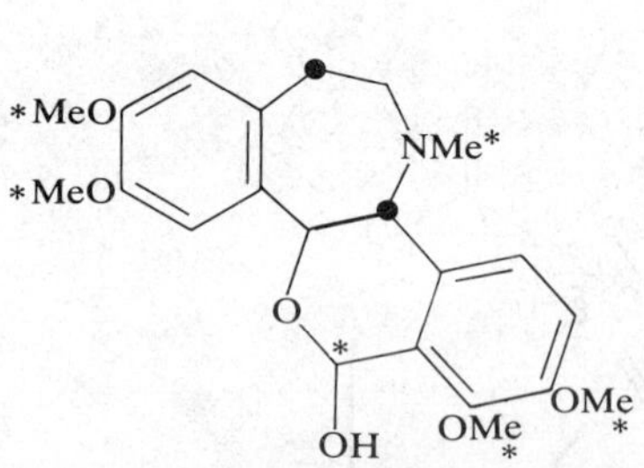

(**212**) Alpinigenine

quaternary nitrogen atom in the necessary bond-breaking [see (**211**)] is thereby indicated.[178]

Corydaline (**213**) and ochotensimine (**214**) can be seen as yet further variants on the protoberberine theme. The more obvious correlation with corydaline (**213**) has been substantiated by the incorporation of reticuline (**145**).[179] Further evidence consistent with both alkaloids being modified protoberberines comes from the labelling patterns produced by [*Me*-^{14}C]methionine and [3-^{14}C]tyrosine (see Scheme 26).[180] In particular, methionine labelling of C-8 (corydaline) and C-9 (ochotensimine) is that expected for the ostensible 'berberine-bridge' carbons. The point at which *C*-methylation occurs in a corydaline precursor is speculative, as is the skeletal rearrangement necessary for the formation of ochotensimine, although chemical analogy for this does exist.

30.1.4.6 *Erythrina* alkaloids

The structures of these alkaloids, *e.g.* erythraline (**219**), suggest an unusual biogenesis. This is certainly true for the late stages of biosynthesis, but the early stages involve formation of the ubiquitous isoquinoline skeleton. Most interestingly, the biosynthesis of some *Dicentra* aporphines (Section 30.1.4.3) involves the same isoquinoline and dienone intermediates (compare Schemes 22 and 27).

Equal labelling of C-8 and C-10 of β-erythroidine (**220**) by [2-^{14}C]tyrosine [as (**114**)] (no incorporation of phenylalanine) is consistent with an intermediate of the type (**221**).[181] This base was, however, poorly incorporated into erythraline (**219**) and erythratine (**218**).[182] On the other hand, the benzylisoquinoline base (*S*)-*N*-norprotosinomenine (**215**) could be firmly defined as a precursor for the *Erythrina* alkaloids.[182–184] This finding is significant and allows proposition of a rational pathway (Scheme 27) to these alkaloids in which (**216**) and erysodienone (**217**) are intermediates. In accord, both were incorporated without fragmentation.[182,184] Moreover, only (−)-erysodienone (**217**), which has the (5*S*)-chirality of the natural alkaloids, is involved in biosynthesis.[185] [4′-*Methoxy*-^{14}C]-*N*-Norprotosinomenine [as (**215**)] gave erythraline (**219**) with an equal distribution of label

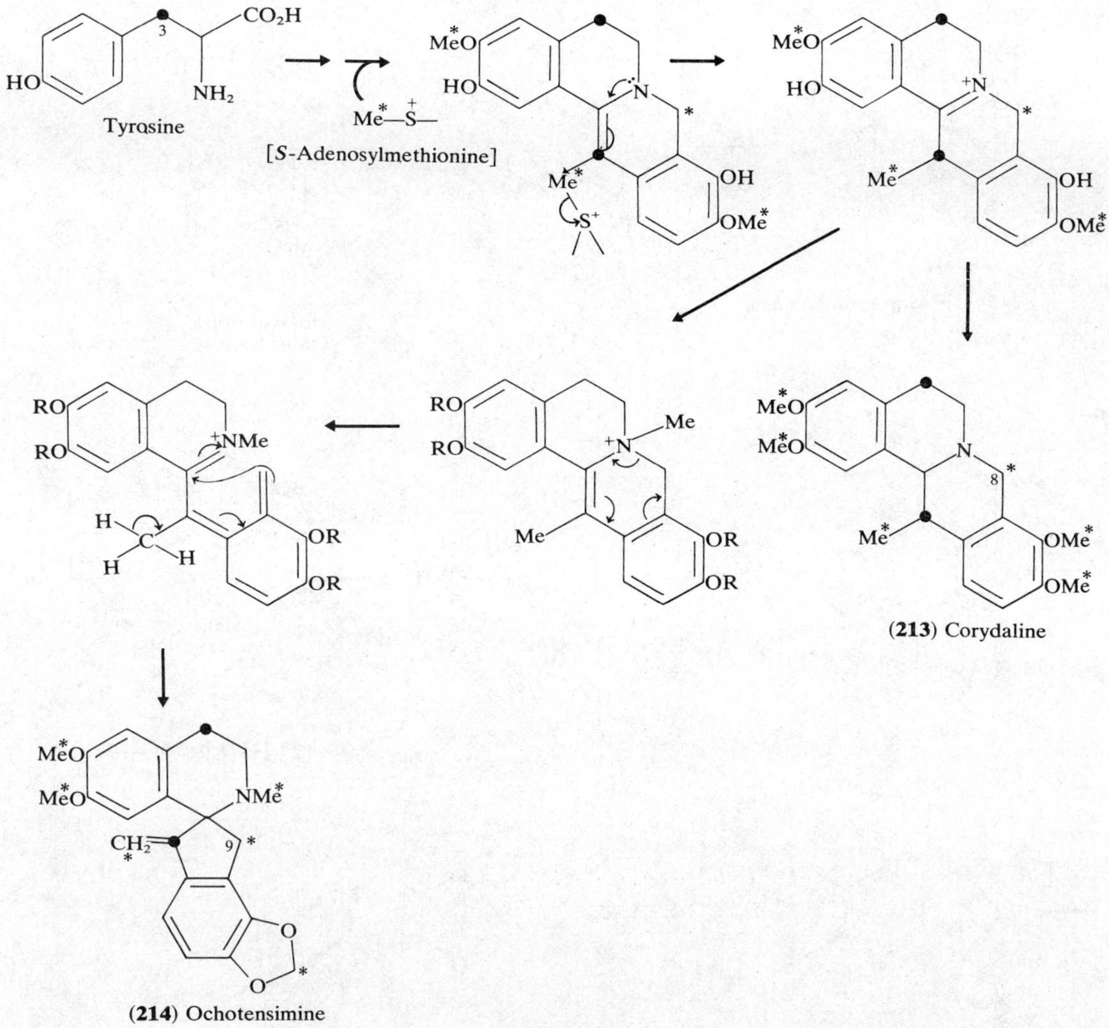

(213) Corydaline

(214) Ochotensimine

SCHEME 26

between methoxy- and methylenedioxy-groups, so clearly the pathway must involve a symmetrical intermediate [as (**216**)][184] and the asymmetry of (*S*)-norprotosinomenine (**215**) is not transferred to (*S*)-erysodienone (**217**).[185]

Results of further experiments establish that erythraline (**219**) is formed from 2-epierythratine [as (**218**)],[182] and that the lactonic erythroidines (**220**) arise from bases [(**222**) and (**223**)] of the erythraline type.[185] This latter transformation obviously involves aromatic ring scission — a rare thing in the biosynthesis of plant bases (*cf.* betalains in Section 30.1.10).

30.1.4.7 Hasubanonine and protostephanine

Hasubanonine (**224**) and protostephanine (**225**), both produced by *Stephania japonica*, might seem to be variations of the benzylisoquinoline skeleton, but attempts to establish this or any other pathway have been frustrated.[186] Many benzylisoquinolines and bisphenethylamines [as (**221**)] have been tested as precursors, with negative results. Positive results have, nevertheless, been obtained with tyrosine, dopa, tyramine, and dopamine. This allows partial definition of the biosynthesis of these alkaloids. They are both

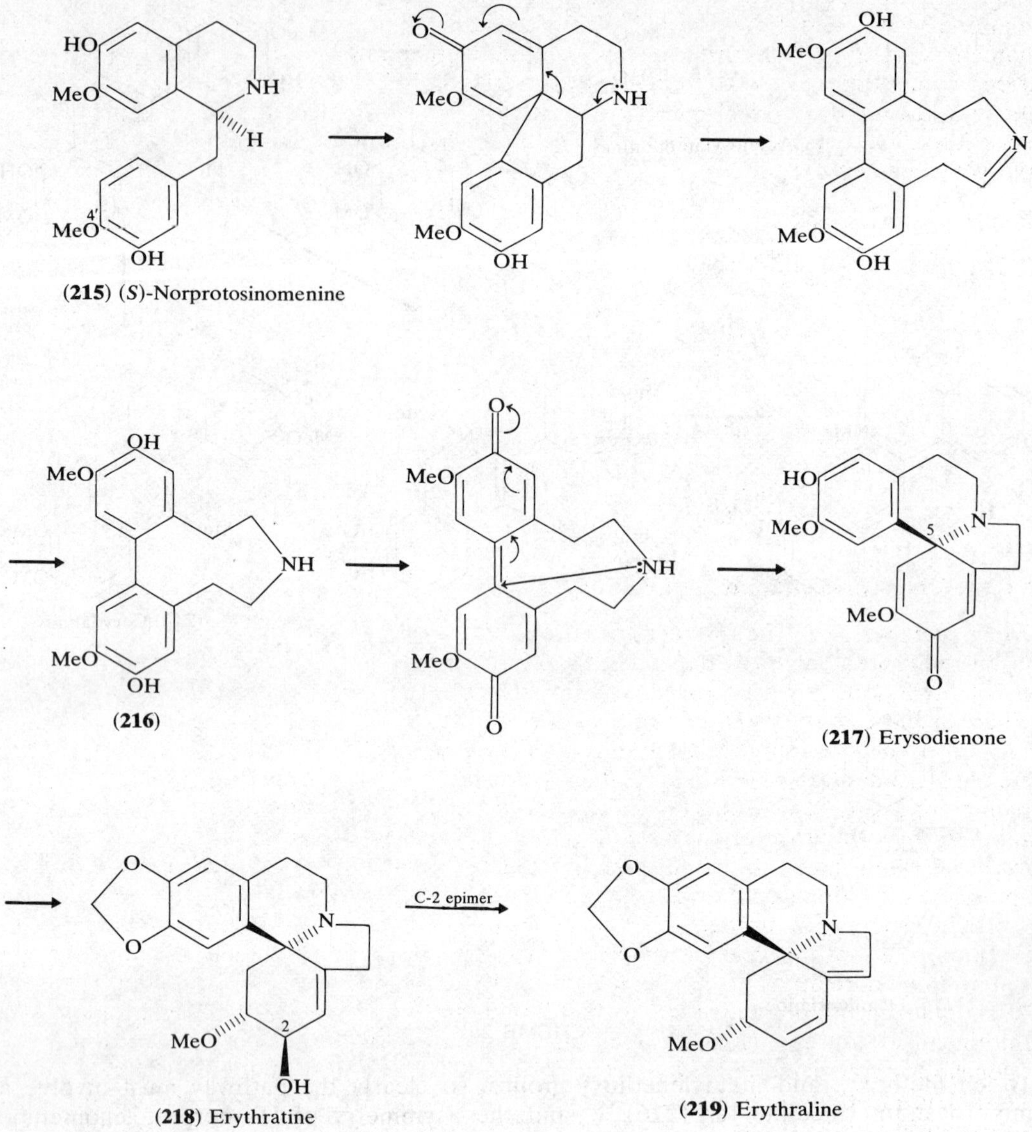

(**215**) (*S*)-Norprotosinomenine

(**216**)

(**217**) Erysodienone

(**218**) Erythratine

(**219**) Erythraline

SCHEME 27

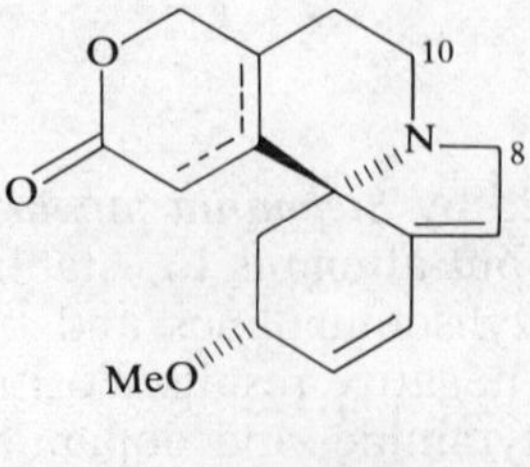

(**220**)

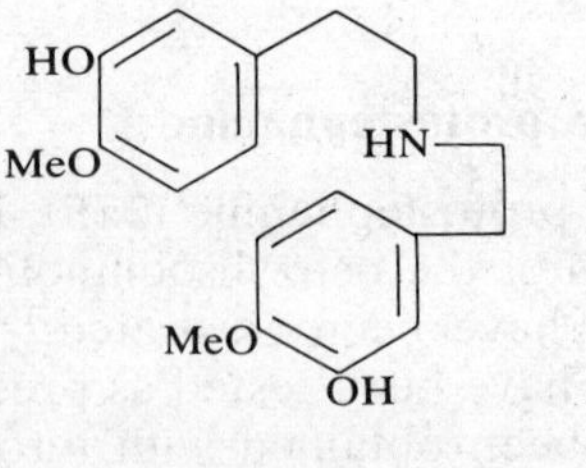

(**221**)

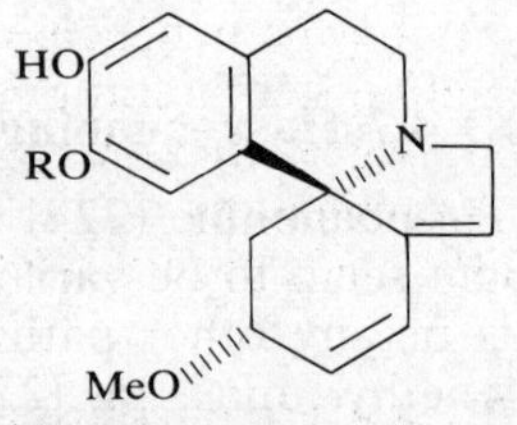

(**222**) R = H
(**223**) R = Me

constructed from C_6–C_2 units, for which tyrosine (**114**) is the source, and dopa (**118**), dopamine (**119**), and tyramine (**115**) only serve as the source of ring C with attached ethanamine residue. This unit is clearly a phenethylamine and the results of feeding various compounds of this type indicate that this unit is trioxygenated but monomethylated, *i.e.* (**122**), on combination with the other C_6–C_2 unit. Although these results suggest a set of possible benzylisoquinoline and bisphenethylamine precursors, no further results have yet been reported.

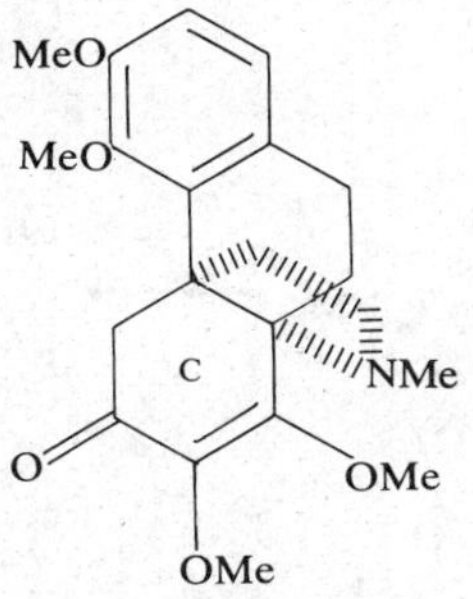

(**224**)

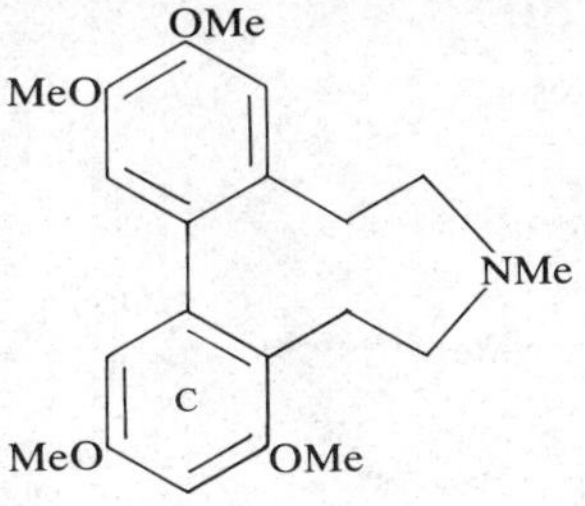

(**225**)

30.1.4.8 Phenethylisoquinoline alkaloids: colchicine

Most remarkable in the unique structure (**232**) for colchicine is the tropolone ring. An unusual biogenesis might be expected, therefore, for this non-basic alkaloid, and although this is true in part, it is also true that the pathway to (**232**) has fairly orthodox features.[187]

Phenylalanine, by way of cinnamic acid, serves in *Colchicum* species as the source of ring A plus carbon atoms 5, 6, and 7.[188] The seven carbons of the tropolone ring are generated from tyrosine with loss of C-1 and C-2, labels from [4′-^{14}C]- and [3-^{14}C]-tyrosine appearing at C-9 and C-12 of (**232**), respectively.[189] It follows that the tropolone ring arises by expansion of the aromatic ring of what was tyrosine, with inclusion of the benzylic carbon atom (a similar ring expansion has been deduced for the fungal tropolones, *e.g.* stipitatonic acid, although the origins lie in acetate[190]).

Further progress in unravelling colchicine biosynthesis depended on an unsuspected structural relationship that between the alkaloid and androcymbine which had been isolated from a relative of *Colchicum.* Assignment was made of the dienone structure (**228**) to androcymbine, which could then be thought of as arising from the phenethylisoquinoline skeleton [as (**226**)], which was unknown at the time. A similar possible biogenesis for colchicine (**232**) followed.[191] Plausibly the transformation of the androcymbine skeleton to that of colchicine could occur by hydroxylation to give (**229**). Homoallylic ring expansion and subsequent minor modification would yield colchicine (Scheme 28).

The crucial test for this hypothesis lay in the examination of compounds of the type (**226**) and (**228**) as colchicine precursors. In the event, *O*-methylandrocymbine (**227**) was a spectacularly good precursor for colchicine and the phenethylisoquinoline called autumnaline [as (**226**)] was an efficient and specific precursor for the alkaloid.[192] (This result has recently been affirmed in an experiment with [^{13}C]autumnaline.[193]) Extended exploration with various phenethylisoquinoline precursors allowed further definition of the pathway as that shown in Scheme 28;[192] only the (*S*)-isomer of autumnaline (**226**) with the same absolute configuration as colchicine (**232**) is involved with biosynthesis and oxidative coupling of this base occurs in a *para–para* sense and not by *ortho–para* coupling.[194]

Examination of the final stages of colchicine biosynthesis showed that demecolcine (**230**) and deacetylcolchicine (**231**) were involved.[194]

30.1.4.9 Homoaporphines

The phenethylisoquinoline skeleton seen above in colchicine makes a more orthodox appearance in the homoaporphines, *e.g.* floramultine (**235**). As their name indicates, a

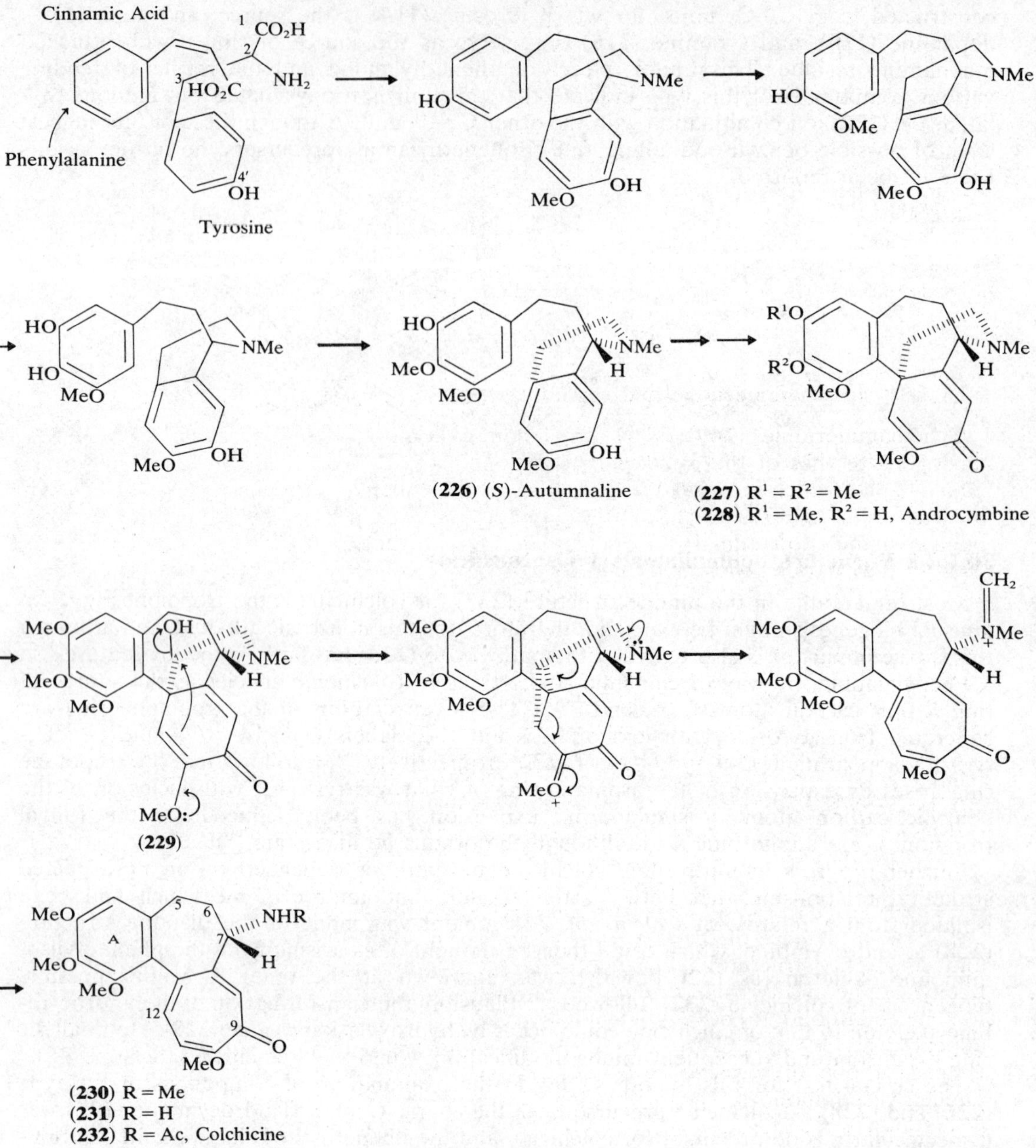

(226) (S)-Autumnaline

(227) $R^1 = R^2 = Me$
(228) $R^1 = Me$, $R^2 = H$, Androcymbine

(229)

(230) R = Me
(231) R = H
(232) R = Ac, Colchicine

SCHEME 28

biogenesis analogous to the aporphine alkaloids (Section 30.1.4.3) is expected, *i.e.* through a phenethylisoquinoline. Accordingly the *Kreysigia multiflora* bases, floramultine (**235**), multifloramine (**236**), and kreysigine (**237**), have been found to derive from autumnaline (**233**), the colchicine precursor (see above). By contrast, the base (**239**) was poorly utilized, confirming that these alkaloids arise by direct oxidative coupling on (**233**), with floramultine (**235**) presumably the first alkaloid to be formed. Dienone intermediates are not involved, but it is interesting to note that the dienone kreysiginone (**238**) and its dihydro-derivative have been isolated from *K. multiflora.* They are manifestly derived from (**234**), which was found to be a further precursor for the homoaporphines in this plant.[195]

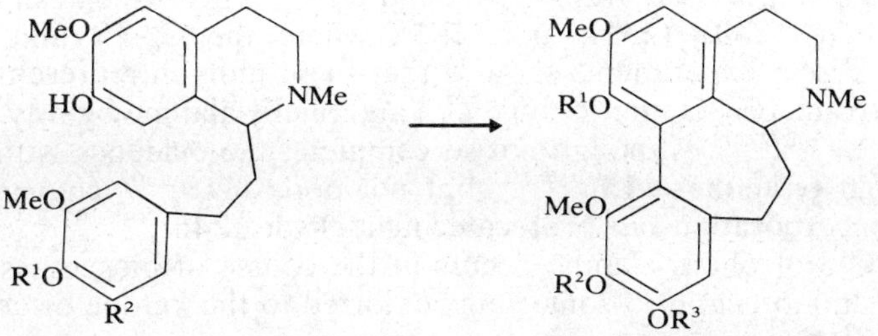

(**233**) $R^1 = Me$, $R^2 = OH$
(**234**) $R^1 = R^2 = H$

(**235**) $R^1 = R^3 = H$, $R^2 = Me$
(**236**) $R^1 = R^2 = H$, $R^3 = Me$
(**237**) $R^1 = H$, $R^2 = R^3 = Me$

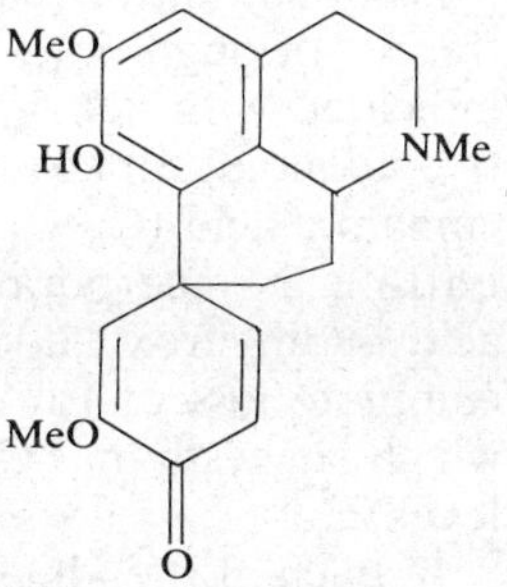

(**238**)

30.1.4.10 Schelhammeridine and cephalotaxine

Schelhammeridine (**240**) is by inspection a homo-*Erythina* alkaloid. A biogenesis analogous to that of the *Erythrina* bases (Section 30.1.4.6) might be expected for this alkaloid, one which in particular passed through a phenethylisoquinoline [as (**233**)]. The incorporations of tyrosine (label from C-2 appeared at the asterisked site in the alkaloid), phenylalanine, dopamine, and cinnamic acid support this hypothesis in a preliminary way[196] (they parallel those obtained for colchicine above).

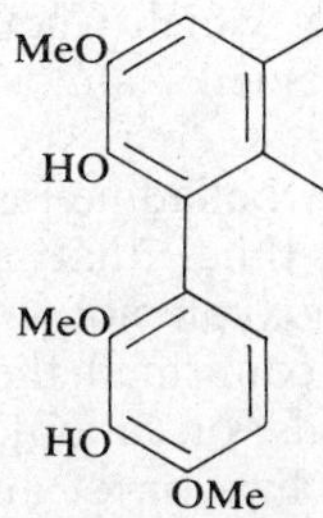

(**239**)

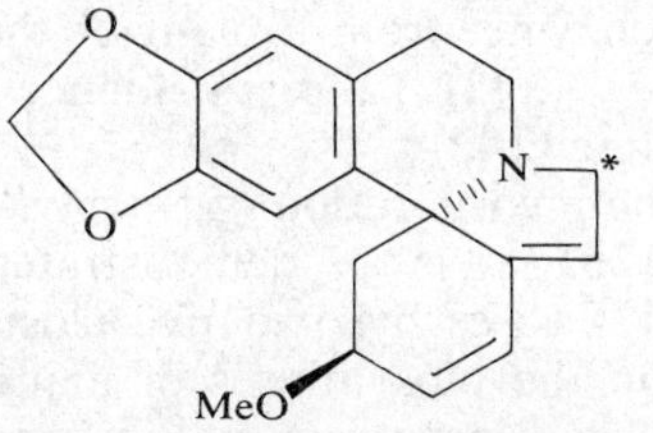

(**240**) Schelhammeridine

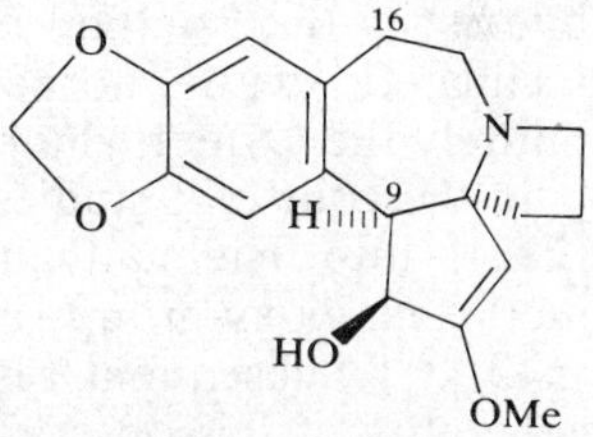

(**241**) Cephalotaxine

The co-occurrence of cephalotaxine (**241**) with alkaloids of the schelhammeridine type and apparent structural kinship suggests a similar origin for cephalotaxine from phenylalanine and tyrosine through a phenethylisoquinoline intermediate. Tyrosine (**114**) was indeed found to be a precursor, but both the labelling pattern (label from C-3 appeared at C-9 and C-16) and the fact that two molecules of tyrosine are involved, instead of one of tyrosine and one of phenylalanine, rules out a pathway involving phenethylisoquinolines.[197]

30.1.5 AMARYLLIDACEAE AND MESEMBRINE ALKALOIDS

30.1.5.1 Amaryllidaceae alkaloids

The structures found amongst the Amaryllidaceae alkaloids are diverse, although they may be classified into three main groups represented by (**248**), (**251**), and (**255**). The quite brilliant recognition[2] that these diverse alkaloids could arise simply by different modes of phenol oxidative coupling (Section 30.1.1) of a molecule of type (**242**) was of central importance to biosynthetic studies in this area, studies which were to prove the correctness of the original concept.

Examination of (**242**) indicates an assembly from C_6–C_2 and C_6–C_1 units, which can be traced through into the alkaloids (**248**), (**251**), and (**255**), where the C_6–C_2 unit is indicated with heavy bonding. Tracer experiments showed that these units in representative alkaloids all arose from tyrosine (C_6–C_2 unit only)[198–201] and phenylalanine by way of cinnamic acid (C_6–C_1 unit only).[198,202,203] Although not so complete, the evidence is that tyrosine is incorporated through tyramine (**115**),[198,202] but not dopa (**118**);[204] cinnamic acid is dihydroxylated before incorporation *via* protocatechualdehyde (**246**);[198,202,203,205] complete loss of tritium from C-3 of phenylalanine occurs in the course of biosynthesis, which suggests that C-3 of the amino-acid is at some stage oxidized to the ketone or acid level.[206]

In each of the alkaloids formed from these various precursors the siting of the label was completely in accord with the hypothetical pathway (Scheme 29), which involves norbelladine (**242**). Important evidence for the participation of (**242**) in biosynthesis came with the demonstration that radioactive norbelladine (**242**) was incorporated intact into lycorine (**248**), haemanthamine (**251**), and galanthamine (**255**), with labelling confined to the expected site in each alkaloid. Thus norbelladine is a precursor for each of the different groups of alkaloids.[198,200,201] Studies with norbelladine derivatives have shown that *O*-methylnorbelladine (**243**) is implicated in the biosynthesis of haemanthamine (**251**), norpluviine (**244**), and galanthine (**257**),[201,204] but not always in that of galanthamine (**255**) (positive results in one plant but negative in another).[201,207] On the other hand, the *N*-methyl derivatives (**247**) and (**249**) are involved only in galanthamine biosynthesis. Also shown in these studies is the uniqueness of the substituents in compounds of the norbelladine type for them to be involved in biosynthesis. Of further interest is the isolation of an enzyme, from a plant of the Amaryllidaceae, which when incubated with norbelladine (**242**) and *S*-adenosylmethionine yielded almost entirely the *O*-methylnorbelladine (**243**).[208]

In the course of studies on the incorporation of [*methyl*-^{14}C]-*O*-methylnorbelladine [as (**243**)] into haemanthamine (**251**) it was demonstrated, for the first time, that a methylenedioxy-group [in (**251**)] arises by oxidative closure of an *o*-methoxyphenol [in (**243**)].[201] Subsequent results on the biosynthesis of many alkaloids have confirmed the generality of this observation. The mechanism may be one involving radicals, or cationic species (Scheme 30). A related oxidative cyclization is apparent in the cyclization of an *N*-methyl group on to an aromatic nucleus in the conversion of an isoquinoline into a protoberberine (Section 30.1.4.5).

It is to be noted that the norbelladine derivatives (**243**) and (**249**), which are intermediates in biosynthesis, have phenolic groups in the appropriate positions for oxidative coupling *ortho* or *para*, consistent with hypothesis[2] (Scheme 29). Appropriate *ortho–para* coupling on (**249**) leads to the dienone (**253**), which as the ether (**254**) formed simply from (**253**) is a naturally occurring alkaloid, narwedine. Moreover, narwedine (**254**) itself is an efficient and specific precursor for galanthamine (**255**). This alkaloid is in turn a precursor for chlidanthine (**256**).[209]

Para–ortho coupling on (**243**) leads through to norpluviine (**244**). This compound is, like (**243**), an efficient precursor for lycorine (**248**), from which the sequence of biosynthesis follows in which the extra hydroxy-group in lycorine arises by allylic oxidation on (**244**).[198] Incorporation of (**243**) [labelled with tritium as shown in (**258**)] into norpluviine (**244**) occurred with retention of two of the four tritium atoms, one of which was located at C-2. The simplest pathway [(a), Scheme 31] between these two bases requires retention of three tritium atoms. Therefore, modification is necessary so that the tritium label at C-11b (norpluviine) can be lost, *e.g.* through (**259**) [path (b)].[204]

The fate of tritium at C-5′ (≡C-3′) of *O*-methylnorbelladine (**243**) on transformation into lycorine (**248**) has been examined in two plants, with completely opposite results. In the first study, tritium from (**243**) appeared in norpluviine (**244**) at C-2 with the β-configuration and was retained in the α-configuration of the lycorine subsequently formed. This indicates that hydroxylation occurs with inversion of configuration, a very unusual course for this biological reaction and the epoxide (**260**) has been proposed as an

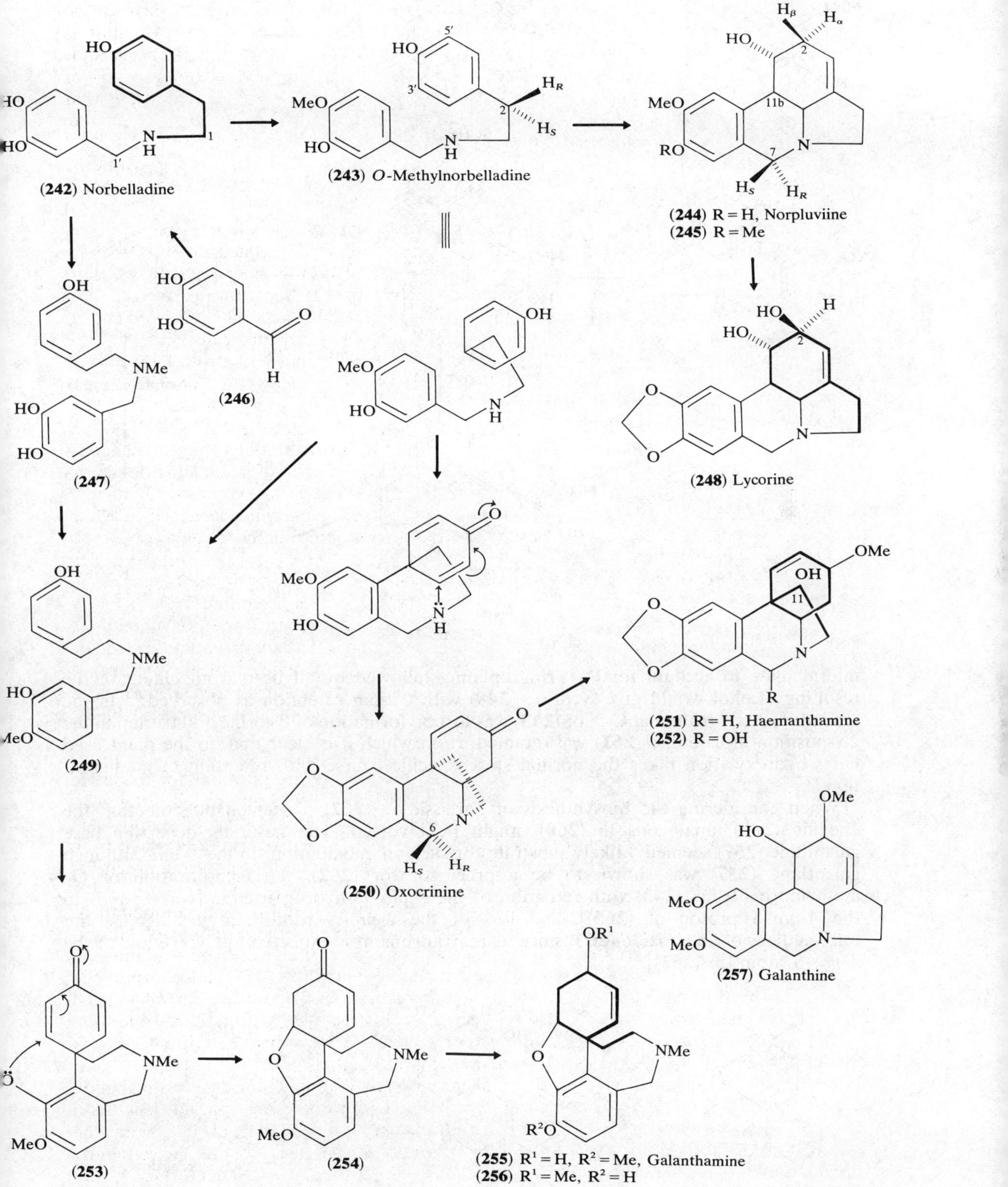
(242) Norbelladine
(243) O-Methylnorbelladine
(244) R = H, Norpluviine
(245) R = Me
(246)
(247)
(248) Lycorine
(249)
(250) Oxocrinine
(251) R = H, Haemanthamine
(252) R = OH
(253)
(254)
(255) R¹ = H, R² = Me, Galanthamine
(256) R¹ = Me, R² = H
(257) Galanthine

SCHEME 29

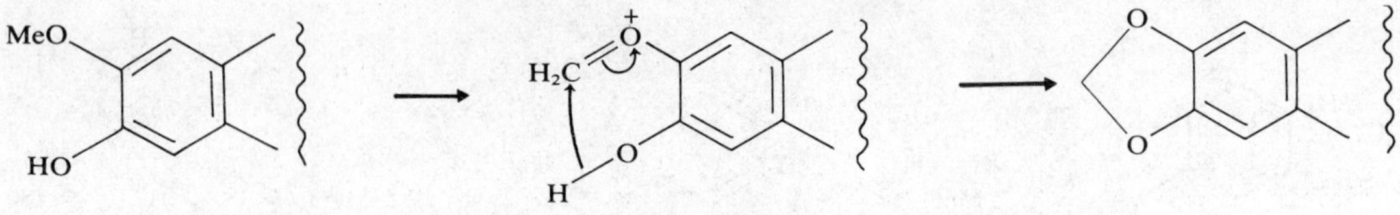

SCHEME 30

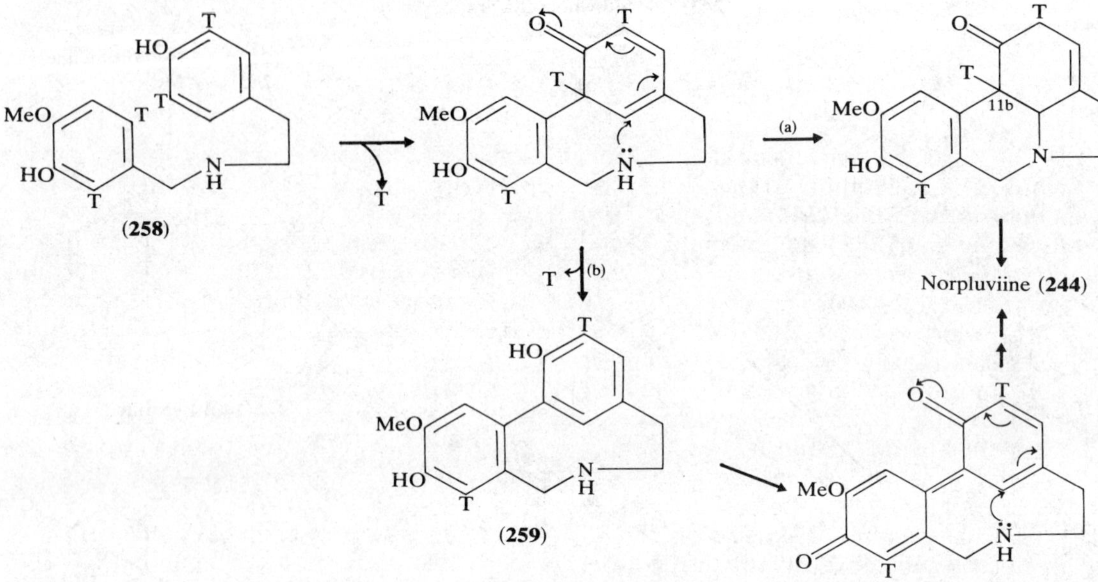

(258) (259)

SCHEME 31

intermediate to account for this; ring opening followed by allylic rearrangement of the resulting alcohol would give lycorine (**248**) with tritium retention as observed.[80] In the second study,[210] tritium at C-5′ of (**243**) was lost on formation of lycorine and tritium in the 2α-position in caranine (**261**) was retained, from which it is clear that, in the plant used here, hydroxylation takes the normal stereospecific course with retention of configuration.[177]

When considering the biosynthesis of narcissidine (**262**) it seemed possible that the epoxide (**263**), analogous to (**260**), might be involved. The naturally occurring base galanthine (**257**) seemed a likely substrate for such an epoxidation. In the event, although galanthine (**257**) was shown to be a precursor for (**262**), the incorporation of *O*-methylnorbelladine (**243**) with retention of the 2-*pro-R* hydrogen atom [corresponds to the 4-*pro-S* proton of (**263**)], and loss of the 2-*pro-S* proton, argues against the intermediacy of an epoxide (**263**) since its rearrangement is expected to involve loss of the 4-*pro-S* atom [in (**263**)].[211]

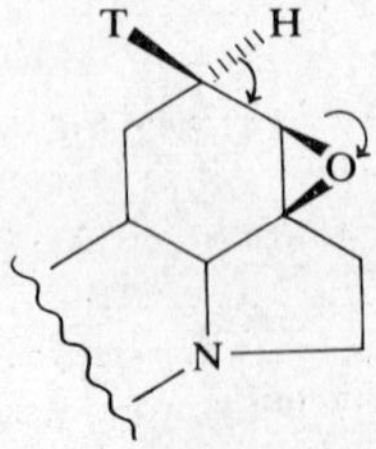

(**260**)

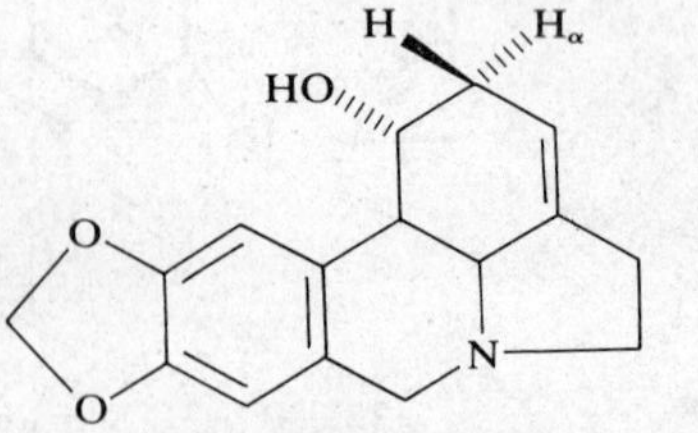

(**261**)

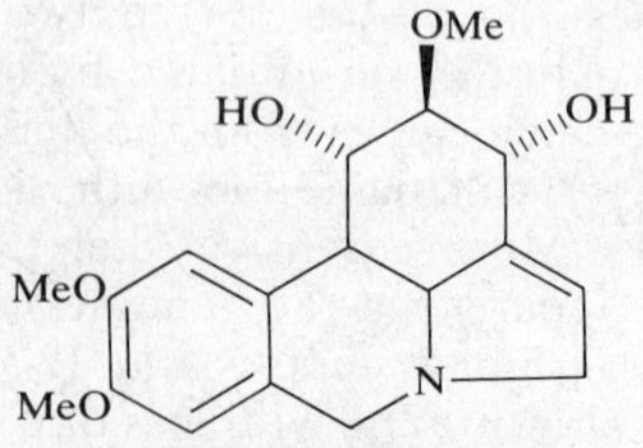

(**262**) Narcissidine

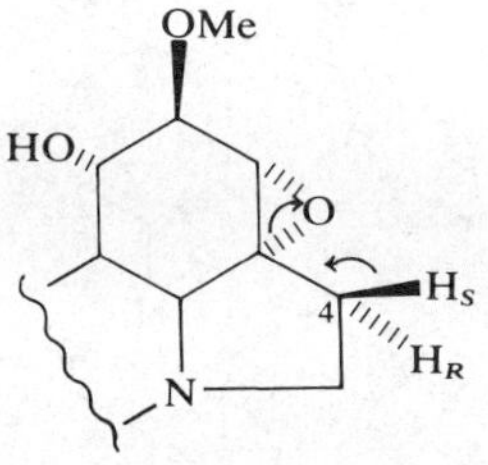

(263)

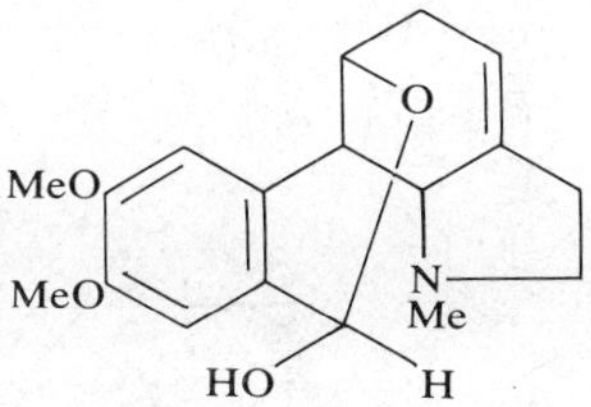

(264) Lycorenine

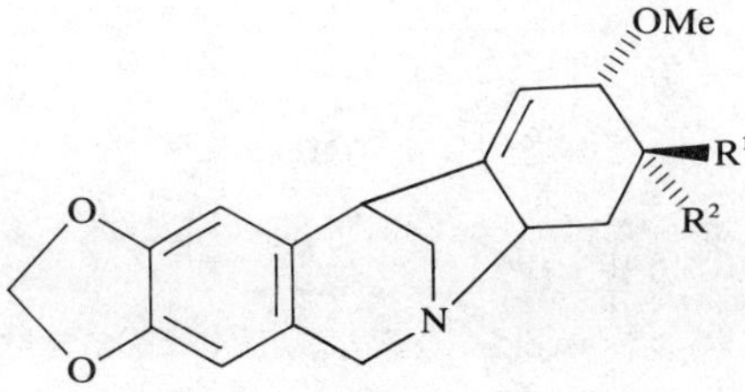

(265) $R^1 = H$, $R^2 = OH$, Manthine
(266) $R^1 = OH$, $R^2 = H$, Montanine

The biological conversion of protocatechualdehyde (**246**) into lycorenine (**264**) formally involves the addition of hydrogen to the aldehyde carbon in the formation of *O*-methylnorbelladine (**243**) and subsequent removal on introduction of a hydroxy-group on to a pluviine (**245**) intermediate leading to lycorenine (**264**). Results of detailed and elegant experiments using tritium-labelled precursors have shown that proton addition and removal is stereospecific and it is the hydrogen atom added initially [to the *re*-face of (**246**)] which is later removed [*i.e.* the 7-*pro-R* hydrogen in (**244**)].[212,213]

The third group of alkaloids which arise from norbelladine (**242**), this time by *para–para* phenol oxidative coupling, is exemplified by haemanthamine (**251**), in the biosynthesis of which, compounds of the type (**250**) are involved. Haemanthamine (**251**) displays an extra hydroxy-group at C-11 which has been shown[214] to arise by a hydroxylation reaction in which normal retention of configuration is observed[177] (see also refs. cited in Ref. 214).

Haemanthamine (**251**) is a precursor for haemanthidine (**252**) and the conversion of protocatechualdehyde (**246**) through the sequence (**243**) → (**251**) → haemanthidine (**252**) involves stereospecific hydrogen addition to, and removal from, what begins as the aldehyde carbonyl, the hydrogen added being later removed.[212] The course of biosynthesis parallels then that observed for lycorenine (**264**) (above).

The chemical conversion of compounds of the haemanthamine (**251**) type into alkaloids of type (**265**) suggests that alkaloids like manthine (**265**) and montanine (**266**) may be biosynthesized in a similar way. The potential relationship between (**251**) and these bases is supported to some extent by the observation[215] that *O*-methylnorbelladine (**243**) is a precursor for (**265**) and (**266**), but in the course of biosynthesis the 2-*pro-S* proton is lost from (**243**), indicating that functionalization at C-11 of a putative intermediate of the (**251**) type occurs with opposite stereochemistry to that observed in the entry of the hydroxy-group in the biosynthesis of haemanthamine (**251**) (above).

The structure of the alkaloid narciclassine (**269**) is such that it can *a priori* arise either *via* a route involving intermediates of the norpluviine (**244**) type or one which involves, for example, oxocrinine (**250**). Tracer experiments have shown that the latter route is utilized and it involves as intermediates oxocrinine (**250**), vittatine (**267**), and those of the crinine [as (**267**)] type with a pseudo-axial hydroxy-group or carbonyl function at C-3 (formation of the methylenedioxy-group is independent of the particular oxidation level at C-3).[216] Loss of the two carbon bridge in the conversion of (**267**) into (**269**) could plausibly be initiated by a retro-Prins reaction on 11-hydroxyvittatine (**268**), support for which comes from the finding that (**268**) is an efficient precursor for narciclassine (**269**).[217] (*cf.* colchicine biosynthesis, Section 30.1.4.8, for a related fragmentation)

More extensive skeletal degradation than that seen in narciclassine biosynthesis is apparent in the generation of ismine (**270**), and tracer experiments have shown that it has a similar genesis *via* (**271**) and oxocrinine (**250**).[218] Incorporation of the former was shown to be with loss of C-12 (it does not become the *N*-methyl group of ismine therefore) and the 6-*pro-R* hydrogen atom. Interestingly, a similar *pro-R* hydrogen atom is lost in the conversion of (**251**) into (**252**) and of (**244**) into (**264**).

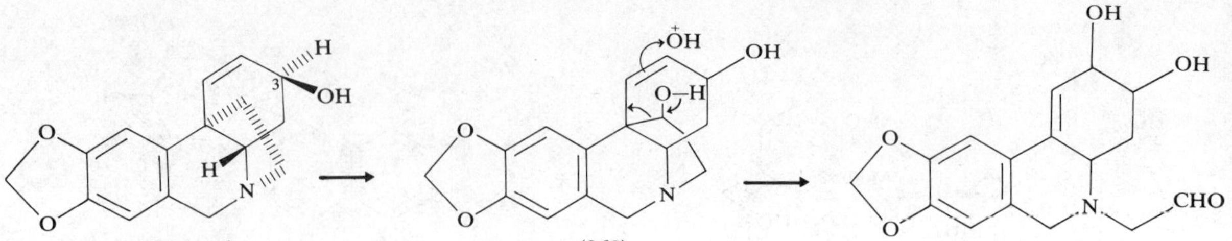

(**267**) Vittatine (**268**)

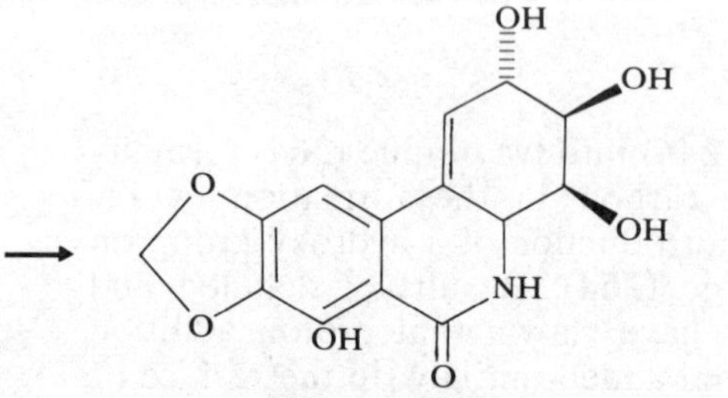

(**269**) Narciclassine

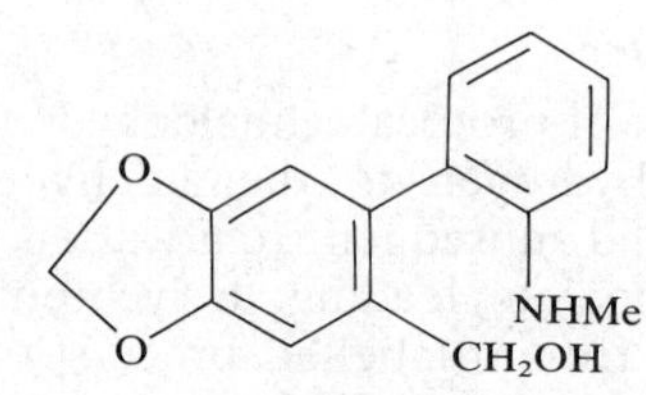

(**270**) Ismine

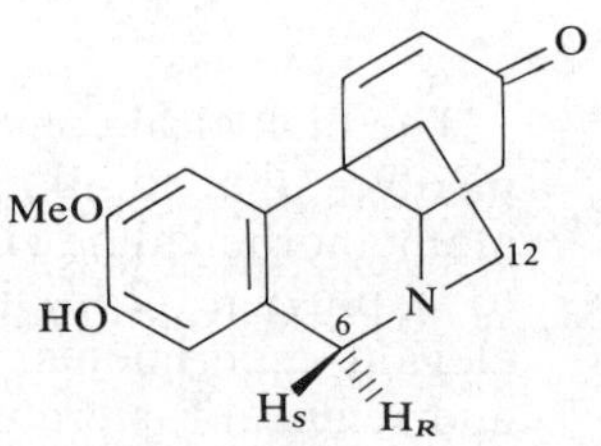

(**271**)

30.1.5.2 Mesembrine alkaloids

The mesembrine alkaloids, exemplified by mesembrine (**272**) itself, show a strong structural kinship with Amaryllidaceae alkaloids of the haemanthamine (**251**) kind. Careful investigation, however, has shown that the only aspect of biosynthesis common to these two groups of alkaloids is their origin in phenylalanine and tyrosine. A variety of derivatives of norbelladine (**242**), the key intermediate in Amaryllidaceae alkaloid biosynthesis, were only incorporated after fragmentation.[219]

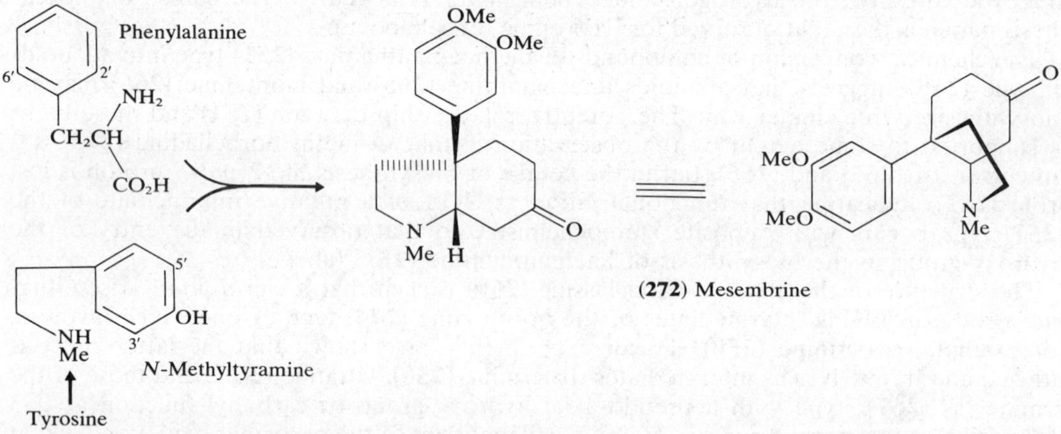

(**272**) Mesembrine

Tyrosine provides a C_6–C_2 unit *via* tyramine and *N*-methyltyramine, which accounts for the octahydroindole moiety in the mesembrine alkaloids.[220,221] The unusual, if not unique, mesembrine C_6 unit derives from the aromatic nucleus of phenylalanine.[220]

Although the steps of biosynthesis which lie beyond phenylalanine and tyrosine/*N*-methyltyramine remain largely unknown, results with tritiated samples of these precursors define in part the course of biosynthesis. Tritium at C-2′ and C-6′ in phenylalanine is retained on formation of mesembrine (**272**). This further excludes Amaryllidaceae intermediates of the crinine [as (**250**)] type, which would result in loss of one tritium atom. Further, bond formation between the phenylalanine and tyrosine derived units must occur at the carbon atom corresponding to C-1′ of phenylalanine. This leads logically to a

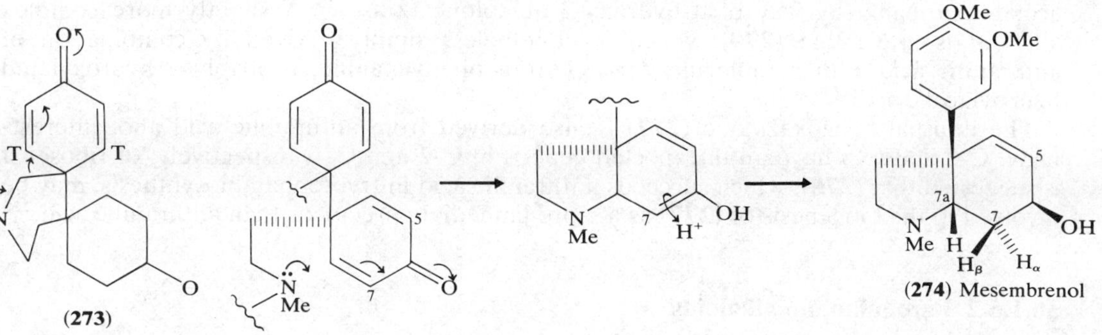

SCHEME 32

dienone of type (**273**), which can undergo fragmentation and aromatization as indicated without tritium loss.[219] Also to be considered is the cyclization at some point of nitrogen on to C-7a [see (**274**)]. Results with tyrosine, tyramine, and *N*-methyltyramine labelled with tritium at C-3′ and C-5′ have bearing on this. These precursors were incorporated with loss of half the label, residual tritium appearing at H-5 and H-7α. The results are consistent with the pathway shown (Scheme 32) involving internal conjugate addition of nitrogen followed by stereospecific α-protonation of the resulting enolate at C-7. Since residual tritium is equally distributed between C-5 and C-7α in mesembrenol (**274**), tritium loss must occur while the two sites are equivalent, *i.e.* before dienone formation.[221]

30.1.6 QUINOLINE AND RELATED ALKALOIDS

30.1.6.1 Simple anthranilic acid derivatives

The alkaloids discussed in Section 30.1.6 all derive in part from anthranilic acid, an intermediate in tryptophan biosynthesis. Damascenine (**275**) is such an alkaloid and, as apparently do other alkaloids in this section, it derives from anthranilic acid without the intermediacy of tryptophan. This follows from the incorporation of label from the carboxy-group of anthranilic acid into the methoxycarbonyl group of (**275**) (incorporation of anthranilic acid *via* tryptophan results in loss of the carboxy-group). 3-Methoxyanthranilic acid is a further intermediate in damascenine biosynthesis.[222]

The simple quinoline echinorine (**276**) has a simple genesis from anthranilic acid and

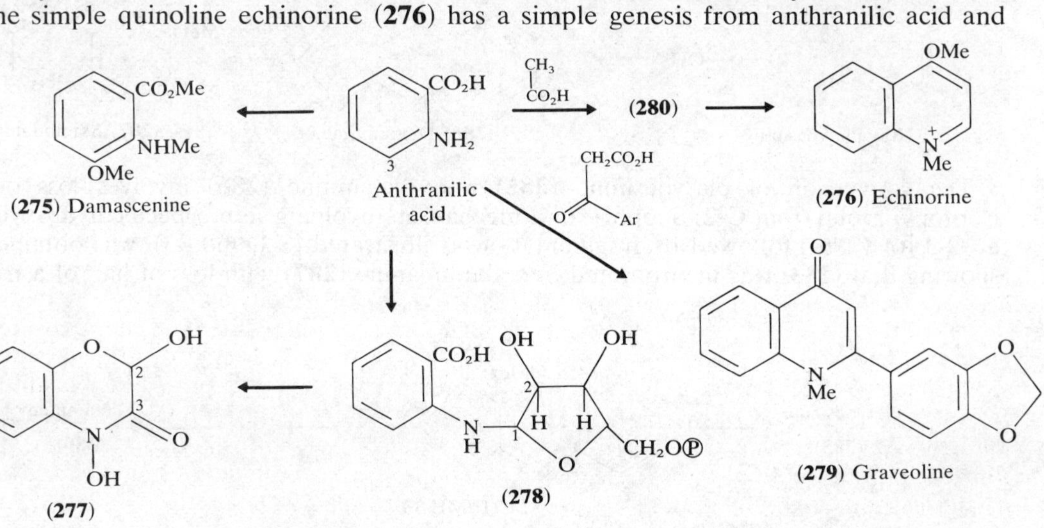

acetate, probably by way of 4-hydroxy-2-quinolone (**280**).[223] A slightly more complex alkaloid is graveoline (**279**), which is nonetheless simply derived by combination of anthranilic acid with a molecule derived from phenylalanine, probably a hydroxylated benzoylacetic acid.[224]

The unusual benzoxazinone (**277**) is also derived from anthranilic acid and, interestingly, C-2 and C-3 have an unexpected genesis in C-2 and C-1, respectively, of ribose. It is suggested that (**278**), which succeeds anthranilic acid in tryptophan biosynthesis, may be involved in the biogenesis of (**277**) as a more immediate precursor than anthranilic acid.[225]

30.1.6.2 Furoquinoline alkaloids

Furoquinoline alkaloids, *e.g.* dictamnine (**286**), are partly derived like echinorine (**276**) from anthranilic acid (but not tryptophan) and acetate, with methionine providing the methyl groups.[226] Carbon atoms 2 and 3 [see (**286**)] were not labelled by these precursors. The clue that they are mevalonoid in origin comes from the observation that bases like (**285**) are naturally occurring. Accordingly, satisfactory and specific incorporations into skimmianine (**287**) were recorded of mevalonic acid and dimethylallyl alcohol (**284**)[227] (label at C-4 and C-5 of mevalonic acid gave rise to alkaloid labelled at C-2 and C-3, respectively).

The results of further experiments established that 4-hydroxy-2-quinolone (**280**), (**282**), and platydesmine (**285**) are intermediates in the biosynthesis of alkaloids like dictamnine.[228,229] Prenylation of the quinoline skeleton may occur as well with a methoxy-group at C-4 as a hydroxy-group for both (**280**) and (**281**), and (**282**) and (**283**) are efficient precursors [(**283**) being proven to be incorporated without loss of its methyl group], but methylation at C-2 of (**281**) prevents utilization in biosynthesis.[229,230]

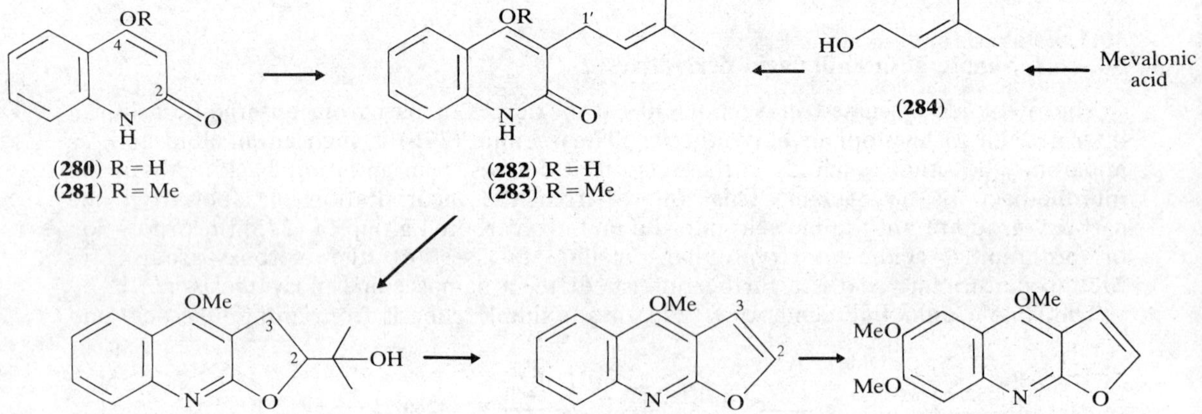

(**280**) R = H
(**281**) R = Me

(**282**) R = H
(**283**) R = Me

(**284**)

(**285**) Platydesmine

(**286**) Dictamnine

(**287**) Skimmianine

The conversion of platydesmine (**285**) into dictamnine (**286**) involves loss of the isopropyl group from C-2. Support for a mechanism involving stereospecific hydroxylation at C-3 [of (**285**)] followed by fragmentation as illustrated (Scheme 33), was obtained by showing that (**283**) was incorporated into skimmianine (**287**) with loss of half of a tritium

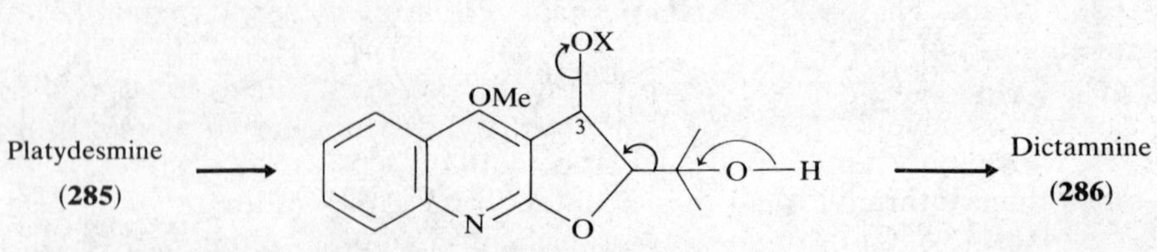

SCHEME 33

label sited at C-1′ [of (**283**)] (hydride abstraction may be substituted for hydroxyl loss in this scheme, but a ketone intermediate is excluded).[230]

The biosynthesis of furocoumarins involves a similar fragmentation reaction and there is a further similarity in the late hydroxylation of the benzene ring: skimmianine (**287**) is derived from dictamnine (**286**) (as are two more complex alkaloids choisyine and evoxine).[231]

Supporting evidence for the biosynthetic pathway to the furoquinoline alkaloids gleaned in whole plants has come from an investigation using cell suspension cultures of *Ruta graveolens*, which produces these alkaloids and edulinine (**288**). In particular, (**280**) was

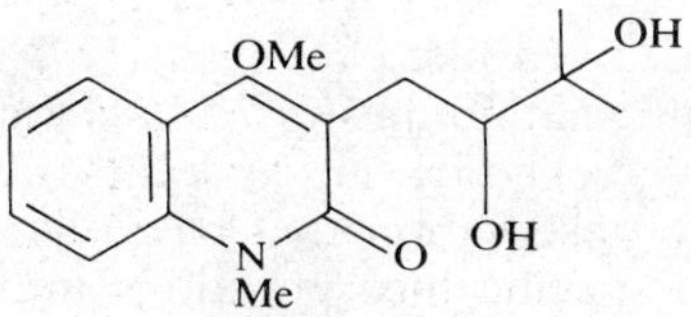

(**288**) Edulinine

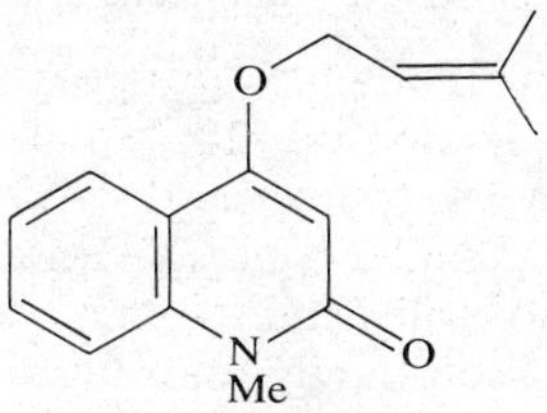

(**289**)

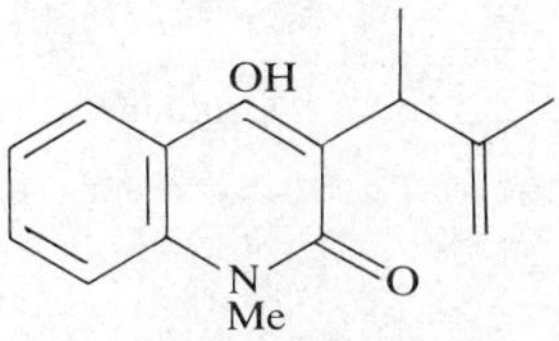

(**290**)

deduced to be at the branch point for the biosynthesis of the two alkaloid types, with *N*-methylation of (**280**) causing diversion from furoquinoline to edulinine biosynthesis, a conclusion which may extend to whole plants. The pathway to edulinine (**288**) also involves the *N*-methyl derivatives of (**282**) and (**283**).[232]

Ravenine (**289**) is the precursor for a different type of alkaloid, ravenoline (**290**). The conversion plausibly occurs through a Claisen rearrangement, which may be applicable to the biosynthesis of other alkaloids.[233]

30.1.6.3 Acridone alkaloids

Acridone alkaloids, *e.g.* melicopicine (**291**), are found in the same plant family (Rutaceae) as the furoquinolines and again anthranilic acid and acetic acid are implicated in biosynthesis, the latter precursor in all probability providing the whole of ring C. *N*-Methylanthranilic acid, 4-hydroxy-2-quinolone (**280**), and its *N*-methyl derivative are also implicated in acridone alkaloid biosynthesis.[234,235]

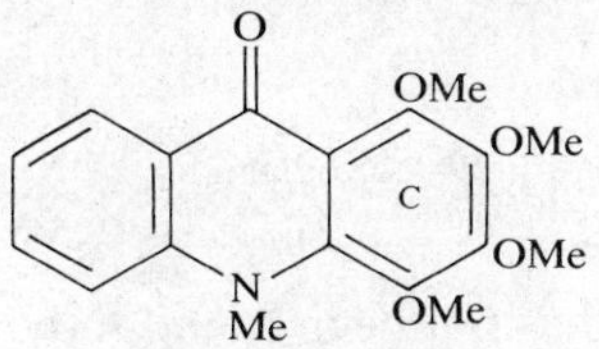

(**291**) Melicopicine

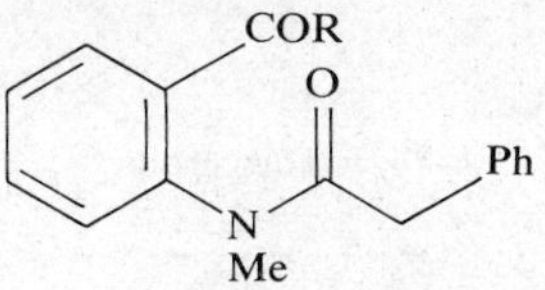

(**292**) R = OH
(**293**) R = NH_2

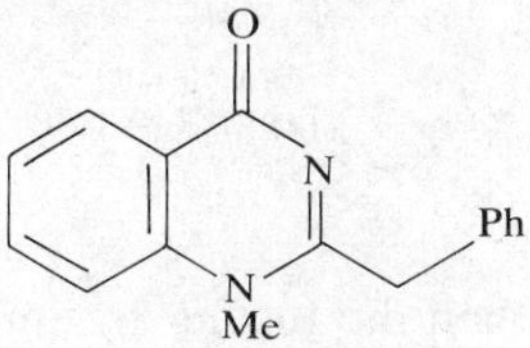

(**294**) Arborine

30.1.6.4 Quinazoline alkaloids

Arborine (**294**) is formed, like graveoline (**279**), from anthranilic acid and phenylalanine, but the nature of the intermediates, apart from *N*-methylanthranilic acid, is not beyond doubt. The best evidence perhaps is that phenylacetic acid, (**292**), and (**293**) are intermediates.[235,236]

More uncertainty is associated with the biosynthesis of peganine (≡vasicine) (**295**) since different pathways have been deduced[237,238] for the alkaloid in two plants from different families, in which only anthranilic acid is common.[239] In the first study, origin of the atoms, other than anthranilic acid, has been deduced to be ornithine (and putrescine)—both [2-^{14}C]- and [5-^{14}C]-ornithine gave specific and equal labelling of C-1 and C-10 of peganine.[237]

In the second study, evidence for the involvement of ornithine and its congeners could not be obtained. Instead, aspartic acid was implicated as a C_2 unit (C-1 and C-2 of peganine) and the remaining two carbon atoms originate from acetate, supported by the intact incorporation of *N*-acetylanthranilic acid.[238] Anthranoylaspartic acid (**296**) is also a fairly specific precursor for peganine, which suggests that the most important precursor may be (**297**).[240]

If both these pathways (Scheme 34) to peganine are valid, and this remains to be seen,

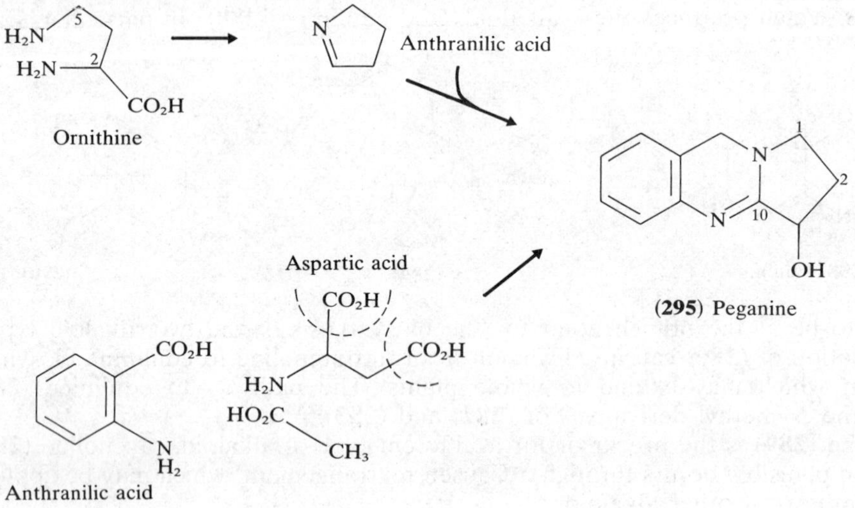

SCHEME 34

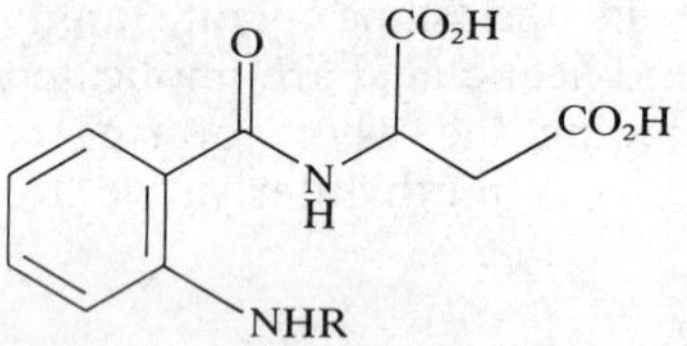

(**296**) R = H
(**297**) R = Ac

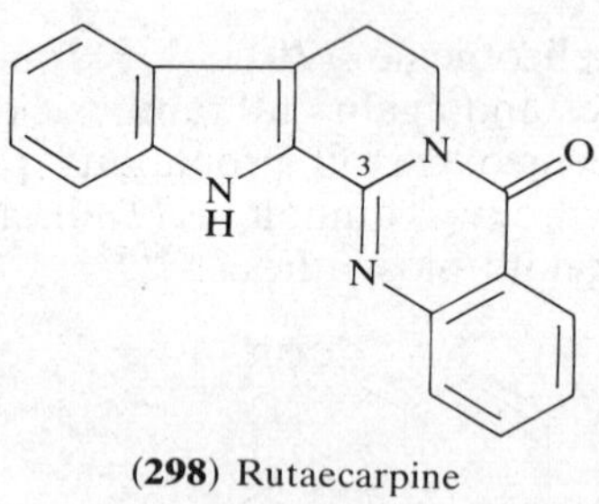

(**298**) Rutaecarpine

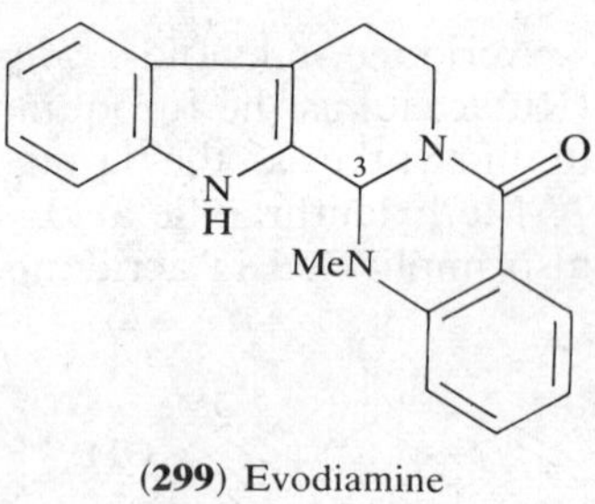

(**299**) Evodiamine

then the results are unique, for so far plant alkaloid biosynthesis has not been found to be species specific (*cf.* gramine biosynthesis, Section 30.1.7.1) unless one includes the radically different pathways to coniine (**61**) and *N*-methylpelletierine (**66**) (Section 30.1.3.1) and the differing pathways to the alkaloids discussed in Section 30.1.3.2.

The origins of the alkaloids rutaecarpine (**298**) and evodiamine (**299**) are more certain: both anthranilic acid and tryptophan are incorporated, with methionine supplying C-3 in both alkaloids in addition to the *N*-methyl group of (**299**).[241] It is a reasonable assumption that C-3 arises in the course of biosynthesis from an *N*-methyl group (*cf.* the origins of the 'berberine bridge' in Section 30.1.4.5).

30.1.7 INDOLE ALKALOIDS

The biosynthesis of Ergot and related alkaloids, which is not covered in this review, has been surveyed recently in detail.[242]

30.1.7.1 Simple indole deravatives

One of the simplest modifications of the important aromatic amino-acid tryptophan (**300**) is gramine (**301**), the indole alkaloid of barley. Study of its biosynthesis is a milestone in plant alkaloid biosynthesis because demonstration[243] that label from [3-^{14}C]tryptophan appeared specifically at the side-chain methylene of gramine was the first of the multitude of such experiments beyond studies of methylation, which form the basis of this review. It was rigorously established in subsequent experiments[244] that the indole nucleus of (**300**) and C-3 together with its two hydrogen atoms were incorporated as an intact unit into (**301**). The side-chain nitrogen atom also appears to arise from tryptophan,[245] and introduction of the methyl groups occurs stepwise from (**302**).[246] These observations are consistent with a plausible mechanism for gramine biosynthesis which involves a pyridoxal-linked adduct (see Scheme 35).[247] The incorporation of the tryptophan, amino-group is in agreement with the model provided that in the plant the nitrogen of (**304**) is transferred to (**303**).

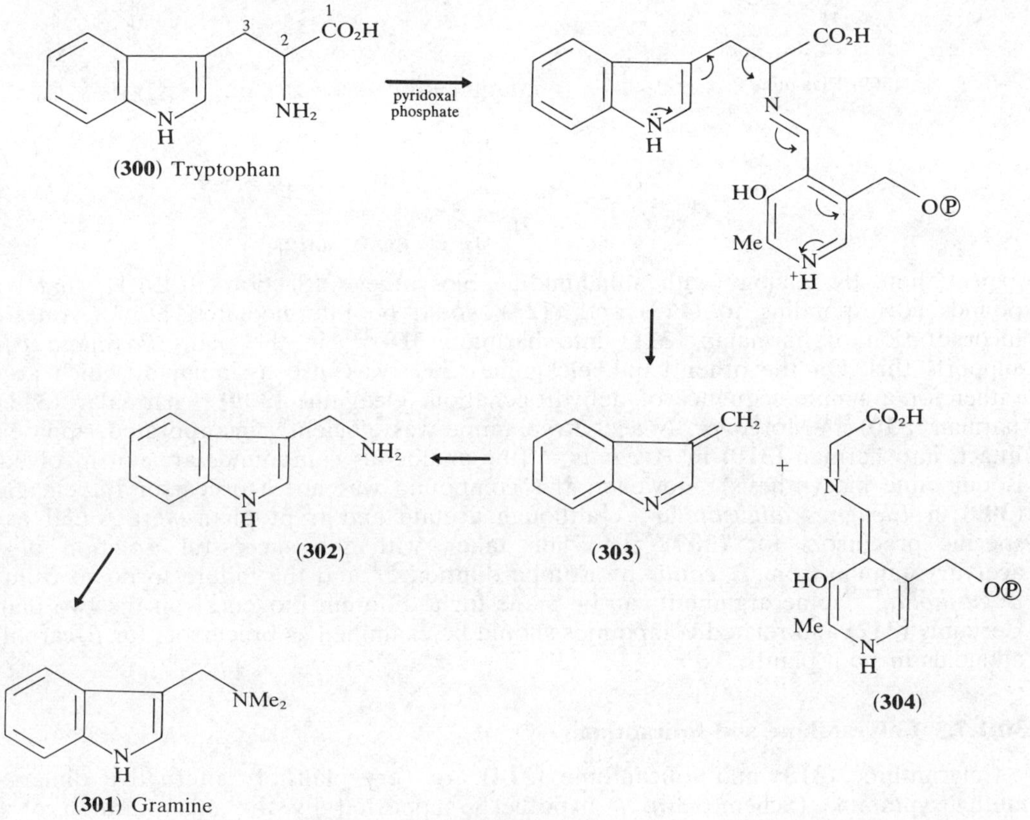

SCHEME 35

Recently it has been shown that the biosynthesis of gramine probably follows the same course in a plant of a different family to barley, from which one may conclude that the mode of gramine biosynthesis is species independent.[248]

Other simple indole derivatives are (**305**) and (**306**), and psilocybin (**308**). The biosynthesis of the latter has been deduced to be: tryptophan (**300**) → tryptamine → *N*-methyltryptamine → *N,N*-dimethyltryptamine → psilocin (**307**) → psilocybin (**308**).[249] The biosynthesis of (**305**) and (**306**) from tryptophan is more complex. Almost all the possible pathways (made up from alternative sequences of decarboxylation, hydroxylation, and *O*- and *N*-methylation) appear to operate.[250] Such results are a caution against the simplistic belief in a single biosynthetic pathway to secondary metabolites.

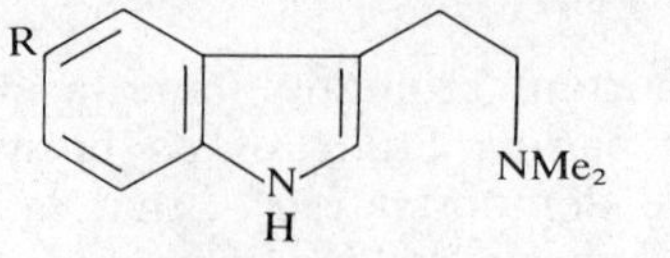

(305) R = H
(306) R = OMe

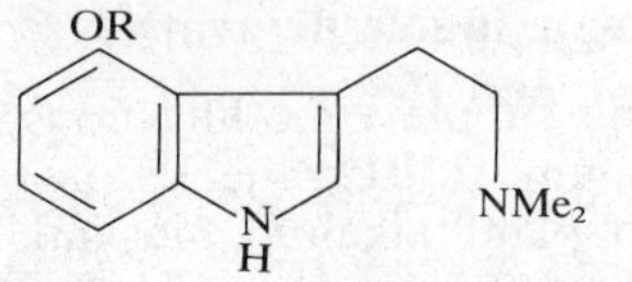

(307) R = H
(308) R = PO_3H_2, Psilocybin

30.1.7.2 β-Carboline alkaloids

There is an apparent structural resemblance between the β-carboline alkaloids, *e.g.* eleagnine (**309**), and the simple isoquinoline alkaloids, *e.g.* anhalonidine (**128**). One might

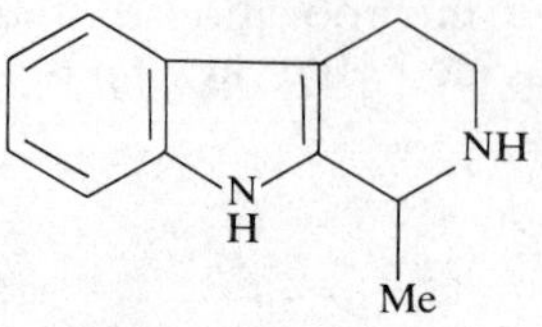

(309) Eleagnine

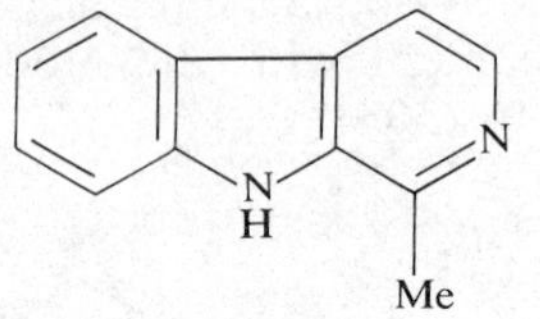

(310) Harman

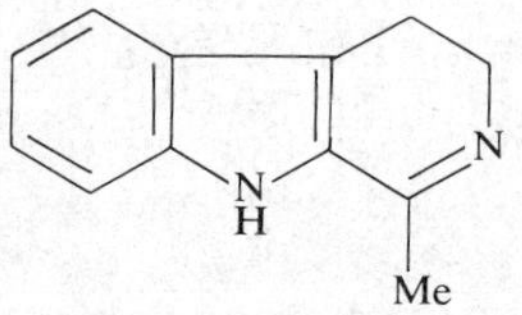

(311) Harmalan

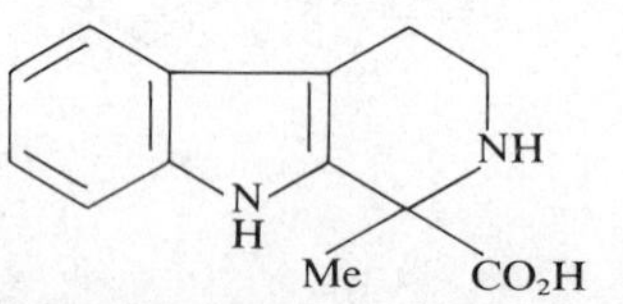

(312)

expect then, by analogy with anhalonidine biosynthesis (Section 30.1.4.1), that compounds corresponding to (**123**) and (**125**) would be intermediates. So far, only the incorporation of harmalan (**311**) into harman (**310**)[251] in the plant *Passiflora edulis* supports this. On the other hand, eleagnine (**309**) was also assimilated, which argues rather for a simple sequence of dehydrogenation: eleagnine (**309**) → harmalan (**311**) → harman (**310**).[251] Moreover, *N*-acetyltryptamine was efficiently incorporated, apparently intact, into harman (**310**) in *P. edulis*[251] (the analogous compounds are not involved in isoquinoline biosynthesis). However, this compound was not a precursor for eleagnine (**309**) in *Eleagnus angustifolia*[127] [although acetate and tryptophan were found to be specific precursors for (**309**)[252]]. When taken with the successful isolation of *N*-acetyltryptamine from *P. edulis* by isotope dilution,[251] and the failure to do so from *E. angustifolia*,[127] some argument can be made for a different biogenesis in the two plants. Certainly (**312**) and related compounds should be examined as precursors for β-carboline alkaloids in both plants.

30.1.7.3 Calycanthine and folicanthine

Calycanthine (**313**) and folicanthine (**314**) are very plausibly alternative dimers of methyltryptamine (Scheme 36), a hypothesis supported by the incorporation of [2-^{14}C]tryptophan [as (**300**)] into both alkaloids;[253,254] in the more simple case of folicanthine (**314**) the label was confined to the asterisked positions, as required.[253]

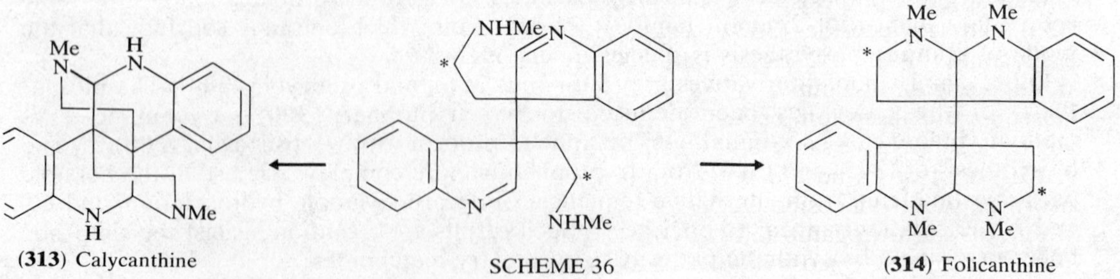

(313) Calycanthine SCHEME 36 **(314)** Folicanthine

30.1.7.4 Terpenoid indole alkaloids: *Corynanthé–Strychnos, Aspidosperma*, and *Iboga* bases

Rich structural variation within the family of terpenoid indole alkaloids brings the total number of bases to over 1000. The invariant tryptamine unit in association with a C_9–C_{10} unit of varying structure serves to identify the family, and where examined the tryptamine moiety has been found to be derived from tryptophan *via* tryptamine.[255–257] The structural variation found in terpenoid indole alkaloids is generally occasioned by variation in the C_9–C_{10} unit, and although at first sight this might seem bewildering, closer inspection reveals three main groups of bases: (a) *Corynanthé–Strychnos, e.g.* ajmalicine (**323**) and akuammicine (**324**) with C_9–C_{10} unit (**319**); (b) *Aspidosperma, e.g.* vindoline (**322**) with C_9–C_{10} unit (**320**); (c) *Iboga, e.g.* catharanthine (**321**) with a third C_9–C_{10} variation (**318**) (where only nine carbon atoms are present in the non-tryptamine residue it is invariably the carbon indicated by the dotted line in (**318**)–(**320**) which is lost).

Speculation on the origins of the ubiquitous C_9–C_{10} units has not stood the experimental test,[8,10] with one exception, namely the proposition that the units have a terpenoid origin.[247,258] It was envisaged that each C_9–C_{10} unit could be derived by scission of a cyclopentane ring of a monoterpene, with further modification as indicated in Scheme 37. Extensive, elegant, and rigorous experimentation has given results which prove the correctness of this hypothesis and which define the detail of biosynthesis. Thus (after early difficulties) it could be shown that the C_9–C_{10} unit of representatives of the three groups of alkaloids were each derived along a pathway involving the normal head-to-tail linkage of two mevalonate units passing through geraniol (**316**)/nerol (**325**)[259,260] and the key cyclic monoterpenoid, loganin (**329**).[261] Not only was loganin (**329**) a specific alkaloid precursor, but its presence and biosynthesis from geraniol in *Vinca rosea*, the plant used for most of the experiments, could also be demonstrated.[262]

In the incorporation of mevalonate it is the (3*R*)-isomer (**315**) and not its enantiomer which is involved,[260] consistent with observations in other systems. Further, incorporation of mevalonate (label at C-2 or C-3′) resulted in loss of identity between the two terminal positions (C-9 and C-10) of the stylized intermediate (**319**), and it is to be noted that a similar observation has been made in studies of the biosynthesis of cyclopentanoid monoterpenes.[263]

Further experiments established that deoxyloganin (**328**) is an intermediate between geraniol (**316**)/nerol (**325**) and loganin;[264] and the C-10 hydroxylated derivatives (**326**) and (**327**) of geraniol and nerol, respectively, are also to be included as earlier biosynthetic intermediates (Scheme 38).[263a,265] Failure of various other derivatives of geraniol and nerol to act as precursors restricts the range of possible intermediates beyond these terpenes, and leads to a plausible mechanism for cyclization (Scheme 39) which accounts for the observation that label passing from mevalonate through C-9/C-10 of (**326**) and (**327**) becomes equally divided between C-10 and C-9 of loganin (**329**) and the equivalent positions in the derived alkaloids.[263a]

Conversion of loganin (**329**) into the indole alkaloids must involve cleavage of the cyclopentane ring [*cf.* (**317**) → (**319**) in Scheme 37], which may be rationalized in terms of the mechanism shown in (**330**), to give secologanin (**331**). This terpene was shown to be a precursor for representative alkaloids and also a natural constituent of *V. rosea*, thus validating the hypothesis.[266] Further consideration of possible pathways leads to vincoside (**333**) and isovincoside (**332**), the products of condensation between tryptamine and secologanin (**331**).[257,267] Both bases were shown to be natural constituents of *V. rosea* and to derive from loganin,[257,268] but it is only vincoside (**333**) which is used for the subsequent elaboration of terpenoid indole alkaloids.[257] This is remarkable because isovincoside has the same configuration at C-3, as, say, (**323**) and (**334**), and accepting that bases of this kind are precursors for those of type (**321**) and (**322**) (see below), the incorporation of vincoside (**333**) must be associated with inversion of this centre.[267] Since tritium at C-5 of loganin (**329**) (corresponds to label at C-3 of vincoside) is retained on formation of post-vincoside alkaloids,[269] epimerization does not involve proton loss from

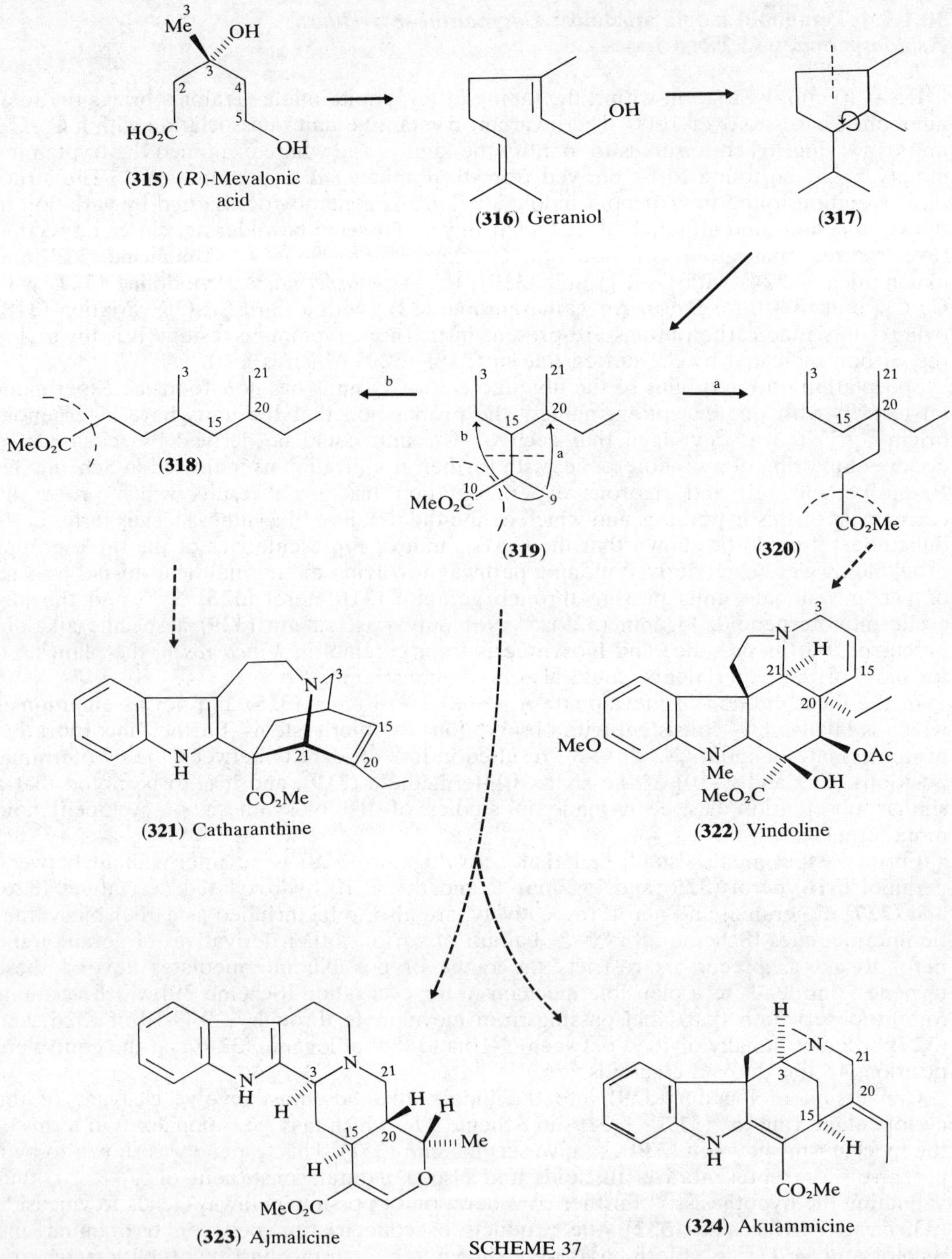

SCHEME 37

C-3 of (**333**) and must therefore arise through C—C or, more likely, C—N cleavage. [Note added in proof: results of recent experiments have shown unequivocally that it is *not* vincoside (**333**) but isovincoside (**332**) which is the true precursor for terpenoid indole alkaloids.[270]]

The formation of the *Corynanthé*-type bases, *e.g.* corynantheine (**335**) and ajmalicine (**323**), from vincoside requires no skeletal rearrangement, and is regarded as involving

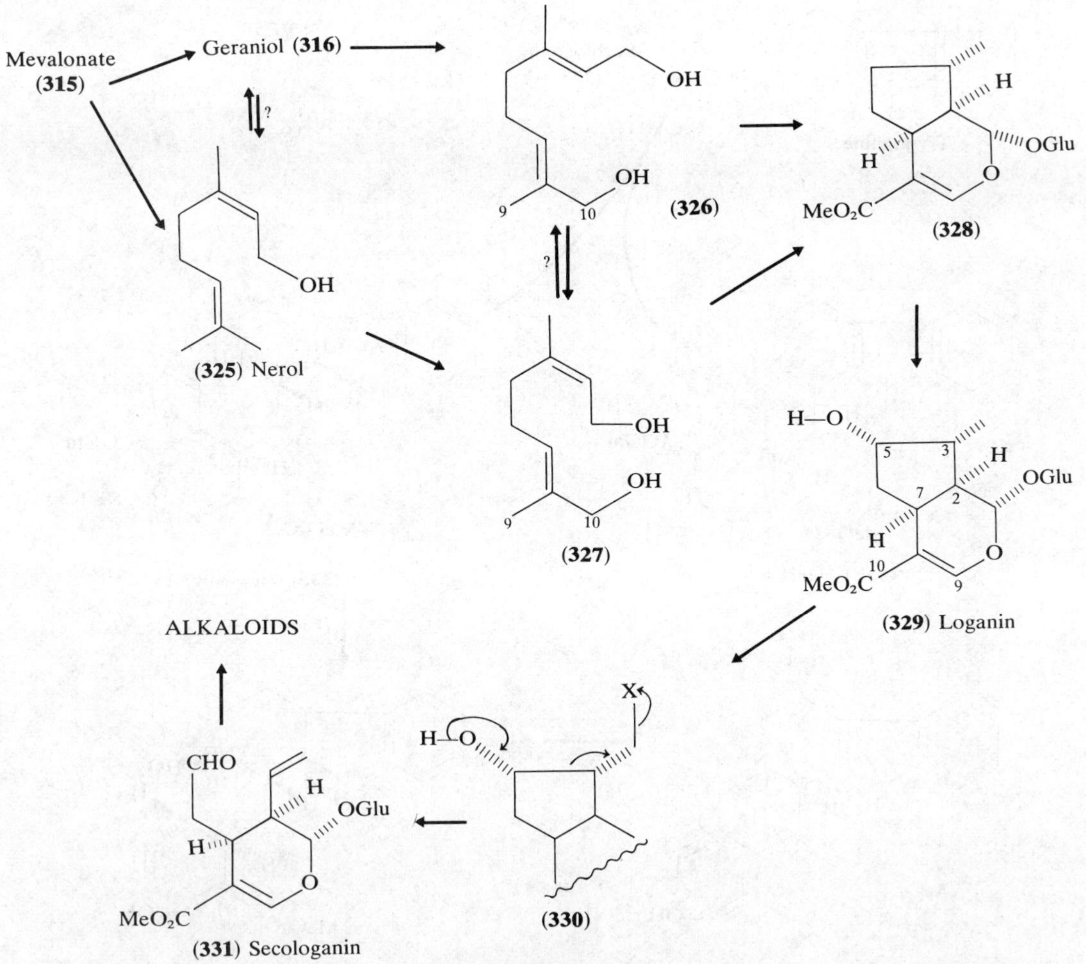

SCHEME 38

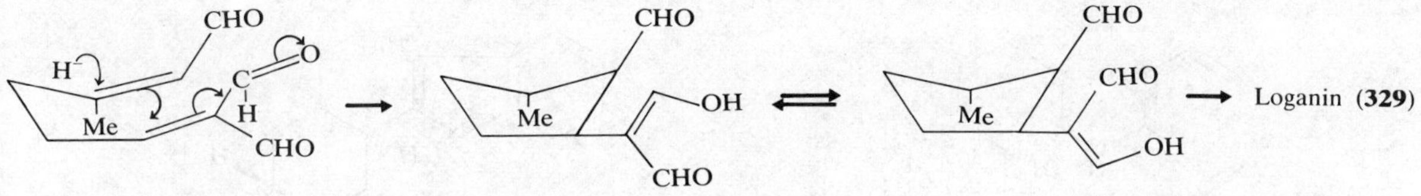

SCHEME 39

enzymatic cleavage of glucose as a first step with plausible following transformations (Scheme 40).[257] It is reasonable to regard this skeletal type as the precursor for alkaloids with rearranged skeleton. This is supported by the observation[271] that alkaloids with *Corynanthé*-type structures appear in *V. rosea* seedlings before bases of the *Aspidosperma* and *Iboga* type, and tracer experiments show that the key intermediate for the alkaloids with rearranged C_9–C_{10} units is the *Corynanthé* base geissoschizine (**334**). It is an intact precursor for representatives of the *Aspidosperma* and *Iboga* group of bases[272] and also for those with the *Strychnos* skeleton, represented by akuammicine (**324**)[272] and strychnine (**339**).[273] [Strychnine has the expected genesis in mevalonate *via* geraniol and in tryptophan, and the unusual C_2 unit (C-22 and C-23) arises from acetate.[274] A search[273]

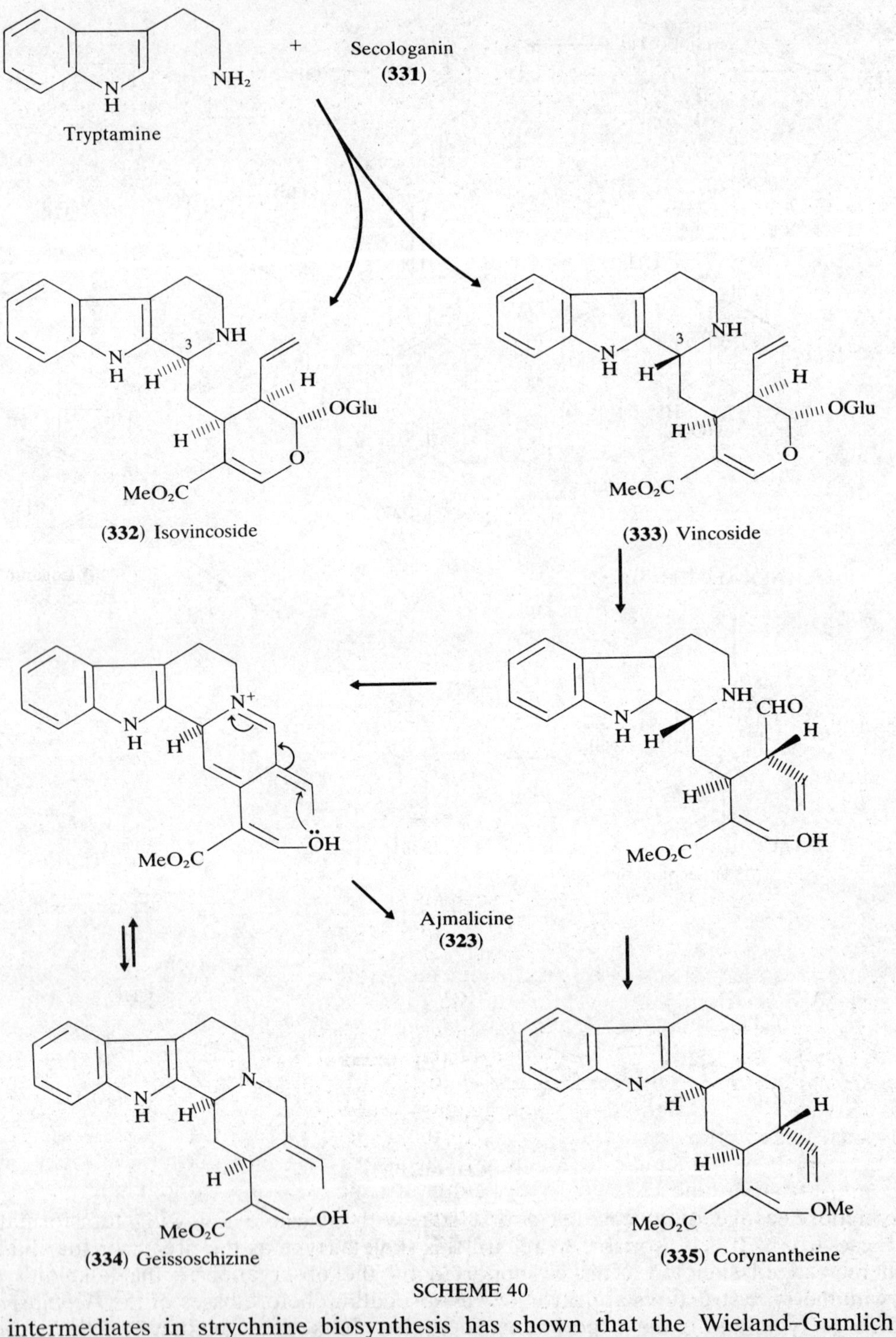

SCHEME 40

for intermediates in strychnine biosynthesis has shown that the Wieland–Gumlich aldehyde (**337**) is a precursor, and negative results with geissoschizal (**340**) and diaboline (**341**) have suggested that (**336**) and (**338**) may be intermediates (Scheme 41).]

From the results of orthodox feeding experiments and experiments where the sequence of alkaloid formation is obtained by noting the appearance of label in individual alkaloids in relation to time, it has been concluded[271,275] that stemmadenine (**343**), tabersonine

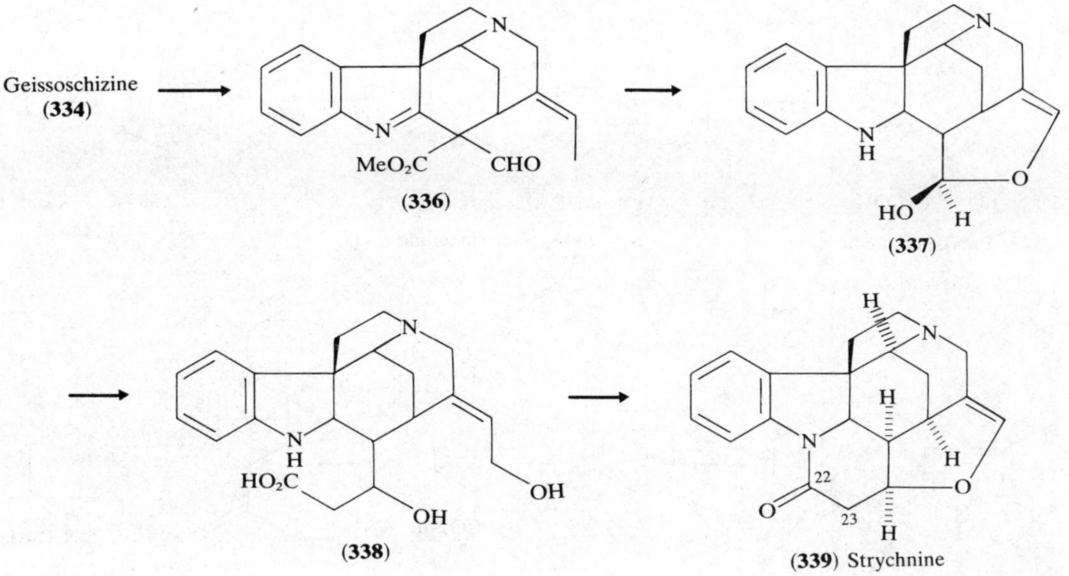

SCHEME 41

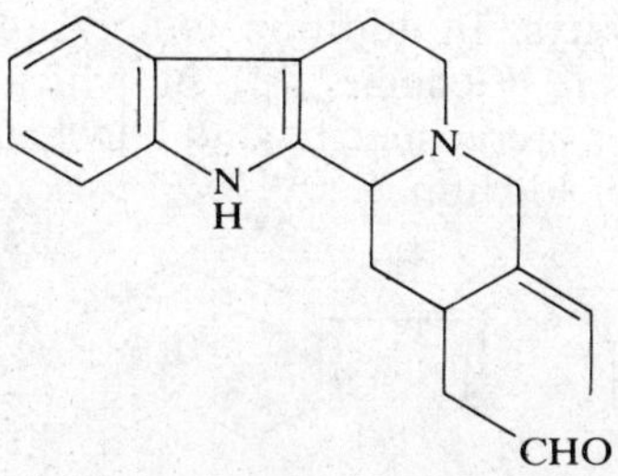

(340)

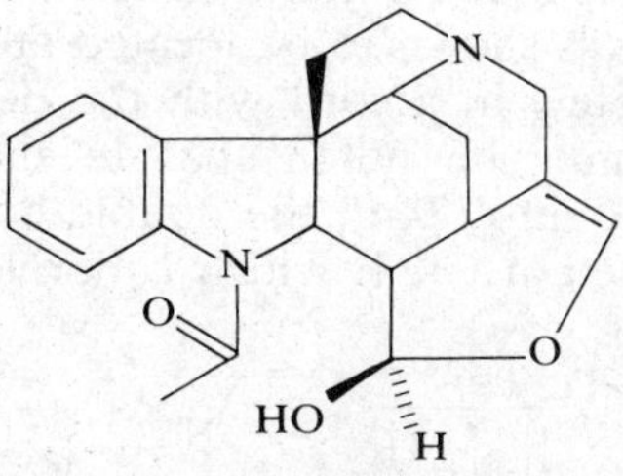

(341)

(**346**), and preakuammicine (**342**) are further biosynthetic intermediates of importance. It has also been concluded that stemmadenine (**343**) is an intermediate in catharanthine (**321**) (*Iboga* type) and vindoline (**322**) (*Aspidosperma* type) biosynthesis, being utilized by way of tabersonine (**346**). The formation of this alkaloid may be rationalized in terms of the pathway shown in Scheme 42. Central to this hypothetical pathway is the enamine (**345**) arrived at by fragmentation of stemmadenine (**343**). Although alternative cyclization paths from this intermediate could give both the *Iboga* type [as (**321**)] and *Aspidosperma* type [as (**346**)], such a simple idea is argued against by the incorporation of tabersonine (**346**) into catharanthine (**321**).[271] The validity of (**345**) as a biosynthetic intermediate is supported by the isolation of simple derivatives of secodine (**347**) from plants and by the specific incorporation of labelled secodine (**347**) into vindoline (**322**); loss of tritium from the dehydropiperidine ring of secodine on transformation into vindoline (**322**) is consistent with involvement of secodine in biosynthesis *via* a more highly oxidized intermediate [as (**345**)].[276]

Only brief reference has been made so far to the use of tritium labelling as an aid to studying the biosynthesis of alkaloids under the present heading. In fact tritium labelling has been quite extensively used here and the results provide a test of the conclusions so far reached. [In the following the numbers in brackets refer to the site of labelling in loganin (**329**).] Experiments were carried out with geraniol (C-1 and C-2), nerol (C-2),

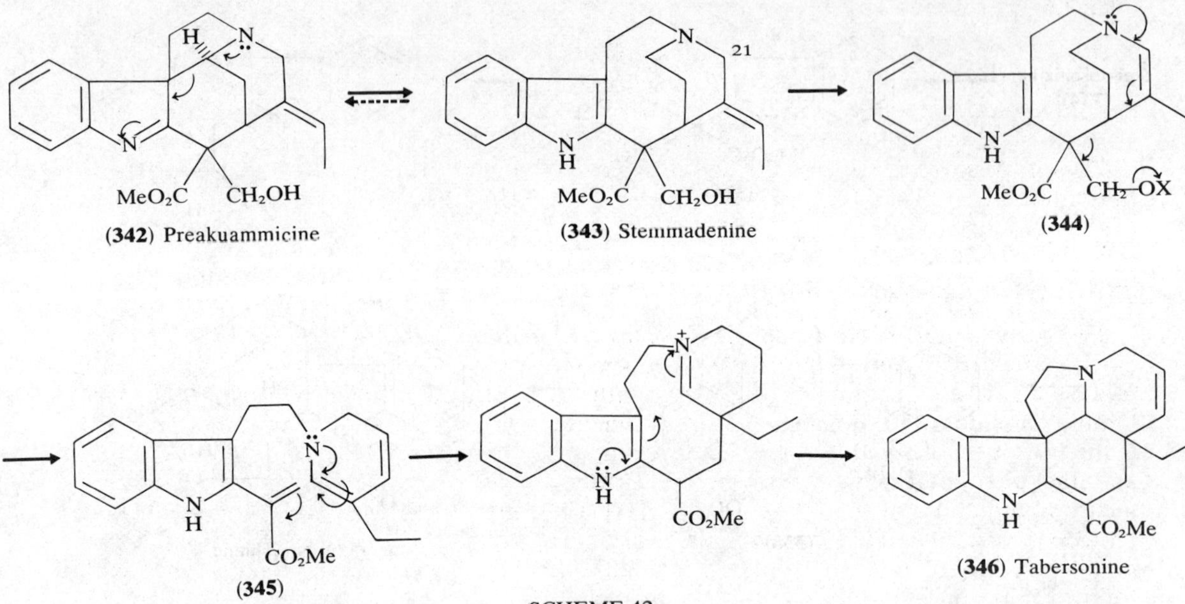

(342) Preakuammicine (343) Stemmadenine (344)

(345) (346) Tabersonine

SCHEME 42

mevalonate (C-2 and C-7), and loganin (C-5). The results are that tritium from C-1, C-5, and C-7 is retained in the formation of each alkaloid type, whereas that at C-2 is lost [the C-5 label was shown to be located at C-3 in (**321**), (**322**), and (**323**)]. These observations are completely in accord with the delineated pathways. In addition to the points made earlier it may be noted that in the model scheme (Scheme 42) in which (**343**) is transformed into (**344**), the double-bond migration is stereospecific and leaves unaffected the hydrogen at C-21, which is derived from C-1 of loganin.[260,269]

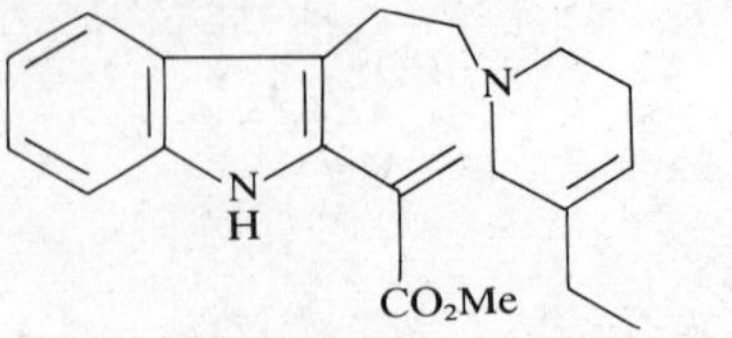

(**347**) Secodine

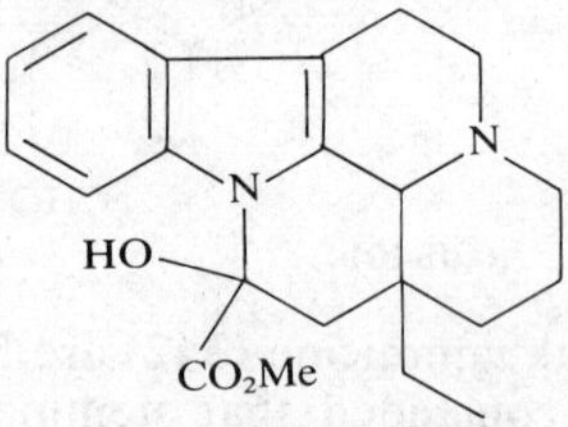

(**348**) Vincamine

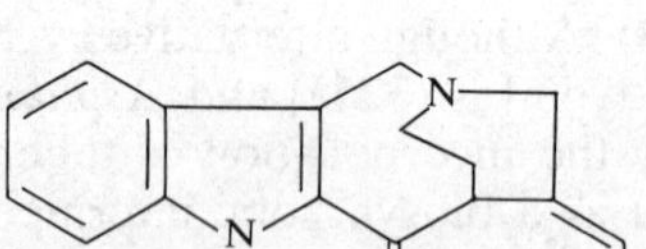

(**349**) Apparicine

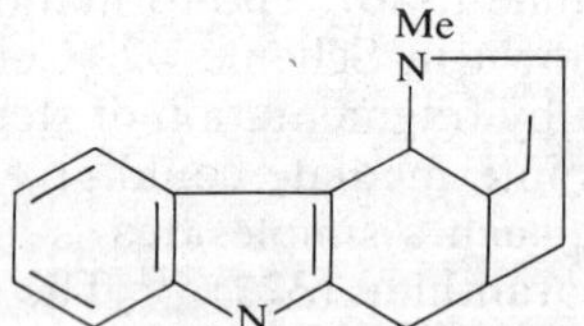

(**350**) Uleine

30.1.7.5 Vincamine

Support for the proposal that vincamine (**348**) is derived from the *Aspidosperma* skeleton has come in a provisional way with the reported incorporations of tryptophan, stemmadenine (**343**), and tabersonine (**346**).[277]

30.1.7.6 Apparicine and uleine

It is clear from the structures for apparicine (**349**) and uleine (**350**) that modification of the normal tryptamine side-chain has occurred in the course of biosynthesis. There is no definitive evidence on uleine biosynthesis,[278] but it appears that apparicine (**349**) does arise from tryptophan with the expected loss of C-2.[279] Stemmadenine (**343**) and secodine (**347**) are precursors, so rearrangement to the apparicine skeleton must be a late step in biosynthesis.[280]

30.1.7.7 *Cinchona* alkaloids

The early stages in the evidently complex biosynthesis of the *Cinchona* alkaloids, *e.g.* cinchonidine (**355**) and quinine (**356**), are similar to those delineated for *Corynanthé* and related alkaloids (Section 30.1.7.4): biosynthesis is proved to involve tryptophan, tryptamine, geraniol (**316**), loganin (**329**), and vincoside (**333**).[259a,281–283] It appears that (**351**) is the last of the *Corynanthé*-type intermediates [(**351**) but not (**352**) was incorporated][282] and the next intermediate may be (**353**)[282] since the corresponding alcohol was not incorporated and half the label from [1-$^3H_2$]tryptamine [labelling site corresponds to C-5 of (**351**)] was retained, thus excluding the corresponding carboxylic acid as an intermediate.[283] There is no information available on the steps of the rearrangement which follow (**353**), but biosynthesis proceeds plausibly along the pathway shown in Scheme 43. Cinchonidinone (**354**) is an important late intermediate in this scheme, and its presence in *Cinchona* plants has been demonstrated, as has its conversion into *Cinchona* alkaloids.[283] Other results indicate the reversibility of the conversion (**354**) → (**355**), the probability of aromatic hydroxylation occurring on (**354**), and the accessible epimerization at C-8 in this ketone.[283]

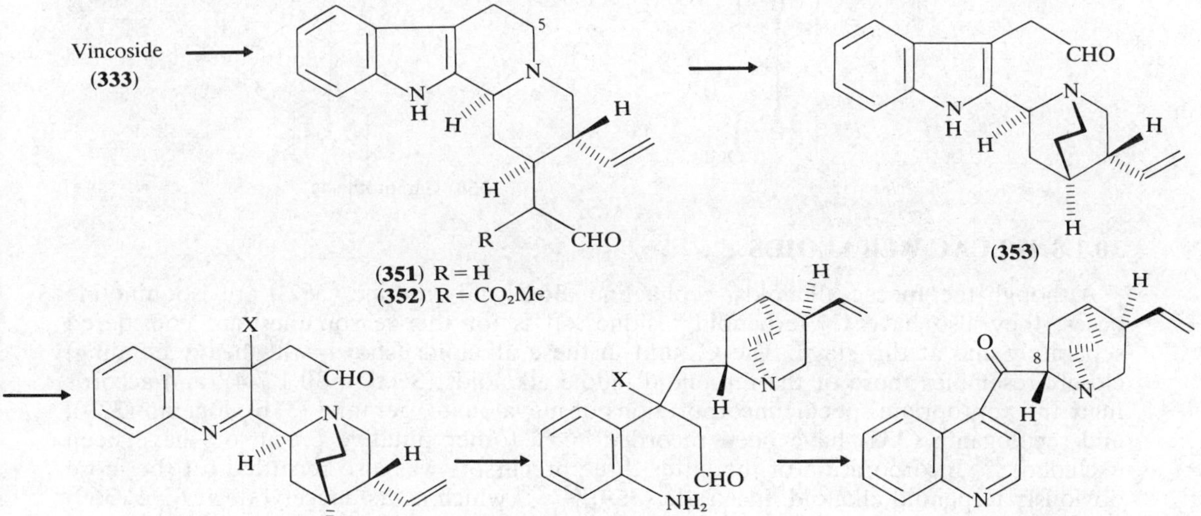

(**351**) R = H
(**352**) R = CO_2Me
(**353**)
(**354**)
(**355**) R = H, Cinchonidine
(**356**) R = OMe, Quinine

SCHEME 43

30.1.7.8 Camptothecin

Following the suggestion that camptothecin (**358**) is derived along a similar pathway to the terpenoid indole alkaloids (Section 30.1.7.4), it has been shown by tracer experiments that the two pathways are the same as far as vincoside/isovincoside.[284,285] In notable contrast to the biosynthesis of terpenoid indole alkaloids, which involves vincoside (**333**),* that of camptothecin continues with the lactam (**357**) derived from isovincoside (**332**) (testing of the epimeric lactams was carried out with ^{13}C labels, one of the first such applications in plant alkaloid biosynthesis).[284]

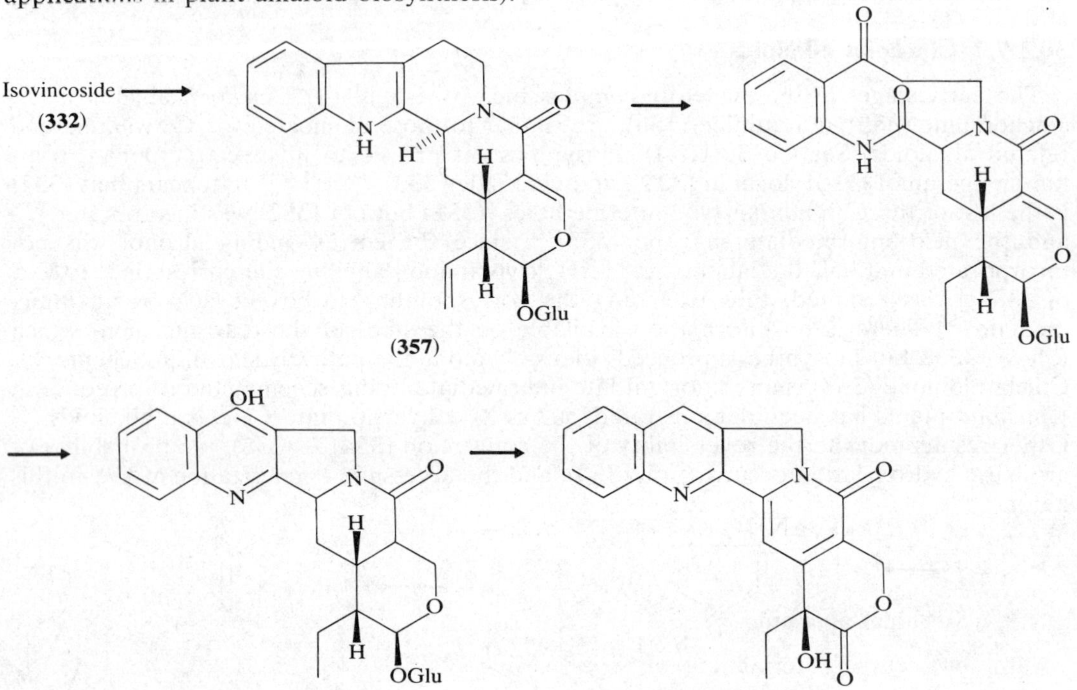

(**358**) Camptothecin

30.1.8 IPECAC ALKALOIDS

Although the Ipecac alkaloids, cephaeline (**361**) and emetine (**362**), are isoquinoline bases, they also have C_9 terpenoid residues. It is for this reason they are considered separately and at this stage. The C_9 unit in these alkaloids (shown with heavy bonding) closely resembles those of the terpenoid indole alkaloids (Section 30.1.7.4), and accordingly the appropriate specific incorporations of mevalonate, geraniol (**316**), loganin (**329**), and secologanin (**331**) have been recorded[286–288] (other putative precursors have been excluded).[287] Incorporation of the latter three precursors was also recorded for the more obviously terpenoid alkaloid, ipecoside (**359**),[286,288] which as its deacetyl derivative (**360**) turns out to be a key intermediate in the biosynthesis of the more complex emetine (**362**) and cephaeline (**361**).[288] Deacetylipecoside (**360**), and not its C-5 epimer, was incorporated into these alkaloids whose configuration at C-11b is remarkably opposite to that of the equivalent site (C-5) in the precursor. Further, retention of tritium from C-5 of loganin (**329**) [corresponds to C-5 in (**360**)] in the derived alkaloids establishes that the epimerization which necessarily occurs in the conversion of (**360**) into (**361**) and (**362**) is not associated with proton (tritium) removal and replacement.[288] The epimerization thus parallels closely that observed for the terpenoid indole alkaloids (Section 30.1.7.4). It is probably associated with C—N bond cleavage, but in this case is not so easily rationalized. The biological significance of such reactions is obscure.

* However, see Ref. 270.

The non-terpenoid fragments of cephaeline (**361**) and emetine (**362**) are manifestly phenethylamine in origin and appropriately (*cf.* Section 30.1.4) tyrosine serves as a specific precursor. The pathway which the combined results indicate is illustrated in Scheme 44.

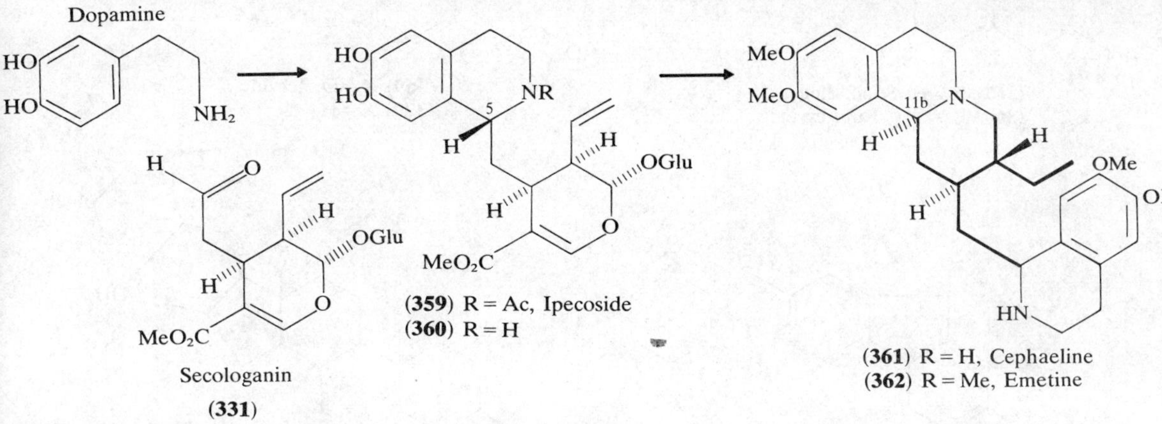

SCHEME 44

30.1.9 STEROIDAL AND TERPENOID ALKALOIDS

30.1.9.1 Steroidal alkaloids

Naturally occurring steroidal alkaloids, *e.g.* (**363**), (**367**), and (**369**), can be thought of as arising by simple modification of the normal steroid nucleus. In accord with this, acetate and mevalonate are precursors for, *e.g.* (**363**) and (**366**),[289,290] cholesterol is a precursor for tomatidine (**366**) and solanidine (**367**) and the *Veratrum* bases jervine (**369**) and veratramine (**371**),[291] and cycloartenol and lanosterol are tentatively identified as precursors for solanidine (**367**), solanocapsine (**372**), and tomatidine (**366**).[292] The quite differently substituted 3-aminopregnenone alkaloids, holaphylline (**374**) and holaphyllamine (**375**), also derive from cholesterol; and pregnenolone (**373**), but not progesterone, appears to be an intermediate.[293]

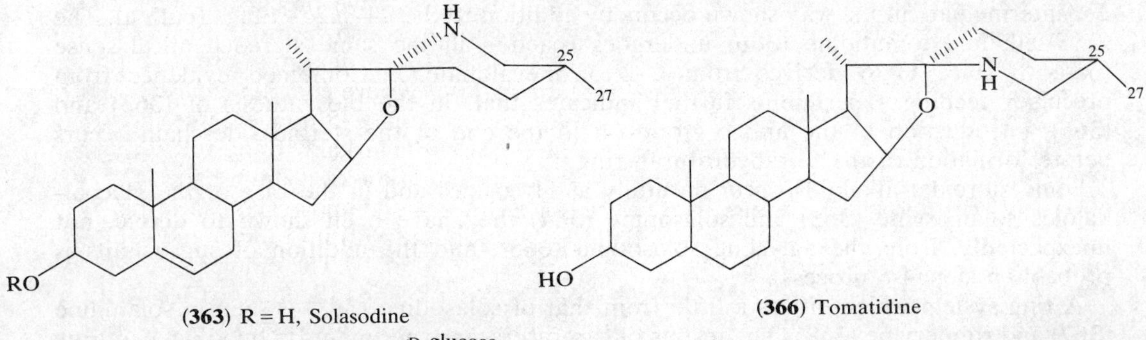

(**363**) R = H, Solasodine
(**364**) R = D-galactose < D-glucose, L-rhamnose
(**365**) R = D-glucose < L-rhamnose, L-rhamnose

(**366**) Tomatidine

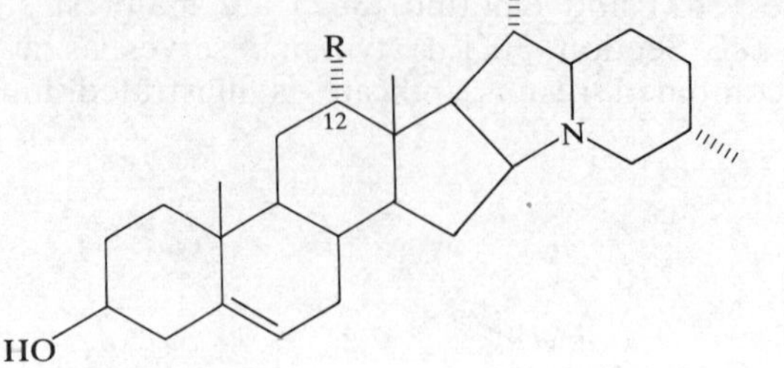

(**367**) R = H, Solanidine
(**368**) R = OH, Rubijervine

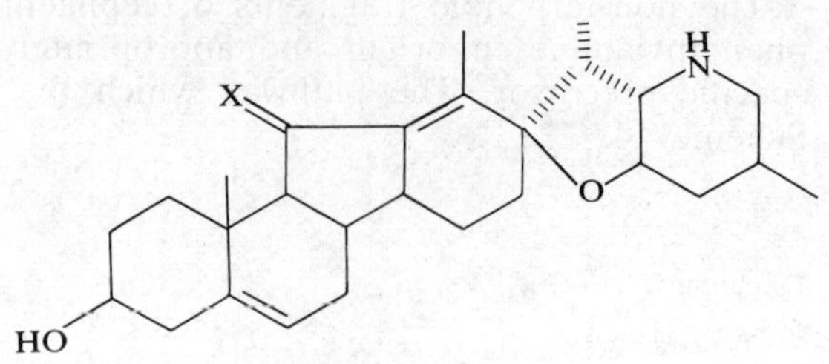

(**369**) X = O, Jervine
(**370**) X = H

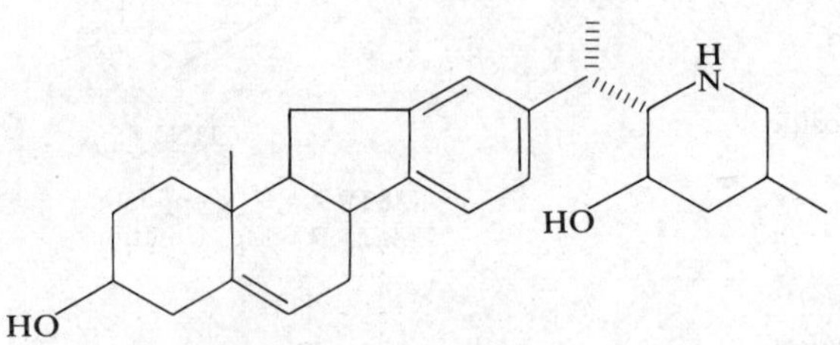

(**371**) Veratramine

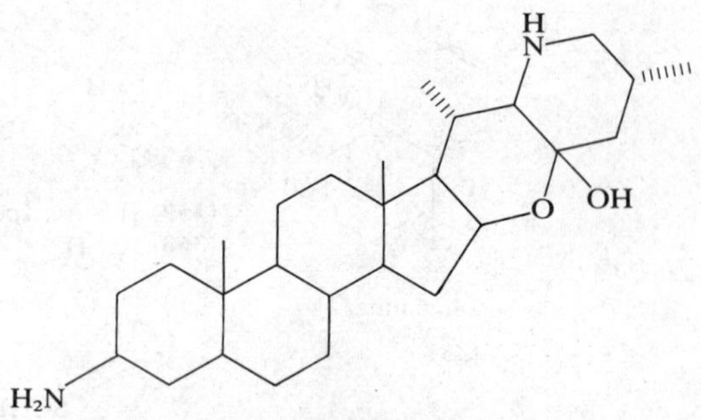

(**372**) Solanocapsine

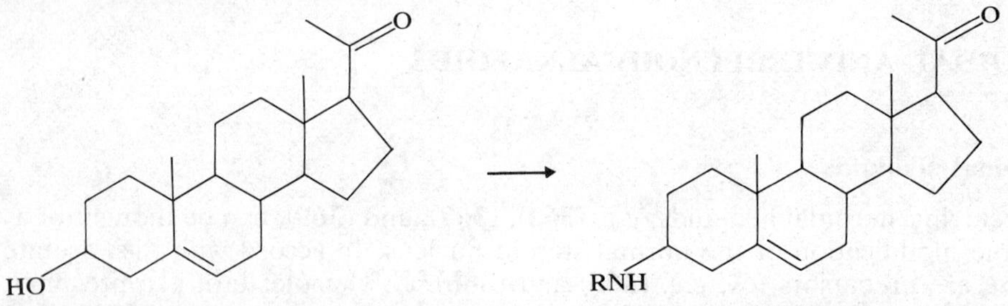

(**373**)

(**374**) R = Me
(**375**) R = H

Modification of the steroidal side-chain which leads to steroidal alkaloids like tomatidine (**366**) and solasodine (**363**) may be thought of as occurring by way of the sequence shown in Scheme 45. As the C-27 methyl group of solasodine (**363**), a (25*R*)-alkaloid, derives from C-2 of mevalonate,[289] it follows that saturation of a Δ^{24}-intermediate in the way shown occurs by addition to the 24-*si*,25-*si* face (path a). The (25*S*)-alkaloid tomatidine (**366**) undergoes reaction in the same stereochemical sense (path b) since C-26 derives from C-2 of mevalonate.[294] Combined evidence from precursor feeding experiments further indicates that, in the biosynthesis of (**363**) and (**366**), introduction of the amino-group on to the end of the steroid side-chain occurs before formation of the tetrahydrofuran ring.[295]

Some steroidal alkaloids occur naturally as glycosides and in the case of the glycoalkaloids solamargine (**365**) and solasonine (**364**) they have been shown to derive, not unexpectedly, from the aglycone solasodine (**363**), and the addition of sugar units is probably a stepwise process.[296]

A ring system which differs a little from that of solasodine (**363**) is seen in solanidine (**367**) and rubijervine (**368**). The results of careful tracer experiments in budding *Veratrum grandiflorum* and in dormant rhizome slices have indicated that the extra ring present in these alkaloids is formed at a late stage of biosynthesis, but interestingly the pathways

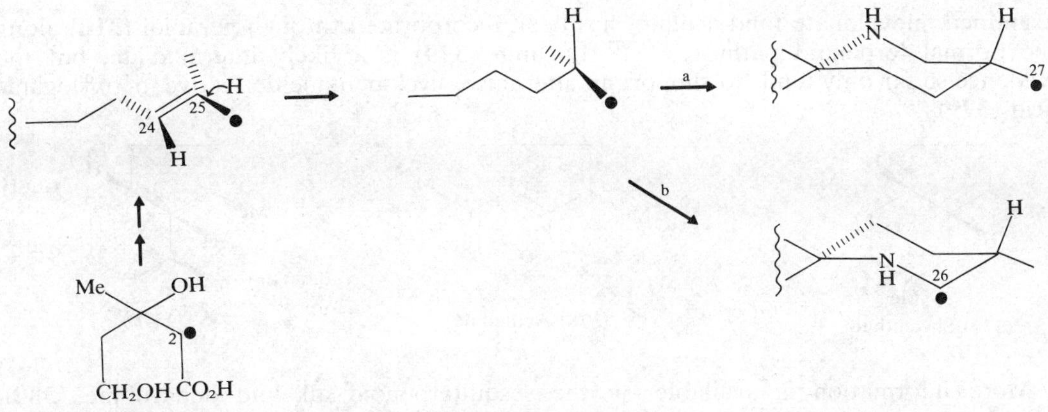

Mevalonic acid SCHEME 45

from verazine (**376**), an early intermediate, to the two quite similar alkaloids is different (see Scheme 46).[297]

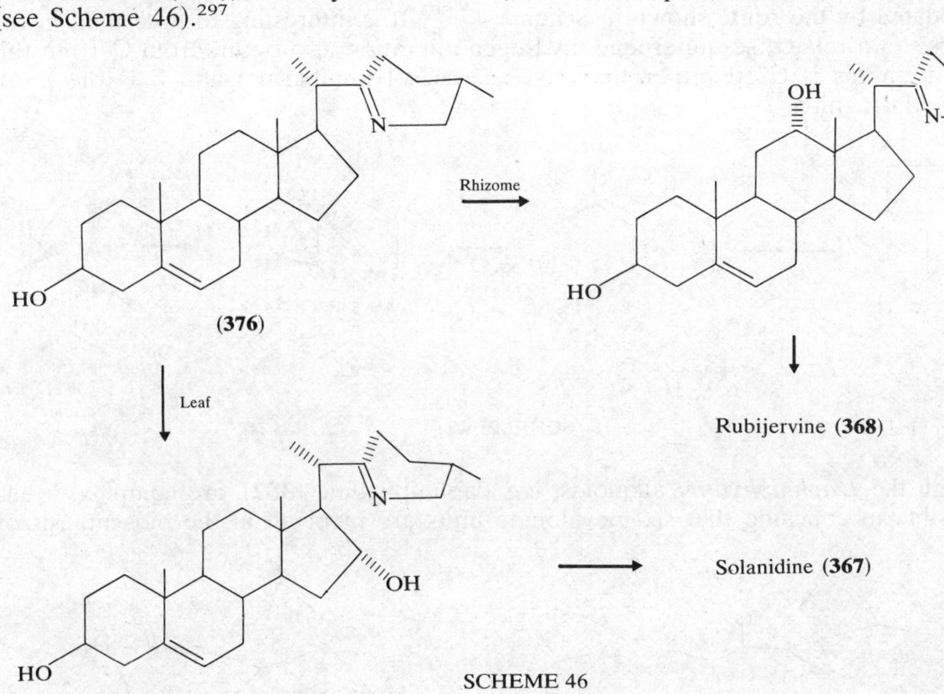

(**376**)

Rubijervine (**368**)

Solanidine (**367**)

SCHEME 46

The results discussed earlier indicate that the interesting C-nor-D-homosteroidal alkaloids, *e.g.* jervine (**369**), originate from the normal steroid skeleton, It is an attractive possibility that the distinctive skeleton of these alkaloids could originate from a solanidine (**367**) derivative with an equatorial hydroxy-group at C-12. Substance is given to this hypothesis by the observation that radioactivity from [1-^{14}C]acetate was transferred to solanidine glycoside in *V. grandiflorum* grown in the dark, and subsequently transferred to jervine (**369**) and veratramine (**371**) on illumination.[298] A further detail is that 11-deoxyjervine (**370**) appears to be an intermediate in jervine biosynthesis.[299]

30.1.9.2 Terpenoid alkaloids

The monoterpenoid C_9–C_{10} unit found in the terpenoid indole alkaloids (Section 30.1.7.4) appears in simpler bases like skytanthine (**377**) and actinidine (**378**), and, where examined, mevalonate (and acetate) has been incorporated through geraniol (**316**) along the normal terpenoid pathway.[300,301] Loganin (**329**) is a likely intermediate, but the evidence so far only tends to support an alternative cyclopentane derivative, deoxyloganic acid (**379**).[301]

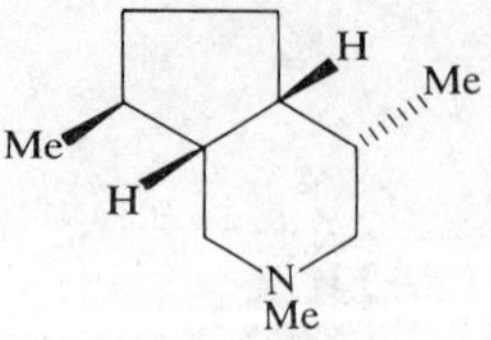

(**377**) Skytanthine

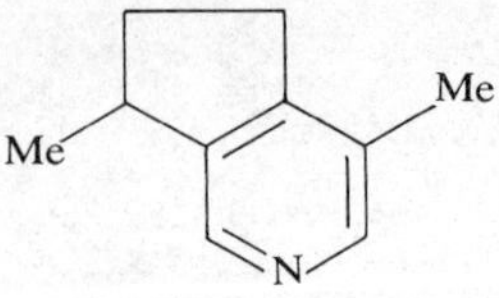

(**378**) Actinidine

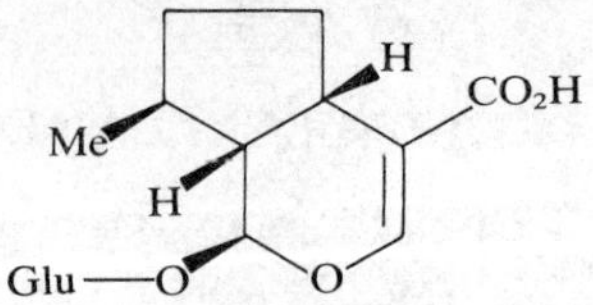

(**379**)

More information is available on the sesquiterpenoid alkaloid dendrobine (**380**). Mevalonate was incorporated[302] by way of 2-*trans*,6-*trans*-farnesol (**381**).[303] Tritium-labelling results establish that in the conversion of (**381**) into dendrobine (**380**) the 1-*pro-R* hydrogen migrates to appear at C-8 of the alkaloid. These results may be accommodated by the route shown in Scheme 47.[303] It is interesting to note that, in the biosynthesis of a related sesquiterpene, hydrogen migration also occurs from C-1 but this time the terminus is C-10 rather than C-11 (farnesol numbering) and it is the *pro-S* hydrogen which shifts.[304]

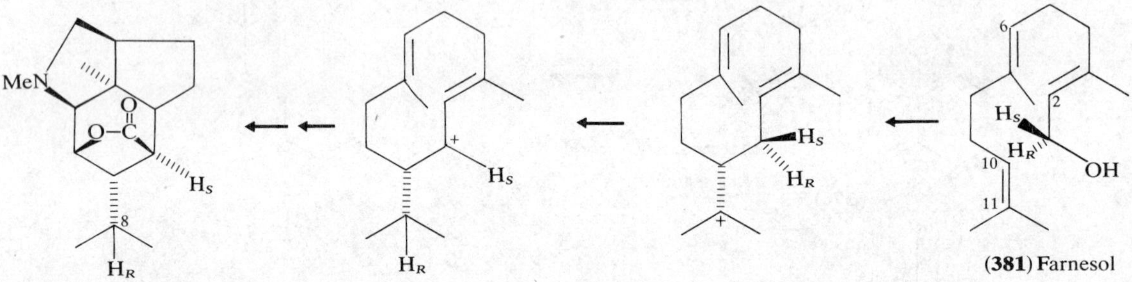

(**381**) Farnesol

(**380**) Dendrobine

SCHEME 47

Although the *Daphniphyllum* alkaloids, *e.g.* daphniphylline (**382**), are complex, it has been possible to conclude that six mevalonate units are involved in the biosynthesis of

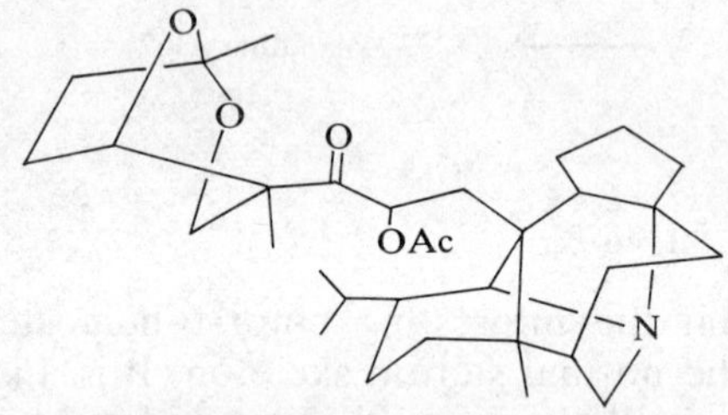

(**382**) Daphniphylline

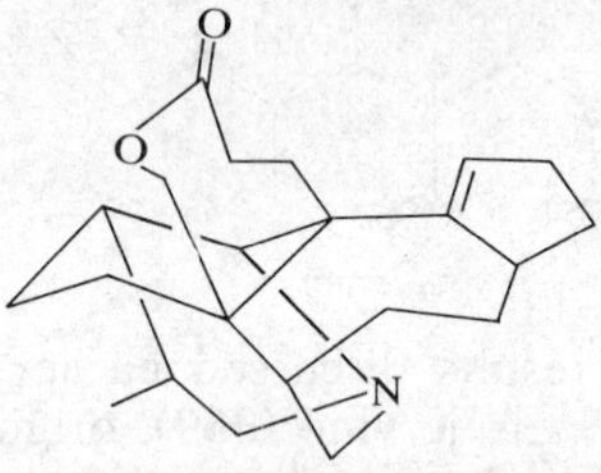

(**383**)

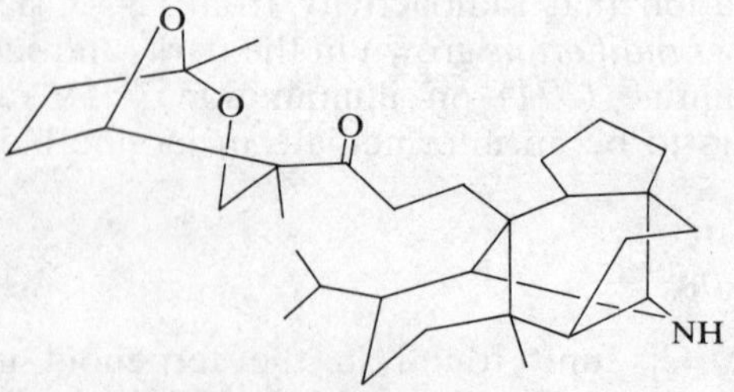

(**384**)

daphniphylline (**382**) and to propose a possible biogenesis.[305] The C_{22} triterpenoid alkaloids like daphnilactone B (**383**) are derived plausibly *via* C_{30} intermediates such as secodaphniphylline (**384**), and this is supported by the results of preliminary experiments with [2-^{14}C]mevalonate.[306]

30.1.10 MISCELLANEOUS ALKALOIDS

Of all the plant bases discussed in Chapter 30.1, none has a so clearly defined function as the betalains, *e.g.* betanin (**390**). These are a group of red-violet and yellow pigments found abundantly and uniquely in plants of the Centrospermae order and which account, for example, for the colours of some flowers.

Investigation of the biosynthesis of betanin (**390**) has revealed that it derives in a unique way from two complete molecules of tyrosine through dopa.[307] As indicated in Scheme 48, incorporation of one of these molecules occurs in a simple way, but the other, which

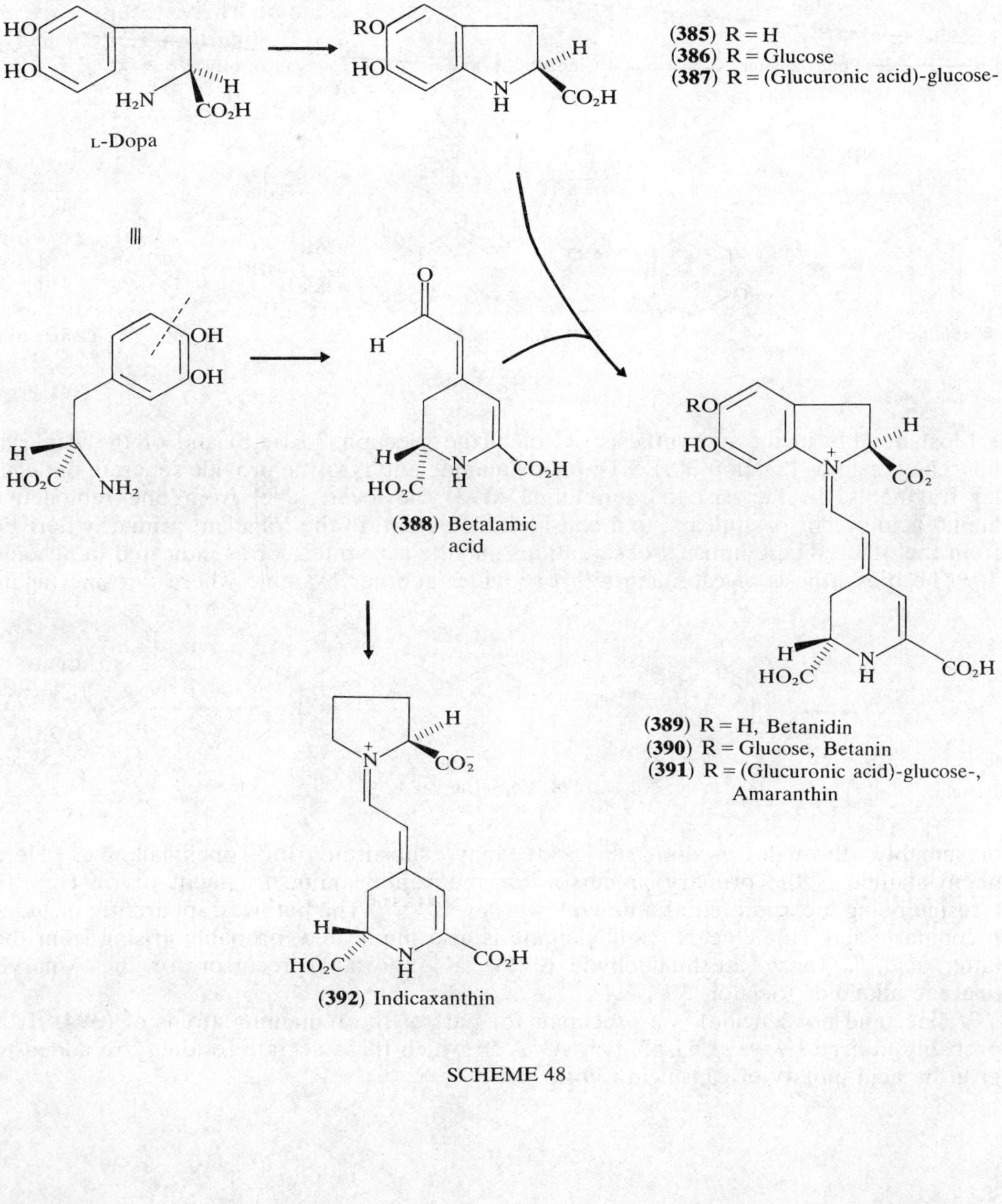

SCHEME 48

provides the betalamic acid grouping (**388**), is incorporated after fragmentation of the aromatic ring; results of tritium-labelling experiments establish that for both betanin (**390**)[308] and indicaxanthin (**392**)[309] cleavage occurs as indicated. Ring cleavage of phenols is fairly generally observed *in vivo* but is rare in alkaloid biosynthesis, the only other documented example being in the *Erythrina* alkaloids (Section 30.1.4.6).

The conversion of betanidin (**389**) into betanin (**390**) establishes that glycosylation is probably the last step of biosynthesis.[310] On the other hand, it appears that the formation of another betalain, amaranthin (**391**), takes a different course, one which involves the sequence (**385**) → (**386**) → (**387**) → (**391**).[311]

Although annuloline (**393**) is a unique oxazole alkaloid, it is derived from the products of tyrosine and phenylalanine metabolism which are involved in the biosynthesis of many other alkaloids, namely tyramine and cinnamic acid.[312] The deduced pathway is illustrated in Scheme 49.

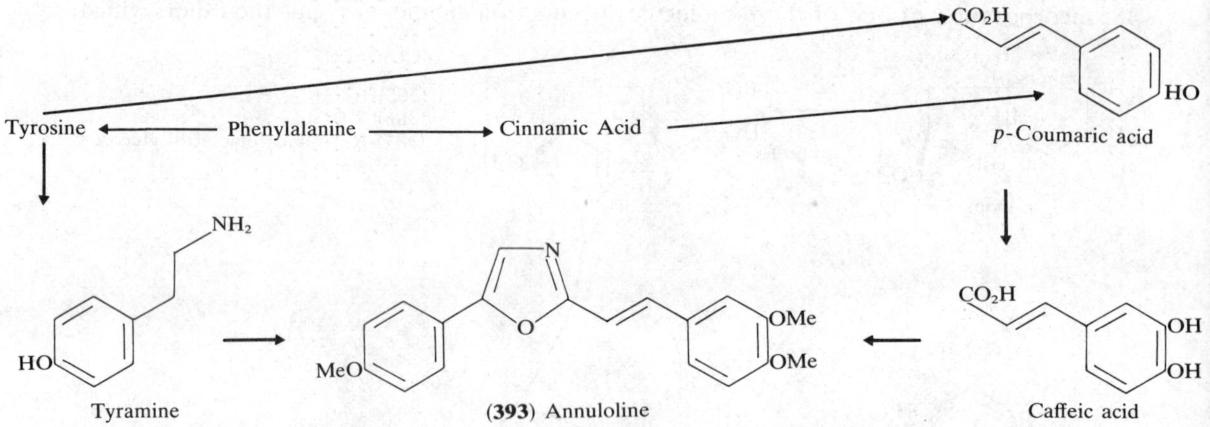

SCHEME 49

Most notably in the biosynthesis of colchicine (Section 30.1.4.8) and of the Amaryllidaceae alkaloids (Section 30.1.5.1), phenylalanine and tyrosine provide separate molecular fragments. In the case of annuloline (**393**), however, label from one radioactive amino-acid precursor appears, to a considerable extent, in the fragment primarily derived from the other. These unusual observations may be accounted for as indicated in Scheme 49. The biosynthesis of capsaicin (**394**) provides another example where tyrosine (again

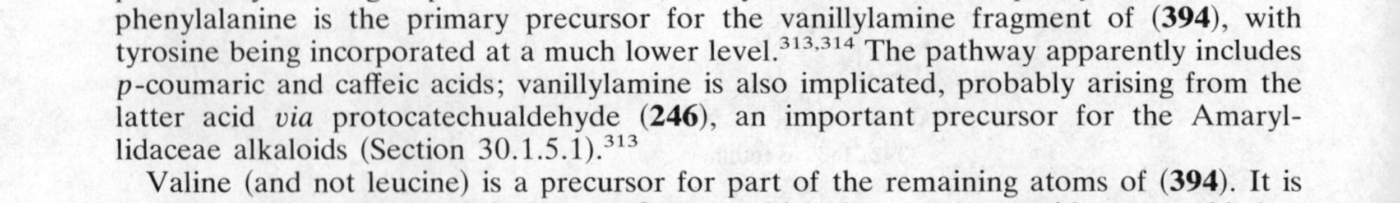

(**394**) Capsaicin

presumably through *p*-coumaric acid) may substitute for phenylalanine. Here phenylalanine is the primary precursor for the vanillylamine fragment of (**394**), with tyrosine being incorporated at a much lower level.[313,314] The pathway apparently includes *p*-coumaric and caffeic acids; vanillylamine is also implicated, probably arising from the latter acid *via* protocatechualdehyde (**246**), an important precursor for the Amaryllidaceae alkaloids (Section 30.1.5.1).[313]

Valine (and not leucine) is a precursor for part of the remaining atoms of (**394**). It is probably utilized by way of isobutyryl-CoA to which three acetate residues are added to give the acid moiety of capsaicin (**394**).[314]

The 1,3-azole system [as (**393**)] is seen again in the unique cactus alkaloid dolichotheline (**395**). The origins of this base are unexceptionally in histidine by way of histamine, with the isovaleryl unit arising from leucine (and valine) through isovaleric acid.[315] Further research[6] on dolichotheline biosynthesis has been within the interesting new area of exploring the ability of the plants to make unnatural alkaloid analogues (see Section 30.1.1).

Shihunine (**397**) is an alkaloid produced by an orchid, *Dendrobium pierardii.* For an alkaloid it has, perhaps not inappropriately, somewhat exotic origins in (**396**).[316] This

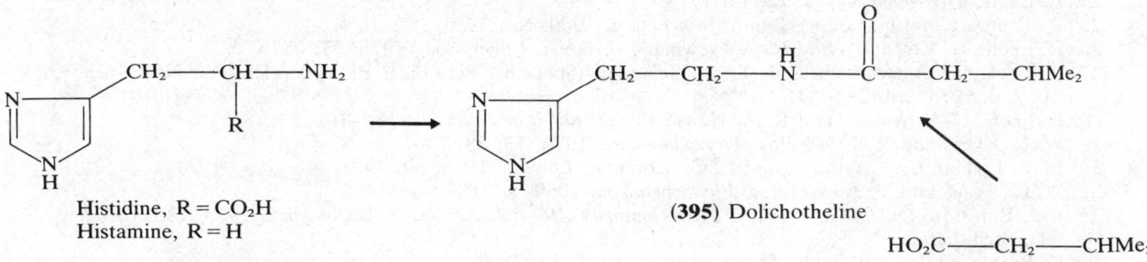

Histidine, R = CO_2H
Histamine, R = H

(**395**) Dolichotheline

Isovaleric acid

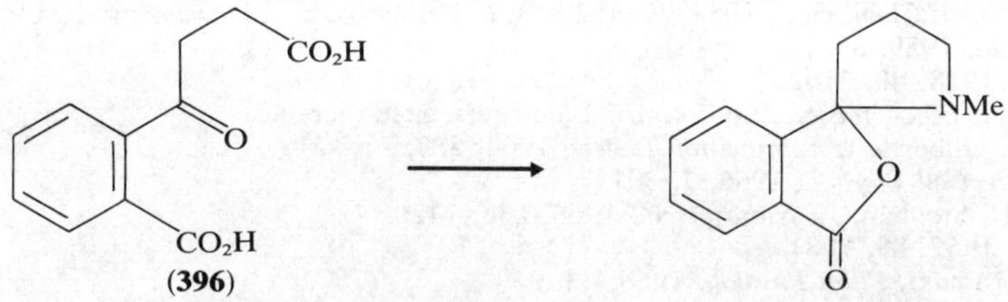

(**396**)

(**397**) Shihunine

compound is an important intermediate in the biosynthesis of several 1,4-naphthoquinones of plant and bacterial origin, being elaborated from the aromatic amino-acid precursor shikimic acid, and it will be singularly interesting to see if shihunine has similar origins. (For a discussion of naphthoquinone biosynthesis, see Chapter 30.3.)

References

1. R. Robinson, 'The Structural Relations of Natural Products', Oxford University Press, 1955.
2. D. H. R. Barton and T. Cohen, in 'Festschrift Dr. A. Stoll', Birkhäuser, Basle, 1957, p. 117.
3. C. Poupat and G. Kunesch, *Compt. rend.*, 1971, **273C**, 433.
4. S. A. Brown, in 'Biosynthesis', ed. T. A. Geissman (Specialist Periodical Reports), The Chemical Society, London, 1972, vol. 1, p. 1.
5. M. L. Rueppel and H. Rapoport, *J. Amer. Chem. Soc.*, 1971, **93,** 7021; E. Leete, G. B. Bodem, and M. F. Manuel, *Phytochemistry*, 1971, **10,** 2687; G. W. Kirby, S. R. Massey, and P. Steinreich, *J.C.S. Perkin I*, 1972, 1642.
6. H. Rosenberg and S. J. Stohs, *Phytochemistry*, 1976, **15,** 501, and refs. cited therein.
7. E. Leete, in 'Biogenesis of Natural Compounds', 2nd edn., ed. P. Bernfeld, Pergamon, Oxford, 1967, p. 953.
8. K. Mothes and H. R. Schütte, 'Biosynthese der Alkaloide', V. E. B. Deutscher Verlag der Wissenschaften, Berlin, 1969.
9. I. D. Spenser, in 'Chemistry of the Alkaloids', ed. S. W. Pelletier, Van Nostrand Reinhold, New York, 1970, p. 669.
10. I. D. Spenser, in 'Comprehensive Biochemistry', ed. M. Florkin and E. H. Stotz, Elsevier, Amsterdam, 1968, vol. 20, p. 231.
11. 'The Alkaloids', ed. J. E. Saxton (vols. 1–5) and M. F. Grundon (vols. 6 and 7) (Specialist Periodical Reports), The Chemical Society, London, 1971–1977.
12. 'Biosynthesis', ed. T. A. Geissman (vols. 1–3) and J. D. Bu'Lock (vol. 4) (Specialist Periodical Reports), The Chemical Society, London, 1972–1976.
13. R. Robinson, *J. Chem. Soc.*, 1917, 762.
14. E. Leete, *J. Amer. Chem. Soc.*, 1962, **84,** 55.
15. E. Leete and S. J. Nelson, *Phytochemistry*, 1969, **8,** 413.

16. H. W. Liebisch, H. R. Schütte, and K. Mothes, *Annalen*, 1963, **668,** 139.
17. H. W. Liebisch, H. Ramin, I. Schöffinius, and H. R. Schütte, *Z. Naturforsch.*, 1965, **20b,** 1183.
18. F. E. Baralle and E. G. Gros, *Chem. Comm.*, 1969, 721, and refs. cited therein; A. Ahmad and E. Leete, *Phytochemistry*, 1970, **9,** 2345.
19. H. W. Liebisch, A. S. Radwan, and H. R. Schütte, *Annalen*, 1969, **721,** 123.
20. J. M. Essery, D. J. McCaldin, and L. Marion, *Phytochemistry*, 1962, **1,** 209, and refs. cited therein.
21. E. F. L. J. Anet, G. K. Hughes, and E. Ritchie, *Austral. J. Sci. Res.*, 1949, **Ser. 2A,** 616.
22. D. G. O'Donovan and M. F. Keogh, *J. Chem. Soc.* (*C*), 1969, 223.
23. F. E. Baralle and E. G. Gros, *Phytochemistry*, 1969, **8,** 849, 853; I. Kaczkowski, H. R. Schütte, and K. Mothes, *Biochim. Biophys. Acta*, 1961, **46,** 588.
24. E. Leete, *Phytochemistry*, 1972, **11,** 1713.
25. A. Romeike and G. Fodor, *Tetrahedron Letters*, 1960, No. 22, 1.
26. E. Leete, N. Kowanko, and R. A. Newmark, *J. Amer. Chem. Soc.*, 1975, **97,** 6826.
27. E. Leete, in 'Biosynthesis', ed. T. A. Geissman (Specialist Periodical Reports), The Chemical Society, London, 1973, vol. 2, p. 115.
28. E. Leete, N. Kowanko, and R. A. Newmark, *Tetrahedron Letters*, 1975, 4103.
29. W. C. Evans and J. G. Woolley, *Phytochemistry*, 1976, **15,** 287.
30. B. V. Prabhu, C. A. Gibson, and L. C. Schramm, *Lloydia*, 1976, **39,** 79.
31. W. C. Evans and V. A. Woolley, *Phytochemistry*, 1969, **8,** 2183.
32. P. J. Beresford and J. G. Woolley, *Phytochemistry*, 1974, **13,** 2143; E. Leete, *ibid.*, 1973, **12,** 2203, and refs. cited therein.
33. K. Basey and J. G. Woolley, *Phytochemistry*, 1973, **12,** 2197.
34. P. J. Beresford and J. G. Woolley, *Phytochemistry*, 1974, **13,** 2511.
35. E. Leete and K. J. Siegfried, *J. Amer. Chem. Soc.*, 1957, **79,** 4529; B. L. Lamberts, L. J. Dewey, and R. U. Byerrum, *Biochim. Biophys. Acta*, 1959, **33,** 22.
36. E. Leete, *J. Amer. Chem. Soc.*, 1958, **80,** 2162.
37. P.-H. L. Wu and R. U. Byerrum, *Biochemistry*, 1965, **4,** 1628, and refs. cited therein.
38. E. Leete, E. G. Gros, and T. J. Gilbertson, *Tetrahedron Letters*, 1964, 587.
39. D. Yoshida and T. Mitake, *Plant Cell Physiol.*, 1966, **7,** 301.
40. H. R. Schütte, W. Maier, and U. Stephan, *Z. Naturforsch.*, 1968, **23b,** 1426.
41. E. Leete, *J. Amer. Chem. Soc.*, 1967, **89,** 7081.
42. S. Mizusaki, T. Kisaki, and E. Tamaki, *Plant Physiol.*, 1968, **43,** 93.
43. S. Mizusaki, Y. Tanabe, M. Noguchi, and E. Tamaki, *Plant Cell Physiol.*, 1971, **12,** 633; *Phytochemistry*, 1972, **11,** 2757.
44. T. J. Gilbertson and E. Leete, *J. Amer. Chem. Soc.*, 1967, **89,** 7085.
45. M. L. Rueppel, B. P. Mundy, and H. Rapoport, *Phytochemistry*, 1974, **13,** 141; for earlier work see refs. cited therein.
46. T. A. Scott and J. P. Glynn, *Phytochemistry*, 1967, **6,** 505; G. M. Frost, K. S. Yang, and G. R. Waller, *J. Biol. Chem.*, 1967, **242,** 887.
47. E. Leete and Y.-Y. Liu, *Phytochemistry*, 1973, **12,** 593, and refs. cited therein.
48. H. R. Zielke, C. M. Reinke, and R. U. Byerrum, *J. Biol. Chem.*, 1969, **244,** 95, and refs. cited therein.
49. N. M. Bale and D. H. G. Crout, *Phytochemistry*, 1975, **14,** 2617.
50. C. A. Hughes, R. Letcher, and F. L. Warren, *J. Chem. Soc.*, 1964, 4974.
51. W. Bottomley and T. A. Geissman, *Phytochemistry*, 1964, **3,** 357.
52. N. M. Davies and D. H. G. Crout, *J.C.S. Perkin I*, 1974, 2079; D. H. G. Crout, N. M. Davies, E. H. Smith, and D. Whitehouse, *ibid.*, 1972, 671.
53. D. J. Robins, N. M. Bale, and D. H. G. Crout, *J.C.S. Perkin I*, 1974, 2082.
54. D. H. G. Crout, *J. Chem. Soc.* (*C*), 1967, 1233; *ibid.*, 1966, 1968.
55. N. B. Mulchandani, S. S. Iyer, and L. P. Badheka, *Phytochemistry*, 1969, **8,** 1931; *ibid.*, 1971, **10,** 1047.
56. R. B. Herbert, F. B. Jackson, and I. T. Nicolson, *J.C.S. Chem. Comm.*, 1976, 865.
57. E. Leete, *J. Amer. Chem. Soc.*, 1964, **86,** 2509.
58. E. Leete and J. Olsen, *J. Amer. Chem. Soc.*, 1972, **94,** 5472.
59. M. F. Roberts, *Phytochemistry*, 1971, **10,** 3057.
60. M. F. Roberts, *Phytochemistry*, 1975, **14,** 2393.
61. J. W. Fairbairn and P. N. Suwal, *Phytochemistry*, 1961, **1,** 38.
62. S. M. C. Dietrich and R. O. Martin, *Biochemistry*, 1969, **8,** 4163.
63. E. Leete and K. N. Juneau, *J. Amer. Chem. Soc.*, 1969, **91,** 5614.
64. E. Leete and R. A. Carver, *J. Org. Chem.*, 1975, **40,** 2151; E. Leete, J. C. Lechleiter, and R. A. Carver, *Tetrahedron Letters*, 1975, 3779.
65. R. N. Gupta and I. D. Spenser, *Phytochemistry*, 1969, **8,** 1937.
66. M. F. Keogh and D. G. O'Donovan, *J. Chem. Soc.* (*C*), 1970, 1792.
67. R. N. Gupta and I. D. Spenser, *Phytochemistry*, 1970, **9,** 2329.
68. R. N. Gupta and I. D. Spenser, *Canad. J. Chem.*, 1967, **45,** 1275.
69. E. Leete, *J. Amer. Chem. Soc.*, 1956, **78,** 3520.
70. E. Leete, E. G. Gros, and T. J. Gilbertson, *J. Amer. Chem. Soc.*, 1964, **86,** 3907.
71. E. Leete and A. R. Friedman, *J. Amer. Chem. Soc.*, 1964, **86,** 1224, and refs. cited therein.
72. E. Leete, *J. Amer. Chem. Soc.*, 1969, **91,** 1697.

73. (a) E. Leistner and I. D. Spenser, *J. Amer. Chem. Soc.*, 1973, **95,** 4715; (b) refs. cited therein.
74. P. Korzan and T. J. Gilbertson, *Phytochemistry*, 1974, **13,** 435.
75. E. Leete and M. R. Chedekel, *Phytochemistry*, 1972, **11,** 2751.
76. E. Leistner and I. D. Spenser, *J.C.S. Chem. Comm.*, 1975, 378.
77. T. J. Gilbertson, *Phytochemistry*, 1972, **11,** 1737; E. Leistner, R. N. Gupta, and I. D. Spenser, *J. Amer. Chem. Soc.*, 1973, **95,** 4040.
78. W. D. Marshall, T. T. Nguyen, D. B. MacLean, and I. D. Spenser, *Canad. J. Chem.*, 1975, **53,** 41.
79. R. N. Gupta and I. D. Spenser, *J. Biol. Chem.*, 1969, **244,** 88.
80. I. T. Bruce and G. W. Kirby, *Chem. Comm.*, 1968, 207.
81. D. G. O'Donovan and P. B. Creedon, *Tetrahedron Letters*, 1971, 1341.
82. E. Leete and A. R. Pinder, *Phytochemistry*, 1972, **11,** 3219.
83. M. F. Keogh and D. G. O'Donovan, *J. Chem. Soc.* (*C*), 1970, 2470.
84. D. G. O'Donovan, D. J. Long, E. Forde, and P. Geary, *J.C.S. Perkin I*, 1975, 415.
85. D. G. O'Donovan and T. Forde, *J. Chem. Soc.* (*C*), 1971, 2889.
86. R. N. Gupta and I. D. Spenser, *Canad. J. Chem.*, 1971, **49,** 384.
87. R. J. Parry, *J.C.S. Chem. Comm.*, 1975, 144; *Tetrahedron Letters*, 1974, 307.
88. U. Sankawa, K. Yamasaki, and Y. Ebizuka, *Tetrahedron Letters*, 1974, 1867.
89. M. W. Golebiewski, P. Horsewood, and I. D. Spenser, *J.C.S. Chem. Comm.*, 1976, 217.
90. E. Leete, *J.C.S. Chem. Comm.*, 1975, 9.
91. D. G. O'Donovan and P. B. Creedon, *J.C.S. Perkin I*, 1974, 2524.
92. D. G. O'Donovan and P. B. Creedon, *J. Chem. Soc.* (*C*), 1971, 1604.
93. H. R. Schütte, K.-L. Kelling, D. Knöfel, and K. Mothes, *Phytochemistry*, 1964, **3,** 249.
94. S. H. Koo, F. Comer, and I. D. Spenser, *Chem. Comm.*, 1970, 897.
95. S. H. Koo, R. N. Gupta, I. D. Spenser, and J. T. Wrobel, *Chem. Comm.*, 1970, 396.
96. M. Castillo, R. N. Gupta, D. B. MacLean, and I. D. Spenser, *Canad. J. Chem.*, 1970, **48,** 1893.
97. M. Castillo, R. N. Gupta, Y. K. Ho, D. B. MacLean, and I. D. Spenser, *Canad. J. Chem.*, 1970, **48,** 2911.
98. Y. K. Ho, R. N. Gupta, D. B. MacLean, and I. D. Spenser, *Canad. J. Chem.*, 1971, **49,** 3352.
99. J.-C. Braekman, R. N. Gupta, D. B. MacLean, and I. D. Spenser, *Canad. J. Chem.*, 1972, **50,** 2591.
100. M. Souček and H. R. Schütte, *Angew. Chem. Internat. Edn.*, 1962, **1,** 597.
101. H. R. Schütte, H. Hindorf, K. Mothes, and G. Hübner, *Annalen*, 1964, **680,** 93.
102. H. R. Schütte, F. Bohlmann, and W. Reusche, *Arch. Pharm.*, 1961, **294,** 610.
103. H. R. Schütte and H. Hindorf, *Z. Naturforsch.*, 1964, **19b,** 855.
104. H. R. Schütte and G. Seelig, *Annalen*, 1968, **711,** 221.
105. H. R. Schütte and D. Knöfel, *Z. Pflanzenphysiol.*, 1968, **59,** 80.
106. E. K. Nowacki and G. R. Waller, *Phytochemistry*, 1975, **14,** 155.
107. E. Nowacki, D. Nowacka, and R. U. Byerrum, *Bull. Acad. Polon. Sci., Ser. Sci. Biol.*, 1966, **14,** 25.
108. H. R. Schütte, E. Nowacki, P. Kovacs, and H. W. Liebisch, *Arch. Pharm.*, 1963, **296,** 438.
109. (a) H. R. Schütte and H. Hindorf, *Annalen*, 1965, **685,** 187; (b) refs. cited therein.
110. Y. D. Cho, R. O. Martin, and J. N. Anderson, *J. Amer. Chem. Soc.*, 1971, **93,** 2087; Y. D. Cho and R. O. Martin, *Canad. J. Biochem.*, 1971, **49,** 971.
111. H. R. Schütte and J. Lehfeldt, *J. prakt. Chem.*, 1964, Ser. 4, **24,** 143.
112. H. R. Schütte, J. Lehfeldt, and H. Hindorf, *Annalen*, 1965, **685,** 194.
113. S. Shibata and U. Sankawa, *Chem. and Ind.* (*London*), 1963, 1161.
114. J. K. Kuschmuradov, D. Gross, and H. R. Schütte, *Phytochemistry*, 1972, **11,** 3441.
115. H. R. Schütte, G. Sandke, and J. Lehfeldt, *Arch. Pharm.*, 1964, **297,** 118.
116. H. P. Tiwari, W. R. Penrose, and I. D. Spencer, *Phytochemistry*, 1976, **6,** 1245, and refs. cited therein.
117. S. R. Johns and L. Marion, *Canad. J. Chem.*, 1966, **44,** 23; D. Gross, P. Banditt, J. W. Kurbatow, and H. R. Schütte, *Z. Pflanzenphysiol.*, 1968, **58,** 410; K. S. Yang and G. R. Waller, *Phytochemistry*, 1965, **4,** 881; and refs. cited therein.
118. R. D. Johnson and G. R. Waller, *Phytochemistry*, 1974, **13,** 1493.
119. S. D. Sastry and G. R. Waller, *Phytochemistry*, 1972, **11,** 2241.
120. H. J. Lee and G. R. Waller, *Phytochemistry*, 1972, **11,** 2233.
121. E. Leete and L. Marion, *Canad. J. Chem.*, 1953, **31,** 126; *ibid.*, 1954, **32,** 646; E. Leete, S. Kirkwood, and L. Marion, *ibid.*, 1952, **30,** 749.
122. J. Lundström, *Acta Pharm. Suecica*, 1971, **8,** 275; *Acta Chem. Scand.*, 1971, **25,** 3489; H. Rosenberg, K. L. Khanna, M. Takido, and A. G. Paul, *Lloydia*, 1969, **32,** 334; A. G. Paul, K. L. Khanna, H. Rosenberg, and M. Takido, *Chem. Comm.*, 1969, 838; and refs. cited therein.
123. G. P. Basmadjian and A. G. Paul, *Lloydia*, 1971, **34,** 91.
124. A. R. Battersby, R. Binks, and R. Huxtable, *Tetrahedron Letters*, 1968, 6111; *ibid.*, 1967, 563.
125. G. J. Kapadia, G. S. Rao, E. Leete, M. B. E. Fayez, Y. N. Vaishnav, and H. M. Fales, *J. Amer. Chem. Soc.*, 1970, **92,** 6943.
126. J. Lundström, *Acta Pharm. Suecica*, 1971, **8,** 485.
127. I. J. McFarlane and M. Slaytor, *Phytochemistry*, 1972, **11,** 229.
128. E. Leete and J. D. Braunstein, *Tetrahedron Letters*, 1969, 451.
129. G. Hahn, L. Barwald, O. Schales, and H. Werner, *Annalen*, 1935, **520,** 107; G. Hahn and F. Rumpf, *Ber.*, 1938, **71,** 2141; G. Hahn and K. Stiehl, *Ber.*, 1936, **69,** 2627.
130. I. J. McFarlane and M. Slaytor, *Phytochemistry*, 1972, **11,** 235.

131. J. G. Bruhn, U. Svensson, and S. Agurell, *Acta Chem. Scand.*, 1970, **24,** 3775.
132. D. G. O'Donovan and H. Horan, *J. Chem. Soc.* (*C*), 1968, 2791; H. R. Schütte and G. Seelig, *Annalen*, 1969, **730,** 186.
133. D. G. O'Donovan and E. Barry, *J.C.S. Perkin I*, 1974, 2528.
134. D. G. O'Donovan and H. Horan, *J. Chem. Soc.* (*C*), 1969, 1737.
135. S. Agurell, I. Granelli, K. Leander, B. Lüning, and J. Rosenblom, *Acta Chem. Scand.*, 1974, **B28,** 239.
136. S. Agurell, I. Granelli, K. Leander, and J. Rosenblom, *Acta Chem. Scand.*, 1974, **B28,** 1175.
137. W. J. Keller, L. A. Spitznagle, and L. R. Brady, *Lloydia*, 1973, **36,** 397; T. A. Wheaton and I. Stewart, *Phytochemistry*, 1969, **8,** 85.
138. K. Yamasaki, T. Tamaki, S. Uzawa, U. Sankawa, and S. Shibata, *Phytochemistry*, 1973, **12,** 2877.
139. S. Tewari, D. S. Bhakuni, and R. S. Kapil, *J.C.S. Chem. Comm.*, 1975, 554.
140. M. L. Wilson and C. J. Coscia, *J. Amer. Chem. Soc.*, 1975, **97,** 431.
141. A. R. Battersby, R. C. F. Jones, and R. Kazlauskas, *Tetrahedron Letters*, 1975, 1873.
142. E. Winterstein and G. Trier, 'Die Alkaloide', Bornträger–Verlag, Berlin, 1910, p. 307.
143. A. R. Battersby and B. J. T. Harper, *J. Chem. Soc.*, 1962, 3526.
144. A. R. Battersby, R. Binks, R. J. Francis, D. J. McCaldin, and H. Ramuz, *J. Chem. Soc.*, 1964, 3600.
145. E. Brochmann-Hanssen, C. Chen, C. R. Chen, H. Chiang, A. Y. Leung, and K. McMurtrey, *J.C.S. Perkin I*, 1975, 1531; H. Uprety, D. S. Bhakuni, and R. S. Kapil, *Phytochemistry*, 1975, **14,** 1535.
146. D. H. R. Barton, G. W. Kirby, and A. Wiechers, *J. Chem. Soc.* (*C*), 1966, 2313.
147. G. Blaschke, G. Waldheim, M. von Schantz, and P. Peura, *Arch. Pharm.*, 1974, **307,** 122; G. Blaschke, *ibid.*, 1970, **303,** 358; *ibid.*, 1968, **301,** 432.
148. E. Brochmann-Hanssen, C.-H. Chen, H.-C. Chiang, and K. McMurtrey, *J.C.S. Chem. Comm.*, 1972, 1269.
149. E. Brochmann-Hanssen, C.-C. Fu, and L. Y. Misconi, *J. Pharm. Sci.*, 1971, **60,** 1880; E. Brochmann-Hanssen, C. H. Chen, H.-C. Chiang, C.-C. Fu, and H. Nemoto, *ibid.*, 1973, **62,** 1291.
150. S. Tewari, D. S. Bhakuni, and R. S. Kapil, *J.C.S. Chem. Comm.*, 1974, 940.
151. A. R. Battersby, R. T. Brown, J. H. Clements, and G. G. Iverach, *Chem. Comm.*, 1965, 230.
152. A. R. Battersby and T. H. Brown, *Chem. Comm.*, 1966, 170.
153. A. R. Battersby, T. J. Brocksom, and R. Ramage, *Chem. Comm.*, 1969, 464.
154. A. R. Battersby, J. L. McHugh, J. Staunton, and M. Todd, *Chem. Comm.*, 1971, 985.
155. L. J. Haynes, K. L. Stuart, D. H. R. Barton, D. S. Bhakuni, and G. W. Kirby, *Chem. Comm.*, 1965, 141.
156. D. H. R. Barton, D. S. Bhakuni, G. M. Chapman, and G. W. Kirby, *J. Chem. Soc.* (*C*), 1967, 2134.
157. D. S. Bhakuni, S. Satish, H. Uprety, and R. S. Kapil, *Phytochemistry*, 1974, **13,** 2767.
158. A. R. Battersby, R. Binks, and B. J. T. Harper, *J. Chem. Soc.*, 1962, 3534; and refs. cited therein.
159. A. R. Battersby and R. J. Francis, *J. Chem. Soc.*, 1964, 4078.
160. E. Leete and J. B. Murrill, *Tetrahedron Letters*, 1964, 147.
161. A. R. Battersby and R. Binks, *Proc. Chem. Soc.*, 1960, 360.
162. D. H. R. Barton, G. W. Kirby, W. Steglich, G. M. Thomas, A. R. Battersby, T. A. Dobson, and H. Ramuz, *J. Chem. Soc.*, 1965, 2423.
163. A. R. Battersby, D. M. Foulkes, and R. Binks, *J. Chem. Soc.*, 1965, 3323.
164. E. Brochmann-Hanssen and B. Nielsen, *Tetrahedron Letters*, 1965, 1271; A. R. Battersby, G. W. Evans, R. O. Martin, M. E. Warren, jun., and H. Rapoport, *ibid.*, p. 1275; E. Brochmann-Hanssen and T. Furuya, *J. Pharm. Sci.*, 1964, **53,** 575.
165. R. O. Martin, M. E. Warren, jun., and H. Rapoport, *J. Amer. Chem. Soc.*, 1964, **86,** 4726.
166. H. I. Parker, G. Blaschke, and H. Rapoport, *J. Amer. Chem. Soc.*, 1972, **94,** 1276, and refs. cited therein; A. R. Battersby and B. J. T. Harper, *Tetrahedron Letters*, 1960, No. 27, p. 21.
167. J. R. Gear and I. D. Spenser, *Canad. J. Chem.*, 1963, **41,** 783; I. Monkovic and I. D. Spenser, *ibid.*, 1965, **43,** 2017.
168. R. N. Gupta and I. D. Spenser, *Canad. J. Chem.*, 1965, **43,** 133.
169. D. H. R. Barton, R. H. Hesse, and G. W. Kirby, *J. Chem. Soc.*, 1965, 6379; A. R. Battersby, R. J. Francis, M. Hirst, and J. Staunton, *Proc. Chem. Soc.*, 1963, 268.
170. A. R. Battersby, R. J. Francis, M. Hirst, E. A. Ruveda, and J. Staunton, *J.C.S. Perkin I*, 1975, 1140.
171. E. Leete, *J. Amer. Chem. Soc.*, 1963, **85,** 473.
172. E. Leete and J. B. Murrill, *Phytochemistry*, 1967, **6,** 231.
173. A. R. Battersby, J. Staunton, H. R. Wiltshire, R. J. Francis, and R. Southgate, *J.C.S. Perkin I*, 1975, 1147.
174. A. R. Battersby, J. Staunton, H. R. Wiltshire, B. J. Bircher, and C. Fuganti, *J.C.S. Perkin I*, 1975, 1162.
175. C. Tani and K. Tagahara, *Chem. Pharm. Bull.* (*Japan*), 1974, **22,** 2457.
176. A. R. Battersby, M. Hirst, D. J. McCaldin, R. Southgate, and J. Staunton, *J. Chem. Soc.* (*C*), 1968, 2163.
177. R. Bentley, 'Molecular Asymmetry in Biology', Academic Press, New York, 1970, vol. 2.
178. H. Rönsch, *European J. Biochem.*, 1972, **28,** 123, and refs. cited therein.
179. G. Blaschke, *Arch. Pharm.*, 1968, **301,** 439.
180. H. L. Holland, M. Castillo, D. B. MacLean, and I. D. Spenser, *Canad. J. Chem.*, 1974, **52,** 2818.
181. E. Leete and A. Ahmad, *J. Amer. Chem. Soc.*, 1966, **88,** 4722.
182. D. H. R. Barton, R. James, G. W. Kirby, D. W. Turner, and D. A. Widdowson, *J. Chem. Soc.* (*C*), 1968, 1529.
183. D. H. R. Barton, C. J. Potter, and D. A. Widdowson, *J.C.S. Perkin I*, 1974, 346.
184. D. H. R. Barton, R. B. Boar, and D. A. Widdowson, *J. Chem. Soc.* (*C*), 1970, 1213.
185. D. H. R. Barton, R. D. Bracho, C. J. Potter, and D. A. Widdowson, *J.C.S. Perkin I*, 1974, 2278.

186. A. R. Battersby, R. C. F. Jones, R. Kazlauskas, C. Poupat, C. W. Thornber, S. Ruchirawat, and J. Staunton, *J.C.S. Chem. Comm.*, 1974, 773.
187. A. R. Battersby, *Pure Appl. Chem.*, 1967, **14,** 117.
188. A. R. Battersby, R. Binks, J. J. Reynolds, and D. A. Yeowell, *J. Chem. Soc.*, 1964, 4257; E. Leete and P. E. Nemeth, *J. Amer. Chem. Soc.*, 1960, **82,** 6055; E. Leete, *ibid.*, 1963, **85,** 3666.
189. A. R. Battersby, T. A. Dobson, D. M. Foulkes, and R. B. Herbert, *J.C.S. Perkin I*, 1972, 1730; E. Leete, *Tetrahedron Letters*, 1965, 333.
190. A. I. Scott and K. J. Wiesner, *J.C.S. Chem. Comm.*, 1972, 1075.
191. A. R. Battersby, R. B. Herbert, L. Pijewska, F. Šantavý and P. Sedmera, *J.C.S. Perkin I*, 1972, 1736.
192. A. R. Battersby, R. B. Herbert, E. McDonald, R. Ramage, and J. H. Clements, *J.C.S. Perkin I*, 1972, 1741.
193. A. R. Battersby, P. W. Sheldrake, and J. A. Milner, *Tetrahedron Letters*, 1974, 3315.
194. A. C. Barker, A. R. Battersby, E. McDonald, R. Ramage, and J. H. Clements, *Chem. Comm.*, 1967, 390.
195. A. R. Battersby, P. Böhler, M. H. G. Munro, and R. Ramage, *J.C.S. Perkin I*, 1974, 1399.
196. A. R. Battersby, E. McDonald, J. A. Milner, S. R. Johns, J. A. Lamberton, and A. A. Sioumis, *Tetrahedron Letters*, 1975, 3419.
197. R. J. Parry and J. M. Schwab, *J. Amer. Chem. Soc.*, 1975, **97,** 2555.
198. A. R. Battersby, R. Binks, S. W. Breuer, H. M. Fales, W. C. Wildman, and R. J. Highet, *J. Chem. Soc.*, 1964, 1595.
199. P. W. Jeffs, *Proc. Chem. Soc.*, 1962, 80; W. C. Wildman, H. M. Fales, and A. R. Battersby, *J. Amer. Chem. Soc.*, 1962, **84,** 681.
200. A. R. Battersby, H. M. Fales, and W. C. Wildman, *J. Amer. Chem. Soc.*, 1961, **83,** 4098.
201. D. H. R. Barton, G. W. Kirby, J. B. Taylor, and G. M. Thomas, *J. Chem. Soc.*, 1963, 4545.
202. R. J. Suhadolnik, A. G. Fischer, and J. Zulalian, *J. Amer. Chem. Soc.*, 1962, **84,** 4348.
203. R. J. Suhadolnik and A. G. Fischer, *Proc. Chem. Soc.*, 1963, 132; R. J. Suhadolnik and J. Zulalian, *ibid.*, p. 216.
204. G. W. Kirby and H. P. Tiwari, *J. Chem. Soc. (C)*, 1966, 676.
205. J. Zulalian and R. J. Suhadolnik, *Proc. Chem. Soc.*, 1964, 422.
206. R. H. Wightman, J. Staunton, A. R. Battersby, and K. R. Hanson, *J.C.S. Perkin I*, 1972, 2355.
207. C. Fuganti, *Chim. Ind. (Milan)*, 1969, **51,** 1254.
208. H. M. Fales, J. Mann, and S. H. Mudd, *J. Amer. Chem. Soc.*, 1963, **85,** 2025.
209. J. G. Bhandarkar and G. W. Kirby, *J. Chem. Soc. (C)*, 1970, 1224.
210. C. Fuganti and M. Mazza, *J.C.S. Chem. Comm.*, 1972, 936.
211. C. Fuganti, D. Ghiringhelli, and P. Grasselli, *J.C.S. Chem. Comm.*, 1974, 350.
212. C. Fuganti and M. Mazza, *Chem. Comm.*, 1971, 1196.
213. C. Fuganti and M. Mazza, *J.C.S. Perkin I*, 1973, 954.
214. A. R. Battersby, J. E. Kelsey, J. Staunton, and K. E. Suckling, *J.C.S. Perkin I*, 1973, 1609; G. W. Kirby and J. Michael, *ibid.*, p. 115.
215. C. Fuganti, D. Ghiringhelli, and P. Grasselli, *J.C.S. Chem. Comm.*, 1973, 430.
216. C. Fuganti, J. Staunton, and A. R. Battersby, *Chem. Comm.*, 1971, 1154; C. Fuganti and M. Mazza, *ibid.*, p. 1388; *J.C.S. Chem. Comm.*, 1972, 239.
217. C. Fuganti, *Gazzetta*, 1973, **103,** 1255.
218. C. Fuganti and M. Mazza, *Chem. Comm.*, 1970, 1466; C. Fuganti, *Tetrahedron Letters*, 1973, 1785.
219. P. W. Jeffs, H. F. Campbell, D. S. Farrier, G. Ganguli, N. H. Martin, and G. Molina, *Phytochemistry*, 1974, **13,** 933.
220. P. W. Jeffs, W. C. Archie, R. L. Hawks, and D. S. Farrier, *J. Amer. Chem. Soc.*, 1971, **93,** 3752.
221. P. W. Jeffs, D. B. Johnson, N. H. Martin, and B. S. Rauckman, *J.C.S. Chem. Comm.*, 1976, 82.
222. D. Munsche and K. Mothes, *Phytochemistry*, 1965, **4,** 705.
223. P. Schröder, in 'Biochemie und Physiologie der Alkaloide', Fourth International Symposium, 1969, ed. K. Mothes, K. Schreiber, and H. R. Schütte, Akademie-Verlag, Berlin, 1972, p. 519; P. Schröder and M. Luckner, *Pharmazie*, 1966, **21,** 642.
224. M. Blaschke-Cobet and M. Luckner, *Phytochemistry*, 1973, **12,** 2393.
225. C. L. Tipton, M.-C. Wang, F. H.-C. Tsao, C. L. Tu, and R. R. Husted, *Phytochemistry*, 1973, **12,** 347; J. E. Reimann and R. U. Byerrum, *Biochemistry*, 1964, **3,** 847.
226. I. Monković, I. D. Spenser, and A. O. Plunkett, *Canad. J. Chem.*, 1967, **45,** 1935; M. Matsuo and Y. Kasida, *Chem. Pharm. Bull. (Japan)*, 1966, **14,** 1108; M. Matsuo, M. Yamasaki, and Y. Kasida, *Biochem. Biophys. Res. Comm.*, 1966, **23,** 679.
227. A. O. Colonna and E. G. Gros, *Phytochemistry*, 1971, **10,** 1515.
228. M. Cobet and M. Luckner, *Phytochemistry*, 1971, **10,** 1031; *European J. Biochem.*, 1968, **4,** 76.
229. J. F. Collins, W. J. Donnelly, M. F. Grundon, and K. J. James, *J.C.S. Perkin I*, 1974, 2177.
230. M. F. Grundon, D. M. Harrison, and C. G. Spyropoulos, *J.C.S. Perkin I*, 1975, 302.
231. M. F. Grundon, D. M. Harrison, and C. G. Spyropoulos, *J.C.S. Perkin I*, 1974, 2181.
232. D. Boulanger, B. K. Bailey, and W. Steck, *Phytochemistry*, 1973, **12,** 2399.
233. T. R. Chamberlain, J. F. Collins, and M. F. Grundon, *Chem. Comm.*, 1969, 1269.
234. R. H. Prager and H. M. Thredgold, *Austral. J. Chem.*, 1969, **22,** 2627.
235. D. Gröger and S. Johne, *Z. Naturforsch.*, 1968, **23b,** 1072.
236. S. Johne, K. Waiblinger and D. Gröger, *European J. Biochem.*, 1970, **15,** 415; D. G. O'Donovan and H. Horan, *J. Chem. Soc. (C)*, 1970, 2466.

237. D. R. Liljegren, *Phytochemistry*, 1971, **10,** 2661.
238. K. Waiblinger, S. Johne, and D. Gröger, *Phytochemistry*, 1972, **11,** 2263; S. Johne and D. Gröger, *ibid.*, 1968, **7,** 429.
239. S. Johne, D. Gröger, and G. Richter, *Arch. Pharm.*, 1968, **301,** 721; D. Gröger, S. Johne, and K. Mothes, *Experientia*, 1965, **21,** 13; D. Gröger and K. Mothes, *Arch. Pharm.*, 1960, **293,** 1049.
240. S. Johne, K. Waiblinger, and D. Gröger, *Pharmazie*, 1973, **28,** 403.
241. M. Yamazaki, A. Ikuta, T. Mori, and T. Kawana, *Tetrahedron Letters*, 1967, 3317; M. Yamazaki and A. Ikuta, *ibid.*, 1966, 3221.
242. H. G. Floss, *Tetrahedron*, 1976, **32,** 873.
243. K. Bowden and L. Marion, *Canad. J. Chem.*, 1951, **29,** 1037.
244. D. Gross, H. Lehman, and H. R. Schütte, *Tetrahedron Letters*, 1971, 4047; D. O'Donovan and E. Leete, *J. Amer. Chem. Soc.*, 1963, **85,** 461; and refs. cited therein.
245. D. Gross, A. Nemeckova, and H. R. Schütte, *Z. Pflanzenphysiol.*, 1967, **57,** 60.
246. B. G. Gower and E. Leete, *J. Amer. Chem. Soc.*, 1963, **85,** 3683.
247. E. Wenkert, *J. Amer. Chem. Soc.*, 1962, **84,** 98.
248. E. Leete, *Phytochemistry*, 1975, **14,** 471.
249. S. Agurell and J. L. G. Nilsson, *Acta Chem. Scand.*, 1968, **22,** 1210.
250. C. Baxter and M. Slaytor, *Phytochemistry*, 1972, **11,** 2767.
251. M. Slaytor and I. J. McFarlane, *Phytochemistry*, 1968, **7,** 605.
252. D. G. O'Donovan and M. F. Kenneally, *J. Chem. Soc. (C)*, 1967, 1109.
253. D. G. O'Donovan and M. F. Keogh, *J. Chem. Soc. (C)*, 1966, 1570.
254. H. R. Schütte and B. Maier, *Arch. Pharm.*, 1965, **298,** 459.
255. E. Leete, A. Ahmad, and I. Kompis, *J. Amer. Chem. Soc.*, 1965, **87,** 4168; Y. Yamasaki and E. Leete, *Tetrahedron Letters*, 1964, 1499; E. Leete, *J. Amer. Chem. Soc.*, 1960, **82,** 6338; *Tetrahedron*, 1961, **14,** 35.
256. J. P. Kutney, W. J. Cretney, J. R. Hadfield, E. S. Hall, V. R. Nelson, and D. C. Wigfield, *J. Amer. Chem. Soc.*, 1968, **90,** 3566.
257. A. R. Battersby, A. R. Burnett, and P. G. Parsons, *J. Chem. Soc. (C)*, 1969, 1193.
258. R. Thomas, *Tetrahedron Letters*, 1961, 544; see also Ref. 187.
259. (a) A. R. Battersby, R. T. Brown, R. S. Kapil, J. A. Knight, J. A. Martin, and A. O. Plunkett, *Chem. Comm.*, 1966, 888; (b) refs. cited therein; T. Money, I. G. Wright, F. McCapra, E. S. Hall, and A. I. Scott, *J. Amer. Chem. Soc.*, 1968, **90,** 4144; D. Gröger, K. Stolle, and K. Mothes, *Arch. Pharm.*, 1967, **300,** 393.
260. A. R. Battersby, J. C. Byrne, R. S. Kapil, J. A. Martin, T. G. Payne, D. Arigoni, and P. Loew, *Chem. Comm.*, 1968, 951.
261. A. R. Battersby, R. T. Brown, R. S. Kapil, J. A. Martin, and A. O. Plunkett, *Chem. Comm.*, 1966, 890; A. R. Battersby, R. S. Kapil, J. A. Martin, and L. Mo, *ibid.*, 1968, 133; P. Loew and D. Arigoni, *ibid.*, p. 137.
262. A. R. Battersby, E. S. Hall, and R. Southgate, *J. Chem. Soc. (C)*, 1969, 721, and refs. cited therein.
263. (a) S. Escher, P. Loew, and D. Arigoni, *Chem. Comm.*, 1970, 823; (b) refs. cited therein.
264. A. R. Battersby, A. R. Burnett, and P. G. Parsons, *Chem. Comm.*, 1970, 826.
265. A. R. Battersby, S. H. Brown, and T. G. Payne, *Chem. Comm.*, 1970, 827.
266. A. R. Battersby, A. R. Burnett, and P. G. Parsons, *J. Chem. Soc. (C)*, 1969, 1187
267. K. T. De Silva, G. N. Smith, and K. E. H. Warren, *Chem. Comm.*, 1971, 905; W. P. Blackstock, R. T. Brown and G. K. Lee, *ibid.*, p. 910.
268. R. T. Brown, G. N. Smith, and K. S. J. Stapleford, *Tetrahedron Letters*, 1968, 4349.
269. A. R. Battersby, A. R. Burnett, E. S. Hall, and P. G. Parsons, *Chem. Comm.*, 1968, 1582; A. R. Battersby and K. H. Gibson, *ibid.*, 1971, 902.
270. M. Rueffer, N. Nagakura, and M. H. Zenk, *Tetrahedron Letters*, 1978, 1593, and refs. cited therein.
271. A. A. Qureshi and A. I. Scott, *Chem. Comm.*, 1968, 948.
272. A. R. Battersby and E. S. Hall, *Chem. Comm.*, 1969, 793; A. I. Scott, P. C. Cherry, and A. A. Qureshi, *J. Amer. Chem. Soc.*, 1969, **91,** 4932.
273. S. I. Heimberger and A. I. Scott, *J.C.S. Chem. Comm.*, 1973, 217.
274. Ch. Schlatter, E. E. Waldner, H. Schmid, W. Maier, and D. Gröger, *Helv. Chim Acta*, 1969, **52,** 776.
275. A. I. Scott, P. B. Reichardt, M. B. Slaytor, and J. G. Sweeny, *Bioorg. Chem.*, 1971, **1,** 157.
276. J. P. Kutney, J. F. Beck, N. J. Eggers, H. W. Hanssen, R. S. Sood, and N. D. Westcott, *J. Amer. Chem. Soc.*, 1971, **93,** 7322.
277. J. P. Kutney, J. F. Beck, V. R. Nelson, and R. S. Sood, *J. Amer. Chem. Soc.*, 1971, **93,** 255.
278. J. P. Kutney, *Heterocycles*, 1976, **4,** 429.
279. J. P. Kutney, V. R. Nelson, and D. C. Wigfield, *J. Amer. Chem. Soc.*, 1969, **91,** 4278.
280. J. P. Kutney, V. R. Nelson, and D. C. Wigfield, *J. Amer. Chem. Soc.*, 1969, **91,** 4279; J. P. Kutney, J. F. Beck, C. Ehret, G. Poulton, R. S. Sood, and N. D. Westcott, *Bioorg. Chem.*, 1971, **1,** 194.
281. N. Kowanko and E. Leete, *J. Amer. Chem. Soc.*, 1962, **84,** 4919; E. Leete and J. N. Wemple, *ibid.*, 1969, **91,** 2698; A. R. Battersby and E. S. Hall, *Chem. Comm.*, 1970, 194.
282. A. R. Battersby and R. J. Parry, *Chem. Comm.*, 1971, 30.
283. A. R. Battersby and R. J. Parry, *Chem. Comm.*, 1971, 31.
284. C. R. Hutchinson, A. H. Heckendorf, P. E. Daddona, E. Hagaman, and E. Wenkert, *J. Amer. Chem. Soc.*, 1974, **96,** 5609.
285. G. M. Sheriha and H. Rapoport, *Phytochemistry*, 1976, **15,** 505.

286. A. R. Battersby and B. Gregory, *Chem. Comm.*, 1968, 134.
287. A. R. Battersby, R. Binks, W. Lawrie, G. V. Parry, and B. R. Webster, *J. Chem. Soc.*, 1965, 7459.
288. A. R. Battersby and R. J. Parry, *Chem. Comm.*, 1971, 901.
289. A. R. Guseva and V. P. Paseshnichenko, *Biochemistry* (*U.S.S.R.*), 1962, **27,** 721.
290. H. Sander and H. Grisebach, *Z. Naturforsch.*, 1958, **13b,** 755.
291. E. Heftmann, E. R. Lieber, and R. D. Bennett, *Phytochemistry*, 1967, **6,** 225; K. Kaneko, H. Mitsuhashi, K. Hirayama, and N. Yoshida, *ibid.*, 1970, **9,** 2489; R. Tschesche and H. Hulpke, *Z. Naturforsch.*, 1966, **21b,** 893; *ibid.*, 1967, **22b,** 791.
292. H. Ripperger, W. Moritz, and K. Schreiber, *Phytochemistry*, 1971, **10,** 2699.
293. R. D. Bennett and E. Heftmann, *Phytochemistry*, 1965, **4,** 873; *Arch. Biochem. Biophys.*, 1965, **112,** 616.
294. F. Ronchetti and G. Russo, *J.C.S. Chem. Comm.*, 1974, 785.
295. F. Ronchetti, G. Russo, G. Ferrara, and G. Vecchio, *Phytochemistry*, 1975, **14,** 2423.
296. D. R. Liljegren, *Phytochemistry*, 1971, **10,** 3061.
297. K. Kaneko, H. Seto, C. Motoki, and H. Mitsuhashi, *Phytochemistry*, 1975, **14,** 1295.
298. K. Kaneko, M. Watanabe, S. Taira, and H. Mitsuhashi, *Phytochemistry*, 1972, **11,** 3199.
299. K. Kaneko, H. Mitsuhashi, K. Hirayama, and S. Ohmori, *Phytochemistry*, 1970, **9,** 2497.
300. H. Auda, G. R. Waller, and E. J. Eisenbraun, *J. Biol. Chem.*, 1967, **242,** 4157; H. Auda, H. R. Juneja, E. J. Eisenbraun, G. R. Waller, W. R. Kays and H. H. Appel, *J. Amer. Chem. Soc.*, 1967, **89,** 2476.
301. D. Gross, W. Berg, and H. R. Schütte, *Biochem. Physiol. Pflanzen*, 1972, **163,** 576.
302. M. Yamazaki, M. Matsuo, and K. Arai, *Chem. Pharm. Bull.* (*Japan*), 1966, **14,** 1058.
303. A. Corbella, P. Gariboldi, G. Jommi, and M. Sisti, *J.C.S. Chem. Comm.*, 1975, 288; A. Corbella, P. Gariboldi, and G. Jommi, *ibid.*, 1973, 729.
304. J. R. Hanson and R. Nyfeler, *J.C.S. Chem. Comm.*, 1975, 171; *ibid.*, p. 824.
305. K. T. Suzuki, S. Okuda, H. Niwa, M. Toda, Y. Hirata, and S. Yamamura, *Tetrahedron Letters*, 1973, 799.
306. H. Niwa, Y. Hirata, K. T. Suzuki and S. Yamamura, *Tetrahedron Letters*, 1973, 2129.
307. H. E. Miller, H. Rösler, A. Wohlpart, H. Wyler, M. E. Wilcox, H. Frohofer, T. J. Mabry, and A. S. Dreiding, *Helv. Chim. Acta*, 1968, **51,** 1470; E. Dunkelblum, H. E. Miller, and A. S. Dreiding, *ibid.*, 1972, **55,** 642; see also, C. Chang, L. Kimler, and T. J. Mabry, *Phytochemistry*, 1974, **13,** 2771.
308. N. Fischer and A. S. Dreiding, *Helv. Chim. Acta*, 1972, **55,** 649.
309. G. Impellizzeri and M. Piattelli *Phytochemistry*, 1972, **11,** 2499.
310. S. Sciuto, G. Oriente, and M. Piattelli, *Phytochemistry*, 1972, **11,** 2259.
311. S. Sciuto, G. Oriente, M. Piattelli G. Impellizzeri, and V. Amico, *Phytochemistry*, 1974, **13,** 947.
312. D. G. O'Donovan and H. Horan, *J. Chem. Soc.* (*C*), 1971, 331.
313. D. J. Bennett and G. W. Kirby, *J. Chem. Soc.* (*C*), 1968, 442.
314. E. Leete and M. C. L. Louden, *J. Amer. Chem. Soc.*, 1968, **90,** 6837.
315. H. Horan and D. G. O'Donovan, *J. Chem. Soc.* (*C*), 1971, 2083; H. Rosenberg and A. G. Paul, *Lloydia*, 1972, **34,** 372.
316. E. Leete and G. B. Bodem, *J.C.S. Chem. Comm.*, 1973, 522.

30.2

Porphyrin, Chlorophyll, and Corrin Biosynthesis

M. AKHTAR AND P. M. JORDAN
University of Southampton

30.2.1 GENERAL INTRODUCTION

The study of porphyrins, chlorophylls, and corrins, particularly in the past 30 years, has challenged the expertise of both organic chemists and biochemists. Even as early as 1880 the recognition of similarities in the spectra of haem and chlorophyll suggested that they were structurally related compounds. In the 1920s the pioneering work of Willstätter[1] and Fischer[2] established the structure of haem, which was confirmed by synthesis as early as 1929. In contrast, chlorophyll, although its structure was known in 1934, was not chemically synthesized until as late as 1960.[3] The structure of vitamin B_{12}, the biologically important corrin derivative, was elucidated in 1955,[4] and its total synthesis accomplished by Woodward[5] and Echenmoser[6] nearly 20 years later marked one of the milestones in modern synthetic organic chemistry.

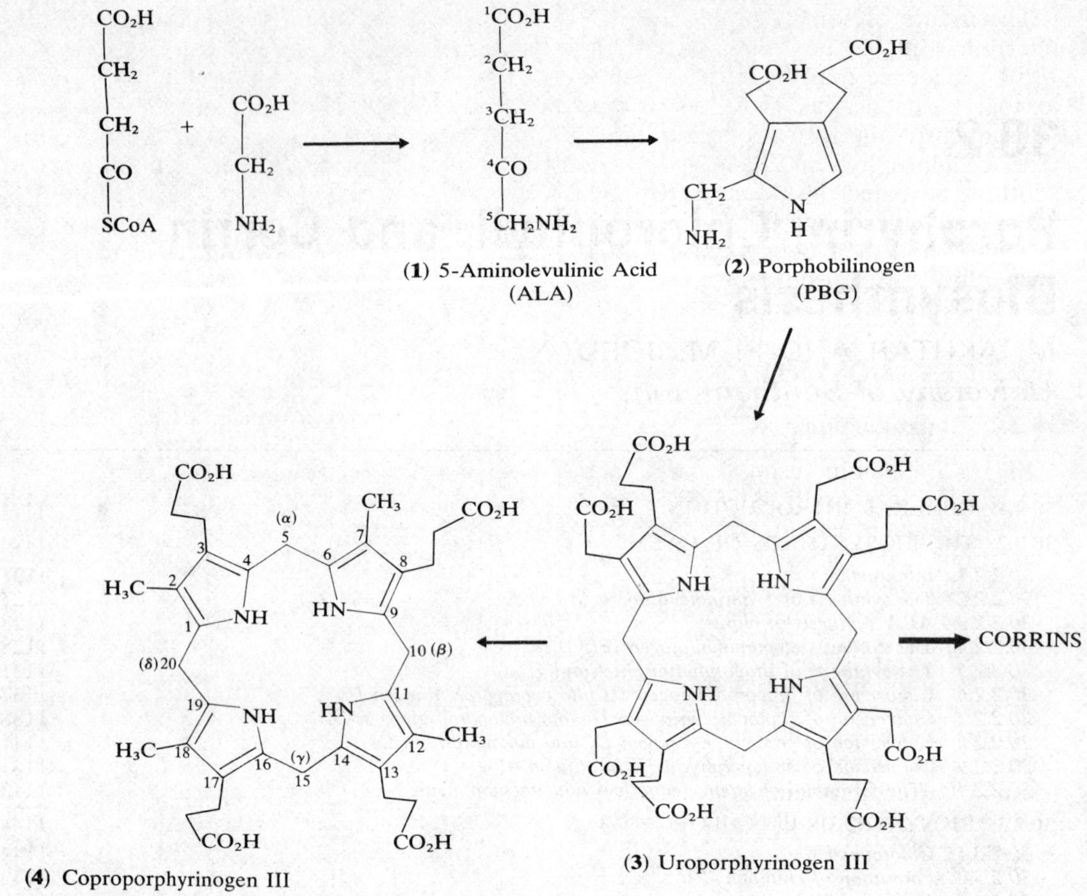

SCHEME 1 Summary pathway for the biosynthesis of haem, chlorophylls, and corrins. Throughout the chapter the 1–20 numbering system shown in structure (**4**) is used. The positions 2, 3, 7, 8, 12, 13, 17, and 18 according to this system correspond to positions 1, 2, 3, 4, 5, 6, 7, and 8 of the 1–8 numbering system

The nature of some of the enzymic intermediates involved in haem biosynthesis was inferred from the now classical labelling studies of Shemin and his colleagues around 1950.[7] Evidence that the early stages of chlorophyll biosynthesis followed a route similar to that of haem was provided by the work of Granick using mutants of *Chlorella*.[8] Subsequently, the formation of corrins from the same initial building blocks that contribute to chlorophyll and haem was demonstrated by Shemin.[9] More recently the work of Scott has revealed the extent to which the biosynthesis of corrin parallels the pathways of haem and chlorophyll.[10] A summary of these biosynthetic pathways is illustrated in Scheme 1 and their details are discussed in three sections: the biosynthesis of haem; the biosynthesis of chlorophylls; and the biosynthesis of corrins.

30.2.2 THE BIOSYNTHESIS OF HAEM

30.2.2.1 Introduction

In 1945, Shemin, using himself as the experimental subject, established that when [^{15}N]glycine was orally administered, the label was effectively incorporated into haem (**6**).[11] Subsequent experiments using avian erythrocyte preparations showed that radioactivity from [2-^{14}C]glycine, but not from [1-^{14}C]glycine, was incorporated into haem.[12,13] The design of a degradation scheme whereby all the carbon atoms of haem could be isolated and their position in the macrocycle unambiguously assigned was a crucial experimental development. In this degradation (Scheme 2), haem was converted into mesoporphyrin IX (**7**), which was then cleaved by chromic oxide to yield ethylmethylmaleimide (**8**) and hematinic acid (**9**). The maleimides were then fragmented and the positions of the carbon atoms originally in haem were delineated. Subsequently, labelling experiments established that [1-^{14}C]- and [2-^{14}C]-acetate were incorporated into haem[13,14] after first being converted by way of the TCA cycle[15] into succinyl-CoA.[16] The biosynthetic origin of all the carbon and nitrogen atoms in haem is thus as shown in (**6a**).

30.2.2.2 The synthesis of 5-aminolevulinic acid (ALA)

The labelling data obtained from incorporation studies with radioactive glycine and acetate, and the knowledge that the carboxyl group of glycine was lost during the biosynthesis of haem, led to the proposal[17] that the condensation of glycine with succinyl-CoA results in a transient intermediate (**10**), which yielded ALA (**1**) on decarboxylation.

This suggestion was confirmed when [5-^{14}C]ALA was chemically synthesized and shown to be incorporated into haem, not only in high yield but with a labelling pattern identical to that of [2-^{14}C]glycine.[18] However, it was not until 1958 that two research groups independently recognized the enzyme ALA synthetase, which converted glycine and succinyl-CoA into ALA in both bacterial[19] and avian preparations.[20] The enzyme has been highly purified from bacterial[21] and mammalian sources[22] and much information about its properties and physiological role is now known.[23]

The enzyme from all sources requires the cofactor pyridoxal phosphate for full catalytic activity. Spectroscopic data[24] indicate that at pH 5 or pH 8.5 the enzyme–pyridoxal phosphate link is largely in the form of a Schiff base (**11**) absorbing at 415 nm. At pH 7.2, where the enzyme is catalytically active, the experimental evidence is more consistent with the presence of a carbinolamine (**12**) or equivalent structure absorbing at 330 nm.

The mechanism by which glycine condenses with succinyl-CoA has been studied[25] using glycine labelled at the important 2-position, where the bond-forming events take place. It is well established that the 2-position of glycine is incorporated into the 5-position of the product ALA (Scheme 3). Accordingly, [2*RS*-$^{3}H_2$]glycine (**13**) together with [2-^{14}C]glycine were transformed into ALA by highly purified ALA synthetase from *Rhodopseudomonas spheroides*. To prevent loss or randomization of any tritium label at C-5, the

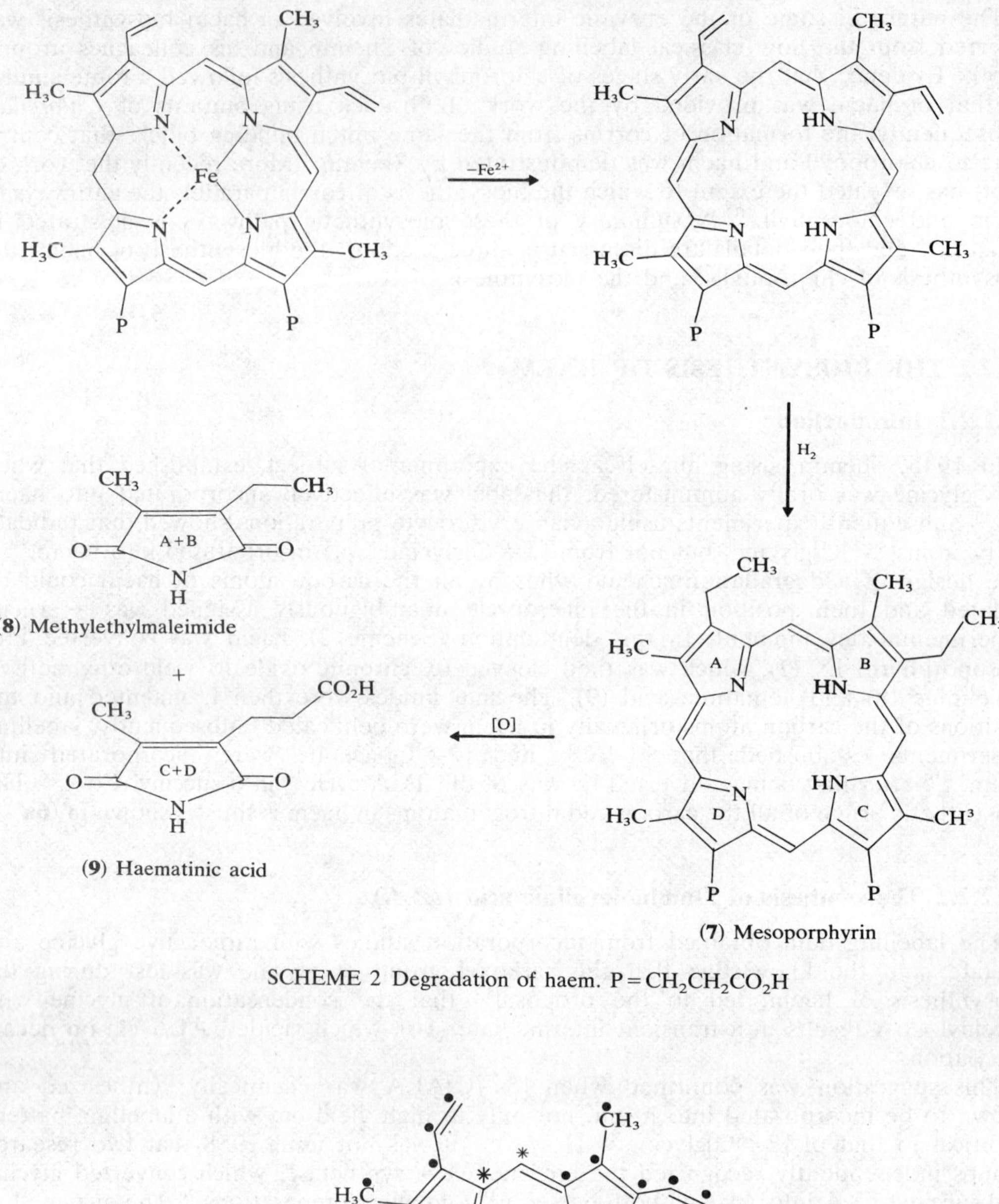

SCHEME 2 Degradation of haem. $P = CH_2CH_2CO_2H$

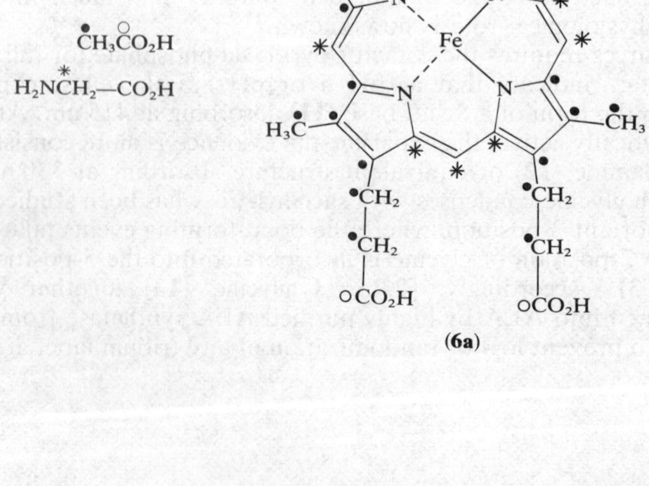

(6a)

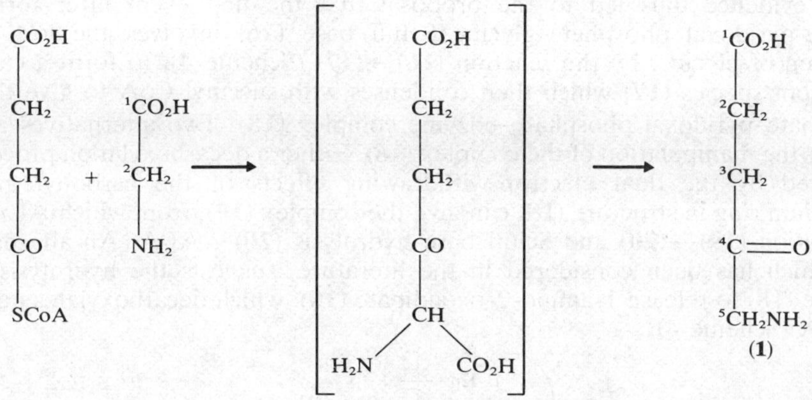

(10) (1)

SCHEME 3 The ALA-synthetase reaction

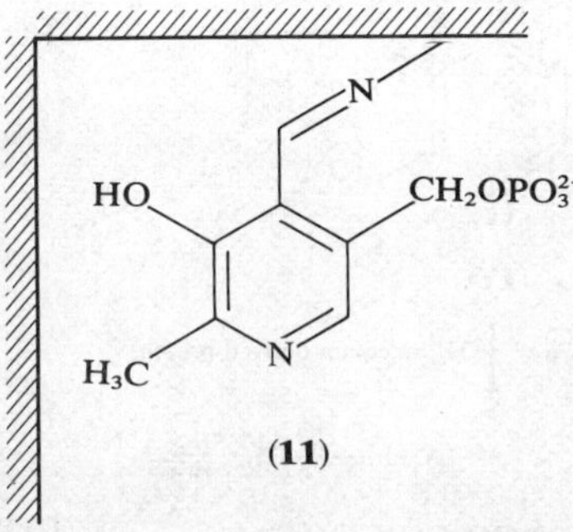

(11)

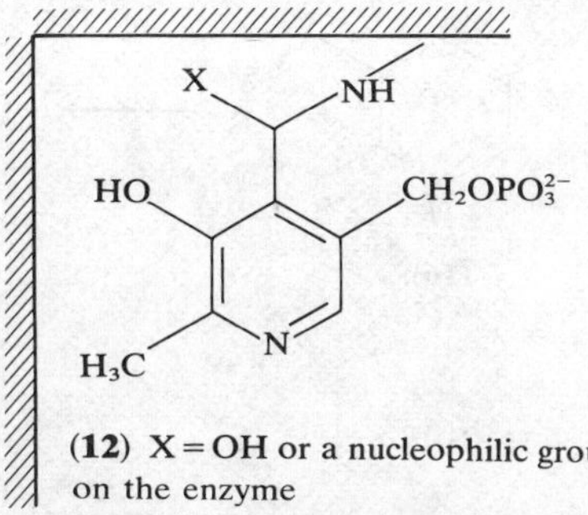

(12) X = OH or a nucleophilic group on the enzyme

ALA was reduced with $NaBH_4$ and the resulting amino-alcohol was cleaved with periodate to yield formaldehyde. Comparison between the $^3H/^{14}C$ ratio of the precursor glycine and the formaldehyde isolated from the degration of this ALA indicated that 50% of the tritium label had been lost during the enzymic conversion. Subsequent experiments using [2*R*-^{3}H]glycine (**14**) and [2*S*-^{3}H]glycine (**15**) revealed that all the tritium radioactivity was incorporated into ALA from [2*S*-^{3}H]glycine, whereas ALA biosynthesized from [2*R*-^{3}H]glycine contained negligible tritium label.[25]

Substrate	*% of the ^3H of Gly at C-5 of ALA*
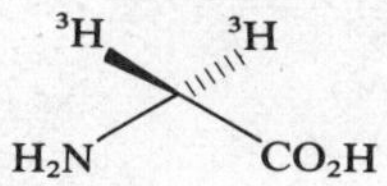 (**13**) [2*RS*-^{3}H]glycine	47%
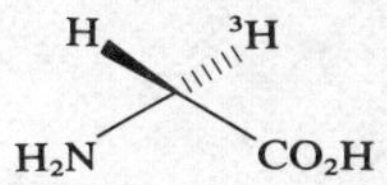 (**14**) [2*R*-^{3}H]glycine	3%
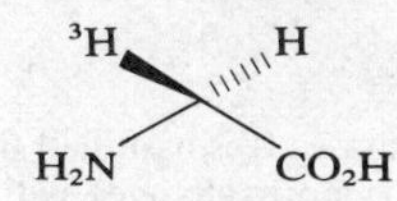 (**15**) [2*S*-^{3}H]glycine	100%

The evidence thus led to the proposal that the first event after formation of the enzyme-pyridoxal phosphate–glycine Schiff base (**16**) involves the loss of the *pro-R* hydrogen of glycine, by the reaction (**16**)→(**17**) (Scheme 4), to form a carbanion or an equivalent species (**17**) which then condenses with succinyl-CoA to give the 1-amino-2-oxoadipate–pyridoxal phosphate–enzyme complex (**18**). Two alternatives are now available for the manipulation of the complex (**18**). Either a decarboxylation process (**18**)→(**19**) promoted by the dual electron-withdrawing effects of the carbonyl group and the pyridinium ring in structure (**18**) can give the complex (**19**), from which ALA is formed by protonation (**19**)→(**20**) and Schiff base hydrolysis (**20**)→ALA. An alternative mechanism, which has been considered in the literature, suggests the hydrolysis of the intermediate (**18**) to release 1-amino-2-oxoadipate (**10**), which decarboxylates non-enzymically to ALA (Scheme 4).

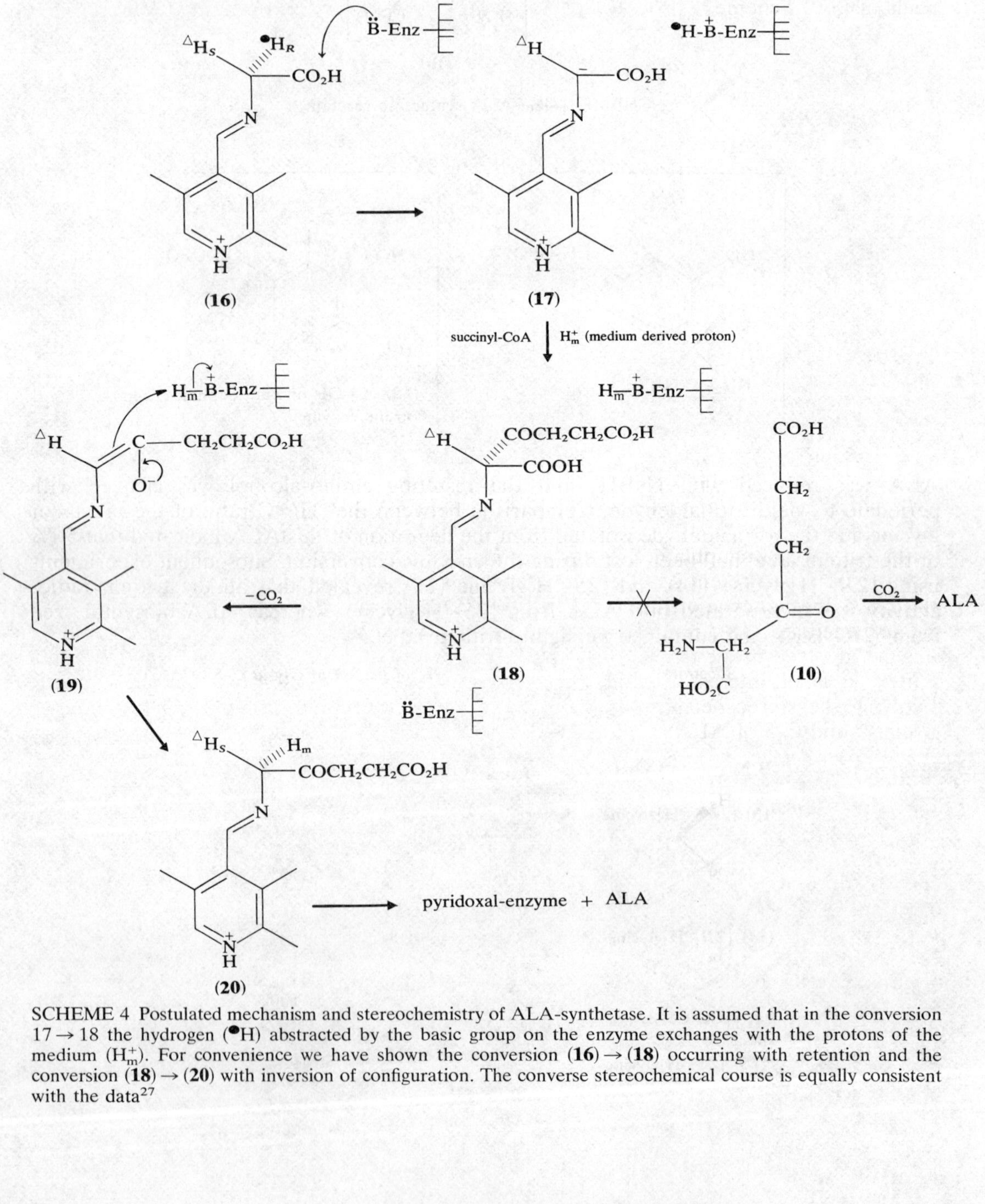

SCHEME 4 Postulated mechanism and stereochemistry of ALA-synthetase. It is assumed that in the conversion 17→18 the hydrogen (•H) abstracted by the basic group on the enzyme exchanges with the protons of the medium (H_m^+). For convenience we have shown the conversion (**16**)→(**18**) occurring with retention and the conversion (**18**)→(**20**) with inversion of configuration. The converse stereochemical course is equally consistent with the data[27]

That it is the former mechanism which operates has been inferred through the determination of the absolute stereochemistry of the pro-chiral centre at C-5 of ALA.[26] This was achieved by the incorporation of [2*RS*-$^3H_2$]glycine (**13**) into ALA followed by the conversion of the latter into porphobilinogen using porphobilinogen synthetase (also known as ALA-dehydratase). The degradation of the porphobilinogen to glycine (Scheme 5) and the subsequent determination of its chirality established that the biosynthetic ALA contained the majority of the radioactive label in the *pro-S* position at C-5 (Scheme 5). This stereoselective formation of ALA could only have occurred if the decarboxylation of the intermediate (**18**) had proceeded on the enzyme surface and the species (**19**) thus generated was subsequently protonated, incorporating a hydrogen derived from the medium to yield ALA by the sequence (**19**) → (**20**) → ALA. The stereochemical information may be interpreted to suggest that the overall reaction mechanism of Scheme 4 proceeds by involving one retention and one inversion.[27]

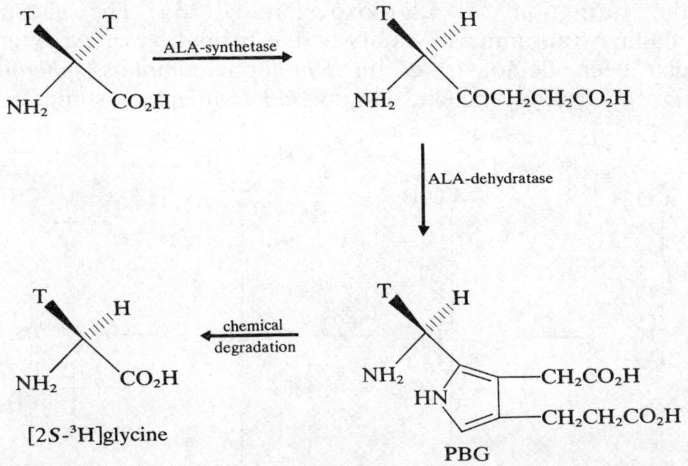

SCHEME 5 Biosynthesis and degradation of porphobilinogen (PBG)

For the sake of simplicity we have assumed here that the condensation (**16**) → (**17**) → (**18**) occurs with a retention of configuration and that the inversion of configuration takes place as a consequence of the fact that the enol (**19**), formed from the decarboxylation of the intermediate (**18**), is protonated to yield (**20**), possibly by the same group on the enzyme involved in the deprotonation of glycine in the reaction (**16**) → (**17**). In the latter context it is of interest to note that ALA synthetase catalyses two partial reactions involving the stereospecific exchange of the *pro-R* hydrogen atom attached to C-2 of glycine[28] and C-5 of ALA,[29] in the absence of succinyl-CoA or CoA as shown in Scheme 5a.

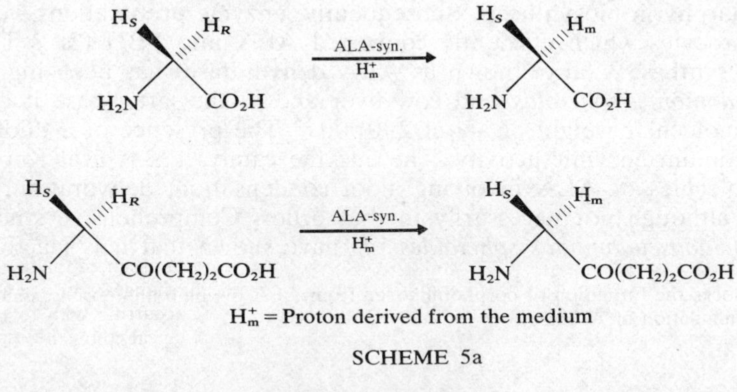

H_m^+ = Proton derived from the medium

SCHEME 5a

30.2.2.3 ALA synthesis in plants

All attempts to isolate ALA synthetase from plants have been unsuccessful and the early observation that radioactive label was incorporated into chlorophyll equally well from either [1-^{14}C]- or [2-^{14}C]-glycine was inconsistent with the orthodox route followed by animals and photosynthetic bacteria. A more comprehensive study[30] in which labelled compounds were fed to preparations of cucumber, barley, or bean plant tissues in the presence of levulinic acid* demonstrated that succinate, acetate, and glycine are far less effectively incorporated into ALA than glutamate (**21**) and related five-carbon compounds such as 2-oxoglutarate (**22**). Moreover, the pattern of labelling supported the intact incorporation of the glutamate carbon skeleton in which the 1-carboxyl of glutamate becomes the 5-position in ALA. It thus appears that an alternative route for the biosynthesis of ALA operates in plants in which glutamate is first converted into 2-oxoglutarate and then two enzymes participate in the conversion of the latter into ALA.[31] The first enzyme requires NADH as a cofactor and catalyses a novel reaction resulting in the formation of 4,5-dioxovalerate (**23**). The second enzyme, 4,5-dioxovalerate: alanine transaminase, catalyses the formation of ALA and pyruvate. This enzyme has also been demonstrated in *Rhodopseudomonas spheroides*, although its metabolic role in this case is unclear. The overall sequence is summarized in Scheme 6.

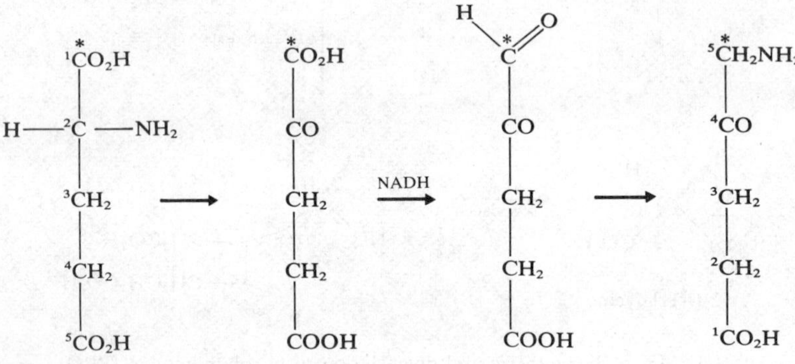

(**21**) (**22**) (**23**) (**1**)

SCHEME 6 The pathway for the biosynthesis of ALA in plants

30.2.2.4 The synthesis of porphobilinogen (PBG)

For a great many years the urine of porphyriac patients had been known to turn dark red when stored for a period of hours. The substance responsible for the formation of the red compound was isolated by Westall[32] and characterized two years later as porphobilinogen, although at this time its significance as an intermediate in porphyrin biosynthesis was not fully appreciated. It was subsequently shown that PBG was incorporated into haem in higher yields than ALA, confirming that porphobilinogen was the missing link in porphyrin biosynthesis. Subsequently, enzyme preparations were isolated from duck erythrocytes which efficiently converted ALA into PBG (**2**).[33] The enzyme porphobilinogen synthase[34] (also known as ALA dehydratase) has been highly purified from *Rhodopseudomonas spheroids* and cow liver and in the latter case it exists as an octamer with a molecular weight of about 290 000.[35] The presence of a thiol reagent is required for maximum enzymic activity. The enzyme catalyses a typical Knorr reaction between two molecules of ALA involving aldol condensation, dehydration, and Schiff base formation, although not necessarily in that order. Comprehensive studies on the enzyme from *Rhodopseudomonas spheroides*[36,37] have shown that it is inactivated when

* Levulinic acid blocks the formation of porphobilinogen from ALA by inhibiting porphobilinogen synthase, thus causing the accumulation of ALA.

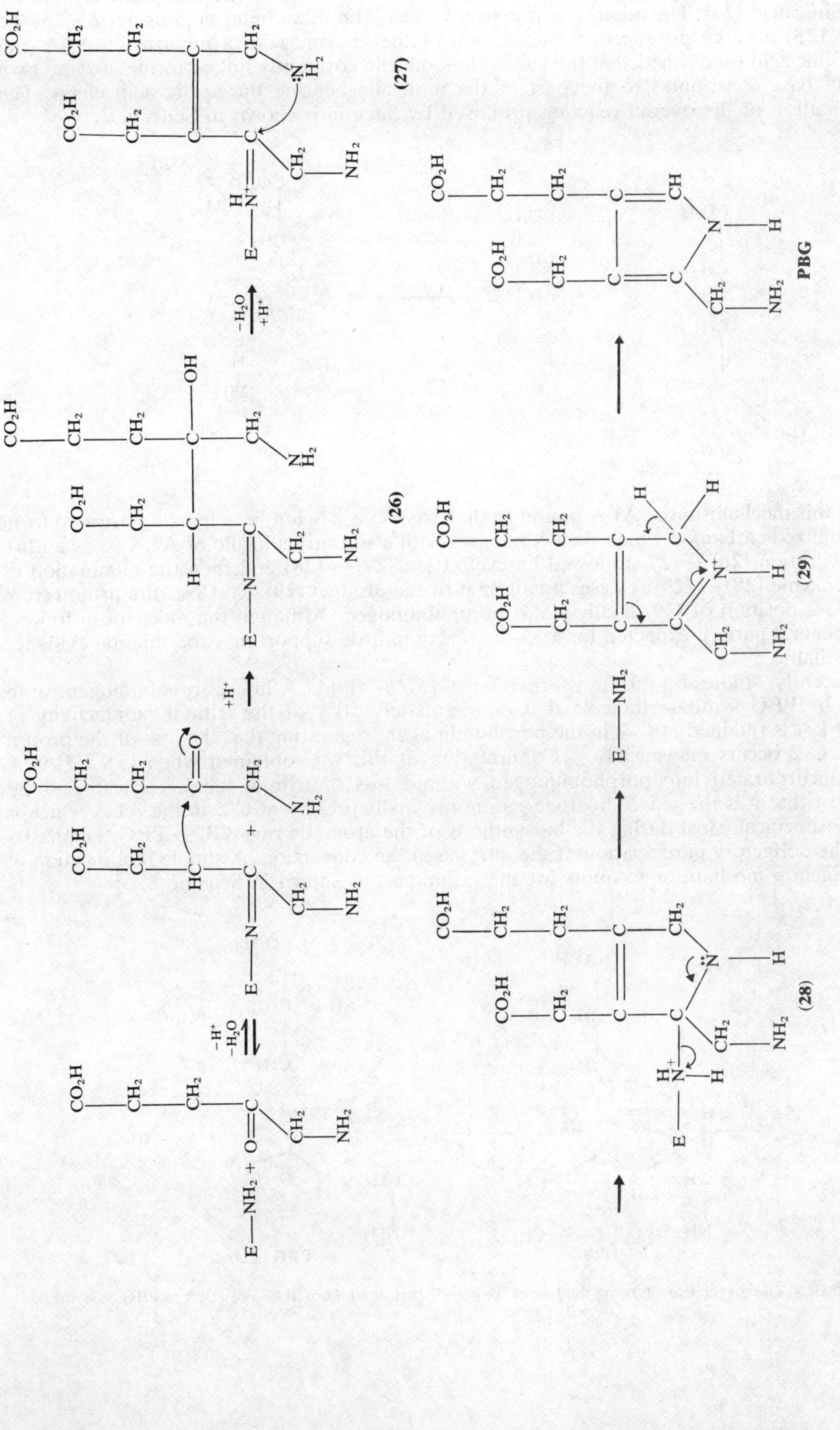

SCHEME 7 The mechanism for the biosynthesis of PBG according to Nandi and Shemin[37]

treated with $NaBH_4$ in the presence of the substrate ALA or the competitive inhibitor levulinic acid (**24**). The additional important observation that heterologous pyrroles of the type (**25**) may be produced by incubation of the enzyme with a mixture of ALA and levulinic acid established that the substrate molecule covalently linked to the enzyme by a Schiff base contributes to the part of the molecule bearing the acetic acid chain. The mechanism of the overall reaction proposed by Shemin is shown in Scheme 7.

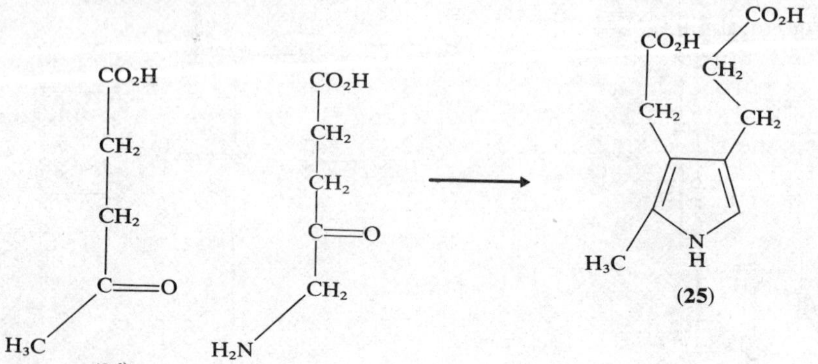

(**24**) (**25**)

In this mechanism the ALA bound to the enzyme as a Schiff base loses a proton to form a stabilized carbanion which then condenses with a second molecule of ALA to give (**26**). Dehydration (**26**) → (**27**) followed by cyclization (**27**) → (**28**) and then the elimination of the enzyme (**28**) → (**29**) releases a tautomer of the product (**29**). The loss of a proton from the C-2 position of (**29**) finally yields porphobilinogen. Although the mechanism follows the general pattern expected for a Knorr reaction, little supporting experimental evidence is available.

Recently, studies on the transformation of [5*RS*-$^3H_2$]ALA into porphobilinogen catalysed by PBG synthase indicate that approximately 50% of the tritium radioactivity of the ALA is retained at C-2 in the porphobilinogen, suggesting that the loss of the proton from C-2 occurs enzymically.[38] Confirmation of this was obtained when [5*S*-^{3}H]ALA was incorporated into porphobilinogen without loss of tritium label.[39] These findings suggest that it is the *pro-R* hydrogen atom originally present at C-5 in the ALA which is stereospecifically lost during the biosynthesis of the aromatic ring (**28**) → PBG, emphasizing the obligatory participation of the enzyme in the conversion. A simple modification of the Shemin mechanism accounts for these findings, as shown in Scheme 8.

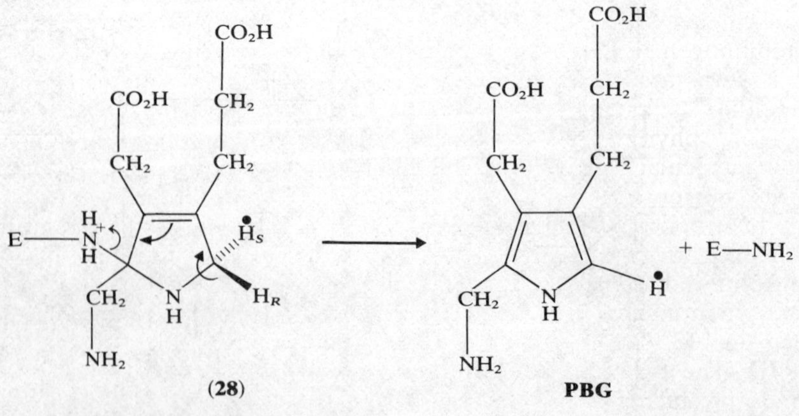

(**28**) **PBG**

SCHEME 8 The mechanism showing the loss of the *pro-R* hydrogen atom at C-5 of ALA in PBG biosynthesis

30.2.2.5 The synthesis of uroporphyrinogen I and III

Of all the problems in porphyrin biosynthesis, the biosynthetic mechanism for the formation of uroporphyrinogen III has attracted the most attention. Although uroporphyrinogen I is formed in a few isolated situations by biological systems, it is uroporphyrinogen III which is utilized exclusively for haem, chlorophyll, and corrin biosynthesis. The other two isomers of uroporphyrinogen, II and IV, are known to arise only from non-enzymic processes. Extensive studies have been made on the chemical reactivity of porphobilinogen and related pyrroles, and on the nature of the products formed by their polymerization under various conditions.[40] The acid-catalysed self-condensation of porphobilinogen yields all four possible isomers (Scheme 9) in the ratio III, IV, II, I of 4:2:1:1; furthermore, the treatment of any of these isomers with acid, under N_2, gives the same equilibrium mixture.[41]

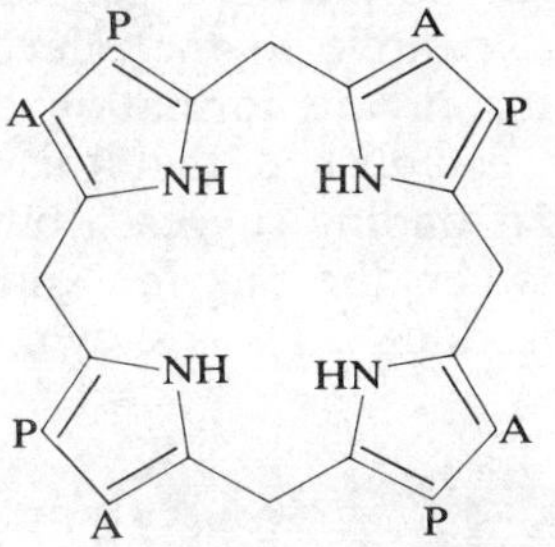

Uroporphyrinogen I

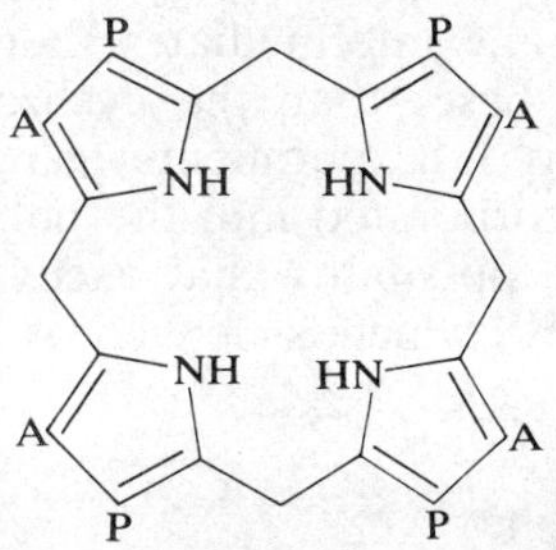

Uroporphyrinogen II

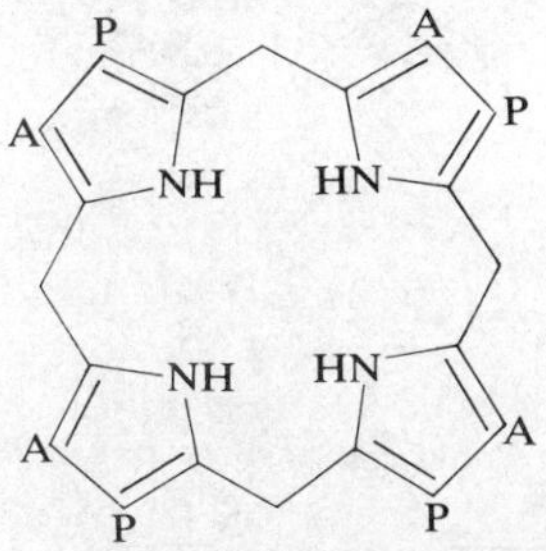

Uroporphyrinogen III

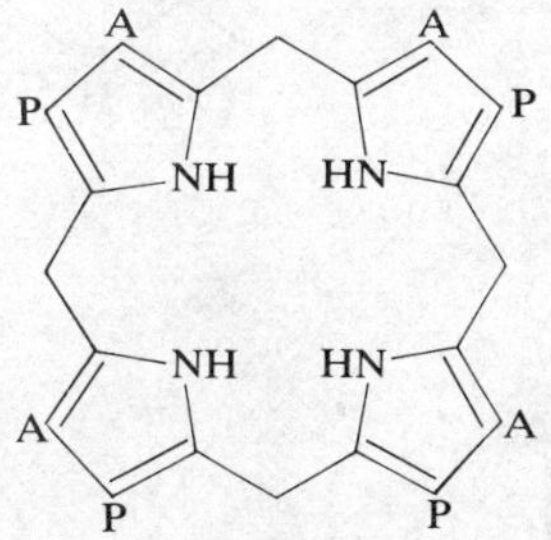

Uroporphyrinogen IV

$A = CH_2CO_2H$; $P = CH_2CH_2CO_2H$

SCHEME 9 Isomers of uroporphyrinogen

Initial experiments demonstrated that the enzymic formation of uroporphyrinogen III from porphobilinogen required the coordinated participation of two enzymes,[42] uroporphyrinogen I synthetase and uroporphyrinogen III co-synthetase. The two enzymes may be isolated either as an impure complex (termed porphobilinogenase) or as separate proteins.[43] Uroporphyrinogen I synthetase, which has been obtained from a variety of sources, has a molecular weight of 36 000 and acting alone converts porphobilinogen only into uroporphyrinogen I.

The study of uroporphyrinogen III co-synthetase has been hindered mainly owing to the lack of a straightforward assay system. This latter enzyme has been partially purified from wheat germ and spinach leaves and has a molecular weight of about 60 000.[44] The co-synthetase when incubated with porphobilinogen displays no apparent catalytic activity. It is also unable to promote the isomerization of uroporphyrinogen I into uroporphyrinogen III. The two separate enzymes may be reassociated to restore the uroporphyrinogen III producing capability and, in the presence of a fixed amount of uroporphyrinogen I synthetase, the formation of uroporphyrinogen III is proportional to the concentration of the co-synthetase. It thus appears that the co-synthetase is able to assert its activity only when it is associated as a complex with uroporphyrinogen I synthetase. In this way it modulates the activity of the uroporphyrinogen I synthetase, steering the polymerization of porphobilinogen towards the production of the III isomer.[45]

Apart from an isolated report,[46] one of the most striking features of either uroporphyrinogen I or III biosynthesis is the absence, under normal conditions, of the formation of any free intermediates. The studies of the mechanism by which uroporphyrinogens are formed was greatly stimulated therefore when polypyrromethanes were identified from incubation of PBG with uroporphyrinogen I synthetase in the presence of high concentrations of ammonia or hydroxylamine.[47]

These bases do not appreciably affect the consumption of porphobilinogen by the enzyme but dramatically inhibit the formation of the product, resulting in the liberation of compounds tentatively characterized as polypyrromethanes[48] (**32**) and (**33**) (or derivatives where NH_2 is replaced by NHOH). The condensation of porphobilinogen to form the pyrromethane intermediates therefore seems much less sensitive to the effects of these inhibitory bases than the cyclization stage, which results in the formation of uroporphyrinogen I. Experiments using [^{14}C]methoxyamine[49] established that this inhibitory base is incorporated into the polypyrromethane (**31**). This finding suggests, but does not prove, the possibility that a covalent linkage exists between the enzyme and an intermediate (**30**) which is cleaved by the inhibitory base to produce (**31**), as shown in Scheme 10.

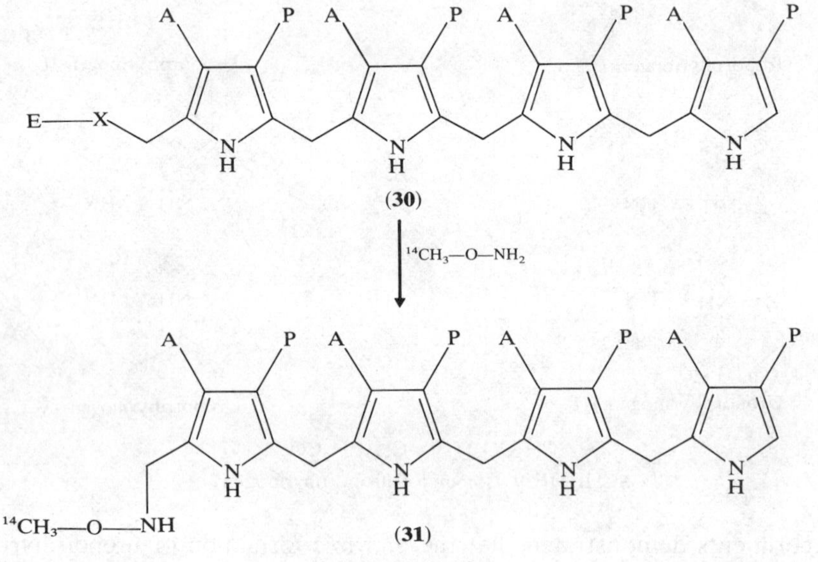

SCHEME 10

The recognition of pyrromethanes as intermediates in the biosynthesis of uroporphyrinogen I heralded the beginning of a period of intense activity in synthetic organic chemistry by the groups of Frydman[50] and Battersby.[51] Major emphasis was placed on the synthesis of various pyrromethanes [(**32**), (**39**)], *etc.*] in anticipation that one of them would act as substrate for the enzymic synthesis of uroporphyrinogen III. The strategy of these workers was based on the many hypotheses proposed for the formation of uroporphyrinogen III which had accumulated over the last 20 years. These hypotheses may be summarized in the form of three broad reaction mechanisms.

The spiro-mechanism. Proposed by Mathewson and Corwin,[52] this mechanism (Scheme 11) involves the formation of a tetrapyrromethane or bilane of the type (**33**) by three successive reactions of porphobilinogen. The bilane (**33**) cyclizes to form the spiro-intermediate (**35**), which on cleavage of the C—C bond yields the isomeric bilane or its

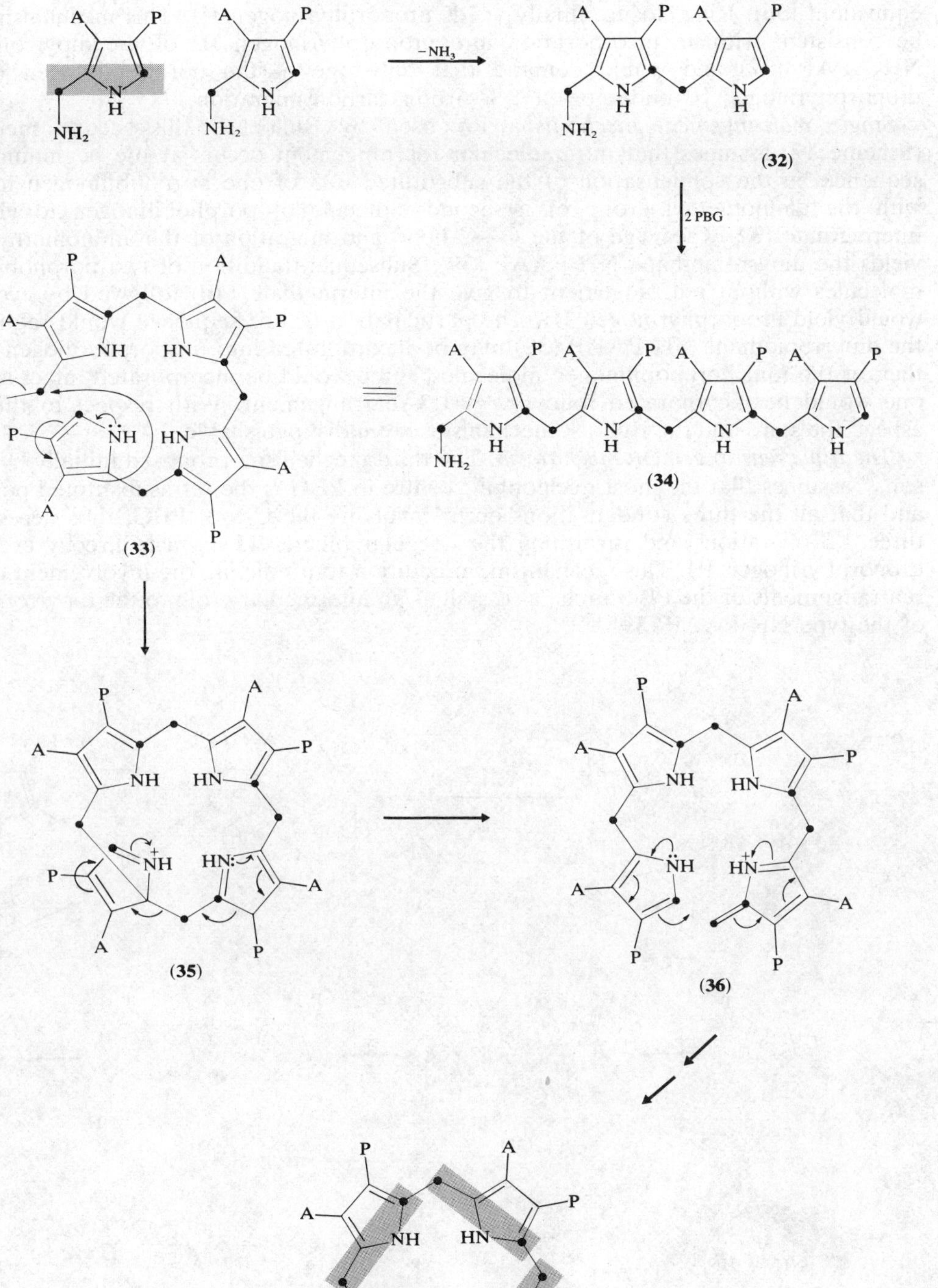

SCHEME 11 Mathewson and Corwin mechanism for the biosynthesis of uroporphyrinogen III. Heavy screen, rearranged PBG unit; light screen, unrearranged PBG units

equivalent (**36**). Ring closure finally yields uroporphyrinogen III. This mechanism would be consistent with an incorporation into uroporphyrinogen III of the dipyrromethane NH_2APAP (**32**) and would demand that only one of the porphobilinogen units of uroporphyrinogen III undergoes a 1,3-carbon–carbon migration.

Single rearrangement mechanism. Proposed by Bullock,[53] this second mechanism (Scheme 12) assumes that intramolecular rearrangement occurs at the beginning of the sequence by the condensation of the substituted C-5 of one porphobilinogen molecule with the aminomethyl group of a second molecule of porphobilinogen to yield the intermediate (**37**). Cleavage of the C—C bond and migration of the aminomethyl group yields the dipyrromethane NH_2PAAP (**39**). Subsequent addition of two porphobilinogen molecules without rearrangement to give the intermediate (**40**) followed by cyclization would yield uroporphyrinogen III. The participation of this sequence would require that the dipyrromethane NH_2PAAP (**39**) may be incorporated into uroporphyrinogen III and that, of the four porphobilinogen molecules, three would be incorporated intact and only one would be incorporated following a 1,3-rearrangement. With respect to this latter aspect the spiro and the Bullock mechanisms are indistinguishable.

The triple rearrangement mechanism. The third mechanism, proposed initially by Robinson,[54] assumes that the most nucleophilic centre in PBG is the tetrasubstituted position-5 and that all the three condensations occur involving the C-5 of PBG, thus necessitating three 1,3-migrations and furnishing the irregular bilane (**41**), which directly cyclizes to uroporphyrinogen III. This mechanism, in addition to predicting the involvement of three rearrangements of the PBG units, also assigns an intermediary role to the dipyrromethane of the type NH_2PAAP (**39**).

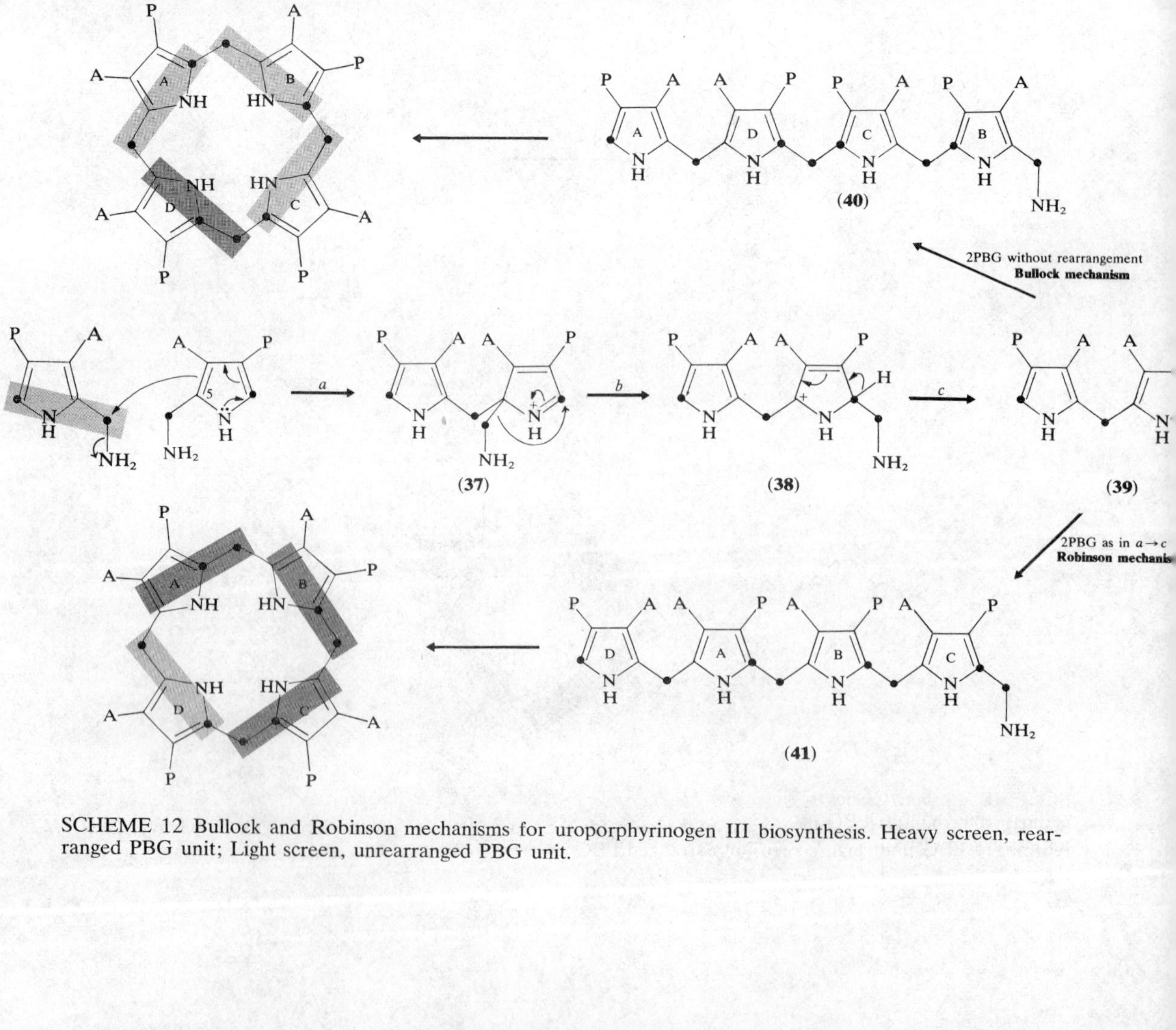

SCHEME 12 Bullock and Robinson mechanisms for uroporphyrinogen III biosynthesis. Heavy screen, rearranged PBG unit; Light screen, unrearranged PBG unit.

The problem of deciding how many intramolecular rearrangements occur is tailor-made for the application of ^{13}C n.m.r. spectroscopy. This technique allows definitive conclusions to be made regarding the position of a ^{13}C probe in a molecule, particularly with respect to other adjacent or nearby ^{13}C atoms. Accordingly, porphobilinogen labelled with ^{13}C at the key positions C-2 and C-11 was prepared enzymically from [5-^{13}C]ALA for use in biosynthetic experiments.[51] A crucial feature of the experimental approach adopted by Battersby and his groups was to five-fold dilute the ^{13}C-labelled porphobilinogen with unlabelled porphobilinogen. The porphobilinogen was then incubated with an avian erythrocyte system to form uroporphyrinogen III. The erythrocyte system also contains the enzymes necessary for the subsequent conversion of the uroporphyrinogen III into protoporphyrin IX. Because of the dilution of the ^{13}C-labelled porphobilinogen, most of the molecules of uroporphyrinogen III produced would have been formed on average from *one* labelled and *three* non-labelled porphobilinogen molecules. As the subsequent enzymic conversion of uroporphyrinogen III into protoporphyrin IX is known to occur without rearrangement of the carbon skeleton, a study of the ^{13}C spectrum of protoporphyrin IX (as the diester) can be extrapolated directly to account for the mechanism of uroporphyrinogen III formation. Examination of the ^{13}C n.m.r. spectrum of the protoporphyrin IX as a mixture of (**42**)–(**45**) showed that signals from the α-, β-, and δ-*meso* carbon atoms (positions 5, 10, and 20) all appeared as doublets split by 5.5 Hz owing to 1,4-coupling of the ^{13}C atom in each case. In contrast, the γ-*meso* signal consisted of a doublet split by 72 Hz expected from the coupling to an adjacent ^{13}C atom. (Scheme 13).

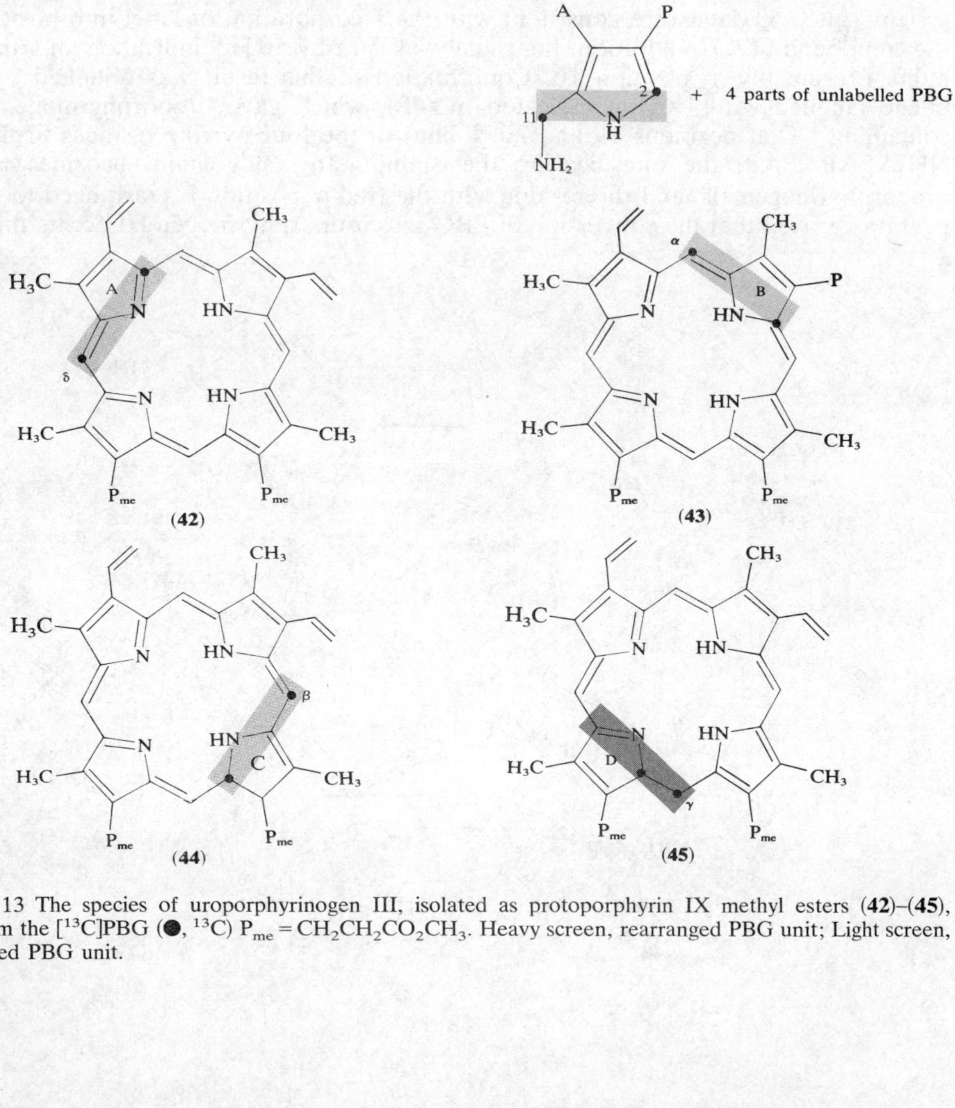

(**42**) (**43**) (**44**) (**45**)

SCHEME 13 The species of uroporphyrinogen III, isolated as protoporphyrin IX methyl esters (**42**)–(**45**), formed from the [^{13}C]PBG (●, ^{13}C) $P_{me} = CH_2CH_2CO_2CH_3$. Heavy screen, rearranged PBG unit; Light screen, unrearranged PBG unit.

From these cumulative results it may be concluded that the A ring together with the δ-*meso* position, the B ring together with α-*meso* position, and the C ring together with the β-*meso* position all represent intact incorporation of porphobilinogen molecules without rearrangement. The porphobilinogen residue forming ring D and the γ-*meso* carbon atom, however, shows evidence that an intramolecular rearrangement has occurred during the formation of the macrocyclic ring. These elegant studies confirm the labelling studies of Shemin and eliminate the triple rearrangement mechanism, although they do not preclude the ingeneous variation of Russell[54b] or allow a distinction between the two remaining mechanisms involving single rearrangements.

The other main approach in the elucidation of the mechanism of uroporphyrinogen III biosynthesis centres on the study of pyrromethanes, particularly of the type NH_2APAP (**32**), implicated in the Mathewson–Corwin mechanism, and NH_2PAAP (**39**), favoured as an intermediate in the Bullock mechanism.[50,51] However, in view of the tendency of these compounds to form uroporphyrinogens non-enzymically, the results from enzymic incubation have not yielded any conclusive information except in a recent report.[54c]

Recent reports by Battersby and the Cambridge group describing experiments with the tetrapyrromethane (or bilane) (**33**) are of considerable interest.[55] It was found that in control experiments, *without the enzyme*, the bilane NH_2APAPAPAP (**33**) was transformed as expected into uroporphyrinogen I. In the *presence* of the combined synthetase/cosynthetase enzyme system the reaction was diverted towards the formation of uroporphyrinogen III at the expense of uroporphyrinogen I. When the bilane labelled with ^{13}C at the positions indicated in structure (**46**) was incubated with the enzyme system[56] the ^{13}C data were consistent with the incorporation of label into position 15 of the compound (**47**). In addition, the signal was split by 50 Hz, indicating an adjacent ^{13}C atom, presumably at position 16. Confirmation of this result was obtained by using a bilane labelled with ^{13}C, as indicated in (**48**), which gave uroporphyrinogen III (**49**) containing ^{13}C at positions 19 and 20.[56] Thus of the four pyrrole residues in the bilane, NH_2APAPAPAP, the one bearing the aminomethyl side-chain, becomes ring A in uroporphyrinogen III, and the residue with the free α-position is rearranged to form ring D. If it is proved that the conversion of PBG into uroporphyrinogen III occurs through the

(**46**) (**47**)

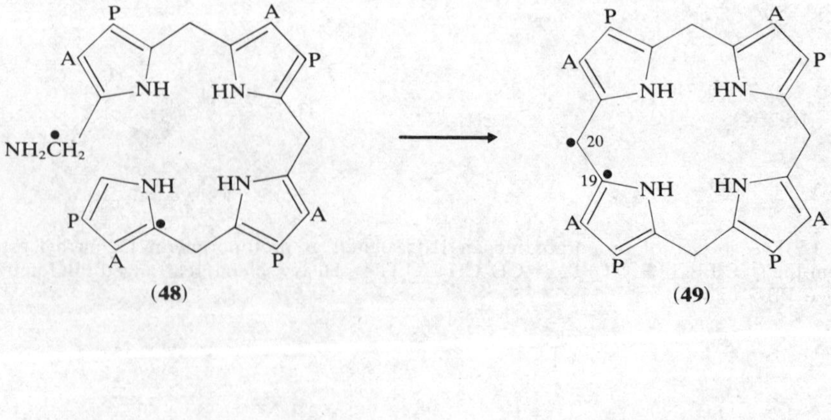

(**48**) (**49**)

intermediacy of the bilane, then these findings in particular[55,56] represent a considerable contribution to the understanding of the uroporphyrinogen III problem. The main conclusions from these studies, that the rearrangement to form the inverted D ring in uroporphyrinogen III occurs after the formation of the bilane (**33**), makes the Bullock mechanism in which rearrangement occurs at an early stage less attractive and favours the Mathewson–Corwin mechanism or one of its variations.

One such possibility is that the rearrangement occurs not *via* the spiro-compound (**35**) but involves the participation of covalent links between intermediates and either one or other (or both) of the enzymes. Whatever the mechanistic route followed, the final stages would be expected to require the intermediacy of the irregular bilane (**34**), its deamino variant (**36**), or an enzyme-linked equivalent.

30.2.2.6 Conversion of uroporphyrinogen III into coproporphyrinogen III

Early workers in the porphyrin field considered that uroporphyrin III was an intermediate in haem biosynthesis. We know now that it is the reduced porphyrin, uroporphyrinogen III, that is the true enzymic intermediate[57] and that the biosynthesis of haem proceeds through the intermediacy of coproporphyrinogen III and protoporphyrinogen IX before the macrocyclic ring is oxidized to a porphyrin (Scheme 1). The conversion of uroporphyrinogen III into coproporphyrinogen III by decarboxylation of the acetic acid side-chains at positions 2, 7, 12, and 18 (Scheme 14) is catalysed by a single enzyme,

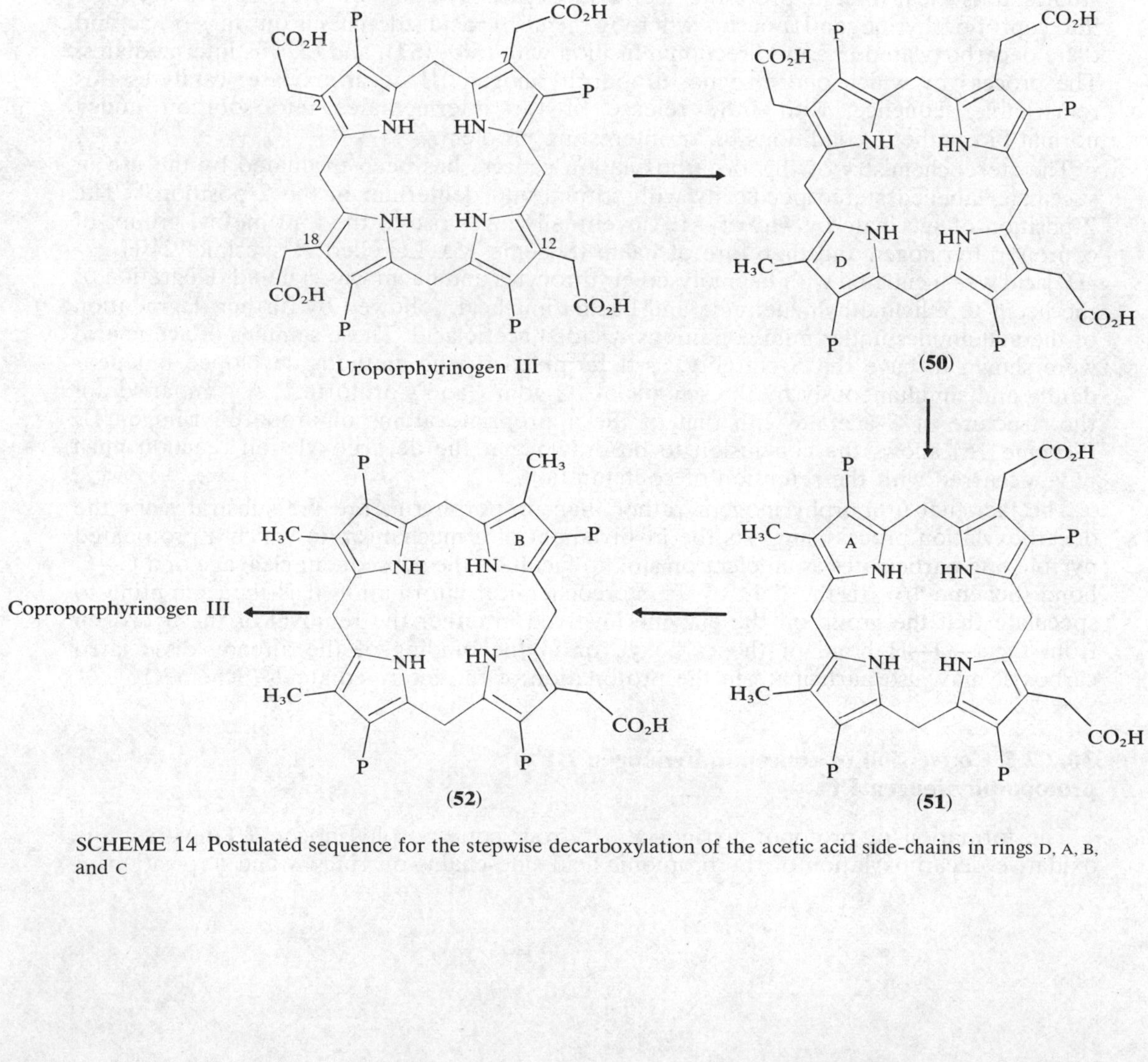

SCHEME 14 Postulated sequence for the stepwise decarboxylation of the acetic acid side-chains in rings D, A, B, and C

uroporphyrinogen III decarboxylase. The enzyme also catalyses the decarboxylation of uroporphyrinogen I, but the II and IV isomers are poor substrates.[58]

Theoretically there are 14 possible intermediates between uroporphyrinogen III and coproporphyrinogen III, containing five, six, or seven carboxylic acid side-chains. Some of these intermediates have been isolated as their porphyrins from the urine of porphyriac patients[59] or from experimental animals treated with the drug hexachlorobenzene.[60] All these intermediates are related to the III series, as shown by their subsequent chemical decarboxylation to coproporphyrin III.

The heptacarboxylic acid porphyrin, phriaporphyrin (**50**), (also known as porphyrin 208 and pseudouroporphyrin) has been isolated from various sources and shown to have the structure (**50**), originating from the decarboxylation of the D-ring acetic acid side-chain of uroporphyrinogen III.[61] There are six possible isomers of the hexacarboxylic acid derivative, formed through the decarboxylation of two of the acetic acid side-chains of uroporphyrinogen III, but only one such isomer has been isolated from the natural sources, rat faeces, and formulated on the basis of its ^1H n.m.r. spectrum and unambiguous synthesis as (**51**).[61]

Four pentacarboxylic acid porphyrins can be derived from uroporphyrinogen III by decarboxylation of three acetic acid side-chains. The knowledge that the hexacarboxylic porphyrin had lost the A- and D-ring acetic acid side-chains left two isomers for firm consideration as intermediates, *i.e.* those where decarboxylation had occurred in the rings A, B, and D or A, C, and D. Distinction between these two main contenders was again provided by chemical synthesis, and subsequent comparison of the ^1H n.m.r. spectra of the synthetic product (**52**) with a biological sample obtained from rat faeces.[61] From these studies it is clear that a preferred stepwise decarboxylation or uroporphyrinogen III into coproporphyrinogen III occurs whereby the acetic acid side-chains on rings D, A, B, and C are decarboxylated in a unidirectional fashion with (**50**), (**51**), and (**52**) as intermediates. The process by which one enzyme, uroporphyrinogen III decarboxylase, catalyses this remarkable sequence, with little release of the intermediates into solution under normal biosynthetic conditions, is an interesting problem.

The stereochemistry of this decarboxylation process has been evaluated by the use of succinate labelled stereospecifically with tritium and deuterium in the 2-position.[62] The 2-position of succinate by way of ALA eventually gives rise to the four methyl groups of coproporphyrinogen and therefore of haem (Scheme 15). Labelled *R*-succinic [2-$^2H_1$, 2-$^3H_1$]acid was incubated with haemolysed erythrocytes and haem was isolated. Degration of the haem to ethylmethylmaleimide and haematinic acid, followed by further degradation of the maleimides under mild conditions, yielded acetic acid. These samples of acetic acid were shown to have the *S*-chirality, as determined by the methods developed independently and simultaneously by the schools of Arigoni[63] and Cornforth.[64] A comparison of the structure of *S*-acetate with that of the appropriate atoms of uroporphyrinogen III (Scheme 15) allows the conclusion to be drawn that the decarboxylation reaction must have occurred with the retention of configuration.[62]

The fact that uroporphyrinogens rather than uroporphyrins are the substrates for the decarboxylation process suggests the involvement of a mechanism in which a protonated pyrrole ring participates as an electron sink to facilitate the subsequent cleavage of a C—C bond (Scheme 16). In the light of the stereochemical information it is then tempting to speculate that the group on the enzyme involved in either the removal of the hydrogen from the —O—H bond of the carboxyl, or in the binding of the already dissociated carboxyl, may also participate in the protonation of the intermediate *b* (Scheme 16).

30.2.2.7 Conversion of coproporphyrinogen III to protoporphyrinogen IX

The formation of protoporphyrinogen IX from coproporphyrinogen III involves the oxidative decarboxylation of the propionic acid side-chains on rings A and B (positions 3

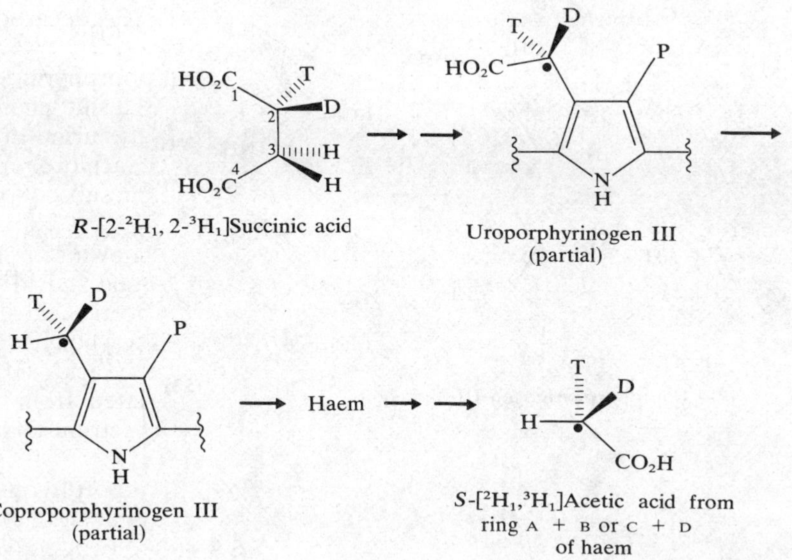

SCHEME 15 The stereochemistry of the porphyrinogen carboxy-lyase reaction

and 8) to vinyl groups (Scheme 17). The enzyme requires molecular oxygen,[65] although other electron acceptors must be utilized in anaerobic organisms.

There are two main considerations regarding the conversion of coproporphyrinogen III into protoporphyrinogen IX. Firstly, the sequence of decarboxylation and the possible involvement of other intermediates and, secondly, the mechanistic route followed during the formation of the vinyl group. In normal metabolism the amount of free porphyrins

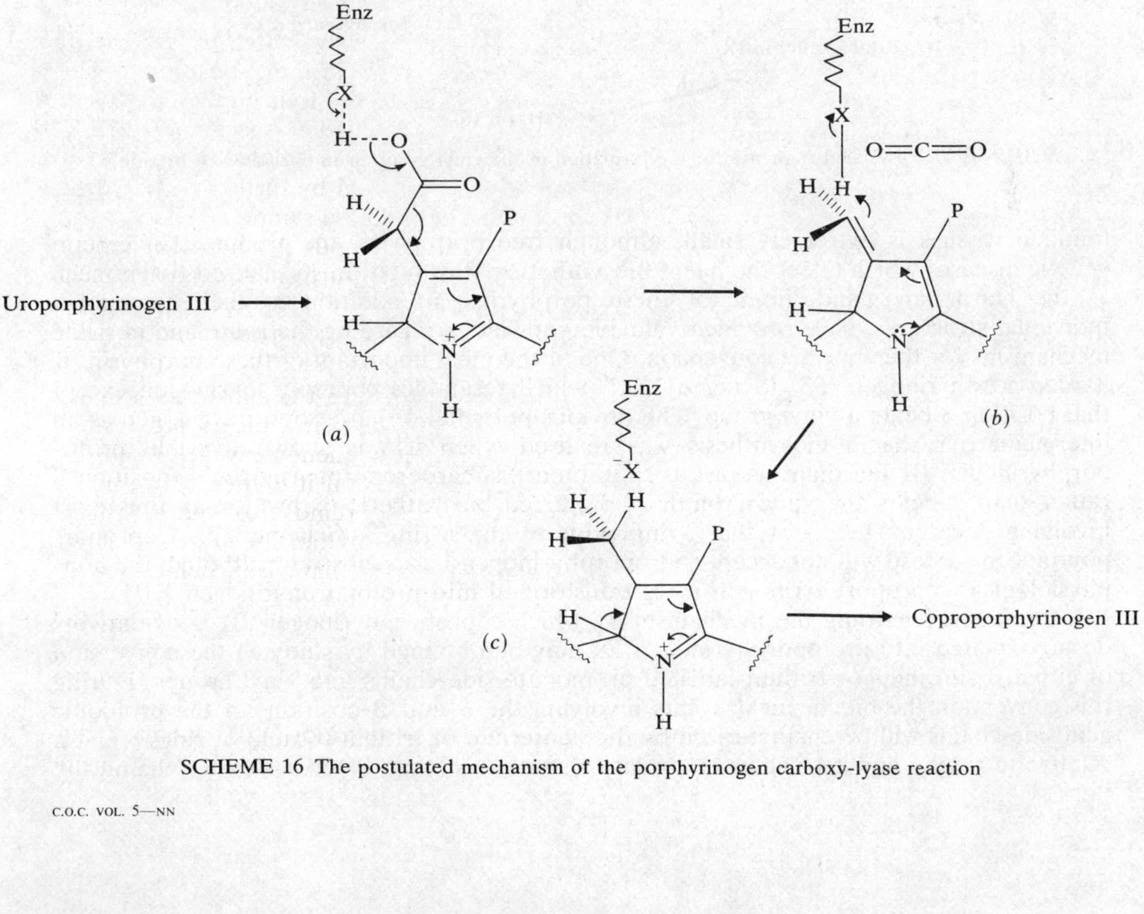

SCHEME 16 The postulated mechanism of the porphyrinogen carboxy-lyase reaction

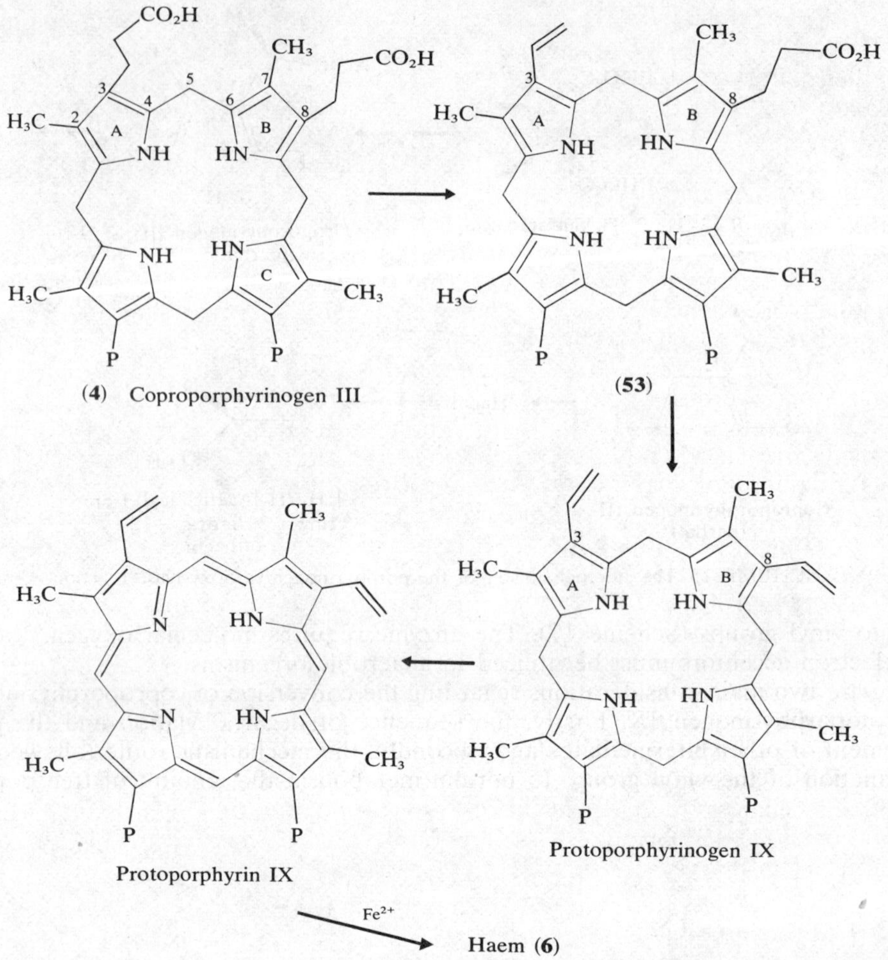

SCHEME 17 Postulated sequence for the formation of the vinyl group in ring A followed by ring B

found in tissues is extremely small, although free porphyrins are produced in certain genetic diseases which affect the haem biosynthetic pathway, or in specialized tissues such as the Harderian gland. Some of these porphyrins, in addition to their interesting individual structures, have provided valuable knowledge regarding the route and possible mechanisms for the enzymic conversion. One of the most important of these porphyrins is harderoporphyrinogen (**53**) (Scheme 17),[66] which resembles coproporphyrinogen except that position 3 bears a vinyl group. The obvious potential of harderoporphyrinogen as an intermediate in haem biosynthesis was realised when it was shown to yield protoporphyrinogen IX ten times as fast as the isomer, isoharderoporphyrinogen[67] (position 8 rather than 3 bears the vinyl group). A preferred biosynthetic pathway may thus exist involving decarboxylation of the A-ring prior to the B-ring[61] (Scheme 17). Coproporphyrinogen oxidase will not accept coproporphyrinogen I as a substrate, although the non-physiological coproporphyrinogen IV is transformed into protoporphyrinogen XIII.

Information regarding the mechanism by which coproporphyrinogen III is oxidatively decarboxylated into protoporphyrinogen IX may be obtained by studying the conversion of either deuterium- or tritium-labelled propionate side-chains into vinyl groups. During this conversion the biochemical events involving the α-and β-positions of the propionic acid side-chains will be characterized by the deuterium or tritium content of rings A and B relative to rings C and D of haem. In order to study this facet, PBG was either chemically synthesized[68] containing deuterium in the side-chain, (**55a**) and (**55b**), or biosynthesized[27,69] from variously tritiated succinates (**54**), as in Scheme 18, and converted into haem with a haemolysed avian erythrocyte preparation. The biosynthesized material was degraded to haematinic acid and ethylmethylmaleimide for the determination of their isotopic content. These experiments showed that the only hydrogen atom removed from the propionate side-chains was from the β-position. That it was the *pro-S* hydrogen which is eliminated was revealed when a sample of PBG stereospecifically labelled with tritium in the side-chain was used as the precursor, as shown in Scheme 18. These findings eliminate mechanisms involving the participation of either β-keto or acrylic acid intermediates in the formation of vinyl groups from the corresponding propionate side-chains. The cumulative evidence presented above is therefore compatible with three closely related mechanistic alternatives, shown in Scheme 19.[70] In mechanism *a* a hydride ion is removed from the β-carbon atom with simultaneous decarboxylation to a vinyl group. The mechanism *b* procceds *via* an oxygen-dependent hydroxylation process in which a hydroxyl group introduced at the β-carbon atom facilitates decarboxylation. Support for the mechanism *b* stems from the observation that synthetic 3-(3-hydroxypropionate)-8-propionate deuteroporphyrinogen IX (**4**, but C-3 substituent = $CH(OH)CH_2CO_2H$) and, to a lesser extent, 8-(3-hydroxypropionate)-3-propionate deuteroporphyrinogen IX (**4**, but C-8 substituent = $CH(OH)CH_2CO_2H$) are both enzymically converted into protoporphyrinogen IX with avian erythrocyte preparations.[45] In the event that either mechanism *a* or *b* operates, the overall reaction must occur *via* an antiperiplanar elimination process[71] (Scheme 18).

Finally, the mechanism *c* predicts an oxidative reaction involving a β-hydrogen atom and the pyrrole N—H bond to give a Schiff base intermediate which, owing to its electron-withdrawing property, aids decarboxylation to give the vinyl group. The last mechanism is similar to several biological decarboxylation reactions catalysed by enzymes linked to pyridoxal phosphate and thiamine pyrophosphate. Although, at present, a choice between these three mechanims (*a, b, c*, Scheme 19) cannot be made, it is clear that the mechanism *b* predicting an O_2-dependent formation of a β-hydroxylated derivative is unlikely to be involved in haem biosynthesis in anaerobic organisms.

30.2.2.8 Conversion of protoporphyrinogen IX to protoporphyrin IX

Although the non-enzymic conversion of the porphyrinogens to porphyrins proceeds rapidly in the presence of oxygen and in the light, nature rarely leaves any reaction to chemistry alone. Since a distinction between enzymic and non-enzymic processes can often be made from a study of kinetic isotope effects, protoporphyrinogen IX was prepared, labelled randomly with tritium at each of the four *meso* positions. Incubation of this substrate with a chicken erythrocyte preparation followed by isolation of protoporphyrin IX revealed that exactly 50% of the tritium label was lost in the conversion, providing strong evidence that the reaction is enzyme catalysed.[72] In a model non-enzymic conversion of coproporphyrinogen III into coproporphyrin III only 4% of the tritium label was lost, indicating that a high isotopic discrimination occurs in the non-enzymic porphyrinogen oxidation. The extracts* which catalyse the oxidation of protoporphyrinogen IX also catalyse the oxidation of protoporphyrinogen XIII and isoharderoporphyrinogen, although at significantly slower rates.

30.2.2.9 Conversion of protoporphyrin IX to haem

The biological insertion of ferrous iron into protoporphyrin IX is catalysed by extracts from a wide variety of sources and is termed ferrochelatase.[73] Most systems seem to have similar properties in that they require anaerobic conditions and a reducing agent such as glutathione. In addition to iron, the enzyme is also able to catalyse the insertion of other

* The enzyme has recently been partially purified from mammalian mitochondria (R. Poulson, *J. Biol. Chem.*, 1976, **251** 3730) and anaerobically grown *E. coli* (J. M. Jacobs and N. J. Jacobs, *Biochem. Biophys. Res. Comm.*, 1977, **78,** 429). The mammalian enzyme utilized molecular oxygen as the oxidant while with the *E. coli* enzyme the final electron acceptor was shown to be fumarate.

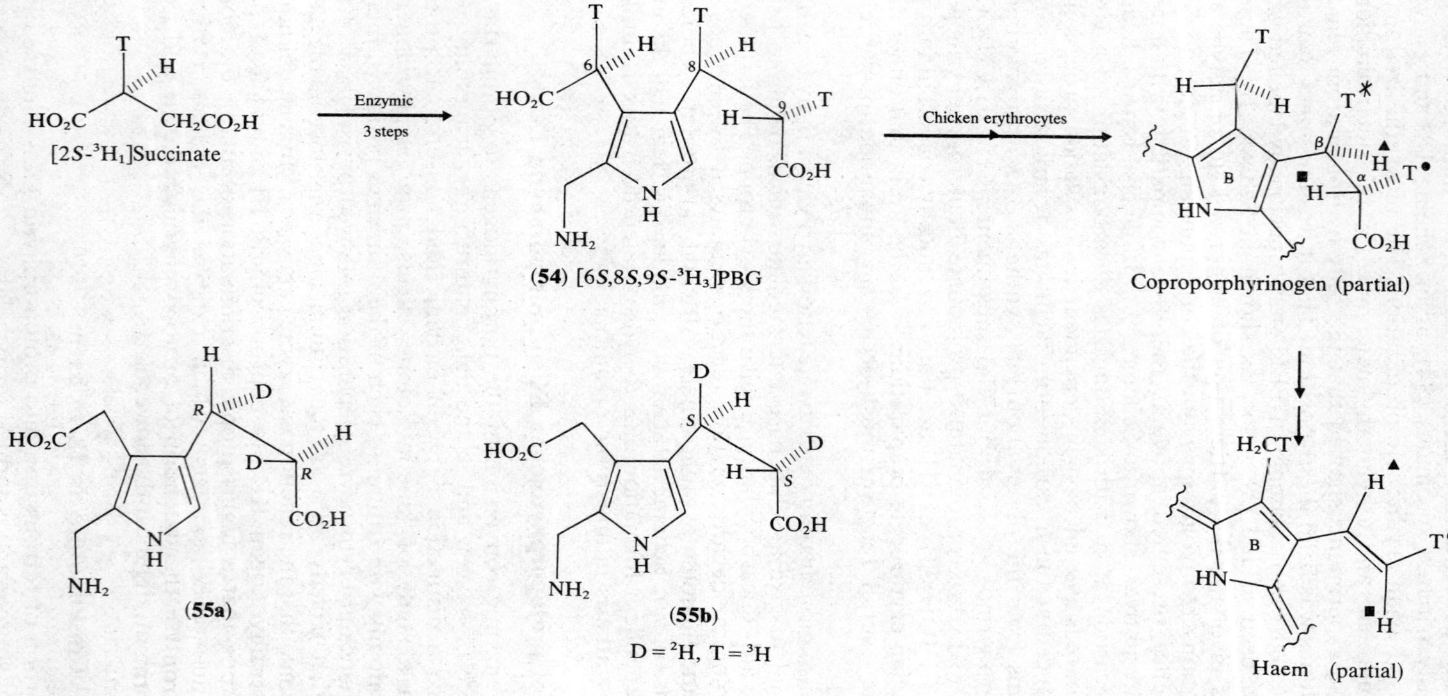

SCHEME 18 The loss of the *pro-S* hydrogen atom from the β-position in vinyl group formation was shown[27] using the enzymically prepared PBG (**54**); this conclusion has recently been confirmed.[124b] That it is an antiperiplanar elimination was proved by Battersby *et al.*[71] using a sample of PBG containing (**55a**) + (**55b**)

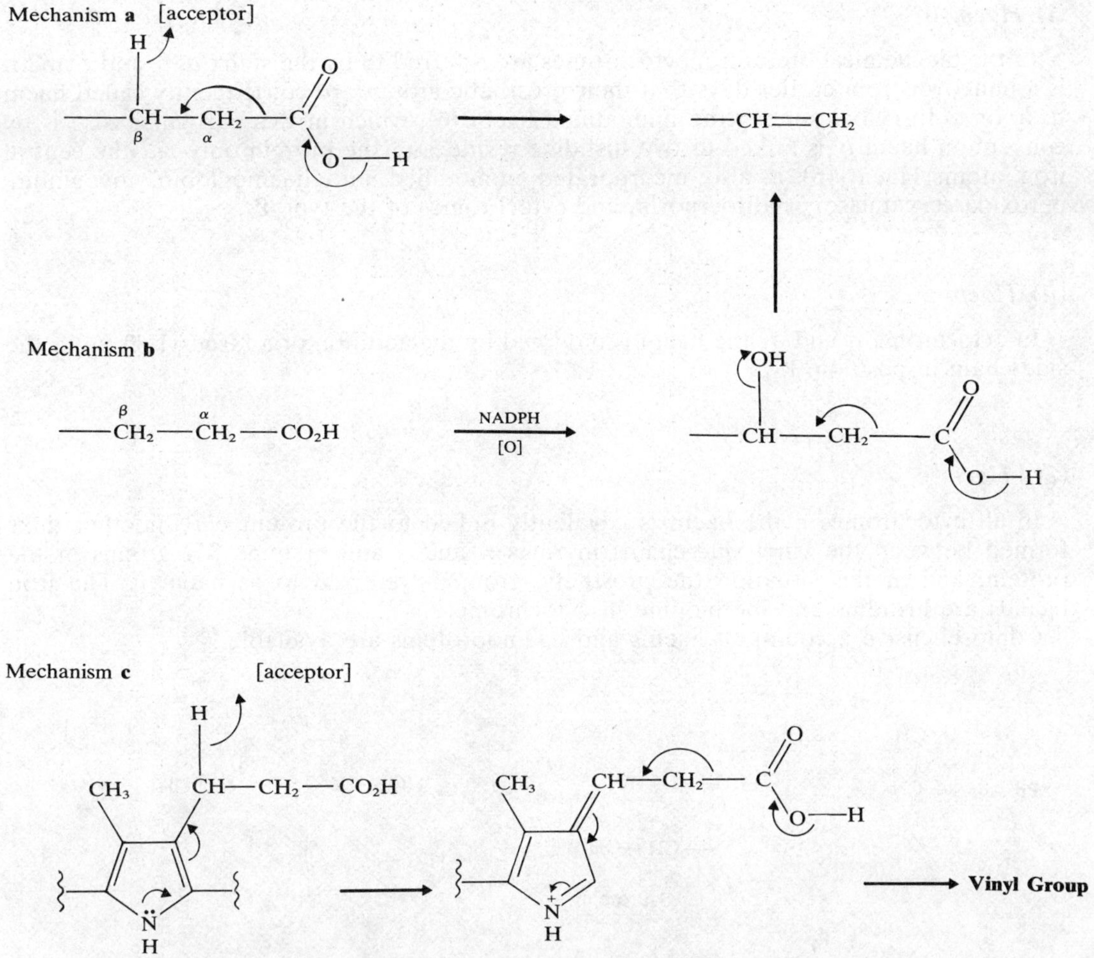

SCHEME 19 Possible mechanisms for the formation of vinyl groups in haem biosynthesis

divalent metals such as cobalt, zinc, copper, manganese, and nickel, although only cobalt and, to a lesser extent, zinc compare with the rate at which iron is inserted.

The structure of the porphyrin ring plays an important role in determining the rate at which the metal is chelated. For instance, the enzyme accepts mesoporphyrin and deuteroporphyrin even more efficiently than protoporphyrin IX. In general, two free carboxyl groups, preferably propionic acid residues, favour the binding to the enzyme.[74]

Iron may be incorporated in protoporphyrin IX non-enzymically by heating with a ferrous salt in acetic acid or pyridine. At physiological pH, however, the incorporation of iron is rather slow. Iron is not incorporated into porphyrinogens in which the conformation of the pyrrole rings is not so well suited to chelation as the planar conjugated porphyrin ring system; zinc, however, is able to chelate porphyrinogens and porphyrins both enzymically and non-enzymically and unwanted formation of zinc complexes interfere in studies on the formation of iron porphyrins.

30.2.2.10 The formation of haem derivatives and haemoproteins

A detailed account of the incorporation of haem into haemoproteins is outside the scope of this chapter; nevertheless a few comments on this aspect are included below.

(*i*) *Haem* b

In the biochemical literature, cytochromes are referred to by the suffix *a*, *b*, and *c* and it is a hangover from earlier days that their prosthetic groups are consequently called haem *a*, *b*, or *c*. In cytochrome *b* the unmodified haem (**6**), which in this particular case is by convention haem *b*, is linked to two histidine residues of the protein only *via* the central iron atom. Haem (**6**) is also incorporated unmodified into haemoglobin, myoglobin, peroxidases, catalase, erythrocruorin, and cytochromes of the type P_{450}.

(*ii*) *Haem* a

In cytochrome *a* and a_3 the haem is modified by the addition of a farnesyl group to the side chain in position 3.

(*iii*) *Haem* c

In all cytochromes *c* the haem is covalently linked to the protein *via* thioether links formed between the vinyl side-chains in rings A and B and cysteine SH groups in the protein, and in this situation the prosthetic group is referred to as haem *c*. The iron ligands are histidine and methionine in cytochrome *c*.

Comprehensive accounts of haems and haemoproteins are available.[75]

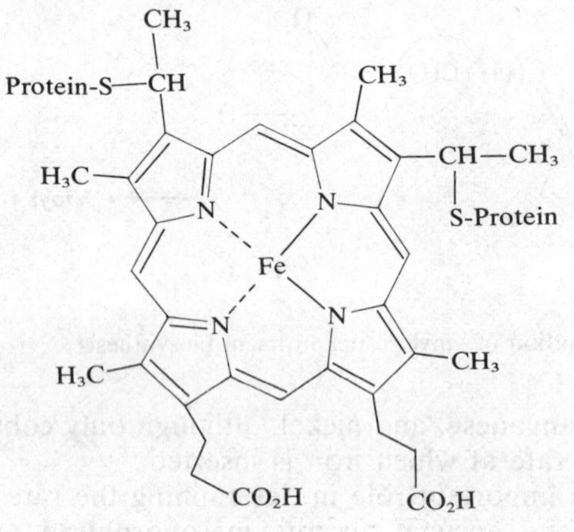

Haem *c*

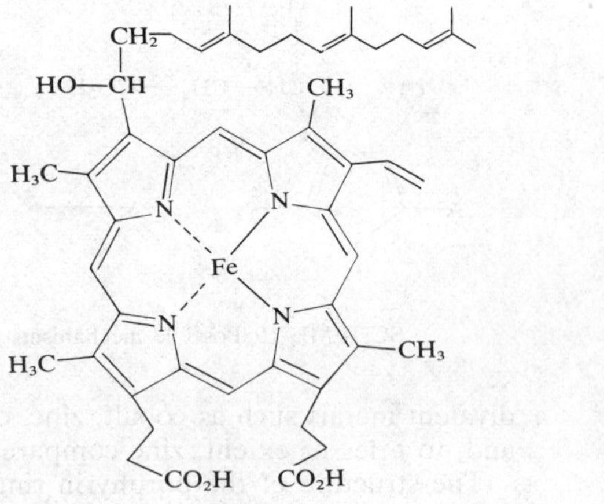

Haem *a*

30.2.3 THE BIOSYNTHESIS OF CHLOROPHYLLS[76,77]

30.2.3.1 Chlorophyll *a*

Our present knowledge of the biosynthesis[77] of chlorophyll *a* (**58**) (Scheme 20) is based mainly on the early experiments of Granick, who showed that mutants of *Chlorella* unable to synthesize chlorophyll *a* accumulated intermediates such as protoporphyrin IX, Mg-protoporphyrin IX (**56**), and Mg-protoporphyrin IX methyl ester (**57**).[8,77–79] Subsequently these and related intermediates were also isolated from barley grown in the dark or barley mutants defective in chlorophyll biosynthesis.[79] These studies led to the proposal that haem and chlorophyll may share a common pathway up to protoporphyrin IX and allowed the early events in chlorophyll biosynthesis to be formulated as shown in Scheme 20.[77–79]

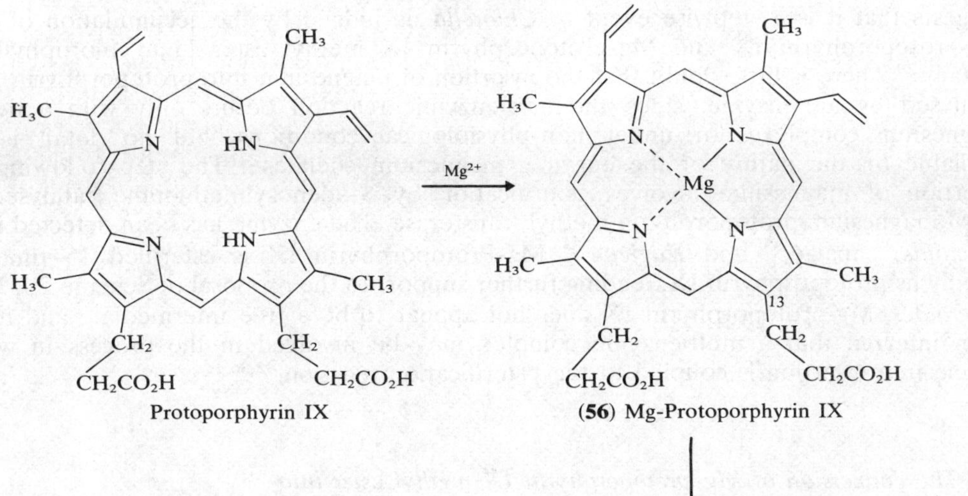

SCHEME 20 Early stages in the biosynthesis of chlorophyll *a*

Recently, the use of cell-free systems from greening cucumber cotyledons[80] and dark-grown barley etioplasts[81] for the incorporation of labelled ALA and protoporphyrin IX into chlorophyll precursors has provided more direct evidence in support of the earlier biogenetic view. The knowledge currently available in the area is summarized under two main headings, involving first the conversion of protoporphyrin IX into Mg-protoporphyrin IX methyl ester and then the conversion of the latter into chlorophyll *a*.

(*i*) *The conversion of protoporphyrin IX into Mg-protoporphyrin IX methyl ester*

The precise sequence in which two of the reactions of Scheme 20 occur, *i.e.* the insertion of Mg into the porphyrin nucleus and the esterification of the propionic acid side-chain at position 13, is not yet unambiguously known, although indirect evidence suggests that it is a stepwise event in *Chlorella* as judged by the accumulation of both Mg-protoporphyrin IX and Mg-protoporphyrin IX methyl ester in a chlorophyll-less mutant.[79] There is little doubt that the insertion of magnesium into protoporphyrin IX is catalysed by an enzyme, since the non-enzymic reaction occurs only with activated magnesium complexes or under non-physiological conditions, but no detail is yet available on the nature of the enzyme, magnesium chelatase. The step following the insertion of magnesium involves esterification by *S*-adenosylmethionine catalysed by SAM magnesium protoporphyrin methyl transferase. The enzyme has been detected in *R. spheroids*,[82] maize,[83] and *Euglena*.[84] Mg-Protoporphyrin IX is esterified 15 times as rapidly as protoporphyrin IX, lending further support to the proposal of Scheme 20. In *R. spheroides*, Mg-protoporphyrin IX does not appear to be a free intermediate and it has been inferred that a multienzyme complex may be involved in the process in which magnesium insertion is coupled to the esterification reaction.[85,86]

(*ii*) *The conversion of Mg-protoporphyrin IX methyl ester into protochlorophyllide* a

En route to chlorophyll *a*, the steps in the biological conversion of Mg-protoporphyrin IX methyl ester involve (a) the elaboration of the isocyclic E-ring from the methyl propionate side-chain at C-13 to give 3,8-divinylphaeoporphyrin a_5 methyl ester (**64a**, Scheme 21); (b) the reduction of the 8-vinyl group of the latter to produce protochlorophyllide *a* (**64b**); (c) the *trans* reduction of the 17–18 double bond in ring D to convert the protochlorophyllide *a* (**64b**) into chlorophyllide *a* (**68**, Scheme 23); and finally (d), the esterification of the C-17 propionate side-chain of (**68**) with phytol to give chlorophyll *a* (Scheme 23).

The sequence in which the first two of these conversions (a and b above) occur is a matter of some controversy[86] at the present time and the salient features of the protagonists' views are briefly mentioned later in this section; however, first we consider the pathway for the construction of the isocyclic ring E. The careful work of Ellsworth and co-workers[87] demonstrated that two mutants of *Chlorella* blocked in chlorophyll *a* biosynthesis accumulated a number of magnesium-tetrapyrrole intermediates which on the basis of mass spectrometric analysis were shown to contain modified side-chains at position 13 and which were formulated[87] as the dehydro (**60**), hydroxy (**61**), and keto (**62**) derivatives. The last (**62**) processes the active methylene group which may be used in a Michael type of addition reaction to produce the five-membered ring E (Scheme 21). A critical examination of the mechanism outlined in Scheme 21 reveals that the intermediate (**63b**) formed by this reaction may furnish chlorophyllide *a* (**68**) by a simple rearrangement of the double bonds, as shown in Scheme 22. However, the burden of evidence in the literature[86] has been interpreted to suggest that the earliest product of the ring closure is a prototype of compound (**64**), implying that the formation of the ring E may be rapidly followed by an oxidative reaction (**63a**→**64a** or **63b**→**64b**).

We now return to the controversy regarding the stage at which the C-8 vinyl group in ring B is reduced. Two pieces of information suggest that the reduction reaction may occur before the construction of the ring E. Firstly, the intermediates of the type (**60b**), (**61b**), (**62b**), which have no ring E, possess[87] already reduced side-chains at position 8. Secondly, it has recently been reported[88] that an enzyme present in the cell-free extract of etiolated wheat seedlings reduces vinyl groups of Mg-protoporphyrin IX methyl ester 20 times more efficiently than that of magnesium 3,8-divinylphaeoporphyrin a_5 methyl ester (**64a**). It is interesting to note that, as expected, the reduction reaction requires the presence of a reduced pyridine nucleotide and in the process one of the hydrogen atoms from C-4 of the pyridine nucleotide is transferred to the side-chain ethyl group.[88]

The converse view, that the vinyl group reduction occurs[86] after the formation of the ring E, is based on the isolation of magnesium 3,8-divinylphaeoporphyrin a_5 methyl ester

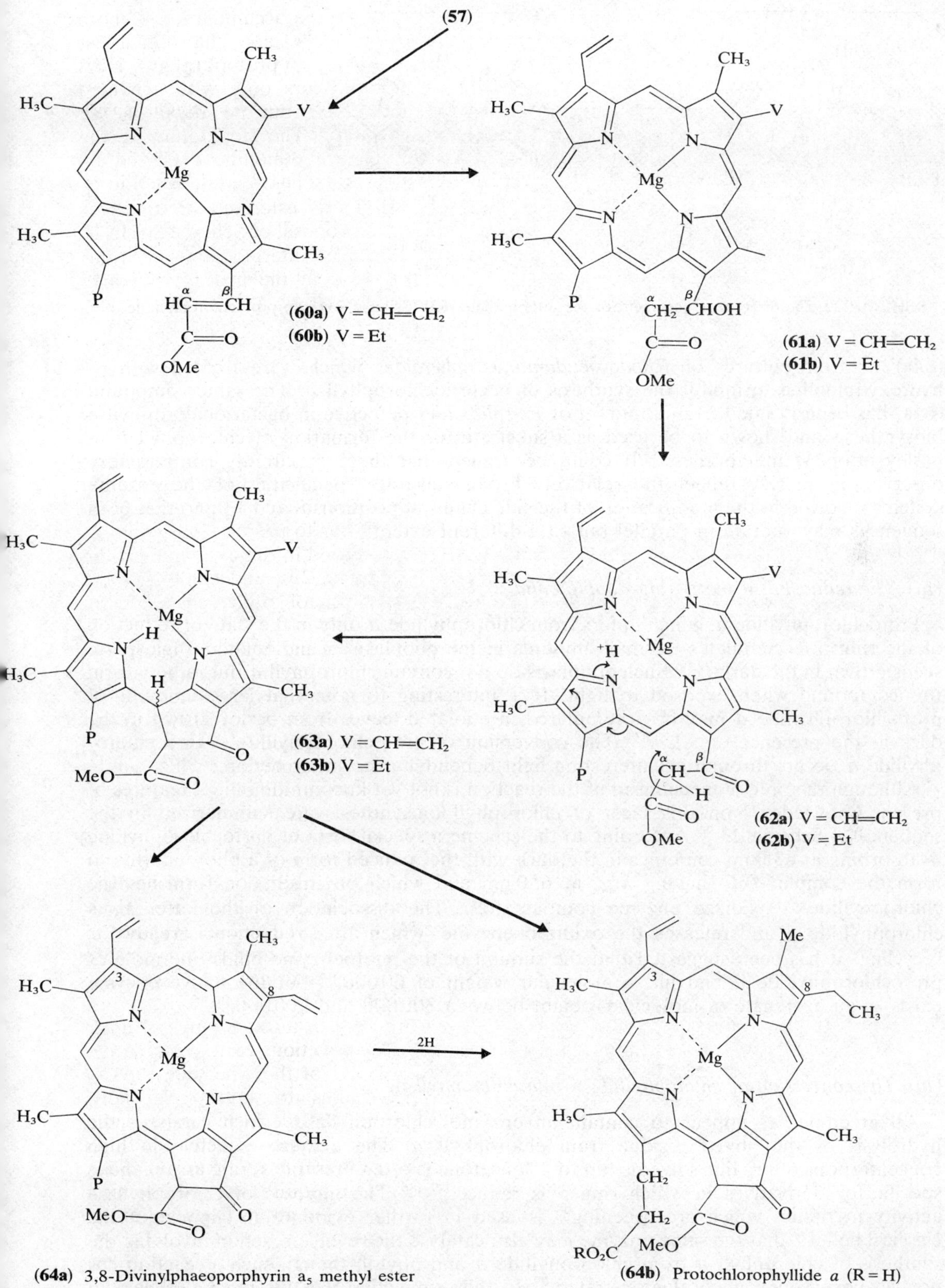

SCHEME 21 The elaboration of the E-ring and the formation of protochlorophyllide *a*

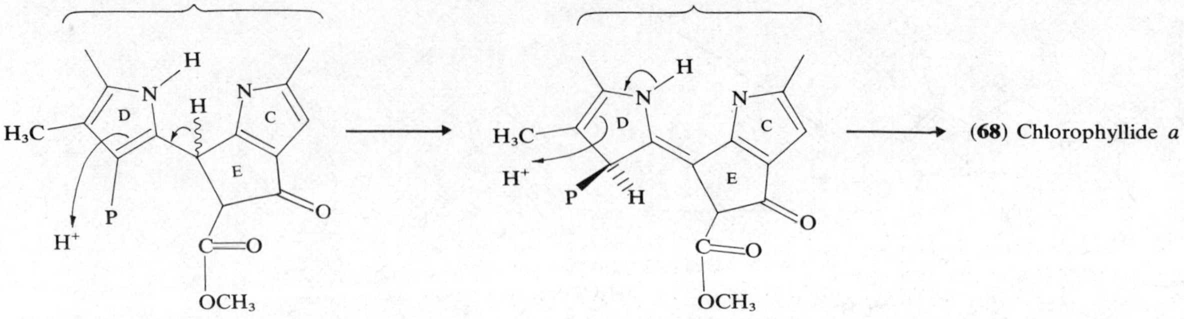

(**63b**) (**65**)

SCHEME 22 The direct rearrangement of the intermediate (**63**) (Scheme 21) into protochlorophyllide *a*

(**64a**)[89,90] from cultures of *Rhodopseudomonas spheroides* which were treated with 8-hydroxyquinoline to inhibit the synthesis of bacteriochlorophyll *a*. The same compound (**64a**) has been isolated from a mutant of *R. spheroides* defective in bacteriochlorophyll *a* biosynthesis and shown to be used as a substrate for the formation of chlorophyll *a* in barley etioplast membranes.[91] It could be argued that these apparently contradictory observations merely reflect the relatively broad substrate specificities of the enzyme systems involved in the modification of the side chains at positions 8 and 13 and that both sequences may operate in parallel but to a different extent.

(*iii*) *The reduction of the* D-*ring double bond*

Protochlorophyllide *a*, which differs from chlorophyllide *a* only in the state of reduction of the ring D, accumulates in small amounts in the etioplasts of germinated angiosperm seeds grown in the dark.[77,86] Such etioplasts do not contain chlorophyll *a* but rapidly form this compound when exposed to light. It is interesting to note that large amounts of protochlorophyllide *a* may be accumulated in etiolated leaves from barley grown in the dark in the presence of ALA.[79] The conversion of protochlorophyllide *a* into chlorophyllide *a* occurs through an interesting light-dependent enzymic reaction.[77,86,92]

Although the precise mechanism of the reaction is not yet known, the salient features of present knowledge[86] on this facet of chlorophyll biosynthesis are summarized in the sequence of Scheme 23.[93] According to the scheme a special form of protochlorophyllide *a* absorbing at 635 nm combines in the dark with the reduced form of a photoenzyme to form the complex (**66**) having λ_{max} at 650 nm and which on irradiation furnishes the chlorophyllide *a*–oxidized enzyme complex (**67**). The dissociation of the latter gives chlorophyllide *a* and releases the oxidized enzyme, which after reduction is ready for recycling. It has been suggested that the subunit of the photoenzyme binds one mole of protochlorophyllide *a* and has a molecular weight of 60 000,[94] but the native enzyme exists as an aggregate of molecular weight between 300 000 and 900 000.[92,95]

(*iv*) *The conversion of chlorophyllide* a *into chlorophyll* a

All green leaves appear to contain an enzyme, chlorophyllase, which catalyses the hydrolysis of the phytyl residue from chlorophyll *a*. The enzyme is active in high concentrations of organic solvents (up to 4% acetone is often present, during assay) and is specific for derivatives in which ring D is reduced.[77,86] The increase of chlorophyllase activity in tissues which are greening[96] is used in further evidence in support of the classical belief[77] that the same enzyme may also catalyse the reverse reaction involving the synthesis of chlorophyll *a* from chlorophyllide *a* and phytol, though such a reaction has not yet been convincingly demonstrated.[97] In this connection it is interesting to bear in

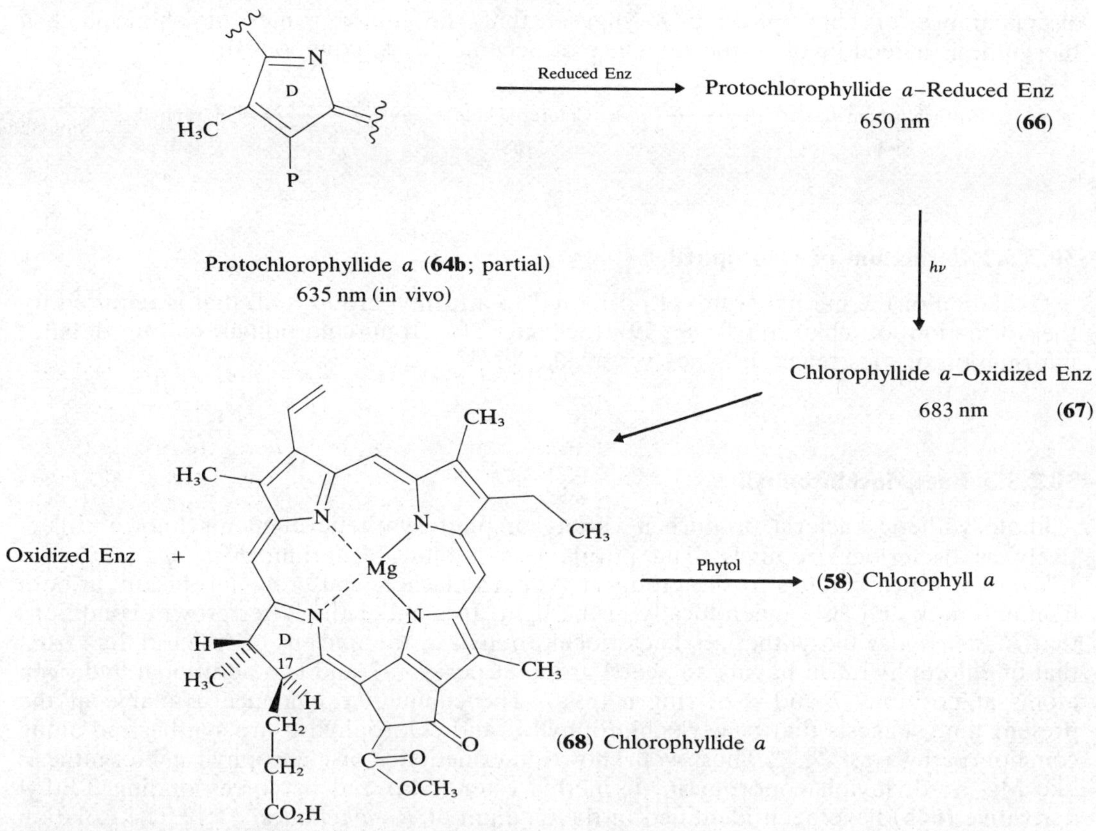

SCHEME 23 The conversion of protochlorophyllide *a* into chlorophyllide *a* and then chlorophyll *a*. The complex of the type (**66**) involving protochlorophyllide *a* and the enzyme responsible for the reduction of ring D is known in the literature as protochlorophyllide-holochrome'

mind that in other areas of biochemistry the hydrolysis and synthesis of ester linkages are usually catalysed by separate enzymes, the biosynthetic enzyme compulsorily requiring the participation of active-ester derivatives as shown in equation (1).

$$R^1\text{—COOH} \xrightarrow{\text{ATP}} R^1\text{—C}(\text{=O})\text{—O—}\textcircled{P} \xrightarrow{R^2\text{–OH}} R^1\text{—COOR}^2 \qquad (1)$$

The presence in etioplasts of varying amounts of protochlorophyll *a* (**64b**; R = phytyl residue) has given rise to another confusing debate[77,80] in the literature and it has been argued that the sequence (equation 2),

$$\underset{(\mathbf{64b})}{\text{Protochlorophyllide } a} + C_{20}\text{ alcohol} \longrightarrow \underset{(\mathbf{64b};\ R = \text{phytyl})}{\text{Protochlorophyll } a} \xrightarrow{h\nu + 2H} \underset{(\mathbf{58})}{\text{Chlorophyll } a} \qquad (2)$$

in which the esterification with a C_{20} isoprenoid alcohol* precedes the reduction of the D-ring may be involved in chlorophyll *a* biosynthesis. In our view the experiments of Wolff and Price[98] in which the formation of chlorophyll *a* was clearly correlated with the

* That the C-20 isoprenoid unit is transferred in the form of a geranylgeranyl moiety is suggested by recent work (W. Rudiger, P. Hedden, H.-P. Kost, and D. J. Chapman, *Biochem. Biophys. Res. Comm.*, 1977, **74,** 1268; S. Schoch, U. Lempert, and W. Rudiger. *Z. Pflanzenphysiol.*, 1977, **83,** 427; also see A. R. Wellburn, *Phytochemistry*, 1970, **9,** 2311).

disappearance of chlorophyllide *a* suggest that the major route for chlorophyll *a* biosynthesis indeed involves the sequence of Scheme 23 as (equation 3):

$$\underset{\textbf{(64b)}}{\text{Protochlorophyllide } a} \longrightarrow \underset{\textbf{(68)}}{\text{Chlorophyllide } a} \xrightarrow[\text{footnote on p. 1149)}]{C_{20}\text{ alcohol (see}} \underset{\textbf{(58)}}{\text{Chlorophyll } a} \quad (3)$$

30.2.3.2 Formation of chlorophyll *b*

Oxidation of the methyl group at position 7 to a formyl group is all that is required for the formation of chlorophyll *b* (**59**) (Scheme 20) from chlorophyll *a*, but detailed information on this reaction is not yet available.

30.2.3.3 Bacteriochlorophyll *a*

Photosynthetic bacteria produce a variety of photosynthetic pigments known collectively as bacteriochlorophylls. The purple non-sulphur bacterium *Rhodopseudomonas spheroides*, which belongs to the group of Athiorhodacaea, can grow aerobically in both light and dark and also anaerobically in the light. It is under the latter growth conditions that *R. spheriodes* biosynthesizes bacteriochlorophyll *a*, the nucleus of which differs from that of chlorophyll *a* in having an acetyl group at position 3 and two additional hydrogen atoms at positions 7 and 8 of ring B (**69a**). The cumulative evidence available at the present time suggests that bacteriochlorophyll *a* and chlorophyll *a* are synthesized along common pathways.[92,99,100] Thus, well-known intermediates of chlorophyll *a* biosynthesis like Mg-3,8-divinylphaeoporphyrin a_5 methyl ester (**64a**) and its corresponding 8-ethyl derivative (**64b**) have been identified in the medium of *R. spheriodes.*[89,90,99] The isolation of 3-devinyl-3-(1-hydroxyethyl)phaeophorbide *a* methyl ester (**70**)[9,101] from the medium

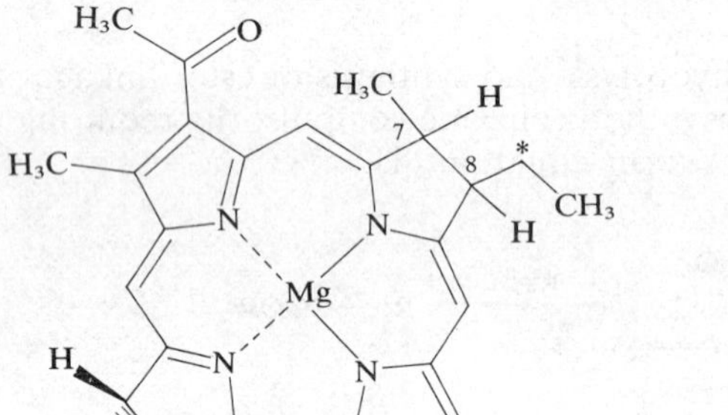

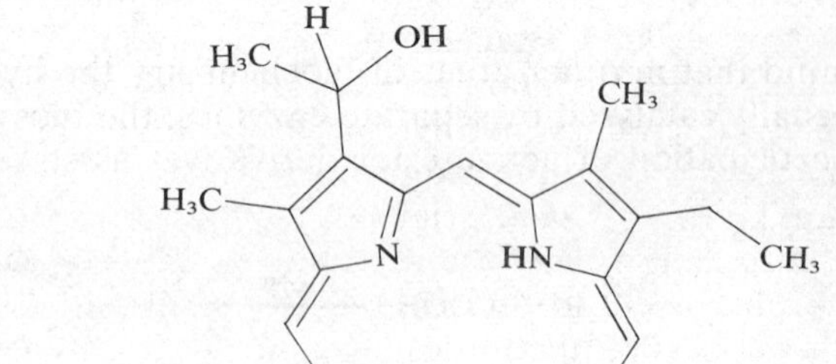

(**70**)

(**69a**) Bacteriochlorophyll *a*
(**69b**) Bacteriochlorophyll *b* = double bond between C-8—C*

when *R. spheroides* is grown in the presence[101] of 8-hydroxyquinoline to inhibit the chlorophyll biosynthesis, is cited as further evidence that bacteriochlorophyll *a* shares the general pathway as for the biogenesis of chlorophyll *a*, except that in the case of bacteriochlorophyll *a*, at some stage in the biosynthesis, the 3-vinyl group of an intermediate is hydrated and oxidized and the ring B is reduced.[102]

Until very recently it had been assumed that bacteriochlorophyll *a*, like chlorophyll *a*, is esterified with phytol on the C-17 propionate group, but this picture has now been changed by the finding that the esterifying group of bacteriochlorophyll isolated from *Rhodospirillum rubrum* is all-*trans*-geranylgeraniol.[103]

30.2.3.4 Bacteriochlorophyll *b*

The revised nuclear structure[104] of bacteriochlorophyll *b* is as shown in (**69b**), but little information is available at the present time regarding its biosynthesis.

30.2.3.5 Bacteriochlorophylls *c* and *d* (*Chlorobium* Chlorophylls)

Members of the green sulphur bacteria (Chlorobacteriaceae) produce a number of closely related pigments which are known as *Chlorobium* chlorophylls and may be subdivided into two main classes, each made up of mixtures of 4–7 different homologues.[105,106] One of these classes comprises bacteriochlorophyll *c* (**72**), absorbing at 660 nm in ether, and the other bacteriochlorophyll *d* (**71**) at 650 nm. In view of their absorption bands these compounds are often referred to as chlorophylls series (660), and chlorophylls series (650), respectively.[105] It is the latter nomenclature which we shall use below. Compared with bacteriochlorophyll *a*, the *Chlorobium* chlorophylls have the following distinguishing features.[105,106] The vinyl side-chain originally at position 3 is hydrated, which presumably occurs through an electrophilic addition mechanism, but not oxidized to the methyl ketone; the methoxycarbonyl group in ring E is absent; and the esterifying alcohol in the *Chlorobium* chlorophylls is all-*trans*-farnesol.[105,106] The most important feature of *Chlorobium* chlorophyll 650 (**71**), however, is the presence of extra methyl groups attached to the 8 and 12 side-chains. The chlorophylls 660 (**72**), in addition, also contain an extra alkyl group[105] at the 20-position (Scheme 24). It is known that these extra methyl groups arise from *S*-adenosylmethionine,[105] but the precise mechanism, and the sequence in which the methyl groups are introduced, is at the present not known. It was originally proposed by Richards and Rapoport[107] that *Chlorobium* chlorophylls follow the pathway common to all chlorophylls up to the stage of Mg-protoporphyrin IX methyl ester, but this view has recently been challenged by Kenner and co-workers on theoretical grounds. These workers[105] have drawn attention to the fact that for *S*-adenosylmethionine to transfer the methyl group, the recipient carbon atom in the side chain must be activated towards electrophilic attack. This they suggest may be achieved through the formation of enols from the corresponding propionate and acetate side-chains as shown in structures (*b*) and (*f*) (Scheme 25). According to this proposal[105] the 'extra' methyl groups required for the elaboration of *Chlorobium* chlorophylls would be introduced at the stage of uroporphyrinogen III or an equivalent species, requiring that the pathway for the *Chlorobium* chlorophyll biosynthesis branches at a much earlier point than protoporphyrin IX in contrast to the biosynthesis of other chlorophylls. In our view the methylation could be conveniently carried out using the vinyl side-chain of an orthodox intermediate in chlorophyll biosynthesis to produce the ethyl, n-propyl, and butyl side chains as shown in Scheme 26. Analogous mechanisms have been shown to operate for the elaboration of a variety of side chains in ergosterol and phytosterol biosynthesis.[108]

Since it is known that hydrogen atoms from the C-12 methyl groups of intermediates of the type (*a*) (Scheme 27) are readily labilized[109] by the reaction $(a) \rightarrow (b)$, there is no reason to believe that a species such as (*b*) generated at the enzyme active-site should not

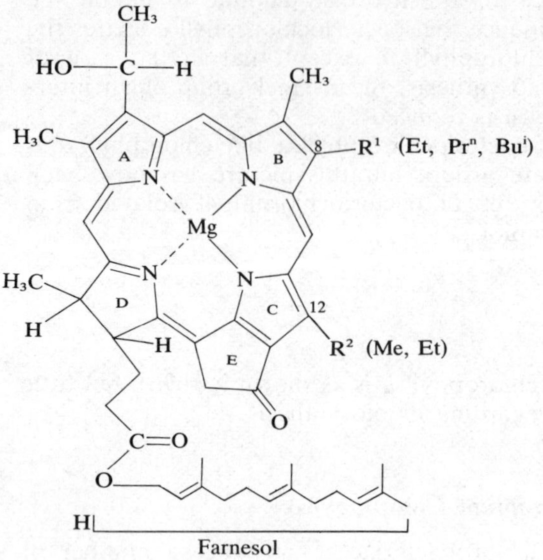

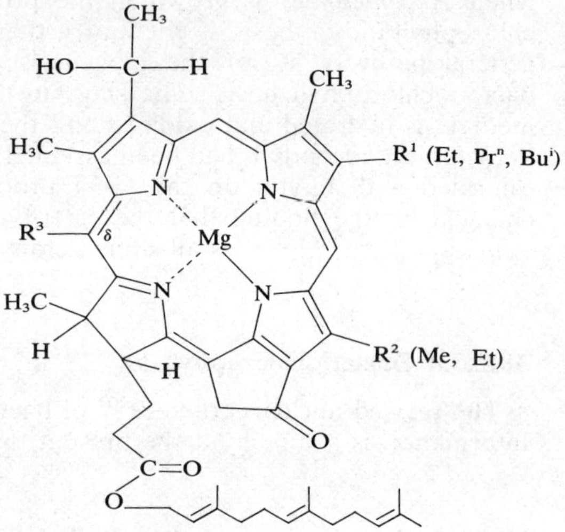

(71) *Chlorobium* Chlorophylls *d* (Chlorophylls-650)

(72) *Chlorobium* Chlorophylls *c* (Chlorophylls-660)

Band	R^1	R^2	R^1	R^2	R^3
1	Bu^i	Et	Bu^i	Et	Et
2	Pr^n	Et	Bu^i	Et	Me
3	Bu^i	Me	Pr^n	Et	Et
4	Et	Et	Pr^n	Et	Me
5	Pr^n	Me	Et	Et	Me
6	Et	Me	Et	Me	Me

SCHEME 24 The structure[105,106] of *Chlorobium* chlorophylls *d* (series 650) and *c* (series 660). The chlorophylls 650 and chlorophylls 660 from strains of *C. thiosulphatophilum* each separate into six bands on celite columns. The structure of these bands in the two series is tabulated above

be a potential site for an electrophilic attack by *S*-adenosylmethionine. Once again the sequence in which the nuclear and side-chain modifications occur in the elaboration of *Chlorobium* chlorophylls is not known at the present time.

30.2.3.6 Bacteriochlorophyll *e*

The structures of bacteriochlorophylls *e* have been elucidated by mass spectrometry and n.m.r. analysis and are shown in (**73**).[110] A distinguishing feature of bacteriochlorophylls *e* is the presence of a formyl group at position 7, a feature incidently found in chlorophyll *b*. Other aspects of the molecule relate[110] to bacteriochlorophylls *c* and *d* in the various forms that the side chains in positions 8 and 12 assume. The absolute stereochemistry[110] of positions 17 and 18 are *S* and the 3-hydroxyethyl group is *R*. Little is known of the route for the biosynthesis of this group of compounds.

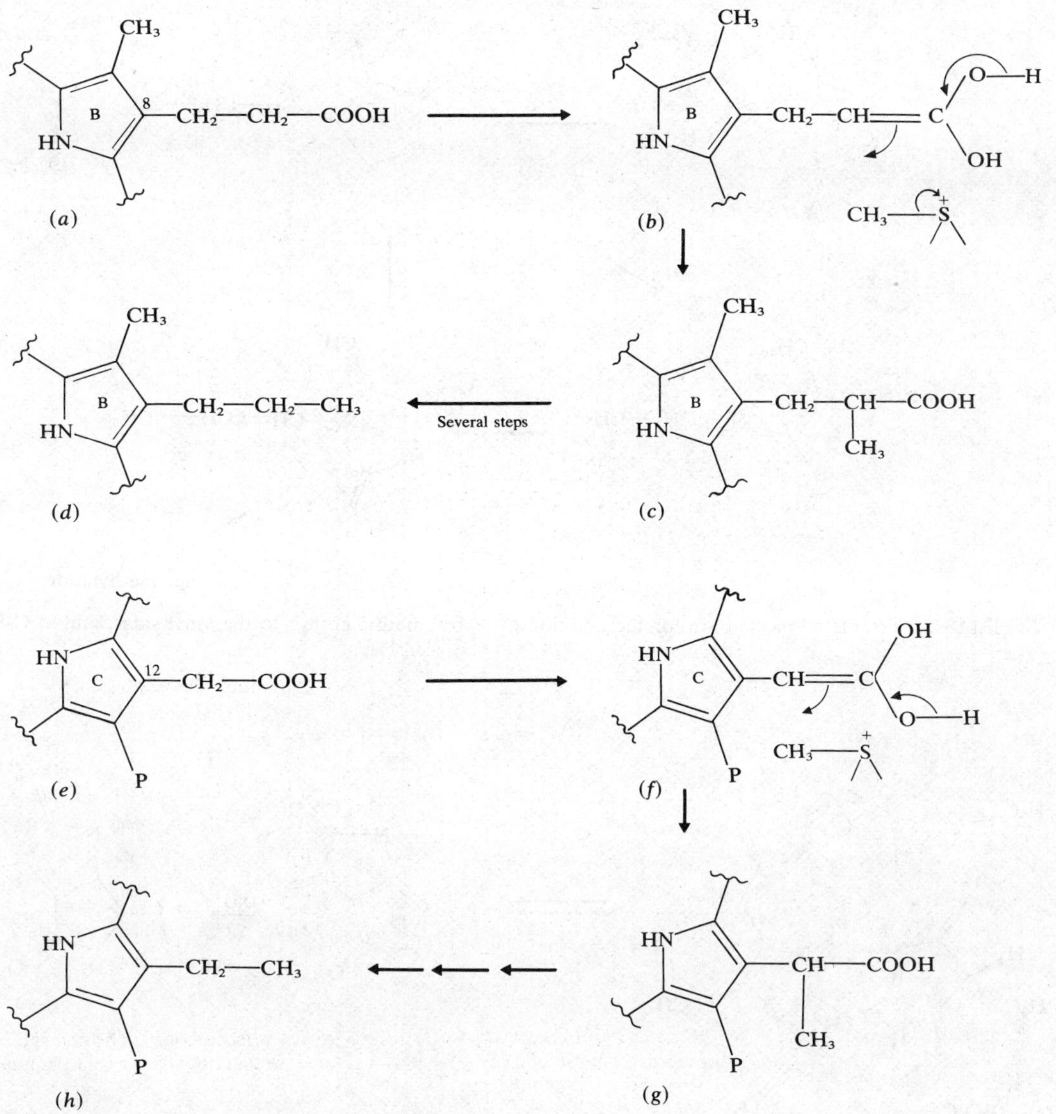

SCHEME 25 The mechanism for the addition of 'extra' methyl groups to the side chain at C-8 ($a \rightarrow d$) and C-12 ($e \rightarrow h$) according to Kenner *et al.*[105] ^{+}S—CH_3 = *S*-adenosylmethionine.

30.2.4 THE BIOSYNTHESIS OF CORRINS

30.2.4.1 Introduction

Vitamin B_{12} (**74**) was isolated independently, and almost simultaneously, in Great Britain[111] and in the United States,[112] and was shown to possess miraculous therapeutic activity for reversing the pathological state produced by pernicious anaemia. The major structural features of the vitamin emerging from the now classical X-ray diffraction studies of Dorothy Hodgkin[113] include the presence of a corrin nucleus containing a central cobalt atom which is linked to the imino nitrogen of an unusual base, dimethylbenzimidazole. The latter is linked to the corrin nucleus *via* a loop comprising ribosyl, phosphoryl, and aminopropanol moieties. All seven carboxyl functions surrounding the

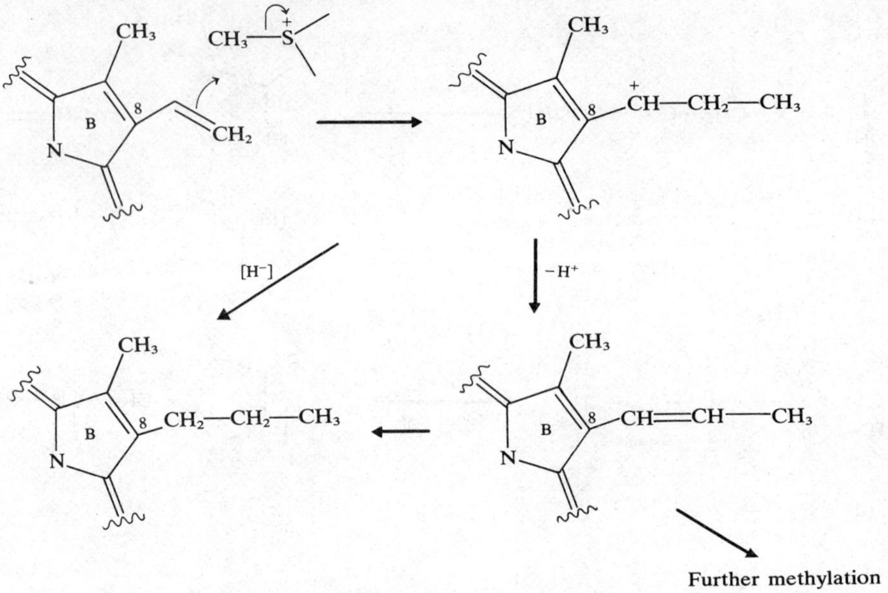

SCHEME 26 Postulated mechanism for the addition of 'extra' methyl groups to the vinyl side-chain at C-8

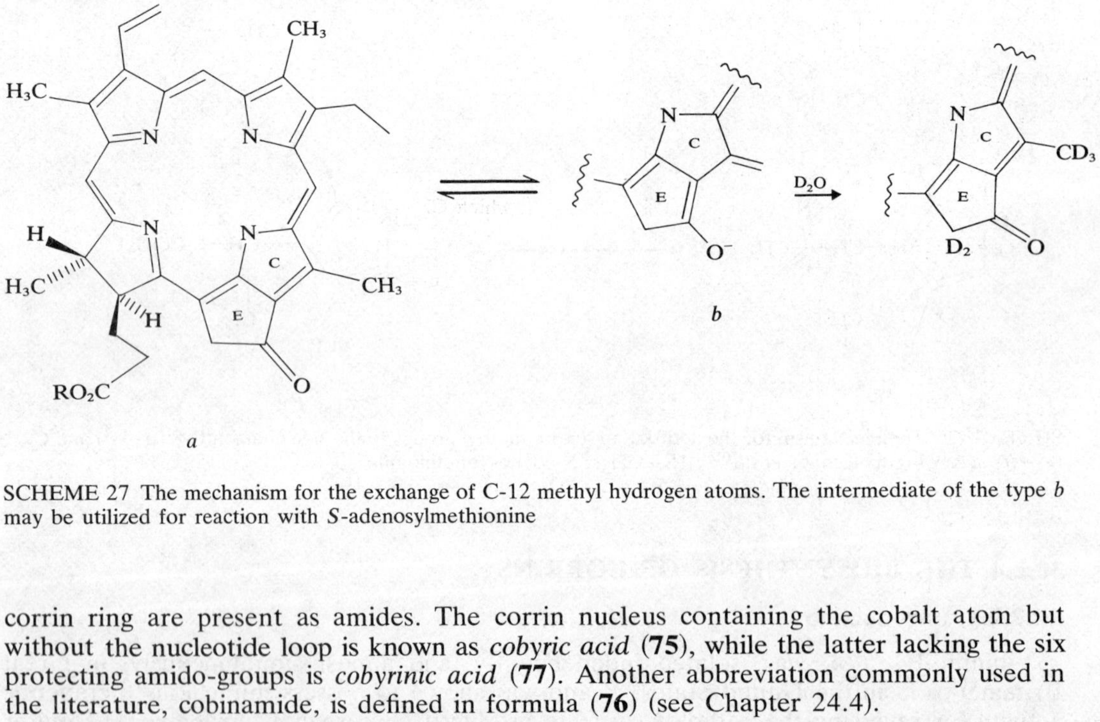

SCHEME 27 The mechanism for the exchange of C-12 methyl hydrogen atoms. The intermediate of the type *b* may be utilized for reaction with *S*-adenosylmethionine

corrin ring are present as amides. The corrin nucleus containing the cobalt atom but without the nucleotide loop is known as *cobyric acid* (**75**), while the latter lacking the six protecting amido-groups is *cobyrinic acid* (**77**). Another abbreviation commonly used in the literature, cobinamide, is defined in formula (**76**) (see Chapter 24.4).

30.2.4.2 The early precursors of the corrin nucleus

The close structural similarities of the corrin to the porphyrin nucleus suggested that the two groups of natural products may share common biosynthetic pathways.[9,115] This

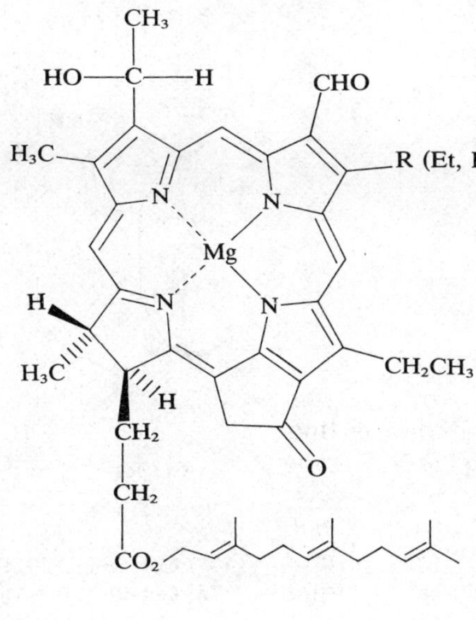

(**73**) Bacteriochlorophyll *e*

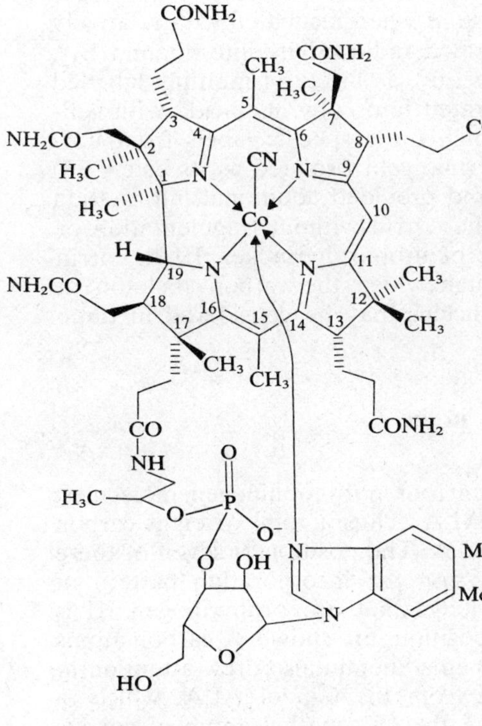

(**74**) Vitamin B_{12}

(**74a**) Dicyanocobalamine; by dissolving (**74**) in 0.1% KCN, (**74a**) is obtained in which CN replaces the nucleotide N

expectation was substantiated by Shemin's group, who in 1957 showed that [^{14}C]aminolevulinic acid (ALA) was efficiently incorporated into vitamin B_{12} by a *Streptomycete*.[114] Two years later the incorporation of the next porphyrin intermediate, porphobilinogen (PBG), into vitamin B_{12} was reported by Schwartz *et al.*[116]

The original proposal of Burnham,[117] that the divergence of the corrin pathway from the haem and chlorophyll pathway occurs after the formation of uroporphyrinogen III, was always attractive. However, early attempts to incorporate labelled uroporphyrinogen

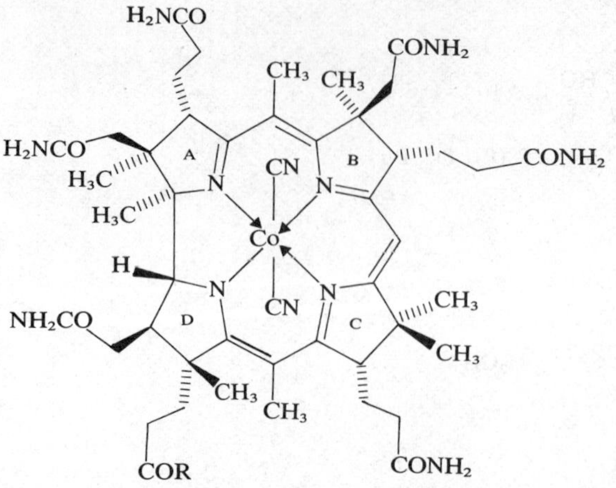

(75) Cobyric acid, R = OH

(76) Cobinamide, R = $NHCH_2CH(Me)OH$

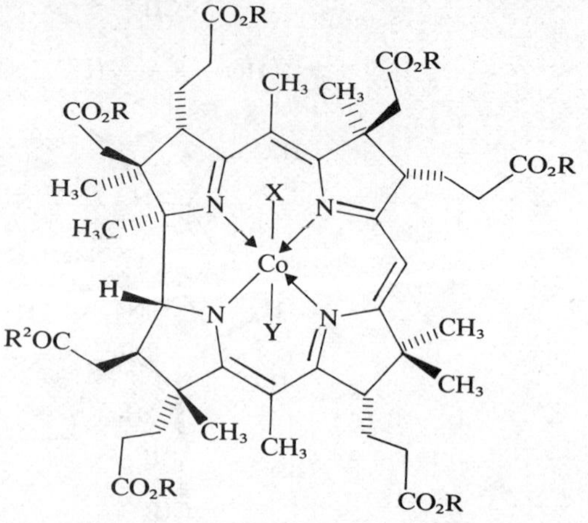

(77) Cobyrinic acid, R = H, X = OH, Y = H_2O
(78) Cobester, R = Me, X = Y = CN

III into corrins were disappointing. The demonstration by Scott's laboratory,[118] that under carefully controlled conditions whole cells of *P. shermanii* when incubated with relatively large amounts of [^{14}C]uroporphyrinogen III incorporated radioactivity into vitamin B_{12}, stimulated further activity in the field.[119–121] Subsequently, a variety of multiply labelled samples of uroporphyrinogen III[119,120] were incorporated into cobyrinic acid* with cell-free preparations of *P. shermanii.* The positions which these specific labels from uroprophyrinogen III occupied in the biosynthetic cobyrinic acid (isolated as its ester, **78**) confirmed the original conclusion of Scott *et al.*[118] and provided additional information that the uroporphyrinogen III is incorporated into the corrin without fragmentation or rearrangement of the four rings.[119,120] With the role of uroporphyrinogen III in corrin biosynthesis securely established, the pathway through which the carbon skeleton of uroporphyrinogen III is converted into the corrin nucleus may be considered in three stages.

30.2.4.3 The status of the C-5 carbon atoms of ALA in the elaboration of the corrin nucleus

The four hetrocyclic rings of corrins originate from four porphobilinogen molecules, which in turn are derived from eight molecules of ALA. Thus a total of eight carbon atoms in the corrin nucleus could arise from C-5 of ALA. The position of seven of these carbon atoms was deduced from Shemin's studies[115] on the incorporation pattern of [5-^{14}C]ALA into vitamin B_{12} and the further knowledge that uroporphyrinogen III is involved in the construction of the corrin ring. These positions are shown by carbon atoms (■, **80**) in Scheme 28. From the ^{14}C-labelling experiments Shemin also drew attention to the possibility that the eighth carbon atom derived from the C-5 of ALA, which in porphyrin(ogen) occupies the δ-position, may become the C-1 methyl group of corrins (**CH_3**, structure **80**). An unambiguous evaluation of this proposition was, however, prohibited at the time, primarily due to the unavailability of suitable degradation methods for the specific isolation of the methyl group (**CH_3**, structure **80**). With the advent of ^{13}C Fourier transform nuclear magnetic resonance (FT n.m.r.) in the late 1960s and the development of improved methods for the incorporation of substrate quantities of

* In cell-free preparations, available at the present time, the biosynthesis stops at the stage of cobyrinic acid.

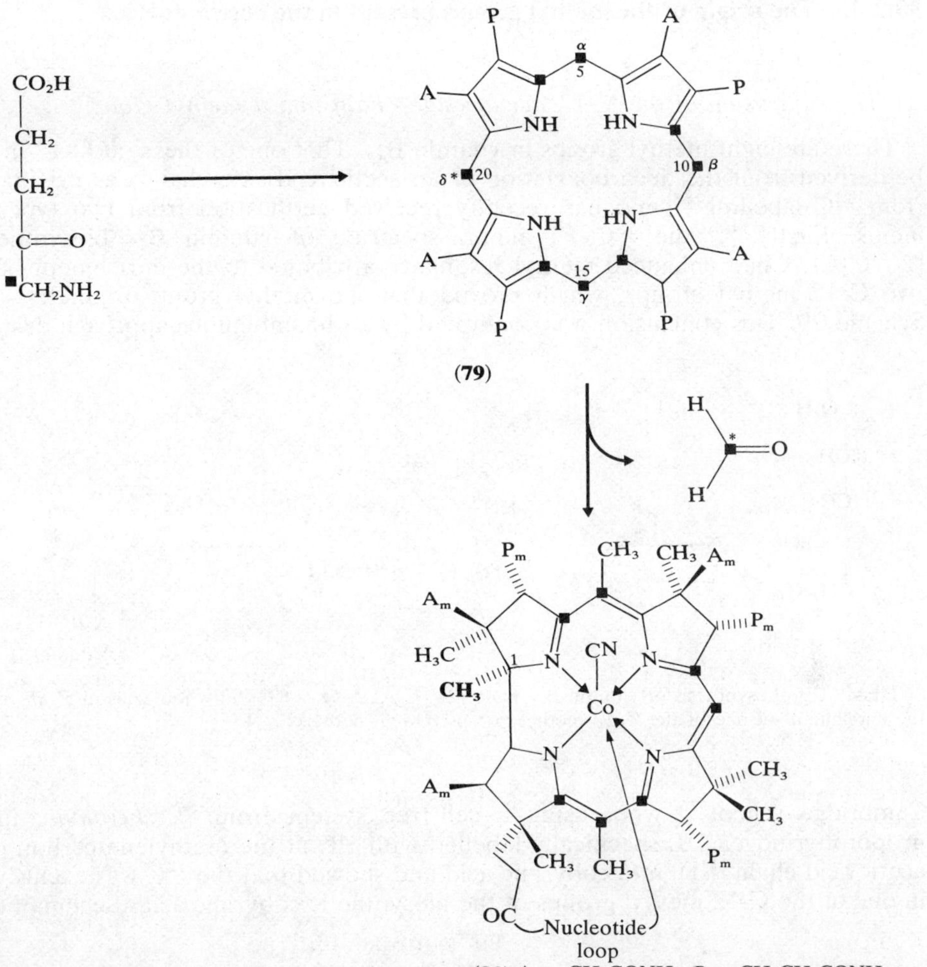

(**80**) $A_m = CH_2CONH_2$; $P_m = CH_2CH_2CONH_2$

SCHEME 28 Biosynthesis of vitamin B_{12} from [5-^{13}C]ALA (■ = ^{13}C), in whole cells of *P. shermanii*, showing the enrichment (■) of the six sp^2 carbon atoms of (**80**) but not the C-1 methyl group, $\mathbf{CH_3}$.

precursors into vitamin B_{12} without dilution from the endogenous pool, the problem was examined almost simultaneously in two laboratories.[122,123]

Administration of [5-^{13}C]ALA to whole cells of *P. shermanii* afforded a sample of vitamin B_{12} which was subjected to ^{13}C FT n.m.r. The rationale for the interpretation of the n.m.r. data involved the consideration that, of the eight carbon atoms under examination, seven atoms (■, **80**) are sp^2, C═C or C═N, and only the critical eighth carbon atom ($\mathbf{CH_3}$) is sp^3. In the spectrum of the biosynthesized material the resonances which appeared between 180–187 and 100–113 p.p.m. were assigned to the seven sp^2 carbon atoms. However, it was the *absence* of the signal in the high-field region of the spectrum, between 25 and 30 p.p.m.,* due to a sp^3 carbon atom which proved that no enrichment of the methyl carbon at C-1 ($\mathbf{CH_3}$, **80**) had occurred. The results allow the conclusion to be drawn[122–124] that one of the —$^{13}CH_2NH_2$ termini of ALA, and hence the corresponding carbon atom of PBG or uroporphyrinogen, has been extruded in the formation of the vitamin. It has recently been shown[125] that the C atom (■) can be trapped as formaldehyde (Scheme 28).

* The expectation that the ^{13}C resonances for sp^3 carbon atoms are between 20–28 p.p.m. is shown by the data in Scheme 31.

30.2.4.4 The origin of the methyl groups present in the corrin nucleus

(i) The conversion of the C-12 *acetate side-chain into a methyl group*

There are eight methyl groups in vitamin B_{12}. That one of these, at C-12 in ring C, may be derived from the decarboxylation of an acetic acid side-chain was originally deduced from ^{14}C labelling[115] and has recently received verification from two types of experiments. Firstly,[119b] the ^{13}C FT n.m.r. spectrum of vitamin B_{12} biosynthesized from [2-^{13}C]ALA had enhanced methyl resonance attributed to the enrichment of one of the two C-12 methyl groups, which proved that the methyl group originates as shown in Scheme 29. This conclusion was confirmed by an unambiguous approach designed by the

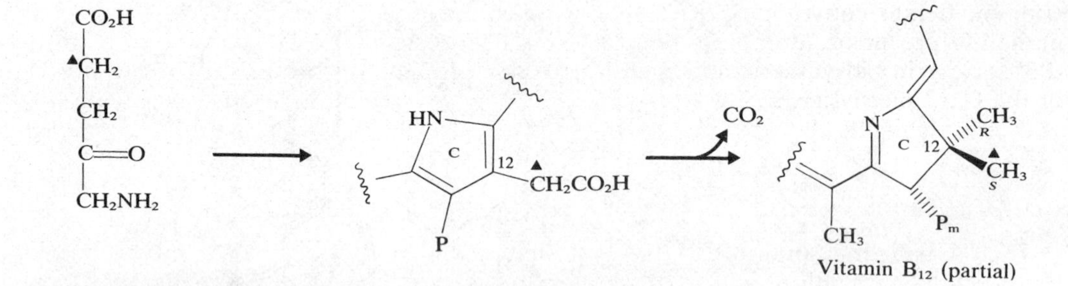

SCHEME 29 Biosynthesis of vitamin B_{12} from [2-^{13}C]ALA (▲ = ^{13}C), in whole cells of *P. shermanii*, showing the enrichment of one of the C-12 methyl groups (*pro-S*-methyl)

Cambridge school[126] who, using a cell-free system from *P. shermanii*, incorporated uroporphyrinogen III, specifically labelled with ^{14}C at the methylene carbon of the C-12 acetic acid chain (**81**), into cobyrinic acid and showed that the ^{14}C was exclusively located in one of the C-12 methyl groups of the biosynthetic cobyrinic acid (Scheme 30). That the

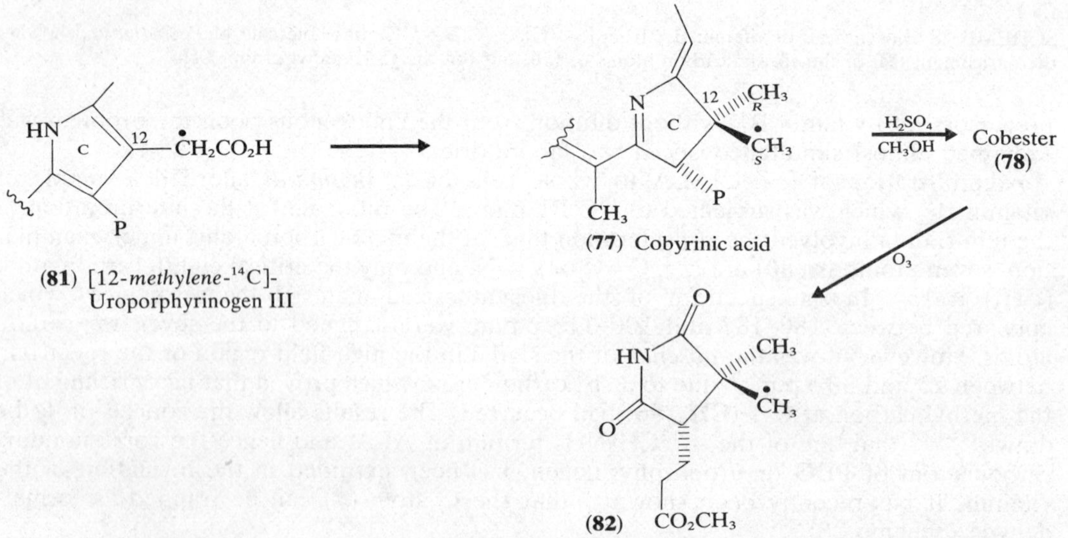

SCHEME 30 The incorporation of specifically labelled uroporphyrinogen III, with broken cells of *P. shermanii*, into cobyrinic acid isolated as cobester and degraded to (**82**) for the isolation of ring C

C-12 methyl group derived from the decarboxylation of the acetic acid side-chain occupies the β-orientation (*pro-S* methyl group) was established by the experiments described in the latter part of this section. Secondly, synthetic 12-decarboxy-uroporphyrinogen III (**83**), in which the C-12 acetic acid side-chain of the ring C of uroporphyrinogen III is replaced by a methyl group, was shown to be incorporated[125] into cobyrinic acid with a cell-free system from *P. shermanii.* Although it has not been unambiguously shown that the C-12 methyl group of the decarboxyuroporphyrinogen III (**83**) occupies C-12 in cobyrinic acid, experiments have been performed[125] which prove that the conversion, (**83**) → cobyrinic acid, occurs without rearrangement or fragmentation of the carbon skeleton. Furthermore, in the conversion of decarboxyuroporphyrinogen III (**83**) into cobyrinic acid, like that of uroporphyrinogen III, the C-20 carbon atom is also removed as formaldehyde.[125] However, decarboxyuroporphyrinogen III cannot be an obligatory intermediate in corrin biosynthesis[126] since the conversion of this compound into cobyrinic acid was an order of magnitude lower[125,126] than that of the authentic precursor, uroporphyrinogen III. The utilization of the decarboxy-derivative (**83**) for corrin biosynthesis nonetheless provides additional evidence for the fact that one of the C-12 methyl groups of vitamin B_{12} is derived from the C-12 acetic acid side-chain and also highlights the relatively broad substrate specificity of the enzymes involved.

(*ii*) *Methionine derived methyl groups*

The ^{14}C-labelling approach[115] used by Shemin and co-workers during the early 1960s provided the basic information on the origin of the extra methyl groups of vitamin B_{12} and in particular established that *at least* six of the seven remaining methyl groups, that is those present at C-2, C-5, C-7, C-12, C-15, and C-17 (see **84**), are derived from methionine. Recent biosynthetic experiments using ^{13}C n.m.r. have confirmed this conclusion and have, in addition, shown that the C-1 methyl group also originates from methionine. The conclusion is based on experiments reported almost simultaneously from Chicago,[127] Yale,[123] and Cambridge,[124] in which [*methyl*-^{13}C]methionine was incorporated by the whole cells of *P. shermanii* into vitamin B_{12}. The latter either[123,127] as its dicyano-derivative (**74a**) or[124] after conversion into the cobester (of the type **78**) when subjected to ^{13}C FT n.m.r. showed seven* well-defined resonances between 19.6–26 p.p.m., attributed to the seven methyl groups at C-1, C-2, C-5, C-7, C-12, C-15, and C-17 in (**84**).

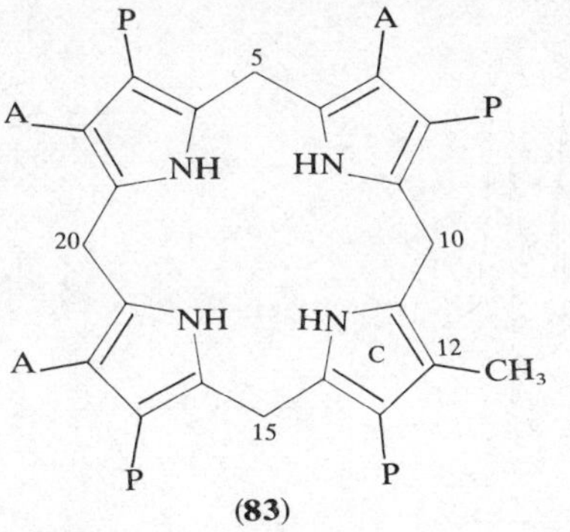

(**83**)

(*iii*) *The absolute stereochemistry of the two* C-12 *methyl groups*

Of the two methyl groups present at C-12 of ring C, one originates from the decarboxylation of an acetic acid side-chain and another from methionine. The assignment of the

* The spectrum of ^{13}C-enriched vitamin B_{12} shows only six signals, but the conversion of this into either dicyanocobalamin or, preferably, cobester simplifies the spectrum and gives improved resolution.

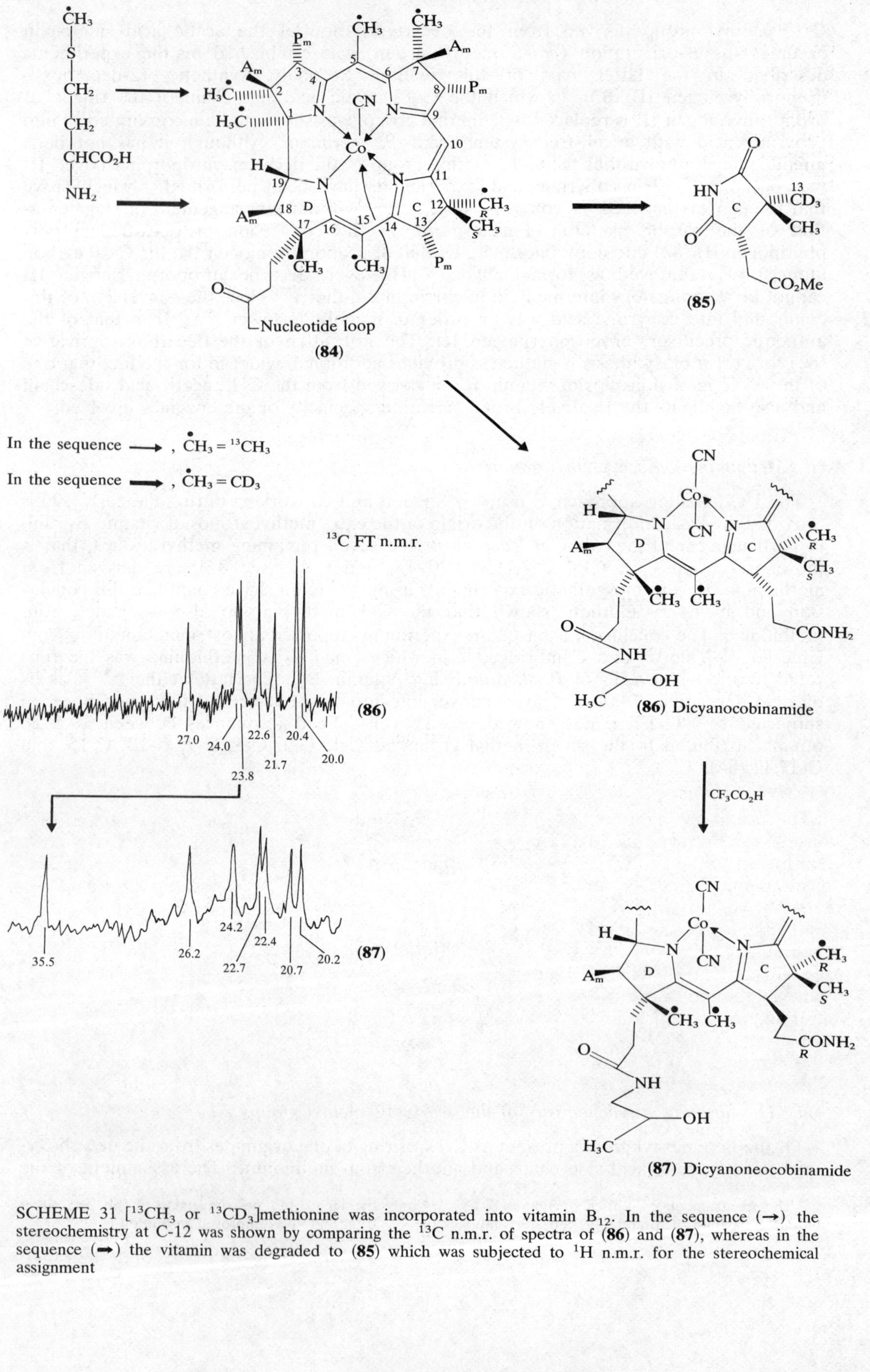

SCHEME 31 [$^{13}CH_3$ or $^{13}CD_3$]methionine was incorporated into vitamin B_{12}. In the sequence (→) the stereochemistry at C-12 was shown by comparing the ^{13}C n.m.r. of spectra of (**86**) and (**87**), whereas in the sequence (➡) the vitamin was degraded to (**85**) which was subjected to ^{1}H n.m.r. for the stereochemical assignment absolute stereochemical configuration to the two methyl groups was approached by considering[128] that the conformation of the C-ring of vitamin B_{12} places the 12α-methyl group *cis* to the axially oriented propionamide side-chain at C-13. This juxtaposition would be predicted to produce a gamma-effect on the ^{13}C chemical shift of the 12α-methyl group which should be lost when the propionamide side-chain is epimerized to the β-configuration as is present in the neo-derivative of the type (**87**). A comparison[119b,128] of the ^{13}C FT n.m.r. spectrum of [*methyl*-^{13}C]methionine-enriched dicyanocobinamide (**86**) with that of its neo-derivative (**87**) produced by CF_3COOH-catalysed isomerization[128b] of the C-13 centre of (**86**) revealed that the signal at 23.8 p.p.m. in dicyanocobinamide (**86**) suffered, through the removal of the gamma effect, a downfield shift to 35.5 p.p.m. in dicyano-neo-cobinamide (Scheme 31, **87**). These results establish that it is the [*methyl*-^{13}C]methionine-derived methyl group which must occupy the α- (or *pro-R*) position at C-12 in the biosynthetic sample.[119b,128] That the methyl group *cis* to the propionamide (the *pro-R* methyl group) is the one derived from methionine was also independently established by Battersby and co-workers.[124] These authors isolated the ring C imide (Scheme 31, **85**) from the degradation of a sample of vitamin B_{12} biosynthesized in the presence of [$^{13}CD_3$]methionine and assigned the absolute configuration to the two methyl groups using the ^1H n.m.r. approach developed by Dubs and Eschenmoser.[129]

*(iv) The sequence of methylation and the role of sirohydrochlorin**

The sequence in which methyl groups from methionine are introduced into the uroporphyrinogen III nucleus during cobyrinic acid biosynthesis is not yet defined with certainty, although a major advancement in this direction has been made through the isolation of a methylated intermediate in Scott's laboratory[130a] at Yale. It was found that a cell-free preparation of *P. shermanii* when incubated with ALA and *S*-adenosylmethionine, but in the absence of cobalt and dimethylbenzimidazole, led to the accumulation of a bis-methylated-uroporphyrinogen III derivative which was shown to be identical with sirohydrochlorin (**88b**)[130a] produced by the removal of iron from sirohaem (**88a**). The latter is a newly discovered prosthetic group of bacterial nitrite and sulphite reductases. The sirohydrochlorin (**88b**) obtained from the above biosynthetic experiment after reduction with sodium amalgam was incorporated into cobyrinic acid in a cell-free system, thus securely placing the new class of derivative as an intermediate between uroporphyrinogen III and cobyrinic acid. Further insight into the nature of intermediates has been provided by other workers.[130b,130c]

(v) The mechanism of the methylation steps

The broad mechanism through which methyl groups from *S*-adenosylmethionine are successively incorporated into uroporphyrinogen III to produce sirohydrochlorin is now considered (Scheme 32). *S*-Adenosylmethionine, which has been suggested to be an electrophilic agent, may be envisaged to attack the electron-rich β-position of one of the four rings of uroporphyrinogen III. For the sake of convenience we have chosen the first methylation to occur in ring A to produce the cation (**89**), which after proton elimination rearranges to give (**90**). A second methylation by an analogous mechanism then furnishes the siro-derivative (**91a** or **91b**). Recent observations showing that the CD_3 groups from [CD_3]methionine were transferred intact[131–133] into vitamin B_{12}, and the elegant demonstration by Arigoni and co-workers[134] using [chiral-*methyl*]methionine that the overall displacement at the methyl group in such transfer reactions occurs with the inversion of configuration, is consistent with the basic tenets of the mechanism proposed in Scheme 32. The mechanism in Scheme 32 makes the further prediction that the dimethylated derivative contains the arrangement of double bonds as shown in structure (**91a** or **91b**). It is thus possible that sirohydrochlorin (**88b**) normally isolated from the biosynthetic experiments[130] is the oxidation product of the true biosynthetic intermediate (**91**), and that it is the latter species which is produced by the chemical reduction of sirohydrochlorin

* This account is based on typescripts kindly made available to us by Professor A. I. Scott in June, 1977. The work has now been published (see Refs. 119b, 125, and A. I. Scott, A. J. Irwin, L. M. Siegel, and J. N. Shoolery, *J. Amer. Chem. Soc.*, 1978, **100,** 317) almost at the same time, as that of an independent study by Professor A. R. Battersby and his colleagues (A. R. Battersby, E. McDonald, M. Thompson, and V. Ya. Bykhovsky, *J.C.S. Chem. Comm.*, 1978, 150).

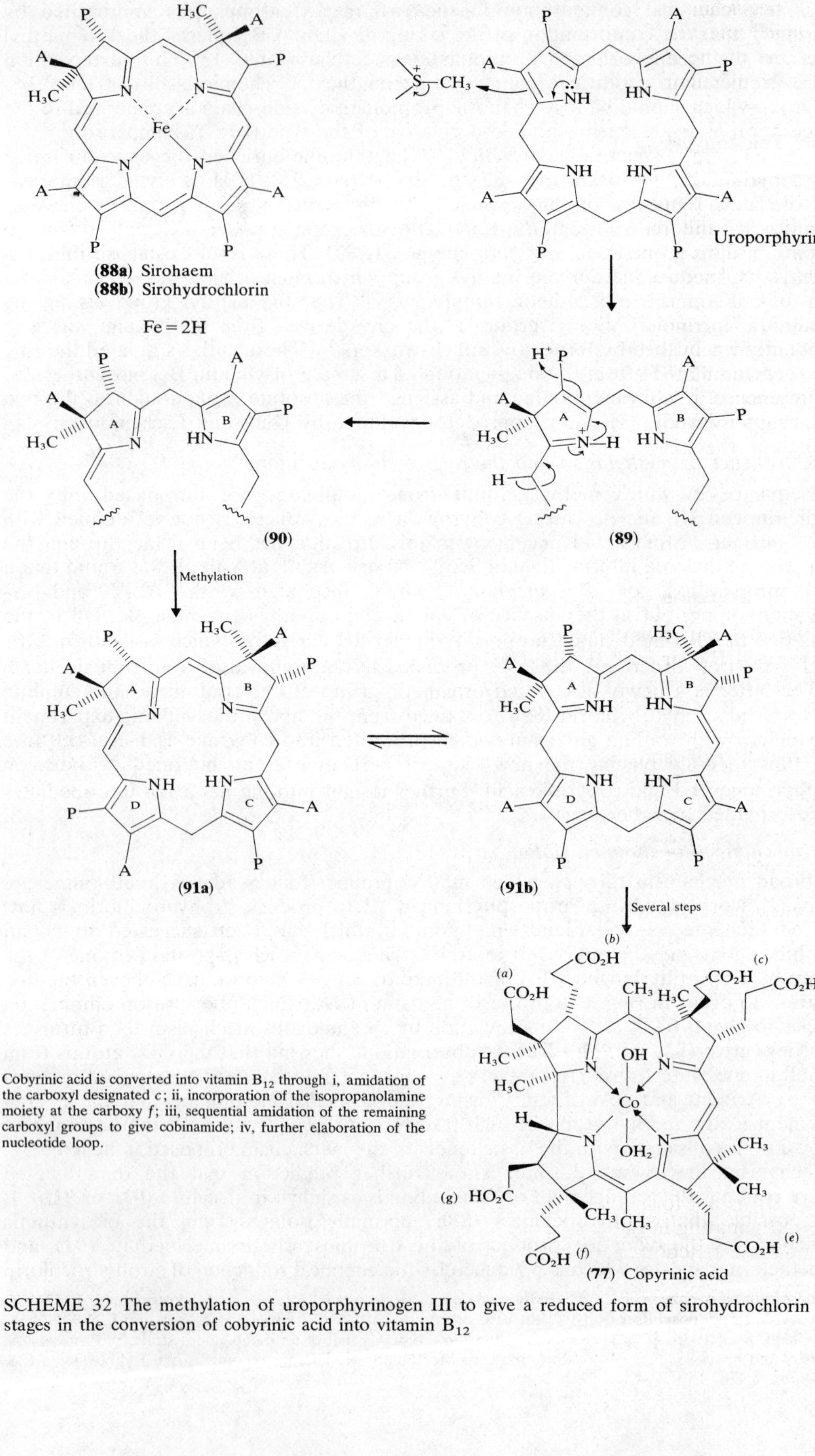

Cobyrinic acid is converted into vitamin B_{12} through i, amidation of the carboxyl designated *c*; ii, incorporation of the isopropanolamine moiety at the carboxy *f*; iii, sequential amidation of the remaining carboxyl groups to give cobinamide; iv, further elaboration of the nucleotide loop.

SCHEME 32 The methylation of uroporphyrinogen III to give a reduced form of sirohydrochlorin (**91**) and stages in the conversion of cobyrinic acid into vitamin B_{12}

and biologically incorporated into cobyrinic acid. A stepwise delineation of the molecular events which intervene between the sirohydrochlorin type of structure and cobyrinic acid remains one of the many unsolved problems in the field of vitamin B_{12} biosynthesis, as indeed is the mechanism and the stage at which cobalt is inserted into the corrin nucleus.

30.2.4.5 The conversion of cobyrinic acid into vitamin B_{12}

That cobyrinic acid occupies a central place in the vitamin B_{12} biosynthetic pathway was one of the most significant discoveries made in this field during the middle 1960s and owes much to the meticulous and painstaking work of the Stuttgart group led by Bernhauer.[135] This group developed many of the chemical and microbiological techniques which found subsequent application in the more recent work described in the preceding sections. It was found that young growing cultures of *P. shermanii* contained a significant amount of a corrinoid compound, shown to be cobyrinic acid, while cultures of *P. shermanii* grown in the presence of cobalt but without the precursor 5,6-dimethylbenzimidazole accumulated a variety of variously amidated derivatives of cobyrinic acid. The precise sequence in which the six carboxyl groups of cobyrinic acid are amidated and the seventh group is elaborated to produce the nucleotide loop of the vitamin is not yet known. Nevertheless, the kinetics of the accumulation of intermediates has led Wagner[135c] to propose the preferred sequence shown in Scheme 32 for the biosynthesis of vitamin B_{12} from cobyrinic acid in *P. shermanii.* The most important assertion made by the work is that the introduction of the isopropanol residue compulsorily requires that one of the carboxyl groups, probably the one designated (*c*) in Scheme 32, is first amidated.

30.2.4.6 Participation of vitamin B_{12} in enzymic reactions

Vitamin B_{12} participates in enzymic reactions after being converted either into 5′-deoxyadenosyl B_{12} or methyl B_{12}. The conversion of vitamin B_{12} into 5′-deoxyadenosyl B_{12}* is catalysed by an enzyme system† which requires a reduced flavin and ATP. The reduced flavin is used for the conversion of Co^{III} into Co^{I}. The two electrons added in the process are then utilized in the next step to form a cobalt–carbon bond through the elimination of triphosphate (Scheme 33).

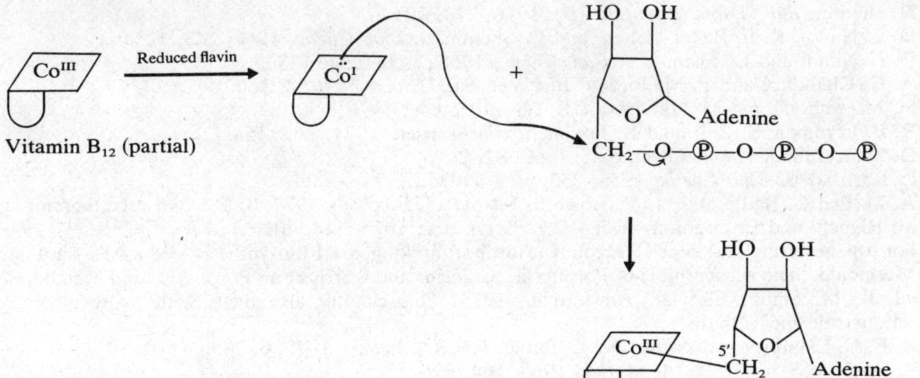

SCHEME 33

Methyl B_{12} is not formed as a free intermediate but is generated at the active site of the enzyme participating in the conversion of homocysteine into methionine. A more detailed review of these reactions is given in Chapter 24.4 and is also available elsewhere.[136,137]

* Also known as coenzyme B_{12} or 5′-deoxyadenosylcobalamin.
† ATP : 5′-deoxyadenosyl corrin transferase.

References

1. R. Willstätter, 'Uber Pflanzenfarbstroffe in Nobel stiftelsen Stockholm Les Prix Nobel en 1914–18', Norstedt, Stockholm, 1920, p. 1.
2. H. Fischer and H. Orth 'Die Chemie des Pyrrols', Akademische Verlagsgesellschaft, Leipzig, 1937.
3. R. B. Woodward, *Angew. Chem.*, 1960, **72,** 651.
4. D. C. Hodgkin, *Proc. Roy. Soc. (A)*, 1955, **288,** 294.
5. R. B. Woodward, *Pure Appl. Chem.*, 1973, **33,** 145.
6. A. Eschenmoser, '23rd Internat. Congress Pure and Applied Chem.', Butterworths, London, 1971, vol. 2, p. 69
7. D. Shemin, *Harvey Lectures*, 1955, **50,** 258, and references cited therein.
8. S. Granick, *Ann. Rev. Plant Physiol*, 1951, **2,** 115.
9. D. Shemin and R. C. Bray, *Ann. N. Y. Acad. Sci.*, 1964, **112,** 615, and references cited therein.
10. A. I. Scott, *Phil. Trans. Roy. Soc. (B)*, 1976, **273,** 303.
11. D. Shemin and D. Rittenberg, *J. Biol. Chem.*, 1945, **159,** 567.
12. I. M. London, D. Shemin and D. Rittenberg, *J. Biol. Chem.*, 1950, **183,** 757.
13. H. M. Muir and A. Neuberger, *Biochem. J.*, 1950, **47,** 97.
14. N. S. Radin, D. Rittenberg, and D. Shemin, *J. Biol. Chem.*, 1950 **184,** 755.
15. D. Shemin and J. Wittenberg, *J. Biol. Chem.*, 1952, **192,** 315.
16. D. Shemin and S. Kumin, *J. Biol. Chem.*, 1952, **198,** 827.
17. D. Shemin and C. S. Russell, *J. Amer. Chem. Soc.*, 1953, **75,** 4873.
18. D. Shemin, C. S. Russell, and T. Abramsky, *J. Biol. Chem.*, 1955, **215,** 613.
19. G. Kikuchi, A. Kumar, P. Talmage, and D. Shemin, *J. Biol. Chem.*, 1958, **233,** 1214.
20. K. D. Gibson, W. G. Laver, and A. Neuberger, *Biochem. J.*, 1958, **70,** 71.
21. G. W. Warnick and B. F. Burnham, *J. Biol. Chem.*, 1971, **246,** 6880.
22. M. J. Whiting and S. Granick, *J. Biol. Chem.*, 1976, **251,** 1340.
23. P. M. Jordan and D. Shemin, 'The Enzymes', 3rd edn., ed. P. Boyer, Academic Press, New York, 1972, vol. 7, p. 339.
24. M. Fanica-Gaignier and J. Clement-Metral, *European J. Biochem.*, 1973, **40,** 13, 19.
25. Z. Zaman, P. M. Jordan, and M. Akhtar, *Biochem. J.*, 1973, **135,** 257.
26. M. M. Abboud, P. M. Jordan, and M. Akhtar, *J.C.S. Chem. Comm.*, 1974, 643.
27. M. Akhtar, M. M. Abboud, G. Barnard, P. M. Jordan, and Z. Zaman, *Phil. Trans. Roy. Soc. (B)*, 1976, **273,** 117.
28. P. M. Jordan and A. Laghai, *Biochem. Soc. Trans.*, 1976, **4,** 52.
29. A. Laghai and P. M. Jordan, *Biochem. Soc. Trans.*, 1977, **5,** 299.
30. S. I. Beale, *Phil. Trans. Roy. Soc. (B)*, 1976, **273,** 99, and references cited therein.
31. J. B. Lohr and H. C. Friedmann, *Biochem. Biophys. Res. Comm.*, 1976, **69,** 908.
32. R. G. Westall, *Nature*, 1952, **170,** 614.
33. R. Schmid and D. Shemin, *J. Amer. Chem. Soc.*, 1955, **77,** 506.
34. D. Shemin, 'The Enzymes', 3rd edn., ed. P. Boyer, Academic Press, New York, 1972, vol. 7, p. 323.
35. D. Shemin, *Phil. Trans. Roy. Soc. (B)*, 1976, **273,** 109.
36. D. L. Nandi, K. F. Baker-Cohen, and D. Shemin, *J. Biol. Chem.*, 1968, **243,** 1224.
37. D. L. Nandi and D. Shemin, *J. Biol. Chem.*, 1968, **243,** 1231, 1236.
38. A. G. Chaudhry and P. M. Jordan, *Biochem. Soc. Trans.*, 1976, **4,** 760.
39. M. M. Abboud and M. Akhtar, *J.C.S. Chem. Comm.*, 1976, 1007.
40. R. B. Frydman, S. Reil, and B. Frydman, *Biochemistry*, 1971, **10,** 1154.
41. D. Mauzerall, *J. Amer. Chem. Soc.*, 1960, **82,** 2605.
42. L. Bogorad, *J. Biol. Chem.*, 1958, **233,** 501, 510, 516.
43. A. M. Del C. Batlle and M. V. Rossetti, *Internat. J. Biochem*, 1977, **8,** 251 *and* refs. therein.
44. M. Higuchi and L. Bogorad, *Ann. N. Y. Acad. Sci.*, 1975, **244,** 401.
45. For a penetrating and concise account of uroporphyrinogen III biosynthesis, see A. R. Battersby and E. McDonald, 'The Biosynthesis of Porphyrins, Chlorins and Corrins', in 'Porphyrins and Metalloporphyrins', ed. K. M. Smith, Elsevier, Amsterdam, 1975. This chapter also deals with many other aspects of tetrapyrrole biosynthesis.
46. E. B. C. Llambias and A. M. Del C. Batlle, *F.E.B.S. Letters*, 1970. **6,** 285.
47. L. Bogorad, *Ann. N. Y. Acad. Sci.*, 1963, **104,** 676.
48. J. Plusec and L. Bogorad, *Biochemistry*, 1970, **9,** 4736.
49. R. C. Davies and A. Neuberger, *Biochem. J.*, 1973, **133,** 471.
50. B. Frydman, R. B. Frydman, A. Valasinas, E. S. Levy, and G. Feinstein, *Phil. Trans. Roy. Soc. (B)*, 1976, **273,** 137.
51. A. R. Battersby and E. McDonald, *Phil. Trans. Roy. Soc. (B)*, 1976, **273,** 161.
52. J. H. Mathewson and A. H. Corwin, *J. Amer. Chem. Soc.*, 1961, **83,** 135.
53. E. Bullock, *Nature*, 1965, **205,** 70.
54. (a) R. Robinson, 'The Structural Relations of Natural Products', Clarendon Press, Oxford, 1955; (b) C. S. Russell, *J. Theoret. Biol.*, 1974, **47,** 145; (c) A. I. Scott, K. S. Ho, M. Kajiwara, and T. Takahashi, *J. Amer. Chem. Soc.*, 1976, **98,** 1589.
55. A. R. Battersby, E. McDonald, D. C. Williams, and H. K. W. Wurziger, *J.C.S. Chem. Comm.*, 1977, 113.

56. A. R. Battersby, C. J. R. Fookes, E. McDonald, and M. J. Meegan, *J.C.S. Chem. Comm.*, 1978, 185.
57. R. A. Neve, R. F. Labbe, and R. A. Aldrich, *J. Amer. Chem. Soc.*, 1956, **78,** 691.
58. D. Mauzerall and S. Granick, *J. Biol. Chem.*, 1958, **232,** 1141.
59. L. C. San Martin de Viale and M. Grinstein, *Biochem. Biophys. Acta*, 1968, **158,** 79.
60. L. C. San Martin de Viale, A. A. Viale, S. Nacht, and M. Grinstein, *Clin. Chim. Acta*, 1970, **28,** 13.
61. A. H. Jackson, H. A. Sancovich, A. M. Ferramola, N. Evans, D. E. Games, S. A. Matlin, G. H. Elder, and S. G. Smith, *Phil. Trans. Roy. Soc.* (*B*), 1976, **273,** 191.
62. G. F. Barnard and M. Akhtar, *J.C.S. Chem. Comm.*, 1975, 495.
63. J. Luthy, J. Rétey, and D. Arigoni, *Nature*, 1969, **221,** 1213.
64. J. W. Cornforth, J. W. Redmond, H. Eggerer, W. Buckel, and C. Gutschow, *Nature*, 1969, **221,** 1212.
65. S. Sano and S. Granick, *J. Biol. Chem.*, 1961, **236,** 1173.
66. G. Y. Kennedy, A. H. Jackson, G. W. Kenner, and C. J. Suckling, *F.E.B.S. Letters*, 1970, **6,** 9; 1970, **7,** 205.
67. J. A. S. Cavaleiro, G. W. Kenner, and K. M. Smith, *J.C.S. Perkin I*, 1973, 183.
68. A. R. Battersby, J. Baldas, J. Collins, D. H. Grayson, K. J. James, and E. McDonald, *J.C.S. Chem. Comm.*, 1972, 1265.
69. Z. Zaman, M. M. Abboud, and M. Akhtar, *J.C.S. Chem. Comm.*, 1972, 1263.
70. Z. Zaman and M. Akhtar, *European J. Biochem.*, 1976, **61,** 215.
71. A. R. Battersby, E. McDonald, H. K. W. Wurziger, and K. J. James, *J.C.S. Chem. Comm.*, 1975, 493.
72. A. H. Jackson, D. E. Games, P. Couch, J. R. Jackson, R. V. Belcher, and S. G. Smith, *Enzyme*, 1974, **17,** 81.
73. H. A. Dailey, Jnr. and J. Lascelles, *Arch. Biochem. Biophys.* 1974, **160,** 523, and refs. cited therein.
74. R. J. Porra and O. T. G. Jones, *Biochem. J.*, 1963, **87,** 181, 186.
75. For a detailed discussion on haems and haem proteins, see 'Inorganic Biochemistry', ed. G. L. Eichhorn, Elsevier, Amsterdam, 1973, vol. 2, part VI, pp. 797–1133.
76. For a detailed review on the structural aspects of chlorophylls, see A. H. Jackson, in 'Chemistry and Biochemistry of Plant Pigments', ed. T. W. Goodwin, Academic Press, New York, 1976, vol. 1, chapter 1.
77. An up-to-date and scholarly account of the biochemistry of chlorophylls is available; L. Bogorad, in 'Chemistry and Biochemistry of Plant Pigments', ed. T. W. Goodwin, Academic Press, New York, 1976, vol. 1, chapter 2.
78. S. Granick, *J. Biol. Chem.*, 1948, **172,** 717; 1948, **175,** 333.
79. S. Granick, *J. Biol. Chem.*, 1961, **236,** 1168.
80. C. A. Rebeiz and P. A. Castelfranco, *Ann. Rev. Plant Physiol.*, 1973, **24,** 129, and references cited therein.
81. R. K. Ellsworth and A. S. Hsing, *Photosynthetica*, 1974, **8,** 228.
82. K. D. Gibson, A. Neuberger, and G. H. Tait, *Biochem. J.*, 1963, **88,** 325.
83. R. J. Radmer and L. Bogorad, *Plant Physiol.*, 1967, **42,** 463.
84. J. G. Ebbon and G. H. Tait, *Biochem. J.*, 1969, **111,** 573.
85. A. Gorchein, *Biochem. J.*, 1973, **134,** 833.
86. O. T. G. Jones, *Phil. Trans. Roy. Soc.* (*B*), 1976, **273,** 207, and references cited therein.
87. R. K. Ellsworth and S. Aronoff, *Arch. Biochem. Biophys.*, 1969, **130,** 374, and references cited therein.
88. R. K. Ellsworth and A. S. Hsing, *Biochim. Biophys. Acta*, 1973, **313,** 119.
89. O. T. G. Jones, *Biochem. J.*, 1963, **88,** 335.
90. O. T. G. Jones, *Biochem. J.*, 1966, **101,** 153.
91. W. T. Griffiths and O. T. G. Jones, *F.E.B.S. Letters*, 1975, **50,** 355.
92. J. H. C. Smith, in 'Comparative Biochemistry of Photoreactive Systems', ed. M. B. Allen, Academic Press, New York, 1960, p. 257.
93. The original proposal of S. Granick and M. Gassman, *Plant Physiol.*, 1970, **45,** 201 as modified in Ref. 77.
94. K. W. Henningsen and A. Kahn, *Plant Physiol.*, 1971, **47,** 685.
95. P. Schopfer and H. W. Siegelman, *Plant Physiol.*, 1968, **43,** 990.
96. M. Holden, *Biochem. J.*, 1961, **78,** 359.
97. A. O. Klein and W. Vishniac, *J. Biol. Chem.*, 1961, **236,** 2544.
98. J. B. Wolff and L. Price, *Arch. Biochem. Biophys.*, 1957, **72,** 293.
99. J. Lascelles, *Biochem. J.*, 1966, **100,** 175.
100. For a review on bacteriochlorophyll, see R. C. Davies, A. Gorchein, A. Neuberger, J. D. Sandy, and G. H. Tait, *Nature*, 1973, **245,** 15.
101. O. T. G. Jones, *Biochem. J.*, 1964, **91,** 572.
102. J. Lascelles and T. Altschuler, *Arch. Mikrobiol.*, 1967, **59,** 204.
103. J. J. Katz, H. H. Strain, A. L. Harkness, M. H. Studier, W. A. Svec, T. R. Janson, and B. T. Cope, *J. Amer. Chem. Soc.*, 1972, **94,** 7938.
104. H. Scheer, W. A. Svec, B. T. Cope, M. H. Studier, R. G. Scott, and J. J. Katz, *J. Amer. Chem. Soc.*, 1974, **96,** 3714.
105. G. W. Kenner, J. Rimmer, K. M. Smith, and J. F. Unsworth, *Phil. Trans. Roy. Soc.* (*B*), 1976, **273,** 255, and references cited therein.
106. A. S. Holt, J. W. Purdie, and J. W. F. Wasley, *Canad. J. Chem.*, 1966, **44,** 88, and references cited therein.
107. W. R. Richards and H. Rapoport, *Biochemistry*, 1967, **6,** 3830.

108. M. Akhtar, P. F. Hunt, and M. A. Parvez, *Biochem. J.*, 1967, **103,** 616; E. Lederer, *Quart. Rev. Chem. Soc.*, 1969, **23,** 453; L. J. Goad, J. R. Lentom, F. F. Knapp and T. W. Goodwin, *Lipids*, 1974, **9,** 382.
109. C. D. Mengler, *Illinois Inst. Technol. N.M.R. Letters*, 1967, **100,** 27, as quoted in Ref. 105.
110. H. Brockmann, *Phil. Trans. Roy. Soc.* (*B*), 1976, **273,** 277, and references cited therein.
111. E. L. Smith and L. F. J. Parker, *Biochem, J.*, 1948, **43,** viii; K. H. Fantes, J. E. Page, L. E. J. Parker, and E. L. Smith, *Proc. Roy. Soc.* (*B*), 1949, **136,** 592.
112. E. L. Rickes, N. G. Brink, F. R. Koniuszy, T. R. Wood, and K. Folkers, *Science*, 1948, **107,** 396.
113. D. C. Hodgkin, J. Kamper, M. Mackay, J. Pickworth, K. N. Trueblood, and J. G. White, *Nature*, 1956, **178,** 64, and references cited therein.
114. J. W. Corcoran and D. Shemin, *Biochim. Biophys. Acta*, 1957, **25,** 661.
115. R. C. Bray and D. Shemin, *J. Biol. Chem.*, 1963, **238,** 1501.
116. S. Schwartz, K. Ikeda, I. M. Miller, and C. J. Watson, *Science*, 1959, **129,** 40.
117. B. F. Burnham, in 'Metabolic Pathways', 3rd edn., ed. D. M. Greenberg, Academic Press, New York, 1969, vol. 3, chapter 18.
118. A. I. Scott, C. A. Townsend, K. Okada, M. Kajiwara, and R. J. Cushley, *J. Amer. Chem. Soc.*, 1972, **94,** 8269.
119. (a) A. I. Scott, N. Georgopapadakou, K. S. Ho, S. Klioze, E. Lee, S. L. Lee, G. H. Temme III, C. A. Townsend, and I. M. Armitage, *J. Amer. Chem. Soc.*, 1975, **97,** 2548; for correction of this paper see *J. Amer. Chem. Soc.*, 1976, **98,** 2371. (b) For a review of the work of Scott's group see, A. I. Scott, *Accounts Chem. Res.*, 1978, **11,** 29.
120. A. R. Battersby, M. Ihara, E. McDonald, F. Satoh, and D. C. Williams, *J.C.S. Chem. Comm.*, 1975, 436.
121. H. Dauner and G. Müller, *Z. physiol. chem.*, 1975, **356,** 1353.
122. C. E. Brown, J. J. Katz, and D. Shemin, *Proc. Nat. Acad. Sci. U.S.A.*, 1972, **69,** 2585.
123. A. I. Scott, C. A. Townsend, K. Okada, M. Kajiwara, R. J. Cushley, and P. J. Whitman, *J. Amer. Chem. Soc.*, 1972, **94,** 8267; 1974, **96,** 8069.
124. (a) A. R. Battersby, M. Ihara, E. McDonald, J. R. Stephenson, and B. T. Golding, *J.C.S. Chem. Comm.*, 1973, 404; A. R. Battersby, M. Ihara, E. McDonald, J. R. Redfern and B. T. Golding, *J.C.S. Perkin I*, 1977, 158, and references cited therein; (b) for a review of the work of Battersby's group, see A. R. Battersby, *Experientia*, 1978, **34,** 1.
125. M. Kajiwara, K. S. Ho, H. Klein, A. I. Scott, A. Gossauer, J. Engel, E. Neumann, and H. Zilch, *Bio-org. Chem.*, 1977, **6,** 397.
126. A. R. Battersby, E. McDonald, R. Hollenstein, M. Ihara, F. Satoh, and D. C. Williams, *J.C.S. Perkin I*, 1977, 166.
127. C. E. Brown, D. Shemin, and J. J. Katz, *J. Biol. Chem.*, 1973, **248,** 8015.
128. (a) A. I. Scott, C. A. Townsend, and R. J. Cushley, *J. Amer. Chem. Soc.*, 1973, **95,** 5759, and references cited therein; (b) R. Bonnett, *Phil. Trans. Roy. Soc.* (*B*), 1976, **273,** 295.
129. P. Dubs and A. Eschenmoser, unpublished results quoted in Ref. 124; P. Dubs, Dissertation No. 4297, E.T.H. Zurich, 1969.
130. (a) A. I. Scott, A. J. Irwin, L. M. Siegel, and J. N. Shoolery, presented at Internat. Symp. Stereochemistry, Kingston, Ontario, June 1976; (b) R. Deeg, H. P. Kriemler, K. H. Bergman, and G. Müller, *Z. physiol. Chem.*, 1977, 358, 399; (c) K. H. Bergman, R. Deeg, K. D. Gneuss, H. P. Kriemler, and G. Müller, *ibid.*, p. 1315.
131. M. Imfeld, C. A. Townsend, and D. Arigoni, *J.C.S. Chem. Comm.*, 1976, 541.
132. A. R. Battersby, R. Hollenstein, E. McDonald, and D. C. Williams, *J.C.S. Chem. Comm.*, 1976, 543.
133. A. I. Scott, M. Kajiwara, T. Takahashi, I. M. Armitage, P. Demou, and D. Petrocine, *J.C.S. Chem. Comm.*, 1976, 544.
134. D. Arigoni, unpublished work.
135. (a) K. Bernhauer, O. Müller, and F. Wagner, *Angew. Chem. Internat. Edn.*, 1964, **3,** 200, and references cited therein; (b) K. Bernhauer, F. Wagner, H. Michna, P. Rapp, and H. Vogelmann: *Z. physiol. Chem.*, 1968, **349,** 1297. (c) F. Wagner, *Ann. Rev. Biochem.*, 1966, **35,** 405.
136. For reviews see: H. A. Barker, *Ann. Rev. Biochem.*, 1972, **41,** 55; M. Akhtar and D. C. Wilton, *Ann. Reports* (*B*), 1970, **67,** 557; 1972, **69,** 140.
137. H. Weissbach and R. T. Taylor, *Vitamins and Hormones*, 1970, **28,** 415.

30.3
Shikimic Acid Metabolites

E. HASLAM
University of Sheffield

30.3.1 THE SHIKIMATE PATHWAY: BIOSYNTHESIS OF CHORISMIC ACID

Although Fischer and Dangschat[1] first noted the close structural similarities between plant phenols such as gallic acid and the alicyclic acids quinic (**4**) and shikimic (**7**) in 1935, and were led on this basis to postulate a biosynthetic relationship, the full significance of this suggestion was not realised until the 1950s, when elegant investigations first by Davis[2] and later by Sprinson[3] and Gibson[4] and their collaborators established the pathway of aromatic amino-acid metabolism in plants and micro-organisms which is now universally referred to as the shikimate pathway. (−)-Shikimic acid (**7**) itself was first isolated as a natural product in 1885 by Eykmann, who obtained it from the aniseed fruit, *Illicium religiosum* Sieb., and its common name in fact derives from the Japanese name for that plant *shikimi-no-ki.* Reviews[2–5] and a monograph[6] described the principal biochemical features and associated metabolites of the shikimate pathway. Whilst plants and micro-organisms are able to synthesize the three aromatic amino-acids, L-phenylalanine (**10**), L-tyrosine (**11**), and L-tryptophan (**12**) using the shikimate pathway, and in plants this pathway also constitutes the principal route for the synthesis of aromatic metabolites, it is significant to note that animals cannot utilize this biochemical pathway for the *de novo* synthesis of these amino-acids from carbohydrate precursors.

The first step in the shikimate pathway (Scheme 1) is an aldol condensation between phosphoenolpyruvate (**2**) and D-erythrose 4-phosphate (**1**) to give 3-deoxy-D-arabinoheptulosonic acid 7-phosphate (**3**; DAHP). From this point a series of six enzyme-catalysed steps leads to chorismate (**9**) *via* shikimate. The main stem of the shikimate pathway terminates at this point, and from chorismate (**9**) there diverge at least five biosynthetic pathways to essential metabolites – the three protein α-amino-acids (L-phenylalanine, L-tyrosine, and L-tryptophan), *p*-aminobenzoic acid (**14**) and the folate coenzymes, and the isoprenoid quinones. The essential biochemical and chemical details of the stages in this common part of the pathway from carbohydrate precursors to

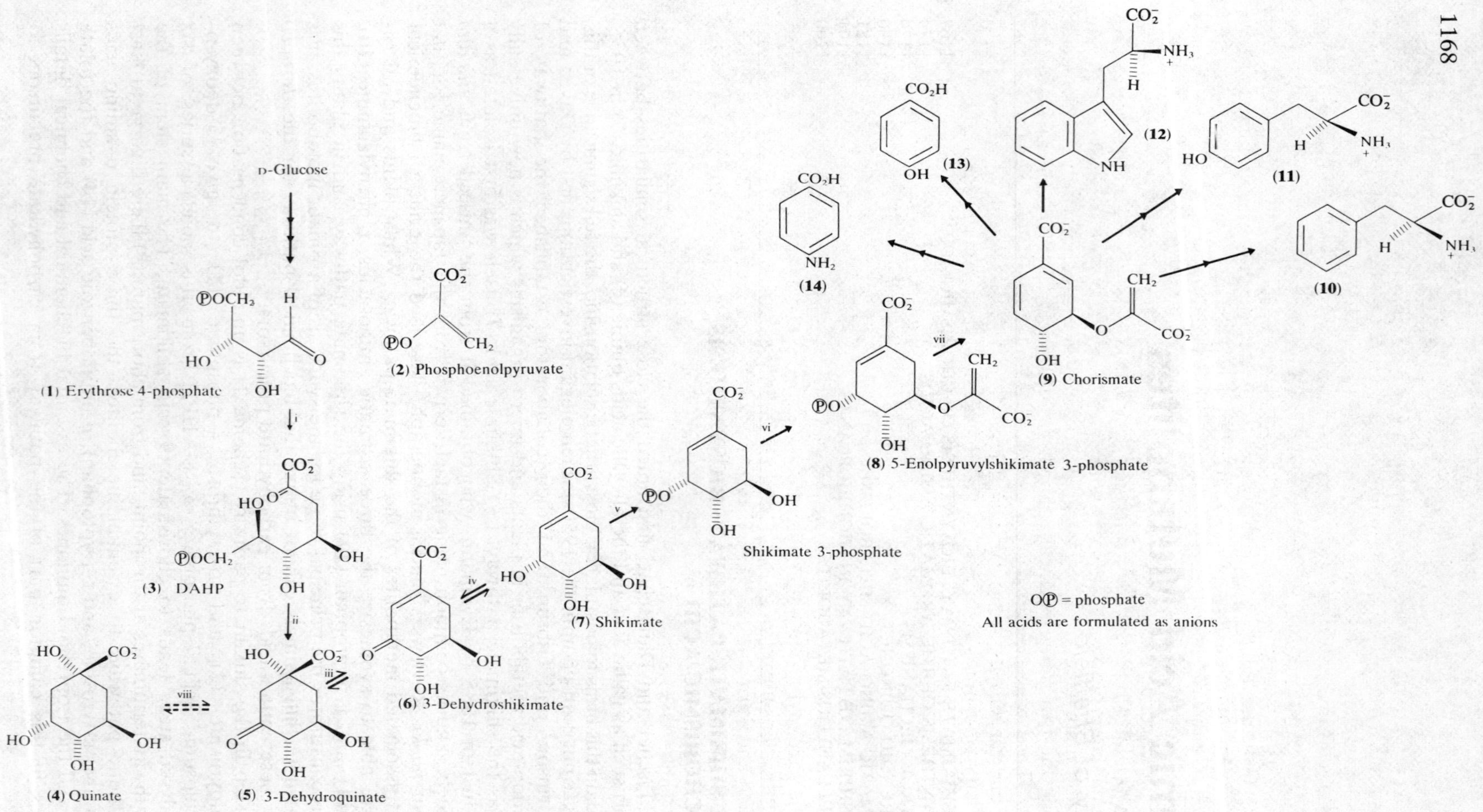

i, DAHP synthetase; ii, 3-dehydroquinate synthetase (NAD^+, Co^{2+}); iii, 3-dehydroquinate dehydratase; iv, 3-dehydroshikimate reductase (NADPH); v, shikimate kinase (ATP); vi, 5-enolpyruvylshikimate 3-phosphate synthetase (PEP); vii, chorismate synthetase; viii, quinate dehydrogenase (NADH).

SCHEME 1 The shikimate pathway

chorismate were determined by the combined use of isotopic tracers, auxotrophic mutants, and enzyme studies. The role of quinic acid (**4**) in the shikimate pathway has been a subject of some debate, but all the available evidence favours the view that, although some micro-organisms (e.g. *Aerobacter aerogenes*) can utilize the acid as a carbon source by incorporation *via* 3-dehydroquinate (**5**) into the pathway, it is not a normal intermediate.

Enzymes which function as catalysts for each of the steps from 3-dehydroquinate (**5**) to chorismate (**9**) in the main stem of the pathway have been isolated from bacteria and characterized. Studies of the details of their mode of action, and in particular their stereochemical features of catalysis, have revealed several distinctive features worthy of note. The initial condensation of phosphoenolpyruvate (**2**; PEP) and D-erythrose 4-phosphate (**1**) catalysed by the enzyme DAHP synthetase was first studied by DeLeo and Sprinson[7] and later Floss and his collaborators.[8,9] The latter group showed that the reaction at the methylene group of the PEP occurred in a predominantly stereospecific manner using the two stereospecifically labelled (^{3}H) samples of PEP (**15**) and (**16**) (see Scheme 2). These were separately converted to samples of [6-^{3}H]shikimic acid, *via* DAHP (**3**), and the shikimic acid was then degraded to (2*RS*)-[3-^{3}H]malic acid (**17**). The stereospecificity of tritium labelling in this product was examined using the fumarase reaction. Only the 2*S*-isomer of malic acid is active in the fumarase reaction and Floss showed by this means that (3*E*)-[3-^{3}H]PEP (**15**) gave predominantly (6*S*)-[6-^{3}H]shikimic acid (**18**) in the enzymic synthesis and that (3*Z*)-[3-^{3}H]PEP (**16**) conversely gave mainly the (6*R*)-[6-^{3}H]shikimic acid (**19**). This work, combined with a knowledge of the absolute stereochemistry of DAHP (**3**) and of (−)-shikimic acid (**7**), therefore defines the *si*-face as the direction of attack at the methylene group of the phosphoenolpyruvate (**2**) and the *re*-face at the carbonyl group of D-erythrose 4-phosphate (**1**). This evidence is accommodated in the mechanism shown in Scheme 3, based on the evidence put forward by Floss[9] and by DeLeo and Sprinson,[7] in which a carbanion is formed at C-3 of phosphoenolpyruvate and that during the condensation this normally attacks the aldehyde C-atom of D-erythrose 4-phosphate to give DAHP (**3**) in the manner shown.

The reversible *cis* elimination and addition of water observed for the enzyme 3-dehydroquinate dehydratase, which catalyses the interconversion of 3-dehydroquinate (**5**) and 3-dehydroshikimate (**6**), represents the first example of this type recorded in enzyme chemistry and it contrasts with other well-known examples of the addition of water to $\alpha\beta$-unsaturated carboxyl systems such as *cis*-aconitate and fumarate. Hanson and Rose[10] studied the reverse reaction in tritiated water and isolated the tritiated 3-dehydroquinic acid by conversion to quinic acid (**4**) using an extract of an *Aerobacter aerogenes* mutant. In addition, a further sample of [2-^{3}H]quinic acid (**4**) was also isolated from the equilibration of quinic acid with the same enzyme system in tritiated water. Both samples of [2-^{3}H]quinic acid were separately degraded to citric acid, which was then equilibrated with aconitate hydratase. Under these conditions both samples of citric acid lost over 95% of their tritium, thus establishing that the stereochemical position of the tritium in the quinic acid was 2*R* and hence that the overall addition of water in the conversion of 3-dehydroshikimate to 3-dehydroquinate was *cis* (Scheme 4).

Similar observations were later made in an independent study of the forward biosynthetic reaction[11] utilizing samples of (2*S*)-[2-^{2}H]-3-dehydroquinate (**20**; 0.6–0.7 $^2H_1$) and (2*R*)-[2-^{2}H]-3-dehydroquinate (**21**; 0.5-0.6 $^2H_1$), which were chemically prepared from quinic and shikimic acid, respectively (Scheme 5).

As a consequence of these observations, Hanson and Rose[10] formulated this reversible addition-elimination reaction as taking place in two distinct steps *via* the enol or enolate anion of 3-dehydroquinate. They postulated that, given this type of mechanism, the overall stereochemistry of the reaction would then be entirely dependent on the relative spatial disposition of catalytic groupings on the enzyme active site. This view has been amplified by later workers[11,12] in the mechanism shown in Scheme 6 in which the 3-dehydroquinate substrate adopts a 'boat' or 'skew boat' conformation in the active site of the enzyme and forms a Schiff base with a free amino-group on the enzyme.

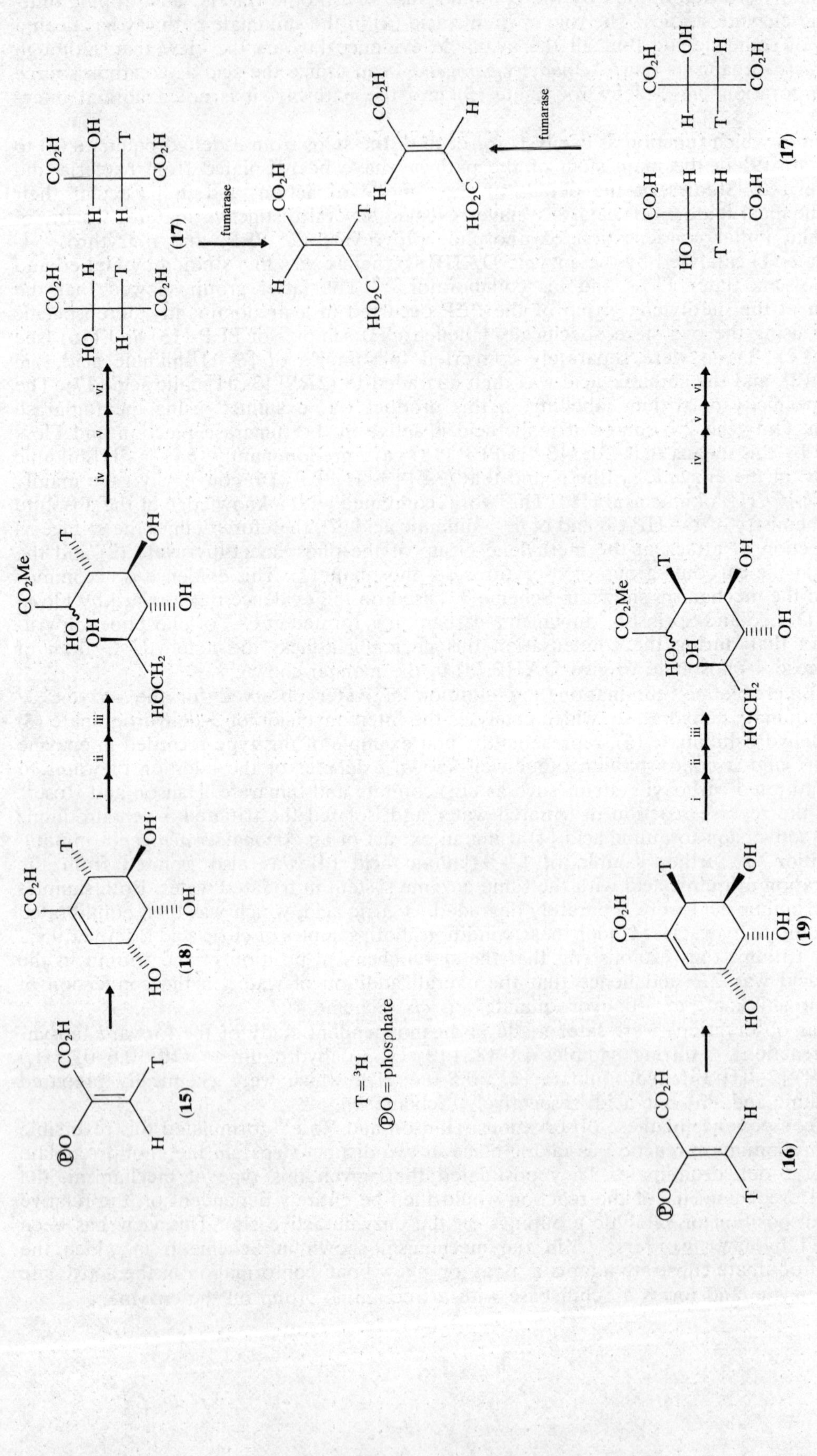

i, MeOH, H^+; ii, O_3; iii, $NaBH_4$; iv, $NaIO_4$; v, Br_2; vi, OH^-.

SCHEME 2

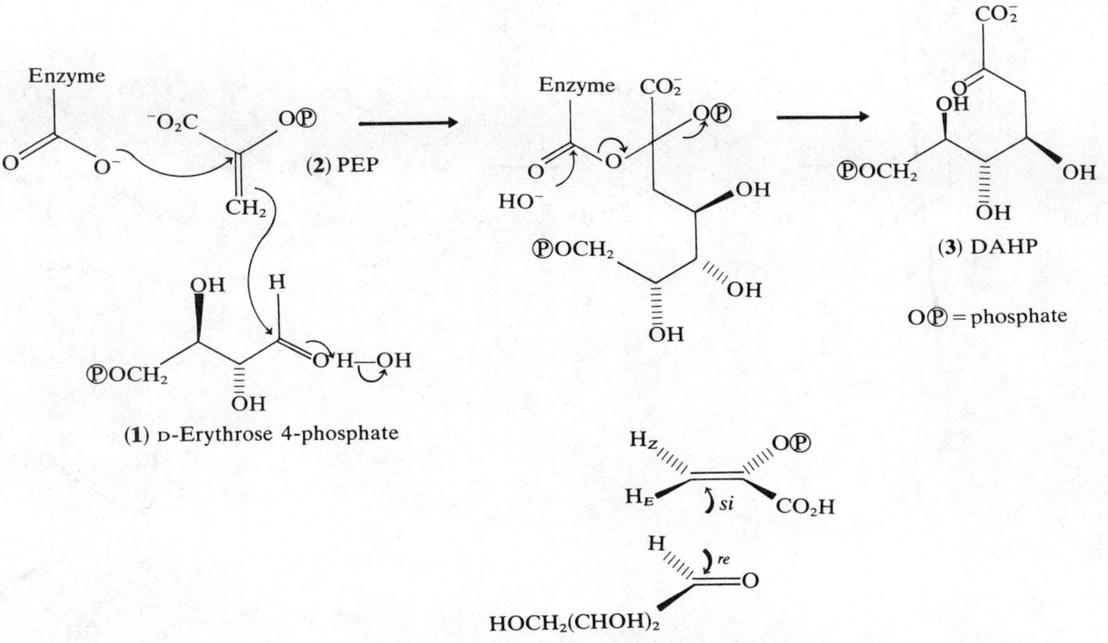

SCHEME 3 DAHP synthetase—a postulated mechanism

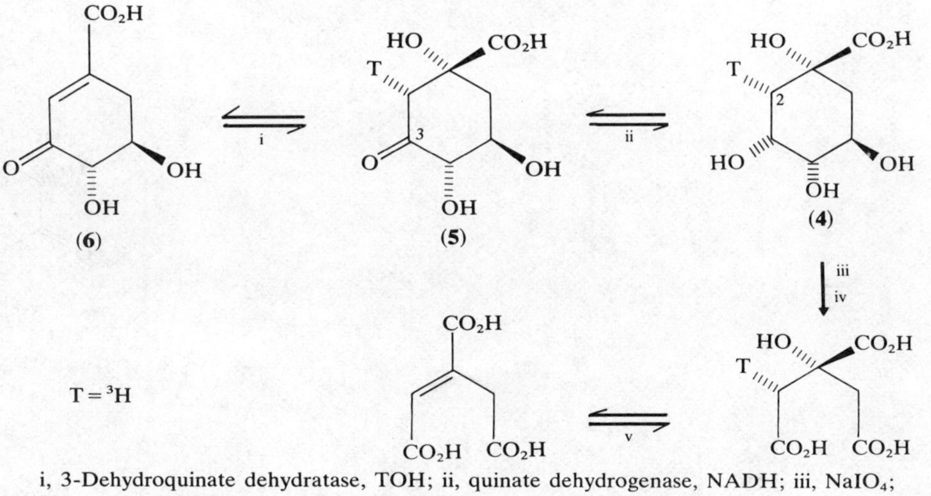

i, 3-Dehydroquinate dehydratase, TOH; ii, quinate dehydrogenase, NADH; iii, $NaIO_4$; iv, Br_2; v, aconitase.

SCHEME 4

Generation of the enolpyruvyl group of 5-enolpyruvylshikimate 3-phosphate (**8**) from phosphoenolpyruvate (**2**) represents an unusual type of biosynthetic reaction. Only one other example is known in primary metabolism–namely in the formation of UDP-*N*-acetylpyruvylglucosamine in the biosynthesis of UDP-*N*-acetylmuramic acid. Based on reactions carried out in tritiated and deuteriated water, Sprinson and his collaborators[13] formulated the reaction broadly as depicted in Scheme 7 where H—A is a proton donor attached to the enzyme active site. Kinetic isotope effects were observed in the protonation and deprotonation steps,[14] but in order to accommodate the random incorporation of isotopic hydrogen into the methylene group of the enolpyruvyl system it was concluded that there is little or no restriction to rotation of the methyl group (**22**) generated in the intermediate stage of the reaction.

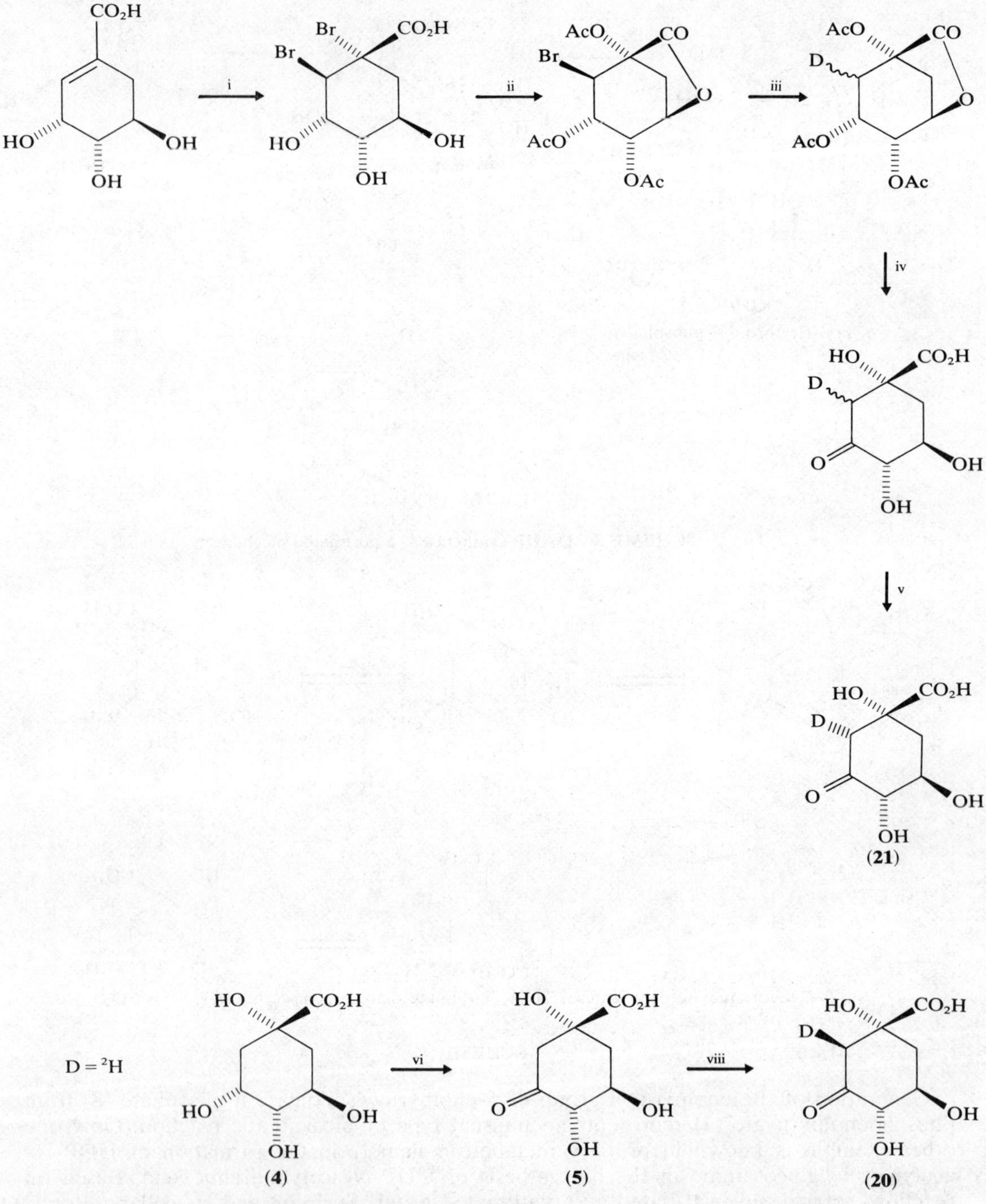

(21) (4) (5) (20)

i, Br_2; ii, Ag_2CO_3, H_2O; Ac_2O; iii, D_2, Pd/C; iv, OH^-, Pt, O_2; v, H_2O, pH 7.0; vi, Pt, O_2; vii, D_2O, pH 7.0.

SCHEME 5 Preparation of 2(*S*)- and 2(*R*)-2-deuterio-3-dehydroquinic acids (**20**) and (**21**)

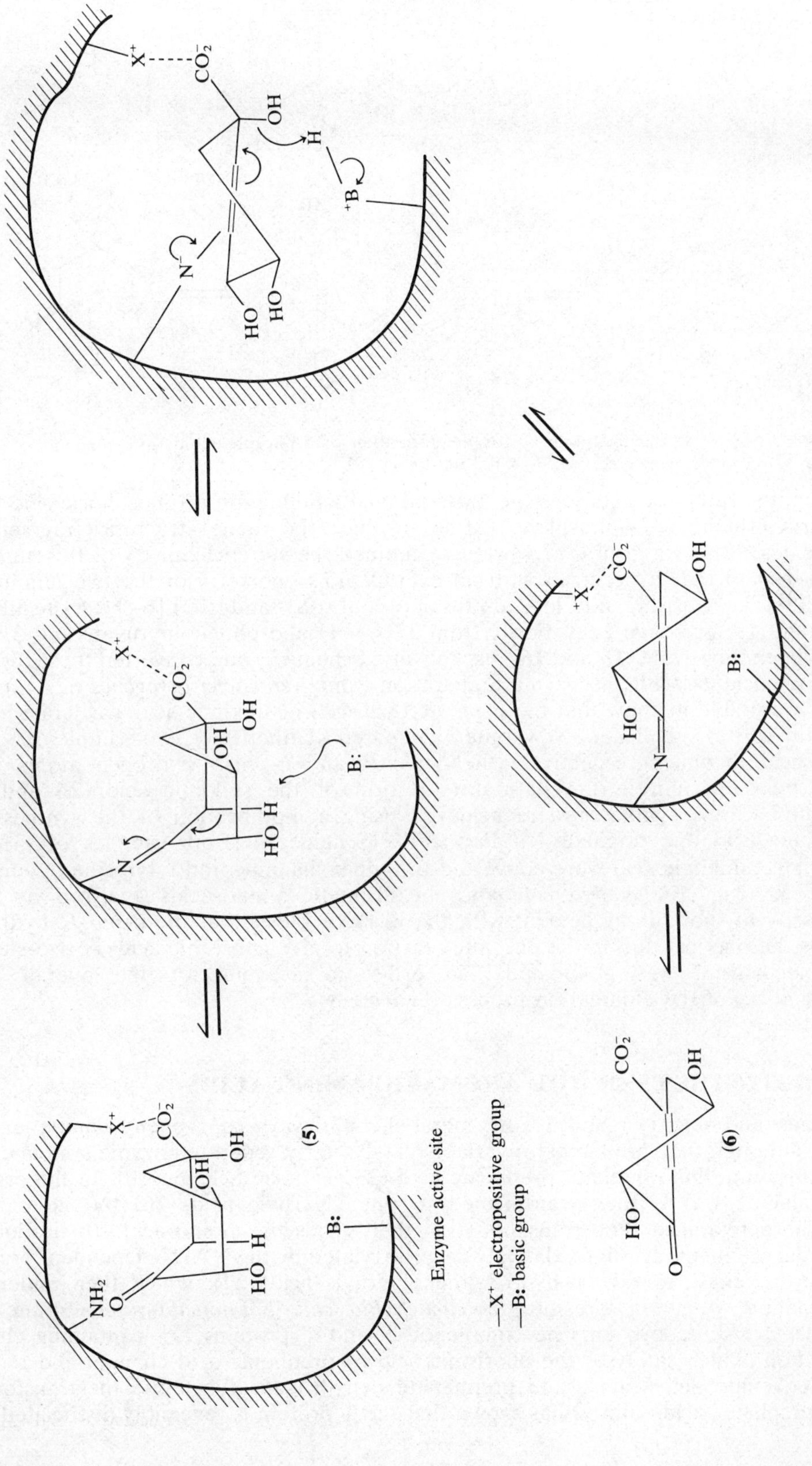

SCHEME 6 3-Dehydroquinate dehydratase—a postulated mechanism

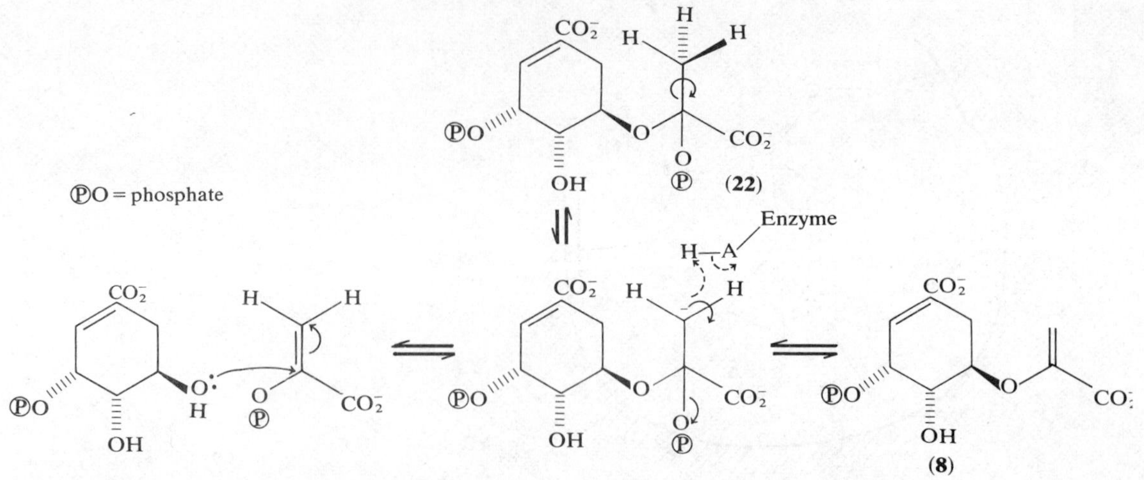

SCHEME 7 Postulated pathway for the formation of 5-phosphoenolpyruvylshikimic acid 3-phosphate (**8**)

Chorismate synthetase catalyses the 1,4-conjugate elimination of phosphoric acid from 5-enolpyruvylshikimate 3-phosphate (**8**) to produce the diene structural fragment of chorismic acid (**9**). Two groups of workers examined the stereochemistry of this elimination and showed that it occurred with an overall *trans* geometry of the two eliminated groups. Onderka, Carroll, and Floss[9] used samples of (6*S*)- and (6*R*)-[6-^{3}H]-(−)-shikimic acid (**18** and **19**) prepared enzymically from 3*E*-[3-^{3}H]phosphoenolpyruvate and 3*Z*-[3-^{3}H]phosphoenolpyruvate (**15** and **16**, respectively; Scheme 2) and converted these directly to chorismic acid (**9**) with an enzyme preparation from *Aerobacter aerogenes* 62.1. In this way they were able to show that the 6-*pro-R* hydrogen in shikimic acid was labile in the elimination process. Hill and Newkome[15] prepared synthetically two samples of (±)-shikimic acid in which respectively the 6-*pro-R* and 6-*pro-S* hydrogen atoms were replaced by deuterium in the natural (−) form of the shikimic acid (**23** and **24**, respectively). These syntheses were achieved using a modification of the synthesis of (±)-shikimic acid due originally to Raphael[15] (Scheme 8). Both samples of racemic 6-deuteriated shikimic acid were converted to L-phenylalanine and L-tyrosine by enzyme preparations from *Escherichia coli*, and the aromatic amino-acids analysed by mass spectroscopy to show, in agreement with the work of Floss, that the 6-*pro-R* hydrogen atom was the one lost during the enzymic reaction to give chorismic acid. A two-step 'X group mechanism' was postulated[15] in order to account for the overall *trans* stereochemistry of the elimination process (Scheme 9).

30.3.2 BIOSYNTHESIS OF THE AROMATIC AMINO-ACIDS

In plants and micro-organisms the metabolic pathways to L-phenylalanine and L-tyrosine, although they are almost invariably catalysed by separate enzyme systems, pass through the same intermediate, prephenic acid (**25**)[16,17] (see Scheme 10). In the case of L-phenylalanine this is then transformed to phenylpyruvic acid (**26**) by the enzyme prephenate dehydratase, and transamination then yields the amino-acid. To produce L-tyrosine the prephenic acid is oxidatively decarboxylated by the NAD$^+$-dependent prephenate dehydrogenase to give *p*-hydroxyphenylpyruvic acid (**27**), which then undergoes transamination to give L-tyrosine. In *Escherichia coli*, *Salmonella typhimurium*, and *Aerobacter aerogenes* two enzyme complexes–P and T proteins[18,19]–containing chorismate mutase (which catalyses the chorismic acid to prephenic acid change) and respectively prephenate dehydratase and prephenate dehydrogenase catalyse the transformations of prephenic acid. Gibson has shown that the T protein is reversibly dissociated into

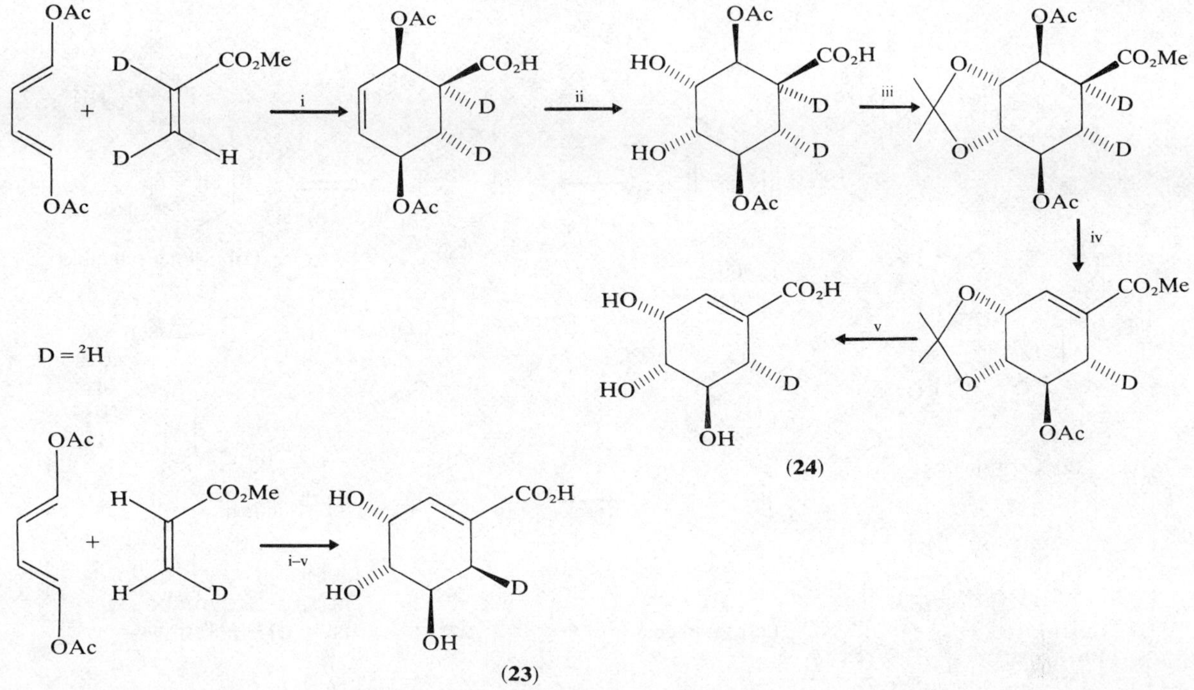

i, Δ; ii, OsO_4; iii, CH_2N_2; iv, MgO, 290°C; v, H^+.

SCHEME 8 Synthesis of 6-deuterioshikimic acids (all formulae show one form of a racemate)

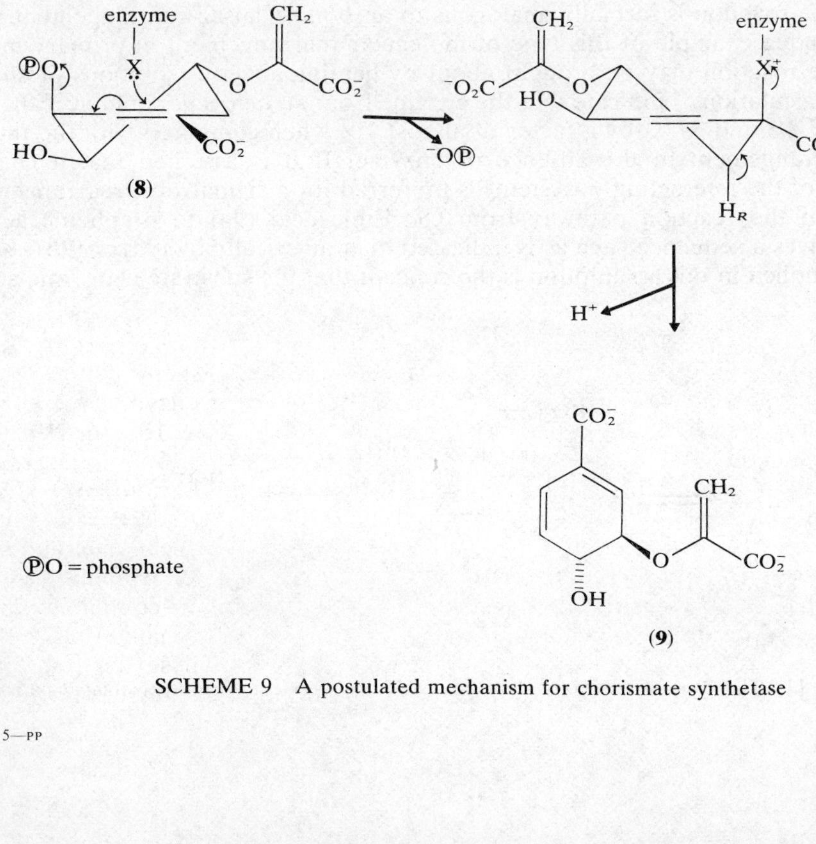

SCHEME 9 A postulated mechanism for chorismate synthetase

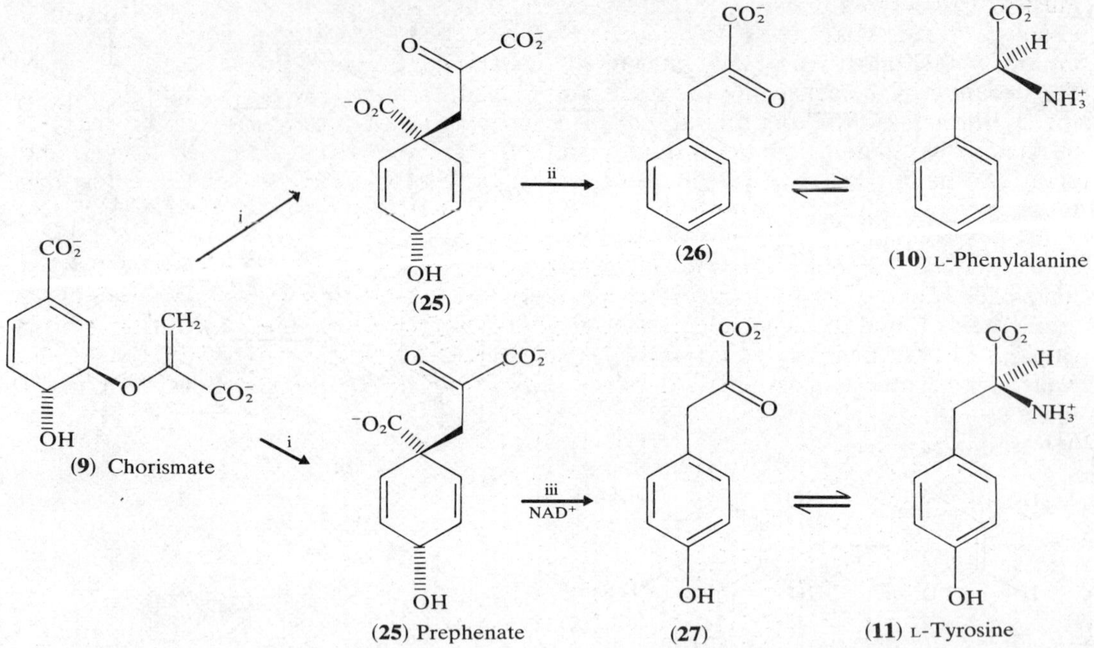

i, chorismate mutase; ii, prephenate dehydratase; iii, prephenate dehydrogenase, NAD^+.

SCHEME 10 The biosynthesis of L-phenylalanine and L-tyrosine from chorismate

two sub-units of approximately equal size. Once separated the individual sub-units are inactive and this behaviour strongly suggests that the active site(s) for expression of the two enzymic activities is only created by the interaction of the two protein sub-units.

The rearrangement of chorismic acid (**9**) to prephenic acid (**25**) is catalysed by chorismate mutase and the reaction is formally analogous to an *ortho*-Claisen rearrangement.[20,21] As such it is a unique example of this type of molecular rearrangement in primary metabolism. The same reaction may be brought about by heating aqueous solutions of chorismic acid (**9**), and calculations[14] indicate that the enzyme from *A. aerogenes* enhances the rate of reaction at pH 7.5 and 37 °C by a factor of almost 10^7 when compared with the thermally activated rearrangement in the absence of enzyme. If it is assumed that a chair-type conformation of the interacting π-systems is preferred for a sigmatropic rearrangement of this type, then the reaction pathway from chorismic acid (**9a**) to prephenic acid (**25**) probably involves a sequence such as is indicated diagramatically by (**9a**) ⇌ (**9b**) → (**25**) in Scheme 11. Implicit in this assumption is the concept that the substrate chorismic acid (**9a**)

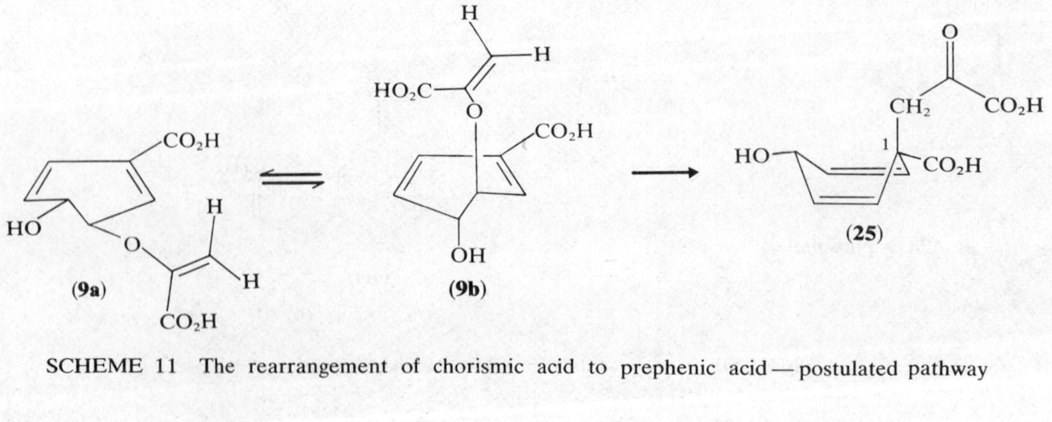

SCHEME 11 The rearrangement of chorismic acid to prephenic acid — postulated pathway

must first undergo an inversion of the conformation of the ring to give (**9b**) and that the enolpyruvyl side-chain must be orientated precisely. The role of the enzyme chorismate mutase is then most reasonably explained[14] in terms of Pauling's original concept[22] that the structure of the active site of the enzyme is complimentary to that of the substrate in the conformation (**9b**), and the subsequent transition state which resembles (**9b**), and that all the intrinsic binding energy is used to stabilize this intermediate and thus decrease the activation energy for the reaction. Suggestions have been advanced to indicate how this binding of the substrate may occur.[14]

In higher plants, particularly members of the families Cruciferae, Resedaceae, Iridaceae, and Curcubitaceae, four *m*-carboxy-substituted aromatic amino-acids **(30)–(33)** have been found.[23,24] These acids are members of what now constitutes a large group of amino-acids found in higher plants that are not normally encountered as constituents of proteins. Their chemistry and biogenesis has been studied by several groups, and pathways of biosynthesis have been adumbrated as shown in Scheme 12. A route has been

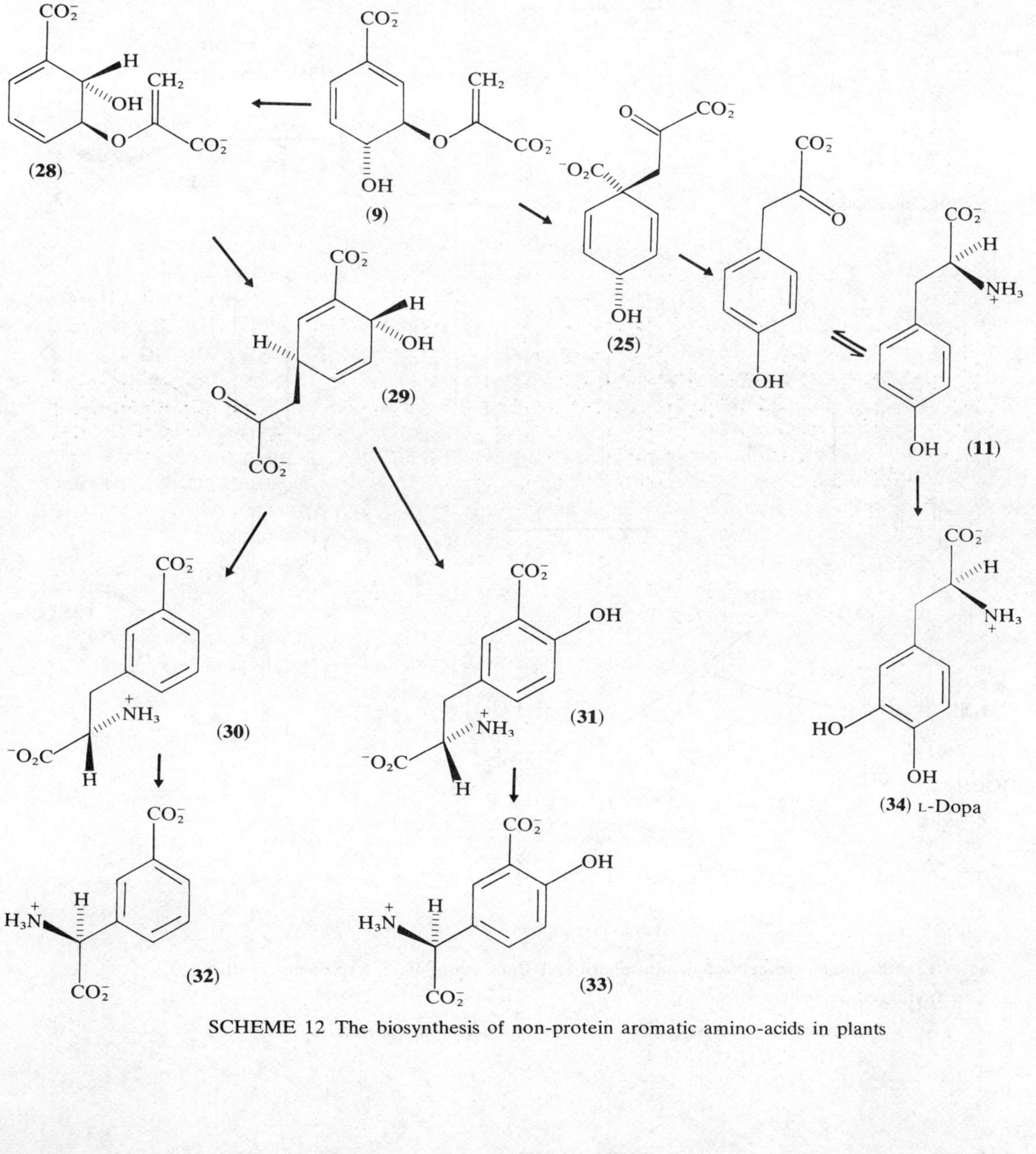

SCHEME 12 The biosynthesis of non-protein aromatic amino-acids in plants

proposed[25] in which isochorismic acid (**28**) is formed from chorismic acid (**9**), and this then undergoes a rearrangement to give (**29**) in a process which is *formally* analogous to the *ortho*-Claisen rearrangement catalysed by chorismate mutase. The amino-acids (**30**) and (**31**) are then derived by processes which in principle are similar to those outlined for L-phenylalanine and L-tyrosine from prephenic acid (**25**). The phenylglycine derivatives (**32**) and (**33**) are presumed to arise by a chain-shortening process involving loss of C-1 from (**30**) and (**31**).

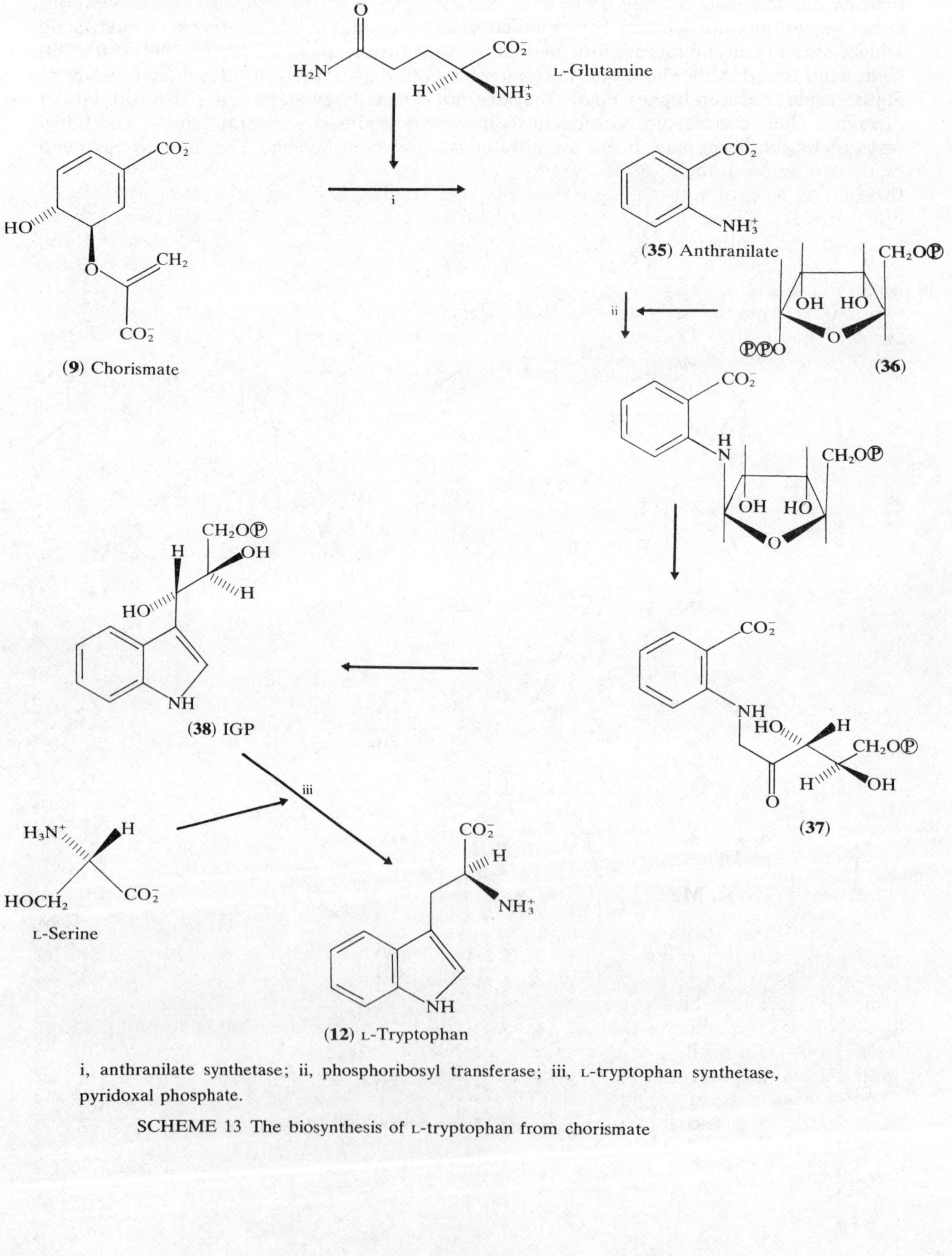

i, anthranilate synthetase; ii, phosphoribosyl transferase; iii, L-tryptophan synthetase, pyridoxal phosphate.

SCHEME 13 The biosynthesis of L-tryptophan from chorismate

3′,4′-Dihydroxy-L-phenylalanine (**34**; L-dopa) is a further important non-protein amino-acid which is found in plants and which has been strongly implicated in several types of alkaloid biosynthesis[26] in plants. Biosynthetic evidence suggests that it is formed by direct hydroxylation of L-tyrosine (**11**) (see Scheme 18). Tissues containing L-dopa when they are damaged or die undergo a characteristic blackening due to the formation of melanin pigments from L-dopa.

The biogenesis of L-tryptophan from chorismate takes place *via* the pathway outlined in Scheme 13. The first step leading to anthranilate (**35**) is a complex one and the mechanism is not yet fully understood.[27] Tracer studies have shown that the amino-group (from the amide nitrogen of L-glutamine) is added to C-2 of chorismate (**9**) (and hence C-6 of shikimate) and that the enolpyruvyl grouping is eliminated, after protonation, as pyruvate. Reaction of anthranilate with 5-phosphoribosyl pyrophosphate (**36**) followed by an Amadori-type rearrangement leads to the deoxyribulose (**37**), from which indoleglycerol phosphate (**38**; IGP) is obtained. Several interesting mechanistic observations have been made in relation to the last stage of L-tryptophan synthesis from IGP (**38**). Early studies of the biosynthesis of L-tryptophan suggested that indole (**39**) was an intermediate in its biosynthesis[28]. However, in *E. coli* the final stage is catalysed by an enzyme complex L-tryptophan synthetase ($\alpha\beta_2\alpha$) composed of two non-identical sub-units (an α form, M.W. 29 500 and a β_2 form, M.W. 108 000).[29,30] Highly purified preparations of the separated α and β_2 sub-units catalyse distinct reactions (equations 1 and 2). The α and β_2 sub-units combine when mixed and as the complex $\alpha\beta_2\alpha$ they catalyse a reaction which is the sum of equations (1) and (2), but in which indole (**39**) – a product of (1) and a substrate for (2) – *cannot be detected as a true intermediate.*

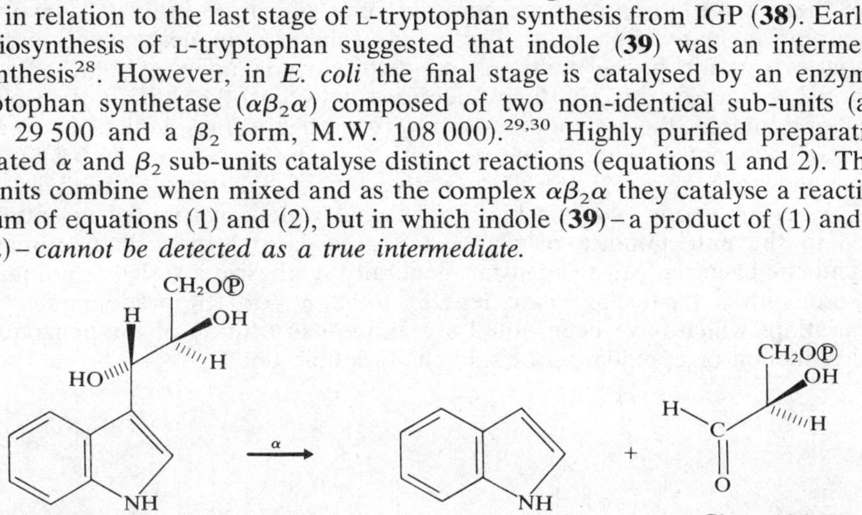

(**38**) IGP (**39**) D-Glyceraldehyde (1)

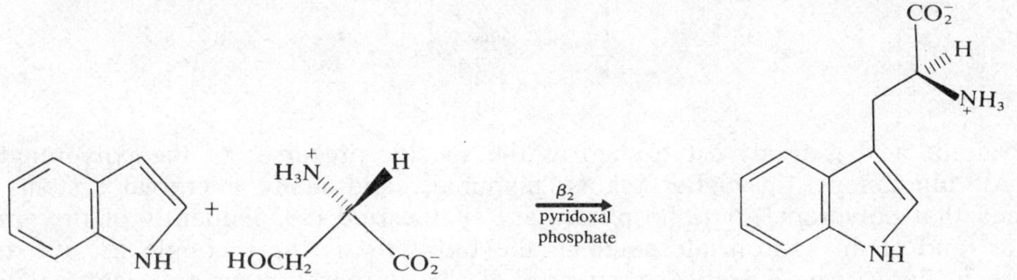

(**39**) L-Serine (**12**) L-Tryptophan (2)

30.3.3 CONTROL MECHANISMS[4,6]

At the enzymic level, control of the shikimate pathway has been examined for several organisms and may occur by the control of enzyme synthesis or by the regulation of enzymic activity. These studies have also revealed the diversity of regulatory mechanisms which exist for the shikimate pathway in different types of organism. In many, control of the synthesis and activity of DAHP synthetase (the first enzyme of the pathway, Scheme 1) is a key factor in the overall metabolic control of the synthesis of the three aromatic α-amino-acids. In several micro-organisms, for example, this enzyme exists in three forms (isoenzymes), each of which is sensitive to regulation (feed-back control) by one of the end-product α-amino-acids, L-phenylalanine, L-tyrosine, or L-tryptophan. The individual enzymes which catalyse the first steps in the separate pathways to the individual amino acids are also subject to similar end-product feed-back control. For the L-phenylalanine and L-tyrosine pathways the end-product inhibition effects are seen most clearly with the prephenate metabolizing enzymes (Scheme 10), as opposed to chorismate mutase with which these enzymes are frequently found complexed. Analogously, in all cases which have been examined in detail, anthranilate synthetase, the first enzyme in the separate L-tryptophan pathway of biosynthesis (Scheme 13), is subject to feed-back inhibition by the end-product L-tryptophan.

30.3.4 ISOPRENOID QUINONES

In addition to its function as a means of synthesizing the three aromatic α-amino-acids, the shikimate pathway also provides a route for the synthesis of other essential metabolites. One of these is the group of isoprenoid quinones which are intimately involved in electron transport in many organisms. These lipid-soluble quinones appear to have a definite spatial orientation in the multi-enzyme respiratory complex of some organisms, and they probably function by 'shuttling' electrons between other different respiratory coenzymes. For example, the ubiquinones are very probably intermediate between the flavoproteins and the cytochromes in the respiratory chain (see Section 24.3.2.3).

The ubiquinones (coenzyme Q) are all derivatives of 5,6-dimethoxy-3-methyl-2-*trans*-polyprenyl-1,4-benzoquinone (**40**; $n = 1$–12). They are widely distributed in nature and are localized in the mitochondria of plants and animals and the cell membranes of non-photosynthetic bacteria. Most organisms generally synthesize a series of ubiquinones in which those with a particular chain length ($n = 8$, 9, or 10) predominate. Other structural variations which have been noted are amino-substitution of the benzoquinone nucleus and reduction or epoxidation of side-chain double-bonds.

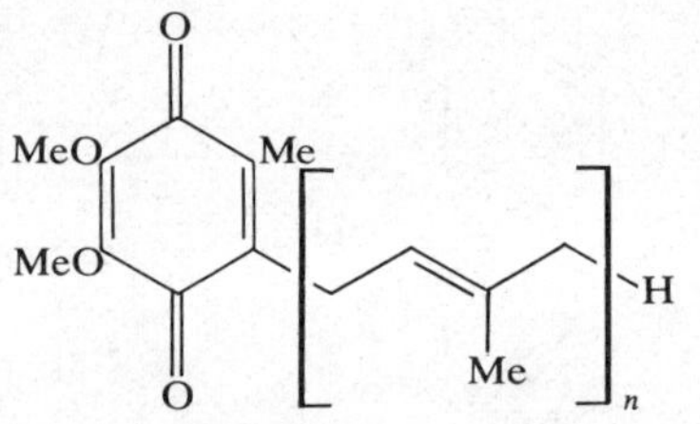

(**40**)

Mevalonic acid is firmly established as the specific precursor of the polyprenyl side-chain of ubiquinones in higher plants, mammals, and many microbial systems. It is assumed that polyprenyl pyrophosphates are synthesized independently of the aromatic nucleus, and at an appropriate stage in the biosynthesis these couple to give an aryl polyprenyl derivative. Evidence in support of this proposal is the frequent isolation of polyprenyl pyrophosphate synthetases from living systems and the co-occurence of ubiquinones with polyprenyl alcohols.[31,32]

p-Hydroxybenzoic acid (**41**) has been shown to be a direct precursor of the benzoquinone ring in a whole range of organisms from mammals and higher plants to bacteria. In bacteria, *p*-hydroxybenzoic acid is formed[33] directly from chorismic acid (**9**) but in mammals it is produced[34] from the aromatic α-amino-acids (L-phenylalanine and L-tyrosine) by oxidative degradation of the aliphatic side-chain. A similar pathway may well operate[35] in higher plants and in algae and fungi.

Two procedures have been utilized to determine the sequence of biochemical reactions leading from *p*-hydroxybenzoic acid and the polyprenyl pyrophosphates to the ubiquinones themselves. The approach pioneered by Folkers and his collaborators[36] has relied on the painstaking detection and isolation of polyprenyl-substituted phenols and quinones in lipid extracts of *Rhodospirillum rubrum*, determination of their structure, and then formulation of a biosynthetic pathway based on conversions between the various polyprenyl aryl systems, which seem eminently reasonable on chemical grounds. The route (Scheme 14) has provided an extremely useful basis for biosynthetic work in this area, and although variations have from time to time been proposed, it agrees closely with the route (Scheme 15) of ubiquinone biosynthesis which Gibson and his collaborators[37,38] have deduced using mutants of *E. coli* and a genetic analysis. The compatability of the evidence from these two quite different approaches suggests that most organisms use very similar if not identical pathways from *p*-hydroxybenzoic acid to the ubiquinones.

The biosynthetic pathway which leads to the formation of the phylloquinones (vitamin K_1, *e.g.* **42**) in higher plants and the menaquinones (vitamin K_2, e.g. **43**) in bacteria bears many similarities to that deduced for the formation of the ubiquinones. The polyprenyl side-chains are derived from mevalonic acid and the nuclear *C*-methyl groups from L-methionine. However, the exact origin of the naphthoquinone nucleus has proved to be an intriguing biosynthetic problem.

Using the normal tracer techniques, (−)-shikimic acid was established as an immediate precursor of the naphthoquinone nuclei of both menaquinones and phylloquinones, and the metabolite is utilized directly as a seven-carbon fragment.[39,40] Conflicting results were obtained, however, by several groups initially to suggest that the carboxyl group contributes equally to each of the two quinone carbonyl functions. More recent work by Baldwin, Snyder, and Rapoport[41] in which [7-^{14}C]shikimic acid was incorporated into menaquinones in *Mycobacterium phlei* has shown that the carboxyl group of shikimic acid contributes *solely* to C-4 in the menaquinone molecule (*e.g.* **43**). The biosynthetic precursor of the three remaining carbon atoms in the quinone nucleus was traced[39] to C-2, C-3, and C-4 of L-glutamic acid, and a route has been proposed[6] (Scheme 16) in which a carbanion is formed at C-2 of the amino-acid *via* α-ketoglutaric acid and its adduct with thiamine pyrophosphate. This may be visualized to condense with (−)-shikimic acid, chorismic acid, or isochorismic acid (**44**; Scheme 16) to yield *o*-succinylbenzoic acid (**45**) as the first aromatic intermediate on the pathway of biosynthesis.[39,42] The remaining stages in the construction of the naphthoquinone nucleus remain to be clarified in detail, although suggestions that 1,4-naphthoquinone and 2-methyl-1,4-naphthoquinone are intermediates have recently been discounted.[39] The same problem also remains to be solved in the biosynthesis of some higher plant naphthoquinones such as lawsone,[43] and these cases are discussed below (Section 30.3.6).

30.3.5 AROMATIC AMINO-ACID METABOLISM

The aromatic amino-acids are the precursors in many organisms for several physiologically active metabolites, and in this section particular reference is made to those compounds which are formed by oxidative pathways in micro-organisms and higher organisms. Schemes 17 and 18 summarize the range of products which fall into these categories for L-tryptophan and for L-phenylalanine and L-tyrosine, respectively. Under appropriate conditions all the amino-acids may, in addition, be metabolized to yield two to four carbon metabolites such as acetate, fumarate, acetoacetate, and succinate.

The aryl hydroxylases are a group of mixed-function oxidases which occur in all types of organism and are of special importance in many of the pathways of oxidative metabolism of the aromatic amino-acids and of other aromatic substrates.[44] Their action, the overall stoichiometry of which is indicated in equation (3), results in the introduction of a phenolic hydroxyl group into the aromatic ring. Various compounds such as ascorbate, pteridine derivatives, cytochromes, and pyridine and flavin nucleotides, besides metal ions (Fe, Cu) may serve as electron donors in the reaction.

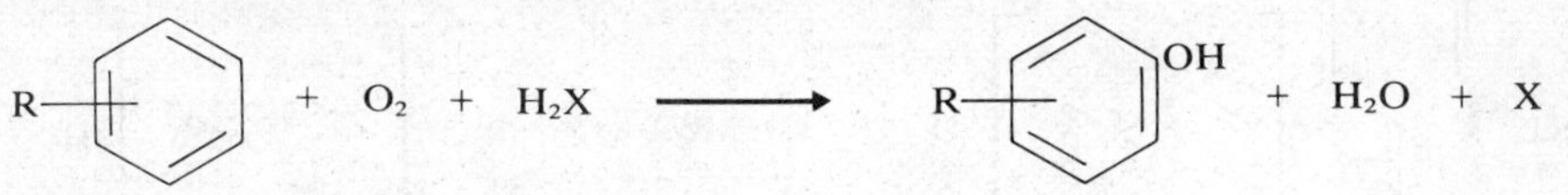

(3)

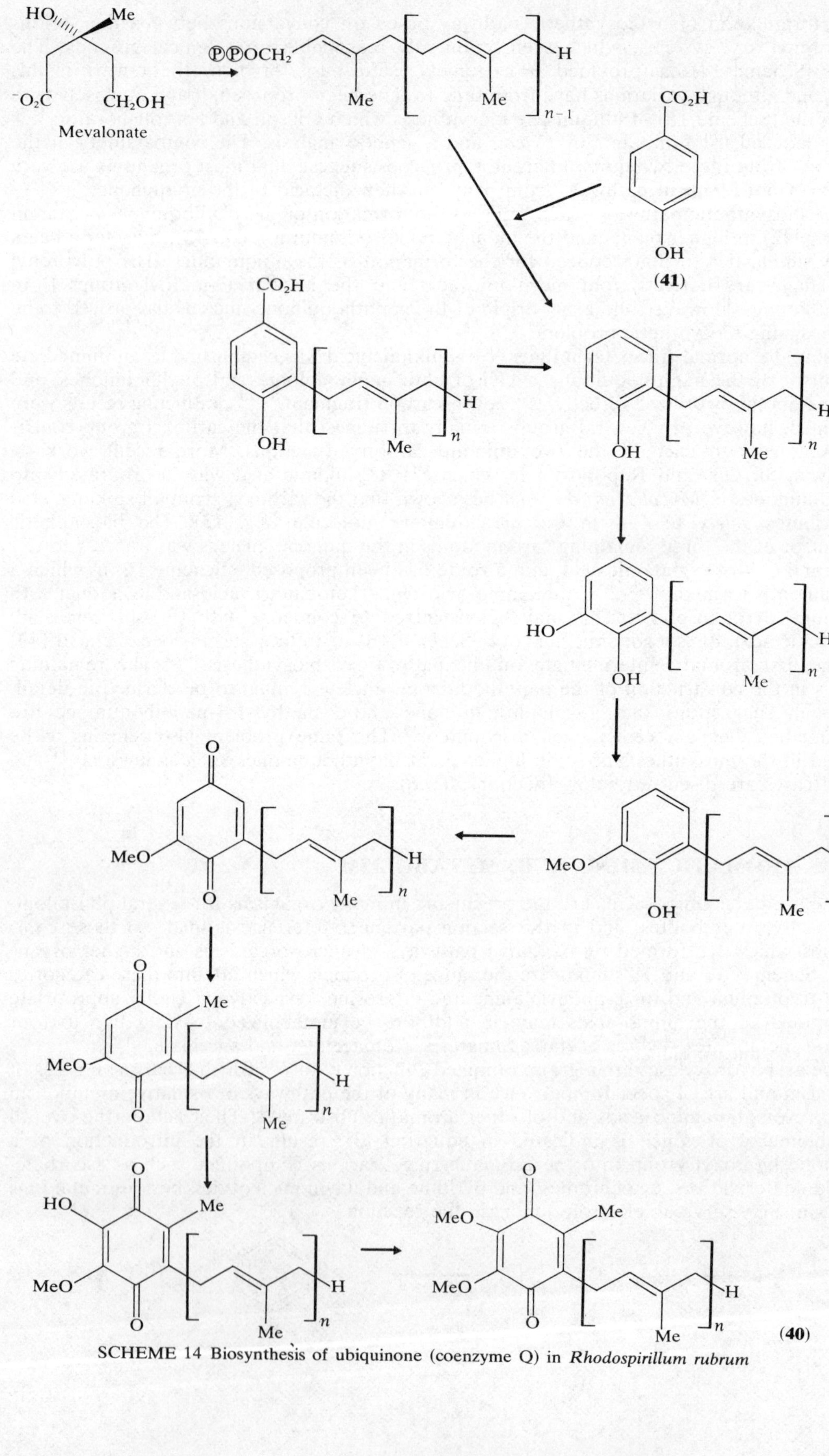

SCHEME 14 Biosynthesis of ubiquinone (coenzyme Q) in *Rhodospirillum rubrum*

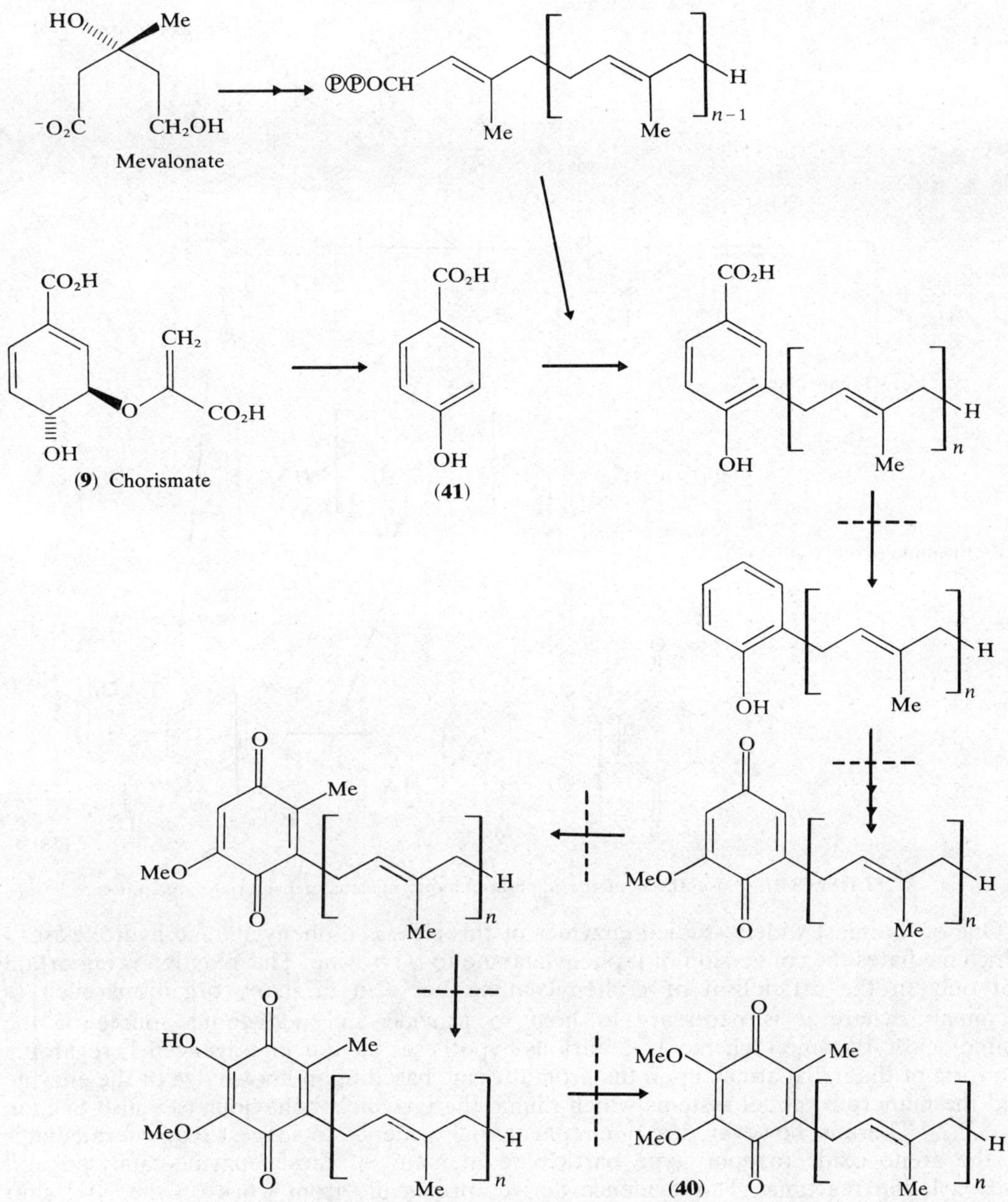

SCHEME 15 15 Biosynthesis of uniquinone in *Escherichia coli.* Mutants which were isolated and identified were blocked in one of the reactions shown

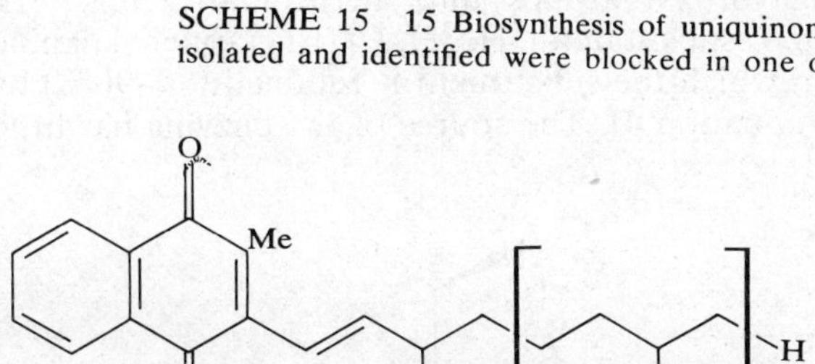

(42)

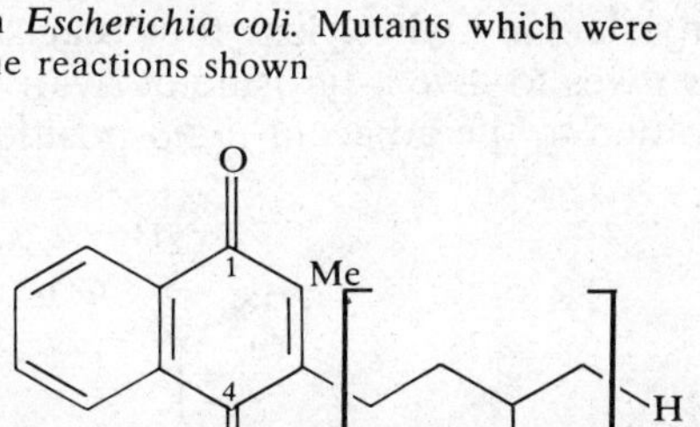

(43)

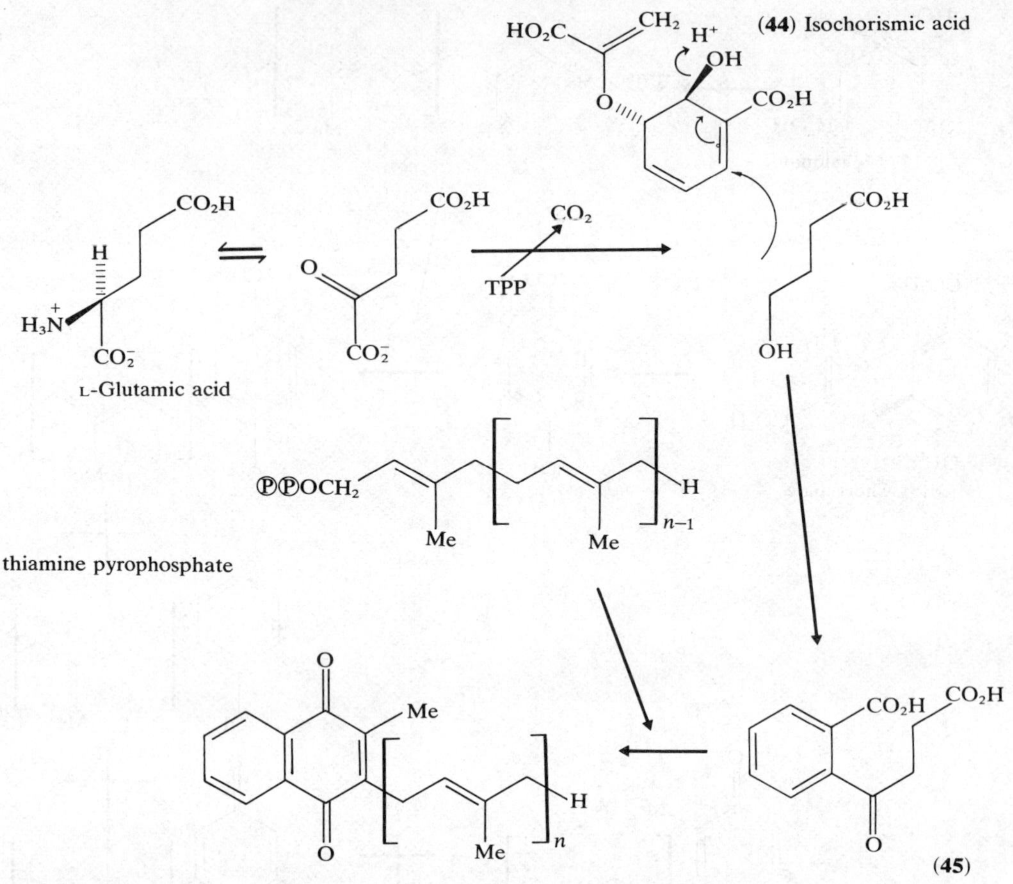

SCHEME 16 Biosynthesis of the naphthoquinone nucleus of bacterial menaquinones

One of the most widely studied enzymes of this class is L-phenylalanine hydroxylase,[45] which mediates the conversion of L-phenylalanine to L-tyrosine. This reaction is important not only in the catabolism of L-phenylalanine but also in many organisms such as mammals, where it is necessary to help to provide an endogenous source of the amino-acid L-tyrosine (Scheme 16). Various hypotheses have been advanced [46] regarding the form of the initial attack upon the aromatic ring based upon knowledge of the enzyme and the numerous model systems which mimic the enzyme's behaviour (see also Section 24.3.2.2). There is, however, strong circumstantial evidence to suggest that intermediates of the arene oxide (oxepin) type participate in many of these enzyme-catalysed aryl hydroxylation reactions. This evidence derives principally from work on the NIH shift which accompanies the introduction of hydroxyl groups into an aromatic ring.[47] L-Phenylalanine hydroxylase thus reacts with 4′-substituted (^2H, ^3H, Cl, Br) L-phenylalanine derivatives to give L-tyrosine derivatives in which the substituent is retained (80–90%) by migration to the adjacent *ortho* positions (equation 4). The source of the enzyme has little

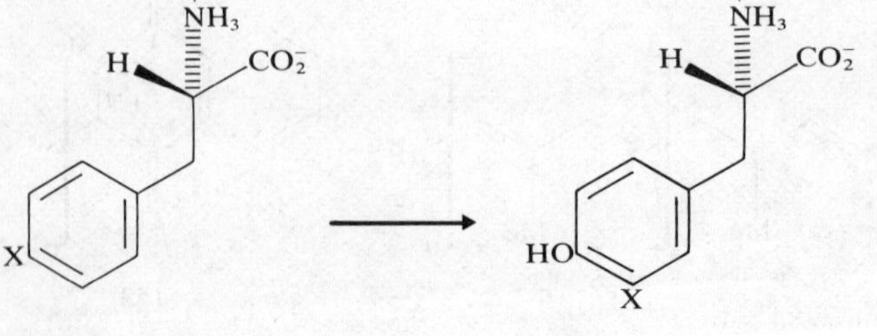

(4)

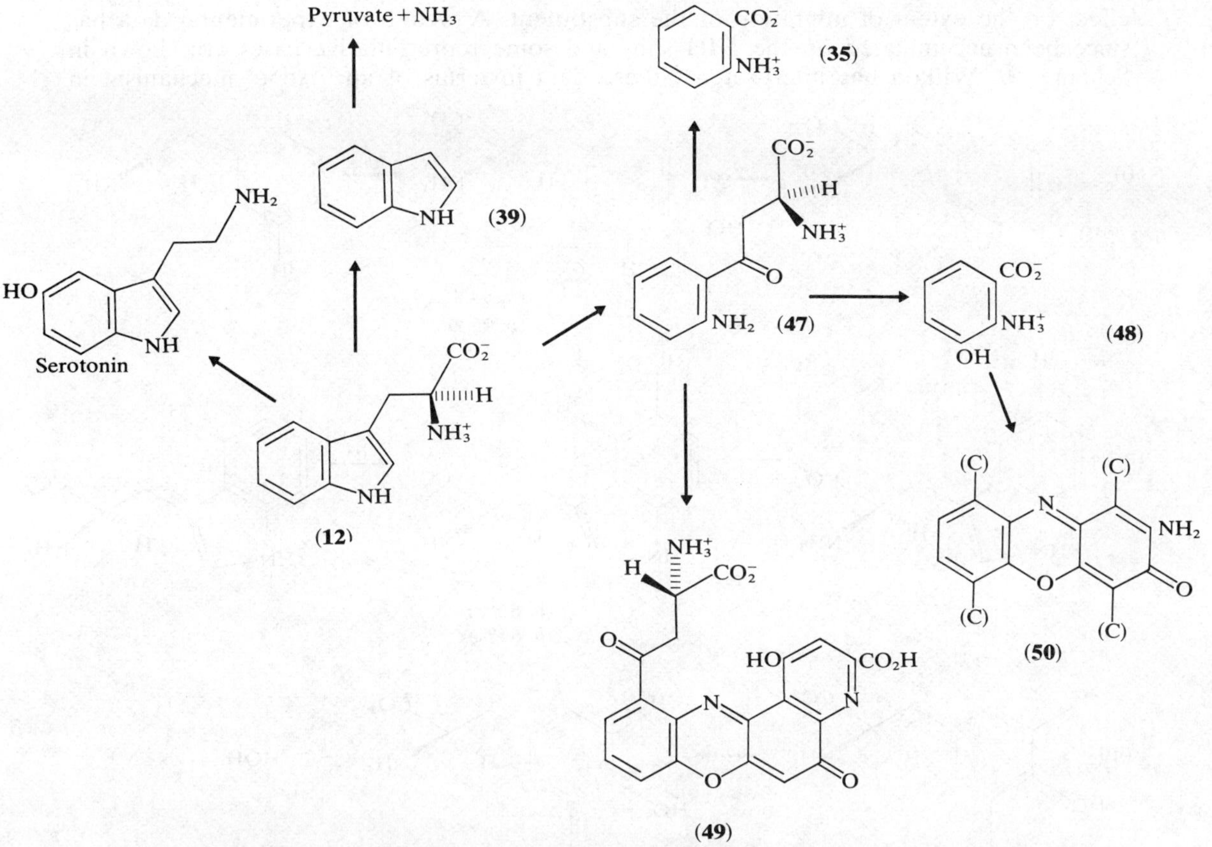

SCHEME 17 Metabolism of L-tryptophan in higher organisms and micro-organisms

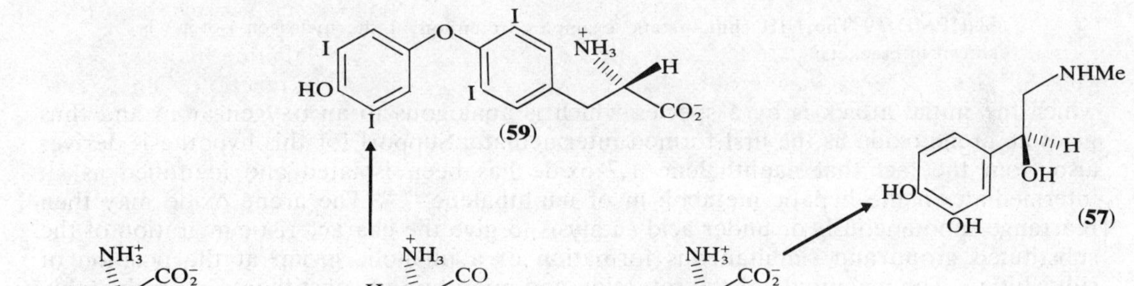

SCHEME 18 Metabolism of L-phenylalanine and L-tyrosine in higher organisms and micro-organisms

effect on the extent of migration of the substituent. A wealth of experimental data has since been accumulated on the NIH shift and some representative cases are shown in Scheme 19. Witkop has interpreted[47] these data in terms of an 'oxene' mechanism in

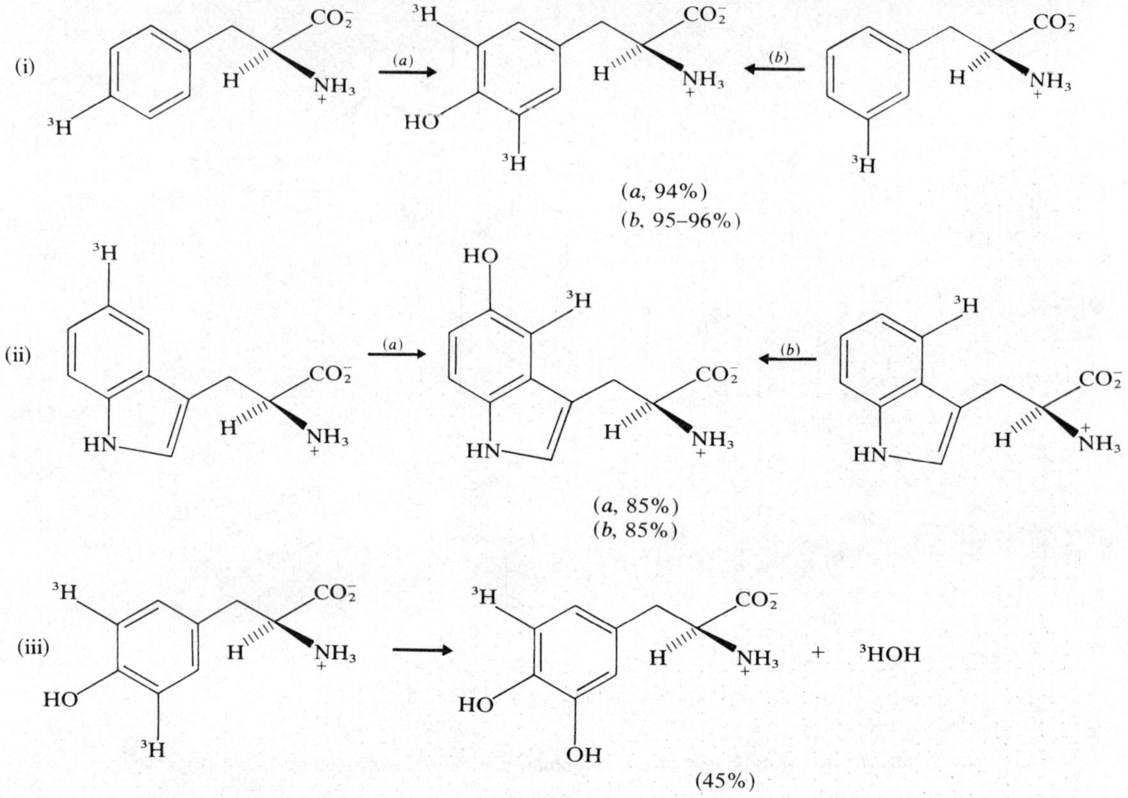

SCHEME 19 The NIH shift—some examples (retention of the hydrogen isotope is shown in brackets)

which the initial attack is by a species which is analogous to an oxygen atom and thus gives an arene oxide as the first formed intermediate. Support for this hypothesis derives also from the fact that naphthalene 1,2-oxide has been isolated and identified as an intermediate in the hepatic metabolism of naphthalene.[48,49] The arene oxide may then rearrange spontaneously or under acid catalysis to give the characteristic migration of the substituted group and simultaneous formation of a phenolic group at the position of substitution. The magnitude of the retention and migration of substituents is considerably influenced by other substituents in the aromatic ring and this has led to the proposal that two pathways (Scheme 20, a and b) exist for the decomposition of the arene oxide.

Kirby and his collaborators[50] measured the isotope effect which accompanied the NIH shift in the hydroxylation of L-phenylalanine to give L-tyrosine. Isotopically labelled [4'-^{3}H]- and [4'-^{3}H, 3',5'-$^2H_2$]-specimens of DL-phenylalanine were converted to L-tyrosine by a strain of *Pseudomonas.* Hydroxylation of the 4'-tritio-precursor proceeded with high (95%) migration and retention of the tritium. However, with the [4'-^{3}H, 3',5'-$^2H_2$]-precursor a lower retention of the tritium atom (74%) was observed. Kirby and his co-workers[50] calculated a k_H/k_D ratio of 10 ± 1 for the process.

The loss of tritium (~55%) by the introduction of a further hydroxyl group into [3',5'-$^3H_2$]-L-tyrosine to give L-dopa[51] using the enzyme L-tyrosine hydroxylase (Figure 19, iii) is readily accounted for by the arene oxide mechanism. In this and similar cases, loss of the hydrogen isotope is clearly the preferred process required to stabilize the intermediate (**46**) derived from the arene oxide (Scheme 21).

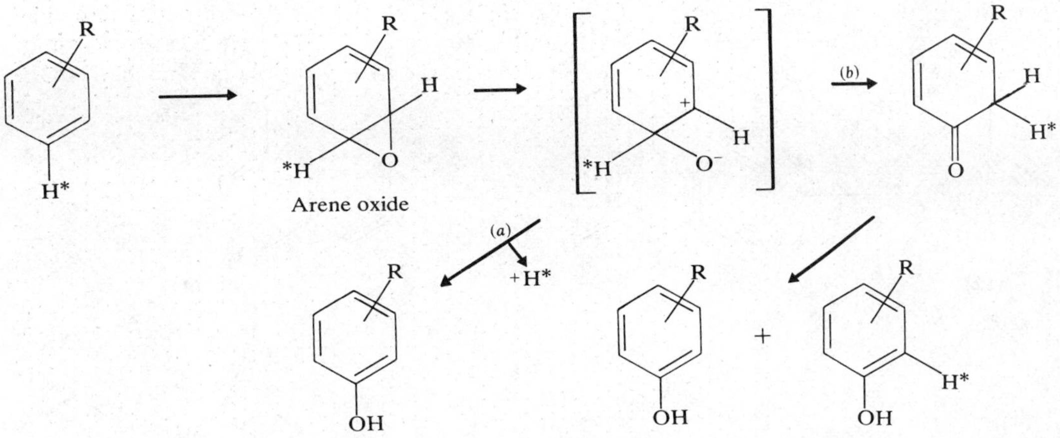

SCHEME 20 Aryl hydroxylation and the NIH shift

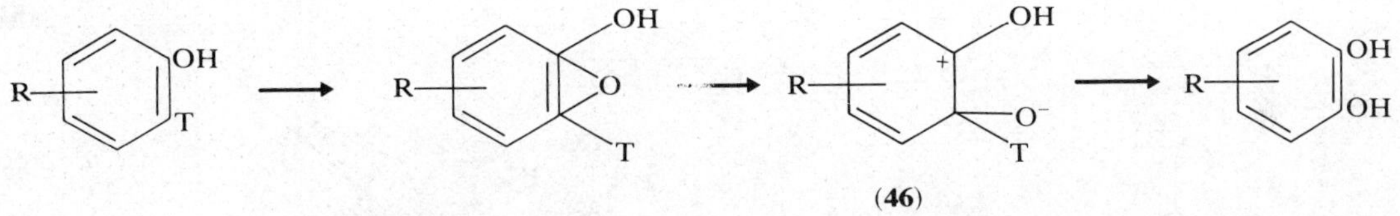

SCHEME 21 Hydroxylation *ortho* to a phenolic group

The first step in the oxidative metabolism of L-tryptophan is to give *N*-formylkynurenine and then kynurenine (**47**) in an enzyme-catalysed oxygenation in which both atoms of a molecule of oxygen are inserted into the C-2,3 bond of the pyrrole ring.[52] Several pathways diverge from kynurenine and in mammals the quantitatively major route is oxidation to carbon dioxide *via* 3-hydroxyanthranilate (**48**). There is also now firm evidence that kynurenine (**47**) and 3-hydroxyanthranilate (**48**) are intermediates in the biosynthesis of the chromophore of the phenoxazinones. Thus Butenandt[53] has shown that the ommochrome eye pigments of *Drosophila* species (*e.g.* xanthommatin, **49**) may be made in a biogenetically patterned synthesis by oxidation of 3-hydroxykynurenine (**51**) with potassium ferricyanide. On this basis, and with tracer experiment support, he proposed that the ommochromes were biosynthesized by oxidative condensation of two molecules of 3-hydroxykynurenine (**51**) (Scheme 22)[53].

Actinocin (**52**) and its derivatives the actinomycins (a family of chromopeptide antibiotics, see Chapter 23.4) can be analogously prepared by chemical oxidation of 3-hydroxy-4-methylanthranilate (**53**), and it has been postulated that this amino-acid is the direct precursor of the phenoxazinone chromophore of the actinomycins *in vivo*[54,55]. Biochemical evidence in support of this hypothesis has been obtained by Katz and Weissbach, and the synthesis of actinocin (**52**) from (**53**) by a cell-free system from *Streptomyces antibioticus* has been achieved.[56] The methyl groups of actinocin are derived[57] from L-methionine, but although it occurs after the formation of 3-hydroxykynurenine (**51**), the stage at which methylation is achieved is not known precisely. Similarly, the point at which the two peptide chains are attached is not entirely clear, although peptide derivatives of 3-hydroxy-4-methylanthranilic acid (**54**) may be smoothly transformed[58] to analogues of actinomycin (**55**) by the phenoxazinone synthetase of *Streptomyces antibioticus* (Scheme 23). The fungal pigment cinnabarinic acid (**56**) and its derivatives are similarly derived[59] by oxidation of 3-hydroxyanthranilic acid (**48**).

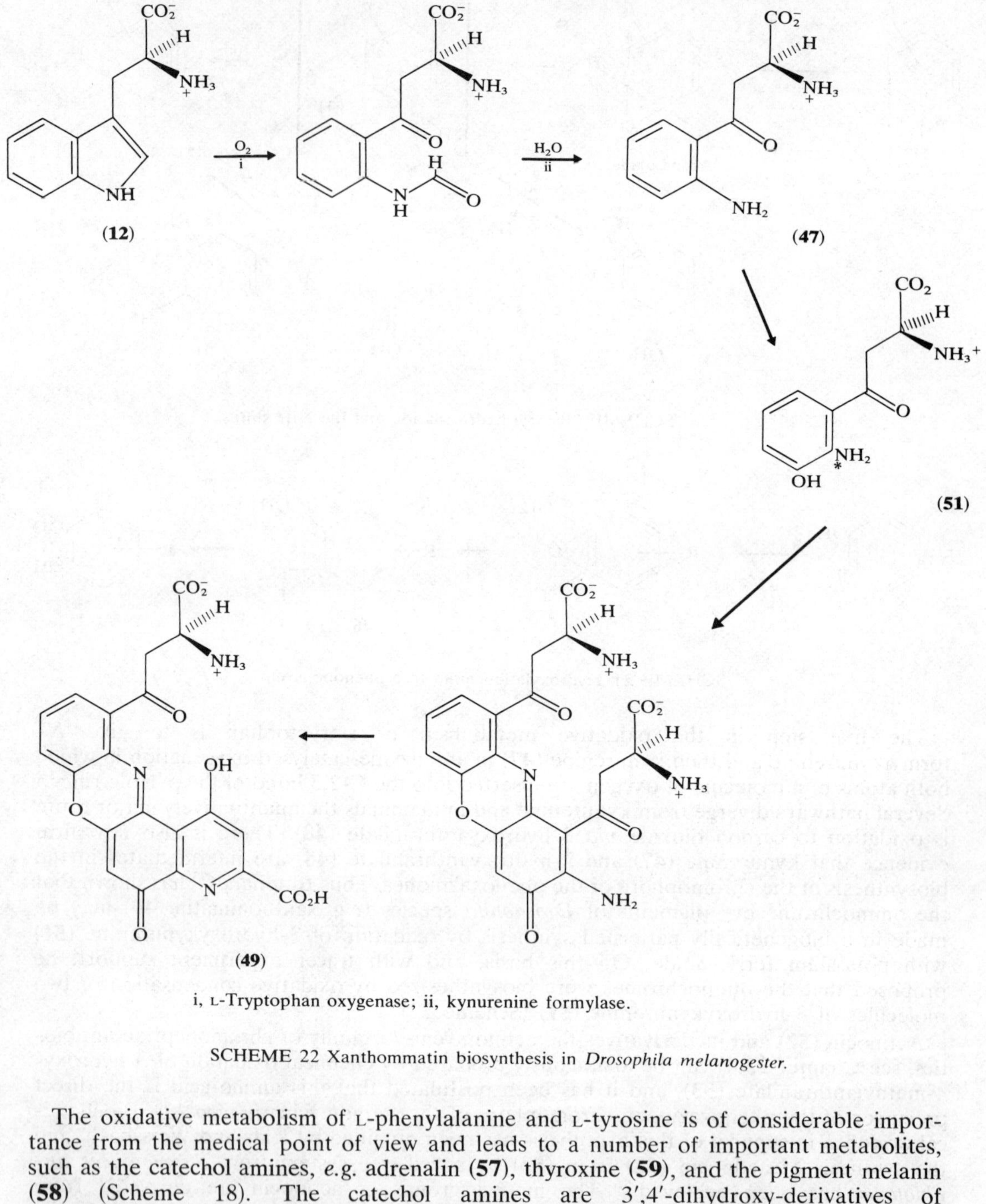

i, L-Tryptophan oxygenase; ii, kynurenine formylase.

SCHEME 22 Xanthommatin biosynthesis in *Drosophila melanogester.*

The oxidative metabolism of L-phenylalanine and L-tyrosine is of considerable importance from the medical point of view and leads to a number of important metabolites, such as the catechol amines, *e.g.* adrenalin (**57**), thyroxine (**59**), and the pigment melanin (**58**) (Scheme 18). The catechol amines are 3′,4′-dihydroxy-derivatives of phenylethylamine and they influence the actions of a wide range of mammalian tissues. The biogenesis of the catechol amines begins with L-tyrosine (Scheme 24) and each enzymic step in the subsequent pathway to adrenalin (**57**) has been identified and fully characterized.[60–63]

The molecule of thyroglobulin is a complex glycoprotein located in the thyroid follicle and serves as a storage vehicle for the two thyroid hormones thyroxine (**59**) and tri-iodothyronine. Considerable evidence has accumulated to suggest[64] that thyroxine is

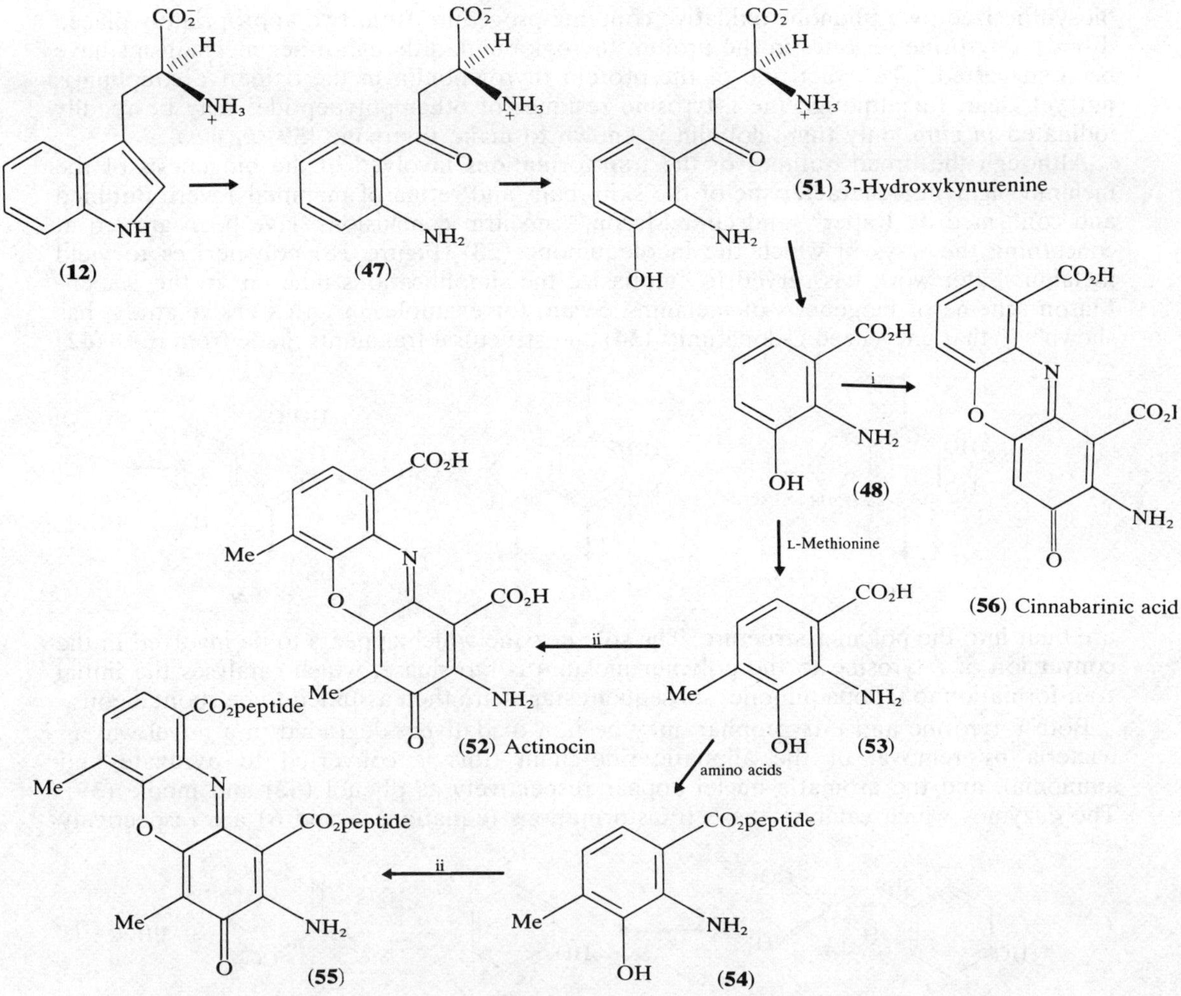

i, cinnabarinate synthetase; ii, phenoxazinone synthetase

SCHEME 23 Phenoxazinone biosynthesis: actinocin, actinomycins, and cinnabarinic acid

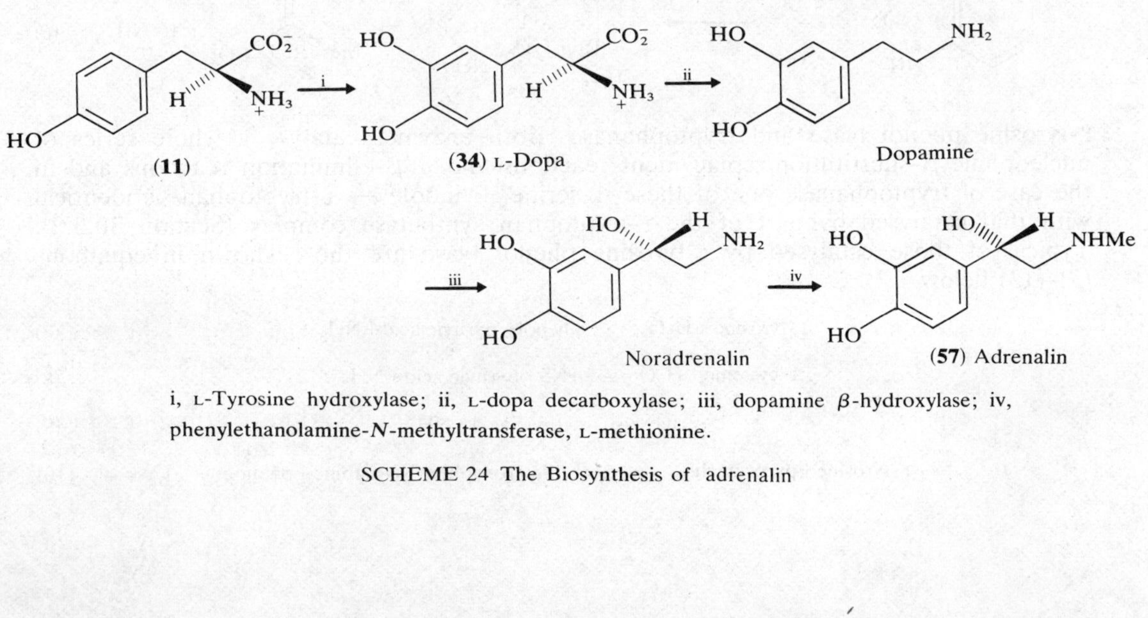

i, L-Tyrosine hydroxylase; ii, L-dopa decarboxylase; iii, dopamine β-hydroxylase; iv, phenylethanolamine-*N*-methyltransferase, L-methionine.

SCHEME 24 The Biosynthesis of adrenalin

biosynthesized by a phenolic oxidative coupling procedure from two appropriately placed di-iodo-L-tyrosine residues in the protein thyroglobulin, although other mechanisms have been suggested. The exact role of the protein thyroglobulin in the oxidative coupling is not yet clear, for although the L-tyrosine residues of other polypeptides may be readily iodinated *in vitro*, only thyroglobulin is known to make thyroxine (**59**) *in vivo*.

Although the broad outlines of the transformations involved in the biogenesis of the melanin pigments (characteristic of the skin, hair, and retina of mammals) were outlined and confirmed by Raper[65] and later Mason,[66] no firm conclusions have been arrived at concerning the ways in which the indolequinone (**58**) (Figure 18) polymerizes to yield melanin. Later work has served to emphasize the simplifications inherent in the Raper–Mason scheme of biogenesis of melanins. Swan, for example, in an extensive study, has shown[67,68] that uncyclized L-dopa units (**34**) and structural fragments made from (**60**)–(**62**)

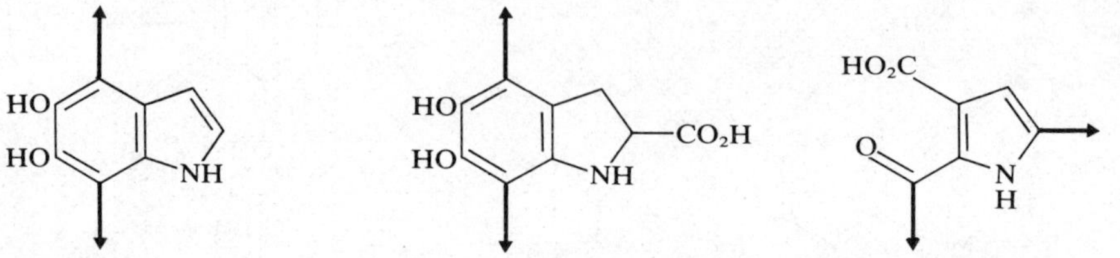

(**60**) (**61**) (**62**)

are built into the polymer structure. The sole enzyme which appears to be involved in the conversion of L-tyrosine to the polymer melanin is tyrosinase, which catalyses the initial transformation to L-dopaquinone; subsequent stages are then assumed to be spontaneous.

Both L-tyrosine and L-tryptophan may be non-oxidatively degraded in a novel way in bacteria by removal of the aliphatic side-chain (this is converted to pyruvate and ammonia), and the aromatic nuclei appear respectively as phenol (**63**) and indole (**39**). The enzymes which catalyse these transformations (equations 5 and 6) are respectively

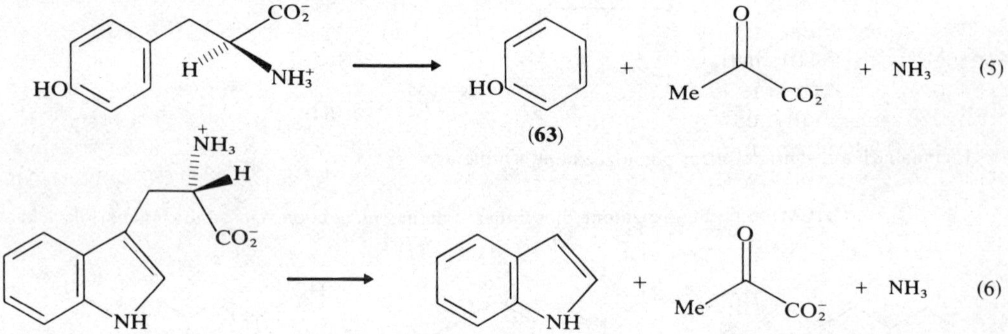

L-tyrosine phenol lyase and tryptophanase. Both enzymes catalyse a whole series of nucleophilic β-substitution/replacement reactions and α,β-elimination reactions and in the case of tryptophanase one of these–L-serine + indole → L-tryptophan–is identical with that catalysed by part of the L-tryptophan synthetase complex (Section 30.3.2). Typical of those catalysed by L-tyrosine phenol lyase are those shown in equations (7)–(13) below.

$$\text{L-tyrosine} + H_2O \rightarrow \text{phenol} + \text{pyruvic acid} + NH_3 \qquad (7)$$

$$\text{L-cysteine} + H_2O \rightarrow H_2S + \text{pyruvic acid} + NH_3 \qquad (8)$$

$$\text{L-serine} \rightarrow \text{pyruvic acid} + NH_3 \qquad (9)$$

$$\text{L-tyrosine} + \text{pyrocatechol} \rightarrow 3',4'\text{-dihydroxyphenyl-L-alanine} + \text{phenol} \qquad (10)$$

S-methyl-L-cysteine + resorcinol → 2′,4′-dihydroxyphenyl-L-alanine + phenol (11)

phenol + pyruvic acid + NH_3 → L-tyrosine + H_2O (12)

pyrocatechol + pyruvic acid + NH_3 → 3′,4′-dihydroxyphenyl-L-alanine (13)

Both enzymes function[69] with the cofactor pyridoxal phosphate and, in the case of tryptophanase, Snell and his colleagues[69] have proposed a mechanism for the reaction in which the required cleavage occurs *via* the formation of a pyridoxal phosphate–tryptophan–metal complex (Scheme 25). A similar mechanism may be written for L-tyrosine phenol lyase.

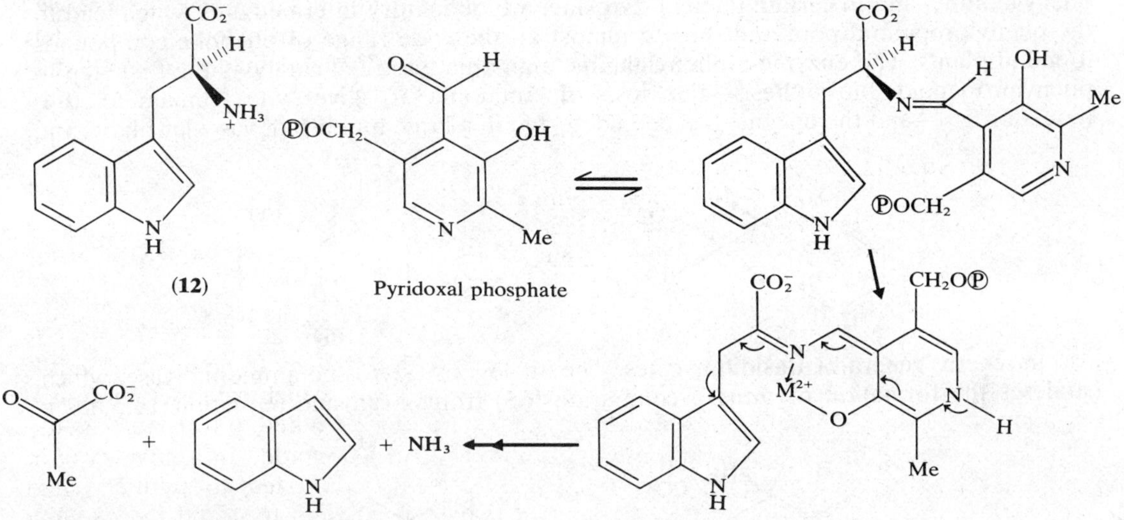

SCHEME 25 Mechanism of the tryptophanase reaction

The stereochemistry of the β-replacement reaction catalysed by L-tryosine phenol lyase (*e.g.* equation 10) has been examined by Hill and his collaborators.[70] They studied the reaction of L-tyrosine (stereospecifically deuteriated at C-3) with resorcinol to give 2′,4′-dihydroxy-L-phenylalanine. The configuration of the deuterium atom at C-3 in the product was determined by ^{1}H n.m.r. spectroscopy and showed that the exchange (equation 14) proceeds with retention of configuration at C-3. Similarly, Fuganti and his

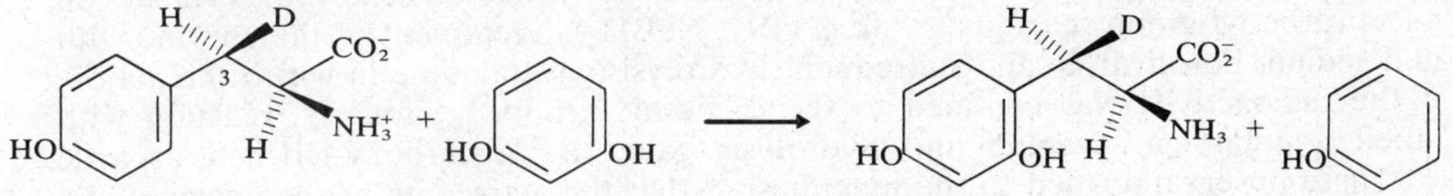
(14)

co-workers[71] have shown the reverse L-tyrosine–phenol lyase reaction itself to proceed with retention of configuration, and Floss[72] has similarly shown the L-tryptophan synthetase reaction (Section 30.3.2) to proceed with retention of configuration at C-3 of the L-serine molecule.

30.3.6 THE SHIKIMATE PATHWAY IN HIGHER PLANTS

Historically, organic chemistry is inextricably linked with the study of compounds extracted from living matter, and one of the most productive areas of enquiry has been that of the chemistry of secondary metabolites – compounds such as terpenes, alkaloids, quinones, and phenols – which many organisms produce but for which no clear biological function has been discerned. Plants contain a wide range of these substances and many are derived biosynthetically from the aromatic amino-acids or from intermediates in the shikimate pathway. Predominant amongst these metabolites are the alkaloids (see Chapter 30.1) and the various plant phenols. Some of the general biogenetic features associated with this second group of compounds and of other miscellaneous shikimic acid derived metabolites found in plants are discussed here.

Plant phenols vary in complexity from that of the structural polymer lignin to simple C_6 phenols themselves. Classifications of plant phenols[6] based on structural comparisons nevertheless show that many contain the C_6–C_3 or phenylpropanoid fragment (cinnamic acids, lignans, lignins, stilbenes, and flavonoids), and it is generally agreed that L-phenylalanine, and in certain plants L-tyrosine, are obligatory intermediates which lead to the phenylpropanoid pool and thus to almost all the wide range of phenolic compounds found in plants. The enzyme L-phenylalanine ammonia lyase[73,74] catalyses the first step in phenylpropanoid biosynthesis—the loss of ammonia to give *trans*-cinnamate (**64**) (equation 15)—and the enzyme has been detected in a large number of vascular plants and

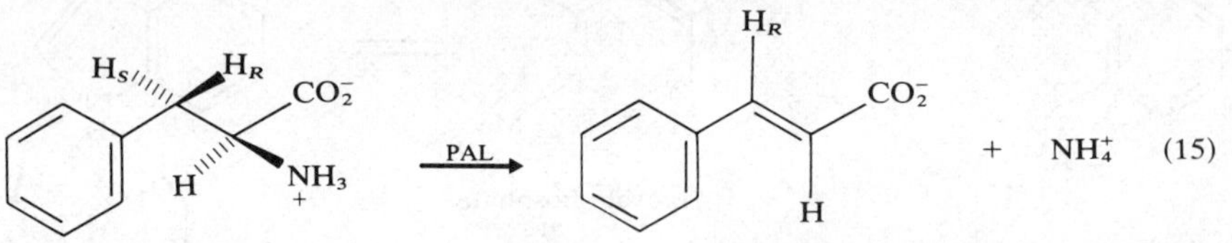

(**64**)

also in several genera of Basidiomycetes. The analogous L-tyrosine ammonia lyase, which catalyses the formation of *trans-p*-coumarate (**65**) from L-tyrosine (equation 16), has a

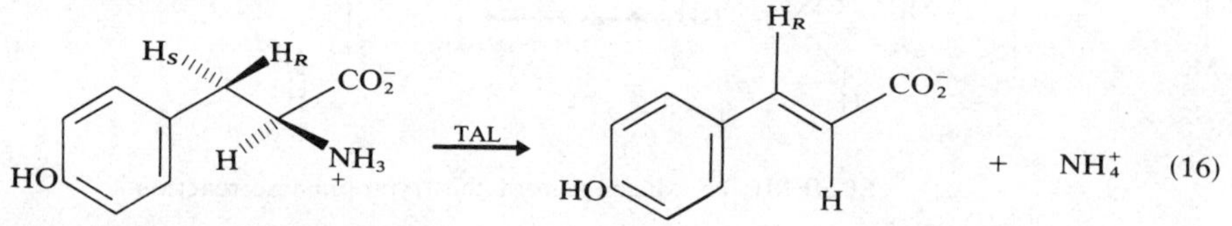

(**65**)

more limited distribution in the plant kingdom[74] and is found most commonly in the Gramineae. It seems reasonable to assume that these enzymes act at a branch point in metabolism to divert the aromatic amino-acids from protein synthesis to the biosynthesis of phenylpropanoid compounds.

L-Phenylalanine ammonia lyase has been isolated from many plant sources[73,74] and extensively purified. Chemical inhibition of enzyme activity may be achieved by the action of typical carbonyl group reagents[75,76] (*e.g.* CN^-, $NaBH_4$). Treatment of the enzyme with tritiated sodium borohydride and subsequent hydrolysis gave alanine in which the majority of the radioactivity was confined to the β-methyl group.[75] Similarly, reaction with ^{14}C-labelled potassium cyanide and hydrolysis gave [β-^{14}C]carboxy-labelled aspartic acid.[76] These observations led to the hypothesis[75] that the active site of the enzyme, like that of other related amino-acid ammonia lyases, contain an α,β-dehydroalanine residue. A mode of action for the enzyme was proposed[75] (Scheme 26) in which the amino-group of the amino-acid first adds to the methylene group of the dehydroalanine. Two groups of workers[77,78] have also shown that the elimination reaction proceeds in a clearly defined stereochemical sense by loss of the *pro*-3*S* hydrogen atom, once again analogous to that of other ammonia lyase reactions.[79]

L-Tyrosine ammonia lyase, when it is found in plants, is generally found in association with L-phenylalanine ammonia lyase but in a much lower amount.[74] In maize it has been shown[80] that a single enzyme is responsible for the elimination of ammonia from both L-phenylalanine and L-tyrosine and that the same active site acts for both substrates. L-Tyrosine ammonia lyase acts in the same highly stereospecific manner as its counterpart, which acts on L-phenylalanine and the reaction proceeds[81] by removal of the *pro*-3*S* hydrogen (equation 16).

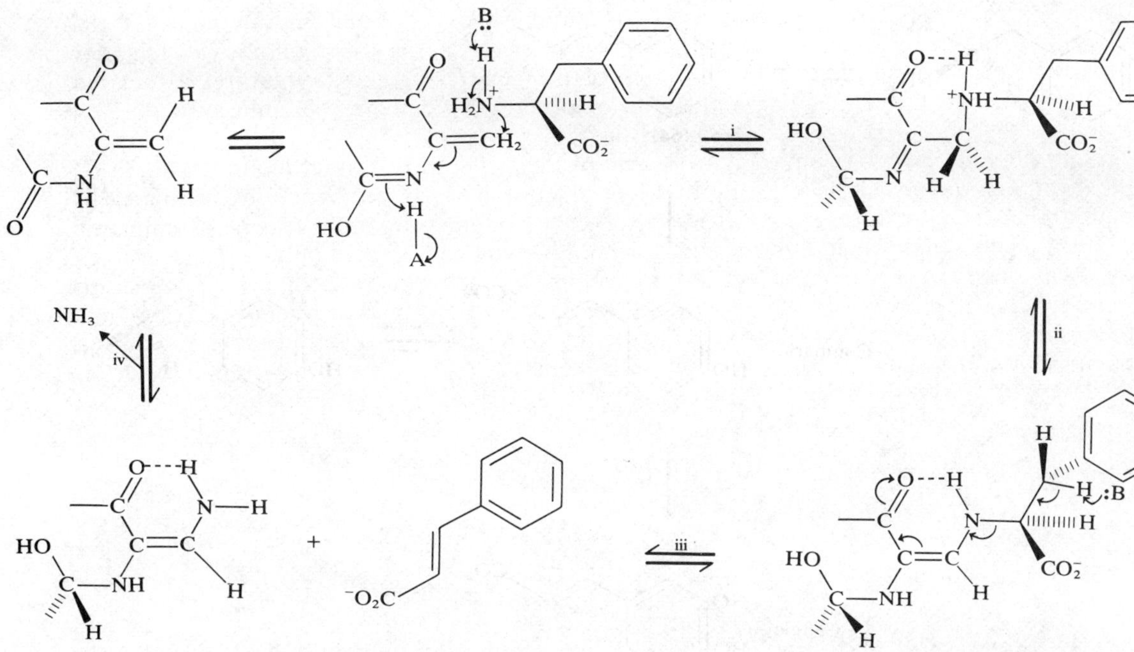

i, Addition of amino-group of L-phenylalanine to dehydroalanine; ii, 1,3-prototropic shift; iii, elimination to generate *trans*-cinnamate and amino-enzyme; iv, regeneration of enzyme and loss of ammonia.

SCHEME 26 Mode of action of L-phenylalanine ammonia lyase

In the majority of higher plants there is ample evidence[82] to support the view that the remaining hydroxycinnamic acids–which either themselves or as activated esters are the principal constituents of the phenolic phenylpropanoid pool–are formed sequentially by aryl hydroxylation and methylation from *trans*-cinnamate (**64**) (Scheme 27). The characteristic NIH shift which accompanies aryl hydroxylation reactions in micro-organisms and higher organisms has been shown to occur with cinnamate hydroxylase,[83] and in *Fagopyrum esculentum* when [4′-^{3}H]-*trans*-cinnamic acid was fed to the plant chlorogenic acid was isolated in which 50% of the tritium was retained.[84] This result is again consistent with the tenets of the NIH hypothesis, namely that introduction of the first hydroxyl group causes an almost quantitative retention and migration of the tritium atom to the 3′-position, but that substitution of the second phenolic group *ortho* to the *p*-hydroxyl group results in loss of the tritium atom at the position of substitution. Similar results have also been demonstrated for phenylpropanoid residues where they are incorporated into more complex phenolic structures.[6,85,86]

The plant lignins are polymers[87,88] formed largely if not entirely from monomers with a C_6–C_3 phenylpropyl alcohol skeleton and corresponding in hydroxylation patterns to the various cinnamic acids (Scheme 27). Conifer or softwood lignins are formed almost exclusively from coniferyl alcohol (**67**) and hardwood lignins from varying proportions of coniferyl and sinapyl (**68**) alcohols, whilst lignins from grasses contain structural fragments not only from (**67**) and (**68**) but also from *p*-coumaryl alcohol (**66**). The necessary reduction of the carboxyl group of the cinnamic acids to give the cinnamyl alcohols may be envisaged as a two-stage process proceeding *via* the coenzyme A ester and exactly analogous to the reduction of β-hydroxy-β-methylglutaric acid to give mevalonic acid (Chapter 29.2). Based on a series of model reactions in which lignin-type polymers were produced by enzymic oxidation (laccase) of coniferyl alcohol (**67**), Freudenberg[87] concluded that this *in vitro* process closely resembled the natural processes of lignification. He

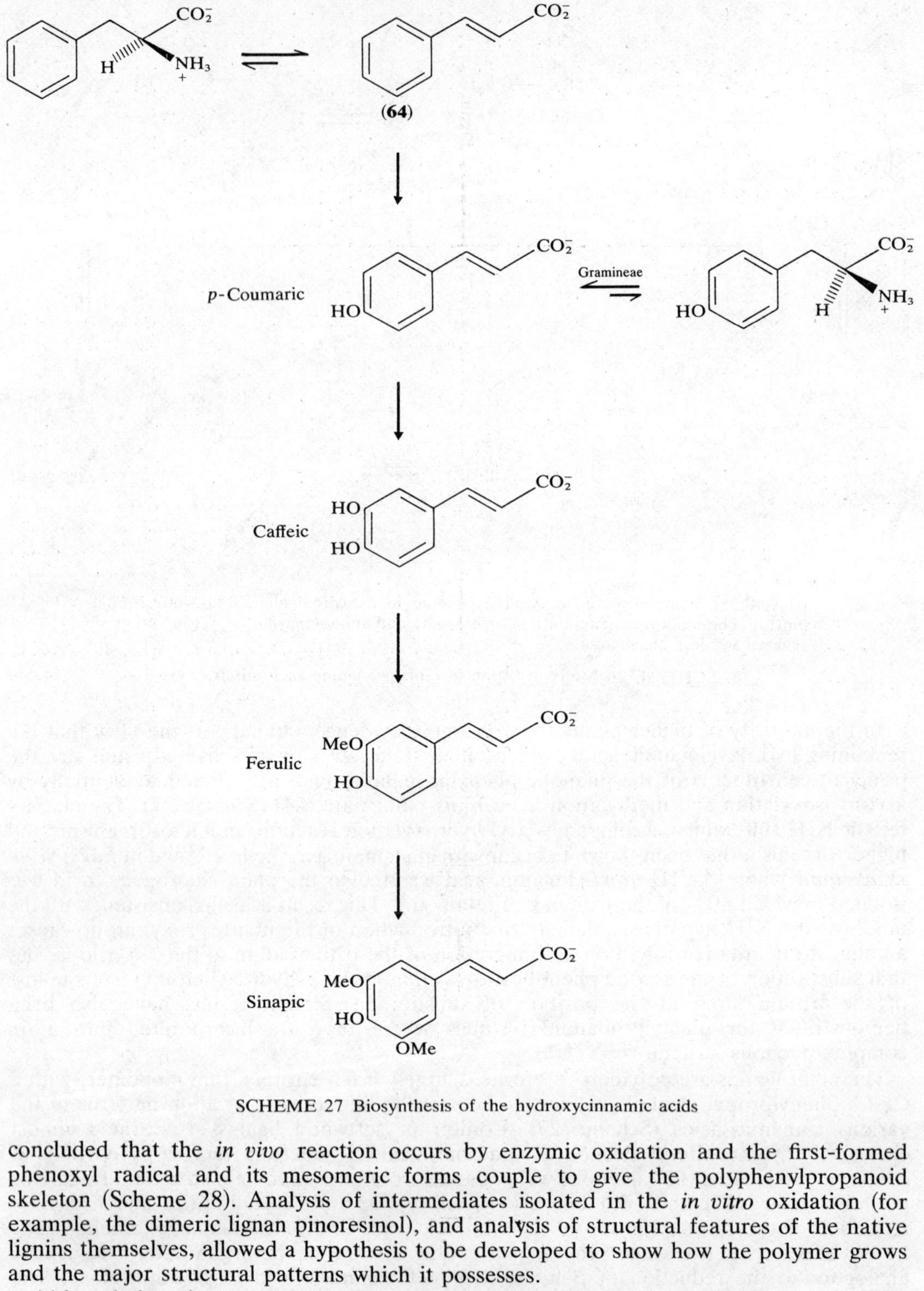

SCHEME 27 Biosynthesis of the hydroxycinnamic acids

concluded that the *in vivo* reaction occurs by enzymic oxidation and the first-formed phenoxyl radical and its mesomeric forms couple to give the polyphenylpropanoid skeleton (Scheme 28). Analysis of intermediates isolated in the *in vitro* oxidation (for example, the dimeric lignan pinoresinol), and analysis of structural features of the native lignins themselves, allowed a hypothesis to be developed to show how the polymer grows and the major structural patterns which it possesses.

Although in micro-organisms several hydroxybenzoic acids have been shown to be derived from non-aromatic intermediates in the shikimate pathway, there is little direct evidence, except in the case of gallic acid, to suggest that similar pathways operate in

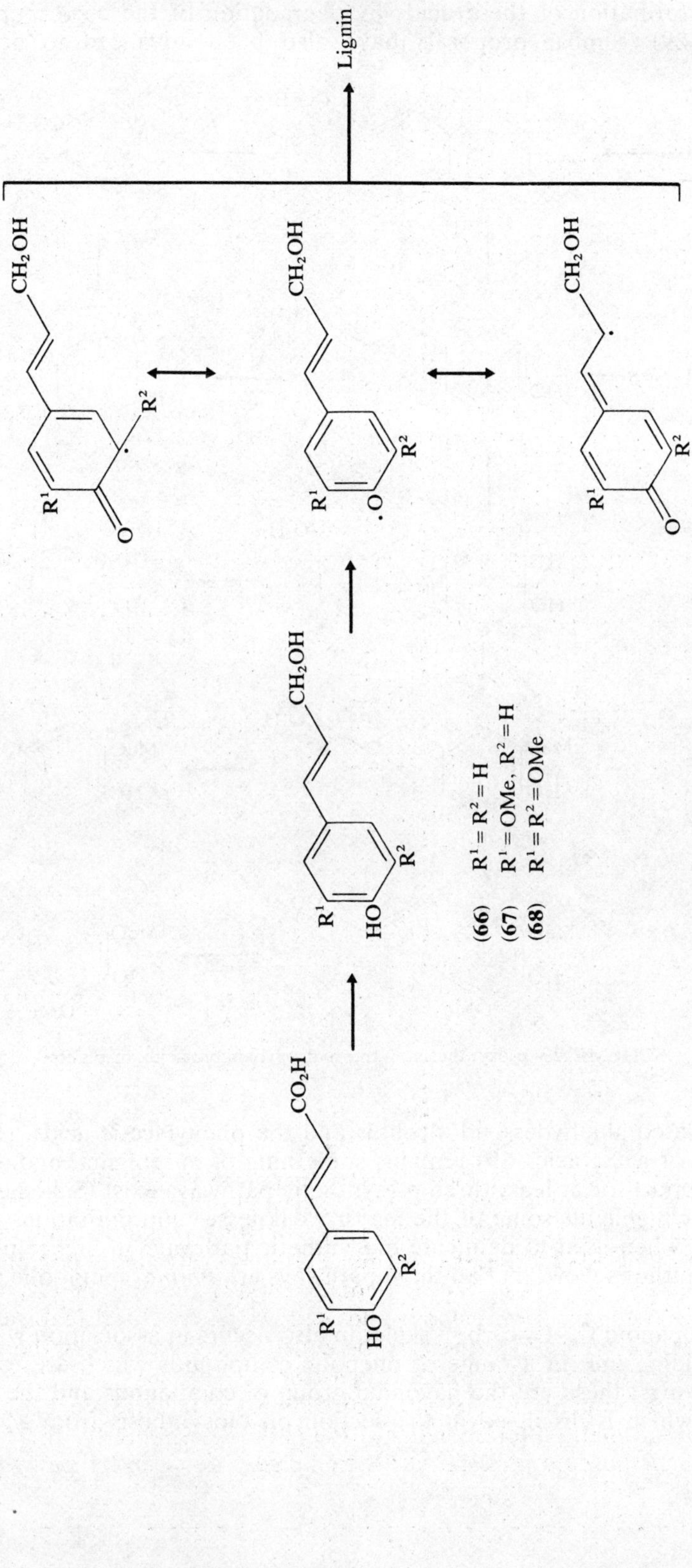

(66) $R^1 = R^2 = H$
(67) $R^1 = OMe$, $R^2 = H$
(68) $R^1 = R^2 = OMe$

SCHEME 28 Biosynthesis of Lignin

higher plants. Geissman and Hinreiner[89] first postulated that β-oxidation of cinnamic acids would form a biosynthetic route to the substituted benzoic acids, and experimental evidence using isotopic tracers has since been accumulated to support this view. Using this work as a basis, Zenk[90] and others have formulated a general metabolic grid biosynthetic scheme for the formation of these acids by degradation of the corresponding cinnamic acids (Scheme 29). Similar proposals have also been advanced to account for the

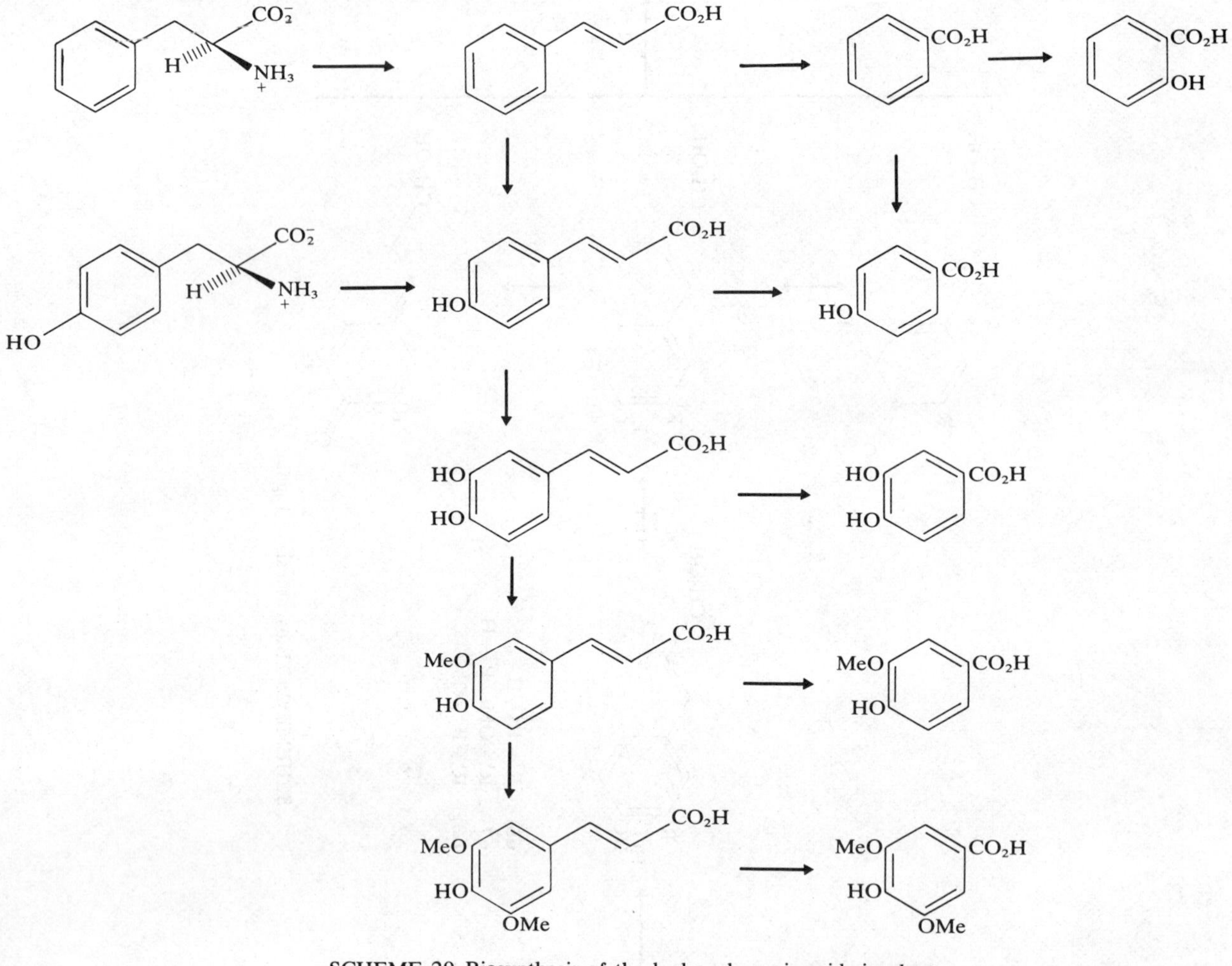

SCHEME 29 Biosynthesis of the hydroxybenzoic acids in plants

biogenesis of related aldehydes and alcohols and the phenylacetic acids.[6] In this context the biosynthesis of gallic acid (**69**) remains something of an enigma, and at the moment experimental support for at least three biosynthetic pathways exist (Scheme 30).[91–93] This case in particular highlights some of the major weaknesses and limitations of the isotopic tracer technique when used to delineate biosynthetic pathways in higher plants. It seems unlikely, in the author's view, that all three pathways are *normal* metabolic routes to gallic acid.

The phenylpropanoid C_6–C_3 carbon skeleton also occurs in association with acetate as a biosynthetic building unit in a range of phenolic compounds which are found in plants. Pre-eminent amongst these are the flavonoid group of compounds and the stilbenes,[6] the biosynthesis of which is discussed in the section on biosynthesis from acetate (Chapter 29.1).

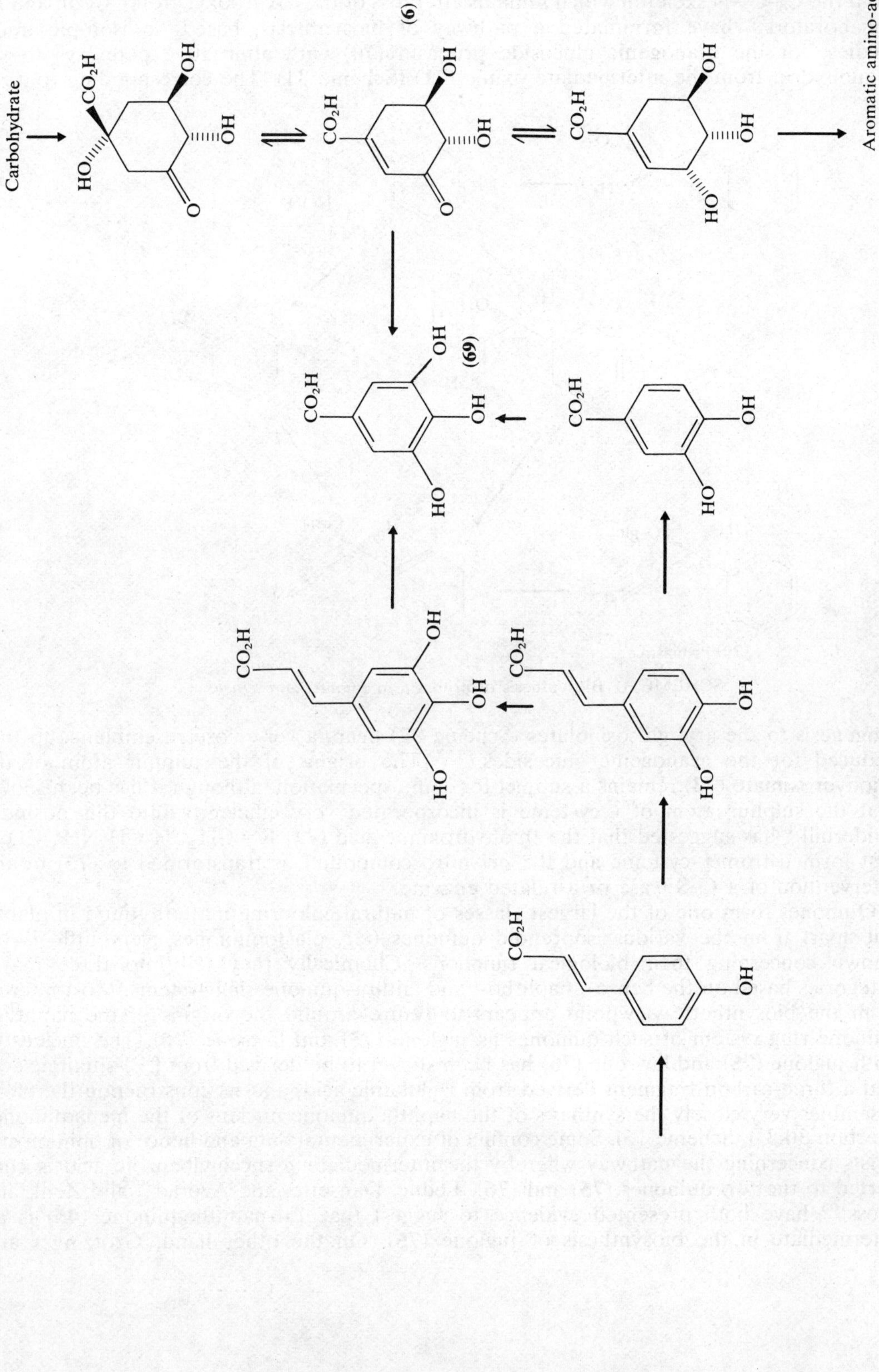

SCHEME 30 Postulated pathways for gallic acid biosynthesis

Plants may also use the shikimate-derived aromatic amino-acids more directly for the synthesis of natural products, and examples of their utilization in this way are in the biosynthesis of the betalain pigments of the Centrospermae (Chapter 30.1), the formation of indoleacetic acid from L-tryptophan,[94] and the biogenesis of aryl glucosinolates and cyanogenic glucosides. The aglycones of the aryl cyanogenetic glucosides are thus derived from the C_6–C_2–N skeleton which remains after loss of the α-carboxyl group. Conn and his collaborators[95] have formulated a pathway of biosynthesis, based on isotopic tracer studies, for the cyanogenic glucoside prunasin (**70**) with alternative pathways to the cyanohydrin from the intermediate oxime (**71**) (Scheme 31). The corresponding route of

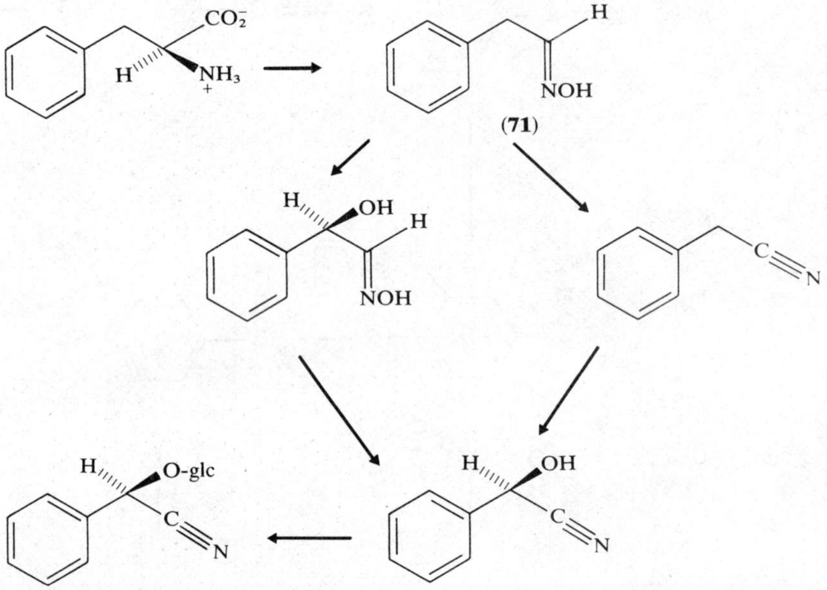

(**70**) Prunasin

SCHEME 31 Biosynthesis of prunasin in *Prunus laurocerasus*

biogenesis to the aryl glucosinolates (Scheme 32) bears a very close resemblance to that deduced for the cyanogenic glucosides.[96,97] The origins of the sulphur atom of the thiohydroximate (**73**) remains a subject for some speculation, although it has been shown that the sulphur atom of L-cysteine is incorporated very efficiently into this position. Underhill[98] has suggested that the thiohydroximic acid (**72**; $R = CH_2CH_2CH(NH_3^+)CO_2^-$), first formed from L-cysteine and the *aci*-nitro compound, is transformed to (**73**) by the intervention of a C–S lyase or a related enzyme.

Quinones form one of the largest classes of natural colouring matters found in plants, but apart from the various isoprenoid quinones (*e.g.* plastoquinones) very little is yet known concerning their biological functions. Chemically they fall into three major categories based on the benzo-, naphtho-, and anthra-quinone ring systems. Most interest from the biosynthetic viewpoint appears to centre around the origins of the naphthoquinone ring system of such quinones as juglone (**75**) and lawsone (**76**). The nucleus of both juglone (**75**) and lawsone (**76**) has been shown to be derived from (−)-shikimic acid and a three-carbon fragment derived from L-glutamic acid, and its construction therefore resembles very closely the synthesis of the naphthoquinone nucleus of the menaquinones (Section 30.3.4, Scheme 16). Some conflict of experimental data and hence of opinion still exists concerning the pathway whereby the intermediate *o*-succinylbenzoic acid is converted to the two quinones (**75**) and (**76**). Leduc, Dansette, and Azerad[99] and Zenk and Floss[100] have both presented evidence to suggest that 1,4-naphthoquinone (**45**) is an intermediate in the biosynthesis of juglone (**75**). On the other hand, Grotzinger and

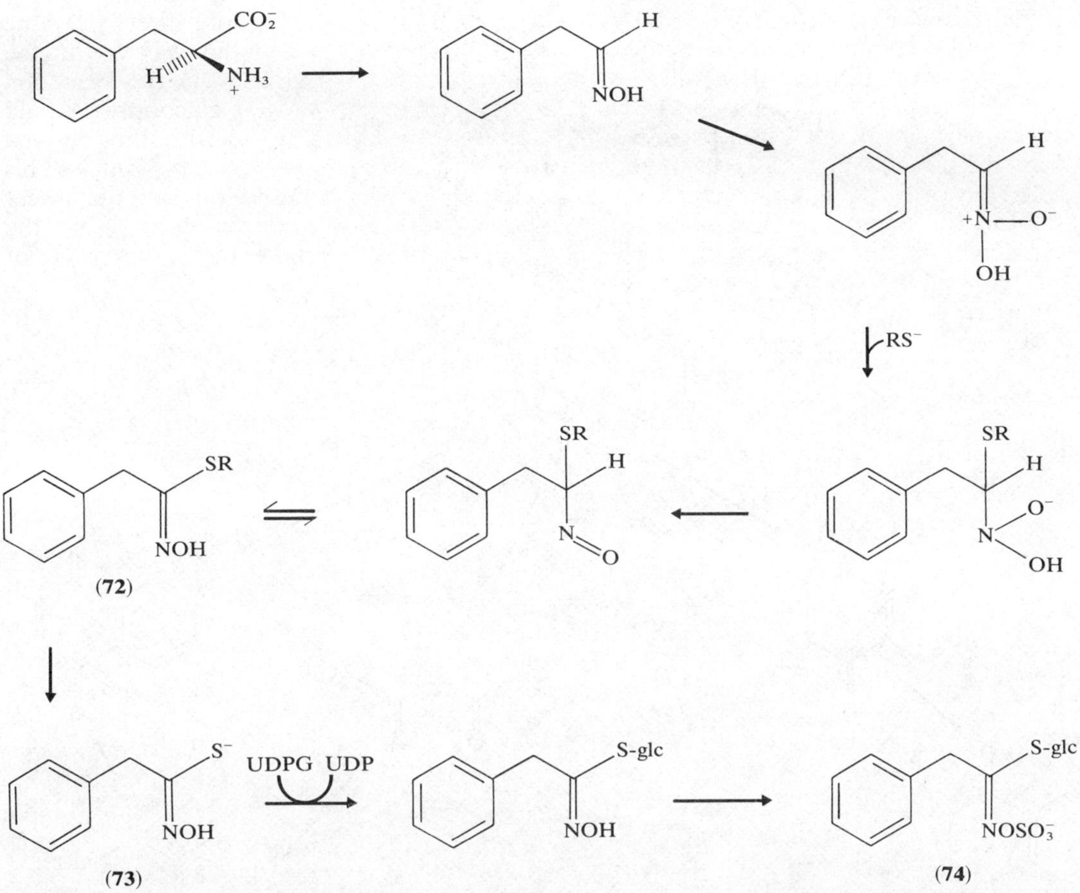

SCHEME 32 Proposed mechanism for the biosynthesis of benzylglucosinolate (**74**; glucotropaeolin) in *Tropaeolum majus*

Campbell[101] have produced data which show that symmetrical intermediates such as 1,4-naphthoquinone (**45**) cannot be involved in the biosynthesis of lawsone (**76**). These results, based on the administration of [3,4-$^{14}C_2$]-L-glutamic acid and sodium [2-^{14}C]acetate, showed that the phenolic hydroxyl group of lawsone is attached to the carbon atom which was originally C-2 in the intermediate *o*-succinylbenzoic acid. On the basis of this evidence, two alternative pathways to lawsone from the intermediate (**45**) were suggested (Scheme 33). Structurally, juglone and lawsone appear to be very closely related naphthoquinones; however, the present evidence suggests that they are derived from the common intermediate (**45**) by quite different pathways.

In the plant kingdom the largest number of quinones is found in the family Rubiaceae, and amongst these are the distinctive anthraquinones alizarin (**78**) and pseudopurpurin (**79**). The experimental evidence which is presently available[102] suggests that these anthraquinones are derived from a naphthalenic precursor and the third aromatic ring is from a branched C_5 chain (mevalonate derived) by cyclization and oxidation. Once again there is a conflict of evidence[103,104] as to the nature of the napthalene intermediate which follows *o*-succinylbenzoic acid (**45**) in the biosynthetic pathway, and evidence both for and against the intermediacy of the symmetrical 1,4-naphthoquinone has been presented for the formation of alizarin (**78**) in *Rubia tinctora.* With the details of these intermediates still to be finally elucidated, the route shown (Scheme 34) represents the outline of the pathway to these anthraquinones.

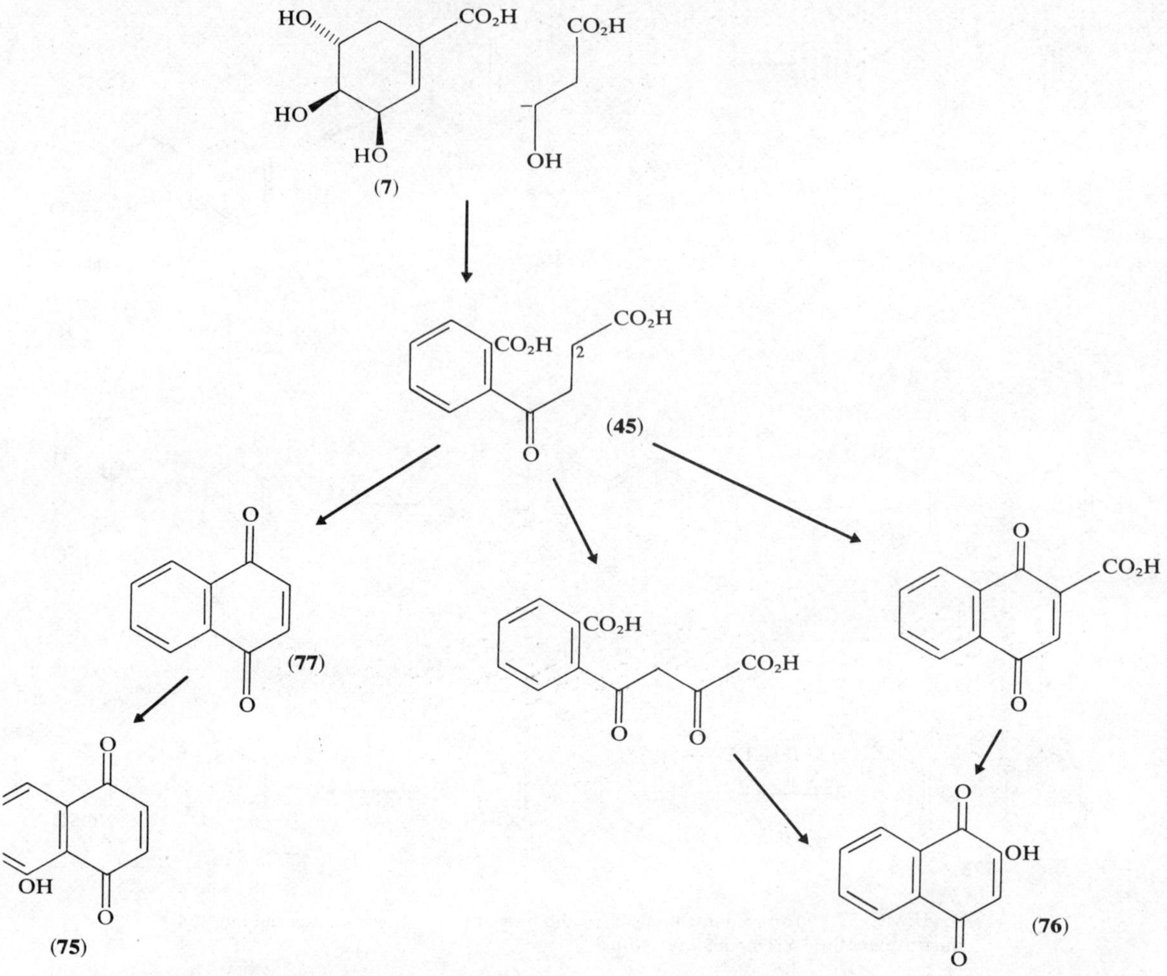

SCHEME 33 Biosynthesis of juglone (**75**) and lawsone (**76**)

30.3.7 MISCELLANEOUS METABOLITES

The shikimate pathway provides a metabolic route not only to the three aromatic α-amino-acids and the isoprenoid quinones, but also to *p*-aminobenzoic acid (**80**), which is an essential structural component of the folate coenzymes (Section 24.3.3.1). The details of this conversion are, however, still not clear. Weiss and Srinivasan[105] have demonstrated that the amide group of L-glutamine is the precursor of the aryl amino-group and evidence has been presented to show that the synthesis of *p*-aminobenzoic acid and anthranilic acid (**35**) from chorismic acid (**9**) are very closely linked in many systems.[106,107] Similar uncertainties surround the biosynthetic origins of the aryl amino-group of *p*-amino-L-phenylalanine (**81**), whose isolation and biosynthesis in *Vigna vexillata* has been reported by Larsen.[108] The same amino-acid (**81**) has been strongly implicated in the biosynthesis of the antibiotic chloramphenicol (**82**) in *Streptomyces venezuelae.*[109]

Considerable interest has surrounded the efforts to discover the mode in which the phenazine ring system is constructed by micro-organisms. Although there is little evidence to support the original suggestion of Carter and Richards[110] that this ring system is formed by coupling of two anthranilic acid molecules, various groups[111,112] have finally concluded that a C_6–C_1–N type of intermediate–related to and derived from chorismic acid (**9**)–is

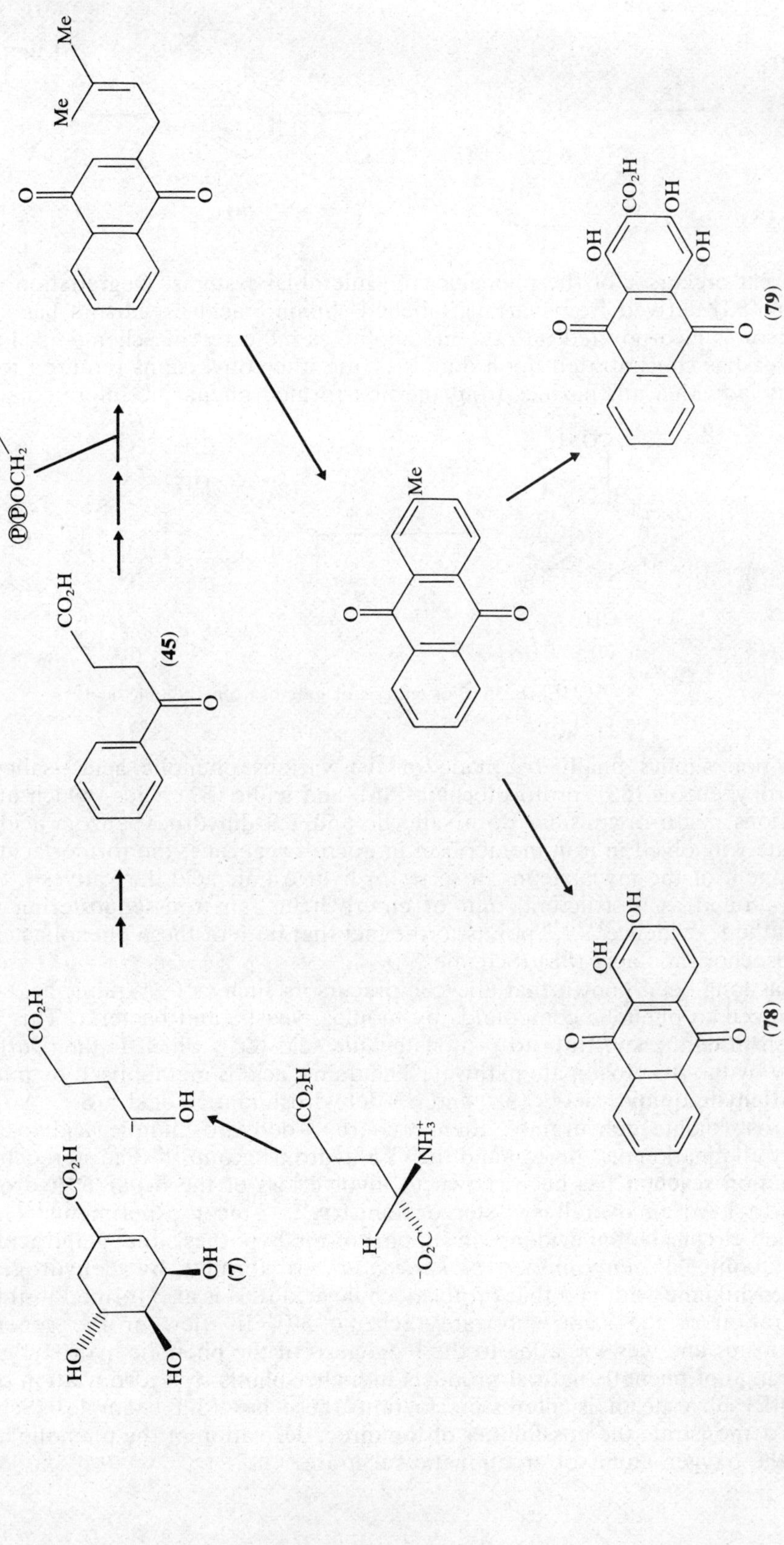

SCHEME 34 Biosynthesis of anthraquinones in the Rubiaceae

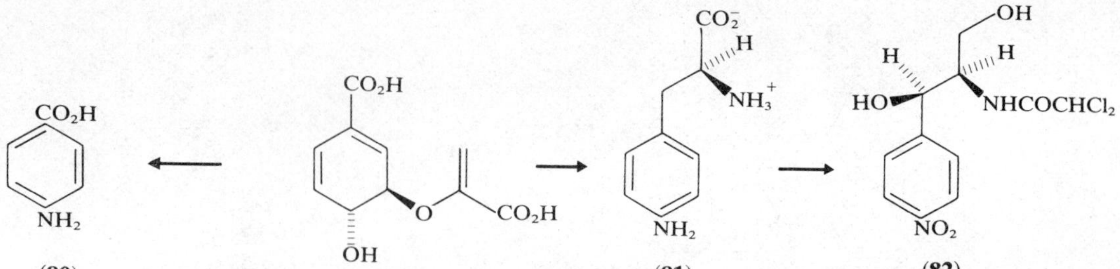

(80) **(9)** **(81)** **(82)**

the direct precursor of the phenazines in microbial systems. Degradation of the pigment iodinin (**83**) derived from various labelled shikimic acid precursors has shown that the precursor is incorporated into the metabolite as indicated in Scheme 35. Further work in this area has concentrated upon details of the interconversions required to elaborate the various individual phenazines from the first-formed phenazine intermediate.[113,114]

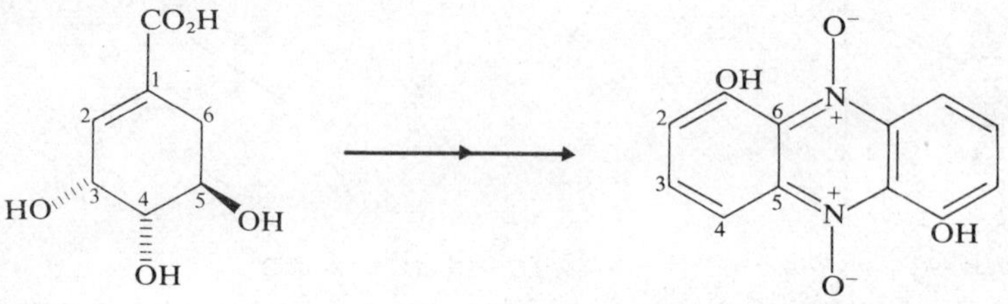

(7) **(83)**

SCHEME 35 Biosynthesis of iodinin from shikimic acid

Mention should finally be made of the various phenolic acids–salicylic (**84**), 2,3-dihydroxybenzoic (**85**), protocatechuic (**86**), and gallic (**87**) acids–which are metabolized in various micro-organisms. Both salicylic and 2,3-dihydroxybenzoic acid appear to be intimately involved in iron metabolism in micro-organisms; the former acid is a structural component of the mycobactins (a series of hydroxamic acid derivatives[115]) and the latter acid is found as a structural unit of enterochelin (an iron-sequestering metabolite[116]). Biosynthetic evidence[117,118] points to the fact that both of these phenolic acids are derived from isochorismic acid (**88**) (Scheme 36).

It has long been known that alicyclic precursors such as (−)-quinic acid (**4**) are readily aromatized to phenolic compounds by moulds, yeasts, and bacteria. This is achieved by catabolism of the substrate to protocatechuic acid (**86**), which is then further degraded, usually by the β-ketoadipate pathway. The quinic acid is metabolized by initial conversion to 3-dehydroquinic acid (**5**) and 3-dehydroshikimic acid (**6**). An enzyme, 3-dehydroshikimate dehydratase, then converts 3-dehydroshikimic acid to protocatechuic acid by elimination of a proton and the C-5 hydroxyl group.[119] The stereochemistry of this elimination reaction has been shown to involve loss of the 6-*pro-R* hydrogen atom and hence to have an overall *syn*-stereochemistry.[120] Tracer experiments[121] have similarly provided circumstantial evidence in support of the hypothesis that gallic acid (**87**), which is a metabolite of *Phycomyces blakesleeanus*, is formed by dehydrogenation of 3-dehydroshikimic acid, and that protocatechuic acid (**86**) is also formed in this organism by dehydration of the same substrate (Scheme 37). In view of the generally accepted mechanisms and views relating to the biogenesis of the phenolic hydroxyl group in a very wide range of phenolic natural products in higher plants by hydroxylation of a preformed aromatic substrate, it is interesting to note these last four examples (Schemes 36, 37), which demonstrate the possibilities of the direct derivation of the phenolic hydroxyl group from the oxygen atoms of an aliphatic substrate.

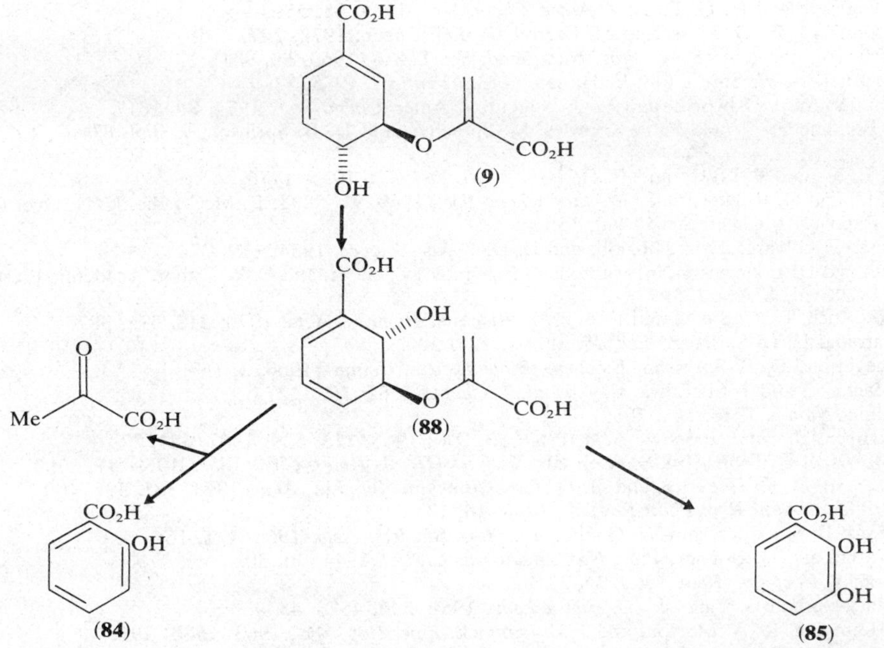

SCHEME 36 Biosynthesis of salicylic acid and 2,3-dihydroxybenzoic acid in bacteria

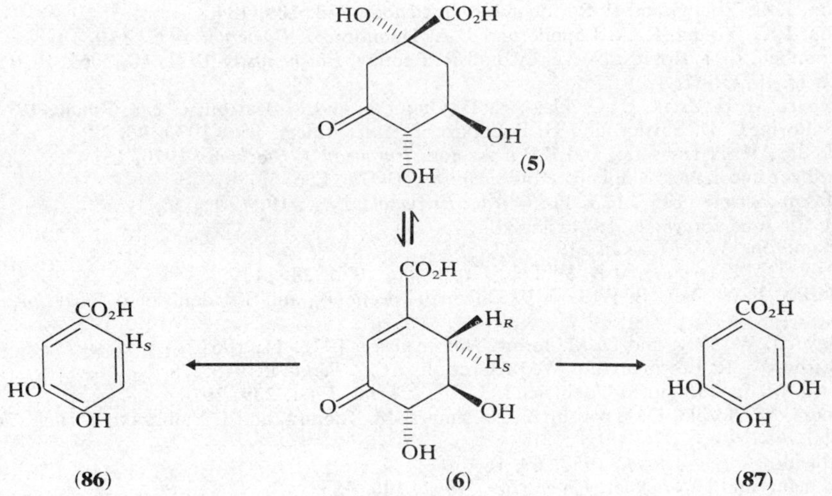

SCHEME 37 Biosynthesis of protocatechuic acid and gallic acid in micro-organisms

References

1. H. O. L. Fischer and G. Dangschat, *Helv. Chim. Acta*, 1935, **18**, 1206.
2. B. D. Davis, *Adv. Enzymol.*, 1955, **16,** 287.
3. D. B. Sprinson, *Adv. Carbohydrate Chem.*, 1960, **15,** 235.
4. F. Gibson and J. Pittard, *Bact. Rev.*, 1968, **32,** 468.
5. B. A. Bohm, *Chem. Rev.*, 1965, **65,** 435.
6. E. Haslam, 'The Shikimate Pathway', Butterworths, London, 1974.
7. A. B. DeLeo and D. B. Sprinson, *Biochem. Biophys. Res. Comm.*, 1968, **32,** 873.

8. D. K. Onderka and H. G. Floss, *J. Amer. Chem. Soc.*, 1969, **91,** 5894.
9. D. K. Onderka, H. G. Floss, and M. Carroll, *J. Biol. Chem.*, 1972, **247,** 736.
10. K. R. Hanson and I. A. Rose, *Proc. Nat. Acad. Sci. U.S.A.* 1963, **50,** 981.
11. M. J. Turner, B. W. Smith, and E. Haslam, *J.C.S. Perkin I,* 1975, 52.
12. J. R. Butler, W. L. Alworth, and M. J. Nugent, *J. Amer. Chem. Soc.*, 1974, **96,** 1617.
13. W. E. Bondinell, J. Vnek, P. F. Knowles, M. Sprecher, and D. B. Sprinson, *J. Biol. Chem.*, 1971, **246,** 6191.
14. R. Ife, L. F. Ball, P. Lowe, and E. Haslam, *J.C.S. Perkin I,* 1976, 1776.
15. R. K. Hill and G. R. Newkome, *J. Amer. Chem. Soc.*, 1969, **91,** 5893; R. McCrindle, K. H. Overton, and R. A. Raphael, *J. Chem. Soc.*, 1960, 1560.
16. U. Weiss, C. Gilvarg, E. S. Mingioli, and B. D. Davis, *Science*, 1954, **119,** 774.
17. J. Dayan and D. E. Sprinson, in 'Methods in Enzymology', ed. H. and C. W. Tabor, Academic Press, New York, 1970, vol. 17A, p. 559.
18. G. L. E. Koch, D. C. Shaw, and F. Gibson, *Biochim. Biophys. Acta*, 1970, **212,** 375, 387.
19. J. Dayan and D. B. Sprinson, *Fed. Proc.*, 1968, **27,** 290.
20. J. G. Levin and D. B. Sprinson, *Biochem. Biophys. Res. Comm.*, 1960, **3,** 157.
21. L. M. Jackman and J. M. Edwards, *Austral. J. Chem.*, 1965, **18,** 1227.
22. L. Pauling, *Nature*, 1948, **161,** 706.
23. P. O. Larsen, *Biochim. Biophys. Acta*, 1964, **93,** 200; 1966, **115**, 529; 1967, **141,** 27.
24. C. J. Morris, J. F. Thompson, S. Asen, and F. Irreverre, *J. Amer. Chem. Soc.*, 1959, **81,** 6069.
25. P. O. Larsen, D. K. Onderka, and H. G. Floss, *Biochim. Biophys. Acta*, 1975, **381,** 397, 409.
26. J. B. Pridham, *Ann. Rev. Plant Physiol.*, 1965, **16,** 13.
27. F. Lingens, B. Sprössler, and W. Goebel, *Biochim. Biophys. Acta*, 1966, **121,** 164.
28. E. L. Tatum and D. Bonner, *Proc. Nat. Acad. Sci. U.S.A.*, 1944, **30,** 30.
29. C. Yanofsky, *Bacteriol. Rev.*, 1960, **24,** 221.
30. D. A. Jackson and C. Yanofsky, *J. Biol. Chem.*, 1969, **244,** 4526, 4539.
31. F. W. Hemming, R. A. Morton, and J. F. Pennock, *Proc. Roy. Soc.*, 1963, **158B,** 291.
32. C. M. Allen, W. Alworth, A. MacRae, and K. Bloch, *J. Biol. Chem.*, 1967, **242,** 1895.
33. F. Gibson and M. I. Gibson, *Biochim. Biophys. Acta*, 1962, **65,** 160.
34. R. E. Olson, *Vitamins and Hormones*, 1966, **24,** 551.
35. G. R. Whistance, D. R. Threlfall, and T. W. Goodwin, *Biochem. J.*, 1967, **105,** 145.
36. P. Friis, G. D. Daves, and K. Folkers, *J. Amer. Chem. Soc.*, 1966, **88,** 4754.
37. F. Gibson, I. G. Young, and P. Stroobant, *J. Bacteriol*, 1972, **109,** 134.
38. F. Gibson, I. G. Young, R. A. Leppik, and J. A. Hamilton, *J. Bacteriol*, 1972, **110,** 18.
39. I. M. Campbell, D. J. Robins, N. Kelsey, and R. Bentley, *Biochemistry* 1971, **10,** 3069; R. Bentley, *Pure Appl. Chem.*, 1975, **41,** 47.
40. K. H. Scharf, M. H. Zenk, H. G. Floss, K. D. Onderka, and M. Carroll, *Chem. Comm.*, 1971, 576.
41. R. M. Baldwin, C. D. Snyder, and H. Rapoport, *J. Amer. Chem. Soc.*, 1973, **95,** 276.
42. M. M. Leduc, M. P. Dansette, and R. G. Azerad, *European J. Biochem.*, 1970, **15,** 428.
43. E. Grotzinger and I. M. Campbell, *Phytochemistry*, 1972, **11,** 675.
44. H. S. Mason, *Science*, 1957, **125,** 1185; *Adv. Enzymol.*, 1957, **19,** 79.
45. S. Kaufman, *Adv. Enzymol.*, 1971, **35,** 245.
46. G. A. Hamilton, *Adv. Enzymol.*, 1969, **32,** 55.
47. J. W. Daly, D. M. Jerina, and B. Witkop, *Experentia*, 1972, **28,** 1129.
48. D. M. Jerina, J. W. Daly, B. Witkop, P. Zaltzman-Nirenberg, and S. Udenfriend, *Biochemistry*, 1970, **9,** 147.
49. D. R. Boyd, J. W. Daly, and D. M. Jerina, *Biochemistry*, 1972, **11,** 1961.
50. G. W. Kirby, W. R. Bowman, and W. R. Gretton, *J.C.S. Perkin I*, 1973,218.
51. T. Nagatsu, M. Levitt, and S. Udenfriend, *J. Biol. Chem.*, 1964, **239,** 2910.
52. Y. Ishimura, M. Nozaki, O. Hayaishi, Y. Nakamura, M. Tamura, and I. Yamazaki, *J. Biol. Chem.*, 1970, **245,** 3593.
53. A. Butenandt, *Angew. Chem.*, 1957, **69,** 16.
54. H. Brockmann and H. Lackner, *Chem. Ber.*, 1967, **100,** 353.
55. H. Weissbach, G. B. Redfield, V. Beaven, and E. Katz, *J. Biol. Chem.* 1965, **240,** 4377.
56. E. Katz and H. Weissbach, *J. Biol. Chem.*, 1963, 238, 666.
57. A. J. Birch, D. W. Cameron, P. W. Holloway, and R. W. Rickards, *Tetrahedron Letters*, 1960, 26; R. B. Herbert, *ibid.*, 1974, 4525.
58. L. Salzman, H. Weissbach, and E. Katz, *Arch. Biochem. Biophys.*, 1969, **130,** 536.
59. N. N. Gerber, *Canad. J. Chem.*, 1968, **46,** 790.
60. H. Blaschko, *J. Physiol.*, 1939, **96,** 50P.
61. T. Nagatsu, M. Levitt, and S. Udenfriend, *Analyt. Biochem.*, 1964, **9,** 122.
62. E. Y. Levin and S. Kaufman, *J. Biol. Chem.*, 1961, **236,** 2043.
63. S. Kaufman and S. Friedman, *J. Biol. Chem.*, 1965, **240,** PC 552.
64. R. V. Pitt-Rivers and R. R. Cavalieri, in 'The Thyroid Gland', ed. R. V. Pitt-Rivers and W. R. Trotter, Butterworths, London, 1964, vol. 1.
65. H. S. Raper, *Biochem. J.*, 1927, **21,** 89; *J. Chem. Soc.*, 1938, 125.
66. H. S. Mason, *J. Biol. Chem.*, 1948, **172,** 83.
67. G. A. Swan and A. Waggott, *J. Chem. Soc.* (*C*), 1970, 1409.

68. G. A. Swan, J. A. G. King, A. Percival, and N. C. Robson, *J. Chem. Soc.* (*C*), 1970, 1419.
69. W. A. Newton, Y. Morino, and E. E. Snell, *J. Biol. Chem.*, 1965, **240,** 1211.
70. S. Sawada, H. Kumagi, H. Yamada, and R. K. Hill, *J. Amer. Chem. Soc.*, 1975, **97,** 4334.
71. C. Fuganti, D. Ghiringhelli, D. Giangrasso, and P. Grasselli, *J.C.S. Chem. Comm.*, 1974, 726.
72. G. E. Syke, R. Potts, and H. G. Floss, *J. Amer. Chem. Soc.*, 1974, **96,** 1593.
73. E. A. Havir and K. R. Hanson, *Biochemistry*, 1968, **7,** 1896, 1904.
74. M. R. Young, G. H. N. Towers, and A. C. Neish, *Canad. J. Bot.*, 1966, **44,** 341.
75. E. A. Havir and K. R. Hanson, *Arch. Biochem. Biophys.*, 1970, **141,** 1.
76. D. S. Hodgins, *Arch. Biochem. Biophys.*, 1972, **149,** 91.
77. K. R. Hanson, R. H. Wightman, J. Staunton, and A. R. Battersby, *J.C.S. Perkin I*, 1972, 2355.
78. R. Ife and E. Haslam, *J. Chem. Soc.* (*C*), 1971, 2818.
79. I. L. Givot, T. A. Smith, and R. H. Abeles, *J. Biol. Chem.*, 1969, **244,** 6341.
80. E. A. Havir, P. D. Reid, and H. V. Marsh, *Plant Physiol.*, 1971, **48,** 130; 1972, **50,** 480.
81. P. G. Strange, J. Staunton, R. H. Wiltshire, A. R. Battersby, K. R. Hanson, and E. A. Havir, *J.C.S. Perkin I*, 1972, 2364.
82. A. C. Neish, in 'Biochemistry of Phenolic Compounds', ed. J. B. Harborne, Academic Press, New York, 1964, p. 294.
83. D. W. Russell, E. E. Conn, A. Sutter, and H. Grisebach, *Biochim. Biophys. Acta*, 1968, **170,** 210.
84. M. H. Zenk and N. Amrhein, *Phytochemistry*, 1969, **8,** 107.
85. A. Sutter and H. Grisebach, *Phytochemistry*, 1969, **8,** 101.
86. E. Haslam, L. J. Porter, D. Jacques, and C. T. Opie, *J.C.S. Perkin I*, 1977, 1637.
87. K. Freudenberg, *Pure Appl. Chem.*, 1962, **5,** 9.
88. T. Higuchi, *Adv. Enzymol.*, 1971, **34,** 207.
89. T. A. Geissman and E. Hinreiner, *Bot. Rev.*, 1952, **18,** 165.
90. M. H. Zenk, *in 'Biosynthesis of Aromatic Compounds'*, ed. G. Billek, Pergamon, Oxford, 1965, p. 45.
91. S. Z. El-Basyouni, D. Chen, R. K. Ibrahim, A. C. Neish, and G. H. N. Towers, *Phytochemistry*, 1964, **3,** 485.
92. M. H. Zenk, *Z. Naturforsch.*, 1964, **19b,** 83.
93. P. M. Dewick and E. Haslam, *Biochem. J.*, 1969, **113,** 537.
94. K. V. Thimann, *J. Biol. Chem.*, 1935, **109,** 279.
95. B. A. Tapper, H. Zilg, and E. E. Conn, *Phytochemistry*, 1972, **11,** 1047.
96. W. E. Underhill, M. D. Chisholm, and L. R. Wetter, *Canad. J. Biochem.*, 1962, **40,** 1505.
97. M. Matsuo, D. F. Kirkland, and W. E. Underhill, *Phytochemistry*, 1972, **11,** 697.
98. L. R. Wetter and W. E. Underhill, *Plant Physiol.*, 1969, **44,** 584.
99. M. M. Leduc, P. M. Dansette, and R. G. Azerad, *European J. Biochem.*, 1970, **15,** 428.
100. K.-H. Scharf, M. H. Zenk, D. K. Onderka, M. Carroll, and H. G. Floss, *Chem. Comm.*, 1971, 576.
101. E. Grotzinger and I. M. Campbell, *Phytochemistry*, 1972, **11,** 675.
102. R. H. Thompson and A. R. Burnett, *J. Chem. Soc.* (*C*), 1967, 2100; 1968, 850, 854, 2437.
103. M. H. Zenk and E. Leistner, *Tetrahedron Letters*, 1968, 861, 1395.
104. E. Leistner, *Phytochemistry*, 1973, **12,** 337.
105. P. R. Srinivasan and B. Weiss, *Biochim. Biophys. Acta*, 1961, **51,** 597.
106. K. H. Altendorf, A. Bacher, and F. Lingens, *Z. Naturforsch.*, 1969, **24b,** 1602.
107. R. A. Jensen, W. H. Holmes, and J. F. Kane *J. Biol. Chem.*, 1972, **247,** 1587.
108. G. A. Dardenne, P. O. Larsen, and E. Wieczorkowska, *Biochim. Biophys. Acta*, 1975, **381,** 416.
109. L. C. Vining, V. S. Malik, and D. W. S. Westlake, *Lloydia*, 1968, **31,** 355.
110. R. E. Carter and J. H. Richards, *J. Amer. Chem. Soc.*, 1961, **83,** 495.
111. U. Hollstein and D. A. McCamey, *J. Org. Chem.*, 1973, **38,** 3415.
112. F. G. Holliman, R. B. Herbert, and J. B. Sheridan, *Tetrahedron Letters*, 1974, 4201.
113. M. E. Flood, R. B. Herbert, and F. G. Holliman, *J.C.S. Perkin I*, 1972, 622.
114. G. S. Hansford, F. G. Holliman, and R. B. Herbert, *J.C.S. Perkin I*, 1972, 103.
115. G. A. Snow, *Bacteriol. Rev.*, 1970, **34,** 99.
116. I. G. O'Brien and F. Gibson, *Biochim. Biophys. Acta*, 1970, **215,** 393.
117. I. G. Young and F. Gibson, *Biochim. Biophys. Acta*, 1969, **177,** 348, 401.
118. B. J. Marshall and C. Ratledge, *Biochim. Biophys. Acta*, 1972, 264, 106.
119. S. R. Gross, *J. Biol. Chem.*, 1958, **233,** 1146.
120. K. H. Scharf, M. H. Zenk, D. K. Onderka, M. Carroll, and H. G. Floss, *Chem. Comm.*, 1971, 765.
121. P. F. Knowles, E. Haslam, and R. D. Haworth, *J. Chem. Soc.*, 1961, 1854.